Meyler's Side Effects of Antimicrobial Drugs

Meyler's Side Effects of Antimicrobial Drugs

Editor

J K Aronson, MA, DPhil, MBChB, FRCP, FBPharmacolS, FFPM (Hon)
Oxford, United Kingdom

ELSEVIER

AMSTERDAM • BOSTON • HEIDELBERG • LONDON • NEW YORK • OXFORD
PARIS • SAN DIEGO • SAN FRANCISCO • SINGAPORE • SYDNEY • TOKYO

Elsevier
Radarweg 29, PO Box 211, 1000 AE Amsterdam, The Netherlands
The Boulevard, Langford Lane, Kidlington, Oxford OX5 1GB, UK
525 B Street, Suite 1900, San Diego, CA 92101-4495, USA

Notice
No responsibility is assumed by the publisher for any injury and/or damage to persons
or property as a matter of products liability, negligence or otherwise, or from any use or operation
of any methods, products, instructions or ideas contained in the material herein. Because of rapid
advances in the medical sciences, in particular, independent verification of diagnoses and drug
dosages should be made

Medicine is an ever-changing field. Standard safety precautions must be followed, but as new
research and clinical experience broaden our knowledge, changes in treatment and drug therapy
may become necessary or appropriate. Readers are advised to check the most current product
information provided by the manufacturer of each drug to be administered to verify the
recommended dose, the method and duration of administrations, and contraindications. It is the
responsibility of the treating physician, relying on experience and knowledge of the patient, to
determine dosages and the best treatment for each individual patient. Neither the publisher nor the
authors assume any liability for any injury and/or damage to persons or property arising from this
publication.

British Library Cataloguing in Publication Data
A catalogue record for this book is available from the British Library

Library of Congress Cataloging in Publication Data
A catalog record for this book is available from the Library of Congress

ISBN: 978-044-453272-5

Typeset by Integra Software Services Pvt. Ltd, Pondicherry, India www.integra-india.com

Contents

Contents

Preface

This volume covers the adverse effects of antimicrobial drugs. The material has been collected from *Meyler's Side Effects of Drugs: The International Encyclopedia of Adverse Drug Reactions and Interactions* (15th edition, 2006, in six volumes), which was itself based on previous editions of *Meyler's Side Effects of Drugs* and *Side Effects of Drugs Annuals*, and from later *Side Effects of Drugs Annuals* (SEDA) 28, 29, and 30. The main contributors of this material were BJ Angus, JK Aronson, VV Banu Rekha, AGC Bauer, T Bicanic, J Blaser, J Buser, A Cerny, C Chiou, N Corti, P Cottagnaud, MD de Jong, S Dittmann, MNG Dukes, CJ Ellis, J Evison, K Fattinger, H Furrer, C Fux, AH Groll, F Hackenberger, R Hoigné, A Imhof, DJ Jeffries, AMM Kaddu, O Koch, H Kolve, M Krause, S Krishna, R Laffer, P Leuenberger, P Magee, IR McNicholl, T Midtvedt, KA Neftel, JN Pande, T Planche, P Reiss, I Ribiero, C Ruef, A Schaffner, M Schlegel, P Schmid, M Schneemann, R Serafino, S Sheehy, OO Simooya, S Swaminathan, C Thurnheer, PJJ van Genderen, PL Vernazza, TJ Walsh, R Walter, CJM Whitty, C Woodrow, J-P Zellweger, AS Zinkernagel, and M Zoppi. For contributors to earlier editions of *Meyler's Side Effects of Drugs* and the *Side Effects of Drugs Annuals*, see http://www.elsevier.com/wps/find/bookseriesdescription.cws_home/BS_SED/description.

A brief history of the Meyler series

Leopold Meyler was a physician who was treated for tuberculosis after the end of the Nazi occupation of The Netherlands. According to Professor Wim Lammers, writing a tribute in Volume VIII (1975), Meyler got a fever from para-aminosalicylic acid, but elsewhere Graham Dukes has written, based on information from Meyler's widow, that it was deafness from dihydrostreptomycin; perhaps it was both. Meyler discovered that there was no single text to which medical practitioners could look for information about unwanted effects of drug therapy; Louis Lewin's text "Die Nebenwirkungen der Arzneimittel" ("The Untoward Effects of Drugs") of 1881 had long been out of print (SEDA-27, xxv-xxix). Meyler therefore determined to make such information available and persuaded the Netherlands publishing firm of Van Gorcum to publish a book, in Dutch, entirely devoted to descriptions of the adverse effects that drugs could cause. He went on to agree with the Elsevier Publishing Company, as it was then called, to prepare and issue an English translation. The first edition of 192 pages (*Schadelijke Nevenwerkingen van Geneesmiddelen*) appeared in 1951 and the English version (*Side Effects of Drugs*) a year later.

The book was a great success, and a few years later Meyler started to publish what he called surveys of unwanted effects of drugs. Each survey covered a period of two to four years. They were labelled as volumes rather than editions, and after Volume IV had been published Meyler could no longer handle the task alone. For subsequent volumes he recruited collaborators, such as Andrew Herxheimer. In September 1973 Meyler died unexpectedly, and Elsevier invited Graham Dukes to take over the editing of Volume VIII.

Dukes persuaded Elsevier that the published literature was too large to be comfortably encompassed in a four-yearly cycle, and he suggested that the volumes should be produced annually instead. The four-yearly volume could then concentrate on providing a complementary critical encyclopaedic survey of the entire field. The first *Side Effects of Drugs Annual* was published in 1977. The first encyclopaedic edition of *Meyler's Side Effects of Drugs*, which appeared in 1980, was labelled the ninth edition, and since then a new encyclopaedic edition has appeared every four years. The 15th edition was published in 2006, in both hard and electronic versions.

Monograph structure

The monographs in this volume are arranged in six sections:

- Antibacterial drugs
- Antiviral drugs, including immunoglobulins and interferons
- Antifungal drugs
- Antiprotozoal and antihelminthic drugs
- Vaccines
- Disinfectants and antiseptics

In each monograph in the Meyler series the information is organized into sections as shown below (although not all the sections are covered in each monograph).

Drug names

Drugs have usually been designated by their recommended or proposed International Non-proprietary Names (rINN or pINN); when these are not available, chemical names have been used. In some cases brand names have been used.

Spelling

For indexing purposes, American spelling has generally been used, e.g. anemia, estrogen, rather than anaemia, oestrogen.

Cross-references

The various editions of *Meyler's Side Effects of Drugs* are cited in the text as SED-13, SED-14, etc; the *Side Effects of Drugs Annuals* are cited as SEDA-1, SEDA-2, etc.

J K Aronson
Oxford, August 2009

Organization of material in monographs in the Meyler series (not all sections are included in each monograph)

General information
Drug studies
  Observational studies
  Comparative studies
  Drug-combination studies
  Placebo-controlled studies
  Systematic reviews
Organs and systems
  Cardiovascular
  Respiratory
  Ear, nose, throat
  Nervous system
  Neuromuscular function
  Sensory systems
  Psychological
  Psychiatric
  Endocrine
  Metabolism
  Nutrition
  Electrolyte balance
  Mineral balance
  Metal metabolism
  Acid-base balance
  Fluid balance
  Hematologic
  Mouth
  Teeth
  Salivary glands
  Gastrointestinal
  Liver
  Biliary tract
  Pancreas
  Urinary tract
  Skin
  Hair
  Nails
  Sweat glands
  Serosae
  Musculoskeletal
  Sexual function
  Reproductive system
  Breasts
  Immunologic
  Autacoids
  Infection risk
  Body temperature

  Multiorgan failure
  Trauma
  Death
Long-term effects
  Drug abuse
  Drug misuse
  Drug tolerance
  Drug resistance
  Drug dependence
  Drug withdrawal
  Genotoxicity
  Cytotoxicity
  Mutagenicity
  Tumorigenicity
Second-generation effects
  Fertility
  Pregnancy
  Teratogenicity
  Fetotoxicity
  Lactation
  Breast feeding
Susceptibility factors
  Genetic factors
  Age
  Sex
  Physiological factors
  Disease
  Other features of the patient
Drug administration
  Drug formulations
  Drug additives
  Drug contamination and adulteration
  Drug dosage regimens
  Drug administration route
  Drug overdose
Interactions
  Drug-drug interactions
  Food-drug interactions
  Drug-device interactions
  Smoking
  Other environmental interactions
Interference with diagnostic tests
Diagnosis of adverse drug reactions
Management of adverse drug reactions
Monitoring therapy
References

ANTIBACTERIAL DRUGS

Alatrofloxacin and trovafloxacin

See also Fluoroquinolones

General Information

Alatrofloxacin is a fluoronaphthyridone that is hydrolysed to the active moiety, trovafloxacin, after intravenous administration. This fourth-generation broad-spectrum fluoroquinolone has activity against Gram-positive, Gram-negative, anerobic, and atypical respiratory pathogens. Because it has significant hepatotoxicity, the list of appropriate indications for trovafloxacin has been restricted.

In a multicenter, double-blind, randomized comparison of trovafloxacin 200 mg and clarithromycin 500 mg bd in 176 subjects with acute exacerbations of chronic bronchitis, the most common adverse effects of trovafloxacin were nausea (5%), dizziness (5%), vomiting (3%), and constipation (3%) (1). Because trovafloxacin is hepatotoxic, the list of appropriate indications has been limited to patients who have at least one of several specified infections, such as nosocomial pneumonia or complicated intra-abdominal infections that are serious and life- or limb-threatening in the physician's judgement.

Trovafloxacin may down-regulate cytokine mRNA transcription in human peripheral blood mononuclear cells stimulated with lipopolysaccharide or lipoteichoic acid (2). Likewise, trovafloxacin inhibited *Salmonella typhimurium* induced production of TNFα, HIV-1 replication, and reactivation of latent HIV-1 in promonocytic U1 cells at concentrations comparable to the plasma and tissue concentrations achieved by therapeutic dosages (3).

Organs and Systems

Cardiovascular

Phlebitis can occur during parenteral administration of trovafloxacin. High concentrations of trovafloxacin (2 mg/ml) significantly reduced intracellular ATP content in cultured endothelial cells and reduced concentrations of ADP, GTP, and GDP (4). These in vitro data suggest that high doses of trovafloxacin are not compatible with maintenance of endothelial cell function and may explain the occurrence of phlebitis. Commercial formulations should be diluted and given into large veins.

Nervous system

Alatrofloxacin can cause seizures (5).

- A 37-year-old Asian man received several antibiotics (including intravenous ceftazidime, gentamicin, meropenem, metronidazole, and vancomycin) postoperatively. After 3 weeks he was given alatrofloxacin 75 mg in 25 ml of dextrose 5% (1.875 mg/ml) and developed generalized clonus. On rechallenge, infusing at half the initial rate, the seizure recurred. A CT scan of the brain was normal.

Seizures are rare but have occurred during treatment with other fluoroquinolones. This is the first report of a case of seizures associated with slow infusion of alatrofloxacin. However, as of 21 June 2000, the manufacturers had received 53 reports of seizures through worldwide postmarketing surveillance. In rat hippocampus slices, trovafloxacin had significant convulsive potential; the underlying mechanism is hitherto incompletely understood.

Trovafloxacin has been associated with diffuse weakness due to a demyelinating polyneuropathy in a patient without an underlying neurological disorder (6).

Sensory systems

Intravitreal trovafloxacin in doses of 50 mg and higher in the pigmented rabbit eye caused retinal and nerve fiber injury; intravitreal doses of 25 mg and lower appear to be safe, with no evidence of ocular toxicity (7).

Hematologic

Alatrofloxacin has been associated with severe leukopenia (8).

- A 79-year-old white man was treated with intravenous alatrofloxacin mesylate 200 mg bd for 5 days. His leukocyte count fell from $10.9 \times 10^9/l$ to $2.2 \times 10^9/l$; the hemoglobin did not change. Alatrofloxacin was withdrawn, and 3 days later the leukocyte count had increased to $11.5 \times 10^9/l$.

The mechanism of trovafloxacin-induced leukopenia is unknown. Nevertheless, since quinolones exert their antibacterial effect by inhibiting bacterial DNA gyrase and since similar topoisomerases are involved in the organization and function of mammalian DNA, it is possible that trovafloxacin acts by modulating bone marrow stem-cell DNA production.

Alatrofloxacin has been associated with severe thrombocytopenia (9).

- A 54-year-old woman was given alatrofloxacin 300 mg intravenously qds and on day 4 developed epistaxis. Her platelet count was $7 \times 10^9/l$, with normal hemoglobin and white blood cell counts. Direct antiglobulin testing showed coating of erythrocytes with polyspecific immunoproteins, and platelet-associated antibody testing was positive for IgM and IgG antibodies. Alatrofloxacin was withdrawn and azithromycin was given instead. She was given methylprednisolone 125 mg intravenously bd and the platelet count fell to $2 \times 10^9/l$ and then rose, reaching $60 \times 10^9/l$ on day 8.

During clinical trials, thrombocytopenia occurred in under 1% of more than 7000 patients who received alatrofloxacin or trovafloxacin.

Liver

More than 100 cases of hepatotoxicity associated with trovafloxacin have been reported to the FDA.

- A 19-year-old woman developed severe acute hepatitis and peripheral eosinophilia during oral trovafloxacin therapy for recurrent sinusitis (10). Liver biopsy showed extensive centrilobular hepatocyte necrosis, probably causing veno-occlusive disease. Clinical and laboratory abnormalities resolved completely after prolonged treatment with steroids.

- A 66-year-old man had taken trovafloxacin 100 mg/day for 4 weeks for refractory chronic sinusitis (11). For several years he had also taken allopurinol, doxepin, hydrochlorothiazide, losartan, metoprolol, and nabumetone. He developed nausea, vomiting, malaise, and abdominal distension. His white cell count was 8000×10^9/l with 16% eosinophils; his serum aspartate transaminase was 537 IU/l, alanine transaminase 841 IU/l, direct bilirubin 17 µmol/l; total bilirubin 27 µmol/l, alkaline phosphatase 111 IU/l; blood urea nitrogen 5 µmol/l; and creatinine 190 µmol/l. Tests for hepatitis A, B, and C were negative. A biopsy of the liver showed centrilobular and focal periportal necrosis and eosinophilic infiltration; the sinusoids were dilated and contained lymphocytes and eosinophils; many hepatocytes were undergoing mitosis. After withdrawal of trovafloxacin and treatment with prednisone, his hepatic and renal function returned to normal, and the eosinophilia gradually resolved.

Skin

The photosensitizing potential of trovafloxacin 200 mg od has been compared with that of ciprofloxacin 500 mg bd, lomefloxacin 400 mg od, and placebo in 48 healthy men (aged 19–45 years) (12). Trovafloxacin had significantly less photosensitizing potential than either ciprofloxacin or lomefloxacin. Photosensitivity seemed to be induced only by wavelengths in the UVA region, was maximal at 24 hours, and had a short-term effect.

Musculoskeletal

Trovafloxacin inhibited growth and extracellular matrix mineralization in MC3T3-E1 osteoblast-like cell cultures (13). The IC_{50} was 0.5 µg/ml, which is below clinically achievable serum concentrations. The authors suggested that the clinical relevance of this observation to bone healing in orthopedic patients should be evaluated.

In rats experimental fractures systemically exposed to trovafloxacin had impaired healing during the early stages of fracture repair (14).

Second-Generation Effects

Teratogenicity

In an ex vivo study, trovafloxacin crossed the human placenta by simple diffusion and neither accumulated in the media nor bound to tissues or accumulated in the placenta (15). This implies that it should have no effects on the fetus if given during pregnancy.

Susceptibility Factors

Age

The pharmacokinetics of a single intravenous dose of alatrofloxacin have been determined in six infants aged 3–12 months and in 14 children aged 2–12 years (16). The peak trovafloxacin concentration at the end of the infusion was 4.3 µg/ml; the volume of distribution at steady state was 1.6 l/kg, clearance 2.5 ml/min/kg, and the half-life 9.8 hours, with no age-related differences. Less than 5% of the administered dose was excreted in the urine over 24 hours.

Other features of the patient

The pharmacokinetics of trovafloxacin after the administration of alatrofloxacin were not substantially altered in seven critically ill patients (three men, four women) with APACHE II scores of 27 (range 15–32) and normal or mildly impaired hepatic function (17).

Monitoring Therapy

In 17 patients aged over 18 years with severe acute community-acquired pneumonia trovafloxacin concentrations were persistently high in the sputum, bronchial secretions, bronchoalveolar lavage fluid, and epithelial lining fluid, with no significant difference between these compartments (18). The authors proposed that measurement of sputum concentrations could be used to monitor the outcome of treatment.

References

1. Sokol WN Jr, Sullivan JG, Acampora MD, Busman TA, Notario GF. A prospective, double-blind, multicenter study comparing clarithromycin extended-release with trovafloxacin in patients with community-acquired pneumonia. Clin Ther 2002;24(4):605–15.
2. Purswani M, Eckert S, Arora H, Johann-Liang R, Noel GJ. The effect of three broad-spectrum antimicrobials on mononuclear cell responses to encapsulated bacteria: evidence for down-regulation of cytokine mRNA transcription by trovafloxacin. J Antimicrob Chemother 2000;46(6):921–9.
3. Gollapudi S, Gupta S, Thadepalli H. Salmonella typhimurium-induced reactivation of latent HIV-1 in promonocytic U1 cells is inhibited by trovafloxacin. Int J Mol Med 2000;5(6):615–8.
4. Armbruster C, Robibaro B, Griesmacher A, Vorbach H. Endothelial cell compatibility of trovafloxacin and levofloxacin for intravenous use. J Antimicrob Chemother 2000;45(4):533–5.
5. Melvani S, Speed BR. Alatrofloxacin-induced seizures during slow intravenous infusion. Ann Pharmacother 2000;34(9):1017–9.
6. Murray CK, Wortmann GW. Trovafloxacin-induced weakness due to a demyelinating polyneuropathy. South Med J 2000;93(5):514–5.
7. Ng EW, Joo MJ, Eong KG, Green WR, O'Brien TP. Ocular toxicity of intravitreal trovafloxacin in the pigmented rabbit. Curr Eye Res 2003;27:387–93.

8. Mitropoulos FA, Angood PB, Rabinovici R. Trovafloxacin-associated leukopenia. Ann Pharmacother 2001;35(1):41–4.
9. Gales BJ, Sulak LB. Severe thrombocytopenia associated with alatrofloxacin. Ann Pharmacother 2000;34(3):330–4.
10. Lazarczyk DA, Goldstein NS, Gordon SC. Trovafloxacin hepatotoxicity. Dig Dis Sci 2001;46(4):925–6.
11. Chen HJ, Bloch KJ, Maclean JA. Acute eosinophilic hepatitis from trovafloxacin. N Engl J Med 2000; 342(5):359–60.
12. Ferguson J, McEwen J, Al-Ajmi H, Purkins L, Colman PJ, Willavize SA. A comparison of the photosensitizing potential of trovafloxacin with that of other quinolones in healthy subjects. J Antimicrob Chemother 2000;45(4):503–9.
13. Holtom PD, Pavkovic SA, Bravos PD, Patzakis MJ, Shepherd LE, Frenkel B. Inhibitory effects of the quinolone antibiotics trovafloxacin, ciprofloxacin, and levofloxacin on osteoblastic cells in vitro. J Orthop Res 2000;18(5):721–7.
14. Perry AC, Prpa B, Rouse MS, Piper KE, Hanssen AD, Steckelberg JM, Patel R. Levofloxacin and trovafloxacin inhibition of experimental fracture-healing. Clin Orthop Relat Res 2003;(414):95–100.
15. Casey B, Bawdon RE. Ex vivo human placental transfer of trovafloxacin. Infect Dis Obstet Gynecol 2000;8(5–6):228–9.
16. Bradley JS, Kearns GL, Reed MD, Capparelli EV, Vincent J. Pharmacokinetics of a fluoronaphthyridone, trovafloxacin (CP 99,219), in infants and children following administration of a single intravenous dose of alatrofloxacin. Antimicrob Agents Chemother 2000; 44(5):1195–9.
17. Olsen KM, Rebuck JA, Weidenbach T, Fish DN. Pharmacokinetics of intravenous trovafloxacin in critically ill adults. Pharmacotherapy 2000;20(4):400–4.
18. Peleman RA, Van De Velde V, Germonpre PR, Fleurinck C, Rosseel MT, Pauwels RA. Trovafloxacin concentrations in airway fluids of patients with severe community-acquired pneumonia. Antimicrob Agents Chemother 2000;44(1):178–80.

Amikacin

See also Aminoglycoside antibiotics

General Information

Amikacin is a semisynthetic derivative of kanamycin with similar pharmacokinetic properties and dosages. It is resistant to many of the bacterial R factor-mediated enzymes that inactivate kanamycin and gentamicin. Noteworthy is its effect against *Pseudomonas aeruginosa* and against most Gram-negative aerobes that are resistant to gentamicin and tobramycin. There are strains of *Staphylococcus aureus* that inactivate amikacin by phosphorylation and adenylation. Ticarcillin or azlocillin plus amikacin is considered one of the most efficacious empiric antibiotic combinations in febrile granulocytopenia in patients with cancer. On a weight basis amikacin is less active than gentamicin, and so the usual dose is 10–20 mg/kg/day. Wherever possible, peak concentrations of 40 µg/ml and troughs of 10 µg/ml should not be exceeded during twice-daily dosing.

Organs and Systems

Cardiovascular

Episodes of hypotension have been attributed to amikacin (1).

• A 68-year-old woman on continuous ambulatory peritoneal dialysis was given amikacin (250 mg/day) and on the third day fainted and had a blood pressure of 90/60 mmHg. On the next 2 days, she had episodes of postural hypotension of between 90/60 and 80/50 mmHg two or three times a day. She was given antibiotics for the next 6 days and felt very bad the entire time, with a blood pressure of 80/50 mmHg. Her condition improved 2 days after withdrawal of amikacin and the episodes of hypotension did not recur.

Respiratory

Amikacin may have been the causative agent in an apneic episode in an infant on peritoneal dialysis (2).

Sensory systems

Ears
• A 43-year-old man, who was receiving hemodialysis through a permanent catheter developed severe irreversible sensorineural hearing loss after using an amikacin–heparin lock for 16 weeks (3). He suddenly developed a high-frequency sensorineural hearing loss of 40 decibels. His condition progressed over 1 week, despite immediate withdrawal of the amikacin–heparin lock, and he developed severe irreversible hearing loss below 80 decibels for both high and low frequencies.

Retinal toxicity can occur when aminoglycosides are given intravitreally for endophthalmitis.

• Preretinal hemorrhages developed in a 58-year old man who was treated with two intravitreal injections of amikacin (0.4 mg) and cefazolin (2.25 mg) 48 hours apart for postoperative endophthalmitis following routine extracapsular cataract extraction (4).
• Macular toxicity followed the use of intravitreal amikacin 0.2 mg for postoperative endophthalmitis in a 69-year-old white woman (5).

Ototoxicity was observed in three of 195 patients who received amikacin (15 mg/kg/day) with either cefepime (2 g bd) or ceftazidime (2 g tds) (6). Two patients had severe loss of hearing, which persisted after drug withdrawal and resulted in permanent disability. The other had mild ototoxicity that required no action and resolved spontaneously.

Olfaction
Olfactory disorders are among the rare adverse effects of antibiotic therapy. Reversible anosmia has been described.

• A 50-year-old man with lymphangitis of the forearm was given intravenous amikacin sulfate 500 mg bd and intravenous co-amoxiclav 1.2 g tds for 5 days (7). Before treatment began, there was no investigation of his nose or sense of smell. However, after a septoplasty some 5 years earlier he thought that his olfaction was completely normal. At the end of the treatment period, he noticed a disturbance in his ability to smell, which led to complete anosmia within a few days. Psychometric examination was compatible with complete anosmia. Some 18 months later, he reported that his sense of smell had largely returned during the previous 6 months.In a guinea-pig model the non-ototoxic dose of amikacin (20 mg/kg/day) administered before the ototoxic dose (400 mg/kg/day) had a statistically significant protective effect on the basal turns of the cochlea, observed histologically (8).

Urinary tract

After administration of the recommended doses of amikacin for 10 days, renal damage probably occurs in less than 10% of cases. Limited data support the view that amikacin is less nephrotoxic than other aminoglycosides, possibly because of lower binding affinity to proximal tubular cells or reduced potential to cause phospholipidosis (SEDA-20, 236). In several prospective randomized studies the liability of amikacin to cause nephrotoxicity was no greater than that of gentamicin or tobramycin (9–11). In a prospective study there was significantly lower nephrotoxicity with amikacin 15 mg/kg/day (4% toxicity) compared with netilmicin 7 mg/kg/day (12%) (12). As with other aminoglycosides, renal toxicity is reversible in most cases (13).

Nephrotoxicity occurred in five of 195 patients who received amikacin (15 mg/kg/day) with either cefepime (2 g bd) or ceftazidime (2 g tds) (6). In two patients the deterioration in renal function was mild and resolved without withdrawal of amikacin. In the three other patients, renal insufficiency necessitated drug withdrawal; two of these patients recovered, but one died with sepsis, and renal function was still abnormal at the time of death.

Long-Term Effects

Drug tolerance

In Spain, an epidemic strain of *Acinetobacter baumannii* with resistance to amikacin was isolated in eight different hospitals (14).

Susceptibility Factors

In an open study in eight young healthy Japanese women, the pharmacokinetics of amikacin were affected by the phase of the menstrual cycle (15). In patients with hematological malignancies, bodyweight, renal function, acute myeloblastic leukemia, and hypoalbuminemia were the most important co-variates for the interindividual variability in amikacin pharmacokinetics (16).

Drug Administration

Drug formulations

The pharmacokinetics and toxicity of liposomal amikacin have been investigated in a patient treated for advanced pulmonary multidrug-resistant tuberculosis (17). The serum concentrations of amikacin obtained with the liposomal formulation were considerably greater than those obtained with the conventional formulation. Liposomal amikacin was well tolerated and led to clinical improvement, but the patient's sputum remained smear- and culture-positive during the treatment period and for 9 months.

Drug administration route

Amikacin has been tested for compatibility with a chlorhexidine-bearing central venous catheter, the ARROWg + ard Blue Plus, and did not cause a substantial increase in chlorhexidine delivery (18). The amount of amikacin sulfate that was delivered was slightly less than the amount in the infusion solution (92%), but this was considered acceptable.

Drug–Drug Interactions

Colistin

In a prospective study in 80 patients with cystic fibrosis and normal renal function, the combination of aminoglycosides with the polymyxin antibiotic colistin may have increased the risk of nephrotoxicity; in a multiple linear regression model there was a strong correlation between the use of aminoglycosides and reduced renal function, which was potentiated by colistin (19).

Penicillins

Amikacin may be inactivated by penicillins. This inactivation occurs not only with a mixture of the agents in solution but also in vivo, particularly in patients with renal insufficiency. Amikacin offers, at least in vitro, the advantage of being much less inactivated than tobramycin or gentamicin (20).

Thalidomide

Thalidomide may potentiate the nephrotoxicity of aminoglycosides. Three patients with refractory multiple myeloma taking thalidomide developed severe renal insufficiency shortly after starting to take amikacin for concurrent infections (21).

References

1. Wanic-Kossowska M, Grzegorzewska AE, Tykarski A. Hypotonia during amikacin administration in a patient treated with continuous ambulatory peritoneal dialysis. Adv Perit Dial 2006;22:69–72.
2. Cano F, Morales M, Delucchi A. Amykacin-related apneic episode in an infant on peritoneal dialysis. Pediatr Nephrol 2000;14(4):357.
3. Saxena AK, Panhotra BR, Naguib M. Sudden irreversible sensory-neural hearing loss in a patient with diabetes receiving amikacin as an antibiotic–heparin lock. Pharmacotherapy 2002;22(1):105–8.

4. Kumar A, Dada T. Preretinal haemorrhages: an unusual manifestation of intravitreal amikacin toxicity. Aust NZ J Ophthalmol 1999;27(6):435–6.

5. Galloway G, Ramsay A, Jordan K, Vivian A. Macular infarction after intravitreal amikacin: mounting evidence against amikacin. Br J Ophthalmol 2002;86(3):359–60.

6. Erman M, Akova M, Akan H, Korten V, Ferhanoglu B, Koksal I, Cetinkaya Y, Uzun O, Unal SFebrile Neutropenia Study Group of Turkey. Comparison of cefepime and ceftazidime in combination with amikacin in the empirical treatment of high-risk patients with febrile neutropenia: a prospective, randomized, multicenter study. Scand J Infect Dis 2001;33(11):827–31.

7. Welge-Luessen A, Wolfensberger M. Reversible anosmia after amikacin therapy. Arch Otolaryngol Head Neck Surg 2003;129:1331–3.

8. Oliveira JA, Canedo DM, Rossato M, Andrade MH. Self-protection against aminoglycoside ototoxicity in guinea pigs. Otolaryngol Head Neck Surg 2004;131(3):271–9.

9. Plaut ME, Schentag JJ, Jusko WJ. Aminoglycoside nephrotoxicity: comparative assessment in critically ill patients. J Med 1979;10(4):257–66.

10. Smith CR, Baughman KL, Edwards CQ, Rogers JF, Lietman PS. Controlled comparison of amikacin and gentamicin. N Engl J Med 1977;296(7):349–53.

11. Feld R, Valdivieso M, Bodey GP, Rodriguez V. Comparison of amikacin and tobramycin in the treatment of infection in patients with cancer. J Infect Dis 1977;135(1):61–6.

12. Noone M, Pomeroy L, Sage R, Noone P. Prospective study of amikacin versus netilmicin in the treatment of severe infection in hospitalized patients. Am J Med 1989;86(6 Pt 2):809–13.

13. Lane AZ, Wright GE, Blair DC. Ototoxicity and nephrotoxicity of amikacin: an overview of phase II and phase III experience in the United States. Am J Med 1977;62(6):911–918.

14. Vila J, Ruiz J, Navia M, Becerril B, Garcia I, Perea S, Lopez-Hernandez I, Alamo I, Ballester F, Planes AM, Martinez-Beltran J, de Anta TJ. Spread of amikacin resistance in Acinetobacter baumannii strains isolated in Spain due to an epidemic strain. J Clin Microbiol 1999;37(3):758–761.

15. Matsuki S, Kotegawa T, Tsutsumi K, Nakamura K, Nakano S. Pharmacokinetic changes of theophylline and amikacin through the menstrual cycle in healthy women. J Clin Pharmacol 1999;39(12):1256–62.

16. Romano S, Fdez de Gatta MM, Calvo MV, Caballero D, Dominguez-Gil A, Lanao JM. Population pharmacokinetics of amikacin in patients with haematological malignancies. J Antimicrob Chemother 1999;44(2):235–42.

17. Whitehead TC, Lovering AM, Cropley IM, Wade P, Davidson RN. Kinetics and toxicity of liposomal and conventional amikacin in a patient with multidrug-resistant tuberculosis. Eur J Clin Microbiol Infect Dis 1998;17(11):794–7.

18. Xu QA, Zhang Y, Trissel LA, Gilbert DL. Adequacy of a new chlorhexidine-bearing polyurethane central venous catheter for administration of 82 selected parenteral drugs. Ann Pharmacother 2000;34(10):1109–16.

19. Al-Aloul M, Miller H, Alapati S, Stockton PA, Ledson MJ, Walshaw MJ. Renal impairment in cystic fibrosis patients due to repeated intravenous aminoglycoside use. Pediatr Pulmonol 2005;39(1):15–20.

20. Meyer RD. Amikacin. Ann Intern Med 1981;95(3):328–32.

21. Montagut C, Bosch F, Villela L, Rosinol L, Blade J. Aminoglycoside-associated severe renal failure in patients with multiple myeloma treated with thalidomide. Leuk Lymphoma 2004;45(8):1711–2.

Aminoglycoside antibiotics

See also Amikacin, Gentamicin, Isepamycin, Kanamycin, Tobramycin

General Information

Eleven aminoglycosides have been, or are still, important in medical practice: amikacin (rINN), gentamicin (pINN), isepamicin (rINN), kanamycin (rINN), neomycin (rINN), netilmicin (rINN), paromomycin (rINN), sisomicin (rINN), streptomycin (rINN) and dihydrostreptomycin (rINN), and tobramycin (rINN). The following aminoglycosides are also covered in separate monographs: amikacin, gentamicin, isepamicin, kanamycin, and tobramycin.

Being chemically similar, the aminoglycosides have many features in common, in particular their mechanism of antibacterial action, a broad antibacterial spectrum, partial or complete cross-resistance, bactericidal action in a slightly alkaline environment, poor absorption from the gastrointestinal tract, elimination by glomerular filtration, nephrotoxicity, ototoxicity, a potential to cause neuromuscular blockade, and partial or complete cross-allergy (1). Aminoglycosides have a moderate capacity for diffusion into bone tissue (2).

The aminoglycosides have probably more than one mechanism of action on bacterial cells. They cause misreading of the RNA code and/or inhibition of the polymerization of amino acids.

All the aminoglycosides have similar patterns of adverse reactions, although there are important differences with regard to their frequency and severity (Table 1).

Strategies for minimizing aminoglycoside toxicity include early bedside detection of cochlear and vestibular dysfunction, which should lead to prompt withdrawal, use of short periods of treatment, dosing intervals of at least 12 hours, monitoring of serum concentrations, and awareness of relative contraindications, such as renal or hepatic dysfunction, old age, hearing impairment, and previous recent aminoglycoside exposure (3).

Observational studies

In an open, randomized, comparative study of the efficacy, safety, and tolerance of two different antibiotic regimens in the treatment of severe community-acquired or nosocomial pneumonia, 84 patients were analysed (4). Half were treated with co-amoxiclav (amoxicillin 2 g and clavulanic acid 200 mg) every 8 hours plus a single-dose of 3–6 mg/kg of an aminoglycoside (netilmicin or gentamicin), and half with piperacillin 4 g and tazobactam

Table 1 Relative adverse effects of aminoglycoside antibiotics on eighth nerve function, the nervous system, and neuromuscular function

Aminoglycoside[1]	Usual parenteral dose (mg/kg/day)[2]	Site of adverse effect[3]			
		Vestibular function	Cochlear function	Nervous system	Neuromuscular function
Amikacin	10–15	+	++	++	+
Gentamicin	3–6	++	++	++	+
Kanamycin	10–20	+	+++	++	+
Netilmicin	3–6	+	+	++	+
Sisomicin	2–4	++	++	++	+
Streptomycin	10–20	+++	+	+	++
Dihydrostreptomycin	10–20	+	+++	+++	++
Tobramycin	3–6	++	++	++	+

[1]Neomycin and paromomycin: only accidental absorption after topical or gastrointestinal use.
[2]In general, older patients require lower doses. Low doses are required during therapy of infectious endocarditis when an aminoglycoside is given with a penicillin. Monitoring serum concentrations should be considered, particularly when using dosing regimens that involve more than one dose a day and high doses for several days.
[3]The number of signs (+) indicates the relative clinical importance of each reaction.

500 mg every 8 hours. The patients were treated for between 48 hours and 21 days. Clinical cure was achieved in 65% of patients with co-amoxiclav/aminoglycoside and in 81% of patients with piperacillin/tazobactam. Cure or improvement was observed in 84 and 90% respectively. Treatment failures were recorded in 14 versus 7%. One patient in each group relapsed. There was only one fatal outcome in the piperacillin/tazobactam group compared with six in the co-amoxiclav/aminoglycoside group. The adverse event rate was non-significantly lower in the piperacillin/tazobactam group. In one patient given piperacillin/tazobactam, there were raised transaminases. In the co-amoxiclav/aminoglycoside group, acute renal insufficiency developed in two patients and possibly drug-related fever in one. Bacteriological efficacy was comparable (92% versus 96%). The authors concluded that piperacillin/tazobactam is highly efficacious in the treatment of severe pneumonia in hospitalized patients and compares favorably with the combination of co-amoxiclav plus an aminoglycoside.

General adverse effects

The main adverse reactions of aminoglycosides consist of kidney damage (often presenting as non-oliguric renal insufficiency) and ototoxicity, including vestibular and/or cochlear dysfunction. Neuromuscular transmission can be inhibited. Hypersensitivity reactions are most frequent after topical use, which should be avoided. Anaphylactic reactions can occur. Tumor-inducing effects have not been reported.

Pharmacoeconomics

The pharmacoeconomic impact of adverse effects of antimicrobial drugs is enormous. Antibacterial drug reactions account for about 25% of adverse drug reactions. The adverse effects profile of an antimicrobial agent can contribute significantly to its overall direct costs (monitoring

costs, prolonged hospitalization due to complications or treatment failures) and indirect costs (quality of life, loss of productivity, time spent by families and patients receiving medical care). In one study an adverse event in a hospitalized patient was associated on average with an excess of 1.9 days in the length of stay, extra costs of $US2262 (1990–93 values), and an almost two-fold increase in the risk of death. In the outpatient setting, adverse drug reactions result in 2–6% of hospitalizations, and most of them were thought to be avoidable if appropriate interventions had been taken. In a review, economic aspects of antibacterial therapy with aminoglycosides have been summarized and critically evaluated (5).

Organs and Systems

Cardiovascular

Anecdotal reports refer to tachycardia, electrocardiographic changes, hypotension, and even cardiac arrest (6). In practice, effects on the cardiovascular system are unlikely to be of any significance.

Respiratory

Severe respiratory depression due to neuromuscular blockade has been observed (7). Bronchospasm can occur as part of a hypersensitivity reaction.

Neuromuscular function

The aminoglycosides have a curare-like action, which can be antagonized by calcium ions and acetylcholinesterase inhibitors (8). The mechanisms include reduced release of acetylcholine prejunctionally and an interaction with the postjunctional acetylcholine receptor-channel complex. While neomycin interacts with the open state of the receptor, streptomycin blocks the receptor (9).

Aminoglycoside-induced neuromuscular blockade can be clinically relevant in patients with respiratory acidosis, in myasthenia gravis, and in other neuromuscular diseases. Severe illness, the simultaneous use of anesthetics, for example in the immediate postoperative phase, and application of the antibiotic to serosal surfaces are predisposing factors (10).

With regard to this effect, neomycin is the most potent member of the group. Several deaths and cases of severe respiratory depression due to neomycin have been reported (11). Severe clinical manifestations are rare in patients treated with aminoglycosides that are administered in low doses, such as gentamicin, netilmicin, and tobramycin. In some cases the paralysis was reversed by prostigmine.

Sensory systems

Eyes
Collagen corneal shields pre-soaked with antibiotics are used as a means of delivering drug to the cornea and anterior chamber of the eye. Gentamicin damages the primate retina, particularly by macular infarction, and amikacin can have a similar effect; a subconjunctival injection of tobramycin causes macular infarction (SEDA-19, 245). This potentially devastating consequence suggests that care must be exercised when contemplating instillation of aminoglycosides directly into the eye. Allergic contact dermatitis causing conjunctivitis and blepharitis has been reported with topical ophthalmic tobramycin (12).

Ears
Ototoxicity is a major adverse effect of aminoglycoside antibiotics (13). They all affect both vestibular and cochlear function, but different members of the family have different relative effects (Table 1).

Differences between different aminoglycosides
Ototoxicity due to amikacin is primarily cochlear; however, in comparisons with equipotent dosages, ototoxicity was of the same order as that caused by gentamicin (14–16).

In 40 patients tobramycin had little effect on audiometric thresholds, but produced a change in the amplitude of the distortion products, currently considered an objective method for rapidly evaluating the functional status of the cochlea (17). In one case, tobramycin caused bilateral high-frequency vestibular toxicity, which subsequently showed clinical and objective evidence of functional recovery (18).

In a quantitative assessment of vestibular hair cells and Scarpa's ganglion cells in 17 temporal bones from 10 individuals with aminoglycoside ototoxicity, streptomycin caused a significant loss of both type I and type II hair cells in all five vestibular sense organs (19). The vestibular ototoxic effects of kanamycin appeared to be similar to those of streptomycin, whereas neomycin did not cause loss of vestibular hair cells. There was no significant loss of Scarpa's ganglion cells.

Incidence
The incidence of ototoxicity due to aminoglycosides varies in different studies, depending on the type of patients treated, the methods used to monitor cochlear and vestibular function, and the aminoglycoside used (20). Clinically recognizable hearing loss and vestibular damage occur in about 2–4% of patients, but pure-tone audiometry, particularly at high frequencies, and electronystagmography show hearing loss and/or vestibular damage in up to 26 and 10% respectively, despite careful dosage adjustment (21). In patients with Pseudomonas endocarditis receiving prolonged high-dose gentamicin, auditory toxicity was found in 44% (22,23).

A review of nearly 10 000 adults suggested rates of 14% for amikacin, 8.6% for gentamicin, and 2.4% for netilmicin. Aminoglycoside toxicity is markedly lower in infants and children, with an incidence of 0–2%. A long duration of treatment and repeated courses or high cumulative doses appear to be critical for ototoxicity, which occurs in high frequency hearing beyond the range of normal speech.

There is a discrepancy between clinical observations, in which very few patients receiving aminoglycosides actually complain of hearing loss, and the reported incidences of ototoxicity in studies of audiometric thresholds. A major reason for this discrepancy relates to the fact that aminoglycosides cause high-frequency hearing loss well before they affect the speech frequency range in which they can be detected by the patient (21).

Gentamicin also damages the vestibular apparatus at a rate of 1.4–3.7%, resulting in vertigo and impaired balance. This effect is reversible in only about 50% from 1 week to 6 months after administration.

In a retrospective study in 81 men and 29 women, hearing loss of 15 decibels at two or more frequencies, or at least 20 decibels at at least one frequency, was found in 18% of patients treated with aminoglycosides (amikacin, kanamycin, and/or streptomycin) (24). In those treated with kanamycin the rate was 16%. Age, sex, treatment duration, total aminoglycoside dose, and first serum creatinine concentration were not associated with hearing loss.

Dose-relatedness
Hearing loss was attributed to repeated exposure to aminoglycosides in 12 of 70 patients with cystic fibrosis (one child) (25). There was a non-linear relation between the number of courses of therapy and the incidence of hearing loss. The severity of loss was not related to the number of courses. Assuming that the risk of hearing loss was independent of each course, the preliminary estimate of the risk was less than 2 per 100 courses.

Presentation
Clinically, cochlear ototoxicity is more frequent and easier to detect than vestibular toxicity; combined defects are relatively rare. Symptoms of cochlear damage include tinnitus, hearing loss, pressure, and sometimes pain in the ear. The manifestations of vestibular toxicity are

dizziness, vertigo, ataxia, and nystagmus. These are often overlooked in severely ill, bed-ridden patients.

Symptoms of ototoxicity can occur within 3–5 days of starting treatment, but most patients with severe damage have received prolonged courses of aminoglycosides. In some cases, hearing loss progresses after the administration of the causative drug has been interrupted. The ototoxicity is reversible in only about 50% of patients. Permanent deafness is often seen in patients with delayed onset of symptoms, progressive deterioration after withdrawal of treatment, and hearing loss of over 25 db (21).

There are interesting differences in the toxicity patterns of aminoglycosides in animals. Gentamicin and tobramycin affect the cochlear and vestibular systems to a similar extent, while amikacin, kanamycin, and neomycin preferentially damage the cochlear and streptomycin the vestibular system. Netilmicin appears to be the least toxic (26,27).

In man, differences in the ototoxic risks of the currently used aminoglycosides are difficult to evaluate (20). There have been no prospective comparisons of more than two drugs using the same criteria in similar patient populations. However, several controlled comparisons of two aminoglycosides are available and provide some information. A survey of 24 such trials showed the following mean frequencies of ototoxicity: gentamicin 7.7%, tobramycin 9.7%, amikacin 13.8%, netilmicin 2.3% (28). There was also a lower incidence of netilmicin-induced inner ear damage compared with tobramycin in two studies (29,30).

Of 20 patients with Dandy's syndrome, 15 had previously been treated with aminoglycosides (13 with gentamicin and 2 with streptomycin), of whom 10 had symptoms of pre-existing chronic nephrosis or transitory renal insufficiency. In all 13 patients who had gentamicin, peripheral vestibular function was destroyed or severely damaged, whereas there was no hearing loss (31).

Mechanism
The mechanism of ototoxicity by aminoglycosides is still not fully clarified. Most of the experimental data have been gained in the guinea-pig model, which seems to resemble man.

Animal studies
Traditionally, toxic damage is considered to be the consequence of drug accumulation in the inner ear fluids (32,33). After a period of reversible functional impairment, destruction of outer hair cells occurs in the basilar turn of the cochlear duct and proceeds to the apex. Similar changes are found in the hair cells of the vestibular system. Gentamicin was detected in the outer hair cells in the cochlea of animals receiving non-ototoxic doses of the drug and continued to increase for several days after withdrawal (34). This was followed by a third and very much slower phase of elimination (estimated half-life 6 months), and in the absence of ototoxicity gentamicin could still be detected in the outer hair cells 11 months after treatment. This may explain why patients who receive several courses of

aminoglycosides in a year may be more susceptible to ototoxicity, and suggests that the cumulative dose (and by implication the duration of therapy) is the more important determinant of cochlear damage. However, it has been questioned whether ototoxicity correlates with plasma perilymph or whole-tissue concentrations of aminoglycosides (33).

The effects of aminoglycosides on the medial efferent system have been assessed in awake guinea-pigs (35,36). The ensemble background activity and its suppression by contralateral acoustic stimulation was used as a tool to study the medial efferent system. A single intramuscular dose of gentamicin 150 mg/kg reduced or abolished the suppressive effect produced by activation of the olivocochlear system by contralateral low-level broadband noise stimulation. This effect was dose-dependent and could be demonstrated ipsilaterally on the compound action potential, otoacoustic emissions, and ensemble background activity of the eighth nerve. Long-term gentamicin treatment (60 mg/kg for 10 days) had no effect, at least before the development of ototoxicity. Single-dose intramuscular netilmicin 150 mg/kg displayed blocking properties similar to gentamicin, although less pronounced, while amikacin 750 mg/kg and neomycin 150 mg/kg had no effect. With tobramycin 150 mg/kg and streptomycin 400 mg/kg a decrease in suppression was usually associated with a reduction of the ensemble background activity measured without acoustic stimulation, which may be a fist sign of alteration to cochlear function. There was no correlation between specificity and degree of aminoglycoside ototoxicity and their action on the medial efferent system.

Possible mechanisms and preventive strategies have also been investigated in pigmented guinea-pigs (37–39). Animals that received alpha-lipoic acid (100 mg/kg/day), a powerful free radical scavenger, in combination with amikacin (450 mg/kg/day intramuscularly) had a less severe rise in compound action potential threshold than animals that received amikacin alone. In a similar study in pigmented guinea-pigs, the iron chelator deferoxamine (150 mg/kg bd for 14 days) produced a significant protective effect against ototoxicity induced by neomycin (100 mg/kg/day for 14 days). The spin trap alpha-phenyl-tert-butyl-nitrone also protected against acute ototopical aminoglycoside ototoxicity in guinea-pigs. These studies have provided further evidence for the hypothesis that aminoglycoside ototoxicity is mediated by the formation of an aminoglycoside-iron complex and reactive oxygen species.

Aminoglycoside-induced ototoxicity may be in part a process that involves the excitatory activation of cochlear NMDA receptors (40). In addition, the uncompetitive NMDA receptor antagonist dizocilpine attenuated the vestibular toxicity of streptomycin in a rat model, further stressing that excitotoxic mechanisms mediated by NMDA receptors also contribute to aminoglycoside-induced vestibular toxicity (41). In two studies in guinea-pigs, nitric oxide and free radicals, demonstrated by the beneficial effect of the antioxidant/free radical scavenger alpha-lipoic acid, have been suggested to be involved in aminoglycoside-induced ototoxicity (38,42). In contrast, insulin, transforming growth factor alpha, or retinoic acid may offer a protective potential against ototoxicity caused by aminoglycosides. However, they do not seem to promote cochlear hair cell repair (43). In guinea-pigs, brain-derived neurotrophic factor, neurotrophin-3, and the iron chelator and antioxidant 2-hydroxybenzoate (salicylate), at concentrations corresponding to anti-inflammatory concentrations in humans, attenuated aminoglycoside-induced ototoxicity (44,45). In rats, concanavalin A attenuated aminoglycoside-induced ototoxicity, and kanamycin increased the expression of the glutamate-aspartate transporter gene in the cochlea, which might play a role in the prevention of secondary deaths of spiral ganglion neurons (46,47). Finally, 4-methylcatechol, an inducer of nerve growth factor synthesis, enhanced spiral ganglion neuron survival after aminoglycoside treatment in mice (48).

In hatched chicks repeatedly injected with kanamycin, afferent innervation of the regenerated hair cells was related more to the recovery of hearing than efferent innervation (49).

In an animal model of ototoxicity, the most severe degeneration in the cristae ampullaris, utricle, and saccule was observed after administration of streptomycin. The severity of the vestibular damage in terms of magnitude was in the order streptomycin > gentamicin > amikacin > netilmicin (50).

Human studies

There is evidence that the site of ototoxic action is the mitochondrial ribosome (51,52). In some countries, such as China, aminoglycoside toxicity is a major cause of deafness. Susceptibility to ototoxicity in these populations appears to be transmitted by women, suggesting mitochondrial inheritance. In Chinese, Japanese, and Arab-Israeli pedigrees a common mutation was found. A point mutation in a highly conserved region of the mitochondrial 12S ribosomal RNA gene was common in all pedigrees with maternally inherited ototoxic deafness (51). A mutation at nucleotide 1555 has been reported to confer susceptibility to aminoglycoside antibiotics, and to cause non-syndromic sensorineural hearing loss. Outside these susceptible families, sporadic cases also have this mutation in increased frequency. In patients bearing this mitochondrial mutation hearing loss was observed after short-term exposure to isepamicin sulfate (53). These findings might create a molecular baseline for preventive screening of patients when aminoglycosides are to be used (SEDA-18, 1) (51).

Differences between sera from patients with resistance or susceptibility to aminoglycoside ototoxicity have been described in vitro (54). Sera from sensitive but not from resistant individuals metabolized aminoglycosides to cytotoxins, whereas no sera were cytotoxic when tested without the addition of aminoglycosides. This effect persisted for up to 1 year after aminoglycoside treatment.

Table 2 Factors that increase susceptibility to the adverse effects of aminoglycosides

Patient factors	Drug-related factors
Prior renal insufficiency	High temperature
Prior abnormal audiogram	Dose (blood concentration exceeding the usual target range)
Age (mainly older patients)	Total cumulative dose
Septicemia	Prolonged duration of therapy (2–3 weeks)
Dehydration	Prior aminoglycoside exposure

Susceptibility factors

Several factors predispose to ototoxic effects (Table 2). Drug-related toxicity is influenced by the quality of prescribing. Overdosage in patients with impaired renal function, unnecessary prolongation of treatment, and the concomitant administration of other potentially ototoxic agents should be avoided. The exact mechanism of increased toxicity in patients with septicemia and a high temperature is not clear; the possible relevance of additive damage by bacterial endotoxins has been discussed (55). Dehydration with hypovolemia is probably the main reason for the increased toxicity experienced when aminoglycosides are given with loop diuretics, but furosemide itself does not seem to be an independent risk factor (56,57). Attempts have been made in animals to protect against ototoxicity by antioxidant therapy (for example glutathione and vitamin C), as well as iron chelators and neurotrophins (58).

Hereditary deafness is a heterogeneous group of disorders, with different patterns of inheritance and due to a multitude of different genes (59,60). The first molecular defect described was the A1555G sequence change in the mitochondrial 12S ribosomal RNA gene. A description of two families from Italy and 19 families from Spain has now suggested that this mutation is not as rare as was initially thought (61,62). The A1555G mutation is important to diagnose, since hearing maternal relatives who are exposed to aminoglycosides may lose their hearing. This predisposition is stressed by the fact that 40 relatives in 12 Spanish families and one relative in an Italian family lost their hearing after aminoglycoside exposure. Since the mutation can easily be screened, any patient with idiopathic sensorineural hearing loss may be screened for this and possible other mutations.

In an Italian family of whom five family members became deaf after aminoglycoside exposure, the nucleotide 961 thymidine deletion associated with a varying number of inserted cytosines in the mitochondrial 12S ribosomal RNA gene was identified as a second pathogenic mutation that could predispose to aminoglycoside ototoxicity (63). Molecular analysis excluded the A1555G mutation in this family.

The A1555G mutation in the human mitochondrial 12S RNA, which has been associated with hearing loss after aminoglycoside administration (64) and has been implicated in maternally inherited hearing loss in the absence of aminoglycoside exposure in some families, can be identified by a simple and rapid method for large-scale screening that uses one-step multiplex allele-specific PCR (65).

Using lymphoblastoid cell lines derived from five deaf and five hearing individuals from an Arab-Israeli family carrying the A1555G mutation, the first direct evidence has been provided that the mitochondrial 12S rRNA carrying the A1555G mutation is the main target of the aminoglycosides (66). This suggests that they exert their detrimental effect through altering mitochondrial protein synthesis, which exacerbates the inherent defect caused by the mutation and reduces the overall translation rate below the minimal level required for normal cellular function.

A second pathogenic mutation that could predispose to aminoglycoside ototoxicity has been identified in an Italian family, of whom five members became deaf after aminoglycoside exposure (63). In the mitochondrial 12S ribosomal RNA gene, the deletion of nucleotide 961 thymidine was associated with a varying number of inserted cytosines. Transient evoked otoacoustic emission has been suggested as a sensitive measure for the early effects of aminoglycosides on the peripheral auditory system and may be useful as a tool for the prevention of permanent ototoxicity (67).

In 87 patients with tuberculosis or non-tuberculous mycobacterial infections randomized to receive intravenous streptomycin, kanamycin, or amikacin, 15 mg/kg/day or 25 mg/kg 3 times per week, the dose and the frequency of administration were not associated with the incidences of ototoxicity (hearing loss determined by audiography) or vestibular toxicity (determined by physical examination) (68). Ototoxicity, which occurred in 32 patients, was associated with older age and with a larger cumulative dose. Vestibular toxicity, which occurred in eight patients, usually resolved. Subjective changes in hearing or balance did not correlate with objective findings.

In a case-control study in 15 children under 33 weeks gestation with significant sensorineural hearing loss and 30 matched controls, the children with sensorineural hearing loss had longer periods of intubation, ventilation, oxygen treatment, and acidosis, and more frequent treatment with dopamine or furosemide (69). However, neither peak nor trough aminoglycoside concentrations, nor duration of jaundice or bilirubin concentration varied between the groups. At 12 months of age, seven of the children with sensorineural hearing loss had evidence of cerebral palsy compared with two of the 30 controls. Therefore, preterm children with sensorineural hearing loss required more intensive care in the perinatal period and developed more neurological complications than controls, and the co-existence of susceptibility factors for hearing loss may be more important than the individual factors themselves.

Diagnosis
To recognize auditory damage at an early stage and avoid severe irreversible toxicity, repeated tests of cochlear and vestibular function should be carried out in all patients needing prolonged aminoglycoside treatment. Pure-tone audiometry at 250–8000 Hz and electronystagmography with caloric testing are the standard methods. The first detectable audiometric changes usually occur in the high-tone range (over 4000 Hz) and then progress to lower frequencies. Hearing loss of more than 15 db is usually considered as evidence of toxicity. Brainstem auditory-evoked potentials have been recommended as a means of monitoring ototoxicity in uncooperative, comatose patients (70,71). This technique is time-consuming and requires some expertise, but may become a useful tool for detecting damage at an early stage. It also provides information on pre-existing changes, which is otherwise rarely available in intensive care patients (71).

Prevention
In rats, ototoxicity caused by gentamicin or tobramycin was ameliorated by melatonin, which did not interfere with the antibiotic action of the aminoglycosides (72). The free radical scavenging agent alpha-lipoic acid has previously been shown to protect against the cochlear adverse effects of systemically administered aminoglycoside antibiotics, and in a recent animal study it also prevented cochlear toxicity after the administration of neomycin 5% directly to the round window membrane over 7 days (73).

Loss of spiral ganglion neurons can be prevented by neurotrophin 3, whereas hair cell damage can be prevented by *N*-methyl-D-aspartate (NMDA) receptor antagonists. In an animal study, an NMDA receptor antagonist (MK801) protected against noise-induced excitotoxicity in the cochlea; in addition, combined treatment with neurotrophin 3 and MK801 had a potent effect in preserving both auditory physiology and morphology against aminoglycoside toxicity induced by amikacin (74).

Metabolism

Aminoglycosides can stimulate the formation of reactive oxygen species (free radicals) both in biological and cell-free systems (75,76).

Electrolyte balance

Aminoglycoside-induced proximal tubular dysfunction, which causes some manifestations of Fanconi's syndrome, including electrolyte abnormalities, is rare.

- A 72-year-old man was treated with ceftriaxone (2 g bd) and gentamicin (80 mg tds) for a severe urinary tract infection (77). On day 5 his serum potassium concentration was 3 mmol/l with a normal serum creatinine and urine examination. Despite treatment with oral potassium chloride plus a high potassium diet, his serum potassium fell to 2.3 mmol/l 4 days later, accompanied by inappropriate kaliuresis, hypouricemia with inappropriate uricosuria, and hypophosphatemia with inappropriate phosphaturia. There was no bicarbonate wasting, but there was proteinuria 1.2 g/day, with a predominance of low molecular weight proteins; in contrast, serum creatinine was normal and creatinine clearance was 78 ml/minute. The aminoglycoside was withdrawn with subsequent progressive improvement in renal proximal tubular function, which normalized 9 days later.

Mineral balance

Hypomagnesemia is a well-recognized adverse effect of treatment with various drugs, including aminoglycosides (SEDA-16, 279). Gentamicin-induced magnesium depletion is most likely to occur in older patients when large doses are used over long periods of time (78). Under these circumstances, serum concentrations and urinary electrolyte losses should be monitored.

Metal metabolism

Aminoglycosides can cause renal magnesium wasting and hypomagnesemia, usually associated with acute renal insufficiency. However, animal studies have shown frequent renal magnesium wasting, even in the absence of renal insufficiency and abnormalities of renal tubular morphology. In 24 patients with cystic fibrosis, treatment with amikacin plus ceftazidime for exacerbation of pulmonary symptoms by *Pseudomonas aeruginosa* resulted in mild hypomagnesemia due to renal magnesium wasting, even in the absence of a significant rise in circulating creatinine and urea concentrations (79).

In five healthy volunteers gentamicin 5 mg/kg caused immediate but transient renal calcium and magnesium wasting (80).

Reversible hypokalemic metabolic alkalosis and hypomagnesemia can occur with gentamicin, and routine monitoring has been recommended (81).

The results of an in vitro study on immortalized mouse distal convoluted tubule cells have suggested that aminoglycosides act through an extracellular polyvalent cation-sensing receptor and that they inhibit hormone-stimulated magnesium absorption in the distal convoluted tubule (82).

In children with cystic fibrosis, aminoglycosides can cause hypomagnesemia due to excessive renal loss (83).

Hematologic

In an in vitro study both gentamicin sulfate and netilmicin sulfate showed competitive inhibition of glucose-6-phosphate dehydrogenase from human erythrocytes, whereas streptomycin sulfate showed non-competitive inhibition (84).

Liver

Certain aminoglycosides affect liver function tests. Increases in alkaline phosphatase after gentamicin and tobramycin have been described.

Urinary tract

Non-oliguric renal insufficiency is a well-known nephrotoxic consequence of aminoglycosides, although reversible tubular damage in the absence of any change in renal function has occasionally been found. Two representative cases of reversible tubular damage due to prolonged aminoglycoside administration have been reported: a patient with a Fanconi-like syndrome of proximal tubular dysfunction and a patient with a syndrome of hypokalemic metabolic alkalosis associated with hypomagnesemia (85). In a survey of the use of antibiotics in a surgical service, aminoglycosides were given to 26 patients, of whom four developed nephrotoxicity (86).

In patients undergoing peritoneal dialysis, intraperitoneal or intravenous aminoglycosides can increase the rapidity of the fall in residual renal function. In an observational, non-randomized study, 72 patients on peritoneal dialysis were followed for 4 years (87). Patients who had been treated for peritonitis with intraperitoneal or intravenous aminoglycosides for more than 3 days had a more rapid fall in renal creatinine clearance and a more rapid fall in daily urine volume than patients without peritonitis or patients treated for peritonitis with antibiotics other than aminoglycosides. However, the use of other nephrotoxins or agents that may improve renal function was not quantified.

In a meta-analysis of randomized, controlled trials in children there was a significantly worse outcome in secondary nephrotoxicity with multiple daily doses of aminoglycosides (16%, 11 of 69 cases) versus once-daily dosing (4.4%, 3 of 69 cases) (88). There were no statistical significant differences in the outcomes of primary nephrotoxicity.

Incidence

Impairment of kidney function in different studies is highly variable. It depends on the study population and the definition of toxicity, and can range from a few percent to more than 30% in severely ill patients (89,90).

In a retrospective study, the rate of nephrotoxicity at the end of treatment with aminoglycosides was 7.5% (kanamycin 4.5%) (24). Patients who developed nephrotoxicity had a longer duration of treatment and received larger total doses.

Susceptibility factors

As in the case of ototoxicity, certain susceptibility factors have been identified for nephrotoxicity (91,92). Total dose, age, abnormal initial creatinine clearance, the 1-hour post-dose and trough aminoglycoside serum concentrations, duration of treatment, and co-existent liver disease were predictors of subsequent kidney damage. However, the clinical significance and utility of predictive nomograms based on such risk factors have been challenged (93). Other factors, such as dehydration, hypovolemia, potassium and magnesium depletion, and sepsis are also important. Gentamicin nephrotoxicity is increased in biliary obstruction (SEDA-20, 235). Nephrotoxicity is potentiated by ACE inhibitors, cephalosporins, ciclosporin, cisplatin, loop diuretics (secondary to volume depletion), methoxyflurane, non-steroidal anti-inflammatory drugs, and any basic amino acid such as lysine. Additional factors, such as previous aminoglycoside exposure, metabolic acidosis, female sex, and diabetes mellitus, are less clearly associated with renal damage (94,95).

In a prospective, non-interventional surveillance study of 249 patients receiving a once-daily aminoglycoside

(17% amikacin and 83% gentamicin), serum creatinine increased by more than 50% in 12%; none developed oliguric renal insufficiency (96). Renal damage correlated significantly with: a high aminoglycoside trough concentration (over 1.1 µg/ml); a hemoglobin concentration below 10 g/dl; hospital admission for more than 7 days before aminoglycoside treatment; and aminoglycoside treatment for more than 11 days.

Mechanism

The mechanism of nephrotoxicity has been studied in vitro, in animals, and in man. The data suggest that accumulation of aminoglycosides in the renal cortex is an important reason for functional damage. In addition, aminoglycoside-induced alterations of glomerular ultrastructure have been described (97). Tissue uptake is predominantly mediated by proximal tubular reabsorption of the filtered drug. The degree of renal injury caused by an aminoglycoside correlates with the amount of drug that accumulates in the renal cortex. However, intrinsic toxicity differs among the aminoglycosides. Endotoxins can increase intracortical accumulation (98). After the drug binds to an amino acid receptor site, there is pinocytotic entry into the tubular cells. Once the drug is within the cell, it persists there for a long time and is liberated only slowly, with a half-life of several days (99,100). Direct toxicity to tubular cells is mainly explained by inhibition of lysosomal activity, with accumulation of lamellar myeloid bodies consisting of phospholipids. Inhibition of lysosomal phospholipids can also be shown in purified liposomes in vitro (101). In renal biopsy specimens, vacuolization of the proximal tubular epithelium, clumped nuclear chromatin, and swollen mitochondria are seen. Patchy tubular cell necrosis, desquamation, and luminal obstruction are found at later stages.

The uptake of aminoglycosides into proximal renal tubular epithelial cells is limited to the luminal cell border and is saturable. Less frequent administration of doses larger than needed for saturation of this uptake may therefore reduce drug accumulation in the renal cortex (102). In vitro and in vivo data have provided evidence that partially reduced oxygen metabolites (superoxide anion, hydrogen peroxide, and hydroxyl radical), which are generated by renal cortical mitochondria, are important mediators of aminoglycoside-induced acute renal insufficiency (103).

A lipopeptide, daptomycin, which has bactericidal activity against Gram-positive bacteria by inhibition of the synthesis of lipoteichonic acid, prevents tobramycin-induced nephrotoxicity in rats. Fourier transformed infrared spectroscopy has been used to monitor the hydrolysis of phosphatidylcholine phospholipase A_2 in the presence of different aminoglycosides and/or daptomycin (104). Among the various aminoglycosides investigated there were major differences, directly related to chemical structure. The number of charges, the size, and the hydrophobicity of the substituents of an aminoglycoside determined the influence on the lag phase, on the maximal rate, and on the final extent of hydrolysis. Daptomycin alone eliminated the initial latency period,

and reduced the maximal extent of hydrolysis. When daptomycin was combined with any of the aminoglycosides, the latency period also disappeared, but the phospholipase activity was higher than with the lipopeptide alone. The strongest activation of phospholipase A_2 activation was observed when daptomycin was combined with gentamicin.

Stress proteins seem to be actively involved in aminoglycoside-induced renal damage. In a study in Wistar rats, subcutaneous gentamicin caused tubular necrosis, followed by tubular regenerative changes and interstitial fibrosis (105). Both the regenerated and phenotypically altered tubulointerstitial cells were found to express heat-shock protein 47 in and around the fibrosis. Increased shedding of tubular membrane components, followed by rapid inductive repair processes with overshoot protein synthesis, can be detected by analysis of tubular marker protein in the urine of rats after administration of aminoglycosides (106).

The role of megalin, a giant endocytic receptor abundantly expressed at the apical membrane of renal proximal tubules, has been discussed as an important factor in the binding and endocytosis of aminoglycosides in proximal tubular cells (107). The authors suggested that inhibition of the uptake process is the most promising approach to prevent aminoglycoside-induced nephrotoxicity.

Prevention

In a prospective, randomized, double-blind study of once-daily versus twice-daily aminoglycoside therapy in 123 patients (of whom 83 received treatment for over 72 hours), once-daily administration of aminoglycosides had a predictably lower probability of causing nephrotoxicity than twice-daily administration. In a Kaplan–Meier analysis, once-daily dosing provided a longer time of administration until the threshold for nephrotoxicity was met. The risk of nephrotoxicity was modulated by the daily AUC for aminoglycoside exposure, and the concurrent use of vancomycin significantly increased the probability of nephrotoxicity and shortened the time to its occurrence (102). In another study, once-daily aminoglycoside in 200 patients was prospectively compared with a retrospectively evaluated group of 100 patients treated with individualized traditional dosing. Patients with a baseline creatinine clearance below 40 ml/minute, meningitis, burns, spinal cord injury, endocarditis, or enterococcal infection were excluded. Clinical and microbial outcome was similar in the two groups. Nephrotoxicity occurred in 7.5% of the patients treated with once-daily aminoglycosides compared with 14% of the others. The cumulative AUC was significantly larger in patients on traditional dosing. Minimum serum concentrations, length of aminoglycoside therapy, and AUC over 24 hours were related to nephrotoxicity. Whereas vancomycin and concurrent nephrotoxic agents were independently related to toxicity, sex and age were not (108).

Careful tailoring of the dose can prevent nephrotoxicity. In 89 critically ill patients with a creatinine clearance over 30 ml/minute who were treated with gentamicin or tobramycin 7 mg/kg/day independent of renal function, with subsequent doses chosen on the basis of the pharmacokinetics of the first dose, signs of renal impairment occurred in 14%; in all survivors renal function recovered completely and hemofiltration was not needed (109).

Presentation and diagnosis

Nephrotoxicity can present as acute tubular necrosis or, more commonly, as gradually evolving non-oliguric renal failure. The time-course of toxicity is variable, but it usually develops only after several days of treatment. Early diagnosis is difficult, since there can be a reduction in glomerular filtration before a significant rise in serum creatinine concentration occurs (110). An increased number of casts in the urinary sediment can also precede the increase in serum creatinine (111). Measurement of phospholipids and urinary enzymes, such as $beta_2$-microglobulin or alanine aminopeptidase, has been proposed as a means of detecting early toxicity (112–114). However, data on these enzymes are not very useful, since they can be increased for various other reasons and raised concentrations do not reliably predict pending renal toxicity. Fortunately, recovery of renal function nearly always follows the withdrawal of aminoglycosides, although serum creatinine concentrations can continue to rise for several days after the last dose.

There is an increase in the urinary output of tubular marker proteins after aminoglycoside administration (106). Determination of N-acetyl-beta-D-glucosaminidase activity in the urine may be used as a screening test to facilitate early detection of the nephrotoxic effect of aminoglycosides (115). Urine chemiluminescence may aid in the detection of neonatal aminoglycoside-induced nephropathy (116).

Relative potency of different aminoglycosides

The nephrotoxicity of the various aminoglycosides has been compared in many animal experiments. The order of relative nephrotoxicity is as follows: neomycin > gentamicin > tobramycin > amikacin > netilmicin (117,118). However, in man conclusive data on the relative toxicity of the various aminoglycosides is still lacking. An analysis of 24 controlled trials showed the following average rates for nephrotoxicity: gentamicin 11%; tobramycin 11.5%; amikacin 8.5%; and netilmicin 2.8% (28). In contrast to this survey, direct comparison in similar patient groups showed no significant differences between the various agents in most trials (14,15,119–123). In fact, the relative advantage of lower nephrotoxicity rates observed with netilmicin in some studies may be limited to administration of low doses. In a prospective study there was significantly lower nephrotoxicity with amikacin 15 mg/kg/day (4% toxicity) compared with netilmicin 7 mg/kg/day (12%) (124). One prospective trial showed a significant advantage of tobramycin over gentamicin (125). However, these findings were not subsequently confirmed (126). Nephrotoxicity is a serious risk with all the currently available aminoglycoside antibiotics, and no drug in this series can be regarded as safe.

Skin

It is generally accepted that antibiotics that are important for systemic use should not be administered topically. This rule applies particularly to the aminoglycoside antibiotics. Even though neomycin and streptomycin are no longer often used systemically, the frequency of sensitization after topical administration of these drugs is particularly high.

Contact dermatitis due to arbekacin has been reported (127). The risk of sensitization by topically administered gentamicin seems to be smaller. Nevertheless, and in order to avoid resistance, topical use should be restricted to life-endangering thermal burns and to severe skin infections in which strains of *Pseudomonas* resistant to other antibiotics are involved.

Immunologic

In a prospective study of the results of skin patch testing in 149 patients who were scheduled for ear surgery, 14% of the patients had a positive skin reaction to one of the aminoglycosides (13% for gentamicin, 13% for neomycin) (128). In 16% of the patients with chronic otitis media and 6.7% of the patients with otosclerosis there was allergy to one of the aminoglycosides commonly found in antibiotic ear-drops. Patients who had previously received more than five courses of antibiotic ear-drops had a greater tendency to develop allergy to the aminoglycosides (35%).

Cross-reactivity between aminoglycoside antibiotics has long been known. Aminoglycoside antibiotics can be categorized in two groups, depending on the aminocycolitol nucleus: streptidine (streptomycin) and deoxystreptamine (neomycin, kanamycin, gentamicin, paromomycin, spectinomycin, and tobramycin). Another antigenic determinant is neosamine, a diamino sugar present in neomycin and, with minor changes, also in paromomycin, kanamycin, tobramycin, amikacin, and isepamicin. Streptomycin shares no common antigenic structures with the other aminoglycoside antibiotics, and cross-sensitivity with streptomycin has not been reported. Acute contact dermatitis was described in a 30-year-old man after rechallenge with gentamicin 80 mg; a patch test was positive for gentamicin, neomycin, and amikacin (129).

Occasional cases of anaphylactic shock have been reported, most of which have been due to streptomycin; other aminoglycosides have rarely been implicated, such as gentamicin (130,131) and neomycin (132).

Long-Term Effects

Drug resistance

During treatment of infections with either Gram-negative or Gram-positive bacteria, resistant subpopulations emerge, unless the peak to MIC ratio is high enough to reduce the bacterial inoculum drastically within a few hours (133). Similarly, rapid emergence of resistant subpopulations has been reported during aminoglycoside treatment in neutropenic animals. The virulence and clinical relevance of the relatively slow-growing resistant subpopulations (small colony variants) have been documented in both animal and clinical studies (134,135). Resistant subpopulations can be detected only by direct plating of specimens on aminoglycoside-containing agar plates, since resistance may be lost within one subculture. Combination therapy of an aminoglycoside plus a beta-lactam antibiotic has been successfully used to prevent selection of resistant subpopulations (136).

In a prospective surveillance study in France, the development of resistance to aminoglycosides and fluoroquinolones among Gram negative bacilli was assessed in 51 patients with a second infection due to Gram negative organisms at least 8 days after a first Gram negative infection (137). Treatment of the first infection with a beta-lactam antibiotic plus an aminoglycoside significantly reduced the susceptibility to amikacin (the most prescribed antibiotic). When the first infection was treated with a beta-lactam plus a fluoroquinolone, susceptibility to ciprofloxacin, pefloxacin, netilmicin, and tobramycin, but not gentamicin and amikacin, fell significantly.

A multidrug efflux system that appears to be a major contributor to intrinsic high-level resistance to aminoglycosides and macrolides has been identified in *Burkholderia pseudomallei* (138).

Out of 1102 consecutive clinical Gram negative blood isolates from Belgium and Luxembourg, 157 were "not susceptible" to aminoglycosides (139). The resistance levels were 5.9% for gentamicin, 7.7% for tobramycin, 7.5% for netilmicin, 2.8% for amikacin, and 1.2% for isepamicin. In a large European multinational study on 7057 Gram negative bacterial isolates, resistance levels were 0.4–3% for amikacin, 2–13% for gentamicin, and 2.5–15% for tobramycin among Enterobacteriaceae; 75% of *Staphylococcus aureus* isolates, but only 21% of enterococcal strains were susceptible to gentamicin (140). In a study from Spain, 9% of 1014 clinical *P. aeruginosa* isolates were resistant to amikacin or tobramycin (141). In the same country, resistance to amikacin rose from 21% to 84% and to tobramycin from 33% to 72% between 1991 and 1996 (142).

In a leading cancer center in Houston, 24% of 758 Gram negative clinical isolates were resistant to tobramycin and 12% were resistant to amikacin (143). In 3144 bacterial isolates causing urinary tract infections in Chile, 74% were identified as *Escherichia coli*; 4.2% of these strains were resistant to gentamicin, and 1.3% were resistant to amikacin (144). In contrast, the resistance levels were 30% and 17% respectively, in the other enterobacterial strains. In Brazil, all isolates of methicillin-resistant *S. aureus* were also resistant to gentamicin, amikacin, kanamycin, neomycin, and tobramycin (145), and 97% of such strains from Spain were resistant to tobramycin (146).

In mice, the bacterial neomycin phosphotransferase gene, which confers resistance to aminoglycoside antibiotics, prevented the loss of hair cells after neomycin treatment (147).

In in vitro susceptibility studies on 99 clinical isolates of *S. aureus*, 68 of 73 strains of methicillin-resistant *S. aureus* (MRSA) and two of 26 strains of methicillin-susceptible *S. aureus* were gentamicin-resistant (148). However, the combination of arbekacin plus vancomycin produced synergistic killing against 12 of 13 gentamicin-resistant MRSA isolates. Combinations of meropenem and aminoglycosides may be effective against strains of *P. aeruginosa* that are resistant to meropenem at clinically relevant concentrations; synergistic effects were observed in combinations that included arbekacin or amikacin (149).

Second-Generation Effects

Teratogenicity

During pregnancy the aminoglycosides cross the placenta and might theoretically be expected to cause otological and perhaps nephrological damage in the fetus. However, no proven cases of intrauterine damage by gentamicin and tobramycin have been recorded. In 135 mother/child pairs exposed to streptomycin during the first 4 months of pregnancy, there was no increase in the risk of any malformation (150). However, streptomycin and dihydrostreptomycin have caused severe otological damage (151).

Using the population-based dataset of the Hungarian Case-Control Surveillance of Congenital Abnormalities (1980–96), which includes 38 151 pregnant women who had newborn infants without any defects and 22 865 pregnant women who had fetuses or newborns with congenital abnormalities, no teratogenic risk of parenteral gentamicin, streptomycin, tobramycin, or oral neomycin was discovered when restricted to structural developmental disturbances (152).

Susceptibility Factors

Age

In a prospective study on the prevalence of hearing impairment in a neonatal intensive care unit population (a total of 942 neonates were screened), aminoglycoside administration did not seem to be an important risk factor for communication-related hearing impairment (153). In almost all cases, another factor was the more probable cause of the hearing loss (dysmorphism, prenatal rubella or cytomegaly, a positive family history of hearing loss, and severe perinatal and postnatal complications).

Other features of the patient

The main susceptibility factors are summarized in Table 2. The most important factors are renal insufficiency, high serum drug concentrations, and a long duration of treatment (55).

Drug Administration

Drug dosage regimens

Twice-daily or thrice-daily administration was for a long time standard in aminoglycoside therapy of systemic bacterial infections in patients with normal renal function. However, in vitro, in vivo, and clinical studies have suggested that once-daily dosing of the same total daily dose might be more beneficial with respect to both efficacy and toxicity (SEDA-21, 265; 154–157).

Less frequent administration of higher doses results in higher peak concentrations. Owing to the pronounced concentration dependence of bacterial killing, higher peaks potentiate the efficacy of the aminoglycosides. The importance of a high ratio of peak concentration to the minimal inhibitory concentration (MIC) has been shown in vitro for both bactericidal activity and prevention of the emergence of resistance (158). Clinically the predictive value of the peak to MIC ratio has been documented for aminoglycoside therapy of Gram-negative bacteremia (159). Although once-daily dosing can result in prolonged periods of subinhibitory concentrations, bacterial regrowth does not occur immediately after the aminoglycoside concentration drops below the MIC. The term "post-antibiotic effect" has been suggested to describe this hit-and-run effect, in which there is persistent suppression of bacterial regrowth after cessation of exposure of bacteria to an active antibiotic. This phenomenon has been observed both in vitro and in vivo (160).

Various in vivo models have been used to study the effect of the dosage regimen on aminoglycoside nephrotoxicity and ototoxicity. Renal uptake of aminoglycosides is not proportional to serum concentration, because of saturation at the high concentrations achieved with once-daily regimens. The degree of renal injury increases the more often the aminoglycoside is given in humans, dogs, rabbits, guinea-pigs, and rats, as long as the total daily dose is kept constant (161,162). Early auditory alterations also occur more often, or to a greater extent, the more often the aminoglycoside is given.

Once-daily regimens have been compared with multiple-daily regimens in at least 27 randomized clinical trials and eight meta-analyses and in a mega-analysis of meta-analyses (163). In all the studies, extended-interval dosing was at least as efficacious as multiple-daily dosing, and may have been slightly better; it was no more toxic, and may have been less nephrotoxic. Extended dosing is also less expensive than multiple-daily dosing and markedly reduces costs. An analysis of 24 randomized clinical trials of amikacin, gentamicin, and netilmicin, including 3181 patients, showed superior results with once-daily regimens with respect to clinical efficacy (90 versus 85%) and bacteriological efficacy (89 versus 83%) (156). There were no statistically significant differences for toxicity. Nevertheless, both nephrotoxicity and ototoxicity occurred less often during once-daily dosing (4.5 versus 5.5% and 4.2 versus 5.8% respectively). Finally, once-daily dosing is more economical, since less nursing time and infusion material are required and drug monitoring can be reduced. In conclusion, amikacin, gentamicin, and netilmicin can be given once a day.

In a randomized trial in 249 patients with suspected or proven serious infections, the safety and efficacy of gentamicin once-daily compared with three-times-a-day was assessed in 175 patients who were treated with ticarcillin–clavulanate combined with gentamicin once-daily or three times daily or with ticarcillin–clavulanate alone (164).

The achievement of protocol-defined peak serum gentamicin concentrations was required for evaluability. There were no significant differences between treatment regimens with respect to clinical or microbiological efficacy; the incidence of nephrotoxicity was similar in the three groups. In a post-hoc analysis, renal function was better preserved in those treated with gentamicin once a day plus ticarcillin + clavulanate than ticarcillin + clavulanate only.

Once-daily amikacin was as effective and safe as twice-daily dosing in a prospective randomized study in 142 adults with systemic infections (165).

In 43 patients, once-daily tobramycin (4 mg/kg/day) was at least as effective and was no more and possibly even less toxic than a twice-daily regimen (166).

The pharmacokinetics of once-daily intravenous tobramycin have been investigated in seven children with cystic fibrosis (167). All responded well. There was one case of transient ototoxicity but no nephrotoxicity.

In a prospective study, only increasing duration of once-daily aminoglycoside therapy was recognized as risk factor for toxicity in 88 patients aged 70 years and over (168).

In controlled trials of once-daily aminoglycoside regimens patients with endocarditis have often been excluded, and therefore few clinical data are available on this topic. Based on pharmacokinetic data in experimental endocarditis and integrative computer modeling, it has been hypothesized that drug penetration into vegetations would be enhanced by once-daily aminoglycoside administration, resulting in a long residence time in the vegetations and a beneficial effect in providing synergistic bactericidal action in combination with another antibiotic (169). With the exception of two early studies using enterococci, results from animal studies have been promising in once-daily aminoglycoside treatment of endocarditis. Only two human studies of once-daily aminoglycoside therapy in streptococcal endocarditis were identified. The regimen was efficacious and safe. The authors concluded that there appeared to be sufficient promising evidence to justify large trials to further investigate once-daily aminoglycoside in the treatment of endocarditis due to viridans streptococci, whereas more preliminary investigations were needed for enterococcal endocarditis.

Once-daily dosing regimens of aminoglycosides are routinely used in critically ill patients with trauma, although there is a marked variability in pharmacokinetics in these patients, eventually leading to prolonged drug-free intervals, and individualized dosing on the basis of at least two serum aminoglycoside concentrations can be recommended when once-daily dosing regimens are chosen (170).

Not all populations of patients have been included or extensively studied in published trials. Therefore, conventional multiple-daily aminoglycoside dosing with individualized monitoring should be used in neonates and children, in patients with moderate to severe renal insufficiency (creatinine clearance below 40 ml/minute), serious burns (over 20% of body surface area), ascites, severe sepsis, endocarditis, and mycobacterial disease, in pregnancy, in patients on dialysis, in invasive *P. aeruginosa* infection in neutropenic patients, and with concomitant administration of other nephrotoxic agents (for example amphotericin, cisplatin, radiocontrast agents, and NSAIDs). Extended-interval aminoglycoside dosing may be safe and efficacious in patients with mild to moderate renal insufficiency (over 40 ml/minute) and febrile neutropenia, especially where the prevalence of *P. aeruginosa* is low. In other serious Gram-negative infections warranting aminoglycoside treatment, extended-interval dosing is strongly suggested (171).

Drug administration route

Intraocular

Aminoglycosides are often used in ophthalmology to treat or to prevent bacterial infections, and they can be toxic to retinal structures if high concentrations are reached in the vitreous. Retinal ultrastructure was examined at various intervals after a single intravitreal injection of 100–4000 µg of gentamicin into rabbit eyes. Three days after injection of 100–500 µg, numerous abnormal lamellar lysosomal inclusions were observed in the retinal pigment epithelium and in macrophages in the subretinal These changes were typical of drug-induced lipid storage and were comparable to inclusions reported in kidney and other tissues as manifestations of gentamicin toxicity. One week after similar injections, focal areas of retinal pigment epithelium necrosis and hyperplasia with disruption of outer segments appeared, but the inner segments and the inner retina were intact. Doses of 800–4000 µg produced a combined picture of lipidosis of the retinal pigment epithelium and macrophages within the first 3 days, with increasing superimposed inner retinal necrosis (172).

Intratympanic

Intratympanic gentamicin therapy has gained some popularity in the treatment of vertigo associated with Menière's disease, as it offers some advantages over traditional surgical treatment. However, although the vestibulotoxic effect of gentamicin is well documented, there is no general agreement about the dose needed to control attacks of vertigo without affecting hearing. In 27 patients treated with small doses of gentamicin delivered via a microcatheter in the round window niche and administered by an electronic micropump, vertigo was effectively controlled; however, the negative effect on hearing was unacceptable (173,174).

Intraperitoneal

Bolus intraperitoneal gentamicin or tobramycin (5 mg/kg ideal body weight) is safe, achieves therapeutic blood concentrations for extended intervals, causes no clinical ototoxicity or vestibular toxicity, is cost-effective, and is convenient for patients and nurses (175).

Drug overdose

Although aminoglycoside antibiotics are dialysable, peritoneal dialysis may not remove aminoglycosides from the blood after overdosage (176). However, hemodialysis is effective (177). In one study in eight patients hemodiafiltration removed more netilmicin than conventional hemodialysis (178).

Drug–Drug Interactions

Amphotericin

Amphotericin prolongs the half-life of aminoglycoside antibiotics (217).

Atracurium dibesilate

Another interaction that has been reported not to occur in man is potentiation by the aminoglycoside antibiotics, gentamicin and tobramycin (218). In animals, however, gentamicin was found to enhance atracurium blockade (219), so further investigation is required to clarify this point.

Cephalosporins

There are many reports of acute renal insufficiency from combined treatment with gentamicin (or another aminoglycoside) and one of the cephalosporins (179–181). The potential nephrotoxic effect of the combinations seems to be related mainly to the nephrotoxic effect of the aminoglycosides. In contrast, there is some evidence, both experimental (182) and clinical (183), that ticarcillin may attenuate the renal toxicity of the aminoglycosides.

Cisplatin/carboplatin

The nephrotoxic and ototoxic effects of cisplatin and carboplatin can be potentiated by concurrent administration of aminoglycosides, as shown in animals (184).

Etacrynic acid

Etacrynic acid potentiates aminoglycoside ototoxicity by facilitating the entry of the antibiotics from the systemic circulation into the endolymph (220). Animal evidence suggests that this effect may be potentiated by glutathione depletion (221). Conversely, neomycin can enhance the penetration of etacrynic acid into the inner ear (222).

Furosemide

Furosemide increases the ototoxic risks of aminoglycoside antibiotics (223,224) by reducing their clearance by about 35% (225); permanent deafness has resulted from the use of this combination.

- A 60-year-old woman developed moderately severe sensorineural hearing loss bilaterally after receiving five doses of gentamicin and one of furosemide 20 mg (226).

The cumulative dose and duration of aminoglycoside therapy are more important than serum concentrations in the development of gentamicin ototoxicity, except when interacting medications such as furosemide are co-administered.

Hippocastanaceae

Beta-aescin can precipitate renal insufficiency when combined with aminoglycoside antibiotics (SEDA-3, 181) (SEDA-9, 190).

Ibuprofen

High-dose ibuprofen can slow the progression of lung disease in patients with cystic fibrosis and is usually well tolerated (SEDA-20, 93). However, transient renal insufficiency developed in four children with cystic fibrosis who were taking maintenance ibuprofen when an intravenous aminoglycoside was added to their regimen to treat an exacerbation of lung disease (227). Ibuprofen should probably be stopped during intravenous aminoglycoside therapy.

Loop diuretics

Among the agents that promote the nephrotoxic effects of the aminoglycosides, the loop diuretics furosemide and etacrynic acid are often mentioned. However, this interaction is by no means clearly established (26). These agents are not nephrotoxic in themselves. This supposed interaction may only be a consequence of sodium and volume depletion. Other types of diuretics, such as mannitol, hydrochlorothiazide, and acetazolamide, do not produce this interaction (26).

- A 60-year-old white woman developed ototoxicity after only 5 days of gentamicin therapy (500 mg, 6.8 mg/kg/day) and one dose of furosemide 20 mg (185).

Loop diuretics greatly potentiate the cochleotoxic effects of aminoglycosides (186). In pigmented guinea-pigs the effects of high-dose topical (10 μl of a 100 mg/ml solution directly on to the round window) or single-dose systemic (100 mg/kg) gentamicin and intracardiac administration of the loop diuretic etacrynic acid (40 mg/kg) on cochlear function have been studied (187). Compound action potentials were elicited at 8 kHz. All animals that received etacrynic acid had an immediate and profound rise in hearing threshold, irrespective of the method of gentamicin administration. The maximum threshold shift occurred within 30 minutes. Animals that received topical gentamicin recovered after etacrynic acid treatment; by day 20 the mean threshold shift was 7 dB. This group did not differ statistically from animals that received etacrynic acid alone. In contrast, animals that received systemic gentamicin initially recovered within 2 hours after etacrynic acid, but subsequently deteriorated over the next 24 hours. The mean threshold shift was 70 dB at day 20. Animals that received topical gentamicin had a temporary shift that resolved within 24–48 hours; by day 20, the mean threshold shift was 7 dB. Animals treated with systemic gentamicin alone did not have hearing loss. This study suggests that the potentiating effect of etacrynic acid on aminoglycoside ototoxicity is only after systematic and not topical aminoglycoside administration. This may be due to an etacrynic acid-induced increase in leakiness of the stria vascularis, thereby facilitating diffusion of aminoglycosides from the systemic circulation into the endolymphatic fluid.

Neuromuscular blocking drugs

The aminoglycosides have a curare-like action, which can be antagonized by calcium ions and acetylcholinesterase inhibitors (8). In patients who require general anesthesia, the effect of muscle relaxants, such as D-tubocurarine, pancuronium, and suxamethonium, can be potentiated by aminoglycosides (188).

Penicillins

There is an in vitro interaction between aminoglycoside antibiotics and carbenicillin or ticarcillin, leading to a significant loss of aminoglycoside antibacterial activity if these antibiotics are mixed in the same infusion bottle (189). The extent of inactivation depends on the penicillin concentration, the contact time, and the temperature. Azlocillin and mezlocillin inactivate aminoglycosides in a similar manner to that described for carbenicillin (190,191). Aminoglycosides should not be mixed with penicillins or cephalosporins in the same infusion bottle.

However, the clinical significance of the presence of both types of antibiotics in the patient is debatable. In patients who received a combination of gentamicin and carbenicillin the measured serum gentamicin concentrations were lower than the pharmacokinetically predicted values (192). This interaction may be especially important in patients with severe renal impairment, in whom long in vivo incubation of these drug combinations takes place before supplemental doses of the aminoglycoside drugs are given.

Vancomycin

Combined use of vancomycin and an aminoglycoside can increase the risk of toxicity (193–195).

Vancomycin is more nephrotoxic than teicoplanin when it is combined with aminoglycosides (228).

Monitoring Therapy

Serum concentrations have often been monitored during multiple-daily dosing of aminoglycosides, particularly when high doses are used and also during prolonged therapy. The main goals of individual dosing and monitoring are to reach high bactericidal drug concentrations and to avoid drug accumulation in the serum, to minimize the risk of toxicity. Since both goals are much more likely to be met with once-daily dosing it may be feasible to reduce monitoring efforts during such dosing regimens. However, more clinical experience needs to be accumulated to establish solid guidelines for monitoring once-daily dosing regimens. Although more clinicians are nowadays switching to once-daily dosing of aminoglycosides, some continue to use multiple-daily regimens. Therefore, two separate concepts of aminoglycoside monitoring have to be considered, depending on the dosing schedule.

In a retrospective case-control study, 2405 patients received aminoglycosides in doses that were decided either by individualized pharmacokinetic monitoring or by the physician (196). Those who received individualized pharmacokinetic monitoring were significantly less likely to develop aminoglycoside-associated nephrotoxicity. Women were also less likely to develop nephrotoxicity. Age 50 years and above, high initial aminoglycoside trough concentration, long duration of therapy, and concurrent therapy with piperacillin, clindamycin, or vancomycin increased the risk.

Monitoring multiple-daily dosing regimens

Peak serum concentrations of gentamicin and tobramycin over 5–7 µg/ml and of amikacin over 20–28 µg/ml are associated with improved survival in patients with septicemia and pneumonia caused by Gram-negative bacteria (197,198). On the other hand, excessive peak concentrations (over 10–12 µg/ml) and trough concentrations (over 2 µg/ml) of gentamicin and tobramycin increase the risk of ototoxicity and nephrotoxicity (199). Dosage requirements to obtain aminoglycoside concentrations in the target range can differ considerably, even in patients with normal renal function (200). Owing to the great individual variability in pharmacokinetics, dosage adjustments with the commonly used nomograms often result in suboptimal or potentially toxic aminoglycoside concentrations (201). An individualized treatment based on serum concentration monitoring is therefore necessary to achieve maximum bactericidal efficacy without a concomitant high risk of adverse reactions. Measurements of peak and trough serum concentrations should be carried out during the first 24–48 hours in all patients and repeated after 3–5 days to detect any tendency to abnormal drug accumulation. However, one must realize that close monitoring alone cannot completely eliminate the danger of ototoxicity and nephrotoxicity, since aminoglycoside drug concentrations progressively increase in renal tissue and in the inner ear with repeated administration, even if optimum serum concentrations are maintained.

The relation between serum concentration of aminoglycosides and the two clinically important adverse effects, ototoxicity and nephrotoxicity, has been debated for many years. Whereas some authors have found a definite relation between the frequency of adverse effects and serum concentrations, others have not (199). This controversy can be partly explained by the pharmacokinetic behavior common to all aminoglycosides, which leads to drug accumulation in deep compartments, particularly in the renal cortex. Assuming that the extent of accumulation is a factor that relates to the frequency of adverse effects, it is not surprising to find a correlation between serum concentrations and toxicity in one group of patients but not in another. The amount accumulated depends not only on the dosing schedule and the serum concentration achieved during treatment, but also on the duration of drug administration. The same concentration in the same patient can be associated with a significantly different amount of drug in the body, depending on whether it was sampled during the second or the tenth day after the beginning of treatment.

In a survey of aminoglycoside treatment in 2022 patients in Saudi Arabia, 8.8%, 18%, and 12% had trough concentrations considered toxic for amikacin, gentamicin, and tobramycin respectively, whereas there were peak serum drug concentrations in the subtherapeutic range in 53%, 50%, and 57% respectively (202). Toxic concentrations were noticed mainly in patients aged over 60 years and in patients in the intensive care unit, coronary care unit, and burn unit.

Monitoring once-daily dosing regimens

In general, both peak and trough concentrations are determined during monitoring of multiple-daily dosing regimens, and doses are subsequently adjusted to achieve the target concentrations. However, peak and trough concentrations do not necessarily offer the most valuable information for dosage adjustments during once-daily dosing. The indication and frequency of monitoring and the timing of serum concentrations within the interval, along with their target ranges, have yet to be established. Different targets have been proposed and used clinically (203–205), including mid-dosage interval plasma concentration monitoring as an estimate of the AUC (206).

In one prospective study 8-hour concentrations have been considered for monitoring as an alternative to the measurement of peaks and troughs (207). In 51 adult patients given an average dose of 400 mg once a day, doses were adjusted during therapy if 8-hour concentrations were not within the target range of 1.5–6.0 µg/ml. Concentrations above or below this target range correlated significantly with nephrotoxicity, 24-hour trough concentrations, and AUCs. Determination of 8-hour concentrations was therefore useful for identifying patients with either low AUCs or an increased risk of nephrotoxicity. In other studies, aminoglycoside serum concentrations were determined in the second half of the dosing interval (203,205,208–210). However, trough concentrations do not allow for extrapolations of concentrations achieved in the period after infusion. Thus, patients with very low peak concentrations due to unusually high volumes of distribution cannot be identified. Similarly, very rapid elimination, as frequently happens in patients with burns or children, may not be noticed.

Depending on the goals of drug monitoring, peak 8-, 12- or 24-hour concentrations might be more important (156). Correlations of efficacy with high peak concentrations, high ratios of peak to MIC, and high AUCs suggest that drug concentrations should be determined during the first part of the dosing interval. In order to minimize toxicity, drug accumulation should be avoided. Therefore early detection of increased trough concentrations is of importance. It has been strongly suggested that the threshold values for troughs during once-daily dosing should be lowered, compared with multiple-daily dosing (208,211). To reduce the cost of aminoglycoside therapy, the indications for and frequency of monitoring serum concentrations should be minimal. Instead, serum creatinine concentrations might be used to monitor renal function. The strategy selected for monitoring aminoglycoside therapy must take a number of factors into account, including the type and severity of infection, the duration of therapy, or the presence of factors associated with increased risk of toxicity. In addition, local factors should be considered for the timing of blood samples, including the sensitivity of the drug assay available (212) and the time required for processing the specimens, in order to modify the subsequent dose, if necessary.

Major considerations in determining the appropriate dose of an aminoglycoside are its volume of distribution and rate of clearance. Concern has been raised regarding the use of the Cockcroft–Gault equation with either actual or ideal body weight, resulting in systematic errors, especially in malnourished patients (213). However, even extended-interval dosing should not obviate the need for monitoring drug concentrations. Even in patients with normal renal function in whom treatment is necessary for over 3 days, mid-interval or trough drug concentrations should be obtained once or twice a week to optimize the dosing regimen (171,214).

Economics of serum concentration monitoring

The economic impact of aminoglycoside toxicity and its prevention through therapeutic drug monitoring have been investigated in a cost-effectiveness study. It was estimated that to offset the cost of providing high level drug monitoring, that is serum drug concentration monitoring with assessment and consultation by trained personnel using computerized resources to determine individualized pharmacokinetic parameters, for the purpose of achieving an optimum dosage regimen, by cost saving due to reducing nephropathy, the service should reduce the risk of nephrotoxicity by 6.6%. Therefore, high-level therapeutic drug monitoring is only cost-justified in populations in which high rates of nephrotoxicity would be expected. Risk factors for a high rate of nephrotoxicity include age, duration of therapy, high drug concentrations, the presence of ascites or liver disease, and the concomitant use of nephrotoxic drugs (215). The economic significance of aminoglycoside peak concentrations has been assessed in 61 febrile neutropenic patients with hematological malignancies. Since the clinical outcome and average infection-related costs depended significantly on peak aminoglycoside concentration, it was concluded that successful pharmacokinetic intervention may save money (216).

A pharmacy-based active therapeutic drug monitoring service has been examined in a prospective study (64). In the patients that were not monitored, the gentamicin dosage regimen was determined by the physician; in the patients that were monitored, the regimen was calculated using a population model and measured serum gentamicin concentrations. This resulted in significantly different mean dosage intervals: 19 hours in the monitored patients and 14 hours in the others. Active monitoring resulted in higher peak concentrations of aminoglycosides, lower trough concentrations, a reduction in the length of hospitalization, a reduced duration of aminoglycoside therapy, reduced nephrotoxicity, and a trend toward reduced mortality that was significant for patients with an infection on admission. Costs were lower with active monitoring. Although all patients treated with an aminoglycoside profited from active monitoring, it was most beneficial in patients who were admitted to the hospital with a suspected or proven Gram negative infection.

References

1. Brewer NS. Antimicrobial agents—Part II. The aminoglycosides: streptomycin, kanamycin, gentamicin, tobramycin, amikacin, neomycin. Mayo Clin Proc 1977;52(11):675–9.
2. Boselli E, Allaouchiche B. Diffusion osseuse des antibiotiques. [Diffusion in bone tissue of antibiotics.] Presse Méd 1999;28(40):2265–76.
3. John JF Jr. What price success? The continuing saga of the toxic:therapeutic ratio in the use of aminoglycoside antibiotics. J Infect Dis 1988;158(1):1–6.
4. Speich R, Imhof E, Vogt M, Grossenbacher M, Zimmerli W. Efficacy, safety, and tolerance of piperacillin/tazobactam compared to co-amoxiclav plus an aminoglycoside in the treatment of severe pneumonia. Eur J Clin Microbiol Infect Dis 1998;17(5):313–7.
5. Beringer PM, Wong-Beringer A, Rho JP. Economic aspects of antibacterial adverse effects. Pharmacoeconomics 1998;13(1 Pt 1):35–49.
6. Martin PD. ECG change associated with streptomycin. Chest 1974;65(4):478.
7. Emery ER. Neuromuscular blocking properties of antibiotics as a cause of post-operative apnoea. Anesthesia 1963;18:57.
8. Adams HR, Mathew BP, Teske RH, Mercer HD. Neuromuscular blocking effects of aminoglycoside antibiotics on fast- and slow-contracting muscles of the cat. Anesth Analg 1976;55(4):500–7.
9. Fiekers JF. Sites and mechanisms of antibiotic-induced neuromuscular block: a pharmacological analysis using quantal content, voltage clamped end-plate currents and single channel analysis. Acta Physiol Pharmacol Ther Latinoam 1999;49(4):242–50.
10. Holtzman JL. Gentamicin and neuromuscular blockade. Ann Intern Med 1976;84(1):55.
11. Dzoljic M, Atanackovic D. Effect of neomycin on smooth muscle. Arch Int Pharmacodyn Ther 1966;162(2):493–6.
12. Caraffini S, Assalve D, Stingeni L, Lisi P. Allergic contact conjunctivitis and blepharitis from tobramycin. Contact Dermatitis 1995;32(3):186–7.
13. Tange RA. Ototoxicity. Adverse Drug React Toxicol Rev 1998;17(2–3):75–89.
14. Feld R, Valdivieso M, Bodey GP, Rodriguez V. Comparison of amikacin and tobramycin in the treatment of infection in patients with cancer. J Infect Dis 1977;135(1):61–6.
15. Barza M, Lauermann MW, Tally FP, Gorbach SL. Prospective, randomized trial of netilmicin and amikacin, with emphasis on eighth-nerve toxicity. Antimicrob Agents Chemother 1980;17(4):707–14.
16. Matz GJ, Lerner SA. Prospective studies of aminoglycoside ototoxicity in adults. In: Lerner SA, Matz GJ, Hawkins JE, editors. Aminoglycoside Ototoxicity. Boston: Little, Brown and Co, 1981:327.
17. Orts Alborch M, Morant Ventura A, Garcia Callejo J, Ferrer Baixauli F, Martinez Beneito MP, Marco Algarra J. Monitorizacion de la ototoxicidad por farmacos con productos de distorsion. [Monitoring drug ototoxicity with distortion products.] Acta Otorrinolaringol Esp 2000;51(5):387–95.

18. Walsh RM, Bath AP, Bance ML. Reversible tobramycin-induced bilateral high-frequency vestibular toxicity. ORL J Otorhinolaryngol Relat Spec 2000;62(3):156–9.
19. Tsuji K, Velazquez-Villasenor L, Rauch SD, Glynn RJ, Wall C 3rd, Merchant SN. Temporal bone studies of the human peripheral vestibular system. Aminoglycoside ototoxicity. Ann Otol Rhinol Laryngol Suppl 2000;181:20–5.
20. Brummett RE, Fox KE. Aminoglycoside-induced hearing loss in humans. Antimicrob Agents Chemother 1989;33(6):797–800.
21. Fee WE Jr. Aminoglycoside ototoxicity in the human. Laryngoscope 1980;90(10 Pt 2 Suppl 24):1–19.
22. Tablan OC, Reyes MP, Rintelmann WF, Lerner AM. Renal and auditory toxicity of high-dose, prolonged therapy with gentamicin and tobramycin in Pseudomonas endocarditis. J Infect Dis 1984;149(2):257–63.
23. Federspil P, Schatzle W, Tiesler E. Pharmakokinetische, histologische und histochemische Untersuchungen zur Ototoxizitat des Gentamicins, Tobramycins und Amikacins. [Pharmacokinetical, histological, and histochemical investigation on the ototoxicity of gentamicin, tobramycin, and amikacin.] Arch Otorhinolaryngol 1977;217(2):147–66.
24. de Jager P, van Altena R. Hearing loss and nephrotoxicity in long-term aminoglycoside treatment in patients with tuberculosis. Int J Tuberc Lung Dis 2002;6(7):622–7.
25. Mulheran M, Degg C, Burr S, Morgan DW, Stableforth DE. Occurrence and risk of cochleotoxicity in cystic fibrosis patients receiving repeated high-dose aminoglycoside therapy. Antimicrob Agents Chemother 2001;45(9):2502–9.
26. Brummett RE, Fox KE. Studies of aminoglycoside ototoxicity in animal models. In: Whelton A, Neu HC, editors. The Aminoglycosides. New York, Basel: Marcel Dekker, 1982:419.
27. Brummett RE, Fox KE, Brown RT, Himes DL. Comparative ototoxic liability of netilmicin and gentamicin. Arch Otolaryngol 1978;104(10):579–84.
28. Cone LA. A survey of prospective, controlled clinical trials of gentamicin, tobramycin, amikacin, and netilmicin. Clin Ther 1982;5(2):155–62.
29. Lerner AM, Reyes MP, Cone LA, Blair DC, Jansen W, Wright GE, Lorber RR. Randomised, controlled trial of the comparative efficacy, auditory toxicity, and nephrotoxicity of tobramycin and netilmicin. Lancet 1983;1(8334):1123–6.
30. Gatell JM, SanMiguel JG, Araujo V, Zamora L, Mana J, Ferrer M, Bonet M, Bohe M, Jimenez de Anta MT. Prospective randomized double-blind comparison of nephrotoxicity and auditory toxicity of tobramycin and netilmicin. Antimicrob Agents Chemother 1984;26(5):766–9.
31. Lange G, Keller R. Beidseitiger Funktionsverlust der peripheren Gleichgewichtsorgane. Beobachtungen zu 20 Fallen von Dandy-Syndrom. [Bilateral malfunction of peripheral vestibular organs. Observations of 20 cases of Dandy syndrome.] Laryngorhinootologie 2000;79(2):77–80.
32. Federspil P. Zur Ototoxizität der Aminoglykosid-Antibiotika. [Ototoxicity of the aminoglycoside antibiotics.] Infection 1976;4(4):239–48.
33. Henley CM 3rd, Schacht J. Pharmacokinetics of aminoglycoside antibiotics in blood, inner-ear fluids and tissues and their relationship to ototoxicity. Audiology 1988;27(3):137–46.
34. Rybak LP. Ototoxicity. Curr Opin Otolaryngol Head Neck Surg 1996;4:302–7.
35. Lima da Costa D, Erre JP, Pehourq F, Aran JM. Aminoglycoside ototoxicity and the medial efferent system: II. Comparison of acute effects of different antibiotics. Audiology 1998;37(3):162–73.
36. Lima da Costa D, Erre JP, Aran JM. Aminoglycoside ototoxicity and the medial efferent system: I. Comparison of acute and chronic gentamicin treatments. Audiology 1998;37(3):151–61.
37. Conlon BJ, Perry BP, Smith DW. Attenuation of neomycin ototoxicity by iron chelation. Laryngoscope 1998;108(2):284–7.
38. Conlon BJ, Aran JM, Erre JP, Smith DW. Attenuation of aminoglycoside-induced cochlear damage with the metabolic antioxidant alpha-lipoic acid. Hear Res 1999;128(1–2):40–4.
39. Hester TO, Jones RO, Clerici WJ. Protection against aminoglycoside otic drop-induced ototoxicity by a spin trap: I. Acute effects. Otolaryngol Head Neck Surg 1998;119(6):581–7.
40. Segal JA, Harris BD, Kustova Y, Basile A, Skolnick P. Aminoglycoside neurotoxicity involves NMDA receptor activation. Brain Res 1999;815(2):270–7.
41. Basile AS, Brichta AM, Harris BD, Morse D, Coling D, Skolnick P. Dizocilpine attenuates streptomycin-induced vestibulotoxicity in rats. Neurosci Lett 1999;265(2):71–4.
42. Nakagawa T, Yamane H, Takayama M, Sunami K, Nakai Y. Involvement of nitric oxide in aminoglycoside vestibulotoxicity in guinea pigs. Neurosci Lett 1999;267(1):57–60.
43. Romand R, Chardin S. Effects of growth factors on the hair cells after ototoxic treatment of the neonatal mammalian cochlea in vitro. Brain Res 1999;825(1–2):46–58.
44. Ruan RS, Leong SK, Mark I, Yeoh KH. Effects of BDNF and NT-3 on hair cell survival in guinea pig cochlea damaged by kanamycin treatment. Neuroreport 1999;10(10):2067–71.
45. Vila J, Ruiz J, Navia M, Becerril B, Garcia I, Perea S, Lopez-Hernandez I, Alamo I, Ballester F, Planes AM, Martinez-Beltran J, de Anta TJ. Spread of amikacin resistance in Acinetobacter baumannii strains isolated in Spain due to an epidemic strain. J Clin Microbiol 1999;37(3):758–61.
46. Matsuda K, Ueda Y, Doi T, Tono T, Haruta A, Toyama K, Komune S. Increase in glutamate-aspartate transporter (GLAST) mRNA during kanamycin-induced cochlear insult in rats. Hear Res 1999;133(1–2):10–6.
47. Zheng JL, Gao WQ. Concanavalin A protects hair cells against gentamicin ototoxicity in rat cochlear explant cultures. J Neurobiol 1999;39(1):29–40.
48. Kimura N, Nishizaki K, Orita Y, Masuda Y. 4-methylcatechol, a potent inducer of nerve growth factor synthesis, protects spiral ganglion neurons from aminoglycoside ototoxicity—preliminary report. Acta Otolaryngol Suppl 1999;540:12–5.
49. Xiang ML, Mu MY, Pao X, Chi FL. The reinnervation of regenerated hair cells in the basilar papilla of chicks after kanamycin ototoxicity. Acta Otolaryngol 2000;120(8):912–21.
50. Selimoglu E, Kalkandelen S, Erdogan F. Comparative vestibulotoxicity of different aminoglycosides in guinea pigs. Yonsei Med J 2003;44:517–22.
51. Fischel-Ghodsian N, Prezant TR, Bu X, Oztas S. Mitochondrial ribosomal RNA gene mutation in a patient with sporadic aminoglycoside ototoxicity. Am J Otolaryngol 1993;14(6):399–403.
52. Jacobs HT. Mitochondrial deafness. Ann Med 1997;29(6):483–91.
53. Usami S, Abe S, Tono T, Komune S, Kimberling WJ, Shinkawa H. Isepamicin sulfate-induced sensorineural

hearing loss in patients with the 1555 A→G mitochondrial mutation. ORL J Otorhinolaryngol Relat Spec 1998;60(3):164–9.

54. Wang S, Bian Q, Liu Z, Feng Y, Lian N, Chen H, Hu C, Dong Y, Cai Z. Capability of serum to convert streptomycin to cytotoxin in patients with aminoglycoside-induced hearing loss. Hear Res 1999;137(1–2):1–7.

55. Moore RD, Smith CR, Lietman PS. Risk factors for the development of auditory toxicity in patients receiving aminoglycosides. J Infect Dis 1984;149(1):23–30.

56. Smith CR, Lietman PS. Effect of furosemide on aminoglycoside-induced nephrotoxicity and auditory toxicity in humans. Antimicrob Agents Chemother 1983;23(1):133–7.

57. Schonenberger U, Streit C, Hoigne R. Nephro- und Ototoxizitat von Aminoglykosid-Antibiotica unter besonderer Berücksichtigung von Gentamicin. [Nephro- and ototoxicity of aminoglycoside-antibiotics, with special reference to gentamicin.] Schweiz Rundsch Med Prax 1981;70(5):169–73.

58. Schacht J. Aminoglycoside ototoxicity: prevention in sight? Otolaryngol Head Neck Surg 1998;118(5):674–7.

59. Hardisty RE, Fleming J, Steel KP. The molecular genetics of inherited deafness—current knowledge and recent advances. J Laryngol Otol 1998;112(5):432–7.

60. Steel KP. Progress in progressive hearing loss. Science 1998;279(5358):1870–1.

61. Casano RA, Bykhovskaya Y, Johnson DF, Hamon M, Torricelli F, Bigozzi M, Fischel-Ghodsian N. Hearing loss due to the mitochondrial A1555G mutation in Italian families. Am J Med Genet 1998;79(5):388–91.

62. Estivill X, Govea N, Barcelo E, Perello E, Badenas C, Romero E, Moral L, Scozzri R, D'Urbano L, Zeviani M, Torroni A. Familial progressive sensorineural deafness is mainly due to the mtDNA A1555G mutation and is enhanced by treatment of aminoglycosides. Am J Hum Genet 1998;62(1):27–35.

63. Casano RA, Johnson DF, Bykhovskaya Y, Torricelli F, Bigozzi M, Fischel-Ghodsian N. Inherited susceptibility to aminoglycoside ototoxicity: genetic heterogeneity and clinical implications. Am J Otolaryngol 1999;20(3):151–6.

64. Hutchin T. Sensorineural hearing loss and the 1555G mitochondrial DNA mutation. Acta Otolaryngol 1999;119(1):48–52.

65. Scrimshaw BJ, Faed JM, Tate WP, Yun K. Rapid identification of an A1555G mutation in human mitochondrial DNA implicated in aminoglycoside-induced ototoxicity. J Hum Genet 1999;44(6):388–90.

66. Guan MX, Fischel-Ghodsian N, Attardi G. A biochemical basis for the inherited susceptibility to aminoglycoside ototoxicity. Hum Mol Genet 2000;9(12):1787–93.

67. Stavroulaki P, Apostolopoulos N, Dinopoulou D, Vossinakis I, Tsakanikos M, Douniadakis D. Otoacoustic emissions—an approach for monitoring aminoglycoside induced ototoxicity in children. Int J Pediatr Otorhinolaryngol 1999;50(3):177–84.

68. Peloquin CA, Berning SE, Nitta AT, Simone PM, Goble M, Huitt GA, Iseman MD, Cook JL, Curran-Everett D. Aminoglycoside toxicity: daily versus thrice-weekly dosing for treatment of mycobacterial diseases. Clin Infect Dis 2004;38(11):1538–44.

69. Marlow ES, Hunt LP, Marlow N. Sensorineural hearing loss and prematurity. Arch Dis Child Fetal Neonatal Ed 2000;82(2):F141–4.

70. Guerit JM, Mahieu P, Houben-Giurgea S, Herbay S. The influence of ototoxic drugs on brainstem auditory evoked potentials in man. Arch Otorhinolaryngol 1981;233(2):189–99.

71. Hotz MA, Allum JH, Kaufmann G, Follath F, Pfaltz CR. Shifts in auditory brainstem response latencies following plasma-level-controlled aminoglycoside therapy. Eur Arch Otorhinolaryngol 1990;247(4):202–5.

72. Lopez-Gonzalez MA, Guerrero JM, Torronteras R, Osuna C, Delgado F. Ototoxicity caused by aminoglycosides is ameliorated by melatonin without interfering with the antibiotic capacity of the drugs. J Pineal Res 2000;28(1):26–33.

73. Conlon BJ, Smith DW. Topical aminoglycoside ototoxicity: attempting to protect the cochlea. Acta Otolaryngol 2000;120(5):596–9.

74. Duan M, Agerman K, Ernfors P, Canlon B. Complementary roles of neurotrophin 3 and a N-methyl-D-aspartate antagonist in the protection of noise and aminoglycoside-induced ototoxicity. Proc Natl Acad Sci USA 2000;97(13):7597–602.

75. Lopez-Gonzalez MA, Delgado F, Lucas M. Aminoglycosides activate oxygen metabolites production in the cochlea of mature and developing rats. Hear Res 1999;136(1–2):165–8.

76. Sha SH, Schacht J. Stimulation of free radical formation by aminoglycoside antibiotics. Hear Res 1999;128(1–2):112–8.

77. Liamis G, Alexandridis G, Bairaktari ET, Elisaf MS. Aminoglycoside-induced metabolic abnormalities. Ann Clin Biochem 2000;37(Pt 4):543–4.

78. Kes P, Reiner Z. Symptomatic hypomagnesemia associated with gentamicin therapy. Magnes Trace Elem 1990;9(1):54–60.

79. von Vigier RO, Truttmann AC, Zindler-Schmocker K, Bettinelli A, Aebischer CC, Wermuth B, Bianchetti MG. Aminoglycosides and renal magnesium homeostasis in humans. Nephrol Dial Transplant 2000;15(6):822–6.

80. Elliott C, Newman N, Madan A. Gentamicin effects on urinary electrolyte excretion in healthy subjects. Clin Pharmacol Ther 2000;67(1):16–21.

81. Shetty AK, Rogers NL, Mannick EE, Aviles DH. Syndrome of hypokalemic metabolic alkalosis and hypomagnesemia associated with gentamicin therapy: case reports. Clin Pediatr (Phila) 2000;39(9):529–33.

82. Kang HS, Kerstan D, Dai L, Ritchie G, Quamme GA. Aminoglycosides inhibit hormone-stimulated Mg^{2+} uptake in mouse distal convoluted tubule cells. Can J Physiol Pharmacol 2000;78(8):595–602.

83. Akbar A, Rees JH, Nyamugunduru G, English MW, Spencer DA, Weller PH. Aminoglycoside-associated hypomagnesaemia in children with cystic fibrosis. Acta Paediatr 1999;88(7):783–5.

84. Ciftci M, Kufrevioglu OI, Gundogdu M, Ozmen II. Effects of some antibiotics on enzyme activity of glucose-6-phosphate dehydrogenase from human erythrocytes. Pharmacol Res 2000;41(1):107–11.

85. Alexandridis G, Liberopoulos E, Elisaf M. Aminoglycoside-induced reversible tubular dysfunction. Pharmacology 2003;67:118–20.

86. English WP, Williams MD. Should aminoglycoside antibiotics be abandoned? Am J Surg 2000;180(6):512–6.

87. Shemin D, Maaz D, St Pierre D, Kahn SI, Chazan JA. Effect of aminoglycoside use on residual renal function in peritoneal dialysis patients. Am J Kidney Dis 1999;34(1):14–20.

88. Contopoulos-Ioannidis DG, Giotis ND, Baliatsa DV, Ioannidis JP. Extended-interval aminoglycoside adminis-

tration for children: a meta-analysis. Pediatrics 2004;114(1):e111-8.

89. Plaut ME, Schentag JJ, Jusko WJ. Aminoglycoside nephrotoxicity: comparative assessment in critically ill patients. J Med 1979;10(4):257–66.

90. Schentag JJ, Plaut ME, Cerra FB. Comparative nephrotoxicity of gentamicin and tobramycin: pharmacokinetic and clinical studies in 201 patients. Antimicrob Agents Chemother 1981;19(5):859–66.

91. Moore RD, Smith CR, Lipsky JJ, Mellits ED, Lietman PS. Risk factors for nephrotoxicity in patients treated with aminoglycosides. Ann Intern Med 1984;100(3):352–7.

92. Sawyers CL, Moore RD, Lerner SA, Smith CR. A model for predicting nephrotoxicity in patients treated with aminoglycosides. J Infect Dis 1986;153(6):1062–8.

93. Lam YW, Arana CJ, Shikuma LR, Rotschafer JC. The clinical utility of a published nomogram to predict aminoglycoside nephrotoxicity. JAMA 1986;255(5):639–42.

94. Thatte L, Vaamonde CA. Drug-induced nephrotoxicity: the crucial role of risk factors. Postgrad Med 1996;100(6):83–487–8, 91 passim.

95. Samaniego-Picota MD, Whelton A. Aminoglycoside-induced nephrotoxicity in cystic fibrosis: a case presentation and review of the literature. Am J Ther 1996;3(3):248–57.

96. Raveh D, Kopyt M, Hite Y, Rudensky B, Sonnenblick M, Yinnon AM. Risk factors for nephrotoxicity in elderly patients receiving once-daily aminoglycosides. QJM 2002;95(5):291–7.

97. Luft FC, Evan AP. Comparative effects of tobramycin and gentamicin on glomerular ultrastructure. J Infect Dis 1980;142(6):910–4.

98. Tardif D, Beauchamp D, Bergeron MG. Influence of endotoxin on the intracortical accumulation kinetics of gentamicin in rats. Antimicrob Agents Chemother 1990;34(4):576–80.

99. Appel GB. Aminoglycoside nephrotoxicity: physiologic studies of the sites of nephron damage. In: Whelton A, Neu HC, editors. The Aminoglycosides. New York, Basel: Marcel Dekker, 1982:269.

100. Whelton A. Renal tubular transport and intrarenal aminoglycoside distribution. In: Whelton A, Neu HC, editors. The Aminoglycosides. New York, Basel: Marcel Dekker, 1982:191.

101. Carlier MB, Laurent G, Claes PJ, Vanderhaeghe HJ, Tulkens PM. Inhibition of lysosomal phospholipases by aminoglycoside antibiotics: in vitro comparative studies. Antimicrob Agents Chemother 1983;23(3):440–9.

102. Rybak MJ, Abate BJ, Kang SL, Ruffing MJ, Lerner SA, Drusano GL. Prospective evaluation of the effect of an aminoglycoside dosing regimen on rates of observed nephrotoxicity and ototoxicity. Antimicrob Agents Chemother 1999;43(7):1549–55.

103. Walker PD, Barri Y, Shah SV. Oxidant mechanisms in gentamicin nephrotoxicity. Ren Fail 1999;21(3–4):433–42.

104. Carrier D, Bou Khalil M, Kealey A. Modulation of phospholipase A2 activity by aminoglycosides and daptomycin: a Fourier transform infrared spectroscopic study. Biochemistry 1998;37(20):7589–97.

105. Cheng M, Razzaque MS, Nazneen A, Taguchi T. Expression of the heat shock protein 47 in gentamicin-treated rat kidneys. Int J Exp Pathol 1998;79(3):125–32.

106. Scherberich JE, Mondorf WA. Nephrotoxic potential of antiinfective drugs as assessed by tissue-specific proteinuria of renal antigens. Int J Clin Pharmacol Ther 1998;36(3):152–8.

107. Nagai J, Takano M. Molecular aspects of renal handling of aminoglycosides and strategies for preventing the nephrotoxicity. Drug Metab Pharmacokinet 2004;19(3):159–70.

108. Murry KR, McKinnon PS, Mitrzyk B, Rybak MJ. Pharmacodynamic characterization of nephrotoxicity associated with once-daily aminoglycoside. Pharmacotherapy 1999;19(11):1252–60.

109. Buijk SE, Mouton JW, Gyssens IC, Verbrugh HA, Bruining HA. Experience with a once-daily dosing program of aminoglycosides in critically ill patients. Intensive Care Med 2002;28(7):936–42.

110. Keys TF, Kurtz SB, Jones JD, Muller SM. Renal toxicity during therapy with gentamicin or tobramycin. Mayo Clin Proc 1981;56(9):556–9.

111. Schentag JJ, Gengo FM, Plaut ME, Danner D, Mangione A, Jusko WJ. Urinary casts as an indicator of renal tubular damage in patients receiving aminoglycosides. Antimicrob Agents Chemother 1979;16(4):468–74.

112. Schentag JJ, Sutfin TA, Plaut ME, Jusko WJ. Early detection of aminoglycoside nephrotoxicity with urinary beta-2-microglobulin. J Med 1978;9(3):201–10.

113. Tulkens PM. Pharmacokinetic and toxicological evaluation of a once-daily regimen versus conventional schedules of netilmicin and amikacin. J Antimicrob Chemother 1991;27(Suppl C):49–61.

114. Mondorf AW. Urinary enzymatic markers of renal damage. In: Whelton A, Neu HC, editors. The Aminoglycosides. New York, Basel: Marcel Dekker, 1982:283.

115. Marchewka Z, Dlugosz A. Enzymes in urine as markers of nephrotoxicity of cytostatic agents and aminoglycoside antibiotics. Int Urol Nephrol 1998;30(3):339–48.

116. Panova LD, Farkhutdinov RR, Akhmadeeva EN. [Urine chemiluminescence in preclinical diagnosis of neonatal drug-induced nephropathy.]Urol Nefrol (Mosk) 1998;(4):25–9.

117. Luft FC, Yum MN, Kleit SA. Comparative nephrotoxicities of netilmicin and gentamicin in rats. Antimicrob Agents Chemother 1976;10(5):845–9.

118. Hottendorf GH, Gordon LL. Comparative low-dose nephrotoxicities of gentamicin, tobramycin, and amikacin. Antimicrob Agents Chemother 1980;18(1):176–81.

119. Smith CR, Baughman KL, Edwards CQ, Rogers JF, Lietman PS. Controlled comparison of amikacin and gentamicin. N Engl J Med 1977;296(7):349–53.

120. Love LJ, Schimpff SC, Hahn DM, Young VM, Standiford HC, Bender JF, Fortner CL, Wiernik PH. Randomized trial of empiric antibiotic therapy with ticarcillin in combination with gentamicin, amikacin or netilmicin in febrile patients with granulocytopenia and cancer. Am J Med 1979;66(4):603–10.

121. Lau WK, Young LS, Black RE, Winston DJ, Linne SR, Weinstein RJ, Hewitt WL. Comparative efficacy and toxicity of amikacin/carbenicillin versus gentamicin/carbenicillin in leukopenic patients: a randomized prospective trial. Am J Med 1977;62(6):959–66.

122. Fong IW, Fenton RS, Bird R. Comparative toxicity of gentamicin versus tobramycin: a randomized prospective study. J Antimicrob Chemother 1981;7(1):81–8.

123. Bock BV, Edelstein PH, Meyer RD. Prospective comparative study of efficacy and toxicity of netilmicin and amikacin. Antimicrob Agents Chemother 1980;17(2):217–25.

124. Noone M, Pomeroy L, Sage R, Noone P. Prospective study of amikacin versus netilmicin in the treatment of severe infection in hospitalized patients. Am J Med 1989;86(6 Pt 2):809–13.

125. Smith CR, Lipsky JJ, Laskin OL, Hellmann DB, Mellits ED, Longstreth J, Lietman PS. Double-blind comparison of the nephrotoxicity and auditory toxicity of gentamicin and tobramycin. N Engl J Med 1980;302(20):1106–9.

126. Matzke GR, Lucarotti RL, Shapiro HS. Controlled comparison of gentamicin and tobramycin nephrotoxicity. Am J Nephrol 1983;3(1):11–7.

127. Akaki T, Dekio S. Contact dermatitis from arbekacin sulfate: report of a case. J Dermatol 2002;29(10):674–5.

128. Yung MW, Rajendra T. Delayed hypersensitivity reaction to topical aminoglycosides in patients undergoing middle ear surgery. Clin Otolaryngol Allied Sci 2002;27(5):365–8.

129. Paniagua MJ, Garcia-Ortega P, Tella R, Gaig P, Richart C. Systemic contact dermatitis to gentamicin. Allergy 2002;57(11):1086–7.

130. Schulze S, Wollina U. Gentamicin-induced anaphylaxis. Allergy 2003;58(1):88–9.

131. Hall FJ. Anaphylaxis after gentamycin. Lancet 1977;2(8035):455.

132. Goh CL. Anaphylaxis from topical neomycin and bacitracin. Australas J Dermatol 1986;27(3):125–6.

133. Blaser J, Stone BB, Zinner SH. Efficacy of intermittent versus continuous administration of netilmicin in a two-compartment in vitro model. Antimicrob Agents Chemother 1985;27(3):343–9.

134. Gerber AU, Craig WA. Aminoglycoside-selected subpopulations of Pseudomonas aeruginosa: characterization and virulence in normal and leukopenic mice. J Lab Clin Med 1982;100(5):671–81.

135. Olson B, Weinstein RA, Nathan C, Chamberlin W, Kabins SA. Occult aminoglycoside resistance in Pseudomonas aeruginosa: epidemiology and implications for therapy and control. J Infect Dis 1985;152(4):769–74.

136. Hilf M, Yu VL, Sharp J, Zuravleff JJ, Korvick JA, Muder RR. Antibiotic therapy for Pseudomonas aeruginosa bacteremia: outcome correlations in a prospective study of 200 patients. Am J Med 1989;87(5):540–6.

137. Mathon L, Decaillot F, Allaouchiche B. Impact de l'antibiothérapie initiale sur l'evolution des résistances aux fluoroquinolones et aux aminosides des bacilles a gram négatif isolés chez des patients de réanimation. [Impact of initial antibiotic therapy on the course of resistance to fluoroquinolones and aminoglycosides in Gram-negative bacilli isolated from intensive care patients.] Ann Fr Anesth Reanim 1999;18(10):1054–60.

138. Moore RA, DeShazer D, Reckseidler S, Weissman A, Woods DE. Efflux-mediated aminoglycoside and macrolide resistance in Burkholderia pseudomallei. Antimicrob Agents Chemother 1999;43(3):465–70.

139. Vanhoof R, Nyssen HJ, Van Bossuyt E, Hannecart-Pokorni EAminoglycoside Resistance Study Group. Aminoglycoside resistance in Gram-negative blood isolates from various hospitals in Belgium and the Grand Duchy of Luxembourg. J Antimicrob Chemother 1999;44(4):483–8.

140. Schmitz FJ, Verhoef J, Fluit AC. Prevalence of aminoglycoside resistance in 20 European university hospitals participating in the European SENTRY Antimicrobial Surveillance Programme. Eur J Clin Microbiol Infect Dis 1999;18(6):414–21.

141. Bouza E, Garcia-Garrote F, Cercenado E, Marin M, Diaz MS. Pseudomonas aeruginosa: a survey of resistance in 136 hospitals in Spain. The Spanish Pseudomonas aeruginosa Study Group. Antimicrob Agents Chemother 1999;43(4):981–2.

142. Ruiz J, Nunez ML, Perez J, Simarro E, Martinez-Campos L, Gomez J. Evolution of resistance among clinical isolates of Acinetobacter over a 6-year period. Eur J Clin Microbiol Infect Dis 1999;18(4):292–5.

143. Jacobson K, Rolston K, Elting L, LeBlanc B, Whimbey E, Ho DH. Susceptibility surveillance among Gram-negative bacilli at a cancer center. Chemotherapy 1999;45(5):325–34.

144. Valdivieso F, Trucco O, Prado V, Diaz MC, Ojeda A. Resistencia a los antimicrobianos en agentes causantes de infeccion del tracto urinario en 11 hospitales chilenos. [Antimicrobial resistance of agents causing urinary tract infections in 11 Chilean hospitals. PRONARES project.] Rev Med Chil 1999;127(9):1033–40.

145. Freitas FI, Guedes-Stehling E, Siqueira-Junior JP. Resistance to gentamicin and related aminoglycosides in Staphylococcus aureus isolated in Brazil. Lett Appl Microbiol 1999;29(3):197–201.

146. del Valle O, Trincado P, Martin MT, Gomez E, Cano A, Vindel A. Prevalencia de Staphylococcus aureus resistentes a meticilina fagotipo 95 en los Hospitales Vall d'Hebron de Barcelona. [The prevalence of methicillin-resistant Staphylococcus aureus phagotype 95 in the Hospitales Vall d'Hebron of Barcelona.] Enferm Infecc Microbiol Clin 1999;17(10):498–505.

147. Dulon D, Ryan AF. The bacterial Neo gene confers neomycin resistance to mammalian cochlear hair cells. Neuroreport 1999;10(6):1189–93.

148. You I, Kariyama R, Zervos MJ, Kumon H, Chow JW. In-vitro activity of arbekacin alone and in combination with vancomycin against gentamicin- and methicillin-resistant Staphylococcus aureus. Diagn Microbiol Infect Dis 2000;36(1):37–41.

149. Nakamura A, Hosoda M, Kato T, Yamada Y, Itoh M, Kanazawa K, Nouda H. Combined effects of meropenem and aminoglycosides on Pseudomonas aeruginosa in vitro. J Antimicrob Chemother 2000;46(6):901–4.

150. Heinonen OP, Slone D, Shapiro S. Birth defects and drugs in pregnancy. In: Kaufmann DW, editor. Antimicrobial and Parasite Agents. Littleton, MA: John Wright, 1982:296 435.

151. Conway N, Birt BD. Streptomycin in pregnancy: effect on the foetal ear. BMJ 1965;5456:260–3.

152. Czeizel AE, Rockenbauer M, Olsen J, Sorensen HT. A teratological study of aminoglycoside antibiotic treatment during pregnancy. Scand J Infect Dis 2000;32(3):309–13.

153. Hess M, Finckh-Kramer U, Bartsch M, Kewitz G, Versmold H, Gross M. Hearing screening in at-risk neonate cohort. Int J Pediatr Otorhinolaryngol 1998;46(1–2):81–9.

154. Mattie H, Craig WA, Pechere JC. Determinants of efficacy and toxicity of aminoglycosides. J Antimicrob Chemother 1989;24(3):281–93.

155. Gilbert DN. Once-daily aminoglycoside therapy. Antimicrob Agents Chemother 1991;35(3):399–405.

156. Blaser J, Konig C. Once-daily dosing of aminoglycosides. Eur J Clin Microbiol Infect Dis 1995;14(12):1029–38.

157. Lacy MK, Nicolau DP, Nightingale CH, Quintiliani R. The pharmacodynamics of aminoglycosides. Clin Infect Dis 1998;27(1):23–7.

158. Blaser J, Stone BB, Groner MC, Zinner SH. Comparative study with enoxacin and netilmicin in a pharmacodynamic model to determine importance of ratio of antibiotic peak concentration to MIC for bactericidal activity and emergence of resistance. Antimicrob Agents Chemother 1987;31(7):1054–60.

159. Moore RD, Lietman PS, Smith CR. Clinical response to aminoglycoside therapy: importance of the ratio of peak concentration to minimal inhibitory concentration. J Infect Dis 1987;155(1):93–9.

160. Vogelman B, Gudmundsson S, Turnidge J, Leggett J, Craig WA. In vivo postantibiotic effect in a thigh infection in neutropenic mice. J Infect Dis 1988;157(2):287–98.

161. Powell SH, Thompson WL, Luthe MA, Stern RC, Grossniklaus DA, Bloxham DD, Groden DL, Jacobs MR, DiScenna AO, Cash HA, Klinger JD. Once-daily vs. continuous aminoglycoside dosing: efficacy and toxicity in animal and clinical studies of gentamicin, netilmicin, and tobramycin. J Infect Dis 1983;147(5):918–32.

162. De Broe ME, Verbist L, Verpooten GA. Influence of dosage schedule on renal cortical accumulation of amikacin and tobramycin in man. J Antimicrob Chemother 1991;27(Suppl C):41–7.

163. Freeman CD, Strayer AH. Mega-analysis of meta-analysis: an examination of meta-analysis with an emphasis on once-daily aminoglycoside comparative trials. Pharmacotherapy 1996;16(6):1093–102.

164. Gilbert DN, Lee BL, Dworkin RJ, Leggett JL, Chambers HF, Modin G, Tauber MG, Sande MA. A randomized comparison of the safety and efficacy of once-daily gentamicin or thrice-daily gentamicin in combination with ticarcillin–clavulanate. Am J Med 1998;105(3):182–91.

165. Karachalios GN, Houpas P, Tziviskou E, Papalimneou V, Georgiou A, Karachaliou I, Halkiadaki D. Prospective randomized study of once-daily versus twice-daily amikacin regimens in patients with systemic infections. Int J Clin Pharmacol Ther 1998;36(10):561–4.

166. Sanchez-Alcaraz A, Vargas A, Quintana MB, Rocher A, Querol JM, Poveda JL, Hermenegildo M. Therapeutic drug monitoring of tobramycin: once-daily versus twice-daily dosage schedules. J Clin Pharm Ther 1998;23(5):367–73.

167. Bragonier R, Brown NM. The pharmacokinetics and toxicity of once-daily tobramycin therapy in children with cystic fibrosis. J Antimicrob Chemother 1998;42(1):103–6.

168. Paterson DL, Robson JM, Wagener MM. Risk factors for toxicity in elderly patients given aminoglycosides once daily. J Gen Intern Med 1998;13(11):735–9.

169. Tam VH, Preston SL, Briceland LL. Once-daily aminoglycosides in the treatment of Gram-positive endocarditis. Ann Pharmacother 1999;33(5):600–6.

170. Barletta JF, Johnson SB, Nix DE, Nix LC, Erstad BL. Population pharmacokinetics of aminoglycosides in critically ill trauma patients on once-daily regimens. J Trauma 2000;49(5):869–72.

171. Gerberding JL. Aminoglycoside dosing: timing is of the essence. Am J Med 1998;105(3):256–8.

172. Havener WH. Ocular Pharmacology. 4th ed.. St Louis: CV Mosby;. 1978.

173. Thomsen J, Charabi S, Tos M. Preliminary results of a new delivery system for gentamicin to the inner ear in patients with Menière's disease. Eur Arch Otorhinolaryngol 2000;257(7):362–5.

174. Quaranta A, Piazza F. Menière's disease: diagnosis and new treatment perspectives. [Mèniere's disease: diagnosis and new treatment perspectives.] Recenti Prog Med 2000;91(1):33–7.

175. Mars RL, Moles K, Pope K, Hargrove P. Use of bolus intraperitoneal aminoglycosides for treating peritonitis in end-stage renal disease patients receiving continuous ambulatory peritoneal dialysis and continuous cycling peritoneal dialysis. Adv Perit Dial 2000;16:280–4.

176. Green FJ, Lavelle KJ, Aronoff GR, Vander Zanden J, Brier GL. Management of amikacin overdose. Am J Kidney Dis 1981;1(2):110–2.

177. Lu CM, James SH, Lien YH. Acute massive gentamicin intoxication in a patient with end-stage renal disease. Am J Kidney Dis 1996;28(5):767–71.

178. Basile C, Di Maggio A, Curino E, Scatizzi A. Pharmacokinetics of netilmicin in hypertonic hemodiafiltration and standard hemodialysis. Clin Nephrol 1985;24(6):305–9.

179. Bailey RR. Renal failure in combined gentamicin and cephalothin therapy. BMJ 1973;2(5869):776–7.

180. Cabanillas F, Burgos RC, Rodriguez C, Baldizon C. Nephrotoxicity of combined cephalothin–gentamicin regimen. Arch Intern Med 1975;135(6):850–2.

181. Tobias JS, Whitehouse JM, Wrigley PF. Severe renal dysfunction after tobramycin/cephalothin therapy. Lancet 1976;1(7956):425.

182. English J, Gilbert DN, Kohlhepp S, Kohnen PW, Mayor G, Houghton DC, Bennett WM. Attenuation of experimental tobramycin nephrotoxicity by ticarcillin. Antimicrob Agents Chemother 1985;27(6):897–902.

183. Wade JC, Schimpff SC, Wiernik PH. Antibiotic combination-associated nephrotoxicity in granulocytopenic patients with cancer. Arch Intern Med 1981;141(13):1789–93.

184. Caston J, Doinel L. Comparative vestibular toxicity of dibekacin, habekacin and cisplatin. Acta Otolaryngol 1987;104(3–4):315–21.

185. Bates DE, Beaumont SJ, Baylis BW. Ototoxicity induced by gentamicin and furosemide. Ann Pharmacother 2002;36(3):446–51.

186. Santucci RA, Krieger JN. Gentamicin for the practicing urologist: review of efficacy, single daily dosing and "switch" therapy. J Urol 2000;163(4):1076–84.

187. Conlon BJ, McSwain SD, Smith DW. Topical gentamicin and ethacrynic acid: effects on cochlear function. Laryngoscope 1998;108(7):1087–9.

188. Burkett L, Bikhazi GB, Thomas KC Jr, Rosenthal DA, Wirta MG, Foldes FF. Mutual potentiation of the neuromuscular effects of antibiotics and relaxants. Anesth Analg 1979;58(2):107–15.

189. Holt HA, Broughall JM, McCarthy M, Reeves DS. Interactions between aminoglycoside antibiotics and carbenicillin or ticarillin. Infection 1976;4(2):107–9.

190. Adam D, Haneder J. Studies on the inactivation of aminoglycoside antibiotics by acylureidopenicillins and piperacillin. Infection 1981;9:182.

191. Henderson JL, Polk RE, Kline BJ. In vitro inactivation of gentamicin, tobramycin, and netilmicin by carbenicillin, azlocillin, or mezlocillin. Am J Hosp Pharm 1981;38(8):1167–70.

192. Thompson MI, Russo ME, Saxon BJ, Atkin-Thor E, Matsen JM. Gentamicin inactivation by piperacillin or carbenicillin in patients with end-stage renal disease. Antimicrob Agents Chemother 1982;21(2):268–73.

193. Rybak MJ, Albrecht LM, Boike SC, Chandrasekar PH. Nephrotoxicity of vancomycin, alone and with an aminoglycoside. J Antimicrob Chemother 1990;25(4):679–87.

194. de Lemos E, Pariat C, Piriou A, Fauconneau B, Courtois P. Variations circadiennes de la nephrotoxicité de l'association vancomycine–gentamicine chez le rat. [Circadian variations in the nephrotoxicity of the

vancomycin–gentamicin combination in rats.] Pathol Biol (Paris) 1991;39(1):12–5.

195. Pauly DJ, Musa DM, Lestico MR, Lindstrom MJ, Hetsko CM. Risk of nephrotoxicity with combination vancomycin–aminoglycoside antibiotic therapy. Pharmacotherapy 1990;10(6):378–82.

196. Streetman DS, Nafziger AN, Destache CJ, Bertino AS Jr. Individualized pharmacokinetic monitoring results in less aminoglycoside-associated nephrotoxicity and fewer associated costs. Pharmacotherapy 2001;21(4):443–51.

197. Moore RD, Smith CR, Lietman PS. The association of aminoglycoside plasma levels with mortality in patients with Gram-negative bacteremia. J Infect Dis 1984;149(3):443–8.

198. Moore RD, Smith CR, Lietman PS. Association of aminoglycoside plasma levels with therapeutic outcome in Gram-negative pneumonia. Am J Med 1984;77(4):657–62.

199. Wenk M, Vozeh S, Follath F. Serum level monitoring of antibacterial drugs. A review. Clin Pharmacokinet 1984;9(6):475–92.

200. Zaske DE, Cipolle RJ, Rotschafer JC, Solem LD, Mosier NR, Strate RG. Gentamicin pharmacokinetics in 1,640 patients: method for control of serum concentrations. Antimicrob Agents Chemother 1982;21(3):407–11.

201. Lesar TS, Rotschafer JC, Strand LM, Solem LD, Zaske DE. Gentamicin dosing errors with four commonly used nomograms. JAMA 1982;248(10):1190–3.

202. Adjepon-Yamoah KK, Al-Homrany M, Bahar Y, Ahmed ME. Aminoglycoside usage and monitoring in a Saudi Arabian teaching hospital; a ten-year laboratory audit. J Clin Pharm Ther 2000;25(4):303–7.

203. Konrad F, Wagner R, Neumeister B, Rommel H, Georgieff M. Studies on drug monitoring in thrice and once daily treatment with aminoglycosides. Intensive Care Med 1993;19(4):215–20.

204. Janknegt R. Aminoglycoside monitoring in the once- or twice-daily era. The Dutch situation considered. Pharm World Sci 1993;15(4):151–5.

205. Parker SE, Davey PG. Practicalities of once-daily aminoglycoside dosing. J Antimicrob Chemother 1993;31(1):4–8.

206. Barclay ML, Kirkpatrick CM, Begg EJ. Once daily aminoglycoside therapy. Is it less toxic than multiple daily doses and how should it be monitored? Clin Pharmacokinet 1999;36(2):89–98.

207. Blaser J, Konig C, Simmen HP, Thurnheer U. Monitoring serum concentrations for once-daily netilmicin dosing regimens. J Antimicrob Chemother 1994;33(2):341–8.

208. MacGowan AP, Reeves DS. Serum monitoring and practicalities of once-daily aminoglycoside dosing. J Antimicrob Chemother 1994;33(2):349–50.

209. Giamarellou H, Yiallouros K, Petrikkos G, Moschovakis E, Vavouraki E, Voutsinas D, Sfikakis P. Comparative kinetics and efficacy of amikacin administered once or twice daily in the treatment of systemic Gram-negative infections. J Antimicrob Chemother 1991;27(Suppl C):73–9.

210. Maller R, Ahrne H, Holmen C, Lausen I, Nilsson LE, Smedjegard JScandinavian Amikacin Once Daily Study Group. Once- versus twice-daily amikacin regimen: efficacy and safety in systemic Gram-negative infections. J Antimicrob Chemother 1993;31(6):939–48.

211. Reeves DS, MacGowan AP. Once-daily aminoglycoside dosing. Lancet 1993;341(8849):895–6.

212. Blaser J, Konig C, Fatio R, Follath F, Cometta A, Glauser M. Multicenter quality control study of amikacin assay for monitoring once-daily dosing regimens. International Antimicrobial Therapy Cooperative Group of the European Organization for Research and Treatment of Cancer. Ther Drug Monit 1995;17(2):133–6.

213. Kotler DP, Sordillo EM. Nutritional status and aminoglycoside dosing. Clin Infect Dis 1998;26(1):249–52.

214. Bailey TC, Reichley RM. Nutritional status and aminoglycoside dosing. Clin Infect Dis 1998;26:251–2.

215. Slaughter RL, Cappelletty DM. Economic impact of aminoglycoside toxicity and its prevention through therapeutic drug monitoring. Pharmacoeconomics 1998;14(4):385–94.

216. Binder L, Schiel X, Binder C, Menke CF, Schuttrumpf S, Armstrong VW, Unterhalt M, Erichsen N, Hiddemann W, Oellerich M. Clinical outcome and economic impact of aminoglycoside peak concentrations in febrile immunocompromised patients with hematologic malignancies. Clin Chem 1998;44(2):408–14.

217. Goren MP, Viar MJ, Shenep JL, Wright RK, Baker DK, Kalwinsky DK. Monitoring serum aminoglycoside concentrations in children with amphotericin B nephrotoxicity. Pediatr Infect Dis J 1988;7(10):698–703.

218. Dupuis JY, Martin R, Tetrault JP. Atracurium and vecuronium interaction with gentamicin and tobramycin. Can J Anaesth 1989;36(4):407–11.

219. Chapple DJ, Clark JS, Hughes R. Interaction between atracurium and drugs used in anaesthesia. Br J Anaesth 1983;55(Suppl 1):S17–22.

220. Conlon BJ, McSwain SD, Smith DW. Topical gentamicin and ethacrynic acid: effects on cochlear function. Laryngoscope 1998;108(7):1087–9.

221. Hoffman DW, Whitworth CA, Jones-King KL, Rybak LP. Potentiation of ototoxicity by glutathione depletion. Ann Otol Rhinol Laryngol 1988;97(1):36–41.

222. Orsulakova A, Schacht J. A biochemical mechanism of the ototoxic interaction between neomycin and ethacrynic acid. Acta Otolaryngol 1982;93(1–2):43–8.

223. Brown CB, Ogg CS, Cameron JS, Bewick M. High dose frusemide in acute reversible intrinsic renal failure. A preliminary communication. Scott Med J 1974;19(Suppl 1):35–9.

224. Thomsen J, Bech P, Szpirt W. Otologic symptoms in chronic renal failure. The possible role of aminoglycoside–furosemide interaction. Arch Otorhinolaryngol 1976;214(1):71–9.

225. Lawson DH, Tilstone WJ, Gray JM, Srivastava PK. Effect of furosemide on the pharmacokinetics of gentamicin in patients. J Clin Pharmacol 1982;22(5–6):254–8.

226. Bates DE, Beaumont SJ, Baylis BW. Ototoxicity induced by gentamicin and furosemide. Ann Pharmacother 2002;36(3):446–51.

227. Kovesi TA, Swartz R, MacDonald N. Transient renal failure due to simultaneous ibuprofen and aminoglycoside therapy in children with cystic fibrosis. N Engl J Med 1998;338(1):65–6.

228. Paganini H, Marin M. Caracteristicas farmacocineticas y espectro antimicrobiano de la teicoplanina. [Pharmacokinetic characteristics and antimicrobial spectrum of teicoplanin.] Medicina (B Aires) 2002;62(Suppl 2):52–5.

Antituberculosis drugs

See also Capreomycin, Clofazimine, Cycloserine, Ethambutol, Ethionamide and protionamide, Isoniazid, Pyrazinamide, Rifamycins, Thiacetazone

General Information

Antituberculosis drugs are classified as first-line and second-line.

First-line drugs

- ethambutol
- isoniazid
- pyrazinamide
- rifampicin
- streptomycin

Second-line drugs

- capreomycin
- clofazimine
- cycloserine
- ethionamide and propionamide
- fluoroquinolones
- kanamycin
- *para*-aminosalicylic acid
- rifabutin
- thiacetazone

As a rule, a regimen of two, three, or four of the five first-line antituberculosis drugs (isoniazid, rifampicin, pyrazinamide, ethambutol, and streptomycin) is used in tuberculosis (1). The 6-month short-course regimen consists of isoniazid, rifampicin, and pyrazinamide for 2 months, followed by isoniazid and rifampicin for 4 months (1). It may be advisable to include ethambutol in the initial phase when isoniazid resistance is suspected or if the prevalence of primary resistance to isoniazid is over 4% in new cases. A 9-month regimen consisting of isoniazid and rifampicin is also highly successful (1). Treatment should always include at least two drugs to which the mycobacteria are susceptible.

Careful monitoring and the addition of pyridoxine to isoniazid have reduced the number of adverse drug effects in tuberculosis. Awareness of potentially severe hepatotoxic reactions is vital, because hepatic failure may be a devastating and often fatal condition. Fulminant hepatic failure caused by rifampicin, isoniazid, or both has been described (2).

Treatment problems that can arise are mainly of two types: adverse reactions (collateral, toxic, or hypersusceptibility reactions), and initial or acquired resistance of *Mycobacterium tuberculosis*, *Mycobacterium bovis*, or non-tuberculous mycobacteria to one or more of the antituberculosis drugs. The latter probably only occurs when the patient has not taken the full combination or the full doses of the drugs all the time. Combination formulations are thus particularly useful. Multidrug-resistant tuberculosis, defined as resistance against at least isoniazid and rifampicin, is the most clinically relevant form of resistance to treatment worldwide.

Observational studies

An increasing number of patients with multidrug-resistant tuberculosis are being treated with second-line drugs worldwide, often in places with poor resources. The number of drugs used is large (4–9) and treatment is prolonged (1–2 years). There is justifiable concern over patients' tolerance of such regimens and their adverse effects, which determine adherence to treatment. Treatment has to be individualized according to the WHO guidelines for a DOTS-plus strategy.

It is therefore encouraging to read a report from Lima, Peru, where 60 patients from a shanty town tolerated a median of eight antituberculosis drugs fairly well for a median duration of 20 months (3). All received a parenteral aminoglycoside daily for 6 months, cycloserine, and a fluoroquinolone, and most also took *para*-aminosalicylic acid and ethionamide. Of 60 patients, 23 took clofazimine, 23 pyrazinamide, 25 isoniazid, and 3 rifampicin. Commonly encountered adverse effects included dermatological effects, including bronzing of the skin (many of these patients were taking clofazimine and fluoroquinolones), depression, anxiety, and peripheral neuropathy. All complained of mild gastritis. There were no cases of serious hepatic or renal toxicity. This may have been because only a few patients took rifampicin. Absence of eighth nerve toxicity was striking, and can be attributed to close monitoring of patients by physicians with experience of DOTS-plus regimens.

In a similar report from Turkey, adverse reactions to drugs led to withdrawal of one or more drugs in 62 of 158 patients (39%) (4). Outcomes were favorable and cultures became negative in 95% of the patients within 2 months.

The authors of an observational study in 367 HIV-infected patients with 372 episodes of culture-confirmed tuberculosis analysed the factors that complicate antituberculosis therapy (5). In 25% there was hepatic disease at the time of the diagnosis of tuberculosis or during antituberculosis therapy, and there were rises in serum transaminases to at least twice the upper limits of the reference ranges during the first month of antituberculosis therapy in 116 (31%) of the episodes. The most commonly reported adverse effects were rash (28%), nausea (26%), leukopenia or neutropenia (20%), diarrhea (19%), vomiting (19%), and raised temperature (17%). There was co-prescription of rifampicin with medications that interact with rifampicin during 270 episodes (72%).

General adverse reactions

Adverse reactions are often due to the combined effects of two or more drugs used simultaneously (6). Hypersusceptibility reactions can occur even to more than one agent. The incidence of adverse reactions to drugs used in the treatment of tuberculosis is higher in elderly patients, who are more likely to have intercurrent illnesses and a lower lean body mass than younger patients. In two studies from Hong Kong in patients being treated for tuberculosis with rifampicin, the incidence of adverse reactions was higher with regimens containing rifampicin;

furthermore, patients taking rifampicin had a higher steady-state plasma concentration of isoniazid (1,7).

Some simple rules about which drugs are more likely to cause which reactions reflect the principle that the most probable causative agent (or agents) must be stopped.

Nervous and sensory systems

Ethambutol is the most likely drug to cause visual disturbances. Isoniazid is associated with polyneuritis and reactions of the central nervous system. Streptomycin can cause eighth nerve toxicity.

Liver

The hepatotoxic potential of isoniazid, pyrazinamide, and rifampicin during antituberculosis chemotherapy has been reviewed (8). Hepatotoxicity is consistently the most common serious adverse reaction in patients taking antituberculosis drugs. Hepatic necrosis is the most important adverse effect of first-line antituberculosis drug therapy (2). Asymptomatic rises in transaminases are common and are not by themselves justification for withdrawing medication, since they settle spontaneously in most patients while treatment continues. All patients taking antituberculosis drugs should be told to report all new illnesses, especially when associated with vomiting. Hepatitis B carriers were no more likely to react adversely to antituberculosis drugs than non-carriers (9).

If liver damage occurs, isoniazid is probably an important factor and it should be stopped before rifampicin or pyrazinamide (10). Prediction of hepatotoxicity is possible (11). In a case-control study of 60 patients in India, conducted in order to identify features predicting hepatotoxicity, the body mass index was significantly lower (17.2 kg/m^2) in patients who experienced hepatotoxicity than in controls (19.5 kg/m^2) (12).

Urinary tract

In cases of renal insufficiency, streptomycin and ethambutol or second-line antituberculosis drugs with renal toxicity should be immediately withdrawn.

Skin

Adverse reactions to all types of medication are more common in HIV-positive individuals; skin reactions are especially frequent and often severe. Thiacetazone is well recognized as a cause of severe reactions, some of them fatal, but even in combination antituberculosis drug regimens that exclude thiacetazone, the incidence of adverse skin reactions is much higher in HIV-positive than HIV-negative patients: 23% against 1% in one study from Cameroon (13).

Immunologic

In allergic reactions, the drug most probably responsible can be difficult to identify, since the same kind of reaction can occur independently of the chemical nature of the drug. For evaluation of allergic drug reactions, the analysis of time relations (duration of exposure, reaction time, drug-free interval before re-exposure) is extremely important. Particularly in allergic reactions to rifampicin, intermittent treatment or re-exposure after a drug free-

interval favors sensitization and occurrence. Depending on the severity of the adverse effects, one, two, or all drugs must be stopped until the adverse reaction has completely disappeared. The use of second-line antituberculosis drugs may sometimes be necessary. In patients with drug fever or common rashes, specific desensitization may be attempted, at least with isoniazid (14). In more severe reactions, with anaphylactic shock, agranulocytosis, thrombocytopenia, toxic epidermal necrolysis, or Stevens–Johnson syndrome, specific desensitization should not be considered and the drug should be discarded from the combination.

Organs and Systems

Liver

Hepatotoxicity is the most important adverse effect of antituberculosis drug therapy. Isoniazid, rifampicin, and pyrazinamide are the main culprits. There is wide variability in the risk of hepatotoxic reactions reported from different parts of the world or in different populations (for example African-American women in the postpartum period) (SEDA-24, 353).

The American Thoracic Society has issued a statement on the hepatotoxicity of antituberculosis drugs (15). The liver has a central role in drug metabolism and detoxification, and is consequently vulnerable to injury. The pathogenesis and types of drug-induced liver injury range from hepatic adaptation to hepatocellular injury. Systematic steps for preventing and managing liver damage include patient and regimen selection to optimize the benefits to harm balance, staff and patient education, ready access to care for patients, good communication among providers, and judicious use of clinical and biochemical monitoring. During treatment of latent tuberculosis, alanine transaminase monitoring is recommended for those who chronically take alcohol or concomitant hepatotoxic drugs, have viral hepatitis or other pre-existing liver disease or abnormal baseline transaminase activity, have had prior isoniazid-induced hepatitis, are pregnant, or are within 3 months post-partum. Treatment should be withdrawn and a modified or alternative regimen used for those with raised transaminase activity more than three times the upper limit of normal in the presence of hepatitis symptoms and/or jaundice, or five times the upper limit in the absence of symptoms.

In a retrospective comparison of isoniazid for 9 months (n = 770) and rifampicin for 4 months (n = 1379) for latent tuberculosis the percentages of patients who completed 80% or more of their prescribed treatment were 53% and 72% respectively (16). Clinically recognized adverse reactions resulted in permanent treatment withdrawal in 4.6% and 1.9% respectively. Clinically recognized hepatotoxicity was more common with isoniazid (1.8%) than rifampicin (0.08%).

Risks

The risks of hepatotoxicity differ with different antituberculosis drug combinations.

Rifampicin plus isoniazid

There are mild to moderate increases in liver transaminases during treatment with rifampicin plus isoniazid in most patients. However, biochemical hepatitis is diagnosed when transaminase activities increase to more than four times the upper limit of the reference ranges on two occasions at least 1 week apart, or more than five times on any single occasion. This calls for withdrawal of all potentially hepatotoxic drugs (rifampicin, isoniazid, and pyrazinamide) until the enzymes return to the reference ranges. During this period, streptomycin plus ethambutol, with or without cycloserine and a fluoroquinolone, is recommended in seriously ill patients.

Hepatotoxicity is generally considered to be rare among children who receive antituberculosis drugs. However, a report from Japan has suggested that this might not be the case, at least in Asia (17). The authors noted high activities of transaminases, more than five times the upper end of the reference range, in eight of 99 children aged 0–16 years who received various combinations of drugs, including rifampicin and isoniazid; 18 children were excluded because of baseline abnormalities in liver function. Age under 5 years and pyrazinamide in the drug regimen were risk factors for hepatotoxicity. There have been few other reports of the risk of hepatotoxicity in children.

The efficacy of weekly rifapentine + isoniazid has been compared with daily rifampicin + pyrazinamide in preventing tuberculosis in 399 household contacts of patients with pulmonary tuberculosis in Brazil (18). The median age was 34 years and the median weight 63 kg; 60% were female and only one patient was HIV infected. The trial was halted by the investigators before completion because of unanticipated hepatotoxicity in the rifampicin + pyrazinamide arm. Of 193 participants who received rifampicin + pyrazinamide 20 had grade 3 or 4 hepatotoxicity, compared with two of 206 participants who received rifapentine + isoniazid. There were no hospitalizations or deaths due to hepatotoxicity, and all the participants' liver enzymes returned to normal during follow-up. During follow-up, there were four cases of active tuberculosis, three in the rifapentine + isoniazid group and one in the rifampicin + pyrazinamide group. The authors concluded that rifapentine + isoniazid was better tolerated than rifampicin + pyrazinamide and was associated with good protection against tuberculosis.

Pyrazinamide plus rifampicin

There is continuing concern about the hepatotoxicity of the combination of pyrazinamide with rifampicin for the treatment of latent pulmonary tuberculosis. Among 148 who were given the combination for 2 months, grade 3 hepatotoxicity (transaminases more than 5–20 times the upper limit of the reference range) and grade 4 hepatotoxicity (transaminases more than 20 times the upper limit of the reference range) were reported in 10 and four patients respectively (19). The risk of hepatotoxicity was associated with female sex (OR = 4.1; 95% CI = 1.2, 14) and presumed recent infection (OR = 14.3; 95% CI = 1.8, 115). The investigators recommended caution

in using the combination of pyrazinamide with rifampicin in populations in whom its safety has not been established.

Others consider that this combination is useful for high-risk, traditionally non-adherent patients, such as alcoholics and the homeless, but have also emphasized the need for careful monitoring for toxicity (20). This suggestion is based on the presumption that the combination regimen, although toxic, is more likely to be completed by high-risk patients than 6 months of isoniazid alone. However, others observed similar completion rates for the two regimens in a multicenter study (61% and 57%) (21), and the safety and cost-effectiveness of this combination in the treatment of latent tuberculosis has been questioned (22).

Two cases of severe and fatal hepatitis have been reported to the CDC from New York and Georgia among patients taking rifampicin and pyrazinamide for latent tuberculosis (23). Between February and August 2001, another 21 patients had severe hepatotoxicity following treatment with rifampicin plus pyrazinamide for 2 months for latent tuberculosis, as reported to the CDC; five died of fulminant hepatic failure, two of whom had recovered from isoniazid-induced hepatitis (24). This report led to revision of previous guidelines of the American Thoracic Society. According to the revised guidelines, isoniazid for 9 months is the preferred treatment for latent tuberculosis infection in HIV-negative subjects, followed by isoniazid for 6 months or rifampicin for 4 months. Rifampicin plus pyrazinamide for 2 months can be used with caution in these patients, especially if they are taking other medications that are associated with liver injury, and those who drink a lot of alcohol. This combination is not recommended for those with underlying liver disease or who have had isoniazid-associated liver damage. Liver function tests should be measured at baseline and at 2-weekly intervals thereafter for 6 weeks. Only two non-fatal cases of severe liver injury have been reported to the CDC since the publication of revised guidelines as of November 2002 (25).

Data on patients who died or were hospitalized for liver disease within 1 month after taking rifampicin + pyrazinamide for latent tuberculosis have been reported (26). Liver injury was attributable to rifampicin + pyrazinamide in 50 patients reported, of whom 12 died. Rifampicin + pyrazinamide was the likeliest cause of liver injury in 47 patients. The median age was 44 (range 17–73) years and 32 (64%) were men. Seven of 43 patients tested had hepatitis C virus antibodies, one of 45 had chronic hepatitis B, three of 22 had positive HIV serology, 34 of 48 had excess alcohol use, and 33 of 50 were taking other hepatotoxic medications. Six patients, two of whom died, had no predictors of liver disease. Patients who died were older and had taken more medications than did those who recovered. Of 31 who were monitored according to guidelines, nine died. Death was predicted by age and the use of other medications, but none of the co-factors was a promising predictor of severe liver injury.

Susceptibility factors

Acetylator status and other genetic factors, old age, pre-existing hepatic dysfunction, alcoholism, co-infection with hepatitis virus, and malnutrition are important potential susceptibility factors for the development of liver damage in patients taking antituberculosis drugs (27), but there have been inconsistent findings with regard to some of these risk factors in different studies.

Genetic factors

The issue of hepatic dysfunction and acetylator status during treatment with isoniazid plus rifampicin has been re-examined in 77 Japanese patients with pulmonary tuberculosis (28). There was a marked increase in the risk of hepatotoxicity amongst slow acetylator NAT2* genotypes (a combination of mutant alleles) compared with the rapid acetylator genotype (homozygous NAT2*4). Using Taylor's series analysis it can be calculated that the relative risk was 28 (95% CI = 4.1, 192). Despite a small sample size (seven slow acetylator genotypes, 42 intermediate, and 28 rapid) the relative risk was highly significant, which is not surprising if all seven of the slow acetylators and only one of the 28 rapid acetylators developed hepatotoxicity.

A unique feature of this study was the determination of acetylator status by genotyping rather than phenotyping. There is generally good concordance between the two methods, but in the presence of hepatic dysfunction the phenotype assessment may not reflect the genotype. Furthermore, 42 of the 77 patients were assigned to the intermediate acetylator genotype, based on heterozygosity for NAT2*4 and a mutant allele. However, phenotyping by estimation of concentrations of metabolites of the commonly used probes does not consistently result in identification of intermediate acetylators. The Japanese are mostly fast acetylators (~90%) compared with Caucasians or Indians (40–50%). It is unlikely, however, that the observed association between slow acetylator genotype and the high risk of hepatic dysfunction was affected by any of these considerations. The dose of isoniazid was rather large (~8 mg/kg/day) and this may have increased the risk of hepatotoxicity. Furthermore, hepatotoxicity was defined as an increase in transaminases to one and a half times the top of the reference range. This degree of hepatic dysfunction is not uncommon in patients taking antituberculosis drugs, and is no indication for withdrawal or modification of treatment. It is not possible to assess the risk of severe hepatotoxicity during treatment, owing to lack of detailed information in the published report.

A case-control study has suggested that there is also an increased risk of antituberculosis drug-induced hepatotoxicity in individuals with a glutathione-S-transferase M1 "null" mutation (29). Reduced glutathione transferase activity could theoretically predispose individuals to adverse effects of toxic metabolites and xenobiotics. These observations need to be confirmed.

Old age

Old age and the presence of hepatic dysfunction on baseline evaluation are the most consistent predictors of hepatotoxicity during antituberculosis therapy. Workers from Florida have reported a five-fold increase in the likelihood of drug-induced hepatotoxicity in patients who are hepatitis C-positive and a four-fold increase in patients who are HIV-positive, compared with seronegative patients treated for tuberculosis (30). In all, 134 patients taking antituberculosis drugs were monitored for drug-induced hepatotoxicity, defined as an increase in aspartate transaminase and/or alanine transaminase activity from normal to at least three times normal and/or an increase in bilirubin above normal. Of the 22 patients who developed drug-induced hepatotoxicity, only six developed drug-induced hepatotoxicity on re-introduction of treatment after an interval in which the abnormalities had resolved. Four of the six had liver biopsies, which showed active inflammation, attributed (at least in part) to hepatitis C. These were then treated with interferon alfa, with improvement of liver chemistry. On improvement, antituberculosis drug therapy was successfully re-introduced in the form of isoniazid and rifabutin, the latter being considered to be less hepatotoxic than rifampicin. A report from Hong Kong has suggested that the risk is much greater in hepatitis B virus carriers taking antituberculosis drugs (31). Even after excluding patients who had raised baseline alanine transaminase activity or with HbeAg seroconversion during the phase of hepatic dysfunction, the risk of hepatotoxicity was still significantly higher in hepatitis B carriers taking antituberculosis drugs compared with non-carriers (26% versus 8.8%). These observations are of considerable importance in regions of the world in which the prevalence of hepatitis B infection as well as tuberculosis is high, such as South-East Asia and sub-Saharan Africa.

Liver transplantation

Enhanced hepatotoxicity of conventional antituberculosis regimens has been reported in recipients of orthotopic liver transplants, which is not unexpected, because of bouts of organ rejection (32). The authors recommended ofloxacin for these patients on the basis of favorable outcome in six cases. A conventional antituberculosis induction regimen was used initially until hepatotoxicity developed in all six patients. Thereafter they were treated with a combination of ofloxacin and ethambutol, with apparent cure in all. It should be noted that most of the patients took isoniazid + rifampicin for almost 2 months, which is the usual period when hepatotoxic reactions occur. Perhaps one should evaluate substitution of rifampicin with ofloxacin from the very beginning in order to minimize hepatotoxicity, as well as interference with ciclosporin leading to graft rejection noted in an earlier study (33).

Prevention

One of the most important predictors of hepatotoxicity during antituberculosis drug therapy is an abnormal liver function test at baseline. It is reasonable to avoid potentially hepatotoxic drugs in the management of patients with pre-existing liver disease.

The use of ofloxacin instead of rifampicin in antituberculosis drug regimens for patients with underlying chronic liver disease has been reported to be associated with a significantly lower risk of hepatotoxicity (34). Similar observations have been reported among carriers of hepatitis B and liver transplant recipients by other investigators.

Management

There are four issues related to the management of patients who develop hepatotoxicity during treatment with antituberculosis drugs:

1. whatthe preferred treatment regimen should be for patients with significantly abnormal liver functions at baseline;
2. when treatment should be stopped/modified if hepatic dysfunction develops;
3. what antituberculosis treatment, if any, should be used until liver function improves;
4. what a safe regimen is for re-treatment of these patients.

Several herbal products have been claimed to mitigate drug-induced hepatitis caused by antituberculosis agents. However, few of them have undergone rigorous randomized controlled trials.

- Glycyrrhizin is widely used in Japan for the treatment of chronic hepatitis, but in a non-randomized trial in 24 patients who developed drug-induced hepatitis while undergoing antituberculosis chemotherapy, there was no difference in the time required for recovery between the patients who were treated with or without intravenous glycyrrhizin 40 ml/day (35).
- *Moringa oleifera*, commonly known as "drumstick," has been mentioned in the treatment of various illnesses in Indian folk medicine, and an ethanolic extract of the leaves had a hepatoprotective effect in a rat model of antituberculosis drug-induced liver injury (36).
- Russian investigators have reported a hepatoprotective effect of a plant product "Galstena" in a rat model and have extended their observations to a clinical trial, with favorable results; however, few data were given in this report (37).

It cannot be denied that there is a need for continuing research in this area, but it is necessary to undertake well-conducted scientific studies before claims of hepatoprotective effects of herbal products can be accepted.

There is a lack of consensus on the best re-treatment protocol for patients who develop hepatotoxicity during treatment with standard antituberculosis agents. Investigators from Turkey have reported a high risk of recurrence of hepatitis (in six of 25 patients) on re-introduction of all drugs in full doses after recovery from hepatitis (38). This risk was less when rifampicin and isoniazid were re-introduced sequentially in increasing doses and when pyrazinamide was replaced by streptomycin.

- A 19-year-old woman with ovarian and peritoneal tuberculosis was given isoniazid 300 mg/day, rifampicin 600 mg/day, ethambutol 1500 mg/day, and pyrazinamide 1500 mg/day (39). On the third day she developed fatigue, malaise, and jaundice. Other causes were ruled out and the antituberculosis drug therapy was immediately withdrawn. The next day she became unconscious and icteric and did not respond to verbal or painful stimuli. Her serum aspartate transaminase activity was 1301 IU/l, alanine transaminase 1332 IU/l, total bilirubin 10.5 mg/dl, and prothrombin time 71 seconds. She underwent living-related partial liver transplantation and was then given non-hepatotoxic antituberculosis drugs and low-dose immunosuppressants.

The authors found four similar published cases and concluded that liver transplantation is feasible and effective for antituberculosis drug-induced hepatic failure.

Immunologic

Treatment of tuberculosis can be associated with a transient worsening of clinical symptoms a few days to weeks after the start of therapy. This scenario was described over 30 years ago and was designated as a "paradoxical reaction" (40). The most common paradoxical reactions include worsening of lymphadenopathy, increasing central nervous system lesions, and worsening respiratory symptoms (41). The pathogenesis of this syndrome, the so-called immune reconstitution syndrome, is thought to be reconstitution of a cell-mediated immune response to mycobacterial antigens that were either absent, suppressed, or dysregulated in untreated tuberculosis (42,43).

Immune reconstitution disease is more common in HIV-infected patients with tuberculosis, especially those who are receiving antiretroviral drug therapy. This entity needs to be more widely recognized, as it can present challenging diagnostic and management scenarios for practitioners (44,45). In one prospective study of HIV-infected patients with tuberculosis, immune reconstitution disease was reported among 36% of those who were taking antiretroviral drugs and 7% of those who were taking antituberculosis drugs only ([46]). After the start of antiretroviral therapy, there is reconstitution of antigen-specific responses to a number of pathogens, and this presumably contributes to the amplified inflammatory responses (47). Immune reconstitution disease needs to be distinguished from treatment failure, progression of drug resistant tuberculosis, drug reactions, and other HIV-related complications that can mimic immune reconstitution disease.

Immune reconstitution disease secondary to tuberculosis most commonly presents with fever, lymphadenopathy, and worsening respiratory symptoms. Most cases have been reported within the first 2 months of antiretroviral therapy, with a median duration of HIV therapy of 4 weeks. A CD4+ cell count of under 50×10^6/l before HIV therapy is a significant risk factor for the development of immune reconstitution disease. There is usually a rise in

CD4+ cell count when immune reconstitution disease develops, but that is not a prerequisite for the diagnosis (48).

In most cases, the management of suspected immune reconstitution disease includes continuation of tuberculosis and antiretroviral drug therapy, close observation, evaluation of other possible intercurrent illnesses, and review of adherence to HIV medications. In the presence of severe lymphadenopathy, worsening respiratory status, or worrisome central nervous system symptoms, a course of glucocorticoids can be tried (44). In extreme cases, antiretroviral drugs may need to be withheld. While there have been no deaths reported from tuberculosis-associated immune reconstitution disease, morbidity from this syndrome can be severe.

Long-Term Effects

Drug tolerance

Primary and secondary bacterial resistance to antituberculosis drugs represents a major problem. This can be demonstrated by resistance tests, if available. When testing is not done, primary or secondary resistance can only be suspected when drug treatment fails. The incidence of bacterial resistance varies enormously from country to country and from population to population. In Tanzania (49), where the WHO and the International Union Against Tuberculosis and Lung Diseases (IUATLD) has developed antituberculosis programs, primary resistance to isoniazid and/or streptomycin is found in 10% of cases (50). In contrast, South-East Asian and African patients harbor resistant bacteria markedly more often (SED-10, 572); resistance tests therefore remain mandatory for good epidemiological and therapeutic control. Recent reports from the WHO drug resistance surveillance network have allowed an overview of the current situation worldwide and facilitate the choice of medication according to the origin of the patients (51).

Multidrug-resistant tuberculosis generally results from inadequate therapy or lack of compliance with therapy. A strain of mycobacteria is called resistant when it is insensitive to one of the first-line drugs. It is called multiresistant when it is insensitive to both isoniazid and rifampicin. In this case other antituberculosis drugs may also be ineffective (52). In practice, at least two second-line antituberculosis drugs, selected on the basis of individual drug susceptibility, are given in combination with a fluoroquinolone (53).

Drug malabsorption may contribute to the emergence of acquired drug resistance. It has been described in HIV-infected patients with advanced disease (54), and also in immunocompetent patients (55). Thus, in addition to the use of directly observed therapy to ensure compliance, it is advisable to monitor antimycobacterial drug concentrations routinely in such patients. Practical proposals for the choice of antituberculosis drugs in special circumstances, including drug resistance, have been made (56).

Susceptibility Factors

Genetic

Differences in drug response between individuals can be due to the occurrence of genetic polymorphisms in drug metabolizing enzymes (57). Because of genetic variation in drug metabolizing capacity, a predisposed individual may experience:

1. lack of efficacy at a normal drug dose, requiring a higher dose to achieve the expected therapeutic response;
2. a much larger effect at the usual dose, leading to adverse effects.

Acetylation

Drug metabolizing enzymes are responsible for degradation of drugs and environmental pollutants and are important determinants of drug action. An example is the polymorphism in acetylation that is mediated by N-acetyltransferase isoenzymes NAT1 and NAT2 in the liver (58). More than 25 NAT2 genotypes and about 20 NAT1 genotypes have been reported. Based on NAT2 phenotype, individuals are characterized as rapid, intermediate, or slow acetylators. Isoniazid and some sulfonamides, such as sulfadimidine, are typical substrates for NAT2, while NAT1 metabolizes para-aminosalicylic acid and para-aminobenzoic acid. Caffeine is metabolized by both enzymes. There is significant variation in the distributions of various phenotypes in different parts of the world. The Inuit and Japanese have the lowest rates for slow acetylators (about 10%), while in India it is high at around 60%. This has implications for the metabolism and detoxification of isoniazid. Slow acetylators have an increased risk of peripheral neuropathy during therapy with isoniazid while rapid acetylators are more likely to have treatment failure and relapse if they take isoniazid twice weekly. Hepatotoxic reactions may be more common in slow acetylators, who also have an increased susceptibility to phenytoin toxicity. There is an increased risk of lupus-like syndrome among slow acetylators who take isoniazid, hydralazine, or procainamide.

Oxidation

The cytochrome P450 (CYP) mono-oxygenase system of enzymes is responsible for the major portion of drug metabolism in humans. Among the numerous P450 subtypes, CYP2D6, CYP3A4/5, CYP1A2, CYP2E1, CYP2C9, and CYP2C19 play important roles in genetically determined responses to a broad spectrum of drugs. If the in vitro clearance of a drug is largely mediated by a single polymorphically expressed or allelic variant, poor metabolizers will be characterized by disparate pharmacokinetics (for example high plasma AUCs and/or prolonged half-lives). About 40% of human CYP-dependent drug metabolism is carried out by enzymes that are polymorphically distributed.

Combined polymorphisms

There has been a recent evaluation of whether polymorphism of the CYP2E1 gene is associated with the development of antituberculosis drug-induced hepatitis (59). The CYP2E1 and NAT2 genotypes were determined using PCR. Patients with the homozygous wild genotype CYP2E1 c1/c1 had a very significantly higher risk of hepatotoxicity (20%, OR = 2.52) than those with the mutant allele c2 (9%). When the CYP2E1 genotype was combined with acetylator status, the risk of hepatotoxicity increased from 3.94 for CYP2E1 c1/c1 plus rapid acetylator status to 7.43 for CYP2E1 c1/c1 plus slow acetylator status. The authors concluded that CYP2E1 genetic polymorphism may be associated with susceptibility to antituberculosis drug-induced hepatitis, even after adjustment for age.

Such genetic variations may account for the variability in reported adverse effects due to antituberculosis drugs in different populations. Thus, drug metabolism is affected by genetic factors impacting enzyme activity as well as external and internal factors, such as age, sex, environment, diet, and drug-drug interactions.

Future prospects

The study of pharmacogenetic differences holds the potential to improve therapeutic effectiveness and limit the adverse effects of available drugs, in other words to individualize medicine. Pharmacogenetics can also provide substantial efficiency in clinical research by facilitating the conduct of smaller clinical trials by targeting groups of patients with similar genetic backgrounds. Pharmacogenomics will also play a role in the development of new clinical entities in order to reduce adverse drug reactions.

Other factors

In a retrospective study of the adverse effects of antituberculosis drugs and associated susceptibility factors in patients with active tuberculosis admitted to the Respiratory Ward of a University Clinic between 1984 and 2001, 95 of 1149 patients (8.3%) had adverse effects (60). The frequency of adverse drug reactions increased from 0.6% at ages under 20 years to 5.2% at ages 20–40. There were no sex or age differences between patients who did and did not have adverse effects. There were asymptomatic liver function disturbances in 56 patients (4.9%) but the rate of hepatotoxicity was only 2.4%. There were no age or sex differences among those who had hepatotoxicity and who had not. The major adverse effects were ototoxicity (1.7%), hepatotoxicity (0.8%), neuropsychiatric effects (0.7%), and hyperuricemia (0.6%). The authors concluded that severe adverse reactions to antituberculosis drugs are relatively common, especially among patients hospitalized for pulmonary tuberculosis.

Current management of tuberculosis involves taking at least four drugs and a minimum treatment duration of 6 months. Moreover, the emergence of newer strains of tuberculosis-like multidrug resistant organisms (MDR-TB) and extremely resistant organisms (XDR-TB) has focused attention on the development of new antituberculosis drugs. There is now a pipeline of new compounds or classes of compounds that are being specifically tested for their potential effectiveness in the treatment of tuberculosis.

Newer Antituberculosis Drugs

Drug targets

The most important aim in drug development is the identification of novel drug targets involved in vital aspects of bacterial growth, metabolism, and viability, the inactivation of which will lead to bacterial death or inability to persist (61). The Mycobacterium tuberculosis genome sequence and mycobacterial genetic tools, such as transposon mutagenesis and signature-tagged mutagenesis, have been used to identify genes essential for growth of the organism in vitro and its survival in vivo (62,63). Inhibition of the host tissue liquefaction process, which facilitates reduced transmission, represents a novel approach to the design of new drugs (62). Targeted knockout of specific genes, whose disruption leads to non-viability of the bacilli, is a valuable approach to identifying essential gene products involved in mycobacterial persistence. Some enzymes, including isocitrate lyase (ICL), PcaA (a methyltransferase involved in the modification of mycolic acid), RelA (ppGpp synthase), and DosR (controlling a 48-gene regulon involved in mycobacterial survival under hypoxic conditions) have been identified as targets for the development of drugs to kill persistent bacilli (64,65,66,67). Energy production pathways, such as the electron transport chain, glycolytic pathways (like the Embden–Meyerhof pathway), and fermentation pathways, could be good targets for drug development (61). In identifying drugs that kill persistent organisms and thereby shortening the duration of treatment, novel drug screens that mimic in vivo conditions in lesions (i.e. acidic pH and hypoxia) and act against old stationary-phase non-growing bacilli could be important (68,69). In addition, drug combination screens could be performed to identify drugs that have synergistic effects. The systems biology approach, which proposes using multiple compounds that hit multiple targets in different pathways to achieve the desired outcome, can be used for identifying novel drug combinations against tuberculosis (61).

In the growing pipeline of potential new antituberculosis drugs there are currently seven novel compounds that are not yet approved for the treatment of tuberculosis and are in various stages of clinical development (70). The most advanced of these are the fluoroquinolones, specifically gatifloxacin and moxifloxacin, which are currently being evaluated in phase 2 and 3 clinical trials. Two other compounds (TMC207 and OPC67683) have completed phase 1 clinical trials and early bactericidal activity (EBA) studies, PA824 has completed its phase 1 program, and two other compounds (LL3858 and SQ109) are currently being evaluated in phase 1 clinical studies.

Diarylquinoline (TMC207)

A novel diarylquinoline TMC207 (previously referred to as "R207910") is being developed by Tibotec, a Johnson & Johnson subsidiary. It has many characteristics, both in vitro and in vivo, that make it a very attractive antituberculosis drug candidate. It has very potent in vitro activity against both multidrug-resistant and drug-susceptible strains of *Mycobacterium tuberculosis* (71). The target for diarylquinoline has been proposed to be mycobacterial F1F0 proton ATP synthase, which is a new drug target in mycobacteria. TMC207 was more active than isoniazid and rifampicin in a mouse model and shortened therapy from 4 months to 2 months in mice with established infection. In phase 1 studies in humans tolerability was good and the pharmacokinetics were linear over the dose range studied. In a phase 1 study of multiple ascending doses there was accumulation, with increases in the AUC by a factor of about two between day 1 and day 14. The "effective half-life" was of the order of 24 hours. TMC207 is about to enter a randomized phase 2 study in a population of patients with MDR-TB (70).

Nitroimidazoles (PA824 and OPC67683)

Two nitroimidazoles are currently in clinical development, the nitroimidazo-oxazine PA824, which is being developed by the TB Alliance, and the dihydroimidazo-oxazole OPC67683, which is being developed by Otsuka Pharmaceutical (70).

PA824 has an MIC as low as 0.015–0.250 µg/ml against drug sensitive and multidrug resistant *Mycobacterium tuberculosis* (72). PA824 is a prodrug that requires activation by a bacterial F420-depedent glucose-6-phosphate dehydrogenase (Fgd) and nitroreductase to activate components that then inhibit bacterial mycolic acid and protein synthesis (72). Pharmacokinetic studies of PA824 in rats have shown that it has excellent tissue penetration (73). In animals, PA824 was active against non-growing bacilli, even in microaerophilic conditions. and its activity is comparable to that of isoniazid, rifampicin, and moxifloxacin (72). PA824 had bactericidal activity in mice in the first 2 months of treatment and also in the continuation phase, which suggests that it has significant activity against non-growing persistent bacilli in vivo (73).

OPC67683 is extremely potent in vitro and in vivo against Mycobacterium tuberculosis (74). In a mouse model of chronic infection, OPC67683 was more efficacious that currently used antituberculosis drugs. The effective plasma concentration was 100 mg/l, which was achieved with an oral dose of 0.625 mg/kg. OPC67683 showed no cross-resistance with any of the currently used antituberculosis drugs. MICs against multiple clinically isolated tuberculosis strains were of the order of 6 mg/l (75).

New fluoroquinolones

C-8-methoxy-FQ, moxifloxacin, and gatifloxacin have a longer half-life and are more active against M. tuberculosis than the older quinolones. Moxifloxacin, in combination with rifampicin and pyrazinamide, killed tubercle bacilli in mice more effectively than the standard regimen of isoniazid + rifampicin + pyrazinamide and achieve stable cures in 4 months without relapse (76,77). Moxifloxacin has early bactericidal activity against tubercle bacilli comparable to that of isoniazid and was well tolerated in a preliminary human study (78). In a clinical trial conducted by the TB Trials Consortium in patients with pulmonary tuberculosis substitution of moxifloxacin for ethambutol did not influence 2-month sputum culture status but did result in a higher frequency of negative cultures at earlier times, which suggests that moxifloxacin has good sterilizing activity (79). There is a current trial by the TB Trials Consortium, in which moxifloxacin replaces isoniazid in patients with pulmonary tuberculosis. In addition, at the Tuberculosis Research Centre in Chennai, patients are being recruited for a randomized clinical study of the efficacy and tolerability of 3- and 4 month regimens containing moxifloxacin.

Pyrrole (LL3858)

Pyrrole LL3858, developed by Lupin, is being evaluated in a multidose phase 1 trial in healthy volunteers in India. It has submicromolar MICs and was be very active in a mouse model of tuberculosis. In combination with currently used antituberculosis drugs, LL3858 is reported to sterilize the lungs and spleens of infected animals more quickly than conventional therapy (80).

Diamine (SQ109)

The most recent compound to enter phase 1 clinical trials for tuberculosis is SQ109, which is being developed by Sequella. In in vitro and mouse in vivo studies the MIC against Mycobacterium tuberculosis was 100–630 mg/l (81). SQ109 has recently entered phase 1 studies in human volunteers.

References

1. Bass JB Jr, Farer LS, Hopewell PC, O'Brien R, Jacobs RF, Ruben F, Snider DE Jr, Thornton G. Treatment of tuberculosis and tuberculosis infection in adults and children. American Thoracic Society and The Centers for Disease Control and Prevention. Am J Respir Crit Care Med 1994;149(5):1359–74.
2. Mitchell I, Wendon J, Fitt S, Williams R. Anti-tuberculous therapy and acute liver failure. Lancet 1995;345(8949):555–6.
3. Furin JJ, Mitnick CD, Shin SS, Bayona J, Becerra MC, Singler JM, Alcantara F, Castanieda C, Sanchez E, Acha J, Farmer PE, Kim JY. Occurrence of serious adverse

effects in patients receiving community-based therapy for multidrug-resistant tuberculosis. Int J Tuberc Lung Dis 2001;5(7):648–55.

4. Tahaoglu K, Torun T, Sevim T, Atac G, Kir A, Karasulu L, Ozmen I, Kapakli N. The treatment of multidrug-resistant tuberculosis in Turkey. N Engl J Med 2001;345(3):170–4.

5. Dworkin MS, Adams MR, Cohn DL, Davidson AJ, Buskin S, Horwitch C, Morse A, Sackoff J, Thompson M, Wotring L, McCombs SB, Jones JL. Factors that complicate the treatment of tuberculosis in HIV-infected patients. J Acquir Immune Defic Syndr 2005;39:464–70.

6. Ormerod LP, Horsfield N. Frequency and type of reactions to antituberculosis drugs: observation in routine treatment. Tubercle Lung Dis 1966;77:37.

7. Combs DL, O'Brien RJ, Geiter LJ. USPHS Tuberculosis Short-Course Chemotherapy Trial 21: effectiveness, toxicity, and acceptability. The report of final results. Ann Intern Med 1990;112(6):397–406.

8. Yew WW, Leung CC. Antituberculosis drugs and hepatotoxicity. Respirology 2006;11:699–707.

9. Hwang SJ, Wu JC, Lee CN, Yen FS, Lu CL, Lin TP, Lee SD. A prospective clinical study of isoniazid–rifampicin–pyrazinamide–induced liver injury in an area endemic for hepatitis B. J Gastroenterol Hepatol 1997;12(1):87–91.

10. Schaberg T, Rebhan K, Lode H. Risk factors for side-effects of isoniazid, rifampin and pyrazinamide in patients hospitalized for pulmonary tuberculosis. Eur Respir J 1996;9(10):2026–30.

11. Dossing M, Wilcke JT, Askgaard DS, Nybo B. Liver injury during antituberculosis treatment: an 11-year study. Tuber Lung Dis 1996;77(4):335–40.

12. Singh J, Arora A, Garg PK, Thakur VS, Pande JN, Tandon RK. Antituberculosis treatment-induced hepatotoxicity: role of predictive factors. Postgrad Med J 1995;71(836):359–62.

13. Kuaban C, Bercion R, Koulla-Shiro S. HIV seroprevalence rate and incidence of adverse skin reactions in adults with pulmonary tuberculosis receiving thiacetazone free antituberculosis treatment in Yaounde, Cameroon. East Afr Med J 1997;74(8):474–7.

14. Hoigne R. Allergische Erkrankungen. In: Stucki P, Hess T, editors. Hadorn, Lehrbuch der Therapie. 7th ed.. Berne-Stuttgart-Vienna: Verlag Hans Huber, 1983:155.

15. Saukkonen JJ, Cohn DL, Jasmer RM, Schenker S, Jereb JA, Nolan CM, Peloquin CA, Gordin FM, Nunes D, Strader DB, Bernardo J, Venkataramanan R, Sterling TR An official ATS statement: hepatotoxicity of antituberculosis therapy. Am J Respir Crit Care Med 2006;174:935–52.

16. Page KR, Sifakis F, Montes de Oca R, Cronin WA, Doherty MC, Federline L, Bur S, Walsh T, Karney W, Milman J, Baruch N, Adelakun A, Dorman SE. Improved adherence and less toxicity with rifampin vs isoniazid for treatment of latent tuberculosis: a retrospective study. Arch Intern Med 2006;166:1863–70.

17. Ohkawa K, Hashiguchi M, Ohno K, Kiuchi C, Takahashi S, Kondo S, Echizen H, Ogata H. Risk factors for antituberculous chemotherapy-induced hepatotoxicity in Japanese pediatric patients. Clin Pharmacol Ther 2002;72(2):220–6.

18. Schechter M, Zajdenverg R, Falco G, Barnes G L, Faulhaber JC, Coberly JS, Moore RD, Chaisson RE. Weekly rifapentine/isoniazid or daily rifampin/pyrazinamide for latent tuberculosis in household contacts. Am J Respir Crit Care Med 2006;173:922–6.

19. Lee AM, Mennone JZ, Jones RC, Paul WS. Risk factors for hepatotoxicity associated with rifampin and pyrazinamide

for the treatment of latent tuberculosis infection: experience from three public health tuberculosis clinics. Int J Tuberc Lung Dis 2002;6(11):995–1000.

20. Stout JE, Engemann JJ, Cheng AC, Fortenberry ER, Hamilton CD. Safety of 2 months of rifampin and pyrazinamide for treatment of latent tuberculosis. Am J Respir Crit Care Med 2003;167(6):824–7.

21. Jasmer RM, Saukkonen JJ, Blumberg HM, Daley CL, Bernardo J, Vittinghoff E, King MD, Kawamura LM, Hopewell PC. Short-Course Rifampin and Pyrazinamide for Tuberculosis Infection (SCRIPT) Study Investigators. Short-course rifampin and pyrazinamide compared with isoniazid for latent tuberculosis infection: a multicenter clinical trial. Ann Intern Med 2002;137(8):640–7.

22. Jasmer RM, Daley CL. Rifampin and pyrazinamide for treatment of latent tuberculosis infection: is it safe? Am J Respir Crit Care Med 2003;167(6):809–10.

23. Centers for Disease Control and Prevention (CDC). Fatal and severe hepatitis associated with rifampin and pyrazinamide for the treatment of latent tuberculosis infection—New York and Georgia, 2000. MMWR Morb Mortal Wkly Rep 2001;50(15):289–91.

24. Centers for Disease Control and Prevention (CDC). Update: fatal and severe liver injuries associated with rifampin and pyrazinamide for latent tuberculosis infection, and revisions in American Thoracic Society/CDC recommendations—United States, 2001. MMWR Morb Mortal Wkly Rep 2001;50(34):733–5.

25. Centers for Disease Control and Prevention (CDC). Update: fatal and severe liver injuries associated with rifampin and pyrazinamide treatment for latent tuberculosis infection. MMWR Morb Mortal Wkly Rep 2002;51(44):998–9.

26. Ijaz K, Jereb JA, Lambert LA, Bower WA, Spradling PR, McElroy PD, Iademarco MF, Navin TR, Castro KG. Severe or fatal liver injury in 50 patients in the United States taking rifampin and pyrazinamide for latent tuberculosis infection. Clin Infect Dis 2006;42:346–55.

27. Pande JN, Singh SP, Khilnani GC, Khilnani S, Tandon RK. Risk factors for hepatotoxicity from antituberculosis drugs: a case-control study. Thorax 1996;51(2):132–6.

28. Ohno M, Yamaguchi I, Yamamoto I, Fukuda T, Yokota S, Maekura R, Ito M, Yamamoto Y, Ogura T, Maeda K, Komuta K, Igarashi T, Azuma J. Slow N-acetyltransferase 2 genotype affects the incidence of isoniazid and rifampicin-induced hepatotoxicity. Int J Tuberc Lung Dis 2000;4(3):256–61.

29. Roy B, Chowdhury A, Kundu S, Santra A, Dey B, Chakraborty M, Majumder PP. Increased risk of antituberculosis drug-induced hepatotoxicity in individuals with glutathione S-transferase M1 'null' mutation. J Gastroenterol Hepatol 2001;16(9):1033–7.

30. Ungo JR, Jones D, Ashkin D, Hollender ES, Bernstein D, Albanese AP, Pitchenik AE. Antituberculosis drug-induced hepatotoxicity. The role of hepatitis C virus and the human immunodeficiency virus. Am J Respir Crit Care Med 1998;157(6 Pt 1):1871–6.

31. Wong WM, Wu PC, Yuen MF, Cheng CC, Yew WW, Wong PC, Tam CM, Leung CC, Lai CL. Antituberculosis drug-related liver dysfunction in chronic hepatitis B infection. Hepatology 2000;31(1):201–6.

32. Meyers BR, Papanicolaou GA, Sheiner P, Emre S, Miller C. Tuberculosis in orthotopic liver transplant patients: increased toxicity of recommended agents; cure of disseminated infection with nonconventional regimens. Transplantation 2000;69(1):64–9.

33. Aguado JM, Herrero JA, Gavalda J, Torre-Cisneros J, Blanes M, Rufi G, Moreno A, Gurgui M, Hayek M, Lumbreras C, Cantarell C. Clinical presentation and outcome of tuberculosis in kidney, liver, and heart transplant recipients in Spain. Spanish Transplantation Infection Study Group, GESITRA. Transplantation 1997;63(9):1278–86.

34. Saigal S, Agarwal SR, Nandeesh HP, Sarin SK. Safety of an ofloxacin-based antitubercular regimen for the treatment of tuberculosis in patients with underlying chronic liver disease: a preliminary report. J Gastroenterol Hepatol 2001;16(9):1028–32.

35. Miyazawa N, Takahashi H, Yoshiike Y, Ogura T, Watanuki Y, Sato M, Kakemizu N, Yamakawa Y, U CH, Goto H, Odagiri S. [Effect of glycyrrhizin on anti-tuberculosis drug-induced hepatitis.]Kekkaku 2003;78(1):15–9.

36. Pari L, Kumar NA. Hepatoprotective activity of Moringa oleifera on antitubercular drug-induced liver damage in rats. J Med Food 2002;5(3):171–7.

37. Katikova OIu, Asanov BM, Vize-Khripunova MA, Burba EN, Ruzov VI. [Use of the plant hepatoprotector Galstena tuberculostatics-induced hepatic lesions: experimental and clinical study.]Probl Tuberk 2002;(4):32–6.

38. Tahaoglu K, Atac G, Sevim T, Tarun T, Yazicioglu O, Horzum G, Gemci I, Ongel A, Kapakli N, Aksoy E. The management of anti-tuberculosis drug-induced hepatotoxicity. Int J Tuberc Lung Dis 2001;5(1):65–9.

39. Idilman R, Ersoz S, Coban S, Kumbasar O, Bozkaya H. Antituberculous therapy-induced fulminant hepatic failure: successful treatment with liver transplantation and nonstandard antituberculous therapy. Liver Transplant 2006;12(9):1427–30.

40. Smith, H. Paradoxical responses during the chemotherapy of tuberculosis. J Infect 1987;15(1):1–3.

41. Cheng VC, Ho PL, Lee RA, Chan KS, Chan KK, Woo PC, Lau SK, Yuen KY. Clinical spectrum of paradoxical deterioration during antituberculosis therapy in non-HIV-infected patients. Eur J Clin Microbiol Infect Dis 2002;21(11):803–9.

42. Wendland T, Furrer H, Vernazza PL, Frutig K, Christen A, Matter L, Malinverni R, Pichler WJ. HAART in HIV-infected patients: restoration of antigen-specific CD4 T-cell responses in vitro is correlated with CD4 memory T-cell reconstitution, whereas improvement in delayed type hypersensitivity is related to a decrease in viraemia. AIDS 1999;13(14):1857–62.

43. Stone SF, Price P, French MA. Immune restoration disease: a consequence of dysregulated immune responses after HAART. Curr HIV Res 2004;2(3):235–42.

44. Breton G, Duval X, Estellat C, Poaletti X, Bonnet D, Mvondo Mvondo D, Longuet P, Leport C, Vilde JL. Determinants of immune reconstitution inflammatory syndrome in HIV type 1-infected patients with tuberculosis after initiation of antiretroviral therapy. Clin Infect Dis 2004;39(11):1709–12.

45. Lawn SD, Bekker LG, Miller RF. Immune reconstitution disease associated with mycobacterial infections in HIV-infected individuals receiving antiretrovirals. Lancet Infect Dis 2005;5(6):361–73.

46. Narita M, Ashkin D, Hollender ES, Pitchenik AE. Paradoxical worsening of tuberculosis following antiretroviral therapy in patients with AIDS. Am J Respir Crit Care Med 1998;158(1):157–61.

47. Shelburne SA 3rd, Hamill RJ, Rodriguez-Barradas MC, Greenberg SB, Atmar RL, Musher DW, Gathe JC Jr, Visnegarwala F, Trautner BW. Immune reconstitution inflammatory syndrome: emergence of a unique syndrome during highly active antiretroviral therapy. Medicine (Baltimore) 2002;81(3):213–27.

48. Carcelain G, Debre P, Autran B. Reconstitution of CD4+ T lymphocytes in HIV-infected individuals following antiretroviral therapy. Curr Opin Immunol 2001;13(4):483–8.

49. Tanzanian/British Medical Research Council Collaborative Study. Tuberculosis in Tanzania—a national survey of newly notified cases. Tubercle 1985;66(3):161–78.

50. Glassroth J, Robins AG, Snider DE Jr. Tuberculosis in the 1980s. N Engl J Med 1980;302(26):1441–50.

51. WHO Global Tuberculosis Programme, Geneva. Antituberculosis drug resistance in the world. The WHO/IUATLD Global Project on Anti-tuberculosis Drug Resistance Surveillance, 1994–1997.

52. Yew WW, Chau CH. Drug-resistant tuberculosis in the 1990s. Eur Respir J 1995;8(7):1184–92.

53. Iseman MD. Treatment of multidrug-resistant tuberculosis. N Engl J Med 1993;329(11):784–91.

54. Peloquin CA, MacPhee AA, Berning SE. Malabsorption of antimycobacterial medications. N Engl J Med 1993;329(15):1122–3.

55. Turner M, McGowan C, Nardell E, Haskal R. Serum drug levels in tuberculosis patients. Am J Respir Crit Care Med 1994;149:A102.

56. Des Prez RM, Heim CR. Mycobacterium tuberculosis. In: Mandell GL, Douglas Jr. RG Jr, Bennett JE, editors. Principles and Practice of Infectious Diseases. 3rd ed.. New York: Churchill Livingstone, 1990:1877.

57. Srivastava P. Drug metabolism and individualized medicine. Curr Drug Metab 2003;4:33–44.

58. Pande JN, Pande A, Singh SPN. Acetylator status, drug metabolism and disease. Natl Med J India 2003;16:24–6.

59. Huang YS, Chern HD, Su WJ, Wu JC, Chang SC, Chiang CH, Chang FY and Lee SD. Cytochrome P450 2E1 genotype and the susceptibility to antituberculosis drug-induced hepatitis. Hepatology 2003;37:924–30.

60. Gülbay BE, Gürkan OU, Yildiz OA, Onen ZP, Erkekol FO, Baççioğlu A, Acican T. Side effects due to primary antituberculosis drugs during the initial phase of therapy in 1149 hospitalized patients for tuberculosis. Respir Med 2006;100:1834–42.

61. Zhang Y, Post-Martens K, Denkin S. New drug candidates and therapeutic targets for tuberculosis therapy. Drug Discov Today 2006;11(1-2):21–7.

62. Lamichhane G, Zignol M, Blades NJ, Geiman DE, Dougherty A, Grosset J, Broman KW, Bishai WR. A postgenomic method for predicting essential genes at subsaturation levels of mutagenesis: application to Mycobacterium tuberculosis. Proc Natl Acad Sci USA 2003;100:7213–8.

63. Sassetti CM, Rubin EJ. Genetic requirements for mycobacterial survival during infection. Proc Natl Acad Sci USA 2003;100:1298–9.

64. McKinney JD, Höner zu Bentrup K, Muñoz-Elías EJ, Miczak A, Chen B, Chan WT, Swenson D, Sacchettini JC, Jacobs WR Jr, Russell DG. Persistence of Mycobacterium tuberculosis in macrophages and mice requires the glyoxylate shunt enzyme isocitrate lyase. Nature 2000;406(6797):735–8.

65. Glickman MS, Cox JS, Jacobs WR. A novel mycolic acid cyclopropane synthetase is required for cording, persistence, and virulence of Mycobacterium tuberculosis. Mol Cell 2000;5:717–27.

66. Dahl JL, Kraus CN, Boshoff HI, Doan B, Foley K, Avarbock D, Kaplan G, Mizrahi V, Rubin H, Barry CE 3rd. The role of RelMtb-mediated adaptation to stationary phase in long-term

persistence of *Mycobacterium tuberculosis* in mice. Proc Natl Acad Sci USA 2003;100(17):10026–31.

67. Dong PH, Guinn KM, Harrell MI, Liao R, Voskuil MI, Tompa M, Schoolnik GK, Sherman DR. Rv3133c/dosR is a transcription factor that mediates the hypoxic response of *Mycobacterium tuberculosis*. Mol Microbiol 2003;48:833–43.

68. Zhang Y. The magic bullets and tuberculosis drug targets. Annu Rev Pharmacol Toxicol 2005;45:529–64.

69. Mitchison DA. The search for new sterilizing anti-tuberculosis drugs. Front Biosci 2004;9:1059–72.

70. Spigelman MK. New tuberculosis therapeutics: a growing pipeline. J Infect Dis 2007;196:S28-34.

71. Andries K, Verhasselt P, Guillemont J, Go"hlmann HWH, Neefs JM., Winkler H, Van Gestel J. A diarylquinoline drug active on the ATP synthase of *Mycobacterium tuberculosis*. Science 2005;307:223–7.

72. Stover CK, Warrener P, VanDevanter DR, Sherman DR, Arain TM, Langhorne MH, Anderson SW, Towell JA, Yuan Y, McMurray DN, Kreiswirth BN, Barry CE, Baker WR. A small-molecule nitroimidazopyran drug candidate for the treatment of tuberculosis. Nature 2000;405:962–6.

73. Tyagi S, Nuermberger E, Yoshimatsu T, Williams K, Rosenthal I, Lounis N, Bishai W, Grosset J. Bactericidal activity of the nitroimidazopyran PA-824 in the murine model of tuberculosis. Antimicrob Agents Chemother 2005;49(6):2289–93.

74. Matsumoto M, Hshizume H, Tomishige T, Kawasaki M. In vitro and in vivo efficacy of novel antituberculous candidate OPC-67683 (abstract F-1462). In: Program and abstracts of the 45th Interscience Conference on Antimicrobial Agents and Chemotherapy (Washington, DC). Washington, DC: American Society for Microbiology, 2005:204.

75. Sasaki H, Haraguchi Y, Itotani M, Kuroda H, Hashizume H, Tomishige T, Kawasaki M, Matsumoto M, Komatsu M, Tsubouchi H. Synthesis and antituberculosis activity of a novel series of optically active 6-nitro-2,3-dihydroimidazo(2,1-b)oxazoles. J Med Chem 2006;49(26):7854–60.

76. Nuermberger EL, Yoshimatsu T, Tyagi S, O'Brien RJ, Vernon AN, Chaisson RE, Bishai WR, Grosset JH. Moxifloxacin-containing regimen greatly reduces time to culture conversion in murine tuberculosis. Am J Respir Crit Care Med 2004;169:421–6.

77. Nuermberger EL, Yoshimatsu T, Tyagi S, Williams K, Rosenthal I, O'Brien RJ, Vernon AA, Chaisson RE, Bishai WR, Grosset JH. Moxifloxacin-containing regimens of reduced duration produce a stable cure in murine tuberculosis. Am J Respir Crit Care Med 2004;170:1131–4.

78. Pletz MW, Roux AD, Roth A, Neumann KH, Mauch H, Lode H. Early bactericidal activity of moxifloxacin in treatment of pulmonary tuberculosis: a prospective, randomized study. Antimicrob Agents Chemother 2004;48:780–2.

79. Burman WJ, Goldberg S, Johnson JL, Muzanye G, Engle M, Mosher AW, Choudhri S, Daley CL, Munsiff SS, Zhao Z, Vernon A, Chaisson RE. Moxifloxacin versus ethambutol in the first 2 months of treatment for pulmonary tuberculosis. Am J Respir Crit Care Med 2006;174(3):331–8.

80. Sinha RK, Arora SK, Sinha N, Modak VM. In vivo activity of LL4858 against Mycobacterium tuberculosis (abstract F-1116). In: Program and abstracts of the 44th Interscience Conference on Antimicrobial Agents and Chemotherapy (Washington, DC). Washington, DC: American Society for Microbiology, 2004:212.

81. Jia L, Tomaszewski JE, Hanrahan C, Coward L, Noker P, Gorman G, Nikonenko B, Protopopova M. Pharmacodynamics and pharmacokinetics of SQ109, a new diamine-based antitubercular drug. Br J Pharmacol 2005;144(1):80-7.

Azithromycin

See also Macrolide antibiotics

General Information

The overall adverse reactions rate for azithromycin is 0.7% (1). Only diarrhea, nausea, and abdominal pain occurred in over 1% of patients. Other adverse effects that were reported from clinical trials in adults are palpitation, angina, dyspepsia, flatus, vomiting, melena, jaundice, vaginal candidiasis, vaginitis, nephritis, dizziness, headache, vertigo, somnolence, and fatigue. Azithromycin has also been reported to cause angioedema and photosensitivity, intrahepatic cholestasis, hypersensitivity syndrome, toxic pustuloderma, and irreversible deafness after low-dose use. It can cause a maculopapular eruption when given to patients with infectious mononucleosis. It can also cause contact dermatitis.

Observational studies

In 3995 patients who took azithromycin 1.5 g in divided doses over 5 days or who took 1 g as a single dose for urethritis/cervicitis adverse events occurred in 12% (2). In patients over 65 years the rate was 9.3%, and in children under 14 years of age it was 5.4%. The most common adverse effects were gastrointestinal (9.6%); central nervous system and peripheral nervous system effects were reported in 1.3%. Overall, 59% of the adverse events were considered mild, 34% moderate, and only 6% severe, involving mainly the gastrointestinal tract. Adverse events resulted in withdrawal in 0.7% of patients, lower than the rate reported with other macrolides. Treatment-related rises in liver enzymes were uncommon (under 2%), as was leukopenia (1.1–1.5%).

Phase II/III clinical trials in the USA have yielded data on 1928 children aged 6 months to 15 years who took azithromycin for infections that included acute otitis media ($n = 1150$) and streptococcal pharyngitis ($n = 754$) (3). Most took a 5-day course of azithromycin (5–12 mg/kg/day). There were adverse effects in 190 patients (9.9%): diarrhea (3.1%), vomiting (2.5%), abdominal pain (1.9%), loose stools (1%), and rash (2.5%). In three comparisons with co-amoxiclav, the overall incidence of adverse effects was significantly lower with azithromycin (7.7 versus 29%), with withdrawal rates of 0.3 versus 3.6%. However, the incidence of adverse effects was significantly greater with azithromycin than with penicillin V in comparisons in patients with streptococcal pharyngitis (13 versus 6.7%). In conclusion, it appears that the safety and tolerability of azithromycin is similar in children and adults.

In an open study, children with end-stage lung disease or chronic airflow limitation unresponsive to conventional therapy were treated with long-term azithromycin. Seven children (mean age 12 years), all of whom were colonized with *Pseudomonas aeruginosa* and who took azithromycin for more than 3 months, were studied. There was a significant improvement in FVC and FEV_1 (4). The mechanism whereby azithromycin works is unknown, but it may be other than antibacterial. It has been hypothesized that the effect may be due to upregulation of a P glycoprotein, a member of the family of multidrug resistant proteins, since erythromycin upregulates P glycoprotein expression in a monkey model. Multidrug resistance (MDR) is homologous to CFTR, and previous in vitro experiments have shown that the MDR and CFTR genes can complement each other (5). However, direct proof of this hypothesis is lacking at the moment.

Of 42 adult HIV-positive patients with confirmed or presumed acute toxoplasmic encephalitis who received azithromycin 900, 1200, or 1500 mg/day plus pyrimethamine, 28 responded to therapy during the induction period (6). Six patients withdrew during the induction period because of reversible toxic effects (three with raised liver enzymes, two with hearing loss, one with neutropenia). Treatment-terminating adverse events occurred most often among the patients who took 1500 mg/day.

In an open, prospective trial gingival hyperplasia due to ciclosporin was successfully treated with azithromycin 250 mg/day for 5 days in 30 of 35 patients, who reported esthetic satisfaction and disappearance of bleeding and pain (7). There was no change in ciclosporin concentration or renal function after azithromycin.

Comparative studies

The tolerability of azithromycin oral suspension, 10 mg/kg od for 3 days, has been assessed in children in a review of 16 multicenter studies (8). Of 2425 patients, 1213 received azithromycin and 1212 received other drugs. The incidence of treatment-related adverse events was significantly lower in those who took azithromycin, while withdrawal rates were similar. There were significantly fewer gastrointestinal events with azithromycin and their duration was significantly shorter.

Co-amoxiclav
In a multicenter, parallel-group, double-blind trial in 420 evaluable patients aged 6 months to 16 years with community-acquired pneumonia, the therapeutic effect of azithromycin (once-daily for 5 days) was similar to that of co-amoxiclav in children under 5 years and to that of erythromycin tds for 10 days. Treatment-related adverse events occurred in 11% of those given azithromycin and 31% in the comparator group (9).

Azithromycin (500 mg/day for 3 days) has been used to treat acute periapical abscesses (10). Of 150 patients treated with azithromycin 18 reported a total of 26 adverse events. Slightly more (24 out of 153) treated with co-amoxiclav reported 34 adverse events, but this difference did not reach statistical significance. Most of the adverse events (44/60) were gastrointestinal, mostly diarrhea or abdominal pain. There were no significant differences between the two groups in the severity of adverse events or in the number of withdrawals because of adverse events.

Fluoroquinolones
In a multicenter, open, randomized comparison of levofloxacin 500 mg/day orally or intravenously and azithromycin 500 mg/day intravenously for up to 2 days plus ceftriaxone 1 g/day intravenously for 2 days in 236 patients, the most common drug-related adverse events in those given azithromycin were diarrhea (4.2%), vein disorders (2.5%), and pruritus (1.7%) (11).

In a randomized, double-blind comparison, single doses of azithromycin 1000 mg and levofloxacin 500 mg were used for travelers' diarrhea (12). The most common adverse events in those given azithromycin were mild abdominal pain (20%) and fecal urgency (13%). There were also one case each of anxiety and a transient rash.

Other macrolides
The incidence of disseminated MAC infection has increased dramatically with the AIDS epidemic. Treatment regimens for patients with a positive culture for MAC from a sterile site should include two or more drugs, including clarithromycin. Prophylaxis against disseminated MAC should be considered for patients with a CD4 cell count of less than $50 \times 10^6/l$ (13). In a randomized, open trial in 37 patients with HIV-associated disseminated MAC infection, treatment with clarithromycin + ethambutol produced more rapid resolution of bacteremia, and was more effective at sterilization of blood cultures after 16 weeks than azithromycin + ethambutol (14).

Tetracyclines
Compared with tetracycline, azithromycin had a favorable short-term effect on childhood morbidity in a mass trial for trachoma in rural Gambian villages, and adverse effects were limited (15).

Treatment of facial comedonic and papulopustular acne with azithromycin (500 mg/day for 4 days in four cycles every 10 days) may be at least as effective as minocycline (100 mg/day for 6 weeks). Both were well tolerated and mild adverse effects were reported in 10% of patients given azithromycin and 12% of those given minocycline (16).

Placebo-controlled studies

In 169 patients with acute infective rhinitis, azithromycin (500 mg/day for 3 days) resulted in a better cure rate after 11 days than placebo; however, after 25 days the results for both improvement and cure were equal (17).

In a randomized, double-blind, placebo-controlled multicenter trial in 174 HIV-infected patients with CD4 cell counts of under $100 \times 10^6/l$, azithromycin (1200 mg once a week) was safe and effective in preventing disseminated MAC infection, death due to MAC infection, and respiratory tract infections (18).

In a triple-masked, randomized, placebo-controlled study in 1867 women, prophylaxis with azithromycin 500 mg 1 hour before IUCD insertion did not affect the rate of IUCD removal, the frequency of medical attention after insertion, or the risk of upper genital tract infection at 90 days (19). Women were at low risk of sexually transmitted disease according to self-reported medical history. Gastrointestinal adverse effects were infrequent (3% azithromycin; 2% placebo). Fewer women taking azithromycin (0.7%) than those taking placebo (1.3%)

were treated with antibiotics for pelvic tenderness; however, this difference was not statistically significant. Since cervical infections increase the risk of pelvic infection in women who use IUCDs, generalization of these results may be difficult (20).

In a meta-analysis of randomized, controlled trials of 3–5 days of azithromycin or other antibiotics that are typically given in longer courses for upper respiratory tract infections, there were no significant differences in bacteriological outcomes (21). Azithromycin was withdrawn because of adverse events in only 37 (0.8%) of 4870 patients.

Three well-designed randomized controlled trials have demonstrated a small but significant improvement in respiratory function with azithromycin compared with placebo (22). Mild adverse events (wheeze, diarrhea, and nausea) were significantly increased in one trial. Daily azithromycin (250 mg/day) as prophylaxis against malaria in adults was well-tolerated during 20 weeks in 148 patients, but azithromycin recipients complained more frequently of heartburn, paresthesia, and itching than those taking doxycycline (23). There was no evidence of hearing loss or hematologic, hepatic, or renal toxicity.

Organs and Systems

Cardiovascular

Azithromycin can prolong the QT_c interval (24,25), although in a prospective study in 47 healthy subjects, azithromycin (3 g total dose given during 5 days) resulted in only a small, non-significant prolongation of the QT interval (<<Ruuskanen S135).

- A 65-year-old man with idiopathic dilated cardiomyopathy developed significant prolongation of the QT interval after taking azithromycin for 2 days for a community-acquired pneumonia. Three days after withdrawal the QT interval returned to normal.Azithromycin can cause life-threatening bradycardia (26).
- A 9-month-old infant who was inadvertently given azithromycin 50 mg/kg, taken from floor stock, instead of the prescribed ceftriaxone, became unresponsive and pulseless. The initial heart rhythm observed when cardiopulmonary resuscitation was started was a broad-complex bradycardia, with a prolonged rate-corrected QT interval and complete heart block. She was resuscitated with adrenaline and atropine but suffered severe anoxic encephalopathy.Polymorphous ventricular tachycardia has been attributed to azithromycin (27).
- A 51-year-old woman took azithromycin 500 mg for an upper respiratory tract infection shortly after a dose of over-the-counter pseudoephedrine. Two hours later she had two syncopal events due to polymorphous ventricular tachycardia without QT interval prolongation. Azithromycin was withdrawn and the ventricular tachycardia abated after 10 hours. She was symptom-free 1 year later.

Torsade de pointes and cardiorespiratory arrest have been reported in a patient with congenital long QT syndrome who took azithromycin (28). In a prospective study of 47 previously healthy people, there was a modest statistically insignificant prolongation of the QT_c interval without clinical consequences after the end of a course of azithromycin 3 g/day for 5 days (29).

Sensory systems

Ears

Azithromycin can cause ototoxicity. In one study, 8 (17%) of 46 HIV-positive patients had probable ($n = 6$) or possible ($n = 2$) ototoxicity with azithromycin (30). The effects were hearing loss (88%), tinnitus (37%), plugged ears (37%), and vertigo (25%), developing at a mean of 7.6 weeks (1.5–20 weeks) after the start of long-term azithromycin therapy for *Mycobacterium avium* infection. The symptoms resolved in a mean of 4.9 weeks (2–11 weeks) after withdrawal.

Sensorineural hearing loss has been attributed to azithromycin (31).

- A 35-year old Caucasian man with AIDS and multiple opportunistic infections, including *Mycobacterium kansasii* and *Mycobacterium avium* complex (MAC) disease developed moderate to severe primary sensorineural hearing loss after 4–5 months of therapy with oral azithromycin 500 mg/day. Other medications included ethambutol, isoniazid, rifabutin, ciprofloxacin, co-trimoxazole, fluconazole, zidovudine (later switched to stavudine), lamivudine, indinavir, methadone, modified-release oral morphine, pseudoephedrine, diphenhydramine, megestrol acetate, trazodone, sorbitol, salbutamol by metered-dose inhaler and nebulizer, ipratropium, and oral morphine solution as needed. Significant improvement of the hearing impairment was documented 3 weeks after drug withdrawal.

A literature review identified several cases of ototoxicity in HIV-positive patients treated with azithromycin for *M. avium* complex infection. In four series, 14–41% of such patients had some degree of hearing loss. However, some patients were also taking other potentially ototoxic drugs, which may have contributed to the high frequency of hearing loss reported. Hearing loss improved markedly after withdrawal of azithromycin. Hearing loss may be more common and probably more severe with high-dose azithromycin than with high-dose clarithromycin.

- A 47-year-old woman who had a left lung transplantation 3 months earlier and who was taking ticarcillin + clavulanate and aztreonam for sinusitis, was given co-trimoxazole, ticarcillin + clavulanate, azithromycin (500 mg/day intravenously), and ganciclovir for presumed pneumonia (32). Other drugs included tacrolimus, mycophenolate, prednisone, lansoprazole, diltiazem, itraconazole, warfarin, alendronate, ipratropium bromide, folic acid, and nystatin. The next day, rimantadine and vancomycin were added, and co-trimoxazole was reduced. A neurological examination to assess symptoms of peripheral neuropathy noted no hearing deficit. On day 3, vancomycin, ticarcillin + clavulanate, and ganciclovir were withdrawn. On the fifth day, mild tinnitus and reduced hearing developed and gradually progressed to complete deafness. After eight doses, azithromycin was withdrawn, and 20 days later her hearing was back to baseline.

Low-dose exposure to azithromycin has been associated with irreversible sensorineural hearing loss in otherwise healthy subjects (33). Even a single oral dose of azithromycin altered the conjunctival bacterial flora of children from a trachoma endemic area (34). However, the clinical significance is not yet clear.

Psychological, psychiatric

Azithromycin can cause delirium (35) and did so in two elderly patients who took 500 mg initially followed by 250 mg/day (36).

Hematologic

The effects of combining azithromycin and rifabutin have been studied in 50 subjects with or without HIV infection, of whom 19 took azithromycin 1200 mg/day and rifabutin 600 mg/day, and 31 took azithromycin 600 mg/day and rifabutin 300 mg/day (37). Neutropenia was the most common adverse event, in 33 of 50 subjects. Low-grade nausea, diarrhea, fatigue, and headache were also common, and most subjects had more than one type of event. There was no significant pharmacokinetic interaction between the two drugs.

Gastrointestinal

In a review of 12 clinical studies most of the adverse events in those taking azithromycin affected the gastrointestinal system, and were reported in 138 (8.5%) azithromycin-treated patients (38). Abdominal pain, diarrhea, nausea, and vomiting were the most frequently reported gastrointestinal adverse events.

Gastrointestinal symptoms were the most common adverse effects reported in a trial of azithromycin in disseminated *Mycobacterium avium* complex in 62 patients with AIDS (39). Erythromycin is a motilin receptor agonist (40–42). This mechanism may be at least partly responsible for the gastrointestinal adverse effects of macrolides. Azithromycin may act on gastrointestinal motility in a similar way to erythromycin, as it produces a significant increase in postprandial antral motility (43).

In a randomized, double-blind, placebo-controlled trial in 186 patients with reactive arthritis treated with oral azithromycin for 13 weeks, there were more adverse events with azithromycin than placebo; the adverse events in those taking azithromycin were most often gastrointestinal (44).

In an open, non-comparative study in 35 patients with acne vulgaris azithromycin 500 mg thrice weekly for 12 weeks caused heartburn and nausea in four patients (45).

Liver

Liver damage has been attributed to azithromycin (46).

- A 75-year-old woman took azithromycin 500 mg/day for 3 days and developed jaundice, fatigue, and diarrhea. She had a history of allergy to penicillin, morphine, and pethidine. Transaminases, bilirubin, and alkaline phosphatase were raised. Liver biopsy showed centrilobular necrosis consistent with a drug reaction. Two months later her liver function tests had returned to baseline.

Severe liver damage can be caused by intravenous infusion of azithromycin (47).

Azithromycin can cause intrahepatic cholestasis (48).

- A 33-year-old woman and a 72-year-old man developed cholestasis after they had taken a 5-day course of azithromycin. The woman was given colestyramine and underwent six courses of plasmapheresis; 2 months later, her total bilirubin and serum transaminases were back to normal (49). After withdrawal of azithromycin, the man's symptoms resolved within 1 month and his liver enzymes returned to normal (50).

Urinary tract

A 14-year-old Caucasian girl developed acute interstitial nephritis after taking azithromycin (1.5 g total or 6 mg/kg/day) for 5 days (51).

Skin

In a double-blind, placebo-controlled trial of azithromycin (750 mg loading dose followed by 250 mg/day) in malaria prophylaxis in 300 patients, the most important adverse event was a maculopapular rash (52).

- A 19-year-old man with infectious mononucleosis developed a maculopapular, non-pruritic rash after one dose of azithromycin 500 mg (53).

In a phase II, randomized, double-blind, treatment-controlled study comparison of topical 2% azithromycin versus 2% erythromycin in 20 subjects with moderate inflammatory acne and 20 with rosacea, the number of inflammatory lesions was reduced by 2% erythromycin in both acne and rosacea (54). Azithromycin was not as effective in rosacea. There were no significant adverse events in those with acne. In five patients with rosacea, there was transient irritation.

After 10 days treatment with azithromycin for an upper respiratory tract infection, a 62-year-old woman developed Stevens–Johnson syndrome, which improved during treatment with corticosteroids after few days (55).

In a randomized, multicenter, double-blind study *rash* was reported in 2.5% of patients after a single dose of azithromycin for acute otitis media (56).

- A 5-year-old child developed Stevens–Johnson syndrome 3 days after starting a course of azithromycin and improved after treatment with a glucocorticoid for 28 days (57).

Immunologic

Occupational allergic contact dermatitis has been attributed to azithromycin (58).

- A 32-year-old pharmaceutical worker had been loading reactors at three different stages of azithromycin synthesis for the past 3 years and had been exposed to airborne powders. He wore overalls and latex gloves. His symptoms had persisted for 1 year in the form of pruritus, erythema, vesicles, and scaling of the face and forearms. A positive patch test and a positive workplace challenge were considered reliable in the diagnosis of occupational allergic contact dermatitis induced by azithromycin. After transfer to another work station that excluded exposure to azithromycin, he had no further work-related symptoms.

  Airborne allergic contact dermatitis from azithromycin occurred in two workers employed in the blending department of a pharmaceutical company; both cases were confirmed by patch testing (59).

- A 5-year-old boy developed Stevens–Johnson syndrome after taking azithromycin for 3 days. He had oral pain and skin eruptions with bullae (60). He was given betamethasone, panipenem/betamipron, and ulinastatin and gradually improved over 27 days.

Leukocytoclastic vasculitis occurred in an 8-month-old boy who was treated with azithromycin (61). Fever and erythematous lesions on the legs, feet, arms, buttocks, and face were seen on the third day of treatment. After withdrawal of azithromycin, body temperature returned to normal within 3 days; the skin lesions began to fade on the next day and disappeared within 3 days.

Hypersensitivity to azithromycin has been reported (62).

- A 79-year-old man developed fever, mental changes, a rash, acute renal insufficiency, and hepatitis after he had completed a 5-day course of oral azithromycin (500 mg initially then 250 mg/day). With intravenous hydration only, his fever abated and his urinary output and renal and hepatic function returned to normal over the next 4 days. His mental status improved significantly. The skin rash was followed by extensive desquamation.

Azithromycin has been associated with Churg–Strauss syndrome in a patient with atopy (63).

Second-Generation Effects

Pregnancy

In two randomized trials in pregnant women with cervical *Chlamydia trachomatis* infection, women were randomized to oral amoxicillin 500 mg tds for 7 days or oral azithromycin 1 g in a single dose (64,65). The two drugs had similar efficacy. Adverse effects were common in both groups: 40% of those who took azithromycin reported moderate to severe gastrointestinal adverse effects compared with 17% of those who took amoxicillin.

Teratogenicity

In a case-control study in 123 patients gestational exposure to azithromycin was not associated with an increase in the rate of major malformations above the baseline of 1–3% (66).

Susceptibility Factors

Renal disease

In eight patients a single 500 mg oral dose of azithromycin was not substantially removed by continuous ambulatory peritoneal dialysis in the absence of peritonitis. Azithromycin cannot be recommended for widespread use in CAPD at present. However, the successful use of azithromycin in treating peritonitis, perhaps because of an intracellular drug transport mechanism, has been reported (67).

Drug–Drug Interactions

Interactions with azithromycin have been reviewed (<<Scheinfeld 350). It does not interact with the P450 complex and thus has few drug interactions. Even so, the package insert suggests caution in using it with medications with P450 interactions.

Antacids

Co-administration of antacids reduced the peak concentration of azithromycin; however, the extent of absorption was not significantly reduced (<<Scheinfeld 350). Azithromycin should be given at least 1 hour before or 2 hours after antacids. Antacids containing aluminium and magnesium reduce peak serum concentrations, but the total extent of azithromycin absorption is not altered (68).

Antihistamines

The effects of azithromycin 250 mg/day on the pharmacokinetics of desloratadine 5 mg/day and fexofenadine 60 mg bd have been studied in a parallel-group, third-party-blind, multiple-dose, randomized, placebo-controlled study (69). There were small increases (under 15%) in the mean plasma concentrations of desloratadine. In contrast, peak fexofenadine concentrations were increased by 69% and the AUC by 67%. There were no changes in the electrocardiogram.

Carbamazepine

A retrospective analysis of 3995 patients treated with azithromycin did not show any pharmacokinetic interactions in patients who were also taking various other drugs, including carbamazepine (2,68).

Cardiac glycosides

- In a 31-month-old boy with Down's syndrome and Fallot's tetralogy during a 5-day course of azithromycin 5 mg/kg/day the serum digoxin concentration rose and the child had anorexia, diarrhea, and second-degree atrioventricular block with junctional extra beats (84).

The mechanism was not investigated, but it is likely that azithromycin affects *Eubacterium lentum*. When azithromycin is used concomitantly with digoxin, serum digoxin concentrations need to be monitored.

Ciclosporin

When azithromycin is used concomitantly with ciclosporin, blood ciclosporin concentrations need to be monitored (70). Anecdotal evidence suggests that azithromycin is effective for ciclosporin-induced gingival hyperplasia in recipients of solid organ transplants. Two heart transplant recipients insidiously developed gingival hyperplasia, probably because of immunosuppression with ciclosporin, which was successfully treated with azithromycin 250 mg/day for 10 days (71). The concomitant use of azithromycin (500 mg/day for 3 days) with ciclosporin in eight stable renal transplant patients produced only a 7% increase in the AUC of ciclosporin and a 19% increase in peak plasma concentration, effects that are not likely to be clinically significant (72).

Cimetidine

A retrospective analysis of 3995 patients treated with azithromycin did not show any pharmacokinetic interactions in patients who were also taking various other drugs, including cimetidine (2,68).

Cytochrome P450

Azithromycin has a 15-membered ring and does not induce or inhibit cytochrome P450 in rats (73).

Methylprednisolone

A retrospective analysis of 3995 patients treated with azithromycin did not show any pharmacokinetic interactions in patients who were also taking various other drugs, including methylprednisolone (2,68).

Midazolam

In an open, randomized, crossover, pharmacokinetic and pharmacodynamic study in 12 healthy volunteers who took clarithromycin 250 mg bd for 5 days, azithromycin 500 mg/day for 3 days, or no pretreatment, followed by a single dose of midazolam (15 mg), clarithromycin increased the AUC of midazolam by over 3.5 times and the mean duration of sleep from 135 to 281 minutes (74). In contrast, there was no change with azithromycin, suggesting that it is much safer for co-administration with midazolam.

Piroxicam

In 66 patients undergoing oral surgery, treatment with azithromycin impaired the periodontal disposition of piroxicam (75).

Rifamycins

An interaction involving azithromycin with rifabutin, and less commonly rifampicin, was observed in patients with MAC infections (76).

Terfenadine

The potential interaction of azithromycin with terfenadine has been evaluated in a randomized, placebo-controlled study in 24 patients who took terfenadine plus azithromycin or terfenadine plus placebo (77). Azithromycin did not alter the pharmacokinetics of the active carboxylate metabolite of terfenadine or the effect of terfenadine on the QT interval.

Theophylline and other xanthines

A retrospective analysis of 3995 patients treated with azithromycin did not show any pharmacokinetic interactions in patients who were also taking various other drugs, including theophylline (2,68).

In two double-blind, randomized, placebo-controlled studies there was no inhibition of the metabolism of theophylline by azithromycin (78,79). However, there has been a report of reduced theophylline concentrations after withdrawal of azithromycin (80). The authors concluded that the mechanism of interaction was best explained by concomitant induction and inhibition of theophylline metabolism by azithromycin, followed by increased availability of unbound enzyme sites as azithromycin was cleared from the system.

Warfarin

A retrospective analysis of 3995 patients treated with azithromycin did not show any pharmacokinetic interactions in patients who were also taking various other drugs, including warfarin (2,68)

- An 83-year-old African American man stabilized on warfarin took azithromycin 500 mg on one day; the prothrombin time rose and normalized 3 days after azithromycin was withdrawn (81).

Zidovudine

Zidovudine does not affect azithromycin concentrations and azithromycin does not affect zidovudine concentrations (82).

Food–Drug Interactions

Owing to interference by food (83), azithromycin should be given at least 1 hour before or 2 hours after food.

References

1. Scheinfeld NS, Tutrone WD, Torres O, Weinberg JM. Macrolides in dermatology. Dis Mon 2004;50(7):350–68.
2. Peters DH, Friedel HA, McTavish D. Azithromycin. A review of its antimicrobial activity, pharmacokinetic properties and clinical efficacy. Drugs 1992;44(5):750–99.
3. Hopkins SJ, Williams D. Clinical tolerability and safety of azithromycin in children. Pediatr Infect Dis J 1995;14(Suppl):S67–71.
4. Jaffe A, Francis J, Rosenthal M, Bush A. Long-term azithromycin may improve lung function in children with cystic fibrosis. Lancet 1998;351(9100):420.

5. Altschuler EL. Azithromycin, the multidrug-resistant protein, and cystic fibrosis. Lancet 1998;351(9111):1286.

6. Jacobson JM, Hafner R, Remington J, Farthing C, Holden-Wiltse J, Bosler EM, Harris C, Jayaweera DT, Roque C, Luft BJ. ACTG 156 Study Team. Dose-escalation, phase I/II study of azithromycin and pyrimethamine for the treatment of toxoplasmic encephalitis in AIDS. AIDS 2001;15(5):583–9.

7. Citterio F, Di Pinto A, Borzi MT, Scata MC, Foco M, Pozzetto U, Castagneto M. Azithromycin treatment of gingival hyperplasia in kidney transplant recipients is effective and safe. Transplant Proc 2001;33(3):2134–5.

8. Treadway G, Reisman A. Tolerability of 3-day, once-daily azithromycin suspension versus standard treatments for community-acquired paediatric infectious diseases. Int J Antimicrob Agents 2001;18(5):427–31.

9. Harris JA, Kolokathis A, Campbell M, Cassell GH, Hammerschlag MR. Safety and efficacy of azithromycin in the treatment of community-acquired pneumonia in children. Pediatr Infect Dis J 1998;17(10):865–71.

10. Adriaenssen CF. Comparison of the efficacy, safety and tolerability of azithromycin and co-amoxiclav in the treatment of acute periapical abscesses. J Int Med Res 1998;26(5):257–65.

11. Frank E, Liu J, Kinasewitz G, Moran GJ, Oross MP, Olson WH, Reichl V, Freitag S, Bahal N, Wiesinger BA, Tennenberg A, Kahn JB. A multicenter, open-label, randomized comparison of levofloxacin and azithromycin plus ceftriaxone in hospitalized adults with moderate to severe community-acquired pneumonia. Clin Ther 2002;24(8):1292–308.

12. Adachi JA, Ericsson CD, Jiang ZD, DuPont MW, Martinez-Sandoval F, Knirsch C, DuPont HL. Azithromycin found to be comparable to levofloxacin for the treatment of US travelers with acute diarrhea acquired in Mexico. Clin Infect Dis 2003; 37: 1165-71.

13. Faris MA, Raasch RH, Hopfer RL, Butts JD. Treatment and prophylaxis of disseminated *Mycobacterium avium* complex in HIV-infected individuals. Ann Pharmacother 1998;32(5):564–73.

14. Ward TT, Rimland D, Kauffman C, Huycke M, Evans TG, Heifets L. Randomized, open-label trial of azithromycin plus ethambutol vs. clarithromycin plus ethambutol as therapy for *Mycobacterium avium* complex bacteremia in patients with human immunodeficiency virus infection. Veterans Affairs HIV Research Consortium. Clin Infect Dis 1998;27(5):1278–85.

15. Whitty CJ, Glasgow KW, Sadiq ST, Mabey DC, Bailey R. Impact of community-based mass treatment for trachoma with oral azithromycin on general morbidity in Gambian children. Pediatr Infect Dis J 1999;18(11):955–8.

16. Gruber F, Grubisic-Greblo H, Kastelan M, Brajac I, Lenkovic M, Zamolo G. Azithromycin compared with minocycline in the treatment of acne comedonica and papulopustulosa. J Chemother 1998;10(6):469–73.

17. Haye R, Lingaas E, Hoivik HO, Odegard T. Azithromycin versus placebo in acute infectious rhinitis with clinical symptoms but without radiological signs of maxillary sinusitis. Eur J Clin Microbiol Infect Dis 1998;17(5): 309–12.

18. Oldfield EC 3rd, Fessel WJ, Dunne MW, Dickinson G, Wallace MR, Byrne W, Chung R, Wagner KF, Paparello SF, Craig DB, Melcher G, Zajdowicz M, Williams RF, Kelly JW, Zelasky M, Heifets LB, Berman JD. Once weekly azithromycin therapy for prevention of *Mycobacterium avium* complex infection in patients with AIDS: a randomized, double-blind, placebo-controlled multicenter trial. Clin Infect Dis 1998;26(3):611–9.

19. Walsh T, Grimes D, Frezieres R, Nelson A, Bernstein L, Coulson A, Bernstein GIUD Study Group. Randomised controlled trial of prophylactic antibiotics before insertion of intrauterine devices. Lancet 1998;351(9108):1005–8.

20. Coggins C, Sloan NL. Prophylactic antibiotics before insertion of intrauterine devices. Lancet 1998;351(9120):1962–3.

21. Ioannidis JP, Contopoulos-Ioannidis DG, Chew P, Lau J. Meta-analysis of randomized controlled trials on the comparative efficacy and safety of azithromycin against other antibiotics for upper respiratory tract infections. J Antimicrob Chemother 2001;48(5):677–89.

22. Southern KW, Barker PM. Azithromycin for cystic fibrosis. Eur Respir J 2004;24(5):834–8.

23. Ruuskanen O. Safety and tolerability of azithromycin in pediatric infectious diseases: 2003 update. Pediatr Infect Dis J 2004;23(2 Suppl):S135-9.

24. Matsunaga N, Oki Y, Prigollini A. A case of QT-interval prolongation precipitated by azithromycin. N Z Med J 2003;116: U666.

25. Russo V, Puzio G, Siniscalchi N. Azithromycin-induced QT prolongation in elderly patient. Acta Biomed 2006;77(1):30–2.

26. Tilelli JA, Smith KM, Pettignano R. Life-threatening bradyarrhythmia after massive azithromycin overdose. Pharmacotherapy 2006;26(1):147–50.

27. Kim MH, Berkowitz C, Trohman RG. Polymorphic ventricular tachycardia with a normal QT interval following azithromycin. Pacing Clin Electrophysiol 2005;28(11):1221–2.

28. Arellano-Rodrigo E, Garcia A, Mont L, Roque M. Torsade de pointes y parada cardiorrespiratoria inducida pot azitromicina en una paciente con sindrome de QT largo congenito. [Torsade de pointes and cardiorespiratory arrest induced by azithromycin in a patient with congenital long QT syndrome.] Med Clin (Barc) 2001;117(3):118–9.

29. Strle F, Maraspin V. Is azithromycin treatment associated with prolongation of the Q-Tc interval? Wien Klin Wochenschr 2002;114(10–11):396–9.

30. Tseng AL, Dolovich L, Salit IE. Azithromycin-related ototoxicity in patients infected with human immunodeficiency virus. Clin Infect Dis 1997;24(1):76–7.

31. Lo SH, Kotabe S, Mitsunaga L. Azithromycin-induced hearing loss. Am J Health Syst Pharm 1999;56(4):380–3.

32. Bizjak ED, Haug MT 3rd, Schilz RJ, Sarodia BD, Dresing JM. Intravenous azithromycin-induced ototoxicity. Pharmacotherapy 1999;19(2):245–8.

33. Mamikoglu B, Mamikoglu O. Irreversible sensorineural hearing loss as a result of azithromycin ototoxicity. A case report. Ann Otol Rhinol Laryngol 2001;110(1):102.

34. Chern KC, Shrestha SK, Cevallos V, Dhami HL, Tiwari P, Chern L, Whitcher JP, Lietman TM. Alterations in the conjunctival bacterial flora following a single dose of azithromycin in a trachoma endemic area. Br J Ophthalmol 1999;83(12):1332–5.

35. Sirois F. Delirium associé à l'azithromycine. [Delirium associated with azithromycin administration.] Can J Psychiatry 2002;47(6):585–6.

36. Cone LA, Padilla L, Potts BE. Delirium in the elderly resulting from azithromycin therapy. Surg Neurol 2003;59:509–11.

37. Hafner R, Bethel J, Standiford HC, Follansbee S, Cohn DL, Polk RE, Mole L, Raasch R, Kumar P, Mushatt D, Drusano GDATRI 001B Study Group. Tolerance and pharmacokinetic interactions of rifabutin and azithromycin. Antimicrob Agents Chemother 2001;45(5):1572–7.

38. Treadway G, Pontani D, Reisman A. The safety of azithromycin in the treatment of adults with community-acquired respiratory tract infections. Int J Antimicrob Agents 2002;19(3):189–94.

39. Koletar SL, Berry AJ, Cynamon MH, Jacobson J, Currier JS, MacGregor RR, Dunne MW, Williams DJ. Azithromycin as treatment for disseminated *Mycobacterium avium* complex in AIDS patients. Antimicrob Agents Chemother 1999;43(12):2869–72.

40. Lin HC, Sanders SL, Gu YG, Doty JE. Erythromycin accelerates solid emptying at the expense of gastric sieving. Dig Dis Sci 1994;39(1):124–8.

41. Hasler WL, Heldsinger A, Chung OY. Erythromycin contracts rabbit colon myocytes via occupation of motilin receptors. Am J Physiol 1992;262(1 Pt 1):G50–5.

42. Kaufman HS, Ahrendt SA, Pitt HA, Lillemoe KD. The effect of erythromycin on motility of the duodenum, sphincter of Oddi, and gallbladder in the prairie dog. Surgery 1993;114(3):543–8.

43. Sifrim D, Matsuo H, Janssens J, Vantrappen G. Comparison of the effects of midecamycin acetate and azithromycin on gastrointestinal motility in man. Drugs Exp Clin Res 1994;20(3):121–6.

44. Kvien TK, Gaston JS, Bardin T, Butrimiene I, Dijkmans BA, Leirisalo-Repo M, Solakov P, Altwegg M, Mowinckel P, Plan PA, Vischer T. Three month treatment of reactive arthritis with azithromycin: a EULAR double blind, placebo controlled study. Ann Rheum Dis 2004;63(9):1113–9.

45. Kapadia N, Talib A. Acne treated successfully with azithromycin. Int J Dermatol 2004;43(10):766–7.

46. Baciewicz AM, Al-Nimr A, Whelan P. Azithromycin-induced hepatotoxicity. Am J Med 2005;118(12):1438–9.

47. An N, Gui XM, Wang YH. [A case of severe liver damage caused by intravenous infusion of azithromycin.] Zhonghua Er Ke Za Zhi 2006;44(4):313.

48. Longo G, Valenti C, Gandini G, Ferrara L, Bertesi M, Emilia G. Azithromycin-induced intrahepatic cholestasis. Am J Med 1997;102(2):217–8.

49. Suriawinata A, Min AD. A 33 year old woman with jaundice after azithromycin use. Semin Liver Dis 2002;22(2):207–10.

50. Chandrupatla S, Demetris AJ, Rabinovitz M. Azithromycin-induced intrahepatic cholestasis. Dig Dis Sci 2002;47(10):2186–8.

51. Soni N, Harrington JW, Weiss R, Chander P, Vyas S. Recurrent acute interstitial nephritis induced by azithromycin. Pediatr Infect Dis J 2004;23(10):965–6.

52. Taylor WR, Richie TL, Fryauff DJ, Ohrt C, Picarima H, Tang D, Murphy GS, Widjaja H, Braitman D, Tjitra E, Ganjar A, Jones TR, Basri H, Berman J. Tolerability of azithromycin as malaria prophylaxis in adults in Northeast Papua, Indonesia. Antimicrob Agents Chemother 2003;47:2199–203.

53. Dakdouki GK, Obeid KH, Kanj SS. Azithromycin-induced rash in infectious mononucleosis. Scand J Infect Dis 2002;34(12):939–41.

54. McHugh RC, Rice A, Sangha ND, McCarty MA, Utterback R, Rohrback JM, Osborne BE, Fleischer AB Jr, Feldman SR. A topical azithromycin preparation for the treatment of acne vulgaris and rosacea. J Dermatolog Treat 2004;15(5):295–302.

55. Brkljacic N, Gracin S, Prkacin I, Sabljar-Matovinovic M, Mrzljak A, Nemet Z. Stevens–Johnson syndrome as an unusual adverse effect of azithromycin. Acta Dermatovenereol Croat 2006;14(1):40–5.

56. Arguedas A, Emparanza P, Schwartz RH, Soley C, Guevara S, de Caprariis PJ, Espinoza G. A randomized, multicenter, double blind, double dummy trial of single dose azithromycin versus high dose amoxicillin for treatment of uncomplicated acute otitis media. Pediatr Infect Dis J 2005;24(2):153–61.

57. Schmutz JL, Barbaud A, Trechot P. [Azithromycin and Stevens–Johnson syndrome]. Ann Dermatol Venereol 2005;132(8-9 Pt 1):728.

58. Milkovic-Kraus S, Kanceljak-Macan B. Occupational airborne allergic contact dermatitis from azithromycin. Contact Dermatitis 2001;45(3):184.

59. Mimesh S, Pratt M. Occupational airborne allergic contact dermatitis from azithromycin. Contact Dermatitis 2004;51(3):151.

60. Aihara Y, Ito S, Kobayashi Y, Aihara M. Stevens–Johnson syndrome associated with azithromycin followed by transient reactivation of Herpes simplex virus infection. Allergy 2004;59(1):118.

61. Odemis E, Kalyoncu M, Okten A, Yildiz K. Azithromycin-induced leukocytoclastic vasculitis. J Rheumatol 2003;30:2292.

62. Cascaval RI, Lancaster DJ. Hypersensitivity syndrome associated with azithromycin. Am J Med 2001;110(4):330–1.

63. Hubner C, Dietz A, Stremmel W, Stiehl A, Andrassy H. Macrolide-induced Churg–Strauss syndrome in a patient with atopy. Lancet 1997;350(9077):563.

64. Kacmar J, Cheh E, Montagno A, Peipert JF. A randomized trial of azithromycin versus amoxicillin for the treatment of *Chlamydia trachomatis* in pregnancy. Infect Dis Obstet Gynecol 2001;9(4):197–202.

65. Jacobson GF, Autry AM, Kirby RS, Liverman EM, Motley RU. A randomized controlled trial comparing amoxicillin and azithromycin for the treatment of *Chlamydia trachomatis* in pregnancy. Am J Obstet Gynecol 2001;184(7):1352–4discussion 1354–6.

66. Sarkar M, Woodland C, Koren G, Einarson AR. Pregnancy outcome following gestational exposure to azithromycin. BMC Pregnancy Childbirth 2006;618.

67. Kent JR, Almond MK, Dhillon S. Azithromycin: an assessment of its pharmacokinetics and therapeutic potential in CAPD. Perit Dial Int 2001;21(4):372–7.

68. Hopkins S. Clinical toleration and safety of azithromycin. Am J Med 1991;91(3A):S40–5.

69. Gupta S, Banfield C, Kantesaria B, Marino M, Clement R, Affrime M, Batra V. Pharmacokinetic and safety profile of desloratadine and fexofenadine when coadministered with azithromycin: a randomized, placebo-controlled, parallel-group study. Clin Ther 2001;23(3):451–66.

70. Ljutic D, Rumboldt Z. Possible interaction between azithromycin and cyclosporin: a case report. Nephron 1995;70(1):130.

71. Strachan D, Burton I, Pearson GJ. Is oral azithromycin effective for the treatment of cyclosporine-induced gingival hyperplasia in cardiac transplant recipients? J Clin Pharm Ther 2003;28:329–38.

72. Bachmann K, Jauregui L, Chandra R, Thakker K. Influence of a 3-day regimen of azithromycin on the disposition kinetics of cyclosporine A in stable renal transplant patients. Pharmacol Res 2003;47:549–54.

73. Yeates RA, Laufen H, Zimmermann T. Interaction between midazolam and clarithromycin: comparison with azithromycin. Int J Clin Pharmacol Ther 1996;34(9):400–5.

74. Amacher DE, Schomaker SJ, Retsema JA. Comparison of the effects of the new azalide antibiotic, azithromycin, and erythromycin estolate on rat liver cytochrome P-450. Antimicrob Agents Chemother 1991;35(6):1186–90.

75. Malizia T, Batoni G, Ghelardi E, Baschiera F, Graziani F, Blandizzi C, Gabriele M, Campa M, Del Tacca M, Senesi S. Interaction between piroxicam and azithromycin during

distribution to human periodontal tissues. J Periodontol 2001;72(9):1151–6.

76. Griffith DE, Brown BA, Girard WM, Wallace RJ Jr. Adverse events associated with high-dose rifabutin in macrolide-containing regimens for the treatment of *Mycobacterium avium* complex lung disease. Clin Infect Dis 1995;21(3):594–8.

77. Harris S, Hilligoss DM, Colangelo PM, Eller M, Okerholm R. Azithromycin and terfenadine: lack of drug interaction. Clin Pharmacol Ther 1995;58(3):310–5.

78. Gardner M, Coates P, Hilligoss D, Henry E. Lack of effect of azithromycin on the pharmacokinetics of theophylline in man. In: Proceedings of the Mediterranean Congress of ChemotherapyAthens 1992:327.

79. Clauzel A, Visier S, Michel F. Efficacy and safety of azithromycin in lower respiratory tract infections. Eur Resp J 1990;3(Suppl 10):89.

80. Pollak PT, Slayter KL. Reduced serum theophylline concentrations after discontinuation of azithromycin: evidence for an unusual interaction. Pharmacotherapy 1997;17(4):827–9.

81. Rao KB, Pallaki M, Tolbert SR, Hornick TR. Enhanced hypoprothrombinemia with warfarin due to azithromycin. Ann Pharmacother 2004;38(6):982–5.

82. Chave JP, Munafo A, Chatton JY, Dayer P, Glauser MP, Biollaz J. Once-a-week azithromycin in AIDS patients: tolerability, kinetics, and effects on zidovudine disposition. Antimicrob Agents Chemother 1992;36(5):1013–8.

83. Schmidt LE, Dalhoff K. Food–drug interactions. Drugs 2002;62(10):1481–502.

84. Ten Eick AP, Sallee D, Preminger T, Weiss A, Reed MD. Possible drug interaction between digoxin and azithromycin in a young child. Clin Drug Invest 2000;20:61–4.

Bacitracin

General Information

Bacitracin has mostly dermatological uses. It can cause allergic reactions of the delayed type. A shock-like picture after local application has occurred in a hypersensitive individual (1). Since bacitracin is nephrotoxic, it should not be given intraperitoneally to patients with renal impairment.

Observational studies

In a randomized study of the effects of a triple antibiotic ointment (polymyxin B + bacitracin + neomycin) and simple gauze-type dressings on scarring of dermabrasion wounds, the ointment was superior to the simple dressing in minimizing scarring; the beneficial effect on pigmentary changes was especially pronounced (2).

Organs and Systems

Sensory systems

Inadvertent injection of bacitracin ointment into the orbit can cause a postoperative orbital compartment syndrome.

- Acute proptosis, chemosis, reduced vision, and ophthalmoplegia occurred after endoscopic sinus surgery in a 73-year-old woman (3). The orbit was tense and the intraocular pressure was 54 mmHg. The presence of bacitracin ointment was established by computed tomography.

Immunologic

Bacitracin is one of the most important clinical allergens (4). Anaphylaxis rarely occurs after topical administration of bacitracin ointment (5,6).

- A 45-year-old man developed a near-fatal anaphylactic reaction after he applied bacitracin ointment to an excoriated area on his foot. He had had a similar, but less severe, episode 4 years earlier. IgE antibodies to bacitracin were positive.

- A 24-year-old man injured in a motorcycle accident was treated with viscous lidocaine and bacitracin zinc ointment for extensive abrasions on the extremities. Five minutes later, he developed symptoms of severe anaphylaxis and required adrenaline, antihistamines, intravenous fluids, and glucocorticoids. Two weeks later, only the prick test to bacitracin zinc ointment was positive.

Anaphylaxis has also been reported after bacitracin nasal packing (7).

- A 48-year-old man underwent uneventful septorhinoplasty, after which his right nostril was packed with 6 ft of vaseline gauze placed in the finger of a latex glove coated with bacitracin ointment. Within seconds, his oxygen saturation fell from 97 to 94% (and increased to 97% with 100% oxygen), but blood pressure and heart rate were unchanged. After the left nostril had been packed, his oxygen saturation fell to 89%, no pulse wave was registered and an electrocardiogram showed a heart rate of 39/minute with first-degree atrioventricular block, and the blood pressure was not obtainable by non-invasive measurement. Cardiopulmonary resuscitation was successful, but the patient remained intubated for 2 days because of concerns about facial and upper airway edema. Later he gave a history of an episode of irritation and swelling after nasal application of polymyxin B and bacitracin ointment 2–3 weeks before surgery. Skin prick testing was positive for bacitracin but negative for latex, polymyxin, cefazolin, and saline.

Cases of anaphylaxis have also been reported after bacitracin irrigation (8,9).

- A 65-year-old man undergoing elective sternal debridement and rewiring was given a prophylactic infusion of vancomycin 1 g preoperatively. Anesthesia was induced with thiopental, suxamethonium, and fentanyl, and maintained with fentanyl, vecuronium, and isoflurane. A few minutes after wound irrigation with bacitracin (about 25 U/ml), his blood pressure fell precipitously, necessitating intravenous fluids and adrenaline. His face and arms were flushed. Afterwards, he reported having had a rash several years before after the use of an over-the-counter ointment composed of polymyxin B, bacitracin, and neomycin.

- A 9-year-old child with a repaired myelomeningocele and congenital hydrocephalus who had undergone four previous shunt revisions in the past had two episodes of anaphylaxis during insertion of the ventriculoperitoneal shunt. The shunt tubing had been soaked in a solution of bacitracin 2500 U/ml. A skin prick test was positive for bacitracin.

Thiomersal, a mercury derivative of thiosalicylic acid, is a preservative used in several types of consumer products, including cosmetics, ophthalmic and otolaryngological medications, and vaccines. In a retrospective study in 574 patients, people who were allergic to thiomersal were more likely to be allergic to bacitracin (10).

Long-Term Effects

Drug resistance

Although there is evidence of acquired resistance to bacitracin in enterococci and staphylococci isolated from animals (11), a review found no evidence that the prevalence of such resistance has increased over time or in relation to the use of bacitracin in man or in animals (in which bacitracin is used as a growth promoter) (12).

Drug Administration

Drug contamination

Commercial bacitracin comprises more than 30 different substances, but the major antibiotic isoforms A and B account for about 60% of the mixture. An impurity has been identified in some but not all bacitracin lots (13). The impurity is a powerful subtilisin-type protease capable of cleaving many proteins, including protein disulfide isomerase, myosin, and a variety of artificial substrates. Investigators using bacitracin are therefore reminded to determine whether their bacitracin is contaminated by a protease. If it is, careful reinterpretation of the results or retesting with an enzyme-free bacitracin reagent may be warranted.

Interference with Diagnostic Tests

Semen detection

In forensic medicine, the detection of semen may be critical, and identification due to Wood's lamp-induced fluorescence has been suggested to be helpful. In a study to investigate whether semen can be distinguished from other products, none of 41 physicians was able to differentiate semen from other products using a Wood's lamp; however, some ointments and creams that contained bacitracin were mistaken for semen (14).

References

1. Kanof NB. Bacitracin and tyrothricin. Med Clin North Am 1970;54(5):1291–3.
2. Berger RS, Pappert AS, Van Zile PS, Cetnarowski WE. A newly formulated topical triple-antibiotic ointment minimizes scarring. Cutis 2000;65(6):401–4.
3. Castro E, Seeley M, Kosmorsky G, Foster JA. Orbital compartment syndrome caused by intraorbital bacitracin ointment after endoscopic sinus surgery. Am J Ophthalmol 2000;130(3):376–8.
4. Maouad M, Fleischer AB Jr, Sherertz EF, Feldman SR. Significance–prevalence index number: a reinterpretation and enhancement of data from the North American contact dermatitis group. J Am Acad Dermatol 1999;41(4):573–6.
5. Lin FL, Woodmansee D, Patterson R. Near-fatal anaphylaxis to topical bacitracin ointment. J Allergy Clin Immunol 1998;101(1 Pt 1):136–7.
6. Saryan JA, Dammin TC, Bouras AE. Anaphylaxis to topical bacitracin zinc ointment. Am J Emerg Med 1998;16(5):512–3.
7. Gall R, Blakley B, Warrington R, Bell DD. Intraoperative anaphylactic shock from bacitracin nasal packing after septorhinoplasty. Anesthesiology 1999;91(5):1545–7.
8. Blas M, Briesacher KS, Lobato EB. Bacitracin irrigation: a cause of anaphylaxis in the operating room. Anesth Analg 2000;91(4):1027–8.
9. Carver ED, Braude BM, Atkinson AR, Gold M. Anaphylaxis during insertion of a ventriculoperitoneal shunt. Anesthesiology 2000;93(2):578–9.
10. Suneja T, Belsito DV. Thimerosal in the detection of clinically relevant allergic contact reactions. J Am Acad Dermatol 2001;45(1):23–7.
11. Butaye P, Devriese LA, Haesebrouck F. Phenotypic distinction in *Enterococcus faecium* and *Enterococcus faecalis* strains between susceptibility and resistance to growth-enhancing antibiotics. Antimicrob Agents Chemother 1999;43(10):2569–70.
12. Phillips I. The use of bacitracin as a growth promoter in animals produces no risk to human health. J Antimicrob Chemother 1999;44(6):725–8.
13. Rogelj S, Reiter KJ, Kesner L, Li M, Essex D. Enzyme destruction by a protease contaminant in bacitracin. Biochem Biophys Res Commun 2000;273(3):829–32.
14. Santucci KA, Nelson DG, McQuillen KK, Duffy SJ, Linakis JG. Wood's lamp utility in the identification of semen. Pediatrics 1999;104(6):1342–4.

Beta-lactam antibiotics

See also Carbapenems, Cephalosporins, Monobactams, Penicillins

General Information

The beta-lactam antibiotics still comprise roughly half of the antibiotic market worldwide. The common structure that defines the whole family of beta-lactam antibiotics is the four-membered, highly reactive beta-lactam ring, which is essential for antimicrobial activity (1). The following simplifying classification is practical:

1. penicillins
2. cephalosporins
3. monobactams (containing no second ring system besides the beta-lactam ring)
4. carbapenems

In addition, beta-lactamase inhibitors also contain the beta-lactam structure.

The crucial event that initiates the antimicrobial effects of beta-lactam antibiotics is binding to and inhibition of bacterial enzymes located in the cell membrane, the so-called penicillin-binding proteins (2). This happens by covalent binding, through opening of the beta-lactam ring. Enzyme activities of penicillin-binding proteins are involved in the last steps of bacterial cell wall (peptidoglycan) synthesis, and their inhibition halts cell growth, causing cell death and lysis (3). Beta-lactamases are genetically and structurally closely related to penicillin-binding proteins.

Despite their chemical diversity, their adverse effects profiles share various common aspects. There are several reasons why beta-lactam antibiotics belonging to different classes can cause comparable reactions. Besides the beta-lactam ring, other structural similarities (for example side chains) or antimicrobial activity can be relevant. However, the incidence of a given reaction, and in particular instances also the severity, varies among beta-lactam classes.

Incidence and cause–effect relations

It is difficult to establish clearly the incidence and cause–effect relations of many reactions and hence to identify patients at risk. The following factors are important:

1. The range of recommended daily doses varies by more than an order of magnitude, according to clinical need. Hence, the incidence of some collateral and toxic reactions varies greatly among different populations.
2. Combinations of beta-lactam antibiotics with antimicrobial drugs from other molecular classes are often used, especially in severe infection.
3. The spectrum of potential beta-lactam-antibiotic-induced reactions is especially broad, and in most cases no test procedure is available to distinguish beta-lactam antibiotics from other causes of a reaction, in particular from the consequences of the treated infection.

Relation to dose

Many reactions to beta-lactam antibiotics are clearly not immune mediated. These include bleeding disorders, neurotoxicity, and most cases of diarrhea. In addition, many reactions, the pathogenesis of which is still being discussed, clearly depend on the daily and the cumulative dose of beta-lactam antibiotics and hence the duration of treatment. Although the rare, but well-understood, immune hemolysis after penicillin is seen mostly with high-dose and long-term treatment, dose dependency and time dependency point to direct toxicity rather than to immunological mechanisms. Indeed, direct toxic effects

of beta-lactam antibiotics on eukaryotic cells and specific interactions with receptor proteins and enzymes have been shown (4) and may underlie particular reactions.

There are three lines of evidence that beta-lactam antibiotics cause a variety of reactions by toxic mechanisms:

1. Certain reactions are overwhelmingly reported to be dose-dependent and time-dependent.
2. Particular compounds cause adverse effects with unexpectedly high frequencies in certain circumstances (for example cystic fibrosis, bacterial endocarditis, and osteomyelitis) that require particularly high doses and prolonged treatment.
3. Beta-lactam antibiotics affect a variety of cultured eukaryotic cells.

For other reactions, the underlying mechanisms are less clear. The body of individual reports and some published series suggest that their incidence increases disproportionately with prolonged, high-dosage treatment, that is, with accumulation. This is particularly the case in the following reactions:

- severe neutropenia up to total agranulocytosis, as observed with virtually all beta-lactam antibiotics.
- acute interstitial nephritis, seen with methicillin but more rarely also with other beta-lactams, for example penicillin G.
- one type of hepatitis induced by isoxazolyl penicillins;
- varying combinations of symptoms positively referred to or not as "serum sickness-like syndromes."

There have been reports of high overall frequencies of adverse effects after the use of very high cumulative doses of beta-lactam antibiotics in healthy volunteers and in patients with, for example, chronic osteomyelitis, pulmonary exacerbations in cystic fibrosis, and infective endocarditis (5–9). In one series, 23% of patients treated with an average cumulative dose of carbenicillin of 925 g and 68% of those treated with ureidopenicillins 329 g developed adverse effects, including rash, fever, leukopenia, eosinophilia, thrombocytopenia, and hepatic damage, requiring change of therapy in 52% of cases in the latter group (10). Another study included a total of 292 treatment courses with five different beta-lactams for infective endocarditis (6). With a treatment duration of 9 days or less, drug was withdrawn in only 3% because of adverse reactions. However, treatment courses ranging from 10 days to 6 weeks were associated with adverse reactions in 33%, one-quarter of which consisted of neutropenia. Fourteen of 44 patients receiving piperacillin up to 900 mg/kg/day for acute pulmonary exacerbations in cystic fibrosis developed a syndrome that resembled serum sickness; the symptoms were mainly fever, malaise, anorexia, eosinophilia, and rashes (8). The reaction occurred after a minimum of 9 days and the frequency of symptoms was dose-related. All patients who developed the reaction were re-admitted at 4–28 months after the initial episode and in every case re-exposure to piperacillin did not evoke the reaction.

The dose-relation of reactions to piperacillin in patients with cystic fibrosis has created a debate about its usefulness in this condition (11–15). However, comparable dose-related patterns and frequencies of adverse effects were found in other patients treated with piperacillin (15) and with other beta-lactam antibiotics (16), as well as in patients with both cystic fibrosis and other conditions (6,10,16). Three later studies showed that piperacillin more often caused fever, rash, and other reactions per treatment course in patients with cystic fibrosis compared with a large variety of other beta-lactam or non-beta-lactam antibiotics (17–19). Of particular interest is a study in which volunteers who took high doses of cefalotin or cefapirin for up to 4 weeks developed comparable syndromes, with an overall incidence of adverse effects of 100% (9). Despite these astonishingly high frequencies, these reactions were predominantly regarded as being allergic, although their pathogenesis was mostly unclear.

Thus, a disproportionately high frequency of apparently unrelated adverse effects occurs in a relatively small group of patients, those needing high-dose prolonged treatment, who are at particular risk.

Mechanisms

Degradation products spontaneously formed in aqueous solutions, for example culture media, rather than the parent molecules themselves, may be responsible for the observed effects (4). Antiproliferative activities were generally more pronounced with cephalosporins than with penicillins, while monobactams appear to be practically free from such effects. Carbapenems have not been thoroughly studied in this respect, and some data on clavulanic acid and two other beta-lactamase inhibitors do not clearly reflect the same kind of toxicity as observed with penicillins and cephalosporins (20).

The selectivity of beta-lactam antibiotics for bacterial target proteins is not absolute. A specific interaction of modified cephalosporins with mammalian serine proteases has been shown (21) and the affinity of various penicillins for the benzodiazepine receptor may be part of the chain of events leading to neurotoxicity (22) However, most intriguing are observations made in proliferating cultured cells. Biological effects associated with proliferation were dose-dependently inhibited by a large array of beta-lactam antibiotics in a variety of cells from both man and animals (4,23). Resting cells, on the other hand, were not susceptible, even to very high concentrations.

The clinical impact of the inhibitory effects of beta-lactam antibiotics on proliferating eukaryotic cells is as yet unknown, and formal proof of a correlation with toxicity in patients is lacking. However, there are reasons for considering this type of toxicity as the cause of neutropenia and thrombocytopenia (SEDA-13, 230). In dogs, high-dose cefonicid and cefazedone for up to several months caused bone marrow damage, resembling the findings in clinical cases of neutropenia, which could explain peripheral cytopenias (24,25). In addition, mild thrombocytopenia and reticulocytopenia, which have been concomitantly found respectively in 30 and 17% of cases of neutropenia (26), are also paralleled by results in

dogs. On the other hand, in the same dogs, IgG associated with erythrocytes, neutrophils, and platelets was found after high-dose treatment with cefazedone (27) Antigranulocyte IgG antibodies in beta-lactam-induced neutropenia have also been described in man (28–30). However, the relevance of these findings is unclear, since high cumulative doses of beta-lactams often induce beta-lactam-specific IgG antibodies in patients with and without adverse effects (6,31). Newer data from human and animal cell culture investigations suggest that ceftazidime-induced myelosuppression could be the consequence of multiple effects on various myeloid and non-myeloid cells in the bone marrow (32–34). They also give hints of a more rational basis for using G-CSF or other cytokines in beta-lactam-antibiotic-induced neutropenia (32,34). Hence, there is still controversy about whether beta-lactam antibiotics can cause neutropenia by both toxic and immunological mechanisms and how both mechanisms could act in concert with each other.

For evaluation of local tolerability, human peritoneal cells (35), human osteoblasts (36), and human as well as animal endothelial cells (37) have been studied in culture. The type of toxicity and rank efficacy among various compounds were congruent with the results from earlier studies on other cells (4,23). The clinical relevance of these data remains to be established.

The Jarisch–Herxheimer reaction

The Jarisch–Herxheimer reaction is a systemic reaction that occurs hours after initial treatment of spirochete infections, such as syphilis, leptospirosis, Lyme disease, and relapsing fever, and presents with fever, rigors, hypotension, and flushing (38,39). In patients with syphilis the reaction is more frequent in secondary syphilis and can cause additional manifestations, such as flare-up of cutaneous lesions, sudden aneurysmal dilatation of the aortic arch (40), and angina pectoris or acute coronary occlusion (SED-8, 559). It can easily be mistaken for a drug-induced hypersensitivity reaction. The underlying mechanism is initiated by antibiotic-induced release of spirochete-derived pyrogens. Transient rises in TNF, IL-6, and IL-8 have been detected (41). The role of TNF-alpha in the pathogenesis of the Jarisch–Herxheimer reaction is further underscored by the observation that in patients undergoing penicillin treatment for louse-borne relapsing fever, pretreatment with anti-TNF antibody Fab fragments partially protected against the reaction (42). The reaction lasts 12–24 hours and can be alleviated by aspirin. Alternatively, prednisone can be used and is recommended as adjunctive treatment of symptomatic cardiovascular syphilis or neurosyphilis.

Organs and Systems

Respiratory

Allergic bronchospasm can principally be a consequence of IgE antibody-mediated allergy to all beta-lactam antibiotics.

Nervous system

The neurotoxic effects of beta-lactam antibiotics have been reviewed (43). Since the first observation of convulsions after intraventricular administration of penicillin more than 50 years ago (44), neurotoxicity has been attributed to most beta-lactam antibiotics. Its manifestations are considered to be the consequence of GABAergic inhibition (45,46)and include clear epileptic manifestations as well as more atypical reactions, such as asterixis, drowsiness, and hallucinations. Epileptogenic activity of beta-lactam antibiotics has also been documented in animals and in brain slices in vitro (47). With penicillins and cephalosporins, integrity of the beta-lactam ring is a prerequisite, and epileptogenic activity is extinguished by beta-lactamase (48,49). However, this may not be true of the carbapenems, the neurotoxicity of which is differently related to their structure (50). However, clinical manifestations are always clearly dose-dependent, and brain tissue concentrations appear to be more relevant than CSF or blood concentrations (47). Accordingly, the major risk factor is impaired renal function, particularly when it is not recognized. Other risk factors are age (very young or very old), meningitis, intraventricular therapy, and a history of epilepsy (51).

The neurotoxic potential differs considerably among the various beta-lactam antibiotics, and experimental models have been developed for investigating this (52,53). Currently, imipenem + cilastatin appears to cause the highest frequency of neurotoxic effects (54,55) and the above-mentioned risk factors have been particularly confirmed with this compound (SEDA-18, 261) (56). Quinolone antibiotics, which themselves are proconvulsant, can potentiate excitation of the central nervous system by beta-lactam antibiotics, at least in animals (57,58).

Tardive seizures in psychiatric patients undergoing electroconvulsive therapy and receiving a beta-lactam have been reported (59).

- A 62-year-old man undergoing ECT developed pneumonia and was given piperacillin 2 g/day + tazobactam. After 5 days, and after his third ECT session, he had generalized tonic–clonic convulsions. Electroencephalography showed no focal abnormalities and other examinations, including MRI scans, laboratory tests, and cerebrospinal fluid examination, were all negative. Piperacillin was withdrawn. He had recurrent seizures during the next 2 days and gradually improved over the next weeks.
- A 24-year-old man undergoing ECT developed a urinary tract infection and was given cefotiam 2 g/day intravenously for 5 days. One day later and after his third ECT session, he had recurrent attacks of generalized tonic–clonic seizure. Electroencephalography showed no focal seizure activity and MRI and laboratory findings were normal. ECT was stopped and he gradually improved. Four weeks later he had ECT again without subsequent seizures.

Reviewing the literature, the authors found a case of seizures in a patient receiving ECT who was given ciprofloxacin (60). The epileptogenic effect of ciprofloxacin is thought to be mediated through suppression of the inhibitory function of GABA, as is that of some beta-lactams. In mice piperacillin and cefotiam inhibit GABA receptor function, inducing convulsions (61).

Sensory systems

In vitro, methicillin and ceftazidime in high concentrations produced toxic effects on corneal and endothelial cells of the eye (62,63).

Metabolism

Pivaloyl-containing compounds (baccefuconam, cefetamet pivoxil, cefteram pivoxil, pivampicillin, pivmecillinam) can significantly increase urinary carnitine excretion (64,65). These compounds are esterified prodrugs, which become effective only after the release of pivalic acid, which in turn is esterified with carnitine. Carnitine loss induced by pivaloyl-containing beta-lactams was first described in children and can produce symptoms similar to other types of carnitine deficiency, for example secondary to organic acidurias (64). Carnitine is essential for the transport of fatty acids through the mitochondrial membrane for beta-oxidation. Consequences of its deficiency include skeletal damage, cardiomyopathy, hypoglycemia and reduced ketogenesis, encephalopathy, hepatomegaly, and Reye-like syndromes (66).

The administration of pivaloyl-conjugated beta-lactam antibiotics to healthy volunteers for 54 days reduced mean serum carnitine 10-fold and muscle carnitine, as measured per non-collagen protein, more than 2-fold (66). Long-term treatment of children for 12–37 months to prevent urinary tract infection resulted in serum carnitine concentrations of 0.9–3.6 µmol/l (reference range 23–60 µmol/l). In four cases, muscle carnitine was 0.6–1.4 µmol/g non-collagen protein (reference range 7.1–19) (67).

Although oral carnitine aided the elimination of the pivaloyl moiety, its simultaneous use did not fully compensate for the adverse metabolic effects of pivaloyl-containing beta-lactams (68,69). The consequences of pivaloyl-induced carnitine loss seem to be generally reversible. But as long as the risk of pivaloyl-induced urinary loss of carnitine and particular risk factors are not better defined, it is prudent to use pivaloyl-containing prodrugs only in short-term treatment.

Electrolyte balance

Since beta-lactam antibiotics contain sodium or potassium, they can cause or at least aggravate electrolyte disturbances when given in sufficiently high doses. The most frequent manifestations are hypernatremia and hypokalemia. The sodium content of injectable beta-lactam antibiotics per gram of active compound varies by up to a factor of three (70).

Hematologic

Neutropenia

Neutropenia due to beta-lactam antibiotics has been reviewed (71). It usually occurs after high-dose therapy lasting more than 10 days, and the frequency rises with cumulative dose. It is often preceded by a fever or rash, usually lasts less than 10 days, and is uncommonly associated with infectious complications or death. Although any beta-lactam can cause neutropenia, there seems to be a high incidence associated with the prolonged use of cefepime or piperacillin + tazobactam.

While in large series of several thousands of patients, neutropenia has generally been reported as an adverse effect in under 0.1–1.0% (SEDA-13, 212), an overview in 1985 estimated that neutropenia (neutrophil count below $1.0 \times 10^9/l$) occurs in up to 15% of all patients treated with high-dose intravenous beta-lactam antibiotics for more than 10 days (26). In subsequent series of patients treated for several weeks with various beta-lactam antibiotics, up to 25% developed neutropenia (5,21,72–74).

In one series, 22 of 128 patients receiving cloxacillin for staphylococcal infections became neutropenic (72). Neutropenia appeared, on average, 23 days after the start of therapy. The same authors, in a somewhat bigger population, found neutropenia in 1.1% of patients who received cumulative doses of oxacillin below 150 g, but in 43% (22 of 51) who received more than 150 g (5). Similarly, in 132 patients, cefapirin in a cumulative dose of less than 90 g did not cause neutropenia, but did in 26% (five of 19) of those who used higher total doses (21).

In addition, for a given compound, higher daily doses increase the risk of neutropenia. In one study, seven of 14 patients became neutropenic with a mean dose of penicillin G of 17 g/day after 9–23 days (73), while in another study only 12 of 193 patients developed neutropenia with a mean dose of 11 g/day for an average duration of 20 days (74). A considerable extension of the aforementioned study (73) corroborated this: neutropenia occurred in 35% of those treated with a mean daily dose of 17 g of penicillin G for an average of 23 days, while it was found in only 8% of those who received 12 g for 22 days (6).

Epidemiological studies (7,75) as well as single cases of severe neutropenia observed with newer compounds have invariably confirmed the dose- and time-dependent pattern described above. For example, cefepime, a fourth-generation cephalosporin, possibly or probably caused neutropenia in only 0.2% of 3314 treatment courses, while 7.1% of those who received cefepime for several weeks developed neutropenia (76). Accordingly, high-dose cefepime (150 mg/kg/day) was given for 7–10 days to 43 children for bacterial meningitis without causing neutropenia (77), while there were two cases in adults after total doses of 112 g (over 28 days) and 120 g (30 days) respectively (78).

It is therefore not surprising that after consecutive or simultaneous treatment with more than one beta-lactam antibiotic, neutropenia is similarly observed, suggesting additive toxicity (6,79,80).

There is so far no clear evidence about the different risks of different compounds. The data best fit the assumption that the risk of neutropenia correlates with the cumulative dose, or probably more precisely with the area under the serum concentration versus time curve (AUC). Hence, renal insufficiency is a potential risk factor. In addition, beta-lactam-antibiotic-induced leukopenia has been associated with hepatic dysfunction (81).

Recovery in most cases is rapid and uneventful. In patients who were re-exposed to the same or other beta-lactam antibiotics, there was similar dependence of neutropenia on the duration of treatment and the cumulative dose (26). Whether the use of hemopoietic growth factors, and in particular G-CSF, is useful is unclear. There are case reports of positive clinical effects (82–84). However, the recovery time in these reports did not differ from that observed in a large population of untreated patients (26). Theoretically, early use of growth factors could even be counterproductive, since some toxic effects of beta-lactam antibiotics on bone marrow cells appear to be related to the S-phase of the cell cycle (4). On the other hand, G-CSF maintained the proliferative activity of bone marrow cells exposed to ceftazidime in vitro, if it was added at the beginning of the culture process (32).

Neutropenia is accompanied by fever, eosinophilia, and/or a rash in more than 80% of cases.

Hemolytic anemia

Immune hemolytic anemia was originally described with penicillin G, but subsequently also with other penicillins and cephalosporins. It is usually seen during treatment with very high doses after the so-called "drug absorption" mechanism. The beta-lactam antibiotic binds covalently to the erythrocyte surface, forming complete antigens, which can in turn bind drug-specific circulating IgG antibody. Typically, direct and indirect Coombs' tests are positive, but complement is not activated (85–87). Rarely, other immunological mechanisms have been observed, for example the so-called "innocent bystander" type of hemolysis (87), in which complement can be detected on the erythrocyte surface. Some cephalosporins, clavulanic acid, and imipenem + cilastatin can cause positive direct antiglobulin tests (88). The phenomenon is due to non-specific serum protein absorption on to the erythrocyte membrane and is not related to immune hemolytic processes. Detection of non-immunologically bound serum proteins is improved if the reagents used include additional anti-albumin activity (89). The phenomenon is a known source of difficulties in evaluating suspected immune hemolysis or routine cross-matching of blood products (90). The true frequency of the phenomenon is unclear, since it has not been positively sought.

Thrombocytosis

Thrombocytosis is frequently mentioned as an adverse effect of beta-lactam antibiotics. However, it has been suggested that this reflects healing from infection rather than toxicity (91).

Eosinophilia

Virtually all beta-lactam antibiotics can cause eosinophilia, either isolated or in the context of very different reactions.

Bleeding disorders

Treatment with beta-lactam antibiotics can result in impaired hemostasis and bleeding (92). The true incidence of bleeding is difficult to assess, since many non-antibiotic factors can be involved, such as malnutrition with vitamin K depletion (93), renal insufficiency (94), and serious infection (95). Cancer, the use of cytotoxic drugs, and surgery have made conclusive interpretation of coagulation disorders difficult (96). Between the different beta-lactam antibiotics, the reported incidence of clinical relevant bleeding varies widely, and was highest with moxalactam (22% of patients), now withdrawn (SED-12, 625). With other cephalosporins, bleeding was observed with frequencies ranging from 2.7% (cefazolin/cefalotin) to 8.2% (cefoxitime) (97). Two basic mechanisms have been proposed.

Altered coagulation

Both direct inhibition of the hepatic production of vitamin K-dependent clotting factors and alterations in the intestinal flora, with subsequent reduction of microbial supply of vitamin K, have been implicated (98,99). The relative role of either mechanism is difficult to assess, but experimental support for the flora theory is weak (100,101).

Several of the cephalosporins that contain either a non-substituted *N*-methylthiotetrazole (NMTT) side chain, such as cefamandole, cefamazole, cefmenoxime, cefmetazole, cefoperazone, cefotetan, and moxalactam, as well as a substituted NMTT side chain (ceforanide, ceforicid, or cefotiam), or the structurally similar *N*-methylthiotriazine ring in ceftriaxone and the 2-methyl-1,2,4-thiadiazole-5-thiol (MTD) ring of cefazolin interfere with vitamin K-dependent clotting factor synthesis in the liver (factors II, VII, IX, and X). The molecular mechanism involves dose-dependent inhibition of microsomal carboxylase function, as shown in animals (102), and inhibition of the epoxide reductase system in both animals and man (103–106). Cefoxitin, a non-NMTT compound, was implicated significantly more often than the NMTT-containing compounds cefamandole and cefoperazone (107).

The NMTT must leave the parent antibiotic to inhibit the carboxylation reaction (108). The NMTT molecule leaves the parent cephalosporin either during spontaneous hydrolysis in the blood or during nucleophilic cleavage of the beta-lactam ring by intestinal bacteria, and is reabsorbed from the gut into the portal circulation (109). Studies in healthy volunteers show compound-related differences in the ability of NMTT antibiotics to generate free NMTT, reflecting drug-specific differences in susceptibility to in vitro hydrolysis or differences in gut NMTT production, which may be a function of biliary excretion of the drug (110).

Altered platelet numbers and function

Platelet dysfunction occurs dose-dependently with carbenicillin, ticarcillin, and, infrequently, other broad-spectrum penicillins (111), but the NMTT cephalosporin moxalactam has also been associated with altered platelet function in both healthy subjects and in patients treated with standard regimens (112–116). In contrast, clinical studies including cefotaxime, ceftizoxime, cefoperazone, and ceftracone did not show platelet dysfunction attributable to these compounds (115–117). There is evidence that beta-lactam-antibiotic-induced platelet dysfunction is at least partially irreversible (118).

From a practical point of view it can be concluded that:

1. the use of cephalosporins containing an NMTT side chain is associated with a risk of dose-dependent inhibition of vitamin K-dependent clotting factor synthesis.
2. platelet dysfunction occurs primarily with the broad-spectrum penicillins, but the NMTT cephalosporins, notably moxalactam, have also been implicated; monitoring of bleeding time should be considered in patients at risk (bleeding history, clinical bleeding, concomitant thrombocytopenia, or the use of other drugs known to interfere with platelet function.
3. the presence of non-antibiotic factors, such as therapy with vitamin K antagonists or NSAIDs, renal insufficiency, hepatic dysfunction, impaired gastrointestinal function, and malnutrition, can increase the risk of bleeding in cephalosporin-treated patients; close monitoring of homeostasis (prothrombin time, bleeding time), as well as prophylactic supplementation with vitamin K or, if necessary, therapeutic administration of fresh-frozen plasma and/or platelets is warranted according to the clinical context.

Gastrointestinal

Gastrointestinal upsets, nausea, and vomiting have been observed with virtually all beta-lactam antibiotics, both oral and parenteral. Even when comparing analogous applications and doses, no particular risk can be clearly ascribed to a given compound. Acute hemorrhagic colitis without pseudomembrane formation has been described after treatment with various penicillins and cephalosporins (SEDA-21, 261).

Antibiotic-induced diarrhea

There are three types of antibiotic-induced diarrhea:

- simple diarrhea due to altered bowel flora; this is quite common, for example it occurs in about 8% of patients who take ampicillin (119).
- diarrhea due to loss of bowel flora and overgrowth of *Clostridium difficile*, with toxin production; this is much less common.
- a rare form of diarrhea that is due to allergy.

Almost all antibacterial agents have been observed to cause diarrhea in a variable proportion of patients (120,121). The proportion depends not only on the antibiotic, but also on the clinical setting (in-patient/out-patient), age, race, and the definition of diarrhea. Severe colonic inflammation develops in a variable proportion of cases, and in some cases pseudomembranous colitis occurs (122–127). Since 1977, much evidence has accumulated that the most important causative agent in antibiotic-associated diarrhea is an anaerobic, Gram-positive, toxin-producing bacterium, *C. difficile* (128–130).

Pseudomembranous colitis was known before the introduction of antimicrobial agents and can still occur without previous antibiotic use, for example after antineoplastic chemotherapy (131) or even spontaneously. However, the number of cases has increased dramatically since antibiotics began to be used (132). Patients treated with lincomycin or clindamycin, cephalosporins, penicillinase-resistant penicillins, or combinations of several antibiotics are at especially high risk (133–136). A low risk is usually associated with sulfonamides, co-trimoxazole, chloramphenicol, and tetracyclines (122). Although few data have yet been published on this subject for the quinolones, they seldom seem to cause diarrhea and pseudomembranous colitis (137).

Presentation

In pseudomembranous colitis the stools are generally watery, with occult blood loss, which is seldom gross. Common findings include abdominal pain, cramps, fever, and leukocytosis. Especially severe forms can run such a rapid course that diarrhea does not occur; they present with symptoms of severe toxicity and shock (138). As a rare complication, marked dilatation of the colon and paralytic ileus can develop, that is, toxic megacolon.

Pseudomembranes are described as initially punctuate creamy to yellow plaques, 0.2–2.0 cm in size, which may be confluent, with "skip areas" of edematous mucosa. Histologically they are composed of fibrin, mucous, necrotic epithelial cells, and leukocytes.

An acute colitis, different from pseudomembranous colitis, was observed in five patients taking penicillin and penicillin derivatives (139). There was considerable rectal bleeding. The radiographic findings were those of ischemic colitis (spasm, transverse ridging, "thumbprinting," and punctuate ulceration). On sigmoidoscopy and biopsy, the mucosa was normal, except for an inflammatory cell infiltration in one case. Conservative treatment resulted in rapid remission.

Occurrence and frequency

Clostridium difficile has been isolated in 11–33% of patients with antibiotic-associated diarrhea, 60–75% of patients with antibiotic-associated colitis, and 96–100% of patients with pseudomembranous colitis (123,140,141). However, about 2% of the adult population are asymptomatic carriers (133). Primary symptomless colonization with *C. difficile* reduces the risk of antibiotic-associated diarrhea (142). Infants up to 2 years seem to be refractory to pseudomembranous colitis, although a high percentage may be carriers of *C. difficile* (141,143). The reasons for this are unknown. It has been speculated that infants lack receptors for the toxin.

There have been several reports of frequent diarrhea in patients treated with combinations of ampicillin or amoxicillin with beta-lactamase inhibitors, such as sulbactam or clavulanic acid (144–147). A double-blind crossover study in healthy volunteers showed disturbances of small bowel motility after oral co-amoxiclav (148).

The appearance of pseudomembranous colitis in clusters of patients (149–152) may explain the wide variation in occurrence, and suggests that the disease may result from cross-contamination among patients rendered susceptible by antibiotic treatment. This is especially true for epidemic outbreaks in hospitals, where the disease may be considered a nosocomial infection favored by serious illness, frequent and prolonged use of broad-spectrum antibiotics (especially cephalosporins), and poor compliance with the rules of hospital hygiene (153). In such an epidemic, a variable proportion of patients will harbor the organism as asymptomatic carriers. An additional possible explanation for the large differences in reported frequencies may be the use of different methods of detection and differences in the definition of the disease. If colonoscopy was routinely performed in all patients with diarrhea taking clindamycin, pseudomembranous colitis was found in as many as 10% (154).

Although the first antibiotics reported to cause pseudomembranous colitis were lincomycin and clindamycin, the disease was later described with all other antimicrobial drugs, even topically applied (155). Vancomycin (156) and metronidazole (157), which may be used as specific treatments, have also been implicated.

Susceptibility factors

Besides the type of antibiotic therapy, other factors such as the age of the patient, the severity of the underlying disease, colonic stasis, cytostatic therapy, surgical interventions, and gastrointestinal manipulations are predisposing factors for antibiotic-associated colitis (158–162).

It is still not established if there is a correlation between toxin production or genotype of the *C. difficile* and the clinical manifestations of the infection (163,164). Although hospital-acquired antibiotic-associated colitis is by far the major problem, community-acquired diarrhea associated with *C. difficile* has also been described (165).

Mechanism

Clostridium difficile produces two well-characterized toxins (130,166)—toxin A, an enterotoxin, and toxin B, an extremely potent cytotoxin—which are thought to be responsible for the disease. The toxigenicity of toxins A and B varies between different strains of *C. difficile* and seems to correlate with symptomatic disease (167). Pseudomembranes were found in a higher percentage of patients with stools positive for cytotoxin than in patients whose stools were positive for *C. difficile*, but toxin-negative (159) Although there is also a high association with *C. difficile* (about 20% are toxin-positive) in antibiotic-associated diarrhea without pseudomembranes, it is possible that this microorganism plays no pathogenic role in some of these usually milder forms of the disease. In these cases the diarrhea may be due to impaired metabolism of carbohydrates, altered fatty acid profiles, or the composition and deconjugation of bile acids by quantitatively and qualitatively altered fecal flora (120,121,141).

Diagnosis

The diagnosis of antibiotic-related colitis should be considered in any patient with severe diarrhea during or within 4–6 weeks after antibiotic therapy. The single best diagnostic procedure is sigmoidoscopy, although in a number of cases the typical pseudomembranous lesions may be seen only above the rectosigmoid area (168). Radiographic investigations (barium enema and air contrast) may show typical findings, but are dangerous in advanced cases and should be avoided. Computerized tomography showed typical but not pathognomonic patterns in two patients (169).

Clostridium difficile can be cultured from the stool, and toxins A and B can be assessed by different techniques (122). The most accurate method is still a cytotoxin tissue culture assay. This detects the cytopathic effect of cytotoxin B, which can be neutralized by *Clostridium sordellii* antitoxin, but it takes 24–48 hours to show a result. Alternative tests that produce faster results have been developed. A latex agglutination test lacks sensitivity and specificity, and does not distinguish toxigenic from non-toxigenic strains. An enzyme immunoassay for toxin A may be an acceptable alternative to the cell cytotoxin assay and the results are rapidly available. A dot immunobinding assay has not yet been extensively studied (170).

Management

Therapy consists of withdrawal of the antibiotic when diarrhea occurs and replacement of fluid and electrolyte losses. In less severe cases of antibiotic-associated diarrhea, no further treatment is needed. However, in patients with pseudomembranous colitis, a more intensive approach is usually required. When a toxic syndrome develops, fluid losses within the bowel can be very large. In these cases, a central venous line offers the chance to measure central venous pressure. Usually there is also loss of serum proteins and in some cases blood, which need appropriate replacement. In the rare cases with fulminant colitis and toxic megacolon, surgical intervention may be necessary (171,172).

In pseudomembranous colitis (typical endoscopic findings, positive test for *C. difficile* or its toxin), the preferred treatment is oral metronidazole, 250 mg qds or 500 mg tds (126,173). Metronidazole is as effective as vancomycin 125–250 mg qds, which is significantly more expensive (174). Oral bacitracin 25 000 U qds (175) and oral teicoplanin (176) are acceptable alternatives.

Relapses are similarly frequent after treatment with metronidazole and vancomycin (122). In 189 adult patients, a first relapse occurred in up to 24% and a second relapse in 46% (175). Relapse may be due to sporulation of *C. difficile* and not to the development of resistance. Relapses usually respond to further courses of the initial treatment. Some alternative treatments have been proposed for repeatedly relapsing cases, including the combination of vancomycin with rifampicin for 10 days (177).

The role of anion exchange resins (colestyramine and colestipol), which bind *C. difficile* toxin, is still controversial (178). If ion exchange resins are given at all, they should not be given together with vancomycin, because they also bind the antibiotic (179). Attempts to restore the intestinal flora with *Lactobacillus GG* (180), or with fecal enemas (181) from healthy volunteers have shown some favorable results in less severe cases. However, esthetic and infectious concerns may be an obstacle. It also has been suggested that treatment with *Saccharomyces boulardii* may help prevent the development of antibiotic-associated diarrhea (182). Its value in the prevention and treatment of relapses has still to be demonstrated. Antimotility agents have been associated with an increased incidence of antibiotic-related diarrhea and can worsen symptoms when the disease is already established (183). They should therefore be avoided.

There is little evidence that re-exposure to the same antibiotic that caused pseudomembranous colitis confers a further risk for relapse. Still, it would be wise to avoid the antibiotics that are most often related to pseudomembranous colitis in a patient who has had this complication.

Liver

Increases in serum transaminases and alkaline phosphatase, largely without additional symptoms, have been reported with the majority of beta-lactam antibiotics. With different compounds the estimated frequencies vary by up to a factor of 10. However, the frequency also depends on patient-related factors; in one study only a minority of transaminase increases could not be explained by factors other than antibiotic treatment (91).

More severe liver disease, presenting as hepatitis and/or intrahepatic cholestasis, has been seen with beta-lactam antibiotics of various classes, the isoxazolyl penicillins being most frequently involved. Co-amoxiclav has repeatedly been associated with cholestatic hepatitis.

Hepatitis is accompanied by fever, eosinophilia, and/or a rash in more than 80% of cases. This hints at the possibility of overlapping pathogenetic steps and sheds some doubt on the reliability of these accompanying symptoms as indicators of immune-mediated reactions, for example serum sickness-like syndromes.

One type of hepatitis is mainly associated with oxacillin (184,185). Eight of 54 patients developed this reaction after a mean cumulative dose of oxacillin 157 g (186).

Prolonged duration of treatment and increasing age were risk factors for flucloxacillin-induced jaundice (187), and cholestatic liver injury has been described most often with flucloxacillin (188,189) and other isoxazolylpenicillins (190). Whether cholestatic hepatitis after the combination of amoxicillin with clavulanic acid (co-amoxiclav) is related to one of these categories is not yet clear.

Urinary tract

Methicillin-induced acute interstitial nephritis follows a similar pattern of dose-dependence and time-dependence to that of neutropenia (191,192). This reaction occurred in 16% of all children treated with high-dose methicillin

(193). Nephritis occurred after a mean of 17 days and a mean cumulative dose of 120 g.

With other beta-lactams, mainly penicillin G, acute interstitial nephritis is rare, but it can follow the same pattern (194).

Nephritis is accompanied by fever, eosinophilia, and/or a rash in more than 80% of cases. This hints at the possibility of overlapping pathogenetic steps and sheds some doubt on the reliability of these accompanying symptoms as indicators of immune-mediated reactions, for example serum sickness-like syndromes.

Acute renal insufficiency, with or without skin rash and eosinophilia, has been reported with various beta-lactam antibiotics, most often with methicillin. Hence, the designation "methicillin-nephritis" is still sometimes used. The pathogenesis is largely unknown and is different from the nephrotoxicity of older cephalosporins (cefaloridine and cefalotin).

Skin

Rashes are among the most common adverse reactions to drugs in general and occur in 2–3% of hospitalized patients (195). Most distinct mucocutaneous reactions that can be induced by drugs have been associated with the use of individual beta-lactam antibiotics. These reactions include urticaria, angioedema, maculopapular rash, fixed drug eruption, erythema multiforme, Stevens–Johnson syndrome, toxic epidermal necrolysis, allergic vasculitis, serum sickness-like syndrome, eczematous lesions, pruritus, and stomatitis (196–199). The maculopapular rash, starting on the trunk or areas of pressure or trauma, is more frequent than all other skin manifestations together (195,200). Involvement of mucous membranes, palms, and soles is variable; the eruption can be associated with moderate to severe pruritus and fever. In addition, an indistinguishable rash often accompanies various reactions in other organs.

Pustular drug eruptions due to penicillin (201), amoxicillin (202), ampicillin (203), bacampicillin (204), cefazolin (205,206), cefradine (207), cefalexin (208), cefaclor (209), or imipenem + cilastin (210) seem to form a distinct clinical entity that has to be differentiated from pustular psoriasis, which can be drug-induced as well (210). A history of drug exposure, rapid disappearance of the eruption after the drug is stopped, and eosinophils in the inflammatory infiltrate argue in favor of pustular drug eruptions.

In patients with mononucleosis, aminopenicillins, and, less so, cephalosporins evoke rashes in a much higher percentage than usual (211). The incidence of rashes in infectious mononucleosis without antibiotics is 3–15%, compared with 40–100% with ampicillin. The underlying mechanism is speculative.

Baboon syndrome

Historically the condition called "baboon syndrome" was described as a special entity of a mild cutaneous erythema after exposure to so-called type IV allergens, such as nickel, mercury, immunoglobulins, and several drugs, such as aminophylline, heparin, terbinafine, and some antibacterial drugs. It is located on the buttocks and anogenital area, reminiscent of the red buttocks of baboons. It has recently been proposed that this term be replaced by the acronym SDRIFE (symmetrical drug-related intertriginous and flexural exanthem). The condition has been described in a child who received cefadroxil (212) and in another who received amoxicillin (213).

Immune hemolytic anemia

Drug-induced hemolytic anemia, which is uncommon, is characterized by a sudden fall in hemoglobin concentration. The full range of causative drugs has not been fully evaluated, but beta-lactams are often involved. This view has been verified in a 20-year retrospective analysis of 73 patients with drug-dependent antibodies to 23 different drugs from an immune hematological reference laboratory in the USA (214). Cephalosporins were at the top of the list (n = 37), followed by penicillins and/or penicillin derivatives (n = 12), non-steroidal anti-inflammatory drugs (n = 11), and others (n = 13).

Immunologic

Allergic reactions to beta-lactams are the most frequent immunological drug reactions (215,216). The first cases were described soon after the introduction of penicillin G.

The pathogenesis of many presumably immunologically mediated reactions to beta-lactam antibiotics is still unknown. Reliable and standardized tests to predict hypersensitivity only exist for a minority of allergic reactions, that is, IgE-mediated reactions. The matter is further complicated by the fact that beta-lactams can readily induce immune responses that by themselves do not necessarily result in disease. This is the case, for example, when antierythrocyte antibodies directed against beta-lactam bound to the erythrocyte surface are formed. This biological property (immunogenicity) has to be distinguished from allergenicity, that is, immune responses causing disease.

Cross-reactivity

Cross-reactivity, that is, hypersensitivity reactions initially induced by one compound but triggered by another, is an important and as yet unresolved problem, complicated by the fact that beta-lactams undergo structural modifications after administration, and that different parts of the molecule (such as the nucleus or side chains) can be involved. Data from cross-exposed patients (skin tests or drug challenge) suggest a high degree of cross-reactivity between compounds belonging to the same class and between the penicillins and carbapenems, but a low degree of cross-reactivity between penicillins and cephalosporins and between monobactams and the other beta-lactams.

For some decades, as several new beta-lactams came on the market, the classical dogma was that cross-reactivity between different penicillins was quite high, side-chain specific responses were negligible, in vivo tests with major and minor determinants would provide the diagnosis in more than 90% of cases, and the oral route was not relevant in inducing allergic reactions (26).

Now views may have changed, exemplified by a study on the frequency of anaphylactic reactions and cross-reactivity in 1170 children with suspected immediate allergic reactions to cephalosporins and/or penicillins (217). In vivo skin tests and challenges and in vitro tests for specific IgE showed that 58% were skin or challenge test positive; among them, 94% were positive to penicillins and 36% were positive to cephalosporins. The frequency of positive reactions with in vivo testing was 36–88% for penicillins and 0.3–29% for cephalosporins; 32% of the children who were allergic to penicillin cross-reacted to some cephalosporin. In contrast, if a child was allergic to a cephalosporin, the frequency of positive reactions to penicillin was 84%. The cross-reactivity among different generations of cephalosporins was 0–69%, and was highest with first- and second-generation cephalosporins and zero with third-generation cephalosporins. No information was given about the medication histories of these children and one cannot conclude that any beta-lactams is less allergenic than another. However, the results suggest a great degree of specificity and cross-reactivity to beta-lactams in children.

A leading Spanish group has stressed that allergy to beta-lactams has become a more complicated problem, because of the contribution of different chemical structures in inducing clinically relevant specificities (<<Gomez 216). The incidences of atopy and allergy in children are increasing all around Europe. Many theories have been brought forward, but the reasons are still unknown. The increased complexity of allergic reactions to beta-lactams might in fact reflect this increase in allergy.

Cross-reactivity with cephalosporins

Penicillins and other beta-lactam antibiotics, such as cephalosporins and carbapenems, are still by far the most widely used antibiotics for common infections. However, they are also by far the most common cause of drug allergy, especially IgE-mediated reactions, such as urticaria and anaphylaxis. Over the years, an important question has been cross-reactivity between the different classes of beta-lactams, and there is still confusion about whether it is safe to give a cephalosporin to a patient who is allergic to penicillin. This can lead to either underestimation or overestimation of the risks. In both cases, there can be negative consequences for the patient. It has recently been reported that six of 12 fatal anaphylactic reactions occurred after the first dose of cephalosporin (218). Three of the six patients were known to be allergic to amoxicillin and one was allergic to benzylpenicillin. On the other hand, because of the fear of cross-reactivity, the most common therapeutic approach to patients who are allergic to penicillin is to select antibiotics that do not contain a beta-lactam ring. However, reduced effectiveness, increased antimicrobial resistance, and higher costs are major drawbacks of this policy (219).

Cross-reactivity to cephalosporins in patients with well documented severe penicillin allergy has been studied in 128 consecutive patients who had an anaphylactic reaction (n = 81) or urticaria (n = 47) and had positive skin tests for at least one of the penicillin reagents tested (penicilloyl-polylysine, minor determinant mixture, benzylpenicillin) (220). They were skin tested with cefamandole, cefotaxime, ceftazidime, ceftriaxone, cefuroxime, and cephalothin. Patients with negative results with four cephalosporins were challenged with cefuroxime axetil and ceftriaxone. Fourteen of the 128 patients (11%) had positive skin tests for one or more of the cephalosporins tested. Challenge with a cephalosporin was refused by 22 patients. All 101 patients with negative results on skin tests tolerated cefuroxime axetil and ceftriaxone.

The authors stated that their data confirmed the advisability of avoiding cephalosporins in patients with positive skin tests for penicillins.

However, it should be emphasized that cross-reactivity can be very specific and not easy to predict, as shown in a recent report (221).

- A 3-year-old girl developed an anaphylactic reaction within 10 min after intravenous administration of the second dose (750 mg) of ceftriaxone for a urinary tract infection (222). After treatment with anti-shock therapy, her symptoms were considerably reduced within 1 hour and completely resolved after 12 hours. She had no previous history of drug allergy or atopy and no family history of allergic disease. Six months later she underwent prick and intradermal skin tests with standard concentrations of penicilloyl-polylysine, minor determinant mixture, penicillin G, penicillin V, ampicillin, amoxicillin, cefaclor, cefotaxime, ceftabuten, and cefalexin. Assays for specific IgE were performed for penicilloyl G and penicilloyl V. The patient had positive responses to prick testing with ceftriaxone and ampicillin. All assays for specific IgE were negative. When both in vivo and in vitro tests were negative, single-blind, placebo-controlled challenges with progressively increasing amounts of the respective drugs were performed. Challenges were positive only to cefalexin.

Taken together, these results suggest the presence of antigenic determinants unrelated to the beta-lactam ring but at the same time not completely side-chain specific. The patient reacted to cefalexin and ampicillin but not to cefaclor, all drugs with identical side-chains. In fact, the difference between cefalexin and cefaclor is in only one functional group (a methyl group instead of a chlorine at the R2 position of the thiazolidine ring). Whatever the mechanisms might be, this is the first case of immediate hypersensitivity to ceftriaxone and cross-reactivity with cefalexin and ampicillin in a child.

There are few data about the cross-reactivity of the carbapenems (imipenem–cilastatin, meropenem, ertapenem) with other beta-lactams and between the three compounds in the group. In an early report it was stated that about 50% of patients with penicillin allergy also reacted to imipenem–cilastatin (223). However, the statement was based on skin testing of only 20 patients.

A more recent retrospective review of patients undergoing bone marrow transplantation found about 10% cross-reactivity with imipenem–cilastatin in patients with self-reported or confirmed penicillin allergy (224).

There has been a large retrospective study of the comparative incidence of cross-reactivity associated with a carbapenem (imipenem–cilastatin or meropenem) in patients with and without penicillin allergy (n = 100 and n = 111 respectively) (225). Of those with reported or documented penicillin allergy 11% had an allergic-type reaction to a carbapenem, 5.2 times greater than the risk in patients who were reportedly not allergic to penicillin. There was no difference in the occurrence of allergic-type reactions between the two carbapenems. The authors concluded that "clinicians should be cautious whenever imipenem-cilastatin or meropenem is administered to patients who are allergic to penicillins".

Thus, cross-reactivity between carbapenems and penicillins is well established. Cross-reactivity between the various carbapenems is still a more open question. In animals, anti-meropenin antibodies raised in rabbits and guinea pigs had cross-reacted weakly with imipenem (226). In one report of anaphylaxis in a patient treated with imipenem-cilastatin, skin testing 2 weeks after the reaction was positive for imipenem–cilastatin and for imipenem alone, but negative for cilastatin alone and for meropenem (227). A similar case was recently published.

- A 41-year-old woman was admitted to an intensive care unit with postoperative septicemia and was given imipenem–cilastatin (dosage not stated) for a presumed intra-abdominal infection (228). Within 48 hours of starting imipenem-cilastatin, a large erythematous maculopapular rash with areas of urticaria appeared in several places on her body. Imipenem-cilastatin was withdrawn and within some days the skin eruption faded. However, she developed several other complications, and after 10 weeks in the intensive care unit several abdominal abscesses had to be drained subcutaneously. Cultures from these abscesses showed Gram negative rods and Gram positive cocci, resistant to nearly all commercially available antibiotics. One of the few antibiotics that had some activity against both groups of bacteria was imipenem. It was believed that the benefit of using meropenem outweighed the possible risks, and she received challenge doses of meropenem, which she tolerated well, followed by a full 14-day course without skin eruptions. About 7 days after the completion of the course of meropenem her clinical status had not improved significantly, her family opted to withdraw care, and she died. The authors reported that they did not recommend indiscriminate use of meropenem in patients with a history of imipenem allergy, and that it should only be considered if clinical circumstances demand a carbapenem antibiotic and then only after a dose-challenging regimen.

Possible cross-reactivity between penicillins and cephalosporins was discussed in SEDA-28 (p. 261). There are now new data. In a meta-analysis, using Medline and EMBASE databases and the key words "cephalosporin", "penicillin", and "cross-sensitivity" for the years 1960–2005, 219 articles were found, of which nine served as sources for an evidence-based analysis (229). First-generation cephalosporins have cross-allergy with penicillins, but cross-allergy is negligible with second- and third-generation cephalosporins; this is also most probably true for fourth-generation cephalosporins (230). The value of emphasizing the role of chemical structure has been underlined in a case report.

- A 39-year-old woman developed anaphylactic shock a few minutes after taking a tablet of cefuroxime axetil 500 mg (231). Skin tests confirmed that she was allergic to cefuroxime, and the reaction was defined as probable according to the Naranjo probability scale. A structure-activity relation study was performed using skin testing. She was sensitized to beta-lactam antibiotics with a methoxylimino group, but not to similar compounds that lack this chemical group (for example amoxicillin, penicillin G, and penicillin V). Intravenous amoxicillin was well tolerated.

The authors suggested that an approach based on structure-activity relations could help physicians and pharmacists in advising patients with allergic reactions.

A similar approach to studying the tolerability of other beta-lactams has been used in patients with a history of aminopenicillin-induced rash (232). Skin testing was followed by oral challenge to identify beta-lactams that patients with confirmed delayed-type non-immunoglobulin E-mediated allergic hypersensitivity to aminopenicillins could tolerate. Of 71 patients, 69 tolerated cephalosporins without an aminobenzyl side-chain (such as cefpodoxime or cefixime) and 51 also tolerated phenoxymethylpenicillin. The authors concluded that skin tests and drug challenge tests can help in determining individual cross-reactivity. Beta-lactam-specific IgE has good specificity but poor sensitivity (233). More sensitive methods should therefore be developed.

In a survey of 83 patients who reported penicillin allergy and were given a cephalosporin, seven had an adverse reaction (234). Six of these reported a definite history of an immediate reaction to penicillin, including urticaria. Of 62 patients who reported that their penicillin reaction had been delayed, probable, or unknown, only one had a reaction to a cephalosporin. The risk of a reaction was highest with second-generation cephalosporins and least with fourth-generation cephalosporins. The presence of an aminobenzyl ring in the cephalosporin molecule increased the risk.

There have been reports of presumed IgE-mediated hypersensitivity reactions of individual cephalosporins without cross-reactivity to other beta-lactam antibiotics.

- A 51-year-old woman had anaphylactic shock 10 minutes after an intravenous dose of cefodizime 1 g, having tolerated intramuscular cefodizime 11 months before (235). Skin tests with major and minor penicillin determinants, amoxicillin, ampicillin, benzylpenicillin, cefamandole, cefotaxime, ceftriaxone, ceftazidime, and cefuroxime were all negative. Cefodizime produced wheal, maximum diameter 10 mm, surrounded by erythema after 20 minutes.
- A 3-year-old girl had an anaphylactic reaction 20 minutes taking a second oral dose of ceftibuten 135 mg,

which she had tolerated 6 months before (236). Prick and intradermal skin tests with standard concentration of major and minor determinants, amoxicillin, ampicillin, penicillin G, penicillin V, cefaclor, cefotaxime, ceftriaxone, and cephalexin were all negative. There was a positive response to prick testing with ceftibuten. In both cases, although the authors did not detect specific IgE to cefodizime they attributed this to the fact that the tests were carried out some time after the reactions, and they suggested that the reactions were IgE-mediated. They also suggested that the lack of cross-reactivity with other beta-lactams emphasized the role of the IgE epitope present on R1 side-chain in inducing immediate hypersensitivity.

Tables 1 and 2 show penicillins and cephalosporins that are structurally similar and for which cross-reactivity is more likely. The author of an extensive review of the evidence reached the following conclusions (237):

- If a patient has a reaction to a penicillin or cephalosporin that was not IgE-mediated and was not serious, it is safe to give repeated courses of that antibiotic and related antibiotics; only IgE-mediated reactions are likely to become more severe with time and to result in anaphylaxis.
- If the rash is non-urticarial and non-pruritic, it is almost certain that it is not IgE-mediated and the risk of recurrence of the same rash with repeated courses of the same antibiotic is not increased; in uncertain cases, elective penicillin skin testing is advisable.
- If the patient has a history that is consistent with a severe IgE-mediated reaction to a penicillin, cephalosporins with a similar 7-position side chain on the beta-lactam ring (Table 1) should be used with caution.
- If the allergic reaction followed administration of ampicillin or amoxicillin, cephalosporins with a similar side chain (Table 1) should be used with caution.
- Other cephalosporins with different side-chains are not more likely to produce allergic reactions in penicillin-allergic patients than among non-allergic patients.
- A cephalosporin can be given to a patient who has had a non-IgE-mediated adverse reaction to a penicillin, i.e. a type II, III, IV, or unclassified reaction, such as hemolytic anemia, serum sickness, contact dermatitis, or a morbilliform or maculopapular rash; in uncertain cases, elective penicillin skin testing is advisable.

- When patients give a history of penicillin allergy, it is advisable to question this information, because very often the drug was not actually taken or a recognized non-immunological adverse effect (for example vomiting, diarrhea, or a non-specific rash) occurred.
- Penicillin skin testing can be useful to identify allergic patients, and testing is about 60% predictive of clinical hypersensitivity.
- Cephalosporins cause allergic or immune-mediated reactions among about 1–3% of patients, even if they are not allergic to penicillins.
- The incidence of allergic reactions to cephalosporins among penicillin-allergic patients, attributable to cross-reactive antibodies, varies with the side-chain similarity of the cephalosporin to the penicillin.
- For first-generation cephalosporins, the risk is 0.4%; for cefuroxime, cefpodoxime, and cefdinir the risk is nearly zero.
- A patient who has an allergic reaction to a specific cephalosporin probably should not use that cephalosporin again; however, the risk of a drug reaction when a different cephalosporin is given appears to be very low or non-existent if the side-chains of the drugs are not similar.
- Penicillin skin testing is not predictive of cephalosporin allergy unless the side-chain of the penicillin is similar to the side-chain of the cephalosporin.

In a retrospective review of 101 patients who underwent penicillin skin tests, 92 had a negative result and five had a positive result; in four the test result was indeterminate (238). Of patients with negative skin tests 49% were given a penicillin-based drug and 48% a cephalosporin; there were no serious adverse reactions. There was a 96% reduction in the use of vancomycin and a 96% (23/24) reduction in the use of fluoroquinolones in patients with negative skin tests.

Skin tests with commercially available haptens to major and minor determinants (benzylpenicilloyl poly-l-lysine and a mixture of minor determinants), penicillin G, injectable amoxicillin, and ampicillin and cephalosporins, if they have been incriminated by the patient, have typically been used in the diagnosis of allergy to beta-lactam antibiotics. However, both benzylpenicilloyl poly-l-lysine and mixtures of minor determinants have been withdrawn commercially in most countries. The likely effect of this withdrawal on the diagnosis of beta-lactam allergy has

Table 1 Structural similarities in the 7-position side-chains of some penicillins and cephalosporins

Similar structures			Dissimilar structures	
Related	Related	Related	Not Related	
Penicillin G	Amoxicillin	Cefepime	Cefazolin	Cefamandole
Cefoxitin	Ampicillin	Cefetamet	Cefdinir	Cefixime
Cephaloridine	Cefaclor	Cefotaxime	Cefonicid	Cefmetazole
Cephalothin	Cefadroxil	Cefpirome	Cefoperazone	Cefotiam
	Cefatrizine	Cefpodoxime	Cefotetan	Ceftazidime
	Cefprozil	Cefteram	Cefsulodin	Ceftibuten
	Cephalexin	Ceftizoxime	Cefuroxime	Cephapirin
	Cephradine	Ceftriaxone		Moxalactam

Table 2 Structural similarities in the 3-position side-chains of some cephalosporins

Similar structures							Dissimilar structures
Related	Related	Related	Related	Related	Related	Related	Unrelated
Cephradine	Cefmetazole	Cefdinir	Cephapirin	Ceftazidime	Cefuroxime	Ceftibuten	Cefatrizine
Cefadroxil	Cefoperazone	Cefixime	Cefotaxime	Cefsulodin	Cefoxitin	Ceftizoxime	Cefotiam
Cephalexin	Cefotetan		Cephalothin				Cefpodoxime
	Cefamandole						Cefprozil
							Ceftibuten
							Ceftriaxone
							Cefonicid
							Cefepime
							Cefotiam
							Cefazolin
							Cephaloridine
							Cefaclor

been retrospectively analysed in 824 patients (mean age 37 years, 254 men and 570 women) (239). There was a positive skin test response in 136 (16.5%), of whom six (4.4% of those with positive skin test responses) had positive skin test responses to benzylpenicilloyl poly-l-lysine only, nine (6.6%) to a mixture of minor determinants only, and five (3.7%) to both without a positive reaction to other beta-lactams. There were positive skin tests to other beta-lactams in 116 (85.3%), of whom about 30% of those with positive skin test responses also had positive responses to benzylpenicilloyl poly-l-lysine, a mixture of minor determinants, or both. Thus, had benzylpenicilloyl poly-l-lysine and a mixture of minor determinants not been available, the diagnosis would not have been made in 20 patients, who could not otherwise have been identified. In another study it was estimated that a misdiagnosis would have occurred in 47% of 463 patients in the absence of benzylpenicilloyl poly-l-lysine and a mixture of minor determinants (240). Although there were differences between the results of these two studies, both sets of authors concluded that major and minor determinants are necessary in the investigation of allergy to beta-lactams.

Mechanisms
Drug allergy or hypersensitivity represents an acquired capacity of the organism to mount an immunologically mediated reaction to a compound. This ultimately involves covalent or exceptionally non-covalent binding to and modification of host molecules (presumably proteins) by the drug, to which the host becomes sensitized (induction phase). Re-exposure to the sensitizing drug can trigger a series of immunological effector mechanisms (effector phase). These can be defined as pathways of inflammation or tissue injury, but they also represent mechanisms of immune protection from infectious agents.

Traditionally, the classification scheme defined by Gell and Coombs (241) distinguishes four types of reactions:

- type I reactions, which are IgE-mediated immediate hypersensitivity reactions.

- type II reactions, which are mediated by cytotoxic IgM and/or IgG.
- type III reactions, which are mediated by immune complexes.
- type IV reactions, which are cell-mediated hypersensitivity responses.

However, this classification fails to account for the complex and sequential involvement of several cell types and mediators in the immune response, as recognized today (242).

IgE-antibody-mediated adverse reactions
IgE-antibody-mediated hypersensitivity can serve as a paradigm to demonstrate some important features of beta-lactam hypersensitivity. Beta-lactams are small molecules that have to combine with a host macromolecule to be recognized by the immune system. In the case of penicillin, this reaction involves coupling of reactive degradation products to a protein-containing carrier (243). There are several degradation pathways, which result in the formation of reactive compounds, most importantly penicilloyl (244), also called the major determinant. Other less abundant degradation products include penilloate, benzylpenicilloate, and benzylpenilloate, the so-called minor determinants.

The complex contains haptens, often multiple, coupled to a protein-containing carrier molecule, and can induce T cell-dependent B cell activation, leading to the formation of antihapten antibodies. The mechanisms that govern the selection of the different immunoglobulin isotypes are reviewed elsewhere (242).

The time required for sensitization is called "latency" and is variable, depending on factors such as route of exposure, hapten dose, and chemical reactivity of the drug, as well as on genetic and acquired host factors. The period between the last exposure to the drug and the first appearance of symptoms has been termed the "reaction time." It is part of the clinical description of an adverse event and may help to attribute it to a specific drug (SED-12, 594).

Once sensitivity has been established, that is, once hapten-specific IgE-producing B cells have been formed, exposure to even small amounts of hapten can induce a cascade of events that lead to immediate reactions, such as anaphylaxis (245). Briefly, preformed IgE antibodies to drug determinants recognize the hapten-carrier complex and fix to the surface of mast cells or basophils, triggering the release of a series of mediators, such as histamine, neutral proteases, biologically active arachidonic acid products, and cytokines. This ultimately leads to a clinical spectrum that ranges from a mild local reaction to anaphylactic shock.

Non-IgE-antibody-mediated immunological reactions

Modification of erythrocyte surface components due to binding of beta-lactams or their metabolic products is thought to be the cause of the formation of antierythrocyte antibodies and the development of a positive Coombs' test implicated in the development of immune hemolytic anemia (246). About 3% of patients receiving large doses of intravenous penicillin (10–20 million units/day) will develop a positive direct Coombs' test (247). However, only a small fraction of Coombs' positive patients will develop frank hemolytic anemia (248). Antibody-coated erythrocytes are probably eliminated by the reticuloendothelial system (extravascular hemolysis) (249), or less often by complement-mediated intravascular erythrocyte destruction (250). Another mechanism implicates circulating immune complexes (anti-beta-lactam antibody/beta-lactam complexes), resulting in erythrocyte elimination by an "innocent bystander" mechanism (87). Similar mechanisms have been implicated in thrombocytopenia associated with beta-lactam antibiotics (251,252).

Contact dermatitis was often observed when penicillin was used in topical formulations and still continues to be described in cases of occupational exposure to beta-lactams (253,254). The underlying mechanism is thought to involve chemical modification of antigen-presenting cells in the epidermis, leading to sensitization of drug-specific T cells (255,256).

The underlying mechanism of a series of clinical entities associated with beta-lactams, such as maculopapular rash, drug fever, eosinophilia, serum sickness-like disease, vesicular and bullous skin reactions, erythema nodosum, and acute interstitial nephritis, is suspected to be immunological but is still largely unknown.

Reactions specific to side chains

Side chain–specific allergic reactions to beta-lactams are a steadily increasing problem (SEDA-21, 260) (257–259). Apart from epitopes generated by the beta-lactam nucleus, side chains attached to it can serve as additional epitopes recognized by the host immune response. Side chain–specific antibodies can be detected in patients who are allergic to beta-lactam antibiotics, even in the absence of reactivity to the mother compound. The clinical importance of this is debated. Serious anaphylactic reactions to amoxicillin occurred in three patients who tolerated benzylpenicillin (260). The phenomenon is mostly relevant for patients given semisynthetic penicillins, cephalosporins, carbapenems, and monobactams: compounds derived from each of these classes of drugs share certain side chains that may be cross-recognized by preformed antibody. Diagnosis of side chain–specific allergy requires a panel of diagnostic tools available only at selected research centers.

Pseudoallergy

The term "hypersensitivity" includes both immunoallergic and pseudoallergic reactions. Immunoallergic reactions occur when highly specific mechanisms involving immunological memory and recognition are involved. Pseudoallergic reactions are reactions that mimic immunoallergic reactions, but in which a specific immune-mediated mechanism is not involved. The so-called ampicillin rash is an example of a pseudoallergic reaction. Some major mediators of pseudoallergic reactions have been reviewed (261). The roles of newer putative mechanisms, involving cytokines, kinins, and other host-derived substances, remain to be ascertained. Most important is the fact that currently there are no standardized and validated animal models for predicting pseudoallergic reactions (262).

Animal models

Guinea pigs have been used for years in studies of systemic anaphylaxis. However, variations in predictability and sensitivity limit their value (263,264). Passive cutaneous anaphylaxis is another guinea pig model, but it is no more sensitive than systemic anaphylaxis (263). Respiratory sensitization, resulting in IgE-mediated immediate hypersensitivity has been investigated in mice and guinea pigs (265,266). Most often highly reactive chemicals have been used, and the models are of limited value in testing antimicrobial drugs.

Contact sensitizers have been studied in guinea pigs and mice, and it has been stated that "these models can reasonably identify the majority of human contact sensitizers" (262).

The best approach to induce a specific immune response against substances of low molecular weights is to use hapten-carrier conjugates. This method is of value in assessing the potential for cross-reactivity between closely related compounds, such as beta-lactam antibiotics (267).

It is obvious that the development of new animal models, for example transgenic and knock-out mice, should create new possibilities for predicting the sensitizing potential of new antimicrobials. The sad fact that hypersensitivity reactions are among the most commonly occurring adverse effects when antibiotics are used underlines the urgent need for research efforts in academia and industry (262).

Incidence

The true incidence of immunologically mediated reactions to beta-lactam antibiotics is hard to evaluate. This is mainly because of problems associated with the case definition of hypersensitivity reactions. The pathogenetic mechanism for a significant number of reactions presumed to be immunological in nature has not yet been conclusively determined. Furthermore, studies that address the incidences of adverse reactions face the problem of dealing with heterogeneous patient populations, treated with different types of beta-lactams, and administered by diverse routes in various dosages. The issue can be illustrated by reviewing data derived from four pharmacoepidemiological studies.

1. The International Rheumatic Fever Study, a prospective multicenter study that recorded allergic reactions, defined as hypotension, dyspnea, pruritus, urticaria, angioedema, arthralgia, and maculopapular rash in 1790 patients treated with monthly intramuscular benzathine penicillin for prophylaxis of rheumatic fever (32 430 injections during 2736 patient years). There was a 3.2% case incidence of allergic reactions and a 0.2% case incidence of anaphylaxis (12/100 000 injections), including one death (0.05%, equivalent to 3.1/100 000 injections) (268).

2. A large national study by venereal disease clinics in the USA, including four cooperative surveys conducted at 5-year intervals (1954, 1959, 1964, and 1969). The study included data from 94 655 patients unselected with regard to a history of penicillin allergy. The frequency of anaphylaxis was 0.055%, including one death (269).

3. A retrospective analysis of allergic reactions (drug-induced fever and rash) in 90 adults with cystic fibrosis, of whom 26 developed probable allergic reactions to parenteral beta lactams. There was drug-induced fever in 54 and skin reactions in 28 of 897 treatment courses (6 and 3.1% respectively). There was one case of non-fatal anaphylaxis. The numbers of allergic reactions per number of patients receiving specific antibiotics were: carbenicillin 4/56, mezlocillin 7/42, piperacillin 11/31, ticarcillin 1/20, cefazolin 0/24, ceftazidime 1/35, imipenem + cilastatin 4/16, and nafcillin 3/36 (17).

4. The Boston Collaborative Drug Surveillance Program. In this classic study in in-patients during 1966–82, beta-lactams headed the list of drugs causing skin reactions, presumably allergic. The overall reaction rate (the number of drug-related skin reactions per 1000 treated patients) was 51 for amoxicillin, 42 for ampicillin, 29 for semisynthetic penicillins, 16 for penicillin G, and 13 for cephalosporins (18,199).

These four sets of data illustrate a spectrum of diverse settings of beta-lactam administration: single-dose parenteral use (in venereal disease clinics), intermittent parenteral use (in rheumatic fever), and continuous high-dose parenteral use (in cystic fibrosis). Factors other than route of administration and dosing, such as drug history, underlying disease, co-administered drugs, and the risk profile of a particular compound, will be important in assessing the risk of giving a beta-lactam to a particular patient, as discussed in more detail below.

Table 3 contains a list of presumably immunologically mediated effects of beta-lactams, according to their estimated frequencies. The mechanisms of most of these reactions are not completely understood, which implies that some of the entities listed may be due to non-immunological mechanisms. The frequencies of the various adverse effects vary among different beta-lactams and depend on additional factors, discussed below. A compilation of reported frequencies of occurrence related to different compounds has been published and was used and extended to prepare Table 1 (200).

Presentation

The requirement for sensitization explains why a drug may be administered for a variable length of time without adverse effects. Once the organism is sensitized, the manifestation of hypersensitivity will depend on the route and dose of the allergen as well as the type of effector mechanism involved, preformed IgE being the most rapid, others evolving more slowly, typically over days. Generally much less drug is required to trigger a hypersensitivity reaction in a sensitized subject than for induction. Anaphylactic reactions have been described after ingestion of meat from penicillin-fed animals or after sexual intercourse in a penicillin-sensitive patient (304,305).

The time for sensitization to occur is often difficult to establish in a patient who develops symptoms during continuous therapy. A classification scheme that distinguishes between immediate, accelerated, and late reactions is of limited clinical use, since it will only allow distinction between IgE-mediated reactions (that is, rapid reactions) and non-IgE-mediated reactions (that is, more slowly evolving reactions) in the setting of re-exposure of a sensitized subject (85).

In contrast to hypersensitivity, other adverse reactions do not require sensitization and require similar doses of drug for recurrence. A special case is a syndrome (Hoigné's syndrome) that resembles an immediate allergic reaction combined with hallucinations, aggressive behavior, anxiety, and auditory and visual disturbances, which has been described after intramuscular procaine penicillin and benzathine penicillin. It is probably due to accidental intravascular injection and results from microembolism of the penicillin depot formulation (306–310).

Susceptibility factors

Several factors that influence hypersensitivity have been recognized and reviewed (311).

Patient-related factors

Patient-related factors include an increased incidence of allergic reactions to beta-lactams in patients with systemic lupus erythematosus (312) but not with atopic diseases (313). Genetic factors that influence drug metabolism and excretion, as well as the underlying disease of the patient and host immune reactivity, are likely to modulate the risk and severity of hypersensitivity reactions.

A history of a prior penicillin reaction increases the risk of a subsequent exposure. A classic study showed a frequency of allergic reactions to penicillin of 0.62% (155 of 24 906 treatment courses) in patients without a history of penicillin allergy compared with 13% (10 of 78 treatment courses) in patients with a history of penicillin allergy

Table 3 Presumably immunologically mediated adverse reactions to beta-lactams

Adverse reaction	References
Expected in one or more of 100 treatment courses	
Maculopapular rash[a]	(199,200,270)
Expected once in 100–1000 treatment courses	
Urticaria, angioedema	(85,200,270,271)
Drug fever [b]	(17,272)
Eosinophilia[c]	(273–275)
Expected once in 1000–10 000 treatment courses	
Anaphylactic shock	(245,276,277)
Bronchospasm and acute severe dyspnea	(278,279)
Thrombocytopenia	(97,280)
Serum sickness-like disease	(281–284)
Vasculitis	
Expected less than once in 10 000 treatment courses	
Hemolytic anemia	(86,87,285–288)
Vesicular and bullous skin reactions (including Stevens–Johnson syndrome and toxic epidermal necrolysis)	(289,295)
Erythema multiforme[d]	(292,296)
Erythema nodosum[e]	(292,296)
Interstitial nephritis[f]	(297)
Observed after occupational exposure	
Contact sensitivity	(254,298)
Anaphylaxis	(245)
Asthma, pneumonitis	(299–303)

[a]Occurs with all beta-lactams; more often with aminopenicillins, penicillinase-resistant penicillins, and anti-*Pseudomonas* penicillins
[b]Occurs with all beta-lactams; probably more often with piperacillin + tazobactam and aztreonam
[c]Occurs with all beta-lactams; probably more often with meticillin, nafcillin, oxacillin, second- and third-generation cephalosporins, aztreonam, and imipenem
[d]Occurs with all beta-lactams; probably more often with penicillins G and V, antipseudomonal penicillins, cefaclor, cefadroxil, cefalexin, loracarbef, aztreonam, and imipenem
[e]Occurs probably with all beta-lactams, but more often with penicillins G and V, cefuroxime, cefoperazone, cefaclor, and imipenem
[f]Occurs probably with all beta-lactams; well documented for methicillin

(269). Reaction rates are higher in patients with a history that suggests IgE-mediated reactions (314).

Patients with chronic lymphatic leukemia or with concurrent infection with Epstein–Barr virus or HIV have an increased frequency of ampicillin- and amoxicillin-associated rashes (315).

Drug-related factors
Drug dosage, mode of administration, and duration of treatment probably influence the frequency of allergic reactions. Topical administration has been associated with a high incidence of sensitization, in contrast to a low incidence with the oral route. For IgE-mediated reactions, a frequent and intermittent course of treatment is more likely to cause allergy than a prolonged course without a drug-free interval (243). High doses of parenteral beta-lactams are usually required for the induction of penicillin-induced hemolytic anemia (247). Similarly, it is likely that the high dosages of beta-lactams used in patients with cystic fibrosis result in a high incidence of drug fever (17).

Co-administration of beta-blockers has been associated with an increased risk of severe allergic drug reactions and reduces the effect of adrenaline in the immediate treatment of anaphylactic shock. The mechanism involves changes in the regulation of anaphylactic mediators (316).

Evidence that allopurinol potentiates skin reactions to ampicillin is controversial (198,317).

Long-Term Effects

Drug resistance
The introduction of penicillin G more than 50 years ago was one of the milestones in the treatment of infectious diseases, leading to a drastic reduction in mortality from severe infections (318–320). Two years later the emergence of the first penicillin-resistant *Staphylococcus aureus* rapidly cooled clinicians' enthusiasm. Since then microorganisms have developed various mechanisms to survive antibiotic pressure, including the following:

1. Modification of the targets of beta-lactam antibiotics, that is, the penicillin-binding proteins (321,322), resulting in a reduced affinity of the antibiotic.
2. The synthesis of new penicillin-binding proteins with very low affinity for the antibiotic, providing a high degree of resistance.
3. The production and secretion into the periplasmic space of beta-lactamases, that is, enzymes sharing structural analogies with the penicillin-binding proteins without fulfilling any function in the cell wall synthesis but hydrolysing and inactivating the beta-

lactam ring. The genes encoding for these beta-lactamases are usually located on a plasmid but can also be anchored in the bacterial genome (323,324).

4. Structural modification of porines, proteins that form channels in the outer membrane of Gram-negative bacteria, preventing the antibiotics from reaching the penicillin-binding proteins by impairing their penetration through channels of the outer membrane (325,326).

All of these mechanisms are mainly based on interbacterial exchange of DNA or on point mutations (327), the predominant mechanisms for genetic exchange being transformation, transduction, and conjugation (328). Transformation is the simplest way to transfer DNA to another bacterium, provided that this is ready to accept foreign DNA (= competent bacteria). This mechanism of genetic transfer is mostly used by several human pathogens: *Streptococcus pneumoniae*, *Hemophilus influenzae*, *Neisseria meningitidis*, *Neisseria gonorrhoeae*, and *Bacillus subtilis*. The transduction needs a bacteriophage (a virus) as a vector to inject DNA into a bacterium. This elaborate system is limited by the restricted specificity of the vectors for few microorganisms. The most sophisticated system is conjugation. This mechanism was first observed in *E. coli* in 1952 by Hayes, who described the first transfer of genetic material (F-Factor, or fertility factor) by conjugation (329). Microorganisms use conjugation to transfer several types of genetic material, either extrachromosomal (that is, plasmids) or intrachromosomal (that is, transposons).

In summary, microorganisms can exchange and acquire genetic material in order to adapt to a changing environment, the antibiotic pressure. Furthermore, DNA exchange between prokaryotes and eukaryotic cells (yeasts) and some plants has been observed (330). A review has addressed in detail the problem of emergence and spread of resistance among clinical isolates (331). Here we shall only briefly discuss as examples the development of resistance by two major pathogens, *S. pneumoniae* and *S. aureus*.

One of the most striking features in microbiology is the rapid emergence and worldwide spread of penicillin-resistant *S. pneumoniae* (pneumococci), mainly due to the uncontrolled use of penicillin in certain countries (332). The first cases of penicillin-resistant pneumococci were reported in the 1960s in New Guinea and Australia. Penicillin-resistant pneumococci have now been registered in all continents. The highest rate was reported in 1989 in Hungary, amounting to 57% of all clinical isolates (333). Until now, pneumococci have not acquired the genes encoding for beta-lactamases. Accordingly, the underlying mechanism of penicillin resistance is structural modification of penicillin-binding proteins, leading to reduced affinity to the penicillin molecule (see above). These modifications of the penicillin-binding protein usually require several genetic steps. Moreover, horizontal transfer of pieces of penicillin-binding protein genes has been described between *Streptococcus mitis* and *Streptococcus pneumoniae* (334). Such modifications of penicillin-binding proteins lead to reduced affinity of these enzymes for their natural substrates (disaccharide-pentapeptides) (335) and eventually to the synthesis of a structurally different cell wall harboring more branched peptides. This is the biological price that pneumococci pay to survive antibiotic pressure (336). Usually in an epidemic area a few clones of penicillin-resistant pneumococci are responsible for the majority of the registered cases (337). Furthermore, DNA polymorphism analysis has shown that isolates have been imported from one continent to another, causing new epidemics (338–341).

Another major pathogen *S. aureus* has developed two different mechanisms of resistance:

- synthesis of beta-lactamase (nowadays more than 80% of *S. aureus* secrete beta-lactamase).
- the emergence of so-called methicillin-resistant *S. aureus* (MRSA), which can grow even in the presence of high concentrations of methicillin (up to 800 μg/ml).

The unique feature of MRSA is based on the acquisition of a low-affinity penicillin-binding protein for beta-lactam antibiotic molecules (the penicillin-binding protein 2A), which allows the bacteria to carry on synthesis of its cell wall, whereas the other penicillin-binding proteins are already inactivated by the high concentration of methicillin or other beta-lactamase-resistant beta-lactam antibiotics.

The origin of the penicillin-binding protein 2A is a matter of debate. Until now the only antibiotics that inhibit MRSA are the glycopeptides, such as vancomycin. A nightmare scenario would be the transfer of vancomycin resistance from enterococci to MRSA, which would cause a major epidemiological and therapeutic problem in the treatment of staphylococcal infections. There have been a few cases of vancomycin-resistant coagulase-negative *S. aureus* (342,343), but none of vancomycin-resistant MRSA, although transfer of vancomycin resistance to staphylococci has been achieved experimentally (344).

The continuous spread of resistance among clinical isolates, especially of multiresistant microorganisms represents a unique challenge in the treatment of infectious diseases. The detection of asymptomatic carriers of pneumococci, especially young children in day care centers, makes early detection even more difficult (345). Uncontrolled use of antibiotics in agriculture selects multiresistant fecal flora in animals, and meat can be contaminated by imperfect processing.

The problem of increasing resistance of micro-organisms is a major worldwide issue that necessitates close collaboration of clinicians, epidemiologists, and basic research laboratories. Newer and fast diagnostic tools (such as the polymerase chain reaction), routinely introduced into clinical laboratories, and better understanding at the molecular level of mechanisms of resistance of microorganisms are key prerequisites for the prevention of further spread of resistant microorganisms. Additional measures, for example broad use of vaccines and restrictions on antibiotic use imposed by health authorities, will require global cooperation and will have to be addressed by international organizations such as the WHO.

Second-Generation Effects

Teratogenicity

Since the days of the thalidomide disaster about 40 years ago, resulting in the birth of some thousands of malformed babies, it has been well recognized that drugs taken by pregnant mothers can have severe adverse effects on their unborn children. A consequence of the thalidomide disaster was worldwide awareness that drugs can cause congenital malformations and the necessity to investigate this possibility in animals. Since thalidomide, around 30 drugs have been proven to be teratogenic, not all of which are currently in clinical use (346). For most drugs, however, safety in pregnancy has still to be established. With the risk of teratogenicity and dysmorphogenesis ever present, clinicians are in general very cautious in prescribing drugs for pregnant women. Despite this, over 60% of pregnant women consume therapeutic agents not directly related to their pregnancy, and it has been estimated that about 5% of birth defects are caused by maternal drug therapy (347).

Even if a drug is generally recognized as being safe after animal experiments, it is wise to be suspicious when giving it to a pregnant woman. One major obstacle in evaluating safety in humans is the sample size required to reach sound conclusions. For example, in Europe, neural tube defects and cleft lips both occur with a prevalence of around seven per 10 000 live births (348). It has been calculated that for an uncommon drug exposure (that is, a frequency of under one per 1000 pregnant women and a background malformation prevalence of 0.001), one would have to monitor more than 1 000 000 births in order to detect a teratogenic effect, even though the relative risk associated with the drug might be as high as 20 (that is, a 20-fold increased risk of a particular malformation). In contrast, for formulations that are commonly used in pregnancy (for example by 2% of women, as was the case with thalidomide) and that are associated with an extremely high relative risk (such as 175), 1000 births would be sufficient to detect the teratogenic potential, even when the background prevalence of the malformation was as low as 0.0024 (349).

Another issue that has to be taken into consideration is the temporal relation between drug exposure and the effect on the embryo or fetus (350). Exposure to harmful drugs in the 2 weeks after conception usually leads to abortion, which may not be noticed. In the next 6–7 weeks the embryo is assumed to be extremely sensitive to teratogens (351). However, different organs and systems may be susceptible to teratogens at different times during this period. Therefore, in order to link drug use during pregnancy to a congenital malformation, drug intake must have taken place when the organ or organ system was sensitive to its harmful effects (350). It goes without saying that exact information on the timing of exposure is crucial.

In an ideal world, no drug would become available before it had been thoroughly tested for safety and effectiveness in a randomized, double-blind, placebo-controlled trial in pregnant women (352). However, because of ethical concerns about the welfare of the mother and fetus, pregnant women are traditionally excluded from drug trials. Therefore, usage is most often based on indirect measures of safety, such as in vitro studies and animal models. However, the thalidomide affair reminds us of the potential inadequacy of animal models.

In reality, most information about the safety of antimicrobial drugs in pregnancy comes from a history of long-term use with no reported adverse outcomes. As has been emphasized (352) most practitioners are happy to prescribe penicillin and its derivatives although there are no data from formal trials. However, there are data that show that penicillin V is safe during pregnancy (353). The study took place in Hungary between 1980 and 1996. The case group consisted of 22 865 malformed infants or fetuses, of whom 173 (0.8%) had mothers who had taken penicillin V during pregnancy. Two control neonates without malformations were matched with every case according to sex, week of birth, and the district of the parent's residence. Of the 38 151 infants in the control group, 218 had been treated with penicillin V. This difference was explained mainly by recall bias and confounders, because there was no difference in the adjusted odds ratio for medically documented phenoxymethylpenicillin treatment during the second and third months of gestation, that is, during the critical period for most major congenital abnormalities in case-matched control pairs. Thus, treatment with oral phenoxymethylpenicillin during pregnancy presents very little, if any, teratogenic risk.

There have also been two studies of the teratogenic potential of other penicillins. In the first study, in 791 women who had redeemed a prescription for pivampicillin during their first pregnancy, birth outcomes (malformations, pre-term delivery, and low birth weight) were matched with similar outcomes in 7472 reference pregnancies in which the mother had not redeemed any prescription for pivampicillin during pregnancy (354). There were no significant effects of pivampicillin. In the second study, in 78 women who took cefuroxime axetil during pregnancy, none of the 13 women who were treated in the first trimester gave birth to a malformed child, but one baby with hip dysplasia was found among 20 babies from mothers treated in their second trimester, and there was one case of hypospadias and one of imperforate anus in 47 children of mothers treated in the third trimester (355). The authors correctly concluded that the number of patients who had taken cefuroxime in the first trimester of pregnancy was small, and that cefuroxime should be used with caution in the early months of pregnancy.

Susceptibility Factors

Renal disease

Renal insufficiency is a risk factor for the toxic effects of the beta-lactams (356,357), including neurotoxic reactions (358), inhibition of platelet aggregation (359), and to some extent interaction with vitamin K-dependent synthesis of coagulation factors (360).

Drug Administration

Drug contamination

The administration of piperacillin + tazobactam or co-amoxiclav can result in detectable amounts of *Aspergillus* galactomannan antigen in the plasma (SEDA-29, 248), since some beta-lactam antibiotics may contain galactomannan. In a study of 39 batches of four different beta-lactam antibiotics galactomannan was not detected in nine batches of piperacillin, whereas it was detected in all 10 batches of amoxicillin and co-amoxiclav and in six of 10 batches of piperacillin + tazobactam (361). Within each four groups of drugs, all the batches came from the same company. The real size of this problem is not known, but it is large enough for regulatory agencies to start taking action.

Drug–Drug Interactions

Antacids

Antacids increase gastric pH and can result in impaired dissolution of some cephalosporins (362–364).

Anticoagulants

Some beta-lactam antibiotics impair coagulation by inhibiting hepatic and intestinal vitamin K production and impairing platelet function.

- A patient suffered significant postoperative bleeding 4 days after dental surgery in a patient taking amoxicillin, despite the use of a tranexamic acid (4.8%) mouth rinse to control hemostasis (365).

Interactions with drugs that affect coagulation and platelet function must therefore be borne in mind.

- A 58-year-old woman developed a raised INR and microscopic hematuria while taking warfarin and co-amoxiclav (366). This was attributed to an interaction of the two drugs.

In contrast, some penicillinase-resistant penicillins (dicloxacillin, nafcillin) provoke resistance to warfarin, lasting for up to 3 weeks after withdrawal of the antibiotic (367,368).

- A patient experienced the effects of interactions of warfarin with nafcillin and dicloxacillin (369). During co-administration of nafcillin, warfarin doses were increased to as much as 4.5 times the previous amounts needed to provide adequate anticoagulation. During co-administration of dicloxacillin, warfarin doses gradually fell, but still stabilized at a higher maintenance dose than before.
- A 41-year-old man taking warfarin 22 mg/week with a prothrombin time of 20.7 seconds was given dicloxacillin 500 mg qds for 10 days (367). The prothrombin time and S- and R-warfarin concentrations fell by 17, 25, and 20% respectively after 5 days. In a retrospective review of seven other patients, the mean prothrombin time fell

by 17% (range 11–26%) within 4 days of starting dicloxacillin.

This type of interaction may be due to induction of warfarin metabolism.

Digoxin

In 10% of patients taking digoxin, there is inactivation of up to 40% of the drug before absorption, by intestinal *Eubacterium lentum*. This can be reversed by antibiotics (370,371). The lack of effects of some beta-lactam antibiotics on serum digoxin concentrations in one study (372) might have been due to the small sample size or resistance of the bacteria.

Gentamicin

Gentamicin and other aminoglycosides have increased activity when they are combined with beta-lactams, resulting in increased bacterial aminoglycoside uptake (405). The proposed mechanism of synergism is damage to the cell membrane by the beta-lactam, followed by improved diffusion of gentamicin across the outer bacterial membrane. A second type of synergism, pharmacodynamic synergism, occurs when high serum concentrations of aminoglycosides cause efficient bacterial killing, resulting in reduced bacterial concentrations, which are more effectively eliminated by beta-lactams, as they work more efficiently against lower bacterial concentrations. The action of gentamicin is inhibited by some antimicrobials, which are bacteriostatic rather than bactericidal; for example antagonism occurs with macrolides, tetracycline and doxycycline, and chloramphenicol. The clinical significance of this antagonism is unknown.

There are many reports of acute renal insufficiency from combined treatment with gentamicin and one of the cephalosporins (406–408). The potential nephrotoxic effect of this combination seems to be related mainly to the nephrotoxic effect of gentamicin.

Neuromuscular blocking drugs

Penicillins G and V (409) have been reported to cause neuromuscular block in animal preparations, but only at exceptionally high doses. Calcium is effective in reversal. The acylaminopenicillins augment vecuronium-induced blockade (410). Possible "re-curarization" with piperacillin was successfully reversed by neostigmine (411).

Nifedipine

An active dipeptide transport system that depends on hydrogen ions takes up non-ester amino-beta-lactams (penicillin, amoxicillin, and oral first-generation cephalosporins) (373–375) and specific cephalosporins that lack the alpha-amino group (cefixime, ceftibuten, cefdinir, cefprozil) (376,377). Nifedipine increases amoxicillin and cefixime absorption, probably by stimulating the dipeptide transport system, since the serum concentrations of passively absorbed drugs and intestinal blood flow did not change (378–380).

Oral contraceptive steroids

There have been anecdotal reports that oral antibiotics reduced the efficacy of oral contraceptives. The proposed mechanism is that the antibiotics reduce the amount of gut bacteria that normally deconjugate excreted estrogens prior to reabsorption. However, in small formal studies this effect has not been confirmed (381–383). It is nevertheless possible that this is the mechanism, but that it occurs in too few women to be detected in small formal studies.

As early as 1971, it was observed that there was an increased incidence of intermenstrual bleeding in women who were reliably taking oral contraceptives and rifampicin for tuberculosis (384). Subsequent studies showed that pregnancy was a possible adverse effect of the combined use of antibiotics and oral contraceptives, the most commonly involved antimicrobial drugs being ampicillin, co-trimoxazole, and tetracyclines (385). In an analysis based on reports to the Committee on Safety of Medicines in the UK between 1968 and 1984, 63 pregnancies were identified in women taking this combination of drugs (381). Tetracyclines and penicillin were the drugs most often involved. However, the question of under-reporting was underlined. Over the years, there have been a few reports of the possibility of ineffectiveness of oral contraceptives during the use of antimicrobial drugs.

- A healthy 21-year-old woman, using an oral contraceptive, became pregnant while taking minocycline 100 mg/day for acne (386).

At least three different mechanisms have been proposed.

1. Antibiotics, for example rifampicin (SEDA-8, 256), interfere with the hepatic metabolism of the compounds in oral contraceptives.
2. Antibiotics increase gastric emptying and small intestinal motility. This in turn alters gastrointestinal absorption (and reabsorption) of oral contraceptives (probably both estrogens and progestogens). Increased gastrointestinal motility has been best studied in relation to macrolide antibiotics (SEDA-18, 269). However, there is no evidence of reduced efficacy of oral contraceptives through an effect of macrolides on gastrointestinal motility, which is in any case an unlikely mechanism.
3. Antibiotics interfere with the normal flora in the gastrointestinal tract, thereby also interfering with the normal enterohepatic circulation of the compounds that are included in oral contraceptives. This is a complex story. Oral contraceptives contain estrogens and progestogens. There are several different derivatives in each group and the amounts of each compound vary from formulation to formulation. However, most (if not all) of the compounds have an enterohepatic circulation—they are excreted into the bile, usually as conjugates, and are reabsorbed from the intestinal tract. The efficacy of their absorption and reabsorption depends on physicochemical factors in the gut (such as pH, conjugated versus unconjugated compounds, and adsorption to microbes and dietary constituents). In addition to this basic physiological circulation, intestinal microbes interact with the compounds, most often by splitting conjugates, but also by acting on double bonds, hydroxyl groups, and keto groups, present in the original compounds. These derivatives may have different absorption rates from the original ones. Partly because of their complexity, these microbial interactions have been very little studied. However, it is well known that microbes interfere with the enterohepatic circulation of other molecules with similar structures, such as androgens (SEDA-13, 215) (SEDA-16, 262) and bile acids (387). A similar effect can take place in patients with gastrointestinal bacterial infections. In addition, and at least from a theoretical point of view, an effect on this enterohepatic circulation may also occur in individuals with a high load of living microbes, as is the case when ingesting probiotics (products that contain live bacteria such as Bifidobacterium and Lactobacillus). This possibility has not been adequately addressed by any of the companies that sell tons of probiotics to fertile women taking oral contraceptives.

Given the problems of under-reporting, the lack of proper investigations, and the possibility of an unwanted pregnancy, the concern can be reduced to a practical one: what should women be told? An excellent recommendation has been published in the UK (388): "Your doctor has prescribed antibiotics, which are necessary to treat [your] infection. However, the antibiotics can interfere with the pill and make it less effective. This means that you may not be protected from pregnancy if you have sex, even though you are taking the pill. You are advised that you should take precautions, that is use of a condom, during the time you are taking the antibiotics, and for 7 days after completing the course of antibiotics ···" For the time being, this is the best advice. However, more investigations in this important field are needed.

Phenobarbital

There was an unexpectedly high frequency of adverse effects in a pediatric intensive care unit with the combination of high-dose phenobarbital and beta-lactam antibiotics, mainly cefotaxime (389). The reactions, which mostly affected the skin and blood, were only rarely reproduced by a single component, suggesting an interaction. However, these findings have not been confirmed, and their impact is unclear.

Probenecid

Probenecid inhibits the tubular resorption of anions and inhibits the renal excretion of most beta-lactam antibiotics (390,391).

Vecuronium

There are various conflicting reports about acute interactions of beta-lactam antibiotics, especially acylaminopenicillins (apalcillin, azlocillin, mezlocillin, piperacillin), with vecuronium, leading to prolongation of muscle blockade. Reports of clinically relevant effects (392–394) conflict with reports of no effect (395).

Food–Drug Interactions

Co-administration of acid-labile beta-lactams, such as penicillin and ampicillin, with food reduces their systemic availability by lowering gastric pH and delaying gastric emptying (396).

Food increases the systemic availability of some cephalosporin prodrug esters, possibly by improving dissolution or blocking premature hydrolysis (397,398).

Interference with Diagnostic Tests

Aminoglycosides

Plasma aminoglycoside concentrations can be falsely low because they are inactivated by penicillins and cephalosporins. This effect occurred in plasma stored for 24 hours or longer before measurement (399–401).

Ciclosporin measurement

A retrospective study found an increased risk of ciclosporin-associated early nephrotoxicity in nafcillin-treated patients, despite the fact that ciclosporin concentrations were not different from controls (402). Possible interference of nafcillin with ciclosporin measurement, giving rise to falsely low concentrations, was considered as a possible explanation.

Coombs' test

There are often false-positive antiglobulin tests by non-immunologically bound serum proteins, especially with cephalosporins, clavulanic acid, and imipenem + cilastin. This is a source of difficulty in cross-matching blood products.

Glycosuria

Urine samples containing beta-lactams should be tested for glucose by the glucose oxidase method, since falsely high values are observed with the copper reduction method (403,404).

References

1. In: Morin RB, Gorman M, editors. The Chemistry and Biology of the Beta-lactam Antibiotics. New York: Academic Press, 1982:1–3.
2. Waxman DJ, Strominger JL. Penicillin-binding proteins and the mechanism of action of beta-lactam antibiotics. Annu Rev Biochem 1983;52:825–69.
3. Frere JM, Joris B. Penicillin-sensitive enzymes in peptidoglycan biosynthesis. Crit Rev Microbiol 1985;11(4):299–396.
4. Neftel KA, Hafkemeyer P, Cottagnoud P, Eich G, Hübscher U. Did evolutionary forerunners of betalactam antibiotics bind to nucleid acid replication enzymes? In: 50 Years of Penicillin Application. Berlin, Prague: Technische Universität Berlin, 1993:394.
5. Rello J, Gatell JM, Miro JM, Martinez JA, Soriano E, Garcia San Miguel J. Effectos secundarios associados a la cloxacillina. [Secondary effects associated with cloxacillin.] Med Clin (Barc) 1987;89(15):631–3.
6. Olaison L, Belin L, Hogevik H, Alestig K. Incidence of beta-lactam-induced delayed hypersensitivity and neutropenia during treatment of infective endocarditis. Arch Intern Med 1999;159(6):607–15.
7. Himelright IM, Keerasuntonpong A, McReynolds JA, Smith EA, Abell E, Smith RJ, Baddour LM. Gender predilection of antibiotic-induced granulocytopenia in outpatients with septic arthritis or osteomyelitis. Infect Dis Clin Pract 1997;6:183.
8. Reed MD, Stern RC, Myers CM, Klinger JD, Yamashita TS, Blumer JL. Therapeutic evaluation of piperacillin for acute pulmonary exacerbations in cystic fibrosis. Pediatr Pulmonol 1987;3(2):101–9.
9. Sanders WE Jr, Johnson JE 3rd, Taggart JG. Adverse reactions to cephalothin and cephapirin. Uniform occurrence on prolonged intravenous administration of high doses. N Engl J Med 1974;290(8):424–9.
10. Lang R, Lishner M, Ravid M. Adverse reactions to prolonged treatment with high doses of carbenicillin and ureidopenicillins. Rev Infect Dis 1991;13(1):68–72.
11. Stead RJ, Kennedy HG, Hodson ME, Batten JC. Adverse reactions to piperacillin in cystic fibrosis. Lancet 1984;1(8381):857–8.
12. Strandvik B. Adverse reactions to piperacillin in patients with cystic fibrosis. Lancet 1984;1(8390):1362.
13. Brock PG, Roach M. Adverse reactions to piperacillin in cystic fibrosis. Lancet 1984;1(8385):1070–1.
14. McDonnell TJ, FitzGerald MX. Cystic fibrosis and penicillin hypersensitivity. Lancet 1984;1(8389):1301–2.
15. Stead RJ, Kennedy HG, Hodson ME, Batten JC. Adverse reactions to piperacillin in adults with cystic fibrosis. Thorax 1985;40(3):184–6.
16. Koch C, Hjelt K, Pedersen SS, Jensen ET, Jensen T, Lanng S, Valerius NH, Pedersen M, Hoiby N. Retrospective clinical study of hypersensitivity reactions to aztreonam and six other beta-lactam antibiotics in cystic fibrosis patients receiving multiple treatment courses. Rev Infect Dis 1991;13(Suppl 7):S608–11.
17. Pleasants RA, Walker TR, Samuelson WM. Allergic reactions to parenteral beta-lactam antibiotics in patients with cystic fibrosis. Chest 1994;106(4):1124–8.
18. Wills R, Henry RL, Francis JL. Antibiotic hypersensitivity reactions in cystic fibrosis. J Paediatr Child Health 1998;34(4):325–9.
19. Mallon P, Murphy P, Elborn S. Fever associated with intravenous antibiotics in adults with cystic fibrosis. Lancet 1997;350(9092):1676–7.
20. Yamabe S, Adachi K, Watanabe M, Ueda S. The effects of three beta-lactamase inhibitors: YTR830H, sulbactam and clavulanic acid on the growth of human cells in culture. Chemioterapia 1987;6(5):337–40.
21. Vidal Pan C, Gonzalez Quintela A, Roman Garcia J, Millan I, Martin Martin F, Moya Mir M. Cephapirin-induced neutropenia. Chemotherapy 1989;35(6):449–53.
22. Antoniadis A, Muller WE, Wollert U. Benzodiazepine receptor interactions may be involved in the neurotoxicity of various penicillin derivatives. Ann Neurol 1980;8(1):71–3.
23. Neftel KA, Hubscher U. Effects of beta-lactam antibiotics on proliferating eucaryotic cells. Antimicrob Agents Chemother 1987;31(11):1657–61.

24. Bloom JC, Lewis HB, Sellers TS, Deldar A. The hematologic effects of cefonicid and cefazedone in the dog: a potential model of cephalosporin hematotoxicity in man. Toxicol Appl Pharmacol 1987;90(1):135–42.

25. Deldar A, Lewis H, Bloom J, Weiss L. Cephalosporin-induced changes in the ultrastructure of canine bone marrow. Vet Pathol 1988;25(3):211–8.

26. Neftel KA, Hauser SP, Muller MR. Inhibition of granulopoiesis in vivo and in vitro by beta-lactam antibiotics. J Infect Dis 1985;152(1):90–8.

27. Bloom JC, Thiem PA, Sellers TS, Deldar A, Lewis HB. Cephalosporin-induced immune cytopenia in the dog: demonstration of erythrocyte-, neutrophil-, and platelet-associated IgG following treatment with cefazedone. Am J Hematol 1988;28(2):71–8.

28. Rouveix B, Lassoued K, Regnier B. Neutropénies induites par les bétalactamines: mécanisme toxique ou immun?. [Beta lactam-induced neutropenia: toxic or immune mechanism?.] Therapie 1988;43(6):489–92.

29. Murphy MF, Riordan T, Minchinton RM, Chapman JF, Amess JA, Shaw EJ, Waters AH. Demonstration of an immune-mediated mechanism of penicillin-induced neutropenia and thrombocytopenia. Br J Haematol 1983;55(1):155–60.

30. Murphy MF, Metcalfe P, Grint PC, Green AR, Knowles S, Amess JA, Waters AH. Cephalosporin-induced immune neutropenia. Br J Haematol 1985;59(1):9–14.

31. Lee D, Dewdney JM, Edwards RG, Neftel KA, Walti M. Measurement of specific IgG antibody levels in serum of patients on regimes comprising high total dose beta-lactam therapy. Int Arch Allergy Appl Immunol 1986;79(4):344–8.

32. Charak BS, Brown EG, Mazumder A. Role of granulocyte colony-stimulating factor in preventing ceftazidime-induced myelosuppression in vitro. Bone Marrow Transplant 1995;15(5):749–55.

33. Hauser SP, Udupa KB, Lipschitz DA. Murine marrow stromal response to myelotoxic agents in vitro. Br J Haematol 1996;95(4):596–604.

34. Hauser SP, Allewelt MC, Lipschitz DA. Effects of myelotoxic agents on cytokine production in murine long-term bone marrow cultures. Stem Cells 1998;16(4):261–70.

35. Yen CJ, Tsai TJ, Chen HS, Fang CC, Yang CC, Lee PH, Lin RH, Tsai KS, Hung KY, Yen TS. Effects of intraperitoneal antibiotics on human peritoneal mesothelial cell growth. Nephron 1996;74(4):694–700.

36. Edin ML, Miclau T, Lester GE, Lindsey RW, Dahners LE. Effect of cefazolin and vancomycin on osteoblasts in vitro. Clin Orthop Relat Res 1996;(333):245–51.

37. Lanbeck P, Paulsen O. Cytotoxic effects of four antibiotics on endothelial cells. Pharmacol Toxicol 1995;77(6):365–70.

38. Friedland JS, Warrell DA. The Jarisch–Herxheimer reaction in leptospirosis: possible pathogenesis and review. Rev Infect Dis 1991;13(2):207–10.

39. Maloy AL, Black RD, Segurola RJ Jr. Lyme disease complicated by the Jarisch–Herxheimer reaction. J Emerg Med 1998;16(3):437–8.

40. Young EJ, Weingarten NM, Baughn RE, Duncan WC. Studies on the pathogenesis of the Jarisch–Herxheimer reaction: development of an animal model and evidence against a role for classical endotoxin. J Infect Dis 1982;146(5):606–15.

41. Negussie Y, Remick DG, DeForge LE, Kunkel SL, Eynon A, Griffin GE. Detection of plasma tumor necrosis factor, interleukins 6, and 8 during the Jarisch–Herxheimer reaction of relapsing fever. J Exp Med 1992;175(5):1207–12.

42. Fekade D, Knox K, Hussein K, Melka A, Lalloo DG, Coxon RE, Warrell DA. Prevention of Jarisch–Herxheimer reactions by treatment with antibodies against tumor necrosis factor alpha. N Engl J Med 1996;335(5):311–5.

43. Chow KM, Hui AC, Szeto CC. Neurotoxicity induced by beta-lactam antibiotics: from bench to bedside. Eur J Clin Microbiol Infect Dis 2005;24(10):649–53.

44. Johnson HC, Walker A. Convulsive factor in commercial penicillin. Arch Surg 1945;50:69.

45. Macdonald RL, Barker JL. Pentylenetetrazol and penicillin are selective antagonists of GABA-mediated post-synaptic inhibition in cultured mammalian neurones. Nature 1977;267(5613):720–1.

46. Chow P, Mathers D. Convulsant doses of penicillin shorten the lifetime of GABA-induced channels in cultured central neurones. Br J Pharmacol 1986;88(3):541–7.

47. Schliamser SE, Cars O, Norrby SR. Neurotoxicity of beta-lactam antibiotics: predisposing factors and pathogenesis. J Antimicrob Chemother 1991;27(4):405–25.

48. Gutnick MJ, Prince DA. Penicillinase and the convulsant action of penicillin. Neurology 1971;21(7):759–64.

49. Sobotka P, Safanda J. The epileptogenic action of penicillins: structure–activity relationship. J Mol Med 1976;1:151.

50. Sunagawa M, Nouda H. [Neurotoxicity of carbapenem compounds and other beta-lactam antibiotics.]Jpn J Antibiot 1996;49(1):1–16.

51. Barrons RW, Murray KM, Richey RM. Populations at risk for penicillin-induced seizures. Ann Pharmacother 1992;26(1):26–9.

52. Grondahl TO, Langmoen IA. Epileptogenic effect of antibiotic drugs. J Neurosurg 1993;78(6):938–43.

53. De Sarro A, Ammendola D, Zappala M, Grasso S, De Sarro GB. Relationship between structure and convulsant properties of some beta-lactam antibiotics following intra-cerebroventricular microinjection in rats. Antimicrob Agents Chemother 1995;39(1):232–7.

54. Winston DJ, Ho WG, Bruckner DA, Champlin RE. Beta-lactam antibiotic therapy in febrile granulocytopenic patients. A randomized trial comparing cefoperazone plus piperacillin, ceftazidime plus piperacillin, and imipenem alone. Ann Intern Med 1991;115(11):849–59.

55. Rolston KV, Berkey P, Bodey GP, Anaissie EJ, Khardori NM, Joshi JH, Keating MJ, Holmes FA, Cabanillas FF, Elting L. A comparison of imipenem to ceftazidime with or without amikacin as empiric therapy in febrile neutropenic patients. Arch Intern Med 1992;152(2):283–91.

56. Pestotnik SL, Classen DC, Evans RS, Stevens LE, Burke JP. Prospective surveillance of imipenem/cilastatin use and associated seizures using a hospital information system. Ann Pharmacother 1993;27(4):497–501.

57. De Sarro A, Zappala M, Chimirri A, Grasso S, De Sarro GB. Quinolones potentiate cefazolin-induced seizures in DBA/2 mice. Antimicrob Agents Chemother 1993;37(7):1497–503.

58. De Sarro A, Ammendola D, De Sarro G. Effects of some quinolones on imipenem-induced seizures in DBA/2 mice. Gen Pharmacol 1994;25(2):369–79.

59. Saito T, Nakamura M, Watari M, Isse K. Radive seizure and antibiotics: case reports and review of literature J ECT 2008;24(4):275–6.

60. Kisa C, Yildirim SG, Aydemir C, Cebeci S, Goka E. Prolonged electroconvulsive therapy seizure in a patient taking ciprofloxacin. J ECT 2005;21(1):43–44.

61. Sugimoto M, Uchida I, Mashimoto T, Yamazaki S, Hatano K, Ikeda P, Mochizuki Y, Terai T, Matsuoka N. Evidence for the involvement of GABA(A) receptor blockade in convulsion induced by cephalosporin. Neuropharmacology 2003;43:304–14.

62. Berry M, Gurung A, Easty DL. Toxicity of antibiotics and antifungals on cultured human corneal cells: effect of mixing, exposure and concentration. Eye 1995;9(Pt 1):110–5.

63. Duch-Samper AM, Capdevila C, Menezo JL, Hurtado-Sarrio M. Endothelial toxicity of ceftazidime in anterior chamber irrigation solution. Exp Eye Res 1996;63(6):739–45.

64. Holme E, Greter J, Jacobson CE, Lindstedt S, Nordin I, Kristiansson B, Jodal U. Carnitine deficiency induced by pivampicillin and pivmecillinam therapy. Lancet 1989;2(8661):469–73.

65. Melegh B, Kerner J, Bieber LL. Pivampicillin-promoted excretion of pivaloylcarnitine in humans. Biochem Pharmacol 1987;36(20):3405–9.

66. Abrahamsson K, Eriksson BO, Holme E, Jodal U, Jonsson A, Lindstedt S. Pivalic acid-induced carnitine deficiency and physical exercise in humans. Metabolism 1996;45(12):1501–7.

67. Holme E, Jodal U, Linstedt S, Nordin I. Effects of pivalic acid-containing prodrugs on carnitine homeostasis and on response to fasting in children. Scand J Clin Lab Invest 1992;52(5):361–72.

68. Nakashima M, Kosuge K, Ishii I, Ohtsubo M. [Influence of multiple-dose administration of cefetamet pivoxil on blood and urinary concentrations of carnitine and effects of simultaneous administration of carnitine with cefetamet pivoxil.]Jpn J Antibiot 1996;49(10):966–79.

69. Melegh B, Pap M, Molnar D, Masszi G, Kopcsanyi G. Carnitine administration ameliorates the changes in energy metabolism caused by short-term pivampicillin medication. Eur J Pediatr 1997;156(10):795–9.

70. Baron DN, Hamilton-Miller JM, Brumfitt W. Sodium content of injectable beta-lactam antibiotics. Lancet 1984;1(8386):1113–4.

71. Peralta G, Sanchez-Santiago MB. Neutropenia secundaria a betalactámicos. Una vieja compañera olvidada. [Beta-lactam-induced neutropenia. An old forgotten companion.] Enferm Infecc Microbiol Clin 2005;23(8):485–91.

72. Gatell JM, Rello J, Miro JM, Martinez JA, Soriano E, SanMiguel Garcia J. Cloxacillin-induced neutropenia. J Infect Dis 1986;154(2):372.

73. Olaison L, Alestig K. A prospective study of neutropenia induced by high doses of beta-lactam antibiotics. J Antimicrob Chemother 1990;25(3):449–53.

74. Neftel KA, Walti M, Schulthess HK, Gubler J. Adverse reactions following intravenous penicillin-G relate to degradation of the drug in vitro. Klin Wochenschr 1984;62(1):25–9.

75. Vial T, Pofilet C, Pham E, Payen C, Evreux JC. Agranulocytoses aiguës médicamenteuses: expérience du Centre Régional de Pharmacovigilance de Lyon sur 7 ans. [Acute drug-induced agranulocytosis: experience of the Regional Center of Pharmacovigilance of Lyon over 7 years.] Therapie 1996;51(5):508–15.

76. Neu HC. Safety of cefepime: a new extended-spectrum parenteral cephalosporin. Am J Med 1996;100(6A):S68–75.

77. Saez-Llorens X, Castano E, Garcia R, Baez C, Perez M, Tejeira F, McCracken GH Jr. Prospective randomized comparison of cefepime and cefotaxime for treatment of bacterial meningitis in infants and children. Antimicrob Agents Chemother 1995;39(4):937–40.

78. Dahlgren AF. Two cases of possible cefepime-induced neutropenia. Am J Health Syst Pharm 1997;54(22):2621–2.

79. Gerber L, Wing EJ. Life-threatening neutropenia secondary to piperacillin/tazobactam therapy. Clin Infect Dis 1995;21(4):1047–8.

80. Wilson C, Greenhood G, Remington JS, Vosti KL. Neutropenia after consecutive treatment courses with nafcillin and piperacillin. Lancet 1979;1(8126):1150.

81. Oldfield EC 3rd. Leukopenia associated with the use of beta-lactam antibiotics in patients with hepatic dysfunction. Am J Gastroenterol 1994;89(8):1263–4.

82. Ramos Fernandez de Soria R, Martin Nunez G, Sanchez Gil F. Agranulocitosis inducida por drogas. Rapida recuperacion con el uso precoz de G-CSF. [Agranulocytosis induced by drugs. Rapid recovery with the early use of G-CSF.] Sangre (Barc) 1994;39(2):145–6.

83. Bradford CR, Ong EL, Hendrick DJ, Saunders PW. Use of colony stimulating factors for the treatment of drug-induced agranulocytosis. Br J Haematol 1993;84(1):182–3.

84. Borgbjerg BM, Hovgaard D, Laursen JB, Aldershvile J. Granulocyte colony stimulating factor in neutropenic patients with infective endocarditis. Heart 1998;79(1):93–5.

85. Levine BB, Redmond AP, Fellner MJ, Voss HE, Levytska V. Penicillin allergy and the heterogenous immune responses of man to benzylpenicillin. J Clin Invest 1966;45(12):1895–906.

86. Petz LD, Fudenberg HH. Coombs-positive hemolytic anemia caused by penicillin administration. N Engl J Med 1966;274(4):171–8.

87. Funicella T, Weinger RS, Moake JL, Spruell M, Rossen RD. Penicillin-induced immunohemolytic anemia associated with circulating immune complexes. Am J Hematol 1977;3:219–23.

88. Garratty G. Review: Immune hemolytic anemia and/or positive direct antiglobulin tests caused by drugs. Immunohematol 1994;10(2):41–50.

89. Petz LD, Garratty G. Acquired Immune Hemolytic AnemiasNew York: Churchill Livingstone;. 1980.

90. Williams ME, Thomas D, Harman CP, Mintz PD, Donowitz GR. Positive direct antiglobulin tests due to clavulanic acid. Antimicrob Agents Chemother 1985;27(1):125–7.

91. Norrby SR. Side effects of cephalosporins. Drugs 1987;34(Suppl 2):105–20.

92. Mellerup MT, Bruun NE, Nielsen JD. [Increased bleeding tendency induced by beta-lactam antibiotics.] Ugeskr Laeger 2005;167(25-31):2790–1.

93. Barza M, Furie B, Brown AE, Furie BC. Defects in vitamin K-dependent carboxylation associated with moxalactam treatment. J Infect Dis 1986;153(6):1166–9.

94. Andrassy K, Koderisch J. An open study on hemostasis in 20 patients with normal and impaired renal function treated with cefotetan alone or combined with tobramycin. In: Abstracts, 15th International Congress of Chemotherapy. Istanbul 1987;327.

95. Conly JM, Ramotar K, Chubb H, Bow EJ, Louie TJ. Hypoprothrombinemia in febrile, neutropenic patients with cancer: association with antimicrobial suppression of intestinal microflora. J Infect Dis 1984;150(2):202–12.

96. Holt J. Hypoprothrombinemia and bleeding diathesis associated with cefotetan therapy in surgical patients. Arch Surg 1988;123(4):523.

97. Hicks MJ, Flaitz CM. The role of antibiotics in platelet dysfunction and coagulopathy. Int J Antimicrob Agents 1993;2:129.

98. Shirakawa H, Komai M, Kimura S. Antibiotic-induced vitamin K deficiency and the role of the presence of intestinal flora. Int J Vitam Nutr Res 1990;60(3):245–51.

99. Williams KJ, Bax RP, Brown H, Machin SJ. Antibiotic treatment and associated prolonged prothrombin time. J Clin Pathol 1991;44(9):738–41.

100. Lipsky JJ. Antibiotic-associated hypoprothrombinaemia. J Antimicrob Chemother 1988;21(3):281–300.

101. Sattler FR, Weitekamp MR, Sayegh A, Ballard JO. Impaired hemostasis caused by beta-lactam antibiotics. Am J Surg 1988;155(5A):30–9.

102. Lipsky JJ. Mechanism of the inhibition of the gamma-carboxylation of glutamic acid by N-methylthiotetrazole-containing antibiotics. Proc Natl Acad Sci USA 1984;81(9):2893–7.

103. Suttie JW, Engelke JA, McTigue J. Effect of N-methyl-thiotetrazole on rat liver microsomal vitamin K-dependent carboxylation. Biochem Pharmacol 1986;35(14):2429–33.

104. Uchida K, Yoshida T, Komeno T. Mechanism for hypo-prothrombinemia caused by N-methyltetrazolethiol (NNTT)-containing antibiotics. Abstracts, 15th International Congress of Chemotherapy. Istanbul 1987;1153.

105. Shearer MJ, Bechtold H, Andrassy K, Koderisch J, McCarthy PT, Trenk D, Jahnchen E, Ritz E. Mechanism of cephalosporin-induced hypoprothrombinemia: relation to cephalosporin side chain, vitamin K metabolism, and vitamin K status. J Clin Pharmacol 1988;28(1):88–95.

106. Jones P, Bodey GP, Rolston K, Fainstein V, Riccardi S. Cefoperazone plus mezlocillin for empiric therapy of febrile cancer patients. Am J Med 1988;85(1A):3–8.

107. Brown RB, Klar J, Lemeshow S, Teres D, Pastides H, Sands M. Enhanced bleeding with cefoxitin or moxalactam. Statistical analysis within a defined population of 1493 patients. Arch Intern Med 1986;146(11):2159–64.

108. Boyd DB, Lunn WH. Electronic structures of cephalosporins and penicillins. 9. Departure of a leaving group in cephalosporins. J Med Chem 1979;22(7):778–84.

109. Mizojiri K, Norikura R, Takashima A, Tanaka H, Yoshimori T, Inazawa K, Yukawa T, Okabe H, Sugeno K. Disposition of moxalactam and N-methyltetrazolethiol in rats and monkeys. Antimicrob Agents Chemother 1987;31(8):1169–76.

110. Schentag JJ, Welage LS, Williams JS, Wilton JH, Adelman MH, Rigan D, Grasela TH. Kinetics and action of N-methylthiotetrazole in volunteers and patients. Population-based clinical comparisons of antibiotics with and without this moiety. Am J Surg 1988;155(5A):40–4.

111. Fletcher C, Pearson C, Choi SC, Duma RJ, Evans HJ, Qureshi GD. In vitro comparison of antiplatelet effects of beta-lactam penicillins. J Lab Clin Med 1986;108(3):217–23.

112. Bang NU, Tessler SS, Heidenreich RO, Marks CA, Mattler LE. Effects of moxalactam on blood coagulation and platelet function. Rev Infect Dis 1982;4(Suppl):S546–54.

113. Weitekamp MR, Aber RC. Prolonged bleeding times and bleeding diathesis associated with moxalactam administration. JAMA 1983;249(1):69–71.

114. Weitekamp MR, Caputo GM, Al-Mondhiry HA, Aber RC. The effects of latamoxef, cefotaxime, and cefoperazone on platelet function and coagulation in normal volunteers. J Antimicrob Chemother 1985;16(1):95–101.

115. Weitekamp MR, Holmes P, Walker ME. A double blind study on the effects of cefoperazone (CPZ), ceftizoxime (CTZ), moxalactam (MOX) on platelet function and prothrombin time in normal volunteers. In: Abstracts, 25th Interscience Conference on Antimicrobial Agents and Chemotherapy. Minnesota: Minneapolis, 1985:959.

116. Fass RJ, Copelan EA, Brandt JT, Moeschberger ML, Ashton JJ. Platelet-mediated bleeding caused by broad-spectrum penicillins. J Infect Dis 1987;155(6):1242–8.

117. Norrby R, Foord RD, Hedlund P. Clinical and pharmaco-kinetic studies on cefuroxime. J Antimicrob Chemother 1977;3(4):355–62.

118. Burroughs SF, Johnson GJ. Beta-lactam antibiotic-induced platelet dysfunction: evidence for irreversible inhibition of platelet activation in vitro and in vivo after prolonged exposure to penicillin. Blood 1990;75(7):1473–1480.

119. Knudsen ET, Harding JW. A multicentre comparative trial of talampicillin and ampicillin in general practice. Br J Clin Pract 1975;29(10):255–64.

120. Ewe K. Diarrhoea and constipation. Baillieres Clin Gastroenterol 1988;2(2):353–84.

121. Hooker KD, DiPiro JT. Effect of antimicrobial therapy on bowel flora. Clin Pharm 1988;7(12):878–88.

122. Bartlett JG. Antibiotic-associated diarrhea. Clin Infect Dis 1992;15(4):573–81.

123. Kelly CP, Pothoulakis C, LaMont JT. Clostridium difficile colitis. N Engl J Med 1994;330(4):257–62.

124. George WL. Antimicrobial agent-associated colitis and diarrhea: historical background and clinical aspects. Rev Infect Dis 1984;6(Suppl 1):S208–13.

125. Talbot RW, Walker RC, Beart RW Jr. Changing epidemiology, diagnosis, and treatment of Clostridium difficile toxin-associated colitis. Br J Surg 1986;73(6):457–60.

126. Hogenauer C, Hammer HF, Krejs GJ, Reisinger EC. Mechanisms and management of antibiotic-associated diarrhea. Clin Infect Dis 1998;27(4):702–10.

127. Johnson S, Gerding DN. Clostridium difficile-associated diarrhea. Clin Infect Dis 1998;26(5):1027–34.

128. Larson HE, Price AB. Pseudomembranous colitis: Presence of clostridial toxin. Lancet 1977;2(8052–8053):1312–4.

129. Borriello SP. 12th C. L. Oakley lecture. Pathogenesis of Clostridium difficile infection of the gut. J Med Microbiol 1990;33(4):207–15.

130. Bartlett JG. Clostridium difficile: history of its role as an enteric pathogen and the current state of knowledge about the organism. Clin Infect Dis 1994;18(Suppl 4):S265–72.

131. Anand A, Glatt AE. Clostridium difficile infection associated with antineoplastic chemotherapy: a review. Clin Infect Dis 1993;17(1):109–13.

132. Bartlett JG. Antibiotic-associated pseudomembranous colitis. Rev Infect Dis 1979;1(3):530–9.

133. Aronsson B, Mollby R, Nord CE. Antimicrobial agents and Clostridium difficile in acute enteric disease: epidemiological data from Sweden, 1980–1982. J Infect Dis 1985;151(3):476–81.

134. Aronsson B, Mollby R, Nord CE. Clostridium difficile and antibiotic associated diarrhoea in Sweden. Scand J Infect Dis Suppl 1982;35:53–8.

135. Fekety R, Shah AB. Diagnosis and treatment of Clostridium difficile colitis. JAMA 1993;269(1):71–5.

136. Barbut F, Corthier G, Charpak Y, Cerf M, Monteil H, Fosse T, Trevoux A, De Barbeyrac B, Boussougant Y, Tigaud S, Tytgat F, Sedallian A, Duborgel S, Collignon A, Le Guern ME, Bernasconi P, Petit JC. Prevalence and pathogenicity of Clostridium difficile in hospitalized patients. A French multicenter study. Arch Intern Med 1996;156(13):1449–54.

137. Zehnder D, Kunzi UP, Maibach R, Zoppi M, Halter F, Neftel KA, Muller U, Galeazzi RL, Hess T, Hoigne R. Die Häufigkeit der Antibiotika-assoziierten Kolitis bei hospitalisierten Patienten der Jahre 1974–1991 im 'Comprehensive Hospital Drug Monitoring' Bern/St. Gallen. [Frequency of antibiotics-associated colitis in hospitalized patients in 1974–1991 in "Comprehensive Hospital Drug Monitoring," Bern/St. Gallen.] Schweiz Med Wochenschr 1995;125(14):676–83.

138. Burke GW, Wilson ME, Mehrez IO. Absence of diarrhea in toxic megacolon complicating *Clostridium difficile* pseudomembranous colitis. Am J Gastroenterol 1988;83(3):304–7.

139. Toffler RB, Pingoud EG, Burrell MI. Acute colitis related to penicillin and penicillin derivatives. Lancet 1978;2(8092 Pt 1):707–9.

140. Finegold SM. Clinical considerations in the diagnosis of antimicrobial agent-associated gastroenteritis. Diagn Microbiol Infect Dis 1986;4(Suppl 3):S87–91.

141. Viscidi R, Willey S, Bartlett JG. Isolation rates and toxigenic potential of *Clostridium difficile* isolates from various patient populations. Gastroenterology 1981;81(1):5–9.

142. Shim JK, Johnson S, Samore MH, Bliss DZ, Gerding DN. Primary symptomless colonisation by *Clostridium difficile* and decreased risk of subsequent diarrhoea. Lancet 1998;351(9103):633–6.

143. Mardh PA, Helin I, Colleen I, Oberg M, Holst E. *Clostridium difficile* toxin in faecal specimens of healthy children and children with diarrhoea. Acta Paediatr Scand 1982;71(2):275–8.

144. Pitts NE, Gilbert GS, Knirsch AK, Noguchi Y. Worldwide clinical experience with sultamicillin. APMIS Suppl 1989;5:23–34.

145. McLinn SE, Moskal M, Goldfarb J, Bodor F, Aronovitz G, Schwartz R, Self P, Ossi MJ. Comparison of cefuroxime axetil and amoxicillin–clavulanate suspensions in treatment of acute otitis media with effusion in children. Antimicrob Agents Chemother 1994;38(2):315–8.

146. Todd PA, Benfield P. Amoxicillin/clavulanic acid. An update of its antibacterial activity, pharmacokinetic properties and therapeutic use. Drugs 1990;39(2):264–307.

147. Friedel HA, Campoli-Richards DM, Goa KL. Sultamicillin. A review of its antibacterial activity, pharmacokinetic properties and therapeutic use. Drugs 1989;37(4):491–522.

148. Caron F, Ducrotte P, Lerebours E, Colin R, Humbert G, Denis P. Effects of amoxicillin–clavulanate combination on the motility of the small intestine in human beings. Antimicrob Agents Chemother 1991;35(6):1085–8.

149. Cerquetti M, Pantosti A, Gentile G, D'Ambrosio F, Mastrantonio P. Epidemie ospedaliere di diarrea da *Clostridium difficile*: dimostrazione di infezione crociata mediante tecniche di tipizzazione. [Hospital epidemic of *Clostridium difficile* diarrhea: demonstration of cross-infection using a typing technic.] Ann Ist Super Sanita 1989;25(2):327–32.

150. McFarland LV, Surawicz CM, Stamm WE. Risk factors for *Clostridium difficile* carriage and *C. difficile*-associated diarrhea in a cohort of hospitalized patients. J Infect Dis 1990;162(3):678–84.

151. Nolan NP, Kelly CP, Humphreys JF, Cooney C, O'Connor R, Walsh TN, Weir DG, O'Briain DS. An epidemic of pseudomembranous colitis: importance of person to person spread. Gut 1987;28(11):1467–73.

152. Impallomeni M, Galletly NP, Wort SJ, Starr JM, Rogers TR. Increased risk of diarrhoea caused by *Clostridium difficile* in elderly patients receiving cefotaxime. BMJ 1995;311(7016):1345–6.

153. Starr JM, Rogers TR, Impallomeni M. Hospital-acquired *Clostridium difficile* diarrhoea and herd immunity. Lancet 1997;349(9049):426–8.

154. Tedesco FJ. Clindamycin and colitis: a review. J Infect Dis 1977;135(Suppl):S95–8.

155. Milstone EB, McDonald AJ, Scholhamer CF Jr. Pseudomembranous colitis after topical application of clindamycin. Arch Dermatol 1981;117(3):154–5.

156. Hecht JR, Olinger EJ. *Clostridium difficile* colitis secondary to intravenous vancomycin. Dig Dis Sci 1989;34(1):148–9.

157. Saginur R, Hawley CR, Bartlett JG. Colitis associated with metronidazole therapy. J Infect Dis 1980;141(6):772–4.

158. Brown E, Talbot GH, Axelrod P, Provencher M, Hoegg C. Risk factors for *Clostridium difficile* toxin-associated diarrhea. Infect Control Hosp Epidemiol 1990;11(6):283–90.

159. Gerding DN, Olson MM, Peterson LR, Teasley DG, Gebhard RL, Schwartz ML, Lee JT Jr. *Clostridium difficile*-associated diarrhea and colitis in adults. A prospective case-controlled epidemiologic study. Arch Intern Med 1986;146(1):95–100.

160. Church JM, Fazio VW. A role for colonic statis in the pathogenesis of disease related to *Clostridium difficile*. Dis Colon Rectum 1986;146:95.

161. Pierce PF Jr, Wilson R, Silva J Jr, Garagusi VF, Rifkin GD, Fekety R, Nunez-Montiel O, Dowell VR Jr, Hughes JM. Antibiotic-associated pseudomembranous colitis: an epidemiologic investigation of a cluster of cases. J Infect Dis 1982;145(2):269–74.

162. de Lalla F, Privitera G, Ortisi G, Rizzardini G, Santoro D, Pagano A, Rinaldi E, Scarpellini P. Third generation cephalosporins as a risk factor for *Clostridium difficile*-associated disease: a four-year survey in a general hospital. J Antimicrob Chemother 1989;23(4):623–31.

163. Cheng SH, Lu JJ, Young TG, Perng CL, Chi WM. *Clostridium difficile*-associated diseases: comparison of symptomatic infection versus carriage on the basis of risk factors, toxin production, and genotyping results. Clin Infect Dis 1997;25(1):157–8.

164. Samore M, Killgore G, Johnson S, Goodman R, Shim J, Venkataraman L, Sambol S, DeGirolami P, Tenover F, Arbeit R, Gerding D. Multicenter typing comparison of sporadic and outbreak *Clostridium difficile* isolates from geographically diverse hospitals. J Infect Dis 1997;176(5):1233–8.

165. Hirschhorn LR, Trnka Y, Onderdonk A, Lee ML, Platt R. Epidemiology of community-acquired *Clostridium difficile*-associated diarrhea. J Infect Dis 1994;169(1):127–33.

166. Lyerly DM, Krivan HC, Wilkins TD. *Clostridium difficile*: its disease and toxins. Clin Microbiol Rev 1988;1(1):1–18.

167. Wren B, Heard SR, Tabaqchali S. Association between production of toxins A and B and types of *Clostridium difficile*. J Clin Pathol 1987;40(12):1397–401.

168. Seppala K, Hjelt L, Sipponen P. Colonoscopy in the diagnosis of antibiotic-associated colitis. A prospective study. Scand J Gastroenterol 1981;16(4):465–8.

169. Mukai JK, Janower ML. Diagnosis of pseudomembranous colitis by computed tomography: a report of two patients. Can Assoc Radiol J 1987;38(1):62–3.

170. Woods GL, Iwwen PC. Comparison of a dot immunobinding assay, latex agglutination, and cytotoxin assay for laboratory diagnosis of *Clostridium difficile*-associated diarrhea. J Clin Microbiol 1990;28(5):855–7.

171. Morris JB, Zollinger RM Jr, Stellato TA. Role of surgery in antibiotic-induced pseudomembranous enterocolitis. Am J Surg 1990;160(5):535–9.

172. Van Ness MM, Cattau EL Jr. Fulminant colitis complicating antibiotic-associated pseudomembranous colitis: case report and review of the clinical manifestations and treatment. Am J Gastroenterol 1987;82(4):374–7.

173. Wenisch C, Parschalk B, Hasenhundl M, Hirschl AM, Graninger W. Comparison of vancomycin, teicoplanin, metronidazole, and fusidic acid for the treatment of *Clostridium difficile*-associated diarrhea. Clin Infect Dis 1996;22(5):813–8.

174. Teasley DG, Gerding DN, Olson MM, Peterson LR, Gebhard RL, Schwartz MJ, Lee JT Jr. Prospective randomised trial of metronidazole versus vancomycin for *Clostridium-difficile*-associated diarrhoea and colitis. Lancet 1983;2(8358):1043–6.

175. Bartlett JG. Treatment of antibiotic-associated pseudomembranous colitis. Rev Infect Dis 1984;6(Suppl 1):S235–41.

176. de Lalla F, Santoro D, Rinaldi E, Suter F, Cruciani M, Guaglianone MH, Rizzardini G, Pellegata G. Teicoplanin in the treatment of infections by staphylococci, *Clostridium difficile* and other Gram-positive bacteria. J Antimicrob Chemother 1989;23(1):131–42.

177. Buggy BP, Fekety R, Silva J Jr. Therapy of relapsing *Clostridium difficile*-associated diarrhea and colitis with the combination of vancomycin and rifampin. J Clin Gastroenterol 1987;9(2):155–9.

178. Ariano RE, Zhanel GG, Harding GK. The role of anion-exchange resins in the treatment of antibiotic-associated pseudomembranous colitis. CMAJ 1990;142(10):1049–51.

179. Taylor NS, Bartlett JG. Binding of *Clostridium difficile* cytotoxin and vancomycin by anion-exchange resins. J Infect Dis 1980;141(1):92–7.

180. Gorbach SL, Chang TW, Goldin B. Successful treatment of relapsing *Clostridium difficile* colitis with *Lactobacillus* GG. Lancet 1987;2(8574):1519.

181. Bowden TA Jr, Mansberger AR Jr, Lykins LE. Pseudomembraneous enterocolitis: mechanism for restoring floral homeostasis. Am Surg 1981;47(4):178–83.

182. McFarland LV, Surawicz CM, Greenberg RN, Elmer GW, Moyer KA, Melcher SA, Bowen KE, Cox JL. Prevention of beta-lactam-associated diarrhea by *Saccharomyces boulardii* compared with placebo. Am J Gastroenterol 1995;90(3):439–48.

183. Novak E, Lee JG, Seckman CE, Phillips JP, DiSanto AR. Unfavorable effect of atropine-diphenoxylate (Lomotil) therapy in lincomycin-caused diarrhea. JAMA 1976;235(14):1451–4.

184. Olans RN, Weiner LB. Reversible oxacillin hepatotoxicity. J Pediatr 1976;89(5):835–8.

185. Michelson PA. Reversible high dose oxacillin-associated liver injury. Can J Hosp Pharm 1981;34:83.

186. Onorato IM, Axelrod JL. Hepatitis from intravenous high-dose oxacillin therapy: findings in an adult inpatient population. Ann Intern Med 1978;89(4):497–500.

187. Fairley CK, McNeil JJ, Desmond P, Smallwood R, Young H, Forbes A, Purcell P, Boyd I. Risk factors for development of flucloxacillin associated jaundice. BMJ 1993;306(6872):233–5.

188. Turner IB, Eckstein RP, Riley JW, Lunzer MR. Prolonged hepatic cholestasis after flucloxacillin therapy. Med J Aust 1989;151(11–12):701–5.

189. Devereaux BM, Crawford DH, Purcell P, Powell LW, Roeser HP. Flucloxacillin associated cholestatic hepatitis.

An Australian and Swedish epidemic? Eur J Clin Pharmacol 1995;49(1–2):81–5.

190. Kleinman MS, Presberg JE. Cholestatic hepatitis after dicloxacillin-sodium therapy. J Clin Gastroenterol 1986;8(1):77–8.

191. Ditlove J, Weidmann P, Bernstein M, Massry SG. Methicillin nephritis. Medicine (Baltimore) 1977;56(6):483–91.

192. Galpin JE, Shinaberger JH, Stanley TM, Blumenkrantz MJ, Bayer AS, Friedman GS, Montgomerie JZ, Guze LB, Coburn JW, Glassock RJ. Acute interstitial nephritis due to methicillin. Am J Med 1978;65(5):756–65.

193. Sanjad SA, Haddad GG, Nassar VH. Nephropathy, an underestimated complication of methicillin therapy. J Pediatr 1974;84(6):873–7.

194. Neftel KA. Verträglichkeit der hochdosierten Therapie mit Betalactam-Antibiotika-Pathogenese der Nebenwirkungen insbesondere der Neutropenie. Fortschr Antimikr Antineoplast Chemother 1984;3–1:71.

195. Bigby M, Jick S, Jick H, Arndt K. Drug-induced cutaneous reactions. A report from the Boston Collaborative Drug Surveillance Program on 15,438 consecutive inpatients, 1975 to 1982. JAMA 1986;256(24):3358–63.

196. Zurcher K, Krebs A. Cutaneous Drug ReactionsBasel: Karger-Verlag;. 1991.

197. Stubb S, Heikkila H, Kauppinen K. Cutaneous reactions to drugs: a series of in-patients during a five-year period. Acta Derm Venereol 1994;74(4):289–91.

198. Hoigne R, Sonntag MR, Zoppi M, Hess T, Maibach R, Fritschy D. Occurrence of exanthema in relation to aminopenicillin preparations and allopurinol. N Engl J Med 1987;316(19):1217.

199. Arndt KA, Jick H. Rates of cutaneous reactions to drugs. A report from the Boston Collaborative Drug Surveillance Program. JAMA 1976;235(9):918–23.

200. Hunziker T, Hoigné RV, Kuenzi UP, et al. Comprehensive hospital drug monitoring (CHDM), the adverse skin reactions, a 20-years survey. Pharmacoepidemiology 1995;4(Suppl 1):S13.

201. Katz M, Seidenbaum M, Weinrauch L. Penicillin-induced generalized pustular psoriasis. J Am Acad Dermatol 1987;17(5 Pt 2):918–20.

202. Prieto A, de Barrio M, Lopez-Saez P, Baeza ML, de Benito V, Olalde S. Recurrent localized pustular eruption induced by amoxicillin. Allergy 1997;52(7):777–8.

203. Beylot C, Bioulac P, Doutre MS. Pustuloses exanthématiques aiguës généralisées. [Acute generalized exanthematic pustuloses (four cases).] Ann Dermatol Venereol 1980;107(1–2):37–48.

204. Isogai Z, Sunohara A, Tsuji T. Pustular drug eruption due to bacampicilin hydrochloride in a patient with psoriasis. J Dermatol 1998;25(9):612–5.

205. Stough D, Guin JD, Baker GF, Haynie L. Pustular eruptions following administration of cefazolin: a possible interaction with methyldopa. J Am Acad Dermatol 1987;16(5 Pt 1):1051–2.

206. Fayol J, Bernard P, Bonnetblanc JM. Pustular eruption following administration of cefazolin: a second case report. J Am Acad Dermatol 1988;19(3):571.

207. Kalb RE, Grossman ME. Pustular eruption following administration of cephradine. Cutis 1986;38(1):58–60.

208. Jackson H, Vion B, Levy PM. Generalized eruptive pustular drug rash due to cephalexin. Dermatologica 1988;177(5):292–4.

209. Ogoshi M, Yamada Y, Tani M. Acute generalized exanthematic pustulosis induced by cefaclor and acetazolamide. Dermatology 1992;184(2):142–4.

210. Spencer JM, Silvers DN, Grossman ME. Pustular eruption after drug exposure: is it pustular psoriasis or a pustular drug eruption? Br J Dermatol 1994;130(4):514–9.

211. McCloskey GL, Massa MC. Cephalexin rash in infectious mononucleosis. Cutis 1997;59(5):251–4.

212. Dhingra B, Grover C. Baboon syndrome. Indian Pediatr 2007; 44(12): 937.

213. Handisurya A, Stingi G, Wöhrl S. SDRIFE (baboon syndrome) induced by penicillin Clin Exp Dermatol 2008;34:355–7.

214. Johnson ST, Fueger JT, Gottschall JL One center's experience: the serology and drugs associated with drug-induced immune hemolytic anemia—a new paradigm. Transfusion 2007;47(4):697–702.

215. Romano A, Gueant-Rodriguez RM, Viola M, Amoghly F, Gaeta F, Nicolas JP, Gueant JL. Diagnosing immediate reactions to cephalosporins. Clin Exp Allergy 2005;35:1234–42.

216. Gomez MB, Torres MJ, Mayorga C, Perez-Inestrosa E, Suau R, Montanez MI, Juarez C. Immediate allergic reactions to betalactams: facts and controversies. Curr Opin Allergy Clin Immunol 2004;4:261–6.

217. Atanaskovic-Markovic M, Circovic Velickovic T, Gavrovic-Jankulovic M, Vuckovic O, Nestorivic B. Immediate allergic reactions to cephaloporins and penicillins and their cross-reactivity in children. Pediatr Allergy Immunol 2005;16:341–7.

218. Pumphrey RS, Davis S. Under-reporting of antibiotic anaphylaxis may put patients at risk. Lancet 1999.353,1157-8.

219. Kelkar PS, Li JT. Cephalosporin allergy. New Engl J Medicine 2001;345:804–9.

220. Romano A, Gueant-Rodrique RM, Viola M, Gueant JL, Cross-reactivity and tolerability of cephalosporins in patients with immediate hypersensitivity to penicillins. Ann Intern Med 2004;141:16–22.

221. Atanaskovic-Markovic M, Gavrovic-Jankulovic M, Cirkovic Velickovic T, Vuckovic O, Todoric D. Type-I hypersensitivity to ceftriaxone and cross-reactivity with cefalexin and ampicillin. Allergy 2003;58:537–8.

222. Romano A, Quaratino D, Venemalm L, Torres MJ, Venuti A, Blanca M. A case of IgE-mediated hypersensitivity to ceftriaxone. J Allergy Clin Immunol 1999;104:1113–4.

223. Saxon A, Adelman DC, Patel A, Hajdu R, Calandra GB. Imipenem cross-reactivity with penicillins in humans. J Allergy Clin Immunol 1988;88:213–7.

224. McConnell SA, Penzal SR, Warmack TS, Anaisse EJ, Gibbins PO. Incidence of imipenem hypersensitivity reactions in febrile neutropenic bone marrow transplant patients with a history of penicillin allergy. Clin Infect Dis 2000;31:1512–4.

225. Prescott WA, DePestel DD, Ellis JJ, Regal RE. Incidence of carbapenem- associated allergic-type reactions among patients with versus patients without a reported penicillin allergy. Clin Infect Dis 2004;38:1101–7.

226. Nakanish T, Kohda A, Kato T, Appleford DJA, Pulsford AH. Antigenicity tests of meropenem. Chemotherapy (Tokyo) 1992;40:251–7.

227. Chen Z, Baus X, Kutscha-Lissberg F, Merget R. IgE-mediated anaphylactic reaction to imipenem. Allergy 2000;55:92–3.

228. Bauer SL, Wall GC, Skoglund KJ, Peters LK. Lack of cross-reactivity to meropenem in a patient with an allergy to imipenem–cilastatin. J Allergy Clin Immunol 2004;113:173–5.

229. Pichichero ME, Casey JR. Safe use of selected cephalosporins in penicillin-allergic patients: a meta-analysis. Otolaryngol Head Neck Surg 2007;136:340–7.

230. Moreno E, Davila I, Laffond E, Macias E, Isodoro M, Ruiz A, Lorente F. Selective immediate hypersensitivity to cefepime. J Investig Allergol Clin Immunol 2007;17:52–4.

231. Hasdenteufel F, Luyasu S, Renaudin JM, Trechot P, Kanny G. Anaphylactic chock associated with cefuroxime axetil: structure-activity relationships. Ann Pharmacother 2007;41:1069–72.

232. Trcka J, Seitz CS, Bröcker EB, Gross GE, Trautmann A. Aminopenicillin-induced exanthema allows treatment with certain cephalosporins or phenoxymethylpenicillin. J Antimicrob Chemother 2007;60:107–11.

233. Lucena Fontaine C, Mayorga C, Bouscuet PJ Arnoux B, Torres MJ, Bianca M, Demoly P. Relevance of the determination of serum-specific IgE antibodies in the diagnosis of immediate beta-lactam allergy. Allergy 2007;62:47–52.

234. Fonacier L, Hirschberg R, Gerson S. Adverse drug reactions to a cephalosporins in hospitalized patients with a history of penicillin allergy. Allergy Asthma Proc 2005;26(2):135–41.

235. Romano A, Viola M, Gueant-Rodriguez RM, Valluzzi RL, Gueant JL. Selective immediate hypersensitivity to cefodizime. Allergy 2005;60(12):1545–6.

236. Atanasković-Marković M, Cirković Velicković T, Gavrović-Jankulović M, Ivanovski P, Nestorović B. A case of selective IgE-mediated hypersensitivity to ceftibuten. Allergy 2005;60(11).1454.

237. Pichichero ME. A review of evidence supporting the American Academy of Pediatrics recommendation for prescribing cephalosporin antibiotics for penicillin-allergic patients. Pediatrics 2005;115(4):1048–57.

238. Nadarajah K, Green GR, Naglak M. Clinical outcomes of penicillin skin testing. Ann Allergy Asthma Immunol 2005;95(6):541–5.

239. Bousquet PJ, Co-Minh HB, Arnoux B, Daures JP, Demoly P.Importance of mixture of minor determinants and benzylpenicilloyl poly-L-lysine skin testing in the diagnosis of beta-lactam allergy. J Allergy Clin Immunol 2005;115(6):1314–6.

240. Matheu V, Pérez-Rodriguez E, Sánchez-Machin I, de la Torre F, García-Robaina JC. Major and minor determinants are high-performance skin tests in β-lactam allergy diagnosis. J Allergy Clin Immunol 2005;116(5):1167–8.

241. Gell PGH, Coombs RRA. Classification of allergic reactions responsible for clinical hypersensitivity and disease. In: Gell PGH, Coombs RRA, Lachmann PJ, editors. Clinical Aspects of Immunology. Oxford: Blackwell Scientific Publications, 1975:251–4.

242. Plaut M, Zimmerman EM. Allergy and mechanisms of hypersensitivity. In: Paul WE, editor. Fundamental Immunology. 3rd ed.. New York: Raven Press, 1993:1399.

243. De Weck AL. Pharmacologic and immunochemical mechanisms of drug hypersensitivity. Immunol Allergy Clin North Am 1991;11:461.

244. Lafaye P, Lapresle C. Fixation of penicilloyl groups to albumin and appearance of anti-penicilloyl antibodies in penicillin-treated patients. J Clin Invest 1988;82(1):7–12.

245. Bochner BS, Lichtenstein LM. Anaphylaxis. N Engl J Med 1991;324(25):1785–90.

246. Levine B, Redmond A. Immunochemical mechanisms of penicillin induced Coombs positivity and hemolytic anemia in man. Int Arch Allergy Appl Immunol 1967;31(6):594–606.

247. Abraham GN, Petz LD, Fudenberg HH. Immunohaematological cross-allergenicity between penicillin and cephalothin in humans. Clin Exp Immunol 1968;3(4):343–57.

248. Garratty G, Petz LD. Drug-induced immune hemolytic anemia. Am J Med 1975;58(3):398–407.

249. Worlledge SM. Immune drug-induced hemolytic anemias. Semin Hematol 1973;10(4):327–44.

250. Kerr RO, Cardamone J, Dalmasso AP, Kaplan ME. Two mechanisms of erythrocyte destruction in penicillin-induced hemolytic anemia. N Engl J Med 1972;287(26):1322–5.

251. Christie DJ, Lennon SS, Drew RL, Swinehart CD. Cefotetan-induced immunologic thrombocytopenia. Br J Haematol 1988;70(4):423–6.

252. Gharpure V, O'Connell B, Schiffer CA. Mezlocillin-induced thrombocytopenia. Ann Intern Med 1993;119(8):862.

253. Moller NE, Nielsen B, von Wurden K. Contact dermatitis to semisynthetic penicillins in factory workers. Contact Dermatitis 1986;14(5):307–11.

254. Tadokoro K, Niimi N, Ohtoshi T, Nakajima K, Takafuji S, Onodera K, Suzuki S, Muranaka M. Cefotiam-induced IgE-mediated occupational contact anaphylaxis of nurses; case reports, RAST analysis, and a review of the literature. Clin Exp Allergy 1994;24(2):127–33.

255. Hertl M, Geisel J, Boecker C, Merk HF. Selective generation of CD8+ T cell clones from the peripheral blood of patients with cutaneous reactions to beta-lactam antibiotics. Br J Dermatol 1993;128(6):619–26.

256. Scheper RJ, von Blomberg BM. Immunoregulation of T cell-mediated skin hypersensitivity. Arch Toxicol Suppl 1994;16:63–70.

257. Adkinson NF Jr. Beta-lactam crossreactivity. Clin Exp Allergy 1998;28(Suppl 4):37–40.

258. Bolzacchini E, Meinardi S, Orlandi M, Rindone B. 'In vivo' models of hapten generation. Clin Exp Allergy 1998;28(Suppl 4):83–6.

259. Perez Pimiento A, Gomez Martinez M, Minguez Mena A, Trampal Gonzalez A, de Paz Arranz S, Rodriguez Mosquera M. Aztreonam and ceftazidime: evidence of in vivo cross allergenicity. Allergy 1998;53(6):624–5.

260. Blanca M, Perez E, Garcia J, Miranda A, Fernandez J, Vega JM, Terrados S, Avila M, Martin A, Suau R. Anaphylaxis to amoxycillin but good tolerance for benzyl penicillin. In vivo and in vitro studies of specific IgE antibodies. Allergy 1988;43(7):508–10.

261. Dejarnatt AC, Grant JA. Basic mechanisms of anaphylaxis and anaphylactoid reactions. Immunol Allergy Clin North Am 1992;12:33–46.

262. Choquet-Kastylevsky G, Descotes J. Value of animal models for predicting hypersensitivity reactions to medicinal products. Toxicology 1998;129(1):27–35.

263. Chazal I, Verdier F, Virat M, Descotes J. Prediction of drug induced immediate hypersensitivity in guinea-pigs. Toxicol In Vitro 1994;8:1045–9.

264. Nagami K, Matsumoto H, Maki E, Motegi K, Aoyagi K, Naruse S, Samura K, Losos GJ, Ikemoto F. Experimental methods for immunization and challenge in antigenicity studies in guinea pigs. J Toxicol Sci 1995;20(5):579–94.

265. Sarlo K, Karol MH. Guinea pig predictive tests for allergy. In: Dean JH, Luster MI, Munson AE, Kiber I, editors. Immunotoxicology and Immunopharmacology. 2nd ed.. New York: Raven Press, 1994:703–20.

266. Hilton J, Dearman RJ, Boylett MS, Fielding I, Basketter DA, Kimber I. The mouse IgE test for the identification of potential chemical respiratory allergens:

267. Saxon A, Swabb EA, Adkinson NF Jr. Investigation into the immunologic cross-reactivity of aztreonam with other beta-lactam antibiotics. Am J Med 1985;78(2A):19–26.

268. International Rheumatic Fever Study Group. Allergic reactions to long-term benzathine penicillin prophylaxis for rheumatic fever. Lancet 1991;337(8753):1308–10.

269. Rudolph AH, Price EV. Penicillin reactions among patients in venereal disease clinics. A national survey. JAMA 1973;223(5):499–501.

270. Shepherd GM. Allergy to beta-lactam antibiotics. Immunol Allergy Clin North Am 1991;11:611.

271. Hantson P, de Coninck B, Horn JL, Mahieu P. Immediate hypersensitivity to aztreonam and imipenem. BMJ 1991;302(6771):294–5.

272. Mackowiak PA, LeMaistre CF. Drug fever: a critical appraisal of conventional concepts. An analysis of 51 episodes in two Dallas hospitals and 97 episodes reported in the English literature. Ann Intern Med 1987;106(5):728–33.

273. Calandra GB, Wang C, Aziz M, Brown KR. The safety profile of imipenem/cilastatin: worldwide clinical experience based on 3470 patients. J Antimicrob Chemother 1986;18(Suppl E):193–202.

274. Sanders CV, Greenberg RN, Marier RL. Cefamandole and cefoxitin. Ann Intern Med 1985;103(1):70–8.

275. Swabb EA. Review of the clinical pharmacology of the monobactam antibiotic aztreonam. Am J Med 1985;78(2A):11–8.

276. Delage C, Irey NS. Anaphylactic deaths: a clinicopathologic study of 43 cases. J Forensic Sci 1972;17(4):525–40.

277. Weiss ME, Adkinson NF. Immediate hypersensitivity reactions to penicillin and related antibiotics. Clin Allergy 1988;18(6):515–40.

278. Hoigné RV, Braunschweig S, Zehnder D, et al. Pharmacoepidemiology 1994;4(Suppl 1):S90.

279. Hoigné R, Jaeger MD, Hess T, Wymann R, Muller U, Galeazzi R, Maibach R, Kunzi UP. Akute schwere Dyspnoea als Medikamentennebenwirkung. [Acute severe dyspnea as a side effect of drugs. Report from the CHDM (Comprehensive Hospital Drug Monitoring).] Schweiz Med Wochenschr 1990;120(34):1211–6.

280. Adkinson NF Jr. Immunogenicity and cross-allergenicity of aztreonam. Am J Med 1990;88(3C):S12–5.

281. Platt R, Dreis MW, Kennedy DL, Kuritsky JN. Serum sickness-like reactions to amoxicillin, cefaclor, cephalexin, and trimethoprim–sulfamethoxazole. J Infect Dis 1988;158(2):474–7.

282. Levine LR. Quantitative comparison of adverse reactions to cefaclor vs. amoxicillin in a surveillance study. Pediatr Infect Dis 1985;4(4):358–61.

283. Moskovitz BL. Clinical adverse effects during ceftriaxone therapy. Am J Med 1984;77(4C):84–8.

284. Stricker BH, Tijssen JG. Serum sickness-like reactions to cefaclor. J Clin Epidemiol 1992;45(10):1177–84.

285. White JM, Brown DL, Hepner GW, Worlledge SM. Penicillin-induced haemolytic anaemia. BMJ 1968;3(609):26–9.

286. Tuffs L, Manoharan A. Flucloxacillin-induced haemolytic anaemia. Med J Aust 1986;144(10):559–60.

287. Garratty G, Postoway N, Schwellenbach J, McMahill PC. A fatal case of ceftriaxone (Rocephin)-induced hemolytic anemia associated with intravascular immune hemolysis. Transfusion 1991;31(2):176–9.

considerations of stability and controls. J Appl Toxicol 1996;16(2):165–70.

288. Chambers LA, Donovan LM, Kruskall MS. Ceftazidime-induced hemolysis in a patient with drug-dependent antibodies reactive by immune complex and drug adsorption mechanisms. Am J Clin Pathol 1991;95(3):393–6.

289. Fellner MJ. Adverse reactions to penicillin and related drugs. Clin Dermatol 1986;4(1):133–41.

290. Schopf E, Stuhmer A, Rzany B, Victor N, Zentgraf R, Kapp JF. Toxic epidermal necrolysis and Stevens–Johnson syndrome. An epidemiologic study from West Germany. Arch Dermatol 1991;127(6):839–42.

291. Wakelin SH, Allen J, Zhou S, Wojnarowska F. Drug-induced linear IgA disease with antibodies to collagen VII. Br J Dermatol 1998;138(2):310–4.

292. Fellner MJ, Mark AS. Penicillin- and ampicillin-induced pemphigus vulgaris. Int J Dermatol 1980;19(7):392–3.

293. Manders SM, Heymann WR. Acute generalized exanthemic pustulosis. Cutis 1994;54(3):194–6.

294. McDonald BJ, Singer JW, Bianco JA. Toxic epidermal necrolysis possibly linked to aztreonam in bone marrow transplant patients. Ann Pharmacother 1992;26(1):34–5.

295. Brenner S, Wolf R, Ruocco V. Drug-induced pemphigus. I. A survey. Clin Dermatol 1993;11(4):501–5.

296. Blacker KL, Stern RS, Wintroub BU. Cutaneous reactions to drugs. In: Fitzpatrick TB, editor. Dermatology in General Medicine. 4th ed. New York: McGraw-Hill, 1993:1783.

297. Murray KM, Keane WR. Review of drug-induced acute interstitial nephritis. Pharmacotherapy 1992;12(6):462–7.

298. Schulz KH, Schopf E, Wex O. Allergische Berufsekzeme durch Ampicillin. [Allergic occupational eczemas caused by ampicillin.] Berufsdermatosen 1970;18(3):132–43.

299. Davies RJ, Hendrick DJ, Pepys J. Asthma due to inhaled chemical agents: ampicillin, benzyl penicillin, 6 amino penicillanic acid and related substances. Clin Allergy 1974;4(3):227–47.

300. Wengrower D, Tzfoni EE, Drenger B, Leitersdorf E. Erythroderma and pneumonitis induced by penicillin? Respiration 1986;50(4):301–3.

301. de Hoyos A, Holness DL, Tarlo SM. Hypersensitivity pneumonitis and airways hyperreactivity induced by occupational exposure to penicillin. Chest 1993;103(1):303–4.

302. Stenton SC, Dennis JH, Hendrick DJ. Occupational asthma due to ceftazidime. Eur Respir J 1995;8(8):1421–3.

303. Moscato G, Galdi E, Scibilia J, Dellabianca A, Omodeo P, Vittadini G, Biscaldi GP. Occupational asthma, rhinitis and urticaria due to piperacillin sodium in a pharmaceutical worker. Eur Respir J 1995;8(3):467–9.

304. Kanny G, Puygrenier J, Beaudoin E, Moneret-Vautrin DA. Choc anaphylactique alimentaire: implication des residues de pénicilline. [Alimentary anaphylactic shock: implication of penicillin residues.] Allerg Immunol (Paris) 1994;26(5):181–3.

305. Green RL, Green MA. Postcoital urticaria in a penicillin-sensitive patient. Possible seminal transfer of penicillin. JAMA 1985;254(4):531.

306. Hoigne R. Akute Nebenreaktionen auf Penicillinpräparate. [Acute side-reactions to penicillin preparations.] Acta Med Scand 1962;171:201–8.

307. Silber TJ, D'Angelo L. Psychosis and seizures following the injection of penicillin G procaine. Hoigné's syndrome. Am J Dis Child 1985;139(4):335–7.

308. Kraus SJ, Green RL. Pseudoanaphylactic reactions with procaine penicillin. Cutis 1976;17(4):765–7.

309. Kryst L, Wanyura H. Hoigné's syndrome—its course and symptomatology. J Maxillofac Surg 1979;7(4):320–6.

310. Tompsett R. Pseudoanaphylactic reactions to procaine penicillin G. Arch Intern Med 1967;120(5):565–7.

311. Van Arsdel PP Jr. Classification of risk factors for drug allergy. Immunol Allergy Clin North Am 1991;11:475.

312. Petri M, Albritton J. Antibiotic allergy in systemic lupus erythematosus: a case control study. J Rheumatol 1992;20:399.

313. Capaul R, Maibach R, Kuenzi UP, et al. Atopy, bronchial asthma and previous adverse drug reactions (ADRs): risk factors for ADRs? Post Marketing Surveillance 1993;7:331.

314. Green GR, Rosenblum AH, Sweet LC. Evaluation of penicillin hypersensitivity: value of clinical history and skin testing with penicilloyl-polylysine and penicillin G. A cooperative prospective study of the penicillin study group of the American Academy of Allergy. J Allergy Clin Immunol 1977;60(6):339–45.

315. Battegay M, Opravil M, Wuthrich B, Luthy R. Rash with amoxycillin–clavulanate therapy in HIV-infected patients. Lancet 1989;2(8671):1100.

316. Toogood JH. Risk of anaphylaxis in patients receiving beta-blocker drugs. J Allergy Clin Immunol 1988;81(1):1–5.

317. Jick H, Porter JB. Potentiation of ampicillin skin reactions by allopurinol or hyperuricemia. J Clin Pharmacol 1981;21(10):456–8.

318. Fleming A. On the bactericidal action of cultures of a Penicillium with a special reference to their use in the isolation of B. influenzae. Br J Exp Pathol 1929;10:226.

319. Abraham EP. Further observation on penicillin. Lancet 1941;1:177.

320. Florey HW. Penicillins in war wounds. A report from the Mediterranean. Lancet 1943;2:742.

321. Tomasz A. Penicillin-binding proteins and the antibacterial effectiveness of beta-lactam antibiotics. Rev Infect Dis 1986;8(Suppl 3):S260–78.

322. Georgopapadakou NH. Penicillin-binding proteins and bacterial resistance to beta-lactams. Antimicrob Agents Chemother 1993;37(10):2045–53.

323. Ghuysen JM. Serine beta-lactamases and penicillin-binding proteins. Annu Rev Microbiol 1991;45:37–67.

324. Philippon A, Labia R, Jacoby G. Extended-spectrum beta-lactamases. Antimicrob Agents Chemother 1989;33(8):1131–6.

325. Nikaido H. Prevention of drug access to bacterial targets: permeability barriers and active efflux. Science 1994;264(5157):382–8.

326. Livermore DM. Interplay of impermeability and chromosomal beta-lactamase activity in imipenem-resistant Pseudomonas aeruginosa. Antimicrob Agents Chemother 1992;36(9):2046–8.

327. Moreillon P. La résistance bactérienne aux antibiotiques. [Bacterial resistance to antibiotics.] Schweiz Med Wochenschr 1995;125(23):1151–61.

328. Watson JD, Hopkins NH, Roberts JW, et al. The genetic systems provided by E. coli and its viruses. In: Gillen JR, editor. Molecular Biology of the Gene. Menlo Park: Benjamin/Cummings, 1987:176.

329. Hayes W. Recombination in Bact. coli K 12; unidirectional transfer of genetic material. Nature 1952;169(4290):118–9.

330. Amabile-Cuevas CF, Chicurel ME. Bacterial plasmids and gene flux. Cell 1992;70(2):189–99.

331. Neu HC. The crisis in antibiotic resistance. Science 1992;257(5073):1064–73.

332. Friedland IR, McCracken GH Jr. Management of infections caused by antibiotic-resistant Streptococcus pneumoniae. N Engl J Med 1994;331(6):377–82.

333. Marton A, Gulyas M, Munoz R, Tomasz A. Extremely high incidence of antibiotic resistance in clinical isolates

of *Streptococcus pneumoniae* in Hungary. J Infect Dis 1991;163(3):542–8.

334. Coffey TJ, Dowson CG, Daniels M, Spratt BG. Genetics and molecular biology of beta-lactam-resistant pneumococci. Microb Drug Resist 1995;1(1):29–34.

335. Tomasz A. Multiple-antibiotic-resistant pathogenic bacteria. A report on the Rockefeller University Workshop. N Engl J Med 1994;330(17):1247–51.

336. Garcia-Bustos J, Tomasz A. A biological price of antibiotic resistance: major changes in the peptidoglycan structure of penicillin-resistant pneumococci. Proc Natl Acad Sci USA 1990;87(14):5415–9.

337. Appelbaum PC. Antimicrobial resistance in *Streptococcus pneumoniae*: an overview. Clin Infect Dis 1992;15(1):77–83.

338. Barnes DM, Whittier S, Gilligan PH, Soares S, Tomasz A, Henderson FW. Transmission of multidrug-resistant serotype 23F *Streptococcus pneumoniae* in group day care: evidence suggesting capsular transformation of the resistant strain in vivo. J Infect Dis 1995;171(4):890–6.

339. Munoz R, Coffey TJ, Daniels M, Dowson CG, Laible G, Casal J, Hakenbeck R, Jacobs M, Musser JM, Spratt BG, et al. Intercontinental spread of a multiresistant clone of serotype 23F *Streptococcus pneumoniae*. J Infect Dis 1991;164(2):302–6.

340. Soares S, Kristinsson KG, Musser JM, Tomasz A. Evidence for the introduction of a multiresistant clone of serotype 6B Streptococcus pneumoniae from Spain to Iceland in the late 1980s. J Infect Dis 1993;168(1):158–63.

341. Tomasz A. The pneumococcus at the gates. N Engl J Med 1995;333(8):514–5.

342. Cherubin CE, Corrado ML, Sierra MF, Gombert ME, Shulman M. Susceptibility of Gram-positive cocci to various antibiotics, including cefotaxime, moxalactam, and N-formimidoyl thienamycin. Antimicrob Agents Chemother 1981;20(4):553–5.

343. Schwalbe RS, Stapleton JT, Gilligan PH. Emergence of vancomycin resistance in coagulase-negative staphylococci. N Engl J Med 1987;316(15):927–31.

344. Noble WC, Virani Z, Cree RG. Co-transfer of vancomycin and other resistance genes from *Enterococcus faecalis* NCTC 12201 to *Staphylococcus aureus*. FEMS Microbiol Lett 1992;72(2):195–8.

345. Reichler MR, Allphin AA, Breiman RF, Schreiber JR, Arnold JE, McDougal LK, Facklam RR, Boxerbaum B, May D, Walton RO, et al. The spread of multiply resistant *Streptococcus pneumoniae* at a day care center in Ohio. J Infect Dis 1992;166(6):1346–53.

346. Koren G, Pastuszak A, Ito S. Drugs in pregnancy. N Engl J Med 1998;338(16):1128–37.

347. Rao JM, Arulappu R. Drug use in pregnancy: how to avoid problems. Drugs 1981;22(5):409–14.

348. EUROCAT Working Group. EUROCAT Report 7:15 years of surveillance of congenital anomalies in Europe 1980–1994Brussels: Scientific Institute of Public Health—Louis Pasteur;. 1997.

349. Khoury MJ, Holtzman NA. On the ability of birth defects monitoring to detect new teratogens. Am J Epidemiol 1987;126(1):136–43.

350. Irl C, Hasford J. Assessing the safety of drugs in pregnancy: the role of prospective cohort studies. Drug Saf 2000;22(3):169–77.

351. Lenz W. Kindliche Missbildungen nach Medikamenten während der Gravidität. Dtsch Med Wochenschr 1961;86:2555–6.

352. Weller TM, Rees EN. Antibacterial use in pregnancy. Drug Saf 2000;22(5):335–8.

353. Czeizel AE, Rockenbauer M, Olsen J, Sorensen HT. Oral phenoxymethylpenicillin treatment during pregnancy. Results of a population-based Hungarian case-control study. Arch Gynecol Obstet 2000;263(4):178–81.

354. Larsen H, Nielsen GL, Sorensen HT, Moller M, Olsen J, Schonheyder HC. A follow-up study of birth outcome in users of pivampicillin during pregnancy. Acta Obstet Gynecol Scand 2000;79(5):379–83.

355. Manka W, Solowiow R, Okrzeja D. Assessment of infant development during an 18-month follow-up after treatment of infections in pregnant women with cefuroxime axetil. Drug Saf 2000;22(1):83–8.

356. Fossieck B Jr, Parker RH. Neurotoxicity during intravenous infusion of penicillin. A review. J Clin Pharmacol 1974;14(10):504–12.

357. Andrassy K, Weischedel E, Ritz E, Andrassy T. Bleeding in uremic patients after carbenicillin. Thromb Haemost 1976;36(1):115–26.

358. Schliamser SE, Bolander H, Kourtopoulos H, Norrby SR. Neurotoxicity of benzylpenicillin: correlation to concentrations in serum, cerebrospinal fluid and brain tissue fluid in rabbits. J Antimicrob Chemother 1988;21(3):365–72.

359. Bang NU, Kammer RB. Hematologic complications associated with betalactam antibiotics. Rev Infect Dis 1983;5(Suppl):380.

360. Sattler FR, Weitekamp MR, Ballard JO. Potential for bleeding with the new beta-lactam antibiotics. Ann Intern Med 1986;105(6):924–31.

361. Aubry A, Porcher R, Bottero J, Touratier S, Leblanc T, Brethon B, Rousselit P, Raffoux E, Menotti J, Derouin F, Ribaud P, Sulahian A. Occurrence and kinetics of false-positive *Aspergillus galactomannan test results following treatment with beta-lactam antibiotics in patients with hematological disorder. J Clin Microbial 2006;44:389–94.*

362. Hughes GS, Heald DL, Barker KB, Patel RK, Spillers CR, Watts KC, Batts DH, Euler AR. The effects of gastric pH and food on the pharmacokinetics of a new oral cephalosporin, cefpodoxime proxetil. Clin Pharmacol Ther 1989;46(6):674–85.

363. Saathoff N, Lode H, Neider K, Depperman KM, Borner K, Koeppe P. Pharmacokinetics of cefpodoxime proxetil and interactions with an antacid and an H2 receptor antagonist. Antimicrob Agents Chemother 1992;36(4):796–800.

364. Blouin RA, Kneer J, Ambros RJ, Stoeckel K. Influence of antacid and ranitidine on the pharmacokinetics of oral cefetamet pivoxil. Antimicrob Agents Chemother 1990;34(9):1744–8.

365. Bandrowsky T, Vorono AA, Borris TJ, Marcantoni HW. Amoxicillin-related postextraction bleeding in an anticoagulated patient with tranexamic acid rinses. Oral Surg Oral Med Oral Pathol Oral Radiol Endod 1996;82(6):610–2.

366. Davydov L, Yermolnik M, Cuni LJ. Warfarin and amoxicillin/clavulanate drug interaction. Ann Pharmacother 2003;37(3):367–70.

367. Mailloux AT, Gidal BE, Sorkness CA. Potential interaction between warfarin and dicloxacillin. Ann Pharmacother 1996;30(12):1402–7.

368. Heilker GM, Fowler JW Jr, Self TH. Possible nafcillin–warfarin interaction. Arch Intern Med 1994;154(7): 822–824.

369. Taylor AT, Pritchard DC, Goldstein AO, Fletcher JL Jr. Continuation of warfarin–nafcillin interaction during dicloxacillin therapy. J Fam Pract 1994;39(2):182–5.

370. Lindenbaum J, Rund DG, Butler VP Jr, Tse-Eng D, Saha JR. Inactivation of digoxin by the gut flora: reversal by antibiotic therapy. N Engl J Med 1981;305(14):789–94.

371. Saha JR, Butler VP Jr, Neu HC, Lindenbaum J. Digoxin-inactivating bacteria: identification in human gut flora. Science 1983;220(4594):325–7.

372. Rhodes KM, Brown SN. Do the penicillin antibiotics interact with digoxin? Eur J Clin Pharmacol 1994;46(5):479–80.

373. Ganapathy ME, Prasad PD, Mackenzie B, Ganapathy V, Leibach FH. Interaction of anionic cephalosporins with the intestinal and renal peptide transporters PEPT 1 and PEPT 2. Biochim Biophys Acta 1997;1324(2):296–308.

374. Matsumoto S, Saito H, Inui K. Transcellular transport of oral cephalosporins in human intestinal epithelial cells, Caco-2: interaction with dipeptide transport systems in apical and basolateral membranes. J Pharmacol Exp Ther 1994;270(2):498–504.

375. Sugawara M, Iseki K, Miyazaki K, Shiroto H, Kondo Y, Uchino J. Transport characteristics of ceftibuten, cefixime and cephalexin across human jejunal brush-border membrane. J Pharm Pharmacol 1991;43(12):882–4.

376. Dantzig AH, Duckworth DC, Tabas LB. Transport mechanisms responsible for the absorption of loracarbef, cefixime, and cefuroxime axetil into human intestinal Caco-2 cells. Biochim Biophys Acta 1994;1191(1):7–13.

377. Winstanley PA, Orme ML. The effects of food on drug bioavailability. Br J Clin Pharmacol 1989;28(6):621–8.

378. Westphal JF, Trouvin JH, Deslandes A, Carbon C. Nifedipine enhances amoxicillin absorption kinetics and bioavailability in humans. J Pharmacol Exp Ther 1990;255(1):312–7.

379. Duverne C, Bouten A, Deslandes A, Westphal JF, Trouvin JH, Farinotti R, Carbon C. Modification of cefixime bioavailability by nifedipine in humans: involvement of the dipeptide carrier system. Antimicrob Agents Chemother 1992;36(11):2462–7.

380. Deslandes A, Camus F, Lacroix C, Carbon C, Farinotti R. Effects of nifedipine and diltiazem on pharmacokinetics of cefpodoxime following its oral administration. Antimicrob Agents Chemother 1996;40(12):2879–81.

381. Back DJ, Grimmer SF, Orme ML, Proudlove C, Mann RD, Breckenridge AM. Evaluation of Committee on Safety of Medicines yellow card reports on oral contraceptive-drug interactions with anticonvulsants and antibiotics. Br J Clin Pharmacol 1988;25(5):527–32.

382. Orme ML, Back DJ. Factors affecting the enterohepatic circulation of oral contraceptive steroids. Am J Obstet Gynecol 1990;163(6 Pt 2):2146–52.

383. Hanker JP. Gastrointestinal disease and oral contraception. Am J Obstet Gynecol 1990;163(6 Pt 2):2204–7.

384. Reimers D, Jezek A. Rifampicin und andere Antituberkulotika bei gleichzeitiger oraler Kontrazeption. [The simultaneous use of rifampicin and other antitubercular agents with oral contraceptives.] Prax Pneumol 1971;25(5):255–62.

385. Szoka PR, Edgren RA. Drug interactions with oral contraceptives: compilation and analysis of an adverse experience report database. Fertil Steril 1988;49(5 Suppl 2):S31–8.

386. de Groot AC, Eshuis H, Stricker BH. Ineffectiviteit van orale anticonceptie tijdens gebruik van minocycline. [Inefficacy of oral contraception during use of minocycline.] Ned Tijdschr Geneeskd 1990;134(25):1227–9.

387. Midtvedt T. Microbial functional activities. In: Hanson LA, Yolken RH, editors. Probiotics, Other Nutritional Factors, and Intestinal Microflora. Philadelphia: Lippincott-Raven, 1999:79–96.

388. Mastrantonio M, Minhas H, Gammon A. Antibiotics, the pill, and pregnancy. J Accid Emerg Med 1999;16(4):268–70.

389. Harder S, Schneider W, Bae ZU, Bock U, Zielen S. Unerwünschte Arzneimittel-reaktionen bei gleichzeitiger Gabe von hochdosiertem Phenobarbital und Betalaktam-Antibiotika. [Undesirable drug reactions in simultaneous administration of high-dosage phenobarbital and beta-lactam antibiotics.] Klin Padiatr 1990;202(6):404–7.

390. Young DS. Effects of Drugs on Clinical Laboratory Tests. 3rd ed.. Washington: AACC Press;. 1990.

391. Garton AM, Rennie RP, Gilpin J, Marrelli M, Shafran SD. Comparison of dose doubling with probenecid for sustaining serum cefuroxime levels. J Antimicrob Chemother 1997;40(6):903–6.

392. Singh YN, Harvey AL, Marshall IG. Antibiotic-induced paralysis of the mouse phrenic nerve-hemidiaphragm preparation, and reversibility by calcium and by neostigmine. Anesthesiology 1978;48(6):418–24.

393. Segredo V, Caldwell JE, Matthay MA, Sharma ML, Gruenke LD, Miller RD. Persistent paralysis in critically ill patients after long-term administration of vecuronium. N Engl J Med 1992;327(8):524–8.

394. Tryba M. Wirkungsverstarkung nicht-depolarisierender Muskelrelaxantien durch Acylaminopencilline. Untersuchungen am Beispiel von Vecuronium. [Potentiation of the effect of non-depolarizing muscle relaxants by acylaminopenicillins. Studies on the example of vecuronium.] Anaesthesist 1985;34(12):651–5.

395. Condon RE, Munshi CA, Arfman RC. Interaction of vecuronium with piperacillin or cefoxitin evaluated in a prospective, randomized, double-blind clinical trial. Am Surg 1995;61(5):403–6.

396. Welling PG. Interactions affecting drug absorption. Clin Pharmacokinet 1984;9(5):404–34.

397. Finn A, Straughn A, Meyer M, Chubb J. Effect of dose and food on the bioavailability of cefuroxime axetil. Biopharm Drug Dispos 1987;8(6):519–26.

398. Sommers DK, van Wyk M, Moncrieff J, Schoeman HS. Influence of food and reduced gastric acidity on the bioavailability of bacampicillin and cefuroxime axetil. Br J Clin Pharmacol 1984;18(4):535–9.

399. Tindula RJ, Ambrose PJ, Harralson AF. Aminoglycoside inactivation by penicillins and cephalosporins and its impact on drug-level monitoring. Drug Intell Clin Pharm 1983;17(12):906–8.

400. Pickering LK, Rutherford I. Effect of concentration and time upon inactivation of tobramycin, gentamicin, netilmicin and amikacin by azlocillin, carbenicillin, mecillinam, mezlocillin and piperacillin. J Pharmacol Exp Ther 1981;217(2):345–9.

401. Blair DC, Duggan DO, Schroeder ET. Inactivation of amikacin and gentamicin by carbenicillin in patients with end-stage renal failure. Antimicrob Agents Chemother 1982;22(3):376–9.

402. Jahansouz F, Kriett JM, Smith CM, Jamieson SW. Potentiation of cyclosporine nephrotoxicity by nafcillin in lung transplant recipients. Transplantation 1993;55(5):1045–8.

403. LeBel M, Paone RP, Lewis GP. Effect of ten new beta-lactam antibiotics on urine glucose test methods. Drug Intell Clin Pharm 1984;18(7–8):617–20.

404. Kowalsky SF, Wishnoff FG. Evaluation of potential interaction of new cephalosporins with Clinitest. Am J Hosp Pharm 1982;39(9):1499–501.

Beta-lactamase inhibitors

General Information

Beta-lactamases are genetically and structurally closely related to penicillin-binding proteins. Their production by bacteria is a major mechanism of resistance to the action of beta-lactam antibiotics. Drugs have therefore been developed that inhibit beta-lactamase, as a way of overcoming this resistance (SEDA-20, 229). They are beta-lactam compounds with particularly high affinities for beta-lactamases (1,2), which therefore act as competitive inhibitors of beta-lactamases. Beta-lactamase inhibitors have no important antimicrobial activity and are only given in combination with an antimicrobial beta-lactam.

The following beta-lactamase inhibitors are in use:

- clavulanic acid (rINN), produced naturally by *Streptomyces clavuligerus*, used in combination with amoxicillin or ticarcillin; the combination clavulanic acid + amoxicillin is also called co-amoxiclav (BAN);
- sulbactam (rINN), a halogenated derivative of penicillanic acid, used in combination with ampicillin or cefoperazone;
- tazobactam (rINN), a halogenated derivative of penicillanic acid, used in combination with piperacillin.

In order to improve absorption, sulbactam has also been bound to ampicillin in the single molecule sultamicillin, which is hydrolysed to the active components after absorption.

An inherent obstacle in evaluating the adverse effects of beta-lactamase inhibitors is that they are only co-administered with antimicrobial beta-lactams, the doses of which are usually several times higher. For the most part, combinations of beta-lactam antibiotics with beta-lactamase inhibitors produce the adverse effects of the individual drugs. However, it is not always possible to say whether an adverse effect is due to one drug alone or to the combination. For example, cholestatic hepatitis occurs more often with co-amoxiclav than with ampicillin alone; however, this could be because it is primarily due to clavulanic acid or because the combination somehow increases the risk.

Organs and Systems

Respiratory

Interstitial pneumonitis has been attributed to ampicillin + sulbactam in two Japanese patients (3).

Psychological, psychiatric

Behavioral changes occurred in four children aged 1.5–10.5 years, taking co-amoxiclav (4).

Hematologic

Parallel to a well-known phenomenon seen with cephalosporins, clavulanic acid can be associated with a positive direct antiglobulin test. In three patients antibiotic courses, including intravenous ticarcillin + clavulanic acid, were associated with positive direct antiglobulin tests in over 50% of cases (5–7). Corresponding observations were made in patients taking ampicillin + sulbactam (8). In vitro studies showed that clavulanic acid and sulbactam caused non-immunological absorption of plasma proteins on to the erythrocyte surface (5,9). There seems to be no clinical impact of this phenomenon, but it can interfere with cross-matching of blood products or with the investigation of true hemolysis.

Reversible bone marrow suppression after high-dose piperacillin + tazobactam was seen in an underweight woman and was thought to be a dose-dependent and piperacillin-related effect (10).

Gastrointestinal

Two cases of hemorrhagic colitis, apparently not related to *Clostridium difficile*, have been reported after co-amoxiclav (11,12). However, the same type of colitis has repeatedly been observed with aminopenicillins alone (12).

Co-amoxiclav often causes diarrhea and other gastrointestinal problems. Oral administration was associated with motor disturbances of the small intestine (13).

Liver

Cholestatic hepatitis

> DoTS classification
> Dose-relation: hypersusceptibility effect
> Time-course: immediate
> Susceptibility factors: genetic (DRB1*1501*DRB5-DRB5*0101-DQB1*0602 haplotypes); age over 55 years; male sex; duration of treatment

Co-amoxiclav can cause cholestatic hepatitis. The first report appeared in 1988 (14), since when several hundreds of cases have been reported, for example to health authorities (15), and over 100 cases have been described in detail (16–31). Clavulanic acid is instrumental, either alone or in combination with amoxicillin, since the risk of acute liver injury is much smaller with amoxicillin alone (32).

The relative contributions of amoxicillin and clavulanate to co-amoxiclav-induced hepatotoxicity are incompletely understood. In patients with co-amoxiclav hepatotoxicity, previous use of amoxicillin and rechallenge with amoxicillin were both uneventful, pointing to clavulanic acid as the more likely culprit (16). In a report from the UK, the incidence of liver injury with amoxicillin alone was 0.3 per 10 000 prescriptions versus 1.7 with co-amoxiclav (32). The risk increased after multiple use and with increasing age to 1 per 1000 prescriptions of co-amoxiclav. The main message is that the combination should be used with caution in elderly patients. A patient who has had documented hepatotoxicity related to co-amoxiclav should be well informed about this adverse drug reaction and any future use should be prohibited.

Ticarcillin + clavulanic acid, which is only used intravenously, has also been reported to induce a similar syndrome (33,34) and can also aggravate pre-existing hepatitis (35). One cholestatic reaction has been reported with intravenous sulbactam + ampicillin (36).

The importance of taking a careful history in patients in whom drug-related hepatitis is suspected has been underlined (37).

Incidence

Considering the large number of patients who take this very widely used combination, the risk of hepatitis was initially estimated to be very low, probably below 1/100 000 (23). Newer data, however, have suggested a risk of 1/10 000 or higher (32). In a retrospective cohort study of family practitioners' records, with a high proportion of mild cases, there was a rate of 1 per 4449 prescriptions (32). If this is the case, it is wise to reserve co-amoxiclav (and maybe also ticarcillin + clavulanate) for use in infections caused by strains producing beta-lactamases that can destroy amoxicillin (or ticarcillin).

The syndrome is practically unknown in children; only one pediatric case has been reported so far, in a 4-year-old boy with spherocytosis (16) and causality was disputed (17).

Presentation

The hepatitis usually develops acutely, although an interval of up to 4 weeks between the end of treatment and the first signs of hepatitis is frequent and can prevent rapid diagnosis. Although the clinical effects can be impressive, the condition is usually reversible within 4–6 weeks.

- A 40-year-old woman with a history of chronic sinusitis and asthma developed nausea, vomiting, abdominal pain, and diarrhea (38). Six weeks before, she had taken a 10-day course of co-amoxiclav for acute sinusitis. Her transaminase activities were markedly increased, as was total bilirubin. All drugs were withdrawn, her symptoms progressively improved, and she was discharged without a clearly identified cause of her illness. Liver function tests normalized completely within a few weeks. Two months later she had another episode of acute sinusitis and was again given co-amoxiclav. A few days later she developed nausea, vomiting, a skin rash, abdominal pain, and reduced appetite. Her alanine transaminase activity was 199 U/l (0–65), aspartate transaminase 99 U/l (0–60), and alkaline phosphatase 362 U/l (50–180). The antibiotic was withdrawn, and she completely recovered in 2 weeks and had normal liver function tests over the next several months.
- A 33-month-old boy took co-amoxiclav (dose not stated) for 10 days for otitis media (39). He had taken it twice before. One day after completing the course he developed a rash over his entire body, followed 3 days later by lethargy, jaundice, pale stools, and pruritus. The jaundice persisted, the liver was markedly enlarged, and all liver function tests were abnormal. Tests for known viral and metabolic causes of cholestasis were negative. A percutaneous liver biopsy showed centrilobular cholestasis "consistent with a drug reaction." He was given ursodeoxycholic acid (30 mg/kg/day) and vitamins A, D, and K; later prednisolone was added. However, his jaundice persisted, as did severe pruritus. He also developed extensive xanthomatosis and failure to grow. A liver transplantation was successfully performed 8 months after the onset of symptoms. His explanted liver had features of biliary cirrhosis, with ductular proliferation and ductopenia.

One fatal outcome was described in a patient who was also taking ethinylestradiol, which can itself cause cholestasis (40).

In isolated cases, Stevens–Johnson syndrome together with cholestasis and bone marrow aplasia have been associated with either amoxicillin alone (41) or with co-amoxiclav (42).

Histological features

Liver biopsy shows predominantly centrilobular or panlobular cholestasis, and occasionally granulomatous hepatitis (43).

Mechanism

The mechanism of the syndrome and its possible relation to the liver injury that other beta-lactams, particularly isoxazolylpenicillins, can cause are unclear. A slight eosinophilia has been seen in many cases (26,43,44) and some of the sporadic cases of rechallenge were positive (16). However, there is no other evidence supporting an immunoallergic basis. Hepatic accumulation or biliary secretion of clavulanic acid or its metabolites have not been demonstrated (45).

Two other theories have been proposed (46). One is based on the metabolic formation of neo-antigens, and subsequent recognition of these antigens as foreign by the immune system. This "immune allergic hypothesis" is supported by the strong association with an HLA class II haplotype. The authors argued that "HLA class II molecules are required for antigen presentation to CD4-positive T cells. HLA alleles may differ by a little as a single codon, and one amino acid residue difference at a critical site in the resulting polypeptide may be functionally significant, determining not only the affinity with which a given antigen is presented but also the interaction of the HLA peptide complex with the T cell receptor." Their other theory was that the liver disease may arise through linkage with another gene on chromosome 6p. This "linked-gene" hypothesis, they proposed, may explain why jaundice is rare after treatment with co-amoxiclav, although this particular HLA haplotype is common in Northern Europe, where co-amoxiclav is commonly prescribed.

Whatever the mechanisms might be, at present it is reasonable to look on clavulanate as the main contributor to the development of hepatotoxicity with co-amoxiclav. Hepatotoxicity has also been reported with clavulanate plus ticarcillin (34). So far, however, there has been no genetic evaluation of patients with hepatotoxic reactions after therapy with clavulanate and ticarcillin.

Susceptibility factors

Increasing age (over 55 years), male sex, and duration of treatment are risk factors (26,32,47), while drug dose and route of administration, other medications, previous drug allergies, or prior use of co-amoxiclav were not significantly associated with the reaction.

The importance of HLA antigens in the pathogenesis of some liver diseases (autoimmune hepatitis, primary biliary cirrhosis, primary cholangitis) is also reasonably well established (48), and there is a significant association between co-amoxiclav-induced liver damage and an HLA haplotype (49,50). HLA-class antigens have been investigated in 35 patients with biopsy-documented liver damage due to co-amoxiclav and 300 controls (volunteer bone marrow donors) (46). HLA-A and HLA-B were typed using allo-antisera and HLA-DBR and HLA-DWB were typed by PCR. The patients with hepatitis were characterized by a higher frequency of the DRB1*1501*DRB5-DRB5*0101-DQB1*0602 haplotype (57 versus 12% in controls). Patients with that haplotype tended to have a cholestatic rather than a hepatocellular type of hepatitis. However, these data also suggest that other factors must act concurrently. These factors may involve heterogeneity of the formed antigens and/or polymorphism of the T cell receptor. A reaction to (unknown) metabolites of clavulanic acid also has to be kept in mind. The authors suggested that "metabolic factors may play a greater role in the pathogenesis of hepatocellular cases, whereas immunological factors may be more involved in the pathogenesis of cholestatic ones." This HLA association has been confirmed in a study in which there was an increased frequency of homozygous status for this haplotype (51). This might reflect population differences and the small sample size in both studies (46). It is reasonable to assume that HLA characterization will be implemented in a future diagnostic armamentarium.

Urinary tract

The combination of piperacillin + tazobactam was thought to have caused acute interstitial nephritis in a 51-year-old woman (52). It remains open whether the combination or one of the components was the culprit.

Immunologic

Clavulanic acid has a very low immunogenic and allergenic potential in animals. The possible impact of its co-administration with other beta-lactam antibiotics is unknown (53). Two patients with IgE-mediated hypersensitivity to oral co-amoxiclav and positive skin tests for clavulanic acid, but not for penicillins, both tolerated oral amoxicillin. One patient was also challenged with clavulanic acid and developed urticaria, conjunctivitis, and bronchial obstruction (54). Since co-amoxiclav has been widely used since its introduction in 1981, the frequency of hypersensitivity reactions is low.

Although there have been many reports of allergic reactions and cross-reactivity to penicillins, there have been few reports of allergic reactions to clavulanic acid, which contains a beta-lactam ring, but differs from penicillin G and penicillin V in its second ring, which is an oxazolidine instead of a thiazolidine ring [55]. Clavulanic acid shares this ring with cloxacillin.

- A 46-year-old woman with a history of allergy to co-amoxiclav was challenged with a single tablet; 30 minutes later she developed itching, wheals, and flares, which started on her groins and armpits but then spread over her entire body within a few minutes [56]. Radioallergosorbent tests for specific IgE to penicillin V, penicillin G, amoxicillin, and ampicillin were all negative. Skin prick tests and intradermal tests to amoxicillin, penicillin major determinant (penicilloyl polylysine), and penicillin minor determinant were also negative. Oral challenge with progressively increasing doses of amoxicillin up to 500 mg gave no reactions. However, 10 minutes after she took a half a capsule of co-amoxiclav 500/125 mg she developed itching, wheals, and flares. Her symptoms were relieved by adrenaline, a glucocorticoid, and an antihistamine. A histamine release test was negative for amoxicillin, but positive for co-amoxiclav and clavulanic acid alone. Skin prick tests were positive with the combination, but negative with cloxacillin, tazobactam, and sulbactam. Specific IgE determination 2 months later was positive with penicillin V, but negative with penicillin G, amoxicillin, and ampicillin. Skin prick tests to amoxicillin, penicillin major and minor determinants, and penicillin V were all negative. She refused oral challenge with clavulanic acid.

In this case, allergy to clavulanic acid was demonstrated by specific IgE antibody concentrations, skin prick tests, a histamine release test, and oral challenge. The results of the skin prick tests with cloxacillin and the other beta-lactamase inhibitors almost certainly excluded the oxazolidine ring as the epitope responsible for sensitization, and this is consistent with previous reports of clavulanic acid allergy, in most of which type I sensitization has been demonstrated by skin prick tests [57], although delayed reactions have also been described [58]. The authors also suggested that sensitization to penicillin V, which occurred in this case, might have been due to repeated exposure during the months of evaluation.

Allergy to clavulanic acid has been reported in 10 children aged 4–12 years [59]. The diagnosis was based on the confirmation of an IgE-mediated etiology by an positive oral challenge test with clavulanic acid and negative tests, including skin tests with amoxicillin, ampicillin, penicillins G and V, and cefaclor. Allergic contact dermatitis has also been attributed to clavulanic acid [60]. The clinical data available on sulbactam and tazobactam are still limited and do not allow an assessment of the frequency and pattern of associated hypersensitivity reactions (61).

Body temperature

A woman immediately developed hyperpyrexia up to 40°C after a first dose of intravenous ampicillin + sulbactam, having previously tolerated ampicillin alone for 10 days. Hyperpyrexia was repeatedly observed after six more doses of sulbactam (62).

Interference with Diagnostic Tests

Leukocyte dipstick test

Clavulanic acid caused false-positive dipstick tests for leukocytes; sulbactam and tazobactam did not (63).

References

1. Rolinson GN. Evolution of beta-lactamase inhibitors. Rev Infect Dis 1991;13(Suppl 9):S727–32.
2. Hoover JRE. Betalactam antibiotics: structure-activity relationships. In: Demain AL, Solomon NA, editors. Antibiotics Containing the Betalactam-Structure, Part II. Berlin: Springer-Verlag, 1983:119.
3. Miyashita N, Nakajima M, Kuroki M, Kawabata S, Hashiguchi K, Niki Y, Kawane H, Matsushima T. [Sulbactam/ampicillin-induced pneumonitis.]Nihon Kokyuki Gakkai Zasshi 1998;36(8):684–9.
4. Macknin ML. Behavioral changes after amoxicillin–clavulanate. Pediatr Infect Dis J 1987;6(9):873–4.
5. Williams ME, Thomas D, Harman CP, Mintz PD, Donowitz GR. Positive direct antiglobulin tests due to clavulanic acid. Antimicrob Agents Chemother 1985;27(1):125–7.
6. Finegold SM, Johnson CC. Lower respiratory tract infection. Am J Med 1985;79(5B):73–7.
7. Blanchard M, Oppliger R, Bucher U. Positiver direkter Coombs-Test bei akuten Leukämien und anderen Hämoblastosen: Zusammenhang mit clavulansäurehaltigen Antibiotika?. [Positive direct Coombs' test in acute leukemias and other hemoblastoses: relation to clavulanic acid-containing antibiotics?.] Schweiz Med Wochenschr 1989;119(2):39–45.
8. Lutz P, Dzik W. Very high incidence of a positive direct antiglobulin test (+DAT) in patients receiving Unasyn. Transfusion 1992;32:23.
9. Garratty G, Arndt PA. Positive direct antiglobulin tests and haemolytic anaemia following therapy with beta-lactamase inhibitor containing drugs may be associated with nonimmunologic adsorption of protein onto red blood cells. Br J Haematol 1998;100(4):777–83.
10. Ruiz-Irastorza G, Barreiro G, Aguirre C. Reversible bone marrow depression by high-dose piperacillin/tazobactam. Br J Haematol 1996;95(4):611–2.
11. Klotz F, Barthet M, Perreard M. A propos d'un cas de colite aiguë hémorragique après la prise orale d'Augmentin. [A case of acute hemorrhagic colitis after oral ingestion of Augmentin.] Ann Med Interne (Paris) 1990;141(3):276.
12. Heer M, Sulser H, Hany A. Segmentale, hämorrhagische Kolitis nach Amoxicillin-Therapie. [Segmental hämorrhagic colitis following amoxicillin therapy.] Schweiz Med Wochenschr 1989;119(21):733–5.
13. Caron F, Ducrotte P, Lerebours E, Colin R, Humbert G, Denis P. Effects of amoxicillin–clavulanate combination on the motility of the small intestine in human beings. Antimicrob Agents Chemother 1991;35(6):1085–8.
14. van den Broek JW, Buennemeyer BL, Stricker BH. Cholestatische hepatitis door de combinatie amoxicilline en clavulaanzuur (Augmentin). [Cholestatic hepatitis caused by a combination of amoxicillin and clavulanic acid (Augmentin).] Ned Tijdschr Geneeskd 1988;132(32):1495–7.
15. Thomson JA, Fairley CK, McNeil JJ, Purcell P. Augmentin-associated jaundice. Med J Aust 1994;160(11):733–4.
16. Stricker BH, Van den Broek JW, Keuning J, Eberhardt W, Houben HG, Johnson M, Blok AP. Cholestatic hepatitis due to antibacterial combination of amoxicillin and clavulanic acid (Augmentin). Dig Dis Sci 1989;34(10):1576–80.
17. Reddy KR, Brillant P, Schiff ER. Amoxicillin–clavulanate potassium-associated cholestasis. Gastroenterology 1989;96(4):1135–41.
18. Reddy KR, Schiff ER. Hepatitis and Augmentin. Dig Dis Sci 1990;35(8):1045–6.
19. Verhamme M, Ramboer C, Van de Bruaene P, Inderadjaja N. Cholestatic hepatitis due to an amoxycillin/clavulanic acid preparation. J Hepatol 1989;9(2):260–4.
20. Dowsett JF, Gillow T, Heagerty A, Radcliffe M, Toadi R, Isle I, Russell RC. Amoxycillin/clavulanic acid (Augmentin)-induced intrahepatic cholestasis. Dig Dis Sci 1989;34(8):1290–3.
21. Schneider JE, Kleinman MS, Kupiec JW. Cholestatic hepatitis after therapy with amoxicillin/clavulanate potassium. NY State J Med 1989;89(6):355–6.
22. Pelletier G, Ink O, Fabre M, Hagege H. Hépatite cholestatique probablement due à l'association d'amoxilline et d'acide clavulanique. [Hepatic cholestasis probably due to the combination of amoxicillin and clavulanic acid.] Gastroenterol Clin Biol 1990;14(6–7):601.
23. Larrey D, Vial T, Micaleff A, Babany G, Morichau-Beauchant M, Michel H, Benhamou JP. Hepatitis associated with amoxycillin–clavulanic acid combination report of 15 cases. Gut 1992;33(3):368–71.
24. Hanssens M, Mast A, Van Maele V, Pauwels W. Cholestatische icterus door amoxicilline–clavulaanzuur bij 4 patienten. [Cholestatic jaundice caused by amoxicillin–clavulanic acid in 4 patients.] Ned Tijdschr Geneeskd 1994;138(29):1481–3.
25. Wong FS, Ryan J, Dabkowski P, Dudley FJ, Sewell RB, Smallwood RA. Augmentin-induced jaundice. Med J Aust 1991;154(10):698–701.
26. Alexander P, Roskams T, Van Steenbergen W, Peetermans W, Desmet V, Yap SH. Intrahepatic cholestasis induced by amoxicillin/clavulanic acid (Augmentin): a report on two cases. Acta Clin Belg 1991;46(5):327–32.
27. Maggini M, Raschetti R, Agostinis L, Cattaruzzi C, Troncon MG, Simon G. Use of amoxicillin and amoxicillin-clavulanic acid and hospitalization for acute liver injury. Ann Ist Super Sanita 1999;35(3):429–33.
28. Soza A, Riquelme F, Alvarez M, Duarte I, Glasinovic JC, Arrese M. Hepatotoxicidad por amoxicilina/acido clavulanico: caso clinico. [Hepatotoxicity by amoxicillin/clavulanic acid: case report.] Rev Med Chil 1999;127(12):1487–91.
29. Ma C, Bayliff CD, Ponich T. Amoxicillin–clavuanic acid-induced hepatotoxicity. Can J Hosp Pharm 1999;52:30–2.
30. Richardet JP, Mallat A, Zafrani ES, Blazquez M, Bognel JC, Campillo B. Prolonged cholestasis with ductopenia after administration of amoxicillin/clavulanic acid. Dig Dis Sci 1999;44(10):1997–2000.
31. Limauro DL, Chan-Tompkins NH, Carter RW, Brodmerkel GJ Jr, Agrawal RM. Amoxicillin/clavulanate-associated hepatic failure with progression to Stevens–Johnson syndrome. Ann Pharmacother 1999;33(5):560–4.
32. Garcia Rodriguez LA, Stricker BH, Zimmerman HJ. Risk of acute liver injury associated with the combination of

amoxicillin and clavulanic acid. Arch Intern Med 1996;156(12):1327–32.

33. Ryan J, Dudley FJ. Cholestasis with ticarcillin–potassium clavulanate (Timentin). Med J Aust 1992;156(4):291.
34. Sweet JM, Jones MP. Intrahepatic cholestasis due to ticarcillin–clavulanate. Am J Gastroenterol 1995;90(4):675–6.
35. Van der Auwera P, Legrand JC. Ticarcillin–clavulanic acid therapy in severe infections. Drugs Exp Clin Res 1985;11(11):805–13.
36. Lode H, Springsklee M. Klinische Ergebnisse mit Sulbactam/ Ampicillin in einer multizentrischen Studie an 425 Patienten. [Clinical results with sulbactam/ampicillin in a multicenter study of 425 patients.] Med Klin (Munich) 1989;84(5):236–41.
37. Aithal PG, Day CP. The natural history of histologically proved drug induced liver disease. Gut 1999;44(5):731–5.
38. Nathani MG, Mutchnick MG, Tynes DJ, Ehrinpreis MN. An unusual case of amoxicillin/clavulanic acid-related hepatotoxicity. Am J Gastroenterol 1998;93(8):1363–5.
39. Chawla A, Kahn E, Yunis EJ, Daum F. Rapidly progressive cholestasis: an unusual reaction to amoxicillin/clavulanic acid therapy in a child. J Pediatr 2000;136(1):121–3.
40. Hebbard GS, Smith KG, Gibson PR, Bhathal PS. Augmentin-induced jaundice with a fatal outcome. Med J Aust 1992;156(4):285–6.
41. Cavanzo FJ, Garcia CF, Botero RC. Chronic cholestasis, paucity of bile ducts, red cell aplasia, and the Stevens–Johnson syndrome. An ampicillin-associated case. Gastroenterology 1990;99(3):854–6.
42. Escallier F, Dalac S, Caillot D, Boulitrop C, Collet E, Lambert D. Erythème polymorphe, aplasie, hépatite cholestatique au cours d'un traitement par Augmentin (amoxicilline acide clavulanique). [Erythema multiforme, aplasia, cholestatic hepatitis during treatment with Augmentin (amoxicillin + clavulanic acid).] Rev Med Interne 1990;11(1):73–5.
43. Silvain C, Fort E, Levillain P, Labat-Labourdette J, Beauchant M. Granulomatous hepatitis due to combination of amoxicillin and clavulanic acid. Dig Dis Sci 1992;37(1):150–2.
44. Belknap MK, McClelland KJ. Cholestatic hepatitis associated with amoxicillin–clavulanate. Wis Med J 1993;92(5):241–2.
45. Reading C, Slocombe B. Augmentin: clavulanate-potentiated amoxicillin. In: Queener SF, Webber JA, Queener SW, editors. Betalactam Antibiotics for Clinical Use. New York: Marcel Dekker, 1986:527.
46. Hautekeete ML, Horsmans Y, Van Waeyenberge C, Demanet C, Henrion J, Verbist L, Brenard R, Sempoux C, Michielsen PP, Yap PS, Rahier J, Geubel AP. HLA association of amoxicillin–clavulanate-induced hepatitis. Gastroenterology 1999;117(5):1181–6.
47. Thomson JA, Fairley CK, Ugoni AM, Forbes AB, Purcell PM, Desmond PV, Smallwood RA, McNeil JJ. Risk factors for the development of amoxycillin–clavulanic acid associated jaundice. Med J Aust 1995;162(12):638–40.
48. Berson A, Freneaux E, Larrey D, Lepage V, Douay C, Mallet C, Fromenty B, Benhamou JP, Pessayre D. Possible role of HLA in hepatotoxicity. An exploratory study in 71 patients with drug-induced idiosyncratic hepatitis. J Hepatol 1994;20(3):336–42.
49. Van Waeryenberge C, Hautekeete ML, Horsmans Y, Demanet C, et al. Amoxycillin–clavulanate-induced hepatitis is linked to the DRB11501-DRB50101-DQB10602 haplotype but not to DPB antigens. Europ J Immunogen 1998;25(Suppl 1):66.
50. Donaldson PT, Underhill JA, Clare M, O'Donohue J, et al. Is there a genetic basis for Augmentin associated jaundice?

A link with HLA DRB11501-DQA10102-DQB10602 haplotype. Hepatology 1998;28:256A.
51. O'Donohue J, Oien KA, Donaldson P, Underhill J, Clare M, MacSween RN, Mills PR. Co-amoxiclav jaundice: clinical and histological features and HLA class II association. Gut 2000;47(5):717–20.
52. Pill MW, O'Neill CV, Chapman MM, Singh AK. Suspected acute interstitial nephritis induced by piperacillin-tazobactam. Pharmacotherapy 1997;17(1):166–9.
53. Edwards RG, Dewdney JM, Dobrzanski RJ, Lee D. Immunogenicity and allergenicity studies on two beta-lactam structures, a clavam, clavulanic acid, and a carbapenem: structure-activity relationships. Int Arch Allergy Appl Immunol 1988;85(2):184–9.
54. Fernandez-Rivas M, Perez Carral C, Cuevas M, Marti C, Moral A, Senent CJ. Selective allergic reactions to clavulanic acid. J Allergy Clin Immunol 1995;95(3):748–50.
55. Edwards RG, Dewdney JM, Dobrzanski RJ, Lee D. Immunogenicity and allergenicity studies on two beta-lactam structures, a clavam, clavulanic acid, and a carbapenem: structure-activity relationships. Int Arch Allergy Appl Immunol 1988; 85: 184-9.
56. Gonzales de Olano D, Losada PA, Caballer BdeL, Vazquez Gonzales AC, Diegues Pastor MC, Cuevas Agustin M. Selective sensitization to clavulanic acid and penicillin V. J Investig Allergol Clin Immunol 2007; 17: 119-21.
57. Raison-Peyron N, Messaad D, Bousquet J, Demoli P. Selective immediate hypersensitivity to clavulanic acid. Ann Pharmacother 2003; 37: 1146-7.
58. Kamphof WC, Rustemeyer T, Bruyzeel DP. Sensitization to clavulanic acid in Augmentin. Contact Dermatitis 2002; 47: 47.
59. Tortajada GM, Ferrer FA, Gracia AM, Clement PA, Garia ME, Tallion GM. Hypersensitivity to clavulanic aid in children. Allergol Immunopathol 2008;36:308–10.
60. Kim YH, Ko JY, Kim YS, Ro YS. A case of allergic contact dermatitis to clavulanic acid. Contact Dermatitis 2008;59(6):378–379.
61. Wilson SE, Nord CE. Clinical trials of extended spectrum penicillin/beta-lactamase inhibitors in the treatment of intra-abdominal infections. European and North American experience. Am J Surg 1995;169(Suppl 5A):S21–6.
62. Olivencia-Yurvati AH, Sanders SP. Sulbactam-induced hyperpyrexia. Arch Intern Med 1990;150(9):1961.
63. Beer JH, Vogt A, Neftel K, Cottagnoud P. False positive results for leucocytes in urine dipstick test with common antibiotics. BMJ 1996;313(7048):25.

Capreomycin

See also Antituberculosis drugs

General Information

Capreomycin has been abandoned for the treatment of tuberculosis and replaced by first-choice drugs. In rare cases, it has been administered in infections with non-tuberculous mycobacteria when there is multiple drug resistance to the first-line antituberculosis drugs (1).

Organs and Systems

Electrolyte balance

Marked renal loss of sodium, chloride, potassium, and magnesium with progressive metabolic alkalosis and hyper-reninemia have been reported (2).

Urinary tract

A Bartter-like syndrome has been reported in a 25-year-old man taking prolonged capreomycin for drug-resistant pulmonary tuberculosis (2).

Drug–Drug Interactions

Streptomycin

Capreomycin should never be combined with streptomycin or other aminoglycosides because of nephrotoxicity and ototoxicity (3).

References

1. Mandell GL, Sande MA. Antimicrobial agents: drugs used in the chemotherapy of tuberculosis and leprosy. In: Goodman Gilman A, Rall TW, Nies AS, Taylor P, editors. Goodman and Gilman's The Pharmacological Basis of Therapeutics. 8th ed.. New York: Pergamon Press, 1990:1146 Chapter 49.
2. Steiner RW, Omachi AS. A Bartter's-like syndrome from capreomycin, and a similar gentamicin tubulopathy. Am J Kidney Dis 1986;7(3):245–9.
3. Hugues FC, Moore N, Julien D. Les interactions observées avec les médicaments antituberculeux. [Interactions of antitubercular drugs.] Rev Pneumol Clin 1988;44(6):278–85.

Carbapenems

See also Beta-lactam antibiotics

General Information

Carbapenems differ from penicillins and cephalosporins by a methylene substitution for sulfur in the five-membered beta-ring structure. Imipenem and meropenem belong to this class of compounds.

In the last 25 years, various natural carbapenems have been discovered (1). However, their potential is limited by chemical instability. Imipenem (*N*-formimidoylthienamycin), the first carbapenem in use, is therefore a stabilized synthetic compound. To overcome a second difficulty, namely inactivation by a kidney dehydropeptidase, imipenem has to be combined with cilastatin, a competitive inhibitor of that enzyme. Meropenem has better stability in the presence of renal dehydropeptidase I (2). The antibacterial spectrum of carbapenems is among the broadest of all beta-lactam antibiotics, and they have good stability against many beta-lactamases.

General adverse reactions

The safety profile of the carbapenems is comparable to that of other beta-lactam antibiotics, in particular with regard to laboratory abnormalities, the most common ones being those related to liver function (3,4). In patients with pre-existing nervous system disease or who take dosages above the recommended limits (for example in renal impairment) seizures appear to be more common with imipenem + cilastatin.

Organs and Systems

Nervous system

Seizures associated with imipenem + cilastatin have repeatedly been reported (5–7). As with other beta-lactam antibiotics, it is difficult to assess clearly the cause of a seizure in patients with a cluster of other predisposing factors for neurotoxicity (8) and hence to reach clear estimates of frequency. In a review of 1754 patients there was a similar incidence of seizures with imipenem + cilastatin as with other antibiotic regimens usually containing another beta-lactam (9). In rabbits imipenem + cilastatin and another carbapenems were more neurotoxic than benzylpenicillin (10). In mice, ataxia and seizures were seen, with much lower blood concentrations of imipenem than cefotaxime or benzylpenicillin (1900 µg/ml versus 3400 µg/ml and 5800 µg/ml) (11). In mice imipenem also lowered the convulsive threshold of pentetrazol (pentylenetetrazole) more than cefazolin or two other carbapenems (12). Cilastatin alone was not proconvulsant, but it increased the effects of co-administered imipenem.

Imipenem is a more common cause of seizures than other beta-lactam antibiotics, particularly when high doses are given (13–15). In one study, seven of 21 children developed seizure activity while receiving imipenem + cilastatin for bacterial meningitis, a recognized risk factor (13). However, computer-assisted monitoring of imipenem + cilastatin dosages in relation to renal function resulted in a reduced incidence of seizures (16).

In animals, meropenem (17) and other carbapenems (18,19) were less epileptogenic than imipenem. In 403 children there was no meropenem-associated neurotoxicity (20) and meropenem was well tolerated in children with bacterial meningitis (21). In summary, a larger dose range of meropenem than imipenem appears to be tolerated, but when strictly observing known risk factors for seizure propensity the difference between the two compounds is very small (22,23).

Sensory systems

Taste alterations were seen in some early patients who were treated with carbapenems (24); these observations have not subsequently been confirmed.

Hematologic

As with some cephalosporins and clavulanic acid, the Coombs' test was positive in a number of patients taking carbapenems, but without hemolysis (25).

Mouth and teeth

Yellowish-brown staining of the teeth was related to imipenem in several cases (26,27). Staining was mostly removable with dental assistance.

Urinary tract

In animals, the tubular toxicity of imipenem was completely prevented by cilastatin. Accordingly, definite nephrotoxicity of this combination has not been documented in patients (25) or in healthy volunteers (28). The cilastatin component may even reduce the nephrotoxic effects of ciclosporin after kidney transplantation (29) or bone marrow transplantation (30).

Skin

Of all the drugs that have been implicated in drug-induced toxic epidermal necrolysis, antimicrobial drugs account for 29–42% (now more than 100 in number) (31,32), and almost all antimicrobial drugs have been implicated, including meropenem (33).

- A 75-year-old woman developed acute pneumonia. She was first given oral co-amoxiclav, fluconazole, and ciprofloxacin for 10 days, followed by cefotaxime and amikacin, both intravenously. Six days later, she developed a progressive erythematous rash, soon involving 40% of her body surface. The antibiotics were withdrawn and she was rehydrated and given intravenous immunoglobulins 0.75 g/kg for 5 days, but not systemic glucocorticoids. She then developed severe septic shock because of a combination of two very resistant bacterial strains and was given meropenem 1 g and teicoplanin 800 mg bd. However, within 2 days her skin lesions recurred, extending to previously uninvolved skin areas and including over 60% of her body. A biopsy showed typical features of toxic epidermal necrolysis. Meropenem was withdrawn and replaced by aztreonam. However, she died 5 days later.

Imipenem, which is related to meropenem, has also been reported to cause toxic epidermal necrolysis (34). The authors stated that to the best of their knowledge, this was the first report of a possible cross-reaction between two classes of antibiotics in causing toxic epidermal necrolysis. The time between first administration and the occurrence of epidermal necrolysis is considerably shorter in recurrence or provocation testing (35,36). They also claimed that it is likely that the beta-lactam ring is responsible for this hypersensitivity reaction, citing the evidence that the patient had been given amoxicillin 15 days before the cephalosporin, and that could have served as the sensitizing event. They did not discuss whether aztreonam, a monobactam, also could have caused a cross-reaction; however, it has been involved in two cases of fatal toxic epidermal necrolysis (37).

A pustular rash, as repeatedly observed with cephalosporins, has been described in one case (38). The frequencies of rash, urticaria, and pruritus were similar to those seen with other beta-lactam antibiotics (25).

Meropenem can cause occupational allergic contact dermatitis (39).

- A 45-year-old nurse presented with a 6-month history of recurrent periorbital erythema with itching and runny eyes. Each episode lasted about 5–6 days and settled on withdrawal from her place at work, where she handled a large number of drugs, including meropenem. She was patch-tested with the European standard series, and was negative to all except meropenem. She had complete remission after quitting her workplace.

The authors assumed that the localization to the periorbital region was due to involuntary contact with her hands or airborne contact.

Immunologic

There was a high degree of cross-reactivity between imipenem determinants, analogous to the penicillin determinants in penicillin-allergic patients. Nine of twenty patients with positive penicillin skin tests had positive skin reactions to analogous imipenem determinants (40). In view of this appreciable cross-reactivity, imipenem should not be given to patients with penicillin allergy.

Immediate hypersensitivity related to imipenem has been reported in a patient allergic to penicillin and aztreonam (41).

Drug–Drug Interactions

Ciclosporin

The addition of imipenem + cilastatin to long-term ciclosporin provoked seizures in patients without a corresponding history (42). In general, nervous system toxicity occurred shortly after the start of the antibiotic therapy and drug concentrations were stable, suggesting a synergistic effect.

Ciclosporin-induced acute nephrotoxicity was reduced by cilastatin, an inhibitor of active tubular resorption (30,43). Reduced renal parenchymal accumulation of the drug may account for this effect. Serum ciclosporin concentrations were unchanged or insignificantly reduced. A similar protective effect against tacrolimus is possible, but unproven (44).

GABA receptor antagonists

Carbapenems (imipenem more than meropenem) are believed to increase central nervous system excitation by inhibition of GABA binding to receptors. Combinations with other GABA inhibiting drugs such as theophylline or quinolones have been reported to provoke seizures (45,46).

Valproic acid

Interactions of carbapenems with valproic acid are so common that the Ministry of Health and Welfare in Japan has banned co-administration.

Two carbapenems, meropenem and panipenem + beta-mipron, reduced serum valproic acid concentrations, and increased the risk of seizures (47,48). The mechanism of the interaction was unclear. Accelerated renal elimination of valproic acid is probable, since the almost immediate fall in serum drug concentrations and low protein binding of carbapenems argue against enzyme induction or protein binding displacement (49).

There have been many reports of seizures in epileptic patients due to lowered plasma concentrations of valproic acid, including a child with a neurodegenerative disorder and epilepsy, in whom valproate serum concentrations fell rapidly on two occasions when meropenem was used (50).

- A 21-year-old woman was given valproic acid 1000 mg/day as a continuous intravenous infusion for generalized tonic–clonic seizures (serum valproate concentration 53 μg/ml) (51). After 13 days she was given intravenous meropenem 1 g tds, and 2 days later, when she was afebrile, had numerous myoclonic episodes involving her arms and face (valproate concentration 42 μg/ml). The dose of valproic acid was increased to 2880 mg/day. Two days later, she had a generalized tonic–clonic seizure and despite the increased dosage the valproate concentration was 7 μg/ml. The dose of valproic acid was increased to 3600 mg/day, but the serum concentration remained below 10 μg/ml. Meropenem was withdrawn and the serum valproate concentration rose to 52 μg/ml.

There are several possible mechanisms for this interaction (52,53,54), such as reduced plasma protein binding and inhibition of the enterohepatic circulation of valproic acid and valproic acid glucuronide. The results of a retrospective study in 39 patients suggested that the most critical mechanism is inhibition by carbapenems of the hydrolysis of valproic acid glucuronide in the liver (55). There was an interaction of meropenem with valproic acid in all patients, leading to a fall in valproic acid concentration by an average of 66% within 24 hours. The authors concluded that to avoid neurological deterioration, meropenem and valproate should not be co-administered. This may also be true for all carbapenems, although data on doripenem are still lacking.

Interference with Diagnostic Tests

Leukocyte urine dipstick test

Imipenem caused positive dipstick tests for leukocytes in patients with agranulocytosis and normal urinary sediments. This phenomenon was reproducible in vitro with imipenem, meropenem, and clavulanic acid. Sulbactam, tazobactam, three penicillins, three cephalosporins, and the basic structures of penicillins, cephalosporins, and monobactams tested negative (56).

References

1. Birnbaum J, Kahan FM, Kropp H, MacDonald JS. Carbapenems, a new class of beta-lactam antibiotics. Discovery and development of imipenem/cilastatin. Am J Med 1985;78(6A):3–21.
2. Fukasawa M, Sumita Y, Harabe ET, Tanio T, Nouda H, Kohzuki T, Okuda T, Matsumura H, Sunagawa M. Stability of meropenem and effect of 1 beta-methyl substitution on its stability in the presence of renal dehydropeptidase I. Antimicrob Agents Chemother 1992;36(7):1577–9.
3. Calandra GB, Brown KR, Grad LC, Ahonkhai VI, Wang C, Aziz MA. Review of adverse experiences and tolerability in the first 2,516 patients treated with imipenem/cilastatin. Am J Med 1985;78(6A):73–8.
4. Ahonkhai VI, Cyhan GM, Wilson SE, Brown KR. Imipenem–cilastatin in pediatric patients: an overview of safety and efficacy in studies conducted in the United States. Pediatr Infect Dis J 1989;8(11):740–4.
5. Brotherton TJ, Kelber RL. Seizure-like activity associated with imipenem. Clin Pharm 1984;3(5):536–40.
6. Job ML, Dretler RH. Seizure activity with imipenem therapy: incidence and risk factors. DICP 1990;24(5):467–9.
7. Tse CS, Hernandez Vera F, Desai DV. Seizure-like activity associated with imipenem–cilastatin. Drug Intell Clin Pharm 1987;21(7–8):659–60.
8. Schliamser SE, Cars O, Norrby SR. Neurotoxicity of beta-lactam antibiotics: predisposing factors and pathogenesis. J Antimicrob Chemother 1991;27(4):405–25.
9. Calandra G, Lydick E, Carrigan J, Weiss L, Guess H. Factors predisposing to seizures in seriously ill infected patients receiving antibiotics: experience with imipenem/cilastatin. Am J Med 1988;84(5):911–8.
10. Schliamser SE, Broholm KA, Liljedahl AL, Norrby SR. Comparative neurotoxicity of benzylpenicillin, imipenem/cilastatin and FCE 22101, a new injectible penem. J Antimicrob Chemother 1988;22(5):687–95.
11. Eng RH, Munsif AN, Yangco BG, Smith SM, Chmel H. Seizure propensity with imipenem. Arch Intern Med 1989;149(8):1881–3.
12. Williams PD, Bennett DB, Comereski CR. Animal model for evaluating the convulsive liability of beta-lactam antibiotics. Antimicrob Agents Chemother 1988;32(5):758–60.
13. Wong VK, Wright HT Jr, Ross LA, Mason WH, Inderlied CB, Kim KS. Imipenem/cilastatin treatment of bacterial meningitis in children. Pediatr Infect Dis J 1991;10(2):122–5.
14. Winston DJ, Ho WG, Bruckner DA, Champlin RE. Beta-lactam antibiotic therapy in febrile granulocytopenic patients. A randomized trial comparing cefoperazone plus piperacillin, ceftazidime plus piperacillin, and imipenem alone. Ann Intern Med 1991;115(11):849–59.
15. Leo RJ, Ballow CH. Seizure activity associated with imipenem use: clinical case reports and review of the literature. DICP 1991;25(4):351–4.
16. Pestotnik SL, Classen DC, Evans RS, Stevens LE, Burke JP. Prospective surveillance of imipenem/cilastatin use and associated seizures using a hospital information system. Ann Pharmacother 1993;27(4):497–501.
17. Patel JB, Giles RE. Meropenem: evidence of lack of proconvulsive tendency in mice. J Antimicrob Chemother 1989;24(Suppl A):307–9.
18. Kurihara A, Hisaoka M, Mikuni N, Kamoshida K. Neurotoxicity of panipenem/betamipron, a new carbapenem, in rabbits: correlation to concentration in central nervous system. J Pharmacobiodyn 1992;15(7):325–32.

19. Sunagawa M, Matsumura H, Fukasawa M. Structure-activity relationships of carbapenem and penem compounds for the convulsive property. J Antibiot (Tokyo) 1992;45(12):1983–5.

20. Fujii R, Yoshioka H, Fujita K, Maruyama S, Sakata H, Inyaku F, Chiba S, Tsutsumi H, Wagatsuma Y, Fukushima N, et al. [Pharmacokinetic and clinical studies with meropenem in the pediatric field. Pediatric Study Group of Meropenem.] Jpn J Antibiot 1992;45(6):697–717.

21. Klugman KP, Dagan R. Randomized comparison of meropenem with cefotaxime for treatment of bacterial meningitis. Meropenem Meningitis Study Group. Antimicrob Agents Chemother 1995;39(5):1140–6.

22. Norrby SR, Faulkner KL, Newell PA. Differentiating meropenem and imipenem/cilastatin. Inf Dis in Clin Pract 1997;6:291.

23. Norrby SR, Gildon KM. Safety profile of meropenem: a review of nearly 5,000 patients treated with meropenem. Scand J Infect Dis 1999;31(1):3–10.

24. Freimer EH, Donabedian H, Raeder R, Ribner BS. Empirical use of imipenem as the sole antibiotic in the treatment of serious infections. J Antimicrob Chemother 1985;16(4):499–507.

25. Calandra GB, Wang C, Aziz M, Brown KR. The safety profile of imipenem/cilastatin: worldwide clinical experience based on 3470 patients. J Antimicrob Chemother 1986;18(Suppl E):193–202.

26. Ku C, O'Neill D. Imipenem induced dental stains. Can J Hosp Pharm 1994;47:288.

27. Scanlon N, Wilsher M, Kolbe J. Imipenem induced dental staining. Aust NZ J Med 1997;27(2):190.

28. Drusano GL, Standiford HC, Bustamante CI, Rivera G, Forrest A, Leslie J, Tatem B, Delaportas D, Schimpff SC. Safety and tolerability of multiple doses of imipenem/cilastatin. Clin Pharmacol Ther 1985;37(5):539–43.

29. Hammer C, Thies JC, Mraz W, Mihatsch M. Reduction of cyclosporin (CSA) nephrotoxicity by imipenem/cilastatin after kidney transplantation in rats. Transplant Proc 1989;21(1 Pt 1):931.

30. Gruss E, Tomas JF, Bernis C, Rodriguez F, Traver JA, Fernandez-Ranada JM. Nephroprotective effect of cilastatin in allogeneic bone marrow transplantation. Results from a retrospective analysis. Bone Marrow Transplant 1996;18(4):761–5.

31. Roujeau JC, Guillaume JC, Fabre JP, Penso D, Flechet ML, Girre JP. Toxic epidermal necrolysis (Lyell syndrome). Incidence and drug etiology in France, 1981–1985. Arch Dermatol 1990;126(1):37–42.

32. Schopf E, Stuhmer A, Rzany B, Victor N, Zentgraf R, Kapp JF. Toxic epidermal necrolysis and Stevens–Johnson syndrome. An epidemiologic study from West Germany. Arch Dermatol 1991;127(6):839–42.

33. Paquet P, Jacob E, Damas P, Pierard GE. Recurrent fatal drug-induced toxic epidermal necrolysis (Lyell's syndrome) after putative beta-lactam cross-reactivity: case report and scrutiny of antibiotic imputability. Crit Care Med 2002;30(11):2580–3.

34. Brand R, Rohr JB. Toxic epidermal necrolysis in Western Australia. Australas J Dermatol 2000;41(1):31–3.

35. Dreyfuss DA, Gottlieb LJ, Wilkerson DK, Parsons RW, Krizek TJ. Survival after a second episode of toxic epidermal necrolysis. Ann Plast Surg 1988;20(2):146–7.

36. Roujeau JC, Chosidow O, Saiag P, Guillaume JC. Toxic epidermal necrolysis (Lyell syndrome). J Am Acad Dermatol 1990;23(6 Pt 1):1039–58.

37. McDonald BJ, Singer JW, Bianco JA. Toxic epidermal necrolysis possibly linked to aztreonam in bone marrow transplant patients. Ann Pharmacother 1992;26(1):34–5.

38. Escallier F, Dalac S, Foucher JL, Lorcerie B, Lucet A, Lambert D. Pustulose exanthématique aiguë généralisée: imputabilité à l'imipénème (Tienam). [Acute generalized exanthematic pustulosis. Imputability of imipenem (Tienam).] Ann Dermatol Venereol 1989;116(5):407–9.

39. Yesudian PD, King CM. Occupational allergic contact dermatitis from meropenem. Contact Dermatitis 2001;45(1):53.

40. Saxon A, Adelman DC, Patel A, Hajdu R, Calandra GB. Imipenem cross-reactivity with penicillin in humans. J Allergy Clin Immunol 1988;82(2):213–7.

41. Hantson P, de Coninck B, Horn JL, Mahieu P. Immediate hypersensitivity to aztreonam and imipenem. BMJ 1991;302(6771):294–5.

42. Bosmuller C, Steurer W, Konigsrainer A, Willeit J, Margreiter R. Increased risk of central nervous system toxicity in patients treated with ciclosporin and imipenem/cilastatin. Nephron 1991;58(3):362–4.

43. Carmellini M, Frosini F, Filipponi F, Boggi U, Mosca F. Effect of cilastatin on cyclosporine-induced acute nephrotoxicity in kidney transplant recipients. Transplantation 1997;64(1):164–6.

44. Paterson DL, Singh N. Interactions between tacrolimus and antimicrobial agents. Clin Infect Dis 1997;25(6):1430–40.

45. De Sarro A, Ammendola D, De Sarro G. Effects of some quinolones on imipenem-induced seizures in DBA/2 mice. Gen Pharmacol 1994;25(2):369–79.

46. Semel JD, Allen N. Seizures in patients simultaneously receiving theophylline and imipenem or ciprofloxacin or metronidazole. South Med J 1991;84(4):465–8.

47. De Turck BJ, Diltoer MW, Cornelis PJ, Maes V, Spapen HD, Camu F, Huyghens LP. Lowering of plasma valproic acid concentrations during concomitant therapy with meropenem and amikacin. J Antimicrob Chemother 1998;42(4):563–4.

48. Yamagata T, Momoi MY, Murai K, Ikematsu K, Suwa K, Sakamoto K, Fujimura A. Panipenem–betamipron and decreases in serum valproic acid concentration. Ther Drug Monit 1998;20(4):396–400.

49. Nagai K, Shimizu T, Togo A, Takeya M, Yokomizo Y, Sakata Y, Matsuishi T, Kato H. Decrease in serum levels of valproic acid during treatment with a new carbapenem, panipenem/betamipron. J Antimicrob Chemother 1997;39(2):295–6.

50. Santucci M, Parmeggiani A, Riva R. Seizure worsening caused by decreased serum valproate during meropenem therapy. J Child Neurol 2005; 20(5): 456–7.

51. Coves-Orts FJ, Borrás-Blasco J, Navarro-Ruiz A, Murcia-López A, Palacios-Ortega F. Acute seizures due to a probable interaction between valproic acid and meropenem. Ann Pharmacother 2005; 39(3): 533–7.

52. Hobara N, Kameya H, Hokama N, Ohshiro S. Altered pharmacokinetics of sodium valproate by simultaneous administration of imipenem/cilastatin sodium. Jpn J Hosp Pharm 1998; 24: 464–72.

53. Kojima S, Nadai M, Kitaichi K, Wang L, Nabeshima T, Hasegawa T. Possible mechanism by which the carbapenem antibiotic panipenem decreases the concentration of valproic acid in plasma in rats. Antimicrob Agents Chemother 1998; 42: 3136–40.

54. Nakajima Y, Mizobuchi M, Nakamura M, Takagi H, Inagaki H, I, Kominami G, Koike M, Yamaguchi T. Mechanism of the drug interaction between valproic acid and carbapenem antibiotics in monkeys and rats. Drug Metab Dispos 2004: 32: 1381–91.

55. Spriet L Goyens J, Meersseman W, Wilmer A, Willems L, van Paesschen W. Interaction between valproate and meropenem: a retrospective study. Ann Pharmacol 2007; 41: 1130-6.
56. Beer JH, Vogt A, Neftel K, Cottagnoud P. False positive results for leucocytes in urine dipstick test with common antibiotics. BMJ 1996;313(7048):25.

Cephalosporins

See also Beta-lactam antibiotics

General Information

The cephalosporins represent a family of beta-lactam antibiotics originally derived from the naturally occurring cephalosporin C. Isolation of the cephalosporin C nucleus, 7-aminocephalosporanic acid, made it possible to introduce new groups into this molecule to obtain the current variety of compounds (1). Cephalosporins vary widely in their antibacterial properties, beta-lactamase stability, and pharmacokinetic behavior, but there is as yet no unequivocal classification (2). For reasons more practical than pharmacological, cephalosporins are often classified into first-, second-, third- and fourth-generation compounds, as shown in Table 1.

General adverse effects

Reactions that parallel those observed with penicillins include local reactions to parenteral administration, epileptogenicity, effects on sodium and potassium balance, autoimmune hemolytic anemia, neutropenia, thrombocytopenia, and altered platelet function.

More specifically associated with cephalosporins are false-positive Coombs' tests (also seen with clavulanic acid and imipenem + cilastatin), impaired vitamin K-dependent clotting factor synthesis with cephalosporins that contain the N-methylthiotetrazole side chain, biliary sludge formation with ceftriaxone, tubular nephrotoxicity of some older compounds (cefaloridine, cefaloglycin, and cefalotin), and

Table 1 Some cephalosporins and cephamycins (all rINNs except where stated)

Orally active cephalosporins	Injectable cephalosporins
Cefaclor[2] (pINN)	Cefamandole[2] (pINN)
Cefadroxil[3] (pINN)	Cefazolin[3] (pINN)
Cefalexin[2] (pINN)	Cefotaxime[3]
Cefetamet[3]	Cefoxitin[2]
Cefixime[3]	Ceftazidime[3]
*Cefradine[1]	Ceftizoxime[3]
*Cefuroxime[2]	Ceftriaxone[3]
	Cefuroxime[2]
	Latamoxef[3] (pINN)

[1]First-generation drugs
[2]Second-generation drugs
[3]Third-generation drugs
*Can also be given by injection

disulfiram-like interactions with alcohol. Anaphylactic shock and other IgE antibody-mediated reactions are rare, but analogous to those experienced with the penicillins. Sufficiently reliable tests to predict or prove these reactions are still lacking for the cephalosporins. Other hypersensitivity reactions include acute interstitial nephritis, the majority of drug-inducible mucocutaneous manifestations, and various combinations of symptoms often referred to as "serum sickness-like reactions."

Organs and Systems

Cardiovascular

Angina and myocardial infarction can occur in the absence of angiographically stenosed coronary arteries, because of arterial vasospasm during a drug-induced allergic reaction. This rare condition is called the Kounis syndrome, and it has been reported in a 70-year-old woman after intravenous cefuroxime (3). The authors suggested that individuals in whom there is increased mast cell degranulation may be more susceptible to this effect. In isolated cases, cefalotin (4) and cefaclor (5) have been suspected to cause hypersensitivity myocarditis.

Respiratory

Diffuse pulmonary inflammation, as documented by gallium scanning in one case, was possibly caused by ceftriaxone (6). Cefotiam and in another case cefotiam followed by ceftazidime have been suspected to have caused pulmonary hypersensitivity (7,8).

Cefotetan-induced hiccup recurred after each cefotetan infusion and disappeared immediately after withdrawal; it did not recur after administration of another antibiotic (9). Intractable hiccups in a boy were attributed to ceftriaxone-induced pseudolithiasis (10).

Nervous system

Since penicillin G-induced convulsions were observed in 1945 (11), many investigators have studied the convulsive effects of the beta-lactams, the mechanism(s) of action, and structure-activity relations. The convulsive effects of seven different cephalosporins have been studied both in vivo and in vitro, by intracerebroventricular administration in mice, labelled ^{3}H-muscimol binding in mouse brain synaptosomes, and inhibition assays in *Xenopus* oocytes (12). The rank orders of convulsive effects were cefazolin > cefoselis > cefotiam > cefpirome > cefepime> ceftazidime > cefozopran. The authors suggested that the strong correlation between this effect and affinity for muscimol binding sites implies that most cephalosporins inhibit GABA-mediated transmission by binding to some specific subunits of the GABA receptors. They speculated that cefazolin and cefoselis might recognize a different molecular motif on GABA receptors from that detected by other cephalosporins. However, they also emphasized that while these results are suggestive, they do not completely clarify the convulsive risk of these cephalosporins and that intensive pharmacokinetic and pharmacodynamic studies will be required to predict the potential convulsive risk of these antibiotics.

In rats, intracerebroventricular injection of various cephalosporins produced markedly different responses in epileptogenic potential (13), later confirmed with a total of 15 compounds (14). Compounds with two heterocyclic rings at both position 3 and position 7 of the cephalosporin molecule, for example ceftriaxone, cefoperazone, and ceftazidime, were even more epileptogenic than benzylpenicillin, while others, with only one heterocyclic ring at position 7, for example cefotaxime and cefonicid, were less potent. Cefazolin, a tetrazole derivative, similar to the convulsant phenyltetrazole, was most potent.

The convulsive effects of the cephalosporins may be caused by suppression of inhibitory neurotransmission via modulation of $GABA_A$ receptors (15). Furthermore, radioligand binding studies show that cephalosporins inhibit GABA binding (16). Consistent with this hypothesis, positive modulators of these receptors, such as benzodiazepines and barbiturates, can prevent or treat cephalosporin-induced convulsions.

Clinical reports (17,18) and experimental reports (19) continue to appear. The following two case reports can be taken as brief reminders

- A 60-year-old man with diabetes mellitus and end-stage renal disease on hemodialysis was given cefepime 2 g/day for a supposed pneumonia (20). Five days later he became confused and agitated and had visual and auditory hallucinations. His symptoms did not improve with further dialysis over 2 days, and cefepime was withdrawn. Within 1 day he started to improve, and 2 days later had regained his baseline mental status. He had had hallucinations and confusion after taking ceftazidime a year before.
- A 66-year-old woman with acute myeloid leukemia had a fever on third day of the initial chemotherapy cycle (21). She was given empirical antibiotic treatment with cefepime 2 g every 8 hours and 10 days later developed acute renal insufficiency and altered consciousness (Glasgow coma scale 6). An electroencephalogram showed generalized spike and sharp wave activity compatible with non-convulsive status. Cefepime was withdrawn and the epileptiform activity disappeared with clonazepam. She regained consciousness 48 hours after cefepime withdrawal.

The authors emphasized that neurotoxic symptoms are not uncommon in patients receiving cephalosporins and that non-convulsive status epilepticus is often underdiagnosed.

Neurotoxicity has been reported with intracerebroventricular cefazolin (22) and with systemic cefazolin (23–25), cefepime (26), cefotaxime (27–29), ceftazidime (30,31), and cefuroxime (32). Ceftazidime also caused truncal asterixis (33) and absence status and toxic hallucinations (34). Even the least epileptogenic of 15 cephalosporins, namely cefonicid (14), caused seizures (35), although this effect was disputed (36). As expected, seizures after systemic treatment were predominant in uremic patients, and neurotoxicity has been associated with intraperitoneal ceftazidime therapy in a patient undergoing CAPD (37). Of practical interest was a patient treated with intravenous cefmetazole in whom the CSF concentration was twice as high as the corresponding blood concentration (236 versus 103 µg/ml) (38). Uremia may have contributed to this unusual distribution pattern.

- A 66-year-old woman had several episodes of recurrent aseptic meningitis associated with cefalexin, cefazolin, and ceftazidime (39). Intrathecal ceftazidime-specific IgG antibodies were isolated and a skin test with cefazolin provoked recurrence of the aseptic meningitis.

Two patients developed status epilepticus during treatment with cefepime for *Pseudomonas aeruginosa* sepsis (40).

- A 44-year-old man, who had previously had bilateral lung transplantation and who was on hemodialysis for chronic renal insufficiency, was given cefepime 2 g/day. Within 24 hours he started to become confused and developed diffuse hyper-reflexia. Two days later an electroencephalogram showed nearly continuous, generalized sharp-wave/slow-wave activity. After lorazepam 2 mg the status epilepticus resolved, but he remained confused. A follow-up electroencephalogram showed recurrence of generalized sharp-wave activity. Cefepime was withdrawn, and within hours he rapidly recovered his mental status. An electroencephalogram showed absence of epileptiform discharges.
- A 28-year-old woman with a thoracic spina bifida was given cefepime 1 g bd for an infection with *P. aeruginosa*. After a time (not stated) an electroencephalogram showed a continuous generalized spike and wave pattern. She was given lorazepam 2 mg, which resulted in resolution. Cefepime was withdrawn and she promptly recovered.

In the first case the dose was inappropriate for the degree of renal impairment and in the second case the dose was inappropriate for the patient's body weight. The authors underlined the importance of giving cefepime with great care, especially to patients with renal impairment and low body weight.

Five patients developed severe symptoms after receiving cefepime (41). The patients, three men and two women, aged 16–75 years, received 2 g/day ($n = 3$) 4 g/day ($n = 1$) or 9 g/day ($n = 1$). The symptoms started 12–16 days after the start of therapy. In all cases, the initial neurological symptoms (disorientation, confusion, and reduced consciousness) were progressive and were attributed to the infection. Facial or multifocal myoclonic movements occurred subsequently and were rapidly followed by convulsive or non-convulsive status epilepticus. The dose of cefipime had not been adjusted for renal function in any of the patients. Cefepime serum concentrations were measured in three cases, and were 72, 73, and 134 µg/ml. All the patients underwent hemodialysis, and the serum concentrations of cefepime fell to 4.3, 21, and 25 µg/ml respectively. In the other two patients, the serum concentrations after dialysis were 14 and 54 µg/ml, suggesting high concentrations before dialysis. There was complete recovery in four of the patients. One, a 73-year-old woman, died of multiorgan failure with refractory status epilepticus and coma. The authors referred to four other reports of cefepime-induced generalized seizures. They seriously questioned whether the actual frequency of this complication might not be underestimated, owing to insufficient knowledge, underreporting, and/or lack of specificity of clinical features.

- An 82-year-old man on chronic hemodialysis had pneumonia, for which he was given intravenous cefepime 1 g/day (42). After 4 days he developed a seizure and cefepime was withdrawn. Hemodialysis was started and his conscious level improved. On the next day, after a second hemodialysis, he recovered completely.

There have been many other reports of neurotoxicity and neuropsychiatric adverse effects of cefepime in patients with impaired kidney function (43, 44, 45, 46, 47). It might be wise to follow a recommendation given in a recent review: "Since reports of neurotoxic effects and of an increased mortality have created some concerns regarding its safety·····.. caution in the use of cefepime should be adopted until new evidence on (its) safety is available" (48).

Seven patients developed reversible cefepime-induced encephalopathy with a peculiar electroencephalographic pattern, characterized by semiperiodic diffuse triphasic waves (49). Abnormal electroencephalography was also found in a US patient (50). Most often, these types of adverse effects occur in patents with reduced renal func tion (<<Lam 1169,51), but they can occur in patients with normal renal function (52). In a 79-year-old patient with normal renal function subtle mental status changes during cefepime therapy were shown by electroencephalography to be due to non-convulsive status epilepticus (53). In three other patients non-convulsive status epilepticus due to cefepime was associated with varying degrees of renal impairment (54). Cefepime toxicity should be suspected whenever a patient taking the drug has a change in mental status or myoclonus.

Pseudotumor cerebri has been attributed to ceftriaxone (55).

Psychological, psychiatric

It has long been known that intramuscular procaine penicillin can cause some peculiar psychological adverse reactions, and that other penicillin derivatives, such as amoxicillin, can cause psychiatric reactions, such as hallucinations (SEDA-21, 259). In a report from the Netherlands, neuropsychiatric symptoms occurred in six patients who received cefepime for febrile neutropenia (56). The patients, two men and four women, aged 32–75 years, received 6 g/day ($n = 5$) or 3 g/day ($n = 1$). The symptoms started 1–5 days after the first dose and varied from nightmares, anxiety, agitation, and visual and auditory hallucinations to coma and seizures. After withdrawal of cefepime, they recovered within 1–5 days. The causality between their neuropsychiatric symptoms and cefepime was considered as probable (WHO criteria)

because of the temporal relation, lack of other causal neurological explanations, and positive rechallenge in five patients.

The mechanisms of these adverse effects are unknown, although it might be of value to take a closer look at the theory that drug-induced limbic kindling may be the principal pathogenic factor (SEDA-21, 259).

Metabolism

Although possible interference with the metabolism of carnitine by pivaloylmethyl-esterified beta-lactams is a matter of concern (SEDA-21, 260), new similar prodrug derivatives of cephalosporins continue to be marketed, as do reports that they can be given to healthy volunteers without concern (57).

However, taking a closer look at the data, it is evident that healthy volunteers lost around 10% of their body stores of carnitine within 2 weeks of being given antibiotics containing pivalic acid (58). The authors emphasized that prolonged used of such drugs might result in profound carnitine depletion and that this depletion might be associated with clinical sequelae.

Valproate also causes urinary loss of carnitine (SEDA-12, 209), most probably by a different mechanism than pivalic acid (59). However, the combination can rapidly cause serious adverse effects (60).

- A 72-year-old woman taking valproate as monotherapy for her epilepsy developed a urinary tract infection and was given pivmecillinam 600 mg/day. During the next few days she became stuporose; her serum ammonia concentration was high (113 mmol/l) but liver function was normal. Pivmecillinam and valproate were withdrawn and she recovered rapidly.

The authors recommended caution when treating patients taking valproate with pivmecillinam because of the risk of hyperammonemic encephalopathy. It seems reasonable to assume that this caution should include all beta-lactams that incorporate pivalic acid.

However, there may be another mechanism by which cephalosporins can interfere with carnitine metabolism. Cephalosporins with a quaternary nitrogen (cefepime, cefluprenam, cefoselide, and cefaloridine) compete with carnitine for renal reabsorption due to OCNT2, a major member of the family of organic cationic transporters (61). Mutations in the OCNT2gene are responsible for the genetic disorder primary systemic carnitine deficiency (62,63). Since carnitine and the cephalosporins mentioned above compete for the same substrate-binding site on OCTN2, it is likely that such mutations will interfere with the pharmacokinetics of these drugs. Consequently these cephalosporins should not be given to patients with such mutations.

Hematologic

Since the days when chloramphenicol was more commonly used, it has been recognized that many antimicrobial drug are associated with severe blood dyscrasias, such as aplastic anemia, neutropenia, agranulocytosis, thrombocytopenia, and hemolytic anemia. Information on this association has come predominantly from case series and hospital surveys (64–66). Some evidence can be extracted from population-based studies that have focused on aplastic anemia and agranulocytosis and their association with many drugs, including antimicrobial drugs (67,68). The incidence rates of blood dyscrasias in the general population have been estimated in a cohort study with a nested case-control analysis, using data from a General Practice Research Database in Spain (69). The study population consisted of 822 048 patients aged 5–69 years who received at least one prescription (in all 1 507 307 prescriptions) for an antimicrobial drug during January 1994 to September 1998. The main outcome measure was a diagnosis of neutropenia, agranulocytosis, hemolytic anemia, thrombocytopenia, pancytopenia, or aplastic anemia. The incidence was 3.3 per 100 000 person-years in the general population. Users of antimicrobial drugs had a relative risk (RR), adjusted for age and sex, of 4.4, and patients who took more than one class of antimicrobial drug had a relative risk of 29. Among individual antimicrobial drugs, the greatest risk was with cephalosporins (RR = 14), followed by the sulfonamides (RR = 7.6) and penicillins (RR = 3.1).

Hemolytic anemia

Drug-induced immune hemolytic anemia has been reviewed (70). Twelve cephalosporins have been implicated, the most common being cefotetan and ceftriaxone. Autoimmune hemolytic anemia has rarely been reported with the older cephalosporins, including cefalexin (71), cefalotin (72,73), cefazolin (74), and cefaloridine (75). The main laboratory findings correspond to the "drug adsorption" mechanism classically found in benzylpenicillin-induced immune hemolysis. Antibodies cross-reacting with cefalotin and benzylpenicillin were found in both benzylpenicillin-induced and cefalotin-induced hemolysis (74,76) Cases have also been reported with cefamandole (77), cefalexin (78), ceftriaxone (79), cefotaxime (80,81), cefotetan (82,83) and ceftazidime (84).

Cefotetan-induced hemolytic anemia has been discussed in a review of 35 cases, eight of which were discussed in greater detail (85). All eight cases were associated with the prophylactic use of cefotetan in gynecological or obstetric procedures. The patients had received 1–4 doses of cefotetan, and they left hospital in good shape, but all were readmitted with hemolytic anemia within 2 weeks after the last dose. All needed several blood transfusions and two underwent plasmapheresis twice. They all survived, but the authors underlined the seriousness of this adverse effect. Of 43 cases of drug-induced hemolytic anemia that had been referred to their laboratory in the previous 8 years, 35 had been caused by cefotetan and three by ceftriaxone; 11 had a fatal outcome, eight and three caused by cefotetan and ceftriaxone respectively.

Hemolytic anemia has also been attributed to ceftizoxime (86).

- A 76-year-old Japanese man, who had been given 23 courses of intravenous antibiotic therapy over 2 years for chronic bronchitis, was given intravenous ceftizoxime 1 g/day. He developed anaphylactic shock and

hemolysis. Despite very extensive therapy he died 2 weeks later.

The patient's serum was tested for antibodies against five penicillins and 30 different cephems (that is all types of cephalosporins), using protocols to detect drug adsorption as well as immune-complex mechanisms. His serum contained an IgM antibody that formed immune complexes with 10 of the 30 cephems. The 10 drugs were classified as oxime-type cephalosporins, that is they had a common structural formula at the C7 position on 7-aminocephalosporinic acid. This antibody did not show any cross-reactivity with five kinds of penicillins (ampicillin, aspoxicillin, carbenicillin, piperacillin, sulbenicillin). The authors asked a difficult question: Why did anaphylactic shock accompany acute hemolysis? Their answer was that the complex of ceftizoxime with IgM anti-ceftizoxime might act like anti-A or anti-B. This hypothesis will surely be further tested. In the meantime, it would be wise not to use the newer cephalosporins too freely.

Ceftriaxone has been associated with autoimmune hemolytic anemia, erythroblastocytopenia, and acute hepatitis (87). The ceftriaxone in this case was given intravenously and not orally, as erroneously published (written communication from the authors). Other cases of hemolysis have been reported after ceftriaxone (88).

- A 38-year-old man with no known disease rapidly developed disturbed consciousness 4 days after having taken co-amoxiclav for sinusitis (89). On admission, he was given intravenous ceftriaxone 2 g bd for purulent meningitis. On the 10th day of therapy, he developed icterus and hemolytic anemia. Despite vigorous resuscitative efforts, he died with evidence of multiple organ failure on day 5.
- A 48-year-old woman who had been given ceftriaxone 2 g/day intravenously for 7 days for Lyme disease developed severe hemolytic anemia (90). She had previously been given ceftriaxone twice without any adverse effects. An immune complex mechanism was suggested.
- A 14-year-old girl, perinatally infected with HIV, had a medical history of recurrent infections that had been treated with several antibiotics, including ceftriaxone. She was given ceftriaxone (60 mg/kg intravenously) for pneumonia and 30 minutes later complained of severe back pain, became nauseated, vomited, and collapsed. Despite intensive medical care she died within a few hours with massive intravascular hemolysis and disseminated intravascular coagulopathy. Autopsy was refused.

Of 10 patients with hemolysis due to ceftriaxone, seven died, six of them children (79,88,91–93).

- A 5-year-old girl, who had taken co-trimoxazole prophylaxis for recurrent urinary tract infections since the age of 1 year, received ceftriaxone intramuscularly for 7 days at the age of 5 years, uneventfully (94). Six months later she was given intramuscular ceftriaxone 50 mg/kg/day and amikacin 20 mg/kg for a new urinary tract infection. After 3 days she became unexpectedly ill and had a generalized seizure 30 minutes after the administration of both drugs. A day later, her seizures recurred, she rapidly became worse, and she was referred to an ICU for ventilatory support. There, she had two cardiac arrests and was resuscitated successfully. Her hematocrit and reticulocyte count were eight and 0.2% respectively. A direct antiglobulin test was strongly positive. She was given methylprednisone 5 mg/kg/day and three units of red blood cells. The direct antiglobulin test became negative on day 3, and the dose of methylprednisone was reduced to 3 mg/kg/day but then had to be increased again to 5 mg/kg/day because of recurring hemolysis. The steroid was withdrawn uneventfully after 8 weeks. She remained well and had no neurological deficit.

The authors stated that in children, hemolysis usually starts within minutes to some few hours after the administration of the drug, whereas in adults it starts after a period of days.

Hemolytic anemia associated with ceftriaxone-dependent antibodies has been described in a patient with hemoglobin SC disease (95) and in one with sickle cell anemia (96). In the former it was fatal.

Mechanism
In addition to the "drug-adsorption" mechanism, the findings in some instances of anemia associated with cephalosporins were consistent with concomitant formation of autoantibodies (82) or the so-called "innocent bystander" mechanism, leading to acute intravascular hemolysis, one with ceftriaxone being fatal (79). After this, another six cases were reported, of which five were fatal (88,91,92,97–99). All were due to ceftriaxone and all were in children aged 2–14 years, who were immunocompromised and/or had a hematological disease. All had been previously exposed to ceftriaxone and in some instances also to other cephalosporins. The hemolysis occurred abruptly within 5–54 minutes. However, it is not settled whether the risk reflected by those observations is specific to ceftriaxone or occurs with all cephalosporins. Ceftriaxone has a large share of the cephalosporin market, and three of the reports (92,98,99) were stimulated by a foregoing one (91), which may point to publication bias. Cross-reactivity of ceftriaxone antibodies with other cephalosporins was not studied. The significance of the underlying immunological or hematological diseases and the youth of the patients is also uncertain. Cefalotin and other cephalosporins can cause a positive direct antiglobulin test (100,101). This phenomenon is due to non-specific serum protein adsorption on to the erythrocyte membrane and is not related to immune hemolytic processes. Detection of non-immunologically bound serum proteins improves if reagents used in the direct antiglobulin test include additional antialbumin activity (102). The phenomenon is a known source of difficulties in evaluating suspected immune hemolysis or in routine cross-matching of blood products (103). The true frequency with many individual cephalosporins is unclear, since it has not been positively sought. However, it may depend on daily doses and in particular on the duration of treatment. For example, in a study with cefepime there were positive direct antiglobulin tests in 43% (104). The mean duration of treatment was 19 days and positive direct antiglobulin tests were principally seen in patients taking

long-term treatment for osteomyelitis (mean duration of treatment 32 days). On the other hand, ceftazidime given for 9 days induced positive direct antiglobulin tests in 12% of patients (105). Positive direct antiglobulin tests can turn negative again while treatment with cephalosporins continues (104).

In a case of hemolysis associated with ceftriaxone the causative antibodies appeared to be stimulated solely by a degradation product of ceftriaxone (93).

- A 16-year-old girl with craniofacial dysplasia was given ceftriaxone 4 g/day for pneumococcal meningitis. On the seventh day of therapy she developed neck muscle spasms, dizziness, and tachypnea immediately after the administration of ceftriaxone. On the next day, the same symptoms occurred about 30 minutes after the end of the infusion of ceftriaxone. Her plasma was red and her hemoglobin had fallen to 2.4 g/dl. A direct antiglobulin test was positive only for C3d and no ceftriaxone-dependent antibodies were detected. Her serum reacted strongly with erythrocytes in the presence of ex vivo antigen related to ceftriaxone (urine samples from patients receiving ceftriaxone). All therapeutic attempts were unsuccessful and she developed renal insufficiency and died.

The authors concluded that this was the first reported case in which the causative antibodies appeared to be stimulated solely by a degradation product of ceftriaxone. Unfortunately, they were not able to characterize the degradation products. They ended their report by advising that degradation products should be taken into account in all suspected cases of drug-dependent hemocytopenia in which the antibody remains undetectable.

Neutropenia
Virtually all cephalosporins can cause neutropenia and agranulocytosis (106). This has been associated with cefapirin (107), cefepime (108), cefmenoxime (109), cefmetazole (110), ceftriaxone (111–114), moxalactam (115), and others. All of these cases were seen after high cumulative doses given in one treatment course. In one series, cefapirin-induced neutropenia occurred in five of 19 patients who took a total of 90 g or more, but not in 113 patients who took smaller cumulative doses (107). It has not been settled whether toxic mechanisms, immunological mechanisms, or both are involved.

Agranulocytosis has been attributed to cefoperazone in a patient with end-stage renal insufficiency (116).

Thrombocytopenia
Thrombocytopenia has rarely been reported, always associated with cefalotin (117–119). In one case there were drug-dependent antibodies. In two other cases the role of drug-dependent antibodies was further evaluated. In one case the antibodies only reacted with platelets in the presence of exogenous cefotetan, but not with cefotetan-coated platelets (120). In another case associated with cefamandole, antibodies cross-reacted with two cephalosporins that had a thiomethyltetrazole group at position 3 but not with other cephalosporins (121). In an additional

case, cefuroxime has been implicated (122). In about one-third of cases with cephalosporin-induced neutropenia, slight concomitant thrombocytopenia has been found (106).

Impaired hemostasis

As with other beta-lactam antibiotics, cephalosporins can cause impaired hemostasis and bleeding by altering coagulation and platelet function.

Pseudolymphoma

Pseudolymphoma leukemia (Sézary syndrome) has been reported to occur mainly after the use of phenytoin and other anticonvulsant drugs (123,124), but cefixime has also been implicated (125).

• A 48-year-old woman developed fever and cough, for which she was given oral cefixime. After 48 hours she developed an itchy diffuse erythematous maculopapular rash all over her body. Her white blood cell count was $20 \times 10^9/l$, with 8% eosinophils; 5 days later it was $12.5 \times 10^9/l$, with atypical, so-called Sézary-like, cells and 64% eosinophils. A bone marrow aspirate showed 20% atypical lymphocytes with a Sézary-like appearance, having large cerebriform nuclei. An increase in eosinophil precursors was also noted. Cefixime was withdrawn, and as she continued to be febrile with a rash and her eosinophil count increased to 72% she was given corticosteroids. Her fever fell to 37.5°C, the rash improved, and her white cell count fell. At follow-up 12 months later her leukocytosis and eosinophilia had completely resolved and she did not have any other problems.

The authors stressed that, like patients with phenytoin-induced pseudolymphoma, this patient will need long-term follow-up to differentiate true prelymphoma. This case can be taken as a reminder that every new beta-lactam coming on the market may have an adverse effects profile of its own.

Mouth

Three children receiving cefaclor developed intraoral ulcers covered with a thick pseudomembrane together with various skin lesions, suggesting a viral disease at first sight (126).

Gastrointestinal

Acute hemorrhagic colitis without pseudomembranes has been reported after oral cefuroxime (SEDA 21, 261). In mice, some cephalosporins accelerated gastric emptying, in some instances even more effectively than erythromycin or metoclopramide (127). The relevance of this to the gastrointestinal adverse effects of cephalosporins, such as nausea and vomiting, is uncertain.

Cefdinir has been reported to cause red stools in a child (128).

• A 7-month-old girl, with a history of apnea of prematurity, developed red-colored stools after taking oral cefdinir 100 mg/day for 6 days for recurrent otitis media. The stools were guaiacum negative on two separate tests. Cefdinir was withdrawn and she had had no further episodes of red stools.

The incidence of red stools in children taking cefdinir is not known, although estimates vary from 1% (129) to 10 % (130). The few reports have so far been limited to children. However, the possibility that it also may occur in adults can by no means be excluded. The mechanism is not known, but it seems to occur most often in children who are also taking iron supplements. Despite the mild nature of this effect, it can create distress and confusion among both parents and physicians.

Liver

As with all beta-lactam antibiotics, slight increases in serum transaminases and alkaline phosphatase have been reported with cephalosporins. However, in contrast to the isoxazolylpenicillins and co-amoxiclav, only very sporadic cases with more severe liver disease have occurred (131–135). Very prolonged cholestasis was reported in isolated cases with cefmenoxime (136) and ceftibuten (137).

A rare case of hepatitis has been attributed to cefprozil (138).

Biliary tract

Ceftriaxone is eliminated partly in the bile, partly in the urine and can precipitate in the gallbladder (139), causing "biliary sludge" or biliary pseudolithiasis. This effect of ceftriaxone can occur at doses as low as 50 mg/kg/day (140) but is rarely symptomatic. However, it can occasionally result in cholelithiasis, as in the case of a 7-year-old boy (141). This adverse effect has been reviewed in the context of a 53-year-old man who was given intravenous ceftriaxone 2 g every 12 hours and became jaundiced after 7 days (142). Ultrasonography showed biliary sludge and cholelithiasis without cholecystitis. Ceftriaxone was withdrawn, the jaundice subsided, and the liver function test results normalized within 14 days. The condition is defined by the presence of low amplitude echoes with absent postacoustic shadows in the gallbladder on ultrasonography and has also been called "biliary pseudolithiasis" (143). Findings consistent with the presence of cholelithiasis (high amplitude echoes with acoustic shadowing) have also been reported (144).

The frequency of biliary sludge in 37 children treated for serious infections was 40% and was lower in adults (about 20%) (143,145,146). The condition usually runs a benign course and ultrasound becomes normal after withdrawal of ceftriaxone within a mean of 15 days (143). However, clinical evidence of cholecystitis has been reported (147). Gallstones containing mostly ceftriaxone have been described (148). In a case-control study of patients treated with ceftriaxone for Lyme disease, more serious biliary complications (cholecystitis and cholelithiasis) were observed in 2%. Age over 18 years and female sex were risk factors (149).

Of 50 children who were given ceftriaxone 13 developed abnormal biliary sonography suggestive of biliary sludge (so-called pseudolithiasis) (150). Susceptibility was independent of age, weight, and sex. In another similar study, 19 of 33 children developed ultrasonographic biliary sludge within 10 days of treatment with intravenous ceftriaxone 100 mg/kg/day; all were asymptomatic (151). Ceftriaxone does not seem to predispose to subsequent gallbladder stone formation, as assessed 6 and 12 months later (152). The pathogenesis relates to ceftriaxone's high rate of biliary excretion and the subsequent formation of calcium-containing precipitates (153). Thus, apart from the risk factors mentioned above, the risk of biliary sludge may increase with the use of high doses, rapid bolus injection, gall bladder stasis, and renal insufficiency associated with enhanced biliary excretion. Routine ultrasound scans are not required in patients taking ceftriaxone. In the presence of symptoms possibly due to ceftriaxone-associated sludge or cholelithiasis, confirmed by ultrasonography, the drug should be withdrawn. Surgery is mostly unnecessary.

In neurosurgical patients the symptoms of biliary pseudolithiasis can mimic those of raised intracranial pressure (154).

Pancreas

Ceftriaxone has been reported to have caused pancreatitis (155).

- A 13-year-old boy received long-term intravenous ceftriaxone after surgical drainage of a right frontal subdural empyema secondary to sinusitis. After about 5 weeks he developed abdominal pain with profuse emesis; his serum amylase was 1133 U/l and lipase 3528 U/l. Abdominal ultrasound showed cholelithiasis, and he had an uncomplicated cholecystectomy. The material in the gallbladder was 100% ceftriaxone.

The authors ended by stating that patients taking long-term ceftriaxone may be at risk of cholelithiasis and pancreatitis. Ultrasound screening might be useful for monitoring such patients.

Urinary tract

Two of the older cephalosporins, cefaloridine and cefalotin, are nephrotoxic and have often caused renal dysfunction (156,157). The nephrotoxicity of cefaloridine is related to its unusual renal transport, resulting in higher intracellular concentrations in the proximal tubular cells than with other cephalosporins (158). Since less toxic cephalosporins are available, the clinical use of cefaloridine is no longer justified.

Cefalotin can cause two types of renal disease in man (157): acute tubular necrosis, similar to that seen with cefaloridine, although less often; and acute interstitial nephritis, often accompanied by a rash, fever, or eosinophilia, resembling the same disorder as that caused by methicillin.

Beta-lactam nephrotoxicity has been reviewed, particularly considering structure-activity relations (158). Acute proximal tubular necrosis as a consequence of beta-lactam toxicity develops in proportion to:

- uptake by the cupular cell
- acylation inactivation of mitochondrial substrate transporters

- in the case of cefaloridine, lipid peroxidation.

For example, cefaloglycin, which is no longer used, is highly nephrotoxic because of its reactivity, while cefalexin, being comparably sequestered in the tubular cells, is not (158).

Broad clinical use of second- and third-generation cephalosporins has not produced clear evidence of significant nephrotoxicity. In two groups of 10 and 17 patients treated with ceftazidime 3 or 4 g/day, falls in glomerular filtration rate of about 10 ml/minute were seen after treatment courses of 4–9 days or 7–31 days respectively (159,160). For many other cephalosporins comparable data are lacking. Furthermore, an increase in aminoglycoside nephrotoxicity have only been documented with cefalotin and cefaloridine when used in combination (156,161). Nevertheless, adjustment of cephalosporin dosages in renal insufficiency is justified, and acute tubular necrosis can occur with very high doses of cephalosporins other than cefaloridine and cefalotin (162). Isolated cases of acute interstitial nephritis with newer compounds have been related to cefaclor (163), cefamandole (164), cefotaxime (165,166), cefotetan (167), cefoxitin (168), ceftazidime (169), ceftriaxone (170), and cefuroxime (171). Ceftriaxone, on the other hand, protects against tobramycin nephrotoxicity in rats (172). In addition to biliary sludge and stone formation with ceftriaxone, occasional patients developing urinary stones have been described (173,174).

In a report of acute tubulointerstitial nephritis in a patient receiving cefdinir, it was stated that among 41 kinds of cephalosporins available in Japan, this complication has been reported with 12 (175).

- A 58-year-old woman developed acute renal insufficiency soon after a 7-day course of cefdinir 300 mg/day for acute bronchitis. Renal histology showed tubular atrophy and interstitial fibrosis accompanied by moderate lymphocyte infiltration. A lymphocyte stimulation test with cefdinir was positive. Serum creatinine concentrations continued to rise even after withdrawal of cefdinir, but steroid therapy normalized renal function.

The outcome of drug-induced acute tubulointerstitial nephritis can be life-long dialysis or renal transplantation if it is not adequately treated in time. The authors therefore concluded that it is important to look out for this complication when using any cephalosporin.

Nephrolithiasis and ceftriaxone

Ceftriaxone can cause renal stones (175,176,177). All the patients had some form of renal impairment with either anuria, a raised serum creatinine, or a dilated collecting system on imaging. One required extracorporeal shockwave lithotripsy and one required nephrostomy.

- A 14-year-old boy with severe sinusitis complicated by an epidural abscess was given intravenous ceftriaxone 4 g/day and metronidazole (178). On day 8 he developed colicky abdominal pain and vomiting. His serum creatinine had risen from 70 to 420 µmol/l. His urine output fell and he had anuria for 24 hours. A CT scan of the abdomen showed high-density material in his gallbladder and in the collecting system of both kidneys and throughout both ureters. The ceftriaxone was withdrawn. On day 9 bilateral urethral stents were placed at cystoscopy and proteinaceous toothpaste-like material was found in both ureters. After another 3 weeks a CT scan showed complete resolution of the biliary pseudolithiasis and the material in the ureters, and the serum creatinine had returned to 60 μmol/l.

Nephrolithiasis has also been attributed to ceftriaxone in a child with acute pyelonephritis (179) and in another with septic arthritis (180). In the latter the stones were composed of 90% ceftriaxone; in other words this is a between-the-eyes adverse effect of type 1 (181).

In a prospective study, 51 children with various infections received ceftriaxone (27 received 50 mg/kg/day and 24 received 100 mg/kg/day) (182). Abdominal ultrasonography after treatment showed renal stones up to 2 mm in size in four of them. Comparison of those with and without stones showed no significant differences with respect to age, sex distribution, or duration of treatment. The renal stones resolved spontaneously in three of the four cases, but were still present in one 7 months after ceftriaxone treatment.

The possible relation to daily dose or length of therapy has caused some discussion. However, nephrolithiasis has been observed in children who received less than the recommended dose (183,184), and symptoms from the kidney have been reported after 3 days of ceftriaxone therapy (185). Thus, on the background of an increasing number of reports, it seems prudent to underscore the need to reiterate the indications for using ceftriaxone in daily practice and also the need for prompt clinical and laboratory monitoring with regard to nephrolithiasis in selected patients (<<Avci 1153), using described monitoring and selection procedures (<<Avci 1069). Abdominal ultrasonography before and after ceftriaxone therapy might be wise.

Taking all reports together, it is easy to accept the authors' suggestion that nephrolithiasis secondary to ceftriaxone is generally more serious than the biliary complications , since intervention to relieve renal obstruction is often necessary. In patients receiving ceftriaxone, any impairment in renal function should be taken as a warning signal.

Skin

Generalized pustular eruptions, histologically presenting as leukocytoclastic vasculitis with neutrophils forming subcorneal pustules, have been reported with different cephalosporins, such as cefaclor (186), cefazolin (187,188), cefalexin (189), and cefradine (190).

Isolated cases of the following syndromes have been reported more recently in connection with cephalosporins:

- pseudolymphoma leukemia syndrome after cefixime presenting with itchy rash, atypical Sézary-like cells, and lymphadenopathy (125);

- pemphigus vulgaris with cefadroxil (191) and cefalexin (192);
- erythema multiforme with cefotaxime (193) and cefalotin (194);
- adult linear IgA disease associated with an erythema multiforme-like reaction with cefamandole (195);
- Stevens–Johnson syndrome with cefalexin (196);
- toxic epidermal necrolysis with cefsulodin (197) and cefazolin (198);
- photosensitivity due to intravenous ceftazidime in a patient with cystic fibrosis (199);
- fixed drug eruption with cefalexin (200);
- occupational contact allergy with ceftiofur (201);
- a rash from cefprozil in a patient with infectious mononucleosis (202).

The frequencies of rashes have been retrospectively investigated in 5923 children (203). All the children who developed a rash after treatment with one or more of the commonly used oral antibiotics were identified, 472 in all. Significantly more rashes were documented with cefaclor (4.79%) compared with sulfonamides (3.46%), penicillins (2.72%), and other cephalosporins (1.04%). Based on the numbers of patients for whom the antibiotics were prescribed, the frequencies of rashes were 12.3% for cefaclor, 8.5% for sulfonamides, 7.4% for penicillins, and 2.6% for other cephalosporins.

Telangiectasiae in a light-exposed distribution have rarely been reported with cephalosporins.

- An otherwise healthy 57-year-old man developed telangiectatic skin lesions after receiving intramuscular cefotaxime 1 g bd for 7 days for a urinary tract infection (204). He had used no other medications, and there was no history of photosensitivity or rosacea. He had several asymptomatic telangiectasiae widely distributed on the forehead and on the backs of both hands. Antinuclear antibodies and antineutrophil cytoplasmic antibodies were negative. Skin biopsy showed dilated capillaries without signs of vasculitis. A light provocation test produced telangiectatic lesions at 36 hours. Because of the relation between the administration of cefotaxime and the onset of the telangiectasiae, confirmed by light testing, cefotaxime was withdrawn, with progressive improvement and complete resolution after 2 months; rechallenge was not performed.

Iatrogenic telangiectasis is a poorly understood dermatological adverse effect of several drugs, including cephalosporins (199,205). Telangiectasiae localized to light-exposed areas, as in this case, have been described with some calcium channel blockers (206,207).

Fixed drug eruptions are unusual and are characterized by recurrent site-specific lesions each time the drug thought to be responsible is given. A ceftriaxone-associated fixed drug eruption has been reported (208). Topical provocation with ceftriaxone, both at previously affected sites and in unaffected skin, was negative, whereas re-exposure to ceftriaxone 1 g intravenously confirmed the diagnosis. The patient tolerated cefazolin and amoxicillin, suggesting that the sensitizing portion was not the beta-lactam ring.

Immunologic

Type I reactions

Immediate hypersensitivity reactions, mediated through IgE antibodies to cephalosporin determinants, are a major factor limiting their use. Early cases of anaphylaxis to cephalosporins were probably due to contamination with trace amounts of penicillin (209). These studies may therefore have over-reported cross-sensitivity.

Hypersensitivity reactions to cephalosporins have been reported in 148 children who had such reactions, mainly to cefaclor and ceftriaxone; 105 had non-immediate effects (occurring after 1 hour—mostly urticarial eruptions and maculopapular rashes) and 43 had immediate effects (occurring within 1 hour—anaphylactic shock, urticaria, and/or angioedema, and erythema) (210). None of the non-immediate reactors had positive results in patch tests and/or delayed skin tests and only one had positive responses to penicillin skin-tests reagents. Of 104 children with negative results, 95 underwent challenges, which 94 tolerated. The other reacted to cefaclor pediatric suspension, but tolerated a challenge with a cefaclor capsule. In the 43 children with immediate reactions, there was IgE-mediated hypersensitivity in 34 (79%). The authors concluded that "extremely few non-immediate manifestations with cephalosporin therapy are actually hypersensitivity reactions, whereas most immediate reactions to cephalosporins are IgE-mediated".

Typical anaphylactic reactions to beta-lactams usually involve the skin and the respiratory and cardiovascular systems, and commonly present with urticaria, angioedema, dyspnea, wheeze, dizziness, syncope, and hypotension. Gastrointestinal symptoms, including nausea, vomiting, diarrhea, and abdominal cramp, can occur, but mucosal lesions are rarely detected or indeed sought.

In a retrospective study the frequency of systemic anaphylaxis to cefaloridine, cefalotin, or cefalexin was two out of 9388 patients (0.02%) without a history of penicillin allergy, and two out of 450 patients (0.4%) with a history of penicillin allergy (211). In the first group, two of the 1983 patients treated with cefalotin accounted for the adverse event.

Two of 178 prospective patients, of whom 151 had a history of penicillin allergy but were negative on penicillin skin testing, had reactions to a cephalosporin (212). There were 27 who had a positive penicillin skin test but did not react to a cephalosporin. Similar results were found by others (213).

However, a history of penicillin allergy is often vague, and many studies have suggested that it is an unreliable indicator, which has been confirmed (214). In 62 penicillin skin-test-positive patients, cephalosporins produced only one reaction of mild urticaria and bronchospasm (215).

Primary cephalosporin allergy in patients not allergic to penicillin has been reported, but the exact frequency is not known (216,217). The true incidence of allergic reactions may differ among the cephalosporins. Several reports implicating particular compounds have been published (218–222).

An accurate molecular definition of cephalosporin allergy is not currently available. Relevant determinants of cephalosporin-induced anaphylaxis may not reside in the bicyclic core, but rather in the side chain (223,224).

Neither in vitro tests nor skin tests reliably predict cephalosporin allergy (225). The true frequency of allergic reactions in penicillin-allergic patients exposed to cephalosporins has been estimated to be 1 or 2% (226). Nevertheless, when there is a history of penicillin anaphylaxis or other severe IgE-mediated reactions, it is wise to avoid cephalosporins.

- A 27-year-old man visited a local clinic complaining of fatigue and arthralgia (227). He was given paracetamol, cefaclor, and diclofenac, and 20 minutes later developed generalized itching, oral dysesthesia, and dyspnea, followed by hypotension and loss of consciousness. He was successfully treated with adrenaline and an intravenous glucocorticoid, but 15 hours after admission developed epigastric pain, nausea, and vomiting. Gastroduodenoscopy showed generalized ulcers with white exudates and petechiae from the bulb to the second part of the duodenum; histology of the duodenal ulcers showed eosinophilic infiltration. His symptoms resolved with a proton pump inhibitor. Total IgE was 160 IU/ml. Specific IgE antibodies were negative to 11 common allergens. Aspirin intolerance was excluded by a negative inhalation challenge with sodium tolmetin and oral challenges with paracetamol and diclofenac were also negative.
- In preparation for an operation a patient received cefazolin 2 g intravenously (228). The patient had no known drug allergies and had previously received intravenous cefazolin intraoperatively 2 months before without any problems, but 45 seconds later, while fully awake, developed shortness of breath, became unconscious, and had a cardiac arrest. Successful resuscitation included endotracheal intubation, external cardiac compression, electric defibrillation, and multiple large doses of adrenaline, atropine, and sodium bicarbonate.
- A patient developed a systemic anaphylactic reaction to a subconjunctival injection of cefazolin (229). The patient did well, but required intensive therapy, including airway intubation.
- A 50-year-old man developed fever, headache, nausea, vomiting, myalgia, arthralgia, and a pruritic skin rash 4 days after the start of a course of ceftriaxone 4 g/day for 21 days for meningitis (230). His temperature was 38.8°C, heart rate 110/minute, and blood pressure 120/70 mmHg. He had extensive erythroderma, petechiae, and desquamation, edema on the face, and painful subconjunctival hyperemia and hemorrhages. His leukocyte count was $24 \times 10^9/l$ with 8% eosinophils, the C-reactive protein 98 mg/l, and the erythrocyte sedimentation rate 30 mm/hour. All other investigations were normal or negative. After treatment with prednisone 48 mg/day for 3 days the syndrome resolved.

An acute, life-threatening, anaphylactic reaction has been described in a child who received his first intravenous injection of ceftriaxone (231).

- A 3-year-old boy developed a high fever and a petechial rash. In the past he had been treated four times with amoxicillin for upper respiratory tract infections without allergic reactions. At presentation he had multiple petechiae over the trunk and limbs. There were no signs of meningeal irritation. He was given intravenous ceftriaxone 100 mg/kg and after 1 minute developed excitation and a generalized papular urticarial rash. His heart rate increased to 160/minute and the blood pressure was not measurable. He was given subcutaneous adrenaline 0.15 mg plus intravenous clemastine fumarate 2 mg, dexamethasone 3 mg, and fluids. Within 15 minutes his circulation was restored and the urticarial rash abated. Instead of ceftriaxone, he was given chloramphenicol for 7 days, and no further allergic reaction was observed. *Neisseria meningitidis*, sensitive to chloramphenicol and ceftriaxone, was cultured from his blood and spinal fluid. He was discharged well 12 days after admission. One month later, skin tests for ceftriaxone and benzylpenicillin were negative, as was a test for ceftriaxone-specific IgE. Because hypersensitivity could not be demonstrated, a controlled intravenous challenge with ceftriaxone 100 mg/kg was performed, and 20 seconds later there was again excitation and a generalized papular urticarial rash. He was treated as before and recovered within 15 minutes.

According to the authors, anaphylaxis after a single injection of ceftriaxone without previous exposure to the drug is very rare, and they referred to only one previous report (232). However, despite the fact that hypersensitivity could not be demonstrated by skin testing or the presence of ceftriaxone-specific IgE, the outcome of the challenge to ceftriaxone very clearly pointed to an anaphylactic reaction.

An anaphylactic reaction to a subconjunctival injection of cefazolin has been described.

- A 70-year-old white woman with a history of penicillin allergy underwent an uncomplicated eye operation, at the end of which she received a subconjunctival injection of cefazolin 50 mg (233). About 90 minutes postoperatively she developed acute respiratory distress and an erythematous macular rash over the face, neck, chest, forearms, and lateral thighs. The rash became urticarial without pruritus. She was given bronchodilators and intravenous glucocorticoids, with little improvement, but repeated subcutaneous injections of adrenaline gave some improvement in breathing. However, she had to be intubated, and was weaned from the ventilator only after some hours.

The main lesson from this case is that a small amount of a beta-lactam antibiotic anywhere in the body can cause life-threatening anaphylaxis.

Type III reactions

Serum sickness was first described by von Pirquet and Schick in 1905 and was regarded as a syndrome resulting from the administration of heterologous serum or other foreign proteins. The immunopathology of classic serum sickness results from antigen-antibody complex formation with a foreign protein as the antigen. Characteristic symptoms include fever, cutaneous eruptions, edema, arthralgias, and lymphadenopathy. The incidence of classical serum sickness has fallen secondary to the refinement of foreign proteins. However, a serum sickness-like reaction that is clinically similar to classical serum sickness can result from the administration of a number of non-protein drugs, such as tetracyclines, penicillins, cephalosporins (234). The reaction typically occurs within 1 month of the start of therapy and resolves after withdrawal.

Most cephalosporins have sporadically been reported to cause reactions closely resembling classical serum sickness (235,236), and intriguingly comparable syndromes were seen in the majority of a series of volunteers treated for up to 4 weeks with high doses of cefalotin and cefapirin (237). There is so far no clear evidence of drug-specific antibodies. The outcome has always been benign. Earlier suspicions that cefaclor increases the risk of serum sickness-like syndromes (238) have subsequently been confirmed (239–244). The true incidence is still a matter of debate. Data suggest that it is several-fold higher than with other cephalosporins or comparator antimicrobials, such as amoxicillin-containing products, penicillin, flucloxacillin, trimethoprim, and others (245,246). More children are affected, which might at least in part reflect preferential use of cefaclor in this age group (247,248).

Serum sickness has been attributed to cefaclor (249).

- A 2-year-old girl was given cefaclor for an infection, having been given it on one previous occasion without any adverse effects. However, on this occasion, after 6 days she developed a rash, edema, and joint swelling. Her symptoms were thought to be due to serum sickness caused by cefaclor, which was withdrawn. The rash resolved within 2 days and the arthralgia settled within 1 week.

It is still uncertain whether cefaclor-associated serum sickness-like reactions correspond to true hypersensitivity. It has indeed been suggested that they may result from inherited defects in the metabolism of reactive intermediates and may be a unique adverse reaction requiring biotransformation of the parent drug (250). A lymphocyte-based cytotoxicity assay has shown that cefaclor metabolites, as generated by murine hepatic microsomes, mediated cytotoxicity among patients with cefaclor-induced serum sickness-like disease, but not with other immediate or delayed hypersensitivity reactions (251). The positive tests were not shared with loracarbef, a carbacephem with a methylene (instead of a sulfur) substituent at position 5, but otherwise structurally identical to cefaclor. It is not known whether the chemical degradation of cefaclor in vivo (252) contributes to the pathophysiology of serum sickness-like reactions.

Drug reactions with eosinophilia and systemic symptoms

Drug reactions with eosinophilia and systemic symptoms (DRESS) can occur 3–6 weeks after drug administration. Although the clinical and biological manifestations of DRESS are characteristic, the diagnosis can be missed because of the time gap between drug consumption and the onset of symptoms and signs. Although the syndrome is rare, many drugs have been implicated, often in connection with a viral infection (253). Three reports have also implicated cefixime (254), cefadroxil (255), and ceftriaxone (256).

Second-Generation Effects

Teratogenicity

Sporadic reports of findings in cell cultures (257) and animals (258) have suggested potential second-generation effects of cephalosporins; however, no such effects are known in man.

Lactation

The safety of cefuroxime during lactation has been studied prospectively in 38 women (259). Women taking cefalexin (n = 11) were used as controls and were matched for indication, duration of treatment, and age. The frequencies of adverse effects in the infants were not significantly different (OR = 0.92; 95% CI = 0.94, 1.06). All adverse effects were minor and self-limiting and did not necessitate interruption of breast-feeding.

Drug Administration

Drug administration route

Thrombophlebitis is a common reaction to the administration of cephalosporins into peripheral veins. The use of buffered solutions mitigated the reaction with cefalotin (260). Published trials have mainly compared older cephalosporins, but the overall results are still contradictory (261,262). Pain and inflammatory reactions after intramuscular injection are also common. Ceftriaxone is probably given more often intramuscularly now than any other cephalosporin. Its local tolerability does not differ from that of other compounds (263).

Drug–Drug Interactions

ACE inhibitors

Intestinal absorption of beta-lactams occurs at least in part by an active mechanism involving a dipeptide carrier, and this pathway can result in interactions with dipeptides and tripeptides (264,265), which reduce the rate of absorption of the beta-lactams. In particular, angiotensin-converting enzyme (ACE) inhibitors, which have an oligopeptide structure, are absorbed by the same carrier (266) and interact with beta-lactams in isolated rat intestine (267). However, there might be a second site of interaction between ACE inhibitors and beta-lactams. Both groups of substances are excreted by the renal anionic transport system, and concomitant administration of both drugs sometimes results in pronounced inhibition of the elimination of beta-lactams (268). In the case of

cefalexin, it may not lead to toxic effects. However, when more toxic beta-lactams are used, the possibility of this interaction has to be kept in mind.

In rats, ACE inhibitors reduced renal cefdinir excretion, probably by competition at the tubular anionic carrier (269).

Alcohol

Cephalosporins that contain a methyltetrazole-thiol side chain (as in cefamandole, cefmenoxime, cefmetazole, cefonizid, cefoperazone, ceforanide, cefotetan, cefotiam, latamoxef, and moxalactam) and chemically similar structures (cefepime, ceftriaxone) co-administered with alcohol can produce a disulfiram-like syndrome by inhibiting aldehyde dehydrogenase (263,270,271).

Transient toxic reactions to cefepime and alcohol in the inner ear have been reported (272). Alcohol intolerance needs some time to build up and has been described for up to 5 days after the ingestion of the anti biotic (273,274). It has to be kept in mind that many pharmacological formulations contain alcohol.

Sudden but reversible bilateral labyrinthine hearing loss occurred after cefepime therapy for one-sided otitis and simultaneous intake of alcohol (272). An abortive form of the acetaldehyde syndrome was assumed.

Aminoglycoside antibiotics

There are many reports of acute renal insufficiency from combined treatment with gentamicin (or another aminoglycoside) and one of the cephalosporins (275–277). The potential nephrotoxic effect of the combinations seems to be related mainly to the nephrotoxic effect of the aminoglycosides. In contrast, there is some evidence, both experimental (278) and clinical (279), that ticarcillin may attenuate the renal toxicity of the aminoglycosides.

Calcium salts

Biliary sludge and nephrolithiasis are well known adverse effects of ceftriaxone and animal and in vitro studies of the physiochemical basis for this have suggested that when the solubility product of the ceftriaxone–calcium salt is exceeded, precipitation can occur. In France four deaths occurred in neonates who received ceftriaxone and calcium simultaneously through the same intravenous line (280). Precipitates were observed in the intravenous line or pulmonary or renal tissue. The mechanism of death was assumed to be that calcium/ceftriaxone insoluble particles had occluded small arteries and capillaries, in both the lungs and kidneys. In a fatal case in a neonate, ceftriaxone was given at different times and sites. The FDA in the USA subsequently identified four additional cases (three fatal), and crystals were found in the lungs of one patient.

The manufacturer and the FDA have now recommended that ceftriaxone should not be given to neonates under 4 weeks of age. In addition, solutions containing ceftriaxone and calcium should not be co-administered within 48 hours of each other (based on the half-life of ceftriaxone), regardless of patient age (281).

However, the last statement has given rise to some debate (282, 283). Ceftriaxone has been on the market for more than 20 years, and it has been suggested that one need not be concerned about adults. As the calcium/ceftriaxone interaction is a physicochemical interaction, occurring in vitro as well as in vivo, it seems reasonable to assume that microcrystals might be found in adults. Neonates, given the anatomical structure of their pulmonary and renal vasculatures, might be most vulnerable to this problem. However, especially in elderly people, who have reduced vascular capacity, it might be wise to take a safe approach and not to administered calcium salts within 48 hours after the last dose of ceftriaxone. A panel of experts has provided a summary; ceftriaxone should be avoided or significantly minimized in neonates (especially those treated concomitantly with intravenous calcium solutions and those with hyperbilirubinemia) and potentially restricted in the geriatric population who are being treated concomitantly with intravenous calcium (284).

Ciclosporin

Ceftriaxone reportedly increases ciclosporin blood concentrations (285).

Iron

There was a more than 10-fold reduction in AUC when cefdinir was given concomitantly with iron (286). Since the formation of a chelation complex is believed to account for this effect, similar interactions with aluminium- and magnesium-containing antacids are likely.

Metronidazole

The use of high doses of metronidazole in combination with cefamandole and clindamycin has been associated with encephalopathy (SEDA-12, 705).

Vancomycin

Ceftazidime formed microprecipitates when it was mixed with vancomycin (287).

Verapamil

Competitive albumin binding of drugs with high serum protein affinity can increase pharmacologically active unbound concentrations and enhance the metabolism of low clearance drugs. Acute verapamil toxicity with ceftriaxone and clindamycin may be explained by this mechanism (288).

Interference with Diagnostic Tests

Galactose absorption

Some cephalosporins non-competitively inhibited the active absorption of D-galactose by reducing the activity of Na/K-ATPase (289). Theoretically, this effect might interfere with galactose tolerance testing.

Jaffee Test

Certain cephalosporins (cefoxitin, cefpirome, cefacetrile, cefaloglycin, cefaloridine, cefalotin) react with alkaline picrate solution, forming a chromogen with the same spectrum of absorbance as that formed by creatinine and alkaline picrate, producing falsely high serum creatinine values (290,291). This reaction occurs at therapeutic concentrations. Other cephalosporins (cefamandole, cefazolin, cefoperazone, cefotaxime, ceftazidime, ceftriaxone, cefuroxime, latamoxef) do not interfere with this assay (292).

References

1. Abraham EP. Cephalosporins 1945–1986. Drugs 1987;34(Suppl 2):1–14.
2. Williams JD. Classification of cephalosporins. Drugs 1987;34(Suppl 2):15–22.
3. Mazarakis A, Koutsojannis CM, Kounis NG, Alexopoulos D. Cefuroxime-induced coronary artery spasm manifesting as Kounis syndrome. Acta Cardiol 2005; 60(3): 341-5.
4. Burke AP, Saenger J, Mullick F, Virmani R. Hypersensitivity myocarditis. Arch Pathol Lab Med 1991;115(8):764–9.
5. Beghetti M, Wilson GJ, Bohn D, Benson L. Hypersensitivity myocarditis caused by an allergic reaction to cefaclor. J Pediatr 1998;132(1):172–3.
6. Krasnow AZ, McNamara M, Akhtar R, Holanders E, Collier BD, Isitman AT. Cephalosporin-induced diffuse pulmonary inflammation depicted by Ga-67 scintigraphy. Clin Nucl Med 1989;14(5):379–80.
7. Irie M, Teshima H, Matsuura T, Sogawa H, Kihara H, Kubo C, Nakagawa T. [Pulmonary infiltration with eosinophilia possibly induced by cefotiam in a case of steroid-dependent asthma.]Nihon Kyobu Shikkan Gakkai Zasshi 1990;28(10):1353–8.
8. Suzuki K, Inagaki T, Adachi S, Matsuura T, Yamamoto T. [A case of ceftazidime-induced pneumonitis.]Nihon Kyobu Shikkan Gakkai Zasshi 1993;31(4):512–6.
9. Morris JT, McAllister CK. Cefotetan-induced singultus. Ann Intern Med 1992;116(6):522–3.
10. Bonioli E, Bellini C, Toma P. Pseudolithiasis and intractable hiccups in a boy receiving ceftriaxone. N Engl J Med 1994;331(22):1532.
11. Johnson HC, Walker AE. Intraventricular penicillin: a note of warning. J Am Med Assoc 1945;127:212–9.
12. Sugimoto M, Uchida I, Mashimo T, Yamazaki S, Hatano K, Ikeda F, Mochizuki Y, Terai T, Matsuoka N. Evidence for the involvement of GABA A receptor induced by cephalosporins. Neuropharmacology 2003;45:304–14.
13. De Sarro A, De Sarro GB, Ascioti C, Nistico G. Epileptogenic activity of some beta-lactam derivatives: structure–activity relationship. Neuropharmacology 1989;28(4):359–65.
14. De Sarro A, Ammendola D, Zappala M, Grasso S, De Sarro GB. Relationship between structure and convulsant properties of some beta-lactam antibiotics following intracerebroventricular microinjection in rats. Antimicrob Agents Chemother 1995;39(1):232–7.
15. Fujimoto M, Munakata M, Akaike N. Dual mechanisms of GABA-A response inhibition by beta-lactam antibiotics in the pyramidal neurones of the rat cerebral cortex. Br J Pharmacol 1995;116:3014–20.

16. Hori S, Kurioka M, Matsuda M, Shimada J. Inhibitory effect of cephalosporins on gamma-aminobutyric acid receptor binding in rat synaptic membranes. Antimicrob Agents Chemother 1985;27:650–1.

17. Chow KM, Szeto CC, Hui AC, Wong TY, Li PK. Retrospective review of neurotoxicity induced by cefepime and ceftazidime. Pharmacotherapy 2003;23:369–73.

18. Primavera A, Cocito L, Audenino D. Nonconvulsive status epilepticus during cephalosporin therapy. Neuropsychobiology 2004;49:218–22.

19. Zivanovic D, Stanojlovic O, Stojanovic J, Susic V. Induction of audiogenic seizures in imipenem/cilastatin-treated rats. Epilepsy Behav 2004;5:151–8.

20. Dakdouki GK, Al-Alwar GN. Cefepime-induced encephalopathy Int J Inf Dis. 2004;8:59–61.

21. Abanades S, Nolla J. Reversible coma secondary to cefepime neurotoxicity. Ann Pharmacother 2004; 38:606–8.

22. Manzella JP, Paul RL, Butler IL. CNS toxicity associated with intraventricular injection of cefazolin. Report of three cases. J Neurosurg 1988;68(6):970–1.

23. Herd AM, Ross CA, Bhattacharya SK. Acute confusional state with postoperative intravenous cefazolin. BMJ 1989;299(6695):393–4.

24. Josse S, Godin M, Fillastre JP. Cefazolin-induced encephalopathy in a uraemic patient. Nephron 1987;45(1):72.

25. Geyer J, Hoffler D, Demers HG, Niemeyer R. Cephalosporin-induced encephalopathy in uremic patients. Nephron 1988;48(3):237.

26. Chetaille E, Hary L, de Cagny B, Gras-Champel V, Decocq G, Andrejak M. Crises convulsives associées à un surdosage en céfépime. [Convulsive crisis associated with an overdose of cefepime.] Therapie 1998;53(2):167–8.

27. Vincent JP, Dervanian P, Bodak A. Encéphalopathie sous céfotaxime. Une observation chez un sujet âgé insuffisant rénal. [Encephalopathy caused by cefotaxime. A case in an aged patient with renal failure.] Ann Med Interne (Paris) 1989;140(4):322.

28. Pascual J, Liano F, Ortuno J. Cefotaxime-induced encephalopathy in an uremic patient. Nephron 1990;54(1):92.

29. Wroe SJ, Ellershaw JE, Whittaker JA, Richens A. Focal motor status epilepticus following treatment with azlocillin and cefotaxime. Med Toxicol 1987;2(3):233–4.

30. Douglas MA, Quandt CM, Stanley DA. Ceftazidime-induced encephalopathy in a patient with renal impairment. Arch Neurol 1988;45(9):936–7.

31. Hoffler D, Demers HG, Niemeyer R. Neurotoxizität moderner Cefalosporine. [Neurotoxicity of modern cephalosporins.] Dtsch Med Wochenschr 1986;111(5):197–8.

32. Herishanu YO, Zlotnik M, Mostoslavsky M, Podgaietski M, Frisher S, Wirguin I. Cefuroxime-induced encephalopathy. Neurology 1998;50(6):1873–5.

33. Hillsley RE, Massey EW. Truncal asterixis associated with ceftazidime, a third-generation cephalosporin. Neurology 1991;41(12):2008.

34. Jackson GD, Berkovic SF. Ceftazidime encephalopathy: absence status and toxic hallucinations. J Neurol Neurosurg Psychiatry 1992;55(4):333–4.

35. Tse CS, Madura AJ, Vera FH. Suspected cefonicid-induced seizure. Clin Pharm 1986;5(8):629.

36. Higbee M, Ramsey R, Swenson E. Cefonicid-induced seizure. Clin Pharm 1987;6(4):271–2.

37. Lye WC, Leong SO. Neurotoxicity associated with intra-peritoneal ceftazidime therapy in a CAPD patient. Perit Dial Int 1994;14(4):408–9.

38. Uchihara T, Tsukagoshi H. Myoclonic activity associated with cefmetazole, with a review of neurotoxicity of cephalosporins. Clin Neurol Neurosurg 1988;90(4):369–71.

39. Creel GB, Hurtt M. Cephalosporin-induced recurrent aseptic meningitis. Ann Neurol 1995;37(6):815–7.

40. Dixit S, Kurle P, Buyan-Dent L, Sheth RD. Status epilepticus associated with cefepime. Neurology 2000;54(11):2153–5.

41. Chatellier D, Jourdain M, Mangalaboyi J, Ader F, Chopin C, Derambure P, Fourrier F. Cefepime-induced neurotoxicity: an underestimated complication of antibiotherapy in patients with acute renal failure. Intensive Care Med 2002;28(2):214–7.

42. Ferrara N, Abete P, Giordano M, Ferrara P, Carnovale V, Leosco D, Beneduce F, Ciarambino T, Varricchio M, Rengo F. Neurotoxicity induced by cefepime in a very old hemodialysis patient. Clin Nephrol 2003;59(5):388–90.

43. Garces EO, Andrade de Anzambuja MF, da Silva D, Bragatti JA, Jacoby T, Saldanha Thomé F. Renal failure is a risk factor for cefepime-induced encephalopathy. J Nephrol 2008;21(4):526–34.

44. Sonck J, Laurey G, Verbeelen D. The neurotoxicity and safety of treatment with cefepime in patients with renal failure. Nephrol Dial Transplant 2008;23:966–70.

45. Parotte MC, Krzesinski JM. Le cas clinique du mois. Antibiotiques et patient dialyse: trois cas de toxicite neurologique du cefepime. [Clinical case of the month. Antibiotics and hemodialysis: three cases of neurotoxicity from cefepime.] Rev Med Liège 2008;63(3):119–21.

46. Bresson J, Paugam-Burtz C, Josserand J, Bardin C, Mantz J, Pease S. Cefepime overdosage with neurotoxicity recovered by high-volume haemofiltration. J Antimicrob Chemother 2008;62:849–50.

47. De Silva DA, Pan ABS, Lim SH. Cefepime-indused encephalopathy with triphasic waves in three Asian patients. Ann Acad Med 2007;6:450–1.

48. Drago L, De Vecchi E The safety of cefipime in the treatment of infections. Exp Opin Drug Saf 2007;7:377–87.

49. Bragatti JA, Rossato R, Ziomkowski S, Kliemann FA. Encefalopatia induzida por cefepime: achados clinicos e eletroencefalograficos em sete pacientes. [Cefepime-induced encephalopathy: clinical and electroencephalographic features in seven patients.] Arq Neuropsiquiatr 2005; 63(1): 87–92.

50. Lam S, Gomolin IH. Ceepime neurotoxicity: case report, pharmacokinetic considerations, and literature review. Pharmacotherapy 2006; 26: 1169–74.

51. Lin CM, Chen YM, Po HL, Hseuh IH. Acute neurological deficit caused by cefipime: a case report and review of literature. Acta Neurol Taiwan 2006; 15: 279–72.

52. Capparelli FJ, Diaz MF, Hiavnika A, Wainsztein NA, Leigurda R, Del Castillo ME. Cefepime and cefixime-induced encephalopathy in a patient with normal renal function. Neurology 2005; 65: 1840.

53. Maganti R, Jolin D, Rishi D, Biswas A. Nonconvulsive status epilepticus due to cefepime in a patient with normal renal function. Epilepsy Behav 2006; 8(1): 312–4.

54. Fernandez-Torre JL, Martinez-Martinez M, Gonzalez-Rato J, Maestro I, Alonso I, Rodrigo E, Horcajada JP. Cephalosporin-induced nonconvulsive status epilepticus: clinical and electroencephalographic features. Epilepsia 2005; 46(9): 1550–2.

55. Goenaga Sánchez MA, Sánchez Haya E, Martínez Soroa I, Millet Sampedro M, Garde Orbáiz C. [Pseudotumor cerebri probably due to ceftriaxone.] An Med Interna 2005; 22(3): 147–8.

56. Diemont W, MacKenzie M, Schaap N, Goverde G, Van Heereveld H, Hekster Y, Van Grootheest K. Neuropsychiatric symptoms during cefepime treatment. Pharm World Sci 2001;23:36.

57. Brass EP, Mayer MD, Mulford DJ, Stickler TK, Hoppel CHL. Impact on carnitine homeostasis of short-term treatment with the pivalate prodrug cefditoren pivoxil. Clin Pharmacol Ther 2003;73:338–47.

58. Abrahamsson K, Melander M, Eriksson BO, Holme E, Jodal U, Jonasson A. Transient reduction of human left ventricular mass in carnitine depletion induced by antibiotics containing pivalic acid. Br Heart J 1995;74:656–9.

59. Melegh B, Kerner J, Jaszai V, Bieber L. Differential excretion of xenobiotic acylesters of carnitine due to administration of pivampicillin and valproate. Biochem Med Metabol Biol 1990;43:30–8.

60. Lokrantz CM. Eriksson B, Rosen I, Asztely F. Hyperammonemic encephalopathy induced by a combination of valproate and pivmecillinam Acta Neurol Scand 2004;109:297–301.

61. Ganapapathy ME, Huang W, Rajan DP, Carter AL, Sugawara M, Isek K, Leiback FH, Ganapathy V. Beta-lactam antibiotics as substrates for OCTN2, an organic cation/carnitine transporter. J Biol Chem 2000;275:1699–707.

62. Nezu J, Tamai I, Oku A, Ohashi R, Yabuuchi H, Hashimoto N, Nikaido H, Say Y, Koizumi A, Shoji Y, Takada G, Matsuishi T, Yoshino M, Kato A, Ohura T, Tsujimoto G, Hayakawa J, Shimane M, Tsuji A. Primary systemic carnitine deficiency is caused by mutations in a gene encoding sodium ion-dependent carnitine transporter, Nature Genet 1999;21:91–4.

63. Burwinkel B, Kreuder J, Schweitzer S, Vorgerd M, Gempel K, Gerbitz KD, Kilimann MW. Carnitine transporter OCTN2 mutations in systemic primary carnitine deficiency: a novel Arg169Gln mutation and a recurrent Arg282ter mutation associated with an unconventional splicing abnormality. Biochem Biophys Res Commun 1999;261:484–7.

64. George JN, Raskob GE, Shah SR, Rizvi MA, Hamilton SA, Osborne S, Vondracek T. Drug-induced thrombocytopenia: a systematic review of published case reports. Ann Intern Med 1998;129(11):886–90.

65. Wright MS. Drug-induced hemolytic anemias: increasing complications to therapeutic interventions. Clin Lab Sci 1999;12(2):115–8.

66. Arneborn P, Palmblad J. Drug-induced neutropenia—a survey for Stockholm 1973–1978. Acta Med Scand 1982;212(5):289–92.

67. Baumelou E, Guiguet M, Mary JY. Epidemiology of aplastic anemia in France: a case-control study. I. Medical history and medication use. The French Cooperative Group for Epidemiological Study of Aplastic Anemia. Blood 1993;81(6):1471–8.

68. International Agranulocytosis and Aplastic Anemia Study Group. Anti-infective drug use in relation to the risk of agranulocytosis and aplastic anemia. Arch Intern Med 1989;149(5):1036–40.

69. Huerta C, Garcia Rodriguez LA. Risk of clinical blood dyscrasia in a cohort of antibiotic users. Pharmacotherapy 2002;22(5):630–6.

70. Arndt PA, Garratty G. The changing spectrum of drug-induced immune hemolytic anemia. Semin Hematol 2005; 42(3): 137–44.

71. Forbes CD, Craig JA, Mitchell R, McNicol GP. Acute intravascular haemolysis associated with cephalexin therapy. Postgrad Med J 1972;48(557):186–8.

72. Gralnick HR, McGinniss M, Elton W, McCurdy P. Hemolytic anemia associated with cephalothin. JAMA 1971;217(9):1193–7.

73. Jeannet M, Bloch A, Dayer JM, Farquet JJ, Girard JP, Cruchaud A. Cephalothin-induced immune hemolytic anemia. Acta Haematol 1976;55(2):109–17.

74. Moake JL, Butler CF, Hewell GM, Cheek J, Spruell MA. Hemolysis induced by cefazolin and cephalothin in a patient with penicillin sensitivity. Transfusion 1978;18(3):369–73.

75. Kaplan K, Reisberg B, Weinstein L. Cephaloridine. Studies of therapeutic activity and untoward effects. Arch Intern Med 1968;121(1):17–23.

76. Nesmith LW, Davis JW. Hemolytic anemia caused by penicillin. Report of a case in which antipenicillin antibodies cross-reacted with cephalothin sodium. JAMA 1968;203(1):27–30.

77. Branch DR, Berkowitz LR, Becker RL, Robinson J, Martin M, Gallagher MT, Petz LD. Extravascular hemolysis following the administration of cefamandole. Am J Hematol 1985;18(2):213–9.

78. Manoharan A, Kot T. Cephalexin-induced haemolytic anaemia. Med J Aust 1987;147(4):202.

79. Garratty G, Postoway N, Schwellenbach J, McMahill PC. A fatal case of ceftriaxone (Rocephin)-induced hemolytic anemia associated with intravascular immune hemolysis. Transfusion 1991;31(2):176–9.

80. Shulman IA, Arndt PA, McGehee W, Garratty G. Cefotaxime-induced immune hemolytic anemia due to antibodies reacting in vitro by more than one mechanism. Transfusion 1990;30(3):263–6.

81. Salama A, Gottsche B, Schleiffer T, Mueller-Eckhardt C. 'Immune complex' mediated intravascular hemolysis due to IgM cephalosporin-dependent antibody. Transfusion 1987;27(6):460–3.

82. Chenoweth CE, Judd WJ, Steiner EA, Kauffman CA. Cefotetan-induced immune hemolytic anemia. Clin Infect Dis 1992;15(5):863–5.

83. Eckrich RJ, Fox S, Mallory D. Cefotetan-induced immune hemolytic anemia due to the drug-adsorption mechanism. Immunohematol 1994;10(2):51–4.

84. Chambers LA, Donovan LM, Kruskall MS. Ceftazidime-induced hemolysis in a patient with drug-dependent antibodies reactive by immune complex and drug adsorption mechanisms. Am J Clin Pathol 1991;95(3):393–6.

85. Garratty G, Leger RM, Arndt PA. Severe immune hemolytic anemia associated with prophylactic use of cefotetan in obstetric and gynecologic procedures. Am J Obstet Gynecol 1999;181(1):103–4.

86. Endoh T, Yagihashi A, Sasaki M, Watanabe N. Ceftizoxime-induced hemolysis due to immune complexes: case report and determination of the epitope responsible for immune complex-mediated hemolysis. Transfusion 1999;39(3):306–9.

87. Longo F, Hastier P, Buckley MJ, Chichmanian RM, Delmont JP. Acute hepatitis, autoimmune hemolytic anemia, and erythroblastocytopenia induced by ceftriaxone. Am J Gastroenterol 1998;93(5):836–7.

88. Moallem HJ, Garratty G, Wakeham M, Dial S, Oligario A, Gondi A, Rao SP, Fikrig S. Ceftriaxone-related fatal hemolysis in an adolescent with perinatally acquired human immunodeficiency virus infection. J Pediatr 1998;133(2):279–81.

89. Punar M, Ozsut H, Eraksoy H, Calangu S, Dilmener M. An adult case of fatal hemolysis induced by ceftriaxone. Clin Microbiol Infect 1999;5(9):585–6.

90. Maraspin V, Lotric-Furlan S, Strle F. Ceftriaxone associated hemolysis. Wien Klin Wochenschr 1999;111(9):368–70.

91. Bernini JC, Mustafa MM, Sutor LJ, Buchanan GR. Fatal hemolysis induced by ceftriaxone in a child with sickle cell anemia. J Pediatr 1995;126(5 Pt 1):813–5.

92. Scimeca PG, Weinblatt ME, Boxer R. Hemolysis after treatment with ceftriaxone. J Pediatr 1996;128(1):163.

93. Meyer O, Hackstein H, Hoppe B, Gobel FJ, Bein G, Salama A. Fatal immune haemolysis due to a degradation product of ceftriaxone. Br J Haematol 1999;105(4):1084–5.

94. Citak A, Garratty G, Ucsel R, Karabocuoglu M, Uzel N. Ceftriaxone-induced haemolytic anaemia in a child with no immune deficiency or haematological disease. J Paediatr Child Health 2002;38(2):209–10.

95. Bell MJ, Stockwell DC, Luban NL, Shirey RS, Shaak L, Ness PM, Wong EC. Ceftriaxone-induced hemolytic anemia and hepatitis in an adolescent with hemoglobin SC disease. Pediatr Crit Care Med 2005; 6(3): 363-6.

96. Corso M, Ravindranath TM. Albuterol-induced myocardial ischemia in sickle cell anemia after hemolysis from ceftriaxone administration. Pediatr Emerg Care 2005; 21(2): 99-101.

97. Lo G, Higginbottom P. Ceftriaxone induced hemolytic anemia. Transfusion 1993;33(Suppl):25S.

98. Lascari AD, Amyot K. Fatal hemolysis caused by ceftriaxone. J Pediatr 1995;126(5 Pt 1):816–7.

99. Borgna-Pignatti C, Bezzi TM, Reverberi R. Fatal ceftriaxone induced hemolysis in a child with acquired immunodeficiency syndrome. Pediatr Infect Dis J 1995;14(12):1116–7.

100. Molthan L, Reidenberg MM, Eichman MF. Positive direct Coombs tests due to cephalothin. N Engl J Med 1967;277(3):123 5.

101. Garratty G. Review: Immune hemolytic anemia and/or positive direct antiglobulin tests caused by drugs. Immunohematol 1994;10(2):41–50.

102. Petz LD, Garratty G. Acquired Immune Hemolytic AnemiasNew York: Churchill Livingstone;. 1980.

103. Williams ME, Thomas D, Harman CP, Mintz PD, Donowitz GR. Positive direct antiglobulin tests due to clavulanic acid. Antimicrob Agents Chemother 1985;27(1):125–7.

104. Jauregui L, Matzke D, Scott M, Minns P, Hageage G. Cefepime as treatment for osteomyelitis and other severe bacterial infections. J Antimicrob Chemother 1993;32(Suppl B):141–9.

105. Joshi M, Bernstein J, Solomkin J, Wester BA, Kuye OPiperacillin/Tazobactam Nosocomial Pneumonia Study Group. Piperacillin/tazobactam plus tobramycin versus ceftazidime plus tobramycin for the treatment of patients with nosocomial lower respiratory tract infection. J Antimicrob Chemother 1999;43(3):389–97.

106. Neftel KA, Hauser SP, Muller MR. Inhibition of granulopoiesis in vivo and in vitro by beta-lactam antibiotics. J Infect Dis 1985;152(1):90–8.

107. Vidal Pan C, Gonzalez Quintela A, Roman Garcia J, Millan I, Martin Martin F, Moya Mir M. Cephapirin-induced neutropenia. Chemotherapy 1989;35(6):449–53.

108. Dahlgren AF. Two cases of possible cefepime-induced neutropenia. Am J Health Syst Pharm 1997;54(22):2621–2.

109. Lucht F, Guy C, Perrot JL, et al. Agranulocytose à la cefménoxime. Therapie 1988;43:506.

110. Sugimoto M, Saito K, Hashimoto M, Horie S, Wakabayashi Y, Hirose S, Murata Y, Kobayashi M.

[Antibiotics-induced agranulocytosis. Patient's IgG inhibits a GM colony formation.]Rinsho Ketsueki 1989;30(5):768–73.

111. Rey D, Martin T, Albert A, Pasquali JL. Ceftriaxone-induced granulopenia related to a peculiar mechanism of granulopoiesis inhibition. Am J Med 1989;87(5):591–2.

112. Baciewicz AM, Skiest DJ, Weinshel EL. Ceftriaxone-associated neutropenia. Drug Intell Clin Pharm 1988;22(10):826–7.

113. Becq-Giraudon B, Cazenave F, Breux JP. Agranulocytose aiguë réversible au cours d'un traitement par la ceftriaxone. [Acute reversible agranulocytosis during ceftriaxone treatment.] Pathol Biol (Paris) 1986;34(5):534–5.

114. Osterwalder P, Stocker D. Agranulozytose. [Agranulocytosis.] Schweiz Rundsch Med Prax 1998;87(33):1030–3.

115. Miyano T, Kawauchi K, Suto Y, Yokoyama M, Sato Y. [Latamoxef-induced agranulocytosis—direct inhibition of CFU-C by in vitro colony assay.]Rinsho Ketsueki 1988;29(2):174–9.

116. Balkarova OV, Fomin VV, Kozlovskaia LV. [Agranulocytosis due to cefoperazon treatment in a patient with terminal renal failure.] Ter Arkh 2005; 77(6): 76-8.

117. Gralnick HR, McGinniss M, Halterman R. Thrombocytopenia with sodium cephalothin therapy. Ann Intern Med 1972;77(3):401–4.

118. Sheiman L, Spielvogel AR, Horowitz HI. Thrombocytopenia caused by cephalothin sodium. Occurrence in a penicillin-sensitive individual. JAMA 1968;203(8):601–3.

119. Naraqi S, Raiser M. Nonrecurrence of cephalothin-associated granulocytopenia and thrombocytopenia. J Infect Dis 1982;145(2):281.

120. Christie DJ, Lennon SS, Drew RL, Swinehart CD. Cefotetan-induced immunologic thrombocytopenia. Br J Haematol 1988;70(4):423–6.

121. Lown JA, Barr AL. Immune thrombocytopenia induced by cephalosporins specific for thiomethyltetrazole side chain. J Clin Pathol 1987;40(6):700–1.

122. Aitken P, Zaidi SM. Cefuroxime-induced thrombocytopenia? Postgrad Med J 1996;72(854):757–8.

123. Rosenthal CJ, Noguera CA, Coppola A, Kapelner SN. Pseudolymphoma with mycosis fungoides manifestations, hyperresponsiveness to diphenylhydantoin, and lymphocyte disregulation. Cancer 1982;49(11):2305–14.

124. D'Incan M, Souteyrand P, Bignon YJ, Fonck Y, Roger H. Hydantoin-induced cutaneous pseudolymphoma with clinical, pathologic, and immunologic aspects of Sézary syndrome. Arch Dermatol 1992;128(10):1371–4.

125. Jabbar A, Siddique T. A case of pseudolymphoma leukaemia syndrome following cefixime. Br J Haematol 1998;101(1):209.

126. Blignaut E. Cefaclor associated with intra-oral ulceration. S Afr Med J 1990;77(8):426–7.

127. Kuo WH, Wadwa KS, Ferris CD. Cephalosporin antibiotics accelerate gastric emptying in mice. Dig Dis Sci 1998;43(8):1690–4.

128. Graves R, Weaver SP. Cefdinir-associated "bloody stools" in an infant. J Am Board Fam Med 2008;21:246-8P.

129. Gilbert DN, Moellering RC, Eliopoulos GM, Sande MA. The Standford Guide to Antimicrobial Therapy 2007. 17th edition. Speryville (VA): Antimicrobial Inc, 2007:87.

130. Bowlware KL, McCracken GH Jr, Lozano-Hernandez J, Ghaffar F. Cefdinir pharmacokinetics and tolerability in children receiving 25 mg/kg once daily. Pediatr Infect Dis J 2006;25(3):208–10.

131. Horsmans Y, Larrey D, Pessayre D, Benhamou JP. Hépatotoxicité des médicaments anti-infectieux. [Hepatotoxicity of antimicrobial agents.] Gastroenterol Clin Biol 1990;14(12):911–24.

132. Vial T, Biour M, Descotes J, Trepo C. Antibiotic-associated hepatitis: update from 1990. Ann Pharmacother 1997;31(2):204–20.

133. Ammann R, Neftel K, Hardmeier T, Reinhardt M. Cephalosporin-induced cholestatic jaundice. Lancet 1982;2(8293):336–7.

134. Kojima N, Kumamoto I, Masumoto T, Onji M. A case report of drug-induced allergic hepatitis probably due to the N-methyltetrazolethiol group cephalosporin. Arerugi 1994;43(3):511–4.

135. Combe C, Banas B, Zoller WG, Manns MP, Schlondorff D. Antibiotikumreduzierte prolongierte Cholestase: Verdacht der Auslösung durch Ceftibuten. [Antibiotic-induced prolonged cholestasis: suspected induction by ceftibuten.] Z Gastroenterol 1996;34(7):434–7.

136. Kajii N, Matsuda S, Okazaki M, Nishizaki Y, Ohmura R, Fukumoto Y, Harada T. [A case of granulomatous hepatitis caused by administration of antibiotics.]Nippon Shokakibyo Gakkai Zasshi 1993;90(3):710–4.

137. Schaad UB, Wedgwood-Krucko J, Tschaeppeler H. Reversible ceftriaxone-associated biliary pseudolithiasis in children. Lancet 1988;2(8625):1411–3.

138. Bilici A, Karaduman M, Cankir Z. A rare case of hepatitis associated with cefprozil therapy. Scan J Infect Dis 2007;39:190–2.

139. Schaad UB, Tschappeler H, Lentze MJ. Transient formation of precipitations in the gallbladder associated with ceftriaxone therapy. Pediatr Infect Dis 1986;5:708–10.

140. Avci Z, Karadag A, Odemis E, Catal F. Re: Pediatric ceftriaxone nephrolithiasis. J Urol 2005; 174(3): 1153.

141. Costa DL, Barbosa MD, Barbosa MT. [Cholelithiasis associated with the use of ceftriaxone.] Rev Soc Bras Med Trop 2005; 38(6): 521-3.

142. Bickford CL, Spencer AP. Biliary sludge and hyperbilirubinemia associated with ceftriaxone in an adult: case report and review of the literature. Pharmacotherapy 2005; 25(10): 1389-95.

143. Sahni PS, Patel PJ, Kolawole TM, Malabarey T, Chowdhury D, el-Rashed Gorish M. Ultrasound of ceftriaxone-associated reversible cholelithiasis. Eur J Radiol 1994;18(2):142–5.

144. Heim-Duthoy KL, Caperton EM, Pollock R, Matzke GR, Enthoven D, Peterson PK. Apparent biliary pseudolithiasis during ceftriaxone therapy. Antimicrob Agents Chemother 1990;34(6):1146–9.

145. Pigrau C, Pahissa A, Gropper S, Sureda D, Martinez Vazquez JM. Ceftriaxone-associated biliary pseudolithiasis in adults. Lancet 1989;2(8655):165.

146. Jacobs RF. Ceftriaxone-associated cholecystitis. Pediatr Infect Dis J 1988;7(6):434–6.

147. Lopez AJ, O'Keefe P, Morrissey M, Pickleman J. Ceftriaxone-induced cholelithiasis. Ann Intern Med 1991;115(9):712–4.

148. Genese C, Finelli L, Parkin W, Spitalny KC. From the Centers for Disease Control and Prevention. Ceftriaxone-associated biliary complications of treatment of suspected disseminated Lyme disease—New Jersey, 1990–1992. JAMA 1993;269(8):979–80.

149. Cometta A, Gallot-Lavallee-Villars S, Iten A, Cantoni L, Anderegg A, Gonvers JJ, Glauser MP. Incidence of gallbladder lithiasis after ceftriaxone treatment. J Antimicrob Chemother 1990;25(4):689–95.

150. Ceran C, Oztoprak I, Cankorkmaz L, Gumuş C, Yildiz T, Koyluoglu G. Ceftriaxone-associated biliary pseudolithiasis in paediatric surgical patients. Int J Antimicrob Agents 2005; 25(3): 256-9.

151. Ozturk A, Kaya M, Zeyrek D, Ozturk E, Kat N, Ziylan SZ. Ultrasonographic findings in ceftriaxone: associated biliary sludge and pseudolithiasis in children. Acta Radiol 2005; 46(1): 112-6.

152. Shiffman ML, Keith FB, Moore EW. Pathogenesis of ceftriaxone-associated biliary sludge. In vitro studies of calcium-ceftriaxone binding and solubility. Gastroenterology 1990;99(6):1772–8.

153. Maranan MC, Gerber SI, Miller GG. Gallstone pancreatitis caused by ceftriaxone. Pediatr Infect Dis J 1998;17(7):662–3.

154. Evliyaoglu C, Kizartici T, Bademci G, Unal B, Keskil S. Ceftriaxone-induced symptomatic pseudolithiasis mimicking ICP elevation. Zentralbl Neurochir 2005; 66(2): 92-4.

155. Zhanel GG. Cephalosporin-induced nephrotoxicity: does it exist? DICP 1990;24(3):262–5.

156. Foord RD. Cephaloridine, cephalothin and the kidney. J Antimicrob Chemother 1975;1(Suppl 3):119–33.

157. Tune BM. Nephrotoxicity of beta-lactam antibiotics: mechanisms and strategies for prevention. Pediatr Nephrol 1997;11(6):768–72.

158. Norrby SR, Burman LA, Linderholm H, Trollfors B. Ceftazidime: pharmacokinetics in patients and effects on the renal function. J Antimicrob Chemother 1982;10(3):199–206.

159. Alestig K, Trollfors B, Andersson R, Olaison L, Suurkula M, Norrby SR. Ceftazidime and renal function. J Antimicrob Chemother 1984;13(2):177–81.

160. Rankin GO, Sutherland CH. Nephrotoxicity of aminoglycosides and cephalosporins in combination. Adverse Drug React Acute Poisoning Rev 1989;8(2):73–88.

161. Lentnek AL, Rosenworcel E, Kidd L. Acute tubular necrosis following high-dose cefamandole therapy for Hemophilus parainfluenzae endocarditis. Am J Med Sci 1981;281(3):164–8.

162. Pommer W, Krause PH, Berg PA, Neumayer HH, Mihatsch MJ, Molzahn M. Acute interstitial nephritis and non-oliguric renal failure after cefaclor treatment. Klin Wochenschr 1986;64(6):290–3.

163. Csanyi P, Rado JP, Hormay M. Acute renal failure due to cephamandole. BMJ (Clin Res Ed) 1988;296(6619):433.

164. Grcevska L, Polenakovic M. Second attack of acute tubulointerstitionephritis induced by cefataxim and pregnancy. Nephron 1996;72(2):354–5.

165. al Shohaib S, Satti MS, Abunijem Z. Acute interstitial nephritis due to cefotaxime. Nephron 1996;73(4):725.

166. Nguyen VD, Nagelberg H, Agarwal BN. Acute interstitial nephritis associated with cefotetan therapy. Am J Kidney Dis 1990;16(3):259–61.

167. Toll LL, Lee M, Sharifi R. Cefoxitin-induced interstitial nephritis. South Med J 1987;80(2):274–5.

168. Ladagnous JF, Rousseau JM, Gaucher A, Saissy JM, Pitti R. Néphrite interstitielle aiguë: rôle de la ceftazidime. [Acute interstitial nephritis. Role of ceftazidime.] Ann Fr Anesth Reanim 1996;15(5):677–80.

169. Mancini S, Iacovoni R, Fierimonte V, Spaini A, Parisi MC, Pichi A. Nefrite interstiziale da farmaci. Descrizione di un caso clinico. [Drug-induced interstitial nephritis. A case report.] Minerva Pediatr 1994;46(12):557–60.

170. Goddard JK, Janning SW, Gass JS, Wilson RF. Cefuroxime-induced acute renal failure. Pharmacotherapy 1994;14(4):488–91.

171. Beauchamp D, Theriault G, Grenier L, Gourde P, Perron S, Bergeron Y, Fontaine L, Bergeron MG. Ceftriaxone protects against tobramycin nephrotoxicity. Antimicrob Agents Chemother 1994;38(4):750–6.

172. Schaad UB, Suter S, Gianella-Borradori A, Pfenninger J, Auckenthaler R, Bernath O, Cheseaux JJ, Wedgwood J. A comparison of ceftriaxone and cefuroxime for the treatment of bacterial meningitis in children. N Engl J Med 1990;322(3):141–7.

173. Cochat P, Cochat N, Jouvenet M, Floret D, Wright C, Martin X, Vallon JJ, David L. Ceftriaxone-associated nephrolithiasis. Nephrol Dial Transplant 1990;5(11):974–6.

174. Diemont W, MacKenzie M, Schaap N, Goverde G, van Heereveld H, Hekster Y, van Grootheest K. Neuropsychiatric symptoms during cefepime treatment. Pharm World Sci 2001;23(1):36.

175. Shaad UB, Tschaeppler H,Lentze MJ. Transient formation of precipitations in the gallbladder associated with ceftriaxone therapy. Pediatric Infect Dis 1986;5:708–10.

176. De More RA, Egberts ACG, Schroder CH. Ceftriaxone-associated nephrolithiasis and biliary pseudolithiasis. Eur J Pediatr 1999;158:975–7.

177. Grasberger H, Otto B, Loeschke K. Ceftriaxone-associated nephrolithiasis. Ann Pharmacother 2000;34:1076–7.

178. Prince JS, Senac MO. Ceftriaxone-associated nephrolithiasis and biliary pseudolithiasis in a child. Pediatr Radiol 2003;33:648–51.

179. Tasic V, Sofijanova A, Avramoski V. Nephrolithiasis in a child with acute pyelonephritis. Ceftriaxone-induced nephrolithiasis and biliary pseudolithiasis. Pediatr Nephrol 2005; 20(10): 1510-1, 1512-3.

180. Gargollo PC, Barnewolt CE, Diamond DA. Pediatric ceftriaxone nephrolithiasis. J Urol 2005; 173(2): 577-8.

181. Aronson JK, Hauben M. Anecdotes that provide definitive evidence. BMJ 2006; 333(7581): 1267-9.

182. Avci Z, Koktener A, Uras N, Catal F, Karadag A, Tekin O, Degirmencioglu H, Baskin A. Nephrolithiasis associated with cetritraxone therapy: a prospective study in 51 children. Arch Dis Child 2004;89:1069–72.

183. Gargollo P, Barnevolt CE, Diamond DA. Pediatric ceftriaxone nephrolithiasis J Urol 2005;173:577–8.

184. Avci Z, Karadag A. Pediatric cefrixone nephrolithiasis. J Urol 2005;174:1153.

185. Acun C, Oktay Erdem L, Sogut A, Zuhal Erdem C, Tomac N, Gundogdo S. Ceftriaxone-induced biliary pseudolithiasis and urinary bladder sludge. Pediatr Int. 2004;46:368–70.

186. Ogoshi M, Yamada Y, Tani M. Acute generalized exanthematic pustulosis induced by cefaclor and acetazolamide. Dermatology 1992;184(2):142–4.

187. Stough D, Guin JD, Baker GF, Haynie L. Pustular eruptions following administration of cefazolin: a possible interaction with methyldopa. J Am Acad Dermatol 1987;16(5 Pt 1):1051–2.

188. Fayol J, Bernard P, Bonnetblanc JM. Pustular eruption following administration of cefazolin: a second case report. J Am Acad Dermatol 1988;19(3):571.

189. Jackson H, Vion B, Levy PM. Generalized eruptive pustular drug rash due to cephalexin. Dermatologica 1988;177(5):292–4.

190. Kalb RE, Grossman ME. Pustular eruption following administration of cephradine. Cutis 1986;38(1):58–60.

191. Wilson JP, Koren JF, Daniel RC 3rd, Chapman SW. Cefadroxil-induced ampicillin-exacerbated pemphigus vulgaris: case report and review of the literature. Drug Intell Clin Pharm 1986;20(3):219–23.

192. Wolf R, Dechner E, Ophir J, Brenner S. Cephalexin. A nonthiol drug that may induce pemphigus vulgaris. Int J Dermatol 1991;30(3):213–5.

193. Green ST, Natarajan S, Campbell JC. Erythema multiforme following cefotaxime therapy. Postgrad Med J 1986;62(727):415.

194. Munoz D, Del Pozo MD, Audicana M, Fernandez E, Fernandez De Corres LF. Erythema-multiforme-like eruption from antibiotics of 3 different groups. Contact Dermatitis 1996;34(3):227–8.

195. Argenyi ZB, Bergfeld WF, Valenzuela R, McMahon JT, Taylor JS. Adult linear IgA disease associated with an erythema multiforme-like drug reaction. Cleve Clin J Med 1987;54(5):445–50.

196. Murray KM, Camp MS. Cephalexin-induced Stevens–Johnson syndrome. Ann Pharmacother 1992;26(10):1230–3.

197. Okano M, Kitano Y, Ohzono K. Toxic epidermal necrolysis due to cephem. Int J Dermatol 1988;27(3):183–4.

198. Julsrud ME. Toxic epidermal necrolysis. J Foot Ankle Surg 1994;33(3):255–9.

199. Vinks SA, Heijerman HG, de Jonge P, Bakker W. Photosensitivity due to ambulatory intravenous ceftazidime in cystic fibrosis patient. Lancet 1993;341(8854):1221–2.

200. Baran R, Perrin C. Fixed-drug eruption presenting as an acute paronychia. Dr J Dermatol 1991;125(6):592–5.

201. Garcia F, Juste S, Garces MM, Carretero P, Blanco J, Herrero D, Perez R. Occupational allergic contact dermatitis from ceftiofur without cross-sensitivity. Contact Dermatitis 1998;39(5):260.

202. Baciewicz AM, Chandra R. Cefprozil-induced rash in infectious mononucleosis. Ann Pharmacother 2005; 39(5): 974-5.

203. Ibia EO, Schwartz RH, Wiedermann BL. Antibiotic rashes in children: a survey in a private practice setting. Arch Dermatol 2000;136(7):849–54.

204. Borgia F, Vaccaro M, Guarneri F, Cannavo SP. Photodistributed telangiectasia following use of cefotaxime. Br J Dermatol 2000;143(3):674–5.

205. Flax SH, Uhle P. Photo recall-like phenomenon following the use of cefazolin and gentamicin sulfate. Cutis 1990;46(1):59–61.

206. Collins P, Ferguson J. Photodistributed nifedipine-induced facial telangiectasia. Br J Dermatol 1993;129(5):630–3.

207. Basarab T, Yu R, Jones RR. Calcium antagonist-induced photo-exposed telangiectasia. Br J Dermatol 1997;136(6):974–5.

208. Ozkaya E, Mirzoyeva L, Jhaish MS. Ceftriaxone-induced fixed drug eruption: first report. Am J Clin Dermatol 2008;9:345–7.

209. Pedersen-Bjergaard J. Cephalotin in the treatment of penicillin-sensitive patients. Acta Allergol 1967;22:299.

210. Romano A, Gaeta F, Valluzi RL, Alonzi C, Viola M, Bousquet P. Diagnosing hypersensitivity reactions to cephalosporins in children. Pediatrics 2008;122:521–7.

211. Petz LD. Immunologic reactions of humans to cephalosporins. Postgrad Med J 1971;47(Suppl):64–9.

212. Solley GO, Gleich GJ, Van Dellen RG. Penicillin allergy: clinical experience with a battery of skin-test reagents. J Allergy Clin Immunol 1982;69(2):238–44.

213. Van Arsdel PP Jr, Miller S. Antimicrobial treatment of patients with a penicillin allergy history. J Allergy Clin Immunol 1990;85:188.

214. Surtees SJ, Stockton MG, Gietzen TW. Allergy to penicillin: fable or fact? BMJ 1991;302(6784):1051–2.
215. Saxon A, Beall GN, Rohr AS, Adelman DC. Immediate hypersensitivity reactions to beta-lactam antibiotics. Ann Intern Med 1987;107(2):204–15.
216. Abraham GN, Petz LD, Fudenberg HH. Cephalothin hypersensitivity associated with anti-cephalothin antibodies. Int Arch Allergy Appl Immunol 1968;34(1):65–74.
217. Ong R, Sullivan T. Detection and characterization of human IgE to cephalosporin determinants. J Allergy Clin Immunol 1988;81:222.
218. Nishioka K, Katayama I, Kobayashi Y, Takijiri C. Anaphylaxis due to cefaclor hypersensitivity. J Dermatol 1986;13(3):226–7.
219. Hama R, Mori K. High incidence of anaphylactic reactions to cefaclor. Lancet 1988;1(8598):1331.
220. Levine LR. Quantitative comparison of adverse reactions to cefaclor vs. amoxicillin in a surveillance study. Pediatr Infect Dis 1985;4(4):358–61.
221. Bloomberg RJ. Cefotetan-induced anaphylaxis. Am J Obstet Gynecol 1988;159(1):125–6.
222. Hashimoto Y, Soeda A, Takarada M, Tanioka H. Anaphylaxis to moxalactam: report of a case. J Oral Maxillofac Surg 1990;48(9):1004–6.
223. Blanca M, Fernandez J, Miranda A, Terrados S, Torres MJ, Vega JM, Avila MJ, Perez E, Garcia JJ, Suau R. Cross-reactivity between penicillins and cephalosporins: clinical and immunologic studies. J Allergy Clin Immunol 1989;83(2 Pt 1):381–5.
224. Anderson JA. Cross-sensitivity to cephalosporins in patients allergic to penicillin. Pediatr Infect Dis 1986;5(5):557–61.
225. Saxon A, Beall GN, Rohr AS, et al. Immediate hypersensitivity reactions to beta-lactam antibiotics. Urology 1988;31(Suppl):14.
226. Saxon A. Antibiotic choices for the penicillin-allergic patient. Postgrad Med 1988;83(4):135–8141–2, 147–8.
227. Shirai T, Mori M, Uotani T, Chida K. Gastrointestinal disorders in anaphylaxis. Intern Med 2007; 46(6): 315-6.
228. Kuczkowski KM. Anaphylaxis and anesthesia. Minerva Anestesiol 2004;71:55.
229. Berrocal AM, Schuman JS. Subconjungtival cephalosporin anaphylaxis. Ophthalmic Surg Lasers 2001;32:79–80.
230. Akcam FZ, Aygun FO, Akkaya VB. DRESS like severe drug rash with eosinophilia, atypic lymphocytosis and fever secondary to ceftriaxone. J Infect 2006; 53(2): e51-3.
231. Ernst MR, van Dijken PJ, Kabel PJ, Draaisma JM. Anaphylaxis after first exposure to ceftriaxone. Acta Paediatr 2002;91(3):355–6.
232. Romano A, Piunti E, Di Fonso M, Viola M, Venuti A, Venemalm L. Selective immediate hypersensitivity to ceftriaxone. Allergy 2000;55(4):415–6.
233. Berrocal AM, Schuman JS. Subconjunctival cephalosporin anaphylaxis. Ophthalmic Surg Lasers 2001;32(1):79–80.
234. Mannik M. Serum sickness and pathophysiology of immune complexes. In: Rich RR, Fleisher TA, Schwartz BD, editors. Clinical Immunology, Principles and Practice. St Louis: Mosby Year Book Inc, 1996:1062–71.
235. Lowery N, Kearns GL, Young RA, Wheeler JG. Serum sickness-like reactions associated with cefprozil therapy. J Pediatr 1994;125(2):325–8.
236. Plantin P, Milochau P, Dubois D. Nefrite interstiziale da farmaci. Descrizione di un caso clinico. [Drug-induced serum sickness after ingestion of cefatrizine. First reported case.] Presse Méd 1992;21(40):1915.

237. Sanders WE Jr, Johnson JE 3rd, Taggart JG. Adverse reactions to cephalothin and cephapirin. Uniform occurrence on prolonged intravenous administration of high doses. N Engl J Med 1974;290(8):424–9.
238. Murray DL, Singer DA, Singer AB, Veldman JP. Cefaclor—a cluster of adverse reactions. N Engl J Med 1980;303(17):1003.
239. Platt R, Dreis MW, Kennedy DL, Kuritsky JN. Serum sickness-like reactions to amoxicillin, cefaclor, cephalexin, and trimethoprim–sulfamethoxazole. J Infect Dis 1988;158(2):474–7.
240. Heckbert SR, Stryker WS, Coltin KL, Manson JE, Platt R. Serum sickness in children after antibiotic exposure: estimates of occurrence and morbidity in a health maintenance organization population. Am J Epidemiol 1990;132(2):336–42.
241. Stricker BH, Tijssen JG. Serum sickness-like reactions to cefaclor. J Clin Epidemiol 1992;45(10):1177–84.
242. Vial T, Pont J, Pham E, Rabilloud M, Descotes J. Cefaclor-associated serum sickness-like disease: eight cases and review of the literature. Ann Pharmacother 1992;26(7–8):910–4.
243. Hebert AA, Sigman ES, Levy ML. Serum sickness-like reactions from cefaclor in children. J Am Acad Dermatol 1991;25(5 Pt 1):805–8.
244. Parra FM, Igea JM, Martin JA, Alonso MD, Lezaun A, Sainz T. Serum sickness-like syndrome associated with cefaclor therapy. Allergy 1992;47(4 Pt 2):439–40.
245. Parshuram CS, Phillips RJ. Retrospective review of antibiotic-associated serum sickness in children presenting to a paediatric emergency department. Med J Aust 1998;169(2):116.
246. Martin J, Abbott G. Serum sickness like illness and antimicrobials in children. NZ Med J 1995;108(997):123–4.
247. Boyd IW. Cefaclor-associated serum sickness. Med J Aust 1998;169(8):443–4.
248. Anonymous. Canadian Adverse Drug Reaction Newsletter. Can Med Assoc J 1996;155:913.
249. Isaacs D. Serum sickness-like reaction to cefaclor. J Paediatr Child Health 2001;37(3):298–9.
250. Kearns GL, Wheeler JG, Childress SH, Letzig LG. Serum sickness-like reactions to cefaclor: role of hepatic metabolism and individual susceptibility. J Pediatr 1994;125(5 Pt 1):805–11.
251. Kearns GL, Wheeler JG, Rieder MJ, Reid J. Serum sickness-like reaction to cefaclor: lack of in vitro cross-reactivity with loracarbef. Clin Pharmacol Ther 1998;63(6):686–93.
252. Sourgens H, Derendorf H, Schifferer H. Pharmacokinetic profile of cefaclor. Int J Clin Pharmacol Ther 1997;35(9):374–80.
253. Eshki M, Allanore L, Musette P, Milpied B, Grange A, Guillaume JC, Chosidow O, Guillot I, Paradis v, Joly P, Crickx B, Ranger-Rogez S, Descamps V. Twelve-year analysis of severe cases of drug reaction with eosinophilia and systemic symptoms. Arch Dermatol 2009;145: 67-72.
254. Garnier A, El Marabet El H, Kwon T, Peuchmaur M, Mourier O, Baudouin V, Deschênes G, Loirat C. Acute renal failure in a 3-year-old child as part of the drug reaction with eosinophilia and systemic symptoms (DRESS) syndrome following hepatitis A. Pediatr Nephrol 2008;23(4):667–9.
255. Suswardana HM, Yadani BA, Pudjiayi SR, Indrastut N. DRESS syndrome from cefadroxil confirmed by positive patch test. Allergy 2009;62:1216–7.

256. Akcam FZ, Aygun FO, Akkaya VB. DRESS like severe drug rash with eosinophilia, atypic lymphocytosis and fever secondary to ceftriaxone. J Infect 2006;53(2):e51-3.

257. Jaju M, Jaju M, Ahuja YR. Effect of cephaloridine on human chromosomes in vitro in lymphocyte cultures. Mutat Res 1982;101(1):57–66.

258. Hoover DM, Buening MK, Tamura RN, Steinberger E. Effects of cefamandole on spermatogenic development of young CD rats. Fundam Appl Toxicol 1989;13(4):737–46.

259. Benyamini L, Merlob P, Stahl B, Braunstein R, Bortnik O, Bulkowstein M, Zimmerman D, Berkovitch M. The safety of amoxicillin/clavulanic acid and cefuroxime during lactation. Ther Drug Monit 2005; 27(4): 499-502.

260. Berger S, Ernst EC, Barza M. Comparative incidence of phlebitis due to buffered cephalothin, cephapirin, and cefamandole. Antimicrob Agents Chemother 1976;9(4):575–9.

261. Cole DR. Double-blind comparison of phlebitis associated with cefazolin and cephalothin. Int J Clin Pharmacol Biopharm 1976;14(1):75–7.

262. Browning MC, Tune BM. Toxicology of Betalactam Antibiotics. In: Demain AL, Solomon NA, editors. Antibiotics Containing the Betalactam Structure, Part II. Berlin: Springer-Verlag, 1983:371.

263. Moskovitz BL. Clinical adverse effects during ceftriaxone therapy. Am J Med 1984;77(4C):84–8.

264. Sugawara M, Toda T, Iseki K, Miyazaki K, Shiroto H, Kondo Y, Uchino J. Transport characteristics of cephalosporin antibiotics across intestinal brush-border membrane in man, rat and rabbit. J Pharm Pharmacol 1992;44(12):968–72.

265. Dantzig AH, Bergin L. Uptake of the cephalosporin, cephalexin, by a dipeptide transport carrier in the human intestinal cell line, Caco-2. Biochim Biophys Acta 1990;1027(3):211–7.

266. Friedman DI, Amidon GL. Intestinal absorption mechanism of dipeptide angiotensin converting enzyme inhibitors of the lysyl-proline type: lisinopril and SQ 29,852. J Pharm Sci 1989;78(12):995–8.

267. Hu M, Amidon GL. Passive and carrier-mediated intestinal absorption components of captopril. J Pharm Sci 1988;77(12):1007–11.

268. Padoin C, Tod M, Perret G, Petitjean O. Analysis of the pharmacokinetic interaction between cephalexin and quinapril by a nonlinear mixed-effect model. Antimicrob Agents Chemother 1998;42(6):1463–9.

269. Jacolot A, Tod M, Petitjean O. Pharmacokinetic interaction between cefdinir and two angiotensin-converting enzyme inhibitors in rats. Antimicrob Agents Chemother 1996;40(4):979–82.

270. Kitson TM. The effect of cephalosporin antibiotics on alcohol metabolism: a review. Alcohol 1987;4(3):143–8.

271. Norrby SR. Side effects of cephalosporins. Drugs 1987;34(Suppl 2):105–20.

272. Klemm E, Mross C. Akute beidseitige Innenohrschwerhörigkeit durch Cefepim und Alkohol. [Sudden bilateral labyrinthine hearing loss due to cefepime and alcohol.] Otorhinolaryngol Nova 1997;7/1:45.

273. Bailey RR. Renal failure in combined gentamicin and cephalothin therapy. BMJ 1973;2(5869):776–7.

274. Cabanillas F, Burgos RC, Rodriguez C, Baldizon C. Nephrotoxicity of combined cephalothin–gentamicin regimen. Arch Intern Med 1975;135(6):850–2.

275. Tobias JS, Whitehouse JM, Wrigley PF. Severe renal dysfunction after tobramycin/cephalothin therapy. Lancet 1976;1(7956):425.

276. English J, Gilbert DN, Kohlhepp S, Kohnen PW, Mayor G, Houghton DC, Bennett WM. Attenuation of experimental tobramycin nephrotoxicity by ticarcillin. Antimicrob Agents Chemother 1985;27(6):897–902.

277. Wade JC, Schimpff SC, Wiernik PH. Antibiotic combination-associated nephrotoxicity in granulocytopenic patients with cancer. Arch Intern Med 1981;141(13):1789–93.

278. Buening MK, Wold JS. Ethanol–moxalactam interactions in vivo. Rev Infect Dis 1982;4(Suppl):S555–63.

279. Heizmann WR, Trautmann M, Marre R. Antiinfektiöse Chemotherapie. Stuttgart: Wissenschaftliche Verlags gesellschaft 1996;176.

280. Agence française de sécurité sanitaire des produit de santé. Ceftriaxone et incompatibilités physico-chimiques, particulièrement avec les solutions contenant des sels de calcium. Modifications des Résumés des Caractéristiques du Produit. http://www.afssaps.fr/Infos-de-securite/Lettres-aux-professionnels-de-sante/Ceftriaxone-et-incompatibi-lites-physico-chimiques-particulierement-avec-les-solu-tions-contenant-des-sels-de-calcium-Modifications-des-Resumes-des-Caracteristiques-du-Produit/(language)/fre-FR. 30 November 2006.

281. US Food and Drug Administration. Ceftriaxone (marketed as Rocephin) Information. http://www.fda.gov/Drugs/DrugSafety/PostmarketDrugSafetyInformationforPatientsandProviders/ucm109103.htm.

282. Rapp RP, Kuhn R. Clinical pharmaceutics and calcium ceftriaxone. Ann Pharmacother 2007;41(12):2072.

283. Gin AS, Wheaton H, Dalton B. Comment. Clinical pharmaceutics and calcium ceftriaxone. Ann Pharmacother 2007;2(3):450–1.

284. Monte SV, Prescott WA, Johnson KK, Kuhman L, Paladino JA. Safety of ceftriaxone at extremes of age. Expert Opin Drug Saf 2008;7(5):515–23.

285. Soto Alvarez J, Sacristan Del Castillo JA, Alsar Ortiz MJ. Interaction between ciclosporin and ceftriaxone. Nephron 1991;59(4):681–2.

286. Stoeckel K, Hayton WL, Edwards DJ. Clinical pharmacokinetics of oral cephalosporins. Antibiot Chemother 1995;47:34–71.

287. Fiscella RG. Physical incompatibility of vancomycin and ceftazidime for intravitreal injection. Arch Ophthalmol 1993;111(6):730.

288. Kishore K, Raina A, Misra V, Jonas E. Acute verapamil toxicity in a patient with chronic toxicity: possible interaction with ceftriaxone and clindamycin. Ann Pharmacother 1993;27(7–8):877–80.

289. Idoate I, Mendizabal MV, Urdaneta E, Larralde J. Interactions of cephradine and cefaclor with the intestinal absorption of D-galactose. J Pharm Pharmacol 1996;48(6):645–50.

290. Swain RR, Briggs SL. Positive interference with the Jaffe reaction by cephalosporin antibiotics. Clin Chem 1977;23(7):1340–2.

291. Grotsch H, Hajdu P. Interference by the new antibiotic cefpirome and other cephalosporins in clinical laboratory tests, with special regard to the "Jaffe" reaction. J Clin Chem Clin Biochem 1987;25(1):49–52.

292. Kroll MH, Elin RJ. Mechanism of cefoxitin and cephalothin interference with the Jaffe method for creatinine. Clin Chem 1983;29(12):2044–8.

Chloramphenicol

See also Thiamphenicol

General Information

Chloramphenicol is one of the older broad-spectrum antibiotics. It was introduced in 1948 and grew in popularity because of its high antimicrobial activity against a wide range of Gram-positive and Gram-negative bacteria, *Rickettsiae*, *Chlamydia*, and *Mycoplasma* species. It is particularly useful in infections caused by *Salmonella typhi* and *Haemophilus influenzae*. It is mainly bacteriostatic. It readily crosses tissue barriers and diffuses rapidly into nearly all tissues and body fluids.

The main route of elimination of chloramphenicol is metabolic transformation by glucuronidation. The microbiologically inactive metabolites are excreted rapidly and only a small proportion of unchanged drug is excreted in the urine. The usual daily dose is 50 mg/kg for adults and children over 2 months. The total dose should not exceed 3.0–3.5 g/70 kg. The statement that neither the dose nor the interval of chloramphenicol administration needs to be adjusted in patients with significant renal dysfunction (1) probably has to be modified in view of recent findings (2).

By 1950 it became evident that chloramphenicol could cause serious and fatal blood dyscrasias. Its use has therefore steadily fallen during the past 50 years. Since the risk of serious chloramphenicol toxicity is so small (1: 18 000 or probably less) it is of more than historical interest. There are still many areas in which its benefits outweigh its risks. These include:

- typhoid and paratyphoid fever;
- other septic forms of *Salmonella* infections;
- meningitis due to *H. influenzae*, *Streptococcus pneumoniae*, and *Neisseria meningitidis* when the patient is allergic to beta-lactam antibiotics or when the strains (*H. influenzae*, Enterobacteriaceae) are resistant to aminopenicillins and cephalosporins;
- brain abscess;
- serious infections caused by *Bacteroides fragilis* (as an alternative to clindamycin or metronidazole).

Since chloramphenicol is still one of the cheapest antibiotics, this list of indications is longer in developing countries, where chloramphenicol may be more readily available than newer expensive antibiotics. However, most infections can be readily, safely, and effectively treated with alternative drugs. Therefore, the role of chloramphenicol in the treatment of infectious diseases is likely to diminish further.

Chloramphenicol and its metabolites act primarily on the 50S ribosomal subunit, with suppression of the activity of the enzyme peptidyltransferase. It inhibits mitochondrial membrane protein synthesis, leading to suppression of mitochondrial respiration and ultimately to cessation of cell proliferation (3). Analogous mechanisms may operate in the production of the reversible type of bone marrow depression, which is the most prominent toxic effect in patients taking chloramphenicol. Its potency to induce toxic effects on mitochondria in maturing or rapidly proliferating eukaryotic cells is very close to that for inhibiting prokaryotic cells (bacteria and blue-green algae). However, little progress has been made in elucidating the pathogenesis of irreversible bone marrow aplasia (3–5).

Observational studies

The response to chloramphenicol has been assessed in cases of bacteremia due to vancomycin-resistant enterococci, of whom 65% received chloramphenicol (6). Among those in whom a response could be assessed, 61% had a clinical response and 79% had a microbiological response. Mortality was non-significantly lower in patients treated with chloramphenicol. In cases with central line-related bacteremia, there was no difference in mortality among those treated with chloramphenicol, line removal, or both. No adverse effect could be definitely attributed to chloramphenicol.

General adverse effects

Chloramphenicol has been associated with two serious but rare toxic effects, each with a high mortality. One is the "gray syndrome," vasomotor collapse in neonates caused by excessive parenteral doses. The second is bone marrow aplasia, which is a hypersusceptibility reaction. Prolonged use can result in neuropathies. Mild gastrointestinal disturbances are common. Chloramphenicol can cause a Jarisch–Herxheimer reaction, for example in patients with louse-borne relapsing fever (7). Hypersensitivity reactions are commonly mild and more frequent with topical use (allergic contact dermatitis, rashes, glossitis). The late, severe type of bone marrow reaction may be of allergic origin. Tumor-inducing effects have not been described; a statement that chloramphenicol might cause cancer in the fetus appears to have been purely speculative.

Organs and Systems

Cardiovascular

The "gray syndrome" is the term given to the vasomotor collapse that occurs in neonates who are given excessive parenteral doses of chloramphenicol. The syndrome is characterized by an ashen gray, cyanotic color of the skin, a fall in body temperature, vomiting, a protuberant abdomen, refusal to suck, irregular and rapid respiration, and lethargy. It is mainly seen in newborn infants, particularly when premature. It usually begins 2–9 days after the start of treatment.

Inadequate glucuronyl transferase activity combined with reduced glomerular filtration in the neonatal period is responsible for a longer half-life and accumulation of the drug. In addition, the potency of chloramphenicol to inhibit protein synthesis is higher in proliferating cells and tissues. The most important abnormality seems to be respiratory deficiency of mitochondria, due, for example to suppressed synthesis of cytochrome oxidase. The dosage should be adjusted according to the age of the neonate, and blood concentrations should be monitored. In most cases of gray syndrome, the daily dose of chloramphenicol has been higher than 25 mg/kg (8,9). Occasionally, treatment of older children and teenagers with large doses of chloramphenicol (about 100 mg/kg) has resulted in a similar form of vasomotor collapse (10).

Nervous system

Peripheral neuropathy has been seen after prolonged courses of chloramphenicol (11).

Retrobulbar optic neuritis and polyneuritis have been attributed to prolonged chloramphenicol therapy (12).

Sensory systems

Eyes
Optic neuropathy has been seen after prolonged courses of chloramphenicol (13). Alterations in color perception and optic neuropathy, in some cases resulting in optic atrophy and blindness, have been observed, especially in children with cystic fibrosis receiving relatively high doses for many months (12,14). Most of these complications were reversible and were attributed to a deficiency of B vitamins.

Ears
Local application of chloramphenicol can cause hearing defects. Asymmetrical hearing loss with lowered perception of high tones has been documented after treatment of chronic bilateral otitis media with chloramphenicol powder (15). Propylene glycol is often used as a vehicle for chloramphenicol ear-drops, and ototoxicity may be due to chloramphenicol and/or propylene glycol, which is itself strongly ototoxic. Ototoxic effects can also occur after systemic drug administration (16).

Hematologic

The first death resulting from bone marrow aplasia induced by chloramphenicol eye-drops was described in 1955 (17). Chloramphenicol causes two types of bone marrow damage (18).

- A frequent, early, dose-related, reversible suppression of the formation of erythrocytes, thrombocytes, and granulocytes (early toxicity).
- A rare, late type of bone marrow aplasia, a hypersusceptibility reaction, which is generally irreversible, and has a high mortality rate (aplastic anemia) (19–21).

Chloramphenicol inhibits mRNA translation by the 70S ribosomes of prokaryotes, but does not affect 80S eukaryotic ribosomes. Most mitochondrial proteins are encoded by nuclear DNA and are imported into the organelles from the cytosol where they are synthesized. Mitochondria retain the capacity to translate, on their own ribosomes, a few proteins encoded by the mitochondrial genome. True to its prokaryotic heritage, mitochondrial ribosomes are similar to those of bacteria, meaning that chloramphenicol inhibits protein synthesis by these ribosomes. Chloramphenicol-induced anemia is believed to result from this inhibition (22). Chloramphenicol can also cause apoptosis in purified human bone marrow CD34+ cells (23).

Dose-related bone marrow suppression
The early, dose-related type of chloramphenicol toxicity is usually seen after the second week of treatment, and is characterized by inhibited proliferation of erythroid cells and reduced incorporation of iron into heme. The clinical correlates in the peripheral blood are anemia, reticulocytopenia, normoblastosis, and a shift to early erythrocyte forms. The plasma iron concentration is increased. Early erythroid forms and granulocyte precursors show cytoplasmic vacuolation. After withdrawal, complete recovery is the rule. Leukopenia and thrombocytopenia are less frequent.

Although there is no evidence that these abnormalities progress to frank bone marrow aplasia, continuation of chloramphenicol after the appearance of early toxicity is thought to be hazardous. Pre-existing liver damage (for example due to infectious hepatitis or alcoholism) and impaired kidney function can lead to reduced elimination of chloramphenicol and its metabolites, thereby aggravating marrow toxicity. As a rule, this is not the irreversible type.

Aplastic anemia
Although bone marrow aplasia has not been related with certainty to either the daily or the total dose of chloramphenicol or to the sex or age of the patients, it has occurred almost exclusively in individuals who were taking prolonged therapy, particularly if they were exposed to the drug on more than one occasion (24). The condition is rare, occurring about once in every 18 000–50 000 subjects in various countries. These variations may in part depend on ethnic factors (25,26). For example, there have been very few cases reported in blacks (27). Bone marrow aplasia due to chloramphenicol has usually resulted in aplastic anemia with pancytopenia; other forms, such as red cell hypoplasia, selective leukopenia, or thrombocytopenia, are less common.

When bone marrow aplasia was complete, the fatality rate approached 100%. As a rule, it has been found that the longer the interval between the last dose of chloramphenicol and the appearance of the first sign of a blood dyscrasia, the more severe the resulting aplasia. Nearly all patients in whom the interval was longer than 2 months died as a result of this complication. However, fatal aplastic anemia can also occur shortly after normal doses of chloramphenicol (28).

The pathogenesis of bone marrow aplasia after chloramphenicol is still uncertain. Compared with normal cells, bone marrow aspirates from patients with bone marrow aplasia are relatively resistant to the toxic effects of chloramphenicol in vitro. This has been explained by the hypothesis that during treatment with chloramphenicol, chloramphenicol-sensitive cells were eliminated, leaving behind only a chloramphenicol-insensitive population of blood cell precursors with poor proliferative capacity (29). Chloramphenicol can induce apoptosis in purified human bone marrow CD34+ cells; however, there was no protection from a variety of antioxidants on chloramphenicol-induced suppression of burst-forming unit erythroid and colony-forming unit granulocyte/monocyte in vitro (30). In contrast, a caspase inhibitor ameliorated the apoptotic-inducing effects of chloramphenicol.

Since thiamphenicol, which causes very few cases of aplastic anemia, differs from chloramphenicol by substitution of the para-nitro group by a methylsulfonyl group, interest has been focused on the para-nitro group and metabolites of that part of the molecule, nitrosochloramphenicol and chloramphenicol hydroxylamine. In human bone marrow, nitrosochloramphenicol inhibited DNA synthesis at 10% of the concentration of chloramphenicol required for the same effect, and proliferation of myeloid progenitors was irreversibly inhibited. The covalent binding of nitrosochloramphenicol to marrow cells was 15 times greater than that of chloramphenicol (31). This has lent support to the hypothesis that abnormal metabolism may contribute to the susceptibility to bone marrow aplasia. The production of reduced derivatives by intestinal microbes may contribute to toxicity, but oral administration of chloramphenicol is not essential for the development of aplastic anemia (32). There is evidence that genetic predisposition may play a role (33,34). The wide geographical variations in the incidence of aplastic anemia may also reflect environmental factors.

For many years it had been said that there were no cases of aplastic anemia after parenteral administration of chloramphenicol; however, a few cases of aplastic anemia have been reported (35). There have also been reports of bone marrow hypoplasia after the use of chloramphenicol eye-drops (36,37).

There is controversy about the risk of aplastic anemia with topical chloramphenicol. In a prospective case-control surveillance of aplastic anemia in a population of patients who had taken chloramphenicol for a total of 67.2 million person-years, 145 patients with aplastic anemia and 1226 controls were analysed. Three patients and five controls had been exposed to topical chloramphenicol, but two had also been exposed to other known causes of aplastic anemia. Based on these findings, an association between ocular chloramphenicol and aplastic anemia could not be excluded, but the risk was less than one per million treatment courses (38). In another study, a review of the literature identified seven cases of idiosyncratic hemopoietic reactions associated with topical chloramphenicol. However, the authors failed to find an association between the epidemiology of acquired aplastic anemia and topical chloramphenicol. Furthermore, after topical therapy they failed to detect serum accumulation of chloramphenicol by high performance liquid chromatography. They concluded that these findings support the view that topical chloramphenicol was not a risk factor for dose-related bone marrow toxicity and that calls for abolition of treatment with topical chloramphenicol based on current data are not supported (39).

In a study using general practitioner-based computerized data, 442 543 patients were identified who received 674 148 prescriptions for chloramphenicol eye-drops. Among these patients, there were three with severe hematological toxicity and one with mild transient leukopenia. The causal link between topical chloramphenicol and hematological toxicity was not further evaluated in detail (40).

Chloramphenicol eye drops are used to treat bacterial conjunctivitis and infections of the external eye. Among more than 200 million ocular chloramphenicol products dispensed in the UK in the past 10 years, only 11 cases of suspected chloramphenicol-induced blood dyscrasias (none fatal) were reported to the Committee on Safety of Medicines (41). However, under-reporting in this way, which may be as low as 6% (42), suggests that the number of incidents could have been as many as 200. Over 40 cases of blood dyscrasias or aplastic anemia after the use of topical ocular chloramphenicol have been reported in the literature or to the National Registry of Drug-Induced Ocular Side Effects (Casey Eye Institute, Portland, Oregon, USA). Although there was no proof of causality, we feel, based on WHO criteria, that the association in these cases was probable. The medical literature contains many papers both for and against the use of chloramphenicol in ophthalmology.

Leukemia

In a small fraction of patients who survive the chronic type of bone marrow damage, myeloblastic leukemia develops (43,44). In most instances this complication has appeared within a few months of the diagnosis of aplasia and was considered to be a sequel of chloramphenicol treatment. Sometimes the delay was shorter. The majority were either children or adults aged 50–70 years.

The occurrence of acute leukemia has been studied in relation to preceding use of drugs (before the 12 months preceding the diagnosis) in a case-control study of 202 patients aged over 15 years with a diagnosis of acute leukemia and age- and sex-matched controls (45). Among users of chloramphenicol or thiamphenicol the odds ratio for any use was 1.1 (0.6–2.2) whereas the odds ratio for high doses was 1.8 (0.6–5.3). Other systemic antibiotics showed no substantial relation with the occurrence of leukemia.

Paroxysmal nocturnal hemoglobinuria

Paroxysmal nocturnal hemoglobinuria occurred after 6 weeks of oral chloramphenicol therapy for treatment of a hip prosthesis infection with *Bacillus fragilis* in a 78-year-old man (46).

Gastrointestinal

Mild gastrointestinal disturbances are common in patients taking chloramphenicol. In 51 children with Mediterranean spotted fever randomized for 7 days to either clarithromycin, 15 mg/kg/day orally in two divided doses, or chloramphenicol, 50 mg/kg/day orally in four divided doses, the two drugs were equally well tolerated and there were no major adverse effects; there was vomiting in two patients treated with clarithromycin and in one treated with chloramphenicol (47). None of the patients required drug withdrawal.

Skin

Hypersensitivity occurs about four times more often after topical than after oral use (48). In fact, there has been a continuous increase in chloramphenicol hypersensitivity, owing to the use of dermatological formulations (49). Allergic contact dermatitis and macular or vesicular skin rashes are usually limited to skin areas previously exposed to the drug. Contact conjunctivitis has also been reported.

A case of a facial contact dermatitis due to chloramphenicol with cross-sensitivity to thiamphenicol has been reported (50).

Immunologic

Systemic reactions with collapse, bronchospasm, angioedema, and urticaria occur rarely (51,52).

Infection risk

The number and types of microorganisms that constitute the normal microflora of the alimentary, respiratory, and genital tracts change during therapy with chloramphenicol. Superinfections can then develop with *Staphylococcus aureus*, *Pseudomonas*, *Proteus*, and fungi. The changes in intestinal flora may be partly responsible for a reduction in the synthesis of vitamin K-dependent clotting factors, especially in patients with severe illnesses and malnutrition or during the administration of oral anticoagulants.

Long-Term Effects

Drug tolerance

There have been reports of chloramphenicol-resistant *H. influenzae* from various countries, but there have been few cases (53). Outbreaks of chloramphenicol-resistant *S. typhi* have been observed in several countries (54,55).

High-level chloramphenicol-resistant strains of *N. meningitidis* serogroup B were isolated from 11 patients in Vietnam and one patient in France. Resistance was due to the presence of the catP gene on a truncated transposon that has lost mobility because of internal deletions (56).

Salmonella typhimurium DT104 is usually resistant to ampicillin, chloramphenicol, streptomycin, sulfonamides, and tetracycline. An outbreak of 25 culture-confirmed cases of multidrug-resistant *S. typhimurium* DT104 has been identified in Denmark (57). The strain was resistant to the above-mentioned antibiotics and nalidixic acid and had reduced susceptibility to fluoroquinolones. A swineherd was identified as the primary source (57). The DT104 strain was also found in cases of salmonellosis in Washington State, and soft cheese made with unpasteurized milk was identified as an important vehicle of its transmission (58).

A high rate of resistance of non-typhoid *Salmonella* to commonly used antimicrobial agents was found in Taiwan; 67% of the isolated strains were resistant to chloramphenicol (59).

Streptococcus pneumoniae was isolated in 30% of 40 HIV-infected and 50% of 162 HIV-negative children living in a Romanian orphanage (60). Multidrug-resistant streptococci were highly prevalent, and 21% of the isolates were resistant to chloramphenicol.

Esterases in serum from rabbits and to a lesser extent humans can convert diacetyl chloramphenicol back to an active antibiotic. Therefore, in vitro findings may not accurately reflect the level of chloramphenicol resistance by chloramphenicol acetyltransferase-bearing bacteria in vivo, when growth media supplemented with serum are used (61).

The flo gene that confers resistance to chloramphenicol and the veterinary antibiotic florfenicol have previously been identified in *Photobacterium piscicida* and *Salmonella enterica serovar typhimurium* DT104 (62). Florfenicol-resistant isolates of *Escherichia coli* were tested and found to contain large flo-positive plasmids, suggesting that several of these isolates may have a chromosomal flo gene. The *E. coli* flo gene also specifies non-enzymatic cross-resistance to both florfenicol and chloramphenicol (63). Florfenicol resistance has emerged among veterinary isolates of *E. coli* incriminated in bovine diarrhea.

Second-Generation Effects

Teratogenicity

In the large population-based dataset of the Hungarian Case Control Surveillance of Congenital Abnormalities, of 38 151 pregnant women who had babies without any defects and 22 865 pregnant women who had neonates or fetuses with congenital abnormalities, 51 and 52 had been treated with oral chloramphenicol respectively. Treatment during early pregnancy presented little, if any, teratogenic risk to the fetus (64).

Fetotoxicity

Chloramphenicol penetrates the fetal circulation and should therefore be avoided during the last phase of pregnancy (65,66). The gray syndrome has been observed in babies born to mothers who had received chloramphenicol in the final stage of pregnancy.

Lactation

Chloramphenicol has been found in relatively large amounts in breast milk. It should therefore be avoided during breast-feeding (65,66).

Susceptibility Factors

Age

The pharmacokinetics of chloramphenicol in children have been reviewed (67).

In children, a high cumulative dose seems to be an important risk factor. As leukopenia can occur in the early phase of treatment, a complete blood count every third day is recommended (SEDA-15, 267).

Renal disease

Impaired kidney function, with reduced clearance of chloramphenicol, may be a risk factor for toxicity.

Hepatic disease

Liver damage, with reduced clearance of chloramphenicol, may be a risk factor for toxicity.

Other features of the patient

In a retrospective study of 30 consecutive children with sepsis treated with oral chloramphenicol, weight, albumin, and white blood cell count were the most important determinants for chloramphenicol distribution volume, whereas age, white blood cell count, and serum creatinine were most important for drug clearance (68). A pre-existing blood dyscrasia is generally considered to be an absolute contraindication to the use of chloramphenicol.

Drug Administration

Drug administration route

In 1993 the American National Register of Drug Induced Ocular Side Effects received reports of 23 patients with blood dyscrasias that could have been related to topical ocular administration of chloramphenicol (69).

Of the two types of bone marrow toxicity that chloramphenicol can cause, it may cause the late type only in genetically predisposed patients. The overall risk of aplastic anemia after oral administration of chloramphenicol is 1:30 000 to 1:50 000, which is 13 times greater than the risk of idiopathic aplastic anemia in the population as a whole. Since topical administration achieves systemic effects by absorption through the conjunctival membrane or through drainage down the lacrimal duct, with eventual absorption from the gastrointestinal tract, the risk may be similar to that after oral administration. However, based on two case-control studies and a cohort study, the incidence of blood dyscrasias due to chloramphenicol eye-drops was estimated to be somewhat lower, namely 1:100 000 treated patients (40,70).

It is difficult to justify subjecting patients to this small potential risk, in view of the availability of other antibiotics for use in the eye. In the USA the Physician's Desk Reference emphasizes with repeated warnings the importance of not using ocular chloramphenicol unless there is no alternative, and this warning should be respected on both sides of the Atlantic (71,72).

Allergic reactions to chloramphenicol eye-drops include conjunctivitis, keratitis, and palpebral and periocular eczema (50).

Erythema multiforme caused by local treatment with chloramphenicol eye-drops has been described (50). The possible role of an allergic mechanism in this reaction was suggested, based on a positive mast cell degranulation test (73).

Drug–Drug Interactions

Drug interactions with chloramphenicol have been reviewed (74). Concurrent use of rifampicin or phenobarbital, both inducers of hepatic microsomal enzymes, can reduce serum concentrations of chloramphenicol. Concomitant administration of phenytoin can cause accumulation of chloramphenicol in serum to toxic concentrations; toxic concentrations of phenytoin also have been reported. Other drugs that are metabolized in the liver, such as paracetamol, isoniazid, and theophylline, can have similar effects.

Ciclosporin

A possible interaction between ciclosporin and chloramphenicol has been observed (75).

- A morbidly obese 17-year-old Hispanic girl, who had a cadaveric renal transplantation 5 years before, took ciclosporin and prednisone for stabilization. She was treated with chloramphenicol 875 mg qds and ceftazidime 2 g tds for vancomycin-resistant enterococcal sinusitis. There was a substantial and sustained increase in ciclosporin concentrations after chloramphenicol was added. Normalization was achieved after withdrawal of chloramphenicol.

Cyclophosphamide

Chloramphenicol inhibits the biotransformation of cyclophosphamide (76).

Cytochrome P450

Chloramphenicol can interfere with the elimination of drugs that are inactivated by hepatic metabolism, probably through a mechanism involving inhibition of microsomal enzymes. The mechanism has been claimed to be inactivation of microsomal enzymes via an intermediate reactive metabolite that binds covalently to the protein moiety of cytochrome P450 (76). Assuming such a mechanism, chloramphenicol would be expected to interact with the metabolism of other drugs dealt with by cytochrome P450.

Paracetamol (acetaminophen)

Paracetamol altered the pharmacokinetics of chloramphenicol in some studies but not in others.

In six adults the half-life of chloramphenicol, 1 g intravenously was increased from 3.3 to 15 hours by paracetamol 100 mg intravenously (77).

In contrast, in five children aged 2.5–5 years paracetamol 50 mg/kg/day for several days significantly lowered the C_{max} of chloramphenicol, increased its apparent volume of distribution and clearance, and slightly shortened its half-life (78).

Other studies have failed to show any interaction in children (79) or adults (80).

Phenobarbital

Phenobarbital can increase the rate of chloramphenicol metabolism and so lead to abnormally low serum chloramphenicol concentrations. In 17 children receiving chloramphenicol succinate alone, mean peak and trough serum concentrations were 25 and 13 µg/ml respectively. In six patients phenobarbital reduced these concentrations to 17 and 7.5 µg/ml respectively (81).

Phenytoin

The ability of chloramphenicol to inhibit the biotransformation of phenytoin is well established and clinically relevant (82).

Conversely, phenytoin can increase the rate of chloramphenicol metabolism and so lead to abnormally low serum chloramphenicol concentrations, as has been anecdotally reported (83). However, contrary to expectation, in another study serum chloramphenicol concentrations rose during concomitant phenytoin therapy (81).

Rifampicin

Rifampicin can increase the rate of chloramphenicol metabolism and so lead to abnormally low serum chloramphenicol concentrations (84,85).

Tacrolimus

There is a rapid and severe interaction between chloramphenicol and tacrolimus, with greatly increased tacrolimus concentrations during co-administration, and a rapid fall after chloramphenicol withdrawal. This was reported in a 56-year-old man with a cadaveric kidney–pancreas transplant (86).

Inhibition of tacrolimus clearance has been observed in an adolescent renal transplant recipient who was treated with standard doses of chloramphenicol for vancomycin-resistant enterococci. Toxic concentrations of tacrolimus were observed on the second day of chloramphenicol treatment, requiring an 83% reduction in the dose of tacrolimus (87).

- A 47-year-old white man with a cadaveric liver transplant took chloramphenicol for a urinary tract infection due to a vancomycin-resistant *Enterococcus* and inadvertently received 1850 mg qds (roughly twice the maximum recommended dose) (88). On day 4 he had a 12-hour trough tacrolimus concentration of over 60 ng/ml, and complained of fatigue, lethargy, headache, and tremor, symptoms consistent with tacrolimus toxicity.

It was suggested that the underlying mechanism might be inhibition of CYP3A4 by chloramphenicol.

Tolbutamide

Chloramphenicol inhibits the biotransformation of tolbutamide (76).

Warfarin

An interaction of warfarin with ocular chloramphenicol (5 mg/ml; 1 drop qds in each eye), which led to an increase in INR, has been suspected in an 83-year-old white woman (89). The authors suggested that the effect may be due to chloramphenicol inhibition of hepatic microsomal CYP2C9, since the pharmacologically active enantiomer S-warfarin is metabolized by this enzyme).

References

1. Bennett WM, Aronoff GR, Morrison G, Golper TA, Pulliam J, Wolfson M, Singer I. Drug prescribing in renal failure: dosing guidelines for adults. Am J Kidney Dis 1983;3(3):155–93.
2. Phelps SJ, Tsiu W, Barrett FF, Nahata MC, Disessa TG, Stidham G, Roy S 3rd. Chloramphenicol-induced cardiovascular collapse in an anephric patient. Pediatr Infect Dis J 1987;6(3):285–8.
3. Yunis AA. Effects of chloramphenicol on erythropoiesis. In: Dimitrov NV, Nodine JH, editors. Drugs and Hematological Reactions. London, New York: Grune and Stratton, 1974:133.
4. Keiser G. Co-operative study of patients treated with thiamphenicol. Comparative study of patients treated with chloramphenicol and thiamphenicol. Postgrad Med J 1974;50(Suppl 5):143–5.
5. Polin HB, Plaut ME. Chloramphenicol. NY State J Med 1977;77(3):378–81.
6. Lautenbach E, Schuster MG, Bilker WB, Brennan PJ. The role of chloramphenicol in the treatment of bloodstream infection due to vancomycin-resistant enterococcus. Clin Infect Dis 1998;27(5):1259–65.
7. Perine PL, Teklu B. Antibiotic treatment of louse-borne relapsing fever in Ethiopia: a report of 377 cases. Am J Trop Med Hyg 1983;32(5):1096–100.
8. Lietman PS. Chloramphenicol and the neonate—1979 view. Clin Perinatol 1979;6(1):151–62.
9. Nahata MC. Lack of predictability of chloramphenicol toxicity in paediatric patients. J Clin Pharm Ther 1989;14(4):297–303.
10. Brown RT. Chloramphenicol toxicity in an adolescent. J Adolesc Health Care 1982;3(1):53–5.
11. Bomb BS, Bedi HK. Chloramphenicol-induced peripheral neuropathy (a case report). J Assoc Physicians India 1974;22(8):623–5.
12. Murayama E, Miyakawa T, Sumiyoshi S, Deshimaru M, Sugita K. [Retrobulbar optic neuritis and polyneuritis due to prolonged chloramphenicol therapy.]Rinsho Shinkeigaku 1973;13(4):213–20.
13. Venegas-Francke P, Fruns-Quintana M, Oporto-Caroca M. Neuritis optica bilateral por cloranfenicol. [Bilateral optic neuritis caused by chloramphenicol.] Rev Neurol 2000;31(7):699–700.
14. Beyer CR. Chloramphenicol-induced acute bilateral optic neuritis in cystic fibrosis. J Pediatr Ophthalmol Strabismus 1978;15:291.
15. Anonymous. Ear drops and iatrogenic deafness. Med J Aust 1975;2(16):626.
16. Iqbal SM, Srivatsav CB. Chloramphenicol ototoxicity. A case report. J Laryngol Otol 1984;98(5):523–5.
17. Rosenthal RL, Blackman A. Bone-marrow hypoplasia following the use of chloramphenicol eye drops. JAMA 1965;191:136–7.

18. Yunis AA. Chloramphenicol toxicity: 25 years of research. Am J Med 1989;87(3N):N44–8.
19. Turton JA, Andrews CM, Havard AC, Robinson S, York M, Williams TC, Gibson FM. Haemotoxicity of thiamphenicol in the BALB/c mouse and Wistar Hanover rat. Food Chem Toxicol 2002;40(12):1849–61.
20. Holdiness MR. Management of cutaneous erythrasma. Drugs 2002;62(8):1131–41.
21. Lam RF, Lai JS, Ng JS, Rao SK, Law RW, Lam DS. Topical chloramphenicol for eye infections. Hong Kong Med J 2002;8(1):44–7.
22. Alcindor T, Bridges KR. Sideroblastic anaemias. Br J Haematol 2002;116(4):733–43.
23. Ahmed SG, Ibrahim UA. Bone marrow morphological features in anaemic patients with acquired immune deficiency syndrome in Nigeria. Niger Postgrad Med J 2001;8(3):112–5.
24. Najean Y, Guerin MN, Chomienne C. Etiology of acquired aplastic anemia. In: Najean Y, Tognoni G, Yunis AA, editors. Safety Problems Related to Chloramphenicol and Thiamphenicol Therapy. New York: Raven Press, 1981:61.
25. Wallerstein RO, Condit PK, Kasper CK, Brown JW, Morrison FR. Statewide study of chloramphenicol therapy and fatal aplastic anemia. JAMA 1969;208(11):2045–50.
26. Mary JY, Baumelou E, Guiguet M. Epidemiology of aplastic anemia in France: a prospective multicentric study. The French Cooperative Group for Epidemiological Study of Aplastic Anemia. Blood 1990;75(8):1646–53.
27. Froese EA. Chloramphenicol-associated aplastic anemia: its occurrence in Africans and with parenteral administration. Cent Afr J Med 1978;24:58.
28. Daum RS, Cohen DL, Smith AL. Fatal aplastic anemia following apparent "dose-related" chloramphenicol toxicity. J Pediatr 1979;94(3):403–6.
29. Yunis AA, Mayan DR, Arimura GK. Comparative metabolic effects of chloramphenicol and thiamphenicol in mammalian cells. J Pediatr 1974;94:403.
30. Kong CT, Holt DE, Ma SK, Lie AK, Chan LC. Effects of antioxidants and a caspase inhibitor on chloramphenicol-induced toxicity of human bone marrow and HL-60 cells. Hum Exp Toxicol 2000;19(9):503–10.
31. Murray TR, Downey KM, Yunis AA. Chloramphenicol-mediated DNA damage and its possible role in the inhibitory effects of chloramphenicol on DNA synthesis. J Lab Clin Med 1983;102(6):926–32.
32. Chaplin S. Bone marrow depression due to mianserin, phenylbutazone, oxyphenbutazone, and chloramphenicol—Part II. Adverse Drug React Acute Poisoning Rev 1986;5(3):181–96.
33. Nagao T, Mauer AM. Concordance for drug-induced aplastic anemia in identical twins. N Engl J Med 1969;281(1):7–11.
34. Yunis AA. Differential in-vitro toxicity of chloramphenicol, nitroso-chloramphenicol, and thiamphenicol. Sex Transm Dis 1984;11(Suppl 4):340–2.
35. Fink TJ, Gump DW. Chloramphenicol: an impatient study of use and abuse. J Infect Dis 1978;138(5):690–4.
36. West BC, DeVault GA Jr, Clement JC, Williams DM. Aplastic anemia associated with parenteral chloramphenicol: review of 10 cases, including the second case of possible increased risk with cimetidine. Rev Infect Dis 1988;10(5):1048–51.
37. Brodsky E, Biger Y, Zeidan Z, Schneider M. Topical application of chloramphenicol eye ointment followed by fatal bone marrow aplasia. Isr J Med Sci 1989;25(1):54.
38. Laporte JR, Vidal X, Ballarin E, Ibanez L. Possible association between ocular chloramphenicol and aplastic anaemia—the absolute risk is very low. Br J Clin Pharmacol 1998;46(2):181–4.
39. Walker S, Diaper CJ, Bowman R, Sweeney G, Seal DV, Kirkness CM. Lack of evidence for systemic toxicity following topical chloramphenicol use. Eye 1998;12(Pt 5):875–9.
40. Lancaster T, Swart AM, Jick H. Risk of serious haematological toxicity with use of chloramphenicol eye drops in a British general practice database. BMJ 1998;316(7132):667.
41. Fraunfelder FW, Fraunfelder FT. Scientific challenges in postmarketing surveillance of ocular adverse drug reactions. Am J Ophthalmol 2007;143(1):145–9.
42. Smith CC, Bennett PM, Pearce HM, Harrison PI, Reynolds DJ, Aronson JK, Grahame-Smith DG. Adverse drug reactions in a hospital general medical unit meriting notification to the Committee on Safety of Medicines. Br J Clin Pharmacol 1996;42(4):423–9.
43. Baumelou E, Najean Y. Why still prescribe chloramphenicol in 1983? Comparison of the clinical and biological hematologic effects of chloramphenicol and thiamphenicol. Blut 1983;47(6):317–20.
44. Shu XO, Gao YT, Linet MS, Brinton LA, Gao RN, Jin F, Fraumeni JF Jr. Chloramphenicol use and childhood leukaemia in Shanghai. Lancet 1987;2(8565):934–7.
45. Traversa G, Menniti-Ippolito F, Da Cas R, Mele A, Pulsoni A, Mandelli F. Drug use and acute leukemia. Pharmacoepidemiol. Drug Saf 1998;7(2):113–23.
46. Diskin C. Paroxysmal nocturnal hemoglobinuria after chloramphenicol therapy. Mayo Clin Proc 2005;80(10):1392, 1394.
47. Cascio A, Colomba C, Di Rosa D, Salsa L, di Martino L, Titone L. Efficacy and safety of clarithromycin as treatment for Mediterranean spotted fever in children: a randomized controlled trial. Clin Infect Dis 2001;33(3):409–11.
48. Forck G. Häufigkeit und Bedeutung von Chloramphenicol-Allergien. [Incidence and significance of chloramphenicol hypersensitivities with respect to different ways of sensitization.] Dtsch Med Wochenschr 1971;96(4):161–5.
49. van Joost T, Dikland W, Stolz E, Prens E. Sensitization to chloramphenicol; a persistent problem. Contact Dermatitis 1986;14(3):176–8.
50. Le Coz CJ, Santinelli F. Facial contact dermatitis from chloramphenicol with cross-sensitivity to thiamphenicol. Contact Dermatitis 1998;38(2):108–9.
51. Palchick BA, Funk EA, McEntire JE, Hamory BH. Anaphylaxis due to chloramphenicol. Am J Med Sci 1984;288(1):43–5.
52. Liphshitz I, Loewenstein A. Anaphylactic reaction following application of chloramphenicol eye ointment. Br J Ophthalmol 1991;75(1):64.
53. Kinmonth AL, Storrs CN, Mitchell RG. Meningitis due to chloramphenicol-resistant *Haemophilus influenzae* type b. BMJ 1978;1(6114):694.
54. Butler T, Linh NN, Arnold K, Pollack M. Chloramphenicol-resistant typhoid fever in Vietnam associated with R factor. Lancet 1973;2(7836):983–5.
55. Cherubin CE, Neu HC, Rahal JJ, Sabath LD. Emergence of resistance to chloramphenicol in *Salmonella*. J Infect Dis 1977;135(5):807–12.
56. Galimand M, Gerbaud G, Guibourdenche M, Riou JY, Courvalin P. High-level chloramphenicol resistance in *Neisseria meningitidis*. N Engl J Med 1998;339(13):868–74.
57. Villar RG, Macek MD, Simons S, Hayes PS, Goldoft MJ, Lewis JH, Rowan LL, Hursh D, Patnode M, Mead PS. Investigation of multidrug-resistant *Salmonella* serotype typhimurium DT104 infections linked to raw-milk cheese in Washington State. JAMA 1999;281(19):1811–6.

58. Molbak K, Baggesen DL, Aarestrup FM, Ebbesen JM, Engberg J, Frydendahl K, Gerner-Smidt P, Petersen AM, Wegener HC. An outbreak of multidrug-resistant, quinolone-resistant *Salmonella enterica* serotype typhimurium DT104. N Engl J Med 1999;341(19):1420–5.

59. Chen YH, Chen TP, Tsai JJ, Hwang KP, Lu PL, Cheng HH, Peng CF. Epidemiological study of human salmonellosis during 1991–1996 in southern Taiwan. Kaohsiung J Med Sci 1999;15(3):127–36.

60. Leibovitz E, Dragomir C, Sfartz S, Porat N, Yagupsky P, Jica S, Florescu L, Dagan R. Nasopharyngeal carriage of multidrug-resistant *Streptococcus pneumoniae* in institutionalized HIV-infected and HIV-negative children in northeastern Romania. Int J Infect Dis 1999;3(4):211–5.

61. Sohaskey CD, Barbour AG. Esterases in serum-containing growth media counteract chloramphenicol acetyltransferase activity in vitro. Antimicrob Agents Chemother 1999;43(3):655–60.

62. Cloeckaert A, Sidi Boumedine K, Flaujac G, Imberechts H, D'Hooghe I, Chaslus-Dancla E. Occurrence of a *Salmonella enterica* serovar typhimurium DT104-like antibiotic resistance gene cluster including the floR gene in *S. enterica* serovar agona. Antimicrob Agents Chemother 2000;44(5):1359–61.

63. White DG, Hudson C, Maurer JJ, Ayers S, Zhao S, Lee MD, Bolton L, Foley T, Sherwood J. Characterization of chloramphenicol and florfenicol resistance in *Escherichia coli* associated with bovine diarrhea. J Clin Microbiol 2000;38(12):4593–8.

64. Czeizel AE, Rockenbauer M, Sorensen HT, Olsen J. A population-based case-control teratologic study of oral chloramphenicol treatment during pregnancy. Eur J Epidemiol 2000;16(4):323–7.

65. Havelka J, Frankova A. Prispevek K Vedlejsim ucinkum chloramfenikolu u nororozencu. [Adverse effects of chloramphenicol in newborn infants.] Cesk Pediatr 1972;27(1):31–3.

66. Kunz J, Schreiner WE. Breitspektrumantibiotika. In: Kunz J, Schreiner WE, editors. Pharmakotherapie während Schwangerschaft und Stillperiode. Stuttgart, New York: Thieme Verlag, 1982:28.

67. Anonymous. Chloramphenicol. Nurs Times 2004;100(48):33.

68. Lugo Goytia G, Lares-Asseff I, Perez Guille MG, Perez AG, Mejia CL. Relationship between clinical and biologic variables and chloramphenicol pharmacokinetic parameters in pediatric patients with sepsis. Ann Pharmacother 2000;34(3):393–7.

69. Fraunfelder FT, Morgan RL, Yunis AA. Blood dyscrasias and topical ophthalmic chloramphenicol. Am J Ophthalmol 1993;115(6):812–3.

70. Wiholm BE, Kelly JP, Kaufman D, Issaragrisil S, Levy M, Anderson T, Shapiro S. Relation of aplastic anaemia to use of chloramphenicol eye drops in two international case-control studies. BMJ 1998;316(7132):666.

71. Doona M, Walsh JB. Use of chloramphenicol as topical eye medication: time to cry halt? BMJ 1995;310(6989):1217–8.

72. Fraunfelder FT. In: Drug-induced Ocular Side Effects. 4th ed.. Baltimore, MD: Williams and Wilkins, 1996:23–6.

73. Amichai B, Grunwald MH, Halevy S. Erythema multiforme resulting from chloramphenicol in eye drops: confirmation by mast cell degranulation test. Ann Ophthalmol Glaucoma 1998;30:225–7.

74. Balbi HJ. Chloramphenicol: a review. Pediatr Rev 2004;25(8):284–8.

75. Bui L, Huang DD. Possible interaction between cyclosporine and chloramphenicol. Ann Pharmacother 1999;33(2):252–3.

76. Halpert J, Naslund B, Betner I. Suicide inactivation of rat liver cytochrome P-450 by chloramphenicol in vivo and in vitro. Mol Pharmacol 1983;23(2):445–52.

77. Buchanan N, Moodley GP. Interaction between chloramphenicol and paracetamol. BMJ 1979;2(6185):307–8.

78. Spika JS, Davis DJ, Martin SR, Beharry K, Rex J, Aranda JV. Interaction between chloramphenicol and acetaminophen. Arch Dis Child 1986;61(11):1121–4.

79. Kearns GL, Bocchini JA Jr, Brown RD, Cotter DL, Wilson JT. Absence of a pharmacokinetic interaction between chloramphenicol and acetaminophen in children. J Pediatr 1985;107(1):134–9.

80. Stein CM, Thornhill DP, Neill P, Nyazema NZ. Lack of effect of paracetamol on the pharmacokinetics of chloramphenicol. Br J Clin Pharmacol 1989;27(2):262–4.

81. Krasinski K, Kusmiesz H, Nelson JD. Pharmacologic interactions among chloramphenicol, phenytoin and phenobarbital. Pediatr Infect Dis 1982;1(4):232–5.

82. Rose JQ, Choi HK, Schentag JJ, Kinkel WR, Jusko WJ. Intoxication caused by interaction of chloramphenicol and phenytoin. JAMA 1977;237(24):2630–1.

83. Powell DA, Nahata MC, Durrell DC, Glazer JP, Hilty MD. Interactions among chloramphenicol, phenytoin, and phenobarbital in a pediatric patient. J Pediatr 1981;98(6):1001–3.

84. Prober CG. Effect of rifampin on chloramphenicol levels. N Engl J Med 1985;312(12):788–9.

85. Kelly HW, Couch RC, Davis RL, Cushing AH, Knott R. Interaction of chloramphenicol and rifampin. J Pediatr 1988;112(5):817–20.

86. Bakri R, Breen C, Maclean D, Taylor J, Goldsmith D. Serious interaction between tacrolimus FK506 and chloramphenicol in a kidney-pancreas transplant recipient. Transpl Int 2003;16:441–3.

87. Schulman SL, Shaw LM, Jabs K, Leonard MB, Brayman KL. Interaction between tacrolimus and chloramphenicol in a renal transplant recipient. Transplantation 1998;65(10):1397–8.

88. Taber DJ, Dupuis RE, Hollar KD, Strzalka AL, Johnson MW. Drug–drug interaction between chloramphenicol and tacrolimus in a liver transplant recipient. Transplant Proc 2000;32(3):660–2.

89. Leone R, Ghiotto E, Conforti A, Velo G. Potential interaction between warfarin and ocular chloramphenicol. Ann Pharmacother 1999;33(1):114.

Ciprofloxacin

See also Fluoroquinolones

General Information

Ciprofloxacin is a fluoroquinolone antibacterial drug with a wider spectrum of activity than nalidixic acid.

Observational studies

Antimicrobial prophylaxis to prevent inhalational anthrax has been recommended for people potentially exposed to *Bacillus anthracis* as a result of recent bioterrorist attacks. Of 3428 people taking ciprofloxacin,

666 (19%) reported severe nausea, vomiting, diarrhea, or abdominal pain; 484 (14%) reported fainting, light-headedness, or dizziness; 250 (7%) reported heartburn or acid reflux; and 216 (6%) reported rashes, hives, or an itchy skin. Of those taking ciprofloxacin, 287 (8%) stopped taking it, 116 (3%) because of adverse events, 27 (1%) because of fear of possible adverse events, and 28 (1%) because they "did not think it was needed" (1–4).

The imaging of inflammation/infection with ^{99m}Tm-labeled ciprofloxacin in 96 patients had a sensitivity of 81% and specificity of 87%. The positive and negative predictive values were 90 and 75% respectively. No adverse effects were reported (5).

In a post-marketing surveillance study in 3859 hospitalized patients with urinary tract infections, 136 (3.5%) had 181 adverse events; 106 of them (2.8%) had 138 adverse events that were possibly or probably related to ciprofloxacin (6). The most frequent adverse events that were considered drug-related were diarrhea (1.7%), rash (0.3%), nausea (0.2%), allergic reactions (0.2%), pruritus (0.2%), vomiting (0.1%), and abnormal tests of liver function (0.1%). There were severe adverse events in 26 patients (0.7%).

Comparative studies

In a multicenter, double-blind study of 234 patients with acute bacterial exacerbations of chronic bronchitis, ciprofloxacin (500 mg bd) was associated with a trend toward a longer infection-free interval and a significantly higher bacteriological eradication rate compared with clarithromycin (500 mg bd) after 14 days (7).

In a double-blind study, ciprofloxacin (500 mg bd) was associated with an infection-free interval and clinical response that were similar to those achieved with cefuroxime axetil (500 mg bd), but the bacteriological eradication rate associated with ciprofloxacin was significantly higher (8).

Organs and Systems

Cardiovascular

Ciprofloxacin causes prolongation of the QT interval (9,10).

The risk of ciprofloxacin-associated torsade de pointes (11) appears to be low, but caution is still warranted when it is given in the presence of pre-existing QT prolongation.

- A 70-year-old woman who was taking azathioprine, olanzapine, and valsartan developed marked QT interval prolongation after the intravenous administration of ciprofloxacin 800 mg/day. Ciprofloxacin was immediately replaced by a third-generation cephalosporin, and during the next few days the QT interval gradually returned to normal (12).
- Torsade de pointes and QT interval prolongation occurred when intravenous ciprofloxacin was given to a 22-year-old healthy man with pneumonia (13).

- Cardiac arrest temporally related to ciprofloxacin occurred in two women (aged 44 and 67 years) when they developed marked QT_c interval prolongation (590 and 680 ms) within 24 hours of ciprofloxacin administration, with recurrent syncope and documented torsade de pointes requiring defibrillation (14). The QTc interval normalized after withdrawal of ciprofloxacin.
- A 76-year-old man with acute on chronic renal insufficiency, developed torsade de pointes in combination with hypocalcemia, triggered by hemodialysis (15). The QT interval prolongation was corrected by treating the hypocalcemia.

Respiratory

Ciprofloxacin can cause interstitial pneumonitis with acute respiratory failure.

- A 68-year-old woman developed severe respiratory failure six days after taking ciprofloxacin. C-reactive protein rose she had an intermittent high fever, dyspnea, and severe hypoxemia (16). Bronchoalveolar lavage was consistent with hypersensitivity pneumonitis. Her symptoms rapidly improved with systemic glucocorticoid therapy.

Nervous system

Headache was recorded in 8% of patients taking ciprofloxacin (17)

Ciprofloxacin can cause confusion and general seizures (18,19). and can exacerbate the risk of seizures in epilepsy.

- A 65-year-old woman on peritoneal dialysis had recurrent generalized tonic-clonic seizures while taking Ciproxin ear drops (ciprofloxacin 2 mg, hydrocortisone 10 mg) for otitis media. She had a seizure-free period of 9 months after she stopped using the eardrops, despite tapering of the dose of sodium valproate (20).

Ciprofloxacin might have prolonged a seizure associated with electroconvulsive therapy in a patient with a urinary tract infection (21).

- A 43-year old woman developed a subacute confusional state and involuntary movements after being given ciprofloxacin, cephalexin, and co-amoxiclav after an insect bite (22). The antibiotics were withdrawn. She later had two generalized tonic–clonic seizures.

Ciprofloxacin can cause propriospinal myoclonus by inhibiting gamma-aminobutyric acid metabolism (23).

Ciprofloxacin can cause facial dyskinesia (24,25).

Two cases of generalized painful dysesthesia associated with ciprofloxacin have been reported (26).

Sensory systems

Eyes

Bilateral acute visual loss, possibly due to toxic optic neuropathy, was observed after 4 weeks of treatment with ciprofloxacin 1.5 g/day and improved after withdrawal (27).

Ciprofloxacin 0.3% ophthalmic drops can cause micro-precipitates of pure ciprofloxacin in the corneal epithelium (28). In four corneal transplantation patients treated preoperatively with ciprofloxacin 0.3% ophthalmic drops, there were microprecipitates associated with damaged corneal epithelium in two patients; another developed a macroprecipitate in a corneal ulcer (29). The crystalline precipitates were pure ciprofloxacin.

Precipitation of topical ciprofloxacin on the corneal surface can delay recovery from viral ocular surface infection (30)

- A 27-year-old man with an ophthalmic infection with adenovirus type 3 had worse eye symptoms within a week of treatment with 0.3% ciprofloxacin ophthalmic solution. The ocular symptoms resolved after ciprofloxacin had been withdrawn and white corneal precipitates had been scraped off.

Ears

Topical 0.2% ciprofloxacin (0.2 ml od for 7 days) did not significantly affect the auditory brainstem response thresholds of guinea pigs, whereas 4% gentamicin (0.2 ml od for 7 days) resulted in total hearing loss (31).

Topical 0.2% ciprofloxacin solution was effective and well tolerated in 232 patients with chronic suppurative otitis media; the most frequently reported adverse events were pruritus, stinging, and earache. Audiometric tests did not show changes attributable to ciprofloxacin (32).

In children with tympanic membrane perforation, topical ciprofloxacin caused no signs of local intolerance or ototoxicity and did not result in significant serum concentrations (33).

Psychological, psychiatric

The administration of ciprofloxacin has been associated with psychosis (34,35) and hypoactive delirium (36,37).

- A 27-year-old woman developed an acute psychotic reaction following the use of ciprofloxacin eye-drops (1 drop hourly to each eye) (38).
- A 28-year-old white man with a history of primary sclerosing cholangitis and controlled ulcerative colitis developed signs of suppurative cholangitis and was treated with intravenous metronidazole 500 mg bd, intravenous cefazolin 1 g 8-hourly, intravenous ciprofloxacin 400 mg bd, Gravol 25–50 mg 4-hourly, urso-deoxycholic acid 1500 mg bd, and folate 1 mg/day (39). On day 1, sulfasalazine 300 mg tds was added and increased on the next day to 1000 mg tds. He improved quickly. On day 5, all medications were discontinued, except ciprofloxacin 500 mg bd. The next day, he became agitated and violent. He was highly irritable, expansive, and grandiose, wanting to publish a book and run a religious mission. He was given piperacillin + tazobactam 4.5 g qds, thiamine 100 mg, and aciclovir 600 mg 8-hourly. Ciprofloxacin was withdrawn, and he was given intramuscular haloperidol 5 mg every 4–6 hours as required and sublingual lorazepam 1 mg 8-hourly. He was then given olanzapine 5 mg bd, the dose of haloperidol was reduced to 2.5 mg 2-hourly

(maximum 10 mg/day), and all other medications were withdrawn. On the next day, he had increased delusional thought content and expansive mood. By day 15, he was alert and oriented, with no delusions or persisting hostility, and with insight into his previous mental disturbance.

- A 45-year-old man without previous mental disorders developed an acute delusional parasitosis after taking ciprofloxacin. He had a complete sustained remission within a few days after withdrawal of ciprofloxacin, without the use of any other medication (40).
- Psychosis occurred in a 32-year-old woman who was taking ciprofloxacin for multidrug resistant tuberculosis; the symptoms resolved within 48 hours after the ciprofloxacin was withdrawn (41).

Hematologic

Ciprofloxacin has been associated with hemolysis in combination with a severe skin reaction in a young adult (42).

Fatal ciprofloxacin-associated thrombotic thrombocytopenic purpura (43) and thrombocytopenia (44) have been reported.

Treatment with ciprofloxacin and piperacillin/tazobactam was associated with thrombocytosis in a 50-year-old man (45).

In two patients with cystic fibrosis the INR and activated partial thromboplastin time rose after ciprofloxacin treatment and returned to normal after withdrawal (46).

Gastrointestinal

In a randomized, double-blind comparison of prulifloxacin 600 mg/day and ciprofloxacin 500 mg bd in 235 patients with acute exacerbations of chronic bronchitis, the most common treatment-related adverse event was gastric pain of mild or moderate intensity, reported in 8.5% of the patients taking prulifloxacin and 6.8% of those taking ciprofloxacin (47).

Ciprofloxacin has been associated with diarrhea due to *Clostridium difficile* (48–50). In 27 patients the only significant risk factor for nosocomial *C. difficile*-associated diarrhea was the use of ciprofloxacin (51).

Liver

Ciprofloxacin causes a mild reversible rise in liver enzymes in 2–3% of patients. Acute hepatitis is rare, but has been reported in a 32-year-old man (52).

There is increasing evidence that ciprofloxacin can cause severe liver damage, and 14 cases have been reported.

- A 22-year-old man took ciprofloxacin 500 mg/day and developed acute liver failure 14 days later (53). Liver biopsy showed extensive hepatocellular necrosis. His symptoms resolved after glucocorticoid therapy.
- Severe liver injury was associated with ciprofloxacin 500 mg bd in a 79-year-old woman with a Gram negative infection; she developed a metabolic acidosis 48 hours after the first dose and her symptoms resolved after withdrawal (54).

Pancreas

A report has suggested that ciprofloxacin can cause pancreatitis (55).

Urinary tract

Ciprofloxacin can cause acute interstitial nephritis, as in the case of an 81-year-old man with diabetic nephropathy (56).

- A 77-year-old man developed interstitial nephritis after taking ciprofloxacin for an infected knee-prosthesis; withdrawal led to recovery of renal function (57).

Acute renal failure secondary to ciprofloxacin crystal nephropathy has been reported.

- A 90-year-old woman had surgery for a perforated duodenal ulcer and was given ciprofloxacin 750 mg bd for a respiratory infection (58). After 8 days her serum creatinine rose to 140 µmol/l (1.6 mg/dl) and she developed oliguria. The fractional sodium excretion was 2%, the urine pH was 6.5, and the urine sediment showed granular casts, epithelial cells, and needle-shaped birefringent crystals in round conglomerates. There was no eosinophilia and no urinary leukocytes. Intravenous fluids had no effect on urine output. Ciprofloxacin was withdrawn and the oliguria resolved over the next day. The serum creatinine peaked at 336 µmol/l (3.8 mg/dl) and returned to baseline within 10 days.
- A 58-year-old woman developed acute renal insufficiency after taking ciprofloxacin 1 g without any other identifiable risk factor (59). Renal biopsy showed no evidence of interstitial nephritis, but there were tubular lesions and deposits of a brown-yellowish substance, which was identified as ciprofloxacin salt. The outcome was favorable.

Two cases of acute renal insufficiency due to necrotizing vasculitis associated with ciprofloxacin were reported in elderly patients (60). In two patients who underwent high-dose chemotherapy with autologous stem cell rescue, acute renal insufficiency developed while they were taking prophylactic ciprofloxacin; withdrawal resulted in prompt reversal of renal insufficiency (61). Ciprofloxacin-induced acute renal insufficiency has been reported in cancer patients undergoing high-dose chemotherapy and autologous stem cell rescue (61). A case report has suggested that ciprofloxacin overdose can lead to acute renal insufficiency characterized by acute tubular necrosis with distal nephron apoptosis (62).

- An 18-year-old woman with cystic fibrosis had pronounced impairment of renal function after taking oral ciprofloxacin 750 mg tds (30 mg/kg/day) for 3 weeks; withdrawal led to normalization of renal function within 10 days (63).

Hemolytic-uremic syndrome has been attributed to ciprofloxacin (64).

- A 53-year-old white man was given chemotherapy for acute lymphoblastic leukemia and after 4 weeks recovered his blood cell count but developed a fever and was given oral ciprofloxacin 500 mg bd. After four doses he developed the typical features of hemolytic–uremic syndrome with microangiopathic hemolytic anemia. The ciprofloxacin was withdrawn, and he received five sessions of plasma exchange. He recovered completely.

Bilateral hydronephrosis and acute renal insufficiency due to urinary tract stones predominantly composed of ciprofloxacin has been reported (65).

Skin

Ciprofloxacin can cause a fixed drug eruption (66,67), purpuric skin lesions (67,68), bullous pemphigoid (69), cutaneous vasculitis (68–70), and ultraviolet recall-like phenomenon (71).

Two cases of Stevens–Johnson syndrome and one of toxic epidermal necrolysis associated with ciprofloxacin have been reported (72,73).

- A 93-year-old lady died after developing extensive skin lesions following treatment with ciprofloxacin (74).

Photosensitivity reactions reported with ciprofloxacin mimic those of sunburn, with erythema and edema in the milder forms and painful blistering with subsequent peeling when severe. Purpuric eruptions during treatment with ciprofloxacin have been rarely reported.

- A 30-year-old man took a 15-day course of ciprofloxacin 500 mg bd and after exposure to the sun developed an itchy eruption consisting of erythematous, petechial lesions located on the anterior aspect of his thighs and legs, clearly delimited by his bathing suit. The lesions cleared completely after withdrawal of the drug and treatment with topical clobetasol (75).

An 80-year-old woman taking ciprofloxacin developed acute generalized exanthematous pustulosis mimicking a bullous drug eruption (76).

Musculoskeletal

Tendinopathy

Ciprofloxacin can cause partial or complete tendon rupture (77,78). In two cases there was bilateral Achilles tendon rupture (79, 80). Of 72 lung transplant recipients who received ciprofloxacin, 20 had Achilles tendon involvement (tendinitis 15, rupture 5) (81). Tendon rupture occurred at a lower dosage of ciprofloxacin than tendinitis and the mean recovery duration was significantly longer.

- Achilles tendon rupture without any sudden pain occurred in a 45-year-old female runner who developed bilateral tendinopathy of the Achilles tendon after repeated treatment with ciprofloxacin; histological analysis showed cystic changes with focal necrosis (82).
- A 35-year-old man developed discomfort in his left medial thigh 3 days after taking ciprofloxacin 500 mg bd for pneumonia; rupture of the adductor longus tendon was confirmed by MRI (83).

- A 35-year-old postal worker who had previously taken ciprofloxacin 1.5 g/day for about 10 days because of concerns about a possible anthrax bacillus infection while working in the post office had a tendon rupture (84).

Ciprofloxacin chelates magnesium, with a resulting 60% reduction in fibroblast activity. This is important in all enzymatic reactions involved in the production of type-1 collagen, which is the main component of tendons.

Arthropathy
The available data suggest that the incidence of arthrotoxicity in children taking ciprofloxacin is the same as in adults; the use of other fluoroquinolones is too rare to obtain clear information about the risks in children (85). In 12 children with sickle cell disease treated successfully for acute osteomyelitis with oral ciprofloxacin, transient bilateral Achilles tendon tendinitis occurred in one 5-year old (86). Another case was reported in a hemodialysis patient with a ciprofloxacin-associated Achilles tendon rupture (87).

In 75 children with typhoid fever, aged under 6 years (mean age 32 months), ciprofloxacin had no adverse effects on growth or joints (88). In another study only 2 of 219 children treated with ciprofloxacin developed arthropathy, in one case transiently (89). In a necropsy study on children treated with ciprofloxacin 20–40 mg/kg/day for an average of 148 days, there were no chondrotoxic effects; however, synovial membranes showed signs of subacute synovitis, which had not been noted in life (90).

In children being treated with ciprofloxacin the rates of arthralgia and quinolone-induced cartilage toxicity were low (91). Episodes of arthralgia were mostly reversible, based on published surveillance data. Recent data from Bayer's ciprofloxacin clinical trials database showed that the incidence of arthralgia in children did not differ between ciprofloxacin and non-quinolone antimicrobial drugs.

Data on more than 1500 children treated with ciprofloxacin suggest that the safety profile of ciprofloxacin in children and adolescents is similar to the profile in adults (92). Adverse events, mostly involving the gastrointestinal tract, were noted in 5–15% of patients. Reversible arthralgia occurred in 36 of 1113 patients, but there was no radiographic evidence of cartilage damage.

An acute reversible arthropathy has been described in a child with cancer treated with a short course of ciprofloxacin for febrile neutropenia (93).

Immunologic

Anaphylactoid reactions occurred in 3 of about 3200 students who took ciprofloxacin 500 mg for chemoprophylaxis of meningococcal meningitis; two had no history of atopic illness (94). Additional adverse reactions were mild skin rashes in three students and nausea and vomiting in two.

- A non-IgE-mediated anaphylactic (anaphylactoid) reaction occurred in a 79-year-old following a first-time exposure to intravenous ciprofloxacin (400 mg) (95).

Ciprofloxacin has been associated with leukocytoclastic vasculitis and other types of necrotizing vasculitis (96). Ciprofloxacin-induced ANCA-negative cutaneous and renal vasculitis has been described in a 50-year-old man with a history of hypertension who developed gross hematuria after taking ciprofloxacin for 10 days (97).

- A 15-year-old girl had an anaphylactoid reaction, with angioedema, shock, and loss of consciousness, within a few minutes of taking an oral dose of ciprofloxacin. She regained consciousness within a few hours of treatment with fluid resuscitation (60 ml/kg), hydrocortisone, and diphenhydramine (98).

Angioimmunoblastic lymphadenopathy is a rare disorder characterized by generalized lymphadenopathy, fever, hepatosplenomegaly, immune hemolytic anemia, and polyclonal hypergammaglobulinemia. Biopsy-proven angioimmunoblastic lymphadenopathy has been reported in a 79-year-old man who had received ciprofloxacin (99).

A Jarisch–Herxheimer reaction to ciprofloxacin has been reported (100).

- A 14-year-old girl developed tachycardia, hypotension, and disseminated intravascular coagulation after her first dose of oral ciprofloxacin 500 mg for presumed pyelonephritis. A peripheral blood smear showed spirochetes consistent with *Borrelia* species.

Susceptibility Factors

Age

Children
In 36 premature infants, delivered at 25–35 weeks and with birth weights of 750–2050 g, ciprofloxacin (13.8 mg/kg/day in two or three divided doses for 3–20 days) had good efficacy in 66% of cases (101). Thrombocytopenia (five cases), raised transaminases (three cases), hyperbilirubinemia (three cases), and raised creatinine concentration (two patients) were reported as adverse events; one child developed femoral osteitis.

In a Russian study of children with cystic fibrosis, the adverse effects of ciprofloxacin were chiefly gastrointestinal (nausea, stomach pain, diarrhea) and increased transaminase activity (102). One episode of arthrotoxicity was transient. There were no negative effects on growth and no chondrotoxicity.

Oral ciprofloxacin (10 mg/kg bd) was as safe and effective as intramuscular ceftriaxone (50 mg/kg/day) in the treatment of acute invasive diarrhea in 201 children (aged 6 months to 10 years) (103). Possible drug-related adverse events occurred in 8% and were mild and transient. Joints were normal during and after the completion of therapy in all patients.

Elderly people
In a retrospective analysis there were no clinically important differences in the safety profile of ciprofloxacin in

patients aged under or over 65 years (104). The incidence of drug-related adverse events was higher in those under 65 years (25%) than in those aged 65 years or more (17%); the most common adverse events affected the gastrointestinal and central nervous systems.

Other features of the patient

Ten patients with peripheral arterial occlusive disease were scheduled to undergo elective percutaneous transluminal angioplasty after a single dose of ciprofloxacin 400 mg (105). Antibiotic concentrations were significantly reduced in ischemic lesions compared with healthy adipose tissue. However, improvement of arterial blood flow in the affected limb was associated with increased cure rates of soft tissue infections.

The pharmacokinetics of intravenous ciprofloxacin have been studied in intensive care unit patients during continuous venovenous hemofiltration ($n = 5$) or hemodiafiltration ($n = 5$) (106). Ciprofloxacin clearance was not altered. A dosage of 400 mg/day was sufficient to maintain effective drug plasma concentrations in patients undergoing continuous renal replacement therapy.

Drug Administration

Drug administration route

In a comparison of intravenous and oral ciprofloxacin in children, treatment associated adverse events were reported in 11% of children taking oral ciprofloxacin, compared with 19% of the children who were treated intravenously (107). In 31 children (1.5%) arthralgia occurred, but it was generally mild to moderate and resolved spontaneously.

In a prospective multicenter trial in 624 patients with presumed bacterial keratitis who were treated with topical ciprofloxacin 0.3% solution 95 (15%) developed a white corneal precipitate during ciprofloxacin therapy, and 473 (76%) began within the first 3 days of treatment (108). Older patients treated with topical ciprofloxacin for bacterial keratitis have a higher risk of corneal deposition.

Drug overdose

A patient developed acute renal insufficiency after ciprofloxacin overdose. This was mediated by tubulointerstitial nephritis with distal nephron apoptosis, as evidenced by renal biopsy (62).

Drug–Drug Interactions

Ciclosporin

In 42 patients who had received a kidney transplant, cases were treated with ciprofloxacin in the first 1–6 months after transplantation, and matched controls (two per case) were not. The proportion of cases with at least one episode of biopsy-proven rejection 1–3 months after transplantation (45%) was significantly higher than in the controls (19%). The authors speculated that ciprofloxacin increases rejection rates in renal transplant recipients by antagonizing ciclosporin-dependent inhibition of interleukin-2 production (109).

Clozapine

Ciprofloxacin can alter plasma clozapine concentrations, perhaps by inhibition of cytochrome P450 enzymes (110).

A possible pharmacokinetic interaction between ciprofloxacin, which inhibits CYP1A2, and clozapine, with moderately increased serum concentrations of clozapine, has been reported (129).

Cyclophosphamide

In eight patients with non-Hodgkin's lymphoma ciprofloxacin co-administration with cyclophosphamide resulted in a significantly lower exposure to cyclophosphamide and in a lower ratio of the AUCs of cyclophosphamide and its active metabolite 4-hydroxycyclophosphamide (111). The authors postulated that ciprofloxacin inhibited CYP3A2, which is involved in the conversion of cyclophosphamide to 4- hydroxycyclophosphamide.

Didanosine

Didanosine one enteric-coated capsule/day (400 mg/day) did not affect the absorption of ciprofloxacin in 16 patients (112).

Glibenclamide

Hypoglycemia and raised serum concentrations of glibenclamide, which is metabolized by CYP2C9, occurred after treatment with ciprofloxacin for 1 week in a patient taking long-term glibenclamide (113).

- A man with diabetes taking glibenclamide was given ciprofloxacin and developed prolonged hypoglycemia, which persisted for over 24 hours (114).

Insulin

Hypoglycemia occurred in a patient treated with insulin and ciprofloxacin 500 mg bd (44).

Methadone

Ciprofloxacin, given to a patient who had been successfully treated with methadone for more than 6 years, caused profound sedation, confusion, and respiratory depression (115). This may have been due to inhibition of CYP1A2 and CYP3A4, two of the isozymes involved in the metabolism of methadone.

Methotrexate

Methotrexate elimination can be delayed by ciprofloxacin. Two adolescents with malignant diseases had reduced elimination of methotrexate (12 g/m^2 4-hourly) when they took ciprofloxacin 500 mg bd (116).

Olanzapine

The plasma concentration of olanzapine doubled in a patient who also took ciprofloxacin, a potent inhibitor of

CYP1A2 (130). The magnitude of the interaction was surprising, because available data suggest that CYP1A2-mediated oxidation of olanzapine accounts for only a small portion of the biotransformation of olanzapine relative to glucuronidation.

Oral contraceptives

Some antibiotics can reduce the efficacy of oral contraceptives. However, there is pharmacokinetic evidence that plasma concentrations of oral contraceptive steroids are unchanged by co-administration of ciprofloxacin (117). Furthermore, ciprofloxacin (500 mg bd) did not interfere with the ovarian suppression produced by the oral contraceptive Marvelon (30 micrograms of ethinylestradiol plus 150 micrograms of desogestrel) in 24 healthy women in a randomized, double-blind, placebo-controlled, crossover trial (118).

Phenazopyridine

In a double-blind, crossover, randomized trial in 24 healthy men oral co-administration of phenazopyridine increased the systemic availability of a single dose of ciprofloxacin (119).

Phenytoin

Ciprofloxacin may interact with phenytoin reducing phenytoin concentrations (120).

- A lower than expected phenytoin serum concentration has been measured in a 78-year-old white woman with a grade III astrocytoma of the right parieto-occipital region treated with ciprofloxacin (500 mg bd) (121).

Increased renal excretion has been suggested to be at least partly responsible for the increased clearance.

Besides this kinetic interaction, the possible epileptogenic potential of ciprofloxacin itself may contribute to the development of seizure activity.

Probenecid

The renal excretion of ciprofloxacin was reduced and plasma concentrations increased by probenecid (122).

Quinidine

Serum quinidine concentrations rose during concomitant administration of ciprofloxacin (123). The authors speculated that the mechanism was inhibition of cytochrome P450 by ciprofloxacin.

Rifamycins

Rifampicin-induced lupus-like syndrome is associated with combination therapy with ciprofloxacin, since rifampicin is metabolized by (among others) CYP3A4, which is inhibited by ciprofloxacin, and combined usage may lead to higher rifampicin blood concentrations (124).

Ropinirole

During co-administration of ciprofloxacin with ropinirole in 12 patients there was an increase in the plasma ropinirole concentration, which is metabolized by CYP1A2 (125).

Teicoplanin

Teicoplanin should be administered separately from ciprofloxacin, since precipitation has been provoked by concomitant parenteral administration (131).

Tizanidine

In a retrospective survey of medical records obtained from 1165 patients there were eight patients who took tizanidine and ciprofloxacin (126). None was taking medications that inhibit CYP1A2, besides ciprofloxacin. There were adverse effects attributed to tizanidine in three patients (low heart rate in two, low body temperature in one, low blood pressure in two, drowsiness in two, and confusion in one). In one patient ciprofloxacin was withdrawn because of confusion.

- A 45-year-old Japanese woman taking tizanidine 3 mg/day was given ciprofloxacin 400 mg/day for 7 days. On day 2 she complained of drowsiness and had a low blood pressure (92/54 mmHg). She recovered immediately after withdrawal of ciprofloxacin. The doses of her other medications remained unchanged during this period.

Warfarin

Ciprofloxacin can occasionally cause an exaggerated hypoprothrombinemic response and bleeding in patients taking warfarin. In 66 patients (median age 72 years, range 36–94), the mean time to detection of the coagulopathy after ciprofloxacin challenge was 5.5 days (127). Hospitalization was reported in 15 cases, bleeding in 25, and death in one. The median INR was 10.0. Patients in their seventh decade and those requiring polypharmacy were most at risk.

Food–Drug Interactions

The systemic availability of ciprofloxacin is reduced by 30–36% when it is taken with dairy products (128).

References

1. Centers for Disease Control and Prevention (CDC). Update: adverse events associated with anthrax prophylaxis among postal employees—New Jersey, New York City, and the District of Columbia metropolitan area, 2001. MMWR Morb Mortal Wkly Rep 2001;50(47):1051–4.
2. Centers for Disease Control and Prevention (CDC). Update: Investigation of bioterrorism-related anthrax and adverse events from antimicrobial prophylaxis. MMWR Morb Mortal Wkly Rep 2001;50(44):973–6.
3. The Centers for Disease Control and Prevention. Investigation of bioterrorism-related anthrax and adverse events from antimicrobial prophylaxis. JAMA 2001;286(20):2536–7.
4. The Centers for Disease Control and Prevention. Update: investigation of bioterrorism-related anthrax and interim

guidelines for exposure management and antimicrobial therapy, October 2001. JAMA 2001;286(18):2226–32.

5. Sundram FX, Wong WY, Ang ES, Goh AS, Ng DC, Yu S. Evaluation of technetium-99m ciprofloxacin (Infecton) in the imaging of infection. Ann Acad Med Singapore 2000;29(6):699–703.

6. Naber KG, Landen H. Rapid resolution of symptoms with ciprofloxacin therapy in 3859 hospitalised patients with urinary tract infection. Int J Antimicrob Agents 2004;23 Suppl 1:S35-40.

7. Chodosh S, Schreurs A, Siami G, Barkman HW Jr, Anzueto A, Shan M, Moesker H, Stack T, Kowalsky SThe Bronchitis Study Group. Efficacy of oral ciprofloxacin vs. clarithromycin for treatment of acute bacterial exacerbations of chronic bronchitis. Clin Infect Dis 1998;27(4):730–8.

8. Chodosh S, McCarty J, Farkas S, Drehobl M, Tosiello R, Shan M, Aneiro L, Kowalsky SThe Bronchitis Study Group. Randomized, double-blind study of ciprofloxacin and cefuroxime axetil for treatment of acute bacterial exacerbations of chronic bronchitis. Clin Infect Dis 1998;27(4):722–9.

9. Owens RC Jr, Ambrose PG. Torsades de pointes associated with fluoroquinolones. Pharmacotherapy 2002;22(5):663–8discussion 668–72.

10. Singh H, Kishore K, Gupta MS, Khetarpal S, Jain S, Mangla M. Ciprofloxacin-induced QTc prolongation. J Assoc Physicians India 2002;50:430–1.

11. Owens RC Jr. QT prolongation with antimicrobial agents: understanding the significance. Drugs 2004;64(10):1091–124.

12. Letsas KP, Sideris A, Kounas SP, Efremidis M, Korantzopoulos P, Kardaras F. Drug-induced QT interval prolongation after ciprofloxacin administration in a patient receiving olanzapine. Int J Cardiol 2006;109(2):273–4.

13. Flanagan MC, Mitchell ES, Haigney MC. Ciprofloxacin-induced torsade de pointes. Int J Cardiol 2006;113(2):239–41.

14. Prabhakar M, Krahn AD. Ciprofloxacin-induced acquired long QT syndrome. Heart Rhythm 2004; 1(5):624–6.

15. Daya SK, Gowda RM, Khan IA. Ciprofloxacin- and hypocalcemia-induced torsade de pointes triggered by hemodialysis. Am J Ther 2004;11(1):77–9.

16. Steiger D, Bubendorf L, Oberholzer M, Tamm M, Leuppi JD. Ciprofloxacin-induced acute interstitial pneumonitis. Eur Respir J 2004;23(1):172–4.

17. McCarty JM, Richard G, Huck W, Tucker RM, Tosiello RL, Shan M, Heyd A, Echols RM. A randomized trial of short-course ciprofloxacin, ofloxacin, or trimethoprim/sulfamethoxazole for the treatment of acute urinary tract infection in women. Ciprofloxacin Urinary Tract Infection Group. Am J Med 1999;106(3):292–9.

18. Tattevin P, Messiaen T, Pras V, Ronco P, Biour M. Confusion and general seizures following ciprofloxacin administration. Nephrol Dial Transplant 1998;13(10):2712–3.

19. Kushner JM, Peckman HJ, Snyder CR. Seizures associated with fluoroquinolones. Ann Pharmacother 2001;35(10):1194–8.

20. Orr CF, Rowe DB. Eardrop attacks: seizures triggered by ciprofloxacin eardrops. Med J Aust 2003;178:343.

21. Kisa C, Yildirim SG, Aydemir C, Cebeci S, Goka E. Prolonged electroconvulsive therapy seizure in a patient taking ciprofloxacin. J ECT 2005;21(1):43–4.

22. Azar S, Ramjiani A, Van Gerpen JA. Ciprofloxacin-induced chorea. Mov Disord 2005;20(4):513–4.

23. Post B, Koelman JH, Tijssen MA. Propriospinal myoclonus after treatment with ciprofloxacin. Mov Disord 2004;19(5):595–7.

24. Lee CH, Cheung RT, Chan TM. Ciprofloxacin-induced oral facial dyskinesia in a patient with normal liver and renal function. Hosp Med 2000;61(2):142–3.

25. MacLeod W. Case report: severe neurologic reaction to ciprofloxacin. Can Fam Physician 2001;47:553–5.

26. Zehnder D, Hoigne R, Neftel KA, Sieber R. Painful dysaesthesia with ciprofloxacin. BMJ 1995;311(7014):1204.

27. Vrabec TR, Sergott RC, Jaeger EA, Savino PJ, Bosley TM. Reversible visual loss in a patient receiving high-dose ciprofloxacin hydrochloride (Cipro). Ophthalmology 1990;97(6):707–10.

28. Madhavan HN, Rao SK. Ciprofloxacin precipitates in the corneal epithelium. J Cataract Refract Surg 2002;28(6):909.

29. Eiferman RA, Snyder JP, Nordquist RE. Ciprofloxacin microprecipitates and macroprecipitates in the human corneal epithelium. J Cataract Refract Surg 2001;27(10):1701–2.

30. Patwardhan A, Khan M. Topical ciprofloxacin can delay recovery from viral ocular surface infection. J R Soc Med 2005;98(6):274–5.

31. Ikiz AO, Serbetcioglu B, Guneri EA, Sutay S, Ceryan K. Investigation of topical ciprofloxacin ototoxicity in guinea pigs. Acta Otolaryngol 1998;118(6):808–12.

32. Miro N. Controlled multicenter study on chronic suppurative otitis media treated with topical applications of ciprofloxacin 0.2% solution in single-dose containers or combination of polymyxin B, neomycin, and hydrocortisone suspension Otolaryngol Head Neck Surg 2000;123(5):617–23.

33. Claros P, Sabater F, Claros A Jr, Claros A. Determinacion de niveles plasmaticos de ciprofloxacino en niños tratados con ciprofloxacino topico al 0.2% en presencia de perforacion timpanica. [Determination of plasma ciprofloxacin levels in children treated with 0.2% topical ciprofloxacin for tympanic perforation.] Acta Otorrinolaringol Esp 2000;51(2):97–9.

34. Zabala S, Gascon A, Bartolome C, Castiella J, Juyol M. Ciprofloxacino y psicosis aguda. [Ciprofloxacin and acute psychosis.] Enferm Infecc Microbiol Clin 1998;16(1):42.

35. James EA, Demian AZ. Acute psychosis in a trauma patient due to ciprofloxacin. Postgrad Med J 1998;74(869):189–90.

36. Grassi L, Biancosino B, Pavanati M, Agostini M, Manfredini R. Depression or hypoactive delirium? A report of ciprofloxacin-induced mental disorder in a patient with chronic obstructive pulmonary disease. Psychother Psychosom 2001;70(1):58–9.

37. Imani K, Druart F, Glibert A, Morin T. Troubles neuropsychiatriques induits par la ciprofloxacine. [Neuropsychiatric disorders induced by ciprofloxacin.] Presse Méd 2001;30(27):1356.

38. Tripathi A, Chen SI, O'Sullivan S. Acute psychosis following the use of topical ciprofloxacin. Arch Ophthalmol 2002;120(5):665–6.

39. Bhalerao S, Talsky A, Hansen K, Kingstone E, Schroeder B, Karim Z, Fung I. Ciprofloxacin-induced manic episode. Psychosomatics 2006;47(6):539–40.

40. Steinert T, Studemund H. Acute delusional parasitosis under treatment with ciprofloxacin. Pharmacopsychiatry 2006;39(4):159–60.

41. Norra C, Skobel E, Breuer C, Haase G, Hanrath P, Hoff P. Ciprofloxacin-induced acute psychosis in a patient with

multidrug-resistant tuberculosis. Eur Psychiatry 2003;18:262–3.

42. Kundu AK. Ciprofloxacin-induced severe cutaneous reaction and haemolysis in a young adult. J Assoc Physicians India 2000;48(6):649–50.

43. Mouraux A, Gille M, Pieret F, Declercq I. Purpura thrombotique thrombocytopenique fulminant au decour d'un traitement pas ciprofloxacine. [Fulminant thrombotic thrombocytopenic purpura in the course of ciprofloxacin therapy.] Rev Neurol (Paris) 2002;158(11):1115–7.

44. Kljucar S, Rost KL, Landen H. Ciprofloxacin in der Therapie des nosobomialen pneumonie: Eine Anwendungs beobachtung bei 676 patienten. [Ciprofloxacin in the treatment of hospital-acquired pneumonia: a surveillance study in 676 patients.] Pneumologie 2002;56(10):599–604.

45. Finsterer J, Kotzailias N. Thrombocytosis under ciprofloxacin and tazobactam/piperacillin. Platelets 2003;14:329–31.

46. Moore GC, Redfern J, Shiach CR, Webb K, Jones AM. Coagulopathy in two patients with cystic fibrosis treated with ciprofloxacin. J Cyst Fibros 2007;6(3):209–11.

47. Grassi C, Salvatori E, Rosignoli MT, Dionisio PPrulifloxacin Study Group. Randomized, double-blind study of prulifloxacin versus ciprofloxacin in patients with acute exacerbations of chronic bronchitis. Respiration 2002;69(3):217–22.

48. Fernandez de la Puebla Gimenez RA, Lechuga Varona MT, Garcia Sanchez E. Colitis seudomembranosa por ciprofloxacino. [Pseudomembranous colitis due to ciprofloxacin.] Med Clin (Barc) 1998;111(7):278–9.

49. Zabala Lopez S, Iglesias Quiros E, Gonzalez Heras S, Martinez Navarro C, Gambaro Royo B. Diarrea asociada a Clostridium difficile secundaria al uso de ciprofloxacino, complicando un primer brote de enfermedad inflamatoria intestinal. [Diarrhea associated with Clostridium difficile secondary to the use of ciprofloxacin, complicating a first occurrence of intestinal inflammatory disease.] Rev Esp Enferm Dig 2000;92(8):539–40.

50. Thomas C, Golledge CL, Riley TV. Ciprofloxacin and Clostridium difficile-associated diarrhea. Infect Control Hosp Epidemiol 2002;23(11):637–8.

51. Yip C, Loeb M, Salama S, Moss L, Olde J. Quinolone use as a risk factor for nosocomial Clostridium difficile-associated diarrhea. Infect Control Hosp Epidemiol 2001;22(9):572–5.

52. Contreras MA, Luna R, Mulero J, Andreu JL. Severe ciprofloxacin-induced acute hepatitis. Eur J Clin Microbiol Infect Dis 2001;20(6):434–5.

53. Zimpfer A, Propst A, Mikuz G, Vogel W, Terracciano L, Stadlmann S. Ciprofloxacin-induced acute liver injury: case report and review of literature. Virchows Arch 2004;444(1):87–9.

54. Goetz M, Galle PR, Schwarting A. Non-fatal acute liver injury possibly related to high-dose ciprofloxacin. Eur J Clin Microbiol Infect Dis 2003;22:294–6.

55. Mann S, Thillainayagam A. Is ciprofloxacin a new cause of acute pancreatitis? J Clin Gastroenterol 2000;31(4):.

56. Fogo AB. Quiz page. Diabetic nephropathy with superimposed acute interstitial nephritis. Am J Kidney Dis 2003;41:A47, E1-3.

57. Mannaerts L, Van der Wurff AA, Wolfhagen FH. Interstitiele nefritis toegeschreven aan het gebruik van piperacilline–tazobactam en van ciprofloxacine. [Interstitial nephritis attributed to treatment with piperacillin–tazobactam and with ciprofloxacin.] Ned Tijdschr Geneeskd 2006;150(14):804–7.

58. Sedlacek M, Suriawinata AA, Schoolwerth A, Remillard BD. Ciprofloxacin crystal nephropathy—a 'new' cause of acute renal failure. Nephrol Dial Transplant 2006;21(8):2339–40.

59. Montagnac R, Briat C, Schillinger F, Sartelet H, Birembaut P, Daudon M. [Fluoroquinolone induced acute renal failure. General review about a case report with crystalluria due to ciprofloxacin.] Nephrol Ther 2005;1(1):44–51.

60. Shih DJ, Korbet SM, Rydel JJ, Schwartz MM. Renal vasculitis associated with ciprofloxacin. Am J Kidney Dis 1995;26(3):516–9.

61. Raja N, Miller WE, McMillan R, Mason JR. Ciprofloxacin-associated acute renal failure in patients undergoing high-dose chemotherapy and autologous stem cell rescue. Bone Marrow Transplant 1998;21(12):1283–4.

62. Dharnidharka VR, Nadeau K, Cannon CL, Harris HW, Rosen S. Ciprofloxacin overdose: acute renal failure with prominent apoptotic changes. Am J Kidney Dis 1998;31(4):710–2.

63. Bald M, Ratjen F, Nikolaizik W, Wingen AM. Ciprofloxacin-induced acute renal failure in a patient with cystic fibrosis. Pediatr Infect Dis J 2001;20(3):320–1.

64. Allan DS, Thompson CM, Barr RM, Clark WF, Chin-Yee IH. Ciprofloxacin-associated hemolytic–uremic syndrome. Ann Pharmacother 2002;36(6):1000–2.

65. Chopra N, Fine PL, Price B, Atlas I. Bilateral hydronephrosis from ciprofloxacin induced crystalluria and stone formation. J Urol 2000;164(2):438.

66. Maquirriain Gorriz MT, Merino Munoz F, Tres Belzuncgui JC, Sangros Gonzalez FJ. Erupcion fija por farmacos inducida por ciprofloxacino. [Fixed drug eruption induced by ciprofloxacin.] Aten Primaria 1998;21(8):585–6.

67. Rodriguez-Morales A, Llamazares AA, Benito RP, Cocera CM. Fixed drug eruption from quinolones with a positive lesional patch test to ciprofloxacin. Contact Dermatitis 2001;44(4):255.

68. Pons R, Escutia B. Vasculitis por ciprofloxacino con afectacion cutanea y renal. [Ciprofloxacin-induced vasculitis with cutaneous and renal involvement.] Nefrologia 2001;21(2):209–12.

69. Kimyai-Asadi A, Usman A, Nousari HC. Ciprofloxacin-induced bullous pemphigoid. J Am Acad Dermatol 2000;42(5 Pt 1):847.

70. Perez Vazquez A, Gutierrez Perez B, Carreter de Granda E, Zuniga Perez-Lemaur M, Conde Yague R. Vasculitis cutanea por ciprofloxacino. [Cutaneous vasculitis caused by ciprofloxacin.] An Med Interna 2000;17(4):225.

71. Terzano C, Taurino AE, Peona V. Nebulized tobramycin in patients with chronic respiratory infections during clinical evolution of Wegener's granulomatosis. Eur Rev Med Pharmacol Sci 2001;5(4):131–8.

72. Jongen-Lavrencic M, Schneeberger PM, Van der Hoeven JG. Ciprofloxacin-induced toxic epidermal necrolysis in a patient with systemic lupus erythematosus. Infection 2003;31:428–9.

73. Hallgren J, Tengvall-Linder M, Persson M, Wahlgren CF. Stevens–Johnson syndrome associated with ciprofloxacin: a review of adverse cutaneous events reported in Sweden as associated with this drug. J Am Acad Dermatol 2003;49 Suppl 5:S267-9.

74. Mandal B, Steward M, Singh S, Jones H. Ciprofloxacin-induced toxic epidermal necrolysis (TEN) in a nonagenarian: a case report. Age Ageing 2004;33(4):405–6.

75. Urbina F, Barrios M, Sudy E. Photolocalized purpura during ciprofloxacin therapy. Photodermatol Photoimmunol Photomed 2006;22(2):111–2.

76. Hausermann P, Scherer K, Weber M, Bircher AJ. Ciprofloxacin-induced acute generalized exanthematous pustulosis mimicking bullous drug eruption confirmed by a positive patch test. Dermatology 2005;211(3):277–80.

77. Ozaras R, Mert A, Tahan V, Uraz S, Ozaydin I, Yilmaz MH, Ozaras N. Ciprofloxacin and Achilles' tendon rupture: a causal relationship. Clin Rheumatol 2003;22:500–1.

78. Khaliq Y, Zhanel GG. Fluoroquinolone-associated tendinopathy: a critical review of the literature. Clin Infect Dis 2003;36:1404–10.

79. Shortt P, Wilson R, Erskine I. Tendinitis: the Achilles heel of quinolones! Emerg Med J 2006;23(12):e63.

80. Palin SL, Gough SC. Rupture of the Achilles tendon associated with ciprofloxacin. Diabet Med 2006;23(12):1386–7.

81. Chhajed PN, Plit ML, Hopkins PM, Malouf MA, Glanville AR. Achilles tendon disease in lung transplant recipients: association with ciprofloxacin. Eur Respir J 2002;19(3):469–71.

82. Petersen W, Laprell H. Die "schleichende" Ruptur der Achillessehne nach Ciprofloxacin induzierter Tendopathie. Ein Fallbericht. [Insidious rupture of the Achilles tendon after ciprofloxacin-induced tendopathy. A case report.] Unfallchirurg 1998;101(9):731–4.

83. Mouzopoulos G, Stamatakos M, Vasiliadis G, Skandalakis P. Rupture of adductor longus tendon due to ciprofloxacin. Acta Orthop Belg 2005;71(6):743–5.

84. Fama U, Irace S, Frati R, de Gado F, Scuderi N. Is it a real risk to take ciprofloxacin? Plast Reconstr Surg 2004;114(1):267.

85. Gendrel D, Moulin F. Fluoroquinolones in paediatrics. Paediatr Drugs 2001;3(5):365–77.

86. Gbadoe AD, Dogba A, Dagnra AY, Atakouma Y, Tekou H, Assimadi JK. Osteomyelites aïgues de l'enfant drepanocytaire en zone tropicale: interêt de l'utilisation des fluoroquinolones par voie orale. [Acute osteomyelitis in the child with sickle cell disease in a tropical zone: value of oral fluoroquinolones.] Arch Pediatr 2001;8(12):1305–10.

87. Malaguti M, Triolo L, Biagini M. Ciprofloxacin-associated Achilles tendon rupture in a hemodialysis patient. J Nephrol 2001;14(5):431–2.

88. Doherty CP, Saha SK, Cutting WA. Typhoid fever, ciprofloxacin and growth in young children. Ann Trop Paediatr 2000;20(4):297–303.

89. Singh UK, Sinha RK, Prasad B, Chakrabarti B, Sharma SK. Ciprofloxacin in children: is arthropathy a limitation? Indian J Pediatr 2000;67(5):386–7.

90. Postnikov SS, Nazhimov VP, Semykin SIu, Kapranov NI. [Comparative morphological analysis of the articular cartilage, epiphyseal plate, spongy bone, and synovial membrane of the knee joint in children treated and not treated with ciprofloxacin.]Antibiot Khimioter 2000; 45(11):9–13.

91. Grady R. Safety profile of quinolone antibiotics in the pediatric population. Pediatr Infect Dis J 2003;22:1128–32.

92. Kubin R. Safety and efficacy of ciprofloxacin in paediatric patients—review. Infection 1993;21(6):413–21.

93. Mullen CA, Petropoulos D, Rytting M, Jeha S, Zipf T, Roberts WM, Rolston KV. Acute reversible arthropathy in a pediatric patient with cancer treated with a short course of ciprofloxacin for febrile neutropenia. J Pediatr Hematol Oncol 1998;20(5):516–7.

94. Burke P, Burne SR. Allergy associated with ciprofloxacin. BMJ 2000;320(7236):679.

95. Ho DY, Song JC, Wang CC. Anaphylactoid reaction to ciprofloxacin. Ann Pharmacother 2003;37:1018–23.

96. Sable D, Murakawa GJ. Quinolones in dermatology. Dis Mon 2004; 50(7): 381-94.

97. Storsley L, Geldenhuys L. Ciprofloxacin-induced ANCA-negative cutaneous and renal vasculitis—resolution with drug withdrawal. Nephrol Dial Transplant 2007;22(2): 660-1.

98. Kothur K, Singh M, Dayal D. Ciprofloxacin-induced anaphylactoid reaction. Eur J Pediatr 2006;165(8):573–4.

99. Knoops L, van den Neste E, Hamels J, Theate I, Mineur P. Angioimmunoblastic lymphadenopathy following ciprofloxacin administration. Acta Clin Belg 2002;57(2):71–3.

100. Webster G, Schiffman JD, Dosanjh AS, Amieva MR, Gans HA, Sectish TC. Jarisch–Herxheimer reaction associated with ciprofloxacin administration for tick-borne relapsing fever. Pediatr Infect Dis J 2002;21(6):571–3.

101. Wlazlowski J, Krzyzanska-Oberbek A, Sikora JP, Chlebna-Sokol D. Use of the quinolones in treatment of severe bacterial infections in premature infants. Acta Pol Pharm 2000;57(Suppl):28–31.

102. Postnikov SS, Semykin SIu, Kapranov NI, Perederko LV, Polikarpova SV. [The efficacy and safety of ciprofloxacin in treating children with mucoviscidosis.]Antibiot Khimioter 2000;45(4):14–7.

103. Leibovitz E, Janco J, Piglansky L, Press J, Yagupsky P, Reinhart H, Yaniv I, Dagan R. Oral ciprofloxacin vs. intramuscular ceftriaxone as empiric treatment of acute invasive diarrhea in children. Pediatr Infect Dis J 2000;19(11):1060–7.

104. Heyd A, Haverstock D. Retrospective analysis of the safety profile of oral and intravenous ciprofloxacin in a geriatric population. Clin Ther 2000;22(10):1239–50.

105. Joukhadar C, Klein N, Frossard M, Minar E, Stass H, Lackner E, Herrmann M, Riedmuller E, Muller M. Angioplasty increases target site concentrations of ciprofloxacin in patients with peripheral arterial occlusive disease. Clin Pharmacol Ther 2001;70(6):532–9.

106. Malone RS, Fish DN, Abraham E, Teitelbaum I. Pharmacokinetics of levofloxacin and ciprofloxacin during continuous renal replacement therapy in critically ill patients. Antimicrob Agents Chemother 2001;45(10):2949–54.

107. Hampel B, Hullmann R, Schmidt H. Ciprofloxacin in pediatrics: worldwide clinical experience based on compassionate use—safety report. Pediatr Infect Dis J 1997;16(1):127–9.

108. Wilhelmus KR, Abshire RL. Corneal ciprofloxacin precipitation during bacterial keratitis. Am J Ophthalmol 2003;136:1032–7.

109. Wrishko RE, Levine M, Primmett DR, Kim S, Partovi N, Lewis S, Landsberg D, Keown PA. Investigation of a possible interaction between ciprofloxacin and cyclosporine in renal transplant patients. Transplantation 1997;64(7): 996–999.

110. Joos AA. Pharmakologische Interaktionen von Antibiotika und Psychopharmaka. [Pharmacologic interactions of antibiotics and psychotropic drugs.] Psychiatr Prax 1998;25(2):57–60.

111. Afsharian P, Mollgard L, Hassan Z, Xie H, Kimby E, Hassan M. The effect of ciprofloxacin on cyclophosphamide pharmacokinetics in patients with non-Hodgkin lymphoma. Eur J Haematol 2005;75(3):206–11.

112. Jablonowski H. Didanosin als Kapsel. Bewahrtes Medikament in neuer Form. [Didanosine as a capsule. A reliable drug in a new dosage form.] MMW Fortschr Med 2001;143(Suppl 1):92–5.
113. Roberge RJ, Kaplan R, Frank R, Fore C. Glyburide–ciprofloxacin interaction with resistant hypoglycemia. Ann Emerg Med 2000;36(2):160–3.
114. Lin G, Hays DP, Spillane L. Refractory hypoglycemia from ciprofloxacin and glyburide interaction. J Toxicol Clin Toxicol 2004;42(3):295–7.
115. Herrlin K, Segerdahl M, Gustafsson LL, Kalso E. Methadone, ciprofloxacin, and adverse drug reactions. Lancet 2000;356(9247):2069–70.
116. Dalle JH, Auvrignon A, Vassal G, Leverger G, Kalifa C. Interaction methotrexate–ciprofloxacine: à propos de deux cas d'intoxication sévère. [Methotrexate–ciprofloxacin interaction: report of two cases of severe intoxication.] Arch Pediatr 2001;8(10):1078–81.
117. Archer JS, Archer DF. Oral contraceptive efficacy and antibiotic interaction: a myth debunked. J Am Acad Dermatol 2002;46(6):917–23.
118. Scholten PC, Droppert RM, Zwinkels MG, Moesker HL, Nauta JJ, Hoepelman IM. No interaction between ciprofloxacin and an oral contraceptive. Antimicrob Agents Chemother 1998;42(12):3266–8.
119. Marcelin-Jimenez G, Angeles AP, Martinez-Rossier L, Fernandez SA. Ciprofloxacin bioavailability is enhanced by oral co-administration with phenazopyridine: a pharmacokinetic study in a Mexican population. Clin Drug Investig 2006;26(6):323–8.
120. Otero MJ, Moran D, Valverde MP. Interaction between phenytoin and ciprofloxacin. Ann Pharmacother 1999;33(2):251–2.
121. McLeod R, Trinkle R. Comment: unexpectedly low phenytoin concentration in a patient receiving ciprofloxacin. Ann Pharmacother 1998;32(10):1110–1.
122. Jaehde U, Sorgel F, Reiter A, Sigl G, Naber KG, Schunack W. Effect of probenecid on the distribution and elimination of ciprofloxacin in humans. Clin Pharmacol Ther 1995;58(5):532–41.
123. Teppo AM, Haltia K, Wager O. Immunoelectrophoretic "tailing" of albumin line due to albumin-IgG antibody complexes: a side effect of nitrofurantoin treatment? Scand J Immunol 1976;5(3):249–61.
124. Patel GK, Anstey AV. Rifampicin-induced lupus erythematosus. Clin Exp Dermatol 2001;26(3):260–2.
125. Kaye CM, Nicholls B. Clinical pharmacokinetics of ropinirole. Clin Pharmacokinet 2000;39(4):243–54.
126. Momo K, Homma M, Kohda Y, Ohkoshi N, Yoshizawa T, Tamaoka A. Drug interaction of tizanidine and ciprofloxacin: case report. Clin Pharmacol Ther 2006;80(6):717–9.
127. Ellis RJ, Mayo MS, Bodensteiner DM. Ciprofloxacin–warfarin coagulopathy: a case series. Am J Hematol 2000;63(1):28–31.
128. Schmidt LE, Dalhoff K. Food–drug interactions. Drugs 2002;62(10):1481–502.
129. Raaska K, Neuvonen PJ. Ciprofloxacin increases serum clozapine and N-desmethylclozapine: a study in patients with schizophrenia. Eur J Clin Pharmacol 2000;56(8):585–9.
130. Markowitz JS, DeVane CL. Suspected ciprofloxacin inhibition of olanzapine resulting in increased plasma concentration. J Clin Psychopharmacol 1999;19(3):289–91.
131. Jim LK. Physical and chemical compatibility of intravenous ciprofloxacin with other drugs. Ann Pharmacother 1993;27(6):704–7.

Clarithromycin

See also Macrolide antibiotics

General Information

Clarithromycin is a commonly used macrolide antibiotic and is a regular part of regimens for the eradication of *Helicobacter pylori*, often in combination with a nitroimidazole antibiotic as well, in addition to a proton pump inhibitor. Variable rates of adverse events (4–30%) have been reported with clarithromycin.

The most common adverse effects of clarithromycin in clinical trials were diarrhea and abnormal taste, both of which occurred in 6% of patients [1]. Other events that occurred less often were nausea (3%), dyspepsia (2%), abdominal pain (2%), and headache (2%). Clarithromycin can also cause fixed drug eruptions, hypersensitivity reactions, and leukocytoclastic vasculitis. It can also cause mania, especially when given with glucocorticoids.

Comparative studies

In a double-blind, multicenter trial in 328 patients with *H. pylori* infection and non-ulcer dyspepsia, omeprazole 20 mg bd, amoxicillin 1 g bd, and clarithromycin 500 mg bd were compared with omeprazole alone. The rate of success and quality of life were similar in both groups. There were no serious adverse events. However, there were 12 withdrawals in the group given omeprazole and antibiotics and two in the group given omeprazole alone. Diarrhea occurred in 63 patients in those given omeprazole and antibiotics and in ten patients given omeprazole alone (2). In another double-blind, placebo-controlled trial eradication of *H. pylori* (omeprazole 20 mg, amoxicillin 1 g, and clarithromycin 500 mg bd) in long-term users of NSAIDs with past or current peptic ulcer or troublesome dyspepsia led to impaired healing of gastric ulcers and did not affect the rate of peptic ulcers or dyspepsia over 6 months (3).

Beta-lactam antibiotics

In a multicenter, double-blind, randomized comparison of cefprozil, 500 mg bd for 5 days and clarithromycin 500 mg bd for 10 days in 295 subjects with an acute exacerbation of chronic bronchitis, the most common adverse effects of clarithromycin were nausea (8%), diarrhea (12%), taste disturbance (8%), and dry mouth (5%) (4).

Clarithromycin (250 mg bd for 10 days) was as effective as cefuroxime axetil (250 mg bd for 10 days) in the treatment of acute maxillary sinusitis in a randomized, double-blind, multicenter study in 370 patients; 10% of patients in each group had adverse events (5).

Other macrolides

The incidence of disseminated *Mycobacterium avium* complex (MAC) infection has increased dramatically with the AIDS epidemic. Treatment regimens for patients with a positive culture for MAC from a sterile site should include two or more drugs, including clarithromycin. Prophylaxis against disseminated MAC should be considered for patients with a CD4 cell count of less than $50 \times 10^6/l$ (6). In a randomized, open trial in 37 patients with HIV-associated disseminated MAC infection, treatment with clarithromycin + ethambutol produced more rapid resolution of bacteremia, and was more effective at sterilization of blood cultures after 16 weeks than azithromycin + ethambutol (7).

In a direct comparison of clarithromycin with erythromycin stearate, the rate of adverse events was 19% in 96 patients taking clarithromycin and 35% in 112 patients taking erythromycin (8). Most of the adverse events associated with clarithromycin affect the gastrointestinal tract (7%).

In a prospective, single-blind, randomized study of a 7-day course of clarithromycin (7.5 mg/kg bd) and a 14-day course of erythromycin (13.3 mg/kg tds) in 153 children with pertussis, the incidence of treatment-emergent drug-related adverse events was significantly higher with erythromycin than with clarithromycin (62 versus 45%) (9). Three subjects given erythromycin withdrew prematurely because of adverse events: one because of a rash; one with vomiting and diarrhea; and one with vomiting, abdominal pain, and rash.

Quinolones

In a multicenter, double-blind, randomized comparison of trovafloxacin 200 mg and clarithromycin 500 mg bd in 176 subjects with acute exacerbations of chronic bronchitis, the most common adverse effects of clarithromycin were nausea (3%), diarrhea (4%), and taste disturbances (4%) (10).

Tetracyclines

Clarithromycin (0.75–2 g/day), minocycline (200 mg/day), and clofazimine (100 mg/day) for 15 months were investigated as treatment of MAC lung disease in 30 HIV-negative patients. Eight patients did not complete the study owing to deviations from protocol or adverse effects. Persistently negative cultures were found in 14 of the other patients. There were three cases of hepatic disturbances and three of ototoxicity, which required a reduction in clarithromycin dosage after a short interruption of treatment (11).

Organs and Systems

Cardiovascular

Clarithromycin caused modest but significant prolongation of the QT_c interval in 28 children with respiratory tract infections after 24 hours of treatment (mean prolongation 22 ms; 95% CI = 14, 30); in seven cases the QT_c interval was over 440 ms, and in one case it was over 460 ms [12].

- QT interval prolongation and a ventricular dysrhythmia occurred in an HIV-positive 30-year-old man at the start of intravenous clarithromycin therapy 500 mg 12-hourly (13).

Intravenous clarithromycin caused thrombophlebitis in four patients when it was given inappropriately as a rapid bolus injection instead of a short infusion; the manufacturers have received other reports of similar reactions, even with infusions, but the incidence seems to be considerably lower than with erythromycin (14). In a prospective, non-randomized study, phlebitis occurred in 15 of 19 patients treated with intravenous erythromycin (incidence rate of 0.40 episodes/patient-day) and in 19 of 25 patients treated with intravenous clarithromycin (0.35 episodes/patient-day) (15).

Respiratory

Bronchospasm with clarithromycin occurred in a 44-year-old woman who had no history of respiratory allergies but had had adverse drug reactions to general and regional anesthetics and to ceftriaxone (16). After the administration of a quarter of the therapeutic dose the patient had dyspnea, cough, and bronchospasm throughout the lung.

Pulmonary infiltration with eosinophilia occurred in a 17-year old youth who took clarithromycin 500 mg bd for 7 days; clarithromycin was immediately withdrawn, and he quickly recovered [17].

Nervous system

Adverse events on the nervous system due to clarithromycin have been observed in 3% of patients.

- Progressive loss of strength and difficulty in swallowing and eye opening after the first dose of clarithromycin (2 g/day) occurred in a patient with cerebral toxoplasmosis and AIDS (18). This myasthenic syndrome resolved within 6 hours of withdrawal of clarithromycin and administration of pyridostigmine.

The authors postulated that this adverse effect may have been the consequence of neuromuscular blockade, through inhibition of the presynaptic release of acetylcholine.

Sensory systems

Eyes

Topical clarithromycin can cause self-resolving corneal deposits (19).

Ears

Tinnitus has been attributed to clarithromycin.

- A 50-year-old man developed tinnitus 9 days after he started to take clarithromycin 500 mg bd for a peptic ulcer [20]. The clarithromycin was withdrawn, but the patient continued to take his other medications (amoxicillin, and lansoprazole). The tinnitus resolved after 2 days and he did not have any other symptoms.

An 81-year-old woman developed irreversible sensorineural hearing loss in the right ear 3 days after starting low-dose clarithromycin for an acute exacerbation of chronic obstructive pulmonary disease [21].

- Ototoxicity was attributed to clarithromycin in a 76-year-old man 4 days after he started to take clarithromycin for atypical pulmonary tuberculosis (22). When the clarithromycin was withdrawn his hearing improved subjectively, but it worsened again on re-exposure.

Taste

Abnormal taste developed in 17 of 175 patients treated with clarithromycin 250 mg bd for 10 days for community-acquired pneumonia, compared with 3 of 167 patients treated with sparfloxacin (23). Mild to moderate gastro-intestinal disturbances were the most common adverse events and were reported in 13 and 11% respectively.

Psychological, psychiatric

Two patients, a man aged 74 and a woman aged 56 years, developed delirium after taking clarithromycin (24).

- A 63-year-old woman took clarithromycin for a community-acquired pneumonia and had an episode of agitation and delirium; she recovered when the clarithromycin was stopped [25].

Two patients (aged 21 and 33 years) with late-stage AIDS had acute psychoses shortly after taking clarithromycin (2 g/day) for MAC bacteremia (26). In both cases the psychosis resolved on withdrawal but recurred on rechallenge. In one case treatment with azithromycin was well tolerated.

Of cases of mania attributed to antibiotics and reported to the WHO, 28% were due to clarithromycin (27).

- A 77-year-old man who was HIV-negative developed mania after 6 days treatment with clarithromycin 1 g/day for a soft tissue infection; his mental state resolved on withdrawal (28).
- A 53-year-old Canadian lawyer taking long-term fluox-etine and nitrazepam developed a frank psychosis 1–3 days after starting to take clarithromycin 500 mg/day for a chest infection (29). His symptoms resolved on withdrawal of all three drugs, and did not recur with erythromycin or when fluoxetine and nitrazepam were restarted in the absence of antibiotics.

The symptoms may have been due to a direct effect of clarithromycin or else inhibition of hepatic cytochrome P450 metabolism, leading to fluoxetine toxicity.

Clarithromycin occasionally causes hallucinations.

- Visual hallucinations with marked anxiety and nervousness occurred after the second dose of oral clarithromycin 500 mg in a 32-year-old woman (30). Clarithromycin was withdrawn and the symptoms disappeared a few hours later.
- Visual hallucinations developed in a 56-year-old man with chronic renal insufficiency and underlying alumi-nium intoxication maintained on peritoneal dialysis 24 hours after he started to take clarithromycin 500 mg bd for a chest infection, and resolved completely 3 days after withdrawal (31).

Hematologic

- Thrombotic thrombocytopenic purpura was reported in a 42-year-old man with no past medical history after he had just completed a 30-day course of clarithromycin 250 mg bd (32).

Gastrointestinal

Erythromycin acts as a motilin receptor agonist (33–35). This mechanism may be at least partly responsible for the gastrointestinal adverse effects of macrolides. Clarithromycin may act on gastrointestinal motility in a similar way. In dogs, clarithromycin caused contractions and discomfort, as did erythromycin (36). In healthy volunteers, oral clarithromycin 250 mg bd caused a statistically significant increase in the number of postprandial antral contractions and antral motility (37). A single oral dose of clarithromycin 3000 mg resulted in severe abdominal pain within 1 hour of administration in two patients (38).

In a randomized comparison of oral telithromycin 800 mg/day for 5 or 7 days and oral clarithromycin 500 mg bd for 10 days in 575 patients, 257 (45%) had at least one treatment-related adverse event [39]. The most frequent adverse effects were gastrointestinal (diarrhea, nausea, and dysgeusia). None of the patients taking clarithromycin withdrew because of gastrointestinal events.

Based on observations made in dogs and rabbits, clarithromycin is significantly less potent than azithromycin and erythromycin as an agonist for stimulation of smooth muscle contraction (40). Therefore, a lower rate of gastro-intestinal adverse events would be expected with clarithromycin.

Pseudomembranous colitis is relatively rarely seen with macrolides, but has been reported with clarithromycin (SED-12, 597) (41,42).

Liver

Abnormal liver function tests and hepatomegaly have been described with clarithromycin; in 4291 patients, the frequency of increased alanine transaminase activity was 5% (43). Clarithromycin was also associated with chole-static hepatitis (36). To date, at least nine cases of hepa-totoxicity have been described in HIV-negative patients taking clarithromycin 1–2 g/day for chronic lung disease due to *M. avium* or *Mycobacterium abscessus* (44). The pattern of liver enzyme abnormality was primarily chole-static, and the patients were typically elderly (all but one aged over 60 years), or of low weight. Only three patients were symptomatic, and the liver function abnormalities resolved on withdrawal. Subsequent rechallenge was successful in four patients, unsuccessful in one, and not performed in four. There was some dispute as to whether toxicity was dose-related or not, but it is wise to recommend that elderly patients should receive an initial daily dose of 1 g in this disease setting.

Although cholestatic hepatitis has been typically described in association with erythromycin, newer macrolides are not totally free of this risk. A gradual increase in bilirubin and transaminases has been reported during treatment of a *Mycobacterium chelonae* infection with clarithromycin. These alterations were quickly reversible after withdrawal, but re-appeared on re-exposure to clarithromycin 1 g (45).

- Fatal drug-induced cholestasis associated with clarithromycin 500 mg bd for 3 days has been reported in a 59-year-old woman with diabetes mellitus and chronic renal insufficiency (46).
- Severe progressively worsening cholestatic liver disease occurred in a 15-year-old girl after a 15-day course of clarithromycin 500 mg/day and nimesulide 100 mg/day [47]. Prednisone was started after 20 days after which there was prompt improvement. After 2 months she became symptom-free with normal liver function test results.

Fulminant hepatic failure can occasionally occur.

- Fulminant liver failure was reported in a 58-year-old white woman while she was taking clarithromycin for pneumonia. She recovered spontaneously within a few days after drug withdrawal [48].

Pancreas

Acute pancreatitis was reported in a 63-year-old woman and in an 84-year-old woman taking clarithromycin [49,50].

Urinary tract

Drug-induced interstitial nephritis represents 4% of all cases studied histologically, with an estimated frequency of 8%. Granulomatous interstitial nephritis occurred in a 38-year-old woman who was given clarithromycin for pharyngitis [51].

- Idiosyncratic drug-induced fatal fulminant hepatic failure has been reported in a 40-year-old woman with end-stage renal insufficiency taking clarithromycin 500 mg bd (52).

Skin

Clarithromycin has been associated with fixed drug eruptions and hypersensitivity reactions (53,54). In one case a clarithromycin-induced fixed drug eruption was reproduced by oral provocation, whereas patch tests on both unaffected and residual pigmented skin were negative (55).

- A 31-year-old woman developed Stevens–Johnson syndrome after she had taken oral erythromycin 333 mg tds for otitis media (56). After two doses she developed oral ulcers, tongue swelling, and a generalized erythematous rash. The diagnosis was confirmed histologically. She recovered slowly after withdrawal of erythromycin.
- Roxithromycin-induced generalized urticaria and tachycardia with a positive prick test and a cross-reaction to erythromycin and clarithromycin has been reported in a 31-year-old woman (57).

- A 68-year-old woman developed a fixed drug eruption on the tongue after taking clarithromycin for 4 days for an acute upper respiratory infection [58]. The erythematous violaceous patch over the posterior dorsum of the tongue resolved 15 days after withdrawal of the drug.
- A 29-year-old woman with tonsillitis was given clarithromycin and developed a pruritic rash after 48 hours [59]. Despite switching to amoxicillin her condition worsened, with the appearance of bullous lesions on her abdomen, a fever of 40°C, peeling of 70% of the body surface, and mucosal involvement. Stevens–Johnson syndrome was diagnosed. She recovered with residual scars on the skin and ophthalmic lesions.

Clarithromycin can cause phototoxicity (60)

Immunologic

Henoch–Schönlein purpura occurred 4 days after a 48-year-old white man started to take clarithromycin 500 mg /day; after a few days clarithromycin was withdrawn and his symptoms quickly resolved [61].

Drug Administration

Drug formulations

In a multicenter, double blind, randomized comparison of a 5-day course of clarithromycin extended-release 500 mg/day or clarithromycin immediate-release 250 mg bd in 706 subjects with acute bacterial exacerbations, the incidence of drug-related adverse events was 7% in the extended-release group and 5% in the immediate-release group. The only drug-related adverse events that occurred with an incidence of over 1% in either treatment group were abdominal pain (1% in each treatment group), diarrhea (2% for the extended-release formulation and 1% for the immediate-release formulation), and taste disturbances (1% for both formulations) [62,63].

Drug–Drug Interactions

Drug interactions with clarithromycin have been reviewed [1, 64]. Clarithromycin causes less inhibition of cytochrome P450 than erythromycin, but co-administration increases the serum concentrations of carbamazepine, ciclosporin, terfenadine, and theophylline.

Amprenavir

A pharmacokinetic study has shown a minor interaction of amprenavir with clarithromycin in healthy men (65). The mean AUC, $C_{max.ss}$ and $C_{min.ss}$ of amprenavir increased by 18, 15, and 39% respectively. Amprenavir had no effect on the AUC of clarithromycin, but the median $t_{max.ss}$ increased by 2 hours, renal clearance increased by 34%, and the AUC for 14-(R)-hydroxyclarithromycin fell by 35%. These effects were felt not to be clinically important and dosage adjustment was not recommended.

Antifungal imidazoles

In three patients with pulmonary MAC and aspergillosis infections, itraconazole was suggested to increase the plasma concentration of clarithromycin as well as the clarithromycin:14-hydroxyclarithromycin ratio (66). This effect may have been due to inhibition of CYP3A4 by itraconazole.

Antihistamines

Toxic effects of terfenadine and astemizole have been reported in patients taking concomitant macrolides, especially clarithromycin (67–70), typically resulting in prolongation of the QT interval and cardiac dysrhythmias (torsade de pointes) (71).

Cabergoline

In 10 healthy men and seven patients with Parkinson's disease, co-administration of cabergoline with clarithromycin increased the blood concentration of cabergoline [72].

Carbamazepine

Clarithromycin can inhibit the metabolism of other drugs.

- An 85-year old woman, who was taking carbamazepine, fluoxetine, a benzodiazepine, an ACE inhibitor, a statin, and a coumarin anticoagulant, developed confusion and asterixis, with raised liver enzymes and an increased international normalized ratio (INR) after taking clarithromycin for 10 days for a respiratory infection [73].

This presentation was associated with carbamazepine and coumarin toxicity caused by an interaction with clarithromycin.

Cardiac glycosides

Clarithromycin has been reported to cause digoxin toxicity (74). Two different mechanisms are involved, inhibition of the renal excretion of digoxin and alteration of intestinal flora, which reduces the presystemic hydrolysis of digoxin.

- A 70-year-old woman taking digoxin for atrial fibrillation developed nausea, vomiting, and dizziness 2 days after starting to take clarithromycin (75). Her serum digoxin concentration was 3.9 ng/ml (target range 0.5–2.0).
- In a 72-year-old woman taking digoxin 0.25 mg/day, the addition of clarithromycin caused a rise in the serum digoxin concentration to 4.6 ng/ml (76).

In two other cases in which clarithromycin increased serum digoxin concentrations there was an associated reduction in the rate of renal digoxin clearance, which may be another mechanism for this interaction (77). The authors hypothesized that clarithromycin inhibited P glycoprotein. This was supported by the observation of a concentration-dependent effect of clarithromycin on in vitro transcellular transport of digoxin.

There is anecdotal evidence that this interaction may be clinically important. Digoxin toxicity occurred in a patient taking clarithromycin (78).

Cisapride

Concurrent administration of clarithromycin and cisapride prolongs the QT_c interval by a mean effect of 25 msec [79], with a risk of ventricular dysrhythmias (80,81). Clarithromycin increases serum concentrations of cisapride (82). This potentially dangerous interaction can result in QT interval prolongation and dysrhythmias such as torsade de pointes.

- Torsade de pointes occurred in a 77-year-old woman taking cisapride and clarithromycin (83).

Warnings have been issued by the manufacturers to avoid concomitant administration.

Colchicine

An interaction of clarithromycin with colchicine has been described [84].

- A 76-year-old man, who had taken colchicine 1.5 mg/day for 6 years for familial Mediterranean fever, was given clarithromycin, amoxicillin, and omeprazole for *Helicobacter pylori*-associated gastritis for 7 days. He developed fever, abdominal pain, and diarrhea 3 days later, dehydration, pancytopenia, metabolic acidosis, and increased lipase activity on day 8, alopecia 2 weeks later. He recovered fully after rehydration and reduction in the dosage of colchicine to 0.5 mg/day. The previous dosage was then reinstituted without adverse effects.

The authors suggested that clarithromycin reduced the biliary excretion of colchicine by inhibiting P glycoprotein.

- Fatal colchicine intoxication occurred in a 67-year-old man who had taken clarithromycin 500 mg bd for 4 days (85).

Didanosine

When clarithromycin was administered with didanosine in seven adults with HIV, there was a 40% increase in the AUC of didanosine, a difference that could be clinically relevant [1].

Digoxin

A clinically important interaction between digoxin and clarithromycin has been suggested. Digoxin concentrations increased during concomitant administration of clarithromycin in eight patients, and this effect was related to the dose of clarithromycin. The percentage increase in digoxin concentrations after the usual oral dose of clarithromycin (400 mg/day) is about 70%. The cause of this interaction could be increased oral systemic availability and reduced non-glomerular renal clearance of digoxin, probably by inhibition of intestinal and renal P glycoproteins [86,87].

Disulfiram

Fatal toxic epidermal necrolysis and fulminant hepatitis occurred shortly after the start of treatment with clarithromycin in a 47-year-old man who was taking disulfiram (88).

- A 47-year-old man with a history of chronic alcoholism took disulfiram 250 mg/day for 1 month. He then took clarithromycin 500 mg bd and paracetamol 500 mg tds and 1 week later noticed non-pruritic cutaneous maculopapular lesions on his legs, extending to the rest of his body, excluding the palms and soles. Previous drug therapy was withdrawn. A skin biopsy showed toxic epidermal necrolysis. During the next several days the skin lesions worsened. Cutaneous blisters became evident, initially covering less than 10% of the body surface, but then extending all over the body. The serum bilirubin concentration was 359 µmol/l (direct bilirubin 213 µmol/l), the partial thromboplastin time longer than 200 seconds, and the prothrombin time 26 seconds. He developed septic shock and, despite supportive measures, died.

Ergot alkaloids

In patients with ergotamine toxicity, vasoconstriction can lead to frank ischemia. Clarithromycin interferes with ergotamine metabolism.

- A 41-year-old woman developed worsening lower leg pain, pallor, and a sensation of coolness aggravated by exertion; there was severe vasospasm in the legs (89). She had taken a caffeine + ergotamine formulation for migraine for many years and had recently been given clarithromycin 500 mg bd for flu-like symptoms.

Fluconazole

The effects of fluconazole and clarithromycin on the pharmacokinetics of rifabutin and 25-O-desacetylrifabutin have been studied in 10 HIV-infected patients who were given rifabutin 300 mg qds in addition to fluconazole 200 mg qds and clarithromycin 500 mg qds (90). There was a 76% increase in the plasma AUC of rifabutin when either fluconazole or clarithromycin was given alone and a 152% increase when both drugs were given together. The authors concluded that patients should be monitored for adverse effects of rifabutin when it is co-administered with fluconazole or clarithromycin.

HIV nucleoside reverse transcriptase inhibitors

Clarithromycin reduced the peak concentration and AUC of zidovudine at steady state by about 12% (38), possibly as a result of reduced zidovudine absorption (91). However, if the two drugs were taken at least 2 hours apart, the pharmacokinetics of zidovudine were unaffected.

In 12 HIV-positive patients there was no statistically significant difference in concentrations of didanosine when clarithromycin was added (92).

Itraconazole

In an open phase II trial in 17 patients with cystic fibrosis taking itraconazole the highest itraconazole concentrations were observed in a patient who had concurrently taken clarithromycin [93]. This may have resulted from competitive inhibition of metabolism by CYP3A4.

Meglitinides

Clarithromycin, an inhibitor of CYP3A4, increases the plasma concentrations and the effect of repaglinide; this can enhance the blood glucose lowering effect and increase the risk of hypoglycemia (94).

Midazolam

In an open, randomized, crossover, pharmacokinetic and pharmacodynamic study in 12 healthy volunteers who took clarithromycin 250 mg bd for 5 days, azithromycin 500 mg/day for 3 days, or no pretreatment, followed by a single dose of midazolam (15 mg), clarithromycin increased the AUC of midazolam by over 3.5 times and the mean duration of sleep from 135 to 281 minutes (95). In contrast, there was no change with azithromycin, suggesting that it is much safer for co-administration with midazolam.

Nifedipine

In a 77-year-old man a life-threatening pharmacokinetic interaction of clarithromycin with nifedipine was suspected as the cause of vasodilatory shock [96].

Omeprazole

In 21 healthy volunteers, clarithromycin (400 mg bd) for 3 days before omeprazole (20 mg/day) significantly inhibited the metabolism of omeprazole (97).

Paroxetine

Serotonin syndrome has been reported in a patient taking paroxetine and Clarithromycin [98].

- A 36-year-old woman who had taken paroxetine 10 mg/day for mild anxiety for 3 months developed an upper respiratory tract infection and stopped taking paroxetine. Three days later she was given clarithromycin 500 mg bd and after the fourth dose developed acute ocular clonus, akathisia, and fever. She received intravenous fluids and lorazepam and clarithromycin was withdrawn; the syndrome abated within 24 hours.

Pimozide

Clarithromycin inhibits the metabolism of pimozide, pimozide plasma concentrations increase, and there is an increased risk of cardiotoxicity through prolongation of the QT interval and fatal ventricular dysrhythmias. In 12 healthy volunteers given oral pimozide 6 mg after 5 days of treatment with clarithromycin (500 mg bd) or placebo, pimozide significantly prolonged the QT_c interval in the first 20 hours in both groups (maximum changes in QT_c 16 and 13 ms respectively) (99).

Rifamycins

Clarithromycin is one of the core drugs for MAC infections in both HIV-infected and non-infected patients. For this indication, doses of up to 2000 mg/day are used, typically in combination with other drugs.

The interaction of clarithromycin with the rifamycins is complex. Clarithromycin inhibits CYP3A4, while both rifampicin and rifabutin induce P450 cytochromes, including CYP3A4, resulting in enhanced metabolism of drugs. The changes in serum concentrations of clarithromycin and its metabolite in the presence of the enzyme inducers rifampicin and rifabutin suggest that metabolism of clarithromycin by CYP3A4 is increased (100).

After the addition of rifampicin, peak serum concentrations of clarithromycin fell markedly, from a mean of 5.8–2.5 µg/ml (100). At the same time the ratio of the serum concentrations of clarithromycin and its 14-OH metabolite was reversed from 3.3:1 to 1:2.7. There were similar, although less marked, changes after the addition of rifabutin 600 mg/day to a regimen that included clarithromycin 1000 mg/day.

Whether these changes in serum clarithromycin concentrations are relevant to its antimicrobial activity is unknown, since prediction of clinical efficacy based on serum concentrations of clarithromycin is probably not justified, given that the macrolides accumulate to a large degree in tissues and macrophages.

In patients with MAC infections taking rifabutin or rifampicin the addition of clarithromycin resulted in rifamycin-related adverse events in 77% of patients (101). These included uveitis (102–105), especially at rifabutin doses of 600 mg/day or more, neutropenia, nausea, vomiting, diarrhea, and abnormal liver enzyme activities. In addition, diffuse polyarthralgia (19%) was observed. Since inhibition of cytochrome P450 by clarithromycin can interfere with rifabutin metabolism, as illustrated by a report of a significant increase in the AUC of rifabutin during treatment with clarithromycin (101), the authors recommended using rifabutin in a dosage of 300 mg/day in regimens that include a macrolide.

The effects of fluconazole and clarithromycin on the pharmacokinetics of rifabutin and 25-O-desacetylrifabutin have been studied in ten HIV-infected patients who were given rifabutin 300 mg qds in addition to fluconazole 200 mg qds and clarithromycin 500 mg qds (106). There was a 76% increase in the plasma AUC of rifabutin when either fluconazole or clarithromycin was given alone and a 152% increase when both drugs were given together. The authors concluded that patients should be monitored for adverse effects of rifabutin when it is co-administered with fluconazole or clarithromycin.

Severe interactions have been observed when rifabutin and clarithromycin were given simultaneously (107). The mean concentrations of rifabutin and 25-O-desacetyl-rifabutin in healthy subjects who took clarithromycin and rifabutin concomitantly were respectively more than 4 times and 37 times greater than the concentrations recorded when rifabutin was administered alone. Neutropenia was detected in 14 of 18 subjects taking rifabutin. Myalgia and high fever were also common.

Physicians should be aware that recommended prophylactic doses of rifabutin can be associated with severe neutropenia within 2 weeks after the start of therapy, and all patients taking rifabutin, especially with clarithromycin, should be monitored carefully for neutropenia.

Statius

Lovastatin has been reported to interact with clarithromycin (SEDA-14, 1531; 108), and a similar reaction has been observed with simvastatin (109).

- A 64-year-old African-American man developed worsening renal insufficiency, raised creatine kinase activity, diffuse muscle pain, and severe muscle weakness. He had been taking simvastatin for about 6 months and clarithromycin for sinusitis for about 3 weeks. He was treated aggressively with intravenous hydration, sodium bicarbonate, and hemodialysis. A muscle biopsy showed necrotizing myopathy secondary to a toxin. He continued to receive intermittent hemodialysis until he died from infectious complications 3 months after admission.

Sulfonylureas

- Severe hypoglycemia occurred in two elderly men with type 2 diabetes mellitus and mild to moderate impaired renal function, who took clarithromycin 1000 mg/day for respiratory infections, in addition to a sulfonylurea (glibenclamide 5 mg/day in one case and glipizide 15 mg/day in the other) (110). Both developed severe hypoglycemia within 48 hours of starting clarithromycin.

Clarithromycin inhibits cytochrome CYP3A4. It should be used carefully in patients with diabetes and reduced renal function taking oral drugs.

Tacrolimus

Clarithromycin can increase the steady-state concentrations of drugs that depend primarily on CYP3A metabolism.

- Steady-state tacrolimus concentrations rose in a 32-year-old African-American man who took clarithromycin 500 mg bd for 4 days (111).
- In two women aged 37 and 69, acute and reversible tacrolimus nephrotoxicity developed after the addition of clarithromycin for an upper respiratory tract infection (112).

Theophylline and other xanthines

Inhibition of cytochrome P450 activity by clarithromycin affects the metabolism of theophylline. However, the results of several studies of the effect of clarithromycin on theophylline concentrations are conflicting. While the total body clearance of theophylline fell (113) and plasma theophylline concentrations increased by 18% (114), mean theophylline concentrations remained within the target range (114). Based on these data it is wise to monitor serum theophylline concentrations in patients taking high dosages of theophylline or in patients with theophylline concentrations in the upper target range who start to take clarithromycin (43).

Verapamil

- A 53-year-old woman had periods of dizziness and episodes of fainting when she stood up 24 hours after having been given clarithromycin for an acute exacerbation of chronic obstructive pulmonary disease and verapamil for atrial fibrillation (109). One day later she developed severe hypotension and bradycardia. Since her symptoms matched those of severe verapamil overdosage, the drug was withdrawn and her condition improved within two days.
- A 77-year-old hypertensive woman receiving both verapamil and propranolol for hypertrophic cardiomyopathy and paroxysmal atrial fibrillation developed symptomatic bradycardia within 2–4 days of initiation of both erythromycin or clarithromycin on two separate occasions, for pneumonia and sinusitis respectively (116).

Warfarin

Increases in International Normalized Ratio (INR) have been detected in patients who have previously been stabilized on warfarin when they were simultaneously given clarithromycin. In one case this caused a suprachoroidal hemorrhage (117).

See also above under Carbamazepine.

Zidovudine

Clarithromycin has an unpredictable effect on the absorption of zidovudine; blood concentrations may rise or fall (118,119).

References

1. Scheinfeld NS, Tutrone WD, Torres O, Weinberg JM. Macrolides in dermatology. Dis Mon 2004;50(7):350–68.
2. Blum AL, Talley NJ, O'Morain C, van Zanten SV, Labenz J, Stolte M, Louw JA, Stubberod A, Theodors A, Sundin M, Bolling-Sternevald E, Junghard O. Lack of effect of treating *Helicobacter pylori* infection in patients with nonulcer dyspepsia. Omeprazole plus Clarithromycin and Amoxicillin Effect One Year after Treatment (OCAY) Study Group. N Engl J Med 1998;339(26): 1875–81.
3. Hawkey CJ, Tulassay Z, Szczepanski L, van Rensburg CJ, Filipowicz-Sosnowska A, Lanas A, Wason CM, Peacock RA, Gillon KR. Randomised controlled trial of *Helicobacter pylori* eradication in patients on non-steroidal anti-inflammatory drugs: HELP NSAIDs study. Helicobacter Eradication for Lesion Prevention. Lancet 1998;352(9133):1016–21.
4. McCarty JM, Pierce PF. Five days of cefprozil versus 10 days of clarithromycin in the treatment of an acute exacerbation of chronic bronchitis. Ann Allergy Asthma Immunol 2001;87(4):327–34.
5. Stefansson P, Jacovides A, Jablonicky P, Sedani S, Staley H. Cefuroxime axetil versus clarithromycin in the treatment of acute maxillary sinusitis. Rhinology 1998;36(4):173–8.
6. Faris MA, Raasch RH, Hopfer RL, Butts JD. Treatment and prophylaxis of disseminated *Mycobacterium avium* complex in HIV-infected individuals. Ann Pharmacother 1998;32(5):564–73.
7. Ward TT, Rimland D, Kauffman C, Huycke M, Evans TG, Heifets L. Randomized, open-label trial of azithromycin plus ethambutol vs. clarithromycin plus ethambutol as therapy for *Mycobacterium avium* complex bacteremia in patients with human immunodeficiency virus infection. Veterans Affairs HIV Research Consortium. Clin Infect Dis 1998;27(5):1278–85.
8. Anderson G, Esmonde TS, Coles S, Macklin J, Carnegie C. A comparative safety and efficacy study of clarithromycin and erythromycin stearate in community-acquired pneumonia. J Antimicrob Chemother 1991;27(Suppl A):117–24.
9. Lebel MH, Mehra S. Efficacy and safety of clarithromycin versus erythromycin for the treatment of pertussis: a prospective, randomized, single blind trial. Pediatr Infect Dis J 2001;20(12):1149–54.
10. Sokol WN Jr, Sullivan JG, Acampora MD, Busman TA, Notario GF. A prospective, double-blind, multicenter study comparing clarithromycin extended-release with trovafloxacin in patients with community-acquired pneumonia. Clin Ther 2002;24(4):605–15.
11. Roussel G, Igual J. Clarithromycin with minocycline and clofazimine for *Mycobacterium avium intracellulare* complex lung disease in patients without the acquired immune deficiency syndrome. GETIM. Groupe d'Etude et de Traitement des Infections a Mycobacteries. Int J Tuberc Lung Dis 1998;2(6):462–70.
12. Germanakis I, Galanakis E, Parthenakis F, Vardas PE, Kalmanti M. Clarithromycin treatment and QT prolongation in childhood. Acta Paediatr 2006;95(12):1694–6.
13. Vallejo Camazon N, Rodriguez Pardo D, Sanchez Hidalgo A, Tornos Mas MP, Ribera E, Soler Soler J. Taquicardia Ventricular y QT largo asociades a la administracion de claritromycina en un paciente afectado de infeccion por el VIH. [Ventricular tachycardia and long QT associated with clarithromycin administration in a patient with HIV infection.] Rev Esp Cardiol 2002;55(8):878–81.
14. Cousins D, Upton D. Beware bolus clarithromycin. Pharm Pract 1996;4:443–5.
15. de Dios Garcia-Diaz J, Santolaya Perrin R, Paz Martinez Ortega M, Moreno-Vazquez M. Flebitis relacionada con la administracion intravenosa de antibioticos macrolidos. Estudio comparativo de eritromicina y claritromicina. [Phlebitis due to intravenous administration of macrolide antibiotics. A comparative study of erythromycin versus clarithromycin.] Med Clin (Barc) 2001;116(4):133–5.
16. Gangemi S, Ricciardi L, Fedele R, Isola S, Purello-D'Ambrosio F. Immediate reaction to clarithromycin. Allergol Immunopathol (Madr) 2001;29(1):31–2.
17. Terzano C, Petroianni A. Clarithromycin and pulmonary infiltration with eosinophilia. BMJ 2003;326:1377–8.
18. Pijpers E, van Rijswijk RE, Takx-Kohlen B, Schrey G. A clarithromycin-induced myasthenic syndrome. Clin Infect Dis 1996;22(1):175–6.
19. Tyagi AK, Kayarkar VV, McDonnell PJ. An unreported side effect of topical clarithromycin when used successfully to treat *Mycobacterium avium-intracellulare* keratitis. Cornea 1999;18(5):606–7.
20. Uzun C. Tinnitus due to clarithromycin. J Laryngol Otol 2003;117:1006–7.

21. Coulston J, Balaratnam N. Irreversible sensorineural hearing loss due to clarithromycin. Postgrad Med J 2005;81(951):58–9.

22. Kolkman W, Groeneveld JH, Baur HJ, Verschuur HP. Door claritromycine geinduceerde ototoxiciteite. [Ototoxicity induced by clarithromycin.] Ned Tijdschr Geneeskd 2002;146(37):1743–5.

23. Ramirez J, Unowsky J, Talbot GH, Zhang H, Townsend L. Sparfloxacin versus clarithromycin in the treatment of community-acquired pneumonia. Clin Ther 1999;21(1):103–17.

24. Pijlman AH, Kuck EM, van Puijenbroek EP, Hoekstra JB. Acuut delies, waarschijulijk uitgelokt door claritromycine. [Acute delirium, probably precipitated by clarithromycin.] Ned Tijdschr Geneeskd 2001;145(5):225–8.

25. Vicente de Vera C, Garcia M, Pifarre Teixido R, Barbe F. Delirium induced by clarithromycin in a patient with community-acquired pneumonia. Eur Respir J 2006;28(3):671–2.

26. Nightingale SD, Koster FT, Mertz GJ, Loss SD. Clarithromycin-induced mania in two patients with AIDS. Clin Infect Dis 1995;20(6):1563–4.

27. Abouesh A, Stone C, Hobbs WR. Antimicrobial-induced mania (antibiomania): a review of spontaneous reports. J Clin Psychopharmacol 2002;22(1):71–81.

28. Cone LA, Sneider RA, Nazemi R, Dietrich EJ. Mania due to clarithromycin therapy in a patient who was not infected with human immunodeficiency virus. Clin Infect Dis 1996;22(3):595–6.

29. Pollak PT, Sketris IS, MacKenzie SL, Hewlett TJ. Delirium probably induced by clarithromycin in a patient receiving fluoxetine. Ann Pharmacother 1995;29(5):486–8.

30. Jimenez-Pulido SB, Navarro-Ruiz A, Sendra P, Martinez-Ramirez M, Garcia-Motos C, Montesinos-Ros A. Hallucinations with therapeutic doses of clarithromycin. Int J Clin Pharmacol Ther 2002;40(1):20–2.

31. Steinman MA, Steinman TI. Clarithromycin-associated visual hallucinations in a patient with chronic renal failure on continuous ambulatory peritoneal dialysis. Am J Kidney Dis 1996;27(1):143–6.

32. Alexopoulou A, Dourakis SP, Kaloterakis A. Thrombotic thrombocytopenic purpura in a patient treated with clarithromycin. Eur J Haematol 2002;69(3):191–2.

33. Lin HC, Sanders SL, Gu YG, Doty JE. Erythromycin accelerates solid emptying at the expense of gastric sieving. Dig Dis Sci 1994;39(1):124–8.

34. Hasler WL, Heldsinger A, Chung OY. Erythromycin contracts rabbit colon myocytes via occupation of motilin receptors. Am J Physiol 1992;262(1 Pt 1):G50–5.

35. Kaufman HS, Ahrendt SA, Pitt HA, Lillemoe KD. The effect of erythromycin on motility of the duodenum, sphincter of Oddi, and gallbladder in the prairie dog. Surgery 1993;114(3):543–8.

36. Nakayoshi T, Izumi M, Tatsuta K. Effects of macrolide antibiotics on gastrointestinal motility in fasting and digestive states. Drugs Exp Clin Res 1992;18(4):103–9.

37. Sifrim D, Janssens J, Vantrappen G. Comparison of the effects of midecamycin acetate and clarithromycin on gastrointestinal motility in man. Drugs Exp Clin Res 1992;18(8):337–42.

38. Polis M, Haneiwich S, Kovacs J, et al. Dose escalation study to determine the safety, maximally tolerated dose and pharmacokinetics of clarithromycin with zidovudine in HIV-infected patients. In: Interscience Conference on Antimicrobial Agents and ChemotherapyAmerican Society for Microbiology, 1991:327.

39. Tellier G, Niederman MS, Nusrat R, Patel M, Lavin B. Clinical and bacteriological efficacy and safety of 5 and 7 day regimens of telithromycin once daily compared with a 10 day regimen of clarithromycin twice daily in patients with mild to moderate community-acquired pneumonia. J Antimicrob Chemother 2004;54(2):515–23.

40. Nellans H, Petersen A. Stimulation of gastrointestinal motility: clarithromycin significantly less potent than azithromycin. In: Seventh International Congress of ChemotherapyBerlin 1991:327.

41. Teare JP, Booth JC, Brown JL, Martin J, Thomas HC. Pseudomembranous colitis following clarithromycin therapy. Eur J Gastroenterol Hepatol 1995;7(3):275–7.

42. Gantz NM, Zawacki JK, Dickerson WJ, Bartlett JG. Pseudomembranous colitis associated with erythromycin. Ann Intern Med 1979;91(6):866–7.

43. Peters DH, Clissold SP. Clarithromycin. A review of its antimicrobial activity, pharmacokinetic properties and therapeutic potential. Drugs 1992;44(1):117–64.

44. Brown BA, Wallace RJ Jr, Griffith DE, Girard W. Clarithromycin-induced hepatotoxicity. Clin Infect Dis 1995;20(4):1073–4.

45. Yew WW, Chau CH, Lee J, Leung CW. Cholestatic hepatitis in a patient who received clarithromycin therapy for a *Mycobacterium chelonae* lung infection. Clin Infect Dis 1994;18(6):1025–6.

46. Fox JC, Szyjkowski RS, Sanderson SO, Levine RA. Progressive cholestatic liver disease associated with clarithromycin treatment. J Clin Pharmacol 2002;42(6):676–80.

47. Giannattasio A, D'Ambrosi M, Volpicelli M, Iorio R. Steroid therapy for a case of severe drug-induced cholestasis. Ann Pharmacother 2006;40(6):1196–9.

48. Tietz A, Heim MH, Eriksson U, Marsch S, Terracciano L, Krahenbuhl S. Fulminant liver failure associated with clarithromycin. Ann Pharmacother 2003;37:57–60.

49. Schouwenberg BJ, Deinum J. Acute pancreatitis after a course of clarithromycin. Neth J Med 2003;61:266–7.

50. Rassiat E, Michiels C, Jouve JL, Sgro C, Faivre J, Hillon P. Pancreatite aigue après prise de clarithromycine et de beta-methasone. Gastroenterol Clin Biol 2003;27:123.

51. Audimoolam VK, Bhandari S. Clarithromycin-induced granulomatous tubulointerstitial nephritis. Nephrol Dial Transplant 2006;21(9):2654–5.

52. Christopher K, Hyatt PA, Horkan C, Yodice PC. Clarithromycin use preceding fulminant hepatic failure. Am J Gastroenterol 2002;97(2):489–90.

53. Rosina P, Chieregato C, Schena D. Fixed drug eruption from clarithromycin. Contact Dermatitis 1998;38(2):105.

54. Igea JM, Lazaro M. Hypersensitivity reaction to clarithromycin. Allergy 1998;53(1):107–9.

55. Hamamoto Y, Ohmura A, Kinoshita E, Muto M. Fixed drug eruption due to clarithromycin. Clin Exp Dermatol 2001;26(1):48–9.

56. Sullivan S, Harger B, Cleary JD. Stevens–Johnson syndrome secondary to erythromycin. Ann Pharmacother 1999;33(12):1369.

57. Kruppa A, Scharffetter-Kochanek K, Krieg T, Hunzelmann N. Immediate reaction to roxithromycin and prick test cross-sensitization to erythromycin and clarithromycin. Dermatology 1998;196(3):335–6.

58. Alonso JC, Melgosa AC, Gonzalo MJ, Garcia CM. Fixed drug eruption on the tongue due to clarithromycin. Contact Dermatitis 2005;53(2):121–2.

59. Khaldi N, Miras A, Gromb S. Toxic epidermal necrolysis and clarithromycin. Can J Clin Pharmacol 2005;12(3):e264-8.

60. Parkash P, Gupta SK, Kumar S. Phototoxic reaction due to clarithromycin. J Assoc Physicians India 2002;50:1192–3.

61. Borras-Blasco J, Enriquez R, Amoros F, Cabezuelo JB, Navarro-Ruiz A, Perez M, Fernandez J. Henoch–Schönlein purpura associated with clarithromycin. Case report and review of literature. Int J Clin Pharmacol Ther 2003;41:213–16.

62. Darkes MJ, Perry CM. Clarithromycin extended-release tablet: a review of its use in the management of respiratory tract infections. Am J Respir Med 2003;2:175–201.

63. Nalepa P, Dobryniewska M, Busman T, Notario G. Short-course therapy of acute bacterial exacerbation of chronic bronchitis: a double-blind, randomized, multicenter comparison of extended-release versus immediate-release clarithromycin. Curr Med Res Opin 2003;19:411–20.

64. Brown CS, Farmer RG, Soberman JE, Eichner SF. Pharmacokinetic factors in the adverse cardiovascular effects of antipsychotic drugs. Clin Pharmacokinet 2004;43(1):33–56.

65. Brophy DF, Israel DS, Pastor A, Gillotin C, Chittick GE, Symonds WT, Lou Y, Sadler BM, Polk RE. Pharmacokinetic interaction between amprenavir and clarithromycin in healthy male volunteers. Antimicrob Agents Chemother 2000;44(4):978–84.

66. Auclair B, Berning SE, Huitt GA, Peloquin CA. Potential interaction between itraconazole and clarithromycin. Pharmacotherapy 1999;19(12):1439–44.

67. Tran HT. Torsades de pointes induced by nonantiarrhythmic drugs. Conn Med 1994;58(5):291–5.

68. Zechnich AD, Hedges JR, Eiselt-Proteau D, Haxby D. Possible interactions with terfenadine or astemizole. West J Med 1994;160(4):321–5.

69. Jurima-Romet M, Crawford K, Cyr T, Inaba T. Terfenadine metabolism in human liver. In vitro inhibition by macrolide antibiotics and azole antifungals. Drug Metab Dispos 1994;22(6):849–57.

70. Honig P, Wortham D, Zamani K, Cantilena L. Comparison of the effect of the macrolide antibiotics erythromycin, clarithromycin and azithromycin on terfenadine steady-state pharmacokinetics and electrocardiographic parameters. Drug Invest 1994;7:148.

71. Botstein P. Is QT interval prolongation harmful? A regulatory perspective. Am J Cardiol 1993;72(6):B50–2.

72. Nakatsuka A, Nagai M, Yabe H, Nishikawa N, Nomura T, Moritoyo H, Moritoyo T, Nomoto M. Effect of clarithromycin on the pharmacokinetics of cabergoline in healthy controls and in patients with Parkinson's disease. J Pharmacol Sci 2006;100(1):59–64.

73. Leclercq V, Lacaille S, Delpierre S, Karoubi E, Legrain S. [Avoidable adverse event: carbamazepine encephalopathy when introducing clarithromycin.] Rev Med Interne 2005;26(10):835–6.

74. Ford A, Smith LC, Baltch AL, Smith RP. Clarithromycin-induced digoxin toxicity in a patient with AIDS. Clin Infect Dis 1995;21(4):1051–2.

75. Xu H, Rashkow A. Clarithromycin-induced digoxin toxicity: a case report and a review of the literature. Conn Med 2001;65(9):527–9.

76. Gooderham MJ, Bolli P, Fernandez PG. Concomitant digoxin toxicity and warfarin interaction in a patient receiving clarithromycin. Ann Pharmacother 1999;33(7–8):796–9.

77. Wakasugi H, Yano I, Ito T, Hashida T, Futami T, Nohara R, Sasayama S, Inui K. Effect of clarithromycin on renal excretion of digoxin: interaction with P-glycoprotein. Clin Pharmacol Ther 1998;64(1):123–8.

78. Trivedi S, Hyman J, Lichstein E. Clarithromycin and digoxin toxicity. Ann Intern Med 1998;128(7):604.

79. Owens RC Jr, Nolin TD. Antimicrobial-associated QT interval prolongation: pointes of interest. Clin Infect Dis 2006;43(12):1603–11.

80. Evans ME, Feola DJ, Rapp RP. Polymyxin B sulfate and colistin: old antibiotics for emerging multiresistant Gram-negative bacteria. Ann Pharmacother 1999;33(9):960–7.

81. Tonini M, De Ponti F, Di Nucci A, Crema F. Review article: cardiac adverse effects of gastrointestinal prokinetics. Aliment Pharmacol Ther 1999;13(12):1585–91.

82. Wysowski DK, Bacsanyi J. Cisapride and fatal arrhythmia. N Engl J Med 1996;335(4):290–1.

83. Piquette RK. Torsade de pointes induced by cisapride/clarithromycin interaction. Ann Pharmacother 1999;33(1):22–6.

84. Rollot F, Pajot O, Chauvelot-Moachon L, Nazal EM, Kelaidi C, Blanche P. Acute colchicine intoxication during clarithromycin administration. Ann Pharmacother 2004;38(12):2074–7.

85. Dogukan A, Oymak FS, Taskapan H, Guven M, Tokgoz B, Utas C. Acute fatal colchicine intoxication in a patient on continuous ambulatory peritoneal dialysis (CAPD). Possible role of clarithromycin administration. Clin Nephrol 2001;55(2):181–2.

86. Rengelshausen J, Goggelmann C, Burhenne J, Riedel KD, Ludwig J, Weiss J, Mikus G, Walter-Sack I, Haefeli WE. Contribution of increased oral bioavailability and reduced nonglomerular renal clearance of digoxin to the digoxin–clarithromycin interaction. Br J Clin Pharmacol 2003;56:32–8.

87. Tanaka H, Matsumoto K, Ueno K, Kodama M, Yoneda K, Katayama Y, Miyatake K. Effect of clarithromycin on steady-state digoxin concentrations. Ann Pharmacother 2003;37:178–81.

88. Masia M, Gutierrez F, Jimeno A, Navarro A, Borras J, Matarredona J, Martin-Hidalgo A. Fulminant hepatitis and fatal toxic epidermal necrolysis (Lyell disease) coincident with clarithromycin administration in an alcoholic patient receiving disulfiram therapy. Arch Intern Med 2002;162(4):474–6.

89. Ausband SC, Goodman PE. An unusual case of clarithromycin associated ergotism. J Emerg Med 2001;21(4):411–3.

90. Jordan MK, Polis MA, Kelly G, Narang PK, Masur H, Piscitelli SC. Effects of fluconazole and clarithromycin on rifabutin and 25-O-desacetylrifabutin pharmacokinetics. Antimicrob Agents Chemother 2000;44(8):2170–2.

91. Amsden GW. Macrolides versus azalides: a drug interaction update. Ann Pharmacother 1995;29(9):906–17.

92. Gillum JG, Bruzzese VL, Israel DS, Kaplowitz LG, Polk RE. Effect of clarithromycin on the pharmacokinetics of 2′,3′-dideoxyinosine in patients who are seropositive for human immunodeficiency virus. Clin Infect Dis 1996;22(4):716–8.

93. Conway SP, Etherington C, Peckham DG, Brownlee KG, Whitehead A, Cunliffe H. Pharmacokinetics and safety of itraconazole in patients with cystic fibrosis. J Antimicrob Chemother 2004;53(5):841–7.

94. Niemi M, Neuvonen PJ, Kivisto KT. The cytochrome P4503A4 inhibitor clarithromycin increases the plasma concentrations and effects of repaglinide. Clin Pharmacol Ther 2001;70(1):58–65.

95. Yeates RA, Laufen H, Zimmermann T. Interaction between midazolam and clarithromycin: comparison with azithromycin. Int J Clin Pharmacol Ther 1996;34(9):400–5.

96. Geronimo-Pardo M, Cuartero-del-Pozo AB, Jimenez-Vizuete JM, Cortinas-Saez M, Peyro-Garcia R. Clarithromycin–nifedipine interaction as possible cause of vasodilatory shock. Ann Pharmacother 2005;39(3):538–42.

97. Furuta T, Ohashi K, Kobayashi K, Iida I, Yoshida H, Shirai N, Takashima M, Kosuge K, Hanai H, Chiba K, Ishizaki T, Kaneko E. Effects of clarithromycin on the metabolism of omeprazole in relation to CYP2C19 genotype status in humans. Clin Pharmacol Ther 1999;66(3):265–74.

98. Jaber BL, Lobon LF, Madias NE. The serotonin syndrome complicating co-prescription of paroxetine and clarithromycin. Am J Med 2006;119(4):e3.

99. Desta Z, Kerbusch T, Flockhart DA. Effect of clarithromycin on the pharmacokinetics and pharmacodynamics of pimozide in healthy poor and extensive metabolizers of cytochrome P450 2D6 (CYP2D6). Clin Pharmacol Ther 1999;65(1):10–20.

100. Wallace RJ Jr, Brown BA, Griffith DE, Girard W, Tanaka K. Reduced serum levels of clarithromycin in patients treated with multidrug regimens including rifampin or rifabutin for *Mycobacterium avium–M. intracellulare* infection. J Infect Dis 1995;171(3):747–50.

101. Griffith DE, Brown BA, Girard WM, Wallace RJ Jr. Adverse events associated with high-dose rifabutin in macrolide-containing regimens for the treatment of Mycobacterium avium complex lung disease. Clin Infect Dis 1995;21(3):594–8.

102. Fuller JD, Stanfield LE, Craven DE. Rifabutin prophylaxis and uveitis. N Engl J Med 1994;330(18):1315–6.

103. Shafran SD, Deschenes J, Miller M, Phillips P, Toma E. Uveitis and pseudojaundice during a regimen of clarithromycin, rifabutin, and ethambutol. MAC Study Group of the Canadian HIV Trials Network. N Engl J Med 1994,330(6):438–9.

104. Frank MO, Graham MB, Wispelway B. Rifabutin and uveitis. N Engl J Med 1994;330(12):868.

105. Havlir D, Torriani F, Dube M. Uveitis associated with rifabutin prophylaxis. Ann Intern Med 1994;121(7):510–2.

106. Jordan MK, Polis MA, Kelly G, Narang PK, Masur H, Piscitelli SC. Effects of fluconazole and clarithromycin on rifabutin and 25-O-desacetylrifabutin pharmacokinetics. Antimicrob Agents Chemother 2000;44(8):2170–2.

107. Rubinstein E. Comparative safety of the different macrolides. Int J Antimicrob Agents 2001;18(Suppl 1):S71–6.

108. Grunden JW, Fisher KA. Lovastatin-induced rhabdomyolysis possibly associated with clarithromycin and azithromycin. Ann Pharmacother 1997;31(7–8):859–63.

109. Lee AJ, Maddix DS. Rhabdomyolysis secondary to a drug interaction between simvastatin and clarithromycin. Ann Pharmacother 2001;35(1):26–31.

110. Bussing R, Gende A. Severe hypoglycemia from clarithromycin–sulfonylurea drug interaction. Diabetes Care 2002;25(9):1659–61.

111. Ibrahim RB, Abella EM, Chandrasekar PH. Tacrolimus-clarithromycin interaction in a patient receiving bone marrow transplantation. Ann Pharmacother 2002;36(12):1971–2.

112. Gomez G, Alvarez ML, Errasti P, Lavilla FJ, Garcia N, Ballester B, Garcia I, Purroy A. Acute tacrolimus nephrotoxicity in renal transplant patients treated with clarithromycin. Transplant Proc 1999;31(6):2250–1.

113. Niki Y, Nakajima M, Tsukiyama K, et al. Effect of TE-031(A-56268), a new oral macrolide antibiotic on serum theophylline concentration. Chemotherapy 1988;36:515.

114. Ruf F, Chu S, Sonders R, Sennello L. Effect of multiple doses of clarithromycin on the pharmacokinetics of theophylline. In: International Conference on Antimicrobial Agents and Chemotherapy. Atlanta, Georgia, USA: American Society for Microbiology, 1990:327.

115. Kaeser YA, Brunner F, Drewe J, Haefeli WE. Severe hypotension and bradycardia associated with verapamil and clarithromycin. Am J Health Syst Pharm 1998;55(22):2417–8.

116. Steenbergen JA, Stauffer VL. Potential macrolide interaction with verapamil. Ann Pharmacother 1998;32(3):387–8.

117. Dandekar SS, Laidlaw DA. Suprachoroidal haemorrhage after addition of clarithromycin to warfarin. J R Soc Med 2001;94(11):583–4.

118. Gustavson LE, Chu SY, Mackenthun A, Gupta MS, Craft JC. Drug interaction between clarithromycin and oral zidovudine in HIV-1 infected patients. Clin Pharmacol Ther 1993;53:163.

119. Vance E, Watson-Bitar M, Gustavson L, Kazanjian P. Pharmacokinetics of clarithromycin and zidovudine in patients with AIDS. Antimicrob Agents Chemother 1995;39(6):1355–60.

Clofazimine

See also Antituberculosis drugs

General Information

Clofazimine is weakly bactericidal against *Mycobacterium leprae*. It is active in chronic skin ulcers (Buruli ulcer) and partly against *Mycobacterium avium* intracellulare. The usual adult dosage is 50–100 mg/day. At higher doses, its anti-inflammatory effect seems to prevent the development of acute reactions, such as erythema nodosum leprosum.

Clofazimine is a strongly lipophilic dye and accumulates in tissues, especially fat, bile, macrophages, the reticuloendothelial system, and skin. This is the basis of adverse reactions, including skin discoloration (1). Lymphedema (2), diminished sweating, and reduced tearing have been observed (3).

Observational studies

In 84 patients with leprosy who took clofazimine, the most common adverse effect was a dark red skin pigmentation of varying intensity, which appeared within 10 weeks of the start of therapy (4). The intensity of the colour was proportional to the density of the infiltration. There was ichthyosis in 56 patients. Adverse effects such as anorexia, diarrhea, lymph gland enlargement, liver enlargement, corneal drying, and loss of weight were self-correcting, but nine patients had severe gastrointestinal effects (severe abdominal pain, vomiting, and diarrhea).

Organs and Systems

Respiratory

Clofazimine crystals were observed in macrophages in the respiratory alveoli of a patient with AIDS; this is considered harmless (5).

Nervous system

Intraneural deposition of a ceroid-like pigment has been seen after treatment of lepromatous leprosy with clofazimine (6). This pigment does not affect the healing process. Treatment can be continued, provided that the dose is not too high.

Sensory systems

In 76 patients taking multidrug therapy, including clofazimine, for at least 6 months, 46% had conjunctival deposition, 53% had deposition in the cornea, and crystals were found in the tears of 32% (7). Conjunctival pigmentation has been reported, as well as reversible linear brownish corneal streaks. Two cases of macular pigmentation have also been described (SEDA-5, 294).

Gastrointestinal

Nausea, vomiting, diarrhea, abdominal pain, and anorexia have been reported with clofazimine (8).

Clofazimine can accumulate and precipitate in tissues, such as the wall of the small bowel, after prolonged administration. Enteropathy can develop if crystals are stored in the lamina propria of the jejunal mucosa and the mesenteric lymph nodes. These effects depend on the dosage and duration of therapy. When this complication is suspected jejunal biopsy is indicated. At laparotomy, all organs can have an orange–yellow color (SEDA-9, 272) (1,9). On drug withdrawal, the enteropathy progressively improves. Recognized for the first time in 1967, only 14 cases of this complication have so far appeared in the English-language literature (10). Acute or chronic abdominal pain was the main symptom. In most patients, the diagnosis was made after exploratory laparotomy. Barium meal follow-through or CT scanning of the abdomen showed mucosal thickening in the small intestine. Mesenteric lymph node enlargement was present in the index case. Characteristic eosinophilic clofazimine crystals were demonstrated in the histiocytes of all patients except three. The authors proposed the term "clofazimine-induced crystal-storage histiocytosis" to emphasize causes other than B cell neoplasms for crystal-storage histiocytosis. Awareness of this complication of clofazimine can avoid unnecessary surgical exploration.

Liver

Clofazimine can inhibit the liver damage that is associated with lepromatous leprosy and the leprosy reaction; it has only minimal or no deleterious effects on liver function (11).

Skin

Pigmentation and purple skin discoloration are the most frequent adverse effects of clofazimine. It is therefore unacceptable to most light-skinned patients (12).

Ichthyosis is very frequent at dosages of clofazimine over 100 mg/day (SEDA-8, 290).

Nails

Nail changes, such as brown discoloration of the nail plate and onycholysis, have been described in patients taking high doses (300 mg/day) of clofazimine (13). Clofazimine crystals were demonstrated in the nails and nail beds.

References

1. Merrett MN, King RW, Farrell KE, Zeimer H, Guli E. Orange/black discolouration of the bowel (at laparotomy) due to clofazimine. Aust NZ J Surg 1990;60(8):638–9.
2. Oommen T. Clofazimine-induced lymphoedema. Lepr Rev 1990;61(3):289.
3. Braude AL, Davis Ch E, Fierer J. In: Infectious Diseases and Medical Microbiology. 2nd ed.. Philadelphia: WB Saunders, 1966:1171.
4. Ramu G, Iyer GG. Side effects of clofazimine therapy. Lepr India 1976;48(Suppl 4):722–31.
5. Sandler ED, Ng VL, Hadley WK. Clofazimine crystals in alveolar macrophages from a patient with the acquired immunodeficiency syndrome. Arch Pathol Lab Med 1992;116(5):541–3.
6. McDougall AC, Jones RL. Intra-neural ceroid-like pigment following the treatment of lepromatous leprosy with clofazimine (B663; Lamprene). J Neurol Neurosurg Psychiatry 1981;44(2):116–20.
7. Kaur I, Ram J, Kumar B, Kaur S, Sharma VK. Effect of clofazimine on eye in multibacillary leprosy. Indian J Lepr 1990;62(1):87–90.
8. Lal S, Garg BR, Hameedulla A. Gastro-intestinal side effects of clofazimine. Lepr India 1981;53(2):285–8.
9. Jost JL, Venencie PY, Cortez A, Orieux G, Debbasch L, Chomette G, Puissant A, Vayre P. Entéropathie à la clofazimine. [Enteropathy caused by clofazimine.] J Chir (Paris) 1986;123(1):7–9.
10. Sukpanichnant S, Hargrove NS, Kachintorn U, Manatsathit S, Chanchairujira T, Siritanaratkul N, Akaraviputh T, Thakerngpol K. Clofazimine-induced crystal-storing histiocytosis producing chronic abdominal pain in a leprosy patient. Am J Surg Pathol 2000;24(1):129–35.
11. Bulakh PM, Kowale CN, Ranade SM, Burte NP, Chandorkar AG. The effect of clofazimine on liver function tests in lepra reaction (ENL). Lepr India 1983;55(4):714–8.
12. Burte NP, Chandorkar AG, Muley MP, Balsara JJ, Bulakh PM. Clofazimine in lepra (ENL) reaction, one year clinical trial. Lepr India 1983;55(2):265–77.
13. Dixit VB, Chaudhary SD, Jain VK. Clofazimine induced nail changes. Indian J Lepr 1989;61(4):476–8.

Cycloserine

See also Antituberculosis drugs

General Information

Cycloserine is an aminoisoxazolidone that shows no cross-resistance with other tuberculostatic agents (1). Because of its high toxicity, it should only be used when micro-organisms are resistant to other drugs (relapse or primary resistance). It is usually given orally.

Organs and Systems

Nervous system

Cycloserine can cause headache, somnolence, and tremor (2). In mice, L-cycloserine protects against auditory-invoked seizures by binding to pyridoxal phosphate in the presence of zinc ions (3). The addition of a fluoroquinolone can increase the risk of nervous system effects (4).

Psychological, psychiatric

Cycloserine can cause altered mood, cognitive deterioration, dysarthria, confusion, and even psychotic crises (2).

Susceptibility Factors

Renal disease

Cycloserine accumulates to toxic concentrations in patients with renal insufficiency (5).

References

1. Peloquin CA. Pharmacology of the antimycobacterial drugs. Med Clin North Am 1933;77(6):1253–62.
2. Mitchell RS, Lester W. Clinical experience with cycloserine in the treatment of tuberculosis. Scand J Respir Dis Suppl 1970;71:94–108.
3. Chung SH, Johnson MS, Gronenborn AM. L-cycloserine: a potent anticonvulsant. Epilepsia 1984;25(3):353–62.
4. Berning SE. The role of fluoroquinolones in tuberculosis today. Drugs 2001;61(1):9–18.
5. Wareska W, Klott M, Izdebska-Makosa Z. Wydalenie cykloseryny w przypadkach schorze'n nerkowych. [Excretion of cycloserine in cases of kidney diseases.] Gruzlica 1963;31:664–8.

Dapsone and analogues

General Information

Dapsone is 4,4-diaminodiphenylsulfone (DDS, avlosulfone, disulfone) (SEDA-17, 352). It is a bacteriostatic antileprosy drug with a sulfonamide-like structure. The dosage should be 50–100 mg/day in adults (1). In children aged 3–5 years, the dosage should be reduced according to weight to, for example, 25 mg/day (2).

Studies of patients with borderline leprosy have suggested that dapsone has mild immunosuppressive effects (SEDA-18, 30). Thus, it has been given with some success to patients with dermatitis herpetiformis, subcorneal pustular dermatosis, bullous dermatoses, relapsing polychondritis, thrombocytopenic purpura (3), giant cell arteritis (4), rheumatoid arthritis, and systemic lupus erythematosus (5).

Dapsone is an alternative drug for *Pneumocystis jiroveci* pneumonia prophylaxis in individuals who cannot tolerate co-trimoxazole, and although this has become less of a problem since the advent of highly active antiretroviral drug therapy in countries able to afford these regimens, further data on the toxicity of dapsone in children are welcome. In a multicenter study from the USA daily and weekly dapsone regimens have been compared in 94 HIV-infected children, monitoring hematological and liver toxicity and the incidence of skin rashes, *P. jiroveci* pneumonia, or death (6). They concluded that the weekly regimen produced less hematological toxicity, but that this advantage was offset by a trend toward higher breakthrough rates of *P. jiroveci* pneumonia.

Dapsone analogues

The sulfone acedapsone (rINN) (4,4'-diacetyldiaminodiphenylsulfone, DADDS) is the diacetyl derivative of dapsone with a long half-life (7). However, its plasma concentrations are much lower than those of dapsone and it could enhance the emergence of resistant strains of *Mycobacterium leprae*. Its adverse effects are similar to those of dapsone, which it can replace if gastrointestinal symptoms become severe. It is available in an enteric-coated formulation, given in a dosage of 330 mg/day.

General adverse effects

The adverse effects of dapsone have been comprehensively reviewed (8). The most common untoward effect is hemolysis of varying degree; it is usually mild, except in patients with glucose-6-phosphate dehydrogenase (G6PD) deficiency. Methemoglobinemia and Heinz body formation also occur and methemoglobinemia can be a problem at doses over 200 mg/day. Agranulocytosis is rare but potentially fatal. Mild gastrointestinal complaints and neurological effects of dapsone are not uncommon. Under 0.5% of the patients taking prolonged dapsone therapy develop the dapsone syndrome (SEDA-16, 347). Other rare adverse effects include peripheral neuropathy, psychosis, hepatitis, nephritic syndrome, and renal papillary necrosis. Hypoalbuminemia has been seen after prolonged dapsone therapy of dermatitis herpetiformis (9). Erythema nodosum leprosum may be due to an immune complex mechanism; the antigen is provided by the bacteria and their degeneration products, possibly as a kind of Jarisch–Herxheimer reaction (10). Anaphylactic shock and tachycardia are rare. Rashes, serious cutaneous reactions, and erythema nodosum may have an immunological basis. Tumor-inducing effects have not been reported.

In France, dapsone is available in combination with ferrous oxalate as Disulone (Aventis), and in 1983–98, 249 adverse reactions were reported to French pharmacovigilance centers, mainly blood dyscrasias (often neutropenia and agranulocytosis, rarely hemolysis and anemia) (11). Five patients died; three of them had septicemia secondary to agranulocytosis. There were 29 cases of dapsone syndrome, 39 skin reactions, 27 cases of liver damage, and 27 cases of neurological and psychiatric adverse effects. Patients taking dapsone need to be under close medical supervision for early recognition of adverse reactions.

Organs and Systems

Cardiovascular

Signs of heart failure with edema, ascites, and severe hypoalbuminemia have been described in the treatment of dermatitis herpetiformis with dapsone (9).

Respiratory

Two cases of pulmonary eosinophilia attributed to dapsone were reported in 1998. Four cases had previously been reported, in which the fixed combination of dapsone and pyrimethamine (a malaria prophylactic) had been implicated, but only one previous report had implicated dapsone alone.

- A French woman with chronic urticaria (12) and an Indian man with lepromatous leprosy (13) developed fever, wheezing, and breathlessness. Both had peripheral eosinophilia and chest X-rays showed infiltrates. The woman's symptoms began 2 weeks after she started to take dapsone but recurred a few hours after a subsequent rechallenge. The man's symptoms occurred a few hours after each daily dose. Symptoms in both cases resolved within a few days of stopping dapsone.

Nervous system

Dapsone-induced neuropathy is not common (14), in spite of its widespread use in a variety of unrelated disorders. It has not been reported in patients with leprosy, but it would be easy to miss, since worsening neuropathy would readily be attributed to the underlying disease. Neuropathy has not been reported in patients with leprosy taking the usually recommended dosage of 100 mg/day. Isolated cases of dapsone-induced peripheral neuropathy, including motor and minor sensory defects, have been published (15,16). The clinical characteristics include a motor neuropathy affecting the extremities with onset within 5 years after the start of dapsone therapy in doses of over 300 mg/day. Complete recovery from the neuropathy almost always occurs after the dose is reduced or the drug is withdrawn.

Infection with *M. leprae* affects the peripheral nerves and the dermis, causing an accumulation of macrophages and other immune cells at the infected site (17).

Sensory systems

Ocular toxicity is rare with dapsone. Reduced visual acuity has been described after overdosage (18).

Psychological, psychiatric

Dapsone-induced psychosis has rarely been reported (SEDA-15, 331) (19).

Hematologic

The most common adverse effect of dapsone therapy is hemolysis (SEDA-15, 331). In varying degrees, it develops in nearly all patients taking dapsone in dosages of 200–300 mg/day. Even with the usual dose of 100 mg/day in normal persons and 50 mg/day in patients with G6PD deficiency, there can be some degree of hemolysis (20,21). Red cell survival can be reduced by dapsone, depending on the dose of the sulfone and its oxidizing activity, but hemolytic anemia generally does not occur without a preexisting disorder of the erythrocytes or bone marrow. The hemolysis can be so severe that manifestations of hypoxia become striking (20).

The medical records of 12 patients with ocular cicatricial disease, of whom 11, mean age 72 (range 55–89) years, were treated with dapsone 50 mg bd as first line therapy for 19 (range 1–60) months, were reviewed (22). Six had reticulocytosis, including four with clinically significant hemolytic anemia. These patients were not checked for glucose-6-phosphate hydrogenase (G6PD) deficiency, which is a susceptibility factor for hemolytic anemia.

- Mild hemolytic anemia occurred in a breast-fed infant and his mother, who had continuously been taking 100–150 mg/day; dapsone and its metabolite, monoacetyldapsone, were identified in the infant's serum (SEDA-8, 290).

Dapsone is an oxidant and can trigger methemoglobinemia and delayed sulfhemoglobinemia (SEDA-8, 289) (23). Methemoglobinemia remains subclinical in most cases. However, it can be accurately inferred by a discrepancy between the oxygen saturation measured by pulse oximetry and the concentration of oxygen in the arterial blood (24). Methemoglobinemia at concentrations over 10% produces a visible lavender-colored cyanosis, and concentrations over 35% result in weakness and shortness of breath. Methemoglobinemia can be minimized by the administration of an antioxidant, such as vitamin C or vitamin E (25). Methemoglobinemia disappears after withdrawal of dapsone (SEDA-10, 273).

- A patient who had been taking dapsone inappropriately instead of an antispasmodic that had been prescribed for a spinal condition, because of incorrect labeling in a pharmacy, developed methemoglobinemia (26). Methylene blue was given intravenously and may have contributed to the severe hemolysis that followed.
- An immunocompromised child with chronic immune thrombocytopenic purpura developed a fever, cough, perioral cyanosis, bilateral lower lobe coarse crackles, and a low O_2 saturation by pulse oximetry (89%). His

medications included prednisone and rituximab for chronic immune thrombocytopenic purpura, and dapsone for *Pneumocystis jiroveci* pneumonia prophylaxis. The methemoglobin concentration was 9.6%.

The frequency of methemoglobinemia and reduced cytochrome b5 reductase (Cb5r) activity have been studied in 15 children with acute lymphoblastic leukemia taking dapsone prophylaxis for *Pneumocystis jiroveci* and 10 taking co-trimoxazole (27). At a mean of 6.6 weeks, three children taking dapsone developed symptomatic methemoglobinemia, defined as increased concentrations of methemoglobin in association with hypoxemia (oxygen saturation <95%). The other 12 were all asymptomatic. All the controls had normal methemoglobin concentrations. Cb5 reductase activities in two of the three symptomatic children were below 50% of normal, suggesting heterozygosity, which was confirmed by assessment of their parent's Cb5 reductase activities. Heterozygosity for Cb5 reductase deficiency can predispose to methemoglobinemia, and such patients tend to be symptomatic at low methemoglobin concentrations. Heterozygotes should therefore be monitored closely for the possibility of methemoglobinemia.

Agranulocytosis is a rare complication of dapsone treatment (28), and a fatal case has been reported:

- A 60-year-old man took 100 mg dapsone daily for leprosy and after 3 weeks developed pneumonia and agranulocytosis (29). He died a few hours after diagnosis despite parenteral antibiotics.

Dapsone was the suspected cause of agranulocytosis. However, the patient had also received concomitant rifampicin, clofazimine, and prednisolone.

Aplastic anemia is a rare complication of dapsone, but a fatal case has been reported (30).

- A 23-year-old Brazilian man was given dapsone, clofazimine, and rifampicin for lepromatous leprosy. After 10 months he had a severe episode of epistaxis associated with pancytopenia (hematocrit 8.3%, white cell count 1.3 x 109/l, 28% neutrophils, platelet count 5 x 109/l). Aplastic anemia was confirmed by a bone marrow biopsy. Despite numerous transfusions of platelets and packed red cell concentrates, he died from bleeding and nosocomial infection.

Late-onset aplastic anemia, as in this case, does not depend on the dose and duration of exposure to dapsone and is irreversible. Therefore, periodic hematological monitoring is recommended and patients should be educated about suggestive symptoms.

Pure red cell aplasia has been attributed to dapsone (31).

- A 75-year-old man with type 2 diabetes taking glibenclamide developed granuloma annulare and was given dapsone 100 mg/day. A full blood count was normal, but 4 weeks later his hemoglobin was 3.6 g/dl, reticulocyte count 0.54%, MCV 109 fl, serum iron 27 µg/l, and serum ferritin 711 ng/l. Vitamin B_{12} and folic acid concentrations were normal. A bone marrow aspirate

showed a normocellular marrow with profound erythroid hypoplasia. Pure red cell aplasia was diagnosed and linked to dapsone, which was withdrawn. After blood transfusion the hematological profile gradually returned to normal by day 8.

Liver

Jaundice and hepatitis can occur as part of the sulfone syndrome with dapsone (32). Previous liver damage can predispose to serious hepatic or other adverse effects.

Pancreas

A case of pancreatitis associated with sulfone syndrome has been documented; a second case has now been reported (33).

- An 87-year-old white man developed acute abdominal pain 4 weeks after starting to take dapsone 100 mg/day. He had raised serum amylase and lipase activities. His symptoms resolved when dapsone was withdrawn and recurred when it was re-introduced 4 months later.

Urinary tract

Renal insufficiency can be associated with severe hemolysis due to dapsone (34).

Skin

Pruritus and various forms of rash can occur with dapsone. In 17 cases of dapsone syndrome, the symptoms developed on average 27 days after the start of treatment. The skin lesions took the form of erythematous papules and plaques (n = 13), eczematous lesions (n = 4), and associated bullous lesions (n = 2) (35). The other manifestations were: fever (n = 16), pruritus (n = 15), lymphadenopathy (n = 14), hepatomegaly (n = 10), icterus and oral erosions (n = 5 each), photosensitivity (n = 4), and splenomegaly (n = 2).

Serious cutaneous reactions, such as exfoliative dermatitis, toxic epidermal necrolysis, and erythema multiforme bullosum are extremely rare. Erythema nodosum leprosum has been described during dapsone therapy, mostly in the lepromatous type of leprosy (10). If erythema nodosum develops before the start of therapy, the drug should be withheld until the reaction has disappeared. Severe erythema nodosum can be controlled by short-term glucocorticoid therapy. Desensitization to dapsone in patients with hypersensitivity reactions has been proposed (36).

Immunologic

The sulfones occasionally exacerbate lepromatous leprosy, the so-called sulfone syndrome or dapsone syndrome, which resembles acute infectious mononucleosis (SEDA-16, 347) (37), and can develop 3–6 weeks after the start of treatment in malnourished patients. It appears to be an allergic reaction. It includes fever, malaise, pruritus, exfoliative dermatitis, photosensitivity, polyarthritis (38), jaundice and even hepatic necrosis, hepatosplenomegaly, lymphadenopathy, methemoglobinemia, and anemia. The syndrome is accompanied by the formation of atypical T lymphocytes with markedly increased spontaneous thymidine uptake (SEDA-8, 289). The full syndrome is probably rare, but it is important to recognize its partial expression (5,39,40). It has been suggested that it has become more common since the introduction of multidrug therapy (41), especially with rifampicin plus dapsone. The syndrome usually resolves rapidly after withdrawal of dapsone and with glucocorticoid treatment (for example prednisolone 30–60 mg/day). However, it can also end in a fatal allergic reaction (SEDA-12, 259).

Hypersensitivity to dapsone, called the "sulfone syndrome", manifests as fever, malaise, exfoliative dermatitis, raised transaminase activities, lymphadenopathy, methemoglobinemia, and hemolytic anemia. The sulfone syndrome can occur at doses of 50–300 mg/day. In all cases the syndrome occurs within 2 months of the start of therapy.

- A 42-year-old HIV-infected African-American man taking dapsone prophylaxis for *Pneumocystis jiroveci* pneumonia developed fever, lymphadenopathy, exfoliative dermatitis, hepatitis, and methemoglobinemia 4 weeks after starting to take dapsone. His symptoms and laboratory abnormalities completely resolved on withdrawal of dapsone (42).
- A similar syndrome occurred in a 55-year-old Caucasian woman with urticarial vasculitis, who developed near-fatal sulfone syndrome with hepatic and renal failure (43).
- A 51-year old woman, with a history of sulfonamide allergy, received a matched unrelated hemopoietic stem cell transplant for acute myelogenous leukemia (44). Dapsone 100 mg/day was started on day 28 for *Pneumocystis jiroveci* prophylaxis. One month later she developed hepatitis, hemolytic anemia, fever, methemoglobinemia of 8%, and exfoliative dermatitis.
- A 12-year-old girl was admitted to hospital 24 days after starting to take the World Health Organization (WHO) recommended multidrug therapy for leprosy (dapsone, clofazimine, and rifampicin) (45). She had intense jaundice, generalized lymphadenopathy, hepatosplenomegaly, oral lesions, conjunctivitis, a morbilliform rash, and edema of the face, ankles, and hands. Her hemoglobin was 8.4 g/dl, the white blood cell count $15.7 \times 10^9/l$, and the platelet count $100 \times 10^9/l$. Her INR was 49 and there was an increase in liver enzymes. She later deteriorated and developed exfoliative dermatitis, shock, and acute renal and hepatic insufficiency. She was given adrenergic drugs, fluids, blood, and antibiotics and made a good recovery. She was later given clofazimine and rifampicin without adverse effects.

Hypersensitivity pneumonitis is an immunologically mediated lung disease characterized by dyspnea, fever, cough, chest pain, and hypoxemia. The characteristic radiographic findings include bilateral reticular infiltrates. This type of pneumonitis is acute and unrelated to the dose or duration of therapy with the incriminated drug. Bronchoalveolar lavage may show a predominance of eosinophils or lymphocytes, with an increased CD8/CD4 ratio. Inciting antigens include those of birds, fungi, chemicals, and drugs. Adverse drug reactions related to antimicrobial use are common in patients with HIV infection. In particular, sulfonamides, including co-trimoxazole, commonly cause adverse reactions, but sulfonamide-induced pulmonary complications are rare.

In HIV-negative patients with ulcerative colitis, sulfa-related drugs such as sulfasalazine have been reported to cause hypersensitivity pneumonitis with eosinophilia. However, hypersensitivity pneumonitis has rarely been reported in patients infected with HIV, and this can be attributed to the low absolute numbers of CD4 and CD8 lymphocytes and reduced function of CD4 cells. The differential diagnosis of interstitial pneumonitis in patients with HIV is extensive and includes *Pneumocystis jiroveci* pneumonia, other infectious causes such as cytomegalovirus and mycobacteria, and non-infectious causes such as neoplasms, eosinophilic pneumonia, and non-specific interstitial pneumonia (46).

- A 40 year old Native American woman with newly diagnosed AIDS developed osteomyelitis and was given vancomycin and co-trimoxazole prophylaxis for *Pneumocystis* pneumonia when her CD4 count was reported as $140 \times 10^6/l$ (47). She developed a drug rash and was subsequently switched to dapsone, but 4 days later developed a maculopapular rash along with a non-productive cough, dyspnea, and hypoxemia. There were scattered rales and a chest X-ray showed new diffuse interstitial pulmonary infiltrates. A complete blood count showed eosinophilia. She was treated with clindamycin, primaquine, and prednisone, and the dapsone and vancomycin were withdrawn. No evidence for *Pneumocystis* pneumonia was found and treatment for that was withdrawn after 1 week, after which she continued to improve. A chest X-ray 10 days after dapsone had been withdrawn showed resolution of the interstitial infiltrates.

Although it is extremely unusual, drug-induced hypersensitivity pneumonitis should be considered in the differential diagnosis of interstitial pneumonitis in patients with HIV.

Anaphylactic shock and tachycardia are among the most severe allergic reactions to dapsone (41).

Second-Generation Effects

Pregnancy

The use of antileprosy drugs during pregnancy depends upon the severity of the disease and the relative need for treatment. Even though untoward effects on the fetus have not been reported, treatment of leprosy during pregnancy probably predisposes to erythema nodosum leprosum; if this occurs during pregnancy, clofazimine is considered the best drug available (48).

Lactation

Dapsone passes into the breast milk, with a milk to plasma AUC ratio of 0.22:0.45 (49). Assuming a daily milk ingestion of 1 liter by the infant, the maximum percentage of the maternal dose in milk was 14% over 9 days. Usually no adverse effects are noted in the newborn, unless there is G6PD deficiency.

Susceptibility Factors

Genetic factors

Glucose-6-phosphate dehydrogenase deficiency is a risk factor for hemolytic anemia (50–52).

To what extent phenotype (fast versus slow acetylators) affects the metabolism of dapsone is controversial. Since dapsone is acetylated in the liver, an effect might be anticipated (SEDA-4, 217) (53), but has not been demonstrated.

Hepatic disease

Previous liver damage predisposes to adverse effects during dapsone therapy (54).

Other features of the patient

Malnutrition predisposes to adverse effects during dapsone therapy (55).

Drug Administration

Drug administration route

Aqueous or oily suspensions of dapsone for intramuscular injection can be used to ensure adherence and to prevent resistance due to irregular drug intake. The therapeutic efficacy of intramuscular formulations lasts 3–4 weeks (56).

Drug overdose

Several reports of accidental poisoning with dapsone in children have appeared since the early 1980s. Two reports (57,58) have emphasized the persistence of the problem, although the number of childhood cases has fallen over the years. Poisoning with dapsone results in cyanosis due to methemoglobinemia, vomiting, mental confusion, tachycardia, and dyspnea. It has been suggested that treatment with multiple doses of activated charcoal may be sufficient for less severely poisoned children (methemoglobin concentration below 30%) and that a single dose of methylthioninium chloride (methylene blue) should be given to those with higher concentrations (58).

Drug–Drug Interactions

Clofazimine

In 28 patients with lepromatous leprosy, clofazimine did not influence the urinary excretion of dapsone, except in one case (59).

Fluconazole

Dapsone hydroxylamine formation is thought cause high rates of adverse reactions to dapsone in HIV-infected individuals. The effect of fluconazole on hydroxylamine formation in individuals with HIV infection has been investigated in 23 HIV-infected subjects (60). Fluconazole reduced the AUC by 49%, the percentage of the dose excreted in the urine in 24 hours by 53%, and the formation clearance of the hydroxylamine by 55%. This inhibition of in vivo hydroxylamine formation was quantitatively consistent with that predicted from human liver microsomal experiments.

Primaquine

The combination of primaquine with clindamycin is used as second choice in the treatment or prevention of *P. jiroveci* pneumonia. If the patient has been treated immediately beforehand with dapsone, methemoglobinemia can result, especially in patients infected with HIV (SEDA-21, 296).

Probenecid

Probenecid increases plasma dapsone concentrations by inhibiting its renal clearance (61).

Rifamycins

When rifampicin was given, dapsone blood concentrations were lowered and urinary excretion was increased during the first 2 days; however, blood concentrations remained in the therapeutic range (62).

The interaction of rifampicin with dapsone has been reviewed (63). In seven patients with leprosy, rifampicin shortened the half-life of dapsone by 50% (64).

Rifabutin had no effect on the plasma AUC of hydroxylamine or the percent excreted in the urine in 24 hours but increased the formation clearance by 92% (60). Dapsone clearance was increased by rifabutin and rifabutin plus fluconazole (67% and 38% respectively) but was unaffected by fluconazole or clarithromycin. Hydroxylamine production was unaffected by clarithromycin. On the basis of these data, and assuming that exposure to dapsone hydroxylamine determines dapsone toxicity, co-administration of fluconazole should reduce the rate of adverse reactions to dapsone in people with HIV infection and rifabutin and clarithromycin will have no effect. When dapsone is given in combination with rifabutin, dapsone dosage adjustment may be necessary.

References

1. Garg SK, Kumar B, Bakaya V, Lal R, Shukla VK, Kaur S. Plasma dapsone and its metabolite monoacetyldapsone levels in leprotic patients. Int J Clin Pharmacol Ther Toxicol 1988;26(11):552–4.
2. Thangaraj RH, Yawalkar SJ. Leprosy for Medical Practitioners and Paramedical Workers. 2nd ed.. Basel: Ciba-Geigy;. 1987.

3. Godeau B, Oksenhendler E, Bierling P. Dapsone for auto-immune thrombocytopenic purpura. Am J Hematol 1993;44(1):70–2.

4. Liozon F, Vidal E, Barrier JH. Dapsone in giant cell arteritis treatment. Eur J Intern Med 1993;4:207.

5. Kraus A, Jakez J, Palacios A. Dapsone induced sulfone syndrome and systemic lupus exacerbation. J Rheumatol 1992;19(1):178–80.

6. McIntosh K, Cooper E, Xu J, Mirochnick M, Lindsey J, Jacobus D, Mofenson L, Yogev R, Spector SA, Sullivan JL, Sacks H, Kovacs A, Nachman S, Sleasman J, Bonagura V, McNamara J. Toxicity and efficacy of daily vs. weekly dapsone for prevention of *Pneumocystis carinii* pneumonia in children infected with human immunodeficiency virus. ACTG 179 Study Team. AIDS Clinical Trials Group. Pediatr Infect Dis J 1999;18(5):432–9.

7. George J, Balakrishnan S. Blood dapsone levels in leprosy patients treated with acedapsone. Indian J Lepr 1986;58(3):401–6.

8. Zhu YI, Stiller MJ. Dapsone and sulfones in dermatology: overview and update. J Am Acad Dermatol 2001;45(3):420–34.

9. Cowan RE, Wright JT. Dapsone and severe hypoalbuminaemia in dermatitis herpetiformis. Br J Dermatol 1981;104(2):201–4.

10. Somorin AO. Erythema nodosum leprosum in Nigeria. Int J Dermatol 1975;14(9):664–6.

11. Benedetti-Bardet C, Guy C, Boudignat O, Regnier-Zerbib A, Ollagnier M. Centres Regionaux de Pharmacovigilance. Effets indésirables de la Disulone: résultants de l'enquête française de pharmacovigilance. [Adverse effects of Disulone; results of the France pharmacovigilance inquiry. Regional Centers of Pharmacovigilance.] Therapie 2001;56(3):295–9.

12. Jaffuel D, Lebel B, Hillaire-Buys D, Pene J, Godard P, Michel FB, Blayac JP, Bousquet J, Demolyi P. Eosinophilic pneumonia induced by dapsone. BMJ 1998;317(7152):181.

13. Arunthathi S, Raju S. Dapsone-induced pulmonary eosinophilia without cutaneous allergic manifestations—an unusual encounter—a case report. Acta Leprol 1998;11(2):3–5.

14. Saqueton AC, Lorincz AL, Vick NA, Hamer RD. Dapsone and peripheral motor neuropathy. Arch Dermatol 1969;100(2):214–7.

15. Waldinger TP, Siegle RJ, Weber W, Voorhees JJ. Dapsone-induced peripheral neuropathy. Case report and review. Arch Dermatol 1984;120(3):356–9.

16. Ahrens EM, Meckler RJ, Callen JP. Dapsone-induced peripheral neuropathy. Int J Dermatol 1986;25(5):314–6.

17. Kaplan G, Cohn ZA. The immunobiology of leprosy. Int Rev Exp Pathol 1986;28:45–78.

18. Alexander TA, Raju R, Kuriakose T, Cherian AM. Presumed DDS ocular toxicity. Indian J Ophthalmol 1989;37(3):150–1.

19. Balkrishna, Bhatia MS. Dapsone induced psychosis. J Indian Med Assoc 1989;87(5):120–1.

20. Mandell GL, Sande MA. Antimicrobial agents: drugs used in the chemotherapy of tuberculosis and leprosy. In: Goodman Gilman A, Rall TW, Nies AS, Taylor P, editors. Goodman and Gilman's The Pharmacological Basis of Therapeutics. 8th ed.. New York: Pergamon Press, 1990:1146 Chapter 49.

21. Byrd SR, Gelber RH. Effect of dapsone on haemoglobin concentration in patients with leprosy. Lepr Rev 1991;62(2):171–8.

22. Wertheim MS, Males JJ, Cook SD, Tole DM. Dapsone induced haemolytic anaemia in patients treated for ocular cicatricial pemphigoid. Br J Ophthalmol 2006;90:516.

23. Schiff DE, Roberts WD, Sue Y-J. Methemoglobinemia associated with dapsone therapy in a child with pneumonia and chronic immune thrombocytopenic purpura. J Pediatr Hematol Oncol 2006;28:395–8.

24. Trillo RA Jr, Aukburg S. Dapsone-induced methemoglobinemia and pulse oximetry. Anesthesiology 1992;77(3):594–6.

25. Prussick R, Ali MA, Rosenthal D, Guyatt G. The protective effect of vitamin E on the hemolysis associated with dapsone treatment in patients with dermatitis herpetiformis. Arch Dermatol 1992;128(2):210–3.

26. Southgate HJ, Masterson R. Lessons to be learned: a case study approach: prolonged methaemoglobinaemia due to inadvertent dapsone poisoning; treatment with methylene blue and exchange transfusion. J R Soc Health 1999;119(1):52–5.

27. Williams S, MacDonald P, Hoyer JD, Barr RD, Athale UH. Methemoglobinemia in children with acute lymphoblastic leukemia (ALL) receiving dapsone for *Pneumocystis carinii* pneumonia (PCP) prophylaxis: a correlation with cytochrome b5 reductase (Cb5R) enzyme levels. Pediatr Blood Cancer 2005;44:55–62.

28. Braude AL, Davis Ch E, Fierer J. In: Infectious Diseases and Medical Microbiology. 2nd ed.. Philadelphia: WB Saunders, 1966:1171.

29. Bhat RM, Radhakrishnan K. A case report of fatal dapsone-induced agranulocytosis in an Indian mid-borderline leprosy patient. Lepr Rev 2003;74:167–70.

30. Goulart IM, Reis AC, De Rezende TM, Borges AS, Ferreira MS, Nishioka SA. Aplastic anaemia associated with multidrug therapy (dapsone, rifampicin and clofazimine) in a patient with lepromatous leprosy. Lepr Rev 2005;76:167–9.

31. Borrás-Blasco J, Conesa-García V, Navarro-Ruiz A, Devesa P, Matarredona J. Pure red cell aplasia associated with dapsone therapy. Ann Pharmacother 2005; 39(6):1137–8.

32. Tomecki KJ, Catalano CJ. Dapsone hypersensitivity. The sulfone syndrome revisited. Arch Dermatol 1981;117(1):38–9.

33. Jha SH, Reddy JA, Dave JK. Dapsone-induced acute pancreatitis. Ann Pharmacother 2003;37:1438–40.

34. Chugh KS, Singhal PC, Sharma BK, Mahakur AC, Pal Y, Datta BN, Das KC. Acute renal failure due to intravascular hemolysis in the North Indian patients. Am J Med Sci 1977;274(2):139–46.

35. Kumar RH, Kumar MV, Thappa DM. Dapsone syndrome—a five year retrospective analysis. Indian J Lepr 1998;70(3):271–6.

36. Browne SG. Desensitization for dapsone dermatitis. BMJ 1963;5358:664–6.

37. Chan HL, Lee KO. Tonsillar membrane in the DDS (dapsone) syndrome. Int J Dermatol 1991;30(3):216–7.

38. Pavithran K. Dapsone syndrome with polyarthritis: a case report. Indian J Lepr 1990;62(2):230–2.

39. Johnson DA, Cattau EL Jr, Kuritsky JN, Zimmerman HJ. Liver involvement in the sulfone syndrome. Arch Intern Med 1986;146(5):875–7.

40. Mohle-Boetani J, Akula SK, Holodniy M, Katzenstein D, Garcia G. The sulfone syndrome in a patient receiving dapsone prophylaxis for *Pneumocystis carinii* pneumonia. West J Med 1992;156(3):303–6.

41. Richardus JH, Smith TC. Increased incidence in leprosy of hypersensitivity reactions to dapsone after

introduction of multidrug therapy. Lepr Rev 1989;60(4):267–73.

42. Lee KB, Nashed TB. Dapsone-induced sulfone syndrome. Ann Pharmacother 2003;37:1044–6.

43. Leslie KS, Gaffney K, Ross CN, Ridley S, Barker TH, Garoich JJ. A near fatal case of the dapsone hypersensitivity syndrome in a patient with utricarial vasculitis. Clin Exp Dermatol 2003;28:496–8.

44. Abidi MH, Kozlowski JR, Ibrahim RB, Peres E. The sulfone syndrome secondary to dapsone prophylaxis in a patient undergoing unrelated hematopoietic stem cell transplantation. Hematol Oncol 2006;24:164–5.

45. Bucaretchi F, Vicente DC, Pereira RM, Tresold T. Dapsone hypersensitivity in an adolescent during treatment for leprosy. Rev Inst Med Trop S Paulo 2004;46 (6):331–4.

46. Sattler F, Nichols L, Hirano L, Hiti A, Hofman F, Hughlett C, Zeng L, Boylen CT, Koss M. Nonspecific interstitial pneumonitis mimicking Pneumocystis carinii pneumonia. Am J Respir Crit Care Med 1997;156: 912–7.

47. Tobin-D'Angelo MJ, Hoteit MA, Brown KV, Ray SM, King MD. Dapsone-induced hypersensitivity pneumonitis mimicking Pneumocystis carinii pneumonia in a patient with AIDS. Am J Med Sci 2004;327(3):163–5.

48. Duncan ME, Pearson JM. The association of pregnancy and leprosy—III. Erythema nodosum leprosum in pregnancy and lactation. Lepr Rev 1984;55(2):129–42.

49. Edstein MD, Veenendaal JR, Newman K, Hyslop R. Excretion of chloroquine, dapsone and pyrimethamine in human milk. Br J Clin Pharmacol 1986;22(6):733–5.

50. Editorial. Adverse reactions to dapsone. Lancet 1981;2(8239):184–5.

51. Menezes S, Rege VL, Sehgal VN. Dapsone haemolysis in leprosy. A preliminary report. Lepr India 1981;53(1).63–9.

52. Halmekoski J, Mattila MJ, Mustakallio KK. Metabolism and haemolytic effect of dapsone and its metabolites in man. Med Biol 1978;56(4):216–21.

53. Peters JH, Gordon GR, Levy L, Strokan MA, Jacobson RR, Enna CD, Kirchheimer WF. Metabolic disposition of dapsone in patients with dapsone-resistant leprosy. Am J Trop Med Hyg 1974;23(2):222–30.

54. Goette DK. Dapsone-induced hepatic changes. Arch Dermatol 1977;113(11):1616–7.

55. Gawkrodger DJ, Ferguson A, Barnetson RS. Nutritional status in patients with dermatitis herpetiformis. Am J Clin Nutr 1988;48(2):355–60.

56. Modderman ES, Huikeshoven H, Zuidema J, Leiker DL, Merkus FW. Intramuscular injection of dapsone in therapy for leprosy: a new approach. Int J Clin Pharmacol Ther Toxicol 1982;20(2):51–6.

57. Carrazza MZ, Carrazza FR, Oga S. Clinical and laboratory parameters in dapsone acute intoxication. Rev Saude Publica 2000;34(4):396–401.

58. Bucaretchi F, Miglioli L, Baracat EC, Madureira PR, Capitani EM, Vieira RJ. Exposicao aguda a dapsona e metemoglobinemia em criancas: tratamento com doses multiplas de carvao ativado. [Acute dapsone exposure and methemoglobinemia in children: treatment with multiple doses of activated charcoal with or without the administration of methylene blue.] J Pediatr (Rio J) 2000;76(4):290–4.

59. Grabosz JA, Wheate HW. Effect of clofazimine on the urinary excretion of DDS (dapsone). Int J Lepr Other Mycobact Dis 1975;43(1):61–2.

60. Winter HR, Trapnell CB, Slattery JT, Jacobson M, Greenspan DL, Hooton TM, Unadkat JD. The effect of clarithromycin, fluconazole, and rifabutin on dapsone hydroxylamine formation in individuals with human immunodeficiency virus infection (AACTG 283). Clin Pharmacol Ther 2004;76:579–87.

61. Goodwin CS, Sparell G. Inhibition of dapsone excretion by probenecid. Lancet 1969;2(7626):884–5.

62. Balakrishnan, Seshadri PS. Drug interactions—the influence of rifampicin and clofazimine on the urinary excretion of DDS. Lepr India 1981;53(1):17–22.

63. Baciewicz AM, Self TH. Rifampin drug interactions. Arch Intern Med 1984;144(8):1667–71.

64. Krishna DR, Appa Rao AVN, Ramanakar TV, Prabhakar MC. Pharmacokinetic interaction between dapsone and rifampicin (rifampin) in leprosy patients. Drug Dev Ind Pharm 1986;12:443–9.

Daptomycin

General Information

Daptomycin is a novel lipopeptide antibiotic, an inhibitor of lipoteichoic acid synthesis, with potent bactericidal activity against most clinically important Gram-positive bacteria, including resistant strains (1,2), such as meticillin-resistant *Staphylococcus aureus* and vancomycin-resistant enterococci and was the first inhibitor of lipoteichoic acid synthesis.

Daptomycin was initially developed in the late 1980s and early 1990s (3) but was ultimately shelved owing to concerns regarding adverse effects, in particular drug-induced myopathy (4,5). In clinical trials, the most common adverse effects included gastrointestinal disorders, for example constipation (6%), nausea (6%), and diarrhea (5%), injection site reactions (6%), headache (5%), and rash (4%) (6,7,8).Most of the reported adverse effects of daptomycin have been mild to moderate in intensity.

Phase III study results have suggested no difference in efficacy or tolerability between daptomycin 4 mg/kg/day intravenously and vancomycin or semisynthetic penicillins for complicated skin and skin-structure infections; however, specific adverse effects were not discussed (9).

Daptomycin has linear pharmacokinetics at doses of 0.5–6 mg/kg, but in 20% of adults given 8 mg/kg intravenously there is non-linear accumulation. Daptomycin is highly protein-bound and is distributed mainly in the extracellular fluid. In a study of daptomycin concentrations collected from serum and inflammatory blisters, the mean concentrations at 1 and 2 hours were 9.4 and 14.5 μg/ml respectively (35).

When daptomycin 4 mg/kg was given intravenously to seven healthy men the mean peak concentrations in plasma and inflammatory fluid were 78 and 28 μg/ml respectively; the mean half-lives were 7.7 and 13 hours respectively; the overall penetration of total drug into the inflammatory fluid was 68% and the mean urinary recovery over 24 hours was 60% (10).

Placebo-controlled studies

Three cohorts of 12 healthy subjects each were given daptomycin 10 mg/kg or placebo once daily for 14 days, daptomycin 12 mg/kg or placebo once daily for 14 days, or daptomycin 6 or 8 mg/kg once daily for 4 days. Daptomycin did not cause electrocardiographic abnormalities or electrophysiological evidence of muscle or nerve toxicity (11).

Organs and Systems

Nervous system

Daptomycin has been associated with neuropathy in a few cases in phase 2 clinical studies (12). The dose in these studies (3 mg/kg bd) was higher than in phase 3 studies. Pooled laboratory data showed no differences in hematological measurements, blood chemistry, or hepatobiliary function between daptomycin and comparative antibiotics.

In a randomized trial, 120 patients with *Staphylococcus aureus* bacteremia with or without endocarditis received daptomycin 6 mg/kg/day intravenously and 115 received initial low-dose gentamicin plus either an antistaphylococcal penicillin or vancomycin (13). Eleven of the 120 patients who received daptomycin (9.2%) had adverse events related to the peripheral nervous system (e.g. paresthesia, dysesthesia, and peripheral neuropathies) compared with two of 116 patients (1.7%) who received standard therapy. All of the events were classified as mild to moderate in intensity; most were short-lived and resolved during continued treatment.

Musculoskeletal

Daptomycin may have adverse effects on skeletal muscle, since it increases serum creatine kinase activity. To find the dosing regimen that has the least effects on skeletal muscle, dogs were given repeated intravenous daptomycin every 24 hours or every 8 hours for 20 days (14). The results suggested that adverse effects on skeletal muscle are primarily related to dosing frequency but not peak plasma concentrations, and once-daily administration appeared to minimize the potential for daptomycin-related skeletal-muscle effects, possibly by allowing more time between doses for repair of subclinical effects.

In phase 3 trials, rises in creatine kinase activity occurred in 15 of 534 (2.8%) daptomycin-treated patients compared with 10 of 558 (1.8%) comparator-treated patients (15). In 13 cases the rises reversed within about 7–10 days after withdrawal. In phase III clinical trials the reported incidence was 2.8% (16,17).

People receiving daptomycin should be monitored for muscle pain or weakness, and creatine kinase activity should be monitored weekly. Daptomycin should be withdrawn in patients who develop an otherwise unexplained myopathy with raised creatine kinase activity (over 5 times the upper limit of the reference range) or an isolated marked increase (over 10 times). Patients with abnormal findings who do not meet these criteria should be monitored closely, especially if they are taking other agents that can cause muscle damage, such as statins (12). Rhabdomyolysis has also been described (18).

- A 52-year-old man developed a severe myopathy after taking daptomycin 500 mg/day (6.5 mg/kg/day) for osteomyelitis (19). He had also taken simvastatin, which was withdrawn when he started to take daptomycin. The baseline creatine kinase activity was 102 U/l (reference range 25–220). Nine days into the course, he developed generalized muscle pain and weakness, progressing to the point where he could not get out of bed. The creatine kinase activity was 20 771 U/l, urinalysis was negative for blood, and serum creatinine was 80 µmol/l. Daptomycin was withdrawn and he was admitted to the intensive care unit for close monitoring and hydration. He gradually recovered his muscle strength over the next 48 hours, with resolution of pain, although the creatine kinase activity was still 2700 U/l on day 14. Two weeks later the creatine kinase activity had normalized and all his muscle symptoms had resolved.
- A 45-year-old woman with refractory acute myeloid leukemia was given intravenous daptomycin 550 mg (6 mg/kg) every 24 hours because of bacteremia due to vancomycin-resistant enterococci. She was neutropenic (white cell count <400 x 106/l; absolute neutrophil count 10 x 106/l). Her creatine kinase activity was 108 U/l at baseline, but it gradually increased over the next 7 days, and on day 10 of daptomycin therapy had risen to 996 U/l. The blood urea nitrogen concentration was 26 mmol/l (73 mg/dl) and the serum creatinine 168 µmol/l (1.9 mg/dl). Myoglobin was present in a high concentration in the urine. Daptomycin was withdrawn and she recovered over a period of 2 weeks with hydration.
- A 53-year-old African–American woman with diabetes mellitus, hypertension, and peripheral vascular disease was given daptomycin 360 mg/day (6 mg/kg/day) intravenously and after 10 days developed generalized muscle weakness progressing to the point where she could not walk or even get out of bed, followed by non-oliguric acute renal failure with a serum creatinine rising to 239 µmol/l from a baseline of 80 µmol/l. The serum creatine kinase activity was raised at 21 243 U/l, as were transaminase and lactate dehydrogenase activities. Urinalysis was positive for red blood cells, hemoglobin, and myoglobin. Daptomycin was withdrawn and she was given intravenous fluids with urine alkalinization. She progressively recovered.
- A 26-year-old African American woman with methicillin-resistant *Staphylococcus aureus* endocarditis was given intravenous daptomycin 6 mg/kg/day for 14 days. She developed muscle aches and pains with only a minor rise in creatine kinase (492 U/l); both resolved after daptomycin was withdrawn (20).

Another case of daptomycin-related raised serum creatine kinase activity was reported in a randomized comparison of daptomycin 4 mg/kg every 24 hours intravenously, vancomycin, or a semi-synthetic penicillin (21). The patient had arm pain and weakness during the second week of treatment with daptomycin. Daptomycin was withdrawn and all the clinical and laboratory abnormalities rapidly resolved. This was one of two cases (0.4%) of rises in creatine kinase activity leading to withdrawal of daptomycin among all 534 patients treated in phase 3 studies of daptomycin.

In a randomized trial, 120 patients with *Staphylococcus aureus* bacteremia with or without endocarditis received daptomycin 6 mg/kg/day intravenously and 115 received initial low-dose gentamicin plus either an antistaphylococcal penicillin or vancomycin (166). Of the 120 who received daptomycin, 62 (52%) had a serious adverse event. Rises in creatine kinase activity were significantly more common in the daptomycin group than the standard-therapy group (6.7% versus 0.9%). Among patients with normal baseline creatine kinase activity, there were rises in activity in 23 of 92 patients who received daptomycin compared with 12 of 96 patients who received standard therapy (25% versus 13%). Raised CK activity led to withdrawal of treatment in 3 of 120 patients treated with daptomycin (2.5%). In 20 of the 24 patients who received daptomycin and who had increased creatine kinase activity at baseline (83%), the activity returned to the reference range during treatment (n =18) or after treatment (n = 2). Three patients had small rises in creatine kinase (range 114 to 451 U/l) throughout the course of daptomycin therapy.

Long-Term Effects

Drug tolerance

The activity of daptomycin against both vancomycin-sensitive and vancomycin-resistant *Enterococcus faecalis* was greater than that of quinupristin + dalfopristin (22). Daptomycin was as active as quinupristin + dalfopristin but more active than linezolid. At concentrations four times the MIC, daptomycin and vancomycin achieved 99.9% killing of methicillin-resistant *Staphylococcus aureus* after 8 hours, which was greater than the killing seen with linezolid and quinupristin + dalfopristin. However, the antibacterial activity of daptomycin strongly depended on the calcium concentration of the medium.

Susceptibility Factors

Age

The pharmacokinetics of daptomycin were evaluated after a single 30-minute intravenous infusion of 4 mg/kg

in young adults (18–30 years) and elderly volunteers (75 years and over) (23). There were increases in AUC and half-life with increased age. Total and renal clearances both fell with increasing age. There were no statistically significant differences between the two groups in C_{max} and V_{ss}. These results show that changes in the pharmacokinetics of daptomycin in the elderly are attributable to changes in renal function, rather than age per se.

Sex

In a double-blind pharmacokinetic study in 24 healthy subjects given three doses of daptomycin (4, 6, and 8 mg/kg/day for 7–14 days there was no difference between the sexes (24).

Obesity

The pharmacokinetics of daptomycin 4 mg/kg have been studied in moderately obese adults (BMI over 25 and under 40 kg/m^2), morbidly obese adults (BMI 40 kg/m^2 and over), and a non-obese control group matched for sex, age, and renal function (25). The terminal plasma half-life, the fraction of the dose excreted unchanged in the urine, and renal clearance were not affected by obesity. The volume of distribution and total plasma clearance were higher in the obese subjects. The rates of change of volume and clearance with increasing BMI were greater when they were expressed in absolute terms than when they were normalized for total or ideal body weight. This suggests that increases in body mass associated with obesity are proportionally higher than the corresponding increases in volume and clearance. C_{max} and AUC were respectively 25% and 30% higher in the obese subjects.

Renal disease

Daptomycin is eliminated primarily by renal excretion (about 54%) (12). Data from subjects in nine phase 1 (n = 153) and six phase 2/3 (n = 129) clinical trials were combined to identify factors that contribute to interindividual variability in daptomycin pharmacokinetics (26). Daptomycin clearance varied linearly with the estimated creatinine clearance. Daptomycin clearance in dialysis subjects was about one-third of that in healthy subjects (0.27 versus 0.81 l/hour). Daptomycin clearance in women was 80% that in men; however, in clinical trials, the outcome was not affected by sex. This suggests that modified dosing regimens are indicated for patients with severe renal disease and for those undergoing dialysis. In such patients the dosage interval should be 48 hours.

Liver disease

Dosage adjustment does not appear to be necessary in mild to moderate hepatic impairment. The use of daptomycin in patients with severe hepatic impairment has not been assessed (9).

In a single-dose, parallel-design, matched-control study in subjects aged 18–80 years with moderately impaired hepatic function (Child-Pugh Class B, n = 10) the pharmacokinetics were similar to those in control subjects (27).

Drug-Drug Interactions

No drug-drug interactions have been reported with daptomycin. Nevertheless, the manufacturer recommends giving consideration to temporarily discontinuing HMG CoA reductase inhibitors while patients take daptomycin because of the potential confusion that could occur if a patient taking an HMG CoA reductase inhibitor were to experience myopathy or an increase in CK activity (7).

References

1. Stephenson J. Researchers describe latest strategies to combat antibiotic-resistant microbes. JAMA 2001;285(18):2317–8.
2. Anonymous. Daptomycin. Cidecin, Dapcin, LY 146032. Drugs R D 2002;3(1):33–9.
3. Anonymous. First in a new class of antibiotics. FDA Consum 2003;37:4.
4. Shoemaker DM, Simou J, Roland WE. A review of daptomycin for injection (Cubicin) in the treatment of complicated skin and skin structure infections. Ther Clin Risk Manag 2006;2(2):169–74.
5. Lee SY, Fan HW, Kuti JL, Nicolau DP. Update on daptomycin: the first approved lipopeptide antibiotic. Expert Opin Pharmacother 2006;7(10):1381–97.
6. Rybak MJ. The efficacy and safety of daptomycin: first in a new class of antibiotics for Gram-positive bacteria. Clin Microbiol Infect 2006;12 Suppl 124-32.
7. Tedesco KL, Rybak MJ. Daptomycin. Pharmacotherapy 2004;24(1):41–57.
8. Carpenter CF, Chambers HF. Daptomycin: another novel agent for treating infections due to drug-resistant gram-positive pathogens. Clin Infect Dis 2004;38(7):994–1000.
9. Jeu L, Fung HB. Daptomycin: a cyclic lipopeptide antimicrobial agent. Clin Ther 2004;26(11):1728–57.
10. Wise R, Gee T, Andrews JM, Dvorchik B, Marshall G. Pharmacokinetics and inflammatory fluid penetration of intravenous daptomycin in volunteers. Antimicrob Agents Chemother 2002;46(1):31–3.
11. Benvenuto M, Benziger DP, Yankelev S, Vigliani G. Pharmacokinetics and tolerability of daptomycin at doses up to 12 milligrams per kilogram of body weight once daily in healthy volunteers. Antimicrob Agents Chemother 2006;50(10):3245–9.
12. Stein GE. Safety of newer parenteral antibiotics. Clin Infect Dis 2005;41 Suppl 5:S293-302.
13. Fowler VG, Jr., Boucher HW, Corey GR, Abrutyn E, Karchmer AW, Rupp ME, Levine DP, Chambers HF, Tally FP, Vigliani GA, Cabell CH, Link AS, DeMeyer I, Filler SG, Zervos M, Cook P, Parsonnet J, Bernstein JM, Price CS, Forrest GN, Fätkenheuer G, Gareca M, Rehm SJ, Brodt HR, Tice A, Cosgrove SE. Daptomycin versus standard therapy for bacteremia and endocarditis caused by *Staphylococcus aureus*. N Engl J Med 2006;355(7):653–65.
14. Oleson FB Jr, Berman CL, Kirkpatrick JB, Regan KS, Lai JJ, Tally FP. Once-daily dosing in dogs optimizes daptomycin safety. Antimicrob Agents Chemother 2000;44(11):2948–53.
15. Oberholzer CM, Caserta MT. Antimicrobial update: daptomycin. Pediatr Infect Dis J 2005;24(10):919–20.
16. Papadopoulos S, Ball AM, Liewer SE, Martin CA, Winstead PS, Murphy BS. Rhabdomyolysis during therapy with daptomycin. Clin Infect Dis 2006;42(12):e108-10.
17. Fenton C, Keating GM, Curran MP. Daptomycin. Drugs 2004;64(4):445-55; discussion 457-8.
18. Kazory A, Dibadj K, Weiner ID. Rhabdomyolysis and acute renal failure in a patient treated with daptomycin. J Antimicrob Chemother 2006;57(3):578–9.
19. Echevarria K, Datta P, Cadena J, Lewis JS, 2nd. Severe myopathy and possible hepatotoxicity related to daptomycin. J Antimicrob Chemother 2005;55(4):599–600.
20. Veligandla SR, Louie KR, Malesker MA, Smith PW. Muscle pain associated with daptomycin. Ann Pharmacother 2004;38(11):1860–2.
21. Lipsky BA, Stoutenburgh U. Daptomycin for treating infected diabetic foot ulcers: evidence from a randomized, controlled trial comparing daptomycin with vancomycin or semi-synthetic penicillins for complicated skin and skin-structure infections. J Antimicrob Chemother 2005;55(2):240–5.
22. Rybak MJ, Hershberger E, Moldovan T, Grucz RG. In vitro activities of daptomycin, vancomycin, linezolid, and quinupristin–dalfopristin against staphylococci and enterococci, including vancomycin-intermediate and -resistant strains. Antimicrob Agents Chemother 2000;44(4):1062–6.
23. Dvorchik BH, Damphousse D. The pharmacokinetics of daptomycin in moderately obese, morbidly obese, and matched nonobese subjects. J Clin Pharmacol 2005;45(1):48–56.
24. Dvorchik BH, Brazier D, DeBruin MF, Arbeit RD. Daptomycin pharmacokinetics and safety following administration of escalating doses once daily to healthy subjects. Antimicrob Agents Chemother 2003;47:1318–23.
25. Dvorchik B. Moderate liver impairment has no influence on daptomycin pharmacokinetics. J Clin Pharmacol 2004;44(7):715–22.
26. Dvorchik B, Arbeit RD, Chung J, Liu S, Knebel W, Kastrissios H. Population pharmacokinetics of daptomycin. Antimicrob Agents Chemother 2004;48(8):2799–807.
27. Dvorchik B, Damphousse D. Single-dose pharmacokinetics of daptomycin in young and geriatric volunteers. J Clin Pharmacol 2004;44(6):612-20.

Dirithromycin

See also Macrolide antibiotics

General Information

Adverse effects of dirithromycin were studied in 4263 patients (1). There was abdominal pain in 5.6%, diarrhea in 5.0%, and nausea in 4.9%. Headache was relatively common (4.5%). In 63% of cases the adverse events were considered mild, in 31% moderate, and in 6.3% severe. Adverse events resulted in withdrawal in 3.1%, mainly because of gastrointestinal symptoms such as nausea and abdominal pain.

In a randomized, investigator-blinded, parallel-group trial in acute exacerbations of chronic obstructive pulmonary disease, dirithromycin 500 mg/day was well tolerated; the most frequent adverse events were chest pain and paresthesia (2).

Organs and Systems

Respiratory

Adverse events involving the respiratory system (dyspnea, increased cough, or asthma) were reported in about 2% of patients taking dirithromycin (1).

Drug–Drug Interactions

Ciclosporin

Dirithromycin has a small effect on ciclosporin concentrations but to a clinically insignificant extent (3).

Oral contraceptives

Oral dirithromycin reduced the mean ethinylestradiol 24-hour AUC and increased its oral clearance in women using an oral contraceptive (4). However, since there was no effect on inhibition of ovulation, the clinical importance of this interaction may be negligible.

Theophylline and other xanthines

Drug interactions with dirithromycin have rarely been reported, since it does not interact with cytochrome P450. Nevertheless, in 13 healthy volunteers 500 mg/day for 10 days caused a significant 18% fall in the average steady-state plasma theophylline concentration and a 26% fall in peak concentration, with a 14–15% increase in clearance (5). In contrast, the steady-state pharmacokinetics of theophylline did not change in 14 patients with chronic obstructive airways disease who took dirithromycin for 10 days (6).

References

1. Brogden RN, Peters DH. Dirithromycin. A review of its antimicrobial activity, pharmacokinetic properties and therapeutic efficacy. Drugs 1994;48(4):599–616.
2. Castaldo RS, Celli BR, Gomez F, LaVallee N, Souhrada J, Hanrahan JP. A comparison of 5-day courses of dirithromycin and azithromycin in the treatment of acute exacerbations of chronic obstructive pulmonary disease. Clin Ther 2003;25:542–57.
3. Bachmann K, Sullivan TJ, Reese JH, Jauregui L, Miller K, Scott M, Sides GD, Shapiro R. The influence of dirithromycin on the pharmacokinetics of cyclosporine in healthy subjects and in renal transplant patients. Am J Ther 1995;2(7):490–8.
4. Wermeling DP, Chandler MH, Sides GD, Collins D, Muse KN. Dirithromycin increases ethinyl estradiol clearance without allowing ovulation. Obstet Gynecol 1995;86(1):78–84.
5. Bachmann K, Nunlee M, Martin M, Sullivan T, Jauregui L, DeSante K, Sides GD. Changes in the steady-state pharmacokinetics of theophylline during treatment with dirithromycin. J Clin Pharmacol 1990;30(11):1001–5.
6. Bachmann K, Jauregui L, Sides G, Sullivan TJ. Steady-state pharmacokinetics of theophylline in COPD patients treated with dirithromycin. J Clin Pharmacol 1993;33(9):861–5.

Doxycycline

See also Tetracyclines

General Information

Doxycycline and minocycline are more lipophilic tetracyclines. They are well absorbed after oral administration. Their half-lives are 16–18 hours. Their higher affinity for fatty tissues improves their effectiveness and changes their adverse effects profile. Local gastrointestinal irritation and disturbance of the intestinal bacterial flora occur less often than with the more hydrophilic drugs, which have to be given in higher oral doses for sufficient absorption.

Nevertheless, their toxic effects are similar to those of other tetracyclines and arise from accumulation in fatty tissues. Accumulation in a third compartment and the resulting long half-life may contribute to an increased incidence of various toxic adverse effects during long-term treatment, even if lower daily doses are used. This seems also to be the case for pigmentation disorders and possibly for neurological disturbances (1).

Minocycline and doxycycline are predominantly eliminated by the liver and biliary tract (70–90%). Therefore, no change in dose is needed in patients with impaired renal function. However, it should be considered that hepatic elimination of doxycycline or minocycline might be accelerated by co-administration of agents that induce hepatic enzymes.

Placebo-controlled studies

In a randomized, placebo-controlled study in 150 patients with chronic meibomian gland dysfunction, doxycycline 20 or 200 mg bd both increased tear production and reduced symptoms (2). However, the high-dose group reported adverse effects more often than the low-dose group (18 versus 8, 39% versus 17%).

Doxycycline is currently used in the treatment of filariasis to deplete *Wolbachia* endosymbionts, which leads to inhibition of worm development, embryogenesis, fertility, and viability in onchocerciasis. In a double-blind, placebo-controlled trial of doxycycline 200 mg/day for 3 weeks, followed by standard antifilarial treatment (albendazole 400 mg + ivermectin 150 micrograms/kg) at month 4 in 44 individuals from the western region of Ghana with *Wuchereria bancrofti* infections adverse reactions to standard antifilarial treatment were similar in frequency between doxycycline and placebo (3). Moderate adverse reactions (increased body temperature and/or chills, papular rash, itching, and headache) occurred only with placebo followed by standard antifilarial treatment and were recorded in three of 17 patients.

The severity of the adverse reactions was associated with microfilaremia, Wolbachia bacteria in the plasma, and proinflammatory cytokines in the plasma, in line with previous observations that suggest that adverse events after antifilarial treatment are related to the release of parasite-related antigens into the circulation.

Organs and Systems

Nervous system

Peripheral neuropathy from doxycycline has been reported (4).

- A 61-year-old doctor with recurrent bronchopneumonia took two courses of doxycycline. During the first course he had persistent numbness in his feet. During the second course, a few months later, he noticed after only 2 or 3 days that the numbness accelerated markedly and was associated with a low threshold for muscle cramps in the feet. He stopped taking doxycycline and during the following weeks noticed slight improvement. However, some symptoms persisted and he had neurological investigations. A wide range of clinical and laboratory tests showed no cause for his neuropathy.

A search for an association between doxycycline and polyneuropathy failed to identify any documented cases. An inquiry to the Swedish Adverse Drug Reactions Advisory Committee elicited information about three cases of "paresthesia," two cases of "sensitivity disturbance," and one case of "neuropathy." The last was a man who had had pain and paresthesia in the feet, arms, and face after taking doxycycline 100 mg/day for 2 weeks for prostatitis. The symptoms began to wane 1 week after treatment was stopped, and disappeared completely 1 week later.

The idiopathic form of benign intracranial hypertension typically occurs in obese women in their 30s and 40s. Intracranial hypertension has been described in two patients taking doxycycline (5).

- A slightly overweight 21-year-old woman presented with headaches and blurred vision after taking doxycycline 100 mg/day for malaria prophylaxis during a 3-week vacation. She had severe papilledema with hemorrhages and cotton wool spots. Lumbar puncture showed an increased opening pressure of 52 cm of fluid, and intracranial hypertension was diagnosed. Doxycycline was withdrawn, and her symptoms gradually resolved.
- A slightly overweight 19-year-old woman presented with vomiting, headache, and blurred vision after taking doxycycline 100 mg/day for malaria prophylaxis over about 4 months. Her vision was severely reduced and her visual fields constricted. She had papilledema with hemorrhages and cotton wool spots. Lumbar puncture showed an increased cerebrospinal fluid pressure of over 40 cm of fluid. Intracranial hypertension was diagnosed and doxycycline was withdrawn. Her symptoms stabilized and the disc swelling resolved, but optic

atrophy developed. Colour vision and visual fields remained poor, with an estimated 70% loss of vision.

The authors concluded that doxycycline should be prescribed with caution to overweight women of childbearing age or with a history of idiopathic intracranial hypertension and that awareness of this adverse drug reaction in travellers should be increased to allow prompt withdrawal of the causing drug and appropriate medical therapy in affected individuals.

Metabolism

Doxycycline can cause hypoglycemia.

- A 70-year-old man with type 2 diabetes mellitus presented with sudden confusion, which rapidly progressed to loss of consciousness (6). The only drug he had taken during the previous 2 months was doxycycline (100 mg/day), which he had taken for 5 days for an upper respiratory tract infection. Urine tests for sulfonylureas were negative. Routine hematological and biochemical tests and an electrocardiogram were normal. He improved with intravenous glucose and withdrawal of doxycycline and had no further episodes of hypoglycemia over the next 3 months.

Plasma insulin was not measured in this case, so the mechanism of hypoglycemia was unclear.

Hypoglycemia has also been attributed to doxycycline in a non-diabetic patient (7).

Teeth

Four patients with brucellosis developed yellow-brown discoloration of permanent teeth after taking doxycycline 200 mg/day for 30–45 days (8). In all cases, the staining completely resolved and the teeth recovered their original color after abrasive dental cleaning. The authors suggested that staining of the permanent dentition associated with doxycycline may be much commoner than reported, especially if the drug is taken during the summer months, and that strict avoidance of sunlight exposure during high-dose, long-term doxycycline therapy may prevent this complication.

Gastrointestinal

Esophageal ulcers have been described in association with oral doxycycline or tetracycline. They can occur in children (9) and can respond to proton pump inhibitors and sucralfate (10). Acute onset of substernal burning pain and dysphagia was noted within hours of taking the drug (11–15). Remaining parts of the ingested capsule were identified by esophagoscopy.

Thirty centers for pharmacovigilance in France have reported 81 cases of esophageal damage after treatment with tetracyclines collected between 1985 and 1992 (16). There were 64 ulcers, eight cases of dysphagia, and nine of esophagitis. Most (96%) of the cases were caused by doxycycline and 73% of the patients were female, mean age 29 years. Prescriptions were for dermatological (54%), urogenital (23%), and ENT diseases. In one patient, a 71-year-old man, an esophagobronchial fistulation required esophagectomy. In 92% the drugs were not taken correctly, that is at bedtime or without a sufficient quantity of fluid. Treatment with sucralfate 1 g tds did not change the outcome of tetracycline-induced esophageal ulcers (17).

Esophagitis in children has been reported.

- In a 15-year-old girl esophageal ulceration was attributed to doxycycline 100 mg/day (18).
- A 46-year-old woman with rosacea took doxycycline 50 mg/day and after 48 hours developed retrosternal chest pain and dysphagia due to two benign esophageal ulcers, which responded to omeprazole and sucralfate (19).
- A 12-year-old boy developed lower chest pain, having taken doxycycline for 7 days for presumed epididymitis. He had a normal chest X-ray and electrocardiogram and no signs of infection (20). Doxycycline was withdrawn, but the chest pain persisted. After 4 days endoscopy showed two very large ulcerated craters measuring about 5 × 12 mm in the distal esophagus.
- A 15-year-old girl developed chest pain and difficulty in swallowing after having taken five doses of doxycycline (20). Physical examination was normal and she had a normal chest X-ray and electrocardiogram. Doxycycline was withdrawn and she was given sucralfate 10 ml tds, omeprazole having been ineffective. Gastroscopy was not performed. Her pain began to improve 2 days later.

The authors stated that it is important to inquire about the use of tetracyclines in children with chest pain, and to consider them as a possible cause of the pain.

Esophageal ulceration occurred in two adults taking doxycycline as malaria chemoprophylaxis (21).

- A 20-year-old woman, who had been taking doxycycline malaria prophylaxis, took a doxycycline capsule (dose not given) before going to bed and awoke hours later with the feeling that the capsule was stuck in her esophagus. Over the next 4 days she developed worsening dysphagia. Esophagoscopy showed an esophageal ulcer over 20% of the esophageal surface. She was treated with ranitidine and sucralfate and improved over the next 2 days.
- A 27-year-old man with an 8-day history of dysphagia and retrosternal pain was taking doxycycline prophylaxis and occasional terfenadine (doses not stated). He recalled no problems with taking any of his doxycycline prophylaxis. He had an esophagoscopy, which showed a 1 cm esophageal ulcer. He improved with ranitidine.

This is a rare complication—of all cases of esophageal ulcers attributed to drugs, doxycycline has been implicated in 1 in about 4000.

Of 1442 patients with chronic prostatitis due to *Ureaplasma urealyticum* 63 were randomized to azithromycin 4.5 g over 3 weeks or doxycycline 100 mg bd for 21 days; five patients who took doxycycline had nausea (22).

Liver

Doxycycline-induced liver injury has been reported (23). The patient took oral doxycycline 200 mg/day for 8 days and had markedly altered liver function. The liver enzyme activities normalized only 109 days after withdrawal.

Skin

Fixed drug eruptions have been attributed to doxycycline (24).

Annular and semicircular actinic granulomata have been described in two patients taking doxycycline 100 mg/day for 1-5 months for malaria prophylaxis (25). Both had had phototoxicity and severe sunburn after starting doxycycline.

Sweet's syndrome has been attributed to doxycycline (26).

Nails

Discoloration of the nails is rarely attributed to doxycycline and has previously been reported as being painful (27,28).

- An 11-year-old-boy developed painless brown discoloration of the nails after taking doxycycline 100 mg/day for 15 days; the thumbs were particularly affected (29). The discoloration resolved within 1 month after withdrawal.

Immunologic

Renal small-vessel vasculitis related to doxycycline has been reported (30).

Second-Generation Effects

Teratogenicity

Doxycycline has not been shown to be teratogenic (31).

Susceptibility Factors

Renal disease

Doxycycline is almost completely eliminated via the liver and the biliary tract and is therefore safe in patients with pre-existing renal insufficiency. However, for intravenous administration, doxycycline is solubilized with polyvinylpyrrolidone, the clearance of which is less than that of doxycycline. In patients with a serum creatinine concentration of more than 250 µmol/l (3 mg/dl), it is therefore advisable to limit the duration of treatment to a few days.

Drug Administration

Drug administration route

When intravenous tetracycline became no longer available, many centers began to use doxycycline as a sclerosant. In one review (32), chest pain was the most frequent adverse event with doxycycline, occurring in about 40% of the 60 patients in whom it had been used, and fever occurred in about 7%. In a more recent controlled trial in 106 patients treated with either doxycycline or bleomycin, there was chest pain in 20% of the patients treated with doxycycline, and nausea in one patient (33).

Drug–Drug Interactions

Methotrexate

Doxycycline can be added to the long list of drugs (SEDA-18, 262) (SEDA-23, 253) that can interact with methotrexate (34).

- A 17-year-old girl with a femoral osteosarcoma received her 11th cycle of methotrexate and simultaneously oral doxycycline 100 mg bd for a palpebral abscess. As in previous cycles, pharmacokinetic monitoring of methotrexate was performed. On this occasion the half-life of methotrexate was more than doubled. She developed hematological and gastroenterological toxicity.

The authors recommended that in patients receiving methotrexate an alternative to doxycycline should be used.

Warfarin

There has been a report of bleeding and prolonged international normalized ratio in a 69-year-old man given warfarin and doxycycline (35).

References

1. Lander CM. Minocycline-induced benign intracranial hypertension. Clin Exp Neurol 1989;26:161–7.
2. Yoo SE, Lee DC, Chang MH. The effect of low-dose doxycycline therapy in chronic meibomian gland dysfunction. Korean J Ophthalmol 2005;19(4):258–63.
3. Turner JD, Mand S, Debrah AY, Muehlfeld J, Pfarr K, McGarry HF, Adjei O, Taylor MJ, Hoerauf A. A randomized, double-blind clinical trial of a 3-week course of doxycycline plus albendazole and ivermectin for the treatment of Wuchereria bancrofti infection. Clin Infect Dis 2006;42:1081–89.
4. Olsson R. Can doxycycline cause polyneuropathy? J Intern Med 2002;251(4):361–2.
5. Lochhead J, Elston JS. Doxycycline induced intracranial hypertension. BMJ 2003;326:641–2.
6. Odeh M, Oliven A. Doxycycline-induced hypoglycemia. J Clin Pharmacol 2000;40(10):1173–4.
7. Basaria S, Braga M, Moore WT. Doxycycline-induced hypoglycemia in a nondiabetic young man. South Med J 2002;95(11):1353–4.
8. Ayaslioglu E, Erkek E, Oba AA, Cebecioğlu E. Doxycycline-induced staining of permanent adult dentition. Aust Dent J 2005;50(4):273–5.
9. Pociello Almiñana N, Vilar Escrigas P, Luaces Cubells C. Esofagitis por doxiciclina. A propósito de dos casos. An Pediatr (Barc) 2005;62(2):171–3.
10. Grgurević I, Marusić S, Banić M, Buljevac M, Crncevic-Urek M, Cabrijan Z, Kardumz D, Kujundzić M, Lesnjaković I, Tadić M, Loncar B. [Doxycycline induced esophageal ulcers: report of two cases and review of the literature.] Lijec Vjesn 2005;127(11-12):285–7.
11. Baeriswyl G, Bengoa J, de Peyer R, Loizeau E. Importance des ulcérations médicamenteuses dans les lésions endoscopiques de l'oesophage. [Importance of drug-induced ulceration in endoscopic lesions of the esophagus.] Schweiz Med Wochenschr Suppl 1985;19:6–9.
12. Bonavina L, DeMeester TR, McChesney L, Schwizer W, Albertucci M, Bailey RT. Drug-induced esophageal strictures. Ann Surg 1987;206(2):173–83.
13. Zijnen-Suyker MP, Hazenberg BP. Oesophagusbeschadiging door doxycycline. [Esophageal lesions caused by doxycycline.] Ned Tijdschr Geneeskd 1981;125(35):1407–10.
14. Schneider R. Doxycycline esophageal ulcers. Am J Dig Dis 1977;22(9):805–7.
15. Crowson TD, Head LH, Ferrante WA. Esophageal ulcers associated with tetracycline therapy. JAMA 1976;235(25):2747–8.
16. Champel V, Jonville-Bera AP, Bera F, Autret E. Les tétracyclines peuvent être responsables d'ulcérations oesophagiennes si leur prise est incorrecte. Rev Prat Med Gen 1998;12:9–10.
17. Huizar JF, Podolsky I, Goldberg J. Ulceras esofagicas inducidas por doxiciclina. [Doxycycline-induced esophageal ulcers.] Rev Gastroenterol Mex 1998;63(2):101–5.
18. Passalidou P, Giudicelli H, Moreigne M, Khalfi A. [Doxycycline-induced esophageal ulceration.] Arch Pediatr 2006;13(1):90–1.
19. Fernándeza MA. Úlcera esofágica y doxiciclina. Aten Primaria 2005;36:224.
20. Palmer KM, Selbst SM, Shaffer S, Proujansky R. Pediatric chest pain induced by tetracycline ingestion. Pediatr Emerg Care 1999;15(3):200–1.
21. Morris TJ, Davis TP. Doxycycline-induced esophageal ulceration in the U.S. Military service Mil Med 2000;165(4):316–9.
22. Skerk V, Mareković I, Markovinović L, Begovac J, Skerk V, Barsić N, Majdak-Gluhinić V. Comparative randomized pilot study of azithromycin and doxycycline efficacy and tolerability in the treatment of prostate infection caused by *Ureaplasma urealyticum*. Chemotherapy 2006;52(1):9–11.
23. Bjornsson E, Lindberg J, Olsson R. Liver reactions to oral low-dose tetracyclines. Scand J Gastroenterol 1997;32(4):390–5.
24. Alanko K. Topical provocation of fixed drug eruption. A study of 30 patients. Contact Dermatitis 1994;31(1):25–7.
25. Lim DS, Triscott J. O'Brien's actinic granuloma in association with prolonged doxycycline phototoxicity. Australas J Dermatol 2003;44:67–70.
26. Jamet A, Lagarce L, Le Clec'h C, Croué A, Hoareau F, Diquet B, Laine-Cessac P. Doxycycline-induced Sweet's syndrome Eur J Dermatol 2008;18(5):595–6.
27. Coffin SE, Puck J. Painful discoloration of the fingernails in a 15-year-old boy. Pediatr Infect Dis J 1993;12: 702–703; 706.
28. Yong CK, Prendiville J, Peacock DL, Wong LT, Davidson AG. An unusual presentation of doxycycline-induced photosensitivity. Pediatrics 2000;106:E13.
29. Akcam M, Artan R, Akcam FZ, Yilmaz A. Nail discoloration induced by doxycycline. Pediatr Infect Dis J 2005;24(9):845–6.
30. Goland S, Kazarsky R, Kagan A, Huszar M, Abend I, Malnick SDH. Renal vasculitis associated with doxycycline. J Pharm Technol 2001;17:220–2.

31. Czeizel AE, Rockenbauer M. Teratogenic study of doxycycline. Obstet Gynecol 1997;89(4):524–8.
32. Walker-Renard PB, Vaughan LM, Sahn SA. Chemical pleurodesis for malignant pleural effusions. Ann Intern Med 1994;120(1):56–64.
33. Patz EF Jr, McAdams HP, Erasmus JJ, Goodman PC, Culhane DK, Gilkeson RC, Herndon J. Sclerotherapy for malignant pleural effusions: a prospective randomized trial of bleomycin vs doxycycline with small-bore catheter drainage. Chest 1998;113(5):1305–11.
34. Tortajada-Ituren JJ, Ordovas-Baines JP, Llopis-Salvia P, Jimenez-Torres NV. High-dose methotrexate–doxycycline interaction. Ann Pharmacother 1999;33(7–8):804–8.
35. Baciewicz AM, Bal BS. Bleeding associated with doxycycline and warfarin treatment. Arch Intern Med 2001;161(9):1231.

Erythromycin

See also Macrolide antibiotics

General Information

For some decades erythromycin was the only macrolide antibiotic available, but with the development of new macrolides with remarkable pharmacokinetic and safety features (1), it has met fierce competition and has, at least in some health-care systems, lost its place as the most important macrolide.

Patients who took erythromycin during its premarket clinical trials reported that abdominal pain, anorexia, diarrhea, nausea, and vomiting were the most common adverse effects. There have also been reports of cardiac conduction abnormalities, allergic reactions, skin eruptions, and reversible hearing loss. Erythromycin can cause cholestatic hepatitis, with nausea, vomiting, abdominal pain, jaundice, fever, liver function abnormalities, and occasionally eosinophilia.

Comparative studies

In a direct comparison of clarithromycin with erythromycin stearate, the rate of adverse events was 19% in 96 patients taking clarithromycin and 35% in 112 patients taking erythromycin (2). Most of the adverse events associated with clarithromycin affect the gastrointestinal tract (7%).

In a prospective, single-blind, randomized study of a 7-day course of clarithromycin (7.5 mg/kg bd) and a 14-day course of erythromycin (13.3 mg/kg tds) in 153 children with pertussis, the incidence of treatment-emergent drug-related adverse events was significantly higher with erythromycin than with clarithromycin (62 versus 45%) (3). Three subjects given erythromycin withdrew prematurely because of adverse events: one because of a rash; one with vomiting and diarrhea; and one with vomiting, abdominal pain, and rash.

In a double-blind, randomized, multicenter trial in 302 children, a 10-day course of erythromycin estolate (40 mg/kg/day in two doses) was as safe and effective as amoxicillin (50 mg/kg/day in two doses) in acute otitis media. Treatment-related adverse events occurred in 5.3% of patients given erythromycin and in 7.3% of patients given amoxicillin (4).

General adverse effects

Erythromycin is relatively well tolerated, with the exception of gastrointestinal adverse effects. Cholestasis resulting from the use of all forms of erythromycin is virtually the only serious effect. However, local irritation (affecting the gastrointestinal system, the muscles, or the veins, depending on the route of administration) is common. Erythromycin can increase serum theophylline concentrations and occasionally causes theophylline toxicity. Hypersensitivity reactions are rare, unless cholestasis is to be regarded as allergic. They probably have clinical effects in under 0.5% of treated patients, and consist mainly of maculopapular rashes, pruritus, urticaria, and angioedema; anaphylaxis and acute respiratory distress have also been reported. Fixed drug eruptions, urticaria, and Stevens–Johnson syndrome have also been reported.

Organs and Systems

Cardiovascular

Erythromycin has antidysrhythmic properties similar to those of Class IA antidysrhythmic drugs, and causes an increase in atrial and ventricular refractory periods. This is only likely to be a problem in patients with heart disease or in those who are receiving drugs that delay ventricular repolarization (5). High-doses intravenously have caused ventricular fibrillation and torsade de pointes (6). Each episode of dysrhythmia, QT interval prolongation, and myocardial dysfunction occurred 1–1.5 hours after erythromycin infusion and resolved after withdrawal.

In an FDA database analysis, 346 cases of cardiac dysrhythmias associated with erythromycin were identified. There was a preponderance of women, as there was among those with life-threatening ventricular dysrhythmias and deaths after intravenous erythromycin lactobionate. A sex difference in cardiac repolarization response to erythromycin is a potential contributing factor, since in an in vitro experiment on rabbit hearts, erythromycin caused significantly greater QT prolongation in female than in male hearts (7).

In the Tennessee Medicaid cohort, during 1 249 943 person-years of follow-up there were 1476 cases of sudden death from cardiac causes; the multivariate adjusted rate of sudden death from cardiac causes among patients currently using erythromycin was twice as high (8). There was no significant increase in the risk of sudden death among former users of erythromycin. The adjusted rate of sudden death from cardiac causes was five times as high among those who concurrently used CYP3A inhibitors and erythromycin as among those who had used neither CYP3A inhibitors nor any of the study antibiotic medications.

In 35 women and 28 men erythromycin caused QT interval prolongation after the first few doses of erythromycin (9). Similarly, in a prospective, comparative study in 19 patients with uncomplicated community-acquired pneumonia, a single dose of intravenous erythromycin 500 mg increased the heart rate and prolonged the QT interval. These effects were seen after 15 minutes of infusion and disappeared 5 minutes after the infusion had been stopped (10).

Owing to prolongation of the QT interval, a newborn with congenital AV block developed ventricular extra beats and non-sustained ventricular tachycardia after intravenous erythromycin; the QT interval normalized after withdrawal (11).

- Intravenous erythromycin (1 g 6-hourly by intravenous infusion over 30 minutes) resulted in QT interval prolongation, ventricular fibrillation, and torsade de pointes in a 32-year-old woman (6).

Torsade de pointes occurred in 16 of 23 patients who received intravenous erythromycin 3-4 g/day and 3 of 23 who received oral erythromycin 1.5-2 g/day (12). There were marked differences in xposure between the two regimens, with typical peak erythromycin concentrations of 30 µg/ml after intravenous administration versus 2–4 µg/ml after oral administration. Non-clinical models have shown a concentration-related effect of erythromycin on action potential duration over a wide range of concentrations.

Intravenous administration of erythromycin into peripheral veins relatively commonly causes thrombophlebitis, although the lactobionate form of erythromycin may be less irritating to veins than other parenteral forms (13,14). In a prospective study of 550 patients with 1386 peripheral venous catheters, the incidence of phlebitis was 19% with antibiotics and 8.8% without; erythromycin was associated with an increased risk (15).

Respiratory

Adverse effects involving the respiratory system were reported in 2% of patients taking erythromycin stearate (2).

A beneficial effect of erythromycin on sputum volume has been reported in a patient with severe airways obstruction due to bronchorrhea (16).

Nervous system

In patients without neuromuscular disease erythromycin caused subclinical loss of motor unit contractions, which improved with intravenous edrophonium or neostigmine (17).

Sensory systems

Ototoxicity, resulting in hearing loss, and usually reversible, has been reported in patients treated with erythromycin lactobionate 4 g/day or more or large oral doses of erythromycin estolate (18). Ototoxic reactions have also been seen after the use of esters of erythromycin, such as the ethylsuccinate, stearate, and propionate (19). High parenteral doses of erythromycin have resulted in

transient perceptive deafness (20–23). Since renal and hepatic disease was a prominent feature in these patients, ototoxicity was thought to result from high blood concentrations (24). Recovery occurred within a few days after withdrawal. The phenomenon differs from the permanent type of ototoxicity caused by aminoglycosides. Erythromycin should not be given together with other potentially ototoxic drugs and hearing acuity should be monitored during erythromycin therapy, especially in the elderly. Acute psychotic reactions have been related to ototoxicity and high-dose erythromycin therapy (25).

Psychological, psychiatric

Erythromycin has been associated with complications such as confusion, paranoia, visual hallucinations, fear, lack of control, and nightmares. These suspected psychiatric adverse effects were seen within 12–48 hours of starting therapy with conventional doses. Such complications may even be under-reported (25–28).

Gastrointestinal

Erythromycin is a motilin receptor agonist (29–31). This mechanism may be at least partly responsible for the gastrointestinal adverse effects of macrolides.

The prokinetic effect of erythromycin has been investigated in healthy volunteers, in whom a dose of 3 mg/kg seemed to have the largest prokinetic effect (32)

Pylorospasm and hypertrophic pyloric stenosis is associated with early postnatal erythromycin exposure and has been observed in neonates after 1–2 days of oral erythromycin therapy (33). The prominent gastrokinetic properties of erythromycin have been postulated as the mechanism (34).

- Pyloric stenosis has also been reported in a boy born at 23 weeks gestation, weight 690 g, after treatment of the child with three doses of oral erythromycin 10 mg/kg/day (35).

The use of erythromycin in postexposure prophylaxis for pertussis in 200 infants was followed by an increased number of cases of infantile hypertrophic pyloric stenosis, and all seven cases had taken erythromycin prophylactically (36). A case review and cohort study supported these preliminary findings (37). In a retrospective study in 314 029 children, very early exposure to erythromycin (at 3–13 days of life) was associated with a nearly eight-fold increased risk of pyloric stenosis (38). There was no increased risk in infants exposed to erythromycin after 13 days of life or in infants exposed to antibiotics other than erythromycin.

Intravenous erythromycin should be restricted to as few patients as possible. It can cause severe abdominal cramps, probably by a direct action on smooth muscle (39–41).

Liver

Erythromycin can cause two different types of liver damage (42,43). Administration of erythromycin as base or salt can be followed in 0–10% of cases by apparently benign increases in serum transaminases, which may or may not recur on rechallenge. In children, raised transaminases were noted at dosages of 40 mg/kg/day but not 20 mg/kg/day.

Cholestatic hepatitis, which is associated primarily with erythromycin estolate, can be caused by all forms of erythromycin, including the base, estolate, ethylsuccinate, propionate, and stearate (42,44). Although it was originally speculated that a hypersensitivity reaction to the estolate ester rather than to the erythromycin itself was responsible for this adverse reaction (45), erythromycin does inhibit bile flow (46). Most probably the differences in hepatotoxicity between the various erythromycin derivatives are of a quantitative rather than a qualitative nature (47,48), perhaps because of better intestinal absorption of the estolate. Potentially severe but rare cholestatic liver injury occurs in perhaps up to 2–4% of treated patients. Erythromycin-induced cholestasis is rare in children under the age of 12 years, but has occurred in infants at 6 weeks of age, in whom it can mimic acute cholecystitis, biliary atresia, or neonatal hepatitis (49).

- A young woman developed severe cholestasis and jaundice after taking erythromycin stearate (50). A second severe episode of jaundice and malaise occurred after treatment with erythromycin succinate 2 years later pointing to erythromycin itself as the culprit.
- After two doses of erythromycin ethylsuccinate, following unsuccessful treatment with penicillin for a respiratory illness, a 10-year-old previously healthy girl developed liver damage (51). Liver biopsy showed moderate panlobular parenchymatous degeneration with cholestasis due to numerous intracellular and intraductal bile plugs. She recovered completely 6–8 weeks after withdrawal of erythromycin.

The syndrome generally starts 10–14 days after the start of therapy, but earlier after re-exposure, sometimes within 12–24 hours (52). At all ages it often begins with abdominal pain, nausea, vomiting, pyrexia, pruritus, and jaundice; fever, rash, leukocytosis, raised serum transaminases, and eosinophilia can also occur. However, it can be ushered in with severe acute upper abdominal pain or right subcostal tenderness, simulating an acute abdomen, or it can resemble obstructive jaundice. Serum bilirubin, alkaline phosphatase, and transaminases are raised. Histological examination typically shows intrahepatic cholestasis and periportal inflammatory infiltration, with lymphocytes, neutrophils, and disproportionate numbers of eosinophils (53). These histological findings could be interpreted as reflecting a hypersensitivity reaction, but this hypothesis has been rejected (46). if erythromycin is promptly withdrawn, the clinical signs often improve rapidly, although prolonged jaundice has been reported.

Urinary tract

Acute renal insufficiency has been observed in a patient with Henoch–Schönlein syndrome (54). Another case presented as interstitial nephritis with acute renal insufficiency (55).

Skin

Topical erythromycin in benzoylperoxide, marketed for acne treatment, must be compounded by a pharmacist and requires subsequent refrigeration, warranting the development of alternative formulations. In a double-blind, parallel-group, multicenter study in 327 patients, a single-use erythromycin in benzoylperoxide combination package was compared with the vehicle alone and the original, reconstituted formulation packaged in a jar. Dry skin was the most frequently reported skin-related adverse event; it occurred in 3.2% of patients who used the preformulated erythromycin in benzoylperoxide and 5.0% of those who used the reconstituted erythromycin in benzoylperoxide (56).

A fixed drug eruption due to erythromycin has been observed (57). In another case skin tests with erythromycin were positive for the immediate and or delayed types of hypersensitivity (58).

- A 27-year-old woman developed urticaria 30 minutes after taking a single dose of erythromycin (59). An identical episode had occurred three years before. Erythromycin-specific IgE was detected in her serum by radioimmunoassay.
- Stevens–Johnson syndrome developed in a 64-year-old man who took erythromycin stearate for non-specific upper respiratory tract symptoms (60). After four doses of 250 mg, he developed a fever and typical lesions in the mouth and conjunctivae and on the lips. He was treated with prednisolone and recovered rapidly.

Second-Generation Effects

Teratogenicity

In a retrospective study there was no evidence of an increased risk of pyloric stenosis among infants born to mothers exposed to erythromycin during pregnancy (61).

Drug–Drug Interactions

Antihistamines

The association of ventricular dysrhythmias with co-administration of erythromycin and terfenadine (62) is thought to be due to inhibition of CYP3A4 by erythromycin. This combination should be avoided.

Erythromycin increases concentrations of astemizole (63). This combination should be avoided.

A double-blind, crossover study of the potential interaction between erythromycin and loratadine in 24 healthy volunteers showed that the AUCs of loratadine and its metabolite descarboethoxyloratadine were increased 40% and 46% respectively by erythromycin, but with no discernible effect on the QT interval (64).

Combined desloratadine 7.5 mg/day plus erythromycin 500 mg qds in 24 healthy volunteers was well tolerated and had no clinically important electrocardiographic effects (65). Although co-administration of erythromycin slightly increased plasma concentrations of desloratadine, this change did not correlate with prolongation of the QT interval, and there was no toxicity.

Benzodiazepines

Erythromycin can increase concentrations of midazolam and triazolam by inhibition of CYP3A4, and dosage reductions of 50% have been proposed if concomitant therapy is unavoidable (63).

Bupropion

Co-administration of modified-release erythromycin 999 mg in an adolescent taking methylphenidate and bupropion was associated with an acute myocardial infarction (66). Co-administration of erythromycin may have inhibited CYP3A4-dependent metabolism of bupropion, leading to raised plasma bupropion concentrations and excessive vasospasm due to the sympathetic effect of methylphenidate and bupropion.

Buspirone

Co-administration of erythromycin with the anxiolytic drug buspirone increased the plasma concentration of buspirone (67).

Erythromycin, an inhibitor of CYP3A, can increase buspirone concentrations (SEDA-22, 39).

Carbamazepine

Erythromycin can cause acute carbamazepine intoxication, probably by inhibiting its hepatic metabolism (68). Erythromycin may also directly inhibit the conversion of carbamazepine to its epoxide. In a controlled study of the effects of erythromycin on carbamazepine pharmacokinetics in healthy volunteers, the clearance of a single dose of carbamazepine was reduced by 19% during erythromycin treatment (69). In contrast, the single-dose pharmacokinetics of phenytoin were not affected by erythromycin (70,71). After withdrawal of the macrolide, carbamazepine concentrations quickly return to normal (72). If co-administration of erythromycin and carbamazepine cannot be avoided, a dosage reduction of carbamazepine of around 25% should be considered, with careful monitoring of serum concentrations (63).

Cardiac glycosides

The interaction of erythromycin with digoxin (73,74) is probably due to inhibition of its presystemic metabolism by inhibition of the growth of *Eubacterium glenum*.

- Digoxin toxicity occurred in a neonate who was also given erythromycin (75). She had bradycardia and coupled extra beats. Digoxin and erythromycin were withdrawn and she was given antidigoxin antibodies. Her plasma digoxin concentration, which had previously been 1.8 ng/ml, had risen to 8.0 ng/ml.

Cilostazol

When erythromycin was given to 16 healthy non-smoking male volunteers taking cilostazol, the C_{max} and AUC of cilostazol increased significantly by 47 and 73%

respectively, and the unbound clearance of the major metabolite of cilostazol, OPC-13015, fell by about 50% (76).

Cisapride

- QT interval prolongation and torsade de pointes occurred after co-administration of cisapride and erythromycin 500 mg qds for 1 week in a 47-year-old woman (77).

Clozapine

Increased clozapine serum concentrations have been reported with erythromycin (78,79) and can cause adverse effects (SEDA-21, 55). However, in 12 healthy men who took a single dose of clozapine 12.5 mg alone or in combination with a daily dose of erythromycin 1.5 g, the metabolism of clozapine was not altered (80). This confirms that CYP3A4 is a relatively minor pathway for clozapine metabolism, in contrast to CYP1A2.

In a case of neutropenia the authors suggested that an interaction of clozapine with erythromycin had been the precipitating factor (81).

Disopyramide

Disopyramide altered protein binding of erythromycin and this resulted in increased plasma erythromycin concentrations in vitro (82). The interaction between erythromycin and disopyramide was potentially fatal in two cases (83).

Ergot alkaloids

The interaction of erythromycin with ergotamine or dihydroergotamine can cause ergotism, sometimes leading to gangrene, by inhibition of the metabolism of the ergopeptides (84).

Felbamate

Erythromycin (333 mg tds for 10 days) had no effect on the pharmacokinetics of felbamate (3.0 or 3.6 g/d) used as monotherapy in epilepsy (85).

Felodipine

The metabolism of felodipine is inhibited by erythromycin (SEDA-17, 239).

Halofantrine

Erythromycin inhibited halofantrine metabolism in vitro, suggesting that increased cardiotoxicity might be clinically important (86).

Ketoconazole

If erythromycin and ketoconazole, both CYP3A4 inhibitors, are taken in combination, there will be an even more dramatic effect on the metabolism of other drugs, such as terfenadine and astemizole, midazolam and triazolam, and ciclosporin.

Lidocaine

Erythromycin can increase the plasma concentration and toxicity of oral lidocaine, as shown in a crossover study in nine volunteers who took erythromycin orally (500 mg tds) for 4 days and 1 mg/kg of oral lidocaine on day 4 (87).

The effects of erythromycin, an inhibitor of CYP3A4, on the pharmacokinetics of lidocaine have been studied in nine healthy volunteers. Steady-state oral erythromycin had no effect on the plasma concentration versus time curve of lidocaine after intravenous administration, but erythromycin increased the plasma concentrations of the major metabolite of lidocaine, MEGX (88). It is not clear what the interpretation of these results is, particularly since the authors did not study enough subjects to detect what might have been small but significant changes in various disposition parameters of lidocaine and did not report unbound concentrations of lidocaine or its metabolites. However, whatever the pharmacokinetic explanation, the clinical relevance is that one would expect that erythromycin would potentiate the toxic effects of lidocaine that are mediated by MEGX.

Losartan

Because losartan is metabolized by CYP2C9 and CYP3A4 to an active metabolite, E3174, which has greater antihypertensive activity than the parent compound, the effects of co-administration of losartan 50 mg/day and erythromycin, a moderate inhibitor of CYP3A4, have been investigated in a well-designed study in healthy volunteers. There was no significant effect of erythromycin.

Mosapride

The co-administration of mosapride 15 mg/day with erythromycin 1200 mg/day did not affect the electrocardiogram in healthy men, indicating a reduced likelihood of severe clinical adverse events, such as QT interval prolongation and torsade de pointes (89).

Oral anticoagulants

Erythromycin can interact with warfarin, resulting in a modest increase in blood concentrations and a rise in the prothrombin time of around 8% (90).

Erythromycin may potentiate acenocoumarol anticoagulant treatment, as reported in a 68-year-old man on stable anticoagulation with acenocoumarol 3 mg/day who took erythromycin ethylsuccinate 1.5 g/day (91).

Quetiapine

Co-administration of erythromycin tds increased exposure to quetiapine 200 mg bd significantly in Chinese patients by inhibition of the CYP3A4-dependent metabolism of quetiapine; the C_{max} rose by 68% and the AUC by 129% (92). Intra individual variability was large and so a fixed modification of the dosage regimen could not be recommended.

Quinidine

Erythromycin reduced the total clearance of quinidine, reduced its partial clearance by 3-hydroxylation, and increased its maximal serum concentration in an open study in 30 healthy young volunteers (93).

Quinine

Erythromycin is a competitive in vitro inhibitor of quinine 3-hydroxylation and may therefore interact with quinine (94).

Saquinavir

In 11 healthy men, erythromycin 250 mg qds increased the AUC of saquinavir (given as a soft gel capsule 1200 mg tds) by 69% when both were given for 7 days (95).

Drug interactions leading to the serotonin syndrome usually result from pharmacodynamic mechanisms. However, the antibiotic erythromycin may have precipitated the serotonin syndrome in a patient taking sertraline by a pharmacokinetic mechanism (96).

- A 12-year-old boy, who had taken sertraline 37.5 mg/day for 5 weeks for obsessive-compulsive disorder, started to take erythromycin 200 mg bd. Within 4 days he began to feel anxious; this was followed over the next 10 days by panic, restlessness, irritability, tremulousness, and confusion. These symptoms resolved within 72 hours of withdrawal of sertraline and erythromycin.

The authors proposed that in this case erythromycin had inhibited sertraline metabolism by inhibiting CYP3A. This could have led to increased concentrations of sertraline and signs of serotonin toxicity. Unfortunately sertraline concentrations were not measured to confirm this suggestion.

Somatostatin and analogues

Pretreatment with octreotide enhanced the gastric prokinetic effects of erythromycin in eight healthy subjects, suggesting that octreotide may be clinically useful in patients with tachyphylaxis to this effect of erythromycin (97).

Statins

It has been proposed that the risk of myotoxicity increases when statins are prescribed concurrently with erythromycin (98). There are no data for any pharmacokinetic interaction with fluvastatin or pravastatin, but as in the case of simvastatin the major route of metabolism of these drugs is by CYP3A4 and there is potential for an adverse interaction.

Atorvastatin
When erythromycin was co-administered with atorvastatin, the mean C_{max} and AUC of atorvastatin increased by more than 30% (5,99).

Lovastatin

Rhabdomyolysis with or without renal impairment has been reported in patients taking both erythromycin and lovastatin (100). The exact mechanism is unknown, but lovastatin is extensively metabolized by CYP3A4 and its metabolism may therefore be inhibited by erythromycin. The manufacturers have advised that careful monitoring is required when these two drugs are given together.

Rhabdomyolysis due to a short course of erythromycin in a 73-year-old man who had taken lovastatin for 7 years was accompanied by signs of multiple organ toxicity so severe as to mimic sepsis (101).

Simvastatin

A case-control analysis of 7405 cases and 28 327 controls suggested that concomitant use of simvastatin and erythromycin is associated with an increased risk of cataract (102). Studies in dogs have shown that some statins are associated with cataract when given in excessive doses (103).

Theophylline and other xanthines

Erythromycin can increase the serum concentrations of theophylline by 20–25%. However, patients with an average serum concentration of theophylline under 15 µg/ml will probably only experience a small increase in their serum theophylline concentration during erythromycin therapy, whereas patients with steady-state concentrations above 15 µg/ml deserve careful monitoring and close observation for symptoms of theophylline toxicity during treatment with erythromycin (104,105).

Tiagabine

Both tiagabine and erythromycin are metabolized by cytochrome P450. In an open, crossover study in 13 healthy volunteers, tiagabine (4 mg bd) and erythromycin (500 mg bd) were co-administered for 4 days (106). Maximum plasma concentration, AUC, and half-life of tiagabine were comparable when tiagabine was administered alone or in combination with erythromycin. The t_{max} was prolonged after administration with erythromycin in women; this effect may be due to a differential effect of erythromycin on gastric emptying. The interpretation of these findings is limited by the rather low doses of tiagabine used in the study and the short time of co-administration.

Triazolam

Erythromycin can increase concentrations of triazolam by inhibition of CYP3A4, and dosage reductions of 50% have been proposed if concomitant therapy is unavoidable (107).

Verapamil

- A 79-year-old white woman developed extreme fatigue and dizziness (108). Her heart rate was 40/minute and her blood pressure 80/40 mmHg. An electrocardiogram showed complete atrioventricular block, an escape rhythm at 50/minute, and QT_c interval prolongation to 583 milliseconds. This event was attributed to concomitant treatment with verapamil 480 mg/day and erythromycin 2000 mg/day, which had been prescribed 1 week before admission.

This is the first report of complete AV block and prolongation of the QT interval after co-administration of erythromycin and verapamil, both of which are principally metabolized by CYP3A4. Both drugs are potent inhibitors of CYP3A4 and P-glycoprotein, which may be the basis of this interaction.

The effects of a combination of erythromycin and verapamil on the pharmacokinetics of a single dose of simvastatin have been studied in a randomized, double-blind, cross-over study in 12 healthy volunteers simultaneously taking the three drugs. Both erythromycin and verapamil interacted with simvastatin, producing significant increases in the serum concentrations of simvastatin and its active metabolite simvastatin acid. The mean C_{max} of active simvastatin acid was increased about five-fold and the AUC_{0-24} four-fold by erythromycin; verapamil increased the C_{max} of simvastatin acid 3.4-fold and the AUC_{0-24} 2.8-fold. There was a substantial interindividual variation in the extent of these interactions. Concomitant use of erythromycin, verapamil, and simvastatin should be avoided (109).

Vinca alkaloids

Toward the end of a phase I study of vinblastine plus oral ciclosporin to reverse multidrug resistance, three patients also received erythromycin to raise their ciclosporin concentrations; all developed severe toxicity consistent with a much higher dose of vinblastine than was actually given (110).

Voriconazole

No interaction of voriconazole with erythromycin was reported in an open, randomized, parallel-group study in 30 healthy men who took oral voriconazole 200 mg bd for 14 days plus either erythromycin 1 g bd on days 8-14 or azithromycin 500 mg/day on days 12–14 (111).

Ximelagatran

There was a pharmacokinetic interaction between erythromycin and ximelagatran in 16 healthy volunteers (mean age 24 years; seven men and nine women) (112). They took a single dose of ximelagatran 36 mg on day 1, erythromycin 500 mg tds on days 2–5, and a single dose of ximelagatran 36 mg plus erythromycin 500 mg on day 6. The ratio of AUCs of ximelagatran + erythromycin to ximelagatran alone was 1.82 (95% CI = 1.64, 2.01) and the ratio of Cmax values of melagatran, the active form of ximelagatran, was 1.74 (1.52, 2.00). Erythromycin inhibited P glycoprotein-mediated transport of both ximelagatran and melagatran in vitro and reduced the biliary excretion of melagatran in rat.

Zaleplon

Erythromycin increases plasma concentrations of the Z drugs and increases their sedative effects (113); this occurs to a lesser extent than the similar effect on benzodiazepines that are exclusively metabolized by CYP3A4.

References

1. Schlossberg D. Azithromycin and clarithromycin. Med Clin North Am 1995;79(4):803–15.
2. Anderson G, Esmonde TS, Coles S, Macklin J, Carnegie C. A comparative safety and efficacy study of clarithromycin and erythromycin stearate in community-acquired pneumonia. J Antimicrob Chemother 1991;27(Suppl A):117–24.
3. Lebel MH, Mehra S. Efficacy and safety of clarithromycin versus erythromycin for the treatment of pertussis: a prospective, randomized, single blind trial. Pediatr Infect Dis J 2001;20(12):1149–54.
4. Scholz H, Noack RAOM Study Group. Multicenter, randomized, double-blind comparison of erythromycin estolate versus amoxicillin for the treatment of acute otitis media in children. Eur J Clin Microbiol Infect Dis 1998;17(7):470–8.
5. Rubinstein E. Comparative safety of the different macrolides. Int J Antimicrob Agents 2001;18(Suppl 1):S71–6.
6. Orban Z, MacDonald LL, Peters MA, Guslits B. Erythromycin-induced cardiac toxicity. Am J Cardiol 1995;75(12):859–61.
7. Drici MD, Knollmann BC, Wang WX, Woosley RL. Cardiac actions of erythromycin: influence of female sex. JAMA 1998;280(20):1774–6.
8. Ray WA, Murray KT, Meredith S, Narasimhulu SS, Hall K, Stein CM. Oral erythromycin and the risk of sudden death from cardiac causes. N Engl J Med 2004;351(11):1089–96.
9. Kdesh A, McPherson CA, Yaylali Y, Yasick D, Bradley K, Manthous CA. Effect of erythromycin on myocardial repolarization in patients with community-acquired pneumonia. South Med J 1999;92(12):1178–82.
10. Mishra A, Friedman HS, Sinha AK. The effects of erythromycin on the electrocardiogram. Chest 1999;115(4):983–6.
11. Brixius B, Lindinger A, Baghai A, Limbach HG, Hoffmann W. Ventrikuläre Tachykardie nach Erythromycin-Gabe bei einem Neugeborenen mit angeborenem AV-Block. [Ventricular tachycardia after erythromycin administration in a newborn with congenital AV-block.] Klin Padiatr 1999;211(6):465–8.
12. Nakatsuka A, Nagai M, Yabe H, Nishikawa N, Nomura T, Moritoyo H, Moritoyo T, Nomoto M. Effect of clarithromycin on the pharmacokinetics of cabergoline in healthy controls and in patients with Parkinson's disease. J Pharmacol Sci 2006;100(1):59–64.
13. Washington JA 2nd, Wilson WR. Erythromycin: a microbial and clinical perspective after 30 years of clinical use (1). Mayo Clin Proc 1985;60(3):189–203.
14. Washington JA 2nd, Wilson WR. Erythromycin: a microbial and clinical perspective after 30 years of clinical use (2). Mayo Clin Proc 1985;60(4):271–8.
15. Lanbeck P, Odenholt I, Paulsen O. Antibiotics differ in their tendency to cause infusion phlebitis: a prospective observational study. Scand J Infect Dis 2002;34(7):512–9.
16. Marom ZM, Goswami SK. Respiratory mucus hypersecretion (bronchorrhea): a case discussion—possible mechanisms(s) and treatment. J Allergy Clin Immunol 1991;87(6):1050–5.
17. Herishanu Y, Taustein I. The electromyographic changes by antibiotics. A preliminary study. Confin Neurol 1971;33(1):41–5.
18. Eckman MR, Johnson T, Riess R. Partial deafness after erythromycin. N Engl J Med 1975;292(12):649.
19. Schweitzer VG, Olson NR. Ototoxic effect of erythromycin therapy. Arch Otolaryngol 1984;110(4):258–60.
20. Brummett RE, Fox KE. Vancomycin- and erythromycin-induced hearing loss in humans. Antimicrob Agents Chemother 1989;33(6):791–6.
21. Haydon RC, Thelin JW, Davis WE. Erythromycin ototoxicity: analysis and conclusions based on 22 case reports. Otolaryngol Head Neck Surg 1984;92(6):678–84.
22. Huang MY, Schacht J. Drug-induced ototoxicity. Pathogenesis and prevention. Med Toxicol Adverse Drug Exp 1989;4(6):452–67.
23. Agusti C, Ferran F, Gea J, Picado C. Ototoxic reaction to erythromycin. Arch Intern Med 1991;151(2):380.
24. Quinnan GV Jr, McCabe WR. Ototoxicity of erythromycin. Lancet 1978;1(8074):1160–1.
25. Umstead GS, Neumann KH. Erythromycin ototoxicity and acute psychotic reaction in cancer patients with hepatic dysfunction. Arch Intern Med 1986;146(5):897–9.
26. Black RJ, Dawson TA. Erythromycin and nightmares. BMJ (Clin Res Ed) 1988;296(6628):1070.
27. Williams NR. Erythromycin: a case of nightmares. BMJ (Clin Res Ed) 1988;296(6616):214.
28. Murdoch J. Psychiatric complications of erythromycin and clindamycin. Can J Hosp Pharm 1988;41:277.
29. Lin HC, Sanders SL, Gu YG, Doty JE. Erythromycin accelerates solid emptying at the expense of gastric sieving. Dig Dis Sci 1994;39(1):124–6.
30. Hasler WL, Heldsinger A, Chung OY. Erythromycin contracts rabbit colon myocytes via occupation of motilin receptors. Am J Physiol 1992;262(1 Pt 1):G50–5.
31. Kaufman HS, Ahrendt SA, Pitt HA, Lillemoe KD. The effect of erythromycin on motility of the duodenum, sphincter of Oddi, and gallbladder in the prairie dog. Surgery 1993;114(3):543–8.
32. Boivin MA, Carey MC, Levy H. Erythromycin accelerates gastric emptying in a dose-response manner in healthy subjects. Pharmacotherapy 2003;23:5–8.
33. SanFilippo A. Infantile hypertrophic pyloric stenosis related to ingestion of erythromycine estolate: A report of five cases. J Pediatr Surg 1976;11(2):177–80.
34. Hauben M, Amsden GW. The association of erythromycin and infantile hypertrophic pyloric stenosis: causal or coincidental? Drug Saf 2002;25(13):929–42.
35. Shiima Y, Tsukahara H, Kobata R, Hayakawa K, Hiraoka M, Mayumi M. Erythromycin in ELBW infants. J Pediatr 2002;141(2):297–8.
36. Centers for Disease Control and Prevention (CDC). Hypertrophic pyloric stenosis in infants following pertussis prophylaxis with erythromycin—Knoxville, Tennessee, 1999. MMWR Morb Mortal Wkly Rep 1999;48(49):1117–20.
37. Honein MA, Paulozzi LJ, Himelright IM, Lee B, Cragan JD, Patterson L, Correa A, Hall S, Erickson JD. Infantile hypertrophic pyloric stenosis after pertussis prophylaxis with erythromcyin: a case review and cohort study. Lancet 1999;354(9196):2101–5.

38. Cooper WO, Griffin MR, Arbogast P, Hickson GB, Gautam S, Ray WA. Very early exposure to erythromycin and infantile hypertrophic pyloric stenosis. Arch Pediatr Adolesc Med 2002;156(7):647–50.

39. Tomomasa T, Kuroume T, Arai H, Wakabayashi K, Itoh Z. Erythromycin induces migrating motor complex in human gastrointestinal tract. Dig Dis Sci 1986;31(2):157–61.

40. Omura S, Tsuzuki K, Sunazuka T, Marui S, Toyoda H, Inatomi N, Itoh Z. Macrolides with gastrointestinal motor stimulating activity. J Med Chem 1987;30(11):1941–3.

41. Lehtola J, Jauhonen P, Kesaniemi A, Wikberg R, Gordin A. Effect of erythromycin on the oro-caecal transit time in man. Eur J Clin Pharmacol 1990;39(6):555–8.

42. Braun P. Hepatotoxicity of erythromycin. J Infect Dis 1969;119(3):300–6.

43. Ginsburg CM, Eichenwald HF. Erythromycin: a review of its uses in pediatric practice. J Pediatr 1976;89(6):872–84.

44. Ginsburg CM. A prospective study on the incidence of liver function abnormalities in children receiving erythromycin estolate, erythromycin ethylsuccinate or penicillin V for treatment of pneumonia. Pediatr Infect Dis 1986;5(1):151–3.

45. Tolman KG, Sannella JJ, Freston JW. Chemical structure of erythromycin and hepatotoxicity. Ann Intern Med 1974;81(1):58–60.

46. Lee WM. Drug-induced hepatotoxicity. N Engl J Med 1995;333(17):1118–27.

47. Funck-Brentano C, Pessayre D, Benhamou JP. Hepatites dues à divers derivés de l'érythromycine. [Hepatitis caused by various derivatives of erythromycin.] Gastroenterol Clin Biol 1983;7(4):362–9.

48. Inman WH, Rawson NS. Erythromycin estolate and jaundice. BMJ (Clin Res Ed) 1983;286(6382):1954–5.

49. Krowchuk D, Seashore JH. Complete biliary obstruction due to erythromycin estolate administration in an infant. Pediatrics 1979;64(6):956–8.

50. Horn S, Aglas F, Horina JH. Cholestasis and liver cell damage due to hypersensitivity to erythromycin stearate—recurrence following therapy with erythromycin succinate. Wien Klin Wochenschr 1999;111(2):76–7.

51. Karthik SV, Casson D. Erythromycin-associated cholestatic hepatitis and liver dysfunction in children: the British experience. J Clin Gastroenterol 2005;39(8):743-4.

52. Eichenwald HF. Adverse reactions to erythromycin. Pediatr Infect Dis 1986;5(1):147–50.

53. Lunzer MR, Huang SN, Ward KM, Sherlock S. Jaundice due to erythromycin estolate. Gastroenterology 1975;68(5 Pt 1):1284–91.

54. Handa SP. The Schönlein-Henoch syndrome: Glomerulonephritis following erythromycin. Sth Med J (Bgham Ala) 1972;65:917.

55. Rosenfeld J, Gura V, Boner G, Ben-Bassat M, Livni E. Interstitial nephritis with acute renal failure after erythromycin. BMJ (Clin Res Ed) 1983;286(6369):938–9.

56. Thiboutot D, Jarratt M, Rich P, Rist T, Rodriguez D, Levy S. A randomized, parallel, vehicle-controlled comparison of two erythromycin/benzoyl peroxide preparations for acne vulgaris. Clin Ther 2002;24(5):773–85.

57. Pigatto PD, Riboldi A, Riva F, Altomare GF. Fixed drug eruption to erythromycin. Acta Dermatol Venereol 1984;64(3):272–3.

58. van Ketel WG. Immediate- and delayed-type allergy to erythromycin. Contact Dermatitis 1976;2(6):363–4.

59. Pascual C, Crespo JF, Quiralte J, Lopez C, Wheeler G, Martin-Esteban M. In vitro detection of specific IgE antibodies to erythromycin. J Allergy Clin Immunol 1995;95(3):668–71.

60. Pandha HS, Dunn PJ. Stevens–Johnson syndrome associated with erythromycin therapy. NZ Med J 1995;108(992):13.

61. Hussain N, Herson VC. Erythromycin use during pregnancy in relation to pyloric stenosis. Am J Obstet Gynecol 2002;187(3):821–2author reply 822.

62. Wynn RL. Erythromycin and ketoconazole (Nizoral) associated with terfenadine (Seldane)-induced ventricular arrhythmias. Gen Dent 1993;41(1):27–9.

63. Amsden GW. Macrolides versus azalides: a drug interaction update. Ann Pharmacother 1995;29(9):906–17.

64. Brannan MD, Reidenberg P, Radwanski E, Shneyer L, Lin CC, Cayen MN, Affrime MB. Loratadine administered concomitantly with erythromycin: pharmacokinetic and electrocardiographic evaluations. Clin Pharmacol Ther 1995;58(3):269–78.

65. Banfield C, Hunt T, Reyderman L, Statkevich P, Padhi D, Affrime M. Lack of clinically relevant interaction between desloratadine and erythromycin. Clin Pharmacokinet 2002;41(Suppl 1):29–35.

66. George AK, Kunwar AR, Awasthi A. Acute myocardial infarction in a young male on methylphenidate, bupropion, and erythromycin. J Child Adolesc Psychopharmacol 2005;15(4):693–5.

67. Mahmood I, Sahajwalla C. Clinical pharmacokinetics and pharmacodynamics of buspirone, an anxiolytic drug. Clin Pharmacokinet 1999;36(4):277–87.

68. Hedrick R, Williams F, Morin R, Lamb WA, Cate JC 4th. Carbamazepine–erythromycin interaction leading to carbamazepine toxicity in four epileptic children. Ther Drug Monit 1983;5(4):405–7.

69. Wong YY, Ludden TM, Bell RD. Effect of erythromycin on carbamazepine kinetics. Clin Pharmacol Ther 1983;33(4):460–4.

70. Bachmann K, Schwartz JI, Forney RB Jr, Jauregui L. Single dose phenytoin clearance during erythromycin treatment. Res Commun Chem Pathol Pharmacol 1984;46(2):207–17.

71. Milne RW, Coulthard K, Nation RL, Penna AC, Roberts G, Sansom LN. Lack of effect of erythromycin on the pharmacokinetics of single oral doses of phenytoin. Br J Clin Pharmacol 1988;26(3):330–3.

72. Wroblewski BA, Singer WD, Whyte J. Carbamazepine–erythromycin interaction. Case studies and clinical significance. JAMA 1986;255(9):1165–7.

73. Suri A, Forbes WP, Bramer SL. Effects of CYP3A inhibition on the metabolism of cilostazol. Clin Pharmacokinet 1999;37(Suppl 2):61–8.

74. Kyrmizakis DE, Chimona TS, Kanoupakis EM, Papadakis CE, Velegrakis GA, Helidonis ES. QT prolongation and torsades de pointes associated with concurrent use of cisapride and erythromycin. Am J Otolaryngol 2002;23(5):303–7.

75. Thomsen MS, Groes L, Agerso H, Kruse T. Lack of pharmacokinetic interaction between tiagabine and erythromycin. J Clin Pharmacol 1998;38(11):1051–6.

76. Lindenbaum J, Rund DG, Butler VP Jr, Tse-Eng D, Saha JR. Inactivation of digoxin by the gut flora: reversal by antibiotic therapy. N Engl J Med 1981;305(14):789–94.

77. Morton MR, Cooper JW. Erythromycin-induced digoxin toxicity. DICP 1989;23(9):668–70.

78. Taylor D. Pharmacokinetic interactions involving clozapine. Br J Psychiatry 1997;171:109–12.

79. Cohen LG, Chesley S, Eugenio L, Flood JG, Fisch J, Goff DC. Erythromycin-induced clozapine toxic reaction. Arch Intern Med 1996;156(6):675–7.

80. Hagg S, Spigset O, Mjorndal T, Granberg K, Persbo-Lundqvist G, Dahlqvist R. Absence of interaction between erythromycin and a single dose of clozapine. Eur J Clin Pharmacol 1999;55(3):221–6.

81. Usiskin SI, Nicolson R, Lenane M, Rapoport JL. Retreatment with clozapine after erythromycin-induced neutropenia. Am J Psychiatry 2000;157(6):1021.

82. Zini R, Fournet MP, Barre J, Tremblay D, Tillement JP. In vitro study of roxithromycin binding to serum proteins and erythrocytes in man. Br J Clin Pract 1987;42(Suppl 5):54.

83. Ragosta M, Weihl AC, Rosenfeld LE. Potentially fatal interaction between erythromycin and disopyramide. Am J Med 1989;86(4):465–6.

84. Eadie MJ. Clinically significant drug interactions with agents specific for migraine attacks. CNS Drugs 2001;15(2):105–18.

85. Sachdeo RC, Narang-Sachdeo S, Montgomery PA, Shumaker RC, Perhach JL, Lyness WH, Rosenberg A. Evaluation of the potential interaction between felbamate and erythromycin in patients with epilepsy. J Clin Pharmacol 1998;38(2):184–90.

86. Baune B, Flinois JP, Furlan V, Gimenez F, Taburet AM, Becquemont L, Farinotti R. Halofantrine metabolism in microsomes in man: major role of CYP 3A4 and CYP 3A5. J Pharm Pharmacol 1999;51(4):419–26.

87. Isohanni MH, Neuvonen PJ, Olkkola KT. Effect of erythromycin and itraconazole on the pharmacokinetics of oral lignocaine. Pharmacol Toxicol 1999;84(3):143–6.

88. Isohanni MH, Neuvonen PJ, Palkama VJ, Olkkola KT. Effect of erythromycin and itraconazole on the pharmacokinetics of intravenous lignocaine. Eur J Clin Pharmacol 1998;54(7):561–5.

89. Katoh T, Saitoh H, Ohno N, Tateno M, Nakamura T, Dendo I, Kobayashi S, Nagasawa K. Drug interaction between mosapride and erythromycin without electrocardiographic changes. Jpn Heart J 2003;44:225–34.

90. Bachmann K, Schwartz JI, Forney R Jr, Frogameni A, Jauregui LE. The effect of erythromycin on the disposition kinetics of warfarin. Pharmacology 1984;28(3):171–6.

91. Grau E, Real E, Pastor E. Macrolides and oral anticoagulants: a dangerous association. Acta Haematol 1999;102(2):113–4.

92. Li KY, Li X, Cheng ZN, Zhang BK, Peng WX, Li HD. Effect of erythromycin on metabolism of quetiapine in Chinese suffering from schizophrenia. Eur J Clin Pharmacol 2005;60(11):791–5.

93. Damkier P, Hansen LL, Brosen K. Effect of diclofenac, disulfiram, itraconazole, grapefruit juice and erythromycin on the pharmacokinetics of quinidine. Br J Clin Pharmacol 1999;48(6):829–38.

94. Zhao XJ, Ishizaki T. A further interaction study of quinine with clinically important drugs by human liver microsomes: determinations of inhibition constant (Ki) and type of inhibition. Eur J Drug Metab Pharmacokinet 1999;24(3):272–8.

95. Grub S, Bryson H, Goggin T, Ludin E, Jorga K. The interaction of saquinavir (soft gelatin capsule) with ketoconazole, erythromycin and rifampicin: comparison of the effect in healthy volunteers and in HIV-infected patients. Eur J Clin Pharmacol 2001;57(2):115–21.

96. Lee DO, Lee CD. Serotonin syndrome in a child associated with erythromycin and sertraline. Pharmacotherapy 1999;19(7):894–6.

97. Athanasakis E, Chrysos E, Zoras OJ, Tsiaoussis J, Karkavitsas N, Xynos E. Octreotide enhances the accelerating effect of erythromycin on gastric emptying in healthy subjects. Aliment Pharmacol Ther 2002;16(8):1563–70.

98. Prieto JC. El perfil de seguridad de las estatinas. [Safety profile of statins.] Rev Med Chil 2001;129(11):1237–40.

99. Williams D, Feely J. Pharmacokinetic–pharmacodynamic drug interactions with HMG-CoA reductase inhibitors. Clin Pharmacokinet 2002;41(5):343–70.

100. Garnett WR. Interactions with hydroxymethylglutaryl-coenzyme: a reductase inhibitors. Am J Health Syst Pharm 1995;52(15):1639–45.

101. Wong PW, Dillard TA, Kroenke K. Multiple organ toxicity from addition of erythromycin to long-term lovastatin therapy. South Med J 1998;91(2):202–5.

102. Schlienger RG, Haefeli WE, Jick H, Meier CR. Risk of cataract in patients treated with statins. Arch Intern Med 2001;161(16):2021–6.

103. Bernini F, Poli A, Paoletti R. Safety of HMG-CoA reductase inhibitors: focus on atorvastatin. Cardiovasc Drugs Ther 2001;15(3):211–8.

104. Zarowitz BJ, Szefler SJ, Lasezkay GM. Effect of erythromycin base on theophylline kinetics. Clin Pharmacol Ther 1981;29(5):601–5.

105. Paulsen O, Hoglund P, Nilsson LG, Bengtsson HI. The interaction of erythromycin with theophylline. Eur J Clin Pharmacol 1987;32(5):493–8.

106. Coudray S, Janoly A, Belkacem-Kahlouli A, Bourhis Y, Bleyzac N, Bourgeois J, Putet G, Aulagner G. Erythromycin-induced digoxin toxicity in a neonatal intensive care unit. J Pharm Clin 2001;20:129–31.

107. Amsden GW. Macrolides versus azalides: a drug interaction update. Ann Pharmacother 1995;29(9):906–17.

108. Goldschmidt N, Azaz-Livshits T, Gotsman, Nir-Paz R, Ben-Yehuda A, Muszkat M. Compound cardiac toxicity of oral erythromycin and verapamil. Ann Pharmacother 2001;35(11):1396–9.

109. Kantola T, Kivisto KT, Neuvonen PJ. Erythromycin and verapamil considerably increase serum simvastatin and simvastatin acid concentrations. Clin Pharmacol Ther 1998;64(2):177–82.

110. Tobe SW, Siu LL, Jamal SA, Skorecki KL, Murphy GF, Warner E. Vinblastine and erythromycin: an unrecognized serious drug interaction. Cancer Chemother Pharmacol 1995;35(3):188–90.

111. Purkins L, Wood N, Ghahramani P, Kleinermans D, Layton G, Nichols D. No clinically significant effect of erythromycin or azithromycin on the pharmacokinetics of voriconazole in healthy male volunteers. Br J Clin Pharmacol 2003;56 Suppl 1:30–6.

112. Eriksson UG, Dorani H, Karlsson J, Fritsch H, Hoffmann KJ, Olsson L, Sarich TC, Wall U, Schutzer KM. Influence of erythromycin on the pharmacokinetics of ximelagatran may involve inhibition of P-glycoprotein-mediated excretion. Drug Metab Dispos 2006;34(5):775–82.

113. Hesse LM, von Moltke LL, Greenblatt DJ. Clinically important drug interactions with zopiclone, zolpidem and zaleplon. CNS Drugs 2003;17(7):513–32.

Ethambutol

See also Antituberculosis drugs

General Information

Ethambutol is tuberculostatic and acts against *Mycobacterium tuberculosis* and *Mycobacterium kansasii* as well as some strains of *Mycobacterium avium* complex. It has no effect on other bacteria. The sensitivities of non-tuberculous mycobacteria are variable. Ethambutol suppresses the growth of most isoniazid-resistant and streptomycin-resistant tubercle bacilli (1).

The adverse effects of ethambutol are mainly seen in patients taking very high doses, that is over 25 mg/kg/day. Dosages over 25 mg/kg/day should be administered for only about 2 weeks at the beginning of therapy. Treatment can then be continued with 15 mg/kg/day. If initial problems arise and ethambutol is essential, the daily dose should not exceed 10 mg/kg.

About 80% of an oral dose of ethambutol is absorbed from the gastrointestinal tract. Peak plasma concentrations are reached in 2–4 hours and the half-life is 3–4 hours. Within 24 hours, about 60% is excreted unchanged in the urine. In patients with renal insufficiency, the dose has to be altered (2). There is no indication for ethambutol monotherapy.

General adverse effects

Visual disturbances are the most common adverse effects and also the most important. These include diminished visual acuity, retrobulbar neuritis, retinal pigment displacement, and (rarely) hemorrhages. Gastrointestinal symptoms (abdominal pain or vomiting), and headache, dizziness, mental confusion, and hallucinations are all rarely seen. Adverse effects are more frequent in elderly patients and patients with alcoholism, diabetes, or renal insufficiency. Stevens–Johnson syndrome, toxic epidermal necrolysis (3), purpura-like vasculitis, acute thrombogenic purpura, joint pain, drug fever, tachycardia, and leukopenia have been attributed to allergy. As these reactions often arise during combined treatment with other tuberculostatic drugs, it is difficult or impossible to determine which drug is responsible. Tumor-inducing and teratogenic effects have not been described.

Organs and Systems

Respiratory

Acute lung injury has been attributed to ethambutol (4).

- A 78 year old woman with active pulmonary tuberculosis was given isoniazid, rifampicin, ethambutol, and pyrazinamide. After 10 days she developed a fever, dyspnea, left-sided pleuritic chest pain, and hypoxemia. A chest CT scan showed bilateral pleural effusions. A biopsy specimen from the anterior basal segment of the right lower lobe showed diffuse alveolar epithelial damage, focal hemorrhage, and alveolar wall thickening, with moderate infiltration of inflammatory cells. Drug-induced acute lung injury was suspected and the antituberculosis drugs were withdrawn after 34 days. The fever subsided and the dyspnea and general weakness improved. The chest X-ray showed significant clearing of the infiltration with reduced pleural effusions but persistent atelectasis of the middle lobe in the right lung 1 month later .Isoniazid, rifampicin, and pyrazinamide were restarted 14, 19,and 30 days later respectively and there was no recurrence. Rechallenge was not undertaken.

The authors concluded that the lung injury was due to ethambutol.

Nervous system

Peripheral neuropathy can precede or accompany ocular damage from ethambutol. These symptoms can serve as warning of impending eye damage (5). Loss of sensitivity, with numbness and tingling of the fingers, are relatively rare adverse effects (6). Electroneuromyography in ethambutol-induced neuropathy has confirmed that elderly patients are at increased risk (7). Sensory changes are more severe than motor dysfunction (6).

Sensory systems

Optic neuropathy, which is primarily retrobulbar, is the most important adverse effect of ethambutol and takes two forms central and peripheral. The commoner is axial neuritis (central type), which involves the papillomacular bundle and results in reduced visual acuity, cecocentral scotoma, and blue-yellow color vision impairment. In periaxial neuritis (peripheral type) there is peripheral visual field loss, especially bitemporal defects, with sparing of visual acuity and red-green color vision impairment (8). Rarely damage to the retina and macula has also been reported (9) (10).

Presentation

Ocular symptoms begin with bilateral progressive blurred vision or defects in color vision. However, some individuals are asymptomatic and abnormalities are detected only by tests of vision. A central scotoma is the most common visual field defect. Dyschromatopsia in the form of red-green color changes may be the earliest sign. Fundoscopy is usually normal.

Ethambutol can reportedly precipitate symptoms in Leber's hereditary optic neuropathy (11).

- A 70-year-old woman with tuberculosis took ethambutol, rifampicin, isoniazid, and pyrazinamide. She had marked reduction in visual acuity in both eyes after 3 months of treatment and ethambutol was withdrawn. Her corrected visual acuity was 0.03 in both eyes. There was no hyperemia, swelling of the optic disc, or capillary dilatation in either eye. Centrocecal scotomas were found bilaterally. After 1 month her visual acuity had further reduced to 0.01 and the scotomas had enlarged.

Genetic analysis revealed a point mutation in mitochondrial DNA 11778.

The authors concluded that ethambutol could be a risk factor for Leber's hereditary optic neuropathy and found three similar cases in a literature review.

Mechanism

The exact mechanism of ethambutol-induced ocular toxicity is not known.

- The earliest onset of visual disturbance occurred in a 26-year-old man 3 days after beginning combined treatment including 15 mg/kg/day ethambutol, suggesting an idiosyncratic reaction (12).

However, there is an association with low serum zinc concentrations and reduced renal function (13) (14). Biochemical research has shown the importance of zinc metabolism in the retina (20). Zinc is found in high concentrations in the choroid, the retina, and especially the ganglion cells. Retinol dehydrogenase, a zinc-containing enzyme, interferes with the transformation of retinol (vitamin A_1), which is essential for color sensation and conal vision. Furthermore, zinc is involved in the biosynthesis of the specific transport of retinol from the liver to the effector cells. Ethambutol is a chelating agent and makes zinc unavailable for axoplasmic transport, provoking optic or retrobulbar neuritis (15). Patients with zinc blood concentrations below 0.7 µg/ml (reference range 0.9–1.0 µg/ml) before the use of ethambutol are at high risk of ocular disturbances (16).

Dose relation

Ocular toxicity due to ethambutol is dose related in the therapeutic range. At doses of over 50 mg/kg/day, over 40% of adults develop toxicity, compared with 0–3% at a dose of 15 mg/kg/day (17).

Dosages not over 25 mg/kg/day during the first 2 months of treatment and 15 mg/kg/day thereafter are generally accepted as adequate (18). At a dosage of 15 mg/kg/day, which should be regarded as a maximum for maintenance therapy, ocular toxicity developed in only 1.6% of patients (19). Advanced age (20), renal insufficiency, and diabetes can enhance ocular damage.

Time course

The mean interval between the onset of therapy and the adverse effects is as short as 1.5 months or as long as 12 months after the start of therapy (17).

The onset of visual loss can be sudden and dramatic, with color vision defects in the red–green or blue–yellow spectra, as well as variable field defects. In acute cases, disc edema is accompanied by splinter hemorrhages. Retrobulbar neuritis with ethambutol can be predominantly axial, presenting with reduced visual acuity and central scotoma, or periaxial, with peripheral field defects. In non-acute types the fundi and discs appear normal (21). Visual defects can be unilateral or bilateral.

Diagnosi

Visual-evoked potential tests, such as flash electroretinography, flash and pattern visual-evoked responses, flicker fusion thresholds, and visual field perimetry, are reported to be the most reliable methods for early detection of ocular abnormalities (19,15,22). The routine use of visual-evoked potentials in the systematic follow-up of ethambutol-treated patients has been recommended (23).

Color vision and visual acuity should be examined before beginning ethambutol and every 2–4 weeks during treatment, using color tables and reading tests (SED-9, 525). Some recommend computerized perimetry of the central visual fields (SED-10, 581) (24,25). In patients with diabetes with retinopathy monthly monitoring of the fundi and visual acuity are mandatory. Early childhood, when visual examination is difficult or impossible, is a major indication for electroretinography. As a rule, ethambutol is not recommended for children under 5 years of age, because of the difficulty in reliably testing their visual acuity, unless the severity of tuberculosis makes the use of ethambutol necessary (23).

There was abnormal multifocal electroretinography in two cases of presumed ethambutol toxicity (26).

- An 81-year-old woman with Parkinson's disease and hypothyroidism developed gradual, bilateral, painless, progressive visual loss over 2–3 months. Her medications included clarithromycin, clopidogrel, oxybutynin, omeprazole, ciprofloxacin, levothyroxine, carbidopa + levodopa, salbutamol, paracetamol, and ethambutol 400 mg bd. She had been taking ethambutol for 4 months when she noticed the visual changes. Her visual acuity was 20/250 in the right eye and 20/100 in the left eye. The pupils were equal and reactive without a relative afferent pupillary defect. Ophthalmoscopy showed minimal retinal pigment epithelial changes in the macula and minimal temporal pallor of the optic discs. Goldmann visual fields showed a dense centrocecal scotoma with breakout to the superior temporal periphery. Ethambutol was withdrawn. The amplitudes of multifocal electroretinography were reduced by 1.1–4.5 standard deviations at different locations, and the change was greatest in the centre of the visual fields. Over the next 2 months, her visual acuity gradually improved to 20/80 (right eye) and 20/50 (left eye) and after 10 months 20/70 and 20/40; the visual fields showed markedly improved central scotomas.
- A 60-year-old woman with *Mycobacterium avium intracellulare* pulmonary infection was given rifampicin 600 mg/day and ethambutol 1200 mg/day. After 9 months she developed reduced vision centrally in both eyes, which she described as a "shadow". Her visual acuity was 20/25. Multifocal electroretinography showed a moderate diffuse reduction in amplitudes by 2.1–4.2 standard deviations. Ethambutol was continued, but after 1 month there was continued subjective visual loss and visual acuity was 20/50 with a new central scotoma. Multifocal electroretinography showed a reduction in amplitudes greater than 2 standard deviations and disproportionately in the centre of the field.

Ethambutol was withdrawn and her visual functions returned to normal within 6 months.

Multifocal electroretinography is useful in detecting retinal dysfunction caused by damage to photoreceptors or bipolar cells or their connections. The authors concluded that abnormal electroretinography with diffuse depression and additional central field depression could have contributed to the reduction in visual field sensitivity, in particular the central scotomas. This gives further evidence that ethambutol can produce toxicity in layers of the retina deep to the ganglion layer in addition to optic nerve toxicity.

Outcome
If ethambutol is not withdrawn when visual symptoms occur, optic atrophy or permanent blindness can occur. Patients therefore have to be instructed to interrupt treatment if they experience any visual abnormality. Hydroxocobalamin can accelerate recovery (27).

Ocular toxicity in elderly people
Of 299 patients, mean age 63 years, taking ethambutol-containing multidrug therapy for Mycobacterium avium complex lung disease for a mean period of 16 months, 42% consulted an ophthalmologist and 10% stopped taking ethambutol at least temporarily (28). Ethambutol-associated ocular toxicity was present only in patients who took daily treatment (6%), while none of the patients who took ethambutol intermittently reported this complication. Baseline ocular function was restored on withdrawal of ethambutol.

Two elderly women taking ethambutol for atypical mycobacterial disease developed reduced visual acuity after 10 and 14 months (29). At the start both developed severe bilateral visual loss with bitemporal hemianopia-like visual field disturbance, and ethambutol was withdrawn. However, there was no disc pallor or nerve fiber layer defect (NFLD) at any time during the follow-up period, and they both had good recovery of visual function at 18 months and 19 months after the onset of optic neuropathy. The authors concluded that in patients with severe ethambutol optic neuropathy, as long as disc pallor and NFLD are not observed, good visual recovery can be expected even if severe visual loss persists for a long time.

Ocular toxicity in children
There has been considerable reluctance to use ethambutol in young children, because of the potentially serious nature of ocular complications. Most international and national guidelines recommend that ethambutol should not be given to children younger than 5 or 7 years of age.

In a recent review of several studies that carefully evaluated significant numbers of children taking ethambutol 15–30 mg/kg/day there was no evidence of ocular toxicity. Moreover, the cases reported in those with tuberculous meningitis could not be attributed to the drug, as the disease itself would often have been responsible. In only two of 3811 cases (0.05%) was ethambutol stopped because of fears of poorly documented ocular toxicity.

These results endorse the safety of ethambutol in children of all ages.

Metabolism
Blood urate concentrations can be increased because of reduced excretion of uric acid in patients taking ethambutol (30). This is probably enhanced by combined treatment with isoniazid and pyridoxine. Special attention should be paid when tuberculostatic drug combinations include pyrazinamide. However, severe untoward clinical effects are rare, except in patients with gout or renal insufficiency (2,31).

Hematologic
Acute thrombocytopenia, probably due to an immunological mechanism, has been described in a single patient (32).

Liver
Jaundice and non-icteric liver disturbances probably occur in under 0.1% of patients, as long as isoniazid and rifampicin are not given simultaneously (33).

Urinary tract
Ethambutol-induced acute renal insufficiency is very rare; only three cases of tubulointerstitial nephritis have previously been reported. In those cases, renal function deteriorated only after administration of antituberculosis drug treatment over several months, and liver function tests were normal.

- A 33 year old Korean man who had been treated for pulmonary tuberculosis 5 years before with rifampicin, pyrazinamide, isoniazid, and ethambutol, developed a recurrence and the same four medications were started (34). One day later he noted jaundice, abdominal pain, and oliguria. He had abnormal liver and renal function tests (blood urea nitrogen 26 mmol/l (72 mg/dl), serum creatinine 400 μmol/l (5.2 mg/dl), total bilirubin 240 μmol/l (14 mg/dl), aspartate transaminase 170 IU/l, alanine transaminase 28 IU/l, proteinuria 3+ with many red and white blood cells in the urine sediment). The anuria persisted for 4 days. On days 8 and 15, isoniazid and rifampicin were reintroduced without any problems. On day 24, ethambutol was reintroduced, and that evening he developed abdominal pain, fever, and general weakness. The serum creatinine concentration increased to 8.6 mg/dl and his urine output fell; hemodialysis was initiated. A renal biopsy showed features characteristic of acute tubular necrosis. He completely recovered renal function after treatment with rifampicin, isoniazid, pyrazinamide, and cycloserine.

In this case it was suggested that the renal damage was due to a toxic rather than an allergic effect, producing a tubular lesion and interstitial nephritis.

Musculoskeletal

Crystal arthropathy, which is a rare and poorly recognized adverse effect of pyrazinamide and ethambutol, has been reported in a patient taking both drugs (35).

- A 76-year-old woman was given rifampicin, ethambutol, isoniazid and pyrazinamide for tuberculosis of the cervical spine and after 3 weeks developed an acute monoarthropathy of the right wrist associated with pyrexia. The wrist was swollen, erythematous, boggy, and tender. Active movement was impaired by pain and passive movement was limited. Radiography showed chondrocalcinosis. The CRP had risen to 600 g/l, there was a leukocytosis, and the serum uric acid concentration was 580 μmol/l, having been normal on admission. From the joint 0.5 ml of "creamy" pus was aspirated. Synovial fluid analysis showed 50% neutrophils and 50% lymphocytes. The fluid was negative for Gram staining organisms and acid-fast bacilli. Polarized light microscopy showed urate and pyrophosphate crystals. She was treated with diclofenac sodium for 5 days and her symptoms subsided. She completed her course of antituberculosis drug therapy.

A metabolite of pyrazinamide, pyrazinoic acid, inhibits the renal tubular secretion of uric acid and ethambutol also reduces its renal clearance. The resulting hyperuricemia can arthralgias and very rarely arthritis.

Second-Generation Effects

Pregnancy

Ethambutol in combination with isoniazid in pregnancy has been suspected of causing one case of gastrointestinal malformation, but other factors may have been implicated (SEDA-12, 256).

Teratogenicity

Fetal damage has not been attributed to ethambutol (36).

Fetotoxicity

Toxic effects in the newborn have not been attributed to ethambutol (37).

Lactation

No adverse effects during lactation have been attributed to ethambutol (37).

Susceptibility Factors

Age

The reason for the low toxicity observed in children could be attributed to differences in drug pharmacokinetics between children and adults.

In adults, peak ethambutol concentrations after daily doses of 25 and 50 mg/kg were 5 and 10 μg/ml respectively (38). Serum concentrations were proportional to the dose. Less than 10% of the administered dose was present in

the serum after 24 hours and there was no evidence of accumulation of the drug over more than 3 months.

There have been few studies of the pharmacokinetics of ethambutol in children. Serum concentrations in children taking doses of 15–35 mg/kg were lower than those in adults after similar doses (39). Furthermore, serum ethambutol concentrations were lower in younger than in older children. The serum concentrations reached in adults and children given similar doses of ethambutol were clearly different, suggesting that in order to achieve serum concentrations equivalent to those reached in adults given 15 mg/kg a child would require a dose of 25 mg/kg or more. However, factors such as the ratio of extracellular to intracellular and total body water, biotransformation, and elimination have also to be considered in interpreting the results. The authors speculated that delayed absorption of ethambutol could be the reason, and slow and incomplete absorption of ethambutol has been observed in children.

Renal disease

In patients with impaired glomerular filtration or any other renal dysfunction, ethambutol dosages have to be altered (40).

Drug Administration

Drug dosage regimens

In the absence of overt ocular toxicity in children aged from under 1 year to 18 years, who had taken ethambutol 15–30 mg/kg/day the following dosage regimens have been recommended (41):

- Daily treatment—20 mg/kg (range 15–25 mg/kg) for children of all ages. Increasing the dose beyond this range to compensate for deficiencies in serum concentrations might increase the risk of ethambutol ocular toxicity.
- Intermittent treatment—30 mg/kg (range 20–35 mg/kg) three times weekly or 45 mg/kg (range 40–50 mg/kg) twice weekly (as currently recommended for adults).

Just as in adults, care should be taken to establish that a child does not have renal disease, as this could lead to exposure to unacceptably high serum concentrations of ethambutol.

Drug–Drug Interactions

Antacids

In 13 patients with tuberculosis, aluminium hydroxide significantly lowered serum ethambutol concentrations during the first 4 hours after a dose; in healthy volunteers, both aluminium hydroxide and glycopyrrhonium, alone or in combination, reduced ethambutol absorption (42).

Isoniazid

The risk of optic neuropathy due to ethambutol may be increased when it is combined with isoniazid.

- A 40-year-old patient underwent unsuccessful cadaver kidney transplantation and was treated with ethambutol and isoniazid (43). Bilateral retrobulbar neuropathy with an unusual central bitemporal hemianopic scotoma developed and ethambutol was withdrawn, but there was only a small improvement. When isoniazid was also withdrawn, there was dramatic improvement in visual acuity.

In another patient taking isoniazid and ethambutol, optic neuropathy improved only when both drugs were withdrawn (44).

Isoniazid 300 mg/day orally for 7 days increased serum ethambutol concentrations at 4, 6, and 8 hours after a daily dose of 20 mg/kg; the cumulative percentage dose excreted was significantly reduced at 4, 6, and 24 hours (45). In another study from the same center, ethambutol 20 mg/kg did not alter the pharmacokinetics of a single dose of isoniazid 300 mg in 10 patients with tuberculosis (46).

References

1. Dickinson JM, Aber VR, Mitchison DA. Bactericidal activity of streptomycin, isoniazid, rifampin, ethambutol, and pyrazinamide alone and in combination against *Mycobacterium tuberculosis*. Am Rev Respir Dis 1977;116(4):627–35.
2. Mandell GL, Sande MA. Antimicrobial agents: drugs used in the chemotherapy of tuberculosis and leprosy. In: Goodman Gilman A, Rall TW, Nies AS, Taylor P, editors. Goodman and Gilman's The Pharmacological Basis of Therapeutics. 8th ed.. New York: Pergamon Press, 1990:1146 Chapter 49.
3. Pegram PS Jr, Mountz JD, O'Bar PR. Ethambutol-induced toxic epidermal necrolysis. Arch Intern Med 1981;141(12):1677–8.
4. Choi WI, June Jeon Y, Kwon KY, Ko SM, Keum DY, Kwon Park C. Acute lung injury caused by ethambutol. Respir Med Extra 2006;2:55–7.
5. Nair VS, LeBrun M, Kass I. Peripheral neuropathy associated with ethambutol. Chest 1980;77(1):98–100.
6. Takeuchi H, Takahashi M, Tarui S, Sanagi S, Takenaka H. Peripheral nerve conduction function in patients treated with antituberculotic agents, with special reference to ethambutol and isoniazid. Folia Psychiatr Neurol Jpn 1980;34(1):57–64.
7. Takeuchi H, Takahashi M, Kang J, Ueno S, Tarui S, Nakao Y, Otori T. Ethambutol neurophathy: clinical and electroneuromyographic studies. Folia Psychiatr Neurol Jpn 1980;34(1):45–55.
8. Schild HS, Fox BC. Rapid-onset reversible ocular toxicity from ethambutol therapy. Am J Med 1991;90:404–6.
9. Kakisu Y, Adachi-Usami E, Mizota A. Pattern electroretinogram and visual evoked cortical potential in ethambutol optic neuropathy. Doc Ophthalmol 1987;67:327–34.
10. Lai TYY, Chan W-M, Lam DSC, Lim E. Multifocal electroretinogram demonstrated macular toxicity associated with ethambutol related optic neuropathy Br J Ophthalmol 2005;89(6):774–5.
11. Ikeda A, Ikeda T, Ikeda N, Kawakami Y, Mimura O. Leber's hereditary optic neuropathy precipitated by ethambutol. Jap J Ophthalmol 2006;50:280–3.
12. Karnik AM, Al-Shamali MA, Fenech FF. A case of ocular toxicity to ethambutol—an idiosyncratic reaction? Postgrad Med J 1985;61(719):811–3.
13. Jhamaria JP, Rajput VS, Luhadia SK, Bansal PP, Ved ML, Gandhi VC, Rajput VS. Ocular toxicity of ethambutol and its correlation with serum zinc levels. Lung India 1989;7(4):183–5.
14. Ji-Tseng Fang, Yung-Chang Chen, Ming-Yang Chang. Ethambutol-induced optic neuritis in patients with end stage renal disease on hemodialysis: two case reports and literature review. Renal Fail 2004;2(26):89–93.
15. Yolton DP. Nutritional effects of zinc on ocular and systemic physiology. J Am Optom Assoc 1981;52(5):409–14.
16. Delacoux E, Moreau Y, Godefroy A, Evstigneeff T. Prévention de la toxicité oculaire de l'éthambutol: intérêt de la zincémie et de l'analyse du sens chromatique. [Prevention of ocular toxicity of ethambutol: study of zincaemia and chromatic analysis.] J Fr Ophtalmol 1978;1(3):191–6.
17. Donald PR, Maher D, Maritz JS, Qazi S. Ethambutol dosage for the treatment of children: literature review and recommendations. Int J Tuberc Lung Dis 2006;10:1318–30.
18. Otori T. Drug-induced ocular side effects. Asian Med J 1981;24:141.
19. Garrett CR. Optic neuritis in a patient on ethambutol and isoniazid evaluated by visual evoked potentials: case report. Mil Med 1985;150(1):43–6.
20. Cole A, May PM, Williams DR. Metal binding by pharmaceuticals. Part 1. Copper(II) and zinc(II) interactions following ethambutol administration. Agents Actions 1981;11(3):296–305.
21. Baciewicz AM, Self TH, Bekemeyer WB. Update on rifampin drug interactions. Arch Intern Med 1987;147(3):565–8.
22. Williams DE. Visual electrophysiology and psychophysics in chronic alcoholics and in patients on tuberculostatic chemotherapy. Am J Optom Physiol Opt 1984;61(9):576–85.
23. Trebucq A. Should ethambutol be recommended for routine treatment of tuberculosis in children? A review of the literature. Int J Tuberc Lung Dis 1997;1(1):12–5.
24. Trau R, Salu P, Jonckheere P, Leysen R. Early diagnosis of Myambutol (ethambutol) ocular toxicity by electrophysiological examination. Bull Soc Belge Ophtalmol 1981;193:201–12.
25. Gramer E, Jeschke R, Krieglstein GK. Zur computergesteuerten Gesichtsfeldkontrolle bei Kindern mit Ethambutol-Medikation. [Computerized perimetry in infants treated with ethambutol.] Klin Padiatr 1982;194(1):52–5.
26. Kardon RH, Morrisey MC, Lee AG. Abnormal multifocal electroretinogram (mfERG) in ethambutol toxicity. Semin Ophthalmol 2006;21(4):215–22.
27. Guerra R, Casu L. Hydroxycobalamin for ethambutol-induced optic neuropathy. Lancet 1981;2(8256):1176.
28. Griffith DE, Brown-Elliott BA, Shepherd S, McLarty J, Griffith L, Wallace Jr RJ. Ethambutol ocular toxicity in treatment regimens for Mycobacterium avium complex lung disease. Am J Respir Crit Care Med 2005;172:250–3.
29. Takada Ritsuko, Takagi Mineo, Oshima Akira, Miki Atsushi, Usui Tomoaki, Hasegawa Shigeru, Abe Haruki. Delayed visual recovery from severe ethambutol optic

neuropathy in two patients with atypical mycobacterium infection. Neuro-Ophthalmology 2005;29:187–93.

30. Khanna BK, Gupta VP, Singh MP. Ethambutol-induced hyperuricaemia. Tubercle 1984;65(3):195–9.

31. Khanna BK. Acute gouty arthritis following ethambutol therapy. Br J Dis Chest 1980;74(4):409–10.

32. Rabinovitz M, Pitlik SD, Halevy J, Rosenfeld JB. Ethambutol-induced thrombocytopenia. Chest 1982;81(6):765–6.

33. Ansari MM, Beg MH, Haleem S. Hepatitis in patients with surgical complications of pulmonary tuberculosis. Indian J Chest Dis Allied Sci 1991;33(3):133–8.

34. Kwon SH, Kim JH, Yang JO, Lee E-Y, Hong SY. Ethambutol-induced acute renal failure. Nephrol Dial Transplant 2004;19:1335–6.

35. Abraham S, Mitchell A, Cope A. Anti-tuberculous therapy-induced crystal arthropathy. Rheumatology 2006;45:1173–4.

36. Lewit T, Nebel L, Terracina S, Karman S. Ethambutol in pregnancy: observations on embryogenesis. Chest 1974;66(1):25–6.

37. Kunz J, Schreiner WE. Pharmakotherapie während Schwangerschaft und StillperiodeStuttgart-New York: G Thieme;. 1982.

38. Place VA, Thomas JP. Clinical pharmacology of ethambutol. Am Rev Respir Dis 1963;87:901–4.

39. Zhu M, Burman WJ, Starke JR, Stambaugh JJ, Steiner P, Bulpitt AE, Ashkin D, Auclair B, Berning SE, Jelliffe RW, Jaresko GS, Peloquin CA. Pharmacokinetics of ethambutol in children and adults with tuberculosis. Int J Tuberc Lung Dis 2004;8:1360–7..

40. Strauss I, Erhardt F. Ethambutol absorption, excretion and dosage in patients with renal tuberculosis. Chemotherapy 1970;15(3):148–57.

41. World Health Organization. Ethambutol efficacy and toxicity. Literature review and recommendation for daily and intermittent dosage in children. Geneva, Switzerland: WHO, 2006.

42. Mattila MJ, Linnoila M, Seppala T, Koskinen R. Effect of aluminium hydroxide and glycopyrrhonium on the absorption of ethambutol and alcohol in man. Br J Clin Pharmacol 1978;5(2):161–6.

43. Karmon G, Savir H, Zevin D, Levi J. Bilateral optic neuropathy due to combined ethambutol and isoniazid treatment. Ann Ophthalmol 1979;11(7):1013–7.

44. Jimenez-Lucho VE, del Busto R, Odel J. Isoniazid and ethambutol as a cause of optic neuropathy. Eur J Respir Dis 1987;71(1):42–5.

45. Singhal KC, Rathi R, Varshney DP, Kishore K. Serum concentration and urinary excretion of ethambutol administered alone and in combination with isoniazid in patients of pulmonary tuberculosis. Indian J Physiol Pharmacol 1985;29(4):223–6.

46. Singhal KC, Varshney DP, Rathi R, Kishore K, Varshney SC. Serum concentration of isoniazid administered with & without ethambutol in pulmonary tuberculosis patients. Indian J Med Res 1986;83:360–2.

Ethionamide and protionamide

See also Antituberculosis drugs

General Information

Ethionamide is a synthetic derivative of thio-isonicotinamide. The initial oral dosage for adults is 250 mg/day, slowly increasing up to 15–20 mg/kg/day (maximum 1 g/day).

Protionamide is a pyridine derivative of ethionamide. The dosage is 250–300 mg/day.

Ethionamide and protionamide have often proved to be effective in non-tuberculous mycobacterial infections. Acute rheumatic symptoms and difficulty in the management of diabetes have been reported (1).

Organs and Systems

Nervous system

Mental depression, weakness, drowsiness, and hypotension are not rare in patients taking ethionamide or protionamide (1,2). Other neurological reactions include diplopia, olfactory disturbances, metallic taste, dizziness, paresthesia, headache, and tremor (3).

Psychological, psychiatric

Psychotic reactions have been described in patients taking ethionamide (4,5) and may be exacerbated by alcohol (6).

Endocrine

Goitrous hypothyroidism has rarely been described in patients taking ethionamide (7), with recovery after withdrawal (8). Ethionamide inhibits both the uptake of iodine and its incorporation into trichloroacetic acid-precipitable protein (9).

Metabolism

Thionamides lower the blood glucose concentration and also suppress appetite, which influences carbohydrate intake (10).

Gastrointestinal

Nausea, vomiting, and anorexia reflect gastric intolerance, and ethionamide is best taken with meals (1,2). Protionamide has better gastric tolerance, but patients with a history of stomach ulcer are susceptible to gastric problems (11).

Skin

Severe allergic skin reactions, alopecia, acne, and purpura can occur in patients taking ethionamide (1,12). Other adverse effects include acneiform eruptions, photodermatitis, and hair loss (13).

Sexual function

Gynecomastia, impotence, and menorrhagia have been observed in patients taking ethionamide (7,14).

Second-Generation Effects

Teratogenicity

Ethionamide is teratogenic (15,16), and neither ethionamide nor protionamide should be administered during the first trimester of pregnancy.

Susceptibility Factors

Hepatic disease

Since hepatitis occurs in about 5% of patients treated with ethionamide or protionamide, frequent monitoring of liver function is compulsory (17).

Drug–Drug Interactions

Rifamycins

When thionamides are used in combination with rifampicin, hepatotoxicity is more common and severe (18,19). There was a 13% incidence of hepatotoxicity in patients with multibacillary leprosy treated with dapsone, rifampicin, and protionamide 10 mg/kg/day, and a 17% incidence in 110 patients treated with dapsone, rifampicin, and protionamide 5 mg/kg/day; however, although the lower dose of protionamide did not reduce the incidence of hepatotoxicity, it did reduce its severity (20). Protionamide does not affect the pharmacokinetics of rifampicin (21).

References

1. Council on Drugs. Ethionamide. A tuberculostatic agent (Trecator). JAMA 1964;187(7):527.
2. Schwartz WS. Comparison of ethionamide with işoniazid in original treatment cases of pulmonary tuberculosis. XIV. A report of the Veterans Administration—Armed Forces cooperative study. Am Rev Respir Dis 1966;93(5):685–92.
3. Verbist L, Prignot J, Cosemans J, Gyselen A. Tolerance to ethionamid and PAS in original treatment of tuberculous patients. Scand J Respir Dis 1966;47(4):225–35.
4. Narang RK. Acute psychotic reaction probably caused by ethionamide. Tubercle 1972;53(2):137–8.
5. Sharma GS, Gupta PK, Jain NK, Shanker A, Nanawati V. Toxic psychosis to isoniazid and ethionamide in a patient with pulmonary tuberculosis. Tubercle 1979;60(3):171–2.
6. Lansdown FS, Beran M, Litwak T. Psychotoxic reaction during ethionamide therapy. Am Rev Respir Dis 1967;95(6):1053–5.
7. Moulding T, Fraser R. Hypothyroidism related to ethionamide. Am Rev Respir Dis 1970;101(1):90–4.
8. Drucker D, Eggo MC, Salit IE, Burrow GN. Ethionamide-induced goitrous hypothyroidism. Ann Intern Med 1984;100(6):837–9.
9. Danan G, Pessayre D, Larrey D, Benhamou JP. Pyrazinamide fulminant hepatitis: an old hepatotoxin strikes again. Lancet 1981;2(8254):1056–7.
10. Filla E, Comenale D. La terapia con PAS, rifamicina ed etionamide per via endovenosa e sue ripercussioni sul ricambio glucidico in diabetici affetti da tubercolosi polmonare. [Therapy with PAS, rifamycin and ethionamide administered intravenously and its effects on glucide metabolism in diabetics affected by pulmonary tuberculosis.] Minerva Med 1965;56(103):4570–3.
11. Samtsov VS. [Treatment of patients with pulmonary tuberculosis, complicated by peptic ulcer, with ethionamide and prothionamide.]Probl Tuberk 1973;51(7):67–70.
12. Desmons MT. Aspect pseudolupique provoqué par l'éthionamide (Trecator). Bull Soc Franc Derm Syph 1973;80:168.
13. Holdiness MR. Adverse cutaneous reactions to antituberculosis drugs. Int J Dermatol 1985;24(5):280–5.
14. Hussey HH. Editorial: Gynecomastia. JAMA 1974;228(11):1423.
15. Potworowska M. Leczenie etionamidem a ciaza. Gruzlica Chor Pluc 1966;34(4):345.
16. Jentgens H. Ethionamid und teratogene Wirkung. [Ethionamide and teratogenic effect.] Prax Pneumol 1968;22(11):699–704.
17. Mitchell I, Wendon J, Fitt S, Williams R. Anti-tuberculous therapy and acute liver failure. Lancet 1995;345(8949):555–6.
18. Braude AL, Davis Ch E, Fierer J. In: Infectious Diseases and Medical Microbiology. 2nd ed.. Philadelphia: WB Saunders, 1966:1171.
19. Pattyn SR, Janssens L, Bourland J, Saylan T, Davies EM, Grillone S, Feracci C. Hepatotoxicity of the combination of rifampin–ethionamide in the treatment of multibacillary leprosy. Int J Lepr Other Mycobact Dis 1984;52(1):1–6.
20. Cartel JL, Naudillon Y, Artus JC, Grosset JH. Hepatotoxicity of the daily combination of 5 mg/kg prothionamide + 10 mg/kg rifampin Int J Lepr Other Mycobact Dis 1985;53(1):15–8.
21. Mathur A, Venkatesan K, Girdhar BK, Bharadwaj VP, Girdhar A, Bagga AK. A study of drug interactions in leprosy—1. Effect of simultaneous administration of prothionamide on metabolic disposition of rifampicin and dapsone. Lepr Rev 1986;57(1):33–7.

Fluoroquinolones

See also Alatrofloxacin and trovafloxacin, Ciprofloxacin, Garenoxacin, Gatifloxacin, Gemifloxacin, Grepafloxacin, Levofloxacin, Lomefloxacin, Moxifloxacin, Norfloxacin, Ofloxacin, Pazufloxacin, Pefloxacin, Sitafloxacin, Sparfloxacin, Tosufloxacin

General Information

Following the introduction of the first quinolone antibiotic (nalidixic acid, see separate monograph), structural modifications to the basic quinolone and naphthyridone nucleus and to the side-chains produced fluoroquinolones with improved coverage of bacterial pathogens, with high activity against Gram-negative species and a number of atypical pathogens, and good-to-moderate activity against Gram-positive species. However, despite their broad spectrum and clinical success, defects became evident, and compounds developed in recent years have targeted improvements in pharmacokinetic properties (improved systemic availability, once-daily dosing), greater activity against Gram-positive cocci and anerobes, activity against fluoroquinolone-resistant strains, and better coverage of non-fermenting Gram-negative species (1–4).

However, owing to adverse effects (including severe anaphylaxis, QT interval prolongation, and potential cardiotoxicity), several fluoroquinolones have had to be withdrawn (for example temafloxacin and grepafloxacin) or strictly limited in their use (for example trovafloxacin) after marketing. A serious idiosyncratic reaction profile is possibly related to the immunologically reactive 1-difluorophenyl substituent that characterizes temafloxacin, trovafloxacin, and tosufloxacin (5).

The following fluoroquinolones are covered in separate monographs: alatrofloxacin and trovafloxacin, ciprofloxacin, garenoxacin, gatifloxacin, gemifloxacin, grepafloxacin, levofloxacin, lomefloxacin, moxifloxacin, norfloxacin, ofloxacin, pazufloxacin, pefloxacin, prulifloxacin, sitafloxacin, sparfloxacin, and tosufloxacin (all rINNs).

Fluoroquinolones are administered by several routes (6). The duration of therapy can vary from a single dose to several weeks or months. Owing to their adverse effects (including severe anaphylaxis, QT interval prolongation, and potential cardiotoxicity), several fluoroquinolones have had to be withdrawn (for example temafloxacin and grepafloxacin) or strictly limited in their use (for example trovafloxacin) after marketing (7). A serious idiosyncratic reaction profile may be related to the immunologically reactive 1-difluorophenyl substituent that characterizes temafloxacin, trovafloxacin, and tosufloxacin (1).

Withdrawal of fluoroquinolones

Owing to adverse effects (including severe anaphylaxis, QT interval prolongation, and potential cardiotoxicity), several fluoroquinolones have had to be withdrawn (for example temafloxacin and grepafloxacin) or strictly limited in their uses (for example trovafloxacin) after marketing. A serious idiosyncratic reaction profile is possibly related to the immunologically reactive 1-difluorophenyl substituent that characterizes temafloxacin, tosufloxacin, and trovafloxacin (8).

The withdrawal of temafloxacin in 1992, only 6 months after its introduction, followed the observation of serious adverse events that were labeled the "temafloxacin syndrome" (9). Adverse effects, including hemolysis, renal dysfunction, coagulopathy, and hepatic dysfunction, were estimated to occur in one in 3500 patients. For comparison, incidence rates for these adverse events were about one in 17 000 patients treated with ciprofloxacin and one in 33 000 patients treated with ofloxacin.

In 1999 trovafloxacin was withdrawn after reports of lethal hepatic damage. Before this development, several studies had been published describing impressive clinical efficacy coupled with an almost flawless safety profile (10,11).

Grepafloxacin was withdrawn from the market in 1999 because of its adverse cardiovascular events, which included dysrhythmias (12).

Clinafloxacin has been withdrawn because of phototoxicity and hypoglycemia and sparfloxacin because of phototoxicity (13).

These experiences demonstrate the need to be vigilant about any untoward events associated with large-scale use of new fluoroquinolones.

Comparative studies

In a crossover, randomized study, 16 fasted volunteers (8 men, 8 women) took single oral doses of gemifloxacin 320 mg and ofloxacin 400 mg on two separate occasions, in order to assess urinary excretion (14). Urine concentrations of ofloxacin were higher than those of gemifloxacin. There were no adverse effects.

Quinolones have been compared retrospectively in 11 controlled trials in acute cystitis in 7535 women (15). Photosensitivity reactions were more common with sparfloxacin than ofloxacin. Adverse cutaneous events and photosensitivity causing withdrawal were more common with lomefloxacin than ofloxacin. Nervous system adverse events and insomnia were reported more frequently with rufloxacin and enoxacin than with pefloxacin or ciprofloxacin.

General adverse effects

In spite of widely publicized negative experiences, most fluoroquinolones are safe, and adverse events are, compared with other groups of antimicrobial agents, relatively rare (16). The rates of overall adverse events were similar in several comparisons of individual quinolones among themselves or with other antimicrobial agents. In one multicenter, double-blind, randomized study drug-related adverse events were reported by 8.9% of those taking levofloxacin and 8.2% of those taking ciprofloxacin for 7–10 days (17). A meta-analysis of data from 20 phase II and III studies in 4926 patients treated with moxifloxacin showed that adverse events led to withdrawal of treatment in 3.8% of patients (18). In a comparison of grepafloxacin and ciprofloxacin for exacerbations of chronic bronchitis withdrawal was precipitated by adverse events in 3% of 624 patients; there was no difference between ciprofloxacin and grepafloxacin (19). The most frequent causes of withdrawal were nausea, vomiting, dysgeusia, dizziness, and diarrhea. In smaller studies, adverse events are occasionally observed more often. In a prospective, double-blind comparison of ciprofloxacin, ofloxacin, and co-trimoxazole for short-course treatment of acute urinary tract infections, there were drug-related adverse events in 26% of patients treated with ciprofloxacin and 34% of patients treated with ofloxacin (20).

If analysis of the frequency of adverse events is based on prescription event monitoring (PEM), the rates of adverse events are usually substantially lower. In one study, over 11 000 patients taking each antibiotic were monitored (21). Among the fluoroquinolones, ciprofloxacin, norfloxacin, and ofloxacin were used. Adverse events resulted in withdrawal of norfloxacin or ofloxacin in under 1% of patients (21).

The main systems affected by adverse effects of the quinolones are the skin, liver, and nervous system. The best-known adverse effect is phototoxicity, the risk of which varies markedly among the quinolones; lomefloxacin and sparfloxacin carry a particularly high risk. The development of phototoxicity is based on an interaction between light and the drug. Neurotoxicity also occurs, with marked variation of incidence between the various compounds. Hypersensitivity reactions to quinolones are rare, and include anaphylactic shock and anaphylactoid reactions. Organ-specific reactions attributed to hypersensitivity involve the liver and kidneys. If hypersensitivity reactions occur, switching from one quinolone compound to another is probably not advisable, since there is cross-reactivity.

The risk of neoplastic disease is minimal, even during long-term use, but may be increased by exposure to UVA light.

Using electron spin resonance spectroscopy and spin trapping, ciprofloxacin has been shown to cause free radical production in a dose- and time-dependent manner; the authors suggested that this effect may contribute to drug-related adverse effects, including phototoxicity and cartilage defects (22).

The most common drug interactions include malabsorption interactions associated with multivalent cations and CYP450 interactions (23).

Pharmacoeconomics

The pharmacoeconomic impact of adverse effects of antimicrobial drugs is enormous. Antibacterial drug reactions account for about 25% of adverse drug reactions. The adverse effects profile of an antimicrobial agent can contribute significantly to its overall direct costs (monitoring costs, prolonged hospitalization due to complications or treatment failures) and indirect costs (quality of life, loss of productivity, time spent by families and patients receiving medical care). In one study an adverse event in a hospitalized patient was associated on average with an excess of 1.9 days in the length of stay, extra costs of $US2262 (1990–93 values), and an almost two-fold increase in the risk of death. In the outpatient setting, adverse drug reactions result in 2–6% of hospitalizations, and most of them were thought to be avoidable if appropriate interventions had been taken. In a review, economic aspects of antibacterial therapy with fluoroquinolones have been summarized and critically evaluated (24).

Organs and Systems

Cardiovascular

grepafloxacin = moxifloxacin > ciprofloxacin (25). In guinea-pig ventricular myocardium sparfloxacin prolonged the action potential duration by about 8% at 10 µmol/l and 41% at 100 µmol/l (26). Gatifloxacin, grepafloxacin, and moxifloxacin were less potent, but prolonged the action potential duration at 100 µmol/l by about 13%, 24%, and 25% respectively. In contrast, ciprofloxacin, gemifloxacin, levofloxacin, sitafloxacin, tosufloxacin, and trovafloxacin had little or no effect on the action potential at concentrations as high as 100 µmol/l.

Preclinical and clinical trial data and data from phase IV studies have shown that levofloxacin, moxifloxacin, and gatifloxacin cause prolongation of the QT interval, but that the potential for torsade de pointes is rare and is influenced by several independent variables (for example concurrent administration of class Ia and III antidysrhythmic agents) (27). There is a moderate increase in the QT interval associated with sparfloxacin, averaging 3%, and the few serious adverse cardiovascular events that have been reported during postmarketing surveillance all occurred in patients with underlying heart disease (28).

In patients taking quinolones (ciprofloxacin 11 477, enoxacin 2790, ofloxacin 11 033, and norfloxacin 11 110; mean ages 49–57 years) there was no evidence of drug-induced dysrhythmias associated with enoxacin within 42 days of drug administration (29). Of the other quinolones, atrial fibrillation was reported most often within 42 days of ciprofloxacin administration, with no change in event rate over that time. The crude rate of palpitation did not change significantly with ciprofloxacin, norfloxacin, or ofloxacin. Syncope and tachycardia were also reported with ciprofloxacin and ofloxacin. There was no evidence of drug-induced hepatic dysfunction within 42 days of drug administration with any of the quinolones used.

In a retrospective database analysis 25 cases of torsade de pointes associated with ciprofloxacin ($n = 2$), ofloxacin ($n = 2$), levofloxacin ($n = 13$), and gatifloxacin ($n = 8$) were identified in the USA (30). Ciprofloxacin was associated with a significantly lower rate of torsade de pointes (0.3 cases/10 million prescriptions) than levofloxacin (5.4/10 million) or gatifloxacin (27/10 million). When the analysis was limited to the first 16 months after initial approval of the drug, the rates for levofloxacin (16/10 million) and gatifloxacin (27/10 million) were similar.

Nervous system

The main central nervous system adverse effects of the fluoroquinolones include dizziness, convulsions, psychosis, and insomnia; levofloxacin, moxifloxacin, and ofloxacin are reportedly least likely to cause these effects, based on a study of European and international data from about 130 million prescriptions (31).

A review has suggested that fluoroquinolone-associated peripheral nervous system events are mild and short-term (32). Among 60 courses of quinolones in 45 patients (levofloxacin 33 courses, ciprofloxacin 11 courses, ofloxacin 6 courses, lomefloxacin 1 course, trovafloxacin 1 course; in eight cases the same antibiotic was prescribed twice) there were 36 severe events that typically involved multiple organ systems. The symptoms lasted more than 3 months in 71% of cases and more than 1 year in 58%. The onset of the adverse events in the 45 patients was usually rapid: 15 events began within 24 hours of the start of treatment, 26 within 72 hours, and 38 within 1 week.

Dizziness is not rare in patients taking fluoroquinolones, and was observed in 2.8% of patients taking moxifloxacin (18), in 8 and 9% of patients taking grepafloxacin

400 mg/day and 600 mg/day, and in 6% of patients taking ciprofloxacin (19). Prescription event monitoring found markedly lower rates of this adverse event during treatment with ciprofloxacin, norfloxacin, or ofloxacin (21).

Headache was recorded in 8% of patients taking ciprofloxacin and 9% taking ofloxacin during short-course treatment of urinary tract infections (20). Similar rates are reported for grepafloxacin (19).

In two patients with myasthenia gravis, ciprofloxacin and norfloxacin exacerbated the symptoms (33).

Encephalopathy with unconsciousness has been reported in a 48-year-old woman with Machado–Joseph disease after the administration of fleroxacin (200 mg/day) for 3 days (34). She recovered after withdrawal. The serum and cerebrospinal fluid concentrations of fleroxacin were within normal limits.

Seizures

Quinolone antibiotics vary in their ability to cause seizures. Trovafloxacin has the greatest potential and levofloxacin possibly the least. Ciprofloxacin can cause confusion and general seizures (35,36). Seizures also occurred in patients taking ofloxacin (37) and levofloxacin (36). However, this must be rare, since there have only been isolated case reports.

Among over 30 000 patients treated with ciprofloxacin, norfloxacin, or ofloxacin, no seizures were detected during prescription event monitoring (21). The risk of seizures during treatment with individual quinolones is currently unknown. Electrophysiological field potentials in animals are affected to varying degrees by different quinolones; the smallest effect was observed with ofloxacin, followed by ciprofloxacin and nalidixic acid, whereas there was an increasing excitatory effect with clinafloxacin, enoxacin, fleroxacin, lomefloxacin, moxifloxacin, and trovafloxacin (38). The pathophysiological basis for the triggering of seizures probably lies in the binding of fluoroquinolones to $GABA_A$ receptors in the brain, blocking the natural ligand GABA; this results in nervous system stimulation (39). Binding to this receptor is strongly influenced by the side chain in the 7-position; quinolones with bulky moieties, such as temafloxacin and sparfloxacin, bind less efficiently to GABA receptors.

Sensory systems

Taste

Dysgeusia is reported with many quinolones. This was observed more often with higher doses of grepafloxacin (600 mg/day; 17%) than with lower doses (400 mg/day; 9%) (40). In another study there was a similar dose–response relation: 13% of patients taking 400 mg/day and 27% of those taking 600 mg/day (19).

In 135 unmedicated young volunteers, 13 elderly volunteers and 14 unmedicated HIV-infected patients, enoxacin applied to the tongue was described as metallic by young subjects, but bitter by elderly subjects; leflxacin and ofloxacin were described as bitter (41).

Psychological, psychiatric

Isolated cases of depression (20) and psychosis have been described in temporal association with fluoroquinolones (42). In a retrospective study the data on fluoroquinolones and other antibacterial drugs, rufloxacin was associated with a reporting rate of 221 reports/daily defined dose/1000 inhabitants/day, and the most frequent were psychiatric disorders (43).

Metabolism

There was a higher rate of hyperglycemia with gatifloxacin or levofloxacin compared with ceftriaxone in a retrospective chart review of 17 000 patients (44). Sulfonylurea therapy was identified as an independent risk factor for hypoglycemia.

Hematologic

In over 33 000 patients treated with ciprofloxacin, norfloxacin, or ofloxacin, no case of hemolytic anemia was discovered by prescription event monitoring (21). However, ciprofloxacin has been associated with hemolysis in combination with a severe skin reaction in a young adult (45).

Leukopenia has been observed in 0.1–0.7% of patients given fluoroquinolones in Japan and eosinophilia was observed in 0.5–2.2% of these patients (46). Leukopenia was generally mild and reversible after dosage reduction or withdrawal (47,48).

Anemia, thrombocytopenia, and thrombocytosis have only rarely been reported (49–52).

Gastrointestinal

Adverse gastrointestinal events are not uncommon during treatment with fluoroquinolones. There may be some dose-dependency, since with grepafloxacin 600 mg/day the rates of the following adverse events were noticeably higher than with 400 mg/day: nausea (15 versus 11%), vomiting (6 versus 1%), and diarrhea (4 versus 3%) (40). In a randomized, double-blind comparison of prulifloxacin 600 mg/day and ciprofloxacin 500 mg bd in 235 patients with acute exacerbations of chronic bronchitis, the most common treatment-related adverse event was gastric pain of mild or moderate intensity, reported in 8.5% of the patients taking prulifloxacin and 6.8% of those taking ciprofloxacin (53).

Liver

Transient rises in serum transaminase and serum alkaline phosphatase activities have been observed with all fluoroquinolones. This occurred in 0.9–4.3% of patients in Japan. In the vast majority of the cases this alteration was self-limited and reversible without withdrawal of the drug (46).

Urinary tract

In a Medline search to investigate the incidence and features of fluoroquinolone nephrotoxicity only primarily case reports and temporally related events could be identified (54). Ciprofloxacin was associated with an increased risk of renal insufficiency, probably because it has been in use longer and more widely than the newer agents. Raised serum creatinine or blood urea nitrogen concentrations have been observed in 0.1–0.7% of patients (46).

Single cases of reversible, acute non-oliguric or oliguric renal insufficiency, probably due to tubulointerstitial nephritis, have been reported (55,56), as well as isolated cases of hematuria (57). The newer fluoroquinolones have only rarely been associated with nephrotoxicity, with an estimated incidence of 0.4–0.8%.

Rare cases of possibly drug-related crystalluria have been described in patients taking fluoroquinolones (58–60).

Skin

Skin rashes are relatively common with fluoroquinolones. A retrospective cohort study in patients in general practice in the Netherlands focused on the use of antibacterial agents and the occurrence of adverse cutaneous events covered 469 505 consultations with 87 475 patients, of whom 13 679 received prescriptions for antibiotics (61). After adjustment for age, sex, and co-medications, the incidence density ratio (incidence density per 1000 exposed days) for various groups of antibacterial agents was as follows: tetracyclines 1.0, macrolides 1.1, fluoroquinolones 2.8, penicillins 2.9, and co-trimoxazole 4.4 (61). No details of the types of skin reactions were given, and it is therefore possible that phototoxic events were included. Compared with other studies, the reported rate of antibiotic-associated adverse cutaneous events in this outpatient population was rather low.

Exposure to light causes photosesitivity reactions in patients taking quinolones and fluoroquinolones. In guinea-pigs the phototxic potencies were: enoxacin, lomeflox > ofloxacin > nalidixic acid, tosufloxacin > norfloxacin, ciprofloxacin (62). Photosensitivity reactions to ofloxacin may be initiated by oxygen radicals and/or by ofloxacin radicals acting as haptens (63). In a mouse model pefloxacin and ciprofloxacin augmented the effect of ultraviolet A by increasing sunburn and apoptosis, depleting Langerhans cells, and suppressing local immune responses (64).

A combination of primary ear swelling analysis and cell counting of ear-draining lymph nodes after UV irradiation in mice was fast and highly predictive of the risks of photosensitization and photoirritancy of fluoroquinolones, depending on the route of exposure (oral or dermal) and may therefore be good tools for preclinical risk assessment in terms of discriminating photoreactions (65).

Tosufloxacin has been associated with a fixed drug eruption (66).

In a phase III, randomized, investigator-blinded comparison of the safety and efficacy of clinafloxacin with those of piperacillin/tazobactam in the treatment of adults with severe skin and soft-tissue infections, four of 84 patients randomized to clinafloxacin developed phototoxicity (67).

The incidence of cutaneous hypersensitivity reactions (erythema, pruritus, urticaria, rash) with quinolones is low (0.4–2.2%) (68).

Photosensitivity

Photosensitivity is a common adverse effect of the fluoroquinolones. It results from an abnormal reaction of the skin to natural or artificial light sources, usually associated with the UVA part of the electromagnetic spectrum (315–400 nm), mediated by the absorption of light energy into fluoroquinolones, followed by degradation of the molecule and formation of cytotoxic photoproducts (69). The chemical structure at position 8 in the quinolone ring probably determines the phototoxic potential, since the introduction of a methoxy group at this position markedly reduces the phototoxicity of individual drugs (70).

The early fluoroquinolones, that is ciprofloxacin, norfloxacin, and ofloxacin, are mild photosensitizers (71–75). Clinically relevant photosensitivity was generally only observed when high dosages were used and the patient was exposed to large amounts of sunlight. The phototoxic potential of lomefloxacin, sparfloxacin, and fleroxacin is far greater than that of earlier compounds (76–81). However, even with fluoroquinolones that are reported to have a relatively high rate of phototoxicity, the incidence is low. In a Japanese study of 4276 patients, photosensitivity was found in 44 (1.03%) and was typically not severe (82). Patients with co-morbidity or those over 60 years were at higher risk. The risk may be further increased by a long duration of treatment.

Experience reported to date suggests that for gatifloxacin, gemifloxacin, and moxifloxacin phototoxicity occurs at a lower rate than with widely used fluoroquinolones such as ciprofloxacin and levofloxacin (83).

Enoxacin 200 mg tds caused significant phototoxicity in healthy volunteers (84). The photosensitizing effect of grepafloxacin is relatively weak and similar to that of ciprofloxacin (40). In one study, one of 207 patients taking grepafloxacin 400 mg/day and six of 204 taking 600 mg/day developed phototoxicity (19). Moxifloxacin is not phototoxic (85). Data obtained in albino mice have suggested that the phototoxic potential of sitafloxacin is milder than that of lomefloxacin or sparfloxacin (86).

Musculoskeletal

Tendinopathy

Tendinopathy and partial or complete tendon rupture as adverse events of fluoroquinolones have been reported during or shortly after the use of fluoroquinolones (87–91). In the Achilles tendon tendonitis starts with swelling and inflammation of the tendon followed by pain. In 50% of cases, it is bilateral. The onset is at 1–152 days. Tendon rupture has been observed as late as 120 days after treatment. Pefloxacin and ofloxacin have been implicated, as has ciprofloxacin (92,93). In six patients taking fluoroquinolones risk factors included renal insufficiency, glucocorticoid therapy, secondary hyperparathyroidism, advanced age, and diabetes mellitus (94). Cases have also been reported among immunocompromised renal transplant recipients (95).

In 42 spontaneous reports of fluoroquinolone-associated tendon disorders, 32 patients had tendinitis, 24 bilaterally, and 10 had a tendon rupture; most affected the Achilles tendon (96). The median age was 68 years and there was a male predominance. In 16 cases ofloxacin was implicated, in 13 ciprofloxacin, in eight norfloxacin, and in five pefloxacin. The delay between the start of treatment and the appearance of the first symptoms was 1–510 (median 6) days. Most patients recovered within 2 months after withdrawal, but 26% had not yet recovered at follow-up.

In a retrospective analysis, quinolone arthropathy developed during the first 3 weeks, and depended on the patient's age and history (97). It resolved fully within 7 days to 3 months after drug withdrawal.

In a study of fibroblast metabolism in vitro, ciprofloxacin stimulated matrix-degrading protease activity and inhibited fibroblast metabolism; these effects may both contribute to the tendinopathy that is associated with fluoroquinolones (98).

Animal studies show that the propensity to induce tendon lesions varies among fluoroquinolones. Fleroxacin, pefloxacin, levofloxacin, and ofloxacin were the most toxic, while enoxacin, norfloxacin, and ciprofloxacin had little or no effect (99). The structure of the substituent at position 7 seems to be a determining factor. Quinolones with toxic effects in animals share a methylpiperadinyl substituent at the position 7 of the fluoroquinolone core structure, while this position is occupied by a piperadinyl substituent in ciprofloxacin, norfloxacin, and enoxacin (99). It is recommended that quinolones be discontinued at the first sign of tendon pain or inflammation and that the patient should not exercise until the diagnosis of tendonitis can be confidently excluded.

The quinolones are toxic to immature cartilage and also affect the epiphyseal growth plate. The mechanism of cartilage toxicity is chelation and depletion of magnesium in chondrocytes (100). In rodents, pefloxacin (400 mg/kg for several days) caused oxidative damage to the type I collagen in the Achilles tendon; these alterations were identical to those observed in experimental tendinous ischemia and a reperfusion model (101). Oxidative damage was prevented by the co-administration of N-acetylcysteine (150 mg/kg).

Arthropathy

Arthropathy has been reported with various fluoroquinolones, particularly in children. In a Russian study in children with cystic fibrosis five were withdrawn (four taking ciprofloxacin and one taking pefloxacin) (102). Two had an arthropathy that was drug- and age-dependent. Quinolone-induced arthropathy was more common with pefloxacin and occurred only in children over 10 years old with a history of joint problems. The arthropathy fully

recovered within 7 days to 3 months and there was no cartilage damage.

In Wistar rats treated with ciprofloxacin, there was poor healing of experimental fractures during the early stages of repair, suggesting that fluoroquinolones may compromise the clinical course of fracture healing (103).

Sexual function

In a large study vaginitis and vulvitis were detected by prescription event monitoring in a significantly higher proportion of women taking fluoroquinolones than women taking azithromycin or cefixime (21).

Immunologic

Fluoroquinolones have immunomodulatory effects, at least partly at the gene transcription level, demonstrated by inhibition of cytokine (IL-1α, TNF-α, IL-6, and IL-8) mRNA and cytokine (IL-1α and IL-1β) concentrations by grepafloxacin (1–30 mg/l) in vitro (104).

The frequency of fluoroquinolone-associated anaphylaxis has been estimated at 1.8–23 per 10 million days of treatment, based on spontaneous reports (105). All fluoroquinolone-associated cases of IgE- and non-IgE-mediated anaphylactic reactions spontaneously reported to the Federal Institute for Drugs and Medical Devices in Germany between 1 January 1993 and 31 December 2004 were identified and assessed with regard to the correctness of the diagnosis. In 166 of 204 cases, the diagnosis of anaphylaxis and a causal relation with the drug were considered at least possible. Moxifloxacin, levofloxacin, ciprofloxacin, and ofloxacin respectively accounted for 90 (54%), 25 (15%), 21 (13%), and 16 (10%) of the 166 cases; the corresponding reporting rates per 1 million defined daily doses based on crude estimates of exposure were 3.3, 0.6, 0.2, and 0.2. Anaphylaxis after the first dose or within the first 3 days was reported in 71 of 166 (43%) cases, but there was no information about prior exposure with this or any other fluoroquinolone. In 21 (13%) of the 166 cases, the reaction occurred within the first 3 days and it was stated that the particular fluoroquinolone had never been taken before.

Anaphylactic shock associated with cinoxacin was reported in three patients by the Netherlands Center for Monitoring of Adverse Reactions to Drugs (106). Another 17 cases were reported to the WHO Collaborating Center for International Drug Monitoring. In some cases the reaction was observed immediately after the first dose of a repeat cycle of treatment. Anaphylactoid reactions to ciprofloxacin have been reported in patients with cystic fibrosis (107–109).

Organ-specific reactions attributed to hypersensitivity involve the liver and kidneys. In one instance centrilobular hepatic necrosis developed during treatment of a urinary tract infection with ciprofloxacin (110). Among 14 cases of drug-induced allergic nephritis, two were associated with quinolones (111). Isolated case reports of allergic nephropathy associated with norfloxacin and ciprofloxacin therapy suggest that this type of reaction is probably very rare (112), since the authors were able to find only 28 other reported cases. If hypersensitivity reactions occur, switching from one quinolone compound to another is probably not advisable, since oral provocation using different agents reproduced the initially observed hypersensitivity reaction (113). This clinical observation is supported by the results of in vitro lymphocyte transformation tests, which were positive with ofloxacin in two patients with allergy to ciprofloxacin (114).

Infection risk

Ciprofloxacin, norfloxacin, and ofloxacin increased gastrointestinal colonization by *Candida albicans* in 17 patients (115). Ciprofloxacin caused the largest increase, but the difference was not statistically significant.

In a retrospective cohort study of patients hospitalized in a Canadian teaching hospital during January 2003 to June 2004 there were 7421 episodes of care in 5619 patients, who were observed until they developed *Clostridium difficile*-associated diarrhea, or died, or for 60 days after discharge (116). Fluoroquinolones were the antibiotics that were most strongly associated with *Clostridium difficile*-associated diarrhea (adjusted hazard ratio = 3.44; 95% CI = 2.65, 4.47). Almost one-quarter of all in-patients received quinolones, for which the population-attributable fraction of *Clostridium difficile*-associated diarrhea was 36%. All three generations of cephalosporins, macrolides, clindamycin, and intravenous beta-lactam/beta-lactamase inhibitors were intermediate-risk antibiotics, with similar hazard ratios (1.56–1.89).

Long-Term Effects

Drug tolerance

To study the impact on the normal intestinal microflora, gatifloxacin was given to 18 healthy volunteers (400 mg/day orally). In the aerobic intestinal microflora *E. coli* strains were eliminated or strongly suppressed and the number of enterococci fell significantly, while the number of staphylococci increased. In the anerobic microflora the numbers of *Clostridia* and fusobacteria fell significantly. The microflora normalized 40 days after the gatifloxacin withdrawal. No selection or overgrowth of resistant bacterial strains or yeasts occurred (117).

Salmonella typhimurium DT104 is usually resistant to ampicillin, chloramphenicol, streptomycin, sulfonamides, and tetracycline. An outbreak of 25 culture-confirmed cases of multidrug-resistant *Salmonella typhimurium* DT104 has been identified in Denmark. The strain was resistant to the abovementioned antibiotics and nalidixic acid and had reduced susceptibility to fluoroquinolones. A swineherd was identified as the primary source (118). The DT104 strain was also found in cases of salmonellosis in Washington State, and soft cheese made with unpasteurized milk was identified as an important vehicle of its transmission (119).

From 8419 worldwide clinical isolates of *Streptococcus pneumoniae* associated with lower respiratory tract or blood infections obtained from 519 geographically distinct hospital laboratories during 1997–98, 69 had reduced susceptibility or resistance to fluoroquinolones. Only mutations in parC and gyrA (especially in combination), but not in gyrB or parE, contributed significantly to resistance. Efflux is probably crucial in reduced susceptibility for new hydrophilic fluoroquinolones (120).

In an in vitro study, ciprofloxacin, grepafloxacin, levofloxacin, moxifloxacin, ofloxacin, and sparfloxacin had similar good activity against *Haemophilus influenzae* and *Moraxella catarrhalis* (121). Against *S. pneumoniae* (irrespective of the strain's susceptibility to penicillin), grepafloxacin, levofloxacin, moxifloxacin, and sparfloxacin had better activity than ciprofloxacin and ofloxacin.

Clinafloxacin, moxifloxacin, sparfloxacin, and trovafloxacin were significantly more active in vitro than ciprofloxacin and levofloxacin against *Stenotrophomonas maltophilia*, a microorganism with inherent resistance to many antibiotics; new-generation quinolones may become very useful in the treatment of certain severe or life-threatening infectious conditions due to this bacterium (122).

In over 90 000 routine samples of *Escherichia coli* in five Dutch laboratories during 1989–98 resistance to norfloxacin increased from 1.3% to 5.8% (123). In addition, multiresistance, defined as resistance to norfloxacin and at least two other antibiotics (from a group consisting of amoxicillin, trimethoprim, and nitrofurantoin), increased from 0.5% to 4.0%.

Mutagenicity

In vitro lomefloxacin photochemically produced oxidative DNA damage, an effect known to be of mutagenic potential (124). This may be the basis of the photochemical mutagenicity and photochemical carcinogenicity of quinolones.

Tumorigenicity

Results of carcinogenicity studies have suggested that the risk of neoplastic disease is minimal, even during long-term use (125). However, the risk may be increased by exposure to UVA light. Skin tumors, only a minority of them malignant, developed in mice after treatment with various quinolones for up to 78 weeks and exposure to UV light. It therefore appears that fluoroquinolones have the potential to enhance the UVA-induced phototumorigenic effect (126).

Second-Generation Effects

Teratogenicity

Possible adverse outcomes were investigated in the newborns of 38 mothers who had received quinolones during pregnancy between 1989 and 1992. The majority (35 women) took norfloxacin or ciprofloxacin for urinary tract infections during the first trimester of pregnancy. There were no malformations in the quinolone group, whereas one child in the control group had a ventricular septal defect. There were no differences between the groups in the achievement of developmental milestones or in the musculoskeletal system (127). However, the authors of the study pointed out that the duration of follow-up may have been too short to draw firm conclusions regarding the safety of quinolones during pregnancy. In addition, the size of the study population was probably inadequate for statistically meaningful comparisons.

There was no increased risk of overall malformations after prenatal exposure to fluoroquinolones in a database cohort study of 217 women before or during pregnancy (128). However, there were more cases of bone malformations in the children of women who had been exposed before or during the first 30 days of pregnancy.

Treatment with fluoroquinolones during embryogenesis was not associated with an increased risk of major malformations in a multicenter, prospective, controlled trial in 400 women (129). There was a higher rate of therapeutic abortions in quinolone-exposed women. This may have been explained by the misperception of a major risk related to quinolones during pregnancy.

In pregnant rats fed with norfloxacin, there was DNA fragmentation in fetal tissues (16). Even though this effect was attributed to general toxicity rather than genotoxicity, quinolones should be avoided during pregnancy (130).

Lactation

Given the potential for untoward effects on the cartilage of breast-feeding infants, fluoroquinolones should be avoided by nursing mothers.

Susceptibility Factors

Genetic factors

The pharmacokinetics of fleroxacin (200 mg intravenously or 200 mg orally) have been studied in 19 Nigerian men. C_{max} and AUC were 3–4 fold lower than previously reported after identical doses, but systemic availability profile was as previously reported (131).

Age

Children
The use of fluoroquinolones in children is controversial. Fluoroquinolones have greatly facilitated the management of exacerbations of pulmonary infections in cystic fibrosis. Apart from this indication, they have rarely been used in children for other indications, such as urinary tract infection, prophylaxis in neutropenia, gastrointestinal infections, and nervous system infections (132). Thorough reviews of published experience suggest that fluoroquinolones can be used without increased risks of short-term or long-term adverse events. A study of adverse events within 45 days of receiving a prescription for ciprofloxacin was remarkable for the fact that among over 1700 children, no cases of newly diagnosed acute arthritis or other serious disturbances of liver or kidney function were

recorded (133). One patient with hemolytic–uremic syndrome had an exacerbation that was possibly caused by ciprofloxacin.

Renal disease

Based on their predominant renal elimination, dosage adjustment is necessary in the presence of renal disease for ciprofloxacin, gatifloxacin, levofloxacin, and sitafloxacin (134).

Hepatic disease

The penetration of routinely used fluoroquinolones into ascitic fluid after intravenous administration has been studied in patients with uncompensated hepatic cirrhosis (135). Three patients received three doses of ciprofloxacin 200 mg, six received three doses of ciprofloxacin 300 mg, seven received three doses of pefloxacin 400 mg, and six received three doses of ofloxacin 400 mg. Pefloxacin and ofloxacin produced serum and ascitic fluid concentrations above the MICs of the common pathogens that cause spontaneous bacterial peritonitis, and the authors concluded that they could be given to cirrhotic patients in dosage regimens similar to those in patients with normal hepatic function.

Other features of the patient

Given the well-described risk of phototoxicity associated with fluoroquinolones, sunlight and direct exposure to the sun are risk factors. Patients therefore need to be instructed about protective measures.

In patients with leg ischemia fleroxacin diffused into both ischemic and non-ischemic tissues (bone, subcutaneous fat, muscle, and tendons) after a 400 mg intravenous dose (136). Since the maximum antibiotic concentrations were lower than the MICs of various relevant pathogens, the dose used for perioperative prophylaxis in these patients should be increased to 800 mg.

Drug–Drug Interactions

Antacids

Concurrent treatment with antacids reduces the oral absorption of many quinolones, such as ciprofloxacin and enoxacin (137), moxifloxacin (18), norfloxacin (138), ofloxacin (139), and sparfloxacin (140).

Caffeine

CYP1A2 participates in the metabolism of both enoxacin and caffeine, and inhibition of caffeine metabolism by enoxacin can cause adverse effects (141).

In 24 healthy volunteers, 12 men and 12 women, the women had significantly different caffeine pharmacokinetics in the presence of ciprofloxacin and fleroxacin compared with the men (142). There were also significant differences between the sexes in the pharmacokinetics of ciprofloxacin and fleroxacin in the presence of caffeine.

The differences were in part due to different body weights.

Clinafloxacin 200 or 400 mg reduced the mean clearance of caffeine by 84% (143).

Cimetidine

Cimetidine reduces the metabolic clearance of both enoxacin and pefloxacin by about 20% (144,145).

Digoxin

The pharmacokinetics of digoxin were not altered by sparfloxacin (146).

Drugs that prolong the QT interval

In order to prevent any untoward effect on cardiac repolarization, agents that prolong the QT interval, such as terfenadine, astemizole, erythromycin, and agents with Class Ia and Class III antidysrhythmic activity, should not be used together with sparfloxacin (147).

Furosemide

The renal excretion of lomefloxacin and ofloxacin was reduced and plasma concentrations increased by furosemide, resulting in higher concentrations (148).

Iron

All of the fluoroquinolones interact with polyvalent cations, and their systematic availability is reduced by 50% when they are co-administered with iron compounds; ciprofloxacin and moxifloxacin (18) are more affected than gemifloxacin or levofloxacin (149).

Ferrous salts reduce the systemic availability of fluoroquinolones (137), including sparfloxacin (150).

NSAIDs

Co-administration of fenbufen and fluoroquinolones has been associated with seizures (151). The structure at the 7-position greatly affects the risk of NSAID-potentiated nervous system effects. Fluoroquinolones with unsubstituted piperazinyl rings (ciprofloxacin, enoxacin, and norfloxacin) have a strong interaction with NSAIDs (152). The increased risk of seizures is not caused by increased serum concentrations of fluoroquinolones, since their pharmacokinetics are not altered by NSAIDs (153). The mechanism has been suggested to be facilitation by fenbufen of the fluoroquinolone-induced inhibition of $GABA_A$ receptor function in the hippocampus and frontal cortex (154).

Probenecid

Probenecid increases the serum concentrations of cinoxacin (155), enoxacin (156), and nalidixic acid (157) probably by inhibiting their renal tubular secretion.

Sucralfate

Sucralfate markedly reduces the systemic availability of fluoroquinolones (137,158,159).

Theophylline

Fluoroquinolones are potent inhibitors of hepatic cytochrome P450 isozymes (160). They inhibit theophylline metabolism, and accumulation of theophylline has led to seizures (161). Theophylline clearance is reduced by about 10% by norfloxacin, 30% by ciprofloxacin, and 70% by enoxacin (162–172). In a comparison of grepafloxacin (400 and 600 mg/day) and ciprofloxacin, increased theophylline concentrations associated with clinical symptoms were found with both doses of grepafloxacin but not in patients taking ciprofloxacin (19). The dosage of theophylline should be reduced when a quinolone is given.

No interaction has been reported with levofloxacin (149) or sparfloxacin (147,171).

Warfarin

The anticoagulant effect of warfarin was altered by ciprofloxacin (172) but not by sparfloxacin (159) or moxifloxacin (18).

Interference with Diagnostic Tests

Opioid immunoassays

The reactivity of 13 quinolones and fluoroquinolones (ciprofloxacin, clinafloxacin, enoxacin, gatifloxacin, levofloxacin, lomefloxacin, moxifloxacin, nalidixic acid, norfloxacin, ofloxacin, pefloxacin, sparfloxacin, and trovafloxacin) with five commercial opiate immunoassays has been tested in vitro and in three healthy volunteers (173). In vitro, levofloxacin and ofloxacin (using Abbott AxSYM, CEDIA, EMIT II, and Roche OnLine assays), pefloxacin (using CEDIA, EMIT II, and Roche OnLine assays), enoxacin (using CEDIA and EMIT II assays), gatifloxacin (using EMIT II assay), and ciprofloxacin, lomefloxacin, moxifloxacin, and norfloxacin (using Roche OnLine assay) cross-reacted and cause a positive test result for opiates. Clinafloxacin, nalidixic acid, sparfloxacin, and trovafloxacin did not cross-react in any of the assays. A single dose of levofloxacin 500 mg caused a false-positive test result using the EMIT II assay within 2 hours for as long as 22 hours in all three volunteers. Ofloxacin (a single dose of 400 mg) produced a similar pattern. Detectable opiate activity in the urine was seen for more than 30 hours with both levofloxacin and ofloxacin.

References

1. Ball P. New antibiotics for community-acquired lower respiratory tract infections: improved activity at a cost? Int J Antimicrob Agents 2000;16(3):263–72.
2. Hooper DC. The fluoroquinolones after ciprofloxacin and ofloxacin. Curr Clin Top Infect Dis 2000;20:63–91.
3. O'Donnell JA, Gelone SP. Fluoroquinolones. Infect Dis Clin North Am 2000;14(2):489–513.
4. King DE, Malone R, Lilley SH. New classification and update on the quinolone antibiotics. Am Fam Physician 2000;61(9):2741–8.

5. Ball P. Adverse drug reactions: implications for the development of fluoroquinolones. J Antimicrob Chemother 2003;51 Suppl 1:21–7.

6. Hooper DC. Expanding uses of fluoroquinolones: opportunities and challenges. Ann Intern Med 1998;129(11):908–10.

7. Bertino J Jr, Fish D. The safety profile of the fluoroquinolones. Clin Ther 2000;22(7):798–817.

8. Rubinstein E. History of quinolones and their side effects. Chemotherapy 2001;47(Suppl 3):3–8discussion 44–8.

9. Lietman PS. Fluoroquinolone toxicities. An update. Drugs 1995;49(Suppl 2):159–63.

10. Williams DJ, Hopkins S. Safety and tolerability of intravenous-to-oral treatment and single-dose intravenous or oral prophylaxis with trovafloxacin. Am J Surg 1998;176(Suppl 6A):S74–9.

11. Donahue PE, Smith DL, Yellin AE, Mintz SJ, Bur F, Luke DR. Trovafloxacin in the treatment of intra-abdominal infections: results of a double-blind, multicenter comparison with imipenem/cilastatin. Trovafloxacin Surgical Group. Am J Surg 1998;176(Suppl 6A):S53–61.

12. Gibaldi M. Grepafloxacin withdrawn from market. Drug Ther Topics Suppl 2000;29:6.

13. Zhanel GG, Ennis K, Vercaigne L, Walkty A, Gin AS, Embil J, Smith H, Hoban DJ. A critical review of the fluoroquinolones: focus on respiratory infections. Drugs 2002;62(1):13–59.

14. Barker PJ, Sheehan R, Teillol-Foo M, Palmgren AC, Nord CE. Impact of gemifloxacin on the normal human intestinal microflora. J Chemother 2001;13(1):47–51.

15. Rafalsky V, Andreeva I, Rjabkova E. Quinolones for uncomplicated acute cystitis in women. Cochrane Database Syst Rev 2006;3CD003597.

16. Norrby SR, Lietman PS. Safety and tolerability of fluoroquinolones. Drugs 1993;45(Suppl 3):59–64.

17. Nicodemo AC, Robledo JA, Jasovich A, Neto W. A multicentre, double-blind, randomised study comparing the efficacy and safety of oral levofloxacin versus ciprofloxacin in the treatment of uncomplicated skin and skin structure infections. Int J Clin Pract 1998;52(2):69–74.

18. Balfour JA, Wiseman LR. Moxifloxacin. Drugs 1999;57(3):363–73.

19. Chodosh S, Lakshminarayan S, Swarz H, Breisch S. Efficacy and safety of a 10-day course of 400 or 600 milligrams of grepafloxacin once daily for treatment of acute bacterial exacerbations of chronic bronchitis: comparison with a 10-day course of 500 milligrams of ciprofloxacin twice daily. Antimicrob Agents Chemother 1998;42(1):114–20.

20. McCarty JM, Richard G, Huck W, Tucker RM, Tosiello RL, Shan M, Heyd A, Echols RM. A randomized trial of short-course ciprofloxacin, ofloxacin, or trimethoprim/sulfamethoxazole for the treatment of acute urinary tract infection in women. Ciprofloxacin Urinary Tract Infection Group. Am J Med 1999;106(3):292–9.

21. Wilton LV, Pearce GL, Mann RD. A comparison of ciprofloxacin, norfloxacin, ofloxacin, azithromycin and cefixime examined by observational cohort studies. Br J Clin Pharmacol 1996;41(4):277–84.

22. Gurbay A, Gonthier B, Daveloose D, Favier A, Hincal F. Microsomal metabolism of ciprofloxacin generates free radicals. Free Radic Biol Med 2001;30(10):1118–21.

23. Berning SE. The role of fluoroquinolones in tuberculosis today. Drugs 2001;61(1):9–18.

24. Beringer PM, Wong-Beringer A, Rho JP. Economic aspects of antibacterial adverse effects. Pharmacoeconomics 1998;13(1 Pt 1):35–49.

25. Patmore L, Fraser S, Mair D, Templeton A. Effects of sparfloxacin, grepafloxacin, moxifloxacin, and ciprofloxacin on cardiac action potential duration. Eur J Pharmacol 2000;406(3):449–52.

26. Hagiwara T, Satoh S, Kasai Y, Takasuna K. A comparative study of the fluoroquinolone antibacterial agents on the action potential duration in guinea pig ventricular myocardia. Jpn J Pharmacol 2001;87(3):231–4.

27. Owens RC Jr, Ambrose PG. Torsades de pointes associated with fluoroquinolones. Pharmacotherapy 2002;22(5):663–8discussion 668–72.

28. Jaillon P, Morganroth J, Brumpt I, Talbot G. Overview of electrocardiographic and cardiovascular safety data for sparfloxacin. Sparfloxacin Safety Group. J Antimicrob Chemother 1996;37(Suppl A):161–7.

29. Clark DW, Layton D, Wilton LV, Pearce GL, Shakir SA. Profiles of hepatic and dysrhythmic cardiovascular events following use of fluoroquinolone antibacterials: experience from large cohorts from the Drug Safety Research Unit Prescription-Event Monitoring database. Drug Saf 2001;24(15):1143–54.

30. Frothingham R. Rates of torsades de pointes associated with ciprofloxacin, ofloxacin, levofloxacin, gatifloxacin, and moxifloxacin. Pharmacotherapy 2001;21(12):1468–72.

31. Carbon C. Comparison of side effects of levofloxacin versus other fluoroquinolones. Chemotherapy 2001;47(Suppl 3):9–14discussion 44–8.

32. Cohen JS. Peripheral neuropathy associated with fluoroquinolones. Ann Pharmacother 2001;35(12):1540–7.

33. Moore B, Safani M, Keesey J. Possible exacerbation of myasthenia gravis by ciprofloxacin. Lancet 1988;1(8590):882.

34. Kimura M, Fujiyama J, Nagai A, Hirayama M, Kuriyama M. [Encephalopathy induced by fleroxacin in a patient with Machado–Joseph disease.]Rinsho Shinkeigaku 1998;38(9):846–8.

35. Tattevin P, Messiaen T, Pras V, Ronco P, Biour M. Confusion and general seizures following ciprofloxacin administration. Nephrol Dial Transplant 1998;13(10):2712–3.

36. Kushner JM, Peckman HJ, Snyder CR. Seizures associated with fluoroquinolones. Ann Pharmacother 2001;35(10):1194–8.

37. Walton GD, Hon JK, Mulpur TG. Ofloxacin-induced seizure. Ann Pharmacother 1997;31(12):1475–7.

38. Schmuck G, Schurmann A, Schluter G. Determination of the excitatory potencies of fluoroquinolones in the central nervous system by an in vitro model. Antimicrob Agents Chemother 1998;42(7):1831–6.

39. Bryskier A, Chantot JF. Classification and structure–activity relationships of fluoroquinolones. Drugs 1995;49(Suppl 2):16–28.

40. Stahlmann R, Schwabe R. Safety profile of grepafloxacin compared with other fluoroquinolones. J Antimicrob Chemother 1997;40(Suppl A):83–92.

41. Schiffman SS, Zervakis J, Westall HL, Graham BG, Metz A, Bennett JL, Heald AE. Effect of antimicrobial and anti-inflammatory medications on the sense of taste. Physiol Behav 2000;69(4–5):413–24.

42. Reeves RR. Ciprofloxacin-induced psychosis. Ann Pharmacother 1992;26(7–8):930–1.

43. Leone R, Venegoni M, Motola D, Moretti U, Piazzetta V, Cocci A, Resi D, Mozzo F, Velo G, Burzilleri L, Montanaro N, Conforti A. Adverse drug reactions related to the use of fluoroquinolone antimicrobials: an analysis of

spontaneous reports and fluoroquinolone consumption data from three Italian regions. Drug Saf 2003;26:109–20.

44. Mohr JF, McKinnon PS, Peymann PJ, Kenton I, Septimus E, Okhuysen PC. A retrospective, comparative evaluation of dysglycemias in hospitalized patients receiving gatifloxacin, levofloxacin, ciprofloxacin, or ceftriaxone. Pharmacotherapy 2005;25(10):1303–9.

45. Kundu AK. Ciprofloxacin-induced severe cutaneous reaction and haemolysis in a young adult. J Assoc Physicians India 2000;48(6):649–50.

46. Shimada J, Hori S. Adverse effects of fluoroquinolones. Prog Drug Res 1992;38:133–43.

47. Eron LJ, Harvey L, Hixon DL, Poretz DM. Ciprofloxacin therapy of infections caused by *Pseudomonas aeruginosa* and other resistant bacteria. Antimicrob Agents Chemother 1985;28(2):308–10.

48. Patoia L, Guerciolini R, Menichetti F, Bucaneve G, Del Favero A. Norfloxacin and neutropenia. Ann Intern Med 1987;107(5):788–9.

49. Wang C, Sabbaj J, Corrado M, Hoagland V. World-wide clinical experience with norfloxacin: efficacy and safety. Scand J Infect Dis Suppl 1986;48:81–9.

50. Ball P. Ciprofloxacin: an overview of adverse experiences. J Antimicrob Chemother 1986;18(Suppl D):187–93.

51. Sawada M, Nakamura S, Yamada A, Kobayashi T, Okada S. Phase IV study and post-marketing surveillance of ofloxacin in Japan. Chemotherapy 1991;37(2):134–42.

52. Simon J, Guyot A. Pefloxacin: safety in man. J Antimicrob Chemother 1990;26(Suppl B):215–8.

53. Grassi C, Salvatori E, Rosignoli MT, Dionisio PPrulifloxacin Study Group. Randomized, double-blind study of prulifloxacin versus ciprofloxacin in patients with acute exacerbations of chronic bronchitis. Respiration 2002;69(3):217–22.

54. Lomaestro BM. Fluoroquinolone-induced renal failure. Drug Saf 2000;22(6):479–85.

55. Hootkins R, Fenves AZ, Stephens MK. Acute renal failure secondary to oral ciprofloxacin therapy: a presentation of three cases and a review of the literature. Clin Nephrol 1989;32(2):75–8.

56. Hatton J, Haagensen D. Renal dysfunction associated with ciprofloxacin. Pharmacotherapy 1990;10(5):337–40.

57. Rastogi S, Atkinson JL, McCarthy JT. Allergic nephropathy associated with ciprofloxacin. Mayo Clin Proc 1990;65(7):987–9.

58. Swanson BN, Boppana VK, Vlasses PH, Rotmensch HH, Ferguson RK. Norfloxacin disposition after sequentially increasing oral doses. Antimicrob Agents Chemother 1983;23(2):284–8.

59. Schaeffer AJ. Multiclinic study of norfloxacin for treatment of urinary tract infections. Am J Med 1987;82(6B):53–8.

60. Campoli-Richards DM, Monk JP, Price A, Benfield P, Todd PA, Ward A. Ciprofloxacin. A review of its antibacterial activity, pharmacokinetic properties and therapeutic use. Drugs 1988;35(4):373–447.

61. van der Linden PD, van der Lei J, Vlug AE, Stricker BH. Skin reactions to antibacterial agents in general practice. J Clin Epidemiol 1998;51(8):703–8.

62. Horio T, Miyauchi H, Asada Y, Aoki Y, Harada M. Phototoxicity and photoallergenicity of quinolones in guinea pigs. J Dermatol Sci 1994;7(2):130–5.

63. Navaratnam S, Claridge J. Primary photophysical properties of ofloxacin. Photochem Photobiol 2000;72(3):283–90.

64. Sun YW, Heo EP, Cho YH, Bark KM, Yoon TJ, Kim TH. Pefloxacin and ciprofloxacin increase UVA-induced edema and immune suppression. Photodermatol Photoimmunol Photomed 2001;17(4):172–7.

65. Blotz A, Michel L, Moysan A, Blumel J, Dubertret L, Ahr HJ, Vohr HW. Analyses of cutaneous fluoroquinolones photoreactivity using the integrated model for the differentiation of skin reactions. J Photochem Photobiol B 2000;58(1):46–53.

66. Sangen Y, Kawada A, Asai M, Aragane Y, Yudate T, Tezuka T. Fixed drug eruption induced by tosufloxacin tosilate. Contact Dermatitis 2000;42(5):285.

67. Siami FS, LaFleur BJ, Siami GA. Clinafloxacin versus piperacillin/tazobactam in the treatment of severe skin and soft-tissue infections in adults at a Veterans Affairs medical center. Clin Ther 2002;24(1):59–72.

68. Sable D, Murakawa GJ. Quinolones in dermatology. Dis Mon 2004;50(7):381–94.

69. Matsumoto M, Kojima K, Nagano H, Matsubara S, Yokota T. Photostability and biological activity of fluoroquinolones substituted at the 8 position after UV irradiation. Antimicrob Agents Chemother 1992;36(8):1715–9.

70. Marutani K, Matsumoto M, Otabe Y, Nagamuta M, Tanaka K, Miyoshi A, Hasegawa T, Nagano H, Matsubara S, Kamide R, et al. Reduced phototoxicity of a fluoroquinolone antibacterial agent with a methoxy group at the 8 position in mice irradiated with long-wavelength UV light. Antimicrob Agents Chemother 1993;37(10):2217–23.

71. Wainwright NJ, Collins P, Ferguson J. Photosensitivity associated with antibacterial agents. Drug Saf 1993;9(6):437–40.

72. Ferguson J, Johnson BE. Clinical and laboratory studies of the photosensitizing potential of norfloxacin, a 4-quinolone broad-spectrum antibiotic. Br J Dermatol 1993;128(3):285–95.

73. Jensen T, Pedersen SS, Nielsen CH, Hoiby N, Koch C. The efficacy and safety of ciprofloxacin and ofloxacin in chronic *Pseudomonas aeruginosa* infection in cystic fibrosis. J Antimicrob Chemother 1987;20(4):585–94.

74. Jungst G, Mohr R. Side effects of ofloxacin in clinical trials and in postmarketing surveillance. Drugs 1987;34(Suppl 1):144–9.

75. Scheife RT, Cramer WR, Decker EL. Photosensitizing potential of ofloxacin. Int J Dermatol 1993;32(6):413–6.

76. Lopitaux R, Hermet R, Sirot J, Filiu P, Terver S. Tolérance de la péfloxacine au cours du traitement d'une série d'infections ostéo-articulaires. [Tolerance to pefloxacine during treatment of a series of osteoarticular infections. 36 cases.] Therapie 1985;40(5):349–52.

77. Stahlmann R. Safety profile of the quinolones. J Antimicrob Chemother 1990;26(Suppl D):31–44.

78. Bowie WR, Willetts V, Jewesson PJ. Adverse reactions in a dose-ranging study with a new long-acting fluoroquinolone, fleroxacin. Antimicrob Agents Chemother 1989;33(10):1778–82.

79. Cohen JB, Bergstresser PR. Inadvertent phototoxicity from home tanning equipment. Arch Dermatol 1994;130(6):804–6.

80. Correia O, Delgado L, Barros MA. Bullous photodermatosis after lomefloxacin. Arch Dermatol 1994;130(6):808–9.

81. Kurumaji Y, Shono M. Scarified photopatch testing in lomefloxacin photosensitivity. Contact Dermatitis 1992;26(1):5–10.

82. Arata J, Horio T, Soejima R, Ohara K. Photosensitivity reactions caused by lomefloxacin hydrochloride: a

multicenter survey. Antimicrob Agents Chemother 1998;42(12):3141–5.

83. Saravolatz LD, Leggett J. Gatifloxacin, gemifloxacin, and moxifloxacin: the role of 3 newer fluoroquinolones. Clin Infect Dis 2003;37:1210–5.

84. Dawe RS, Ibbotson SH, Sanderson JB, Thomson EM, Ferguson J. A randomized controlled trial (volunteer study) of sitafloxacin, enoxacin, levofloxacin and sparfloxacin phototoxicity. Br J Dermatol 2003;149:1232–41.

85. Man I, Murphy J, Ferguson J. Fluoroquinolone phototoxicity: a comparison of moxifloxacin and lomefloxacin in normal volunteers. J Antimicrob Chemother 1999;43(Suppl B):77–82.

86. Shimoda K, Ikeda T, Okawara S, Kato M. Possible relationship between phototoxicity and photodegradation of sitafloxacin, a quinolone antibacterial agent, in the auricular skin of albino mice. Toxicol Sci 2000;56(2):290–6.

87. Ribard P, Audisio F, Kahn MF, De Bandt M, Jorgensen C, Hayem G, Meyer O, Palazzo E. Seven Achilles tendinitis including 3 complicated by rupture during fluoroquinolone therapy. J Rheumatol 1992;19(9):1479–81.

88. Blanco Andres C, Bravo Toledo R. Tendinitis bilateral secundaria a ciprofloxacino. [Bilateral tendinitis caused by ciprofloxacin.] Aten Primaria 1998;21(3):184–5.

89. West MB, Gow P. Ciprofloxacin, bilateral Achilles tendonitis and unilateral tendon rupture—a case report. NZ Med J 1998;111(1058):18–9.

90. Stahlmann R. Clinical toxicological aspects of fluoroquinolones. Toxicol Lett 2002;127(1–3):269–77.

91. Burstein GR, Berman SM, Blumer JL, Moran JS. Ciprofloxacin for the treatment of uncomplicated gonorrhea infection in adolescents: does the benefit outweigh the risk? Clin Infect Dis 2002;35(Suppl 2):S191–9.

92. Casparian JM, Luchi M, Moffat RE, Hinthorn D. Quinolones and tendon ruptures. South Med J 2000;93(5):488–91.

93. Saint F, Gueguen G, Biserte J, Fontaine C, Mazeman E. Rupture du ligament patellaire un mois après traitement par fluoroquinolone. [Rupture of the patellar ligament one month after treatment with fluoroquinolone.] Rev Chir Orthop Reparatrice Appar Mot 2000;86(5):495–7.

94. Gabutti L, Stoller R, Marti HP. Fluoroquinolone als Ursache von Tendinopathien. [Fluoroquinolones as etiology of tendinopathy.] Ther Umsch 1998;55(9):558–61.

95. Donck JB, Segaert MF, Vanrenterghem YF. Fluoroquinolones and Achilles tendinopathy in renal transplant recipients. Transplantation 1994;58(6):736–7.

96. van der Linden PD, van Puijenbroek EP, Feenstra J, Veld BA, Sturkenboom MC, Herings RM, Leufkens HG, Stricker BH. Tendon disorders attributed to fluoroquinolones: a study on 42 spontaneous reports in the period 1988 to 1998. Arthritis Rheum 2001;45(3):235–9.

97. Postnikov SS, Semykin Slu, Nazhimov VP, Novichkova GA. Mesto pefloksatsina v lechenii bol'nykh mukovistsidozom. [On fluoroquinolones treatment safety in children (clinical, morphological and catamnesis data).] Antibiot Khimioter 2002;47(9):14–7.

98. Williams RJ 3rd, Attia E, Wickiewicz TL, Hannafin JA. The effect of ciprofloxacin on tendon, paratenon, and capsular fibroblast metabolism. Am J Sports Med 2000;28(3):364–9.

99. Kashida Y, Kato M. Characterization of fluoroquinolone-induced Achilles tendon toxicity in rats: comparison of toxicities of 10 fluoroquinolones and effects of anti-inflammatory compounds. Antimicrob Agents Chemother 1997;41(11):2389–93.

100. Stahlmann R, Schwabe R, Pfister K, Lozo E, Shakibaei M, Vormann J. Supplementation with magnesium and tocopherol diminishes quinolone-induced chondrotoxicity in immature rats. Drugs 1999;58 Suppl 2:393–4.

101. Simonin MA, Gegout-Pottie P, Minn A, Gillet P, Netter P, Terlain B. Pefloxacin-induced achilles tendon toxicity in rodents: biochemical changes in proteoglycan synthesis and oxidative damage to collagen. Antimicrob Agents Chemother 2000;44(4):867–72.

102. Postnikov SS. Sravnitel'naia effektivnost' i bezopasnost' tsiprofloksatsina, ofloksatsina i pefloksatsina pri lechenii infektsii dykhatel'nykh putei u detei s mukovitsidozom. [Comparative efficacy and safety of ciprofloxacin, ofloxacin, and pefloxacin in treatment of respiratory infections in children with cystic fibrosis.] Antibiot Khimioter 2001;46(3):16–20.

103. Huddleston PM, Steckelberg JM, Hanssen AD, Rouse MS, Bolander ME, Patel R. Ciprofloxacin inhibition of experimental fracture healing. J Bone Joint Surg Am 2000;82(2):161–73.

104. Ono Y, Ohmoto Y, Ono K, Sakata Y, Murata K. Effect of grepafloxacin on cytokine production in vitro. J Antimicrob Chemother 2000;46(1):91–4.

105. Sachs B, Riegel S, Seebeck J, Beier R, Schichler D, Barger A, Merk HF, Erdmann S. Fluoroquinolone-associated anaphylaxis in spontaneous adverse drug reaction reports in Germany: differences in reporting rates between individual fluoroquinolones and occurrence after first-ever use. Drug Saf 2006;29(11):1087–100.

106. Stricker BH, Slagboom G, Demaeseneer R, Slootmaekers V, Thijs I, Olsson S. Anaphylactic reactions to cinoxacin. BMJ 1988;297(6661):1434–5.

107. Davis H, McGoodwin E, Reed TG. Anaphylactoid reactions reported after treatment with ciprofloxacin. Ann Intern Med 1989;111(12):1041–3.

108. Kennedy CA, Goetz MB, Mathisen GE. Ciprofloxacin-induced anaphylactoid reactions in patients infected with the human immunodeficiency virus. West J Med 1990;153(5):563–4.

109. Miller MS, Gaido F, Rourk MH Jr, Spock A. Anaphylactoid reactions to ciprofloxacin in cystic fibrosis patients. Pediatr Infect Dis J 1991;10(2):164–5.

110. Grassmick BK, Lehr VT, Sundareson AS. Fulminant hepatic failure possibly related to ciprofloxacin. Ann Pharmacother 1992;26(5):636–9.

111. Shibasaki T, Ishimoto F, Sakai O, Joh K, Aizawa S. Clinical characterization of drug-induced allergic nephritis. Am J Nephrol 1991;11(3):174–80.

112. Hadimeri H, Almroth G, Cederbrant K, Enestrom S, Hultman P, Lindell A. Allergic nephropathy associated with norfloxacin and ciprofloxacin therapy. Report of two cases and review of the literature. Scand J Urol Nephrol 1997;31(5):481–5.

113. Davila I, Diez ML, Quirce S, Fraj J, De La Hoz B, Lazaro M. Cross-reactivity between quinolones. Report of three cases. Allergy 1993;48(5):388–90.

114. Ronnau AC, Sachs B, von Schmiedeberg S, Hunzelmann N, Ruzicka T, Gleichmann E, Schuppe HC. Cutaneous adverse reaction to ciprofloxacin: demonstration of specific lymphocyte proliferation and cross-reactivity to ofloxacin in vitro. Acta Derm Venereol 1997;77(4):285–8.

115. Mavromanolakis E, Maraki S, Cranidis A, Tselentis Y, Kontoyiannis DP, Samonis G. The impact of norfloxacin, ciprofloxacin and ofloxacin on human gut colonization by *Candida albicans*. Scand J Infect Dis 2001;33(6):477–8.

116. Pepin J, Saheb N, Coulombe MA, Alary ME, Corriveau MP, Authier S, Leblanc M, Rivard G, Bettez M, Primeau V, Nguyen M, Jocob CE, Lanthier L. Emergence of fluoroquinolones as the prodominant risk factor for *Clostridium difficile*-associated diarrhea: a cohort study during an epidemic in Quebec. Clin Infect Dis 2005;41(9):1254–60.

117. Edlund C, Nord CE. Ecological effect of gatifloxacin on the normal human intestinal microflora. J Chemother 1999;11(1):50–3.

118. Molbak K, Baggesen DL, Aarestrup FM, Ebbesen JM, Engberg J, Frydendahl K, Gerner-smidt P, Petersen AM, Wegener HC. An outbreak of multidrug-resistant, quinolone-resistant Salmonella enterica serotype typhimurium DT104. N Engl J Med 1999;341(19):1420–5.

119. Villar RG, Macek MD, Simons S, Hayes PS, Goldoft MJ, Lewis JH, Rowan LL, Hursh D, Patnode M, Mead PS. Investigation of multidurg-resistant Salmonella serotype typhimurium DT104 infections linked to raw-milk cheese in Washington State. JAMA 1999;281(19):1811–6.

120. Jones ME, Sahm DF, Martin N, Scheuring S, Heisig P, Thornsberry C, Kohrer K, Schmitz FJ. Prevalence of gyrA, gyrB, parC, and parE mutations in clinical isolates of *Streptococcus pneumoniae* with decreased susceptibilities to different fluoroquinolones and originating from worldwide surveillance studies during the 1997–1998 respiratory season. Antimicrob Agents Chemother 2000;44(2):462–6.

121. Esposito S, Noviello S, Ianniello F. Comparative in vitro activity of older and newer fluoroquinolones against respiratory tract pathogens. Chemotherapy 2000;46(5):309–14.

122. Weiss K, Restieri C, De Carolis E, Laverdiere M, Guay H. Comparative activity of new quinolones against 326 clinical isolates of *Stenotrophomonas maltophilia*. J Antimicrob Chemother 2000;45(3):363–5.

123. Goettsch W, van Pelt W, Nagelkerke N, Hendrix MG, Buiting AG, Petit PL, Sabbe LJ, van Griethuysen AJ, de Neeling AJ. Increasing resistance to fluoroquinolones in *Escherichia coli* from urinary tract infections in the netherlands. J Antimicrob Chemother 2000;46(2):223–8.

124. Jeffrey AM, Shao L, Brendler-Schwaab SY, Schluter G, Williams GM. Photochemical mutagenicity of phototoxic and photochemically carcinogenic fluoroquinolones in comparison with the photostable moxifloxacin. Arch Toxicol 2000;74(9):555–9.

125. Fort FL. Mutagenicity of quinolone antibacterials. Drug Saf 1992;7(3):214–22.

126. Klecak G, Urbach F, Urwyler H. Fluoroquinolone antibacterials enhance UVA-induced skin tumors. J Photochem Photobiol B 1997;37(3):174–81.

127. Berkovitch M, Pastuszak A, Gazarian M, Lewis M, Koren G. Safety of the new quinolones in pregnancy. Obstet Gynecol 1994;84(4):535–8.

128. Wogelius P, Norgaard M, Gislum M, Pedersen L, Schonheyder HC, Sorensen HT. Further analysis of the risk of adverse birth outcome after maternal use of fluoroquinolones. Int J Antimicrob Agents 2005;26(4):323–6.

129. Loebstein R, Addis A, Ho E, Andreou R, Sage S, Donnenfeld AE, Schick B, Bonati M, Moretti M, Lalkin A, Pastuszak A, Koren G. Pregnancy outcome following gestational exposure to fluoroquinolones: a multicenter prospective controlled study. Antimicrob Agents Chemother 1998;42(6):1336–9.

130. Cukierski MA, Prahalada S, Zacchei AG, Peter CP, Rodgers JD, Hess DL, Cukierski MJ, Tarantal AF, Nyland T, Robertson RT, et al. Embryotoxicity studies of norfloxacin in cynomolgus monkeys: I. Teratology studies and norfloxacin plasma concentration in pregnant and nonpregnant monkeys. Teratology 1989;39(1):39–52.

131. Chukwuani CM, Coker HA, Oduola AM, Sowunmi A, Ifudu ND. Bioavailability of ciprofloxacin and fleroxacin: results of a preliminary investigation in healthy adult Nigerian male volunteers. Biol Pharm Bull 2000;23(8):968–72.

132. Dagan R. Fluoroquinolones in paediatrics—1995. Drugs 1995;49(Suppl 2):92–9.

133. Jick S. Ciprofloxacin safety in a pediatric population. Pediatr Infect Dis J 1997;16(1):130–3.

134. Aminimanizani A, Beringer P, Jelliffe R. Comparative pharmacokinetics and pharmacodynamics of the newer fluoroquinolone antibacterials. Clin Pharmacokinet 2001;40(3):169–87.

135. Sambatakou H, Giamarellos-Bourboulis EJ, Galanakis N, Giamarellou H. Pharmacokinetics of fluoroquinolones in uncompensated cirrhosis: the significance of penetration in the ascitic fluid. Int J Antimicrob Agents 2001;18(5):441–4.

136. Miglioli PA, Kafka R, Bonatti H, Fraedrich G, Allerberger F, Schoeffel U. Fleroxacin uptake in ischaemic limb tissue. Acta Microbiol Immunol Hung 2001;48(1):11–5.

137. Deppermann KM, Lode H. Fluoroquinolones: interaction profile during enteral absorption. Drugs 1993;45(Suppl 3):65–72.

138. Cordoba-Diaz M, Cordoba-Borrego M, Cordoba-Diaz D. Influence of pharmacotechnical design on the interaction and availability of norfloxacin in directly compressed tablets with certain antacids. Drug Dev Ind Pharm 2000;26(2):159–66.

139. Flor S, Guay DR, Opsahl JA, Tack K, Matzke GR. Effects of magnesium–aluminum hydroxide and calcium carbonate antacids on bioavailability of ofloxacin. Antimicrob Agents Chemother 1990;34(12):2436–8.

140. Johnson RD, Dorr MB, Talbot GH, Caille G. Effect of Maalox on the oral absorption of sparfloxacin. Clin Ther 1998;20(6):1149–58.

141. Carrillo JA, Benitez J. Clinically significant pharmacokinetic interactions between dietary caffeine and medications. Clin Pharmacokinet 2000;39(2):127–53.

142. Kim MK, Nightingale C, Nicolau D. Influence of sex on the pharmacokinetic interaction of fleroxacin and ciprofloxacin with caffeine. Clin Pharmacokinet 2003;42(11):985–96.

143. Randinitis EJ, Alvey CW, Koup JR, Rausch G, Abel R, Bron NJ, Hounslow NJ, Vassos AB, Sedman AJ. Drug interactions with clinafloxacin. Antimicrob Agents Chemother 2001;45(9):2543–52.

144. Misiak PM, Eldon MA, Toothaker RD, Sedman AJ. Effects of oral cimetidine or ranitidine on the pharmacokinetics of intravenous enoxacin. J Clin Pharmacol 1993;33(1):53–6.

145. Bressolle F, Goncalves F, Gouby A, Galtier M. Pefloxacin clinical pharmacokinetics. Clin Pharmacokinet 1994;27(6):418–46.

146. Johnson R, Wilson J, Talbot G. The effect of sparfloxacin on the steady-state pharmacokinetics of digoxin in healthy male volunteers. Pharm Res 1994;11(Suppl):429.

147. Goa KL, Bryson HM, Markham A. Sparfloxacin. A review of its antibacterial activity, pharmacokinetic properties, clinical efficacy and tolerability in lower respiratory tract infections. Drugs 1997;53(4):700–25.

148. Sudoh T, Fujimura A, Shiga T, Sasaki M, Harada K, Tateishi T, Ohashi K, Ebihara A. Renal clearance of

lomefloxacin is decreased by furosemide. Eur J Clin Pharmacol 1994;46(3):267–9.

149. Lode H. Evidence of different profiles of side effects and drug–drug interactions among the quinolones—the pharmacokinetic standpoint. Chemotherapy 2001;47(Suppl 3):24–31discussion 44–8.

150. Kanemitsu K, Hori S, Yanagawa A, Shimada J. Effect of ferrous sulfate on the absorption of sparfloxacin in healthy volunteers and rats. Drugs 1995;49(Suppl 2):352–6.

151. Morita H, Maemura K, Sakai Y, Kaneda Y. [A case of convulsion, loss of consciousness and subsequent acute renal failure caused by enoxacin and fenbufen.]Nippon Naika Gakkai Zasshi 1988;77(5):744–5.

152. Furuhama K, Akahane K, Iawara K, Takayama S. Interaction of the new quinolone antibacterial agent levofloxacin with fenbufen in mice. Arzneimittelforschung 1992;43(3A):406–8.

153. Fillastre JP, Leroy A, Borsa-Lebas F, Etienne I, Gy C, Humbert G. Effects of ketoprofen (NSAID) on the pharmacokinetics of pefloxacin and ofloxacin in healthy volunteers. Drugs Exp Clin Res 1992;18(11–12):487–92.

154. Motomura M, Kataoka Y, Takeo G, Shibayama K, Ohishi K, Nakamura T, Niwa M, Tsujihata M, Nagataki S. Hippocampus and frontal cortex are the potential mediatory sites for convulsions induced by new quinolones and non-steroidal anti-inflammatory drugs. Int J Clin Pharmacol Ther Toxicol 1991;29(6):223–7.

155. Rodriguez N, Madsen PO, Welling PG. Influence of probenecid on serum levels and urinary excertion of cinoxacin. Antimicrob Agents Chemother 1979;15(3):465–9.

156. Wijnands WJ, Vree TB, Baars AM, van Herwaarden CL. Pharmacokinetics of enoxacin and its penetration into bronchial secretions and lung tissue. J Antimicrob Chemother 1988;21(Suppl B):67–77.

157. Vree TB, Van den Biggelaar-Martea M, Van Ewijk-Beneken Kolmer EW, Hekster YA. Probenecid inhibits the renal clearance and renal glucuronidation of nalidixic acid. A pilot experiment. Pharm World Sci 1993;15(4):165–70.

158. Zix JA, Geerdes-Fenge HF, Rau M, Vockler J, Borner K, Koeppe P, Lode H. Pharmacokinetics of sparfloxacin and interaction with cisapride and sucralfate. Antimicrob Agents Chemother 1997;41(8):1668–72.

159. Lee LJ, Hafkin B, Lee ID, Hoh J, Dix R. Effects of food and sucralfate on a single oral dose of 500 milligrams of levofloxacin in healthy subjects. Antimicrob Agents Chemother 1997;41(10):2196–200.

160. Polk RE. Drug–drug interactions with ciprofloxacin and other fluoroquinolones. Am J Med 1989;87(5A):S76–81.

161. Grasela TH Jr, Dreis MW. An evaluation of the quinolone–theophylline interaction using the Food and Drug Administration spontaneous reporting system. Arch Intern Med 1992;152(3):617–21.

162. Takagi K, Hasegawa T, Yamaki K, Suzuki R, Watanabe T, Satake T. Interaction between theophylline and enoxacin. Int J Clin Pharmacol Ther Toxicol 1988;26(6):288–92.

163. Beckmann J, Elsasser W, Gundert-Remy U, Hertrampf R. Enoxacin—a potent inhibitor of theophylline metabolism. Eur J Clin Pharmacol 1987;33(3):227–30.

164. Wijnands WJ, Vree TB, van Herwaarden CL. The influence of quinolone derivatives on theophylline clearance. Br J Clin Pharmacol 1986;22(6):677–83.

165. Nix DE, DeVito JM, Whitbread MA, Schentag JJ. Effect of multiple dose oral ciprofloxacin on the pharmacokinetics of theophylline and indocyanine green. J Antimicrob Chemother 1987;19(2):263–9.

166. Schwartz J, Jauregui L, Lettieri J, Bachmann K. Impact of ciprofloxacin on theophylline clearance and steady-state concentrations in serum. Antimicrob Agents Chemother 1988;32(1):75–7.

167. Bowles SK, Popovski Z, Rybak MJ, Beckman HB, Edwards DJ. Effect of norfloxacin on theophylline pharmacokinetics at steady state. Antimicrob Agents Chemother 1988;32(4):510–2.

168. Ho G, Tierney MG, Dales RE. Evaluation of the effect of norfloxacin on the pharmacokinetics of theophylline. Clin Pharmacol Ther 1988;44(1):35–8.

169. Davis RL, Kelly HW, Quenzer RW, Standefer J, Steinberg B, Gallegos J. Effect of norfloxacin on theophylline metabolism. Antimicrob Agents Chemother 1989;33(2):212–4.

170. Prince RA, Casabar E, Adair CG, Wexler DB, Lettieri J, Kasik JE. Effect of quinolone antimicrobials on theophylline pharmacokinetics. J Clin Pharmacol 1989;29(7):650–4.

171. Takagi K, Yamaki K, Nadai M, Kuzuya T, Hasegawa T. Effect of a new quinolone, sparfloxacin, on the pharmacokinetics of theophylline in asthmatic patients. Antimicrob Agents Chemother 1991;35(6):1137–41.

172. Ellis RJ, Mayo MS, Bodensteiner DM. Ciprofloxacin–warfarin coagulopathy: a case series. Am J Hematol 2000;63(1):28–31.

173. Baden LR, Horowitz G, Jacoby H, Eliopoulos GM. Quinolones and false-positive urine screening for opiates by immunoassay technology. JAMA 2001; 286(24):3115–9.

Fosfomycin

General Information

Fosfomycin is a broad-spectrum antibiotic used to treat uncomplicated lower urinary tract infections. It penetrates interstitial space fluids of soft tissues well and reaches concentrations sufficient to substantially inhibit the growth of relevant bacteria at the target site (1).

Fosfomycin has relatively low toxicity. Its penetration into tissues, including bones and joints, and into the cerebrospinal fluid is good. When given orally (2–3 g/day), it can produce gastrointestinal distress; when injected intramuscularly, it can cause local pain. Fosfomycin is recommended in daily doses of 4–16 g intravenously for the treatment of severe infections resistant to other commonly used antibiotics. Fosfomycin diffuses moderately well into bone tissue (2).

During systemic administration, rises in the activities of serum transaminases and lactate dehydrogenase, skin reactions, eosinophilia, and altered vision have been noted (3).

Comparative studies

In a multicenter trial in 749 ambulatory women aged at least 12 years with an acute uncomplicated urinary tract infection, a single dose of fosfomycin tromethamine 3 g had an equivalent bacteriological and clinical cure rate as a 7-day course of nitrofurantoin (4). Adverse events were reported by 5.3% of fosfomycin-treated patients (versus 5.6%). The most common adverse effects were diarrhea (2.4%), vaginitis (1.8%), and nausea (0.8%), and 1.9% of fosfomycin-treated patients were withdrawn owing to adverse events (versus 4.3%).

Organs and Systems

Nervous system

Fosfomycin concentrations in brain interstitium were measured in two patients after the intravenous administration of 4 g (5). Brain C_{max} values were above MIC for relevant pathogens, such as *Streptococcus pneumoniae* and *Neisseria meningitidis*. Variability in brain penetration might be explained by the degree to which the integrity of the blood–brain barrier is disrupted by the underlying disease.

Hematologic

The effects of fosfomycin on several indices of neutrophil function have been studied in vitro (6). Phagocytosis was unaffected but there was enhanced bactericidal ability, increased intracellular calcium concentrations, raised extracellular reactive oxygen intermediate production, and reduced chemotaxis; fosfomycin did not affect intracellular oxygen intermediate production or chemokinesis.

Gastrointestinal

The most frequent adverse effects of oral fosfomycin are nausea (1%) and dyspepsia [7,8].

Liver

Increased transaminase activities occurred in 0.3% of patients treated with fosfomycin [9]. Fosfomycin-induced recurrent liver toxicity occurred in a patient with cystic fibrosis (10).

Urinary tract

In Japanese children the use of fosfomycin within the first two days for treatment of illness caused by *Escherichia coli* O157 reduced the risk of subsequent hemolytic–uremic syndrome (11). However, treatment with oral fluoroquinolones may be preferable (12).

In rats, co-administration of fosfomycin significantly reduced the nephrotoxicity of vancomycin (13).

Skin

Urticaria occurred in 0.3% of patients treated with fosfomycin [7].

Immunologic

The immunomodulatory effect of fosfomycin may in part be explained by an effect on cytokine production, as shown in mice in vivo (14).

Infection risk

Bacterial biofilms develop on a number of living and inert surfaces within the urinary tract, producing chronic intractable urinary tract infections. Combination therapy with fosfomycin and a fluoroquinolone (or a fluoroquinolone and a macrolide) may be the most effective regimen available at present. Nevertheless, management of the local urinary condition and removal of the local underlying disease are the most effective approaches for treating urinary biofilm infection (15).

Long-Term Effects

Drug tolerance

There is a serious reduction in the susceptibility of strains of *E. coli* to amoxicillin (due to R-TEM enzymes), to co-trimoxazole, and to trimethoprim. However, fosfomycin trometamol remains highly active against urinary *Enterobacteriaceae*, and over 90% of *E. coli* are susceptible (16). Fosfomycin is as active as fluoroquinolones or better for treating intestinal infections caused by *Salmonella* species, pathogenic *E. coli*, *Campylobacter* species, and *Shigella* species (17).

Drug–Drug Interactions

Cisplatin

Fosfomycin is both otoprotective and nephroprotective against cisplatin-induced toxicity, without inhibiting the tumoricidal activity of cisplatin (18). Mice treated with cisplatin and fosfomycin also survived longer than animals treated with cisplatin alone, probably owing to lessening of immediate cisplatin systemic toxicity (19).

Vancomycin

Co-administration of fosfomycin may reduce the nephrotoxicity of vancomycin, as suggested by a study in rats (20).

References

1. Frossard M, Joukhadar C, Erovic BM, Dittrich P, Mrass PE, Van Houte M, Burgmann H, Georgopoulos A, Muller M. Distribution and antimicrobial activity of fosfomycin in the interstitial fluid of human soft tissues. Antimicrob Agents Chemother 2000;44(10):2728–32.
2. Boselli E, Allaouchiche B. Diffusion osseuse des antibiotiques. [Diffusion in bone tissue of antibiotics.] Presse Méd 1999;28(40):2265–76.
3. Reports of fosfomycin (76 laboratory and clinical studies in Japanese, with summaries in English). Chemotherapy (Tokyo) 1975;23:1649.
4. Stein GE. Comparison of single-dose fosfomycin and a 7-day course of nitrofurantoin in female patients with uncomplicated urinary tract infection. Clin Ther 1999;21(11):1864–72.
5. Brunner M, Reinprecht A, Illievich U, Spiss CK, Dittrich P, van Houte M, Muller M. Penetration of fosfomycin into the

parenchyma of human brain: a case study in three patients. Br J Clin Pharmacol 2002;54(5):548–50.

6. Krause R, Patruta S, Daxbock F, Fladerer P, Wenisch C. The effect of fosfomycin on neutrophil function. J Antimicrob Chemother 2001;47(2):141–6.
7. Gobernado M. Fosfomicina. [Fosfomycin.] Rev Esp Quimioter 2003;16:15–40.
8. Lobel B. Short term therapy for uncomplicated urinary tract infection today. Clinical outcome upholds the theories. Int J Antimicrob Agents 2003;22 Suppl 2:85–7.
9. Gobernado M. Fosfomicina. Rev Esp Quimioter 2003; 16:15–40.
10. Durupt S, Josserand RN, Sibille M, Durieu I. Acute, recurrent fosfomycin-induced liver toxicity in an adult patient with cystic fibrosis. Scand J Infect Dis 2001;33(5):391–2.
11. Ikeda K, Ida O, Kimoto K, Takatorige T, Nakanishi N, Tatara K. Effect of early fosfomycin treatment on prevention of hemolytic uremic syndrome accompanying Escherichia coli O157:H7 infection. Clin Nephrol 1999;52(6):357–62.
12. Shiomi M, Togawa M, Fujita K, Murata R. Effect of early oral fluoroquinolones in hemorrhagic colitis due to Escherichia coli O157:H7. Pediatr Int 1999;41(2):228–32.
13. Nakamura T, Kokuryo T, Hashimoto Y, Inui KI. Effects of fosfomycin and imipenem–cilastatin on the nephrotoxicity of vancomycin and cisplatin in rats. J Pharm Pharmacol 1999;51(2):227–32.
14. Matsumoto T, Tateda K, Miyazaki S, Furuya N, Ohno A, Ishii Y, Hirakata Y, Yamaguchi K. Fosfomycin alters lipopolysaccharide-induced inflammatory cytokine production in mice. Antimicrob Agents Chemother 1999;43(3):697–8.
15. Kumon H. Management of biofilm infections in the urinary tract. World J Surg 2000;24(10):1193–6.
16. Chomarat M. Resistance of bacteria in urinary tract infections. Int J Antimicrob Agents 2000;16(4):483–7.
17. Fukuyama M, Furuhata K, Oonaka K, Hara T, Sunakawa K. [Antibacterial activity of fosfomycin against the causative bacteria isolated from bacterial enteritis.]Jpn J Antibiot 2000;53(7):522–31.
18. Sakamoto M, Kaga K, Kamio T. Extended high-frequency ototoxicity induced by the first administration of cisplatin. Otolaryngol Head Neck Surg 2000;122(6):828–33.
19. Tandy JR, Tandy RD, Farris P, Truelson JM. In vivo interaction of cis-platinum and fosfomycin on squamous cell carcinoma. Laryngoscope 2000;110(7):1222–4.
20. Nakamura T, Kokuryo T, Hashimoto Y, Inui KI. Effects of fosfomycin and imipenem-cilastatin on the nephrotoxicity of vancomycin and cisplatin in rats. J Pharm Pharmacol 1999;51(2):227–32.

Fosmidomycin

General Information

Fosmidomycin is an antimicrobial drug that acts by inhibiting 1-deoxy-D-xylulose 5-phosphate reductoisomerase, a key enzyme of the non-mevalonate pathway of isoprenoid biosynthesis. It inhibits the synthesis of isoprenoids by Plasmodium falciparum and suppresses the growth of multidrug-resistant strains in vitro (1).

In an open, uncontrolled study, fosmidomycin was administered for 3–5 days (1.2 g every 8 hours) to 27 adults with malaria, of whom 16 reported possibly drug-related adverse events (2). The most frequent adverse events were headache, weakness, myalgia, abdominal pain, and loose stools. There were two cases of raised alanine transaminase activity.

Observational studies

In an open, uncontrolled trial in 26 patients, fosmidomycin 1200 mg every 8 hours for 7 days caused at least one adverse reaction in 20 subjects. None was serious and all were mild or moderate in intensity. Ten adverse events were categorized as being possibly related to the drug, and of these gastrointestinal events were the most frequent: loose stools (n = 3), diarrhea (n = 2), and flatulence (n = 2). There were increases in alanine transaminase activity in two subjects and dizziness in one (3).

In a randomized, controlled, open study of fosmidomycin combined with clindamycin (n = 12; 30 and 5 mg/kg respectively every 12 hours for 5 days), compared with fosmidomycin alone (n = 12; 30 mg/kg every 12 hours for 5 days) and clindamycin alone (n = 12; 5 mg/kg every 12 hours for 5 days) for the clearance of asymptomatic Plasmodium falciparum infections in schoolchildren in Gabon there were no serious adverse event (4). Gastrointestinal adverse events (eight cases of abdominal pain, seven of diarrhea, four of vomiting, two of loose stools, and two of nausea) were the most frequently reported adverse events. All gastrointestinal adverse events were judged to be possibly or probably related to the treatment. Nevertheless, all gastrointestinal adverse events were self-limiting and intervention was not required. There were no significant differences in the incidences of gastrointestinal adverse events between the groups.

In 52 children with P. falciparum malaria in Gabon who were given an oral combination of fosmidomycin + clindamycin (30 mg/kg and 10 mg/kg) there were no serious adverse events (5). Of 64 adverse events, 26 mild and three moderate adverse events were judged to be either possibly or probably related to the study drugs. The most frequent adverse events were gastrointestinal events (29 events, all mild), mostly loose stools (n = 13) and abdominal pain (n = 9). There was no difference in the distribution of gastrointestinal adverse events among the cohorts of different treatment durations.

Hematologic

In a single-arm study of an oral 3-day fixed-ratio combination of fosmidomycin 30 mg/kg and clindamycin 10 mg/kg every 12 hours for uncomplicated Plasmodium falciparum malaria in 51 children aged 1–14 years, one serious adverse event was reported in a boy aged 2.7 years who developed severe anemia on day 2 (6). His hemoglobin concentration fell from 9.2 g/dl on admission to 4.5 g/dl on day 2. He was given parenteral quinine but no blood transfusion. He made an uneventful recovery. In the whole group mean hemoglobin concentration, mean leukocyte count, and mean alanine transaminase activity fell significantly from day 0 to day 2. In seven patients (including the patient described above) the hemoglobin concentration fell by 2 g/dl but remained above 8 g/dl in all but two patients and subsequently recovered by day 28. The glucose-6-phosphate dehydrogenase status of these patients was not determined. The neutrophil counts in two of three patients with grade I neutropenia on admission recovered slightly by day 2 but fell further in the third patient from 1200 to 700 x 10^6/l, resulting in grade II neutropenia on day 2. Seven other patients aged 6–9 years developed transient asymptomatic neutropenia on day 2 (six grade I and one grade II). In four other patients, the neutrophil count fell below 1500 x 10^6/l (grade I) between days 7 and 14. Neutrophil counts returned to normal values by day 7 in all patients with day 2 neutropenia, without intervention. Neutropenia did not occur in children under 6 years old.

Susceptibility Factors

Age

In 50 children with *Plasmodium falciparum* malaria given consecutively shortened regimens of artesunate + fosmidomycin (1–2 mg/kg and 30 mg/kg respectively every 12 hours) the most frequent adverse events before day 7 were gastrointestinal (7/12), mostly abdominal pain (n = 5), but not diarrhea or loose stools (7). However, there were two cases of transient rises in alanine transaminase activity, in an 8-year-old girl (up to 96 U/l on day 2 from 24 U/l on admission) and in a 10-year-old boy (up to 157 U/l from 26 U/l).

References

1. Wiesner J, Borrmann S, Jomaa H. Fosmidomycin for the treatment of malaria. Parasitol Res 2003;90 Suppl 2:S71-6.
2. Missinou MA, Borrmann S, Schindler A, Issifou S, Adegnika AA, Matsiegui PB, Binder R, Lell B, Wiesner J, Baranek T, Jomaa H, Kremsner PG. Fosmidomycin for malaria. Lancet 2002;360(9349):1941–2.
3. Lell B, Ruangweerayut R, Wiesner J, Missinou MA, Schindler A, Baranek T, Hintz M, Hutchinson D, Jomaa H, Kremsner PG. Fosmidomycin, a novel chemotherapeutic agent for malaria. Antimicrob Agents Chemother 2003;47:735–8.
4. Borrmann S, Adegnika AA, Matsiegui PB, Issifou S, Schindler A, Mawili-Mboumba DP, Baranek T, Wiesner J, Jomaa H, Kremsner PG. Fosmidomycin–clindamycin for Plasmodium falciparum Infections in African children. J Infect Dis 2004;189(5):901–8.
5. Borrmann S, Issifou S, Esser G, Adegnika AA, Ramharter M, Matsiegui PB, Oyakhirome S, Mawili-Mboumba DP, Missinou MA, Kun JF, Jomaa H, Kremsner PG. Fosmidomycin-clindamycin for the treatment of *Plasmodium falciparum* malaria. J Infect Dis 2004;190(9):1534–40.
6. Borrmann S, Lundgren I, Oyakhirome S, Impouma B, Matsiegui PB, Adegnika AA, Issifou S, Kun JF, Hutchinson D, Wiesner J, Jomaa H, Kremsner PG. Fosmidomycin plus clindamycin for treatment of pediatric patients aged 1 to 14 years with *Plasmodium falciparum* malaria. Antimicrob Agents Chemother 2006;50(8):2713–8.
7. Borrmann S, Adegnika AA, Moussavou F, Oyakhirome S, Esser G, Matsiegui PB, Ramharter M, Lundgren I, Kombila M, Issifou S, Hutchinson D, Wiesner J, Jomaa H, Kremsner PG. Short-course regimens of artesunate–fosmidomycin in treatment of uncomplicated *Plasmodium falciparum malaria*. Antimicrob Agents Chemother 2005;49(9):3749–54.

Fusidic acid

General Information

Fusidic acid is the best-known representative of a group of antibiotics with steroid structures, which are eliminated primarily by biliary excretion as microbiologically inactive metabolites. The antibacterial action of fusidic acid is bacteriostatic, although it can be bactericidal at higher concentrations. It exerts its antibacterial effect by inhibiting protein synthesis, but the exact mechanism by which this inhibition occurs has not yet been elucidated. It has a narrow antibiotic spectrum, with a bacteriostatic or slow bactericidal effect mainly directed at both coagulase-negative staphylococci and *Staphylococcus aureus* 1. Fusidic acid may also have a role as a clinically useful suppressor of immunoinflammatory processes 2.

Fusidic acid is adequately absorbed from the gastrointestinal tract, but has also been used topically. It has the important property of good tissue penetration, including entry into bones and joints, but does not reach the cerebrospinal fluid. Elimination of fusidic acid is primarily by non-renal mechanisms, and a proportion is metabolized to several breakdown products detectable in bile. Systemic clearance is increased by hypoalbuminemia, reduced by severe cholestasis, and is unchanged in renal insufficiency (3).

General adverse effects

The most common adverse effects of fusidic acid are minor and are related to the gastrointestinal tract (discomfort, diarrhea) (4). Rare adverse events include granulocytopenia, thrombocytopenia, venous spasm, and skin reactions 5. Fusidic acid has detergent properties and can cause hemolysis when injected intravenously or can induce tissue damage when given intramuscularly. However, its systemic toxicity is relatively low.

Bacterial resistance has been an obstacle to the widespread use of fusidic acid; it develops in a single step in vitro and has also been observed in patients, particularly during prolonged administration (6).

Organs and Systems

Hematologic

The hematological adverse effects of fusidic acid, such as granulocytopenia and thrombocytopenia, have been rarely reported. Two cases of fusidic acid-induced leukopenia and thrombocytopenia after 2 weeks of fusidic acid treatment have been reported; in both cases, the abnormality resolved in 3-6 days after withdrawal of fusidic acid (7).

There have been six reports of fusidic acid-induced sideroblastic anemia after treatment with fusidic acid for 32–190 (mean 81) days; four required repeated blood transfusions (8). After fusidic acid withdrawal in five patients, there was complete recovery. In one patient, rechallenge with fusidic acid resulted in recurrence of anemia and resolution resolved after definitive withdrawal.

Gastrointestinal

The major adverse effects of fusidic acid are mild gastrointestinal discomfort and diarrhea (9). Oral fusidic acid can cause *Klebsiella oxytoca*-associated colitis (10).

Liver

Fusidic acid is chemically very similar to bile acids and hence competes with them for elimination and metabolism. A patient with a past history of alcohol-induced cirrhosis developed cholestatic jaundice after taking oral fusidic acid for 2 days (11).

Urinary tract

Severe hypocalcemia and acute renal insufficiency developed in two diabetic patients who took oral fusidic acid 500 mg tds (12).

Skin

Acanthosis nigricans has been associated with fusidic acid (13).

Sensitization to sodium fusidate is rare, and most often found in patients with stasis dermatitis or atopic dermatitis.

- Allergic contact dermatitis has been reported in a 26-year-old Korean woman after treatment of an abrasion on the left knee with fusidic acid ointment and povidone iodine (Betadine) (14). The diagnosis was confirmed by a strongly positive patch test for sodium fusidate.

Immunologic

Fusidic acid may have an immunomodulatory effect that seems to be partly mediated by suppression of cytokine production. The effect on several diseases has been investigated, but it remains to be characterized more in detail whether there is any therapeutic usefulness (5).

Positive patch tests occurred in three of 1119 patients who had used topical fusidic acid (15). In the second part of this study, all cases of positive patch tests to fusidic acid over the previous 20 years were reviewed; the average frequency was 1.62 patch-tested patients per year (1.45%) of those who were patch-tested.

Long-Term Effects

Drug tolerance

Mechanisms of resistance to fusidic acid include alterations in elongation factor G (whose inhibition at the ribosome level mediates the effect of fusidic acid), altered drug permeability, binding by chloramphenicol acetyltransferase type 1, and enhanced efflux (16).

Most studies show low levels of resistance in staphylococci, with a slightly higher rate in methicillin-resistant *S. aureus* (MRSA) (16). This has been confirmed in recent studies. Only 1 of 106 isolates of methicillin-resistant *S. aureus* was also resistant to fusidic acid (17). Only very little resistance to fusidic acid has been reported from Wales (18). However, clonal spread of fusidic acid-resistant methicillin-resistant *S. aureus* has been documented in Norway (19). A resistance rate of about 50% has been found in coagulase-negative staphylococci isolated in eight Swedish intensive care units between 1995 and 1997 (20). A high rate of resistance to fusidic acid was also found in bacteria that were cultured from blood after dental treatment (21).

Pathogenic nocardia may inactivate fusidic acid (22). In an in vitro study, salicylate and related compounds increased the MIC in strains of *S. aureus* both resistant and susceptible to fusidic acid (23).

In a prospective randomized trial, oral fusidic acid alone (500 mg tds for 7 days) failed to eradicate methicillin-resistant *S. aureus* colonization but resulted in the emergence of fusidic acid-resistant strains (24).

In an in vitro susceptibility study of 170 clinical isolates of *Mycobacterium tuberculosis* to fusidic acid, 19 isolates were resistant to at least one first-line antituberculosis drug (25). In all, 1.8% of the isolates were resistant to fusidic acid. Fusidic acid can be a potential supplementary drug for the treatment of infections due to multidrug-resistant strains of *M. tuberculosis*.

Susceptibility Factors

Other features of the patient

The half-life of fusidic acid was shortened in ten patients with severe burns treated with 500 mg tds intravenously (26). This effect may be explained by translesional extrahepatic clearance and an increase in hepatic clearance.

Drug Administration

Drug dosage regimens

In a multicenter, randomized, double-blind, parallel-group comparison of a new regimen of fusidic acid suspension (10 mg/kg bd) against a standard regimen (17 mg/kg tds) in 213 children with skin and soft tissue infections, adverse drug reactions were significantly less common with the new twice-daily regimen (27). This was because of a significantly higher incidence of gastrointestinal reactions with the higher dose. Adverse drug reactions caused or contributed to treatment withdrawal in three children given twice-daily doses and in 12 children given thrice-daily doses, and were the primary reasons for stopping treatment in one and in eight children respectively.

Drug–Drug Interactions

HIV protease inhibitors

In a 32-year-old man infected with HIV plasma concentrations of ritonavir, saquinavir, and fusidic acid were significantly raised when these drugs were administered in combination, possibly from mutual inhibition of metabolism (28).

Phenazone

Activation of the CYP450 enzyme system, demonstrated by an increase in the metabolism of phenazone (antipyrine), has been found in 30 HIV-positive drug abusers (29). This activating effect was demonstrated after 28 but not after 14 days of treatment with fusidic acid 500 mg/day, suggesting time dependency.

Statins

Acute rhabdomyolysis has been attributed to an interaction of fusidic acid with simvastatin.

Acute rhabdomyolysis has been reported in a 71-year-old man taking simvastatin (40 mg/day) plus fusidic acid 250 mg tds for an infection due to methicillin-resistant *Staphylococcus aureus* 30.

- A 66-year-old kidney transplant recipient developed a gangrenous lesion on the left foot infected with *S. aureus* and *Escherichia coli* (31). He was given ciprofloxacin and clindamycin for 6 weeks and then fusidic acid 1500 mg/day for 2 weeks. He became ill, with myalgia and no active movement of his legs, and rhabdomyolysis was established by laboratory tests. He had also taken atorvastatin 10 mg/day and he slowly recovered after withdrawal of both atorvastatin and fusidic acid.

In a model of staphylococcal meningitis in rabbits, there was antagonism between methicillin and fusidic acid (32).

References

1. Collignon P, Turnidge J. Fusidic acid in vitro activity. Int J Antimicrob Agents 1999;12(Suppl 2):S45–58.
2. Bendtzen K, Diamant M, Faber V. Fusidic acid, an immunosuppressive drug with functions similar to cyclosporin A. Cytokine 1990;2(6):423–9.
3. Turnidge J. Fusidic acid pharmacology, pharmacokinetics and pharmacodynamics. Int J Antimicrob Agents 1999;12(Suppl 2):S23–34.
4. Johnston GA. Treatment of bullous impetigo and the staphylococcal scalded skin syndrome in infants. Expert Rev Anti Infect Ther 2004;2(3):439–46.
5. Christiansen K. Fusidic acid adverse drug reactions. Int J Antimicrob Agents 1999;12(Suppl 2):S3–9.
6. Shanson DC. Clinical relevance of resistance to fusidic acid in Staphylococcus aureus. J Antimicrob Chemother 1990;25(Suppl B):15–21.
7. Liao YM, Chiu CF, Ho MW, Hsueh CT. Fusidic acid-induced leukopenia and thrombocytopenia. J Chin Med Assoc 2003;66:429–32.
8. Vial T, Grignon M, Daumont M, Guy C, Zenut M, Germain ML, Jaubert J, Ruivard M, Guyotat D, Descotes J. Sideroblastic anaemia during fusidic acid treatment. Eur J Haematol 2004;72(5):358–60.
9. Garcia-Rodriguez JA, Gutierrez Zufiaurre N, Munoz Bellido JL. Acido fusidico. Rev Esp Quimioter 2003;16:161–71.
10. Seksik P, Galula G, Maury E, Levy VG, Offenstadt G. [Klebsiella oxytoca-associated colitis after oral administration of fusidic acid.]Gastroenterol Clin Biol 2000;24(5):587–588.
11. Carbonell N, Thabut D, Podevin P, Biour M, Serfaty L, Poupon R. Ictère choléstatique induit par la prise d'acide fucidique chez un patient cirphotique. [Cholestatic icterus induced by the administration of fusidic acid in a cirrhotic patient.] Presse Méd 2002;31(23):1083–4.
12. Biswas M, Owen K, Jones MK. Hypocalcaemia during fusidic acid therapy. J R Soc Med 2002;95(2):91–3.
13. Cairo F, Rubino I, Rotundo R, Prato GP, Ficarra G. Oral acanthosis nigricans as a marker of internal malignancy. A case report. J Periodontol 2001;72(9):1271–5.
14. Lee AY, Joo HJ, Oh JG, Kim YG. Allergic contact dermatitis from sodium fusidate with no underlying dermatosis. Contact Dermatitis 2000;42(1):53.
15. Morris SD, Rycroft RJ, White IR, Wakelin SH, McFadden JP. Comparative frequency of patch test reactions to topical antibiotics. Br J Dermatol 2002;146(6):1047–51.
16. Turnidge J, Collignon P. Resistance to fusidic acid. Int J Antimicrob Agents 1999;12(Suppl 2):S35–44.
17. Liu CP, Lee CM, Su SC, Li YT. Susceptibility testing and clinical effect of fusidic acid in oxacillin-resistant Staphylococcus aureus infections. J Microbiol Immunol Infect 1999;32(3):194–8.
18. Morgan M, Salmon R, Keppie N, Evans-Williams D, Hosein I, Looker DN. All Wales surveillance of methicillin-resistant Staphylococcus aureus (MRSA): the first year's results. J Hosp Infect 1999;41(3):173–9.
19. Andersen BM, Bergh K, Steinbakk M, Syversen G, Magnaes B, Dalen H, Bruun JN. A Norwegian nosocomial outbreak of methicillin-resistant Staphylococcus aureus resistant to fusidic acid and susceptible to other antistaphylococcal agents. J Hosp Infect 1999;41(2):123–32.
20. Erlandsson CM, Hanberger H, Eliasson I, Hoffmann M, Isaksson B, Lindgren S, Nilsson LE, Soren L, Walther SM. Surveillance of antibiotic resistance in ICUs in southeastern Sweden. ICU Study Group of the South East of Sweden. Acta Anaesthesiol Scand 1999;43(8):815–20.
21. Messini M, Skourti I, Markopulos E, Koutsia-Carouzou C, Kyriakopoulou E, Kostaki S, Lambraki D, Georgopoulos A. Bacteremia after dental treatment in mentally handicapped people. J Clin Periodontol 1999;26(7):469–73.
22. Harada K, Tomita K, Fujii K, Sato N, Uchida H, Yazawa K, Mikami Y. Inactivation of fusidic acid by pathogenic Nocardia. J Antibiot (Tokyo) 1999;52(3):335–9.

23. Price CT, O'Brien FG, Shelton BP, Warmington JR, Grubb WB, Gustafson JE. Effects of salicylate and related compounds on fusidic acid MICs in *Staphylococcus aureus*. J Antimicrob Chemother 1999;44(1):57–64.
24. Chang SC, Hsieh SM, Chen ML, Sheng WH, Chen YC. Oral fusidic acid fails to eradicate methicillin-resistant *Staphylococcus aureus* colonization and results in emergence of fusidic acid-resistant strains. Diagn Microbiol Infect Dis 2000;36(2):131–6.
25. Cicek-Saydam C, Cavusoglu C, Burhanoglu D, Hilmioglu S, Ozkalay N, Bilgic A. In vitro susceptibility of *Mycobacterium tuberculosis* to fusidic acid. Clin Microbiol Infect 2001;7(12):700–2.
26. Lesne-Hulin A, Bourget P, Le Bever H, Carsin H. Pharmacocinetique de l'acide fusidique administre au sujet gravement brule infecte. [Pharmacokinetics of fusidic acid in patients with seriously infected burns.] Pathol Biol (Paris) 1999;47(5):486–90.
27. Torok E, Somogyi T, Rutkai K, Iglesias L, Bielsa I. Fusidic acid suspension twice daily: a new treatment schedule for skin and soft tissue infection in children, with improved tolerability. J Dermatolog Treat 2004;15(3):158–63.
28. Reimann G, Barthel B, Rockstroh JK, Spatz D, Brockmeyer NH. Effect of fusidic acid on the hepatic cytochrome P450 enzyme system. Int J Clin Pharmacol Ther 1999;37(11):562–6.
29. Khaliq Y, Gallicano K, Leger R, Foster B, Badley A. A drug interaction between fusidic acid and a combination of ritonavir and saquinavir. Br J Clin Pharmacol 2000;50(1):82–3.
30. Yuen SL, McGarity B. Rhabdomyolysis secondary to interaction of fusidic acid and simvastatin. Med J Aust 2003;179:172.
31. Wenisch C, Krause R, Fladerer P, El Menjawi T, Pohanka E. Acute rhabdomyolysis after atorvastatin and fusidic acid therapy. Am J Med 2000;109(1):78.
32. Ostergaard C, Yieng-Kow RV, Knudsen JD, Frimodt-Moller N, Espersen F. Evaluation of fusidic acid in therapy of experimental *Staphylococcus aureus* meningitis. J Antimicrob Chemother 2003;51:1301–5.

Garenoxacin

See also Fluoroquinolones

General Information

Garenoxacin (T-3811ME, BMS-284756) is a novel des-F(6) quinolone that is effective in vitro against a wide range of clinically important pathogens, including Gram-positive and Gram-negative aerobes and anaerobes (1).

Placebo-controlled studies

In a randomized, double-blind, placebo-controlled, dose escalation trial in 40 healthy subjects receiving garenoxacin 100-1200 mg/day, the most common adverse events in the subjects who received garenoxacin were headache (23%), pharyngitis (17%), dizziness (13%), and a white exudate (13%). There was no relation between the dose of garenoxacin and either the type or the frequency of adverse events [2].

Organs and Systems

Nervous system

In rodents the effects of garenoxacin on the central nervous system were weaker than those of norfloxacin, ciprofloxacin, sitafloxacin, and trovafloxacin [3]. Garenoxacin may therefore have a low potential for central nervous system adverse reactions.

Musculoskeletal

In dogs garenoxacin concentrations in plasma and joint tissue were higher than those of ciprofloxacin and norfloxacin (4). However, the articular toxicity of garenoxacin was much less than that of the other two quinolones.

Second-Generation Effects

Lactation

Excretion of garenfloxacin into breast milk has been studied in six lactating women taking garenoxacin 600 mg/day [5]. Garenoxacin exposure in breast milk was minimal—a mean of 0.07% of the administered dose was recovered within 120 hours. Garenoxacin was undetectable in the breast milk of most subjects within 84 hours of dosing. A nursing infant whose mother had taken a single oral dose of garenoxacin could 600 mg would theoretically be exposed to 0.42 mg of garenoxacin (0.105 mg/kg/day for a 4-kg infant over 5 days of nursing). If extrapolated to a 14-day course of garenoxacin 600 mg/day, total exposure would be about 5.88 mg.

References

1. Wise R, Gee T, Marshall G, Andrews JM. Single-dose pharmacokinetics and penetration of BMS 284756 into an inflammatory exudate. Antimicrob Agents Chemother 2002;46(1):242–4.
2. Gajjar DA, Bello A, Ge Z, Christopher L, Grasela DM. Multiple-dose safety and pharmacokinetics of oral garenoxacin in healthy subjects. Antimicrob Agents Chemother 2003;47:2256–63.
3. Nagai A, Miyazaki M, Morita T, Furubo S, Kizawa K, Fukumoto H, Sanzen T, Hayakawa H, Kawamura Y. Comparative articular toxicity of garenoxacin, a novel quinolone antimicrobial agent, in juvenile beagle dogs. J Toxicol Sci 2002;27(3):219–28.
4. Nakamura T, Fukuda H, Morita Y, Soumi K, Kawamura Y. Pharmacological evaluation of garenoxacin, a novel des-F(6)-quinolone antimicrobial agent: effects on the central nervous system. J Toxicol Sci 2003;28:35–45.
5. Amsden GW, Nicolau DP, Whitaker AM, Maglio D, Bello A, Russo R, Barros A Jr, Gajjar DA. Characterization of the penetration of garenoxacin into the breast milk of lactating women. J Clin Pharmacol 2004;44(2):188–92.

Gatifloxacin

See also Fluoroquinolones

General Information

Gatifloxacin is an 8-methoxyfluoroquinolone with enhanced activity against Gram-positive, atypical agents, and some anerobes, and broad-spectrum activity against Gram-negative bacteria (1–4). It is bactericidal and produces a post-antibiotic effect in Gram-positive and Gram-negative bacteria. Gatifloxacin is well absorbed from the gastrointestinal tract (oral availability almost 100%), and concomitant administration of a continental breakfast, 1050 kcal, had no effect on its availability (5). The standard dose is 400 mg od and both oral and intravenous formulations are available (6,7).

Since gatifloxacin has a high oral systemic availability (96%), oral and intravenous formulations are bioequivalent and interchangeable (8). It has a large volume of distribution (about 1.8 l/kg), low protein binding (about 20%), broad tissue distribution, and is primarily excreted unchanged in the urine (over 80%) (7). After daily repeated administration, there was predictable modest accumulation; steady state concentrations were reached after the third dose (9).

The in vitro antibacterial spectrum of gatifloxacin has been tested against a variety of clinically important microorganisms (10). It is two to four times more potent than ciprofloxacin and ofloxacin against staphylococci, streptococci, pneumococci, and enterococci. However, it is two times less potent than ciprofloxacin, but the same as or two times more potent than ofloxacin against *Enterobacteriaceae*. Gatifloxacin and ofloxacin have similar antipseudomonal activity, while ciprofloxacin is two to eight times more potent. Gatifloxacin is highly potent against *Hemophilus influenzae*, *Legionella* species, and *Helicobacter pylori*, and also has activity against *Bacteroides fragilis* and *Clostridium difficile*. Like other quinolones, it has poor activity against *Mycobacterium avium* intracellulare, but is 8–16 times more potent against *Mycobacterium tuberculosis*.

Intravenous gatifloxacin can cause dose-related local reactions (8).

Gatifloxacin does not interact with drugs metabolized by the CYP450 enzyme family, as assessed in 14 healthy adult men using midazolam as a probe (11).

Comparative studies

In a double-blind, randomized study in 102 adult women who took either gatifloxacin 400 mg as a single dose or 200 mg/day for 3 days or ciprofloxacin 250 mg bd for 3 days, the frequencies of treatment-related adverse events were 17%, 19%, and 14% respectively (12). The most common adverse effects were nausea (4.6%, 5.4%, and 5.0%), headache (1.9%, 1.9%, and 1.4%), dizziness (2.2%, 1.3%, and 1.4%), and diarrhea (1.1%, <1%, and 1.4%).

Organs and Systems

Cardiovascular

Although early studies suggested that gatifloxacin has little effect on the QT interval of the electrocardiogram (7,13), clinical trial data and data from phase IV studies have shown that it prolongs the QT interval (14). Four cases of gatifloxacin-associated cardiac toxicity have been reported in patients with known risk factors for this adverse event (15).

- A 95-year-old woman took gatifloxacin and developed recurrent episodes of torsade de pointes on the 4th day of treatment and one hour after infusion (16).

Nervous system

In patients taking gatifloxacin there have been reports of myoclonus and generalized seizures (17) and ataxia and generalized seizures (18). Status epilepticus has been attributed to gatifloxacin 400 mg/day in a 79-year-old woman with a non-febrile urinary tract infection with Escherichia coli (19)

Of five patients who developed suspected drug-induced *aseptic meningitis*, one had been treated for *Salmonella* gastroenteritis with ciprofloxacin (20). Symptoms of nervous system disease, including psychomotor excitement and deterioration of auto and allopsychical orientation, occurred after 4 days. She had a concomitant urticarial rash and considerable swelling of the face. Ciprofloxacin was immediately withdrawn and her symptoms improved.

Sensory systems

In phase III trials of gatifloxacin, the most frequently reported adverse events were conjunctival irritation, increased lacrimation, keratitis, and papillary conjunctivitis, which occurred in 5–10% of patients with bacterial conjunctivitis (21). Chemosis, conjunctival hemorrhage, dry eyes, discharge, eye irritation, eye pain, eyelid edema, headache, red eye, reduced visual acuity, and taste disturbance were reported by 1–4% of patients. Treatment-related adverse events were reported by 3.8% of those who took twice-daily doses and 1.9% of those who took doses four times a day.

Gatifloxacin can cause intrastromal macroscopic crystalline deposits through a compromised corneal epithelium.

- An 85-year-old man developed faint crystal-like white precipitates in the mid-peripheral stroma of his left cornea 3 weeks after starting to use 0.3% gatifloxacin eye-drops (22).

Psychiatric

Psychiatric adverse effects occur at a rate of 2?4%, causing headache (2–4% of patients), dizziness (2–3%), and other symptoms (under 1%), including confusion, agitation, insomnia, depression, somnolence, vertigo, lightheadedness, and tremors. Seizures are rare. Some quinolones displace GABA or compete with GABA binding at receptor sites in the nervous system. Substitution of compounds containing 7-piperazinyl or 7-pyrrolidinyl , such as gatifloxacin, gemifloxacin, and moxifloxacin, is associated with reduced seizure-causing potential. Administration of non-steroidal anti-inflammatory drugs concurrently with certain quinolones has been linked to an increase in the possibility of seizures (23).

- Delirium occurred in a 69-year-old white man with a history of depression, non-insulin-dependent diabetes mellitus, hypertension, and atherosclerotic disease who was treated with intravenous gatifloxacin 400 mg/day (24). After the first dose of gatifloxacin he had numerous hallucinations and the symptoms got worse after each dose. After withdrawal no further hallucinations occurred.
Gatifloxacin-induced hallucinations have been described (25).
- An otherwise-healthy 19-year-old young military recruit developed a community-acquired pneumonia, for which he was given intravenous gatifloxacin 400 mg/day. He denied a past psychiatric history, but did have a brother with bipolar affective disorder. After 2 days he complained of hallucinations and had paranoia. No changes in his medications were made for 2 days, and he continued to have intermittent hallucinations. The pneumonia did not improve and piperacillin/tazobactam 3.375 g 6-hourly was added. The gatifloxacin was then withdrawn, and his hallucinations resolved with a time-course consistent with the half-life of gatifloxacin, 7–14 hours.

A psychosis attributed to gatifloxacin occurred in an elderly woman with dementia (26).

Endocrine

Gatifloxacin can cause either hyperglycemia or hypoglycemia. Hyperglycemia was reported in three patients taking gatifloxacin (200 mg/day) and hypoglycemia in one patient (27,28,29). The mechanism of hypoglycemia, shown in vitro, is increased insulin secretion by inhibition of pancreatic beta-cell K_{ATP} channels (30).

Metabolism

Gatifloxacin was well tolerated in patients with non-insulin-dependent diabetes mellitus maintained with diet and exercise (31). It had no significant effect on glucose homeostasis, beta cell function, or long-term fasting serum glucose concentrations, but it caused a brief increase in serum insulin concentrations.

However, in two population-based nested case-control studies of 788 patients who were treated for hypoglycemia within 30 days of antibiotic therapy, gatifloxacin was associated with an increased risk of hypoglycemia (adjusted OR = 4.3; 95% CI = 2.9, 6.3) (32). Levofloxacin was also associated with a slightly increased risk (adjusted OR = 1.5; 95% CI = 1.2, 2.0), but there was no such risk with moxifloxacin or ciprofloxacin. In 470 patients treated for hyperglycemia within 30 days of antibiotic therapy, gatifloxacin was associated with a considerably increased risk of hyperglycemia (adjusted

OR = 17; 95% CI = 10, 27), but there was no risk with the other antibiotics.

In a review of hypoglycemia or hyperglycemia attributed to fluoroquinolones, 93% of all episodes of dysglycemia were due to gatifloxacin; only 11% of reports were associated with levofloxacin and only 10% of reports associated with moxifloxacin involved either hypoglycemia or hyperglycemia (33). Analysis of the data suggested that significantly more episodes of hypoglycemia were associated with gatifloxacin than with either levofloxacin or moxifloxacin.

Gatifloxacin-induced hypoglycemia is associated with concomitant use of sulfonylureas, and usually occurs immediately after drug administration. It stimulates insulin secretion in mouse pancreatic beta-cells and the effect is additive with glibenclamide (34). On the other hand, long-term gatifloxacin treatment reduces islet cell insulin contents by inhibiting insulin biosynthesis, which may be associated with gatifloxacin-induced hyperglycemia, which often takes several days to develop. Withdrawal of gatifloxacin results in improved insulin secretion.

The manufacturer of gatifloxacin has changed the labelling to strengthen an existing warning about hypoglycemia and hyperglycemia, adding a contraindication to its use in people with diabetes, and providing information identifying other susceptibility factors for low or high blood glucose concentrations, including advanced age, kidney malfunction, and taking glucose-altering medications while taking gatifloxacin (35, 36, 37, 38).

- Hyperglycemia occurred in an 86-year-old man who took (39) gatifloxacin 400 mg/day for suspected pneumonia. After 4 days his mean blood glucose concentration increased from 7.4 mmol/l (133 mg/dl) to 30 mmol/l (537 mg/dl). Although he had glycosuria, he did not complain of symptoms of hyperglycemia, such as polyuria, polyphagia, or polydipsia. The hyperglycemia resolved after gatifloxacin was withdrawn and he had received regular insulin 15 U subcutaneously over 5 hours.

Another 15 cases of dysglycemia associated with gatifloxacin have been described (40, 41, 42).

In an observational study of spontaneous adverse events reports gatifloxacin was associated with much higher rates of *hypoglycemia* and *hyperglycemia* (477 reports per 10^7 retail prescriptions) compared with ciprofloxacin (4 reports), levofloxacin (11 reports), and moxifloxacin (36 reports) (43).

In four pivotal studies in 867 children, low concentrations of fasting and non-fasting glucose occurred more often in children taking gatifloxacin (5.8%) than in those taking co-amoxiclav (2.5%) (44).

Gatifloxacin 600 mg intravenously has been associated with a serious episode of hypoglycemia after cardiopulmonary bypass (45).

Severe hyperglycemia occurred in a non-diabetic patient with progressive renal dysfunction who took gatifloxacin 200 mg/day for 9 days (46). Gatifloxacin was withdrawn and blood glucose concentrations returned to normal within several days after intensive treatment with insulin.

Gastrointestinal

Diarrhea and nausea were the most common adverse events in several clinical studies of gatifloxacin (4,47–49).

Liver

Gatifloxacin can cause hepatitis (50,51).

- A 44-year-old woman developed acute hepatitis while taking gatifloxacin for chronic sinusitis (52). After 5 days she developed nausea, lethargy, and abdominal pain, all of which progressed over the next few days. Liver function tests were abnormal, and the bilirubin peaked at 161 µmol/l. A percutaneous liver biopsy showed acute hepatitis with eosinophilic infiltrates, consistent with drug-induced hepatitis.

Susceptibility Factors

Age

In a pharmacokinetic study in 111 children taking gatifloxacin oral suspension or tablets 5, 10, or 15 mg/kg, the most common adverse events were vomiting (17%), headache (3%), and rash (4%) (53). Three had transiently increased liver function tests, up to seven times the upper limit of the reference range.

Gatifloxacin can be administered without dose modification in the elderly (54).

Sex

The effects of sex on the pharmacokinetics of a single oral dose of gatifloxacin 400 mg have been investigated in 12 healthy subjects (6 men, 6 women) (55). The women had a significantly smaller volume of distribution. However, there was no significant difference when normalized for total body weight or lean body weight. Given the smaller volume of distribution, women may have slightly higher maximum concentrations, but these differences are unlikely to have clinical significance. Gatifloxacin can be administered without dose modification in women (56).

Renal impairment

Gatifloxacin is excreted in the urine unchanged, via glomerular filtration alone, resulting in 80- 95% recovery of an administered dose by this route (26). Doses should therefore be reduced in renal insufficiency.

Hepatic disease

Gatifloxacin can be administered without dose modification in patients with hepatic impairment (56).

Drug–Drug Interactions

Calcium carbonate

Reduced absorption of quinolones by divalent and trivalent cations such as aluminium, magnesium, calcium, iron, and zinc occurs by the formation of insoluble chelates.

- A 77-year old woman took gatifloxacin 400 mg/day for pneumonia together with calcium carbonate 500 mg and a multivitamin formulation and failed to improve (57). The time of administration of gatifloxacin was changed and her symptoms improved 2 days later. However, plasma concentrations of gatifloxacin were not measured.

Oral hypoglycemic drugs

Four cases of severe persistent hypoglycemia due to gatifloxacin 400 mg/day have been reported in adults with diabetes mellitus who were taking oral hypoglycemic agents (repaglinide, glibenclamide, glimepiride) (58,59).

Oxycodone

Oral oxycodone and gatifloxacin can be co-administered without a significant reduction in systemic availability (60).

Sulfonylureas

During the post-marketing period, reports to the manufacturers of gatifloxacin of hypoglycemia and hyperglycemia caused a revision of the package insert to include warnings of serious disturbances in glucose metabolism. Hyperglycemia was reported in two patients taking gatifloxacin 200 mg/day and glipizide. Hypoglycemia occurred in a 68-year-old woman who was also taking glibenclamide and gatifloxacin (61).

Warfarin

Enhanced hypoprothrombinemia has been reported when gatifloxacin was given with warfarin (62). Anticoagulation has been studied retrospectively in patients taking stable dosages of warfarin who started taking levofloxacin (n = 54) or gatifloxacin (n = 38), by reviewing general patient databases and electronic medical records (63). The INR was above 4 in one patient taking levofloxacin and in eight patients taking gatifloxacin. The authors concluded that close monitoring of INR is warranted when gatifloxacin is added.

Interference with Diagnostic Tests

Gatifloxacin can interfere with urinary screening for opiates (64).

- A 48-year-old man was given gatifloxacin for a urinary tract infection. While he was taking the antibiotic, two urine screens were positive for opiates; the results of previous screens had been negative. A test using a different assay method was negative and 2 weeks after completing the course of gatifloxacin the urine screen was again negative for opiates.

The mechanism by which gatifloxacin cross-reacted with the opiate immunoassay was not known.

References

1. Fish DN, North DS. Gatifloxacin, an advanced 8-methoxy fluoroquinolone. Pharmacotherapy 2001;21(1):35–59.
2. Naber CK, Steghafner M, Kinzig-Schippers M, Sauber C, Sorgel F, Stahlberg HJ, Naber KG. Concentrations of gatifloxacin in plasma and urine and penetration into prostatic and seminal fluid, ejaculate, and sperm cells after single oral administrations of 400 milligrams to volunteers. Antimicrob Agents Chemother 2001;45(1):293–7.
3. Honeybourne D, Banerjee D, Andrews J, Wise R. Concentrations of gatifloxacin in plasma and pulmonary compartments following a single 400 mg oral dose in patients undergoing fibre-optic bronchoscopy J Antimicrob Chemother 2001;48(1):63–6.
4. Perry CM, Ormrod D, Hurst M, Onrust SV. Gatifloxacin: a review of its use in the management of bacterial infections. Drugs 2002;62(1):169–207.
5. Mignot A, Guillaume M, Gohler K, Stahlberg HJ. Oral bioavailability of gatifloxacin in healthy volunteers under fasting and fed conditions. Chemotherapy 2002;48(3):111–5.
6. Blondeau JM. Gatifloxacin: a new fluoroquinolone. Expert Opin Investig Drugs 2000;9(8):1877–95.
7. Grasela DM. Clinical pharmacology of gatifloxacin, a new fluoroquinolone. Clin Infect Dis 2000;31(Suppl 2):S51–8.
8. LaCreta FP, Kaul S, Kollia GD, Duncan G, Randall DM, Grasela DM. Interchangeability of 400-mg intravenous and oral gatifloxacin in healthy adults. Pharmacotherapy 2000;20(6 Pt 2):S59–66.
9. Gajjar DA, LaCreta FP, Uderman HD, Kollia GD, Duncan G, Birkhofer MJ, Grasela DM. A dose-escalation study of the safety, tolerability, and pharmacokinetics of intravenous gatifloxacin in healthy adult men. Pharmacotherapy 2000;20(6 Pt 2):S49–58.
10. Fung-Tomc J, Minassian B, Kolek B, Washo T, Huczko E, Bonner D. In vitro antibacterial spectrum of a new broad-spectrum 8-methoxy fluoroquinolone, gatifloxacin. J Antimicrob Chemother 2000;45(4):437–46.
11. Grasela DM, LaCreta FP, Kollia GD, Randall DM, Uderman HD. Open-label, nonrandomized study of the effects of gatifloxacin on the pharmacokinetics of midazolam in healthy male volunteers. Pharmacotherapy 2000;20(3):330–5.
12. Naber KG, Allin DM, Clarysse L, Haworth DA, James IG, Raini C, Schneider H, Wall A, Weitz P, Hopkins G, Ankel-Fuchs D. Gatifloxacin 400 mg as a single shot or 200 mg once daily for 3 days is as effective as ciprofloxacin 250 mg twice daily for the treatment of patients with uncomplicated urinary tract infections. Int J Antimicrob Agents 2004;23(6):596–605.
13. Iannini PB, Circiumaru I. Gatifloxacin-induced QTc prolongation and ventricular tachycardia. Pharmacotherapy 2001;21(3):361–2.
14. Owens RC Jr, Ambrose PG. Torsades de pointes associated with fluoroquinolones. Pharmacotherapy 2002;22(5):663–8discussion 668–72.
15. Bertino JS Jr, Owens RC Jr, Carnes TD, Iannini PB. Gatifloxacin-associated corrected QT interval prolongation, torsades de pointes, and ventricular fibrillation in patients with known risk factors. Clin Infect Dis 2002;34(6):861–3.
16. Fteha A, Fteha E, Haq S, Kozer L, Saul B, Kassotis J. Gatifloxacin induced torsades de pointes. Pacing Clin Electrophysiol 2004;27(10):1449–50.
17. Marinella MA. Myoclonus and generalized seizures associated with gatifloxacin treatment. Arch Intern Med 2001;161(18):2261–2.
18. Mohan N, Menon K, Rao PG. Oral gatifloxacin-induced ataxia. Am J Health Syst Pharm 2002;59(19):1894.
19. Koussa SF, Chahine SL, Samaha EI, Riachi MA. Generalized status epilepticus possibly induced by gatifloxacin. Eur J Neurol 2006;13(6):671–2.
20. Kepa L, Oczko-Grzesik B, Stolarz W, Sobala-Szczygiel B. Drug-induced aseptic meningitis in suspected central nervous system infections. J Clin Neurosci 2005;12(5):562–4.
21. Olson R. Zymar as an ocular therapeutic agent. Int Ophthalmol Clin 2006;46(4):73–84.
22. Awwad ST, Haddad W, Wang MX, Parmar D, Conger D, Cavanagh HD. Corneal intrastromal gatifloxacin crystal deposits after penetrating keratoplasty. Eye Contact Lens 2004;30(3):169–72.
23. Saravolatz LD, Leggett J. Gatifloxacin, gemifloxacin, and moxifloxacin: the role of 3 newer fluoroquinolones. Clin Infect Dis 2003;37:1210–15.
24. Sumner CL, Elliott RL. Delirium associated with gatifloxacin. Psychosomatics 2003;44:85–6.
25. Adams M, Tavakoli H. Gatifloxacin-induced hallucinations in a 19-year-old man. Psychosomatics 2006;47(4):360.
26. Satyanarayana S, Campbell B. Gatifloxacin-induced delirium and psychosis in an elderly demented woman. J Am Geriatr Soc 2006;54(5):871.
27. Happe MR, Mulhall BP, Maydonovitch CL, Holtzmuller KC. Gatifloxacin-induced hyperglycemia. Ann Intern Med 2004;141(12):968–9.
28. Khovidhunkit W, Sunthornyothin S. Hypoglycemia, hyperglycemia, and gatifloxacin. Ann Intern Med 2004;141(12):969.
29. Arce FC, Bhasin RS, Pasmantier R. Severe hyperglycemia during gatifloxacin therapy in patients without diabetes. Endocr Pract 2004;10(1):40–4.
30. Saraya A, Yokokura M, Gonoi T, Seino S. Effects of fluoroquinolones on insulin secretion and beta-cell ATP-sensitive K+ channels. Eur J Pharmacol 2004;497(1):111–7.
31. Gajjar DA, LaCreta FP, Kollia GD, Stolz RR, Berger S, Smith WB, Swingle M, Grasela DM. Effect of multiple-dose gatifloxacin or ciprofloxacin on glucose homeostasis and insulin production in patients with noninsulin-dependent diabetes mellitus maintained with diet and exercise. Pharmacotherapy 2000;20(6 Pt 2):S76–86.
32. Park-Wyllie LY, Juurlink DN, Kopp A, Shah BR, Stukel TA, Stumpo C, Dresser L, Low DE, Mamdani MM. Outpatient gatifloxacin therapy and dysglycemia in older adults. N Engl J Med 2006;354(13):1352–61.
33. Tailor SA, Simor AE, Cornish W, Phillips EJ, Knowles S, Rachlis A. Glucose homeostasis abnormalities and gatifloxacin. Clin Infect Dis 2006;42(6):895.
34. Yamada C, Nagashima K, Takahashi A, Ueno H, Kawasaki Y, Yamada Y, Seino Y, Inagaki N. Gatifloxacin acutely stimulates insulin secretion and chronically suppresses insulin biosynthesis. Eur J Pharmacol 2006;553(1-3):67–72.
35. Anonymous. Stronger warning for antibiotic Tequin. FDA Consum 2006;40(3):4.
36. Kesavadev J, Rasheed SA. Gatifloxacin induced abnormalities in glucose homeostasis in a patient on glimepiride. J Assoc Physicians India 2006;54951-2.
37. Ittner KP. Gatifloxacin and dysglycemia in older adults. N Engl J Med 2006;354(25):2725-6; author reply 2725-6.
38. Gurwitz JH. Serious adverse drug effects—seeing the trees through the forest. N Engl J Med 2006;354(13):1413–5.
39. Yip C, Lee AJ. Gatifloxacin-induced hyperglycemia: a case report and summary of the current literature. Clin Ther 2006;28(11):1857–66.

40. Talwalkar PG. Gatifloxacin-induced severe hypoglycemia in a patient with type 2 diabetes mellitus. J Assoc Physicians India 2006;54953-4.

41. Zvonar R. Gatifloxacin-induced dysglycemia. Am J Health Syst Pharm 2006;63(21):2087–92.

42. Bobba RK, Arsura EL. Hyperglycemia in an elderly diabetic patient: drug-drug or drug-disease interaction? South Med J 2006;99(1):94–5.

43. Frothingham R. Glucose homeostasis abnormalities associated with use of gatifloxacin. Clin Infect Dis 2005;41(9):1269–76.

44. Pichichero ME, Arguedas A, Dagan R, Sher L, Saez-Llorens X, Hamed K, Echols R. Safety and efficacy of gatifloxacin therapy for children with recurrent acute otitis media (AOM) and/or AOM treatment failure. Clin Infect Dis 2005;41(4):470–8.

45. Brogan SE, Cahalan MK. Gatifloxacin as a possible cause of serious postoperative hypoglycemia. Anesth Analg 2005;101(3):635–6.

46. Blommel AL, Lutes RA. Severe hyperglycemia during renally adjusted gatifloxacin therapy. Ann Pharmacother 2005;39(7-8):1349–52.

47. Jones RN, Andes DR, Mandell LA, Gothelf S, Ehrhardt AF, Nicholson SC. Gatifloxacin used for therapy of outpatient community-acquired pneumonia caused by *Streptococcus pneumoniae*. Diagn Microbiol Infect Dis 2002;44(1):93–100.

48. Nicholson SC, Wilson WR, Naughton BJ, Gothelf S, Webb CD. Efficacy and safety of gatifloxacin in elderly outpatients with community-acquired pneumonia. Diagn Microbiol Infect Dis 2002;44(1):117–25.

49. Nicholson SC, High KP, Gothelf S, Webb CD. Gatifloxacin in community-based treatment of acute respiratory tract infections in the elderly. Diagn Microbiol Infect Dis 2002;44(1):109–16.

50. Gotfried M, Quinn TC, Gothelf S, Wikler MA, Webb CD, Nicholson SC. Oral gatifloxacin in outpatient community-acquired pneumonia: results from TeqCES, a community-based, open-label, multicenter study. Diagn Microbiol Infect Dis 2002;44(1):85–91.

51. Nicholson SC, Webb CD, Moellering RC Jr. Antimicrobial-associated acute hepatitis. Pharmacotherapy 2002;22(6):794–6discussion 796–7.

52. Henann NE, Zambie MF. Gatifloxacin-associated acute hepatitis. Pharmacotherapy 2001;21(12):1579–82.

53. Capparelli EV, Reed MD, Bradley JS, Kearns GL, Jacobs RF, Damle BD, Blumer JL, Grasela DM. Pharmacokinetics of gatifloxacin in infants and children. Antimicrob Agents Chemother 2005;49(3):1106–12.

54. LaCreta FP, Kollia GD, Duncan G, Behr D, Grasela DM. Age and gender effects on the pharmacokinetics of gatifloxacin. Pharmacotherapy 2000;20(6 Pt 2):S67–75.

55. Zhang X, Overholser BR, Kays MB, Sowinski KM. Gatifloxacin pharmacokinetics in healthy men and women. J Clin Pharmacol 2006;46(10):1154–62.

56. Grasela DM, Christofalo B, Kollia GD, Duncan G, Noveck R, Manning JA Jr, LaCreta FP. Safety and pharmacokinetics of a single oral dose of gatifloxacin in patients with moderate to severe hepatic impairment. Pharmacotherapy 2000;20(6 Pt 2):S87–94.

57. Mallet L, Huang A. Coadministration of gatifloxacin and multivitamin preparation containing minerals: potential treatment failure in an elderly patient. Ann Pharmacother 2005;39(1):150–2.

58. Baker SE, Hangii MC. Possible gatifloxacin-induced hypoglycemia. Ann Pharmacother 2002;36(11):1722–6.

59. Menzies DJ, Dorsainvil PA, Cunha BA, Johnson DH. Severe and persistent hypoglycemia due to gatifloxacin interaction with oral hypoglycemic agents. Am J Med 2002;113(3):232–4.

60. Grant EM, Nicolau DR, Nightingale C, Quintiliani R. Minimal interaction between gatifloxacin and oxycodone. J Clin Pharmacol 2002;42(8):928–32.

61. Biggs WS. Hypoglycemia and hyperglycemia associated with gatifloxacin use in elderly patients. J Am Board Fam Pract 2003;16:455–7.

62. Artymowicz RJ, Cino BJ, Rossi JG, Walker JL, Moore S. Possible interaction between gatifloxacin and warfarin. Am J Health Syst Pharm 2002;59(12):1205–6.

63. Mathews S, Cole J, Ryono RA. Anticoagulation-related outcomes in patients receiving warfarin after starting levofloxacin or gatifloxacin. Pharmacotherapy 2006;26(10):1446–52.

64. Straley CM, Cecil EJ, Herriman MP. Gatifloxacin interference with opiate urine drug screen. Pharmacotherapy 2006;26(3):435–9.

Gemifloxacin

See also Fluoroquinolones

General Information

Gemifloxacin is a fluoroquinolone that has enhanced affinity for topoisomerase. Compared with other fluoroquinolones, gemifloxacin was the most potent against penicillin-intermediate and penicillin-resistant pneumococci, methicillin-susceptible and methicillin-resistant *Staphylococcus epidermidis* isolates, and coagulase-negative staphylococci (1,2). It has excellent activity against *Haemophilus influenzae* and *Moraxella catarrhalis* and is unaffected by beta-lactamases. It is generally two-fold less active than ciprofloxacin against most *Enterobacteriaceae* (3). Atypical respiratory pathogens (*Legionella*, *Mycoplasma*, and *Chlamydia* species) and *Neisseria gonorrheae* are highly susceptible (4).

The pharmacokinetics of oral gemifloxacin have been characterized in healthy male volunteers (5). About 20–30% of the dose was excreted unchanged in the urine. The renal clearance was 160 ml/minute on average after single and multiple doses, which was slightly greater than the accepted glomerular filtration rate. There were no adverse effects.

Observational studies

In phase II trials oral gemifloxacin 320 mg/day produced bacteriological responses in 94% of patients with acute exacerbations of chronic bronchitis (6–8) and in 95% of patients with uncomplicated urinary tract infections. Adverse events included nausea, abdominal pain, headache, and a mild rash in both patients and healthy volunteers.

After a single dose of 20–800 mg of gemifloxacin, there were no significant changes in clinical chemistry, hematology, or urinalysis, vital signs, or 12-lead electrocardiograms in healthy men, irrespective of dose (9).

Gemifloxacin 320 mg od and trovafloxacin 200 mg od have been compared in 571 patients with community-acquired pneumonia in a multicenter, double-blind, parallel-group, randomized study (10). Gemifloxacin was slightly more effective (88%) than trovafloxacin (81%). Gemifloxacin was well tolerated and the incidence of transient liver function abnormalities was very low.

The effect of oral gemifloxacin 320 mg for 7 days on the human intestinal microflora has been investigated in 10 healthy subjects (11). The numbers of enterobacteria, enterococci, and streptococci were reduced. No other aerobic microorganisms were affected. The numbers of anerobic cocci and lactobacilli were reduced. The microflora normalized 49 days after withdrawal. There was no selection or overgrowth of resistant bacterial strains or yeasts.

Organs and Systems

Cardiovascular

In 16 trials worldwide, gemifloxacin has been reported to produce small, non-significant QT interval prolongation (12).

Hematologic

Gemifloxacin reached intracellular concentrations in human polymorphonuclear leukocytes eight times higher than extracellular concentrations. Uptake was rapid, reversible, and non-saturable and was affected by environmental temperature, cell viability, and membrane stimuli (13).

Gastrointestinal

In a randomized, double-blind, multicenter comparison of a 5-day course of gemifloxacin 320 mg/day with a standard 7-day regimen of clarithromycin 500 mg bd in 712 patients with acute exacerbations of chronic bronchitis, the most frequently reported gemifloxacin-related adverse events were diarrhea (5.1%) and nausea (4.3%) (14).

Liver

Gemifloxacin was generally well tolerated in a pharmacokinetic study, although one subject was withdrawn after 6 days at 640 mg because of mild, transient rises in alanine transaminase and aspartate transaminase not associated with signs or symptoms (5).

Mild reversible rises in liver enzymes have been reported in 4.1% of patients taking gemifloxacin (12).

Skin

Gemifloxacin can cause mild phototoxicity (15) and its potential to cause mild phototoxicity appears similar to that of ciprofloxacin. The abnormal responses occur within the ultraviolet A region (335–365 nm) and clear 48 hours after withdrawal (12,16).

The risk of rashes with gemifloxacin is quite high (3%), especially in women under 40 years of age who take it for more than 7 days (17). More serious skin reactions, such as Stevens–Johnson syndrome, toxic epidermal necrolysis, or eosinophilic dermatosis, have not been reported (18).

The risk is greater in women aged over 40 years and those who take gemifloxacin for more than 7 days.

Drug–Drug Interactions

Drug interactions with gemifloxacin have been reviewed (12). Gemifloxacin has lower systemic availability if it is co-administered with multivalent cations, such as aluminium, magnesium, or iron salts. In one study, gemifloxacin systemic availability was reduced by 85% if it was taken 10 minutes before Maalox (which contains aluminium hydroxide, magnesium hydroxide, and simethicone), but did not change if taken 2 hours before or 3 hours after. Sucralfate reduced gemifloxacin's systemic availability by as much as 53% when the fluoroquinolone was taken 3 hours after sucralfate; however, gemifloxacin systemic availability did not significantly change if it was administered 2 hours before sucralfate. Ferrous sulfate did not clinically significantly alter the systemic availability or maximum plasma concentration of gemifloxacin if the agents were taken 2 hours apart.

Calcium carbonate

Simultaneous co-administration of calcium carbonate 1000 mg in 16 healthy volunteers reduced the systemic availability of gemifloxacin 320 mg (19). C_{max} fell by 21% and AUC by 17%. Administration of calcium either 2 hours before or 2 hours after gemifloxacin had no effect.

Iron

In an open, randomized, single-dose, five-way crossover study of the effects of ferrous sulfate on the systemic availability of gemifloxacin, there were no changes when gemifloxacin was given at least 2 hours before or at least 3 hours after ferrous sulfate (9).

Sucralfate

In an open, randomized, single-dose, five-way crossover study of the effects of sucralfate on the systemic availability of gemifloxacin, there were no changes when gemifloxacin was given at least 2 hours before sucralfate (9).

Food–Drug Interactions

Food had a minor and clinically insignificant effect on the systemic availability of gemifloxacin (320 and 640 mg) (20). In another study in healthy men and women a high-fat meal slightly reduced gemifloxacin AUC (by 3%) and C_{max} (by 12%) and caused a slight delay in t_{max} of 0.75 and 0.21 hours after doses of 320 and 640 mg respectively (12). These changes were not clinically significant.

References

1. Hardy D, Amsterdam D, Mandell LA, Rotstein C. Comparative in vitro activities of ciprofloxacin, gemifloxacin, grepafloxacin, moxifloxacin, ofloxacin, sparfloxacin, trovafloxacin, and other antimicrobial agents against

bloodstream isolates of gram-positive cocci. Antimicrob Agents Chemother 2000;44(3):802–5.

2. Jones RN, Pfaller MA, Erwin ME. Evaluation of gemifloxacin (SB-265805, LB20304a): in vitro activity against over 6000 gram-positive pathogens from diverse geographic areas. Int J Antimicrob Agents 2000;15(3): 227–230.

3. Marchese A, Debbia EA, Schito GC. Comparative in vitro potency of gemifloxacin against European respiratory tract pathogens isolated in the Alexander Project. J Antimicrob Chemother 2000;46(Suppl T1):11–5.

4. Berron S, Vazquez JA, Gimenez MJ, de la Fuente L, Aguilar L. In vitro susceptibilities of 400 Spanish isolates of *Neisseria gonorrhoeae* to gemifloxacin and 11 other antimicrobial agents. Antimicrob Agents Chemother 2000;44(9):2543–4.

5. Allen A, Bygate E, Vousden M, Oliver S, Johnson M, Ward C, Cheon A, Choo YS, Kim I. Multiple-dose pharmacokinetics and tolerability of gemifloxacin administered orally to healthy volunteers. Antimicrob Agents Chemother 2001;45(2):540–5.

6. File T, Schlemmer B, Garau J, Lode H, Lynch S, Young CThe 070 Clinical Study group. Gemifloxacin versus amoxicillin/clavulanate in the treatment of acute exacerbations of chronic bronchitis. J Chemother 2000;12(4):314–25.

7. Hong CY. Discovery of gemifloxacin (Factive, LB20304a): a quinolone of a new generation. Farmaco 2001;56(1–2):41–4.

8. Ball P, File TM, Twynholm M, Henkel T. Efficacy and safety of gemifloxacin 320 mg once-daily for 7 days in the treatment of adult lower respiratory tract infections Int J Antimicrob Agents 2001;18(1):19–27.

9. Allen A, Bygate E, Oliver S, Johnson M, Ward C, Cheon AJ, Choo YS, Kim IC. Pharmacokinetics and tolerability of gemifloxacin (SB-265805) after administration of single oral doses to healthy volunteers. Antimicrob Agents Chemother 2000;44(6):1604–8.

10. Naber CK, Hammer M, Kinzig-Schippers M, Sauber C, Sorgel F, Bygate EA, Fairless AJ, Machka K, Naber KG. Urinary excretion and bactericidal activities of gemifloxacin and ofloxacin after a single oral dose in healthy volunteers. Antimicrob Agents Chemother 2001;45(12):3524–30.

11. File TM Jr, Schlemmer B, Garau J, Cupo M, Young C049 Clinical Study Group. Efficacy and safety of gemifloxacin in the treatment of community-acquired pneumonia: a randomized, double-blind comparison with trovafloxacin. J Antimicrob Chemother 2001;48(1):67–74.

12. Yoo BK, Triller DM, Yong CS, Lodise TP. Gemifloxacin: a new fluoroquinolone approved for treatment of respiratory infections. Ann Pharmacother 2004;38(7-8):1226–35.

13. Garcia I, Pascual A, Ballesta S, Joyanes P, Perea EJ. Intracellular penetration and activity of gemifloxacin in human polymorphonuclear leukocytes. Antimicrob Agents Chemother 2000;44(11):3193–5.

14. Wilson R, Schentag JJ, Ball P, Mandell L068 Study Group. A comparison of gemifloxacin and clarithromycin in acute exacerbations of chronic bronchitis and long-term clinical outcomes. Clin Ther 2002;24(4):639–52.

15. Lowe MN, Lamb HM. Gemifloxacin. Drugs 2000;59(5):1137–47.

16. File TM Jr, Tillotson GS. Gemifloxacin: a new, potent fluoroquinolone for the therapy of lower respiratory tract infections. Expert Rev Anti Infect Ther 2004;2(6):831–43.

17. Saravolatz LD, Leggett J. Gatifloxacin, gemifloxacin, and moxifloxacin: the role of 3 newer fluoroquinolones. Clin Infect Dis 2003;37:1210–15.

18. Mandell LA, Iannini PB, Tillotson GS. Respiratory fluoroquinolones: differences in the details. Clin Infect Dis 2004;38:1331–2.

19. Pletz MW, Petzold P, Allen A, Burkhardt O, Lode H. Effect of calcium carbonate on bioavailability of orally administered gemifloxacin. Antimicrob Agents Chemother 2003;47:2158–60.

20. Allen A, Bygate E, Clark D, Lewis A, Pay V. The effect of food on the bioavailability of oral gemifloxacin in healthy volunteers. Int J Antimicrob Agents 2000;16(1):45–50.

Gentamicin

See also Aminoglycoside antibiotics

General Information

Gentamicin is well established for the treatment of several bacterial infections, especially those caused by Gram-negative bacteria, including *Pseudomonas aeruginosa*, *Klebsiella* species, and *Serratia marcescens*. In adults, it is usually given in daily doses of 240–360 mg.

Observational studies

In 17 patients with suspected postoperative endophthalmitis treated with 0.2 mg vancomycin and 0.05 mg gentamicin intravitreally, there were adequate intravitreal vancomycin and gentamicin concentrations for over a week; there were no adverse effects (1).

Intratympanic gentamicin therapy has gained some popularity in the treatment of vertigo associated with Menière's disease, as it offers some advantages over traditional surgical treatment. In a 2-year follow-up of 15 patients with Menière's disease, gentamicin solution 0.5 ml (20 mg/ml) injected intratympanically once a week minimized the risk of hearing loss in the treated ear, allowing complete control of vertigo in eight patients after two doses and in 14 patients after four doses (2).

In an open, randomized, controlled trial, once-daily and thrice-daily gentamicin were compared in 173 children aged 1 month to 12 years; there was no nephrotoxicity or ototoxicity (3). Daily doses of gentamicin in both groups were 7.5 mg/kg (under 5 years old), 6.0 mg/kg (5–10 years old), and 4.5 mg/kg (over 10 years old).

Organs and Systems

Respiratory

Acute respiratory failure with near-fatal bronchoconstriction has been reported in an adult with bronchiectasis and chronic *P. aeruginosa* airways colonization immediately after the first inhalation of a commercially available gentamicin solution (4).

Nervous system

Intraventricular gentamicin can cause aseptic meningitis (5).

Sensory systems

Eyes

Gentamicin can induce sclerochorioretinal necrosis after subconjunctival injection (6).

An intraocular injection of a high dose of gentamicin given by mistake led to severe retinal damage, which was at first misdiagnosed as central retinal artery occlusion; the damage was completely reversed by vitrectomy (7).

Ears

In patients with acute bacterial conjunctivitis there were adverse drug reactions in 4 of 103 treated with gentamicin (8). The adverse effects included redness, itching, and burning, and none was serious.

The frequency of aminoglycoside-associated hearing loss is 2–45%. Since gentamicin-induced ototoxicity in most cases only involves vestibular function, the symptoms are easily overlooked in severely ill patients who are unable to sit. If diagnosed early, the vestibular damage is usually reversible. In some cases, severe long-term disability has been described (9). Six patients presented with unilateral vestibulotoxicity after systemic gentamicin therapy (10). All had ataxia and oscillopsia, but none had a history of vertigo. The authors suggested that a subacute course of vestibulotoxicity with time for compensation or asymmetrical recovery of vestibular function after bilateral vestibular loss could have explained the lack of vertigo in these patients.

The risk of ototoxicity from gentamicin in children is probably less than in adults. In many studies of serious neonatal infections treated with gentamicin there have been very few cases that have provided unequivocal evidence of gentamicin-induced ototoxicity. Gentamicin can be an excellent drug in neonatal sepsis, and its potential toxicity should not preclude its use when it is needed.

Gentamicin ear-drops can cause serious adverse effects (for example vertigo, imbalance, ataxia, oscillating vision, hearing loss, and tinnitus) when they are used by patients with perforated tympanic membranes or tympanostomy tubes (11).

The symptom complex known as visual vestibular mismatch can be caused by peripheral vestibular disease. In a retrospective study of 28 patients with Menière's disease, 17 had visual vestibular mismatch; gentamicin therapy increased the number of positive answers (12).

In a retrospective analysis of 85 patients treated with intratympanic gentamicin, using a fixed-dose regimen of 26.7 mg/ml tds on 4 consecutive days, hearing loss occurred in 26% of individuals (13).

The characteristics of ototoxicity of topical gentamicin have been studied retrospectively in 16 patients (14). All used ear-drops for more than 7 days before the development of symptoms, and all had some degree of vestibulotoxicity, but only one had a worsening of cochlear reserve. Even if the tympanic membrane is intact, one should hesitate to use gentamicin in ear-drops or in other topical forms for the treatment of otitis media.

In two animal studies methylcobalamin or dimethylsulfoxide inhibited the ototoxic adverse effects of gentamicin (15,16).

Intratympanic injections of gentamicin 27 mg/ml were performed at weekly intervals in 71 patients with Menière's disease (17). Vertigo was controlled by gentamicin instillation in 83%. Two years after treatment, there was *hearing loss* as a result of the gentamicin injections in only 11 patients.

Gentamicin ototoxicity presents with gait imbalance and oscillopsia, but only rarely with hearing loss and vertigo. Sinusoidal rotational stimuli with a high rate of acceleration, such as the bedside head-thrust test or rotational step changes in velocity, are useful in diagnosing bilateral vestibulopathy. In a retrospective review of the quantitative vestibular function testing results in 35 patients with imbalance, 33 of whom had oscillopsia, three reported a noticeable change in hearing and five reported vertigo; 15 had renal insufficiency at the time of gentamicin administration (18). Those with pre-existing peripheral neuropathy compensated poorly.

In a review of the literature 54 patients with vestibular toxicity attributed to gentamicin were found, 24 of whom had associated auditory toxicity (19). A mutation in the mitochondrial genome A1555G can result in hypersensitivity to the ototoxic effects of the aminoglycosides compared with non-carriers of the mutation, which suggests that there may be an additional intrinsic sensitivity to ototoxicity in certain individuals.

The pathogenesis of Menière's disease is associated with a disorder of ionic homeostasis, the pathological correlate being endolymphatic hydrops. Despite uncertainty about its mode of action, it is accepted wisdom that intratympanic gentamicin has a definite therapeutic role in the control of symptoms in patients who fail to respond to other medical therapy. In a retrospective review of 56 patients undergoing intratympanic gentamicin treatment for Menière's disease there was a 21% rate of significant hearing loss, defined as greater than 10 dB, with an average loss of 19 dB; however, there was an overall significant improvement in vertigo symptoms of 81% (20).

In a retrospective case series from previously published prospective and retrospective studies of vestibular function in 25 patients receiving gentamicin, patients with vestibulotoxic reactions to gentamicin therapy had little additional hearing loss than the general population (21).

The authors of a literature review found two cases of hearing loss attributable to gentamicin and 14 patients with tympanic membrane perforations who had ototoxicity from Garasone (gentamicin sulfate 0.3%, betamethasone, sodium sulfate) (22).

The effect of aspirin in preventing gentamicin-induced hearing loss has been studied in a double-blind, randomized, placebo-controlled trial in 195 patients (23). They received gentamicin 80–160 mg bd by intravenous infusion (generally for 5–7 days) and were randomly assigned to aspirin 3 g/day divided into three doses (n = 89) or placebo (n = 106) for 14 days. The incidence of hearing loss in the placebo group was within the anticipated range (13%) but in the aspirin group it was significantly lower (3%). The efficacy of gentamicin was not affected by aspirin. However, gastric symptoms were more common in the aspirin group and three patients were removed from the study because of gastric bleeding.

In guinea-pigs treated with gentamicin (100 mg/kg/day intramuscularly) alone or gentamicin (100 mg/kg/day intramuscularly) plus alpha-tocopherol (100 mg/kg/day intramuscularly) for 2 weeks, both hearing loss and vestibular dysfunction induced by gentamicin were significantly attenuated by alpha-tocopherol (24).

Psychological, psychiatric

There are several case reports of acute toxic psychoses due to gentamicin (25).

Mineral balance

Gentamicin-induced magnesium depletion is most likely to occur in older patients when large doses are used over long periods of time (26). Under these circumstances, serum concentrations and urinary electrolyte losses should be monitored.

In a prospective study in 659 neonates who received gentamicin for more than 4 days, the incidence of hypocalcemia was five times higher after the dosage was changed from 2.5 mg/kg bd to 4 mg/kg/day (27).

Liver

Increases in alkaline phosphatase after gentamicin have been described (28).

Urinary tract

A full course of gentamicin therapy causes nephrotoxicity in 1–55% of patients. Two types of gentamicin-induced nephrotoxicity are recognized: (1) a gradual reduction in creatinine clearance, occurring after about 2 weeks, in about 5–10% of patients receiving the drug in full doses, the reduction being rapidly reversible in most cases as soon as gentamicin is withdrawn; (2) acute renal insufficiency due to tubular necrosis, usually associated with oliguria lasting 10–12 days, followed by a diuretic phase; this type of nephrotoxicity occurs far less often than the first type.

The following order of relative nephrotoxicity has been found in many animal experiments: neomycin > gentamicin > tobramycin > amikacin > netilmicin (29,30). However, in humans, conclusive data regarding the relative toxicity of the various aminoglycosides are still lacking. An analysis of 24 controlled trials showed the following average rates for nephrotoxicity: gentamicin 11%; tobramycin 11.5%; amikacin 8.5%; and netilmicin 2.8% (31). In contrast to this survey, direct comparison in similar patient groups showed no significant differences between the various agents in

most trials (32–38). In fact, the relative advantage of lower nephrotoxicity rates observed with netilmicin in some studies may be limited to administration of low doses. One prospective trial showed a significant advantage of tobramycin over gentamicin (39). However, these findings could subsequently not be confirmed (40). The risk of gentamicin nephrotoxicity is increased in biliary obstruction (SEDA-20, 235).

In a few cases, gentamicin nephrotoxicity was associated with a Fanconi syndrome, with raised serum enzymes activities in the urine. Among these, muramidase seemed to be especially useful in checking for proximal tubular dysfunction (41).

Gentamicin is of considerable value in the management of sepsis in immunosuppressed patients and renal transplant recipients. Although it has been suggested that gentamicin should be avoided in such patients because of potential renal toxicity in the allograft (42,43), experienced physicians have felt that gentamicin may be given, provided the dosage schedule is adapted to the degree of allograft function and that blood concentrations are monitored.

After a full course of gentamicin 1–55% of patients have nephrotoxicity. The increased serum creatinine concentration peaks on day 6 of therapy and is reversible in most cases within 30 days. Nephrotoxicity appears to be more common among patients with pre-existing renal impairment, longer treatment duration (over 7 days), repeated courses of aminoglycosides, and after the co-administration of other nephrotoxic drugs (for example amphotericin, cisplatin, daunorubicin, furosemide, and vancomycin). Animal studies have suggested that hydrocortisone, angiotensin converting enzyme inhibitors, and hypercalcemia can also increase aminoglycoside nephrotoxicity, whereas acetazolamide, bicarbonate, ceftriaxone, lithium, magnesium, melatonin, piperacillin, polyaspartic acid, pyridoxal-5'-phosphate, and a high protein diet may be protective (44,45).

In 87 patients with intertrochanteric hip fractures, preoperative antibiotic prophylaxis (gentamicin 240 mg and dicloxacillin 2 g) had no significant effect on wound infections; however, there were 16 reversible cases of nephrotoxicity and 1 irreversible case among patients who received antibiotic prophylaxis, compared with only 4 cases of reversible kidney damage among 76 patients who did not receive antibiotics (46).

- Acute renal insufficiency occurred in an 83-year-old woman after two-stage revision of an infected knee prosthesis with gentamicin-impregnated beads and block spacers (47). The combined use of beads and a cement block spacer, both gentamicin impregnated, may have caused this severe complication.
- A 43-year-old black woman with a 13-year history of lupus developed severe acute tubular necrosis secondary to gentamicin (48).

Since serum creatinine does not accurately reflect renal function in patients with spinal cord injury, dosage regimens of gentamicin should be individualized, based on age, sex, weight, height, the level of spinal cord injury, and renal function (49).

In animals melatonin (50,51) or l-carnitine (52) protected the kidneys against oxidative damage and the nephrotoxic effect of gentamicin.

Nephrocalcinosis occurred in 16 of 101 babies born at less than 32 weeks gestation (53). Multivariate analysis showed that the strongest predictors of nephrocalcinosis were duration of ventilation, toxic gentamicin/vancomycin concentrations, low fluid intake, and male sex.

In a retrospective review of 744 patients who were dose-individualized with gentamicin once daily, in those patients in whom nephrotoxicity was predicted from a change in gentamicin clearance, this change occurred on average 3 days before the change in creatinine clearance (54).

Agents that can augment aminoglycoside-induced nephrotoxicity (for example calcium channel blockers and nephrotoxic agents such as ciclosporin) should not be combined with these antibiotics. However, antioxidant drugs, especially the natural antioxidants, seem to have the most potential for clinical use. Of these natural antioxidants, melatonin seems to be the most promising in abating nephrotoxicity (55,56,57).

Four women presented with a Bartter-like syndrome, with hypokalemia, metabolic alkalosis, hypomagnesemia, hypermagnesiuria, hypocalcemia, and hypercalciuria, after receiving gentamicin 1.2–2.6 g (58). The syndrome lasted for 2–6 weeks after withdrawal of gentamicin.

Aminoglycoside-induced renal tubular dysfunction can be divided into Fanconi-like syndrome and Bartter-like syndrome. Rarely they can occur together.

- A 66-year-old Chinese man developed Fanconi-like syndrome, Bartter-like syndrome, distal renal tubular acidosis, acute renal failure, and deafness after receiving a large dose of gentamicin, which was considered the major risk factor (59). The dose of gentamicin was 4.9 mg/kg/day, higher than the calculated dose of 3.75 mg/kg/day based on estimated creatinine clearance.
 Impaired mitochondria and ATP generation might participate in the mechanism of renal tubular dysfunction due to aminoglycosides.
- A 53-year-old man who received intermittent intravenous gentamicin (320–560 mg/day) for a total of 4 months to a total cumulative dose of 9.4 g developed Fanconi syndrome, with profound hypophosphatemia, hypocalcemia, hyperphosphaturia, and aminoaciduria (60). The electrolyte disturbances persisted until gentamicin was withdrawn and recurred after rechallenge.

Acute tubular necrosis occurred in an adolescent with cystic fibrosis receiving intravenous gentamicin and ceftazidime (61A).

Skin

Erythema multiforme developed in a 4-year-old girl after treatment with topical aural gentamicin sulfate (0.3%) plus hydrocortisone acetate (1%) prescribed for otorrhea (62).

Musculoskeletal

In human osteoblast-like cells in vitro, gentamicin, at high concentrations (100 µg/ml and above), as achieved after topical application, inhibited cell proliferation; It may therefore be detrimental to repair in vivo (63).

Immunologic

Allergic contact dermatitis due to gentamicin is rare in patients with eyelid dermatitis, but it can occur.

- A 55-year-old housewife developed pruritic, erythematous, scaly plaques on the eyelids, spreading in a few days periorbitally after treatment with gentamicin eyedrops (Colircusi Gentamicina) (64). A positive patch test reaction to kanamycin, to which the patient had not been previously exposed, suggested cross-reactivity. Systemic allergy to gentamicin, including drug-induced cytopenias and rashes, is very rare.
- Systemic anaphylaxis after intravenous gentamicin has been reported in a 70-year-old woman (65). During the first injection of gentamicin she developed anaphylaxis and was treated in intensive care for 5 days. A few weeks after full recovery, open patch testing showed type IV sensitization to gentamicin sulfate.

Death

In a randomized trial in infants with Gram-negative meningitis and ventriculitis, the use of intraventricular antibiotics in addition to intravenous antibiotics resulted in a three-fold increased relative risk of death compared with standard treatment with intravenous antibiotics alone (66). The authors speculated that the increased case-fatality rate in those who were given intraventricular antibiotics could have been related to the procedure or to a direct toxic effect of gentamicin.

Second-Generation Effects

Teratogenicity

During pregnancy the aminoglycosides cross the placenta and they might theoretically be expected to cause otological and perhaps nephrological damage to the fetus. However, no proven cases of intrauterine damage by gentamicin have been recorded.

Susceptibility Factors

Age

In premature neonates, gentamicin clearance depends on gestational age, with a cut-off at 30 weeks: younger neonates have lower gentamicin clearance, a slightly higher volume of distribution, and a longer half-life compared with the older neonates (67). Loading doses of 3.7 and 3.5 mg/kg followed by maintenance doses of 2.8 mg/kg/day and 2.6 mg/kg/18 hours have been recommended for younger and older neonates respectively.

Drug Administration

Drug contamination

Several patients developed severe shaking chills, often accompanied by fever, tachycardia, and/or a significant reduction in systolic blood pressure within 3 hours of receiving intravenous once-daily dosing regimens of gentamicin produced by Fujisawa USA, Inc (Deerfield, Illinois). Investigations showed that gentamicin formulations that contain concentrations of endotoxin that are within the USP standards may deliver amounts of endotoxin that are above the threshold for pyrogenic reactions with once-daily dosing (68).

Of 155 patients (38% men, mean age 41 years) with pyrogenic reactions due to gentamicin, 81% received once-daily dosing (70% in a dose of 5–7 mg/kg) and 10% received a conventional dose (3 mg/kg in three divided doses (69)). Reactions typically occurred within 3 hours after infusion (98%) and lasted for less than 3 hours (96%). Patients reported chills, shaking, or shivering (75%), rigors (23%), fever (68%), tachycardia (17%), hypertension or hypotension (17%), and respiratory symptoms (47%). More serious reactions also occurred, including cyanosis (4%), oxygen saturation below 80% (7%), and pulmonary edema (one patient); 8% had severe reactions leading to hospitalization (with intubation, resuscitation, or admission to the intensive care unit in five cases), but none died. An FDA investigation showed that 10% of gentamicin lots tested had raised endotoxin concentrations, and an additional 4% of the lots would have exposed a patient to concentrations above the acceptable threshold with once-daily dosing. The two products implicated in these clusters involved the same supplier of bulk gentamicin; inadequacies in manufacturing practices had led to an increase in overall impurities (70,71).

Drug dosage regimens

Once-daily regimens are appealing for cost saving and may have a therapeutic advantage and reduced toxicity. A prolonged distribution time has been noted with high-dose gentamicin (7 mg/kg). Higher doses are used for extended-interval aminoglycoside therapy. In 12 healthy volunteers receiving extended-interval high-dose gentamicin, sampling within 90 minutes after the start of the infusion provided information that led to overestimation of peak serum concentration/minimum inhibitory concentration and inaccurate pharmacokinetic calculations (72).

Although once-daily dosing appears to be effective in limited studies in children, its role in Gram-positive coccal endocarditis, in individuals with neutropenia or cystic fibrosis, and in individuals with altered volumes of distribution remains uncertain (73).

In 18 patients receiving empirical therapy for CAPD-related peritonitis, once-daily intraperitoneal gentamicin (0.6 mg/kg) had less therapeutic benefit and peak serum gentamicin concentrations were lower than the suggested value of 4 µg/ml, whereas trough serum gentamicin concentrations were higher than the minimum toxic concentration; dialysate gentamicin concentrations were higher than therapeutic concentrations for only 4.75 hours of each day (74).

In 45 neonates (75) and 123 older children (76) gentamicin 4–5 mg/kg once a day produced peak concentrations associated with greater efficacy and trough concentrations associated with less toxicity than 2.0–2.5 mg/kg bd.

In febrile neutropenic episodes after intensive chemotherapy, once-daily gentamicin (7 mg/kg/day) in combination with azlocillin was more effective than a multiple-daily dosing regimen, but the incidence of toxicity was low overall and was slightly but not significantly higher in the once-daily group (77).

For external otitis, therapeutic local antibiotic concentrations can be achieved by giving gentamicin ear-drops twice daily; more frequent administration is not needed (78).

Drug administration route

Oral administration of gentamicin in low dosages to reduce the intestinal flora is rarely practiced, although it is probably as effective as neomycin. In the presence of intestinal mucosal inflammation more than 10% of the dose can be absorbed.

Endotracheal administration of gentamicin can be used in patients with a tracheotomy. This route of administration does not produce toxic plasma concentrations, but in patients with renal insufficiency the absorption of a certain amount via the respiratory tract should be taken into account.

Topical application of gentamicin to large areas of burns has caused ototoxic effects, ranging from mild to severe loss of hearing, with decrease of vestibular function (79). Positional vertigo occurred in one patient (80). In another case a woman complained of tinnitus each time she treated her paronychia with gentamicin cream (81).

Intrathecal gentamicin can cause neurotoxic lesions (82).

Increased penetration of gentamicin through a thin sclera can lead to toxic concentrations of the drug in a localized area adjacent to the site of injection, as has been shown in rabbits (83). These toxic effects are also influenced by the degree of pigmentation and acute inflammation. In a case of accidental injection of gentamicin 20 mg into the vitreous in a 70-year-old man during vitrectomy, no toxic signs occurred after the operation (84).

There is uncertainty about the risk of ototoxicity after the topical administration of aminoglycosides into the ear when the indication was appropriate (85,86). Nine cases of iatrogenic topical vestibulotoxicity have been reported (87). All had used ear-drops containing gentamicin sulfate and betamethasone sodium phosphate for prolonged periods. Toxicity was primarily vestibular rather than cochlear. Although compensation occurred in unilateral cases, the disability in bilateral cases was typically severe and often resulted in litigation.

Drug overdose

Although aminoglycoside antibiotics are dialysable, peritoneal dialysis may not remove them from the blood after overdosage. However, hemodialysis is effective (88).

Drug–Drug Interactions

Beta-lactam antibiotics

Gentamicin and other aminoglycosides have increased activity when they are combined with beta-lactams, resulting in increased bacterial aminoglycoside uptake (46). The proposed mechanism of synergism is damage to the cell membrane by the beta-lactam, followed by improved diffusion of gentamicin across the outer bacterial membrane. A second type of synergism, pharmacodynamic synergism, occurs when high serum concentrations of aminoglycosides cause efficient bacterial killing, resulting in reduced bacterial concentrations, which are more effectively eliminated by beta-lactams, as they work more efficiently against lower bacterial concentrations. The action of gentamicin is inhibited by some antimicrobials, which are bacteriostatic rather than bactericidal; for example antagonism occurs with macrolides, tetracycline and doxycycline, and chloramphenicol. The clinical significance of this antagonism is unknown.

There are many reports of acute renal insufficiency from combined treatment with gentamicin and one of the cephalosporins (89–91). The potential nephrotoxic effect of this combination seems to be related mainly to the nephrotoxic effect of gentamicin.

COX-2-selective inhibitors

In animals the combination of COX-2-selective inhibitors with gentamicin posed no additional risk of nephrotoxic renal insufficiency (92).

Methylthioninium chloride

Gentamicin is synergistic with methylthioninium chloride (methylene blue) in vitro against *P. aeruginosa* (93).

Metronidazole

In guinea-pigs, metronidazole augmented gentamicin-induced ototoxicity, determined by the measurement of compound action potentials (94).

NSAIDs

Aminoglycosides used in combination with non-steroidal anti-inflammatory drugs can be associated with renal insufficiency.

- An adolescent with cystic fibrosis developed renal insufficiency and severe vestibular toxicity after treatment with gentamicin and standard-dose ibuprofen (95). A low intravascular volume was a possible contributing factor.

Management of Adverse Drug Reactions

In animals there was biochemical and histopathological evidence that chelerythrine, a commonly used protein kinase C inhibitor, reduced gentamicin-induced kidney damage (96).

Monitoring Therapy

During multiple-daily dosing regimens peak serum concentrations of gentamicin over 5–7 µg/ml are associated with improved survival in patients with septicemia and pneumonia caused by Gram-negative bacteria (97,98). On the other hand, excessive peak concentrations (over 10–12 µg/ml) and trough concentrations (over 2 µg/ml) of gentamicin increase the risk of ototoxicity and nephrotoxicity (99).

Salivary sampling is of potential interest in monitoring drug therapy, especially in children. Although there was no correlation between serum gentamicin concentrations and salivary concentrations when gentamicin was given two or three times daily in children with uncomplicated infections, there was a good correlation after once-daily dosing (100).

References

1. Gan IM, van Dissel JT, Beekhuis WH, Swart W, van Meurs JC. Intravitreal vancomycin and gentamicin concentrations in patients with postoperative endophthalmitis. Br J Ophthalmol 2001;85(11):1289–93.
2. Quaranta A, Scaringi A, Aloidi A, Quaranta N, Salonna I. Intratympanic therapy for Menière's disease: effect of administration of low concentration of gentamicin. Acta Otolaryngol 2001;121(3):387–92.
3. Carapetis JR, Jaquiery AL, Buttery JP, Starr M, Cranswick NE, Kohn S, Hogg GG, Woods S, Grimwood K. Randomized, controlled trial comparing once daily and three times daily gentamicin in children with urinary tract infections. Pediatr Infect Dis J 2001;20(3):240–6.
4. Melani AS, Di Gregorio A. Acute respiratory failure due to gentamicin aerosolization. Monaldi Arch Chest Dis 1998;53(3):274–6.
5. Haase KK, Lapointe M, Haines SJ. Aseptic meningitis after intraventricular administration of gentamicin. Pharmacotherapy 2001;21(1):103–7.
6. Salati C, Migliorati G, Brusini P. Scleroretinal necrosis after a subconjunctival injection of gentamicin in a patient with a surgically repaired episcleral retinal detachment. Eur J Ophthalmol 2004;14(6):575–7.
7. Zhang J, Li M, Lin X. [A case report of injecting gentamicin intraocularly by mistake being misdiagnosed as central retinal artery occlusion]. Yan Ke Xue Bao 2005;21(2):88–91.
8. Papa V, Aragona P, Scuderi AC, Blanco AR, Zola P, Di BA, Santocono M, Milazzo G. Treatment of acute bacterial conjunctivitis with topical netilmicin. Cornea 2002;21(1):43–7.
9. Dayal VS, Chait GE, Fenton SS. Gentamicin vestibulotoxicity. Long term disability. Ann Otol Rhinol Laryngol 1979;88(1 Pt 1):36–9.

10. Waterston JA, Halmagyi GM. Unilateral vestibulotoxicity due to systemic gentamicin therapy. Acta Otolaryngol 1998;118(4):474–8.

11. Wooltorton E. Ototoxic effects from gentamicin ear drops. CMAJ 2002;167(1):56.

12. Longridge NS, Mallinson AI, Denton A. Visual vestibular mismatch in patients treated with intratympanic gentamicin for Meniere's disease. J Otolaryngol 2002;31(1):5–8.

13. Kaplan DM, Nedzelski JM, Al-Abidi A, Chen JM, Shipp DB. Hearing loss following intratympanic instillation of gentamicin for the treatment of unilateral Menière's disease. J Otolaryngol 2002;31(2):106–11.

14. Bath AP, Walsh RM, Bance ML, Rutka JA. Ototoxicity of topical gentamicin preparations. Laryngoscope 1999;109(7 Pt 1):1088–93.

15. Ali BH, Mousa HM. Effect of dimethyl sulfoxide on gentamicin-induced nephrotoxicity in rats. Hum Exp Toxicol 2001;20(4):199–203.

16. Jin X, Jin X, Sheng X. Methylcobalamin as antagonist to transient ototoxic action of gentamicin. Acta Otolaryngol 2001;121(3):351–4.

17. Martin E, Perez N. Hearing loss after intratympanic gentamicin therapy for unilateral Meniere's disease. Otol Neurotol 2003;24:800–6.

18. Ishiyama G, Ishiyama A, Kerber K, Baloh RW. Gentamicin ototoxicity: clinical features and the effect on the human vestibulo-ocular reflex. Acta Otolaryngol 2006;126(10):1057–61.

19. Pappas S, Nikolopoulos TP, Korres S, Papacharalampous G, Tzangarulakis A, Ferekidis E. Topical antibiotic ear drops: are they safe? Int J Clin Pract 2006;60(9):1115–9.

20. Flanagan S, Mukherjee P, Tonkin J. Outcomes in the use of intra-tympanic gentamicin in the treatment of Ménière's disease. J Laryngol Otol 2006;120(2):98–102.

21. Dobie RA, Black FO, Pezsnecker SC, Stallings VL. Hearing loss in patients with vestibulotoxic reactions to gentamicin therapy. Arch Otolaryngol Head Neck Surg 2006;132(3):253–7.

22. Wanic-Kossowska M, Grzegorzewska AE, Tykarski A. Hypotonia during amikacin administration in a patient treated with continuous ambulatory peritoneal dialysis. Adv Perit Dial 2006;22:69–72.

23. Sha SH, Qiu JH, Schacht J. Aspirin to prevent gentamicin-induced hearing loss. N Engl J Med 2006;354(17):1856–1.

24. Sergi B, Fetoni AR, Ferraresi A, Troiani D, Azzena GB, Paludetti G, Maurizi M. The role of antioxidants in protection from ototoxic drugs. Acta Otolaryngol Suppl 2004(552):42–5.

25. Kane FJ Jr, Byrd G. Acute toxic psychosis associated with gentamicin therapy. South Med J 1975;68(10):1283–5.

26. Kes P, Reiner Z. Symptomatic hypomagnesemia associated with gentamicin therapy. Magnes Trace Elem 1990;9(1):54–60.

27. Jackson GL, Sendelbach DM, Stehel EK, Baum M, Manning MD, Engle WD. Association of hypocalcemia with a change in gentamicin administration in neonates. Pediatr Nephrol 2003;18:653–6.

28. Mor F, Leibovici L, Cohen O, Wysenbeek AJ. Prospective evaluation of liver function tests in patients treated with aminoglycosides. DICP 1990;24(2):135–7.

29. Luft FC, Yum MN, Kleit SA. Comparative nephrotoxicities of netilmicin and gentamicin in rats. Antimicrob Agents Chemother 1976;10(5):845–9.

30. Hottendorf GH, Gordon LL. Comparative low-dose nephrotoxicities of gentamicin, tobramycin, and amikacin. Antimicrob Agents Chemother 1980;18(1):176–81.

31. Cone LA. A survey of prospective, controlled clinical trials of gentamicin, tobramycin, amikacin, and netilmicin. Clin Ther 1982;5(2):155–62.

32. Smith CR, Baughman KL, Edwards CQ, Rogers JF, Lietman PS. Controlled comparison of amikacin and gentamicin. N Engl J Med 1977;296(7):349–53.

33. Feld R, Valdivieso M, Bodey GP, Rodriguez V. Comparison of amikacin and tobramycin in the treatment of infection in patients with cancer. J Infect Dis 1977;135(1):61–6.

34. Love LJ, Schimpff SC, Hahn DM, Young VM, Standiford HC, Bender JF, Fortner CL, Wiernik PH. Randomized trial of empiric antibiotic therapy with ticarcillin in combination with gentamicin, amikacin or netilmicin in febrile patients with granulocytopenia and cancer. Am J Med 1979;66(4):603–10.

35. Lau WK, Young LS, Black RE, Winston DJ, Linne SR, Weinstein RJ, Hewitt WL. Comparative efficacy and toxicity of amikacin/carbenicillin versus gentamicin/carbenicillin in leukopenic patients: a randomized prospective trail. Am J Med 1977;62(6):959–66.

36. Fong IW, Fenton RS, Bird R. Comparative toxicity of gentamicin versus tobramycin: a randomized prospective study. J Antimicrob Chemother 1981;7(1):81–8.

37. Bock BV, Edelstein PH, Meyer RD. Prospective comparative study of efficacy and toxicity of netilmicin and amikacin. Antimicrob Agents Chemother 1980;17(2):217–25.

38. Barza M, Lauermann MW, Tally FP, Gorbach SL. Prospective, randomized trial of netilmicin and amikacin, with emphasis on eighth-nerve toxicity. Antimicrob Agents Chemother 1980;17(4):707–14.

39. Smith CR, Lipsky JJ, Laskin OL, Hellmann DB, Mellits ED, Longstreth J, Lietman PS. Double-blind comparison of the nephrotoxicity and auditory toxicity of gentamicin and tobramycin. N Engl J Med 1980;302(20):1106–9.

40. Matzke GR, Lucarotti RL, Shapiro HS. Controlled comparison of gentamicin and tobramycin nephrotoxicity. Am J Nephrol 1983;3(1):11–7.

41. Russo JC, Adelman RD. Gentamicin-induced Fanconi syndrome. J Pediatr 1980;96(1):151–3.

42. Wellwood JM, Tighe JR. Proceedings: Evidence of gentamicin nephrotoxicity in patients with renal allografts. Br J Surg 1975;62(2):156.

43. Termeer A, Hoitsma AJ, Koene RA. Severe nephrotoxicity caused by the combined use of gentamicin and cyclosporine in renal allograft recipients. Transplantation 1986;42(2):220–1.

44. Santucci RA, Krieger JN. Gentamicin for the practicing urologist: review of efficacy, single daily dosing and "switch" therapy. J Urol 2000;163(4):1076–84.

45. Ozbek E, Turkoz Y, Sahna E, Ozugurlu F, Mizrak B, Ozbek M. Melatonin administration prevents the nephrotoxicity induced by gentamicin. BJU Int 2000;85(6):742–6.

46. Solgaard L, Tuxoe JI, Mafi M, Due Olsen S, Toftgaard Jensen T. Nephrotoxicity by dicloxacillin and gentamicin in 163 patients with intertrochanteric hip fractures. Int Orthop 2000;24(3):155–7.

47. van Raaij TM, Visser LE, Vulto AG, Verhaar JA. Acute renal failure after local gentamicin treatment in an infected total knee arthroplasty. J Arthroplasty 2002;17(7):948–50.

48. Fogo AB. Quiz page. Mesangial lupus nephritis, WHO IIb, and severe acute tubular necrosis, secondary to gentamycin toxicity. Am J Kidney Dis 2002;40(5):xlix.

49. Vaidyanathan S, Watt JW, Singh G, Soni BM, Sett P. Dosage of once-daily gentamicin in spinal cord injury patients. Spinal Cord 2000;38(3):197–8.

50. Sener G, Sehirli AO, Altunbas HZ, Ersoy Y, Paskaloglu K, Arbak S, Ayanoglu-Dulger G. Melatonin protects against gentamicin-induced nephrotoxicity in rats. J Pineal Res 2002;32(4):231–6.

51. Reiter RJ, Tan DX, Sainz RM, Mayo JC, Lopez-Burillo S. Melatonin: reducing the toxicity and increasing the efficacy of drugs. J Pharm Pharmacol 2002;54(10):1299–321.

52. Kopple JD, Ding H, Letoha A, Ivanyi B, Qing DP, Dux L, Wang HY, Sonkodi S. L-carnitine ameliorates gentamicin-induced renal injury in rats. Nephrol Dial Transplant 2002;17(12):2122–31.

53. Narendra A, White MP, Rolton HA, Alloub ZI, Wilkinson G, McColl JH, Beattie J. Nephrocalcinosis in preterm babies. Arch Dis Child Fetal Neonatal Ed 2001;85(3):F207–13.

54. Kirkpatrick CM, Duffull SB, Begg EJ, Frampton C. The use of a change in gentamicin clearance as an early predictor of gentamicin-induced nephrotoxicity. Ther Drug Monit 2003;25:623–30.

55. Ali BH, Al-Qarawi AA, Haroun EM, Mousa HM. The effect of treatment with gum Arabic on gentamicin nephrotoxicity in rats: a preliminary study. Renal Fail 2003;25:15–20.

56. Ali BH. Agents ameliorating or augmenting experimental gentamicin nephrotoxicity: some recent research. Food Chem Toxicol 2003;41:1447–52.

57. Maldonado PD, Barrera D, Medina-Campos ON, Hernandez-Pando R, Ibarra-Rubio ME, Pedraza Chavorri J. Aged garlic extract attenuates gentamicin induced renal damage and oxidative stress in rats. Life Sci 2003;73:2543–56.

58. Chou CL, Chen YH, Chau T, Lin SH. Acquired Bartter-like syndrome associated with gentamicin administration. Am J Med Sci 2005;329(3):144–9.

59. Hung CC, Guh JY, Kuo MC, Lai YH, Chen HC. Gentamicin-induced diffuse renal tubular dysfunction. Nephrol Dial Transplant 2006;21(2):547–8.

60. Ghiculescu RA, Kubler PA. Aminoglycoside-associated Fanconi syndrome. Am J Kidney Dis 2006;48(6):e89-93.

61. Kennedy SE, Henry RL, Rosenberg AR. Antibiotic-related renal failure and cystic fibrosis. J Paediatr Child Health 2005;41(7):382–3.

62. Siddiq MA. Erythema multiforme after application of aural Gentisone HC drops. J Laryngol Otol 1999;113(11):1002–3.

63. Isefuku S, Joyner CJ, Simpson AH. Gentamicin may have an adverse effect on osteogenesis. J Orthop Trauma 2003;17:212–6.

64. Sanchez-Perez J, Lopez MP, De Vega Haro JM, Garcia-Diez A. Allergic contact dermatitis from gentamicin in eyedrops, with cross-reactivity to kanamycin but not neomycin. Contact Dermatitis 2001;44(1):54.

65. Schulze S, Wollina U. Gentamicin-induced anaphylaxis. Allergy 2003;58:88–9.

66. Shah S, Ohlsson A, Shah V. Intraventricular antibiotics for bacterial meningitis in neonates. Cochrane Database Syst Rev 2004(4):CD004496.

67. Rocha MJ, Almeida AM, Afonso E, Martins V, Santos J, Leitao F, Falcao AC. The kinetic profile of gentamicin in premature neonates. J Pharm Pharmacol 2000;52(9):1091–7.

68. Centers for Disease Control and Prevention (CDC). Endotoxin-like reactions associated with intravenous gentamicin—California, 1998. MMWR Morb Mortal Wkly Rep 1998;47(41):877–80.

69. Fanning MM, Wassel R, Piazza-Hepp T. Pyrogenic reactions to gentamicin therapy. N Engl J Med 2000;343(22):1658–9.

70. Chuck SK, Raber SR, Rodvold KA, Areff D. National survey of extended-interval aminoglycoside dosing. Clin Infect Dis 2000;30(3):433–9.

71. Buchholz U, Richards C, Murthy R, Arduino M, Pon D, Schwartz W, Fontanilla E, Pegues C, Boghossian N, Peterson C, Kool J, Mascola L, Jarvis WR. Pyrogenic reactions associated with single daily dosing of intravenous gentamicin. Infect Control Hosp Epidemiol 2000;21(12):771–4.

72. McNamara DR, Nafziger AN, Menhinick AM, Bertino JS Jr. A dose-ranging study of gentamicin pharmacokinetics: implications for extended interval aminoglycoside therapy. J Clin Pharmacol 2001;41(4):374–7.

73. The ARDS Network. Ketoconazole for early treatment of acute lung injury and acute respiratory distress syndrome: a randomized controlled trial. JAMA 2000;283(15):1995–2002.

74. Tosukhowong T, Eiam-Ong S, Thamutok K, Wittayalertpanya S, Na Ayudhya DP. Pharmacokinetics of intraperitoneal cefazolin and gentamicin in empiric therapy of peritonitis in continuous ambulatory peritoneal dialysis patients. Perit Dial Int 2001;21(6):587–94.

75. Chotigeat U, Narongsanti A, Ayudhya DP. Gentamicin in neonatal infection: once versus twice daily dosage. J Med Assoc Thai 2001;84(8):1109–15.

76. Robinson RF, Nahata MC. Safety of intravenous bolus administration of gentamicin in pediatric patients. Ann Pharmacother 2001;35(11):1327–31.

77. Bakri FE, Pallett A, Smith AG, Duncombe AS. Once-daily versus multiple-daily gentamicin in empirical antibiotic therapy of febrile neutropenia following intensive chemotherapy. J Antimicrob Chemother 2000;45(3):383–6.

78. Rakover Y, Smuskovitz A, Colodner R, Keness Y, Rosen G. Duration of antibacterial effectiveness of gentamicin ear drops in external otitis. J Laryngol Otol 2000;114(11):827–9.

79. Dayal VS, Smith EL, McCain WG. Cochlear and vestibular gentamicin toxicity. A clinical study of systemic and topical usage. Arch Otolaryngol 1974;100(5):338–40.

80. Lelliever WC. Topical gentamicin-induced positional vertigo. Otolaryngol Head Neck Surg 1985;93(4):553–5.

81. Drake TE. Letter: Reaction to gentamicin sulfate cream. Arch Dermatol 1974;110(4):638.

82. Watanabe I, Hodges GR, Dworzack DL, Kepes JJ, Duensing GF. Neurotoxicity of intrathecal gentamicin: a case report and experimental study. Ann Neurol 1978;4(6):564–72.

83. Loewenstein A, Zemel E, Vered Y, Lazar M, Perlman I. Retinal toxicity of gentamicin after subconjunctival injection performed adjacent to thinned sclera. Ophthalmology 2001;108(4):759–64.

84. Burgansky Z, Rock T, Bartov E. Inadvertent intravitreal gentamicin injection. Eur J Ophthalmol 2002;12(2):138–40.

85. Indudharan R. Ototopic aminoglycosides and ototoxicity. J Otolaryngol 1998;27(3):182.

86. Walby P, Stewart R, Kerr AG. Aminoglycoside ear drop ototoxicity: a topical dilemma? Clin Otolaryngol Allied Sci 1998;23(4):289–90.

87. Marais J, Rutka JA. Ototoxicity and topical eardrops. Clin Otolaryngol Allied Sci 1998;23(4):360–7.

88. Lu CM, James SH, Lien YH. Acute massive gentamicin intoxication in a patient with end-stage renal disease. Am J Kidney Dis 1996;28(5):767–71.

89. Bailey RR. Renal failure in combined gentamicin and cephalothin therapy. BMJ 1973;2(5869):776–7.

90. Cabanillas F, Burgos RC, Rodriguez C, Baldizon C. Nephrotoxicity of combined cephalothin–gentamicin regimen. Arch Intern Med 1975;135(6):850–2.

91. Tobias JS, Whitehouse JM, Wrigley PF. Letter: Severe renal dysfunction after tobramycin/cephalothin therapy. Lancet 1976;1(7956):425.

92. Hosaka EM, Santos OF, Seguro AC, Vattimo MF. Effect of cyclooxygenase inhibitors on gentamicin-induced nephrotoxicity in rats. Braz J Med Biol Res 2004;37(7):979–85.

93. Gunics G, Motohashi N, Amaral L, Farkas S, Molnar J. Interaction between antibiotics and non-conventional antibiotics on bacteria. Int J Antimicrob Agents 2000;14(3):239–42.

94. Riggs LC, Shofner WP, Shah AR, Young MR, Hain TC, Matz GJ. Ototoxicity resulting from combined administration of metronidazole and gentamicin. Am J Otol 1999;20(4):430–4.

95. Scott CS, Retsch-Bogart GZ, Henry MM. Renal failure and vestibular toxicity in an adolescent with cystic fibrosis receiving gentamicin and standard-dose ibuprofen. Pediatr Pulmonol 2001;31(4):314–6.

96. Parlakpinar H, Tasdemir S, Polat A, Bay-Karabulut A, Vardi N, Ucar M, Yanilmaz M, Kavakli A, Acet A. Protective effect of chelerythrine on gentamicin-induced nephrotoxicity. Cell Biochem Funct 2006; 24(1):41–8.

97. Moore RD, Smith CR, Lietman PS. The association of aminoglycoside plasma levels with mortality in patients with Gram-negative bacteremia. J Infect Dis 1984;149(3):443–8.

98. Moore RD, Smith CR, Lietman PS. Association of aminoglycoside plasma levels with therapeutic outcome in Gram-negative pneumonia. Am J Med 1984;77(4):657–62.

99. Wenk M, Vozeh S, Follath F. Serum level monitoring of antibacterial drugs. A review. Clin Pharmacokinet 1984;9(6):475–92.

100. Berkovitch M, Goldman M, Silverman R, Chen-Levi Z, Greenberg R, Marcus O, Lahat E. Therapeutic drug monitoring of once daily gentamicin in serum and saliva of children. Eur J Pediatr 2000;159(9):697–8.

Glycopeptides

General Information

The glycopeptides antibiotics include avoparcin, dalbavancin, oritavancin, teicoplanin, and vancomycin; the last two are covered in separate monographs.

Dalbavancin

Dalbavancin (BI 397) is a semi-synthetic derivative of A40926, a glycopeptide structurally related to teicoplanin (1,2). It has in vitro activity against a variety of Gram-positive pathogens (3). Dalbavancin is more active against *Streptococcus pneumoniae* than conventional glycopeptides are. Compared with other available agents it has favorable minimum inhibitory concentration ranges against meticillin-susceptible and meticillin-resistant *Staphylococcus aureus*. It is highly protein bound (>90%), which contributes to its prolonged half-life of 149300 hours, because of which once-weekly dosing strategies have been used in clinical trials. However, it is not more active than teicoplanin against enterococci harboring the VanA phenotype of resistance to glycopeptides. Dalbavancin is also characterized by a marked bactericidal character and synergism with penicillin.

Preclinical studies in rats and dogs have shown that dalbavancin is well tolerated, even at dosages several-fold higher than those likely to be used clinically. In Phase I, II, and III clinical trials, dalbavancin was well tolerated, without evidence of dose- or duration-related adverse effects. It is important to note that clinical and pharmacokinetic studies published to date have excluded patients with a history of glycopeptide hypersensitivity, and there is no information about the incidence of cross-reactivity (4).

Oritavancin

Oritavancin (LY333328) is a novel glycopeptide antimicrobial agent that was obtained by reductive alkylation with 4'chloro-biphenylcarboxaldehyde of the natural glycopeptide chloroeremomycin, which differs from vancomycin by the addition of a 4-epi-vancosamine sugar and the replacement of the vancosamine by a 4-epi-vancosamine. Although oritavancin has a general spectrum of activity comparable to that of vancomycin, it offers considerable advantages in terms of intrinsic activity (especially against streptococci), and is insensitive to resistance mechanisms in staphylococci and enterococci. Because the binding affinities of vancomycin and oritavancin to free D-Ala-D-Ala and D-Ala-D-Lac are of the same order of magnitude, the difference in their activity has been attributed to cooperative interactions that can occur between the drug and both types of precursors in situ (1).

Oritavancin has a long half-life (195 hours), allowing once-a-day administration. Studies on cultured macrophages show that it accumulates slowly (by an endocytic process) in lysosomes, from which its efflux is extremely slow. This explains why it is bactericidal against intracellular forms of staphylococci or enterococci, but not against cytosolic bacteria, such as *Listeria monocytogenes*. In healthy volunteers oritavancin reached high concentrations not only in epithelial lining fluid but also in alveolar macrophages (1).

Observational studies

In a single-dose, open, non-controlled, dose-escalation study in 11 healthy subjects oritavancin was infused intravenously over 1 hour in doses of 0.020.325 mg/kg; four subjects each received a single dose of 0.5 mg/kg (5). The pharmacokinetics of oritavancin were linear. Renal clearance was 0.457 ml/minute, and less than 5% and 1% of administered drug were recovered in the urine and feces respectively after 7 days.

Comparative studies

Dalbavancin has been compared with vancomycin in an open, multicenter, phase 2 study in 67 patients with catheter-related septicemia (6). Intravenous dalbavancin was given to 33 patients as a single dose of 1000 mg followed by a dose of 500 mg 1 week later. The most common medication related adverse events were diarrhea (7%), constipation (6%), fever (6%), and oral candidiasis (4%).

Organs and Systems

Liver

In five of 11 healthy subjects there were asymptomatic and transient rises in hepatic transaminase activities in a dose-escalation study (5).

Long-Term Effects

Drug resistance

Over the years increased antimicrobial drug resistance, which has been observed all over the world, is the most serious adverse effect of this type of drug. The reason is very simple: it reduces the effectiveness of antimicrobial drug treatment of infectious diseases everywhere and in all organisms (humans, animals, fishes, plants), thereby leading to increased morbidity and mortality, as well as increased costs. In addition, when we have to switch to another type of drug, simply because of resistance, the other drug has often an increased spectrum of adverse effects.

Resistance is certainly not a new problem. Penicillinase, an enzyme that destroys penicillins, was described as early as in 1940 (7), before penicillin had been used clinically, and instances occurred in patients with infections caused by staphylococci soon after its introduction. Strains of Shigella that were resistant to several antimicrobial drugs, i.e. that expressed multiresistance, were described in Japan in the mid-1950s, and new principles for the transfer of resistance genes from one species to another, first from Shigella to E. coli, were described soon after (8). Since then, terms such as conjugation, transformation, transduction, and horizontal transfer have become part of the everyday language of those who combat infections.

It is now well established that antimicrobial drug usage and microbial resistance in one area of use may have, and most often are shown to have, consequences for the profile of resistance in other areas. Microbes are everywhere, and they share genetic material with each other freely and frequently. And when addressing the problem of antimicrobial drug resistance, its occurrence, causes, consequences, and prevention, a global approach is necessary. In other words, we have to focus on usage in all areas, human medicine, veterinary medicine, agriculture, fish farming, and so on.

The sad story of the glycopeptide avoparcin, which chemically resembles vancomycin, deserves to be underlined. From the end of the 1970s, avoparcin came into use in several countries as a growth promotor in husbandry, including poultry. In the 1980s, some instances of vancomycin-resistant enterococci were described in humans (9,10), although nobody associated this new type of resistance with the use of avoparcin. However, in the mid-1990s it was clearly shown in Germany, i.e. in a country in which several tons of avoparcin had been used as a growth promoter, that vancomycin-resistant enterococci were being isolated far more frequently on farms in which avoparcin had been used than from farms in which it had not (11), and the correlation between the use of avoparcin and vancomycin-resistant enterococci was soon firmly established. However, it took more than 2 years, and several meetings, before avoparcin was prohibited in all countries in the European Union. More than 2 years later, the manufacturer withdrew avoparcin from the market, but by that time the problem was well established. Between 1988 and 1993, vancomycin-resistant enterococci were isolated in 30 different UK hospitals, and the number of hospitals increased from 1 in 1988 to 18 in 1993 (12). The problem is still with us, 10 years after the ban of avoparcin, in farmers and on their farms (13). The mechanisms behind this persistence are still not clearly understood (14).

The good news, although now somewhat old, is that several countries, especially in Europe, have implemented monitoring programmes for antimicrobial drug resistance and antimicrobial drug usage in human and veterinary medicine. The EU supported this approach (Copenhagen Recommendation) some years ago. Several international organizations, such as the World Health Organization (WHO), the Food and Agriculture Organization (FAO), and the World Animal Health Organization (OIE), have emphasized the importance of monitoring antimicrobial drug resistance and usage and they have published several reports and recommendations.

This is good, but it is certainly not enough. The use of antimicrobial drugs continue to increase and so does the emergence of resistant organisms from all over the world; Iran, China, the USA, travellers or birds, it doesn t matter (15,16).

Among the many things that could be done, the most important is the need for a new strategy among regulatory agencies when new antimicrobial drugs are going to be introduced. In Europe and in the USA, the Food and Drug Administration (FDA) and the European Medicines Agency (EMEA) have relatively vague requirements for environmental documentation. In principle, the rules are based on toxicological rather than biological principles. But the spread of antimicrobial drug resistance is a biological event of great concern and one that must be solved according to biological principles, including prudent use of old and new antibiotics. It is also remarkable that most agencies have reduced the chance of creating rules for the prudent use of antibiotics. If a drug is as good and safe as its comparators, it has to be registered for the indications for which it has been shown to be active. So far, not a single strain of streptococcus group A has developed resistance to penicillin V, and more potent beta-lactams should be saved for other indications.

Drug Administration

Drug dosage regimens

Dalbavancin has such a long half-life that its plasma concentration exceeds the minimal bactericidal concentrations (MICs) of target organisms even at 1 week after administration of a single dose of 1000 mg; however, free concentrations are close to the MICs. Dalbavancin is therefore currently being evaluated in a once-a-week dosage regimen. Pilot phase II trials have shown excellent clinical efficacy (over 90%) in patients who receive dalbavancin 1000 mg on day 1 and 500 mg on day 8 for skin and soft tissue infections or catheter-related bloodstream infections by Gram-positive organisms (1,17).

References

1. Van Bambeke F. Glycopeptides in clinical development: pharmacological profile and clinical perspectives. Curr Opin Pharmacol 2004;4(5):471–8.
2. Guay DR. Dalbavancin: an investigational glycopeptide. Expert Rev Anti Infect Ther 2004;2(6):845–52.
3. Pope SD, Roecker AM. Dalbavancin: a novel lipoglycopeptide antibacterial. Pharmacotherapy 2006;26(7):908–18.
4. Lin SW, Carver PL, DePestel DD. Dalbavancin: a new option for the treatment of Gram-positive infections. Ann Pharmacother 2006;40(3):449–60.
5. Bhavnani SM, Owen JS, Loutit JS, Porter SB, Ambrose PG. Pharmacokinetics, safety, and tolerability of ascending single intravenous doses of oritavancin administered to healthy human subjects. Diagn Microbiol Infect Dis 2004;50(2):95–102.
6. Raad I, Darouiche R, Vazquez J, Lentnek A, Hachem R, Hanna H, Goldstein B, Henkel T, Seltzer E. Efficacy and safety of weekly dalbavancin therapy for catheter-related bloodstream infection caused by Gram-positive pathogens. Clin Infect Dis 2005;40(3):374–80.
7. Abraham EP, Chain E. An enzyme from bacteria able to destroy penicillin. Nature 1940;106:837.
8. Watanabe T. Infective heredity of multiple drug resistance in bacteria. Bact Rev 1963;27:87–115.
9. Leclercq R, Derlot E, Duval J, Courvalin P. Plasmid-mediated resistance to vancomycin and teicoplanin in Enterococcus faecium. N Engl J Med 1988;319:157–61.
10. Uttley A, Collins CH, Naidoo J, George RC. Vancomycin-resistant enterococci. Lancet 1988;1:57–8.
11. Klare I, Heier H, Claus H, Reissbrodt R, Witte W. VanA-mediated high-level glycopeptide resistance in Enterococcus faecium from animal husbandry. FEMS Microbiol Lett 1995;125(2-3):165-71; erratum in 127(3):273.
12. Woodford N, Johnsen AP, Morrison D, Speller DC. Current perspective on glycopeptide resistance Clin Microbiol Rev 1995;8:585–615.
13. Johnsen PJ, Osterhus JI, Sletvold H, Sorum M, Kruse H, Nielsen K, Simonsen GS, Sundsfjord A. Persistence of animal and human glycopeptide-resistant enterococci on two Norwegian poultry farms formerly exposed to avoparcin is associated with a widespread plasmid-mediated vanA element within a polyclonal Enterococcus faecium population. Appl Environment Microbiol 2005;71:159–68.
14. Johnsen PJ, Simonsen GS, Olsvik O, Midtvedt T, Sundsfjord A. Stability, persistence, and evolution on plasmid-encoded VanA glycopeptide resistance in enterococci in the absence of antibiotic selection in vitro and in gnotobiotic mice. Microbiol Drug Resist 2002;8:161–70.
15. Wang H, Dzink-Fox JL, Chen M, Levy SB. Genetic characterization of highly fluoroquinolone-resistant clinical Escherichia coli strains from China: role of acrR mutations. Antimicrob Agents Chemother 2001;45:1515–21.
16. Moniri R, Dastegholi K. Fluoroquinolone-resistant Escherichia coli isolated from healthy broilers with previous exposure to fluoroquinolone: is there a link? Microb Ecol Health Dis 2005;17:69–74.
17. Leighton A, Gottlieb AB, Dorr MB, Jabes D, Mosconi G, VanSaders C, Mroszczak EJ, Campbell KC, Kelly E. Tolerability, pharmacokinetics, and serum bactericidal activity of intravenous dalbavancin in healthy volunteers. Antimicrob Agents Chemother 2004;48(3):940–5.

Glycylcyclines

General Information

The glycylcyclines are semisynthetic analogues of earlier tetracyclines, the first of which is tigecycline. Tetracyclines have been in use for more than 50 years, and in most species the levels of antibacterial resistance are high and still increasing. There is therefore a need for new derivatives. The earliest glycylcyclines had a dimethyl-glycylamido group at the C-9 position of the basic tetracycline molecule. The structure-activity relations of the tetracyclines and glycylcyclines are discussed extensively elsewhere (1). The clinical status of tigecycline has been summarized (2). The adverse effects of tigecycline have been reviewed; nausea and vomiting are the main adverse effects and can be managed with antiemetics (3,4,5).

Comparative studies

In a double-blind study in 1116 adults with complicated skin infections, in which tigecycline was compared with vancomycin + aztreonam, the main adverse effects of tigecycline were *nausea* (35%), vomiting (20%), diarrhea (8.5%), and dyspepsia (3.7%); tigecycline also prolonged the activated partial thromboplastin time and the prothrombin time in about 3% and caused increased transaminase activities in 12% (6). It is not clear whether two other report from the same group of similar study in 573 and 543 patients overlap with this (7,8), but the adverse effects of tigecycline were similar in the three studies.

Organs and Systems

Gastrointestinal

In a pooled analysis of two phase 3, double-blind comparisons of intravenous tigecycline and imipenem + cilastatin in 1642 adults with complicated intra-abdominal infections, nausea (24% tigecycline, 19% imipenem + cilastatin), vomiting (19% tigecycline, 14% imipenem + cilastatin), and diarrhea (14% tigecycline, 13% imipenem + cilastatin) were the most common adverse events (9).

In one of those studies, intravenous tigecycline (100 mg initially then 50 mg every 12 hours) and imipenem + cilastatin (500 + 500 mg 6-hourly or adjusted for renal dysfunction) for 5 to 14 days was compared in 825 patients (10). Tigecycline was as efficacious as imipenem + cilastatin but caused more adverse effects, of which nausea (31% versus 5%), vomiting (26% versus 19%), and diarrhea (21% versus 19%) were the most common.

In a multicenter study there was nausea in 22% of the patients who received a low dose of tigecycline (25 mg/day intravenously) and in 38% of those who received a high dose (50 mg/day intravenously) (11). Vomiting occurred in 13% and 19% respectively. It was not stated whether the patients received antiemetics. It has also been found that gastrointestinal adverse effects are perceived to be more common with oral than with intravenous regimens (2). There was a similar high incidence of gastrointestinal disturbances in a phase I study (2). An interesting hypothesis is that tigecycline might cause the release of gastrointestinal serotonin, which then causes nausea and vomiting. However, whatever the mechanisms might be, it seems reasonable to assume that before tigecycline can be used clinically, the mechanisms behind these very common adverse effects have to be elucidated.

Drug-Drug Interactions

Ciclosporin

Tigecycline does not undergo extensive metabolism. It works independently of the cytochrome P-450 enzyme system and therefore does not affect medications metabolized by these enzymes. Biliary and fecal excretion account for 59% of the administered dose, 32% is excreted renally, and 22% is excreted unchanged in the urine (12). However, interactions between tigecycline and other drugs that are handled by the liver have been reported (13).

- A 61-year old woman with a urinary tract infection 5 years after renal transplantation was given meropenem and then tigecycline 100mg/day intravenously. She was also taking oral ciclosporin 120 mg/day. On the third day of tigecycline therapy, the whole-blood trough concentration of ciclosporin was about 600 ng/ml, about double the concentration found before tigecycline was given. Her serum creatinine concentration rose from around 100 μmol/l to a peak around 170 μmol/l. Ciclosporin was omitted for 1 day and the dosage was reduced by 50%. During the entire period of tigecycline therapy no other potentially interacting drugs were administered.

The authors suggested that ciclosporin concentrations should be monitored during treatment with tigecycline. It was subsequently suggested that both drugs might depend on the P glycoprotein efflux transporter system in the small intestine (14). This led to the conclusion that intravenous tigecycline could alter the pharmacokinetics of many other drugs. Even though this has not be shown to be so, it would be wise not to give tigecycline to patients who are taking other drugs that might be handled by P glycoprotein. When monitoring ciclosporin in such cases intralymphocytic concentrations, if available, might be helpful, since lymphocytes express P glycoprotein (15).

Food-Drug Interactions

In two pharmacokinetic studies of tigecycline food increased the maximum tolerated single dose from 100 to 200 mg (16).

References

1. Zhanel GG, Homenuik K, Nichol K, Noreddin A, Vercaigne L, Embil J, Gin A, Karlowsky JA, Hoban DJ. The glycylcyclines: a comparative review with the tetracyclines Drug 2004;64:63–88.
2. Garrison MW, Neumiller JJ, Setter SM. Tigecycline. An investigational glycylcycline antimicrobial with activity against resistant Gram-positive organisms. Clin Ther 2005;27:12–22.
3. Garrison MW, Neumiller JJ, Setter SM. Tigecycline: an investigational glycylcycline antimicrobial with activity against resistant gram-positive organisms. Clin Ther 2005;27(1):12–22.
4. Rello J. Pharmacokinetics, pharmacodynamics, safety and tolerability of tigecycline. J Chemother 2005;17 Suppl 1:12–22.
5. Scheinfeld N. Tigecycline: a review of a new glycylcycline antibiotic. J Dermatolog Treat 2005;16(4):207–12.
6. Ellis-Grosse EJ, Babinchak T, Dartois N, Rose G, Loh E; Tigecycline 300 cSSSI Study Group. The efficacy and safety of tigecycline in the treatment of skin and skin-structure infections: results of 2 double-blind phase 3 comparison

studies with vancomycinaztreonam. Clin Infect Dis 2005;41 Suppl 5:S341-53.

7. Sacchidanand S, Penn RL, Embil JM, Campos ME, Curcio D, Ellis-Grosse E, Loh E, Rose G. Efficacy and safety of tigecycline monotherapy compared with vancomycin plus aztreonam in patients with complicated skin and skin structure infections: results from a phase 3, randomized, double-blind trial. Int J Infect Dis 2005;9(5):251–61.

8. Breedt J, Teras J, Gardovskis J, Maritz FJ, Vaasna T, Ross DP, Gioud-Paquet M, Dartois N, Ellis-Grosse EJ, Loh E; Tigecycline 305 cSSSI Study Group. Safety and efficacy of tigecycline in treatment of skin and skin structure infections: results of a double-blind phase 3 comparison study with vancomycinaztreonam. Antimicrob Agents Chemother 2005;49(11):4658–66.

9. Babinchak T, Ellis-Grosse E, Dartois N, Rose GM, Loh E; Tigecycline 301 Study Group; Tigecycline 306 Study Group. The efficacy and safety of tigecycline for the treatment of complicated intra-abdominal infections: analysis of pooled clinical trial data. Clin Infect Dis 2005;41 Suppl 5:S354-67.

10. Oliva ME, Rekha A, Yellin A, Pasternak J, Campos M, Rose GM, Babinchak T, Ellis-Grosse EJ, Loh E; 301 Study Group. A multicenter trial of the efficacy and safety of tigecycline versus imipenem/cilastatin in patients with complicated intra-abdominal infections [Study ID Numbers: 3074A1-301-WW; ClinicalTrials.gov Identifier: NCT00081744]. BMC Infect Dis 2005;5:88.

11. Postier RG, Green SL, Klein SR, Ellis-Grosse EJ, Loh E. Tigecycline 200 study group: results of a multicenter, randomized, open-label efficacy and safety study of two doses of tigecycline for complicated skin and skin-structure infections in hospitalized patients. Clin Ther 2004;26:704–14.

12. Bhattacharya M, Parakh A, Narang M. Tigecycline. Drug Rev 2009;55:65–8.

13. Stumf AN, Schmidt C, Hiddeman W, Gerbitz A. High serum concentration of ciclosporin related to administration of tigecycline. Eur J Clin Pharmacol 2009;65:101–3.

14. Srinivas NR. Tigecycline and cyclosporine interactionan interesting case of biliary-excreted drug enhancing the oral bioavailability of cyclosporine. Eur J Clin Pharmacol. 2009;65(5):543–4.

15. Falck P, Asberg A, Gulseth H, Bremer S, Akhlaghi F, Reuset J, Pfeffer P, Hartmann A, Midtvedt K. Declining intra-cellular T-lymphocyte concentration of cyclosporine precedes acute rejection in kidney transplant recipients. Transplantation 2008;85:179–84.

16. Muralidharan G, Micalizzi M, Speth J, Raible D, Troy S. Pharmacokinetics of tigecycline after single and multiple doses in healthy subjects. Antimicrob Agents Chemother 2005;49(1):220–9.

Grepafloxacin

See also Fluoroquinolones

General Information

Grepafloxacin is a synthetic fluoroquinolone antibiotic with extensive tissue distribution and strong antibacterial activity in vivo (1,2). However, it was withdrawn in 1999 because of its adverse cardiovascular effects, which included dysrhythmias (3–5).

Observational studies

In phase II and III trials of grepafloxacin in a total of more than 3000 patients, the most common adverse events with grepafloxacin 400 or 600 mg were gastrointestinal, such as nausea, vomiting, and diarrhea. Significantly more patients reported a mild unpleasant metallic taste with grepafloxacin than with ciprofloxacin, but under 1% of patients withdrew because of this. Headache occurred significantly more often with ciprofloxacin than grepafloxacin. In a study of more than 9000 patients, only 2.3% reported adverse events (nausea 0.8%; gastrointestinal symptoms 0.4%; dizziness 0.3%; photosensitization 0.04%). Rarely, an unpleasant taste has been reported as an adverse event in spontaneous reports (6).

Organs and Systems

Cardiovascular

Grepafloxacin has been removed from the US market because of deaths as a result of torsade de pointes (7).

Nervous system

Headache was recorded in patients taking grepafloxacin, at rates similar to those reported with ciprofloxacin and ofloxacin (8–9%), during short-course treatment of urinary tract infections (8).

Gastrointestinal

With grepafloxacin 600 mg/day the rates of the following adverse events were noticeably higher than with 400 mg/day: nausea (15 versus 11%), vomiting (6 versus 1%), and diarrhea (4 versus 3%) (9).

References

1. Suzuki T, Kato Y, Sasabe H, Itose M, Miyamoto G, Sugiyama Y. Mechanism for the tissue distribution of grepafloxacin, a fluoroquinolone antibiotic, in rats. Drug Metab Dispos 2002;30(12):1393–9.

2. Yamamoto H, Koizumi T, Hirota M, Kaneki T, Ogasawara H, Yamazaki Y, Fujimoto K, Kubo K. Lung tissue distribution after intravenous administration of grepafloxacin: comparative study with levofloxacin. Jpn J Pharmacol 2002;88(1):63–8.

3. Gibaldi M. Grepafloxacin withdrawn from market. Drug Ther Topics Suppl 2000;29:6.

4. Carbon C. Comparison of side effects of levofloxacin versus other fluoroquinolones. Chemotherapy 2001;47(Suppl 3):9–14discussion 44–8.

5. Zhanel GG, Ennis K, Vercaigne L, Walkty A, Gin AS, Embil J, Smith H, Hoban DJ. A critical review of the fluoroquinolones: focus on respiratory infections. Drugs 2002;62(1):13–59.

6. Goldblatt EL, Dohar J, Nozza RJ, Nielsen RW, Goldberg T, Sidman JD, Seidlin M. Topical ofloxacin versus systemic amoxicillin/clavulanate in purulent otorrhea in children with tympanostomy tubes. Int J Pediatr Otorhinolaryngol 1998;46(1–2):91–101.

7. Amankwa K, Krishnan SC, Tisdale JE. Torsades de pointes associated with fluoroquinolones: importance of concomitant risk factors. Clin Pharmacol Ther 2004;75(3):242–7.

8. Chodosh S, Lakshminarayan S, Swarz H, Breisch S. Efficacy and safety of a 10-day course of 400 or 600 milligrams of grepafloxacin once daily for treatment of acute bacterial exacerbations of chronic bronchitis: comparison with a 10-day course of 500 milligrams of ciprofloxacin twice daily. Antimicrob Agents Chemother 1998; 42(1):114–20.

9. Stahlmann R, Schwabe R. Safety profile of grepafloxacin compared with other fluoroquinolones. J Antimicrob Chemother 1997;40(Suppl A):83–92.

Helicobacter pylori eradication regimens

General Information

Drugs and regimens for *Helicobacter pylori* eradication have been reviewed (1,2). The major factor in choosing an antibiotic regimen is the pattern of antibiotic resistance in the community. Generally two antimicrobial drugs plus bismuth or ranitidine or a proton pump inhibitor such as omeprazole are required to achieve a cure rate of over 90% and avoid the resistance that occurs when clarithromycin or metronidazole is the single antimicrobial used. In regions where metronidazole and clarithromycin resistance is common initial quadruple therapy with bismuth, metronidazole, tetracycline, and a proton pump inhibitor is recommended. In general, higher doses and longer duration of therapy are associated with better outcomes. Although therapy for 1 week has become accepted in first-line regimens, therapy for 2 weeks is better when treating nitroimidazole-resistant or clarithromycin-susceptible strains.

Adverse effects of these regimens are related to the individual drugs used.

Combination regimens

Bismuth biskalcitrate + metronidazole + tetracycline + omeprazole

In an international multicenter trial this 10-day regimen, which overcomes metronidazole resistance, was effective and well tolerated (3). Eradication rates were 95% (38/40) for strains resistant to metronidazole and 99% (75/76) for strains sensitive to metronidazole.

Esomeprazole + clarithromycin + metronidazole

Esomeprazole 20 mg bd + clarithromycin 500 mg bd + metronidazole 400 mg bd was well tolerated and of similar efficacy to a regimen of omeprazole 20 mg bd + clarithromycin + metronidazole in 72 patients (4). Adverse effects were diarrhea, taste disturbance, and headache; all were mild and self limiting.

Esomeprazole + amoxicillin + gatifloxacin

In a pilot study (n = 30) of sequential therapy with esomeprazole 120 mg/day + amoxicillin 3000 mg/day for 12 days, adding gatifloxacin 400 mg/day on days 6–12, there was an eradication rate of 80%; 13 patients reported adverse effects, of which diarrhea was the most common (5).

Esomeprazole + tetracycline + metronidazole

A 10-day regimen of esomeprazole 40 mg qds, tetracycline 500 mg qds, and metronidazole 500 mg qds has been studied in 20 patients with penicillin allergy, duodenal ulcer disease, and *Helicobacter pylori* infection (6). In 17 this was the first course of treatment and in three there had been prior treatment failure. Baseline and follow-up endoscopy was performed after 30 days or more of treatment for urease tests and biopsies for *Helicobacter pylori*. Eradication rates by intention to treat were 85% for first-time treatment and 100% for prior failure. Endoscopy was normal in 70% of all cases at follow-up and 85% and 100% of the patients had healed erosive gastritis and duodenal ulcers respectively. Three patients reported oral candidiasis which required drug withdrawal at days 6, 7, and 9. The symptoms resolved with fluconazole 100 mg/day and nystatin lozenges. In all three *Helicobacter pylori* was eradicated. One patient reported anorexia, nausea, and diarrhea. The combination of esomeprazole, tetracycline, and metronidazole is effective and well tolerated in patients with penicillin allergy.

Lansoprazole + amoxicillin + clarithromycin

In 55 patients *Helicobacter pylori* eradication with lansoprazole + amoxicillin + clarithromycin (LAC) was compared with LAC and subsequently lying in the left lateral position and LAC plus cetraxate (7). Adverse events were more common in the group treated with LAC alone (70%). There were two serious adverse events, one case of diarrhea in the LAC plus cetraxate group and one case of abdominal pain in the LAC group. Common adverse events were diarrhea, taste disturbances, nausea, constipation, eczema, headache, abdominal pain, and indigestion.

Omeprazole + amoxicillin + clarithromycin

In a comparative study (8) of two *Helicobacter pylori* eradication regimens using omeprazole 40 mg/day + amoxicillin 1500 mg/day and two different doses of clarithromycin (400 and 800 mg/day, n = 143 and 145 respectively) there were more adverse events in the higher dose group (52% versus 47%). The most common adverse events were diarrhea (33 and 49 patients respectively) and altered taste (3 and 18 patients respectively).

Adverse effects of triple therapy (clarithromycin 500 mg bd + amoxicillin 1 g bd + omeprazole 20 mg/day or esomeprazole 40 mg/day) for *Helicobacter pylori* eradication occurred in about 30% of 200 patients. Common adverse effects included nausea and vomiting, constipation and diarrhea, metallic and bitter taste sensations, headache, and mild abdominal pain (9).

In 582 patients the most common adverse effects of the most popular eradication regimen (omeprazole + amoxicillin + clarithromycin) were mild only; they included bitter taste, diarrhea, dizziness, and malaise (10).

When a low-dose eradication regimen of omeprazole + amoxicillin (1.5 g/day) + clarithromycin (800 mg/day) was compared with a high-dose regimen of omeprazole + amoxicillin (2 g/day) + clarithromycin (1 g/day) in 225 patients, 1.8% of patients in the low-dose group withdrew because of adverse effects (*sleep disorder and vertigo*) and 0.9% of patients in the high-dose group withdrew because of diarrhea (11). There were adverse events in 66% of the low-dose group and 61% of the high-dose group. These comprised diarrhea, bitter taste, and reflux symptoms. There were four severe adverse events, all in the low-dose group; these were esophageal cancer, pyelonephritis, myelopathy, and erythrocytosis and bronchitis, which were not thought to be due to the trial drugs.

Omeprazole + azithromycin + amoxicillin or tinidazole
In randomized comparison of omeprazole + azithromycin + either amoxicillin or tinidazole, both regimens were only partially successful at eradicating *Helicobacter pylori* (eradication rates 63% and 71% respectively) (12). The adverse effects with both regimens were negligible.

Omeprazole + clarithromycin + tinidazole
Omeprazole + clarithromycin + tinidazole for 7 days has been compared with the same regimen followed by omeprazole or placebo for 21 days, in a double-blind, placebo-controlled study in 103 patients with *Helicobacter pylori* positive peptic ulcer disease (13). There was no difference in ulcer healing (95% versus 96%) or rate of eradication of *Helicobacter pylori* (84% versus 83%) at 16 weeks. At 1 year all the patients were free from both peptic ulcer disease and *Helicobacter pylori* infection.

Pantoprazole + amoxicillin + rifabutin
In 130 patients who had failed eradication therapy a regimen of low-dose rifabutin 150 mg/day, pantoprazole 240 mg/day, and amoxicillin 3000 or 4500 mg/day for (14) 12 days was highly effective, with an eradication rate of 91% when amoxicillin 3000 mg/day was used and 97% when 4500 mg/day was use. All reported adverse events (n = 52) were mild and non-serious, the most common being diarrhea (8), nausea (5), and abdominal pain (5).

Rabeprazole + amoxicillin + clarithromycin
Rabeprazole 10 mg bd + amoxicillin 1 g bd + clarithromycin 500 mg bd, given for 1 week to 180 patients, was successful in 166 (92%) (15). Ulcer healing occurred by 4 weeks (with no continuation of proton pump inhibitor beyond the 1 week triple therapy), in 42/44 patients (96%) in whom eradication therapy was successful. Ulcer healing did not occur in the three patients in whom the bacteria were not eradicated. The authors commented that this underlines the significant role of *Helicobacter pylori* in ulcer pathogenesis. Adverse effects were bitter taste (39%), giddiness (8%), abdominal pain (5%), lethargy (5%), loose bowel motions (8%), and rash (1%). Triple therapy was not completed in 2% of the patients because of adverse effects.

In an open pilot study of eradication therapy using rabeprazole 40 mg/day + amoxicillin 2000 mg/day + clarithromycin 1000 mg/day in 111 patients there was a cure rate of 85% (16). Of these 111 patients, 76 reported adverse events; the most common was a metallic taste in the mouth (44%) followed by abdominal pain in 15% (70% mild, 30% moderate).

Helicobacter pylori eradication therapy was evaluated in 458 patients using rabeprazole 20 mg/day + amoxicillin + clarithromycin for either 7 or 10 days (17). There was no significant difference in adverse effects between the two groups. The most common adverse effects were a metallic taste sensation (19%), oral mucositis (19%), and diarrhea (7%). Symptoms were limited to the duration of treatment.

Rabeprazole + amoxicillin + metronidazole
As second-line therapy in Japanese patients who remained *Helicobacter pylori* positive after a proton pump inhibitor + amoxicillin + clarithromycin), a regimen of rabeprazole 20 mg bd + amoxicillin 750 mg bd + metronidazole 250 mg bd for 1 week produced an eradication rate of 97% (61/63) for metronidazole-sensitive strains and 82% (18/22) for metronidazole-resistant strains (18).

Rabeprazole + amoxicillin + metronidazole was well tolerated when used for *H pylori* eradication in 120 patients in a randomized controlled study (19). Adverse effects occurred equally in those given rabeprazole 10 mg or 20 mg (30%). Common adverse effects included diarrhea (17%), dizziness (10%), and nausea (3%).

Rabeprazole + clarithromycin + metronidazole
Eradication therapy with rabeprazole + clarithromycin + metronidazole was well tolerated; only two of 10 patients in an observational study reported a metallic taste alone (20).

Rabeprazole + levofloxacin + furazolidone
Twelve patients with duodenal ulcer who had previously been treated unsuccessfully with two regimens containing clarithromycin and metronidazole, were given a daily combination of rabeprazole 20 mg,

levofloxacin 500 mg, and furazolidone 200 mg as a single dose for 10 days (21). Two patients discontinued treatment because of nausea and vomiting. After 90 days of treatment 10 patients were left in the study, in six of whom culture of gastric tissue fragments was obtained: 100% and 83% of the samples showed sensitivity to furazolidone and levofloxacin respectively. Eradication rates for *Helicobacter pylori* as determined per protocol and intention to treat were 100 and 83%.

Rabeprazole + levofloxacin + rifabutin
Rabeprazole 20 mg bd + levofloxacin 500 mg/day + rifabutin 300 mg/day for 7 days has been compared with rabeprazole 20 mg bd, metronidazole 400 mg tds, bismuth subcitrate 120 mg qds, and tetracycline 500 mg qds for 7 days (22). All 109 patients had previously failed *Helicobacter pylori* eradication therapy. There was significant resistance to metronidazole and clarithromycin in patients who had previously received these drugs. The two regimens were as efficacious as second-line eradication treatments, with eradication rates of 91% in the triple-therapy group and 92% in the quadruple-therapy group.

Rabeprazole + levofloxacin + tinidazole
Duration of treatment has implications for adherence and consequently efficacy. In a pilot study of 169 patients (23) comparing 4 and 7 days of eradication therapy with rabeprazole 40 mg/day + levofloxacin 1000 mg/day + tinidazole 1000 mg/day efficacy was very similar for both durations of treatment (94% versus 95% respectively). The 4-day treatment group, however, had significantly fewer adverse events (13% versus 29%) suggesting that it may be a preferable method for eradicating *Helicobacter pylori*.

Ranitidine bismuth citrate + amoxicillin + clarithromycin
There are few validated treatment regimens for eradicating *Helicobacter pylori* in children. In 206 children (median age 12 years) 4-day and 7-day eradication regimens with ranitidine bismuth citrate + amoxicillin + clarithromycin were investigated (24). The 7-day regimen was significantly more efficacious (89% eradication versus 78%). There were no significant differences in adverse effects, the most common of which was blackening of stools due to bismuth citrate; loose stools, dyspepsia, and headache were also frequently noted.

Comparative studies

The MACH-2 study has assessed the role of omeprazole in triple therapy in 539 patients with duodenal ulcers associated with *H. pylori* (25). The addition of omeprazole resulted in significantly higher eradication rates (over 90%) than antibiotics alone (amoxicillin plus

clarithromycin about 25%; clarithromycin plus metronidazole 70%), and reduced the impact of primary resistance to metronidazole. About one-third of the patients who took amoxicillin reported diarrhea/loose stools. The frequency of taste disturbance was dose-dependent with clarithromycin. Increased liver enzymes were more commonly reported in those taking metronidazole. The addition of omeprazole did not increase the frequency of reported adverse effects.

The DU-MACH study assessed the efficacy of two omeprazole-based triple therapies (omeprazole, amoxicillin, clarithromycin versus omeprazole, metronidazole, clarithromycin) given for 1 week to 149 patients for eradicating *H. pylori*, healing duodenal ulcers, and preventing ulcer relapse over 6 months after treatment (26). Both regimens achieved high eradication rates (about 90%) and were well tolerated. Adverse effects were similar in the two groups, and included diarrhea, taste disturbance, headache, nausea, and dyspepsia.

Ranitidine 300 mg bd and omeprazole 20 mg bd have been compared as components of triple therapies (combining them with either amoxicillin plus clarithromycin or amoxicillin plus metronidazole) in 320 patients with *H. pylori* (27). Omeprazole and ranitidine combined with two antibiotics for 1 week were equally effective in eradicating *H. pylori*. This result questions the role of profound acid suppression in eradication. There was no difference in the reported adverse effects, which included nausea, vomiting, diarrhea, metallic taste, skin rashes, and headache.

In a similar study in 221 patients with peptic ulcer disease associated with *H. pylori*, rabeprazole has been compared with omeprazole and lansoprazole (combining them with amoxicillin plus clarithromycin for 1 week) (28). Rabeprazole was as effective as omeprazole and lansoprazole in eradicating *H. pylori* (84–88% each). There were no differences in reported adverse events. Common adverse effects were soft stools, glossitis, taste disturbances, and skin rashes.

Dual therapy (omeprazole plus clarithromycin) for 2 weeks has been compared with triple therapy (omeprazole plus amoxicillin and clarithromycin) for 1 week in the eradication of *H. pylori* in 145 patients with duodenal ulcers (29). Triple therapy was significantly more effective in eradicating *H. pylori* (71 versus 48%). There were no significant differences in compliance or adverse effects. The most frequent adverse effects were metallic taste and nausea in the dual-therapy group and metallic taste, mild abdominal pain, and diarrhea in the triple-therapy group.

Quadruple therapy (omeprazole, amoxicillin, roxithromycin, and metronidazole for 1 week) has been studied in an open trial in 169 patients with *H. pylori* (30). This regimen achieved an eradication rate of 92%. It was also beneficial in patients infected with pretreatment resistant strains to the antibiotics, in which cases the eradication rates achieved (over 90%) were similar to eradication rates in patients infected with sensitive strains. Compliance was good and there was only one serious adverse effect, ana-phylaxis, probably due to amoxicillin. Frequent adverse effects were abdominal distension (10%), glossitis (9%), and diarrhea (8%).

Sucralfate 1 g tds in combination with amoxicillin 500 mg tds and clarithromycin 400 mg bd for 2 weeks was as effective as a combination of lansoprazole 30 mg bd plus amoxicillin 500 mg tds and clarithromy-cin 400 mg bd for 2 weeks for *H. pylori* eradication in a randomized, multicenter trial in 150 patients (31). There was no significant difference in adverse effects between the two groups. Diarrhea, abdominal pain, glossitis, and taste disturbance were the adverse effects commonly reported.

In an open trial, 7-day triple therapy with omeprazole 30 mg bd, amoxicillin 500 mg tds, and clarithromycin 400 mg bd was safe and effective in eradicating *H. pylori* in 12 of 13 patients undergoing hemodialysis (32). There were adverse effects in two patients (compared with three of 27 patients not undergoing hemodialysis) and treat-ment had to be discontinued in one, owing to severe nausea and vomiting.

In a randomized study of the treatment options for patients allergic to penicillin the regimens studied were: omeprazole + clarithromycin + metronidazole for 7 days (group 1, n = 12); ranitidine + bismuth + tetracycline + metronidazole for 7 days (group 2, n = 17); omeprazole + rifabutin + clarithromycin for 10 days (group 3, n = 9); and levofloxacin + clarithro-mycin + omeprazole for 10 days (group 4, n = 2) (33). Group 1 tolerated the treatment best—only 17% of patients had adverse events, these being nausea and diarrhea. In group 2 adverse events occurred in 53%, including nausea, heartburn, diarrhea, and abdominal pain. In group 3 there was a drop-out rate of 33%, primarily because of the adverse events experienced by 89%, including arthralgia (55%) and myelosuppres-sion (44%). Group 4 had an adverse incident rate of 50%, including anorexia and arthralgia.

Four different combinations for *Helicobacter pylori* eradication have been studied in patients with penicillin allergy (34). The regimens comprised:

- first-line (n = 12)—metronidazole, omeprazole, and clarithromycin for 7 days;
- second-line (n = 17)—metronidazole, ranitidine bis-muth citrate, and tetracycline for 7 days;
- third-line (n = 9)—rifabutin, clarithromycin, and ome-prazole for 10 days;
- fourth-line (n = 12)—levofloxacin, clarithromycin, and omeprazole for 10 days.

The outcome measure was a negative (13)urea breath test 8 weeks after completion of treatment. The eradica-tion rates for the four groups were 58%, 47%, 11%, and 100% for the four treatments. Compliance was generally good apart from the rifabutin-based regimen, which had adverse effects in 89% of the patients, including four cases of myelotoxicity.

Moxifloxacin-based regimens have been compared with comparable clarithromycin-based therapies in 320 patients with *Helicobacter pylori* (35). The regimens com-prised:

- moxifloxacin, amoxicillin, and esomeprazole;
- moxifloxacin, tinidazole, and esomeprazole;
- clarithromycin, amoxicillin, and esomeprazole;
- clarithromycin, tinidazole, and esomeprazole.

Each group had 80 patients. Eradication rates were 89%, 92%, 78%, and 79%. Taste disturbance and bloating were less frequent in those who took moxifloxacin than in those who took the other treatments.

Failure of *Helicobacter pylori* eradication in childhood is common. The efficacy of amoxicillin, bismuth subci-trate, and omeprazole, with either nifuratel (n = 37) or furazolidone (n = 39) has been evaluated in 76 children aged 12–16 years who had failed one attempt at eradica-tion using a metronidazole-containing triple regimen (36). Eradication rates were 33/37 with nifuratel and 34/39 with furazolidone. More patients who took furazolidone com-plained of severe anorexia than those who took nifuratel (eight versus one).

Failure of first-line eradication therapy is a common problem in areas with high levels of resistance to metro-nidazole and clarithromycin. In a randomized controlled trial in 106 such patients there was comparable efficacy between a quadruple regimen containing metronidazole 1200 mg/day + tetracycline 200 mg/day + bismuth subci-trate 480 mg/day + lansoprazole 60 mg/day and a triple regimen of lansoprazole 60 mg/day + amoxicillin 2000 mg/day + levofloxacin 1000 mg/day (37). The quad-ruple therapy had an eradication rate of 71%, the triple therapy 57%. The two regimens were associated with similar rates of adverse events (21 and 18 patients respec-tively), most commonly nausea, vomiting, diarrhea, and dizziness.

Bismuth salts are commonly recommended as a com-ponent of rescue therapy for failed *Helicobacter* eradica-tion therapy but are not available world wide. In a prospective study of 93 patients who had failed eradica-tion therapy esomeprazole + clarithromycin + tetracy-cline + metronidazole was compared with esomeprazole + bismuth subcitrate + tetracycline + metronidazole (38). There were similar eradication rates in the two groups (74% versus 77%). There were adverse events in 57% and 36% respectively. This included significantly more taste disturbance (28 versus 16 patients) and vomiting (12 versus 4) in the former, and three versus one patients withdrew because of adverse effects. Good adherence to therapy independently pre-dicted treatment success in these groups (OR = 22).

In a randomized study of patients who had failed first-line eradication therapy, rabeprazole 40 mg/day + amox-icillin 2000 mg/day was less effective than rabeprazole 20 mg/day + amoxicillin 1500 mg/day + metronidazole 500 mg/day (74% versus 97%) (39). Adverse events were similar in the two groups, the most common being diarrhea (n = 4), abdominal bloating (2), and heartburn (3).

In a pilot study of the use of rifaximin in eradication therapy in 48 patients, there were limited rates of eradication (58% with esomeprazole + clarithromycin, 42% with esomeprazole + levofloxacin) despite high degrees of adherence (40). Adverse events were very similar in the two groups (13/24 and 10/24 respectively), the most common being taste disturbance.

Placebo-controlled studies

The effect of eradicating *H. pylori* in 20 patients with chronic idiopathic urticaria has been assessed in a randomized, placebo-controlled trial (41). After 7 days of treatment with omeprazole 20 mg bd, amoxicillin 500 mg bd, and clarithromycin 500 mg bd *H. pylori* was eradicated in nine of 10 patients (compared with three of 10 in the placebo group) and there was a significant improvement in urticaria in four of these nine (compared with one of the seven in whom *H. pylori* was not eradicated). No serious adverse effects were reported in either treatment group.

In a randomized, controlled trial in 120 patients supplementation with inactivated *Lactobacillus acidophilus* tds significantly improved the efficacy of a standard 7-day regimen with rabeprazole 20 mg bd, clarithromycin 250 mg tds, and amoxicillin 500 mg tds (42). There was no significant difference in adverse effects between the two groups. Those reported were abdominal pain, nausea, and diarrhea.

In a randomized, placebo-controlled trial in 60 healthy, asymptomatic subjects who screened positive for *H. pylori*, supplementation with *Lactobacillus* GG twice daily for 14 days significantly reduced the adverse effects (diarrhea, nausea, taste disturbance) and improved the overall tolerability of a standard 7-day eradication regimen consisting of rabeprazole 20 mg bd, clarithromycin 500 mg bd, and tinidazole 500 mg bd (43).

Organs and Systems

Gastrointestinal

Pseudomembranous colitis has been reported in an 86-year-old woman with non-ulcer dyspepsia a few days after she had taken triple eradication therapy (omeprazole 20 mg bd, metronidazole 400 mg tds, and clarithromycin 500 mg bd); she recovered after treatment with oral vancomycin (44).

Susceptibility Factors

Age

The efficacy of triple therapy with bismuth plus amoxicillin and metronidazole has been assessed in an open trial in 26 children with duodenal ulcers associated with *H. pylori* (45). *Helicobacter pylori* was eradicated in 25 children and the ulcer was healed in 24. During a mean follow-up of nearly 2 years, the annual ulcer relapse rate was 9% (compared with 56% in historical controls, in whom the infection was not eradicated). Adverse events were reported by 13% of the children, and included diarrhea, dizziness, nausea, and vomiting.

Triple therapy with bismuth plus amoxicillin and metronidazole for 2 weeks has been compared with dual therapy with omeprazole plus amoxicillin for 2 weeks in 126 patients over the age of 60 years, who were *H. pylori* positive and had functional dyspepsia (46). Eradication rates were similar in the two groups 2 months after the end of therapy (66% with triple therapy and 64% with dual therapy), and there was a significant reduction in dyspeptic symptoms in patients in whom *H. pylori* was eradicated compared with those in whom infection persisted. Adverse effects were reported by 15% of patients who took triple therapy (nausea and metallic taste) compared with 4% of those who took dual therapy (nausea and headache). In all cases no adverse effect was severe enough to interrupt therapy.

Drug Administration

Drug dosage regimens

The effect of adding adherence-enhancing measures to triple therapy with omeprazole plus amoxicillin and metronidazole for 10 days has been studied in 119 Australian patients with *H. pylori* infection (47). The adherence-enhancing measures were:

- the provision of the medication in a dose-dispensing unit;
- the use of a medication chart;
- the provision of an information sheet about the treatment;
- a reminder by telephone 2 days after the start of therapy.

Ten days of triple therapy was effective for *H. pylori* eradication (85–90%) and patient adherence was excellent in both groups (97%). Attempts to improve adherence had no impact on outcome and adverse effects were common and were significantly associated with treatment failure and poor adherence to therapy in both groups. There were no significant differences in the adverse effects profiles. Female sex predicted more adverse effects. Two patients reported severe adverse effects, 19 reported moderate adverse effects, and 75 reported mild adverse effects. Common adverse effects included anorexia and nausea (35%), vomiting (7%), constipation (16%), diarrhea/flatulence (37%), abdominal pain/cramps (27%), a metallic taste (44%), headache/dizziness (48%), rash/itch (3%), lethargy/confusion (9%), and insomnia/agitation (4%). The authors attributed the unusually high prevalence of adverse effects to comprehensive reporting.

References

1. Goodwin CS. Antimicrobial treatment of *Helicobacter pylori* infection. Clin Infect Dis 1997;25(5):1023–6.
2. Nakajima S, Graham DY, Hattori T, Bamba T. Strategy for treatment of *Helicobacter pylori* infection in adults. II. Practical policy in 2000. Curr Pharm Des 2000;6(15):1515–29.
3. O'Morain C, Borody T, Farley A, De Boer WA, Dallaire C, Schuman R, Piotrowski J, Fallone CA, Tytgat G, Megraud F, Spenard J. Efficacy and safety of single-triple capsules of bismuth biskalcitrate, metronidazole and tetracycline, given with omeprazole, for the eradication of Helicobacter pylori: an international multi-centre study. Aliment Pharmacol Ther 2003;17:415–20.
4. Miehlke S, Schneider-Brachert W, Bastlein E, Ebert S, Kirsch C, Haferland C, Buchners M, Neumeyer M, Vieth M, Stolte M, Lehn N, Bayerdorffer E. Metronidazole-based one week triple therapy with clarithromycin and metronidazole is effective in eradicating Helicobacter pylori in the absence of antimicrobial resistance. Aliment Pharmacol Ther 2003;18:799–804.
5. Graham DY, Abudayyeh S, El-Zimaity HMT, Hoffman J, Reddy R, Opekun AR. Sequential therapy using high-dose esomeprazole–amoxicillin followed by gatifloxacin for *Helicobacter pylori* infection. Aliment Pharmacol Ther 2006;24:845–50.
6. Rodríguez-Torres M, Salgado- Mercado R, Ríos-Bedoya CF, Aponte-Rivera E, Marxuach-Cuétara AM, Rodríguez-Orengo JF, Fernández-Carbia A. High eradication rates of *Helicobacter pylori* infection with second-line combination of esomeprazole, tetracycline, and metronidazole in patients allergic to penicillin. Dig Dis Sci 2005;50(4):634–9.
7. Kubota K, Shimizu N, Nozaki K, Takeshita Y, Ueda T, Imamura K, Hiki N, Yamaguchi H, Shimoyama S, MaFune K, Kaminishi M. Efficacy of triple therapy plus cetraxate for the *Helicobacter pylori* eradication in partial gastrectomy patients. Dig Dis Sci2005;50(5):842–6.
8. Higuchi K, Maekawa T, Nakagawa K, Chouno S, Hayakumo T, Tomono N, Orino A, Tanimura H, Asahina K, Matsuura N, Endo M, Hirano M, Sakamoto C, Inomoto T, Arakawa T. Efficacy and safety of *Helicobacter pylori* eradication therapy with omeprazole, amoxicillin and high-and low-dose clarithromycin in Japanese patients. Clin Drug Invest 2006;26(7):403–4.
9. Sheu B, Kao A, Cheng H, Hunag S, Chen T, Lu C, Wu J. Esomeprazole 40 mg twice daily in triple therapy and the efficacy of *Helicobacter pylori* eradication related to CYP2C19 metabolism. Aliment Pharmacol Ther 2005;21:283–8.
10. Wong W, Xiao S, Hu P, Wang W, Gu Q, Huang J, Xia H, Wu S, Li C, Chen M, Lai K, Chan C, Lam S, Wong B. Standard treatment for *Helicobacter pylori* infection is suboptimal in non ulcer dyspepsia compared with duodenal ulcer in Chinese. Aliment Pharmacol Ther 2005;21(1):73–81.
11. Kuwayama H, Luk G, Yoshida S, Nakamura T, Kubo M, Uemura N, Harasawa S, Kaise M, Sanuki E, Haruma K, Inoue M, Shimatani T, Meino H, Kawanishi M, Watanabe H, Nakashima M, Nakazawa S. Efficacy of a low dose omeprazole based triple therapy regimen for *Helicobacter pylori* eradication independent of cytochrome P450 genotype. Clin Drug Investig 2005;25(5):293–305.
12. Anagnostopoulos GK, Kostopoulos P, Margantinis G, Tsiakos S, Arvanitidis D. Omeprazole plus azithromycin and either amoxicillin or tinidazole for eradication of Helicobacter pylori infection. J Clin Gastroenterol 2003;36:325–8.
13. Marzio L, Cellini L, Angelucci D. Triple therapy for 7 days vs. triple therapy for 7 days plus omeprazole for 21 days in treatment of active duodenal ulcer with Helicobacter pylori infection. A double blind placebo controlled trial. Dig Liver Dis 2003;35:20–3.
14. Borody TJ, Pang G, Wettstein AR, Clancy R, Herdman K, Surace R, Llorente R, Ng C. Efficacy and safety of rifabutin-containing 'rescue therapy' for resistant *Helicobacter pylori* infection. Aliment Pharmacol Ther 2006;23:481–8.
15. Goh KL, Rosaida MS, Salem O, Cheau PL, Ranjeev P, Tan YM, Rosmawati M, Chin SC. Efficacy of a 1-week rabeprazole triple therapy for eradicating Helicobacter pylori and ulcer healing: an "in-clinical-practice" study. J Dig Dis 2003;4:204–8.
16. Cardenas VM, Graham DY, El-Zimaity HM, Opekun AR, Campos A, Chavez A, Guerrero L. Rabeprazole containing triple therapy to eradicate *Helicobacter pylori* infection on the Texas–Mexican border. Aliment Pharmacol Ther 2006;23:295–301.
17. Calvet X, Ducons J, Bujanda L, Bory F, Monserrat A, Gisbert J; Hp Study Group of the Asociación Española de Gastroenterología. Seven versus ten days of rabeprazole triple therapy for *Helicobacter pylori* eradication: a multicenter randomised trial. Am J Gastroenterol 2005;100(8):1696–701.
18. Mukrakami K, Sato R, Okimoto T, Nasu M, Fujioka T, Kodama M, Kagawa J. Efficacy of triple therapy comprising rabeprazole, amoxicillin and metronidazole for second-line Helicobacter pylori eradication in Japan and the influence of metronidazole resistance. Aliment Pharmacol Ther 2003;17:119–23.
19. Wong W, Huang J, Xia H, Fung F, Tong T, Cheung K, Ho V, Lai K, Chan C, Chan A, Hui C, Lam S, Wong B. Low dose rabeprazole, amoxicillin and metronidazole triple therapy for the treatment of *Helicobacter pylori* infection in Chinese patients. J Gastroenterol Hepatol 2005;20:935–40.
20. Giannini E, Malfatti F, Botta F, Polegato S, Testa E, Fumagalli A, Mamone M, Savarino V, Testa R. Influence of 1 week *Helicobacter pylori* eradication therapy with rabeprazole, clarithromycin, and metronidazole on 13C-aminopyrine breath test. Dig Dis Sci 2005;50(7):1207–13.
21. Coelho L, Moretsohn L, Viera W, Gallo M, Passos M, Cindr J, Cerqueira M, Vitiello L, Ribeiro M, Mendonca S, Pedrazzoli-Junior J, Castro L. New once daily, highly effective rescue triple therapy after multiple *Helicobacter pylori* treatment failure: a pilot study. Aliment Pharmacol Ther 2005;21:783–7.
22. Wong WM, Gu O, Lam SK, Fung FMY, Lai KC, Hu WHC, Yee YK, Chan CK, Xia HHX, Yuen MF, Wong BCY. Randomised controlled study of rabeprazole, levofloxacin and rifabutin triple therapy vs. quadruple therapy as a second-line treatment for Helicobacter pylori infection. Aliment Pharmacol Ther 2003;17: 553-60.
23. Giannini EG, Bilardi C, Dulbecco P, Mamone M, Santi ML, Testa R, Mansi C, Savarino V. Can *Helicobacter pylori* eradication regimens be shortened in clinical practice? An open-label, randomized, pilot study of 4 and 7-day triple therapy with rabeprazole, high-dose levofloxacin, and tinidazole. J Clin Gastroenterol 2006;40:515–20.

24. Tam YH, Yeung CK, Lee KH. Seven-day is more effective than 4-day ranitidine bismuth citrate-based triple therapy in eradication of *Helicobacter pylori* in children: a prospective randomized study. Aliment Pharmacol Ther 2006;24:81–6.

25. Lind T, Megraud F, Unge P, Bayerdorffer E, O'morain C, Spiller R, Veldhuyzen Van Zanten S, Bardhan KD, Hellblom M, Wrangstadh M, Zeijlon L, Cederberg C. The MACH2 study: role of omeprazole in eradication of *Helicobacter pylori* with 1-week triple therapies. Gastroenterology 1999;116(2):248–53.

26. Zanten SJ, Bradette M, Farley A, Leddin D, Lind T, Unge P, Bayerdorffer E, Spiller RC, O'Morain C, Sipponen P, Wrangstadh M, Zeijlon L, Sinclair P. The DU-MACH study: eradication of *Helicobacter pylori* and ulcer healing in patients with acute duodenal ulcer using omeprazole based triple therapy. Aliment Pharmacol Ther 1999;13(3):289–95.

27. Savarino V, Zentilin P, Bisso G, Pivari M, Mele MR, Mela GS, Mansi C, Vigneri S, Termini R, Celle G. Head-to-head comparison of 1-week triple regimens combining ranitidine or omeprazole with two antibiotics to eradicate *Helicobacter pylori*. Aliment Pharmacol Ther 1999;13(5):643–9.

28. Miwa H, Ohkura R, Murai T, Sato K, Nagahara A, Hirai S, Watanabe S, Sato N. Impact of rabeprazole, a new proton pump inhibitor, in triple therapy for *Helicobacter pylori* infection—comparison with omeprazole and lansoprazole. Aliment Pharmacol Ther 1999;13(6):741–6.

29. Calvet X, Lopez-Lorente M, Cubells M, Bare M, Galvez E, Molina E. Two-week dual vs. one-week triple therapy for cure of *Helicobacter pylori* infection in primary care: a multicentre, randomized trial. Aliment Pharmacol Ther 1999;13(6):781–6.

30. Okada M, Nishimura H, Kawashima M, Okabe N, Maeda K, Seo M, Ohkuma K, Takata T. A new quadruple therapy for *Helicobacter pylori*: influence of resistant strains on treatment outcome. Aliment Pharmacol Ther 1999;13(6):769–74.

31. Adachi K, Ishihara S, Hashimoto T, Hirakawa K, Niigaki M, Takashima T, Kaji T, Kawamura A, Sato H, Okuyama T, Watanabe M, Kinoshita Y. Efficacy of sucralfate for *Helicobacter pylori* eradication triple therapy in comparison with a lansoprazole-based regimen. Aliment Pharmacol Ther 2000;14(7):919–22.

32. Tsukada K, Miyazaki T, Katoh H, Masuda N, Ojima H, Fukai Y, Nakajima M, Manda R, Fukuchi M, Kuwano H, Tsukada O. Seven-day triple therapy with omeprazole, amoxycillin and clarithromycin for *Helicobacter pylori* infection in haemodialysis patients. Scand J Gastroenterol 2002;37(11):1265–8.

33. Gisbert J, Gisbert J, Marcos S, Olivares D, Pajares J. *Helicobacter pylori* first line treatment and rescue options in patients allergic to penicillin. Aliment Pharmacol Ther 2005;22:1041–6.

34. Gisbert JP, Gisbert JL, Marcos S, Olivares D, Pajares JM. *Helicobacter pylori* first line treatment and rescue operations in patients allergic to penicillin. Aliment Pharmacol Ther 2005;22 (10):1041–6.

35. Nista EC, Candelli M, Zocco MA, Cazzato IA, Cremonini F, Ojetti V, Santoro M, Finizio R, Pignataro G, Cammarota G, Gasbarrini G, Gasbarrini A. Moxifloxacin-based

36. Nijevitch AA, Shcherbakov PL, Sataev VU, Khasanov R SH, Al Khashash R, Tuygunov MM. *Helicobacter pylori* eradication in childhood after failure on initial treatment: advantage of quadruple therapy with nifuratel to furazolidine. Aliment Pharmacol Ther 2005;22(9):881–7.

37. Wong WM, Gu Q, Chu KM, Yee YK, Fung FM, Tong TS, Chan AO, Lai KC, Chan CK, Wong BC. Lansoprazole, levofloxacin and amoxicillin triple therapy vs. quadruple therapy as second-line treatment of resistant *Helicobacter pylori* infection. Aliment Pharmacol Ther 2006;23:421–7.

38. Wu DC, Hsu PI, Chen A, Lai KH, Tsay FW, Wu CJ, Lo GH, Wu JY, Wu IC, Wang WM, Tseng HH. Randomized comparison of two rescue therapies for *Helicobacter pylori* infection. Eur J Clin Invest 2006;36 (11):803–9.

39. Kawai T, Kawakami K, Kataoka M, Taira S, Itoi T, Moriyaus F, Takagi Y, Aoki T, Rimbara E, Noguchi N, Sasatsu M. Comparison of efficacies of dual therapy and triple therapy using rabeprazole in second-line eradication of *Helicobacter pylori* in Japan. Aliment Pharmacol Ther 2006;24 (Suppl 4):16–22.

40. Gasbarrini A, Lauritano EC, Nista EC, Candelli M, Gabrielli M, Santoro M, Zocco MA, Cazzato A, Finizio R, Ojetti V, Cammarota G, Gasbarrini G. Rifaximin-based regimens for eradication of *Helicobacter pylori*: a pilot study. Dig Dis 2006;24:195–200.

41. Gaig P, Garcia-Ortega P, Enrique E, Papo M, Quer JC, Richard C. Efficacy of the eradication of *Helicobacter pylori* infection in patients with chronic urticaria. A placebo-controlled double blind study. Allergol Immunopathol (Madr) 2002;30(5):255–8.

42. Canducci F, Armuzzi A, Cremonini F, Cammarota G, Bartolozzi F, Pola P, Gasbarrini G, Gasbarrini A. A lyophilized and inactivated culture of *Lactobacillus acidophilus* increases *Helicobacter pylori* eradication rates. Aliment Pharmacol Ther 2000;14(12):1625–9.

43. Armuzzi A, Cremonini F, Bartolozzi F, Canducci F, Candelli M, Ojetti V, Cammarota G, Anti M, De Lorenzo A, Pola P, Gasbarrini G, Gasbarrini A. The effect of oral administration of *Lactobacillus GG* on antibiotic-associated gastrointestinal side-effects during *Helicobacter pylori* eradication therapy. Aliment Pharmacol Ther 2001;15(2):163–9.

44. Harsch IA, Hahn EG, Konturek PC. Pseudomembranous colitis after eradication of *Helicobacter pylori* infection with a triple therapy. Med Sci Monit 2001;7(4):751–4.

45. Huang FC, Chang MH, Hsu HY, Lee PI, Shun CT. Long-term follow-up of duodenal ulcer in children before and after eradication of *Helicobacter pylori*. J Pediatr Gastroenterol Nutr 1999;28(1):76–80.

46. Catalano F, Branciforte G, Brogna A, Bentivegna C, Luca S, Terranova R, Michalos A, Dawson BK, Chodash HB. *Helicobacter pylori*-positive functional dyspepsia in elderly patients: comparison of two treatments. Dig Dis Sci 1999;44(5):863–7.

47. Henry A, Batey RG. Enhancing compliance not a prerequisite for effective eradication of *Helicobacter pylori*: the HelP Study. Am J Gastroenterol 1999;94(3):811–5.

Isepamycin

See also Aminoglycoside antibiotics

General Information

Isepamicin is similar to amikacin but has better activity against strains that produce type I 6'-acetyltransferase. It can cause nephrotoxicity, vestibular toxicity, and ototoxicity. However, it is one of the less toxic of the aminoglycosides (1). The antibacterial spectrum of isepamicin includes *Enterobacteriaceae* and staphylococci; anaerobes, *Neisseriae*, and streptococci are resistant (1). Isepamicin was as effective and safe as amikacin in the treatment of acute pyelonephritis in children and might prove an advantageous alternative in areas with a high incidence of resistance to other aminoglycosides (2).

Isepamicin is given intravenously or intramuscularly in a dosage of 15 mg/kg/day or 7.5 mg/kg bd. It is not bound to plasma proteins, it distributes in extracellular fluids, and it enters some cells (outer hair cells, kidney cortex) by an active transport mechanism (1); the transference of isepamicin to the bone marrow is excellent (3). Isepamicin is not metabolized and is renally excreted with a half-life of 2–3 hours in adults with normal renal function. Its clearance is reduced in neonates, and a dose of 7.5 mg/kg/day is recommended in children younger than 16 days. Its clearance is also reduced in elderly people, but no dosage adjustment is required. In patients with chronic renal impairment, isepamicin clearance is proportional to creatinine clearance.

The bone tissue penetration of isepamicin has been studied in an open, non-comparative study, and the results compared with microbiologic data to estimate the clinical efficacy of isepamicin in bone infections (4). In 12 subjects of similar age, body weight, height, and creatinine clearance, who were undergoing elective total hip replacement, a single parenteral dose of isepamicin 15 mg/kg achieved concentrations in both cancellous and cortical bone tissue greater than the minimum concentrations required to inhibit the growth of 90% of strains of most of the susceptible pathogens commonly involved in bone infections.

Organs and Systems

Sensory systems

Hereditary deafness is a heterogeneous group of disorders, with different patterns of inheritance and due to a multitude of different genes (5,6). The first molecular defect described was the A1555G sequence change in the mitochondrial 12S ribosomal RNA gene. Two cases of hearing loss after short-term exposure to isepamicin sulfate, an aminoglycoside with milder adverse effects than other aminoglycosides, have also been described in patients with the A1555G mutation (7).

Urinary tract

Fleroxacin had a protective effect on isepamicin-induced nephrotoxicity in rats (8).

Drug Administration

Drug dosage regimens

In a randomized, multicenter study 236 patients were randomly assigned to isepamicin in a calculated dose or a loading dose of 25 mg/kg (9). The calculated dose was estimated using a specific population model with a Bayesian method, including information about age, weight, height, sex, and serum creatinine. The Bayesian method was significantly more accurate and performed particularly well in ventilated patients and severely ill patients, compared with a loading dose of 25 mg/kg, in obtaining a first isepamicin peak concentration of 80 micrograms/ml in patients in an intensive care unit.

References

1. Tod M, Padoin C, Petitjean O. Clinical pharmacokinetics and pharmacodynamics of isepamicin. Clin Pharmacokinet 2000;38(3):205–23.
2. Kafetzis DA, Maltezou HC, Mavrikou M, Siafas C, Paraskakis I, Delis D, Bartsokas C. Isepamicin versus amikacin for the treatment of acute pyelonephritis in children. Int J Antimicrob Agents 2000;14(1):51–5.
3. Shibata Y, Midorikawa K, Naito M, Yatsunami M, Hamada K. [Concentration of isepamicin sulfate in bone marrow blood.]Jpn J Antibiot 2000;53(9):609–13.
4. Boselli E, Breilh D, Bel JC, Debon R, Saux MC, Chassard D, Allaouchiche B. Diffusion of isepamicin into cancellous and cortical bone tissue. J Chemother 2002;14(4):361–5.
5. Hardisty RE, Fleming J, Steel KP. The molecular genetics of inherited deafness—current knowledge and recent advances. J Laryngol Otol 1998;112(5):432–7.
6. Steel KP. Progress in progressive hearing loss. Science 1998;279(5358):1870–1.
7. Usami S, Abe S, Tono T, Komune S, Kimberling WJ, Shinkawa H. Isepamicin sulfate-induced sensorineural hearing loss in patients with the 1555 A→G mitochondrial mutation. ORL J Otorhinolaryngol Relat Spec 1998;60(3):164–9.
8. Yazaki T, Yoshiyama Y, Wong P, Beauchamp D, Kanke M. Protective effect of fleroxacin against the nephrotoxicity of isepamicin in rats. Biol Pharm Bull 2002;25(4):516–9.
9. Krishnan RS, Lewis AT, Kass JS, Hsu S. Ultraviolet recall-like phenomenon occurring after piperacillin, tobramycin, and ciprofloxacin therapy. J Am Acad Dermatol 2001;44(6):1045–7.

Isoniazid

See also Antituberculosis drugs

General Information

Isoniazid is the hydrazide of isonicotinic acid. It is a first-line drug for therapy and prophylaxis of tuberculosis. It is bactericidal for rapidly dividing mycobacteria, but bacteriostatic for "resting bacilli". Among non-tuberculous mycobacteria, only a few strains, such as *Mycobacterium kansasii*, are susceptible. As a rule, sensitivity should always be tested in vitro, since the minimum inhibitory concentration varies greatly.

Isoniazid diffuses rapidly into body fluids and cells, including cerebrospinal fluid. It is as effective against bacteria growing within cells as it is against bacteria in culture media. There is no cross-resistance between isoniazid and other antituberculosis drugs.

The daily dose of isoniazid is 5 mg/kg, with a maximum of 300 mg/day in adults with normal liver and kidney function. In children, 8–10 mg/kg/day is an appropriate dosage, with a maximum daily dose of 300 mg, since the metabolism of isoniazid in children is rapid. Untoward effects of isoniazid as a single antituberculosis drug can be evaluated in preventive tuberculosis therapy, since curative regimens usually consist of multiple drugs.

Pharmacokinetics

After oral administration, isoniazid reaches a peak plasma concentration of 3–5 micrograms/ml within 1–2 hours. It equilibrates into all body fluids and tissues and 75–95% is excreted in the urine within 24 hours. The most important urinary metabolites are products of acetylation (acetylisoniazid) and hydrolysis (isonicotinic acid). Isonicotinyl glycine, isonicotinyl hydrazones, and *N*-methylisoniazid appear in only small amounts. The rate of acetylation of isoniazid significantly alters its plasma concentrations and half-life. The mean half-life of isoniazid in rapid acetylators is about 70 minutes and in slow acetylators 3 hours.

General adverse reactions

In a survey of more than 2000 patients treated with isoniazid, the most frequent adverse effects were rash (2%), fever (1.2%), jaundice (0.6%), and peripheral neuropathy (0.2%). Seizures can also occur. Isoniazid prolongs the half-life of rifampicin and other drugs metabolized by the liver. Morbilliform, maculopapular, purpuric, and urticarial rashes, with or without fever, are considered to be of allergic origin. Hematological adverse effects consist of agranulocytosis, thrombocytopenia, pure red cell aplasia, and eosinophilia (SEDA-9, 268). Dyspnea, with thoracic pain, cough, fever, and eosinophilia, as well as micronodular densities in the chest X-ray, may also be due to an immunological process (1). Vasculitis associated with antinuclear antibodies has been observed, as well as arthritic symptoms (2). Liver injury (mainly hepatocellular) has been considered another form of hypersensitivity reaction, but usually occurs during combined antituberculosis treatment with rifampicin in combination with anticonvulsants or halothane, or in association with alcoholism. Some are of the opinion that the main factor that induces liver damage is the fast acetylator phenotype (3), while others have found no correlation (4). No increase in cancer deaths was observed in a series of 338 women treated with isoniazid for pulmonary tuberculosis (5). Isoniazid is a potent inhibitor of hepatic enzymes and interferes with the metabolism of many drugs (SEDA-8, 287) (SEDA-11, 271).

Organs and Systems

Respiratory

Symptoms suggestive of bronchial obstruction occur only very rarely during isoniazid treatment (1).

Nervous system

Untoward neurological effects (peripheral neuropathy and focal seizures) occur in 2% of patients if pyridoxine is not given and in 12–20% taking higher doses of isoniazid (10–15 mg/kg) (6).

Peripheral neuropathy due to isoniazid results from inhibition of the formation of the co-enzyme form of vitamin B6, pyridoxine. Numbness or tingling of the extremities in the "glove and stocking" distribution can occur early during treatment with isoniazid when pyridoxine is not given. Neuropathy was noted by several authors in up to 20% of malnourished patients and fast acetylators taking isoniazid in doses over 5–6 mg/kg. Only 2% of patients taking 5 mg/kg of isoniazid plus pyridoxine developed neuropathy. Symptoms generally consist of hyperesthesia, reduced vibration and position sense, and exaggerated or reduced tendon reflexes, but ataxia, muscle weakness, and even paralysis can develop (7). Shoulder–arm syndrome has also been described.

Neurohistology shows disappearance of synaptic vesicles, mitochondrial swelling or condensation, and fragmentation of axon endings. Alterations in the lumbar and sacral spinal ganglia and the spinal cord have also been reported (8).

The addition of pyridoxine to usual doses of isoniazid of 5 mg/kg/day in adults and 8–10 mg/kg/day in children markedly reduces neurotoxicity (9). Adherence to therapy is improved by prescribing combined tablets containing 20 mg of pyridoxine per 100 mg of isoniazid. In otherwise healthy people, prescription of pyridoxine is not mandatory. However, it should be routinely given to malnourished patients and those predisposed to neuropathy (for example pregnant women, elderly people, and people with diabetes, alcoholism, or uremia) (7).

Pellagra-associated encephalopathy has been suspected as an adverse effect of isoniazid administration in several patients with tuberculosis. Deficiency of niacin (nicotinic acid) is characterized by dermatitis, diarrhea, and dementia. Other symptoms can occur, such as seizures, hallucinations, spasticity, and glossitis. Pellagra induced by isoniazid is promoted by malnutrition or a vegetarian diet with low intake of the nicotinamide precursors tryptophan and nicotinic acid. Specific supplementation is essential (10).

Acute isoniazid intoxication is rare and is classically identified by a triad of signs and symptoms—seizures refractory to standard anticonvulsants, metabolic acidosis, and coma. The initial picture often mimic encephalitis. Laboratory findings often include metabolic acidosis, hyperglycemia, hypokalemia, glycosuria, and ketonuria, mimicking diabetic ketoacidosis. The nervous system effects, particularly seizures, result from the ability of isoniazid to react with pyridoxine, a co-factor for the production of gamma-aminobutyric acid (GABA).

Isoniazid and pyridoxine form a compound that is rapidly excreted in the urine, resulting in acute GABA deficiency. Status epilepticus after overdose of isoniazid has been reported (11).

- A previously healthy 10-year old was noticed to have noisy irregular breathing during sleep. She vomited repeatedly and was agitated and not fully responsive to commands. Her only regular medication was isoniazid 400 mg/day because of recent contact with active tuberculosis. On the assumption that the child had taken an overdose of isoniazid, she was given activated charcoal via nasogastric tube and pyridoxine 5 mg/kg and sodium bicarbonate intravenously. There was rapid improvement and she recovered full consciousness within a few hours.

Sensory systems

Optic neuritis and atrophy have occurred in patients taking isoniazid (12).

Psychological, psychiatric

Isoniazid can cause neuropsychiatric syndromes, including euphoria, transient impairment of memory, separation of ideas and reality, loss of self-control, psychoses (9), and obsessive-compulsive neurosis (13). Isoniazid should be used with caution in patients with pre-existing psychoses, as it can cause relapse of paranoid schizophrenia (14). Patients on chronic dialysis appear to be vulnerable to neurological adverse drug reactions, because of abnormal metabolism of uremic toxins. It is therefore recommended that a higher than usual dose of pyridoxine be given to patients on dialysis taking isoniazid (15,16).

Endocrine

Cushing's syndrome, gynecomastia, amenorrhea, and precocious puberty have been regarded as reflecting the enzyme-inhibitory activity of isoniazid, with resulting derangement of hormone metabolism in the liver.

Metabolism

Isoniazid can cause transient hyperglycemia in overdose (17).

In five volunteers taking isoniazid 15 mg/kg/day over a period of 6 weeks, serum cholesterol concentrations were reduced (SEDA-9, 268).

Hematologic

Agranulocytosis, thrombocytopenia, hemolytic anemia, sideroblastic anemia (18), pure red cell aplasia (19), methemoglobinemia, and eosinophilia can occur exceptionally during isoniazid treatment (SEDA-9, 268) (20). An acquired coagulation factor XIII inhibitor developed in a patient taking isoniazid and resulted in a bleeding disorder (21).

Gastrointestinal

No severe gastrointestinal adverse effects of isoniazid have been observed; symptoms are limited to such as epigastric distress, gastric burning, and dry mouth. Of 814 patients with pulmonary tuberculosis treated for 9 months with isoniazid and rifampicin, 18 had minor symptoms of gastrointestinal intolerance (22). Nausea can occur, particularly if the drug is taken before breakfast or in combination with other antituberculosis drugs; 5.5% of 912 patients taking a fixed tablet combination of isoniazid, protionamide, and diaphenyl sulfone plus rifampicin had sensations of fullness, nausea, or vomiting (23).

Liver

Abnormal liver function is the most commonly described adverse effect of isoniazid (24). It may be related to acetylator phenotype.

Hepatotoxicity caused considerable alarm after an episode in Washington DC, when 19 of 2231 government employees given isoniazid prophylaxis developed clinical signs of liver disease within 6 months; 13 were severely jaundiced and 2 died (24).

In a series of 13 838 individuals treated prophylactically, 114 developed overt hepatic disease (SED-9, 574) (25). There were 13 deaths, submassive necrosis in 9 cases, and massive necrosis in 4. The 20 other patients from whom hepatic tissue was available included 16 with moderately severe acute hepatocellular injury (four with a mixed hepatocellular-cholestatic pattern), and four with chronic hepatic damage (one with cirrhosis). The effects of liver disease before the onset of jaundice included vague digestive and "viral" disease-like complaints, some with and some without gastrointestinal symptoms. Jaundice was the presenting complaint in 10%, fever and rash were reported in under 4%, and there was a modest eosinophilia in about 10%. Liver damage was recognized during the first month of administration in 15% and during the second month in a further 31%. The other 54% had taken isoniazid for 2–11 months. Although liver damage in these patients may have been related to isoniazid, some probably had viral hepatitis.

In a study of isoniazid prophylactic therapy in Seattle, USA, only 11 patients of 11 141 had hepatotoxic reactions (26). The rate was 0.1% of those starting and 0.15% of those completing the course of therapy. The duration of therapy was not stated, but 10 of the 11 episodes occurred within 3 months of starting.

The mechanism responsible for hepatotoxicity is not known. A metabolite of isoniazid, acetylhydrazine, causes hepatic damage and may play a role. Alcoholic hepatitis is an aggravating factor. The sensitivity of chronic carriers of hepatitis B virus is controversial (27). In an investigation in which the urinary metabolic profile of isoniazid was assayed in patients who developed isoniazid-related liver damage, it was impossible to predict which patients would be susceptible. It was also impossible to show that rifampicin plays a significant role in inducing liver damage when added to isoniazid (28,29). Nevertheless, it is conceivable that enzyme induction by rifampicin alters the metabolism of isoniazid. Further risk factors are alcoholism, malnutrition, diabetes, previous liver damage, renal insufficiency, and drug abuse. Age is certainly another risk factor: hepatic damage seems to be rare in patients under the age of 20; the incidence of liver toxicity is 0.3% between the ages of 20 and 34, and increases with age to 1.2% between 34 and 49 and 2.3% over the age of 50 (9). Even in prophylaxis with isoniazid monotherapy, adverse effects are more frequent in patients aged over 35 years (30).

Liver damage usually appears 1–2 months after the start of therapy. In children, raised liver enzymes are common during the first few months of treatment, but withdrawal is seldom necessary. A careful watch should be kept for early symptoms of isoniazid-induced hepatitis, such as malaise, fatigue, nausea, and epigastric distress. The dangers of continuing isoniazid after the onset of symptoms of toxicity have been highlighted (31). The earliest symptoms of isoniazid toxicity should be clearly described to the patient, particularly to hepatitis B carriers, who may be more susceptible to hepatotoxicity (27).

It is advisable to measure the liver enzymes aspartate transaminase and alanine transaminase before treatment and monthly thereafter, for as long as isoniazid administration lasts. Isoniazid should be withdrawn if the transaminases increase to over five times normal (9).

Drug users have multiple risk factors for hepatotoxicity, including chronic infection with hepatitis C. In a prospective study among drug users the only two factors that were independently associated with isoniazid hepatotoxicity were excessive alcohol consumption (OR = 4.2, 95% CI = 1.6, 10.8) and a high baseline alanine transaminase activity (OR = 4.3, 95% CI = 1.6, 11.4) (32). The presence of hepatitis C antibodies by itself did not confer an additional risk of toxicity, which was observed in 20 of 415 patients (4.8%). In another study, among 3788 patients taking isoniazid in a US county tuberculosis clinic, 673 (18%) had one or more adverse effects, including 10 (0.3%) with isoniazid-associated liver damage (33). There were higher rates with increasing age, but the overall rates were not alarming, even in individuals with risk factors. However, the greatest obstacle to the successful use of isoniazid preventive therapy is not toxicity but low completion rates.

There have been no clinical trials of the benefits and harms of routinely monitoring liver function tests for all patients taking isoniazid for latent tuberculosis infection. Data from two case series have suggested that routine liver function test monitoring leads to withdrawal of isoniazid prophylaxis from about 6% of patients because of abnormal results (34,35). This is 10–60 times the hepatitis rate found in case series using a symptom-based monitoring strategy (36,37,38,39). In a pooled analysis of more than 200 000 patients taking isoniazid prophylaxis there was a hepatitis rate of 1.2% and only two deaths. There is an increased risk of hepatitis in advancing age; in one series the rates were 3/1000 in those aged 20–34 years, 12/1000 in those aged 35–49 years, and 23/1000 in those aged 50–64 years. The CDC and ATS joint guidelines for treatment of LTBI state that baseline laboratory testing is not routinely indicated, even for those over 35 years but may be considered for patients who are taking other hepatotoxic drugs or have chronic medical conditions. Baseline measurements of bilirubin and aspartate transaminase or alanine transaminase along with monthly monitoring of liver function tests are recommended for patients with pre-existing liver disease, patients with HIV infection, pregnant or postpartum women, and alcoholics. Isoniazid should be withdrawn if liver function tests exceed 5 times the upper limit of the reference range or three times the upper limit if the patient is symptomatic.

Hepatotoxicity, defined as aspartate transaminase activity more than five times the upper limit of the reference range, associated with the treatment of latent tuberculosis has been evaluated over 7 years in a retrospective study in adults (40). Of 3377 patients taking isoniazid, 19 had high aspartate transaminase activities (5.6 per 1000 patients). Only one of the 19 had prodromal symptoms associated with hepatotoxicity. After 1, 3, and 6 months of therapy the numbers of hepatotoxic events per 1000 patients were 2.75, 7.20, and 4.10 respectively. Age over 49 years and a baseline aspartate transaminase activity above the upper limit of normal were susceptibility factors. Moderate to severe hepatotoxicity often occurs without symptoms, which emphasizes the value of transaminase monitoring.

Pancreas

Acute pancreatitis has been attributed to isoniazid (41,42).

- A 42-year-old Asian man developed clinical, biochemical, and imaging features of acute pancreatitis 11 days after starting to take rifampicin, isoniazid, and pyrazinamide for spinal tuberculosis (43). He had no history of excessive alcohol or other drug therapy. He improved after withdrawal of all drugs, but the pancreatitis recurred on reintroduction of isoniazid and resolved after withdrawal.

Causality here was difficult to establish. Pancreatitis has rarely been seen in patients who have taken an overdose of isoniazid. Rifampicin, also used in this case, is more likely to cause pancreatitis in usual doses.

Urinary tract

Isoniazid can rarely cause renal damage (44).

Urinary retention is a rare complication of isoniazid overdose (45).

Skin

Morbilliform, maculopapular, or urticarial rashes have been observed in up to 2% of patients (9). In one study, acneiform eruptions developed in only 11 cases (1.4%), including 0.5% of patients taking isoniazid, 1.5% of patients taking rifampicin, and 0.6% of patients taking ethambutol (46).

Acquired cutis laxa has been mentioned in a single child taking isoniazid, but it was probably coincidental (47).

Erythema nodosum and purpura have been described in patients taking isoniazid (48).

Pellagra has been reported in a patient taking isoniazid, despite concomitant pyridoxine prophylaxis (49).

- A 52-year-old man developed pellagra with a classical photosensitive distribution after taking isoniazid for 14 months; the first 6 months being treatment for possible tuberculous meningitis, the rest to provide antituberculosis protection while glucocorticoids were given for possible neurosarcoidosis. He was said to take alcohol only occasionally and took pyridoxine throughout the entire period of treatment. He improved rapidly on withdrawal of isoniazid and supplementation with nicotinamide.

The authors noted that isoniazid-induced pellagra was first described in 1956, shortly after the introduction of isoniazid for the treatment of tuberculosis, but that subsequent reports have been very uncommon.

Isoniazid has also been reported to cause subepidermal blistering (50).

- A 63-year-old man developed bullous lesions on the trunk and limbs and a six-fold rise in liver enzymes 15 days after treatment for tuberculosis. His skin recovered and his liver enzymes returned to normal within 2 weeks of withdrawal of all treatment, but the abnormalities recurred when treatment was resumed. At this point a skin biopsy showed subepidermal blistering. Once again withdrawal of treatment led to improvement, and when treatment was resumed without isoniazid the improvement continued.

Desensitization to isoniazid has been attempted in some patients with drug fever or rashes. A procedure of rush desensitization over a few days or a week can be used, starting with 1 mg orally and increasing the dosage every second day (51) or even every few hours.

Musculoskeletal

Arthritic symptoms, with back pain, bilateral proximal interphalangeal joint involvement, arthralgia of the knees, elbows, and wrists, and the so-called "shoulder–arm" syndrome with cervicobrachial neuralgia can occur (9).

Rhabdomyolysis occurred in subjects with isoniazid poisoning in a retrospective analysis of 270 patients seen over a 5-year period at the Phillipine General Hospital in Manilla (52). Skeletal muscle creatine kinase activity was raised in 31 of the 52 evaluable subjects who had taken more than 2.4 g/day of isoniazid. Creatine kinase activity peaked on days 5–6 and fell thereafter. Two patients developed acute renal insufficiency and required dialysis. Seizures occurred in all patients, and their duration, but not their frequency, correlated with raised creatine kinase activity. However, it is likely that factors other than seizures contribute to rhabdomyolysis in patients with isoniazid poisoning.

Immunologic

The expansion or new development of tuberculous lesions during ultimately successful therapy has been termed a "paradoxical response." It is most often reported in relation to intracranial tuberculomata, but is probably most common in tuberculous lymphadenopathy. It is also described in tuberculous pleurisy and in parenchymal lung disease. In most cases, the problem eventually settles, but sometimes glucocorticoid therapy is used empirically.

Anaphylaxis secondary to prophylactic isoniazid has been reported (53).

- A 40-year-old Hispanic man with a 20-mm positive PPD reaction was given isoniazid 900mg and pyridoxine 50 mg twice weekly. One month later, he developed chest pain and nausea 1 hour after taking isoniazid. This was repeated 1 week later after another dose, when he also had a tachycardia, a low blood pressure, and oxygen desaturation. He recovered after treatment for an anaphylactic reaction.

The hydrazine metabolite of isoniazid may be the cause of allergy in susceptible patients.

A lupus-like syndrome or vasculitis, with arthritis, rheumatic pain, fever, pleurisy, and leukopenia, has been reported in patients taking isoniazid (48,54). Tests for antinuclear antibodies are useful to distinguish idiopathic systemic lupus erythematosus and drug-induced lupus-like syndromes. However, many patients taking isoniazid have antinuclear antibodies, usually without signs or symptoms of systemic lupus erythematosus. Isoniazid can also exacerbate pre-existing systemic lupus erythematosus (48,54). Long-term glucocorticoid treatment may be necessary if symptoms persist after withdrawal of isoniazid.

- Two Japanese patients developed pleural effusions while taking antituberculosis therapy and were believed to have isoniazid-induced lupus-like syndrome (55). This diagnosis was based on the presence of antinuclear antibody in the effusate, and in one patient, a positive lymphocyte stimulation test using isoniazid; in the other patient it was negative. Both had moderately strongly positive serum antinuclear antibodies (1:160). In the first patient, the effusion disappeared 2 weeks after withdrawal of isoniazid; in the other treatment was continued but the effusion nevertheless resolved in 10 weeks.

It is worth checking for evidence of lupus-like syndrome in patients with paradoxical responses to antituberculosis therapy but it remains to be seen how many cases would be explained by it.

Long-Term Effects

Mutagenicity

Isoniazid is not mutagenic (SEDA-6, 276), but patients taking combined isoniazid and rifampicin therapy for 3–10 months developed an increased rate of chromosomal aberrations in peripheral blood lymphocytes (SEDA-9, 276). However, this effect is not known to have clinical consequences.

Second-Generation Effects

Fertility

Isoniazid has no effect on fertility (SEDA-6, 276).

Pregnancy

Isoniazid is the safest antituberculosis drug for use during pregnancy (56,57). During pregnancy, isoniazid with ethambutol is considered to be the preferred combination (56).

Teratogenicity

Isoniazid diffuses into blastocytes, but no teratogenic effects have been reported (57).

Lactation

Isoniazid is the safest antituberculosis drug for use during lactation (56,57). However, as isoniazid passes into breast milk, a watch must be kept for adverse effects in the infant when a nursing mother is taking isoniazid (58). However, the serum concentrations that occur in children have no therapeutic effect and cannot be considered as a form of preventive chemotherapy in infants of mothers with active tuberculosis.

Susceptibility Factors

Genetic factors

People are divided genetically into rapid and slow acetylators of isoniazid, depending on the amount of N-acetyltransferase they have. The rate of acetylation of isoniazid depends on race, but is not influenced by sex or by age after childhood. In Eskimos and Japanese, fast acetylators predominate. Slow acetylation is the predominant phenotype in most Scandinavians, Jews, and North African Caucasians (9). Slow acetylators are homozygous for an autosomal recessive gene, while fast acetylators are homozygous or heterozygous for a dominant gene. In the case of isoniazid (which has a low hepatic extraction ratio) the rate of acetylation significantly alters its plasma concentrations and half-life. The mean half-life of isoniazid in rapid acetylators is about 70 minutes and in slow acetylators 3 hours. Despite this, there is probably no difference in the effectiveness of isoniazid in the two phenotypes. The relation between isoniazid toxicity and acetylator status continues to be discussed (3,4,6).

Drug Administration

Drug overdose

Acute poisoning with isoniazid in children (59) and adults (13,27,45) causes recurrent seizures, profound metabolic acidosis, coma, and even death. In adults, toxicity can occur with the acute ingestion of as little as 1.5 g of isoniazid. Doses larger than 30 mg/kg often produce seizures and 80–150 mg/kg or more can be rapidly fatal. The first signs and symptoms of isoniazid toxicity usually appear 0.5–2.0 hours after ingestion, by which time peak absorption occurs (60), and include nausea, vomiting, slurred speech, dizziness, tachycardia, and urinary retention, followed by stupor, coma, and recurrent tonic-clonic seizures. The seizures are often refractory to antiepileptic

therapy. However, pyridoxine, 20–1500 mg per gram of isoniazid ingested, usually eliminates seizure activity and helps to correct the metabolic acidosis (61).

Two reports highlight the fact that isoniazid poisoning should be considered in the differential diagnosis of adults or children who present with intractable seizures or an encephalopathy-like syndrome (62,63). Slow acetylators of the drug are at greater risk because of higher peak concentrations. Isoniazid produces a deficiency of endogenous pyridoxine, by both increasing the renal excretion of pyridoxine and inhibiting its action. This causes a rapid fall in nervous system GABA (γ-aminobutyric acid) because of non-availability of pyridoxal phosphate, thereby lowering the intrinsic seizure threshold. The symptoms of isoniazid intoxication are nausea, vomiting, fever, skin rashes, ataxia, speech disorders, and altered consciousness, followed by seizures and coma. The specific antidote is pyridoxine (Vitamin B_6) which should be given intravenously.

Drug–Drug Interactions

Antacids

Absorption of isoniazid can be inhibited by aluminium-containing antacids (64).

Barbiturates

Barbiturates which induce liver drug-metabolizing enzymes increase the rate of metabolism of isoniazid, rifampicin, or combined antituberculosis treatment, since they may result in accelerated breakdown of these drugs (65).

Benzodiazepines

In healthy volunteers, the half-life of triazolam was prolonged from 2.5 to 3.3 hours when it was given with isoniazid (SEDA-9, 267). However, isoniazid did not affect the pharmacokinetics of oxazepam.

Beta-adrenoceptor antagonists

Lipophilic beta-adrenoceptor antagonists are metabolized to varying degrees by oxidation by liver microsomal cytochrome P450 (for example propranolol by CYP1A2 and CYP2D6 and metoprolol by CYP2D6). They can therefore reduce the clearance and increase the steady-state plasma concentrations of other drugs that undergo similar metabolism, potentiating their effects. Drugs that interact in this way include isoniazid (66).

Carbamazepine

Patients taking carbamazepine can develop signs of carbamazepine intoxication; serum concentrations of carbamazepine are increased by isoniazid, probably through enzyme inhibition (67).

Disulfiram

In seven patients taking isoniazid 0.6–1.0 g/day for at least 1 month, disulfiram 0.5 g/day caused dizziness, disorientation, staggering, insomnia, irritability, listlessness, lethargy, and in one case hypomania (68). This interaction may have been due to inhibition of isoniazid metabolism by disulfiram or to a complex interaction with dopamine metabolism.

Ethambutol

The risk of optic neuropathy due to ethambutol may be increased when it is combined with isoniazid.

- A 40-year-old patient underwent unsuccessful cadaver kidney transplantation and was treated with ethambutol and isoniazid (28). Bilateral retrobulbar neuropathy with an unusual central bitemporal hemianopic scotoma developed and ethambutol was withdrawn, but there was only a small improvement. When isoniazid was also withdrawn, there was dramatic improvement in visual acuity.

In another patient taking isoniazid and ethambutol, optic neuropathy improved only when both drugs were withdrawn (76).

Isoniazid 300 mg/day orally for 7 days increased serum ethambutol concentrations at 4, 6, and 8 hours after a daily dose of 20 mg/kg; the cumulative percentage dose excreted was significantly reduced at 4, 6, and 24 hours (77). In another study from the same center, ethambutol 20 mg/kg did not alter the pharmacokinetics of a single dose of isoniazid 300 mg in 10 patients with tuberculosis (78).

Ethosuximide

In one case, isoniazid increased blood concentrations of the antiepileptic drug ethosuximide and caused hiccups, anorexia, nausea, vomiting, insomnia, and an acute psychosis (SEDA-9, 267) (69). This may have been due to inhibition of ethosuximide metabolism by isoniazid.

Ketoconazole

Combined administration of ketoconazole with isoniazid can lead to increased concentrations of the latter, and there are possibly also alterations in ketoconazole concentrations (SED-12, 678) (79).

Monoamine oxidase inhibitors

Isoniazid inhibits monoamine oxidase, and hence reduces tyramine metabolism; this effect is enhanced by co-administration of other monoamine oxidase inhibitors (70).

Isoniazid inhibits MAO, and during isoniazid treatment the ingestion of several kinds of cheese can cause flushing, palpitation, tachycardia, and increased blood pressure (80). Similar symptoms occur with isoniazid after the ingestion of skipjack fish (*Thunnidae*) (80). The symptoms, notably headache, palpitation, erythema, redness of the eyes, itching, diarrhea, and wheezing, are thought to be caused mainly by the high histamine content of this fish. The undesirable effects that occur when MAO inhibitors are taken together with isoniazid resemble the symptoms seen after the simultaneous ingestion of these foods.

Phenazone

Isoniazid significantly reduces the hepatic clearance of phenazone (antipyrine) by about 40% (SEDA-9, 267).

Phenytoin

Isoniazid inhibits the parahydroxylation of phenytoin. Symptoms of phenytoin intoxication, such as dizziness, incoordination, or excessive sedation, can occur, particularly in patients who are slow acetylators (SEDA-11, 271). Dosages of phenytoin should be adjusted according to plasma concentrations (71).

Stavudine

A five-fold increase in the risk of distal sensory neuropathy has been reported in patients taking stavudine plus isoniazid (55 versus 11%) compared with patients taking stavudine without isoniazid (72). In nine of 12 patients, the neuropathy resolved on changing antiretroviral drugs. Peripheral neuropathy is a distressing complication during treatment with stavudine and the risk is considerably increased with co-administration of isoniazid. This combination of drugs should be avoided, if possible, in patients with tuberculosis and AIDS.

A distal sensory neuropathy is the commonest neurological complication in HIV-infected individuals, and has been documented in up to 30% of patients with AIDS. There is evidence from a retrospective case-note review in 30 individuals that co-administration of stavudine and isoniazid increases the incidence of distal sensory neuropathy. Of 22 patients taking stavudine in combination with other drugs, all took isoniazid for tuberculosis and 12 developed a distal sensory neuropathy, with a median time to onset of 5 months (81). Those taking stavudine alone had an incidence of 11%.

Thiacetazone

Thiacetazone combined with isoniazid (and possibly thioacetazone alone) has been reported to damage chromosomes in cultured human lymphocytes (82).

Triazolam

Isoniazid is a potent hepatic enzyme inhibitor and interferes with the metabolism of many drugs (SEDA-8, 287) (SEDA-11, 271). In healthy volunteers the half-life of triazolam was prolonged from 2.5 to 3.3 hours when it was given with isoniazid, whereas isoniazid did not affect the kinetics of oxazepam (SEDA-9, 267).

Food–Drug Interactions

Tyramine-containing foods

Adverse reactions during treatment with isoniazid have been noted after ingestion of several kinds of cheese (73,74). Flushing, palpitation, tachycardia, and increased blood pressure have been observed 0.5–2 hours after cheese. The symptoms generally disappear within 2–4 hours. Interference by isoniazid with monoamine oxidase, and hence with tyramine metabolism, has been incriminated.

Histamine-containing foods

Headache, palpitation, erythema, redness of the eyes, itching, diarrhea, and wheezing can occur with isoniazid after skipjack fish (*Thunnidae*) (73,75), probably because of the high histamine content of this fish.

References

1. Schelling JL. Dyspnées médicamenteuses. [Medicamentous dyspnea.] Ther Umsch 1981;38(2):163–5.
2. Ueda Y, Fujita K, Kohno K, Ichinose K, Fukushima H, Nakatomi M. [A case of isoniazid-induced lupus.]Kekkaku 1989;64(10):613–9.
3. Mitchell JR, Thorgeirsson UP, Black M, Timbrell JA, Snodgrass WR, Potter WZ, Jollow DJ, Keiser HR. Increased incidence of isoniazid hepatitis in rapid acetylators: possible relation to hydrazine metabolites. Clin Pharmacol Ther 1975;18(1):70–9.
4. Gurumurthy P, Krishnamurthy MS, Nazareth O, Parthasarathy R, Sarma GR, Somasundaram PR, Tripathy SP, Ellard GA. Lack of relationship between hepatic toxicity and acetylator phenotype in three thousand South Indian patients during treatment with isoniazid for tuberculosis. Am Rev Respir Dis 1984;129(1):58–61.
5. Boice JD, Fraumeni JF Jr. Late effects following isoniazid therapy. Am J Public Health 1980;70(9):987–9.
6. Mitchell I, Wendon J, Fitt S, Williams R. Anti-tuberculous therapy and acute liver failure. Lancet 1995;345(8949):555–6.
7. Snider DE Jr. Pyridoxine supplementation during isoniazid therapy. Tubercle 1980;61(4):191–6.
8. Schröder JM. Zur Pathogenese der Isoniazid-Neuropathie. I. Eine feinstrukturelle Differenzierung gegenüber der Wallerschen Degeneration. [The pathogenesis of INH-neuropathy. I. Fine structural similarities and differences to Wallerian degeneration.] Acta Neuropathol (Berl) 1970;16(4):301–23.
9. Mandell GL, Sande MA. Antimicrobial agents: drugs used in the chemotherapy of tuberculosis and leprosy. In: Goodman Gilman A, Rall TW, Nies AS, Taylor P, editors. Goodman and Gilman's The Pharmacological Basis of Therapeutics. 8th ed.. New York: Pergamon Press, 1990:1201 Chapter 49.
10. Ishii N, Nishihara Y. Pellagra encephalopathy among tuberculous patients: its relation to isoniazid therapy. J Neurol Neurosurg Psychiatry 1985;48(7):628–34.
11. Tibussek D, Mayatepek E, Distelmaier F, Rosenbaum T. Status epilepticus due to attempted suicide with isoniazid. Eur J Pediatr 2006;165(2):136–7.
12. Leibold JE. Drugs having a toxic effect on the optic nerve. Int Ophthalmol Clin 1971;11(2):137–57.
13. Bhatia MS. Isoniazid-induced obsessive compulsive neurosis. J Clin Psychiatry 1990;51(9):387.
14. Bernardo M, Gatell JM, Parellada E. Acute exacerbation of chronic schizophrenia in a patient treated with antituberculosis drugs. Am J Psychiatry 1991;148(10):1402.
15. Siskind MS, Thienemann D, Kirlin L. Isoniazid-induced neurotoxicity in chronic dialysis patients: report of three cases and a review of the literature. Nephron 1993;64(2):303–6.
16. Cheung WC, Lo CY, Lo WK, Ip M, Cheng IK. Isoniazid induced encephalopathy in dialysis patients. Tuber Lung Dis 1993;74(2):136–9.
17. Ferner RE. Drug-induced diabetes. Baillière's Clin Endocrinol Metab 1992;6(4):849–66.
18. Sharp RA, Lowe JG, Johnston RN. Anti-tuberculous drugs and sideroblastic anaemia. Br J Clin Pract 1990;44(12):706–7.
19. Johnsson R, Lommi J. A case of isoniazid-induced red cell aplasia. Respir Med 1990;84(2):171–4.
20. Goldman AL, Braman SS. Isoniazid: a review with emphasis on adverse effects. Chest 1972;62(1):71–7.
21. Krumdieck R, Shaw DR, Huang ST, Poon MC, Rustagi PK. Hemorrhagic disorder due to an isoniazid-associated acquired factor XIII inhibitor in a patient with Waldenstrom's macroglobulinemia. Am J Med 1991;90(5):639–45.
22. Dutt AK, Moers D, Stead WW. Undesirable side effects of isoniazid and rifampin in largely twice-weekly short-course chemotherapy for tuberculosis. Am Rev Respir Dis 1983;128(3):419–24.
23. Hoose C, Eberhardt K, Hartmann W, Wosniok W. Kurz- und Langzeitergebnisse der Tuberkulosetherapie mit einer fixen Kombination aus Isoniazid, Prothionamid und Diaphenylsulfon (IPD) in Verbindung mit Rifampicin. [Short- and long-term results of tuberculosis therapy with a fixed combination of isoniazide, prothionamide and diaminodiphenylsulfone combined with rifampicin.] Pneumologie 1990;44(Suppl 1):458–9.
24. Snider DE Jr, Caras GJ. Isoniazid-associated hepatitis deaths: a review of available information. Am Rev Respir Dis 1992;145(2 Pt 1):494–7.
25. Black M, Mitchell JR, Zimmerman HJ, Ishak KG, Epler GR. Isoniazid-associated hepatitis in 114 patients. Gastroenterology 1975;69(2):289–302.
26. Nolan CM, Goldberg SV, Buskin SE. Hepatotoxicity associated with isoniazid preventive therapy: a 7-year survey from a public health tuberculosis clinic. JAMA 1999;281(11):1014–8.
27. Amarapurkar DN, Prabhudesai PP, Kalro RH, Desai HG. Antituberculosis drug induced hepatitis and HBsAg carriers. Tuber Lung Dis 1993;74(3):215–6.
28. Gangadharam PR. Isoniazid, rifampin, and hepatotoxicity. Am Rev Respir Dis 1986;133(6):963–5.
29. Timbrell JA, Wright JM. Urinary metabolic profile of isoniazid in patients who develop isoniazid-related liver damage. Hum Toxicol 1984;3(6):485–95.
30. Mandell GL, Sande MA. Antimicrobial agents: drugs used in the chemotherapy of tuberculosis and leprosy. In: Goodman Gilman A, Goodman LS, Rall TW, et al., editors. The Pharmacological Basis of Therapeutics. 7th ed.. New York: Macmillan Publishing, 1985:1201.
31. Halpern M, Meyers B, Miller C, Bodenheimer H, Thung SN, Adler J, Toth D, Cohen D, Baccardo L, Diferdinando G, Birkhead A. from the Centers for Disease Control and Prevention. Severe isoniazid-associated hepatitis—New York, 1991–1993. JAMA 1993;270(7):809.
32. Fernandez-Villar A, Sopena B, Vasquez R, Ulloa F, Fluiters E, Mosteiro M, Martinez-Vasquez C, Pineiro L. Isoniazid hepatotoxicity among drug users: the role of hepatitis C. Clin Infect Dis 2003;36:293–8.
33. Lobue PA, Moser KS. Use of isoniazid for latent tuberculosis infection in a public health clinic. Am J Resp Crit Care Med 2003;168:443–7.
34. Byrd RB, Horn BR, Solomon DA, Griggs GA. Toxic effects of isoniazid in tuberculosis chemoprophylaxis. Role of biochemical monitoring in 1,000 patients. JAMA 1979;241:1239–41.
35. Stuart RL, Wilson J, Grayson ML. Isoniazid toxicity in health care workers. Clin Infect Dis 1999;28:895–97.

36. Kopanoff DE, Snider DE Jr, Caras GJ. Isoniazid-related hepatitis: a U.S. Public Health Service cooperative surveillance study. Am Rev Res Dis 1978;117:991–1001.

37. International Union Against Tuberculosis Committee on Prophylaxis. Efficacy of various durations of isoniazid preventive therapy for tuberculosis: five years of follow-up in the IUAT trial. Bull World Health Org 1982;60:555–64.

38. Dash LA, Comstock GW, Flynn JP. Isoniazid preventive therapy: retrospect and prospect. Am Rev Respir Dis 1980;121:1039–44.

39. Nolan CM, Goldberg SV, Buskin SE. Hepatotoxicity associated with isoniazid preventive therapy: a 7-year survey from a public health tuberculosis clinic. JAMA 1999;281:1014–8.

40. Fountain FF, Tolley E, Chrisman CR, Self TH. Isoniazid hepatotoxicity associated with treatment of latent tuberculosis infection: a 7-year evaluation from a public health tuberculosis clinic. Chest 2005;128:116–23.

41. Chan KL, Chan HS, Lui SF, Lai KN. Recurrent acute pancreatitis induced by isoniazid. Tuber Lung Dis 1994;75(5):383–5.

42. Rabassa AA, Trey G, Shukla U, Samo T, Anand BS. Isoniazid-induced acute pancreatitis. Ann Intern Med 1994;121(6):433–4.

43. Stephenson I, Wiselka MJ, Qualie MJ. Acute pancreatitis induced by isoniazid in the treatment of tuberculosis. Am J Gastroenterol 2001;96(7):2271–2.

44. Trainin EB, Turin RD, Gomez-Leon G. Acute renal insufficiency complicating isoniazid therapy. Int J Pediatr Nephrol 1981;2(1):53–4.

45. Romero JA, Kuczler FJ Jr. Isoniazid overdose: recognition and management. Am Fam Physician 1998;57(4):749–52.

46. Sharma RP, Kathari AK, Sharma NK. Acneiform eruptions and anti-tubercular drugs. Indian J Dermatol Venereol Leprol 1995;61:26.

47. Koch SE, Williams ML. Acquired cutis laxa: case report and review of disorders of elastolysis. Pediatr Dermatol 1985;2(4):282–8.

48. Rothfield NF, Bierer WF, Garfield JW. Isoniazid induction of antinuclear antibodies. A prospective study. Ann Intern Med 1978;88(5):650–2.

49. Darvay A, Basarab T, McGregor JM, Russell-Jones R. Isoniazid induced pellagra despite pyridoxine supplementation. Clin Exp Dermatol 1999;24(3):167–9.

50. Scheid P, Kanny G, Trechot P, Rosner V, Menard O, Vignaud JM, Anthoine D, Martinet Y. Isoniazid-induced bullous skin reaction. Allergy 1999;54(3):294–6.

51. De Weck AL. Approaches to prevention and treatment of drug allergy. In: Turk JL, Parker D, editors. Drugs and Immune Responsiveness, 13. Baltimore: University Park Press, 1979:211.

52. Panganiban LR, Makalinao IR, Corte-Maramba NP. Rhabdomyolysis in isoniazid poisoning. J Toxicol Clin Toxicol 2001;39(2):143–51.

53. Hoigne R, Biedermann HP, Naegeli HR. INH-induzierter systemischer Lupus Erythematodes: 2 Beobachtungen mit Reexposition. Schweiz Med Wochenschr 1975;105(50):1726.

54. Crook MJ. Isoniazid induced anaphylaxis. J Clin Pharmacol 2003;43:545–6.

55. Hiraoka K, Nagata N, Kawajiri T, Suzuki K, Kurokawa S, Kido M, Sakamoto N. Paradoxical pleural response to antituberculous chemotherapy and isoniazid-induced lupus. Review and report of two cases. Respiration 1998;65(2):152–5.

56. Des Prez RM, Heim CR. Mycobacterium tuberculosis. In: Mandell GL, Douglas Jr. RG Jr, Bennett JE, editors.

57. Principles and Practice of Infectious Diseases. 3rd ed.. New York: Churchill Livingstone, 1990:1877.

57. Kunz J, Schreiner WE. Pharmakotherapie während Schwangerschaft und StillperiodeStuttgart-New York: G Thieme;. 1982.

58. Olive G. Interactions médicamenteuses chez le nouveau né. In: Comptes Rendus. 25e Congrès de l'Association des Pédiatres de Langue Française, Tunis, 1978. Paris: Expansion Scientifique Française 1978;57.

59. Miller J, Robinson A, Percy AK. Acute isoniazid poisoning in childhood. Am J Dis Child 1980;134(3):290–2.

60. Shah BR, Santucci K, Sinert R, Steiner P. Acute isoniazid neurotoxicity in an urban hospital. Pediatrics 1995;95(5):700–4.

61. Gilhotra R, Malik SK, Singh S, Sharma BK. Acute isoniazid toxicity—report of 2 cases and review of literature. Int J Clin Pharmacol Ther Toxicol 1987;25(5):259–61.

62. Caksen H, Odabas D, Erol M, Anlar O, Tuncer O, Atas B. Do not overlook acute isoniazid poisoning in children with status epilepticus. J Child Neurol 2003;18:142–3.

63. Maw G, Aitken P. Isoniazid overdose: a case series, literature review and survey of antidote availability. Clin Drug Invest 2003;23:479–85.

64. Hurwitz A, Schlozman DL. Effects of antacids on gastrointestinal absorption of isoniazid in rat and man. Am Rev Respir Dis 1974;109(1):41–7.

65. Duroux P. Surveillance et accidents de la chimiothérapie antituberculeuse. [Surveillance and complications of antituberculosis chemotherapy.] Rev Prat 1979;29(33):2681–90.

66. Santoso B. Impairment of isoniazid clearance by propranolol. Int J Clin Pharmacol Ther Toxicol 1985;23(3):134–6.

67. Valsalan VC, Cooper GL. Carbamazepine intoxication caused by interaction with isoniazid. BMJ (Clin Res Ed) 1982;285(6337):261–2.

68. Whittington HG, Grey L. Possible interaction between disulfiram and isoniazid. Am J Psychiatry 1969;125(12):1725–9.

69. van Wieringen A, Vrijlandt CM. Ethosuximide intoxication caused by interaction with isoniazid. Neurology 1983;33(9):1227–8.

70. DiMartini A. Isoniazid, tricyclics and the "cheese reaction". Int Clin Psychopharmacol 1995;10(3):197–8.

71. Miller RR, Porter J, Greenblatt DJ. Clinical importance of the interaction of phenytoin and isoniazid: a report from the Boston Collaborative Drug Surveillance Program. Chest 1979;75(3):356–8.

72. Breen RA, Lipman MC, Johnson MA. Increased incidence of peripheral neuropathy with co-administration of stavudine and isoniazid in HIV-infected individuals. AIDS 2000;14(5):615.

73. Hauser MJ, Baier H. Interactions of isoniazid with foods. Drug Intell Clin Pharm 1982;16(7–8):617–8.

74. Lejonc JL, Schaeffer A, Brochard P, Portos JL. Hypertension artérielle paroxystique provoquée sous isoniazide par l'ingestion de Gruyère: deux cas. [Paroxystic hypertension after ingestion of Gruyère cheese during isoniazide treatment: a report on two cases.] Ann Med Interne (Paris) 1980;131(6):346–8.

75. Uragoda CG. Histamine poisoning in tuberculous patients after ingestion of tuna fish. Am Rev Respir Dis 1980;121(1):157–9.

76. Jimenez-Lucho VE, del Busto R, Odel J. Isoniazid and ethambutol as a cause of optic neuropathy. Eur J Respir Dis 1987;71(1):42–5.

77. Singhal KC, Rathi R, Varshney DP, Kishore K. Serum concentration and urinary excretion of ethambutol administered alone and in combination with isoniazid in patients of

pulmonary tuberculosis. Indian J Physiol Pharmacol 1985;29(4):223-6.

78. Singhal KC, Varshney DP, Rathi R, Kishore K, Varshney SC. Serum concentration of isoniazid administered with & without ethambutol in pulmonary tuberculosis patients. Indian J Med Res 1986;83:360-2.
79. Bickers DR. Antifungal therapy: potential interactions with other classes of drugs. J Am Acad Dermatol 1994;31(3 Pt 2):S87-90.
80. Hauser MJ, Baier H. Interactions of isoniazid with foods. Drug Intell Clin Pharm 1982;16(7-8):617-8.
81. Breen RA, Lipman MC, Johnson MA. Increased incidence of peripheral neuropathy with co-administration of stavudine and isoniazid in HIV-infected individuals. AIDS 2000;14(5):615.
82. Ahuja YR, Jaju M, Jaju M. Chromosome-damaging action of isoniazid and thiacetazone on human lymphocyte cultures in vivo. Hum Genet 1981;57(3):321-2.

Josamycin

See also Macrolide antibiotics

General Information

In a randomized open study, 325 children aged 2–15 years with acute tonsillitis and a positive test for *Streptococcus pyogenes* antigen were treated with josamycin 25 mg/kg bd for 5 days, or penicillin 50 000–100 000 IU/day for 10 days; in five patients taking josamycin treatment was withdrawn because of gastrointestinal adverse events (nausea/vomiting) (1).

Organs and Systems

Liver

Cholestatic hepatitis has been attributed to josamycin (2).

Biliary tract

In female rats josamycin caused bile duct proliferation at a high dose of 1460 mg/kg (3).

Skin

Delayed-type hypersensitivity to josamycin has been reported in a 32-year-old woman, 4 hours after a full dose of josamycin (one tablet of 500 mg) with a generalized maculopapular rash, which increased in intensity during 24 hour and regressed over 1 week (4). She had previously had generalized erythema on the second day of treatment with josamycin 2 years before.

References

1. Portier H, Bourrillon A, Lucht F, Choutet P, Gehanno P, Meziane L, Bingen E. Groupe d'etude de pathologie infectieuse pediatrique. [Treatment of acute group A beta-hemolytic streptococcal tonsillitis in children with a 5-day course of josamycin.] Arch Pediatr 2001;8(7):700-6.

2. Lavin I, Mundi JL, Trillo C, Trapero A, Fernandez R, Lopez MA, Cervilla E, Quintero D, Palacios A. [Cholestatic hepatitis by josamycin.]Gastroenterol Hepatol 1999;22(3):160.
3. Kasahara K, Nishikawa A, Furukawa F, Ikezaki S, Tanakamaru Z, Lee IS, Imazawa T, Hirose M. A chronic toxicity study of josamycin in F344 rats. Food Chem Toxicol 2002;40(7):1017-22.
4. Freymond N, Catelain A, Cousin F, Nicolas JF. Skin delayed-type hypersensitivity to josamycin. Allergy 2003;58:1319-20.

Kanamycin

See also Aminoglycoside antibiotics

General Information

Kanamycin can be used for short-term treatment of severe infections caused by susceptible strains (for example *Escherichia coli*, *Proteus* species, *Enterobacter aerogenes*, *Klebsiella pneumoniae*, *Serratia marcescens*, and *Mima-Herellea*) that are resistant to other less ototoxic aminoglycosides. It is not indicated for long-term therapy, for example in tuberculosis.

Organs and Systems

Neuromuscular function

Like the other aminoglycosides, kanamycin has neuromuscular blocking properties, particularly if given directly into the peritoneum (1). However, it seems to be less dangerous in this respect than neomycin or streptomycin.

Sensory systems

Kanamycin causes mainly cochlear damage (2). After prolonged administration (for example 1 g for periods of 30–180 days) the frequency of this adverse reaction is higher than 40%. Vestibular toxicity occurs in less than 10% of cases treated with usual doses and is generally reversible soon after withdrawal.

Antioxidant therapy protects against aminoglycoside-induced ototoxicity in animals. In adult mice receiving concurrent treatment with kanamycin (700 mg/kg bd for 15 days), extracts of *Salviae miltiorrhizae* significantly attenuated auditory threshold shifts induced by kanamycin (approximately 50 dB) but did not reduce the serum concentrations or antibacterial efficacy of kanamycin (3).

Gastrointestinal

Presurgical bowel preparation with oral kanamycin is seldom practiced and can be followed by an intestinal malabsorption syndrome (4). Only negligible amounts of kanamycin are absorbed through an intact intestinal mucosa, but increased systemic availability and potential toxicity can result from the presence of ulcerated or denuded areas.

Urinary tract

Kanamycin causes very little nephrotoxicity after short courses of treatment with daily doses of less than 15 mg/kg; if total doses of 30 g or more are given, the incidence of renal damage can be 50% or higher (1).

Immunologic

Sensitization (rash, drug fever) after parenteral administration of kanamycin is less frequent than with streptomycin. Anaphylaxis has only rarely been described. Cross-allergy with the other aminoglycosides is frequent (5).

References

1. Finegold SM. Toxicity of kanamycin in adults. Ann NY Acad Sci 1966;132(2):942–56.
2. Hinojosa R, Nelson EG, Lerner SA, Redleaf MI, Schramm DR. Aminoglycoside ototoxicity: a human temporal bone study. Laryngoscope 2001;111(10):1797–805.
3. Wang AM, Sha SH, Lesniak W, Schacht J. Tanshinone (*Salviae miltiorrhizae extract*) preparations attenuate aminoglycoside-induced free radical formation in vitro and ototoxicity in vivo. Antimicrob Agents Chemother 2003;47:1836–41.
4. Lees GM, Percy WH. Antibiotic-associated colitis: an in vitro investigation of the effects of antibiotics on intestinal motility. Br J Pharmacol 1981;73(2):535–47.
5. Chung CW, Carson TR. Cross-sensitivity of common aminoglycoside antibiotics. Arch Dermatol 1976;112(8):1101–7.

Ketolides

General Information

The ketolides cethromycin (ABT-773) and telithromycin are semisynthetic derivatives of 14-membered-ring macrolide antibiotics, members of the macrolide-lincosamide-streptogramin-B, MLS(B), class of antimicrobials, but they differ from macrolides by having a 3-keto group instead of an L-cladinose group on the erythronolide A ring (1–3). In a study of the efficacy of the ketolides HMR 3004 (previously RU 004) and telithromycin (HMR 3647, previously 66 647) against beta-lactamase-producing *Haemophilus influenzae*, the in vitro activity of both ketolides against 30 clinical isolates was comparable to that of azithromycin and superior to clarithromycin and erythromycin A (4). Compared with macrolides, it has increased affinity for the binding sites on domains II and V of the 50S ribosomal subunit. Furthermore, HMR 3004 and HMR 3647 were both active in a murine model of experimental pneumonia, as assessed by pulmonary clearance of beta-lactamase producing *H. influenzae* (5).

Telithromycin has a good spectrum of activity against respiratory pathogens, including penicillin-resistant and erythromycin-resistant pneumococci, intracellular bacteria, and atypical bacteria. It has an absolute oral availability of 57% in young and elderly subjects (6), and the rate and extent of absorption is unaffected by food (7). It penetrates rapidly into bronchopulmonary, tonsillar, sinus, and middle ear tissues/fluids, achieves high concentrations at sites of infection, and concentrates within polymorphonuclear neutrophils. In a dosage of 800 mg/day it is well tolerated across all patient populations, and adverse events, most commonly diarrhea, nausea, dizziness, and vomiting, are generally mild to moderate in intensity and seldom lead to treatment withdrawal (8–10),11,12,13,14,15,16.

Observational studies

In an open multicenter study in 432 patients with community-acquired pneumonia given telithromycin 800 mg/day for 7 days, diarrhea was reported in 8.1% and nausea in 5.8% (17). Six patients discontinued treatment because of allergic reactions, abdominal pain, vomiting, vertigo, or increased alanine transaminase activity, which were considered possibly related to telithromycin. Two patients had non-significant prolongation of the QT interval.

Placebo-controlled studies

In three randomized, double blind, multicenter studies in 609 patients who took telithromycin 800 mg/day for 5 days for exacerbations of chronic bronchitis, diarrhea and nausea occurred in 6.4% and 4.8% respectively (18). One patient had QT_c interval prolongation from 488 ms to 531 ms, which returned to baseline after the end of therapy.

Organs and Systems

Cardiovascular

Telithromycin causes prolongation of the QT interval, especially in elderly patients with predisposing conditions (2), but not in healthy men and women (19). Torsade de pointes has been reported (20).

Sensory systems

In a prospective, double-blind, parallel-group, randomized, multicenter study, 349 adults with acute bacterial rhinosinusitis were randomized to oral telithromycin 800 mg/day for 5 days or oral moxifloxacin 400 mg/day for 10 days; blurred vision was a clinically significant adverse event in three patients taking telithromycin (21).

Neuromuscular

Telithromycin has been implicated in exacerbation or unmasking of myasthenia gravis. There has been a report of two cases, including a summary of eight other suspected cases notified to the French pharmacovigilance system, highlighting a potentially life-threatening risk of telithromycin treatment in patients with myasthenia (22). An important common feature was that in seven cases the symptoms occurred within 2 hours of the first dose, notably in cases of severe exacerbation. Another case has been described, in a 46-year-old woman who had an exacerbation of myasthenia gravis while taking telithromycin (23).

Gastrointestinal

In a randomized study of a 5-day or 7-day regimen of oral telithromycin 800 mg/day and a 10-day regimen of oral clarithromycin 500 mg bd in 575 randomized patients, 257 patients (45%) had at least one adverse effect (24). The most common adverse events were gastrointestinal (diarrhea, nausea, and dysgeusia). With telithromycin, all cases of diarrhea, nausea, and dysgeusia were mild or moderate in intensity and none was severe or categorized as serious. Two cases of diarrhea—one in each telithromycin group—resulted in withdrawal.

In another double-blind, parallel-group, randomized, multicenter study, 349 adults with acute bacterial rhinosinusitis were randomized to oral telithromycin 800 mg/day for 5 days or oral moxifloxacin 400 mg/day for 10 days, 35% of the patients who received telithromycin had one or more adverse effect; most were mild to moderate in intensity (22). There was at least one severe adverse event in 5.2% and 2.3% of patients treated with telithromycin and moxifloxacin respectively. The most commonly reported adverse events that were assessed as being possibly related to the drug were diarrhea (8.1%) and nausea (5.8%). The drugs had to be withdrawn in 2.9% of patients because of adverse events.

Liver

Telithromycin has been associated with rare cases of serious liver injury and liver failure, with four reported deaths and one liver transplantation (25,26).

Urinary tract

Telithromycin-induced severe acute interstitial nephritis has been reported (27).

- An 18-year-old man received telithromycin 800 mg/day and after 3 days his serum creatinine concentration was 140 µmol/l. There was no rash or lymphadenopathy. Urinalysis showed aseptic leukocyturia, proteinuria, and glycosuria. One week later, the serum creatinine level rose to 407 µmol/l. A percutaneous renal biopsy showed severe interstitial edema, marked diffuse infiltration with lymphocytes, and some eosinophils. The glomeruli were normal. He was treated with pulse methylprednisolone 500 mg/day for 3 days followed by oral methylprednisolone 32 mg/day, and within 2 weeks the serum creatinine concentration fell to 80 µmol/l.

Skin

- An 18-year-old woman with infectious mononucleosis developed a pruritic rash after treatment with co-amoxiclav (28). Treatment was switched to telithromycin, and the rash started to resolve but worsened again, with the development of additional headache and joint pains. The antibiotic was withdrawn and the symptoms and the rash resolved with supportive therapy with antihistamines and methylprednisolone.

Musculoskeletal

Telithromycin can exacerbate myasthenia gravis in patients with pre-existing myasthenia gravis (29).

Long-Term Effects

Drug tolerance

In subjects receiving oral telithromycin (800 mg/day for 10 days), high drug concentrations were detected in the saliva indicating a good therapeutic profile for throat infections. Quantitative ecological disturbances in the normal microflora during administration of telithromycin were moderate, and no overgrowth of yeasts or *Clostridium difficile* occurred. However, resistant bacterial strains emerged (30).

The in vitro activity of cethromycin was very similar to that of telithromycin, with an MIC_{90} of 0.5 mg/l or less for all bacteria examined, except methicillin-resistant *Staphylococcus aureus*, *Enterococcus faecalis*, *Enterococcus faecium*, *H. influenzae*, and *Bacteroides* species. However, the antichlamydial activity of cethromycin was greater than that of telithromycin (31).

Susceptibility Factors

Renal disease

The pharmacokinetic profiles of single and repeated oral doses of telithromycin 800 mg/day were comparable in patients with mild to moderate renal impairment and healthy subjects (32). However, in patients with severe renal impairment dosage adjustment should be considered.

Hepatic impairment

The pharmacokinetic profiles of single and repeated oral doses of telithromycin 800 mg/day were comparable in patients with hepatic impairment and healthy subjects with normal renal function (33).

Drug–Drug Interactions

Telithromycin inhibits CYP-3A4. It is contraindicated in patients taking cisapride, pimozide, or class IA and class III antidysrhythmic agents in whom it can prolong the QT interval (4).

Digoxin

Telithromycin increases plasma digoxin concentrations and can cause signs and symptoms of toxicity (34).

- A 58-year-old woman, who had been taking digoxin 0.25 mg/day for more than 35 years for palpitation after mitral valve repair, was given telithromycin for acute bronchitis. After 6 days she complained of general malaise after three episodes of syncope over the previous 2 days. The plasma digoxin concentration was raised and electrocardiography showed several non-specific repolarization anomalies.

The authors attributed this interaction to inhibition of renal P glycoprotein by telithromycin.

Verapamil

Telithromycin is a substrate of CYP3A4 and has been reported to interact with verapamil (35).

- A 76-year-old white woman taking verapamil 180 mg/day for hypertension was given telithromycin 800 mg/day for 2 days and suddenly developed shortness of breath and weakness and was profoundly hypotensive and bradycardic, with a systolic blood pressure of 50–60 mm Hg and a heart rate of 30/minute. Telithromycin was withdrawn and within 72 hours her blood pressure and heart rate returned to normal.

Warfarin

Telithromycin can interact with warfarin, resulting in a raised international normalized ratio (INR).

- A 73-year-old white man taking warfarin developed a raised INR and mild hemoptysis 5 days after he started to take telithromycin 800 mg/day (36).

Cytokines

Cytokines modify phagocyte activity and may interfere with the immunomodulating properties of antibacterial agents. In an in vitro study, TNF-α and GM-CSF reduced the inhibitory effect of telithromycin on oxidant production by polymorphonuclear neutrophils, suggesting an effect of telithromycin downstream of the priming effect of cytokines. In addition, TNF-α and GM-CSF moderately impaired the uptake of telithromycin by polymorphonuclear neutrophils; the inhibitory effect of these two cytokines seemed to be related to the activation of the p38 mitogen-activated protein kinase (37).

References

1. Nilius AM, Ma Z. Ketolides: the future of the macrolides? Curr Opin Pharmacol 2002;2(5):493–500.
2. Shain CS, Amsden GW. Telithromycin: the first of the ketolides. Ann Pharmacother 2002;36(3):452–64.
3. Van Rensburg DJ, Matthews PA, Leroy B. Efficacy and safety of telithromycin in community-acquired pneumonia. Curr Med Res Opin 2002;18(7):397–400.
4. Turner M, Corey GR, Abrutyn E. Telithromycin. Ann Intern Med 2006;144(6):447–8.
5. Piper KE, Rouse MS, Steckelberg JM, Wilson WR, Patel R. Ketolide treatment of *Haemophilus influenzae* experimental pneumonia. Antimicrob Agents Chemother 1999;43(3):708–10.
6. Perret C, Lenfant B, Weinling E, Wessels DH, Scholtz HE, Montay G, Sultan E. Pharmacokinetics and absolute oral bioavailability of an 800-mg oral dose of telithromycin in healthy young and elderly volunteers. Chemotherapy 2002;48(5):217–23.
7. Bhargava V, Lenfant B, Perret C, Pascual MH, Sultan E, Montay G. Lack of effect of food on the bioavailability of a new ketolide antibacterial, telithromycin. Scand J Infect Dis 2002;34(11):823–6.
8. Balfour JA, Figgitt DP. Telithromycin. Drugs 2001;61(6):815–29.
9. Baltch AL, Smith RP, Ritz WJ, Franke MA, Michelsen PB. Antibacterial effect of telithromycin (HMR 3647) and comparative antibiotics against intracellular *Legionella pneumophila*. J Antimicrob Chemother 2000;46(1):51–5.
10. Bearden DT, Neuhauser MM, Garey KW. Telithromycin: an oral ketolide for respiratory infections. Pharmacotherapy 2001;21(10):1204–22.
11. Mikamo H, Ninomiya M, Tamaya T. Penetration of oral telithromycin into female genital tissues. J Infect Chemother 2003;9:358–60.
12. Carbon C. A pooled analysis of telithromycin in the treatment of community-acquired respiratory tract infections in adults. Infection 2003;31:308–17.
13. Gehanno P, Sultan E, Passot V, Nabet P, Danon J, Romanet P, Attal P. Telithromycin (HMR 3647) achieves high and sustained concentrations in tonsils of patients undergoing tonsillectomy. Int J Antimicrob Agents 2003;21:441–5.
14. Spiers KM, Zervos MJ. Telithromycin. Expert Rev Anti Infect Ther 2004;2(5):685–93.
15. Wellington K, Noble S. Telithromycin. Drugs 2004;64(15):1683-94; discussion 1695-6.
16. Reinert RR. Clinical efficacy of ketolides in the treatment of respiratory tract infections. J Antimicrob Chemother 2004;53(6):918–27.
17. Fogarty CM, Patel TC, Dunbar LM, Leroy BP. Efficacy and safety of telithromycin 800 mg once daily for 7 days in community-acquired pneumonia: an open-label, multicenter study. BMC Infect Dis 2005;5(1):43.
18. Fogarty C, Zervos M, Tellier G, Aubier M, Rangaraju M, Nusrat R. Telithromycin for the treatment of acute exacerbations of chronic bronchitis. Int J Clin Pract 2005;59(3):296–305.
19. Demolis JL, Vacheron F, Cardus S, Funck-Brentano C. Effect of single and repeated oral doses of telithromycin on cardiac QT interval in healthy subjects. Clin Pharmacol Ther 2003;73:242–52.
20. Owens RC Jr. QT prolongation with antimicrobial agents: understanding the significance. Drugs 2004;64(10):1091–124.
21. Ferguson BJ, Guzzetta RV, Spector SL, Hadley JA. Efficacy and safety of oral telithromycin once daily for 5 days versus moxifloxacin once daily for 10 days in the treatment of acute bacterial rhinosinusitis. Otolaryngol Head Neck Surg 2004;131(3):207–14.
22. Perrot X, Bernard N, Vial C, Antoine JC, Laurent H, Vial T, Confavreux C, Vukusic S. Myasthenia gravis exacerbation or unmasking associated with telithromycin treatment. Neurology 2006;67(12):2256–8.
23. Jennett AM, Bali D, Jasti P, Shah B, Browning LA. Telithromycin and myasthenic crisis. Clin Infect Dis 2006;43(12):1621–2.
24. Tellier G, Niederman MS, Nusrat R, Patel M, Lavin B. Clinical and bacteriological efficacy and safety of 5 and 7 day regimens of telithromycin once daily compared with a 10 day regimen of clarithromycin twice daily in patients with mild to moderate community-acquired pneumonia. J Antimicrob Chemother 2004;54(2):515–23.
25. Anonymous. New safety information on Ketek. FDA Consumer 2006;40(5):8.
26. Clay KD, Hanson JS, Pope SD, Rissmiller RW, Purdum PP 3rd, Banks PM. Brief communication: severe hepatotoxicity of telithromycin: three case reports and literature review. Ann Intern Med 2006;144(6):415–20.

27. Tintillier M, Kirch L, Almpanis C, Cosyns JP, Pochet JM, Cuvelier C. Telithromycin-induced acute interstitial nephritis: a first case report. Am J Kidney Dis 2004;44(2):e25–27.
28. Wargo KA, McConnell V, Jennings M. Amoxicillin/telithromycin-induced rash in infectious mononucleosis. Ann Pharmacother 2005;39(9):1577.
29. Nieman RB, Sharma K, Edelberg H, Caffe SE. Telithromycin and myasthenia gravis. Clin Infect Dis 2003;37:1579.
30. Edlund C, Alvan G, Barkholt L, Vacheron F, Nord CE. Pharmacokinetics and comparative effects of telithromycin (HMR 3647) and clarithromycin on the oropharyngeal and intestinal microflora. J Antimicrob Chemother 2000;46(5): 741–9.
31. Andrews JM, Weller TM, Ashby JP, Walker RM, Wise R. The in vitro activity of ABT773, a new ketolide antimicrobial agent. J Antimicrob Chemother 2000;46(6):1017–22.
32. Shi J, Montay G, Chapel S, Hardy P, Barrett JS, Sack M, Marbury T, Swan SK, Vargas R, Leclerc V, Leroy B, Bhargava VO. Pharmacokinetics and safety of the ketolide telithromycin in patients with renal impairment. J Clin Pharmacol 2004;44(3):234–44.
33. Cantalloube C, Bhargava V, Sultan E, Vacheron F, Batista I, Montay G. Pharmacokinetics of the ketolide telithromycin after single and repeated doses in patients with hepatic impairment. Int J Antimicrob Agents 2003;22:112–21.
34. Nenciu LM, Laberge P, Thirion DJ. Telithromycin-induced digoxin toxicity and electrocardiographic changes. Pharmacotherapy 2006;26(6):872–6.
35. Reed M, Wall GC, Shah NP, Heun JM, Hicklin GA. Verapamil toxicity resulting from a probable interaction with telithromycin. Ann Pharmacother 2005;39(2):357–60.
36. Kolilekas L, Anagnostopoulos GK, Lampaditis I, Eleftheriadis I. Potential interaction between telithromycin and warfarin. Ann Pharmacother 2004;38(9):1424–7.
37. Vazifeh D, Bryskier A, Labro MT. Effect of proinflammatory cytokines on the interplay between roxithromycin, HMR 3647, or HMR 3004 and human polymorphonuclear neutrophils. Antimicrob Agents Chemother 2000;44(3): 511–21.

Levofloxacin

See also Fluoroquinolones

General Information

Levofloxacin, the levorotatory (*S*)-enantiomer of the racemate ofloxacin, is an oral and parenteral fluoroquinolone that has bactericidal activity against a wide spectrum of Gram-negative and Gram-positive bacilli (including *Streptococcus pneumoniae*), as well as atypical respiratory pathogens.

In patients with meningitis, levofloxacin penetration in cerebrospinal fluid and the liquor-to-plasma ratio was assessed at 2 hours after dosing in five patients with spontaneous acute bacterial meningitis. Cerebrospinal fluid levofloxacin concentration at 2 hours after dosing was 2.0 µg/ml, and the liquor-to-plasma ratio at 2 hours after dosing was 0.35 (1).

Observational studies

In 10 patients who took levofloxacin 500 mg/day and rifampicin 600 mg/day for 2–6 months, there were no adverse reactions in 46% of patients, occasional digestive symptoms in 40%, and mild diarrhea in 13%; these patients also took unspecified anti-inflammatory drugs (2). There was sleeplessness in 6% but neither tendinitis nor changes in liver function.

In a prospective, multicenter open trial, 313 patients with clinical signs and symptoms of bacterial infections of the respiratory tract, skin, or urinary tract were treated with levofloxacin (3). Of these, 134 patients had a pathogen recovered from the primary infection site and had an MIC of the pathogen to levofloxacin determined. Levofloxacin generated clinical and microbiological response rates of about 95%. These response rates included pathogens such as *Streptococcus pneumoniae* and *Staphylococcus aureus*. In a logistic regression analysis, the clinical outcome was predicted by the ratio of peak plasma concentration to MIC and site of infection. Microbiological eradication was predicted by the peak concentration/MIC ratio. Both clinical and microbiological outcomes were most likely to be favorable if the peak concentration/MIC ratio was at least 12.

Of 17 individuals with suspected latent multidrug-resistant tuberculosis treated with pyrazinamide and levofloxacin, 11 developed musculoskeletal adverse effects related to therapy, 5 had nervous system effects, and 15 had raised liver enzymes, uric acid, or creatinine kinase (4).

Comparative studies

In comparative trials involving commonly used regimens, levofloxacin had equivalent if not greater activity in the treatment of community-acquired pneumonia, acute bacterial exacerbations of chronic bronchitis, acute bacterial sinusitis, acute pyelonephritis, and complicated urinary tract infection (5).

General adverse effects

The most frequently reported adverse events of levofloxacin are nausea and diarrhea; compared with some other quinolones it has a low photosensitizing potential, and clinically significant cardiac and hepatic adverse events are rare (6).

The adverse effects rates of levofloxacin are 1.3% for nausea, 0.1% for anxiety, 0.3% for insomnia, and 0.1% for headache. No levofloxacin-related adverse events were reported at a rate higher than 1.3%, and most were less common. High-dose levofloxacin (750 mg) was also well tolerated. Surveillance data reported low adverse event rates: nausea 0.8%, rash 0.5%, abdominal pain 0.4%, and diarrhea, dizziness, and vomiting 0.3%. The adverse drug reactions rate for levofloxacin is still one of the lowest of any fluoroquinolone, at 2% compared with 2–10% for other fluoroquinolones (7–10).

Organs and Systems

Cardiovascular

Preclinical and clinical trial data and data from phase IV studies have suggested that levofloxacin causes prolongation of the QT interval (11), for example in healthy volunteers who took levofloxacin 1000 mg compared with placebo (12). There were cardiovascular problems in 1 in 15 million prescriptions compared with 1–3% of patients taking sparfloxacin, who had QT_c prolongation to over 500 ms. Polymorphous ventricular tachycardia with a normal QT interval has been associated with oral levofloxacin in the absence of other causes (7,10,13,14).

Among 23 patients who took levofloxacin 500 mg/day there was prolongation of the QT_c interval by more than 30 ms in four patients and 60 ms in two patients (15). There was absolute QT interval prolongation to over 500 ms in four patients, one of whom developed torsade de pointes.

- A 65-year-old woman had torsade de pointes after receiving levofloxacin 250 mg/day intravenously for 3 days (16).

Phlebitis can occur during parenteral administration of levofloxacin. High concentrations of levofloxacin (5 mg/ml) significantly reduced intracellular ATP content in cultured endothelial cells and reduced ADP, GTP, and GDP concentrations (16). These in vitro data suggest that high doses of levofloxacin are not compatible with maintenance of endothelial cell function and may explain the occurrence of phlebitis. Commercial formulations should be diluted and given into large veins.

Respiratory

Eosinophilic pneumonia complicated by bronchial asthma has been attributed to levofloxacin (18).

- A 76-year-old woman took levofloxacin for a productive cough with non-segmental infiltration in both lung fields. She developed eosinophilia in both the peripheral blood (24%) and the sputum (10%), airflow limitation, hypoxemia, and increased airway responsiveness to methacholine. Bronchoalveolar lavage fluid showed increased total cells and a 55% increase in eosinophils, and the CD4/CD8 ratio was reduced to 0.8. Histological features included increased infiltration of eosinophils in the alveolar and interstitial compartments and goblet cell metaplasia. Levofloxacin was withdrawn, and her symptoms improved without steroid therapy. A leukocyte migration test for levofloxacin was weakly positive.

Pneumonitis has been attributed to levofloxacin and kampo medicine (19).

- A 55 year old woman developed dyspnea on exertion. She had hypoxemia and restrictive ventilatory impairment. Chest x-rays showed diffuse patchy infiltration in both lungs. She gradually improved after withdrawal of the drugs. Transbronchial lung biopsy specimens showed lymphocyte dominant infiltration in the alveolar septa and Masson bodies. Lymphocyte stimulation

tests were positive for levofloxacin and shin-i-seihai-to, a kampo medicine.

Nervous system

Levofloxacin can cause seizures (20). In one study convulsions occurred in two per million prescriptions (10,21).

- A 75-year-old white woman was given oral levofloxacin (500 mg on day 1 followed by 250 mg/day) for ischemic toes (20). After three doses she had a seizure. One month later, she was challenged with ciprofloxacin 400 mg intravenously every 12 hours and again had a seizure.
- A 74-year-old white woman was given oral levofloxacin 500 mg/day for bacterial pneumonia and had a seizure after five doses (20).
- A 75-year-old man with Alzheimer's disease, chronic renal insufficiency, and seizures was given levofloxacin 500 mg/day for 1 week for a urinary tract infection and developed increased seizure activity over 3 days; the seizure activity ceased after withdrawal of levofloxacin (22).
- A 73-year old man with community-acquired pneumonia took oral levofloxacin after completing treatment with intravenous ceftriaxone and 2 days later became disoriented and confused; the symptoms regressed completely after withdrawal of levofloxacin (23).

A myasthenic crisis and respiratory depression occurred in a 45-year-old man 36 hours administration of levofloxacin (24).

Sensory systems

Taste disturbance occurred in less than three per million prescriptions of levofloxacin (10).

Metabolism

Hypoglycemia has uncommonly been reported with levofloxacin (25). It appears to occur most often in elderly patients with type 2 diabetes mellitus who are taking oral hypoglycemic drugs.

- A 64-year-old woman with type 2 diabetes treated by diet was given levofloxacin 500 mg/day for a urinary infection and pneumonia (26). After 2 days she became lethargic and disoriented. Her blood glucose concentration was 1.8 mmol/l (32 mg/dl) and the simultaneous blood insulin concentration was 6.7 (reference range 5–25) IU/ml. A 30% dextrose infusion was started, and she regained consciousness without any neurological deficit. Although she did not take any insulin during the dextrose infusion, her blood glucose concentrations were 3 and 3.4 mmol/l at 1 hour and 4 hours. On day 3 of levofloxacin treatment, she had another episode of hypoglycemia. Levofloxacin was withdrawn and she was given piperacillin + tazobactam instead. There were no further episodes of hypoglycemia.
- A 65-year-old woman who took levofloxacin 500 mg/day had episodes of hypoglycemia, which stopped spontaneously 6 days after the end of therapy (27).
- A 78-year-old, non-diabetic man had steroid-induced hyperglycemia, which was treated with glibenclamide; 12 hours after taking levofloxacin 500 mg orally he

developed severe hypoglycemia, resulting in anoxic brain injury (28).

In two population-based, nested case-control studies in 788 patients who had been treated for hypoglycemia within 30 days after antibiotic therapy, levofloxacin was associated with a slightly increased risk (adjusted OR = 1.5; 95% CI = 1.2 to 2.0) (37).

The effect of levofloxacin on serum glucose concentrations has been investigated in rats (29). The serum glucose concentration fell after injection of levofloxacin 100 mg/kg, and increased after levofloxacin 300 mg/kg. Adrenaline and histamine concentrations rose after levofloxacin 300 mg/kg. Diphenhydramine prevented the hyperglycemia induced by levofloxacin 300 mg/kg.

Hematologic

Pancytopenia occurred 9 days after treatment of pelvic inflammatory disease with levofloxacin 500 mg/day and resolved after withdrawal (30).

Levofloxacin can cause coagulopathies. Three patients who took oral levofloxacin 500 mg/day for 3 days for urinary tract infections developed prolonged prothrombin times (31). One later developed acquired von Willebrand's syndrome during surgery. The coagulopathies were successfully corrected preoperatively with parenteral vitamin K. The patient with acquired von Willebrand syndrome required multiple transfusions.

- A 70-year-old woman developed pseudothrombocytopenia after taking levofloxacin and ceftriaxone for acute bronchitis (32).

Autoimmune hemolytic anemia due to levofloxacin is extremely rare, but potentially fatal. However, hemolytic anemia was reported in an 82-year-old white man 3 days after the end of a course of levofloxacin 500 mg/day for cellulitis (33).

Gastrointestinal

Of 48 patients taking pyrazinamide 30 mg/kg/day plus levofloxacin 500 mg/day for 1 year, 27 discontinued therapy within 4 months owing to adverse events. Gastrointestinal intolerance was the major adverse event that resulted in early withdrawal (34).

In a double-blind, randomized, controlled trial in 394 patients who received either intravenous/oral moxifloxacin 400 mg/day) or intravenous/oral levofloxacin 500 mg/day) for 7–14 days, the rate of treatment-related adverse events due to any cause was significantly higher in those who received moxifloxacin, although the rates of drug-related and serious adverse events were similar in the two treatment arms (35). There was no difference between treatments with regard to the rate of premature discontinuation because of an adverse event; 17 of 35 of these events were considered to be drug related (10 in the moxifloxacin arm and 7 in the levofloxacin arm). Serious adverse events were reported in 23% of patients in both treatment groups. The most commonly reported adverse event in the levofloxacin arm was diarrhea (5%).

Levofloxacin can cause pseudomembranous colitis due to *Clostridium difficile* (36).

Liver

Levofloxacin can cause liver injury. Two patients with renal insufficiency developed acute hepatitis, which resolved on withdrawal (37,38). Another case of hepatitis after treatment with levofloxacin was reported in a 22-year-old woman who took amoxicillin 2 g/day for 12 days before levofloxacin 500 mg/day) (39).

In a study based on European and international data from about 130 million prescriptions, the adverse effects profile of levofloxacin was compared with that of other fluoroquinolones; there was a low rate of hepatic abnormalities (1/650 000) (7). However, two cases of severe acute liver toxicity were reported in patients who had received intravenous levofloxacin (40,41).

Acute fulminant fatal hepatic failure has been reported in a patient taking levofloxacin (42).

- A 55-year-old woman with asymptomatic hepatitis B infection was given oral levofloxacin 500 mg/day for 10 days for an upper respiratory tract infection. On day 12 she developed mixed hepatocellular and cholestatic liver damage, with transaminases 20 tomes the upper limit of the reference range. Despite supportive treatment she died 12 weeks later.

Pancreas

Two case reports have suggested that levofloxacin can cause pancreatitis (43).

Urinary tract

Levofloxacin can cause tubulointerstitial nephritis (44). A case of nephrotoxicity and purpura associated with levofloxacin has also been reported; allergic interstitial nephritis or vasculitis was believed to be the underlying pathologic process (45).

- A 73-year-old white man took levofloxacin for a lower urinary tract infection for 3 days and developed palpable purpura and erythematous skin lesions over the lower limbs and trunk, with a markedly reduced urine output. Serum creatinine was 560 μmol/l (6.4 mg/dl). Levofloxacin was withdrawn, and prednisone, furosemide, and intravenous fluids were given. The patient recovered fully over the next 4 weeks.

Granulomatous interstitial nephritis with associated granulomatous vasculitis was reported in a 47-year-old woman who took levofloxacin 250 mg/day for a urinary tract infection (46).

Levofloxacin can cause interstitial nephritis and vasculitis (47).

Skin

In a double-blind, randomized study in 30 healthy adults oral levofloxacin (500 mg/day for 5 days) had a low photosensitizing potential (48), as it did in preclinical animal studies and postmarketing surveillance (49). In preclinical studies levofloxacin was 20 times less photo-toxic than sparfloxacin. Phototoxicity occurs in only 1 in 1.8 million cases.

Levofloxacin can cause a rash similar to the ampicillin rash in patients with infectious mononucleosis (50).

- A 78-year-old woman developed a rash with blistering 2 days after completing a course of levofloxacin (51). The rash progressed to toxic epidermal necrolysis in 7 days. She was treated with intravenous fluids and wound dressings. Her condition improved and she was discharged after 22 days.

Cases of toxic epidermal necrolysis have been reported in patients taking levofloxacin.

- An 87-year-old woman developed a pruritic rash on her trunk and limbs 3 hours after taking a single dose of levofloxacin followed by a generalized seizure (52). A skin biopsy confirmed the diagnosis of toxic epidermal necrolysis. She had developed a rash after taking cipro-floxacin several years before, which had probably sensitized her to levofloxacin.
- A 15-year-old boy developed fatal toxic epidermal necrolysis, with a pruritic rash, fever, and joint pains, after taking levofloxacin 500 mg/day for 9 days (53). The exfoliation progressed to involve 80% of the body surface and he died despite supportive treatment.

Musculoskeletal

Tendinopathy is a class-related adverse effect of the fluor-oquinolones, with old age, and renal dysfunction as susceptibility factors. Several cases have been attributed to levofloxacin (53,54–66). Of four patients with

Achilles tendinitis two were on chronic dialysis, one was a kidney transplant recipient, and one had chronic vasculitis (67). In all four cases, tendinitis had an acute onset with bilateral involvement and was incapacitating. In three cases the onset was early during levofloxacin treatment and in one case it began 10 days after the end of treatment. All the patients recovered completely after 3–8 weeks.

Old age, renal dysfunction, and concomitant corticos-teroid therapy are predisposing risk factors (68,69,70). Tendon rupture occurred in less than four per million prescriptions (10).

In rats systemically exposed to levofloxacin, experimental fractures healed less well during the early stages of fracture repair (71).

Rhabdomyolysis has been attributed to levofloxacin.

- A 19-year-old man developed rhabdomyolysis with swelling and weakness of the arms and creatine kinase activity up to 16 546 U/l having taken ofloxacin and then levofloxacin for periorbital cellulitis; the symptoms improved after withdrawal of levofloxacin (72).
- After receiving a second cadaveric-donor renal transplant and after 10 days of therapy with levofloxacin 250 mg/day a 68-year old woman developed worsening myalgia and difficulty in walking (73). Her proximal leg muscles were tender. Her creatine kinase activity was up to 6200 U/l and there was myoglobinuria.

Levofloxacin, simvastatin, and colchicine were withdrawn. Intravenous hydration was maintained for 5 days. The serum creatine kinase activity fell and her muscle pain and tenderness disappeared after 14 days.

It was not clear in the second case what the relative contributions of the three drugs were.

Immunologic

Anaphylactic and anaphylactoid reactions are rare adverse events after the administration of fluoroquino-lones (about 0.46–1.2 per 100 000 patients).

- On two occasions a 49-year-old asthmatic woman who took levofloxacin for a chest infection developed worse respiratory distress, requiring intubation (74). The second reaction was accompanied by a marked skin reaction.

Four patients developed presumed hypersensitivity reactions to levofloxacin (75), including a 31-year-old man with a type I immunological mechanism, who developed an itching micropapular exanthema, facial and scrotal edema, and generalized erythema 30 minutes after an oral dose of levofloxacin 500 mg. Sensitivity to levoflox-acin was confirmed by a positive skin prick test.

- A 33-year old Japanese man developed a severe anaphylactic reaction 10 minutes after taking his first dose of levofloxacin with mefenamic acid and L-carbocys-teine orally; patch tests only showed a reaction to levo-floxacin (76).

An in vitro study in rat peritoneal mast cells showed that levofloxacin-mediated release of histamine may be closely linked to activation of pertussis toxin-sensitive G proteins (77).

Susceptibility Factors

The pharmacokinetics of intravenous levofloxacin have been studied in intensive care unit patients during continuous venovenous hemofiltration or hemodiafiltration (78,79). Levofloxacin clearance was substantially increased during both types of continuous renal replacement therapy. Levofloxacin 250 mg/day maintained effective plasma drug concentrations in these patients.

Drug–Drug Interactions

Acecainide

Levofloxacin and ciprofloxacin reduced the renal clearance of N-acetylprocainamide (acecainide) (80).

Calcium carbonate

In patients with cystic fibrosis, but not healthy volunteers, calcium carbonate reduced the C_{max} of levofloxacin by 19% and increased the tmax by 37% (81). The AUC was unaffected, but the authors concluded that multivalent cations should be taken at a separate time from oral levofloxacin in patients with cystic fibrosis.

Chinese medicines

Chinese medicines did not influence the systemic availability or renal excretion of levofloxacin (82).

Efavirenz

Levofloxacin pharmacokinetics in HIV-positive patients were not altered by steady-state treatment with efavirenz (83).

HIV protease inhibitors

Levofloxacin pharmacokinetics in HIV-positive patients were not altered by steady-state treatment with nelfinavir (83).

Immunosuppressants

Concomitant treatment with levofloxacin and ciclosporin or tacrolimus increased the AUC of ciclosporin and tacrolimus by 25%, probably by inhibition of hepatic metabolism by levofloxacin (89).

Lithium

Co-administration with levofloxacin can cause severe lithium toxicity; the authors did not discuss the mechanism (85).

Procainamide

Levofloxacin and ciprofloxacin reduced the renal clearance of procainamide (80).

Theophylline

Theophylline clearance was reduced by levofloxacin plus clarithromycin in a 59-year-old Japanese man, who had stimulation, insomnia, and tachycardia due to theophylline toxicity (86).

The mechanism was probably inhibition of theophylline metabolism by CYP1A2 and CYP3A4.

Vancomycin

Changes in pharmacokinetics were studied in male Wistar rats when intravenous vancomycin 100 mg/kg and levofloxacin 20 mg/kg were administered together (90). There was an increase in the AUC and half-life of vancomycin. There was also an increase in the AUC and a delay in the t_{max} of levofloxacin, but no effect on C_{max}; these data suggested delayed absorption of levofloxacin. Concomitant administration had no effect on the correlation between serum and hepatic tissue concentrations of levofloxacin, but it markedly reduced the correlation between the serum and renal tissue concentrations of vancomycin. Vancomycin increased serum creatinine concentrations 8 hours after administration. However, there was no difference in animals who received monotherapy compared with animals who received combined therapy. The authors suggested the cautious use of a combination of levofloxacin and vancomycin and advised monitoring blood concentrations of vancomycin in such cases.

Warfarin

Enhanced hypoprothrombinemia has been reported when levofloxacin was given with warfarin (87–89), although in another study co-administration of warfarin with levofloxacin did not affect the INR compared with warfarin alone (91).

References

1. Villani P, Viale P, Signorini L, Cadeo B, Marchetti F, Villani A, Fiocchi C, Regazzi MB, Carosi G. Pharmacokinetic evaluation of oral levofloxacin in human immunodeficiency virus-infected subjects receiving concomitant antiretroviral therapy. Antimicrob Agents Chemother 2001;45(7):2160–2.
2. Ortega M, Soriano A, Garcia S, Almela M, Alvarez JL, Tomas X, Mensa J, Soriano E. Perfil de tolerabilidad y seguridad de levofloxacinoen tratamientos prolongados. [Tolerability and safety of levofloxacinin long-term treatment.] Rev Esp Quimioter 2000;13(3):263–6.
3. Preston SL, Drusano GL, Berman AL, Fowler CL, Chow AT, Dornseif B, Reichl V, Natarajan J, Corrado M. Pharmacodynamics of levofloxacin: a new paradigm for early clinical trials. JAMA 1998;279(2):125–9.
4. Papastavros T, Dolovich LR, Holbrook A, Whitehead L, Loeb M. Adverse events associated with pyrazinamide and levofloxacin in the treatment of latent multidrug-resistant tuberculosis. CMAJ 2002;167(2):131–6.
5. Wimer SM, Schoonover L, Garrison MW. Levofloxacin: a therapeutic review. Clin Ther 1998;20(6):1049–70.
6. Croom KF, Goa KL. Levofloxacin: a review of its use in the treatment of bacterial infections in the United States. Drugs 2003;63:2769–802.
7. Carbon C. Comparison of side effects of levofloxacin versus other fluoroquinolones. Chemotherapy 2001;47(Suppl 3):9–14discussion 44–8.
8. Rossi C, Sternon J. Les fluoroquinolones de troisième et quatrième generations. [Third and fourth generation fluoroquinolones.] Rev Med Brux 2001;22(5):443–56.
9. Chow AT, Fowler C, Williams RR, Morgan N, Kaminski S, Natarajan J. Safety and pharmacokinetics of multiple 750-milligram doses of intravenous levofloxacin in healthy volunteers. Antimicrob Agents Chemother 2001;45(7):2122–5.
10. Kahn JB. Latest industry information on the safety profile of levofloxacin in the US. Chemotherapy 2001;47(Suppl 3):32–7discussion 44–8.
11. Owens RC Jr, Ambrose PG. Torsades de pointes associated with fluoroquinolones. Pharmacotherapy 2002;22(5):663–8discussion 668–72.
12. Noel GJ, Natarajan J, Chien S, Hunt TL, Goodman DB, Abels R. Effects of three fluoroquinolones on QT interval in healthy adults after single doses. Clin Pharmacol Ther 2003;73:292–303.
13. Scotton PG, Pea F, Giobbia M, Baraldo M, Vaglia A, Furlanut M. Cerebrospinal fluid penetration of levofloxacin in patients with spontaneous acute bacterial meningitis. Clin Infect Dis 2001;33(9):e109–11.
14. Paltoo B, O'Donoghue S, Mousavi MS. Levofloxacin induced polymorphic ventricular tachycardia with normal QT interval. Pacing Clin Electrophysiol 2001;24(5):895–7.
15. Carbon C. Tolérance de la lévofloxacine, dossier clinique et données de pharmacovigilance. [Levofloxacin adverse effects, data from clinical trials and pharmacovigilance.] Therapie 2001;56(1):35–40.

16. Amankwa K, Krishnan SC, Tisdale JE. Torsades de pointes associated with fluoroquinolones: importance of concomitant risk factors. Clin Pharmacol Ther 2004;75(3):242–7.

17. Armbruster C, Robibaro B, Griesmacher A, Vorbach H. Endothelial cell compatibility of trovafloxacin and levofloxacin for intravenous use. J Antimicrob Chemother 2000;45(4):533–5.

18. Fujimori K, Shimatsu Y, Suzuki E, Arakawa M, Gejyo F. [Levofloxacin-induced eosinophilic pneumonia complicated by bronchial asthma.]Nihon Kokyuki Gakkai Zasshi 2000;38(5):385–90.

19. Tohyama M, Arakaki N, Tamaki K, Shimoji T. [A case of drug-induced pneumonitis due to levofloxacin and kampo medicine.] Nihon Kokyuki Gakkai Zasshi 2006;44(12):951–6.

20. Kushner JM, Peckman HJ, Snyder CR. Seizures associated with fluoroquinolones. Ann Pharmacother 2001;35(10):1194–8.

21. Pedros A, Emilio Gomez J, Angel Navarro L, Tomas A. Levofloxacino y sindrome confusional agudo. [Levofloxacin and acute confusional syndrome.] Med Clin (Barc) 2002;119(1):38–9.

22. Bird SB, Orr PG, Mazzola JL, Brush DE, Boyer EW. Levofloxacin-related seizure activity in a patient with Alzheimer's disease: assessment of potential risk factors. J Clin Psychopharmacol 2005;25(3):287–8.

23. Hakko E, Mete B, Ozaras R, Tabak F, Ozturk R, Mert A. Levofloxacin-induced delirium. Clin Neurol Neurosurg 2005;107(2):158–9.

24. Gunduz A, Turedi S, Kalkan A, Nuhoglu I. Levofloxacin induced myasthenia crisis. Emerg Med J 2006;23(8):662.

25. Friedrich LV, Dougherty R. Fatal hypoglycemia associated with levofloxacin. Pharmacotherapy 2004;24(12):1807–12.

26. Kanbay M, Aydogan T, Bozalan R, Isik A, Uz B, Kaya A, Akcay A. A Rare but serious side effect of levofloxacin: hypoglycemia in a geriatric patient. Diabetes Care 2006;29(7):1716–7.

27. Wang S, Rizvi AA. Levofloxacin-induced hypoglycemia in a nondiabetic patient. Am J Med Sci 2006;331(6):334–5.

28. Lawrence KR, Adra M, Keir C. Hypoglycemia-induced anoxic brain injury possibly associated with levofloxacin. J Infect 2006;52(6):e177-80.

29. Ishiwata Y, Itoga Y, Yasuhara M. Effect of levofloxacin on serum glucose concentration in rats. Eur J Pharmacol 2006;551(1-3):168–74.

30. Deng JY, Tovar JM. Pancytopenia with levofloxacin therapy for pelvic inflammatory disease in an otherwise healthy young patient. Ann Pharmacother 2006;40(9):1692–3.

31. Psarros T, Trammell T, Morrill K, Giller C, Morgan H, Allen B. Abnormal coagulation studies associated with levofloxacin. Report of three cases. J Neurosurg 2004;100(4):710–2.

32. Kinoshita Y, Yamane T, Kamimoto A, Oku H, Iwata Y, Kobayashi T, Hino M. [A case of pseudothrombocytopenia during antibiotic administration]. Rinsho Byori 2004;52(2):120–3.

33. Oh YR, Carr-Lopez SM, Probasco JM, Crawley PG. Levofloxacin-induced autoimmune hemolytic anemia. Ann Pharmacother 2003;37:1010–3.

34. Fleisch F, Hartmann K, Kuhn M. Fluoroquinolone-induced tendinopathy: also occurring with levofloxacin. Infection 2000;28(4):256–7.

35. Anzueto A, Niederman MS, Pearle J, Restrepo MI, Heyder A, Choudhri SH. Community-Acquired Pneumonia Recovery in the Elderly (CAPRIE):efficacy and safety of moxifloxacin therapy versus that of levofloxacin therapy. Clin Infect Dis 2006;42(1):73–81.

36. Casado Burgos E, Vinas Ponce G, Lauzurica Valdemoros R, Olive Marques A. Tendinitis por levofloxacino. [Levofloxacin-induced tendinitis.] Med Clin (Barc) 2000;114(8):319.

37. Schwalm JD, Lee CH. Acute hepatitis associated with oral levofloxacin therapy in a hemodialysis patient. Can Med Assoc J 2003;168:847–8.

38. Heluwaert F, Roblin X, Duffournet V, Capony P, Martin D. A propos d'un cas d'hepatite aigue medicamenteuse mixte après traitement par amoxicilline et levofloxacine. Rev Med Interne 2003;24:841–3.

39. Airey K, Koller E. Acute hepatitis associated with levofloxacin in a patient with renal insufficiency. Can Med Assoc J 2003;169:755.

40. Gates GA. Safety of ofloxacin otic and other ototopical treatments in animal models and in humans. Pediatr Infect Dis J 2001;20(1):104–7discussion 120–2.

41. Karim A, Ahmed S, Rossoff LJ, Siddiqui RK, Steinberg HN. Possible levofloxacin-induced acute hepatocellular injury in a patient with chronic obstructive lung disease. Clin Infect Dis 2001;33(12):2088–90.

42. Coban S, Ceydilek B, Ekiz F, Erden E, Soykan I. Levofloxacin-induced acute fulminant hepatic failure in a patient with chronic hepatitis B infection. Ann Pharmacother 2005;39(10):1737–40.

43. Spahr L, Rubbia-Brandt L, Marinescu O, Armenian B, Hadengue A. Acute fatal hepatitis related to levofloxacin. J Hepatol 2001;35(2):308–9.

44. Wood ML, Schlessinger S. Levaquin induced acute tubulointerstitial nephritis—two case reports. J Miss State Med Assoc 2002;43(4):116–7.

45. Famularo G, De Simone C. Nephrotoxicity and purpura associated with levofloxacin. Ann Pharmacother 2002;36(9):1380–2.

46. Ramalakshmi S, Bastacky S, Johnson JP. Levofloxacin-induced granulomatous interstitial nephritis. Am J Kidney Dis 2003;41:E7.

47. Zaigraykin N, Kovalev J, Elias N, Naschitz JE. Levofloxacin-induced interstitial nephritis and vasculitis in an elderly woman. Isr Med Assoc J 2006;8(10):726–7.

48. Boccumini LE, Fowler CL, Campbell TA, Puertolas LF, Kaidbey KH. Photoreaction potential of orally administered levofloxacin in healthy subjects. Ann Pharmacother 2000;34(4):453–8.

49. Mennecier D, Thiolet C, Bredin C, Potier V, Vergeau B, Farret O. Pancreatite aiguë survenant après la prise de lévofloxacine et de methylprédnisolone. [Acute pancreatitis after treatment by levofloxacin and methylprednisolone.] Gastroenterol Clin Biol 2001;25(10):921–2.

50. Paily R. Quinolone drug rash in a patient with infectious mononucleosis. J Dermatol 2000;27(6):405–6.

51. Digwood-Lettieri S, Reilly KJ, Haith LR Jr, Patton ML, Guilday RJ, Cawley MJ, Ackerman BH. Levofloxacin-induced toxic epidermal necrolysis in an elderly patient. Pharmacotherapy 2002;22(6):789–93.

52. Christie MJ, Wong K, Ting RH, Tam PY, Sikaneta TG. Generalized seizure and toxic epidermal necrolysis following levofloxacin exposure. Ann Pharmacother 2005;39(5):953–5.

53. Islam AF, Rahman MD. Levofloxacin-induced fatal toxic epidermal necrolysis. Ann Pharmacother 2005;39(6):1136–7.

54. Vergara Fernandez I. Afeccion musculotendinosa secundaria a levofloxacino: revision a proposito de un caso. [Muscle and tendon problems as a side-effect of levofloxacine: review of a case]. Aten Primaria 2004;33(4):214.

55. Gomez Rodriguez N, Ibanez Ruan J, Gonzalez Perez M. Tendonitis aquilea bilateral y levofloxacino. [Achilles bilateral tendonitis and levofloxacin]. An Med Interna 2004;21(3):154.

56. Kowatari K, Nakashima K, Ono A, Yoshihara M, Amano M, Toh S. Levofloxacin-induced bilateral Achilles tendon rupture: a case report and review of the literature. J Orthop Sci 2004;9(2):186–90.

57. Braun D, Petitpain N, Cosserat F, Loeuille D, Bitar S, Gillet P, Trechot P. Rupture of multiple tendons after levofloxacin therapy. Joint Bone Spine 2004;71(6):586–7.

58. Aouam K, El Aidli S, Daghfous R, Kastalli S, Lakhal M, Loueslati MH, Belkahia C. La tendinite achiléenne à l'ofloxacine malgré l'absence de prédisposition. [Ofloxacin-induced Achilles tendinitis in the absence of a predisposition]. Thérapie 2004;59(6):653–5.

59. Sanchez Munoz LA, Sanjuan Portugal FJ, Naya Machado J, Castiella Herrero J. Levofloxacino y rotura bilateral del tendon de Aquiles con evolucion fatal. [Levofloxacin-induced Achilles tendon rupture with fatal outcome.] An Med Interna 2006;23(2):102.

60. Filippucci E, Farina A, Bartolucci F, Spallacci C, Busilacchi P, Grassi W. Tendinopatia dell'Achilleo da levofloxacina: dalle immagini alla diagnosi. Reumatismo 2003;55:267–9.

61. Gold L, Igra H. Levofloxacin-induced tendon rupture: a case report and review of the literature. J Am Board Fam Pract 2003;16:458–60.

62. Melhus A, Apelqvist J, Larsson J, Eneroth M. Levofloxacin-associated Achilles tendon rupture and tendinopathy. Scand J Infect Dis 2003;35:768–70.

63. De La Red G, Mejia JC, Cervera R, Llado A, Mensa J, Font J. Bilateral Achilles tendinitis with spontaneous rupture induced by levofloxacin in a patient with systemic sclerosis. Clin Rheumatol 2003;22:367–8.

64. Mathis AS, Chan V, Gryszkiewicz M, Adamson RT, Friedman GS. Levofloxacin associated Achilles tendon rupture. Ann Pharmacother 2003;37:1014–7.

65. Cebrian P, Manjon P, Caba P. Ultrasonography of non-traumatic rupture of the Achilles tendon secondary to levofloxacin. Foot Ankle Int 2003;24:122–4.

66. Bernacer L, Artigues A, Serrano A. Levofloxacino y rotura espontanea bilateral del tendon de Aquiles. Med Clin (Barc) 2003;120:78–9.

67. Lou HX, Shullo MA, McKaveney TP. Limited tolerability of levofloxacin and pyrazinamide for multidrug-resistant tuberculosis prophylaxis in a solid organ transplant population. Pharmacotherapy 2002;22(6):701–4.

68. Ozawa TT, Valadez T. *Clostridium difficile* infection associated with levofloxacin treatment. Tenn Med 2002;95(3):113–5.

69. Aros C, Flores C, Mezzano S. Tendinitis aquiliana asociada al uso de levofloxacino: comunicacion de cuatro casos. [Achilles tendinitis associated with levofloxacin: report of 4 cases.] Rev Med Chil 2002;130(11):1277–81.

70. Leone R, Venegoni M, Motola D, Moretti U, Piazzetta V, Cocci A, Resi D, Mozzo F, Velo G, Burzilleri L, Montanaro N, Conforti A. Adverse drug reactions related to the use of fluoroquinolone antimicrobials: an analysis of spontaneous reports and fluoroquinolone consumption data from three Italian regions. Drug Saf 2003;26:109–20.

71. Perry AC, Prpa B, Rouse MS, Piper KE, Hanssen AD, Steckelberg JM, Patel R. Levofloxacin and trovafloxacin inhibition of experimental fracture-healing. Clin Orthop Relat Res 2003;(414):95–100.

72. Hsiao SH, Chang CM, Tsao CJ, Lee YY, Hsu MY, Wu TJ. Acute rhabdomyolysis associated with ofloxacin/levofloxacin therapy. Ann Pharmacother 2005;39(1):146–9.

73. Korzets A, Gafter U, Dicker D, Herman M, Ori Y. Levofloxacin and rhabdomyolysis in a renal transplant patient. Nephrol Dial Transplant 2006;21(11):3304–5.

74. Smythe MA, Cappelletty DM. Anaphylactoid reaction to levofloxacin. Pharmacotherapy 2000;20(12):1520–3.

75. Gonzalez-Mancebo E, Fernandez-Rivas M. Immediate hypersensitivity to levofloxacin diagnosed through skin prick test. Ann Pharmacother 2004;38(2):354.

76. Takahama H, Tsutsumi Y, Kubota Y. Anaphylaxis due to levofloxacin. Int J Dermatol 2005;44(9):789–90.

77. Mori K, Maru C, Takasuna K, Furuhama K. Mechanism of histamine release induced by levofloxacin, a fluoroquinolone antibacterial agent. Eur J Pharmacol 2000;394(1):51–5.

78. Malone RS, Fish DN, Abraham E, Teitelbaum I. Pharmacokinetics of levofloxacin and ciprofloxacin during continuous renal replacement therapy in critically ill patients. Antimicrob Agents Chemother 2001;45(10):2949–54.

79. Yagawa K. Latest industry information on the safety profile of levofloxacin in Japan. Chemotherapy 2001;47(Suppl 3):38–43discussion 44–8.

80. Bauer LA, Black DJ, Lill JS, Garrison J, Raisys VA, Hooton TM. Levofloxacin and ciprofloxacin decrease procainamide and N-acetylprocainamide renal clearances. Antimicrob Agents Chemother 2005;49(4):1649–51.

81. Pai MP, Allen SE, Amsden GW. Altered steady state pharmacokinetics of levofloxacin in adult cystic fibrosis patients receiving calcium carbonate. J Cyst Fibros 2006;5(3):153–7.

82. Iunda IF, Kushniruk IuI. [Functional state of the testis after the use of certain antibiotics and nitrofuran preparations.]Antibiotiki 1975;(9):843–6.

83. Nakamura H, Ohtsuka T, Enomoto H, Hasegawa A, Kawana H, Kuriyama T, Ohmori S, Kitada M. Effect of levofloxacin on theophylline clearance during theophylline and clarithromycin combination therapy. Ann Pharmacother 2001;35(6):691–3.

84. Federico S, Carrano R, Capone D, Gentile A, Palmiero G, Basile V. Pharmacokinetic interaction between levofloxacin and ciclosporin or tacrolimus in kidney transplant recipients: ciclosporin, tacrolimus and levofloxacin in renal transplantation. Clin Pharmacokin 2006;45(2):169–75.

85. Takahashi H, Higuchi H, Shimizu T. Severe lithium toxicity induced by combined levofloxacin administration. J Clin Psychiatry 2000;61(12):949–50.

86. Gheno G, Cinetto L. Levofloxacin–warfarin interaction. Eur J Clin Pharmacol 2001;57(5):427.

87. Jones CB, Fugate SE. Levofloxacin and warfarin interaction. Ann Pharmacother 2002;36(10):1554–7.

88. Hansen E, Bucher M, Jakob W, Lemberger P, Kees F. Pharmacokinetics of levofloxacin during continuous venovenous hemofiltration. Intensive Care Med 2001;27(2):371–5.

89. Ravnan SL, Locke C. Levofloxacin and warfarin interaction. Pharmacotherapy 2001;21(7):884–5.

90. Mori H, Nakajima T, Nakayama A, Yamori M, Izushi F, Gomita Y. Interaction between levofloxacin and vancomycin in rats—study of serum and organ levels. Chemotherapy 1998;44(3):181–9.

91. Yamreudeewong W, Lower DL, Kilpatrick DM, Enlow AM, Burrows MM, Greenwood MC. Effect of levofloxacin coadministration on the international normalized ratios during warfarin therapy. Pharmacotherapy 2003;23:333–8.

Lincosamides

General Information

The two established members of the group of antibiotics known as the lincosamides, lincomycin and its semisynthetic derivative clindamycin, have a narrow antibacterial spectrum involving mostly Gram-positive species and some obligate anerobes, such as *Bacteroides*. Like chloramphenicol and erythromycin, they combine with a subunit of bacterial ribosomes and interfere with protein synthesis.

Whereas oral lincomycin has a systemic availability of about 40%, which may be further compromised by food, clindamycin is absorbed from the gastrointestinal tract about 90–100%. Both are eliminated mainly by hepatic metabolism and biliary excretion.

Observational studies

Long-term oral clindamycin therapy has been successfully used in a 36-year-old woman with late-stage AIDS who presented with disseminated, nodular cutaneous lesions and underlying osteomyelitis due to a microsporidial infection with an Encephalitozoon-like species (1).

In 123 patients with uncomplicated falciparum malaria, intravenous clindamycin 5 mg/kg every 8 hours and quinine 8 mg/kg every 8 hours for 3 days, 12 patients had minor adverse effects, including transient hypoacusis (4%), nausea (3.2%), transient hypoglycemia (1%), anxiety (0.8%), diarrhea (0.8%), and transient rash (0.8%). Treatment was withdrawn in two cases on day 2 because one patient had severe diarrhea and the other had intense abdominal pain. In both of these, the clindamycin was withdrawn and the patients completed therapy with a 7-day course of quinine and remained well (2).

Comparative studies

In a multicenter, double-blind, randomized trial in 87 patients, clindamycin + primaquine was compared with co-trimoxazole as therapy for AIDS-related *Pneumocystis jiroveci* pneumonia; efficacy was similar. In patients with a PaO_2 under 70 mmHg, clindamycin + primaquine was associated with fewer adverse events and less glucocorticoid use, but more rashes (3).

In a prospective, open, randomized trial clindamycin (600 mg tds) and quinine (650 mg tds) were compared with atovaquone (750 mg bd) plus azithromycin (500 mg on day 1 followed by 250 mg/day) in 58 patients with non-life-threatening babesiosis (4). Bacterial response was complete 3 months after the end of treatment. Adverse effects were reported by 72% of those who received clindamycin and quinine compared with 15% of those who received atovaquone and azithromycin. The most common adverse effects with clindamycin and quinine were tinnitus (39%), diarrhea (33%), and impaired hearing (28%); the symptoms had resolved in 73% of the patients assigned to clindamycin/quinine 3 months after the start of therapy and in 100% after 6 months.

In 233 women with bacterial vaginosis, a 3-day regimen of clindamycin (intravaginal ovules, 100 mg/day) was as effective as a 7-day regimen of oral metronidazole (500 mg bd) and better tolerated (5). Treatment-related adverse events were reported more often with metronidazole, and systemic symptoms, such as nausea and taste disturbance, accounted for most of the difference between the groups.

Placebo-controlled studies

In a 10-week, multicenter, double-blind study 480 patients with moderate to moderately severe acne were randomized to receive twice-daily 5% benzoyl peroxide, 1% clindamycin, 5% benzoyl peroxide plus 1% clindamycin, or vehicle, all topically (6). There were significantly greater reductions in the numbers of inflammatory and total lesions in patients who used combination therapy compared with those who used any of its three individual components. The most frequent adverse effect, dry skin, occurred to a similar extent with the combination and with benzoyl peroxide alone.

General adverse effects

The direct toxicity of the lincosamides is relatively low (SED-7, 389) (7). The adverse effects of clindamycin may be well below 1%. In a tertiary care center, adverse reactions to clindamycin were reported in 0.47% of 3896 courses, and in half of these events an effect of other medications could not be excluded (8). However, clindamycin has not been given in as high doses as lincomycin.

The most common adverse effect is diarrhea, which occurs in as many as 10–20% of patients. The most serious gastrointestinal complication is colitis due to *Clostridium difficile*, which occurs with about equal frequency after oral and parenteral treatment (9,10). Skin rashes, urticaria, and angioedema have been reported with lincomycin, but are rare. In contrast, maculopapular and pruritic eruptions occur after 1–2 weeks of treatment in up to 10% of patients taking clindamycin. Clindamycin has also been incriminated in one case of Stevens–Johnson syndrome and in one of anaphylaxis; in the latter, hemagglutinating antibodies were found against clindamycin and lincomycin (11). Leukocytoclastic angiitis associated with clindamycin is rare. If a similar angiitis can also occur in the colon, the question arises whether some cases of antibiotic-associated colitis might be caused by the drug itself rather than by bacterial toxins. Tumor-inducing effects have not been reported.

Organs and Systems

Cardiovascular

Rapid intravenous infusion of large doses of lincomycin (600 mg in 5–10 minutes) can cause flushing and a sensation of warmth for about 10 minutes.

- A patient who received 200 mg/kg of lincomycin experienced nausea, vomiting, hypotension, dyspnea, and electrocardiographic changes for 20 minutes (SED-7, 389) (12).

Rapid intravenous infusion of lincomycin 1–2 g can cause phlebitis.

Clindamycin can prolong the QT interval and cause ventricular fibrillation (13). Cardiac arrest associated with rapid intravenous administration of clindamycin has been reported (SEDA-8, 258) (14).

Respiratory

The benzyl alcohol component of clindamycin injection can cause "gasping syndrome" in premature neonates.

- A baby boy born at 24 weeks gestation weighing 710 g was given intravenous clindamycin (15). The first two doses were given without problem, but after the third and fourth doses he had profound desaturation and chest splinting, requiring resuscitation. The clindamycin was withdrawn and he had no obvious sequelae.

Nervous system

Clindamycin, either alone or in combination with neuromuscular blocking drugs or aminoglycosides, has been associated with neuromuscular blockade.

- A 58-year-old woman who was accidentally overdosed with 2400 mg (40 mg/kg) of clindamycin during anesthesia (with tubocurarine and suxamethonium induction) developed prolonged neuromuscular blockade, unresponsive to intravenous calcium or cholinesterase inhibitors (edrophonium and neostigmine), and required assisted ventilation for 11 hours (16). It seems likely that the large dose of clindamycin used in this case produced sustained neuromuscular blockade in the absence of non-depolarizing relaxants and after full recovery from suxamethonium.

Neuromuscular function

Lincosamide antibiotics can produce neuromuscular block postjunctionally by interacting with the open state of the acetylcholine receptor-channel complex. Lipophilicity, rather than stereochemistry of the molecule, is important for open-channel blockade affecting primarily the "off" rate of channel blocking (17). Neuromuscular blockade after clindamycin has been reported (18).

- A 44-year-old woman with mild asthma and mitral valve prolapse was given intravenous clindamycin 300 mg, methylprednisolone 125 mg, and midazolam 2 mg 30 minutes before surgery. Suxamethonium 120 mg was given and general anesthesia was induced with fentanyl 0.1 mg and propofol 120 mg, and maintained with 60% nitrous oxide in oxygen, desflurane, and a single dose of propofol 80 mg. Five hours after uneventful surgery and about 20 minutes after another intravenous dose of clindamycin 600 mg, she complained of profound weakness and had bilateral ptosis, difficulty in speaking, and rapid shallow respiration. Her weakness rapidly became more profound, and she was given neostigmine 4 mg and glycopyrrolate 0.8 mg, after which her muscle strength returned to normal. Electromyography showed no evidence of neuromuscular disease, and acetylcholine receptor antibodies were negative.

An in vitro study has shown a direct effect of clindamycin on nicotinic but not muscarinic acetylcholine receptors (19). This may explain improvement of tremor on treatment with clindamycin in a patient with Parkinson's disease. On each of three occasions, the tremor almost completely disappeared shortly after the start of therapy with clindamycin but reappeared within 1–3 days after withdrawal.

Sensory systems

In a prospective study subconjunctival injections of clindamycin did not produce any general adverse effects (20). However, conjunctival inflammation and keratitis were observed in one of 13 cases, caused by an error in the administered concentration of clindamycin, which was too high.

Hematologic

Granulocytopenia and thrombocytopenia have been described in a few patients taking lincosamides. However, a cause-and-effect relation has not been unequivocally established.

Lymphadenitis has been attributed to clindamycin (21).

- A 54-year-old woman with paraplegia due to spina bifida was treated with clindamycin, ciprofloxacin, and gentamicin for osteomyelitis of the ischium and acetabulum, but developed painful swelling of lymph nodes in the neck, which she said occurred with each dose of clindamycin. Withdrawal of gentamicin had no effect, but withdrawal of clindamycin resolved the adenitis over 7 days; the ciprofloxacin was continued.

Gastrointestinal

Clindamycin can cause esophageal disorders, including inflammation, strictures, bleeding, and ulceration (22) (23).

Drug-induced esophagitis is rare, accounting for about 1% of all cases of esophagitis. An incidence of 3.9 in 100 000 has been reported. After the first description, there have been more than 250 observations, with more than 50 different drugs. Among those, the principal antibiotics included tetracyclines (doxycycline, metacycline, minocycline, oxytetracycline, and tetracycline), penicillins (amoxicillin, cloxacillin, penicillin V, and pivmecillinam), clindamycin (24), co-trimoxazole, erythromycin, lincomycin, spiramycin, and tinidazole. Doxycycline alone was involved in one-third of all cases. Risk factors included prolonged esophageal passage, due to motility disorders, stenosis, cardiomegaly, the formulation, supine position during drug ingestion, and failure to use liquid to wash down the tablet. Direct toxic effects of the drug (pH, accumulation in epithelial cells, non-uniform dispersion) also seem to contribute to the development of drug-induced esophagitis (25). Clindamycin should be taken with a meal or followed by a glass of water (26).

The most prominent adverse reaction of the lincosamides is diarrhea, which varies from mildly loose bowel movements to life-threatening pseudomembranous colitis (see monograph on Beta-lactam antibiotics). Almost all antimicrobial drugs have been associated with severe diarrhea and colitis; however, lincomycin and clindamycin have been particularly incriminated. The incidence of clindamycin-induced diarrhea in hospital is 23%. Diarrhea resolves promptly after withdrawal in most cases. It seems to be dose-related and may result from a direct action on the intestinal mucosa. Severe colitis due to *C. difficile* is not dose-related and occurs in 0.01–10% of recipients. Clustering of cases in time and place suggests the possibility of cross-infection. Even low doses of clindamycin, in some cases after topical administration, can cause marked alterations in several intestinal functions related to bowel flora (27). There was reduced susceptibility of *C. difficile* to clindamycin in 80% of French isolates in 1997 (28). Lincomycin was among the antibiotics that were most often associated with the development of antibiotic-associated diarrhea in a Turkish study of 154 patients; other associated antibiotics were azithromycin and ampicillin (29).

In a one-year retrospective study at a tertiary hospital in Spain, 17% of 148 episodes of diarrhea associated with *C. difficile* developed after therapy with clindamycin (30). The possible association of toxin-positive *C. difficile*-induced colitis and the use of clindamycin phosphate vaginal cream for bacterial vaginosis has been reported in a 25-year-old white woman postpartum (31).

In a prospective study patients treated with antibiotics, including clindamycin, for 3 days had a significantly lower frequency of antibiotic-associated diarrhea than those treated for longer periods (32).

Restricting the use of clindamycin has been successful in terminating outbreaks of *C. difficile* diarrhea associated with its use (33). Between 1989 and 1992, outbreaks of diarrhea due to a clindamycin-resistant strain of *C. difficile* occurred in different parts of the USA. Resistance was mediated by the ermB gene. The use of clindamycin was a specific risk factor for diarrhea due to this strain (34).

- A young otherwise healthy nurse developed severe diarrhea and vomiting, profuse ascites, pleural effusion, abdominal tenderness, peritoneal irritation, and systemic toxicity 10 days after taking oral clindamycin for a dental infection (35). Although the assay for *C. difficile* was repeatedly negative, features compatible with pseudomembranous colitis were seen at sigmoidoscopy, and the diagnosis was confirmed histologically.

Topical application of clindamycin to the skin has been used in acne vulgaris. However, percutaneous absorption can occur (SEDA-8, 160) and several cases of diarrhea have been reported, including cases of pseudomembranous colitis (36).

Clindamycin can rarely cause antibiotic-associated diarrhea after short-term use as vaginal cream (37).

Liver

Since the lincosamides are eliminated by biliary excretion, toxicity would be expected in patients with liver disease. High doses of clindamycin may be hepatotoxic (38). Abnormal liver function tests during treatment with lincomycin are rare, and only in patients who had taken large doses (over 4 g/day) for more than 3 weeks (7). In another series, intravenous lincomycin 4–18 g/day was not associated with renal or hepatic toxicity (SED-7, 388).

Skin

The most common adverse events in patients using clindamycin/benzoyl peroxide were dry skin, peeling, erythema, and rash (39,40). However, withdrawal rates due to adverse events were low (0–0.8%).

- A 57-year-old Caucasian woman with a history of ocular toxoplasmosis, treated with intravitreal clindamycin (1 mg/0.1 ml) and dexamethasone (0.4 mg/0.1 ml), developed a generalized erythematous macular rash over the scalp, face, arms, thighs, and trunk 2 days after the start of treatment (41).

Precautions are necessary to avoid ultraviolet radiation after taking photoreactive drugs (42). Metabolism of lincomycin can lead to the formation of reactive oxygen species and cause tissue injury and damage to various cellular macromolecules, which can result in phototoxicity. Typical photosensitivity with a maculopapular eruption has been observed with lincomycin in two patients treated intramuscularly 7.

- Toxic epidermal necrolysis has been described in a 50-year-old insulin-dependent diabetic who received clindamycin (300 mg 8-hourly) (43). On the seventh day, he developed flu-like symptoms, fever, and an erythematous rash, associated with sloughing of 30% of the body surface after a further 4 days (when Nikolsky's sign became positive). He was given oral methylprednisolone 32 mg and eventually made a full recovery.

Lincomycin can cause acne rosacea (44).

A fixed drug eruption has been reported in a patient taking clindamycin 13.

- In a 38-year-old non-atopic man, a generalized pruriginous maculopapular eruption with lip edema and facial erythema developed after 10 days of treatment with oral clindamycin phosphate (300 mg qds) and amoxicillin (500 mg qds) for bronchopneumonia (27). A patch test was positive 2 months later for clindamycin phosphate but negative for penicillin, amoxicillin, ampicillin, and erythromycin. Prick tests and intradermal tests were all negative. Oral rechallenge with clindamycin phosphate 300 mg was positive.

In 300 subjects with acne in a multicenter, randomized, investigator-blinded comparison of adapalene gel 0.1% plus clindamycin topical solution 1% versus clindamycin topical solution 1% alone, nine subjects (four in the combination group and five in the monotherapy group) had 19 local adverse events, mostly erythema (n = 8) followed by scaling (n = 3), burning (n = 3), pruritus (n = 2), and stinging, papules, and pustules (n = 1 each) (45). Seven patients withdrew from the study during the initial treatment phase owing to local adverse events; all had erythema in combination with burning, scaling, pruritus, or papules and pustules, five in the monotherapy group and two in the combination group.

In a retrospective study of drug-induced reactions among children with various ear, nose, and throat (ENT) infections, 7 of 62 reported cases with skin reactions were attributed to clindamycin (46).

Acute, generalized, exanthematous pustulosis has been associated with clindamycin (47).

Acute generalized exanthematous pustulosis (AGEP) involves numerous non-follicular sterile pustular lesions associated with fever above 38°C, neutrophilic leukocytosis, an intensely pruritic rash, and in the later stages desquamation. AGEP was associated with clindamycin in a 38-year-old woman who was also undergoing therapy for systemic lupus erythematosus (48) and in two elderly patients taking clindamycin 300 mg qds (49).

- Erythroderma occurred in 73-year-old woman who developed intense pruritus, malaise, and chills 7 days after treatment with intravenous clindamycin (600 mg 6-hourly). Two years later she had a similar reaction with more rapid onset (48 hours) after treatment with aztreonam (500 mg 8-hourly) (50).

In a retrospective study of drug reactions in children with various ear, nose, and throat infections, 7 of 62 reported cases of skin reactions were attributed to clindamycin (51).

Immunologic

Although the risk of drug hypersensitivity is increased in patients with AIDS, clindamycin hypersensitivity has been considered to be relatively uncommon, despite its widespread use, with rash developing in about 9% of patients. However, in a retrospective survey of 50 patients with AIDS recruited in a European multicenter study of treatment for *Toxoplasma* encephalitis, the incidence of rash in 26 patients given pyrimethamine plus clindamycin was 58%, compared with 75% in those given pyrimethamine plus sulfadiazine, a non-significant difference (52). Treatment was initially continued throughout the duration of hypersensitivity, and was tolerated in all patients taking pyrimethamine plus clindamycin, but had to be withdrawn in half of those taking pyrimethamine plus sulfadiazine. Stevens–Johnson syndrome developed in two patients and fatal toxic epidermal necrolysis in one. Thus, the continuation of treatment despite a rash was more likely to succeed with pyrimethamine plus clindamycin but was potentially hazardous with pyrimethamine plus sulfadiazine.

- A 47-year-old man with buccal cancer who was undergoing radical neck dissection developed life-threatening anaphylactic shock after receiving an intravenous infusion of clindamycin 600 mg in 30 ml of lactated Ringer's solution over about 10 minutes through a peripheral vein (53). Within 3 minutes after the end of the infusion he developed bronchospasm, hypotension (systolic blood pressure under 40 mmHg), tachycardia, and wheals and erythema over almost all of the body but made an uneventful recovery with standard treatment.

In a prospective study, true-positive patch tests were seen in four of six patients with known clindamycin hypersensitivity, while 22 healthy controls were negative; there was one false positive and one false negative reaction (54).

Successful desensitization has been described in a 35-year-old woman who developed a generalized rash after taking clindamycin (600 mg 6-hourly) and pyrimethamine for 12 days for AIDS-associated cerebral toxoplasmosis; the rash resolved after withdrawal of clindamycin (55). Subsequent oral rechallenge was performed (without pretreatment with glucocorticoids or antihistamines), starting with three doses of 20 mg on day 1, 40 mg on day 2, 80 mg on day 3, and so on, until a dose of 600 mg qds was reached on day 7. A transient rash lasting 5 hours developed after the second dose of 600 mg. She remained free from adverse reactions for the duration of follow-up (13 months).

Infection risk

Superinfection with resistant strains of *Pseudomonas*, *Proteus*, or staphylococci has been observed with lincosamides. Suppression of *Bacteroides* in the intestinal flora may be related to the proliferation of *C. difficile*, which is important in causing pseudomembranous colitis. Excessive growth of *Candida* on the skin occurred when lincomycin was applied topically (7).

Long-Term Effects

Drug resistance

Staphylococcus aureus, pneumococci, Group A streptococci, and viridans streptococci acquire resistance in vitro to the lincosamides regularly, easily, and quickly (SED-8, 638). In endometrial cultures taken after clindamycin therapy, the occurrence of clindamycin-resistant anaerobic bacteria was significantly higher than before therapy (56). Their similar mechanism of action has been used to explain cross-resistance between the macrolide antibiotics and the lincosamides. Among erythromycin-resistant staphylococci, 50% of the isolated strains were also resistant to lincomycin (SED-6, 304). In patients who need long-term suppressive therapy, but who are allergic to penicillin, the development of such combined resistance of the oropharyngeal flora can be a serious clinical problem.

The erythromycin ribosomal methylase (erm) genes encode 23S ribosomal RNA methylases. This modification results in reduced binding of all known macrolides, lincosamides, and streptogramin B to the ribosome (MLS resistance). Novel triazine-containing methyltransferase inhibitors that may reverse erm-mediated resistance are under development (57).

The MLS resistance was found in four of 137 consecutive clinical isolates of *Streptococcus pyogenes* (58). In two, both ermB and ermTR genes were present. In the other two, these genes were not identified, suggesting a new mechanism of high-level resistance to these antibiotics. Erm genes were detected in 45% of 173 strains of *Streptococcus pneumoniae* isolated from surveillance studies in day-care centers in Central Italy (59). From 387 clinical strains of erythromycin-resistant strains of *S. pyogenes* isolated in Italian laboratories from 1995 to 1998, 31% were assigned to the inducible and 17% to the constitutive MLS resistance phenotype (60). Resistance to erythromycin increased to 33% in 1997 among community-acquired isolates of *S. pneumoniae* from central Italy (61). Most carried an erm gene and were also resistant to clindamycin. Regulatory regions located upstream of the erm genes were amplified and sequenced in clinical isolates of enterococci and streptococci with either inducible or constitutive resistance. Expression of constitutive resistance in two strains of *S. pneumoniae* and *Enterococcus faecalis* could be accounted for by a large deletion or a DNA duplication within the regulatory regions respectively (62). In 294 macrolide-, lincosamide-, and/or streptogramin-resistant clinical isolates of *S. aureus* and coagulase-negative staphylococci isolated in 1995 from 32 French hospitals, ermA or ermC genes were found in 88% (63). Genes related to linA/linA' and conferring resistance to lincomycin were detected in one strain of *S. aureus* and 7 strains of coagulase-negative staphylococci.

A resistance gene, named linB, which encodes a lincosamide nucleotidyltransferase catalysing the adenylation of the hydroxyl group in position 3 of the molecules of lincomycin and clindamycin has been characterized in a clinical isolate of *Enterococcus faecium* (64). Expression of linB was also observed in *Escherichia coli* and *S. aureus*, and the spread of this gene in other clinical isolates of *E. faecium* has been suggested.

In *Brachyspira hyodysenteriae* (formerly *Serpulina hyodysenteriae*) isolates, resistance to clindamycin was associated with a transversion mutation in the nucleotide position homologous with position 2058 of the *E. coli* 23S rRNA gene (65).

There was a significant increase in the rate of resistance in clinical strains of *Bacteroides fragilis* from Canada (20% in 1997) (66). This trend was also observed in a prospective multicenter survey from the USA (67).

Among *P. acnes* isolated from acne lesions, a resistance level of 4% (of 50 strains) and 6.8% (of 70 strains) were found in Japanese and German studies respectively (68,69).

Of 302 clinical isolates of *S. pyogenes* from Portugal, 108 were resistant to erythromycin, and 86 also had a constitutive resistance to clindamycin (MLSB phenotype) (70). Four isolates had a phenotype characterized by low-level erythromycin resistance and high-level clindamycin resistance. In another European study of 286 *S. pneumoniae* strains, 7% were resistant to penicillin, and 35% were also resistant to clindamycin (71). Of 3205 group A streptococcal strains from Canada, only 18 and 2 strains respectively showed inducible and constitutive resistance to clindamycin (72). Among 180 strains of the *Streptococcus milleri* group isolated in Spain, 17% were resistant to clindamycin (73).

An increase in resistance to clindamycin has been found in group B streptococcal strains causing neonatal infection (74).

Second-Generation Effects

Pregnancy

In pregnant women with bacterial vaginosis, a 7-day regimen of 2% vaginal clindamycin cream was effective and did not alter the rates of preterm deliveries or peripartum infections (75,76).

In a randomized study in pregnant women with abnormal vaginal flora and bacterial vaginosis, clindamycin 300 mg bd for 5 days reduced the rate of late miscarriage and spontaneous preterm birth. Adverse effects included any gastrointestinal upset (n=5), nausea, vomiting, diarrhea, abdominal pains, or a combination of these), rashes (n=1); vulvo-vaginal candidiasis (n = 1), and headache (n = 4) (77).

Fetotoxicity

Although lincomycin penetrates the fetal circulation, no fetal abnormalities were related to lincomycin administration in 302 women who had completed a course of lincomycin therapy (500 mg every 6 hours) for 1 week; all three trimesters of pregnancy were included (78).

Drug Administration

Drug formulations

Whereas only about 4% of clindamycin from a vaginal cream was systemically absorbed, systemic absorption was in the range of 30% when clindamycin was intravaginally administered as a phosphate ovule (79).

Drug–Drug Interactions

Ciclosporin

In two patients who had lung transplants and were taking an immunosuppressive regimen that included ciclosporin, the addition of clindamycin (600 mg tds) resulted in a significant reduction in ciclosporin concentrations (80).

Ganglion blocking agents

Clindamycin may potentiate the effects of some ganglion blocking agents (81).

Macrolides

Antagonism between the lincosamides and the macrolide antibiotics has been observed in vitro and was explained by binding to the same subunit of bacterial ribosomes (82). This mechanism of bacteriostatic action also suggests that the lincosamides might prevent the bactericidal action of the penicillins and the cephalosporins.

Neuromuscular blocking drugs

The lincosamides have prejunctional and postjunctional effects, the principal action probably being on the muscle. This blockade is difficult to reverse with cholinesterase inhibitors or calcium.

Clindamycin and lincomycin potentiate the action of non-depolarizing neuromuscular blocking drugs, such as pancuronium and D-tubocurarine. The lincosamide-induced block cannot be reliably reversed pharmacologically (83).

Paclitaxel

In 16 patients with advanced ovarian cancer clindamycin altered paclitaxel pharmacokinetics (46). However, the changes were minimal and of questionable clinical relevance.

References

1. Kester KE, Turiansky GW, McEvoy PL. Nodular cutaneous microsporidiosis in a patient with AIDS and successful treatment with long-term oral clindamycin therapy. Ann Intern Med 1998;128(11):911–4.
2. Adehossi E, Parola P, Foucault C, Delmont J, Brouqui P, Badiaga S, Ranque S. Three-day quinine-clindamycin treatment of uncomplicated falciparum malaria imported from the tropics. Antimicrob Agents Chemother 2003;47:1173.
3. Toma E, Thorne A, Singer J, Raboud J, Lemieux C, Trottier S, Bergeron MG, Tsoukas C, Falutz J, Lalonde R, Gaudreau C, Therrien RCTN-PCP Study Group. Clindamycin with primaquine vs. trimethoprim–sulfamethoxazole therapy for mild and moderately severe *Pneumocystis carinii* pneumonia in patients with AIDS: a multicenter, double-blind, randomized trial (CTN 004). Clin Infect Dis 1998;27(3):524–30.
4. Krause PJ, Lepore T, Sikand VK, Gadbaw J Jr, Burke G, Telford SR 3rd, Brassard P, Pearl D, Azlanzadeh J, Christianson D, McGrath D, Spielman A. Atovaquone and azithromycin for the treatment of babesiosis. N Engl J Med 2000;343(20):1454–8.
5. Paavonen J, Mangioni C, Martin MA, Wajszczuk CP. Vaginal clindamycin and oral metronidazole for bacterial vaginosis: a randomized trial. Obstet Gynecol 2000;96(2):256–60.
6. Leyden JJ, Berger RS, Dunlap FE, Ellis CN, Connolly MA, Levy SF. Comparison of the efficacy and safety of a combination topical gel formulation of benzoyl peroxide and clindamycin with benzoyl peroxide, clindamycin and vehicle gel in the treatments of acne vulgaris. Am J Clin Dermatol 2001;2(1):33–9.
7. Herrell WE. Considerations of toxicity of lincomycin. In: Herrell WE, editor. Lincomycin. Chicago: Modern Scientific Publications, 1969:147.
8. Mazur N, Greenberger PA, Regalado J. Clindamycin hypersensitivity appears to be rare. Ann Allergy Asthma Immunol 1999;82(5):443–5.
9. Klainer AS. Clindamycin. Med Clin North Am 1987;71(6):1169–75.
10. Zehnder D, Kunzi UP, Maibach R, Zoppi M, Halter F, Neftel KA, Muller U, Galeazzi RL, Hess T, Hoigne R. Die Häufigkeit der Antibiotika-assoziierten Kolitis bei hospitalisierten Patienten der Jahre 1974–1991 im "Comprehensive Hospital Drug Monitoring" Bern/St. Gallen. [Frequency of antibiotics-associated colitis in hospitalized patients in 1974–1991 in "Comprehensive Hospital Drug Monitoring", Bern/St. Gallen.] Schweiz Med Wochenschr 1995;125(14):676–83.
11. Lochmann O, Kohout P, Vymola F. Anaphylactic shock following the administration of clindamycin. J Hyg Epidemiol Microbiol Immunol 1977;21(4):441–7.
12. Vacek V, Tesarova-Magrova J, Stafova J. Prevention of adverse reactions in therapy with high doses of lincomycin. Arzneimittelforschung 1970;20(1):99–101.
13. Gabel A, Schymik G, Mehmel HC. Ventricular fibrillation due to long QT syndrome probably caused by clindamycin. Am J Cardiol 1999;83(5):813–5A11.
14. Aucoin P, Beckner RR, Gantz NM. Clindamycin-induced cardiac arrest. South Med J 1982;75(6):768.
15. Hall CM, Milligan DW, Berrington J. Probable adverse reaction to a pharmaceutical excipient. Arch Dis Child Fetal Neonatal Ed 2004;89(2):F184.
16. al Ahdal O, Bevan DR. Clindamycin-induced neuromuscular blockade. Can J Anaesth 1995;42(7):614–7.
17. Fiekers JF. Sites and mechanisms of antibiotic-induced neuromuscular block: a pharmacological analysis using quantal content, voltage clamped end-plate currents and single channel analysis. Acta Physiol Pharmacol Ther Latinoam 1999;49(4):242–50.
18. Best JA, Marashi AH, Pollan LD. Neuromuscular blockade after clindamycin administration: a case report. J Oral Maxillofac Surg 1999;57(5):600–3.
19. Schulze J, Toepfer M, Schroff KC, Aschhoff S, Remien J, Muller-Felber W, Endres S. Clindamycin and nicotinic neuromuscular transmission. Lancet 1999;354(9192):1792–3.
20. Ben Zina Z, Abid D, Kharrat W, Chaker N, Aloulou K, Chaabouni M. Intérêt du traitement par clindamycine en sous conjonctivale dans les retinochoroïdites toxoplasmiques. [Interest in treatment with subconjunctival clindamycin in toxoplasmic retinochoroiditis.] Tunis Med 2001;79(3):157–60.
21. Southern PM Jr. Lymphadenitis associated with the administration of clindamycin. Am J Med 1997;103(2):164–5.
22. Petersen KU, Jaspersen D. Medication-induced oesophageal disorders. Expert Opin Drug Saf 2003;2:495–507.
23. Rivera Vaquerizo PA, Santisteban Lopez Y, Blasco Colmenarejo M, Vicente Gutierrez M, Garcia Garcia V, Perez-Flores R. Clindamycin-induced esophageal ulcer. Rev Esp Enferm Dig 2004;96(2):143–5.
24. Jaspersen D. Drug-induced oesophageal disorders: pathogenesis, incidence, prevention and management. Drug Saf 2000;22(3):237–49.
25. Zerbib F. Les oesophagites médicamenteuses. Hepato-Gastro 1998;5:115–20.
26. Bott SJ, McCallum RW. Medication-induced oesophageal injury. Survey of the literature. Med Toxicol 1986;1(6):449–57.
27. Midtvedt T, Carlstedt-Duke B, Hoverstad T, Lingaas E, Norin E, Saxerholt H, Steinbakk M. Influence of peroral antibiotics upon the biotransformatory activity of the intestinal microflora in healthy subjects. Eur J Clin Invest 1986;16(1):11–7.
28. Barbut F, Decre D, Burghoffer B, Lesage D, Delisle F, Lalande V, Delmee M, Avesani V, Sano N, Coudert C, Petit JC. Antimicrobial susceptibilities and serogroups of clinical strains of *Clostridium difficile* isolated in France in 1991 and 1997. Antimicrob Agents Chemother 1999;43(11):2607–11.
29. Gorenek L, Dizer U, Besirbellioglu B, Eyigun CP, Hacibektasoglu A, Van Thiel DH. The diagnosis and treatment of *Clostridium difficile* in antibiotic-associated diarrhea. Hepatogastroenterology 1999;46(25):343–8.
30. Barreiro PM, Pintor E, Rosario Buron M, Diaz B, Valverde J, de la Torre F. Diarrea asociada a Clostridium

difficile. Estudio retrospectiro a un ano en un hospital terciario. [Diarrhea associated with *Clostridium difficile*. One-year retrospective study at a tertiary hospital.] Enferm Infecc Microbiol Clin 1998;16(8):359–63.

31. Meadowcroft AM, Diaz PR, Latham GS. *Clostridium difficile* toxin-induced colitis after use of clindamycin phosphate vaginal cream. Ann Pharmacother 1998;32(3):309–11.

32. Wistrom J, Norrby SR, Myhre EB, Eriksson S, Granstrom G, Lagergren L, Englund G, Nord CE, Svenungsson B. Frequency of antibiotic-associated diarrhoea in 2462 antibiotic-treated hospitalized patients: a prospective study. J Antimicrob Chemother 2001;47(1):43–50.

33. Samore MH. Epidemiology of nosocomial *Clostridium difficile* diarrhoea. J Hosp Infect 1999;43(Suppl):S183–90.

34. Johnson S, Samore MH, Farrow KA, Killgore GE, Tenover FC, Lyras D, Rood JI, DeGirolami P, Baltch AL, Rafferty ME, Pear SM, Gerding DN. Epidemics of diarrhea caused by a clindamycin-resistant strain of *Clostridium difficile* in four hospitals. N Engl J Med 1999;341(22):1645–51.

35. Boaz A, Dan M, Charuzi I, Landau O, Aloni Y, Kyzer S. Pseudomembranous colitis: report of a severe case with unusual clinical signs in a young nurse. Dis Colon Rectum 2000;43(2):264–6.

36. Parry MF, Rha CK. Pseudomembranous colitis caused by topical clindamycin phosphate. Arch Dermatol 1986;122(5):583–4.

37. Vikenes K, Lund-Tonnesen S, Schreiner A. Clostridium difficile-associated diarrhea after short term vaginal administration of clindamycin. Am J Gastroenterol 1999;94(7):1969–70.

38. Gray JE, Purmalis A, Purmalis B, Mathews J. Ultrastructural studies of the hepatic changes brought about by clindamycin and erythromycin in animals. Toxicol Appl Pharmacol 1971;19(2):217–33.

39. Warner GT, Plosker GL. Clindamycin/benzoyl peroxide gel: a review of its use in the management of acne. Am J Clin Dermatol 2002;3(5):349–60.

40. Weiss JW, Shavin J, Davis M. Preliminary results of a non-randomized, multicenter, open-label study of patient satisfaction after treatment with combination benzoyl peroxide/clindamycin topical gel for mild to moderate acne. Clin Ther 2002;24(10):1706–17.

41. Kim P, Younan N, Coroneo MT. Hypersensitivity reaction to intravitreal clindamycin therapy. Clin Experiment Ophthalmol 2002;30(2):147–8.

42. Ray RS, Mehrotra S, Shankar U, Babu GS, Joshi PC, Hans RK. Evaluation of UV-induced superoxide radical generation potential of some common antibiotics. Drug Chem Toxicol 2001;24(2):191–200.

43. Paquet P, Schaaf-Lafontaine N, Pierard GE. Toxic epidermal necrolysis following clindamycin treatment. Br J Dermatol 1995;132(4):665–6.

44. de Kort WJ, de Groot AC. Clindamycin allergy presenting as rosacea. Contact Dermatitis 1989;20(1):72–3.

45. Zhang JZ, Li LF, Tu YT, Zheng J. A successful maintenance approach in inflammatory acne with adapalene gel 0.1% after an initial treatment in combination with clindamycin topical solution 1% or after monotherapy with clindamycin topical solution 1%. J Dermatolog Treat 2004;15(6):372–8.

46. Rallis E, Balatsouras DG, Kouskoukis C, Verros C, Homsioglou E. Drug eruptions in children with ENT infections. Int J Pediatr Otorhinolaryngol 2006;70(1):53–7.

47. Schwab RA, Vogel PS, Warschaw KE. Clindamycin-induced acute generalized exanthematous pustulosis. Cutis 2000;65(6):391–3.

48. Kapoor R, Flynn C, Heald PW, Kapoor JR. Acute generalized exanthematous pustulosis induced by clindamycin. Arch Dermatol 2006;142(8):1080–1.

49. Valois M, Phillips EJ, Shear NH, Knowles SR. Clindamycin-associated acute generalized exanthematous pustulosis. Contact Dermatitis 2003;48:169.

50. Gonzalo-Garijo MA, de Argila D. Erythroderma due to aztreonam and clindamycin. J Investig Allergol Clin Immunol 2006;16(3):210–1.

51. Rallis E, Balatsouras DG, Kouskoukis C, Verros C, Homsioglou E. Drug eruptions in children with ENT infections. Int J Pediatr Otorhinolaryngol 2006;70(1):53–7.

52. Caumes E, Bocquet H, Guermonprez G, Rogeaux O, Bricaire F, Katlama C, Gentilini M. Adverse cutaneous reactions to pyrimethamine/sulfadiazine and pyrimethamine/clindamycin in patients with AIDS and toxoplasmic encephalitis. Clin Infect Dis 1995;21(3):656–8.

53. Chiou CS, Lin SM, Lin SP, Chang WG, Chan KH, Ting CK. Clindamycin-induced anaphylactic shock during general anesthesia. J Chin Med Assoc 2006;69(11):549–51.

54. Lammintausta K, Tokola R, Kalimo K. Cutaneous adverse reactions to clindamycin: results of skin tests and oral exposure. Br J Dermatol 2002;146(4):643–8.

55. Marcos C, Sopena B, Luna I, Gonzalez R, de la Fuente J, Martinez-Vazquez C. Clindamycin desensitization in an AIDS patient. AIDS 1995;9(10):1201–2.

56. Ohm-Smith MJ, Sweet RL, Hadley WK. Occurrence of clindamycin-resistant anaerobic bacteria isolated from cultures taken following clindamycin therapy. Antimicrob Agents Chemother 1986;30(1):11–4.

57. Hajduk PJ, Dinges J, Schkeryantz JM, Janowick D, Kaminski M, Tufano M, Augeri DJ, Petros A, Nienaber V, Zhong P, Hammond R, Coen M, Beutel B, Katz L, Fesik SW. Novel inhibitors of Erm methyltransferases from NMR and parallel synthesis. J Med Chem 1999;42(19):3852–9.

58. Portillo A, Lantero M, Gastanares MJ, Ruiz-Larrea F, Torres C. Macrolide resistance phenotypes and mechanisms of resistance in *Streptococcus pyogenes* in La Rioja, Spain. Int J Antimicrob Agents 1999;13(2):137–40.

59. Latini L, Ronchetti MP, Merolla R, Merolla R, Guglielmi F, Bajaksouzian S, Villa MP, Jacobs MR, Ronchetti R. Prevalence of mefE, erm and tet(M) genes in *Streptococcus pneumoniae* strains from Central Italy. Int J Antimicrob Agents 1999;13(1):29–33.

60. Giovanetti E, Montanari MP, Mingoia M, Varaldo PE. Phenotypes and genotypes of erythromycin-resistant *Streptococcus pyogenes* strains in Italy and heterogeneity of inducibly resistant strains. Antimicrob Agents Chemother 1999;43(8):1935–40.

61. Oster P, Zanchi A, Cresti S, Lattanzi M, Montagnani F, Cellesi C, Rossolini GM. Patterns of macrolide resistance determinants among community-acquired *Streptococcus pneumoniae* isolates over a 5-year period of decreased macrolide susceptibility rates. Antimicrob Agents Chemother 1999;43(10):2510–2.

62. Rosato A, Vicarini H, Leclercq R. Inducible or constitutive expression of resistance in clinical isolates of streptococci and enterococci cross-resistant to erythromycin and lincomycin. J Antimicrob Chemother 1999;43(4):559–62.

63. Lina G, Quaglia A, Revexdy ME, Lederq R, Vandenesch F, Etienne J. Distribution of genes encoding resistance to masrolides, lincosamedes, and streptogramins among staphylococci. Antimicrob Agents Chemother 1999;43(5):1062–6.

64. Bozdogan B, Berrezouga L, Kuo MS, Yurek DA, Farley KA, Stockman BJ, Leclercq R. A new resistance gene, linB, conferring resistance to lincosamides by nucleotidylation in *Enterococcus faecium* HM1025. Antimicrob Agents Chemother 1999;43(4):925–9.

65. Karlsson M, Fellstrom C, Heldtander MU, Johansson KE, Franklin A. Genetic basis of macrolide and lincosamide resistance in *Brachyspira* (*Serpulina*) hyodysenteriae. FEMS Microbiol Lett 1999;172(2):255–60.

66. Labbe AC, Bourgault AM, Vincelette J, Turgeon PL, Lamothe F. Trends in antimicrobial resistance among clinical isolates of the *Bacteroides fragilis* group from 1992 to 1997 in Montreal, Canada. Antimicrob Agents Chemother 1999;43(10):2517–9.

67. Snydman DR, Jacobus NV, McDermott LA, Supran S, Cuchural GJ Jr, Finegold S, Harrell L, Hecht DW, Iannini P, Jenkins S, Pierson C, Rihs J, Gorbach SL. Multicenter study of in vitro susceptibility of the *Bacteroides fragilis* group, 1995 to 1996, with comparison of resistance trends from 1990 to 1996. Antimicrob Agents Chemother 1999;43(10):2417–22.

68. Fluhr JW, Gloor M, Dietz P, Hoffler U. In Vitro activity of 6 antimicrobials against propionibacteria isolates from untreated acne papulopustulosa. Zentralbl Bakteriol 1999;289(1):53–61.

69. Kurokawa I, Nishijima S, Kawabata S. Antimicrobial susceptibility of *Propionibacterium acnes* isolated from acne vulgaris. Eur J Dermatol 1999;9(1):25–8.

70. Melo-Cristino J, Fernandes ML. *Streptococcus pyogenes* isolated in Portugal: macrolide resistance phenotypes and correlation with T types. Portuguese Surveillance Group for the Study of Respiratory Pathogens. Microb Drug Resist 1999;5(3):219–25.

71. Fluit AC, Schmitz FJ, Jones ME, Acar J, Gupta R, Verhoef J. Antimicrobial resistance among community-acquired pneumonia isolates in Europe: first results from the SENTRY antimicrobial surveillance program 1997. SENTRY Participants Group. Int J Infect Dis 1999;3(3):153–6.

72. De Azavedo JC, Yeung RH, Bast DJ, Duncan CL, Borgia SB, Low DE. Prevalence and mechanisms of macrolide resistance in clinical isolates of group A streptococci from Ontario, Canada. Antimicrob Agents Chemother 1999;43(9):2144–7.

73. Limia A, Jimenez ML, Alarcon T, Lopez-Brea M. Five-year analysis of antimicrobial susceptibility of the *Streptococcus milleri* group. Eur J Clin Microbiol Infect Dis 1999;18(6):440–4.

74. Morales WJ, Dickey SS, Bornick P, Lim DV. Change in antibiotic resistance of group B streptococcus: impact on intrapartum management. Am J Obstet Gynecol 1999;181(2):310–4.

75. Kekki M, Kurki T, Pelkonen J, Kurkinen-Raty M, Cacciatore B, Paavonen J. Vaginal clindamycin in preventing preterm birth and peripartal infections in asymptomatic women with bacterial vaginosis: a randomized, controlled trial. Obstet Gynecol 2001;97(5 Pt 1):643–8.

76. Vermeulen GM, van Zwet AA, Bruinse HW. Changes in the vaginal flora after two percent clindamycin vaginal cream in women at high risk of spontaneous preterm birth. BJOG 2001;108(7):697–700.

77. Ugwumadu A, Manyonda I, Reid F, Hay P. Effect of early oral clindamycin on late miscarriage and preterm delivery in asymptomatic women with abnormal vaginal flora and bacterial vaginosis: a randomised controlled trial. Lancet 2003;361:983–8.

78. Mickal A, Dildy GA, Miller HJ. Lincomycin in the treatment of cervicitis and vagini in pregnancy. South Med J 1966;59(5):567–70.

79. Borin MT, Ryan KK, Hopkins NK. Systemic absorption of clindamycin after intravaginal administration of clindamycin phosphate ovule or cream. J Clin Pharmacol 1999;39(8):805–10.

80. Thurnheer R, Laube I, Speich R. Possible interaction between clindamycin and cyclosporin. BMJ 1999;319(7203):163.

81. Konopka LM, Neel DS, Parsons RL. Clindamycin-induced alteration of ganglionic function. II. Effect of nicotinic receptor-channel function. Brain Res 1988;458(2):278–84.

82. Ruiz NM, Ramirez-Ronda CH. Tetracyclines, macrolides, lincosamides & chloramphenicol. Bol Asoc Med P R 1990;82(1):8–17.

83. Marshall IG, Henderson F. Drug interactions at the neuromuscular junction. Clin Anaesthesiol 1985;3:261.

Lomefloxacin

See also Fluoroquinolones

General Information

Lomefloxacin is a fluoroquinolone antibacterial drug with actions and uses similar to those of ciprofloxacin.

Organs and Systems

Gastrointestinal

In a multicenter, prospective, randomized study of oral lomefloxacin 400 mg/day in 182 patients with chronic bacterial prostatitis, the most frequent adverse events were gastrointestinal disorders (1).

Skin

In a retrospective study the data on fluoroquinolones and other antibacterial drugs were obtained from a spontaneous reporting system database. Lomefloxacin was associated with the reporting rate from 196 reports/daily defined dose/1000 inhabitants/day, and the most frequent were phototoxic reactions (2).

Photosensitivity was found in 44 (1.03%) of 4276 patients treated with lomefloxacin in Japan. Most cases were not severe and improved after withdrawal. Risk factors for a sensitivity reaction were age over 60 years with concomitant diseases and complications, total amount of lomefloxacin over 20 g, treatment for longer than 30 days, and previous treatment with a quinolone (3).

In eight patients (mean age 69 years) with eczematous or acute sunburn-like lesions in photo-exposed areas, who took lomefloxacin for 1 week to several months, phototoxicity appeared to be the main mechanism of photosensitivity, particularly in older patients with concomitant diseases and long-term use of the drug (4).

Long-Term Effects

Mutagenicity

Lomefloxacin was a weak clastogen in mouse bone marrow cells and non-mutagenic in germ cells (5).

Drug-Drug Interactions

Fluconazole

There have been reports that fluconazole can prolong the QT interval, and co-administration of fluconazole with levofloxacin may further increase this risk (6).

References

1. Naber KGEuropean Lomefloxacin Prostatitis Study Group. Lomefloxacin versus ciprofloxacin in the treatment of chronic bacterial prostatitis. Int J Antimicrob Agents 2002;20(1):18–27.
2. Leone R, Venegoni M, Motola D, Moretti U, Piazzetta V, Cocci A, Resi D, Mozzo F, Velo G, Burzilleri L, Montanaro N, Conforti A. Adverse drug reactions related to the use of fluoroquinolone antimicrobials: an analysis of spontaneous reports and fluoroquinolone consumption data from three Italian regions. Drug Saf 2003;26:109–20.
3. Arata J, Horio T, Soejima R, Ohara K. Photosensitivity reactions caused by lomefloxacin hydrochloride: a multicenter survey. Antimicrob Agents Chemother 1998;42(12):3141–5.
4. Oliveira HS, Goncalo M, Figueiredo AC. Photosensitivity to lomefloxacin. A clinical and photobiological study. Photodermatol Photoimmunol Photomed 2000;16(3):116–20.
5. Singh AC, Kumar M, Jha AM. Genotoxicity of lomefloxacin—an antibacterial drug in somatic and germ cells of Swiss albino mice in vivo. Mutat Res 2003;535:35–42.
6. Gandhi PJ, Menezes PA, Vu HT, Rivera AL, Ramaswamy K. Fluconazole- and levofloxacin-induced torsades de pointes in an intensive care unit patient. Am J Health-Syst Pharm 2003;60:2479–83.

Macrolide antibiotics

General Information

The basic structure of the macrolide antibiotics is characterized by a lactonic cycle with two osidic chains, and they are classified according to the number of carbon atoms in the cycle: 14-membered macrolides (for example clarithromycin, dirithromycin, erythromycin, roxithromycin, troleandomycin), 15-membered macrolides (for example azithromycin), and 16-membered macrolides (for example josamycin, midecamycin, spiramycin).

The following macrolides are covered in separate monographs: azithromycin, clarithromycin, dirithromycin, erythromycin, josamycin, midecamycin, rokitamycin, roxithromycin, spiramycin, and troleandomycin (all rINNs).

The antibacterial activity of the macrolides is based on interference with protein synthesis by combining with a subunit of bacterial ribosomes. This generally results in a bacteriostatic effect. They cover a broad range of pathogens. Most are active against Gram-positive cocci and some Gram-negative bacteria, including *Campylobacter* and *Haemophilus*, but not most *Enterobacteriaceae*. However, they are very active against *Mycoplasma*, *Ureaplasma*, *Chlamydia*, *Legionella*, and *Coxiella*. Their activity against *Mycobacterium avium* complex, *Cryptosporidia*, and *Toxoplasma* is variable (1). The macrolide antibiotics have established themselves in the treatment of community-acquired infections. In addition, some have found new indications in the treatment of opportunistic infections in HIV-infected patients (1).

Actions and uses in non-infective conditions

Erythromycin is a motilin receptor agonist (2–4). Azithromycin also produced a significant increase in postprandial antral motility (5). Clarithromycin is also prokinetic, as shown in 16 patients with functional dyspepsia and *Helicobacter pylori* gastritis (6). For this reason, the macrolides have been used to treat gastroparesis (7).

Azithromycin may be useful in reducing ciclosporin-induced gingival hyperplasia in renal transplant recipients (8).

A beneficial effect of low doses of clarithromycin on sputum rheology has been reported in patients with chronic pulmonary diseases, such as chronic bronchitis (9).

In a murine model of virus-induced lung injury, erythromycin significantly improved survival rate (10). This may be explained by inhibition of inflammatory-cell responses and suppression of nitric oxide overproduction in the lungs of the virus-infected mice.

In 202 patients with unstable angina pectoris, roxithromycin prevented death and re-infarction for at least 6 months after initial treatment (11). However, these findings could not be confirmed in another study in 302 patients with coronary heart disease and a seropositive reaction to *Chlamydia pneumoniae* who were treated with azithromycin. While global tests of markers of inflammation improved, there were no differences in antibody titers and clinical events (12).

General adverse effects

Macrolides are in general well tolerated and are used over wide dosage ranges. The rates of adverse reactions are dose-related, as exemplified by the rate of adverse reactions to clarithromycin, which was higher with 2000 or 4000 mg in two doses per day than with 500 or 1000 mg/day (1). Severe toxicity is very rarely observed with macrolides. Most adverse reactions are rated as either mild or moderate, regardless of the macrolide used. Among 245 patients who were hospitalized because of toxic epidermal necrolysis or Stevens–Johnson syndrome, 6 patients had had exposure to macrolides within the week before the onset of the illness (13). The following antibiotics were used: erythromycin, roxithromycin (14), and spiramycin (14). The results of a case-control study suggest that macrolides do not pose an excess risk of this toxic complication (13). Anaphylactic reactions to macrolides are exceedingly uncommon, but anaphylaxis and acute respiratory distress have been reported (15,16). Skin tests with erythromycin were positive for the immediate and/or delayed types of hypersensitivity (17). Tumor-inducing effects have not been reported.

Table 1 and Table 2 summarize the frequencies of adverse reactions and premature withdrawal of the most widely used macrolides.

Pharmacoeconomics

The pharmacoeconomic impact of adverse effects of antimicrobial drugs is enormous. Antibacterial drug reactions account for about 25% of adverse drug reactions. The adverse effects profile of an antimicrobial agent can contribute significantly to its overall direct costs (monitoring costs, prolonged hospitalization due to complications or treatment failures) and indirect costs (quality of life, loss of productivity, time spent by families and patients receiving medical care). In one study an adverse event in a hospitalized patient was associated on average with an excess of 1.9 days in the length of stay, extra costs of $US2262 (1990–3 values), and an almost two-fold increase in the risk of death. In the outpatient setting, adverse drug reactions result in 2–6% of hospitalizations, and most of them were thought to be avoidable if appropriate interventions had been taken. In a review, economic aspects of

Table 1 Frequencies of adverse reactions to macrolide antibiotics

Drug (number of patients)	Adverse reactions (%)	Premature withdrawal (%)	Reference
Erythromycin (112)	33	18.7	(18)
Clarithromycin (4291)	19.6	3.5*	(19)
Roxithromycin (2917)	4.1	0.9	(20)
Azithromycin (3995)	12	0.7	(21)
Dirithromycin (2825)	33.1	3.1	(22)

*at high doses (2000 mg/day) withdrawal rate 22%.

Table 2 Adverse effects of macrolide antibiotics

Drug	Gastro-intestinal (%)	Nervous system (%)	Skin(%)	Reference
Erythromycin (n = 112)	27	4	1	(18)
Clarithromycin (n = 96)	6	2	0	(19)
Roxithromycin (n = 2917)	3.9	0.4	0.7	(20)
Azithromycin (n = 3995)	9.6	1.3	0.6	(21)
Dirithromycin (n = 4263)	5.6	4.5	0.4	(22)

antibacterial therapy with macrolides have been summarized and critically evaluated (23).

Organs and Systems

Cardiovascular

Of the currently available antimicrobial classes, the macrolides appear to be associated with the greatest degree of QT interval prolongation and risk of torsade de pointes (24). Macrolides with at least one published report of torsade de pointes include azithromycin, clarithromycin, erythromycin, roxithromycin, spiramycin, and troleandomycin. However, cardiovascular reactions are rare if macrolide antibiotics are used in the absence of susceptibility factors, which include drug interactions, increasing age, female sex, concomitant diseases, and co-morbidity (25).

Respiratory

In an animal model, acute lung injury was inhibited by pretreatment with clarithromycin or roxithromycin, which significantly ameliorated bleomycin-induced increases in the total cell and neutrophil counts in bronchoalveolar lavage fluids and wet lung weight (26). Pretreatment with clarithromycin or roxithromycin also suppressed inflammatory cell infiltration and interstitial lung edema. Pretreatment with azithromycin was much less effective.

Nervous system

Table 2 lists the rates of adverse events affecting the nervous system attributed to erythromycin and newer macrolides.

Sensory systems

Eyes
Bilateral ischemic optic neuropathy can develop secondary to macrolides (27).

Ears
The cochlear toxicity of systemic macrolides, azithromycin, clarithromycin, and erythromycin, has been investigated in guinea pigs by measuring transiently evoked otoacoustic emissions (28). A single intravenous dose of erythromycin 125 mg/kg caused no change in evoked otoacoustic emissions, whereas oral azithromycin 45 mg/kg and intravenous clarithromycin 75 mg/kg reversibly reduced the emission response. This could have been caused by transient dysfunction of the outer hair cells.

Taste
The bitterness of 18 different antibiotic and antiviral drug formulations was evaluated using an artificial taste sensor (29). Seven of the formulations had a bitterness intensity exceeding 1.0 in gustatory sensation tests and were therefore assumed to have an unpleasant taste to children. In the case of three macrolide antibiotic formulations containing erythromycin, clarithromycin, and azithromycin, the bitterness intensities of suspensions in acidic sports

drinks were much higher than the corresponding scores of suspensions in water.

Psychological, psychiatric

Although there is no evidence that neuropsychiatric complications of macrolides develop more readily in uremic patients, several factors may predispose toward these adverse effects, such as reduced drug clearance, altered plasma protein binding, different penetration of drug across the blood–brain barrier, and an increased propensity for drug interactions.

Two women, aged 49 and 50 years, developed altered mental status a few days after starting to take clarithromycin for eradication of *H. pylori* (30). There was incoherent speech with perseveration, inability to sustain attention, impaired ability to comprehend, coprolalia, euphoria, restlessness, visual hallucinations, anxiety, and inappropriate affect. Similarly, in three cases, a 46-year-old man, a 39-year-old woman, and a 4-year-old boy, treatment with clarithromycin was followed by nervous system and psychiatric symptoms that included euphoria, insomnia, aggressive behavior, hyperactivity, and emotional lability (31).

Hematologic

Clarithromycin and roxithromycin slightly inhibited the down-regulation of L-selectin expression on neutrophils induced by interleukin-8 stimulation (32). Furthermore, clarithromycin strongly inhibited the interleukin-8-induced up-regulation of the expression of Mac-1, an adhesion molecule, on neutrophils.

Hematological changes with macrolides are very rare. Isolated instances of neutropenia are occasionally reported (33).

Gastrointestinal

The gastrointestinal adverse effects are the most common untoward effects of the macrolides (Table 2). Nausea and vomiting associated with abdominal pain and occasionally diarrhea can be minor and transitory or, in a small percentage of patients, become severe enough to result in premature withdrawal. The rate of these adverse effects varies among the different antibiotics. In general, newer macrolides, such as azithromycin, clarithromycin, or roxithromycin, are better tolerated and cause fewer adverse effects than erythromycin.

Erythromycin is a motilin receptor agonist (2–4). This mechanism may be at least partly responsible for the gastrointestinal adverse effects of macrolides.

Based on observations in dogs and rabbits, clarithromycin is significantly less potent than azithromycin and erythromycin as an agonist for stimulation of smooth muscle contraction (34). A lower rate of gastrointestinal adverse events would therefore be expected with clarithromycin than with azithromycin (Table 2). Since most of these data were compiled from several studies, and since most have not been obtained by direct comparison of the various macrolides in single studies, the rates should be interpreted with caution. They most probably provide

only an approximate indication of the rate of adverse events. Small differences in rates between individual macrolides will in most cases not be clinically useful indicators of the true risk for the occurrence of adverse events.

In contrast to the macrolides mentioned above, macrolides with a 16-membered lactone ring (acetylspiramycin, josamycin, leucomycin, midecamycin, rokitamycin, spiramycin, tylosin) have little if any motor-stimulating effects (35,36).

Drug-induced esophagitis is rare, accounting for about 1% of all cases of esophagitis. An incidence of 3.9 in 100 000 has been reported. After the first description, there have been more than 250 observations, with more than 50 different drugs. Among those, the principal antibiotics included tetracyclines (doxycycline, metacycline, minocycline, oxytetracycline, and tetracycline), penicillins (amoxicillin, cloxacillin, penicillin V, and pivmecillinam), clindamycin, co-trimoxazole, erythromycin, lincomycin, spiramycin, and tinidazole. Doxycycline alone was involved in one-third of all cases. Risk factors included prolonged esophageal passage, due to motility disorders, stenosis, cardiomegaly, the formulation, supine position during drug ingestion, and failure to use liquid to wash down the tablet. Direct toxic effects of the drug (pH, accumulation in epithelial cells, non-uniform dispersion) also seem to contribute to the development of drug-induced esophagitis (37).

Liver

Erythromycin can cause two different types of liver damage (38,39), benign increases in serum transaminases, which may or may not recur on rechallenge, and cholestatic hepatitis. Reports of intrahepatic cholestasis with azithromycin (40), clarithromycin (41,42), and josamycin (43) suggest that the newer macrolides are not free of this adverse effect, although the relative risks compared with erythromycin are unclear. Similar involvement of the liver has been seen with the ester of triacetyloleandomycin, but not with the unesterified antibiotic.

Macrolides such as josamycin, midecamycin, and spiramycin, which do not form stable complexes with cytochrome P450, rarely if ever cause cholestatic hepatitis.

Urinary tract

- A 77-year-old man taking regular captopril, furosemide, salbutamol inhaler, vitamin C, and nasal beclomethasone dipropionate took clarithromycin 250 mg bd for sinusitis and bronchitis and 5 days later developed abdominal pain and intermittent fever (44). Laboratory findings included raised serum creatinine and blood urea nitrogen, aspartate transaminase, amylase, lactate dehydrogenase, and creatine kinase (not of the MB isoenzyme). The cause of the non-oliguric renal insufficiency was diagnosed by renal biopsy as interstitial nephritis with eosinophilic infiltrates. During the course of illness he also developed thrombocytopenia.

Skin

Skin rashes and fixed drug eruptions can occur during treatment with various macrolides but are rare (under 1%) (17).

Immunologic

Immunomodulatory effects of macrolides have been repeatedly reported; for example suppression of the release of chemotactic mediators may be important for the clinical effect of roxithromycin in patients with chronic lower respiratory tract infections (45). Both clarithromycin and azithromycin altered cytokine production in human monocytes in vitro (46).

The suppressive activity of macrolide antibiotics on pro-inflammatory cytokine production has also been shown in human peripheral blood monocytes, in which roxithromycin inhibited the in vitro production of interleukin-1 beta and tumor necrosis factor alpha (47). It also suppressed cytokine production after a prolonged pretreatment period in mice. In another mouse model both roxithromycin and clarithromycin inhibited angiogenesis and enhanced the antitumor activity of some cytotoxic agents, suggesting a beneficial effect when combined with such drugs against solid tumors (48,49). Furthermore, growth suppression of human fibroblasts by roxithromycin has been demonstrated both in vitro and in vivo (50).

Together, these anti-inflammatory effects result in improved pulmonary functions and fewer airway infections (51,52,53,54), and for this reason the macrolides are sometimes used after lung transplantation.

Azithromycin has been associated with Churg–Strauss syndrome in a patient with atopy (55).

- A 46-year-old man with asthma was treated with oral roxithromycin 300 mg/day for 5 days for purulent rhinitis and 2 weeks later developed arthritis, mononeuritis multiplex, eosinophilia (64%), eosinophilic infiltrations in the bone marrow, raised IgE concentrations, and transient pulmonary infiltrates. Churg–Strauss syndrome was diagnosed.

A similar course of disease had occurred 1 year before, after the administration of azithromycin (56).

Leukocytoclastic vasculitis associated with clarithromycin has been reported in an 83-year-old woman who was treated for pneumonia. All her symptoms resolved after withdrawal and a short course of glucocorticoids (57).

- Henoch–Schönlein purpura developed in an 84-year-old Indian woman 10 days after she started to take clarithromycin (250 mg bd) for pneumonia (58). She was otherwise healthy and taking no regular medications. Histology confirmed a leukocytoclastic vasculitis of superficial vessels, with extravasation of erythrocytes, and direct immunofluorescence showed immunoglobulin A in superficial dermal vessels. Treatment with prednisone (1 mg/kg/day) was required. Most of the symptoms and signs resolved within a few days, but renal function remained impaired.

The authors identified two previous case reports of clarithromycin-induced leukocytoclastic vasculitis.

- A 39-year-old man developed acute angioedema and urticaria 6 hours after taking erythromycin base 500 mg in enteric-coated pellets for acute sinusitis (59). He remembered having taken erythromycin once before without any problem. He had no known allergies and was taking no regular medications, but he had had chemotherapy for non-Hodgkin's lymphoma several years earlier.

The authors identified five previous reports of erythromycin-associated urticarial reactions. However, it was not possible to exclude a reaction to the ingredients of the coated pellets.

Long-Term Effects

Drug tolerance

An area of increasing concern and clinical importance is the increasing macrolide resistance that has been reported over the last several years with some of the common pathogens, particularly *Streptococcus pneumoniae*, group A streptococci, and *Haemophilus influenzae*, and may result in failure of therapy of pneumonia, pharyngitis, and skin infections (60). High rates of resistance of several groups of streptococci to macrolides have been reported from all parts of the world (61–70).

Resistance to erythromycin can develop rapidly and is usually associated with bacterial cross-resistance to the other macrolide antibiotics, and also to the chemically unrelated lincomycins. Resistance has been detected in strains of staphylococci, Group A hemolytic streptococci, viridans streptococci, *Streptococcus pyogenes*, *Neisseria gonorrhoeae*, *Bacteroides fragilis*, and *Clostridium difficile* (71,72). It has tended to occur in hospitals, where either erythromycin or the lincosamides were used extensively, but can also result from multiple drug resistance when other antibiotics are used. Subinhibitory concentrations of erythromycin can cause resistance in staphylococci.

In combination with proton pump inhibitors and other antibiotics, macrolides are still successfully used for the eradication of *H. pylori* infection (73,74). However, resistance of *H. pylori* to macrolides has emerged in a number of countries. The first case of *H. pylori* resistance to clarithromycin has now also been documented in Denmark and follows increased use of this macrolide in eradication regimens (75).

Clinical isolates of *N. gonorrhoeae* with reduced susceptibility to azithromycin are commonly found in Uruguay, and one of the mechanisms involved included mutations in the mtrR gene (76).

Resistance of *H. pylori* to clarithromycin appears to have increased in proportion to clarithromycin use. Clarithromycin resistance arises through mutations that lead to base changes in 23S ribosomal RNA subunits. A rapid PCR hybridization assay with a sensitivity of 97% for the detection of clarithromycin resistance of strains of *H. pylori* has been described (77). Resistance to clarithromycin has a serious impact on the efficiency of eradicating regimens that include clarithromycin (78–80). The reported incidences of primary resistance of *H. pylori* to clarithromycin are 6.1% in the USA, 8% in Austria, 8.7% in Bulgaria, 9.5% in Japan, 10% in Spain, 11% in France, 13% in Nigeria, and 23% in Italy (81–88).

Among strains of *Enterococcus faecium*, resistance against tylosin was mainly detected in strains from poultry, but also in some strains from pork. Among strains of *E. faecium* and *Enterococcus faecalis* isolated from pigs and poultry in Denmark, resistance to tylosin was often observed among isolates from places in which these antimicrobials had been widely used, but rarely among isolates from places in which their use had been limited (89).

Second-Generation Effects

Teratogenicity

In a large case-control surveillance study from Hungary in 38 151 pregnant women, oral erythromycin during pregnancy did not present a detectable teratogenic risk to the fetus (90).

Susceptibility Factors

Cystic fibrosis

In a Cochrane database review there was a significant increase in mild adverse events (nausea, diarrhea, and wheezing) in patients with cystic fibrosis treated with macrolides (91).

Sjögren's syndrome

Allergic reactions to antimicrobials are frequent in patients with Sjögren's syndrome. They are especially susceptible to reactions to penicillins, cephalosporins, and sulfonamides, but reactions to macrolides and tetracyclines also seem to be over-represented in these patients (92).

Drug–Drug Interactions

Alfentanil

The metabolism of alfentanil, a potent short-acting narcotic, is inhibited by macrolide antibiotics, resulting in significant changes in half-life and clearance (93).

Antihistamines

Toxic effects of terfenadine and astemizole have been reported in patients taking concomitant macrolides, especially clarithromycin (94–96,123), typically resulting in prolongation of the QT interval and cardiac dysrhythmias (torsade de pointes) (118). The potential interaction of azithromycin with terfenadine has been evaluated in a randomized, placebo-controlled study in 24 patients who took terfenadine plus azithromycin or terfenadine plus placebo (97). However, azithromycin did not alter the pharmacokinetics of the active carboxylate metabolite of

terfenadine or the effect of terfenadine on the QT_c interval.

Only a modest increase in QT_c interval has been observed with concomitant use of erythromycin and the antihistamine ebastine (98).

Clarithromycin (500 mg bd for 10 days) significantly increased the steady-state maximum plasma concentration and the steady-state AUC of loratadine (10 mg/day for 10 days) (99). In contrast, the addition of loratadine did not affect the steady-state pharmacokinetics of clarithromycin or its active metabolite, 14(R)-hydroxyclarithromycin. No QT_c interval exceeded 439 ms in any subject.

Benzodiazepines

Interactions of macrolides with triazolam and midazolam are clinically important. Increased serum concentration, AUC, and half-life, and reduced clearance have been documented (100,101,125). These changes can result in clinical effects, such as prolonged psychomotor impairment, amnesia, or loss of consciousness (102). No interactions between ciprofloxacin and diazepam have been reported (103).

In a randomized, double-blind, pharmacokinetic–pharmacodynamic study, 12 volunteers took placebo or triazolam 0.125 mg orally, together with placebo, azithromycin, erythromycin, or clarithromycin. The apparent oral clearance of triazolam was significantly reduced by erythromycin and clarithromycin. The peak plasma concentration was correspondingly increased, and the half-life was prolonged. The effects of triazolam on dynamic measures were nearly identical when triazolam was given with placebo or azithromycin, but benzodiazepine agonist effects were enhanced by erythromycin and clarithromycin (205).

Erythromycin can increase concentrations of triazolam by inhibition of CYP3A4, and dosage reductions of 50% have been proposed if concomitant therapy is unavoidable (206).

CYP3A4 is mainly responsible for catalyzing the hydroxylation of miocamycin metabolites, which can alter the metabolism of concomitantly administered drugs by the formation of a metabolic intermediate complex with CYP450 or by competitive inhibition of CYP450 (207). This can cause excessive sedation with benzodiazepines such as triazolam.

Carbamazepine

Significant increases in serum carbamazepine concentrations due to reduced clearance (104) and prolonged half-life (105,129,130) can result in confusion, somnolence, ataxia, vertigo, nausea, and vomiting in patients taking macrolides (107,131,132). Toxicity can occur rapidly after addition of the macrolide and abate quickly on withdrawal (133). However, a retrospective analysis of 3995 patients treated with azithromycin did not show any pharmacokinetic interaction in patients who were also taking carbamazepine (134,135).

- Carbamazepine toxicity, with dizziness, lethargy, and nystagmus, developed in a 17-year-old boy two days after he started to take clarithromycin 500 mg/day (136). His serum carbamazepine concentration (previously acceptable) rose, but returned to within the target range after withdrawal of clarithromycin.

Ciclosporin

The pharmacokinetics of ciclosporin can be altered by macrolides. Commonly observed changes include increases in ciclosporin AUC and peak plasma concentration and reductions in the time to peak and clearance (109,111,137,138). Ciclosporin concentrations should therefore be monitored, to minimize the risk of toxicity in patients taking certain macrolides. When azithromycin is used concomitantly with ciclosporin, blood ciclosporin concentrations need to be monitored (139).

Cytochrome P450

The frequency and pattern of drug interactions with macrolides is influenced by the chemical structure of the individual macrolide. The most important mechanism determining many drug interactions is the effect of macrolides on the hepatic cytochrome P450, which oxidizes macrolides after binding of the drug to oxidized (Fe_3) cytochrome P450 (133). Binding of the macrolide to group IIIA cytochrome P450 can have one of several consequences: oxidation can result in the formation of a stable iron-metabolite complex, causing either induction of cytochrome P450 and increased metabolism of the antibiotic, or inactivation of cytochrome P450 and inhibition of drug metabolism (133).

Structural properties of the macrolides determine whether the drug metabolizing enzyme is induced or inactivated. Macrolides with a 14-membered lactone ring have a greater potential to inhibit cytochrome P450 than bulkier macrolides with 15-membered or 16-membered lactone rings. It is therefore useful to group the macrolides according to their molecular structure and their potential for drug interactions via cytochrome P450 (Table 3).

The differences in interactions of different macrolides with cytochrome P450 are marked. Troleandomycin is a more potent inhibitor of microsomal drug metabolism

Table 3 Molecular structures of macrolides and the extent of their interactions with cytochrome P450

Degree of interaction	Number of carbon atoms in macrolide ring		
	14	15	16
High	Troleandomycin Erythromycin		
Low	Flurithromycin Clarithromycin Roxithromycin		Josamycin Midecamycin Miocamycin
Not incriminated	Dirithromycin	Azithromycin	Rokitamycin Spiramycin

than erythromycin, while josamycin, midecamycin, and spiramycin have not so far been incriminated (140,141). Azithromycin does not induce or inhibit cytochrome P450 in rats (142).

The occurrence of an interaction in a particular patient is difficult to predict. Hepatic concentrations of CYP3A4, the activity of which can be estimated by an erythromycin breath test (140), vary at least ten-fold among patients (143,144). The variability is compounded by the fact that, in addition to the liver, mucosal cells of the small intestine commonly express CYP3A4, where it is responsible for significant first-pass metabolism of orally administered substrates (151).

Drugs that participate in major CYP3A4 interactions with macrolides are listed in Table 4. Other drugs that may be affected include alprazolam (145), the water-soluble artemisinin analogue artelinic acid (146), gallopamil (147), lovastatin (138), nefazodone (149), and risperidone (150). Pharmacokinetic interactions with clarithromycin, erythromycin, and troleandomycin have been reviewed (151).

Digoxin

The interaction between macrolides and digoxin is not a consequence of altered cytochrome P450 activity. Digoxin is metabolized in the gastrointestinal tract by *Eubacterium lentum* in the bowel flora of about 10% of patients (113). The direct antibacterial effect of macrolides reduces digoxin metabolism and increases its systemic availability (141). The resultant increased digoxin plasma concentrations are associated with severe nausea, vomiting, and dysrhythmias (110,152,153).

A concomitant clarithromycin-associated interaction with digoxin and warfarin has been reported (154).

- A 72-year-old man was taking regular digoxin (0.25 mg/day) and warfarin (22.5 mg/week) for chronic atrial fibrillation; other medications included furosemide, enalapril, glyceryl trinitrate paste, and a beclomethasone inhaler. He developed signs of digoxin intoxication after taking clarithromycin (500 mg tds) for 12 days and a raised serum digoxin concentration was confirmed. In addition, the INR was well above the target range (7.3 instead of 2.0–3.0).

An interaction with CYP450 enzymes was thought to be involved in the interaction with warfarin, and alteration of gut flora may have been involved in digoxin intoxication. Macrolides have also been suggested to interact with digoxin by inhibiting P glycoprotein (SEDA-26, 200;148). However, in a study of this mechanism in nine healthy Japanese men, clarithromycin 200 mg bd and erythromycin 200 mg qds did not alter the plasma concentration versus time curve of a single intravenous dose of digoxin 0.5 mg, but increased its renal clearance (198). This contrasts with an observation of reduced renal clearance of digoxin in two patients taking clarithromycin (SEDA-23, 194). That inhibition of renal P glycoprotein may not reduce the renal clearance of digoxin has also been suggested by studies with talinolol (SEDA-25, 172) and atorvastatin (199). Other transport mechanisms for digoxin, including the organic anion-transporting

Table 4   Summary of major interactions with macrolides

Interacting drug	Azithromycin	Clarithromycin	Erythromycin	Roxithromycin	Reference
Alfentanil	ND	ND	++	ND	(106)
Carbamazepine	–	++	++	–	(107,108)
Ciclosporin	+	ND	++	(+)	(109,110)
Digoxin	ND	++	++	ND	(111,112)
Disopyramide	ND	ND	++	(+)	(113,114)
Methylprednisolone	–	ND	+	ND	(115)
Oral contraceptives	ND	–	ND	–	(116,117)
Rifamycins	+?	++	ND	ND	(118–122)
Terfenadine, astemizole	ND	++	++	ND	(123)
Theophylline	–	+	++	(+)	(31)(108)(124)
Triazolam, midazolam	–	ND	++	(+)	(125)
Warfarin	–	–	++	–	(126,127)
Zidovudine	–	+	ND	ND	(128)

++ clinically relevant
+ potentially clinically relevant
(+) probably insignificant interaction
– documented lack of interaction
ND no published data
? insufficient information to judge relevance

polypeptides, have not been well studied and may play a role in digoxin disposition and hence drug interactions.

Azithromycin

- In a 31-month-old boy with Down's syndrome and Fallot's tetralogy during a 5-day course of azithromycin 5 mg/kg/day the serum digoxin concentration rose and the child had anorexia, diarrhea, and second-degree atrioventricular block with junctional extra beats (156).

The mechanism was not investigated, but it is likely that azithromycin affects *Eubacterium lentum*. When azithromycin is used concomitantly with digoxin, serum digoxin concentrations need to be monitored (139).

Clarithromycin

- In a 72-year-old woman taking digoxin 0.25 mg/day, the addition of clarithromycin caused a rise in the serum digoxin concentration to 4.6 ng/ml (154).

In two other cases in which clarithromycin increased serum digoxin concentrations there was an associated reduction in the rate of renal digoxin clearance, which may be another mechanism for this interaction (157). The authors hypothesized that clarithromycin inhibited P glycoprotein. This was supported by the observation of a concentration-dependent effect of clarithromycin on in vitro transcellular transport of digoxin.

There is anecdotal evidence that this interaction may be clinically important. Digoxin toxicity occurred in a patient taking clarithromycin (158).

Erythromycin

- Digoxin toxicity occurred in a neonate who was also given erythromycin (159). She had bradycardia and coupled extra beats. Digoxin and erythromycin were withdrawn and she was given antidigoxin antibodies.

Her plasma digoxin concentration, which had previously been 1.8 ng/ml, had risen to 8.0 ng/ml.

Disopyramide

Severe cardiac dysrhythmias and major hypoglycemia have occurred in patients taking disopyramide with some macrolide antibiotics, especially erythromycin and clarithromycin (160).

Some of the macrolide antibiotics have been reported to inhibit the clearance of disopyramide (SEDA-21, 200) (SEDA-22, 207), resulting in serious dysrhythmias or hypoglycemia. The mechanism of this interaction is presumed to be inhibition of dealkylation of disopyramide to its major metabolite, mono-*N*-dealkyldisopyramide. For example, in human liver microsomes the macrolide antibiotic troleandomycin significantly inhibited the mono-*N*-dealkylation of disopyramide enantiomers by inhibition of CYP3A4 (200). This interaction can result in serious dysrhythmias or other adverse effects of disopyramide.

- A 76-year-old woman developed torsade de pointes 5 days after starting to take clarithromycin 200 mg bd in addition to disopyramide 100 mg tds (201). Her serum potassium concentration was 2.8 mmol/l and the QT_c interval was prolonged to 0.71 seconds. The plasma disopyramide concentration was in the usual target range (3.2 µg/ml). The disopyramide and clarithromycin were withheld and potassium was given; 14 hours later the serum potassium concentration was 4.3 mmol/l and there was no further dysrhythmia, despite prolongation of the QT_c interval to 0.67 seconds, falling to 0.45 seconds 10 days later.

- A 35-year-old woman taking disopyramide phosphate modified-release capsules 150 mg qds was given azithromycin 500 mg initially and 250 mg/day thereafter (202). In 11 days she developed malaise, light-headedness, and urinary retention. After the insertion of a urinary catheter she developed a monomorphic ventricular tachycardia with left bundle branch block. She was successfully cardioverted and the electrocardiogram showed a markedly prolonged QT interval of 560 ms and T wave inversion in the anterolateral leads. Her serum disopyramide concentration, which had previously been 2.6 µg/ml, was 11 µg/ml.

- In a 59-year-old man taking disopyramide 50 mg/day the addition of clarithromycin 600 mg/day caused hypoglycemia, and the serum disopyramide concentration rose from 1.5 to 8.0 µg/ml (203). The ratio of plasma insulin concentration to blood glucose concentration was greatly increased, suggesting that hypersecretion of insulin was responsible, confirming the likelihood that the hypoglycemia was due to disopyramide intoxication secondary to inhibition of its metabolism by clarithromycin. There was also slight prolongation of the QT_c interval, but no cardiac dysrhythmias. After withdrawal of clarithromycin and disopyramide both the blood glucose concentration and the QT_c interval returned to normal.
- An 86-year-old woman presented with severe hypoglycemia after clarithromycin 500 mg/day had been added for 3 days to her other therapy, which included disopyramide 500 mg/day (204). The hypoglycemia resolved completely after withdrawal of disopyramide.

Clarithromycin

- In a 76-year old woman, severe prolongation of the QT_c interval and self-terminating torsade de pointes were induced by the combined use of clarithromycin and disopyramide; concomitant hypokalemia may have contributed to the cardiac dysrhythmia (161).
- Symptomatic hypoglycemia developed in a 59-year-old man treated with a combination of disopyramide (50 mg/day) and clarithromycin (600 mg/day) (162).

Additional investigations in the second case suggested enhanced insulin secretion induced by toxic disopyramide concentrations as the probable mechanism.

Erythromycin

Disopyramide alters the protein binding of erythromycin, and this results in increased plasma concentrations in vitro (163). The interaction between erythromycin and disopyramide was potentially fatal in two cases (114).

Roxithromycin

Disopyramide alters the protein binding of roxithromycin, and this results in increased plasma concentrations in vitro (163). However, this effect has not been observed with roxithromycin in vivo.

Glucocorticoids

Some macrolides have a dose- and time-related effect on methylprednisolone elimination, resulting in a prolonged half-life and reduced clearance (164). These changes were considered advantageous (steroid sparing) in patients with asthma (165,166). However, a retrospective analysis of 3995 patients treated with azithromycin did not show any pharmacokinetic interaction in patients who were also taking methylprednisolone (134,135).

In an open study in six adults with mild to moderate asthma, clarithromycin significantly reduced methylprednisolone clearance, thereby increasing the risk of steroid-induced adverse effects (167). In contrast, prednisolone clearance and mean prednisolone plasma concentrations were unaffected. Frank psychosis due to combined therapy with prednisone and clarithromycin has been reported (168).

Lincomycins

Antagonism in vitro has been described between macrolides and lincomycins (169). Competition for similar binding sites at bacterial ribosomes probably underlies this phenomenon.

Oral contraceptives

Most reports associating cholestatic jaundice with the concomitant use of oral contraceptives and macrolides have implicated troleandomycin (170–173). The exact mechanism of this interaction has not been established. Troleandomycin possibly inhibits the metabolism of estrogens and progestogens (140). In most patients jaundice persisted for more than a month after the antibiotic was withdrawn.

In a non-blinded study in 20 healthy women using Ortho Novum 7/7/7–28, oral dirithromycin 500 mg/day for 14 days produced a small (7.6%) but statistically significant reduction in the mean ethinylestradiol 24-hour AUC and an increase in apparent oral clearance (174). No woman ovulated or had compromised effectiveness of the oral contraceptive in this small series, so the clinical importance of this interaction may be negligible.

Risperidone

Drugs that inhibit CYP3A, such as the macrolides, can significantly alter risperidone concentrations, especially in patients with CYP2D6 deficiency (150).

SSRIs

- A 12-year-old boy developed the serotonin syndrome, which is normally associated with the interaction of two or more serotonergic agents, after the co-administration of erythromycin and sertraline.

This could have been due to erythromycin-induced inhibition of sertraline metabolism by CYP3A (175).

Statins

Macrolide antibiotics inhibit the metabolism of statins that are metabolized by CYP3A4 (atorvastatin, cerivastatin, lovastatin, simvastatin) (176). This interaction can cause myopathy and rhabdomyolysis, particularly in patients with renal insufficiency or in those who are concurrently taking medications associated with myopathy. Rhabdomyolysis is a rare disorder that mostly results from adverse drug effects. Hydroxymethylglutaryl-coenzyme A (HMG-CoA) reductase inhibitors (statins) are among the most important drugs causing this disorder, and the risk is increased when gemfibrozil or macrolide antibiotics are concurrently administered. The interaction usually develops within 2 weeks of combination therapy with macrolide antibiotics. Treatment includes drug withdrawal, intravenous fluids, urine alkalinization, and in severe cases dialysis.

Atorvastatin

In 12 healthy volunteers, erythromycin increased both the maximal plasma concentration and AUC of co-administered atorvastatin (177).

Lovastatin

Based on case reports or small studies, the potential for drug interactions between macrolides and lovastatin should be considered (178).

- Rhabdomyolysis, acute renal insufficiency, pancreatitis, ileus, livedo reticularis, and raised transaminase activities developed in a patient who had taken lovastatin for 7 years and took erythromycin before a dental procedure.
- Rhabdomyolysis requiring short-term dialysis occurred in a 57-year-old man with chronic renal insufficiency treated with lovastatin 40 mg/day, gemfibrozil 600 mg bd, and clarithromycin 500 mg tds (179).

In a review of the literature, three other reported instances of erythromycin and lovastatin interaction presenting with rhabdomyolysis, raised transaminase activities, and acute renal insufficiency were identified (180).

Simvastatin

Erythromycin interacts with simvastatin, probably by inhibiting its metabolism by CYP3A4. In a randomized, double-blind crossover study in 12 healthy volunteers, erythromycin significantly increased mean peak serum concentration and AUC for both unchanged simvastatin and its active metabolite simvastatin acid. However, there was extensive interindividual variability in the extent of this interaction (181).

- A 64-year-old African-American man developed rhabdomyolysis resulting from concomitant use of clarithromycin and simvastatin (182).

Theophylline and other xanthines

The most frequent effects of macrolides on theophylline pharmacokinetics are increased half-life and serum theophylline concentration and reduced clearance (140). The interaction with theophylline is mainly seen with higher doses of macrolides and can result in theophylline toxicity (115).

The effect of macrolides on serum theophylline concentration and clearance has been investigated in 53 patients with moderate asthma treated with theophylline (400 mg/day) in a randomized trial reference. Erythromycin (500 mg bd) and roxithromycin (150 mg bd), but not clarithromycin (250 mg bd) or azithromycin

(250 mg bd), caused increased serum theophylline concentrations and reduced clearance.

However, a retrospective analysis of 3995 patients treated with azithromycin did not show any pharmacokinetic interaction in patients who were also taking theophylline (134,135). Furthermore, in two double-blind, placebo-controlled, randomized studies azithromycin did not inhibit the metabolism of theophylline (183,184). However, there has been a report of reduced theophylline concentrations after withdrawal of azithromycin (185). The authors concluded that the mechanism of interaction was best explained by concomitant induction and inhibition of theophylline metabolism by azithromycin, followed by increased availability of unbound enzyme sites as azithromycin was cleared from the system.

Clarithromycin increases the serum concentration of theophylline by inhibiting CYP450 enzymes.

- In a 72-year-old man, the administration of clarithromycin (400 mg/day) plus modified-release theophylline (200 mg/day) resulted in acute rhabdomyolysis with renal insufficiency requiring hemodialysis after 5 days of combined treatment (186).

Verapamil

- A 77-year-old woman taking verapamil and propranolol for hypertrophic cardiomyopathy and paroxysmal atrial fibrillation developed symptomatic bradycardia on two separate occasions within days of taking either erythromycin or clarithromycin. The proposed mechanism of the interaction was inhibition of cytochrome P450, since verapamil is a substrate of an isoenzyme that is inhibited by some macrolides (187,188).

Severe hypotension and bradycardia after combined therapy with verapamil and clarithromycin has been reported in another case (189).

Verapamil is both a substrate and an inhibitor of CYP3A4, which is inhibited by clarithromycin and erythromycin. Giving these macrolide antibiotics during verapamil therapy is likely to reduce the first-pass metabolism of verapamil, increase its systemic availability, and impair its elimination. In patients taking this combination, verapamil should be started in a low dosage and its hemodynamic effects should be monitored closely.

- A 53-year-old woman had periods of dizziness and episodes of fainting when she stood up 24 hours after having been given clarithromycin for an acute exacerbation of chronic obstructive pulmonary disease and verapamil for atrial fibrillation (215). One day later she developed severe hypotension and bradycardia. Since her symptoms matched those of severe verapamil overdosage, the drug was withdrawn and her condition improved within two days.
- A 77-year-old hypertensive woman receiving both verapamil and propranolol for hypertrophic cardiomyopathy and paroxysmal atrial fibrillation developed symptomatic bradycardia within 2–4 days of initiation of both erythromycin or clarithromycin on two separate occasions, for pneumonia and sinusitis respectively.

- A 79-year-old white woman developed extreme fatigue and dizziness (210). Her heart rate was 40/minute and her blood pressure 80/40 mmHg. An electrocardiogram showed complete atrioventricular block, an escape rhythm at 50/minute, and QT_c interval prolongation to 583 milliseconds. This event was attributed to concomitant treatment with verapamil 480 mg/day and erythromycin 2000 mg/day, which had been prescribed 1 week before admission.

This is the first report of complete AV block and prolongation of the QT interval after co-administration of erythromycin and verapamil, both of which are principally metabolized by CYP3A4. Both drugs are potent inhibitors of CYP3A4 and P-glycoprotein, which may be the basis of this interaction.

The effects of a combination of erythromycin and verapamil on the pharmacokinetics of a single dose of simvastatin have been studied in a randomized, double-blind, cross-over study in 12 healthy volunteers simultaneously taking the three drugs. Both erythromycin and verapamil interacted with simvastatin, producing significant increases in the serum concentrations of simvastatin and its active metabolite simvastatin acid. The mean C_{max} of active simvastatin acid was increased about five-fold and the AUC_{0-24} four-fold by erythromycin; verapamil increased the C_{max} of simvastatin acid 3.4-fold and the AUC_{0-24} 2.8-fold. There was a substantial interindividual variation in the extent of these interactions. Concomitant use of erythromycin, verapamil, and simvastatin should be avoided (211).

Warfarin

Azithromycin can increase the effect of warfarin (190), perhaps by inhibiting its metabolism. Despite only modest effects of macrolides on serum warfarin concentrations and increases in prothrombin time (126), morbidity caused by hemorrhage may be significant, as illustrated by several case reports (191–195). It is likely that this interaction is potentiated by other factors, such as old age or dietary restrictions (126). However, a retrospective analysis of 3995 patients treated with azithromycin did not show any pharmacokinetic interaction in patients who were also taking warfarin (134).

- A 71-year old woman, who was taking maintenance warfarin for a prosthetic heart valve, took a 5-day course of azithromycin, and 6 days later her INR rose from 2.5 to 3.5–15 (196).

A concomitant clarithromycin-associated interaction with digoxin and warfarin has been reported (154).

- A 72-year-old man was taking regular digoxin (0.25 mg/day) and warfarin (22.5 mg/week) for chronic atrial fibrillation; other medications included furosemide, enalapril, glyceryl trinitrate paste, and a beclomethasone inhaler. He developed signs of digoxin intoxication after taking clarithromycin (500 mg tds) for 12 days and a raised serum digoxin concentration was

confirmed. In addition, the INR was well above the target range (7.3 instead of 2.0–3.0).

An interaction with CYP450 enzymes was thought to be involved in the interaction with warfarin, and alteration of gut flora may have been involved in digoxin intoxication. Inhibition of P-glycoprotein by clarithromycin may also have contributed to the interaction with digoxin (155).

Zafirlukast

Although zafirlukast inhibits CYP3A, it did not have any significant pharmacokinetic effect on single doses of azithromycin or clarithromycin in 12 healthy volunteers (197).

References

1. Barradell LB, Plosker GL, McTavish D. Clarithromycin. A review of its pharmacological properties and therapeutic use in Mycobacterium avium-intracellulare complex infection in patients with acquired immune deficiency syndrome. Drugs 1993;46(2):289–312.
2. Lin HC, Sanders SL, Gu YG, Doty JE. Erythromycin accelerates solid emptying at the expense of gastric sieving. Dig Dis Sci 1994;39(1):124–8.
3. Hasler WL, Heldsinger A, Chung OY. Erythromycin contracts rabbit colon myocytes via occupation of motilin receptors. Am J Physiol 1992;262(1 Pt 1):G50–5.
4. Kaufman HS, Ahrendt SA, Pitt HA, Lillemoe KD. The effect of erythromycin on motility of the duodenum, sphincter of Oddi, and gallbladder in the prairie dog. Surgery 1993;114(3):543–8.
5. Sifrim D, Matsuo H, Janssens J, Vantrappen G. Comparison of the effects of midecamycin acetate and azithromycin on gastrointestinal motility in man. Drugs Exp Clin Res 1994;20(3):121–6.
6. Bortolotti M, Mari C, Brunelli F, Sarti P, Miglioli M. Effect of intravenous clarithromycin on interdigestive gastroduodenal motility of patients with functional dyspepsia and *Helicobacter pylori* gastritis. Dig Dis Sci 1999;44(12):2439–42.
7. Vandenplas Y, Hauser B, Salvatore S. Current pharmacological treatment of gastroparesis. Expert Opin Pharmacother 2004;5(11):2251–4.
8. Nash MM, Zaltzman JS. Efficacy of azithromycin in the treatment of cyclosporine-induced gingival hyperplasia in renal transplant recipients. Transplantation 1998;65(12):1611–5.
9. Tamaoki J, Takeyama K, Tagaya E, Konno K. Effect of clarithromycin on sputum production and its rheological properties in chronic respiratory tract infections. Antimicrob Agents Chemother 1995;39(8):1688–90.
10. Sato K, Suga M, Akaike T, Fujii S, Muranaka H, Doi T, Maeda H, Ando M. Therapeutic effect of erythromycin on influenza virus-induced lung injury in mice. Am J Respir Crit Care Med 1998;157(3 Pt 1):853–7.
11. Gurfinkel E, Bozovich G, Beck E, Testa E, Livellara B, Mautner B. Treatment with the antibiotic roxithromycin in patients with acute non-Q-wave coronary syndromes. The final report of the ROXIS Study. Eur Heart J 1999;20(2):121–7.

12. Anderson JL, Muhlestein JB, Carlquist J, Allen A, Trehan S, Nielson C, Hall S, Brady J, Egger M, Horne B, Lim T. Randomized secondary prevention trial of azithromycin in patients with coronary artery disease and serological evidence for *Chlamydia pneumoniae* infection: The Azithromycin in Coronary Artery Disease: Elimination of Myocardial Infection with Chlamydia (ACADEMIC) study. Circulation 1999;99(12):1540–7.

13. Roujeau JC, Kelly JP, Naldi L, Rzany B, Stern RS, Anderson T, Auquier A, Bastuji-Garin S, Correia O, Locati F, et al. Medication use and the risk of Stevens–Johnson syndrome or toxic epidermal necrolysis. N Engl J Med 1995;333(24):1600–7.

14. Schlossberg D. Azithromycin and clarithromycin. Med Clin North Am 1995;79(4):803–15.

15. Periti P, Mazzei T, Mini E, Novelli A. Adverse effects of macrolide antibacterials. Drug Saf 1993;9(5):346–64.

16. Slater JE. Hypersensitivity to macrolide antibiotics. Ann Allergy 1991;66(3):193–5.

17. van Ketel WG. Immediate- and delayed-type allergy to erythromycin. Contact Dermatitis 1976;2(6):363–4.

18. Orban Z, MacDonald LL, Peters MA, Guslits B. Erythromycin-induced cardiac toxicity. Am J Cardiol 1995;75(12):859–61.

19. Schoenenberger RA, Haefeli WE, Weiss P, Ritz RF. Association of intravenous erythromycin and potentially fatal ventricular tachycardia with Q-T prolongation (torsades de pointes). BMJ 1990;300(6736):1375–6.

20. Bartkowski RR, McDonnell TE. Prolonged alfentanil effect following erythromycin administration. Anesthesiology 1990;73(3):566–8.

21. Kafetzis DA, Blanc F. Efficacy and safety of roxithromycin in treating paediatric patients. A European multicentre study. J Antimicrob Chemother 1987;20(Suppl B):171–7.

22. Anderson G, Esmonde TS, Coles S, Macklin J, Carnegie C. A comparative safety and efficacy study of clarithromycin and erythromycin stearate in community-acquired pneumonia. J Antimicrob Chemother 1991;27(Suppl A):117–24.

23. Beringer PM, Wong-Beringer A, Rho JP. Economic aspects of antibacterial adverse effects. Pharmacoeconomics 1998;13(1 Pt 1):35–49.

24. Owens RC Jr. QT prolongation with antimicrobial agents: understanding the significance. Drugs 2004;64(10):1091–124.

25. Shaffer D, Singer S, Korvick J, Honig P. Concomitant risk factors in reports of torsades de pointes associated with macrolide use: review of the United States Food and Drug Administration Adverse Event Reporting System. Clin Infect Dis 2002;35(2):197–200.

26. Kawashima M, Yatsunami J, Fukuno Y, Nagata M, Tominaga M, Hayashi S. Inhibitory effects of 14-membered ring macrolide antibiotics on bleomycin-induced acute lung injury. Lung 2002;180(2):73–89.

27. Sommer S, Delemazure B, Wagner M, Xenard L, Rozot P. Neuropathie optique ischémique bilatéral secondaire à un ergotisme aigu. [Bilateral ischemic optic neuropathy secondary to acute ergotism.] J Fr Ophtalmol 1998;21(2):123–5.

28. Uzun C, Koten M, Adali MK, Yorulmaz F, Yagiz R, Karasalihoglu AR. Reversible ototoxic effect of azithromycin and clarithromycin on transiently evoked otoacoustic emissions in guinea pigs. J Laryngol Otol 2001;115(8):622–8.

29. Ishizaka T, Miyanaga Y, Mukai J, Asaka K, Nakai Y, Tsuji E, Uchida T. Bitterness evaluation of medicines for pediatric use by a taste sensor. Chem Pharm Bull (Tokyo) 2004;52(8):943–8.

30. Gomez-Gil E, Garcia F, Pintor L, Martinez JA, Mensa J, de Pablo J. Clarithromycin-induced acute psychoses in peptic ulcer disease. Eur J Clin Microbiol Infect Dis 1999;18(1):70–1.

31. Geiderman JM. Central nervous system disturbances following clarithromycin ingestion. Clin Infect Dis 1999;29(2):464–5.

32. Enomoto F, Andou I, Nagaoka I, Ichikawa G. Effect of new macrolides on the expression of adhesion molecules on neutrophils in chronic sinusitis. Auris Nasus Larynx 2002;29(3):267–9.

33. Usiskin SI, Nicolson R, Lenane M, Rapoport JL. Retreatment with clozapine after erythromycin-induced neutropenia. Am J Psychiatry 2000;157(6):1021.

34. Nellans H, Petersen A. Stimulation of gastrointestinal motility: clarithromycin significantly less potent than azithromycin. In: Seventh International Congress of ChemotherapyBerlin 1991:1201.

35. Peeters TL. Erythromycin and other macrolides as prokinetic agents. Gastroenterology 1993;105(6):1886–99.

36. Nakayoshi T, Izumi M, Tatsuta K. Effects of macrolide antibiotics on gastrointestinal motility in fasting and digestive states. Drugs Exp Clin Res 1992;18(4):103–9.

37. Zerbib F. Les oesophagites médicamenteuses. Hepato-Gastro 1998;5:115–20.

38. Braun P. Hepatotoxicity of erythromycin. J Infect Dis 1969;119(3):300–6.

39. Ginsburg CM, Eichenwald HF. Erythromycin: a review of its uses in pediatric practice. J Pediatr 1976;89(6):872–84.

40. Longo G, Valenti C, Gandini G, Ferrara L, Bertesi M, Emilia G. Azithromycin-induced intrahepatic cholestasis. Am J Med 1997;102(2):217–8.

41. Stahlman R, Lode H. Macrolides: tolerability and interactions with other drugs. Anti-infective Drugs Chemother 1996;14:155–62.

42. Yew WW, Chau CH, Lee J, Leung CW. Cholestatic hepatitis in a patient who received clarithromycin therapy for a Mycobacterium chelonae lung infection. Clin Infect Dis 1994;18(6):1025–6.

43. Lavin I, Mundi JL, Trillo C, Trapero A, Fernandez R, Lopez MA, Cervilla E, Quintero D, Palacios A. Hepatitis colestasica por josamicina. [Cholestatic hepatitis from josamycin.] Gastroenterol Hepatol 1999;22(3):160.

44. Baylor P, Williams K. Interstitial nephritis, thrombocytopenia, hepatitis, and elevated serum amylase levels in a patient receiving clarithromycin therapy. Clin Infect Dis 1999;29(5):1350–1.

45. Nakamura H, Fujishima S, Inoue T, Ohkubo Y, Soejima K, Waki Y, Mori M, Urano T, Sakamaki F, Tasaka S, Ishizaka A, Kanazawa M, Yamaguchi K. Clinical and immunoregulatory effects of roxithromycin therapy for chronic respiratory tract infection. Eur Respir J 1999;13(6):1371–9.

46. Khan AA, Slifer TR, Araujo FG, Remington JS. Effect of clarithromycin and azithromycin on production of cytokines by human monocytes. Int J Antimicrob Agents 1999;11(2):121–32.

47. Suzaki H, Asano K, Ohki S, Kanai K, Mizutani T, Hisamitsu T. Suppressive activity of a macrolide antibiotic, roxithromycin, on pro-inflammatory cytokine production in vitro and in vivo. Mediators Inflamm 1999;8(4–5):199–204.

48. Yatsunami J, Fukuno Y, Nagata M, Tsuruta N, Aoki S, Tominaga M, Kawashima M, Taniguchi S, Hayashi S.

Roxithromycin and clarithromycin, 14-membered ring macrolides, potentiate the antitumor activity of cytotoxic agents against mouse B16 melanoma cells. Cancer Lett 1999;147(1–2):17–24.

49. Yatsunami J, Fukuno Y, Nagata M, Tominaga M, Aoki S, Tsuruta N, Kawashima M, Taniguchi S, Hayashi S. Antiangiogenic and antitumor effects of 14-membered ring macrolides on mouse B16 melanoma cells. Clin Exp Metastasis 1999;17(4):361–7.

50. Nonaka M, Pawankar R, Tomiyama S, Yagi T. A macrolide antibiotic, roxithromycin, inhibits the growth of nasal polyp fibroblasts. Am J Rhinol 1999;13(4):267–72.

51. Tamaoki J, Kadota J, Takizawa H. Clinical implications of the immunomodulatory effects of macrolides. Am J Med 2004;117 Suppl 9A:5S-11S.

52. Siddiqui J. Immunomodulatory effects of macrolides: implications for practicing clinicians. Am J Med 2004;117 Suppl 9A:26S-29S.

53. Amsden GW. Anti-inflammatory effects of macrolides—an underappreciated benefit in the treatment of community-acquired respiratory tract infections and chronic inflammatory pulmonary conditions? J Antimicrob Chemother 2005;55(1):10–21.

54. Kanazawa S, Nomura S, Muramatsu M, Yamaguchi K, Fukuhara S. Azithromycin and bronchiolitis obliterans. Am J Respir Crit Care Med 2004;169(5):654-5; author reply 655.

55. Hubner C, Dietz A, Stremmel W, Stiehl A, Andrassy H. Macrolide-induced Churg–Strauss syndrome in a patient with atopy. Lancet 1997;350(9077):563.

56. Dietz A, Hubner C, Andrassy K. Makrolid–Antibiotika indurierte Vaskulitis (Churg–Strauss syndrome). [Macrolide antibiotic-induced vasculitis (Churg–Strauss syndrome).] Laryngorhinootologie 1998;77(2):111–4.

57. Gavura SR, Nusinowitz S. Leukocytoclastic vasculitis associated with clarithromycin. Ann Pharmacother 1998;32(5):543–5.

58. Goldberg EI, Shoji T, Sapadin AN. Henoch–Schönlein purpura induced by clarithromycin. Int J Dermatol 1999;38(9):706–8.

59. Gallardo MA, Thomas I. Hypersensitivity reaction to erythromycin. Cutis 1999;64(6):375–6.

60. Alvarez-Elcoro S, Enzler MJ. The macrolides: erythromycin, clarithromycin, and azithromycin. Mayo Clin Proc 1999;74(6):613–34.

61. Oster P, Zanchi A, Cresti S, Lattanzi M, Montagnani F, Cellesi C, Rossolini GM. Patterns of macrolide resistance determinants among community-acquired Streptococcus pneumoniae isolates over a 5-year period of decreased macrolide susceptibility rates. Antimicrob Agents Chemother 1999;43(10):2510–2.

62. Melo-Cristino J, Fernandes ML. Streptococcus pyogenes isolated in Portugal: macrolide resistance phenotypes and correlation with T types. Portuguese Surveillance Group for the Study of Respiratory Pathogens. Microb Drug Resist 1999;5(3):219–25.

63. De Azavedo JC, Yeung RH, Bast DJ, Duncan CL, Borgia SB, Low DE. Prevalence and mechanisms of macrolide resistance in clinical isolates of group A streptococci from Ontario, Canada. Antimicrob Agents Chemother 1999;43(9):2144–7.

64. Morales WJ, Dickey SS, Bornick P, Lim DV. Change in antibiotic resistance of group B streptococcus: impact on intrapartum management. Am J Obstet Gynecol 1999;181(2):310–4.

65. Hsueh PR, Teng LJ, Lee LN, Yang PC, Ho SW, Luh KT. Extremely high incidence of macrolide and trimethoprim-sulfamethoxazole resistance among clinical isolates of Streptococcus pneumoniae in Taiwan. J Clin Microbiol 1999;37(4):897–901.

66. Jacobs MR, Bajaksouzian S, Zilles A, Lin G, Pankuch GA, Appelbaum PC. Susceptibilities of Streptococcus pneumoniae and Haemophilus influenzae to 10 oral antimicrobial agents based on pharmacodynamic parameters: 1997 U.S. Surveillance study Antimicrob Agents Chemother 1999;43(8):1901–8.

67. Lopez B, Cima MD, Vazquez F, Fenoll A, Gutierrez J, Fidalgo C, Caicoya M, Mendez FJ. Epidemiological study of Streptococcus pneumoniae carriers in healthy primary-school children. Eur J Clin Microbiol Infect Dis 1999;18(11):771–6.

68. Varaldo PE, Debbia EA, Nicoletti G, Pavesio D, Ripa S, Schito GC, Tempera GArtemis–Italy Study Group. Nationwide survey in Italy of treatment of Streptococcus pyogenes pharyngitis in children: influence of macrolide resistance on clinical and microbiological outcomes. Clin Infect Dis 1999;29(4):869–73.

69. Wisplinghoff H, Reinert RR, Cornely O, Seifert H. Molecular relationships and antimicrobial susceptibilities of viridans group streptococci isolated from blood of neutropenic cancer patients. J Clin Microbiol 1999;37(6):1876–80.

70. York MK, Gibbs L, Perdreau-Remington F, Brooks GF. Characterization of antimicrobial resistance in Streptococcus pyogenes isolates from the San Francisco Bay area of northern California. J Clin Microbiol 1999;37(6):1727–31.

71. Washington JA 2nd, Wilson WR. Erythromycin: a microbial and clinical perspective after 30 years of clinical use (1). Mayo Clin Proc 1985;60(3):189–203.

72. Washington JA 2nd, Wilson WR. Erythromycin: a microbial and clinical perspective after 30 years of clinical use (2). Mayo Clin Proc 1985;60(4):271–8.

73. Pohle T, Stoll R, Kirchner T, Heep M, Lehn N, Bock H, Domschke W. Eradication of Helicobacter pylori with lansoprazole, roxithromycin and metronidazole—an open pilot study. Aliment Pharmacol Ther 1998;12(12):1273–8.

74. Laine L, Suchower L, Frantz J, Connors A, Neil G. Twice-daily, 10-day triple therapy with omeprazole, amoxicillin, and clarithromycin for Helicobacter pylori eradication in duodenal ulcer disease: results of three multicenter, double-blind, United States trials. Am J Gastroenterol 1998;93(11):2106–12.

75. Petersen AM, Schradieck W, Krogfelt KA. Helicobacter pylori-resistens over for clarithromycin. [Resistance of Helicobacter pylori to clarithomycin.] Ugeskr Laeger 1998;160(23):3412–3.

76. Barbut F, Decre D, Burghoffer B, Lesage D, Delisle F, Lalande V, Delmee M, Avesani V, Sano N, Coudert C, Petit JC. Antimicrobial susceptibilities and serogroups of clinical strains of Clostridium difficile isolated in France in 1991 and 1997. Antimicrob Agents Chemother 1999;43(11):2607–11.

77. Gibson JR, Saunders NA, Burke B, Owen RJ. Novel method for rapid determination of clarithromycin sensitivity in Helicobacter pylori. J Clin Microbiol 1999;37(11):3746–8.

78. Bazzoli F, Berretti D, De Luca L, Nicolini G, Pozzato P, Fossi S, Zagari M. What can be learnt from the new data about antibiotic resistance? Are there any practical clinical

consequences of *Helicobacter pylori* antibiotic resistance? Eur J Gastroenterol Hepatol 1999;11(Suppl 2):S39–42.

79. Ellenrieder V, Boeck W, Richter C, Marre R, Adler G, Glasbrenner B. Prevalence of resistance to clarithromycin and its clinical impact on the efficacy of *Helicobacter pylori* eradication. Scand J Gastroenterol 1999;34(8):750–6.

80. Megraud F. Resistance of *Helicobacter pylori* to antibiotics: the main limitation of current proton-pump inhibitor triple therapy. Eur J Gastroenterol Hepatol 1999;11(Suppl 2):S35–7.

81. Boyanova L, Spassova Z, Krastev Z, Petrov S, Stancheva I, Docheva J, Mitov I, Koumanova R. Characteristics and trends in macrolide resistance among *Helicobacter pylori* strains isolated in Bulgaria over four years. Diagn Microbiol Infect Dis 1999;34(4):309–13.

82. Ducons JA, Santolaria S, Guirao R, Ferrero M, Montoro M, Gomollon F. Impact of clarithromycin resistance on the effectiveness of a regimen for *Helicobacter pylori*: a prospective study of 1-week lansoprazole, amoxycillin and clarithromycin in active peptic ulcer. Aliment Pharmacol Ther 1999;13(6):775–80.

83. Gschwantler M, Dragosics B, Schutze K, Wurzer H, Hirschl AM, Pasching E, Wimmer M, Klimpfinger M, Oberhuber G, Brandstatter G, Hentschel E, Weiss W. Famotidine versus omeprazole in combination with clarithromycin and metronidazole for eradication of *Helicobacter pylori*—a randomized, controlled trial. Aliment Pharmacol Ther 1999;13(8):1063–9.

84. Lamouliatte H, Samoyeau R, De Mascarel A, Megraud F. Double vs. single dose of pantoprazole in combination with clarithromycin and amoxycillin for 7 days, in eradication of *Helicobacter pylori* in patients with non-ulcer dyspepsia. Aliment Pharmacol Ther 1999;13(11):1523–30.

85. Murakami K, Kimoto M. [Antibiotic-resistant *H. pylori* strains in the last ten years in Japan.]Nippon Rinsho 1999;57(1):81–6.

86. Osato MS, Reddy R, Graham DY. Metronidazole and clarithromycin resistance amongst *Helicobacter pylori* isolates from a large metropolitan hospital in the United States. Int J Antimicrob Agents 1999;12(4):341–7.

87. Realdi G, Dore MP, Piana A, Atzei A, Carta M, Cugia L, Manca A, Are BM, Massarelli G, Mura I, Maida A, Graham DY. Pretreatment antibiotic resistance in *Helicobacter pylori* infection: results of three randomized controlled studies. Helicobacter 1999;4(2):106–12.

88. Ani AE, Malu AO, Onah JA, Queiroz DM, Kirschner G, Rocha GA. Antimicrobial susceptibility test of *Helicobacter pylori* isolated from Jos, Nigeria. Trans R Soc Trop Med Hyg 1999;93(6):659–61.

89. Aarestrup FM, Seyfarth AM, Emborg HD, Pedersen K, Hendriksen RS, Bager F. Effect of abolishment of the use of antimicrobial agents for growth promotion on occurrence of antimicrobial resistance in fecal enterococci from food animals in Denmark. Antimicrob Agents Chemother 2001;45(7):2054–9.

90. Czeizel AE, Rockenbauer M, Sorensen HT, Olsen J. A population-based case-control teratologic study of oral erythromycin treatment during pregnancy. Reprod Toxicol 1999;13(6):531–6.

91. Southern KW, Barker PM, Solis A. Macrolide antibiotics for cystic fibrosis. Cochrane Database Syst Rev 2004(2):CD002203.

92. Antonen JA, Markula KP, Pertovaara MI, Pasternack AI. Adverse drug reactions in Sjögren's syndrome. Frequent allergic reactions and a specific trimethoprim-associated systemic reaction. Scand J Rheumatol 1999;28(3):157–9.

93. Bartkowski RR, Goldberg ME, Larijani GE, Boerner T. Inhibition of alfentanil metabolism by erythromycin. Clin Pharmacol Ther 1989;46(1):99–102.

94. Tran HT. Torsades de pointes induced by nonantiarrhythmic drugs. Conn Med 1994;58(5):291–5.

95. Zechnich AD, Hedges JR, Eiselt-Proteau D, Haxby D. Possible interactions with terfenadine or astemizole. West J Med 1994;160(4):321–5.

96. Honig P, Wortham D, Zamani K, Cantilena L. Comparison of the effect of the macrolide antibiotics erythromycin, clarithromycin and azithromycin on terfenadine steady-state pharmacokinetics and electrocardiographic parameters. Drug Invest 1994;7:148.

97. Harris S, Hilligoss DM, Colangelo PM, Eller M, Okerholm R. Azithromycin and terfenadine: lack of drug interaction. Clin Pharmacol Ther 1995;58(3):310–5.

98. Moss AJ, Chaikin P, Garcia JD, Gillen M, Roberts DJ, Morganroth J. A review of the cardiac systemic side-effects of antihistamines: ebastine. Clin Exp Allergy 1999;29(Suppl 3):200–5.

99. Carr RA, Edmonds A, Shi H, Locke CS, Gustavson LE, Craft JC, Harris SI, Palmer R. Steady-state pharmacokinetics and electrocardiographic pharmacodynamics of clarithromycin and loratadine after individual or concomitant administration. Antimicrob Agents Chemother 1998;42(5):1176–80.

100. Warot D, Bergougnan L, Lamiable D, Berlin I, Bensimon G, Danjou P, Puech AJ. Troleandomycin–triazolam interaction in healthy volunteers: pharmacokinetic and psychometric evaluation. Eur J Clin Pharmacol 1987;32(4):389–93.

101. Phillips JP, Antal EJ, Smith RB. A pharmacokinetic drug interaction between erythromycin and triazolam. J Clin Psychopharmacol 1986;6(5):297–9.

102. Hiller A, Olkkola KT, Isohanni P, Saarnivaara L. Unconsciousness associated with midazolam and erythromycin. Br J Anaesth 1990;65(6):826–8.

103. Nelson WO, Bunge RG. The effect of therapeutic dosages of nitrofurantoin (Furadantin) upon spermatogenesis in man. J Urol 1957;77(2):275–81.

104. Wong YY, Ludden TM, Bell RD. Effect of erythromycin on carbamazepine kinetics. Clin Pharmacol Ther 1983;33(4):460–4.

105. Albin H, Vincon G, Pehourcq F, Dangoumau J. Influence de la josamycine sur la pharmacokinétique de la carbamazépine. [Effect of josamycin on the pharmacokinetics of carbamazepine.] Therapie 1982;37(5):563–6.

106. Amsden GW. Macrolides versus azalides: a drug interaction update. Ann Pharmacother 1995;29(9):906–17.

107. Mesdjian E, Dravet C, Cenraud B, Roger J. Carbamazepine intoxication due to triacetyloleandomycin administration in epileptic patients. Epilepsia 1980;21(5):489–96.

108. Jensen CW, Flechner SM, Van Buren CT, Frazier OH, Cooley DA, Lorber MI, Kahan BD. Exacerbation of cyclosporine toxicity by concomitant administration of erythromycin. Transplantation 1987;43(2):263–70.

109. Yee GC, McGuire TR. Pharmacokinetic drug interactions with cyclosporin. Clin Pharmacokinet 1990;19(4):319–321990;19(5):400–15.

110. Herishanu Y, Taustein I. The electromyographic changes by antibiotics. A preliminary study. Confin Neurol 1971;33(1):41–5.

111. Couet W, Istin B, Seniuta P, Morel D, Potaux L, Fourtillan JB. Effect of ponsinomycin on cyclosporin pharmacokinetics. Eur J Clin Pharmacol 1990;39(2):165–7.

112. Maxwell DL, Gilmour-White SK, Hall MR. Digoxin toxicity due to interaction of digoxin with erythromycin. BMJ 1989;298(6673):572.

113. Lindenbaum J, Rund DG, Butler VP Jr, Tse-Eng D, Saha JR. Inactivation of digoxin by the gut flora: reversal by antibiotic therapy. N Engl J Med 1981;305(14):789–94.

114. Ragosta M, Weihl AC, Rosenfeld LE. Potentially fatal interaction between erythromycin and disopyramide. Am J Med 1989;86(4):465–6.

115. Parish RA, Haulman NJ, Burns RM. Interaction of theophylline with erythromycin base in a patient with seizure activity. Pediatrics 1983;72(6):828–30.

116. Back DJ, Tija J, Martin C, Millar E, Salmon P, Orme M. The interaction between clarithromycin and combined oral-contraceptive steroids. J Pharm Med 1991;2:81–7.

117. Meyer B, Muller F, Wessels P, Maree J. A model to detect interactions between roxithromycin and oral contraceptives. Clin Pharmacol Ther 1990;47(6):671–4.

118. Botstein P. Is QT interval prolongation harmful? A regulatory perspective. Am J Cardiol 1993;72(6):B50–2.

119. Griffith DE, Brown BA, Girard WM, Wallace RJ Jr. Adverse events associated with high-dose rifabutin in macrolide-containing regimens for the treatment of *Mycobacterium avium* complex lung disease. Clin Infect Dis 1995;21(3):594–8.

120. Fuller JD, Stanfield LE, Craven DE. Rifabutin prophylaxis and uveitis. N Engl J Med 1994;330(18):1315–6.

121. Shafran SD, Deschenes J, Miller M, Phillips P, Toma EMAC Study Group of the Canadian HIV Trials Network. Uveitis and pseudojaundice during a regimen of clarithromycin, rifabutin, and ethambutol. N Engl J Med 1994;330(6):438–9.

122. Frank MO, Graham MB, Wispelway B. Rifabutin and uveitis. N Engl J Med 1994;330(12):868.

123. Jurima Romet M, Crawford K, Cyr T, Inaba T. Terfenadine metabolism in human liver. In vitro inhibition by macrolide antibiotics and azole antifungals. Drug Metab Dispos 1994;22(6):849–57.

124. Garnett WR. Interactions with hydroxymethylglutaryl-coenzyme A reductase inhibitors. Am J Health Syst Pharm 1995;52(15):1639–45.

125. Gascon MP, Dayer P, Waldvogel F. Les interactions médicamenteuses du midazolam. [Drug interactions with midazolam.] Schweiz Med Wochenschr 1989;119(50):1834–6.

126. Weibert RT, Lorentz SM, Townsend RJ, Cook CE, Klauber MR, Jagger PI. Effect of erythromycin in patients receiving long-term warfarin therapy. Clin Pharm 1989;8(3):210–4.

127. Agache P, Amblard P, Moulin G, Barriere H, Texier L, Beylot C, Bergoend H. Roxithromycin in skin and soft tissue infections. J Antimicrob Chemother 1987;20(Suppl B):153–6.

128. Nightingale SD, Koster FT, Mertz GJ, Loss SD. Clarithromycin-induced mania in two patients with AIDS. Clin Infect Dis 1995;20(6):1563–4.

129. Couet W, Istin B, Ingrand I, Girault J, Fourtillan JB. Effect of ponsinomycin on single-dose kinetics and metabolism of carbamazepine. Ther Drug Monit 1990;12(2):144–9.

130. Barzaghi N, Gatti G, Crema F, Faja A, Monteleone M, Amione C, Leone L, Perucca E. Effect of flurithromycin, a new macrolide antibiotic, on carbamazepine disposition in normal subjects. Int J Clin Pharmacol Res 1988;8(2):101–5.

131. Berrettini WH. A case of erythromycin-induced carbamazepine toxicity. J Clin Psychiatry 1986;47(3):147.

132. Carranco E, Kareus J, Co S, Peak V, Al-Rajeh S. Carbamazepine toxicity induced by concurrent erythromycin therapy. Arch Neurol 1985;42(2):187–8.

133. Periti P, Mazzei T, Mini E, Novelli A. Pharmacokinetic drug interactions of macrolides. Clin Pharmacokinet 1992;23(2):106–31.

134. Peters DH, Friedel HA, McTavish D. Azithromycin. A review of its antimicrobial activity, pharmacokinetic properties and clinical efficacy. Drugs 1992;44(5):750–99.

135. Hopkins S. Clinical toleration and safety of azithromycin. Am J Med 1991;91(3A):S40–5.

136. Stafstrom CE, Nohria V, Loganbill H, Nahouraii R, Boustany RM, DeLong GR. Erythromycin-induced carbamazepine toxicity: a continuing problem. Arch Pediatr Adolesc Med 1995;149(1):99–101.

137. Wadhwa NK, Schroeder TJ, O'Flaherty E, Pesce AJ, Myre SA, Munda R, First MR. Interaction between erythromycin and cyclosporine in a kidney and pancreas allograft recipient. Ther Drug Monit 1987;9(1):123–5.

138. Azanza J, Catalan M, Alvarez P, Honorato J, Herreros J, Llorens R. Possible interaction between cyclosporine and josamycin. J Heart Transplant 1990;9(3 Pt 1):265–6.

139. Ljutic D, Rumboldt Z. Possible interaction between azithromycin and cyclosporin: a case report. Nephron 1995;70(1):130.

140. Watkins PB, Murray SA, Winkelman LG, Heuman DM, Wrighton SA, Guzelian PS. Erythromycin breath test as an assay of glucocorticoid-inducible liver cytochromes P-450. Studies in rats and patients. J Clin Invest 1989;83(2):688–97.

141. Ludden TM. Pharmacokinetic interactions of the macrolide antibiotics. Clin Pharmacokinet 1985;10(1):63–79.

142. Amacher DE, Schomaker SJ, Retsema JA. Comparison of the effects of the new azalide antibiotic, azithromycin, and erythromycin estolate on rat liver cytochrome P-450. Antimicrob Agents Chemother 1991;35(6):1186–90.

143. Lown K, Kolars J, Turgeon K, Merion R, Wrighton SA, Watkins PB. The erythromycin breath test selectively measures P450IIIA in patients with severe liver disease. Clin Pharmacol Ther 1992;51(3):229–38.

144. Lown KS, Kolars JC, Thummel KE, Barnett JL, Kunze KL, Wrighton SA, Watkins PB. Interpatient heterogeneity in expression of CYP3A4 and CYP3A5 in small bowel. Lack of prediction by the erythromycin breath test. Drug Metab Dispos 1994;22(6):947–55.

145. Gorski JC, Jones DR, Hamman MA, Wrighton SA, Hall SD. Biotransformation of alprazolam by members of the human cytochrome P4503A subfamily. Xenobiotica 1999;29(9):931–44.

146. Grace JM, Skanchy DJ, Aguilar AJ. Metabolism of artelinic acid to dihydroqinqhaosu by human liver cytochrome P4503A. Xenobiotica 1999;29(7):703–17.

147. Suzuki A, Iida I, Tanaka F, Akimoto M, Fukushima K, Tani M, Ishizaki T, Chiba K. Identification of human cytochrome P-450 isoforms involved in metabolism of R(+)- and S(-)-gallopamil: utility of in vitro disappearance rate. Drug Metab Dispos 1999;27(11):1254–9.

148. Jacobsen W, Kirchner G, Hallensleben K, Mancinelli L, Deters M, Hackbarth I, Benet LZ, Sewing KF, Christians U. Comparison of cytochrome P-450-dependent metabolism and drug interactions of the 3-hydroxy-3-methylglutaryl-CoA reductase inhibitors lovastatin and pravastatin in the liver. Drug Metab Dispos 1999;27(2):173–9.

149. von Moltke LL, Greenblatt DJ, Granda BW, Grassi JM, Schmider J, Harmatz JS, Shader RI. Nefazodone, meta-chlorophenylpiperazine, and their metabolites in vitro: cytochromes mediating transformation, and P450-3A4 inhibitory actions. Psychopharmacology (Berl) 1999;145(1):113–22.

150. Bork JA, Rogers T, Wedlund PJ, de Leon J. A pilot study on risperidone metabolism: the role of cytochromes P450 2D6 and 3A. J Clin Psychiatry 1999;60(7):469–76.

151. Rubinstein E. Comparative safety of the different macrolides. Int J Antimicrob Agents 2001;18(Suppl 1):S71–6.

152. Friedman HS, Bonventre MV. Erythromycin-induced digoxin toxicity. Chest 1982;82(2):202.

153. Ford A, Smith LC, Baltch AL, Smith RP. Clarithromycin-induced digoxin toxicity in a patient with AIDS. Clin Infect Dis 1995;21(4):1051–2.

154. Gooderham MJ, Bolli P, Fernandez PG. Concomitant digoxin toxicity and warfarin interaction in a patient receiving clarithromycin. Ann Pharmacother 1999;33(7–8):796–9.

155. Juurlink DN, Ito S. Comment: clarithromycin–digoxin interaction. Ann Pharmacother 1999;33(12):1375–6.

156. Ten Eick AP, Sallee D, Preminger T, Weiss A, Reed MD. Possible drug interaction between digoxin and azithromycin in a young child. Clin Drug Invest 2000;20:61–4.

157. Wakasugi H, Yano I, Ito T, Hashida T, Futami T, Nohara R, Sasayama S, Inui K. Effect of clarithromycin on renal excretion of digoxin: interaction with P-glycoprotein. Clin Pharmacol Ther 1998;64(1):123–8.

158. Trivedi S, Hyman J, Lichstein E. Clarithromycin and digoxin toxicity. Ann Intern Med 1998;128(7):604.

159. Coudray S, Janoly A, Belkacem-Kahlouli A, Bourhis Y, Bleyzac N, Bourgeois J, Putet G, Aulagner G. Erythromycin-induced digoxin toxicity in a neonatal intensive care unit. J Pharm Clin 2001;20:129–31.

160. Anonymous. Disopyramide: interactions with marcolide antibiotics. Prescrire Int 2001;10(55):151.

161. Hayashi Y, Ikeda U, Hashimoto T, Watanabe T, Mitsuhashi T, Shimada K. Torsades de pointes ventricular tachycardia induced by clarithromycin and disopyramide in the presence of hypokalemia. Pacing Clin Electrophysiol 1999;22(4 Pt 1):672–4.

162. Iida H, Morita T, Suzuki E, Iwasawa K, Toyo-oka T, Nakajima T. Hypoglycemia induced by interaction between clarithromycin and disopyramide. Jpn Heart J 1999;40(1):91–6.

163. Zini R, Fournet M, Barre J, et al. In vitro study of roxithromycin binding to serum proteins and erythrocytes in man. Br J Clin Pract 1987;42(Suppl 5):54.

164. LaForce CF, Szefler SJ, Miller MF, Ebling W, Brenner M. Inhibition of methylprednisolone elimination in the presence of erythromycin therapy. J Allergy Clin Immunol 1983;72(1):34–9.

165. Spector S, Katz F, Farr R. Troleandomycin: effectiveness in steroid-dependent asthma and bronchitis. J Allergy Clin Immunol 1974;54:367.

166. Zeiger RS, Schatz M, Sperling W, Simon RA, Stevenson DD. Efficacy of troleandomycin in outpatients with severe, corticosteroid-dependent asthma. J Allergy Clin Immunol 1980;66(6):438–46.

167. Fost DA, Leung DY, Martin RJ, Brown EE, Szefler SJ, Spahn JD. Inhibition of methylprednisolone elimination in the presence of clarithromycin therapy. J Allergy Clin Immunol 1999;103(6):1031–5.

168. Finkenbine RD, Frye MD. Case of psychosis due to prednisone–clarithromycin interaction. Gen Hosp Psychiatry 1998;20(5):325–6.

169. Ruiz NM, Ramirez-Ronda CH. Tetracyclines, macrolides, lincosamides & chloramphenicol. Bol Asoc Med P R 1990;82(1):8–17.

170. Claudel S, Euvrard P, Bory R, Chavaillon A, Paliard P. Cholestase intrahépatique apres aasociation triacetylo-leandomycine–estroprogestatif. [Intra-hepatic cholestasis after taking a triacetyloleandomycin–estroprogestational combination.] Nouv Presse Méd 1979;8(14):1182.

171. Fevery J, Van Steenbergen W, Desmet V, Deruyttere M, De Groote J. Severe intrahepatic cholestasis due to the combined intake of oral contraceptives and triacetyloleandomycin. Acta Clin Belg 1983;38(4):242–5.

172. Haber I, Hubens H. Cholestatic jaundice after triacetyloleandomycin and oral contraceptives. The diagnostic value of gamma-glutamyl transpeptidase. Acta Gastroenterol Belg 1980;43(11–12):475–82.

173. Miguet JP, Vuitton D, Pessayre D, Allemand H, Metreau JM, Poupon R, Capron JP, Blanc F. Jaundice from troleandomycin and oral contraceptives. Ann Intern Med 1980;92(3):434.

174. Wermeling DP, Chandler MH, Sides GD, Collins D, Muse KN. Dirithromycin increases ethinyl estradiol clearance without allowing ovulation. Obstet Gynecol 1995;86(1):78–84.

175. Shimada N, Omuro H, Saka S, Ebihara I, Koide H. [A case of acute renal failure with rhabdomyolysis caused by the interaction of theophylline and clarithromycin.]Nippon Jinzo Gakkai Shi 1999;41(4):460–3.

176. Einarson TR, Metge CJ, Iskedjian M, Mukherjee J. An examination of the effect of cytochrome P450 drug interactions of hydroxymethylglutaryl-coenzyme A reductase inhibitors on health care utilization: a Canadian population-based study. Clin Ther 2002;24(12):2126–36.

177. Siedlik PH, Olson SC, Yang BB, Stern RH. Erythromycin coadministration increases plasma atorvastatin concentrations. J Clin Pharmacol 1999;39(5):501–4.

178. Ayanian JZ, Fuchs CS, Stone RM. Lovastatin and rhabdomyolysis. Ann Intern Med 1988;109(8):682–3.

179. Landesman KA, Stozek M, Freeman NJ. Rhabdomyolysis associated with the combined use of hydroxymethylglutaryl-coenzyme A reductase inhibitors with gemfibrozil and macrolide antibiotics. Conn Med 1999;63(8):455–7.

180. Wong PW, Dillard TA, Kroenke K. Multiple organ toxicity from addition of erythromycin to long-term lovastatin therapy. South Med J 1998;91(2):202–5.

181. Kantola T, Kivisto KT, Neuvonen PJ. Erythromycin and verapamil considerably increase serum simvastatin and simvastatin acid concentrations. Clin Pharmacol Ther 1998;64(2):177–82.

182. Lee AJ, Maddix DS. Rhabdomyolysis secondary to a drug interaction between simvastatin and clarithromycin. Ann Pharmacother 2001;35(1):26–31.

183. Lee DO, Lee CD. Serotonin syndrome in a child associated with erythromycin and sertraline. Pharmacotherapy 1999;19(7):894–6.

184. Gardner M, Coates P, Hilligoss D, Henry E. Lack of effect of azithromycin on the pharmacokinetics of theophylline in man. In: Proceedings of the Mediterranean Congress of ChemotherapyAthens 1992:1201.

185. Clauzel A, Visier S, Michel F. Efficacy safety of azithromycin in lower respiratory tract infections. Eur Resp J 1990;3(Suppl 10):89.

186. Pollak PT, Slayter KL. Reduced serum theophylline concentrations after discontinuation of azithromycin: evidence for an unusual interaction. Pharmacotherapy 1997;17(4):827–9.

187. Arola O, Peltonen R, Rossi T. Arthritis, uveitis, and Stevens-Johnson syndrome induced by trimethoprim. Lancet 1998;351(9109):1102.

188. Steenbergen JA, Stauffer VL. Potential macrolide interaction with verapamil. Ann Pharmacother 1998;32(3):387–8.

189. Kaeser YA, Brunner F, Drewe J, Haefeli WE. Severe hypotension and bradycardia associated with verapamil and clarithromycin. Am J Health Syst Pharm 1998;55(22):2417–8.

190. Woldtvedt BR, Cahoon CL, Bradley LA, Miller SJ. Possible increased anticoagulation effect of warfarin induced by azithromycin. Ann Pharmacother 1998;32(2):269–70.

191. Bartle WR. Possible warfarin-erythromycin interaction. Arch Intern Med 1980;140(7):985–7.

192. Grau E, Fontcuberta J, Felez J. Erythromycin-oral anticoagulants interaction. Arch Intern Med 1986;146(8):1639.

193. Husserl FE. Erythromycin-warfarin interaction. Arch Intern Med 1983;143(9):18311836.

194. Sato RI, Gray DR, Brown SE. Warfarin interaction with erythromycin. Arch Intern Med 1984;144(12):2413–4.

195. Schwartz JI, Bachmann KA. Erythromycin-warfarin interaction. Arch Intern Med 1984;144(10):2094.

196. Foster DR, Milan NL. Potential interaction between azithromycin and warfarin. Pharmacotherapy 1999;19(7):902–8.

197. Garey KW, Peloquin CA, Godo PG, Nafziger AN, Amsden GW. Lack of effect of zafirlukast on the pharmacokinetics of azithromycin, clarithromycin, and 14-hydroxyclarithromycin in healthy volunteers. Antimicrob Agents Chemother 1999;43(5):1152–5.

198. Tsutsumi K, Kotegawa T, Kuranari M, Otani Y, Morimoto T, Matsuki S, Nakano S. The effect of erythromycin and clarithromycin on the pharmacokinetics of intravenous digoxin in healthy volunteers. J Clin Pharmacol 2002;42(10):1159–64.

199. Boyd RA, Stern RH, Stewart BH, Wu X, Reyner EL, Zegarac EA, Randinitis EJ, Whitfield L. Atorvastatin coadministration may increase digoxin concentrations by inhibition of intestinal P-glycoprotein-mediated secretion. J Clin Pharmacol 2000;40(1):91–8.

200. Echizen H, Tanizaki M, Tatsuno J, Chiba K, Berwick T, Tani M, Gonzalez FJ, Ishizaki T. Identification of CYP3A4 as the enzyme involved in the mono-N-dealkylation of disopyramide enantiomers in humans. Drug Metab Dispos 2000;28(8):937–44.

201. Hayashi Y, Ikeda U, Hashimoto T, Watanabe T, Mitsuhashi T, Shimada K. Torsades de pointes ventricular tachycardia induced by clarithromycin and disopyramide in the presence of hypokalemia. Pacing Clin Electrophysiol 1999;22(4 Pt 1):672–4.

202. Granowitz EV, Tabor KJ, Kirchhoffer JB. Potentially fatal interaction between azithromycin and disopyramide. Pacing Clin Electrophysiol 2000;23(9):1433–5.

203. Iida H, Morita T, Suzuki E, Iwasawa K, Toyo-oka T, Nakajima T. Hypoglycemia induced by interaction between clarithromycin and disopyramide. Jpn Heart J 1999;40(1):91–6.

204. Morlet-Barla N, Narbonne H, Vialettes B. Hypoglycémie grave et récidivante secondaire à l'interaction disopyramide–clarithromicine. [Severe hypoglycemia and recurrence caused by disopyramide–clarithromycin interaction.] Presse Méd 2000;29(24):1351.

205. Laux G. Aktueller stand der Behandlung mit Benzodiazepinen. [Current status of treatment with benzodiazepines.] Nervenarzt 1995;66(5):311–22.

206. Amsden GW. Macrolides versus azalides: a drug interaction update. Ann Pharmacother 1995;29(9):906–17.

207. Rubinstein E. Comparative safety of the different macrolides. Int J Antimicrob Agents 2001;18(Suppl 1):S71–6.

208. Kaeser YA, Brunner F, Drewe J, Haefeli WE. Severe hypotension and bradycardia associated with verapamil and clarithromycin. Am J Health Syst Pharm 1998;55(22):2417–8.

209. Steenbergen JA, Stauffer VL. Potential macrolide interaction with verapamil. Ann Pharmacother 1998;32(3):387–8.

210. Goldschmidt N, Azaz-Livshits T, Gotsman, Nir-Paz R, Ben-Yehuda A, Muszkat M. Compound cardiac toxicity of oral erythromycin and verapamil. Ann Pharmacother 2001;35(11):1396–9.

211. Kantola T, Kivisto KT, Neuvonen PJ. Erythromycin and verapamil considerably increase serum simvastatin and simvastatin acid concentrations. Clin Pharmacol Ther 1998;64(2):177–82.

Midecamycin

See also Macrolide antibiotics

General Information

The reported rates of adverse events among 1565 patients treated with midecamycin 900–1800 mg/day are similar to the rates with other macrolides (1). The adverse events profile is also similar, but precise data are not available. In 12 053 Japanese children under 15 years, the rate of adverse events was 0.54% (2). There were adverse events in 0.97% of very young children (under 1 year). These adverse events affected mainly the skin and the gastrointestinal tract. However, this very low rate of adverse events must be interpreted with caution, since lower dosages (21–40 mg/kg/day) were used than are typically prescribed (50 mg/kg/day) (1).

Some drug interactions can be caused by weak induction of cytochrome P450 by midecamycin (3).

Drug–Drug Interactions

Carbamazepine

The half-life of carbamazepine given as a single daily dose (13% increase) and the AUC were significantly affected (4), whereas the trough concentrations and mean AUCs in nine patients treated with carbamazepine 400, 600, or 1200 mg/day increased only minimally (5).

Ciclosporin

Midecamycin 1600 mg/day increased the steady-state concentration of ciclosporin two-fold (6).

Nicoumalone

Midecamycin did not significantly change the pharmacokinetics of acenocoumarol (nicoumalone) in six volunteers (1).

Theophylline

The pharmacokinetics of theophylline were not significantly affected by midecamycin (7).

References

1. Holliday SM, Faulds D. Miocamycin. A review of its antimicrobial activity, pharmacokinetic properties and therapeutic potential. Drugs 1993;46(4):720–45.
2. Mayama T, Maruyama K, Nakazawa T, Iida M. A survey of the side effects of midecamycin acetate (Miocamycin) dry syrup after marketing. Int J Clin Pharmacol Ther Toxicol 1990;28(6):245–50.
3. Periti P, Mazzei T, Mini E, Novelli A. Pharmacokinetic drug interactions of macrolides. Clin Pharmacokinet 1992;23(2):106–31.
4. Couet W, Istin B, Ingrand I, Girault J, Fourtillan JB. Effect of ponsinomycin on single-dose kinetics and metabolism of carbamazepine. Ther Drug Monit 1990;12(2):144–9.
5. Zagnoni P, De Luca M, Casini A. Carbamazepine–miocamycin interaction. Epilepsia 1991;32(Suppl 1):28.
6. Couet W, Istin B, Seniuta P, Morel D, Potaux L, Fourtillan JB. Effect of ponsinomycin on cyclosporin pharmacokinetics. Eur J Clin Pharmacol 1990;39(2):165–7.
7. Couet W, Ingrand I, Reigner B, Girault J, Bizouard J, Fourtillan JB. Lack of effect of ponsinomycin on the plasma pharmacokinetics of theophylline. Eur J Clin Pharmacol 1989;37(1):101–4.

Minocycline

See also Tetracyclines

General Information

Minocycline, in common with doxycycline, is more lipophilic than other tetracyclines. It is well absorbed after oral administration. Its half-life is 16–18 hours. Its higher affinity for fatty tissues improves its effectiveness and changes its adverse effects profile. Local gastrointestinal irritation and disturbance of the intestinal bacterial flora occur less often than with the more hydrophilic drugs, which have to be given in higher oral doses for sufficient absorption.

Nevertheless, its toxic effects are similar to those of other tetracyclines and arise from accumulation in fatty tissues. Accumulation in a third compartment and the resulting long half-life may contribute to an increased incidence of various toxic adverse effects during long-term treatment, even if lower daily doses are used. This seems also to be the case for pigmentation disorders and possibly for neurological disturbances (1).

Minocycline is predominantly eliminated by the liver and biliary tract (70–90%). Therefore, no change in dose is needed in patients with impaired renal function. However, it should be considered that hepatic elimination of doxycycline might be accelerated by co-administration of agents that induce hepatic enzymes.

Adverse effects of minocycline are reported far more often than adverse effects of other tetracycline derivatives. Whatever the mechanisms underlying this larger number of reports might be, they continue to appear. With increasing use of minocycline in acne and other conditions, adverse reactions may become increasingly common; early recognition is important to prevent further deterioration, to hasten recovery, and to avoid invasive investigations and treatment (2). The authors of this review recommended that safer alternatives be considered in the treatment of acne.

Organs and Systems

Respiratory

Minocycline and nicotinamide therapy for bullous pemphigoid have been associated with severe pneumonitis (3).

Pleurocarditis and eosinophilic pneumonia have been associated with minocycline in a 37-year-old man (4).

Nervous system

The syndrome of pseudotumor cerebri consists of symptoms and signs of raised intracranial pressure in the absence of neuroimaging or cerebrospinal fluid abnormalities. Most cases are idiopathic, but several drugs have been implicated as causative or contributory.

The first description of minocycline-related pseudotumor cerebri was published in 1978 (5). Further data have been published concerning 12 patients who developed pseudotumor cerebri after taking standard doses of minocycline for acne vulgaris (6). Nine developed symptoms within 8 weeks of starting minocycline therapy. Minocycline was withdrawn, and none of the patients developed recurrences for at least 1 year afterwards. However, three patients had substantial residual visual field loss.

- Intracranial hypertension has been attributed to lithium and minocycline in a child (7).
- A 15-year-old woman developed worsening bilateral headache and a perception of intracranial noise (8). She had begun taking minocycline for acne several days before the onset of the headache. There was papilledema with loss of physiological cupping, indistinct disc margins, and small retinal hemorrhages bilaterally. Visual acuity was reduced in her right eye (29/60) but her visual fields were full. She had also bilateral sixth nerve palsies on extreme lateral gaze. Minocycline was withdrawn and she was given acetazolamide 250 mg bd. One month later fundoscopy showed indistinct disc margins and she had episodes of blurred vision; acetazolamide was continued.

Of 243 consecutive patients with a diagnosis of pseudotumor cerebri, 18 had a concurrent history of minocycline treatment (9). The mean duration of minocycline treatment before diagnosis was 2.73 months.

It seems reasonable to assume that the mechanisms may be similar to that postulated for other tetracyclines, which reduce cerebrospinal fluid absorption, possibly by an effect on cyclic adenosine monophosphate at the arachnoid villi (10). Minocycline crosses the blood–brain barrier more effectively than other tetracyclines, because of its greater lipid solubility. Therefore, a physician who prescribes minocycline should keep his eye on the patient's eyes.

For some years it has been known that minocycline has biological effects that are completely separate and distinct from its antimicrobial actions, and in recent years it has been claimed to have neuroprotective effects in animal models of ischemic injury (11,12). Ischemic brain injury involves secondary inflammatory responses that contribute significantly to the clinical outcome (13). In some animal models of other neurological disorders, such as Parkinson's disease, Huntington's disease, autoimmune encephalomyelitis, and amyotrophic lateral sclerosis, minocycline has also been claimed to have some protective effect (14,15). These neuroprotective effects have been attributed to many factors, such as inhibition of inducible nitric oxide synthase (iNOS), caspase 1, caspase 3, and metalloproteases.

However, there are clouds in the skies. In an extensive study minocycline ameliorated brain injury in developing rats, but it increased injury in developing mice (16). This detrimental effect in mice was consistent across different regions (cortex, striatum, and thalamus), with both single and multiple injection protocols and with both moderate and high doses. The mechanism of the contrasting effects in mice and rats remained to be elucidated. The authors warned that caution is warranted before clinical use in infants who have suffered hypoxic–ischemic brain injury. Their data in mice suggest that using minocycline in infants with hypoxia–ischemia could cause more severe brain injury.

The authors of an editorial comment on these findings stressed that it was striking that the inconclusive, negative, or deleterious effects of minocycline have come to light only very recently, for which they suggested several explanations, one of which was the difficulty in getting negative findings published, and they proposed that leading journals should start sections dedicated to the communication of inconclusive, negative, or deleterious results (17).

There is experimental evidence that minocycline may be neuroprotective in Parkinson's disease (18).

Psychiatric

Transient depersonalization symptoms have been attributed to by minocycline (19) and may have been caused by increased intracranial pressure.

Endocrine

A distinctive but rare adverse effect of minocycline is black pigmentation of the thyroid gland, of which about 30 cases have been described (20). It is generally harmless, but can occasionally cause harm (21). Based upon two cases, the authors reviewed 28 previous reports, of which 11 (39%) had been found to harbour papillary carcinoma, strongly suggesting an increased incidence of

thyroid cancer in these pigmented glands. They referred to an old theory that the pigment is formed by oxidation of minocycline by the enzyme thyroid peroxidase (22). Minocycline is stable in the presence of thyroid peroxidase unless an iodide substrate is added, which accelerates both the oxidation of minocycline and the production of a reactive intermediate benzoquinone iminium ionized product. This process in turn produces competitive inhibition of the coupling of tyrosyl residues with thyroglobulin, a necessary step in the production of thyroid hormone.

Whatever the mechanism may be, the high incidence of associated papillary thyroid cancer mandates at a minimum that one ask about the use of minocycline in any patient who has an enlarged thyroid. The authors recommended that if a patient has taken minocycline in the past, biopsy and possibly removal of the thyroid gland is advisable.

Mouth and teeth

Staining of permanent teeth by tetracyclines takes place during tooth development by well-documented mechanisms. Tetracycline forms a complex with calcium orthophosphate during calcification, which then darkens with exposure to light. With minocycline, however, the staining occurs after eruption in previously normal-colored, fully mineralized adult teeth. For example, adult-onset tooth discoloration coincident with minocycline administration occurred in four of 72 patients (23).

There are at least four theories about the mechanisms of this adverse effect.

1. That minocycline attaches to the acquired pellicle glycoproteins (the "extrinsic theory") (24). This in turn etches the enamel, and cycles of demineralization and remineralization occur. It oxidizes on exposure to air or as a result of bacterial activity, and the final product is insoluble black quinine.
2. That minocycline, bound to plasma proteins, is deposited in collagen-rich tissues, such as pulp, teeth, and bones (the "intrinsic" theory). Minocycline is then slowly oxidized over time with exposure to light (25,26).
3. That a breakdown product of minocycline chelates with iron to form insoluble complexes (23,27).
4. That minocycline is deposited in dentine during dentinogenesis, accelerating or changing the process (28).

Whatever the mechanism(s) might be, staining of adult dentition occurs in 3–6% of patients who take long-term minocycline 100 mg/day (24,29). The onset of discoloration can occur at any time from 1 month to many years after the start of treatment (27,29).

The importance of avoiding permanent staining of the teeth cannot be over-emphasized, as many patients with acne are already prone to negative psychological effects. Minocycline should therefore be prescribed only with great care.

Minocycline can cause discoloration of the soft tissues of the mouth, but this has often been attributed to staining of the underlying bone.

- A 45-year-old Caucasian woman developed pigmentation of the gums, lips, and nail beds after taking minocycline 100 mg bd for 6 months (30). Biopsies from the gums and lips showed increased melanin and melanocytes in the epithelium and melanin and melanophages in the connective tissues. Nine months after withdrawal of minocycline there was marked reduction in the pigmentation.

Liver

Hepatotoxicity associated with minocycline has been reviewed, covering data reported to the WHO Centre for International Drug Monitoring, which had recorded 8025 reactions to minocycline, of which 493 were reactions involving the liver (31). The authors stated that fields available to define indications for use, time of treatment, and outcome subsequent to the reactions were seldom completed. They therefore concentrated on more complete records in patients known to have used minocycline for acne. Patients taking minocycline for reasons other than acne or those given intravenous minocycline were excluded. Altogether, 65 patients were then commented on; 58% were women and 94% were aged under 40 years. Briefly, two types of hepatic reactions were recognized: autoimmune hepatitis associated with lupus-like symptoms occurring after 1 year or more of exposure to minocycline, and hypersensitivity reactions associated with eosinophilia and exfoliative dermatitis occurring within 35 days of therapy.

The authors stated that they did not have any clear information about the absolute and relative risks of hepatitis, whether these were hypersensitivity reactions or autoimmune hepatitis, in patients taking minocycline for varying lengths of time, and that a study of the comparative rates of hepatitis in people exposed to minocycline compared with those not exposed is required. In the meantime, new reports of severe hepatic reactions to minocycline continue to appear (32–34), including one case of autoimmune hepatitis requiring liver transplantation in a woman who had used minocycline 50–200 mg/day for 3 years (34). Another case of liver transplantation has previously been reported in patients with hepatic failure after minocycline therapy (35).

Some patients consider drugs to treat a common dermatological disease such as acne vulgaris to be cosmetics rather than medications. Safer alternatives than minocycline should be considered in the treatment of acne.

Pancreas

Minocycline has been associated with acute pancreatitis (36).

Skin

Pigmentation of the skin (37) and skin-related structures, such as the as nail beds (38), teeth (SEDA-26, 266), oral mucosa (39) sclerae (40), and conjunctival cysts (41), is a well-documented adverse effect of minocycline. The bones of the oral cavity are probably the most frequently affected sites of pigmentation. On the skin, the blue-black pigmentation develops most frequently on the shins, ankles, and arms. Other patterns include pigmentation that either is generalized and symmetrical or develops at sites of inflammation. The pigmentation is often permanent when sites other than the skin and oral mucosa are involved (42,43). The pigment is a product of an oxidation reaction. In an experimental rat study, pigmentation of the thyroid gland was prevented by ascorbic acid (25). Laser treatment was successfully used in two cases (44,45).

The incidence of skin discoloration from minocycline varies from 2.4% to almost 15% (46,47). According to a well-accepted classification there are three distinct clinical pictures. Type I is blue-black pigmentation confined to sites of scarring or inflammation on the face. Type II is blue-grey circumscribed pigmentation of normal skin of the lower legs and forearms. Type III is diffuse muddy brown pigmentation of normal skin accentuated in sun-exposed areas. A fourth type has been described (48).

- Two 22-year-old men with acne were given minocycline. The first took minocycline 100 mg/day for 1 month, estimated cumulative dose 3 g, 2 years before presentation). The second took minocycline in two different periods before presentation, estimated dose 13.5 g. Both had blue-grey pigmentation confined to acne scars on the back, whereas scars on the face and chest were unaffected, and there was no hyperpigmentation in other areas. Histology showed pigment within dendritic cells and extracellularly throughout the dermis. Histochemistry identified a calcium-containing melanin-like substance. Electron microscopy showed electron-dense granules, both free and membrane bound, within macrophages and some other cells in the dermis. Energy-dispersive X-ray analysis confirmed the presence of calcium. The patients were followed for 3 years and their pigmentation did not change.

Two patients developed minocycline-induced pigmentation of the soft tissue of the palate, confirmed by biopsy (49). Minocycline-induced hyperpigmentation can even take place in scars (50) and, more rarely, in subcutaneous fat (51).

- A 15-year-old girl with no significant past medical history developed bilateral leg discoloration. The lesions were not painful but were vaguely sensitive to pressure and touch. There was a faint bluish discoloration under the tongue, on the alveolar surface of the gums, and on the hard palate. She had taken minocycline twice daily for acne vulgaris for more than 1 year. A punch biopsy showed normal epidermis and dermis, but deep in the subcutaneous fat there were macrophages and multinuclear giant cells containing brown-greenish pigment in their cytoplasm. A Fontana-Masson stain showed that the pigment was melanin or a melanin-like substance. A Perls stain for iron was negative and a Von Kossa stain was negative for calcium. Minocycline was withdrawn, and after 5 years only subtle hyperpigmentation remained on her legs.

According to the authors, this was the first case of minocycline-induced pigmentation with pigment exclusively localized to subcutaneous fat.

Fixed drug eruptions are characterized by solitary or multiple, round or oval, erythematous patches of variable size; some of them develop into bullae or superficial erosions. Characteristically they appear in the same site each time the responsible drug is given, usually 30 minutes to 8 hours after administration, and with itching or burning as the first symptoms. Over a period of 1–2 weeks they fade, often with crusting and scaling, followed by hyperpigmentation, which can persist for months.

Antimicrobial drugs, especially co-trimoxazole, ampicillin, and tetracyclines, can cause fixed drug eruptions (52), and cases have been attributed to minocycline (53). There is usually no cross-reactivity (54). However, a case of fixed drug eruption has been reported after minocycline in a patient with a previous eruption due to doxycycline (55).

- A 48-year-old man with urethritis took doxycycline 100 mg bd and developed a fixed drug eruption on the glans penis. Doxycycline was withdrawn and the eruption healed with symptomatic treatment. Eight months later he took minocycline 100 mg/day for rosacea. A few hours after having taken the first tablet he became aware of oval erythematous patches on the glans, prepuce, and scrotum. The patches rapidly became violaceous and developed into superficial erosions. He stopped taking minocycline and the lesions faded.

According to the authors, this was the first report of a fixed drug eruption after minocycline in a patient with a previous eruption due to doxycycline. They advised clinicians to be aware of this potential cross-reactivity when prescribing these two drugs.

A generalized pustular eruption was reported in a patient with acne treated with minocycline (56). Skin prick tests with minocycline were positive at 48 hours.

Sweet's syndrome is an acute febrile neutrophilic dermatosis marked by non-pruritic, painful, reddish nodules, most often on the head and neck, the chest, and/or the arms, accompanied by fever, arthralgias, and leukocytosis. Histopathology shows a diffuse dermal neutrophilic infiltration. The pathogenesis is not understood, but there is always prompt resolution of the symptoms and lesions with glucocorticoid therapy. Sweet's syndrome due to minocycline was first reported in 1991 (57), since when several reports have appeared (58,59).

The authors of a systematic review concluded that granulocyte colony-stimulation factor (G-CSF), tretinoin (all-trans retinoic acid, ATRA), and vaccines met two of three criteria for an association with drug-induced Sweet's syndrome (60). There were sufficient data for an association with G-CSF and tretinoin and plausible pharmacological mechanisms. Vaccines met the qualitative criteria and also had a plausible pharmacological mechanism. For minocycline, however, the authors concluded that the evidence was of high quality, but the quantity of evidence was small and a plausible pharmacological mechanism was lacking.

Long-term minocycline often results in *pigmentation* of the skin, nails, bones (61), thyroid, mouth, and eyes; on the skin, the blue-black pigmentation develops most often on the shins, ankles, and arms. In one 15-year-old girl bilateral discoloration on the legs was limited to the subcutaneous adipose tissues and completely spared the dermis; unusually for this form of hyperpigmentation, there was a negative stain for iron (62).

Musculoskeletal

An acute myopathy has been attributed to minocycline.

- A 17-year-old youth, who had taken minocycline 100 mg/day for acne 15 days before admission, abruptly developed diffuse myalgia (63). He had increased creatine kinase, aspartate transaminase, alanine transaminase, lactate dehydrogenase, aldolase, alkaline phosphatase, and gamma-glutamyltransferase activities. Other baseline laboratory results were normal. Minocycline was withdrawn. After 1 month the enzyme activities returned to normal and his symptoms resolved.
- A 20-year-old black professional ballet dancer, with no significant past medical history or other drug use, took minocycline for 3 weeks to treat acne and developed arthralgias and facial swelling (64). She stopped taking minocycline and started to take dexamethasone and loratadine for what she described as "an allergic reaction". However, after about 1 month she was admitted to hospital with erythema and edema around her eyes. Her muscle enzyme activities were increased and a tentative diagnosis of rhabdomyolysis was made.

Immunologic

Immunoallergic reactions have been reported with minocycline and include lupus-like syndrome, autoimmune hepatitis, eosinophilic pneumonia, hypersensitivity syndrome, a serum sickness-like illness (65), and Sweet's syndrome (SEDA-21, 262) (SEDA-22, 271). Over 60 minocycline-induced cases of lupus-like syndrome and 24 cases of minocycline-induced autoimmune hepatitis were found in a review of the literature (66). In 13 patients, both disorders co-existed. These patients had symmetrical polyarthralgia/polyarthritis, raised liver enzymes, and positive antinuclear antibodies; they were also generally antihistone-negative, and only two patients had p-ANCA antibodies. Minocycline-related lupus can also occur in adolescents (67).

Genes may also play a part in minocycline-related autoimmune phenomena. All of 13 patients with pANCA positive titers and minocycline-related lupus were either HLA-DR4 or HLA-DR2 positive (68). Four had also cutaneous involvement. However, biopsies were not performed, and a diagnosis of polyarteritis nodosa could not be made.

Minocycline-induced immunological syndromes, such as cutaneous polyarteritis nodosa and lupus-like syndrome, tend to occur in young women, usually after a prolonged period of minocycline therapy. Several cases of minocycline-induced lupus-like syndrome have also been described after restarting treatment. The results of laboratory tests are often similar in minocycline-induced polyarteritis nodosa and lupus-like syndrome, i.e. positive ANA titers, normal concentrations of complement, absence of antibodies to native DNA, and the presence of perinuclear antineutrophil cytoplasmatic antibodies (pANCA).

The mechanisms underlying these autoimmune manifestations after minocycline therapy are not well understood. Hypotheses include reduced production of free radicals (69) inhibition of phospholipase A2 (70), and altered expressing of tumor necrosis factor and interferon-gamma (71). It has been postulated that minocycline or some of its metabolites may interact with the immune system, and that either drug or metabolites can act as haptens and induce an immune response against myeloperoxidases, enzymes that are involved in the metabolism of minocycline (72). It may be more than a coincidence that other drugs that can cause a lupus-like syndrome, such as hydralazine and procainamide, are also substrates for myeloperoxidases (73,74).

It seems wise to avoid using minocycline as much as possible, advice that seems to have to be repeated over and over again (75).

Lupus-like syndrome

Drug-induced lupus is a well-known phenomenon, although the mechanisms are unclear. The diagnostic features should include no prior history of systemic lupus erythematosus (SLE) before the start of treatment, at least one clinical feature of SLE, a positive antinuclear antibody during sustained drug therapy, and dramatic symptomatic improvement after drug withdrawal. Since 1970, at least 49 drugs have been reported to be associated with drug-related lupus (76), of which hydralazine and procainamide have been claimed to the most commonly implicated. However, new reports on minocycline-induced lupus continue to appear (77–79) and it is possible that minocycline is now the most common cause.

In 1999, Schrodt and Callen were the first to describe cutaneous polyarteritis nodosa in a 15-year-old girl after 9 months of minocycline therapy for acne vulgaris (80).

- An 18-year-old girl used minocycline for acne for 1 year at age 15 ([81A]). At age 18, she had a flare-up and took minocycline 100 mg/day. About 4–6 weeks later she developed flu-like symptoms, tender pink nodules on her legs, and a raised antinuclear antibody (ANA) titer (1: 240) with a nuclear pattern. Her aspartate and alanine transaminase activities were slightly raised. Extractable nuclear antigen, antinative DNA antibodies, anticardiolipin antibodies, and lupus anticoagulant were all negative, and complement C3 and C4 concentrations were normal. A punch biopsy of one of the nodules showed a medium-sized artery surrounded by a mixed infiltrate containing predominantly lymphocytes and histiocytes and occasional eosinophils, with disruption of the vessel and fibrin deposition. Minocycline was withdrawn and her symptoms

gradually abated. About 2 months later her ANA titer was 1: 80 and it became normal after another 2 months. After 8 months all of her symptoms had resolved.
- In a 17-year-old boy minocycline-associated lupus-like syndrome presented with a polyarthropathy, odynophagia, and a flu-like illness (82).
- A 19-year-old woman, who was taking an oral contraceptive and minocycline 100 mg bd, abruptly developed a nodular rash, with multiple violaceous, tender, subcutaneous nodules, 0.5–2 cm in diameter, over her lower legs (83). The overlying skin was slightly warm but non-tender. The lesions resolved after treatment with a glucocorticoid for 1 week but then recurred with bilateral ankle pain, stiffness, and swelling. Perinuclear antineutrophilic cytoplasmic antibodies (pANCA) were positive in a titer of 1:256. A skin biopsy showed necrotizing vasculitis of the small vessels of the dermis, panniculitis characterized by vascular wall neutrophil infiltration, hyalinizing necrosis, and intravascular thrombi, and granuloma formation characterized by epithelioid giant cells.

In a retrospective nested case-control study in 27 688 young patients with acne, 7136 had used minocycline, of whom 29 had the lupus-like syndrome (84). Minocycline was associated with an 8.5-fold risk of the lupus-like syndrome. The effect was greater in longer-term users, but the absolute risk of developing lupus-like syndrome seemed to be relatively low.

Minocycline-induced lupus usually occurs some months, or even years, after the start of therapy, and it usually resolves when the drug is withdrawn. The diagnosis can easily be overlooked, especially in patients with rheumatoid arthritis (85). It would always be wise to follow the recommendation that a patient's antinuclear antibody be checked before starting minocycline and when drug-induced lupus is suspected.

In a retrospective review of drug safety databases, minocycline was the only tetracycline derivative that caused drug-induced lupus (SEDA-22, 268) (86). The authors proposed that the propensity of minocycline to cause drug-induced lupus may be due to the presence of a functional group that is easily oxidized to a reactive metabolite. However, the chemically modified tetracycline CMT-3, which has also reportedly caused a lupus-like syndrome (87), lacks this group, so another theory is needed.

- A 16-year-old girl, who had taken minocycline for acne for more than 2 years, developed a severe cough with paroxysms (88). She had also a recent history of joint pain with swelling and stiffness, fever, general weakness, and weight loss of 9 kg. She had been treated as an outpatient for presumed pneumonia with multiple antibiotics, but developed progressive dyspnea. Pulmonary lupus was suspected, and minocycline was withdrawn. She was treated with an initial 3-day course of intravenous methylprednisolone 20 mg tds, and then prednisone 40 mg bd for 2 weeks. She improved very rapidly, and the prednisone was gradually reduced over 7 weeks.

According to the authors, the patient fulfilled all the criteria for a diagnosis of drug-induced lupus-like syndrome, that is no history of lupus erythematosus before minocycline therapy, the presence of antinuclear antibodies, at least one clinical feature of lupus erythematosus, and prompt recovery after withdrawal of minocycline. She also had positive antihistone antibodies, compatible with drug-induced lupus-like syndrome.

- A 54-year-old woman with a 2-week history of low-grade fever, dry cough, and dyspnea was given levofloxacin for a presumed community-acquired pneumonia (89). Five days later she developed severe respiratory failure and was mechanically ventilated and given antibiotics (imipenem and clarithromycin). Microbiological examination of tracheobronchial aspirates was negative for pathogenic organisms, as were serological tests for common agents of atypical pneumonia. She progressively improved and was taken off the ventilator after 6 days and discharged about 10 days later, but 14 days later was readmitted with rapidly progressive pulmonary failure requiring mechanical ventilation. It then transpired that 2 weeks before the first episode of respiratory failure, she had started to take oral minocycline for acne vulgaris and had started to take it again 24 hours before the second episode. The minocycline was stopped and she was given intravenous methylprednisolone. She improved rapidly, and for 12 months after minocycline withdrawal she remained free of respiratory symptoms.

It is a good rule of thumb that patients who develop minocycline-induced lupus should never be rechallenged with minocycline, as symptoms tend to recur (78,90,91). If minocycline is used to treat rheumatoid arthritis, the patient should be followed very carefully, as worsening arthritis may be erroneously attributed to the underlying disease. It should also be emphasized that in the search for a cause of joint pains, minocycline should always be considered (92).

Autoimmune hepatitis
In a review of relevant American and European literature, hepatitis and drug-induced lupus were reported in 66 cases after minocycline therapy, mostly after long-term treatment for acne (93).

- Minocycline-induced hepatitis with antinuclear, antimitochondrial, and antismooth muscle antibodies has been reported in a 19-year-old black West Indian woman who had been treated for acne for two years with oral minocycline (50 mg/day) and topical benzoyl peroxide (5%) (94).
- Seven patients (17–22 years old) developed symptoms of arthralgia and arthritis after having taken minocycline 50–100 mg bd for 6–36 months for acne vulgaris (95). Increased titers of perinuclear ANCA were detected in all seven, five had fluorescent antinuclear antibodies, two had antihistone autoantibodies, and one had anticardiolipin antibodies. Symptoms resolved in five patients on withdrawal; the other

two were treated with corticosteroids and also achieved remissions.

Autoimmune hepatitis has also been reported (96).

- Three adolescents taking therapeutic doses of minocycline for 12–20 months met the 1993 International Autoimmune Hepatitis Group criteria for autoimmune hepatitis. All had hypogammaglobulinemia and positive antinuclear antibody and antismooth muscle antibody titers. Two underwent liver biopsy that showed severe chronic lymphoplasmocytic inflammation, necrosis, and fibrosis. All other causes of liver disease were excluded. One patient had resolution of symptoms after withdrawal of the drug, while two required immunosuppressive therapy.

Necrotizing vasculitis of the skin and uterine cervix has been reported in a patient taking minocycline (97).

- A 35-year-old woman developed an asymptomatic, erythematous, subcutaneous nodule on the anterior aspect of her left leg. She had been taking minocycline 50–100 mg bd for acne for 2 years, and her only other medication was an oral contraceptive. A presumptive diagnosis of erythema nodosum was made. The nodule was injected twice with triamcinolone acetonide and she was given indometacin 50 mg tds but continued to take minocycline. Some weeks later, after a routine gynecological examination, pathological examination of a cervical biopsy showed necrotizing arteritis of the cervix. A biopsy of the pretibial nodule showed a necrotizing vasculitis with fibrinoid necrosis of the vessel walls and thrombosis. Her serum transaminase activities were increased, and an antinuclear antibody test was positive at 1:60 with a nucleolar pattern. Minocycline was withdrawn, and over the next 3 months her liver function tests normalized, the antinuclear antibody became negative, and the pretibial nodule and uterine cervix vasculitis resolved. There was no recurrence over the next 2 years.

Serum sickness
A serum sickness-like reaction clinically similar to classical serum sickness can result from the administration of a number of non-protein drugs, such as tetracyclines, penicillins, cephalosporins (98). Minocycline has been reported to cause serum sickness (65,99).

- A 16-year-old girl taking minocycline for acne developed typical symptoms of serum sickness after 14 days. Minocycline was withdrawn and her symptoms resolved after a short course of systemic steroids for 5 days.
- An 18-year-old woman developed typical symptoms of serum sickness after taking minocycline 50 mg/day for 10 days for acne. Her symptoms resolved gradually after minocycline was withdrawn.

The authors reviewed some other cases of serum sickness after the use of minocycline. They suggested that the syndrome is under-reported, either because of unawareness that it can be an adverse effect or lack of willingness of physicians to document the event.

Polyarteritis nodosa

A few cases of cutaneous polyarteritis nodosa have been described (100,101,102,103).

Two cases of biopsy-proven cutaneous polyarteritis nodosa with positive perinuclear antineutrophilic cytoplasmic antibodies have been reported with long-term use of minocycline for acne vulgaris (50–100 mg/day for 44 months and 100 mg/day for 65 months) (104). In one of the cases, involvement was not restricted to medium-size vessels alone. In both cases the vasculitis disappeared after a short course of prednisone (40 mg/day) and withdrawal of minocycline. Rechallenge was not performed.

- A 19-year-old woman taking an oral contraceptive abruptly developed a nodular rash on the legs (105). She had taken minocycline 100 mg bd for the past 15 months for pustular acne. There were no cardiac, respiratory, gastrointestinal, musculoskeletal, or neurological symptoms. She had multiple, tender, subcutaneous nodular lesions, 0.5–2 cm in diameter throughout her lower legs. She was given a glucocorticoid and her skin lesions resolved for 1 week and then recurred with bilateral ankle pain, stiffness, and swelling. Perinuclear antineutrophilic cytoplasmic antibodies (pANCA) were positive (titre 1: 256). A skin biopsy showed a necrotizing vasculitis of the small vessels in the dermis and panniculitis with vascular well neutrophilic infiltration, hyalinizing necrosis, and intravascular thrombi. She was given prednisolone 20 mg/day and minocycline was withdrawn. She responded well and 4 months later was free of lesions; the pANCA titer fell to 1: 64.

The authors proposed that six of the following seven criteria should be fulfilled for a diagnosis of minocycline-induced polyarteritis nodosa:

- minocycline use for more than 12 months;
- skin manifestations, including livedo reticularis and/or subcutaneous nodules;
- arthritis and/or myalgias and/or neuropathy in the distribution of the rash;
- lack of systemic organ involvement;
- a skin biopsy with necrotizing vasculitis of small and/or medium-sized vessels
- positive pANCA;
- improvement after withdrawal of minocycline.

This report adds to the growing body of literature regarding drug-induced vasculitis, and, as usual, the mechanisms are not known.

It seems reasonable to assume that minocycline-induced cutaneous polyarteritis nodosa is more common than reported. In some cases the patients stopped taking minocycline themselves and in all cases rapid improvement of symptoms followed withdrawal of medication.

Drug reaction with eosinophilia and systemic symptoms

Drug reaction with eosinophilia and systemic symptoms (DRESS) has most commonly been associated with anticonvulsive drugs, but has been reported with other drug, including minocycline (106,107,108)

- A 15-year-old woman developed a fever and a diffuse erythematous skin eruption after taking minocycline 100 mg/day for 4 weeks for acne vulgaris. Minocycline was withdrawn and she was given oral antihistamines for 1 week followed by 4 days of prednisone in escalating doses from 10 mg/day to 40 mg bd. Despite treatment, she developed progressive erythroderma associated with pruritus, facial swelling, pharyngitis, and diffuse lymphadenopathy. Her white blood cell count was raised dominated by eosinophils (14%). Her serum transaminase activities rose. Viral studies were all negative, thyroid function tests were normal, and antithyroid antibodies were negative. She was treated with corticosteroids, and her cutaneous symptoms, hepatitis, eosinophilia, and leukocytosis gradually improved. She was subsequently found to have autoimmune thyroiditis (Graves disease) and type 1 diabetes.

According to the authors, this is the first reported case of type 1 diabetes following minocycline-induced DRESS. At the time when diabetes was diagnosed, she had raised antinuclear (ANA), anti-Smith, and anti-Sjögren syndrome-A (SS-A/ro) antibody titers. Screening for other organ-specific autoantibodies, including those associated with pernicious anemia, celiac disease, and Addison disease, was negative.

Susceptibility Factors

Renal disease

Minocycline is almost completely eliminated via the liver and the biliary tract and is therefore safe in patients with pre-existing renal insufficiency.

Drug Administration

Drug administration route

The use of minocycline as an agent for inducing chemical pleurodesis has been reported in only a few patients and there is little information regarding its adverse effects. In a retrospective study, 51 patients with primary spontaneous pneumothorax treated by video-assisted thoracoscopic surgery followed by instillation of minocycline hydrochloride 7 mg/kg into the pleural space through a thoracostomy tube were compared with 51 who underwent surgery only (109). Chest pain was a common complaint after minocycline pleurodesis, but the total doses of analgesics were comparable. The rate of prolonged air leaks was significantly lower after minocycline (7 versus 18%). Patients treated with minocycline had shorter periods of postoperative chest drainage and hospitalization. The ipsilateral recurrence rate was also significantly lower (2.9 versus 9.8%).

Drug–Drug Interactions

Phenothiazines

Black galactorrhea was described in a breastfeeding woman during treatment with minocycline and a phenothiazine (110).

References

1. Lander CM. Minocycline-induced benign intracranial hypertension. Clin Exp Neurol 1989;26:161–7.
2. Somech R, Arav-Boger R, Assia A, Spirer Z, Jurgenson U. Complications of minocycline therapy for acne vulgaris: case reports and review of the literature. Pediatr Dermatol 1999;16(6):469–72.
3. Hara H, Fujitsuka A, Morishima C, Kurihara N, Yamaguchi Z, Morishima T. Severe drug-induced pneumonitis associated with minocycline and nicotinamide therapy of a bullous pemphigoid. Acta Derm Venereol 1998;78(5):393–4.
4. Hidalgo Correas FJ, de Andrés Morera S, Ramallal Jiménez de Llano M, Garrote Martínez FJ, García Díaz B. Pleurocarditis y neumonía eosinofílica inducida por minociclina: a propósito de un caso. Farm Hosp 2005;29(2):145–7.
5. Monaco F, Agnetti V, Mutani R. Benign intracranial hypertension after minocycline therapy. Eur Neurol 1978;17(1):48–9.
6. Chiu AM, Chuenkongkaew WL, Cornblath WT, Trobe JD, Digre KB, Dotan SA, Musson KH, Eggenberger ER. Minocycline treatment and pseudotumor cerebri syndrome. Am J Ophthalmol 1998;126(1):116–21.
7. Jonnalagadda J, Saito E, Kafantaris V, Lithium, minocycline, and pseudotumor cerebri. J Am Acad Child Adolesc Psychiatry 2005;44(3):209.
8. Cellucci T, Lee L, Juurlink DN. The headache of teenage acne. Can Med Assoc J 2004;volume:170–1.
9. Kesler A, Goldhammer Y, Hadayer A, Pianka P. The outcome of pseudotumor cerebri induced by tetracycline therapy. Arch Neurol Scand 2004;110:408–11.
10. Walters B, Gubbay S. Tetracycline and benign intracranial hypertension. BMJ (Clin Res Ed) 1981;282(6271):1240.
11. Yrjanheikki J, Keinanen R, Pellikka M, Hokfelt T, Koistinaho J. Tetracyclines inhibit microglial activation and are neuroprotective in global brain ischemia. Proc Natl Acad Sci USA 1998;95:15769–74.
12. Yrjanheikki J, Tikka T, Keinanen R, Goldsteins G, Chan PH, Koistinaho J. A tetracycline derivate, minocycline, reduces inflammation and protect against focal cerebral ischemia with a wide therapeutic window. Proc Natl Soc Acad Sci USA 1999;96:13496–500.
13. Johnston MV, Trescher WH, Ishida A, Nakajima W. Neurobiology of hypoxic-ischemic injury in the developing brain. Pediatr Res 2001;49:735–41.
14. Wee Yong V, Wells J, Guakini F, Casha S, Power C, Metz LM. The promise of minocycline in neurology. Lancet 2004;3:744–51.
15. Stirling DP, Koochesfahani KM, Steeves JD, Tetzlaff W. Minocycline as a neuroprotective agents. Neuroscientist 2005;11:308–22.
16. Tsuji M, Wilson MA, Lange MS, Johnston MV. Minocycline worsens hypoxic–ischemic brain injury in a neonatal mouse model. Exp Neurol 2004;189:58–65.
17. Diguet E, Gross CE, Tison F, Bezard E. Rise and fall of minocycline in neuroprotection: need to promote publication of negative results. Exp Neurol 2004;189:1–4.
18. Thomas M, Le WD. Minocycline: neuroprotective mechanisms in Parkinson's disease. Curr Pharm Des 2004;10:679–86.
19. Cohen PR. Medication-associated depersonalizing symptoms: report of transient depersonalization symptoms induced by minocycline. South Med J 2004;97:70–3.
20. Tsokos M, Schroder S. Black thyroid. Report of an autopsy case. Int J Legal Med 2005;3:1–3.
21. Birkedal C, Tapscott WJ, Giadrosich K, Spence RK, Sperling D. Minocycline-induced black thyroid gland: medical curiosity or a marker for papillary cancer? Curr Surg 2001;58:471–1.
22. Doerge DR, Divi RL, Deck J, Taurog A. Mechanism for the anti-thyroid action of minocycline. Chem Res Toxicol 1997;10:49–58.
23. Poliak SC, DiGiovanna JJ, Gross EG, Gantt G, Peck GL. Minocycline-associated tooth discoloration in young adults. JAMA 1985;254(20):2930–2.
24. Berger RS, Mandel EB, Hayes TJ, Grimwood RR. Minocycline staining of the oral cavity. J Am Acad Dermatol 1989;21(6):1300–1.
25. Bowles WH. Protection against minocycline pigment formation by ascorbic acid (vitamin C). J Esthet Dent 1998;10(4):182–6.
26. Bowles WH, Bokmeyer TJ. Staining of adult teeth by minocycline: binding of minocycline by specific proteins. J Esthet Dent 1997;9(1):30–4.
27. Rosen T, Hoffmann TJ. Minocycline-induced discoloration of the permanent teeth. J Am Acad Dermatol 1989;21(3 Pt 1):569.
28. Good ML, Hussey DL. Minocycline: stain devil? Br J Dermatol 2003;149(2):237–9.
29. Westbury LW, Najera A. Minocycline-induced intraoral pharmacogenic pigmentation: case reports and review of the literature. J Periodontol 1997;68(1):84–91.
30. LaPorta VN, Nikitakis NG, Sindler AJ, Reynolds MA. Minocycline-associated intra-oral soft-tissue pigmentation: clinicopathologic correlations and review. J Clin Periodontol 2005;32(2):119–22.
31. Lawrenson RA, Seaman HE, Sundstrom A, Williams TJ, Farmer RD. Liver damage associated with minocycline use in acne: a systematic review of the published literature and pharmacovigilance data. Drug Saf 2000;23(4):333–49.
32. Kettaneh A, Fain O, Ziol M, Lejeune F, Eclache-Saudreau V, Biaggi A, Guettier-Bouttier C, Thomas M. Minocycline-induced systemic adverse reaction with liver and bone marrow granulomas and Sezary-like cells. Am J Med 2000;108(4):353–4.
33. Nietsch HH, Libman BS, Pansze TW, Eicher JN, Reeves JR, Krawitt EL. Minocycline-induced hepatitis. Am J Gastroenterol 2000;95(10):2993–5.
34. Pohle T, Menzel J, Domschke W. Minocycline and fulminant hepatic failure necessitating liver transplantation. Am J Gastroenterol 2000;95(2):560–1.
35. Boudreaux JP, Hayes DH, Mizrahi S, Hussey J, Regenstein F, Balart L. Fulminant hepatic failure, hepatorenal syndrome, and necrotizing pancreatitis after minocycline hepatotoxicity. Transplant Proc 1993;25(2):1873.
36. Chetaille E, Delcenserie R, Yzet T, Decocq G, Biour M, Andrejak M. Imputabilité de la minocycline dans deux observations de pancréatite aigue. [Minocycline involvement in two cases of acute pancreatitis.] Gastroenterol Clin Biol 1998;22(5):555–6.

37. Fenske, NA. Millns JL. Cutaneous pigmentation due to minocycline hydrochloride. J Am Acad Dermatol 1980;3:308–10.
38. Gordon G, Sparano BM, Iatropolos MJ. Hyperpigmentation of the skin associated with minocycline therapy. Arch Dermatol 1985;121:619–23.
39. Siller G., Tod M, Savage NV. Minocycline induced oral pigmentation. Am Acad Dermatol 1994;30:350–4.
40. Angeloni VL, Salasche SJ, Ortiz R. Nail, skin and scleral pigmentation induced by minocycline. Cutis 1980;40:229–33.
41. Messner E, Font RL, Sheldon G, Murphy D. Pigmented conjunctival cysts following tetracycline/minocycline therapy. Histochemical and electron micrographic observations. Ophthalmology 1983;90:1442–8.
42. Eisen D, Hakim MD. Minocycline-induced pigmentation. Incidence, prevention and management. Drug Saf 1998;18(6):431–40.
43. Cockings JM, Savage NW. Minocycline and oral pigmentation. Aust Dent J 1998;43(1):14–6.
44. Greve B, Schonermark MP, Raulin C. Minocycline-induced hyperpigmentation: treatment with the Q-switched Nd:YAG laser. Lasers Surg Med 1998;22(4):223–7.
45. Karrer S, Szeimies RM, Pfau A, Schroder J, Stolz W, Landthaler M. Minozyklin-induzierte Hyperpigmentierung. [Minocycline-induced hyperpigmentation.] Hautarzt 1998;49(3):219–23.
46. Goulden V, Glass D, Cunliffe WJ. Safety of long-term high dose minocycline in the treatment of acne. Br J Dermatol 1996;134:693–5.
47. Dwyer CM, Cuddihy AM, Kerr REI. Skin pigmentation due to minocycline treatment of facial dermatosis. Br J Dermatol 1993;129:158–62.
48. Mouton RW, Joordan HF, Schneider JW. A new type of minocycline-induced cutaneous hyperpigmentation. Clin Exp Dermatol 2004;29:8–14.
49. Treister N, Magalnick D, Woo SB. Oral mucosal pigmentation secondary to minocycline therapy: report of two cases and a review of the literature. Oral Surg Oral Med Oral Pathol Oral Radiol Endod 2004;97:718–25.
50. Patterson JW, Wilson B, Wick MR, Heath C. Hyperpigmented scar due to minocycline therapy. Cutis 2004;74:293–8.
51. Rahman Z, Lazova R, Antaya RJ. Minocycline hyperpigemtaion isolated to the subcutaneous fat. J Cutan Pathol 2005;32:516–9.
52. Gaffoor PM, George WM. Fixed drug eruptions occurring on the male genitals. Cutis 1990;45(4):242–4.
53. LePaw MI. Fixed drug eruption due to minocycline—report of one case. J Am Acad Dermatol 1983;8(2):263–4.
54. Bargman H. Lack of cross-sensitivity between tetracycline, doxycycline, and minocycline with regard to fixed drug sensitivity to tetracycline. J Am Acad Dermatol 1984;11(5 Pt 1):900–2.
55. Correia O, Delgado L, Polonia J. Genital fixed drug eruption: cross-reactivity between doxycycline and minocycline. Clin Exp Dermatol 1999;24(2):137.
56. Antunes A, Davril A, Trechot P, Grandidier M, Truchetet F, Cuny JF. Syndrome d'hypersensibilité a la minocycline. [Minocycline hypersensitivity syndrome.] Ann Dermatol Venereol 1999;126(6–7):518–21.
57. Mensing H, Kowalzick L, Acute febrile neutrophilic dermatosis (Sweet's syndrome) caused by minocycline. Dermatologica 1991;182:43–6.
58. Thibaut MJ, Bilick RC, Srolovitz H. Minocycline-induced Sweet's syndrome. J Am Acad Dermatol 1992;27:801–4.
59. Khan Durani B, Jappe U. Drug-induced Sweet's syndrome in acne caused by different tetracyclines: case report and review of the literature. Br J Dermatol 2002;147:558–62.
60. Thompson DF, Montarella KE. Drug-induced Sweet's syndrome. Ann Pharmacother 2007;41:802–11.
61. Hepburn MJ, Dooley DP, Hayda RA. Minocycline-induced black bone disease. Orthopedics 2005;28(5):501–2.
62. Rahman Z, Lazova R, Antaya RJ. Minocycline hyperpigmentation isolated to the subcutaneous fat. J Cutan Pathol 2005;32(7):516–9.
63. Narvaez J, Vilaseca-Momplet J. Severe acute myopathy induced by minocycline, Am J Med 2004;116:282–3.
64. Rahman Z, Weinberg J, Scheinfeld N. Minocycline hypersensitivity syndrome manifesting with rhabdomyolysis. Int J Dermatol 2002;41:430.
65. Puyana J, Urena V, Quirce S, Fernandez-Rivas M, Cuevas M, Fraj J. Serum sickness-like syndrome associated with minocycline therapy. Allergy 1990;45(4):313–5.
66. Angulo JM, Sigal LH, Espinoza LR. Coexistent minocycline-induced systemic lupus erythematosus and autoimmune hepatitis. Semin Arthritis Rheum 1998;28(3):187–92.
67. Akin E, Miller LC, Tucker LB. Minocycline-induced lupus in adolescents. Pediatrics 1998;101(5):926.
68. Dunphy J, Oliver M, Rand AL, Lovell CR, McHugh NJ. Antineutrophil cytoplasmic antibodies and HLA class II alleles in minocycline-induced lupus-like syndrome. Br J Dermatol 2000;14:461–7.
69. Miyachi U, Yoshioka A, Imamura S, Niwa Y. Effect of antibiotics on the generation of reactive oxygen species. J Invest Dermatol 1986;86:449–53.
70. Pruzanski W, Greenwald RA, Street IP, Laliberte F, Stefanski E, Vadas P. Inhibition of the enzymatic activity of phospholipase A2 by minocycline and doxycycline. Biochem Pharmacol 1992;44:1165–70.
71. Kloppenburg M, Brinkman BM, de Rooij-Dick HHH, Miltenburg AM, Faha MR, Breedveld FC, Dijkmans BA, Verwilj C. The tetracycline derivative minocycline differentially affects cytokine production by monocytes and T lymphocytes. Antimicrob Agent Chemother 1996;40:934–40.
72. Angulo JM, Sigal LH, Espinoza LR. Coexistent minocycline-induced systemic lupus erythematosus and autoimmune hepatitis. Semin Arthritis Rheum 1998;28:187–92.
73. Hughes GRV. Recent development in drug-associated systemic lupus erythematosus. Adverse Drug React Bull 1987;123:460–3.
74. Yamamoto K, Kawanishi S. Free radical production and site-specific DNA damage induced by hydralazine in the presence of metal ions of peroxidase/hydrogen peroxide. Biochem Pharmacol 1991;41:905–13.
75. McManus P, Iheanacho I. Don't use minocycline as first line oral antibiotic in acne. BMJ 2007;334(7585):154.
76. Hess E. Drug-related lupus. N Engl J Med 1988;318(22):1460–2.
77. Gordon MM, Porter D. Minocycline induced lupus: case series in the West of Scotland. J Rheumatol 2001;28(5):1004–6.
78. Lawson TM, Amos N, Bulgen D, Williams BD. Minocycline-induced lupus: clinical features and response to rechallenge. Rheumatology (Oxford) 2001;40(3):329–35.
79. Graham LE, Bell AL. Minocycline-associated lupus-like syndrome with ulnar neuropathy and antiphospholipid antibody. Clin Rheumatol 2001;20(1):67–9.

80. Schrodt BJ, Callen JP. Polyarteritis nodosa attributable to minocycline treatment for acne vulgaris. Pediatrics 1999;103:146–9.

81. Tehrani R, Nash-Goelitz A, Adams E, Dahiya M, Eilers D. Minocycline-induced cutaneous polyarteritis nodosa. J Clin Rheumatol 2007;13(3):146–9.

82. Leydet H, Armingeat T, Pham T, Lafforgue P. Lupus induit par la minocycline. Rev Med Interne 2006;27(1):72–5.

83. Culver B, Itkin A, Pischel K. Case report and review of minocycline-induced cutaneous polyarteritis nodosa. Arthritis Rheum 2005;53(3):468–70.

84. Sturkenboom MC, Meier CR, Jick H, Stricker BH. Minocycline and lupuslike syndrome in acne patients. Arch Intern Med 1999;159(5):493–7.

85. Marzo-Ortega H, Misbah S, Emery P. Minocycline induced autoimmune disease in rheumatoid arthritis: a missed diagnosis? J Rheumatol 2001;28(2):377–8.

86. Shapiro LE, Knowles SR, Shear NH. Comparative safety of tetracycline, minocycline, and doxycycline. Arch Dermatol 1997;133(10):1224–30.

87. Ghate JV, Turner ML, Rudek MA, Figg WD, Dahut W, Dyer V, Pluda JM, Reed E. Drug-induced lupus associated with COL-3: report of 3 cases. Arch Dermatol 2001;137(4):471–4.

88. Christodoulou CS, Emmanuel P, Ray RA, Good RA, Schnapf BM, Cawkwell GD. Respiratory distress due to minocycline-induced pulmonary lupus. Chest 1999;115(5):1471–3.

89. Oddo M, Liaudet L, Lepori M, Broccard AF, Schaller MD. Relapsing acute respiratory failure induced by minocycline. Chest 2003;123(6):2146–8.

90. Singer SJ, Piazza-Hepp TD, Girardi LS, Moledina NR. Lupuslike reaction associated with minocycline. JAMA 1997;277(4):295–6.

91. Masson C, Chevailler A, Pascaretti C, Legrand E, Bregeon C, Audran M. Minocycline related lupus. J Rheumatol 1996;23(12):2160–1.

92. Bonnotte B, Gresset AC, Chauffert B, Courtois JM, Martin F, Collet E, Sgro C, Lorcerie B. Symptomes évocateurs de maladie de système chez des patients prenant du chlorhydrate de minocycline. [Early signs of systemic disease in patients taking minocycline chlorhydrate.] Presse Méd 1999;28(21):1105–8.

93. Elkayam O, Yaron M, Caspi D. Minocycline-induced autoimmune syndromes: an overview. Semin Arthritis Rheum 1999;28(6):392–7.

94. Pavese P, Sarrot-Reynauld F, Bonadona A, Massot C. Réaction immuno-allergique avec hépatite induite par la minocycline. [Immunoallergic reaction with hepatitis induced by minocycline.] Ann Med Interne (Paris) 1998;149(8):521–3.

95. Elkayam O, Levartovsky D, Brautbar C, Yaron M, Burke M, Vardinon N, Caspi D. Clinical and immunological study of 7 patients with minocycline-induced autoimmune phenomena. Am J Med 1998;105(6):484–7.

96. Teitelbaum JE, Perez-Atayde AR, Cohen M, Bousvaros A, Jonas MM. Minocycline-related autoimmune hepatitis: case series and literature review. Arch Pediatr Adolesc Med 1998;152(11):1132–6.

97. Schrodt BJ, Kulp-Shorten CL, Callen JP. Necrotizing vasculitis of the skin and uterine cervix associated with minocycline therapy for acne vulgaris. South Med J 1999;92(5):502–4.

98. Mannik M. Serum sickness and pathophysiology of immune complexes. In: Rich RR, Fleisher TA, Schwartz BD, editors. Clinical Immunology, Principles and Practice. St Louis: Mosby Year Book Inc, 1996:1062–71.

99. Malakar S, Dhar S, Shah Malakar R. Is serum sickness an uncommon adverse effect of minocycline treatment? Arch Dermatol 2001;137(1):100–1.

100. Schrodt BJ, Callen JP. Polyarteritis nodosa attributable to minocycline treatment for acne vulgaris. Pediatrics 1999;103:503–4.

101. Schaffer JV, Davidson DM, McNiff JM, Bolognia JL. Perinuclear antineutrophilic cytoplasmatic antibody-positive cutaneous polyarteritis nodosa associated with minocycline therapy for acne vulgaris. J Am Acad Dermatol 2001;44:198–206.

102. Pelletier F, Puzenat E, Blanc D, Faivre B, Humbert P, Aubin F. Minocycline-induced cutaneous polyarteritis nodosa with antineutrophilic cytoplasmatic antibodies. Eur J Dermatol 2003;13:396–8.

103. Sakai H, Komatsu S, Matsuo S, Lizuka A. Two cases of minocycline-induced vasculitis. Arerugi [Jpn J Allergol] 2002;51:1153–8.

104. Schaffer JV, Davidson DM, McNiff JM, Bolognia JL. Perinuclear antineutrophilic cytoplasmic antibody-positive cutaneous polyarteritis nodosa associated with minocycline therapy for acne vulgaris. J Am Acad Dermatol 2001;44(2):198–206.

105. Culver B, Itkin A, Pischel K. Case report and review of minocycline-induced cutaneous polyarteritis nodosa. Arthritis Rheum 2005;53:468–70.

106. Peyrière H, Dereure O, Breton H, Demoly P, Cociglio M, Blayac JP, Hillaire-Buys D; Network of the French Pharmacovigilance Centers. Variability in the clinical pattern of cutaneous side-effects of drugs with system symptoms: does a DRESS syndrome really exists? Br J Dermatol 2006;155(2):422–8.

107. Maubec E, Wolkenstein P, Loriot MA, Wechler J, Mulot C, Beaune P, Revuz J, Roujeau JC. Minocycline-induced DRESS: evidence for accumulation of the culprit drug. Dermatology 2008;216:200–4.

108. Brown RJ, Rother KI, Artman H, MercurionMG, Wang R, Looney J, Cowen EW. Minocycline-induced drug hypersensitivity syndrome followed by multiple autoimmune sequelae Arch Dermatol 2009;145:63–6.

109. Chen JS, Hsu HH, Kuo SW, Tsai PR, Chen RJ, Lee JM, Lee YC. Effects of additional minocycline pleurodesis after thoracoscopic procedures for primary spontaneous pneumothorax. Chest 2004;125(1):50–5.

110. Basler RS, Lynch PJ. Black galactorrhea as a consequence of minocycline and phenothiazine therapy. Arch Dermatol 1985;121(3):417–8.

Monobactams

See also Beta-lactam antibiotics

General Information

Two independents groups isolated the first N-thiolated beta-lactams 25 years ago from natural sources. They were given the name "monobactams" because they have a flexible monocyclic ring and lack a carboxylic acid moiety, with no second ring fused to the beta-lactam ring (1,2). However, the highly active compounds of this group do, as a rule, contain the aminothiazole side chain, known to be associated with good antimicrobial activity and beta-lactamase stability in cephalosporins (3). Aztreonam, marketed in 1984, was the first monobactam.

Other compounds, such as carumonam, pirazmonam, and tigemonam, have been developed (4), but so far only aztreonam is widely used. Monobactams are almost exclusively active against Gram-negative bacteria, and their clinical usefulness lies particularly in their antipseudomonal activity.

Like 6-aminopenicillanic acid for the penicillins and 7-aminocephalosporanic acid for the cephalosporins, 3-aminomonobactamic acid is the basic compound for the monobactam family.

In clinical trials the safety profile of aztreonam was similar to or better than that of other more conventional beta-lactam antibiotics, in adult patients with both normal renal function (5,6) and impaired renal function (7). These findings have been confirmed in children (8).

This group has become of increasing interest, because of its effects on mammalian cells. A number of monobactams have been synthesized and their potential as anticancer agents has come to light (9). Some of these derivatives can inducing apoptosis in a wide array of tumor cell types, with little effect on normal cells. Thus, an undesirable side effect on mammalian cells may be turned into a desirable effect on cancer cells.

Organs and Systems

Hematologic

In a child with typhoid fever, severe leukopenia and neutropenia necessitated withdrawal of aztreonam (150 mg/kg/day) on the 10th day, followed by recovery (10). In another case there was myelosuppression (mainly neutropenia) possibly caused by aztreonam after 10 days of treatment; it resolved within 1 week of withdrawal (11). However, acute reversible neutropenia has been otherwise reported in under 0.1% of patients taking aztreonam (8).

Three of twenty-eight patients taking high-dose aztreonam (300 mg/kg/day) for cystic fibrosis developed mild and transient thrombocytopenia (12).

Liver

The frequency of liver function test abnormalities in patients taking monobactams does not differ from the frequencies in patients taking other beta-lactam antibiotics (3,8,12).

Urinary tract

- Acute renal insufficiency associated with a skin rash and eosinophilia has been reported in a 70-year-old man after 9 days of treatment with aztreonam (13).

However, significant nephrotoxicity has not been documented in trials.

Skin

Aztreonam was the suspected cause of toxic epidermal necrolysis in two patients undergoing bone marrow transplantations (14). Graft-versus-host disease seemed a less likely diagnosis of the cutaneous manifestations.

The term erythroderma or generalized exfoliative dermatitis is used to describe any inflammatory skin disease that involves the whole or most of the skin surface. There is usually a poor correlation between the clinical presentation and the histological findings, because the specific cutaneous changes of a dermatosis or a drug reaction are obscured by non-specific changes caused by the erythroderma itself. Erythroderma is uncommon but can be life-threatening.

- A 73-year-old woman was given intravenous clindamycin 600 mg 6-hourly and 7 days later developed generalized erythema with intense pruritus, malaise, and chills (15). The eruptions started on the trunk and spread rapidly over the entire surface of the skin, including the palms and soles. Clindamycin was withdrawn and she was given dexchlorpheniramine and methylprednisolone. She improved slowly, and her symptoms
- A 73-year-old woman was given intravenous clindamycin 600 mg 6-hourly and 7 days later developed generalized erythema with intense pruritus, malaise, and chills (16). The eruptions started on the trunk and spread rapidly over the entire surface of the skin, including the palms and soles. Clindamycin was withdrawn and she was given dexchlorpheniramine and methylprednisolone. She improved slowly, and her symptoms resolved in 12 days. She was given aztreonam and ceftriaxone, which were well tolerated. Two years later, she was given aztreonam 500 mg 8-hourly for a urinary tract infection and developed the same symptoms and signs. Serial skin prick tests and intradermal tests with aztreonam and clindamycin on different days produced delayed reactions (aztreonam at 6?8 hours and clindamycin at 12 hours).

On the basis of the skin tests and the history, the authors identified aztreonam and clindamycin as the cause of these skin reactions. As there are no structural analogies between aztreonam and clindamycin, coincidental sensitization seems more likely than cross-reactivity.

Immunologic

Aztreonam has minor immunogenicity in animals and was associated with a 2% incidence of all presumably immunologically mediated drug reactions in early phase I and II trials (17).

When *Escherichia coli* was co-cultured with mouse peritoneal macrophages and exposed to aztreonam and ceftazidime, there was enhanced secretion of TNF-alpha; imipenem did not do this (18). Both aztreonam and ceftazidime enhanced LPS release from *E. coli* while imipenem did not, consistent with the observed differences in TNF-alpha release. All three antibiotics increased *E. coli*-induced expression of inducible nitric oxide synthase (iNOS), as assessed by both mRNA and protein.

Negligible cross-reactivity has been reported in both animal and human studies involving hapten inhibition, skin tests, and treatment of penicillin-allergic patients with therapeutic doses of aztreonam (17,19–24). Aztreonam therefore seems to be a safe alternative for patients with penicillin allergy. However, the numbers of safely treated patients reported are still small, and immediate type hypersensitivity to aztreonam has been reported in patients with penicillin allergy (25–28).

In a study from Italy all patients over 14 years of age, with a clinical history of immediate reactions to any beta-lactam, and with positive immediate-type skin tests and/or positive specific IgE antibodies to any beta-lactam were studied for their ability to tolerate aztreonam (29). There were 45 patients (27 women and 18 men) with positive skin tests and/or IgE who underwent a skin test with aztreonam. All had negative skin tests and all were asymptomatic after an intramuscular challenge with aztreonam. The authors claimed that their data confirm the general opinion of poor cross-reactivity between beta-lactams and aztreonam. Immediate-type skin tests with aztreonam might represent a simple and rapid diagnostic tool for establishing tolerability in beta-lactam allergic patients who urgently need this type of drugs.

Several cephalosporins, for example ceftazidime, have the same aminothiazole side chain as aztreonam. Sensitization with either drug involving side chain-specific antibodies may therefore predispose to allergy to the other. However, clinical data on this problem are currently not available.

References

1. Ukada A, Kitano K, Kintaka K, Muroi M, Asai M. Sulfazecin and isosulfazecin, novel beta-lactam antibiotics of bacterial origin. Nature 1981;289:590–1.
2 Sykes RB, Cimarusti CM, Bonner DP, Bush K, Floyd DM, Georgopapadakou NH, Koster WM, Liu WC, Parker WL, Principe PA, Rathnum ML, Slusarchyk WA, Trejo WH. Monocyclic beta-lactam antibiotics produced by bacteria. Nature 1981;291:489–91.
3. Hoover JRE. Betalactam antibiotics: structure-activity relationships. In: Demain AL, Solomon NA, editors. Antibiotics Containing the Betalactam-Structure, Part II. Berlin: Springer-Verlag, 1983:119.
4. Kirrstetter R, Durckheimer W. Development of new beta-lactam antibiotics derived from natural and synthetic sources. Pharmazie 1989;44(3):177–85.
5. Newman TJ, Dreslinski GR, Tadros SS. Safety profile of aztreonam in clinical trials. Rev Infect Dis 1985;7(Suppl 4):S648–55.
6. Scully BE, Neu HC. Use of aztreonam in the treatment of serious infections due to multiresistant Gram-negative organisms, including Pseudomonas aeruginosa. Am J Med 1985;78(2):251–61.
7. Sattler FR, Schramm M, Swabb EA. Safety of aztreonam and SQ 26,992 in elderly patients with renal insufficiency. Rev Infect Dis 1985;7(Suppl 4):S622–7.
8. Chartrand SA. Safety and toxicity profile of aztreonam. Pediatr Infect Dis J 1989;8(Suppl 9):S120–3.
9. Kuhn D, Coates C, Daniel K, Chen D, Bhuiyan M, Kaz A,Turos E, Dou QP. Beta-lactams and their potential use as novel anticancer chemotherapeutics chemotherapeutics drugs. Frontiers Biosci 2004;9:2605–17.
10. Tanaka-Kido J, Ortega L, Santos JI. Comparative efficacies of aztreonam and chloramphenicol in children with typhoid fever. Pediatr Infect Dis J 1990;9(1):44–8.
11. Dallal MM, Czachor JS. Aztreonam-induced myelosuppression during treatment of Pseudomonas aeruginosa pneumonia. DICP 1991;25(6):594–7.
12. Schaad UB, Wedgwood-Krucko J, Guenin K, Buehlmann U, Kraemer R. Antipseudomonal therapy in cystic fibrosis: aztreonam and amikacin versus ceftazidime and amikacin administered intravenously followed by oral ciprofloxacin. Eur J Clin Microbiol Infect Dis 1989;8(10):858–65.
13. Pazmino P. Acute renal failure, skin rash, and eosinophilia associated with aztreonam. Am J Nephrol 1988;8(1):68–70.
14. McDonald BJ, Singer JW, Bianco JA. Toxic epidermal necrolysis possibly linked to aztreonam in bone marrow transplant patients. Ann Pharmacother 1992;26(1):34–5.
15. Gonzalo-Garilo MA, de Argila D. Erythroderma due to aztreonam and clindamycin. J Investig Allergol Clin Immunol 2006; 16: 210-1.
16. Gonzalo-Garilo MA, de Argila D. Erythroderma due to aztreonam and clindamycin. J Investig Allergol Clin Immunol 2006; 16: 210-1.
17. Adkinson NF Jr. Immunogenicity and cross-allergenicity of aztreonam. Am J Med 1990;88(3C):S12–5discussion S38–42.
18. Cui W, Lei MG, Silverstein R, Morrison DC. Differential modulation of the induction of inflammatory mediators by antibiotics in mouse macrophages in response to viable Gram-positive and Gram-negative bacteria. J Endotoxin Res 2003;9(4):225–36.
19. Adkinson NF Jr, Swabb EA, Sugerman AA. Immunology of the monobactam aztreonam. Antimicrob Agents Chemother 1984;25(1):93–7.
20. Saxon A, Beall GN, Rohr AS, et al. Immediate hypersensitivity reactions to beta-lactam antibiotics. Urology 1988;31(Suppl):14.
21. Saxon A, Beall GN, Rohr AS, Adelman DC. Immediate hypersensitivity reactions to beta-lactam antibiotics. Ann Intern Med 1987;107(2):204–15.
22. Adkinson NF Jr, Wheeler B Jr, Swabb EA Jr. Clinical tolerance of the monobactam aztreonam in penicillin-allergic subjects. In: Proceedings, 14th International Congress of Chemotherapy 1985:1201 Abstract WS-26–4. Kyoto, Japan.
23. Loria RC, Finnerty N, Wedner HJ. Successful use of aztreonam in a patient who failed oral penicillin desensitization. J Allergy Clin Immunol 1989;83(4):735–7.
24. Jensen T, Koch C, Pedersen SS, Hoiby N. Aztreonam for cystic fibrosis patients who are hypersensitive to other beta-lactams. Lancet 1987;1(8545):1319–20.
25. Hantson P, de Coninck B, Horn JL, Mahieu P. Immediate hypersensitivity to aztreonam and imipenem. BMJ 1991;302(6771):294–5.
26. Iglesias Cadarso A, Saez Jimenez SA, Vidal Pan C, Rodriguez Mosquera M. Aztreonam-induced anaphylaxis. Lancet 1990;336(8717):746–7.
27. Soto Alvarez J, Sacristan del Castillo JA, Sampedro Garcia I, Alsar Ortiz MJ. Immediate hypersensitivity to aztreonam. Lancet 1990;335(8697):1094.
28. Moss RB, McClelland E, Williams RR, Hilman BC, Rubio T, Adkinson NF. Evaluation of the immunologic cross-reactivity of aztreonam in patients with cystic fibrosis

who are allergic to penicillin and/or cephalosporin antibiotics. Rev Infect Dis 1991;13(Suppl 7):S598–607.

29 Patriarca G, Schiavino D, Lombardo C, Altomonte G, De Cinti M, Buonomo A, Nucera E. Tolerability of aztreonam in patients with IgE-mediated hypersensitivity to beta-lactams. Int. J Immunopath Pharmacol 2008;21(2):75–9.

Moxifloxacin

See also Fluoroquinolones

General Information

Moxifloxacin is an 8-methoxyquinolone with enhanced potency against important Gram-positive pathogens, notably *Streptococcus pneumoniae* (penicillin-resistant and penicillin-susceptible strains), and class activity against Gram-negative bacteria. Its activity is not affected by beta-lactamases. Moxifloxacin may therefore represent a promising alternative for treatment of respiratory tract infections (1).

Dosage adjustment is not required for patients of advanced age or those with renal or mild hepatic impairment (2).

Drug interactions with moxifloxacin have been reviewed (3–6).

Observational studies

In a phase I trial, single-dose pharmacokinetics of moxifloxacin have been reported after oral administration of 50–800 mg in 45 healthy Caucasian men. Moxifloxacin was well tolerated (7). There were no serious adverse events, dropouts, or deaths. Only weakness was reported more often in the active treatment group. Other adverse events in subjects taking the active treatment included *Herpes simplex* labialis and an ear disorder. There were no changes in laboratory parameters, electrocardiograms, electroencephalograms, or findings on physical examination. Mean maximum concentrations of moxifloxacin in plasma ranged from 0.29 µg/ml (50 mg dose) to 4.73 µg/ml (800 mg dose) and were reached 0.4–4 hours after drug administration (MICs at which 90% of isolates of penicillin-resistant *Streptococcus pneumoniae* were inhibited were below 0.125 µg/ml). Plasma concentrations fell in a biphasic manner: within 4–5 hours, they fell to 30–55% of the C_{max}. A terminal half-life of 11–14 hours accounted for most of the AUC. Protein binding was about 48%. There was partial tubular reabsorption. No major active metabolites were detected. Concentrations in saliva were higher than in the plasma during the absorption phase, whereas in the terminal phase there was a constant saliva:plasma concentration ratio of 0.5–1.

In an open, randomized, crossover study, the absolute systemic availability of a single 100 mg dose of moxifloxacin was 0.92 in 10 healthy men (mean age 29 years) (8). There was no evidence of active tubular secretion. Both the oral and intravenous formulations were well tolerated, with five reported possible or probable drug-related adverse events, including headache, nausea, and localized urticaria.

In a prospective, uncontrolled, unblinded, phase III trial in 254 patients with community-acquired pneumonia diagnosed by culture or serologically, moxifloxacin (400 mg/day orally for 10 days) produced a bacteriological response of 91% (9). Drug-related adverse events were reported in 33% of patients; nausea (9%), diarrhea (6%), and dizziness (4%) were the most common adverse events.

Moxifloxacin (400 mg/day orally for 7 days) in 12 healthy men significantly reduced the normal oropharyngeal microflora (alpha-hemolytic streptococci and *Neisseriae*), whereas the number of Gram-negative anerobic bacteria increased markedly; however, no new colonizing moxifloxacin-resistant strains were isolated (10). Moxifloxacin caused a significant reduction in enterococci and enterobacteria, while the numbers of staphylococci, streptococci, *Bacillus* species, and *Candida albicans* were unaffected. There was no impact on peptostreptococci, lactobacilli, *Veillonella*, *Bacteroides*, or fusobacteria, but bifidobacteria and *Clostridia* decreased during moxifloxacin administration. The microflora normalized after 35 days (11).

In a multicenter, prospective, double-blind, randomized trial in 455 patients with community-acquired pneumonia oral moxifloxacin 200 mg/day or 400 mg/day for 10 days resulted in a bacteriological response of 91% (12). Most adverse events, possibly or probably related to moxifloxacin, were generally mild or moderate and were mostly related to the digestive system: diarrhea, nausea, and abdominal pain with 200 mg/day; diarrhea, liver function abnormalities, and nausea with 400 mg /day. The drugs were withdrawn because of adverse events in seven of 229 patients taking 200 mg/day, and 11 of 224 taking 400 mg/day.

In a surveillance study in 18 374 patients taking moxifloxacin 400 mg/day for either 5 or 10 days, 18% had adverse events drug-related is 14% (13). Subgroup analyses of this study have also been published (14,15,16). The most common drug-related adverse events were nausea (5.3%), diarrhea (2.2%), and dizziness (2.0%). There was no clinical evidence of increased risk of cardiac dysrhythmias with moxifloxacin treatment.

In a prospective, multicenter trial in 216 patients with acute maxillary sinusitis, oral moxifloxacin 400 mg/day for 7 days resulted in a bacteriological response of 93%. Drug-related adverse events were reported in 13% of moxifloxacin-treated patients and included abdominal pain (2.4%), nausea (2.4%), and diarrhea (1.2%) (17).

Organs and Systems

Cardiovascular

Moxifloxacin blocks the rapid-component delayed-rectifier potassium channel in the heart, and thus prolongs the QT_c interval by 6 minutes after oral administration and 12 minutes after intravenous administration (18). Moxifloxacin carries a greater risk of QT interval prolongation than ciprofloxacin, levofloxacin, and ofloxacin (19), and although the risk of moxifloxacin-induced torsade de pointes is expected to be minimal when the drug is given in the recommended dosage (400 mg/day) (20), moxifloxacin should be used with caution in patients with prodysrhythmic conditions and avoided in patients taking antidysrhythmic drugs, such as quinidine, procainamide, amiodarone, and sotalol (3).

- A man developed a hypertensive crisis and transient left bundle branch block with QT interval prolongation after taking moxifloxacin (21).
- Sinus tachycardia (120/minute) associated with moxifloxacin has been reported in a 49-year-old man about 45 minutes after he took a single 400 mg dose of moxifloxacin (22).

The underlying mechanism may have been vasodilatation, either directly or indirectly, owing to release of histamine, with reflex tachycardia. These effects have been described for other fluoroquinolones.

One case of torsade de pointes has been reported in association with moxifloxacin (23)).

In a prospective observational uncontrolled and monitored study in 13 578 patients with respiratory tract infection moxifloxacin 400 mg/day was associated with 1046 adverse events; there were reports about 678 patients (5%), including 854 drug-related events in 564 patients (4.15%) (24). Of these 1046 adverse events, 95 in 62 patients (0.46%) were serious. Of 34 cardiac adverse events, 25 in 19 patients (0.14%) were thought to have been drug-related: palpitation (n = 13), tachycardia (n = 4), malaise (n = 4), vertigo (n = 3) and pallor (n = 1). All the adverse events were transient and had favorable outcomes.

Nervous system

Syncope has been reported after the use of moxifloxacin (25).

Moxifloxacin has been reported to cause coma (26).

- A 56 year old woman with Child–Pugh A grade alcoholic cirrhosis became comatose after taking oral moxifloxacin 400 mg/day for 8 days. After intravenous treatment with gammahydroxybutyrate (GHB) she became conscious and oriented.

Quinolones inhibit the binding of GABA to its receptor and gammahydroxybutyrate is structurally similar to GABA.

Sensory systems

In in vitro studies in rabbit eyes (27), monkey eyes, and human eyes (28) high concentrations of moxifloxacin ophthalmic solution 0.5% demonstrated a favorable margin of safety in ocular and extraocular tissues and in corneal tissue, which has the highest exposure to the commercial formulation.

However, in clinical studies of corneal wound healing with the use of moxifloxacin ophthalmic solution 0.5%, the findings were inconsistent (29). In two cases sterile corneal ulcers persisted after several weeks of therapy with topical moxifloxacin 0.5% and resolved when antibiotic therapy was changed (30).

Metabolism

In a pooled analysis of 30 (26 controlled, 4 uncontrolled) prospective, phase II/III studies of oral or intravenous moxifloxacin in 8474 subjects no drug-related hypoglycemic events were reported (31).

Gastrointestinal

The most common adverse effects of moxifloxacin are gastrointestinal disturbances (nausea, diarrhea), which are usually mild to moderate in intensity and do not force patients to discontinue treatment (32,33). The frequency of nausea and vomiting is higher than the frequency of diarrhea and abdominal pain (34). Moxifloxacin caused nausea in 7.2% and diarrhea in 5.7% of patients (32).

In a prospective pharmacokinetic study in 12 healthy men the most frequent adverse events, possibly or probably related to moxifloxacin, were generally mild or moderate and were mostly diarrhea, nausea, and abdominal pain (35).

In a double-blind, parallel-group, randomized, multicenter study, 349 adults with acute bacterial rhinosinusitis were randomized to oral telithromycin (800 mg/day for 5 days) or oral moxifloxacin (400 mg/day for 10 days); 28% of those who took moxifloxacin had one or more treatment-related adverse events, most of which were mild to moderate, and there was one or more severe adverse event in 2.3% (36). The most commonly reported adverse effects assessed, in 2% or more of the population, were diarrhea (1.7%) and nausea (4.5%). There was withdrawal in 3.4%.

In a double-blind, randomized, controlled trial in 394 patients who received either intravenous/oral moxifloxacin 400 mg/day or intravenous/oral levofloxacin 500 mg/day for 7–14 days, the rate of adverse events attributed to any cause was significantly higher with moxifloxacin, although the rates of drug-related and serious adverse events were similar in the two treatment arms (37). There was no difference between treatments with regard to the rate of premature withdrawal owing to an adverse event; 17 of 35 events were considered to be drug-related (10 with moxifloxacin and 7 with levofloxacin). There were serious adverse events in 23% of the patients in both treatment groups. The most common adverse event with moxifloxacin was diarrhea (5.6%).

In 36 patients treated with moxifloxacin 400 mg/day orally for resistant tuberculosis, the mean duration of treatment was 6.3 months (38). Twelve patients reported at least one adverse effect due to moxifloxacin, mostly gastrointestinal (n = 8), general (n = 5), and central nervous system (n = 3). In four patients the drug was withdrawn because of major adverse events; there were no irreversible or fatal events.

- A 22-year-old woman developed *Clostridium difficile*-associated diarrhea after taking moxifloxacin 400 mg/day) for 5 days; metronidazole was begun, and the diarrhea resolved with continued moxifloxacin administration (39).

Liver

- A 69-year-old man who took moxifloxacin 400 mg/day for 5 days developed cholestasis (33).

Skin

Moxifloxacin has a low propensity for causing phototoxic reactions relative to other fluoroquinolones (2,3).

Musculoskeletal

In a double-blind, parallel-group, randomized, multicenter study, 349 adults with acute bacterial rhinosinusitis were randomized to oral telithromycin (800 mg/day for 5 days) or oral moxifloxacin (400 mg/day for 10 days) (36). One patient who took moxifloxacin withdrew because of acute Achilles tendonitis.

- A 65-year-old woman developed tendinitis at the insertion of the musculus brachioradialis tendon after taking oral moxifloxacin 400 mg/day for 2 day (40). Despite immediate withdrawal, immobilization of the arm, and local anti-inflammatory treatment, complete clinical resolution took 5 weeks.

Immunologic

In vitro, moxifloxacin has immunomodulatory activity through its capacity to alter the secretion of IL1-α and TNF-α by human monocytes (41).

Hypersensitivity reactions can occur with moxifloxacin.

- A 23-year-old woman developed a moxifloxacin-associated drug hypersensitivity syndrome associated with toxic epidermal necrolysis and fulminant fatal hepatic failure (42). She had no known drug allergies or prior treatment with fluoroquinolones. After 3 days she developed nausea, vomiting, and abdominal pain, and a morbilliform eruption localized to her abdomen. Moxifloxacin was withdrawn. She had increased liver enzymes, fever, and lymphadenopathy. She was given high-dose intravenous immunoglobulin 1g/kg/day for 3 days. However, she required intubation for respiratory support, multipressor therapy for worsening hypotension, and continuous venovenous hemofiltration for acute renal insufficiency. A liver biopsy showed acute hepatitis with hepatocyte necrosis, ^{c}onsistent with a drug reaction or infectious process. Her hepatic function deteriorated further, and despite an orthotopic liver transplant she died 14 days after the start of treatment with moxifloxacin.

Although moxifloxacin is chemically different from other fluoroquinolones, in six patients with hypersensitivity to different fluoroquinolones there was cross-reactivity between moxifloxacin, ciprofloxacin, levofloxacin, and ofloxacin, confirmed by skin tests (43).

Moxifloxacin can cause anaphylactic reactions (44).

A case of simultaneous drug allergies has been reported (45).

- A 32-year-old woman had a generalized urticaria 15 minutes after taking co-amoxiclav and 1 year later developed a non-pruritic micropapular rash some hours after taking moxifloxacin 400 mg.

Susceptibility Factors

In 10 patients with peritonitis who required surgery and drainage of the abdominal cavity who were given a single intravenous infusion of moxifloxacin 400 mg over 1 hour, plasma moxifloxacin concentrations fell from a geometric mean of 3.61 mg/l at 1 hour to 0.36 mg/l at 24 hours (46). Concentrations in peritoneal exudate were highest at 2 hours after the start of the infusion, reached a geometric mean of 3.32 mg/l, and fell to a geometric mean of 0.69 mg/l at 24 hours. The exudate/plasma concentration ratio rose from 1.45 at 2 hours to 1.91 at 24 hours. The AUC tended to be greater in the exudate; the time to peak concentrations was longer in the exudate than in the plasma, as were the half-life and mean residence time.

Drug–Drug Interactions

Antacids

The systemic availability of moxifloxacin is markedly reduced by the co-administration of antacids that contain magnesium or aluminium, unless administration occurs 2 hours before or 4 hours after moxifloxacin (32,47). An interval of 2 hours before or 4 hours after taking antacids ensures that the effect of the interaction is not clinically relevant.

Antidiabetic drugs

In a pooled analysis of 30 (26 controlled, 4 uncontrolled) prospective, phase II/III moxifloxacin studies in 8474 subjects, co-administration of oral antidiabetic drugs did not change the rate of blood glucose increases or decreases in patients with diabetes (37). In five moxifloxacin postmarketing studies (46 130 subjects) there were no episodes of hypoglycemia and two non-drug-related episodes.

Digoxin

There was no clinically relevant effect of moxifloxacin on the pharmacokinetics of digoxin in combination steady-state conditions (4).

Epoetin

- A 76-year-old man with low-risk myelodysplastic syndrome had a major erythroid response to combination therapy with epoetin and moxifloxacin (48). The immunomodulatory effects of moxifloxacin may have explained the synergy with epoetin.

Iron

Iron salts impair the absorption of moxifloxacin (47).

Probenecid

Concomitant administration of probenecid did not affect the elimination of moxifloxacin (4).

Ranitidine

The systemic availability of moxifloxacin was not affected by concurrent administration of ranitidine (4).

Sucralfate

Drugs that contain sucralfate impair the absorption of moxifloxacin (47).

Theophylline

Moxifloxacin did not affect the pharmacokinetics of theophylline or vice versa (49).

Warfarin

Moxifloxacin had no effect on the pharmacokinetics of warfarin in combination steady-state conditions (4). Combination therapy of moxifloxacin 400 mg/day with warfarin can prolong the INR, as reported in three elderly patients; healthy volunteers did not experience this interaction (50).

In five cases an interaction of moxifloxacin with warfarin was suspected to be the cause of a raised international normalized ratio (INR), with clinically significant hemorrhage in one; in three cases the interaction was assessed as probable and in two as possible (51).

Food–Drug Interactions

The effect of food on the pharmacokinetics of moxifloxacin was not clinically important (52). In another study dairy products had no effect on moxifloxacin kinetics (53).

References

1. Esposito S, Noviello S, Ianniello F. Comparative in vitro activity of older and newer fluoroquinolones against respiratory tract pathogens. Chemotherapy 2000;46(5):309–14.
2. Balfour JA, Lamb HM. Moxifloxacin: a review of its clinical potential in the management of community-acquired respiratory tract infections. Drugs 2000;59(1):115–39.
3. Culley CM, Lacy MK, Klutman N, Edwards B. Moxifloxacin: clinical efficacy and safety. Am J Health Syst Pharm 2001;58(5):379–88.
4. Stass H, Kubitza D. Profile of moxifloxacin drug interactions. Clin Infect Dis 2001;32(Suppl 1):S47–50.
5. Stass H, Kubitza D, Schuhly U. Pharmacokinetics, safety and tolerability of moxifloxacin, a novel 8-methoxyfluoroquinolone, after repeated oral administration. Clin Pharmacokinet 2001;40(Suppl 1):1–9.
6. Muijsers RB, Jarvis B. Moxifloxacin in uncomplicated skin and skin structure infections. Drugs 2002;62(6):967–73discussion 974–5.
7. Stass H, Dalhoff A, Kubitza D, Schuhly U. Pharmacokinetics, safety, and tolerability of ascending single doses of moxifloxacin, a new 8-methoxy quinolone, administered to healthy subjects. Antimicrob Agents Chemother 1998;42(8):2060–5.
8. Ballow C, Lettieri J, Agarwal V, Liu P, Stass H, Sullivan JT. Absolute bioavailability of moxifloxacin. Clin Ther 1999;21(3):513–22.
9. Patel T, Pearl J, Williams J, Haverstock D, Church DCommunity Acquired Pneumonia Study Group. Efficacy and safety of ten day moxifloxacin 400 mg once daily in the treatment of patients with community-acquired pneumonia Respir Med 2000;94(2):97–105.
10. Beyer G, Hiemer-Bau M, Ziege S, Edlund C, Lode H, Nord CE. Impact of moxifloxacin versus clarithromycin on normal oropharyngeal microflora. Eur J Clin Microbiol Infect Dis 2000;19(7):548–50.
11. Edlund C, Beyer G, Hiemer-Bau M, Ziege S, Lode H, Nord CE. Comparative effects of moxifloxacin and clarithromycin on the normal intestinal microflora. Scand J Infect Dis 2000;32(1):81–5.
12. Hoeffken G, Meyer HP, Winter J, Verhoef LCAP1 Study Group. The efficacy and safety of two oral moxifloxacin regimens compared to oral clarithromycin in the treatment of community-acquired pneumonia. Respir Med 2001;95(7):553–64.
13. Faich GA, Morganroth J, Whitehouse AB, Brar JS, Arcuri P, Kowalsky SF, Haverstock DC, Celesk RA, Church DA. Clinical experience with moxifloxacin in patients with respiratory tract infections. Ann Pharmacother 2004;38(5):749–54.
14. Ball P, Stahlmann R, Kubin R, Choudhri S, Owens R. Safety profile of oral and intravenous moxifloxacin: cumulative data from clinical trials and postmarketing studies. Clin Ther 2004;26(7):940–50.
15. Katz E, Larsen LS, Fogarty CM, Hamed K, Song J, Choudhri S. Safety and efficacy of sequential i.v. to p.o. moxifloxacin versus conventional combination therapies for the treatment of community-acquired pneumonia in patients requiring initial i.v. therapy. J Emerg Med 2004;27(4):395–405.
16. Johnson P, Cihon C, Herrington J, Choudhri S. Efficacy and tolerability of moxifloxacin in the treatment of acute bacterial sinusitis caused by penicillin-resistant Streptococcus pneumoniae: a pooled analysis. Clin Ther 2004;26(2):224–31.
17. Gehanno P, Berche P, Perrin A. Moxifloxacin in the treatment of acute maxillary sinusitis after first-line treatment failure and acute sinusitis with high risk of complications. J Int Med Res 2003;31:434–47.
18. White CM, Grant EM, Quintiliani R. Moxifloxacin does increase the corrected QT interval. Clin Infect Dis 2001;33(8):1441–4.
19. Moxifloxacin: new preparation. A me-too with more cardiac risks. Prescrire Int 2002;11(62):168–9.
20. Demolis JL, Kubitza D, Tenneze L, Funck-Brentano C. Effect of a single oral dose of moxifloxacin (400 mg and 800 mg) on ventricular repolarization in healthy subjects Clin Pharmacol Ther 2000;68(6):658–66.
21. Salvador Garcia Morillo J, Stiefel Garcia-Junco P, Vallejo Maroto I, Carneado de la Fuente J. Crisis hipertensiva y bloqueo transitorio de rama izquierda con prolongacion del intervalo qt asociados a moxifloxacino. [Hypertensive crisis and transitory left brunch block with QT interval prolongation associated to moxifloxacin.] Med Clin (Barc) 2001;117(5):198–9.
22. Siepmann M, Kirch W. Drug points: tachycardia associated with moxifloxacin. BMJ 2001;322(7277):23.

23. Saravolatz LD, Leggett J. Gatifloxacin, gemifloxacin, and moxifloxacin: the role of 3 newer fluoroquinolones. Clin Infect Dis 2003;37:1210–5.

24. Veyssier P, Voirot P, Begaud B, Funck-Brentano C. Tolerance cardiaque de la moxifloxacine: expérience clinique issue d'une large étude observationnelle française en pratique médicale usuelle (étude IMMEDIAT). [Cardiac tolerance of moxifloxacin: clinical experience from a large observational French study in usual medical practice (IMMEDIAT study) .] Med Mal Infect 2006;36(10):505–12.

25. Carrion Valero F, Facila Rubio L, Marin Pardo J. Sincope tras la administracion de moxifloxacino. [Syncope after administration of moxifloxacin.] Arch Bronconeumol 2000;36(10):603–4.

26. Koehler G, Haimann A, Laferl H, Wenisch C. Rapid reversible coma with intravenous gamma-hydroxybutyrate in a moxifloxacin-treated patient. Clin Drug Investig 2005;25(8):551–4.

27. Aydin E, Kazi AA, Peyman GA, Esfahani MR. Intravitreal toxicity of moxifloxacin. Retina 2006;26(2):187–90.

28. Donaldson KE, Marangon FB, Schatz L, Venkatraman AS, Alfonso EC. The effect of moxifloxacin on the normal human cornea. Curr Med Res Opin 2006;22(10):2073–80.

29. O'Brien TP. Evidence-based review of moxifloxacin. Int Ophthalmol Clin 2006;46(4):61–72.

30. Walter K, Tyler ME. Severe corneal toxicity after topical fluoroquinolone therapy: report of two cases. Cornea 2006;25(7):855–7.

31. Gavin JR, 3rd, Kubin R, Choudhri S, Kubitza D, Himmel H, Gross R, Meyer JM. Moxifloxacin and glucose homeostasis: a pooled-analysis of the evidence from clinical and postmarketing studies. Drug Saf 2004;27(9):671–86.

32. Balfour JA, Wiseman LR. Moxifloxacin. Drugs 1999;57(3):363–73.

33. Soto S, Lopez-Roses L, Avila S, Lancho A, Gonzalez A, Santos E, Urraca B. Moxifloxacin-induced acute liver injury. Am J Gastroenterol 2002;97(7):1853–4.

34. Wilton LV, Pearce GL, Mann RD. A comparison of ciprofloxacin, norfloxacin, ofloxacin, azithromycin and cefixime examined by observational cohort studies. Br J Clin Pharmacol 1996;41(4):277–84.

35. Burkhardt O, Borner K, Stass H, Beyer G, Allewelt M, Nord CE, Lode H. Single- and multiple-dose pharmacokinetics of oral moxifloxacin and clarithromycin, and concentrations in serum, saliva and faeces. Scand J Infect Dis 2002;34(12):898–903.

36. Ferguson BJ, Guzzetta RV, Spector SL, Hadley JA. Efficacy and safety of oral telithromycin once daily for 5 days versus moxifloxacin once daily for 10 days in the treatment of acute bacterial rhinosinusitis. Otolaryngol Head Neck Surg 2004;131(3):207–14.

37. Anzueto A, Niederman MS, Pearle J, Restrepo MI, Heyder A, Choudhri SH. Community-Acquired Pneumonia Recovery in the Elderly (CAPRIE): efficacy and safety of moxifloxacin therapy versus that of levofloxacin therapy. Clin Infect Dis 2006;42(1):73–81.

38. Codecasa LR, Ferrara G, Ferrarese M, Morandi MA, Penati V, Lacchini C, Vaccarino P, Migliori GB. Long-term moxifloxacin in complicated tuberculosis patients with adverse reactions or resistance to first line drugs. Respir Med 2006;100(9):1566–72.

39. Carroll DN. Moxifloxacin-induced Clostridium difficile-associated diarrhea. Pharmacotherapy 2003;23:1517–9.

40. Burkhardt O, Kohnlein T, Pap T, Welte T. Recurrent tendinitis after treatment with two different fluoroquinolones. Scand J Infect Dis 2004;36(4):315–6.

41. Araujo FG, Slifer TL, Remington JS. Effect of moxifloxacin on secretion of cytokines by human monocytes stimulated with lipopolysaccharide. Clin Microbiol Infect 2002;8(1):26–30.

42. Nori S, Nebesio C, Brashear R, Travers JB. Moxifloxacin-associated drug hypersensitivity syndrome with toxic epidermal necrolysis and fulminant hepatic failure. Arch Dermatol 2004;140(12):1537–8.

43. Gonzalez I, Lobera T, Blasco A, del Pozo MD. Immediate hypersensitivity to quinolones: moxifloxacin cross-reactivity. J Investig Allergol Clin Immunol 2005;15(2):146–9.

44. Aleman AM, Quirce S, Cuesta J, Novalbos A, Sastre J. Anaphylactoid reaction caused by moxifloxacin. J Investig Allergol Clin Immunol 2002;12(1):67–8.

45. Gonzalez-Mancebo E, Cuevas M, Gonzalez Gonzalez E, Lara Catedra C, Dolores Alonso M. Simultaneous drug allergies. Allergy 2002;57(10):963–4.

46. Stass H, Rink AD, Delesen H, Kubitza D, Vestweber KH. Pharmacokinetics and peritoneal penetration of moxifloxacin in peritonitis. J Antimicrob Chemother 2006;58(3):693–6.

48. Stass H, Kubitza D. Effects of iron supplements on the oral bioavailability of moxifloxacin, a novel 8-methoxyfluoroquinolone, in humans. Clin Pharmacokinet 2001;40(Suppl 1):57–62.

49. Fragasso A, Mannarella C, Sacco A. Response to erythropoietin and moxifloxacin in a patient with myelodysplastic syndrome non-respondent to erythropoietin alone. Eur J Intern Med 2002;13(8):521–3.

50. O'Connor KA, O'Mahony D. The interaction of moxifloxacin and warfarin in three elderly patients. Eur J Intern Med 2003;14:255–7.

51. Elbe DH, Chang SW. Moxifloxacin-warfarin interaction: a series of five case reports. Ann Pharmacother 2005;39(2):361–4.

52. Stass H, Kubitza D. Lack of pharmacokinetic interaction between moxifloxacin, a novel 8-methoxyfluoroquinolone, and theophylline. Clin Pharmacokinet 2001;40(Suppl 1):63–70.

53. Lettieri J, Vargas R, Agarwal V, Liu P. Effect of food on the pharmacokinetics of a single oral dose of moxifloxacin 400mg in healthy male volunteers. Clin Pharmacokinet 2001;40(Suppl 1):19–25.

54. Stass H, Kubitza D. Effects of dairy products on the oral bioavailability of moxifloxacin, a novel 8-methoxyfluoroquinolone, in healthy volunteers. Clin Pharmacokinet 2001;40(Suppl 1):33–8.

Mupirocin

General Information

Mupirocin is an antibiotic that inhibits bacterial protein synthesis by binding to isoleucyl transfer RNA synthetase. It is primarily bacteriostatic at low concentrations, but is bactericidal in the high concentrations and can have activity against organisms reported to be relatively resistant in vitro.

In a randomized, double-blind, placebo-controlled trial, 97 of 2012 patients treated with mupirocin reported adverse effects, such as rhinorrhea and itching at the site of application (1).

Spontaneous rupture of polyurethane peritoneal catheter has been attributed to mupirocin ointment (2).

- An 84-year-old man with diabetic nephropathy and end-stage renal disease began continuous ambulatory peritoneal dialysis and over the next year had four episodes of exit-site infection and peritonitis and used mupirocin ointment. The exit-site catheter became dilated and during an episode of infection for which he used mupirocin on 6 successive days, a longitudinal rupture developed in the peritoneal catheter, which was removed. The peritoneal liquid contained *Escherichia coli* and *Proteus mirabilis* and the catheter tip contained *E. coli* and *Enterobacter cloacae*. He was treated with ciprofloxacin, without complications, and after 1 month a new peritoneal catheter was inserted.

Organs and Systems

Skin

Mupirocin ointment can occasionally cause allergic contact dermatitis (3).

Long-Term Effects

Drug tolerance

Resistant *Staphylococcus aureus* can emerge during treatment with mupirocin.

In 149 patients undergoing chronic peritoneal dialysis, *S. aureus* was isolated from 26 patients, 25 from the nares, axillae, or groins and one from the catheter exit site (4). So-called "high-level" mupirocin-resistant organisms (minimum inhibitory concentration 256 mg/ml or more) were isolated from four patients. Methicillin-resistant organisms (MRSA) were not detected. One patient with high-level mupirocin-resistant organisms had peritonitis due to *S. aureus*, resulting in treatment failure and catheter loss.

Of 36 patients undergoing continuous ambulatory peritoneal dialysis (mean age 55 years; 21 men), who had been applying mupirocin to the catheter exit site once weekly for an average of 3.1 years before the start of the study, three were nasal carriers of *S. aureus*, and there was only one mupirocin-resistant organism (5). Once-weekly application of mupirocin at catheter exit sites led to comparable rates of colonization by mupirocin-resistant *S. aureus* as did thrice-weekly or more frequent application.

References

1. Perl TM, Cullen JJ, Wenzel RP, Zimmerman MB, Pfaller MA, Sheppard D, Twombley J, French PP, Herwaldt LA. Mupirocin And The Risk Of Staphylococcus Aureus Study Team. Intranasal mupirocin to prevent postoperative *Staphylococcus aureus* infections. N Engl J Med 2002;346(24):1871-7.
2. Riu S, Ruiz CG, Martinez-Vea A, Peralta C, Oliver JA. Spontaneous rupture of polyurethane peritoneal catheter. A possible deleterious effect of mupirocin ointment. Nephrol Dial Transplant 1998;13(7):1870-1.
3. Zappi EG, Brancaccio RR. Allergic contact dermatitis from mupirocin ointment. J Am Acad Dermatol 1997;36(2 Pt 1):266.
4. Annigeri R, Conly J, Vas S, Dedier H, Prakashan KP, Bargman JM, Jassal V, Oreopoulos D. Emergence of mupirocin-resistant *Staphylococcus aureus* in chronic peritoneal dialysis patients using mupirocin prophylaxis to prevent exit-site infection. Perit Dial Int 2001;21(6):554-9.
5. Cavdar C, Atay T, Zeybel M, Celik A, Ozder A, Yildiz S, Gulay Z, Camsari T. Emergence of resistance in staphylococci after long-term mupirocin application in patients on continuous ambulatory peritoneal dialysis. Adv Perit Dial 2004;20:67-70.

Nalidixic acid

General Information

Nalidixic acid, the first quinolone, first introduced in 1962, is now only rarely used and has been supplanted by the fluoroquinolones. It is almost completely absorbed from the gastrointestinal tract and is rapidly eliminated by the kidneys, resulting in urinary concentrations 4-6 times higher than plasma concentrations. There are better drugs to treat urinary tract infections.

General adverse effects

The adverse effects of nalidixic acid tend to be toxic rather than allergic. Nervous system toxicity is predominant, including disturbances of sensory perception and benign intracranial hypertension, which is nearly exclusively found in babies and young children. Gastrointestinal toxicity and skin reactions also occur. Immunological and hypersensitivity reactions mainly affect the skin. Drug fever is rare. Hematological and hepatic reactions are very rare. Tumor-inducing effects have not been reported but require further study, since nalidixic acid, as a DNA gyrase inhibitor, damages DNA (1,2). Fluorinated derivatives, but not nalidixic acid, induced unscheduled DNA synthesis in vitro but not in vivo. While specific fluoroquinolones, such as CP-115,953, have cytotoxic effects on eukaryotic cancer cells, the fluoroquinolones designed specifically as antimicrobial agents have extremely low toxicity against eukaryotic cells. This is due to a marked reduction in the affinity for eukaryotic topoisomerases versus prokaryotic topoisomerases (3,4).

Organs and Systems

Nervous system

Benign intracranial hypertension (pseudotumor cerebri) mainly affects babies, especially during the first 3 months of life. Occasionally even older children can be affected, especially when inordinately high doses are used (5). Very rarely, it occurs in adults with renal insufficiency. In infancy, impaired nalidixic acid elimination (due to underdeveloped glucuronidation), overdosage, or prolonged treatment may be responsible. Metabolic acidosis is usually important in adults (6).

The syndrome includes headache, nausea and vomiting, dizziness, tinnitus, papilledema, and visual disturbances caused by scotoma or optic nerve damage. In babies, the first symptoms appear during the first 3 days of treatment; while in older children and adults symptoms may not appear before a second or even later exposure. In adults, peripheral paresis or severe pyramidal and extrapyramidal symptoms, occasionally followed by a transitory psychosis, have been observed (7).

After withdrawal, most symptoms usually subside quickly. Papillary edema and ocular palsy recede more slowly, the latter not always completely.

Functional short-lasting phenomena, such as sensations of over-bright lights, blurred vision, altered color perception, or difficulty in focusing, are dose-dependent and have been observed in about 7% of cases (8). They begin 30–60 minutes after drug intake and subside within 20 minutes to 3 hours.

Convulsions without benign intracranial hypertension have been reported (9). Affected patients usually had pre-existing cerebral disease or had taken very high doses. Normal doses in otherwise healthy persons can exceptionally provoke convulsions.

Uncharacteristic general symptoms, such as headache, dizziness, drowsiness, insomnia, or restlessness, can occur (10).

Sensory systems

Occasional cases of tinnitus have been attributed to nalidixic acid (8).

Psychological, psychiatric

Disturbances of body perception, hallucinations, confusion, confabulation, and depression can rarely occur with nalidixic acid (11).

Endocrine

Hyperglycemia associated with convulsions has been observed after the use of high single doses of nalidixic acid (7,12).

Metabolism

Nalidixic acid can cause metabolic acidosis in infants (13). This has also been seen in older children and adults with renal insufficiency and can result from disturbed lactate metabolism. Extreme overdosage can cause metabolic acidosis in subjects with normal renal function (12,14).

Hematologic

Rarely, hemolytic anemia associated with glucose-6-phosphate dehydrogenase deficiency has been precipitated by nalidixic acid (15,16).

Rarely hemolytic anemia in neonates has been attributed to nalidixic acid in breast milk (15).

Gastrointestinal

Nausea, vomiting, and occasionally diarrhea or abdominal pain occur in about 8% of patients taking nalidixic acid (17).

Liver

Cholestatic hepatitis has been reported in patients taking nalidixic acid (18).

Skin

Skin reactions occur in about 5% of patients taking nalidixic acid (19), including urticaria, erythematous or maculopapular rashes, isolated pruritus, purpura, lesions resembling pityriasis rosea, erythema multiforme, or exfoliative dermatitis. Except for the latter, they usually run a benign course.

Phototoxic reactions occur in animals given nalidixic acid after exposure to light (20) and have been reported in patients taking nalidixic acid, with blistering (21,22). Sometimes the lesions develop only several days after nalidixic acid withdrawal. The eruptions can persist for several months after withdrawal.

Musculoskeletal

Large doses of nalidixic acid cause degenerative inflammatory damage in the large weight-bearing joints in animals, and occasionally arthritis, arthralgia, or myalgia occur in humans (17). However, three retrospective controlled studies showed no evidence of nalidixic acid-associated arthropathy in children (23,24).

Long-Term Effects

Drug tolerance

Salmonella typhimurium DT104 is usually resistant to ampicillin, chloramphenicol, streptomycin, sulfonamides, and tetracycline. An outbreak of 25 culture-confirmed cases of multidrug-resistant *S. typhimurium* DT104 has been identified in Denmark (25). The strain was resistant to the above-mentioned antibiotics and nalidixic acid and had reduced susceptibility to fluoroquinolones. A swineherd was identified as the primary source (26). The DT104 strain was also found in cases of salmonellosis in Washington State, and soft cheese made with unpasteurized milk was identified as an important vehicle of its transmission (27).

Mutagenicity

The genotoxic effects of nalidixic acid (400 mg bd for 10 days) and metronidazole (250 mg tds for 10 days) have been investigated in a prospective randomized study in 20 patients with *Trichomonas vaginalis* infections (28). Evaluation was by the sister-chromatid exchange test, in which an increased number of exchanges in lymphocytes reflects mutagenic action. Metronidazole had no effect but there was a significant increase with nalidixic acid.

Second-Generation Effects

Teratogenicity

In a case-control surveillance of congenital abnormalities in 22 865 women who had neonates or fetuses with congenital abnormalities, and 38 151 pregnant women who had neonates without any defects, treatment with nalidixic acid during pregnancy was associated with an increased risk of pyloric stenosis (29). Nalidixic acid should be avoided during pregnancy.

Lactation

Very small amounts of nalidixic acid pass into the breast milk. Use during lactation should therefore also be avoided. Rare cases of hemolytic anemia in the newborn have been related to nalidixic acid ingested through breast milk (15).

Drug Administration

Drug overdose

In 18 children with nalidixic acid intoxication, most of whom were aged under 1 year, the clinical effects were neurological disorders of alertness, hypertensive cranial syndrome, and neuronal damage, some had a metabolic acidosis (30). Treatment included gastric lavage, correction of acid-base balance, and control of convulsions.

- A woman survived ingestion of nalidixic acid 32 g despite developing lactic acidosis, hyperglycemia, convulsions, and abnormal behavior (12). The maximum recorded plasma concentration of nalidixic acid was 185 µg/ml and the half-life was 3.2 hours. Carboxynalidixic acid was demonstrated in the plasma.
- A woman developed a severe metabolic acidosis and coma after taking nalidixic acid 28 g (14). She was given sodium bicarbonate 600 mmol and developed a respiratory alkalosis with secondary tetany. She recovered consciousness 9 hours later and the acid–base disturbance resolved after 60 hours.

Drug–Drug Interactions

Nitrofurantoin

Antagonism in antibacterial efficacy between nitrofurantoin and nalidixic acid has been observed (35).

Urine alkalinizing compounds

Alkalinization of the urine increases the excretion of nalidixic acid (31).

Warfarin

Nalidixic acid can displace warfarin from its plasma albumin binding sites, briefly enhancing its anticoagulant effect (32).

Interference with Diagnostic Tests

Glucose measurements

If reducing agents are given with nalidixic acid, glycosuria and hyperglycemia can be mimicked by the presence of nalidixic and glucuronic acid compounds in blood and urine (33).

Ketosteroids

Urine concentrations of C17-ketosteroids (but not of C17-hydroxysteroids) can be falsely raised by nalidixic acid (34).

References

1. McCoy EC, Petrullo LA, Rosenkranz HS. Non-mutagenic genotoxicants: novobiocin and nalidixic acid, 2 inhibitors of DNA gyrase. Mutat Res 1980;79(1):33–43.
2. Wright HT, Nurse KC, Goldstein DJ. Nalidixic acid, oxolinic acid, and novobiocin inhibit yeast glycyl- and leucyl-transfer RNA synthetases. Science 1981;213(4506):455–6.
3. Stratton C. The safety profile of fluoroquinolones. Antimicrob Infect Dis Newslett 1998;17.57–60.
4. Elsea SH, Osheroff N, Nitiss JL. Cytotoxicity of quinolones toward eukaryotic cells. Identification of topoisomerase II as the primary cellular target for the quinolone CP-115,953 in yeast. J Biol Chem 1992;267(19):13150–3.
5. Riyaz A, Aboobacker CM, Sreelatha PR. Nalidixic acid induced pseudotumour cerebri in children. J Indian Med Assoc 1998;96(10):308314.
6. Mobbs JP, Balant L, Revillard C, Favre H. Effets secondaires de l'acide nalidixique chez une patiente atteinte d'insuffisance rénale sévère: étude clinique et proposition d'un modèle pharmacocinétique. [Side effects of nalidixic acid in a patient with severe renal failure. Clinical study and proposal of a pharmacokinetic model.] Schweiz Med Wochenschr 1977;107(9):300–6.
7. Islam MA, Sreedharan T. Convulsions, hyperglycemia and glycosuria from overdose of nalidixic acid. JAMA 1965;192:1100–1.
8. Fish DN. Fluoroquinolone adverse effects and drug interactions. Pharmacotherapy 2001;21(10 Pt 2):S253–72.
9. Fraser AG, Harrower AD. Convulsions and hyperglycaemia associated with nalidixic acid. BMJ 1977;2(6101):1518.
10. Bhatt RV. Management of recurrent urinary infection in females. Role of a new antibacterial agent. Indian J Surg 1975;373Oct/Nov/Dec.
11. Mougeot G, Hugues FC. Bouffées délirantes confuso–oniriques provoquées par l'acide nalidixique. [Acute confusional–hallucinatory states caused by nalidixic acid.] Nouv Presse Méd 1980;9(7):455.
12. Leslie PJ, Cregeen RJ, Proudfoot AT. Lactic acidosis, hyperglycaemia and convulsions following nalidixic acid overdosage. Hum Toxicol 1984;3(3):239–43.

13. Suganthi AR, Ramanan AS, Pandit N, Yeshwanth M. Severe metabolic acidosis in nalidixic acid overdosage. Indian Pediatr 1993;30(8):1025–6.

14. Nogue S, Bertran A, Mas A, Nadal P, Anguita A, Milla J. Metabolic acidosis and coma due to an overdose of nalidixic acid. Intensive Care Med 1979;5(3):141–2.

15. Belton EM, Jones RV. Haemolytic anaemia due to nalidixic acid. Lancet 1965;2(7414):691.

16. Gilbertson C, Jones DR. Haemolytic anaemia with nalidixic acid. BMJ 1972;4(838):493.

17. Gleckman R, Alvarez S, Joubert DW, Matthews SJ. Drug therapy reviews: nalidixic acid. Am J Hosp Pharm 1979;36(8):1071–6.

18. Rached-Mohassel MA, Hanjani AA, Nik-Akhtar B, Chariat C, Tabibi M. Hépatite choléstatique induite par l'administration prolongée de l'acide nalidixique (rapport d'un cas). [Cholestatic hepatitis induced by prolonged administration of nalidixic acid (case report).] Acta Med Iran 1974;17(1–2):47–56.

19. Alexander S, Forman L. Which of the drugs caused the rash? Or the value of the lymphocyte transformation test in eruptions caused by nalidixic acid. Br J Dermatol 1971;84(5):429–34.

20. Horio T, Miyauchi H, Asada Y, Aoki Y, Harada M. Phototoxicity and photoallergenicity of quinolones in guinea pigs. J Dermatol Sci 1994;7(2):130–5.

21. Zelickson AS. Phototoxic reaction with nalidixic acid. JAMA 1964;190:556–7.

22. Birkett DA, Garretts M, Stevenson CJ. Phototoxic bullous eruptions due to nalidixic acid. Br J Dermatol 1969;81(5):342–4.

23. Schaad UB, Wedgwood-Krucko J. Nalidixic acid in children: retrospective matched controlled study for cartilage toxicity. Infection 1987;15(3):165–8.

24. Adam D. Use of quinolones in pediatric patients. Rev Infect Dis 1989;11(Suppl 5):S1113–6.

25. Leone R, Ghiotto E, Conforti A, Velo G. Potential interaction between warfarin and ocular chloramphenicol. Ann Pharmacother 1999;33(1):114.

26. Bui L, Huang DD. Possible interaction between cyclosporine and chloramphenicol. Ann Pharmacother 1999;33(2):252–3.

27. Sohaskey CD, Barbour AG. Esterases in serum-containing growth media counteract chloramphenicol acetyltransferase activity in vitro. Antimicrob Agents Chemother 1999;43(3):655–60.

28. Akyol D, Mungan T, Baltaci V. A comparative study of genotoxic effects in the treatment of Trichomonas vaginalis infection: metronidazole or nalidixic acid. Arch Gynecol Obstet 2000;264(1):20–3.

29. Czeizel AE, Sorensen HT, Rockenbauer M, Olsen J. A population-based case-control teratologic study of nalidixic acid. Int J Gynaecol Obstet 2001;73(3):221–8.

30. Games Eternod J, Juarez Aragon G, Martinez Garcia MC, Ochoa Grijalva P, Palacios Trevino JL. Intoxicacion por acido nalidixico en niños. [Poisoning by nalidixic acid in children.] Bol Med Hosp Infant Mex 1980;37(5):963–9.

31. Portmann GA, McChesney EW, Stander H, Moore WE. Pharmacokinetic model for nalidixic acid in man. II. Parameters for absorption, metabolism, and elimination. J Pharm Sci 1966;55(1):72–8.

32. Hoffbrand BI. Interaction of nalidixic acid and warfarin. BMJ 1974;2(920):666.

33. Klumpp TG. Nalidixic acid··· false-positive glycosuria and hyperglycemia. JAMA 1965;193:746.

34. Llerena O, Pearson OH. Interference of nalidixic acid in urinary 17-ketosteroid determinations. N Engl J Med 1968;279(18):983–4.

35. Stille W, Ostner KH. Antagonismus Nitrofurantoin–Nalidixinsäure. [Nitrofurantoin–Nalidixic acid antagonism.] Klin Wochenschr 1966;44(3):155–6.

Nitrofurantoin

General Information

Nitrofurantoin is a synthetic nitrofuran that belongs to a group of organic substances characterized by a heterocyclic ring consisting of four carbon atoms and one oxygen atom (1). Its mechanism of action is not understood. Nitrofurans enter bacterial cells and interact with several enzymes; thereby inhibiting bacterial growth. Nitrofurantoin is active against many Gram-positive and Gram-negative bacteria. It is particularly useful in treating urinary pathogens, including *Escherichia coli*, *Klebsiella* species, and *Proteus* species. However, *Pseudomonas aeruginosa* is almost always resistant.

After almost complete absorption from the gastrointestinal tract, nitrofurantoin produces high urinary concentrations. Blood and tissue concentrations are low. Nitrofurantoin has been used almost exclusively in the treatment and prophylaxis of urinary tract infections (1). Because of the severity of its adverse effects it should not be used as first choice.

General adverse effects

Harmless gastrointestinal adverse effects are most frequent with nitrofurantoin. Polyneuropathy occurs mainly in patients with renal insufficiency, in whom nitrofurantoin is contraindicated. Rarely, hemolytic anemia and hepatitis occur. Adverse events that are believed to be due to hypersensitivity include rashes, generalized urticaria, and acute pulmonary reactions. Although there are some in vitro data suggesting that nitrofurantoin may be a mutagen, a carcinogenic effect has not been proven in vivo. Teratogenic effects of nitrofurantoin have not been reported. High-dose oral nitrofurantoin may influence spermatogenesis and sperm motility.

Frequency

Drug-related adverse events are rare with nitrofurantoin and occur in fewer patients than with co-trimoxazole or trimethoprim, for example (1–3). The frequency of certain adverse reactions varies in different geographical areas (4,5). Lung reactions are more prevalent in Scandinavia and South Africa than in the UK, whereas polyneuropathies or gastrointestinal reactions are more frequent in the UK than in Sweden. These discrepancies are unexplained.

Organs and Systems

Cardiovascular

Except for cardiovascular collapse in anaphylactic shock, adverse cardiovascular events seem to be extremely rare with nitrofurantoin (4,5). In experimental animals, cardiotoxic effects have been described (6).

Respiratory

Acute respiratory reactions to nitrofurantoin include dyspnea, cough, interstitial pneumonitis, and pleural effusion, while interstitial pneumonitis and fibrosis are common chronic reactions (7). Nitrofurantoin causes acute lung injury more often than any other drug. Since the first well-documented case of an acute lung reaction in 1962 (8), several hundred further observations have been published (5,9).

The frequency of acute severe pulmonary disease has been estimated to be one in every 5000 first administrations (10,11,12) and include dyspnea, cough, interstitial pneumonitis, and pleural effusion; interstitial pneumonitis and fibrosis are common chronic reactions (13). Acute respiratory reactions to nitrofurantoin include dyspnea, cough, interstitial pneumonitis, and pleural effusion; interstitial pneumonitis and fibrosis are common chronic reactions (14,15,16,17).

Women aged 40–50 years are mainly affected. The acute lung reactions are not dose-related, and sensitization occurs at the earliest 1–2 weeks after the onset of exposure during the first course of therapy. Symptoms develop 2–10 hours after administration and consist of severe dyspnea, tachypnea, non-productive cough, high fever (usually with chills), cyanosis, and chest pain. Occasionally, arthralgia, backache or headache, vomiting, rash, collapse, and anaphylactic shock accompany the pulmonary symptoms 18. Lung findings include dense crackles or moist râles, predominantly at the lung bases. X-ray examination may be normal, but more often shows bilateral interstitial lower lobe infiltrates, often with pleural effusion. In one case, transient reverse ventilation-perfusion mismatch was documented by scintigraphy (18). Initially the leukocyte count is normal or raised, with neutrophilia and lymphopenia. Later, eosinophilia is common. When nitrofurantoin is withdrawn, clinical symptoms subside rapidly, usually within 1–3 days (5). However, minor X-ray changes can still be found 2 months later. Re-exposure to nitrofurantoin 50 mg re-induces the syndrome. Single cases of death due to heart failure have been reported in debilitated patients.

In a retrospective analysis of 18 patients with chronic nitrofurantoin-induced lung disease (median age 72 years, range 47–90; 17 women) the median daily dosage of nitrofurantoin was 100 mg (range 50–200) (19). The onset of symptoms occurred after a median duration of 23 months (range 10–144). Although the clinical and radiological presentation is non-specific, there are characteristic features on CT that are distinctly different from those seen in idiopathic pulmonary fibrosis. Most patients with chronic nitrofurantoin-induced lung disease will improve, either on withdrawal of nitrofurantoin alone or with the addition of a glucocorticoid. However, it is not uncommon for residual infiltrates to persist in the long term.

Acute lung reactions to nitrofurantoin are extremely rare in children (20). Lung tissue findings in acute reactions have shown minor vasculitis, granulomatous vasculitis (hypersensitivity angiitis), proliferation of endothelial cells, and empty alveoli (21). Rapidly progressing bronchiolitis obliterans with organizing pneumonia (BOOP) has been reported (22).

Chronic lung reactions are 10–20 times less frequent than acute reactions and mainly involve older patients. Reactions serious enough to require hospitalization occur in one out of 750 long-term users (23). Acute reactions do not seem to predispose to the later occurrence of chronic reactions. During long-term treatment, dyspnea and usually a non-productive cough without fever develop (23). Restrictive respiratory impairment is common. X-rays show interstitial infiltrations, often in the middle and basal lung regions. Fibrotic changes, alveolar exudates, and pleural effusions are rare. Histologically, chronic interstitial pneumonitis with varying degrees of fibrosis most often is found, and in some instances desquamative alveolitis (23–27).

Several mechanisms for the adverse lung effects of nitrofurantoin have been proposed (28–32). The pathogenesis of the acute lung reactions may be allergic (type III reactions) (26). However, there is also evidence that a cytotoxic immune mechanism (type II reactions), cell-mediated immunity (type IV reactions) (2), or direct toxic injury to lung tissue through the production of oxygen radicals (3,4) may be involved (28,29,30,31). In chronic lung reactions, the causative role of nitrofurantoin is less evident. It is supported by analogy and by the clinical course. Sometimes the skin test is positive, even in chronic lung reactions. Lymphocyte transformation tests give variable results. A polyclonal hypergammaglobulinemia is always present, with IgG predominating. Precipitating serum antibodies have not been found. Recent data have supported a toxic pathogenesis similar to that of the herbicide paraquat.

After nitrofurantoin withdrawal, the clinical symptoms regress rapidly. However, in most cases X-ray findings abate slowly and resolution is incomplete in at least 50% of patients. Occasional deaths due to cardiopulmonary failure have also been reported. The therapeutic benefit of corticosteroids is controversial, but the bulk of experience and anecdotal reports suggest that they are useful.

Atypical courses have been rarely described (15). A mixed type of reaction can occur: after an initial short fever peak, the patient becomes either afebrile or subfebrile, despite continuing to take nitrofurantoin and unabated activity of the lung process, or a typical chronic reaction converts to a typical acute reaction on re-exposure to nitrofurantoin after withdrawal. Acute reactions can occur without clinical symptoms, and can be recognized only on X-ray. Single cases of pulmonary hemorrhage, eosinophilic pneumonia, and interstitial giant cell pneumonia have also been reported (34,35).

In nitrofurantoin-induced pulmonary toxicity, in which high-resolution computed tomography initially showed a widespread reticular pattern and associated distortion of the lung parenchyma, thought to represent established and irreversible fibrosis, follow-up scans after drug withdrawal nevertheless showed resolution of pulmonary changes (36). These findings have been corroborated by a report of two middle-aged women who developed respiratory symptoms after prolonged treatment with nitrofurantoin (37). Both had impaired lung function and abnormal CT scans, and lung biopsies showed features compatible with bronchiolitis obliterans organizing pneumonia. Their condition improved when nitrofurantoin was withdrawn and glucocorticoid treatment was given.

In a patient with nitrofurantoin-induced pulmonary toxicity, in whom high resolution CT scans initially showed a widespread reticular pattern and associated distortion of the lung parenchyma, thought to represent established and irreversible fibrosis, follow-up CT scans after withdrawal of the drug showed resolution of the pulmonary changes.

- An 82-year-old woman developed a productive cough after having taken nitrofurantoin 50 mg/day for 4 years (38). She had impaired lung function and abnormal CT scans, and lung biopsies showed features compatible with bronchiolitis obliterans organizing pneumonia. The condition improved when nitrofurantoin was withdrawn and she was given a glucocorticoid.
- In a 58-year-old woman who took nitrofurantoin 100 mg/day for 11 months, pulmonary toxicity occurred, with bilateral interstitial infiltrates in the lower zones of the chest X-ray and loss of lung volume (39). High resolution CT scans showed ground-glass opacification in the mid-thoracic region, with patchy fibrosis and traction bronchiectasis. After withdrawal of nitrofurantoin and administration of prednisone, a chest X-ray 3 months later showed resolution of the pulmonary changes.

Nervous system

More than 140 cases of toxic polyneuropathy have been reported. The frequency depends on dose, tissue concentration, and renal function: in up to 90% of cases polyneuropathy occurred in patients with renal insufficiency (40). Symptoms usually start 9–45 days (at the earliest 3 days) after beginning nitrofurantoin. The neuropathy starts peripherally, predominantly affects the limbs, and remains more severe distally. Initially, there is sensory loss with paresthesia. Later, motor loss develops, often with severe muscle atrophy. As a rule, no further deterioration occurs after withdrawal of nitrofurantoin, and there may be total regression (34% of cases) or partial regression (45% of cases) (40). In some severe cases there is residual disability. The motor loss resolves more slowly and less completely than the sensory impairment. Single cases of retrobulbar optic neuritis, lateral rectus muscle palsy, and facial nerve palsy have been reported (41).

The lesions comprise degeneration of the myelin sheath of the nerves and nerve roots, with degeneration of the corresponding anterior horn cells and muscle fibers. The pathogenesis is unclear. Impairment of glutathione reductase has been considered. Even in healthy people, nitrofurantoin 400 mg/day for 2 weeks causes a significant increase in motor nerve conduction time. If strict attention is paid to the contraindication of renal insufficiency, the risk of polyneuropathy can be reduced. Careful controls for the initial symptoms of paresthesia can prevent the development of severe disablement.

While taking nitrofurantoin after urinary tract surgery, a 10-year-old girl developed diplopia and ptosis (42). A sleep test confirmed ocular myasthenia. Her signs and symptoms resolved after drug withdrawal.

Single cases of benign intracranial hypertension (pseudotumor cerebri), with and without ocular palsy, have been reported (43,44). Uncharacteristic general symptoms, with dizziness, cephalalgia, or drowsiness, are more frequent.

Trigeminal neuralgia (45) and cerebellar symptoms (46) have been attributed to nitrofurantoin.

Sensory systems

One case of crystalline retinopathy has been associated with long-term nitrofurantoin (47).

Psychological, psychiatric

Rarely, concomitant dysphoric, euphoric, or even psychotic reactions have been reported in patients taking nitrofurantoin (1).

Metabolism

One case of hyperlactatemic metabolic acidosis together with hemolytic anemia due to glucose-6-phosphate dehydrogenase deficiency has been reported (48).

Hematologic

Some cases of hemolytic anemia associated with glucose 6-phosphate dehydrogenase deficiency have been described in patients taking nitrofurantoin (49).

Single cases of nitrofurantoin-induced (or enhanced) hemolytic anemia have occurred in patients with deficiencies of other erythrocyte enzymes (enolase or glutathione peroxidase), as have isolated cases of methemoglobinemia (50).

Nitrofurantoin produces oxidant stress and cellular damage by different mechanisms (51). It can disturb folate metabolism, leading to a megaloblastic component in pre-existing (mostly hemolytic) anemia, which responds to folic acid treatment.

There have been single cases of thrombocytopenia and a severe hemorrhagic diathesis with deficiency of coagulation factors due to a nitrofurantoin-induced hepatic disorder (52). Furthermore, nitrofurantoin experimentally inhibits ADP-induced platelet aggregation (53).

Allergic agranulocytosis or neutropenia have been proven in only a few cases (54,55). Pancytopenia is also rarely seen (5).

- A 74-year-old white man developed agranulocytosis after taking nitrofurantoin 100 mg qds for 5 days (56). The total white blood cell count and granulocyte count fell to $1.9 \times 10^9/l$ and $0.5 \times 10^9/l$ respectively. Nitrofurantoin was withdrawn and 2 days later the total white blood cell count and granulocyte count rose to $2.5 \times 10^9/l$ and $1.1 \times 10^9/l$ respectively.

Salivary glands

Some cases of parotitis, rarely proven by rechallenge, have been associated with nitrofurantoin (57,58). The symptoms disappear after withdrawal.

Gastrointestinal

Gastrointestinal symptoms are the most common adverse effects of nitrofurantoin. Nausea and anorexia have been most often reported (55,60), whereas abdominal pain and diarrhea are rare. These effects are dose-related and usually harmless. The manifestations occur mostly after absorption of the drug and are mediated by the central nervous system. Measures that delay absorption, such as sugar coating or the use of a macrocrystalline form of the drug, reduce these adverse effects (61).

Liver

Hepatic reactions to nitrofurantoin are rare, and different forms can be distinguished (62–64). They can be associated with fatal liver necrosis (65,66,67). A 67-year-old woman developed fulminant hepatitis after taking nitrofurantoin (68)

- A 73-year-old white man who had taken nitrofurantoin 50 mg/day for 5 years developed combined hepatic and pulmonary toxicity after taking fluconazole 150 mg/week for onychomycosis (69). Two months after starting fluconazole, his hepatic enzymes rose to 5 times the upper limits of normal. In addition, he reported fatigue, dyspnea on exertion, pleuritic pain, burning tracheal pain, and a cough. Chest X-rays showed bilateral pulmonary disease consistent with nitrofurantoin toxicity. Both drugs were discontinued. The hepatic and pulmonary toxicity resolved on withdrawal of both drugs and use of inhaled glucocorticoids (fluticasone 880 micrograms/day).

Either drug could have caused the hepatic damage in this case, and it is possible that a pharmacokinetic interaction with fluconazole precipitated nitrofurantoin-induced toxicity.

Data from The Netherlands suggest that acute reactions are more common than chronic ones. Acute hepatic reactions may be hepatocellular or cholestatic. In the vast majority of subjects, symptoms appear within the first 6 weeks of nitrofurantoin treatment, and in half of the patients they occur within the first week of treatment. Jaundice is most common, followed by abdominal pain,

malaise, and nausea. Hepatomegaly has been reported in nearly 50% of cases, fever in 30–65%, eosinophilia in 15–50%, and a rash in 12–60%. An immunological pathogenesis has been proposed (70), but experimental data point to a toxic mechanism involving the formation of glutathione-protein mixed disulfides and/or protein alkylation (71). The prognosis after withdrawal of the drug is good, but fatal courses have been reported when the drug is continued or, rarely, even after it has been discontinued.

Chronic active hepatitis, icteric or anicteric, has been described, almost always in women taking long-term nitrofurantoin. Most patients develop symptoms after a period of about 6 months of nitrofurantoin use. Fever (0–24%), rash (0–3%), and eosinophilia (9–23%) occur rarely compared with acute cases. There is hepatomegaly in 30–60% of chronic cases (64). Sometimes a broad spectrum of autoimmune reactions, a lupus-like syndrome, or mild cholestasis can be present (72,73). Some of these cases have occurred in combination with lung reactions of the protracted acute or chronic type, or in patients with ascites and liver cirrhosis. The clinical symptoms usually improve after withdrawal, but a few cases with extensive hepatocellular necrosis have necessitated liver transplantation or ended fatally (74–77). The histological changes can persist. Re-exposure to nitrofurantoin has reproduced the pathological liver tests.

Granulomatous hepatitis, with a rash or isolated increases in serum transaminase activities, can occur in patients with lung reactions (78). In the protracted acute and in chronic lung reactions liver injury (such as chronic active hepatitis) is more frequent than in acute reactions. Such cases usually show a broad spectrum of serological autoimmune reactions (lupus-like syndrome) (79).

Pancreas

A few cases of nitrofurantoin-induced acute pancreatitis, confirmed by rechallenge, have been reported (80,81).

Urinary tract

Nitrofurantoin-induced crystalluria, leading to obstruction of indwelling catheters, has been described in a few patients (82).

In rare cases, interstitial nephritis has been observed (83,84).

Skin

Allergic skin reactions occur in 1–2% of patients who take nitrofurantoin and comprise about 21% of all adverse reactions to nitrofurantoin (5,85). They often occur with other reactions, such as drug fever, lung, or liver reactions. The lesions can present as pruritus, as macular, maculopapular, or vesicular rashes, urticaria, angioedema, or erythema multiforme (86). The frequency of serious cutaneous reactions (erythema multiforme, Stevens–Johnson syndrome, or toxic epidermal necrolysis) after nitrofurantoin has been estimated to be 7 cases per 100 000 exposed individuals (85).

Sweet's syndrome has been observed in association with a 7-day course of nitrofurantoin (87).

Transitory alopecia that has been reported in a few cases is dose-related (88).

Stevens–Johnson syndrome and toxic epidermal necrolysis are more common in the setting of a compromised immune system.

- Stevens–Johnson syndrome occurred in a 24-year-old HIV-positive pregnant woman who took nitrofurantoin for 2 days (89).

Immunologic

About 20 cases of a lupus-like syndrome have been described, mostly in the Scandinavian literature. The clinical picture consisted of arthralgia or, rarely, exacerbation of a pre-existing rheumatoid arthritis and generalized lymphadenopathy, mostly associated with chronic lung and/or liver reactions, such as chronic active hepatitis (27,90,91).

- A 71-year-old woman developed a lupus-like syndrome associated with hepatitis after taking nitrofurantoin for 5 years; her symptoms rapidly improved with systemic glucocorticoid therapy (92). The symptoms resolved after withdrawal of nitrofurantoin and the use of systemic corticosteroids.

In patients with the lupus-like syndrome at least two immunological tests (antinuclear factor, rheumatoid factor, Coombs' test, and antibodies against smooth muscle, thyroglobulin, thyroid cell cytoplasm, or glomeruli) were positive. The lymphocyte transformation test was always positive. However, the LE cell phenomenon was always negative. As in other allergic reactions to nitrofurantoin, circulating albumin IgG complexes were found by immunoelectrophoresis, with tailing of the albumin line (93). The syndrome regresses after withdrawal.

Long-Term Effects

Drug tolerance

Nitrofurantoin is effective against Enterobacteriaceae; the rates of resistance were below 2% in a single-center study (94) and 3.5% in a multicenter study (95).

Mutagenicity

In vitro, nitrofurantoin acts as a mutagen by inhibiting DNA synthetase. In human fibroblast cultures it damages DNA (96). Treatment with nitrofurantoin for 12 months caused a significant increase in chromosome aberrations and sister chromatid exchanges in the lymphocytes of 69 children (97).

Tumorigenicity

Some studies have shown similar metabolism of nitrofurantoin and other (carcinogenic) nitrofurans. Formation of carcinogenic nitrofurantoin metabolites is therefore possible (98). However, a carcinogenic effect has not been proven for nitrofurantoin (99).

Second-Generation Effects

Fertility

High-dose oral nitrofurantoin (10 mg/kg/day) transiently reduces the sperm count in 30% of patients (100). This is due to arrest of maturation. Depression of sperm motility or ejaculate volume can also occur at lower doses (101).

Teratogenicity

Only very small amounts of nitrofurantoin cross the placenta, and teratogenic effects are not known (102). Even when it is used in the first trimester, there have been no associated fetal malformations (103). No teratogenic effects have been associated with nitrofurantoin in Denmark, Finland, Norway, and Sweden (104), or in a Hungarian case-control survey (105).

Fetotoxicity

Hemolytic anemia occurred during the first hours of life in a full-term neonate whose mother had taken nitrofurantoin during the last month of pregnancy (106). It may therefore be wise not to prescribe nitrofurantoin at the end of pregnancy.

Lactation

Nitrofurantoin is actively transported into human milk, achieving concentrations greatly exceeding those in serum. Concern is warranted for suckling infants under 1 month old or for infants with a high frequency of glucose-6-phosphate dehydrogenase deficiency or sensitivity to nitrofurantoin (107).

Susceptibility Factors

Genetic factors

Glucose-6-phosphate dehydrogenase deficiency is decisive for the development of hemolytic anemia in patients taking nitrofurantoin.

Age

Nitrofurantoin is contraindicated during the last trimester of pregnancy and in neonates, because hemolytic anemia can result from immature enzyme systems (106).

Adverse reactions in children related to nitrofurantoin have been reviewed (108). They are gastrointestinal disturbances (4.4/100 person-years) (109,110), skin reactions (2–3%), pulmonary toxicity (9 patients) (111,112), hepatotoxicity (12 patients and 3 deaths) (113,114,115,116), hematological toxicity (12 patients) (117,118,119,120), neurotoxicity (117,121,122), and an increased rate of sister chromatid exchanges (123,124).

Renal disease

The main risk factor for toxic reactions to nitrofurantoin, especially polyneuropathy and gastrointestinal symptoms, is impaired renal function (40).

Drug–Drug Interactions

Estrogens

Nitrofurantoin reduces the enterohepatic circulation of estrogens (125).

Nalidixic acid

Antagonism in antibacterial efficacy between nitrofurantoin and nalidixic acid has been observed (126).

Pyridoxine

Pyridoxine accelerates the renal elimination of nitrofurantoin (127).

Interference with Diagnostic Tests

Glucose measurements

Nitrofurantoin can produce spurious positive urine glucose concentrations or raised blood glucose concentrations if reducing reagents are used (128).

References

1. D'Arcy PF. Nitrofurantoin. Drug Intell Clin Pharm 1985;19(7–8):540–7.
2. Spencer RC, Moseley DJ, Greensmith MJ. Nitrofurantoin modified release versus trimethoprim or co-trimoxazole in the treatment of uncomplicated urinary tract infection in general practice. J Antimicrob Chemother 1994;33(Suppl A):121–9.
3. Cunha BA Nitrofurantoin —current concepts. Urology 1988;32(1):67–71.
4. Lubbers P. Allergische Reaktion gegen Furadantin. Dtsch Med Wochenschr 1962;87:2209.
5. Holmberg L, Boman G, Bottiger LE, Eriksson B, Spross R, Wessling A. Adverse reactions to nitrofurantoin. Analysis of 921 reports. Am J Med 1980;69(5):733–8.
6. Biel B, Younes M, Brasch H. Cardiotoxic effects of nitrofurantoin and tertiary butylhydroperoxide in vitro: are oxygen radicals involved? Pharmacol Toxicol 1993;72(1):50–5.
7. Ben-Noun L. Drug-induced respiratory disorders: incidence, prevention and management. Drug Saf 2000;23(2):143–64.
8. Israel HL, Diamond P. Recurrent pulmonary infiltration and pleural effusion due to nitrofurantoin sensitivity. N Engl J Med 1962;266:1024.
9. Chudnofsky CR, Otten EJ. Acute pulmonary toxicity to nitrofurantoin. J Emerg Med 1989;7(1):15–9.
10. Jick SS, Jick H, Walker AM, Hunter JR. Hospitalizations for pulmonary reactions following nitrofurantoin use. Chest 1989;96(3):512–5.
11. Williams EM, Triller DM. Recurrent acute nitrofurantoin-induced pulmonary toxicity. Pharmacotherapy 2006;26(5):713–8.
12. Arya SC, Agarwal N, Agarwal S. Nitrofurantoin: an effective and ignored antimicrobial. Int J Antimicrob Agents 2006;27(4):354–5.
13. Anonymous. Pulmonary adverse effects of nitrofurantoin. Prescrire Int 2003;12:23.
14. Daba MH, El-Tahir KE, Al-Arifi MN, Gubara OA. Drug-induced pulmonary fibrosis. Saudi Med J 2004;25(6):700–6.
15. Epler GR. Drug-induced bronchiolitis obliterans organizing pneumonia. Clin Chest Med 2004;25(1):89–94.
16. Huggins JT, Sahn SA. Drug-induced pleural disease. Clin Chest Med 2004;25(1):141–53.
17. Bidad K, Harries-Jones R. Nitrofurantoin lung injury. Age Ageing 2004;33(4):414–5.
18. Basoglu T, Erkan L, Canbaz F, Bernay I, Onen T, Sahin M, Furtun F, Yalin T. Transient reverse ventilation–perfusion mismatch in acute pulmonary nitrofurantoin reaction. Ann Nucl Med 1997;11(3):271–4.
19. Mendez JL, Nadrous HF, Hartman TE, Ryu JH. Chronic nitrofurantoin-induced lung disease. Mayo Clin Proc 2005;80(10):1298–302.
20. Fauroux B, Tournier G. Toxicité pulmonaire des drogues chez l'enfant. Med Infant 1990;97:289.
21. Taskinen E, Tukiainen P, Sovijarvi AR. Nitrofurantoin-induced alterations in pulmonary tissue. A report on five patients with acute or subacute reactions. Acta Pathol Microbiol Scand [A] 1977;85(5):713–20.
22. Cohen AJ, King TE Jr, Downey GP. Rapidly progressive bronchiolitis obliterans with organizing pneumonia. Am J Respir Crit Care Med 1994;149(6):1670–5.
23. Rosenow EC 3rd, DeRemee RA, Dines DE. Chronic nitrofurantoin pulmonary reaction. Report of 5 cases. N Engl J Med 1968;279(23):1258–62.
24. Smith GJ. The histopathology of pulmonary reactions to drugs. Clin Chest Med 1990;11(1):95 117.
25. Bone RC, Wolfe J, Sobonya RE, Kerby GR, Stechschulte D, Ruth WE, Welch M. Desquamative interstitial pneumonia following long-term nitrofurantoin therapy. Am J Med 1976;60(5):697–701.
26. Muller U, Abbuhl K, Bisig J, Baumgartner H, Muhlberger F, Scherrer M, Hoigne R, Ueberempfindlichkeitsreaktionen der Lunge auf Nitrofurantoin. [Hypersensitivity reactions of the lung to nitrofurantoin.] Schweiz Med Wochenschr 1970;100(51):2206–12.
27. Back O, Lundgren R, Wiman LG. Nitrofurantoin-induced pulmonary fibrosis and lupus syndrome. Lancet 1974;1(7863):930.
28. Martin WJ 2nd. Nitrofurantoin. Potential direct and indirect mechanisms of lung injury. Chest 1983;83(Suppl 5):S51–2.
29. Pearsall HR, Ewalt J, Tsoi MS, Sumida S, Backus D, Winterbauer RH, Webb DR, Jones H. Nitrofurantoin lung sensitivity: report of a case with prolonged nitrofurantoin lymphocyte sensitivity and interaction of nitrofurantoin-stimulated lymphocytes with alveolar cells. J Lab Clin Med 1974;83(5):728–37.
30. Larsson S, Cronberg S, Denneberg T, Ohlsson NM. Pulmonary reaction to nitrofurantoin. Scand J Respir Dis 1973;54(2):103–10.
31. Boyd MR, Catignani GL, Sasame HA, Mitchell JR, Stiko AW. Acute pulmonary injury in rats by nitrofurantoin and modification by vitamin E, dietary fat, and oxygen. Am Rev Respir Dis 1979;120(1):93–9.
32. Sasame HA, Boyd MR. Superoxide and hydrogen peroxide production and NADPH oxidation stimulated by nitrofurantoin in lung microsomes: possible implications for toxicity. Life Sci 1979;24(12):1091–6.
33. Back O, Liden S, Ahlstedt S. Adverse reactions to nitrofurantoin in relation to cellular and humoral immune responses. Clin Exp Immunol 1977;28(3):400–6.

34. Bucknall CE, Adamson MR, Banham SW. Non fatal pulmonary haemorrhage associated with nitrofurantoin. Thorax 1987;42(6):475–6.

35. Magee F, Wright JL, Chan N, Currie W, Karr G, Hogg J, Thurlbeck WM. Two unusual pathological reactions to nitrofurantoin: case reports. Histopathology 1986;10(7):701–6.

36. Sheehan RE, Wells AU, Milne DG, Hansell DM. Nitrofurantoin-induced lung disease: two cases demonstrating resolution of apparently irreversible CT abnormalities. J Comput Assist Tomogr 2000;24(2):259–61.

37. Cameron RJ, Kolbe J, Wilsher ML, Lambie N. Bronchiolitis obliterans organising pneumonia associated with the use of nitrofurantoin. Thorax 2000;55(3):249–51.

38. Fawcett IW, Ibrahim NB. BOOP associated with nitrofurantoin. Thorax 2001;56(2):161.

39. Hadi HA, Arnold AG. Malaise, weight loss, and respiratory symptoms. Postgrad Med J 2002;78(915):5558.

40. Toole JF, Parrish ML. Nitrofurantoin polyneuropathy. Neurology 1973;23(5):554–9.

41. Thomson RG, James OF. Seventh-nerve palsy and hepatitis associated with nitrofurantoin. Hum Toxicol 1986;5(6):387–8.

42. Wasserman BN, Chronister TE, Stark BI, Saran BR. Ocular myasthenia and nitrofurantoin. Am J Ophthalmol 2000;130(4):531–3.

43. Mushet GR. Pseudotumor and nitrofurantoin therapy. Arch Neurol 1977;34(4):257.

44. Korzets A, Rathaus M, Chen B, Bernheim J. Pseudotumor cerebri and nitrofurantoin. Drug Intell Clin Pharm 1988;22(4):345.

45. Herishanu Y, Milwitzki A. Trigeminal neuralgia induced by nitrofurantoin treatment. J Neurol 1977;216(1):77–8.

46. Graebner RW, Herskowitz A. Cerebellar toxic effects from nitrofurantoin. Arch Neurol 1973;29(3):195–6.

47. Ibanez HE, Williams DF, Boniuk I. Crystalline retinopathy associated with long-term nitrofurantoin therapy. Arch Ophthalmol 1994;112(3):304–5.

48. Lavelle KJ, Atkinson KF, Kleit SA. Hyperlactatemia and hemolysis in G6PD deficiency after nitrofurantoin ingestion. Am J Med Sci 1976;272(2):201–4.

49. Herman J, Ben-Meir S. Overt hemolysis in patients with glucose-6-phosphate dehydrogenase deficiency: a survey in general practice. Isr J Med Sci 1975;11(4):340–6.

50. Gerok W, Waller HD. Hämoglobinbildung durch Furadantin. [Hemoglobin formation by furadantin (N-[5-nitro-2-furfuryliden]-1-aminohydantoin).] Dtsch Med Wochenschr 1956;81(43):1707–9.

51. Novak RF, Kharasch ED, Wendel NK. Nitrofurantoin-stimulated proteolysis in human erythrocytes: a novel index of toxic insult by nitroaromatics. J Pharmacol Exp Ther 1988;247(2):439–44.

52. Murphy KJ, Innis MD. Hepatic disorder and severe bleeding diathesis following nitrofurantoin ingestion. JAMA 1968;204(5):396–7.

53. Rossi EC, Levin NW. Inhibition of primary ADP-induced platelet aggregation in normal subjects after administration of nitrofurantoin (Furadantin). J Clin Invest 1973;52(10):2457–67.

54. Palva IP, Lehmola U. Agranulocytosis caused by nitrofurantoin. Acta Med Scand 1973;194(6):575–6.

55. Carroll DA. Nitrofurantoin induced neutropenia: case report. Mil Med 1984;149(10):570–1.

56. Roberts AD, Neelamegam M. Agranulocytosis associated with nitrofurantoin therapy. Ann Pharmacother 2005;39(1):198.

57. Thompson DF. Drug-induced parotitis. J Clin Pharm Ther 1993;18(4):255–8.

58. Gervilla-Cano J, Otal-Bareche J, Torres-Justribo M, Duran-Rabes J. [Nitrofurantoin associated parotiditis.] Med Clin (Barc) 2005;125(13):519.

59. Brumfitt W, Hamilton-Miller JM. Efficacy and safety profile of long-term nitrofurantoin in urinary infections: 18 years' experience. J Antimicrob Chemother 1998;42(3):363–71.

60. Uhari M, Nuutinen M, Turtinen J. Adverse reactions in children during long-term antimicrobial therapy. Pediatr Infect Dis J 1996;15(5):404–8.

61. Kalowski S, Radford N, Kincaid-Smith P. Crystalline and macrocrystalline nitrofurantoin in the treatment of urinary-tract infection. N Engl J Med 1974;290(7):385–7.

62. Strohscheer H, Wegener HH. Nitrofurantoin-induzierte, granulomatöse Hepatitis. [Nitrofurantoin-induced granulomatous hepatitis.] MMW Munch Med Wochenschr 1977;119(47):1535–6.

63. Zimmerman HJ. Update of hepatotoxicity due to classes of drugs in common clinical use: non-steroidal drugs, anti-inflammatory drugs, antibiotics, antihypertensives, and cardiac and psychotropic agents. Semin Liver Dis 1990;10(4):322–38.

64. Stricker BH, Blok AP, Claas FH, Van Parys GE, Desmet VJ. Hepatic injury associated with the use of nitrofurans: a clinicopathological study of 52 reported cases. Hepatology 1988;8(3):599–606.

65. Edoute Y, Karmon Y, Roguin A, Ben-Ami H. Fatal liver necrosis associated with the use of nitrofurantoin. Isr Med Assoc J 2001;3(5):382–3.

66. Amit G, Cohen P, Ackerman Z. Nitrofurantoin-induced chronic active hepatitis. Isr Med Assoc J 2002;4(3):184–6.

67. Thiim M, Friedman LS. Hepatotoxicity of antibiotics and antifungals. Clin Liver Dis 2003;7:381-99, vi-vii.

68. Peedikayil MC, Dahhan TI, Al Ashgar HI. Nitrofurantoin-induced fulminant hepatitis mimicking autoimmune hepatitis. Ann Pharmacother 2006;40(10):1888–9.

69. Linnebur SA, Parnes BL. Pulmonary and hepatic toxicity due to nitrofurantoin and fluconazole treatment. Ann Pharmacother 2004;38(4):612–6.

70. Kelly BD, Heneghan MA, Bennani F, Connolly CE, O'Gorman TA. Nitrofurantoin-induced hepatotoxicity mediated by CD8+ T cells. Am J Gastroenterol 1998;93(5):819–21.

71. Silva JM, Khan S, O'Brien PJ. Molecular mechanisms of nitrofurantoin-induced hepatocyte toxicity in aerobic versus hypoxic conditions. Arch Biochem Biophys 1993;305(2):362–9.

72. Black M, Rabin L, Schatz N. Nitrofurantoin-induced chronic active hepatitis. Ann Intern Med 1980;92(1):62–4.

73. Fagrell B, Strandberg I, Wengle B. A nitrofurantoin-induced disorder simulating chronic active hepatitis. A case report. Acta Med Scand 1976;199(3):237–9.

74. Sharp JR, Ishak KG, Zimmerman HJ. Chronic active hepatitis and severe hepatic necrosis associated with nitrofurantoin. Ann Intern Med 1980;92(1):14–9.

75. Hebert MF, Roberts JP. Endstage liver disease associated with nitrofurantoin requiring liver transplantation. Ann Pharmacother 1993;27(10):1193–4.

76. Mulberg AE, Bell LM. Fatal cholestatic hepatitis and multisystem failure associated with nitrofurantoin. J Pediatr Gastroenterol Nutr 1993;17(3):307–9.

77. Schattner A, Von der Walde J, Kozak N, Sokolovskaya N, Knobler H. Nitrofurantoin-induced immune-mediated lung and liver disease. Am J Med Sci 1999;317(5):336–40.

78. Sippel PJ, Agger WA. Nitrofurantoin-induced granulomatous hepatitis. Urology 1981;18(2):177–8.

79. Liu ZX, Kaplowitz N. Immune-mediated drug-induced liver disease. Clin Liver Dis 2002;6(3):755–74.

80. Christophe JL. Pancreatitis induced by nitrofurantoin. Gut 1994;35(5):712–3.

81. Mouallem M, Sirotin T, Farfel Z. Nitrofurantoin-induced pancreatitis. Isr Med Assoc J 2003;5:754–5.

82. Macdonald JB, Macdonald ET. Nitrofurantoin crystalluria. BMJ 1976;2(6043):1044–5.

83. Kahn SR. Acute interstitial nephritis associated with nitrofurantoin. Lancet 1996;348(9035):1177–8.

84. Korzets Z, Elis A, Bernheim J, Bernheim J. Acute granulomatous interstitial nephritis due to nitrofurantoin. Nephrol Dial Transplant 1994;9(6):713–5.

85. Chan HL, Stern RS, Arndt KA, Langlois J, Jick SS, Jick H, Walker AM. The incidence of erythema multiforme, Stevens–Johnson syndrome, and toxic epidermal necrolysis. A population-based study with particular reference to reactions caused by drugs among outpatients. Arch Dermatol 1990;126(1):43–7.

86. Chapman JA. An unusual nitrofurantoin-induced drug reaction. Ann Allergy 1986;56(1):16–8.

87. Retief CR, Malkinson FD. Nitrofurantoin-associated Sweet's syndrome. Cutis 1999;63(3):177–9.

88. Johnson SH 3rd, Marshall M Jr. Prophylactic treatment of chronic urinary tract infection with nitrofurantoin: one to five year follow up studies. J Urol 1959;82(1):162–4.

89. Shilad A, Predanic M, Perni SC, Houlihan C, Principe D. Human immunodeficiency virus, pregnancy, and Stevens-Johnson syndrome. Obstet Gynecol 2005;105(5 Pt 2):1254–6.

90. Stratton MA. Drug-induced systemic lupus erythematosus. Clin Pharm 1985;4(6):657–63.

91. Selroos O, Edgren J. Lupus-like syndrome associated with pulmonary reaction to nitrofurantoin. Report of three cases. Acta Med Scand 1975;197(1–2):125–9.

92. Salle V, Lafon B, Smail A, Cevallos R, Chatelain D, Andrejak M, Ducroix JP. [Nitrofurantoin-induced lupus-like syndrome associated with hepatitis.] Rev Med Interne 2006;27(4):344–6.

93. Teppo AM, Haltia K, Wager O. Immunoelectrophoretic "tailing" of albumin line due to albumin-IgG antibody complexes: a side effect of nitrofurantoin treatment? Scand J Immunol 1976;5(3):249–61.

94. Kahlmeter G. The ECO.SENS Project: a prospective, multinational, multicentre epidemiological survey of the prevalence and antimicrobial susceptibility of urinary tract pathogens—interim report J Antimicrob Chemother 2000;46(Suppl 1):15–22.

95. Newell A, Riley P, Rodgers M. Resistance patterns of urinary tract infections diagnosed in a genitourinary medicine clinic. Int J STD AIDS 2000;11(8):499–500.

96. Hirsch-Kauffmann M, Herrlich P, Schweiger M. Nitrofurantoin damages DNA of human cells. Klin Wochenschr 1978;56(8):405–7.

97. Slapsyte G, Jankauskiene A, Mierauskiene J, Lazutka JR. Cytogenetic analysis of peripheral blood lymphocytes of children treated with nitrofurantoin for recurrent urinary tract infection. Mutagenesis 2002;17(1):31–5.

98. Boyd MR, Stiko AW, Sasame HA. Metabolic activation of nitrofurantoin—possible implications for carcinogenesis. Biochem Pharmacol 1979;28(5):601–6.

99. Hasegawa R, Murasaki G, St John MK, Zenser TV, Cohen SM. Evaluation of nitrofurantoin on the two stages of urinary bladder carcinogenesis in the rat. Toxicology 1990;62(3):333–47.

100. Nelson WO, Bunge RG. The effect of therapeutic dosages of nitrofurantoin (Furadantin) upon spermatogenesis in man. J Urol 1957;77(2):275–81.

101. Iunda IF, Kushniruk IuI. [Functional state of the testis after the use of certain antibiotics and nitrofuran preparations.]Antibiotiki 1975;9:843–6.

102. Olshan AF, Faustman EM. Nitrosatable drug exposure during pregnancy and adverse pregnancy outcome. Int J Epidemiol 1989;18(4):891–9.

103. Ben David S, Einarson T, Ben David Y, Nulman I, Pastuszak A, Koren G. The safety of nitrofurantoin during the first trimester of pregnancy: meta-analysis. Fundam Clin Pharmacol 1995;9(5):503–7.

104. Christensen B. Which antibiotics are appropriate for treating bacteriuria in pregnancy? J Antimicrob Chemother 2000;46(Suppl 1):29–34.

105. Czeizel AE, Rockenbauer M, Sorensen HT, Olsen J. Nitrofurantoin and congenital abnormalities. Eur J Obstet Gynecol Reprod Biol 2001;95(1):119–26.

106. Bruel H, Guillemant V, Saladin-Thiron C, Chabrolle JP, Lahary A, Poinsot J. Anémie hémolytique chez un nouveau-né après prise maternelle de nitrofurantoine en fin de grossesse. [Hemolytic anemia in a newborn after maternal treatment with nitrofurantoin at the end of pregnancy.] Arch Pediatr 2000;7(7):745–7.

107. Gerk PM, Kuhn RJ, Desai NS, McNamara PJ. Active transport of nitrofurantoin into human milk. Pharmacotherapy 2001;21(6):669–75.

108. Karpman E, Kurzrock EA. Adverse reactions of nitrofurantoin, trimethoprim and sulfamethoxazole in children. J Urol 2004;172(2):448–53.

109. Holmberg L, Boman G, Bottiger LE, Eriksson B, Spross R, Wessling A. Adverse reactions to nitrofurantoin. Analysis of 921 reports. Am J Med 1980;69:733–8.

110. Koch-Weser J, Sidel VW, Dexter M, Parish C, Finer DC, Kanarek P. Adverse reactions to sulfosoxazole, sulfamethoxazole, and nitrofurantoin. Manifestations and specific reaction rates during 2,118 courses of therapy. Arch Intern Med 1971;128:399–404.

111. Rantala H, Kirvela O, Anttolainen I. Nitrofurantoin lung in a child. Lancet 1979;2:799–800.

112. Holmberg L, Boman G. Pulmonary reactions to nitrofurantoin. 447 cases reported to the Swedish Adverse Drug Reaction Committee 1966-1976. Eur J Respir Dis 1981;62:180–9.

113. Coraggio MJ, Gross T P, Roscelli JD. Nitrofurantoin toxicity in children. Pediatr Infect Dis J 1989;8:163–6.

114. Mulberg AE, Bell LM. Fatal cholestatic hepatitis and multisystem failure associated with nitrofurantoin. J Pediatr Gastroenterol Nutr 1993;17:307–9.

115. Stricker BH, Blok AP, Claas FH, Van Parys GE, Desmet VJ. Hepatic injury associated with the use of nitrofurans: a clinicopathological study of 52 reported cases. Hepatology 1988;8:599–606.

116. Berry WR, Warren GH, Reichen J. Nitrofurantoin-induced cholestatic hepatitis from cow's milk in a teenaged boy. West J Med 1984;140:278–80.

117. Uhari M, Nuutinen M, Turtinen J. Adverse reactions in children during long-term antimicrobial therapy. Pediatr Infect Dis J 1996;15:404–8.

118. Coraggio MJ, Gross TP, Roscelli JD. Nitrofurantoin toxicity in children. Pediatr Infect Dis J 1989;8:163–6.

119. Gait JE. Hemolytic reactions to nitrofurantoin in patients with glucose-6-phosphate dehydrogenase deficiency: theory and practice. DICP 1990;24:1210–3.

120. Bruel H, Guillemant V, Saladin-Thiron C, Chabrolle JP, Lahary A, Poinsot J. Hemolytic anemia in a newborn after maternal treatment with nitrofurantoin at the end of pregnancy. Arch Pediatr 2000;7:745–7.

121. Toole JF, Parrish M L. Nitrofurantoin polyneuropathy. Neurology 1973;23:554–9.

122. Sharma DB, James A. Benign intracranial hypertension associated with nitrofurantoin therapy. Br Med J 1974;4:771.

123. Steineck G, Wiholm BE, Gerhardsson de Verdier M. Acetaminophen, some other drugs, some diseases and the risk of transitional cell carcinoma. A population-based case-control study. Acta Oncol 1995;34:741–8.

124. Slapsyte G, Jankauskiene A, Mierauskiene J, Lazutka JR. Cytogenetic analysis of peripheral blood lymphocytes of children treated with nitrofurantoin for recurrent urinary tract infection. Mutagenesis 2002;17:31–5.

125. Revaz C, Goldenberg B, Achtari H. Etude critique de nouvelles méthodes contraceptives. [Critical study of new contraceptive methods.] Schweiz Med Wochenschr 1971;101(4):127–30concl.

126. Stille W, Ostner KH. Antagonismus Nitrofurantoin–Nalidixinsäure. [Nitrofurantoin–Nalidixic acid antagonism.] Klin Wochenschr 1966;44(3):155–6.

127. Matthews A, Heise H. Die Erhöhung der Nitrofurantoin (Nifuratin) Ausscheidung durch Vitamin B6. Dtsch Gesundheitswes 1973;28:716.

128. Wright LA, Foster MG. Effect of some commonly prescribed drugs on certain chemistry tests. Clin Biochem 1980;13(6):249–52.

Norfloxacin

See also Fluoroquinolones

General Information

Norfloxacin is a fluoroquinolone antibacterial drug with properties similar to those of ciprofloxacin, although less potent in vitro.

Norfloxacin inhibits CYP1A2 and can therefore enhance the effects of other drugs by reducing their clearance (1).

Comparative studies

In a double-blind, multicenter study 171 patients who had acute pyelonephritis were given intravenous cefuroxime for 2–3 days, followed by ceftibuten 200 mg bd or norfloxacin 400 mg bd for 10 days (2). There were fewer bacterial relapses after oral norfloxacin than ceftibuten. Adverse events were reported by 47% of the patients taking ceftibuten and by 38% of those taking norfloxacin. This difference was not significant, but diarrhea or loose stools occurred more often with ceftibuten.

Organs and Systems

Sensory systems

A corneal ulcer associated with deposits of norfloxacin in the right eye has been reported in a 40-year-old man with right trigeminal and facial nerve palsies and reduced tear secretion. He stopped using norfloxacin ophthalmic solution and recovered (3).

Psychological, psychiatric

There have been reports of hallucinations with norfloxacin (4).

Hematologic

• Eosinophilia occurred in a 35-year-old man with alcoholic cirrhosis taking norfloxacin 400 mg bd for prophylaxis of spontaneous bacterial peritonitis (5).

Gastrointestinal

In a prospective study in women with urinary tract infections treated with pivmecillinam 400 mg bd for 3 days (*n* = 483) or norfloxacin 400 mg bd for 3 days (*n* = 471), 36% of the pivmecillinam-treated patients and 39% of the norfloxacin-treated patients reported adverse events (6). Gastrointestinal symptoms were most frequent. Of the patients who took norfloxacin 4.3% had vaginal candidiasis.

Liver

Acute hepatitis has been reported with norfloxacin (7,8).

Pancreas

Norfloxacin can cause pancreatitis (9).

Urinary tract

Acute interstitial nephritis, probably related to norfloxacin, has been reported (10).

• A 38-year-old woman took norfloxacin (300 mg/day) and tiaramide hydrochloride (300 mg tds) for an infection with *Mycoplasma pneumoniae*. One day after the start of treatment, her symptoms of cough and fever worsened and she developed lumbago and hematuria. The diagnosis was confirmed by percutaneous renal biopsy. She slowly improved without specific treatment. Lymphocyte stimulation tests were negative, but rechallenge with norfloxacin was followed by bilateral lumbago.

The authors identified two previous reports of acute interstitial nephritis associated with norfloxacin.

Skin

Acute febrile neutrophilic dermatosis (Sweet's syndrome) is uncommon and rarely related to antibiotic treatment.

• A 66-year-old man with advanced prostate adenocarcinoma took norfloxacin 400 mg bd and developed a fever, myalgias, and a skin eruption consisting of painful red papules, with confluent plaques and some vesicles, affecting the palms, feet, and lumbar region (11). After withdrawal of norfloxacin and glucocorticoid therapy he rapidly improved and the skin lesions resolved after one week.

Toxic epidermal necrolysis has been attributed to moxifloxacin.

- A 40-year-old man developed toxic epidermal necrolysis 10 days after the end of a 14-day course of norfloxacin 800 mg/day (12). He had typical cutaneous and mucous lesions and recovered after treatment with oral prednisolone and fluids.

Musculoskeletal

The incidence of tendinopathy in patients treated with norfloxacin is 10/10 000 (13).

Myalgia has been attributed to norfloxacin (14).

- A 33-year-old man developed myalgia and rhabdomyolysis while taking norfloxacin for cystitis. He complained of general muscle fatigue, tendon disorders, and articular pain. When norfloxacin was withdrawn, his symptoms abated, with persistence of slight myalgia for 10 days.

Drug Administration

Drug formulations

The stability of norfloxacin in suspensions prepared from two brands of film-coated tablets has been studied (15). The vehicle consisted of tragacanth, saccharin sodium, sorbitol solution, glycerol, parabens concentrate, peppermint spirit, purified water, and syrup USP, yielding a final concentration of norfloxacin of 20 mg/ml. The resulting suspensions were chemically stable for 28 days when stored in amber glass bottles at room temperature.

Drug–Drug Interactions

Antacids

Norfloxacin can interact with antacids containing aluminium or magnesium salts, by complexation, reducing its solubility and therefore its absorption; this can result in therapeutic failure. In an in vitro study, dissolution rates were markedly reduced in the presence of all antacids studied; however, this phenomenon was practically avoided when a disintegrant (sodium starch glycolate or crospovidone) was included in the tablet (16).

Ciclosporin

In children undergoing renal transplantation and receiving ciclosporin, the mean daily dose of ciclosporin needed to maintain adequate trough concentrations fell when patients were treated with norfloxacin for urinary tract infections (17).

Mycophenolic acid

The AUC of mycophenolic acid after a single dose was reduced by 10% when norfloxacin was given concomitantly (18).

Oral contraceptives

Plasma concentrations of oral contraceptive steroids were unchanged by norfloxacin (19).

Food–Drug Interactions

The systemic availability of norfloxacin was reduced by 38–52% by dairy products (20).

References

1. Pea F, Furlanut M. Pharmacokinetic aspects of treating infections in the intensive care unit: focus on drug interactions. Clin Pharmacokinet 2001;40(11):833–68.
2. Cronberg S, Banke S, Bergman B, Boman H, Eilard T, Elbel E, Hugo-Persson M, Johansson E, Kuylenstierna N, Lanbeck P, Lindblom A, Paulsen O, Schonbeck C, Walder M, Wieslander P. Fewer bacterial relapses after oral treatment with norfloxacin than with ceftibuten in acute pyelonephritis initially treated with intravenous cefuroxime. Scand J Infect Dis 2001;33(5):339–43.
3. Konishi M, Yamada M, Mashima Y. Corneal ulcer associated with deposits of norfloxacin. Am J Ophthalmol 1998;125(2):258–60.
4. Kundu AK. Norfloxacin-induced hallucination—an unusual CNS toxicity of 4-fluoroquinolones. J Assoc Physicians India 2000;48(9):944.
5. Mofredj A, Boudjema H, Cadranel JF. Norfloxacin-induced eosinophilia in a cirrhotic patient. Ann Pharmacother 2002;36(6):1107–8.
6. Nicolle LE, Madsen KS, Debeeck GO, Blochlinger E, Borrild N, Bru JP, Mckinnon C, O'Doherty B, Spiegel W, Van Balen FA, Menday P. Three days of pivmecillinam or norfloxacin for treatment of acute uncomplicated urinary infection in women. Scand J Infect Dis 2002;34(7):487–92.
7. Lopez-Navidad A, Domingo P, Cadafalch J, Farrerons J. Norfloxacin-induced hepatotoxicity. J Hepatol 1990;11(2):277–8.
8. Bjornsson E, Olsson R, Remotti H. Norfloxacin-induced eosinophilic necrotizing granulomatous hepatitis. Am J Gastroenterol 2000;95(12):3662–4.
9. Drabo YJ, Niakara A, Ouedraogo H. Pancreatite aiguë secondaire à une prise de norfloxacine. [Acute pancreatitis secondary to administration or norfloxacin.] Ann Fr Anesth Reanim 2002;21(1):68–9.
10. Nakamura M, Ohishi A, Aosaki N, Hamaguchi K. Norfloxacin-induced acute interstitial nephritis. Nephron 2000;86(2):204–5.
11. Aguiar-Bujanda D, Aguiar-Morales J, Bohn-Sarmiento U. Sweet's syndrome associated with norfloxacin in a prostate cancer patient. QJM 2004;97(1):55–6.
12. Sahin MT, Ozturkcan S, Inanir I, Filiz EE. Norfloxacin-induced toxic epidermal necrolysis. Ann Pharmacother 2005;39(4):768–70.
13. Shah P. Treten Sehnenschaden bei Chinolon-Gabe auf? Dtsch Med Wochenschr 2003;128:2214.
14. Guis S, Jouglard J, Kozak-Ribbens G, Figarella-Branger D, Vanuxem D, Pellissier JF, Cozzone PJ. Malignant hyperthermia susceptibility revealed by myalgia and rhabdomyolysis during fluoroquinolone treatment. J Rheumatol 2001;28(6):1405–6.
15. Boonme P, Phadoongsombut N, Phoomborplub P, Viriyasom S. Stability of extemporaneous norfloxacin suspension. Drug Dev Ind Pharm 2000;26(7):777–9.

16. Cordoba-Diaz M, Cordoba-Borrego M, Cordoba-Diaz D. Influence of pharmacotechnical design on the interaction and availability of norfloxacin in directly compressed tablets with certain antacids. Drug Dev Ind Pharm 2000;26(2):159–66.
17. McLellan RA, Drobitch RK, McLellan H, Acott PD, Crocker JF, Renton KW. Norfloxacin interferes with cyclosporine disposition in pediatric patients undergoing renal transplantation. Clin Pharmacol Ther 1995;58(3):322–7.
18. Naderer OJ, Dupuis RE, Heinzen EL, Wiwattanawongsa K, Johnson MW, Smith PC. The influence of norfloxacin and metronidazole on the disposition of mycophenolate mofetil. J Clin Pharmacol 2005;45(2):219–26.
19. Archer JS, Archer DF. Oral contraceptive efficacy and antibiotic interaction: a myth debunked. J Am Acad Dermatol 2002;46(6):917–23.
20. Schmidt LE, Dalhoff K. Food–drug interactions. Drugs 2002;62(10):1481–502.

Novobiocin

General Information

Novobiocin is an antimicrobial drug that is structurally related to the coumarins. It is active against Gram-positive bacteria, such as *Staphylococcus aureus* (including meticillin-resistant strains) and other staphylococci; the enterococci are usually resistant. Some Gram-negative organisms, including *Haemophilus influenzae* and *Neisseria* species, are also susceptible, as are some strains of *Proteus*, but most of the Enterobacteriaceae are resistant. Its action is primarily bacteriostatic, although it may be bactericidal against more sensitive species at high concentrations. It inhibits DNA gyrase and is effective in eliminating plasmids, but resistance to novobiocin develops readily both in vitro and during therapy.

Although novobiocin has been used alone or with other drugs, such as rifampicin or tetracycline, in the treatment of infections due to staphylococci and other susceptible organisms, it has largely been superseded by other drugs because of the problems of resistance and toxicity.

Organs and Systems

Hematologic

Reports of drug-induced thrombocytopenia have been systematically reviewed (1). Among the 98 different drugs described in 561 articles the following antibiotics were found with level I (definite) evidence: co-trimoxazole, rifampicin, vancomycin, sulfisoxazole, cefalothin, piperacillin, methicillin, novobiocin. Drugs with level II (probable) evidence were oxytetracycline and ampicillin.

Drug–Drug Interactions

Cytotoxic drugs

Novobiocin potentiates the activity of etoposide in vitro by increasing intracellular accumulation of etoposide. A phase I trial showed that the maximum tolerated dose of novobiocin was 7 g/m^2/day when it was given in combination with etoposide; dose-limiting adverse effects consisted of neutropenic fever and reversible hyperbilirubinemia. Novobiocin did not augment the toxic effects of etoposide on the bone marrow or the gastrointestinal mucosa, and nausea, a dose-limiting adverse effect in other trials of novobiocin, was well controlled by 5-HT3 receptor antagonists (2).

However, in a phase II study of the combination of novobiocin and high-dose cyclophosphamide and thiotepa, followed by autologous bone marrow support in women with chemosensitive advanced breast cancer, there was no significant increase in progression-free survival and overall survival compared with historical controls treated with high-dose alkylating drugs alone (3).

Lactoferrin

Lactoferrin can potentiate the activity of novobiocin against some strains of *Escherichia coli* strains (4).

Sodium polyphosphate

The permeabilizer sodium polyphosphate may enhance the activity of novobiocin against *Pseudomonas aeruginosa* (5).

Vincristine

Novobiocin potentiated the anti-angiogenic effect of vincristine in vitro (6).

References

1. George JN, Raskob GE, Shah SR, Rizvi MA, Hamilton SA, Osborne S, Vondracek T. Drug-induced thrombocytopenia: a systematic review of published case reports. Ann Intern Med 1998;129(11):886–90.
2. Murren JR, DiStasio SA, Lorico A, McKeon A, Zuhowski EG, Egorin MJ, Sartorelli AC, Rappa G. Phase I and pharmacokinetic study of novobiocin in combination with VP-16 in patients with refractory malignancies. Cancer J 2000;6(4):256–65.
3. Hahm HA, Armstrong DK, Chen TL, Grochow L, Passos-Coelho J, Goodman SN, Davidson NE, Kennedy MJ. Novobiocin in combination with high-dose chemotherapy for the treatment of advanced breast cancer: a phase 2 study. Biol Blood Marrow Transplant 2000;6(3A):335–43.
4. Sanchez MS, Watts JL. Enhancement of the activity of novobiocin against Escherichia coli by lactoferrin. J Dairy Sci 1999;82(3):494–9.
5. Ayres HM, Furr JR, Russell AD. Effect of permeabilizers on antibiotic sensitivity of Pseudomonas aeruginosa. Lett Appl Microbiol 1999;28(1):13–6.

6. Yang J, Jiang M, Zhen YS. Novobiocin inhibits angiogenesis and shows synergistic effect with vincristine. Yao Xue Xue Bao 2003; 38: 731-4.

Ofloxacin

See also Fluoroquinolones

General Information

Ofloxacin is a fluoroquinolone antibacterial drug similar to ciprofloxacin.

Comparative studies

The safety and efficacy of topical ofloxacin ear-drops 0.3% (0.25 ml bd) have been compared with that of co-amoxiclav oral suspension (40 mg/kg/day) for acute otitis media in 286 children aged 1–12 years with tympanostomy tubes in place. Topical ofloxacin was as effective as and better tolerated than systemic therapy with co-amoxiclav. Treatment-related adverse event rates were 31% for co-amoxiclav and 6% for ofloxacin (1).

Organs and Systems

Nervous system

Headache was recorded in 9% taking ofloxacin during short-course treatment of urinary tract infections (2).

Seizures have occurred in patients taking ofloxacin (3).

A Tourette-like syndrome developed in a 71-year-old patient, temporally related to ofloxacin; spitting, profuse swearing, echolalia, echopraxia, orofacial and limb automatisms, and hypersalivation all resolved completely after withdrawal (4).

Orofacial dyskinesia with other fluoroquinolones has been rarely reported. However, ofloxacin-induced orofacial dyskinesia has been reported after treatment with ofloxacin 400 mg/day for 3 days in a 43-year-old man (5).

Sensory systems

Eyes

Topical ofloxacin in the treatment of bacterial keratitis caused corneal deposits in six cases (6).

Ears

Compared with aminoglycosides, which caused significant loss of hair cells in the basal turn of the cochlea, ofloxacin caused no loss of hair cells in rats, even at concentrations higher than are achieved clinically (7). Moreover, auditory brainstem testing showed no change in auditory thresholds in the ofloxacin-treated animals, whereas neomycin-treated animals showed substantial threshold shifts. In human studies, topical ofloxacin 0.3% had no demonstrable adverse effects on middle ear or cochlear function and was not associated with any changes in hearing.

Psychiatric

Ofloxacin can cause serious psychiatric adverse effects, particularly in those with a past psychiatric history (8).

Endocrine

Ofloxacin 200 mg bd caused diabetes insipidus in a 25-year-old man (9).

Salivary glands

Ofloxacin can impair salivary gland function. In rat parotid and submandibular glands intraperitoneal ofloxacin (20, 40, and 80 mg/kg) reduced flow rate, amylase activity, total protein and calcium concentrations; in parotid saliva, sodium and potassium were increased (10). Sodium and potassium concentrations were also increased by a dose of 80 mg/kg in submandibular saliva. Possible mechanisms of these effects include altered intracellular cAMP and calcium concentrations and suppression of DNA, RNA, and protein synthesis in acinar cells.

Liver

Ofloxacin can cause fatal hepatic failure (11).

Urinary tract

Acute renal insufficiency has been caused by ofloxacin (12).

Antimicrobial chemotherapy against diseases caused by *Escherichia coli* producing Shiga toxin has been implicated as a risk factor for progression to the hemolytic-uremic syndrome.

- A 75-year-old woman took ofloxacin and developed the hemolytic-uremic syndrome (13).

In an in vitro study the addition of ofloxacin to a cell culture increased toxin activity by more than 200-fold (14).

Skin

Ofloxacin has been implicated in a case of toxic epidermal necrolysis (15).

- A 75-year-old white man took 24 g of ofloxacin over 51 days for epididymitis. He had a severe skin reaction, diagnosed as toxic epidermal necrolysis, and died from complications.

Sweet's syndrome has been associated with ofloxacin 400 mg/day in a 40-year-old woman with Crohn's disease (16).

- A 40-year-old woman developed Sweet's syndrome after taking ofloxacin for 3 days for watery bloody diarrhea in the course of Crohn's disease (17). She had a fever (38.6°C), rash, abdominal tenderness, and painful joints. Her symptoms resolved spontaneously 3 days after withdrawal of ofloxacin.

Musculoskeletal

Achilles tendon rupture has been attributed to ofloxacin and a short course of prednisolone (18).

A 19-year-old man who took ofloxacin 800 mg/day for 3 days developed rhabdomyolysis (19).

Drug Administration

Drug formulations

Chitosan, a positively charged polysaccharide, increases the precorneal residence time of ophthalmic formulations that contain active compounds. Two chitosan products of high molecular weights (1350 and 1930 kDa) and low degrees of deacetylation (50%) significantly increased antibiotic availability compared with controls (20). The duration of efficacy of ofloxacin was significantly increased from about 25 minutes to 46 minutes by the chitosan of higher molecular weight.

Drug–Drug Interactions

Antacids

The systemic availability of ofloxacin is reduced by aluminium-containing antacids, but not by antacids containing magnesium or calcium (21).

Glucocorticoids

The antibacterial activity of polymorphonuclear leukocytes is based on the production of superoxide anions and H_2O_2 in the respiratory burst. Combinations of an antibacterial agent and an anti-inflammatory drug are commonly used in immunosuppressed patients whose respiratory burst of polymorphonuclear leukocytes is impaired. An in vitro study has shown that the combination of a glucocorticoid (dexamethasone, methylprednisolone, betamethasone, hydrocortisone) with ofloxacin 10 mg/ml neutralizes the inhibitory effect of the former on the respiratory burst (22).

NSAIDs

The antibacterial activity of polymorphonuclear leukocytes is based on the production of superoxide anions and H_2O_2 in the respiratory burst. Combinations of an antibacterial agent and an anti-inflammatory drug are commonly used in immunosuppressed patients whose respiratory burst of polymorphonuclear leukocytes is impaired. An in vitro study has shown that the combination of an anti-inflammatory drug (phenylbutazone or acetylsalicylic acid) with ofloxacin 10 mg/ml neutralizes the inhibitory effect of the former on the respiratory burst (22).

Oral contraceptives

Plasma concentrations of oral contraceptive steroids were unchanged by ofloxacin (23).

Procainamide

The renal clearance of procainamide is inhibited by ofloxacin (24).

References

1. Goldblatt EL, Dohar J, Nozza RJ, Nielsen RW, Goldberg T, Sidman JD, Seidlin M. Topical ofloxacin versus systemic amoxicillin/clavulanate in purulent otorrhea in children with tympanostomy tubes. Int J Pediatr Otorhinolaryngol 1998;46(1–2):91–101.
2. McCarty JM, Richard G, Huck W, Tucker RM, Tosiello RL, Shan M, Heyd A, Echols RM. A randomized trial of short-course ciprofloxacin, ofloxacin, or trimethoprim/sulfamethoxazole for the treatment of acute urinary tract infection in women. Ciprofloxacin Urinary Tract Infection Group. Am J Med 1999;106(3):292–9.
3. Walton GD, Hon JK, Mulpur TG. Ofloxacin-induced seizure. Ann Pharmacother 1997;31(12):1475–7.
4. Thomas RJ, Reagan DR. Association of a Tourette-like syndrome with ofloxacin. Ann Pharmacother 1996;30(2):138–41.
5. De Bleecker JL, Vervaet VL, De Sarro A. Reversible orofacial dyskinesia after ofloxacin treatment. Mov Disord 2004;19(6):731–2.
6. Mitra A, Tsesmetzoglou E, McElvanney A. Corneal deposits and topical ofloxacin—the effect of polypharmacy in the management of microbial keratitis. Eye 2007;21(3):410–2.
7. Marinucci P. Risk of torsades de pointes with non-cardiac drugs. Grapefruit juice is source of potentially life threatening adverse drug reactions. BMJ 2001;322(7277):47.
8. Hall CE, Keegan H, Rogstad KE. Psychiatric side effects of ofloxacin used in the treatment of pelvic inflammatory disease. Int J STD AIDS 2003;14:636–7.
9. Bharani A, Kumar H. Drug points: Diabetes inspidus induced by ofloxacin. BMJ 2001;323(7312):547.
10. Abdollahi M, Isazadeh Z. Inhibition of rat parotid and submandibular gland functions by ofloxacin, a fluoroquinolone antibiotic. Fundam Clin Pharmacol 2001;15(5):307–11.
11. Gonzalez Carro P, Huidobro ML, Zabala AP, Vicente EM. Fatal subfulminant hepatic failure with ofloxacin. Am J Gastroenterol 2000;95(6):1606.
12. Espiritu J, Walton T. Acute renal failure due to ofloxacin. W V Med J 1995;91(1):16.
13. Miedouge M, Hacini J, Grimont F, Watine J. Shiga toxin-producing Escherichia coli urinary tract infection associated with hemolytic–uremic syndrome in an adult and possible adverse effect of ofloxacin therapy. Clin Infect Dis 2000;30(2):395–6.
14. Kimmitt PT, Harwood CR, Barer MR. Induction of type 2 Shiga toxin synthesis in Escherichia coli O157 by 4-quinolones. Lancet 1999;353(9164):1588–9.
15. Melde SL. Ofloxacin: a probable cause of toxic epidermal necrolysis. Ann Pharmacother 2001;35(11):1388–90.
16. Ozdemir D, Korkmaz U, Sahin I, Sencan I, Kavak A, Kucukbayrak A, Cakir S. Ofloxacin induced Sweet's syndrome in a patient with Crohn's disease. J Infect 2006;52(5):e155-7.
17. Ozdemir D, Korkmaz U, Sahin I, Sencan I, Kavak A, Kucukbayrak A, Cakir S. Ofloxacin induced Sweet's syndrome in a patient with Crohn's disease. J Infect 2006;52(5):e155-7.
18. Vaucher N, Mosquet B, Levast M. Rupture du tendon d'Achille lors d'un traitement par solution auriculaire d'ofloxacine precédée d'une courte cure orale de prednisolone. [Achilles tendon rupture during ofloxacin treatment and a short course of prednisolone.] Presse Med 2006;35(9 Pt 1):1271–2.
19. Hsiao SH, Chang CM, Tsao CJ, Lee YY, Hsu MY, Wu TJ. Acute rhabdomyolysis associated with ofloxacin/levofloxacin therapy. Ann Pharmacother 2005;39(1):146–9.

20. Felt O, Baeyens V, Buri P, Gurny R. Delivery of antibiotics to the eye using a positively charged polysaccharide as vehicle. AAPS PharmSci 2001;3(4):E34.

21. Flor S, Guay DR, Opsahl JA, Tack K, Matzke GR. Effects of magnesium–aluminum hydroxide and calcium carbonate antacids on bioavailability of ofloxacin. Antimicrob Agents Chemother 1990;34(12):2436–8.

22. Cabrera E, Velert MM, Orero A, Martinez P, Canton E. Efecto de los antiinflamatorios, solos y asociados con oflox-acino, en la explosion respiratoria de los leucocitos polimor-fonucleares humanos. [Effect of anti-inflammatory drugs, alone and combined with ofloxacin, on the respiratory burst of human polymorphonuclear leukocytes.] Rev Esp Quimioter 2001;14(2):165–71.

23. Archer JS, Archer DF. Oral contraceptive efficacy and antibiotic interaction: a myth debunked. J Am Acad Dermatol 2002;46(6):917–23.

24. Martin DE, Shen J, Griener J, Raasch R, Patterson JH, Cascio W. Effects of ofloxacin on the pharmacokinetics and pharmacodynamics of procainamide. J Clin Pharmacol 1996;36(1):85–91.

Oxazolidinones

General Information

The oxazolidinones represent the first truly new class of antibacterial agents to reach the marketplace in several decades (1,2,3,4). They have a unique mechanism of action, involving inhibition of the first step in protein synthesis. This unique mechanism of action makes cross resistance with other antimicrobial agents unlikely. The first marketed member of the class, linezolid, has inhibitory activity against a broad range of Gram-positive bacteria, including meticillin-resistant *Staphylococcus aureus* (MRSA), glycopeptide-intermediate *Staphylococcus aureus* (GISA), vancomycin-resistant enterococci (VRE), and penicillin-resistant *Streptococcus pneumoniae*. It also has activity against certain anerobes, including *Clostridium perfringens*, *Clostridium difficile*, *Peptostreptococcus* species, and *Bacteroides fragilis*. The most frequently reported adverse effects are diarrhea, headache, nausea and vomiting, insomnia, constipation, rash, and dizziness; thrombocytopenia has also been documented in a few patients (about 2%) (5,6).

In controlled phase III studies, linezolid was as effective as vancomycin in patients with infections caused by MRSA and VRE. It is effective both intravenously and orally. Although technically classified as bacteriostatic against a number of pathogens in vitro, linezolid behaves in vivo like a bactericidal antibiotic.

Linezolid is completely absorbed and can be given orally or intravenously; it is cleared by both renal and hepatic routes, but dosage adjustments are not needed in moderate renal or hepatic insufficiency and are not affected by age or sex. Linezolid is metabolized via morpholine ring oxidation, which is independent of CYP450; linezolid is therefore unlikely to interact with medications that induce or inhibit CYP450 enzymes. Plasma linezolid trough concentrations after a 1-hour infusion of 600 mg bd were 0.54–5.3 µg/ml and CSF linezolid trough concentrations were 1.46–7.0 µg/ml; the ratio between CSF and plasma linezolid trough concentrations always exceeded 1 (mean, 1.6; range 1.2–2.3) (7). There was good rapid penetration of linezolid into bone, fat, and muscle (8,9). The mean peak plasma concentration of linezolid was 18.3 µg/ml in six healthy men who took 600 mg orally every 12 hours for five doses. There was good penetration into inflammatory fluid (10,11).

General adverse effects

Linezolid is generally well tolerated (12–14). The most frequently reported adverse events were diarrhea, headache, nausea and vomiting, insomnia, constipation, rash, and dizziness. Thrombocytopenia was also documented in few patients (about 2%). In a phase II, open, multicenter study of intravenous linezolid followed by oral linezolid suspension, both in a dose of 10 mg/kg every 12 hours in 66 children, the most common adverse effects were diarrhea (10%), neutropenia (6.4%), and raised alanine transaminase activity (6.4%) (15).

In children the common adverse events have been similar to those found in adults, although thrombocytopenia has not been as common (16).

Observational studies

In a prospective, multicenter, open, non-comparative, non-randomized trial In patients with serious Gram positive infections, who received linezolid 600 mg bd, the overall adverse event rate was 18%. The most common adverse effects were increased liver function tests, rash, and gastrointestinal disturbances. Three patients required withdrawal of therapy for rash, one for raised liver function tests, and one for thrombocytopenia (17).

In a retrospective analysis in 44 patients taking linezolid (mean duration 29; range 8–185 days), the clinical cure rate was 73%; 28 had adverse reactions (thrombocytopenia, n = 13; anemia, n = 7; gastrointestinal, n = 12; peripheral neuropathy, n = 1; serotonin syndrome, n = 1) (18).

In a retrospective analysis in 20 patients receiving linezolid for orthopedic infections, eight developed reversible myelosuppression, one had irreversible peripheral neuropathy, and two withdrew because of pancytopenia or urticaria (19).

In a retrospective analysis in 42 patients receiving linezolid for osteomyelitis, the clinical cure rate was 55% in the 20 patients who received therapy for at least 6 weeks. Adverse events included gastrointestinal disturbances (15%), thrombocytopenia (10%), anemia (10%), neutropenia (5%), and rash (5%) (20).

Organs and Systems

Cardiovascular

In a placebo-controlled, crossover study in 12 healthy men who took one oral dose of linezolid 600 mg or a placebo tablet followed by an intravenous tyramine pressor test until the systolic blood pressure increased by at least 30 mmHg above baseline, there was a significant difference in the pressor response to intravenous tyramine between linezolid and placebo (21).

Severe bradycardia with an increased blood pressure has been attributed to linezolid (22).

- A 49-year-old woman with cancer of the biliary tree developed a fever and jaundice. Dilatation of the right biliary tract was confirmed by ultrasound and nuclear magnetic resonance. She was given ceftazidime 4 g/day and oral linezolid 600 mg/day. Unexpectedly, 2 days later her blood pressure rose to 170/90 mmHg and was associated with severe bradycardia 37–40/minute. Linezolid was withdrawn. The pulse rate became normal after 48 hours and the blood pressure fell.

The mechanism underlying this effect is unknown, but it may have been related to the fact that linezolid is a monoamine oxidase inhibitor.

Nervous and sensory systems

Long-term use of linezolid can be associated with severe peripheral and optic neuropathy. Eight patients with multidrug-resistant tuberculosis were treated with linezolid 600 mg/day plus at least four other drugs (23). Cultures became negative in all patients in an average of 82 days. Four patients developed a peripheral neuropathy, two developed an optic neuropathy, and one developed anemia. Although the optic neuropathy resolved after withdrawal of linezolid the peripheral neuropathy persisted. Three patients stopped taking linezolid after 7–9 months, two because of adverse effects and one for economic reasons.

There have been several reports of optic and peripheral neuropathies (24,25).

- Peripheral and optic neuropathy occurred in a 76-year-old man after he had taken linezolid for about 6 months (26).
- A 65-year-old man who had taken linezolid 600 mg bd for 1 year because of an infection with a multidrug-resistant Staphylococcus aureus (MRSA) developed bilateral gradual loss of vision over 2 months. Linezolid was withdrawn and his visual acuities returned to normal over 15 months.
- A 40-year-old monocular woman with pyoderma gangrenosum who was taking long-term corticosteroids developed progressive, painless, loss of central vision in her remaining eye over 2 weeks after taking ciprofloxacin and linezolid 600 mg bd for recurrent meticillin-resistant Staphylococcus aureus (MRSA) infections for 6 months. At the same time she developed distal numbness and paresthesia. Linezolid was withdrawn and 1 month later her color vision had improved to 8/14. Four months later, there was 70% subjective improvement, a visual acuity of 20/40, normalization of the visual field, and mild nerve pallor. The

peripheral numbness and tingling did not improve. Normal vision persisted 18 months later.

- A 66-year-old man developed progressive, painless, bilateral loss of vision and peripheral numbness after taking rifampicin and linezolid 600 mg bd for 5 months. Both drugs were withdrawn and 3 months later his visual acuity was 20/25 in the right eye and 20/20 in the left and there was resolution of the disc edema. The peripheral numbness persisted.
- A 79-year-old woman developed progressive, painless loss of vision, more in the left eye than the right over 1 month after taking linezolid for 10 months. There was also an axonal neuropathy, which was attributed to linezolid, which was withdrawn. Three months later, she reported subjective visual improvement, and the acuity in the left eye was 20/50. However, the acuity in the right amblyopic eye was unchanged. Six months later, the acuity in the left eye was 20/30. Color vision and Goldmann visual field improved. The neuropathy was unchanged.

In most cases, optic neuropathies resolved after stopping linezolid but peripheral neuropathies did not; the duration of therapy seems to be the most important factor (27).

- A 46-year-old white woman developed a severe, painful peripheral neuropathy while taking oral linezolid 600 mg bd for 6 months (28). Nerve conduction studies showed a sensorimotor axonal neuropathy. Extensive assessment did not show alternative explanations for her neuropathy. When she died 1 month after withdrawal of linezolid, the neuropathy had not resolved.
- A 27-year-old white woman took linezolid 600 mg bd for 6 months and developed paresthesia in her extremities, numbness of the legs below the knee, and intermittent sharp pain in both feet (29). There was peripheral sensory loss in the "glove and stocking" distribution. Nerve conduction studies showed sensory motor axonal neuropathy. Linezolid was withdrawn and 5 months later she reported no pain. However, nerve conduction studies showed that the peripheral neuropathy persisted.

Other cases of linezolid-induced peripheral neurotoxicity have been reported (30,31).

- A 70-year-old man took linezolid for 10–12 days and complained of numbness in his toes. Nerve conduction studies showed sensory motor axonal damage. During the next 5 months he develop progressive perineal anesthesia without fecal incontinence or urinary tract disorder. Spinal cord MRI and nerve biopsy were normal. The hemoglobin was 8.5 g/dl. Linezolid was withdrawn after 24 weeks of therapy. The perineal anesthesia disappeared after 8 months, but the neuropathy persisted.
- A 45-year-old woman complained of bilateral, predominantly left-sided paresthesia of the toes after taking linezolid for 6 months. One week later, linezolid was withdrawn because of a glove-and-stocking neuropathy without abnormal ankle reflexes. Nerve conduction studies were compatible with a sensory axonal neuropathy. Two weeks after withdrawal of linezolid, she

reported subjective improvement of the sensory perception. After 3 months minor bilateral paresthesia persisted.

- A 38-year-old man presented with bilateral isolated lower limb dysesthesia after taking linezolid for 2 months. Linezolid was withdrawn and voriconazole was continued while the neurological signs improved. Spinal MRI and lumbar puncture were normal; a search for antineuronal antibodies in serum and CSF was negative. After 6 months the paresthesia persisted.
- A 42-year-old woman developed a severe irreversible sensory polyneuropathy after a 6-month course of linezolid.

Bell's palsy has been reported in a 49-year-old man after 3 weeks of linezolid therapy; the symptoms recurred after rechallenge (32).

Sensory systems

Two patients undergoing long-term treatment (both 11 months) with linezolid for pneumonia had reduced visual acuity, dyschromatopsia, and cecocentral scotomata characteristic of toxic optic neuropathy (33). Visual function slowly recovered 3–4 months after withdrawal of linezolid.

- A 56-year-old man developed a toxic optic neuropathy after long-term treatment with linezolid 600 mg bd for 12 months then 600 mg/day for 44 months (34). Linezolid was withdrawn and he noted subjective visual improvement within several weeks.
- A 27-year-old woman developed an optic and peripheral neuropathy after taking linezolid for 154 days for osteomyelitis (35). A glucocorticoid exacerbated the visual loss. Withdrawal of linezolid resulted in marked improvement.

Metabolism

Hyperlactatemia and metabolic acidosis are adverse effects of linezolid that could be related to impaired mitochondrial function.

- Linezolid-induced lactic acidosis occurred in a 70-year-old man during the first 7 days of treatment with linezolid (36). Mitochondria were studied from three patients, in whom weakness and hyperlactatemia developed during therapy with linezolid (37). The results suggested that linezolid interferes with mitochondrial protein synthesis, probably because of similarities between bacterial and mitochondrial ribosomes.
- Linezolid induced hyperlactatemia occurred in a 59-year-old patient during treatment with linezolid after liver transplantation (38).
- A 63-year-old woman developed an optic neuropathy, encephalopathy, skeletal myopathy, lactic acidosis, and renal failure after taking linezolid for 4 months (39). Mitochondrial respiratory chain enzyme activity was reduced in affected tissues, without ultrastructural mitochondrial abnormalities and without mutations or depletion of mtDNA. In experimental animals, line-

zolid caused a dose- and time-related reduction in the activity of respiratory chain complexes containing mtDNA-encoded subunits and reduced the amount of protein in these complexes, whereas the amount of mtDNA was normal.

These results provide direct evidence that linezolid inhibits mitochondrial protein synthesis, with potentially severe clinical consequences.

Hematologic

Linezolid has been associated with myelosuppression, including anemia, leukopenia, pancytopenia, and thrombocytopenia. It is recommended that complete blood counts be monitored weekly in patients who take linezolid, especially those who take it for more than 2 weeks (26,40,41). It is recommended that complete blood counts be monitored weekly in patients who take linezolid, especially those who take it for more than 2 weeks. In children, thrombocytopenia is less common; however, the complete blood count should be monitored weekly while they are taking linezolid.

The mechanism of the anemia has been described and is thought to be inhibition of mitochondrial respiration. It can be managed relatively easily with transfusions. The thrombocytopenia is progressive and may require drug withdrawal; a mechanism for this effect has not been described. A bone marrow biopsy in a patient who developed thrombocytopenia 7 days after starting to take linezolid showed adequate numbers of normal-looking megakaryocytes. This finding alone argues against marrow suppression and supports an immune-mediated mechanism of platelet destruction (42).

Reversible myellosuppression (43) appears to be related to the duration of therapy, with a higher risk after more than 2 consecutive weeks of treatment (44). Myelosuppression with red cell hypoplasia has been reported in three patients taking linezolid 600 mg bd. The bone marrow changes were similar to those seen in reversible chloramphenicol toxicity. Another patient had sideroblastic anemia after taking linezolid for 2 months (45,46).

- Red cell hypoplasia and thrombocytopenia occurred in a 66-year-old man who took linezolid 600 mg bd (47).
- Reversible pure red blood cell aplasia occurred in a 52-year-old black man who had taken linezolid for 8 weeks (48).

Pancytopenia in two cases was reported with linezolid 600 mg bd (49).

In 19 patients there was thrombocytopenia in six of those who had taken it for more than 10 days; gastrointestinal bleeding was observed in one patient and four required platelet transfusions (50). Of 71 patients who took linezolid for 1–44 days, 48 took it for more than 5 days; among those 48, thrombocytopenia, with a 32–89% reduction in platelet count, occurred in 23; the platelet count fell to below $100 \times 10^9/l$ in nine (51).

In a retrospective study patients taking linezolid with lower pre-treatment hematological values were at greater risk not only of anemia but also thrombocytopenia (52).

In a retrospective case-control study of linezolid in 91 patients with end-stage renal disease, 28 of whom were receiving hemodialysis, and patients with non-end-stage renal disease, severe thrombocytopenia and anemia were significantly more frequent in the former: 79% versus 43 and 71% versus 37 respectively (53). Survival analysis for the development of thrombocytopenia or death showed significant differences between the two groups.

In children, thrombocytopenia is less common; however, the complete blood count should be monitored weekly while children are taking linezolid. Vitamin B_6 50 mg/day may prevent or modify the course of linezolid-associated cytopenias. Two patients developed dyserythropoietic anemia, strikingly similar to chloramphenicol-associated myelotoxicity, after taking linezolid for 25–28 days (54).

In a randomized, controlled, open, multicenter study in 430 patients with suspected or proven Gram-positive infections, patients received oral linezolid 600 mg every 12 hours or intravenous or intramuscular teicoplanin for up to 28 days (28). Five patients who were treated with linezolid had hematological adverse effects rated as moderate or severe, including hemolysis, anemia, leukopenia, and thrombocytopenia.

Immunologic

Linezolid-related leukocytoclastic vasculitis occurred in a 68-year-old man after 7 days of treatment with linezolid 600 mg bd (55).

Susceptibility Factors

Age

In single-dose pharmacokinetic studies children had a greater plasma clearance (0.34 l/hour/kg for children aged 3 months to 16 years) than adults (0.10 l/hour/kg) (16).

The pharmacokinetics of linezolid in children have been studied in four clinical trials in over 180 patients, from preterm neonates to 18-year-olds (4). Linezolid clearance is greater in children than in adults and greater in children aged under 12 years. Based on these studies, the recommended dosage of linezolid is 10 mg/kg every 12 hours for children over 12 years (maximum dose 600 mg every 12 hours) and 10 mg/kg every 8 hours for children under 12 years of age. Based on limited pharmacokinetic data in neonates, linezolid clearance appears to be relatively reduced in premature infants under 34 weeks gestation and under 7 days of age; these patients should receive 10 mg/kg every 12 hours, increasing to 10 mg/kg every 8 hours in all neonates beyond the first week of life, regardless of gestational age.

In children the common adverse effects of linezolid have been similar to those seen in adults (diarrhea, vomiting, loose stools, and nausea); however, thrombocytopenia has not been as common (56). In clinical trials of linezolid involving over 950 children, the most common

adverse events were diarrhea, headache, vomiting, nausea, raised serum transaminase activities, and rash (6.5–11% of patients) (41).

Renal and hepatic impairment

Linezolid is cleared by renal and hepatic routes, but dosage adjustments are not needed in moderate renal or hepatic insufficiency (40,57).

In 20 critically ill patients linezolid was significantly eliminated by continuous venovenous hemofiltration (58). The total clearance was 25% higher and the trough serum concentration was 50% lower than in normal conditions. There was large interindividual variability, and the authors concluded that the standard dosage of 600 mg 12-hourly might be ineffective in some patients receiving continuous venovenous hemofiltration; that is, it might be ineffective for the least-susceptible pathogens that have an MIC of >4 mg/l. They also concluded that increasing the dosage to 600 mg 8-hourly might be warranted in selected patients to assure optimal antibacterial activity. Both of the main metabolites of linezolid, PNU-142300 and PNU-142585, accumulated significantly in anuric patients receiving continuous venovenous hemofiltration. The clinical relevance of these inactive metabolites is unknown. However, because prolonged use of linezolid might be associated with severe hemopoietic and neurological adverse effects, special attention has to be paid to the potential toxicity associated with linezolid accumulation.

In a 33-year-old man linezolid was not cleared significantly by continuous venovenous hemofiltration, which contributed 22 ml/min of clearance to the total clearance (59).

In a prospective, single-dose pharmacokinetic study in patients with acute renal insufficiency who received 600 mg of linezolid intravenously and were treated with renal replacement therapy, serum concentrations averaged 12 mg/l and fell at the end to below 4 mg/l in three of eight patients on hemodialysis, three of five patients on sustained low-efficiency dialysis, and two patients on continuous venovenous hemofiltration (60). Mean removal of the drug was 194 mg with hemodialysis (32% of the dose administered), 205 mg with sustained low-efficiency dialysis (34%), and 75 mg (12%) and 105 (18%) mg following the continuous venovenous hemofiltration sessions.

Cystic fibrosis

In 12 adults with cystic fibrosis given a single intravenous dose of linezolid 600 mg over 30 minutes the pharmacokinetics, while variable, with half-lives of 1.76–8.36 hours, were similar to those previously described in other populations (61).

Drug Administration

Drug dosage regimens

Linezolid crosses the blood–retina barrier in non-inflamed eyes. The vitreous concentration in 12 adults after a single 600 mg oral dose rose exponentially with time and 33% of the late group achieved sufficient MIC_{90}

concentrations for the common pathogens found in postoperative endophthalmitis (62). The vitreous linezolid concentration correlated strongly with the interpolated serum concentration. Adequate concentrations might therefore be achieved with an altered dosage regimen to achieve higher serum steady-state concentrations.

Drug–Drug Interactions

Adrenoceptor agonists

An enhanced pressor response has been seen in patients taking linezolid and certain adrenergic agents, including phenylpropanolamine and pseudoephedrine, and the doses of these drugs should be reduced in patients taking linezolid (63).

Bupropion

An interaction between linezolid and bupropion has been reported to cause severe intermittent intraoperative hypertension in a 57-year-old man (64).

Diphenhydramine

- A 56-year-old white man developed delirium with visual and auditory hallucinations and erratic, aggressive behavior, presumably caused by the combination of diphenhydramine 300 mg/day and linezolid 600 mg every 12 hours (65). This persisted for 3 days after diphenhydramine was withdrawn.

Hydroxyzine

Encephalopathy in a 74-year-old woman was attributed to an interaction of linezolid with hydroxyzine (73[A]).

Monoamine oxidase inhibitors

Because the original oxazolidinones, including linezolid, are monoamine oxidase inhibitors, particular attention has been paid to the question of whether adverse interactions with drugs that are metabolized by monoamine oxidase would occur in patients taking linezolid (124) (66).

There is no evidence of interactions of linezolid with dextromethorphan or pethidine.

Salbutamol

There is no evidence of interactions of linezolid with oral or inhaled salbutamol.

Selective serotonin re-uptake inhibitors (SSRIs)

The use of linezolid with medications that increase concentrations of serotonin in the central nervous system can result in serotonin toxicity (67,68,69). Serotonin syndrome due to an interaction of linezolid with sertraline has been described (70). There have been cases of serotonin syndrome due to interaction of linezolid with venlafaxine (71,72), one case in a patient taking linezolid, amitriptyline, and paroxetine (73), and one in a patient

taking linezolid, citalopram, and mirtazapine (74). In a retrospective study 12 patients (mean age 53 years) were found with linezolid-associated serotonin syndrome (75). All had taken linezolid concomitantly with an SSRI. The onset of syndrome was 9.5 days after the introduction of linezolid and was directly correlated with age. The symptoms resolved in 2.9 days. Citalopram was associated with delayed resolution. There was a trend towards a longer resolution time the longer the half-life of the interacting drug. An interaction of duloxetine 60 mg/day with noradrenaline and linezolid caused serotonin syndrome in a 55-year-old man (67).

In a retrospective chart review of 72 inpatients at the Mayo Clinic, 52 received concomitant therapy with linezolid and an SSRI or venlafaxine within 14 days, and 20 did not but did receive linezolid and an SSRI separately within 14 days of each other (76,77). Overall, only two patients had a high probability of serotonin syndrome, and in both cases the symptoms rapidly resolved on withdrawal of the SSRI.

Serotonin syndrome was reported in a 56-year-old white woman who received intravenous linezolid shortly after withdrawal of a selective serotonin reuptake inhibitor, paroxetine (78).

The serotonin syndrome was reported in a 45-year-old white man who received intravenous linezolid (600 mg 12 hourly) and sertraline (79).

Venlafaxine

Linezolid interacts with selective serotonin reuptake inhibitors and sympathomimetic drugs, resulting in serotonin syndrome. Serotonin syndrome due to an interaction of linezolid and venlafaxine has been reported (61,80).

- An 85-year-old man with an infected hip joint prosthesis was given ciprofloxacin 750 mg bd, rifampicin 300 mg bd, and linezolid 600 mg bd (81). He was also taking venlafaxine 150 mg/day for depression and 20 days later became confused and disoriented, with disturbance of the sleep–wake cycle, and was intermittently aggressive; 4 days later he had widespread increased tone, generalized myoclonic jerks, and extensor plantar reflexes. Linezolid and venlafaxine were withdrawn. Within 2 days he had recovered his usual level of mental functioning.

Vitamin C

In an open study in 28 healthy volunteers the pharmacokinetics of linezolid 600 mg/day were not affected by concomitant administration of vitamin C 1000 mg/day) or vitamin E 800 IU/day (82).

Management of Adverse Drug Reactions

Two patients with disseminated *Mycobacterium abscessus* infections who took linezolid for at least 9 months developed cytopenias and one also developed a peripheral neuropathy (83). Because continuing linezolid therapy was required, oral vitamin B$_6$ 50 mg/day was administered in an attempt to mitigate the cytopenia. In both patients the cytopenia resolved after administration of vitamin B_6 and stabilized during prolonged linezolid therapy, although the peripheral neuropathy did not.

A 41-year-old woman who was allergic to oxazolidinones was successfully desensitized with oral administration of an intravenous formulation of linezolid (84).

References

1. Stevens DL, Dotter B, Madaras-Kelly K. A review of linezolid: the first oxazolidinone antibiotic. Expert Rev Anti Infect Ther 2004;2(1):51–9.

2. Harwood PJ, Giannoudis PV. The safety and efficacy of linezolid in orthopaedic practice for the treatment of infection due to antibiotic-resistant organisms. Expert Opin Drug Saf 2004;3(5):405–14.

3. Bozdogan B, Appelbaum PC. Oxazolidinones: activity, mode of action, and mechanism of resistance. Int J Antimicrob Agents 2004;23(2):113–9.

4. Wilcox MH. Update on linezolid: the first oxazolidinone antibiotic. Expert Opin Pharmacother 2005;6(13):2315–26.

5. Evans GA. The oxazolidinones. Curr Infect Dis Rep 2002;4(1):17–27.

6. Moise PA, Forrest A, Birmingham MC, Schentag JJ. The efficacy and safety of linezolid as treatment for *Staphylococcus aureus* infections in compassionate use patients who are intolerant of, or who have failed to respond to, vancomycin. J Antimicrob Chemother 2002;50(6):1017–26.

7. Villani P, Regazzi MB, Marubbi F, Viale P, Pagani L, Cristini F, Cadeo B, Carosi G, Bergomi R. Cerebrospinal fluid linezolid concentrations in postneurosurgical central nervous system infections. Antimicrob Agents Chemother 2002;46(3):936–7.

8. Lovering AM, Zhang J, Bannister GC, Lankester BJ, Brown JH, Narendra G, MacGowan AP. Penetration of linezolid into bone, fat, muscle and haematoma of patients undergoing routine hip replacement. J Antimicrob Chemother 2002;50(1):73–7.

9. Rana B, Butcher I, Grigoris P, Murnaghan C, Seaton RA, Tobin CM. Linezolid penetration into osteo-articular tissues. J Antimicrob Chemother 2002;50(5):747–50.

10. Gee T, Ellis R, Marshall G, Andrews J, Ashby J, Wise R. Pharmacokinetics and tissue penetration of linezolid following multiple oral doses. Antimicrob Agents Chemother 2001;45(6):1843–6.

11. Perry CM, Jarvis B. Linezolid: a review of its use in the management of serious Gram-positive infections. Drugs 2001;61(4):525–51.

12. Marchese A, Schito GC. The oxazolidinones as a new family of antimicrobial agent. Clin Microbiol Infect 2001;7(Suppl 4):66–74.

13. Diekema DJ, Jones RN. Oxazolidinone antibiotics. Lancet 2001;358(9297):1975–82.

14. Zurenko GE, Gibson JK, Shinabarger DL, Aristoff PA, Ford CW, Tarpley WG. Oxazolidinones: a new class of antibacterials. Curr Opin Pharmacol 2001;1(5):470–6.

15. Kaplan SL, Patterson L, Edwards KM, Azimi PH, Bradley JS, Blumer JL, Tan TQ, Lobeck FG, Anderson DCLinezolid Pediatric Pheumonia Study Group. Pharmacia and Upjohn. Linezolid for the treatment of community-acquired pneumonia in hospitalized children. Linezolid Pediatric Pneumonia Study Group. Pediatr Infect Dis J 2001;20(5):488–94.

16. Kaplan SL. Use of linezolid in children. Pediatr Infect Dis J 2002;21(9):870–2.

17. Smith PF, Birmingham MC, Noskin GA, Meagher AK, Forrest A, Rayner CR, Schentag JJ. Safety, efficacy and pharmacokinetics of linezolid for treatment of resistant Gram-positive infections in cancer patients with neutropenia. Ann Oncol 2003;14:795–801.

18. Bishop E, Melvani S, Howden BP, Charles PG, Grayson ML. Good clinical outcomes but high rates of adverse reactions during linezolid therapy for serious infections: a proposed protocol for monitoring therapy in complex patients. Antimicrob Agents Chemother 2006;50(4):1599–602.

19. Razonable RR, Osmon DR, Steckelberg JM. Linezolid therapy for orthopedic infections. Mayo Clin Proc 2004;79(9):1137–44.

20. Aneziokoro CO, Cannon JP, Pachucki CT, Lentino JR. The effectiveness and safety of oral linezolid for the primary and secondary treatment of osteomyelitis. J Chemother 2005;17(6):643–50.

21. Cantarini MV, Painter CJ, Gilmore EM, Bolger C, Watkins CL, Hughes AM. Effect of oral linezolid on the pressor response to intravenous tyramine. Br J Clin Pharmacol 2004;58(5):470–5.

22. Tartarone A, Gallucci G, Iodice G, Romano G, Coccaro M, Vigliotti ML, Mele G, Matera R, Di Renzo N. Linezolid-induced bradycardia: a case report. Int J Antimicrob Agents 2004;23(4):412–3.

23. Park IN, Hong SB, Oh YM, Kim MN, Lim CM, Lee SD, Koh Y, Kim WS, Kim DS, Kim WD, Shim TS. Efficacy and tolerability of daily-half dose linezolid in patients with intractable multidrug-resistant tuberculosis. J Antimicrob Chemother 2006;58(3):701–4.

24. Giannopoulos N, Salam T, Pollock WS. Visual side effects after prolonged MRSA treatment. Eye 2007;21(4):556–62.

25. Rucker JC, Hamilton SR, Bardenstein D, Isada CM, Lee MS. Linezolid-associated toxic optic neuropathy. Neurology 2006;66(4):595–8.

26. Corallo CE, Paull AE. Linezolid-induced neuropathy. Med J Aust 2002;177(6):332.

27. Rho JP, Sia IG, Crum BA, Dekutoski MB, Trousdale RT. Linezolid-associated peripheral neuropathy. Mayo Clin Proc 2004;79(7):927–30.

28. Bressler AM, Zimmer SM, Gilmore JL, Somani J. Peripheral neuropathy associated with prolonged use of linezolid. Lancet Infect Dis 2004;4(8):528–31.

29. Legout L, Senneville E, Gomel JJ, Yazdanpanah Y, Mouton Y. Linezolid-induced neuropathy. Clin Infect Dis 2004;38(5):767–8.

30. Ferry T, Ponceau B, Simon M, Issartel B, Petiot P, Boibieux A, Biron F, Chidiac C, Peyramond D. Possibly linezolid-induced peripheral and central neurotoxicity: report of four cases. Infection 2005;33(3):151–4.

31. Zivkovic SA, Lacomis D. Severe sensory neuropathy associated with long-term linezolid use. Neurology 2005;64(5):926–7.

32. Thai XC, Bruno-Murtha LA. Bell's palsy associated with linezolid therapy: case report and review of neuropathic adverse events. Pharmacotherapy 2006;26(8):1183–9.

33. McKinley SH, Foroozan R. Optic neuropathy associated with linezolid treatment. J Neuroophthalmol 2005;25(1):18–21.

34. Kulkarni K, Del Priore LV. Linezolid induced toxic optic neuropathy. Br J Ophthalmol 2005;89(12):1664–5.

35. Saijo T, Hayashi K, Yamada H, Wakakura M. Linezolid-induced optic neuropathy. Am J Ophthalmol 2005;139(6):1114–6.
36. Kopterides P, Papadomichelakis E, Armaganidis A. Linezolid use associated with lactic acidosis. Scand J Infect Dis 2005;37(2):153–4.
37. Soriano A, Miro O, Mensa J. Mitochondrial toxicity associated with linezolid. N Engl J Med 2005;353(21):2305–6.
38. Pea F, Scudeller L, Lugano M, Baccarani U, Pavan F, Tavio M, Furlanut M, Rocca GD, Bresadola F, Viale P. Hyperlactacidemia potentially due to linezolid overexposure in a liver transplant recipient. Clin Infect Dis 2006;42(3):434–5.
39. De Vriese AS, Coster RV, Smet J, Seneca S, Lovering A, Van Haute LL, Vanopdenbosch LJ, Martin JJ, Groote CC, Vandecasteele S, Boelaert JR. Linezolid-induced inhibition of mitochondrial protein synthesis. Clin Infect Dis 2006;42(8):1111–7.
40. Moellering RC. Linezolid: the first oxazolidinone antimicrobial. Ann Intern Med 2003;138:135–42.
41. Tan TQ. Update on the use of linezolid: a pediatric perspective. Pediatr Infect Dis J 2004;23(10):955–6.
42. Bernstein WB, Trotta RF, Rector JT, Tjaden JA, Barile AJ. Mechanisms for linezolid-induced anemia and thrombocytopenia. Ann Pharmacother 2003;37:517–20.
43. Gerson SL, Kaplan SL, Bruss JB, Le V, Arellano FM, Hafkin B, Kuter DJ. Hematologic effects of linezolid: summary of clinical experience. Antimicrob Agents Chemother 2002;46(8):2723–6.
44. Hau T. Efficacy and safety of linezolid in the treatment of skin and soft tissue infections. Eur J Clin Microbiol Infect Dis 2002;21(7):491–8.
45. Lawyer MC, Lawyer EZ. Linezolid and reversible myelosuppression. JAMA 2001;286(16):1974.
46. Green SL, Maddox JC, Huttenbach ED. Linezolid and reversible myelosuppression. JAMA 2001;285(10):1291.
47. Waldrep TW, Skiest DJ. Linezolid-induced anemia and thrombocytopenia. Pharmacotherapy 2002;22(1):109–12.
48. Monson T, Schichman SA, Zent CS. Linezolid-induced pure red blood cell aplasia. Clin Infect Dis 2002;35(3):E29–31.
49. Halpern M. Linezolid-induced pancytopenia. Clin Infect Dis 2002;35(3):347–8.
50. Attassi K, Hershberger E, Alam R, Zervos MJ. Thrombocytopenia associated with linezolid therapy. Clin Infect Dis 2002;34(5):695–8.
51. Orrick JJ, Johns T, Janelle J, Ramphal R. Thrombocytopenia secondary to linezolid administration: what is the risk? Clin Infect Dis 2002;35(3):348–9.
52. Grau S, Morales-Molina JA, Mateu-de Antonio J, Marin-Casino M, Alvarez-Lerma F. Linezolid: low pre-treatment platelet values could increase the risk of thrombocytopenia. J Antimicrob Chemother 2005;56(2):440–1.
53. Wu VC, Wang YT, Wang CY, Tsai IJ, Wu KD, Hwang JJ, Hsueh PR. High frequency of linezolid-associated thrombocytopenia and anemia among patients with end-stage renal disease. Clin Infect Dis 2006;42(1):66–72.
54. Dawson MA, Davis A, Elliott P, Cole-Sinclair M. Linezolid-induced dyserythropoiesis: chloramphenicol toxicity revisited. Intern Med J 2005;35(10):626–8.
55. Saez de la Fuente J, Escobar Rodriguez I, Perpina Zarco C, Bartolome Colussi M. [Linezolid-related leukocytoclastic vasculitis.] Med Clin (Barc) 2005;124(16):639.
56. Kaplan SL, Deville JG, Yogev R, Morfin MR, Wu E, Adler S, Edge-Padbury B, Naberhuis-Stehouwer S, Bruss JB. Linezolid versus vancomycin for treatment of resistant Gram-positive infections in children. Pediatr Infect Dis J 2003;22:677–86.
57. Pigrau C. Oxazolidinonas y glucopeptidos. Enferm Infecc Microbiol Clin 2003;21:157–64.
58. Meyer B, Thalhammer F. Linezolid and continuous venovenous hemofiltration. Clin Infect Dis 2006;42(3):435–6;author reply 437-8.
59. Mauro LS, Peloquin CA, Schmude K, Assaly R, Malhotra D. Clearance of linezolid via continuous venovenous hemodiafiltration. Am J Kidney Dis 2006;47(6):e83-6.
60. Fiaccadori E, Maggiore U, Rotelli C, Giacosa R, Parenti E, Picetti E, Sagripanti S, Manini P, Andreoli R, Cabassi A. Removal of linezolid by conventional intermittent hemodialysis, sustained low-efficiency dialysis, or continuous venovenous hemofiltration in patients with acute renal failure. Crit Care Med 2004;32(12):2437–42.
61. Bosso JA, Flume PA, Gray SL. Linezolid pharmacokinetics in adult patients with cystic fibrosis. Antimicrob Agents Chemother 2004;48(1):281–4.
62. Ciulla TA, Comer GM, Peloquin C, Wheeler J. Human vitreous distribution of linezolid after a single oral dose. Retina 2005;25(5):619–24.
63. Hendershot PE. Antal EJ. Welshman IR. Batts DH. Hopkins NK. Linezolid: pharmacokinetic and pharmacodynamic evaluation of coadministration with pseudoephedrine HCl, phenylpropanolamine HCl, and dextromethorpan HBr. J Clin Pharmacol 2001;41(5):563–72.
64. Marcucci C, Sandson NB, Dunlap JA. Linezolid–bupropion interaction as possible etiology of severe intermittent intraoperative hypertension? Anesthesiology 2004;101(6):1487–8.
65. Serio RN. Acute delirium associated with combined diphenhydramine and linezolid use. Ann Pharmacother 2004;38(1):62–5.
66. Stein GE. Safety of newer parenteral antibiotics. Clin Infect Dis 2005;41 Suppl 5:S293 302.
67. Strouse TB, Kerrihard TN, Forscher CA, Zakowski P. Serotonin syndrome precipitated by linezolid in a medically ill patient on duloxetine. J Clin Psychopharmacol 2006;26(6):681–3.
68. Lawrence KR, Adra M, Gillman PK. Serotonin toxicity associated with the use of linezolid: a review of postmarketing data. Clin Infect Dis 2006;42(11):1578–83.
69. Huang V, Gortney JS. Risk of serotonin syndrome with concomitant administration of linezolid and serotonin agonists. Pharmacotherapy 2006;26(12):1784–93.
70. Clark DB, Andrus MR, Byrd DC. Drug interactions between linezolid and selective serotonin reuptake inhibitors: case report involving sertraline and review of the literature. Pharmacotherapy 2006;26(2):269–76.
71. Jones SL, Athan E, O'Brien D. Serotonin syndrome due to co-administration of linezolid and venlafaxine. J Antimicrob Chemother 2004;54(1):289–90.
72. Bergeron L, Boule M, Perreault S. Serotonin toxicity associated with concomitant use of linezolid. Ann Pharmacother 2005;39(5):956–61.
73. Morales-Molina JA, Mateu-de Antonio J, Grau Cerrato S, Marin Casino M. Probable sindrome serotoninergico por interaccion entre amitriptilina, paroxetina y linezolid. [Probable serotoninergic syndrome from an interaction between amitryptiline, paroxetine, and linezolid.] Farm Hosp 2005;29(4):292–3.
74. DeBellis RJ, Schaefer OP, Liquori M, Volturo GA. Linezolid-associated serotonin syndrome after concomitant treatment with citalopram and mirtazepine in a critically ill

bone marrow transplant recipient. J Intensive Care Med 2005;20(6):351–3.

75. Morales-Molina JA, Mateu-de Antonio J, Marin-Casino M, Grau S. Linezolid-associated serotonin syndrome: what we can learn from cases reported so far. J Antimicrob Chemother 2005;56(6):1176–8.

76. Taylor JJ, Wilson JW, Estes LL. Linezolid and serotonergic drug interactions: a retrospective survey. Clin Infect Dis 2006;43(2):180–7.

77. Taylor JJ, Estes LL, Wilson JW. Linezolid and serotonergic drug interactions. Clin Infect Dis 2006;43(10):1371.

78. Wigen CL, Goetz MB. Serotonin syndrome and linezolid. Clin Infect Dis 2002;34(12):1651–2.

79. Lavery S, Ravi H, McDaniel WW, Pushkin YR. Linezolid and serotonin syndrome. Psychosomatics 2001;42(5):432–4.

80. Bruel H, Guillemant V, Saladin-Thiron C, Chabrolle JP, Lahary A, Poinsot J. Hemolytic anemia in a newborn after maternal treatment with nitrofurantoin at the end of pregnancy. Arch Pediatr 2000;7:745–7.

81. Thomas CR, Rosenberg M, Blythe V, Meyer WJ, 3rd. Serotonin syndrome and linezolid. J Am Acad Child Adolesc Psychiatry 2004;43(7):790.

82. Gordi T, Tan LH, Hong C, Hopkins NJ, Francom SF, Slatter JG, Antal EJ. The pharmacokinetics of linezolid are not affected by concomitant intake of the antioxidant vitamins C and E. J Clin Pharmacol 2003;43:1161–7.

83. Spellberg B, Yoo T, Bayer AS. Reversal of linezolid-associated cytopenias, but not peripheral neuropathy, by administration of vitamin B_6. J Antimicrob Chemother 2004;54(4):832–5.

84. Cawley MJ, Lipka O. Intravenous linezolid administered orally: a novel desensitization strategy. Pharmacotherapy 2006;26(4):563–8.

Pazufloxacin

See also Fluoroquinolones

General Information

Pazufloxacin is an injectable quinolone antibiotic with bactericidal effect against cephalosporin-resistant, carbapenem-resistant, and aminoglycoside-resistant strains of bacteria.

Organs and Systems

Cardiovascular

Pazufloxacin 3–30 mg/kg intravenously had a low potential for QT interval prolongation in an animal model (1).

Skin

In in vivo studies in animal models of phototoxicity, pazufloxacin was less potent than nalidixic acid, ofloxacin, ciprofloxacin, or sparfloxacin, and there was no photoallergenicity (2).

Drug–Drug Interactions

Theophylline

Co-administration of pazufloxacin and theophylline has been studied in rats (3). Pazufloxacin reduced the clearance of theophylline by about 25%. In seven healthy volunteers taking modified-release theophylline, intravenous pazufloxacin mesilate increased serum theophylline concentrations; analysis of the urinary excretion of theophylline and its metabolites suggested that CYP1A2 had been inhibited (3). Theophylline concentrations need to be monitored if pazufloxacin is co-administered.

References

1. Fukuda H, Morita Y, Shiotani N, Mizuo M, Komae N. [Effect of pazufloxacin mesilate, a new quinolone antibacterial agent, for intravenous use on QT interval]. Jpn J Antibiot 2004;57(4):404–12.

2. Nagasawa M, Nakamura S, Miyazaki M, Nojima Y, Hayakawa H, Kawamura Y. [Phototoxicity studies of pazufloxacin mesilate, a novel parenteral quinolone antimicrobial agent—in vitro and in vivo studies.]Jpn J Antibiot 2002;55(3):259–69.

3. Niki Y, Watanabe S, Yoshida K, Miyashita N, Nakajima M, Matsushima T. Effect of pazufloxacin mesilate on the serum concentration of theophylline. J Infect Chemother 2002;8(1):33–6.

Pefloxacin

See also Fluoroquinolones

General Information

Pefloxacin is a fluoroquinolone antibiotic that inhibits *Plasmodium falciparum* in vitro. It is effective against *Plasmodium yoelii* infections in mice.

Observational studies

In humans pefloxacin was tested in a dosage of 400 mg every 12 hours for 3 days against chloroquine-resistant *P. falciparum* infections in Madagascar, and proved successful in 9 out of 22 cases; seven further cases responded at first but recrudescence followed. The investigators suggested that pefloxacin should be used as a complementary drug rather than as a primary antimalarial drug.

There was a significant reduction in proteinuria in ten children with idiopathic nephrotic syndrome after pefloxacin therapy (mean dose 2–4.6 mg/kg/day for 4–8 weeks) (1). All had received a course of cyclophosphamide at least 6 months before. One patient discontinued pefloxacin within 2 weeks because of nausea and vomiting, one complained of arthralgia, and one developed nail discoloration.

The efficacy and safety of pefloxacin, 15–20 mg/kg bd for 14–28 days in combination with ceftazidime and amikacin, have been investigated in 21 children (aged 7–16 years) with mucoviscidosis or aplastic anemia (2). Combined therapy had good clinical efficacy. Arthropathy developed frequently and children at risk were over 10 years old and had a history of allergies.

In a retrospective study the data on fluoroquinolones and other antibacterial drugs were obtained from a spontaneous reporting system database, pefloxacin was associated with the highest reporting rate (982 reports/daily defined dose/1000 inhabitants/day), and the most frequent were musculoskeletal disorders (3).

Drug–Drug Interactions

Rifampicin

In five healthy volunteers who took a single dose of rifampicin 600 mg plus pefloxacin 500 mg, the excretion rate of rifampicin significantly increased after pefloxacin co-administration. Competition between rifampicin and pefloxacin for liver clearance favors pefloxacin and increases the tubular secretion of rifampicin in the kidney (4).

General adverse effects

The adverse effects of pefloxacin are those of the fluoroquinolones, which are generally well tolerated. Gastrointestinal complaints occur in some 3–6%, and have included (in declining order of frequency) nausea, abdominal discomfort, vomiting, and diarrhea. Colitis due to *Clostridium difficile* infection has been reported infrequently. Nervous system effects have been less common, but headache, dizziness, agitation, sleep disturbances, and more rarely seizures, delirium, and hallucinations have been reported. Allergic reactions are infrequent. Photosensitivity reactions have been reported. The quinolones cause cartilage erosion in weight-bearing joints in young growing animals, and the long-term use of pefloxacin in growing children, such as would be required for prophylaxis, should be avoided. In bacteria that have developed resistance to fluoroquinolones, cross-resistance with chemically unrelated antibiotics, such as tetracycline and chloramphenicol, is possible.

Organs and Systems

Musculoskeletal

In vitro, pefloxacin was more toxic to tendons than ofloxacin, ciprofloxacin, or levofloxacin (5). In rodents, pefloxacin (400 mg/kg for several days) caused oxidative damage to the type I collagen in the Achilles tendon; these alterations were identical to those observed in experimental tendinous ischemia and a reperfusion model (6). Oxidative damage was prevented by the co-administration of *N*-acetylcysteine (150 mg/kg). Several cases of rupture of the Achilles tendon have been reported during or shortly after the use of fluoroquinolones, including five case in which pefloxacin was used (7).

The efficacy and safety of pefloxacin, 15–20 mg/kg bd for 14–28 days in combination with ceftazidime and amikacin, have been investigated in 21 children (aged 7–16 years) with mucoviscidosis or aplastic anemia (2). Combined therapy had good clinical efficacy. Arthropathy developed frequently and children at risk were over 10 years old and had a history of allergies.

In patients with mucoviscidosis treated with pefloxacin the most common adverse event was arthropathy, and the symptoms disappeared 3 days to 3 months after drug withdrawal (8).

Immunologic

Pefloxacin in suprabactericidal concentrations (2.0 mg/ml and 0.4 mg/ml) markedly suppressed T lymphocyte proliferation in blast transformation; 0.08 mg/ml did not (9). Pefloxacin in a maximal effective dose (200 mg/kg) suppressed delayed hypersensitivity skin reactions in mice.

Susceptibility Factors

Renal disease

In moderate impairment of renal function, prolongation of the dosage interval of pefloxacin should be considered; pefloxacin should be avoided in severe renal impairment (10).

Drug–Drug Interactions

Rifampicin

In five healthy volunteers who took a single dose of rifampicin 600 mg plus pefloxacin 500 mg, the excretion rate of rifampicin significantly increased after pefloxacin co-administration. Competition between rifampicin and pefloxacin for liver clearance favors pefloxacin and increases the tubular secretion of rifampicin in the kidney (4).

Theophylline

Pefloxacin inhibits theophylline metabolism (11).

References

1. Sharma RK, Sahu KM, Gulati S, Gupta A. Pefloxacin in steroid dependent and resistant idiopathic nephrotic syndrome. J Nephrol 2000;13(4):271–4.
2. Postnikov SS, Semykin SIu, Kapranov NI, Perederko LV, Polikarpova SV, Khamidullina KF. [Evaluation of tolerance and efficacy of pefloxacin in the treatment and prevention of severe infections in children with mucoviscidosis and aplastic anemia.]Antibiot Khimioter 2000;45(8):25–30.
3. Leone R, Venegoni M, Motola D, Moretti U, Piazzetta V, Cocci A, Resi D, Mozzo F, Velo G, Burzilleri L, Montanaro N, Conforti A. Adverse drug reactions related to the use of fluoroquinolone antimicrobials: an analysis of spontaneous reports and fluoroquinolone consumption data from three Italian regions. Drug Saf 2003;26:109–20.
4. Pouzaud F, Rat P, Cambourieu C, Nourry H, Warnet JM. [Tenotoxic potential of fluoroquinolones in the choice of

surgical antibiotic prophylaxis in ophthalmology.]J Fr Ophtalmol 2002;25(9):921–6.

5. Simonin MA, Gegout-Pottie P, Minn A, Gillet P, Netter P, Terlain B. Pefloxacin-induced achilles tendon toxicity in rodents: biochemical changes in proteoglycan synthesis and oxidative damage to collagen. Antimicrob Agents Chemother 2000;44(4):867–72.

6. Ribard P, Audisio F, Kahn MF, De Bandt M, Jorgensen C, Hayem G, Meyer O, Palazzo E. Seven Achilles tendinitis including 3 complicated by rupture during fluoroquinolone therapy. J Rheumatol 1992;19(9):1479–81.

7. Postnikov SS, Semykin SIu, Polikarpova SV, Nazhimov VP. [Pefloxacin in the treatment of patients with mucoviscidosis.]Antibiot Khimioter 2002;47(4):13–5.

8. Artsimovich NG, Nastoiashchaia NN, Navashin PS. [Effect of pefloxacin on immune response.]Antibiot Khimioter 2001;46(4):11–2.

9. Ma L, Sun C, Wu J. [Study of pefloxacin concentration in blood and sputum in the aged pneumonia patients with impairment of renal function.]Zhonghua Jie He He Hu Xi Za Zhi 2001;24(10):596–8.

10. Orisakwe OE, Agbasi PU, Ofoefule SI, Ilondu NA, Afonne OJ, Anusiem CA, Ilo CE, Maduka SO. Effect of pefloxacin on the urinary excretion of rifampicin. Am J Ther 2004;11(1):13–6.

11. Brouwers JR. Drug interactions with quinolone antibacterials. Drug Saf 1992;7(4):268–81.

Penicillins

See also Beta-lactam antibiotics

General Information

The basic structure of the penicillins consists of a thiazolidine ring, the beta-lactam ring, and a side chain. The beta-lactam ring is essential for antibacterial activity. The side chain determines in large part the antibacterial spectrum and pharmacological properties of a particular penicillin. The rapid emergence of bacteria, particularly *Staphylococcus aureus*, that produce beta-lactamases (penicillinase) has been partly countered by the development of compounds that resist hydrolysis by beta-lactamases and compounds that are more active than penicillin G against Gram-negative species. This has led to the production of many semisynthetic penicillins, the first of which was meticillin, active against beta-lactamase-producing *S. aureus*; followed by ampicillin, active against selected Gram-negative bacilli; carbenicillin, which has activity against *Pseudomonas aeruginosa*; and subsequently many agents with different pharmacological and antimicrobial properties. Some of the more important penicillins are listed in Table 1. Another method of combating beta-lactamase-producing organisms has been the development of beta-lactamase inhibitors.

Clinical experience with penicillins, especially penicillin G and the aminopenicillins, is extensive. These substances are rarely toxic, even when they are given in an extended range of dosages, making them invaluable for use in pregnant women and children. Their major limitation is their propensity to cause allergic reactions.

Use in non-infective conditions

In spite of the fact that most trials of antibiotics in pregnancy have shown that antibiotic administration prolongs pregnancy, the mechanism is unclear. However, it is well established that several antibiotics can alter intracellular calcium concentrations (1–3) or inhibit some enzymes, including various phospholipases (4). It is also well established that bacterial products, such as phospholipases and endotoxins, can stimulate prostaglandin biosynthesis and release by the human amnion (5). Therefore, because prostaglandin biosynthesis depends on the action of phospholipase A2, a calcium-dependent enzyme (4), it has been hypothesized that antibiotics that interfere with phospholipase A2 might affect prostaglandin biosynthesis and release by the amnion. This hypothesis has been tested by evaluating the effect of ampicillin on the release of prostaglandin E from human amnion (6). The results were clear: ampicillin dose-dependently inhibited the release of prostaglandin E from human amnion in vitro. Moreover, ampicillin reversibly counteracted the rise in prostaglandin E induced by arachidonic acid or oxytocin. The authors concluded that inhibition of prostaglandin E release from amnion is a mechanism whereby ampicillin might prevent some cases of premature delivery, even in the absence of infection.

General adverse effects

Non-allergic reactions to penicillins occur mainly with high doses or are related to renal insufficiency. They consist of convulsions or electrolyte disturbances, with hyperkalemia or sodium retention. The penicillinase-resistant and broad-spectrum penicillins can cause specific adverse reactions. Leukopenia, agranulocytosis, liver

Table 1 Some penicillins (all rINNs except where stated)

Beta-lactamase sensitive penicillins	Broad-spectrum penicillins	Beta-lactamase resistant penicillins	Antipseudomonal penicillins
Benzylpenicillin (penicillin G)	Ampicillin	Flucloxacillin	Piperacillin
Benzylpenicillin benethamine	Amoxicillin	Meticillin (pINN)	Ticarcillin (pINN)
Penicillin G procaine	Azlocillin	Oxacillin	
Phenoxymethylpenicillin (penicillin V)	Mezlocillin	Cloxacillin	
	Piperacillin	Dicloxacillin	

damage, and some cases of nephropathy are considered to be due to toxic rather than to allergic mechanisms. Diarrhea is a common complication, whereas severe antibiotic-associated colitis is rare. Hypersensitivity reactions are of great importance and are dealt with in the monograph on beta-lactam antibiotics. They range from mostly harmless skin reactions to life-threatening immediate reactions, including anaphylactic shock, acute bronchial obstruction, and severe skin reactions. Tumor-inducing effects have not been described.

Organs and Systems

Respiratory

Bronchospasm may be a consequence of penicillin allergy (7–12). Acute severe dyspnea with cyanosis has also been observed without symptoms of bronchial obstruction or pulmonary edema (13). Specific mechanisms for such cases have yet to be identified.

Allergic pneumonitis and transient eosinophilic pulmonary infiltrate (Loeffler's syndrome) are rare. These syndromes have also been observed with penicillin hypersensitivity (14–16). In one case, an alveolar allergic reaction, probably due to ampicillin, showed features of an adult respiratory distress syndrome (17).

Nervous system

High doses of penicillins, in the order of several million units/day of penicillin G, can produce myoclonic jerks, hyper-reflexia, seizures, or coma. Drowsiness and hallucinations can occur occasionally (18–20). Such reactions are due to a direct toxic effect and are more likely with high concentrations, as seen with intravenous administration (21,22) and with cardiopulmonary bypass in open-heart surgery (23,24).

Myasthenia gravis can be aggravated by ampicillin (25), a reaction that is well described with aminoglycosides and some other antibiotics.

Intrathecal instillation of more than 10 000 units of penicillin and the topical application of high concentrations of penicillin to the nervous system, especially the brain, during surgery have produced comparable reactions (26). All penicillin formulations can produce this kind of reaction.

Benign intracranial hypertension is an extremely rare reaction to penicillins, possibly allergic (27).

Metabolism

Lipoatrophy can occur after the injection of some drugs, including penicillin (28).

- A 2-year-old boy developed a non-tender, hypopigmented, atrophic patch measuring about 2 × 6 cm on his right buttock. He had been well until 5 months before, when he had received an injection of penicillin into the right buttock.

The incidence of this adverse effect is unknown, as is the mechanism.

In six healthy subjects, ampicillin caused an increase in urinary uric acid excretion; this effect was attributed to competition for active renal tubular reabsorption of urate (SEDA-13, 212).

Electrolyte balance

Potassium penicillin G can significantly alter potassium balance when given in very high doses; 20 million units of potassium penicillin G contains about 30 mmol of potassium, and in patients with renal insufficiency this amount can decisively aggravate potentially lethal hyperkalemia. Similarly, large doses of sodium penicillin G, carbenicillin, or ticarcillin can cause hypernatremia (29,30).

High doses of sodium penicillin can cause urinary potassium loss, presumably by acting as a non-absorbable anion in the distal tubule (30). Apparently by analogous mechanisms a variety of semisynthetic penicillins, including carbenicillin, cloxacillin, mezlocillin, nafcillin, piperacillin, and ticarcillin, caused hypokalemia, mainly in severely ill patients (30–36).

Urinary loss of potassium and interstitial nephritis are well-recognized adverse effects of piperacillin. Since patients in ICU may have increased risks of renal complications, serum electrolyte concentrations have been measured in 43 patients before and after piperacillin administration and in 40 patients who were given other antibiotics (37). The groups were comparable in regard to age and severity of disease and all had normal serum creatinine concentrations before the study. Serum concentrations of magnesium, potassium, and, to a lesser degree, calcium fell significantly 36 hours after the start of therapy in patients who were given piperacillin, but not in patients who were given other antibiotics. The fall was most pronounced in the subgroup of patients who were also given furosemide. The authors concluded that treatment with piperacillin can cause or aggravate electrolyte disorders and tubular dysfunction in ICU patients, even when serum creatinine is normal and that the mechanism is probably exacerbation of pre-existing tubular dysfunction. Serum concentrations of electrolytes, including magnesium, should be regularly monitored and, if necessary, supplements should be given to patients in ICU who are receiving piperacillin. This may hold true for all patients receiving piperacillin.

Hematologic

Since the days when chloramphenicol was more commonly used, it has been recognized that many antimicrobial drug are associated with severe blood dyscrasias, such as aplastic anemia, neutropenia, agranulocytosis, thrombocytopenia, and hemolytic anemia. Information on this association has come predominantly from case series and hospital surveys (38–40). Some evidence can be extracted from population-based studies that have focused on aplastic anemia and agranulocytosis and their association with many drugs, including antimicrobial drugs (41,42). The incidence rates of blood dyscrasias in the general population have been estimated in a cohort study with a nested case-control analysis, using data from a General Practice Research Database in Spain (43). The study population consisted of 822 048 patients aged 5–69 years who received at least one prescription (in all 1 507 307 prescriptions) for an antimicrobial drug during January 1994 to September 1998. The main outcome measure was a diagnosis of neutropenia, agranulocytosis, hemolytic anemia, thrombocytopenia, pancytopenia, or aplastic anemia. The incidence was 3.3 per 100 000 person-years in the general population. Users of antimicrobial drugs had a relative risk (RR), adjusted for age and sex, of 4.4, and patients who took more than one class of antimicrobial drug had a relative risk of 29. Among individual antimicrobial drugs, the greatest risk was with cephalosporins (RR = 14), followed by the sulfonamides (RR = 7.6) and penicillins (RR = 3.1).

An immunologically induced hemolytic anemia due to penicillin or its congeners occurs but is rare (44–47). It typically occurs during treatment with high doses (over 10 million units/day) of penicillin for more than 2 weeks (44,45,48). The dose- and time-dependence of this reaction appear to be explained by the underlying mechanism. During penicillin treatment the erythrocytes are normally coated with penicillin, thereby forming a penicilloyl bond on their surface (49). A drug-specific IgG antibody is directed against the complete antigen, that is the penicillin–erythrocyte complex, and can be shown in direct and indirect Coombs' tests. Clinical hemolysis therefore requires both sufficient coating of erythrocytes and high anti-penicilloyl IgG titers.

Besides this "hapten" or "penicillin-type" of drug-induced hemolysis, a second less frequent mechanism, the so-called "innocent bystander" mechanism can occur (46,49,50). Penicillin–antibody complexes are only loosely bound to erythrocytes and activate complement, which can be detected on the erythrocyte surface with the complement antiglobulin test ("complement" or "non-gamma" type). This mechanism plays a part in immune hemolytic anemias due to various drugs other than penicillins. The hemolytic reaction can continue for weeks after withdrawal of penicillin, that is as long as sufficient penicillin-coated erythrocytes and specific antibodies remain in circulation.

That any penicillin derivative can cause hemolytic anemia is emphasized by the following case (51).

- A 34-year-old woman with cystic fibrosis took piperacillin 6 g tds for respiratory distress. She had no known allergies, although she had previously had pruritus while taking ceftazidime and tingling in her hands with azlocillin. She had completed courses of amoxicillin and flucloxacillin without adverse effects. After about 2 weeks she complained of headache and nausea and passed pink urine. A diagnosis of hemolytic anemia was established and piperacillin was withdrawn. She was given a blood transfusion, prednisone, and folic acid, with good effect.

Various antibiotics, including azlocillin, aztreonam, cefuroxime, ceftazidime, chloramphenicol, colistin, flucloxacillin, gentamicin, imipenem, meropenem, piperacillin, tazobactam, temocillin, and ticarcillin, were incubated with this patient's serum. Only piperacillin

and piperacillin + tazobactam caused agglutination in an indirect agglutination test. The authors concluded that the hemolytic anemia had been caused by piperacillin.

Penicillin has been rarely suspected to cause hemolytic–uremic syndrome (52).

Agranulocytosis and leukopenia are discussed in the monograph on beta-lactam antibiotics. Over the years, several cases of neutropenia after treatment with piperacillin with or without tazobactam have been described, especially in children (53–55).

- A 77-year-old man with chronic obstructive lung disease and pneumonia received piperacillin 4 g + tazobactam 0.5 g every 6 hours (56). The neutrophil count gradually fell to zero after 24 days. Piperacillin + tazobactam was withdrawn and lenograstim was given. Within 4 days, the number of neutrophils started to increase. Lenograstim was withdrawn, and the number of neutrophils returned to normal within a week. He made a full recovery.

Whether the rate of hematological adverse effects is higher for piperacillin + tazobactam than for other penicillins is unclear, as is the question of whether neutropenia is more frequent in children than in adults or more frequent in patients with cystic fibrosis than in patients with other conditions (51). It would anyway be wise to follow patients treated with piperacillin, either alone or in combination with tazobactam, with particular attention.

Thrombocytopenia with penicillins has very rarely been reported (57–61). In two cases with mezlocillin (62) and piperacillin + azobactam (63) antibodies became attached to the platelets in the presence of the incriminated drug. The second of these cases is of particular interest, since drug-dependent antibodies were found in the presence of piperacillin but not tazobactam.

- A young woman developed microangiopathic hemolysis and thrombocytopenia in temporal relation to three separate courses of penicillin or ampicillin (64).

Bleeding disorders with penicillins are discussed in the monograph on beta-lactam antibiotics.

Disseminated intravascular coagulation has been reported during long-term administration of piperacillin (65).

- A 51-year-old man was given piperacillin 2 g bd for osteomyelitis. After close to 4 weeks he developed acute renal insufficiency and superior mesenteric venous thrombosis. His coagulation profile showed disseminated intravascular coagulation. Withdrawal of piperacillin and anticoagulation therapy resulted in clinical improvement and normalization of the laboratory data.

Piperacillin + tazobactam has been associated with neutropenia (66).

- A 19-year-old man had an attack of idiopathic acute pancreatitis and developed a large pseudocyst in the body and tail of the pancreas. He had an endoscopic cystogastrotomy, and 2 weeks later developed a high fever and a leukocyte count of 9.3 x 10^9/l. Blockage of the stent and infection in the cyst was suspected. Pus from the cyst showed a mixed growth dominated by *Pseudomonas aeruginosa*, susceptible to piperacillin, and he was given 8 g/day. His fever responded within 3 days, but 2 weeks later he again developed a fever and was given piperacillin + tazobactam 13.6 g/day. He became afebrile after 2 days. However, after 5 days he developed a neutropenia (lowest value 0.58. x 10^9/l) and thrombocytopenia (72 x 10^9/l). The antibiotics were withdrawn and his blood count returned to normal within 6 days.

The authors referred to several other cases of bone marrow suppression after therapy with piperacillin and/or piperacillin + tazobactam. Bone marrow suppression occurred in patients who had received a cumulative dose of piperacillin + tazobactam of 4929 mg/kg, i.e. 4372 mg/kg of piperacillin. Their patient had received a cumulative dose of piperacillin of 3547 mg/kg.

Teeth

Effects on the mineralization of teeth in children are usually associated with tetracyclines. The possibility that other antibiotics may also be involved has been investigated (67). The authors investigated molar incisor hypomineralization in 141 schoolchildren and recorded the use of antibiotics before 4 years of age. There was molar incisor hypomineralization in 16%, most commonly in those who had taken amoxicillin during their first year of life (OR=2.06; 95% CI = 1.01, 4.17) compared with children who had not used antibiotics. In mouse F18 cells exposed to amoxicillin there was an altered pattern of mineralization.

Gastrointestinal

Antibiotic-induced colitis and diarrhea and non-specific gastrointestinal symptoms are discussed in the monograph on beta-lactam antibiotics.

Liver

Penicillin-induced hepatotoxicity may not be as uncommon as has been thought. There have been three reviews. The first was a comparison of the assessment of drug-induced liver injury obtained by two different methods, the Council for International Organizations of Medical Sciences (CIOMS) scale and the Maria & Victorino (M&V) clinical scale (68). Three independent experts evaluated 215 cases of hepatotoxicity reported using a structured reporting form. There was absolute agreement between the two scales in 18% of cases, but there was no agreement in cases of fulminant hepatitis or death. The authors concluded that the CIOMS instrument is more likely to lead to a conclusion compatible with the specialist's empirical approach.

In the second review some syndromes of drug-induced cholestasis were outlined, with lists of typical examples of which drugs cause what (69). The authors stated that the treatment of drug-induced cholestasis is largely supportive and that the offending drug should be withdrawn immediately.

In the third review the authors' intention was to give new insights into basic mechanism of bile secretion and cholestasis (70). Some drug-induced forms of cholestasis appear to be associated with certain HLA class II haplotypes in patients taking co-amoxiclav (SEDA-24; 276). Whether or not this holds true for hepatotoxicity due to other beta-lactam antibiotics is not known.

In a study in Southwest England over 66 months during 1998?2004, 800 patients presented to a jaundice referral system serving a community of 400 000 (71). Of these, 28 cases were related to drugs (mean age 69 years, 17 men), most often antimicrobial drugs (n = 21). Co-amoxiclav (n = 9) and flucloxacillin (n = 7) were the main culprits, with incidence rates per 100 000 prescriptions of 9.91 and 3.60 respectively. Jaundice due to co-amoxiclav was more common in elderly men (age 65 years; M:F = 7:2). The authors suggested that an alternative to co-amoxiclav should be used if possible in men over the age of 60 years.

Co-amoxiclav

Transient rises in serum transaminases are not uncommon after the use of co-amoxiclav, and hepatic dysfunction with jaundice can also occur. Since this effect is thought to be largely due to the clavulanic acid that co-amoxiclav contains rather than the amoxicillin, it is covered in the monograph on beta-lactamase inhibitors.

The hepatotoxic profile of co-amoxiclav has been further evaluated in a prospective study from Spain (72). In data from all cases of hepatotoxicity reported to the Spanish Registry, co-amoxiclav was implicated in 69 patients (36 men, mean age 56 years), representing 14% of all reported cases. The predominant pattern of lesion was hepatocellular damage, and the mean time lapse between the start of therapy and the onset of jaundice was 16 days. Multiple logistic regression analysis identified advanced age as being associated with the cholestatic/mixed type of injury. There was an unfavourable outcome in 7% of the patients. The lesson is the same as that from England: co-amoxiclav should be used with care in elderly men.

Isoxazolyl penicillins

Flucloxacillin is the most important cause of antimicrobial drug-induced hepatotoxicity in various countries (73–75). The risk has been estimated in some countries to be in the range of 1 in 10 000 to 1 in 30 000 prescriptions (76–78). The hepatic injury is often severe and deaths have occurred (77). The course can be prolonged. Cholestasis is the most frequent and prominent feature and the so-called "vanishing bile duct syndrome" can develop. Female sex, increasing age, and duration of therapy are risk factors (79). High daily doses increase the risk (78).

Chronic hepatitis has been reported in a patient with a history of flucloxacillin-induced hepatitis (80).

- A 55-year-old woman with psoriasis was treated with oral 5-methoxypsoralen and UVA photochemotherapy. After 40 treatments over 5 months she became unwell and complained of headaches, nausea, and abdominal pain. Laboratory tests confirmed a diagnosis of hepatitis. Six years earlier she had had flucloxacillin-induced hepatitis.

Hepatitis after 5-methoxypsoralen is supposedly very rare, and the authors gave only one reference, although there have been several more reports after the use of 8-methoxypsoralen. Without discussing possible mechanisms underlying this difference, it might be wise to remember that a previous history of drug-induced hepatitis should be a reminder of the need to consider hepatotoxic reactions in any patient who develops unexplained symptoms while using another drug.

The other isoxazolyl penicillins, that is cloxacillin, dicloxacillin, and oxacillin, can cause similar hepatotoxicity (81–86). However, it is not known whether the incidence is as high as with flucloxacillin. Nor is it known whether the clearly dose-dependent "oxacillin hepatitis" (87–89) is an identical reaction.

In isolated cases, glucocorticoids (81) and ursodeoxycholic acid (90) apparently improved the outcome of hepatitis related to isoxazolyl penicillins.

Other penicillins

Other penicillins have been only very rarely associated with hepatotoxicity. There are isolated reports involving, among others, penicillin G (91), penicillin V (92), ampicillin and amoxicillin (93,94), carbenicillin (95), and nafcillin (96).

- A 28-year-old woman developed upper abdominal pain, weakness, and dark urine 5 days after a single injection of benzylpenicillin 2 million units for suspected streptococcal pharyngitis (97). Liver dysfunction persisted for up to 18 months.

The authors rated the likelihood that benzylpenicillin had caused cholestasis as probable and referred to three previous reports in which penicillin was claimed to cause hepatotoxicity.

- A 20-year-old man with abdominal trauma received a single dose of piperacillin (1 g) followed by nine doses of imipenem + cilastatin (500 mg tds for 3 days) and 2 weeks later developed jaundice, fatigue, and pruritus (98). A liver biopsy showed centrilobular cholestasis, portal infiltration with eosinophils, and cholangitis. Lymphocyte transformation tests for piperacillin and imipenem/cilastatin were positive, suggesting an immunological mechanism. He made a full clinical and biochemical recovery after 3 months.

The authors concluded that short-term therapy with piperacillin, imipenem + cilastatin, or the combination could cause the same type of liver damage as described with co-amoxiclav and antistaphylococcal penicillins.

Pancreas

The list of agents associated with pancreatitis is long and diverse and is growing. Drugs such as glucocorticoids, estrogens, diuretics, and cancer chemotherapeutic agents have all been implicated, as have various antibiotics, including tetracyclines, rifampicin, and isoniazid (99). Cases of pancreatitis have been reported after the administration of ampicillin (100) and a penicillin derivative (101).

- A 7-year-old boy developed epigastric pain, nausea, and vomiting, starting 10 days after a course of oral penicillin (dose and derivative not stated). His serum amylase activity was 1260 U/l and lipase 528 U/l; electrolytes and liver functions tests were within the reference ranges. Ultrasonography showed a normal liver, spleen, and gallbladder, but his pancreas was diffusely enlarged.

Urinary tract

Acute tubulointerstitial nephritis has been reported in relation to various penicillins, including penicillin G, ampicillin, and amoxicillin (102–106), dicloxacillin (107), meticillin (108–113), nafcillin (114,115), oxacillin (111), and piperacillin (116–118). However, meticillin is the prototype that has caused this reaction more often than any other beta-lactam.

Meticillin-induced acute interstitial nephritis follows a similar pattern of dose-dependence and time-dependence to that of neutropenia (108,109). This reaction occurred in 16% of all children treated with high-dose meticillin (119). Nephritis occurred after a mean of 17 days and a mean cumulative dose of 120 g.

Fever and hematuria (macroscopic or microscopic) are the dominating symptoms. Rash, eosinophilia in the blood, possibly eosinophiluria, and signs of non-oliguric renal insufficiency can occur but are not always present (111). Acute anuric/oliguric renal insufficiency is rare.

There have been reports of renal dysfunction in patients taking nafcillin (120). During 1 year four patients developed acute kidney damage, which the authors attributed to nafcillin-induced interstitial nephritis (121). Similar damage was also described in a patient receiving flucloxacillin (122). In such cases withdrawal of the antibiotic is a must. Referral should be made to a nephrologist, as renal biopsy and corticosteroids may be necessary.

Hemorrhagic cystitis has been observed in connection with meticillin (123) and carbenicillin (124).

A very high dose of amoxicillin (250 mg/kg/day for *Listeria monocytogenes* meningitis) combined with an aminoglycoside led to crystal-induced acute renal insufficiency after 14 days (125).

Drug-induced nephrolithiasis, often seen during the sulfonamide era, is nowadays rare, especially in patients taking beta-lactams. However, it can still occur with penicillins (126).

- A 48-year-old woman with pneumococcal meningitis developed acute oliguric renal insufficiency after taking high-dose amoxicillin (320 mg/kg/day) for 4 days. Amoxicillin crystallization was documented by infrared spectrometry. The outcome was favorable after dosage reduction, a single hemodialysis, and adequate hydration.

As was true for the sulfonamides, the risk of crystalluria due to penicillins is increased by high doses, a low urinary pH, and low urine output.

Skin

Skin reactions are the commonest adverse effects of therapeutically administered penicillins (127). Penicillin-contaminated milk or meat can cause itching or generalized skin reactions (128) or even anaphylaxis (129,130).

Incidence

The overall annual incidence of severe erythema multiforme (toxic epidermal necrolysis and Stevens–Johnson syndrome) is about one case per million, antibiotics being involved in 30–40% (131,132). The clinical differentiation between these syndromes can be difficult (133). Allergic contact dermatitis is usually caused by topical drugs, but is also seen in connection with ingestion, injection, or inhalation (134–136).

The increased frequency of contact eczema due to cloxacillin and bacampicillin may be because they are intensely irritant and lipophilic (137).

Mechanisms

Mechanisms of non-immediate reactions are unclear; but may be immunological and non-immunological. Delayed reactions of the IgE type are known (138). Aminopenicillins seem to be an important cause of non-immediate reactions (139–141). The morbilliform rash that begins 1–10 days after amoxicillin can be caused by a delayed cell mediated immune reaction (142) as can fixed drug eruptions (143,144), toxic epidermal necrolysis (145–147), bullous erythroderma (148), and contact eczema (149). Investigation of these disorders should include delayed readings of skin tests (142). In patients with chronic urticaria, penicillin allergy was demonstrated by cutaneous tests.

Presentation

In contrast to other drugs, penicillin-induced skin reactions can occur after more than 1 week of therapy (60). The typical presentation is a maculopapular, erythematous, symmetrically disposed rash on the legs, buttocks, and trunk.

In very rare cases, certain penicillins can cause pemphigus vulgaris (150) or pemphigoid-like reactions (151,152).

Linear IgA disease is an acquired subepidermal bullous disease characterized by linear deposits of IgA at the cutaneous basement membrane zone and by circulating IgA anti-basement membrane antibodies. Although the cause of linear IgA disease is usually unknown, a few cases have been reported to follow ingestion of drugs, especially vancomycin and diclofenac. A patient with penicillin G-induced linear IgA disease who had circulating IgA antibodies showed specificity against type VII collagen (153).

- A previously fit, 76-year-old man developed pneumococcal pneumonia and acute confusion. He was given oxygen, digoxin, furosemide, and penicillin G 9.6 g/day. Because his symptoms of infection continued he was then given higher doses of penicillin, together with intravenous dexamethasone, and his condition slowly improved. After 10 days of treatment with penicillin (cumulative dose 125 g), he developed a maculopapular truncal eruption compatible with a drug rash. Penicillin was withdrawn and the eruption faded over several days, but 1 week later he developed a localized blistering eruption with tense clear bullae and erosions on the penis, scrotum, and inner thighs. This became generalized, affecting most of the body, and he developed large erosions over pressure-bearing sites, oral ulcers, and hemorrhagic nasal crusting. He was given oral prednisolone. His blistering abated within a month, steroid therapy was withdrawn after 3 months, and his disease remained in remission at follow-up 12 months later. Histology of the affected skin showed subepidermal bullae and a mixed inflammatory infiltrate in the dermis. Direct immunofluorescence showed linear IgA deposition along the basement membrane. Anti-basement membrane antibodies were demonstrated by indirect immunofluorescence and were identified by Western blotting to be against a 250 kDa antigen in dermal extracts. Monoclonal antibodies to collagen VII co-migrated to the same spot.

Collagen VII is the major target antigen of epidermolysis bullosa acquisita (154). Consequently, it is open for discussion whether such patients should be classified as having IgA epidermolysis bullosa acquisita or collagen VII linear IgA disease. The authors stated that their patient did not have the clinical phenotype of epidermolysis bullosa acquisita, and the diagnosis of drug-induced collagen VII linear IgA disease seems to have been well validated (99).

Allergic contact urticaria has been attributed to amoxicillin (155).

- A 40-year-old male nurse developed facial angioedema, dyspnea, rhinoconjunctivitis, dysphonia, and dysphagia immediately after opening a sachet containing amoxicillin and clavulanic acid (156). Skin prick tests were positive for both amoxicillin and ampicillin, and an open test with amoxicillin resulted in a severe immediate-type reaction with large localized wheals and erythema at 10 minutes. Six months later, when he was asymptomatic, erythema was observed during open tests with ampicillin 5%.

Consort urticaria has been attributed to penicillins and may occur more often than recognized.

- A 22-year-old woman had labial urticaria with oropharyngeal edema some minutes after kissing her boyfriend, who had taken amoxicillin some minutes before kissing her (157). A few months before, she had had generalized urticaria several minutes after taking amoxicillin. A prick test with amoxicillin was positive and a similar test with penicillin G was negative. Total serum IgE was 90 kU/l and a RAST test for amoxicillin was positive (4.74 kU/l).

- Two episodes of urticarial angioedema occurred in a 45-year-old woman (158). The first episode occurred 1 hour after she had taken a fifth dose of bacampicillin 1200 mg. In the second episode, she had mild itching and edema of the lips and moderate cutaneous itching and swelling about 30 minutes after making love with her husband, who was taking bacampicillin 1200 mg bd and had taken a tablet about 2 hours before. He had used a condom as contraception, and so the only contact between their mucosae was by kissing. Her symptoms disappeared 2 hours after she took cetirizine 10 mg. Some months later, her husband took placebo or bacampicillin 120, 360, or 520 mg on different days, and 2 hours after taking the tablets kissed his wife. She developed mild intraoral itching and itching and wheals on the face and arms 20 minutes after kissing her husband after he had taken bacampicillin 360 mg.

Kissing can cause an allergic reaction if one of the lovers is sensitized to a compound that has just been taken by the other. This holds true for both drugs and food (159,160). Whether a similar reaction can occur if the lovers have intercourse without using a condom has neither been reported nor investigated. However, allergic reactions to penicillin during in vitro fertilization and intrauterine insemination are possible, and the authors recommended that in patients who are penicillin-sensitive, penicillin should not be used during transfer of gamete and embryo for assisted reproductive procedures (161).

Pustular drug eruptions due to penicillin (162), amoxicillin (163), ampicillin (164), bacampicillin (165), or imipenem + cilastin (166) seem to form a distinct clinical entity that has to be differentiated from pustular psoriasis, which can be drug-induced as well (166). A history of drug exposure, rapid disappearance of the eruption after the drug is stopped, and eosinophils in the inflammatory infiltrate argue in favor of pustular drug eruptions.

Acute generalized exanthematous pustulosis (AGEP) has been reported in a patient taking piperacillin + sodium tazobactam (167).

Pseudoallergic reactions are reactions that mimic immunoallergic reactions, but in which a specific immune-mediated mechanism is not involved. The so-called "ampicillin rash," which also occurs with amoxicillin, another aminopenicillin, is an example of a pseudoallergic reaction, which looks like a typical type III allergic rash, but usually occurs at 7–12 days after the start of administration rather than 3–10 days (168). Some major mediators of pseudoallergic reactions have been reviewed (169). The roles of newer putative mechanisms, involving cytokines, kinins, and other host-derived substances, remain to be ascertained. Most important is the fact that currently there are no standardized and validated animal models for predicting pseudoallergic reactions (170). In patients with infectious mononucleosis, the aminopenicillins evoke rashes in a much higher percentage than usual (171). The incidence of rashes in infectious mononucleosis without antibiotics is 3–15%, compared with 40–100% with ampicillin. The underlying mechanism is speculative.

Although several reports have described fixed rashes due to amoxicillin, palmar exfoliation has rarely been described (172).

- Five patients (one man, four women, aged 30–72 years) developed intense palmar rashes and itching during treatment with amoxicillin (doses not stated). All the episodes began after several days of treatment with amoxicillin, either alone or in combination with clavulanic acid. Three of the patients had repeated episodes, and the interval between treatment and onset was shorter each time (down to 5 hours on the third occasion). In all cases the rash was followed by exfoliation and cleared in 7–10 days without residual lesions. Skin prick tests, intradermal tests, and patch tests were performed with several beta-lactams, including amoxicillin, and all were negative. A challenge test with amoxicillin was performed in one patient, and the erythema recurred in 3–4 hours. All five patients tolerated cefuroxime and ceftazidime. Cefalexin was given to one patient only, and palmar exfoliative erythema developed a few days later.

Pseudoporphyria is a bullous disorder of the skin that is clinically and histologically similar to porphyria cutanea tarda, but with normal porphyrin metabolism (173). Several groups of drugs have been implicated, including tetracyclines and nalidixic acid (174,175), but never before beta-lactams.

- A 24-year-old African American woman with a history of systemic lupus erythematosus, end-stage renal disease requiring hemodialysis, autoimmune hemolytic anemia, and idiopathic thrombocytopenia developed a chronic tubo-ovarian abscess and was given intravenous ampicillin + sulbactam (dosage not stated) (176). After 7 days, she developed pruritus, an increased white blood cell count with an eosinophilia of 29%. The ampicillin/sulbactam was withdrawn and intravenous cefepime + oral metronidazole were substituted. Six days later, she developed discrete, pruritic, non-inflammatory, tense bullae on the forehead and cheeks. Laboratory studies showed an increase in the eosinophil count to 41%. Cefepime was withdrawn. Biopsies showed lesions similar to porphyria cutanea tarda. She recovered within 2 weeks.

The authors stated that it was not clear which of the two beta-lactams had caused the bullous eruption. However, since her trouble began when she was taking ampicillin + sulbactam therapy, they thought that that combination was the more likely culprit.

Diagnosis

A maximum of 20% of subjects with a history of allergy-like reactions after administration of a penicillin antibiotic have positive skin or RAST tests (177–179). Tests using benzylpenicillin derivatives or semisynthetic penicillins can almost double positive test results (180,181). Patients with a positive history but negative skin tests run a 1–3% risk of an IgE-mediated reaction and 60% of test-positive patients had evidence of an immediate reaction, including urticaria and angioedema (177).

Management
Penicillins should be avoided in any patient who gives a history of a skin reaction or anaphylaxis to any penicillin derivative. To prevent mild skin reactions becoming severe when they occur, it is advisable to withdraw the culprit antibiotic not only when a type I reaction is suspected but in all kinds of common rashes, in view of a possible epidermolytic process. A diet free of dairy products was curative in 30 of 70 patients with positive tests (182).

Immunologic

Type I reactions
Anaphylactic shock can occur, even after oral administration of penicillin and skin testing. However, anaphylactic shock is less common after oral than parenteral administration (183). In one study the incidence of anaphylactic shock was 0.04% of all patients treated with penicillin (7). It is also low in patients receiving long-term benzathine penicillin (1.2 million units every 4 weeks). Four episodes of anaphylaxis occurred in 0.012% of injections (1.2 reactions to 10 000 injections) (184). Anaphylactic shock resulting in death occurred in 0.002% of all patients treated with penicillin (7) and in 0.003% of those treated with benzathine penicillin (184).

In nearly half of the cases, the course of anaphylactic shock, especially that induced by penicillin and other small molecular substances, is that of a cardiovascular reaction without any other effects suggestive of an allergic mechanism (185–187). There is an extensive list of articles on anaphylactic shock to penicillins (7–10,185, 186,188,189). General anesthesia does not inhibit the development of anaphylactic shock in penicillin allergy (190).

Diagnosis
The two most important elements in the evaluation of an individual for the presence or absence of beta-lactam hypersensitivity are the drug history and skin tests. Other diagnostic tools, such as measurement of drug-specific antibodies and lymphocyte transformation tests, are investigational or restricted to specialized laboratories. Standardized and widely used protocols for skin testing only exist for the penicillins and allow assessment of IgE-mediated hypersensitivity. The most commonly used reagents are penicilloyl-polylysine (PPL, which contains multiple penicilloyl molecules coupled to a polylysine carrier) and fresh penicillin followed by minor determinant mixtures (MDM), containing penicilloate, benzylpenicilloate, and benzylpenilloate (191). A survey conducted among members of the American Academy of Allergy and Immunology reported the use of penicilloyl-polylysine and fresh penicillin by 86% and minor determinant mixtures by 40% of those responding to the questionnaire (192).

Skin tests are first applied as a prick test for safety. In the absence of a local or systemic reaction, an intradermal test is performed and interpreted as described elsewhere (193,194). Experience with skin testing in penicillin allergy has been reviewed (188,195). Properly performed sequential testing is considered a safe procedure, and only an estimated 1% or less of penicillin allergic patients will have systemic symptoms while undergoing skin tests. However, at least three deaths have been reported with both epicutaneous and intradermal testing (196).

In a collaborative study in the National Institute of Allergy and Infectious Diseases (NIAID), hospitalized patients were tested with major and minor skin test reagents in order to assess the predictive value of skin testing. Among 600 history-negative patients, 568 had negative skin tests and none had a reaction to penicillin. Among 726 history-positive patients, 566 had a negative skin test and received penicillin, seven of whom (1.2%) had a possibly IgE-mediated reaction. Nine of the 167 patients with positive skin tests were exposed to penicillin, two of whom had reactions compatible with IgE-mediated reactions. These data suggest that overall, 99% of patients with negative skin tests to penicilloyl-polylysine and minor determinant mixtures can safely receive penicillin. A history of a previous reaction slightly increases the risk of an adverse reaction, to 1.2%. Most positive skin tests were detected with penicilloyl-polylysine with or without minor determinant mixtures, and a further 16% reacted to minor determinant mixtures alone (178).

In another study in an outpatient clinic for sexually transmitted diseases, 5063 consecutive patients were tested with penicilloyl-polylysine with and without minor determinant mixtures (179). The role of the history of a previous penicillin reaction was emphasized in this study: 1.7% of history-negative subjects had a positive skin test; in contrast, 7.1% of history-positive patients had a positive skin test, and a previous history of anaphylaxis or urticaria was associated with positive skin tests in 17% and 12% respectively. Penicillin was safe in more than 99% of patients with a negative history and a negative skin test. Reactions were more common (2.9%) in patients with a positive history and a negative skin test. The reactions were mild and self-limiting. Two patients with a history of severe IgE-mediated reaction had mild anaphylactic reactions.

Relatively safe doses for skin testing, provided that one begins with a prick test, are 25 nmol/ml of penicilloyl-polylysine and purified benzylpenicillin. Positive skin tests of the immediate type with penicilloyl-polylysine are usually obtained 2 weeks to 3 months after the clinical reaction (197).

The safety of such an approach has been challenged in a description of three patients who were negative in skin tests with penicilloyl-polylysine and minor determinant mixtures and who tolerated therapeutic doses of benzylpenicillin, but reacted to amoxicillin (185). In an extension of that study, 177 patients who were allergic to beta-lactams were identified using the clinical history, a skin test panel including penicilloyl-polylysine, and minor determinant mixtures, as well as ampicillin and amoxicillin and drug-specific radio-allergosorbent tests. Fifty-four patients (31%) tolerated penicillin G but reacted to amoxicillin with anaphylaxis, urticaria, or angioedema. Skin tests with penicilloyl-polylysine and minor determinant mixtures failed to detect those patients, but tests with amoxicillin were positive in 63% (180).

Canadian data have partly confirmed these findings (181). Benzylpenicillin derivatives and semisynthetic penicillins were applied to 112 patients with a history of an allergic reaction to penicillins. The tests were positive in 21 patients (19%), of whom 10 reacted against the semisynthetic penicillin reagents only. Reports of subjects allergic to flucloxacillin (198), cloxacillin (199), and cefadroxil (200), but not penicillin, lend further support to the concept of side chain-specific allergic reactions (see the monograph on beta-lactam antibiotics).

Management

Fearing penicillin anaphylaxis, many clinicians overdiagnose penicillin allergy in patients who have not had a true allergic reaction. Consequently, penicillins are withheld from many patients who could safely receive them. This was the background to a study whose objectives were to determine the likelihood of true penicillin allergy, taking into consideration the clinical history, and to evaluate the diagnostic value added by appropriate skin testing (202). The authors searched MEDLINE for relevant English-language articles dated 1966 to October 2000. Bibliographies were searched to identify additional articles. Original articles describing the precision of skin tests in the diagnosis of penicillin allergy were included, and studies that did not use both minor and major determinants were excluded; 14 studies met the inclusion criteria. At least three authors independently reviewed and abstracted the data from all the articles and reached a consensus about any discrepancies. Some of their conclusions are worth remembering:

- 80–90% of all patients who report penicillin allergy have negative skin tests, suggesting that penicillins are withheld from many patients who could safely receive them;
- patients who develop a rash while taking penicillins should not be automatically labeled as allergic without considering other possibilities, including rashes due to ampicillin distinct from allergic rashes, and rashes caused by the infection being treated or by other drugs;
- for patients with a history of immediate (type I) penicillin allergy who have a compelling need for penicillin, skin testing should be performed;
- at least 98% of patients with positive histories of penicillin allergy and negative skin test results can tolerate penicillin without any sequelae.

Desensitization

Patients with a history of penicillin allergy should undergo skin testing with both penicilloyl-polylysine and minor determinant mixtures. Patients with positive skin tests should be treated with another immunologically unrelated compound or should undergo desensitization. The management of patients with a negative skin test but a history of a severe IgE-mediated reaction has to be individualized; options include the use of an alternative compound, desensitization, or the controlled administration of a test dose.

Acute drug desensitization is commonly described as the process by which a drug-allergic individual is converted from a highly sensitive state to a state in which the drug is tolerated. The procedure involves cautious administration of incremental doses of the drug over a short period of time (hours to a few days). In the past it has mainly been considered to be of value in patients in whom IgE antibodies to a particular drug are known or assumed to exist and no alternative treatment agent is available. In clinical practice, most of the desensitization protocols have involved penicillins (202). However, the principle has been applied successfully to other agents as well (203,204), including other antibiotics, insulin, chemotherapeutic agents, vaccines, heterologous sera, and other proteins.

Mechanism

It has been stated that in patients with penicillin-specific IgE antibodies who underwent successful penicillin desensitization, the data suggest that anti-specific, mast cell desensitization is responsible for the tolerant state and that mediator depletion plays no role (202). Additionally, the clinical observation that wheal-and-flare skin responses to penicillin often become negative with successful desensitization, while IgE responses to other antigens remain unchanged, also supports an involvement of an antigen-specific mechanism. Furthermore, both clinical reactivity and skin-test reactivity return within a few days, unless a tolerant state is maintained by continued drug administration. The author stressed that these findings show that the desensitized state depends on the continuous presence of antigen and that clinical sensitivity returns rapidly in the absence of antigen.

However, the underlying mechanisms responsible for the antigen-specific desensitized state are still unclear. It has been hypothesized that IgE receptor aggregation may generate counter-regulatory forces that, instead of causing cell activation, actually extinguish activating signals (203). The key point is that during desensitization, the drug is introduced very slowly and the drug concentration rises gradually. The slow rate of possible receptor aggregation caused by the gradual increase in drug concentration, along with suppression of cellular activation signals, may lead to antigen-specific desensitization and clinical tolerance. It has also been long thought that during desensitization, univalent drug-hapten protein conjugates are formed and may act by inhibiting the cross-linking of drug-specific IgE molecules on mast cells. It is slightly surprising that this prospect has not come into routine therapy.

Procedure

Beta-lactam desensitization should be done in an intensive care unit and any concomitant risk factors for anaphylaxis, such as use of beta-blockers should be corrected. Protocols based on incremental use of the drug orally or parenterally have been described (202,205). The oral route is preferable and is associated with a lower incidence of adverse events, but mild transient reactions are frequent (183,206,207). Pregnant women with limited antibiotic choices have been treated with immunotherapy (208). Repeated administration will maintain a state of anergy, which is often lost after withdrawal (209). At the conclusion of therapy, patients must be informed that after withdrawal, they may once again become allergic to penicillin, with a new reaction to the first subsequent application (209).

Desensitization is not effective in non-IgE-mediated reactions and should therefore not be attempted, for example in cases of serum sickness-like syndromes or Stevens–Johnson syndrome.

Treatment of acute anaphylaxis

For acute anaphylaxis, immediate treatment is essential, with adrenaline followed by intravenous histamine H_1 receptor antagonists, glucocorticoids, fluids, and electrolytes. In view of the frequency of cardiac dysrhythmias and conduction disturbances in patients with anaphylactic shock, they should immediately be monitored (210,211).

Type III reactions

Serum sickness was first described by von Pirquet and Schick in 1905 and was regarded as a syndrome resulting from the administration of heterologous serum or other foreign proteins. The immunopathology of classic serum sickness results from antigen–antibody complex formation with a foreign protein as the antigen. Characteristic symptoms include fever, cutaneous eruptions, edema, arthralgia, and lymphadenopathy. The incidence of classical serum sickness has fallen secondary to the refinement of foreign proteins. However, a serum sickness-like reaction that is clinically similar to classical serum sickness can result from the administration of a number of non-protein drugs, such as tetracyclines, penicillins, and cephalosporins (212). The reaction typically occurs within 1 month of the start of therapy and resolves after withdrawal.

Serum sickness has been associated with penicillins (213).

- A 39-year-old woman developed the characteristic symptoms for serum sickness, having completed a 5-day course of amoxicillin for a perilingual infection 1 week before. She was treated with prednisone 60 mg/day and diphenhydramine 25–50 mg tds. Her symptoms gradually resolved and the prednisolone dose was tapered over 2 weeks.
- A 29-year-old woman complained of fever, rash, throat and facial swelling, abdominal pain, and increasing joint pain, leaving her wheelchair-bound. Her symptoms started a week after she had completed a 10-day course of penicillin V for a dental abscess. She was given oral methylprednisolone, 40 mg every 6 hours,

and over the next few days her symptoms gradually resolved.
- A 29-year-old woman developed symptoms of serum-sickness 2 weeks after completing a 21-day course of co-amoxiclav for sinusitis. She responded to prednisone 40 mg/day.

The authors suggested that serum sickness may be more common than has previously been described and that the reaction may be under-reported or unrecognized.

Second-Generation Effects

Teratogenicity

Early animal studies suggested that malformations could be caused by penicillins. However, this was not confirmed in later, more extensive evaluations (214). Penicillin G, ampicillin, and probably most other penicillins can be safely used in pregnant women and children. Experience with the newer semisynthetic penicillins is not extensive enough to allow definite conclusions regarding their safety for mother and fetus during pregnancy.

Fetotoxicity

Increased antenatal administration of ampicillin for the prevention of neonatal group B streptococcal disease may be responsible for the higher incidence of more severe neonatal sepsis with ampicillin-resistant non-group B streptococci, without a change in the overall infection rate (215,216).

Lactation

The transfer of penicillins (ampicillin, pivampicillin, phenoxymethylpenicillin) into the breast milk of nursing mothers with puerperal mastitis to the breastfed infant is minimal (217). The safety of co-amoxiclav during lactation has been studied prospectively in 67 women (218). Women taking amoxicillin (n = 40) were used as controls and were matched for indication, duration of treatment, and age. There were adverse effects in 15 infants and the frequency increased with dosage. In the control group three infants had adverse effects (RR = 2.99; 95% CI = 0.92, 9.7). All the adverse effects were minor and self-limiting and did not necessitate interruption of breast-feeding. The risk of adverse drug reactions due to penicillins is therefore negligible, unless the infant has penicillin allergy (SEDA-14, 215).

Susceptibility Factors

Renal disease

Toxic effects of penicillins can occur in patients with renal insufficiency if the dosage is not altered (219).

Other features of the patient

Toxic reactions can occur in patients with pre-existing cerebral damage or during cardiopulmonary bypass.

Embolic reactions can have severe consequences in patients with pre-existing cardiac or pulmonary disease.

Drug Administration

Drug formulations

Embolic-toxic reactions to penicillin depot formulations were first described in patients with syphilis (220). The symptoms include fear of death, confusion, acoustic and visual hallucinations, and possibly palpitation, tachycardia, and cyanosis (SEDA-8, 559) (187,220–224). Generalized seizures or twitching of the limbs have been observed in children and adults (223,225–229). As a rule, the symptoms abate and disappear within several minutes to an hour. They rarely persist for up to 24 hours. If a cardiovascular reaction with a fall in blood pressure occurs simultaneously with typical symptoms, a combination with anaphylactic shock must be considered (230,231).

Such reactions have been called "pseudo-anaphylactic reactions" or "acute non-allergic reactions" (186,221–223,232–236), "panic attack syndrome," and "acute psychotic reactions" (225,237). In several countries, the term "Hoigné syndrome" is used.

The frequency of such reactions is about 1–3 reactions per 1000 intramuscular injections of penicillin G procaine, the usual dose being about 0.6–1.2 million units (187,226,238–240). Eight of 920 patients with venereal diseases had a definite toxic-like reaction with a dose of 4.8 million units of penicillin G procaine, corresponding to about one in 120 patients (241). In a series of 7700 intramuscular injections with only 400 000 units of penicillin G procaine, there was not one episode (234).

The mechanism is probably embolic (184,220,232,235,238), as has been shown in one case at autopsy, in which emboli of benzathine penicillin crystals were found in the lungs (232).

Some reports suggest that the procaine component may be especially important (226,235). Plasma procaine esterase activity was low in patients with systemic toxic reactions (226). The same symptoms occurred in three patients after erroneous administration of penicillin G procaine by intravenous infusion (235), but also in two after procaine-free antihistamine penicillin was injected intramuscularly (185,238). However, observations of similar symptoms with a procaine-free antihistamine penicillin argue against a central role of the procaine component (SED-8, 560; 187,230,233,234,238).

Drug administration route

Intramuscular injection of high doses of depot formulations of penicillins can lead to painful swelling, especially when over 600 000–1 000 000 units are given at a single site (242,243). Such reactions occurred in two of 878 patients (0.2%) with intramuscular penicillin G procaine (244). Arthus phenomenon seems to be rare (134,245).

Nicolau syndrome (embolia cutis medicamentosa) is a very rare complication of intramuscular injections, in which there is extensive necrosis of the injected skin area, perhaps due to accidental intra-arterial and/or para-arterial injection (246). It usually occurs in children: in a review of 102 patients, 80 were under 12 years of age (247). Complications can include everything from an ischemic syndrome with local necrosis of the skin, subcutaneous tissue, and muscle, often combined with vascular and nervous system involvement, intestinal and renal hemorrhage, necrosis of the entire leg, and even paraplegia from spinal cord damage (248–254). Necrosis of the forearm has been described in two patients after inadvertent intra-arterial administration of dicloxacillin (255).

Special emphasis should be put on the precautionary measures to be taken when injecting long-acting penicillins or other drugs in crystalline suspensions intramuscularly.

Repeated intramuscular injections into the thighs of newborns and infants can cause severe and widespread muscular contractures of the quadriceps femoris (SEDA-8, 560). In some cases, penicillin was implicated.

Drug–Drug Interactions

Allopurinol

The risk of rashes caused by aminopenicillins does not seem to be increased by parallel treatment with allopurinol (256), as had been suggested before (257).

Amikacin

Amikacin may be inactivated by penicillins. This inactivation occurs not only with a mixture of the agents in solution but also in vivo, particularly in patients with renal insufficiency. Amikacin offers, at least in vitro, the advantage of being much less inactivated than tobramycin or gentamicin (291).

Aminoglycosides

High doses of parenteral penicillin can inactivate aminoglycosides (258). In patients receiving low doses of aminoglycosides because of reduced renal function this can be clinically important (259,260). Parenteral administration of these drugs in neonatal dosages does not seem to produce relevant inactivation, and so temporal separation of the infusions is not required (261).

Piperacillin protected against aminoglycoside nephrotoxicity without reducing its blood concentration; this was possibly a protective effect of co-administered mineral salts (262).

Aminoglycoside antibiotics

There is an in vitro interaction between aminoglycoside antibiotics and carbenicillin or ticarcillin, leading to a significant loss of aminoglycoside antibacterial activity if these antibiotics are mixed in the same infusion bottle (292). The extent of inactivation depends on the penicillin concentration, the contact time, and the temperature. Azlocillin and mezlocillin inactivate aminoglycosides in a similar manner to that described for carbenicillin

(293,294). Aminoglycosides should not be mixed with penicillins or cephalosporins in the same infusion bottle.

However, the clinical significance of the presence of both types of antibiotics in the patient is debatable. In patients who received a combination of gentamicin and carbenicillin the measured serum gentamicin concentrations were lower than the pharmacokinetically predicted values (295). This interaction may be especially important in patients with severe renal impairment, in whom long in vivo incubation of these drug combinations takes place before supplemental doses of the aminoglycoside drugs are given.

Ciclosporin

In a study in lung transplant recipients, ciclosporin nephrotoxicity was potentiated by nafcillin (116).

Methotrexate

Beta-lactams are weak organic acids that compete with the renal tubular secretion of methotrexate and its metabolites and reduce their clearance, leading to methotrexate toxicity (263,264). Consecutive aplastic crises have been described, particularly in patients with impaired renal clearance (263,265,267). In contrast, co-administration of flucloxacillin in another study produced a significant but not clinically important reduction in methotrexate AUC (267).

An interaction of piperacillin + tazobactam with methotrexate has been described (268).

- A 50-year-old woman with Burkitt's lymphoma was given one cycle of chemotherapy with IVAC (ifosfamide, etoposide, and high-dose cytarabine) and on day 10 developed febrile neutropenia and cavitating pneumonia due to *Pseudomonas aeruginosa*. She was given piperacillin + tazobactam (dose not stated) empirically. She then had a cycle of CPDOX-M (cyclophosphamide, doxorubicin, vincristine, and high-dose methotrexate), during which a drug interaction was suspected. Cytotoxic methotrexate concentrations were sustained for 8 days and did not fall below 0.05 µmol/l until piperacillin + tazobactam was withdrawn. During a second cycle of CPDOX-M, piperacillin + tazobactam was not administered and the serum methotrexate concentrations were lower. Her serum creatinine concentration was the same during the two cycles, and concurrent medications, aside from piperacillin + tazobactam, did not differ. Methotrexate total body clearance fell to 3% of normal in the presence of piperacillin + tazobactam.

This interaction has been reported only once before (269). However, it has been forecast from experiments in monkeys and rabbits. In *Rhesus* and *Cynomolgus* monkeys, penicillin and methotrexate share a common secretory system in the kidney and penicillin blocks methotrexate secretion (270). Piperacillin blocks the renal excretion of methotrexate and its main metabolite 7-hydroxymethotrexate in rabbits by an effect on the tubular transport mechanism for organic acids (271).

The more basic interactions between piperacillin and methotrexate and its major metabolite 7-hydroxymethotrexate have been studied in rabbits (272). The interaction was mainly caused by reduced renal clearance of both methotrexate and its metabolite. The authors concluded that renal function in patients taking this combination should be monitored, with adequate fluid intake, especially in elderly patients, because dehydration may accelerate the occurrence of toxicity.

In general, it would be wise to think of possible interactions when the combination of piperacillin + tazobactam is given with weak organic acids that are excreted renally.

Phenylbutazone

Phenylbutazone interferes with the tubular excretion of penicillins (296).

Phenytoin

Competitive albumin binding of drugs with high serum protein affinity can increase pharmacologically active unbound concentrations and enhance the metabolism of low clearance drugs. In vitro data suggest a significant increase in unbound phenytoin concentration by high doses of oxacillin, especially with hypoalbuminemia or uremia (273).

Warfarin

Over the years, interactions between warfarin and penicillins have been infrequently reported. Most of them have been related to use of nafcillin, followed by dicloxacillin. In one case a patient first took nafcillin then dicloxacillin (274).

- A 39-year-old man, with a history of deep vein thrombosis and septic arthritis was stabilized on warfarin (32 g/weekly) and cefazolin (275[AR]). When nafcillin 2 g six times a day was given instead of cefazolin, his INR started to fall and his warfarin dosage had to be increased to a maximum of 88 g/weekly to achieved the target INR. After nafcillin was withdrawn his warfarin requirements slowly fell over several weeks to a maintenance dose of 42–48 mg/week.

The authors thought that enzyme induction by nafcillin was the most likely explanation for this interaction. They described the clinical course as follows: the usual onset is within 1 week after starting nafcillin and the offset is usually evident 4 weeks after withdrawal.

Interference with Diagnostic Tests

Aspergillus galactomanan antigen

It is clinically well established that the presence of *Aspergillus* galactomanan antigen is useful in the early detection of invasive aspergillosis, but it is also well recognized that this test can yield false positive results. This has been reported in patients taking piperacillin +

tazobactam and co-amoxiclav (SEDA-29, 248). An increasing number of reports continue to appear (276, 277, 278, 279).

In a multicenter prospective Spanish four-year study 85 patients who had received liver transplants had 414 samples tested for the presence of galactomanan antigens using a commercially available test (280). The mean number of samples per patient (among those who could be assessed) was 4.8. There were 40 false-positive results (9.6%), corresponding to 28 patients. The frequency of false-positive results in samples obtained during the first week after transplantation was 36% (27 of 75), falling to zero in the fourth week. Multivariate analysis showed that prophylaxis with ampicillin was the only independent factor associated with a false-positive result. The authors investigated three vials of three different beta-lactams, and obtained a positive galactomanan test in four of six different vials, in three of six vials containing piperacillin + tazobactam, and in none of six vials containing cefotaxime. They underlined the fact that ampicillin is routinely used prophylactically for 2–5 days after transplantation, whereas piperacillin + tazobactam is rarely used. They also mentioned that the epitope detected in their test is not exclusively present in *Aspergillus* species, but can also be found in *Penicillium* species. Obviously, some producers have still a way to go until their products are good enough.

The detection of *Aspergillus* galactomannan antigen in plasma is widely used in the diagnosis of invasive aspergillosis. In the last 2 years several authors have reported the presence of this antigen in patients without aspergillosis but who were taking piperacillin + tazobactam (281,282,283,284,285) although others did not find it (286). Seeking an explanation for this discordance, an Italian group has investigated 30 randomly selected batches of piperacillin + tazobactam; 26 of them were positive for the *Aspergillus* galactomannan antigen (287). Another combination, amoxicillin + clavulanic acid (co-amoxiclav), has been found to give false positive results in this test (288,289). It seems reasonable to assume that the companies are working on this problem. In the meantime, it might be wise from a clinical point of view to remember the old lesson that a positive laboratory result is not the same as a disease in your patient. False positive results challenge your clinical judgement.

Pseudoproteinuria

Patients who take penicillin G or ureidopenicillin derivatives in doses over 5 g/day develop pseudoproteinuria. Proteinuria should be evaluated by a bromphenol blue test (Albustix) or after urine dialysis (SEDA-3, 219).

17-Ketosteroids

High-dosage penicillin produces abnormally high concentrations of 17-ketogenic steroids in the blood and high concentrations of 17-ketosteroid in the urine (290).

References

1. Cloutier MM, Guernsey L, Sha'afi RI. Duramycin increases intracellular calcium in airway epithelium. Membr Biochem 1993;10(2):107–18.
2. Burroughs SF, Johnson GJ. Beta-lactam antibiotics inhibit agonist-stimulated platelet calcium influx. Thromb Haemost 1993;69(5):503–8.
3. Bird SD, Walker RJ, Hubbard MJ. Altered free calcium transients in pig kidney cells (LLC-PK1) cultured with penicillin/streptomycin. In Vitro Cell Dev Biol Anim 1994;30A(7):420–4.
4. Verheij HM, Slotboom AJ, de Haas GH. Structure and function of phospholipase A2. Rev Physiol Biochem Pharmacol 1981;91:91–203.
5. Romero R, Mazor M, Wu YK, Sirtori M, Oyarzun E, Mitchell MD, Hobbins JC. Infection in the pathogenesis of preterm labor. Semin Perinatol 1988;12(4):262–79.
6. Vesce F, Buzzi M, Ferretti ME, Pavan B, Bianciotto A, Jorizzo G, Biondi C. Inhibition of amniotic prostaglandin E release by ampicillin. Am J Obstet Gynecol 1998;178(4):759–64.
7. Idsoe O, Guthe T, Willcox RR, de Weck AL. Art und Ausmass der Penizillinnebenwirkungen unter besonderer Berücksichtigung von 151 Todesfällen nach anaphylaktischem Schock. [Nature and extent of penicillin side effects with special reference to 151 fatal cases after anaphylactic shock.] Schweiz Med Wochenschr 1969;99(33):1190–7contd.
8. Capaul R, Maibach R, Kunzi UP, et al. Atopy, bronchial asthma and previous adverse drug reactions (ADRs): risk factors for ADRs? Post Marketing Surveillance 1993;7:331.
9. Bertelsen K, Dalgaard JB. Penicillindodsfald. 16 secerede Danske tilfaelde. [Death due to penicillin. 16 Danish cases with autopsies.] Nord Med 1965;73:173–7.
10. Hoffman DR, Hudson P, Carlyle SJ, Massello W 3rd. Three cases of fatal anaphylaxis to antibiotics in patients with prior histories of allergy to the drug. Ann Allergy 1989;62(2):91–3.
11. Davies RJ, Hendrick DJ, Pepys J. Asthma due to inhaled chemical agents: ampicillin, benzyl penicillin, 6 amino penicillanic acid and related substances. Clin Allergy 1974;4(3):227–47.
12. Hoigné R, Braunschweig S, Zehnder D, et al. Drug-induced bronchial asthma attack: epidemiological aspects (communication of CHDM Berne/St. Gallen, Switzerland). Pharmacoepidemiol Drug Saf 1994;3:S90.
13. Hoigne R, Jaeger MD, Hess T, Wymann R, Muller U, Galeazzi R, Maibach R, Kunzi UP. Akute schwere Dyspnoe als Medikamentennebenwirkung. [Acute severe dyspnea as a side effect of drugs. Report from the CHDM (Comprehensive Hospital Drug Monitoring).] Schweiz Med Wochenschr 1990;120(34):1211–6.
14. Reichlin S, Loveless MH, Kane EG. Loeffler's syndrome following penicillin therapy. Ann Intern Med 1953;38(1):113–20.
15. Wengrower D, Tzfoni EE, Drenger B, Leitersdorf E. Erythroderma and pneumonitis induced by penicillin? Respiration 1986;50(4):301–3.
16. de Hoyos A, Holness DL, Tarlo SM. Hypersensitivity pneumonitis and airways hyperreactivity induced by occupational exposure to penicillin. Chest 1993;103(1):303–4.
17. Poe RH, Condemi JJ, Weinstein SS, Schuster RJ. Adult respiratory distress syndrome related to ampicillin sensitivity. Chest 1980;77(3):449–51.

18. New PS, Wells CE. Cerebral toxicity associated with massive intravenous penicillin therapy. Neurology 1965;15(11):1053–8.

19. Nicholls PJ. Neurotoxicity of penicillin. J Antimicrob Chemother 1980;6(2):161–5.

20. Schliamser SE, Bolander H, Kourtopoulos H, Norrby SR. Neurotoxicity of benzylpenicillin: correlation to concentrations in serum, cerebrospinal fluid and brain tissue fluid in rabbits. J Antimicrob Chemother 1988;21(3):365–72.

21. Boston Collaborative Drug Surveillance Program. Drug-induced convulsions. Lancet 1972;2(7779):677–9.

22. Smith H, Lerner PI, Weinstein L. Neurotoxicity and "massive" intravenous therapy with penicillin. A study of possible predisposing factors. Arch Intern Med 1967;120(1):47–53.

23. Currie TT, Hayward NJ, Westlake G, Williams J. Epilepsy in cardiopulmonary bypass patients receiving large intravenous doses of penicillin. J Thorac Cardiovasc Surg 1971;62(1):1–6.

24. Seamans KB, Gloor P, Dobell RA, Wyant JD. Penicillin-induced seizures during cardiopulmonary bypass. A clinical and electroencephalographic study. N Engl J Med 1968;278(16):861–8.

25. Argov Z, Brenner T, Abramsky O. Ampicillin may aggravate clinical and experimental myasthenia gravis. Arch Neurol 1986;43(3):255–6.

26. Reuling JR, Cramer C. Intrathecal penicillin. JAMA 1947;134:16.

27. Schmitt BD, Krivit W. Benign intracranial hypertension associated with a delayed penicillin reaction. Pediatrics 1969;43(1):50–3.

28. Kuperman-Beade M, Laude TA. Partial lipoatrophy in a child. Pediatr Dermatol 2000;17(4):302–3.

29. Wright AJ, Wilkowske CJ. The penicillins. Mayo Clin Proc 1987;62(9):806–20.

30. Brunner FP, Frick PG. Hypokalaemia, metabolic alkalosis, and hypernatraemia due to "massive" sodium penicillin therapy. BMJ 1968;4(630):550–2.

31. Mohr JA, Clark RM, Waack TC, Whang R. Nafcillin-associated hypokalemia. JAMA 1979;242(6):544.

32. Wade JC, Schimpff SC, Newman KA, Fortner CL, Standiford HC, Wiernik PH. Piperacillin or ticarcillin plus amikacin. A double-blind prospective comparison of empiric antibiotic therapy for febrile granulocytopenic cancer patients. Am J Med 1981;71(6):983–90.

33. Rotstein C, Cimino M, Winkey K, Cesari C, Fenner J. Cefoperazone plus piperacillin versus mezlocillin plus tobramycin as empiric therapy for febrile episodes in neutropenic patients. Am J Med 1988;85(1A):36–43.

34. Kibbler CC, Prentice HG, Sage RJ, Hoffbrand AV, Brenner MK, Mannan P, Warner P, Bhamra A, Noone P. A comparison of double beta-lactam combinations with netilmicin/ureidopenicillin regimens in the empirical therapy of febrile neutropenic patients. J Antimicrob Chemother 1989;23(5):759–71.

35. Arevalo A, Mateos F, Otero MJ, Fuertes A. Hipopotasemia inducida por cloxacilina. [Hypopotassemia induced by cloxacillin.] Rev Clin Esp 1996;196(7):494–5.

36. Garcia Diaz B, Plaza S, Garcia Benayas E, Santos D. Hipopotasemia por cloxacilina: un nuevo caso. [Hypopotassemia caused by cloxacillin: a new case.] Rev Clin Esp 1997;197(11):792–3.

37. Polderman KH, Girbes AR. Piperacillin-induced magnesium and potassium loss in intensive care unit patients. Intensive Care Med 2002;28(4):520–2.

38. George JN, Raskob GE, Shah SR, Rizvi MA, Hamilton SA, Osborne S, Vondracek T. Drug-induced thrombocytopenia: a systematic review of published case reports. Ann Intern Med 1998;129(11):886–90.

39. Wright MS. Drug-induced hemolytic anemias: increasing complications to therapeutic interventions. Clin Lab Sci 1999;12(2):115–8.

40. Arneborn P, Palmblad J. Drug-induced neutropenia—a survey for Stockholm 1973–1978. Acta Med Scand 1982;212(5):289–92.

41. Baumelou E, Guiguet M, Mary JY. Epidemiology of aplastic anemia in France: a case-control study. I. Medical history and medication use. The French Cooperative Group for Epidemiological Study of Aplastic Anemia. Blood 1993;81(6):1471–8.

42. International Agranulocytosis and Aplastic Anemia Study Group. Anti-infective drug use in relation to the risk of agranulocytosis and aplastic anemia. Arch Intern Med 1989;149(5):1036–40.

43. Huerta C, Garcia Rodriguez LA. Risk of clinical blood dyscrasia in a cohort of antibiotic users. Pharmacotherapy 2002;22(5):630–6.

44. Petz LD, Fudenberg HH. Coombs-positive hemolytic anemia caused by penicillin administration. N Engl J Med 1966;274(4):171–8.

45. White JM, Brown DL, Hepner GW, Worlledge SM. Penicillin-induced haemolytic anaemia. BMJ 1968;3(609):26–9.

46. Funicella T, Weinger RS, Moake JL, Spruell M, Rossen RD. Penicillin-induced immunohemolytic anemia associated with circulating immune complexes. Am J Hematol 1977;3:219–23.

47. Tuffs L, Manoharan A. Flucloxacillin-induced haemolytic anaemia. Med J Aust 1986;144(10):559–60.

48. Spath P, Garratty G, Petz LD. Immunhämatologische Reaktionen bei Penizillinbehandlung. [Immunohematologic reactions during treatment with penicillin.] Schweiz Med Wochenschr 1973;103(10):383–8.

49. Kerr RO, Cardamone J, Dalmasso AP, Kaplan ME. Two mechanisms of erythrocyte destruction in penicillin-induced hemolytic anemia. N Engl J Med 1972;287(26):1322–5.

50. Harris JW. Studies on the mechanism of a drug-induced hemolytic anemia. J Lab Clin Med 1956;47(5):760–75.

51. Thickett KM, Wildman MJ, Fegan CD, Stableforth DE. Haemolytic anaemia following treatment with piperacillin in a patient with cystic fibrosis. J Antimicrob Chemother 1999;43(3):435–6.

52. Brandslund I, Petersen PH, Strunge P, Hole P, Worth V. Haemolytic uraemic syndrome and accumulation of haemoglobin–haptoglobin complexes in plasma in serum sickness caused by penicillin drugs. Haemostasis 1980;9(4):193–203.

53. Gerber L, Wing EJ. Life-threatening neutropenia secondary to piperacillin/tazobactam therapy. Clin Infect Dis 1995;21(4):1047–8.

54. Ruiz-Irastorza G, Barreiro G, Aguirre C. Reversible bone marrow depression by high-dose piperacillin/tazobactam. Br J Haematol 1996;95(4):611–2.

55. Reichardt P, Handrick W, Linke A, Schille R, Kiess W. Leukocytopenia, thrombocytopenia and fever related to

piperacillin/tazobactam treatment—a retrospective analysis in 38 children with cystic fibrosis. Infection 1999;27(6):355–6.

56. Ortega Garcia MP, Guevara Serrano J, Gil Gomez I, Iglesias Iglesias AA, Fernandez Villalba EM. Neutropenia reversibile secundaria al tratamiento con piperacillina/tazobactam. Atencion Pharmaceutica 2002;4:44–8.

57. Schiffer CA, Weinstein HJ, Wiernik PH. Methicillin-associated thrombocytopenia. Ann Intern Med 1976;85(3):338–9.

58. Lee M, Sharifi R. Severe thrombocytopenia due to apalcillin. Urol Int 1987;42(4):313–5.

59. Brocks AP. Thrombocytopenia during treatment with ampicillin. Lancet 1974;2:273.

60. Hsi YJ, Kuo HY, Ouyang A. Thrombocytopenia following administration of penicillin. Report of a case. Chin Med J 1966;85(4):249–51.

61. Olivera E, Lakhani P, Watanakunakorn C. Isolated severe thrombocytopenia and bleeding caused by piperacillin. Scand J Infect Dis 1992;24(6):815–7.

62. Gharpure V, O'Connell B, Schiffer CA. Mezlocillin-induced thrombocytopenia. Ann Intern Med 1993;119(8):862.

63. Perez-Vazquez A, Pastor JM, Riancho JA. Immune thrombocytopenia caused by piperacillin/tazobactam. Clin Infect Dis 1998;27(3):650–1.

64. Parker JC, Barrett DA 2nd. Microangiopathic hemolysis and thrombocytopenia related to penicillin drugs. Arch Intern Med 1971;127(3):474–7.

65. Miyazaki H, Yanagitani S, Matsumoto T, Yoshida K, Amoh Y, Watanabe T, Kubota Y, Inoue K. Hypercoagulopathy with piperacillin administration in osteomyelitis. Intern Med 2000;39(5):424–7.

66. Kumar A, Choudhuri G, Aggarwal, R. Piperacillin induced bone marrow suppression: a case report. BMC Clin Pharmacol 2003:3.

67. Laisi S, Ess A, Sahlberg C, Arvio P, Lukinmaa PL, Alaluusua S. Amoxicillin may cause molar incisor hypomineralization. J Dent Res 2009;88(2):132–6.

68. Lucena MI, Camargo R, Andrade RJ, Perez-Sanchez CJ, Sanchez De La Cuesta F. Comparison of two clinical scales for causality assessment in hepatotoxicity. Hepatology 2001;33(1):123–30.

69. Chitturi S, Farrell GC. Drug-induced cholestasis. Semin Gastrointest Dis 2001;12(2):113–24.

70. Trauner M, Boyer JL. Cholestatic syndromes. Curr Opin Gastroenterol 2001;17:242–56.

71. Hussaini SH, O'Brien CS, Despott EJ, Dalton HR. Antibiotic therapy; a major cause of drug-induced jaundice in southwest England. Eur J Gastroenterol Hepatol 2007; 19: 15-20.

72. Lucena MI, Andrade RJ, Fernandez MC, Pachkoria K, Pelaez G, Duran JA, Villar M, Rodrigo L, Romero-Gomez M, Planas R, Barriocanal A, Costa J, Guarner C, Blanco S, Navarro JM, Pons F, Castiella A, Avila S; Spanish Group for the Study of Drug-Induced Liver Disease (Grupo de Estudio para las Hepatopatias Asociadas a Medicamentos (GEHAM)). Determinants of the clinical expression of amoxicillin–clavulanate hepatotoxicity: a prospective series from Spain. Hepatology 2006; 44(4): 850-6.

73. George DK, Crawford DH. Antibacterial-induced hepatotoxicity. Incidence, prevention and management. Drug Saf 1996;15(1):79–85.

74. Farrell GC. Drug-induced hepatic injury. J Gastroenterol Hepatol 1997;12(9–10):S242–50.

75. Pillans PI. Drug associated hepatic reactions in New Zealand: 21 years experience. NZ Med J 1996;109(1028):315–9.

76. Derby LE, Jick H, Henry DA, Dean AD. Cholestatic hepatitis associated with flucloxacillin. Med J Aust 1993;158(9):596–600.

77. Devereaux BM, Crawford DH, Purcell P, Powell LW, Roeser HP. Flucloxacillin associated cholestatic hepatitis. An Australian and Swedish epidemic? Eur J Clin Pharmacol 1995;49(1–2):81–5.

78. Olsson R, Wiholm BE, Sand C, Zettergren L, Hultcrantz R, Myrhed M. Liver damage from flucloxacillin, cloxacillin and dicloxacillin. J Hepatol 1992;15(1–2):154–61.

79. Fairley CK, McNeil JJ, Desmond P, Smallwood R, Young H, Forbes A, Purcell P, Boyd I. Risk factors for development of flucloxacillin associated jaundice. BMJ 1993;306(6872):233–5.

80. Stephens RB, Cooper A. Hepatitis from 5-methoxypsoralen occurring in a patient with previous flucloxacillin hepatitis. Australas J Dermatol 1999;40(4):217–9.

81. Goland S, Malnick SD, Gratz R, Feldberg E, Geltner D, Sthoeger ZM. Severe cholestatic hepatitis following cloxacillin treatment. Postgrad Med J 1998;74(867):59–60.

82. Barrio J, Castiella A, Cosme A, Lopez P, Fernandez J, Arenas JI. Hepatotoxicidad pox cloxacilina. [Hepatotoxicity caused by cloxacillin.] Rev Esp Enferm Dig 1997;89(7):559–60.

83. Konikoff F, Alcalay J, Halevy J. Clocaxcillin-induced cholestatic jaundice. Am J Gastroenterol 1987;482–3.

84. Siegmund JB, Tarshis AM. Prolonged jaundice after dicloxacillin therapy. Am J Gastroenterol 1993;88(8):1299–300.

85. Tauris P, Jorgensen NF, Petersen CM, Albertsen K. Prolonged severe cholestasis induced by oxacillin derivatives. A report on two cases. Acta Med Scand 1985;217(5):567–9.

86. Kleinman MS, Presberg JE. Cholestatic hepatitis after dicloxacillin-sodium therapy. J Clin Gastroenterol 1986;8(1):77–8.

87. Olans RN, Weiner LB. Reversible oxacillin hepatotoxicity. J Pediatr 1976;89(5):835–8.

88. Michelson PA. Reversible high dose oxacillin-associated liver injury. Can J Hosp Pharm 1981;34:83.

89. Onorato IM, Axelrod JL. Hepatitis from intravenous high-dose oxacillin therapy: findings in an adult inpatient population. Ann Intern Med 1978;89(4):497–500.

90. Piotrowicz A, Polkey M, Wilkinson M. Ursodeoxycholic acid for the treatment of flucloxacillin-associated cholestasis. J Hepatol 1995;22(1):119–20.

91. Bauer TM, Bircher AJ. Drug-induced hepatocellular liver injury due to benzylpenicillin with evidence of lymphocyte sensitization. J Hepatol 1997;26(2):429–32.

92. Onate J, Montejo M, Aguirrebengoa K, Ruiz-Irastorza G, Gonzalez de Zarate P, Aguirre C. Hepatotoxicity associated with penicillin V therapy. Clin Infect Dis 1995;20(2):474–5.

93. Davies MH, Harrison RF, Elias E, Hubscher SG. Antibiotic-associated acute vanishing bile duct syndrome: a pattern associated with severe, prolonged, intrahepatic cholestasis. J Hepatol 1994;20(1):112–6.

94. Anderson CS, Nicholls J, Rowland R, LaBrooy JT. Hepatic granulomas: a 15-year experience in the Royal Adelaide Hospital. Med J Aust 1988;148(2):71–4.

95. Wilson FM, Belamaric J, Lauter CB, Lerner AM. Anicteric carbenicillin hepatitis. Eight episodes in four patients. JAMA 1975;232(8):818–21.

96. Presti ME, Janney CG, Neuschwander-Tetri BA. Nafcillin-associated hepatotoxicity. Report of a case and review of the literature. Dig Dis Sci 1996;41(1):180–4.

97. Andrade RJ, Guilarte J, Salmeron FJ, Lucena MI, Bellot V. Benzylpenicillin-induced prolonged cholestasis. Ann Pharmacother 2001;35(6):783–4.

98. Quattropani C, Schneider M, Helbling A, Zimmermann A, Krahenbuhl S. Cholangiopathy after short-term administration of piperacillin and imipenem/cilastatin. Liver 2001;21(3):213–6.

99. Marshall JB. Acute pancreatitis. A review with an emphasis on new developments. Arch Intern Med 1993;153(10):1185–98.

100 Hanline MH Jr. Acute pancreatitis caused by ampicillin. South Med J 1987;80(8):1069.

101 Sammett D, Greben C, Sayeed-Shah U. Acute pancreatitis caused by penicillin. Dig Dis Sci 1998;43(8):1778–83.

102. Ruley EJ, Lisi LM. Interstitial nephritis and renal failure due to ampicillin. J Pediatr 1974;84(6):878–82.

103. Tannenberg AM, Wicher KJ, Rose NR. Ampicillin nephropathy. JAMA 1971;218(3):449.

104. Kleinknecht D, Vanhille P, Morel-Maroger L, Kanfer A, Lemaitre V, Mery JP, Laederich J, Callard P. Acute interstitial nephritis due to drug hypersensitivity. An up-to-date review with a report of 19 cases. Adv Nephrol Necker Hosp 1983;12:277–308.

105. Gilbert DN, Gourley R, d'Agostino A, Goodnight SH Jr, Worthen H. Interstitial nephritis due to methicillin, penicillin and ampicillin. Ann Allergy 1970;28(8):378–85.

106. Dharnidharka VR, Rosen S, Somers MJ. Acute interstitial nephritis presenting as presumed minimal change nephrotic syndrome. Pediatr Nephrol 1998;12(7):576–8.

107. Hedstrom SA, Hybbinette CH. Nephrotoxicity in isoxazolylpenicillin prophylaxis in hip surgery. Acta Orthop Scand 1988;59(2):144–7.

108. Ditlove J, Weidmann P, Bernstein M, Massry SG. Methicillin nephritis. Medicine (Baltimore) 1977;56(6):483–91.

109. Galpin JE, Shinaberger JH, Stanley TM, Blumenkrantz MJ, Bayer AS, Friedman GS, Montgomerie JZ, Guze LB, Coburn JW, Glassock RJ. Acute interstitial nephritis due to methicillin. Am J Med 1978;65(5):756–65.

110. Baldwin DS, Levine BB, McCluskey RT, Gallo GR. Renal failure and interstitial nephritis due to penicillin and methicillin. N Engl J Med 1968;279(23):1245–52.

111. Appel GB. A decade of penicillin related acute interstitial nephritis—more questions than answers. Clin Nephrol 1980;13(4):151–4.

112. Woodroffe AJ, Thomson NM, Meadows R, Lawrence JR. Nephropathy associated with methicillin administration. Aust NZ J Med 1974;4(3):256–61.

113. Hansen ES, Tauris P. Methicillin-induced nephropathy. A case with linear deposition of IgG and C3 on the tubular-basement-membrane. Acta Pathol Microbiol Scand [A] 1976;84(5):440–2.

114. Parry MF, Ball WD, Conte JE Jr, et al. Nafcillin nephritis. JAMA 1973;225:178.

115. Jahansouz F, Kriett JM, Smith CM, Jamieson SW. Potentiation of cyclosporine nephrotoxicity by nafcillin in lung transplant recipients. Transplantation 1993;55(5):1045–8.

116. Dorner O, Piper C, Dienes HP, et al. Akute interstitielle Nephritis nach Piperacillin. Klin Wochenschr 1988;67:682.

117. Soto J, Bosch JM, Alsar Ortiz MJ, Moreno MJ, Gonzalez JD, Diaz JM. Piperacillin-induced acute interstitial nephritis. Nephron 1993;65(1):154–5.

118. Tanaka H, Waga S, Kakizaki Y, Tateyama T, Koda M, Yokoyama M. Acute tubulointerstitial nephritis associated with piperacillin therapy in a boy with glomerulonephritis. Acta Paediatr Jpn 1997;39(6):698–700.

119. Sanjad SA, Haddad GG, Nassar VH. Nephropathy, an underestimated complication of methicillin therapy. J Pediatr 1974;84(6):873–7.

120. Lestico MR, Vick KE, Hietsko CM. Hepatic and renal dysfunction following nafcillin administration Ann Pharmacother 1992;26:985–90.

121. Hoppes T, Prikis M, Segal A. Four cases of nafcillin-associated acute interstitial nephritis in one institution. Nat Clin Pract Nephrol 2007;3:456–61.

122. Xu B, Murray M. Flucloxacillin-induced acute renal failure. Aust Fam Physician 2008;37:1009–11.

123. Bracis R, Sanders CV, Gilbert DN. Methicillin hemorrhagic cystitis. Antimicrob Agents Chemother 1977;12(3): 438–439.

124. Moller NE. Carbenicillin-induced haemorrhagic cystitis. Lancet 1978;2(8096):946.

125. Boursas M, Benhassine L, Kempf J, Petit B, Vuillemin F. Insuffisance rénale obstructive par cristallurie à l'amoxicilline. [Obstructive renal insufficiency caused by amoxicillin crystalluria.] Ann Fr Anesth Reanim 1997;16(7):908–10.

126. Boffa JJ, De Preneuf H, Bouadma L, Daudon M, Pallot JL. Insuffisance rénale aiguë par cristallisation d'amoxicilline. [Acute renal failure after amoxicillin crystallization.] Presse Méd 2000;29(13):699–701.

127. De Weck AL. Penicillins and cephalosporins. In: Allergic Reactions to Drugs. Heidelberg: Springer-Verlag, 1983:423.

128. Lindemayr H, Knobler R, Kraft D, Baumgartner W. Challenge of penicillin-allergic volunteers with penicillin-contaminated meat. Allergy 1981;36(7):471–8.

129. Schwartz HJ, Sher TH. Anaphylaxis to penicillin in a frozen dinner. Ann Allergy 1984;52(5):342–3.

130. Tscheuschner I. Anaphylaktische Reaktion auf Penicillin nach Genuss von Schweinefleisch. [Penicillin anaphylaxis following pork consumption.] Z Haut Geschlechtskr 1972;47(14):591–2.

131. Schopf E, Stuhmer A, Rzany B, Victor N, Zentgraf R, Kapp JF. Toxic epidermal necrolysis and Stevens–Johnson syndrome. An epidemiologic study from West Germany. Arch Dermatol 1991;127(6):839–42.

132. Roujeau JC, Guillaume JC, Fabre JP, Penso D, Flechet ML, Girre JP. Toxic epidermal necrolysis (Lyell syndrome). Incidence and drug etiology in France, 1981–1985. Arch Dermatol 1990;126(1):37–42.

133. Bastuji-Garin S, Rzany B, Stern RS, Shear NH, Naldi L, Roujeau JC. Clinical classification of cases of toxic epidermal necrolysis, Stevens–Johnson syndrome, and erythema multiforme. Arch Dermatol 1993;129(1):92–6.

134. Fellner MJ. Adverse reactions to penicillin and related drugs. Clin Dermatol 1986;4(1):133–41.

135. Schulz KH, Schopf E, Wex O. Allergische Berufsekzeme durch Ampicillin. [Allergic occupational eczemas caused by ampicillin.] Berufsdermatosen 1970;18(3):132–43.

136. Calkin JM, Maibach HI. Delayed hypersensitivity drug reactions diagnosed by patch testing. Contact Dermatitis 1993;29(5):223–33.

137. Kristofferson A, Ahlstedt S, Enander I. Contact sensitivity in guinea pigs to different penicillins. Int Arch Allergy Appl Immunol 1982;69(4):316–21.
138. Hoigné R, D'Andrea Jaeger M, Wymann R, et al. Time pattern of allergic reactions to drugs. In: Weber E, Lawson DH, Hoigné R, editors. Risk Factors for Adverse Drug Reactions. Agents Actions 69 1990:316–21.
139. de Haan P, Bruynzeel DP, van Ketel WG. Onset of penicillin rashes: relation between type of penicillin administered and type of immune reactivity. Allergy 1986;41(1):75–8.
140. Dolovich J, Ruhno J, Sauder DN, Ahlstedt S, Hargreave FE. Isolated late cutaneous skin test response to ampicillin: a distinct entity. J Allergy Clin Immunol 1988;82(4):676–9.
141. Vega JM, Blanca M, Carmona MJ, Garcia J, Claros A, Juarez C, Moya MC. Delayed allergic reactions to beta-lactams. Four cases with intolerance to amoxicillin or ampicillin and good tolerance to penicillin G and V. Allergy 1991;46(2):154–7.
142. Barbaud AM, Bene MC, Schmutz JL, Ehlinger A, Weber M, Faure GC. Role of delayed cellular hypersensitivity and adhesion molecules in amoxicillin-induced morbilliform rashes. Arch Dermatol 1997;133(4):481–6.
143. Shiohara T, Nickoloff BJ, Sagawa Y, Gomi T, Nagashima M. Fixed drug eruption. Expression of epidermal keratinocyte intercellular adhesion molecule-1 (ICAM-1). Arch Dermatol 1989;125(10):1371–6.
144. Jimenez I, Anton E, Picans I, Sanchez I, Quinones MD, Jerez J. Fixed drug eruption from amoxycillin. Allergol Immunopathol (Madr) 1997;25(5):247–8.
145. Surbled M, Lejus C, Milpied B, Pannier M, Souron R. Syndrome de Lyell consecutif a l'administration d'amoxicilline chez un enfant de 2 ans. [Lyell syndrome after amoxicillin administration in a 2 year old child.] Ann Fr Anesth Reanim 1996;15(7):1095–8.
146. Miyauchi H, Hosokawa H, Akaeda T, Iba H, Asada Y. T-cell subsets in drug-induced toxic epidermal necrolysis. Possible pathogenic mechanism induced by CD8-positive T cells. Arch Dermatol 1991;127(6):851–5.
147. Correia O, Delgado L, Ramos JP, Resende C, Torrinha JA. Cutaneous T-cell recruitment in toxic epidermal necrolysis. Further evidence of CD8+ lymphocyte involvement. Arch Dermatol 1993;129(4):466–8.
148. Hertl M, Bohlen H, Jugert F, Boecker C, Knaup R, Merk HF. Predominance of epidermal CD8+ T lymphocytes in bullous cutaneous reactions caused by beta-lactam antibiotics. J Invest Dermatol 1993;101(6):794–9.
149. Stejskal VD, Forsbeck M, Olin R. Side-chain-specific lymphocyte responses in workers with occupational allergy induced by penicillins. Int Arch Allergy Appl Immunol 1987;82(3–4):461–4.
150. Fellner MJ, Mark AS. Penicillin- and ampicillin-induced pemphigus vulgaris. Int J Dermatol 1980;19(7):392–3.
151. Miralles J, Barnadas MA, Baselga E, Gelpi C, Rodriguez JL, de Moragas JM. Bullous pemphigoid-like lesions induced by amoxicillin. Int J Dermatol 1997;36(1):42–7.
152. Wakelin SH, Allen J, Wojnarowska F. Drug-induced bullous pemphigoid with dermal fluorescence on salt-split skin. J Eur Acad Dermatol Venereol 1996;7:266.
153. Wakelin SH, Allen J, Zhou S, Wojnarowska F. Drug-induced linear IgA disease with antibodies to collagen VII. Br J Dermatol 1998;138(2):310–4.
154. Woodley DT, Burgeson RE, Lunstrum G, Bruckner-Tuderman L, Reese MJ, Briggaman RA. Epidermolysis bullosa acquisita antigen is the globular carboxyl terminus of type VII procollagen. J Clin Invest 1988;81(3):683–7.
155. Gamboa P, Jauregui I, Urrutia I. Occupational sensitization to aminopenicillins with oral tolerance to penicillin V. Contact Dermatitis 1995;32(1):48–9.
156. Conde-Salazar L, Guimaraens D, Gonzalez MA, Mancebo E. Occupational allergic contact urticaria from amoxicillin. Contact Dermatitis 2001;45(2):109.
157. Petavy-Catala C, Machet L, Vaillant L. Consort contact urticaria due to amoxycillin. Contact Dermatitis 2001;44(4):251.
158. Liccardi G, Gilder J, D'Amato M, D'Amato G. Drug allergy transmitted by passionate kissing. Lancet 2002;359(9318):1700.
159. Wuthrich B. Oral allergy syndrome to apple after a lover's kiss. Allergy 1997;52(2):235–6.
160. Wuthrich B, Dascher M, Borelli S. Kiss-induced allergy to peanut. Allergy 2001;56(9):913.
161. Smith YR, Hurd WW, Menge AC, Sanders GM, Ansbacher R, Randolph JF Jr. Allergic reactions to penicillin during in vitro fertilization and intrauterine insemination. Fertil Steril 1992;58(4):847–9.
162. Katz M, Seidenbaum M, Weinrauch L. Penicillin-induced generalized pustular psoriasis. J Am Acad Dermatol 1987;17(5 Pt 2):918–20.
163. Prieto A, de Barrio M, Lopez-Saez P, Baeza ML, de Benito V, Olalde S. Recurrent localized pustular eruption induced by amoxicillin. Allergy 1997;52(7):777–8.
164. Beylot C, Bioulac P, Doutre MS. Pustuloses exanthématiques aiguës généralisées. A propos de 4 cas. [Acute generalized exanthematic pustuloses (four cases).] Ann Dermatol Venereol 1980;107(1–2):37–48.
165. Isogai Z, Sunohara A, Tsuji T. Pustular drug eruption due to bacampicillin hydrochloride in a patient with psoriasis. J Dermatol 1998;25(9):612–5.
166. Spencer JM, Silvers DN, Grossman ME. Pustular eruption after drug exposure: is it pustular psoriasis or a pustular drug eruption? Br J Dermatol 1994;130(4):514–9.
167. Grieco T, Cantisani C, Innocenzi D, Bottoni U, Calvieri S. Acute generalized exanthematous pustulosis caused by piperacillin/tazobactam. J Am Acad Dermatol 2005;52(4):732-3.
168. Beckmann H. Exantheme unter der Behandlung mit Ampicillin. [Exanthemas during ampicillin treatment.] Munch Med Wochenschr 1971;113(43):1423–9.
169. Dejarnatt AC, Grant JA. Basic mechanisms of anaphylaxis and anaphylactoid reactions. Immunol Allergy Clin North Am 1992;12:33–46.
170. Choquet-Kastylevsky G, Descotes J. Value of animal models for predicting hypersensitivity reactions to medicinal products. Toxicology 1998;129(1):27–35.
171. McCloskey GL, Massa MC. Cephalexin rash in infectious mononucleosis. Cutis 1997;59(5):251–4.
172. Gastaminza G, Audicana MT, Fernandez E, Anda M, Ansotegui IJ. Palmar exfoliative exanthema to amoxicillin. Allergy 2000;55(5):510–1.
173. Green J, Manders S. Pseudoporphyria. J Am Acad Dermatol 2001;44:100–8.
174. Epstein J, Tuffanelli D, Seibert J, Epstein W. Porphyria-like cutaneous changes induced by tetracycline hydrochloride photosensitization. Arch Dermatol 1976;112:661–6.
175. Birkett D, Garrettts M, Stevenson C. Phototoxic bullous eruptions due to nalidixic acid. Br J Dermatol 1969;81:342–4.

176. Phung TL. Piplin CA, Tahan SR, Chiu DS. Beta-lactam antibiotic-induced pseudoporphyria. J Am Acad Dermatol 2004;51:80–2.
177. Terrados S, Blanca M, Garcia J, Vega J, Torres MJ, Carmona MJ, Miranda A, Moya M, Juarez C, Fernandez J. Nonimmediate reactions to betalactams: prevalence and role of the different penicillins. Allergy 1995;50(7):563–7.
178. Sogn DD, Evans R 3rd, Shepherd GM, Casale TB, Condemi J, Greenberger PA, Kohler PF, Saxon A, Summers RJ, VanArsdel PP Jr, et al. Results of the National Institute of Allergy and Infectious Diseases Collaborative Clinical Trial to test the predictive value of skin testing with major and minor penicillin derivatives in hospitalized adults. Arch Intern Med 1992;152(5):1025–32.
179. Gadde J, Spence M, Wheeler B, Adkinson NF Jr. Clinical experience with penicillin skin testing in a large inner-city STD clinic. JAMA 1993;270(20):2456–63.
180. Vega JM, Blanca M, Garcia JJ, Carmona MJ, Miranda A, Perez-Estrada M, Fernandez S, Acebes JM, Terrados S. Immediate allergic reactions to amoxicillin. Allergy 1994;49(5):317–22.
181. Silviu-Dan F, McPhillips S, Warrington RJ. The frequency of skin test reactions to side-chain penicillin determinants. J Allergy Clin Immunol 1993;91(3):694–701.
182. Boonk WJ, van Ketel WG. Chronische urticaria, penicilline-allergie en melkprodukten in de voeding. [Chronic urticaria, penicillin allergy and dairy products in the diet.] Ned Tijdschr Geneeskd 1980;124(42):1771–3.
183. Bochner DS, Lichtenstein LM. Anaphylaxis. N Engl J Med 1991;324(25):1785–90.
184. Markowitz M, Kaplan E, Cuttica R, et alInternational Rheumatic Fever Study Group. Allergic reactions to long-term benzathine penicillin prophylaxis for rheumatic fever. Lancet 1991;337(8753):1308–10
185. Blanca M, Perez E, Garcia J, Miranda A, Fernandez J, Vega JM, Terrados S, Avila M, Martin A, Suau R. Anaphylaxis to amoxycillin but good tolerance for benzyl penicillin. In vivo and in vitro studies of specific IgE antibodies. Allergy 1988;43(7):508–10.
186. Hunziker I, Kunzi UP, Braunschweig S, Zehnder D, Hoigné R. Comprehensive hospital drug monitoring (CHDM): adverse skin reactions, a 20-year survey. Allergy 1997;52(4):388–93.
187. Hoigne R. Akute Nebenreaktionen auf Penicillinpräparate. [Acute side-reactions to penicillin preparations.] Acta Med Scand 1962;171:201–8.
188. Lin RY. A perspective on penicillin allergy. Arch Intern Med 1992;152(5):930–7.
189. Spark RP. Fatal anaphylaxis due to oral penicillin. Am J Clin Pathol 1971;56(3):407–11.
190. Cullen DJ. Severe anaphylactic reaction to penicillin during halothane anaesthesia. A case report. Br J Anaesth 1971;43(4):410–2.
191. Macy E, Richter PK, Falkoff R, Zeiger R. Skin testing with penicilloate and penilloate prepared by an improved method: amoxicillin oral challenge in patients with negative skin test responses to penicillin reagents. J Allergy Clin Immunol 1997;100(5):586–91.
192. Wickern GM, Nish WA, Bitner AS, Freeman TM. Allergy to beta-lactams: a survey of current practices. J Allergy Clin Immunol 1994;94(4):725–31.
193. Levine BB, Redmond AP, Fellner MJ, Voss HE, Levytska V. Penicillin allergy and the heterogenous immune responses of man to benzylpenicillin. J Clin Invest 1966;45(12):1895–906.
194. VanArsdel PP Jr, Larson EB. Diagnostic tests for patients with suspected allergic disease. Utility and limitations. Ann Intern Med 1989;110(4):304–12.
195. Barbaud A, Reichert-Penetrat S, Trechot P, Jacquin-Petit MA, Ehlinger A, Noirez V, Faure GC, Schmutz JL, Bene MC. The use of skin testing in the investigation of cutaneous adverse drug reactions. Br J Dermatol 1998;139(1):49–58.
196. Ressler C, Mendelson LM. Skin test for diagnosis of penicillin allergy—current status. Ann Allergy 1987;59(3):167–70.
197. Erffmeyer JE. Adverse reactions to penicillin. Ann Allergy 1981;47(4):288–300.
198. Baldo BA, Pham NH, Weiner J. Detection and side-chain specificity of IgE antibodies to flucloxacillin in allergic subjects. J Mol Recognit 1995;8(3):171–7.
199. Torres MJ, Blanca M, Fernandez J, Esteban A, Moreno F, Vega JM, Garcia J. Selective allergic reaction to oral cloxacillin. Clin Exp Allergy 1996;26(1):108–11.
200. Sastre J, Quijano LD, Novalbos A, Hernandez G, Cuesta J, de las Heras M, Lluch M, Fernandez M. Clinical cross-reactivity between amoxicillin and cephadroxil in patients allergic to amoxicillin and with good tolerance of penicillin. Allergy 1996;51(6):383–6.
201. Salkind AR, Cuddy PG, Foxworth JW. The rational clinical examination. Is this patient allergic to penicillin? An evidence-based analysis of the likelihood of penicillin allergy. JAMA 2001;285(19):2498–505.
202. Gruchalla RS. Acute drug desensitization. Clin Exp Allergy 1998;28(Suppl 4):63–4.
203. Sullivan TJ. Drug Allergy. In: Middleton Jr. E Jr, editor. Allergy—Principles and Practice. 4th ed.CV Mosby Co, 1993:1726.
204. Tidwell BH, Clearly JD, Lorenz KR. Antimicrobial desensitization: a review of published protocols. Hosp Pharm 1997;32:1362–70.
205. Sullivan TJ. Management of patients allergic to antimicrobial drugs. Allergy Proc 1991;12(6):361–4.
206. Sullivan TJ, Yecics LD, Shatz GS, Parker CW, Wedner HJ. Desensitization of patients allergic to penicillin using orally administered beta-lactam antibiotics. J Allergy Clin Immunol 1982;69(3):275–82.
207. Chisholm CA, Katz VL, McDonald TL, Bowes WA Jr. Penicillin desensitization in the treatment of syphilis during pregnancy. Am J Perinatol 1997;14(9):553–4.
208. Wendel GD Jr, Stark BJ, Jamison RB, Molina RD, Sullivan TJ. Penicillin allergy and desensitization in serious infections during pregnancy. N Engl J Med 1985;312(19):1229–32.
209. Naclerio R, Mizrahi EA, Adkinson NF Jr. Immunologic observations during desensitization and maintenance of clinical tolerance to penicillin. J Allergy Clin Immunol 1983;71(3):294–301.
210. Booth BH, Patterson R. Electrocardiographic changes during human anaphylaxis. JAMA 1970;211(4):627–31.
211. Petsas AA, Kotler MN. Electrocardiographic changes associated with penicillin anaphylaxis. Chest 1973;64(1):66–9.
212. Mannik M. Serum sickness and pathophysiology of immune complexes. In: Rich RR, Fleisher TA, Schwartz BD, editors. Clinical Immunology, Principles and Practice. St Louis: Mosby Year Book Inc, 1996:1062–71.
213. Tatum AJ, Ditto AM, Patterson R. Severe serum sickness-like reaction to oral penicillin drugs: three case reports. Ann Allergy Asthma Immunol 2001;86(3):330–4.

214. Heinonen OP, Slone D, Shapiro S. Antimicrobial and antiparasitic agents. In: Birth Defects and Drugs in Pregnancy. 4th ed.. Boston: John Wright PSG, 1982:296.

215. Joseph TA, Pyati SP, Jacobs N. Neonatal early-onset *Escherichia coli* disease. The effect of intrapartum ampicillin. Arch Pediatr Adolesc Med 1998;152(1):35–40.

216. Towers CV, Carr MH, Padilla G, Asrat T. Potential consequences of widespread antepartal use of ampicillin. Am J Obstet Gynecol 1998;179(4):879–83.

217. Matheson I, Samseth M, Sande HA. Ampicillin in breast milk during puerperal infections. Eur J Clin Pharmacol 1988;34(6):657–9.

218. Benyamini L, Merlob P, Stahl B, Braunstein R, Bortnik O, Bulkowstein M, Zimmerman D, Berkovitch M. The safety of amoxicillin/clavulanic acid and cefuroxime during lactation. Ther Drug Monit 2005;27(4):499–502.

219. Manian FA, Stone WJ, Alford RH. Adverse antibiotic effects associated with renal insufficiency. Rev Infect Dis 1990;12(2):236–49.

220. Batchelor RC, Horne GO, Rogerson HL. An unusual reaction to procaine penicillin in aqueous suspension. Lancet 1951;2(5):195–8.

221. Hoigné R, Schoch K. Anaphylaktischer Schock und akute nichtallergische Reaktionen nach Procain-Penicillin. [Anaphylactic shock and acute nonallergic reactions following procaine-penicillin.] Schweiz Med Wochenschr 1959;89:1350–6.

222. Dry J, Leynadier F, Damecour C, Pradalier A, Herman D. Réaction pseudo-anaphylactique à la procaine-pénicilline G. Trois cas de syndrome de Hoigné. [Pseudo-anaphylactic reaction to procaine-penicillin G. 3 cases of Hoigné's syndrome.] Nouv Presse Med 1976;5(22):1401–3.

223. Schmied C, Schmied E, Vogel J, Saurat JH. Syndrome de Hoigné ou réaction pseudo-anaphylactique à la procaine pénicilline G: un classique d'actualité. [Hoigné's syndrome or pseudo-anaphylactic reaction to procaine penicillin G: a still current classic.] Schweiz Med Wochenschr 1990;120(29):1045–9.

224. Lewis GW. Acute immediate reactions to penicillin. BMJ 1957;(5028):1151–2.

225. Silber TJ, D'Angelo L. Psychosis and seizures following the injection of penicillin G procaine. Hoigné's syndrome. Am J Dis Child 1985;139(4):335–7.

226. Downham TF 2nd, Cawley RA, Salley SO, Dal Santo G. Systemic toxic reactions to procaine penicillin G. Sex Transm Dis 1978;5(1):4–9.

227. Silber TJ, D'Angelo LJ. Panic attack following injection of aqueous procaine penicillin G (Hoigné syndrome). J Pediatr 1985;107(2):314–5.

228. Berger H, Juchinka H, Tomczyk D, et al. Pseudo-anaphylactic syndrome after procaine penicillin in children. In: Abstracts, 10th Jubilee Congress. Bialystok: Polish Neurology Society, 1977:82.

229. Menke HE, Pepplinkhuizen L. Acute non-allergic reaction to aqueous procaine penicillin. Lancet 1974;2(7882):723–4.

230. Hoigné R, Krebs A. Kombinierte anaphylaktische und embolisch-toxische Reaktion durch akzidentelle intravaskuläre Injektion von Procain-Penicillin. [Combined anaphylactic and embolic-toxic reaction caused by the accidental intravascular injection of procaine penicillin.] Schweiz Med Wochenschr 1964;94:610–4.

231. Kryst L, Wanyura H. Hoigné's syndrome—its course and symptomatology. J Maxillofac Surg 1979;7(4):320–6.

232. Ernst G, Reuter E. Nicht-allergische tödliche Zwischenfälle nach depot-Penicillin. Beitrag zur Pathogenese und Prophylaxe. [Nonallergic fatal incidents following depot penicillin. Pathogenesis and prevention.] Dtsch Med Wochenschr 1970;95(12):618.

233. Bornemann K, Schulz E, Heinecker R. Akute, nicht-allergische Reaktionen nach i.m. Gabe von Clemizol-Penicillin G und Streptomycin. [Acute, non-allergic reactions following i.m. administration of clemizole-penicillin G and streptomycin.] Munch Med Wochenschr 1966;108(15):834–7.

234. Bredt J. Akute nicht-allergische Reaktionen bei Anwendung von Depot-Penicillin. [Acute non-allergic reactions in the use of depot-penicillin.] Dtsch Med Wochenschr 1965;90:1559–63.

235. Galpin JE, Chow AW, Yoshikawa TT, Guze LB. "Pseudoanaphylactic" reactions from inadvertent infusion of procaine penicillin G. Ann Intern Med 1974;81(3):358–359.

236. Kraus SJ, Green RL. Pseudoanaphylactic reactions with procaine penicillin. Cutis 1976;17(4):765–7.

237. Ilechukwu ST. Acute psychotic reactions and stress response syndromes following intramuscular aqueous procaine penicillin. Br J Psychiatry 1990;156:554–9.

238. Clauberg G. Wiederbelebung bei embolisch-toxischer Komplikation. [Resuscitation in embolic and toxic complication caused by intravascular administration of a depot-penicillin.] Anaesthesist 1966;15(8):284–5.

239. Utley PM, Lucas JB, Billings TE. Acute psychotic reactions to aqueous procaine penicillin. South Med J 1966;59(11):1271–4.

240. Randazzo SD, DiPrima G. Psicosi allucinatoria acuta da penicillina-procaina in sospensione acquosa. [Acute hallucinatory psychoses caused by procaine penicillin in aqueous suspension.] Minerva Dermatol 1959;34(6):422–8.

241. Green RL, Lewis JE, Kraus SJ, Frederickson EL. Elevated plasma procaine concentrations after administration of procaine penicillin G. N Engl J Med 1974;291(5):223–6.

242. Fishman LS, Hewitt WL. The natural penicillins. Med Clin North Am 1970;54(5):1081–99.

243. Lloyd-Roberts GC, Thomas TG. The etiology of quadriceps contracture in children. J Bone Joint Surg Br 1964;46:498–517.

244. Greenblatt DJ, Allen MD. Intramuscular injection-site complications. JAMA 1978;240(6):542–4.

245. Girard JP, Zawodnik S. Diagnostic procedures in drug allergy. In: De Weck AL, Bundgard H, editors. Handbook Exp Pharmacol 63. Heidelberg: Springer-Verlag, 1983:207.

246. Nicolau S. Dermite livédoïde et gangréneuse de la fesse, consécutive aux injections intra-musculaires, dans la syphilis: à propos d'un cas d'embolie artérielle bismuthique. Ann Mal Vener 1925;20:321.

247. Saputo V, Bruni G. La sindrome di Nicolau da preparati di penicillina: analisi della letteratura alla ricercá di potenziali fattori di rischio. [Nicolau syndrome caused by penicillin preparations: review of the literature in search for potential risk factors.] Pediatr Med Chir 1998;20(2):105–23.

248. Schanzer H, Gribetz I, Jacobson JH 2nd. Accidental intra-arterial injection of penicillin G. A preventable catastrophe. JAMA 1979;242(12):1289–90.

249. Vivell O, Hennewig J. Infarktähnliche Nekrosen nach intramuskulärer Injektion von Antibiotika. Padiatr Prax 1963;2:415.

250. Friederiszick FK. Embolien während intramuskulärer Penicillinbehandlung. Klin Wochenschr 1949;27:173.

251. Deutsch J. Schwere lokale Reaktion nach Benzathin-Penizillin. Ein Beitrag zum Nicolau-Syndrom (Dermatitis livedoides). [Severe local reaction to benzathine penicillin.

A contribution to the Nicolau syndrome (dermatitis live-doides).] Dtsch Gesundheitsw 1966;21(51):2433–7.

252. Gerbeaux J, Couvreur J, Lajouanine P, Canet J, Bonvallet. Sur deux cas d'ischémie étendue transitoire après injection intramusculaire de benzathine-pénicilline chez l'enfant. [On 2 cases of transitory extensive ischemia after intramuscular injection of benzathine penicillin in children.] Presse Méd 1966;74(7):299–302.

253. Muller-Vahl H. Adverse reactions after intramuscular injections. Lancet 1983;1(8332):1050.

254. Shaw EB. Transverse myelitis from injection of penicillin. Am J Dis Child 1966;111:548.

255. Ehringer H, Fischer M, Holzner JH, Imhof H, Kubiena K, Lechner K, Pichler H, Schnack H, Seidl K, Staudacher M. Gangrän nach versehentlicher intraaerterieller Injektion von Dicloxacillin. [Gangrene following erroneous intra-arterial injection of dicloxacillin.] Dtsch Med Wochenschr 1971;96(26):1127–30.

256. Hoigné R, Sonntag MR, Zoppi M, Hess T, Maibach R, Fritschy D. Occurrence of exanthema in relation to aminopenicillin preparations and allopurinol. N Engl J Med 1987;316(19):1217.

257. Jick H, Porter JB. Potentiation of ampicillin skin reactions by allopurinol or hyperuricemia. J Clin Pharmacol 1981;21(10):456–8.

258. Henderson JL, Polk RE, Kline BJ. In vitro inactivation of gentamicin, tobramycin, and netilmicin by carbenicillin, azlocillin, or mezlocillin. Am J Hosp Pharm 1981;38(8):1167–70.

259. Thompson MI, Russo ME, Saxon BJ, Atkin-Thor E, Matsen JM. Gentamicin inactivation by piperacillin or carbenicillin in patients with end-stage renal disease. Antimicrob Agents Chemother 1982;21(2):268–73.

260. Halstenson CE, Hirata CA, Heim-Duthoy KL, Abraham PA, Matzke GR. Effect of concomitant administration of piperacillin on the dispositions of netilmicin and tobramycin in patients with end-stage renal disease. Antimicrob Agents Chemother 1990;34(1):128–33.

261. Daly JS, Dodge RA, Glew RH, Keroack MA, Bednarek FJ, Whalen M. Effect of time and temperature on inactivation of aminoglycosides by ampicillin at neonatal dosages. J Perinatol 1997;17(1):42–5.

262. Sabra R, Branch RA. Role of sodium in protection by extended-spectrum penicillins against tobramycin-induced nephrotoxicity. Antimicrob Agents Chemother 1990;34(6):1020–5.

263. Ronchera CL, Hernandez T, Peris JE, Torres F, Granero L, Jimenez NV, Pla JM. Pharmacokinetic interaction between high-dose methotrexate and amoxycillin. Ther Drug Monit 1993;15(5):375–9.

264. Yamamoto K, Sawada Y, Matsushita Y, Moriwaki K, Bessho F, Iga T. Delayed elimination of methotrexate associated with piperacillin administration. Ann Pharmacother 1997;31(10):1261–2.

265. Mayall B, Poggi G, Parkin JD. Neutropenia due to low-dose methotrexate therapy for psoriasis and rheumatoid arthritis may be fatal. Med J Aust 1991;155(7):480–4.

266. Dawson JK, Abernethy VE, Lynch MP. Methotrexate and penicillin interaction. Br J Rheumatol 1998;37(7):807.

267. Herrick AL, Grennan DM, Griffen K, Aarons L, Gifford LA. Lack of interaction between flucloxacillin and methotrexate in patients with rheumatoid arthritis. Br J Clin Pharmacol 1996;41(3):223–7.

268. Zarychanski R, Wlodarczyk K, Ariano R, Bow E. Pharmacokinetic interaction between methotrexate and piperacillin/tazobactam resulting in prolonged toxic concentrations of methotrexate. J Antimicrob Chemother 2006; 58: 228-30.

269. Yamamoto K, Sawada Y, Matsushita Y, Moriwaki K, Bessho F, Iga T. Delayed elimination of methotrexate associated with piperacillin administration. Ann Pharmacother 1997; 31: 1261-2.

270. Williams WM, Chen TS, Huang KC. Effect of penicillin on the renal tubular secretion of methotrexate in the monkey. Cancer Res 1984; 44: 1913-7.

271. Iven H, Brasch H. Influence of the antibiotics piperacillin, doxycycline and tobramycin on the pharmacokinetics of methotrexate in rabbits. Cancer Chemother Pharmacol 1986; 17: 218-22.

272. Najjar TA, Abou-Auda HS, Ghilzai NM. Influence of piperacillin on the pharmacokinetics of methotrexate and 7-hydroxymethotrexate. Cancer Chemother Pharmacol 1998;42(5):423–8.

273. Dasgupta A, Sperelakis A, Mason A, Dean R. Phenytoin–oxacillin interactions in normal and uremic sera. Pharmacotherapy 1997;17(2):375–8.

274. Taylor At, Pritchard DC, Goldstein AO, Fletcher JL Jr. Continuation of warfarin–nafcillin interaction during dicloxacillin therapy. J Fam Pract 1994;39:182–5.

275. Kim KY, Frey RJ, Epplen K, Foruhari F. Interaction between warfarin and nafcillin: case report and review of the literature. Pharmacotherapy 2007;27:1467–70.

276. Alhambra A, Cuétara MS, Ortiz MC, Moreno JM, del Pelacio A, Pontón J, del Palacio A. False positive galactomannan results in adult hematological patients treated with piperacillin–tazobactam. Rev Iberoam Micol 2007;24(2):106–12.

277. Penack O, Rempf P, Graf B, Thiel E, Blau IW. False-positive *Aspergillus* antigen testing due to application of piperacillin/tazobactam—is it still an issue? *Diagn Microbiol Infect Dise* 2008;60(1):117–120.

278. Zandijck E, Mewis A, Magerman K, Cartuyveis R. False-positive results by the platelia *Aspergillus* galactomannan antigen test for patents treated with amoxicillin–clavulanate. *Clin Vaccine Immunol* 2008;15(7):1132–3.

279. Orlopp K, von Lilienfeld-Toal M, Marklein G, Reiffert SM, Welter A, Hahn-Ast C, Purr I, Gorschlüter M, Molitor E, Glasmacher A. False positivity of the *Aspegillus galactomannan platelia ELISA because of piperacillin/tazobactam treatment: does it represent a clinical problem? J Antimicrob Chemother* 2008;2(5):1109–12.

280. Fortún J, Martín-Dávila P, Alvarez ME, Norman F, Sánchez-Sousa A, Gajate L, Bárcena R, Nuño SJ, Moreno S. False-positive results of *Aspergillus galactomannan antigenemia in liver transplant recipients. Transplantation* 2009;87(2):256–60.

281. Adam O, Auperin A, Wilquin F. Bourhis JH, Gashot B, Chachaty E. Treatment with piperacillin/tazobactam and false-positive Aspergillus galactomannan antigen test results for patients with haematological malignancies. Clin Infect 2004;38:917–20.

282. Sing N, Obman A, Husain S, Aspinall S, Mictner S, Stout JE. Reactivity of Platelia Aspergillus galactomannan antigen with piperacillin–tazobactam: clinical implications based on achievable concentrations in serum. Antimicrob Agents Chemother 2004;48:1989–92.

283. Sulahian A, Touratier S, Ribaud P. False positive test for Aspergillus antigemia related to concominant administration of piperacillin and tazobactam. N Engl J Med 2003;349:2366–7.

284. Visculi C, Machetti M, Caooellano P, Bucci B, Bruzzi P, van Lint MT, Bacigalupo A. False-positive galactomannan Platelia Aspergillus test results for patients receiving piperacillin–tazobactam. Clin Infect Dis 2004;38:913–5.

285. Walsh TJ, Shoham S, Petraitiene R, Sein T, Schaufele R, Kelaher A, Murray H, Mya-San C, Vacher J, Petraitis V. Detection of galactomannan antigemia in patients receiving piperacillin–tazobactam and correlations between in vitro, in vivo and clinical properties of the drug–antigen interaction. J Clin Microbiol 2004;42:4744–8.

286. Penack O, Schwartz S, Thiel E, Wolfgang Blau I. Lack of evidence that false-positive Aspergillus galactomannan antigen test results are due to treatment with piperacillin–tazobactam. Clin Infect Dis 2004;39:1401–2.

287. Machetti M, Furfaro E, Viscoli C. Galactomannan in piperacillin–tazobactam: how much and to what extent? Antimicrob Agents Chemother 2005;49:3984–5.

288. Mattei D, Rapezzi D, Mordini N, Cuda F, Nigro CL, Mussi M, Arnelli A, Cagnassi S, Gallamini A, Cligo J. False-positive Aspergillus fumigatus galactomannan enzyme-linked immunoabsorbent assay results in vivo during amoxicillin–clavulanic acid treatment. J Clin Microbiol 2004;42:5362–3.

289. Metan G, Durusu M, Uzun O. False positivity for Aspergillus antigemia with amoxicillin–clavulanic acid. J Clin Microbiol 2005;43:2548–9.

290. Bower BF, McComb R, Ruderman M. Effect of penicillin on urinary 17-ketogenic and 17-ketosteroid excretion. N Engl J Med 1967;277(10):530–2.

291. Meyer RD. Amikacin. Ann Intern Med 1981;95(3):328–32.

292. Holt HA, Broughall JM, McCarthy M, Reeves DS. Interactions between aminoglycoside antibiotics and carbenicillin or ticarillin. Infection 1976;4(2):107–9.

293. Adam D, Haneder J. Studies on the inactivation of aminoglycoside antibiotics by acylureidopenicillins and piperacillin. Infection 1981;9:182.

294. Henderson JL, Polk RE, Kline BJ. In vitro inactivation of gentamicin, tobramycin, and netilmicin by carbenicillin, azlocillin, or mezlocillin. Am J Hosp Pharm 1981;38(8):1167–70.

295. Thompson MI, Russo ME, Saxon BJ, Atkin-Thor E, Matsen JM. Gentamicin inactivation by piperacillin or carbenicillin in patients with end-stage renal disease. Antimicrob Agents Chemother 1982;21(2):268–73.

296. In: Martindale: The Extra Pharmacopoeia. 28th ed.. London: The Pharmaceutical Press, 1983:273.

Polymyxins

General Information

The polymyxins are antibacterial agents that are produced from different strains of *Bacillus polymyxa*. Because of their poor tissue distribution and their substantial nephrotoxicity and neurotoxicity, they are mainly restricted to topical use. However, they can be considered for serious systemic infections caused by multidrug-resistant Gram-negative bacteria (1–4).

The polymyxins that are used clinically are polymyxin B (rINN) and colistin (pINN, formerly known as polymyxin E). Colistimethate sodium (rINN, also called colistin methanesulfonate and colistin sulfomethate sodium) is prepared from colistin.

The polymyxins are cationic, basic, and amphipathic polypeptides that interact with lipopolysaccharides in the outer membrane of bacilli. They potently neutralize endotoxin, reduce blood endotoxin concentrations in patients with septic shock during direct hemoperfusion over immobilized polymyxin B fibers, and are bactericidal for many Gram-negative rods, even in resting bacteria. Alteration of the cell wall is also thought to be the mechanism of damage to renal epithelia and to the nervous system.

The polymyxins are effective against Gram-negative bacteria, with the exception of *Proteus* and *Neisseria* (*Branhamella*). They have in the past been used particularly to treat infections due to *Pseudomonas*, including inhalation therapy in patients with cystic fibrosis (5). They are prescribed in mg or units; 1 mg of polymyxin B corresponds to 10 000 units and 1 mg of colistin corresponds to 20 000–30 000 units.

After intravenous colistimethate (5 mg/kg/day) the CSF concentration was 25% of the serum concentration in a patient with meningitis (6). Colistin sulfate is administered orally for bowel decontamination and is used topically as a powder for the treatment of bacterial skin infections; colistimethate sodium (also called colistin methanesulfate, pentasodium colistimethanesulfate, colistin sulfamethate, and colistin sulfonyl methate) is given intravenously and intramuscularly. Both formulations have been used in aerosols. For life-threatening meningitis due to such organisms, polymyxins can also be given intrathecally as adjunctive therapy (7). Polymyxins are also used in regimens of selective decontamination of the digestive tract (8). However, colistimethate sodium is associated with fewer adverse effects, such as chest tightness, throat irritation, and cough, than colistin sulfate (9).

Observational studies

In a randomized study of the effects of a triple antibiotic ointment (polymyxin B + bacitracin + neomycin) and simple gauze-type dressings on scarring of dermabrasion wounds, the ointment was superior to the simple dressing in minimizing scarring; the beneficial effect on pigmentary changes was especially pronounced (10).

General adverse effects

Even in patients with normal renal function, adverse reactions have occurred in up to 25%, contributing to death in 5% (11). At therapeutically equivalent doses, suggestions of differences in nephrotoxicity or neurotoxicity between polymyxin B and colistin are not convincing. In view of their potential for adverse effects, the polymyxins have now been largely replaced by other antibacterial drugs.

Organs and Systems

Cardiovascular

Transient hypotension occurred in a 62-year-old man after simultaneous administration of intravenous colistin 2 million IU and aerosolized colistin 2 million IU for the management of multidrug-resistant Pseudomonas aeruginosa (12). Two possible explanations were considered. First, colistin is a relatively potent stimulator of degranulation of rat mast cells, and histamine released from mast cells reduces blood pressure. Secondly, colistin stimulates the activity of pseudomonal elastase, which might play a role in causing hypotension.

Respiratory

The polymyxins are bronchial irritants, probably by histamine release (13). Treatment with aerosolized colistin can be complicated by bronchoconstriction and chest tightness (14). This reaction can be very rapid, but treatment with aerosolized beta$_2$-adrenoceptor agonists before starting treatment with aerosolized colistin can prevent it (15).

In 58 children with bronchoconstriction in response to nebulized colistin, FEV$_1$ was significantly reduced for 15 minutes (16). In 20 children the reduction was greater than 10% from baseline FEV$_1$ and was still at that level in five children after 30 minutes. Subjective assessment, baseline FEV$_1$, and serum IgE did not distinguish susceptible children.

In 62 children with cystic fibrosis there was increased dyspnea in seven and pharyngitis in three in response to nebulized colistimethate (80 mg dissolved in 3 ml of preservative-free isotonic saline, by inhalation twice daily for 4 weeks) (17).

Nine patients with cystic fibrosis chronically infected with *Pseudomonas aeruginosa* participated in a double-blind, randomized, crossover comparison of nebulized colistin sulfate or colistin sulfomethate (18). Nebulized colistin sulfate caused a significantly larger mean reduction in lung function than nebulized colistin sulfomethate. In three patients, there was a reduction in FEV$_1$ of 10% or more. Seven patients were not able to complete the course of nebulized colistin sulfate because of throat irritation and severe cough.

Colistin caused bronchospasm in 20 patients with cystic fibrosis chronically infected with *Pseudomonas aeruginosa* in a placebo-controlled clinical trial with a crossover design testing colistin 75 mg in 4 ml of saline solution and a placebo solution of the same osmolarity using a breath-enhanced nebulizer for administration (19). However, treatment with inhaled beta$_2$-adrenoceptor agonists before the start of treatment can prevent bronchoconstriction (20).

Nervous system

During treatment with any of the polymyxins, neurotoxicity can occur in up to 7% of patients with normal renal function. Circumoral paresthesia, vasomotor instability, ataxia, dizziness, convulsions of varying severity, and apnea have been reported. Of 31 patients with cystic fibrosis 21 had one or more adverse effects attributed to colistin (21). The most common reactions involved reversible neurological effects, including oral and perioral paresthesia ($n = 16$), headache ($n = 5$), and lower limb weakness ($n = 5$). All of these effects, although bothersome, were benign and reversible. There was no relation between the occurrence of any colistin-associated adverse effect and plasma colistin concentration or colistin pharmacokinetics.

Paresthesia occurs in about 27% and 7.3% of patients receiving intravenous and intramuscular colistimethate sodium respectively. This is associated with dizziness, muscle weakness, which can be generalized, facial and peripheral paresthesia, partial deafness, visual disturbances, vertigo, confusion, hallucinations, seizures, ataxia, and neuromuscular blockade. The last of these usually produces a myasthenia-like clinical syndrome, as well as apnea due to respiratory muscle paralysis (14).

Meningeal irritation rarely occurs in daily doses of 50 000 units (5 mg) given intrathecally, but higher doses can cause a stiff neck with CSF pleocytosis (22) and sometimes convulsions (20).

Neuromuscular function

In animals the polymyxins can cause neuromuscular blockade similar to that observed with the aminoglycosides, aggravated by curare, ether, and suxamethonium, and antagonized by calcium. Neuromuscular blockade induced by polymyxins has been attributed to a presynaptic action, through blockade of the release of acetylcholine into the synaptic gap (14). This can be noted first as fatiguability 1–26 hours after dosing, and can progress to severe muscular weakness, including respiratory paralysis (13). This complication has been reported both in neurologically normal subjects exposed to high plasma concentrations of the polymyxins, and also in some individuals with concentrations that were considered to be in the target range. Particularly at risk are patients with myasthenia gravis, who may require increased doses of neostigmine.

In patients with chronic pulmonary disease, polymyxin-induced neuromuscular block can result in fatal apnea. Finally, after anesthesia involving muscle relaxants the polymyxins can cause relapse of muscle weakness and inadequate ventilation (23).

Effective treatment of polymyxin-induced neuromuscular blockade requires awareness of the complication, with appropriate supervision and immediate ventilatory support, if required. Calcium gluconate and neostigmine are not of proven efficacy and should not be relied on (13).

Sensory systems

The polymyxins are occasionally ototoxic (24).

Electrolyte balance

Hyponatremia, hypokalemia, and hypocalcemia, with corresponding clinical manifestations, have occurred in patients treated for 3 weeks with doses of polymyxin B over 2 g/m^2 body surface (25,26). These abnormalities were interpreted as consequences of polymyxin-induced nephrotoxicity. Hyperchloremia and a negative anion gap seem to result from the polycationic properties of polymyxin B (27).

Urinary tract

Adverse reactions involving the kidneys occur in about 20% of patients receiving polymyxins (11). The potential for kidney damage seems to be related to age. Whereas in neonates and young infants colistimethate 20 mg/kg may be well tolerated, children over 2 years should not receive more than 10 mg/kg/day and adults even less.

Nephrotoxicity occurs more often in patients with preexisting impairment of renal function. The incidence of nephrotoxicity attributable to colistin was 36% in patients with pre-existing acute or chronic renal disease and 20% in a study of 288 patients. Additionally, in three studies intravenous colistimethate sodium was given for the treatment of patients with Gram-negative bacterial infections, 10.5% of patients had prolonged increase of blood urea nitrogen levels, 26% of patients experienced renal impairment during therapy, and 50% had a fall in creatinine clearance and an increase in serum creatinine levels. Another interesting finding was the relatively high number of case reports that were published in the old literature reporting patients who experienced acute renal failure during treatment with colistimethate sodium. However, is that in most of these cases the total daily dose of colistimethate sodium was considerably higher compared to the currently recommended dose. Of note, polymyxin B was reported in the old literature to be associated with a relatively increased incidence of toxicity compared to colistimethate sodium. However, these data were not verified in two recent studies that showed that the incidence of nephrotoxicity was 14% and 10% among patients receiving polymyxin B therapy. It has been suggested that the toxicity of polymyxins may be partly due to their D-amino acid content and fatty acid component. The proposed mechanism by which polymyxin B induces nephrotoxic events is by increasing membrane permeability, resulting in an increased influx of cations, anions, and water, leading to cell swelling and lysis. Renal toxicity associated with the use of polymyxins is considered to be dose-dependent (14,28). Doses must be adjusted in patients with renal insufficiency, because colistin is excreted principally by the kidneys, and raised blood concentrations can further impair renal function (29).

In a retrospective study in 60 patients taking polymyxin B 1.5–2.5 mg/kg/day for treatment of multiresistant bacteria, the development of renal insufficiency was independent of the daily and cumulative doses of polymyxin B and the length of treatment, but was significantly associated with age. Overall mortality was 20%, but it increased to 57% in those who developed renal insufficiency (30).

In a prospective, observational, cohort study in 21 patients who received intravenous colistin for at least 7 days (median daily dose 17.7 mg and median duration of treatment 15 days), three patients developed nephrotoxicity (31). The cumulative dose of colistin correlated with the difference in serum creatinine concentrations between the start and end of treatment.

- A 57-year-old man developed acute renal insufficiency after receiving colistin 250 mg intravenously every 6 hours for 4 days (31).
- A 35-year-old man who received colistin 6 MU intravenously divided into three daily doses had no adverse effects. However, during two subsequent courses 1 month and 4 months later he developed acute renal insufficiency (32).

In both of these episodes, renal function returned to normal values within 3–5 days after colistin withdrawal and despite the continuation of all other drugs.

In an observational retrospective cohort study of 17 patients who received intravenous colistin for more than 4 weeks, 19 courses of prolonged treatment were identified (33). The mean duration of administration was 43 days, and the mean cumulative dose was 190 million IU. The median creatinine concentration rose by 220 μmol/l during treatment compared with baseline but returned to near baseline at the end of treatment.

Intermittent proteinuria was observed on urinalysis in 14 of 31 patients with cystic fibrosis, and one patient developed reversible, colistin-induced nephrotoxicity (14). There was no relation between the occurrence of any colistin-associated adverse effect and plasma colistin concentration or colistin pharmacokinetics.

Serum colistin concentrations and toxicity appear to be a function of glomerular filtration rate (34). Renal toxicity is dose-related and is usually reversible after early withdrawal. However, there are a few published reports of irreversible nephrotoxicity after withdrawal of colistin (19).

Skin

Hypersensitivity reactions, rash, urticaria, generalized itching, and fever can occur during therapy with colistin, the incidence of allergic reactions being 2% (20).

- A 27-year-old woman developed an allergic contact dermatitis after having both ears pierced followed by prophylactic application of an ointment containing colistin sulfate and bacitracin (35). The eruption responded to a glucocorticoid ointment. Patch tests showed positive reactions to colistin sulfate and bacitracin.

Topical polymyxin B was the predominant allergen in patients who underwent patch-testing for evaluation of eczema of the external ear canal (36).

Musculoskeletal

Of 23 critically ill patients treated with colistin, one developed diffuse muscular weakness on day 10 of treatment; the symptom resolved within 1 week of withdrawal of colistin (37).

Immunologic

Compared with their toxic effects, allergic reactions to the polymyxins are relatively unimportant. Nevertheless, drug fever and maculopapular eruptions and other skin lesions have been observed in few patients (38,39).

In 145 patients with eczema of the external ear canal, allergic contact dermatitis was diagnosed in one-third; topical therapeutic agents, especially neomycin sulfate and probably polymyxin B, were the dominating allergens (36).

Long-Term Effects

Drug resistance

The Gram-negative organism *Burkholderia pseudomallei* is the pathogen that causes melioidosis. This bacterium is intrinsically resistant to the killing action of cationic antimicrobial peptides, including polymyxins. An in vitro study has now identified genetic loci that are associated with resistance to polymyxins in a virulent clinical isolate (40). In patients with cystic fibrosis *Pseudomonas aeruginosa* synthesized specific lipid A structures containing palmitate and aminoarabinose, which were associated with resistance to cationic antibiotics and increased inflammatory responses (1).

A two-component regulatory system has been characterized that is involved in the resistance of *P. aeruginosa* in response to external magnesium concentrations (41). Similarly, the PmrA/PmrB two-component system of *Salmonella enterica* that mediates modifications in the lipopolysaccharide, resulting in resistance to polymyxins, has been characterized in more detail (42). The *pqaB* locus affected polymyxin B resistance in *Salmonella typhi* (43).

All hemolytic and cytotoxic *Aeromonas* species that have been isolated from water samples from various sources were resistant to polymyxin B (44). Resistance to polymyxin B was also found in *Vibrio vulnificus* (45).

Resistance to colistin has been analysed in 44 adults with cystic fibrosis treated with inhaled colistin. Five developed polymyxin resistance (46). After therapy *P. aeruginosa* became sensitive to polymyxins within a few months, enabling the reintroduction of colistin for antibacterial treatment.

Second-Generation Effects

Teratogenicity

Colistin, and probably also polymyxin B, crosses the placenta (47). Although there is no evidence that the polymyxins are teratogenic, they should be avoided in pregnancy.

Susceptibility Factors

Renal impairment

Intravenous colistin methanesulfonate is converted in vivo to colistin, and these two compounds have substantially different pharmacokinetics, antibacterial activities, and adverse effects. Patients who are currently receiving colistin methanesulfonate are often in intensive care units, have multiple organ dysfunctions, and receive renal replacement therapy.

- A 53-year-old woman weighing 110 kg undergoing continuous venovenous hemodiafiltration was given intravenous colistin methanesulfonate 150 mg (equivalent to 2.46 mg/kg ideal body weight) every 24 hours, reduced to 150 mg/48 hours after 14 days (48). Each dose was given as a 30-minute infusion. The maximum concentration of colistin in plasma occurred 30 minutes after completion of the infusion, consistent with relatively rapid conversion to colistin. The terminal plasma half-lives of colistin methanesulfonate and colistin were 6.83 and 7.52 hours respectively. The total clearance of colistin methanesulfonate was 49 ml/minute and its volume of distribution was 11 liters. From 0 to 8 hours after dosing, 20% of the dose was recovered in the dialysate as colistin methanesulfonate and 6.9% as colistin. The hemodiafiltration clearances of colistin methanesulfonate and colistin were similar (11 and 12 ml/minute).

Drug Administration

Drug formulations

Dry powder inhalation may be an alternative to nebulization of colistin in the treatment of chest infections in patients with cystic fibrosis. Colistin 25 mg of dry by dry powder inhaler (prototype Twincer) has been compared with 158 mg by nebulization in 10 patients with cystic fibrosis a randomized crossover study (49). The dry powder was well tolerated and there was no significant reduction in FEV1. The relative availability of colistin by dry powder inhalation was about 140% based on actual dose. Similar results were reported in 10 healthy volunteers (50).

A food additive (flavored BMI-60) may help to mask the bitter taste of polymyxin B sulfate tablets (51).

Drug administration route

Bolus intravenous colistin (160 mg in 10 ml of saline tds) has been studied in a phase I open study during acute respiratory exacerbations in adults with cystic fibrosis and chronic *P. aeruginosa* infection; patients without total indwelling venous access systems had mild to moderate injection-like pain (52).

Intramuscular injection of the sulfates of polymyxin or colistin often causes pain at the site of injection; with sodium colistimethate this adverse reaction is largely absent (53).

In a study of the intrathecal administration of colistin adverse events were not reported (54). This may be an effective alternative treatment of bacterial meningitis caused by multidrug-resistant Gram-negative rods.

During sepsis, toxins (for example released from bacteria) can cause shock, disseminated intravascular coagulation, multiorgan dysfunction, and death. Apheresis may be a way of reducing the amounts of toxins and other harmful compounds in the circulation, and polymyxin B may serve as an adsorber. In three patients with septic shock, direct hemoperfusion using a polymyxin B-immobilized fiber column was carried out after antibacterial and antishock therapy. As a result, cardiovascular instabilities improved without increasing the supply of catecholamines (55). Furthermore, in seven patients with endotoxic shock after laparotomy undergoing hemoperfusion with the polymyxin B-immobilized fiber, there was an early increase in urine volume, attributable to increased glomerular filtration independent of systemic hemodynamic factors (56).

Drug overdose

Overdoses are mainly reported with colistimethate sodium (14). There is no antidote for polymyxin overdose. Management requires early withdrawal of the medication and appropriate supportive treatment. In the presence of established acute renal failure, hemodialysis and peritoneal dialysis can only manage renal complications, since they have little influence on the elimination of polymyxins (57). If apnea occurs, mechanical ventilatory support is needed. Exchange transfusion has been proposed (58,59).

Drug–Drug Interactions

Cefalothin

Co-administration of sodium cefalothin and polymyxins can increase the risk of neurotoxicity (14).

Neuromuscular blocking drugs

There may be difficulty in reversing neuromuscular blockade if polymyxin is given in combination with neuromuscular blocking drugs (60).

The polymyxins probably produce a predominantly postjunctional effect (via ion channel block) and reduce muscle contractility. The block is difficult to reverse, calcium being only partly successful. Neostigmine has been reported to increase blockade produced by polymyxin B and colistin; in such cases 4-aminopyridine might be helpful.

References

1. Ernst RK, Yi EC, Guo L, Lim KB, Burns JL, Hackett M, Miller SI. Specific lipopolysaccharide found in cystic fibrosis airway *Pseudomonas aeruginosa*. Science 1999;286(5444):1561–5.
2. Asanuma Y, Furuya T, Tanaka J, Sato T, Shibata S, Koyama K. The application of immobilized polymyxin B fiber in the treatment of septic shock associated with severe acute pancreatitis: report of two cases. Surg Today 1999;29(11):1177–82.
3. Nakamura T, Ebihara I, Shoji H, Ushiyama C, Suzuki S, Koide H. Treatment with polymyxin B-immobilized fiber reduces platelet activation in septic shock patients: decrease in plasma levels of soluble P-selectin, platelet factor 4 and beta-thromboglobulin. Inflamm Res 1999;48(4):171–5.
4. Hellman J, Warren HS. Antiendotoxin strategies. Infect Dis Clin North Am 1999;13(2):371–86ix.
5. Mordasini C, Aebischer CC, Schoch OD. Zur inhalativen Antibiotika-Therapie bei Patienten mit zystischer Fibrose und *Pseudomonas*-Befall. [Inhalational antibiotic therapy in patients with cystic fibrosis and *Pseudomonas* infection.] Schweiz Med Wochenschr 1997;127(21):905–10.
6. Jimenez-Mejias ME, Pichardo-Guerrero C, Marquez-Rivas FJ, Martin-Lozano D, Prados T, Pachon J. Cerebrospinal fluid penetration and pharmacokinetic/pharmacodynamic parameters of intravenously administered colistin in a case of multidrug-resistant *Acinetobacter baumannii* meningitis. Eur J Clin Microbiol Infect Dis 2002;21(3):212–4.
7. Segal-Maurer S, Mariano N, Qavi A, Urban C, Rahal JJ Jr. Successful treatment of ceftazidime-resistant *Klebsiella pneumoniae* ventriculitis with intravenous meropenem and intraventricular polymyxin B: case report and review. Clin Infect Dis 1999;28(5):1134–8.
8. Rommes JH, Zandstra DF, van Saene HK. Selectieve darmdecontaminatie voorkomt sterfte bij intensive-carepatienten. [Selective decontamination of the digestive tract reduces mortality in intensive care patients.] Ned Tijdschr Geneeskd 1999;143(12):602–6.
9. Michalopoulos A, Kasiakou SK, Falagas ME. The significance of different formulations of aerosolized colistin. Crit Care 2005;9(4):417–8.
10. Berger RS, Pappert AS, Van Zile PS, Cetnarowski WE. A newly formulated topical triple-antibiotic ointment minimizes scarring. Cutis 2000;65(6):401–4.
11. Koch-Weser J, Sidel VW, Federman EB, Kanarek P, Finer DC, Eaton AE. Adverse effects of sodium colistimethate. Manifestations and specific reaction rates during 317 courses of therapy. Ann Intern Med 1970;72(6):857–68.
12. Hakeam HA, Almohaizeie AM. Hypotension following treatment with aerosolized colistin in a patient with multidrug-resistant *Pseudomonas aeruginosa*. Ann Pharmacother 2006;40(9):1677–80.
13. Lindesmith LA, Baines RD Jr, Bigelow DB, Petty TL. Reversible respiratory paralysis associated with polymyxin therapy. Ann Intern Med 1968;68(2):318–27.
14. Falagas ME, Kasiakou SK. Toxicity of polymyxins: a systematic review of the evidence from old and recent studies. Crit Care 2006;10(1):R27.
15. Falagas ME, Kasiakou SK, Tsiodras S, Michalopoulos A. The use of intravenous and aerosolized polymyxins for the treatment of infections in critically ill patients: a review of the recent literature. Clin Med Res 2006;4(2):138–46.
16. Cunningham S, Prasad A, Collyer L, Carr S, Lynn IB, Wallis C. Bronchoconstriction following nebulised colistin in cystic fibrosis. Arch Dis Child 2001;84(5):432–3.
17. Nikolaizik WH, Trociewicz K, Ratjen F. Bronchial reactions to the inhalation of high-dose tobramycin in cystic fibrosis. Eur Respir J 2002;20(1):122–6.
18. Westerman EM, Le Brun PP, Touw DJ, Frijlink HW, Heijerman HG. Effect of nebulized colistin sulphate and colistin sulphomethate on lung function in patients with cystic fibrosis: a pilot study. J Cyst Fibros 2004;3(1):23–8.

19. Alothman GA, Ho B, Alsaadi MM, Ho SL, O'Drowsky L, Louca E, Coates AL. Bronchial constriction and inhaled colistin in cystic fibrosis. Chest 2005;127(2):522–9.

20. Falagas ME, Kasiakou SK. Colistin: the revival of polymyxins for the management of multidrug-resistant gram-negative bacterial infections. Clin Infect Dis 2005;40(9):1333–41.

21. Reed MD, Stern RC, O'Riordan MA, Blumer JL. The pharmacokinetics of colistin in patients with cystic fibrosis. J Clin Pharmacol 2001;41(6):645–54.

22. Everett ED, Strausbaugh LJ. Antimicrobial agents and the central nervous system. Neurosurgery 1980;6(6):691–714.

23. Sobek V. Arrest of respiration induced by polypeptide antibiotics. Arzneimittelforschung 1982;32(3):235–7.

24. Linder TE, Zwicky S, Brandle P. Ototoxicity of ear drops: a clinical perspective. Am J Otol 1995;16(5):653–7.

25. O'Regan S, Carson S, Chesney RW, Drummond KN. Electrolyte and acid-base disturbances in the management of leukemia. Blood 1977;49(3):345–53.

26. Rodriguez V, Green S, Bodey GP. Serum electrolyte abnormalities associated with the administration of polymyxin B in febrile leukemic patients. Clin Pharmacol Ther 1970;11(1):106–11.

27. O'Connor DT, Stone RA. Hyperchloremia and negative anion gap associated with polymyxin B administration. Arch Intern Med 1978;138(3):478–80.

28. Linden PK, Paterson DL. Parenteral and inhaled colistin for treatment of ventilator-associated pneumonia. Clin Infect Dis 2006;43 Suppl 2S:89–94.

29. Stein A, Raoult D. Colistin: an antimicrobial for the 21st century? Clin Infect Dis 2002;35(7):901–2.

30. Ouderkirk JP, Nord JA, Turett GS, Kislak JW. Polymyxin B nephrotoxicity and efficacy against nosocomial infections caused by multiresistant gram-negative bacteria. Antimicrob Agents Chemother 2003;47:2659–62.

31. Falagas ME, Fragoulis KN, Kasiakou SK, Sermaidis GJ, Michalopoulos A. Nephrotoxicity of intravenous colistin: a prospective evaluation. Int J Antimicrob Agents 2005;26(6):504–7.

32. Kallel H, Hamida CB, Ksibi H, Bahloul M, Hergafi L, Chaari A, Chelly H, Bouaziz M. Suspected acute interstitial nephritis induced by colistin. J Nephrol 2005;18(3):323–6.

33. Falagas ME, Rizos M, Bliziotis IA, Rellos K, Kasiakou SK, Michalopoulos A. Toxicity after prolonged (more than four weeks) administration of intravenous colistin. BMC Infect Dis 2005;5(1):1.

34. Daram SR, Gogia S, Bastani B. Colistin-associated acute renal failure: revisited. South Med J 2005;98(2):257–8.

35. Sowa J, Tsuruta D, Kobayashi H, Ishii M. Allergic contact dermatitis caused by colistin sulfate and bacitracin. Contact Dermatitis 2005;53(3):175–6.

36. Hillen U, Geier J, Goos M. Kontaktallergien bei Patienten mit Ekzemen des ausseren Gehorgangs. Ergebnisse des Informationsverbundes Dermatologischer Kliniken und der Deutschen Kontaktallergie-Gruppe. [Contact allergies in patients with eczema of the external ear canal. Results of the Information Network of Dermatological Clinics and the German Contact Allergy Group.] Hautarzt 2000;51(4):239–43.

37. Linden PK, Kusne S, Coley K, Fontes P, Kramer DJ, Paterson D. Use of parenteral colistin for the treatment of serious infection due to antimicrobial-resistant Pseudomonas aeruginosa. Clin Infect Dis 2003;37:e154-60.

38. Sasaki S, Mitsuhashi Y, Kondo S. Contact dermatitis due to sodium colistimethate. J Dermatol 1998;25(6):415–7.

39. Zehnder D, Kunzi UP, Maibach R, Zoppi M, Halter F, Neftel KA, Muller U, Galeazzi RL, Hess T, Hoigne R. Die Häufigkeit der Antibiotika-assoziierten Kolitis bei hospitalisierten Patienten der Jahre 1974–1991 im "Comprehensive Hospital Drug Monitoring" Bern/St. Gallen. [Frequency of antibiotics-associated colitis in hospitalized patients in 1974–1991 in "Comprehensive Hospital Drug Monitoring", Bern/St. Gallen.] Schweiz Med Wochenschr 1995;125(14):676–83.

40. Burtnick MN, Woods DE. Isolation of polymyxin B-susceptible mutants of Burkholderia pseudomallei and molecular characterization of genetic loci involved in polymyxin B resistance. Antimicrob Agents Chemother 1999;43(11):2648–56.

41. Macfarlane EL, Kwasnicka A, Ochs MM, Hancock RE. PhoP-PhoQ homologues in Pseudomonas aeruginosa regulate expression of the outer-membrane protein OprH and polymyxin B resistance. Mol Microbiol 1999;34(2):305–16.

42. Wosten MM, Groisman EA. Molecular characterization of the PmrA regulon. J Biol Chem 1999;274(38):27185–90.

43. Baker SJ, Gunn JS, Morona R. The Salmonella typhi melittin resistance gene pqaB affects intracellular growth in PMA-differentiated U937 cells, polymyxin B resistance and lipopolysaccharide. Microbiology 1999;145(Pt 2):367–78.

44. Alavandi SV, Subashini MS, Ananthan S. Occurrence of haemolytic & cytotoxic Aeromonas species in domestic water supplies in Chennai. Indian J Med Res 1999;110:50–5.

45. Ghinsberg RC, Dror R, Nitzan Y. Isolation of Vibrio vulnificus from sea water and sand along the Dan region coast of the Mediterranean. Microbios 1999;97(386):7–17.

46. Tamm M, Eich C, Frei R, Gilgen S, Breitenbucher A, Mordasini C. Inhalatives Colistin bei zystischer fibrose. [Inhaled colistin in cystic fibrosis.] Schweiz Med Wochenschr 2000;130(39):1366–72.

47. MacAulay MA, Charles D, Burgess FM. Placental transmission of colistimethate. Clin Pharmacol Ther 1967;8(4):578–86.

48. Li J, Rayner CR, Nation RL, Deans R, Boots R, Widdecombe N, Douglas A, Lipman J. Pharmacokinetics of colistin methanesulfonate and colistin in a critically ill patient receiving continuous venovenous hemodiafiltration. Antimicrob Agents Chemother 2005;49(11):4814–5.

49. Westerman EM, De Boer AH, Le Brun PP, Touw DJ, Roldaan AC, Frijlink HW, Heijerman HG. Dry powder inhalation of colistin in cystic fibrosis patients: a single dose pilot study. J Cyst Fibros 2007;6(4):284–92.

50. Westerman EM, de Boer AH, Le Brun PP, Touw DJ, Frijlink HW, Heijerman HG. Dry powder inhalation of colistin sulphomethate in healthy volunteers: a pilot study. Int J Pharm 2007;335(1-2):41–5.

51. Saito M, Hoshi M, Igarashi A, Ogata H, Edo K. The marked inhibition of the bitter taste of polymyxin B sulfate and trimethoprim × sulfamethoxazole by flavored BMI-60 in pediatric patients. Biol Pharm Bull 1999;22(9):997–8.

52. Conway SP, Etherington C, Munday J, Goldman MH, Strong JJ, Wootton M. Safety and tolerability of bolus intravenous colistin in acute respiratory exacerbations in adults with cystic fibrosis. Ann Pharmacother 2000;34(11):1238–42.

53. Kucers A, Bennett NM. In: Polymixins. Philadelphia: Lippincott, 1987:905–17.

54. Vasen W, Desmery P, Ilutovich S, Di Martino A. Intrathecal use of colistin. J Clin Microbiol 2000;38(9):3523.

55. Yuasa J, Naya Y, Tanaka M, Amakasu M, Yamaguchi K. [Clinical experiences of endotoxin removal columns in septic shock due to urosepsis: report of three cases.]Hinyokika Kiyo 2000;46(11):819–22.
56. Terawaki H, Kasai K, Kobayashi H, Hirano K, Hamaguchi A, Kase Y, Horiguchi T, Yokoyama K, Yamamoto H, Nakayama M, Kawaguchi Y, Hosoya T. [A study on the mechanism of enhanced diuresis following direct hemoperfusion with polymyxin B-immobilized fiber.]Nippon Jinzo Gakkai Shi 2000;42(5):359–64.
57. Goodwin NJ, Friedman EA. The effects of renal impairment, peritoneal dialysis, and hemodialysis on serum sodium colistimethate levels. Ann Intern Med 1968;68(5):984–94.
58. Hoeprich PD. The polymyxins. Med Clin North Am 1970;54(5):1257–65.
59. Brown JM, Dorman DC, Roy LP. Acute renal failure due to overdosage of colistin. Med J Aust 1970;2(20):923–4.
60. Cammu G. Interactions of neuromuscular blocking drugs. Acta Anaesthesiol Belg 2001;52(4):357–63.

Pyrazinamide

See also Antituberculosis drugs

General Information

Pyrazinamide is a pyrazine analogue of nicotinamide. It is bactericidal for *Mycobacterium tuberculosis* in an acid environment and within macrophages (1). Regimens that include pyrazinamide produce significantly more rapid rates of sputum conversion than any other combination. Pyrazinamide is therefore especially appropriate in the initial phase of treatment. In the 6-month regimen of the American Thoracic Society, pyrazinamide was used together with isoniazid and rifampicin for the first 2 months (2).

Pyrazinamide is distributed throughout the body. Peak plasma concentrations are reached 2 hours after oral administration. Excretion is primarily by glomerular filtration. Serum concentrations are generally 30–50 µg/ml with daily doses of 20–25 mg/kg. The maximum daily dose should not exceed 3 g, regardless of weight. At a pH of 5.5, the minimal inhibitory concentration of pyrazinamide for *Mycobacterium tuberculosis* is 20 µg/ml (1).

Observational studies

The combination of pyrazinamide plus levofloxacin is first-line treatment for multidrug-resistant latent tuberculosis. In 17 Canadian patients there were important adverse reactions affecting the musculoskeletal and central nervous systems; hyperuricemia, gastrointestinal effects, and dermatological effects were also common (3). This combination may be used with careful monitoring for adverse effects.

General adverse effects

Most of the adverse effects of pyrazinamide are toxic effects. Reactions involving the liver, hyperuricemia

with and without gout, and symptoms of pellagra have particularly been recognized. Fever and urticaria are described. Sideroblastic anemia, thrombocytopenia, anorexia, nausea and vomiting, dysuria, malaise, and aggravation of peptic ulcer can occur (SEDA-13, 261). Allergic reactions and tumor-inducing effects have not been reported.

Organs and Systems

Cardiovascular

Acute symptomatic hypertension consistently followed the administration of pyrazinamide to a 65-year-old woman with pulmonary tuberculosis (4).

Metabolism

Pyrazinamide interferes with the renal excretion of urate, resulting in hyperuricemia. Acute episodes of gout or arthralgia have occurred. Arthralgia responds better to NSAIDs than to uricosuric drugs (5).

Pyrazinamide can aggravate porphyria (1).

Liver

Liver damage is the most common adverse effect of pyrazinamide (6). It varies from asymptomatic alteration of liver function detectable only by laboratory tests, through a mild syndrome characterized by fever, anorexia, malaise, liver tenderness, hepatomegaly, and splenomegaly, to more serious reactions with clinical jaundice, and finally the rare form with progressive acute yellow atrophy and death. As most patients take a combined regimen of pyrazinamide with isoniazid and rifampicin, it is difficult to determine which of the three drugs causes the hepatotoxicity; it could be due to a combined effect (7). As with isoniazid and rifampicin, hepatic function should initially be monitored every few weeks.

Skin

Pyrazinamide can cause pellagra. However, prophylactic nicotinamide is not generally recommended. In cases of pellagra, a dosage of 300 mg/day should be given (8).

In one case, erythema multiforme and urticaria occurred together after administration of pyrazinamide, and there were circulating immune complexes (9).

Photosensitization has been rarely described in patients taking pyrazinamide (10).

Susceptibility Factors

In a retrospective analysis of 430 patients with tuberculosis at a chest center between 1990 and 1999, the incidence of all major adverse effects was 1.48 per 100 person-months of exposure (95% CI = 1.31, 1.61) for pyrazinamide compared with 0.49 (0.42, 0.55) for isoniazid, 0.43 (0.37, 0.49) for rifampicin, and 0.07 (0.04, 0.1) for ethambutol [11]. The occurrence of any major adverse effect was associated with female sex, age over 60 years, birthplace in Asia, and HIV-positive status. The incidence of pyrazinamide-induced hepatotoxicity and rash during treatment for active tuberculosis was substantially higher than with other first-line antituberculosis drugs.

Age

Although pyrazinamide is a part of conventional combination therapy for children with tuberculosis, as in adults (12), there is little published information on its safety. In those with raised transaminases before the start of therapy there was no increase during therapy, and the activities normalized in all patients after its conclusion.

Hepatic disease

Pyrazinamide should be avoided in patients with liver disease and porphyria (1). Liver function tests should be repeated at frequent intervals during the entire period of treatment.

Other features of the patient

Pyrazinamide should be used with extreme caution in patients with a history of gout, especially in elderly people, in whom urinary urate stones can cause renal insufficiency.

In patients with hemoptysis, the possibility that pyrazinamide may have an adverse effect on blood clotting time or vascular integrity should be borne in mind (13,14).

References

1. Mandell GL, Sande MA. Antimicrobial agents: drugs used in the chemotherapy of tuberculosis and leprosy. In: Goodman Gilman A, Rall TW, Nies AS, Taylor P, editors. Goodman and Gilman's The Pharmacological Basis of Therapeutics. 8th ed.. New York: Pergamon Press, 1990:1146 Chapter 49.
2. Bass JB Jr, Farer LS, Hopewell PC, O'Brien R, Jacobs RF, Ruben F, Snider DE Jr, Thornton G. Treatment of tuberculosis and tuberculosis infection in adults and children. American Thoracic Society and The Centers for Disease Control and Prevention. Am J Respir Crit Care Med 1994;149(5):1359–74.
3. Papastavros T, Dolovich LR, Holbrook A, Whitehead L, Loeb M. Adverse events associated with pyrazinamide and levofloxacin in the treatment of latent multidrug-resistant tuberculosis. CMAJ 2002;167(2):131–6.
4. Goldberg J, Moreno F, Barbara J. Acute hypertension as an adverse effect of pyrazinamide. JAMA 1997;277(17):1356.
5. Patel AM, McKeon J. Avoidance and management of adverse reactions to antituberculosis drugs. Drug Saf 1995;12(1):1–25.
6. Danan G, Pessayre D, Larrey D, Benhamou JP. Pyrazinamide fulminant hepatitis: an old hepatotoxin strikes again. Lancet 1981;2(8254):1056–7.
7. Pretet S, Perdrizet S. La toxicité du pyrazinamide dans les traitements antituberculeux. [Toxicity of pyrazinamide in antituberculous treatments.] Rev Fr Mal Respir 1980;8(4):307–30.
8. Jorgensen J. Pellagra probably due to pyrazinamide: development during combined chemotherapy of tuberculosis. Int J Dermatol 1983;22(1):44–5.
9. Perdu D, Lavaud F, Prevost A, Deschamps F, Cambie MP, Bongrain E, Barhoum K, Kalis B. Erythema multiforme due to pyrazinamide. Allergy 1996;51(5):340–2.
10. Chan SL. Chemotherapy of tuberculosis. In: Davies PDO, editor. Clinical Tuberculosis. London: Chapman and Hall, 1994:141.
11. Yee D, Valiquette C, Pelletier M, Parisiea I, Rocher I, Menzies D. Incidence of serious side effects from first line antituberculosis drugs among patients treated for active tuberculosis. Am J Respir Crit Care Med 2003; 167: 1472-7.
12. Britisch Medical Association and Royal Pharmaceutical Society of Great Britain. Antituberculous drugs. Br Natl Formulary 1998;35:160.
13. Jenner PJ, Ellard GA, Allan WG, Singh D, Girling DJ, Nunn AJ. Serum uric acid concentrations and arthralgia among patients treated with pyrazinamide-containing regimens in Hong Kong and Singapore. Tubercle 1981;62(3):175–9.
14. Lukaszczyk E, Radecki A, Ignasiak J. Wplyw pobierania pyrazinamidu na uklad krzepniecia krwi i fibrynolize. [Effect of pyrazinamide on the blood coagulation system and fibrinolysis.] Gruzlica 1970;38(3):229–37.

Rifamycins

See also Antituberculosis drugs

General Information

The rifamycins are a group of macrocyclic antibiotics that were originally derived from a culture of *Streptomyces mediterranei*, the products of which were named after the French film "Rififi chez les hommes" (director Jules Dassin).

Rifampicin

Rifampicin (rINN) is a semisynthetic derivative of rifamycin B. By suppressing the initiation of chain formation in RNA synthesis, it inhibits the DNA-dependent RNA polymerase of mycobacteria and other microorganisms (1). It inhibits the growth of most Gram-positive and many Gram-negative bacteria: *Escherichia coli*, some strains of *Pseudomonas*, *Proteus*, *Klebsiella*, *Staphylococcus aureus*, *Neisseria meningitidis*, *Hemophilus influenzae*, and *Legionella* species. *Mycobacterium tuberculosis*, *Mycobacterium kansasii*, *Mycobacterium scrofulaceum*, and *Mycobacterium intracellulare* are suppressed with increasing concentrations between 0.005 and 4 μg/ml. Rifampicin is bactericidal for *Mycobacterium leprae* in a concentration of less than 1 μg/ml. *Mycobacterium fortuitum* is not susceptible. Primary resistance is very rare, but secondary resistance among mycobacteria and meningococci develops very rapidly when rifampicin is used as a single drug (1).

In a prospective, randomized trial in 12 high-risk adult patients, the use of polyurethane, triple-lumen, central venous catheters impregnated with minocycline and rifampicin (on both the luminal and external surfaces) is associated with a lower rate of infection than the use of catheters impregnated with chlorhexidine and silver sulfadiazine (on the external surface only) (2). There were low rates of catheter-related bloodstream infection (0.3%) and catheter colonization (7.9%) with the use of catheters impregnated with minocycline and rifampicin. This favorable result may be explained either by differences in the coating of the catheters (internal and external surfaces versus external surface only), by microbiological advantages of minocycline and rifampicin over chlorhexidine and silver sulfadiazine, or even by an effect unrelated to the antibacterial activity. The additional costs of preventing an infection and a death would be about $3125 and $12 500 respectively (3).

Pharmacokinetics

Rifampicin is distributed to nearly all organs and body fluids in adequate antibacterial concentrations. The rifamycins are concentrated several-fold in alveolar macrophages (4). This explains their efficacy against intracellular bacteria. Peak concentrations of up to 7 µg/ml are reached within 2–4 hours after a dose of 600 mg before a meal. After gastrointestinal absorption, rifampicin is quickly eliminated in the bile, following enterohepatic circulation. It is also excreted in various fluids, and causes orange–red discoloration of the urine, feces, saliva, tears, sputum, and sweat (SED-10, 576) (5). Rifampicin is progressively deacetylated. Its half-life varies between 2.5 and 5 hours and is prolonged to various degrees by isoniazid and hepatic disorders (1). There is a progressive shortening of the half-life of rifampicin by about 40% during the first 14 days of treatment, owing to self-induction of the hepatic microsomal enzymes that metabolize it (6).

Because rifampicin is a potent inducer of hepatic microsomal drug-metabolizing enzymes, it has been implicated in reducing the effectiveness of many drugs that are metabolized in the liver (SEDA-21, 313). Interactions continue to be recognized, reflecting the extension of the use of rifampicin from an antituberculosis agent to an antistaphylococcal drug, particularly useful in the treatment of methicillin-resistant *Staphylococcus aureus* (MRSA) and when prostheses are infected.

Comparative studies

The efficacy and safety of rifampicin + pyrazinamide versus isoniazid for the prevention of tuberculosis among people with or without HIV infection has been evaluated in a meta-analysis of three trials in HIV infected patients and three in HIV non-infected persons (7). The rates of tuberculosis and mortality were similar in the two groups, whether the subjects were HIV infected or not. However, both subgroup analyses showed a higher incidence of all severe adverse events in those who took rifampicin + pyrazinamide among non-HIV-infected persons (29% versus 7 %) (see also Liver below).

General adverse effects

Rifampicin given in usual doses (for example 10 mg/kg/day) is well tolerated and causes adverse effects in only

about 4% of patients. Adverse reactions are predominantly hepatic and allergic. Gastrointestinal symptoms are generally transient. Risk factors are age, alcoholism, and hepatic disorders (1). As a potent microsomal enzyme inducer, rifampicin shortens the half-lives of many other drugs (1). This effect occurs after about 7 days and persists for a few days after withdrawal. Allergic reactions can cause rashes (in 0.8%), fever, a flu-like illness (malaise, headache), eosinophilia, and much less often hemolytic anemia, hemoglobinuria, and kidney damage with acute renal insufficiency. These reactions occur especially during intermittent treatment (less than twice weekly) or after re-introduction of rifampicin. Anaphylactoid reactions to rifampicin have been described in HIV-positive patients (8). Light-chain proteinuria and concomitant kidney damage are attributed to an immunological process (SEDA-10, 273). Rifampicin antibodies have been found using an antiglobulin test (9). Hemolysis as an adverse effect seems to be mainly of the immune-complex type, exceptionally of the IgG-antibody type (10). The drug-induced lupus-like syndrome has been linked to rifampicin in a few cases. Tumor-inducing effects and chromosome aberrations have not been noted with rifampicin or during the combined use of isoniazid and rifampicin (SEDA-9, 276).

A flu-like illness, with fever, headache, malaise, and bone pain, can occur shortly after the administration of rifampicin, and was observed in a man who had taken rifampicin 600 mg monthly for multibacillary leprosy (11). However, the reaction usually occurs with higher doses given weekly or twice weekly. The usual procedure is to reduce the dose or increase the frequency of treatment. Antipyretic drugs can be used to provide symptomatic relief.

Intermittent rifampicin therapy introduces risks of hematological and renal adverse reactions, probably through immunological mechanisms. Restarting rifampicin after a drug-free interval has to be carefully guided using small initial dosages of about 75 mg/day and increasing to a final dosage of about 500–600 mg/day. It is essential to monitor blood counts, coagulation factors, and kidney function. The authors of a comprehensive review of the adverse effects of rifampicin (12) have made the point that these are likely to increase, because reactions such as hemolysis, thrombocytopenia, and flu-like syndromes are more likely in patients who take rifampicin intermittently, a pattern of treatment that is becoming recognized as the best way to manage tuberculosis in developing countries and in patients everywhere whose compliance cannot be relied upon, since it is impractical to administer directly observed therapy (DOTS) more often than two or three times a week.

Some drug interactions with rifampicin have been reviewed (SEDA-21, 313; SEDA-24, 354).

Rifabutin

Rifabutin (rINN), a spiropiperidyl derivative of rifamycin, is more effective in vitro against *M. tuberculosis* than rifampicin. It has a longer half-life, better tissue penetration, and causes less enzyme induction than rifampicin. Rifabutin is well absorbed from the gastrointestinal tract and reaches a peak serum concentration of about 0.5 µg/ml 4 hours after a single dose of 300 mg.

A 300 mg dose of rifabutin is usually well tolerated. Adverse effects include neutropenia, thrombocytopenia, rash, and gastrointestinal disturbances (nausea, flatulence). Myositis (13) and uveitis (14) are rarely observed. The drug-induced lupus-like syndrome has been linked in a few cases with rifampicin and rifabutin.

In a multicenter study from the National Institute of Allergy and Infectious Diseases in the USA, azithromycin (600 mg/day) plus rifabutin (300 mg/day) was poorly tolerated by 31 patients with or without HIV infection (15). Gastrointestinal symptoms and neutropenia were the major adverse effects. There was no significant pharmacokinetic interaction between the two drugs.

Rifamycin SV

Rifamycin SV (rINN), a semisynthetic macrocyclic antibiotic derived from natural rifamycin B, has been used in the therapy of tuberculosis and in some European countries as a topical antibiotic. Anaphylaxis has been reported after systemic administration, and rarely after topical application.

Organs and Systems

Cardiovascular

Shock and a flu-like illness (fever, chills, and myalgia) have been observed most often in patients taking intermittent rifampicin, dosages over 1000 mg/day, or on restarting treatment (1,16). Shock and cerebral infarction have been reported in an HIV-positive patient after re-exposure to rifampicin (17).

Local thrombophlebitis can occur during prolonged intravenous administration (18).

The incidence of venous thromboembolic complications (deep vein thrombosis and pulmonary embolism) in patients being treated for tuberculosis has been reported (19). In all, 1237 patients (mean age 44 years, 66% men, 41% immigrants, 75% new cases, 92% pulmonary cases, and 16% multidrug resistant cases) were followed up for a mean of 49 days after diagnosis and the immediate starting of an appropriate antituberculosis regimen. Five patients (0.4%) developed a proximal deep vein thrombosis after a mean interval of 20 days, complicated in two cases by pulmonary embolism. There were two other cases of pulmonary embolism without deep vein thrombosis in the first week of treatment. All cases of venous thromboembolism occurred among new cases of pulmonary tuberculosis, with rifampicin as part of the initial standardized treatment regimen. All except one case of venous thromboembolism occurred while the patients were in hospital in the absence of venous thromboprophylaxis. The presence of a hypercoagulable state among patients with tuberculosis was postulated as a consequence of a raised plasma fibrinogen, reactive thrombocytosis, direct endothelial damage promoted by the tubercle bacillus, and the use of rifampicin.

Respiratory

Respiratory symptoms from rifampicin are very rare. They can be part of a flu-like illness with bronchial obstruction (20,21).

Rifampicin-induced pneumonitis is rare.

- An 81-year-old man with smear- and culture-positive pulmonary tuberculosis developed clinical and radiological features of localized interstitial pneumonitis 1 week after starting to take rifampicin, isoniazid, and ethambutol (22). The bronchoalveolar lavage fluid contained 83% lymphocytes with a CD4/CD8 ratio of 10.5. Antituberculosis treatment was withheld and he was treated with methylprednisolone for 3 days because of progressive respiratory failure. A drug lymphocyte stimulation test showed a high stimulation index with rifampicin (370%). He was subsequently treated with streptomycin instead of rifampicin. Re-challenge with rifampicin was not undertaken.

Sudden clinical and radiological worsening during treatment for pulmonary tuberculosis may be due to bronchogenic spread of infection, immune reconstitution producing a paradoxical reaction, drug-induced hypersensitivity pneumonitis, or other unrelated causes, such as pulmonary embolism. Allergic pneumonitis with rifampicin is most unusual, as is the high CD4/CD8 ratio observed in the bronchoalveolar lavage fluid.

Nervous system

Rifampicin is widely used in tuberculous and meningococcal meningitis, since it passes into the cerebrospinal fluid (23). Rifampicin-induced neurological effects include drowsiness, headache, dizziness, ataxia, generalized numbness, pain in the extremities, muscular weakness, confusion, inability to concentrate, delusions, disorientation, hallucinations, and agitation (1,24).

Sensory systems

Drug-induced uveitis is rare. Antibiotics that have been implicated include rifabutin and sulfonamides. Furthermore, nearly all antibiotics injected intracamerally have been reported to produce uveitis (25).

In one study, the most important adverse effect of rifabutin was uveitis, which occurred in 24 of 63 patients taking rifabutin 600 mg/day and three of 53 patients taking 300 mg/day (26). No patients taking quadruple therapy developed uveitis (27). Initially, uveitis was thought to occur only with dosages of rifabutin over 1200 mg/day. However, a review of 54 cases has shown that it can occur at dosages of 300–600 mg/day and is more likely to occur in patients with a low body mass (28). The patients presented with uveitis 2 weeks to 7 months after starting therapy. In all cases they were also taking fluconazole and clarithromycin, drugs that inhibit cytochrome P450 drug metabolism, leading to increased blood concentrations of rifabutin. Rifabutin is less likely to cause uveitis when it is used in combination with azithromycin 500 mg/day than with clarithromycin. If rifabutin is combined with clarithromycin 1 g/day the dose of rifabutin should be limited to 300 mg/day—advice which has been endorsed by the UK Medicines Control Agency (29).

Topical administration of a glucocorticoid and a cycloplegic drug (such as atropine) is suitable as initial treatment. Withdrawal of causative drugs is not always necessary (30).

Endocrine

In patients taking glucocorticoids for Addison's disease, rifampicin may necessitate an increase in glucocorticoid dosage. Thus, incipient adrenal insufficiency can be unmasked by rifampicin (SEDA-13, 261). The phenomenon is due to liver enzyme induction (31).

A significantly raised concentration of TSH during therapy with rifampicin has been reported in a man taking levothyroxine; TSH concentrations returned to baseline 9 days after withdrawal of rifampicin (32).

Rifampicin-induced hypothyroidism has been reported in three euthyroid patients (33).

- A 62-year-old man with recurrent non-Hodgkin's lymphoma developed pulmonary tuberculosis, for which he received rifampicin. Within 2 weeks, his thyrotropin (TSH) concentration increased to 170 mU/l and the serum concentrations of thyroxine (T_4) and triiodothyronine (T_3) fell to 24 µg/l and 180 ng/l respectively. He was given thyroxine. After the course of rifampicin therapy had been completed, thyroxine was withdrawn and he remained euthyroid for 4 years.
- A 66-year-old woman with tuberculous peritonitis was given rifampicin and developed hypothyroidism (thyrotropin concentration 12.5 mU/l, T_4 48 µg/l, T_3 8.7 ng/l). She was given thyroxine for 3 months. Hypothyroidism developed again, and thyroxine was resumed for the duration of the course of rifampicin therapy and then withdrawn, after which she remained euthyroid for 42 months.
- A 56-year-old woman with liver abscesses and tuberculous lymphadenitis was given rifampicin and 2 weeks later developed a raised thyrotropin concentration of 21 mU/ml, for which she was given thyroxine. The hypothyroidism resolved on withdrawal of rifampicin. However on re-starting rifampicin she developed hypothyroidism within 4 weeks. She was again given thyroxine, which was withdrawn on completion of the course of rifampicin. She remained euthyroid for 12 months.

Hypothyroidism developed within 2 weeks of rifampicin therapy in these patients and resolved when it was withdrawn. Rifampicin increases thyroxine clearance, possibly by enhancing hepatic thyroxine metabolism and the biliary excretion of iodothyronine conjugates. In healthy volunteers rifampicin reduces circulating thyroid hormone concentrations without affecting thyrotropin, suggesting that rifampicin directly reduces thyroid hormone concentrations.

Metabolism

The combination of rifampicin and isoniazid reduces serum concentrations of 25-hydroxycholecalciferol. Rifampicin acts by induction of an enzyme that promotes conversion of 25-hydroxycholecalciferol to an inactive metabolite, and isoniazid acts by inhibiting 25-hydroxylation and 1-hydroxylation (SEDA-14, 258). Children or pregnant women with tuberculosis have increased calcium requirements independent of rifampicin administration (34). In 132 children of Afro-Asian origin there was a significant increase in serum alkaline phosphatase activity. This was more pronounced in patients taking both isoniazid and rifampicin than with isoniazid alone (35). The rise in alkaline phosphatase could reflect an effect on either liver or bone. The possibility of a link between this effect and osteomalacia is unclear.

Rifampicin-induced porphyria cutanea tarda has been described in one case, combined with altered liver function (36).

Hematologic

Hemolysis (37,38), agranulocytosis (39), leukopenia (40), and thrombocytopenia (41) have been reported in patients taking rifampicin (SED-10, 578) and constitute contraindications to continuation of therapy.

Hemorrhagic states have been induced by rifampicin in pregnant women and their offspring because of drug-induced hepatic breakdown of vitamin K (42).

Disseminated intravascular coagulation has been attributed to rifampicin (43).

- A 46-year-old woman died of severe disseminated intravascular coagulation after her third monthly dose of rifampicin, given with daily dapsone for the treatment of leprosy (44).

Erythrocytes

Isolated cases of massive hemolysis, with or without renal insufficiency, have been observed in patients taking rifampicin (45–47). Whereas rifampicin given repeatedly and in the usual dose is quite safe, it can cause intravascular hemolysis when given intermittently (48).

Antibodies to rifampicin have been found in several studies, with a positive Coombs' test in the presence of the drug (SEDA-5, 291), more often of the immunocomplex type (49,50) than with IgG or IgM antibodies (46,47).

- In one case, hemolysis started after the second dose of rifampicin, and the patient's blood contained rifampicin-dependent IgG and IgM antibodies, which caused erythrocyte lysis through an interaction with the I antigen on the erythrocyte surface (38). This antigen is also expressed on renal tubular epithelium and the hemolysis was accompanied by acute renal insufficiency.

Intermittent or interrupted treatment appears to predispose to this complication.

Leukocytes

There were four cases of agranulocytosis due to antituberculosis drugs (rifampicin, isoniazid, ethambutol, streptomycin, or pyrazinamide) among about 6400 patients who underwent chemotherapy from 1981 to 2002; the incidence rate of agranulocytosis was estimated at 0.06% (39).

In 140 patients who took Rimapen (Orion, Finland) there were 11 cases of leukopenia, while in 132 patients who took Rimactan (Ciba-Geigy, Switzerland) there was only one case, a statistically significant difference (40).

Platelets

Reports of drug-induced thrombocytopenia have been systematically reviewed (51). Among the 98 different drugs described in 561 articles, the following antibiotics were found with level I (definite) evidence: co-trimoxazole, rifampicin, vancomycin, sulfisoxazole, cefalothin, piperacillin, methicillin, novobiocin. Drugs with level II (probable) evidence were oxytetracycline and ampicillin. There is an increased frequency of thrombocytopenia with intermittent therapy (43).

- A 40-year-old man with multiple anesthetic plaques due to Hansen's multibacillary disease was given WHO multidrug therapy consisting of once-a-month supervised rifampicin 600 mg and clofazimine 300 mg along with unsupervised dapsone 100 mg/day and clofazimine 50 mg/day for 1 year (52). After 2 months he developed multiple ecchymoses and bleeding from the gums. His hemoglobin was 9.2 g/dl, total white cell count 10.5×10^9/l, platelet count 17×10^9/l; the bleeding and clotting times were normal. Dapsone was withdrawn and rifampicin and clofazimine were continued. However, within 24 hours he developed malaise and fatigue, with purpuric spots and ecchymoses all over the body. Although other hematological parameters were normal, the platelet count was 10×10^9/l. Rifampicin was withdrawn and he was given minocycline with clofazimine and dapsone. The platelet count improved in 3 weeks and the purpura and ecchymoses resolved within 1 week.

Rifampicin-induced thrombocytopenia is rare, but even a single monthly supervised dose can cause life-threatening thrombocytopenia.

Immune thrombocytopenia during rifampicin therapy has been attributed to drug-dependent binding of an IgG antibody to platelets; the binding epitope of the antibody was found in the glycoprotein Ib/IX complex (41).

Gastrointestinal

Nausea, vomiting, epigastric pain, diarrhea, loss of appetite, abdominal cramps, and meteorism are often restricted to the beginning of rifampicin therapy. However, gastric burning can oblige some patients to take the drug after meals. Hemorrhage from gastric erosions is a rare complication (53).

- Tablet-associated esophagitis has been reported in a 70-year-old white man on the fourth day of antibiotic therapy with vancomycin, gentamicin, and oral rifampicin for *Staphylococcus epidermidis* prosthetic valve endocarditis (54).

The authors noted that age, bedridden state, gastroesophageal reflux disease, simultaneous administration of several medications, and nasopharyngeal obstruction may have increased the risk of esophagitis. They found a second case of tablet-associated esophagitis caused by rifampicin in their review of the published literature.

Histologically confirmed pseudomembranous colitis has been reported in patients taking rifampicin. Bacteriology showed mainly *Clostridium difficile* resistant to rifampicin and several other antibiotics. Withdrawal of rifampicin and the use of vancomycin has been helpful (SEDA-6, 275) (SEDA-7, 310) (55).

Liver

Rifampicin is rarely used as monotherapy. The risk of hepatotoxicity appears to be very low in patients with normal liver function, especially if rifampicin is given continuously. When given with isoniazid, rifampicin can cause a fulminant liver reaction. This may be attributable to enhancement of isoniazid hepatotoxicity as a result of enzyme induction by rifampicin. In some cases, jaundice occurred within 6–10 days after beginning isoniazid plus rifampicin (56). High serum transaminase activities, disturbances of consciousness, and centrilobular necrosis were found. All the patients recovered.

Rifampicin + pyrazinamide for 2 months (n = 153) has been compared with a 6-month course of isoniazid (n = 199) for latent tuberculosis in HIV-negative contacts of patients with infectious pulmonary tuberculosis (57). Treatment was withdrawn because of hepatotoxicity (transaminases over 5 times the upper limit of normal) in 10% of contacts who took rifampicin + pyrazinamide and in 2.5% of those who took isoniazid. This higher than expected rate of hepatotoxicity led to premature termination of the study. There were no cases of severe or fatal liver injury. Liver function tests normalized after withdrawal of treatment. The authors concluded that the use of rifampicin + pyrazinamide should only be considered when other regimens are unsuitable and that intensive monitoring of liver function is feasible.

Hepatotoxicity of combined therapy for leprosy has been reported in 39 patients treated with dapsone, protionamide, and rifampicin. There were similar findings in 50 patients treated with dapsone, clofazimine, rifampicin, and protionamide. Deaths probably related to the drugs occurred in both groups after 3–4 months of treatment (58). The drug responsible for liver injury may have been protionamide, although rifampicin administered simultaneously could also have contributed (58).

Among 50 000 patients treated prophylactically with rifampicin, there were 16 deaths associated with jaundice (0.03%) (SED-10, 578) (59,60).

Raised transaminases in children commonly cause withdrawal of therapy (61).

Transaminase activities and other liver function tests should be measured weekly in cases with liver dysfunction and every 4 weeks in patients with no known liver disease.

Severe itching is often a distressing symptom in patients with primary biliary cirrhosis and in cholestatic jaundice due to other causes. Rifampicin has been recommended for controlling this symptom, even though it is known to be hepatotoxic. Rifampicin-induced hepatitis has been reported in three of 41 patients with primary biliary cirrhosis (7.3%; 95% CI = 2.5, 19) (62). This risk is greater than the risk of hepatitis during rifampicin monotherapy for latent tuberculosis. Pre-existing liver disease is a recognized risk factor for drug-induced hepatitis, and these patients need to be monitored carefully during rifampicin therapy.

Biliary tract

Total serum bile acid concentrations were raised in 72% of 61 patients treated with rifampicin and isoniazid; in some patients, the concentrations were as much as 40 times above normal, but in only four was the serum bilirubin raised (63).

Urinary tract

Acute renal insufficiency is a rare life-threatening complication of rifampicin. The Tuberculosis Research Centre in India has treated more than 8000 patients with pulmonary and extrapulmonary tuberculosis, including three who developed rifampicin-induced acute renal insufficiency (64).

- A 25-year-old man with sputum-positive pulmonary tuberculosis was given a daily rifampicin-containing regimen and responded with negative sputum smears and cultures by the end of treatment. However, he relapsed after 5 months. Antituberculosis treatment was re-started with thrice-weekly isoniazid 600 mg, rifampicin 450 mg, ethambutol 1200 mg, and pyrazinamide 1500 mg. After 20 days, he complained of vomiting, anorexia, fever, and oliguria. His blood urea concentration was 62 mmol/l and serum creatinine 1529 µmol/l. Abdomen ultrasonography showed normal kidneys with increased cortical echoes. He was treated with three sessions of peritoneal dialysis, along with salt, protein, and fluid restriction. He refused renal biopsy. His renal function improved and after 4 weeks his blood urea was 15 mmol/l and serum creatinine 88 µmol/l. He was then successfully treated with isoniazid 300 mg/day, ethambutol 800 mg/day, and pyrazinamide 1500 mg/day
- A 14-year-old boy with a brain tuberculoma was treated successfully with a rifampicin-containing daily regimen. He developed sputum-positive pulmonary tuberculosis 12 years later and was given thrice-weekly isoniazid 600 mg, rifampicin 450 mg, ethambutol 1200 mg and pyrazinamide 1500 mg. After 10 days he developed low back pain, anorexia, and vomiting. His blood urea concentration was 27 mmol/l and serum creatinine 356 µmol/l. Renal biopsy suggested immune complex deposition in the interstitium and blood

vessels. Rifampicin was withdrawn, and isoniazid, ethambutol, and pyrazinamide were continued. Renal insufficiency was managed with peritoneal dialysis and salt, protein, and fluid restriction. His renal function returned to normal after 5 days (blood urea 9 mmol/l and serum creatinine 88 µmol/l).
- A 25-year-old man had taken irregular treatment with a rifampicin-containing regimen for 5 months about 1 year before being treated with thrice-weekly isoniazid 600 mg, rifampicin 450 mg, ethambutol 1200 mg, and pyrazinamide 1500 mg. After 10 days he complained of oliguria, facial puffiness, pedal edema, and vomiting. His blood urea concentration was 168 mg/dl and serum creatinine 1132 µmol/l. He refused renal biopsy. Rifampicin was withdrawn. With peritoneal dialysis and supportive measures, his renal function returned to normal in 1 month (blood urea 14 mmol/l and serum creatinine 88 µmol/l).

All three patients reported here had previously taken daily rifampicin-containing regimens. They developed acute renal insufficiency after re-treatment with an intermittent rifampicin regimen after 5 months to 11 years, which manifested within 10–20 days of starting re-treatment. Although rifampicin-dependent antibody titers in these patients were not measured, the sequence of events was highly suggestive of rifampicin-induced acute renal insufficiency. On withdrawal of rifampicin and dialysis and supportive care renal function returned to normal.

The mechanism postulated for immune-induced rifampicin nephrotoxicity is that antibodies accumulate during the antigen-free interval when there is a gap in treatment or during an intermittent dosage regimen. When rifampicin is re-administered there is an intense immune reaction. Immune complexes are deposited in the blood vessels or interstitium and cause glomerular endotheliosis, leading to tubular injury, thereby impairing renal function.

Skin

Pruritus, rashes, and urticaria have been reported. Rifampicin-induced rashes often resolve spontaneously, even without drug withdrawal, although in others they can be severe and can be accompanied by systemic symptoms, in some cases amounting to anaphylaxis (12).

Acne occurs more often in patients taking rifampicin and isoniazid combinations than those without rifampicin (SEDA-9, 270).

- A patient developed pemphigus foliaceus induced by rifampicin and improved after withdrawal (78).
- Another patient developed an exacerbation of pemphigus vulgaris during rifampicin therapy, which improved after withdrawal (79).

Severe bullous reactions have been associated with subepidermal detachment typical of toxic epidermal necrolysis (80).

After intravenous rifampicin, a severe cutaneous hypersensitivity reaction has been described, similar to the pattern known as the red man syndrome seen after rapid infusion of vancomycin. The reaction responded to a histamine H_1 receptor antagonist (81).

Rifampicin can cause a fixed drug eruption (82).

Rifampicin-induced generalized exanthematous pustulosis (AGEP) has been reported (83).

- A 79-year-old woman with meticillin-sensitive *Staphylococcus aureus* endocarditis was given vancomycin and rifampicin. Within 48 hours she developed a severe generalized erythematous maculopapular rash, which worsened over the next 3 days. The rash was pruritic and non-tender and Nikolsky's sign was negative. A biopsy of one of the pustules showed histological features consistent with acute generalized exanthematous pustulosis. Given the intensity of the rash she was given a 5-day course of oral steroids (prednisolone 50 mg/day for 2days then 25mg/day). Over the next 7 days the rash resolved. She was given linezolid and successfully completed treatment.

AGEP is usually caused by systemic medications with a typical onset within 48 hours of the start of treatment and less than 24 hour in 50% of cases. The most commonly implicated drugs are antibiotics, particularly penicillins and macrolides. On withdrawal of the offending drug, the rash typically resolves within 2 weeks without further complications.

Musculoskeletal

Of 26 patients who received rifabutin 600 mg/day in combination with ethambutol, streptomycin, and either clarithromycin (500 mg bd; $n - 15$) or azithromycin (600 mg/day; $n = 11$), there were rifabutin-related adverse events in 20; these included a diffuse polyarthralgia syndrome in 5 (84).

Immunologic

There is no evidence that rifampicin causes clinically significant deleterious effects on the immune system in humans (85), whereas it can cause immunosuppression in animals (86). Rifampicin partially suppresses cutaneous hypersensitivity to tuberculin and T cell function (87). In 33 patients with leprosy treated with a rifampicin drug combination, a flu-like illness or antibodies to rifampicin-conjugated proteins were not observed (88).

A possible explanation of the association of allergic reactions with intermittent therapy is that during daily regimens the antigen–antibody complexes are continuously cleared from the plasma without reaching a critical concentration, whereas in intermittent regimens, antibody titers can increase markedly during the drug-free days. This is supported by the observation that anti-rifampicin antibodies, measured by the indirect Coombs' test, developed more commonly during intermittent than during daily therapy and that antibodies may disappear from the serum when patients change from intermittent to daily regimens.

Three patients taking rifampicin developed immediate urticarial reactions (89). Only intradermal tests at a dilution of at least 1:10 000 (concentration of rifampicin approximately 6 mg/l) gave true positive results and in vitro tests (IgE, LTT, and CAST) did not correctly identify hypersensitive patients. All three patients were successfully desensitized with rifampicin using to a 7-day protocol. Rifampicin immediate hypersensitivity, which is rare, can be diagnosed by intradermal skin tests. In vitro tests did not contribute to the diagnosis and so an IgE-mediated mechanism remains to be proven.

Severe anaphylaxis has been reported in two patients with infected wounds that had been treated with topical rifamycin for several months (90). There was urticaria, angioedema, and hypotension in one case, and urticaria, wheezing, dyspnea, and hypotensive shock in the other. In both cases, prick tests with 10% rifamycin solution were positive, while there were no positive reactions in 20 controls.

- A 36-year-old woman developed generalized urticaria during a second course of treatment with rifamycin eye-drops within a month and a 49-year-old man had systemic urticaria, bronchospasm, and hypotension shortly after his surgical wound had been washed with a solution of rifamycin (91). Both patients had positive skin prick tests to rifamycin (1 mg/ml) when tested several weeks after the acute episode, while 10 healthy volunteers had negative tests. The woman also had a positive skin prick test to rifampicin 2 mg/ml, although she had never taken it before.

An HIV-infected patient who developed an anaphylactic reaction to rifampicin tolerated treatment with rifabutin without any adverse event.

The lupus-like syndrome has been reported in seven patients, six of them women who were taking rifampicin ($n = 4$) or rifabutin ($n = 3$) in standard dosages for mycobacterial infections (92). None was HIV-1 positive, none was also taking isoniazid, and although they were taking other antimycobacterial drugs, their symptoms disappeared after withdrawal of the rifamycin alone. All had two or more episodes of fever, malaise, myalgia, and arthralgia, and all had positive antinuclear antibodies. All were also taking either ciprofloxacin or clarithromycin, and the authors speculated that these drugs, which are cytochrome P_{450} enzyme inhibitors, could have increased the serum concentrations of the rifamycins.

Desensitization protocols can be helpful in patients who have had anaphylactic reactions, and the detection of IgE antibodies to rifampicin may be helpful in clarifying pathogenesis. A switch to a daily regimen, when administration was previously intermittent, may allow resumption of rifampicin without further problems. In 35 HIV-positive patients with previous allergic reactions to rifampicin, oral desensitization was safe and allowed the reintroduction of rifampicin in 60% of cases (93). However, the flu-like syndrome, hemolytic and thrombocytopenic crises, and acute renal insufficiency are not IgE-mediated, and when rifampicin is thought to be indispensable, a course of treatment may be completed under glucocorticoid cover. Four patients with reactions to rifampicin, one with rash, fever, and lymphadenopathy and one with hepatitis, completed courses of antituberculosis therapy for nervous system infections under glucocorticoid cover (94).

Second-Generation Effects

Pregnancy

Rifampicin is currently recommended by the WHO for the treatment of tuberculosis during pregnancy. On the other hand, drug companies advise against the use of rifampicin during the first 3 months of pregnancy, even though deleterious effects on the fetus have not been confirmed in man.

Induction of hepatic microsomal enzymes by rifampicin is believed to be the cause of vitamin K deficiency in pregnancy, leading to hemorrhagic disturbances in pregnant women and their neonates. Prophylactic vitamin K should therefore be given to all mothers and their offspring when the mother has taken rifampicin during late pregnancy. Blood coagulation tests should be done on both (SEDA-8, 288).

Teratogenicity

Rifampicin crosses the placenta, and although teratogenicity is uncertain, it may not be as innocuous during pregnancy as isoniazid or ethambutol (SEDA-6, 277) (95).

Lactation

Lactation is not a contraindication to rifampicin; only small amounts pass into the milk with no relevant effects on the newborn (96).

Susceptibility Factors

Renal disease

Impaired renal function has to be taken into consideration when calculating the dosage of rifampicin (97).

Hepatic disease

Pre-existing liver damage has to be taken into consideration when calculating the dosage of rifampicin (98).

HIV/AIDS

There are conflicting reports about the absorption and systemic availability of antituberculosis drugs in patients with HIV/AIDS. The effects of sex and AIDS status on the absorption of rifampicin have been investigated in 10 men and women in San Francisco with and without AIDS (99). The mean CD4 count in men and women with AIDS was 265 x 10^6/l, and all were taking antiretroviral drugs. Rifampicin was measured in plasma, epithelial lining fluid, and alveolar cells after 5 days of treatment with rifampicin 600 mg/day. Plasma and alveolar cell concentrations were not significantly different and both were greater than concentrations in the epithelial lining fluid and were adequate to inhibit *Mycobacterium tuberculosis*. There were no differences in plasma rifampicin concentrations between those with and without AIDS. The authors concluded that the absorption of oral rifampicin was not affected by sex or AIDS.

However, in a study in Chennai there was significant malabsorption of rifampicin (reduced C_{max} and AUC and increased clearance), isoniazid, and other antituberculosis drugs in patients with advanced HIV disease both with and without tuberculosis (mean CD4 cell counts 60 and 98 x 10^6/l respectively) (100). The systemic availability of isoniazid was affected more in rapid acetylators. This study was performed in patients with more advanced disease, who had diarrhea and cryptosporidial infection, and who were not taking antiretroviral drugs. The difference between studies could have been due to differences in patient profiles, associated infections, stage of the disease, acetylator profile, and other genetic differences. Monitoring the plasma concentrations of antituberculosis drugs should therefore be considered for patients with advanced HIV disease who have suboptimal response to antituberculosis treatment. Further studies are required to assess whether increasing the dosages of antituberculosis drugs can help overcome the effect of malabsorption and to correlate plasma drug concentrations with treatment outcome and the emergence of mycobacterial drug resistance in HIV co-infected patients with tuberculosis.

Alcoholism

Chronic alcoholism is an important risk factor for adverse effects of the rifamycins. In 79 consecutive patients taking rifampicin in combination with isoniazid and another drug, there was a high incidence of acute clinical liver disease; about half of the patients were advanced alcoholics and almost all the cases of hepatitis came from this group (101). Most of those with pretreatment abnormalities of liver function had abnormalities in liver biopsies, not attributable to alcohol. In one patient, active chronic hepatitis was attributed to rifampicin.

However, in 531 eligible patients enrolled in a US Public Health Service Cooperative Trial of Short-Course Chemotherapy of Pulmonary Tuberculosis, of whom 58% were classified as alcoholic, although the alcoholics had more abnormal concentrations of aspartate transaminase before and during therapy, there was no significant difference between the alcoholics and non-alcoholics in the incidence of adverse reactions, including hepatotoxic reactions (102). The authors concluded that in the absence of clinically significant and persistent pretreatment abnormalities of hepatic function tests, rifampicin and isoniazid are not contraindicated in patients categorized as alcoholic.

Vitamin D and calcium disorders

In patients with suspected disorders of vitamin D and calcium metabolism, caution has to be taken (SEDA-9, 269).

Drug Administration

Drug overdose

Severe overdosage has been observed in inadvertent administration of an excessive dose of rifampicin in children (103) and also in suicide attempts (104,105). It can also occur when there is impaired hepatic function or severe renal insufficiency. Symptoms are nausea, vomiting, headache, abdominal pain, diarrhea, and pruritus.

Drug–Drug Interactions

Because it induces the cytochrome P450 enzyme system in the liver and intestine, rifampicin can produce many clinically important drug interactions.

Amiodarone

Rifampicin has been reported to reduce the effects of amiodarone (106).

- A 33-year-old woman was given rifampicin to suppress an MRSA infection of a pacing system that could not be removed. She was already taking amiodarone which, with the pacing system, was intended to manage her complex dysrhythmias. The introduction of antibiotic therapy was followed by an increase in bouts of palpitation and in shocks from her defibrillator. Her amiodarone concentrations had fallen and returned to the target range, with disappearance of her symptoms, when the rifampicin was withdrawn.

The authors discussed the possible reasons for this interaction, including a reduction in systemic availability of amiodarone or induction of metabolism by rifampicin.

Amprenavir

Co-administration of amprenavir with rifampicin and rifabutin should be avoided (135).

Antacids

Antacids, particularly aluminium hydroxide gel, magnesium trisilicate, and sodium bicarbonate, reduce the systemic availability of rifampicin (107).

Antidiabetic drugs

The effects of rifampicin on the pharmacokinetics of gliclazide (108) and nateglinide (109) have been investigated in two different studies in healthy subjects. Rifampicin reduced the plasma concentration of both drugs; the mean AUC for gliclazide was reduced by 70% and for nateglinide by 24%. Rifampicin may reduce the blood glucose-lowering effect of these drugs.

Antiretroviral drugs

The recommendations of the Center for Disease Control and Prevention in Atlanta regarding the use of antituberculosis drugs in combination with antiretroviral drugs have been published (110). In general, the use of rifampicin in patients taking protease inhibitors is contraindicated, except in the following circumstances:

- patients taking the NNRTI efavirenz and two NRTIs;
- patients taking the protease inhibitor ritonavir and one or more NRTIs;
- patients taking a combination of the two protease inhibitors ritonavir and saquinavir.

However, Roche Pharma have recommended that rifampicin should not be used in patients who are taking ritonavir + saquinavir, because of the results of a study in which 11 of 28 patients who took rifampicin 300 mg/day with ritonavir 100 mg bd + saquiniavir 1000 mg bd developed severe hepatotoxicity [http://www.fda.gov/medwatch/safety/2005/Saquinavir-Rifampin_deardoc_Feb05].

The use of rifampicin with the protease inhibitors indinavir, nelfinavir, and amprenavir is contraindicated. However, these agents can be used with rifabutin after appropriate dosage reduction. Failure to reduce the dosage of rifabutin can result in toxic manifestations, such arthralgia and uveitis.

Rifamycins can be used with the NNRTIs nevirapine or efavirenz, but not with delavirdine.

Data on drug pharmacokinetics and drug interactions in patients taking treatment for HIV infection and tuberculosis are scanty, and the current recommendations are almost certain to be modified in the near future. Furthermore, it may be prudent to monitor rifampicin concentrations in the event of intolerance or adverse drug reactions.

Atazanavir + ritonavir

Rifampicin is contraindicated in combination with most HIV protease inhibitors. The pharmacokinetics of atazanavir in three regimens with atazanavir, ritonavir, and rifampicin has been evaluated in 71 healthy subjects, 53 of whom took atazanavir 400 mg od on days 1–6 followed by atazanavir 300 mg and ritonavir 100 mg on days 7–16 (n –52) (111). These subjects were then randomized to one of three regimens on days 17–26: atazanavir 300 + ritonavir 100 + rifampicin 600 (n =17), atazanavir 300 + ritonavir 200 + rifampicin 600 (n =17), or atazanavir 400 + ritonavir 200 + rifampicin 600 od (n =14) (all in mg/day). Another 18 subjects underwent procedures identical to those who received study drug on days 1–16, except that they did not receive the study drug or have a pharmacokinetic evaluation on days 1–16. This group took rifampicin 600 alone on days 17–26 and underwent both trough and serial (day 26) pharmacokinetic sampling. With atazanavir 400 + ritonavir 200 + rifampicin 600, the atazanavir AUCs were comparable, but the C_{min} values were lower than with atazanavir 400 alone. Atazanavir exposure was substantially reduced with the other rifampicin-containing regimens relative to atazanavir 400 alone and with all regimens relative to atazanavir 300 + ritonavir 100. Rifampicin and des-rifampicin exposures were 1.6–2.5 times higher than with rifampicin 600 alone. The authors concluded that atazanavir and rifampicin should not be co-administered in the doses studied, although atazanavir with rifampicin was safe.

Atenolol

The effect of rifampicin on the pharmacokinetics of atenolol has been studied in healthy volunteers (112). Rifampicin reduced the mean AUC of atenolol to 81% and increased renal clearance to 109%. Rifampicin pretreatment reduced the peak plasma concentration (C_{max}), AUC_{0-33h}, and the amount of atenolol excreted to 85%,

81%, and 86% of the respective placebo values. The average heart rate and diastolic blood pressure were slightly higher after rifampicin than after placebo. Thus, although the inducing effect of rifampicin may not have been at its maximum by day 6, it has only a minor effect on the pharmacokinetics of atenolol, evidenced by a slight reduction in its systemic availability.

Azithromycin

An interaction involving azithromycin with rifabutin, and less commonly rifampicin, was observed in patients with MAC infections (136).

Barbiturates

Barbiturates, which induce liver drug-metabolizing enzymes, increase the rate of metabolism of rifampicin (113).

Cardiac glycosides

In eight healthy volunteers rifampicin reduced digoxin plasma concentrations (114), mirroring clinical experience. The authors hypothesized that this effect may be brought about by induction of P-glycoprotein, which increases excretion of digoxin into the gut lumen; however, renal digoxin clearance, which is also mediated by P-glycoprotein, was not affected. In these healthy volunteers, rifampicin reduced digoxin concentrations by about 50%, so the interaction is likely to be clinically important.

Digitoxin plasma concentrations can be reduced by rifampicin, and the dosage should be adjusted according to the serum concentration (SEDA-10, 272).

Ciclosporin

Increased dosages of ciclosporin are recommended when patients take rifampicin (SEDA-21, 314). In one study it was necessary to increase the dosage of ciclosporin from 225 to 800 mg/day during treatment with rifampicin and for 7 days afterwards in order to maintain therapeutic concentrations (115). Low ciclosporin blood concentrations and acute graft rejection have been observed in a renal transplant recipient during prophylactic rifampicin therapy (116).

Ciprofloxacin

Rifampicin-induced lupus-like syndrome is associated with combination therapy with ciprofloxacin, since rifampicin is metabolized by (among others) CYP3A4, which is inhibited by ciprofloxacin, and combined usage may lead to higher rifampicin blood concentrations (137).

Clarithromycin

Severe interactions have been observed when rifabutin and clarithromycin were given simultaneously (117). The mean concentrations of rifabutin and 25-*O*-desacetyl-rifabutin in healthy subjects who took clarithromycin and rifabutin concomitantly were respectively more than 4 times and 37 times greater than the concentrations

recorded when rifabutin was administered alone. Neutropenia was detected in 14 of 18 subjects taking rifabutin. Myalgia and high fever were also common. In another study, clarithromycin increased the AUC of rifabutin by 76% (118). Physicians should be aware that recommended prophylactic doses of rifabutin can be associated with severe neutropenia within 2 weeks after the start of therapy, and all patients taking rifabutin, especially with clarithromycin, should be monitored carefully for neutropenia.

Clarithromycin is one of the core drugs for MAC infections in both HIV-infected and non-infected patients. For this indication, doses of up to 2000 mg/day are used, typically in combination with other drugs.

The interaction of clarithromycin with the rifamycins is complex. Clarithromycin inhibits CYP3A4, while both rifampicin and rifabutin induce P450 cytochromes, including CYP3A4, resulting in enhanced metabolism of drugs. The changes in serum concentrations of clarithromycin and its metabolite in the presence of the enzyme inducers rifampicin and rifabutin suggest that metabolism of clarithromycin by CYP3A4 is increased (138).

After the addition of rifampicin, peak serum concentrations of clarithromycin fell markedly, from a mean of 5.8–2.5 µg/ml (138). At the same time the ratio of the serum concentrations of clarithromycin and its 14-OH metabolite was reversed from 3.3:1 to 1:2.7. There were similar, although less marked, changes after the addition of rifabutin 600 mg/day to a regimen that included clarithromycin 1000 mg/day.

Whether these changes in serum clarithromycin concentrations are relevant to its antimicrobial activity is unknown, since prediction of clinical efficacy based on serum concentrations of clarithromycin is probably not justified, given that the macrolides accumulate to a large degree in tissues and macrophages.

In patients with MAC infections taking rifabutin or rifampicin the addition of clarithromycin resulted in rifamycin-related adverse events in 77% of patients (139). These included uveitis (64–86), especially at rifabutin doses of 600 mg/day or more, neutropenia, nausea, vomiting, diarrhea, and abnormal liver enzyme activities. In addition, diffuse polyarthralgia (19%) was observed. Since inhibition of cytochrome P450 by clarithromycin can interfere with rifabutin metabolism, as illustrated by a report of a significant increase in the AUC of rifabutin during treatment with clarithromycin (139), the authors recommended using rifabutin in a dosage of 300 mg/day in regimens that include a macrolide.

The effects of fluconazole and clarithromycin on the pharmacokinetics of rifabutin and 25-*O*-desacetylrifabutin have been studied in ten HIV-infected patients who were given rifabutin 300 mg qds in addition to fluconazole 200 mg qds and clarithromycin 500 mg qds (144). There was a 76% increase in the plasma AUC of rifabutin when either fluconazole or clarithromycin was given alone and a 152% increase when both drugs were given together. The authors concluded that patients should be monitored for adverse effects of rifabutin when it is co-administered with fluconazole or clarithromycin.

Clofibrate

The interaction of rifampicin with clofibrate has been reviewed (119). Rifampicin induces the metabolism of clofibrate, and in five subjects rifampicin 600 mg/day caused a 35% fall in steady-state serum clofibrate concentrations (120).

Clonidine

Rifampicin 600 mg/day had no effect on the pharmacokinetics of clonidine 0.4 mg/day in six healthy subjects (119,121).

Clozapine

Rifampicin can reduce plasma concentrations of clozapine and exacerbate psychotic symptoms (121).

Contrast media

Excretion of contrast media through the bile can be impaired by rifampicin (1,77).

Corticosteroids

Rifampicin enhances the catabolism of many glucocorticoids. For example, it increases the plasma clearance of prednisolone by 45% and may reduce the amount of drug available to the tissues by as much as 66% (SEDA-8, 288). Glucocorticoid therapy for concomitant diseases should therefore be adjusted in the light of plasma concentrations and clinical effects during rifampicin treatment (SEDA-10, 272). For example, rejection of a kidney transplant occurred after 7 weeks of rifampicin therapy (123).

Co-trimoxazole

In patients taking co-trimoxazole, rifampicin in a standard oral antituberculosis regimen (600 mg/day plus isoniazid 300 mg/day and pyrazinamide 30 mg/kg for more than 12 days) reduced the steady-state AUC of trimethoprim and sulfamethoxazole by 47 and 23% respectively (124). The same authors had previously reported reduced efficacy of co-trimoxazole in the prevention of toxoplasmosis in HIV-infected subjects (125). The reduction in prophylactic efficiency was more pronounced in subjects who took a single double-strength tablet.

There have been no reports on the risk of *Pneumocystis jiroveci* pneumonia in patients taking rifampicin plus co-trimoxazole prophylaxis. Until such time as more data are available, it is prudent to use double-strength co-trimoxazole tablets twice daily for prophylaxis of toxoplasmosis and *P. jiroveci* pneumonia in patients taking concomitant rifampicin.

Coumarin anticoagulants

Coumarin anticoagulants undergo accelerated metabolism in patients taking rifampicin (126). Rifampicin greatly increases the clearance of warfarin by inducing hepatic enzymes, and difficulty in achieving an optimal INR despite an increase in the dose of warfarin has been highlighted (127). In patients taking rifampicin, warfarin should be started in dosages of 20–30 mg/day for two days, subsequently adjusting the dose according to the prothrombin time; a daily maintenance dose of 20 mg may be required (128) and close monitoring is required during the initial weeks to achieve a therapeutic INR (129). Offset of the interaction when rifampicin is withdrawn can take as long as 4–5 weeks, during which time the dose of warfarin should be reduced gradually to avoid a high INR and bleeding complications.

Dapsone

The interaction of rifampicin with dapsone has been reviewed (106). In seven patients with leprosy, rifampicin shortened the half-life of dapsone by 50% (130).

Dapsone and analogues

When rifampicin was given, dapsone blood concentrations were lowered and urinary excretion was increased during the first 2 days; however, blood concentrations remained in the therapeutic range (145).

Diazepam

The interaction of rifampicin with diazepam has been reviewed (106). The mean half-life of diazepam fell from 58 to 14 hours in seven patients who took isoniazid, rifampicin, and ethambutol (131).

Efavirenz

Rifampicin induces the metabolism of efavirenz. In 24 patients (21 men, 3 women; mean age 37 years) with HIV infection and tuberculosis the C_{max}, C_{min}, and AUC of efavirenz interval fell by 24%, 25%, and 22% respectively in the presence of rifampicin (146).

Ethionamide and protionamide

When thionamides are used in combination with rifampicin, hepatotoxicity is more common and severe (147,148). There was a 13% incidence of hepatotoxicity in patients with multibacillary leprosy treated with dapsone, rifampicin, and protionamide 10 mg/kg/day, and a 17% incidence in 110 patients treated with dapsone, rifampicin, and protionamide 5 mg/kg/day; however, although the lower dose of protionamide did not reduce the incidence of hepatotoxicity, it did reduce its severity (149). Protionamide does not affect the pharmacokinetics of rifampicin (150).

Fluconazole

Fluconazole increased the AUC of rifabutin by 76% (105).

The combination of rifampicin with fluconazole has insignificant effects (SED-12, 682) (151,152,153).

Haloperidol

Blood concentrations of haloperidol are reduced during rifampicin administration, owing to shortening of the half-life of haloperidol (SEDA-12, 258) (132).

Hormonal contraceptives—oral

Increased intermenstrual or breakthrough bleeding and pregnancy have been reported in women taking rifampicin in conjunction with contraceptive steroids (154). There is evidence that rifampicin increases the rate of metabolism of both the estrogenic and progestogenic components of oral contraceptives through hepatic microsomal enzyme induction (155) involving the same CYP isozyme as that induced by anticonvulsants, CYP3A. A four-fold increase in the rate of steroid metabolism has been shown; it is therefore unwise for women taking rifampicin to rely on steroid contraception and an alternative method should be used.

Itraconazole

Rifampicin 600 mg/day for 14 days had a very strong inducing effect on the metabolism of a single dose of itraconazole 200 mg, indicating that these two drugs should not be used concomitantly (156).

Ketoconazole

Ketoconazole interacts with rifampicin, and the serum concentrations of both drugs are reduced (SEDA-9, 269).

Serum concentrations of ketoconazole are reduced by concomitant use of drugs that induce hepatic microsomal enzymes, such as rifampicin. There may at the same time be a change in rifampicin serum concentration (SED-12, 678) (157,158,159).

Lopinavir and ritonavir

In a pilot, non-randomized study, HIV-infected patients with tuberculosis were treated with regimens containing rifampicin and ritonavir (160). Despite the effects on CYP3A4, there was no significant interaction; plasma ritonavir concentrations remained sufficiently high and rifampicin concentrations did not rise to the toxic range.

Methadone

Methadone metabolism is increased and a withdrawal syndrome has been reported (1).

Mexiletine

The interaction of rifampicin with mexiletine has been reviewed (106). The mean half-life of a single dose of mexiletine 400 mg fell from 8.5 to 5 hours in eight healthy subjects who took rifampicin 600 mg/day for 10 days (133).

Mycophenolate mofetil

In a prospective, open, non-randomized, controlled study of the interaction of rifampicin with mycophenolate mofetil in eight stable renal allograft recipients total MPA $AUC_{0\rightarrow12}$ fell by 18% after rifampicin co-administration, partly attributable to induction of MPA glucuronidation (134). This was mainly because of a 33% reduction in total MPA $AUC_{6\rightarrow12}$, attributable to reduced enterohepatic recirculation. Free MPA $AUC_{6\rightarrow12}$ fell by 22%. Total MPAG and AcMPAG $AUC_{0\rightarrow12}$ increased by 34 and 193% respectively attributable to altered in MRP2-mediated transport of MPAG and AcMPAG.

In a heart–lung transplant recipient long-term rifampicin caused a more than two-fold reduction in dose-corrected mycophenolic acid exposure; subsequent withdrawal of rifampicin resulted in reversal of these changes after 2 weeks of washout (135). The effect of rifampicin on the metabolism of mycophenolate mofetil may be explained by simultaneous induction of renal, hepatic, and gastrointestinal uridine diphosphate-glucuronosyl transferases and organic anion transporters, with subsequent inhibition of enterohepatic recirculation.

Nevirapine

In 140 HIV-infected Thai patients randomized nevirapine 400 mg/day as part of highly active antiretroviral therapy with or without rifampicin, mean plasma nevirapine concentrations at weeks 8 and 12 were 5.40 mg/l and 6.56 mg/l respectively, attributable to enzyme induction by rifampicin (136). However, there was no significant difference in CD4+ cell count at 24 weeks, plasma alanine transaminase activity at 8 and 12 weeks, or the incidences of nevirapine-associated rashes (7.0% versus 8.6%).

The effects of rifampicin 450/600 mg/day for 1 week on the steady-state pharmacokinetics of nevirapine 200 mg bd have been studied in 13 HIV-infected patients (137). Rifampicin caused a 42% reduction in C_{max}, a 53% reduction in C_{min}, and a 46% reduction in AUC of nevirapine. Increasing the dose of nevirapine to 300 mg/day in seven patients increased the C_{min} to within the target range without exceeding the toxic concentration of 12 µg/ml.

Nifedipine

The elimination of nifedipine is enhanced by rifampicin (SEDA-14, 259).

Oral contraceptives

The serum concentrations of ethinylestradiol and similar steroids used as contraceptives are reduced during rifampicin therapy; it is therefore not surprising that unexpected pregnancies have occurred (138). The effects of rifampicin have been compared with those of rifabutin (increasingly used for the treatment of atypical mycobacterial infections) in a double-blind, crossover study in 12 women who were taking a stable oral contraceptive regimen containing ethinylestradiol and norethindrone (139). Rifampicin had the greater effect in reducing the mean AUC, but none of the subjects ovulated during the cycle in which either of the rifamycins was administered.

Oral hypoglycemic agents

Oral hypoglycemic agents (glibenclamide, gliclazide, repaglinide, and rosiglitazone) undergo accelerated metabolism in patients taking rifampicin (113).

The addition of rifampicin to glimepiride (metabolized by CYP2C9) produced only a modest reduction in AUC and no significant effect on blood glucose (140).

Phenazone

Rifampicin significantly reduces the half-life of phenazone (antipyrine) (141).

Probenecid

Probenecid can increase the serum concentration of rifampicin; this can reduce costs and hepatotoxicity in long-term therapy (142). Interactions of rifampicin with probenecid have been reviewed (106).

Propranolol

The metabolism of propranolol is enhanced by rifampicin (SEDA-9, 269).

Quinidine

Quinidine plasma concentrations can be lowered by rifampicin because of enzyme induction (143).

Quinine

The effect of adding rifampicin to quinine has been assessed in adults with uncomplicated falciparum malaria in Thailand (144). Although parasite clearance times were shorted in patients who took quinine + rifampicin (mean 70 versus 82 hours), recrudescence rates were five times higher than those obtained with quinine alone (15/23, 65% versus 3/25, 12%). Rifampicin increases the metabolic clearance of quinine and reduces cure rates. Rifampicin should not be combined with quinine and doses of quinine should probably be increased in patients who are already taking rifampicin.

Repaglinide

Co-administration of rifampicin with repaglinide considerably lowers the concentration of repaglinide and alters its therapeutic effect in diabetes mellitus (145). In nine healthy volunteers, the maximum reduction in blood glucose after a single 0.5 mg dose of repaglinide fell from 1.6 to 1.0 mmol/l after pretreatment with rifampicin for 5 days. This presumably occurred by induction of CYP3A4.

Ritonavir/saquinavir

The efficacy and safety of concomitant use of rifampicin and regimens containing ritonavir/saquinavir (400/400 mg bd) have been studied in HIV-positive patients with tuberculosis (146). Of 20 patients 15 withdrew mainly because of adverse reactions. Therapeutic concentrations of the drugs were achieved in the other five patients, with reduction of viral load

Saquinavir

In 14 healthy men, rifampicin 600 mg/day reduced the AUC of saquinavir (given as a soft gel capsule 1200 mg tds) by 46% when both were given for 14 days; this

reduction can be counteracted by the addition of ritonavir (161).

Simvastatin

Rifampicin greatly reduced the plasma concentrations of simvastatin and simvastatin acid in 10 healthy volunteers in a randomized, crossover study (162). Because the half-life of simvastatin was not affected by rifampicin, induction of CYP3A4-mediated first-pass metabolism of simvastatin in the intestine and liver probably explains this interaction.

Statins

In healthy subjects taking rifampicin the C_{max} and AUC of simvastatin were greatly reduced (147), presumably by enzyme induction. It is likely that other HMG-CoA reductase inhibitors, including lovastatin and atorvastatin, also have clinically significant interactions with rifampicin.

Tacrolimus

An interaction of rifampicin with tacrolimus has been documented (148,149).

- A 61-year-old Chinese renal transplant recipient, who had taken tacrolimus for 1 year after an episode of rejection had failed to respond to ciclosporin, took rifampicin and 12 days later had an episode of biopsy-proven graft rejection, associated with very low serum tacrolimus concentrations.
- A 50-year-old woman, a renal transplant recipient, developed a brain abscess due to *Nocardia otitidiscaviarum* after craniotomy and was given meropenem and rifampicin. The dose of tacrolimus had to be increased three-fold to maintain adequate trough concentrations.

Tenofovir disoproxil fumarate

Rifampicin has major pharmacokinetic interactions with HIV protease inhibitors and non-nucleoside reverse transcriptase inhibitors (NNRTI), which complicates the management of those who are co-infected with HIV and tuberculosis. However, a pharmacokinetic study in 24 healthy subjects has shown no interaction of tenofovir 300 mg, an NNRTI, and rifampicin 600 mg (150). The 95% confidence intervals for AUC and C_{min} were 0.84–0.92 and 0.80–0.91 respectively, while for C_{max} the confidence interval was 0.78–0.90, suggesting pharmacokinetic equivalence when tenofovir was given with or without rifampicin. One patient had a grade 3 rise in hepatic enzymes, which led to withdrawal of therapy.

Terbinafine

Rifampicin 600 mg/day reduced terbinafine concentrations by about 50% by enzyme induction (163).

Theophylline

Rifampicin increases the metabolic clearance of theophylline by 45%, especially after intravenous use (151,152).

Thyroxine

Induction of cytochrome P_{450} may be the explanation for a modest increase in the clearance of levothyroxine (10–15%) which may necessitate an increase in dosage. In two cases, rifampicin led to significant increases in TSH concentrations in patients being treated for hypothyroidism (32).

Trimethoprim

The interaction of rifampicin with trimethoprim has been reviewed (106). Rifampicin has a small effect on the clearance of trimethoprim but it is not clinically significant (153,154).

Trimethoprim and co-trimoxazole

In 10 HIV-positive patients who had been taking one double-strength tablet of co-trimoxazole daily for more than 1 month, the concentrations of trimethoprim and sulfamethoxazole in serum were significantly reduced after the administration of rifampicin (164).

Zidovudine

Rifampicin, a well-known enzyme inducer, increased the metabolism of zidovudine, and the effect persisted for 2 weeks after rifampicin had been withdrawn (165).

Interference with Diagnostic Tests

von Jaksch test

Rifampicin metabolites in urine can cause a false positive test for melanin when assessed with the von Jaksch test (155).

References

1. Mandell GL, Sande MA. Antimicrobial agents: drugs used in the chemotherapy of tuberculosis and leprosy. In: Goodman Gilman A, Rall TW, Nies AS, Taylor P, editors. 8th ed.Goodman and Gilman's The Pharmacological Basis of Therapeutics . New York: Pergamon Press, 1990:1146 Chapter 49 Rothfield NF, Bierer WF, Garfield JW. Isoniazid induction of antinuclear antibodies. A prospective study. Ann Intern Med 1978;88(5):650–2.
2. Darouiche RO, Raad II, Heard SO, Thornby JI, Wenker OC, Gabrielli A, Berg J, Khardori N, Hanna H, Hachem R, Harris RL, Mayhall GCatheter Study Group. A comparison of two antimicrobial-impregnated central venous catheters. N Engl J Med 1999;340(1):1–8.
3. Wenzel RP, Edmond MB. The evolving technology of venous access. N Engl J Med 1999;340(1):48–50.
4. Hand WL, Boozer RM, King-Thompson NL. Antibiotic uptake by alveolar macrophages of smokers. Antimicrob Agents Chemother 1985;27(1):42–5.
5. Newton RW, Forrest AR. Rifampicin overdosage—"the red man syndrome". Scott Med J 1975;20(2):55–6.
6. Immanuel C, Jayasankar K, Narayana AS, Santha T, Sundaram V, Sarma GR. Induction of rifampicin metabolism during treatment of tuberculous patients with daily and fully intermittent regimens containing the drug. Indian J Chest Dis Allied Sci 1989;31(4):251–7.
7. Gao XF, Wang L, Liu GJ, Wen J, Sun X, Xie Y, Li YP. Rifampicin plus pyrazinamide versus isoniazid for treating latent tuberculosis infection: a meta-analysis. Int J Tuberc Lung Dis 2006;10:1080–90.
8. Wurtz RM, Abrams D, Becker S, Jacobson MA, Mass MM, Marks SH. Anaphylactoid drug reactions to ciprofloxacin and rifampicin in HIV-infected patients. Lancet 1989;1(8644):955–6.
9. Pujet JC, Homberg JC, Decroix G. Sensitivity to rifampicin: incidence, mechanism, and prevention. BMJ 1974;2(916):415–8.
10. Stevens E, Bloemmen F, Mbuyi JM, Gyselen A, Mattson K, Hellström H, Riska N. Aspects immunologiques des reactions secondaires a la rifampicine. [Immunological aspects of reactions caused by rifampicin (proceedings).] Bull Int Union Tuberc 1979;54(2):179–80.
11. Vaz M, Jacob AJ, Rajendran A. "Flu" syndrome on once monthly rifampicin: a case report. Lepr Rev 1989;60(4):300–2.
12. Martinez E, Collazos J, Mayo J. Hypersensitivity reactions to rifampin. Pathogenetic mechanisms, clinical manifestations, management strategies, and review of the anaphylactic-like reactions. Medicine (Baltimore) 1999;78(6):361–9.
13. Masur H. Recommendations on prophylaxis and therapy for disseminated Mycobacterium avium complex disease in patients infected with the human immunodeficiency virus. Public Health Service Task Force on Prophylaxis and Therapy for Mycobacterium avium Complex. N Engl J Med 1993;329(12):898–904.
14. Havlir D, Torriani F, Dube M. Uveitis associated with rifabutin prophylaxis. Ann Intern Med 1994;121(7):510–2.
15. Hafner R, Bethel J, Standiford HC, Follansbee S, Cohn DL, Polk RE, Mole L, Raasch R, Kumar P, Mushatt D, Drusano GDATRI 001B Study Group. Tolerance and pharmacokinetic interactions of rifabutin and azithromycin. Antimicrob Agents Chemother 2001;45(5):1572–7.
16. Ramachandran A, Bhatia VN. Rifampicin induced shock—a case report. Indian J Lepr 1990;62(2):228–9.
17. Martinez E, Collazos J, Mayo J. Shock and cerebral infarct after rifampin re-exposure in a patient infected with human immunodeficiency virus. Clin Infect Dis 1998;27(5):1329–30.
18. Kissling M, Xilinas M. Rimactan parenteral formulation in clinical use. J Int Med Res 1981;9(6):459–69.
19. Ambrosetti M, Ferrarese M, Codecasa LR, Besozzi G, Sarassi P, Viggiani P, Migliori GB. Incidence of venous thromboembolism in tuberculosis patients. Respiration 2006;73:396.
20. Singapore Tuberculosis Service and British Medical Research Council. Controlled trial of intermittent regimens of rifampicin plus isoniazid for pulmonary tuberculosis in Singapore. Lancet 1975;2(7945):1105–9.
21. Anastasatu C, Bungeteanu G, Sibila S. The intermittent chemotherapy of tuberculosis with rifampicin regimens on ambulatory basis. Scand J Respir Dis Suppl 1973;84:136–9.
22. Kunichika N, Miyahara N, Kotani K, Takeyama H, Harada M, Tanimoto M. Pneumonitis induced by rifampicin. Thorax 2002;57(11):1000–1.
23. Artaza A, Gallofre M, Arboix M, Laporte JR. Niveles de la rifampicina en liquido cefalorraquideo en cuadros de inflammacion meningea. [Rifampicin levels in cerebrospinal fluid in meningeal inflammation.] Arch Farmacol Toxicol 1983;9(1):121–4.

24. Pratt TH. Rifampin-induced organic brain syndrome. JAMA 1979;241(22):2421–2.

25. Moorthy RS, Valluri S, Jampol LM. Drug-induced uveitis. Surv Ophthalmol 1998;42(6):557–70.

26. Kelleher P, Helbert M, Sweeney J, Anderson J, Parkin J, Pinching A. Uveitis associated with rifabutin and macrolide therapy for Mycobacterium avium intracellulare infection in AIDS patients. Genitourin Med 1996;72(6):419–21.

27. Shafran SD, Singer J, Zarowny DP, Phillips P, Salit I, Walmsley SL, Fong IW, Gill MJ, Rachlis AR, Lalonde RG, Fanning MM, Tsoukas CMCanadian HIV Trials Network Protocol 010 Study Group. A comparison of two regimens for the treatment of Mycobacterium avium complex bacteremia in AIDS: rifabutin, ethambutol, and clarithromycin versus rifampin, ethambutol, clofazimine, and ciprofloxacin. N Engl J Med 1996;335(6):377–83.

28. Tseng AL, Walmsley SL. Rifabutin-associated uveitis. Ann Pharmacother 1995;29(11):1149–55.

29. Committee on Safety of Medicines and the Medicines Control Agency. Revised indication and drug interaction of rifabutin. Curr Prob Pharmacovig 1997;23:14.

30. Anonymous. Drug-induced uveitis can usually be easily managed. Drugs Ther Perspect 1998;11:11–4.

31. Kyriazopoulou V, Parparousi O, Vagenakis AG. Rifampicin-induced adrenal crisis in Addisonian patients receiving corticosteroid replacement therapy. J Clin Endocrinol Metab 1984;59(6):1204–6.

32. Nolan SR, Self TH, Norwood JM. Interaction between rifampin and levothyroxine. South Med J 1999;92(5):529–31.

33. Nobuyuki Takasu, Masaki Takara, Ichiro Komiya. Rifampicin induced hypothyroidism in patients with Hashimoto's thyroiditis. N Engl J Med 2005;352(5):518–9.

34. Williams SE, Wardman AG, Taylor GA, Peacock M, Cooke NJ. Long term study of the effect of rifampicin and isoniazid on vitamin D metabolism. Tubercle 1985;66(1):49–54.

35. Toppet M, Vainsel M, Cantraine F, Franckson M. Evolution de la phosphatase alcaline sérique sous traitement d'isoniazide et de rifampicine. [Course of serum alkaline phosphatase during treatment with isoniazid and rifampicin.] Arch Fr Pediatr 1985;42(2):79–80.

36. Millar JW. Rifampicin-induced porphyria cutanea tarda. Br J Dis Chest 1980;74(4):405–8.

37. Conen D, Blumberg A, Weber S, Schubothe H. Hämolytische Krise und akutes Nierenversagen unter Rifampicin. [Hemolytic crisis and acute kidney failure from rifampicin.] Schweiz Med Wochenschr 1979;109(15):558–62.

38. De Vriese AS, Robbrecht DL, Vanholder RC, Vogelaers DP, Lameire NH. Rifampicin-associated acute renal failure: pathophysiologic, immunologic, and clinical features. Am J Kidney Dis 1998;31(1):108–15.

39. Shishido Y, Nagayama N, Masuda K, Baba M, Tamura A, Nagai H, Akagawa S, Kawabe Y, Machida K, Kurashima A, Komatsu H, Yotsumoto H. [Agranulocytosis due to anti-tuberculosis drugs including isoniazid (INH) and rifampicin (RFP)—a report of four cases and review of the literature.]Kekkaku 2003;78(11):683–9.

40. van Assendelft AH. Leucopenia caused by two rifampicin preparations. Eur J Respir Dis 1984;65(4):251–8.

41. Pereira J, Hidalgo P, Ocqueteau M, Blacutt M, Marchesse M, Nien Y, Letelier L, Mezzano D. Glycoprotein Ib/IX complex is the target in rifampicin-induced immune thrombocytopenia. Br J Haematol 2000;110(4):907–10.

42. Chouraqui JP, Bessard G, Favier M, Kolodie L, Rambaud P. Hémorragie par avitaminose K chez la femme enceinte et le nouveau-né. [Haemorrhage associated with vitamin K deficiency in pregnant women and newborns. Relationship with rifampicin therapy in two cases.] Therapie 1982;37(4):447–50.

43. Ip M, Cheng KP, Cheung WC. Disseminated intravascular coagulopathy associated with rifampicin. Tubercle 1991;72(4):291–3.

44. Souza CS, Alberto FL, Foss NT. Disseminated intravascular coagulopathy as an adverse reaction to intermittent rifampin schedule in the treatment of leprosy. Int J Lepr Other Mycobact Dis 1997;65(3):366–71.

45. van Assendelft AH. Leucopenia in rifampicin chemotherapy. J Antimicrob Chemother 1985;16(3):407–8.

46. Diamond JR, Tahan SR. IgG-mediated intravascular hemolysis and nonoliguric acute renal failure complicating discontinuous rifampicin administration. Nephron 1984;38(1):62–4.

47. Tahan SR, Diamond JR, Blank JM, Horan RF. Acute hemolysis and renal failure with rifampicin-dependent antibodies after discontinuous administration. Transfusion 1985;25(2):124–7.

48. Criel A, Verwilghen RL. Intravascular haemolysis and renal failure caused by intermittent rifampicin treatment. Blut 1980;40(2):147–50.

49. Hoigne R, Biedermann HP, Naegeli HR. INH-induzierter systemischer Lupus Erythematodes: 2 Beobachtungen mit Reexposition. Schweiz Med Wochenschr 1975;105(50):1726.

50. Rothfield NF, Bierer WF, Garfield JW. Isoniazid induction of antinuclear antibodies. A prospective study. Ann Intern Med 1978;88(5):650–2.

51. George JN, Raskob GE, Shah SR, Rizvi MA, Hamilton SA, Osborne S, Vondracek T. Drug-induced thrombocytopenia: a systematic review of published case reports. Ann Intern Med 1998;129(11):886–90.

52. Sudip D, Kumar RA, Arunasis M. Rifampicin induced thrombocytopenia. Ind J Dermatol 2006;51:222.

53. Zargar SA, Thapa BR, Sahni A, Mehta S. Rifampicin-induced upper gastrointestinal bleeding. Postgrad Med J 1990;66(774):310–1.

54. Smith SJ, Lee AJ, Maddix DS, Chow AW. Pill-induced esophagitis caused by oral rifampin. Ann Pharmacother 1999;33(1):27–31.

55. Moriarty HJ, Scobie BA. Pseudomembranous colitis in a patient on rifampicin and ethambutol. NZ Med J 1980;91(658):294–5.

56. Pessayre D. Present views on isoniazid and isoniazid-rifampicin hepatitis. Agressologie 1982;23(A):13–5.

57. Tortajada C, Martínez-Lacasa J, Sánchez F, Jiménez-Fuentes A, De Souza ML, García JF, Martínez JA, Caylà JA. Is the combination of pyrazinamide plus rifampicin safe for treating latent tuberculosis infection in persons not infected by the human immunodeficiency virus? Int J Tuberc Lung Dis 2005;9:276–81.

58. Ji BH, Chen JK, Wang CM, Xia GA. Hepatotoxicity of combined therapy with rifampicin and daily prothionamide for leprosy. Lepr Rev 1984;55(3):283–9.

59. Mandell GL, Sande MA. Antimicrobial agents: drugs used in the chemotherapy of tuberculosis and leprosy. In: Goodman LS, Gilman A, editors. The Pharmacological Basis of Therapeutics. 6th ed.. New York: Macmillan Publishing Co Inc, 1980:1200.

60. Mitchell I, Wendon J, Fitt S, Williams R. Anti-tuberculous therapy and acute liver failure. Lancet 1995;345(8949):555–6.

61. Linna O, Uhari M. Hepatotoxicity of rifampicin and isoniazid in children treated for tuberculosis. Eur J Pediatr 1980;134(3):227–9.

62. Prince MI, Burt AD, Jones DE. Hepatitis and liver dysfunction with rifampicin therapy for pruritus in primary biliary cirrhosis. Gut 2002;50(3):436–9.

63. Berg JD, Pandov HI, Sammons HG. Serum total bile acid levels in patients receiving rifampicin and isoniazid. Ann Clin Biochem 1984;21(Pt 3):218–22.

64. Banu Rekha VV, Santha T, Jawahar MS. Rifampicin induced renal toxicity in the re-treatment of patients with pulmonary TB—case report. JAPI 2005;53:811–3.

65. Kar HK, Roy RG. Reversible acute renal failure due to monthly administration of rifampicin in a leprosy patient. Indian J Lepr 1984;56(4):835–9.

66. Mauri JM, Fort J, Bartolome J, Camps J, Capdevila L, Morlans M, Martin-Vega C, Piera L. Antirifampicin antibodies in acute rifampicin-associated renal failure. Nephron 1982;31(2):177–9.

67. Warrington RJ, Hogg GR, Paraskevas F, Tse KS. Insidious rifampin-associated renal failure with light-chain proteinuria. Arch Intern Med 1977;137(7):927–30.

68. Winter RJ, Banks RA, Collins CM, Hoffbrand BI. Rifampicin induced light chain proteinuria and renal failure. Thorax 1984;39(12):952–3.

69. Cohn JR, Fye DL, Sills JM, Francos GC. Rifampicin-induced renal failure. Tubercle 1985;66(4):289–93.

70. Covic A, Goldsmith DJ, Segall L, Stoicescu C, Lungu S, Volovat C, Covic M. Rifampicin-induced acute renal failure: a series of 60 patients. Nephrol Dial Transplant 1998;13(4):924–9.

71. Covic A, Gusbeth-Tatomir P, Tarevici Z, Mihaescu T, Covic M. Insuficientă renală acută post-rifampicină—complicatie redutabilă însă putin cunoscută a tratamentulvi tuberculostatic. [Post-rifampicin acute renal failure—serious, but seldom recognized complication of the anti-tuberculosis treatment.] Pneumologia 2001;50(4):225–31.

72. Munteanu L, Golea O, Nicolicioiu M, Tudorache V. Particularitatile insuficientei renale acute (IRA) la bolnavii tratati cu rifampicină. [Specific features of acute renal failure in patients treated with rifampicin.] Pneumologia 2002;51(1):15–20.

73. Prakash J, Kumar NS, Saxena RK, Verma U. Acute renal failure complicating rifampicin therapy. J Assoc Physicians India 2001;49:877–80.

74. Muthukumar T, Jayakumar M, Fernando EM, Muthusethupathi MA. Acute renal failure due to rifampicin: a study of 25 patients. Am J Kidney Dis 2002;40(4):690–6.

75. Bassilios N, Vantelon C, Baumelou A, Deray G. Continuous rifampicin administration inducing acute renal failure. Nephrol Dial Transplant 2001;16(1):190–1.

76. Ogata H, Kubo M, Tamaki K, Hirakata H, Okuda S, Fujishima M. Crescentic glomerulonephritis due to rifampin treatment in a patient with pulmonary atypical mycobacteriosis. Nephron 1998;78(3):319–22.

77. Yoshioka K, Satake N, Kasamatsu Y, Nakamura Y, Shikata N. Rapidly progressive glomerulonephritis due to rifampicin therapy. Nephron 2002;90(1):116–8.

78. Lee CW, Lim JH, Kang HJ. Pemphigus foliaceus induced by rifampicin. Br J Dermatol 1984;111(5):619–22.

79. Miyagawa S, Yamashina Y, Okuchi T, Konoike Y, Kano T, Sakamoto K. Exacerbation of pemphigus by rifampicin. Br J Dermatol 1986;114(6):729–32.

80. Prazuck T, Fisch A, Simonnet F, Noat G. Lyell's syndrome associated with rifampicin therapy of tuberculosis in an AIDS patient. Scand J Infect Dis 1990;22(5):629.

81. Nahata MC, Fan-Havard P, Barson WJ, Bartkowski HM, Kosnik EJ. Pharmacokinetics, cerebrospinal fluid concentration, and safety of intravenous rifampin in pediatric patients undergoing shunt placements. Eur J Clin Pharmacol 1990;38(5):515–7.

82. John SS. Fixed drug eruption due to rifampin. Lepr Rev 1998;69(4):397–9.

83. Azad A, Connelly N. Case of rifampicin-induced acute generalized exanthematous pustulosis. Int Med J 2006;36:619–20.

84. Griffith DE, Brown BA, Girard WM, Wallace RJ Jr. Adverse events associated with high-dose rifabutin in macrolide-containing regimens for the treatment of Mycobacterium avium complex lung disease. Clin Infect Dis 1995;21(3):594–8.

85. Farr B, Mandell GL. Rifampin. Med Clin North Am 1982;66(1):157–68.

86. Bassi L, Di Berardino L, Arioli V, Silvestri LG, Ligniere EL. Conditions for immunosuppression by rifampicin. J Infect Dis 1973;128(6):736–44.

87. Dickinson JM, Aber VR, Mitchison DA. Bactericidal activity of streptomycin, isoniazid, rifampin, ethambutol, and pyrazinamide alone and in combination against Mycobacterium tuberculosis:. Am Rev Respir Dis 1977;116(4):627–35.

88. Rook GA. Absence from sera from normal individuals or from rifampin-treated leprosy patients (THELEP trials) of antibody to rifamycin-protein or rifamycin-membrane conjugates. Int J Lepr Other Mycobact Dis 1985;53(1):22–7.

89. Buergin S, Scherer K, Hausermann P, Bircher AJ. Immediate hypersensitivity to rifampicin in 3 patients: diagnostic procedures and induction of clinical tolerance. Int Arch Allergy Immunol 2006;140:20–6.

90. Baciewicz AM, Self TH, Bekemeyer WB. Update on rifampin drug interactions. Arch Intern Med 1987;147(3):565–8.

91. Garcia F, Blanco J, Carretero P, Herrero D, Juste S, Garces M, Perez R, Fuentes M. Anaphylactic reactions to topical rifamycin. Allergy 1999;54(5):527–8.

92. Berning SE, Iseman MD. Rifamycin-induced lupus syndrome. Lancet 1997;349(9064):1521–2.

93. Arrizabalaga J, Casas A, Camino X, Iribarren JA, Rodriguez Arrondo F, Von Wichmann MA. Utilidad de la desensibilizacion a rifampicina en el tratamiento de enfermedades producidas por micobacterias en pacientes con SIDA. [The usefulness of the desensitization to rifampin in the treatment of mycobacterial disease in patients with AIDS.] Med Clin (Barc) 1998;111(3):103–4.

94. Morris H, Muckerjee J, Akhtar S, Abdullahi L, Harrison M, Scott G. Use of corticosteroids to suppress drug toxicity in complicated tuberculosis. J Infect 1999;39(3):237–40.

95. Siskind MS, Thienemann D, Kirlin L. Isoniazid-induced neurotoxicity in chronic dialysis patients: report of three cases and a review of the literature. Nephron 1993;64(2):303–6.

96. Paumgartner G. Medikamente in der Muttermilch. Pharma-Kritik (Switzerland) 1979;1:53.

97. Acocella G. Clinical pharmacokinetics of rifampicin. Clin Pharmacokinet 1978;3(2):108–27.

98. Bergamini N, Fowst G, Rifamycin SV. A review. Arzneimittelforschung 1965;15(Suppl 8):951–1002.

99. Conte JE Jr, Golden JA, Kipps JE, Lin ET, Zurlinden E. Effect of sex and AIDS status on the plasma and intrapulmonary pharmacokinetics of rifampicin. Clin Pharmacokinet 2004;43(6):395–404.

100. Gurumurthy P, Ramachandran G, Hemanth Kumar AK, Rajasekaran S, Padmapriyadarsini C, Swaminathan S, Bhagavathy S, Venkatesan P, Sekar L, Mahilmaran A, Ravichandran N, Paramesh P. Decreased bioavailability of rifampicin and other antituberculosis drugs in patients with advanced human immunodeficiency virus disease. Antimicrob Agents Chemother 2004;48:4473–6.

101. Thompson JE. The effect of rifampicin on liver morphology in tuberculous alcoholics. Aust NZ J Med 1976;6(2):111–6.

102. Cross FS, Long MW, Banner AS, Snider DE Jr. Rifampin-isoniazid therapy of alcoholic and nonalcoholic tuberculous patients in a U.S. Public Health Service Cooperative Therapy Trial Am Rev Respir Dis 1980;122(2):349–53.

103. Bolan G, Laurie RE, Broome CV. Red man syndrome: inadvertent administration of an excessive dose of rifampin to children in a day-care center. Pediatrics 1986;77(5):633–5.

104. Broadwell RO, Broadwell SD, Comer PB. Suicide by rifampin overdose. JAMA 1978;240(21):2283–4.

105. Meisel S, Brower R. Rifampin: a suicidal dose. Ann Intern Med 1980;92(2 Pt 1):262–3.

106. Zarembski DG, Fischer SA, Santucci PA, Porter MT, Costanzo MR, Trohman RG. Impact of rifampin on serum amiodarone concentrations in a patient with congenital heart disease. Pharmacotherapy 1999;19(2):249–51.

107. Khalil SAH, El-Khordagui LK, El-Gholmy ZA. Effect of antacids on oral absorption of rifampicin. Int J Pharm 1984;20:99.

108. Park JY, Kim KA, Park PW, Part CW, Shin JG. Effect of rifampin on pharmacokinetics and pharmacodynamics of gliclazide. Clin Pharmacol Ther 2003;74:334–40.

109. Niemi M, Backman JT, Neuvonen M, Neuvonen PJ. Effect of rifampicin on the pharmacokinetics and pharmacodynamics of nateglinide in healthy subjects. J Clin Pharmacol 2003;56:427–32.

110. Centers for Disease Control and Prevention (CDC). Updated guidelines for the use of rifabutin or rifampin for the treatment and prevention of tuberculosis among HIV-infected patients taking protease inhibitors or nonnucleoside reverse transcriptase inhibitors. MMWR Morb Mortal Wkly Rep 2000;49(9):185–9.

111. Burger DM, Agarwala S, Child M, Been-Tiktak A, Wang Y, Bertz R. Effect of rifampin on steady-state pharmacokinetics of atazanavir with ritonavir in healthy volunteers. Antimicrob Agents Chemother 2006;50(10):3336–42.

112. Lilja JJ, Juntti-Patinen L, Neuvonen PJ. Effect of rifampicin on the pharmacokinetics of atenolol. Basic Clin Pharmacol Toxicol 2006;98:555–8.

113. Duroux P. Surveillance et accidents de la chimiothérapie antituberculeuse. [Surveillance and complications of antituberculosis chemotherapy.] Rev Prat 1979;29(33):2681–90.

114. Greiner B, Eichelbaum M, Fritz P, Kreichgauer HP, von Richter O, Zundler J, Kroemer HK. The role of intestinal P-glycoprotein in the interaction of digoxin and rifampin. J Clin Invest 1999;104(2):147–53.

115. Capone D, Aiello C, Santoro GA, Gentile A, Stanziale P, D'Alessandro R, Imperatore P, Basile V. Drug interaction between cyclosporine and two antimicrobial agents, josamycin and rifampicin, in organ-transplanted patients. Int J Clin Pharmacol Res 1996;16(2–3):73–6.

116. Offermann G, Keller F, Molzahn M. Low cyclosporin A blood levels and acute graft rejection in a renal transplant recipient during rifampin treatment. Am J Nephrol 1985;5(5):385–7.

117. Rubinstein E. Comparative safety of the different macrolides. Int J Antimicrob Agents 2001;18(Suppl 1):S71–6.

118. Jordan MK, Polis MA, Kelly G, Narang PK, Masur H, Piscitelli SC. Effects of fluconazole and clarithromycin on rifabutin and 25-O-desacetylrifabutin pharmacokinetics. Antimicrob Agents Chemother 2000;44(8):2170–2.

119. Baciewicz AM, Self TH. Rifampin drug interactions. Arch Intern Med 1984;144(8):1667–71.

120. Houin G, Tillement JP. Clofibrate and enzymatic induction in man. Int J Clin Pharmacol Biopharm 1978;16(4):150–4.

121. Affrime MB, Lowenthal DT, Rufo M. Failure of rifampin to induce the metabolism of clonidine in normal volunteers. Drug Intell Clin Pharm 1981;15(12):964–6.

122. Joos AA, Frank UG, Kaschka WP. Pharmacokinetic interaction of clozapine and rifampicin in a forensic patient with an atypical mycobacterial infection. J Clin Psychopharmacol 1998;18(1):83–5.

123. Pallardo L, Moreno R, Garcia Martinez J, et al. Rechado agudo tardio del injerto renal inducido por rifampicina. Nefrologia 1987;7:93.

124. Ribera E, Pou L, Fernandez-Sola A, Campos F, Lopez RM, Ocana I, Ruiz I, Pahissa A. Rifampin reduces concentrations of trimethoprim and sulfamethoxazole in serum in human immunodeficiency virus-infected patients. Antimicrob Agents Chemother 2001;45(11):3238–41.

125. Ribera E, Fernandez-Sola A, Juste C, Rovira A, Romero FJ, Armadans-Gil L, Ruiz I, Ocana I, Pahissa A. Comparison of high and low doses of trimethoprim-sulfamethoxazole for primary prevention of toxoplasmic encephalitis in human immunodeficiency virus-infected patients. Clin Infect Dis 1999;29(6):1461–6.

126. Held H. Interaktion von Rifampicin mit Phenprocoumon: Beobachtungen bei tuberkulosekranken Patienten. [Interaction of rifampicin with phenprocoumon.] Dtsch Med Wochenschr 1979;104(37):1311–4.

127. Lee CR, Thrasher KA. Difficulties in anticoagulation management during coadministration of warfarin and rifampin. Pharmacotherapy 2001;21(10):1240–6.

128. Casner PR. Inability to attain oral anticoagulation: warfarin-rifampin interaction revisited. South Med J 1996;89(12):1200–3.

129. Cropp JS, Bussey HI. A review of enzyme induction of warfarin metabolism with recommendations for patient management. Pharmacotherapy 1997;17(5):917–28.

130. Krishna DR, Appa Rao AVN, Ramanakar TV, Prabhakar MC. Pharmacokinetic interaction between dapsone and rifampicin (rifampin) in leprosy patients. Drug Dev Ind Pharm 1986;12:443–9.

131. Ochs HR, Greenblatt DJ, Roberts GM, Dengler HJ. Diazepam interaction with antituberculosis drugs. Clin Pharmacol Ther 1981;29(5):671–8.

132. Takeda M, Nishinuma K, Yamashita S, Matsubayashi T, Tanino S, Nishimura T. Serum haloperidol levels of schizophrenics receiving treatment for tuberculosis. Clin Neuropharmacol 1986;9(4):386–97.

133. Pentikainen PJ, Koivula IH, Hiltunen HA. Effect of rifampicin treatment on the kinetics of mexiletine. Eur J Clin Pharmacol 1982;23(3):261–6.

134. Naesens M, Kuypers DRJ, Streit F, Armstrong VW, Oellerich M, Verbeke K, Vanrenterghem Y. Rifampin induces alterations in mycophenolic acid glucuronidation and elimination: implications for drug exposure in renal allograft recipients. Clin Pharmacol Ther 2006;80:509–21.

135. Kuypers DRJ, Verleden G, Naesens M, Vanrenterghem Y. Drug interaction between mycophenolate mofetil and rifampin: possible induction of uridine diphosphate-glucuronosyltransferase. Clin Pharmacol Ther 2005;78:81–8.

136. Manosuthi W, Sungkanuparph S, Thakkinstian A, Rattanasiri S, Chaovavanich A, Prasithsirikul W, Likanonsakul S, Ruxrungtham K. Plasma nevirapine levels and 24-week efficacy in HIV-infected patients receiving nevirapine-based highly active antiretroviral therapy with or without rifampicin. Clin Infect Dis 2006;43:253–5.

137. Ramachandran G, Hemanthkumar AK, Rajasekaran S, Padmapriyadarsini C, Narendran G, Sukumar B, Sathishnarayan S, Raja K, Kumaraswami V, Swaminathan S. Increasing nevirapine dose can overcome reduced bioavailability due to rifampicin coadministration. J Acquir Immune Defic Syndr 2006;42(1):36–41.

138. Skolnick JL, Stoler BS, Katz DB, Anderson WH. Rifampin, oral contraceptives, and pregnancy. JAMA 1976;236(12):1382.

139. Barditch-Crovo P, Trapnell CB, Ette E, Zacur HA, Coresh J, Rocco LE, Hendrix CW, Flexner C. The effects of rifampin and rifabutin on the pharmacokinetics and pharmacodynamics of a combination oral contraceptive. Clin Pharmacol Ther 1999;65(4):428–38.

140. Niemi M, Kivisto KT, Backman JT, Neuvonen PJ. Effect of rifampicin on the pharmacokinetics and pharmacodynamics of glimepiride. Br J Clin Pharmacol 2000;50(6):591–5.

141. Teunissen MW, Bakker W, Meerburg-Van der Torren JE, Breimer DD. Influence of rifampicin treatment on antipyrine clearance and metabolite formation in patients with tuberculosis. Br J Clin Pharmacol 1984;18(5):701–6.

142. Pankaj R, Lal S, Rao RS. Effect of probenecid on serum rifampicin levels. Indian J Lepr 1985;57(2):329–33.

143. Damkier P, Hansen LL, Brosen K. Rifampicin treatment greatly increases the apparent oral clearance of quinidine. Pharmacol Toxicol 1999;85(6):257–62.

144. Pukrittayakamee S, Prakongpan S, Wancoimotruk S, Clemens R, Looareesuwan S, White NK. Adverse effect of rifampicin on quinine efficacy in uncomplicated falciparum malaria. Antimicrob Agents Chemother 2003;47:1509–13.

145. Niemi M, Backman JT, Neuvonen M, Neuvonen PJ, Kivisto KT. Rifampin decreases the plasma concentrations and effects of repaglinide. Clin Pharmacol Ther 2000;68(5):495–500.

146. Rolla VC, da Silva Vieira MA, Pereira Pinto D, Lourenco MC, de Jesus C da S, Goncalves Morgado M, Ferreira Filho M, Werneck-Barroso E. Safety, efficacy and pharmacokinetics of ritonavir 400 mg/saquinavir 400 mg twice daily plus rifampicin combined therapy in HIV patients with tuberculosis. Clin Drug Invest 2006;26:469–79.

147. Kyrklund C, Backman JT, Kivisto KT, Neuvonen M, Laitila J, Neuvonen PJ. Rifampin greatly reduces plasma simvastatin and simvastatin acid concentrations. Clin Pharmacol Ther 2000;68(6):592–7.

148. Chenhsu RY, Loong CC, Chou MH, Lin MF, Yang WC. Renal allograft dysfunction associated with rifampin–tacrolimus interaction. Ann Pharmacother 2000;34(1):27–31.

149. Hartmann A, Halvorsen CE, Jenssen T, Bjorneklett A, Brekke IB, Bakke SJ, Hirschberg H, Tonjum T, Gaustad P. Intracerebral abscess caused by Nocardia otitidiscaviarum in a renal transplant patient—cured by evacuation plus antibiotic therapy. Nephron 2000;86(1):79–83.

150. Droste JAH, Verweij-Van Wissen CPWGM, Kearney BP, Buffels R, VanHorssen PJ, Hekster YA, Burger DM. Pharmacokinetic study of tenofovir disoproxil fumarate combined with rifampin in healthy volunteers. Antimicrob Agents Chemother 2005;49:680–4.

151. Powell-Jackson PR, Jamieson AP, Gray BJ, Moxham J, Williams R. Effect of rifampicin administration on theophylline pharmacokinetics in humans. Am Rev Respir Dis 1985;131(6):939–40.

152. Ahn HC, Yang JH, Lee HB, Rhee YK, Lee YC. Effect of combined therapy of oral anti-tubercular agents on theophylline pharmacokinetics. Int J Tuberc Lung Dis 2000;4(8):784–7.

153. Buniva G, Palminteri R, Berti M. Kinetics of a rifampicin–trimethoprim combination. Int J Clin Pharmacol Biopharm 1979;17(6):256–9.

154. Emmerson AM, Gruneberg RN, Johnson ES. The pharmacokinetics in man of a combination of rifampicin and trimethoprim. J Antimicrob Chemother 1978;4(6):523–31.

155. Altundag MK, Barista I. False-positive urine melanin pigment reaction caused by rifampin. Ann Pharmacother 1998;32(5):610.

Rokitamycin

See also Macrolide antibiotics

General Information

Rokitamycin is a semisynthetic 16-membered ring macrolide. It is more hydrophobic, and has better bacterial uptake and slower release, more cohesive ribosomal binding, and a longer post-antibiotic effect than other 14-, 15- and 16-membered ring macrolides (1).

Organs and Systems

Immunologic

- Churg–Strauss syndrome has been reported in an 18-year-old woman taking cysteinyl leukotriene receptor antagonists and oral rokitamicin 400 mg bd for 10 days (2).

References

1. Braga PC. Rokitamycin: bacterial resistance to a 16-membered ring macrolide differs from that to 14- and 15-membered ring macrolides. J Chemother 2002;14(2):115–31.

2. Richeldi L, Rossi G, Ruggieri MP, Corbetta L, Fabbri LM. Churg–Strauss syndrome in a case of asthma. Allergy 2002;57(7):647–8.

Roxithromycin

See also Macrolide antibiotics

General Information

Roxithromycin is a macrolide antibiotic with actions and uses similar to those of erythromycin.

Observational studies

In 2917 adults, adverse effects of roxithromycin occurred in 4.1% at a dosage of 150 mg/day (1). Nausea (1.3%), abdominal pain (1.2%), and diarrhea (0.8%) were the most frequently reported events, whereas rash, vomiting, headache, dizziness, pruritus, urticaria, and constipation were reported only rarely. Treatment had to be withdrawn in 0.9% of the patients because of adverse effects.

In an uncontrolled study in 24 HIV-infected patients, roxithromycin (300 mg bd for 4 weeks) was effective against cryptosporidial diarrhea (2). The most limiting adverse effects were abdominal pain (two patients), raised hepatic enzymes (two patients), and abdominal pain with raised hepatic enzymes (one patient). Minor symptoms occurred in nine patients.

In 304 infants and children under 14 years adverse effects occurred in 6.9% (3). Treatment was withdrawn in 10 children (two with vomiting, two diarrhea, and six rashes).

In 480 elderly patients (over 65 years), there were adverse events in 3.1% and treatment was withdrawn in 1.9%. The gastrointestinal tract was most often affected, whereas the laboratory changes (increases in bilirubin, aspartate transaminase, alanine transaminase, and alkaline phosphatase) were seen in under 0.7%.

Organs and Systems

Cardiovascular

The dysrhythmogenic effect of macrolide antibiotics is well known.

- A 6-year-old girl with complex cyanotic heart disease developed torsade de pointes after taking roxithromycin 10 mg/kg/day (4).
- An 83-year-old woman developed torsade de pointes after taking oral roxithromycin 300 mg/day for 7 days (5).
- Torsade de pointes has been reported in an 83-year-old man who developed severe prolongation of the QT interval after taking roxithromycin 300 mg/day for 4 days (6).

Respiratory

Eosinophilic pneumonia of acute onset has been attributed to roxithromycin (7).

- A 21-year-old woman developed a fever of 39°C, a generalized pruritic macular rash, odynophagia, and intense weakness, in conjunction with respiratory difficulties 8 days after starting to take roxithromycin 150 mg bd. Her leukocyte count was 15.4 × 10⁹/l with 9.8% eosinophils and an erythrocyte sedimentation rate of 32 mm/hour. Peripheral infiltrates were evident on the chest X-ray and CT scan, with multiple areas of consolidation and an air bronchogram, mainly peripherally distributed. Bronchoalveolar lavage showed 50% eosinophils. She improved with glucocorticoids and 6 months later was free of respiratory symptoms; a chest X-ray was normal.

Roxithromycin can cause a hypersensitivity pneumonitis (8).

Endocrine

Roxithromycin may have anti-androgenic effects. In an in vitro model roxithromycin 5 µg/ml suppressed androgen receptor transcriptional activity by 21% (9).

Hematologic

In one study lymphopenia or eosinophilia were observed in two of 37 patients (10). However, this finding could not be confirmed in other studies.

Gastrointestinal

During long-term use of roxithromycin 300 mg/day for 2–66 months in nine patients with chronic diffuse sclerosing osteomyelitis of the mandible, diarrhea and stomach discomfort occurred in one case and liver dysfunction in another (11).

Liver

- Cholestatic and hepatocellular liver damage occurred in a previously healthy 20-year-old woman after she took roxithromycin 150 mg bd for 4 days (12).
- Hepatic failure occurred in a previously healthy 5-year-old boy after he was given roxithromycin 50 mg bd for 5 days (13). He developed a non-pruritic, non-urticarial, erythematous, maculopapular, generalized rash, and occasional vomiting. Three days later he became jaundiced, and after 8 days underwent liver transplantation.

Pancreas

Acute pancreatitis occurred in a 58-year-old man 2 days after he took roxithromycin 300 mg/day and betamethasone (4 mg/day) (14).

Urinary tract

Roxithromycin can cause severe liver toxicity, but nephrotoxic effects have not been established.

- A 73-year-old woman developed acute renal insufficiency and hepatotoxicity after taking roxithromycin for 2 days (15). Renal and hepatic function returned to normal on the sixth day after roxithromycin withdrawal.

Skin

Allergic dermatitis with characteristic distribution pattern called "baboon syndrome" has been reported (16).

- A 58-year-old man developed a pruritic skin eruption after he had taken roxithromycin 300 mg/day for 3 days. Large erythematous plaques covered his buttocks. Roxithromycin was immediately withdrawn and following treatment with oral antihistamines and topical corticosteroids the rash resolved within a few days.

Urticarial rashes have been attributed to roxithromycin.

- Roxithromycin-induced generalized urticaria and tachycardia with a positive prick test and a cross-reaction to erythromycin and clarithromycin has been reported in a 31-year-old woman (17).
- Angioedema and urticaria occurred in a 22-year-old woman a few hours after a second dose of roxithromycin 150 mg for a sore throat (18). The lesions subsided within 12 hours of drug withdrawal and there was no relapse after 3 months of follow-up. A skin prick test was positive for roxithromycin (1 mg/ml) and negative for erythromycin and clarithromycin in the same concentrations.

Nails

A 22-year-old office worker developed onycholysis after taking roxithromycin 150 mg bd for 8 weeks for chronic sinusitis (19).

Immunologic

Roxithromycin had an immunomodulatory action on peripheral blood mononuclear cells in patients with psoriasis (20). The anti-inflammatory activity of roxithromycin is due to reduced production of proinflammatory mediators, cytokines, and co-stimulatory molecules, as has been shown in animal studies (21).

Second-Generation Effects

Teratogenicity

In a case-control study, roxithromycin was not a major teratogen (22).

Drug–Drug Interactions

Antacids

Roxithromycin did not interact with antacids containing hydroxides of aluminium and magnesium (1).

Carbamazepine

Roxithromycin did not alter the pharmacokinetics of carbamazepine (23) or interact with warfarin, ranitidine, or antacids containing hydroxides of aluminium and magnesium (1).

Cyclophosphamide

Roxithromycin inhibits CYP3A4, which is involved in cyclophosphamide metabolism, and inhibits P-glycoprotein (24).

- A 66-year old man developed hepatic veno-occlusive disease while taking immunosuppressive doses of cyclophosphamide 100 mg/day and roxithromycin 600 mg/day; after all drugs were withdrawn he recovered within 2 weeks (25).

In vitro the combination of cyclophosphamide and roxithromycin, but not the individual compounds, was toxic to endothelial cells by inducing apoptosis.

Hormonal contraceptives—oral

Roxithromycin does not interfere with the pharmacokinetics of oral contraceptives (26).

Macrolide antibiotics

Disopyramide alters the protein binding of roxithromycin, and this results in increased plasma concentrations in vitro (29). However, this effect has not been observed with roxithromycin in vivo.

Ranitidine

Roxithromycin did not interact with ranitidine (1).

Statins

In an open, randomized, crossover study in 12 healthy volunteers, roxithromycin did not alter the pharmacokinetics of lovastatin in such a way that dosage adjustment of lovastatin should be necessary during co-administration (27).

- Myopathy occurred in a 73-year-old woman taking simvastatin 80 mg/day and gemfibrozil 600 mg bd 4 days after she started to take roxithromycin 300 mg/day (28).

Theophylline

Roxithromycin altered the pharmacokinetics of theophylline, mainly increasing the C_{max}, prolonging the half-life, and increasing the renal clearance (23). However, while these changes were statistically significant, they were considered clinically irrelevant. There was no effect of roxithromycin on trough concentrations of theophylline.

Warfarin

Roxithromycin did not interact with warfarin (1).

References

1. Young RA, Gonzalez JP, Sorkin EM. Roxithromycin. A review of its antibacterial activity, pharmacokinetic properties and clinical efficacy. Drugs 1989;37(1):8–41.
2. Sprinz E, Mallman R, Barcellos S, Silbert S, Schestatsky G, Bem David D. AIDS-related cryptosporidial diarrhoea: an open study with roxithromycin. J Antimicrob Chemother 1998;41(Suppl B):85–91.

3. Kafetzis DA, Blanc F. Efficacy and safety of roxithromycin in treating paediatric patients. A European multicentre study. J Antimicrob Chemother 1987;20(Suppl B):171–7.
4. Promphan W, Khongphatthanayothin A, Horchaiprasit K, Benjacholamas V. Roxithromycin induced torsade de pointes in a patient with complex congenital heart disease and complete atrioventricular block. Pacing Clin Electrophysiol 2003;26:1424–6.
5. Justo D, Mardi T, Zeltser D. Roxithromycin-induced torsades de pointes. Eur J Intern Med 2004;15(5):326–327.
6. Haffner S, Lapp H, Thurmann PA. Unerwunschi te Arzneimittel wirkungen Der konkrete Fall. [Adverse drug reactions—case report.] Dtsch Med Wochenschr 200210;127(19):1021.
7. Perez-Castrillon JL, Jimenez-Garcia R, Martin-Escudero JC, Velasco C. Roxithromycin-induced eosinophilic pneumonia. Ann Pharmacother 2002;36(11):1808–9.
8. Chew GY, Hawkins CA, Cherian M, Hurwitz M. Roxithromycin induced hypersensitivity pneumonitis. Pathology 2006;38(5):475–7.
9. Inui S, Nakao T, Itami S. Modulation of androgen receptor transcriptional activity by anti-acne reagents. J Dermatol Sci 2004;36(2):97–101.
10. Agache P, Amblard P, Moulin G, Barriere H, Texier L, Beylot C, Bergoend H. Roxithromycin in skin and soft tissue infections. J Antimicrob Chemother 1987;20(Suppl B):153–6.
11. Yoshii T, Nishimura H, Yoshikawa T, Furudoi S, Yoshioka A, Takenono I, Ohtsuka Y, Komori T. Therapeutic possibilities of long-term roxithromycin treatment for chronic diffuse sclerosing osteomyelitis of the mandible. J Antimicrob Chemother 2001;47(5):631–7.
12. Hartleb M, Biernat L, Kochel A. Drug-induced liver damage—a three-year study of patients from one gastroenterological department. Med Sci Monit 2002;8(4):CR292–6.
13. Easton-Carter KL, Hardikar W, Smith AL. Possible roxithromycin-induced fulminant hepatic failure in a child. Pharmacotherapy 2001;21(7):867–70.
14. Renkes P, Petitpain N, Cosserat F, Bangratz S, Trechot P. Can roxithromycin and betamethasone induce acute pancreatitis? A case report. JOP 2003;4:184–6.
15. Akcay A, Kanbay M, Sezer S, Ozdemir FN. Acute renal failure and hepatotoxicity associated with roxithromycin. Ann Pharmacother 2004;38(4):721–2.
16. Amichai B, Grunwald MH. Baboon syndrome following oral roxithromycin. Clin Exp Dermatol 2002;27(6):523.
17. Kruppa A, Scharffetter-Kochanek K, Krieg T, Hunzelmann N. Immediate reaction to roxithromycin and prick test cross-sensitization to erythromycin and clarithromycin. Dermatology 1998;196(3):335–6.
18. Gurvinder SK, Tham P, Kanwar AJ. Roxithromycin induced acute urticaria. Allergy 2002;57(3):262.
19. Sharma NL, Mahajan VK. Onycholysis: an unusual side effect of roxithromycin. Indian J Dermatol Venereol Leprol 2005;71(1):49–50.
20. Ohshima A, Takigawa M, Tokura Y. CD8+ cell changes in psoriasis associated with roxithromycin-induced clinical improvement. Eur J Dermatol 2001;11(5):410–5.
21. Shimane T, Asano K, Suzuki M, Hisamitsu T, Suzaki H. Influence of a macrolide antibiotic, roxithromycin, on mast cell growth and activation in vitro. Mediators Inflamm 2001;10(6):323–32.
22. Chun JY, Han JY, Ahn HK, Choi JS, Koong MK, Nava-Ocampo AA, Koren G. Fetal outcome following roxithromycin exposure in early pregnancy. J Matern Fetal Neonatal Med 2006;19(3):189–92.
23. Saint-Salvi B, Tremblay D, Surjus A, Lefebvre MA. A study of the interaction of roxithromycin with theophylline and carbamazepine. J Antimicrob Chemother 1987;20(Suppl B):121–9.
24. Kaufmann P, Haschke M, Torok M, Beltinger J, Bogman K, Wenk M, Terracciano L, Krahenbuhl S. Mechanisms of venooclusive disease resulting from the combination of cyclophosphamide and roxithromycin. Ther Drug Monit 2006;28(6):766–74.
25. Beltinger J, Haschke M, Kaufmann P, Michot M, Terracciano L, Krahenbuhl S. Hepatic veno-occlusive disease associated with immunosuppressive cyclophosphamide dosing and roxithromycin. Ann Pharmacother 2006;40(4):767–70.
26. Archer JS, Archer DF. Oral contraceptive efficacy and antibiotic interaction: a myth debunked. J Am Acad Dermatol 2002;46(6):917–23.
27. Bucher M, Mair G, Kees F. Effect of roxithromycin on the pharmacokinetics of lovastatin in volunteers. Eur J Clin Pharmacol 2002;57(11):787–91.
28. Huynh T, Cordato D, Yang F, Choy T, Johnstone K, Bagnall F, Hitchens N, Dunn R. HMG CoA reductase-inhibitor-related myopathy and the influence of drug interactions. Intern Med J 2002;32(9–10):486–90.
29. Archer JS, Archer DF. Oral contraceptive efficacy and antibiotic interaction: a myth debunked. J Am Acad Dermatol 2002;46(6):917–23.

Sitafloxacin

See also Fluoroquinolones

General Information

Sitafloxacin is a quinolone antibiotic that is effective against methicillin-resistant *Staphylococcus aureus*.

Organs and Systems

Liver

In a phase II, open, multicenter, randomized study sitafloxacin 400 mg/day caused mild transient increases in alanine transaminase and alkaline phosphatase in 69 patients but no effects on other enzymes (1).

Skin

Data obtained in albino mice have suggested that the phototoxic potential of sitafloxacin is milder than that of lomefloxacin or sparfloxacin (2).

In a randomized study in 40 Caucasian volunteers 100 mg bd sitafloxacin was associated with a mild degree of phototoxicity (3).

Drug Administration

Drug formulations

After oral administration sitafloxacin 500 mg is rapidly absorbed, with a systemic availability of 89% (4). By 48 hours, about 61% is excreted unchanged in the urine after oral administration and about 75% after intravenous administration. For both routes, the high renal clearance of sitafloxacin implies active tubular secretion.

References

1. Feldman C, White H, O'Grady J, Flitcroft A, Briggs A, Richards G. An open, randomised, multi-centre study comparing the safety and efficacy of sitafloxacin and imipenem/cilastatin in the intravenous treatment of hospitalised patients with pneumonia. Int J Antimicrob Agents 2001;17(3):177–88.
2. Shimoda K, Ikeda T, Okawara S, Kato M. Possible relationship between phototoxicity and photodegradation of sitafloxacin, a quinolone antibacterial agent, in the auricular skin of albino mice. Toxicol Sci 2000;56(2):290–6.
3. Dawe RS, Ibbotson SH, Sanderson JB, Thomson EM, Ferguson J. A randomized controlled trial (volunteer study) of sitafloxacin, enoxacin, levofloxacin and sparfloxacin phototoxicity. Br J Dermatol 2003;149:1232–41.
4. O'Grady J, Briggs A, Atarashi S, Kobayashi H, Smith RL, Ward J, Ward C, Milatovic D. Pharmacokinetics and absolute bioavailability of sitafloxacin, a new fluoroquinolone antibiotic, in healthy male and female Caucasian subjects. Xenobiotica 2001;31(11):811–22.

Sparfloxacin

See also Fluoroquinolones

General Information

Sparfloxacin is a fluoroquinolone with activity against the major respiratory pathogens and atypical pathogens that cause pneumonia. Photosensitivity, nausea, and diarrhea have been the most common adverse events reported in trials, and sparfloxacin is contraindicated in patients with QT interval prolongation (1).

In an in vivo study in rats, endotoxin reduced the biliary excretion of sparfloxacin and its glucuronide, probably owing to impairment of hepatobiliary transport systems and renal handling (2).

Observational studies

The efficacy and safety of sparfloxacin (400 mg loading dose followed by 200 mg/day for 10 days) in the treatment of acute bacterial maxillary sinusitis have been evaluated in 253 patients (3). The overall success rate was 92%. The majority of adverse events were mild or moderate and the most frequent were photosensitivity reactions, headache, nausea, and diarrhea.

Comparative studies

Analysis of safety data from six multicenter phase III trials in 1585 patients comparing sparfloxacin with standard drugs (cefaclor, ciprofloxacin, clarithromycin, erythromycin, and ofloxacin) showed that 25% of patients treated with sparfloxacin and 28% of patients treated with the other drugs had at least one adverse event considered to be related to the study medication (4). Photosensitivity reactions were seen in 7.4% of treatment episodes with sparfloxacin, compared with 0.5% of episodes with the other drugs. Gastrointestinal reactions, insomnia, and alterations in taste were more common in patients taking the comparator drugs. The mean change from baseline QT interval corrected for heart rate (QT_c) was significantly greater with sparfloxacin, but no ventricular dysrhythmias were detected. Study medication was withdrawn in 8.9% of patients taking the comparator drugs and in 6.6% of those taking sparfloxacin.

Organs and Systems

Cardiovascular

Some quinolones can prolong the QT interval, with a risk of cardiac dysrhythmias.

Sparfloxacin causes greater prolongation of the QT interval than other quinolones (5), as has been shown in in vitro comparisons. Compared with grepafloxacin, moxifloxacin, and ciprofloxacin, sparfloxacin caused the greatest prolongation of the action potential duration (6), and in an in vitro comparison of sparfloxacin, ciprofloxacin, gatifloxacin, gemifloxacin, grepafloxacin, levofloxacin, moxifloxacin, sitafloxacin, tosufloxacin, and trovafloxacin, sparfloxacin caused the largest increase in QT interval (7). In an in vivo study in conscious dogs with stable idioventricular automaticity and chronic complete atrioventricular block, oral sparfloxacin 60 mg/kg caused torsade de pointes, leading to ventricular fibrillation within 24 hours, while 6 mg/kg did not (8). In halothane-anesthetized dogs, intravenous sparfloxacin 0.3 mg/kg prolonged the effective refractory period, and an extra 3.0 mg/kg reduced the heart rate and prolonged the effective refractory period and ventricular repolarization phase to a similar extent, suggesting that a backward shift of the relative repolarization period during the cardiac cycle may be the mechanism responsible for the dysrhythmogenic effect of sparfloxacin. Data from in vitro electrophysiological studies of the effect of sparfloxacin, ofloxacin, and levofloxacin on repolarization of rabbit Purkinje fibers indicate that the prolongation of the action potential is observed only with sparfloxacin (9).

In a double-blind, randomized, placebo-controlled, crossover study of a single oral dose of sparfloxacin in 15 healthy volunteers, prolongation of the QT interval was about 4% greater with sparfloxacin than with placebo (10). An independent Safety Board concluded from the results of various phase I and phase II studies that the increase in the QT_c interval associated with sparfloxacin is moderate, averaging 3%, and that the few serious adverse cardiovascular events that have been reported during postmarketing surveillance all occurred in patients with underlying heart disease (11).

In 25 patients taking sparfloxacin for multiresistant tuberculosis, there were six cases of moderate prolongation of the electrocardiographic QT interval (30–40 ms compared to baseline) without clinical symptoms (12).

- A 37-year-old woman who was taking sparfloxacin as part of modified antituberculosis drug therapy developed torsade de pointes (13).

In rabbits intravenous mexiletine 3 mg/kg reduced the electrical vulnerability of the heart during sparfloxacin overdose and may be a pharmacological strategy against the drug-induced long QT syndrome (14). Care should therefore be taken when sparfloxacin is co-prescribed with other drugs that prolong the QT interval (15).

Sensory systems

Eyes
Sparfloxacin can be deposited in the corneae, but there is no effect on eyesight (16).

Taste
Abnormal taste developed in 17 of 175 patients treated with clarithromycin 250 mg bd for 10 days for community-acquired pneumonia, compared with three of 167 patients treated with sparfloxacin (17). Mild to moderate gastrointestinal disturbances were the most common adverse events and were reported in 13 and 11% respectively.

Gastrointestinal

In 16 patients with pulmonary tuberculosis who were treated with sparfloxacin 200 mg/day with isoniazid and para-aminosalicylic acid, one withdrew because of severe gastrointestinal intolerance (18).

Skin

During the first 9 months of marketing of sparfloxacin, 371 severe phototoxic reactions were reported to the French pharmacovigilance system or the manufacturers, reporting rate of 0.4 per thousand treated patients (about 4–25 times that reported with other fluoroquinolones) (19).

Phototoxicity occurred in four of nine patients with multidrug-resistant tuberculosis treated with a combination of sparfloxacin (400 mg/day), ethionamide, and kanamycin (initially for 3–4 months) (20). It occurred after several months of treatment and was presumably due to the sparfloxacin. Sparfloxacin more commonly causes photosensitivity than ciprofloxacin, levofloxacin, or ofloxacin, which are also effective antituberculosis drugs (19,21). Skin reactions can be severe enough to require withdrawal of sparfloxacin in some patients. In view of this, other fluoroquinolones, particularly ofloxacin, are preferable to sparfloxacin in the management of multidrug-resistant tuberculosis.

In 25 patientws taking sparfloxacin for multidrug-resistant tuberculosis, there were five mild phototoxic reactions (12).

Nails

Blue-black discoloration of the nails occurred in three patients taking sparfloxacin; the discoloration gradually moved distally and disappeared within 6 months (22).

During treatment for tuberculosis with sparfloxacin, streptomycin, ethambutol, and pyrazinamide a 36-year-old man developed an exaggerated sunburn-like rash with painful onycholysis of the fingernails and toenails; after withdrawal of sparfloxacin and substitution of rifampicin, the symptoms improved (23).

Long-Term Effects

Mutagenicity

DNA damage produced by sparfloxacin and UVA in retinal pigment epithelial cells in vitro was remedied by antioxidants, suggesting a possible in vivo strategy for preventing or minimizing retinal damage in humans (24).

Susceptibility Factors

Patients with a history of cardiac disease and patients taking other medications with effects on cardiac repolarization should not be given sparfloxacin, in order to prevent any cumulative inhibitory effect on cardiac repolarization.

Drug–Drug Interactions

Antacids

The absorption of sparfloxacin is impaired by antacids containing aluminium or magnesium (25). In an open, single-dose (400 mg), randomized, four-way crossover study in 20 male volunteers (aged 18–38 years), Maalox (30 ml) given 4 hours after sparfloxacin did not cause a statistically significant reduction in the rate and extent of sparfloxacin absorption. In contrast, Maalox given 2 hours before or 2 hours after sparfloxacin did reduce its absorption: AUC fell by 23 and 17% respectively and mean C_{max} by 29 and 13% (26).

In an vitro study of the effect of some antacids on the systemic availability of sparfloxacin, the release of sparfloxacin from tablets in the presence of sodium bicarbonate, calcium hydroxide, calcium carbonate, aluminium hydroxide, magnesium hydroxide, magnesium carbonate, magnesium trisilicate, and magaldrate was markedly reduced (27). However, magaldrate and calcium carbonate had relatively higher adsorption capacities in simulated gastric juice and magnesium trisilicate and calcium hydroxide in simulated intestinal juice.

Cisapride

Concomitant administration of cisapride increased the rate of absorption and increased the peak concentration of sparfloxacin without having a significant effect on systemic availability (28,29).

Sucralfate

Co-administration of sucralfate leads to a 44% reduction in the systemic availability of sparfloxacin (29). In eight healthy Japanese, the absorption of sparfloxacin (300 mg orally) was reduced when sucralfate (1.5 g orally) was administered concurrently or 2 hours after sparfloxacin, but not 4 hours after sparfloxacin (30).

References

1. Schentag JJ. Sparfloxacin: a review. Clin Ther 2000;22(4):372–87.
2. Nadai M, Zhao YL, Wang L, Nishio Y, Takagi K, Kitaichi K, Takagi K, Yoshizumi H, Hasegawa T. Endotoxin impairs biliary transport of sparfloxacin and its glucuronide in rats. Eur J Pharmacol 2001;432(1):99–105.
3. Garrison N, Spector S, Buffington D, Stafford C, Granito K, Zhang H, Talbot GH. Sparfloxacin for the treatment of acute bacterial maxillary sinusitis documented by sinus puncture. Ann Allergy Asthma Immunol 2000;84(1):63–71.
4. Lipsky BA, Dorr MB, Magner DJ, Talbot GH. Safety profile of sparfloxacin, a new fluoroquinolone antibiotic. Clin Ther 1999;21(1):148–59.
5. Stahlmann R. Clinical toxicological aspects of fluoroquinolones. Toxicol Lett 2002;127(1–3):269–77.
6. Patmore L, Fraser S, Mair D, Templeton A. Effects of sparfloxacin, grepafloxacin, moxifloxacin, and ciprofloxacin on cardiac action potential duration. Eur J Pharmacol 2000;406(3):449–52.
7. Hagiwara T, Satoh S, Kasai Y, Takasuna K. A comparative study of the fluoroquinolone antibacterial agents on the action potential duration in guinea pig ventricular myocardia. Jpn J Pharmacol 2001;87(3):231–4.
8. Chiba K, Sugiyama A, Satoh Y, Shiina H, Hashimoto K. Proarrhythmic effects of fluoroquinolone antibacterial agents: in vivo effects as physiologic substrate for torsades. Toxicol Appl Pharmacol 2000;169(1):8–16.
9. Adamantidis MM, Dumotier BM, Caron JF, Bordet R. Sparfloxacin but not levofloxacin or ofloxacin prolongs cardiac repolarization in rabbit Purkinje fibers. Fundam Clin Pharmacol 1998;12(1):70–6.
10. Demolis JL, Charransol A, Funck-Brentano C, Jaillon P. Effects of a single oral dose of sparfloxacin on ventricular repolarization in healthy volunteers. Br J Clin Pharmacol 1996;41(6):499–503.
11. Jaillon P, Morganroth J, Brumpt I, Talbot G. Overview of electrocardiographic and cardiovascular safety data for sparfloxacin. Sparfloxacin Safety Group. J Antimicrob Chemother 1996;37(Suppl A):161–7.
12. Lubasch A, Erbes R, Mauch H, Lode H. Sparfloxacin in the treatment of drug resistant tuberculosis or intolerance of first line therapy. Eur Respir J 2001;17(4):641–6.
13. Kakar A, Byotra SP. Torsade de pointes probably induced by sparfloxacin. J Assoc Physicians India 2002;50:1077–8.
14. Takahara A, Sugiyama A, Satoh Y, Hashimoto K. Effects of mexiletine on the canine model of sparfloxacin-induced long QT syndrome. Eur J Pharmacol 2003;476:115–22.
15. Sable D, Murakawa GJ. Quinolones in dermatology. Dis Mon 2004;50(7):381–94.
16. Gokhale NS. Sparfloxacin corneal deposits. Indian J Ophthalmol 2004;52(1):79.
17. Ramirez J, Unowsky J, Talbot GH, Zhang H, Townsend L. Sparfloxacin versus clarithromycin in the treatment of community-acquired pneumonia. Clin Ther 1999;21(1):103–17.
18. Rao S. Uncontrolled trial of sparfloxacin in retreatment of pulmonary tuberculosis. Respirology 2004;9(3):402–5.
19. Pierfitte C, Royer RJ, Moore N, Begaud B. The link between sunshine and phototoxicity of sparfloxacin. Br J Clin Pharmacol 2000;49(6):609–12.
20. Singla R, Gupta S, Gupta R, Arora VK. Efficacy and safety of sparfloxacin in combination with kanamycin and

21. Stahlmann R, Lode H. Toxicity of quinolones. Drugs 1999;58(Suppl 2):37–42.
22. Mahajan VK, Sharma NL. Photo-onycholysis due to sparfloxacin. Australas J Dermatol 2005;46(2):104–5.
23. Raad I, Darouiche R, Vazquez J, Lentnek A, Hachem R, Hanna H, Goldstein B, Henkel T, Seltzesr E. Efficacy and safety of weekly dalbavancin therapy for catheter-related bloodstream infection caused by Gram-positive pathogens. Clin Infect Dis 2005;40(3):374–80.
24. Verna LK, Holman SA, Lee VC, Hoh J. UVA-induced oxidative damage in retinal pigment epithelial cells after H_2O_2 or sparfloxacin exposure. Cell Biol Toxicol 2000;16(5):303–12.
25. Shiba K, Hori S, Shimada J. Interaction between oral sparfloxacin and antacid in normal volunteers. Eur J Clin Microbiol Infect Dis 1991;(Special issue):583–5.
26. Johnson RD, Dorr MB, Talbot GH, Caille G. Effect of Maalox on the oral absorption of sparfloxacin. Clin Ther 1998;20(6):1149–58.
27. Hussain F, Arayne MS, Sultana N. Interactions between sparfloxacin and antacids— dissolution and adsorption studies. Pak J Pharm Sci 2006;19(1):16–21.
28. Goa KL, Bryson HM, Markham A. Sparfloxacin. A review of its antibacterial activity, pharmacokinetic properties, clinical efficacy and tolerability in lower respiratory tract infections. Drugs 1997;53(4):700–25.
29. Zix JA, Geerdes-Fenge HF, Rau M, Vockler J, Borner K, Koeppe P, Lode H. Pharmacokinetics of sparfloxacin and interaction with cisapride and sucralfate. Antimicrob Agents Chemother 1997;41(8):1668–72.
30. Kamberi M, Nakashima H, Ogawa K, Oda N, Nakano S. The effect of staggered dosing of sucralfate on oral bioavailability of sparfloxacin. Br J Clin Pharmacol 2000;49(2):98–103.

Spectinomycin

General Information

Spectinomycin is an aminocyclitol antibiotic, distinct from the aminoglycosides. It has been used mainly in single-dose treatment of gonorrhea (2 g intramuscularly), and for treating organisms with multiple antibiotic resistance (1–3). Apart from induration at the site of injection, tolerance in otherwise healthy individuals has been good, even with doses of 8 g/day for 21 days. Allergic reactions have been observed, but are rare. Spectinomycin-resistant gonococci have been found (4,5).

Long-Term Effects

Drug tolerance

Aminoglycoside adenylyltransferase (aadA) genes mediate resistance to streptomycin and spectinomycin. In a pathogenic porcine *Escherichia coli*, the novel aaA5 has now been described integrated with a trimethoprim resistance gene (6). The aadA6 gene has been sequenced from *Pseudomonas aeruginosa* and can confer high-level resistance to spectinomycin in *E. coli* (7). All strains of multidrug resistant *Vibrio cholerae* O1 El Tor isolated in Albania were resistant to spectinomycin, and resistance to this antibiotic was mediated by the aadA1 gene cassette located in the bacterial chromosome within a class 1 integron (8). Another class 1 integron that contained aadA2 was found in pathogenic *Salmonella typhimurium* isolates from France and was also chromosomally located (9). An aadA gene has also been described in *Enterococcus faecalis* (10).

Resistance of *E. coli* isolates from chicken in Saudia Arabia was 96% compared with 71% among human isolates (11). Serotyping showed an overlap between human and chicken strains, suggesting that these animals may serve as a resistance pool for humans.

References

1. Stolz E, Zwart HG, Michel MF. Sensitivity to ampicillin, penicillin, and tetracyline of gonococci in Rotterdam. Br J Vener Dis 1974;50(3):202–7.
2. Levy B, Brown J, Fowler W. Spectinomycin in gonorrhoea. Br J Clin Pract 1974;28(5):174–6.
3. Novak E, Gray JE, Pfeifer RT. Animal and human tolerance of high-dose intramuscular therapy with spectinomycin. J Infect Dis 1974;130(1):50–5.
4. Zenilman JM, Nims LJ, Menegus MA, Nolte F, Knapp JS. Spectinomycin-resistant gonococcal infections in the United States, 1985–1986. J Infect Dis 1987;156(6):1002–4.
5. Boslego JW, Tramont EC, Takafuji ET, Diniega BM, Mitchell BS, Small JW, Khan WN, Stein DC. Effect of spectinomycin use on the prevalence of spectinomycin-resistant and of penicillinase-producing *Neisseria gonorrhoeae*. N Engl J Med 1987;317(5):272–8.
6. Sandvang D. Novel streptomycin and spectinomycin resistance gene as a gene cassette within a class 1 integron isolated from *Escherichia coli*. Antimicrob Agents Chemother 1999;43(12):3036–8.
7. Naas T, Poirel L, Nordmann P. Molecular characterisation of In51, a class 1 integron containing a novel aminoglycoside adenylyltransferase gene cassette, aadA6, in *Pseudomonas aeruginosa*. Biochim Biophys Acta 1999;1489(2–3):445–51.
8. Falbo V, Carattoli A, Tosini F, Pezzella C, Dionisi AM, Luzzi I. Antibiotic resistance conferred by a conjugative plasmid and a class I integron in *Vibrio cholerae* O1 El Tor strains isolated in Albania and Italy. Antimicrob Agents Chemother 1999;43(3):693–6.
9. Poirel L, Guibert M, Bellais S, Naas T, Nordmann P. Integron- and carbenicillinase-mediated reduced susceptibility to amoxicillin–clavulanic acid in isolates of multidrug-resistant *Salmonella enterica* serotype typhimurium DT104 from French patients. Antimicrob Agents Chemother 1999;43(5):1098–104.
10. Clark NC, Olsvik O, Swenson JM, Spiegel CA, Tenover FC. Detection of a streptomycin/spectinomycin adenylyltransferase gene (aadA) in *Enterococcus faecalis*. Antimicrob Agents Chemother 1999;43(1):157–60.
11. Al-Ghamdi MS, El-Morsy F, Al-Mustafa ZH, Al-Ramadhan M, Hanif M. Antibiotic resistance of *Escherichia coli* isolated from poultry workers, patients and chicken in the eastern province of Saudi Arabia. Trop Med Int Health 1999;4(4):278–83.

Spiramycin

See also Macrolide antibiotics

General Information

Spiramycin is a macrolide antibiotic with activity primarily against *Staphylococcus aureus*, beta-hemolytic streptococci, and viridans streptococci. Since spiramycin is retained in bone and the salivary glands and reaches prolonged high concentrations in the saliva, it is used mainly by dentists and otorhinolaryngologists. Because of high concentration in the tonsillar lymphoid tissue it has been proposed that spiramycin be used in the prevention of meningococcal infections. It does not reach the cerebrospinal fluid. Its toxicity is relatively low. Adverse effects include nausea, vomiting, diarrhea, and skin rashes (1).

Organs and Systems

Cardiovascular

QT interval prolongation (2) and torsade de pointes (3) occur rarely in patients taking spiramycin.

Hematologic

Hematological toxicity, including bone marrow suppression (4) and hemolysis (5), has been observed, especially during combined treatment with spiramycin and pyrimethamine for toxoplasmosis.

Liver

Cholestatic hepatitis occurred in a patient taking spiramycin (6).

Skin

In a review of 207 cases of serious acute generalized exanthematous pustulosis, spiramycin was the causal drug in five cases (7).

Immunologic

A peculiar hypersensitivity reaction was reported in an employee of a pharmaceutical company who developed attacks of sneezing, coughing, and breathlessness while working with spiramycin; he had immediate positive skin prick tests to spiramycin and developed blood eosinophilia during asthma attacks (8).

Henoch–Schönlein purpura occurred in a patient taking spiramycin (9).

Drug–Drug Interactions

Neuroleptic drugs

Acute dystonia was observed during spiramycin therapy in a patient who was being treated with neuroleptic drugs (10).

References

1. Johnson RH, Rozanis J, Schofield ID, Haq MS. The effect of spiramycin on plaque accumulation and gingivitis. Dent J 1978;44(10):456–60.
2. Stramba-Badiale M, Guffanti S, Porta N, Frediani M, Beria G, Colnaghi C. QT interval prolongation and cardiac arrest during antibiotic therapy with spiramycin in a newborn infant. Am Heart J 1993;126(3 Pt 1):740–2.
3. Verdun F, Mansourati J, Jobic Y, Bouquin V, Munier S, Guillo P, Pages Y, Boschat J, Blanc JJ. Torsades de pointes sous traitement par spiramycine et mequitazine. A propos d'un cas. [Torsades de pointe with spiramycine and metiquazine therapy. Apropos of a case.] Arch Mal Coeur Vaiss 1997;90(1):103–6.
4. Switala I, Dufour P, Ducloy AS, Vinatier D, Bernardi C, Monnier JC, Plantier I, Fortier B. Aplasie medullaire au cours du traitement d'une toxoplasmose congenitale lors d'une grossesse gemellaire. [Medullary aplasia during treatment for congenital toxoplasmosis in a twin pregnancy.] J Gynecol Obstet Biol Reprod (Paris) 1993;22(5):513–6.
5. Sarma PS. Oxidative haemolysis after spiramycin. Postgrad Med J 1997;73(864):686–7.
6. Denie C, Henrion J, Schapira M, Schmitz A, Heller FR. Spiramycin-induced cholestatic hepatitis. J Hepatol 1992;16(3):386.
7. Saissi EH, Beau-Salinas F, Jonville-Bera AP, Lorette G, Autret-Leca E. Medicaments associés a la survenue d'une pustulose exanthematique aigue generalisée. Ann Dermatol Venéreol 2003;130:612–8.
8. Davies RJ, Pepys J. Asthma due to inhaled chemical agents—the macrolide antibiotic spiramycin. Clin Allergy 1975;5(1):99–107.
9. Valero Prieto I, Calvo Catala J, Hortelano Martinez E, Abril L, de Medrano V, Glez-Cruz Cervellera MI, Herrera Ballester A, Orti E. Purpura de schönlein–Henoch asociada a espiramicina y con importantes manifestaciones digestivas. [Schoenlein–Henoch purpura associated with spiramycin and with important digestive manifestations.] Rev Esp Enferm Dig 1994;85(1):47–9.
10. Benazzi F. Spiramycin-associated acute dystonia during neuroleptic treatment. Can J Psychiatry 1997;42(6):665–6.

Streptogramins

General Information

Pristinamycin

Pristinamycin (rINN) is a mixture of water-insoluble streptogramin A (pristinamycin IIA) and streptogramin B (pristinamycin IB), derived from *Streptomyces pristinaespiralis*. The former is a group A streptogramin (a polyunsaturated macrolactone) and the latter is a group B streptogramin (a peptidic macrolactone or depsipeptide). Separately, group A and group B streptogramins are bacteriostatic, by reversible binding to the 50S subunit of 70S bacterial ribosomes. Together, however, streptogramins from each group are synergic and bactericidal.

Pristinamycin is active against the main bacteria that cause respiratory tract infections and could be useful in the treatment of acute or recurrent sinusitis, community-acquired pneumonia, or periodontal infection with *Actinobacillus actinomycetemcomitans* (1–5).

Quinupristin (rINN) and dalfopristin (rINN) are two semisynthetic pristinamycin derivatives that are given in combination parenterally. The combination has a broad spectrum of activity against Gram-positive bacteria, including multidrug resistant bacilli. A comprehensive review has been published (6). Quinupristin + dalfopristin can be used to treat macrolide-resistant streptococci, staphylococcal infections after failure of conventional therapy, or vancomycin-resistant *Enterococcus faecium* (and probably *Enterococcus raffinosus*), but not vancomycin-resistant *Enterococcus faecalis*, *Enterococcus avium*, *Enterococcus casseliflavus*, or *Enterococcus gallinarum* (7–19). Adverse effects include arthralgia, myalgia, and pain at the infusion site (20).

Quinupristin/dalfopristin

Quinupristin/dalfopristin is a semisynthetic injectable streptogramin combination. Each component has bacteriostatic activity against staphylococci and streptococci (21). Its adverse effects include arthralgia, myalgia, and pain at the infusion site (22). Several challenges to its use have emerged (23). First, a substantial number of patients have developed myalgia, in many cases severe enough to necessitate withdrawal of therapy. Some patients who had continue therapy did so only with the co-administration of opioid analgesia. Secondly, quinupristin/dalfopristin can cause severe venous irritation, necessitating administration through a central venous catheter. Thirdly, resistance of vancomycin-resistant enterococci to quinupristin/dalfopristin emerged fairly rapidly after its clinical introduction. Finally, quinupristin/dalfopristin interacts with drugs metabolized by CYP isoenzymes, such as calcineurin inhibitors used in transplantation.

Quinupristin/dalfopristin has minimal oral absorption and is administered intravenously as a fixed 30:70 ratio of quinupristin to dalfopristin (24). A linear relationship has been observed between the dose administered and maximum plasma concentrations. Single-dose administration of 7.5 mg/kg produced a maximal plasma concentration of 2.3–2.7 mg/l for quinupristin and 6.1–8.2 mg/l for dalfopristin. The AUC obtained with the same dose was 2.7–3.3 and 6.5–7.7 h.mg/l for quinupristin and dalfopristin, respectively. Repeated administration results in 13–21% increases in maximum plasma concentrations and 21–26% increases in area under the curve of the concentration for both quinupristin and dalfopristin. Quinupristin and dalfopristin exhibit steady-state volumes of distribution of 0.46–0.54 and 0.24–0.30 L/kg, respectively. Quinupristin exhibits higher protein binding

(55–78%) than dalfopristin (11–26%), though both entities distribute well into tissues. Concentrations exceeding those in blood have been reported for the kidney, liver, spleen, salivary glands, and white blood cells of primates. Extravascular penetration, as measured in blister fluid, is 40–80%. Both quinupristin and dalfopristin are extensively metabolized via non-enzymatic reactions. Quinupristin is conjugated to form two active compounds, a cysteine moiety and a glutathione moiety. Dalfopristin is hydrolysed to the active metabolite pristinamycin. Quinupristin/dalfopristin is excreted primarily in the feces (75–77%), with lesser renal excretion (15–19%). The elimination half-lives of quinupristin and dalfopristin are similar, and are 0.7–1.3 hours after single doses. The metabolites have slightly longer half-lives, ranging from 1.2 to 1.8 hours. With repeated doses, plasma clearance of quinupristin and dalfopristin is reduced by approximately 20% compared with single doses, resulting in clearances of 0.7–0.8 L/h/kg. Saturable protein binding has been hypothesized as a causative mechanism.

The adverse effects of quinupristin/dalfopristin include infusion-site reactions (42%), infusion-site pain (40%), edema (17%), arthralgias (47%), myalgias (6%), gastrointestinal effects (3–5%), rash (2.5%), headache (1.6%), pruritus (1.5%), and hyperbilirubinemia (25%) (25,26).

Virginiamycin

Virginiamycin is a streptogramin that has been used in animals. Its use has been linked with the selection of quinupristin/dalfopristin-resistant *Enterococcus faecium* (27). Since streptogramins have been used infrequently in human medicine, an animal origin of resistance has been suggested, and spread of resistance via the food chain to humans is probable (28).

Observational studies

In a small, open, phase II pilot study, low-dose quinupristin + dalfopristin (5 mg/kg intravenously every 8 hours; $n = 11$) and high-dose quinupristin + dalfopristin (7.5 mg/kg intravenously every 8 hours; $n = 15$) were compared with vancomycin (1 g intravenously every 12 hours; $n = 13$) in the treatment of catheter-related staphylococcal bacteremia (29). In patients with a baseline pathogen, the outcome was comparable in all groups. Adverse clinical events in the quinupristin + dalfopristin group consisted of arm and chest pain, chills, fever, arthritis, and phlebitis or pain at the injection site; quinupristin + dalfopristin was withdrawn in 12% of patients (compared with 15% of vancomycin-treated patients).

In healthy volunteers receiving quinupristin + dalfopristin, 7.5 mg/kg infused over 1 hour bd, mean fecal antibiotic concentrations were 291 and 42 µg/g of feces for quinupristin and dalfopristin respectively by the fifth day of treatment (30).

In a prospective study in 53 patients with methicillin-resistant *Staphylococcus aureus* treated with pristinamycin 0.5–1.5 g tds, adverse effects included diarrhea or loose stools (n = 8) and one possible rash. Another patient had an infection with *Clostridium difficile* more than 1 month after therapy (31).

In a retrospective chart review of 27 patients with osteoarticular infections, there were adverse effects of pristinamycin in eight cases; seven were gastrointestinal disturbances and there was one instance of an allergic rash, which required drug withdrawal (32).

The safety profile of quinupristin/dalfopristin has been evaluated in more than 2000 patients (33). The most common adverse effect was pain and inflammation at the infusion site. However, treatment withdrawal was reported in under 10% of patients. Other adverse effects include arthralgias (9%) and myalgias (6%), which have led to the withdrawal of quinupristin/dalfopristin in one-third to a half of affected patients. Other common systemic adverse events have been nausea (4.6%), diarrhea (2.7%), vomiting (2.7%), rash (2.5%), headache (1.6%), pruritus (1.5%), and pain (1.5%). Liver function abnormalities have occurred in about 1% of patients who received quinupristin/dalfopristin. However, these effects have generally been mild and transient. No significant effects on renal function have been reported and bone marrow toxicity has been rare.

Comparative studies

The pooled results of two multicenter, phase III, randomized comparisons of quinupristin + dalfopristin (7.5 mg/kg intravenously every 12 hours; $n = 450$) with established comparators (cefazolin, oxacillin, vancomycin; $n = 443$) for complicated skin and skin structure infections have been reported (34). The success rate was equivalent in the two groups: 63% of patients given quinupristin + dalfopristin (versus 54%) reported at least one adverse event, most commonly nausea, vomiting, rash, pain, or pruritus. Although most of the adverse events were mild to moderate, the drug was withdrawn in 19% of patients treated with quinupristin + dalfopristin (versus 4.7%) owing to an untoward event. Adverse venous events (atrophy, edema, hemorrhage, hypersensitivity, inflammation, thrombophlebitis, pain) were reported by 66% of patients treated with quinupristin + dalfopristin (versus 28%) and required withdrawal of the drug in 12% of these patients (versus 2%).

General adverse effects

The adverse effects of quinupristin + dalfopristin include arthralgia, myalgias, reversible rises in serum alkaline phosphatase, itching, diarrhea, vomiting, and pain at the infusion site; adverse effects occurred in 2.5–4.6% of patients (35–42).

Organs and Systems

Electrolyte balance

The combination of quinupristin + dalfopristin has been associated with hyponatremia, probably secondary to inappropriate secretion of antidiuretic hormone (43).

- A 67-year-old woman with peripheral neuropathy, IgM paraproteinemia, and chronic obstructive pulmonary disease developed dyspnea and hyponatremia, and a small-cell lung cancer was diagnosed. She was given 3 cycles of chemotherapy (etoposide, cyclophosphamide, and doxorubicin). Her serum sodium concentrations normalized. On day 103, she was given quinupristin + dalfopristin (7.5 mg/kg every 8 hours) because of septicemia with vancomycin-resistant *E. faecium*. The serum sodium concentration thereafter fell (day 110: 117 mmol/l; serum osmolarity: 268 mosm/l; urine osmolarity: 426 mosm/l). Therapy was withdrawn on day 111, and the sodium concentration gradually returned to normal.

Hematologic

Quinupristin/dalfopristin is rarely associated with hematological adverse events.

- A 44-year-old African American man developed reticulocytopenia after receiving intravenous quinupristin/dalfopristin 550 mg (7.5 mg/kg) every 8 hours for 23 days for a hip joint infection and became anemic (hemoglobin 6.5 g/dl) (44). The results of iron studies were normal and the reticulocyte count was low at 0.22%.

Quinupristin/dalfopristin was withdrawn and vancomycin started. The hemoglobin rose.

Reticulocytopenia has been attributed to quinupristin/dalfopristin ([45A]).

- A 52-year-old woman developed a reticulocytopenia after receiving quinupristin/dalfopristin 750 mg intravenously every 8 hours for 4 weeks. Before therapy, the hemoglobin concentration was 13 g/dl and it fell to 6.8 g/dl, with an absolute reticulocyte count of 2.4×10^9/l (reference range 28.4–150). Quinupristin/dalfopristin was withdrawn and doxycycline 100 mg orally bd was started. Four days later, the absolute reticulocyte count was 146×10^9/l. She completed her course of doxycycline without further complications.

Gastrointestinal

In a multicenter, open, randomized trial in 204 patients with erysipelas treated with either oral pristinamycin 1 g tds or intravenous then oral penicillin, adverse events related to treatment were significantly more common with pristinamycin; they were mostly mild or moderate and mainly involved the gastrointestinal tract (46).

Pristinamycin can cause pseudomembranous colitis.

- An 85-year-old woman developed a severe illness (severe diarrhea and vomiting, abdominal tenderness, peritoneal irritation, and systemic toxicity) 8 days after receiving pristinamycin 3 g/day for 10 days (47). An assay for *Clostridium difficile* was positive. She was treated with metronidazole and her symptoms resolved after 72 hours.

Liver

In 25 liver transplant recipients who received intravenous quinupristin + dalfopristin 7.5 mg/kg 8-hourly for

vancomycin-resistant *E. faecium* infection, hyperbilirubinemia developed, but there was no evidence of drug-specific histological injury (48).

Skin

In a retrospective study of children taking quinupristin + dalfopristin, rash (2%) was the most frequent adverse event, but only one patient with rash discontinued quinupristin + dalfopristin because of it (42).

Patch tests can be used to confirm pristinamycin-induced drug eruptions. In 11 patients with cutaneous drug reactions after oral pristinamycin, patch tests (with pristinamycin diluted to 20% aqueous and 20% petrolatum) were performed 1–3 months after treatment with pristinamycin (49). There were positive patch tests in nine, and there was no relapse of the drug eruption during patch-testing; 30 control patients had negative tests.

Of 207 cases of serious acute generalized exanthematous pustulosis, pristinamycin was the causal drug in 18 (50).

Occupational airborne contact dermatitis from pristinamycin has been reported (51).

- A 23-year-old man was exposed to pristinamycin powder while working for a pharmaceutical company. He used rubber mask and gloves as protective equipment, but developed a pruritic dermatitis involving the neck, eyelids, and cheeks, characterized by erythema, edema, and scales. The skin lesions improved during holidays and with topical corticosteroids, and recurred when he restarted work. Patch testing with pristinamycin 1%, 5% , and 10% in equal volume mixtures of water and alcohol were negative at 20 minutes but positive at 2 and 4 days. Control tests in 10 patients were negative.

In 29 patients with cutaneous adverse drug reactions due to pristinamycin, skin tests were positive in 27 cases, patch tests in 20 cases, and prick tests in three of nine cases; intradermal tests with dalfopristin + quinupristin were positive in four of five cases (52). There were cross-reactions between pristinamycin and virginiamycin in nine of 22 and dalfopristin–quinupristin in seven of eight cases. It is advisable to avoid all streptogramins in patients with cutaneous adverse drug reactions due to pristinamycin.

Toxic epidermal necrolysis has been reported in a patient receiving pristinamycin (53).

- A 75-year-old man developed a rash 4 hours after the start of treatment with pristinamycin. Extensive epidermal detachment was noted 48 hours later, with a positive Nikolsky's sign, and it progressed to cover 40% of the body surface. Skin biopsy confirmed toxic epidermal necrolysis, with subepidermal blistering and numerous necrotic keratinocytes. Re-epithelialization occurred in 3 weeks. Two weeks later pristinamycin was given again and 2 hours later he developed a high fever with a generalized rash and large blisters associated with erosions in the buccal and ophthalmic mucosa. The blisters progressively covered the entire body surface. A skin biopsy confirmed toxic epidermal necrolysis. He died of multiorgan failure.

Sweet's syndrome has been reported after treatment with quinupristin/dalfopristin in a 63-year-old woman; the symptoms and cutaneous lesions rapidly resolved after withdrawal (54).

Mean antibiotic concentrations of pristinamycin in dermal interstitial fluid (from suction bullae) are low; nevertheless, the concentrations achieved should theoretically inhibit the growth of group A streptococci (55).

Musculoskeletal

Of seven patients with end-stage renal insufficiency who received quinupristin + dalfopristin, two developed myalgias (56).

Of 32 patients who received quinupristin + dalfopristin, at least 15 developed arthralgias and/or myalgias (57).

Arthralgias and myalgias have been reported in a retrospective study in 11 children (58). Quinupristin/dalfopristin was begun in a dose of 7.5 mg/kg 8-hourly intravenously. Treatment was planned for at least 14 days, or longer if there was no clinical improvement. Seven children completed the complete course without any adverse effects. In four children, treatment was stopped because of myalgia and arthralgia, and the median time to stopping treatment was 7 (range 7–35) days. All four children needed analgesics; three required additional narcotic analgesics, of whom two required continuous morphine infusion and intensive input from the pain team. In one child, in whom vancomycin-resistant enterococci were isolated from the bile, mechanical ventilation was considered in order that treatment could be maintained pain-free; this child died of overwhelming sepsis after withdrawal of the antibiotic. Of the other three, two were treated with linezolid and one completed the course of quinupristin/dalfopristin while receiving mechanical ventilation for respiratory deterioration.

In 56 patients treated with quinupristin/dalfopristin 7.5 mg/kg every 8 hours for a mean duration of 12 (range 2–52) days, there were myalgias/arthralgias in 20 (59). These patients had significantly higher activities of alkaline phosphatase (mean 319 IU/l) during the midterm therapy cycle compared with patients without any joint or muscular pains (mean 216 IU/l); three had more than five times normal activity, which did not occur in any of the patients who did not develop myalgias/arthralgias. All the myalgias/arthralgias resolved after withdrawal of quinupristin/dalfopristin.

Immunologic

The combination of quinupristin + dalfopristin reduces cytokine production in stimulated monocytes from healthy volunteers, suggesting significant immunomodulatory activity (60).

Long-Term Effects

Drug resistance

The relative frequency of resistance to macrolides, lincosamides, and streptogramins (MLS resistance) has been assessed in a series of 2091 staphylococcus isolates collected during a 3-week period in 1995 in 32 French hospitals. A total of 294 strains (144 *Staphylococcus aureus*, 150 coagulase-negative staphylococci) exhibited resistance to at least one of these groups of antibiotics. Resistance to pristinamycin was phenotypically detected in ten *S. aureus* strains (seven isolates from the same hospital and possibly of the same clone) and three coagulase-negative staphylococcus isolates. It was associated with resistance to type A streptogramins encoded by vat or vatB genes and was associated with the erm genes (61).

A short report demonstrated the absence of a reliable correlation between killing kinetics and normal laboratory tests for pristinamycin susceptibility testing of some pneumococci (62). Eight selected multiresistant clinical isolates and two reference pristinamycin-resistant *Streptococcus pneumoniae* strains were studied. Disk diffusion susceptibility and MICs were determined by the agar dilution method, and all clinical isolates appeared to be susceptible to pristinamycin, whereas the two reference strains were not. In contrast, time-kill experiments identified a limited bactericidal effect of pristinamycin in three clinical and both reference strains. These three strains had been classified as pristinamycin-resistant by the Vitek-II system, which uses a kinetic turbidimetric measurement of bacterial growth. Epidemiological information is hindered by the use of highly selected strains for the study.

The in vitro activity of pristinamycin has been evaluated in 200 isolates of *S. pneumoniae* strains with various degrees of susceptibility to penicillin G and erythromycin (63). All the strains were susceptible to pristinamycin, irrespective of their susceptibility to penicillin G or erythromycin.

After 5 days administration of quinupristin + dalfopristin 7.5 mg/kg infused over 1 hour bd, the fecal microflora in 20 healthy volunteers increased significantly during treatment and returned within 12 weeks to baseline concentrations after the end of treatment. There were anerobes and enterococci resistant to erythromycin or to quinupristin + dalfopristin, but glycopeptide-resistant enterococci did not emerge (30).

The relevance of resistance to virginiamycin

Virginiamycin is a streptogramin that has been used in animals as a growth-promoter. Increasing use of virginiamycin has been associated with a high rate of resistance to quinupristin + dalfopristin (64). Acquired resistance to virginiamycin, which is active against Gram-positive lactic acid-producing bacteria, can be detected in *E. faecium* and *E. faecalis* strains isolated from animals and food (65), including strains from poultry, in which high resistance against virginiamycin was found in Belgium (66). Streptogramin resistance is associated with the resistance genes vatA and vatG in *E. faecium* of both animals and man (64). Because virginiamycin has been used in animals and streptogramins have been used infrequently in man, an animal origin of resistance has been suggested, and spread of resistance via the food chain to humans is probable (67).

Cytotoxicity

Infusion phlebitis is a common problem with quinupristin/dalfopristin. Cultured murine fibroblasts and immortalized human endothelial cells were exposed to quinupristin/dalfopristin, erythromycin, and levofloxacin at clinically relevant concentrations to assess their toxic potential in two cytotoxicity assays (68). Quinupristin/dalfopristin and erythromycin had concentration-related cytotoxic effects in both cell cultures. Cytotoxic effects in both cell cultures occurred in the following order: quinupristin/dalfopristin > erythromycin >> levofloxacin. This ranking correlates well with the frequency of local adverse effects observed with the infusion of these antibiotics.

Susceptibility Factors

Renal disease

The disposition of quinupristin, dalfopristin, and their primary metabolites was similar in eight non-infected patients on peritoneal dialysis compared with eight matched healthy volunteers after the administration of quinupristin + dalfopristin 7.5 mg/kg (69). Since dialysis clearance was insignificant, the combination was thought to be inadequate for the therapy of dialysis-associated peritonitis.

The pharmacokinetics of a single intravenous injection of quinupristin + dalfopristin (7.5 mg/kg over 1 hour) have been assessed in 13 patients with severe chronic renal insufficiency (creatinine clearance 6–28 ml/minute/1.73 m^2) (70). Although the mean peak plasma drug concentration and AUC of quinupristin plus its active derivatives and of both unchanged dalfopristin and dalfopristin plus its active derivatives were about 1.3–1.4 times higher than in healthy volunteers, the authors concluded that no formal reduction in the dosage of quinupristin + dalfopristin is necessary in patients with chronic renal insufficiency.

Hepatic impairment

In a retrospective study in 50 patients who took quinupristin/dalfopristin, significant risk factors for arthralgia or myalgia were chronic liver disease, liver transplantation, raised bilirubin concentration at baseline, major surgery, and the concomitant use of either mycophenolate or ciclosporin; female sex was a less strong factor (71).

Drug–Drug Interactions

Quinupristin + dalfopristin can inhibit the metabolism of agents that are metabolized by CYP3A4 (72).

Ciclosporin

Quinupristin/dalfopristin inhibits CYP3A4, resulting in multiple drug interactions (24). Ciclosporin AUC increased by 5–222% when co-administered with quinupristin/dalfopristin. Patients taking drugs that are substrates of CYP3A4 should be carefully monitored.

Monitoring Drug Therapy

Quinupristin/dalfopristin are active and are converted to active metabolites that contribute to the antimicrobial activity of the formulation (33). The half-life of quinupristin and its metabolites is about 3 hours and the half-life of dalfopristin and its metabolites is about 1 hour. About 75% of a dose is eliminated by the gastrointestinal tract, and urinary excretion accounts for 15% of a dose of quinupristin and 19% of a dose of dalfopristin. The peak plasma concentration after a single intravenous dose of 7.5 mg/kg is 2.3 mg/l for quinupristin and 6.5 mg/l for dalfopristin. Both components penetrate well into interstitial fluid. There is a prolonged post-antibiotic effect of about 10 hours for *Staphylococcus aureus* and 9 hours for *Streptococcus pneumoniae*.

References

1. Leclercq R. Activité in vitro de la pristinamycine sur les germes respiratoires. [In vitro activity of pristinamycin on respiratory bacteria.] Presse Méd 1999;28(Suppl 1):6–9.
2. Klossek JM, Mayaud C. Conclusions: quelle stratégie antibiotique dans les infections respiratoires de l'adulte?. [Conclusion: what is the choice of antibiotics in adult respiratory tract infections?.] Presse Méd 1999;28(Suppl 1):16–8.
3. Poirier R. La place de la pristinamycine dans les pneumopathies aiguës communautaires de l'adulte. [Pristinamycin in the treatment of acute communicable pneumopathies in adults.] Presse Méd 1999;28(Suppl 1):13–5.
4. Pessey JJ. Place de la pristinamycine dans le traitement des sinusites aiguës de l'adulte en ville. [Pristinamycin in the outpatient treatment of acute sinusitis in adults.] Presse Méd 1999;28(Suppl 1):10–2.
5. Madinier IM, Fosse TB, Hitzig C, Charbit Y, Hannoun LR. Resistance profile survey of 50 periodontal strains of Actinobacillus actinomyectomcomitans. J Periodontol 1999;70(8):888–92.
6. Lamb HM, Figgitt DP, Faulds D. Quinupristin/dalfopristin: a review of its use in the management of serious gram-positive infections. Drugs 1999;58(6):1061–97.
7. Betriu C, Redondo M, Palau ML, Sanchez A, Gomez M, Culebras E, Boloix A, Picazo JJ. Comparative in vitro activities of linezolid, quinupristin–dalfopristin, moxifloxacin, and trovafloxacin against erythromycin-susceptible and -resistant streptococci. Antimicrob Agents Chemother 2000;44(7):1838–41.
8. Elsner HA, Sobottka I, Feucht HH, Harps E, Haun C, Mack D, Ganschow R, Laufs R, Kaulfers PM. Nosocomial

outbreak of vancomycin-resistant *Enterococcus faecium* at a German university pediatric hospital. Int J Hyg Environ Health 2000;203(2):147–52.

9. von Eiff C, Reinert RR, Kresken M, Brauers J, Hafner D, Peters G. Nationwide German multicenter study on prevalence of antibiotic resistance in staphylococcal bloodstream isolates and comparative in vitro activities of quinupristin–dalfopristin. J Clin Microbiol 2000;38(8):2819–23.

10. Johnson AP, Warner M, Hallas G, Livermore DM. Susceptibility to quinupristin/dalfopristin and other antibiotics of vancomycin-resistant enterococci from the UK, 1997 to mid-1999. J Antimicrob Chemother 2000;46(1):125–8.

11. McGeer AJ, Low DE. Vancomycin-resistant enterococci. Semin Respir Infect 2000;15(4):314–26.

12. Levison ME, Mallela S. Increasing antimicrobial resistance: therapeutic implications for enterococcal infections. Curr Infect Dis Rep 2000;2(5):417–23.

13. Bergogne-Berezin E. Resistances et nouvelles stratégies antibiotiques. Nouveaux antibiotiques antistaphylococciques. [Resistance and new antibiotic strategies. New anti-staphylococcal antibiotics.] Presse Méd 2000;29(37):2023–7.

14. Livermore DM. Antibiotic resistance in staphylococci. Int J Antimicrob Agents 2000;16(Suppl 1):S3–S10.

15. Bhavnani SM, Ballow CH. New agents for Gram-positive bacteria. Curr Opin Microbiol 2000;3(5):528–34.

16. Bush K, Macielag M. New approaches in the treatment of bacterial infections. Curr Opin Chem Biol 2000;4(4):433–9.

17. Jones RN. Perspectives on the development of new antimicrobial agents for resistant gram-positive pathogens. Braz J Infect Dis 2000;4(1):1–8.

18. Murray BE. Problems and perils of vancomycin resistant enterococci. Braz J Infect Dis 2000;4(1):9–14.

19. Lundstrom TS, Sobel JD. Antibiotics for Gram-positive bacterial infections. Vancomycin, teicoplanin, quinupristin/dalfopristin, and linezolid. Infect Dis Clin North Am 2000;14(2):463–74.

20. Delgado G Jr, Nouhausei MM, Bearden DT, Danziger LH. Quinupristin–dalfopristin: an overview. Pharmacotherapy 2000;20(12):1469–85.

21. Brown J, Freeman BB, 3rd. Combining quinupristin/dalfopristin with other agents for resistant infections. Ann Pharmacother 2004;38(4):677–85.

22. Eliopoulos GM. Quinupristin–dalfopristin and linezolid: evidence and opinion. Clin Infect Dis 2003;36:473–81.

23. Paterson DL. Clinical experience with recently approved antibiotics. Curr Opin Pharmacol 2006;6(5):486–90.

24. Bearden DT. Clinical pharmacokinetics of quinupristin/dalfopristin. Clin Pharmacokinet 2004;43(4):239–52.

25. Moellering RC, Linden PK, Reinhardt J, Blumberg EA, Bompart F, Talbot GH. The efficacy and safety of quinupristin–dalfopristin for the treatment of infections caused by vancomycin-resistant *Enterococcus faecium*. J Antimicrob Chemother 1999;44:251–61.

26. Nichols RL, Graham DR, Barriere SL, Rodgers A, Wilson SE, Zervos M, Dunn DL, Kreter B. Treatment of hospitalized patients with complicated gram-positive skin and skin structure infections: two randomized, multicentre studies of quinupristin/dalfopristin versus cefazolin, oxacillin or vancomycin. Synercid Skin and Skin Structure Infection Group. Antimicrob Chemother 1999;44:263–73.

27. Kieke AL, Borchardt MA, Kieke BA, Spencer SK, Vandermause MF, Smith KE, Jawahir SL, Belongia EA. Use of streptogramin growth promoters in poultry and isolation of streptogramin-resistant *Enterococcus faecium* from humans. J Infect Dis 2006;194(9):1200–8.

28. Hershberger E, Oprea SF, Donabedian SM, Perri M, Bozigar P, Bartlett P, Zervos MJ. Epidemiology of antimicrobial resistance in enterococci of animal origin. J Antimicrob Chemother 2005;55(1):127–30.

29. Raad I, Bompart F, Hachem R. Prospective, randomized dose-ranging open phase II pilot study of quinupristin/dalfopristin versus vancomycin in the treatment of catheter-related staphylococcal bacteremia. Eur J Clin Microbiol Infect Dis 1999;18(3):199–202.

30. Scanvic-Hameg A, Chachaty E, Rey J, Pousson C, Ozoux ML, Brunel E, Andremont A. Impact of quinupristin/dalfopristin (RP59500) on the faecal microflora in healthy volunteers. J Antimicrob Chemother 2002;49(1):135–9.

31. Dancer SJ, Robb A, Crawford A, Morrison D. Oral streptogramins in the management of patients with methicillin-resistant *Staphylococcus aureus* (MRSA) infections. J Antimicrob Chemother 2003;51:731–5.

32. Ng J, Gosbell IB. Successful oral pristinamycin therapy for osteoarticular infections due to methicillin-resistant Staphylococcus aureus (MRSA) and other Staphylococcus spp. J Antimicrob Chemother 2005;55(6):1008–12.

33. Al-Tatari H, Abdel-Haq N, Chearskul P, Asmar B. Antibiotics for treatment of resistant Gram-positive coccal infections. Indian J Pediatr 2006;73(4):323–34.

34. Nichols RL, Graham DR, Barriere SL, Rodgers A, Wilson SE, Zervos M, Dunn DL, Kreter B. Treatment of hospitalized patients with complicated Gram-positive skin and skin structure infections: two randomized, multicentre studies of quinupristin/dalfopristin versus cefazolin, oxacillin or vancomycin. Synercid Skin and Skin Structure Infection Group. J Antimicrob Chemother 1999;44(2):263–73.

35. Raad I, Hachem R, Hanna H, Girgawy E, Rolston K, Whimbey E, Husni R, Bodey G. Treatment of vancomycin resistant enterococcal infections in the immunocompromised host: quinupristin–dalfopristin in combination with minocycline. Antimicrob Agents Chemother 2001;45(11):3202–4.

36. Rehm SJ, Graham DR, Srinath L, Prokocimer P, Richard MP, Talbot GH. Successful administration of quinupristin/dalfopristin in the outpatient setting. J Antimicrob Chemother 2001;47(5):639–45.

37. Linden PK, Moellering RC Jr, Wood CA, Rehm SJ, Flaherty J, Bompart F, Talbot GHSynercid Emergency-Use Study Group. Treatment of vancomycin-resistant *Enterococcus faecium* infections with quinupristin/dalfopristin. Clin Infect Dis 2001;33(11):1816–23.

38. Verma A, Dhawan A, Philpott-Howard J, Rela M, Heaton N, Vergani GM, Wade J. Glycopeptide-resistant *Enterococcus faecium* infections in paediatric liver transplant recipients: safety and clinical efficacy of quinupristin/dalfopristin. J Antimicrob Chemother 2001;47(1):105–108.

39. Allington DR, Rivey MP. Quinupristin/dalfopristin: a therapeutic review. Clin Ther 2001;23(1):24–44.

40. Bonfiglio G, Furneri PM. Novel streptogramin antibiotics. Expert Opin Investig Drugs 2001;10(2):185–98.

41. Blondeau JM, Sanche SE. Quinupristin/dalfopristin. Expert Opin Pharmacother 2002;3(9):1341–64.

42. Loeffler AM, Drew RH, Perfect JR, Grethe NI, Stephens JW, Gray SL, Talbot GH. Safety and efficacy of quinupristin/dalfopristin for treatment of invasive Gram-positive infections in pediatric patients. Pediatr Infect Dis J 2002;21(10):950–6.

43. Cole RP, Roberts WD, Cheng MD. Hyponatremia associated with quinupristin–dalfopristin. Ann Intern Med 2000;133(6):485.

44 Evans PC, Almas JP, Criddle FJ, 3rd. Anemia and reversible reticulocytopenia associated with extended quinupristin/dalfopristin. Ann Pharmacother 2004;38(4):720–1.

45 Chan-Tack KM, Mehta S. Quinupristin–dalfopristin-induced reticulocytopenic anemia. South Med J 2005;98(12):1226–7.

46. Bernard P, Chosidow O, Vaillant LFrench Erysipelas Study Group. Oral pristinamycin versus standard penicillin regimen to treat erysipelas in adults: randomised, non-inferiority, open trial. BMJ 2002;325(7369):864.

47. Gavazzi G, Barnoud R, Lamloum M, Coume M, Fillipi M, Debray M, Couturier P, Franco A. Colite pseudomembraneuse après pristinamycine. [Pseudomembranous colitis after pristinamycin.] Rev Med Interne 2001;22(7):672–3.

48. Linden PK, Bompart F, Gray S, Talbot GH. Hyperbilirubinemia during quinupristin–dalfopristin therapy in liver transplant recipients: correlation with available liver biopsy results. Pharmacotherapy 2001;21(6):661–8.

49. Mayence C, Dompmartin A, Verneuil L, Michel M, Leroy D. Value of patch tests in pristinamycin-induced drug eruptions. Contact Dermatitis 1999;40(3):161–2.

50. Saissi EH, Beau-Salinas F, Jonville-Bera AP, Lorette G, Autret-Leca E. Medicaments associés a la survenue d'une pustulose exanthematique aiguë generalisée. Ann Dermatol Venéreol 2003;130:612–8.

51. Blancas-Espinosa R, Conde-Salazar L, Perez-Hortet C. Occupational airborne contact dermatitis from pristinamycin. Contact Dermatitis 2006;54(1):63–5.

52. Barbaud A, Trechot P, Weber-Muller F, Ulrich G, Commun N, Schmutz JL. Drug skin tests in cutaneous adverse drug reactions to pristinamycin: 29 cases with a study of cross-reactions between synergistins. Contact Dermatitis 2004;50(1):22–6.

53. Chanques G, Girard C, Pinzani V, Jaber S. Fatal pristinamycin-induced toxic epidermal necrolysis (Lyell's syndrome):difficulties in attributing causal association in the polymedicated intensive care unit patient. Acta Anaesthesiol Scand 2005;49(5):721–2.

54. Choi HS, Kim HJ, Lee TH, Lee SH, Lee TW, Ihm CG, Kim MJ. Quinupristin/dalfopristin-induced Sweet's syndrome. Korean J Intern Med 2003;18:187–90.

55. Vaillant L, Le Guellec C, Jehl F, Barruet R, Sorensen H, Roiron R, Autret-Leca E, Lorette G. Diffusions comparées de l'acide fusidique, de l'oxacilline et de la pristinamycine dans le liquide interstitiel dermique après administration orale repetée. [Comparative diffusion of fusidic acid, oxacillin, and pristinamycin in dermal interstitial fluid after repeated oral administration.] Ann Dermatol Venereol 2000;127(1):33–9.

56. Olsen KM, Rebuck JA, Rupp ME. Arthralgias and myalgias related to quinupristin–dalfopristin administration. Clin Infect Dis 2001;32(4):e83–6.

57. Khan AA, Slifer TR, Araujo FG, Remington JS. Effect of quinupristin/dalfopristin on production of cytokines by human monocytes. J Infect Dis 2000;182(1):356–8.

58. Gupte G, Jyothi S, Beath SV, Kelly DA. Quinupristin–dalfopristin use in children is associated with arthralgias and myalgias. Pediatr Infect Dis J 2006;25(3):281.

59. Raad I, Hachem R, Hanna H. Relationship between myalgias/arthralgias occurring in patients receiving quinupristin/dalfopristin and biliary dysfunction. J Antimicrob Chemother 2004;53(6):1105–8.

60. Schwenger V, Mundlein E, Dagrosa EE, Fahr AM, Zeier M, Mikus G, Andrassy K. Treatment of life-threatening multiresistant staphylococcal and enterococcal infections in patients with end-stage renal failure with quinupristin/dalfopristin: preliminary report. Infection 2002;30(5):257–61.

61. Lina G, Quaglia A, Reverdy ME, Leclercq R, Vandenesch F, Etienne J. Distribution of genes encoding resistance to macrolides, lincosamides, and streptogramins among staphylococci. Antimicrob Agents Chemother 1999;43(5):1062–6.

62. Schlegel L, Sissia G, Fremaux A, Geslin P. Diminished killing of pneumococci by pristinamycin demonstrated by time-kill studies. Antimicrob Agents Chemother 1999;43(8):2099–100.

63. Lozniewski A, Lion C, Mory F, Weber M. Comparison of the in vitro activity of pristinamycin and quinupristin/dalfopristin against *Streptococcus pneumoniae*. Pathol Biol (Paris) 2000;48(5):463–6.

64. Hayes JR, McIntosh AC, Qaiyumi S, Johnson JA, English LL, Carr LE, Wagner DD, Joseph SW. High-frequency recovery of quinupristin-dalfopristin-resistant *Enterococcus faecium* isolates from the poultry production environment. J Clin Microbiol 2001;39(6):2298–9.

65. Butaye P, Devriese LA, Haesebrouck F. Phenotypic distinction in *Enterococcus faecium* and *Enterococcus faecalis* strains between susceptibility and resistance to growth-enhancing antibiotics. Antimicrob Agents Chemother 1999;43(10):2569–70.

66. Butaye P, Devriese LA, Haesebrouck F. Differences in antibiotic resistance patterns of *Enterococcus faecalis* and *Enterococcus faecium* strains isolated from farm and pet animals. Antimicrob Agents Chemother 2001;45(5):1374–1378.

67. Voegel LP. Path of drug resistance from farm to clinic. Science 2002;295(5555):625.

68 Kruse M, Kilic B, Flick B, Stahlmann R. Effect of quinupristin/dalfopristin on 3T3 and Eahy926 cells in vitro in comparison to other antimicrobial agents with the potential to induce infusion phlebitis. Arch Toxicol 2007;81(6):447–52.

69. Johnson CA, Taylor CA 3rd, Zimmerman SW, Bridson WE, Chevalier P, Pasquier O, Baybutt RI. Pharmacokinetics of quinupristin–dalfopristin in continuous ambulatory peritoneal dialysis patients. Antimicrob Agents Chemother 1999;43(1):152–6.

70. Chevalier P, Rey J, Pasquier O, Leclerc V, Baguet JC, Meyrier A, Harding N, Montay G. Pharmacokinetics of quinupristin/dalfopristin in patients with severe chronic renal insufficiency. Clin Pharmacokinet 2000;39(1):77–84.

71 Carver PL, Whang E, VandenBussche HL, Kauffman CA, Malani PN. Risk factors for arthralgias or myalgias associated with quinupristin?dalfopristin therapy. Pharmacotherapy 2003;23:159–64.

72. Fost DA, Leung DY, Martin RJ, Brown EE, Szefler SJ, Spahn JD. Inhibition of methylprednisolone elimination in the presence of clarithromycin therapy. J Allergy Clin Immunol 1999;103(6):1031–5.

Sulfonamides

See also Trimethoprim and co-trimoxazole

General Information

The term "sulfonamide" is generic for derivatives of para-aminobenzenesulfonamide (sulfanilamide), which is similar in structure to para-aminobenzoic acid (PABA), a cofactor required by bacteria for folic acid synthesis. The sulfonamides act by competitive inhibition of the incorporation of PABA into tetrahydropteroic acid. Sulfonamides have a higher affinity for the microbial enzyme tetrahydropteroic acid synthetase than the natural substrate PABA. Sulfonamides have a wide range of antimicrobial activity against both Gram-positive and Gram-negative organisms (1). In therapeutic dosages they have only a bacteriostatic effect, and as single agents therefore have a limited role in drug therapy (1). Sulfonamides have been combined with trimethoprim or trimethoprim analogues, since such combinations result in a bactericidal effect (2). Adverse reactions during the administration of sulfamethoxazole + trimethoprim (co-trimoxazole, BAN) can be due to either compound. Although sulfonamides are thought to cause adverse reactions more often than trimethoprim, the culprit in the individual patient can only be determined by re-exposure to the individual agents.

Based on their pharmacological properties and clinical uses, sulfonamides can be classified into four groups (1):

1. short- or medium-acting sulfonamides;
2. long-acting sulfonamides;
3. topical sulfonamides;
4. sulfonamide derivatives used for inflammatory bowel disease.

Short- or medium-acting sulfonamides include the earliest varieties of azosulfonamides (Prontosil, Neoprontosil), sulfapyridine (Dagenan), sulfathiazole (Cibazol), sulfanilamide, and sulfadiazine. With the exception of sulfadiazine, these compounds are no longer used. Sulfadiazine and more recent compounds, including sulfafurazole (sulfisoxazole), sulfamethoxazole, sulfametrole, sulfacitine, and sulfamethizole, are rapidly absorbed and rapidly eliminated. Compared with the older generation they are more soluble, less toxic, and probably less allergenic. Sulfamethoxazole is a medium-acting sulfonamide that is often combined with trimethoprim (as co-trimoxazole). Long-acting sulfonamides include sulfametoxydiazine, sulfadimethoxine, and other compounds, of which many are no longer available, as they were associated with severe hypersensitivity reactions. Although these compounds have the advantage of long administration intervals, their long half-lives (over 100 hours) can be deleterious in case of adverse reactions. A long-acting drug that is still widely used is sulfadoxine (N-(5,6-dimethoxy-4-pyrimidinyl)-sulfanilamide). It is primarily used in combination with pyrimethamine (Fansidar) for the treatment and prophylaxis of malaria.

The topical use of sulfonamides has been discouraged, because of the high risk of sensitization. Nevertheless, sulfacetamide and sulfadicramide are still used topically for eye infections. Topical silver sulfadiazine (see monograph on Silver) is widely and successfully used in treating burns and leg ulcers (3,4). Since sulfonamides are easily absorbed through skin lesions, the same adverse reactions can occur as after systemic use.

A sulfonamide derivative with special use is sulfasalazine (salicylazosulfapyridine), which has been widely used to treat ulcerative colitis and regional ileitis (Crohn's disease). It is a compound of sulfapyridine and 5-aminosalicylate, linked by a diazo bond. Sulfasalazine is broken down in the large bowel to sulfapyridine, which is absorbed systemically, and 5-aminosalicylate, which reaches high concentrations in the feces (5). Sulfasalazine is not used for the antibacterial properties of the sulfapyridine, but for the local anti-inflammatory effect of 5-aminosalicylate in the gut. Because most of the adverse reactions are thought to be due to the absorbed sulfapyridine, the combination has largely been replaced in clinical practice by newer drugs that contain only 5-aminosalicylic acid (mesalazine), such as mesalazine itself and diazosalicylate (olsalazine) (5). The aminosalicylates are covered in a separate monograph.

Observational studies

In an open trial in 25 patients treated with sulfamethoxypyridazine (1 g/day) for mucous membrane pemphigoid unresponsive to topical steroid treatment, 12% of patients were withdrawn, 4% because of allergic reactions, the others because of significant hemolysis (6).

In children with acute uncomplicated *Plasmodium falciparum* malaria, pyrimethamine + sulfadoxine (25 mg + 500 mg) and artesunate (4 mg/kg) were well tolerated, and no adverse reactions attributable to treatment were recorded (7).

Comparative studies

In a randomized, open, multicenter study in 46 patients with sight-threatening ocular toxoplasmosis, those who took pyrimethamine + sulfadiazine had significantly more adverse events, especially malaise, than those who took pyrimethamine + azithromycin (8).

General adverse effects

The frequency and severity of the adverse effects of sulfonamides correspond to those seen with other antibacterial agents (2–5%). Dose-related effects, which tend to be more troublesome than serious, include gastrointestinal symptoms, headache, and drowsiness. Crystalluria can occur, but urinary obstruction is rare. Hematological adverse effects due to folic acid antagonism occur primarily in combination with trimethoprim. Hemolytic anemia occurs in patients with enzyme deficiencies and abnormal hemoglobins. Hypersensitivity is thought to be the mechanism of many adverse effects of the sulfonamides. They can be life-threatening, and immediate withdrawal is recommended. The most important reactions include anaphylactic shock, a serum sickness-like syndrome, systemic vasculitis, severe skin reactions (Stevens–Johnson syndrome and toxic epidermal necrolysis), pneumonitis, hepatitis, and pancytopenia. Sulfonamides should not be used in the third trimester of pregnancy. In premature infants, they displace bilirubin from plasma albumin and can cause kernicterus. Carcinogenicity has not been reported with the use of sulfonamides.

Sulfanilamide and the history of adverse drug reactions

The first major drug catastrophe in the 20th century history of the public control of drugs occurred in 1937 in the USA. A pharmacist introduced Elixir Sulfanilamide, which consisted of sulfanilamide dissolved in diethylene glycol. It had been tested for flavor, appearance, and fragrance, but not for safety. After taking the drug, over 100 patients died in severe pain; many were children, who were given Elixir Sulfanilamide for sore throats and coughs. Public outrage created support for proposed legislation to reinforce the public control of drugs that was pending in the US Congress (9). This led to the US 1938 Food, Drug, and Cosmetic Act, which is still the country's legal foundation for the public control of drugs and devices intended for use in the diagnosis, cure, mitigation, treatment, or prevention of disease in humans or animals. It has been a model for similar legislation in many other countries.

The 1938 Food, Drug, and Cosmetic Act prohibited traffic in new drugs, unless they were safe for use under the conditions of use prescribed on their labels. The Act also explicitly required the labeling of drug products with adequate directions for use.

The burden of proof of harm of new drugs was laid on the Federal Food and Drug Agency (FDA). Companies that wanted to manufacture and sell new drugs in interstate commerce had to investigate their safety and report to the FDA. Unless the FDA, within a specified period of time, found that the safety of a drug had not been established, the company could proceed with its marketing. The FDA was also authorized to remove from the market any drug it could prove to be unsafe (10).

The US Supreme Court also established in 1941, in a legal case over drug adulteration, that responsible individuals in a company can be held personally accountable for the quality of the products manufactured by the company, and that distributors of pharmaceuticals are responsible for the quality of their products, even if they are manufactured elsewhere (11).

After World War II, the pharmaceuticals market changed radically, as many companies started industrial production of drugs that had previously been manufactured in pharmacies. Announcements of new industrially produced drugs were hailed as part of technological advancement, as significant a sign of progress as the launching of satellites and putting a man on the moon. However, public safeguards against the risks of drugs remained unchanged in most countries. Thus, control of the effects of drugs largely lay in the hands of the manufacturers, even though the responsibility for taking precautions rested with pharmacists and doctors.

Organs and Systems

Cardiovascular

Of 98 patients with drug-induced long QT interval, one taking sulfamethoxazole carried a single-nucleotide polymorphism (SNP; found in about 1.6% of the general population) in KCNE2, which encodes MinK-related peptide 1 (MiRP1), a subunit of the cardiac potassium channel I_{Kr} (12). Channels with the SNP were normal at baseline but were inhibited by sulfamethoxazole at therapeutic concentrations, which did not affect wild-type channels.

Cardiovascular reactions can be due to sulfonamide myocarditis or systemic vascular collapse, owing to severe adverse events such as widespread skin disease. Sulfonamide myocarditis has been described in relation to earlier sulfonamides and occurs in combination with other hypersensitivity reactions (13).

Respiratory

Respiratory reactions to sulfonamides include migratory pulmonary infiltrates, chronic pneumonia, asthma, and pulmonary angiitis. These reactions are thought to be mainly due to hypersensitivity, although the precise mechanisms are not well understood (14–16). The link to the drug has been proven in most cases by recurrence after re-exposure to the same sulfonamide or to co-trimoxazole.

Pulmonary reactions have been described with most sulfonamides. Pyrimethamine + sulfadoxine, used in malaria prophylaxis and treatment, also rarely causes pulmonary reactions (17–19). The sulfapyridine moiety of sulfasalazine, used in inflammatory bowel disease, can produce adverse pulmonary reactions (20).

The time between the last drug exposure and the first clinical symptoms varies from hours to a few days, and the lung pathology disappears in most patients within a few days after withdrawal. Most commonly, pulmonary involvement presents with fever, dyspnea, cough, and shortness of breath. Clinical examination reveals râles in the lungs, and there may be pulmonary infiltrates in the chest X-ray. Pulmonary function tests may show bronchial obstruction (18–21), and arterial blood gases show hypoxemia (19,20). Whereas bronchial obstruction is probably an immediate reaction (type I), pulmonary infiltrates may correspond to a type III reaction, similar to the mechanism responsible for extrinsic allergic alveolitis (20,21). Eosinophilia is present in 8–58% of cases (14,16,18,19,21,22). Histologically, the lung tissue is infiltrated by inflammatory cells, and in most cases the alveoli contain numerous macrophages and eosinophils in a protein-rich edema fluid.

Based on the predominant symptoms and their duration, four categories of sulfonamide-related pulmonary hypersensitivity reactions can be distinguished (23–25):

1. transient or migratory pulmonary infiltrations associated with eosinophilia (Loeffler's syndrome) (22,26–28)
2. chronic eosinophilic pneumonia (23,27)
3. asthma with pulmonary eosinophilia (14–16,21)
4. allergic angiitis with pulmonary involvement (25).

In the first three of these, the adverse reaction is limited to the lung, whereas in the fourth the lung involvement is part of a systemic reaction. Syndromes such as allergic granulomatosis and angiitis (Churg–Strauss syndrome) or Wegener's granulomatosis are not associated with the use of sulfonamides (25).

Nervous system

Neurological disturbances that have been attributed to sulfonamides include polyneuritis, neuritis, and optic neuritis (29,30). Tremor and ataxia have been described with co-trimoxazole (31,32). Aseptic meningitis can be separately caused by sulfonamides and trimethoprim. The occurrence of meningitis has been verified in most patients, with recurrence on re-exposure (33–42).

Sensory systems

Eyes
Drug induced uveitis is rare. Antibiotics that have been implicated include rifabutin and sulfonamides. Furthermore, nearly all antibiotics injected intracamerally have been reported to produce uveitis (43). Topical administration of a corticosteroid and a cycloplegic (such as atropine) is suitable as initial treatment. Withdrawal of causative drugs is not always necessary (44).

Transient myopia can be caused by topical or systematic sulfonamides (45–47).

Corneal ring formation has been described simultaneously with an erythematous skin rash in a patient known to have skin hypersensitivity (48).

Taste
In unmedicated young and elderly volunteers and HIV-infected patients, sulfamethoxazole applied to the tongue was described as sour by young subjects but bitter by elderly subjects (49).

Psychological, psychiatric

Headache, drowsiness, lowered mental acuity, and other psychiatric effects can be caused by sulfonamides (50). However, these adverse effects are rare, and the causative role of the drug is usually not clearly established.

Metabolism

Several sulfonamides, including co-trimoxazole in high doses, can produce hyperchloremic metabolic acidosis. This has even been seen in patients with extensive burns receiving topical mafenide (51). Mafenide (Sulfamylon) and its metabolite *para*-sulfamoylbenzoic acid inhibit carbonic acid anhydrase, resulting in reduced reabsorption of bicarbonate and thus bicarbonate wasting.

Electrolyte balance

Although co-trimoxazole in therapeutic doses can cause hyperkalemia (52), it is thought to be caused by the potassium-sparing effects of trimethoprim (53).

Hematologic

Sulfonamides have adverse effects on all bone marrow–derived cell lines. The resulting disturbances include hemolytic anemia, folate deficiency anemia, neutropenia, thrombocytopenia, and pancytopenia. While adverse effects on erythrocytes are rare, the rates of leukopenia, neutropenia, and thrombocytopenia are highly variable. In a hospital drug monitoring program, leukopenia or neutropenia occurred in 0.4% of 1809 patients treated with co-trimoxazole (54), and thrombocytopenia of mild-to-moderate degree in 0.1% (54,55), similar to figures recorded in other studies (56,57). Pancytopenia is an extremely rare form of adverse reaction to sulfonamides (58).

There were similar findings in children (59). It is generally believed that trimethoprim is primarily responsible for neutropenia due to co-trimoxazole.

Severe neutropenia often occurs in HIV-infected patients taking co-trimoxazole (60). It seems to be due to either folic acid deficiency or immunological mechanisms.

Erythrocytes
Sulfonamides rarely have adverse effects on erythrocytes. However, there are various mechanisms by which sulfonamide-induced hemolytic anemia can occur (61):

- abnormally high blood concentrations, due to large doses or reduced excretion of the drug in patients with renal disease (62)
- acquired hypersusceptibility, as reflected by the development of a positive Coombs' test (63,64)
- genetically determined abnormalities of erythrocyte metabolism, for example deficiency of glucose-6-phosphate dehydrogenase or of diaphorase (65,66)
- the presence of an abnormal, so-called "unstable", hemoglobin in the erythrocyte, for example hemoglobin Zürich (67,68), hemoglobin Torino (69), hemoglobin Hasharon (70), and hemoglobins H and M (66).

Simple and readily available in vitro methods have been used to demonstrate the pathogenetic mechanisms, including Coombs' test, Harris's test (71), a quantitative assay or screening for glucose-6-phosphate dehydrogenase activity after recovery (72,73), a test for Heinz bodies, the buffered isopropanol technique (74) to detect abnormal hemoglobins, and hemoglobin electrophoresis (61,66). The direct antiglobulin (Coombs') test can be negative in spite of an immune mechanism. If such a mechanism is suspected and the direct Coombs' test is negative, the indirect Coombs' test on the patient's serum with the addition of the suspected sensitizing agent can be of diagnostic value (75). Heinz bodies in the erythrocytes can be important for early differentiation of a sulfonamide-induced reaction, which could further progress to hemolytic anemia (76). This result can also be of help in distinguishing this from other kinds of anemia.

Sulfonamides are not directly associated with folate deficiency and megaloblastic anemias. Sulfasalazine can affect the absorption of folates, but inflammatory bowel disease can also be responsible for reduced folate absorption. Only in combination with trimethoprim are sulfonamides thought to deplete folate stores in patients with pre-existing deficiency of folate or vitamin B_{12} (77).

Leukocytes

Since the days when chloramphenicol was more commonly used, it has been recognized that many antimicrobial drug are associated with severe blood dyscrasias, such as aplastic anemia, neutropenia, agranulocytosis, thrombocytopenia, and hemolytic anemia. Information on this association has come predominantly from case series and hospital surveys (78–80). Some evidence can be extracted from population-based studies that have focused on aplastic anemia and agranulocytosis and their association with many drugs, including antimicrobial drugs (81,82). The incidence rates of blood dyscrasias in the general population have been estimated in a cohort study with a nested case-control analysis, using data from a General Practice Research Database in Spain (83). The study population consisted of 822 048 patients aged 5–69 years who received at least one prescription (in all 1 507 307 prescriptions) for an antimicrobial drug during January 1994 to September 1998. The main outcome measure was a diagnosis of neutropenia, agranulocytosis, hemolytic anemia, thrombocytopenia, pancytopenia, or aplastic anemia. The incidence was 3.3 per 100 000 person-years in the general population. Users of antimicrobial drugs had a relative risk (RR), adjusted for age and sex, of 4.4, and patients who took more than one class of antimicrobial drug had a relative risk of 29. Among individual antimicrobial drugs, the greatest risk was with cephalosporins (RR = 14), followed by the sulfonamides (RR = 7.6) and penicillins (RR = 3.1).

Agranulocytosis was not infrequent in the early sulfonamide era. The first cases were observed in association with sulfanilamide (84,85), Prontosil (84), sulfapyridine (85,86), sulfathiazole (87), sulfadiazine (87), and sulfasalazine (88). Even with topical silver sulfadiazine, agranulocytosis as a consequence of systemic absorption has been reported (89).

Special observations in patients with agranulocytosis favor an immunological/allergic mechanism rather than a toxic one. Several points justify this view:

1. the sulfonamide is well tolerated by most patients during the initial phase of treatment
2. sulfonamide concentrations in the serum, when determined, are not particularly high in patients with hematological complications
3. in some patients, skin rash, fever, and arthritis start concomitantly with or even before the appearance of leukopenia or agranulocytosis
4. re-exposure to a single dose can be followed by a second episode of severe agranulocytosis

5. an agglutinin for leukocytes has been identified in patients' serum shortly after withdrawal of the drug (86)
6. using in vitro techniques, positive reactions to the drug with the lymphocyte transformation test or inhibition of colony growth in bone marrow have been found (90,91); however, the results of lymphocyte transformation tests must be interpreted with caution—sometimes they are positive in patients who have been exposed to the drug without any evidence of a hypersensitivity reaction.

Platelets

Reports of drug-induced thrombocytopenia have been systematically reviewed (78). Among the 98 different drugs described in 561 articles the following antibiotics were found with level I (definite) evidence: co-trimoxazole, rifampicin, vancomycin, sulfisoxazole, cefalotin, piperacillin, methicillin, novobiocin. Drugs with level II (probable) evidence were oxytetracycline and ampicillin.

In another retrospective analysis of drug-induced thrombocytopenia reported to the Danish Committee on Adverse Drug Reactions, 192 cases caused by the most frequently reported drugs were included and analysed (92). There were pronounced drug-specific differences in the clinical appearance. Early thrombocytopenia was characteristic of cases caused by sulfonamides and co-trimoxazole. These drugs also often caused hemorrhage. Accompanying leukopenia was observed in some cases associated with co-trimoxazole. There were no patient-specific factors responsible for the heterogeneity of the clinical appearance, and factors related to the physician seemed to be of little significance.

Acute thrombocytopenia is rarely associated with the newer sulfonamides (4,93,94). The structurally related sulfonylureas and thiazide diuretics can also cause allergic thrombocytopenia (95). Although some in vitro tests have been reported to predict the occurrence of thrombocytopenia, none of these has been used routinely (96,97). Furthermore, a negative test result with a drug does not definitely exclude it as the responsible allergen.

Salivary glands

Salivary gland enlargement on repeated exposure to sulfafurazole (sulfisoxazole) has been described (98).

Gastrointestinal

Nausea, vomiting, and anorexia occur in a few patients taking sulfonamides (1). They are usually related to dosage, the disposition of the individual patient, and how the question concerning adverse effects is asked.

Liver

Increased activities of alanine transaminase and aspartate transaminase to over five times the upper limit of the reference range were reported in a randomized trial of sulfadiazine in toxoplasmic encephalitis (99).

Co-trimoxazole-induced hepatitis has been repeatedly reported. However, trimethoprim alone can also cause acute liver injury (100). Three forms of liver injury may be related to sulfonamides:

1. hepatitis of the hepatocellular type (101–105)
2. hepatitis of the mixed hepatocellular type accompanied by cholestatic features (106)
3. chronic active hepatitis, possibly leading to cirrhosis (107).

The number of cases of sulfonamide hepatitis published annually fell markedly after 1947, with the introduction of the newer short-acting derivatives (106).

Children can also be affected by drug-induced liver disease (108).

Hitherto, the connection with a sulfonamide has always been investigated by administering a test dose. Immunological in vitro methods that show sensitization to the drug, for example the lymphocyte transformation test, are of limited value. In some patients the hepatic injury develops in connection with a general reaction, such as serum sickness-like syndrome, generalized vasculitis, or rash (109,110). In patients with hypersensitivity, re-exposure can result in generalized malaise, nausea, back pain, and chills within one to several hours (103,107). However, symptoms can be delayed for as long as several days (106). Daily monitoring of liver function on re-exposure seems to be important, since subjective signs can be absent despite rising activities of serum transaminases and alkaline phosphatase (102).

Even in patients with chronic active hepatitis, the histopathology of the liver damage was indistinguishable from non-drug-induced pathology. The degree of piecemeal necrosis usually varies from one area to another. Antinuclear factor and lupus erythematosus factor were positive in some cases (107). Early recognition of drug-induced liver disease is of great importance, since liver injury can be completely reversible after withdrawal.

Pancreas

Pancreatitis has been attributed to sulfonamides. Sulfamethizole (34) and sulfasalazine (111) have been implicated by re-exposure. The 5-aminosalicylic acid moiety of sulfasalazine may also be responsible (112,113).

Urinary tract

Renal complications that occur in relation to sulfonamide administration include crystalluria, tubular necrosis, interstitial nephritis, and glomerular lesions as part of a vasculitis syndrome.

Sulfonamides and their metabolites are excreted in large amounts in the urine. They are relatively insoluble in the acid environment and tend to precipitate in the collecting tubules, calyces, and pelvis of the kidney, and possibly in the ureters. The course is typically benign but adequate hydration and alkalinization may be required (114). Nephrocalcinosis can cause hematuria, renal colic, or acute renal insufficiency (115). Urinary obstruction with anuria/oliguria was seen primarily with the earlier, less soluble sulfonamides. With the newer and more soluble sulfonamides crystal formation is rare, as is acute

renal insufficiency due to other mechanisms. During recent years renal complications have been seen more often in patients with AIDS, because of the use of large doses of sulfonamides combined with trimethoprim against infection with *Pneumocystis jiroveci* (formerly *Pneumocystis carinii*) or *Toxoplasma* encephalitis. Reduced fluid intake and low urinary pH favor crystal formation, and so both adequate fluid intake (about 2 l/day for adults) and urine alkalinization are encouraged when larger doses of sulfonamides are used (60,115–120). For the diagnosis of sulfonamide crystalluria, the Lignin test is recommended. At room temperature crystals can even be found in the urine of patients taking sulfamethoxazole, which is readily soluble (121).

- A 48-year-old man with untreated HIV infection developed confusion and dyspnea. He had a history of ischemic heart disease and hepatitis C infection (122). His CD4 count was 50×10^6/l (reference range 400–1320). He was found to have *Pneumocystis jiroveci* pneumonia and cerebral toxoplasmosis and was given oral co-trimoxazole (320/1600 qds) for the Pneumocystis pneumonia and sulfadiazine (1.5 g qds) for the toxoplasmosis. Baseline renal function was normal. Later, the co-trimoxazole was withdrawn because of concurrent sulfadiazine treatment. On day 7, he developed macroscopic hematuria and profuse crystalluria. Renal function was normal, but 2 days later his creatinine rose to 250 µmol/l. Despite vigorous intravenous hydration the serum creatinine increased to 401 µmol/l. Renal tract ultrasound was normal but morphological examination of crystals confirmed the presence of a sulfonamide. Sulfadiazine was withdrawn and he recovered uneventfully within 1 month.

Sulfadiazine is a weak acid that will precipitate as crystals in the tubular lumen at a urine pH below 5.5; patients taking doses over 4 g/day should maintain a high oral fluid intake or receive adequate intravenous hydration.

Other renal complications reported with sulfonamides are:

- acute tubular necrosis or tubulointerstitial nephritis (117,123);
- interstitial nephritis (124), in some cases combined with granulomatous lesions (125,126);
- acute vasculitis (127);
- acute renal insufficiency in association with a serum sickness-like syndrome, generalized vasculitis, or rashes in combination with hepatic damage (109).

Acute anuria or oliguria is often the first symptom, not only in patients with tubular necrosis or tubulointerstitial nephritis, but also in those with allergic vasculitis. Non-oliguric renal insufficiency can also occur. It is not yet clear whether tubular necrosis in association with sulfonamides is a toxic, collateral, or hypersusceptibility reaction. The unstable hydroxylamine metabolites of some sulfonamides can act as direct renal toxins.

In a French analysis of 22 510 urinary calculi performed by infrared spectroscopy, drug-induced urolithiasis was divided into two categories: first, stones with drugs physically embedded ($n = 238$; 1.0%), notably indinavir monohydrate ($n = 126$; 53%), followed by triamterene ($n = 43$; 18%), sulfonamides ($n = 29$; 12%), and amorphous silica ($n = 24$; 10%); secondly, metabolic nephrolithiasis induced by drugs ($n = 140$; 0.6%), involving mainly calcium/vitamin D supplementation ($n = 56$; 40%) and carbonic anhydrase inhibitors ($n = 33$; 24%) (128). Drug-induced stones are responsible for about 1.6% of all calculi in France. Physical analysis and a thorough drug history are important elements in the diagnosis.

Skin

Rashes are common during sulfonamide administration, and the rate increases with duration of therapy. Maculopapular reactions are most common and occur in about 1–3% of patients (129–133). In a survey of 5923 pediatric records, 3.46% of prescriptions for sulfonamides were followed by the development of a rash, although none was severe enough to require hospitalization (134).

Urticaria, fixed drug eruptions (132,135–138), erythema nodosum (139), photosensitivity reactions (140), and generalized skin reactions involving light-exposed areas (140–142) are less common. Topical silver sulfadiazine cause local reactions, consisting of rash, pruritus, or a burning sensation in 2.5% of patients (4,51).

- Fixed drug eruption has been described in a patient treated with the non-thiazide sulfonamide diuretic indapamide; an oral challenge test showed cross-reactivity to sulfamethoxazole and sulfadiazine (143).

Generalized cutaneous depigmentation after sulfamide therapy occurred in a 41-year-old man (144). Melanocytes were not seen on electron microscopy, but there were clear cells with the characteristics of Langerhans cells along the basal layer.

Other skin eruptions seen with sulfonamides include erythema multiforme, vesicular and bullous rashes, and exfoliative dermatitis (145). In erythema multiforme, linear depositions of IgA at the dermoepidermal junction have been suggested to play a pathogenic role (133).

- Linear IgA dermatosis with erythema multiforme-like clinical features has been reported in a 19-year-old man several days after completion of a 5-day-course treatment with sulfadimethoxine (500 mg bd) for a flu-like syndrome (135). Treatment with methylprednisolone (150 mg) with gradual dosage reduction was started. Slow improvement was followed by a flare-up after reduction to 80 mg/day. Therapy was changed to dapsone 100 mg/day, and there was a dramatic improvement.

The most severe skin reactions associated with sulfonamides are the severe forms of erythema multiforme, Stevens–Johnson syndrome, and toxic epidermal necrolysis (145–153). In a study from Cameroon, eight of ten patients with toxic epidermal necrolysis had taken sulfonamides (five sulfadoxine, three sulfamethoxazole); two patients died after taking sulfadoxine (154).

Mortality in drug-induced toxic epidermal necrolysis has been estimated to be about 20–30% (155,156), and in Stevens–Johnson syndrome 1–10% (33,147–149,152). Some severe skin reactions start with a maculopapular rash or generalized erythema. The culprit drug is often either a long-acting formulation or a short-acting drug that has been continued over a long period. In both toxic epidermal necrolysis and Stevens–Johnson syndrome immediate withdrawal of the sulfonamide and all other non-essential drugs is required, as well as adequate supportive therapy with fluids, proteins, and electrolytes, in order to prevent renal insufficiency and respiratory distress syndrome (155,156). Occasionally, toxic epidermal necrolysis must be distinguished from staphylococcal scalded skin syndrome (Lyell's syndrome) by histology. In toxic epidermal necrolysis, there is subepidermal cleavage of the skin at the level of the basal cells, resulting in full-thickness denudation, whereas in scalded skin syndrome the split occurs in the upper epidermis near the granular layer just beneath the stratum corneum (157).

The role of sulfonamides as an etiological factor in the Stevens–Johnson syndrome is extremely difficult to evaluate, except for patients with re-exposure or in situations where the drug was given prophylactically for meningitis (147) or pneumonia (148,149,152,153). In the first epidemiological study in 1968, 100 000 individuals were given prophylactic sulfadoxine (Fanasil) and 997 (1.0%) had skin reactions (147). Of these, about 100 had severe reactions, such as erythroderma with jaundice, Stevens–Johnson syndrome, or toxic epidermal necrolysis; 11 died from these complications, that is about one in 10 000 patients treated with the probably causative drug. It is not known how many would have had similar skin reactions unrelated to the drug. However, the benefit to risk balance of meningitis prophylaxis clearly favored the use of sulfonamides (147). A second report (148) showed an incidence of three cases of Stevens–Johnson syndrome in 480 healthy, newly recruited Bantu mineworkers treated prophylactically with sulfadimethoxine. In a third epidemiological study in Mozambique in 1981, 149 000 inhabitants in one town were given a single dose of sulfadoxine as mass prophylaxis in an attempt to stem an outbreak of cholera (149); 22 patients with typical Stevens–Johnson syndrome were admitted to hospital over 18 days; three died.

In one case toxic epidermal necrolysis caused by co-trimoxazole improved with high-dose methylprednisolone (158). However, previous studies of the use of glucocorticoids in toxic epidermal necrolysis have given contradictory results.

The combination of pyrimethamine and sulfadoxine, used for prophylaxis of chloroquine-resistant malaria (*Plasmodium falciparum*), causes severe skin reactions in one per 5000–8000 users, with fatal reactions in one per 11 000–25 000 users (159). Even at the stage of early rash or generalized erythema, this therapy should be withdrawn (130).

In an open prospective study in 95 HIV-infected patients with successfully treated *Pneumocystis jiroveci* pneumonia, pyrimethamine + sulfadoxine (25/500 mg) was given twice weekly to prevent relapse (160). There were allergic skin reactions in 16 patients, resulting in permanent withdrawal in six. Two patients developed serious adverse reactions (Stevens–Johnson syndrome), both of whom had continued to take prophylaxis despite progressive hypersensitivity reactions.

Most of the cutaneous adverse reactions to sulfonamides are associated with increased in vitro reactivity to sulfonamide metabolites, such as unstable hydroxylamines (161,162). In some cases glutathione deficiency has been proposed as a major mechanism. This seems to be important in patients with AIDS, in whom glutathione deficiency is frequent, and in whom skin rashes are much more common than in other patients (161,163). A predominance of slow acetylator phenotype has also been observed among patients with sulfonamide hypersensitivity reactions, and an association with the phenotypes HLA-A29, B-12, and DR-7 in patients with bullous cutaneous reactions (162,164–166).

A mechanism for generalized drug-induced delayed skin reactions to sulfamethoxazole may be perforin-mediated killing of keratinocytes by drug-specific CD4+ lymphocytes (167). The requirement of interferon gamma pretreatment of keratinocytes for efficient specific killing might explain the increased frequency of drug allergies in generalized viral infections such as HIV, when interferon gamma concentrations are raised.

Immunologic

Sulfa allergy refers to a specific hypersensitivity response to a group of chemicals containing a sulfonamide moiety covalently bound to a benzene ring; drugs structurally similar to sulfonamides may cross-react, for example sulfonylureas, thiazides, and furosemide (168). Sulfa allergy is most consistent with an immune-mediated reaction with delayed onset, 7–14 days after the start of therapy, characterized by fever, rash, and eosinophilia. IgG antibodies may be present and directed against proteins in the endoplasmic reticulum (about 80% of patients) or against the drug covalently bound to protein (about 5% of patients). High-dose methylprednisolone sodium succinate (250 mg every 6 hours for 48 hours) may not only alleviate the signs but also markedly attenuate the antibody response, as reported in a 19-year-old man (169).

Hypersusceptibility to sulfonamides has been proposed to be the mechanism for many adverse reactions, including anaphylactic shock, serum sickness-like syndrome, systemic allergic vasculitis, drug fever (up to 1–2% in some series), lupus-like syndrome, myocarditis, pulmonary infiltrates, interstitial nephritis, aseptic meningitis, hepatotoxicity, blood dyscrasias (agranulocytosis, thrombocytopenia, eosinophilia, pancytopenia), and a wide variety of skin reactions (urticaria, erythema nodosum, erythema multiforme, erythroderma, toxic epidermal necrolysis, and photosensitivity).

Urticarial and maculopapular rashes are the most frequent adverse reactions to sulfonamides after gastrointestinal symptoms. Although hypersusceptibility is suspected to be the mechanism for these adverse effects, type I allergic reactions, which are induced by IgE

antibodies, have been confirmed only rarely. It appears that with the older sulfonamides severe reactions were more frequent. In some patients who have immediate hypersensitivity reactions to sulfonamides, IgE has been found that can bind to an N4-sulfonamidoyl determinant (N4-SM) (170).

Prediction
It is desirable to predict hypersusceptibility reactions to sulfonamides. IgE-induced in vitro reactions to sulfonamides have mainly been studied in the last 15 years (170–172). A lymphocyte toxicity assay showed a positive result in about 70% of patients with a maculopapular rash, an urticarial reaction, or erythema multiforme (173). This biochemical test determines the percent of cell death due to toxic metabolites. The same in vitro reaction using the hydroxylamine metabolite of sulfamethoxazole gave significantly different results in six patients with fever and skin rash with or without hepatitis than in control patients (161). Unfortunately, in most adverse reactions it is not known whether the reaction is dose-related or allergic. Individual differences in metabolism predispose to idiosyncratic reactions, for example sulfonamides are metabolized by N-acetylation (mediated by a genetically determined polymorphic enzyme) and oxidation to potentially toxic metabolites (165,166). Fever and rash were observed significantly more often in slow than in fast acetylators (165,166). Systemic glutathione deficiency, with a consequently reduced capacity to scavenge such toxic metabolites, might contribute to these adverse reactions, particularly in patients with AIDS (174,175). In a child with dihydropteridine reductase deficiency, a variant of phenylketonuria, adverse drug reactions occurred to co-trimoxazole (176). Unfortunately, there are no reliable in vitro tests to predict idiosyncratic reactions in vivo (161,162,165,166,175).

Cross-reactivity
The immunogenicity of sulfonamide antimicrobials may be due to the presence of an arylamine group at the N4 position of the sulfonamide molecule. Thus, allergic cross-reactions between different sulfonamides can occur. Therefore, in cases of known hypersusceptibility to a specific sulfonamide exposure to other sulfonamides should be avoided. Cross-reactions can even occur with para-aminosalicylic acid and local anesthetics of the procaine type; however, the real frequency of these cross-sensitivities is not known and their significance is undetermined. It should be noted that as many as 50% of patients with rash have recovered in spite of continued treatment with the same drug (177), and even agranulocytosis did not occur after later re-exposures to the causative agent (163).

Susceptibility factors
In an in vitro study, plasma from HIV-positive patients was less able to detoxify nitrososulfamethoxazole than control plasma, suggesting that a disturbance in redox balance in HIV-positive patients may alter metabolic

detoxification capacity, thereby predisposing to sulfamethoxazole hypersensitivity (178).

Types of reaction
Type I reactions
Anaphylactic shock occurs rarely with sulfonamides (161,170,171,179,180). Anaphylaxis to a central venous catheter (ARROWg + ard Blue Catheter) coated with chlorhexidine and sulfadiazine has been reported in a 50-year-old man (181).

Type III reactions
A serum sickness-like syndrome has been observed during sulfonamide administration. This diagnosis should be limited to patients with at least three of the symptoms of classical serum sickness, that is fever, rash, allergic arthritis, lymphadenopathy, and possibly leukopenia or neutropenia. Histologically, severe serum sickness-like syndrome seems to correspond to an allergic vasculitis (127,182). Most of the descriptions of serum sickness-like syndrome with histopathological documentation have been associated with older sulfonamides that are no longer used (183). In some severe forms of serum sickness-like syndrome, the reaction can be complicated by a number of unusual organ manifestations, including plasmacytosis, lymphocytosis, monoclonal gammopathy (184,185), interstitial myocarditis (13,33), allergic pneumonitis, nephropathy, liver damage, and nervous system disorders (127,182).

Lupus-like syndromes
Sulfonamides can cause three different clinical and biological syndromes similar or identical to systemic lupus erythematosus (186,187):

1. exacerbation of pre-existing lupus erythematosus
2. triggering of lupus erythematosus in a susceptible patient
3. serum sickness-like syndrome resembling lupus erythematosus clinically and serologically.

There may be positive LE cells and antinuclear factors. In exacerbation or triggering of lupus erythematosus, two pathogenetic mechanisms may be involved:

1. a reaction to the pharmacological properties of the drug, such as occurs with other drugs, such as hydralazine, diphenylhydantoin, procainamide, isoniazid, and practolol (186–191);
2. a hypersensitivity reaction (187,192,193).

In type I reactions, exposure time, and especially re-exposure time, are usually longer than 1–2 months. In type II reactions, exposure is more variable, lasting from hours to days or up to 1–2 months (186–188). Some patients with ulcerative colitis have developed arthropathy, possibly polyserositis, hematological abnormalities, and even loss of consciousness with positive LE cell and antinuclear antibody tests during treatment with sulfasalazine (188,189).

Diagnosis
No diagnostic tests are available to confirm sulfonamide hypersensitivity, and while avoidance of the drug is generally appropriate when a previous hypersensitivity reaction is suspected, desensitization protocols are available for use in HIV patients in whom *Pneumocystis jiroveci* pneumonia prophylaxis or treatment is indicated (194).

Desensitization
Desensitization has been tried with sulfonamides and especially co-trimoxazole. Desensitization with the combination seems to be essential in patients with AIDS, since co-trimoxazole is the first choice against *Pneumocystis jiroveci* pneumonia and toxoplasmosis. Desensitization is successful in 75% of patients with AIDS (195–197). However, the procedure is not completely safe and even anaphylactic shock can occur (171).

Body temperature
Drug fever due to sulfonamides is usually accompanied by a skin reaction; however, fever without other manifestations can occur (165,166).

Long-Term Effects
Drug resistance
Salmonella typhimurium DT104 is usually resistant to ampicillin, chloramphenicol, streptomycin, sulfonamides, and tetracycline. An outbreak of 25 culture-confirmed cases of multidrug-resistant *S. typhimurium* DT104 has been identified in Denmark (198). The strain was resistant to the abovementioned antibiotics and nalidixic acid and had reduced susceptibility to fluoroquinolones. A swineherd was identified as the primary source (198). The DT104 strain was also found in cases of salmonellosis in Washington State, and soft cheese made with unpasteurized milk was identified as an important vehicle of its transmission (199).

Second-Generation Effects
Fertility
Male infertility with oligospermia has been reported during treatment with sulfasalazine (200,201). However, inflammatory bowel disease can also affect the maturation of spermatozoa (202).

Teratogenicity
The sulfonamides appear to have little if any effect on early human development. This is indicated by the absence of case reports or epidemiological survey data during pregnancy. In one study of 50 282 mother–child pairs, 1455 were exposed to sulfonamides during the first 4 months; there was no increase in the relative risk of any malformation (203,204).

Malaria during pregnancy is associated with an increased risk of severe anemia and babies of low birth weight. Effective intermittent therapy with pyrimethamine + sulfadoxine reduces parasitemia and severe anemia and improves birth weight in areas in which

Plasmodium falciparum is sensitive to this combination. In an open, prospective trial in 287 pregnant women in the Gambia who were exposed to a single dose of a combination of artesunate and pyrimethamine + sulfadoxine there was no evidence of a teratogenic or otherwise harmful effect (205).

Fetotoxicity

Sulfonamides should not be given to pregnant women in the third trimester of pregnancy. They can displace bilirubin from plasma albumin and cause kernicterus (bilirubin encephalopathy) (206–209). For the same reason, the administration of sulfonamides to lactating women or premature infants should be avoided. Successful treatment of neonatal hyperbilirubinemia with higher bilirubin concentrations has been established using exchange transfusion and phototherapy.

Susceptibility Factors

Genetic factors

The acetylator phenotype of a patient can affect the frequency and severity of adverse reactions to drugs that are metabolized by acetylation (66,165,166).

In patients with porphyria, sulfonamides should not be used (210).

Other features of the patient

Although patients with HIV infection are more likely to develop generalized skin reactions to sulfonamides, they can be used for prophylaxis and therapy.

Relative contraindications to sulfonamides are systemic lupus erythematosus and a known predisposition to lupus-like reactions. Allergic reactions to antimicrobials are frequent in patients with Sjögren's syndrome. They are especially susceptible to reactions to penicillins, cephalosporins, and sulfonamides, but reactions to macrolides and tetracyclines also seem to be over-represented in these patients (211).

Drug Administration

Drug administration route

Sulfacetamide sodium (Albucid) in solutions stronger than 5% can cause burning and stinging when applied to the eyes, but this brief discomfort is usually tolerated without serious complaints. Sulfacetamide still compares favorably with newer antibiotics since it is effective against superficial ocular infections caused by a variety of microorganisms. However, serious allergic reactions can develop after ocular treatment (212).

The sulfonamides have a bacteriostatic rather than a bactericidal action. Many local anesthetics used in the eye are esters of *para*-aminobenzoic acid, and such drugs will interfere with the action of sulfonamides. Thus, to obtain the maximum effect from instillation of sulfonamide eyedrops, these drugs should not be used until the effect of the local anesthesia disappears.

Drug–Drug Interactions

Alkalis

Urine alkalinization increases the urinary excretion of sulfonamides (213,214).

Barbiturates

Sulfafurazole enhances the anesthetic effect of short-acting intravenous barbiturates, by competitive displacement from binding sites on plasma albumin (215).

CYP2C9

The inhibitory effect of sulfonamides on tolbutamide metabolism is mediated by CYP2C9, and therefore other drugs with narrow therapeutic ranges that are metabolized by CYP2C9, such as phenytoin and warfarin, deserve attention when certain sulfonamides (for example sulfaphenazole, sulfadiazine, sulfamethizole, and sulfafurazole) are co-administered (216).

Phenylbutazone

Phenylbutazone can displace sulfonamides from protein-binding sites (SED-9, 144) (226).

Phenytoin

Attention is warranted when certain sulfonamides (for example sulfaphenazole, sulfadiazine, sulfamethizole, and sulfafurazole) are co-administered with CYP2C9 substrates with narrow therapeutic ranges, such as phenytoin.

Sulfonylureas

Hypoglycemia, often during the first hours of combining the two drugs, is the result of an important interaction between sulfonylureas and sulfonamides (217–220). For example, the half-life of tolbutamide was increased from 9.5 to 29 hours by chronic sulfaphenazole and from 9.2 to 26 hours by a single dose of sulfaphenazole (221). Interference by sulfonamides with the protein binding of sulfonylureas may contribute.

Most reports of this interaction have described hypoglycemia with tolbutamide in combination with sulfaphenazole (217,218,221,222), sulfafurazole (218), or co-trimoxazole (220,223). The inhibitory effect of sulfonamides on tolbutamide metabolism is mediated by CYP2C9 (216).

Chlorpropamide produces the same interaction (224).

The combination of gliclazide, fluconazole, and sulfamethoxazole can cause severe hypoglycemia (225).

Tolbutamide is mainly metabolized by CYP2C9, which also has a role in the metabolism of sulfonamides. Of various sulfonamides, sulfaphenazole had the largest inhibitory effect on the metabolism of tolbutamide in vitro (227). This gives a theoretical basis for being careful when tolbutamide and sulfonamides are co-administered.

Warfarin

Attention is warranted when certain sulfonamides (for example sulfaphenazole, sulfadiazine, sulfamethizole, and sulfafurazole) are co-administered with CYP2C9 substrates with narrow therapeutic ranges, such as warfarin (215).

Interference with Diagnostic Tests

Folate measurement

Sulfonamides, in contrast to trimethoprim, do not interfere with the microbiological assay of folate (77).

Theophylline

Sulfamethoxazole distorts the results of high performance liquid chromatography used for detection of theophylline plasma concentrations; the antibiotic should be withdrawn 24 hours before the procedure (SEDA-6, 7).

References

1. Zinner SH, Mayer KH. Basic principles in the diagnosis and management of infectious diseases: sulfonamides and trimethoprim. In: Mandell GL, Douglas RG, Bennett JE, editors. Principles and Practice of Infectious Diseases. 4th ed.. Edinburgh: Churchill Livingstone, 1996:354.
2. Bushby SR, Hitchings GH. Trimethoprim, a sulphonamide potentiator. Br J Pharmacol Chemother 1968;33(1):72–90.
3. Ballin JC. Evaluation of a new topical agent for burn therapy. Silver sulfadiazine (silvadene). JAMA 1974;230(8):1184–5.
4. Lowbury EJ, Babb JR, Bridges K, Jackson DM. Topical chemoprophylaxis with silver sulphadiazine and silver nitrate chlorhexidine creams: emergence of sulphonamide-resistant Gram-negative bacilli. BMJ 1976;1(6008):493–6.
5. Sutherland LR, May GR, Shaffer EA. Sulfasalazine revisited: a meta-analysis of 5-aminosalicylic acid in the treatment of ulcerative colitis. Ann Intern Med 1993;118(7):540–9.
6. Thornhill M, Pemberton M, Buchanan J, Theaker E. An open clinical trial of sulphamethoxypyridazine in the treatment of mucous membrane pemphigoid. Br J Dermatol 2000;143(1):117–26.
7. von Seidlein L, Milligan P, Pinder M, Bojang K, Anyalebechi C, Gosling R, Coleman R, Ude JI, Sadiq A, Duraisingh M, Warhurst D, Alloueche A, Targett G, McAdam K, Greenwood B, Walraven G, Olliaro P, Doherty T. Efficacy of artesunate plus pyrimethamine-sulphadoxine for uncomplicated malaria in Gambian children: a double-blind, randomised, controlled trial. Lancet 2000;355(9201):352–7.
8. Bosch-Driessen LH, Verbraak FD, Suttorp-Schulten MS, van Ruyven RL, Klok AM, Hoyng CB, Rothova A. A prospective, randomized trial of pyrimethamine and azithromycin vs pyrimethamine and sulfadiazine for the treatment of ocular toxoplasmosis. Am J Ophthalmol 2002;134(1):34–40.
9. US Congress. House Committee on Interstate and Foreign Commerce and Its Subcommittee on Public Health and Environment, 1974. In: A Brief Legislative History of the Food, Drug, and Cosmetic Act. Washington DC: US Government Printing Offices, 1974:1–4 Committee Print No. 14.
10. USA, 52 Stat. 1040, 75th Congress, 3rd session, 25 June, 1938.
11. USA vs Dotterweich, 1941.
12. Sesti F, Abbott GW, Wei J, Murray KT, Saksena S, Schwartz PJ, Priori SG, Roden DM, George AL Jr, Goldstein SA. A common polymorphism associated with antibiotic-induced cardiac arrhythmia. Proc Natl Acad Sci USA 2000;97(19):10613–8.
13. French AJ, Weller CV. Interstitial myocarditis following the clinical and experimental use of sulfonamide drugs. Am J Pathol 1942;18:109.
14. Jones GR, Malone DN. Sulphasalazine induced lung disease. Thorax 1972;27(6):713–7.
15. Thomas P, Seaton A, Edwards J. Respiratory disease due to sulphasalazine. Clin Allergy 1974;4(1):41–7.
16. Scherpenisse J, van der Valk PD, van den Bosch JM, van Hees PA, Nadorp JH. Olsalazine as an alternative therapy in a patient with sulfasalazine-induced eosinophilic pneumonia. J Clin Gastroenterol 1988;10(2):218–20.
17. Svanbom M, Rombo L, Gustafsson L. Unusual pulmonary reaction during short term prophylaxis with pyrimethamine-sulfadoxine (Fansidar). BMJ (Clin Res Ed) 1984;288(6434):1876.
18. Berliner S, Neeman A, Shoenfeld Y, Eldar M, Rousso I, Kadish U, Pinkhas J. Salazopyrin-induced eosinophilic pneumonia. Respiration 1980;39(2):119–20.
19. Tydd TF. Sulphasalazine lung. Med J Aust 1976;1(16):570–3.
20. Wang KK, Bowyer BA, Fleming CR, Schroeder KW. Pulmonary infiltrates and eosinophilia associated with sulfasalazine. Mayo Clin Proc 1984;59(5):343–6.
21. Klinghoffer JF. Löffler's syndrome following use of a vaginal cream. Ann Intern Med 1954;40(2):343–50.
22. Fiegenberg DS, Weiss H, Kirshman H. Migratory pneumonia with eosinophilia associated with sulfonamide administration. Arch Intern Med 1967;120(1):85–9.
23. Crofton JW, Livingstone JL, Oswald NC, Roberts AT. Pulmonary eosinophilia. Thorax 1952;7(1):1–35.
24. Reeder WH, Goodrich BE. Pulmonary infiltration with eosinophilia (PIE syndrome). Ann Intern Med 1952;36(5):1217–40.
25. Chumbley LC, Harrison EG Jr, DeRemee RA. Allergic granulomatosis and angiitis (Churg–Strauss syndrome). Report and analysis of 30 cases. Mayo Clin Proc 1977;52(8):477–84.
26. Loeffler W. Ueber flüchtige Lungenilfiltrate (mit Eosinophilie). Beitr Klin Tuberk 1932;79:368.
27. Ellis RV, McKinlay CA. Allergic pneumonia. J Lab Clin Med 1941;26:1427.
28. Von Meyenburg H. Das eosinophile Lungenilfilträt: pathologische Anatomie und Pathogenese. Schweiz Med Wochenschr 1942;72:809.
29. Plogge H. Ueber zentrale und periphere nervöse Schäden nach Eubasinummedikation. Dtsch Z Nervenheilkd 1940;151:205.
30. Bucy PC. Toxic optic neuritis resulting from sulfanilamide. JAMA 1937;109:1007.
31. Borucki MJ, Matzke DS, Pollard RB. Tremor induced by trimethoprim-sulfamethoxazole in patients with the acquired immunodeficiency syndrome (AIDS). Ann Intern Med 1988;109(1):77–8.
32. Liu LX, Seward SJ, Crumpacker CS. Intravenous trimethoprim–sulfamethoxazole and ataxia. Ann Intern Med 1986;104(3):448.

33. Whalstrom B, Nystrom-Rosander C, Aberg H, Friman G. [Recurrent meningitis and perimyocarditis after trimethoprim.]Lakartidningen 1982;79(51):4854–5.

34. Barrett PV, Thier SO. Meningitis and pancreatitis associated with sulfamethizole. N Engl J Med 1963;268:36–7.

35. Haas EJ. Trimethoprim–sulfamethoxazole: another cause of recurrent meningitis. JAMA 1984;252(3):346.

36. Kremer I, Ritz R, Brunner F. Aseptic meningitis as an adverse effect of co-trimoxazole. N Engl J Med 1983;308(24):1481.

37. Auxier GG. Aseptic meningitis associated with administration of trimethoprim and sulfamethoxazole. Am J Dis Child 1990;144(2):144–5.

38. Biosca M, de la Figuera M, Garcia-Bragado F, Sampol G. Aseptic meningitis due to trimethoprim–sulfamethoxazole. J Neurol Neurosurg Psychiatry 1986;49(3):332–3.

39. Joffe AM, Farley JD, Linden D, Goldsand G. Trimethoprim–sulfamethoxazole-associated aseptic meningitis: case reports and review of the literature. Am J Med 1989;87(3):332–8.

40. Carlson J, Wiholm BE. Trimethoprim associated aseptic meningitis. Scand J Infect Dis 1987;19(6):687–91.

41. Gordon MF, Allon M, Coyle PK. Drug-induced meningitis. Neurology 1990;40(1):163–4.

42. Derbes SJ. Trimethoprim-induced aseptic meningitis. JAMA 1984;252(20):2865–6.

43. Moorthy RS, Valluri S, Jampol LM. Drug-induced uveitis. Surv Ophthalmol 1998;42(6):557–70.

44. Anonymous. Drug-induced uveitis can usually be easily managed. Drugs Ther Perspect 1998;11:11–4.

45. Bovino JA, Marcus DF. The mechanism of transient myopia induced by sulfonamide therapy. Am J Ophthalmol 1982;94(1):99–102.

46. Hook SR, Holladay JT, Prager TC, Goosey JD. Transient myopia induced by sulfonamides. Am J Ophthalmol 1986;101(4):495–6.

47. Carlberg O. Zur Genese der Sulfonamidmyopie. Acta Ophthalmol 1942;20:275.

48. Gutt L, Feder JM, Feder RS, Grammer LC, Shaughnessy MA, Patterson R. Corneal ring formation after exposure to sulfamethoxazole. Case reports. Arch Ophthalmol 1988;106(6):726–7.

49. Schiffman SS, Zervakis J, Westall HL, Graham BG, Metz A, Bennett JL, Heald AE. Effect of antimicrobial and anti-inflammatory medications on the sense of taste. Physiol Behav 2000;69(4–5):413–24.

50. Wade A, Reynolds JE. In: Sulfonamides. London: The Pharmaceutical Press, 1982:1457.

51. White MG, Asch MJ. Acid-base effects of topical mafenide acetate in the burned patient. N Engl J Med 1971;284(23):1281–6.

52. Alappan R, Buller GK, Perazella MA. Trimethoprim–sulfamethoxazole therapy in outpatients: is hyperkalemia a significant problem? Am J Nephrol 1999;19(3):389–94.

53. Marinella MA. Trimethoprim-induced hyperkalemia: An analysis of reported cases. Gerontology 1999;45(4):209–12.

54. Baumgartner A, Hoigné R, Müller U, et al. Medikamentöse Schäden des Blutbildes: Erfahrungen aus dem Komprehensiven Spital-Drug-Monitoring Bern, 1974–1979. Schweiz Med Wochenschr 1982;112:1530.

55. Muller U. Hämatologische Nebenwirkungen von Medikamenten. [Hematologic side effects of drugs.] Ther Umsch 1987;44(12):942–8.

56. Havas L, Fernex M, Lenox-Smith I. The clinical efficacy and tolerance of co-trimoxazole (Bactrim; Septrim). Clin Trials J 1973;3:81.

57. Hoigne R, Klein U, Muller U. Results of four-week course of therapy of urinary tract infections: a comparative study using trimethoprim with sulfamethoxazole (Bactrim; Roche) and trimethoprim alone. In: Hejzlar M, Semonsky M, Masak S, editors. Advances in Antimicrobial and Antineoplastic Chemotherapy. Munchen-Berlin-Wien: Urban and Schwarzenberg, 1972:1283.

58. Scott JL, Cartwright GE, Wintrobe MM. Acquired aplastic anemia: an analysis of thirty-nine cases and review of the pertinent literature. Medicine (Baltimore) 1959;38(2):119–72.

59. Jick SS, Jick H, Habakangas JA, Dinan BJ. Co-trimoxazole toxicity in children. Lancet 1984;2(8403):631.

60. Malinverni R, Blatter M. Ambulante Therapie und Prophylaxe der haufigsten HIV-assoziierten opportunistischen Infektionen. [Ambulatory therapy and prevention of the most frequent HIV-associated opportunistic infections.] Schweiz Med Wochenschr 1991;121(34):1194–204.

61. Zinkham WH. Unstable hemoglobins and the selective hemolytic action of sulfonamides. Arch Intern Med 1977;137(10):1365–6.

62. de Leeuw N, Shapiro L, Lowenstein L. Drug-induced hemolytic anemia. Ann Intern Med 1963;58:592–607.

63. Worlledge SM. Immune drug-induced haemolytic anemias. Semin Hematol 1969;6(2):181–200.

64. Fishman FL, Baron JM, Orlina A. Non-oxidative hemolysis due to salicylazosulfapyridine: evidence for an immune mechanism. Gastroenterology 1973;64:727.

65. Cohen SM, Rosenthal DS, Karp PJ. Ulcerative colitis and erythrocyte GD deficiency. Salicylazosulfapyridine-provoked hemolysis. JAMA 1968;205(7):528–30.

66. Meyer UA. Drugs in special patient groups: Clinical importance of genetics in drug effects. In: Melmon KL, Morelli HF, Hoffman BB, Nierenberg DW, editors. Nelmon and Morelli's Clinical Pharmacology, Basic Principles in Therapeutics. 3rd ed.. New York-St Louis-San Francisco, etc: McGraw-Hill, 1992:875.

67. Frick PG, Hitzig WH, Stauffer U. Das Hämoglobin Zürich-Syndrom. [Hemoglobin-Zurich syndrome.] Schweiz Med Wochenschr 1961;91:1203–5.

68. Hitzig WH, Frick PG, Betke K, Huisman TH. Hämoglobin Zürich: eine neue Hämoglobinanomalie mit Sulfonamid-induzierter Innenkörperanämie. [Hemoglobin Zurich: a new hemoglobin anomaly with sulfonamide-induced inclusion body anemia.] Helv Paediatr Acta 1960;15:499–514.

69. Beretta A, Prato V, Gallo E, Lehmann H. Haemoglobin Torino—alpha-43 (CD1) phenylalanine replaced by valine. Nature 1968;217(133):1016–8.

70. Adams JG, Heller P, Abramson RK, Vaithianathan T. Sulfonamide-induced hemolytic anemia and hemoglobin Hasharon. Arch Intern Med 1977;137(10):1449–51.

71. Harris JW. Studies on the mechanism of a drug-induced hemolytic anemia. J Lab Clin Med 1956;47(5):760–75.

72. Beutler E. In: Red cell metabolism. Edinburgh/London: Churchill-Livingstone, 1986:16.

73. Gaetani GD, Mareni C, Ravazzolo R, Salvidio E. Haemolytic effect of two sulphonamides evaluated by a new method. Br J Haematol 1976;32(2):183–91.

74. Huisman TH. In: Hemoglobinopathies. Edinburgh/London: Churchill-Livingstone, 1986:15.

75. Shinton NK, Wilson C. Autoimmune haemolytic anaemia due to phenacetin and p-aminosalicylic acid. Lancet 1960;1:226.

76. Lyonnais J. Production de corps de Heinz associée à la prise de salicylazosulfapyridine. [Production of Heinz

bodies after administration of salicylaszosulfapyridine.] Union Med Can 1976;105(2):203–5.

77. Streeter AM, Shum HY, O'Neill BJ. The effect of drugs on the microbiological assay of serum folic acid and vitamin B12 levels. Med J Aust 1970;1(18):900–1.

78. George JN, Raskob GE, Shah SR, Rizvi MA, Hamilton SA, Osborne S, Vondracek T. Drug-induced thrombocytopenia: a systematic review of published case reports. Ann Intern Med 1998;129(11):886–90.

79. Wright MS. Drug-induced hemolytic anemias: increasing complications to therapeutic interventions. Clin Lab Sci 1999;12(2):115–8.

80. Arneborn P, Palmblad J. Drug-induced neutropenia—a survey for Stockholm 1973–1978. Acta Med Scand 1982;212(5):289–92.

81. Baumelou E, Guiguet M, Mary JY. Epidemiology of aplastic anemia in France: a case-control study. I. Medical history and medication use. The French Cooperative Group for Epidemiological Study of Aplastic Anemia. Blood 1993;81(6):1471–8.

82. Anonymous. Anti-infective drug use in relation to the risk of agranulocytosis and aplastic anemia. A report from the International Agranulocytosis and Aplastic Anemia Study. Arch Intern Med 1989;149(5):1036–40.

83. Huerta C, Garcia Rodriguez LA. Risk of clinical blood dyscrasia in a cohort of antibiotic users. Pharmacotherapy 2002;22(5):630–6.

84. Johnston FD. Granulocytopenia following the administration of sulphanilamide compounds. Lancet 1938;2:1044.

85. Rinkoff SS, Spring M. Toxic depression of the myeloid elements following therapy with the sulfonamides: report of 8 cases. Ann Intern Med 1941;15:89.

86. Moeschlin S. Immunological granulocytopenia and agranulocytosis; clinical aspects. Sang 1955;26(1):32–51.

87. Rios Sanchez I, Duarte L, Sanchez Medal L. Agramulocitosis. Analisis de 29 episodes en 19 pacientes. [Agranulocytosis. Analysis of 29 episodes in 19 patients.] Rev Invest Clin 1971;23(1):29–42.

88. Ritz ND, Fisher MJ. Agranulocytosis due to administration of salicylazosulfapyridine (azulfidine). JAMA 1960;172:237.

89. Jarrett F, Ellerbe S, Demling R. Acute leukopenia during topical burn therapy with silver sulfadiazine. Am J Surg 1978;135(6):818–9.

90. Maurer LH, Andrews P, Rueckert F, McIntyre OR. Lymphocyte transformation observed in Sulfamylon agranulocytosis. Plast Reconstr Surg 1970;46(5):458–62.

91. Rhodes EG, Ball J, Franklin IM. Amodiaquine induced agranulocytosis: inhibition of colony growth in bone marrow by antimalarial agents. BMJ (Clin Res Ed) 1986;292(6522):717–8.

92. Pedersen-Bjergaard U, Andersen M, Hansen PB. Drug-specific characteristics of thrombocytopenia caused by non-cytotoxic drugs. Eur J Clin Pharmacol 1998;54(9–10):701–6.

93. Bottiger LE, Westerholm B. Thrombocytopenia. II. Drug-induced thrombocytopenia. Acta Med Scand 1972;191(6):541–8.

94. Gremse DA, Bancroft J, Moyer MS. Sulfasalazine hypersensitivity with hepatotoxicity, thrombocytopenia, and erythroid hypoplasia. J Pediatr Gastroenterol Nutr 1989;9(2):261–3.

95. Bottiger LE, Westerholm B. Thrombocytopenia. I. Incidence and aetiology. Acta Med Scand 1972;191(6):535–40.

96. Kelton JG, Meltzer D, Moore J, Giles AR, Wilson WE, Barr R, Hirsh J, Neame PB, Powers PJ, Walker I, Bianchi F, Carter CJ. Drug-induced thrombocytopenia is associated with increased binding of IgG to platelets both in vivo and in vitro. Blood 1981;58(3):524–9.

97. Kiefel V, Santoso S, Schmidt S, Salama A, Mueller-Eckhardt C. Metabolite-specific (IgG) and drug-specific antibodies (IgG, IgM) in two cases of trimethoprim–sulfamethoxazole-induced immune thrombocytopenia. Transfusion 1987;27(3):262–5.

98. Nidus BD, Field M, Rammelkamp CH Jr. Salivary gland enlargement caused by sulfisoxazole. Ann Intern Med 1965;63(4):663–5.

99. Chirgwin K, Hafner R, Leport C, Remington J, Andersen J, Bosler EM, Roque C, Rajicic N, McAuliffe V, Morlat P, Jayaweera DT, Vilde JL, Luft BJ. Randomized phase II trial of atovaquone with pyrimethamine or sulfadiazine for treatment of toxoplasmic encephalitis in patients with acquired immunodeficiency syndrome: ACTG 237/ANRS 039 Study. AIDS Clinical Trials Group 237/Agence Nationale de Recherche sur le SIDA, Essai 039. Clin Infect Dis 2002;34(9):1243–50.

100. Vial T, Biour M, Descotes J, Trepo C. Antibiotic-associated hepatitis: update from 1990. Ann Pharmacother 1997;31(2):204–20.

101. Fries J, Siragenian R. Sulfonamide hepatitis. Report of a case due to sulfamethoxazole and sulfisoxazole. N Engl J Med 1966;274(2):95–7.

102. Kaufman SF. A rare complication of sulfadimethoxine (Madribon) therapy. Calif Med 1967;107(4):344–5.

103. Konttinen A, Perasalo JO, Eisalo A. Sulfonamide hepatitis. Acta Med Scand 1972;191(5):389–91.

104. Konttinen A. Hepatotoxicity of sulphamethoxyridazine. BMJ 1972;2(806):168.

105. Sotolongo RP, Neefe LI, Rudzki C, Ishak KG. Hypersensitivity reaction to sulfasalazine with severe hepatotoxicity. Gastroenterology 1978;75(1):95–9.

106. Dujovne CA, Chan CH, Zimmerman HJ. Sulfonamide hepatic injury. Review of the literature and report of a case due to sulfamethoxazole. N Engl J Med 1967;277(15):785–8.

107. Tonder M, Nordoy A, Elgjo. Sulfonamide-induced chronic liver disease. Scand J Gastroenterol 1974;9(1):93–6.

108. Gutman LT. The use of trimethoprim–sulfamethoxazole in children: a review of adverse reactions and indications. Pediatr Infect Dis 1984;3(4):349–57.

109. Chester AC, Diamond LH, Schreiner GE. Hypersensitivity to salicylazosulfapyridine: renal and hepatic toxic reactions. Arch Intern Med 1978;138(7):1138–9.

110. Shaw DJ, Jacobs RP. Simultaneous occurrence of toxic hepatitis and Stevens–Johnson syndrome following therapy with sulfisoxazole and sulfamethoxazole. Johns Hopkins Med J 1970;126(3):130–3.

111. Block MB, Genant HK, Kirsner JB. Pancreatitis as an adverse reaction to salicylazosulfapyridine. N Engl J Med 1970;282(7):380–2.

112. Suryapranata H, De Vries H, et al. Pancreatitis associated with sulphasalazine. BMJ 1986;292:732.

113. Deprez P, Descamps C, Fiasse R. Pancreatitis induced by 5-aminosalicylic acid. Lancet 1989;2(8660):445–6.

114. Crespo M, Quereda C, Pascual J, Rivera M, Clemente L, Cano T. Patterns of sulfadiazine acute nephrotoxicity. Clin Nephrol 2000;54(1):68–72.

115. Perazella MA. Crystal-induced acute renal failure. Am J Med 1999;106(4):459–65.

116. Christin S, Baumelou A, Bahri S, Ben Hmida M, Deray G, Jacobs C. Acute renal failure due to sulfadiazine in patients with AIDS. Nephron 1990;55(2):233–4.
117. Miller MA, Gallicano K, Dascal A, Mendelson J. Sulfadiazine urolithiasis during antitoxoplasma therapy. Drug Invest 1993;5:334.
118. Simon DI, Brosius FC 3rd, Rothstein DM. Sulfadiazine crystalluria revisited. The treatment of *Toxoplasma* encephalitis in patients with acquired immunodeficiency syndrome. Arch Intern Med 1990;150(11):2379–84.
119. Furrer H, von Overbeck J, Jaeger P, Hess B. Sulfadiazin-Nephrolithiasis und Nephropathie. [Sulfadiazine nephrolithiasis and nephropathy.] Schweiz Med Wochenschr 1994;124(46):2100–5.
120. Craig WA, Kunin CM. Trimethoprim–sulfamethoxazole: pharmacodynamic effects of urinary pH and impaired renal function. Studies in humans. Ann Intern Med 1973;78(4):491–7.
121. Carbone LG, Bendixen B, Appel GB. Sulfadiazine-associated obstructive nephropathy occurring in a patient with the acquired immunodeficiency syndrome. Am J Kidney Dis 1988;12(1):72–5.
122 Swaminathan S, Pollett S, Loblay R, Macleod C. Sulfonamide crystals and acute renal failure. Int Med J 2006;36:399–402.
123. Robson M, Levi J, Dolberg L, Rosenfeld JB. Acute tubulo-interstitial nephritis following sulfadiazine therapy. Isr J Med Sci 1970;6(4):561–6.
124. Baker SB, Williams RT. Acute interstitial nephritis due to drug sensitivity. BMJ 1963;5346:1655–8.
125. Pusey CD, Saltissi D, Bloodworth L, Rainford DJ, Christie JL. Drug associated acute interstitial nephritis: clinical and pathological features and the response to high dose steroid therapy. Q J Med 1983;52(206):194–211.
126. Cryst C, Hammar SP. Acute granulomatous interstitial nephritis due to co-trimoxazole. Am J Nephrol 1988;8(6):483–8.
127. Van Rijssel TG, Meyler L. Necrotizing generalized arteritis due to the use of sulfonamide drugs. Acta Med Scand 1948;132:251.
128. Cohen-Solal F, Abdelmoula J, Hoarau MP, Jungers P, Lacour B, Daudon M. Les lithiases urinaires d'origine medicamenteuse. [Urinary lithiasis of medical origin.] Therapie 2001;56(6):743–50.
129. Bigby M, Jick S, Jick H, Arndt K. Drug-induced cutaneous reactions. A report from the Boston Collaborative Drug Surveillance Program on 15,438 consecutive inpatients, 1975 to 1982. JAMA 1986;256(24):3358–63.
130. Sonntag MR, Zoppi M, Fritschy D, Maibach R, Stocker F, Sollberger J, Buchli W, Hess T, Hoigne R. Exantheme unter häufig angewandten Antibiotika und anti-bakteriellen Chemotherapeutika (Penicilline, speziell Aminopenicilline, Cephalosporine und Cotrimoxazol) sowie Allopurinol. [Exanthema during frequent use of antibiotics and antibacterial drugs (penicillin, especially aminopenicillin, cephalosporin and cotrimoxazole) as well as allopurinol. Results of The Berne Comprehensive Hospital Drug Monitoring Program.] Schweiz Med Wochenschr 1986;116(5):142–5.
131. Arndt KA, Jick H. Rates of cutaneous reactions to drugs. A report from the Boston Collaborative Drug Surveillance Program. JAMA 1976;235(9):918–23.
132. Gomez B, Sastre J, Azofra J, Sastre A. Fixed drug eruption. Allergol Immunopathol (Madr) 1985;13(2):87–91.
133. Ibia EO, Schwartz RH, Wiedermann BL. Antibiotic rashes in children: a survey in a private practice setting. Arch Dermatol 2000;136(7):849–54.
134. Tonev S, Vasileva S, Kadurina M. Depot sulfonamid associated linear IgA bullous dermatosis with erythema multiforme-like clinical features. J Eur Acad Dermatol Venereol 1998;11(2):165–8.
135. Sehgal VN, Rege VL, Kharangate VN. Fixed drug eruptions caused by medications: a report from India. Int J Dermatol 1978;17(1):78–81.
136. Rollof SI. Erythema nodosum in association with sulphathiazole in children; a clinical investigation with special reference to primary tuberculosis. Acta Tuberc Scand Suppl 1950;24:1–215.
137. Kuokkanen K. Drug eruptions. A series of 464 cases in the Department of Dermatology, University of Turku, Finland, during 1966–1970. Acta Allergol 1972;27(5):407–38.
138. Harber LC, Bickers DR, Armstrong RB, et al. Drug photosensitivity: phototoxic and photoallergic mechanisms. Semin Dermatol 1982;1:183.
139. De Barrio M, Tornero P, Zubeldia JM, Sierra Z, Matheu V, Herrero T. Fixed drug eruption induced by indapamide. Cross-reactivity with sulfonamides. J Investig Allergol Clin Immunol 1998;8(4):253–5.
140. Martinez-Ruiz E, Ortega C, Calduch L, Molina I, Montesinos E, Revert A, Carda C, Navarro V, Jorda E. Generalized cutaneous depigmentation following sulfamide-induced drug eruption. Dermatology 2000;201(3):252–4.
141. Connor EE. Sulfonamide antibiotics. Prim Care Update Ob Gyns 1998;5:32–5.
142. Bergoend H, Loffler A, Amar R, Maleville J. Réactions cutanées survenues au cours de la prophylaxie de masse de la méningite cérébrospinale par un sulfamide longrétard (à propos de 997 cas). [Cutaneous reactions appearing during the mass prophylaxis of cerebrospinal meningitis with a long-delayed action sulfonamide (apropos of 997 cases).] Ann Dermatol Syphiligr (Paris) 1968;95(5):481–90.
143. Taylor GM. Stevens–Johnson syndrome following the use of an ultra-long-acting sulphonamide. S Afr Med J 1968;42(20):501–3.
144. Hernborg A. Stevens–Johnson syndrome after mass prophylaxis with sulfadoxine for cholera in Mozambique. Lancet 1985;2(8463):1072–3.
145. Cohlan SQ. Erythema multiforme exudativum associated with use of sulfamethoxypyridazine. JAMA 1960;173:799–800.
146. Gottschalk HR, Stone OJ. Stevens–Johnson syndrome from ophthalmic sulfonamide. Arch Dermatol 1976;112(4):513–4.
147. Moussala M, Beharcohen F, Dighiero P, Renard G. Le syndrome de Lyell et ses manifestations ophtalmologiques en milieu camerounais. [Lyell's syndrome and its ophthalmologic manifestations in Cameroon.] J Fr Ophtalmol 2000;23(3):229–37.
148. Hoigne R. Interne Manifestationen und Labor-befunde beim Lyell-Syndrom. In: Braun-Falco O, Bandmann HJ, editors. Das Lyell-Syndrom. Bern-Stuttgart-Wien: Verlag H Huber; 1970:27.
149. Revuz J, Roujeau JC, Guillaume JC, Penso D, Touraine R. Treatment of toxic epidermal necrolysis. Creteil's experience. Arch Dermatol 1987;123(9):1156–8.
150. Amon RB, Dimond RL. Toxic epidermal necrolysis. Rapid differentiation between staphylococcal- and drug-induced disease. Arch Dermatol 1975;111(11):1433–7.
151. Soylu H, Akkol N, Erduran E, Aslan Y, Gunes Z, Yildiran A. Co-trimoxazole-induced toxic epidermal necrolysis treated with high dose methylprednisolone. Ann Med Sci 2000;9:38–40.

152. Miller KD, Lobel HO, Satriale RF, Kuritsky JN, Stern R, Campbell CC. Severe cutaneous reactions among American travelers using pyrimethamine–sulfadoxine (Fansidar) for malaria prophylaxis. Am J Trop Med Hyg 1986;35(3):451–8.

153. Schurmann D, Bergmann F, Albrecht H, Padberg J, Grunewald T, Behnsch M, Grobusch M, Vallee M, Wunsche T, Ruf B, Suttorp N. Twice-weekly pyrimethamine-sulfadoxine effectively prevents *Pneumocystis carinii* pneumonia relapse and toxoplasmic encephalitis in patients with AIDS. J Infect 2001;42(1):8–15.

154. Shear NH, Rieder MJ, Spielberg SP, et al. Hypersensitivity reactions to sulfonamide antibiotics are mediated by a hydroxylamine metabolite. Clin Res 1987;35:717.

155. Shear NH, Spielberg SP. In vitro evaluation of a toxic metabolite of sulfadiazine. Can J Physiol Pharmacol 1985;63(11):1370–2.

156. Nixon N, Eckert JF, Holmesk B. The treatment of agranulocytosis with sulfadiazine. Am J Med Sci 1943;206:713.

157. Roujeau JC, Bracq C, Huyn NT, Chaussalet E, Raffin C, Duedari N. HLA phenotypes and bullous cutaneous reactions to drugs. Tissue Antigens 1986;28(4):251–4.

158. Shear NH, Spielberg SP, Grant DM, Tang BK, Kalow W. Differences in metabolism of sulfonamides predisposing to idiosyncratic toxicity. Ann Intern Med 1986;105(2):179–84.

159. Rieder MJ, Shear NH, Kanee A, Tang BK, Spielberg SP. Prominence of slow acetylator phenotype among patients with sulfonamide hypersensitivity reactions. Clin Pharmacol Ther 1991;49(1):13–7.

160. Schnyder B, Frutig K, Mauri-Hellweg D, Limat A, Yawalkar N, Pichler WJ. T cell-mediated cytotoxicity against keratinocytes in sulfamethoxazol-induced skin reaction. Clin Exp Allergy 1998;28(11):1412–7.

161. Dwenger CS. 'Sulpha' hypersensitivity. Anaesthesia 2000;55(2):200–1.

162. Bedard K, Smith S, Cribb A. Sequential assessment of an antidrug antibody response in a patient with a systemic delayed-onset sulphonamide hypersensitivity syndrome reaction. Br J Dermatol 2000;142(2):253–8.

163. Carrington DM, Earl HS, Sullivan TJ. Studies of human IgE to a sulfonamide determinant. J Allergy Clin Immunol 1987;79(3):442–7.

164. Sher MR, Suchar C, Lockey RF. Anaphylactic shock induced by oral desensitization to trimethoprim/sulfmethoxazole. J Allergy Immunol 1986;77:133.

165. Gruchalla RS, Sullivan TJ. Detection of human IgE to sulfamethoxazole by skin testing with sulfamethoxazoyl-poly-L-tyrosine. J Allergy Clin Immunol 1991;88(5):784–92.

166. Ghajar BM, Naranjo CA, Shear NH, Lanctot KL. Improving the accuracy of the differential diagnosis of idiosyncratic adverse drug reactions (IADRs): skin eruptions and sulfonamides. Clin Pharmacol Ther 1990;47(2):127.

167. Delomenie C, Mathelier-Fusade P, Longuemaux S, Rozenbaum W, Leynadier F, Krishnamoorthy R, Dupret JM. Glutathione S-transferase (GSTM1) null genotype and sulphonamide intolerance in acquired immunodeficiency syndrome. Pharmacogenetics 1997;7(6):519–20.

168. Coopman SA, Johnson RA, Platt R, Stern RS. Cutaneous disease and drug reactions in HIV infection. N Engl J Med 1993;328(23):1670–4.

169. Woody RC, Brewster MA. Adverse effects of trimethoprim–sulfamethoxazole in a child with dihydropteridine reductase deficiency. Dev Med Child Neurol 1990;32(7):639–42.

170. Kreuz W, Gungor T, Lotz C, Funk M, Kornhuber B. "Treating through" hypersensitivity to cotrimoxazole in children with HIV infection. Lancet 1990;336(8713):508–9.

171. Naisbitt DJ, Vilar FJ, Stalford AC, Wilkins EG, Pirmohamed M, Park BK. Plasma cysteine deficiency and decreased reduction of nitrososulfamethoxazole with HIV infection. AIDS Res Hum Retroviruses 2000;16(18):1929–38.

172. Binns PM. Anaphylaxis after oral sulphadiazine; two reactions in the same patient within eight days. Lancet 1958;1(7013):194–5.

173. Reichmann J. Anaphylaktischer Schock durch intravenöse Sulfonamidapplikation mit letalem Ausgang. [Anaphylactic shock caused by intravenous sulfonamide application with fatal outcome.] Dtsch Gesundheitsw 1960;15:1139–41.

174. Stephens R, Mythen M, Kallis P, Davies DW, Egner W, Rickards A. Two episodes of life-threatening anaphylaxis in the same patient to a chlorhexidine–sulphadiazine-coated central venous catheter. Br J Anaesth 2001;87(2):306–8.

175. Rich AR. Additional evidence of the role of hypersensitivity in the etiology of periarteritis nodosa. Bull Johns Hopkins Hosp 1942;71:375.

176. Zeek PM, Smith CC, Weeter JC. Studies on periarteritis nodosa. III. Differentiation between vascular lesions of periarteritis nodosa and of hypersensitivity. Am J Pathol 1948;24:889.

177. Delage C, Lagace R. Maladie sérique avec hyperplasie ganglionnaire pseudo-lymphomateuse secondaire à la prise de salicylazosulfapyridine. [Serum sickness with pseudolymphomatous lymph node hyperplasia caused by salicylazosulfapyridine.] Union Med Can 1975;104(4):579–84.

178. Han T, Chawla PL, Sokal JE. Sulfapyridine-induced serum-sickness-like syndrome associated with plasmacytosis, lymphocytosis and multiclonal gamma-globulinopathy. N Engl J Med 1969;280(10):547–8.

179. Lee SL, Rivero I, Siegel M. Activation of systemic lupus erythematosus by drugs. Arch Intern Med 1966;117(5):620–6.

180. Hoigne R, Biedermann HP, Naegeli HR. INH-induzierter systemischer Lupus erythematodes: 2. Beobachtungen mit Reexposition. Schweiz Med Wochenschr 1975;105:1726.

181. Clementz GL, Dolin BJ. Sulfasalazine-induced lupus erythematosus. Am J Med 1988;84(3 Pt 1):535–8.

182. Griffiths ID, Kane SP. Sulphasalazine-induced lupus syndrome in ulcerative colitis. BMJ 1977;2(6096):1188–9.

183. Hess E. Drug-related lupus. N Engl J Med 1988;318(22):1460–2.

184. Cohen P, Gardner FH. Sulfonamide reactions in systemic lupus erythematosus. JAMA 1966;197(10):817–9.

185. Honey M. Systemic lupus erythematosus presenting with sulphonamide hypersensitivity reaction. BMJ 1956;(4978):1272–5.

186. Tilles SA. Practical issues in the management of hypersensitivity reactions: sulfonamides. South Med J 2001;94(8):817–24.

187. Torgovnick J, Arsura E. Desensitization to sulfonamides in patients with HIV infection. Am J Med 1990;88(5):548–9.

188. Papakonstantinou G, Fuessl H, Hehlmann R. Trimethoprim-sulfamethoxazole desensitization in AIDS. Klin Wochenschr 1988;66(8):351–3.

189. Villar RG, Macek MD, Simons S, Hayes PS, Goldoft MJ, Lewis JH, Rowan LL, Hursh D, Patnode M, Mead PS. Investigation of multidrug-resistant *Salmonella* serotype typhimurium DT104 infections linked to raw-milk cheese in Washington State. JAMA 1999;281(19):1811–6.
190. Molbak K, Baggesen DL, Aarestrup FM, Ebbesen JM, Engberg J, Frydendahl K, Gerner-Smidt P, Petersen AM, Wegener HC. An outbreak of multidrug-resistant, quinolone-resistant *Salmonella enterica* serotype typhimurium DT104. N Engl J Med 1999;341(19):1420–5.
191. Levi AJ, Toovey S, Hudson E. Male infertility due to sulphasalazine. Gastroenterology 1981;80:1208.
192. Tobias R, Sapire KE, Coetzee T, Marks IN. Male infertility due to sulphasalazine. Postgrad Med J 1982;58(676):102–3.
193. Karbach U, Ewe K, Schramm P. Samenqualität bei Patienten mit Morbus Crohn. [Quality of semen in patients with Crohn's disease.] Z Gastroenterol 1982;20(6):314–20.
194. Heinonen OP, Slone D, Shapiro S. Antimicrobial and antiparasitic agents. In: Heinonen OP, Slone D, Shapiro S, editors. Birth Defects and Drugs in Pregnancy. 4th ed.. Boston-Bristol-London: John Wright PSG Inc, 1982:296.
195. Karkinen-Jaaskelainen, Saxen L. Maternal influenza, drug consumption, and congenital defects of the central nervous system. Am J Obstet Gynecol 1974;118(6):815–8.
196. Deen JL, von Seidlein L, Pinder M, Walraven GE, Greenwood BM. The safety of the combination artesunate and pyrimethamine–sulfadoxine given during pregnancy. Trans R Soc Trop Med Hyg 2001;95(4):424–8.
197. Brodersen R. Prevention of kernicterus, based on recent progress in bilirubin chemistry. Acta Paediatr Scand 1977;66(5):625–34.
198. Diamond I, Schmid R. Experimental bilirubin encephalopathy. The mode of entry of bilirubin-14C into the central nervous system. J Clin Invest 1966;45(5):678–89.
199. Andersen DH, Blanc WA, Crozier DN, Silverman WA. A difference in mortality rate and incidence of kernicterus among premature infants allotted to two prophylactic antibacterial regimens. Pediatrics 1956;18(4):614–25.
200. Wadsworth SJ, Suh B. In vitro displacement of bilirubin by antibiotics and 2-hydroxybenzoylglycine in newborns. Antimicrob Agents Chemother 1988;32(10):1571–5.
201. Peterkin GA, Khan SA. Iatrogenic skin disease. Practitioner 1969;202(207):117–26.
202. Antonen JA, Markula KP, Pertovaara MI, Pasternack AI. Adverse drug reactions in Sjögren's syndrome. Frequent allergic reaction and a specific trimethoption associated systemic reaction. Scand J Rheumatol 1999;28(3):157–9.
203. Hugues FC, Le Jeunne C. Systemic and local tolerability of ophthalmic drug formulations. An update. Drug Saf 1993;8(5):365–80.
204. Hartshorn EA. Drug interaction. Drug Intell 1968;2:174.
205. Kabins SA. Interactions among antibiotics and other drugs. JAMA 1972;219(2):206–12.
206. Csogor SI, Kerek SF. Enhancement of thiopentone anaesthesia by sulphafurazole. Br J Anaesth 1970;42(11):988–90.
207. Komatsu K, Ito K, Nakajima Y, Kanamitsu S, Imaoka S, Funae Y, Green CE, Tyson CA, Shimada N, Sugiyama Y. Prediction of in vivo drug–drug interactions between tolbutamide and various sulfonamides in humans based on in vitro experiments. Drug Metab Dispos 2000;28(4):475–81.
208. Christensen LK, Hansen JM, Kristensen M. Sulphaphenazole-induced hypoglycaemic attacks in tolbutamide-treated diabetic. Lancet 1963;41:1298–301.
209. Soeldner JS, Steinke J. Hypoglycemia in tolbutamide-treated diabetes; Report of two casses with measurement of serum insulin. JAMA 1965;193:398–9.
210. Dubach UC, Bückert A, Raaflaub J. Einfluss von Sulfonamiden auf die blutzuckersenkende Wirkung oraler Antidiabetica. Schweiz Med Wochenschr 1966;44:1483.
211. Wing LM, Miners JO. Cotrimoxazole as an inhibitor of oxidative drug metabolism: effects of trimethoprim and sulphamethoxazole separately and combined on tolbutamide disposition. Br J Clin Pharmacol 1985;20(5):482–5.
212. Pond SM, Birkett DJ, Wade DN. Mechanisms of inhibition of tolbutamide metabolism: phenylbutazone, oxyphenbutazone, sulfaphenazole. Clin Pharmacol Ther 1977;22(5 Pt 1):573–9.
213. Hansen JM, Christensen LK. Drug interactions with oral sulphonylurea hypoglycaemic drugs. Drugs 1977;13(1):24–34.
214. Schattner A, Rimon E, Green L, Coslovsky R, Bentwich Z. Hypoglycemia induced by co-trimoxazole in AIDS. BMJ 1988;297(6650):742.
215. Baciewicz AM, Swafford WB Jr. Hypoglycemia induced by the interaction of chlorpropamide and co-trimoxazole. Drug Intell Clin Pharm 1984;18(4):309–10.
216. Abad S, Moachon L, Blanche P, Bavoux F, Sicard D, Salmon-Ceron D. Possible interaction between gliclazide, fluconazole and sulfamethoxazole resulting in severe hypoglycaemia. Br J Clin Pharmacol 2001;52(4):456–7.
217. Wardell WM. Drug displacement from protein binding: source of the sulphadoxine liberated by phenylbutazone. Br J Pharmacol 1971;43(2):325–34.
218. Dische FE, Wernstedt C, Westermark GT, Westermark P, Pepys MB, Rennie JA, Gilbey SG, Watkins PJ. Insulin as an amyloid-fibril protein at sites of repeated insulin injections in a diabetic patient. Diabetologia 1988;31(3):158–61.
219. Hunziker T, Braunschweig S, Zehnder D, Hoigné R, Kunzi UP. Comprehensive Hospital Drug Monitoring (CHDM) adverse skin reactions, a 20-year survey. Allergy 1997;52(4):388–93.
220. Kauppinen K, Stubb S. Fixed eruptions: causative drugs and challenge tests. Br J Dermatol 1985;112(5):575–8.
221. Pasricha JS. Drugs causing fixed eruptions. Br J Dermatol 1979;100(2):183–5.
222. Epstein JH. Photoallergy. A review. Arch Dermatol 1972;106(5):741–8.
223. Bottiger LE, Strandberg I, Westerholm B. Drug-induced febrile mucocutaneous syndrome with a survey of the literature. Acta Med Scand 1975;198(3):229–33.
224. Lyell A. A review of toxic epidermal necrolysis in Britain. Br J Dermatol 1967;79(12):662–71.
225. Bjornberg A. Fifteen cases of toxic epidermal necrolysis (Lyell). Acta Dermatol Venereol 1973;53(2):149–52.
226. Alarcon-Segovia D. Drug-induced lupus syndromes. Mayo Clin Proc 1969;44(9):664–81.
227. Finegold I. Oral desensitization to trimethoprim–sulfamethoxazole in a patient with acquired immunodeficiency syndrome. J Allergy Clin Immunol 1986;78(5 Pt 1):905–8.

Teicoplanin

General Information

Teicoplanin has been used in the treatment of Gram-positive infections. Its pharmacological properties allow once-daily dosing. Its safety profile has been thoroughly investigated, and it is generally well tolerated. It diffuses well into bone tissue (1).

Teicoplanin is more effectively administered once daily than vancomycin and it can be given intramuscularly or intravenously; it is not absorbed after oral administration. It is 90% bound to plasma proteins and its elimination is primarily by renal excretion, allowing dosage adjustments to be made on the basis of the measured creatinine clearance. A dosing regimen of 12 mg/kg on day 1 followed by 6 mg/kg/day most often results in efficacious serum concentrations but premature neonates and children require higher dosages. However, doses of 10 mg/kg/day are required to achieve adequate bone concentrations, and there is little penetration into cerebrospinal fluid or aqueous or vitreous humors. In fat, concentrations can be subtherapeutic after a dose of 400 mg. Unlike vancomycin, routine drug monitoring is not required, although it may sometimes be useful for predicting the therapeutic effect of the drug.

Observational studies

In a study of the effects of teicoplanin (6–12 mg/day intravenously), 166 of 342 patients reported one or more adverse events, and in 119 they were thought to be associated with teicoplanin (2). Most had some form of hypersensitivity response (fever, rash, chills, or pruritus), and treatment was withdrawn in 59 patients. There was a clinically significant abnormality of liver function tests in 123 patients, but in only 33 was it judged to be possibly associated with teicoplanin. There was a rise in serum creatinine in 28 patients, in five of whom an association with teicoplanin was suggested.

Cell wall agents, such as glycopeptides, have concentration-independent bactericidal activity, and the time over which the antibiotic serum concentration persists over the minimal inhibitory concentration of the pathogen is a main pharmacodynamic determination of the outcome. In a two-way, randomized, open, two-period, crossover study in 10 healthy adults, teicoplanin given in two 200 mg doses intramuscularly produced steady-state trough concentrations even higher than those after once-daily intravenous administration of 400 mg (3). Conversion from intravenous to intramuscular administration may therefore allow better compliance with preserved efficacy. Intramuscular teicoplanin was well tolerated. Adverse effects reported were mild local pain for 2–3 hours, headache, and backache.

In 76 patients receiving long-term teicoplanin for chronic osteomyelitis due to oxacillin-resistant *Staphylococcus aureus*, teicoplanin had to be withdrawn in only one subject because of low-grade fever, muscular pain, and sleeplessness; these adverse effects abated after withdrawal (4).

Comparative studies

In a prospective study 56 patients with suspected or proven severe Gram-positive infections were randomized to receive vancomycin (1 g 12-hourly intravenously) or teicoplanin (200–400 mg/day intravenously or intramuscularly) (5). Clinical and bacteriological cure rates were similar, but vancomycin was associated with significantly greater toxicity (16 vancomycin, 7 teicoplanin), predominantly histamine-associated reactions (15, including two cases of red man syndrome) (SEDA-17, 312), and nephrotoxicity (five with vancomycin compared with one with teicoplanin). Four of the five vancomycin patients who developed nephrotoxicity were receiving netilmicin at the time, some with amphotericin as well.

The overall rate of adverse reactions in comparative trials was 19%, which is comparable with the rate observed with beta-lactams, but lower than the corresponding rate (39%) associated with vancomycin (6).

Only three of 422 patients treated with a single dose of teicoplanin for antimicrobial prophylaxis during hip or knee arthroplasty (400 mg by intravenous bolus at the time of anesthesia) reported adverse events compared with nine of 424 patients treated with five doses of cefazoline: one had nausea and two had erythema (7). Six patients given teicoplanin had surgical wound infections, and 57 had proven or suspected infections involving other body systems.

In another study, a single dose of teicoplanin (6 mg/kg intravenously) was compared with three doses of cefradine plus metronidazole as prophylaxis for vascular surgery. Efficacy was similar with the two regimens. Adverse events were reported by 40 of 136 patients treated with teicoplanin (versus 39 of 139 treated with cefradine plus metronidazole), and 19 (versus 15) were considered to be related to the study drug; however, none was considered definitely related (8).

In a prospective, randomized trial in 34 drug abusers a short course of a combination of a glycopeptide (vancomycin or teicoplanin) with gentamicin was significantly less effective insided endocarditis caused by *S. aureus* than a combination of cloxacillin and gentamicin (9).

Pharmacoeconomics

In a randomized, prospective, cost-effectiveness study both teicoplanin and vancomycin were assessed as second-line therapy in 66 neutropenic patients after the failure of empirical treatment with a combination of piperacillin + tazobactam and amikacin (10). The primary success of second-line therapy was equivalent, and the direct total costs were similar. Acquisition costs per dose were in favor of vancomycin, but costs derived from administering vancomycin and serum concentration monitoring led to similar costs for both regimens. With the exception of the red man syndrome, which occurred in 10% of vancomycin-treated patients but none of the teicoplanin-treated patients, toxicity (renal, liver, and ear toxicity, diarrhea, phlebitis) was also similar.

In a retrospective cost analysis, the records of 527 patients with acute leukemia were studied (11). They had been treated in a multicenter, randomized trial for febrile neutropenia with ceftazidime and amikacin plus either teicoplanin (6 mg/kg in a single dose; n = 275) or vancomycin (30 mg/kg/day in 2 doses; n = 252). Clinical responses were equivalent. Again the higher acquisition costs for teicoplanin were counterbalanced by the lower incidence of adverse events and easier administration, resulting in overall similar costs for both regimens. A total of 8% of patients treated with vancomycin reported adverse events compared with 3.2% of patients treated with teicoplanin. Rashes occurred in 6.0 versus 1.4% respectively. Nephrotoxicity, ototoxicity, fever, and hypotension occurred in very few patients.

General adverse effects

The most common adverse events associated with teicoplanin are hypersensitivity, fever, rash, diarrhea, nephrotoxicity, and thrombocytopenia (12,13). Local reactions at the injection site include pain, redness, or discomfort after intramuscular injection, or phlebitis after intravenous injection. Erythroderma has occurred during infusion of teicoplanin with fever and hypotension. Allergic reactions have been reported with teicoplanin, with cross-reactivity between teicoplanin and vancomycin documented by in vitro studies showing IgE release by basophils in response to stimulation by both vancomycin and teicoplanin. However, known hypersensitivity to vancomycin is not a contraindication to teicoplanin. Tumor-inducing effects have not been reported.

Red man syndrome

The red man syndrome associated with rapid infusion of vancomycin, resulting in histamine release, is very rarely seen with teicoplanin. Teicoplanin can usually be safely administered to patients with a history of red man syndrome due to vancomycin, as has been confirmed in six children treated with teicoplanin for febrile neutropenia and Gram-positive bacteremia (14).

However, case reports have suggested that there can be cross-reactivity between vancomycin and teicoplanin in the induction of this syndrome, although the underlying mechanism remains elusive (15).

- A swinging pyrexia (peaking at 39–40°C during the day but settling overnight) developed after the fifth day of teicoplanin treatment in a 35-year-old man with infective endocarditis due to *Streptococcus mitis*. Two days after stopping the drug he became apyrexial. Therapy with vancomycin had previously had to be interrupted because of an extensive purpuric rash and erythema, which progressed to exfoliative dermatitis and was subsequently identified as red man syndrome secondary to vancomycin.

Organs and Systems

Cardiovascular

Hypotension has been observed as part of the manifestations of a hypersensitivity reaction to teicoplanin (16).

Respiratory

Bronchospasm required withdrawal of teicoplanin in two of 310 patients (17).

Sensory systems

Ototoxicity has been reported during treatment with teicoplanin (18). A fall in high-frequency auditory threshold was observed in one patient treated with teicoplanin for a serious Gram-positive infection (19) and the hearing thresholds of 12 patients who were treated with teicoplanin for severe staphylococcal infections increased slightly but significantly over time (20). However, the causal relation between teicoplanin and alterations in auditory function has not been established in controlled clinical studies (21). Guinea pigs treated for 28 days with a maximum dose of 75 mg/kg/day remained without any evidence of functional, morphological, or histological changes indicative of ototoxicity (22). There were 11 cases of ototoxicity among 3377 patients treated with teicoplanin (21). It can be concluded that teicoplanin-associated ototoxicity is rare.

Hematologic

The overall rate of hematological adverse effects of teicoplanin was 2.2% in 1431 patients participating in a large non-comparative multicenter study (23).

In a randomized, controlled, open, multicenter study in 430 patients with suspected or proven Gram-positive infections linezolid 600 mg every 12 hours was compared with teicoplanin (dose unspecified) for up to 28 days. Three patients who received teicoplanin had hematological adverse effects rated as moderate or severe (hemolysis, anemia, leukopenia, and thrombocytopenia) (24).

A 68-year-old man developed hemolytic anemia after taking teicoplanin (25).

Hemophagocytosis after vancomycin and teicoplanin occurred in a 45-year-old woman (26).

Leukocytes

The incidence of teicoplanin-induced leukopenia is low (0.33%) and the leukopenia is usually reversible.

- A 10-year-old Malay boy with G6PD deficiency received teicoplanin (300 mg/day) for an epidural abscess and 14 days later developed an erythematous, maculopapular, non-pruritic rash over his trunk and upper limbs and a mild fever with chills (27). He had a leukocyte count of 2×10^9/l. Teicoplanin was withdrawn and by 3 days the rash had almost completely subsided. After rechallenge with a single intravenous dose of teicoplanin 300 mg, he developed a fever (39.3°C), chills, and worsening of the rash on his arms. Teicoplanin was again withdrawn. His fever resolved after 4 days and his leukocyte count normalized by 7 days.
- Neutropenia developed in a patient after 20 days of treatment with teicoplanin (28).

In patients undergoing bone marrow transplantation receiving teicoplanin for the treatment of severe sepsis, the duration of neutropenia was not prolonged by teicoplanin (29).

Teicoplanin rarely causes eosinophilia (30).

Platelets

Thrombocytopenia was associated on two consecutive occasions with teicoplanin in a patient with acute myeloid leukemia (31). The exact mechanism for this adverse event is unknown. Teicoplanin does not alter platelet function or blood coagulation (32).

- Severe thrombocytopenia occurred in a 46-year-old white man treated with teicoplanin 6 mg/kg/day for methicillin-resistant *S. aureus* bacteremia (33). The platelet count fell to $25 \times 10^9/l$ on day 8 (baseline 110). After drug withdrawal the platelet count improved within 4 days. The trough teicoplanin concentrations were 17 µg/ml on day 4 and 15.4 µg/ml on day 9.

Teicoplanin caused thrombocytopenia in one of 17 children (34).

In trials in the USA, thrombocytopenia occurred more commonly with teicoplanin than with vancomycin, but this was almost exclusively in patients who received much larger doses than are now recommended.

Gastrointestinal

Adverse events involving the gastrointestinal system are rare with teicoplanin. Diarrhea has been listed among non-specific events observed in 5.1% of patients (23).

Liver

In one study transient rises in liver enzymes occurred in 2.3% of patients taking teicoplanin (17). Other authors have reported abnormal liver enzymes (19,35–37).

Urinary tract

Teicoplanin can be nephrotoxic, but less often than vancomycin (6). However, concomitant aminoglycoside therapy in some patients makes the contribution of the glycopeptide antibiotic difficult to assess. Renal toxicity was observed more often in patients receiving the combination of netilmicin plus vancomycin than in patients treated with netilmicin plus teicoplanin (38). Similar differences in nephrotoxicity between vancomycin and teicoplanin were observed in febrile neutropenic patients receiving tobramycin plus piperacillin (39). There was significant deterioration of renal function when vancomycin and ciclosporin, but not teicoplanin and ciclosporin, were used together; the mechanism responsible for this interaction is unknown (39).

In two of 76 patients receiving long-term teicoplanin a reduction in dosage was required because of reduced renal function, which recovered within 30 days (4).

There was a fall in residual renal function during episodes of peritonitis in a prospective multicenter study in 152 children with peritoneal dialysis-associated peritonitis (40). The study evaluated the efficacy, safety, and clinical acceptance of combination treatment, including a glycopeptide (teicoplanin or vancomycin) and ceftazidime, each given either intermittently or continuously. Irreversible anuria developed in 19% of the patients with residual diuresis. In patients who retained residual renal function, there was a strong trend toward a decreased residual glomerular filtration rate (−11%). There was no difference with respect to the type of glycopeptide or the mode of use. Aminoglycosides were used concomitantly in only 12 patients and did not seem to affect the evolution of residual renal function. Underdosing/non-adherence or overdosing occurred in 8.2% of continuously treated versus 4.9% of intermittently treated patients. Intermittent and continuous intraperitoneal treatment with glycopeptides and ceftazidime were equally efficacious and safe when measured by objective clinical criteria.

In children with febrile neutropenia and Gram-positive bacteremia associated with antineoplastic drug therapy, teicoplanin was significantly less nephrotoxic than vancomycin (14).

The effect of co-administration of fosfomycin on glycopeptide antibiotic-induced nephrotoxicity for 3 days has been investigated in rats (41). Fosfomycin reduced glycopeptide antibiotic-induced nephrotoxicity, as shown by reduced urinary excretion of *N*-acetyl-beta-D-glucosaminidase and fewer histological signs of nephrotoxicity in those treated with a combination of glycopeptide and fosfomycin compared with a glycopeptide alone. In addition, the higher the dose of fosfomycin, the more it reduced urinary *N*-acetyl-beta-D-glucosaminidase activity, suggesting that the role of fosfomycin in alleviating nephrotoxicity is dose-dependent. There was significantly less accumulation of teicoplanin and vancomycin in the renal cortex of rats treated with glycopeptide antibiotics plus fosfomycin compared with the glycopeptide antibiotics alone.

Skin

A pustular skin eruption occurred in a 60-year-old woman who was given teicoplanin 400 mg as a loading dose followed by 200 mg/day (42). Following the sixth injection of teicoplanin she developed a pruritic eruption on her chest, which evolved into a widespread maculopapular eruption with tiny pustules and occasional vesicles on an erythematous base over the face, trunk, and limbs within 2 days. Teicoplanin was withdrawn and she was given prednisolone 60 mg/day. This resulted in improvement of the eruption and relieved the discomfort over the next 7 days. In other cases teicoplanin caused an acute generalized rash and exanthematous pustulosis (27,43).

Vancomycin reportedly caused a severe delayed skin reaction with allergic cross-reactivity between vancomycin and teicoplanin requiring steroid therapy (44).

- A 68-year-old woman who was being treated with vancomycin for *Staphylococcus epidermidis* bacteremia developed pruritus and a generalized maculopapular skin rash after 2 weeks. After a short course of prednisolone, she was given teicoplanin and developed general malaise, fever, conjunctival injection, an extensive rash, and later blisters on the legs, again requiring treatment with steroids.

Musculoskeletal

Teicoplanin has been successfully used to treat bone and joint infections without any adverse reactions affecting bones or joints (45).

Immunologic

Allergic reactions have been reported with teicoplanin. Erythroderma during infusion of teicoplanin with fever and hypotension was described in a single patient. Re-exposure elicited the same reaction (16). Allergic cross-reactivity between teicoplanin and vancomycin has been reported (46,47). This cross-reactivity was documented by in vitro studies showing IgE release by basophils in response to stimulation by both vancomycin and teicoplanin in a further patient who had an allergic reaction to vancomycin (48). In other studies the second drug did not elicit allergic reactions in patients known to be allergic to one of the two compounds (49–51). IgE-mediated allergy to teicoplanin has been described (52).

- A 41-year-old woman had an episode of severe generalized urticaria/angioedema and vomiting about 2 minutes after the start of an intravenous infusion of teicoplanin. The infusion was immediately stopped and she was promptly and successfully treated with intravenous corticosteroids and antihistamines. She had not had a similar episode in the past, but about 4 months before the last administration of a 10-day course of intramuscular teicoplanin had been followed by local pain, hardening, and inflammation in the buttock that lasted for 3 days. A skin prick test with teicoplanin 75 mg/ml did not elicit any skin reaction, but intradermal injection of 0.02 ml of the same solution induced an immediate wheal-and-flare reaction (mean diameter 2 cm) followed by a red, pruritic, slightly infiltrated lesion of the same size that lasted for about 3 days.

Drug hypersensitivity syndrome to both vancomycin and teicoplanin has been reported.

- A 50-year-old man developed hypersensitivity syndrome to both vancomycin and teicoplanin (53). Skin rash, fever, eosinophilia, interstitial pneumonitis, and interstitial nephritis developed after the administration of each drug, and resolved after withdrawing the drugs and treatment with high-dose corticosteroids. Skin patch tests 2 months after recovery showed a weak positive result for vancomycin (with erythema and vesicles) and a doubtful result for teicoplanin (with macular erythema).
- A 47 year old man developed drug hypersensitivity syndrome 2 hours after the final dose of teicoplanin (54). He developed a fever of 38.5°C, generalized lymphadenopathy, and a raised C reactive protein concentration and hepatic transaminases. Teicoplanin was withdrawn and within 24 hours his fever resolved. Over the following week his liver enzymes, C reactive protein concentration, and skin returned to normal.

Leukocytoclastic vasculitis has been attributed to teicoplanin.

- A 76-year-old man receiving intramuscular teicoplanin 400 mg/day for septic arthritis due to *Staphylococcus aureus* developed a vasculitis with cutaneous and renal involvement (55). On day 17, tender purpura appeared in his legs, and there was concomitant impairment of renal function. A skin biopsy showed the typical features of a leukocytoclastic vasculitis. Teicoplanin was withdrawn and he was given flucloxacillin and prednisolone. The purpura resolved and renal function recovered partly.

Based on these small studies and individual case reports one can conclude that allergic reactions to teicoplanin can occur in patients with known allergic reactions to vancomycin, but the frequency of occurrence of this type of cross-reaction appears to be low. Therefore, known hypersensitivity to vancomycin is not a contraindication to the use of teicoplanin.

Local reactions at the injection site, including pain, redness, or discomfort after intramuscular injection, and phlebitis after intravenous injection, occur in about 3% of patients (6).

Long-Term Effects

Drug tolerance

In a study of antibiotic resistance in enterococci from raw meat, there was a high prevalence of glycopeptide-resistant strains (56). Resistance to vancomycin was significantly associated with resistance to teicoplanin, erythromycin, tetracycline, and chloramphenicol.

Susceptibility Factors

Age

The recommended dosage regimen (for example 12 mg/kg on day 1 followed by 6 mg/kg/day; premature neonates and children require higher dosing regimens) most often results in efficacious serum concentrations.

Renal disease

Teicoplanin is more effectively administered once daily than vancomycin and may be given intramuscularly or intravenously, but it is not absorbed after oral administration. Teicoplanin is 90% bound to plasma proteins, and it is primarily eliminated by renal excretion; since clearance is predictably reduced in renal insufficiency, dosage adjustments can be made on the basis of the ratio of impaired clearance to normal clearance. Steady state concentrations are reached more slowly with increasing renal impairment.

The administration of 5 and 10 mg/kg of teicoplanin to seven anuric patients immediately after the end of hemodialysis gave mean values of C_{max} of 63 and 122 µg/ml, mean AUCs of 526 and 1104 hours.µg/ml, mean half-lives of 109 and 107 hours, mean clearance rates of 13 and 12 ml/minute, mean apparent volumes of distribution of 1.68 and 1.68 l/kg, and mean volumes of distribution at steady state of 0.31 and 0.28 l/kg (57). Trough serum concentrations were above 10 µg/ml for 24 hours after the 5 mg/kg dose and for 48 hours after the 10 mg/kg dose. Teicoplanin was not detected in the dialysate. Its concentrations in the arterial and the venous lines of the fistulae were similar.

In patients undergoing continuous ambulatory peritoneal dialysis teicoplanin serum concentrations above 10 µg/ml were detected 24 hours after a single dose of teicoplanin 10 mg/kg; all dialysate concentrations were very low (58).

In patients receiving continuous hemodiafiltration teicoplanin is not very effectively eliminated, but continuous hemodiafiltration can remove a relevant dose of the drug from the circulation by ultrafiltration using a high-flux membrane (59).

Drug Administration

Drug overdose

- A 10-day-old girl with a history of postasphyxia acute renal failure, which recovered within 7 days, was given teicoplanin for sepsis due to *Staphylococcus hominis* (60). The dosage used was erroneously high (20 mg/kg 12-hourly instead of an initial dose of 16 mg/kg followed by 6–8 mg/kg/day), and therapy was suspended after 5 days. She improved, and blood cultures were negative. Serum creatinine concentrations and cystatin C (as an early marker of glomerular damage) remained in the reference ranges. Urinary parameters for tubulotoxicity (*N*-acetyl-beta-D-glucosaminidase, beta$_1$-microglobulin) were higher than in the following days, but remained in the reference ranges. Urinary concentrations of epithelial growth factor were also higher during therapy than afterwards, probably indicating repair activity. Serum concentrations of teicoplanin were not determined.

Drug–Drug Interactions

Aminoglycosides

The risk of nephrotoxicity increases with the concomitant use of other nephrotoxic drugs, especially aminoglycosides (39).

Ciprofloxacin

Teicoplanin should be administered separately from ciprofloxacin, since precipitation has been provoked by concomitant parenteral administration (61).

Colestyramine

Since colestyramine binds teicoplanin in vitro and reduces its activity against *Clostridium difficile* almost completely, there is potential for a clinically important interaction (62).

Enalapril

A possible drug interaction with enalapril has been reported; a patient with diabetes receiving teicoplanin for osteomyelitis developed renal insufficiency requiring dialysis after the addition of enalapril (63).

Monitoring Therapy

There is evidence that predose concentrations of teicoplanin are related to clinical outcome; the evidence is less conclusive for vancomycin. Predose concentrations can be linked to some toxic effects (nephrotoxicity for vancomycin and thrombocytopenia to teicoplanin) (64,65). In severe infections monitoring serum teicoplanin concentrations is indicated to assure adequate trough concentrations (over 10 µg/ml) (66). For the treatment of *S. aureus* endocarditis trough concentrations should exceed 20 µg/ml (66). Inadequately low serum concentrations of teicoplanin may have been responsible for the high failure rate in intravenous drug users treated with teicoplanin for-sided endocarditis (67). The fact that teicoplanin pharmacokinetics are unpredictably variable in intravenous drug users adds further weight to the importance of monitoring serum drug concentrations in this population during the treatment of severe infections such as endocarditis (68). Since the half-life of teicoplanin also varies greatly in patients with various degrees of renal insufficiency, monitoring serum concentrations can be helpful in guiding therapy in some patients with markedly reduced creatinine clearance (69).

References

1. Boselli E, Allaouchiche B. Diffusion osseuse des antibiotiques. [Diffusion in bone tissue of antibiotics.] Presse Méd 1999;28(40):2265–76.
2. LeFrock J, Ristuccia ATeicoplanin Bone and Joint Cooperative Study Group A. Teicoplanin in the treatment of bone and joint infections: An open study. J Infect Chemother 1999;5(1):32–9.
3. Pea F, Furlanut M, Poz D, Baraldo M. Pharmacokinetic profile of two different administration schemes of teicoplanin. Single 400 mg intravenous dose vs double-refracted 200 mg intramuscular doses in healthy volunteers Clin Drug Invest 1999;18:47–55.
4. Testore GP, Uccella I, Sarrecchia C, Mattei A, Impagliazzo A, Sordillo P, Andreoni M. Long-term intramuscular teicoplanin treatment of chronic osteomyelitis due to oxacillin-resistant *Staphylococcus aureus* in outpatients. J Chemother 2000;12(5):412–5.
5. Neville LO, Brumfitt W, Hamilton-Miller JMT, Harding I. Teicoplanin vs vancomycin for the treatment of serious infections: a randomised trial. Int J Antimicrob Agents 1995;5:187–93.
6. Campoli-Richards DM, Brogden RN, Faulds D. Teicoplanin. A review of its antibacterial activity, pharmacokinetic properties and therapeutic potential. Drugs 1990;40(3):449–86.
7. Periti P, Stringa G, Mini EItalian Study Group for Antimicrobial Prophylaxis in Orthopedic Surgery. Comparative multicenter trial of teicoplanin versus

cefazolin for antimicrobial prophylaxis in prosthetic joint implant surgery. Eur J Clin Microbiol Infect Dis 1999;18(2):113–9.

8. Kester RC, Antrum R, Thornton CA, Ramsden CH, Harding I. A comparison of teicoplanin versus cephradine plus metronidazole in the prophylaxis of post-operative infection in vascular surgery. J Hosp Infect 1999;41(3):233–43.

9. Fortun J, Navas E, Martinez-Beltran J, Perez-Molina J, Martin-Davila P, Guerrero A, Moreno S. Short-course therapy forside endocarditis due to *Staphylococcus aureus* in drug abusers: cloxacillin versus glycopeptides in combination with gentamicin. Clin Infect Dis 2001;33(1):120–5.

10. Vazquez L, Encinas MP, Morin LS, Vilches P, Gutierrez N, Garcia-Sanz R, Caballero D, Hurle AD. Randomized prospective study comparing cost-effectiveness of teicoplanin and vancomycin as second-line empiric therapy for infection in neutropenic patients. Haematologica 1999;84(3):231–6.

11. Bucaneve G, Menichetti F, Del Favero A. Cost analysis of 2 empiric antibacterial regimens containing glycopeptides for the treatment of febrile neutropenia in patients with acute leukaemia. Pharmacoeconomics 1999;15(1):85–95.

12. Harding I, Sorgel F. Comparative pharmacokinetics of teicoplanin and vancomycin. J Chemother 2000;12(Suppl 5):15–20.

13. Wilson AP. Clinical pharmacokinetics of teicoplanin. Clin Pharmacokinet 2000;39(3):167–83.

14. Sidi V, Roilides E, Bibashi E, Gompakis N, Tsakiri A, Koliouskas D. Comparison of efficacy and safety of teicoplanin and vancomycin in children with antineoplastic therapy-associated febrile neutropenia and Gram-positive bacteremia. J Chemother 2000;12(4):326–31.

15. Khurana C, de Belder MA. Red-man syndrome after vancomycin: potential cross-reactivity with teicoplanin. Postgrad Med J 1999;75(879):41–3.

16. Paul C, Janier M, Carlet J, Tamion F, Carlotti A, Fichelle JM, Daniel F. [Erythroderma induced by teicoplanin.]Ann Dermatol Venereol 1992;119(9):667–9.

17. Stille W, Sietzen W, Dieterich HA, Fell JJ. Clinical efficacy and safety of teicoplanin. J Antimicrob Chemother 1988;21(Suppl A):69–79.

18. Maher ER, Hollman A, Gruneberg RN. Teicoplanin-induced ototoxicity in Down's syndrome. Lancet 1986;1(8481):613.

19. Bibler MR, Frame PT, Hagler DN, Bode RB, Staneck JL, Thamlikitkul V, Harris JE, Haregewoin A, Bullock WE Jr. Clinical evaluation of efficacy, pharmacokinetics, and safety of teicoplanin for serious Gram-positive infections. Antimicrob Agents Chemother 1987;31(2):207–12.

20. Bonnet RM, Mattie H, de Laat JA, Schoemaker HC, Frijns JH. Clinical ototoxicity of teicoplanin. Ann Otol Rhinol Laryngol 2004;113(4):310–2.

21. Davey PG, Williams AH. A review of the safety profile of teicoplanin. J Antimicrob Chemother 1991;27(Suppl B):69–73.

22. Brummett RE, Fox KE, Warchol M, Himes D. Absence of ototoxicity of teichomycin A2 in guinea pigs. Antimicrob Agents Chemother 1987;31(4):612–3.

23. Lewis P, Garaud JJ, Parenti F. A multicentre open clinical trial of teicoplanin in infections caused by gram-positive bacteria. J Antimicrob Chemother 1988;21(Suppl A):61–7.

24. Wilcox M, Nathwani D, Dryden M. Linezolid compared with teicoplanin for the treatment of suspected or proven Gram-positive infections. J Antimicrob Chemother 2004;53(2):335–44.

25. Coluccio E, Villa MA, Villa E, Morelati F, Revelli N, Paccapelo C, Garratty G, Rebulla P. Immune hemolytic anemia associated with teicoplanin. Transfusion 2004;44(1):73–6.

26. Lambotte O, Costedoat-Chalumeau N, Amoura Z, Piette JC, Cacoub P. Drug-induced hemophagocytosis. Am J Med 2002;112(7):592–3.

27. Wee IY, Oh HM. Teicoplanin-induced neutropenia in a paediatric patient with vertebral osteomyelitis. Scand J Infect Dis 2001;33(2):157–8.

28. Del Favero A, Patoia L, Bucaneve G, Biscarini L, Menichetti F. Leukopenia with neutropenia associated with teicoplanin therapy. DICP 1989;23(1):45–7.

29. Lang E, Schmid J, Fauser AA. A clinical trial on efficacy and safety of teicoplanin in combination with beta-lactams and aminoglycosides in the treatment of severe sepsis of patients undergoing allogeneic/autologous bone marrow transplantation. Br J Haematol 1990;76(Suppl 2):14–8.

30. Del Favero A, Menichetti F, Guerciolini R, Bucaneve G, Baldelli F, Aversa F, Terenzi A, Davis S, Pauluzzi S. Prospective randomized clinical trial of teicoplanin for empiric combined antibiotic therapy in febrile, granulocytopenic acute leukemia patients. Antimicrob Agents Chemother 1987;31(7):1126–9.

31. Terol MJ, Sierra J, Gatell JM, Rozman C. Thrombocytopenia due to use of teicoplanin. Clin Infect Dis 1993;17(5):927.

32. Agnelli G, Longetti M, Guerciolini R, Menichetti F, Grasselli S, Boldrini F, Bucaneve G, Nenci GG, Del Favero A. Effects of the new glycopeptide antibiotic teicoplanin on platelet function and blood coagulation. Antimicrob Agents Chemother 1987;31(10):1609–12.

33. Lambiotte F, Miczck S, Bierent S, Socolovsky C, Chagnon JL. Thrombopénie acquise sous teicoplanine. Med Mal Infect 2000;30:481–2.

34. Sunakawa K, Nonoyama M, Fujii R, Iwai N, Sakata H, Shirai M, Sato T, Kajino M, Toyonaga Y, Sano T, Naito A, Minagawa K, Niida Y, Oda T, Yokozawa M, Asanuma H, Shimura K, Fujimura M, Kitajima H, Fujinami K, Numazaki K, Fujikawa T, Kobayashi Y, Sato Y, Nishimura T, Iwata S, Tsuchihashi N, Oishi T, Matsumoto S, Motohiro T, Osawa M, Sunahara M, Shirakawa S, Nishida H, Takahashi N, Nakano R, Sai N, Iyoda K, Yoshimitsu K, Ogawa K, Okazaki T, Tsukimoto I, Motoyama O, Takada Y, Kawasaki M, Sunaoshi W, Nakamura S, Ueda Y, Kamata M, Kato T, Chiba M, Ouchi K, Sato S, Horiuchi T, Suzuki K, Shimoyama T, Masaki H, Aikyo M, Kawada M, Banba M, Furukawa S, Okada T, Yamaguchi S, Hirota O, Koizumi S, Wada H, Ohta K, Uehara T, Yukitake K, Mori T, Takakuwa S, Matsuyama K. [Pharmacokinetic and clinical studies on teicoplanin for sepsis by methicillin–cephem resistant *Staphylococcus aureus* in the pediatric and neonate field. Jpn J Antibiot 2002;55(5):656–77

35. Bochud-Gabellon I, Regamey C. Teicoplanin, a new antibiotic effective against Gram-positive bacterial infections of the skin and soft tissues. Dermatologica 1988;176(1):29–38.

36. Verhagen C, De Pauw BE. Teicoplanin for therapy of Gram-positive infections in neutropenic patients. Int J Clin Pharmacol Res 1987;7(6):491–8.

37. Webster A, Russell SJ, Souhami RL, Richards JD, Goldstone AH, Gruneberg RN. Use of teicoplanin for Hickman catheter associated staphylococcal infection in immunosuppressed patients. J Hosp Infect 1987;10(1):77–82.

38. Charbonneau P, Garaud J, Aubertin J, et al. Efficiency and safety of teicoplanin plus netilmicin compared to

vancomycin plus netilmicin in the treatment of severe Gram-positive infections 110. 1987.

39. Kureishi A, Jewesson PJ, Rubinger M, Cole CD, Reece DE, Phillips GL, Smith JA, Chow AW. Double-blind comparison of teicoplanin versus vancomycin in febrile neutropenic patients receiving concomitant tobramycin and piperacillin: effect on cyclosporin A-associated nephrotoxicity. Antimicrob Agents Chemother 1991;35(11):2246–52.

40. Schaefer F, Klaus G, Muller-Wiefel DE, Mehls OThe Mid-European Pediatric Peritoneal Dialysis Study Group (MEPPS). Intermittent versus continuous intraperitoneal glycopeptide/ceftazidime treatment in children with peritoneal dialysis-associated peritonitis. J Am Soc Nephrol 1999;10(1):136–45.

41. Yoshiyama Y, Yazaki T, Wong PC, Beauchamp D, Kanke M. The effect of fosfomycin on glycopeptide antibiotic-induced nephrotoxicity in rats. J Infect Chemother 2001;7(4):243–6.

42. Unal S, Ikizoglu G, Demirkan F, Kaya TI. Teicoplanin-induced skin eruption. Int J Dermatol 2002;41(12):948–9.

43. Chu CY, Wu J, Jean SS, Sun CC. Acute generalized exanthematous pustulosis due to teicoplanin. Dermatology 2001;202(2):141–2.

44. Padial MA, Barranco P, Lopez-Serrano C. Erythema multiforme to vancomycin. Allergy 2000;55(12):1201.

45. Weinberg WG. Safety and efficacy of teicoplanin for bone and joint infections: results of a community-based trial. South Med J 1993;86(8):891–7.

46. McElrath MJ, Goldberg D, Neu HC. Allergic cross-reactivity of teicoplanin and vancomycin. Lancet 1986;1(8471):47.

47. Davenport A. Allergic cross-reactivity to teicoplanin and vancomycin. Nephron 1993;63(4):482.

48. Knudsen JD, Pedersen M. IgE-mediated reaction to vancomycin and teicoplanin after treatment with vancomycin. Scand J Infect Dis 1992;24(3):395–6.

49. Smith SR, Cheesbrough JS, Makris M, Davies JM. Teicoplanin administration in patients experiencing reactions to vancomycin. J Antimicrob Chemother 1989;23(5):810–2.

50. Wood G, Whitby M. Teicoplanin in patients who are allergic to vancomycin. Med J Aust 1989;150(11):668.

51. Schlemmer B, Falkman H, Boudjadja A, Jacob L, Le Gall JR. Teicoplanin for patients allergic to vancomycin. N Engl J Med 1988;318(17):1127–8.

52. Asero R. Teicoplanin-induced anaphylaxis. Allergy 2006;61(11):1370.

53. Kwon HS, Chang YS, Jeong YY, Lee SM, Song WJ, Kim HB, Kim YK, Cho SH, Kim YY, Min KU. A case of hypersensitivity syndrome to both vancomycin and teicoplanin. J Korean Med Sci 2006;21(6):1108–10.

54. Perrett CM, McBride SR. Teicoplanin induced drug hypersensitivity syndrome. BMJ 2004;328(7451):1292.

55. Logan SA, Brown M, Davidson RN. Teicoplanin-induced vasculitis with cutaneous and renal involvement. J Infect 2005;51(3):e185-6.

56. Pavia M, Nobile CG, Salpietro L, Angelillo IF. Vancomycin resistance and antibiotic susceptibility of enterococci in raw meat. J Food Prot 2000;63(7):912–5.

57. Papaioannou MG, Marinaki S, Pappas M, Stamatiadis D, Giamarellos-Bourboulis EJ, Giamarellou H, Stathakis C. Pharmacokinetics of teicoplanin in patients undergoing chronic haemodialysis. Int J Antimicrob Agents 2002;19(3):233–6.

58. Stamatiadis D, Papaioannou MG, Giamarellos-Bourboulis EJ, Marinaki S, Giamarellou H, Stathakis CP. Pharmacokinetics of teicoplanin in patients undergoing continuous ambulatory peritoneal dialysis. Perit Dial Int 2003;23:127–31.

59. Yagasaki K, Gando S, Matsuda N, Kameue T, Ishitani T, Hirano T, Iseki K. Pharmacokinetics of teicoplanin in critically ill patients undergoing continuous hemodiafiltration. Intensive Care Med 2003;29:2094–5.

60. Fanos V, Mussap M, Khoory BJ, Vecchini S, Plebani M, Benini D. Renal tolerability of teicoplanin in a case of neonatal overdose. J Chemother 1998;10(5):381–4.

61. Jim LK. Physical and chemical compatibility of intravenous ciprofloxacin with other drugs. Ann Pharmacother 1993;27(6):704–7.

62. Pantosti A, Luzzi I, Cardines R, Gianfrilli P. Comparison of the in vitro activities of teicoplanin and vancomycin against Clostridium difficile and their interactions with cholestyramine. Antimicrob Agents Chemother 1985;28(6):847–8.

63. Frye R, Job M, Dretler R. Teicoplanin nephrotoxicity: first case report. Pharmacotherapy 1990;10:234.

64. Wilson AP. Comparative safety of teicoplanin and vancomycin. Int J Antimicrob Agents 1998;10(2):143–52.

65. MacGowan AP. Pharmacodynamics, pharmacokinetics, and therapeutic drug monitoring of glycopeptides. Ther Drug Monit 1998;20(5):473–7.

66. Brogden RN, Peters DH. Teicoplanin. A reappraisal of its antimicrobial activity, pharmacokinetic properties and therapeutic efficacy. Drugs 1994;47(5):823–54.

67. Fortun J, Perez-Molina JA, Anon MT, Martinez-Beltran J, Loza E, Guerrero A. Right-sided endocarditis caused by Staphylococcus aureus in drug abusers. Antimicrob Agents Chemother 1995;39(2):525–8.

68. Rybak MJ, Lerner SA, Levine DP, Albrecht LM, McNeil PL, Thompson GA, Kenny MT, Yuh L. Teicoplanin pharmacokinetics in intravenous drug abusers being treated for bacterial endocarditis. Antimicrob Agents Chemother 1991;35(4):696–700.

69. Lam YW, Kapusnik-Uner JE, Sachdeva M, Hackbarth C, Gambertoglio JG, Sande MA. The pharmacokinetics of teicoplanin in varying degrees of renal function. Clin Pharmacol Ther 1990;47(5):655–61.

Tetracyclines

See also Doxycycline, Minocycline

General Information

The tetracyclines are a closely related group of antibiotics with comparable pharmacological properties but different pharmacokinetic characteristics. They have both the advantages and disadvantages of broad-spectrum antibiotics. Tetracyclines are effective not only against bacteria and spirochetes, but also against some forms of *Mycoplasma*, *Chlamydia*, and *Rickettsia*, as well as protozoa. They affect multiplying microorganisms by inhibiting ribosomal protein synthesis. Their effect is therefore primarily bacteriostatic rather than bactericidal, depending on the kind of microorganism.

Most of the adverse effects of the tetracyclines depend on the concentration of the antibiotic in the affected organ. The more lipophilic drugs are more potent with regard to their bacteriostatic efficacy and hence usually require daily doses below 1 g.

Use in non-infective conditions

Tetracyclines have many effects on non-infective inflammatory processes (SEDA-24, 278). These actions have been studied in detail and in a wide variety of areas, such as inflammation, proteolysis, angiogenesis, apoptosis, ionophoresis, and bone metabolism. Clinical trials have also been performed in many different non-microbial fields, including rheumatoid arthritis, periodontal disease, myocardial infections, gastric disorders, and experimentally in the treatment of cancers. In all of these disorders, the proposed mechanisms for effects are on the host rather than the microbial side. An excellent review was published in 2006 (1).

Tetracyclines have many effects on cells involved in inflammatory reactions, including inhibition of neutrophilic functions, such as migration, phagocytosis, degranulation, and the production of free oxygen radical (2). Most of these effects are supposed to be due to chelation of divalent ions and can be partly reversed by the addition of calcium ions or zinc ions. Their ability to inhibit synthesis of mitochondrial proteins may be the background for their effects on lymphocytes, such as inhibition of lymphocyte proliferation in response to mitogens, inhibition of interferon gamma production, and inhibition of immunoglobulin production. Enzyme inactivation caused by tetracyclines has been studied in several models. Tetracyclines inhibited gingival collagenolytic activity in diabetic mice and in humans with periodontal disease (3). Subsequent studies showed that tetracyclines inactivated collagenases found in several places in the body (4,5). Most probably, the inhibitory effects of tetracyclines on collagenases are exerted through inactivation of metalloproteases, rather than serine proteases. Tetracycline also has radical-scavenging properties, which may be partly related to the ulcer-healing effect observed in a rat model of gastric mucosal injury (6).

As is to be expected, modified cycline molecules devoid of antibacterial effects are as effective as non-modified molecules. New derivatives with refined anti-inflammatory and enzyme inhibitory (and reduced antimicrobial) effects are being studied in experimental laboratories (7).

So far, minocycline has been the agent most often tested. Because of its assumed neuroprotective, anti-apoptotic, and anti-inflammatory properties, it was forecast several years ago that it could be of value in the therapy of several neurological diseases, including amyotrophic lateral sclerosis, Huntington's disease, multiple sclerosis, Parkinson's disease, and spinal cord injury (8). Several trials have been performed, but few have fulfilled the promises or provided hard facts on which to rely.

Tetracyclines and metalloproteinases

A major target for non-infective indications of the tetracyclines is inhibition of metalloproteinases. The following is a brief summary of what is known about tetracyclines and metalloproteinases, followed by some comments about possible adverse effects.

The matrix metalloproteinases (MMPs) are a family of calcium- and/or zinc-dependent endopeptidases involved in degradation of extracellular matrix and tissue remodelling (9). At least 21 mammalian MMPs have been described. They participate in various biological processes, such as embryonic development, ovulation, angiogenesis, apoptosis, wound healing, and nerve growth.

Under normal conditions, the activity of MMPs is very low and is strongly regulated by natural tissue inhibitors (TIMPs). The TIMPS are a family of four structurally related proteins (TIMP-1, 2, 3, and 4), exerting dual control of the MMPs by inhibiting both their active forms and their activation. In addition, the proteolytic activity of MMPs is inhibited by non-specific protease inhibitors, such as alpha$_2$-macroglobulin and alpha$_1$-antiprotease.

In the presence of specific stimuli, exemplified by cytokines and growth factors, MMPs can be up-regulated. Chronic activation of MMPs, due to an imbalance between the activity of MMPs and TIMPs, can result in excessive degradation of the extracellular matrix and is believed to contribute to the pathogenesis of several diseases, such as rheumatoid arthritis, osteoarthritis, periodontal disease, emphysema, atherosclerosis, skin ulceration, and cancer.

The physiological and pathophysiological roles of MMPs and TIMPs have been extensively studied in knockout mice (9). For most of the MMPs and TIMPs there seem to be significant overlaps in functions, and a deficiency of one enzyme can be compensated for by the presence of others (9,10). Each MMP is encoded by a distinct gene, and about half of the human MMP genes so far discovered are on chromosome 11 (9).

The mere fact that MMPs might be involved in the pathogenesis of several chronic disorders has made this field attractive to numerous pharmaceutical companies. One major approach for controlling abnormal MMP activity has been the use of small molecular weight inhibitors, and several excellent reviews of the design of such inhibitors have been published (9).

Rheumatoid arthritis

In 1942, the Swedish doctor Nana Svartz introduced sulfasalazine into therapy and suggested that it might be useful in rheumatoid arthritis because of its antibacterial activity. Since then, many antimicrobial agents have been tried in the treatment of rheumatoid arthritis, based on the assumption that the disease may be due to an infectious agent. Interest has focused on tetracyclines for the treatment of rheumatoid arthritis (11), reactive arthritis (12), and osteoarthritis (13) and the state of the art has been reviewed (14). The theory that rheumatoid arthritis is due to an infectious agent has been refined to a statement that it may be a gene transfer disease in which some viruses act as vectors (15), thereby excluding a direct effect of antibacterial agents on the disease.

Two further perspectives on the use of tetracyclines in rheumatoid arthritis have been published (16,17). In addition to an effect on matrix metalloproteinases, the authors focused on a potential antiarthritic action of tetracyclines by their effects in the interaction between the generation of nitric oxide, matrix metalloproteinase release, and chondrocyte apoptosis. Both minocycline and doxycycline inhibit the production of nitric oxide from human cartilage and murine macrophages (18) in concentrations that are achieved in vivo. The authors suggested that tetracyclines may have several potential chondroprotective effects: direct inhibition of matrix metalloproteinase activity and, by inhibition of nitric oxide production, further reduction of matrix metalloproteinase activity, reversal of reduced matrix synthesis, and reduced chondrocyte apoptosis.

In double-blind, placebo-controlled studies, minocycline relieved clinical symptoms and reduced some laboratory measures of disease activity in patients with rheumatoid arthritis (19–21). However, the progression of radiographic damage was not significantly reduced and minocycline caused several adverse effects. This led to investigations of the therapeutic effects of doxycycline in patients with rheumatoid arthritis. In 13 patients with moderate rheumatoid arthritis, low-dose doxycycline 20 mg bd reduced the urinary excretion of pyridinoline (22). Pyridinolines are collagen cross-links that are released from joints during cartilage and bone resorption and correlate with the severity of joint destruction (23). In a later study, doxycycline produced a significant reduction in the number of tender joints and significant improvements in disability and behavior in 12 patients with rheumatoid arthritis (24).

However, these studies were not placebo-controlled, and it was possible that the observed effects were due to factors other than doxycycline. Two double-blind, placebo-controlled trials have since been reported (25,26).

In a crossover study, 66 patients took 50 mg doxycycline or placebo twice a day for 12, 24, or 36 weeks (25). Patients' assessments, swollen and tender joint counts, duration of morning stiffness, erythrocyte sedimentation rate, and the so-called modified disease activity score (27) were used as measures of disease activity. Doxycycline had no significant effects on clinical and laboratory measures of disease activity, pyridinoline excretion, or progression of radiographic joint damage. Furthermore, there were adverse effects during treatment with doxycycline and not with placebo. The authors concluded that doxycycline 50 mg bd provided no therapeutic benefit in patients with rheumatoid arthritis.

In another study, a 16-week, randomized, double-blind, placebo-controlled trial, eligible subjects with active seropositive or erosive rheumatoid arthritis were randomly allocated to three treatment groups: doxycycline 200 mg intravenously, azithromycin 250 mg orally, or placebo (26). The primary end-points were changes between baseline and week 4 in the tender joint count, erythrocyte sedimentation rate, and urinary excretion of pyridinoline. The trial was stopped prematurely after 31 patients had been enrolled. Three subjects were withdrawn because of worsening arthritis. There were no significant differences across the groups in any of the three primary clinical end-

points. The authors concluded that doxycycline did not reduce disease activity or collagen cross-link production.

Ten chemically modified tetracyclines (CMTs), minocycline, and doxycycline have been tested for their capacity to inhibit cartilage degradation in vitro (28). CMT-8 was the most active. The authors concluded that by carefully selecting a tetracycline-based MMP inhibitor and controlling dosages it should be possible to inhibit pathologically excessive MMP-8 and/or MMP-13 activity, especially that which causes bone erosion, without affecting the constitutive activity of MMP-1 needed for tissue remodelling and normal host function. They thought that of the CMTs, CMT-8 and to a lesser extent CMT-3 and CMT-7 were the most effective. However, the effects of these CMTs in a full-scale trial in rheumatoid arthritis have not been reported.

Reactive arthritis
It is well established that a number of microorganisms found in the gastrointestinal tract are associated with reactive arthritis and that most of these organisms might be susceptible to tetracyclines. Several investigations have shown clinical effects of tetracyclines in the treatment of reactive arthritis (14). In these cases, a more direct antibacterial effect of the triggering organism (if still present in the patient) might be the mechanism involved.

Osteoarthritis
Tetracyclines have been investigated in experimental osteoarthritis, because metalloproteases are involved in the breakdown of cartilage matrix seen in this condition. Doxycycline reduced the severity of knee osteoarthritis induced in dogs by ligamentous section (29) and also reduced the degradation of type XI collagen exposed to extract of human arthritic cartilage (30).

Periodontal disease
Periodontal disease is caused by microorganisms, but host responses are in large part responsible for destruction of the periodontal support structures. In these patients, pathological activity of host-derived matrix metalloproteinases occurs in response to the bacterial infection in the periodontal tissues, causing destruction of collagen, the primary structural component of periodontal matrix. This in turn leads to gingival recession, pocket formation, and increased tooth mobility. The final outcome is loss of the tooth.

Tetracyclines inhibit collagenolytic activity in gingival tissue (SEDA-23, 256) (31–33). Based on these findings, a large clinical development program was initiated to demonstrate the potential of a sub-antimicrobial dose of doxycycline to augment and maintain the beneficial effects afforded by conventional non-surgical periodontal therapy in adult periodontitis. A summary of these studies has been published (34). Several different dosage regimens and placebo were compared in patients who had had a variety of adjunctive non-surgical procedures. Of the various parameters studied, the following are worthy of mention: collagenase activity in gingival crevicular fluid and gingival specimens, so-called clinical attachment levels, probing pockets depths, bleeding on probing, and subtraction radiographic measurements of alveolar bone height.

Sub-antimicrobial doses of doxycycline reduced collagenase activity in both gingival crevicular fluid and gingival biopsies, augmented and maintained gains in clinical attachment levels and reductions in pockets depths, reduced bleeding on probing, and prevented loss of alveolar bone height. These clinical effects occurred in the absence of any significant effects on the subgingival microflora and without evidence of an increase in the incidence or severity of adverse reactions relative to the controls. The authors proposed that the main mechanism underlying these effects is inhibition of pathologically high matrix metalloproteinase activity in neutrophils (MMP-8) and bone cells (MMP-13).

Tetracyclines designed for topical treatment of periodontal disease have been developed. In general, antimicrobial agents that are directly administered to the periodontal pocket to treat periodontitis should cause little or no damage to host cells. This holds true even when they are used after conventional surgical surface scaling and curettage, because epithelial regeneration is of importance for healing of inflammation.

This is the background for a recent investigation of the cytocidal effect of eight different antimicrobial agents on a human gingival epithelial cell line (NDUSD-1) (35). The order of the agents according to their cytocidal effects (LD- 50) was minocycline > tetracycline > enoxacin > clarithromycin > roxithromycin = ofloxacin > azithromycin > erythromycin. The maximum non-cytocidal concentrations were 0.3 µmol/l of minocycline and 100 µmol/l for erythromycin. In a previous study the cytocidal effect of minocycline on normal human gingival cells in secondary culture was greater than that of tetracycline, and the effect depended on both the intracellular concentrations of the compounds and their persistence in the cells (36).

These findings need to be extrapolated to in vivo conditions, but in the meantime it would be wise not to use minocycline locally for periodontitis.

Acne
Over the years, various tetracyclines have been used in the treatment of acne. Their mechanism of action is not clear, but appears to be not purely antimicrobial, since they reduce chemotaxis of polymorphonuclear leukocytes, modify complement pathways, and inhibit the polymorphonuclear leukocyte chemotactic factor and lipase production in *Propionibacterium acnes* (37). They may also have a direct effect on sebum secretion (38), for example by modification of free fatty acids (39).

Adult respiratory distress syndrome
Adult respiratory distress syndrome (ARDS) has many causes, is associated with severe lung damage, and is characterized by pulmonary edema and hypoxemia. It has a high mortality. The current method of treatment is supportive and there is no specific therapy. This was the

background to a thorough review of the anti-inflammatory properties of tetracyclines in the prevention of acute lung injury (40). The authors ended with an optimistic forecast, that targeting the proteases that cause ARDS with chemically modified tetracyclines may be useful in prevention and treatment. They ended by suggesting that strategies to prevent ARDS should focus on targets downstream from the initial inflammatory signals that provoke the cascade of events.

Breakthrough endometrial bleeding
The endometrial environment in contraceptive users is characterized by increased production of matrix metalloproteases. In an in vitro model doxycycline only moderately and in a cell-specific manner reduced expression of matrix metalloproteases without inducing their activity (41). The clinical relevance of these findings is unclear.

Diabetic proteinuria
Because matrix metalloproteases contribute to the pathogenesis of diabetic proteinuria, 35 patients with overt diabetic nephropathy (proteinuria over 300 mg/day) were given doxycycline 100 mg/day for 2 months (42). There was a statistically significant reduction in proteinuria, and it increased when doxycycline was withdrawn. Further studies are necessary to determine the long-term effects, the optimal dosage, and the optimal duration of treatment.

Experimental autoimmune encephalitis
Experimental autoimmune encephalitis is an antigen-specific T cell-mediated autoimmune disease of the central nervous system, which has long served as an animal model for studying multiple sclerosis. Minocycline in high doses attenuated the development of experimental autoimmune encephalitis (43), supposedly through an effect on matrix metalloproteases 9 (44). However, clinical trials with minocycline in patients with multiple sclerosis have shown variable results. In a study in mice polyethylene-glycol minocycline liposomes given every 5 days were reportedly as effective in ameliorating experimental autoimmune encephalitis as daily intraperitoneal injections of minocycline (45). Whether it will have the same effect in human multiple sclerosis remains to be seen.

Glaucoma
Tetracyclines, especially demeclocycline, are among the most effective ocular hypotensive agents, according to studies in rabbits and cats. The biochemical mechanism of this effect is unknown. The prolonged effect and apparent lack of adverse ocular adverse effects suggest their possible usefulness for treating glaucoma in man (SEDA-15, 260).

Metamfetamine adverse effects
Metamfetamine can be associated with behavioral changes and neurotoxicity. In a recent study, minocycline ameliorated behavioral changes and neurotoxicity in mouse brain dopaminergic terminals after the administration of metamfetamine (46). The authors suggested that

minocycline may be useful in the treatment of several symptoms associated with metamfetamine abuse, although this may be merely a pipe dream.

Metastatic colorectal cancer

In a complicated model of metastatic colorectal cancer in mice, inhibition of matrix metalloproteases was associated with significantly fewer and smaller liver tumors (47). The authors concluded that after resection of colorectal tumors, inhibition of matrix metalloproteases may be clinically beneficial in preventing recurrence.

Rabies

New therapeutic agents are needed for the treatment of rabies encephalitis in humans. This was the background for experimental studies with minocycline, whose results were expressed declaratively in the title: "Therapy with minocycline aggravates experimental rabies in mice" (48). There was no beneficial effect of minocycline in primary neuron cultures and in the mouse model, minocycline aggravated the clinical neurological disease and resulted in a higher mortality (49). They authors ended by stating that "this team recommends that empirical therapy with minocycline be avoided in the management of rabies and viral encephalitis in humans until more information becomes available"

Schizophrenia

Minocycline inhibits inducible nitric oxide synthase (iNOS) and delays disease onset in mouse models of neuropsychiatric disorders. This was the background for a recent Japanese study, in which minocycline 150 mg/day was given for 4 weeks as an open adjunct to antipsychotic medication to 22 patients with schizophrenia (50). Using a "Positive and Negative Syndrome Scale" for schizophrenia, there was a statistically significant clinical improvement, which was maintained at follow-up 4 weeks after the end of minocycline treatment. There were no adverse events. The authors claimed enthusiastically that "augmentation with minocycline may prove to be a valuable strategy for boosting antipsychotic efficacy and for treating schizophrenia".

There are several possible mechanisms for these effects. More than a quarter of a century ago it was shown that even one single dose of doxycycline significantly reduced memory capacity in healthy volunteers (51). Since then, millions of teenagers have used various tetracyclines, including doxycycline and minocycline, for long-term treatment of acne, at the time of life when memory capacity and creativity are assumed to be needed. Additionally, some millions of individuals take such drugs for long-term prophylaxis and treatment of periodontitis. In both cases the therapeutic effects are most probably due to their effects on matrix metalloproteases (MMPs).

Tendinopathy

Increased activity of matrix metalloproteases and subsequent degradation of extracellular matrix are supposed to be implicated in the pathogenesis of tendinopathy. In an experimental study in tendons in rats' tails doxycycline inhibited pericellular matrix generation (52). The authors hypothesized that this could be beneficial in the treatment of tendinopathy in humans. In another study, related to tendon repair in rats, doxycycline at clinically relevant serum concentrations impaired healing processes in experimentally induced Achilles tendon injuries (53). It currently seems reasonable to conclude that matrix metalloprotease inhibitors should be used with great care in sports medicine, since it is uncertain that these results can be extrapolated from rats' tails to athletes' Achilles tendons.

General adverse effects

The adverse effects profiles of oral doxycycline and minocycline have been systematically reviewed (54). Between 1966 and 2003, a total of 130 and 333 adverse events were published in case reports of doxycycline and minocycline respectively. In 24 doxycycline studies (n = 3833) and 11 minocycline studies (n = 788), adverse events occurred in 0–61% and 12–83% respectively. Gastrointestinal adverse events were most common with doxycycline; central nervous system and gastrointestinal adverse events were most common with minocycline. The FDA MedWatch data contained 628 events for doxycycline and 1099 events for minocycline reported in the USA from January 1998 to August 2003, during which time 47 630 000 new prescriptions for doxycycline and 15 234 000 new prescriptions for minocycline were dispensed, yielding event rates of 13 per million for doxycycline and 72 per million for minocycline.

Mild gastrointestinal disturbances are common. A common and nearly unique feature of all tetracyclines is the formation of drug–melanin complexes, resulting in pigment deposition at various sites. Except for enamel defects and presumable disturbances of osteogenesis these deposits do not give rise to abnormalities of organ function. In view of tooth discoloration and enamel hypoplasia, with a tendency to caries formation, tetracyclines should be avoided in children under 8 years of age and in women after the third month of pregnancy. The risk of photosensitivity reactions largely depends on the dose of the drug and the degree of exposure to sunlight. It may be increased in long-term treatment. Skin, nail, and other organ pigmentation often occur, even with low-dose long-term treatment. The syndrome of fatty liver degeneration is rarely encountered today, because risk factors such as pregnancy are respected, and formulations with lower doses are available. Adverse effects can occur with most tetracyclines if the dosage is not altered in patients with renal insufficiency.

Allergic reactions to tetracyclines are less than half as common as allergic reactions to penicillin. For this reason, tetracyclines are alternatives in patients with allergic reactions to other antibiotics. Exceptional observations of anaphylactic shock have been reported (55,56). In a few cases tetracyclines were assumed to be the cause of hypersensitivity myocarditis (57). Pneumonitis with eosinophilia has been described in association with tetracyclines (58). A serum sickness-like syndrome was probably associated with minocycline in a 19-year-old man treated for acne (59). Allergic and toxic reactions may in some cases have been caused by degraded formulations or additives (60). Tumor-inducing effects have not been reported.

Jarisch–Herxheimer reaction

The Jarisch–Herxheimer reaction is common in patients treated with tetracyclines for louse-borne relapsing fever (61). Two forms of reaction are described at the start of tetracycline therapy: (a) fever, rigor, increased respiratory and heart rates, and occasional delirium and coma (62); (b) fever and disseminated intravascular coagulation (63). At about the time the temperature reaches its peak, spirochetes disappear from the peripheral blood (62). Meptazinol, a partial opioid antagonist, reduces the Jarisch–Herxheimer reaction in relapsing fever (64).

Bleeding with thrombocytopenia and signs of intravascular coagulation in patients treated for louse-borne relapsing fever may be due to a Jarisch–Herxheimer reaction mediated by the release of endotoxins from disintegrating spirochetes (61).

Comparisons of individual tetracyclines

The spectra of bacterial effectiveness of the various tetracyclines are fairly similar. However, metabolism and excretion vary. In general, preference is currently given to drugs that can be given in relatively low doses (not above 1.0 g/day).

Tetracycline (chlortetracycline) and oxytetracycline

These tetracyclines are incompletely absorbed from the gastrointestinal tract. Plasma concentrations fall with half-lives of 6–12 hours. They are predominantly excreted by the kidney, extrarenal elimination amounting at most to 10–20%. They have a lower affinity for fat and membranes, which means that higher dosages to achieve therapeutic effectiveness. However, higher dosages can contribute to an increased risk of systemic toxic effects and, as absorption from oral administration is incomplete, also to an increased risk of gastrointestinal adverse reactions.

Demeclocycline (demethylchlortetracycline)

Demeclocycline produces a significantly higher rate of phototoxic skin reactions than other tetracyclines, presumably owing to its long half-life (about 16 hours), resulting in therapeutic plasma concentrations lasting 24–48 hours, and an even longer-lasting accumulation of the drug in membranes and fatty tissues. Demeclocycline is eliminated about 20–40% by glomerular filtration. In patients with liver cirrhosis, cardiac failure, or impaired renal function, it should be used cautiously, because of its pronounced effect on electrolyte and fluid balance (65).

Doxycycline and minocycline

Doxycycline and minocycline are more lipophilic tetracyclines. They are well absorbed after oral administration. Their half-lives are 16–18 hours. Their higher affinity for fatty tissues improves their effectiveness and changes their adverse effects profile. Local gastrointestinal irritation and disturbance of the intestinal bacterial flora occur less often than with the more hydrophilic drugs, which have to be given in higher oral doses for sufficient absorption.

Nevertheless, their toxic effects are similar to those of other tetracyclines and arise from accumulation in fatty tissues. Accumulation in a third compartment and the resulting long half-life may contribute to an increased incidence of various toxic adverse effects during long-term treatment, even if lower daily doses are used. This seems also to be the case for pigmentation disorders and possibly for neurological disturbances (66).

Minocycline and doxycycline are predominantly eliminated by the liver and biliary tract (70–90%). Therefore, no change in dose is needed in patients with impaired renal function. However, it should be considered that hepatic elimination of doxycycline or minocycline might be accelerated by co-administration of agents that induce hepatic enzymes.

Methacycline

Methacycline is rarely used. It has similar efficacy to ampicillin in acute exacerbations of chronic bronchitis.

Chemically modified tetracyclines

As early as the 1960s it was recognized that tetracyclines might inhibit bone growth, supposedly by interference with calcium metabolism. It was not until collagenase was discovered in 1983 (68) that intensive interest in the non-antimicrobial properties of tetracycline-based antibiotics developed. Most of the initial studies were done with doxycycline and minocycline, that is semi-synthetic tetracyclines with widespread medical and dental uses. However, in 1987, a non-antibacterial chemically modified tetracycline (CMT) that inhibited mammalian collagenase activity was described (69). At present, the CMTs (also known as COLs, from their having been introduced by CollaGenes Pharmaceuticals Inc, Newton, PA, USA) comprise a group of at least 10 analogues compounds (CMT-1 to CMT-10) that differ in their specificity and potency in inhibiting MMP. Comparative pharmacokinetic data, obtained from animals, have recently been published, as has a bibliography covering more than 75 papers and abstracts related to the basic biological properties of these compounds (70). Chemically modified tetracyclines are more active in interfering with mammalian processes than the classical tetracyclines, which is the main justification for their use in non-infective problems (70,71). From a qualitative point of view, it might be reasonable to assume similar host-related adverse effects as the classical tetracyclines, while from a quantitative point of view, dosages, duration of therapy, etc., will influence the number of adverse effects. However, the risks might be quite high.

Hematologic

Sideroblastic anemia is characterized by the accumulation of iron in the mitochondria of erythroblasts. In a Phase I study in 35 patients with refractory tumors, eight taking

CMT-3 developed anemia without leukopenia or thrombocytopenia (72). Three of these patients underwent bone-marrow examination and each had ringed sideroblasts. The authors referred to several cases of aplastic anemia, megaloblastic anemia, and hemolytic anemia in which members of the tetracycline family have been implicated. However, they stated that there has been no previous reports of sideroblastic anemia associated with any tetracycline derivative and that the molecular mechanisms by which CMT-3 might cause sideroblastic anemia are unclear.

Immunologic

Based on the strategy that inhibition of angiogenesis is of importance in anticancer therapy, CMT-3 (COL-3) was used in a phase 1 study in 35 patients at the National Institutes of Health in the USA in patients with refractory metastatic cancer (73). The patients received a test dose of CMT-3, followed by pharmacokinetic testing during daily dosing for 7 days. After a few doses, three patients developed symptoms of drug-induced lupus, and the diagnoses were verified after a few days or weeks. CMT-3 was withdrawn and there was improvement.

Bacterial resistance

Prolonged treatment with the classical antibacterial tetracyclines results in bacterial resistance and/or opportunistic fungal infections. It has therefore been forecast, but not so far proven, that chemically modified tetracyclines without antibacterial activity may not cause microbial resistance. However, accepting that the major mechanism by which bacteria get rid of tetracyclines intracellularly is by increasing their efflux via P glycoprotein, the chemically modified tetracyclines might also trigger that mechanism, thereby causing reduced sensitivity of the exposed bacterial population to all types of tetracyclines. Before this possibility has been ruled out, the new chemically modified tetracyclines should be introduced with great care.

Organs and Systems

Cardiovascular

Cardiovascular reactions to tetracyclines have often been associated with other symptoms of hypersensitivity, such as urticaria, angioedema, bronchial obstruction, and arterial hypotension (55,74). Such reactions occurred in patients who had tolerated tetracyclines previously and were therefore considered as anaphylactic.

Respiratory

There have been a few cases of acute bronchial obstruction after the administration of a tetracycline (75).

Pneumonitis with eosinophilia has also been described (58,76).

When pleural tetracycline instillation is used to produce local inflammation for pleurodesis in spontaneous pneumothorax, there are no severe short-term or long-term adverse effects (77).

Nervous system

Tetracyclines rarely cause benign intracranial hypertension. This syndrome has primarily involved children (78,79), although it has also been observed in adults (66,80–87). Re-exposure can result in recurrence (77). As a rule, the symptoms develop within days or occasionally several months after the start of therapy, in one case only after 18 months of tetracycline therapy for acne (66). The syndrome includes headache, nausea and vomiting, dizziness, tinnitus, papilledema, and visual disturbances caused by scotoma or optic nerve damage. Intracranial pressure can be raised up to three-fold. Distinct intracerebral lesions are not visible on CT scan or angiography and the ventricular spaces are not enlarged (80). With the exception of raised intracranial pressure, all findings, including the cell count and protein concentration of the cerebrospinal fluid, are normal. After withdrawal, symptoms resolve over a period of hours to days or occasionally weeks. Increased use of long-term therapy with tetracyclines for acne may contribute to a higher prevalence of this syndrome (88).

- A 22-year-old girl who took tetracycline 500 mg 6-hourly for 3 days developed intracranial hypertension with esotropia (89).
- A 24-year-old woman developed intracranial hypertension after taking oral tetracycline (90). She had severely reduced visual acuity, papilledema, and concentric impaired visual fields. She was treated with acetazolamide and recurrent lumbar punctures and recovered, but without improvement in either acuity or visual fields.

Early studies showed that curare blockade was increased by tetracyclines (91–93). This effect can be antagonized by calcium ions (94). There was a short-term increase in muscular weakness in patients with myasthenia gravis after intravenous tetracycline. The mechanism may be a calcium-antagonizing effect of magnesium ions present in the tetracycline solvent, as the symptoms could be provoked by similar amounts of magnesium alone. The formation of calcium complexes with non-protein-bound tetracycline, thus lowering the serum concentration of free calcium, may also be involved (95).

Sensory systems

Acute transitory myopia was described as an effect of tetracycline therapy, probably due to changes in refractive power (96).

Ocular pigmentation has not previously been attributed to oral tetracycline hydrochloride.

- A 48 year old healthy white man developed green crystals on the conjunctivae of both eyes after taking tetracycline 500 mg/day for several months for acne vulgaris (97). Tetracycline was identified in a conjunctival biopsy.

This was a between-the-eyes adverse effect of type 1 (98)

Endocrine

Hormone production in patients with "black thyroid," who had taken a tetracycline for prolonged periods, was normal (99,100).

Metabolism

Tetracyclines can increase blood urea nitrogen or serum urea concentrations without a corresponding increase in serum creatinine (that is without accompanying renal damage). The mechanism is an excess nitrogen load of metabolic origin accompanied by negative nitrogen balance. This effect is termed "anti-anabolic" (101,102), but is in fact the result of inhibition of protein synthesis, which affects not only microorganisms but to some degree mammalian cells also. Sodium and water depletion, due to the diuretic effect of some tetracyclines, can further enhance uremia (102).

Tetracyclines have been associated with hypoglycemia (103). Insulin doses may have to be reduced (104).

Hematologic

Hematological changes with tetracyclines are extremely rare. However, in individual cases, hemolytic anemia (105), neutropenia or slight leukopenia, thrombocytopenia (106), and even aplastic anemia have all been described. In the light of the fact that such blood changes are often reported without a clear description of their cause, the relation to the drug often remains doubtful, especially if reactions are listed in tabular form and information on specific details or concomitant drug therapy is lacking.

Bleeding with thrombocytopenia and signs of intravascular coagulation in patients treated for louse-borne relapsing fever may be due to a Jarisch–Herxheimer reaction mediated by the release of endotoxins from disintegrating spirochetes (61).

Mouth and teeth

The long-term esthetic results of treating severely stained teeth due to tetracyclines by endodontics and internal bleaching have been assessed in 20 patients and found to be excellent (107). A therapeutic strategy may be bleaching after the preparation for porcelain laminate veneers or night-guard vital bleaching (108,109).

Stomatitis, other signs of irritation of the oropharynx, and a rash in and around the orifices have been described in patients taking tetracyclines. This may partly be considered as mucous membrane manifestations of allergic or toxic origin. Secondary infections by pathogenic organisms, such as *C. albicans*, viruses, or bacteria, should always be considered (110).

Tetracycline-induced staining of the teeth has been treated successfully with a porcelain laminate veneer (111).

Gastrointestinal

Nausea, vomiting, and epigastric burning are the most common adverse effects of tetracyclines. This is also true for the lower-dose formulations. The symptoms are usually mild and seldom necessitate withdrawal. Nausea occurs in 8–15% of patients.

Esophageal damage

Esophageal ulcers have been described in association with oral doxycycline or tetracycline. Acute onset of substernal burning pain and dysphagia was noted within hours of taking the drug (112–116). Remaining parts of the ingested capsule were identified by esophagoscopy.

Thirty centers for pharmacovigilance in France have reported 81 cases of esophageal damage after treatment with tetracyclines collected between 1985 and 1992 (117). There were 64 ulcers, eight cases of dysphagia, and nine of esophagitis. Most (96%) of the cases were caused by doxycycline and 73% of the patients were female, mean age 29 years. Prescriptions were for dermatological (54%), urogenital (23%), and ENT diseases. In one patient, a 71-year-old man, an esophagobronchial fistulation required esophagectomy. In 92% the drugs were not taken correctly, that is at bedtime or without a sufficient quantity of fluid. Treatment with sucralfate 1 g tds did not change the outcome of tetracycline-induced esophageal ulcers (118).

In three studies of more than 600 children with chest pain, none had tetracycline-induced esophagitis as a cause of their pain (119–121).

Patients should not lie down immediately after taking a tetracycline capsule and the formulation should be swallowed with generous quantities of water (112,122).

Liver

A clinical syndrome, often with fatal outcome, has been recognized as a complication of high doses of tetracycline. It is characterized by nausea, vomiting, and spiking fever. Jaundice, acidosis, and uremia are also often present, while hematemesis and melena are only occasionally observed. Some patients die because of refractory hypotension. The histopathological findings are those of diffuse fatty liver degeneration (123–125). Although the syndrome was primarily described in pregnant women, it is not restricted to them. Impaired renal function, and thus reduced renal elimination of tetracyclines, as well as serious infections, increases the risk of this complication (123,126). Since the introduction of lower doses of tetracycline (under 1 g/day), the syndrome of fatty liver degeneration with severe hepatic insufficiency has become rare (124). However, a few reports have described liver reactions in previously healthy individuals with no pre-existing conditions and given usual oral doses of tetracyclines (127,128). Liver enzymes and renal function should be monitored, especially in patients at risk.

A personally experienced case of likely tetracycline-induced liver injury after low-dose tetracycline has been reported (129). The patient took oral doxycycline 200 mg/day for 8 days and had markedly altered liver function.

The liver enzyme activities normalized only 109 days after withdrawal. The authors also reviewed all reports of liver damage to Swedish Adverse Drug Reactions Advisory Committee (SADRAC) in the period 1965–95. There were 23 liver reactions with a suspected causal relation to oral low-dose tetracycline derivatives. A causal relation was considered likely in three cases and possible in eight, giving an incidence of roughly one in 18 million defined daily doses. There were no deaths from these liver reactions, and liver enzyme activities normalized in all cases without any serious clinical consequences. The authors remarked that the frequency of liver reactions resulting from tetracyclines may be somewhat higher, as previous studies in Sweden suggest that only 20–50% of severe adverse reactions are reported to SADRAC (129).

Pancreas

Pancreatitis is a common feature of the clinical syndrome of fatty liver due to tetracyclines, although it has also been observed without overt liver disease in a patient after two separate exposures (130).

Urinary tract

Tetracyclines not uncommonly cause a raised serum urea concentration due to impaired protein synthesis rather than renal damage, However, renal insufficiency can occur, and patients with pre-existing renal insufficiency are particularly likely to develop raised blood urea nitrogen, serum phosphate, and serum sulfate concentrations during treatment with most tetracyclines. These changes may be associated with acidosis and even symptoms of uremia. Renal dysfunction can be missed if diuresis alone is monitored, since non-oliguric renal insufficiency has been reported (131). Interstitial nephritis is extremely rare (132–134).

- A 42-year-old woman with polycystic kidney disease took tetracycline 250 mg qds after undergoing tooth extraction (135). She developed nausea, vomiting, and diarrhea within days and end-stage renal disease within 2 weeks. Hemodialysis was required.

Uremia, which may at least in part be due to a reduction in renal function, has been observed in patients with liver cirrhosis (136–138), cardiac failure, or interstitial nephritis (139), which is extremely rare (132–134).

Most observations suggest that a determining factor for the risk of renal insufficiency is a relative overdose of the drug in patients with pre-existing renal damage (137,138), particularly in those who are already dehydrated (for example due to diuretic therapy). The half-life of tetracyclines is prolonged several-fold in patients with renal dysfunction (140). With normal renal function, 20–70% of an intravenous dose of tetracycline is eliminated by glomerular filtration. Maintenance therapy must be adjusted to the degree of renal impairment.

Acquired Lignac–De Toni–Fanconi syndrome, with polyuria, polydipsia, glycosuria, aminoaciduria, hyperphosphaturia, and hypercalciuria, was described in a number of patients treated with outdated tetracycline formulations. The degeneration products responsible for the toxic action are probably epitetracycline, anhydro-4-epitetracycline, and anhydrotetracycline (141), as similar renal damage was produced in rats with anhydro-4-epitetracycline (142).

A few observations of a nephrogenic diabetes insipidus-like syndrome with demeclocycline (demethylchlortetracycline hydrochloride) and resistance to exogenous vasopressin suggested impairment of renal concentrating function by tubular damage (65). Demeclocycline has therefore been proposed for the treatment of inappropriate secretion of antidiuretic hormone (143–147).

Acute interstitial nephritis leading to acute renal insufficiency has been reported after a single repeated dose of tetracycline (148).

Skin

Pronounced responsiveness of the skin and nails to light is a complication of systemic therapy with tetracyclines. In general, the pathogenesis of skin reactions promoted by sunlight exposure is mediated by phototoxicity, as suggested by the high occurrence rate (up to 9 out of 10 patients), depending on sun exposure and drug dosage. Experimental studies identified a wavelength close to 320 nm within the ultraviolet spectrum as the most potent area for the induction of phototoxic reactions (149). It has been described with various tetracyclines and presumably occurs with all of them. The skin changes resemble those of minor to severe sunburn, with erythema, edema, papules, blisters, and urticaria (150,151). In some circumstances they resemble cutaneous porphyria and have therefore been described as "porphyria-like cutaneous changes" (152).

Pigmentation due to tetracyclines occurs, but the exact frequency is not known. It has been observed with newer and lower-dose formulations as well as with older drugs, and presumably occurs with all tetracyclines. In contrast to phototoxic reactions, pigmentation can appear without light exposure and without inflammation. Pigment deposits were observed in light-exposed and non-exposed skin, the conjunctivae, the oral mucosa, the tongue, and internal organs, such as the thyroid and heart valve endothelium 99,100,153,154. The pigment is presumed to be a drug–melanin–calcium complex and depends on the chelating properties of the tetracycline molecule. Although pigmentation usually tends to persist after withdrawal, it does not provoke any symptoms or organ dysfunction, except for cosmetic problems. A number of cases occurred in patients treated for acne, and may possibly have been due to more common use of long-term treatments (155), although there is no evidence that the risk is increased in these patients.

Allergic skin and mucous membrane reactions occur with a very low incidence of 0.3%, which is about 10-fold lower than the incidence of such reactions in patients treated with other antibiotics (156). A generalized rash can occur during tetracycline treatment. Various forms have been described, including generalized urticaria, maculopapular rash, erythema exudativum, multiforme-like eruptions, and rare cases of Stevens–Johnson syndrome (157–160). Even acute generalized exanthematous pustulosis has been described (161). Fixed drug eruptions due to systemic administration are rare (162–164), as is a serum sickness-like syndrome (59,165). Allergic contact dermatitis induced by local application of a tetracycline is an exception 157,166–168.

Antimicrobial drugs, especially co-trimoxazole, ampicillin, and tetracyclines, can cause fixed drug eruptions (169).

Nails

Pronounced photosensitivity of the skin and nails is a complication of systemic therapy with tetracyclines. Onycholysis is associated with nail discoloration. Dystrophy is usually preceded by photosensitivity reactions of the skin (157). Depending on the degree and area of light exposure, onycholysis may also affect the toe nails (150). In contrast to idiopathic onycholysis, it not only affects women (170).

Musculoskeletal

Deposition of tetracyclines in bone tissue has been demonstrated in animals (171) and man (172). However, whereas osseous tissue in adult patients treated with tetracycline has shown deposits only in areas of repair or remodelling, children's bones contain extensive areas of deposition. Tetracycline deposition in bone has been reported to have an effect on longitudinal bone growth (173). In experimental tissue cultures, osteogenesis was impaired by tetracyclines in concentrations similar to serum concentrations that are associated with a therapeutic effect (that is 1 µg/ml) (174). The deposition of tetracyclines in human bone begins in utero as early as in the first trimester of pregnancy (172). With regular tissue turnover, the deposits disappear.

Infection risk

Monilial infections of the vulva and vagina in temporal association with tetracycline use can occur, even without other predisposing factors such as diabetes, pregnancy, immunodeficiency, and therapy with oral contraceptives and glucocorticoids. They also occur in patients treated with other antibiotics (175). C. albicans is not always the cause of vaginitis. The same is true for balanitis in men (176). If symptoms persist or change in patients with urethritis or adnexitis, this may be due to treatment failure, because of resistance of the causative pathogen (177).

Long-Term Effects

Drug resistance

Resistance to tetracyclines shows marked inter-regional variations and changes rapidly with time. The selection of resistant bacterial strains may be favored by widespread, often prophylactic, use in veterinary medicine and by long-term therapy for acne, periodontal disease, or symptomatic Borrelia infections. Many of the documented cases of resistance are of limited practical significance, since the tetracyclines are merely one of a number of

therapeutic alternatives. The problem may be different when these drugs are the chemotherapeutic agents of first choice, that is in chronic Borrelia infections, especially arthritis due to Lyme disease and pulmonary or bubonic plague due to Yersinia pestis (178).

For infections with Chlamydia or Mycoplasma, effective alternative antibiotics are available. Increased rates of resistant strains of genital Mycoplasma may explain treatment failures (177,179).

A rather high resistance rate of 25–50% is reported for Hemophilus species (180), and this has to be taken into consideration if tetracyclines are given to patients with chronic pulmonary diseases or for the treatment of acute respiratory infections.

For infections with Neisseria gonorrhoea, tetracyclines are not indicated, owing to the higher prevalence of resistance, especially among penicillinase-producing strains (181). For these patients, better alternatives are usually available.

Second-Generation Effects

Teratogenicity

Teratogenic effects of tetracyclines have been demonstrated (182–184), as evidenced by increased rates of intrauterine death, congenital anomalies in general (185), and congenital cataracts (186) in fetuses exposed to tetracyclines. However, it is often impossible to distinguish between the drug and an underlying unidentified viral infection as a cause of the observed abnormalities.

Of 38 151 pregnant women who had babies without any defects (controls), 214 (0.6%) had taken oral oxytetracycline; in contrast, of 22 865 pregnant women who had offspring with congenital abnormalities, 216 (0.9%) had taken oxytetracycline (OR = 1.7; 95% CI = 1.4, 2.0) (187). More women whose babies had congenital abnormalities had taken oxytetracycline in the second month of pregnancy:

- neural-tube defects (OR = 9.7; CI = 2.0, 47);
- cleft palate (OR = 17; CI = 3.5, 84);
- multiple congenital abnormalities (particularly the combination of neural-tube defects and cardiovascular malformations) (OR = 13; CI = 3.8, 44).

The authors mentioned that their previous study had not shown a teratogenic potential of doxycycline (188), but concluded, far more prudently, that all tetracyclines are contraindicated during pregnancy.

Fetotoxicity

Discoloration of the first teeth is particularly likely if a tetracycline is given to the mother after the third month of pregnancy (189). Tetracyclines pass across the placenta and reach therapeutic concentrations in the fetal circulation (190,191).

Tooth discoloration is due to deposition of tetracyclines in the form of calcium complexes in the mineralizing zones of the teeth, and seems to be pathogenically related to the pigmentation of other organs. It occurs when tetracyclines are used during tooth formation. Tetracyclines pass through the placenta and are also found in high concentrations in the breast milk (192). As mineralization of the deciduous teeth takes place from the fourth month of intrauterine life until 1 year after birth and continues for the permanent teeth up to the age of 7–8 years (190), pregnant women after the third month of pregnancy, nursing women, and children under the age of 8 years should not be treated with tetracyclines. Discoloration of the teeth after intrauterine exposure to tetracyclines was observed in up to 50% of children at risk and was especially high when tetracyclines were used during the last trimester (190). Besides its merely cosmetic aspect, tooth discoloration in children is associated with enamel defects and hypoplasia in severe cases (193). Adult-onset tooth discoloration coincident with minocycline administration occurred in four of 72 patients (194).

Susceptibility Factors

Age

Pregnant women after the third month of pregnancy, nursing women, and children under the age of 8 years should not be treated with tetracyclines, because of the risk of discoloration of the teeth. Besides its merely cosmetic aspect, tooth discoloration in children is associated with enamel defects and hypoplasia in severe cases (193).

Renal disease

Tetracyclines are removed by hemodialysis, but significantly less than creatinine or urea (123,140). Severe adverse effects of tetracyclines occur almost exclusively with doses over 1.0 g/day (124) or in the treatment of pyelonephritis with concomitant renal insufficiency.

Drug Administration

Drug administration route

The intravenous formulation of tetracycline was instilled intrapleurally to produce chemical pleurodesis, with good effect, from the 1970s to the 1990s. The adverse effect most commonly seen with tetracycline was chest pain, which was often severe (195,196). In a comprehensive review of intrapleural therapy published in 1994 (197) the incidence of chest pain was estimated at 14%, and fever occurred in 10% of patients. However, the intravenous form of tetracycline has been withdrawn by the manufacturer and so this agent is no longer available for pleurodesis.

Drug–Drug Interactions

Antacids

Co-administration of tetracyclines with antacids or other drugs containing divalent or trivalent cations, such as calcium, magnesium, or iron, is contraindicated.

Tetracyclines form complexes with such cations, which are very poorly or not at all absorbed 96,198,199.

To avoid this interaction, delay of 2–3 hours between the ingestion of tetracycline and the cation is recommended (199). Similarly, reduced systemic availability results from simultaneous intake of abundant quantities of milk or milk products.

Antidiarrheal drugs

Antidiarrheal drugs, such as kaolin-pectin and bismuth products (200,201), impair the absorption of tetracyclines by chelation (see the interaction with antacids in this monograph).

Coumarin anticoagulants

Although antibiotics can inhibit the production of vitamin K in the gut by intestinal bacteria, they do not thereby interfere with the actions of coumarin anticoagulation, since vitamin K that is produced by intestinal bacteria is of less importance than vitamin K that is obtained from dietary sources (202). Nevertheless, there have been sporadic reports that tetracyclines can enhance the effects of coumarin anticoagulants (203–206). The mechanism is not known, but evidence that tetracyclines can reduce the activity of prothrombin (207) suggests an additive pharmacodynamic interaction.

Diuretics

The combination of tetracyclines with diuretics is particularly detrimental to renal function (208). Tetracyclines accumulate in patients with pre-existing renal insufficiency (for example elderly people, even if the serum creatinine is in the reference range) and can cause nausea and vomiting, which causes dehydration and worsens renal function. This is exacerbated by the effects of diuretics.

Iron

See antacids above (198,199).

Methoxyflurane

In patients taking oral or parenteral tetracyclines who undergo methoxyflurane anesthesia, renal insufficiency and oxalate crystal formation in renal tissue was attributed to an interaction between these drugs (209–211). Tetracyclines are therefore not recommended preoperatively.

Risperidone and/or sertraline

An interaction of tetracycline with risperidone and/or sertraline has been described (212).

- A 15-year-old youth with Asperger's syndrome, Tourette's syndrome, and obsessive-compulsive disorder was stabilized on risperidone 1.5 mg bd and sertraline 100 mg od and had marked improvement in his social skills and tics, until he was given tetracycline 250 mg bd for acne. Within 2 weeks his tics were acutely exacerbated with pronounced neck jerking and guttural sounds. The sertraline was increased to 150 mg/day, but the tics did not resolve. The tetracycline was withdrawn after 1 month, and the tics improved within a few weeks.

The authors reviewed the major metabolic pathways of the three drugs used in this case. Tetracycline may have accelerated the hepatic metabolism of risperidone, but tetracycline has not been shown to induce CYP2D6, the major hepatic enzyme involved in the metabolism of risperidone. The authors claimed that it was more likely that tetracycline binds to risperidone or its active metabolite, making them inactive. They supported this suggestion by citing evidence that clozapine, which has similar pharmacology to risperidone, interacts with tetracycline in vitro (213). They also commented on the possibility that the effect could have come from increased concentrations of sertraline, which can increase tics (214). Tetracycline may inhibit the hepatic enzymes that metabolize sertraline; however, adequate evidence is lacking. Another possibility is a protein binding interaction, because all three drugs are highly protein bound. However, the observation that the tics were not worsened by an increase in sertraline dosage makes this possibility less likely. The mere fact that the withdrawal of tetracycline resulted in an improvement in the tics supports an interaction between tetracycline and risperidone.

References

1. Sapardin AN, Fleischmajer R. Tetracyclines: nonantibiotic properties and their clinical implications. Am Acad Dermaol 2006;54:258–65.
2. Midtvedt T, Lingaas E, Melby K. The effect of 13 antimicrobial agents on the elimination phase of phagocytosis in human polymorphonuclear leukocytes. In: Eickenberg HU, Hahn A, Opferkuch W, editors. The Influence of Antibiotics on the Host-Parasite Relationship. Berlin: Springer Verlag, 1982:118–28.
3. Golub LM, Ramamurthy N, McNamara TF, Gomes B, Wolff M, Casino A, Kapoor A, Zambon J, Ciancio S, Schneir M, et al. Tetracyclines inhibit tissue collagenase activity. A new mechanism in the treatment of periodontal disease. J Periodontal Res 1984;19(6):651–5.
4. Greenwald RA, Golub LM, Lavietes B, Ramamurthy NS, Gruber B, Laskin RS, McNamara TF. Tetracyclines inhibit human synovial collagenase in vivo and in vitro. J Rheumatol 1987;14(1):28–32.
5. Lauhio A, Sorsa T, Lindy O, Suomalainen K, Saari H, Golub LM, Konttinen YT. The anticollagenolytic potential of lymecycline in the long-term treatment of reactive arthritis. Arthritis Rheum 1992;35(2):195–8.
6. Suzuki Y, Ishihara M, Segami T, Ito M. Anti-ulcer effects of antioxidants, quercetin, alpha-tocopherol, nifedipine and tetracycline in rats. Jpn J Pharmacol 1998;78(4):435–41.
7. Golub LM, Ramamurthy NS, Llavaneras A, Ryan ME, Lee HM, Liu Y, Bain S, Sorsa T. A chemically modified nonantimicrobial tetracycline (CMT-8) inhibits gingival matrix metalloproteinases, periodontal breakdown, and extra-oral bone loss in ovariectomized rats. Ann NY Acad Sci 1999;878:290–310.
8. Yong VW, Wells J, Giuliani F, Casha S, Power C, Metz LM. The promise of minocycline in neurology. Lancet Neurol 2004;3:744–51.
9. Skiles JW, Gonnella NC, Jeng AY. The design, structure, and therapeutic application of matrix metalloproteinase inhibitors. Curr Med Chem 2001;8(4):425–74.
10. Hidalgo M, Eckhardt SG. Development of matrix metalloproteinase inhibitors in cancer therapy. J Natl Cancer Inst 2001;93(3):178–93.
11. Trentham DE, Dynesius-Trentham RA. Antibiotic therapy for rheumatoid arthritis. Scientific and anecdotal appraisals. Rheum Dis Clin North Am 1995;21(3):817–34.
12. Pott HG, Wittenborg A, Junge-Hulsing G. Long-term antibiotic treatment in reactive arthritis. Lancet 1988;1(8579):245–6.
13. Dejarnatt AC, Grant JA. Basic mechanisms of anaphylaxis and anaphylactoid reactions. Immunol Allergy Clin North Am 1992;12:33–46.
14. Toussirot E, Despaux J, Wendling D. Do minocycline and other tetracyclines have a place in rheumatology? Rev Rhum Engl Ed 1997;64(7–9):474–80.
15. Grubb R, Grubb A, Kjellen L, Lycke E, Aman P. Rheumatoid arthritis—a gene transfer disease. Exp Clin Immunogenet 1999;16(1):1–7.
16. Cooper SM. A perspective on the use of minocycline for rheumatoid arthritis. J Clin Rheumatol 1999;5:233–6.
17. Alarcon GS. Antibiotics for rheumatoid arthritis? Minocycline shows promise in some patients. Postgrad Med 1999;105(4):95–8.
18. Attur MG, Patel RN, Patel PD, Abramson SB, Amin AR. Tetracycline up-regulates COX-2 expression and prostaglandin E2 production independent of its effect on nitric oxide. J Immunol 1999;162(6):3160–7.
19. Kloppenburg M, Breedveld FC, Terwiel JP, Mallee C, Dijkmans BA. Minocycline in active rheumatoid arthritis. A double-blind, placebo-controlled trial. Arthritis Rheum 1994;37(5):629–36.
20. O'Dell JR, Haire CE, Palmer W, Drymalski W, Wees S, Blakely K, Churchill M, Eckhoff PJ, Weaver A, Doud D, Erikson N, Dietz F, Olson R, Maloley P, Klassen LW, Moore GF. Treatment of early rheumatoid arthritis with minocycline or placebo: results of a randomized, double-blind, placebo-controlled trial. Arthritis Rheum 1997;40(5):842–8.
21. Tilley BC, Alarcon GS, Heyse SP, Trentham DE, Neuner R, Kaplan DA, Clegg DO, Leisen JC, Buckley L, Cooper SM, Duncan H, Pillemer SR, Tuttleman M, Fowler SE. Minocycline in rheumatoid arthritis. A 48-week, double-blind, placebo-controlled trial. MIRA Trial Group. Ann Intern Med 1995;122(2):81–9.
22. Greenwald RA, Moak SA, Golub LM. Low dose doxycycline inhibits pyridinoline excretion in selected patients with rheumatoid arthritis. Ann NY Acad Sci 1994;732:419–21.
23. Astbury C, Bird HA, McLaren AM, Robins SP. Urinary excretion of pyridinium crosslinks of collagen correlated with joint damage in arthritis. Br J Rheumatol 1994;33(1):11–5.
24. Nordstrom D, Lindy O, Lauhio A, Sorsa T, Santavirta S, Konttinen YT. Anti-collagenolytic mechanism of action of doxycycline treatment in rheumatoid arthritis. Rheumatol Int 1998;17(5):175–80.
25. van der Laan W, Molenaar E, Ronday K, Verheijen J, Breedveld F, Greenwald R, Dijkmans B, TeKoppele J. Lack of effect of doxycycline on disease activity and joint

damage in patients with rheumatoid arthritis. A double blind, placebo controlled trial. J Rheumatol 2001;28(9):1967–74.

26. St Clair EW, Wilkinson WE, Pisetsky DS, Sexton DJ, Drew R, Kraus VB, Greenwald RA. The effects of intravenous doxycycline therapy for rheumatoid arthritis: a randomized, double-blind, placebo-controlled trial. Arthritis Rheum 2001;44(5):1043–7.

27. Prevoo ML, van't Hof MA, Kuper HH, van Leeuwen MA, van de Putte LB, van Riel PL. Modified disease activity scores that include twenty-eight-joint counts. Development and validation in a prospective longitudinal study of patients with rheumatoid arthritis. Arthritis Rheum 1995;38(1):44–8.

28. Greenwald RA, Golub LM, Ramamurthy NS, Chowdhury M, Moak SA, Sorsa T. In vitro sensitivity of the three mammalian collagenases to tetracycline inhibition: relationship to bone and cartilage degradation. Bone 1998;22(1):33–8.

29. Yu LP Jr, Smith GN Jr, Brandt KD, Myers SL, O'Connor BL, Brandt DA. Reduction of the severity of canine osteoarthritis by prophylactic treatment with oral doxycycline. Arthritis Rheum 1992;35(10):1150–9.

30. Yu LP Jr, Smith GN Jr, Hasty KA, Brandt KD. Doxycycline inhibits type XI collagenolytic activity of extracts from human osteoarthritic cartilage and of gelatinase. J Rheumatol 1991;18(10):1450–2.

31. Golub LM, Ryan ME, Williams RC. Modulation of the host response in the treatment of periodontitis. Dent Today 1998;17(10):102–6.

32. Sorsa T, Mantyla P, Ronka H, Kallio P, Kallis GB, Lundqvist C, Kinane DF, Salo T, Golub LM, Teronen O, Tikanoja S. Scientific basis of a matrix metalloproteinase-8 specific chair-side test for monitoring periodontal and peri-implant health and disease. Ann NY Acad Sci 1999;878:130–40.

33. Garrett S, Johnson L, Drisko CH, Adams DF, Bandt C, Beiswanger B, Bogle G, Donly K, Hallmon WW, Hancock EB, Hanes P, Hawley CE, Kiger R, Killoy W, Mellonig JT, Polson A, Raab FJ, Ryder M, Stoller NH, Wang HL, Wolinsky LE, Evans GH, Harrold CQ, Arnold RM, Southard GL, et al. Two multi-center studies evaluating locally delivered doxycycline hyclate, placebo control, oral hygiene, and scaling and root planing in the treatment of periodontitis. J Periodontol 1999;70(5):490–503.

34. Ashley RA. Clinical trials of a matrix metalloproteinase inhibitor in human periodontal disease. SDD Clinical Research Team. Ann NY Acad Sci 1999;878:335–46.

35. Inoue K, Kumakura S, Uchida M, Tsutsui T. Effects of eight antibacterial agents on cell survival and expression of epithelial cell- or cell-adhesion-related genes in human gingival epithelial cells. J Periodontal Res 2004;39:50–8.

36. Sato H, Tsutsui T. Effect of tetracyclines on cell survival of cultured human gingival keratinocytes and intracellular concentrations of incorporated tetracyclines. J Jap Soc Periodont 1998;40:1–8.

37. Meynadier J, Alirezai M. Systemic antibiotics for acne. Dermatology 1998;196(1):135–9.

38. Del Rosso JQ. A status report on the use of subantimicrobial-dose doxycycline: a review of the biologic and antimicrobial effects of the tetracyclines. Cutis 2004;74(2):118–22.

39. Huber HP, Pflugshaupt C. Akne und freie Fettsäuren im Hauttalg, Beeinflussung durch Doxycyclin. [Acne and free fatty acids in sebum, modification by doxycycline.] Schweiz Rundsch Med Prax 1990;79(20):631–2.

40. Nieman GF, Zerler BR. A role for the anti-inflammatory properties of tetracyclines in the prevention of acute lung injury. Curr Med Chem 2001;8(3):317–25.

41. Li R, Luo X, Archer DF, Chegini N. Doxycycline alters the expression of matrix metalloproteases in the endometrial cells exposed to ovarian steroid and proinflammatory cytokines. J Reprod Immunol 2007;73:118–29.

42. Naini AE, Harandi AA, Moghtaderi J, Bastani B, Amiran A. Doxycycline: a pilot study to reduce diabetic proteinuria. Am J Nephrol 2007;27:269–73.

43. Nessler S, Dodel R, Bittner A, Reuss S, Du Y, Hemmer B, Sommer N. Effect of minocycline in experimental autoimmune encephalomyelitis Ann Neurol 2002;52(5):689–90.

44. Brundula V, Rewcastle NB, Metz, IM Bernard CC, Yong V. Targeting leucocyte MMPs and transmigration: minocycline as a potential therapy for multiple sclerosis. Brain 2002;125:1297–308.

45. Hu W, Metselaar J, Ben LH, Cravens PD, Singh MP, Frohman EM, Eagar TN, Racke MK, Kieseier BC, Stüve O. PEG minocycline-liposomes ameliorate CNS autoimmune disease. PLoS ONE 2009;4(1):e4151.

46. Zhang L, Kitaichi K, Fujimoto Y, Nakayama H, Shimizu E, Iyo M, Hashimoto K. Protective effects of minocycline on behavioural changes and neurotoxicity in mice after administration of methamphetamine. Prog Neuropsychopharmacol Biol Psychiatry 2006;30:1381–9.

47. Nicoud IB, Jones CM, Pierce JM, Earl TM, Matrisian LM, Chari RS, Gordon DL. Warm hepatic ischemia-reperfusion promotes growth of colorectal carcinoma micrometastases in mouse liver via matrix metalloprotease-9 induction. Cancer Res 2007;15:272–8.

48. Jackson AC, Scott CA, Owen J, Weli SC, Rossiter JP. Therapy with minocycline aggravates experimental rabies in mice. J Virol 2007;81:6248–53.

49. Jackson AC, Scott CA, Owen J, Weli SC, Rossite JP. Human rabies therapy: lessons learned from experimental studies in mice. Develop Biol 2008;131:377–85.

50. Miyakaoka T, Yasukawa R, Yasuda H, Hayashida M, Inagaki T, Horiguchi J. Minocycline as adjunctive therapy for schizophrenia: an open-labeled study. Clin Pharmacol 2008;31:287–92.

51. Idzekowski C, Oswald I. Interference with human memory by an antibiotic. Psychopharmacology 1983;79:108–10.

52. Arnoczky SP, Lavagnino M, Egerbacher M, Caballero O, Gardner K. Matrix metalloproteases inhibitors prevent a decrease in the mechanical properties of stress-deprived tendons: an in vitro experimental study. Am J Sport Med 2007;35:763–9.

53. Pasternak B, Fellenius M, Aspenberg P. Doxycycline impairs tendon repair in rats. Acta Orthop Belg 2006;72:756–60.

54. Smith K, Leyden JJ. Safety of doxycycline and minocycline: a systematic review. Clin Ther 2005;27(9):1329–42.

55. Sastre Dominguez J, Sastre Castillo A, Marin Nunez F. Anafilaxia sistémica a tetraciclinas. [Systemic anaphylaxis caused by tetracyclines.] Rev Clin Esp 1984;174(3–4):135–6.

56. Golbert TM, Patterson R, Pruzansky JJ. Systemic allergic reactions to ingested antigens. J Allergy 1969;44(2):96–107.

57. Fenoglio JJ Jr, McAllister HA Jr, Mullick FG. Drug related myocarditis. I. Hypersensitivity myocarditis. Hum Pathol 1981;12(10):900–7.

58. Guillon JM, Joly P, Autran B, Denis M, Akoun G, Debre P, Mayaud C. Minocycline-induced cell-mediated

hypersensitivity pneumonitis. Ann Intern Med 1992;117(6):476–81.

59. Puyana J, Urena V, Quirce S, Fernandez-Rivas M, Cuevas M, Fraj J. Serum sickness-like syndrome associated with minocycline therapy. Allergy 1990;45(4):313–5.

60. Sulkowski SR, Haserick JR. Simulated systemic lupus erythematosus from degraded tetracycline. JAMA 1964;189:152–4.

61. Zein ZA. Louse borne relapsing fever (LBRF): mortality and frequency of Jarisch–Herxheimer reaction. J R Soc Health 1987;107(4):146–7.

62. Bryceson AD, Parry EH, Perine PL, Warrell DA, Vukotich D, Leithead CS. Louse-borne relapsing fever. Q J Med 1970;39(153):129–70.

63. Perine PL, Kidan TG, Warrell DA, Bryceson AD, Parry EH. Bleeding in louse-borne relapsing fever. II. Fibrinolysis following treatment. Trans R Soc Trop Med Hyg 1971;65(6):782–7.

64. Teklu B, Habte-Michael A, Warrell DA, White NJ, Wright DJ. Meptazinol diminishes the Jarisch–Herxheimer reaction of relapsing fever. Lancet 1983;1(8329):835–9.

65. Maxon HR 3rd, Rutsky EA. Vasopressin-resistant diabetes insipidus associated with short-term demethylchlortetracycline (declomycin) therapy. Mil Med 1973;138(8):500–1.

66. Lander CM. Minocycline-induced benign intracranial hypertension. Clin Exp Neurol 1989;26:161–7.

67. Chodosh S, Baigelman W, Medici TC. Methacycline compared with ampicillin in acute bacterial exacerbations of chronic bronchitis. A double-blind cross over study. Chest 1976;69(5):587–92.

68. Golub LM, Lee HM, Lehrer G, Nemiroff A, McNamara TF, Kaplan R, Ramamurthy NS. Minocycline reduces gingival collagenolytic activity during diabetes. Preliminary observations and a proposed new mechanism of action. J Periodontal Res 1983;18(5):516–26.

69. Golub LM, McNamara TF, D'Angelo G, Greenwald RA, Ramamurthy NS. A non-antibacterial chemically-modified tetracycline inhibits mammalian collagenase activity. J Dent Res 1987;66(8):1310–4.

70. Greenwald R, Golub L. Biologic properties of non-antibiotic, chemically modified tetracyclines (CMTs): a structured, annotated bibliography. Curr Med Chem 2001;8(3):237–42.

71. Liu Y, Ramamurthy N, Marecek J, Lee HM, Chen JL, Ryan ME, Rifkin BR, Golub LM. The lipophilicity, pharmacokinetics, and cellular uptake of different chemically-modified tetracyclines (CMTs). Curr Med Chem 2001;8(3):243–52.

72. Rudek MA, Horne M, Figg WD, Dahut W, Dyer V, Pluda JM, Reed E. Reversible sideroblastic anemia associated with the tetracycline analogue COL-3. Am J Hematol 2001;67(1):51–3.

73. Ghate JV, Turner ML, Rudek MA, Figg WD, Dahut W, Dyer V, Pluda JM, Reed E. Drug-induced lupus associated with COL-3: report of 3 cases. Arch Dermatol 2001;137(4):471–4.

74. Pollen RH. Anaphylactoid reaction to orally administered demethylchlortetracycline. N Engl J Med 1964;271:673.

75. Menon MP, Das AK. Tetracycline asthma—a case report. Clin Allergy 1977;7(3):285–90.

76. Sitbon O, Bidel N, Dussopt C, Azarian R, Braud ML, Lebargy F, Fourme T, de Blay F, Piard F, Camus P.

Minocycline pneumonitis and eosinophilia. A report on eight patients. Arch Intern Med 1994;154(14):1633–40.

77. Almind M, Lange P, Viskum K. Spontaneous pneumothorax: comparison of simple drainage, talc pleurodesis, and tetracycline pleurodesis. Thorax 1989;44(8):627–30.

78. Fields JP. Bulging fontanel: a complication of tetracycline therapy in infants. J Pediatr 1961;58:74–6.

79. Maroon JC, Mealy J Jr. Benign intracranial hypertension. Sequel to tetracycline therapy in a child. JAMA 1971;216(9):1479–80.

80. Rush JA. Pseudotumor cerebri: clinical profile and visual outcome in 63 patients. Mayo Clin Proc 1980;55(9):541–6.

81. Bhowmick BK. Benign intracranial hypertension after antibiotic therapy. BMJ 1972;3(817):30.

82. Meacock DJ, Hewer RL. Tetracycline and benign intracranial hypertension. BMJ (Clin Res Ed) 1981;282(6271):1240.

83. Walters BN, Gubbay SS. Tetracycline and benign intracranial hypertension: report of five cases. BMJ (Clin Res Ed) 1981;282(6257):19–20.

84. Pearson MG, Littlewood SM, Bowden AN. Tetracycline and benign intracranial hypertension. BMJ (Clin Res Ed) 1981;282(6263):568–9.

85. Koch-Weser J, Gilmore EB. Benign intracranial hypertension in an adult after tetracycline therapy. JAMA 1967;200(4):345–7.

86. Lubetzki C, Sanson M, Cohen D, Schaison-Cusin M, Lhermitte F, Lyon-Caen O. Hypertension intracrénienne bénigne et minocycline. [Benign intracranial hypertension and minocycline.] Rev Neurol (Paris) 1988;144(3):218–20.

87. Haenggeli ChA, Laufer D. Pseudotumeur cérébrale chez un jeune homme traité aux tétracyclines pour une acne. Schweiz Med Wochenschr 1990;120(Suppl 34):25.

88. Askmark H, Lundberg PO, Olsson S. Drug-related headache. Headache 1989;29(7):441–4.

89. Santos FX, Parolin A, Lindoso EM, Santos FH, Sousa LB. Hipertensão intracraniana com manifestações oculares associada ao uso de tetraciclina: relato de caso. Arq Bras Oftalmol 2005;68(5):701–3.

90. Altinbas A, Hoogstede HA, Bakker SL. [Intracranial hypertension with severe and irreversible reduced acuity and impaired visual fields after oral tetracycline.] Ned Tijdschr Geneeskd 2005;149(34):1908–12.

91. Bezzi G, Gessa GL. Rapporti tra antibiotici e curarismo (III). Tetracicline e curarismo. [Relation between antibiotics and curarism. III. Tetracycline and curarism.] Boll Soc Ital Biol Sper 1960;36:374–5.

92. Baisset A, Lareng L, Puig G. Incidence d'une thérapeutique antibiotique sur la curarisation. In: Comptes-Rendus, XII Congrès Français d'AnesthésiologieMontpellier, 1962:813.

93. Snavely SR, Hodges GR. The neurotoxicity of antibacterial agents. Ann Intern Med 1984;101(1):92–104.

94. Kubikowski P, Szreniawski Z. The mechanism of the neuromuscular blockade by antibiotics. Arch Int Pharmacodyn Ther 1963;146:549–60.

95. Lambs L, Venturini M, Decock-Le Reverend B, Kozlowski H, Berthon G. Metal ion–tetracycline interactions in biological fluids. Part 8. Potentiometric and spectroscopic studies on the formation of Ca(II) and Mg(II) complexes with 4-dedimethylamino-tetracycline and 6-desoxy-6-demethyl-tetracycline. J Inorg Biochem 1988;33(3):193–210.

96. Edwards TS. Transient myopia due to tetracycline. JAMA 1963;186:69–70.

97. Morrison VL, Kikkawa DO, Herndier BG. Tetracycline induced green conjunctival pigment deposits. Br J Ophthalmol 2005;89(10):1372–3.

98. Aronson JK, Hauben M. Anecdotes that provide definitive evidence. BMJ 2006;333(7581):1267-9.

99. Billano RA, Ward WQ, Little WP. Minocycline and black thyroid. JAMA 1983;249(14):1887.

100. Reid JD. The black thyroid associated with minocycline therapy. A local manifestation of a drug-induced lysosome/substrate disorder. Am J Clin Pathol 1983;79(6):738–46.

101. Korkelia J. Antianabolic effect of tetracyclines. Lancet 1971;1(7706):974–5.

102. Morgan T, Ribush N. The effect of oxytetracycline and doxycycline on protein metabolism. Med J Aust 1972;1(2):55–8.

103. Seltzer HS. Drug-induced hypoglycemia. A review based on 473 cases. Diabetes 1972;21(9):955–66.

104. Garbitelli VP. Tetracycline reduces the need for insulin. NY State J Med 1987;87(10):576.

105. Simpson MB, Pryzbylik J, Innis B, Denham MA. Hemolytic anemia after tetracycline therapy. N Engl J Med 1985;312(13):840–2.

106. Kounis NG. Oxytetracycline-induced thrombocytopenic purpura. JAMA 1975;231(7):734–5.

107. Abou-Rass M. Long-term prognosis of intentional endodontics and internal bleaching of tetracycline-stained teeth. Compend Contin Educ Dent 1998;19(10):1034–8 1040–2, 1044.

108. Leonard RH Jr. Efficacy, longevity, side effects, and patient perceptions of nightguard vital bleaching. Compend Contin Educ Dent 1998;19(8):766–70772, 774.

109. Sadan A, Lemon RR. Combining treatment modalities for tetracycline-discolored teeth. Int J Periodontics Restorative Dent 1998;18(6):564–71.

110. Topoll HH, Lange DE, Muller RF. Multiple periodontal abscesses after systemic antibiotic therapy. J Clin Periodontol 1990;17(4):268–72.

111. Chen JH, Shi CX, Wang M, Zhao SJ, Wang H. Clinical evaluation of 546 tetracycline-stained teeth treated with porcelain laminate veneers. J Dent 2005;33(1):3–8.

112. Baeriswyl G, Bengoa J, de Peyer R, Loizeau E. Importance des ulcérations médicamenteuses dans les lésions endoscopiques de l'oesophage. [Importance of drug-induced ulceration in endoscopic lesions of the esophagus.] Schweiz Med Wochenschr Suppl 1985;19:6–9.

113. Bonavina L, DeMeester TR, McChesney L, Schwizer W, Albertucci M, Bailey RT. Drug-induced esophageal strictures. Ann Surg 1987;206(2):173–83.

114. Zijnen-Suyker MP, Hazenberg BP. Oesophagusbeschadiging door doxycycline. [Esophageal lesions caused by doxycycline.] Ned Tijdschr Geneeskd 1981;125(35):1407–10.

115. Schneider R. Doxycycline esophageal ulcers. Am J Dig Dis 1977;22(9):805–7.

116. Crowson TD, Head LH, Ferrante WA. Esophageal ulcers associated with tetracycline therapy. JAMA 1976;235(25):2747–8.

117. Champel V, Jonville-Bera AP, Bera F, Autret E. Les tétracyclines peuvent être responsables d'ulcérations oesophagiennes si leur prise est incorrecte. Rev Prat Med Gen 1998;12:9–10.

118. Huizar JF, Podolsky I, Goldberg J. Ulceras esofagicas inducidas por doxiciclina. [Doxycycline-induced esophageal ulcers.] Rev Gastroenterol Mex 1998;63(2):101–5.

119. Pantell RH, Goodman BW Jr. Adolescent chest pain: a prospective study. Pediatrics 1983;71(6):881–7.

120. Selbst SM, Ruddy RM, Clark BJ, Henretig FM, Santulli T Jr. Pediatric chest pain: a prospective study. Pediatrics 1988;82(3):319–23.

121. Zavaras-Angelidou KA, Weinhouse E, Nelson DB. Review of 180 episodes of chest pain in 134 children. Pediatr Emerg Care 1992;8(4):189–93.

122. Kobler E, Nuesch HJ, Buhler H, Jenny S, Deyhle P. Medikamentös bedingte Oesophagusulzera. [Drug-induced esophageal ulcers.] Schweiz Med Wochenschr 1979;109(32):1180–2.

123. Dowling HF, Lepper MH. Hepatic reactions to tetracycline. JAMA 1964;188:307–9.

124. Peters RL, Edmondson HA, Mikkelsen WP, Tatter D. Tetracycline-induced fatty liver in nonpregnant patients. A report of six cases. Am J Surg 1967;113(5):622–32.

125. Burette A, Finet C, Prigogine T, De Roy G, Deltenre M. Acute hepatic injury associated with minocycline. Arch Intern Med 1984;144(7):1491–2.

126. Burette A, Finet C, Prigogine T, De Roy G, Deltenre M. Acute hepatic injury associated with minocycline. Arch Intern Med 1984;144(7):1491–2.

127. Hunt CM, Washington K. Tetracycline-induced bile duct paucity and prolonged cholestasis. Gastroenterology 1994;107(6):1844–7.

128. Schrumpf E, Nordgard K. Unusual cholestatic hepatotoxicity of doxycycline in a young male. Scand J Gastroenterol 1986;21(Suppl 120):68.

129. Bjornsson E, Lindberg J, Olsson R. Liver reactions to oral low-dose tetracyclines. Scand J Gastroenterol 1997;32(4):390–5.

130. Elmore MF, Rogge JD. Tetracycline-induced pancreatitis. Gastroenterology 1981;81(6):1134–6.

131. Gant NF Jr, Whalley PJ, Baxter CR. Nonoliguric renal failure. Report of a case. Obstet Gynecol 1969;34(5):675–9.

132. Walker RG, Thomson NM, Dowling JP, Ogg CS. Minocycline-induced acute interstitial nephritis. BMJ 1979;1(6162):524.

133. Wilkinson SP, Stewart WK, Spiers EM, Pears J. Protracted systemic illness and interstitial nephritis due to minocycline. Postgrad Med J 1989;65(759):53–6.

134. Murray KM, Keane WR. Review of drug-induced acute interstitial nephritis. Pharmacotherapy 1992;12(6):462–7.

135. Miller CS, McGarity GJ. Tetracycline-induced renal failure after dental treatment. J Am Dent Assoc 2009;140:56–60.

136. Carrilho F, Bosch J, Arroyo V, Mas A, Viver J, Rodes J. Renal failure associated with demeclocycline in cirrhosis. Ann Intern Med 1977;87(2):195–7.

137. Miller PD, Linas SL, Schrier RW. Plasma demeclocycline levels and nephrotoxicity. Correlation in hyponatremic cirrhotic patients. JAMA 1980;243(24):2513–5.

138. Geheb M, Cox M. Renal effects of demeclocycline. JAMA 1980;243(24):2519–20.

139. Zegers de Beyl D, Naeije R, de Troyer A. Demeclocycline treatment of water retention in congestive heart failure. BMJ 1978;1(6115):760.

140. Kunin CM, Rees SB, Merrill JP, Finland M. Persistence of antibiotics in blood of patients with acute renal failure. I. Tetracycline and chlortetracycline. J Clin Invest 1959;38:1487–97.

141. Carey BW. Abnormal urinary findings and Achromycin V. Pediatrics 1963;31:697.

142. Lowe MB, Obst D, Tapp E. Renal damage caused by anhydro 4-EPI-tetracycline. Arch Pathol 1966;81(4):362–4.

143. Cherrill DA, Stote RM, Birge JR, Singer I. Demeclocycline treatment in the syndrome of inappropriate antidiuretic hormone secretion. Ann Intern Med 1975;83(5):654–6.

144. Forrest JN Jr, Cox M, Hong C, Morrison G, Bia M, Singer I. Superiority of demeclocycline over lithium in the treatment of chronic syndrome of inappropriate secretion of antidiuretic hormone. N Engl J Med 1978;298(4):173–7.

145. Singer I, Rotenberg D. Demeclocycline-induced nephrogenic diabetes insipidus. In-vivo and in-vitro studies. Ann Intern Med 1973;79(5):679–83.

146. De Troyer A, Pilloy W, Broeckaert I, Demanet JC. Correction of antidiuresis by demeclocycline. N Engl J Med 1975;293(18):915–8.

147. Troyer A, Pilloy W, Broeckaert I, Demanet JC. Letter: Demeclocycline treatment of water retention in cirrhosis. Ann Intern Med 1976;85(3):336–7.

148. Bihorac A, Ozener C, Akoglu E, Kullu S. Tetracycline-induced acute interstitial nephritis as a cause of acute renal failure. Nephron 1999;81(1):72–5.

149. Jones HE, Lewis CW, Reisner JE. Photosensitive lichenoid eruption associated with demeclocycline. Arch Dermatol 1972;106(1):58–63.

150. Frank SB, Cohen HJ, Minkin W. Photo-onycholysis due to tetracycline hydrochloride and doxycycline. Arch Dermatol 1971;103(5):520–1.

151. Frost P, Weinstein GD, Gomez EC. Phototoxic potential of minocycline and doxycycline. Arch Dermatol 1972;105(5):681–3.

152. Epstein JH, Tuffanelli DL, Seibert JS, Epstein WL. Porphyria-like cutaneous changes induced by tetracycline hydrochloride photosensitization. Arch Dermatol 1976;112(5):661–6.

153. Fenske NA, Millns JL, Greer KE. Minocycline-induced pigmentation at sites of cutaneous inflammation. JAMA 1980;244(10):1103–6.

154. Butler JM, Marks R, Sutherland R. Cutaneous and cardiac valvular pigmentation with minocycline. Clin Exp Dermatol 1985;10(5):432–7.

155. Shum DT, Smout MS, Pace WE, Headington JT. Unusual skin pigmentation from long-term methacycline and minocycline therapy. Arch Dermatol 1986;122(1):17–8.

156. Arndt KA, Jick H. Rates of cutaneous reactions to drugs. A report from the Boston Collaborative Drug Surveillance Program. JAMA 1976;235(9):918–23.

157. Shelley WB, Heaton CL. Minocycline sensitivity. JAMA 1973;224(1):125–6.

158. Fawcett IW, Pepys J. Allergy to a tetracycline preparation—a case report. Clin Allergy 1976;6(3):301–4.

159. Shoji A, Someda Y, Hamada T. Stevens–Johnson syndrome due to minocycline therapy. Arch Dermatol 1987;123(1):18–20.

160. Curley RK, Verbov JL. Stevens–Johnson syndrome due to tetracyclines—a case report (doxycycline) and review of the literature. Clin Exp Dermatol 1987;12(2):124–5.

161. Trueb RM, Burg G. Acute generalized exanthematous pustulosis due to doxycycline. Dermatology 1993;186(1):75–8.

162. Fiumara NJ, Yaqub M. Pigmented penile lesions (fixed drug eruptions) associated with tetracycline therapy for sexually transmitted diseases. Sex Transm Dis 1981;8(1):23–5.

163. Pasricha JS. Drugs causing fixed eruptions. Br J Dermatol 1979;100(2):183–5.

164. Kanwar AJ, Bharija SC, Singh M, Belhaj MS. Ninety-eight fixed drug eruptions with provocation tests. Dermatologica 1988;177(5):274–9.

165. Domz CA, McNamara DH, Holzapfel HF. Tetracycline provocation in lupus erythematosus. Ann Intern Med 1959;50(5):1217–26.

166. Bojs G, Moller H. Eczematous contact allergy to oxytetracycline with cross-sensitivity to other tetracyclines. Berufsdermatosen 1974;22(5):202–8.

167. Mahaur BS, Sharma VK, Kumar B, Kaur S. Prevalence of contact hypersensitivity to common antiseptics, antibacterials and antifungals in normal persons. Indian. J Dermatol Venerol Leprol 1987;53:269.

168. Burton J. A placebo-controlled study to evaluate the efficacy of topical tetracycline and oral tetracycline in the treatment of mild to moderate acne. Dermatology Research Group. J Int Med Res 1990;18(2):94–103.

169. Gaffoor PM, George WM. Fixed drug eruptions occurring on the male genitals. Cutis 1990;45(4):242–4.

170. Samman PD. The nails: onchylosysis. In: Rook A, Wilkinson DS, Ebling FJG, editors. Textbook of Dermatology. Oxford: Blackwell Scientific Publications, 1972:1647.

171. Rall DP, Loo TL, Lane M, Kelly MG. Appearance and persistence of fluorescent material in tumor tissue after tetracycline administration. J Natl Cancer Inst 1957;19(1):79–85.

172. Totterman LE, Saxen L. Incorporation of tetracycline into human foetal bones after maternal drug administration. Acta Obstet Gynecol Scand 1969;48(4):542–9.

173. Cohlan SQ, Bevelander G, Tiamsic T. Growth inhibition of prematures receiving tetracycline. Am J Dis Child 1963;105:453.

174. Kaitila I, Wartiovaara J, Laitinen O, Saxen L. The inhibitory effect of tetracycline on osteogenesis in organ culture. J Embryol Exp Morphol 1970;23(1):185–211.

175. Gilgor RS. Complications of tetracycline therapy for acne. N C Med J 1972;33(4):331–3.

176. Csonka GW, Rosedale N, Walkden L. Balanitis due to fixed drug eruption associated with tetracycline therapy. Br J Vener Dis 1971;47(1):42–4.

177. Elsner P, Hartmann AA, Burg G. Erfahrungen mit der Oxytetrazyklin-Therapie der nicht-gonorrhoischen Urethritis durch *Ureaplasma urealyticum*. [Experiences with oxytetracycline treatment of non-gonorrhea urethritis caused by *Ureaplasma urealyticum*.] Hautarzt 1990;41(2):94–7.

178. Crook LD, Tempest B. Plague. A clinical review of 27 cases. Arch Intern Med 1992;152(6):1253–6.

179. Jones RB, Van der Pol B, Martin DH, Shepard MK. Partial characterization of *Chlamydia trachomatis* isolates resistant to multiple antibiotics. J Infect Dis 1990;162(6):1309–15.

180. Ling JM, Khin-Thi-Oo H, Hui YW, French GL. Antimicrobial susceptibilities of *Haemophilus* species in Hong Kong. J Infect 1989;19(2):135–42.

181. Schlapfer G, Eichmann A. Penicillinaseproduzierende Stamme von N. gonorrhoeae (PPNG) im Raume Zurich, 1981–1988: Häufigkeit antibiotische Empfindlichkeit und Plasmidprofil (3. Mitteilung). [Penicillinase-producing strains of N. gonorrhoeae (PPNG) in the Zurich area, 1981–1988: incidence, antibiotic sensitivity and plasmid profile (3).] Schweiz Med Wochenschr 1990;120(4):92–7.

182. McColl JD, Globus M, Robinson S. Effect of some therapeutic agents on the developing rat fetus. Toxicol Appl Pharmacol 1965;11:409–17.

183. Krejci L, Brettschneider I. Congenital cataract due to tetracycline. Animal experiments and clinical observation. Ophthalmic Paediatr Genet 1983;3:59–60.

184. Mennie AT. Tetracycline and congenital limb abnormalities. BMJ 1962;2:480.

185. Skosyreva AM. [Comparative evaluation of the embryotoxic effect of various antibiotics.]Antibiot Khimioter 1989;34(10):779–82.

186. Krejci L, Brettschneider I. Congenital cataract due to tetracycline. Ophthalmic Paediatr Genet 1983;3:59.

187. Czeizel AE, Rockenbauer M. A population-based case-control teratologic study of oral oxytetracycline treatment during pregnancy. Eur J Obstet Gynecol Reprod Biol 2000;88(1):27–33.

188. Czeizel AE, Rockenbauer M. Teratogenic study of doxycycline. Obstet Gynecol 1997;89(4):524–8.

189. Anthony JR. Effect on deciduous and permanent teeth of tetracycline deposition in utero. Postgrad Med 1970;48(4):165–8.

190. Seeliger HP, Ronde G. Die Wirkung von Tetracyclingaben auf das kindliche Gebiss bei Listeriosebehandlung von Schwangeren. [Effect of tetracycline administration on the child's dentition in the treatment of listeriosis in pregnant women.] Geburtshilfe Frauenheilkd 1968;28(3):209–23.

191. Briggs GG, Freeman RK, Yaffe SJ. Drugs in Pregnancy and Lactation. 2nd ed.. Baltimore, MD: Williams and Wilkins;. 1986.

192. Charles D, Obst D. Placental transmission of antibiotics. J Obstet Gynaecol Br Emp 1954;61(6):750–7.

193. Witkop CJ Jr, Wolf RO. Hypoplasia and intrinsic staining of enamel following tetracycline therapy. JAMA 1963;185:1008–11.

194. Poliak SC, DiGiovanna JJ, Gross EG, Gantt G, Peck GL. Minocycline-associated tooth discoloration in young adults. JAMA 1985;254(20):2930–2.

195. Ruckdeschel JC, Moores D, Lee JY, Einhorn LH, Mandelbaum I, Koeller J, Weiss GR, Losada M, Keller JH. Intrapleural therapy for malignant pleural effusions. A randomized comparison of bleomycin and tetracycline. Chest 1991;100(6):1528–35.

196. Martinez-Moragon E, Aparicio J, Rogado MC, Sanchis J, Sanchis F, Gil-Suay V. Pleurodesis in malignant pleural effusions: a randomized study of tetracycline versus bleomycin. Eur Respir J 1997;10(10):2380–3.

197. Walker-Renard PB, Vaughan LM, Sahn SA. Chemical pleurodesis for malignant pleural effusions. Ann Intern Med 1994;120(1):56–64.

198. Neuvonen PJ, Gothoni G, Hackman R, Bjorksten K. Interference of iron with the absorption of tetracyclines in man. BMJ 1970;4(734):532–4.

199. Gothoni G, Neuvonen PJ, Mattila M, Hackman R. Iron-tetracycline interaction: effect of time interval between the drugs. Acta Med Scand 1972;191(5):409–11.

200. Albert KS, Welch RD, DeSante KA, DiSanto AR. Decreased tetracycline bioavailability caused by a bismuth subsalicylate antidiarrheal mixture. J Pharm Sci 1979;68(5):586–8.

201. Ericsson CD, Feldman S, Pickering LK, Cleary TG. Influence of subsalicylate bismuth on absorption of doxycycline. JAMA 1982;247(16):2266–7.

202. Koch-Weser J, Sellers EM. Drug interactions with coumarin anticoagulants. Part I. N Engl J Med 1971;285:487.

203. Westfall LK, Mintzer DL, Wiser TH. Potentiation of warfarin by tetracycline. Am J Hosp Pharm 1980;37(12):1620.

204. Danos EA. Apparent potentiation of warfarin activity by tetracycline. Clin Pharm 1992;11(9):806–8.

205. Caraco Y, Rubinow A. Enhanced anticoagulant effect of coumarin derivatives induced by doxycycline coadministration. Ann Pharmacother 1992;26(9):1084–6.

206. Chiavazza F, Merialdi A. Sulle interferenze fra dicumarolo ed antibiotici. [Interference between dicoumarol and antibiotics.] Minerva Ginecol 1973;25(11):630–1.

207. Searcy RL, Simms NM, Foreman JA, Bergquist LM. Evaluation of the blood-clotting mechanism in tetracycline-treated patients. Antimicrobial Agents Chemother 1964;10:179–83.

208. Boston Collaborative Drug Surveillance Program. Tetracycline and drug-attributed rises in blood urea nitrogen. JAMA 1972;220(3):377–9.

209. Albers DD, Leverett CL, Sandin JH. Renal failure following prostatovesiculectomy related to methoxyflurane anesthesia and tetracycline—complicated by *Candida infection.* J Urol 1971;106(3):348–50.

210. Kuzucu EY. Methoxyflurane, tetracycline, and renal failure. JAMA 1970;211(7):1162–4.

211. Proctor EA, Barton FL. Polyuric acute renal failure after methoxyflurane and tetracycline. BMJ 1971;4(788):661–2.

212. Steele M, Couturier J. A possible tetracycline–risperidone–sertraline interaction in an adolescent. Can J Clin Pharmacol 1999;6(1):15–7.

213. Csik V, Molnar J. Possible adverse interaction between clozapine and ampicillin in an adolescent with schizophrenia. J Child Adolesc Psychopharmacol 1994;4:123–8.

214. Hauser RA, Zesiewicz TA. Sertraline-induced exacerbation of tics in Tourette's syndrome. Mov Disord 1995;10(5):682–4.

Thiacetazone

See also Antituberculosis drugs

General Information

Thiacetazone (thioacetazone, thiosemicarbazone) was greeted enthusiastically in 1946 as one of the first synthetic agents against tuberculosis. However, its use rapidly diminished with the increasing observation of untoward effects. It is currently rarely used and then only for economical reasons. Dosages should never exceed 200 mg/day.

Adverse effects of thiacetazone include bone marrow depression and hemolytic anemia. It can also cause serious skin rashes. Continuous laboratory and clinical observations are required (1).

Organs and Systems

Hematologic

Adverse effects of thiacetazone include bone marrow depression, with anemia, leukopenia, agranulocytosis, and thrombocytopenia (2–4). Hemolytic anemia has also been described (5).

Gastrointestinal

Anorexia, nausea, and vomiting are not uncommon with thiacetazone (1).

Urinary tract

Hepatotoxicity with jaundice is frequent with thiacetazone (1).

Skin

Adverse effects of thiacetazone on the skin are of various types, mainly erythematous and maculopapular rashes, angioedema, purpura (6), toxic epidermal necrolysis, Stevens–Johnson syndrome, and pigmentation (7).

Of 50 cases of cutaneous adverse effects of thiacetazone, there were 48 cases of erythema multiforme (25 cases of Stevens–Johnson syndrome and 23 of toxic epidermal necrolysis), one case of erythrodermia, and one case of lichenoid toxidermia; there were 20 deaths, 16 with toxic epidermal necrolysis and four with Stevens–Johnson syndrome (8).

In a retrospective study, 38 cases of toxic epidermal necrolysis were observed in Dakar, and were attributed to thiacetazone in 24 cases; 23 died, mainly because of hypovolemic shock during the first week and septic shock during the second (9). Those who died were generally aged over 50 years, had more than 50% skin involvement, and had evolving tuberculosis at the time of presentation or HIV infection. After-effects were vaginal synechia and two cases of blindness.

Immunologic

Cutaneous allergic reactions to thiacetazone are very common in HIV-positive patients, in the order of 20%. Ethambutol should therefore be used instead of thiacetazone in these patients (10).

Susceptibility Factors

Arguments have been advanced for the abandonment of thiacetazone as an antituberculosis drug (11,12), despite its cheapness, on the grounds that it often causes severe skin reactions, some rapidly fatal, in patients infected with HIV-1 (13). The WHO and IUATLD has recommended careful information and surveillance of possible adverse reactions, particularly cutaneous, in patients treated for tuberculosis in such countries and immediate replacement with ethambutol if there are any prodromal signs of toxicity.

Anergy to tuberculin and lymphopenia have been associated with an increased risk of adverse reactions to thiacetazone. In a randomized study of rifampicin- and thiacetazone-containing regimens in HIV-positive adults with pulmonary tuberculosis, eight of 13 patients who developed adverse reactions were tuberculin-anergic, compared with 12 of 77 patients who did not develop adverse reactions (13). An absolute lymphocyte count below $2.0 \times 10^9/l$ was also associated with adverse reactions.

Drug–Drug Interactions

Isoniazid

Thiacetazone combined with isoniazid (and possibly thioacetazone alone) has been reported to damage chromosomes in cultured human lymphocytes (14).

References

1. Mandell GL, Sande MA. Antimicrobial agents: drugs used in the chemotherapy of tuberculosis and leprosy. In: Goodman Gilman A, Rall TW, Nies AS, Taylor P, editors. Goodman and Gilman's The Pharmacological Basis of Therapeutics. 8th ed.. New York: Pergamon Press, 1990:1146 Chapter 49.
2. Chan SL. Chemotherapy of tuberculosis. In: Davies PDO, editor. Clinical Tuberculosis. London: Chapman and Hall, 1994:141.
3. Gupta SK, Bedi RS, Maini VK. Agranulocytosis due to thiacetazone. Indian J Tuberc 1983;30:146.
4. Jaliluddin, Mohsini AA. Fatal aplastic anaemia due to thiacetazone toxicity. J Indian Med Assoc 1981;77(11):176–7.
5. Anonymous. Antileprosy drugs. BMJ 1971;3(767):174–6.
6. Naraqi S, Temu P. Thiacetazone skin reaction in Papua New Guinea. Med J Aust 1980;1(10):480–1.
7. Short GM. Side-effect of thiacetazone. S Afr Med J 1980;58(1):5–6.
8. Diong MT, Ndiaye B, Camara C. Toxidermies au thiacétazone (TB1) dans un service hospitalier à Dakar. [Skin toxicity of thiacetazone (TB1) at a hospital service in Dakar.] Dakar Med 2001;46(1):1–3.
9. Mame Thierno D, On S, Thierno Ndiaye S, Ndiaye B. Syndrome de Lyell au Senegal: responsabilité de la thiacétazone. [Lyell syndrome in Senegal: responsibility of thiacetazone.] Ann Dermatol Venereol 2001;128(12):1305–7.
10. Nunn P, Kibuga D, Gathua S, Brindle R, Imalingat A, Wasunna K, Lucas S, Gilks C, Omwega M, Were J, McAdam K. Cutaneous hypersensitivity reactions due to thiacetazone in HIV-1 seropositive patients treated for tuberculosis. Lancet 1991;337(8742):627–30.
11. Elliott AM, Foster SD. Thiacetazone: time to call a halt? Considerations on the use of thiacetazone in African populations with a high prevalence of human immunodeficiency virus infection. Tuber Lung Dis 1996;77(1):27–9.
12. Rieder HL, Enarson DA. Rebuttal: time to call a halt to emotions in the assessment of thioacetazone. Tuber Lung Dis 1996;77(2):109–11.
13. Okwera A, Johnson JL, Vjecha MJ, Wolski K, Whalen CC, Hom D, Huebner R, Mugerwa RD, Ellner JJ. Risk factors for adverse drug reactions during thiacetazone treatment of pulmonary tuberculosis in human immunodeficiency virus infected adults. Int J Tuberc Lung Dis 1997;1(5):441–5.
14. Ahuja YR, Jaju M, Jaju M. Chromosome-damaging action of isoniazid and thiacetazone on human lymphocyte cultures in vivo. Hum Genet 1981;57(3):321–2.

Thiamphenicol

See also Chloramphenicol

General Information

Thiamphenicol is a semisynthetic derivative of chloramphenicol, an amine derivative of hydrocarbylsulfonylpropandiol. It differs in that the NO_2 group in the para position is replaced by a methylsulfonyl group. The substitution in the para position of the molecule does not influence its effects on either protein or DNA synthesis in any recognizable way. The antibacterial spectrum is almost identical to that of chloramphenicol.

The two drugs are used in similar dosages, although there are large differences in their elimination. Glucuronidation is unimportant for thiamphenicol: over 90% of a therapeutic dose is excreted by the kidneys in unchanged form. The corresponding figure for chloramphenicol is only about 10%. Thus, in contrast to chloramphenicol, the half-life of thiamphenicol is prolonged in patients with reduced renal function in whom accumulation can occur.

Although the mechanism by which chloramphenicol causes aplastic anemia is not fully understood, all the available evidence suggests that the biochemical consequences of thiamphenicol administration and the risk of serious complications differ from those of chloramphenicol (1).

It does not seem unreasonable, therefore, to use thiamphenicol for wider indications than chloramphenicol, provided that the dosage is adjusted according to renal function. However, thiamphenicol has adverse effects on bone marrow, calling for a degree of caution. It is therefore generally recommended that thiamphenicol should not be given for more than 10–14 days.

The most important adverse effects are an immediate dose-related and reversible disturbance of erythropoiesis (2,3) and peripheral neuropathy. In contrast to chloramphenicol, aplastic anemia and the gray syndrome do not seem to occur with thiamphenicol.

Observational studies

Thiamphenicol has been used to treat 1171 patients with chancroid (4). Each patient was given granulated thiamphenicol 5.0 g orally in a single dose. Only 0.89% did not respond. A few patients had adverse effects, including epigastric pain, headache, nausea, and skin rashes; all were mild and of short duration.

Organs and Systems

Respiratory

To date, there have been no reports of rhinitis or asthma caused by thiamphenicol, but three cases of occupational asthma and rhinitis have been attributed to thiamphenicol inhalation; both immunoglobulin E (IgE)-mediated and non-IgE-mediated responses were involved (5).

Nervous system

Several cases of peripheral neuropathy after the administration of thiamphenicol for 3–5 months have been reported and were thought to be due to a vitamin deficiency. In most cases the sensory function of the nerves of the legs was primarily affected. When discovered early, the damage was reversible, but in other cases it tended to persist (6).

Hematologic

There is evidence that thiamphenicol is more potent than chloramphenicol in causing the early dose-related and therefore predictable type of bone marrow toxicity. Many studies have shown that thiamphenicol in a dose of 1.5 g, and to a much lesser extent 0.75 g/day, causes an immediate disturbance of erythropoiesis in almost every case (7). Alterations in the bone marrow become most evident in patients with renal disease and in elderly subjects (probably because of reduced renal function). Leukocytes and thrombocytes are only slightly affected (8).

Most reviewers have been impressed by the fact that the reported cases of peripheral cytopenia were never accompanied by bone marrow aplasia (9). There have been some doubtful case reports of marrow aplasia, in which factors such as advanced age (predisposing to antibiotic accumulation), neoplasia, concurrent treatment (myelotoxic drugs, anticoagulants), and major surgical interventions have to be mentioned. Even if these cases are to be ascribed to thiamphenicol, they fall within the normal spontaneous incidence of aplastic anemia (10–13).

The most severe hematological changes were reported after treatment of typhoid fever and sepsis with relatively high doses of thiamphenicol. These changes have always been reversible on withdrawal. The serum iron concentration returned to normal within a few days and was rapidly followed by reticulocytosis. Experimental results strongly suggest that the toxic effect of thiamphenicol is temporary and that recovery of the bone marrow after drug-induced suppression is complete, with no subsequent risk of myelodysplasia or leukemic transformation (14).

The occurrence of acute leukemia has been studied in relation to preceding use of drugs (before the 12 months preceding the diagnosis) in a case-control study of 202 patients aged over 15 years with a diagnosis of acute leukemia and age- and sex-matched controls (15). Among users of chloramphenicol or thiamphenicol the odds ratio for any use was 1.1 (0.6–2.2) whereas the odds ratio for high doses was 1.8 (0.6–5.3). Other systemic antibiotics showed no substantial relation with the occurrence of leukemia.

Gastrointestinal

Diarrhea, nausea, or constipation occur in under 10% of cases and are usually mild (16).

Skin

Alopecia has been reported in patients with renal insufficiency taking thiamphenicol, and can lead to complete baldness (7). After withdrawal, normal hair growth will recover. Other skin reactions are rare.

Immunologic

Thiamphenicol has immunosuppressive properties, which are ascribed to an effect on immunocompetent cells, rather than on immunoglobulin synthesis (17). In animals, thiamphenicol prolonged the survival of skin homografts (18).

Second-Generation Effects

Fetotoxicity

Although thiamphenicol penetrates the fetal circulation and is distributed evenly in fetal tissues, no fetal abnormalities have been related to thiamphenicol administration during pregnancy (19). Since mitochondrial protein synthesis in the fetal liver is inhibited by the drug concentrations normally attained, repeated administration of thiamphenicol to pregnant women is not recommended.

Susceptibility Factors

Renal disease

Impaired renal function is an important risk factor for toxic effects of thiamphenicol (20).

Other features of the patient

Pre-existing bone marrow dysfunction and prolonged treatment is an important risk factor for toxic effects of thiamphenicol (21).

References

1. Yunis AA. Chloramphenicol: relation of structure to activity and toxicity. Annu Rev Pharmacol Toxicol 1988;28:83–100.
2. In: International Symposium on Chloramphenicol, Thiamphenicol, Known and Unkown Aspects of Drug-Host InteractionsJanuary 10–12, 1973. Sils-Maria, Switzerland 1973:1201.
3. Ferrari V. Salient features of thiamphenicol: review of clinical pharmacokinetics and toxicity. Sex Transm Dis 1984;11(Suppl 4):336–9.
4. Belda Junior W, Siqueira LF, Fagundes LJ. Thiamphenicol in the treatment of chancroid. A study of 1,128 cases. Rev Inst Med Trop Sao Paulo 2000;42(3):133–5.
5. Ye YM, Kim HM, Suh CH, Nahm DH, Park HS. Three cases of occupational asthma induced by thiamphenicol: detection of serum-specific IgE. Allergy 2006;61(3):394–5.
6. Japanese Ministry of Health and Welfare. Information on adverse reactions to drugs. Peripheral nerve damage due to thiamphenicol. Jpn Med Gaz 1977;20:12.
7. Sotto JJ, Simon P, Subtil P, Rozenbaum A, Najean Y, Pecking A. Toxicité hématologique du thiophénicol. [Hematologic toxicity of thiophenicol.] Nouv Presse Méd 1976;5:2163.
8. Moeschlin S, Novotny Z, Koller F, Ruefli P. Zytostatische Nebenwirkungen des Thiamphenicols: Alopezie, reversible Zytopenien. [Cytostatic side effects of thiamphenicol:
alopecia, reversible cytopenia.] Schweiz Med Wochenschr 1974;104(11):384–7.
9. Keiser G. Co-operative study of patients treated with thiamphenicol. Comparative study of patients treated with chloramphenicol and thiamphenicol. Postgrad Med J 1974;50(Suppl 5):143–5.
10. Najean Y, Guerin MN, Chomienne C. Etiology of acquired aplastic anemia. In: Najean Y, Tognoni G, Yunis AA, editors. Safety Problems Related to Chloramphenicol and Thiamphenicol Therapy. New York: Raven Press, 1981:61.
11. De Renzo A, Formisano S, Rotoli B. Bone marrow aplasia and thiamphenicol. Haematologica 1981;66(1):98–104.
12. Keiser G. Toxizitat von Choramphenicol und Thiamphenicol (CAP und TAP). In: Lohr GW, Arnold H, et al., editors. Probleme der Erythrozytopoese, Granulozytopoese und des malignen Melanoms. Berlin, Heidelberg: Springer Verlag, 1978:179.
13. Martinez-Dalmau A, Fernandez MN, Barbolla L. Haematological toxicity of thiamphenicol: analysis of a case with total irreversible bone marrow aplasia and general review of the problem. [Hematologic toxicity of thiamphenicol. Analysis of a case of irreversible total medullary aplasia and general review of the problem.] Sangre (Barc) 1972;17(1):59–66.
14. Yunis AA, Miller AM, Salem Z, Arimura GK. Chloramphenicol toxicity: pathogenetic mechanisms and the role of the p-NO$_2$ in aplastic anemia. Clin Toxicol 1980;17(3):359–73.
15. Traversa G, Menniti-Ippolito F, Da Cas R, Mele A, Pulsoni A, Mandelli F. Drug use and acute leukemia. Pharmacoepidemiol Drug Saf 1998;7(2):113–23.
16. Belda W. O tratamento da uretrite gonococica aguda masculina pelo tianfenicol: uma revisao. Rev Bras Clin Ter 1978;7:375–9.
17. Vindel JA, Khoury B. Inhibition by thiamphenicol of antibody production induced by different antigens. Postgrad Med J 1974;50(Suppl 5):108–10.
18. Ono K, Hattori T, Kusaba A, Inokuchi K. Prolongation of rat heart allograft survival by thiamphenicol. Surgery 1972;71(2):258–61.
19. Nau H, Welsch F, Ulbrich B, Bass R, Lange J. Thiamphenicol during the first trimester of human pregnancy: placental transfer in vivo, placental uptake in vitro, and inhibition of mitochondrial function. Toxicol Appl Pharmacol 1981;60(1):131–41.
20. Oldershausen HF, Menz HP, Hartmann I, Bezler HJ, Ilg R, Burck GC. Serum levels and elimination of thiamphenicol in patients with impaired liver function and with renal failure on dialysis. Postgrad Med J 1974;50(Suppl 5):44–6.
21. Franceschinis R. In: Drug utilization data for chloramphenicol and thiamphenicol in recent years. New York: Raven Press, 1981:81–9.

Tobramycin

See also Aminoglycoside antibiotics

General Information

Tobramycin closely resembles gentamicin in its microbiological and toxicological properties. The two drugs have similar half-lives, peak serum concentrations, lack of protein binding, volumes of distribution, and predominantly renal excretion by glomerular filtration. The main advantage of tobramycin may be its greater intrinsic activity against *Pseudomonas aeruginosa*. Not all bacterial strains resistant to gentamicin are invariably also resistant to tobramycin. Because of its inherent potential for ototoxicity and nephrotoxicity, renal function and eighth nerve function should be closely monitored (1,2).

Observational studies

In an open, randomized, parallel-group, multicenter study with 184 young people with cystic fibrosis treated with inhaled tobramycin, no adverse events were reported (3).

Comparative studies

In a randomized comparison of nebulized tobramycin and nebulized colistin in patients with cystic fibrosis, 26 of 53 patients treated with tobramycin had at least one respiratory adverse event, most commonly pharyngitis (4). In 520 patients, inhaled tobramycin (300 mg bd for three 28-day cycles, each cycle being separated by a 28-day period of no treatment) was compared with placebo. Respiratory function was significantly improved as early as the second week and remained so for the rest of the study, even during periods without aerosol treatment. There was also a parallel reduction in the relative risk of hospitalization, the number of days of hospitalization, and the number of days of intravenous antibiotic treatment (5).

Placebo-controlled studies

In a double-blind, placebo-controlled, study in 30 patients with bronchiectasis, inhaled tobramycin solution (300 mg bd) was associated with *bronchospasm* in three, but not with detectable ototoxicity or nephrotoxicity (6).

Organs and Systems

Respiratory

Nebulized antipseudomonal antibiotic treatment improves lung function and reduces the frequency of exacerbations of infection in patients with cystic fibrosis and Wegener's granulomatosis (7–9), but the significance of development of antibiotic-resistant organisms remains to be determined (10). In 10 healthy adults, the inhalation of tobramycin 80 mg resulted in the deposition of 11.8 mg in the lungs (11). In a double-blind, randomized, placebo-controlled study, inhaled tobramycin significantly reduced sputum density of *P. aeruginosa*. More patients in the treatment group reported increased cough, dyspnea, wheezing, and non-cardiac chest pain, but the symptoms did not limit therapy (12).

Inhalation of the intravenous formulation of tobramycin can cause bronchoconstriction, as has been confirmed in 26 children with mild to moderate cystic fibrosis (13). Nevertheless, while bronchoconstriction did occur, many patients did not have bronchoconstriction in response to the standard intravenous formulation. Bronchospasm within 30 minutes of administration of tobramycin solution for inhalation in patients with cystic fibrosis was transient and similar to that observed after inhalation of placebo (14,15). The risk of bronchoconstriction may further be reduced by pretreatment with salbutamol.

In a comparison of different dosage regimens, inhaled tobramycin caused bronchial obstruction (16). However, after 10 minutes of inhalation, lung function returned to baseline; the effect was independent of dose.

In two placebo-controlled, randomized studies, inhaled tobramycin significantly reduced sputum density of *P. aeruginosa*; however, more patients in the treatment group reported increased dyspnea and wheezing, although the symptoms did not limit therapy (17,18).

Neuromuscular function

The aminoglycosides have a curare-like action, which can be antagonized by calcium ions and prostigmine. In cases requiring general anesthesia, the effect of muscle relaxants such as D-tubocurarine can be potentiated by aminoglycosides.

Aminoglycoside-induced neuromuscular blockade can be clinically relevant in patients with respiratory acidosis, in myasthenia gravis, and in other neuromuscular diseases. Severe illness, the simultaneous use of anesthetics, for example in the immediate postoperative phase, and the application of the antibiotic to serosal surfaces are predisposing factors to be considered (19). Severe clinical manifestations are rare in patients treated with aminoglycosides that are administered in low doses, such as tobramycin.

Sensory systems

Eyes

Subconjunctival injection of tobramycin has caused macular infarction (SEDA-20, 238). This potentially devastating consequence suggests that care must be exercised when contemplating instillation of aminoglycosides directly into the eye.

Allergic contact dermatitis causing conjunctivitis and blepharitis has been reported with topical ophthalmic tobramycin (20).

- A 59-year-old woman developed severe erythema, edema, and exudation in the left eye and eyelid, with ocular pruritus, after treatment with tobramycin 0.3% eye drops for several days after cataract removal (21). Skin patch testing was positive and there was no response to the excipients. The symptoms subsided within 1 month after withdrawal of the eye drops.

Ears

Tobramycin affects the cochlear and vestibular systems to a similar extent as gentamicin. A survey of 24 trials showed the following mean frequencies of ototoxicity: gentamicin 7.7%, tobramycin 9.7%, amikacin 13.8%, netilmicin 2.3% (22). There was also a lower incidence of netilmicin-induced inner-ear damage compared with tobramycin in two studies (23,24).

- A 15-year-old young man with cystic fibrosis, renal failure and a serum concentration of 134 mg/l developed profound sensorineural hearing loss after using inhaled tobramycin 600 mg every 12 hours for 3 weeks (total dose 29 g or 640 mg/kg) (25). Vestibular damage without hearing loss has been reported with tobramycin.
- A 59-year old woman, a lung transplant recipient with sputum positive for *Pseudomonas aeruginosa*, taking tacrolimus and sirolimus, developed acute renal insufficiency and ataxia with bilateral vestibular system paresis but no hearing loss after using inhaled tobramycin 300 mg every 12 hours for 2 weeks (26). After withdrawal, renal function improved but the vestibular symptoms never fully resolved.

Of 60 adult patients with cystic fibrosis randomized to tobramycin, either 10 mg/kg/day or 3.3 mg/kg tds, two patients (one in each group) had bilateral impairment in pure tone audiography after treatment (27).

Histological examination of the temporal bones from two individuals with ototoxicity due to tobramycin showed reductions in the numbers of both ganglion cells and hair cells (28). Spiral ganglion cell loss was not necessarily subadjacent to areas of hair cell loss in cases of aminoglycoside ototoxicity. Instead, there may be a reduction in the number of ganglion cells in segments of the cochlea with normal-appearing hair cells.

Psychological, psychiatric

As with gentamicin, toxic psychoses can occur with tobramycin (29).

Metal metabolism

Repeated courses of intravenous tobramycin can cause hypomagnesemic tetany (30).

Liver

Certain aminoglycosides affect liver function tests. Increases in alkaline phosphatase after tobramycin have been described (31).

Urinary tract

The following order of relative nephrotoxicity was found in animal experiments: neomycin > gentamicin > tobramycin > amikacin > netilmicin (32,33). However, in humans, conclusive data regarding the relative toxicity of the various aminoglycosides are still lacking. An analysis of 24 controlled trials showed the following average rates for nephrotoxicity: gentamicin, 11%; tobramycin, 11.5%; amikacin, 8.5%; and netilmicin, 2.8% (22). In contrast to this survey, direct comparison in similar patient groups showed no significant differences between the various agents in most trials (34–40). One prospective trial showed a significant advantage of tobramycin over gentamicin (41). However, these findings could not

subsequently be confirmed (42). Nephrotoxicity remains a serious risk with all the currently available aminoglycoside antibiotics, and no drug in this series can be regarded as safe.

- Nephrotoxicity due to inhaled tobramycin has been described in a 20-year-old patient with cystic fibrosis who developed acute non-oliguric renal insufficiency after taking inhaled tobramycin 300 mg bd for 1 week; the clinical and renal biopsy findings were consistent with aminoglycoside-induced changes (43).

A 73-year-old woman with COPD developed nephrotoxicity after using inhaled tobramycin (44), as did a 62-year-old Caucasian woman (45), and two lung transplant patients developed renal failure when nebulized tobramycin was started (46).

In a double blind, randomized, controlled trial once-daily versus thrice-daily intravenous tobramycin was compared in 219 patients with cystic fibrosis and pulmonary exacerbations (47). In children once-daily treatment was significantly less nephrotoxic than thrice daily administration (n = 61 versus 64). However, the study was not powered to detect differences in toxicity.

Cement laden with tobramycin and/or vancomycin can cause systemic toxicity, as in two cases of acute renal failure associated with the use of antibiotic-laden cement incorporated in total hip arthroplasties (48).

- Bone cement impregnated with tobramycin and cefazolin, which was placed into an infected total knee arthroplasty, led to renal insufficiency in an 85-year-old man with pre-existing renal impairment (49).

Skin

Ultraviolet recall has been reported after piperacillin, tobramycin, and ciprofloxacin, with a reaction pattern different from that of chemotherapy-induced reactions.

- A 31-year-old woman with a *Pseudomonas* pneumonia was given intravenous ciprofloxacin 400 mg 12-hourly, tobramycin 400 mg/day, and piperacillin 4 g 6-hourly (50). Three days after the initial administration of the intravenous antibiotics, she developed a morbilliform eruption on the sun-exposed areas of her chest, back, and arms, corresponding to sites of intense sunburn that had occurred a month before.

Immunologic

- Hypersensitivity to inhaled tobramycin has been reported in a 9-year-old boy who developed a rash after a course of gentamicin (51). The rash resolved after withdrawal, but returned all over his body when inhaled tobramycin was restarted. He was desensitized using escalating doses of inhaled tobramycin, tolerated the procedure well, and was still using once-a-day tobramycin 9 months after desensitization.

Long-Term Effects

Drug tolerance

Intermittent administration of inhaled tobramycin has been recommended in patients with cystic fibrosis, as it improves pulmonary function, reduces the density of *P. aeruginosa* in sputum, and reduces the risk of hospitalization. The proportion of patients with isolates of *P. aeruginosa* with higher minimal inhibitory concentrations of tobramycin may increase (52). Treatment with inhaled tobramycin does not increase isolation of *Burkholderia cepacia*, *Stenotrophomonas maltophilia*, or *Alcaligenes xylosoxidans*; however, isolation of *Candida albicans* and *Aspergillus* species may increase (53).

With nebulized tobramycin the introduction or selection of resistant bacteria is relatively rare but is a matter of concern (54).

Second-Generation Effects

Teratogenicity

During pregnancy the aminoglycosides cross the placenta and they might theoretically be expected to cause otological and perhaps nephrological damage to the fetus. However, no proven cases of intrauterine damage by tobramycin have been recorded.

Susceptibility Factors

Renal disease

Renal excretion of tobramycin accounts for 90% of the administered dose, and its elimination is prolonged in patients with impaired renal function. The amount of tobramycin eliminated during plasma exchange represented less than 10% of total body stores (55).

Reduced tobramycin clearance can be associated with a normal creatinine clearance in serum concentration-adjusted dosage of once-daily tobramycin therapy in critically ill patients (56).

Other features of the patient

The pharmacokinetics of tobramycin in patients with cystic fibrosis is significantly altered after lung transplantation, and early and close drug monitoring is recommended (57).

Hydrophilic contact lenses may increase the penetration of tobramycin into the aqueous humor (58).

Drug Administration

Drug dosage regimens

The addition of tobramycin reduced the amount of cefuroxime-induced endotoxin released per killed *Escherichia coli* to a level that was even lower than that of tobramycin alone, despite an increased killing rate (59). Increasing concentrations of tobramycin led to a reduction in endotoxin release, pointing to a possible benefit of once-daily dosing regimens.

In an analysis of sera from 60 adults with cystic fibrosis, it was suggested that the potential benefit of achieving a greater peak/MIC with once-daily aminoglycoside administration may be offset by the significantly greater time that the concentration was below the MIC, compared with that achieved with multiple daily dosing regimens (60).

Based on a study of 10 patients with automated peritoneal dialysis, it was recommended that for empirical treatment of dialysis-related peritonitis, the dosage of intermittent intraperitoneal tobramycin must be 1.5 mg/kg for one exchange during the first day and then 0.5 mg/kg thereafter, to reduce the risk of adverse effects (61).

Drug administration route

With combined inhalational and intravenous tobramycin, toxic serum drug concentrations can occur (62).

Wound irrigation

Although the efficacy of peroperative wound irrigation with aminoglycosides has yet to be firmly established in prospective comparative trials, they are often used during intra-abdominal and thoracic surgery. Depending on the amount of drug used and the nature of the operation, it is likely that significant amounts of tobramycin are absorbed (SEDA-20, 237).

Inhalation

Tobramycin has been used extensively in cystic fibrosis because of its effect on *Pseudomonas* organisms. Used as an adjunct to intravenous antibiotics in acute infections, it can lower sputum colony counts. Aerosolized antibiotics over prolonged periods can improve lung function and reduce hospital admissions, but they carry the potential risk of drug toxicity and resistance. In a review of clinical studies it was concluded that there was no evidence of nephrotoxicity or ototoxicity (63), although long-term toxicity studies at higher dosages are awaited.

The efficacy of nebulized tobramycin 300 mg bd for 4 weeks has been studied in a randomized, double-blind, placebo-controlled trial in 74 patients with bronchiectasis and *Pseudomonas* infection, without cystic fibrosis (38). After 4 weeks there was a significant fall in the density of *Pseudomonas* infection in the sputum of the treated group. This correlated with an improvement in general medical condition, as assessed subjectively 2 weeks later. There was no difference in lung function. Tobramycin resistance (MIC over 16 μg/ml) developed in 4 patients in the treated group and 1 patient in the placebo group. Adverse events were reported by 31 of the 37 patients in each arm. There were significantly increased incidences of dyspnea (32%), chest pain (19%), and wheezing (16%) in the treated group compared with placebo (8%, 0%, 0% respectively). Cough increased in 15 patients (41%) in the treated group and 9 (24%) in the placebo group. The investigators felt that the majority of these respiratory adverse events had been related to the drug. They commented that these adverse events did not generally result in withdrawal of the patients from the trial. No more details were given but the apparent adverse effects profile of nebulized tobramycin in this group is of concern.

There was no difference between recipients of tobramycin by inhalation or placebo in serum creatinine concentrations at week 0 or week 20 in clinical trials (14).

In two studies in which nebulized tobramycin 300 mg twice/day was administered, systemic peak concentrations were below 0.2 and 3.62 µg/ml, and trough concentrations were undetectable, making toxicity from this route of administration negligible. However, high concentrations can occur.

- A 19-year-old woman who received a heart transplant was given tobramycin by inhalation for *Acinetobacter baumanii* pneumonia; her serum trough concentrations were toxic (> 2.0 µg/ml) (64). Her risk factors for these toxic concentrations were renal Insufficiency and administration of the drug by positive pressure ventilation.

Topical administration into the ear

The primary target for the topical effects of aminoglycosides in the ear is controversial. It has been generally believed that outer hair cells are primarily affected, followed by loss of inner hair cells; that degeneration starts basally and proceeds towards the apex; and that neural fibers and ganglion cells would degenerate secondarily to loss of hair cells. However, the possibility that spiral ganglion cells may be a primary target has been raised. In a postmortem study of 10 temporal bones from 6 patients (aged 11–40 years) with cystic fibrosis, whose pulmonary infections had been treated with systemic aminoglycosides (mainly tobramycin), in most cases combined with nebulized tobramycin, cytocochleograms were analysed for each bone (65). Four bones showed typical manifestations of aminoglycoside-induced ototoxicity, with loss of hair cells in the lower turns and degeneration of ganglion cells. Six bones showed no loss or scattered loss of hair cells, but there was degeneration of the spiral ganglion cells. This study supports the hypothesis that degeneration of spiral ganglion cells may occur as a primary manifestation in some cases of aminoglycoside-induced ototoxicity.

Drug overdose

Multiple-dose activated charcoal does not increase the elimination of tobramycin and is therefore not recommended for treatment of poisoning (SEDA24.26.74) (66).

Monitoring Therapy

During multiple daily dosing peak serum concentrations of tobramycin over 5–7 µg/ml are associated with improved survival in patients with septicemia and pneumonia caused by Gram-negative bacteria (67,68). On the other hand, excessive peak concentrations (over 10–12 µg/ml) and trough concentrations (over 2 µg/ml) of tobramycin increase the risk of ototoxicity and nephrotoxicity (69).

References

1. Bendush CL, Weber R. Tobramycin sulfate: a summary of worldwide experience from clinical trials. J Infect Dis 1976;134(Suppl):S219–34.
2. Neu HC. Tobramycin: an overview. J Infect Dis 1976;134(Suppl):S3–S19.
3. Murphy TD, Anbar RD, Lester LA, Nasr SZ, Nickerson B, VanDevanter DR, Colin AA. Treatment with tobramycin solution for inhalation reduces hospitalizations in young CF subjects with mild lung disease. Pediatr Pulmonol 2004;38(4):314–20.
4. Hodson ME, Gallagher CG, Govan JR. A randomised clinical trial of nebulised tobramycin or colistin in cystic fibrosis. Eur Respir J 2002;20(3):658–64.
5. Geller DE, Pitlick WH, Nardella PA, Tracewell WG, Ramsey BW. Pharmacokinetics and bioavailability of aerosolized tobramycin in cystic fibrosis. Chest 2002;122(1):219–26.
6. Drobnic ME, Sune P, Montoro JB, Ferrer A, Orriols R. Inhaled tobramycin in non-cystic fibrosis patients with bronchiectasis and chronic bronchial infection with Pseudomonas aeruginosa. Ann Pharmacother 2005;39(1):39–44.
7. Niedzielska G, Katska E. Rodzinne wystepowanie niedosluchu odbiorczego zwiazanego z leczeniem streptomycyna. [Familial occurrence of hearing loss following streptomycin (SM) treatment.] Otolaryngol Pol 2001;55(3):313–5.
8. Bothamley G. Drug treatment for tuberculosis during pregnancy: safety considerations. Drug Saf 2001;24(7).553–65.
9. Govan J. TOBI: reducing the impact of pseudomonal infection. Hosp Med 2002;63(7):421–5.
10. Ryan G, Mukhopadhyay S, Singh M. Nebulised anti-pseudomonal antibiotics for cystic fibrosis (Cochrane Review)Oxford: Update Software;. 2000.
11. Coates AL, Dinh L, MacNeish CF, Rollin T, Gagnon S, Ho SL, Lands LC. Accounting for radioactivity before and after nebulization of tobramycin to insure accuracy of quantification of lung deposition. J Aerosol Med 2000;13(3):169–78.
12. Barker AF, Couch L, Fiel SB, Gotfried MH, Ilowite J, Meyer KC, O'Donnell A, Sahn SA, Smith LJ, Stewart JO, Abuan T, Tully H, Van Dalfsen J, Wells CD, Quan J. Tobramycin solution for inhalation reduces sputum Pseudomonas aeruginosa density in bronchiectasis. Am J Respir Crit Care Med 2000;162(2 Pt 1):481–5.
13. Ramagopal M, Lands LC. Inhaled tobramycin and bronchial hyperactivity in cystic fibrosis. Pediatr Pulmonol 2000;29(5):366–70.
14. Cheer SM, Waugh J, Noble S. Inhaled tobramycin (TOBI):a review of its use in the management of Pseudomonas aeruginosa infections in patients with cystic fibrosis. Drugs 2003;63:2501–20.
15. Ryan G, Mukhopadhyay S, Singh M. Nebulised anti-pseudomonal antibiotics for cystic fibrosis. Cochrane Database Syst Rev 2003;(3):CD001021.
16. Nikolaizik WH, Trociewicz K, Ratjen F. Bronchial reactions to the inhalation of high-dose tobramycin in cystic fibrosis. Eur Respir J 2002;20(1):122–6.
17. Hinojosa R, Nelson EG, Lerner SA, Redleaf MI, Schramm DR. Aminoglycoside ototoxicity: a human temporal bone study. Laryngoscope 2001;111(10):1797–805.
18. Moss RB. Administration of aerosolized antibiotics in cystic fibrosis patients. Chest 2001;120(Suppl 3):S107–13.
19. Holtzman JL. Letter: Gentamicin and neuromuscular blockade. Ann Intern Med 1976;84(1):55.
20. Caraffini S, Assalve D, Stingeni L, Lisi P. Allergic contact conjunctivitis and blepharitis from tobramycin. Contact Dermatitis 1995;32(3):186–7.

21. Gonzalez-Mendiola MR, Balda AG, Delgado MC, Montano PP, De Olano DG, Sanchez-Cano M. Contact allergy from tobramycin eyedrops. Allergy 2005;60(4):527–8.

22. Cone LA. A survey of prospective, controlled clinical trials of gentamicin, tobramycin, amikacin, and netilmicin. Clin Ther 1982;5(2):155–62.

23. Lerner AM, Reyes MP, Cone LA, Blair DC, Jansen W, Wright GE, Lorber RR. Randomised, controlled trial of the comparative efficacy, auditory toxicity, and nephrotoxicity of tobramycin and netilmicin. Lancet 1983;1(8334):1123–6.

24. Gatell JM, SanMiguel JG, Araujo V, Zamora L, Mana J, Ferrer M, Bonet M, Bohe M, Jimenez de Anta MT. Prospective randomized double-blind comparison of nephrotoxicity and auditory toxicity of tobramycin and netilmicin. Antimicrob Agents Chemother 1984;26(5):766–9.

25. Patatanian L. Inhaled tobramycin-associated hearing loss in an adolescent with renal failure. Pediatr Infect Dis J 2006;25(3):276–8.

26. Ahya VN, Doyle AM, Mendez JD, Lipson DA, Christie JD, Blumberg EA, Pochettino A, Nelson L, Bloom RD, Kotloff RM. Renal and vestibular toxicity due to inhaled tobramycin in a lung transplant recipient. J Heart Lung Transplant 2005;24(7):932–5.

27. Whitehead A, Conway SP, Etherington C, Caldwell NA, Setchfield N, Bogle S. Once-daily tobramycin in the treatment of adult patients with cystic fibrosis. Eur Respir J 2002;19(2):303–9.

28. Gardner TB, Hill DR. Treatment of giardiasis. Clin Microbiol Rev 2001;14(1):114–28.

29. McCartney CF, Hatley LH, Kessler JM. Possible tobramycin delirium. JAMA 1982;247(9):1319.

30. Adams JP, Conway SP, Wilson C. Hypomagnesaemic tetany associated with repeated courses of intravenous tobramycin in a patient with cystic fibrosis. Respir Med 1998;92(3):602–4.

31. Martines G, Butturini L, Menozzi I, Restori G, Boiardi L, Bernardi S, Baldassarri P. Amikacin-induced liver toxicity: correlations between biochemical indexes and ultrastructural features in an experimental model. Rev Med Univ Navarra 1988;32(1):41–5.

32. Luft FC, Yum MN, Kleit SA. Comparative nephrotoxicities of netilmicin and gentamicin in rats. Antimicrob Agents Chemother 1976;10(5):845–9.

33. Hottendorf GH, Gordon LL. Comparative low-dose nephrotoxicities of gentamicin, tobramycin, and amikacin. Antimicrob Agents Chemother 1980;18(1):176–81.

34. Smith CR, Baughman KL, Edwards CQ, Rogers JF, Lietman PS. Controlled comparison of amikacin and gentamicin. N Engl J Med 1977;296(7):349–53.

35. Feld R, Valdivieso M, Bodey GP, Rodriguez V. Comparison of amikacin and tobramycin in the treatment of infection in patients with cancer. J Infect Dis 1977;135(1):61–6.

36. Love LJ, Schimpff SC, Hahn DM, Young VM, Standiford HC, Bender JF, Fortner CL, Wiernik PH. Randomized trial of empiric antibiotic therapy with ticarcillin in combination with gentamicin, amikacin or netilmicin in febrile patients with granulocytopenia and cancer. Am J Med 1979;66(4):603–10.

37. Lau WK, Young LS, Black RE, Winston DJ, Linne SR, Weinstein RJ, Hewitt WL. Comparative efficacy and toxicity of amikacin/carbenicillin versus gentamicin/carbenicillin in leukopenic patients: a randomized prospective trail. Am J Med 1977;62(6):959–66.

38. Fong IW, Fenton RS, Bird R. Comparative toxicity of gentamicin versus tobramycin: a randomized prospective study. J Antimicrob Chemother 1981;7(1):81–8.

39. Bock BV, Edelstein PH, Meyer RD. Prospective comparative study of efficacy and toxicity of netilmicin and amikacin. Antimicrob Agents Chemother 1980;17(2):217–25.

40. Barza M, Lauermann MW, Tally FP, Gorbach SL. Prospective, randomized trial of netilmicin and amikacin, with emphasis on eighth-nerve toxicity. Antimicrob Agents Chemother 1980;17(4):707–14.

41. Smith CR, Lipsky JJ, Laskin OL, Hellmann DB, Mellits ED, Longstreth J, Lietman PS. Double-blind comparison of the nephrotoxicity and auditory toxicity of gentamicin and tobramycin. N Engl J Med 1980;302(20):1106–9.

42. Matzke GR, Lucarotti RL, Shapiro HS. Controlled comparison of gentamicin and tobramycin nephrotoxicity. Am J Nephrol 1983;3(1):11–7.

43. Hoffmann IM, Rubin BK, Iskandar SS, Schechter MS, Nagaraj SK, Bitzan MM. Acute renal failure in cystic fibrosis: association with inhaled tobramycin therapy. Pediatr Pulmonol 2002;34(5):375–7.

44. Izquierdo MJ, Gomez-Alamillo C, Ortiz F, Calabia ER, Ruiz JC, de Francisco AL, Arias M. Acute renal failure associated with use of inhaled tobramycin for treatment of chronic airway colonization with *Pseudomonas aeruginosa*. Clin Nephrol 2006;66(6):464–7.

45. Cannella CA, Wilkinson ST. Acute renal failure associated with inhaled tobramycin. Am J Health Syst Pharm 2006;63(19):1858–61.

46. Laporta R, Ussetti P, Carreno MC. Renal toxicity due to inhaled tobramycin in lung transplant recipients. J Heart Lung Transplant 2006;25(5):608.

47. Smyth A, Tan KH, Hyman-Taylor P, Mulheran M, Lewis S, Stableforth D, Knox A. Once versus three-times daily regimens of tobramycin treatment for pulmonary exacerbations of cystic fibrosis—the TOPIC study: a randomised controlled trial. Lancet 2005;365(9459):573–8.

48. Patrick BN, Rivey MP, Allington DR. Acute renal failure associated with vancomycin- and tobramycin-laden cement in total hip arthroplasty. Ann Pharmacother 2006;40(11):2037–42.

49. Curtis JM, Sternhagen V, Batts D. Acute renal failure after placement of tobramycin-impregnated bone cement in an infected total knee arthroplasty. Pharmacotherapy 2005;25(6):876–80.

50. Terzano C, Taurino AE, Peona V. Nebulized tobramycin in patients with chronic respiratory infections during clinical evolution of Wegener's granulomatosis. Eur Rev Med Pharmacol Sci 2001;5(4):131–8.

51. Spigarelli MG, Hurwitz ME, Nasr SZ. Hypersensitivity to inhaled TOBI following reaction to gentamicin. Pediatr Pulmonol 2002;33(4):311–4.

52. Ramsey BW, Pepe MS, Quan JM, Otto KL, Montgomery AB, Williams-Warren J, Vasiljev-K M, Borowitz D, Bowman CM, Marshall BC, Marshall S, Smith ALCystic Fibrosis Inhaled Tobramycin Study Group. Intermittent administration of inhaled tobramycin in patients with cystic fibrosis. N Engl J Med 1999;340(1):23–30.

53. Burns JL, Van Dalfsen JM, Shawar RM, Otto KL, Garber RL, Quan JM, Montgomery AB, Albers GM, Ramsey BW, Smith AL. Effect of chronic intermittent administration of inhaled tobramycin on respiratory microbial flora in patients with cystic fibrosis. J Infect Dis 1999;179(5):1190–6.

54. Sermet-Gaudelus I, Le Cocguic Y, Ferroni A, Clairicia M, Barthe J, Delaunay JP, Brousse V, Lenoir G. Nebulized antibiotics in cystic fibrosis. Paediatr Drugs 2002;4(7):455–67.

55. Kintzel PE, Eastlund T, Calis KA. Extracorporeal removal of antimicrobials during plasmapheresis. J Clin Apheresis 2003;18:194–205.

56. Reimann IR, Meier-Hellmann A, Pfeifer R, Traut T, Schilling A, Stein G, Reinhart K, Hoffmann A. Serumspiegelorientierte Dosierung der einmal täglichen Aminoglykosidtherapie beim kritisch Kranken: Ergebnisse einer prospektiven Untersuchung. [Serum level-adjusted dosage of once-daily aminoglycoside therapy in critical illness: results of a prospective study.] Anasthesiol Intensivmed Notfallmed Schmerzther 1999;34(5):288–95.

57. Moore RA, DeShazer D, Reckseidler S, Weissman A, Woods DE. Efflux-mediated aminoglycoside and macrolide resistance in *Burkholderia pseudomallei*. Antimicrob Agents Chemother 1999;43(3):465–70.

58. Hehl EM, Beck R, Luthard K, Guthoff R, Drewelow B. Improved penetration of aminoglycosides and fluoroquinolones into the aqueous humour of patients by means of Acuvue contact lenses. Eur J Clin Pharmacol 1999;55(4):317–23.

59. Sjolin J, Goscinski G, Lundholm M, Bring J, Odenholt I. Endotoxin release from Escherichia coli after exposure to tobramycin: dose-dependency and reduction in cefuroxime-induced endotoxin release. Clin Microbiol Infect 2000;6(2):74–81.

60. Beringer PM, Vinks AA, Jelliffe RW, Shapiro BJ. Pharmacokinetics of tobramycin in adults with cystic fibrosis: implications for once-daily administration. Antimicrob Agents Chemother 2000;44(4):809–13.

61. Manley HJ, Bailie GR, Frye R, Hess LD, McGoldrick MD. Pharmacokinetics of intermittent intravenous cefazolin and tobramycin in patients treated with automated peritoneal dialysis. J Am Soc Nephrol 2000;11(7):1310–6.

62. Elidemir O, Maciejewski SR, Oermann CM. Falsely elevated serum tobramycin concentrations in cystic fibrosis patients treated with concurrent intravenous and inhaled tobramycin. Pediatr Pulmonol 2000;29(1):43–5.

63. Touw DJ, Brimicombe RW, Hodson ME, Heijerman HG, Bakker W. Inhalation of antibiotics in cystic fibrosis. Eur Respir J 1995;8(9):1594–604.

64. Kahler DA, Schowengerdt KO, Fricker FJ, Mansfield M, Visner GA, Faro A. Toxic serum trough concentrations after administration of nebulized tobramycin. Pharmacotherapy 2003;23:543–5.

65. Sone M, Schachern PA, Paparella MM. Loss of spiral ganglion cells as primary manifestation of aminoglycoside ototoxicity. Hear Res 1998;115(1–2):217–23.

66. American Academy of Clinical ToxicologyEuropean Association of Poisons Centres and Clinical Toxicologists. Position statement and practice guidelines on the use of multi-dose activated charcoal in the treatment of acute poisoning. J Toxicol Clin Toxicol 1999;37(6):731–51.

67. Moore RD, Smith CR, Lietman PS. The association of aminoglycoside plasma levels with mortality in patients with Gram-negative bacteremia. J Infect Dis 1984;149(3):443–8.

68. Moore RD, Smith CR, Lietman PS. Association of aminoglycoside plasma levels with therapeutic outcome in Gram-negative pneumonia. Am J Med 1984;77(4):657–62.

69. Wenk M, Vozeh S, Follath F. Serum level monitoring of antibacterial drugs. A review. Clin Pharmacokinet 1984;9(6):475–92.

Tosufloxacin

See also Fluoroquinolones

General Information

Tosufloxacin is a fluoroquinolone antibacterial with properties similar to those of ciprofloxacin.

Observational studies

In 58 Japanese patients with typhoid fever, 42 with paratyphoid fever, and one with both typhoid fever and paratyphoid fever, almost 80% of whom were treated with tosufloxacin, there were adverse effects (nausea, urticaria, aphthous stomatitis) in 3.6% and raised serum amylase in 8.3% (1). All the adverse reactions resolved with or without a change in drug therapy.

General adverse effects

The most common adverse reactions to tosufloxacin are gastrointestinal disorders, including diarrhea, abdominal discomfort, nausea and vomiting, and skin disorders, including rash and pruritus (2). Central and peripheral nervous system disorders were observed in 0.36% of patients and the most common symptoms were headache and dizziness.

Organs and Systems

Respiratory

- A syndrome of pulmonary infiltration with eosinophilia occurred in an 83-year-old man who was given piperacillin plus tosufloxacin (3).

In vitro blastogenesis of his peripheral blood lymphocytes was strongly enhanced by piperacillin and tosufloxacin, and they generated a large amount of interleukin-5.

Skin

There is a low incidence of phototoxicity with tosufloxacin (2).

Musculoskeletal

Rhabdomyolysis has been reported in 13 patients taking tosufloxacin (2).
An anaphylactic reaction was reported in patient who took tosufloxacin tosilate (4).

Fluoroquinolones can cause acute renal insufficiency because of interstitial nephritis with or without epithelioid granulomas.

- A 63-year-old Japanese developed slowly progressive renal insufficiency caused by crystal-forming chronic interstitial nephritis with non-Langerhans' cell histiocytosis after exposure for 4 years to tosufloxacin tosilate 300 mg/day (5).

Drug–Drug Interactions

Antacids

The C_{max} and AUC of tosufloxacin tosilate 300 mg were significantly reduced by aluminium hydroxide gel fine granules 1 g or magnesium oxide 1 g (2).

References

1. Ohnishi K, Kimura K, Masuda G, Tsunoda T, Obana M, Yoshida H, Goto T, Sakaue Y, Kim YK, Sakamoto M, Sagara H. Oral administration of fluoroquinolones in the treatment of typhoid fever and paratyphoid fever in Japan. Intern Med 2000;39(12):1044–8.
2. Niki Y. Pharmacokinetics and safety assessment of tosufloxacin tosilate. J Infect Chemother 2002;8(1):1–18.
3. Yamamoto T, Tanida T, Ueta E, Kimura T, Doi S, Osaki T. Pulmonary infiltration with eosinophilia (PIE) syndrome induced by antibiotics, PIPC and TFLX during cancer treatment. Oral Oncol 2001;37(5):471–5.
4. Umebayashi Y, Furuta J. Anaphylaxis due to tosufloxacin tosilate. J Dermatol 2003;30:701–2.
5. Okada H, Watanabe Y, Kotaki S, Ikeda N, Takane H, Kanno Y, Sugahara S, Ban S, Nagata M, Suzuki H. An unusual form of crystal-forming chronic interstitial nephritis following long-term exposure to tosufloxacin tosilate. Am J Kidney Dis 2004;44(5):902–7.

Trimethoprim and co-trimoxazole

General Information

Trimethoprim is a 2,4-diamino-5-(3′,4′,5′-trimethoxybenzyl) pyrimidine that inhibits dihydrofolate reductase, the enzyme in folate synthesis after the step that is blocked by sulfonamides (1). Trimethoprim therefore inhibits the conversion of dihydrofolate to tetrahydrofolate. It has been combined with sulfonamides, including sulfamethoxazole, sulfametrol, sulfadiazine, sulfamoxole, and sulfadimidine (2,3).

A widely available fixed combination is co-trimoxazole (Bactrim, Eusaprim, Septrin), which contains trimethoprim and sulfamethoxazole in a ratio of 1:5. Both trimethoprim and sulfamethoxazole have favorable and comparable pharmacokinetics and the combination is bactericidal (4). Synergy between trimethoprim and sulfonamides has conventionally been ascribed to sequential inhibition of dihydropteroate synthetase by sulfonamides (in competition with para-aminobenzoic acid) and of dihydrofolate reductase by trimethoprim (in competition with dihydrofolate). However, sulfonamides in high concentrations also inhibit dihydrofolate reductase. Thus, an initial partial sequential blockade by trimethoprim (inhibition of dihydrofolate reductase) and sulfonamides (inhibition of dihydropteroate synthetase) leads to defective protein synthesis and cytoplasmic damage, which in turn results in marked increases in the uptake of both agents and "double strength" inhibition of dihydrofolate reductase (5).

Trimethoprim is fairly active against a variety of Gram-positive cocci and Gram-negative rods. Established indications for co-trimoxazole are infections of the sinuses, ears, lungs, and urinary tract, and infections due to Salmonella, Nocardia, Brucella, Stenotrophomonas maltophilia, Pneumocystis jiroveci, and Toxoplasma (1,6). Co-trimoxazole is also used in the treatment of Wegener's granulomatosis, for prevention of spontaneous bacterial peritonitis, and in patients with advanced HIV infection for the prophylaxis of opportunistic infections (1,6).

There have been few comparisons of the efficacy of trimethoprim and co-trimoxazole. In uncomplicated urinary tract, bronchopulmonary, and ear infections, no advantage of co-trimoxazole over trimethoprim has been documented (7,8). However, in complicated urinary tract infections most studies have shown better results with co-trimoxazole than with trimethoprim alone (9). Despite the widespread use of co-trimoxazole for about 35 years, bacterial resistance has not emerged as a major problem (10,11).

The adverse effects of co-trimoxazole are mostly ascribed to the sulfamethoxazole component. However, adverse effects can also be caused by trimethoprim or by the combination. It is important to realize that the culprit agent can only be definitely determined by re-exposure of the single drugs in the individual patient. Thus, in clinical practice it may be impossible to determine the causative compound.

Tetroxoprim is another 2,4-diaminopyrimidine derivative with comparable actions to those of trimethoprim. It is exclusively used in combinations with sulfonamides.

Observational studies

Co-trimoxazole has been recommended for prophylaxis against opportunistic infections in HIV-infected individuals living in sub-Saharan Africa. However, despite this recommendation, there has been limited use of co-trimoxazole in these settings, partly because of concerns about hematological toxicity. The incidence of neutropenia in HIV-infected individuals taking co-trimoxazole prophylaxis has been reported (12). In all, 533 HIV-infected individuals taking co-trimoxazole prophylaxis were followed for 1450 person-years and examined for neutropenia, which was graded as at least one absolute neutrophil count: below $1.5 \times 10^9/l$ (severity grade 1), below $1.0 \times 10^9/l$ (grade 2), below $750 \times 10^6/l$ (grade 3); or below $500 \times 10^6/l$ (grade 4). The probability of being free from neutropenia at 48 months was 0.29 (95% CI = 0.23, 0.34) for grade 1; 0.64 (95% CI = 0.60, 0.71) for grade 2; 0.82 (95% CI = 0.77, 0.86) for grade 3; and 0.96 (95% CI = 0.93, 0.99) for grade 4. There was a significant association between a higher rate of neutropenia and a low baseline CD4 count. However, there was no association between any grade of neutropenia and the overall risk of serious morbidity. These results confirms that mild neutropenia is common in patients taking co-trimoxazole prophylaxis, although there is no evidence of serious clinical consequences.

Adverse drug reactions due to co-trimoxazole occur in 6–8% of patients (13). The most common are mild gastrointestinal distress and cutaneous events; a wide range of hematological abnormalities have also been ascribed to co-trimoxazole.

In a randomized comparison of co-trimoxazole (800/160 mg/day of trimethoprim + sulfamethoxazole 5 days a week) versus norfloxacin (400 mg/day) in preventing spontaneous bacterial peritonitis in 54 patients with cirrhosis and ascites, adverse effects occurred only in patients using co-trimoxazole (14). There were five episodes. One patient had a rash, which disappeared spontaneously, two complained of epigastric pain, and the other two had worsening of renal function unattributable to other causes.

General adverse effects

Severe adverse drug reactions with trimethoprim and co-trimoxazole are rare (15–17). This also applies to children (18). The adverse effects of co-trimoxazole correspond to those expected from a sulfonamide (19). In HIV-infected patients, adverse effects of co-trimoxazole are more frequent and more severe (20–22). Hematological disturbances due to co-trimoxazole include mild anemia, leukopenia, and thrombocytopenia, which may be due to folic acid antagonism. Serious metabolic disturbances that are associated with trimethoprim include hyperkalemia and metabolic acidosis. Trimethoprim can cause hypersensitivity reactions. However, with co-trimoxazole, the sulfonamide is generally believed to be more allergenic (15). Generalized skin reactions predominate. Other effects, such as anaphylactic shock, are extremely rare (23–25). Carcinogenicity due to trimethoprim or co-trimoxazole has not been reported.

Organs and Systems

Respiratory

Cases of allergic pneumonitis and pulmonary infiltrates with eosinophilia have been described with co-trimoxazole (26). Such reactions have not been reported with trimethoprim alone. Pulmonary infiltrates due to co-trimoxazole hypersensitivity in patients with AIDS are particularly worrisome, since they mimic progression of underlying opportunistic pulmonary infections.

- Pneumonitis developed after the administration of co-trimoxazole in a patient with intractable ulcerative colitis complicated by *P. jiroveci* pneumonia. This patient had also previously had sulfasalazine-induced pneumonitis (27).
- A 33-year-old man taking co-trimoxazole developed bilateral pulmonary infiltrates and a fever of 39°C after 2 weeks (28). Co-trimoxazole was withdrawn. The fever resolved 6 days later. A lung biopsy showed non-specific interstitial pneumonia. A lymphocyte stimulation test for co-trimoxazole was positive.

Nervous system

Co-trimoxazole can cause tremor (29), which has been described in patients with AIDS taking co-trimoxazole (30,31).

- A 66-year-old man with pulmonary fibrosis and cor pulmonale was given intravenous co-trimoxazole for 3 weeks followed by oral treatment for an infection with *Nocardia farcinica*. On day 3 of oral treatment, he developed a tremor in both arms and legs, exacerbated by trying to stay calm. The symptoms resolved 3 days after drug withdrawal.
- A girl aged 1.5 years, who was taking long-term azathioprine and methylprednisolone for an immunological lung disease, developed severe pneumonia (32). Azathioprine was withdrawn and ciprofloxacin, aciclovir, fluconazole, and co-trimoxazole (7.5 + 37.5 mg/kg/day) were added. *Pneumocystis jiroveci* was grown from bronchoalveolar lavage fluid. The dose of co-trimoxazole was increased to 20 + 100 mg/kg/day. Five days later she developed involuntary movements consisting of diffuse, continuous, high-frequency, low-amplitude tremors associated with choreic movements. Laboratory investigations were within the reference ranges and electroencephalography showed no signs of epilepsy. A CT scan and lumbar puncture were also normal. On day 10 ciprofloxacin, aciclovir, fluconazole, and methylprednisolone were withdrawn but the symptoms persisted. On day 12, co-trimoxazole was withdrawn and all the abnormal movements disappeared within 3 days.

Aseptic meningitis and meningoencephalitis occur with trimethoprim and co-trimoxazole (33–43). The pathogenetic mechanism is still uncertain.

- A 15-year-old boy developed aseptic meningitis while taking trimethoprim 200 mg in the morning and 100 mg in the evening (44).
- An 18-year-old woman with acute non-lymphoblastic leukemia was admitted for a bone marrow transplant (45). She had attained remission with daunorubicin, thioguanine, and cytarabine. A routine lumbar puncture showed a high leukocyte count and meningitis was suspected. She had been taking co-trimoxazole twice daily on Mondays, Tuesdays, and Wednesdays for the past 3 months and no other medications. Co-trimoxazole was withdrawn and 11 days later lumbar puncture results had returned to normal.
- An 80-year-old woman developed severe headache, nausea, vomiting, and rigors 6 hours after the first dose of co-trimoxazole for dysuria (46). Cerebrospinal fluid examination showed an increased leukocyte count of 717/µl with predominance of polymorphonuclear leukocytes. No infectious cause was identified and autoimmune disease was ruled out. The patient had had a similar episode 6 months earlier after treatment

with co-trimoxazole, and therefore the diagnosis of aseptic meningoencephalitis due to co-trimoxazole was made.

Immune complex deposition, immediate hypersensitivity, direct drug toxicity, and induction of anti-tissue antibodies have all been suggested. Interleukin-6 may be an important mediator of trimethoprim-induced aseptic meningitis in some patients. A case with three episodes after re-exposure to co-trimoxazole has been reported (47). Polymorphonuclear or mononuclear cells may preponderate (48). Interleukin-6 may be an important mediator of trimethoprim-induced aseptic meningitis in some patients (49,50).

Ataxia has been described in two patients with AIDS after intravenous use of co-trimoxazole (51).

Extrapyramidal symptoms developed with co-trimoxazole in a girl with dihydropteridine reductase deficiency and rapidly disappeared after withdrawal. This variant of phenylketonuria should be considered in all infants found to have raised phenylalanine concentrations during the neonatal period (52).

Multifocal myoclonus due to co-trimoxazole is very rare.

- A 63-year-old woman was given trimethoprim 20 mg/kg/day and sulfamethoxazole 100 mg/kg/day for an infection with *Nocardia asteroides* (53). Her fever abated, but 4 days later she began to have progressively worsening involuntary movements involving the head and all four limbs. She had multifocal myoclonus and bilateral asterixis, with no other abnormalities. The co-trimoxazole was withdrawn and the involuntary movements abated markedly on the next day; by the fourth day, they had resolved completely.
- A 46-year-old African–American man with AIDS was admitted on two different occasions within 3 weeks with signs and symptoms of meningitis after using co-trimoxazole (54).

Co-trimoxazole can also exacerbate posthypoxic action myoclonus.

- A 58-year-old man had an exacerbation of posthypoxic myoclonus during high-dose intravenous co-trimoxazole treatment for *Pneumocystis jiroveci* pneumonia (55). Three months before the patient had had a hypoxic insult caused by respiratory arrest due to an anaphylactic reaction to antibiotic therapy. He had developed posthypoxic action myoclonus (Lance–Adams syndrome), which was well controlled by oral piracetam. However, after co-trimoxazole 115 mg/kg/day was started, the myoclonus worsened dramatically, resulting in complete disability. Increased doses of piracetam and valproic acid did not significantly benefit him. Only when the dose of co-trimoxazole was reduced to 38 mg/kg/day on day 12 did the myoclonus cease within 2–3 days.

Sensory systems

Eyes

Co-trimoxazole can cause myopia without accommodative spasm (56).

Optic disc and retinal vasculitis is rarely associated with co-trimoxazole.

- A 53-year-old woman with recurrent painful urticarial skin lesions associated with a vasculitis after using co-trimoxazole started to have visual disturbances in both eyes 2 years after the cutaneous manifestations, recurring at 1-year intervals (57). Her ophthalmological findings were consistent with recurrent vasculitis of the optic nerve and retina. Treatment with high-dose corticosteroids and hydroxychloroquine resulted in the resolution of the cutaneous and ocular manifestations.

Uveitis in combination with meningitis (58) and uveitis in combination with arthritis and Stevens–Johnson syndrome (59) have been reported with trimethoprim. Uveitis with retinal hemorrhages has also been described (60).

Taste

In unmedicated young and elderly volunteers and unmedicated HIV-infected patients, trimethoprim applied to the tongue was primarily described as bitter and medicinal (61).

Psychological, psychiatric

Delirium and psychosis have been rarely reported with co-trimoxazole, but are more likely in elderly people (156).

- An acute psychosis occurred in a 46-year-old woman who had taken oral co-trimoxazole (160 + 800 mg bd) for a urinary tract infection (62). She started to have psychotic symptoms with bizarre behavior 10 days after starting therapy. After withdrawal and antipsychotic drug treatment, her mental state resolved to a stable premorbid level within 36 hours.
- A 46-year-old woman without as premorbid history developed agitation, optic and acoustic hallucinations, delusions of persons and situations, lack of concentration, and incoherent thinking. She was alert and well oriented in time, person, and place, but her mood was labile and inappropriate (63). Her husband said that her mental state had gradually deteriorated over the 2 days before presentation. She had been started to take co-trimoxazole 12 days earlier for a urinary tract infection due to Escherichia coli. Co-trimoxazole was withdrawn and she was given lorazepam 2.5 mg and haloperidol 5 mg. Her mental state improved within 36 hours.
- Delirium occurred after treatment with co-trimoxazole in a patient with AIDS; the episode completely resolved within 72 hours of drug withdrawal (64).

Endocrine

In a child, co-trimoxazole was the probable cause of growth failure (65).

Co-trimoxazole has been suggested to have some antithyroid activity. However, whether this effect is due to trimethoprim alone is still unclear (66,67). Co-trimoxazole 27–31 mg/kg bd orally substantially altered serum total T4 and TSH concentrations and neutrophil counts in dogs within as short a time as a few weeks (68).

Co-trimoxazole 14–16 mg/kg orally every 12 hours for 3 weeks in dogs reduced total and free T4 concentrations and increased the TSH concentration, conditions that would be compatible with hypothyroidism (69). However, hypothyroidism has not been attributed to co-trimoxazole in humans.

Metabolism

Co-trimoxazole can cause reversible hypoglycemia, which may be prolonged, particularly in patients with risk factors for hypoglycemia. Common risk factors include compromised renal function, prolonged fasting, malnutrition, and the use of excessive doses. It has been postulated that the sulfonamide mimics the action of sulfonylureas, stimulating pancreatic islet cells to secrete insulin. In elderly people, co-trimoxazole-induced hypoglycemia can cause altered mental state (70,71).

In 14 cases of refractory hypoglycemia complicated by seizures associated with co-trimoxazole, renal insufficiency was a predisposing risk factor in 13 (72). The mean daily dose of co-trimoxazole was 4.5 double-strength tablets per day (trimethoprim 160 mg + sulfamethoxazole 800 mg). Serum insulin concentrations were increased or inappropriately normal in most of those in whom they were measured, suggesting a sulfonylurea-like effect of co-trimoxazole as the mechanism of hypoglycemia. All required intravenous glucose, and six had protracted hypoglycemia.

Metabolic acidosis has been observed in patients with AIDS after intravenous co-trimoxazole (73). The acidosis developed 3–5 days after the start of treatment and had a favorable course. It is likely that the sulfonamide was responsible, because of renal loss of bicarbonate.

Trimethoprim 15 mg/kg/day increased urinary uric acid excretion and reduced the plasma uric acid concentration in five healthy volunteers from 333 μmol/l (5.6 mg/dl) to 226 μmol/l (3.8 mg/dl) (74). In 90 inpatients with hypouricemia co-trimoxazole was identified as the likely cause in four patients (75). However, since the study was limited to patients with hypouricemia and since exposure rates for co-trimoxazole were not reported for hypouricemic or non-hypouricemic patients, no conclusions about the incidence and the relevance of trimethoprim-associated hypouricemia can be made.

Nutrition

A median increase in serum homocysteine of 50% (range 27–333%) was found in seven healthy male volunteers after a 2-week course of trimethoprim 300 mg bd (76). Concomitantly, serum folate concentrations fell significantly. By day 50, baseline values of homocysteine and folate were regained. Since tetrahydrofolate serves as a methyl group carrier in the remethylation of homocysteine to methionine, the inhibitory effect of trimethoprim on dihydrofolate reductase may be most important, but other mechanisms could not be excluded.

Electrolyte balance

Hyperkalemia is a relatively common complication of trimethoprim therapy and occurs at both high and standard dosages. High-doses of co-trimoxazole can cause hyperkalemia (77) by blocking amiloride-sensitive sodium channels in distal nephrons (156). Trimethoprim, but not sulfamethoxazole, reversibly inhibits the amiloride-sensitive apical sodium channels in the cell membranes of the distal nephron. Inhibition of these sodium channels secondarily blocks sodium/potassium adenosine phosphatase-mediated potassium secretion by hyperpolarizing the luminal membrane in the distal nephron. Elderly people appear to be particularly vulnerable (78–81). However, pre-existing renal impairment and concomitant potassium sparing diuretics may also contribute (78). In almost all cases the electrolyte changes resolve on withdrawal. Trimethoprim reduces renal potassium excretion by competitively inhibiting epithelial sodium channels in the distal nephron like triamterene. Higher dosages and pre-existing renal dysfunction are associated with an increased risk of hyperkalemia, as are probably other disturbances in potassium homeostasis, such as hypoaldosteronism and treatment with drugs that impair renal potassium excretion. Conversely, alkalinization of the urine and induction of high urinary flow rates block the antikaliuretic effect of trimethoprim in the distal nephron (82).

In prospective study in 53 patients there were electrolyte disorders in 9.1% and 22% of patients given co-trimoxazole containing a low dose of trimethoprim (less than 80 mg/day) or a standard dose (80?120 mg/day). There were electrolyte disorders in 86% of patients with renal dysfunction, compared with 18% of those with normal renal function. The dose of trimethoprim and the presence of renal dysfunction increased the incidence of electrolyte disorders, with odds ratios of 2.4 and 80 respectively (83). Co-trimoxazole can cause life-threatening hyperkalemia (84,85), even in therapeutic doses. In one study with standard dosages of co-trimoxazole, up to 62% of patients developed a peak serum potassium concentration of over 5.0 mmol/l and 21% a peak concentration of over 5.5 mmol/l (12,78,85). About 20–53% of patients with AIDS taking high-dosage co-trimoxazole for *P. jiroveci* pneumonia also develop hyperkalemia, and around 10% reach dangerously high concentrations (over 6.0 mmol/l). Withdrawal of trimethoprim is often required. Alkalinization of the urine and the induction of high urinary flow rates with intravenous fluids and loop diuretics block the antikaliuretic effect of trimethoprim on distal nephron cells (86,87).

Life-threatening hyperkalemia secondary to the use of standard doses of co-trimoxazole has been reported in two renal transplant recipients who developed end-stage renal disease secondary to familial Mediterranean fever, who may be at increased risk of hyperkalemia because of concurrent renal insufficiency, concomitant use of ciclosporin, and associated tubulointerstitial disease (88). In one patient, underlying adrenal insufficiency might have contributed to the hyperkalemia.

- Hyperkalemia due to high doses of co-trimoxazole (14 mg/kg/day trimethoprim and 70 mg/kg/day sulfamethoxazole) occurred in an HIV-positive patient with *P. jiroveci* pneumonia (89).
- Life-threatening hyperkalemia secondary to the use of standard oral doses of co-trimoxazole (trimethoprim 320 mg/day and sulfamethoxazole 1600 mg/day) occurred in a 77-year-old man with moderate chronic renal insufficiency from diabetic nephropathy (90). In addition to hyperkalemia, he developed severe metabolic acidosis; both resolved on appropriate medical intervention and withdrawal of co-trimoxazole.
- A 41-year-old black man with AIDS and sickle cell anemia was treated on two separate occasions with co-trimoxazole and prednisone 40 mg/day for *P. jiroveci* pneumonia (91). On both occasions he developed a hyperkalemic metabolic acidosis together with renal tubular acidosis after several days of therapy.

Concurrent acidosis in patients with trimethoprim-induced hyperkalemia is uncommon, which could be explained if the action of trimethoprim, like that of amiloride, is limited to the cortical collecting tubule but does not affect the medullary collecting tubule, which has a large capacity to secrete hydrogen ions and may therefore prevent the development of acidosis. Predisposing factors for the rare adverse effect of renal tubular acidosis in this case may have been aldosterone deficiency or resistance, medullary dysfunction of sickle cell anemia, and renal insufficiency. All these factors could contribute to impaired renal handling of secretion of hydrogen ions (91).

According to studies on animals, tetroxoprim, which is structurally similar to trimethoprim, has stronger antikaliuretic effects than trimethoprim. Tetroxoprim-induced hyperkalemia has not been described yet. However, tetroxoprim is only rarely used and dosages are low (92).

Trimethoprim uncommonly causes hyponatremia.

- A 78-year-old woman developed severe symptomatic hyponatremia after treatment with anhydrous theophylline and 6 months later developed hyponatremia after treatment with co-trimoxazole for a presumptive urinary tract infection (93).

Hematologic

Most hematological adverse effects associated with trimethoprim have been reported with co-trimoxazole. These include macrocytic and megaloblastic anemia, aplastic anemia, neutropenia, hypersegmentation of leukocytes, thrombocytopenia, and pancytopenia (15,79–81,94–98,99,156). The hematological adverse effects of co-trimoxazole have been reviewed (100). Sulfonamides alone have not been associated with folate deficiency, but

in combination with trimethoprim they can deplete folate stores in patients with pre-existing deficiency of folate or vitamin B_{12} (101). Treatment with co-trimoxazole can impair the function of mobilized autologous peripheral blood stem cells (102).

In 2622 children taking co-trimoxazole there were no cases of blood dyscrasias. However, only 17% of the patients received more than two refills of the medicine and the actual dose was not reported. In a prospective study of hematological adverse reactions in children after an oral 10-day course of co-trimoxazole 8 mg/kg/day 34% had neutropenia and 12% had thrombocytopenia.

Mild hemopoietic suppression due to co-trimoxazole in an immunocompromised host is common, even with low-dose regimens. This effect is notable in combination with other marrow-suppressive agents (for example azathioprine, ganciclovir, cyclophosphamide, and allopurinol), malnutrition, or infection (CMV and hepatitis C virus) (103).

Atovaquone + azithromycin and co-trimoxazole have been compared in a randomized, double-blind, placebo-controlled trial for 2 years in 366 HIV-infected children aged 3 months to 19 years (104). Grade 3 and grade 4 hematological adverse events during co-trimoxazole therapy were neutropenia (9.3%), thrombocytopenia (8.8%), and anemia (2.2%).

Pancytopenia in the setting of severe drug hypersensitivity syndrome due to co-trimoxazole has been reported (105).

- A 53-year-old woman with congenital dyskeratosis took six co-trimoxazole double-strength tablets a day (trimethoprim 960 mg/day and sulfamethoxazole 4800 mg/day) for *P. jiroveci*, plus folic acid, and after 9 days developed a fever (39°C) accompanied by a morbilliform rash and painful cervical lymphadenopathy. She had a white blood cell count of 1.7×10^9/l, with 0.5×10^9/l neutrophils; hemoglobin 7.3 g/dl; platelets 17×10^9/l; and hemolysis. A myelogram showed extensive hemophagocytosis. Polyvalent intravenous immunoglobulin (1 g/kg/day) was given for 2 days. She became apyrexial within 48 hours, accompanied by marked clinical improvement, and recovered within 1 month.

Erythrocytes

Megaloblastic anemia and aplastic anemia occur in very few patients (97,106,107). They develop particularly in patients who have pre-existing low folic acid concentrations and who are not taking folic acid supplements (98,107–109). Patients with megaloblastic anemia should not be treated with co-trimoxazole. Other predisposing factors include other drugs with antimetabolite properties or anticonvulsants (110). In patients with pre-existing folic acid deficiency or with HIV infection, the administration of folinic acid has been advised, in order to reduce the hematological adverse effects. However, preliminary data suggest that adjunctive folinic acid may be associated with an increased risk of therapeutic failure and death (111).

Severe hemolytic anemia has been associated with co-trimoxazole in a 10-year-old girl without glucose-6-phosphate dehydrogenase (G6PD) deficiency (112).

Although trimethoprim inhibits dihydrofolate reductase in bacteria, it is estimated that an approximately 50 000 times increased concentration of the drug is required to inhibit the human form of this enzyme. Consequently, trimethoprim does not seem to cause megaloblastic changes when used in the treatment of routine infections, although patients with low folate stores undergoing long-term treatment should be followed up for such an effect (250).

Leukocytes

If leukocytes are routinely monitored during treatment (more often than clinically indicated), mild leukopenia is encountered in 0.4–10% of patients taking either co-trimoxazole (16,106,113–115) or trimethoprim alone (116).

Neutropenia is the most frequent adverse effect of co-trimoxazole in sub-Saharan Africa. In a prospective cohort study the incidence of hematological disorders was estimated during the first 6 months of a zidovudine-containing highly-active antiretroviral therapy regimen in 498 sub-Saharan African adults taking co-trimoxazole (117). There was an unexpectedly high incidence of grade 3–4 neutropenia shortly after the introduction of zidovudine. Almost all of the persistent cases disappeared after co-trimoxazole was withdrawn. This suggests an interaction between these two drugs in sub-Saharan Africans.

In a double-blind, randomized, placebo-controlled trial in children aged 1–14 years with HIV infection treated with either co-trimoxazole (240 or 480 mg suspension) or placebo, mean neutrophil count fell by 0.5×10^9/l from baseline to week 4 in the co-trimoxazole group, and remained at $0.6–1.2 \times 10^9$/l below that in the placebo arm from weeks 12 to 72 (118). In total, 16 children in the co-trimoxazole group and seven in the placebo group had a single neutrophil count less than 0.5×10^9/L after baseline.

There were nine cases of co-trimoxazole-induced agranulocytosis in a prospective cohort study of 91 patients. The mean dose was 1400 mg/day (range 800?2400 mg/day) and the mean duration of treatment was 37 days (range 3?17 days). The median age of the patients was 69 years (range 22?92 years) and the male-female ratio was 2 (119).

In a placebo-controlled trial of co-trimoxazole prophylaxis in Côte d'Ivoire, neutropenia was the most frequent short-term adverse effect (120). In 533 HIV-infected adults receiving co-trimoxazole (sulfamethoxazole 800 mg + trimethoprim 160 mg daily), followed-up for 1450 person-years, the probability of remaining free of neutropenia at 48 months was 0.29 (95% CI = 0.23, 0.34) for grade 1 or worse, 0.64 (95% CI = 0.60, 0.71) for grade 2 or worse, 0.82 (95% CI = 0.77, 0.86) for grade 3 or worse, and 0.96 (95% CI = 0.93, 0.99) for grade 4. The only factor that was significantly associated with a higher rate of all grades of neutropenia was a low baseline CD4 count. There was no association between any grade of neutropenia and the global risk of serious morbidity during the study period.

Platelets

Mild thrombocytopenia is quite common with co-trimoxazole, affecting 0.1% to several% of patients treated (8,9,107,113,115–121). In a retrospective review in Denmark, co-trimoxazole was one of the commonest reported causes of drug-induced thrombocytopenia (11% of reports), with the exception of cytotoxic drugs (122). In a similar review in Sweden from 1985 to 1994 of all drug-induced blood dyscrasias, thrombocytopenia occurred most frequently, the major causes being furosemide (17%) and co-trimoxazole (9%) (123). Older patients were most at risk. However, clinically significant thrombocytopenia is only rarely seen. Although folic acid deficiency may contribute to thrombocytopenia, immunological reactions to either the sulfonamide (94,124) or trimethoprim (125) have been proposed (126,127).

Severe, life-threatening thrombocytopenia associated with co-trimoxazole has been reported (128).

- A 54-year-old white woman took a 10-day course of co-trimoxazole (trimethoprim 160 mg, sulfamethoxazole 800 mg) for chronic sinusitis. One day after finishing the course she developed scattered petechiae on both hands and blood blisters in her mouth. She had a low platelet count of 20×10^9/l. Other laboratory tests were normal, except for a raised blood glucose concentration. She was treated successfully with a transfusion of two units of platelets and oral prednisone. Four days after withdrawal of co-trimoxazole her platelet count increased to 110×10^9/l.

Acute thrombocytopenic purpura has been attributed to co-trimoxazole in a 58-year-old man who had taken it for 2 weeks (129).

- A 40-year-old man developed early onset neutropenia, thrombocytopenia, and a generalized rash after taking co-trimoxazole for 5 days (130). Despite thrombocyte transfusions, whole blood transfusions, red cell concentrates, and filgrastim therapy he patient.

Reports of drug-induced thrombocytopenia have been systematically reviewed (131). Among the 98 different drugs described in 561 articles the following antibiotics were found with level I (definite) evidence: co-trimoxazole, rifampicin, vancomycin, sulfisoxazole, cefalothin, piperacillin, methicillin, novobiocin. Drugs with level II (probable) evidence were oxytetracycline and ampicillin.

In another retrospective analysis of drug-induced thrombocytopenia reported to the Danish Committee on Adverse Drug Reactions, 192 cases caused by the most frequently reported drugs were included and analysed (132). There were pronounced drug-specific differences in the clinical appearance. Early thrombocytopenia was characteristic of cases caused by sulfonamides and co-trimoxazole. These drugs also often caused hemorrhage. Accompanying leukopenia was observed in some cases associated with co-trimoxazole. There were no patient-specific factors responsible for the heterogeneity of the clinical appearance, and factors related to the physician seemed to be of little significance.

Methemoglobinemia

Methemoglobinemia is formed by altered oxidative-reduction balance of hemoglobin, and co-trimoxazole may trigger an increase in methemoglobin by interfering with reducing enzymes. Despite its widespread use in HIV infected patients, methemoglobinemia due to co-trimoxazole has not been reported before in this population.

- A 66 year old African American man with advanced HIV infection developed fever, chills, rhinorrhea, and a cough (133). He had a history of coronary artery disease and pulmonary tuberculosis and had had a cholecystectomy 3 weeks previously. He denied drug allergy and was currently taking aspirin, ibuprofen, glyceryl trinitrate by patch, and an unknown antibiotic. His temperature was 38.3°C and there were chest crackles but no other major findings on physical examination. The CD4 count was 66 x 10⁶/l. Chest radiography showed bilateral scarring and calcification from tuberculosis. He was given ceftizoxime initially, followed by co-trimoxazole. On the next day ceftizoxime was replaced with erythromycin and zidovudine, and lamivudine and saquinavir were added. On the fourth day, he developed tachypnea, tachycardia, and hypotension. Blood gases showed a methemoglobin concentration of 25% and he also had tissue hypoxia and an arterial blood pH of 7.22. Co-trimoxazole and all antiretroviral drugs were withdrawn and exchange transfusion performed. His symptoms improved and the methemoglobin concentration fell to 1% the next day. Co-trimoxazole was replaced with pentamidine for the prevention of *Pneumocystis jiroveci* and all other drugs were restarted. A review of his medical records from another hospital showed a similar episode of methemoglobinemia 7 years before, after treatment with co-trimoxazole.

Gastrointestinal

Nausea and possibly vomiting occur in a few to 20% of adult patients taking normal dosages of co-trimoxazole (16,121). With trimethoprim alone in a dose of up to 400 mg/day, gastrointestinal tolerance was better and skin reactions were less frequent than with the combination (3). A review of the data from five different centers has shown that gastrointestinal complaints are less frequent with trimethoprim than with the combination (9).

Drug-induced esophagitis is rare, accounting for about 1% of all cases of esophagitis. An incidence of 3.9 in 100 000 has been reported. After the first description, there have been more than 250 observations, with more than 50 different drugs. Among those, the principal antibiotics included tetracyclines (doxycycline, metacycline, minocycline, oxytetracycline, and tetracycline), penicillins (amoxicillin, cloxacillin, penicillin V, and pivmecillinam), clindamycin, co-trimoxazole, erythromycin, lincomycin, spiramycin, and tinidazole. Doxycycline alone was involved in one-third of all cases. Risk factors included prolonged esophageal passage, due to motility disorders, stenosis, cardiomegaly, the formulation, supine position

during drug ingestion, and failure to use liquid to wash down the tablet. Direct toxic effects of the drug (pH, accumulation in epithelial cells, non-uniform dispersion) also seem to contribute to the development of drug-induced esophagitis (134).

In 163 women with uncomplicated acute lower urinary tract infections included in a multicenter randomized comparison of cefpodoxime proxetil (100 mg bd) or co-trimoxazole (160 + 800 mg bd) for 3 days both antimicrobials were well tolerated, with the exception of one patient in the co-trimoxazole arm who withdrew because of abdominal pain (135).

Antibiotic-associated colitis is only very rarely seen with co-trimoxazole (136,137).

Quantitative stool cultures in two patients who took co-trimoxazole for 2 years showed a substantially suppressed Gram-negative aerobic flora, while *Enterococcus* species and anaerobes were not affected; yeasts were moderately increased (138).

Liver

The incidence of hepatic toxicity with trimethoprim is 1 in 11 000–45 000 treatments (100). Hepatocellular (40%), mixed (40%), and cholestatic (20%) damage are the most commonly reported pathological subtypes in adults. Intrahepatic cholestasis associated with phospholipidosis has been reported with trimethoprim (139). In one case trimethoprim caused cholestatic hepatitis after re-exposure (140).

The same types of liver disease occur with co-trimoxazole as with sulfonamides alone (141–143). Mild rises in serum transaminases and cholestatic hepatotoxicity are well reported, usually starting after a latent period of several weeks, and associated with a rash. There have been very few case reports of fulminant hepatic failure associated with co-trimoxazole.

- A 32-year-old woman developed a pruritic maculopapular rash and fever (144). She had taken a 12-day course of co-trimoxazole that had finished 5 days before and was taking no other drugs. She had normal hematological indices but a raised alkaline phosphatase and aspartate transaminase. Serological testing for Epstein-Barr virus, hepatitis A, B, and C, cytomegalovirus, echo virus, rubella, and measles showed no evidence of recent infection. Her rash improved but her general condition worsened and steroids were started. An abdominal CT scan showed a large liver with a moderate amount of ascites. The aspartate transaminase rose to 1330 IU/l and the prothrombin time increased. She developed progressive liver failure and died while awaiting liver transplant. At autopsy the liver showed signs of massive hepatic necrosis with no other abnormalities.
- Hepato-renal insufficiency combined with pancytopenia followed the administration of co-trimoxazole for 10 days for suspected pyelonephritis in a 48-year-old man (145). Hemodialysis was temporarily required, and renal and liver function and blood counts returned to normal afterwards.

In a retrospective study of suspected hepatic adverse drug reactions with fatal outcomes received by the Swedish Adverse Drug Reactions Advisory Committee (SADRAC) from 1966 to 2002, six patients (median age 55 years, range 17–87; three men) taking co-trimoxazole died because of liver failure (four hepatocellular, one cholestatic, one mixed); the median duration of treatment was 10 (3–22) days (146).

- A previously healthy 23-year-old man developed acute fulminant liver failure after taking co-trimoxazole for 7 days; after 15 days he received a successful orthotopic liver transplant (147).

In a retrospective study of reports of suspected hepatic adverse drug reactions with fatal outcomes received by the Uppsala Monitoring Centre from 1968 to 2003, 52 patients who had taken co-trimoxazole died because of liver failure (148).

Fulminant liver failure has been associated with co-trimoxazole.

- A 23-year-old previously healthy man presented with myalgia and fever after taking co-trimoxazole for 7 days for otitis externa (149). On the next day, he developed a rash on his neck and chest and jaundice, rapidly followed by fulminant liver failure. His liver enzymes peaked on day 3. Total bilirubin was 176 µmol/l (10.3 mg/dl), direct bilirubin 142 µmol/l (8.3 mg/dl), prothrombin time 61 seconds, and INR 7.5. Screening for paracetamol was negative. Blood cultures and laboratory tests for hepatitis A, B, and C, CMV, HIV, and toxoplasmosis were negative. A liver biopsy showed acute hepatitis with focal hepatocellular necrosis; immunohistochemical studies were negative for HSV, hepatitis B, CMV, and acid-fast bacilli; PAS and Giemsa stains for micro-organisms, and Wharton?Starry and Dieterel stains for spirochetes were negative. On day 8 after admission (15 days after taking co-trimoxazole) he underwent successful liver transplantation.

Pancreas

Acute pancreatitis, with relapse on re-exposure, has been observed in a patient taking co-trimoxazole (150).

- A 53-year-old woman repeatedly developed pancreatitis after co-trimoxazole; a causal relation was confirmed by relapse after rechallenge (151).

Urinary tract

Trimethoprim interferes with the tubular secretion of creatinine, causing an increased serum creatinine concentration and a reduced creatinine clearance, without affecting true glomerular filtration rate (152–155,156). Thus, a small increase in serum creatinine concentration at the beginning of treatment is not necessarily indicative of impaired renal function (157,158). Nevertheless, co-trimoxazole may have a direct nephrotoxic effect, mainly in patients with pre-existing renal impairment taking large dosages (152,153,159). This can be prevented by reducing the dose (153,154). If creatinine clearance is reduced to 15–25 ml/minute, the standard dose of co-trimoxazole should be reduced by half after an initial 3 days of treatment with the usual dose.

Renal tubular acidosis due to antimicrobial drugs has been reviewed (77). There was renal tubular dysfunction associated with hyperkalemia in two elderly patients taking high-dose oral co-trimoxazole; there was hyponatremia in both cases. Type 4 renal tubular acidosis has been reported in patients receiving high-dose intravenous therapy, as well as oral treatment and prophylactic doses of co-trimoxazole. There was heterogeneity in these patients, with both adults and children included. The mechanism by which co-trimoxazole causes renal tubular acidosis is not well understood.

Treatment with co-trimoxazole (or other antibiotics) can increase the risk of the hemolytic-uremic syndrome in children with gastrointestinal infections caused by *Escherichia coli* O157:H7 compared with children with no antibiotic treatment (160).

Extremely rare instances of crystalluria followed by renal insufficiency due to obstruction have been described in patients taking trimethoprim (161).

Skin

The skin reactions observed with co-trimoxazole can be due to trimethoprim or to the sulfonamide. Multiple skin reactions have been described, including a maculopapular rash, urticaria, diffuse erythema, morbilliform lesions, erythema multiforme, purpura, and photosensitivity. Severe reactions, including Stevens?Johnson syndrome and toxic epidermal necrolysis, have been rarely reported. Data collected from five different centers have shown that skin rashes are less frequent with trimethoprim than with co-trimoxazole (9).

In a randomized, double-blind, placebo-controlled comparison of atovaquone + azithromycin and co-trimoxazole for 2 years in 366 HIV-infected children aged 3 months to 19 years, 26% of those who took co-trimoxazole had moderate rashes and two had life-threatening rashes (104).

In a randomized, single-blind study in 59 patients with active ocular toxoplasmosis randomly assigned to pyrimethamine + sulfadiazine or co-trimoxazole, adverse reactions were limited to one patient in each treatment group, in both cases a rash (162).

In 105 patients with established fixed drug eruptions co-trimoxazole was the leading causative agent (64%) and the most common lesions were in the genital mucosa (163).

The skin rashes due to trimethoprim and co-trimoxazole are mostly maculopapular and are related to the duration of treatment (164). They occur with a frequency of 1.3–5.9% (121,165–167)

- Generalized erythematous skin eruptions have now been reported in a 20-year-old Japanese woman and a 70-year-old Japanese man taking co-trimoxazole (168). Patch tests showed that trimethoprim alone was responsible for the erythematous popular-type skin eruption in the young woman. The old man's skin responded to both trimethoprim and sulfamethoxazole.

Two cases of case of fixed drug eruption of the penis have been described in patients taking co-trimoxazole (169,170).

In 81 Nigerian patients with fixed drug eruption there was an association with co-trimoxazole in 23 (171). Co-trimoxazole caused mainly genital and oral lesions, with a predilection for mucocutaneous junctions.

A fixed drug eruption with polysensitivity to piroxicam and co-trimoxazole has been reported in a 38-year-old man (172).

- A bullous necrotizing fixed drug eruption was reported in a 3-month-old child who had received oral co-trimoxazole suspension 3 days before the onset of the lesions (173). The infant was given oral prednisolone 0.5 mg/kg for 1 week along with supportive management. The lesions stopped progressing within 2 days and subsided completely in 10 days.
- A 52-year-old white woman developed a fixed drug eruption with polysensitivity caused by co-trimoxazole and tenoxicam (174).

Urticarial vasculitis is characterized by the association of urticarial papules lasting for more than 24 hours with histological cutaneous vasculitis.

- A 30-year-old woman developed urticarial, purpuric, and necrotic cutaneous lesions of the legs after taking oral co-trimoxazole for 2 weeks (175).
- A 70-year-old Asian man developed toxic epidermal necrolysis within a few hours of the first dose of trimethoprim for a urinary tract infection (176). His skin became sore and itchy and he developed large blisters involving the trunk, buttocks, thighs, and perineum (about 60% of the body surface area).

In a randomized, open trial of long-term intermittent co-trimoxazole on recurrences of toxoplasmic retinochoroiditis, four of 54 patients who took a single tablet of co-trimoxazole (trimethoprim 160 mg, sulfamethoxazole 800 mg) withdrew when they developed mild cutaneous erythema that resolved when drug treatment was stopped (177).

Fixed drug eruptions

Fixed drug eruptions have been analysed in 450 patients (178). The ratio of men to women was 10:11. The mean age of the men was 30 years, and that of the women 31 years. In 13% the fixed drug eruption had occurred for the first time, 2.7% had had more than 40 episodes. There was atopy in 11%, and 23% had a positive family history of drug reactions. Co-trimoxazole was the most common cause of fixed drug eruptions. Other antibiotics included tetracycline, metronidazole, amoxicillin, ampicillin, erythromycin, and clindamycin.

Co-trimoxazole was the offender in 75% of 64 cases of fixed drug eruption. The eruption was mainly located on male genitalia, but unusual findings included familial occurrence, symmetrical and asymmetrical non-pigmented lesions, linear lesions, a solitary plaque on the cheek, and wandering lesions (179).

There may be an association between fixed drug eruption and HLA class I antigens. In 42 of 67 patients with fixed drug eruptions caused by co-trimoxazole there were significantly higher frequencies of the A30 antigen and A30 B13 Cw6 haplotype than in 2378 control subjects (180). HLA-B55 (split of B22) was present exclusively in co-trimoxazole fixed drug eruption, and in a higher frequency than in control subjects.

Topical provocation of fixed drug eruption and positive reactions with co-trimoxazole are extremely rare and have never been seen on unaffected skin.

- In a 42-year-old Caucasian woman with histopathologically confirmed fixed drug eruption induced by co-trimoxazole, positive topical provocation by trimethoprim was obtained on both involved and uninvolved skin (181).

Cross-sensitivity to co-trimoxazole and nimesulide has been reported (182).

- A 10-year-old-boy developed an extensive fixed drug eruption when he took co-trimoxazole (trimethoprim 200 mg, sulfamethoxazole 40 mg bd) 8 weeks after having had a fixed drug eruption due to nimesulide. An oral provocation test with co-trimoxazole 1 month later showed reactivation within 12 hours in the form of severe itching and erythema at the sites of the lesions with one-quarter of the dose.

Erythema multiforme

Erythema multiforme and its related syndromes (Stevens–Johnson syndrome and toxic epidermal necrolysis) are well recognized with co-trimoxazole and can be caused by trimethoprim as a single drug (183). A European review identified sulfonamides (and particularly co-trimoxazole) as a major causative factor in Stevens–Johnson syndrome and toxic epidermal necrolysis, the crude relative risk being far in excess of any other implicated drug (184). In a histopathological study of 111 cases from Germany, co-trimoxazole accounted for 22 cases (20%) (185). There were no histological features that could have discriminated between the different causes of this severe reaction.

- A 34-year-old Asian woman developed toxic epidermal necrolysis associated with co-trimoxazole (160/800 mg bd) (186). She recovered completely without serious sequelae.
- Toxic epidermal necrolysis, associated with co-trimoxazole for sinusitis in a 16-year-old woman, was successfully treated with intravenous immunoglobulins (0.4 g/kg for 5 days) (187).
- Toxic epidermal necrolysis, associated with treatment with co-trimoxazole (480/2400 mg/day) for sepsis in a 66-year-old Caucasian man, was successfully treated

with intravenous immunoglobulin (0.75 g/kg for 5 days) (188).

- An 86-year-old man developed severe and extensive toxic epidermal necrolysis within 24 hours of taking co-trimoxazole for a urinary infection (189). He had had an allergic reaction to co-trimoxazole a few years before. He recovered completely without serious sequelae.
- A 50-year-old man took oral co-trimoxazole (trimethoprim 160 mg, sulfamethoxazole 800 mg) and developed toxic epidermal necrolysis 13 days later (190).

In 20 Indian patients (of whom 70% were women) with Steven–Johnson syndrome and ocular involvement, co-trimoxazole was the commonest identifiable risk factor (191). Conjunctival involvement and its sequelae were the major ocular manifestations.

Susceptibility factors

Skin reactions occur with high frequency during co-trimoxazole therapy in patients with AIDS (21,22). Other symptoms in such patients have been described in association with skin eruptions, including fever, neutropenia, thrombocytopenia, and rises in transaminase activities (21,98). The high dosages used to treat *P. jiroveci* in these patients may partly be responsible for the high frequency of adverse effects, since low-dose co-trimoxazole for the prevention of *P. jiroveci* pneumonia reduced the adverse effects from 28 to 9% (192).

Nails

- Co-trimoxazole was suggested to have caused loss of fingernails and toenails in a 3-year-old boy (193). He was initially treated with gentamicin 2 mg/kg intravenously every 8 hours for *E. coli* urinary tract infection, and co-trimoxazole was given once daily in a prophylactic dose. After 2 weeks his fingernails and toenails began to slough. Co-trimoxazole was withdrawn, and within 2 weeks his nails had returned to normal.

Musculoskeletal

Co-trimoxazole can cause rhabdomyolysis in HIV-positive patients.

- A 39-year-old African–American man developed rhabdomyolysis after taking co-trimoxazole for 14 days (two double-strength tablets qds for presumed Pneumocystis jiroveci pneumonia); 6 months after co-trimoxazole withdrawal he had normal laboratory parameters (194).

Immunologic

Co-trimoxazole can cause a drug hypersensitivity syndrome.

- A 24-year-old woman with severe liver failure developed erythema multiforme and thrombocytopenia after the acute onset of hepatotoxicity and after all medications had been withdrawn. All these features resolved

over weeks, but laboratory abnormalities persisted for up to 8 months (195).

In a prospective observational study in 94 patients with reported sulfonamide allergies, co-trimoxazole allergy was reported by 42 patients, while 42 did not recall the drug to which they were allergic (196). The allergy had caused a rash in 59 patients, anaphylaxis in 13, and Stevens–Johnson syndrome in two. The median time since the last reported allergic reaction to a sulfonamide-containing agent was 20 years. Forty patients had been taking a non-antibiotic sulfonamide for an average of 6.2 years as out-patients; in 24 cases the sulfonamide was furosemide, and 16 reported an allergy to co-trimoxazole. Nine patients had received a non-antibiotic sulfonamide as in-patients (range 2–23 days), most commonly furosemide; there were no adverse events before admission or during the in-patient stay.

- An 82-year-old woman with a history of hypersensitivity reactions to co-trimoxazole resulting in angioedema, developed angioedema and severe dysphagia, shortness of breath, and rash after taking valsartan and hydrochlorothiazide for 4 months (197). Valsartan was identified as the most likely cause of the symptoms and was withdrawn; however, she continued to have weekly episodes of angioedema and when hydrochlorothiazide withdrawn the angioedema disappeared. It recurred more severely after reintroduction of hydrochlorothiazide. Hydrochlorothiazide was permanently withdrawn and the angioedema did not recur.

Cross-reactions between sulfonamides and other drugs with a sulfa group have been discussed controversially. Possible cross-sensitivity between sulfamethoxazole and celecoxib has been described in the context of an anaphylactic reaction.

- A 47-year-old woman developed an anaphylactic reaction, with vomiting, flushing, shivering, hypotension, and finally loss of consciousness, 30 minutes after taking celecoxib 200 mg (198). She had had a previous episode of vomiting and vertigo after taking co-trimoxazole. Scratch and patch tests were negative for all substances, but lymphocyte proliferation tests were clearly positive for sulfamethoxazole and celecoxib.

The authors concluded that the patient had either coincidental sensitization to both drugs or real cross-sensitivity between sulfamethoxazole and celecoxib.

Hypersensitivity syndrome in patients taking co-trimoxazole associated with the reactivation of herpesvirus (HHV)-6 infection has been described.

- A 27-year-old man with a history of bronchial asthma, eosinophilic enteritis, and eosinophilic pneumonia developed a fever, skin eruptions, cervical lymphadenopathy, hepatosplenomegaly, atypical lymphocytosis, and eosinophilia 2 weeks after taking co-trimoxazole (199). After withdrawal of co-trimoxazole and the administration of high-dose steroids, the systemic symptoms gradually resolved. During the disease course, the patient had a transient increase in anti-human herpesvirus (HHV)-6 antibody titers and HHV-6 DNA in the peripheral blood, indicating reactivation of a latent HHV-6 infection.

- HHV-6 encephalitis associated with hypersensitivity syndrome induced by co-trimoxazole has been described in a 72-year-old woman (200).
- A 27-year-old man with a history of bronchial asthma, eosinophilic enteritis, and eosinophilic pneumonia developed a fever, rash, cervical lymphadenopathy, hepatosplenomegaly, atypical lymphocytosis, and eosinophilia 2 weeks after receiving co-trimoxazole (201). After withdrawal of co-trimoxazole and the administration of high-dose steroids the symptoms gradually resolved. There was a transient increase in anti-human herpes virus (HHV-6) antibody titers and HHV-6 DNA in the peripheral blood, suggesting reactivation of a latent HHV-6 infection.

Severe adverse reactions after treatment with co-trimoxazole are common in patients with HIV infection and can manifest as generalized sepsis with hemodynamic compromise, pulmonary infiltrates, and raised creatinine and hepatic transaminases. These reactions usually occur in patients who have previously been exposed to co-trimoxazole. A case of a severe systemic response syndrome in an HIV-positive patient without prior exposure to co-trimoxazole has been reported (202).

- A 33-year-old HIV-positive man returned to hospital with aphthous ulcers, and weight loss. He was not taking any medications. He denied using co-trimoxazole. His CD4 count was 90 x 10^9/l and viral load 354 000 copies/ml. A week earlier a course of co-trimoxazole (one double-strength tablet, 160/800 mg, every other day) had been prescribed. He had taken the first tablet a day before coming to hospital. While in the waiting room he had a fever, chills, and rigors. He had a generalized, erythematous, non-blanching rash and a temperature of 39.3°C. His white blood cell count was 3.5 x 10^9/l with 60% segmented neutrophils, hemoglobin 13.1 g/dl, and platelet count 94 x 10^9/l. Liver and renal function tests, urine examination, and chest radiography were unremarkable. He was given intravenous fluids and ceftriaxone. Co-trimoxazole was withdrawn and he recovered in 5 days.

This case illustrates a probable delayed hypersensitivity reaction to co-trimoxazole in a patient without prior exposure.

Immunologic

Anaphylaxis to trimethoprim is not well recognized.

- Within 5 minutes of ingesting one tablet of trimethoprim for a urinary tract infection, a 79-year-old woman with a history of erythromycin allergy felt sweaty, faint, and nauseated (203). There were no signs of hemodynamic instability and the antibiotic was withdrawn. However, her symptoms persisted and she collapsed with loss of consciousness. Her Glasgow Coma Score was 13 (eyes 3/verbal 4/motor 6), she had unequal reactive pupils and a

rash, and was incontinence of urine and faeces. Her blood pressure was 81/50 mmHg, heart rate 57/minute, and respiratory rate 20/minute. She was pale, cool, and clammy, and mildly hypothermic at 34.2°C. There was generalized limb and facial swelling and an erythematous maculopapular rash over the whole body. She recovered with treatment.

Susceptibility Factors

Genetic

There is conflicting evidence about a possible association between slow acetylator phenotype or genotype and sulfamethoxazole hypersensitivity in HIV-infected patients. Therefore, 40 HIV-infected patients (32 of them with sulfamethoxazole hypersensitivity) and 26 healthy controls were genotyped and phenotyped for acetylator status (204). Furthermore, these data were pooled with the data of 56 patients reported in previously published studies addressing the acetylator status in HIV-infected sulfamethoxazole hypersensitive patients. The frequencies of slow acetylator genotype and phenotype did not differ significantly between HIV-infected patients with and without sulfamethoxazole hypersensitivity and healthy controls. However, since the pooled odds ratio of 2.25 had a wide 95% confidence interval (0.45 to 11), an effect of slow acetylator genotype and phenotype can even now not be definitively ruled out.

Age

The adverse effects of trimethoprim in children have been reviewed (100). In children aged 2–15 years taking long-term prophylaxis with trimethoprim blood dyscrasias occurred at a rate of 0.7 events per 100 years at risk.

Skin reactions occur at a rate of 2 events per 100 years at risk in children aged 2–15 years old (100). They are hypersensitivity reactions and they usually occur immediately after exposure or re-exposure, unlike other adverse reactions, which only occur at high doses or after prolonged therapy. The spectrum of cutaneous reactions includes urticaria, maculopapular rash, fixed drug eruption, erythema multiforme, Stevens–Johnson syndrome, and toxic epidermal necrolysis. Urticaria and maculopapular rash are by far the most common skin reactions. Treatment is withdrawal and resolution is usually complete.

Hepatotoxicity related to co-trimoxazole has been suggested to be a hypersensitivity reaction (100). This is supported by the manifestation of a preceding rash or eosinophilia in children. Cutaneous hypersensitivity reactions, such as toxic epidermal necrolysis, are frequently associated with raised liver enzymes in children. The incidence of skin reactions in children under 2 years is 7.4 events per 100 years at risk. This number falls significantly to 1.4 events per 100 years at risk in children aged 2–15 years, presumably due to prior exposure and the exclusion of those at known risk of allergic reactions. The spectrum of cutaneous reactions to co-trimoxazole includes urticaria, maculopapular rashes, fixed drug eruptions, erythema multiforme, Stevens–Johnson syndrome, and toxic epidermal necrolysis. Urticaria and maculopapular rashes are by far the most common.

Immunosuppression

In immunocompetent patients, adverse reactions to co-trimoxazole are most often related to the skin or gastrointestinal tract and occur in about 8% of cases. In contrast, in patients who are HIV positive there is a much higher incidence of adverse reactions, with frequencies as high as 83% in one study.

- A 37-year-old woman with AIDS developed tremor, acute pancreatitis, and raised serum creatinine concentrations while taking co-trimoxazole (15 mg/kg/day) (205).

Pregnancy

There is an association between co-trimoxazole combinations in early pregnancy and several major malformations, such as neural tube defects and cardiovascular defects (206).

Drug-Drug Interactions

Rosiglitazone

Trimethoprim is a selective inhibitor of CYP2C8 in vitro. Rosiglitazone is predominantly metabolized in the liver by CYP2C8. The effect of trimethoprim on rosiglitazone metabolism in vitro has been determined in pooled liver microsomes and in a randomized crossover study in eight healthy subjects, who took a single dose of rosiglitazone 8 mg before and after taking trimethoprim 200 mg bd for 5 days (207). Trimethoprim significantly inhibited the metabolism of rosiglitazone in vitro and in vivo it increased the AUC of rosiglitazone by 31% and the half-life by 27%. There was also a reduction in the plasma concentration of rosiglitazone metabolites. Trimethoprim should therefore be used with caution in patients with type 2 diabetes who are also taking rosiglitazone.

Sirolimus

In 15 renal transplant recipients a single dose of co-trimoxazole did not affect the pharmacokinetics of sirolimus (208).

The sulfonamide component of co-trimoxazole is generally believed to be more allergenic than trimethoprim. However, trimethoprim alone can cause hypersensitivity reactions more commonly than has previously been thought. Most of these reactions present as generalized skin reactions. The hydroxylamine and other metabolites of sulfamethoxazole can bind covalently to proteins because of their chemical reactivity, resulting in the induction of specific adverse immune responses. Therefore, changes in the activity of detoxification pathways are

associated with a greater risk of allergic reactions to sulfonamides. Allergies to sulfonamides, particularly sulfamethoxazole, are more frequent in patients with AIDS, but the reason for this increased risk is not fully understood. No tools are available to predict which patients have a greater risk for developing allergies to sulfonamides. In a small study in HIV-positive patients with hypersensitivity syndrome reaction, the lymphocyte toxicity assay has a strong potential for use as a diagnostic tool to assess co-trimoxazole hypersensitivity (209). Diagnosis is essential to avoid possible progression to severe reactions and readministration of the offending drug.

Anaphylactic shock is rare, but has been reported with co-trimoxazole (23). However, it is possible that this reaction was due to the sulfonamide compound (24). The case histories of 13 patients (12 women, one man, aged 22–68 years) with anaphylactic reactions to trimethoprim alone that were reported to a national drug safety unit have been analysed (23). Nine were classified as probable anaphylaxis. The casual relation between exposure to trimethoprim and anaphylaxis was classified as definite in three reports, possible in four, and probable in six. In one patient, IgE antibodies against trimethoprim were demonstrated.

- Culture-negative arthritis, bilateral uveitis, mucocutaneous Stevens–Johnson syndrome, and eosinophilia developed in a 31-year-old woman after 3 days of therapy with oral trimethoprim 160 mg bd for a lower urinary tract infection (59). At the start of antibiotic therapy, glucocorticoids and local anesthetics were injected into the lateral aspect of the right knee. Recovery was rapid after trimethoprim was withdrawn. Two months later she developed headache, nausea, malaise, and bilateral uveitis after taking trimethoprim again.

With sera from patients with known hypersensitivity to trimethoprim, IgE-specific recognition of three different but related metabolites has been demonstrated, including the entire molecule itself, the 3,4-dimethoxybenzyl group, and the 2,4-diamino-5-(3′,4′-dimethoxybenzyl) pyrimidine group (210). The incidence of adverse reactions in patients with AIDS is higher than in others (15). Most of these effects are not true allergic reactions, but are related to high doses of co-trimoxazole, and include rashes, nausea, and vomiting. They can be reduced in both frequency and severity by corticosteroids, which are often given in moderate to severe *P. jiroveci* pneumonia. The risk of hypersusceptibility reactions (Stevens–Johnson syndrome, neutropenia, hepatotoxicity, aseptic meningitis, thrombocytopenia) is also higher than in other patients. Since many rashes with co-trimoxazole are not necessarily due to allergic mechanisms, a previous rash should not prevent later re-administration. It may be prudent, however, to use a rapid test dose when co-trimoxazole is the treatment of choice (211).

The incidence of adverse reactions to co-trimoxazole in HIV-infected patients is high. Several reports have shown that an incremental increase in drug dosage may allow a significant proportion of patients to tolerate prophylactic dosages of co-trimoxazole. Eight of 14 selected HIV-infected patients (13 men, 1 woman; patients who experienced severe reactions such as anaphylaxis or Stevens–Johnson syndrome were excluded) were successfully desensitized and after a regimen of gradual incremented exposure over 11 days as an outpatient procedure could continue to take co-trimoxazole (212). *N*-acetylcysteine (3 g of a 20% liquid solution bd) did not prevent hypersensitivity reactions to co-trimoxazole in HIV-infected patients (213). Although cross-reactivity can occur, dapsone can be used for patients with mild hypersensitivity reactions to co-trimoxazole for prophylaxis of *P. jiroveci* pneumonia (214).

A syndrome that resembles bacterial sepsis is well recognized in patients with AIDS (215). This reaction can occur within hours of a large dose of co-trimoxazole, but most often occurs on rechallenge.

- A sepsis-like hypersensitivity reaction occurred in 38-year-old HIV-positive Hispanic man after a 14-day course of co-trimoxazole (216).

The mechanism of this unusual reaction is unclear.

Desensitization

In patients who absolutely require further treatment, successful desensitization can be achieved (217). Two patients with chronic granulomatous disease who had previously been intolerant of co-trimoxazole completed a 5-day desensitization protocol with a good clinical outcome (218).

Desensitization can be efficient in a large proportion of patients (88%) using a 5-day protocol, in which co-trimoxazole is administered orally in a granular formulation in increasing doses, beginning with trimethoprim 0.4 mg and sulfamethoxazole 2 mg and doubling the dose every 12 hours until the therapeutic dose is achieved (219). Another dosage regimen (12 doses of increasing amounts of co-trimoxazole at half-hour intervals) resulted in an overall success rate of 91% at 1 month in 44 patients (220). Such tolerance induction protocols can be adopted, even during pregnancy without risk to the mother or to the fetus (221).

Another uncontrolled trial of a 6-day desensitization procedure in 33 cases has been reported (222). The protocol started with a dose of 0.2 mg rising to 800 mg over 6 days and 32 of the subjects successfully completed the course. In addition, 12 of 14 cases were successfully rechallenged with co-trimoxazole. However, this study lacked a clear description of follow-up or the reasons for the selection of subjects for desensitization or re-challenge, and cannot be used as a basis for recommending this desensitization technique.

Oral desensitization to co-trimoxazole was successfully achieved in patients with AIDS suffering from fever, rash, and wheezing due to the drug (223–226). In a randomized study of desensitization with rechallenge in HIV-positive patients with previous adverse effects of co-trimoxazole 73 patients were given a 14-day course of trimethoprim 200 mg/

day (227). Fourteen had adverse reactions to trimethoprim. The remaining 59 subjects were randomized to a 2-day desensitization technique (34 subjects) or rechallenge (25 subjects). There were seven hypersensitivity reactions in both groups. Clearly there is no advantage of this 2-day desensitization technique over rechallenge with co-trimoxazole in HIV-positive individuals.

In a randomized, double-blind study in HIV patients, gradual introduction of co-trimoxazole was associated with significantly fewer adverse drug reactions compared with standard initiation of therapy (228).

Overall it appears that desensitization to co-trimoxazole is safe in the absence of previous serious adverse events, although it is not yet certain whether desensitization is better than re-challenge or indeed what the ideal desensitization method should be.

Infection risk

There have been rare reports of an association between sulfonamide antibiotics and increased severity of rickettsial infections. Sulfonamides do not increase the pathogenicity of *Ehrlichia* species, but a case of human monocytic ehrlichiosis complicated by ARDS has previously been reported in a patient who had taken oral co-trimoxazole. It has been speculated that oral co-trimoxazole, given for acne, may have contributed to the unusual severity of *Ehrlichia chaffeensis* infection that progressed to respiratory failure in a previously healthy 16-year-old boy (229).

Long-Term Effects

Drug tolerance

Bacteria have developed several mechanisms that make them resistant to trimethoprim. Resistance can occur rapidly, and has been reported in Europe, the USA, Asia, and South America, which may account for the fact that trimethoprim is usually used in combination (1,230,231). However, the clinical significance of resistance to trimethoprim has been debated (1,7,19,232).

From a total of 31 319 *Shigella* strains isolated in Israel between 1990 and 1996, the rates of resistance of *Shigella sonnei*, *Shigella flexneri*, and *Shigella boydii* to co-trimoxazole were 94%, 51%, and 62% respectively; the proportion of strains that exhibited multiple drug resistance was higher for *S. sonnei* than for the other serotypes studied (233).

Among 12 045 isolates of *Streptococcus pneumoniae* collected between 1995 and 1998, resistance to co-trimoxazole increased slightly, from 25% to 29% (234).

Mutagenicity

Genotoxic effects of trimethoprim on cultured human lymphocytes have been described (235). Chromosome studies performed in cultures of peripheral blood lymphocytes did not show significant differences before and after treatment. Cytogenetic studies on bone marrow cells from

12 patients with urinary tract infections treated with co-trimoxazole did not show structural chromosomal aberrations; however, there was an increased number of micronuclei in these patients compared with controls (236).

Second-Generation Effects

Fertility

Whether reported cases of impaired male fertility were due to co-trimoxazole is not clear, since most of the men were being treated for underlying urogenital infections, which may have contributed to reduced fertility (237,238).

Teratogenicity

An early review of newer case reports and placebo-controlled trials involving several hundred patients did not show an increase in fetal abnormalities (239). However, the relative risks of cardiovascular defects and oral clefts in infants whose mothers were exposed to dihydrofolate reductase inhibitors, such as trimethoprim, during the second or third month after the last menstrual period, compared with infants whose mothers had no such exposure, are 3.4 (95% CI = 1.8, 6.4) and 2.6 (1.1, 6.1) respectively (240). Multivitamin supplements containing folic acid reduced the adverse effects of dihydrofolate reductase inhibitors. There have been two reports of severe spinal malformations in the fetuses of HIV-positive women treated with combination antiretroviral therapy and co-trimoxazole (241).

Susceptibility Factors

Renal disease

In renal insufficiency, co-trimoxazole should be used with caution, particularly when there is hyperkalemia (66,115).

The use of co-trimoxazole in HIV-positive patients has been associated with a high rate of hypersensitivity reactions (40–80%), attributed to the bioactivation of the sulfonamide component, sulfamethoxazole, to its toxic hydroxylamine and nitroso metabolites. In a study of HIV-positive patients with (n = 56) and without (n = 89) hypersensitivity to co-trimoxazole, functionally significant polymorphisms in the genes coding for enzymes involved in co-trimoxazole metabolism were unlikely to have been major predisposing factors in determining individual susceptibility to co-trimoxazole hypersensitivity (242).

In a randomized, double-blind study in 372 HIV-positive patients with CD4+ counts below 250 $\times$ 10^6/l, gradual initiation of co-trimoxazole treatment was associated with significantly fewer adverse drug reactions compared with standard initiation (243).

The relation between the onset of adverse reactions to co-trimoxazole in HIV-infected patients and the subsequent development of toxoplasmosis, other AIDS-

defining events, and survival has been studied in 592 French patients (244). A low CD4 cell count when co-trimoxazole was introduced was the only factor associated with the onset of adverse reactions. The occurrence of toxoplasmosis and first AIDS-defining events was significantly and independently linked to a low CD4 cell count when co-trimoxazole was introduced and to previous co-trimoxazole withdrawal for adverse events, but not to previous co-trimoxazole withdrawal for reasons other than adverse events, compared with patients who did not stop taking co-trimoxazole. The survival rate was significantly shorter among patients who stopped taking co-trimoxazole for adverse events and for other reasons, compared with patients who continued to take it.

Other features of the patient

Co-trimoxazole should not be given to patients with malnutrition, pregnancy, severe liver damage, megaloblastic anemia, agranulocytosis, or bone marrow failure (15,17,109) fore prbroind page.

Drug–Drug Interactions

Amiloride

Trimethoprim (SED-14, 675) is structurally similar to amiloride and can cause severe hyperkalemia if co-prescribed with potassium-sparing diuretics (266). This is a particularly important interaction in patients with AIDS.

Anticonvulsants

Most interactions of trimethoprim and co-trimoxazole with other drugs are due to folic acid antagonism. This may be more pronounced with co-trimoxazole than with either drug alone. Such interactions have previously been suspected with anticonvulsants, such as barbiturates, phenytoin, and primidone, which themselves produce folic acid deficiency and megaloblastic anemia (88). In order to circumvent the risk of folate deficiency, folic acid or folinic acid can be given. There is some concern that folate replacement may antagonize the desired antimicrobial effect, particularly in some protozoal parasites, but this concern has been debated (89).

Azathioprine

In renal transplant patients, combining azathioprine with co-trimoxazole was followed by neutropenia and thrombocytopenia (115).

Coumarin anticoagulants

Co-trimoxazole can significantly augment the hypoprothrombinemic effect of warfarin (245,246), acenocoumarol (247,248), and phenprocoumon (248).

Cytotoxic drugs

Most interactions of trimethoprim and co-trimoxazole with other drugs are due to folic acid antagonism. This may be more pronounced with co-trimoxazole than with either drug alone. Such interactions have been suspected with cytotoxic drugs (249,250). In order to circumvent the risk of folate deficiency, folic acid or folinic acid can be given. There is some concern that folate replacement may antagonize the desired antimicrobial effect, particularly in some protozoal parasites, but this concern has been debated (111).

In children with acute lymphoblastic leukemia co-trimoxazole inhibited 6-mercaptopurine metabolism (248).

In another study, co-trimoxazole produced a 66% increase in systemic exposure to methotrexate (249).

Prolonged treatment with co-trimoxazole can reduce cyclophosphamide requirements in patients with Wegener's granulomatosis (251).

Ganciclovir

The interaction kinetics and safety profile of oral ganciclovir when co-administered with trimethoprim have been investigated in HIV- and CMV-seropositive patients (267). Trimethoprim significantly reduced the renal clearance of ganciclovir and prolonged its half-life, although the changes are unlikely to be clinically important. Ganciclovir did not alter trimethoprim pharmacokinetics, with the exception of a 13% increase in C_{min}. Ganciclovir was well tolerated when given alone or in this combination.

Glucocorticoids

Prolonged treatment with co-trimoxazole can reduce glucocorticoid requirements in patients with Wegener's granulomatosis (251).

HIV protease inhibitors

In an animal model, indinavir nephrotoxicity was potentiated by co-trimoxazole, but nelfinavir alone or in combination with co-trimoxazole was not nephrotoxic (252).

Immunosuppressants

Most interactions of trimethoprim and co-trimoxazole with other drugs are due to folic acid antagonism. This may be more pronounced with co-trimoxazole than with either drug alone. Such interactions have previously been suspected with immunosuppressant drugs (248,249). In order to circumvent the risk of folate deficiency, folic acid or folinic acid can be given. There is some concern that folate replacement may antagonize the desired antimicrobial effect, particularly in some protozoal parasites, but this concern has been debated (111).

Pronounced nephrotoxicity resulted from the interaction of co-trimoxazole with ciclosporin in patients with a renal transplant (253,254).

Lithium

Trimethoprim has the same effect on the kidney as amiloride, whose combined use with lithium can cause a raised serum lithium concentration.

- The addition of trimethoprim caused severe lithium toxicity in a 40-year-old woman with a schizoaffective disorder; following rehydration, she made a good recovery (255).
- A 40-year-old woman developed nausea, malaise, impaired concentration, trembling, unsteadiness, diarrhea, and muscle spasm in association with a serum lithium concentration of 2.1 mmol/l while taking trimethoprim 300 mg/day (268).
- A 42-year-old woman developed symptoms of lithium toxicity and a raised serum concentration (2.1 mmol/l) while taking trimethoprim (268).

This interaction may be due to an amiloride-like diuretic effect of trimethoprim, causing lithium retention.

Methotrexate

Methotrexate, a folic acid antagonist, is used in the treatment of several disorders. Its major action is inhibition of dihydrofolate reductase, a critical enzyme in intracellular folate metabolism. Co-trimoxazole competes with methotrexate in inhibiting dihydrofolate reductase and further impairs DNA synthesis.

- A fatal case of toxic epidermal necrolysis that involved 90% of the total body surface has been described in a 15-year-old boy with T cell acute lymphoblastic leukemia treated concomitantly with co-trimoxazole and methotrexate (256).

The authors suggested that methotrexate toxicity was precipitated by co-trimoxazole.

- Fatal bone marrow suppression has been reported in an 82-year-old woman who took methotrexate 7.5 mg/week for one year for rheumatoid arthritis without hematological problems. She was given trimethoprim 100 mg/day at first and later 200 mg/day. One week later, she developed severe pancytopenia. The bone marrow failed to recover despite treatment with folinic acid and G-CSF, and she died of bronchopneumonia.

In a literature review, the authors found two other cases of bone marrow suppression after treatment with methotrexate and trimethoprim with full recovery of both. This interaction is also listed in the British National Formulary 1997 (257,258).

Drugs that inhibit folate metabolism increase the likelihood of serious adverse reactions to methotrexate, particularly hematological toxicity. The additional risk of myelosuppression and subsequent severe pancytopenia has been particularly exemplified by the combination of methotrexate and co-trimoxazole (trimethoprim plus sulfamethoxazole) (269). This should also be taken into account in patients taking trimethoprim alone (SEDA-22, 418).

Oral contraceptives

Some antibiotics interfere with the effects of oral contraceptives, putatively by inhibiting bacterial deconjugation in the bowel and thus reducing their reabsorption after biliary excretion. However, in women taking oral contraceptives, short courses of co-trimoxazole are unlikely to cause any adverse effects on contraceptive control (259).

Phenytoin

Massive hepatic necrosis after exposure to phenytoin and co-trimoxazole is rare.

- Acute liver failure has been reported in a 60-year-old woman after concomitant ingestion of phenytoin and co-trimoxazole over 9 days (260). Autopsy showed acute fulminant hepatic failure.

Drug interactions can potentiate the hepatotoxicity of single agents; withdrawal of co-trimoxazole may be needed in the presence of early liver injury.

Procainamide

Trimethoprim inhibits the active tubular secretion of procainamide (261).

The renal clearance of procainamide is inhibited by trimethoprim (270).

Rifamycins

In 10 HIV-positive patients who had been taking one double-strength tablet of co-trimoxazole daily for more than 1 month, the concentrations of trimethoprim and sulfamethoxazole in serum were significantly reduced after the administration of rifampicin (262).

In patients taking co-trimoxazole, rifampicin in a standard oral antituberculosis regimen (600 mg/day plus isoniazid 300 mg/day and pyrazinamide 30 mg/kg for more than 12 days) reduced the steady-state AUC of trimethoprim and sulfamethoxazole by 47 and 23% respectively (271). The same authors had previously reported reduced efficacy of co-trimoxazole in the prevention of toxoplasmosis in HIV-infected subjects (272). The reduction in prophylactic efficiency was more pronounced in subjects who took a single double-strength tablet.

There have been no reports on the risk of *Pneumocystis jiroveci* pneumonia in patients taking rifampicin plus co-trimoxazole prophylaxis. Until such time as more data are available, it is prudent to use double-strength co-trimoxazole tablets twice daily for prophylaxis of toxoplasmosis and *P. jiroveci* pneumonia in patients taking concomitant rifampicin.

The interaction of rifampicin with trimethoprim has been reviewed (273). Rifampicin has a small effect on the clearance of trimethoprim but it is not clinically significant (274,275).

Trimeprazine

The antihistaminic phenothiazine trimeprazine has significant antibacterial activity in vitro and in vivo, and a combination with trimethoprim is highly synergistic, as shown in vivo in Swiss white mice using *Salmonella typhimurium* as the challenge bacterium (263).

Zidovudine

In one pharmacokinetic study in eight HIV-infected subjects, the renal clearance of zidovudine was significantly reduced by trimethoprim (264). The authors concluded that zidovudine dosages may need to be reduced if trimethoprim is given to patients with impairment of liver function or glucuronidation. Zidovudine, on the other hand, did not alter the pharmacokinetics of trimethoprim.

Interference with Diagnostic Tests

Serum folate concentration

Serum folate concentrations should not be measured by the radioisotope method in patients taking trimethoprim (265).

References

1. Zinner SH, Mayer KH. Basic principles in the diagnosis and management of infectious diseases: sulfonamides and trimethoprim. In: Mandell GL, Douglas RG, Bennett JE, editors. Principles and Practice of Infectious Diseases. 4th ed. Edinburgh: Churchill Livingstone, 1996:354.
2. Finland M. Editorial: Combinations of antimicrobial drugs: trimethoprim–sulfamethoxazole. N Engl J Med 1974;291(12):624–7.
3. Garg SK, Ghosh SS, Mathur VS. Comparative pharmacokinetic study of four different sulfonamides in combination with trimethoprim in human volunteers. Int J Clin Pharmacol Ther Toxicol 1986;24(1):23–5.
4. Bushby SR, Hitchings GH. Trimethoprim, a sulphonamide potentiator. Br J Pharmacol Chemother 1968;33(1):72–90.
5. Richards RM, Taylor RB, Zhu ZY. Mechanism for synergism between sulphonamides and trimethoprim clarified. J Pharm Pharmacol 1996;48(9):981–4.
6. Gruneberg RN. The microbiological rationale for the combination of sulphonamides with trimethoprim. J Antimicrob Chemother 1979;5(B):27–36.
7. Brumfitt W, Hamilton-Miller JM. Co-trimoxazole or trimethoprim alone? A viewpoint on their relative place in therapy. Drugs 1982;24(6):453–8.
8. Brumfitt W, Hamilton-Miller JM. Combinations of sulphonamides with diaminopyrimidines: how, when and why? J Chemother 1995;7(2):136–9.
9. Brogden RN, Carmine AA, Heel RC, Speight TM, Avery GS. Trimethoprim: a review of its antibacterial activity, pharmacokinetics and therapeutic use in urinary tract infections. Drugs 1982;23(6):405–30.
10. Dornbusch K, Toivanen P. Effect of trimethoprim or trimethoprim/sulphamethoxazole usage on the emergence of

trimethoprim resistance in urinary tract pathogens. Scand J Infect Dis 1981;13(3):203–10.

11. Wust J, Kayser FH. Die Empfindlichkeit von Bakterieu gegen chemo therapeutika (zürich, 1993). [Sensitivity of bacteria to chemotherapeutic agents (Zurich, 1993).] Schweiz Rundsch Med Prax 1995;84(4):98–105.

12. Toure S, Gabillard D, Inwoley A, Seyler C, Gourvellec G, Anglaret X. Incidence of neutropenia in HIV-infected African adults receiving co-trimoxazole prophylaxis: a 6-year cohort study in Abidjan, Cte d'Ivoire. Trans R Soc Med Hyg 2006;100:785–90.

13. Kocak Z, Hatipoglu CA, Ertem G, Kinikli S, Tufan A, Irmak H, Demiroz AP. Trimethoprim-sulfamethoxazole induced rash and fatal hematologic disorders. J Infect 2006;52(2):e49-52.

14. Alvarez RF, Mattos AA, Correa EB, Cotrim HP, Nascimento TV. Trimethoprim-sulfamethoxazole versus norfloxacin in the prophylaxis of spontaneous bacterial peritonitis in cirrhosis. Arq Gastroenterol 2005;42(4):256–62.

15. Cribb AE, Lee BL, Trepanier LA, Spielberg SP. Adverse reactions to sulphonamide and sulphonamide–trimethoprim antimicrobials: clinical syndromes and pathogenesis. Adverse Drug React Toxicol Rev 1996;15(1):9–50.

16. Jick H. Adverse reactions to trimethoprim–sulfamethoxazole in hospitalized patients. Rev Infect Dis 1982;4(2):426–8.

17. Lawson DH, Jick H. Adverse reactions to co-trimoxazole in hospitalized medical patients. Am J Med Sci 1978;275(1):53–7.

18. Gutman LT. The use of trimethoprim–sulfamethoxazole in children: a review of adverse reactions and indications. Pediatr Infect Dis 1984;3(4):349–57.

19. Martin AJ, Lacey RW. A blind comparison of the efficacy and incidence of unwanted effects of trimethoprim and co-trimoxazole in the treatment of acute infection of the urinary tract in general practice. Br J Clin Pract 1983;37(3):105–11.

20. Coopman SA, Johnson RA, Platt R, Stern RS. Cutaneous disease and drug reactions in HIV infection. N Engl J Med 1993;328(23):1670–4.

21. Gordin FM, Simon GL, Wofsy CB, Mills J. Adverse reactions to trimethoprim–sulfamethoxazole in patients with the acquired immunodeficiency syndrome. Ann Intern Med 1984;100(4):495–9.

22. Mitsuyasu R, Groopman J, Volberding P. Cutaneous reaction to trimethoprim–sulfamethoxazole in patients with AIDS and Kaposi's sarcoma. N Engl J Med 1983;308(25):1535–6.

23. Bijl AM, Van der Klauw MM, Van Vliet AC, Stricker BH. Anaphylactic reactions associated with trimethoprim. Clin Exp Allergy 1998;28(4):510–2.

24. Johnson MP, Goodwin SD, Shands JW Jr. Trimethoprim–sulfamethoxazole anaphylactoid reactions in patients with AIDS: case reports and literature review. Pharmacotherapy 1990;10(6):413–6.

25. Cabanas R, Caballero MT, Vega A, Martin-Esteban M, Pascual C. Anaphylaxis to trimethoprim. J Allergy Clin Immunol 1996;97(1 Pt 1):137–8.

26. Silvestri RC, Jensen WA, Zibrak JD, Alexander RC, Rose RM. Pulmonary infiltrates and hypoxemia in patients with the acquired immunodeficiency syndrome re-exposed to trimethoprim–sulfamethoxazole. Am Rev Respir Dis 1987;136(4):1003–4.

27. Oshitani N, Matsumoto T, Moriyama Y, Kudoh S, Hirata K, Kuroki T. Drug-induced pneumonitis caused by sulfamethoxazole, trimethoprim during treatment of *Pneumocystis carinii* pneumonia in a patient with refractory ulcerative colitis. J Gastroenterol 1998;33(4):578–81.

28. Hashizume T, Numata H, Matsushita K. [Drug-induced pneumonitis caused by sulfamethoxazole–trimethoprim.]Nihon Kokyuki gakkai Zasshi 2001;39(9):664–7.

29. de Arce Borda AM, Goenaga Sanchez MA. Tremblor producids pox trimetoprim–sulfamethoxazole. [Tremor produced by trimethoprim–sulfamethoxazole.] Neurologia 2000;15(6):264–5.

30. Borucki MJ, Matzke DS, Pollard RB. Tremor induced by trimethoprim-sulfamethoxazole in patients with the acquired immunodeficiency syndrome (AIDS). Ann Intern Med 1988;109(1):77–8.

31. Slavik RS, Rybak MJ, Lerner SA. Trimethoprim/sulfamethoxazole-induced tremor in a patient with AIDS. Ann Pharmacother 1998;32(2):189–92.

32. Bua J, Marchetti F, Barbi E, Sarti A, Ventura A. Tremors and chorea induced by trimethoprim-sulfamethoxazole in a child with Pneumocystis pneumonia. Pediatr Infect Dis J 2005;24(4):934–5.

33. Whalström B, Nyström-Rosander C, Âberg H, Friman G. Upprepad meningit och perimyokardit efter intag ar trimetoprim. [Recurrent meningitis and perimyocarditis after trimethoprim.] Lakartidningen 1982;79(51):4854–5.

34. Haas EJ. Trimethoprim–sulfamethoxazole: another cause of recurrent meningitis. JAMA 1984;252(3):346.

35. Kremer I, Ritz R, Brunner F. Aseptic meningitis as an adverse effect of co-trimoxazole. N Engl J Med 1983;308(24):1481.

36. Joffe AM, Farley JD, Linden D, Goldsand G. Trimethoprim–sulfamethoxazole-associated aseptic meningitis: case reports and review of the literature. Am J Med 1989;87(3):332–8.

37. Carlson J, Wiholm BE. Trimethoprim associated aseptic meningitis. Scand J Infect Dis 1987;19(6):687–91.

38. Derbes SJ. Trimethoprim-induced aseptic meningitis. JAMA 1984;252(20):2865–6.

39. Hedlund J, Aurelius E, Andersson J. Recurrent encephalitis due to trimethoprim intake. Scand J Infect Dis 1990;22(1):109–12.

40. Capra C, Monza GM, Meazza G, Ramella G. Trimethoprim–sulfamethoxazole-induced aseptic meningitis: case report and literature review. Intensive Care Med 2000;26(2):212–4.

41. Meng MV, St Lezin M. Trimethoprim–sulfamethoxazole induced recurrent aseptic meningitis. J Urol 2000;164(5):1664–5.

42. Antonen J, Hulkkonen J, Pasternack A, Hurme M. Interleukin 6 may be an important mediator of trimethoprim-induced systemic adverse reaction resembling aseptic meningitis. Arch Intern Med 2000;160(13):2066–7.

43. Andrade A, Hilmas E, Walter C. A rare occurrence of trimethoprim/sulfamethoxazole (TMP/SMX)-induced aseptic meningitis in an older woman. J Am Geriatr Soc 2000;48(11):1537–8.

44. Redman RC 4th, Miller JB, Hood M, DeMaio J. Trimethoprim-induced aseptic meningitis in an adolescent male. Pediatrics 2002;110(2 Pt 1):e26.

45. Therrien R. Possible trimethoprim/sulfamethoxazole induced aseptic meningitis. Ann Pharmacother 2004;38:1863–7.

46. Sharan L. Trimethoprim-sulfamethoxazole-induced aseptic meningoencephalitis. Infect Med 2003;20:19–20.

47. Auxier GG. Aseptic meningitis associated with administration of trimethoprim and sulfamethoxazole. Am J Dis Child 1990;144(2):144–5.

48. Biosca M, de la Figuera M, Garcia-Bragado F, Sampol G. Aseptic meningitis due to trimethoprim–sulfamethoxazole. J Neurol Neurosurg Psychiatry 1986;49(3):332–3.

49. Muller MP, Richardson DC, Walmsley SL. Trimethoprim-sulfamethoxazole induced aseptic meningitis in a renal transplant patient. Clin Nephrol 2001;55(1):80–4.

50. Antonen JA, Saha HH, Hurme M, Pasternack AI. IL-6 may be the key mediator in trimethoprim-induced systemic adverse reaction and aseptic meningitis: a reply to Muller et al. Clin Nephrol 2001;55(6):489–90.

51. Liu LX, Seward SJ, Crumpacker CS. Intravenous trimethoprim–sulfamethoxazole and ataxia. Ann Intern Med 1986;104(3):448.

52. Woody RC, Brewster MA. Adverse effects of trimethoprim–sulfamethoxazole in a child with dihydropteridine reductase deficiency. Dev Med Child Neurol 1990;32(7):639–42.

53. Dib EG, Bernstein S, Benesch C. Multifocal myoclonus induced by trimethoprim-sulfamethoxazole therapy in a patient with Nocardia infection. N Engl J Med 2004;350(1):88–9.

54. Wambulwa C, Bwayo S, Laiyemo AO, Lombardo F. Trimethoprim-sulfamethoxazole-induced aseptic meningitis. J Natl Med Assoc 2005;97(12):1725–8.

55. Jundt F, Lempert T, Dorken B, Pezzutto A. Trimethoprim-sulfamethoxazole exacerbates posthypoxic action myoclonus in a patient with suspicion of Pneumocystis jiroveci infection. Infection 2004;32(3):176–8.

56. Anonymous. Drug-induced myopia. Prescrire Int 2003;12:22–3.

57. Batioglu F, Taner P, Aydintug OT, Heper AO, Ozmert E. Recurrent optic disc and retinal vasculitis in a patient with drug-induced urticarial vasculitis. Cutan Ocul Toxicol 2006;25(4):281–5.

58. Gilroy N, Gottlieb T, Spring P, Peiris O. Trimethoprim-induced aseptic meningitis and uveitis. Lancet 1997;350(9071):112.

59. Arola O, Peltonen R, Rossi T. Arthritis, uveitis, and Stevens–Johnson syndrome induced by trimethoprim. Lancet 1998;351(9109):1102.

60. Kristinsson JK, Hannesson OB, Sveinsson O, Thorleifsson H. Bilateral anterior uveitis and retinal haemorrhages after administration of trimethoprim. Acta Ophthalmol Scand 1997;75(3):314–5.

61. Schiffman SS, Zervakis J, Westall HL, Graham BG, Metz A, Bennett JL, Heald AE. Effect of antimicrobial and anti-inflammatory medications on the sense of taste. Physiol Behav 2000;69(4–5):413–24.

62. Weis S, Karagulle D, Kornhuber J, Bayerlein K. Cotrimoxazole-induced psychosis: a case report and review of literature. Pharmacopsychiatry 2006;39(6):236–7.

63. Weis S, Karagulle D, Kornhuber J, Bayerlein K. Cotrimoxazole-psychosis: a case report and review of literature. Pharmacopsychiatry 2006;39:236–8.

64. Salkind AR. Acute delirium induced by intravenous trimethoprim–sulfamethoxazole therapy in a patient with the acquired immunodeficiency syndrome. Hum Exp Toxicol 2000;19(2):149–51.

65. Murphy JL, Griswold WR, Reznik VM, Mendoza SA. Trimethoprim/sulfamethoxazole-induced renal tubular acidosis. Child Nephrol Urol 1990;10(1):49–50.

66. Cohen HN, Pearson DW, Thomson JA, Ratcliffe WA, Beastall GH. Trimethoprim and thyroid function. Lancet 1981;1(8221):676–7.

67. Smellie JM, Bantock HM, Thompson BD. Co-trimoxazole and the thyroid. Lancet 1982;2(8289):96.

68. Williamson NL, Frank LA, Hnilica KA. Effects of short-term trimethoprim–sulfamethoxazole administration on thyroid function in dogs. J Am Vet Med Assoc 2002;221(6):802–6.

69. Frank LA, Hnilica KA, May ER, Sargent SJ, Davis JA. Effects of sulfamethoxazole-trimethoprim on thyroid function in dogs. Am J Vet Res 2005;66(2):256–9.

70. Fox GN. Trimethoprim–sulfamethoxazole-induced hypoglycemia. J Am Board Fam Pract 2000;13(5):386.

71. Mathews WA, Manint JE, Kleiss J. Trimethoprim–sulfamethoxazole-induced hypoglycemia as a cause of altered mental status in an elderly patient. J Am Board Fam Pract 2000;13(3):211–2.

72. Strevel EL, Kuper A, Gold WL. Severe and protracted hypoglycaemia associated with co-trimoxazole use. Lancet Infect Dis 2006;6(3):178–82.

73. Porras MC, Lecumberri JN, Castrillon JL. Trimethoprim–sulfamethoxazole and metabolic acidosis in HIV-infected patients. Ann Pharmacother 1998;32(2):185–9.

74. Don BR. The effect of trimethoprim on potassium and uric acid metabolism in normal human subjects. Clin Nephrol 2001;55:45–52.

75. Bairaktari ET, Kakafika AI, Pritsivelis N, Hatzidimou KG, Tsianos EV, Seferiadis KI, Elisaf MS. Hypouricemia in individuals admitted to an inpatient hospital-based facility. Am J Kidney Dis 2003;41:1225–32.

76. Smulders YM, de Man AM, Stehouwer CD, Slaats EH. Trimethoprim and fasting plasma homocysteine. Lancet 1998;352(9143):1827–8.

77. Hemstreet BA. Antimicrobial-associated renal tubular acidosis. Ann Pharmacother 2004;38(6):1031–8.

78. Bugge JF. Severe hyperkalaemia induced by trimethoprim in combination with an angiotensin-converting enzyme inhibitor in a patient with transplanted lungs. J Intern Med 1996;240(4):249–51.

79. Velazquez H, Perazella MA, Wright FS, Ellison DH. Renal mechanism of trimethoprim-induced hyperkalemia. Ann Intern Med 1993;119(4):296–301.

80. Reiser IW, Chou SY, Brown MI, Porush JG. Reversal of trimethoprim-induced antikaliuresis. Kidney Int 1996;50(6):2063–9.

81. Eiam-Ong S, Kurtzman NA, Sabatini S. Studies on the mechanism of trimethoprim-induced hyperkalemia. Kidney Int 1996;49(5):1372–8.

82. Schreiber M, Schlanger LE, Chen CB, Lessan-Pezeshki M, Halperin ML, Patnaik A, Ling BN, Kleyman TR. Antikaliuretic action of trimethoprim is minimized by raising urine pH. Kidney Int 1996;49(1):82–7.

83. Mori H, Kuroda Y, Imamura S, Toyoda A, Yoshida I, Kawakami M, Tabei K. Hyponatremia and/or hyperkalemia in patients treated with the standard dose of trimethoprim-sulfamethoxazole. Intern Med 2003;42:665–9.

84. Greenberg S, Reiser IW, Chou SY, Porush JG. Trimethoprim–sulfamethoxazole induces reversible hyperkalemia. Ann Intern Med 1993;119(4):291–5.

85. Alappan R, Perazella MA, Buller GK. Hyperkalemia in hospitalized patients treated with trimethoprim–sulfamethoxazole. Ann Intern Med 1996;124(3):316–20.

86. Perazella MA. Trimethoprim-induced hyperkalaemia: clinical data, mechanism, prevention and management. Drug Saf 2000;22(3):227–36.

87. Gabriels G, Stockem E, Greven J. Potassium-sparing renal effects of trimethoprim and structural analogues. Nephron 2000;86(1):70–8.

88. Koc M, Bihorac A, Ozener CI, Kantarci G, Akoglu E. Severe hyperkalemia in two renal transplant recipients treated with standard dose of trimethoprim–sulfamethoxazole. Am J Kidney Dis 2000;36(3):E18.

89. Brazille P, Benveniste O, Herson S, Cherin P. Une cause meconnue d'hyperkaliemie: le trimethoprime-sulfamethoxazole. [A cause of unexplained hyperkalemia: trimethoprim–sulfamethoxazole.] Rev Med Interne 2001;22(1):82–3.

90. Margassery S, Bastani B. Life threatening hyperkalemia and acidosis secondary to trimethoprim–sulfamethoxazole treatment. J Nephrol 2001;14(5):410–4.

91. Sheehan MT, Wen SF. Hyperkalemic renal tubular acidosis induced by trimethoprim–sulfamethoxazole in an AIDS patient. Clin Nephrol 1998;50(3):188–93.

92. Gabriels G, Stockem E, Greven J. Hyperkaliämie nach Trimethoprim oder Pentamidin. Eine bisher wenig beachtete Nebenwirkung antimikrobieller Therapiemassnahmen bei AIDS-Patienten. [Hyperkalemia after trimethoprim or pentamidine. Until now, a little noticed side effect of antimicrobial therapeutic measures in AIDS patients.] Dtsch Med Wochenschr 1998;123(45):1351–5.

93. Dreiher J, Porath A. Severe hyponatremia induced by theophylline and trimethoprim. Arch Intern Med 2001;161(2):291–2.

94. Kiefel V, Santoso S, Schmidt S, Salama A, Mueller-Eckhardt C. Metabolite-specific (IgG) and drug-specific antibodies (IgG, IgM) in two cases of trimethoprim–sulfamethoxazole-induced immune thrombocytopenia. Transfusion 1987;27(3):262–5.

95. Stricker RB, Goldberg B. AIDS and pure red cell aplasia. Am J Hematol 1997;54(3):264.

96. Keisu M, Wiholm BE, Palmblad J. Trimethoprim–sulphamethoxazole-associated blood dyscrasias. Ten years' experience of the Swedish spontaneous reporting system. J Intern Med 1990;228(4):353–60.

97. Blackwell EA, Hawson GA, Leer J, Bain B. Acute pancytopenia due to megaloblastic arrest in association with co-trimoxazole. Med J Aust 1978;2(1):38–41.

98. Tulloch AL. Pancytopenia in an infant associated with sulfamethoxazole–trimethoprim therapy. J Pediatr 1976;88(3):499–500.

99. Ermis B, Caner I, Karacan M, Olgun H. Haemolytic anaemia secondary to trimethoprim/sulfamethoxazole use. Thromb Haemost 2003;90:158–9.

100. Karpman E, Kurzrock EA. Adverse reactions of nitrofurantoin, trimethoprim and sulfamethoxazole in children. J Urol 2004;172(2):448–53.

101. Epstein JH. Photoallergy. A review. Arch Dermatol 1972;106(5):741–8.

102. Fuchs M, Scheid C, Schulz A, Diehl V, Sohngen D. Trimethoprim/sulfamethoxazole prophylaxis impairs function of mobilised autologous peripheral blood stem cells. Bone Marrow Transplant 2000;26(7):815–6.

103. Fishman JA. Prevention of infection caused by Pneumocystis carinii in transplant recipients. Clin Infect Dis 2001;33(8):1397–405.

104. Hughes WT, Dankner WM, Yogev R, Huang S, Paul ME, Flores MA, Kline MW, Wei LJ. Comparison of atovaquone and azithromycin with trimethoprim-sulfamethoxazole for the prevention of serious bacterial infections in children with HIV infection. Clin Infect Dis 2005;40(1):136–45.

105. Lambotte O, Costedoat-Chalumeau N, Amoura Z, Piette JC, Cacoub P. Drug-induced hemophagocytosis. Am J Med 2002;112(7):592–3.

106. Muller U. Hämatologische Nebenwirkungen von Medikamenten. [Hematologic side effects of drugs.] Ther Umsch 1987;44(12):942–8.

107. Asmar BI, Maqbool S, Dajani AS. Hematologic abnormalities after oral trimethoprim–sulfamethoxazole therapy in children. Am J Dis Child 1981;135(12):1100–3.

108. Deen JL, von Seidlein L, Pinder M, Walraven GE, Greenwood BM. The safety of the combination artesunate and pyrimethamine–sulfadoxine given during pregnancy. Trans R Soc Trop Med Hyg 2001;95(4):424–8.

109. Poskitt EM, Parkin JM. Effect of trimethoprim–sulphamethoxazole combination on folate metabolism in malnourished children. Arch Dis Child 1972;47(254):626–30.

110. Reynolds EH. Anticonvulsants, folic acid, and epilepsy. Lancet 1973;1(7816):1376–8.

111. Safrin S, Lee BL, Sande MA. Adjunctive folinic acid with trimethoprim-sulfamethoxazole for Pneumocystis carinii pneumonia in AIDS patients is associated with an increased risk of therapeutic failure and death. J Infect Dis 1994;170(4):912–7.

112. Ermis B, Caner I, Karacan M, Olgun H. Haemolytic anaemia secondary to trimethoprim/sulfamethoxazole use. Thromb Haemost 2003;90:158–9.

113. Baumgartner A, Hoigné R, Müller U, et al. Medikamentöse Schäden des Blutbildes: Erfahrungen aus dem Komprehensiven Spital-Drug-Monitoring Bern, 1974–1979. Schweiz Med Wochenschr 1982;112:1530.

114. Wang KK, Bowyer BA, Fleming CR, Schroeder KW. Pulmonary infiltrates and eosinophilia associated with sulfasalazine. Mayo Clin Proc 1984;59(5):343–6.

115. Bradley PP, Warden GD, Maxwell JG, Rothstein G. Neutropenia and thrombocytopenia in renal allograft recipients treated with trimethoprim–sulfamethoxazole. Ann Intern Med 1980;93(4):560–2.

116. Hoigne R, Klein U, Muller U. Results of four-week course of therapy of urinary tract infections: a comparative study using trimethoprim with sulfamethoxazole (Bactrim; Roche) and trimethoprim alone. In: Hejzlar M, Semonsky M, Masak S, editors. Advances in Antimicrobial and Antineoplastic Chemotherapy. Munchen-Berlin-Wien: Urban and Schwarzenberg, 1972:1283.

117. Moh R, Danel C, Sorho S, Sauvageot D, Anzian A, Minga A, Gomis OB, Konga C, Inwoley A, Gabillard D, Bissagnene E, Salamon R, Anglaret X. Haematological changes in adults receiving a zidovudine-containing HAART regimen in combination with cotrimoxazole in Cote d'Ivoire. Antivir Ther 2005;10(5):615–24.

118. Chintu C, Bhat GJ, Walker AS, Mulenga V, Sinyinza F, Lishimpi K, Farrelly L, Kaganson N, Zumla A, Gillespie

SH, Nunn AJ, Gibb DM. Co-trimoxazole as prophylaxis against opportunistic infections in HIV-infected Zambian children (CHAP):a double-blind randomised placebo-controlled trial. Lancet 2004;364(9448):1865–71.

119. Andres E, Noel E, Maloisel F. Trimethoprim-sulfamethoxazole-induced life-threatening agranulocytosis. Arch Intern Med 2003;163:1975–6.

120. Toure S, Gabillard D, Inwoley A, Seyler C, Gourvellec G, Anglaret X. Incidence of neutropenia in HIV-infected African adults receiving co-trimoxazole prophylaxis: a 6-year cohort study in Abidjan, Cote d'Ivoire. Trans R Soc Trop Med Hyg 2006;100(8):785–90.

121. Havas L, Fernex M, Lenox-Smith I. The clinical efficacy and tolerance of co-trimoxazole (Bactrim; Septrim). Clin Trials J 1973;3:81.

122. Pedersen-Bjergaard U, Andersen M, Hansen PB. Thrombocytopenia induced by noncytotoxic drugs in Denmark 1968–91. J Intern Med 1996;239(6):509–15.

123. Wiholm BE, Emanuelsson S. Drug-related blood dyscrasias in a Swedish reporting system, 1985–1994. Eur J Haematol Suppl 1996;60:42–6.

124. Barr AL, Whineray M. Immune thrombocytopenia induced by cotrimoxazole. Aust NZ J Med 1980;10(1):54–5.

125. Claas FH, van der Meer JW, Langerak J. Immunological effect of co-trimoxazole on platelets. BMJ 1979;2(6195):898–9.

126. Moeschlin S. Immunological granulocytopenia and agranulocytosis; clinical aspects. Sang 1955;26(1):32–51.

127. Rios Sanchez I, Duarte L, Sanchez Medal L. Agranulocitosis: analisis de 29 episodios en 19 pacientes. [Agranulocytosis. Analysis of 29 episodes in 19 patients.] Rev Invest Clin 1971;23(1):29–42.

128. Yamreudeewong W, Fosnocht BJ, Weixelman JM. Severe thrombocytopenia possibly associated with TMP/SMX therapy. Ann Pharmacother 2002;36(1):78–82.

129. Papaioannides D, Bouropoulos C, Korantzopoulos P. Co-trimoxazole induced acute thrombocytopenic purpura. Emerg Med J 2003;20:E3.

130. Kocak Z, Hatipoglu CA, Ertem G, Kinikli S, Tufan A, Irmak H, Demiroz AP. Trimethoprim-sulfamethoxazole induced rash and fatal hematologic disorders. J Infect 2006;52(2):e49-52.

131. George JN, Raskob GE, Shah SR, Rizvi MA, Hamilton SA, Osborne S, Vondracek T. Drug-induced thrombocytopenia: a systematic review of published case reports. Ann Intern Med 1998;129(11):886–90.

132. Pedersen-Bjergaard U, Andersen M, Hansen PB. Drug-specific characteristics of thrombocytopenia caused by non-cytotoxic drugs. Eur J Clin Pharmacol 1998;54(9–10):701–6.

133. Koirala J. Trimethoprim-sulfamethoxazole induced methemoglobinemia in an HIV-infected patient. Mayo Clin Proc 2004;79:829–30.

134. Zerbib F. Les oesophagites médicamenteuses. Hepato-Gastro 1998;5:115–20.

135. Kavatha D, Giamarellou H, Alexiou Z, Vlachogiannis N, Pentea S, Gozadinos T, Poulakou G, Hatzipapas A, Koratzanis G. Cefpodoxime-proxetil versus trimethoprim-sulfamethoxazole for short-term therapy of uncomplicated acute cystitis in women. Antimicrob Agents Chemother 2003;47:897–900.

136. Cameron A, Thomas M. Pseudomembranous colitis and co-trimoxazole. BMJ 1977;1(6072):1321.

137. Zehnder D, Kunzi UP, Maibach R, Zoppi M, Halter F, Neftel KA, Muller U, Galeazzi RL, Hess T, Hoigne R. Die Häufigkeit der Antibiotika-assoziierten Kolitis bei hospitalisierten Patienten der Jahre 1974–1991 im 'Comprehensive Hospital Drug Monitoring' Bern/St Gallen. [Frequency of antibiotics-associated colitis in hospitalized patients in 1974–1991 in "Comprehensive Hospital Drug Monitoring," Bern/St. Gallen.] Schweiz Med Wochenschr 1995;125(14):676–83.

138. Kofteridis DP, Maraki S, Mixaki I, Mantadakis E, Samonis G. Impact of prolonged treatment with trimethoprim-sulfamethoxazole on the human gut flora. Scand J Infect Dis 2004;36(10):771–2.

139. Munoz SJ, Martinez-Hernandez A, Maddrey WC. Intrahepatic cholestasis and phospholipidosis associated with the use of trimethoprim–sulfamethoxazole. Hepatology 1990;12(2):342–7.

140. Tanner AR. Hepatic cholestasis induced by trimethoprim. BMJ (Clin Res Ed) 1986;293(6554):1072–3.

141. Horak J, Mertl L, Hrabal P. Severe liver injuries due to sulfamethoxazole–trimethoprim and sulfamethoxydiazine. Hepatogastroenterology 1984;31(5):199–200.

142. Ransohoff DF, Jacobs G. Terminal hepatic failure following a small dose of sulfamethoxazole–trimethoprim. Gastroenterology 1981;80(4):816–9.

143. Thies PW, Dull WL. Trimethoprim–sulfamethoxazole-induced cholestatic hepatitis. Inadvertent rechallenge. Arch Intern Med 1984;144(8):1691–2.

144. Tse W, Singer C, Dominick D. Acute fulminant hepatic failure caused by trimethoprim–sulfamethoxazole. Infect Dis Clin Pract 2000;9:302–3.

145. Windecker R, Steffen J, Cascorbi I, Thurmann PA. Co-trimoxazole-induced liver and renal failure. Case report. Eur J Clin Pharmacol 2000;56(2):191–3.

146. Bjornsson E, Jerlstad P, Bergqvist A, Olsson R. Fulminant drug-induced hepatic failure leading to death or liver transplantation in Sweden. Scand J Gastroenterol 2005;40(9):1095–101.

147. Zaman F, Ye G, Abreo KD, Latif S, Zibari GB. Successful orthotopic liver transplantation after trimethoprim-sulfamethoxazole associated fulminant liver failure. Clin Transplant 2003;17:461–4.

148. Bjornsson E, Olsson R. Suspected drug-induced liver fatalities reported to the WHO database. Dig Liver Dis 2006;38(1):33–8.

149. Zaman F, Ye G, Abreo KD, Latif S, Zibari GB. Successful orthotopic liver transplantation after trimethoprim-sulfamethoxazole associated fulminant liver failure. Clin Transplant 2003;17:461–4.

150. Antonow DR. Acute pancreatitis associated with trimethoprim–sulfamethoxazole. Ann Intern Med 1986;104(3):363–5.

151. Versleijen MW, Naber AH, Riksen NP, Wanten GJ, Debruyne FM. Recurrent pancreatitis after trimethoprim-sulfamethoxazole rechallenge. Neth J Med 2005;63(7):275–7.

152. Bailey RR, Little PJ. Deterioration in renal function in association with co-trimoxazole therapy. Med J Aust 1976;1(24):914916.

153. Kalowski S, Nanra RS, Mathew TH, Kincaid-Smith P. Deterioration in renal function in association with co-trimoxazole therapy. Lancet 1973;1(7800):394–7.

154. Horn B, Cottier P. Kreatininkonzentration im Serum vor und unter Behandlung mit Trimethoprim–Sulfamethoxazol.

[Serum creatinine concentration prior to and following trimethoprim–sulfamethoxazole (Bactrim) treatment.] Schweiz Med Wochenschr 1974;104(49):1809–12.

155. Naderer O, Nafziger AN, Bertino JS Jr. Effects of moderate-dose versus high-dose trimethoprim on serum creatinine and creatinine clearance and adverse reactions. Antimicrob Agents Chemother 1997;41(11):2466–70.

156. Masters PA, O'Bryan TA, Zurlo J, Miller DQ, Joshi N. Trimethoprim-sulfamethoxazole revisited. Arch Intern Med 2003;163:402–10.

157. Trollfors B, Wahl M, Alestig K. Co-trimoxazole, creatinine and renal function. J Infect 1980;2(3):221–6.

158. Kainer G, Rosenberg AR. Effect of co-trimoxazole on the glomerular filtration rate of healthy adults. Chemotherapy 1981;27(4):229–32.

159. Craig WA, Kunin CM. Trimethoprim–sulfamethoxazole: pharmacodynamic effects of urinary pH and impaired renal function. Studies in humans. Ann Intern Med 1973;78(4):491–7.

160. Wong CS, Jelacic S, Habeeb RL, Watkins SL, Tarr PI. The risk of the hemolytic–uremic syndrome after antibiotic treatment of *Escherichia coli* O157:H7 infections. N Engl J Med 2000;342(26):1930–6.

161. Siegel WH. Unusual complication of therapy with sulfamethoxazole-trimethoprim. J Urol 1977;117(3):397.

162. Soheilian M, Sadoughi MM, Ghajarnia M, Dehghan MH, Yazdani S, Behboudi H, Anisian A, Peyman GA. Prospective randomized trial of trimethoprim/sulfamethoxazole versus pyrimethamine and sulfadiazine in the treatment of ocular toxoplasmosis. Ophthalmology 2005;112(11):1876–82.

163. Ozkaya-Bayazit E. Specific site involvement in fixed drug eruption. J Am Acad Dermatol 2003;49:1003–7.

164. Hoigne R, Sonntag MR, Zoppi M, Hess T, Maibach R, Fritschy D. Occurrence of exanthema in relation to aminopenicillin preparations and allopurinol. N Engl J Med 1987;316(19):1217.

165. Bigby M, Jick S, Jick H, Arndt K. Drug-induced cutaneous reactions. A report from the Boston Collaborative Drug Surveillance Program on 15,438 consecutive inpatients, 1975 to 1982. JAMA 1986;256(24):3358–63.

166. Sonntag MR, Zoppi M, Fritschy D, Maibach R, Stocker F, Sollberger J, Buchli W, Hess T, Hoigne R. Exantheme unter häufig angewandten Antibiotika und anti-bakteriellen Chemotherapeutika (Penicilline, speziell Aminopenicilline, Cephalosporine und Cotrimoxazol) sowie Allopurinol. [Exanthema during frequent use of antibiotics and antibacterial drugs (penicillin, especially aminopenicillin, cephalosporin and cotrimoxazole) as well as allopurinol. Results of The Berne Comprehensive Hospital Drug Monitoring Program.] Schweiz Med Wochenschr 1986;116(5):142–5.

167. Arndt KA, Jick H. Rates of cutaneous reactions to drugs. A report from the Boston Collaborative Drug Surveillance Program. JAMA 1976;235(9):918–23.

168. Hattori N, Hino H. Generalized erythematous skin eruptions due to trimethoprim itself and co-trimoxazole. J Dermatol 1998;25(4):269–71.

169. Lawrentschuk N, Pan D, Troy A. Fixed drug eruption of the penis secondary to sulfamethoxazole-trimethoprim. Sci World J 2006;62319-22.

170. Rasi A, Khatami A. Unilateral non-pigmenting fixed drug eruption associated with cotrimoxazole. Dermatol Online J 2006;12(6):12.

171. Nnoruka EN, Ikeh VO, Mbah AU. Fixed drug eruption in Nigeria. Int J Dermatol 2006;45(9):1062–5.

172. Ozkaya E. Polysensitivity in fixed drug eruption due to a novel drug combination-independent lesions due to piroxicam and cotrimoxazole. Eur J Dermatol 2006;16(5):591–2.

173. Dogra S, Handa S. Bullous necrotizing fixed drug eruption in an infant. Pediatr Dermatol 2004;21(3):281–2.

174. Ozkaya-Bayazit E. Independent lesions of fixed drug eruption caused by trimethoprim-sulfamethoxazole and tenoxicam in the same patient: a rare case of polysensitivity. J Am Acad Dermatol 2004;51(2 Suppl):S102-4.

175. Feiza BA, Samy F, Asma D, Rym B, Insaf M. [Urticarial vasculitis. A case report after sulfamethoxazole-trimethoprim ingestion.] Tunis Med 2005;83(11):714–6.

176. Mortimer NJ, Bermingham MR, Chapple SJ, Sladden MJ. Fatal adverse drug reaction to trimethoprim. Aust Fam Physician 2005;34(5):345–6.

177. Silveira C, Belfort R Jr, Muccioli C, Holland GN, Victora CG, Horta BL, Yu F, Nussenblatt RB. The effect of long-term intermittent trimethoprim/sulfamethoxazole treatment on recurrences of toxoplasmic retinochoroiditis. Am J Ophthalmol 2002;134(1):41–6.

178. Mahboob A, Haroon TS. Drugs causing fixed eruptions: a study of 450 cases. Int J Dermatol 1998;37(11):833–8.

179. Ozkaya-Bayazit E, Bayazit H, Ozarmagan G. Drug related clinical pattern in fixed drug eruption. Eur J Dermatol 2000;10(4):288–91.

180. Ozkaya-Bayazit E, Akar U. Fixed drug eruption induced by trimethoprim–sulfamethoxazole: evidence for a link to HLA-A30 B13 Cw6 haplotype. J Am Acad Dermatol 2001;45(5):712–7.

181. Ozkaya-Bayazit E, Gungor H. Trimethoprim-induced fixed drug eruption: positive topical provocation on previously involved and uninvolved skin. Contact Dermatitis 1998;39(2):87–8.

182. Sarkar R, Kaur C, Kanwar AJ. Extensive fixed drug eruption to nimesulide with cross-sensitivity to sulfonamides in a child. Pediatr Dermatol 2002;19(6):553–4.

183. Nwokolo C, Byrne L, Misch KJ. Toxic epidermal necrolysis occurring during treatment with trimethoprim alone. BMJ (Clin Res Ed) 1988;296(6627):970.

184. Revuz JE, Roujeau JC. Advances in toxic epidermal necrolysis. Semin Cutan Med Surg 1996;15(4):258–66.

185. Rzany B, Hering O, Mockenhaupt M, Schroder W, Goerttler E, Ring J, Schopf E. Histopathological and epidemiological characteristics of patients with erythema exudativum multiforme major, Stevens–Johnson syndrome and toxic epidermal necrolysis. Br J Dermatol 1996;135(1):6–11.

186. See S, Mumford JM. Trimethoprim/sulfamethoxazole-induced toxic epidermal necrolysis. Ann Pharmacother 2001;35(6):694–7.

187. Magina S, Lisboa C, Goncalves E, Conceicao F, Leal V, Mesquita-Guimaraes J. A case of toxic epidermal necrolysis treated with intravenous immunoglobin. Br J Dermatol 2000;142(1):191–2.

188. Paquet P, Jacob E, Damas P, Pierard GE. Treatment of drug-induced toxic epidermal necrolysis (Lyell's syndrome) with intravenous human immunoglobulins. Burns 2001;27(6):652–5.

189. Lipozencic J, Milavec-Puretic V, Kotrulja L, Tomicic H, Stulhofer Buzina D. Toxic epidermal necrolysis due to cotrimoxazole. J Eur Acad Dermatol Venereol 2002;16(2):182–3.

190. Nassif A, Bensussan A, Dorothee G, Mami-Chouaib F, Bachot N, Bagot M, Boumsell L, Roujeau JC. Drug specific cytotoxic T cells in the skin lesions of a patient with toxic epidermal necrolysis. J Invest Dermatol 2002;118(4):728–33.

191. Pushker N, Tandon R, Vajpayee RB. Stevens-Johnson syndrome in India — risk factors, ocular manifestations and management. Ophthalmologica 2000;214(4):285–8.

192. Wormser GP, Horowitz HW, Duncanson FP, Forseter G, Javaly K, Alampur SK, Gilroy SA, Lenox T, Rappaport A, Nadelman RB. Low-dose intermittent trimethoprim–sulfamethoxazole for prevention of Pneumocystis carinii pneumonia in patients with human immunodeficiency virus infection. Arch Intern Med 1991;151(4):688–92.

193. Canning DA. A suspected case of trimethoprim–sulfamethoxazole-induced loss of fingernails and toenails. J Urol 2000;163(4):1386–7.

194. Walker S, Norwood J, Thornton C, Schaberg D. Trimethoprim-sulfamethoxazole associated rhabdomyolysis in a patient with AIDS: case report and review of the literature. Am J Med Sci 2006;331(6):339–41.

195. Mainra RR, Card SE. Trimethoprim-sulfamethoxazole-associated hepatotoxicity-part of a hypersensitivity syndrome. Can J Clin Pharmacol 2003;10:175–8.

196. Hemstreet BA, Page RL 2nd. Sulfonamide allergies and outcomes related to use of potentially cross-reactive drugs in hospitalized patients. Pharmacotherapy 2006;26(4):551–7.

197. Ruscin JM, Page RL 2nd, Scott J. Hydrochlorothiazide-induced angioedema in a patient allergic to sulfonamide antibiotics: evidence from a case report and a review of the literature. Am J Geriatr Pharmacother 2006;4(4):325–9.

198. Schuster C, Wuthrich B. Anaphylactic drug reaction to celecoxib and sulfamethoxazole: cross reactivity or coincidence? Allergy 2003;58:1072.

199. Morimoto T, Sato T, Matsuoka A, Sakamoto T, Ohta K, Ando T, Ikushima S, Hagiwara K, Matsuno H, Akiyama O, Oritsu M. Trimethoprim-sulfamethoxazole-induced hypersensitivity syndrome associated with reactivation of human herpesvirus-6. Intern Med 2006;45(2):101–5.

200. Mohammedi I, Mausservey C, Hot A, Najioullah F, Kanitakis J, Robert D. Association d'une enc phalite a herpes virus 6 et d'un syndrome d'hypersensibilit m dicamenteuse au trim thoprime-sulfam thoxazole. [Human herpesvirus 6 encephalitis in trimethoprim-sulfamethoxazole-induced hypersensitivity syndrome.] Rev Med Interne 2006;27(6):499–501.

201. Morimoto T, Sato T, Matsuoka A, Sakamoto T, Ohta K, Ando T, Ikushima S, Hagiwara K, Matsuno H, Akiyama O, Oristu M. Trimethoprim-sulfamethoxazole induced hypersensitivity syndrome associated with reactivation of human herpesvirus-6. Int Med 2006;45:101–5.

202. Shanaah A, Mohapatra S. Systemic inflammatory response to trimethoprim-sulfamethoxazole in an HIV-positive patient. Infect Med 2004;21(3):131–2.

203. Nordstrand IA. Anaphylaxis to trimethoprim: an underappreciated risk in acute medical care. Emerg Med Australas 2004;16(1):82–5.

204. Alfirevic A, Stalford AC, Vilar FJ, Wilkins EG, Park BK, Pirmohamed M. Slow acetylator phenotype and genotype

205. Floris-Moore MA, Amodio-Groton MI, Catalano MT. Adverse reactions to trimethoprim/sulfamethoxazole in AIDS. Ann Pharmacother 2003;37:1810–3.

206. Sivojelezova A, Einarson A, Shuhaiber S, Koren G. Trimethoprim-sulfonamide combination therapy in early pregnancy. Can Fam Phys 2003;49:1085–6.

207. Hruska MW, Amico JA, Langaee TY, Ferrell RE, Fitzgerald SM, Frye RF. The effect of trimethoprim on CYP2C8 mediated rosiglitazone metabolism in human liver microsomes and healthy subjects. Br J Clin Pharmacol 2005;59(1):70–9.

208. Bottiger Y, Brattstrom C, Backman L, Claesson K, Burke JT. Trimethoprim-sulphamethoxazole does not affect the pharmacokinetics of sirolimus in renal transplant recipients. Br J Clin Pharmacol 2005;60(5):566–9.

209. Neuman MG, Malkiewicz IM, Phillips EJ, Rachlis AR, Ong D, Yeung E, Shear NH. Monitoring adverse drug reactions to sulfonamide antibiotics in human immunodeficiency virus-infected individuals. Ther Drug Monit 2002;24(6):728–36.

210. Pham NH, Baldo BA, Manfredi M, Zerboni R. Fine structural specificity differences of trimethoprim allergenic determinants. Clin Exp Allergy 1996;26(10):1155–60.

211. Greenberger PA, Patterson R. Management of drug allergy in patients with acquired immunodeficiency syndrome. J Allergy Clin Immunol 1987;79(3):484–8.

212. Theodore CM, Holmes D, Rodgers M, McLean KA. Co-trimoxazole desensitization in HIV-seropositive patients. Int J STD AIDS 1998;9(3):158–61.

213. Walmsley SL, Khorasheh S, Singer J, Djurdjev O, Schlech W, Thompson W, Duperval R, Toma E, Tsoukas C, Senay H, Wells P, Uetrecht J, Shear N, Rachlis A, Fong B, McGreer A, Smaill F, Cohen J, Ford P, Gilmour J, Mackie I, Williams K, Montaner J, Zarowny DCanadian HIV Trials Network 057 Study Group. A randomized trial of N-acetylcysteine for prevention of trimethoprim-sulfamethoxazole hypersensitivity reactions in Pneumocystis carinii pneumonia prophylaxis (CTN 057). J Acquir Immune Defic Syndr Hum Retrovirol 1998;19(5):498–505.

214. Holtzer CD, Flaherty JF Jr, Coleman RL. Cross-reactivity in HIV-infected patients switched from trimethoprim–sulfamethoxazole to dapsone. Pharmacotherapy 1998;18(4):831–5.

215. O'Kane EB, Schneeweiss R. Trimethoprim–sulfamethoxazole-induced sepsis-like syndrome in a patient with AIDS. J Am Board Fam Pract 1996;9(6):448–50.

216. Moran KA, Ales NC, Hemmer PA. Newly diagnosed human immunodeficiency virus after sepsis-like reaction of trimethoprim–sulfamethoxazole. South Med J 2001;94(3):350–2.

217. Choquet-Kastylevsky G, Vial T, Descotes J. Allergic adverse reactions to sulfonamides. Curr Allergy Asthma Rep 2002;2(1):16–25.

218. Hasui M, Kotera F, Tsuji S, Yamamoto A, Taniuchi S, Fujikawa Y, Nakajima M, Yoshioka A, Kobayashi Y. Successful resumption of trimethoprim–sulfamethoxazole after oral desensitisation in patients with chronic granulomatous disease. Eur J Pediatr 2002;161(6):356–7.

219. Yoshizawa S, Yasuoka A, Kikuchi Y, Honda M, Gatanaga H, Tachikawa N, Hirabayashi Y, Oka S. A 5-day course of oral desensitization to trimethoprim/sulfamethoxazole (T/S) in patients with human immunodeficiency virus type-1 infection

who were previously intolerant to T/S. Ann Allergy Asthma Immunol 2000;85(3):241–4.

220. Demoly P, Messaad D, Reynes J, Faucherre V, Bousquet J. Trimethoprim–sulfamethoxazole-graded challenge in HIV-infected patients: long-term follow-up regarding efficacy and safety. J Allergy Clin Immunol 2000;105(3):588–9.

221. Nucera E, Schiavino D, Buonomo A, Del Ninno M, Sun JY, Patriarca G. Tolerance induction to cotrimoxazole. Allergy 2000;55(7):681–2.

222. Lopez-Serrano MC, Moreno-Ancillo A. Drug hypersensitivity reactions in HIV-infected patients. Induction of cotrimoxazole tolerance. Allergol Immunol Clin 2000;15:347–51.

223. Sher MR, Suchar C, Lockey RF. Anaphylactic shock induced by oral desensitization to trimethoprim/sulfmethoxazole. J Allergy Immunol 1986;77:133.

224. Torgovnick J, Arsura E. Desensitization to sulfonamides in patients with HIV infection. Am J Med 1990;88(5):548–9.

225. Finegold I. Oral desensitization to trimethoprim–sulfamethoxazole in a patient with acquired immunodeficiency syndrome. J Allergy Clin Immunol 1986;78(5 Pt 1):905–8.

226. Papakonstantinou G, Fuessl H, Hehlmann R. Trimethoprim–sulfamethoxazole desensitization in AIDS. Klin Wochenschr 1988;66(8):351–3.

227. Bonfanti P, Pusterla L, Parazzini F, Libanore M, Cagni AE, Franzetti M, Faggion I, Landonio S, Quirino T. The effectiveness of desensitization versus rechallenge treatment in HIV-positive patients with previous hypersensitivity to TMP–SMX: a randomized multicentric study. C.I.S.A.I. Group Biomed Pharmacother 2000;54(1):45–9.

228. Leoung GS, Stanford JF, Giordano MF, Stein A, Torres RA, Giffen CA, Wesley M, Sarracco T, Cooper EC, Dratter V, Smith JJ, Frost KR. American Foundation for AIDS Research (amfAR) Community-Based Clinical Trials Network. Trimethoprim–sulfamethoxazole (TMP-SMZ) dose escalation versus direct rechallenge for Pneumocystis carinii pneumonia prophylaxis in human immunodeficiency virus-infected patients with previous adverse reaction to TMP–SMZ. J Infect Dis 2001;184(8):992–7.

229. Peters TR, Edwards KM, Standaert SM. Severe ehrlichiosis in an adolescent taking trimethoprim–sulfamethoxazole. Pediatr Infect Dis J 2000;19(2):170–2.

230. Konttinen A, Perasalo JO, Eisalo A. Sulfonamide hepatitis. Acta Med Scand 1972;191(5):389–91.

231. Huovinen P, Toivanen P. Trimethoprim resistance in Finland after five years' use of plain trimethoprim. BMJ 1980;280(6207):72–4.

232. Turnidge JD. A reappraisal of co-trimoxazole. Med J Aust 1988;148(6):296–305.

233. Mates A, Eyny D, Philo S. Antimicrobial resistance trends in Shigella serogroups isolated in Israel, 1990–1995. Eur J Clin Microbiol Infect Dis 2000;19(2):108–11.

234. Whitney CG, Farley MM, Hadler J, Harrison LH, Lexau C, Reingold A, Lefkowitz L, Cieslak PR, Cetron M, Zell ER, Jorgensen JH, Schuchat A. Active Bacterial Core Surveillance Program of the Emerging Infections Program Network. Increasing prevalence of multidrug-resistant Streptococcus pneumoniae in the United States. N Engl J Med 2000;343(26):1917–24.

235. Abou-Eisha A, Creus A, Marcos R. Genotoxic evaluation of the antimicrobial drug, trimethoprim, in cultured human lymphocytes. Mutat Res 1999;440(2):157–62.

236. Sorensen PJ, Jensen MK. Cytogenetic studies in patients treated with trimethoprim–sulfamethoxazole. Mutat Res 1981;89(1):91–4.

237. Guillebaud J. Sulpha–trimethoprim combinations and male infertility. Lancet 1978;2(8088):523.

238. Murdia A, Mathur V, Kothari LK, Singh KP. Sulpha–trimethoprim combinations and male fertility. Lancet 1978;2(8085):375–6.

239. Brigg GG, Freedman RK, Jaffe SJ. In: A Reference Guide to Fetal and Neontal Risk: Drugs in Pregnancy and Lactation. 3rd ed.. Baltimore-Hong Kong-London-Sydney: Williams and Wilkins, 1990:621.

240. Hernandez-Diaz S, Werler MM, Walker AM, Mitchell AA. Folic acid antagonists during pregnancy and the risk of birth defects. N Engl J Med 2000;343(22):1608–14.

241. Richardson MP, Osrin D, Donaghy S, Brown NA, Hay P, Sharland M. Spinal malformations in the fetuses of HIV infected women receiving combination antiretroviral therapy and co-trimoxazole. Eur J Obstet Gynecol Reprod Biol 2000;93(2):215–7.

242. Pirmohamed M, Alfirevic A, Vilar J, Stalford A, Wilkins EG, Sim E, Park BK. Association analysis of drug metabolizing enzyme gene polymorphisms in HIV-positive patients with co-trimoxazole hypersensitivity. Pharmacogenetics 2000;10(8):705–13.

243. Para MF, Finkelstein D, Becker S, Dohn M, Walawander A, Black JR. Reduced toxicity with gradual initiation of trimethoprim–sulfamethoxazole as primary prophylaxis for Pneumocystis carinii pneumonia: AIDS Clinical Trials Group 268. J Acquir Immune Defic Syndr 2000;24(4):337–43.

244. Rabaud C, Charreau I, Izard S, Raffi F, Meiffredy V, Leport C, Guillemin F, Yeni P, Aboulker JP. Delta trial group. Adverse reactions to cotrimoxazole in HIV-infected patients: predictive factors and subsequent HIV disease progression. Scand J Infect Dis 2001;33(10):759–64.

245. O'Reilly RA, Motley CH. Racemic warfarin and trimethoprim–sulfamethoxazole interaction in humans. Ann Intern Med 1979;91(1):34–6.

246. Chafin CC, Ritter BA, James A, Self TH. Hospital admission due to warfarin potentiation by TMP–SMX. Nurse Pract 2000;25(12):73–5.

247. Penning-van Beest FJ, van Meegen E, Rosendaal FR, Stricker BH. Drug interactions as a cause of overanticoagulation on phenprocoumon or acenocoumarol predominantly concern antibacterial drugs. Clin Pharmacol Ther 2001;69(6):451–7.

248. Visser LE, Penning-van Bees FJ, Kasbergen AA, De Smet PA, Vulto AG, Hofman A, Stricker BH. Overanticoagulation associated with combined use of antibacterial drugs and acenocoumarol or phenprocoumon anticoagulants. Thromb Haemost 2002;88(5):705–10.

249. Rees CA, Lennard L, Lilleyman JS, Maddocks JL. Disturbance of 6-mercaptopurine metabolism by cotrimoxazole in childhood lymphoblastic leukaemia. Cancer Chemother Pharmacol 1984;12(2):87–9.

250. Ferrazzini G, Klein J, Sulh H, Chung D, Griesbrecht E, Koren G. Interaction between trimethoprim–sulfamethoxazole and methotrexate in children with leukemia. J Pediatr 1990;117(5):823–6.

251. Rusterholz D, Schlegel C. Infektiose Aspekte des Morbus Wegener. [Infectious aspects of Wegener's granulomatosis.] Schweiz Med Wochenschr 2000;(Suppl 125):41S–3S.

252. de Araujo M, Seguro AC. Trimethoprim–sulfamethoxazole (TMP/SMX) potentiates indinavir nephrotoxicity. Antivir Ther 2002;7(3):181–4.

253. Ringden O, Myrenfors P, Klintmalm G, Tyden G, Ost L. Nephrotoxicity by co-trimoxazole and cyclosporin in transplanted patients. Lancet 1984;1(8384):1016–7.

254. Thompson JF, Chalmers DH, Hunnisett AG, Wood RF, Morris PJ. Nephrotoxicity of trimethoprim and cotrimoxazole in renal allograft recipients treated with cyclosporine. Transplantation 1983;36(2):204–6.

255. de Vries PL. Lithiumintoxicatie bij gelijktijdig gebruik van trimethoprim. [Lithium intoxication due to simultaneous use of trimethoprim.] Ned Tijdschr Geneeskd 2001;145(11):539–40.

256. Yang CH, Yang LJ, Jaing TH, Chan HL. Toxic epidermal necrolysis following combination of methotrexate and trimethoprim-sulfamethoxazole. Int J Dermatol 2000;39(8):621–3.

257. Steuer A, Gumpel JM. Methotrexate and trimethoprim: a fatal interaction. Br J Rheumatol 1998;37(1):105–6.

258. Richards AJ. Re: Interaction between methotrexate and trimethoprim. Br J Rheumatol 1998;37(7):806.

259. Grimmer SF, Allen WL, Back DJ, Breckenridge AM, Orme M, Tjia J. The effect of cotrimoxazole on oral contraceptive steroids in women. Contraception 1983;28(1):53–9.

260. Ilario MJ, Ruiz JE, Axiotis CA. Acute fulminant hepatic failure in a woman treated with phenytoin and trimethoprim–sulfamethoxazole. Arch Pathol Lab Med 2000;124(12):1800–3.

261. Trujillo TC, Nolan PE. Antiarrhythmic agents: drug interactions of clinical significance. Drug Saf 2000;23(6):509–32.

262. Ribera E, Pou L, Fernandez-Sola A, Campos F, Lopez RM, Ocana I, Ruiz I, Pahissa A. Rifampin reduces concentrations of trimethoprim and sulfamethoxazole in serum in human immunodeficiency virus-infected patients. Antimicrob Agents Chemother 2001;45(11):3238–41.

263. Guha Thakurta A, Mandal SK, Ganguly K, Dastidar SG, Chakrabarty AN. A new powerful antibacterial synergistic combination of trimethoprim and trimeprazine. Acta Microbiol Immunol Hung 2000;47(1):21–8.

264. Lee BL, Safrin S, Makrides V, Gambertoglio JG. Zidovudine, trimethoprim, and dapsone pharmacokinetic interactions in patients with human immunodeficiency virus infection. Antimicrob Agents Chemother 1996;40(5):1231–6.

265. Streeter AM, Shum HY, O'Neill BJ. The effect of drugs on the microbiological assay of serum folic acid and vitamin B12 levels. Med J Aust 1970;1(18):900–1.

266. Perazella MA. Drug-induced hyperkalemia: old culprits and new offenders. Am J Med 2000;109(4):307–14.

267. Jung D, AbdelHameed MH, Hunter J, Teitelbaum P, Dorr A, Griffy K. The pharmacokinetics and safety profile of oral ganciclovir in combination with trimethoprim in HIV- and CMV-seropositive patients. Br J Clin Pharmacol 1999;47(3):255–9.

268. de Vries PL. Lithiumintoxicatie bij gelijktijdig gebruik van trimethoprim. [Lithium intoxication due to simultaneous use of trimethoprim.] Ned Tijdschr Geneeskd 2001;145(11):539–40.

269. Jeurissen ME, Boerbooms AM, van de Putte LB. Pancytopenia and methotrexate with trimethoprim–sulfamethoxazole. Ann Intern Med 1989;111(3):261.

270. Kosoglou T, Rocci ML Jr, Vlasses PH. Trimethoprim alters the disposition of procainamide and N-acetylprocainamide. Clin Pharmacol Ther 1988;44(4):467–77.

271. Ribera E, Pou L, Fernandez-Sola A, Campos F, Lopez RM, Ocana I, Ruiz I, Pahissa A. Rifampin reduces concentrations of trimethoprim and sulfamethoxazole in serum in human immunodeficiency virus-infected patients. Antimicrob Agents Chemother 2001;45(11):3238–41.

272. Ribera E, Fernandez-Sola A, Juste C, Rovira A, Romero FJ, Armadans-Gil L, Ruiz I, Ocana I, Pahissa A. Comparison of high and low doses of trimethoprim-sulfamethoxazole for primary prevention of toxoplasmic encephalitis in human immunodeficiency virus-infected patients. Clin Infect Dis 1999;29(6):1461–6.

273. Baciewicz AM, Self TH. Rifampin drug interactions. Arch Intern Med 1984;144(8):1667–71.

274. Buniva G, Palminteri R, Berti M. Kinetics of a rifampicin-trimethoprim combination. Int J Clin Pharmacol Biopharm 1979;17(6):256–9.

275. Emmerson AM, Gruneberg RN, Johnson ES. The pharmacokinetics in man of a combination of rifampicin and trimethoprim. J Antimicrob Chemother 1978;4(6):523–31.

Troleandomycin

See also Macrolide antibiotics

General Information

Troleandomycin is a macrolide antibiotic with actions and uses similar to those of erythromycin.

Organs and Systems

Endocrine

Serum TSH concentrations are moderately but significantly reduced by troleandomycin compared with josamycin or placebo given over 10 days. At the same time serum estradiol concentration was significantly increased (1).

Liver

Most reports associating cholestatic jaundice with the concomitant use of oral contraceptives and macrolides have implicated troleandomycin (2–5). The exact mechanism of this interaction has not been established. Troleandomycin possibly inhibits the metabolism of estrogens and progestogens (6). In most patients jaundice persisted for more than a month after the antibiotic was withdrawn.

Drug–Drug Interactions

Alprazolam

Troleandomycin may inhibit the metabolism of alprazolam by inhibiting CYP3A4 (7).

Carbamazepine

Troleandomycin can cause carbamazepine toxicity by inhibiting CYP3A4 (8).

Ecabapide

Troleandomycin may inhibit the metabolism of ecabapide, a gastric prokinetic drug, by inhibiting CYP3A4 (9).

Ergot alkaloids

Clinically severe interactions between troleandomycin and ergotamine (10–13), resulting in ischemia of the extremities, probably result from inhibition of the metabolism of the ergopeptides (14,15), with raised serum concentrations of dihydroergotamine (16,17).

Hormonal contraceptives—oral

There have been several reports of hepatic cholestasis in women taking both troleandomycin and oral contraceptives (25). Oxidation of troleandomycin by CYP3A4 produces a derivative (probably a nitrosylated derivative) that binds tightly to the enzyme and thereby causes inactivation. This inhibition is highly selective for CYP3A4, and hepatic accumulation of ethinylestradiol is possible.

Omeprazole

CYP3A4 plays a major part in the formation of omeprazole sulfone, and also contributes to the 5-hydroxylation of omeprazole; both CYP2C19 and CYP3A contribute to the further elimination of 5-hydroxyomeprazole and omeprazole sulfone. In 18 healthy men who took oral omeprazole 20 mg alone or with troleandomycin 500 mg/day for 2 days, the effect of troleandomycin on the metabolism of omeprazole and its two principal metabolites differed between the different phenotypes of CYP2C19 (18). Mean C_{max} and clearance of omeprazole in poor metabolizers were reduced by troleandomycin. The C_{max} and AUC of 5-hydroxyomeprazole in poor metabolizers were significantly reduced by troleandomycin. There were similar effects in heterozygous extensive metabolizers, but in homozygous extensive metabolizers troleandomycin had no effect.

Quinine

Troleandomycin is an in vitro competitive inhibitor of quinine 3-hydroxylation, which is mediated by CYP3A4, and may therefore interact when co-administered with quinine (19).

References

1. Uzzan B, Nicolas P, Perret G, Vassy R, Tod M, Petitjean O. Effects of troleandomycin and josamycin on thyroid hormone and steroid serum levels, liver function tests and microsomal monooxygenases in healthy volunteers: a double blind placebo-controlled study. Fundam Clin Pharmacol 1991;5(6):513–26.
2. Claudel S, Euvrard P, Bory R, Chavaillon A, Paliard P. Cholestase intrahépatique apres aasociation triacetyloleandomycine–estroprogestatif. [Intra-hepatic cholestasis after taking a triacetyloleandomycin–estroprogestational combination.] Nouv Presse Méd 1979;8(14):1182.
3. Fevery J, Van Steenbergen W, Desmet V, Deruyttere M, De Groote J. Severe intrahepatic cholestasis due to the combined intake of oral contraceptives and triacetyloleandomycin. Acta Clin Belg 1983;38(4):242–5.
4. Haber I, Hubens H. Cholestatic jaundice after triacetyloleandomycin and oral contraceptives. The diagnostic value of gamma-glutamyl transpeptidase. Acta Gastroenterol Belg 1980;43(11–12):475–82.
5. Miguet JP, Vuitton D, Pessayre D, Allemand H, Metreau JM, Poupon R, Capron JP, Blanc F. Jaundice from troleandomycin and oral contraceptives. Ann Intern Med 1980;92(3):434.
6. Watkins PB, Murray SA, Winkelman LG, Heuman DM, Wrighton SA, Guzelian PS. Erythromycin breath test as an assay of glucocorticoid-inducible liver cytochromes P-450. Studies in rats and patients. J Clin Invest 1989;83(2):688–97.
7. Gorski JC, Jones DR, Hamman MA, Wrighton SA, Hall SD. Biotransformation of alprazolam by members of the human cytochrome P4503A subfamily. Xenobiotica 1999;29(9):931–44.
8. Pauwels O. Factors contributing to carbamazepine–macrolide interactions. Pharmacol Res 2002;45(4):291–8.
9. Fujimaki Y, Arai N, Inaba T. Identification of cytochromes P450 involved in human liver microsomal metabolism of ecabapide, a prokinetic agent. Xenobiotica 1999;29(12):1273–82.
10. Bacourt F, Couffinhal JC. Ischémie des membres par association dihydroergotamine-triacetyloléandomycin. [Ischemia of the extremities caused by a combination of dihydroergotamine and triacetyloleandomycin. New case report.] Nouv Presse Méd 1978;7(18):1561.
11. Franco A, Bourlard P, Massot C, Lecoeur J, Guidicelli H, Bessard G. Ergotamine par association dihydroergotamine–triacetyloléandomycin. [Acute ergotism caused by dihydroergotamine–triacetyloleandomycin association.] Nouv Presse Méd 1978;7(3):205.
12. Hayton AC. Precipitation of acute ergotism by triacetyloleandomycin. NZ Med J 1969;69(440):42.
13. Matthews NT, Havill JH. Ergotism with therapeutic doses of ergotamine tartrate. NZ Med J 1979;89(638):476–7.
14. Pea F, Furlanut M. Pharmacokinetic aspects of treating infections in the intensive care unit: focus on drug interactions. Clin Pharmacokinet 2001;40(11):833–68.
15. Eadie MJ. Clinically significant drug interactions with agents specific for migraine attacks. CNS Drugs 2001;15(2):105–18.
16. Azria M, Kiechel J, Lavenne D. Contribution à l'étude de l'interaction de la triacetyloléandomycine avec l'ergotamine ou la dihydroergotamine. J Pharmacol 1979;10:431.
17. Martinet M, Kiechel JR. Interaction of dihydroergotamine and triacetyloleandomycin in the minipig. Eur J Drug Metab Pharmacokinet 1983;8(3):261–7.
18. He N, Huang SL, Zhu RH, Tan ZR, Liu J, Zhu B, Zhou HH. Inhibitory effect of troleandomycin on the metabolism of omeprazole is CYP2C19 genotype-dependent. Xenobiotica 2003;33:211–21.
19. Zhao XJ, Ishizaki T. A further interaction study of quinine with clinically important drugs by human liver microsomes: determinations of inhibition constant (Ki) and type of inhibition. Eur J Drug Metab Pharmacokinet 1999;24(3):272–8.

Tylosin

General Information

Tylosin is a polyketide lactone substituted with three deoxyhexose sugars. It is used as an antimicrobial growth promoter in animals. Acquired resistance has been observed in potentially human pathogenic strains isolated from animals (1). Of further concern is the fact that tylosin confers cross-resistance on erythromycin (2).

References

1. Butaye P, Devriese LA, Haesebrouck F. Phenotypic distinction in *Enterococcus faecium* and *Enterococcus faecalis* strains between susceptibility and resistance to growth-enhancing antibiotics. Antimicrob Agents Chemother 1999;43(10):2569-70.
2. Pedersen KB. Some growth promoters in animals do confer antimicrobial resistance in humans. BMJ 1999;318(7190):1076.

Vancomycin

General Information

Vancomycin is a narrow-spectrum glycopeptide antibiotic with potent antistaphylococcal activity. It was developed in the early 1950s. Early formulations contained substantial impurities, which were presumably responsible for some adverse reactions (1). When rapid infusion rates are avoided, vancomycin is rarely associated with serious toxicity. Reviews have suggested that the potential for vancomycin to cause significant ototoxicity or nephrotoxicity has been exaggerated (2,3). Improved manufacturing has resulted in a purer product and fewer toxic effects, but vancomycin is still associated with potentially serious adverse reactions (4).

Vancomycin inhibits bacterial cell wall synthesis and is bactericidal during cell division at therapeutic concentrations. Bacterial resistance to vancomycin has not been an issue during the first decades of its use. More recently, vancomycin-resistant enterococci have been recovered with increasing frequency from hospitalized patients. In some institutions, multidrug-resistant and vancomycin-resistant enterococci have become important nosocomial pathogens, difficult to treat. Vancomycin-resistant enterococcal bacteremia is associated with a poor prognosis. Judicious use of vancomycin and broad-spectrum antibiotics is recommended, and strict infection control measures must be implemented to prevent nosocomial transmission of these organisms (5).

Vancomycin is particularly useful in infections caused by meticillin-resistant or penicillinase-producing staphylococci and diphtheroids, as well as in the prophylaxis of bacterial endocarditis and in the treatment of antibiotic-associated colitis. Sufficient fecal concentrations can be achieved with oral therapy. It is poorly absorbed from the gastrointestinal tract and painful when injected intramuscularly. It diffuses moderately well into bone tissue (6).

Lactoferrin, a protein contained in tears, increases the activity of vancomycin against biofilms of strains of *Staphylococcus epidermidis* and may be therapeutically helpful in the treatment of infections such as endophthalmitis associated with intraocular lenses (7).

Observational studies

In a prospective, observational study of 742 consecutive patients (390 men; mean age 51 years, range 17–86), the incidence, outcomes, and predictive factors of vancomycin-associated toxic effects in general oncology practice were assessed (8). In all, 47% had hematological malignancies, and of the patients with solid tumors, primary urogenital (12%) and breast (10%) tumors were most common. In 72% vancomycin was given in a dosage of 2 g/day, 16% received a prophylactic dose of 1 g/day, and 12% had another regimen not specified. Phlebitis occurred in 3%, predominantly those with recently inserted central venous catheters. All responded promptly to local therapy, and withdrawal of the catheter or vancomycin was never required. Skin rashes occurred in 11%. However, all but four patients concomitantly received a beta-lactam antibiotic. None of the rashes required withdrawal, although the beta-lactam antibiotic was often withdrawn. There was clinical evidence of ototoxicity in 6% of the patients who received other ototoxic drugs and only 3% of patients who were not receiving other ototoxic drugs. There was nephrotoxicity in 17%. Logistic regression was used to derive a model of risk of nephrotoxicity. Factors associated with an increased risk of nephrotoxicity included administration of other mild to moderate nephrotoxic agents or APACHE II scores over 40. Raised serum vancomycin concentrations did not reliably predict subsequent nephrotoxicity. The derived model was prospectively tested in a validation set of 359 patients. Sensitivity of the Nephrotoxicity Risk Index for any nephrotoxicity was 71% and for serious nephrotoxicity 100%. Specificity for any nephrotoxicity was 90% and for serious nephrotoxicity 81%. A main limitation of the study was the lack of patients who received only vancomycin. The concomitant administration of other agents may therefore have confounded the results.

The adverse effects of once-daily or twice-daily vancomycin were not significantly different in 121 hospitalized patients (9). Nephrotoxicity developed in 11 and 7.7% of the patients respectively; hearing loss in 3.2 and 16%; phlebitis in 14 and 23%; and red man syndrome in 14 and 9.6%.

Observational studies

In a randomized, controlled, open study in 44 patients with osteomyelitis treated with vancomycin, either by intermittent infusion (20 mg/kg over 60 minutes every 12 hours) or continuous infusion (40 mg/kg infused over 24 hours), serum creatinine concentration increased significantly by a mean of 0.4 µmol/l/day with intermittent infusion compared with continuous infusion (10). Adverse drug reactions leading to termination of vancomycin therapy were significantly more frequent with intermittent infusion (n = 9) compared with continuous infusion (n = 2). Adverse drug reactions with intermittent infusion were acute renal damage (n = 4) and allergic reactions (n = 2). One patient each had severe neutropenia, catheter phlebitis, and severe depression. With continuous infusion two patients developed catheter phlebitis. With intermittent infusion 57% of the patients were without an adverse drug reaction at week 6, compared with 95% of patients given continuous infusion.

Comparative studies

In a prospective, randomized trial in 34 drug abusers a short course of a combination of a glycopeptide (vancomycin or teicoplanin) with gentamicin was significantly less effective insided endocarditis caused by *Staphylococcus aureus* than a combination of cloxacillin and gentamicin (11).

Use in non-infective conditions

Many parents of children with regressive-onset autism have noted antecedent antibiotic exposure followed by chronic diarrhea. In a subgroup of children, disruption of indigenous gut flora might promote colonization by one or more neurotoxin-producing bacteria, contributing at least in part to their autistic symptoms. In 11 children with regressive-onset autism who had broad-spectrum antimicrobial exposure followed by chronic persistent diarrhea, oral vancomycin resulted in short-term improvement; however, these gains had largely waned at follow-up (12).

Pharmacoeconomics

The pharmacoeconomic impact of adverse effects of antimicrobial drugs is enormous. Antibacterial drug reactions account for about 25% of adverse drug reactions. The adverse effects profile of an antimicrobial agent can contribute significantly to its overall direct costs (monitoring costs, prolonged hospitalization due to complications or treatment failures) and indirect costs (quality of life, loss of productivity, time spent by families and patients receiving medical care). In one study an adverse event in a hospitalized patient was associated on average with an excess of 1.9 days in the length of stay, extra costs of $US2262 (1990–93 values), and an almost two-fold increase in the risk of death. In the outpatient setting, adverse drug reactions result in 2–6% of hospitalizations, and most of them were thought to be avoidable if appropriate interventions had been taken. In a review, economic aspects of antibacterial therapy with vancomycin have been summarized and critically evaluated (13).

In a randomized, prospective, cost-effectiveness study both teicoplanin and vancomycin were assessed as second-line therapy in 66 neutropenic patients after the failure of empirical treatment with a combination of piperacillin + tazobactam and amikacin (14). The primary success of second-line therapy was equivalent, and

the direct total costs were similar. Acquisition costs per dose were in favor of vancomycin, but costs derived from administering vancomycin and serum concentration monitoring led to similar costs for both regimens. With the exception of the red man syndrome, which occurred in 10% of vancomycin-treated patients but none of the teicoplanin-treated patients, toxicity (renal, liver, and ear toxicity, diarrhea, phlebitis) was also similar.

In a retrospective cost analysis, the records of 527 patients with acute leukemia were studied (15). They had been treated in a multicenter, randomized trial for febrile neutropenia with ceftazidime and amikacin plus either teicoplanin (6 mg/kg in a single dose; $n = 275$) or vancomycin (30 mg/kg/day in 2 doses; $n = 252$). Clinical responses were equivalent. Again the higher acquisition costs for teicoplanin were counterbalanced by the lower incidence of adverse events and easier administration, resulting in overall similar costs for both regimens. A total of 8% of patients treated with vancomycin reported adverse events compared with 3.2% of patients treated with teicoplanin. Rashes occurred in 6.0 versus 1.4% respectively. Nephrotoxicity, ototoxicity, fever, and hypotension occurred in very few patients.

Organs and Systems

Red man syndrome

A unique and peculiar adverse reaction related to the rapid infusion of large doses is the so-called red neck or red man syndrome(16,17). It is the most common adverse reaction to vancomycin, characterized by fever, chills, paresthesia, and erythema at the base of the neck and the upper back, and can be followed by a hypotensive episode (18). It is not a true allergic reaction. It seems to be due to vancomycin-induced release of histamine and possibly other vasoactive substances without the involvement of preformed antibodies (19,20). In rat peritoneal mast cells vancomycin provoked histamine release dose-dependently; fosfomycin inhibited this effect (21).

Antihistamines (H_1 receptor antagonists) prevent anaphylactoid reactions to vancomycin (22). The recommended method of administration is by intravenous infusion of a solution of 0.25–0.5% over 60 minutes, and patients have to be monitored closely. A possible red man syndrome has also been associated with systemic absorption of oral vancomycin in a patient with normal renal function (23).

In a prospective, randomized, double-blind, placebo-controlled study in 30 patients who required vancomycin chemoprophylaxis before elective arthroplasty, oral pretreatment with either a histamine H_1 receptor antagonist (diphenhydramine 1 mg/kg) or a histamine H_2 receptor antagonist (cimetidine 4 mg/kg) significantly reduced the histamine-related adverse effects of rapid vancomycin infusion (24).

It has been stressed that near-fatal red man syndrome can also occur after slow administration of vancomycin (25).

- An 83-year-old white man became distressed and had nausea and facial flushing after 10 minutes of a slow infusion of vancomycin (1 g scheduled over 2 hours). Despite interruption, he developed suffocating respiratory distress requiring intubation and mechanical ventilation. Uncharacteristic of anaphylactoid reactions, he became severely hypertensive. Subsequently, his condition rapidly normalized. During the acute event, there was an increase in serum troponin concentration.

- A 45-year-old man developed hypotension, bradycardia, a change in consciousness, and an erythematous macular rash 10 minutes after the slow infusion of 0.1% vancomycin (26). After appropriate management, he recovered well and was discharged on the following day.

In 50 patients, in whom vancomycin 15 mg/kg was continuously infused at a constant rate over 30 minutes, the occurrence of pruritus suggested that systemic vascular resistance was falling, exposing the patient to a risk of hypotension (27). Therapy with a beta-blocker appeared to confer protection against this hemodynamic effect.

Pretreatment with H_1 and H_2 receptor antagonists (diphenhydramine 1 mg/kg and cimetidine 4 mg/kg) intravenously serially over 3 minutes starting 10 minutes before the infusion of vancomycin permitted rapid vancomycin administration (1 g over 10 minutes) in 17 of 19 patients compared with eight of 19 patients treated with placebo in a prospective, randomized, double-blind, placebo-controlled study of patients undergoing elective arthroplasty (28). Hypotension occurred in 2 versus 12 of the patients, and 12 versus 19 of the patients had a rash. Serum histamine concentrations were raised after vancomycin administration in both groups.

Possible synergism between vancomycin and narcotics in the induction of red man syndrome due to histamine release has been suggested in a 19-year-old women treated with fentanyl and vancomycin during orthopedic surgery (29). Evidence from an animal study has suggested that the risk can also be enhanced by the co-administration of muscle relaxants, resulting in enhanced release of histamine (30).

Cardiovascular

Vancomycin can cause phlebitis if infused peripherally (31,32).

Nervous system

It is still controversial whether vancomycin can cause ototoxicity when given alone. However, vancomycin can augment the ototoxicity of aminoglycosides (33). Tinnitus and dizziness have been noted, resolving on withdrawal (34). Hearing loss can be transient or permanent. If vancomycin is combined with aminoglycosides, toxicity may be additive (35).

Sensory systems

Eyes

In 118 consecutive patients undergoing cataract surgery, the use of supplementary prophylactic vancomycin in the irrigating solution during extracapsular lens extraction was significantly associated with an increased incidence of angiographic and clinical cystoid macular edema (36).

Ears

The risk of ototoxicity has been assessed in a prospective study in 16 patients on continuous ambulatory peritoneal dialysis treated with two infusions of vancomycin (30 mg/kg) in 2 liters of peritoneal dialysate administered at 6-day intervals for episodes of peritonitis (37). Patients who were too ill to respond appropriately, those with pre-existing sensorineural hearing loss, those with a narrow auditory canal, those with a discharging ear or perforated tympanic membrane, and those receiving concurrent ototoxic drugs were excluded. The authors found no evidence of ototoxicity (pure-tone audiometry, electronystagmography, and clinical assessment), even with repeated courses of vancomycin. Average serum concentrations were in the target range. No adverse effects were recorded, except for a transient generalized pruritus in one patient after the start of infusion.

Vancomycin can cause irreversible bilateral sensorineural hearing loss.

- A 63-year-old white man was treated with vancomycin 1 g/day for 4 days because of *Corynebacterium jeikeium* meningitis associated with an Ommaya reservoir, and with intrathecal vancomycin 5 mg. He developed difficulty in hearing after the first intrathecal dose and complete hearing loss after the second intrathecal dose. An audiogram showed eighth nerve bilateral sensorineural hearing loss (38).

In a prospective multicenter study in 152 children with peritoneal dialysis-associated peritonitis one patient with vancomycin intoxication had a 20–30 dB increase in hearing threshold (39).

Hearing loss due to vancomycin toxicity has been treated by hemoperfusion and hemodialysis (40).

- A 14-month-old girl with chronic renal insufficiency due to renal dysplasia was empirically treated with ceftazidime and vancomycin for fever. Her calculated creatinine clearance was 10 ml/minute/1.73 m^2. She erroneously received vancomycin 1.5 g in 3 doses 6 hours apart. Her serum creatinine concentration increased and her vancomycin concentrations remained markedly high (338 mg/l 5 hours after the third dose). The half-life of vancomycin was 145 hours. Hearing loss developed. Continued charcoal hemoperfusion and hemodialysis were used to treat the disorder. Thrombocytopenia was noted as significant consequence of hemoperfusion. The patient did not fully recover her previous renal function and became dialysis dependent. The audiogram normalized by 6 months.

Metabolism

Lactic acidosis occurred in a 56-year-old woman who was given intravenous vancomycin 1 g bd for 10 days (41).

Hematologic

Hemophagocytosis after vancomycin and teicoplanin occurred in a 45-year-old woman (42).

Delayed myeloid engraftment due to vancomycin has been described in a 59-year-old man who received a non-myeloablative allogeneic hemopoietic stem cell transplant (43).

Treatment with vancomycin can rarely cause reversible pancytopenia (44).

- Neutropenia occurred in a 48-year-old man 16 days after the start of vancomycin therapy (45). Vancomycin was withdrawn and he was given a granulocyte-colony stimulating factor. He was then rechallenged with a single dose of vancomycin 1 g. His white blood cell count fell to $600 \times 10^9/l$ but returned to normal with continued granulocyte-colony stimulating factor therapy.
- A 43-year-old white woman was given vancomycin for a suspected methicillin-resistant *S. aureus* infection (46). On day 10 she developed a fever, chills, and a disseminated lacy macular rash. Vancomycin was withdrawn but reinstituted 48 hours later. Another 24 hours later, she again developed a fever, chills, and a confluent erythematous rash. Sepsis developed, and ceftazidime was added. There was thrombocytopenia ($118 \times 10^9/l$), and a fall in the white blood count (from 7.6 to $3.5 \times 10^9/l$) and hemoglobin (from 13.8 to 12.2 g/dl). One day later the leukocyte count was $2.6 \times 10^9/l$ and the hemoglobin fell to 10 g/dl. Both drugs were withdrawn. The rash disappeared, the temperature returned to normal, and the blood cell counts completely recovered 4 days later.
- An 87-year-old white woman was treated with vancomycin for an abscess due to methicillin-resistant *S. epidermidis* (47). On day 30, her white blood cell count fell to $0.5 \times 10^9/l$ after a total dose of 55 g of vancomycin. Vancomycin was withdrawn, and the white blood cell count returned to normal within 5 days. After 6 weeks vancomycin was reintroduced, and the white blood cell count fell to $0.325 \times 10^9/l$ after a total dose of 14 g of vancomycin on day 9, but resolved rapidly after vancomycin was withdrawn.

Leukocytes

Vancomycin-induced neutropenia occurs in about 2% of treated patients, and complete recovery takes place in 2–5 days after withdrawal (48,49). It occurs after prolonged treatment with high doses in patients with normal renal function (for example more than 20 days of more than 15–20 mg/kg) or in the prolonged presence of high serum concentrations in patients with severe renal insufficiency (50). After withdrawal of vancomycin, neutropenia disappeared promptly. The bone marrow seemed to be unaffected. Other cases of agranulocytosis related to vancomycin therapy have been reported (51–53).

In a retrospective chart review of 114 patients treated with vancomycin 14 cases of vancomycin-induced neutropenia were identified; four included a reduction in absolute neutrophil count to $500 \times 10^6/l$ or less (54). The mean duration of vancomycin therapy and time to neutropenia were 32 and 26 days respectively. Resolution occurred promptly after withdrawal. The total vancomycin dose and serum concentrations were not related to the neutropenia, suggesting a hypersusceptibility reaction.

- A 64-year-old man who was treated with intravenous vancomycin 1.5 g/day for finger osteomyelitis developed neutropenia after 21 days (55). The absolute neutrophil count reached a nadir of $418 \times 10^6/l$ during vancomycin use and returned to normal 7 days after withdrawal.

The mechanism of neutropenia caused by vancomycin is unclear, but it appears to be immune-mediated (55,56,57).

Neutropenia normally recovers after drug withdrawal, and the neutrophil count can recover when vancomycin is replaced by teicoplanin (58).

- A 35-year-old man was given vancomycin (1.5 g bd) for a wound infection with a coagulase-negative staphylococcus, a peptostreptococcus, and a coryneform Gram-positive bacillus. On day 37 hi white blood cell count was $2.8 \times 10^9/l$ (baseline 5.2) with a low neutrophil count ($0.5 \times 10^9/l$). Vancomycin was withdrawn and 2 days later he was given teicoplanin 400 mg/day, by which time the cell count had increased to $3.5 \times 10^9/l$, neutrophils $0.9 \times 10^9/l$. On day 45, the total white cell count was $4.9 \times 10^9/l$, neutrophils $2.3 \times 10^9/l$. Teicoplanin was continued for 1 month, and the white cell count remained normal.

A direct toxic effect of vancomycin and/or an immune-mediated mechanism (antineutrophil antibodies, sensitized lymphocytes) have been discussed.

- A typical case of vancomycin-induced neutropenia has been reported in a 39-year-old woman with sickle cell SS disease treated with vancomycin for methicillin-resistant coagulase-negative staphylococcal bacteremia (59). However, in addition to neutropenia, the clinical course was defined by a febrile period characteristic of drug fever, with delayed onset and resolution 48 hours after vancomycin was withdrawn. In this case, fluconazole and cefazoline were also administered, but their contribution to neutropenia was judged unlikely (negative rechallenge with fluconazole, withdrawal of cefazoline after less than 5 days of therapy).

Platelets

Thrombocytopenia is a rare adverse effect of vancomycin and may be associated with the presence of vancomycin-dependent antiplatelet IgG antibodies. Reports of drug-induced thrombocytopenia have been systematically reviewed (60). Among the 98 different drugs described in 561 articles the following antibiotics were found with level I (definite) evidence: co-trimoxazole, rifampicin, vancomycin, sulfisoxazole, cefalothin, piperacillin, methicillin, novobiocin. Drugs with level II (probable) evidence were oxytetracycline and ampicillin.

- Vancomycin-associated thrombocytopenia has been reported in a 72-year-old woman who was treated with gentamicin and vancomycin for infectious endocarditis due to *Clostridium pseudodiphtheriticum* (61). On the 4th day of treatment, the platelet count fell and reached a nadir of 14×10^{12}/l on day 7. Two days after withdrawal of vancomycin (day 8) the platelet count began to rise and reached 150×10^{12}/l within 5 days. Vancomycin-dependent antiplatelet IgG antibodies were not detected 10 days after vancomycin.

- Thrombocytopenia also developed in a 72-year-old man who was treated with vancomycin for staphylococcal sepsis after treatment with gemcitabine for metastatic pancreatic cancer (62). On day 12 thrombocytopenia developed, reached a nadir on day 18 (13×10^{12}/l), and did not respond to platelet transfusions. Bone-marrow megakaryocytes were adequate. Platelet-associated IgG and IgM were increased. Vancomycin was withdrawn on day 28, with prompt recovery of the platelet count to 136×10^{12}/l in 10 days.

- Reversible thrombocytopenia occurred in a 50-year-old man with culture-negative infective endocarditis who was given vancomycin after mitral valve replacement (63,64).

A 2-year-old boy developed severe thrombocytopenia during vancomycin therapy; it resolved promptly after withdrawal (65).

Vancomycin was associated with subsequent thrombocytopenia in a retrospective study that included 193 patients in a surgical trauma intensive care unit (66).

Patients undergoing allogeneic hemopoietic stem cell transplantation always require platelet transfusions, but the increase in platelet count is often less than expected. The factors responsible for poor responses to platelet transfusions in this clinical setting are largely unknown. In a prospective study in 87 consecutive transplanted children univariate analysis showed that concomitant therapy with vancomycin is significantly associated with a lower post-transfusion corrected increment in platelet count (67).

Gastrointestinal

Antibiotic-associated diarrhea can develop with any antibacterial agent. Vancomycin has been implicated as a rare cause of diarrhea associated with *Clostridium difficile* (68), despite the fact that vancomycin is often used to treat it.

- Diarrhea developed in a 60-year-old man on chronic hemodialysis after 20 doses of parenteral vancomycin (250 mg at each dialysis) (69). Although culture for *C. difficile* was not performed, latex agglutination was positive for *C. difficile* toxin.

A few other cases of antibiotic-associated diarrhea with vancomycin have been described in association with *C. difficile*. This paradoxical association, although very uncommon, should be considered in patients who develop diarrhea during vancomycin therapy.

Liver

Raised liver enzymes have been attributed to oral vancomycin (70).

- 57-year-old man who took oral vancomycin 125–500 mg/day on five separate occasions had significant rises in alanine transaminase to 371 U/l and aspartate transaminase to 203 U/l. The rises resolved on each occasion after withdrawal of vancomycin.

Urinary tract

The reported rate of nephrotoxicity of vancomycin is 7–16%, but it can reach 35% with concurrent aminoglycoside therapy and is associated with serum concentrations over 40 µg/ml. In a retrospective study in 19 patients (mean age 51 years; 12 women) who had a 50% increase in their normal baseline serum creatinine during vancomycin therapy, there was no correlation between vancomycin serum concentration and serum creatinine (71).

Of 69 neonates (including 8 with peak vancomycin concentrations over 40 µg/ml) with culture-proven *S. aureus* or coagulase-negative staphylococcal septicemia who received vancomycin for more than 3 days, 6 had a doubling of serum creatinine concentration during vancomycin treatment, and all were in the group with peak serum vancomycin concentrations under 40 µg/ml (72).

Nephrotoxicity of vancomycin is often mild and reversible after withdrawal. It occurs in 6–17% of courses (73). Higher frequencies of nephrotoxicity were reported when vancomycin was used in combination with an aminoglycoside, which is consistent with toxicological data obtained in rats (74). Nevertheless, evidence of synergistic toxicity between vancomycin and aminoglycosides is controversial (2,34,73). Endotoxins seem to potentiate the nephrotoxic effect of vancomycin, at least in rats (75).

Acute interstitial nephritis has been reported in a patient receiving vancomycin (76).

- A 64-year-old white man was treated with intravenous cloxacillin 2 g 4-hourly for a *S. aureus* sternal wound infection and osteomyelitis. On day 14, cloxacillin was discontinued because of a fall in renal function. Urinalysis was positive for occasional red blood cells and hyaline casts, but there were no eosinophils. He was given intravenous vancomycin 1.5 g 36-hourly. Renal function gradually improved and the dosing regimen was adjusted to 1.5 g/day. Predose vancomycin serum concentrations were 11.0 and 6.3 mg/l on days 23 and 30 respectively. On day 32, oral ciprofloxacin 500 mg bd was added for suspected sepsis (catheter tip cultures grew *Klebsiella pneumoniae*, although blood cultures were negative). Other medications were enteric-coated aspirin, warfarin, acebutolol, and digoxin. The next day, a progressing maculopapular rash developed and the patient continued to have a fever. Acute interstitial nephritis was confirmed by kidney biopsy. Hemodialysis was required, and he was treated with prednisone and made a gradual recovery. Four months later, he was treated with vancomycin for a relapse of the sternal infection. His eosinophil count rose and peaked at 18%, and there were eosinophils in the urine.

The authors found four other published cases of vancomycin-induced acute interstitial nephritis.

There was a fall in residual renal function during episodes of peritonitis in a prospective multicenter study in 152 children with peritoneal dialysis-associated peritonitis (39). The study evaluated the efficacy, safety, and clinical acceptance of combination treatment, including a glycopeptide (teicoplanin or vancomycin) and ceftazidime, each given either intermittently or continuously. Irreversible anuria developed in 19% of the patients with residual diuresis. In patients who retained residual renal function, there was a strong trend toward a decreased residual glomerular filtration rate (−11%). There was no difference with respect to the type of glycopeptide or the mode of use. Aminoglycosides were used concomitantly in only 12 patients and did not seem to affect the evolution of residual renal function. Underdosing/non-adherence or overdosing occurred in 8.2% of continuously treated versus 4.9% of intermittently treated patients. Intermittent and continuous intraperitoneal treatment with glycopeptides and ceftazidime were equally efficacious and safe when measured by objective clinical criteria.

In children with febrile neutropenia and Gram-positive bacteremia associated with antineoplastic drug therapy, teicoplanin was significantly less nephrotoxic than vancomycin (77).

The effect of co-administration of fosfomycin on glycopeptide antibiotic-induced nephrotoxicity for 3 days has been investigated in rats (78). Fosfomycin reduced glycopeptide antibiotic-induced nephrotoxicity, as shown by reduced urinary excretion of N-acetyl-beta-D-glucosaminidase and fewer histological signs of nephrotoxicity in those treated with a combination of glycopeptide and fosfomycin compared with a glycopeptide alone. In addition, the higher the dose of fosfomycin, the more it reduced urinary N-acetyl-beta-D-glucosaminidase activity, suggesting that the role of fosfomycin in alleviating nephrotoxicity is dose-dependent. There was significantly less accumulation of teicoplanin and vancomycin in the renal cortex of rats treated with glycopeptide antibiotics plus fosfomycin compared with the glycopeptide antibiotics alone.

Antibiotic-laden cement with aminoglycosides and/or vancomycin can cause systemic toxicity. In two cases acute renal failure was associated with the use of antibiotic-laden cement incorporated in total hip arthroplasties (13).

Skin

Epidermolysis bullosa acquisita occurred in a 73-year-old man after a course of vancomycin for 15 days (79).

Fixed drug eruption is an infrequent adverse effect of vancomycin.

- A 14-year-old African American boy developed a fixed drug eruption involving an extensive area of the body

surface on the fourth day of treatment with vancomycin (80).

A fixed drug eruption after the fifth dose of vancomycin 1 g occurred in a 45-year-old white woman (81).

Vancomycin reportedly caused a severe delayed skin reaction with allergic cross-reactivity between vancomycin and teicoplanin requiring steroid therapy (82).

- A 68-year-old woman who was being treated with vancomycin for *S. epidermidis* bacteremia developed pruritus and a generalized maculopapular skin rash after 2 weeks. After a short course of prednisolone, she was given teicoplanin and developed general malaise, fever, conjunctival injection, an extensive rash, and later blisters on the legs, again requiring treatment with steroids.

Erythema multiforme

Erythema multiforme has been attributed to vancomycin (82).

- A 70-year-old woman given vancomycin postoperatively developed fever on day 5 and vancomycin was withdrawn. She then developed a generalized urticarial rash with oral and vaginal erosive lesions on day 6, requiring treatment with steroids and antihistamines. The diagnosis was confirmed by skin biopsy.

Vancomycin can cause the severe forms of erythema multiforme, both Stevens–Johnson syndrome (involvement of 10–30% of the skin surface) (83) and toxic epidermal necrolysis (more than 30% of the skin surface involved); both conditions include mucosal involvement.

- A 53-year-old white woman with liver cirrhosis took vancomycin (1 g bd) for sepsis due to methicillin-resistant *S. aureus* and a catheter-associated infection due to *Enterococcus faecalis*, developed oral and vaginal mucositis and conjunctivitis followed by a maculopapular rash (84). The diagnosis was confirmed by skin biopsy. Vancomycin was replaced by teicoplanin and corticosteroids; the symptoms disappeared within 7 days.
- A 36-year-old white woman with relapsing acute myeloid leukemia took ceftazidime (2 g tds) and aciclovir for febrile neutropenia and *Herpes labialis*. She developed an itchy rash and treatment was changed to imipenem (500 mg qds for 5 days), vancomycin (1 g bd for 3 days), and gentamicin (2 mg/kg for 3 days); chemotherapy included idarubicin, cytarabine, etoposide, ondansetron, and dexamethasone for 3 days. Within a few days the rash developed into blisters and erosions, affecting more than 80% of the skin. The diagnosis was confirmed histologically, and she subsequently died from shock (85).

Linear IgA bullous disease

Linear IgA bullous disease is an autoimmune subepidermal disorder, characterized by a linear deposition of IgA along the blister base, with a predominantly neutrophilic dermal infiltrate (86,87,88,89,90,91). Most often idiopathic, a subset of linear IgA bullous disease is induced by drugs, and intravenous vancomycin is the best-documented drug that triggers it (92). The diagnosis can be confirmed by direct immunofluorescence (93,94). In some cases the blisters resolve only after withdrawal of therapy and in others glucocorticoid therapy is required.

In a review of 17 cases the features were: mean age 65 years; 11 men/6 women; eruption latency 1–15 days; normal trough concentrations (when reported); cutaneous manifestations included bullous, urticarial, erythematous, and targetoid erythema multiforme-like lesions favoring the trunk, extremities, palms, and soles, and sparing the head and neck; six patients had mucosal involvement (95). The gold standard for diagnosis is direct immunofluorescence of the perilesional skin. Circulating antibasement membrane zone antibodies were identified only in a minority of patients. Therapies included drug withdrawal, prednisone, and dapsone. No spontaneous recurrences were reported.

Linear immunoglobulin A bullous disease has been reported in a 69-year-old white woman treated with vancomycin and ciprofloxacin, in a 53-year-old white man with septic shock, and in other cases in which both vancomycin and *Varicella zoster* infection were present as triggers (96–99).

- An 87-year-old white woman treated with vancomycin and phenytoin and a 65-year-old white man with renal insufficiency developed linear immunoglobulin A bullous disease (100,101).
- A 72-year-old woman developed a bullous eruption on her palms, soles, and conjunctivae a day after receiving vancomycin (intravenously and via a cecostomy tube) for staphylococcal sepsis (102). Immunofluorescence of the skin biopsy showed linear IgA deposition along the basement membrane. The lesions cleared over the next 2 weeks after withdrawal of vancomycin.
- A blistering eruption developed in the skin and buccal mucosa of a 79-year-old man after 8 days of vancomycin therapy for an infected leg ulcer (103). Biopsy showed linear IgA deposition along the basement membrane. The lesions cleared spontaneously within two days of withdrawal.
- After 7 days of antibiotic therapy with intravenous vancomycin, gentamicin, and ticarcillin + clavulanate for *Pseudomonas aeruginosa* and *S. aureus* sepsis, confluent erythematous-based vesicobullae developed in a 65-year-old woman with subarachnoid hemorrhage secondary to a ruptured aneurysm (93). Other medications included ranitidine, glyceryl trinitrate, nimodipine, ferrous sulfate, and phenytoin.
- On the tenth day of antibiotic therapy with vancomycin, imipenem, and gentamicin for an infected enterocutaneous fistula, widespread bullae developed in a 60-year-old woman (93).
- In a 71-year-old woman with pneumonia, vesicles and bullae developed on the eighth day of antibiotic therapy that consisted of one dose of intravenous vancomycin followed by a course of nafcillin (94).
- A 52-year-old woman was given vancomycin 1 g intravenously bd and within 12 hours developed a generalized pruritic maculopapular rash (104). Over the next few days, the lesions progressively worsened and transformed into hemorrhagic and non-hemorrhagic vesicles

and bullae. Mucosal surfaces, palms, and soles were spared. The skin lesions completely healed without scarring within 2 weeks of vancomycin withdrawal. There was no recurrence 5 months later.

- A 48 year old woman developed linear immunoglobulin A bullous disease while taking vancomycin and was rechallenged 4 years later with five doses over 5 days (105). On day 1 she received 10 mg and this was increased to 1000 mg on day 5. She had no recurrence.

Sexual function

- Prostatitis due to vancomycin-resistant enterococci has been reported in a 42-year-old liver transplant recipient (106). The organism, *Enterococcus faecium*, was resistant to vancomycin, ampicillin, ciprofloxacin, and doxycycline. Treatment with a combination of rifampicin and nitrofurantoin for 6 weeks resulted in a long-lasting cure.
- Priapism has been reported in a 37-year-old man with a 30-year history of severe diabetes mellitus after two doses of vancomycin 1 g intravenously for treatment of methicillin-resistant *S. aureus* bursitis (107).

The authors noted that an interaction between vancomycin and other medications (cefazolin, aztreonam, ciclosporin, prednisone, mycophenolate mofetil, and co-trimoxazole) could not be ruled out.

Immunologic

Allergic reactions with vancomycin, such as rashes, chills, fever, and eosinophilia, can occur in up to 5% of patients. Severe anaphylactic reactions are rare. Alpha-tryptase is only detected in the blood during systemic anaphylactic reactions, having been released by degranulation from activated mast cells. Plasma tryptase activities were unchanged, independent of increased histamine concentrations, antihistamine pretreatment, and clinical symptoms of anaphylactoid reaction in 40 patients receiving vancomycin (1 g over 10 minutes) before elective arthroplasty (108). The authors conclude that plasma tryptase activities can be used to distinguish chemical from immunological reactions.

Vancomycin anaphylaxis is a major management problem in patients with methicillin-resistant *S. aureus* sepsis. However, desensitization in patients with previous anaphylaxis is possible (109).

- An anaphylactic reaction occurred in a 77-year-old woman 5 minutes after the start of a vancomycin infusion, when she had received only 40 mg (110). She became unconscious and had a severe cardiovascular collapse, from which she was resuscitated with intravenous ephedrine and adrenaline.
- A 47-year-old white woman with end-stage renal disease had had anaphylactoid shock after vancomycin 1 g intravenously infused over 1.5 hours and gentamicin 90 mg 3 years before, despite premedication with diphenhydramine (111). She was treated with doubling doses of vancomycin every 30 minutes for methicillin-resistant *S. epidermidis*. She had no reaction.

The authors of the second report could not exclude that the previous anaphylactoid reaction had not been due to gentamicin, as no specific testing was done. Although successful vancomycin desensitization has been described, this would be the first time in a patient with a history of anaphylactoid reaction.

In some patients the clinical presentation of red man syndrome is identical to that of acute IgE-mediated anaphylaxis. Vancomycin desensitization should therefore be considered for severe red man syndrome reactions that do not respond to premedication and a slower rate of infusion, and in anaphylactic reactions to vancomycin when substitution of another antibiotic is not feasible. Rapid desensitization is preferred, as it is effective in the majority of patients and enables administration of vancomycin within 24 hours. In patients who fail rapid desensitization, a slow desensitization protocol may be tried (112).

Vasculitic rashes have been described rarely. Two case reports have suggested that there may be cross-reactivity between vancomycin and teicoplanin with respect to biopsy-proven leukocytoclastic vasculitis (113). In both cases, vancomycin-induced vasculitis improved after drug withdrawal. Teicoplanin was started and the rash reappeared several days later. In one case the rash faded after teicoplanin had been withdrawn. In the other, teicoplanin was continued, but the rash improved after prednisolone was given.

- A 56-year-old white woman developed a drug-induced hypersensitivity syndrome after receiving intravenous vancomycin 2 g/day for 20 days for methicillin-resistant *Staphylococcus aureus* (114). She developed a fever and a maculopapular rash, which progressed to purpuric erythema multiforme without mucosal lesions, associated with pharyngitis, leukocytosis, and eosinophilia. There was organ involvement, with raised liver enzymes and creatinine. She was given methylprednisolone for 2 month until the laboratory findings and cutaneous lesions normalized.
- In a 45 year old Caucasian woman acute interstitial nephritis and hepatitis did not improve after withdrawal of vancomycin and empirical glucocorticoid administration, but the rash resolved and renal function recovered after a 5-day course of ciclosporin (115).
- A 76-year-old man developed progressive itching, soreness, photophobia, and lacrimation in the left eye 3 days after completing a 2-week course of 5% vancomycin eye-drops after cataract surgery (116). Vancomycin skin tests were positive and his symptoms resolved after treatment with an antihistamine and systemic and topical glucocorticoids.
- A burned child treated with vancomycin developed an anaphylactic reaction (117).

Long-Term Effects

Drug resistance

The increasing rate of antimicrobial drug resistance has been a matter of great concern and is the most serious unwanted effect of these drugs. However, in spite of the fact that the problem is now very well recognized (SEDA-25, 279), it continues to grow.

History

Most antimicrobial drugs are natural products, that is they are produced by micro-organisms such as bacteria or fungi, often found in the soil. In fact, they can be looked upon as nature's regulatory principle for microbial society. Resistance to antimicrobial drugs is therefore a natural phenomenon. Before the introduction of penicillin in the 1940s, resistance to antimicrobial drugs was not a clinical problem. At that time, the large majority of commensal and infectious bacteria associated with infections in man were susceptible.

Over the last six decades, however, increased use of antimicrobial drugs, not only in human medicine, but in other areas, such as veterinary medicine, agriculture, and fish farming, has had an enormous impact on the microbial society. Nearly everywhere, the numbers of susceptible strains have reduced and resistant strains or variants have increased in numbers. It has been repeatedly reported that the susceptibility profile of bacteria in any human compartment, such as the skin, intestine, and respiratory tract, is very different from what it was in the pre-antibiotic era, and even 15 years ago. The same trend is reported from hospitals and homes. Multidrug resistance, that is resistance to several antimicrobial drugs, is commonly found in bacteria that cause infections as well as in commensal organisms.

A few decades ago, it was a common opinion that various compartments in nature have their own flora. As an example, it was claimed that you could use antimicrobial drugs relatively freely in fish farming without increasing the burden of resistance in humans. Now we have learned the lesson. Micro-organisms circulate everywhere, and there is a continuous exchange of strains between all compartments in nature (humans, animals, birds, fish, etc.). Even if a bacterial species is host-specific, the genetic material that codes for resistance is not. In fact, antibiotics have shown that bacteria have great genetic adaptability, in terms of their ability to exchange genetic traits among genera and species which are evolutionarily millions of years apart. Antibiotic resistance genes on plasmids and transposons flow to and from nearly all types of bacteria. Sometimes they leave the plasmid and jump into bacterial chromosomes; sometimes they jump back again.

However, this knowledge is not being heeded everywhere. Small doses of antimicrobial drugs as "growth promoters" are still commonly used, even in countries in which the health authorities should be aware of the problems. It is easy to blame developing countries for using antimicrobial drugs as growth promoters, or for selling antimicrobial drugs over the counter without prescription, but it took the European Community many years before it started to look into the problem of using antimicrobial drugs as growth promoters. The history of cross-resistance between avoparcin and vancomycin may have provided important background for this alteration in attitude.

Avoparcin and vancomycin are glycopeptide antibiotics, large molecules that are produced by a variety of environmental micro-organisms, which may therefore contain genes that code for antimicrobial drug resistance. Both of these drugs are mostly active against Gram-positive bacteria, such as enterococci and staphylococci. In Europe, avoparcin was allowed to be used as an animal food additive in many countries, while the use of vancomycin was limited to humans. Nobody bothered about the possibility of cross-resistance between avoparcin and vancomycin until 10 years ago. After the emergence of vancomycin-resistant enterococci and after more than 2 years of hard lobbying by several groups, avoparcin was withdrawn from the market in the European Community. However, in the meantime, vancomycin-resistant enterococci had become widespread in many European countries.

Instead of focusing on the development of resistance to a specific antimicrobial drug in a specific species, we should focus on the microbial community as a single entity or a "metagenome". Any use of any microbial agent might cause resistance to develop in one or more microbial species. When such genes have first become established, they may float around and be picked up by other species. This approach to the development and spread of resistance can, and should, be applied to the microbial flora in all mammals, as well as in the environment. The consequence of this approach is that we should, every time we prescribe an antimicrobial drug, try to find a drug that hits nothing but the pathogen in the infected organ(s). Of course, this can be difficult and sometimes even close to impossible. But that should not stop us from trying. For example, in nearly all cases, a third-generation cephalosporin for acute pharyngitis caused by Group A streptococci, or of a fluoroquinolone for acute cystitis, are not the best alternatives.

Ecology
In our struggle for prudent use of antimicrobial drugs, the so-called eco-shadow concept represents a challenging way of following alterations in mammalian and environmental ecosystems produced by exposure of these systems to antibiotics (118). In brief, an eco-shadow (ecological shadow) can be defined as any alteration in any ecosystem following exposure of the system to an antimicrobial drug (or any compound that influences the system). The alterations can be of variable length and may involve a variable number of species and functions, including the development and spread of resistance. When possible, the sum of all the alterations caused by an antibiotic can be included in an "eco-shadow index." A large eco-shadow or high eco-shadow index indicates a high tendency to the development of microbial resistance.

At present, it seems that Europe has taken the lead in the fight against antimicrobial drug resistance. At the Fourth European Conference on Antibiotic Resistance, organized by the European Commission and held in Rome in November 2003, it was strongly emphasized that only a multidisciplinary approach, involving all stakeholders—physicians, researchers, industry, politicians, and patients—can overcome the problem of increasing

resistance. It is to be hoped that this problem will be established as a major objective for the forthcoming European Centre for Disease Prevention and Control.

Reports of organisms that are resistant to vancomycin
The first case of methicillin-resistant *S. aureus* with intermediate resistance to vancomycin was documented in May 1996. Additional cases have been described in the USA:

- a 59-year-old man with diabetes mellitus and chronic renal insufficiency suffering from peritonitis;
- a 66-year-old man with diabetes who developed a septicemia after 18 weeks of vancomycin treatment for recurrent methicillin-resistant *S. aureus* bacteremia;
- a 79-year-old man with diabetes and chronic renal insufficiency, who developed intermediate vancomycin-resistance after a 6-week therapy with vancomycin for MRSA bloodstream infection.

In the last case the strain was identical to eight MRSA isolates obtained from hospitals in the New York City metropolitan area, and all eight isolates, but not control isolates, could be transformed in vitro to develop intermediate resistance to vancomycin. Both the presence of glycopeptides and environmental factors, as demonstrated by increased resistance of *S. aureus* to antibiotics in the presence of prosthetic material in animals, can exert selective pressure to develop new resistance mechanisms (119–121).

There are reports of vancomycin-dependent enterococci that grow only in the presence of vancomycin and are best treated by withdrawal of the compound (122).

Vancomycin/glycopeptide intermediately resistant and methicillin-resistant *S. aureus* have been described in Japan, Europe, the Far East, and the USA. Some vancomycin-susceptible strains of *S. aureus* contain subpopulations with intermediate resistance to vancomycin (heterogeneous strains), and these may escape laboratory detection (119,120,123–125).

Numerous risk factors have been identified for infection with vancomycin-resistant enterococci. In one study the severity of mucositis in cancer patients was significantly associated with vancomycin-resistant enterococci (126). Previous vancomycin therapy was also believed to be a risk factor. However, a meta-analysis concluded that the reported strong association between vancomycin treatment and hospital-acquired vancomycin-resistant enterococci results from selection bias, confounding by duration of hospitalization, and publication bias (127).

Glycopeptide resistance in a cluster of three clinical isolates of vancomycin-resistant *Enterococcus faecium* was due to vanD, which was located on the chromosome and was not transferable to other enterococci. These isolates were indistinguishable but differed from the strain in which vanD-mediated resistance has been reported previously (128). A type of acquired glycopeptide resistance, named vanE, has been characterized in *E. faecalis* BM 4405. It results in low-level resistance of vancomycin but susceptibility to teicoplanin (129). Defects in penicillin-binding protein 4 result in a distorted peptidoglycan composition of the cell of vancomycin-resistant and teicoplanin-resistant laboratory mutants of *S. aureus* and are suggested to be part of the mechanism of glycopeptide resistance in these microbes (130).

Vancomycin-dependent enterococci growing only in the presence of vancomycin are best treated by withdrawing vancomycin (122). However, vancomycin-independent revertants can rapidly emerge in vitro, endangering the cure of patients treated by interruption of vancomycin only (131).

In a study of antibiotic resistance in enterococci from raw meat, there was a high prevalence of glycopeptide-resistant strains (132). Resistance to vancomycin was significantly associated with resistance to teicoplanin, erythromycin, tetracycline, and chloramphenicol.

The prevalence of vancomycin-resistant enterococci (VRE) has been investigated in 49 laboratories from 27 European countries, which collected 4208 clinical isolates; 18 vanA and 5 vanB isolates of VRE were identified (133). The prevalence of vanA VRE was highest in the UK (2.7%), while the prevalence of vanB VRE was highest in Slovenia (2%). Most vanA and vanB vancomycin-resistant enterococci were identified as *E. fecium*. A total of 71 isolates containing the vanC gene were identified. The prevalence of vanC VRE was highest in Latvia and Turkey, where rates were 14 and 12% respectively. Two-thirds of these isolates were identified as *Enterococcus gallinarum* and one-third as *Enterococcus casseliflavus*.

In 155 methicillin-resistant strains of *S. aureus* isolated from patients in Thailand, one-point population analysis identified three methicillin-resistant strains, which contained subpopulations of cells with reduced susceptibility to vancomycin (134).

Susceptibility Factors

Age

The pharmacokinetics of vancomycin in neonates is mainly determined by postconceptional age and renal function (135). In neonates, a patent ductus arteriosus and treatment with indometacin or extracorporeal membrane oxygenation leads to an increased volume of distribution and a reduced clearance. Vancomycin-related nephrotoxicity and ototoxicity in neonates is rare, and there is no clear relation to serum concentrations. Recent guidelines have suggested that dosage can be independent of gestational age or postconceptional age in neonates without renal insufficiency (135).

Renal disease

Vancomycin is eliminated almost exclusively by renal excretion. In oliguria 1 g can produce therapeutic plasma concentrations for 10–14 days. Hemodialysis fails to remove vancomycin from the body to any significant extent. If renal function is compromised, even oral therapy with vancomycin can lead to high and potentially toxic serum and CSF drug concentrations (136).

Drug Administration

Drug contamination

Vancomycin underwent spontaneous chemical modification when kept at room temperature at neutral pH in aqueous solutions containing traces of formaldehyde or acetaldehyde (137). In vitro studies on two different strains of bacteria showed that the resulting compounds had reduced antibacterial activity.

Drug administration route

In a multicenter, prospective, randomized comparison of continuous infusion of vancomycin (targeted serum plateau concentrations of 20–25 µg/ml) and intermittent infusions of vancomycin (targeted serum trough concentrations of 10–15 µg/ml) in 119 critically ill patients, microbiological and clinical outcomes and safety were similar (138). Continuous infusion of vancomycin produced the target concentration faster, and fewer samples were required for monitoring. Serum creatinine concentrations and creatinine clearance increased non-significantly from baseline to the end of treatment in the two groups. There was nephrotoxicity in 21 patients (10 of whom received the continuous infusion). Vancomycin given concomitantly with other antibiotics was associated with a significant increase in the serum creatinine concentration, but not when it was given alone. Dialysis was required for three of those who were given intermittent infusions and six of those who were given continuous infusions. Red man syndrome was reported in one patient who was given an intermittent infusion, and phlebitis and fever in two patients who were given continuous infusions.

Drug overdose

Two premature infants received 10-fold overdoses of vancomycin, with resulting peak plasma concentrations over 300 µg/ml (139). Vancomycin was withdrawn and the plasma vancomycin concentration fell to below 40 µg/ml at under 48 hours in one infant and under 72 hours in the other. Although one infant had a transient increase in serum creatinine neither had further evidence of renal, auditory, or other toxicity acutely or in the long term.

- Accidental 10-fold vancomycin overdose in a 6-month-old girl resulted in flushing and a transient rise in serum creatinine concentration, but she recovered completely without specific therapeutic intervention (140).
- Two preterm twins, born in the 35th week of gestation, accidentally received an overdose of vancomycin as an intravenous bolus injection (141). Both were treated from the day of birth with piperacillin (100 mg/kg/day) and cefotaxime (100 mg/kg/day) for congenital pneumonia. On the 6th day of antibiotic therapy, vancomycin (38 and 35 mg/kg respectively) was given over 1 minute. A few minutes later the twins developed flushed faces and trunks and peripheral cyanosis. Other findings were a prolonged capillary refill time, apnea and hypoxemia, bradycardia (40–60/minute), a

fall in systolic blood pressure of 10 mmHg, and metabolic acidosis (pH 7.29 and 7.24 and base excess −10.5 and −10.9 respectively). Metabolic acidosis and the red man syndrome disappeared after 30 minutes. Serum vancomycin concentrations were 32 and 34.5 µg/ml respectively 9 hours after administration. Fundoscopy and repeated otoacoustic emissions were normal. There was minimal reduction in creatinine clearance. Tubular proteinuria, paralleled by increased activity of N-acetyl-beta-D-glucosaminidase in the urine, resolved completely within 1 week.

- Hemodialysis with high-efficiency dialysis membranes resulted in a removal of plasma vancomycin of about 60% (calculated half-life 2 hours) in two children with initial plasma vancomycin concentrations of 238 µg/ml and 182 µg/ml (142).
- In a 6-day-old neonate with a solitary hypodysplastic kidney, suspected sepsis, and acute renal insufficiency, venovenous hemodiafiltration with a high-flux membrane was successfully used to treat a 10-fold overdose of vancomycin (143).

Multiple-dose activated charcoal does not increase the elimination of vancomycin and is therefore not recommended for detoxification of acute poisoning (144).

Drug–Drug Interactions

Aminoglycoside antibiotics

Vancomycin is more nephrotoxic than teicoplanin when it is combined with aminoglycosides (145).

Combined use of vancomycin and an aminoglycoside can increase the risk of toxicity (164–166).

Beta-lactam antibiotics

Vancomycin and ceftazidime are incompatible in vitro because of precipitation due to the alkaline pH of vancomycin relative to ceftazidime. This phenomenon was encountered in two cases of post-traumatic endophthalmitis (146). Immediately on local administration of the antibiotics, which were injected using different needles and syringes for each drug, dense yellow-white precipitates were observed along the needle tract in the vitreous cavity. During follow-up the vitreous opacities gradually disappeared over a period of 2 months, with complete resolution.

Co-administration of imipenem + cilastatin may reduce the nephrotoxicity of vancomycin, as suggested by a study in rats (147).

Cephalosporins

Ceftazidime formed microprecipitates when it was mixed with vancomycin (167).

Dexamethasone

Intravitreal application of dexamethasone may reduce the elimination of intravitreal vancomycin and may enhance this therapy, as suggested by an in vivo study in rabbits (148).

Dobutamine

In critically ill patients, co-administration with dobutamine can enhance vancomycin clearance by increasing cardiac output and renal blood flow and by interacting with the renal anion transport system, increasing glomerular filtration rate and renal tubular secretion (149).

Dopamine

In critically ill patients, co-administration with dopamine can enhance vancomycin clearance by increasing cardiac output and renal blood flow and by interacting with the renal anion transport system, increasing glomerular filtration rate and renal tubular secretion (149).

Fosfomycin

Co-administration of fosfomycin may reduce the nephrotoxicity of vancomycin, as suggested by a study in rats (147).

Furosemide

Preterm infants often receive theophylline for apnea of prematurity and furosemide to improve dynamic lung compliance and to reduce total pulmonary resistance. In 10 preterm infants receiving furosemide plus theophylline, vancomycin was introduced according to the generally recommended dosing schedules but failed to achieve anticipated therapeutic concentrations (150). In nine infants the addition of furosemide to vancomycin plus theophylline resulted in a fall in serum vancomycin concentrations to subtherapeutic within 24 hours, and only in one did vancomycin concentrations remain within the target range. The addition of furosemide reduced the serum creatinine concentration by 24% (range 12–43). These data suggest that furosemide enhances the excretion of vancomycin, and an acute and transient increase in glomerular filtration rate and creatinine clearance may be the explanation.

Ten preterm infants receiving regular theophylline for apnea of prematurity, who subsequently receive vanomycin and furosemide, have been studied (168). When vancomycin was introduced in the infants who were established on furosemide and theophylline, there was a consistent failure to achieve therapeutic concentrations. Starting furosemide in infants who were already receiving vancomycin resulted in falls in serum vancomycin to subtherapeutic concentrations in all but one case. Serum concentrations fell by a mean of 24% (range 12–43%), in the 24 hours after the start of furosemide treatment. Two of the 10 infants had persistence of coagulase-negative staphylococcal sepsis while vancomycin concentrations were suboptimal. While the mechanism of this interaction is uncertain, the changes in serum vancomycin concentration may indicate acute changes in glomerular filtration rate. Preterm infants receiving theophylline and furosemide need shorter vancomycin dosage intervals to avoid therapeutic failure.

Gelatin

Vancomycin is incompatible with gelatin-containing fluids, resulting in precipitation (151).

Levofloxacin

Changes in pharmacokinetics were studied in male Wistar rats when intravenous vancomycin 100 mg/kg and levofloxacin 20 mg/kg were administered together (152). There was an increase in the AUC and half-life of vancomycin. There was also an increase in the AUC and a delay in the t_{max} of levofloxacin, but no effect on C_{max}; these data suggested delayed absorption of levofloxacin. Concomitant administration had no effect on the correlation between serum and hepatic tissue concentrations of levofloxacin, but it markedly reduced the correlation between the serum and renal tissue concentrations of vancomycin. Vancomycin increased serum creatinine concentrations 8 hours after administration. However, there was no difference in animals who received monotherapy compared with animals who received combined therapy. The authors suggested the cautious use of a combination of levofloxacin and vancomycin and advised monitoring blood concentrations of vancomycin in such cases.

Neuromuscular blocking drugs

Some authors stress the fact that vancomycin can interact with anesthetic drugs, particularly muscle relaxants. In the reported cases anaphylactoid reactions were seen, with intense erythema and marked permeability changes (153).

Monitoring Therapy

There is evidence that predose concentrations of teicoplanin are related to clinical outcome; the evidence is less conclusive for vancomycin. Predose concentrations can be linked to some toxic effects (nephrotoxicity for vancomycin and thrombocytopenia to teicoplanin) (SEDA23.26.77) (154,155). Monitoring of serum concentrations has therefore been advocated to reduce potential nephrotoxicity and ototoxicity due to interpatient variability in vancomycin pharmacokinetics. However, routine monitoring is probably unnecessary in most patients (3,156). Clinical settings in which monitoring is wise include combination therapies with an aminoglycoside and treatment of patients with poor or unstable renal function (2). Vancomycin-induced nephrotoxicity is clearly related to drug plasma concentrations: trough vancomycin concentrations over 15 μg/ml were associated with significantly more nephrotoxicity (157).

In a prospective study to determine if standardized vancomycin doses could produce adequate serum concentrations in 25 full-term neonates with sepsis 13 had adequate peak vancomycin serum concentrations (20–40 mg/ml) and one had a peak concentration with a risk of ototoxicity (over 40 μg/ml) (158). Only 12 had adequate trough concentrations (5–10 mg/ml) and 7 had a risk of nephrotoxicity (over 10 μg/ml). There was no significant difference between peak or trough concentrations and good or bad clinical outcomes.

Intravenous vancomycin 30–40 mg/kg/day, divided every 6 hours, was given to167 infants and children without cancer and 42 cancer patients aged 3 months to 17.5 years with normal serum creatinine concentrations and without evidence of renal dysfunction (159). In 93% of the children without cancer, peak serum vancomycin concentrations were in an adequate target range (8–55 μg/ml). However, 10% of the children with cancers had peak serum vancomycin concentrations below 10 μg/ml, and 21% had trough concentrations less than 5 μg/ml; an increased dosage was required to achieve adequate peak concentrations. There were no treatment failures. The authors suggested the following guidelines:

- the serum creatinine concentration should be measured within 24 hours of beginning therapy, to verify normal renal function, and weekly thereafter to detect an eventual reduction in creatinine clearance;
- an increase of serum creatinine of more than 44 μmol/l (0.5 mg/dl) over baseline should be followed by monitoring serum vancomycin concentrations;
- peak and trough vancomycin serum concentrations measurements are not necessary in children without cancer receiving 40 mg/kg/day or less who have normal renal function and do not receive other potentially nephrotoxic drugs; this approach should be individualized, based on factors such as degree of illness, poor clinical response, or persistent positive cultures;
- monitoring should be considered for patients with cancer, neonates, those receiving concurrent nephrotoxic drugs, patients with renal insufficiency, and for dosages over 40 mg/kg/day.

Fluorescence polarization immunoassay determination of serum vancomycin concentrations can result in falsely high vancomycin concentrations in excess of 30–80% in patients with renal dysfunction. This is due to the formation of a non-toxic, non-microbiologically active pseudometabolite, the vancomycin crystalline degradation product (160). A report on a 48-year-old man underlies the significance of resulting underdosing and eventually suboptimal clinical response (161).

Managing Adverse Effects

Vancomycin clearance is minimally altered by hemodialysis using standard cuprophane membranes. However, it is significantly increased using a high-flux polysulfone membrane.

- A 17-year-old anuric woman with end-stage renal insufficiency received a massive overdose of vancomycin (40 mg/kg/day for 8 days) and was treated three times with high-flux hemodiafiltration with a polysulfone membrane (162). The vancomycin concentration fell from 101 to 17 mg/l at the end of the procedure. There were no adverse effects of either vancomycin or hemodiafiltration.
- In a 46-year-old African-American woman with sickle-cell disease there was clinically significant removal of vancomycin during a plasma exchange transfusion (163).

References

1. Griffith RS. Introduction to vancomycin. Rev Infect Dis 1981;3(Suppl):S200–4.

2. Moellering RC Jr. Monitoring serum vancomycin levels: climbing the mountain because it is there? Clin Infect Dis 1994;18(4):544–6.

3. Cantu TG, Yamanaka-Yuen NA, Lietman PS. Serum vancomycin concentrations: reappraisal of their clinical value. Clin Infect Dis 1994;18(4):533–43.

4. Anderson JL, Muhlestein JB, Carlquist J, Allen A, Trehan S, Nielson C, Hall S, Brady J, Egger M, Horne B, Lim T. Randomized secondary prevention trial of azithromycin in patients with coronary artery disease and serological evidence for *Chlamydia pneumoniae* infection: The Azithromycin in Coronary Artery Disease: Elimination of Myocardial Infection with Chlamydia (ACADEMIC) study. Circulation 1999;99(12):1540–7.

5. Hospital Infection Control Practices Advisory Committee (HICPAC). Recommendations for preventing the spread of vancomycin resistance. Am J Infect Control 1995;23(2):87–94.

6. Boselli E, Allaouchiche B. Diffusion osseuse des antibiotiques. [Diffusion in bone tissue of antibiotics.] Presse Méd 1999;28(40):2265–76.

7. Leitch EC, Willcox MD. Lactoferrin increases the susceptibility of *S. epidermidis* biofilms to lysozyme and vancomycin. Curr Eye Res 1999;19(1):12–9.

8. Elting LS, Rubenstein EB, Kurtin D, Rolston KV, Fangtang J, Martin CG, Raad II, Whimbey EE, Manzullo E, Bodey GP. Mississippi mud in the 1990s: risks and outcomes of vancomycin-associated toxicity in general oncology practice. Cancer 1998;83(12):2597–607.

9. Cohen E, Dadashev A, Drucker M, Samra Z, Rubinstein E, Garty M. Once-daily versus twice-daily intravenous administration of vancomycin for infections in hospitalized patients. J Antimicrob Chemother 2002;49(1):155–60.

10. Vuagnat A, Stern R, Lotthe A, Schuhmacher H, Duong M, Hoffmeyer P, Bernard L. High dose vancomycin for osteomyelitis: continuous vs. intermittent infusion. J Clin Pharm Ther 2004;29(4):351–7.

11. Fortun J, Navas E, Martinez-Beltran J, Perez-Molina J, Martin-Davila P, Guerrero A, Moreno S. Short-course therapy forside endocarditis due to Staphylococcus aureus in drug abusers: cloxacillin versus glycopeptides in combination with gentamicin. Clin Infect Dis 2001;33(1):120–5.

12. Sandler RH, Finegold SM, Bolte ER, Buchanan CP, Maxwell AP, Vaisanen ML, Nelson MN, Wexler HM. Short-term benefit from oral vancomycin treatment of regressive-onset autism. J Child Neurol 2000;15(7):429–35.

13. Beringer PM, Wong-Beringer A, Rho JP. Economic aspects of antibacterial adverse effects. Pharmacoeconomics 1998;13(1 Pt 1):35–49.

14. Vazquez L, Encinas MP, Morin LS, Vilches P, Gutierrez N, Garcia-Sanz R, Caballero D, Hurle AD. Randomized prospective study comparing cost-effectiveness of teicoplanin and vancomycin as second-line empiric therapy for infection in neutropenic patients. Haematologica 1999;84(3):231–6.

15. Bucaneve G, Menichetti F, Del Favero A. Cost analysis of 2 empiric antibacterial regimens containing glycopeptides for the treatment of febrile neutropenia in patients with acute leukaemia. Pharmacoeconomics 1999;15(1):85–95.

16. Sivagnanam S, Deleu D. Red man syndrome. Crit Care 2003;7:119–20.

17. Tilles SA, Slatore CG. Hypersensitivity reactions to non-beta-lactam antibiotics. Clin Rev Allergy Immunol 2003;24:221–8.

18. Newfield P, Roizen MF. Hazards of rapid administration of vancomycin. Ann Intern Med 1979;91(4):581.

19. Davis RL, Smith AL, Koup JR. The "red man's syndrome" and slow infusion of vancomycin. Ann Intern Med 1986;104(2):285–6.

20. Healy DP, Sahai JV, Fuller SH, Polk RE. Vancomycin-induced histamine release and "red man syndrome": comparison of 1- and 2-hour infusions. Antimicrob Agents Chemother 1990;34(4):550–4.

21. Toyoguchi T, Ebihara M, Ojima F, Hosoya J, Shoji T, Nakagawa Y. Histamine release induced by antimicrobial agents and effects of antimicrobial agents on vancomycin-induced histamine release from rat peritoneal mast cells. J Pharm Pharmacol 2000;52(3):327–31.

22. Polk RE. Anaphylactoid reactions to glycopeptide antibiotics. J Antimicrob Chemother 1991;27(Suppl B):17–29.

23. Bergeron L, Boucher FD. Possible red-man syndrome associated with systemic absorption of oral vancomycin in a child with normal renal function. Ann Pharmacother 1994;28(5):581–4.

24. Renz CL, Thurn JD, Finn HA, Lynch JP, Moss J. Oral antihistamines reduce the side effects from rapid vancomycin infusion. Anesth Analg 1998;87(3):681–5.

25. Johnson JR, Burke MS, Mahowald ML, Ytterberg SR. Life-threatening reaction to vancomycin given for noninfectious fever. Ann Pharmacother 1999;33(10):1043–5.

26. Hui YL, Yu CC, Ng YT, Lau WM, Hsieh JR, Chung PC. Red Man's syndrome following administration of vancomycin in a patient under spinal anesthesia—a case report. Acta Anaesthesiol Sin 2002;40(3):149–51.

27. Bertolissi M, Bassi F, Cecotti R, Capelli C, Giordano F. Pruritus: a useful sign for predicting the haemodynamic changes that occur following administration of vancomycin. Crit Care 2002;6(3):234–9.

28. Renz CL, Thurn JD, Finn HA, Lynch JP, Moss J. Antihistamine prophylaxis permits rapid vancomycin infusion. Crit Care Med 1999;27(9):1732–7.

29. Aissaoui M, Mathelier-Fusade P, Leynadier F. Le syndrome vancomycine/narcotiques. [The vancomycin/narcotic syndrome.] Presse Méd 1999;28(39):2152.

30. Shuto H, Sueyasu M, Otsuki S, Hara T, Tsuruta Y, Kataoka Y, Oishi R. Potentiation of vancomycin-induced histamine release by muscle relaxants and morphine in rats. Antimicrob Agents Chemother 1999;43(12):2881–4.

31. Garrelts JC, Smith DF Jr, Ast D, LaRocca J, Peterie JD. Phlebitis associated with vancomycin therapy. Clin Pharm 1988;7(10):720–1.

32. Hadaway L, Chamallas SN. Vancomycin: new perspectives on an old drug. J Infus Nurs 2003;26(5):278–84.

33. Brummett RE. Ototoxicity of vancomycin and analogues. Otolaryngol Clin North Am 1993;26(5):821–8.

34. Mellor JA, Kingdom J, Cafferkey M, Keane CT. Vancomycin toxicity: a prospective study. J Antimicrob Chemother 1985;15(6):773–80.

35. Cook FV, Farrar WE Jr. Vancomycin revisited. Ann Intern Med 1978;88(6):813–8.

36. Axer-Siegel R, Stiebel-Kalish H, Rosenblatt I, Strassmann E, Yassur Y, Weinberger D. Cystoid macular edema after cataract surgery with intraocular vancomycin. Ophthalmology 1999;106(9):1660–4.

37. Gendeh BS, Gibb AG, Aziz NS, Kong N, Zahir ZM. Vancomycin administration in continuous ambulatory peritoneal dialysis: the risk of ototoxicity. Otolaryngol Head Neck Surg 1998;118(4):551–8.

38. Klibanov OM, Filicko JE, DeSimone JA, Jr., Tice DS. Sensorineural hearing loss associated with intrathecal vancomycin. Ann Pharmacother 2003;37:61–5.

39. Schaefer F, Klaus G, Muller-Wiefel DE, Mehls OThe Mid-European Pediatric Peritoneal Dialysis Study Group (MEPPS). Intermittent versus continuous intraperitoneal glycopeptide/ceftazidime treatment in children with peritoneal dialysis-associated peritonitis. J Am Soc Nephrol 1999;10(1):136–45.

40. Panzarino VM, Feldstein TJ, Kashtan CE. Charcoal hemoperfusion in a child with vancomycin overdose and chronic renal failure. Pediatr Nephrol 1998;12(1):63–4.

41. Gavazzi C, Stacchiotti S, Cavalletti R, Lodi R. Confusion after antibiotics. Lancet 2001;357(9266):1410.

42. Lambotte O, Costedoat-Chalumeau N, Amoura Z, Piette JC, Cacoub P. Drug-induced hemophagocytosis. Am J Med 2002;112(7):592–3.

43. Kamble RT, Hamadani M, Selby GB. Delayed myeloid engraftment due to vancomycin in allogeneic haematopoietic stem cell transplant recipients. J Antimicrob Chemother 2006;57(4):795–6.

44. Rocha JL, Kondo W, Baptista MI, Da Cunha CA, Martins LT. Uncommon vancomycin-induced side effects. Braz J Infect Dis 2002;6(4):196–200.

45. Schwartz MD. Vancomycin-induced neutropenia in a patient positive for an antineutrophil antibody. Pharmacotherapy 2002;22(6):783–8.

46. Shahar A, Berner Y, Levi S. Fever, rash, and pancytopenia following vancomycin rechallenge in the presence of ceftazidime. Ann Pharmacother 2000;34(2):263–4.

47. Petit N, Rohrbach P, Mathieu P, Cronier B. Neutropénie a la vancomycine confirmée par une épreuve de réintroduction positive. Med Mal Infect 2000;20:665–6.

48. Weitzman SA, Stossel TP. Drug-induced immunological neutropenia. Lancet 1978;1(8073):1068–72.

49. Henry K, Steinberg I, Crossley KB. Vancomycin-induced neutropenia during treatment of osteomyelitis in an outpatient. Drug Intell Clin Pharm 1986;20(10):783–5.

50. Eich G, Neftel KA. Hämatologische Nebenwirkungen der antiinfektiösen Therapie. Infection 1991;19:S1–S35.

51. Adrouny A, Meguerditchian S, Koo CH, Gadallah M, Rasgon S, Idroos M, Oppenheimer E, Glowalla M. Agranulocytosis related to vancomycin therapy. Am J Med 1986;81(6):1059–61.

52. West BC. Vancomycin-induced neutropenia. South Med J 1981;74(10):1255–6.

53. Koo KB, Bachand RL, Chow AW. Vancomycin-induced neutropenia. Drug Intell Clin Pharm 1986;20(10):780–2.

54. Pai MP, Mercier RC, Koster SA. Epidemiology of vancomycin-induced neutropenia in patients receiving home intravenous infusion therapy. Ann Pharmacother 2006;40(2):224–8.

55. Segarra-Newnham M, Tagoff SS. Probable vancomycin-induced neutropenia. Ann Pharmacother 2004;38(11):1855–9.

56. Jo YI, Yoon JH, Shin SY, Chang WC, Kim BK, Jin CJ, Song JO. Agranulocytosis induced by vancomycin in an ESRD patient on CAPD. Korean J Intern Med 2004;19(1):58–61.

57. Young LS. Hematologic effects of linezolid versus vancomycin. Clin Infect Dis 2004;38(8):1065–6.

58. Sanche SE, Dust WN, Shevchuk YM. Vancomycin-induced neutropenia resolves after substitution with teicoplanin. Clin Infect Dis 2000;31(3):824–5.

59. Smith PF, Taylor CT. Vancomycin-induced neutropenia associated with fever: similarities between two immune-mediated drug reactions. Pharmacotherapy 1999;19(2):240–4.

60. George JN, Raskob GE, Shah SR, Rizvi MA, Hamilton SA, Osborne S, Vondracek T. Drug-induced thrombocytopenia: a systematic review of published case reports. Ann Intern Med 1998;129(11):886–90.

61. Kuruppu JC, Le TP, Tuazon CU. Vancomycin-associated thrombocytopenia: case report and review of the literature. Am J Hematol 1999;60(3):249–50.

62. Govindarajan R, Baxter D, Wilson C, Zent C. Vancomycin-induced thrombocytopenia. Am J Hematol 1999;62(2):122–3.

63. Peel RK, Sykes A, Ashmore S, Turney JH, Woodrow G. A case of immune thrombocytopenic purpura from intraperitoneal vancomycin use. Perit Dial Int 2003;23:506–8.

64. Marraffa J, Guharoy R, Duggan D, Rose F, Nazeer S. Vancomycin-induced thrombocytopenia: a case proven with rechallenge. Pharmacotherapy 2003;23:1195–8.

65. Bay A, Oner AF, Dogan M, Caksen H. A child with vancomycin-induced thrombocytopenia. J Emerg Med 2006;30(1):99–100.

66. Cawley MJ, Wittbrodt ET, Boyce EG, Skaar DJ. Potential risk factors associated with thrombocytopenia in a surgical intensive care unit. Pharmacotherapy 1999;19(1):108–13.

67. Balduini CL, Salvaneschi L, Klersy C, Noris P, Mazzucco M, Rizzuto F, Giorgiani G, Perotti C, Stroppa P, Pumpo MD, Nobili B, Locatelli F. Factors influencing post-transfusional platelet increment in pediatric patients given hematopoietic stem cell transplantation. Leukemia 2001;15(12):1885–91.

68. Hurley BW, Nguyen CC. The spectrum of pseudomembranous enterocolitis and antibiotic-associated diarrhea. Arch Intern Med 2002;162(19):2177–84.

69. Schenfeld LA, Pote HH Jr. Diarrhea associated with parenteral vancomycin therapy. Clin Infect Dis 1995;20(6):1578–9.

70. Cadle RM, Mansouri MD, Darouiche RO. Vancomycin-induced elevation of liver enzyme levels. Ann Pharmacother 2006;40(6):1186–9.

71. Colares VS, Oliveira RB, Abdulkader RC. Nephrotoxicity of vancomycin in patients with normal serum creatinine. Nephrol Dial Transplant 2006;21(12):3608.

72. Bhatt-Mehta V, Schumacher RE, Faix RG, Leady M, Brenner T. Lack of vancomycin-associated nephrotoxicity in newborn infants: a case-control study. Pediatrics 1999;103(4):e48.

73. Downs NJ, Neihart RE, Dolezal JM, Hodges GR. Mild nephrotoxicity associated with vancomycin use. Arch Intern Med 1989;149(8):1777–81.

74. Wold JS, Turnipseed SA. Toxicology of vancomycin in laboratory animals. Rev Infect Dis 1981;3(Suppl):S224–9.

75. Ngeleka M, Auclair P, Tardif D, Beauchamp D, Bergeron MG. Intrarenal distribution of vancomycin in endotoxemic rats. Antimicrob Agents Chemother 1989;33(9):1575–9.

76. Wai AO, Lo AM, Abdo A, Marra F. Vancomycin-induced acute interstitial nephritis. Ann Pharmacother 1998;32(11):1160–4.

77. Sidi V, Roilides E, Bibashi E, Gompakis N, Tsakiri A, Koliouskas D. Comparison of efficacy and safety of teicoplanin and vancomycin in children with antineoplastic therapy-associated febrile neutropenia and Gram-positive bacteremia. J Chemother 2000;12(4):326–31.

78. Yoshiyama Y, Yazaki T, Wong PC, Beauchamp D, Kanke M. The effect of fosfomycin on glycopeptide antibiotic-induced nephrotoxicity in rats. J Infect Chemother 2001;7(4):243–6.

79. Delbaldo C, Chen M, Friedli A, Prins C, Desmeules J, Saurat JH, Woodley DT, Borradori L. Drug-induced

epidermolysis bullosa acquisita with antibodies to type VII collagen. J Am Acad Dermatol 2002;46(Suppl 5):S161–4.

80. Gilmore ES, Friedman JS, Morrell DS. Extensive fixed drug eruption secondary to vancomycin. Pediatr Dermatol 2004;21(5):600–2.

81. Joshi PP, Ahmed N. Fixed drug eruption with vancomycin. Ann Pharmacother 2002;36(6):1104.

82. Padial MA, Barranco P, Lopez-Serrano C. Erythema multiforme to vancomycin. Allergy 2000;55(12):1201.

83. Packer J, Olshan AR, Schwarts AB. Prolonged allergic reaction to vancomycin in endstage renal disease. Dialysis Transplant 1987;16:87.

84. Santaeugenia S, Pedro Botet ML, Sabat M, Sopena N, Sadria M. Sindrome de Stevens–Johnson asociado a vancomicina. [Stevens–Johnson syndrome associated with vancomycin.] Rev Esp Quimioter 2000;13(4):425–6.

85. Thestrup-Pedersen K, Hainau B, Al'Eisa A, Al'Fadley A, Hamadah I. Fatal toxic epidermal necrolysis associated with ceftazidine and vancomycin therapy: a report of two cases. Acta Dermatol Venereol 2000;80(4):316–7.

86. Coelho S, Tellechea O, Reis JP, Mariano A, Figueiredo A. Vancomycin-associated linear IgA bullous dermatosis mimicking toxic epidermal necrolysis. Int J Dermatol 2006;45(8):995–6.

87. Navi D, Michael DJ, Fazel N. Drug-induced linear IgA bullous dermatosis. Dermatol Online J 2006;12(5):12.

88. Lesueur A, Lefort C, Gauthier JM, Andrivet P. Dermatose a IgA lineaire associée a la prise de vancomycine. Presse Med 2003;32:1078.

89. Dellavalle RP, Burch JM, Tayal S, Golitz LE, Fitzpatrick JE, Walsh P. Vancomycin-associated linear IgA bullous dermatosis mimicking toxic epidermal necrolysis. J Am Acad Dermatol 2003;48 Suppl 5:S56-7.

90. Solky BA, Pincus L, Horan RF. Vancomycin-induced linear IgA bullous dermatosis: morphology is a key to diagnosis. Cutis 2004;73(1):65–7.

91. Waldman MA, Black DR, Callen JP. Vancomycin-induced linear IgA bullous disease presenting as toxic epidermal necrolysis. Clin Exp Dermatol 2004;29(6):633–6.

92. Danielsen AG, Thomsen K. Vancomycin-induced linear IgA bullous disease. Br J Dermatol 1999;141(4):756–7.

93. Nousari HC, Costarangos C, Anhalt GJ. Vancomycin-associated linear IgA bullous dermatosis. Ann Intern Med 1998;129(6):507–8.

94. Bernstein EF, Schuster M. Linear IgA bullous dermatosis associated with vancomycin. Ann Intern Med 1998;129(6):508–9.

95. Nousari HC, Kimyai-Asadi A, Caeiro JP, Anhalt GJ. Clinical, demographic, and immunohistologic features of vancomycin-induced linear IgA bullous disease of the skin. Report of 2 cases and review of the literature. Medicine (Baltimore) 1999;78(1):1–8.

96. Wiadrowski TP, Reid CM. Drug-induced linear IgA bullous disease following antibiotics. Australas J Dermatol 2001;42(3):196–9.

97. Chang A, Camisa C, Ormsby A. Intertriginous bullae in a 53-year-old man. Arch Dermatol 2001;137(6):815–20.

98. Palmer RA, Ogg G, Allen J, Banerjee A, Ryatt KS, Ratnavel R, Wojnarowska F. Vancomycin-induced linear IgA disease with autoantibodies to BP180 and LAD285. Br J Dermatol 2001;145(5):816–20.

99. Ahkami R, Thomas I. Linear IgA bullous dermatosis associated with vancomycin and disseminated varicella-zoster infection. Cutis 2001;67(5):423–6.

100. Mofid MZ, Costarangos C, Bernstein B, Wong L, Munster A, Nousari HC. Drug-induced linear immunoglobulin A bullous disease that clinically mimics toxic epidermal necrolysis. J Burn Care Rehabil 2000;21(3):246–7.

101. Klein PA, Callen JP. Drug-induced linear IgA bullous dermatosis after vancomycin discontinuance in a patient with renal insufficiency. J Am Acad Dermatol 2000;42(2 Pt 2):316–23.

102. Richards SS, Hall S, Yokel B, Whitmore SE. A bullous eruption in an elderly woman. Vancomycin-associated linear IgA dermatosis (LAD). Arch Dermatol 1995;131(12):1447–51.

103. Geissmann C, Beylot-Barry M, Doutre MS, Beylot C. Drug-induced linear IgA bullous dermatosis. J Am Acad Dermatol 1995;32(2 Pt 1):296.

104. Neughebauer BI, Negron G, Pelton S, Plunkett RW, Beutner EH, Magnussen R. Bullous skin disease: an unusual allergic reaction to vancomycin. Am J Med Sci 2002;323(5):273–8.

105. Joshi S, Scott G, Looney RJ. A successful challenge in a patient with vancomycin-induced linear IgA dermatosis. Ann Allergy Asthma Immunol 2004;93(1):101–3.

106. Taylor SE, Paterson DL, Yu VL. Treatment options for chronic prostatitis due to vancomycin-resistant *Enterococcus faecium*. Eur J Clin Microbiol Infect Dis 1998;17(11):798–800.

107. Czachor JS, Garzaro P, Miller JR. Vancomycin and priapism. N Engl J Med 1998;338(23):1701.

108. Renz CL, Laroche D, Thurn JD, Finn HA, Lynch JP, Thisted R, Moss J. Tryptase levels are not increased during vancomycin-induced anaphylactoid reactions. Anesthesiology 1998;89(3):620–5.

109. Chopra N, Oppenheimer J, Derimanov GS, Fine PL. Vancomycin anaphylaxis and successful desensitization in a patient with end stage renal disease on hemodialysis by maintaining steady antibiotic levels. Ann Allergy Asthma Immunol 2000;84(6):633–5.

110. Duffy BL. Vancomycin reaction during spinal anaesthesia. Anaesth Intensive Care 2002;30(3):364–6.

111. Sorensen SJ, Wise SL, al-Tawfiq JA, Robb JL, Cushing HE. Successful vancomycin desensitization in a patient with end-stage renal disease and anaphylactic shock to vancomycin. Ann Pharmacother 1998;32(10):1020–3.

112. Wazny LD, Daghigh B. Desensitization protocols for vancomycin hypersensitivity. Ann Pharmacother 2001;35(11):1458–64.

113. Marshall C, Street A, Galbraith K. Glycopeptide-induced vasculitis—cross-reactivity between vancomycin and teicoplanin. J Infect 1998;37(1):82–3.

114. Yazganoglu KD, Ozkaya E, Ergin-Ozcan P, Cakar N. Vancomycin-induced drug hypersensitivity syndrome. J Eur Acad Dermatol Venereol 2005;19(5):648–50.

115. Zuliani E, Zwahlen H, Gilliet F, Marone C. Vancomycin-induced hypersensitivity reaction with acute renal failure: resolution following cyclosporine treatment. Clin Nephrol 2005;64(2):155–8.

116. Hwu JJ, Chen KH, Hsu WM, Lai JY, Li YS. Ocular hypersensitivitiy to topical vancomycin in a case of chronic endophthalmitis. Cornea 2005;24(6):754–6.

117. Barret JP. Vancomycin acute anaphylactoid reaction in a paediatric burn. Burns 2004;30(4):388–90.

118. Midtvedt T. In: The ECO-SHADOW concept—a new way of following environmental impacts of antimicrobials. Heidelberg, New York: Springer-Verlag, 2001:230–6.

119. Smith TL, Pearson ML, Wilcox KR, Cruz C, Lancaster MV, Robinson-Dunn B, Tenover FC,

Zervos MJ, Band JD, White E, Jarvis WR, Arduino MJ, Carr JH, Clark N, Hill B, McAllister S, Miller JM, Jennings GGlycopeptide-Intermediate Staphylococcus Aureus Working Group. Emergence of vancomycin resistance in *Staphylococcus aureus*. N Engl J Med 1999;340(7):493–501.

120. Sieradzki K, Roberts RB, Haber SW, Tomasz A. The development of vancomycin resistance in a patient with methicillin-resistant *Staphylococcus aureus* infection. N Engl J Med 1999;340(7):517–23.

121. Waldvogel FA. New resistance in *Staphylococcus aureus*. N Engl J Med 1999;340(7):556–7.

122. Majumdar A, Lipkin GW, Eliott TS, Wheeler DC. Vancomycin-dependent enterococci in a uraemic patient with sclerosing peritonitis. Nephrol Dial Transplant 1999;14(3):765–7.

123. Bierbaum G, Fuchs K, Lenz W, Szekat C, Sahl HG. Presence of *Staphylococcus aureus* with reduced susceptibility to vancomycin in Germany. Eur J Clin Microbiol Infect Dis 1999;18(10):691–6.

124. Endtz HP, van den Braak N, Verbrugh HA, van Belkum A. Vancomycin resistance: status quo and quo vadis. Eur J Clin Microbiol Infect Dis 1999;18(10):683–90.

125. Perl TM. The threat of vancomycin resistance. Am J Med 1999;106(5A):S26–37.

126. Kuehnert MJ, Jernigan JA, Pullen AL, Rimland D, Jarvis WR. Association between mucositis severity and vancomycin-resistant enterococcal bloodstream infection in hospitalized cancer patients. Infect Control Hosp Epidemiol 1999;20(10):660–3.

127. Carmeli Y, Samore MH, Huskins C. The association between antecedent vancomycin treatment and hospital-acquired vancomycin-resistant enterococci: a meta-analysis. Arch Intern Med 1999;159(20):2461–8.

128. Ostrowsky BE, Clark NC, Thauvin-Eliopoulos C, Venkataraman L, Samore MH, Tenover FC, Eliopoulos GM, Moellering RC Jr, Gold HS. A cluster of VanD vancomycin-resistant *Enterococcus faecium*: molecular characterization and clinical epidemiology. J Infect Dis 1999;180(4):1177–85.

129. Fines M, Perichon B, Reynolds P, Sahm DF, Courvalin P. VanE, a new type of acquired glycopeptide resistance in Enterococcus faecalis BM4405. Antimicrob Agents Chemother 1999;43(9):2161–4.

130. Sieradzki K, Pinho MG, Tomasz A. Inactivated pbp4 in highly glycopeptide-resistant laboratory mutants of *Staphylococcus aureus*. J Biol Chem 1999;274(27):18942–6.

131. Van Bambeke F, Chauvel M, Reynolds PE, Fraimow HS, Courvalin P. Vancomycin-dependent *Enterococcus faecalis* clinical isolates and revertant mutants. Antimicrob Agents Chemother 1999;43(1):41–7.

132. Pavia M, Nobile CG, Salpietro L, Angelillo IF. Vancomycin resistance and antibiotic susceptibility of enterococci in raw meat. J Food Prot 2000;63(7):912–5.

133. Schouten MA, Hoogkamp-Korstanje JA, Meis JF, Voss AEuropean VRE Study Group. Prevalence of vancomycin-resistant enterococci in Europe. Eur J Clin Microbiol Infect Dis 2000;19(11):816–22.

134. Trakulsomboon S, Danchaivijitr S, Rongrungruang Y, Dhiraputra C, Susaemgrat W, Ito T, Hiramatsu K. First report of methicillin-resistant *Staphylococcus aureus* with reduced susceptibility to vancomycin in Thailand. J Clin Microbiol 2001;39(2):591–5.

135. de Hoog M, Mouton JW, van den Anker JN. Vancomycin: pharmacokinetics and administration regimens in neonates. Clin Pharmacokinet 2004;43(7):417–40.

136. Thompson CM Jr, Long SS, Gilligan PH, Prebis JW. Absorption of oral vancomycin—possible associated toxicity. Int J Pediatr Nephrol 1983;4(1):1–4.

137. Heck AJ, Bonnici PJ, Breukink E, Morris D, Wills M. Modification and inhibition of vancomycin group antibiotics by formaldehyde and acetaldehyde. Chemistry 2001;7(4):910–6.

138. Wysocki M, Delatour F, Faurisson F, Rauss A, Pean Y, Misset B, Thomas F, Timsit JF, Similowski T, Mentec H, Mier L, Dreyfuss D. Continuous versus intermittent infusion of vancomycin in severe staphylococcal infections: prospective multicenter randomized study. Antimicrob Agents Chemother 2001;45(9):2460–7.

139. Miner LJ, Faix RG. Large vancomycin overdose in two premature infants with minimal toxicity. Am J Perinatol 2004;21(8):433–8.

140. Balen RM, Betts T, Ensom MHH. Vancomycin overdose in 6-month-old girl. Can J Hosp Pharm 2000;53:32–5.

141. Muller D, Hufnagel M, Suttorp M. Accidental overdose of vancomycin in preterm twins. Pediatr Infect Dis J 1999;18(8):744–5.

142. Bunchman TE, Valentini RP, Gardner J, Mottes T, Kudelka T, Maxvold NJ. Treatment of vancomycin overdose using high-efficiency dialysis membranes. Pediatr Nephrol 1999;13(9):773–4.

143. Goebel J, Ananth M, Lewy JE. Hemodiafiltration for vancomycin overdose in a neonate with end-stage renal failure. Pediatr Nephrol 1999;13(5):423–5.

144. American Academy of Clinical Toxicology. European Association of Poisons Centres and Clinical Toxicologists. Position statement and practice guidelines on the use of multi-dose activated charcoal in the treatment of acute poisoning. J Toxicol Clin Toxicol 1999;37(6):731–51.

145. Paganini H, Marin M. Caracteristicas farmacocineticas y espectro antimicrobiano de la teicoplanina. [Pharmacokinetic characteristics and antimicrobial spectrum of teicoplanin.] Medicina (B Aires) 2002;62(Suppl 2):52–5.

146. Lifshitz T, Lapid-Gortzak R, Finkelman Y, Klemperer I. Vancomycin and ceftazidime incompatibility upon intravitreal injection. Br J Ophthalmol 2000;84(1):117–8.

147. Nakamura T, Kokuryo T, Hashimoto Y, Inui KI. Effects of fosfomycin and imipenem-cilastatin on the nephrotoxicity of vancomycin and cisplatin in rats. J Pharm Pharmacol 1999;51(2):227–32.

148. Park SS, Vallar RV, Hong CH, von Gunten S, Ruoff K, D'Amico DJ. Intravitreal dexamethasone effect on intravitreal vancomycin elimination in endophthalmitis. Arch Ophthalmol 1999;117(8):1058–62.

149. Pea F, Porreca L, Baraldo M, Furlanut M. High vancomycin dosage regimens required by intensive care unit patients cotreated with drugs to improve haemodynamics following cardiac surgical procedures. J Antimicrob Chemother 2000;45(3):329–35.

150. Yeung MY, Smyth JP. Concurrent frusemide–theophylline dosing reduces serum vancomycin concentrations in preterm infants. Aust J Hosp Pharm 1999;29:269–72.

151. Ng HP, Koh KF, Tham LS. Vancomycin causes dangerous precipitation when infused with gelatin fluid. Anaesthesia 2000;55(10):1039–40.

152. Mori H, Nakajima T, Nakayama A, Yamori M, Izushi F, Gomita Y. Interaction between levofloxacin and

vancomycin in rats—study of serum and organ levels. Chemotherapy 1998;44(3):181–9.

153. Symons NL, Hobbes AF, Leaver HK. Anaphylactoid reactions to vancomycin during anaesthesia: two clinical reports. Can Anaesth Soc J 1985;32(2):178–81.

154. Wilson AP. Comparative safety of teicoplanin and vancomycin. Int J Antimicrob Agents 1998;10(2):143–52.

155. MacGowan AP. Pharmacodynamics, pharmacokinetics, and therapeutic drug monitoring of glycopeptides. Ther Drug Monit 1998;20(5):473–7.

156. Freeman CD, Quintiliani R, Nightingale CH. Vancomycin therapeutic drug monitoring: is it necessary? Ann Pharmacother 1993;27(5):594–8.

157. Launay-Vacher V, Izzedine H, Mercadal L, Deray G. Clinical review: use of vancomycin in haemodialysis patients. Crit Care 2002;6(4):313–6.

158. Machado JK, Feferbaum R, Diniz EM, Okay TS, Ceccon ME, Costa Vaz FA. Monitoring the treatment of sepsis with vancomycin in term newborn infants. Rev Hosp Clin Fac Med Sao Paulo 2001;56(1):17–24.

159. Thomas MP, Steele RW. Monitoring serum vancomycin concentrations in children: is it necessary? Pediatr Infect Dis J 1998;17(4):351–3.

160. Smith PF, Morse GD. Accuracy of measured vancomycin serum concentrations in patients with end-stage renal disease. Ann Pharmacother 1999;33(12):1329–35.

161. Kim JS, Perkins RJ, Briceland LL, Tobin EH. Clinical significance of falsely elevated vancomycin concentrations in end-stage renal disease. Ann Pharmacother 1999;33(1):116–8.

162. Akil IO, Mir S. Hemodiafiltration for vancomycin overdose in a patient with end-stage renal failure. Pediatr Nephrol 2001;16(12):1019–21.

163. Foral MA, Heineman SM. Vancomycin removal during a plasma exchange transfusion. Ann Pharmacother 2001;35(11):1400–2.

164. Rybak MJ, Albrecht LM, Boike SC, Chandrasekar PH. Nephrotoxicity of vancomycin, alone and with an aminoglycoside. J Antimicrob Chemother 1990;25(4):679–87.

165. de Lemos E, Pariat C, Piriou A, Fauconneau B, Courtois P. Variations circadiennes de la nephrotoxicité de l'association vancomycine–gentamicine chez le rat. [Circadian variations in the nephrotoxicity of the vancomycin–gentamicin combination in rats.] Pathol Biol (Paris) 1991;39(1):12–5.

166. Pauly DJ, Musa DM, Lestico MR, Lindstrom MJ, Hetsko CM. Risk of nephrotoxicity with combination vancomycin–aminoglycoside antibiotic therapy. Pharmacotherapy 1990;10(6):378–82.

167. Fiscella RG. Physical incompatibility of vancomycin and ceftazidime for intravitreal injection. Arch Ophthalmol 1993;111(6):730.

168. Yeung MY, Smyth JP. Concurrent frusemide–theophylline dosing reduces serum vancomycin concentrations in preterm infants. Aust J Hosp Pharm 1999;29:269–72.

ANTIVIRAL DRUGS, INCLUDING IMMUNOGLOBULINS AND INTERFERONS

Abacavir

See also Nucleoside analogue reverse transcriptase inhibitors (NRTIs)

General Information

Abacavir is a guanidine analogue that inhibits HIV reverse transcriptase. In vitro, its potency is similar to that of zidovudine, protease inhibitors, and dual nucleoside combinations. There is evidence that abacavir is effective in reducing viral load and increasing the CD4 count in HIV-infected patients. Viral resistance is not rapidly selected for, but cross-resistance has been shown to other analogues of cytosine and guanidine (didanosine, lamivudine, and zalcitabine).

Abacavir has good oral systemic availability and penetrates the nervous system. It does not interfere with drugs that are metabolized by liver microsomal cytochrome P450 (1). It has no other significant drug interactions and can be administered without food restrictions.

Observational studies

The effects of abacavir have been evaluated in a study in over 13 000 adults who no longer responded to commercially available treatment regimens (2). By month 2 of treatment with abacavir, plasma HIV-1 RNA concentrations fell by at least half a log unit in 31% of patients, and in 5.6% of the patients HIV-1 RNA concentrations fell to under 400 copies/ml. Serious drug-related adverse events were reported by 7.7% of patients. The most common were nausea, skin rash, diarrhea, malaise or fatigue, and fever. About 4.6% of patients had a hypersensitivity reaction that was possibly drug-related.

General adverse effects

The adverse effects of abacavir that have been most often observed in clinical trials are fatigue, nausea and vomiting, abdominal pain, diarrhea, headache, rash, and dyspepsia (3,4). Allergic reactions lead to withdrawal of therapy in about 3% of patients (5). These can be severe, and anaphylaxis has been reported after rechallenge in a patient with an apparent allergic reaction to abacavir (6). It is wise to avoid rechallenge when allergy is suspected (7). In one study nausea and vomiting occurred in 38–57% of patients, headache in 27–41%, malaise and fatigue in 28%, diarrhea in 18–23%, and weakness in 29% (8). There was also one case of agranulocytosis accompanied by a skin rash.

Organs and Systems

Nervous system

Vertigo has been attributed to abacavir (9).

- A 44-year-old African-American developed vertigo, tinnitus in both ears, headache behind the eyes, and left ear pain and hearing loss soon after starting to take abacavir, lamivudine, and stavudine. There was left-sided nystagmus and vestibular tests showed evidence of vestibular impairment. An MRI scan was normal. All the antiretroviral drugs were withdrawn and he improved. When lamivudine and stavudine were restarted, with nevirapine, the vertigo did not recur.

Psychiatric

Psychiatric adverse effects have been reported in a child (10).

- An 11 year old child had mood changes, headaches, anxiety, and bad dreams 1 month after adding abacavir to his antiretroviral regimen and developed major depression. The symptoms completely resolved after abacavir was withdrawn.

Metabolism

While abacavir has been associated with hyperglycemia in individual cases (11), there were no significant effects on blood glucose concentration in clinical trials.

- A 47-year-old man, with normoglycemia and no family history of diabetes mellitus, who was taking highly active antiretroviral therapy, was given abacavir for treatment intensification. He became lethargic and hyperglycemic. Despite metformin and glibenclamide, the hyperglycemia continued. Abacavir was withdrawn, and within 2 weeks his blood glucose concentration returned to baseline and the hypoglycemic drugs were withdrawn.

This patient was also taking hydrochlorothiazide, but the time-course of onset and resolution were consistent with abacavir-induced hyperglycemia.

Immunologic

Abacavir hypersensitivity is well known (12). In an open, multicenter, randomized trial 291 subjects received abacavir 300 mg/day plus lamivudine 150 mg bd, with a non-nucleoside reverse transcriptase inhibitor (efavirenz 600 mg/day), an enhanced protease inhibitor (amprenavir 1200 mg/day + ritonavir 200 mg/day), or a third nucleoside reverse transcriptase inhibitor (stavudine 30 or 40 mg bd) (13). The rate of adverse effects, including abacavir hypersensitivity, was similar in all three arms. Abacavir hypersensitivity reactions occurred in 21 subjects (7.3%).

The risk of allergic reactions to abacavir may be as high as 10% (14). However, the incidence is more usually reported to be 3–5% (15,16). Allergic reactions usually occur within the first 28 days of therapy and rarely thereafter. They are characterized by non-specific complaints suggestive of an upper respiratory tract infection, fever, rash, nausea, and vomiting. Resolution of the symptoms occurs within days of withdrawal. Severe and even fatal reactions to readministration have been observed, and it has been suggested that rechallenge is contraindicated in any patients who have had an allergic reaction (7). However, it is safe to rechallenge patients who have stopped treatment because of other types of adverse reaction. Of 1201 patients treated in clinical trials, 219 interrupted abacavir therapy for reasons other than allergy; on reintroduction there were no cases of allergy or anaphylaxis (17).

In a retrospective observational cohort study of 730 patients who took abacavir + lamivudine + zidovudine there were hypersensitivity reactions to abacavir in 36 (5%) (18).

In the SOLO study 16 of 211 patients (7.6%) had hypersensitivity reactions to abacavir (19).

The susceptibility factors associated with allergic reactions have been sought in an analysis of all protocols conducted by GlaxoSmithKline that involved abacavir exposure for at least 24 weeks with a quality-assured or validated clinical database by 30 June 2000 (n = 5332) (20). There were 197 allergic reactions (3.7%). The risks of allergic reactions were lower in black people (OR = 0.59; 95% CI = 0.38, 0.91) than in other ethnic groups, and in patients who had received previous therapy for HIV-1 infection with other antiretroviral agents (OR = 0.58; 95% CI = 0.44, 0.78) compared with those receiving therapy for the first time.

Genetic factors affecting the immune response to abacavir have been sought in patients who had taken abacavir for more than 6 weeks, 18 with hypersensitivity reactions and 167 without (21). HLA-B*5701 was present in 14 of the 18 patients with abacavir hypersensitivity, and in four of the 167 others (OR = 117; 95% CI = 29, 481). The combination of HLA-DR7 and HLA-DQ3 was found in 13 of the 18 and five of the 167 (OR = 73; CI = 20, 268). HLA-B*5701, HLA-DR7, and HLA-DQ3 were present in combination in 13 of the 18 and none of the 167 (OR = 822; CI = 43, 15 675). Other MHC markers also present on the 57.1 ancestral haplotype to which these three markers belong confirmed the presence of haplotype-specific linkage disequilibrium, and mapped potential susceptibility loci to a region bounded by C4A6 and HLA-C. HLA-B*5701, HLA-DR7, and HLA-DQ3 had a positive predictive value for hypersensitivity of 100%, and a negative predictive value of 97%. The authors concluded that susceptibility to abacavir hypersensitivity is carried on the 57.1 ancestral haplotype and that withholding abacavir from those with HLA-B*5701, HLA-DR7, and HLA-DQ3 should reduce the prevalence of hypersensitivity from 9% to 2.5% without inappropriately denying abacavir to any patient.

In a retrospective case-control study of patients with allergic reactions, HLA-B57 was present in 39 of 84 patients compared with 4 of 113 controls (22). However, there were few women and other ethnic groups in the study, and so these findings relate largely to white men.

In a retrospective Australian study of 200 patients, 16 of whom had hypersensitivity reactions, the occurrence of HLA-B*5701 and Hsp70-Hom M493T alleles predicted abacavir hypersensitivity (23). The positive predictive value was 93.8% and the negative predictive value was 99.5% for HLA-B*5701 in combination with Hsp70-Hom M493T.

In a model of the cost-effectiveness of HLA B*5701 genotyping, pooling published data, routine testing ranged from being a dominant strategy (less expensive and more beneficial) to an incremental cost-effectiveness ratio of about ?23 000 per hypersensitivity reaction avoided (24).

In a prospective study of 260 abacavir-naive individuals, 20 (7.7%) were positive for HLA-B*5701; there were no cases of abacavir hypersensitivity among 148 HLA-B*5701-negative recipients (25).

Nevertheless, hypersensitivity reactions to abacavir can occur in those who do not have HLA B*5701. In a retrospective study in acute HIV infection nine (18%) of 50 individuals treated with abacavir developed suspected hypersensitivity; only two of these nine and no controls were HLA-B*5701 positive (26).

There were hypersensitivity reactions in two of 17 children who were given high-dose abacavir (12 mg/kg bd, maximum 600 mg) for HIV-associated encephalopathy (27).

Fatal outcomes have been described after rechallenge with abacavir. In an analysis of the HIV cohort at Montpellier, early withdrawal of abacavir was studied in 331 patients (28). The rate of hypersensitivity reaction in this retrospective study was higher than in other studies (8.5%) and the role of other drugs (efavirenz, nevirapine) could not be ruled out in one-third of the cases of hypersensitivity reaction.

Intercurrent influenza infection can be difficult to distinguish from hypersensitivity reactions to abacavir. In a comparison of the clinical presentation of 15 patients with abacavir hypersensitivity with culture confirmed influenza A infection in 30 controls, gastrointestinal symptoms were clearly associated with abacavir hypersensitivity in febrile patients while the presence of cough with gastrointestinal symptoms and fever was more suggestive of influenza A (29).

In 16 (8.2%) of 196 patients, hypersensitivity reactions were more frequent in those taking a fixed combination of zidovudine, lamivudine, and abacavir (Trizivir®) than in those taking abacavir in the form of Ziagen® (9.8% versus 0.04%) (30). The authors hypothesized that excipients such as indigo carmine or polyethylene glycol, which are present in Trizivir® but not in Ziagen® might be responsible. However, earlier studies showed a higher frequency of hypersensitivity reactions in patients taking Ziagen®. There was no association with the co-administration of other drugs that commonly cause hypersensitivity reactions, such as co-trimoxazole, nevirapine, or efavirenz.

In the CNA30021 study abacavir 600 mg/day (n = 384) was compared with 300 mg bd (n = 386) in combination with lamivudine 300 mg and efavirenz 600 mg. The rates of adverse events, including abacavir *hypersensitivity reactions*, were similar in the two arms (9% versus 7%) (31).

In a multicenter trial, 128 children were randomly assigned to zidovudine + lamivudine (n = 36), to zidovudine + abacavir (n = 45), or to lamivudine + abacavir

$(n = 47)$ (32). One child had an allergic reaction to abacavir and stopped taking it, as did three with possible reactions.

Drug-Drug Interactions

Antiretroviral drugs

The combination of abacavir + lamivudine + tenofovir in a triple-nucleoside regimen is no longer recommended (33). This is on the basis of a randomized, open, multicenter study of tenofovir disoproxil fumarate versus efavirenz, both administered once daily with the abacavir + lamivudine fixed-dose combination in 340 treatment-naïve subjects. The abacavir + lamivudine + tenofovir arm had an unacceptably high virological failure rate of 49% compared with 5% in the efavirenz arm at 12 weeks and the study was ended prematurely. Only 54% of the patients who failed had the typical K65R and M184V mutations on subsequent genotyping (34). The mechanism for this interaction is still unclear and there is no classical pharmacokinetic interaction. It has been hypothesized that the interaction occurs at the level of the intracellular nucleotide (35).

References

1. Ravitch JR, Bryant BJ, Reese MJ, et al. In vivo and in vitro studies of the potential for drug interactions involving the anti-retroviral 1592 in humans. In: 5th Conference on Retroviruses and Opportunistic InfectionsChicago 1–5 February 1998:230–6 Abstract 634.
2. Kessler HA, Johnson J, Follansbee S, Sension MG, Mildvan D, Sepulveda GE, Bellos NC, Hetherington SV. Abacavir expanded access program for adult patients infected with human immunodeficiency virus type 1. Clin Infect Dis 2002;34(4):535–42.
3. Kumar PN, Sweet DE, McDowell JA, Symonds W, Lou Y, Hetherington S, LaFon S. Safety and pharmacokinetics of abacavir (1592U89) following oral administration of escalating single doses in human immunodeficiency virus type 1-infected adults. Antimicrob Agents Chemother 1999;43(3):603–8.
4. Vernazza PL, Katlama C, Clotet B, et al. Intensification of stable background (SBG) antiretroviral therapy (ART) with Ziagen (ABC,1592). In: 6th Conference on Retroviruses and Opportunistic InfectionsChicago Jan 31–Feb 4 1999:230–6.
5. Staszewski S. Coming therapies: abacavir. Int J Clin Pract Suppl 1999;103:35–8.
6. Walensky RP, Goldberg JH, Daily JP. Anaphylaxis after rechallenge with abacavir. AIDS 1999;13(8):999–1000.
7. Escaut L, Liotier JY, Albengres E, Cheminot N, Vittecoq D. Abacavir rechallenge has to be avoided in case of hypersensitivity reaction. AIDS 1999;13(11):1419–20.
8. Tikhomirov V, Namek K, Hindes R. Agranulocytosis induced by abacavir. AIDS 1999;13(11):1420–1.
9. Fantry LE, Staecker H. Vertigo and abacavir. AIDS Patient Care STDS 2002;16(1):5–7.
10. Soler Palacin P, Aramburo A, Moraga FA, Cabanas MJ, Figueras C. Neuropsychiatric reaction induced by abacavir in a pediatric human immunodeficiency virus-infected patient. Pediatr Infect Dis J 2006;25(4):382.
11. Modest GA, Fuller J, Hetherington SV, Lenhard JM, Powell GS. Abacavir and diabetes. N Engl J Med 2001;344(2):142–4.
12. Herring SJ, Krieger AC. Acute respiratory manifestations of the abacavir hypersensitivity reaction. AIDS 2006;20(2):301–2.
13. Bartlett JA, Johnson J, Herrera G, Sosa N, Rodriguez A, Liao Q, Griffith S, Irlbeck D, Shaefer MS; Clinically Significant Long-Term Antiretroviral Sequential Sequencing Study (CLASS) Team. Long-term results of initial therapy with abacavir and lamivudine combined with efavirenz, amprenavir/ritonavir, or stavudine. J Acquir Immune Defic Syndr 2006;43(3):284–92.
14. Katlama C, Fenske S, Gazzard B, Lazzarin A, Clumeck N, Mallolas J, Lafeuillade A, Mamet JP, Beauvais LAZL30002 European study team. TRIZAL study: switching from successful HAART to Trizivir (abacavir–lamivudine–zidovudine combination tablet): 48 weeks efficacy, safety and adherence results. HIV Med 2003;4(2):79–86.
15. Hervey PS, Perry CM. Abacavir: a review of its clinical potential in patients with HIV infection. Drugs 2000;60(2):447–79.
16. Henry K, Wallace RJ, Bellman PC, Norris D, Fisher RL, Ross LL, Liao Q, Shaefer MSTARGET Study Team. Twice-daily triple nucleoside intensification treatment with lamivudine–zidovudine plus abacavir sustains suppression of human immunodeficiency virus type 1: results of the TARGET Study. J Infect Dis 2001;183(4):571–8.
17. Loeliger AE, Steel H, McGuirk S, Powell WS, Hetherington SV. The abacavir hypersensitivity reaction and interruptions in therapy. AIDS 2001;15(10):1325–6.
18. Gathe JC Jr, Wood R, Sanne I, DeJesus E, Schürmann D, Gladysz A, Garris C, Givens N, Elston R, Yeo J. Long-term (120-Week) antiviral efficacy and tolerability of fosamprenavir/ritonavir once daily in therapy-naive patients with HIV-1 infection: an uncontrolled, open-label, single-arm follow-on study. Clin Ther 2006;28(5):745–54.
19. Reliquet V, Allavena C, François-Brunet C, Perré P, Bellein V, Garré M, May T, Souala F, Besnier JM, Raffi F. Long-term assessment of nevirapine-containing highly active antiretroviral therapy in antiretroviral-naive HIV-infected patients: 3-year follow-up of the VIRGO study. HIV Med 2006;7(7):431–6.
20. Symonds W, Cutrell A, Edwards M, Steel H, Spreen B, Powell G, McGuirk S, Hetherington S. Risk factor analysis of hypersensitivity reactions to abacavir. Clin Ther 2002;24(4):565–73.
21. Mallal S, Nolan D, Witt C, Masel G, Martin AM, Moore C, Sayer D, Castley A, Mamotte C, Maxwell D, James I, Christiansen FT. Association between presence of HLA-B*5701, HLA-DR7, and HLA-DQ3 and hypersensitivity to HIV-1 reverse-transcriptase inhibitor abacavir. Lancet 2002;359(9308):727–32.
22. Hetherington S, Hughes AR, Mosteller M, Shortino D, Baker KL, Spreen W, Lai E, Davies K, Handley A, Dow DJ, Fling ME, Stocum M, Bowman C, Thurmond LM, Roses AD. Genetic variations in HLA-B region and hypersensitivity reactions to abacavir. Lancet 2002;359(9312):1121–2.
23. Martin AM, Nolan D, Gaudieri S, Almeida CA, Nolan R, James I, Carvalho F, Phillips E, Christiansen FT, Purcell AW, McCluskey J, Mallal S. Predisposition to abacavir hypersensitivity conferred by HLA-B*5701 and a haplotypic Hsp70-Hom variant. Proc Natl Acad Sci USA 2004;101(12):4180–5.

24. Hughes DA, Vilar FJ, Ward CC, Alfirevic A, Park BK, Pirmohamed M. Cost-effectiveness analysis of HLA B*5701 genotyping in preventing abacavir hypersensitivity. Pharmacogenetics 2004;14(6):335–42.

25. Rauch A, Nolan D, Martin A, McKinnon E, Almeida C, Mallal S. Prospective genetic screening decreases the incidence of abacavir hypersensitivity reactions in the Western Australian HIV cohort study. Clin Infect Dis 2006;43(1):99–102.

26. Stekler J, Maenza J, Stevens C, Holte S, Malhotra U, McElrath MJ, Corey L, Collier AC. Abacavir hypersensitivity reaction in primary HIV infection. AIDS 2006;20(9):1269–74.

27. Saavedra-Lozano J, Ramos JT, Sanz F, Navarro ML, de José MI, Martín-Fontelos P, Mellado MJ, Leal JA, Rodriguez C, Luque I, Madison SJ, Irlbeck D, Lanier ER, Ramilo O. Salvage therapy with abacavir and other reverse transcriptase inhibitors for human immunodeficiency-associated encephalopathy. Pediatr Infect Dis J 2006;25(12):1142–52.

28. Peyriere H, Guillemin V, Lotthe A, Baillat V, Fabre J, Favier C, Atoui N, Hansel S, Hillaire-Buys D, Reynes J. Reasons for early abacavir discontinuation in HIV-infected patients. Ann Pharmacother 2003;37:1392–7.

29. Keiser P, Nassar N, Skiest D, Andrews C, Yazdani B, White A, Hetherington S. Comparison of symptoms of influenza A with abacavir-associated hypersensitivity reaction. Int J STD AIDS 2003;14:478–81.

30. Parra-Ruiz J, Martinez-Ramirez M, Munoz-Medina L, Serrano-Falcon C, Hernandez-Quero J. Reasons for early abacavir discontinuation in HIV-infected patients. Ann Pharmacother 2004;38(3):512–3.

31. Moyle GJ, DeJesus E, Cahn P, Castillo SA, Zhao H, Gordon DN, Craig C, Scott TR; Ziagen Once-Daily in Antiretroviral Combination Therapy (CNA30021) Study Team. Abacavir once or twice daily combined with once-daily lamivudine and efavirenz for the treatment of antiretroviral-naive HIV-infected adults: results of the Ziagen Once Daily in Antiretroviral Combination Study. J Acquir Immune Defic Syndr 2005;38(4):417–25.

32. Paediatric European Network for Treatment of AIDS (PENTA). Comparison of dual nucleoside-analogue reverse-transcriptase inhibitor regimens with and without nelfinavir in children with HIV-1 who have not previously been treated: the PENTA 5 randomised trial. Lancet 2002;359(9308):733–40.

33. Kuritzkes DR. Less than the sum of its parts: failure of a tenofovir–abacavir–lamivudine triple-nucleoside regimen. J Infect Dis 2005;192(11):1867–8.

34. Gallant JE, Rodriguez AE, Weinberg WG, Young B, Berger DS, Lim ML, Liao Q, Ross L, Johnson J, Shaefer MS; ESS30009 Study. Early virologic nonresponse to tenofovir, abacavir, and lamivudine in HIV-infected antiretroviral-naive subjects. J Infect Dis 2005;192(11):1921–30.

35. Kuritzkes DR. Less than the sum of its parts: failure of a tenofovir–abacavir–lamivudine triple-nucleoside regimen. J Infect Dis 2005;192(11):1867–8.

Aciclovir

General Information

Aciclovir is an acyclic purine nucleoside. Its antiviral activity depends upon intracellular phosphorylation to its triphosphate derivative. Because of its higher affinity for viral thymidine kinase, aciclovir is phosphorylated at a much higher rate by the viral enzyme. Thus, it is almost exclusively active in infected cells, fulfilling one of the selectivity principles of antiviral drugs. In addition, aciclovir triphosphate serves as a better substrate for viral than for host cell DNA polymerase and thereby causes preferential termination of viral DNA synthesis (1).

Aciclovir is active against *Herpes simplex* virus type 1 (HSV-1), HSV-2, *Varicella zoster* virus (VZV), *Herpesvirus simiae*, and to a lesser degree Epstein–Barr virus (EBV). Resistant strains of HSV can arise owing to the emergence of thymidine kinase-deficient mutants. Other forms of resistance patterns are less common (2,3).

Aciclovir is used topically or systemically, orally or intravenously. Its therapeutic potential is most impressive in active parenchymal or systemic HSV infections. The latency stage of the viral infection is not affected. Since the blood–brain barrier is well penetrated, aciclovir is the treatment of choice for HSV encephalitis.

Very few adverse effects, generally of minor importance, have been reported (4). In immunosuppressed patients abnormal liver function, encephalopathy, and myelosuppression have been observed; however, it is unclear at present whether these adverse effects are related to the drug itself or to the underlying disorder (5–7).

Observational studies In a patient satisfaction questionnaire during a comparison of once- and twice-daily oral suppressive therapy with aciclovir for genital herpes, adverse effects were rarely reported (8).

Comparative studies

The effects of aciclovir and valaciclovir for anogenital herpes have been studied in HIV-infected individuals in two controlled trials (9). In the first study, 1062 patients with CD4+ counts over $100 \times 10^6/l$ received valaciclovir or aciclovir for 1 year and were assessed monthly. In the second study, 467 patients were treated episodically for at least 5 days with valaciclovir or aciclovir and were assessed daily. Valaciclovir was as effective as aciclovir for suppression and episodic treatment of herpesvirus infections. Hazard ratios for the time to recurrence with valaciclovir 500 mg bd and 1000 mg od compared with aciclovir were 0.73 (95% CI = 0.50, 1.06) and 1.31 (0.94, 1.82). Valaciclovir 1000 mg bd and aciclovir had similar effects on the duration of infective episodes (HR = 0.92; CI = 0.75, 1.14). The most common adverse events, which occurred at similar rates with all regimens, were diarrhea, headache, infections, rashes, nausea, rhinitis, pharyngitis, abdominal pain, fever, depression, and cough.

Organs and Systems

Nervous system

Neurotoxicity secondary to aciclovir is rare and is associated with high plasma concentrations (SEDA-18, 299), such as result from impaired renal function (10). Although the risk is greatest with intravenous administration, neurotoxicity has previously been noted with oral use.

Symptoms of neurotoxicity, which usually appear within the first 24–72 hours of administration, include tremor, myoclonus, confusion, lethargy, agitation, hallucinations, dysarthria, asterixis, ataxia, hemiparesthesia, and seizures. While aciclovir-induced neurotoxicity is most prevalent with intravenous administration, it has also been reported after oral use in patients with terminal renal insufficiency on hemodialysis.

Neurotoxicity possibly secondary to the topical use of aciclovir has also been described (11).

- A 59-year-old woman on hemodialysis was treated with oral aciclovir 200 mg/day for ophthalmic Herpes zoster. After a few days, an ophthalmic aciclovir cream was started (one application every 6 hours) because of ipsilateral *Herpes* keratitis. After 1 week of combined oral and topical treatment, she became confused, with dysarthria and audiovisual hallucinations. Aciclovir was withdrawn and hemodialysis was initiated. Complete resolution of symptoms was achieved after three hemodialysis sessions in 3 days. Aciclovir plasma concentrations before hemodialysis were high (45 µmol/l) and fell rapidly during hemodialysis.

There is no conclusive evidence for the contribution of the topically administered aciclovir to the high plasma concentrations and subsequent neurotoxicity in this case. However, the authors argued that the existence of high aciclovir plasma concentrations, in spite of careful adjustment of the oral dosage, pointed to significant topical absorption of the drug, especially since the absorption of aciclovir through the skin and mucous membranes may be unpredictable.

Coma has been attributed to oral aciclovir (12).

- A 73-year-old man with acute respiratory failure, presumed to be secondary to amiodarone toxicity, developed sepsis and acute renal insufficiency, and required intermittent hemodialysis. Following a *Herpes simplex* labialis infection he was treated with oral aciclovir (400 mg tds). The next day he became sleepy, disoriented, and agitated. Over the next 48 hours his neurological condition deteriorated and he responded to pain only, had uncoordinated eye movements, tremors, facial and jaw myoclonus, increased reflexes, and hypertonia. After 7 days of aciclovir he became unresponsive and comatose. Aciclovir was withdrawn and hemodialysis carried out more frequently. His neurological status improved over a period of 4 days. Trough plasma concentrations of aciclovir were well above the upper limit of the usual target range.

This appears to be the first case of coma attributable to oral aciclovir. The fact that the patient was receiving oral rather than intravenous aciclovir and was on regular hemodialysis made neurotoxicity unlikely, and this emphasizes the need to be wary of this potentially serious complication in seriously ill elderly patients.

It has been suggested that hemodialysis can be a useful diagnostic tool in the differential diagnosis between aciclovir-induced neurotoxicity and herpes encephalitis, as well as a fast and reliable treatment of drug-induced neurotoxicity (13).

- A previously healthy 59-year-old woman developed right eye pain for 3 days and vesicle formation on her forehead. She received intravenous aciclovir 250 mg every 8 hours, but 48 hours later she became drowsy and lethargic, with incoherent speech and hallucinations. Her blood urea nitrogen concentration rose from 0.71 mmol/l on admission to 2.41 mmol/l, serum creatinine rose from 53.6 to 442 (reference range 51–115) µmol/l and the serum sodium concentration fell from 135 to 121 mmol/l. Electroencephalography showed mild diffuse cortical dysfunction, with more emphasis in the right hemisphere, and regional epileptiform activity in the bilateral frontal and right parietal regions. An MRI scan of her head was normal. Because of deteriorating renal function and the debilitating nature of the neurological symptoms, which were thought to be due to acyclovir toxicity, hemodialysis was initiated. The pre-hemodialysis trough plasma aciclovir concentration was 18 mg/l, compared with peak and trough concentrations of 5.5–13.8 mg/l and 0.2–1 mg/l respectively in adults who receive 5 mg/kg of acyclovir (14). She underwent two 4-hour sessions of hemodialysis over 2 days. The post-hemodialysis plasma aciclovir concentration fell to 3 mg/l. Her conscious level improved and the blood urea nitrogen and serum creatinine concentrations fell to 6.1 mmol/l and 142 µmol/l respectively. She was well by the fifth day.

Reversible neurotoxicity in a 6-month old child after liver transplantation has been reported (15).

- A 6-month-old girl had liver transplantation after liver failure secondary to an enterovirus infection, followed by initial immunosuppression with antilymphocyte serum, azathioprine, and glucocorticoids. Her renal function improved on day 3, and the antilymphocyte serum was switched to ciclosporin 10mg/kg/day. Antimicrobial drug prophylaxis consisted of ticarcillin + clavulanic acid for 2 days, fluconazole 3 mg/kg/day from day 1, and aciclovir 250 mg/m^2 from day 3, increased to 250 mg/m^2 tds on day 5 to prevent *Herpes simplex* infection. On day 5 she became comatose and agitated, with choreoathetoid movements in all four limbs, random eye movements, and loss of eye contact. A brain CT scan was normal and electroencephalography showed slow waves with no evidence of seizures. The cerebrospinal fluid was normal. The trough plasma ciclosporin concentration was 87 ng/ml (below the target range of 200–250 ng/ml). The plasma aciclovir concentration was 4.5 mg/l on day 7, above the recommended target concentration (2.3 mg/l). Aciclovir was withdrawn and the neurological effects resolved.

Psychiatric

Aciclovir-related neuropsychiatric symptoms have previously been associated with increased serum concentrations of the main metabolite of aciclovir, 9-

Valaciclovir

↓ Valaciclovir hydrolase

Aciclovir

↓ Alcohol dehydrogenase

Aciclovir aldehyde

↓ Aldehydede hydrogenase

CMMG

carboxymethoxymethylguanine (CMMG) (Figure 1). In a further study of nine subjects with neuropsychiatric signs and symptoms (confusion, somnolence, hyper-reflexia, myoclonus, hallucinations, incoherence, and unresponsiveness) and 12 asymptomatic subjects, including 10 from a valaciclovir multiple sclerosis trial and two with recurrent herpes encephalitis, CSF aciclovir concentrations were measured. CMMG could only be detected in the CSF of those with neuropsychiatric symptoms and signs (median concentration 1.0, range 0.6–7.0 μmol/l). The concentration of CMMG was below the limit of quantification (<0.5 μmol/l) in the asymptomatic subjects. All those with neuropsychiatric signs and symptoms, except one, had acute renal function impairment or chronic renal failure (16).

High-dose aciclovir can cause disturbed cognition, changes in the level of consciousness, tremor, asterixis,

hallucinations, and psychiatric syndromes. Often, symptoms of aciclovir toxicity are difficult to differentiate from neurological deterioration caused by *Herpes* encephalitis.

- A 67-year-old man developed severe neurotoxicity after a 7-day course of high-dose aciclovir (800 mg/day) for *Herpes zoster* (17). He had severe disturbance of consciousness, stupor, and loss of spontaneous movement but regained consciousness after only two sessions of hemodialysis.

The authors concluded that hemodialysis is highly effective in eliminating aciclovir and that it can help to differentiate aciclovir-induced neurotoxicity from neurological deterioration due to the underlying encephalitis.

In a retrospective analysis of samples sent for analysis of aciclovir concentrations, neuropsychiatric syndromes correlated with concentrations of 9-carboxymethoxymethylguanine (CM MG), the main metabolite of aciclovir (18). Based on a retrospective chart analysis, one psychiatrist unaware of the drug concentrations divided patients in one group with neuropsychological symptoms (n = 49) and one group without (n = 44). By ROC analysis, CM MG was the strongest predictor of neuropsychiatric symptoms. Symptoms included agitation, confusion, pronounced tiredness, lethargy, coma, dysarthria, myoclonus, and hallucinations and started within 1-2 days of aciclovir administration. Using a cut-off value of 11 mmol/l of CM MG, sensitivity and specificity were 91 and 93% respectively. CM MG was by far the best predictor of neuropsychiatric disorders, compared with aciclovir drug concentrations or exposure, serum creatinine or creatinine clearance. Thus, CM MG concentrations might be a useful alternative tool to differentiate conditions associated with *Herpes* infection from adverse effects of aciclovir. This group also reported the effect of hemodialysis. Among 39 patients who recovered from their condition, 12 received hemodialysis and recovered shortly thereafter. Thus, this study supports the preliminary findings reported in the case report above.

Sensory systems

Local application of 3% ophthalmic ointment can cause mild transient stinging. Diffuse, superficial, punctate, non-progressive keratopathy can develop. This quickly resolves after withdrawal (19,20).

Psychological, psychiatric

One report described reversible psychiatric adverse effects in three dialysis patients receiving intravenous aciclovir (8–10 mg/kg/day) (21).

Hematologic

Neutropenia and thrombocytopenia occurred in an 8-year-old boy who was treated with aciclovir 200 mg bd for 5 months for "chronic cold sores" (22). After withdrawal of aciclovir, the absolute neutrophil and platelet counts normalized within days. There was no recurrence of oral herpes lesions during the ensuing month.

Gastrointestinal

In a randomized, double-blind comparison of aciclovir and famciclovir in 55 patients with uncomplicated *Herpes zoster* infections, constipation was the only significantly different adverse reaction, occurring in 11% and 3.7% respectively (23).

Urinary tract

Renal impairment has been associated with the use of intravenous aciclovir. Transient increases in serum creatinine and urea have been observed in 14% of patients treated with bolus injections (24). These are related to crystal formation in the lower renal tubules when the solubility of aciclovir in urine is exceeded. Slow (1-hour) intravenous infusion and adequate hydration are therefore mandatory. Bolus doses are to be avoided. Dosage modifications for patients with renal insufficiency are based on creatinine clearance (4).

- Crystalluria due to aciclovir occurred within 24 hours of the start of therapy with 500 mg 8-hourly in a 4-year-old African-American boy (25). Slow intravenous infusion over 1–2 hours and volume repletion avoids the problem.

Renal toxicity has not been described in infants treated with intravenous aciclovir, 5–10 mg/kg every 8 hours for 5–10 days (26) or in children receiving aciclovir 500 mg/m^2 intravenously (27) or orally (4).

Reversible acute renal insufficiency has been described after intravenous aciclovir, possibly related to changes in vascular resistance (28).

- A 28-year-old immunocompetent man who had typical varicella infection, a serum creatinine concentration of 106 µmol/l, and normal urinalysis and chest X-ray. He was given intravenous aciclovir 8.5 mg/kg (340 mg/m^2) every 8 hours. During the third infusion, he complained of mild pain in both loins. There was proteinuria (++) and hematuria (++) and his serum creatinine increased to 196 µmol/l. There were no crystals in the urine. Although aciclovir was withdrawn, the serum creatinine rose to 266 µmol/l on the second day. After four days renal function returned to normal. Aciclovir 5 mg/kg was then given again for 4 days without any problem. The patient refused renal biopsy.
- A 21 year old immunocompetent man who had a varicella infection was given intravenous aciclovir 10 mg/kg and during the third infusion complained of nausea and severe pain in both renal angles. He had proteinuria (++), hematuria (+), and reduced renal function (serum creatinine 238 µmol/l). There were no crystals in the urine. Although aciclovir was withdrawn, the serum creatinine rose to 406 µmol/l without any alteration in urine output. A renal biopsy showed minimal changes in the cortex area; the glomeruli showed a small increase in the number of mesangial cells, with some vacuolation of the tubular epithelium; the medulla contained small foci of mononuclear cells and a few polymorphonuclear leukocytes in the interstitium; the

tubules showed ischemia and focal epithelial regeneration, with a few epithelial deposits of hemosiderin. Seven days after withdrawal of aciclovir the renal function returned to normal.

The association with aciclovir in both of these cases was not strong.

Reversible renal failure and encephalopathy have been attributed to a higher dose of aciclovir (29).

- A 42 year old immunocompetent man with herpetic encephalitis was given cefotaxime, fosfomycin, metronidazole, and aciclovir 15 mg/kg 8-hourly. His creatinine rose from 90 µmol/l to a peak of 381 µmol/l within 24 days. This was associated with proteinuria but normal renal ultrasonography. It reversed after withdrawal of aciclovir and intravenous fluids.

Skin

Skin reactions to aciclovir are mostly mild and transitory, including pruritus, pain, rashes, contact dermatitis, and photoallergic contact dermatitis. However, serious reactions occasionally occur (30). Antiviral drugs that have been implicated include topical aciclovir, cidofovir, idoxuridine, imiquimod, lamivudine, penciclovir, podophyllin, podophyllotoxin, trifluridine, tromantadine, vidarabine, intralesional and ophthalmic solutions of interferon, intravitreal injections of fomivirsen and foscarnet, and intraocular implants of ganciclovir. Patch-testing in these cases only rarely caused positive reactions to the antiviral drug.

A case of possible aciclovir-induced Stevens–Johnson syndrome has been reported in an HIV-positive patient with mycobacterial disease (31). However, Stevens–Johnson syndrome is associated with *Herpes simplex* infection and can be prevented by aciclovir (32).

Recall phenomenon, classically chemotherapy-induced reactivation of skin damage caused by radiotherapy months or years later, has been attributed to aciclovir (33).

- A 48-year-old woman took aciclovir 800 mg five times a day for thoracic Herpes zoster. After most of the skin lesions had healed at day 7, she developed a generalized cutaneous erythematous maculopapular eruption. On the previously affected dermatome, confluent linear erythema appeared, designated by the authors as a recall phenomenon, classically a chemotherapy-induced reactivation of skin damage caused by radiotherapy months or years later.

Immunologic

Although allergy to aciclovir is unusual, it can occur; in one case it resulted in a skin rash (34).

- A 38-year-old woman of African descent, with a history of atopy and mild asthma, developed a periumbilical, erythematous, maculopapular rash and generalized pruritus after starting aciclovir. The reaction resolved within a few days after withdrawal, recurred when famciclovir was used, and again resolved when famciclovir was withdrawn. She was successfully stabilized on suppressive therapy after a graded challenge with aciclovir four times a day for 5 days.

Cross-reactivity between aciclovir and famciclovir is unusual. Aciclovir desensitization may be a novel method of treating patients with aciclovir allergy.

Contact sensitization to aciclovir is rare, but frequent application to inflamed skin in relapsing *Herpes simplex* may increase the risk of allergy. Severe contact dermatitis in a teenager has been reported.

- A 16-year-old girl with an 11-year history of frequent cold sores developed an erythematous rash and severe contact dermatitis during oral and topical aciclovir therapy (35). Patch tests showed contact sensitization to aciclovir and to the related compound ganciclovir.
- In a 44-year-old woman who used topical aciclovir for genital herpes, aciclovir contact allergy was associated with a systemic contact allergic reaction with an erythematous vesiculobullous eruption in the labial and perioral skin and a rash on the upper trunk and extremities (36). Patch tests were positive to aciclovir, valaciclovir, and ganciclovir, but not to famciclovir.

Pre-existing vesicular edematous cheilitis (probably due to contact allergy to the protecting lip salve) was aggravated after application of Zovirax cream (37). Patch tests to the lip salve were positive, but in addition there were positive photopatch tests to Zovirax cream, but not to its separate constituents.

Second-Generation Effects

Pregnancy

Aciclovir crosses the placenta and ins concentrated in the amniotic fluid but not in the fetus. The International Acyclovir Pregnancy Registry documented birth outcomes of 1246 women exposed to oral or intravenous aciclovir during pregnancy (61% first trimester, 16% second trimester, 23& third trimester) (38). No aciclovir-associated birth defects were reported. The data had the power to exclude a seven-fold increase in the risk of overall birth defects, but could not address the risks of rare and specific defects.

Animal data suggest that aciclovir is probably safe in pregnancy. There are no reports of teratogenicity in humans, and a report of 312 pregnant women exposed to aciclovir showed no increase in the number of birth defects compared with the numbers expected in the general population (39). However, data from larger numbers of human pregnancies are not available to draw reliable conclusions about the safety of aciclovir in pregnancy.

Drug Administration

Drug formulations

A novel method of applying 5% aciclovir cream to cold sores using iontophoresis has been assessed in a randomized placebo-controlled trial in 200 patients with an incipient outbreak of Herpes labialis infection at the erythema or papular/edema lesion stage (40). The median classic lesion healing time was 1.5 days shorter with the active treatment than with the vehicle (113 versus 148 hours). There were 17 adverse events in 12 patients who were given aciclovir, compared with 19 events in 13 patients who were given placebo. The most common adverse event was an electrical sensation (n = 2 versus 1). There were also single instances each of mildly burned skin and mild erythema on the lip, both with aciclovir. In neither case was treatment required.

Drug administration route

Local necrosis and inflammation can occur due to extravasation of the drug at the site of injection (41).

Various local adverse effects of aciclovir eye-drops have been reported, including pruritus, burning sensations, and irritative or allergic conjunctivitis. Persistent superficial punctate keratitis, delayed epithelial healing, and epithelial dysplasia can develop (42).

References

1. Wagstaff AJ, Faulds D, Goa KL. Aciclovir. A reappraisal of its antiviral activity, pharmacokinetic properties and therapeutic efficacy. Drugs 1994;47(1):153–205.
2. Erlich KS, Mills J, Chatis P, Mertz GJ, Busch DF, Follansbee SE, Grant RM, Crumpacker CS. Acyclovir-resistant *Herpes simplex* virus infections in patients with the acquired immunodeficiency syndrome. N Engl J Med 1989;320(5):293–6.
3. Sacks SL, Wanklin RJ, Reece DE, Hicks KA, Tyler KL, Coen DM. Progressive esophagitis from acyclovir-resistant *Herpes simplex*. Clinical roles for DNA polymerase mutants and viral heterogeneity? Ann Intern Med 1989;111(11):893–9.
4. Drucker JL, Tucker WE, Szczech M Jr. Safety studies of acyclovir: preclinical and clinical. In: Baker DA, editor. Acyclovir Therapy for Herpesvirus Infections. New York: Marcel Dekker, 1990:15.
5. Wade JC, Meyers JD. Neurologic symptoms associated with parenteral acyclovir treatment after marrow transplantation. Ann Intern Med 1983;98(6):921–5.
6. Wade JC, Hintz M, McGuffin R, Springmeyer SC, Connor JD, Meyers JD. Treatment of cytomegalovirus pneumonia with high-dose acyclovir. Am J Med 1982;73(1A):249–56.
7. Straus SE, Smith HA, Brickman C, de Miranda P, McLaren C, Keeney RE. Acyclovir for chronic mucocutaneous *Herpes simplex* virus infection in immunosuppressed patients. Ann Intern Med 1982;96(3):270–7.

8. Taback NA, Bradley C. Validation of the genital herpes treatment satisfaction questionnaire (GHerpTSQ) in status and change versions. Qual Life Res 2006; 15(6): 1043-52.

9. Conant MA, Schacker TW, Murphy RL, Gold J, Crutchfield LT, Crooks RJ, Acebes LO, Aiuti F, Akil B, Anderson J, Melville RL, Ballesteros Martin J, Berry A, Weiner M, Black F, Anderson PL, Bockman W, Borelli S, Bradbeer CS, Braffman M, Brandon W, Clark R, Wisniewski T, Bruun JN, Burdge D, Caputo RM, Chateauvert M, LaLonde R, Chiodo F, et alInternational Valaciclovir HSV Study Group. Valaciclovir versus aciclovir for *Herpes simplex* virus infection in HIV-infected individuals: two randomized trials. Int J STD AIDS 2002;13(1):12–21.

10. Ernst ME, Franey RJ. Acyclovir- and ganciclovir-induced neurotoxicity. Ann Pharmacother 1998;32(1):111–3.

11. Gomez Campdera FJ, Verde E, Vozmediano MC, Valderrabano F. More about acyclovir neurotoxicity in patients on haemodialysis. Nephron 1998;78(2):228–9.

12. Rajan GR, Cobb JP, Reiss CK. Acyclovir induced coma in the intensive care unit. Anaesth Intensive Care 2000;28(3):305–7.

13. Hsu CC, Lai TI, Lien WC, Chen WJ, Fang CC. Emergent hemodialysis for acyclovir toxicity. Am J Emerg Med 2005; 23(7): 899-900.

14. Drugspedia. Acyclovir. http://drugspedia.net/prep/24898.html [last accessed 27 October 2007].

15. Chevret L, Debray D, Poulain C, Durand P, Devictor D. Neurological toxicity of acyclovir: report of a case in a six-month-old liver transplant recipient. Pediatr Transplant 2006; 10(5): 632-4.

16. Helldén A, Lycke J, Vander T, Svensson JO, Odar-Cederlöf I, Ståhle L. The aciclovir metabolite CMMG is detectable in the CSF of subjects with neuropsychiatric symptoms during aciclovir and valaciclovir treatment. J Antimicrob Chemother 2006; 57(5): 945-9.

17. Chiang C-K, Fang C-C, Hsu W D, Chu T-S, Tsai T-J. Hemodialysis reverses acyclovir-induced nephrotoxicity and neurotoxicity. Dial Transplant 2003; 32: 624.

18. Hellden A, Odar-Cederlof I, Diener P, Barkholt L, Medin C, Svensson JO, Sawe J, Stahle L. High serum concentrations of the acyclovir main metabolite 9-carboxymethoxymethylguanine in renal failure patients with acyclovir-related neuropsychiatric side effects: an observational study. Nephrol Dial Transplant 2003; 18: 1135-41.

19. McGill J, Tormey P. Use of acyclovir in herpetic ocular infection. Am J Med 1982;73(1A):286–9.

20. Richards DM, Carmine AA, Brogden RN, Heel RC, Speight TM, Avery GS. Acyclovir. A review of its pharmacodynamic properties and therapeutic efficacy. Drugs 1983;26(5):378–438.

21. Tomson CR, Goodship TH, Rodger RS. Psychiatric side-effects of acyclovir in patients with chronic renal failure. Lancet 1985;2(8451):385–6.

22. Grella M, Ofosu JR, Klein BL. Prolonged oral acyclovir administration associated with neutropenia and thrombocytopenia. Am J Emerg Med 1998;16(4):396–8.

23. Shen MC, Lin HH, Lee SS, Chen YS, Chiang PC, Liu YC. Double-blind, randomized, acyclovir-controlled, parallel-group trial comparing the safety and efficacy of famciclovir and acyclovir in patients with uncomplicated herpes zoster. J Microbiol Immunol Infect 2004;37(2):75–81.

24. Brigden D, Rosling AE, Woods NC. Renal function after acyclovir intravenous injection. Am J Med 1982;73(1A):182–5.

25. Blossom AP, Cleary JD, Daley WP. Acyclovir-induced crystalluria. Ann Pharmacother 2002;36(3):526.

26. Yeager AS. Use of acyclovir in premature and term neonates. Am J Med 1982;73(1A):205–9.

27. Blanshard C, Benhamou Y, Dohin E, Lernestedt JO, Gazzard BG, Katlama C. Treatment of AIDS-associated gastrointestinal cytomegalovirus infection with foscarnet and ganciclovir: a randomized comparison. J Infect Dis 1995;172(3):622–8.

28. Bassioukas K, Katopodis K, Christophorou N, Takouli L, Dimou S, Hatzis J. Acute renal failure induced by intravenous acyclovir. J Eur Acad Dermatol Venereol 2006; 20(9): 1151-2.

29. De Deyne S, De la Gastine B, Gras G, Dargere S, Verdon R, Coquerel A. Acute renal failure with acyclovir in a 42-year-old patient without previous renal dysfunction. Rev Med Interne 2006; 27(11): 892-4.

30. Holdiness MR. Contact dermatitis from topical antiviral drugs. Contact Dermatitis 2001;44(5):265–9.

31. Fazal BA, Turett GS, Justman JE, Hall G, Telzak EE. Stevens–Johnson syndrome induced by treatment with acyclovir. Clin Infect Dis 1995;21(4):1038–9.

32. Cheriyan S, Patterson R. Recurrent Stevens–Johnson syndrome secondary to *Herpes simplex*: a follow up on a successful management program. Allergy Asthma Proc 1996;17(2):71–3.

33. Carrasco L, Pastor MA, Izquierdo MJ, Farina MC, Martin L, Fortes J, Requena L. Drug eruption secondary to aciclovir with recall phenomenon in a dermatome previously affected by Herpes zoster. Clin Exp Dermatol. 2002; 27: 132-4.

34. Kawsar M, Parkin JM, Forster G. Graded challenge in an aciclovir allergic patient. Sex Transm Infect 2001;77(3):204–205.

35. Wollenberg A, Baldauf C, Rueff F, Przybilla B. Allergic contact dermatitis and exanthematous drug eruption following aciclovir-cross reaction with ganciclovir. Allergo J 2000,9,96–9.

36. Lammintausta K, Makela L, Kalimo K. Rapid systemic valaciclovir reaction subsequent to aciclovir contact allergy. Contact Dermatitis 2001;45(3):181.

37. Rodriguez WJ, Bui RH, Connor JD, Kim HW, Brandt CD, Parrott RH, Burch B, Mace J. Environmental exposure of primary care personnel to ribavirin aerosol when supervising treatment of infants with respiratory syncytial virus infections. Antimicrob Agents Chemother 1987;31(7):1143–6.

38. Andrews EB, Yankaskas BC, Cordero JF, Schoeffler K, Hampp S. Acyclovir in pregnancy registry: six years' experience. The Acyclovir in Pregnancy Registry Advisory Committee. Obstet Gynecol 1992;79(1):7–13.

39. Stone KM, Reiff-Eldridge R, White AD, Cordero JF, Brown Z, Alexander ER, Andrews EB. Pregnancy outcomes following systemic prenatal acyclovir exposure: Conclusions from the international acyclovir pregnancy registry, 1984-1999. Birth Defects Res A Clin Mol Teratol. 2004;70(4):201–7.

40. Morrel EM, Spruance SL, Goldberg DI. Topical iontophoretic administration of acyclovir for the episodic treatment of Herpes labialis: a randomized, double-blind, placebo-controlled, clinic-initiated trial. Clin Infect Dis 2006; 43(4): 460-7.

41. Sylvester RK, Ogden WB, Draxler CA, Lewis FB. Vesicular eruption. A local complication of concentrated acyclovir infusions. JAMA 1986;255(3):385–6.

42. Ohashi Y. Treatment of herpetic keratitis with acyclovir: benefits and problems. Ophthalmologica 1997;211(Suppl 1):29–32.

Adefovir

General Information

Adefovir is an adenine analogue reverse transcriptase inhibitor. While it has activity against both HIV and hepatitis B, its use in HIV infection is limited by nephrotoxicity due to the high doses needed (1). The dose used for treatment of hepatitis B is about one-tenth that needed to treat HIV infection, so patients with hepatitis B must have co-infection with HIV ruled out before treatment is started.

Observational studies

A total of 49 consecutive lamivudine-resistant hepatitis B e antigen-negative chronic hepatitis B patients were enrolled in a study of the effects of adefovir 10 mg/day plus lamivudine 100 mg/day (2). After 52 weeks all had some hepatitis B virus DNA response and 57% had a complete virological response. There was a biochemical response in 76%. There were no serious adverse events.

Organs and Systems

Liver

In 35 patients co-infected with hepatitis B virus and HIV given adefovir 10 mg/day plus lamivudine 150 mg bd as part of treatment for hepatitis B and followed for 48 weeks, common adverse effects included raised transaminases, particularly alanine transaminase, increased serum creatinine, and increased blood glucose (3). Concerns about the study include the small number of patients and the lack of a comparison group.

Urinary tract

Adefovir is nephrotoxic, particularly at high doses, and the possible mechanism has been investigated in a 39-year-old man, who had severe acute tubular degenerative changes mainly affecting the proximal tubule; the mitochondria were significantly enlarged, possibly as a result of depletion of mitochondrial DNA (4).

The safety profile of the 10-mg dose of adefovir was similar to that of placebo after 48 weeks of treatment in HBeAg-positive patients; there were renal laboratory abnormalities only at the 30-mg dose (5).

References

1. Danta M, Dusheiko G. Adefovir dipivoxil: review of a novel acyclic nucleoside analogue. Int J Clin Pract 2004;58(9):877–86.
2. Vassiliadis T, Nikolaidis N, Giouleme O, Tziomalos K, Grammatikos N, Patsiaoura K, Zezos P, Gkisakis D, Theodoropoulos K, Katsinelos P, Orfanou-Koumerkeridou E, Eugenidis N. Adefovir dipivoxil added to ongoing lamivudine therapy in patients with lamivudine-resistant hepatitis B e antigen-negative chronic hepatitis B. Aliment Pharmacol Ther 2005;21(5):531–7.
3. Benhamou Y, Bochet M, Thibault V, Calvez V, Fievet MH, Vig P, Gibbs CS, Brosgart C, Fry J, Namini H, Katlama C, Poynard T. Safety and efficacy of adefovir dipivoxil in patients co-infected with HIV-1 and lamivudine-resistant hepatitis B virus: an open-label pilot study. Lancet 2001;358(9283):718–23.
4. Tanji N, Tanji K, Kambham N, Markowitz GS, Bell A, D'agati VD. Adefovir nephrotoxicity: possible role of mitochondrial DNA depletion. Hum Pathol 2001;32(7):734–40.
5. Forton D, Karayiannis P. Established and emerging therapies for the treatment of viral hepatitis. Dig Dis 2006;24(1-2):160–73.

Amantadine

General Information

Amantadine is a symmetrical C10 tricyclic amine with an unusual structure (1-adamantanamine hydrochloride). It interferes with virus uncoating (1) by blocking the M2 ion channel, which is needed to affect a pH change that helps to initiate the uncoating process. Most consistent antiviral activity has been observed against influenza A virus, but amantadine has little or no activity against influenza B virus (2). However, influenza A virus can become rapidly resistant to amantadine in vitro (3). Amantadine also promotes the release of dopamine from nerve endings, but may also delay its reuptake into synaptic vesicles.

Doses of 100–200 mg/day are often used and are usually well tolerated. The best-documented adverse reactions are nausea, psychotic episodes (mania, hallucinations, agitation, confusion), restless legs, and convulsions. Minor adverse reactions often resemble those caused by anticholinergic agents, for example blurred vision, dryness of the mouth, insomnia, lethargy, and rash. In rare cases, photosensitization has been described. Oral or aerosol administration can be accompanied by gastrointestinal or minor neurological symptoms, such as insomnia, light-headedness, concentration difficulties, nervousness, dizziness, and headache in individuals taking 200 mg/day (4). These symptoms disappear on withdrawal. More severe but rare complications include convulsions and coma (5,6). Local nasal adverse effects of the aerosolized form can mimic the symptoms of upper respiratory tract infection. In Parkinson's disease, in which doses of 200 mg/day or more have been used, minor adverse effects resembling those caused by anticholinergic agents (for example blurred vision, dry mouth, as well as livedo reticularis, rash, and photosensitization, can occur (SED-13, 874) (7,8).

Organs and Systems

Cardiovascular

Reversible congestive cardiac failure has been attributed to amantadine (9).

Nervous system

Insomnia is common with amantadine (10).

Myoclonus, especially vocal, can occur (11).

- A 48-year-old woman with a 17-year history of Parkinson's disease developed a sensorimotor peripheral neuropathy after taking amantadine 300 mg/day for 8 years (12). She had livedo reticularis after only 1 year of treatment and this had become increasingly extensive. Attempts to withdraw the drug resulted in worsening Parkinsonian symptoms. However, after the neuropathy had been diagnosed, amantadine was withdrawn, with improvement of the neurological symptoms within 6 weeks and complete resolution after 6 months. However, the livedo reticularis was still present 18 months after withdrawal.

This is the first description of peripheral neuropathy in these circumstances, but it suggests that chronic livedo reticularis may be a forerunner of more severe problems.

Three Japanese women aged 78–87 years who had taken amantadine 100–200 mg/day for 1 month to 5 years, in two cases together with co-careldopa, developed multifocal myoclonus and two were confused (13). Amantadine concentrations were high in the two patients in whom they were measured, at over 3000 ng/ml; a concentration over 1000 ng/ml is regarded as toxic. Amantadine was withdrawn and the myoclonus disappeared within 1–2 weeks and did not recur. Cortical myoclonus has also been described with levodopa and bromocriptine, but the mechanism is not known.

Sensory systems

Rarely, amantadine causes visual impairment due to corneal abrasions, local edema, and superficial keratitis (14).

Psychological, psychiatric

The risk of mental complications seems to increase substantially if doses of 200 mg or more are given (15). Amantadine can cause mania and is contraindicated in patients with bipolar affective disorder (16).

- While taking amantadine, an elderly man developed the Othello syndrome, a severe delusion of marital infidelity, as described in Shakespeare's plays *Othello* and *A Winter's Tale*; it abated with drug withdrawal (SEDA-17, 170).
- Amantadine and phenylpropanolamine may have caused intense recurrent déjà vu experiences in a healthy 39-year-old within 24 hours of starting both drugs for influenza (17).

Fluid balance

Resistant edema of the ankles in the absence of cardiac disease has repeatedly been described and well documented by rechallenge (18).

Skin

Livedo reticularis has been stated to occur in as many as 90% of patients; it is very common in female patients especially those with antiphospholipid antibodies (19).

Long-Term Effects

Drug withdrawal

On withdrawal of amantadine after long-term therapy, acute delirium with confusion, disorientation, agitation, and paranoia occurred in three patients (20). Reintroduction of the drug returned the patients to the baseline status.

In four patients, amantadine withdrawal was associated with delirium and confusion (21). The patients, three of them women, were aged 70–83 years and all but one were considered to have early dementia. Amantadine exposure was 1–5 years and the symptoms occurred within a week of drug withdrawal. In all cases they were reversed by reintroduction. The mechanism of the withdrawal reaction is unknown, but it should obviously be borne in mind, especially in elderly patients with long exposure to the drug and with already impaired cognitive function.

Neuroleptic malignant syndrome can occur on withdrawal of amantadine (22,23).

Susceptibility Factors

Renal disease

In a case of end-stage renal insufficiency, amantadine caused coma, though drug plasma concentrations were not greatly increased (SEDA-18, 160).

References

1. Bukrinskaya AG, Vorkunova NK, Narmanbetova RA. Rimantadine hydrochloride blocks the second step of influenza virus uncoating. Arch Virol 1980;66(3):275–82.
2. Oxford JS, Galbraith A. Antiviral activity of amantadine: a review of laboratory and clinical data. Pharmacol Ther 1980;11(1):181–262.
3. Heider H, Adamczyk B, Presber HW, Schroeder C, Feldblum R, Indulen MK. Occurrence of amantadine- and rimantadine-resistant influenza A virus strains during the 1980 epidemic. Acta Virol 1981;25(6):395–400.
4. Van Voris LP, Newell PM. Antivirals for the chemoprophylaxis and treatment of influenza. Semin Respir Infect 1992;7(1):61–70.
5. Bryson YJ, Monahan C, Pollack M, Shields WD. A prospective double-blind study of side effects associated with the administration of amantadine for influenza A virus prophylaxis. J Infect Dis 1980;141(5):543–7.
6. Ing TS, Daugirdas JT, Soung LS, Klawans HL, Mahurkar SD, Hayashi JA, Geis WP, Hano JE. Toxic effects of amantadine in patients with renal failure. Can Med Assoc J 1979;120(6):695–8.
7. Silver DE, Sahs AL. Livedo reticularis in Parkinson's disease patients treated with amantadine hydrochloride. Neurology 1972;22(7):665–9.

8. van den Berg WH, van Ketel WG. Photosensitization by amantadine (Symmetrel). Contact Dermatitis 1983;9(2):165.

9. Vale JA, Maclean KS. Amantadine-induced heart-failure. Lancet 1977;1(8010):548.

10. Huete B, Varona L. Insomnio durante el tratamiento con amantadina. [Insomnia during treatment with amantadine.] Rev Neurol 1997;25(148):2062.

11. Pfeiffer RF. Amantadine-induced "vocal" myoclonus. Mov Disord 1996;11(1):104–6.

12. Shulman LM, Minagar A, Sharma K, Weiner WJ. Amantadine-induced peripheral neuropathy. Neurology 1999;53(8):1862–5.

13. Matsunaga K, Uozumi T, Qingrui L, Hashimoto T, Tsuji S. Amantadine-induced cortical myoclonus. Neurology 2001;56(2):279–80.

14. Nogaki H, Morimatsu M. Superficial punctate keratitis and corneal abrasion due to amantadine hydrochloride. J Neurol 1993;240(6):388–9.

15. Hunter KR. Treatment of parkinsonsim. Prescr J 1976;16:101.

16. Rego MD, Giller EL Jr. Mania secondary to amantadine treatment of neuroleptic-induced hyperprolactinemia. J Clin Psychiatry 1989;50(4):143–4.

17. Taiminen T, Jaaskelainen SK. Intense and recurrent deja vu experiences related to amantadine and phenylpropanolamine in a healthy male. J Clin Neurosci 2001;8(5):460–2.

18. New Approvals/Indications Pediatric indications for metaproterenol. Drug Ther 1992;Feb 22 and 38.

19. Paulson GW, Brandt JT. Amantadine, livedo reticularis, and antiphospholipid antibodies. Clin Neuropharmacol 1995;18(5):466–7.

20. Factor SA, Molho ES, Brown DL. Acute delirium after withdrawal of amantadine in Parkinson's disease. Neurology 1998;50(5):1456–8.

21. Sommer BR, Wise LC, Kraemer HC. Is dopamine administration possibly a risk factor for delirium? Crit Care Med 2002;30(7):1508–11.

22. Ito T, Shibata K, Watanabe A, Akabane J. Neuroleptic malignant syndrome following withdrawal of amantadine in a patient with influenza A encephalopathy. Eur J Pediatr 2001;160(6):401.

23. Simpson DM, Davis GC. Case report of neuroleptic malignant syndrome associated with withdrawal from amantadine. Am J Psychiatry 1984;141(6):796–7.

Amprenavir

See also Protease inhibitors

General Information

Amprenavir is an HIV protease inhibitor with an enzyme inhibitory constant of 0.6 nmol/l, similar to the inhibitory constants of other protease inhibitors. Its in vitro IC_{50} against wild-type clinical HIV isolates is 115 ng/ml. It has a long half-life (7–10 hours). It can be given twice-daily without food restrictions and had high potency when given as monotherapy in dose-finding studies (1). The recommended doses are 1200 mg bd for adults and 20 mg/kg bd or 15 mg/kg tds for children under 13 years of age or adolescents under 50 kg. Capsules and solution do not have equal systemic availability, and the recommended dose for amprenavir oral solution is 1.5 ml/kg bd or 1.1 ml/kg tds. The systemic availability increases with increasing doses. Amprenavir is about 90% bound to $alpha_1$ acid glycoprotein and 40% to albumin. It does not penetrate the brain well, because it is exported by P glycoprotein. It is mainly metabolized by CYP3A4 and its clearance is reduced in liver disease. The clinical pharmacology of amprenavir has been reviewed (2).

The use of amprenavir has been limited to patients who are highly motivated, because of the high capsule burden (16/day).

Fosamprenavir is a prodrug of amprenavir with better systemic availability. In large clinical trials the most common adverse events in patients taking fosamprenavir, with or without ritonavir, plus abacavir and lamivudine were diarrhea, nausea, vomiting, abdominal pain, drug hypersensitivity, and skin rashes (3).

Observational studies

In a monotherapy trial ($n = 37$), adverse effects were frequent but generally mild, and included rash, diarrhea or loose stools, and headache. In general, these adverse effects tend to disappear or weaken in severity within the first 2–4 weeks of treatment.

In a study of the pharmacokinetics of oral amprenavir administered as soft gelatin capsules to 20 HIV-positive children, the most common adverse event was nausea (4). The kinetics supported twice daily dosing with 20 mg/kg.

Comparative studies

In 249 patients fosamprenavir 1400 mg bd (n = 166) was compared with nelfinavir 1250 mg bd (n = 83), each combined with abacavir and lamivudine, the most common adverse event was diarrhea in both groups, but significantly more often with nelfinavir (5). There was increased lipase activity in 8% of the patients who took fosamprenavir but it was without clinical significance. There were rashes in 7% of the patients who took fosamprenavir. In all 14 patients withdrew from the study because of adverse events (9/166 in the fosamprenavir group; 5/83 in the nelfinavir group). There was a slight increase in total cholesterol in both groups. There were increases in LDL cholesterol requiring medical intervention in 18% of the patients in both groups; the ratio of total:HDL cholesterol was only slightly affected.

General adverse effects

The adverse effects of amprenavir in patients treated with combination therapy included nausea, vomiting, diarrhea, epigastric pain, flatulence, paresthesia, headache, rash, and fatigue (6). The contribution of a single drug to the observed adverse effects is difficult to establish. Amprenavir inhibits CYP3A4 to a greater extent than saquinavir, and to a much lesser extent than ritonavir (7). Co-administration with rifampicin and rifabutin should be avoided. Those who take amprenavir have complained of diarrhea, nausea, headache, and fatigue (8). The frequency of diarrhea may be as high as 50%.

Organs and Systems

Nervous system

Neurotoxicity has been attributed to amprenavir (9).

- A 61-year-old man, who had taken various antiretroviral drugs, took amprenavir 750 mg bd and after the first dose had hallucinations, disorientation, tinnitus, and vertigo. The symptoms abated within 2 hours and recurred after the next dose.

Metabolism

In a phase 1, open, randomized, crossover study 42 subjects were assigned to one of six treatment sequences, each consisting of three periods of 14 days, in which subjects were to receive one of the following treatments with a washout period of 21–28 days between treatments:

- fosamprenavir 700 mg bd + ritonavir 100 mg bd (treatment A);
- foasamprenavir 1400 mg bd + ritonavir 100 mg bd (treatment B);
- foasamprenavir 1400 mg bd + ritonavir 200 mg bd (treatment C).

In six subjects there were marked rises in serum transaminases, mostly in those who took treatment C (10). Mean fasting serum triglycerides increased significantly during all treatments. Mean fasting serum cholesterol increased during treatment A and mean LDL fell during treatment B. Mean HDL cholesterol fell during all the treatments. These lipid changes tended to abate or return to baseline during follow up.

Gastrointestinal

Nausea and vomiting were the main adverse effects that led to withdrawal (in 6%) in patients who were given ritonavir-boosted amprenavir 600/100 bd (11).

Drug Administration

Drug formulations

In a study of an NRTI-based regimen combined with amprenavir 1200 mg bd or fosamprenavir 1395 mg bd or 1860 mg bd, all three regimens produced similar concentration time curves (12). The prodrug formulations led to lower maximal plasma concentrations and higher trough concentrations, thereby theoretically reducing the risk of toxic adverse effects. Adverse events reported in these studies were mainly gastrointestinal, with a slight trend to more frequent diarrhea with amprenavir (but the study was not powered to reach statistical significance on this aspect).

Drug–Drug Interactions

Clarithromycin

A pharmacokinetic study has shown a minor interaction of amprenavir with clarithromycin in healthy men (13).

The mean AUC, $C_{max.ss}$ and $C_{min.ss}$ of amprenavir increased by 18, 15, and 39% respectively. Amprenavir had no effect on the AUC of clarithromycin, but the median $t_{max.ss}$ increased by 2 hours, renal clearance increased by 34%, and the AUC for 14-(R)-hydroxyclarithromycin fell by 35%. These effects were felt not to be clinically important and dosage adjustment was not recommended.

Efavirenz

Four HIV-infected children undergoing intense antiretroviral combination therapy were switched to regimens including amprenavir and efavirenz after the failure of other drugs (14). Pharmacokinetic studies suggested that combinations of these drugs can result in suboptimal concentrations of amprenavir. This was evident in two of the children taking amprenavir and efavirenz, in combination with two NRTIs, who had undetectable concentrations of amprenavir within 4 hours of administration. The addition of ritonavir to the combination restored the blood concentrations of amprenavir to those normally recorded (median 3500 ng/ml). The most probable reason for this effect is enhanced metabolism of amprenavir due to induction by efavirenz.

Indinavir

In an open, randomized study of amprenavir combined with indinavir, nelfinavir, and saquinavir (15) the amprenavir AUC increased by 35% when it was combined with indinavir, and indinavir concentrations also fell, suggesting that this protease inhibitor combination should be avoided. There was no significant interaction of amprenavir with nelfinavir.

Lopinavir + ritonavir

mprenavir plus lopinavir + ritonavir as deep salvage treatment has been studied in a prospective single-center study in 22 patients with virological failure after other drugs (including protease inhibitors). Both amprenavir and lopinavir concentrations were reduced unpredictably when the two were combined (16). Because they have a limited cross-resistance profile, both lopinavir and amprenavir may be treatments of choice in such patients. However, because of their pharmacodynamic properties, the extent to which they can be combined is limited. The authors suggested plasma concentration monitoring in order to adjust amprenavir and lopinavir doses according to viral susceptibility.

Methadone

Amprenavir is extensively metabolized by and induces CYP3A4. Plasma methadone concentrations fell by 35% when amprenavir was used (17).

Rifamycins

Co-administration of amprenavir with rifampicin and rifabutin should be avoided (18).

Ritonavir

In attempts to lower the amprenavir capsule burden, low-dose ritonavir has been used as a pharmacokinetic booster. When ritonavir was added to amprenavir, the amprenavir AUC increased 3–4 times (19), which should allow the total daily capsule burden to be reduced. Adverse effects included diarrhea, nausea, paresthesia, rash, increased cholesterol, and increased triglycerides. The frequency of adverse events correlated with the dose of ritonavir.

Saquinavir

In an open, randomized study of amprenavir combined with indinavir, nelfinavir, and saquinavir (15), saquinavir lowered the amprenavir AUC by 32%; amprenavir did not alter the pharmacokinetics of saquinavir.

Food–Drug Interactions

Grapefruit juice

In 12 healthy volunteers, co-administration of a single dose of amprenavir 1200 mg with grapefruit juice slightly reduced the C_{max} compared with water (7.11 versus 9.10 µg/ml) and slightly increased the t_{max} (1.13 versus 0.75 hours), but did not affect the AUC (20). Thus, grapefruit juice has no clinically important effect on amprenavir pharmacokinetics.

References

1. Sadler BM, Hanson CD, Chittick GE, Symonds WT, Roskell NS. Safety and pharmacokinetics of amprenavir (141W94), a human immunodeficiency virus (HIV) type 1 protease inhibitor, following oral administration of single doses to HIV-infected adults. Antimicrob Agents Chemother 1999;43(7):1686–92.
2. Sadler BM, Stein DS. Clinical pharmacology and pharmacokinetics of amprenavir. Ann Pharmacother 2002;36(1):102–18.
3. Chapman TM, Plosker GL, Perry CM. Fosamprenavir: a review of its use in the management of antiretroviral therapy-naive patients with HIV infection. Drugs 2004;64(18):2101–24.
4. Morse GD, Rosenkranz S, Para MF, Segal Y, Difrancesco R, Adams E, Brizz B, Yarasheski KE, Reichman RC. Amprenavir and efavirenz pharmacokinetics before and after the addition of nelfinavir, indinavir, ritonavir, or saquinavir in seronegative individuals. Antimicrob Agents Chemother 2005;49(8):3373–81.
5. Rodriguez-French A, Boghossian J, Gray GE, Nadler JP, Quinones AR, Sepulveda GE, Millard JM, Wannamaker PG. The NEAT study: a 48-week open-label study to compare the antiviral efficacy and safety of GW433908 versus nelfinavir in antiretroviral therapy-naive HIV-1-infected patients. J Acquir Immune Defic Syndr 2004;35(1):22–32.
6. Adkins JC, Faulds D. Amprenavir. Drugs 1998;55(6):837–42.
7. Decker CJ, Laitinen LM, Bridson GW, Raybuck SA, Tung RD, Chaturvedi PR. Metabolism of amprenavir in liver microsomes: role of CYP3A4 inhibition for drug interactions. J Pharm Sci 1998;87(7):803–7.
8. Kost RG, Hurley A, Zhang L, Vesanen M, Talal A, Furlan S, Caldwell P, Johnson J, Smiley L, Ho D, Markowitz M. Open-label phase II trial of amprenavir, abacavir, and fixed-dose zidovudine/lamivudine in newly and chronically HIV-1-infected patients. J Acquir Immune Defic Syndr 2001;26(4):332–9.
9. James CW, McNelis KC, Matalia MD, Cohen DM, Szabo S. Central nervous system toxicity and amprenavir oral solution. Ann Pharmacother 2002;36(1):174.
10. Shelton MJ, Wire MB, Lou Y, Adamkiewicz B, Min SS. Pharmacokinetic and safety evaluation of high-dose combinations of fosamprenavir and ritonavir. Antimicrob Agents Chemother 2006;50(3):928–34.
11. de Mendoza C, Valer L, Ribera E, Barreiro P, Martín-Carbonero L, Ramirez G, Soriano V. Performance of six different ritonavir-boosted protease inhibitor-based regimens in heavily antiretroviral-experienced HIV-infected patients. HIV Clin Trials 2006;7(4):163–71.
12. Wood R, Arasteh K, Stellbrink HJ, Teofilo E, Raffi F, Pollard RB, Eron J, Yeo J, Millard J, Wire MB, Naderer OJ. Six-week randomized controlled trial to compare the tolerabilities, pharmacokinetics, and antiviral activities of GW433908 and amprenavir in human immunodeficiency virus type 1-infected patients. Antimicrob Agents Chemother 2004;48(1):116–23.
13. Brophy DF, Israel DS, Pastor A, Gillotin C, Chittick GE, Symonds WT, Lou Y, Sadler BM, Polk RE. Pharmacokinetic interaction between amprenavir and clarithromycin in healthy male volunteers. Antimicrob Agents Chemother 2000;44(4):978–84.
14. Wintergerst U, Engelhorn C, Kurowski M, Hoffmann F, Notheis G, Belohradsky BH. Pharmacokinetic interaction of amprenavir in combination with efavirenz or delavirdine in HIV-infected children. AIDS 2000;14(12):1866–8.
15. Sadler BM, Gillotin C, Lou Y, Eron JJ, Lang W, Haubrich R, Stein DS. Pharmacokinetic study of human immunodeficiency virus protease inhibitors used in combination with amprenavir. Antimicrob Agents Chemother 2001;45(12):3663–8.
16. De Luca A, Baldini F, Cingolani A, Di Giambenedetto S, Hoetelmans RM, Cauda R. Deep salvage with amprenavir and lopinavir/ritonavir: correlation of pharmacokinetics and drug resistance with pharmacodynamics. J Acquir Immune Defic Syndr 2004;35(4):359–66.
17. Bart PA, Rizzardi PG, Gallant S, Golay KP, Baumann P, Pantaleo G, Eap CB. Methadone blood concentrations are decreased by the administration of abacavir plus amprenavir. Ther Drug Monit 2001;23(5):553–5.
18. Polk RE, Brophy DF, Israel DS, Patron R, Sadler BM, Chittick GE, Symonds WT, Lou Y, Kristoff D, Stein DS. Pharmacokinetic interaction between amprenavir and rifabutin or rifampin in healthy males. Antimicrob Agents Chemother 2001;45(2):502–8.
19. Sadler BM, Gillotin C, Lou Y, Stein DS. Pharmacokinetic and pharmacodynamic study of the human immunodeficiency virus protease inhibitor amprenavir after multiple oral dosing. Antimicrob Agents Chemother 2001;45(1):30–7.
20. Demarles D, Gillotin C, Bonaventure-Paci S, Vincent I, Fosse S, Taburet AM. Single-dose pharmacokinetics of amprenavir coadministered with grapefruit juice. Antimicrob Agents Chemother 2002;46(5):1589–90.

Atazanavir

General Information

Atazanavir is an inhibitor of HIV protease. It is used once daily in a dose of 400 mg or in combination with ritonavir as atazanavir + ritonavir 300 + 100 mg/day (1).

Organs and Systems

Metabolism

In an analysis of a randomized comparison of atazanavir and nelfinavir in 467 patients cardiovascular risk modelling was used to estimate the impact of dyslipidemia (2). Concentrations of total cholesterol and low-density lipoprotein cholesterol increased significantly more among patients who used nelfinavir (24% and 28%) than among those who used atazanavir (4% and 1%). Overall, the relative risk of coronary disease, adjusted for risk status, age, and sex, was increased by 50% for nelfinavir versus atazanavir over the next 10 years in men or women, regardless of the presence or absence of other coronary risk factors.

Liver

In the first phase 2 licensing study the protease inhibitor atazanavir 200 mg, 400 mg, and 500 mg was evaluated against nelfinavir 750 mg tds in combination with a didanosine and stavudine backbone (3). The 400 mg dose of atazanavir was chosen for further development. Most of the adverse events in this study were grade 1–2 and the rates of grade 3–4 were comparable across the regimens. Jaundice was the most prominent adverse effect of atazanavir and was clearly dose related (6%, 6%, 12% for the three dosing arms) and was not observed in the nelfinavir arm. Rises in bilirubin were predominantly due to the unconjugated form. There was no correlation with raised transaminases.

Preclinical data support the hypothesis that atazanavir-associated hyperbilirubinemia is attributable to inhibition of uridine diphosphate glucuronosyltransferase (UDP-GT) 1A1 (4). This is also the apparent mechanism for the reversible rise in bilirubin that occurs with indinavir (5). Grade 3–4 rises in transaminases were significantly more frequent in patients infected with hepatitis B or C (20–40%) than in patients with negative hepatitis B and C serology (2–12%).

In a study of once-a-day ritonavir-boosted atazanavir in 23 children and adolescents total bilirubin increased significantly (to over 18 mmol/l) in 17 children (6). This was attributed to competitive inhibition of uridine diphosphate glucuronyl transferase (UGT) activity by atazanavir.

The most common adverse effect of atazanavir was a rise in total bilirubin concentration (mostly indirect/unconjugated bilirubin): in 26% of those who continued treatment with atazanavir 400 mg/day (n = 139), 44% of those who took atazanavir 600 mg/day (n = 144), and 13% of those who had previously taken nelfinavir followed by atazanavir 400 mg/day (n = 63) (7). There were grade 4 rises in bilirubin (to over five times the upper of limit normal) in 3% of patients taking atazanavir 600 mg/day and in 1% of patients taking atazanavir 400 mg/day. Jaundice was reported more often in those taking atazanavir 600 mg/day compared with those taking 400 mg/day (22% versus 13%).

Drug-Drug Interactions

Nevirapine

In a population pharmacokinetic analysis in 87 patients, mean trough plasma concentrations of atazanavir were 27, 32, and 81% lower in patients taking atazanavir + ritonavir associated with tenofovir, efavirenz, and nevirapine respectively (8). However, these reductions in trough plasma concentrations were not significant, except for those taking atazanavir + ritonavir and nevirapine. The authors concluded that the risk of reduced plasma concentrations of atazanavir due to increased atazanavir clearance is significant when atazanavir is used in combination with nevirapine and that atazanavir concentrations should be monitoring.

Atazanavir interacted with *saquinavir* hard gel 1600 mg/day + low-dose ritonavir 100 mg/day (9). Atazanavir 150, 200, and 300 mg/day significantly increased the C_{max} and AUC of saquinavir by 40–50% and 50–60% respectively. Ritonavir C_{max} and AUC were not affected by atazanavir 150 and 200 mg/day but were increased to 54 and 40% by atazanavir 300 mg/day. This suggests that the dosage of atazanavir should be reduced when it is co-administered with saquinavir and ritonavir, if atazanavir or ritonavir related adverse effects occur.

Tenofovir

The effect of tenofovir on the pharmacokinetics of atazanavir boosted with ritonavir has been evaluated over 6 weeks in 11 heavily pre-treated patients in the setting of salvage therapy (10). Co-administration of tenofovir led to a significant reduction in atazanavir AUC. The mechanism by which tenofovir reduced atazanavir concentrations remains to be elucidated. There was no significant correlation between the pharmacokinetics and the viral response, partly because of the small sample and probably also the high levels of resistance mutations at baseline.

References

1. von Hentig N. Atazanavir/ritonavir: a review of its use in HIV therapy. Drugs Today (Barc) 2008;44(2):103–32.
2. Grover SA, Coupal L, Gilmore N, Mukherjee J. Impact of dyslipidemia associated with Highly Active Antiretroviral Therapy (HAART) on cardiovascular risk and life expectancy. Am J Cardiol 2005;95(5):586–91.
3. Sanne I, Piliero P, Squires K, Thiry A, Schnittman S; AI424-007 Clinical Trial Group. Results of a phase 2 clinical trial at 48 weeks (AI424-007): a dose-ranging, safety, and efficacy comparative trial of atazanavir at three doses in

combination with didanosine and stavudine in antiretro-viral-naive subjects. J Acquir Immune Defic Syndr 2003;32:18–29.

4. O'Mara EM, Mummaneni V, Randall D, et al. Assessment of the effect of uridine diphosphate glucuronosyltransferase (UDP-GT) 1A1 genotype on indirect bilirubin elevations in healthy subjects dosed with BMS-232632. Presented at the 40th Interscience Conference on Antimicrobial Agents and Chemotherapy, Toronto, September 2000. http://www.as-musa.org/memonly/abstracts/abstractsearch.asp.

5. Zucker SD, Qin X, Rouster SD, Yu F, Green RM, Keshavan P, Feinberg J, Sherman KE. Mechanism of indi-navir-induced hyperbilirubinemia. Proc Natl Acad Sci USA 2001;98:12671–6.

6. Macassa E, Delaugerre C, Teglas JP, Jullien V, Tréluyer JM, Veber F, Rouzioux C, Blanche S. Change to a once-daily combination including boosted atazanavir in HIV-1-infected children. Pediatr Infect Dis J 2006;25(9):809–14.

7. Wood R, Phanuphak P, Cahn P, Pokrovskiy V, Rozenbaum W, Pantaleo G, Sension M, Murphy R, Mancini M, Kelleher T, Giordano M. Long-Term Efficacy and Safety of Atazanavir With Stavudine and Lamivudine in Patients Previously Treated With Nelfinavir or Atazanavir. J Acquir Immune Defic Syndr 2004;36(2):684–92.

8. João EC, Calvet GA, Menezes JA, D'Ippolito MM, Cruz ML, Salgado LA, Matos HJ. Nevirapine toxicity in a cohort of HIV-1-infected pregnant women. Am J Obstet Gynecol 2006;194(1):199–202.

9. Boffito M, Maitland D, Dickinson L, Back D, Hill A, Fletcher C, Moyle G, Nelson M, Gazzard B, Pozniak A. Pharmacokinetics of saquinavir hard-gel/ritonavir and ata-zanavir when combined once daily in HIV Type 1-infected individuals administered different atazanavir doses. AIDS Res Hum Retroviruses 2006;22(8):749–56.

10. Taburet AM, Piketty C, Chazallon C, Vincent I, Gerard L, Calvez V, Clavel F, Aboulker JP, Girard PM. Interactions between atazanavir-ritonavir and tenofovir in heavily pre-treated human immunodeficiency virus-infected patients. Antimicrob Agents Chemother 2004;48(6):2091–6.

Cidofovir

General Information

Cidofovir (S-1-3-hydroxy-2-phosphonylmethoxypropylcy-tosine), also known as HPMPC, is a nucleotide analogue with potent activity against cytomegalovirus (CMV) in vitro and in vivo.

Observational studies

Cidofovir has been used to treat 14 patients with cytome-galovirus infection after stem-cell transplantation; no adverse effects were reported in this small study (1).

In an open study of the use of intralesional cidofovir in treating laryngeal papilloma, 14 adults received monthly injections of cidofovir (maximum dose 37.5 mg per injection in 6 ml of saline; mean 22.5 mg) (2). Remission was achieved in all cases with an average of six injections and without additional laryngeal scarring, vocal cord damage, or systemic adverse effects.

Observational studies

Topical intralaryngeal cidofovir has been used as an effective adjunct in treating recurrences of recurrent laryngeal papillomatosis, a benign laryngeal disease caused by human papilloma virus (3). In 10 patients with recurrent papillomatosis, cidofovir was injected with a laryngeal needle during papilloma resection. There were no local or systemic adverse effects, but the median interval between recurrences increased significantly from 102 days before cidofovir to 239 days after treatment.

Organs and Systems

Sensory systems

In a placebo-controlled study of the efficacy of cidofovir 1% eye-drops with and without ciclosporin 1% eye-drops four or ten times a day for acute adenovirus keratocon-junctivitis in 34 patients, the frequency of severe corneal opacities was lower with cidofovir (4). However, cidofovir caused conjunctival pseudomembranes, conjunctivitis, and erythematous inflammation of the skin of the eyelids, and the trial was stopped as a result.

Two cases of iritis shortly after intravenous administration of cidofovir have been reported (5). Intravitreal administration of cidofovir delays the progression of CMV retinitis, but can be associated with reduced intraocular pressure or vitreitis (6). In two cases the drug association with the iritis was demonstrated by withdrawal and rechallenge (7). These patients had also taken probenecid, but it is not clear whether or not that was involved. The evidence is that in such cases the cidofovir should be withdrawn; treatment should be with a mydriatic and glucocorticoids.

In a study of compassionate use of intravenous cidofovir in AIDS patients with CMV retinitis, iritis developed in 21 of 51 individuals (8). The appearance of this inflammatory process did not fit the characteristics of the vitritis associated with immune reconstitution, or with HIV-induced vitritis. The high rate in this cohort (compared with a 5–7% incidence in randomized trials) was associated with severe CMV retinitis, and the authors suggested that breakdown of the blood–ocular barrier in these patients may promote higher intraocular concentrations of cidofovir and thus enhance local toxicity. Previous correlations of prior use of HIV protease inhibitors with iritis were not confirmed in this study, although patients with iritis had better immunological and virological status than those without the disease.

Uveitis (9) and cystoid macular edema (10) are recognized risks associated with cidofovir treatment for cytomegalovirus retinitis. The latter is almost certainly an immune recovery phenomenon, since it is apparently encountered only in eyes with inactive CMV retinitis; the unaffected contralateral side never develops cystoid macular edema. The time range of appearance (3–48 weeks after administration) also makes a direct toxic effect of cidofovir unlikely.

In a retrospective record review of 18 HIV-infected patients (30 eyes) who were being treated with intravenous cidofovir for complicated cytomegalovirus retinitis, eight patients developed anterior uveitis after a median of four (range 2–8) doses of cidofovir or a median of 55 (20–131) days after the start of therapy (11). While they were receiving treatment with cidofovir, none of the patients showed any evidence of progression of CMV retinitis. Five of the eight had symptoms of photophobia and blurred vision at the onset of uveitis, and the other three were asymptomatic. There was no difference in the use of HIV-1 protease inhibitors between the patients who did or did not develop anterior uveitis. Baseline intraocular pressure measurements were available for 11/18 patients. With the introduction of cidofovir there was a fall in mean intraocular pressure, and a trend for this fall to be more pronounced in those who developed anterior uveitis. Withdrawal of cidofovir was necessary in only one patient, after which all symptoms and signs disappeared and vision returned to baseline within 1 month. In two of the seven other patients, cidofovir had to be withdrawn because of nephrotoxicity. In the other five patients, cidofovir was continued and the uveitis was controlled with topical therapy, consisting of corticosteroids with or without a cycloplegic agent.

During long-term follow-up of patients with AIDS treated with parenteral cidofovir for CMV retinitis, the median time to discontinuation for intolerance was 6.6 months (12). Cidofovir-associated uveitis occurred in 10 of 58 patients and ocular hypotony (a 50% fall in intraocular pressure from baseline to below 5 mmHg) occurred at a rate of 0.16 per person-year. There were 51 episodes of proteinuria in 30 of the 58 patients and 82% of these episodes resolved on withdrawal (median time to resolution 20 days). No nephrotoxic events required dialysis.

Intrapleural cidofovir caused uveitis in a patient with pleural effusion lymphoma associated with herpesvirus-8, which usually has a poor prognosis (13).

- A 78-year-old Italian man with primary effusion lymphoma received two cycles of conventional CHOP chemotherapy with substantial improvement. However, the effusions recurred and one cycle of DHAP and a short course of interferon alfa gave no benefit. He was then given intrapleural cidofovir into the right pleural space on five occasions. This was well tolerated except for mild uveitis.

Anterior uveitis has been described in five of 14 kidney transplant patients given cidofovir for polyomavirus-associated nephropathy (14). The mean number of cidofovir doses given was 7 compared with 9 in patients who did not develop uveitis. Creatinine clearance when the nephropathy was diagnosed was lower in the patients with uveitis, and there was a graft survival of 40% versus 100% in the patients who did not have eye involvement. Treatment was withdrawn in all of the affected patients, and topical steroids and cycloplegics were used; there was complete resolution in 80%. The authors recommended that cidofovir should be used with caution in patients with creatinine clearances below 30 ml/minute.

Urinary tract

The main adverse effect associated with intravenous cidofovir is renal tubular damage (15), although it is usually mild. However, cases of acute renal insufficiency leading to end-stage disease have been reported (16,17). Acute renal insufficiency has also been attributed to topical cidofovir (18).

- A 28-year-old bone marrow transplant recipient with chronic renal insufficiency developed genital condylomata resistant to standard therapy. After application of topical cidofovir (1% daily for 5 days, then 4% for 12 days) the lesions improved but local erosions appeared. He developed acute renal insufficiency with features of tubular acidosis on day 19. He recovered after cidofovir withdrawal.

It was not clear whether this effect was due to the cidofovir or to its vehicle, propylene glycol. The authors suggested that topical cidofovir should be avoided on abraded skin and should be carefully monitored.

The risk of renal toxicity was higher in patients with established CMV infections than in those who were taking it prophylactically, and there was no relation between dosage and nephrotoxicity; toxicity occurred early in therapy, usually within the first 3 weeks (19). Proteinuria seems to be an early indicator of this adverse effect. Other laboratory abnormalities associated with nephrotoxicity are glycosuria, reduced serum phosphate, uric acid, and bicarbonate, and an increased serum creatinine. Serious and potentially irreversible nephrotoxicity can generally be prevented by the concomitant administration of intravenous saline and oral probenecid, monitoring the blood and urine immediately before each infusion of cidofovir, and withdrawing the drug for low-threshold increases in either urinary protein or serum creatinine concentrations (20). Concomitant use of other nephrotoxic drugs should be avoided.

In a rare cytomegalovirus comparison trial, a ganciclovir insert (an intraocular implant) plus oral ganciclovir was compared with intravenous cidofovir (21). Based on data from 61 patients (the trial was stopped early owing to low recruitment), cidofovir was associated with a raised serum creatinine concentration of 142 µmol/l (1.6 mg/dl) or greater (0.48 per person-year), 3+ or greater proteinuria (0.29 per person-year), uveitis (0.35 per person-year), and neutropenia (0.11 per person-year).

Prophylactic cidofovir treatment for CMV antigenemia has been investigated in a prospective study in patients with transplants (22). Renal toxicity, occasionally with proteinuria, was reported, as was nausea. The renal toxicity was mild and reversible.

Two children with liver transplants and life-threatening disseminated adenoviral disease recovered after withdrawal of calcineurin inhibitor treatment and the use of cidofovir; there were no dose-limiting adverse effects, apart from a slight transient rise in creatinine concentration and transient proteinuria in one and moderate neutropenia in the other (23).

The use of intravenous cidofovir has been studied in a retrospective analysis of the records of 57 patients, median age 8 years (range 0.5–26), with adenovirus infection after hemopoietic stem cell transplants (24). Cidofovir 5 mg/kg was given by infusion once a week for 2 weeks and then every 2 weeks until three consecutive samples were negative for adenovirus. The patients were hydrated with isotonic saline before and after each infusion. Probenecid was given orally 3 hours before each infusion (1 g/m^2) and then 2 and 8 hours after the infusion (500 mg/m^2). When possible, other nephrotoxins, such as ciclosporin, aminoglycosides, pentamidine, and foscarnet, were avoided, but if the use of such a nephrotoxin was unavoidable, dosage adjustments were made to achieve a low therapeutic concentration of cidofovir. When indicated for antifungal therapy, liposomal amphotericin B was used. In patients with compromised renal function the dosage of cidofovir was changed to 1 mg/kg given on 3 alternate days weekly for 2 weeks, and the same dose was repeated every other week until three consecutive specimens were negative for adenovirus from all previously involved sites. There were no cases of dose-limiting nephrotoxicity.

- A 48-year-old woman with HIV-related leukoencephalopathy was given intravenous aciclovir 500 mg tds for 15 days and had a partial remission. However, she deteriorated further due to infection with herpesvirus-6 and was given intravenous cidofovir 275 mg/week at first and then 275 mg every 2 weeks. Although her clinical condition gradually improved cidofovir had to be withdrawn because of Fanconi's syndrome and replaced with oral valaciclovir (25).

- A 39-year-old African–American HIV- and HBV-positive man with worsening facial verruca vulgaris for several months, a CD4 count of 22 x 10^6/l, and a viral load of 58 000 copies/mm^3 was given intravenous cidofovir 300 mg in isotonic saline weekly for 2 weeks and then fortnightly, with pretreatment with probenecid and isotonic saline hydration (26). Nevertheless, before the sixth cycles he developed mild proteinuria and the cidofovir was withheld. The proteinuria resolved spontaneously within 1 month.

In 26 cases of BK virus allograft nephropathy the rate of graft loss in cidofovir-treated patients was 15% lower than the historic rate of graft loss before the use of cidofovir of 45% (27). Although the viremia cleared in all but one case, and there was a large reduction in urinary viral load in 20 patients, the serum creatinine did not improve in 50% of cases. This was thought to be due to irreversible tissue damage with progression to chronic allograft nephropathy.

Three kidney transplant recipients with BK virus allograft nephropathy and raised serum creatinine concentrations were given infusions of intermediate-dose cidofovir (0.75–1.0 mg/kg) on 7–15 occasions without probenecid and without nephrotoxicity (28). Renal function improved in all three patients and they all had a marked fall in BK viral loads, with complete clearing of viremia in two.

Nephrotoxicity is reduced by concomitant treatment with oral probenecid and intravenous hydration.

Unfortunately, adverse effects of probenecid (nausea, vomiting, fever, headache, and rash) are frequent (in about 50% of patients) and lead to withdrawal in 4?7%. Based on preliminary studies in monkeys, the effect of an alternative probenecid regimen on renal function during cidofovir administration (5 mg/kg in 100 ml isotonic saline over 1 hour) has been examined in 24 HIV-infected individuals with cytomegalovirus retinitis who were randomly assigned to receive one of two probenecid regimens: 2 g 3 hours before cidofovir infusion and 1 g at 2 and 8 hours after (total 4 g) or 2 g 1 hour before the start of the infusion (29). In both regimens, one liter of isotonic saline was infused 1?2 hours before cidofovir and again after completion of the cidofovir infusion. Renal function was similar in the two groups and the serum concentration time curves of cidofovir were virtually superimposable. The authors suggest that the modified regimen of 2 g 1 hour before cidofovir infusion protects patients from renal toxicity as effectively the standard dose.

A previous study with a low dose of probenecid (1 g 3 hours before the cidofovir infusion followed by 0.5 g at 2 and 8 hours after the infusion failed to protect patients from cidofovir-nephrotoxicity (30).

Of 10 pediatric stem cell recipients with systemic adenovirus infections, one-third had a rise in serum creatinine 50% above baseline despite pre- and post-dose intravenous hydration and pre-dose probenecid (31). Renal tubular cell apoptosis is prevented by probenecid in vitro (32).

Skin

Topical cidofovir 1–3% has been successfully used for the treatment of oral warts in three HIV-positive patients with no adverse effects(33). However, ulceration and pain at the site of the lesion, leaving the normal skin unharmed, were reported as the main adverse effect after topical application to high-grade vulval intraepithelial neoplasia in 12 women(34).

Topical cidofovir 1% has also been used in the successful treatment of a refractory verruca in a child with acute lymphoblastic leukemia (35).

- A 9-year-old girl who had had acute lymphoblastic leukemia since the age of 18 months, had maintenance chemotherapy (methotrexate 10–15 mg/day, mercaptopurine 50–90 mg/day, and intrathecal methotrexate with hydrocortisone) reintroduced for a fourth CNS relapse, and developed a verruca on the sole of her right foot. This was treated initially with curettage, but it recurred quickly and grew larger and more painful. Treatment with cryotherapy was ineffective. The verruca continued to enlarge and became more painful. It again failed to respond to gentle cryotherapy. Cidofovir 1% ointment was compounded from the parenteral formulation (Vistide®, Gilead, Foster City, CA, USA) in ointment (Merck and Co, Inc, Whitehouse Station, NJ, USA) and was applied once daily for 6 weeks, by which time the verruca had completely resolved. No adverse effects or local irritation were observed.

Susceptibility Factors

Renal disease

The pharmacokinetics of cidofovir were studied in 24 subjects with varying degrees of renal insufficiency, who were given a single intravenous dose of cidofovir 0.5 mg/kg over 1 hour (36). Those who were not receiving dialysis were given intravenous hydration and concomitant oral probenecid. Mean cidofovir clearance in control subjects (normal renal function; $n = 5$) was 1.7 ml/minute/kg, and this fell markedly with falling renal function. The mean steady-state volume of distribution did not change significantly in those with kidney disease, but the half-life of cidofovir was significantly prolonged in those with severe renal insufficiency. Cidofovir was not significantly cleared during continuous ambulatory peritoneal dialysis, but high-flux hemodialysis resulted in the removal of 52% of the dose. The authors concluded that very considerable aggressive dosage reduction of cidofovir would be necessary in subjects with kidney disease to ensure comparable drug exposure in terms of serum concentrations.

Drug Administration

Drug administration route

Topical application of cidofovir as a 3% cream over 12 months was reportedly efficient and safe in an HIV-positive patient with recalcitrant molluscum contagiosum (37).

References

1. Bosi A, Bartolozzi B, Vannucchi AM, Orsi A, Guidi S, Rossi Ferrini P. Polymerase chain reaction-based "pre-emptive" therapy with cidofovir for cytomegalovirus reactivation in allogeneic hematopoietic stem cells transplantation recipients: a prospective study. Haematologica 2002;87(4):446–7.
2. Bielamowicz S, Villagomez V, Stager SV, Wilson WR. Intralesional cidofovir therapy for laryngeal papilloma in an adult cohort. Laryngoscope 2002;112(4):696–9.
3. Pontes P, Avelino M, Pignatari S, Weckx LL. Effect of local application of cidofovir on the control of recurrences in recurrent laryngeal papillomatosis. Otolaryngol Head Neck Surg 2006;135(1):22–7.
4. Hillenkamp J, Reinhard T, Ross RS, Bohringer D, Cartsburg O, Roggendorf M, De Clercq E, Godehardt E, Sundmacher R. The effects of cidofovir 1% with and without cyclosporin a 1% as a topical treatment of acute adenoviral keratoconjunctivitis: a controlled clinical pilot study. Ophthalmology 2002;109(5):845–50.
5. Tseng AL, Mortimer CB, Salit IE. Iritis associated with intravenous cidofovir. Ann Pharmacother 1999;33(2):167–71.
6. Kirsch LS, Arevalo JF, Chavez de la Paz E, Munguia D, de Clercq E, Freeman WR. Intravitreal cidofovir (HPMPC) treatment of cytomegalovirus retinitis in patients with acquired immune deficiency syndrome. Ophthalmology 1995;102(4):533–42Erratum in: Ophthalmology 1995;102(5):702.
7. Neau D, Renaud-Rougier MB, Viallard JE, Dutronc H, Cazorla C, Ragnaud JM, Dupon M, Lacut JY. Intravenous cidofovir-induced iritis. Clin Infect Dis 1999;28(1):156–7.
8. Berenguer J, Mallolas J, Padilla B, Colmenero M, Santos ISpanish Cidofovir Study Group. Intravenous cidofovir for compassionate use in AIDS patients with cytomegalovirus retinitis. Clin Infect Dis 2000;30(1):182–4.
9. Ambati J, Wynne KB, Angerame MC, Robinson MR. Anterior uveitis associated with intravenous cidofovir use in patients with cytomegalovirus retinitis. Br J Ophthalmol 1999;83(10):1153–8.
10. Kersten AJ, Althaus C, Best J, Sundmacher R. Cystoid macular edema following immune recovery and treatment with cidofovir for cytomegalovirus retinitis. Graefes Arch Clin Exp Ophthalmol 1999;237(11):893–6.
11. Akler ME, Johnson DW, Burman WJ, Johnson SC. Anterior uveitis and hypotony after intravenous cidofovir for the treatment of cytomegalovirus retinitis. Ophthalmology 1998;105(4):651–7.
12. Jabs DA, Freeman WR, Jacobson M, Murphy R, Van Natta ML, Meinert CL. Long-term follow-up of patients with AIDS treated with parenteral cidofovir for cytomegalovirus retinitis: the HPMPC Peripheral Cytomegalovirus Retinitis Trial. The Studies of Ocular Complications of AIDS Research Group in collaboration with the AIDS Clinical Trials Group. AIDS 2000;14(11):1571–81.
13. Halfdanarson TR, Markovic SN, Kalokhe U, Luppi M. A non-chemotherapy treatment of a primary effusion lymphoma: durable remission after intracavitary cidofovir in HIV negative PEL refractory to chemotherapy. Ann Oncol 2006;17(12):1849–50.
14. Lopez V, Sola E, Gutierrez C, Burgos D, Cabello M, García I, Florez P, Lopez J, Gonzalez-Molina M. Anterior uveitis associated with treatment with intravenous cidofovir in kidney transplant patients with BK virus nephropathy. Transplant Proc 2006;38(8):2412–3.
15. Skiest DJ, Duong M, Park S, Wei L, Keiser P. Complications of therapy with intravenous cidofovir: severe nephrotoxicity and anterior uveitis. Infect Dis Clin Pract 1999;83:151–7.
16. Vandercam B, Moreau M, Goffin E, Marot JC, Cosyns JP, Jadoul M. Cidofovir-induced end-stage renal failure. Clin Infect Dis 1999;29(4):948–9.
17. Meier P, Dautheville-Guibal S, Ronco PM, Rossert J. Cidofovir-induced end-stage renal failure. Nephrol Dial Transplant 2002;17(1):148–9.
18. Bienvenu B, Martinez F, Devergie A, Rybojad M, Rivet J, Bellenger P, Morel P, Gluckman E, Lebbe C. Topical use of cidofovir induced acute renal failure. Transplantation 2002;73(4):661–2.
19. Ljungman P, Deliliers GL, Platzbecker U, Matthes-Martin S, Bacigalupo A, Einsele H, Ullmann J, Musso M, Trenschel R, Ribaud P, Bornhauser M, Cesaro S, Crooks B, Dekker A, Gratecos N, Klingebiel T, Tagliaferri E, Ullmann AJ, Wacker P, Cordonnier C. Cidofovir for cytomegalovirus infection and disease in allogeneic stem cell transplant recipients. The Infectious Diseases Working Party of the European Group for Blood and Marrow Transplantation. Blood 2001;97(2):388–92.
20. Lalezari JP, Holland GN, Kramer F, McKinley GF, Kemper CA, Ives DV, Nelson R, Hardy WD, Kuppermann BD, Northfelt DW, Youle M, Johnson M, Lewis RA, Weinberg DV, Simon GL, Wolitz RA, Ruby AE, Stagg RJ, Jaffe HS. Randomized, controlled

study of the safety and efficacy of intravenous cidofovir for the treatment of relapsing cytomegalovirus retinitis in patients with AIDS. J Acquir Immune Defic Syndr Hum Retrovirol 1998;17(4):339–44.

21. Jabs DA. Studies of Ocular Complications of AIDS Research Group. The AIDS Clinical Trials Group. The ganciclovir implant plus oral ganciclovir versus parenteral cidofovir for the treatment of cytomegalovirus retinitis in patients with acquired immunodeficiency syndrome: The Ganciclovir Cidofovir Cytomegalovirus Retinitis Trial. Am J Ophthalmol 2001;131(4):457–67.

22. Platzbecker U, Bandt D, Thiede C, Helwig A, Freiberg-Richter J, Schuler U, Plettig R, Geissler G, Rethwilm A, Ehninger G, Bornhauser M. Successful preemptive cidofovir treatment for CMV antigenemia after dose-reduced conditioning and allogeneic blood stem cell transplantation. Transplantation 2001;71(7):880–5.

23. Wallot MA, Dohna-Schwake C, Auth M, Nadalin S, Fiedler M, Malagó M, Broelsch C, Voit T. Disseminated adenovirus infection with respiratory failure in pediatric liver transplant recipients: impact of intravenous cidofovir and inhaled nitric oxide. Pediatr Transplant 2006;10(1):121–7.

24. Yusuf U, Hale GA, Carr J, Gu Z, Benaim E, Woodard P, Kasow KA, Horwitz EM, Leung W, Srivastava DK, Handgretinger R, Hayden RT. Cidofovir for the treatment of adenoviral infection in pediatric hematopoietic stem cell transplant patients. Transplantation 2006;81(10):1398–404.

25. Astriti M, Zeller V, Boutolleau D, Gautheret-Dejean A, Allen G, Seilhean D, Agut H, Bricaire F, Katlama C, Bossi P. Fatal HHV-6 associated encephalitis in an HIV-1 infected patient treated with cidofovir. J Infect 2006;52(4):237–42.

26. Kottke MD, Parker SR. Intravenous cidofovir-induced resolution of disfiguring cutaneous human papillomavirus infection. J Am Acad Dermatol 2006;55(3):533–6.

27. Josephson MA, Williams JW, Chandraker A, Randhawa PS. Polyomavirus-associated nephropathy: update on antiviral strategies. Transpl Infect Dis 2006;8(2):95–101.

28. Araya CE, Lew JF, Fennell RS 3rd, Neiberger RE, Dharnidharka VR. Intermediate-dose cidofovir without probenecid in the treatment of BK virus allograft nephropathy. Pediatr Transplant 2006;10(1):32–7.

29. Wolf DL, Rodriguez CA, Mucci M, Ingrosso A, Duncan BA, Nickens DJ. Pharmacokinetics and renal effects of cidofovir with a reduced dose of probenecid in HIV-infected patients with cytomegalovirus retinitis. J Clin Pharmacol 2003;43:43–51.

30. Cundy KC, Petty BG, Flaherty J, Fisher PE, Polis MA, Wachsman M, Lietman PS, Lalezari JP, Hitchcock MJ, Jaffe HS. Clinical pharmacokinetics of cidofovir in human immunodeficiency virus-infected patients. Antimicrob Agents Chemother 1995;39:1247–52.

31. Muller WJ, Levin MJ, Shin YK, Robinson C, Quinones R, Malcolm J, Hild E, Gao D, Giller R. Clinical and in vitro evaluation of cidofovir for treatment of adenovirus infection in pediatric hematopoietic stem cell transplant recipients. Clin Infect Dis 2005;41(12):1812–6.

32. Ortiz A, Justo P, Sanz A, Melero R, Caramelo C, Guerrero MF, Strutz F, Muller G, Barat A, Egido J. Tubular cell apoptosis and cidofovir-induced acute renal failure. Antivir Ther 2005;10(1):185–90.

33. Husak R, Zouboulis CC, Sander-Bähr C, Hummel M, Orfanos CE. Refractory human papillomavirus-associated oral warts treated topically with 1–3% cidofovir solutions in human immunodeficiency virus type 1-infected patients. Br J Dermatol 2005;152(3):590–1.

34. Tristram A, Fiander A. Clinical responses to cidofovir applied topically to women with high grade vulval intraepithelial neoplasia. Gynecol Oncol 2005;99(3):652–5.

35. Tobin AM, Cotter M, Irvine AD, Kirby B. Successful treatment of a refractory verruca in a child with acute lymphoblastic leukaemia with topical cidofovir. Br J Dermatol 2005;152(2):386–8.

36 Brody SR, Humphreys MH, Gambertoglio JG, Schoenfeld P, Cundy KC, Aweeka FT. Pharmacokinetics of cidofovir in renal insufficiency and in continuous ambulatory peritoneal dialysis or high-flux hemodialysis. Clin Pharmacol Ther 1999;65(1):21–8.

37. Baxter KF, Highet AS. Topical cidofovir and cryotherapy—combination treatment for recalcitrant molluscum contagiosum in a patient with HIV infection. J Eur Acad Dermatol Venereol 2004;18(2):230–1.

Delavirdine

See also Non-nucleoside reverse transcriptase inhibitors (NNRTIs)

General Information

Delavirdine is a non-nucleoside reverse transcriptase inhibitor, which is dosed three times daily. No food restrictions apply. It is metabolized mainly by CYP3A, and so interactions with other drugs that use this metabolic pathway can occur.

Organs and Systems

Skin

The most frequent adverse effect of delavirdine is a rash, which usually occurs during the first 3 months of therapy in up to 50% of individuals. It is usually mild and resolves spontaneously in most patients or can be treated successfully with a short course of antihistamines. Interruption of treatment is required in less than 4% of patients (1,2).

Immunologic

Severe hypersensitivity reactions, including anaphylaxis, can occur and may necessitate drug withdrawal (3).

Drug–Drug Interactions

Antacids

Antacids interfere with the absorption of delavirdine and should therefore not be given concomitantly (4). Administration of antacids and buffered formulations of didanosine should be separated from that of delavirdine by at least 1 hour.

Didanosine

Didanosine interferes with the absorption of delavirdine and should therefore not be given concomitantly (5).

Enzyme inducers

Concomitant use of inducers of cytochrome P450, such as rifampicin, rifabutin, phenytoin, phenobarbital, or carbamazepine, should be avoided, since they significantly reduce delavirdine plasma concentrations (4).

HIV protease inhibitors

By inhibiting their metabolism, ritonavir potentiates the actions of other protease inhibitors. The addition of delavirdine instead of another NNRTI in three patients taking protease inhibitors plus ritonavir further increased the exposure to the protease inhibitors (6). Combining delavirdine with indinavir removes the food restrictions during indinavir administration (4). The superior virological response observed in antiretroviral regimens containing delavirdine and protease inhibitors has been attributed in part to the pharmacokinetic interaction.

Statins

- Rhabdomyolysis with acute renal insufficiency has been reported in a 63 year-old man who was taking atorvastatin 20 mg/day and delavirdine 400 mg 8-hourly (7).

The authors suggested that delavirdine had inhibited the metabolism of atorvastatin by inhibiting CYP3A4.

References

1. Been-Tiktak AM, Boucher CA, Brun-Vezinet F, Joly V, Mulder JW, Jost J, Cooper DA, Moroni M, Gatell JM, Staszewski S, Colebunders R, Stewart GJ, Hawkins DA, Johnson MA, Parkin JM, Kennedy DH, Hoy JF, Borleffs JC. Efficacy and safety of combination therapy with delavirdine and zidovudine: a European/Australian phase II trial. Int J Antimicrob Agents 1999;11(1):13–21.
2. Friedland GH, Pollard R, Griffith B, Hughes M, Morse G, Bassett R, Freimuth W, Demeter L, Connick E, Nevin T, Hirsch M, Fischl M. Efficacy and safety of delavirdine mesylate with zidovudine and didanosine compared with two-drug combinations of these agents in persons with HIV disease with CD4 counts of 100 to 500 cells/mm³ (ACTG 261). ACTG 261 Team. J Acquir Immune Defic Syndr 1999;21(4):281–92.
3. Mills G, Morgan J, Hales G, Smith D. Acute hypersensitivity with delavirdine. Antivir Ther 1999;4(1):51.
4. Tran JQ, Gerber JG, Kerr BM. Delavirdine: clinical pharmacokinetics and drug interactions. Clin Pharmacokinet 2001;40(3):207–26.
5. Morse GD, Fischl MA, Shelton MJ, Cox SR, Driver M, DeRemer M, Freimuth WW. Single-dose pharmacokinetics of delavirdine mesylate and didanosine in patients with human immunodeficiency virus infection. Antimicrob Agents Chemother 1997;41(1):169–74.
6. Harris M, Alexander C, O'Shaughnessy M, Montaner JS. Delavirdine increases drug exposure of ritonavir-boosted protease inhibitors. AIDS 2002;16(5):798–9.
7. Castro JG, Gutierrez L. Rhabdomyolysis with acute renal failure probably related to the interaction of atorvastatin and delavirdine. Am J Med 2002;112(6):505.

Didanosine

See also Nucleoside analogue reverse transcriptase inhibitors (NRTIs)

General Information

Didanosine (2′,3′-dideoxyinosine, ddI) is a purine analogue reverse transcriptase inhibitor. The major clinical adverse effects reported during the first years of use of didanosine were acute pancreatitis and a painful neuropathic syndrome (due to a peripheral neuropathy), which appeared to be related to both dosage and cumulative dose (SEDA-17, 340) (1). However, the incidence of acute pancreatitis and peripheral neuropathy in these studies was lower than in earlier studies with didanosine (2,3). This may be related to the fact that treatment was started earlier or to the use of lower dosages (200–400 mg/day) in these studies compared with earlier studies. In the latter studies, gastrointestinal symptoms, most notably nausea and vomiting, were the most commonly reported adverse effects in patients taking didanosine. Minor adverse effects include insomnia, headaches, anxiety, irritability, rash, increased plasma uric acid concentration, and increased hepatic transaminase activities combined with a rash (4). There were no toxic effects of didanosine on hematological laboratory indices (5).

Observational studies

The use of didanosine 125–200 mg bd plus interferon alfa-2b in AIDS-associated Kaposi's sarcoma has been studied in 68 patients (6). Withdrawal of didanosine was required in cases of peripheral neuropathy, rises in serum amylase activity, and hypertriglyceridemia.

Comparative studies

In 44 patient with HIV infection, didanosine added to tenofovir was associated with progressive loss of mtDNA content and reduced cyclo-oxygenase activity over 12-months compared with tenofovir alone (7). In another study in 61 individuals didanosine + stavudine exposure was associated with mtDNA depletion in peripheral blood mononuclear cells and subcutaneous fat (8).

Placebo-controlled studies

In a study of 168 patients with virological failure didanosine was compared with placebo in addition to optimized background treatment (9). The incidence of adverse events was similar in the two groups (38% with didanosine and 36% with placebo). Most of the adverse events were gastrointestinal (20%) or affected the nervous system

(14%). Only five patients (4.5%) in the didanosine group and two (3.6%) in the placebo group complained of grade 1–2 diarrhea; no grade 3 diarrhea was reported. One patient in the didanosine group had a grade 2 rise in serum lipase activity, and one in the placebo group had a grade 3 rise.

Organs and Systems

Sensory systems

Eyes

Didanosine has been associated with retinopathy attributed to irreversible loss of retinal pigment epithelium accompanied by partial loss of the choriocapillaris and neurosensory retina in the mid-periphery. In a single case of retinopathy the electrophysiological abnormalities improved after didanosine withdrawal (10). Retinal depigmentation has also been described in children taking didanosine (11).

Ears

Hearing loss attributed to didanosine has been reported in an HIV-infected adult (12).

- A 37-year-old HIV-infected man developed bilateral deafness while taking didanosine (400 mg/day), which had been started about 4 years before. He was also taking azithromycin, ciprofloxacin, and myambutol for about 1 month for a *Mycobacterium avium* infection. Otoscopic examination and tympanometry were normal. Audiometry showed a bilateral sensorineural hearing deficit of 40–60 dB. There were no other neurological abnormalities. An MRI scan of the brain was normal. Didanosine was withdrawn and replaced by alternative antiretroviral agents. All other medications were continued. His hearing improved progressively and returned to normal after 2 months.

In the absence of rechallenge, there was no conclusive evidence for a causative role of didanosine in the development of hearing loss in this case. However, the authors argued that the improvement on discontinuation of didanosine, which is known to cause neuritis, implicated the drug.

Extensive serial audiological studies in an HIV-infected child showed high-frequency hearing loss after 19 months of combined treatment with zidovudine and didanosine (dosages unknown), which were started at 24 months of age (13). Normal tympanograms indicated that this hearing loss was sensorineural.

While no conclusive evidence was given for a causative role of either antiretroviral drug, the authors concluded that children taking antiretroviral therapy need to be monitored for possible ototoxicity.

In contrast, in a prospective observational pilot study to determine whether zidovudine or didanosine, alone or in combination, are associated with sensorineural hearing loss in HIV-infected persons there was no association (14).

Psychiatric

In a comparison of a regimen that contained a protease inhibitor with a switch to a once-daily combination of efavirenz + didanosine + emtricitabine, didanosine was associated with a higher risk of depressive disorder, although this was not significant in a multivariate analysis (15).

Metabolism

In a cross-sectional study the use of a combination of medication containing stavudine and didanosine for longer than 1 year was significantly more often seen in children with lipoatrophy, although the numbers of patients were very small (4 versus 28 in the control group) (16).

Hematologic

Thrombocytopenia has been observed in patients using didanosine (17,18).

Liver

The reports of all suspected hepatic adverse drug reactions with a fatal outcome received by the Uppsala Monitoring Centre between 1968 and 2003 included over 50 cases of didanosine-associated hepatotoxicity (24).

Pancreas

The combination of didanosine and tenofovir has been associated with unexpected reductions in CD4+ T cell counts and an increased risk of pancreatitis (19). A direct toxic effect of didanosine, particularly when given in combination with tenofovir, on the pancreas has been suggested as the cause for an increase in fasting blood glucose concentrations. In a retrospective comparison of patients taking didanosine + tenofovir and either didanosine or tenofovir in combination with other antiretroviral drugs, both hyperglycemia and diabetes mellitus were significantly more common in recipients of didanosine + tenofovir after 12 months compared with patients taking either tenofovir or didanosine alone; the effect was more pronounced in patients taking didanosine 400 mg/day. The recommended dose for didanosine when given in combination with tenofovir was reduced in 2003 to 250 mg/day for subjects weighing more than 60 kg and 200 mg/day in those weighing less than 60 kg.

Urinary tract

Renal insufficiency has been reported with both tenofovir and didanosine.

- A 12-year-old girl developed nephrogenic diabetes insipidus, renal insufficiency, and a Fanconi-like syndrome while taking tenofovir in combination with lopinavir + ritonavir and didanosine (20).

The authors concluded that these symptoms might have been attributable to tenofovir or didanosine or the combination.

In the BMS Study 045, once-daily atazanavir + ritonavir was compared with twice-daily lopinavir + ritonavir, each with tenofovir and one nucleoside reverse transcriptase inhibitor (21). Adverse renal events were infrequent and occurred in five subjects, who were all taking lopinavir + ritonavir in combination with tenofovir and didanosine.

Musculoskeletal

There is no evidence that didanosine contributes to the development of myopathy in patients taking didanosine and zidovudine in whom this adverse event occurs (22).

Multiorgan failure

There has been a report of a patient in whom pancreatitis, fulminant hepatic failure, and persistent lactic acidosis occurred, culminating in death from liver failure (23).

Second-Generation Effects

Pregnancy

Didanosine was thought to be safe when used in pregnancy after 34 weeks gestation to reduce mother-to-child transmission of HIV (24). There were more grade 3 and 4 events when didanosine was given in combination with stavudine (18%) than with didanosine monotherapy (6%). In the same study treatment was continued for 6 weeks in infants after birth. Serious adverse events occurred less often with didanosine monotherapy (13%) than with other drugs (stavudine or stavudine + didanosine or zidovudine), with which serious adverse events occurred in 22–29%. While the lowest mother-to-child transmission rate was seen in the stavudine + didanosine arm, the authors of this study warned that the combination of stavudine + didanosine should be used with caution during pregnancy.

Drug Administration

Drug dosage regimens

In a retrospective analysis of patients in the HIV Outpatient Study who were taking didanosine + tenofovir the rate of adverse events was studied in patients taking high-dose didanosine (400 mg/day) compared with low-dose didanosine (100–250 mg/day). In keeping with previous findings, high-dose didanosine + tenofovir was more often associated with drug-related toxicity, adverse events, and treatment discontinuation than low-dose regimens. The adverse effects included peripheral neuropathy (12% on high-dose didanosine versus 4% on low-dose) and pancreatitis (6% versus 0%). However, few events were observed, and the differences between the two groups were not statistically significant (25).

Drug-Drug Interactions

Tenofovir

The combination of didanosine with tenofovir requires dosage reduction of didanosine from 400 mg/day to 250 mg/day because of increased didanosine plasma concentrations when tenofovir is co-administered. However, there have not previously been data to suggest that

dosage reduction results in appropriate drug exposure in subjects weighing less than 60 kg.

- A patient weighing 48 kg with renal impairment (creatinine clearance 57 ml/minute) developed didanosine toxicity with lactic acidosis and liver failure 3 months after starting a regimen containing tenofovir and didanosine 200 mg/day (26).

In 21 patients weighing under 60 kg who took didanosine 250 mg/day plus tenofovir there was full viral suppression but the mean CD4 cell count fell to 120 x 106/l; no explanation for the reduction in CD4 count was found other than the combined treatment of tenofovir and didanosine (27).

The efficacy of switching from 400 mg/day to 250 mg/day of didanosine (for patients weighing under 60 kg), has been confirmed by measuring intracellular concentrations, but there was no evidence of a kinetic interaction (28). The long intracellular half-lives of the drugs were also noted (7.5 days for tenofovir).

Therefore, in patients with low body weight and/or renal impairment, this drug combination should be used only with extreme caution, even when low doses of didanosine are used, and close clinical and biochemical monitoring is required.

Drug-Food Interactions

It is usually recommended that didanosine be taken fasting. In a retrospective comparison of the effect of taking enteric-coated didanosine with or without food, those who took it with food had a higher virological failure rate (29). However, those who took it with food but with over 80% adherence had similar virological failure rates to those who took it fasting with adherence over 80%.

References

1. Yarchoan R, Mitsuya H, Pluda JM, Marczyk KS, Thomas RV, Hartman NR, Brouwers P, Perno CF, Allain JP, Johns DG, et al. The National Cancer Institute phase I study of 2′,3′-dideoxyinosine administration in adults with AIDS or AIDS-related complex: analysis of activity and toxicity profiles. Rev Infect Dis 1990;12(Suppl 5):S522–S533.
2. Delta Coordinating Committee. Delta: a randomised double-blind controlled trial comparing combinations of zidovudine plus didanosine or zalcitabine with zidovudine alone in HIV-infected individuals. Lancet 1996;348(9023):283–291.
3. Hammer SM, Katzenstein DA, Hughes MD, Gundacker H, Schooley RT, Haubrich RH, Henry WK, Lederman MM, Phair JP, Niu M, Hirsch MS, Merigan TC. A trial comparing nucleoside monotherapy with combination therapy in HIV-infected adults with CD4 cell counts from 200 to 500 per cubic millimeter. AIDS Clinical Trials Group Study 175 Study Team. N Engl J Med 1996;335(15):1081–90.
4. Franssen RM, Meenhorst PL, Koks CH, Beijnen JH. Didanosine, a new antiretroviral drug. A review. Pharm Weekbl Sci 1992;14(5):297–304.
5. Lambert JS, Seidlin M, Reichman RC, Plank CS, Laverty M, Morse GD, Knupp C, McLaren C, Pettinelli C, Valentine FT, et

al. 2′,3′-dideoxyinosine (ddI) in patients with the acquired immunodeficiency syndrome or AIDS-related complex. A phase I trial. N Engl J Med 1990;322(19):1333–40.

6. Krown SE, Li P, Von Roenn JH, Paredes J, Huang J, Testa MA. Efficacy of low-dose interferon with antiretroviral therapy in Kaposi's sarcoma: a randomized phase II AIDS clinical trials group study. J Interferon Cytokine Res 2002;22(3):295–303.

7. Lopez JC, Moreno S, Jimenez-Onate F, Clotet B, Rubio R, Hernandez-Quero J. A cohort study of the food effect on virological failure and treatment discontinuation in patients on HAART containing didanosine enteric-coated capsules (FOODDIe Study). HIV Clin Trials 2006;7(4):155–62.

8. Cherry CL, Nolan D, James IR, McKinnon EJ, Mallal SA, Gahan ME, Lal L, McArthur JC, Wesselingh SL. Tissue-specific associations between mitochondrial DNA levels and current treatment status in HIV-infected individuals. J Acquir Immune Defic Syndr 2006;42(4):435–40.

9. Molina JM, Marcelin AG, Pavie J, Heripret L, De Boever CM, Troccaz M, Leleu G, Calvez V; AI454-176 JAGUAR Study Team. Didanosine in HIV-1-infected patients experiencing failure of antiretroviral therapy: a randomized placebo-controlled trial. J Infect Dis 2005;191(6):840–7.

10. Fernando AI, Anderson OA, Holder GE, Mitchell SM. Didanosine-induced retinopathy in adults can be reversible. Eye 2006;20(12):1435–7.

11. Pizzo PA, Wilfert C. Antiretroviral therapy for infection due to human immunodeficiency virus in children. Clin Infect Dis 1994;19(1):177–96.

12. Vogeser M, Colebunders R, Depraetere K, Van Wanzeele P, Van Gehuchten S. Deafness caused by didanosine. Eur J Clin Microbiol Infect Dis 1998;17(3):214–5.

13. Christensen LA, Morehouse CR, Powell TW, Alchediak T, Silio M. Antiviral therapy in a child with pediatric human immunodeficiency virus (HIV): case study of audiologic findings. J Am Acad Audiol 1998;9(4):292–8.

14. Schouten JT, Lockhart DW, Rees TS, Collier AC, Marra CM. A prospective study of hearing changes after beginning zidovudine or didanosine in HIV-1 treatment-naive people. BMC Infect Dis 2006;6:28.

15. Journot V, Chene G, De Castro N, Rancinan C, Cassuto JP, Allard C, Vildé JL, Sobel A, Garré M, Molina JM; ALIZE Study Group. Use of efavirenz is not associated with a higher risk of depressive disorders: a substudy of the randomized clinical trial ALIZE-ANRS 099. Clin Infect Dis 2006;42(12):1790–9.

16. Hartman K, Verweel G, de Groot R, Hartwig NG. Detection of lipoatrophy in human immunodeficiency virus-1-infected children treated with highly active antiretroviral therapy. Pediatr Infect Dis J 2006;25(5):427–31.

17. Yarchoan R, Perno CF, Thomas RV, Klecker RW, Allain JP, Wills RJ, McAtee N, Fischl MA, Dubinsky R, McNeely MC, et al. Phase I studies of 2′,3′-dideoxycytidine in severe human immunodeficiency virus infection as a single agent and alternating with zidovudine (AZT). Lancet 1988;1(8577):76–81.

18. Dolin R, Lambert JS, Morse GD, Reichman RC, Plank CS, Reid J, Knupp C, McLaren C, Pettinelli C. 2′,3′-Dideoxyinosine in patients with AIDS or AIDS-related complex. Rev Infect Dis 1990;12(Suppl 5):S540–9.

19. García-Benayas T, Rendón AL, Rodríguez-Novóa S, Barrios A, Maida I, Blanco F, Barreiro P, Rivas P, González-Lahoz J, Soriano V. Higher risk of hyperglycemia in HIV-infected patients treated with didanosine plus tenofovir. AIDS Res Hum Retroviruses 2006;22(4):333–7.

20. Hussain S, Khayat A, Tolaymat A, Rathore MH. Nephrotoxicity in a child with perinatal HIV on tenofovir, didanosine and lopinavir/ritonavir. Pediatr Nephrol 2006;21(7):1034–6.

21. Johnson M, Grinsztejn B, Rodriguez C, Coco J, DeJesus E, Lazzarin A, Lichtenstein K, Wirtz V, Rightmire A, Odeshoo L, McLaren C. 96-week comparison of once-daily atazanavir/ritonavir and twice-daily lopinavir/ritonavir in patients with multiple virologic failures. AIDS 2006;20(5):711–8.

22. Pedrol E, Masanes F, Fernandez-Sola J, Cofan M, Casademont J, Grau JM, Urbano-Marquez A. Lack of muscle toxicity with didanosine (ddI). Clinical and experimental studies. J Neurol Sci 1996;138(1-2):42–8.

23. Calegari J, Lorenzana R, Cheyer C, Gang DL, Higgins TL. Lactic acidosis and fulminant hepatic failure in a patient treated with didanosine, nelfinavir and stavudine. Clin Intensive Care 1999;10:61–3.

24. Gray G, Violari A, McIntyre J, Jivkov B, Schnittman S, Reynolds L, Ledeine JM. Antiviral activity of nucleoside analogues during short-course monotherapy or dual therapy: its role in preventing HIV infection in infants. J Acquir Immune Defic Syndr 2006;42(2):169–76.

25. Young B, Weidle PJ, Baker RK, Armon C, Wood KC, Moorman AC, Holmberg SD; HIV Outpatient Study (HOPS) Investigators. Short-term safety and tolerability of didanosine combined with high- versus low-dose tenofovir disproxil fumarate in ambulatory HIV-1-infected persons. AIDS Patient Care STDS 2006;20(4):238–44.

26. Masia M, Gutierrez F, Padilla S, Ramos JM, Pascual J. Didanosine-associated toxicity: a predictable complication of therapy with tenofovir and didanosine? J Acquir Immune Defic Syndr 2004;35(4):427–8.

27. Negredo E, Molto J, Burger D, Viciana P, Ribera E, Paredes R, Juan M, Ruiz L, Puig J, Pruvost A, Grassi J, Masmitja E, Clotet B. Unexpected CD4 cell count decline in patients receiving didanosine and tenofovir-based regimens despite undetectable viral load. AIDS 2004;18(3):459–63.

28. Pruvost A, Negredo E, Benech H, Theodoro F, Puig J, Grau E, Garcia E, Molto J, Grassi J, Clotet B. Measurement of intracellular didanosine and tenofovir phosphorylated metabolites and possible interaction of the two drugs in human immunodeficiency virus-infected patients. Antimicrob Agents Chemother 2005;49(5):1907–14.

29. Lopez JC, Moreno S, Jimenez-Onate F, Clotet B, Rubio R, Hernandez-Quero J. A cohort study of the food effect on virological failure and treatment discontinuation in patients on HAART containing didanosine enteric-coated capsules (FOODDIe Study). HIV Clin Trials 2006;7(4):155–62.

Efavirenz

See also Non-nucleoside reverse transcriptase inhibitors (NNRTIs)

General Information

Efavirenz is a non-nucleoside reverse transcriptase inhibitor with excellent inhibitory activity against HIV-1. Its most frequent adverse effects involve the central nervous system and the skin (1). At the start of therapy, dizziness, insomnia, or fatigue is observed in most patients, and headache and even psychotic reactions have also been observed. A maculopapular rash is seen in about 10%. These adverse effects usually vanish within the first 2–4 weeks of therapy (2). About 1–2% of individuals stop taking efavirenz because of neurological or dermatological adverse events. Administration of efavirenz at bedtime reduces the incidence of severe adverse effects, and the rash can be managed by short-term antihistamines or topical corticosteroids (1). Nausea and vomiting are less often observed than in patients treated with zidovudine, lamivudine, or indinavir.

There is other evidence that efavirenz is in some respects better tolerated than certain of the alternatives used in HIV infection. However, comparisons are difficult, since efavirenz will generally be used with drugs of other types in order to avoid the rapid development of resistance. Combined efavirenz with nucleoside analogues reduces toxicity (3). Efavirenz-induced adverse effects often begin on the first day of therapy and resolve within 14–28 (median 13) days (4).

Because efavirenz is metabolized by cytochrome P450, several clinically significant interactions have been described. Efavirenz induces CYP3A4 (5); there was 55% mean induction at a dose of 400 mg/day and 33% at 200 mg/day. However, no significant interaction was noted with co-administration of nelfinavir, zidovudine, lamivudine, fluconazole, or azithromycin (2).

Observational studies

In 77 HIV-positive subjects randomized to switch from protease inhibitors to nevirapine or efavirenz or to continue taking protease inhibitors, quality of life significantly improved among those who switched (6). In those who took efavirenz there was an increase in gamma-glutamyltransferase activity and three patients interrupted treatment because of central nervous system symptoms. Eight patients withdrew because of adverse events: rashes ($n = 3$), dizziness ($n = 2$), irritability and depression ($n = 1$), depression ($n = 1$), or hepatotoxicity ($n = 1$).

Comparative studies

In an open comparison of efavirenz and indinavir (plus two nucleoside analogues) in predominantly treatment-naive patients efavirenz-based triple therapy provided at least similar antiviral effects over 48 weeks (7). Furthermore, fewer patients discontinued efavirenz-based triple therapy than indinavir-based therapy because of adverse events. Adverse effects associated with efavirenz included a maculopapular rash and central nervous system disturbances (dizziness, vivid dreams, poor concentration, sleep disturbances), which generally occurred (but later resolved) within the first weeks of therapy.

Organs and Systems

Nervous system

In a randomized comparison of an efavirenz-containing regimen and a protease inhibitor-containing regimen, nervous system adverse effects were specifically sought (8). Patients were randomized to two NRTIs plus efavirenz ($n = 51$) or two NRTIs plus one or more protease inhibitors ($n = 49$). The patients who took efavirenz reported the following at week 4: dizziness (66%), abnormal dreaming (48%), light-headedness (37%), and difficulty in sleeping (35%). At week 24, dizziness (13%), abnormal dreaming (18%), light-headedness (13%), difficulty in sleeping (7%), and nervousness (13%) were significantly less common. Irritability, abnormal dreaming, and nervousness persisted at week 48 in 13%, 10%, and 8% respectively.

An efavirenz-containing HAART regimen was associated with altered dreams or sleep disorders in 14% of patients (9). In another study, efavirenz had a modest but persistent impact on the time spent in several key sleep stages (10).

In 740 patients who took zidovudine + lamivudine + efavirenz, efavirenz was commonly associated with insomnia, vertigo, or nightmares, and less often with severe neuropsychiatric syndromes such as delirium (11). In the same study there was one case of febrile rash, two of intolerable pruritus, and one of cytolytic cellulitis attributed to efavirenz.

Insomnia and abnormal dreams are well described adverse effects of efavirenz. There were grade 3 or 4 sleep disturbances in six of 400 patients in the 2NN trial (12). Eleven of 287 patients stopped taking efavirenz within 30 days because of these disturbances (13). There were sleep abnormalities in 13 of 18 patients taking efavirenz, evaluated using the Pittsburgh Sleep Quality Index. These abnormalities have been explained by disturbances in sleep architecture, since ambulatory encephalographic monitoring showed marked sleep fragmentation (14).

Mild and clinically tolerable neuropsychiatric disorders can persist after a mean of 2 years after withdrawal of an efavirenz-based therapy (15).

In a cross-sectional study, 60 patients taking an efavirenz-based approach were compared with 60 patients taking a protease inhibitor-containing regimen for at least 1 year. The mean times on treatment were 91 weeks and 120 weeks respectively. Mild dizziness (22%), sadness (37%), mood changes (27%), irritability (30%), light-headedness (28%), nervousness (30%), impaired concentration (27%), abnormal dreams (48%), and somnolence (25%) were reported more often with efavirenz than the protease inhibitor. Of 60 patients 49 had plasma concentrations in the target range (1.0–4.0 mg/l). Efavirenz plasma concentrations were similar in subjects with and without neuropsychiatric disorders.

In a retrospective study in 134 patients taking efavirenz there were no significant differences in nervous system adverse effects or discontinuation rates between recreational substance (cocaine, ecstasy, cannabis) users and non-users (16).

Psychological, psychiatric

Efavirenz has been associated with psychiatric problems, such as anxiety, depression, and confusion (17,18). In a retrospective study of 1897 patients, dementia and depression were significantly associated with efavirenz compared with other drugs; the respective odds ratios were 4.0 (95% CI = 1.2, 14) and 1.7 (1.0, 3.0) (19). However, those who were given efavirenz were perhaps more ill than those who were not, judging by CD4 counts and opportunistic infections.

Most clinicians tend to avoid efavirenz in patients with a psychiatric history. However, it is important to remember that efavirenz can precipitate sudden and severe psychiatric symptoms in patients with no such history. Three patients developed sudden irritability; excitability with anxiety; and insomnia, confusion, and amnesia (20).

The psychiatric adverse effects of efavirenz correlate with its plasma concentrations. In 130 HIV-infected patients, toxicity was three times more common in patients who had an efavirenz concentration over 4000 ng/ml (21).

In a retrospective study the overall incidence of neuropsychiatric symptoms of any kind in patients taking an efavirenz-containing regimen was 30%; a previous history of *depression* was significantly associated with depressive symptoms (22). However, in a randomized trial efavirenz was not associated with depression significantly more often than a protease inhibitor-containing regimen (23).

In a single-centre, cross-sectional comparison of patients who had used efavirenz as part of their combination antiretroviral regimen for at least 6 months with a matched cohort who were stable on non-efavirenz combination therapy, those who took efavirenz reported higher degrees of severe *stress* and *anxiety* and a higher rate of *unusual dreams* than patients who were not taking efavirenz (24).

- A 40-year-old man had a classic manic episode after taking efavirenz for 2 weeks (25) He had no previously personal or family psychiatric history. The symptoms completely disappeared 1 month after withdrawal of efavirenz.
- A 40-year-old patient with pre-existing depressive symptoms developed suicidal thoughts and schizophrenia-like first-rank symptoms in within the first month of efavirenz administration and attempted suicide (26).

In a questionnaire survey of 152 patients who stopped taking efavirenz 82 did so because of neuropsychiatric symptoms; a history of multiple episodes of depression was associated with efavirenz discontinuation (27).

Metabolism

In a randomized, double-blind study of 327 patients, cholesterol and triglycerides were raised in patients taking efavirenz, although this did not reach statistical significance (4).

In a cross-sectional evaluation of 1018 HIV-infected patients treated with HAART during the previous 12 months in an Italian clinic, isolated hypertriglyceridemia was more common in 183 naive patients taking efavirenz compared with nevirapine, and both hypertriglyceridemia and hypercholesterolemia appeared earlier (28). In the 295 antiretroviral-experienced patients, in whom an NNRTI was introduced for the first time, the frequency of raised triglyceride concentrations was higher and occurred earlier with efavirenz. In the 145 subjects taking

salvage HAART, including an NNRTI plus a protease inhibitor-containing regimen, the rates of hypertriglyceridemia, hypercholesterolemia, and hyperglycemia were greater among patients taking efavirenz compared with nevirapine, and the time to peak metabolic alterations in hypercholesterolemia and hyperglycemia, but not hypertriglyceridemia, were more rapid in the whole efavirenz group. Comparing all of the 324 patients who took efavirenz with the 299 subjects who took nevirapine, the frequencies of raised triglyceride, cholesterol, and glucose concentrations were much higher in those taking efavirenz. There was some grade of lipodystrophy in 207 pretreated patients, but there was appreciable improvement after an NNRTI was introduced in patients taking efavirenz compared with those taking nevirapine.

Nutrition

A man taking efavirenz developed vitamin D deficiency and the authors speculated that this could have been due to CYP450 enzyme induction by efavirenz (29).

Hematologic

A woman who had previously reacted to nevirapine in form of a rash with angioedema and systemic features, developed an immune-mediated hemolytic anemia (30).

In a 48-year-old man who developed neutropenia while taking efavirenz, the neutrophil count recovered after withdrawal (31).

Mouth

Burning mouth syndrome attributed to efavirenz resolved on withdrawal (32).

Liver

In a prospective study of the incidence of severe hepatotoxicity among 312 patients taking efavirenz, hepatitis C and hepatitis B viruses were detected in 7.7% of the patients (33). There was severe hepatotoxicity in 8.0%, but only 50% of the episodes were detected during the first 12 weeks of therapy. The risk was significantly greater among those with chronic viral hepatitis (69% of cases) and those taking concurrent protease inhibitors (82% of cases). However, 84% of patients with chronic hepatitis C or hepatitis B did not have severe hepatotoxicity.

Urinary tract

There has been a report of urolithiasis suspected to be secondary to efavirenz based on the metabolic composition of the stone (34).

Skin

Efavirenz has been associated with UV photosensitivity, which recurred after skin challenge with efavirenz powder (35).

In a retrospective study of 122 patients who had previously stopped taking nevirapine because of a rash and who took efavirenz instead, 10 (8.2%) developed a rash and required withdrawal of efavirenz (36).

Reproductive system

Gynecomastia and breast hypertrophy have been reported in at least three patients. All had had indinavir replaced by efavirenz. No other medications were changed, and gynecomastia was not present before substitution (37). The breast enlargement was generally asymmetrical and painful. Needle biopsy was consistent with gynecomastia. Six other reports have confirmed painful breast hypertrophy in patients taking efavirenz (38).

However, gynecomastia has also been reported in 15 patients taking a variety of antiretroviral drugs (39). The authors suggested that it was due to increased cytokine concentrations following immune restoration and this adverse effect may therefore not be unique to efavirenz among drugs used to treat HIV infection.

Breasts

In a case-control study of 30 men taking antiretroviral drugs who developed gynecomastia and 30 men who did not, the use of efavirenz was significantly associated with gynecomastia (40). In an observational study seven of 193 men taking an efavirenz-based regimen developed gynecomastia during a median follow-up of 18 months compared with none of 164 men taking nevirapine (13).

Gynecomastia was diagnosed in 13/324 patients who took efavirenz, compared with 2/299 subjects taking nevirapine (28).

Immunologic

The incidence of allergic reactions to efavirenz is 10–34%. They usually cause an erythematous maculopapular rash, with or without fever, 1–3 weeks after the start of therapy. Desensitization has been reported (41).

- A 37-year-old HIV-positive white man was given efavirenz, amprenavir, stavudine, lamivudine, and didanosine after failure of a previous regimen. After 8 days he developed a generalized pruritic rash and all the drugs were withdrawn. Two weeks later he was given efavirenz, stavudine, didanosine, lamivudine, and lopinavir, but developed red itchy skin within a day. All the drugs were withdrawn. He was then successfully restarted on stavudine, didanosine, lamivudine, lopinavir, and amprenavir. Desensitization to efavirenz was undertaken, but on day 12 he again developed a rash on the trunk and limbs, which was treated with a topical steroid and diphenhydramine 45 minutes before each dose of efavirenz. The desensitization protocol was continued for another 4 days, and 16 months later he was taking full-dose efavirenz in combination with the other antiretroviral drugs.

Efavirenz can cause an allergic syndrome called the DRESS syndrome (Drug Rash with Eosinophilia and Systemic Symptoms). It is a life-threatening reaction that typically includes a rash, fever, lymphadenopathy, hepatitis, interstitial nephritis, pneumonia, myocarditis, and hematological abnormalities, particularly eosinophilia and a mononucleosis-like atypical lymphocytosis. The DRESS syndrome has been described in an HIV-infected woman taking efavirenz (42).

- A 44-year-old HIV-1 infected woman from the Ivory Coast, who was taking stavudine, lamivudine, efavirenz, and pyrimethamine plus sulfadiazine for *Toxoplasma* encephalitis, developed a maculopapular rash on both arms. The sulfadiazine was withdrawn and clindamycin was added. Ten days later her condition had worsened. Her temperature was 40°C, pulse rate 137/minute, and respiratory rate 26/minute. She had a generalized maculopapular rash without mucosal involvement, moderate abdominal tenderness, hepatomegaly, jaundice, and bilateral crackles. Her white cell count was $16 \times 10^9/l$ with 9% eosinophils and 51% lymphocytes. A chest X-ray showed moderate bilateral interstitial pneumonitis. All drugs were withdrawn and she was given intravenous methylprednisolone. The skin rash and all systemic manifestations resolved within 1 week and HIV treatment was restarted uneventfully with lamivudine, stavudine, and nelfinavir.

Although this syndrome has been described with sulfonamides (which the patient had taken), the fact that her condition worsened after withdrawal of sulfadiazine, and the characteristic timing of the syndrome (2–6 weeks after starting a drug), suggested efavirenz as the cause.

- In a 39-year-old man efavirenz caused a confluent maculopapular rash, and pulmonary interstitial infiltrates without lymphadenopathy (43). The symptoms resolved when efavirenz was withdrawn while other antiretroviral drugs were continued. The patient was rechallenged, and the rash and fever reappeared; however, recurrence of the pulmonary infiltrates was not addressed.
- A patient with HIV infection and disseminated tuberculosis developed an acute, severe, multiorgan hypersensitivity reaction to efavirenz, with acute hepatic, respiratory, and renal failure, in the absence of skin changes or eosinophilia; the symptoms resolved on withdrawal of efavirenz (44).

A leukocytoclastic vasculitis has been attributed to efavirenz (45).

- A 44-year-old man, having taken various antiretroviral drugs, started to take efavirenz; 5 days later he developed palpable purpura on both legs, with pruritus. His white cell count was $14.4 \times 10^9/l$ and a skin biopsy showed a leukocytoclastic vasculitis. Efavirenz was withdrawn and he was given prednisolone for 3 days; the lesions disappeared, leaving only minimal hyperpigmentation.

Second-Generation Effects

Teratogenicity

- A boy, born at full term to a woman who had taken efavirenz, stavudine, and zidovudine before pregnancy and for the first 24 weeks, had a myelomeningocele (46). The baby was HIV-positive.

The authors suggested that the efavirenz had been responsible, based on previous studies in monkeys.

Susceptibility Factors

Genetic

Plasma efavirenz concentrations were increased in 40 of 100 subjects with the polymorphic homozygous genotype 516G>T at the gene encoding cytochrome CYP2B6 and 19% of subjects with the heterozygous genotype; 20% of those with the wild-type genotype had subtherapeutic concentrations of efavirenz (47). The CYP2B6-516 genotype, which is commoner in African–Americans, may help to identify subjects who will have plasma efavirenz concentrations that are outside the usual target range.

Drug–Drug Interactions

Antiretroviral drugs

In patients who were taking the NNRTIs efavirenz and nevirapine, the apparent oral clearance (CL/F) of lopinavir increased by 39% (48). This is in line with an advised 33% dosage increment for lopinavir+ ritonavir when combined with NNRTIs. There was a 41% increase in indinavir clearance when it was co-administered with nevirapine or efavirenz (49).

The interactions of efavirenz, amprenavir, nelfinavir, indinavir, and ritonavir have been measured in 56 seronegative subjects (ACTG 5043) (50). African–American non-Hispanics had higher efavirenz AUCs than white non-Hispanics on day 14. The authors concluded that efavirenz reduced the AUC of amprenavir, but nelfinavir, indinavir, or ritonavir compensated for the induction of amprenavir metabolism by efavirenz. There were rashes in 11% of subjects but none was worse than grade 1. There was no relation between rash and amprenavir plasma concentration, but there were higher efavirenz concentrations in those with rashes. Efavirenz concentrations did not correlate with nervous system symptoms, possibly because those with more severe symptoms dropped out before the efavirenz AUCs were obtained on day 14.

Atazanavir

In a population pharmacokinetic study in 87 patients, atazanavir clearance was significantly increased when efavirenz was co administered with atazanavir + ritonavir (51).

Ciclosporin

Efavirenz causes increased requirements of ciclosporin by inducing CYP3A4.

- A 39-year-old man took ciclosporin 200 mg bd after a kidney transplant (52). The blood ciclosporin concentration was 307 ng/ml. The co-administration of prednisone caused the blood ciclosporin concentration to rise to 372 ng/ml, but it fell to 203 ng/ml when the dose was reduced to 175 mg bd. Efavirenz 600 mg/day, zidovudine 300 mg bd, and lamivudine 150 mg bd were added, and 7 days later the blood ciclosporin concentration fell to 80 ng/ml and later to 50 ng/ml.

HIV protease inhibitors

The effect of multiple-dose efavirenz 600 mg/day on the steady-state pharmacokinetics of the combination of indinavir (800 mg) + low-dose ritonavir (100 mg) bd, in which ritonavir is used to increase indinavir plasma concentrations, has been investigated in 14 healthy men (53). Efavirenz significantly reduced indinavir AUC (25%), C_{max} (17%), and C_{min} (50%). Ritonavir AUC, C_{max}, and C_{min} were reduced by 36%, 34%, and 39% respectively. The authors proposed that efavirenz had induced the activity of CYP3A and concluded that the dose of indinavir or ritonavir may need to be increased to maintain similar indinavir drug concentrations after the addition of efavirenz.

Methadone

Following a report of withdrawal symptoms in three patients taking methadone maintenance therapy, serum methadone concentrations were measured before and after starting efavirenz in one patient (54). Serum methadone concentrations were as follows: (R)-methadone (the active enantiomer) 168 ng/ml and 90 ng/ml before and after efavirenz. The corresponding (S)-methadone concentrations were 100 and 28 ng/ml. The dosage of methadone was increased from 100 to 180 mg/day before the patient's withdrawal symptoms resolved.

- Three patients taking methadone were given efavirenz-containing regimens and started to complain of opioid withdrawal symptoms 3–7 days later (55). In one case the plasma methadone concentration fell from about 170 ng/ml to about 50 ng/ml over 6 days.

Since methadone is partly metabolized by CYP3A4, the interaction of efavirenz with methadone has been investigated prospectively in 11 patients taking stable methadone maintenance therapy (56). Efavirenz reduced the steady-state methadone AUC by 50%. However, patients generally only needed a 22% increase in their methadone dose to eliminate the symptoms of methadone withdrawal, and the full basis of the interaction is not understood.

Rifamycins

Rifampicin induces the metabolism of efavirenz. In 24 patients (21 men, 3 women; mean age 37 years) with HIV infection and tuberculosis the C_{max}, C_{min}, and

AUC of efavirenz interval fell by 24%, 25%, and 22% respectively in the presence of rifampicin (57).

Because of induction of CYP3A4 by efavirenz the dosage of concurrent rifabutin should be increased to 450–600 mg/day according to current recommendations. However, in one case rifabutin concentrations were inadequate even though the dosage of rifabutin was increased to 1350 mg/day (58). The isoniazid concentration was also inadequate. The concentrations of both tuberculostatics became satisfactory when efavirenz was withdrawn.

References

1. Morales-Ramirez J, Tashima K, Hardy D, et al. In: A phase II, multi-center randomized, open label study to compare the antiretroviral activity and tolerability of efavirenz (EFV) + indinavir (IDV), versus EFV + zidovudine (ZDV) + lamivudine (3TC), versus IDV + 3TC at >36 weeks (DMP 266-006)38th ICAAC, San Diego 24–27 September 1998:103 Abstract.
2. Adkins JC, Noble S. Efavirenz. Drugs 1998;56(6):1055–64.
3. Gazzard BG. Efavirenz in the management of HIV infection. Int J Clin Pract 1999;53(1):60–4.
4. Haas DW, Fessel WJ, Delapenha RA, Kessler H, Seekins D, Kaplan M, Ruiz NM, Ploughman LM, Labriola DF, Manion DJ. Therapy with efavirenz plus indinavir in patients with extensive prior nucleoside reverse-transcriptase inhibitor experience: a randomized, double-blind, placebo-controlled trial. J Infect Dis 2001;183(3):392–400.
5. Mouly S, Lown KS, Kornhauser D, Joseph JL, Fiske WD, Benedek IH, Watkins PB. Hepatic but not intestinal CYP3A4 displays dose-dependent induction by efavirenz in humans. Clin Pharmacol Ther 2002;72(1):1–9.
6. Negredo E, Cruz L, Paredes R, Ruiz L, Fumaz CR, Bonjoch A, Gel S, Tuldra A, Balague M, Johnston S, Arno A, Jou A, Tural C, Sirera G, Romeu J, Clotet B. Virological, immunological, and clinical impact of switching from protease inhibitors to nevirapine or to efavirenz in patients with human immunodeficiency virus infection and long-lasting viral suppression. Clin Infect Dis 2002;34(4):504–10.
7. Moyle GJ. Efavirenz: shifting the HAART paradigm in adult HIV-1 infection. Expert Opin Investig Drugs 1999;8(4):473–86.
8. Fumaz CR, Tuldra A, Ferrer MJ, Paredes R, Bonjoch A, Jou T, Negredo E, Romeu J, Sirera G, Tural C, Clotet B. Quality of life, emotional status, and adherence of HIV-1-infected patients treated with efavirenz versus protease inhibitor-containing regimens. J Acquir Immune Defic Syndr 2002;29(3):244–53.
9. Bartlett JA, Johnson J, Herrera G, Sosa N, Rodriguez A, Liao Q, Griffith S, Irlbeck D, Shaefer MS; Clinically Significant Long-Term Antiretroviral Sequential Sequencing Study (CLASS) Team. Long-term results of initial therapy with abacavir and lamivudine combined with efavirenz, amprenavir/ritonavir, or stavudine. J Acquir Immune Defic Syndr 2006;43(3):284–92.
10. Moyle G, Fletcher C, Brown H, Mandalia S, Gazzard B. Changes in sleep quality and brain wave patterns following initiation of an efavirenz-containing triple antiretroviral regimen. HIV Med 2006;7(4):243–7.
11. Cirino CM, Kan VL. Hypokalemia in HIV patients on tenofovir. AIDS 2006;20(12):1671–3.
12. van Leth F, Phanuphak P, Ruxrungtham K, Baraldi E, Miller S, Gazzard B, Cahn P, Lalloo UG, van der Westhuizen IP, Malan DR, Johnson MA, Santos BR, Mulcahy F, Wood R, Levi GC, Reboredo G, Squires K, Cassetti I, Petit D, Raffi F, Katlama C, Murphy RL, Horban A, Dam JP, Hassink E, van Leeuwen R, Robinson P, Wit FW, Lange JM; 2NN Study team. Comparison of first-line antiretroviral therapy with regimens including nevirapine, efavirenz, or both drugs, plus stavudine and lamivudine: a randomised open-label trial, the 2NN Study. Lancet 2004;363(9417):1253–63.
13. Manfredi R, Calza L, Chiodo F. Efavirenz versus nevirapine in current clinical practice: a prospective, open-label observational study. J Acquir Immune Defic Syndr 2004;35(5):492–502.
14. Gallego L, Barreiro P, del Rio R, Gonzalez de Requena D, Rodriguez-Albarino A, Gonzalez-Lahoz J, Soriano V. Analyzing sleep abnormalities in HIV-infected patients treated with Efavirenz. Clin Infect Dis 2004;38(3):430–2.
15. Fumaz CR, Munoz-Moreno JA, Molto J, Negredo E, Ferrer MJ, Sirera G, Perez-Alvarez N, Gomez G, Burger D, Clotet B. Long-term neuropsychiatric disorders on efavirenz-based approaches: quality of life, psychologic issues, and adherence. J Acquir Immune Defic Syndr 2005;38(5):560–5.
16. Faggian F, Lattuada E, Lanzafame M, Antolini D, Concia E, Vento S. Recreational substance use and tolerance of efavirenz in HIV-1 infected patients. AIDS Care 2005;17(7):908–10.
17. Morales-Ramirez J, Tashima K, Hardy D, et al. In: A phase III, multicenter open-label study to compare the antiretroviral activity and tolerability of efavirenz (EFV) + indinavir (IDV), versus EFV + zidovudine (ZDV) + lamivudine (3TC), versus IDV + ZDV + 3TC at 36 weeks (DMP 266-006)38th Interscience Conference on Antimicrobial Agents and Chemotherapy (San Diego). Washington DC: American Society for Microbiology, 1998:I-103.
18. Mayers D, Jemesk J, Eyster E, et al. In: A double-blind, placebo-controlled study to assess the safety, tolerability and antiretroviral activity of efavirenz (EFV DMP 266) in combination with open-label zidovudine (ZDV) and lamivudine (3TC) in HIV infected patients (DMP 266–004)38th Interscience Conference on Antimicrobial Agents and Chemotherapy (San Diego). Washington DC: American Society for Microbiology, 1998:22340.
19. Welch KJ, Morse A. Association between efavirenz and selected psychiatric and neurological conditions. J Infect Dis 2002;185(2):268–9.
20. Peyriere H, Mauboussin JM, Rouanet I, Fabre J, Reynes J, Hillaire-Buys D. Management of sudden psychiatric disorders related to efavirenz. AIDS 2001;15(10):1323–4.
21. Marzolini C, Telenti A, Decosterd LA, Greub G, Biollaz J, Buclin T. Efavirenz plasma levels can predict treatment failure and central nervous system side effects in HIV-1-infected patients. AIDS 2001;15(1):71–5.
22. Boly L, Cafaro V, Dyner T. Depressive symptoms predict increased incidence of neuropsychiatric side effects in patients treated with efavirenz. J Acquir Immune Defic Syndr 2006;42(4):514–5.
23. Fernando AI, Anderson OA, Holder GE, Mitchell SM. Didanosine-induced retinopathy in adults can be reversible. Eye 2006;20(12):1435–7.
24. Rihs TA, Begley K, Smith DE, Sarangapany J, Callaghan A, Kelly M, Post JJ, Gold J. Efavirenz and chronic neuropsychiatric symptoms: a cross-sectional case control study. HIV Med 2006;7(8):544–8.

25. Shah MD, Balderson K. A manic episode associated with efavirenz therapy for HIV infection. AIDS 2003;17:1713–4.
26. Poulsen HD, Lublin HK. Efavirenz-induced psychosis leading to involuntary detention. AIDS 2003;17:451–3.
27. Spire B, Carrieri P, Garzot MA, L'henaff M, Obadia Y. Factors associated with efavirenz discontinuation in a large community-based sample of patients. AIDS Care 2004;16(5):558–64.
28. Manfredi R, Calza L, Chiodo F. An extremely different dysmetabolic profile between the two available nonnucleoside reverse transcriptase inhibitors: efavirenz and nevirapine. J Acquir Immune Defic Syndr 2005;38(2):236–8.
29. Gyllensten K, Josephson F, Lidman K, Saaf M. Severe vitamin D deficiency diagnosed after introduction of antiretroviral therapy including efavirenz in a patient living at latitude 59 degrees N. AIDS 2006;20(14):1906–7.
30. Freercks RJ, Mehta U, Stead DF, Meintjes GA. Haemolytic anaemia associated with efavirenz. AIDS 2006;20(8):1212–3.
31. Healy BJ, Freedman AR. HIV-related neutropaenia exacerbated by efavirenz. HIV Med 2006;7(2):129–31.
32. Borras-Blasco J, Belda A, Rosique-Robles JD, Castera MD, Abad FJ. Burning mouth syndrome due to efavirenz therapy. Ann Pharmacother 2006;40(7-8):1471–2.
33. Sulkowski MS, Thomas DL, Mehta SH, Chaisson RE, Moore RD. Hepatotoxicity associated with nevirapine or efavirenz-containing antiretroviral therapy: role of hepatitis C and B infections. Hepatology 2002;35(1):182–9.
34. Wirth GJ, Teuscher J, Graf JD, Iselin CE. Efavirenz-induced urolithiasis. Urol Res 2006;34(4):288–9.
35. Treudler R, Husak R, Raisova M, Orfanos CE, Tebbe B. Efavirenz-induced photoallergic dermatitis in HIV. AIDS 2001;15(8):1085–6.
36. Manosuthi W, Thongyen S, Chumpathat N, Muangchana K, Sungkanuparph S. Incidence and risk factors of rash associated with efavirenz in HIV-infected patients with preceding nevirapine-associated rash. HIV Med 2006;7(6):378–82.
37. Caso JA, Prieto Jde M, Casas E, Sanz J. Gynecomastia without lipodystrophy syndrome in HIV-infected men treated with efavirenz. AIDS 2001;15(11):1447–8.
38. Mercie P, Viallard JF, Thiebaut R, Faure I, Rispal P, Leng B, Pellegrin JL. Efavirenz-associated breast hypertrophy in HIV-infection patients. AIDS 2001;15(1):126–9.
39. Qazi NA, Morlese JF, King DM, Ahmad RS, Gazzard BG, Nelson MR. Gynaecomastia without lipodystrophy in HIV-1-seropositive patients on efavirenz: an alternative hypothesis. AIDS 2002;16(3):506–7.
40. Mira JA, Lozano F, Santos J, Ramayo E, Terron A, Palacios R, Leon EM, Marquez M, Macias J, Fernandez-Palacin A, Gomez-Mateos J, Pineda JA; Grupo Andaluz para el Estudio de las Enfermedades Infecciosas. Gynaecomastia in HIV-infected men on highly active antiretroviral therapy: association with efavirenz and didanosine treatment. Antivir Ther 2004;9(4):511–7.
41. Phillips EJ, Kuriakose B, Knowles SR. Efavirenz-induced skin eruption and successful desensitization. Ann Pharmacother 2002;36(3):430–2.
42. Bossi P, Colin D, Bricaire F, Caumes E. Hypersensitivity syndrome associated with efavirenz therapy. Clin Infect Dis 2000;30(1):227–8.
43. Behrens GM, Stoll M, Schmidt RE. Pulmonary hypersensitivity reaction induced by efavirenz. Lancet 2001;357(9267):1503–4.
44. Angel-Moreno-Maroto A, Suarez-Castellano L, Hernandez-Cabrera M, Perez-Arellano JL. Severe efavirenz-induced hypersensitivity syndrome (not-DRESS) with acute renal failure. J Infect 2006;52(2):e39-40.
45. Domingo P, Barcelo M. Efavirenz-induced leukocytoclastic vasculitis. Arch Intern Med 2002;162(3):355–6.
46. Fundaro C, Genovese O, Rendeli C, Tamburrini E, Salvaggio E. Myelomeningocele in a child with intrauterine exposure to efavirenz. AIDS 2002;16(2):299–300.
47. Rodriguez-Novoa S, Barreiro P, Rendón A, Jiménez-Nacher I, González-Lahoz J, Soriano V. Influence of 516G>T polymorphisms at the gene encoding the CYP450-2B6 isoenzyme on efavirenz plasma concentrations in HIV-infected subjects. Clin Infect Dis 2005;40(9):1358–61.
48. Crommentuyn KM, Huitema AD, Brinkman K, van der Ende ME, de Wolf F, Beijnen JH; Athena study. Therapeutic drug monitoring of nevirapine reduces pharmacokinetic variability but does not affect toxicity or virologic success in the ATHENA study. J Acquir Immune Defic Syndr 2005;39(2):249–50.
49. Kappelhoff BS, Huitema AD, Sankatsing SU, Meenhorst PL, Van Gorp EC, Mulder JW, Prins JM, Beijnen JH. Population pharmacokinetics of indinavir alone and in combination with ritonavir in HIV-1-infected patients. Br J Clin Pharmacol 2005;60(3):276–86.
50. Morse GD, Rosenkranz S, Para MF, Segal Y, Difrancesco R, Adams E, Brizz B, Yarasheski KE, Reichman RC. Amprenavir and efavirenz pharmacokinetics before and after the addition of nelfinavir, indinavir, ritonavir, or saquinavir in seronegative individuals. Antimicrob Agents Chemother 2005;49(8):3373–81.
51. Dailly E, Tribut O, Tattevin P, Arvieux C, Perré P, Raffi F, Jolliet P. Influence of tenofovir, nevirapine and efavirenz on ritonavir-boosted atazanavir pharmacokinetics in HIV-infected patients. Eur J Clin Pharmacol 2006;62(7):523–6.
52. Tseng A, Nguyen ME, Cardella C, Humar A, Conly J. Probable interaction between efavirenz and cyclosporine. AIDS 2002;16(3):505–6.
53. Aarnoutse RE, Grintjes KJ, Telgt DS, Stek M Jr, Hugen PW, Reiss P, Koopmans PP, Hekster YA, Burger DM. The influence of efavirenz on the pharmacokinetics of a twice-daily combination of indinavir and low-dose ritonavir in healthy volunteers. Clin Pharmacol Ther 2002;71(1):57–67.
54. Marzolini C, Troillet N, Telenti A, Baumann P, Decosterd LA, Eap CB. Efavirenz decreases methadone blood concentrations. AIDS 2000;14(9):1291–2.
55. Boffito M, Rossati A, Reynolds HE, Hoggard PG, Back DJ, Di Perri G. Undefined duration of opiate withdrawal induced by efavirenz in drug users with HIV infection and undergoing chronic methadone treatment. AIDS Res Hum Retroviruses 2002;18(5):341–2.
56. Clarke SM, Mulcahy FM, Tjia J, Reynolds HE, Gibbons SE, Barry MG, Back DJ. The pharmacokinetics of methadone in HIV-positive patients receiving the non-nucleoside reverse transcriptase inhibitor efavirenz. Br J Clin Pharmacol 2001;51(3):213–7.
57. Lopez-Cortes LF, Ruiz-Valderas R, Viciana P, Alarcon-Gonzalez A, Gomez-Mateos J, Leon-Jimenez E, Sarasanacenta M, Lopez-Pua Y, Pachon J. Pharmacokinetic interactions between efavirenz and rifampicin in HIV-infected patients with tuberculosis. Clin Pharmacokinet 2002;41(9):681–90.
58. Edelstein HE, Cuadros Y. Failure of treatment of tuberculous adenitis due to an unexpected drug interaction with rifabutin and efavirenz. AIDS 2004;18(12):1748–9.

Enfuvirtide

Enfuvirtide is the first available fusion inhibitor for the treatment of HIV infection. It has poor oral systemic availability and is given subcutaneously. Adverse effects were reported in a phase 2 trial of enfuvirtide 45 mg, 67.5 mg, and 90 mg bd combined with other antiretroviral drugs (1). Among 52 patients who received enfuvirtide, all but one reported at least one adverse event. However, the frequency of adverse events was similar in the 19 patients in the control arm, with the exception of injection site reactions, which were reported at least once in 36 patients. However, injection site reactions occasioned drug withdrawal in only three patients receiving enfuvirtide.

Hematologic

One child out of 14 who were given enfuvirtide had grade 3 thrombocytopenia at week 65 and this led to withdrawal of the drug (2).

Fluid balance

One child out of 14 treated had grade 3 edema at week 77 and had stop taking the drug (2).

Skin

In the large randomized TORO trials in 661 patients almost all of those who used enfuvirtide had an injection site reaction (3,4). The reactions were mild in about half the patients. These reactions included erythema (in 87%), induration (84%), and nodules and cysts (82%).

Immunologic

Two patients taking enfuvirtide had hypersensitivity reactions in the TORO trials (5). One had a rash, fever, and vomiting, the other had membranoproliferative glomerulonephritis and on rechallenge severe respiratory distress syndrome. Treatment-related eosinophilia occurred in 11% of the patients who took enfuvirtide compared with 2.4% of the controls.

Infection risk

In an updated safety analysis of both TORO trials there was a significantly higher incidence of bacterial pneumonia (4.7 versus 0.3 events per 100 person-years) in those who took enfuvirtide (5).

References

1. Lalezari JP, DeJesus E, Northfelt DW, Richmond G, Wolfe P, Haubrich R, Henry D, Powderly W, Becker S, Thompson M, Valentine F, Wright D, Carlson M, Riddler S, Haas FF, DeMasi R, Sista PR, Salgo M, Delehanty J. A controlled Phase II trial assessing three doses of enfuvirtide (T-20) in combination with abacavir, amprenavir, ritonavir and efavirenz in non-nucleoside reverse transcriptase inhibitor-naive HIV-infected adults. Antivir Ther 2003;8: 279-87.
2. Church JA, Hughes M, Chen J, Palumbo P, Mofenson LM, Delora P, Smith E, Wiznia A, Hawkins E, Sista P, Cunningham CK; Pediatric AIDS Clinical Trials Group P1005 Study Team. Long term tolerability and safety of enfuvirtide for human immunodeficiency virus 1-infected children. Pediatr Infect Dis J 2004;23(8):713–8.
3. Lalezari JP, Henry K, O'Hearn M, Montaner JS, Piliero PJ, Trottier B, Walmsley S, Cohen C, Kuritzkes DR, Eron JJ Jr, Chung J, DeMasi R, Donatacci L, Drobnes C, Delehanty J, Salgo M; TORO 1 Study Group. Enfuvirtide, an HIV-1 fusion inhibitor, for drug-resistant HIV infection in North and South America. N Engl J Med 2003;348(22):2175–85.
4. Lazzarin A, Clotet B, Cooper D, Reynes J, Arasteh K, Nelson M, Katlama C, Stellbrink HJ, Delfraissy JF, Lange J, Huson L, DeMasi R, Wat C, Delehanty J, Drobnes C, Salgo M; TORO 2 Study Group. Efficacy of enfuvirtide in patients infected with drug-resistant HIV-1 in Europe and Australia. N Engl J Med 2003;348(22):2186–95.
5. Lalezari JP, Henry K, O'Hearn M, Montaner JS, Piliero PJ, Trottier B, Walmsley S, Cohen C, Kuritzkes DR, Eron JJ Jr, Chung J, DeMasi R, Donatacci L, Drobnes C, Delehanty J, Salgo M; TORO 1 Study Group. Enfuvirtide, an HIV-1 fusion inhibitor, for drug-resistant HIV infection in North and South America. N Engl J Med 2003;348(22):2175–85.

Famciclovir

General Information

Famciclovir is an oral prodrug of penciclovir, a selective antiviral drug with activity against *Varicella zoster* virus, *Herpes simplex* virus types 1 and 2, and Epstein–Barr virus, as well as human hepatitis B virus.

After oral administration, famciclovir is well absorbed (systemic availability 77%), with little intersubject variability, and is rapidly converted to penciclovir. This compares favorably with aciclovir, the absorption of which is slow and incomplete, with a highly variable systemic availability of only 10–20%.

Observational studies

In a multicenter study oral famciclovir was well tolerated by 60 women who took famciclovir 125 mg tds and 59 women who took 250 mg tds (1).

Comparative studies

In a study of oral famciclovir versus oral aciclovir, designed to demonstrate equivalence of efficacy of the two drugs in the treatment of mucocutaneous *Herpes simplex* infection in HIV-infected individuals, there was no difference in the incidence or nature of adverse effects in the two groups (2). None of the withdrawals from the trial was considered by the investigator to be related to the study medication.

Placebo-controlled studies

In an integrated safety analysis of 1607 patients who had taken famciclovir for the treatment of *Herpes zoster* or genital herpes, famciclovir was extremely well tolerated with an adverse effect profile similar to placebo (3). Headache, nausea, and diarrhea were the most frequently reported adverse events in those taking both famciclovir and placebo.

In an experimental study of *Herpes simplex* labialis, adverse events (diarrhea and nausea) occurred with similar frequency with famciclovir and placebo (4). No laboratory abnormalities were consistently associated with famciclovir.

In a randomized, placebo-controlled study in 455 patients oral famciclovir (125 or 250 mg tds or 250 mg bd) used to suppress recurrent genital *Herpes simplex* infections, the toxicity profile of famciclovir was comparable to placebo (5). The only serious adverse effects reported as being possibly related to famciclovir were raised bilirubin concentration and lipase activity in one patient after 10 months of treatment with famciclovir 125 mg tds. However, these laboratory abnormalities resolved on therapy after 7 days and did not recur during the rest of the study.

Organs and Systems

Gastrointestinal

Adverse effects associated with famciclovir have been collected in over 6000 patients in two postmarketing surveillance studies (6). Only headache, abdominal symptoms, dizziness, vomiting, and diarrhea were associated with the drug. Two prospective trials have confirmed the low frequency of adverse effects, the more common ones being nausea, headache, vomiting, and diarrhea (7,8).

In a randomized, double-blind comparison of aciclovir and famciclovir in 55 patients with uncomplicated *Herpes zoster* infection, constipation was the only significantly different adverse reaction, occurring in 11% and 3.7% respectively (9).

Sexual function

Prolonged administration of high dosages of famciclovir has been associated with reversible dose-dependent adverse effects on testicular function in rats and dogs. However, in a double-blind, placebo-controlled trial in which 34 men with recurrent genital herpes took famciclovir 250 mg bd for 18 weeks, there were no significant effects on sperm production or function (10).

References

1. Sacks SL. Famciclovir suppression of asymptomatic and symptomatic recurrent anogenital herpes simplex virus shedding in women: a randomized, double-blind, double-dummy, placebo-controlled, parallel-group, single-center trial. J Infect Dis 2004;189(8):1341-7.

2. Romanowski B, Aoki FY, Martel AY, Lavender EA, Parsons JE, Saltzman RLCollaborative Famciclovir HIV Study Group. Efficacy and safety of famciclovir for treating mucocutaneous *Herpes simplex* infection in HIV-infected individuals. AIDS 2000;14(9):1211-7.

3. Saltzman R, Jurewicz R, Boon R. Safety of famciclovir in patients with *Herpes zoster* and genital herpes. Antimicrob Agents Chemother 1994;38(10):2454-7.

4. Spruance SL, Rowe NH, Raborn GW, Thibodeau EA, D'Ambrosio JA, Bernstein DI. Peroral famciclovir in the treatment of experimental ultraviolet radiation-induced *Herpes simplex* labialis: a double-blind, dose-ranging, placebo-controlled, multicenter trial. J Infect Dis 1999;179(2):303-10.

5. Diaz-Mitoma F, Sibbald RG, Shafran SD, Boon R, Saltzman RL. Oral famciclovir for the suppression of recurrent genital herpes: a randomized controlled trial. Collaborative Famciclovir Genital Herpes Research Group. JAMA 1998;280(10):887-92.

6. Engst R, Schiewe U, Hobel W, Machka K, Meister W. Famciclovir in treatment of acute *Herpes zoster*: results of two post-marketing surveillance studies in Germany. Acta Derm Venereol 2001;81(1):59-60.

7. Tyring S, Engst R, Corriveau C, Robillard N, Trottier S, Van Slycken S, Crann RA, Locke LA, Saltzman R, Palestine AGCollaborative Famciclovir Ophthalmic Zoster Research Group. Famciclovir for ophthalmic zoster: a randomised aciclovir controlled study. Br J Ophthalmol 2001;85(5):576-81.

8. Manns MP, Neuhaus P, Atkinson GF, Griffin KE, Barnass S, Vollmar J, Yeang Y, Young CLFamciclovir Liver Transplant Study Group. Famciclovir treatment of hepatitis B infection following liver transplantation: a long-term, multi-centre study. Transpl Infect Dis 2001;3(1):16-23.

9. Shen MC, Lin HH, Lee SS, Chen YS, Chiang PC, Liu YC. Double-blind, randomized, acyclovir-controlled, parallel-group trial comparing the safety and efficacy of famciclovir and acyclovir in patients with uncomplicated herpes zoster. J Microbiol Immunol Infect 2004;37(2):75-81.

10. Perry CM, Wagstaff AJ. Famciclovir. A review of its pharmacological properties and therapeutic efficacy in herpesvirus infections. Drugs 1995;50(2):396-415.

Fomivirsen

General Information

Fomivirsen (ISIS 2922) is an antisense oligonucleotide that specifically inhibits replication of human cytomegalovirus. It has been developed for intravitreal administration to treat cytomegalovirus retinitis in patients with AIDS.

Organs and Systems

Sensory systems

The most common adverse effects of fomivirsen reported in clinical trials have been increased intraocular pressure and mild to moderate uveitis (1,2). These events were generally transient or reversible with topical glucocorticoids.

In 150 patients who were given intravitreous fomivirsen 165 micrograms/injection (35 eyes, 30 patients) or two different regimens of 330 micrograms/injection (142 eyes, 110 patients), anterior chamber inflammation and increased intraocular pressure were dose-related and schedule-dependent: 165 micrograms/injection, 4.1 events/patient-year; 330 micrograms/injection, 6.6 events/patient-year (less intense regimen) and 8.4 events/patient-year (more intense regimen) (3). A large number of other ocular adverse events may not have been related to the drug.

In 309 eyes of 238 patients there were two cases of bull's-eye maculopathy, which resolved after withdrawal (4).

Retinal pigment epithelial changes have been noted at doses of fomivirsen that overlap with those used to treat CMV retinitis (2). In one case there was nyctalopia and reduced visual acuity (20/50 OD) in conjunction with mid-peripheral epithelial pigmentation and cotton-wool spots around the perifoveal capillary network (5).

References

1. Perry CM, Balfour JA, Johnson DW, De Clercq E. Fomivirsen. Drugs 1999;57(3):375–81.
2. Freeman WR. Retinal toxic effects associated with intravitreal fomivirsen. Arch Ophthalmol 2001;119(3):458.
3. Boyer DS, Cowen SJ, Danis RP, Diamond JG, Fish RH, Goldstein DA, Jaffe GJ, Lelezari JP, Lieberman RM, Belfort R Jr, Muccioli C, Palestine AG, Perez JE, Territo C, Johnson DW, Mansour SE, Sheppard JD, Mora-Durate J, Chan CK, Andreu-Andreu D, Deschenes JG, deSmet MD, Fisher M, Gastaut J-A, Gazzard BG, Lightman B, Johnson MA, Klauss V, Gumbel H, et alVitravene Study Group. Safety of intravitreous fomivirsen for treatment of cytomegalovirus retinitis in patients with AIDS. Am J Ophthalmol 2002;133(4):484–98.
4. Stone TW, Jaffe GJ. Reversible bull's-eye maculopathy associated with intravitreal fomivirsen therapy for cytomegalovirus retinitis. Am J Ophthalmol 2000;130(2):242–3.
5. Amin HI, Ai E, McDonald HR, Johnson RN. Retinal toxic effects associated with intravitreal fomivirsen. Arch Ophthalmol 2000;118(3):426–7.

Foscarnet

General Information

Foscarnet (trisodium phosphonoformate hexahydrate) is a pyrophosphate analogue that interacts with the enzymatic action of polymerases and inhibits the cleavage of pyrophosphate from the nucleoside triphosphate. Because of this mechanism, the antiviral activity of the drug is broad. Foscarnet is a non-competitive inhibitor of herpesvirus DNA polymerase, hepatitis B virus DNA polymerase, and reverse transcriptases (1). Intravenous foscarnet has been used for the treatment of mucocutaneous disease due to acyclovir-resistant Herpes simplex (2) and for the treatment of severe CMV infection (3,4).

Foscarnet has been shown to be as effective as ganciclovir in the treatment of gastrointestinal CMV and CMV retinitis in patients with AIDS (5). The two drugs differ, however, in their respective toxicity profile, and in a comparison in CMV retinitis, ganciclovir was better tolerated (6). Treatment-limiting adverse effects are renal toxicity, hypocalcemia, and mucosal ulceration.

Organs and Systems

Electrolyte balance

Significant electrolyte abnormalities have been attributed to foscarnet in a bone marrow transplant recipient on parenteral nutrition (7).

- A 39-year-old man with acute myelogenous leukemia developed a fever after allogeneic bone marrow transplantation and was given prophylactic ganciclovir and antibiotics. Parenteral nutrition was started when severe mucositis and diarrhea limited oral nutrition. On the 18th day after transplantation CMV DNA was detected in his blood and he was given foscarnet. His requirements for potassium, calcium, magnesium, and phosphorus increased dramatically, while his sodium requirements fell. Electrolyte depletion occurred within 24 hours and was accompanied by deteriorating renal function (serum creatinine 106–220, reference range 60–125 µmol/l). On withdrawal of foscarnet, the serum creatinine fell within 24 hours and the electrolyte concentrations returned to normal.

Mineral balance

The second most common adverse effect of foscarnet is symptomatic hypocalcemia, which may be responsible for the cardiac dysrhythmias and seizures that occur after acute overdose or excessively rapid infusion of foscarnet. Foscarnet stimulates the release of parathyroid hormone, which raised concerns about long-term administration (8). However, in a study of seven patients receiving a 14-day foscarnet induction regimen, there were no changes in calcium or phosphate metabolism (9).

Recognizing that foscarnet is a potent chelator of divalent cations, and that acute ionized hypocalcemia and hypomagnesemia are common adverse effects, a trial of intravenous magnesium sulfate has been conducted for foscarnet-induced hypocalcemia and hypomagnesemia in 12 AIDS patients with CMV infection (10). Increasing doses of magnesium sulfate reduced or eliminated foscarnet-induced ionized hypomagnesemia but had no discernible effect on ionized hypocalcemia, despite significant increases in serum parathyroid hormone concentrations. On this basis, intravenous supplementation for patients with normal serum magnesium concentrations was not recommended during treatment with foscarnet.

Urinary tract

Alterations in creatinine clearance or acute renal insufficiency occur in 10–20% of patients with AIDS receiving intravenous foscarnet (11), due to acute tubular damage.

Severe renal insufficiency can be prevented in most cases by careful hydration before and during therapy (12). To minimize the residual incidence of nephrotoxicity, the dose of foscarnet should be frequently recalculated, based on the estimated creatinine clearance.

Patients who have undergone renal transplantation and then require treatment for cytomegalovirus infection are at special risk of renal damage due to foscarnet.

● One renal transplant recipient developed the nephrotic syndrome, with microscopic hematuria and non-oliguric acute renal insufficiency within 15 days after starting foscarnet therapy for cytomegalovirus infection (13). A kidney biopsy showed crystals in all glomeruli and in the proximal tubules. The crystals consisted of several forms of foscarnet salts. Renal function and proteinuria nevertheless improved progressively, and a second transplant biopsy 8 months after the first one showed fibrotic organization of half of the glomeruli and of the interstitial tissues, and a reduction in the amount of crystals.

Even half-dose foscarnet together with half-dose ganciclovir for pre-emptive therapy of cytomegalovirus infection in 24 patients resulted in greater nephrotoxicity than full-dose ganciclovir (14). Creatinine retention and electrolyte disturbances led to changes in therapy in six of those given the combination compared with none of those who were given ganciclovir alone.

The renal damage caused by foscarnet can itself alter the drug's kinetics; the dosage needs to be substantially reduced when renal complications occur, and appropriate guidelines have been developed (15).

Skin

Painful oral, penile, and vulvar ulceration can occur during foscarnet therapy (16), most probably due to local accumulation of the drug (SEDA-17, 338). Penile ulcers have been reported to be the reason for discontinuation of foscarnet therapy in up to 10% of patients (17).

Eosinophilic folliculitis is a common skin manifestation associated with HIV infection, but not commonly associated with medications.

● A 38-year-old patient with AIDS was given intravenous foscarnet 400 mg tds and no other medications (18). On the second day, a pruritic, erythematous, maculopapular, urticarial rash was noted on the limbs and trunk. The infusion was stopped and the reaction disappeared. On restarting the infusion, the reaction recurred but did not clear after withdrawal. Histology showed a pattern consistent with eosinophilic folliculitis. After UVB phototherapy for 2 months the folliculitis resolved.

References

1. Oberg B. Molecular basis of foscarnet action in human herpesvirus infections. In: Lopez C, Roizman B, editors. Human Herpesvirus Infections. New York: Raven Press, 1986:141.
2. Chatis PA, Miller CH, Schrager LE, Crumpacker CS. Successful treatment with foscarnet of an acyclovir-resistant mucocutaneous infection with *Herpes simplex* virus in a patient with acquired immunodeficiency syndrome. N Engl J Med 1989;320(5):297–300.
3. Klintmalm G, Lonnqvist B, Oberg B, Gahrton G, Lernestedt JO, Lundgren G, Ringden O, Robert KH, Wahren B, Groth CG. Intravenous foscarnet for the treatment of severe cytomegalovirus infection in allograft recipients. Scand J Infect Dis 1985;17(2):157–63.
4. Jacobson MA, Drew WL, Feinberg J, O'Donnell JJ, Whitmore PV, Miner RD, Parenti D. Foscarnet therapy for ganciclovir-resistant cytomegalovirus retinitis in patients with AIDS. J Infect Dis 1991;163(6):1348–51.
5. Blanshard C, Benhamou Y, Dohin E, Lernestedt JO, Gazzard BG, Katlama C. Treatment of AIDS-associated gastrointestinal cytomegalovirus infection with foscarnet and ganciclovir: a randomized comparison. J Infect Dis 1995;172(3):622–8.
6. Lewis RA, Clogston P, Fainstein V, et al. Morbidity and toxic effects associated with ganciclovir or foscarnet therapy in a randomized cytomegalovirus retinitis trial. Studies of ocular complications of AIDS Research Group, in collaboration with the AIDS Clinical Trials Group. Arch Intern Med 1995;155(1):65–74.
7. Matarese LE, Speerhas R, Seidner DL, Steiger E. Foscarnet-induced electrolyte abnormalities in a bone marrow transplant patient receiving parenteral nutrition. JPEN J Parenter Enteral Nutr 2000;24(3):170–3.
8. Richman DD. HIV and other human retroviruses. In: Galasso GJ, Whitley RJ, Merigan TC, editors. Antiviral Agents and Viral Diseases of Man. New York: Raven Press, 1990:581.
9. Jacobson MA. Maintenance therapy for cytomegalovirus retinitis in patients with acquired immunodeficiency syndrome: foscarnet. Am J Med 1992;92(2A):S26–9.
10. Huycke MM, Naguib MT, Stroemmel MM, Blick K, Monti K, Martin-Munley S, Kaufman C. A double-blind placebo-controlled crossover trial of intravenous magnesium sulfate for foscarnet-induced ionized hypocalcemia and hypomagnesemia in patients with AIDS and cytomegalovirus infection. Antimicrob Agents Chemother 2000;44(8):2143–8.
11. Katlama C, Dohin E, Caumes E, Cochereau-Massin I, Brancon C, Robinet M, Rogeaux O, Dahan R, Gentilini M. Foscarnet induction therapy for cytomegalovirus retinitis in AIDS: comparison of twice-daily and three-times-daily regimens. J Acquir Immune Defic Syndr 1992;5(Suppl 1):S18–24.
12. Deray G, Martinez F, Katlama C, Levaltier B, Beaufils H, Danis M, Rozenheim M, Baumelou A, Dohin E, Gentilini M, et al. Foscarnet nephrotoxicity: mechanism, incidence and prevention. Am J Nephrol 1989;9(4):316–21.
13. Zanetta G, Maurice-Estepa L, Mousson C, Justrabo E, Daudon M, Rifle G, Tanter Y. Foscarnet-induced crystalline glomerulonephritis with nephrotic syndrome and acute renal failure after kidney transplantation. Transplantation 1999;67(10):1376–8.
14. Mattes FM, Hainsworth EG, Geretti AM, Nebbia G, Prentice G, Potter M, Burroughs AK, Sweny P, Hassan-Walker AF, Okwuadi S, Sabin C, Amooty G, Brown VS, Grace SC, Emery VC, Griffiths PD. A randomized, controlled trial comparing ganciclovir to ganciclovir plus foscarnet (each at half dose) for preemptive therapy of cytomegalovirus infection in transplant recipients. J Infect Dis 2004;189(8):1355–61.
15. Aweeka FT, Jacobson MA, Martin-Munley S, Hedman A, Schoenfeld P, Omachi R, Tsunoda S, Gambertoglio JG. Effect of renal disease and hemodialysis on foscarnet

pharmacokinetics and dosing recommendations. J Acquir Immune Defic Syndr Hum Retrovirol 1999;20(4):350–7.

16. Lacey HB, Ness A, Mandal BK. Vulval ulceration associated with foscarnet. Genitourin Med 1992;68(3):182.

17. Moyle G, Barton S, Gazzard BG. Penile ulceration with foscarnet therapy. AIDS 1993;7(1):140–1.

18. Roos TC, Albrecht H. Foscarnet-associated eosinophilic folliculitis in a patient with AIDS. J Am Acad Dermatol 2001;44(3):546–7.

Ganciclovir

General Information

Ganciclovir (dihydroxypropoxymethylguanine) is a nucleoside analogue with antiviral activity in vitro against herpesviruses. Intracellular phosphorylation of ganciclovir to its triphosphate derivative, which acts as a competitive inhibitor of deoxyguanosine triphosphate, leads to the inhibition of viral DNA synthesis. Because its toxicity profile is more favorable than those of foscarnet and cidofovir, it should be considered first-line treatment of life-threatening or sight-threatening cytomegalovirus (CMV) infections in immunocompromised patients. Ganciclovir is administered by infusion or by intravitreal injection (1). Oral ganciclovir as maintenance therapy for CMV retinitis in patients with AIDS has been reviewed (2). Oral ganciclovir has also been the subject of a pilot study in hepatitis B infection (3).

Observational studies

In 261 patients with CMV retinitis given oral ganciclovir 3–6 g/day or intravenous ganciclovir 5 mg/kg/day the most common adverse effects were on the gastrointestinal tract (nausea in 29–43%, vomiting in 17–23%, diarrhea in 33–52%, and flatulence in 2–18%) (4). There were rashes in 9–32%, a low neutrophil count (below 0.5×10^9/l) in 12–16%, a low hemoglobin concentration (below 8 g/dl) in 8–15%, a low platelet count (below 25×10^9/l) in 0–3%, and a raised serum creatinine in 17–27%.

General adverse effects

The proportion of patients in whom ganciclovir therapy is subsequently interrupted or withdrawn because of adverse effects is estimated at 32% (1).

Fever, rash, and abnormal liver function values are each reported to occur in about 2% of ganciclovir recipients (1). Other infrequently reported adverse effects, which may or may not be associated with ganciclovir, include chills, edema, malaise, vomiting, anorexia, diarrhea, dyspnea, reduced blood glucose, alopecia, impaired renal function, inflammation, pain, or phlebitis at the infusion site (1). These effects may also be due to the underlying illness in such patients.

Organs and Systems

Nervous system

Adverse effects of ganciclovir involving the central nervous system occur in about 5% of patients and include confusion, seizures, abnormal thinking, psychosis, hallucinations, nightmares, anxiety, tremor, dysesthesia, ataxia, coma, headache, and somnolence (1,5,6).

- A clear case of encephalopathy has been described in a patient who had received a bone marrow transplant; the problem resolved on withdrawal (7).

Sensory systems

Adverse effects reported in patients receiving intravitreal ganciclovir include foreign body sensation, conjunctival hemorrhage, mild conjunctival scarring, scleral induration, bacterial endophthalmitis, and retinal detachment (1).

For HIV-infected patients who have failed the intravenous CMV treatment options, intravitreal injections may be the last line. In one case only mild iris atrophy was noted, despite high-dose intravitreal ganciclovir (3 mg twice weekly) and high-dose intravitreal foscarnet (2.4 mg twice weekly) (8).

Endophthalmitis with scleral damage has been associated with a ganciclovir implant (9).

- A 39-year-old woman complained of increasing pain and complete loss of vision in the left eye 1 month after insertion of a ganciclovir implant. The eye was enucleated and pathological examination showed a vitreous abscess; the implant suture tab was intrasclerally located.

The authors concluded that intraocular infection had resulted from a foreign body (the surgical implant) in the scleral gap.

Metabolism

Six infants with cholestasis (aged 3–16 weeks) and signs of CMV infection were given intravenous ganciclovir for 3–7 weeks (10). One patient with septo-optic dysplasia and hypothyroidism had episodes of symptomatic hypoglycemia during treatment, which was withdrawn.

Hematologic

The adverse hematological effects of ganciclovir are generally rapidly reversible after withdrawal (1).

Pure red cell aplasia has been reported in one bone marrow transplant recipient (11) and hemolysis has been observed in two other patients (5).

Neutropenia is the most frequent adverse effect of ganciclovir. It usually occurs before a total dose of 200 mg/kg has been given. Other hematological adverse effects include thrombocytopenia, anemia, and lymphopenia. Since ganciclovir maintenance treatment is often necessary, concomitant use of G-CSF is often required in patients with AIDS with ganciclovir-associated neutropenia (12).

In a study of oral ganciclovir in 36 severely immuno-compromised HIV-positive children with CMV disease, toxicity was minimal and manageable and similar to that in adults in controlled trials of the oral formulation (13). About 20% of the children withdrew, mainly as a result of intolerance of the large volume of oral suspension or numerous capsules. As in adults, neutropenia was the main toxic effect and it was successfully treated with G-CSF.

Of 40 patients who had achieved engraftment after allogeneic hemopoietic stem cell transplantation, 23 of whom had high-risk features, including transplant from an HLA-mismatched or unrelated donor, or associated acute graft-versus-host disease, 19 had pre-emptive therapy with ganciclovir in an initial dose of 5 mg/kg/day (4). There were no significant adverse effects attributed to ganciclovir, except low total leukocyte counts (below 0.5×10^9/l) in three patients, lasting 3, 4, and 14 days, while they were treated with granulocyte colony-stimulating factor.

Neutropenia is the dose-limiting adverse effect of ganciclovir. Of 75 of 165 patients undergoing allogeneic peripheral blood stem cell transplantation who developed cytomegalovirus viremia, 58 received low-dose intravenous ganciclovir 5 mg/kg/day for 21 days (14). Only two of 58 patients had a fall in neutrophil count to below 100 x 106/l, one of whom was switched to foscarnet. In no other patient did the neutrophil count fall below 500 x 106/l. Cytomegalovirus disease developed in only five of the 58 patients and six of the 90 patients who did not develop cytomegalovirus viremia. Four of the five patients with cytomegalovirus disease after low-dose ganciclovir died.

Oral ganciclovir has been evaluated in the prevention of cytomegalovirus disease in 219 liver transplant patients (15). After a 2-week induction period with intravenous ganciclovir 6 mg/kg the patients were randomized to receive either oral ganciclovir 1 g every 8 hours (n = 110) or aciclovir 800 mg every 6 hours (n = 109) until 100 days after transplantation. Cytomegalovirus diseases occurred during the first year after transplantation in only one patient given ganciclovir compared with eight of those given aciclovir. Reversible leukopenia (under 3.0 x 10^9/l) was more common with oral ganciclovir (35%) than with aciclovir (18%).

Congenital cytomegalovirus infection is the most frequently identified viral cause of hearing loss and treatment is not well established. The effect of ganciclovir on subsequent hearing loss has been evaluated in a multicenter study in 100 neonates with symptomatic cytomegalovirus infection (16). The patients were randomized to no treatment or intravenous ganciclovir 6 mg/kg 12-hourly for 6 weeks. Only 42 of the patients had available baseline and 6-month data; 21 of 25 ganciclovir recipients and 10 of 17 control patients had normal hearing. None of 25 patients who were given ganciclovir compared with 41% of controls had worse hearing between baseline and 6 months. In the total study population, 63% of those given ganciclovir and 21% of the controls developed grade 3 or 4 neutropenia during the study.

With an overall incidence of 40%, neutropenia is the most frequent adverse effect associated with ganciclovir (17). It occurs less often in transplant patients than in patients with AIDS and generally reverses within 3–7 days after withdrawal.

It has been suggested that ganciclovir is less toxic to myeloid progenitor cells from cord blood than to progenitor cells from bone marrow. In one study 18 of 20 patients who received ganciclovir for cytomegalovirus infection after bone marrow transplantation developed ganciclovir-related neutropenia, compared with 9 of 17 who received ganciclovir after cord blood transplantation (18). This observation was explained by the differential expression of a putative progenitor cell kinase involved in the phosphorylation (i.e. activation) of ganciclovir. Alternatively, the lower incidence of ganciclovir-related neutropenia in cord blood cells might be due to the slow cycling rate of these cells. In patients with cord blood (but not bone marrow) transplants, a creatinine clearance rate of less than 50 ml/minute and the absence of glucocorticoid therapy were associated with a greater incidence of ganciclovir-associated neutropenia after pre-emptive therapy of cytomegalovirus infection.

Second-Generation Effects

Fertility

Animal data suggest that ganciclovir can inhibit spermatogenesis and fertility (19), but in one study there was no significant change in serum gonadotropin hormone concentrations in 32 men during ganciclovir therapy (20).

Drug Administration

Drug dosage regimens

Ganciclovir prophylaxis of cytomegalovirus disease after bone marrow transplantation was equally well tolerated when it was given in a dosage of 5 mg/kg three times weekly intravenously or 3 g/day orally to 58 patients (21).

Pre-emptive oral ganciclovir 1 g every 8 hours (n = 42) or conventional deferred ganciclovir (n = 38) have been compared in 80 renal transplant patients who developed cytomegalovirus antigenemia (pp65) within 8 weeks after transplantation (22). The incidence of cytomegalovirus disease during the first 12 weeks after transplantation was significantly reduced in the pre-emptive group (0% versus 24%). Two patients in each group developed late cytomegalovirus diseases in the 4?12 month period after transplantation and there was no difference in acute rejection rates. There were no differences in serum creatinine until 1 year after transplantation.

Drug administration route

Treatment options for CMV retinitis include the intravitreal insertion of ganciclovir implants. This method of drug delivery has the attraction of avoiding systemic drug toxicity, but the inherent danger of introducing bacterial infection has been highlighted (23).

- A 42-year-old man with AIDS and a history of CMV retinitis developed pain in his right eye and reduced visual acuity 10 days after receiving a ganciclovir intraocular implant into that eye. A therapeutic vitrectomy was performed and a vitreal tap produced frank pus and white fluffy debris. Cultures grew oxacillin-resistant *Staphylococcus aureus* sensitive only to vancomycin, rifampicin, and co-trimoxazole. The ganciclovir implants were removed and he was given a 4-week course of vancomycin and rifampicin. The bacterial endophthalmitis left him blind in his right eye.

Drug–Drug Interactions

Trimethoprim

The interaction kinetics and safety profile of oral ganciclovir when co-administered with trimethoprim have been investigated in HIV- and CMV-seropositive patients (24). Trimethoprim significantly reduced the renal clearance of ganciclovir and prolonged its half-life, although the changes are unlikely to be clinically important. Ganciclovir did not alter trimethoprim pharmacokinetics, with the exception of a 13% increase in C_{min}. Ganciclovir was well tolerated when given alone or in this combination.

Zidovudine

Zidovudine and ganciclovir have overlapping toxicity profiles with respect to adverse hematological effects. Severe life-threatening hematological toxicity has been reported in 82% of patients treated with a combination of zidovudine and ganciclovir (25). The combination of ganciclovir with didanosine was much better tolerated (26).

Food–Drug Interactions

A well-controlled investigation using high doses of oral ganciclovir in HIV- and CMV-seropositive subjects has shown that food markedly increases the systemic availability of ganciclovir. For example, there is a doubling of blood concentrations and AUC if the drug is given within 30 minutes after a meal rather than on an empty stomach (27). Whatever dosage instructions are given, it seems clear that the relation to meal times should be explained, for example with a firm recommendation to take the drug consistently with food.

References

1. Faulds D, Heel RC. Ganciclovir. A review of its antiviral activity, pharmacokinetic properties and therapeutic efficacy in cytomegalovirus infections. Drugs 1990;39(4):597–638.
2. Spector SA, McKinley GF, Lalezari JP, Samo T, Andruczk R, Follansbee S, Sparti PD, Havlir DV, Simpson G, Buhles W, Wong R, Stempien MRoche Cooperative Oral Ganciclovir Study Group. Oral ganciclovir for the prevention of cytomegalovirus disease in persons with AIDS. N Engl J Med 1996;334(23):1491–7.
3. Hadziyannis SJ, Manesis EK, Papakonstantinou A. Oral ganciclovir treatment in chronic hepatitis B virus infection: a pilot study. J Hepatol 1999;31(2):210–4.
4. Kanda Y, Mineishi S, Saito T, Saito A, Ohnishi M, Niiya H, Chizuka A, Nakai K, Takeuchi T, Matsubara H, Makimoto A, Tanosaki R, Kunitoh H, Tobinai K, Takaue Y. Response-oriented preemptive therapy against cytomegalovirus disease with low-dose ganciclovir: a prospective evaluation. Transplantation 2002;73(4):568–72.
5. Thomson MH, Jeffries DJ. Ganciclovir therapy in iatrogenically immunosuppressed patients with cytomegalovirus disease. J Antimicrob Chemother 1989;23(Suppl E):61–70.
6. Collaborative DHPG Treatment Study Group. Treatment of serious cytomegalovirus infections with 9-(1,3-dihydroxy-2-propoxymethyl)guanine in patients with AIDS and other immunodeficiencies. N Engl J Med 1986;314(13):801–5.
7. Sharathkumar A, Shaw P. Ganciclovir-induced encephalopathy in a bone marrow transplant recipient. Bone Marrow Transplant 1999;24(4):421–3.
8. Velez G, Roy CE, Whitcup SM, Chan CC, Robinson MR. High-dose intravitreal ganciclovir and foscarnet for cytomegalovirus retinitis. Am J Ophthalmol 2001;131(3):396–7.
9. Charles NC, Freisberg L. Endophthalmitis associated with extrusion of a ganciclovir implant. Am J Ophthalmol 2002;133(2):273–5.
10. Fischler B, Casswall TH, Malmborg P, Nemeth A. Ganciclovir treatment in infants with cytomegalovirus infection and cholestasis. J Pediatr Gastroenterol Nutr 2002;34(2):154–7.
11. Emanuel D, Cunningham I, Jules-Elysee K, Brochstein JA, Kernan NA, Laver J, Stover D, White DA, Fels A, Polsky B, et al. Cytomegalovirus pneumonia after bone marrow transplantation successfully treated with the combination of ganciclovir and high-dose intravenous immune globulin. Ann Intern Med 1988;109(10):777–82.
12. Hermans P. Haematopoietic growth factors as supportive therapy in HIV-infected patients. AIDS 1995;9(Suppl 2):S9–S14.
13. Frenkel LM, Capparelli EV, Dankner WM, Xu J, Smith IL, Ballow A, Culnane M, Read JS, Thompson M, Mohan KM, Shaver A, Robinson CA, Stempien MJ, Burchett SK, Melvin AJ, Borkowsky W, Petru A, Kovacs A, Yogev R, Goldsmith J, McFarland EJ, Spector SA. Oral ganciclovir in children: pharmacokinetics, safety, tolerance, and antiviral effects. The Pediatric AIDS Clinical Trials Group. J Infect Dis 2000;182(6):1616–24.
14. Vij R, Khoury H, Brown R, Goodnough LT, Devine SM, Blum W, Adkins D, DiPersio JF. Low-dose short-course intravenous ganciclovir as pre-emptive therapy for CMV viremia post allo-PBSC transplantation. Bone Marrow Transplant 2003; 32: 703-7.
15. Winston DJ, Busuttil RW. Randomized controlled trial of oral ganciclovir versus oral acyclovir after induction with intravenous ganciclovir for long-term prophylaxis of cytomegalovirus disease in cytomegalovirus-seropositive liver transplant recipients. Transplantation 2003; 75: 229-33.
16. Kimberlin DW, Lin CY, Sanchez PJ, Demmler GJ, Dankner W, Shelton M, Jacobs RF, Vaudry W, Pass RF, Kiell JM, Soong SJ, Whitley RJ; National Institute of Allergy and Infectious Diseases Collaborative Antiviral Study Group. Effect of ganciclovir therapy on hearing in symptomatic congenital cytomegalovirus disease involving the central nervous system: a randomized, controlled trial. J Pediatr 2003; 143: 16-25.

17. Scott JC, Partovi N, Ensom MH. Ganciclovir in solid organ transplant recipients: is there a role for clinical pharmacokinetic monitoring? Ther Drug Monit 2004;26(1):68–77.

18. Tomonari A, Iseki T, Takahashi S, Ooi J, Yamada T, Takasugi K, Nagamura F, Uchimaru K, Tojo A, Asano S. Ganciclovir-related neutropenia after preemptive therapy for cytomegalovirus infection: comparison between cord blood and bone marrow transplantation. Ann Hematol 2004;83(9):573–7.

19. Faqi AS, Klug A, Merker HJ, Chahoud I. Ganciclovir induces reproductive hazards in male rats after short-term exposure. Hum Exp Toxicol 1997;16(9):505–11.

20. Dieterich DT, Chachoua A, Lafleur F, Worrell C. Ganciclovir treatment of gastrointestinal infections caused by cytomegalovirus in patients with AIDS. Rev Infect Dis 1988;10(Suppl 3):S532–7.

21. Szer J, Durrant S, Schwarer AP, Bradstock KF, Gibson J, Arthur C, To LB, Hughes T, Raunow H. Oral versus intravenous ganciclovir for the prophylaxis of cytomegalovirus disease after allogeneic bone marrow transplantation. Intern Med J 2004;34(3):98–101.

22. Sagedal S, Nordal KP, Hartmann A, Midtvedt K, Foss A, Asberg A, Degre M, Fauchald P, Rollag H. Pre-emptive therapy of CMVpp65 antigen positive renal transplant recipients with oral ganciclovir: a randomized, comparative study. Nephrol Dial Transplant 2003; 18: 1899-908.

23. Rombos Y, Tzanetea R, Konstantopoulos K, Simitzis S, Zervas C, Kyriaki P, Kavouklis M, Aessopos A, Sakellaropoulos N, Karagiorga M, Kalotychou V, Loukopoulos D. Chelation therapy in patients with thalassemia using the orally active iron chelator deferiprone (L1). Haematologica 2000;85(2):115–7.

24. Jung D, AbdelHameed MH, Hunter J, Teitelbaum P, Dorr A, Griffy K. The pharmacokinetics and safety profile of oral ganciclovir in combination with trimethoprim in HIV- and CMV-seropositive patients. Br J Clin Pharmacol 1999;47(3):255–9.

25. Hochster H, Dieterich D, Bozzette S, Reichman RC, Connor JD, Liebes L, Sonke RL, Spector SA, Valentine F, Pettinelli C, et al. Toxicity of combined ganciclovir and zidovudine for cytomegalovirus disease associated with AIDS. An AIDS Clinical Trials Group Study. Ann Intern Med 1990;113(2):111–7.

26. Jacobson MA, Owen W, Campbell J, Brosgart C, Abrams DI. Tolerability of combined ganciclovir and didanosine for the treatment of cytomegalovirus disease associated with AIDS. Clin Infect Dis 1993;16(Suppl 1):S69–73.

27. Jung D, Griffy K, Dorr A. Effect of food on high-dose oral ganciclovir disposition in HIV-positive subjects. J Clin Pharmacol 1999;39(2):161–5.

Idoxuridine

General Information

Idoxuridine, which is active against *Herpes simplex* and *Varicella zoster*, has only been used in topical antiherpetic solutions. Idoxuridine has been used to treat superficial keratitis, but with poor results in deep stromal diseases because of poor solubility. However, idoxuridine is unstable and cannot eliminate the virus from the eye. Drug allergy and toxicity occur in 5–8% of patients.

Drug Administration

Drug administration route

Idoxuridine eye-drops are locally toxic, especially in patients with dry eyes, because of increased concentrations (or reduced tear secretion), and is topically sensitizing. Lacrimal punctum stenosis, lacrimation, follicular conjunctivitis, narrowing of meibomian gland orifices, inhibition of keratocyte mitosis and corneal stromal repair, and reductions in the strength of healing corneal wounds and the rate of epithelial regeneration are observed (1). After prolonged administration many changes in the conjunctival and corneal epithelium can occur, such as conjunctival cicatrization, punctate keratitis, subepithelial and intra-epithelial edema, and corneal opacities. Idoxuridine can also induce the emergence of resistant virus strains (2).

References

1. Havener WH. Ocular Pharmacology. 4th ed.. St Louis: CV Mosby;. 1978.

2. Lass JH, Thoft RA, Dohlman CH. Idoxuridine-induced conjunctival cicatrization. Arch Ophthalmol 1983;101(5):747–50.

Immunoglobulins

General Information

Immunoglobulin preparations are concentrated protein solutions derived from the pooled plasma of adults or animals. They contain specific antibodies in proportion to the infectious and immunization experience of the population from whose plasma they are prepared (1). Large numbers of donors (at least 1000 donors per lot of final product) are used, in order to ensure inclusion of a broad spectrum of antibodies. Intravenous immunoglobulin is also derived from the pooled plasma of adults, but the alcohol-fractionation procedure is modified to a product suitable for intravenous use. The use of intravenous immunoglobulins in selected immunodeficiency and autoimmune diseases has been reviewed (2).

Specific immunoglobulins, termed "hyperimmune globulins," are derived from human donors known to have high titers of the desired antibody. Specific immunoglobulin preparations for use in infectious disease prevention include hepatitis B, rabies, tetanus, *Varicella zoster*, vaccinia, and cytomegalovirus immunoglobulin.

Specific immunoglobulins of animal origin still currently in use in some countries include antirabies immunoglobulin, diphtheria antitoxin, botulinum antitoxin, antivenins, antilymphocyte globulin, and antithymocyte globulin. Horse antisera against diphtheria (and subsequently against tetanus as well) were produced and used in therapy from about the beginning of the 20th century. Antidiphtheria and antitetanus immunoglobulins are now produced almost exclusively as fractions of plasma of human origin. There are, however, a few cases in which xenogeneic antisera are still used, despite their many and often serious adverse effects.

Uses

Immunoglobulins

In addition to immune deficiencies, intravenous immunoglobulins are used in a wide variety of disorders [3,4,5] and are licensed for a number of indications, such as replacement therapy in primary and secondary immune deficiencies, in children with AIDS and recurrent infections, in patients with idiopathic thrombocytopenic purpura, Guillain–Barré syndrome, and Kawasaki disease, and for allogeneic bone marrow transplantation. Intravenous immunoglobulins have been used with success in several other conditions, mainly autoimmune diseases, such as myasthenia gravis, dermatomyositis, multiple sclerosis, and multifocal motor neuropathy.

Preparations of human immunoglobulins given by intramuscular/subcutaneous administration are mainly given to prevent or treat specific diseases, such as rhesus disease (anti-D) or certain viral infections, for example measles, hepatitis A, hepatitis B, rabies, and cytomegalovirus. Polyclonal antilymphocyte preparations (for example antilymphocyte globulin and antithymocyte globulin) have been developed because of evidence that T cells are primarily responsible for rejection of transplants. Indications for treatment with, for example, horse antilymphocyte globulin and/or antithymocyte globulin are very much the same as the indications for the mouse monoclonal anti-CD3 (muromonab), namely acute rejection of transplants, and aplastic anemia (6,7).

Preparations for intravenous administration are mainly used in patients with general immune deficiency states (primary or secondary) or diseases like idiopathic thrombocytopenic purpura (ITP) and autoimmune diseases (8,9). Neurological disorders (for example Guillain–Barré syndrome and chronic demyelinating polyneuropathy) have been treated with intravenous immunoglobulin (10–12).

Current data suggest that high-dose intravenous immunoglobulin has a beneficial effect on the reduction of anti-HLA antibodies with subsequent improvement in renal transplantation in highly HLA-sensitized patients and is effective in the treatment of antibody-mediated rejection episodes. Intravenous immunoglobulin has minor adverse effects in these patients, such as headaches, associated with the infusion rate [13]. It might also be a potential component of remission induction therapy for patients with myeloperoxidase–antineutrophil cytoplasmic antibody-associated rapidly progressive glomerulonephritis. Only one out of 12 patients had transient hypertension and edema of the limbs during intravenous immunoglobulin infusion, which abated after lowering the infusion rate [14].

The efficacy and safety of intravenous immunoglobulin have been reported in primary and secondary immunodeficiency, as well as in some immune disorders such as idiopathic thrombocytopenic purpura, Kawasaki disease, and Guillain–Barré syndrome (15–20). Intravenous immunoglobulin is also used empirically in a variety of several autoimmune diseases (21). Higher dosages of intravenous immunoglobulin are used for autoimmune indications (400–2000 mg/kg) than in immunodeficiency diseases (100–400 mg/kg). Minor adverse effects, such as headache, myalgia, chest discomfort, and fever, occur in 10% of patients (22). The efficacy of intravenous immunoglobulin has been reported in patients with membranous and membranoproliferative lupus nephritis (21). High-dose intravenous immunoglobulin has also been beneficial in myasthenia gravis, multiple sclerosis, and multifocal motor neuropathy (23,24) and in patients with pemphigus vulgaris and bullous pemphigoid unresponsive to conventional immunosuppressive drugs (25,26).

Intravenous immunoglobulin administered to allogeneic bone marrow transplant recipients modifies graft-versus-host disease and prevents interstitial pneumonia and infections (19).

The use of intravenous immunoglobulins in selected immunodeficiency and autoimmune diseases has been reviewed (2).

Mechanisms of action

The effect of immunoglobulin therapy is either protection against microorganisms (antibody substitution, passive immunization), or immunomodulation, or both. The mechanism of the response in non-infectious diseases is still not clear, but in idiopathic thrombocytopenic purpura an early fall in platelet-associated IgG and IgM may be a primary event, due to interference with antibody binding by platelets (27). Several other mechanisms of action of intravenous immunoglobulin in autoimmune diseases have been suggested, such as enhanced suppressor activity, Fc receptor blockade on neutrophils and macrophages (28), inhibition of complement activation and modulation of anti-idiotype responses (29), cytokine modification, neutralization of superantigens, and regulation of T cells and the idiotypic network (21,30). The suppression of polyclonal immunoglobulin biosynthesis induced by high-dose immunoglobulin infusions has also been suggested as a possible mechanism (31).

The mechanisms of action of intravenous immunoglobulin depend on the dose and the pathogenesis of the underlying disease [32]. Intravenous immunoglobulin products (5% or 6%) are widely used for replacement therapy of immunoglobulin in patients with primary and secondary immune deficiencies and for immune modulation in patients with inflammatory and autoimmune diseases. Almost all formulations are produced by cold ethanol precipitation, but several methods are used to treat Cohn fraction II to obtain the different formulations. These products differ slightly, although they are therapeutically equivalent.

General adverse effects

Common adverse reactions are headache, chills, dizziness, fever, flushing, nausea, backache, rash, arthralgia, myalgia, pruritus, hypotension, diarrhea, dyspnea, and chest pain [33,34,35,36,37,38,39,40]. These reactions are sometimes related to the rate of infusion [33,41,42] and can then be reduced or prevented by lowering the infusion rate. NSAIDs, antihistamines, or glucocorticoids can prevent them [43]. Renal failure was associated with high osmolarity sucrose-containing intravenous immunoglobulin formulations, whereas aseptic meningitis seems to be more common in patients with a history of migraine. Severe reactions have included thromboembolism, acute encephalopathy, bleeding disorders, stroke, neutropenia, hyponatremia, cryoglobulinemia, non-infectious hepatitis, and pleural effusion.

Because of the increased use of intravenous immunoglobulin, there has been an increased number of reports of serious adverse events, such as aseptic meningitis, renal insufficiency (mostly transient), transient hyperviscosity syndrome, thromboembolic events including myocardial infarction and stroke, immune hemolysis, disseminated intravascular coagulation (DIC), acute respiratory distress syndrome (ARDS), and transfusion-related acute lung injury (TRALI) [44]. The postulated mechanisms of action are speculative, and remain to be confirmed. These serious adverse events occur particularly in older patients, partly reflecting the increased use of intravenous immunoglobulin in higher dosages (2 g/kg) for anti-inflammatory or immunomodulatory indications, partly because a large proportion of patients with these diseases are significantly older that patients with immune deficiencies (usual doses of 400–600 mg/kg). In primary immune replacement therapy, adverse events such as headache, backache, chills, fever, chest tightness, and shortness of breath, are generally mild and transient and respond to anti-inflammatory/antipyretic drugs or adjustment of the infusion rate.

Occasionally, vasculitis develops in patients treated with intravenous immunoglobulin for rheumatic diseases, Guillain–Barré syndrome, chronic inflammatory demyelinating polyneuropathy, or dys/hypogammaglobulinemia. Autoimmune hemolytic anemia has been reported in several patients with Kawasaki disease and enhancement of the formation of cryoglobulins in individual patients with pre-existing cryoglobulinemia [45].

In the case of an underlying infection, administration of immunoglobulins can lead to febrile reactions, probably due to the formation of immune complexes [33,35].

Additional antiviral steps are used in all formulations to prevent contamination of blood-borne agents, particularly with viruses. Viral transmission of hepatitis viruses and retroviruses has not been reported over the last years. Concern about transmission of variant Creutzfeldt–Jakob disease has not been justified by human-to-human transmission by intravenous immunoglobulin.

Specific immunoglobulins

When administering different lots of the same product of equine rabies immunoglobulin, significant differences in adverse reactions, reflecting differences in production or purification processes and protein content, have been observed (46). It has been concluded in the past that the incidence of reactions to antirabies immunoglobulin is particularly high, but any of these immunoglobulins can cause severe reactions. The WHO has recommended that animal immunoglobulins should be used only after tests to rule out hypersensitivity.

A review of the discovery of antitoxins, the development of antibody formulations, and possible adverse effects has appeared (47).

Intramuscular immunoglobulin

Allerglobuline is a human gammaglobulin formulation, given intramuscularly, that has been reported to have a protective effect against type I allergic diseases and chronic infections of the upper respiratory tract in both adults and children. In 64 patients given allerglobuline, pain and inflammation at the injection site were the most common adverse effects (48). Fever, drowsiness, headache, nausea, back pain, and conjunctivitis occurred in only few patients. One patient had a rash and myalgia after the third injection; when the rash occurred again after the fourth injection, the patient was withdrawn.

Intravenous immunoglobulin

Adverse effects of intravenous immunoglobulin are generally mild and self-limiting (30). Although more common during the first two infusions, reactions appear to be primarily related to the speed of infusion. When infusions are given over a 1–3-hour period, the incidence of reactions is less than 5% and they occur during the transfusion; when infusions are given rapidly, the reactions may appear soon after completing the infusion. In about 10% of cases they occur 30–60 minutes after the start of the infusion.

Reactions include flushing, myalgia, headache, fever, chills and shaking, low backache, nausea and vomiting, diarrhea, chest tightness and shortness of breath, wheezing, changes in blood pressure, tachycardia, and rashes (49,50). Most of the adverse effects are related to the rate of administration and can be attenuated by slowing the rate of the infusion (51) or by prior administration of hydrocortisone and/or an antihistamine. When infusions are given over a 1–3-hour period, the incidence of reactions is less than 5% and they occur during the transfusion; when infusions are given rapidly the reactions may appear soon after completing the infusion. The current high dosage of 0.8–1.0 g/kg intravenous immunoglobulin for 1–2 days (instead of the original dosage regimen of 0.4 g/kg for 5 days) for patients with idiopathic thrombocytopenic purpura is probably associated with an increased risk of adverse effects (52).

The rate of adverse effects associated with intravenous immunoglobulin varies among different studies, which has been attributed to factors such as the indication, the dosage, the infusion rate, and the patient's age (17). In one study, headache and chills were related to a higher dosage.

Of 37 patients with primary hypogammaglobulinemia who received 1235 immunoglobulin infusions, 10 had adverse reactions during 34 infusions (2.8% of all infusions), but only five reactions were moderately severe. The reactions are related to the rate of administration (53); a rate of 5 mg/kg/minute was well tolerated.

Of 56 patients with autoimmune diseases who received high dosages of intravenous immunoglobulin, 20 had at least one adverse effect after one or more courses of treatment (17). The most frequently reported adverse effects were low-grade fever, headache, and chills. The authors concluded that the occurrence of adverse effects with intravenous immunoglobulin was not related to the clinical response to treatment. However, patients who developed adverse effects during the first course of treatment were more at risk of adverse effects during subsequent courses.

Patients with thrombocytopenia generally tolerate intravenous immunoglobulin well (54). In 16 young patients aged 9 months to 22 years with immune-mediated hemocytopenias (13 with childhood immune thrombocytopenic purpura), who received a total of 210 infusions, minimal adverse effects (transient headaches) were experienced during only four infusions, and later infusions were problem-free in three of the four patients (55).

The most frequent adverse effects of intravenous immunoglobulin (Sandoz) for the treatment of various disorders (affecting 1–3% of patients in all) were headache, nausea, vomiting, and fever. Some other mild symptoms have an incidence below 1%, including abdominal pain, diarrhea, fatigue, malaise, dizziness, myalgia, and chest tightness.

Allerglobuline is an intravenous human gammaglobulin formulation that has been reported to have a protective effect against type I allergic diseases and chronic infections of the upper respiratory tract in both adults and children. In 64 patients given allerglobuline, pain and inflammation at the injection site were the most common adverse effects (48). Fever, drowsiness, headache, nausea, back pain, and conjunctivitis occurred in only few patients. One patient had a rash and myalgia after the third injection; when the rash occurred again after the fourth injection, the patient was withdrawn.

Polyclonal antilymphocyte immunoglobulins

Short-term toxicity from polyclonal antilymphocyte immunoglobulins has been particularly marked in patients treated with combined antilymphocyte/antithymocyte globulin preparations from immunized horses (56). Immediate adverse effects include leukopenia and thrombocytopenia, fever, arthralgia, rash, urticaria, hepatotoxicity, hyperglycemia, hypertension, and diarrhea (57). A later adverse effect is serum sickness. Many of the effects may be due to an increase in tumor necrosis factor (58).

The longer-term effects of immunosuppression, and in particular the residual hematological and immunological abnormalities in patients with aplastic anemia treated with antilymphocyte globulin, have been documented: there is toxicity to hemopoietic cells, eventually leading to clonal marrow diseases years after treatment (56). Paroxysmal nocturnal hemoglobinuria, refractory sideroblastic anemia, chronic myelomonocytic leukemia, or acute leukemia can develop 4–10 years after treatment (59).

Observational studies

Experience with monthly, high-dose intravenous immunoglobulin in patients with different connective tissue diseases who failed to respond to standard therapies or for whom immunosuppressive drugs were contraindicated, the success rate was 70 %, without serious adverse effects [60].

Placebo-controlled studies

In a multicenter, double-blind, placebo-controlled, dose-escalating study of the safety and efficacy of an experimental plasma-derived, donor-selected, polyclonal anti-staphylococcal immunoglobulin with high titers of IgG directed against staphylococcal fibrinogen-binding proteins (INH-A21), 2 % of the 505 infants had 13 drug-associated adverse events, of which seven involved changes in vital signs (apnea, tachycardia, bradycardia, hypertension, and temperature changes) [61].

In a multicenter, randomized, double-blind, placebo-controlled trial the clinical effect of intravenous immunoglobulin 90 g on 3 consecutive days, twice at 3-months intervals was analysed in patients with post-polio syndrome [62]. There were no serious adverse drug reactions. Headache was the most frequent adverse effect. Infusion site reactions, gastrointestinal, nervous system, skin, and subcutaneous disorders were more frequent in those who were given intravenous immunoglobulin.

The potential beneficial effect of intravenous immunoglobulin treatment for recurrent spontaneous pregnancy loss has been assessed in five randomized, controlled trials in 250 patients [63]. Intravenous immunoglobulin was not effective. Adverse events included transient rash and fever, but they were not frequent.

Organs and Systems

Cardiovascular

Intravenous immunoglobulin expands the plasma volume and increases blood viscosity, which can lead to volume overload in patients with cardiac insufficiency (64). Stroke, thromboembolic events, and myocardial infarction have been reported after high-dose treatment with intravenous immunoglobulin, which increases plasma viscosity (64–66).

Thromboembolic events can complicate the administration of intravenous immunoglobulins [33,67,68,69,70,71,72,73,74,75]., and include myocardial infarction, stroke, pulmonary embolism, and deep vein thrombosis [76]. Risk factors are obesity, advanced age, immobilization, hypertension, diabetes mellitus, and a history of vascular disease or thrombosis [33]. Stroke has been described in 16 patients, 15 of whom had recognized risk factors [67]. These events appear to be related to hyperviscosity of the blood after intravenous immunoglobulin [33,71]. This has been confirmed by analysis of blood viscosity after intravenous immunoglobulin infusion [71,77]. Patients with high plasma concentrations of albumin and fibrinogen are considered to be more susceptible to thrombotic complications after intravenous immunoglobulin infusion [78].

- A 54-year-old woman developed an acute severe headache, nausea, and difficulty in speech 1 day after receiving intravenous immunoglobulin (21 g) for isolated IgG1 deficiency [79]. She had a transverse sinus thrombosis, at first considered to be a complication of intravenous immunoglobulin. However, she also had primary thrombocythemia, which is also known to cause headache and is associated with a risk of lateral sinus thrombosis.

Diffuse venous thromboembolism was reported in a patient with streptococcal toxic syndrome after two courses of high-dose intravenous immunoglobulin (0.4 g/kg/day over 5 days and 8 days later 0.4 g/kg/day over 4 days) [80].

- A 38-year-old woman with multiple sclerosis developed a deep vein thrombosis after a course of intravenous immunoglobulin 2 g over 2 days in combination with methylprednisolone [81]. She was not immobilized by her multiple sclerosis.

The authors proposed that the combination of intravenous immunoglobulin and methylprednisolone might have been associated with the thrombotic event, because this patient had tolerated several courses of intravenous immunoglobulin without methylprednisolone. They discussed 27 previously reported cases of thrombotic events after intravenous immunoglobulin. In eight cases a glucocorticoid had also been used. In nine cases, there were no data on glucocorticoid usage, but the clinical conditions made it likely that a glucocorticoid had indeed been used.

Thrombosis in elderly patients with an increased risk of thrombosis, such as those with hypertension or previous episodes of infarction, has been described (82). A few cases of thrombosis subsequent to intravenous immunoglobulin have been reported, including myocardial infarction in five patients, stroke in four cases, and spinal cord ischemia in one (83). It has been postulated that these events are induced by platelet activation and increased plasma viscosity (17).

Several cases of thrombosis have been reported after administration of intravenous immunoglobulin (50).

- A 75-year-old man with idiopathic thrombocytopenia purpura who was treated with intravenous immunoglobulin developed recurrent myocardial ischemia (84).
- A 54-year-old woman with idiopathic thrombocytopenic purpura received intravenous immunoglobulin 1 g/kg/day for 2 days and had an ischemic stroke with hemiparesis; 3 days later she had a deep vein thrombosis (85).
- A 33-year-old woman with Evans' syndrome received intravenous immunoglobulin 400 mg/kg/day and developed a deep vein thrombosis after 1 week (85). She was treated with warfarin, and 6 months later received an additional course of intravenous immunoglobulin for recurrent hemolytic anemia; 1 day later she died of pulmonary thromboembolism.

- A 70-year-old woman with polycythemia rubra vera and Guillain–Barré syndrome, but no known risk factors for thrombosis, had a cerebral infarction 10 days after receiving intravenous immunoglobulin; the authors wondered whether there was a relation to the polycythemia vera (86).

In a randomized, controlled study in 56 patients with untreated autoimmune thrombocytopenic purpura, who were treated with intravenous immunoglobulin 0.7 g/kg/day for 3 days, one had a deep vein thrombosis complicated by pulmonary embolism (87). One of 10 children with toxic epidermal necrolysis, for which they were given intravenous immunoglobulin 0.5 g/kg/day, developed a deep vein thrombosis requiring heparin (88). Of the 10 children, this child was the only one who received intravenous immunoglobulin for 7 days instead of the standard 4-day course.

Several causes of immunoglobulin-mediated thrombosis have been postulated, including increased plasma and blood viscosity, platelet activation, cytokine-mediated vasospasm, and contamination with factor IX. In a retrospective study in seven patients with susceptibility factors who developed such reactions the authors suggested that the thromboembolic complications were caused by clotting factors and vasoactive cytokines within specific batches of intravenous immunoglobulin formulations [89].

- Transient hypertension occurred in a patient with dermatomyositis during therapy with intravenous immunoglobulin (90). In the past, his diastolic blood pressure had been 104–106 mm Hg, but he was normotensive with antihypertensive drug medication.

Several mechanisms for this transient hypertension were postulated, for example stimulation of the vascular endothelium to secrete endothelin to inhibit nitric oxide synthesis.

It has been recommended that patients with cardiac diseases should be monitored during intravenous immunoglobulin therapy, because hypertension and cardiac failure have occurred, presumably as a result of fluid overload or electrolyte shifts (91).

Hypotension after treatment with intravenous immunoglobulin is rare and is due to the presence of IgG dimers in some immunoglobulin formulations [92].

Respiratory

Intravenous immunoglobulin can cause transfusion-related acute lung injury (TRALI) [93].

- A 35-year-old woman with idiopathic thrombocytopenic purpura treated with high dosages of intravenous immunoglobulin developed a recurrent lymphocytic pleural effusion (94).
- In a patient with Guillain–Barré syndrome, and a history of ischemic heart disease, intravenous immunoglobulin 400 mg/kg for 5 consecutive days caused severe bronchospasm and hypercapnia after a dose of 12.5 g had been given (20). The complaints disappeared after withdrawal of the intravenous immunoglobulin.

Nervous system

Headache occurred in 25% of all patients who received an infusion of intravenous immunoglobulin, probably related to larger volumes and fluid shifts, protein loads, and infusion rates (95). Intravenous immunoglobulin caused severe headache in 56% of patients without a history of migraine (96). The pathogenesis of this headache is unknown.

Severe headache has also been reported in children with idiopathic thrombocytopenic purpura using intravenous immunoglobulins (52). In a randomized, controlled study of patients with myasthenia gravis, two of six patients who received intravenous immunoglobulin developed severe headache after the initial dose of 2 g/kg (97).

In 14 patients with primary immunodeficiency disease, progressive neurodegeneration occurred and a possible relation to immunoglobulin therapy could not be ruled out (98).

Aseptic meningitis

Aseptic meningitis has been reported after intravenous immunoglobulin [33,34,99]. The risk of aseptic meningitis is higher in patients with a history of migraine [33]. Among 150 patients who received intravenous immunoglobulin 50, 250, or 500 mg/kg for allogeneic stem cell transplantation, two developed reversible aseptic meningitis [42].

Aseptic meningitis is characterized by headache, photophobia, nausea, vomiting, and meningism, and is confirmed by cerebrospinal fluid pleocytosis with 10–90% polymorphonuclear cells, increased concentrations of several proteins, and negative cultures (SEDA-22, 344) (95,100,101). It has been described in 1–15% of patients receiving high dosages of intravenous immunoglobulin (102,103) and especially in subjects with a history of migraine (30,66,102). However, aseptic meningitis has also been reported after a low dose of intravenous immunoglobulin.

Reversible aseptic meningitis has been reported in a patient who received high-dose intravenous immunoglobulin for idiopathic thrombocytopenic purpura [104]. The risk of aseptic meningitis is higher in patients with a history of migraine (SEDA-28, 372) [40].

- A 50-year-old man developed aseptic meningitis after a low dose of intravenous immunoglobulin (3.5 g) (102). A few weeks after the first infusion, he received a second infusion and again developed symptoms of aseptic meningitis.

The authors thought it unlikely that aseptic meningitis had been caused by an allergic reaction. They proposed that the mechanism of aseptic meningitis involved the entry of immunoglobulin molecules into the cerebrospinal fluid, causing an inflammatory reaction.

It has also been suggested that release of histamine, serotonin, and prostaglandins affects the meningeal microvasculature (50).

- Two children with idiopathic thrombocytopenic purpura developed aseptic meningitis after receiving intravenous immunoglobulin 1 g/kg/day, with unusual large numbers of leukocytes in the cerebrospinal fluid (105).

To prevent aseptic meningitis, it has been advised that intravenous immunoglobulin should be infused at a slow rate and that diluted immunoglobulin solutions should be used (100). Aseptic meningitis can be prevented by the administration of propranolol (64,100). In addition, prehydration and an antihistamine have been helpful in some patients (64,100).

Possible mechanisms of aseptic meningitis include hypersensitivity reactions, stabilizing products, cytokines, cerebrovascular sensitivity, and direct meningeal irritation. It has been suggested that it is caused by aggregated immunoglobulin, antibody–antigen complex formation with subsequent complement activation, or stabilizing carbohydrates used during manufacture (49,101).

Cerebrovascular disease

Stroke and ischemic encephalopathy, probably caused by cerebral vasospasm after intravenous immunoglobulin, have been reported as possible complications of intravenous immunoglobulin (106).

- Hemiplegia occurred on the third day of intravenous immunoglobulin therapy in a 58-year-old man with graft-versus-host disease after transfusion of non-leukocyte-depleted erythrocytes (107). The hemiplegia resolved one day after withdrawal of intravenous immunoglobulin.
- Hemiplegia occurred in a child given intravenous immunoglobulin for idiopathic thrombocytopenic purpura (108).

The authors of both reports suggested that the hemiplegia had been caused either by transient hyperviscosity or by vasospasm.

- An 82-year-old woman with chronic inflammatory demyelinating polyneuropathy was treated over 10 years with 86 doses of intravenous immunoglobulin and had a stroke when she was given an 87th dose (increased from 40 to 50 g) over several hours [109].

Analysis of the factors that can contribute to the occurrence of strokes during or after the administration of intravenous immunoglobulin suggests that certain precautions need to be taken. Pre-existing thrombogenic susceptibility factors, such as age, atherosclerosis, and hypercoagulable and hyperviscosity conditions should be considered.

- A 62-year-old woman with hemophagocytic syndrome was treated with intravenous immunoglobulin (20 g/day for 3 days), followed by methylprednisolone (1 mg/kg/day). On day 8 she had an acute hemiplegia due to cerebral infarction.

The authors question the association with intravenous immunoglobulin alone in the occurrence of the thrombotic event, since in most previous cases additional glucocorticoid therapy was given [110].

- A 3-year-old boy with Kawasaki disease developed a cerebral infarction 4 days after high-dose intravenous immunoglobulin 2 g/kg. Concomitant aspirin 50 mg/kg did not prevent the stroke.

Based on numerous reports of thrombosis due to high dose intravenous immunoglobulin in older patients, caution should be taken when treating children with Kawasaki disease [111].

Cerebral vasospasm, cerebral vasculitis, and serum hyperviscosity have been implicated in the pathogenesis of cerebral infarction after intravenous immunoglobulin (112).

Hyperviscosity with pseudohyponatremia has been attributed to high-dose intravenous immunoglobulin in a full-term child with hemolytic disease of the newborn [113].

Two patients with Susac syndrome (a retinocochleocerebral microangiopathy) were treated with intravenous immunoglobulin and methylprednisolone, which resulted in no further relapses [114]. One had immediate and significant improvement in hearing and MRI lesions. There was one potentially serious adverse event—seizures during the intravenous immunoglobulin infusion.

Sensory systems

Annular crystalline keratopathy has been reported in a 6-year-old boy 1 year after treatment with intravenous immunoglobulin 400 mg/kg/day for pyoderma gangrenosum [115]. Although pyoderma gangrenosum is associated with ocular complications, such as peripheral ulcerative keratitis, annular crystalline keratopathy has never been reported. Annular crystalline keratopathy has been associated with paraglobulinemia in monoclonal gammopathies, also occurring several months after immunoglobulin treatment. The authors suggested that the keratopathy could have been related to exogenous hypergammaglobulinemia by intravenous immunoglobulin or is a previously unknown complication of pyoderma gangrenosum.

Metabolism

Blood glucose should be monitored in patients with diabetes mellitus who receive glucose-containing intravenous immunoglobulin (64).

Hematologic

It has been recommended that the blood count be monitored during intravenous immunoglobulin therapy if there is evidence of mild leukopenia, neutropenia, or thrombocytopenia before the first infusion (23).

Erythrocytes

Severe acute hemolysis due to acquisition of red cell alloantibodies from donor serum has been reported (116–118). In other cases, the suggested mechanism of hemolytic anemia after high dosages of intravenous immunoglobulin was the presence of anti-A and/or anti-B antibodies in the plasma product (119).

Various immunoglobulin products contain IgG of molecular weight of a dimer or greater (over 300 kDa), which in the presence of serum can mimic immune complexes and bind to erythrocytes via CR1 (120). Especially in young adults, the immune complex-like moieties in intravenous immunoglobulin bind to erythrocytes and serve as opsonins in the mediation of erythrocyte sequestration, resulting in significant drops in hematocrit and hemoglobin concentrations (120). It has been suggested that the presence of immune complex-like moieties bound to erythrocyte CR1 after intravenous immunoglobulin treatment is correlated with in vivo hemolysis.

- Severe hemolytic anemia after high doses of intravenous immunoglobulin occurred in a 23-year-old woman with polymyositis (121).

High-dose intravenous immunoglobulin has been used effectively in neonates with proven hemolytic disease of the newborn to reduce the need for exchange transfusion and thereby provide more safety [122]. However, hemolytic anemia occurred in a 4-month-old girl with Kawasaki disease after intravenous immunoglobulin, reportedly because of anti-Rhesus (D) antibodies present in a specific batch [123], indicating the importance of recording batch numbers. The FDA has set limits for the amount of anti-D antibody in intravenous immunoglobulin products; in Europe this has not yet been done [124]. Patients whose blood group is AB rhesus positive have a relatively higher risk of hemolytic anemia after intravenous immunoglobulin, even though current intravenous immunoglobulin products contain relatively small amounts of anti-A and anti-B antibodies [125].

Severe intravascular hemolysis, possibly complicated by disseminated intravascular coagulation, is rare but was detected during post-marketing surveillance of WinRho®SDF, an intravenous human rhesus D immunoglobulin, in patients with idiopathic thrombocytopenic purpura [126].

This adverse reaction was probably due to the presence of allohemagglutinins A and B and high molecular weight IgG complexes in the formulation. Specifications in pharmacopeial monographs and product licenses require that intravenous immunoglobulin be free of significant titers of anti-A anti-B antibodies (SEDA-24, 385).

Leukocytes

High-dose immunoglobulin has been reported to cause neutropenia (50,127,128), disseminated intravascular coagulation and serum sickness (129). Neutropenia after intravenous immunoglobulin is frequent and seems to be transient and self-limiting (130). Neutropenia is not dose-related in the therapeutic range of doses.

Transient neutropenia without infectious complications occurred after 20 of 22 infusions of high-dose intravenous immunoglobulin (0.4 g/kg /day for 1-5 days) [131].

The incidence of neutropenia during 110 treatment courses in 104 children (average age 6.5 years) with idiopathic thrombocytopenic purpura has been studied retrospectively [132]. There were 64 courses of intravenous immunoglobulin (0.8–1 g/kg) and 46 treatment courses of anti-D immunoglobulin (50–75 micrograms/kg). Despite similar neutrophil counts at admission, 28% of those who were given intravenous immunoglobulin developed a neutrophil count below 1.5 x 10^9/l compared with none of those who were given anti-D immunoglobulin. The disadvantages of treating idiopathic thrombocytopenic purpura with intravenous immunoglobulin are the occurrence of acute adverse effects: flushing, headaches, chills, nausea, vomiting in 1–15% of recipients, aseptic meningitis in 4–23%, and renal insufficiency and failure in 6% (mostly with sucrose-containing formulations) [133].

It has been suggested that transient neutropenia can be induced by the presence of antineutrophil antibodies present in intravenous immunoglobulin. However, the possibility that immunoglobulin-mediated neutrophil agglutination causes pseudoleukopenia has also been raised (128). Increased leukocyte aggregation in the circulating pool of peripheral blood, induced by intravenous immunoglobulin, is particularly observed in people with hyperfibrinogenemia (128). It has been suggested that leukopenia detected by electronic counting is not necessarily associated with a real reduction in the absolute number of white blood cells in the peripheral blood, but is artefactual (128).

Platelets

In elderly patients, thrombotic events have been described in up to 10% in patients aged over 60, some of them fatal. Thrombotic events may be related to a rapid rise in numbers of circulating platelets (82,134).

Several cases of intravenous immunoglobulin-related thrombosis have been reported (135,136). It can be either venous or arterial (137). It has been suggested that thrombosis can be caused by platelet activation and increased plasma viscosity (136). In patients with vascular risk factors, such as old age, hypertension, and a history of stroke or coronary artery disease, complications, such as myocardial infarction, pulmonary embolism, stroke, and acute spinal cord events, have been described (137). Intravenous immunoglobulin enhances platelet aggregation and the release of adenosine triphosphate in human platelets in vitro. In addition, there is a dose-related increase in plasma viscosity with increasing plasma immunoglobulin concentration (136,137).

Liver

Although transient rises in liver enzymes have been documented, they are not considered to be serious (117).

Fatal hepatic veno-occlusive disease, characterized by hyperbilirubinemia, hepatomegaly, ascites, and weight gain, has been associated with intravenous immunoglobulin administered prophylactically to prevent transplant-related infections (136). To avoid such thrombotic complications, intravenous immunoglobulin should be infused at a slower rate in patients at risk, and high dosages (400–1000 mg/kg) should not be infused.

In 18 patients with Guillain–Barré syndrome standard high-dose intravenous immunoglobulin only has been compared with combined methylprednisolone plus high-dose intravenous immunoglobulin [138]. There were similar mild adverse effects in the two groups. There was altered liver function, the most common adverse effect, in two of those who were given intravenous immunoglobulin alone and in six in the combined treatment group. Headache occurred in two patients in both groups, suggesting that pre-infusion of a glucocorticoid does not prevent headache.

Urinary tract

Renal insufficiency after high dosages of intravenous immunoglobulin has been observed, mostly in patients with pre-existing renal disease (139). Acute renal insufficiency occurred within 7 days after the administration of intravenous immunoglobulin, with a peak at 5 days. About 40% of patients needed dialysis and 15% died despite treatment (all with severe underlying diseases); the mean time to recovery in survivors was about 10 days (140).

The US FDA has issued recommendations on safety precautions that physicians should take to reduce the potential risk of acute renal insufficiency, which appears to be uncommonly associated with infusion of intravenous immunoglobulin products (141,142). Since 1981, the FDA has received over 114 reports of renal dysfunction and/or acute renal insufficiency in patients given immunoglobulin, and a total of 17 deaths have been reported (51). As the problems have been associated with the use of specific immunoglobulin products, the FDA advises, that "physicians should carefully weigh the potential benefits of administering sucrose-containing intravenous immunoglobulin products against the risks of causing renal damage." Other cases of acute renal insufficiency have been described after the intravenous infusion of immunoglobulin (SEDA-22, 345) (126).

The pathophysiology of acute renal insufficiency due to immunoglobulins is probably related to hyperosmolar renal damage, due to sucrose present in 50 ml intravenous formulations (18,66,143–145,146,147,148). Acute renal insufficiency has also been attributed to sucrose in a kidney allograft recipient treated with intravenous immunoglobulin (149). The high solute load in the kidney can cause osmotic damage (51,139). The histopathology of renal tissue shows osmotic tubular injury, tubular vacuolization, and tubulointerstitial infiltrates (150,139). Cytological findings in the urine included the presence of macrophage-like tubular epithelial cells with multivacuolated cytoplasm (151). Formulations of intravenous immunoglobulin with sucrose as a stabilizer should not be used in patients with renal disease (50,128,152).

In a retrospective study of a heterogeneous group of 119 patients receiving intravenous immunoglobulin, two developed irreversible renal insufficiency and six had a rise in serum creatinine. These patients had received high dosages of two different formulations of intravenous immunoglobulin, one containing sucrose 1.76 g per gram of intravenous immunoglobulin; the other 0.5 g per gram. There was no relation between the amount of sucrose in the intravenous immunoglobulin and the development of renal insufficiency (139).

Studies in rats have suggested that after pinocytosis by renal tubular cells, sucrose is incorporated into phagolysosomes (51). The intracellular accumulation of sucrose leads to vacuole formation and cellular swelling. On withdrawal, renal insufficiency resolves in most cases. However, sometimes hemodialysis is necessary (117,149).

Risk factors for this adverse effect are pre-existing renal disease, age over 65 years, dehydration, diabetes mellitus, hypertension, obesity, use of nephrotoxic medications, and a high infusion rate (SEDA-22, 345) (15,18,51,66,144,153,154). To minimize the risk of renal insufficiency, it has been suggested that immunoglobulin should be diluted with hypotonic fluid, that the infusion rate should be reduced, and that dosing intervals should be increased (149). Patients should be adequately hydrated and potent diuretics should be avoided (15).

- Acute renal insufficiency was attributed to intravenous immunoglobulin (35 g) in an overweight 74-year-old woman with diabetes mellitus who had taken nabumetone and furosemide for 2 weeks.

The authors suggested that it was unlikely that the 2.5% glucose stabilizer in this specific intravenous immunoglobulin product had been responsible [154]. They observed an acute increase in colloid osmotic pressure and suggested that this increase was possibly caused by the macromolecular immunoglobulin itself. In all patients receiving immunoglobulin, good hydration is advised [33].

- Myocarditis and acute renal insufficiency occurred in a 59-year-old woman with hypertension and idiopathic thrombocytopenic purpura, which was treated with intravenous immunoglobulin 110 g on the first day and 50 g on the second day [155]. The immunoglobulin contained 2% glucose and glycine as stabilizers.
- Acute renal insufficiency, probably related to intravascular hemolysis, occurred in a 50-year-old Rh(D)-positive woman with idiopathic thrombocytopenic purpura, treated with intravenous anti-Rhesus (D) immunoglobulin 50 micrograms/kg [156].

Isometric tubular vacuolization was associated with the use of low-dose sucrose-containing intravenous immunoglobulin formulations to remove donor-specific antibodies in case of an ABO-incompatible transplanted kidney [157].

Two patients with chronic inflammatory demyelinating polyneuropathy treated with intravenous immunoglobulin developed acute renal insufficiency. The pathology was similar in the two cases, including tubular epithelial cell injury characterized by isometric intracytoplasmic vacuolization and cellular swelling, attributable to substances like sucrose and maltose in intravenous immunoglobulin formulations [158].

High-dose intravenous immunoglobulin pulse therapy delayed disease progression in patients with progressive immunoglobulin A nephropathy, as assessed by glomerular filtration rate, proteinuria, and renal survival time; compared with the control group, renal survival time was prolonged by 3.5 years [159]. One patient developed acute renal insufficiency due to sucrose-containing intravenous immunoglobulin and was therefore switched to a sucrose-free formulation. Another patient had a deep vein thrombosis.

In most cases the acute renal insufficiency is reversible and recovery occurs after 7–15 days (15).

It has been suggested that kidney function should be monitored during intravenous immunoglobulin therapy if it is abnormal before the first infusion (23). In addition, these patients should be adequately hydrated during treatment with intravenous immunoglobulin (91).

Skin

Skin reactions to intravenous immunoglobulin are rare (91,160–162). Other reported reactions include urticaria, maculopapular rashes, petechiae, eczema, and erythema multiforme (50,163).

Subcutaneous immunoglobulins are associated with fewer adverse reactions, but local reactions can occur at the injection site [164]. Local reactions at the infusion site were also observed in 3% of 649 infusions in 108 patients with multiple sclerosis who were given intravenous immunoglobulin [165]. This kind of reaction is probably not specific to patients with multiple sclerosis. Rashes and eczema after intravenous immunoglobulin occurred in a randomized placebo-controlled trial in 318 patients with secondary progressive multiple sclerosis who received monthly intravenous immunoglobulin 1 g/kg [76].

Four cases of severe extensive eczematous skin reactions occurred about 10 days after intravenous immunoglobulin infusion for polyradiculoneuritis [166]. The onset was characterized by dyshydrotic lesions on the palms extending to the rest of the body. A literature review revealed 33 cases of cutaneous eruptions after infusion of intravenous immunoglobulin, with characteristics similar to the four cases. Of these 37 cases, 35 had a neurological or neuromuscular disease, suggesting that neurological disorders, which are often immunological and are often related to viral infections, could predispose patients to this type of adverse reaction.

Other eczematous lesions have been reported.

- A 7-year-old girl with Stevens–Johnson syndrome was given intravenous immunoglobulin 0.5g/kg for 4 days [167]. Six days later she developed a vesicular eczema on her palms and forehead, which resolved after 3 weeks. However, new lesions on the plantar aspects of the feet then occurred.

A 33-year-old patient with multifocal motor neuropathy received intravenous immunoglobulin 0.4 g/kg/day for 5 days and developed vesicular eczema on the fingers and palms; the vesicles progressed to bullae [168]. A slightly scaling erythema developed on the face and scalp. The lesions healed after treatment with topical glucocorticoids. In spite of systemic antihistamines during the following four courses of intravenous immunoglobulin, the vesicular eczema developed but with a markedly milder course. Antihistamine treatment was stopped after the fourth course of intravenous immunoglobulin, resulting in slight aggravation of the eczema, which was transient and did not recur after the next courses.

Pompholyx has been observed 5–7 days from the last day of therapy in three of 23 neurological patients treated with intravenous immunoglobulin (21,169). Pompholyx occurs on the palms of the hands and is characterized by small papules, subdermal vesicles, desquamation, and crusting

Cutaneous vasculitis has been attributed to intravenous immunoglobulin (163)

- A previously healthy woman with progressive bilateral paresthesia, weakness, and facial paralysis after chiropractic manipulation, initially treated with prednisone for 1 week, received a course of intravenous immunoglobulin 2 mg/kg over 2 days [170]. Two days later she developed asymptomatic erythematous papular eruptions on her shin, progressing to her feet, lower legs, hands, and arms. A skin biopsy showed that this was a thrombotic lymphocytic vasculitis which was attributed to intravenous immunoglobulin.
- In a patient with type II mixed cryoglobulinemia, intravenous immunoglobulin caused severe cutaneous vasculitis accompanied by an increased cryocrit (171).

A simple method of mixing the patient's serum in vitro with intravenous immunoglobulin could probably predict this.

Hair

There have been a few cases of alopecia after the concomitant administration of intravenous immunoglobulin and steroid (87,172–175).

Musculoskeletal

Of 217 children with Kawasaki disease retrospectively reviewed five developed a polyarthropathy after intravenous immunoglobulin at mean of 10 days after the onset of fever and at a mean of 5.8 days after defervescence after the final dose of intravenous immunoglobulin [176].

Immunologic

The administration of equine or other immunoglobulins is associated with a considerable risk of adverse effects and can produce virtually any type of early or late hypersensitivity reaction, ranging from asthma and urticaria to serum sickness and fatal anaphylaxis (177–180). Encephalitis (181), myocarditis (182), nephritis (183), and uveitis (184) can all be manifestations of such reactions. In one case, leukocytoclastic vasculitis was attributed to human immunoglobulin (163).

Of systemic reactions to immunoglobulins, those of an inflammatory nature are thought to be due to the presence of small complexes or microaggregates, probably leading to activation of the complement system. The symptoms are usually mild and influenza-like, and generally respond to slowing down or temporary interruption of the infusion. They include headache, hypotension, sweating, chills, fever, nausea, and vasomotor reactions. The frequency of such adverse reactions has fallen, probably because of the improved quality of the preparations.

In 1978, the frequency of adverse reactions was as high as 55% in those receiving intravenous preparations containing 10–13% polymeric immunoglobulin, whereas by 2000, intravenous immunoglobulin preparations caused adverse reactions in only 3–4% (8). The more severe reactions may also be complement-mediated and may be due to a spontaneous activation of the complement system by the immunoglobulin preparation concerned; immunoglobulin aggregates may again be involved. The symptoms may be similar to those of inflammatory reactions, though they can be more severe. Such serious effects, including rare reactions of an anaphylactic type, may occur in as few as one in 6000 cases (185).

IgA deficiency is one of the more commonly encountered genetic disorders, and homozygotic deficiency is present in 0.3–0.03% of Caucasian populations. Such individuals may develop anti-IgA antibodies and have serum-sickness-like symptoms after the first administration of immunoglobulins, sometimes experiencing more severe reactions after repeated injections (186–188). A mild reaction due to sensitization to genetic IgA variants can also occur. Selective IgA deficiency is a contraindication to intravenous immunoglobulin, because of the risk of anaphylactic shock in patients with IgA deficiency and IgE or IgG antibodies against IgA, which react with IgA in intravenous immunoglobulin (18,12,49,50,64,91,103,161).

Anaphylactic reactions occur rarely. In 13 508 infusions in patients who had previously received intravenous immunoglobulin without problems, there were no severe reactions and the authors questioned the need to have adrenaline at home for patients on home treatment [135].

Some reports suggest that anaphylaxis due to intravenous immunoglobulin infusion occurs most often in patients with primary hypogammaglobulinemia (189,190). However, anaphylactic reactions have been seen in two atopic patients with idiopathic thrombocytopenic purpura, and the authors warned that children with atopic disease should not receive intravenous immunoglobulin (191).

In recipients of left ventricular assist devices awaiting cardiac transplantation treated with intravenous immunoglobulin, clinical manifestations of immune complex disease occurred during four of 27 (15%) monthly courses (192). The immune complex disease was characterized by fever, arthralgia, and maculopapular rashes.

Intravascular hemolysis is a rare adverse effect of immunoglobulins, due to the presence of anti-blood group antibodies (193–197).

Intravenous immunoglobulin can interfere with immunization with live virus vaccines, such as measles, mumps, rubella, or varicella [198,199]. Anaphylactic reactions with antibodies against IgA can occur in IgA-deficient patients who receive intravenous immunoglobulin formulations that contain IgA [37].

- A 17-year-old girl with common variable immunodeficiency developed anaphylactic reactions after intravenous immunoglobulin [200]. These reactions were probably due to the total absence of IgG2, IgG3, and IgG4.

Infection risk

Infections with human parvovirus B19 have been reported in patients who were given intravenous immunoglobulin (SEDA-27, 342). Removal of parvovirus B19 from these formulations requires nanofiltration at 15 nm or liquid heating at 60°C for 10 hours [201].

Body temperature

Intravenous immunoglobulin can cause malaise and fever in patients with infections, probably through a temporarily increased titer of antibodies against different pathogenic microorganisms (202).

Multiorgan failure

Multiorgan failure has been described after intravenous infusion of immunoglobulin (203).

- A 3-year-old mentally retarded girl was given intravenous sulfonated immunoglobulin for prophylaxis of measles. After infusion of about 100 mg, she became cyanotic, confused, and tachycardic. Despite hydrocortisone she developed hypotensive shock. Multiorgan failure developed, with symptoms of disseminated intravascular coagulation, acute renal insufficiency, hepatic dysfunction, respiratory distress syndrome, and rhabdomyolysis. After plasma exchange and continuous hemofiltration she recovered without sequelae. The drug-induced lymphocyte stimulation test using gammaglobulin was negative. In addition, serum concentrations of IgA and IgG were normal. However, concentrations of cytokines, such as interleukin-6, TNF-α, and soluble interleukin-2 receptor, were very high. Complement C3, C4, and CH50 were reduced, but C3a was raised.

The authors suggested that an unknown mechanism associated with intravenous immunoglobulin infusion had caused non-specific activation of complement systems accompanied by fulminant hypercytokinemia.

Second-Generation Effects

Teratogenicity

Intravenous immunoglobulin is indicated for idiopathic thrombocytopenic purpura in pregnancy. No fetal abnormalities have been reported (141).

Susceptibility Factors

Because of reports of severe adverse effects, the presence of rheumatoid factor in patients with B cell neoplasia may constitute a contraindication to intravenous immunoglobulin (141).

Autoimmune disease

The risk of reactions to antilymphocyte globulin is increased in patients with autoimmune disease (204). Fever and chills, sometimes with extreme hyperpyrexia, nausea and vomiting, urticaria, and reduced platelet and granulocyte counts were reported after the administration of horse antithymocyte globulin.

Infusion of intravenous immunoglobulin 0.14 g/kg in 17 patients with autoimmune diseases, in whom circulating immunoglobulins had been depleted, was associated with a high incidence of serious adverse effects (161). Treatment was terminated in four patients because of adverse effects, including urticaria, severe hypotension, arthralgia, and chest discomfort.

In two reviews of intravenous immunoglobulin in systemic lupus erythematosus, there was worsening proteinuria and/or a rise in serum creatinine (205,206), whereas others treating similar patients did not detect deterioration (191,207).

Transplant recipients

The adverse effects of rabbit antithymocyte globulin in transplant recipients were pain and erythema at the injection site and in one instance polyarthritis with urticaria (208). Four cases in which malignant lymphoma developed in renal transplant recipients treated with antithymocyte globulin of animal origin have been reported (209).

Drug Administration

Drug formulations

Febrile reactions to intravenous immunoglobulins are sometimes related to an underlying infection [38]. When changing from one formulation to another, the risks of adverse reactions increase. In September 2000, 49 immune-deficient patients switched to a different intravenous immunoglobulin formulation because Intragam was replaced by Intragam P, which had a different manufacturing process. Although neither the patients nor the clinicians knew about the change in formulation, there were increased reports of adverse reactions. The authors suggested that a patient on home therapy should receive the first infusions of a new formulation in hospital [210]. A dose of 2 g/kg over 2 days, as is often used nowadays in autoimmune diseases, seems not to be associated with more adverse reactions than a dose of 2 g/kg over 5 days [40].

Intravenous immunoglobulin products differ in salt and sugar content, pH, and osmolality. These differences have clinical implications. In a retrospective analysis, infusion-related adverse events were monitored during and after the administration of three intravenous immunoglobulin products, Gamimune 10% (n = 76), Polygam (n = 105), and Carimune (n = 98). Five patients (4.7%) in the Polygam group had acute myocardial infarction, eight (8.2%) in the Carimune group developed acute renal insufficiency, whereas none of the patients in the Gamimune group developed either. All three products caused headaches. These results suggest that pre-existing susceptibility factors should be taken into account in the selection of an intravenous immunoglobulin product [211].

In a controlled study with a new 10% formulation the risk of mild adverse reactions was higher than the 5% product (number of infusions: 9.6% versus 2.5%; number of patients with adverse reactions: 17% versus 8.6%) [212]. In a prospective open study in patients with hypogammaglobulinemia and agammaglobulinemia the pharmacokinetics, efficacy, and safety of this formulationwere similar to the results obtained with other formulations [213]. The most commonly reported drug-related adverse events were headache (9%), pyrexia (9%), and urticaria (5%). Its effects were also assessed in patients with primary immune deficiency [214]. A total of 826 infusions were administered, with a median of 13 infusions per subject and a median dose per subject of 436 g. Drug-related adverse events were considered mild or moderate and consisted of headache, fatigue, pyrexia, rigors, and migraine, headache being the most frequent. Two serious adverse events were drug-related—one subject had two episodes of aseptic meningitis.

Intravenous anti-D immunoglobulin (n = 18) has been compared with low-dose intravenous immunoglobulin (n = 16) in children with chronic immune thrombocytopenic purpura [215]. There was intravascular hemolysis in 61% of the patients on day 3 and in 77% on day 7 after anti-D immunoglobulin infusion. Other less frequent adverse effects following anti-D immunoglobulin infusion were low-grade fever (11%) and allergic skin rashes (5.5%). Low-dose intravenous immunoglobulin caused headache, mild fever, and skin rash; one patient developed aseptic meningitis and had seizures 24 hours after the infusion.

Drug contamination

Specifications in pharmacopeial monographs and product licences require intravenous immunoglobulin products to be free from significant amounts of anti-A and anti-B antibodies (216). It has been advised that a specification be adopted that will prevent the use of batches of intravenous immunoglobulin with abnormally high anti-D titers (over 1:64) (216).

Coagulation factors
Activated factor XIa has been demonstrated in samples of reconstituted intravenous immunoglobulin from eight different manufacturers (16). The degree of factor XIa contamination in intravenous immunoglobulin correlated with the manufacturer, suggesting that the purification process can affect residual factor XI concentration. Factor XIa can activate factor IX to factor IXa, and it can be hypothesized that there is a direct correlation between the presence of factor XI in intravenous immunoglobulin products and an increased risk of thrombotic complications.

Transmission of infection
The transmission of viral infections by immunoglobulins has occasionally been suspected. However, intravenous immunoglobulin preparations are considered relatively safe, and there are no reports of transmission of HIV or hepatitis B (82,174). This is probably because of the high degree of viral inactivation of the cold ethanol fractionation and the screening of every donation for several viruses, such as HIV and hepatitis B and C (143). Of 56 patients with autoimmune diseases who received 167 infusions of intravenous immunoglobulin, none developed antibodies to human immunodeficiency virus and hepatitis C virus or hepatitis B surface antigen (17).

In 1994, several cases of hepatitis C were reported, all traceable to a specific product (Gammagard®) (217). However, since the introduction of additional viral inactivation steps, such as solvent detergent, intravenous immunoglobulin is not considered to pose an infectious risk (118,174,218).

There has been some concern about possible transmission of hepatitis G. However, hepatitis G is an enveloped flavivirus with homology to hepatitis C and is probably also inactivated by the current procedures (100).

Although transmission of Creutzfeldt–Jakob disease through blood components and plasma products has never been documented, the FDA has suggested that there is a theoretical possibility that blood products may carry the responsible agent (219).

Transmission of hepatitis B and HIV after infusion of intravenous immunoglobulin has never been reported.

Life-threatening fulminant hepatitis due to transmission of human parvovirus B19 by intravenous immunoglobulin has been reported (220). The manufacturing process of this intravenous immunoglobulin product includes pasteurization (60°C for 10 hours), treatment with polyethylene glycol, ethanol fractionation, and nanofiltration. Removal of the small, non-lipid-enveloped parvovirus B19, which is highly resistant to different virus-reducing and inactivating steps, requires 15 nm nanofiltration.

A novel DNA virus, TT virus, has been implicated as a cause of post-transfusion hepatitis. A high prevalence of TT virus infection has been found in patients who received blood or blood components, such as factor VIII and IX concentrates. However, the PCR for TT virus DNA was negative in all 17 patients with immunodeficiency, who were treated prophylactically with intravenous immunoglobulin, as well as in 15 tested immunoglobulin formulations (221).

Drug dosage regimens

In a double-blind randomized comparison of two different doses of intravenous immunoglobulin in myasthenia gravis (1 and 2 g/kg), there were many adverse events, such as fever, chills, myalgia, headaches, nausea or vomiting, skin reactions, increased serum creatinine concentration, and increased serum transaminases. However, most were minor and self-limiting and occurrence rates were similar between the two groups, except for headache, which was the most frequent adverse event and more common at the higher dose [222].

Drug administration route

Intramuscular, subcutaneous

Adverse effects associated with intramuscular or subcutaneous administration of immunoglobulin are extremely rare, and are usually related to IgA deficiency in the patient or to additives in the preparation (for example preservatives). Adverse effects associated with intravenous immunoglobulin are more frequently seen and may be either local or systemic (223,224). Local reactions are essentially attributable to the technique used and are not specific to the intravenous immunoglobulin.

Of 65 patients with primary immune deficiency diseases who self-administered subcutaneous immunoglobulin infusions at a mean weekly dose of 158 mg/kg for 12 months 63 had adverse effects, attributable in 62 to a high rate of infusion-site reactions such as local swellings, soreness, redness, and induration, most of mild or moderate intensity. Other treatment-related adverse events were headache, nausea, rash, weakness, and gastrointestinal disorders [225].

Immunoglobulin products for subcutaneous administration can be used in patients with poor venous access or in patients with severe adverse reactions after intravenous immunoglobulin [226]. Subcutaneous immunoglobulin can cause local reactions at the infusion site.

Intravenous

Intravenous immunoglobulin is approved for the following indications: primary immune deficiencies, immune-mediated thrombocytopenia, Kawasaki disease, bone marrow transplantation, chronic B cell lymphocytic leukemia, and pediatric HIV infection (219). In addition, intravenous immunoglobulin is a promising immunomodulatory therapy for several neurological diseases, such as Guillain–Barré syndrome, chronic inflammatory demyelinating polyneuropathy, and multifocal motor neuropathy (64). Intravenous immunoglobulin has also been used in the treatment of several other antibody-mediated diseases, such as dermatomyositis, autoimmune neutropenia, pemphigus vulgaris, and pemphigus foliaceus (227–229).

Mild adverse effects of intravenous immunoglobulin, such as headache, chills, nausea, backache, and flushing, occur at a rate of 5–10% (174,219). Most of these reactions are self-limiting and associated with over-rapid infusion (64). In patients with adverse reactions to intravenous immunoglobulin there was a significant rise in plasma IL-6 and thromboxane B_2 (230).

Drug–Drug Interactions

Contrast media

Acute renal insufficiency after intravenous immunoglobulin therapy has been reported in association with the injection of iodinated radiocontrast agents (231). Contrast media and intravenous immunoglobulin formulations containing maltose or sucrose both have toxic effects on renal cells.

Interference with Diagnostic Tests

Blood glucose measurement

Intravenous immunoglobulins containing maltose as a stabilizer can interfere with blood glucose monitoring in systems that use glucose dehydrogenase [232]. There is no interference with systems that use glucose oxidase.

Falsely high blood glucose readings have been attributed to interference by WinRho®SDF, an intravenous human rhesus D immunoglobulin, or other maltose-containing intravenous immunoglobulin products when using systems that are not glucose-specific. The FDA released a safety alert about this issue for all maltose-containing intravenous immunoglobulins [233]. Another case of false hyperglycemia induced by a maltose-containing immunoglobulin solution was reported where the monitoring device using the principle of bioamperometry created a falsely high blood glucose reading by interference from maltose in the blood [234].

Serum sodium measurement

Pseudohyponatremia, a laboratory artefact due to hyperproteinemia, has been observed during intravenous immunoglobulin treatment (235).

- In a 38-year-old man with end-stage renal insufficiency and thrombocytopenia secondary to systemic lupus erythematosus, pseudohyponatremia occurred after treatment with intravenous immunoglobulin 1 g/kg for 2 days (236).

This phenomenon is explained by the fact that intravenous immunoglobulin increases the non-aqueous phase of the plasma, resulting in a relative loss of plasma water volume. Sodium is virtually restricted to serum water, so each volume of plasma measured will contain less sodium and be interpreted as hyponatremia. Using a direct ion-selective electrode avoids this problem.

Diagnosis of Adverse Drug Reactions

The incidences of immediate and delayed adverse events after intravenous immunoglobulin therapy in children have been determined in a prospective study in 33 patients with immunodeficiency and in 25 for immunomodulation of autoimmune and inflammatory diseases [237]. Immediate and delayed adverse events occurred in 6 and 24 children respectively. The most common delayed adverse events were headache (n = 13), fatigue (n = 12), abdominal pain (n = 8), and myalgia (n = 4). In contrast, headache as an immediate adverse event was reported in only 5 patients. Delayed adverse events were the main cause of absence from school and a need for additional therapies and medical consultations.

References

1. Committee on Infectious Diseases. In: Red Book. 2003 Report of the Committee on Infectious Diseases. 26th ed.. Elk Grove Village, IL: American Academy of Pediatrics, 2003:53–66.
2. Pirofsky B, Kinzey DM. Intravenous immune globulins. A review of their uses in selected immunodeficiency and autoimmune diseases. Drugs 1992;43(1):6–14.
3. Kumar A, Teuber SS, Gershwin ME. Intravenous immunoglobulin: striving for appropriate use. Int Arch Allergy Immunol 2006; 140(3): 185-98.
4. Jolles S, Hughes J. Use of IGIV in the treatment of atopic dermatitis, urticaria, scleromyxedema, pyoderma gangrenosum, psoriasis, and pretibial myxedema. Int Immunopharmacol 2006; 6(4): 579-91.
5. Ahmed AR. Use of intravenous immunoglobulin therapy in autoimmune blistering diseases. Int Immunopharmacol 2006; 6(4): 557-78.
6. Clark KR, Forsythe JL, Shenton BK, Lennard TW, Proud G, Taylor RM. Administration of ATG according to the absolute T lymphocyte count during therapy for steroid-resistant rejection. Transpl Int 1993;6(1):18–21.
7. Moore MA, Castro-Malaspina H. Immunosuppression in aplastic anemia—postponing the inevitable? N Engl J Med 1991;324(19):1358–60.
8. Bjorkander J. Antibody Deficiency Syndromes. ThesisSweden: University of Göteborg;. 1985.
9. Hopkins SJ. Sandoglobulin. Drugs Today 1985;21:277.
10. Bril V, Ilse WK, Pearce R, Dhanani A, Sutton D, Kong K. Pilot trial of immunoglobulin versus plasma exchange in patients with Guillain–Barré syndrome. Neurology 1996;46(1):100–3.
11. Otten A, Vermeulen M, Bossuyt PM, Otten A. Intravenous immunoglobulin treatment in neurological diseases. J Neurol Neurosurg Psychiatry 1996;60(4):359–61.
12. van Dijk GW, Notermans NC, Franssen H, Oey PL, Wokke JH. Response to intravenous immunoglobulin treatment in chronic inflammatory demyelinating polyneuropathy with only sensory symptoms. J Neurol 1996;243(4):318–22.
13. Jordan SC, Vo AA, Peng A, Toyoda M, Tyan D. Intravenous gammaglobulin (IVIG): a novel approach to improve transplant rates and outcomes in highly HLA-sensitized patients. Am J Transplant 2006; 6(3): 459-66.
14. Ito-Ihara T, Ono T, Nogaki F, Suyama K, Tanaka M, Yonemoto S, Fukatsu A, Kita T, Suzuki K, Muso E. Clinical efficacy of intravenous immunoglobulin for patients with MPO-ANCA-associated rapidly progressive glomerulonephritis. Nephron Clin Pract 2006; 102(1): c35-42.
15. Gras V, Andrejak M, Decocq G. Acute renal failure associated with intravenous immunoglobulins. Pharmacoepidemiol Drug Saf 1999;8(Suppl 1):S73–8.
16. Wolberg AS, Kon RH, Monroe DM, Hoffman M. Coagulation factor XI is a contaminant in intravenous immunoglobulin preparations. Am J Hematol 2000;65(1):30–4.
17. Sherer Y, Levy Y, Langevitz P, Rauova L, Fabrizzi F, Shoenfeld Y. Adverse effects of intravenous immunoglobulin therapy in 56 patients with autoimmune diseases. Pharmacology 2001;62(3):133–7.
18. Gupta N, Ahmed I, Nissel-Horowitz S, Patel D, Mehrotra B. Intravenous gammglobulin-associated acute renal failure. Am J Hematol 2001;66(2):151–2.
19. Winston DJ, Antin JH, Wolff SN, Bierer BE, Small T, Miller KB, Linker C, Kaizer H, Lazarus HM, Petersen FB, Cowan MJ, Ho WG, Wingard JR, Schiller GJ, Territo MC, Jiao J, Petrarca MA, Tonetta SA. A multicenter, randomized, double-blind comparison of different doses of intravenous immunoglobulin for prevention of graft-versus-host disease and infection after allogeneic bone marrow transplantation. Bone Marrow Transplant 2001;28(2):187–96.
20. Raphael JC, Chevret S, Harboun M, Jars-Guincestre MCFrench Guillain–Barré Syndrome Cooperative Group. Intravenous immune globulins in patients with Guillain–Barré syndrome and contraindications to plasma exchange: 3 days versus 6 days. J Neurol Neurosurg Psychiatry 2001;71(2):235–8.
21. Levy Y, Sherer Y, George J, Rovensky J, Lukac J, Rauova L, Poprac P, Langevitz P, Fabbrizzi F, Shoenfeld Y. Intravenous immunoglobulin treatment of lupus nephritis. Semin Arthritis Rheum 2000;29(5):321–7.
22. Iannaccone S, Sferrazza B, Quattrini A, Smirne S, Ferini-Strambi L. Pompholyx (vesicular eczema) after i.v. immunoglobulin therapy for neurologic disease Neurology 1999;53(5):1154–5.
23. Bajaj NP, Henderson N, Bahl R, Stott K, Clifford-Jones RE. Call for guidelines for monitoring renal function and haematological variables during intravenous infusion of immunoglobulin in neurological patients. J Neurol Neurosurg Psychiatry 2001;71(4):562–3.
24. Hilkevich O, Drory VE, Chapman J, Korczyn AD. The use of intravenous immunoglobulin as maintenance therapy in myasthenia gravis. Clin Neuropharmacol 2001;24(3):173–6.
25. Ahmed AR. Intravenous immunoglobulin therapy for patients with bullous pemphigoid unresponsive to conventional immunosuppressive treatment. J Am Acad Dermatol 2001;45(6):825–35.
26. Ahmed AR. Intravenous immunoglobulin therapy in the treatment of patients with pemphigus vulgaris unresponsive to conventional immunosuppressive treatment. J Am Acad Dermatol 2001;45(5):679–90.
27. Ball S, Zuiable A, Roter BL, Hegde UM. Changes in platelet immunoprotein levels during therapy in adult immune thrombocytopenia. Br J Haematol 1985;60(4):631–3.
28. Yu Z, Lennon VA. Mechanism of intravenous immune globulin therapy in antibody-mediated autoimmune diseases. N Engl J Med 1999;340(3):227–8.
29. Mouthon L, Kaveri SV, Spalter SH, Lacroix-Desmazes S, Lefranc C, Desai R, Kazatchkine MD. Mechanisms of action of intravenous immune globulin in immune-mediated diseases. Clin Exp Immunol 1996;104(Suppl 1):3–9.
30. Boulton-Jones R, Clark P. Intravenous immunoglobulin and sepsis. CPD Infection 2001;2:48–52.
31. Dammacco F, Iodice G, Campobasso N. Treatment of adult patients with idiopathic thrombocytopenic purpura with intravenous immunoglobulin: effects on circulating T cell subsets and PWM-induced antibody synthesis in vitro. Br J Haematol 1986;62(1):125–35.
32. Jolles S, Sewell WA, Misbah SA. Clinical uses of intravenous immunoglobulin. Clin Exp Immunol 2005; 142(1): 1-11.
33. Knezevic-Maramica I, Kruskall MS. Intravenous immune globulins: an update for clinicians. Transfusion 2003;43:1460–80.
34. Chinen J, Shearer WT. Basic and clinical immunology. J Allergy Clin Immunol 2003;111 (3 Suppl):S813-18.
35. Brennan VM, Salome-Bentley NJ, Chapel HM. Prospective audit of adverse reactions occurring in 459 primary antibody-deficient patients receiving intravenous immunoglobulin. Clin Exp Immunol 2003;133:247–51.

36. Stangel M, Kiefer R, Pette M, Smolka MN, Marx P, Gold R. Side effects of intravenous immunoglobulins in neurological autoimmune disorders-a prospective study. J Neurol 2003;250:818–21.

37. Bierling P, Godeau B. Intravenous immunoglobulin and autoimmune thrombocytopenic purpura: 22 years on. Vox Sanguinis 2004;86(1):8–14.

38. Aghamohammadi A, Farhoudi A, Nikzad M, Moin M, Pourpak Z, Rezaei N, Gharagozlou M, Movahedi M, Atarod L, Afshar AA, Bazargan N, Hosseinpoor AR. Adverse reactions of prophylactic intravenous immunoglobulin infusions in Iranian patients with primary immunodeficiency. Ann Allergy Asthma Immunol 2004;92(1):60–4.

39. Bussel JB, Eldor A, Kelton JG, Varon D, Brenner B, Gillis S, Angiolillo A, Kulkarni R, Abshire TC, Kelleher J; IGIV-C in ITP Study Group. IGIV-C, a novel intravenous immunoglobulin: evaluation of safety, efficacy, mechanisms of action, and impact on quality of life. Thromb Haemost 2004;91(4):771–8.

40. Dalakas MC. The use of intravenous immunoglobulin in the treatment of autoimmune neuromuscular diseases: evidence-based indications and safety profile. Pharmacol Ther 2004;102(3):177–93.

41. Wittstock M, Benecke R, Zettl UK. Therapy with intravenous immunoglobulins: complications and side-effects. Eur Neurol 2003;50:172–5.

42. Cordonnier C, Chevret S, Legrand M, Rafi H, Dhedin N, Lehmann B, Bassompierre F, Gluckman E; GREFIG Study Group. Should immunoglobulin therapy be used in allogeneic stem-cell transplantation? A randomized, double-blind, dose effect, placebo-controlled, multicenter trial. Ann Intern Med 2003;139:8–18.

43. Ruetter A, Luger TA. Efficacy and safety of intravenous immunoglobulin for immune-mediated skin disease: current view. Am J Clin Dermatol 2004;5(3):153–60.

44. Durandy A, Wahn V, Petteway S, Gelfand EW. Immunoglobulin replacement therapy in primary antibody deficiency diseases-maximizing success. Int Arch Allergy Immunol 2005; 136(3): 217-29.

45. Steinbrecher A, Berlit P. Intravenous immunoglobulin treatment in vasculitis and connective tissue disorders. J Neurol 2006; 253 Suppl 5: V39-49.

46. Wilde H, Chomchey P, Prakongsri S, Puyaratabandhu P, Chutivongse S. Adverse effects of equine rabies immune gobulin. Vaccine 1989;7(1):10–1.

47. Gronski P, Seiler FR, Schwick HG. Discovery of antitoxins and development of antibody preparations for clinical uses from 1890 to 1990. Mol Immunol 1991;28(12):1321–32.

48. Bunnag C, Dhorranintra B, Jareoncharsri P. Effect of allerglobuline injection on serum immunoglobulin levels in ENT patients. Asian Pac J Allergy Immunol 1991;9(1):45–50.

49. Jolles S, Hughes J, Rustin M. The treatment of atopic dermatitis with adjunctive high-dose intravenous immunoglobulin: a report of three patients and review of the literature. Br J Dermatol 2000;142(3):551–4.

50. Wiles CM, Brown P, Chapel H, Guerrini R, Hughes RA, Martin TD, McCrone P, Newsom-Davis J, Palace J, Rees JH, Rose MR, Scolding N, Webster AD. Intravenous immunoglobulin in neurological disease: a specialist review. J Neurol Neurosurg Psychiatry 2002;72(4):440–8.

51. Haskin JA, Warner DJ, Blank DU. Acute renal failure after large doses of intravenous immune globulin. Ann Pharmacother 1999;33(7–8):800–3.

52. Bolton-Maggs PH. The management of immune thrombocytopenic purpura. Curr Paediatr 2002;12:298–303.

53. Leen CL, Yap PL, Williams PE, McClelland DB. Tolerance of Scottish National Blood Transfusion Service intravenous immunoglobulin in patients with primary hypogammaglobulinaemia: report of 1235 infusions. Scott Med J 1988;33(4):303–6.

54. Imbach P, Barandun S, d'Apuzzo V, Baumgartner C, Hirt A, Morell A, Rossi E, Schoni M, Vest M, Wagner HP. High-dose intravenous gammaglobulin for idiopathic thrombocytopenic purpura in childhood. Lancet 1981;1(8232):1228–31.

55. Kurtzberg J, Friedman HS, Chaffee S, Falletta JM, Kinney TR, Kurlander R, Matthews TJ, Schwartz RS. Efficacy of intravenous gamma globulin in autoimmune-mediated pediatric blood dyscrasias. AM J Med 1987;83(4A):4–9.

56. Kawano Y, Nissen C, Gratwohl A, Wursch A, Speck B. Cytotoxic and stimulatory effects of antilymphocyte globulin (ALG) on hematopoiesis. Blut 1990;60(5):297–300.

57. Frickhofen N, Kaltwasser JP, Schrezenmeier H, Raghavachar A, Vogt HG, Herrmann F, Freund M, Meusers P, Salama A, Heimpel HThe German Aplastic Anemia Study Group. Treatment of aplastic anemia with antilymphocyte globulin and methylprednisolone with or without cyclosporine. N Engl J Med 1991;324(19):1297–304.

58. Debets JM, Leunissen KM, van Hooff HJ, van der Linden CJ, Buurman WA. Evidence of involvement of tumor necrosis factor in adverse reactions during treatment of kidney allograft rejection with antithymocyte globulin. Transplantation 1989;47(3):487–92.

59. de Planque MM, Brand A, Kluin-Nelemans HC, Eernisse JG, van der Burgh F, Natarajan AT, Beverstock GC, Zwaan FE, Willemze R, van Rood JJ. Haematopoietic and immunologic abnormalities in severe aplastic anaemia patients treated with anti-thymocyte globulin. Br J Haematol 1989;71(3):421–30.

60. Kamali S, Cefle A, Sayarlioglu M, Gul A, Inanc M, Ocal L, Aral O, Konice M. Experience with monthly, high-dose, intravenous immunoglobulin therapy in patients with different connective tissue diseases. Rheumatol Int 2005; 25(3): 211-4.

61. Bloom B, Schelonka R, Kueser T, Walker W, Jung E, Kaufman D, Kesler K, Roberson D, Patti J, Hetherington S; INH-A21 Phase II Study Team. Multicenter study to assess safety and efficacy of INH-A21, a donor-selected human staphylococcal immunoglobulin, for prevention of nosocomial infections in very low birth weight infants. Pediatr Infect Dis J 2005; 24(10): 858-66.

62. Gonzalez H, Sunnerhagen KS, Sjoberg I, Kaponides G, Olsson T, Borg K. Intravenous immunoglobulin for post-polio syndrome: a randomised controlled trial. Lancet Neurol 2006; 5(6): 493-500.

63. Intravenous immunoglobulin (IVIG) and recurrent spontaneous pregnancy loss. Fertil Steril 2006; 86(5 Suppl): S226-7.

64. Stangel M, Hartung HP, Marx P, Gold R. Intravenous immunoglobulin treatment of neurological autoimmune diseases. J Neurol Sci 1998;153(2):203–14.

65. Reinhart WH, Berchtold PE. Effect of high-dose intravenous immunoglobulin therapy on blood rheology. Lancet 1992;339(8794):662–4.

66. Machkhas H, Harati Y. Side effects of immunosuppressant therapies used in neurology. Neurol Clin 1998;16(1):171–188.

67. Caress JB, Cartwright MS, Donofrio PD, Peacock JE, Jr. The clinical features of 16 cases of stroke associated with administration of IVIg. Neurology 2003;60:1822–4.

68. Katz KA, Hivnor CM, Geist DE, Shapiro M, Ming ME, Werth VP. Stroke and deep venous thrombosis complicating intravenous immunoglobulin infusions. Arch Dermatol 2003;139:991–3.

69. Mohaupt M, Krueger T, Girardi V, Mansouri Taleghani B. Stroke after high-dose intravenous immunoglobulin. Transfus Med Hemother 2003;30:186–8.

70. Okuda D, Flaster M, Frey J, Sivakumar K. Arterial thrombosis induced by IVIg and its treatment with tPA. Neurology 2003;60:1825–6.

71. Nishikawa M, Ichiyama T, Hasegawa M, Kawasaki K, Matsubara T, Furukawa S. Safety from thromboembolism using intravenous immunoglobulin therapy in Kawasaki disease. Study of whole-blood viscosity. Pediatr Int 2003;45:156–8.

72. Butler KS, Zeitlin DS. Pulmonary embolism associated with intravenous immunoglobulin therapy. Ann Pharmacother 2003;37:1530.

73. Brown HC, Ballas ZK. Acute thromboembolic events associated with intravenous immunoglobulin infusion in antibody-deficient patients. J Allergy Clin Immunol 2003;112:797–9.

74. Dalakas MC. High-dose intravenous immunoglobulin in inflammatory myopathies: experience based on controlled clinical trials. Neurol Sci 2003;24 Suppl 4:S256-9.

75. Hefer D, Jaloudi M. Thromboembolic events as an emerging adverse effect during high-dose intravenous immunoglobulin therapy in elderly patients: a case report and discussion of the relevant literature. Ann Hematol 2005;84(6):411–5.

76. Hommes OR, Sorensen PS, Fazekas F, Enriquez MM, Koelmel HW, Fernandez O, Pozzilli C, O'Connor P. Intravenous immunoglobulin in secondary progressive multiple sclerosis: randomised placebo-controlled trial. Lancet 2004;364(9440):1149–56.

77. Steinberger BA, Ford SM, Coleman TA. Intravenous immunoglobulin therapy results in post-infusional hyperproteinemia, increased serum viscosity, and pseudohyponatremia. Am J Hematol 2003;73:97–100.

78. Ben-Ami R, Barshtein G, Mardi T, Deutch V, Elkayam O, Yedgar S, Berliner S. A synergistic effect of albumin and fibrinogen on immunoglobulin-induced red blood cell aggregation. Am J Physiol Heart Circ Physiol 2003;285:H2663-9.

79. Evangelou N, Littlewood T, Anslow P, Chapel H. Transverse sinus thrombosis and IVIg treatment: a case report and discussion of risk-benefit assessment for immunoglobulin treatment. J Clin Pathol 2003;56:308–9.

80. Geller JL, Hackner D. Diffuse venous thromboemboli associated with IVIg therapy in the treatment of streptococcal toxic shock syndrome: case report and review. Ann Hematol 2005; 84(9): 601-4.

81. Feuillet L, Guedj E, Laksiri N, Philip E, Habib G, Pelletier J, Ali Cherif A. Deep vein thrombosis after intravenous immunoglobulins associated with methylprednisolone. Thromb Haemost 2004;92(3):662–5.

82. Woodruff RK, Grigg AP, Firkin FC, Smith IL. Fatal thrombotic events during treatment of autoimmune thrombocytopenia with intravenous immunoglobulin in elderly patients. Lancet 1986;2(8500):217–8.

83. Alliot C, Rapin JP, Besson M, Bedjaoui F, Messouak D. Pulmonary embolism after intravenous immunoglobulin. J R Soc Med 2001;94(4):187–8.

84. Crouch ED, Watson LE. Intravenous immunoglobulin-related acute coronary syndrome and coronary angiography in idiopathic thrombocytopenic purpura—a case report and literature review. Angiology 2002;53(1):113–7.

85. Emerson GG, Herndon CN, Sreih AG. Thrombotic complications after intravenous immunoglobulin therapy in two patients. Pharmacotherapy 2002;22(12):1638–41.

86. Byrne NP, Henry JC, Herrmann DN, Abdelhalim AN, Shrier DA, Francis CW, Powers JM. Neuropathologic findings in a Guillain–Barré patient with strokes after IVIg therapy. Neurology 2002;59(3):458–61.

87. Godeau B, Chevret S, Varet B, Lefrere F, Zini JM, Bassompierre F, Cheze S, Legouffe E, Hulin C, Grange MJ, Fain O, Bierling PFrench ATIP Study Group. Intravenous immunoglobulin or high-dose methylprednisolone, with or without oral prednisone, for adults with untreated severe autoimmune thrombocytopenic purpura: a randomised, multicentre trial. Lancet 2002;359(9300):23–9.

88. Keohane SG, Kavanagh GM, Gordon PM, Hunter JAA. Transient hypertension during infusion of intravenous gammaglobulin for dermatomyositis. J Dermatol Treat 1999;10:287–8.

89. Vucic S, Chong PS, Dawson KT, Cudkowicz M, Cros D. Thromboembolic complications of intravenous immunoglobulin treatment. Eur Neurol 2004;52(3):141–4.

90. Tristani-Firouzi P, Petersen MJ, Saffle JR, Morris SE, Zone JJ. Treatment of toxic epidermal necrolysis with intravenous immunoglobulin in children. J Am Acad Dermatol 2002;47(4):548–52.

91. Dahl MV, Bridges AG. Intravenous immune globulin: fighting antibodies with antibodies. J Am Acad Dermatol 2001;45(5):775–83.

92. Kroez M, Kanzy EJ, Gronski P, Dickneite G. Hypotension with intravenous immunoglobulin therapy: importance of pH and dimer formation. Biologicals 2003;31:277–86.

93. Holness L, Knippen MA, Simmons L, Lachenbruch PA. Fatalities caused by TRALI. Transfus Med Rev 2004;18(3):184–8.

94. Bolanos-Meade J, Keung YK, Cobos E. Recurrent lymphocytic pleural effusion after intravenous immunoglobulin. Am J Hematol 1999;60(3):248–9.

95. Kishiyama JL, Valacer D, Cunningham-Rundles C, Sperber K, Richmond GW, Abramson S, Glovsky M, Stiehm R, Stocks J, Rosenberg L, Shames RS, Corn B, Shearer WT, Bacot B, DiMaio M, Tonetta S, Adelman DC. A multicenter, randomized, double-blind, placebo-controlled trial of high-dose intravenous immunoglobulin for oral corticosteroid-dependent asthma. Clin Immunol 1999;91(2):126–33.

96. Finkel AG, Howard JF Jr, Mann JD. Successful treatment of headache related to intravenous immunoglobulin with antimigraine medications. Headache 1998;38(4):317–21.

97. Wolfe GI, Barohn RJ, Foster BM, Jackson CE, Kissel JT, Day JW, Thornton CA, Nations SP, Bryan WW, Amato AA, Freimer ML, Parry GJMyasthenia Gravis-IVIG Study Group. Randomized, controlled trial of intravenous immunoglobulin in myasthenia gravis. Muscle Nerve 2002;26(4):549–52.

98. Ziegner UH, Kobayashi RH, Cunningham-Rundles C, Espanol T, Fasth A, Huttenlocher A, Krogstad P, Marthinsen L, Notarangelo LD, Pasic S, Rieger CH, Rudge P, Sankar R, Shigeoka AO, Stiehm ER, Sullivan KE, Webster AD, Ochs HD. Progressive neurodegeneration in patients with primary immunodeficiency disease on IVIG treatment. Clin Immunol 2002;102(1):19–24.

99. Yata J, Nihei K, Ohya T, Hirano Y, Momoi M, Maekawa K, Sakakihara Y; Study Group for Pediatric Guillain-Barr Syndrome. High-dose immunoglobulin therapy for Guillain-Barre syndrome in Japanese children. Pediatr Int 2003;45:543–9.

100. Jolles S, Hill H. Management of aseptic meningitis secondary to intravenous immunoglobulin. BMJ 1998;316(7135):936.

101. Wittstock M, Benecke R, Zettl UK. Therapie mit intravenös applizierten Immunglobulinen (IVIg). Indikationen und Nebenwirkungen. Neurol Rehabil 2000;6:121–4.

102. Attout H, Mallet H, Desmurs H, Berthier S, Gil H, de Wazieres B, Dupond JL. Méningite aseptique au course d'un traitement par immunoglobulines intraveineuses a très faibles doses. [Aseptic meningitis during treatment with very low doses of intravenous immunoglobulins.] Rev Med Interne 1998;19(2):140–1.

103. Al-Ghamdi H, Mustafa MM, Al-Fawaz I, Al-Dowaish A. Acute aseptic meningitis associated with administration of immunoglobulin in children: a case of report and review of the literature. Ann Saudi Med 1999;19:362–4.

104. Mitterer M, Pescosta N, Vogetseder W, Mair M, Coser P. Two episodes of aseptic meningitis during intravenous immunoglobulin therapy of idiopathic thrombocytopenic purpura. Ann Hematol. 1993;67(3):151–2.

105. Obando I, Duran I, Martin-Rosa L, Cano JM, Garcia-Martin FJ. Aseptic meningitis due to administration of intravenous immunoglobulin with an unusually high number of leukocytes in cerebrospinal fluid. Pediatr Emerg Care 2002;18(6):429–32.

106. Sztajzel R, Le Floch-Rohr J, Eggimann P. High-dose intravenous immunoglobulin treatment and cerebral vasospasm: A possible mechanism of ischemic encephalopathy? Eur Neurol 1999;41(3):153–8.

107. Hazouard E, Sauvagnac X, Corcia P, Legras A, Dequin PF, Ginies G. Accident vasculaire transitoire possiblement lié aux immunoglobulines intraveineuses. [Transient vascular accident, possibly related to intravenous immunoglobulins.] Presse Méd 1998;27(4):161.

108. Tsiouris J, Tsiouris N. Hemiplegia as a complication of treatment of childhood immune thrombocytopenic purpura with intravenously administered immunoglobulin. J Pediatr 1998;133(5):717.

109 Alexandrescu DT, Dutcher JP, Hughes JT, Kaplan J, Wiernik PH. Strokes after intravenous gamma globulin: thrombotic phenomenon in patients with risk factors or just coincidence? Am J Hematol 2005; 78(3): 216-20.

110. Feuillet L, Milandre L, Ali CA. Venous and arterial thrombosis following administration of intravenous immunoglobulins. Blood Coagul Fibrinolysis 2006; 17(1): 85.

111. Wada Y, Kamei A, Fujii Y, Ishikawa K, Chida S. Cerebral infarction after high-dose intravenous immunoglobulin therapy for Kawasaki disease. J Pediatr 2006; 148(3): 399-400.

112. Choudhry VP, Mahapatra M, Kashyap R. Immunoglobulin therapy in immunohematological disorders. Indian J Pediatr 1998;65(5):681–90.

113 Tarcan A, Gokmen Z, Dikmenoglu N, Gurakan B. Pseudohyponatraemia and hyperviscosity due to IVIG therapy in a term newborn. Acta Paediatr 2005; 94(4): 509-10.

114. Fox RJ, Costello F, Judkins AR, Galetta SL, Maguire AM, Leonard B, Markowitz CE. Treatment of Susac syndrome with gamma globulin and corticosteroids. J Neurol Sci 2006; 251(1-2): 17-22.

115. Yata J, Nihei K, Ohya T, Hirano Y, Momoi M, Maekawa K, Sakakihara Y; Study Group for Pediatric Guillain-Barre Syndrome. High-dose immunoglobulin therapy for Guillain-Barre syndrome in Japanese children. Pediatr Int 2003;45:543–9.

116. Turner B, Wills AJ. Cerebral infarction complicating intravenous immunoglobulin therapy in a patient with Miller Fisher syndrome. J Neurol Neurosurg Psychiatry 2000;68(6):790–1.

117. Stangel M, Muller M, Marx P. Adverse events during treatment with high-dose intravenous immunoglobulins for neurological disorders. Eur Neurol 1998;40(3):173–4.

118. Wills AJ, Unsworth DJ. A practical approach to the use of intravenous immunoglobulin in neurological disease. Eur Neurol 1998;39(1):3–8.

119. Nakagawa M, Watanabe N, Okuno M, Kondo M, Okagawa H, Taga T. Severe hemolytic anemia following high-dose intravenous immunoglobulin administration in a patient with Kawasaki disease. Am J Hematol 2000;63(3):160–1.

120. Kessary-Shoham H, Levy Y, Shoenfeld Y, Lorber M, Gershon H. In vivo administration of intravenous immunoglobulin (IVIg) can lead to enhanced erythrocyte sequestration. J Autoimmun 1999;13(1):129–35.

121. Ballot-Brossier C, Mortelecque R, Sinegre M, Marceau A, Dauriat G, Courtois F. Poursuite du traitement d'une polymyosite par les immunoglobulines intraveineuses polyvalentes malgré la survenue d'une anémie hémolytique sévère. [Insisting on intravenous polyvalent immunoglobulin therapy in polymyositis in spite of the occurrence of sever hemolytic anemia.] Transfus Clin Biol 2001;8(2):94–9.

122. Gottstein R, Cooke RW. Systematic review of intravenous immunoglobulin in haemolytic disease of the newborn. Arch Dis Child Fetal Neonatal Ed 2003;88:F6-10.

123. Cleary AG, Brown B, Minards J, Sills J, Bolton-Maggs P. Systematic review of intravenous immunoglobulin in haemolytic disease of the newborn. Arch Dis Child Fetal Neonatal Ed 2003;88:F444.

124. Thorpe SJ, Fox BJ, Dolman CD, Lawrence J, Thorpe R. Batches of intravenous immunoglobulin associated with adverse reactions in recipients contain atypically high anti-Rh D activity. Vox Sanguinis 2003;85:80–4.

125. Trifa M, Simon L, Hamza J, Bavoux F, Des Roziers NB. Haemolytic anaemia associated with high dose intravenous immunoglobulin therapy in a child with Guillain-Barr syndrome. Arch Dis Child 2003;88:836–7.

126 Gaines AR. Disseminated intravascular coagulation associated with acute hemoglobinemia or hemoglobinuria following Rh(0)(D) immune globulin intravenous administration for immune thrombocytopenic purpura. Blood 2005; 106(5): 1532-7.

127. Lee YC, Woodfield DG, Douglas R. Clinical usage of intravenous immunoglobulins in Auckland. NZ Med J 1998;111(1060):48–50.

128. Zelster D, Fusman R, Chapman J, Rotstein R, Shapira I, Elkayam O, Eldor A, Arber N, Berliner S. Increased leukocyte aggregation induced by gamma-globulin: a clue to the presence of pseudoleukopenia. Am J Med Sci 2000;320(3):177–82.

129. Comenzo RL, Malachowski ME, Meissner HC, Fulton DR, Berkman EM. Immune hemolysis, disseminated intravascular coagulation, and serum sickness after large doses of immune globulin given intravenously for Kawasaki disease. J Pediatr 1992;120(6):926–8.

130. Berkovitch M, Dolinski G, Tauber T, Aladjem M, Kaplinsky C. Neutropenia as a complication of intravenous immunoglobulin (IVIG) therapy in children with immune thrombocytopenic purpura: common and non-alarming. Int J Immunopharmacol 1999;21(6):411–5.

131. Matsuda M, Hosoda W, Sekijima Y, Hoshi K, Hashimoto T, Itoh S, Ikeda S. Neutropenia as a complication of high-dose intravenous immunoglobulin therapy in adult patients with neuroimmunologic disorders. Clin Neuropharmacol 2003;26:306–11.

132 Niebanck AE, Kwiatkowski JL, Raffini LJ. Neutropenia following IVIG therapy in pediatric patients with immune-mediated thrombocytopenia. J Pediatr Hematol Oncol 2005; 27(3): 145-7.

133 Shad AT, Gonzalez CE, Sandler SG. Treatment of immune thrombocytopenic purpura in children: current concepts. Paediatr Drugs 2005; 7(5): 325-36.

134. Frame WD, Crawford RJ. Thrombotic events after intravenous immunoglobulin. Lancet 1986;2(8504):468.

135. Go RS, Call TG. Deep venous thrombosis of the arm after intravenous immunoglobulin infusion: case report and literature review of intravenous immunoglobulin-related thrombotic complications. Mayo Clin Proc 2000;75(1):83–5.

136. Paran D, Herishanu Y, Elkayam O, Shopin L, Ben-Ami R. Venous and arterial thrombosis following administration of intravenous immunoglobulins. Blood Coagul Fibrinolysis 2005;16(5):313–8.

137. Elkayam O, Paran D, Milo R, Davidovitz Y, Almoznino-Sarafian D, Zeltser D, Yaron M, Caspi D. Acute myocardial infarction associated with high dose intravenous immunoglobulin infusion for autoimmune disorders. A study of four cases. Ann Rheum Dis 2000;59(1):77–80.

138 Odaka M, Tatsumoto M, Hoshiyama E, Hirata K, Yuki N. Side effects of combined therapy of methylprednisolone and intravenous immunoglobulin in Guillain-Barr syndrome. Eur Neurol 2005; 53(4): 194-6.

139. Dilhuydy MS, Delclaux C, De Precigout V, Haramburu F, Roger I, Deminiere C, Mercie P, Pellegrin JL, Aparicio M. Insuffisance rénale aiguë après cure d'immunoglobulines polyvalentes. [Acute renal failure after polyvalent immunoglobulin therapy.] Presse Méd 2000;29(17):942–3.

140. Levy JB, Pusey CD. Nephrotoxicity of intravenous immunoglobulin. QJM 2000;93(11):751–5.

141. Barton JC, Herrera GA, Galla JH, Bertoli LF, Work J, Koopman WJ. Acute cryoglobulinemic renal failure after intravenous infusion of gamma globulin. Am J Med 1987;82(3 Spec No):624–9.

142. Pasatiempo AM, Kroser JA, Rudnick M, Hoffman BI. Acute renal failure after intravenous immunoglobulin therapy. J Rheumatol 1994;21(2):347–9.

143. Jolles S, Hughes J, Whittaker S. Dermatological uses of high-dose intravenous immunoglobulin. Arch Dermatol 1998;134(1):80–6.

144. Stahl M, Schifferli JA. The renal risks of high-dose intravenous immunoglobulin treatment. Nephrol Dial Transplant 1998;13(9):2182–5.

145. Laidlaw S, Bainton R, Wilkie M, Makris M. Acute renal failure in acquired haemophilia following the use of high dose intravenous immunoglobulin. Haemophilia 1999;5(4):270–2.

146. Wolf HH, Davies SV, Borte M, Caulier MT, Williams PE, Bernuth HV, Egner W, Sklenar I, Adams C, Spath P, Morell A, Andresen I. Efficacy, tolerability, safety and pharmacokinetics of a nanofiltered intravenous immunoglobulin: studies in patients with immune thrombocytopenic purpura and primary immunodeficiencies. Vox Sang 2003;84:45–53.

147. Borte M, Davies SV, Touraine JL, Farber CM, Lipsic T, Adams C, Spath P, Bolli R, Morell A, Andresen I. Clinical properties of a novel liquid intravenous immunoglobulin: studies in patients with immune thrombocytopenic purpura and primary immunodeficiencies. Transfus Med Hemother 2004;31:126–34.

148. Joannidis M. Drug-induced renal failure in the ICU. Int J Artif Organs 2004;27(12):1034–42.

149. Tsinalis D, Dickenmann M, Brunner F, Gurke L, Mihatsch M, Nickeleit V. Acute renal failure in a renal allograft recipient treated with intravenous immunoglobulin. Am J Kidney Dis 2002;40(3):667–70.

150. Ahsan N. Intravenous immunoglobulin induced-nephropathy: a complication of IVIG therapy. J Nephrol 1998;11(3):157–61.

151. Khalil M, Shin HJ, Tan A, DuBose TD Jr, Ordonez N, Katz RL. Macrophage like vacuolated renal tubular cells in the urine of a male with osmotic nephrosis associated with intravenous immunoglobulin therapy. A case report. Acta Cytol 2000;44(1):86–90.

152. Sokos DR, Berger M, Lazarus HM. Intravenous immunoglobulin: appropriate indications and uses in hematopoietic stem cell transplantation. Biol Blood Marrow Transplant 2002;8(3):117–30.

153. Sati HI, Ahya R, Watson HG. Incidence and associations of acute renal failure complicating high-dose intravenous immunoglobulin therapy. Br J Haematol 2001;113(2):556–557.

154. Van Zanten AR, Beekhuyzen M, Van der Meer YG, De Gooijer A, Feith GW. Acuut nierfalen na behandeling met intraveneus toegediende immunoglobulinen. Ned Tijdschr Geneeskd 2003;147:307–10.

155. Akhtar I, Bastani B. Acute renal failure and myocarditis associated with intravenous immunoglobulin therapy. Ann Intern Med 2003;139:W65.

156. Chun NS, Savani B, Seder RH, Taplin ME. Acute renal failure after intravenous anti-D immune globulin in an adult with immune thrombocytopenic purpura. Am J Hematol 2003;74:276–9.

157. Haas M, Sonnenday CJ, Cicone JS, Rabb H, Montgomery RA. Isometric tubular epithelial vacuolization in renal allograft biopsy specimens of patients receiving low-dose intravenous immunoglobulin for a positive crossmatch. Transplantation 2004;78(4):549–56.

158. Soares SM, Sethi S. Impairment of renal function after intravenous immunoglobulin. Nephrol Dial Transplant 2006; 21(3): 816-7.

159. Rasche FM, Keller F, Lepper PM, Aymanns C, Karges W, Sailer LC, M ller L, Czock D. High-dose intravenous immunoglobulin pulse therapy in patients with progressive immunoglobulin A nephropathy: a long-term follow-up. Clin Exp Immunol 2006; 146(1): 47-53.

160. Noseworthy JH, O'Brien PC, Weinshenker BG, Weis JA, Petterson TM, Erickson BJ, Windebank AJ, Whisnant JP, Stolp-Smith KA, Harper CM Jr, Low PA, Romme LJ, Johnson M, An KN, Rodriguez M. IV immunoglobulin does not reverse established weakness in MS. Neurology 2000;55(8):1135–43.

161. Schmaldienst S, Mullner M, Goldammer A, Spitzauer S, Banyai S, Horl WH, Derfler K. Intravenous immunoglobulin application following immunoadsorption: benefit or risk in patients with autoimmune diseases? Rheumatology (Oxford) 2001;40(5):513–21.

162. Morikawa M, Yamada H, Kato EH, Shimada S, Kishi T, Yamada T, Kobashi G, Fujimoto S. Massive intravenous immunoglobulin treatment in women with four or more recurrent spontaneous abortions of unexplained etiology: down-regulation of NK cell activity and subsets. Am J Reprod Immunol 2001;46(6):399–404.

163. Howse M, Bindoff L, Carmichael A. Facial vasculitic rash associated with intravenous immunoglobulin. BMJ 1998;317(7168):1291.

164. Berger M. Subcutaneous immunoglobulin replacement in primary immunodeficiencies. Clin Immunol 2004;112(1):1–7.

165. Achiron A, Kishner I, Dolev M, Stern Y, Dulitzky M, Schiff E, Achiron R. Effect of intravenous immunoglobulin treatment on pregnancy and postpartum-related relapses in multiple sclerosis. J Neurol 2004;251(9):1133–7.

166. Vecchietti G, Kerl K, Prins C, Kaya G, Saurat JH, French LE. Severe eczematous skin reaction after high-dose intravenous immunoglobulin infusion: report of 4 cases and review of the literature. Arch Dermatol 2006; 142(2): 213-7.

167. Young PK, Ruggeri SY, Galbraith S, Drolet BA. Vesicular eczema after intravenous immunoglobulin therapy for treatment of Stevens-Johnson syndrome. Arch Dermatol 2006; 142(2): 247-8.

168. Maetzke J, Sperfeld AD, Scharffetter-Kochanek K, Sunderkotter C. Vesicular and bullous eczema in response to intravenous immunoglobulins (IVIG). Allergy 2006; 61(1): 145-6.

169. Peker S, Kuwert C, Paus R, Moll I. Palmar lokalisierte vesikuläre läsionen nach der intravenösen applikation van immunglobulinen. H G Z Hautkr 2000;75:7–8.

170. Bodemer AA, Longley BJ. Thrombotic lymphocytic vasculitic reaction to intravenous immunoglobulin. J Am Acad Dermatol 2006; 55(5 Suppl): S112-3.

171. Yebra M, Barrios Y, Rincon J, Sanjuan I, Diaz-Espada F. Severe cutaneous vasculitis following intravenous infusion of gammaglobulin in a patient with type II mixed cryoglobulinemia. Clin Exp Rheumatol 2002;20(2):225–7.

172. Ballow M. Mechanisms of action of intravenous immunoglobulin therapy and potential use in autoimmune connective tissue diseases. Cancer 1991;68(Suppl 6):1430–6.

173. Chan-Lam D, Fitzsimons EJ, Douglas WS. Alopecia after immunoglobulin infusion. Lancet 1987;1(8547):1436.

174. Silver RM. Management of idiopathic thrombocytopenic purpura in pregnancy. Clin Obstet Gynecol 1998;41(2):436–48.

175. Leclech C, Maillard H, Penisson-Besnier I, Laine-Cessac P, Verret JL. Réaction cutanée inhabituelle après perfusion intraveineuse d'immunoglobulines polyvalentes. [Unusual skin reaction after intravenous infusion of polyvalent immunoglobulins: 3 case reports.] Presse Méd 1999;28(10):531.

176. Lee KY, Oh JH, Han JW, Lee JS, Lee BC. Arthritis in Kawasaki disease after responding to intravenous immunoglobulin treatment. Eur J Pediatr 2005; 164(7): 451-2.

177. Ducluceau R. Les accidents de la sérothérapie antitétanique. Bull Med Leg Toxicol Med 1971;14:26.

178. Charpin J, Louchet E, Gratecos LA. Subsiste-t-il encore des réactions allergiques de la sérothérapie?. [Does an allergic reaction to serotherapy still occur?.] Mars Med 1969;106(3):223–6.

179. Bianchi R, Dappen U, Hoigné R. Der anaphylaktische Schock des Menschen auf artfremdes Serum. Symptomatologie, Prophylaxe und Therapie. [Anaphylactic shock of humans towards foreign species serum. Symptomatology, prevention and therapy.] Helv Chir Acta 1967;34(3):257–73.

180. World Health Organization. The collection, fractionation, quality control, and uses of blood and blood productsGeneva: WHO;. 1981.

181. Delwaide PJ, Radermecker M, Boverie J. Deux cas de neuropathies multiples d'origine sérothérapique avec paralysie phrénique. [Two cases of multiple neuropathies of a serotherapeutic origin with phrenic paralysis.] Acta Neurol Psychiatr Belg 1967;67(6):452–62.

182. Czirner J, Besznyak G. Myokardinfarkt-ähnliches Bild als seltene Komplikation nach Applikation von Tetanus-Antitoxin. [Myocardial infarct-like clinical picture as a rare complication following the application of tetanus antitoxin.] Z Gesamte Inn Med 1969;24(4):119–21.

183. Humphrey JH, White RG. Reactions due to antigen-antibody complexes (nephritis in man). In: Immunology for Students of Medicine. 3rd ed.. Oxford: Blackwell Scientific Publications, 1970:458.

184. Suarez-Lopez J, Sanchez-Salorio M, Sanchez-Lado J. Uveitis exudativa de caracter sero-anafiláctico. Arch Soc Oftalmol Hisp-Am 1965;25:499.

185. Williams PE, Yap PL, Gillon J, Crawford RJ, Galea G, Cuthbertson B. Non-A, non-B hepatitis transmission by intravenous immunoglobulin. Lancet 1988;2(8609):501.

186. Avoy DR. Delayed serum sickness-like transfusion reactions in a multiply transfused patient. Vox Sang 1981;41(4):239–44.

187. Liblau R, Morel E, Bach JF. Autoimmune diseases, IgA deficiency, and intravenous immunoglobulin treatment. Am J Med 1992;93(1):114–5.

188. Nydegger UE. Intravenous immunoglobulin in combination with other prophylactic and therapeutic measures. Transfusion 1992;32(1):72–82.

189. Hachimi-Idrissi S, de Schepper J, de Waele M, Dab I, Otten J. Type III allergic reaction after infusion of immunoglobulins. Lancet 1990;336(8706):55.

190. McCluskey DR, Boyd NA. Anaphylaxis with intravenous gammaglobulin. Lancet 1990;336(8719):874.

191. Myer I, Andler W. Die Behandlung der idiopathischen thrombozytopenischen Purpura. Krankenhausarzt 1987;60:105.

192. John R, Lietz K, Burke E, Ankersmit J, Mancini D, Suciu-Foca N, Edwards N, Rose E, Oz M, Itescu S. Intravenous immunoglobulin reduces anti-HLA alloreactivity and shortens waiting time to cardiac transplantation in highly sensitized left ventricular assist device recipients. Circulation 1999;100(Suppl 19):II229–35.

193. Brox AG, Cournoyer D, Sternbach M, Spurll G. Hemolytic anemia following intravenous gamma globulin administration. Am J Med 1987;82(3 Spec No):633–5.

194. Chapman JF, Murphy MF, Berney SI, Ord J, Metcalfe P, Amess JA, Waters AH. Post-transfusion purpura associated with anti-Baka and anti-PIA2 platelet antibodies and delayed haemolytic transfusion reaction. Vox Sang 1987;52(4):313–7.

195. Lucas GS, Jobbins K, Bloom AL. Intravenous immunoglobulin and blood group antibodies. Lancet 1987;2(8561):742.

196. Ovesen H, Taaning E, Christensen BA. Posttransfusionel purpura (PTP) forarsaget af anti-Zwb (P1P2). [Posttransfusion purpura caused by ant-Zw b (P1(A2)).] Ugeskr Laeger 1986;148(43):2769–70.

197. Slichter SJ. Post-transfusion purpura: response to steroids and association with red blood cell and lymphocytotoxic antibodies. Br J Haematol 1982;50(4):599–605.

198. Burns JC, Glode MP. Kawasaki syndrome. Lancet 2004;364(9433):533–44.

199. Omwandho CO, Gruessner SE, Roberts TK, Tinneberg HR. Intravenous immunoglobulin (IVIG): modes of action in the clinical management of recurrent pregnancy loss (RPL) and selected autoimmune disorders. Clin Chem Lab Med 2004;42(4):359–70.

200. Almeida Barry LR, Forte WC. Common variable immunodeficiency. Allergol Immunopathol (Madr) 2004;32(2):89–91.

201. Yunoki M, Urayama T, Tsujikawa M, Sasaki Y, Abe S, Takechi K, Ikuta K. Inactivation of parvovirus B19 by liquid heating incorporated in the manufacturing process of human intravenous immunoglobulin preparations. Br J Haematol 2005;128(3):401–4.

202. Teeling JL, Bleeker WK, Hack CE, Kuijpers TW. Nieuwe inzichten in het ontstaan van bijwerkingen van intraveneuze immunoglobuline (IVIG)-preparaten. Ned Tijdschr Allergie 2001;1:20–5.

203. Ikeda M, Hamasaki Y, Hataya H, Honda M, Sugai K. Multiorgan failure induced by intravenous immunoglobulin. Acta Paediatr 2000;89(11):1393.

204. Seiffert J, Brendel W, Lob G, et al. Improvement of the compatibility of ALG. Behring Inst Mitt 1972;51:255.

205. Schifferli J, Leski M, Favre H, Imbach P, Nydegger U, Davies K. High-dose intravenous IgG treatment and renal function. Lancet 1991;337(8739):457–8.

206. Woodruff RK, Grigg AP, Firkin FC, Smith IL. Fatal thrombotic events during treatment of autoimmune thrombocytopenia with intravenous immunoglobulin in elderly patients. Lancet 1986;2(8500):217–8.

207. Jayne DR, Davies MJ, Fox CJ, Black CM, Lockwood CM. Treatment of systemic vasculitis with pooled intravenous immunoglobulin. Lancet 1991;337(8750):1137–9.

208. Doney KC, Weiden PL, Storb R, Thomas ED. Treatment of graft-versus-host disease in human allogeneic marrow graft recipients: a randomized trial comparing antithymocyte globulin and corticosteroids. Am J Hematol 1981;11(1):1–8.

209. Kheirbek AO, Molnar ZV, Choudhury A, Geis WP, Daugirdas JT, Hano JE, Ing TS. Malignant lymphoma in a renal transplant recipient treated with antithymocyte globulin. Transplantation 1983;35(3):267–8.

210 Ameratunga R, Sinclair J, Kolbe J. Increased risk of adverse events when changing intravenous immunoglobulin preparations. Clin Exp Immunol 2004;136(1):111–3.

211. Vo AA, Cam V, Toyoda M, Puliyanda DP, Lukovsky M, Bunnapradist S, Peng A, Yang K, Jordan SC. Safety and adverse events profiles of intravenous gammaglobulin products used for immunomodulation: a single-center experience. Clin J Am Soc Nephrol 2006; 1(4): 844-52.

212 Matamoros N, de Gracia J, Hernandez F, Pons J, Alvarez A, Jimenez V. A prospective controlled crossover trial of a new presentation (10% vs. 5%) of a heat-treated intravenous immunoglobulin. Int Immunopharmacol 2005; 5(3): 619-26.

213. Björkander J, Nikoskelainen J, Leibl H, Lanbeck P, Wallvik J, Lumio JT, Braconier JH, Pavlova BG, Birthistle K, Engl W, Walter S, Ehrlich HJ. Prospective open-label study of pharmacokinetics, efficacy and safety of a new 10% liquid intravenous immunoglobulin in patients with hypo- or agammaglobulinemia. Vox Sang 2006; 90(4): 286-93.

214. Church JA, Leibl H, Stein MR, Melamed IR, Rubinstein A, Schneider LC, Wasserman RL, Pavlova BG, Birthistle K, Mancini M, Fritsch S, Patrone L, Moore-Perry K, Ehrlich HJ; US-PID-IGIV 10% -Study Group10. Efficacy, safety and tolerability of a new 10% liquid intravenous immune globulin [IGIV 10%] in patients with primary immunodeficiency. J Clin Immunol 2006; 26(4): 388-95.

215. El Alfy MS, Mokhtar GM, El Laboudy MA, Khalifa AS. Randomized trial of anti-D immunoglobulin versus low-dose intravenous immunoglobulin in the treatment of childhood chronic idiopathic thrombocytopenic purpura. Acta Haematol 2006; 115(1-2): 46-52.

216. Turner CE, Thorpe SJ, Brasher MD, Thorpe R. Anti-Rh D activity of commercial intravenous immunoglobulin preparations. Vox Sang 1999;76(1):55–8.

217. Yap PL. Intravenous immunoglobulin and hepatitis C virus: an overview of transmission episodes with emphasis on manufacturing data. Clin Ther 1996;18(Suppl B):43–58.

218. Chidwick K, Matejtschuk P, Gascoigne E, Briggs N, More JE, Dash CH. Clinical experience with a new solvent detergent-treated intravenous immunoglobulin free of hypotensive effects. Vox Sang 1999;77(4):204–9.

219. Milgrom H. Shortage of intravenous immunoglobulin. Ann Allergy Asthma Immunol 1998;81(2):97–100.

220. Hayakawa F, Imada K, Towatari M, Saito H. Life-threatening human parvovirus B19 infection transmitted by intravenous immune globulin. Br J Haematol 2002;118(4):1187–9.

221. Azzari C, Resti M, Moriondo M, Gambineri E, Rossi ME, Novembre E, Vierucci A. Lack of transmission of TT virus through immunoglobulins. Transfusion 2001;41(12):1505–8.

222 Gajdos P, Tranchant C, Clair B, Bolgert F, Eymard B, Stojkovic T, Attarian S, Chevret S; Myasthenia Gravis Clinical Study Group. Treatment of myasthenia gravis exacerbation with intravenous immunoglobulin: a randomized double-blind clinical trial. Arch Neurol 2005; 62(11): 1689-93.

223. Misbah SA, Chapel HM. Adverse effects of intravenous immunoglobulin. Drug Saf 1993;9(4):254–62.

224. Nydegger UE. Safety and side effects of i.v. immunoglobulin therapy Clin Exp Rheumatol 1996;14(Suppl 15):S53–S57.

225. Ochs HD, Gupta S, Kiessling P, Nicolay U, Berger M. Safety and efficacy of self-administered subcutaneous immunoglobulin in patients with primary immunodeficiency diseases. J Clin Immunol 2006; 26(3): 265-73.

226. Radinsky S, Bonagura VR. Subcutaneous immunoglobulin infusion as an alternative to intravenous immunoglobulin. J Allergy Clin Immunol 2003;112:630–3.

227. Hern S, Harman K, Bhogal BS, Black MM. A severe persistent case of pemphigoid gestationis treated with intravenous immunoglobulins and cyclosporin. Clin Exp Dermatol 1998;23(4):185–8.

228. Enk AH, Knop J. Adjuvante Therapie von Pemphigus vulgaris und Pemphigus foliaceus mit intravenösen Immunoglobulinen. [Adjuvant therapy of pemphigus vulgaris and pemphigus foliaceus with intravenous immunoglobulins.] Hautarzt 1998;49(10):774–6.

229. Colonna L, Cianchini G, Frezzolini A, De Pita O, Di Lella G, Puddu P. Intravenous immunoglobulins for pemphigus vulgaris: adjuvant or first choice therapy? Br J Dermatol 1998;138(6):1102–3.

230. Bagdasarian A, Tonetta S, Harel W, Mamidi R, Uemura Y. IVIG adverse reactions: potential role of cytokines and vasoactive substances. Vox Sang 1998;74(2):74–82.

231. Bassilios N, Mercadal L, Deray G. Immunoglobulin as a risk factor for contrast media nephrotoxicity. Nephrol Dial Transplant 2001;16(7):1513–4.

232. Kannan S, Rowland CH, Hockings GI, Tauchmann PM. Intragam can interfere with blood glucose monitoring. Med J Aust 2004;180(5):251–2.

233 US Food and Drug Administration. Important safety information on interference with blood glucose measurement following use of parenteral maltose/parenteral galactose/oral xylose-containing products. www.fda.gov. 2006.

234 Souza SP, Castro MC, Rodrigues RA, Passos RH, Ianhez LE. False hyperglycemia induced by polyvalent immunoglobulins. Transplantation 2005; 80(4): 542-3.

235. Lawn N, Wijdicks EF, Burritt MF. Intravenous immune globulin and pseudohyponatremia. N Engl J Med 1998;339(9):632.

236. Ng SK. Intravenous immunoglobulin infusion causing pseudohyponatremia. Lupus 1999;8(6):488–90.

237. Singh-Grewal D, Kemp A, Wong M. A prospective study of the immediate and delayed adverse events following intravenous immunoglobulin infusions. Arch Dis Child 2006; 91(8): 651-4.

Indinavir

See also Protease inhibitors

General Information

Indinavir is an HIV protease inhibitor. The recommended dose is 800 mg tds, given without food. However, as for most other protease inhibitors, the use of indinavir has been markedly simplified by co administration of low-dose ritonavir, which allows indinavir to be given twice daily with or without food. However, increased exposure to indinavir also results in increased adverse effects.

Comparative studies

In a randomized comparison of ritonavir-boosted indinavir versus indinavir alone the two regimens resulted in similar efficacy (1). However, because of an increase in adverse effects, drug-withdrawals were more frequent in the boosted indinavir arm (47/161 versus 20/162). The following adverse effects were significantly more common in the boosted arm: nausea/vomiting, diarrhea, nephrolithiasis and hematuria, dry skin, in-grown toenails, rash, and oral paresthesia.

In a randomized, open comparison of ritonavir-boosted saquinavir versus indinavir, adverse events were reported significantly more often in the indinavir/ritonavir arm (2). While there were no differences in hematological, renal, or hepatic toxicity (except increased bilirubin concentrations with indinavir), there were significantly higher lipid concentrations (total cholesterol, LDL cholesterol, and total triglycerides) in the indinavir arm. Adverse effects (renal, dermatological, and gastrointestinal) were more frequent in the indinavir/ritonavir arm.

Organs and Systems

Cardiovascular

A hypertensive crisis caused a secondary reversible posterior leukoencephalopathy in a patient taking indinavir-containing antiretroviral therapy (3).

- A 40-year-old man, who had taken stavudine 30 mg bd, lamivudine 150 mg bd, and indinavir 800 mg qds, developed an occipital headache, nausea, and vomiting. His blood pressure was 220/140 mmHg and he had bilateral papilledema. His blood pressure was controlled and his symptoms disappeared. An MRI scan of the brain showed lesions in the periventricular white matter; the nuclei semiovale and occipital asta were most severely affected. Indinavir was withdrawn and replaced by nelfinavir; his blood pressure returned to normal and the MRI white matter lesions disappeared.

In a retrospective analysis of 198 normotensive patients in a protease inhibitor comparison study, 30% of those who took indinavir developed stage I or worse hypertension, compared with none of the patients who took nelfinavir, ritonavir, or saquinavir (4).

Respiratory

Shock and respiratory failure have been attributed to indinavir (5).

- A 36-year-old HIV-positive man had started to take zidovudine and zalcitabine 9 months earlier together with co-trimoxazole as primary prophylaxis against *Pneumocystis jiroveci*, but switched to indinavir, stavudine, and lamivudine. Two hours after the first dose of indinavir he developed a high fever, generalized myalgia, and malaise and started to vomit. After the second dose he developed shock and cyanosis. A chest X-ray was compatible with adult respiratory distress syndrome. All cultures were negative for bacterial, viral, mycobacterial, and fungal pathogens. He recovered in 6 days and antiretroviral treatment without indinavir was reintroduced without recurrent problems.

The authors suggested that the severe shock and respiratory distress syndrome had been due to an idiosyncratic reaction to indinavir.

Nervous system

There has been a report of painful neuropathy in two patients who took ritonavir and indinavir respectively (6).

Paraparesis due to epidural lipomatosis has been attributed to indinavir (7).

- A 35-year-old man, who had taken indinavir 2400 mg/day, lamivudine 300 mg/day, and stavudine 80 mg/day for 10 months, developed a slowly progressive paraparesis, with sensory disturbances in the legs. An MRI scan was consistent with epidural lipomatosis. On withdrawal of indinavir, the symptoms gradually resolved.

Although indinavir can cause abnormal fat accumulation, this is thought to have been the first report of epidural lipomatosis.

Metabolism

Protease inhibitors are associated with hyperglycemia and possible diabetes mellitus. In a prospective study in 12 patients indinavir caused hyperglycemia and reduced insulin sensitivity (8).

Indinavir causes insulin resistance over 4–8 weeks of treatment. Tissue plasminogen activator and plasminogen activator inhibitor antigen are both markedly increased in patients with impaired glucose tolerance. Both markers have been linked to impaired thrombolysis, which is associated with an increased cardiovascular risk. In 11 patients taking indinavir and 14 taking fosamprenavir for 8 weeks indinavir was associated with a statistically significant increase in fasting plasma glucose concentration and with a 30% reduction in insulin sensitivity; there were no significant changes in lipid profiles (9). Amprenavir had no significant effect on glucose concentrations or insulin sensitivity but increased total cholesterol, triglycerides, and free fatty acids. Surprisingly, both drugs reduced tissue plasminogen activator, possibly reflecting a reduction in HIV-related inflammation as implied by a reduction in TNF-α.

Hematologic

Indinavir has been associated with severe hemolytic anemia (10).

- A 32-year-old Caucasian, who was taking lamivudine, stavudine, and indinavir for HIV infection, presented with pallor following a period of fatigue and headache (11). His hemoglobin was 8.2 g/dl and there were no clinical findings to suggest bleeding. The reticulocyte count was 3.5% and a direct Coombs' test was negative. A diagnosis of hemolytic anemia secondary to indinavir was made, the indinavir was stopped, and he was transfused with concentrated erythrocytes. The other antiretroviral drugs were continued, saquinavir was added, and a normal hemoglobin concentration was maintained.

Indinavir has been associated with thrombocytopenia (12).

- A man suddenly developed severe thrombocytopenia while taking stavudine + lamivudine + indinavir (13). After withdrawal and treatment with glucocorticoids and immunoglobulin, the platelet count recovered and he was treated with stavudine + didanosine + nevirapine. Six months later, after re-exposure to the initial combination, thrombocytopenia immediately occurred and spontaneously recovered after replacing indinavir with efavirenz.

Liver

Indinavir has been compared with abacavir in a randomized equivalence trial in 562 patients who were also taking lamivudine and zidovudine (14). The only significant difference in adverse effects was that there was hyperbilirubinemia in 8% of those taking indinavir and 2% of those taking abacavir. It has been postulated that indinavir-induced hyperbilirubinemia is due to inhibition of bilirubin UDP glucuronyl transferase activity, since it is more common in individuals with Gilbert's syndrome (15).

Biliary tract

Cholelithiasis has been attributed to indinavir. Of three patients who developed cholelithiasis while taking a protease inhibitor, one was taking indinavir (16).

Urinary tract

Indinavir causes nephrolithiasis as a result of precipitation of indinavir crystals in the urinary tract (17,18). It is dose-related and can be prevented by adequate hydration. In 615 patients (18 864 person-years of follow-up) who did not have risk factors for nephrolithiasis, the incidence was 8.6 episodes per 1000 person-years (19). It is more frequent when indinavir is given with ritonavir.

Several reports have suggested that patients using indinavir may also develop a syndrome consisting of back or flank pain, accompanied by crystalluria, renal function abnormalities, and evidence of tubulointerstitial nephritis on renal biopsy, but without obvious renal calculus formation (20–22).

In a French analysis of 22 510 urinary calculi performed by infrared spectroscopy, drug-induced urolithiasis was divided into two categories: first, stones with drugs physically embedded ($n = 238$; 1.0%), notably indinavir monohydrate ($n = 126$; 53%), followed by triamterene ($n = 43$; 18%), sulfonamides ($n = 29$; 12%), and amorphous silica ($n = 24$; 10%); secondly, metabolic nephrolithiasis induced by drugs ($n = 140$; 0.6%), involving mainly calcium + vitamin D supplementation ($n = 56$; 40%) and carbonic anhydrase inhibitors ($n = 33$; 24%) (23). Drug-induced stones are responsible for about 1.6% of all calculi in France. Physical analysis and a thorough drug history are important elements in the diagnosis.

The incidence of urolithiasis with indinavir has been estimated at 9% (24) but, according to some, may be as high as 20%. Indinavir calculi are often radiolucent and may be missed on CT scan rather than by using a contrast medium (which is itself not without risk) (25,26). It may therefore be that in some cases in which other renal complications with indinavir have been described there was in fact undetected renal stone formation. Any patient taking indinavir who develops renal colic should be suspected of having renal stones (27).

In two cases long-term use of indinavir appeared to have been responsible for renal atrophy (28), and again one wonders whether crystals may have been present but radiologically invisible.

Reversible renal insufficiency (which again could have been due to crystalluria) has been reported with indinavir (29).

- A 38-year-old man developed renal insufficiency while taking indinavir. His serum creatinine increased over a period of about 1 year, but urinalysis was persistently normal. A renal biopsy showed marked tubular crystal deposition. The indinavir was withdrawn, and after 2 months his serum creatinine returned to normal.

Since the basic problem in many such cases is probably crystalluria, it should be possible to treat it with rehydration, perhaps supplemented by brief interruption of therapy; this has been the conclusion of a study in which the unwanted renal effects of indinavir were prominent (30). Of 74 individuals infected with HIV-1 and taking indinavir 2.4 g/day orally, 15 had indinavir-related urological adverse effects (19 episodes), most commonly dull flank pain and dysuria. Microhematuria occurred in 16 of the 19 episodes. Four patients had urinary tract distension on ultrasonography as a possible indirect sign of urolithiasis and one passed a stone. In 4 patients treatment had to be stopped permanently, but in the other 11 it was continued. Some patients required dosage reduction and/or interruption of treatment: only conservative therapeutic measures were required, consisting of rehydration (fluid intake of at least 1.5 l/day) and analgesics.

Of 23 indinavir-treated patients with persistent pyuria, four had interstitial nephritis, seven had urothelial inflammation, 10 had both interstitial nephritis and urothelial inflammation, and two had non-specific urinary tract inflammation (31). In all, 21 patients had multinucleated histiocytes identified by cytological testing of urine specimens. Urine abnormalities resolved in all 20 patients who stopped taking indinavir, and pyuria persisted in the other 3. Six patients had raised serum creatinine concentrations, which returned to baseline when indinavir was withdrawn.

The cause of this urolithiasis has been evaluated in 24 patients taking protease inhibitors of whom 14 were taking indinavir, three ritonavir, two nelfinavir, and five other drugs (32). Of the 14 patients taking indinavir, only four had kidney stones that contained indinavir, the others being calcium oxalate and urate stones. Ten patients underwent 24-hour urine collection and 80% had metabolic abnormalities: five had hypocitraturia, four hyperoxaluria, four hypomagnesuria, three hypercalciuria, three supersaturation of calcium oxalate, and two hyperuricosuria. The authors concluded that the development of kidney stones is probably attributable to underlying metabolic abnormalities rather than to the use of specific protease inhibitors.

In a prospective study of 184 patients taking indinavir, there was persistent leukocyturia in 35% of them at least once during a median follow-up of 48 weeks with 3-monthly visits (33). Leukocyturia coincided with slight increases in serum albumin, erythrocyturia, and crystalluria. Leukocyturia was persistent in 24% of patients with repeated follow-up. Persistent leukocyturia was associated with loss of renal function. The risk of renal toxicity was increased in patients with high indinavir plasma concentrations (over 9000 ng/ml) and a urine pH over 5.7.

In summary, these studies suggest that indinavir toxicity is dose-related and they support the use of plasma concentration monitoring in patients taking indinavir. In fact, a pilot study of low-dose indinavir (400 mg plus ritonavir 100 mg bd) has shown markedly increased tolerability of

the drug (34). There was dry skin in only one of 20 patients and it disappeared after dosage reduction to indinavir 200 mg + ritonavir 100 mg bd. In all 20 patients, HIV-RNA remained fully suppressed below 200 cpm/ml.

Kidney stones and gastrointestinal intolerance were the main adverse effects in eight of 35 patients who had to stop taking ritonavir-boosted indinavir (35).

Skin

HIV-positive/AIDS patients using indinavir develop rashes early in treatment, a finding that is familiar with various other drugs used in this condition. This problem has been quantified in a study using data from postmarketing surveillance (36). Of 110 HIV/AIDS patients with a rash, 67% reported that it had occurred within 2 weeks of the start of indinavir therapy. The rash was initially localized in all cases, but in the majority it went on to spread to other body areas, involving all parts of the body in no less than 44%. It was usually pruritic but not accompanied by fever. Relief was often obtained by use of topical antihistamines or oral or topical glucocorticoids. More than half the patients decided to continue therapy despite the rash.

However, cutaneous toxicity can have a major influence on adherence to treatment and can impact adversely on the quality of life. Of 84 patients taking indinavir plus two nucleoside reverse transcriptase inhibitors for 20 months, 48 developed cheilitis, 34 had skin dryness and pruritus, 10 developed asteatotic dermatitis on the trunk, arms, and thighs, and 10 complained of scalp defluvium (37). Severe alopecia was observed in one patient, while six reported that their body hair had become fairer and thinner and shed considerably. Multiple pyogenic granulomas were observed in the toenails of five patients and softening of the nail plate was noted in five. The temporal relation between starting indinavir and the onset of these effects was striking and regression occurred on withdrawal.

Suggested mechanisms for these adverse effects include:

- retinoid-like effects due to homologies of the amino acid sequences of the HIV-1 protease and cytoplasmic retinoic-acid binding protein type 1 (CRABP-1)
- inhibition of CYP3A by indinavir, resulting in reduced oxidation of retinoic acid and hence augmentation of its biological effects.

Hair

Alopecia has been attributed to indinavir in 10 men (38). Of 337 patients who were given indinavir as part of combination antiretroviral therapy with nucleoside analogues, five (1.5%) developed severe alopecia a median of 50 days after starting indinavir. Three had diffuse shedding of hair involving the entire scalp, and two were initially aware of circumscribed circular areas of alopecia resulting in complete severe hair loss. Although indinavir was discontinued in all five cases, there was no regrowth a median of 30 days later (39).

© 2010 Elsevier B.V. All rights reserved.

Nails

Paronychia associated with indinavir has been reported (40,41). In 288 patients in a retrospective cohort study, paronychia was associated with indinavir, with a hazard ratio of 4.7 (42).

Musculoskeletal

Increased bone mineral density has been documented in patients taking indinavir (43) although in another study protease inhibitor-containing regimens caused accelerated loss of bone density (44).

- Widespread osteosclerosis, accompanied by increased serum concentrations of osteocalcin and C telopeptide, developed in a 56-year-old man who took indinavir 800 mg tds for 27 months (45).

This patient had been taking vitamin A 1 mg/day for 20 years and the authors proposed that the indinavir had also had a vitamin A-like effect.

Frozen shoulder has been attributed to indinavir (SEDA-24, 347) (46). In one case it was associated with Dupuytren's contracture and in one case each with arthralgias and tendinitis (47). An adhesive capsulitis seems to be present.

The results of a questionnaire survey of 878 people with HIV infection treated with antiretroviral drugs confirmed the risk of arthralgias in patients taking indinavir. The authors suggested that crystal deposition in joints, analogous to the crystalluria with nephrolithiasis that indinavir and other protease inhibitors can cause, might be responsible.

Six patients taking stavudine + lamivudine + indinavir developed adhesive capsulitis (48). Based on previous evidence the authors suggested that this adverse effect might have been be attributable to indinavir.

Susceptibility Factors

Sex

The pharmacokinetics of indinavir in 220 women and 94 men have been compared; there was no difference (49).

Variability in the systemic availability of indinavir has been studied in 239 patients (50). Concomitant use of ritonavir and female sex both predicted increased availability. The authors suggested changing the recommended dosages of indinavir according to their findings in order to ensure suitable target drug concentrations in most of the population: women 900 mg/day plus low-dose ritonavir bd; men 1700 mg/day plus low-dose ritonavir bd. They also encouraged drug concentration monitoring. However, close attention has to be paid to adverse effects (for example nephrolithiasis) when exposing patients to regimens with higher dosages than have previously been recommended.

Liver disease

In a pharmacokinetic study of low-dose indinavir, all of six patients with chronic viral hepatitis had an indinavir concentration above the target range defined by the study protocol (150?675 ng/ml) despite low-dose indinavir/ritonavir (400/100 mg bd) (51). Altering the dose to 200/100 mg resulted in drug concentrations in the target range while HIV-RNA concentrations remained suppressed below 200 cpm/ml in all patients.

Drug–Drug Interactions

Cannabinoids

The effects of smoked marijuana (3.95% tetrahydrocannabinol; up to three cigarettes per day) and oral dronabinol (2.5 mg tds) on the pharmacokinetics of indinavir 800 mg 8-hourly ($n = 28$) have been evaluated in a randomized, placebo-controlled study in HIV-infected patients (52). On day 14, marijuana reduced the 8-hour AUC of indinavir by 15%, the C_{max} by 14%, and the C_{min} by 34%. However, only the change in C_{max} was significant. Dronabinol had no effects.

Carbamazepine

Failure of combination therapy including indinavir has been attributed to an interaction of indinavir with carbamazepine (53).

- A 48-year-old HIV-positive man taking indinavir, zidovudine, and lamivudine developed Herpes zoster infection, which was treated with famciclovir. Postherpetic neuralgia was treated with carbamazepine, and his plasma indinavir concentration fell substantially. The carbamazepine was withdrawn after 2.5 months and 2 weeks later HIV-RNA was detectable in his plasma (6×10^3/ml). His circulating virus was resistant to lamivudine. With a further increase in viral load, his therapy was changed to nevirapine, didanosine, and stavudine.

This treatment failure, and possibly the resistance to lamivudine, was attributed to induction of drug metabolism by carbamazepine.

Didanosine

The absorption of indinavir was not affected by co-administration of didanosine in an encapsulated enteric bead formulation in 24 patients (54).

Levothyroxine

Inhibition of glucuronyl transferase activity by indinavir has been blamed for an interaction of indinavir with levothyroxine, causing hyperthyroidism (55).

- A 36-year-old woman, who had taken levothyroxine for several years, took stavudine, lamivudine, and indinavir for 1 month and developed nervousness, palpitation, restlessness, weakness, and weight loss. She had an undetectable serum TSH concentration and raised unbound serum T4 and T3. The dose of levothyroxine was reduced to one-third and then to one-sixth of the previous dose, and the thyroid function tests returned to normal.

Phenylpropanolamine

In one case the use of phenylpropanolamine with triple drug therapy for HIV prophylaxis led to a hypertensive crisis (56). As the patient had previously tolerated phenylpropanolamine well, one must suspect that one or more of the anti-HIV drugs (probably indinavir) had interfered with the metabolic breakdown of the phenylpropanolamine.

Ranunculaceae

Golden seal root inhibits various isoforms of cytochrome P450, including CYP3A4 (57). However, in a crossover study, goldenseal root (1140 mg bd for 14 days) had no effect on the pharmacokinetics of a single oral dose of indinavir 800 mg (58).

Rifampicin

In six patients who took rifampicin for 4 days serum concentrations of both indinavir and ritonavir fell markedly (by 87% and 94% respectively) (57). These data strongly suggest that co administration of rifampicin with indinavir, even in the presence of ritonavir, could lead to subtherapeutic drug concentrations; the combination should therefore be discouraged.

Ritonavir

Using ritonavir as a booster may allow indinavir to be given twice daily and with food. In a cohort survey of 100 patients the combination of indinavir plus ritonavir (400 mg/400 mg or 800 mg/100 mg bd) was a safe and effective option to reduce the tablet burden and improve adherence (58).

Co-administration of efavirenz or nevirapine increased the clearance of indinavir by 41%, irrespective of the presence or absence of ritonavir (59). Women had a 48% higher apparent systemic availability of indinavir than men. Population pharmacokinetic modeling supported ritonavir boosting of indinavir.

In a meta-analysis of three studies (total n = 26) and in a randomized study of 16 patients indinavir concentrations were not reduced significantly by the herbal treatment milk thistle (60).

Sildenafil

There is good evidence that indinavir can substantially increase plasma concentrations of sildenafil (61). Since HIV infection commonly leads to erectile dysfunction, the drugs may well be used together and it will then be prudent to use a lower dose of sildenafil.

Silybum marianum (milk thistle)

The pharmacokinetics of indinavir 800 mg 8-hourly have been studied in the presence and absence of milk thistle 175 mg (containing silymarin 153 mg) tds for 3 weeks in an open study in 10 healthy volunteers; milk thistle had no significant effect except a small reduction in $C_{min.ss}$ (62).

Venlafaxine

Venlafaxine 50 mg 8-hourly reduced the AUC of a single dose of indinavir 800 mg by 28% in nine healthy subjects (63).

Diagnosis of Adverse Drug Reactions

The adverse effects of indinavir have been retrospectively evaluated and correlated with indinavir trough concentration in 63 patients taking indinavir + ritonavir 800/100 mg bd (64). The median indinavir trough concentration of 1446 ng/ml was associated with 60% of measured toxicity. Of 49 patients with an indinavir trough concentration over 500 ng/ml, 46 had at least one dosage adjustment; the main reason for dosage adjustment was toxicity (n = 43). The common adverse effects affected the skin (vitamin A-like reactions; n = 39), kidneys (renal colic, nephrolithiasis, renal insufficiency; n = 35), gastrointestinal tract (nausea, vomiting, and diarrhea; n = 32), and liver (n = 14). After dosage adjustment, the median indinavir trough concentration was 459 ng/ml, which was associated with 8% of toxicity. Trough concentrations over 500 ng/ml correlated with increased toxicity. The authors concluded that indinavir trough concentrations below 500 ng/ml are safe, and that an optimal concentration range for indinavir trough concentration could be 150–500 ng/ml in patients taking twice daily indinavir + ritonavir.

References

1. Arnaiz JA, Mallolas J, Podzamczer D, Gerstoft J, Lundgren JD, Cahn P, Fatkenheuer G, D'Arminio-Monforte A, Casiro A, Reiss P, Burger DM, Stek M, Gatell JM; BEST Study Team. Continued indinavir versus switching to indinavir/ritonavir in HIV-infected patients with suppressed viral load. AIDS 2003;17:831–40.
2. Dragsted UB, Gerstoft J, Pedersen C, Peters B, Duran A, Obel N, Castagna A, Cahn P, Clumeck N, Bruun JN, Benetucci J, Hill A, Cassetti I, Vernazza P, Youle M, Fox Z, Lundgren JD; MaxCmin1 Trial Group. Randomized trial to evaluate indinavir/ritonavir versus saquinavir/ritonavir in human immunodeficiency virus type 1-infected patients: the MaxCmin1 Trial. J Infect Dis 2003;188:635–42.
3. Giner V, Fernandez C, Esteban MJ, Galindo MJ, Forner MJ, Guix J, Redon J. Reversible posterior leukoencephalopathy secondary to indinavir-induced hypertensive crisis: a case report. Am J Hypertens 2002;15(5):465–7.
4. Cattelan AM, Trevenzoli M, Sasset L, Rinaldi L, Balasso V, Cadrobbi P. Indinavir and systemic hypertension. AIDS 2001;15(6):805–7.
5. Dieleman JP, in 't Veld B, Borleffs JC, Schreij G. Acute respiratory failure associated with the human immunodeficiency virus (HIV) protease inhibitor indinavir in an HIV-infected patient. Clin Infect Dis 1998;26(4):1012–3.
6. Colebunders R, De Droogh E, Pelgrom Y, Depraetere K, De Jonghe P. Painful hyperaesthesia caused by protease inhibitors? Infection 1998;26(4):250–1.
7. Cersosimo MG, Lasala B, Folgar S, Micheli F. Epidural lipomatosis secondary to indinavir in an HIV-positive patient. Clin Neuropharmacol 2002;25(1):51–4.

8. Dube MP, Edmondson-Melancon H, Qian D, Aqeel R, Johnson D, Buchanan TA. Prospective evaluation of the effect of initiating indinavir-based therapy on insulin sensitivity and B cell function in HIV-infected patients. J Acquir Immune Defic Syndr 2001;27(2):130–4.

9. Young EM, Considine RV, Sattler FR, Deeg MA, Buchanan TA, Degawa-Yamauchi M, Shankar S, Edmondson-Melancon H, Hernandez J, Dube MP. Changes in thrombolytic and inflammatory markers after initiation of indinavir- or amprenavir-based antiretroviral therapy. Cardiovasc Toxicol 2004;4(2):179–86.

10. Morrison-Griffiths S, Newman M, O'Mahony C, Pirmohamed M. Haemolytic anaemia associated with indinavir. Postgrad Med J 1999;75(883):313–5.

11. Watson A. Reversible acute haemolysis associated with indinavir. AIDS 2000;14(4):465–6.

12. Durand JM. Indinavir and thrombocytopenia. AIDS 1999;13(1):148–9.

13. Camino N, Nunez M, Blanco F, Gonzalez-Requena D, Gonzalez-Lahoz J, Soriano V. Indinavir-induced thrombocytopenia. AIDS Patient Care STDS 2003;17:103–4.

14. Staszewski S, Keiser P, Montaner J, Raffi F, Gathe J, Brotas V, Hicks C, Hammer SM, Cooper D, Johnson M, Tortell S, Cutrell A, Thorborn D, Isaacs R, Hetherington S, Steel H, Spreen WCNAAB3005 International Study Team. Abacavir–lamivudine–zidovudine vs indinavir–lamivudine–zidovudine in antiretroviral-naive HIV-infected adults: A randomized equivalence trial. JAMA 2001;285(9):1155–63.

15. Sen S, Jalan R. Is "Gilbert's" the culprit in indinavir-induced hyperbilirubinemia? Hepatology 2002;35(5):1269–70.

16. Siveke JT, Bogner JR. Cholelithiasis possibly induced by protease inhibitors in 3 patients. Clin Infect Dis 2003;36:1498–500.

17. Pietroski NA. Treating HIV with protease inhibitors. Drug efficacy, tolerability, and dosing. Am Druggist 1996;213:50–7.

18. Tsao JW, Kogan SC. Images in clinical medicine. Indinavir crystalluria. N Engl J Med 1999;340(17):1329.

19. Dworkin MS, Wan PT. Indinavir, zidovudine, lamivudine: 3-year follow-up. Ann Intern Med 2001;134(2):165.

20. Kopp JB, Miller KD, Mican JA, Feuerstein IM, Vaughan E, Baker C, Pannell LK, Falloon J. Crystalluria and urinary tract abnormalities associated with indinavir. Ann Intern Med 1997;127(2):119–25.

21. Tashima KT, Horowitz JD, Rosen S. Indinavir nephropathy. N Engl J Med 1997;336(2):138–40.

22. Chen SC, Nankivell BJ, Dwyer DE. Indinavir-induced renal failure. AIDS 1998;12(4):440–1.

23. Cohen-Solal F, Abdelmoula J, Hoarau MP, Jungers P, Lacour B, Daudon M. Les lithiases urinaires d'origine médicamenteuse. [Urinary lithiasis of medical origin.] Therapie 2001;56(6):743–50.

24. Hermieu J, Prevot M, Ravery V, Sauty L, Moulinier F, Delmas V, Bouvet E, Boccon-Gibod L. Urolithiasis and the protease inhibitor indinavir. Eur Urol 1999;35(3):239–241.

25. Schwartz BF, Schenkman N, Armenakas NA, Stoller ML. Imaging characteristics of indinavir calculi. J Urol 1999;161(4):1085–7.

26. Sundaram CP, Saltzman B. Urolithiasis associated with protease inhibitors. J Endourol 1999;13(4):309–12.

27. Kohan AD, Armenakas NA, Fracchia JA. Indinavir urolithiasis: an emerging cause of renal colic in patients with human immunodeficiency virus. J Urol 1999;161(6):1765–8.

28. Hanabusa H, Tagami H, Hataya H. Renal atrophy associated with long-term treatment with indinavir. N Engl J Med 1999;340(5):392–3.

29. Grabe DW, Eisele G, Miller C, Singh J, Stein D. Indinavir-induced nephropathy. Clin Nephrol 1999;51(3):181–3.

30. Hug B, Naef M, Bucher HC, Sponagel L, Lehmann K, Battegay M. Treatment for human immunodeficiency virus with indinavir may cause relevant urological side-effects, effectively treatable by rehydration. BJU Int 1999;84(6):610–4.

31. Sokal E. Lamivudine for the treatment of chronic hepatitis B. Expert Opin Pharmacother 2002;3(3):329–39.

32. Nadler RB, Rubenstein JN, Eggener SE, Loor MM, Smith ND. The etiology of urolithiasis in HIV infected patients. J Urol 2003;169:475–7.

33. Dieleman JP, van Rossum AM, Stricker BC, Sturkenboom MC, de Groot R, Telgt D, Blok WL, Burger DM, Blijenberg BG, Zietse R, Gyssens IC. Persistent leukocyturia and loss of renal function in a prospectively monitored cohort of HIV-infected patients treated with indinavir. J Acquir Immune Defic Syndr 2003;32:135–42.

34. Ghosn J, Lamotte C, Ait-Mohand H, Wirden M, Agher R, Schneider L, Bricaire F, Duvivier C, Calvez V, Peytavin G, Katlama C. Efficacy of a twice-daily antiretroviral regimen containing 100 mg ritonavir/400 mg indinavir in HIV-infected patients. AIDS 2003;17:209–14.

35. Macassa E, Delaugerre C, Teglas JP, Jullien V, Tréluyer JM, Veber F, Rouzioux C, Blanche S. Change to a once-daily combination including boosted atazanavir in HIV-1-infected children. Pediatr Infect Dis J 2006;25(9):809–14.

36. Gajcwski LK, Grimone AJ, Melbourne KM, Vanscoy GJ. Characterization of rash with indinavir in a national patient cohort. Ann Pharmacother 1999;33(1):17–21.

37. Calista D, Boschini A. Cutaneous side effects induced by indinavir. Eur J Dermatol 2000;10(4):292–6.

38. Bouscarat F, Prevot MH, Matheron S. Alopecia associated with indinavir therapy. N Engl J Med 1999;341(8):618.

39. d'Arminio Monforte A, Testa L, Gianotto M, Gori A, Franzetti F, Sollima S, Bini T, Moroni M. Indinavir-related alopecia. AIDS 1998;12(3):328.

40. Dauden E, Pascual-Lopez M, Martinez-Garcia C, Garcia-Diez A. Paronychia and excess granulation tissue of the toes and finger in a patient treated with indinavir. Br J Dermatol 2000;142(5):1063–4.

41. Sass JO, Jakob-Solder B, Heitger A, Tzimas G, Sarcletti M. Paronychia with pyogenic granuloma in a child treated with indinavir: the retinoid-mediated side effect theory revisited. Dermatology 2000;200(1):40–2.

42. Colson AE, Sax PE, Keller MJ, Turk BK, Pettus PT, Platt R, Choo PW. Paronychia in association with indinavir treatment. Clin Infect Dis 2001;32(1):140–3.

43. Nolan D, Upton R, McKinnon E, John M, James I, Adler B, Roff G, Vasikaran S, Mallal S. Stable or increasing bone mineral density in HIV-infected patients treated with nelfinavir or indinavir. AIDS 2001;15(10):1275–80.

44. Tebas P, Powderly WG, Claxton S, Marin D, Tantisiriwat W, Teitelbaum SL, Yarasheski KE. Accelerated bone mineral loss in HIV-infected patients receiving potent antiretroviral therapy. AIDS 2000;14(4):F63–7.

45. Begovac J, Bayer K, Krpan D, Kusec V. Osteosclerosis and periostal new bone formation during indinavir therapy. AIDS 2002;16(5):803–4.

46. Peyriere H, Mauboussin JM, Rouanet I, Rouveroux P, Hillaire-Buys D, Balmes P. Frozen shoulder in HIV patients

treated with indinavir: report of three cases. AIDS 1999;13(16):2305–6.

47. Cooper CL, Parbhakar MA, Angel JB. Hepatotoxicity associated with antiretroviral therapy containing dual versus single protease inhibitors in individuals coinfected with hepatitis C virus and human immunodeficiency virus. Clin Infect Dis 2002;34(9):1259–63.

48. De Ponti A, Vigano MG, Taverna E, Sansone V. Adhesive capsulitis of the shoulder in human immunodeficiency virus-positive patients during highly active antiretroviral therapy. J Shoulder Elbow Surg 2006;15(2):188–90.

49. Burger DM, Siebers MC, Hugen PW, Aarnoutse RE, Hekster YA, Koopmans PP. Pharmacokinetic variability caused by gender: do women have higher indinavir exposure than men? J Acquir Immune Defic Syndr 2002;29(1):101–2.

50. Csajka C, Marzolini C, Fattinger K, Decosterd LA, Telenti A, Biollaz J, Buclin T. Population pharmacokinetics of indinavir in patients infected with human immunodeficiency virus. Antimicrob Agents Chemother 2004;48(9):3226–32.

51. Bossi P, Peytavin G, Lamotte C, Calvez V, Bricaire F, Costagliola D, Katlama C. High indinavir plasma concentrations in HIV-positive patients co-infected with hepatitis B or C virus treated with low doses of indinavir and ritonavir (400/100 mg twice a day) plus two nucleoside reverse transcriptase inhibitors. AIDS 2003;17:1108–10.

52. Kosel BW, Aweeka FT, Benowitz NL, Shade SB, Hilton JF, Lizak PS, Abrams DI. The effects of cannabinoids on the pharmacokinetics of indinavir and nelfinavir. AIDS 2002;16(4):543–50.

53. Hugen PW, Burger DM, Brinkman K, ter Hofstede HJ, Schuurman R, Koopmans PP, Hekster YA. Carbamazepine–indinavir interaction causes antiretroviral therapy failure. Ann Pharmacother 2000;34(4):465–70.

54. Damle BD, Mummaneni V, Kaul S, Knupp C. Lack of effect of simultaneously administered didanosine encapsulated enteric bead formulation (Videx EC) on oral absorption of indinavir, ketoconazole, or ciprofloxacin. Antimicrob Agents Chemother 2002;46(2):385–91.

55. Lanzafame M, Trevenzoli M, Faggian F, Marcati P, Gatti F, Carolo G, Concia E. Interaction between levothyroxine and indinavir in a patient with HIV infection. Infection 2002;30(1):54–5.

56. Khurana V, de la Fuente M, Bradley TP. Hypertensive crisis secondary to phenylpropanolamine interacting with triple-drug therapy for HIV prophylaxis. Am J Med 1999;106(1):118–9.

57. Justesen US, Andersen AB, Klitgaard NA, Brosen K, Gerstoft J, Pedersen C. Pharmacokinetic interaction between rifampin and the combination of indinavir and low-dose ritonavir in HIV-infected patients. Clin Infect Dis 2004;38(3):426–9.

58. Burger DM, Hugen PW, Aarnoutse RE, Dieleman JP, Prins JM, van der Poll T, ten Veen JH, Mulder JW, Meenhorst PL, Blok WL, van der Meer JT, Reiss P, Lange JM. A retrospective, cohort-based survey of patients using twice-daily indinavir + ritonavir combinations: pharmacokinetics, safety, and efficacy J Acquir Immune Defic Syndr 2001;26(3):218–24.

59. Kappelhoff BS, Huitema AD, Sankatsing SU, Meenhorst PL, Van Gorp EC, Mulder JW, Prins JM, Beijnen JH. Population pharmacokinetics of indinavir alone and in combination with ritonavir in HIV-1-infected patients. Br J Clin Pharmacol 2005;60(3):276–86.

60. Mills E, Wilson K, Clarke M, Foster B, Walker S, Rachlis B, DeGroot N, Montori VM, Gold W, Phillips E, Myers S, Gallicano K. Milk thistle and indinavir: a randomized controlled pharmacokinetics study and meta-analysis. Eur J Clin Pharmacol 2005;61(1):1–7.

61. Merry C, Barry MG, Ryan M, Tjia JF, Hennessy M, Eagling VA, Mulcahy F, Back DJ. Interaction of sildenafil and indinavir when co-administered to HIV-positive patients. AIDS 1999;13(15):F101–7.

62. Piscitelli SC, Formentini E, Burstein AH, Alfaro R, Jagannatha S, Falloon J. Effect of milk thistle on the pharmacokinetics of indinavir in healthy volunteers. Pharmacotherapy 2002;22(5):551–6.

63. Levin GM, Nelson LA, DeVane CL, Preston SL, Eisele G, Carson SW. A pharmacokinetic drug–drug interaction study of venlafaxine and indinavir. Psychopharmacol Bull 2001;35(2):62–71.

64. Solas C, Basso S, Poizot-Martin I, Ravaux I, Gallais H, Gastaut JA, Durand A, Lacarelle B. High indinavir Cmin is associated with higher toxicity in patients on indinavir–ritonavir 800/100 mg twice-daily regimen J Acquir Immune Defic Syndr 2002;29(4):374–7.

Inosine pranobex

General Information

Inosine pranobex is a synthetic product, also known as isoprinosine or inosine dimepranol acedobene, with antiviral properties that are assumed to be related to its effect on T cell-mediated immunity rather than to direct antiviral activity. It has been tried in a wide range of viral diseases and also in rheumatoid arthritis (1), multiple sclerosis (2), and alopecia (3). However, clinical trials have mostly shown only modest therapeutic benefit or none at all (1,4), and no specific adverse effects, except for an increase in serum uric acid concentrations (5), reflecting the metabolic pathways of purines (6).

Organs and Systems

Immunologic

Anecdotal reports have attributed aggravation of polymyositis (7) and generalized *Herpes zoster* virus infection to inosine pranobex (8).

Drug–Drug Interactions

Zidovudine

Inosine pranobex increased zidovudine plasma concentrations and half-life (9).

References

1. Brzeski M, Madhok R, Hunter JA, Capell HA. Randomised, double blind, placebo controlled trial of inosine pranobex in rheumatoid arthritis. Ann Rheum Dis 1990;49(5):293–5.

2. Milligan NM, Miller DH, Compston DA. A placebo-controlled trial of isoprinosine in patients with multiple sclerosis. J Neurol Neurosurg Psychiatry 1994;57(2):164–8.
3. Galbraith GM, Thiers BH, Jensen J, Hoehler F. A randomized double-blind study of inosiplex (isoprinosine) therapy in patients with alopecia totalis. J Am Acad Dermatol 1987;16(5 Pt 1):977–83.
4. Kinghorn GR, Woolley PD, Thin RN, De Maubeuge J, Foidart JM, Engst R. Acyclovir vs isoprinosine (immunovir) for suppression of recurrent genital Herpes simplex infection. Genitourin Med 1992;68(5):312–6.
5. Sarciron ME, Delabre I, Walbaum S, Raynaud G, Petavy AF. Effects of multiple doses of isoprinosine on Echinococcus multilocularis metacestodes. Antimicrob Agents Chemother 1992;36(1):191–4.
6. Thorsen S, Pedersen C, Sandstrom E, Petersen CS, Norkrans G, Gerstoft J, Karlsson A, Christensen KC, Hakansson C, Pehrson PO, et alScandinavian Isoprinosine Study Group. One-year follow-up on the safety and efficacy of isoprinosine for human immunodeficiency virus infection. J Intern Med 1992;231(6):607–15.
7. Chuck AJ, Lloyd Jones JK, Dunn NA. Is inosine pranobex contraindicated in autoimmune disease? BMJ (Clin Res Ed) 1988;296(6622):646.
8. Revuz J, Guillaume JC, Roujeau JC, Perroud AM. Généralisation d'un zona chez un sujet non immunodéprime recevant de l'isoprinosine et de la rifamycine S. V. [Generalization of zona in a nonimmunosuppressed patient receiving isoprinosine and rifamycin SV.] Ann Dermatol Venereol 1983;110(6–7):563.
9. De Simone C, Famularo G, Tzantzoglou S, Moretti S, Jirillo E. Inosine pranobex in the treatment of HIV infection: a review. Int J Immunopharmacol 1991;13(Suppl 1):19–27.

Interferons

See also Interferon alfa, Interferon beta, Interferon gamma

General Information

The interferons, first described in 1957, include at least five natural human glycoproteins (alfa, beta, gamma, omega, and tau). Only the first three types are currently used, and they differ both structurally and antigenically. Interferon alfa is produced by macrophages, B cells, and null lymphocytes, interferon beta by fibroblasts, epithelial cells, and macrophages, and interferon gamma from T lymphocytes and macrophages after antigenic or mitogenic stimulation. The interferons share 30–40% of sequence homology and have antiviral and antiproliferative actions. Interferon gamma, produced by activated T cells and natural killer cells, is recognized by a different receptor and acts primarily as an immunoregulatory cytokine.

Uses

The uses of interferons are listed in Table 1.

Two authoritative reviews have outlined the therapeutic potential of interferons (1,2). A wide range of viral diseases or cancers are other candidates for interferon therapy (3–5).

Mechanisms of action

On binding to surface receptors, interferon alfa results in activation of cytoplasmic enzymes affecting messenger RNA translation and protein synthesis (6). The antiviral state takes hours to develop but can persist for days. Besides broad antiviral activity, interferons are of major importance in regulating immunological functions.

General adverse effects

The adverse effects of interferon are multifarious and the natural products seem to be less toxic than the pure synthetic compounds. Influenza-like symptoms with fever, chills, fatigue, myalgia, arthralgia, nausea, and lethargy, starting within 1 week after the start of treatment and lasting 1–7 days, seem to be very common (7,8). Adverse effects also include neurotoxicity (paresthesia, polyneuropathy), hepatic toxicity, renal toxicity, and an increase in eyelash growth (9–13). Neutralizing antibodies can lead to resistance in patients with hairy cell leukemia and chronic myeloid leukemia (14,15). The route of administration influences the provocation of an antibody response, and recombinant interferon beta is more likely to be immunogenic when given subcutaneously or intramuscularly than when given intravenously (16). Raynaud's phenomenon has been described after treatment with interferon alfa (17), and exacerbation of multiple sclerosis has been observed after treatment with interferon gamma.

The most common adverse effects reported in two large multicenter studies were fever (60%), leukopenia (43%), increased serum aspartate transaminase activity (30%), anorexia (30%), thrombocytopenia (25%), fatigue (21%), nausea, and vomiting (17%) (18,19). Compared with subcutaneous administration, intravenous interferon alfa is associated with similar adverse effects of greater severity and frequency (20,21).

Psychological and psychiatric adverse effects of interferons

Interferon alfa can cause depressive symptoms and potentially dangerous psychiatric adverse effects [22]. The information about interferon alfa-induced psychiatric adverse effects and their recommended management has been comprehensively reviewed [23,24,25]. A past history of psychiatric disorders or substance abuse and a controlled psychiatric disease are no longer considered as contraindications to treatment, provided that patients are closely monitored during treatment.

• In a 39-year-old man with chronic hepatitis C, psychotic mania developed within 3 weeks after standard interferon alfa and ribavirin were changed to peginterferon alfa and ribavirin [26].

The authors discussed the management of manic episodes in patients treated with interferon alfa.

Table 1 Names of different types of interferons and their indications

Name	Indications
Interferon alfa	Malignant diseases: hairy cell leukemia, chronic myelogenous leukemia,
Human natural leukocyte interferon alfa	cutaneous T cell lymphoma, follicular lymphoma, multiple myeloma, Kaposi's
(IFN alfa-n3)	sarcoma, diffuse melanoma, renal cell carcinoma, carcinoid tumors
Lymphoblastoid interferon alfa (IFN alfa-n1)	Viral diseases: condylomata acuminata, chronic active hepatitis B and C
Recombinant interferon alfa-2a (rIFN alfa-2a)	
Recombinant interferon alfa-2b (rIFN alfa-2b)	
Recombinant interferon alfa-2c (rIFN alfa,2c)	
Pegylated interferon alfa (peginterferon alfa)	
Interferon beta	Multiple sclerosis
Natural fibroblast	
Recombinant interferon beta (rIFN beta)	
Recombinant interferon beta-1b (rIFN beta-1b)	
Interferon gamma	Chronic granulomatous disease
Recombinant interferon gamma-1b (rIFN gamma-1b)	

According to data obtained from six controlled and 17 non-controlled studies sponsored by the manufacturer of interferon beta-1a, depression was more frequently reported as an adverse event in the treated patients [27]. However, this was not associated with an increase in depression scores in the only study that used appropriate rating scales. There was also no significant difference in the rate of suicide attempts or suicide in patients taking interferon compared with placebo. There were 42 cases of serious and non-serious reports of depression and suicide attempts, including only five cases of suicide, in postmarketing experience covering more than 161 000 patient-years. Overall these data suggest that interferon beta-1a does not produce typical depression.

Cognitive effects

A number of mild to moderate frontal–subcortical brain symptoms, including cognitive and behavioral slowing, apathy, impaired executive function, and reduced memory have also been attributed to interferon alfa. These symptoms may alter the quality of life. Of 30 adults treated with interferon-alfa alone ($n \approx 13$) or in combination with chemotherapy (low-dose cytarabine arabinoside in 15 or hydroxycarbamide in two), 16 had a significant reduction in one or more cognitive tests compared with baseline [28]. The combination with chemotherapy was associated with greater risks of impaired cognitive performance, but these patients also took higher cumulative doses of interferon alfa. Although one-third of patients had depressive symptoms, there was no significant correlation between new depressive symptoms and most of the cognitive decline, suggesting that depression alone did not account for cognitive dysfunction. Very similar marked cognitive impairment, assessed by a battery of computer-assisted psychological tests, was also found in 70 patients with chronic hepatitis C treated with interferon alfa and ribavirin; there was no difference in the quality or intensity of the cognitive changes when comparing standard and pegylated interferon alfa, and no correlation with age or education [29]. An important finding was that cognitive performances returned to pre-treatment values after the end of treatment.

Neurocognitive performance has been studied in 70 patients receiving interferon alfa 2b (pegylated or conventional) and ribavirin, because impairment of concentration is common during antiviral therapy of chronic hepatitis C [30]. Repeated computer-based testing showed significantly increased reaction times. Accuracy measures, reflected by the number of false reactions, were affected only for the working-memory task. Cognitive performance returned to pre-treatment values after the end of therapy. Cognitive impairment was not significantly correlated with the degree of concomitant depression.

Incidence

Psychiatric symptoms have been prospectively examined in 104 patients with chronic hepatitis C, of whom 84 received interferon alfa-2b 15 MU/week and 20 were not treated [31]. The incidence of clinically relevant scores for depression, anxiety, or anger/hostility increased from 23% of patients before interferon alfa to 58% of patients during treatment, and returned to 30% and 19% of patients 4 weeks and 6 months after withdrawal respectively. In contrast, there were no significant changes in the reference group. There were also significantly higher scores in the 40 patients who took concomitant ribavirin. Six patients successfully received antidepressant therapy, but withdrawal because of untreatable psychiatric symptoms was needed in 8.3% of patients, i.e. about half of the patients who had interferon-induced major depressive disorders.

Depressive symptoms were studied in 186 patients with chronic hepatitis C during treatment for 48 weeks with pegylated interferon alfa-2a and pegylated interferon-alfa-2b. Depressive symptoms increased from 53% at baseline to 61% at 12 weeks with pegylated interferon alfa-2a and from 57 to 65% with pegylated interferon alfa-2b. Three patients had life-threatening psychiatric symptoms with pegylated interferon alfa-2a (psychosis and delirium requiring withdrawal of antiviral therapy and admission to a psychiatric unit).

Mechanisms

Several mechanisms involved in the pathogenesis of interferon alfa-induced psychiatric adverse effects have been hypothesized [23], but the mechanisms by which interferon alfa enters the brain to produce neurotransmitter changes are unclear. Interactions of interferon alfa with the central opioid, serotonin, dopamine, and glutamate neurotransmitter systems are probably involved [32]. The possible predominant role of the serotonergic system has again been emphasized in two studies, with evidence of early and significant changes in tryptophan metabolism resulting in persistently low tryptophan concentrations and increased kynurenine concentrations and kynurenine/tryptophan ratio during interferon alfa treatment compared with values obtained before treatment [33] and in an untreated control group [34]. In addition, the fall in tryptophan concentrations correlated with the development and intensity of mood and cognitive symptoms during interferon alfa treatment [<<Capuron 906].

In a prospective study in 48 patients who received adjuvant interferon alfa for malignant melanoma there was a positive correlation between the increase in serum concentration of soluble ICAM-1 and depression scores; the authors suggested that induction of soluble ICAM-1 by interferon alfa increased the permeability of the blood–brain barrier, allowing the drug to enter the brain more readily [35].

The relation between interferon alfa-induced depressive disorders and the viral response to treatment has been examined in 39 patients with hepatitis C infection and no history of active psychiatric disease [36]. After treatment with interferon alfa-2b and ribavirin for 6–12 months, 13 developed major depressive disorders requiring treatment with citalopram. The end-of-treatment response rates and sustained viral response rates were significantly greater in patients who developed a major depressive disorder than in those who did not (62% versus 27% and 39% versus 12% respectively). Despite the small sample, these results suggest that interferon alfa-induced depression occurs at doses or concentrations that are associated with a therapeutic effect (i.e. is a collateral adverse reaction) and that antidepressant therapy could allow patients to take an optimal dose.

The psychiatric effects of low-dose interferon alfa on brain activity have been assessed using functional MRI during a task of visuospatial attention in patients infected with hepatitis C virus [37]. Despite having symptoms of impaired concentration and fatigue, the 10 patients who received interferon alfa had similar task performance and activation of parietal and occipital brain regions to 11 control subjects infected with hepatitis C. However, in contrast to the controls, the patients who received interferon alfa had significant activation in the dorsal part of the anterior cingulate cortex, which correlated highly with the number of task-related errors; there was no such correlation in the controls. Consistent with the role of the anterior cingulate cortex in conflict monitoring, this activation of the anterior cingulate cortex suggests that interferon alfa might increase the processing of conflict or reduce the threshold for conflict detection, thereby signalling the need for mental effort to maintain performance.

Susceptibility factors

As regular psychiatric assessment is not always possible, the identification of easily detectable predictive factors of severe psychiatric disorders may help select which patients should undergo close psychiatric assessment. In 71 patients treated with interferon alfa alone or combined with ribavirin for chronic hepatitis C, female sex, scores on the MADRS at 4 months of treatment, sleep disorders, and prior antidepressant use were independent risk factors of suicidal behavior or depression [38]. This study also suggested that prolonged follow-up is required, as 8% of patients still had suicidal behavior 6 months after the end of treatment.

The most frequent susceptibility factor for interferon alfa-associated depression is a history of mood and anxiety symptoms before treatment. The roles of a previous history of depression, female sex, and high dose or long duration of interferon alfa treatment have been discussed [24]. High doses undoubtedly play a role in the development of depression, but the duration of treatment is not a consistently replicated susceptibility factor. Although it was initially found that a past history of depression might be a susceptibility factor, this has not been convincingly associated with an increased risk of interferon alfa-induced depression in further studies. Similar conclusions were reached in patients with a past history of drug or alcohol abuse, provided that patients remain abstinent during treatment. Female sex and age have also not emerged as consistent susceptibility factors.

Although depressive symptoms are usually ascribed to interferon alfa alone, ribavirin could also contribute to their occurrence. Among 162 patients with chronic hepatitis C treated with peginterferon alfa-2b (1.5 micrograms/kg once weekly) plus weight-based or standard-dose ribavirin and prospectively evaluated for the incidence and severity of depression, 39% developed moderate to severe depressive symptoms and six patients withdrew because of behavioral symptoms, such as depression, anxiety, or fatigue [39]. Depressive symptoms developed after 4 weeks of treatment and maximal depression scores were reached after 24 weeks. Baseline depression scores, a history of major depressive disorders, and higher doses of ribavirin were significant predictive factors of depression during treatment, in contrast to sex, age, or a history of substance abuse.

In 33 patients with chronic hepatitis C, of whom 10 developed major depressive disorders during interferon alfa treatment, there was no relation between changes in thyroid function and the development of depression [40]. No patients with depression had clinical hypothyroidism and the sole patient with overt hypothyroidism had no depressive symptoms.

A number of other factors, including genetic or biological factors, that are possibly associated with the development of psychiatric disorders, have been explored. In a retrospective study of 110 patients with chronic hepatitis C, those with the apolipoprotein E ε4 allele, the inheritance of which may be associated with several neuropsychiatric outcomes, were more likely than those without this allele to have psychiatric referrals and neuropsychiatric symptoms, in particular irritability, anger, anxiety, or other mood symptoms, during interferon alfa treatment [41]. In addition, patients with this allele had neuropsychiatric events sooner than patients without it. In another study in 14 patients with chronic hepatitis C, lower baseline serum activities of propylendopeptidase and dipeptyl peptidase IV, which both play a role in the pathophysiology of major depression, were possible predictors of higher depressive and anxiety scores during interferon alfa treatment [42].. Whether these factors help to predict susceptibility to interferon alfa-induced neuropsychiatric disorders in clinical practice needs to be further investigated.

Management

Some data suggest that selective serotonin reuptake inhibitors (SSRIs) are effective in treating interferon-induced depression, allowing patients to continue treatment [22,24]. Because interferon alfa increases serotonin reuptake and reduces serotonin synthesis, the addition of 5-hydroxytryptophan to a selective serotonin reuptake inhibitor has also been suggested to produce a synergistic response, but controlled trials are awaited [43].

The successful management of psychiatric symptoms during interferon alfa treatment has been detailed in 60 patients with chronic viral hepatitis, treated for 1 year [44]. A new psychiatric disease developed during treatment in 18 patients (depression in 12, predominant irritability in four, and anxiety in two), and five others had major depression at baseline. Based on the potential mechanisms of interferon alfa-induced psychiatric disorders, and depending on the clinical features, a variety of psychopharmacological treatments were used in 12 of these patients, including selective serotonin reuptake inhibitors in four cases, low-dose pre-synaptic D_2 dopamine receptor antagonists (sulpiride and amisulpride) in three, neuroleptic drugs in two, and benzodiazepines or related drugs in three. Only four patients failed to respond to treatment, and one had to be withdrawn from the study because of persistent irritability. During the study, the 12 patients taking psychiatric drugs had significantly less severe psychiatric symptoms than the 11 untreated patients.

Use in patients with pre-existing psychiatric disease

A prospective study in 81 patients with chronic hepatitis C taking interferon alfa-2a and ribavirin compared adherence to treatment, adverse effects, response rates, and dropout rates in 23 patients without any psychiatric history or drug addiction (control group), 16 patients with psychiatric disorders, 21 patients in a methadone substitution programme, and 21 patients with former intravenous drug addiction [45]. There were no significant differences between groups as regards the frequency and severity of new depressive episodes during treatment, but significantly more patients in the psychiatric groups had to take antidepressants before and during interferon alfa treatment. The rate of dropouts was significantly higher in the drug addiction group (43%) compared with the other groups (13–18%). The cause was psychiatric or somatic adverse effects in half of these patients. No patients in the psychiatric or methadone groups had to stop treatment because of psychiatric adverse effects. Although the limited number of patients and heterogeneity in psychiatric diagnoses precluded any definitive conclusion, these results suggested that treatment with interferon alfa and special psychiatric care is possible in at-risk psychiatric groups.

References

1. Galvani D, Griffiths SD, Cawley JC. Interferon for treatment: the dust settles. BMJ (Clin Res Ed) 1988;296(6636):1554–6.
2. Merigan TC. Human interferon as a therapeutic agent: a decade passes. N Engl J Med 1988;318(22):1458–60.
3. Agarwala SS, Kirkwood JM. Interferons in the therapy of solid tumors. Oncology 1994;51(2):129–36.
4. Urabe A. Interferons for the treatment of hematological malignancies. Oncology 1994;51(2):137–41.
5. Dorr RT. Interferon-alpha in malignant and viral diseases. A review. Drugs 1993;45(2):177–211.
6. Stiehm ER, Kronenberg LH, Rosenblatt HM, Bryson Y, Merigan TC. UCLA conference. Interferon: immunobiology and clinical significance. Ann Intern Med 1982;96(1):80–93.
7. Alexander GJ, Brahm J, Fagan EA, Smith HM, Daniels HM, Eddleston AL, Williams R. Loss of HBsAg with interferon therapy in chronic hepatitis B virus infection. Lancet 1987;2(8550):66–9.
8. Giles FJ, Singer CR, Gray AG, Yong KL, Brozovic M, Davies SC, Grant IR, Hoffbrand AV, Machin SJ, Mehta AB, et al. Alpha-interferon therapy for essential thrombocythaemia. Lancet 1988;2(8602):70–2.
9. Korenman J, Baker B, Waggoner J, Everhart JE, Di Bisceglie AM, Hoofnagle JH. Long-term remission of chronic hepatitis B after alpha-interferon therapy. Ann Intern Med 1991;114(8):629–34.
10. Scott GM, Ward RJ, Wright DJ, Robinson JA, Onwubalili JK, Gauci CL. Effects of cloned interferon alpha 2 in normal volunteers: febrile reactions and changes in circulating corticosteroids and trace metals. Antimicrob Agents Chemother 1983;23(4):589–92.
11. Ingimarsson S, Cantell K, Strander H. Side effects of long-term treatment with human leukocyte interferon. J Infect Dis 1979;140(4):560–3.
12. Cheeseman SH, Rubin RH, Stewart JA, Tolkoff-Rubin NE, Cosimi AB, Cantell K, Gilbert J, Winkle S, Herrin JT, Black PH, Russell PS, Hirsch MS. Controlled clinical trial of prophylactic human-leukocyte interferon in renal transplantation. Effects on cytomegalovirus and herpes simplex virus infections. N Engl J Med 1979;300(24):1345–9.
13. Smedley H, Katrak M, Sikora K, Wheeler T. Neurological effects of recombinant human interferon. BMJ (Clin Res Ed) 1983;286(6361):262–4.
14. Inglada L, Porres JC, La Banda F, Mora I, Carreno V. Anti-IFN-alpha titres during interferon therapy. Lancet 1987;2(8574):1521.

15. Steis RG, Smith JW 2nd, Urba WJ, Clark JW, Itri LM, Evans LM, Schoenberger C, Longo DL. Resistance to recombinant interferon alfa-2a in hairy-cell leukemia associated with neutralizing anti-interferon antibodies. N Engl J Med 1988;318(22):1409–13.

16. Konrad MW, Childs AL, Merigan TC, Borden EC. Assessment of the antigenic response in humans to a recombinant mutant interferon beta. J Clin Immunol 1987;7(5):365–75.

17. Roy V, Newland AC. Raynaud's phenomenon and cryoglobulinaemia associated with the use of recombinant human alpha-interferon. Lancet 1988;1(8591):944–5.

18. Taguchi T. Clinical studies of recombinant interferon alfa-2a (Roferon-A) in cancer patients. Cancer 1986;57(Suppl 8):1705–8.

19. Umeda T, Niijima T. Phase II study of alpha interferon on renal cell carcinoma. Summary of three collaborative trials. Cancer 1986;58(6):1231–5.

20. Mirro J, Dow LW, Kalwinsky DK, Dahl GV, Weck P, Whisnant J, Murphy SB. Phase I-II study of continuous-infusion high-dose human lymphoblastoid interferon and the in vitro sensitivity of leukemic progenitors in nonlymphocytic leukemia. Cancer Treat Rep 1986;70(3):363–7.

21. Muss HB, Costanzi JJ, Leavitt R, Williams RD, Kempf RA, Pollard R, Ozer H, Zekan PJ, Grunberg SM, Mitchell MS, et al. Recombinant alfa interferon in renal cell carcinoma: a randomized trial of two routes of administration. J Clin Oncol 1987;5(2):286–91.

22. Neri S, Pulvirenti D, Bertino G. Psychiatric symptoms induced by antiviral therapy in chronic hepatitis C: comparison between interferon-alpha-2a and interferon-alpha-2b. Clin Drug Investig 2006;26(11):655–62.

23. Hauser P. Neuropsychiatric side effects of HCV therapy and their treatment: focus on IFN alpha-induced depression. Gastroenterol Clin North Am 2004;33 (1 Suppl):S35-50.

24. Loftis JM, Hauser P. The phenomenology and treatment of interferon-induced depression. J Affect Disord 2004;82:175–90.

25. Raison CL, Demetrashvili M, Capuron L, Miller AH. Neuropsychiatric adverse effects of interferon-alpha: recognition and management. CNS Drugs 2005;19:105–23.

26. Onyike CU, Bonner JO, Lyketsos CG, Treisman GJ. Mania during treatment of chronic hepatitis C with pegylated interferon and ribavirin. Am J Psychiatry 2004;161:429–35.

27. Patten SB, Francis G, Metz LM, Lopez-Bresnahan M, Chang P, Curtin F. The relationship between depression and interferon beta-1a therapy in patients with multiple sclerosis. Mult Scler 2005;11:175–81.

28. Scheibel RS, Valentine AD, O'Brien S, Meyers CA. Cognitive dysfunction and depression during treatment with interferon-alpha and chemotherapy. J Neuropsychiatry Clin Neurosci 2004;16:185–91.

29. Kraus MR, Schafer A, Wissmann S, Reimer P, Scheurlen M. Neurocognitive changes in patients with hepatitis C receiving interferon alfa-2b and ribavirin. Clin Pharmacol Ther 2005;77:90–100.

30. Kraus MR, Schafer A, Wissmann S, Reimer P, Scheurlen M. Neurocognitive changes in patients with hepatitis C receiving interferon alfa-2b and ribavirin. Clin Pharmacol Ther 2005;77(1):90–100.

31. Kraus MR, Schafer A, Faller H, Csef H, Scheurlen M. Psychiatric symptoms in patients with chronic hepatitis C receiving interferon alfa-2b therapy. J Clin Psychiatry 2003;64:708–14.

32. Schaefer M, Schwaiger M, Pich M, Lieb K, Heinz A. Neurotransmitter changes by interferon-alfa and therapeutic implications. Pharmacopsychiatry 2003;36 Suppl 3:S203-6.

33. Capuron L, Neurauter G, Musselman DL, Lawson DH, Nemeroff CB, Fuchs D, Miller AH. Interferon-alfa-induced changes in tryptophan metabolism. Relationship to depression and paroxetine treatment. Biol Psychiatry 2003;54:906–14.

34. Van Gool AR, Fekkes D, Kruit WH, Mulder PG, Ten Hagen TL, Bannink M, Maes M, Eggermont AM. Serum amino acids, biopterin and neopterin during long-term immunotherapy with interferon-alfa in high-risk melanoma patients. Psychiatry Res 2003;119:125–32.

35. Schaefer M, Horn M, Schmidt F, Schmid-Wendtner MH, Volkenandt M, Ackenheil M, Mueller N, Schwarz MJ. Correlation between sICAM-1 and depressive symptoms during adjuvant treatment of melanoma with interferon-alpha. Brain Behav Immun 2004;18:555–62.

36. Loftis JM, Socherman RE, Howell CD, Whitehead AJ, Hill JA, Dominitz JA, Hauser P. Association of interferon-alpha-induced depression and improved treatment response in patients with hepatitis C. Neurosci Lett 2004;365:87–91.

37. Capuron L, Pagnoni G, Demetrashvili M, Woolwine BJ, Nemeroff CB, Berns GS, Miller AH. Anterior cingulate activation and error processing during interferon-alpha treatment. Biol Psychiatry 2005;58(3):190–6.

38. Gohier B, Goeb JL, Rannou-Dubas K, Fouchard I, Cales P, Garre JB. Hepatitis C, alfa interferon, anxiety and depression disorders: a prospective study of 71 patients. World J Biol Psychiatry 2003;4:115–18.

39. Raison CL, Borisov AS, Broadwell SD, Capuron L, Woolwine BJ, Jacobson IM, Nemeroff CB, Miller AH. Depression during pegylated interferon-alpha plus ribavirin therapy: prevalence and prediction. J Clin Psychiatry 2005;66:41–8.

40. Loftis JM, Wall JM, Linardatos E, Benvenga S, Hauser P. A quantitative assessment of depression and thyroid dysfunction secondary to interferon-alpha therapy in patients with hepatitis C. J Endocrinol Invest 2004;27:RC16-20.

41. Gochee PA, Powell EE, Purdie DM, Pandeya N, Kelemen L, Shorthouse C, Jonsson JR, Kelly B. Association between apolipoprotein E epsilon4 and neuropsychiatric symptoms during interferon alpha treatment for chronic hepatitis C. Psychosomatics 2004;45:49–57.

42. Maes M, Bonaccorso S. Lower activities of serum peptidases predict higher depressive and anxiety levels following interferon-alpha-based immunotherapy in patients with hepatitis C. Acta Psychiatr Scand 2004;109:126–31.

43. Turner EH, Blackwell AD. 5-Hydroxytryptophan plus SSRIs for interferon-induced depression: synergistic mechanisms for normalizing synaptic serotonin. Med Hypotheses 2005;65:138–44.

44. Maddock C, Baita A, Orru MG, Sitzia R, Costa A, Muntoni E, Farci MG, Carpiniello B, Pariante CM. Psychopharmacological treatment of depression, anxiety, irritability and insomnia in patients receiving interferon-alpha: a prospective case series and a discussion of biological mechanisms. J Psychopharmacol 2004;18:41–6.

45. Schaefer M, Schmidt F, Folwaczny C, Lorenz R, Martin G, Schindlbeck N, Heldwein W, Soyka M, Grunze H, Koenig A, Loeschke K. Adherence and mental side effects during hepatitis C treatment with interferon alfa and ribavirin in psychiatric risk groups. Hepatology 2003;37:443–51.

Interferon alfa

See also Interferons

General Information

Interferon alfa is used as purified natural leukocyte or lymphoblastoid human interferon, or as a recombinant DNA preparation. Work to assign the most frequently observed amino acids at each position has led to a so-called consensus interferon alfa (1). Relatively low doses (3–10 MU three times a week) are now being used in most indications, except in AIDS-related Kaposi's sarcoma (up to 30 MU/day). Although the half-life is only 4–5 hours, its biological effects persist for 2–3 days.

The adverse effects of interferon alfa and ribavirin in combination have been reviewed, with particular emphasis on the management of depressive disorders and hematological adverse effects (2).

Observational studies

Considerable efforts have been made to improve the efficacy of interferon alfa in patients with chronic hepatitis C. Currently used regimens, including long-term interferon alfa alone or in combination with ribavirin, produce a sustained response rate of 40–50%. Other possibly effective strategies include a longer duration of treatment, higher fixed doses, and high-dose induction (3). A longer duration of treatment has been evaluated in patients with chronic hepatitis B. In 118 patients, treatment for 32 rather than 16 weeks enhanced the virological response to hepatitis B without increasing the severity of adverse effects, except for hair loss, which was more frequent during prolonged therapy (4).

A wide range of persistent symptoms has been reported during interferon alfa treatment for chronic hepatitis C. An analysis of 222 patients from the USA and France, enrolled in a multicenter trial, suggested that pretreatment symptoms were an important predictor of moderate or severe (defined as debilitating) adverse effects during treatment with interferon alfa (5). Compared with baseline, the incidences of moderate and severe fatigue, myalgia, arthralgia, headache, dry eyes, and dry mouth increased significantly after 6 months. In each case, the development of these debilitating adverse effects was associated with the presence of that symptom at baseline. They were more often reported in patients who received interferon alfa daily than three times weekly, and in US than French patients, suggesting possible differences in cultural attitudes toward illness. There was also increased use of antidepressants during the 6-month survey.

Low daily doses of interferon alfa have been used with small doses of cytarabine in the treatment of early chronic myelogenous leukemia (6). With doses sufficient to obtain a good cytogenetic response (for example 3.7 MU/m^2/day plus cytarabine 7.5 mg/day) toxicity was considered acceptable. There was significant fatigue in 43% of cases, significant neurological changes in 27%, weight loss in 19%, and oral ulcers in 4%.

Comparative studies

Interferon alfa, in combination with ribavirin, is currently first-line therapy for patients with chronic hepatitis C and compensated liver disease, and its use has been extensively reviewed (7). A meta-analysis of trials in patients who were previously non-responsive to interferon alfa alone showed that treatment withdrawal for an adverse event was more frequent in patients who received combination therapy (8.8%) compared with interferon alfa monotherapy (4%) (8). However, treatment withdrawal is more frequent in practice. In a retrospective analysis of 441 consecutive patients treated with interferon alfa and ribavirin, 25% of patients discontinued treatment because of adverse effects (9). The study identified female sex, a dose of interferon alfa above 15 MU/week, and naive patients as independent susceptibility factors for premature withdrawal.

The safety profile of ribavirin plus human leukocyte interferon alfa was more favorable than ribavirin plus interferon alfa-2b in a randomized, controlled study of 423 therapy-naive patients with chronic hepatitis C infection (10). Whereas patients in the two groups achieved similar virological responses to treatment, there were fewer adverse events in those who were given human leukocyte interferon alfa and fewer patients withdrew because of adverse events or laboratory abnormalities compared with interferon alfa-2b (4% versus 11%). In particular, the flu-like syndrome was observed in 52 and 75% of patients respectively, and there were no severe adverse effects in either group. Conversion to human leukocyte interferon alfa may be therefore considered in patients with poor tolerability of recombinant interferon alfa.

The safety of retreatment with leukocyte interferon alfa has been confirmed in 43 patients with chronic hepatitis C intolerant of previous recombinant or lymphoblastoid interferon alfa monotherapy (11). Only five patients had to discontinue interferon alfa owing to the reappearance of the same adverse effect in four and the occurrence of a new adverse effect in one. In six other patients, dosage reduction was required because of the recurrence of the same adverse effect.

Fatigue is a frequent adverse effect that may require dosage reduction during the first months of treatment. The efficacy of L-carnitine on interferon alfa-induced fatigue has been studied in 50 patients with chronic hepatitis C (12). The patients who were given L-carnitine (2 g/day) plus interferon alfa had less severe fatigue and an earlier reduction of physical and mental fatigue levels. It was therefore concluded that carnitine supplementation may ameliorate this adverse effect.

In 50 patients treated with peginterferon alfa-2b and ribavirin for 1 year and carefully examined at each 4–6 week visit by a physician trained in pharmacovigilance, 10 patients developed a serious adverse effect (13). Compared with clinical trials, blurred vision, injection site reactions, and endocrine adverse effects were more frequent, and depression and insomnia were less frequent with peginterferon; hair loss occurred with similar frequency.

Comparisons of different forms of interferon alfa

In a randomized comparison of recombinant interferon alfa-2b and interferon alfa n-3 (9 MU/week for 1 year) in 168 naive patients with chronic hepatitis C, there was no significant difference in clinical outcomes and the incidence or type of adverse effects between the groups (14). There was a non-significant trend toward more severe leukopenia and a higher incidence of severe thyroid disorders in patients who received recombinant interferon alfa-2b.

Pegylated interferon alfa-2a is a modified form of interferon alfa; it produces higher serum concentrations and has greater efficacy. In 1530 patients with chronic hepatitis C, pegylated interferon alfa-2b had a similar profile of adverse effects to unmodified interferon alfa-2b, but with more frequent dose-limiting neutropenia (15). Two other studies have shown that peginterferon alfa-2a once weekly is more effective than unmodified interferon alfa-2a three times weekly in patients with chronic hepatitis C (16,17). The frequency and severity of adverse effects with the two treatments were very similar and were consistent with the known adverse effects of interferon alfa. In one study, a neutrophil count below $0.5 \times 10^9/l$ was more frequent with peginterferon alfa than with unmodified interferon alfa (12/265 versus 4/261), but none of these patients required treatment withdrawal or had serious infections in relation to neutropenia (17). In the other study, the proportion of patients who required dosage modification because of thrombocytopenia was also higher with peginterferon alfa (18 versus 6%), but no patients had clinically significant bleeding disorders (16). Taken together, these studies suggest that pegylated interferon alfa may produce more frequent or more severe hemotoxic effects than unmodified interferon alfa.

In 1530 patients with chronic hepatitis C, pegylated interferon alfa-2b had a similar profile of adverse effects to unmodified interferon alfa-2b, but with more frequent dose-limiting neutropenia (15). No particular adverse effect has emerged since the use of this new formulation.

General adverse effects

The adverse effects of interferon alfa have mostly been reported after systemic administration, as intranasal use was not associated with more frequent adverse effects than placebo (18). Almost all patients treated with interferon alfa have experienced adverse effects, most of which are mild to moderate in intensity and easily manageable without withdrawal of treatment (19). The incidence and profile of adverse effects reported with the available types of interferon alfa are very similar (SEDA-21, 369; SEDA-22, 399), but they differ with the dose, schedule of administration, and the disease. At least 4–5% of patients with chronic hepatitis C had to discontinue treatment because of adverse effects, and dosage reduction was required in 9–22% of those receiving 9–15 MU/week (20). In a large retrospective evaluation of 11 241 patients with chronic viral hepatitis, the incidence of fatal or life-threatening adverse effects related to interferon alfa was one in 1000; events included irreversible

liver failure, severe bone marrow depression, and attempted suicide (21). Overall, severe adverse effects were observed in 1% of patients, and comprised mostly thyroid disorders, neuropsychiatric manifestations, and cutaneous adverse effects. In other studies in patients with chronic hepatitis, the incidence of major adverse effects was 25% in 659 patients (22); in Japan, dosage reduction or withdrawal was necessary in 31% of 987 patients receiving relatively high dosages of interferon alfa (18–70 MU weekly) (23). The safety of interferon alfa in children appears to be similar to that in adults (24). The pathogenesis of most adverse effects observed with interferon alfa is poorly understood, but two main mechanisms are commonly postulated, namely a direct toxic effect or an indirect immune-mediated effect.

During the first days of treatment, virtually all patients have a flu-like syndrome with fever, chills, tachycardia, malaise, headache, arthralgias and myalgias, but tachyphylaxis usually develops after 1–2 weeks of treatment (25). Late febrile reactions are rarely noted (26). Although the severity increases with the dose, the flu-like syndrome is rarely treatment-limiting and it can be partly prevented by the prophylactic administration of paracetamol (acetaminophen). The acute release of fever-promoting factors, for example the eicosanoids, IL-1, and TNF alfa, secondary to interferon alfa is the suggested mechanism.

Although the adverse effects profiles of the currently available formulations of interferon alfa are very similar, patients who have adverse effects with one formulation can be successfully re-treated with another type of interferon alfa. This has been shown in 22 patients in whom lymphoblastoid interferon alfa was withdrawn because of severe adverse effects (leukopenia, thrombocytopenia, thyroid disorders, and psychiatric disturbances) and were successfully re-treated with similar dosages of leukocyte interferon alfa (27). Only one of these patients had severe leukopenia again.

Organs and Systems

Cardiovascular

Hypotension or hypertension, benign sinus or supraventricular tachycardia, and rarely distal cyanosis, have been reported within the first days of treatment in 5–15% of patients receiving high-dose interferon alfa (25). These adverse effects are usually benign, except in high-risk patients with a previous history of dysrhythmias, coronary disease, or cardiac dysfunction.

Cardiac complications

Severe or life-threatening cardiotoxicity is infrequent and mostly reported in the form of a subacute complication in patients with cancer, and in those with pre-existing heart disease or receiving high-dose interferon alfa. Atrioventricular block, life-threatening ventricular dysrhythmias, pericarditis, dilated cardiomyopathy, cardiogenic shock, asymptomatic or symptomatic myocardial ischemia or even infarction, and sudden death have been observed (SED-13, 1091; SEDA-20, 326; SEDA-22, 369) (28). The combination of high-dose interleukin-2 (IL-2) with interferon alfa enhanced cardiovascular complications, namely cardiac ischemia and ventricular dysfunction (29).

Cardiomyopathy has been attributed to interferon alfa in an infant (30).

- A 3-month-old boy was given interferon alfa (2.5–5.5 MU/m^2) for chronic myelogenous leukemia. After 7.5 months he developed progressive respiratory distress, with anorexia, irritability, and nocturnal sweating. A chest X-ray showed cardiomegaly, an echocardiogram showed a markedly dilated left ventricle, and an electrocardiogram showed left ventricular hypertrophy with abnormal repolarization. Viral cultures and serology for cytomegalovirus, parvovirus B19, and enterovirus were negative. Infectious diseases and metabolic disturbances were excluded. Interferon alfa was withdrawn and digoxin, furosemide, and an angiotensin-converting enzyme inhibitor were given. One year later, he was asymptomatic without further cardiac treatment.

Similar, but anecdotal reports were also described in patients without evidence of previous cardiac disease and receiving low-dose interferon alfa (SEDA-20, 326). In chronic viral hepatitis, only seven of 11 241 patients had severe cardiac adverse effects (21). The exact risk of such cardiovascular adverse effects is unknown. In patients with chronic viral hepatitis, cardiovascular test results were not modified when patients were re-examined after at least 6 months of treatment, even where there was an earlier cardiac history (31), but there was a potentially critical reversible reduction in left ventricular ejection of more than 10% in another prospective study (32).

Myocardial dysfunction can completely reverse after withdrawal of interferon alfa and does not exclude further treatment with lower doses (33).

- A 47-year-old man with renal cancer and no previous history of cardiovascular disease developed gradually worsening exertional dyspnea after he had received interferon alfa in a total dose of 990 MU over 5 years. Echocardiography and a myocardial CT scan confirmed a dilated cardiomyopathy, with left ventricle dilatation and diffuse heterogeneous perfusion at rest. He improved after interferon alfa withdrawal and treatment with furosemide, quinapril, and digoxin. Myocardial scintigraphy confirmed normal perfusion. He restarted low-dose interferon alfa (6 MU/week) 1 year later and had no recurrence of congestive heart failure after a 1-year follow-up period.

Acute myocardial infarction occurred about 12 hours after the administration of pegylated interferon alfa-2b 1 microgram/kg in a healthy 76-year-old woman during a single-dose pharmacokinetic study in 24 patients, of whom 18 were over 65 years (34). Although the case

was poorly documented, the close temporal relation strongly suggested a causal role of the drug.

Pulmonary hypertension has previously been reported in one patient and was briefly mentioned as a potential complication of interferon alfa in two other patients who also had multiple ulcers involving the feet or toes (35). Other vascular events including Raynaud's phenomenon, digital ulcerations and gangrene, pulmonary vasculitis, and thrombotic thrombocytopenic purpura are occasionally reported (32).

Pericarditis has been attributed to interferon alfa (36).

- A 40-year-old woman with chronic hepatitis C developed pericarditis without the typical features of a lupus-like syndrome after 4 months of treatment with low-dose interferon alfa-2b in (37). She simultaneously developed a polyneuropathy. Both disorders disappeared after interferon withdrawal.

- A 42-year-old woman with a history of atrioventricular septal defect and atrial fibrillation was given interferon alfa-2b for chronic hepatitis C. Pre-treatment echocardiography showed moderate mitral valve disease and left atrial hypertrophy but no pericardial anomalies. Six hours after her first injection of interferon alfa (3 MU) she had an acute episode of thoracic pain, which disappeared spontaneously within 24 hours. Echocardiography showed a moderate non-constrictive pericardial effusion. Similar symptoms recurred 7 hours after a subsequently reduced dose of interferon alfa (1 MU), and echocardiography again showed a reversible pericardial effusion. Autoimmune diseases, cryoglobulinemia, and myocardial infarction were ruled out. Interferon alfa was withdrawn and no recurrence of pericarditis was noted over 5 years of follow-up.

This seems to have been the first case of direct interferon alfa-induced pericarditis, because lupus-like syndrome and mixed type 2 cryoglobulinemia with vasculitis were deemed to be the direct cause in previously reported cases.

Patients with pre-existing cardiac disease are more likely to develop cardiovascular toxicity while receiving interferon alfa, but these complications are rare. Among 89 patients with chronic hepatitis C, 12-lead electrocardiography monthly during a 12-month treatment period and follow-up for 6-months showed only minimal and non-specific abnormalities in five patients (two had right bundle branch block, one left anterior hemiblock, and two unifocal ventricular extra beats) (38). None of these disorders required treatment withdrawal, and complete non-invasive cardiovascular assessment was normal. Overall, the role of interferon alfa was uncertain and the 5.6% incidence of electrocardiographic abnormalities was suggested to be similar to that expected in the general population. Nevertheless, severe cardiac dysrhythmias are still possible in isolated cases, as illustrated by the development of third-degree atrioventricular block, reversible on withdrawal, in a 57-year-old man with lower limb arteritis but no other cardiovascular disorder (39).

Peripheral vascular complications
Raynaud's phenomenon
Raynaud's phenomenon can occur, particularly in patients with chronic myelogenous leukemia (SEDA-20, 326) (40), and severe cases were complicated by digital necrosis (SEDA-20, 326) (SEDA-21, 369).

In 24 cases of Raynaud's syndrome, interferon alfa was the causative agent in 14, interferon beta in 3, and interferon gamma in 5 (41). There was no consistent delay in onset and the duration of treatment before the occurrence of symptoms ranged from 2 weeks to more than 4 years. The most severe cases were complicated by digital artery occlusion and necrosis requiring amputation. Few patients had other ischemic symptoms, such as myocardial, ophthalmic, central nervous system, or muscular manifestations. Severe Raynaud's phenomenon was also reported in a 5-year-old girl with hepatitis C (42).

Cryoglobulinemia
Although most patients with mild-to-moderate clinical manifestations of hepatitis C virus-associated mixed cryoglobulinemia improved during treatment with interferon alfa, acute worsening of ischemic lesions has been reported in three patients who had prominent cryoglobulinemia-related ischemic manifestations (43). All three had acute progression of pre-existing peripheral ischemia or leg ulcers within the first month of treatment, and transmetatarsal or right toe amputations were required in two. The lesions healed after interferon alfa withdrawal. It was therefore suggested that the anti-angiogenic activity of interferon alfa may also impair revascularization and healing of ischemic lesions in patients with initially severe ischemic manifestations.

Venous thrombosis
Whereas clinically insignificant coagulation abnormalities have been documented in patients receiving high-dose continuous interferon alfa (44), isolated cases of venous thrombosis have been observed (SEDA-20, 329) (45). Interferon alfa can also induce the production of antiphospholipid antibodies (SEDA-20, 329) (SEDA-21, 371). In one study, antiphospholipid antibodies were found in five of 12 patients with melanoma treated with interferon alfa alone or with interferon alfa plus interleukin-2; deep venous thrombosis occurred in four patients with antiphospholipid antibodies (46). Although the underlying neoplasia undoubtedly played a role in the further development of venous thrombosis, the causative role of interferon alfa was suggested by the absence of antiphospholipid antibodies and venous thrombosis in eight patients treated with interleukin-2 alone.

Other vascular complications
Other anecdotal reports included acrocyanosis and peripheral arterial occlusion (SED-13, 1095) (SEDA-22, 400). Although the causal relation is unclear, interferon alfa was considered as a possible cause in the triggering of acute cerebrovascular hemorrhage or ischemic neurological symptoms in few patients (SEDA-21, 370) (SEDA-22, 400). The pathogenic mechanisms of these vascular effects are still unclear; vasculitis, hypercoagulability, vasospasm, a paradoxical anginal effect of interferon alfa, or an underlying cardiovascular disease have all been suggested as underlying processes.

Respiratory
The respiratory adverse effects of interferon alfa include interstitial pneumonitis (47), which is rare. Since the first description of interstitial pneumonitis associated with interferon alfa in Japanese patients who also used the popular "Sho-saiko-t" herbal formulation, similar cases have been described in Western patients, suggesting that interferon alfa can be the sole cause in some patients (SED-13, 1091; SEDA-20, 326; SEDA-21, 370) (48,49).

Interferon alfa was also suspected to be involved in one case of biopsy-proven bronchiolitis obliterans-organizing pneumonia (50). Clinical symptoms of pneumonitis appeared 3–12 weeks after the onset of interferon alfa therapy, and after withdrawal of treatment they usually completely resolved, either spontaneously or after a short course of glucocorticoid treatment. Immune-mediated pulmonary toxicity involving the activation of T cells was considered as a likely mechanism. The uncommon features of bronchiolitis obliterans-organizing pneumonia have been reported in three other patients who received interferon alfa together with ribavirin or cytosine arabinoside (51,52).

In four patients with hematological malignancies who developed symptoms suggestive of pneumonitis (that is a dry cough and dyspnea) after 1 week to 38 months of interferon alfa treatment, there was a marked reduction in carbon monoxide diffusion capacity in all cases, whereas there were pathological findings in ordinary chest X-rays in only two (53). What the authors called ultracardiography and high-resolution CT scanning were suggested to have higher sensitivity in evaluating pulmonary symptoms. In three patients, complete reversal was obtained after interferon alfa withdrawal, either spontaneously or after corticosteroid treatment, although one patient required long-term glucocorticoid treatment.

Among 558 Japanese patients with chronic hepatitis C treated with or without ribavirin, interstitial pneumonitis was found in six (1.1%) patients aged 54-66 years (54). One woman had two reversible episodes of interstitial pneumonitis 9 years apart with two different forms of interferon. Two patients were also taking Sho-saiko-to, a herbal medicine previously involved in interstitial pneumonitis. Serum concentrations of KL-6, a lung epithelium-specific protein, were increased at the onset of interstitial pneumonitis in all cases. Serum KL-6 concentrations were increased before starting interferon alfa in five of the seven episodes and in only three of 48 age-adjusted patients (6%) who did not develop interstitial pneumonitis during interferon alfa treatment. The authors suggested that measurement of serum KL-6 concentrations may be helpful not only in diagnosing interferon alfa-induced interstitial pneumonia, but also before starting interferon alfa for detecting patients who are more likely to develop interstitial pneumonia during treatment. It is also tempting to speculate that increased concentrations of serum KL-6 before treatment indicate subclinical hepatitis C-related interstitial pneumonitis, which can be triggered by subsequent interferon alfa administration.

- A 49-year-old man taking pegylated interferon alfa-2b and ribavirin developed interstitial pneumonitis with subsequent adult respiratory distress syndrome after 2 weeks of treatment (55). He died on day 26 from sepsis and multiorgan failure. At necropsy his lungs were five times the normal weight, with diffuse alveolar damage and advanced fibrosis.

Whether the pharmacokinetics of pegylated interferon alfa accounted for the rapid onset or a more severe outcome is speculative, but prompt withdrawal of treatment is advised in cases of unexplained pulmonary symptoms.

- A 65-year-old man developed bronchiolitis obliterans organizing pneumonia with fulminant respiratory distress 48 hours after starting treatment with interferon alfa, oral cytarabine ocfosfate, and hydroxyurea for chronic myelogenous leukemia (56). He died 5 days later.
- A 60-year-old man developed the first symptoms of an acute interstitial pneumonia during the second week of interferon alfa and 5-fluorouracil treatment for hepatocellular carcinoma (57). He died 3 weeks later.

Although the suspected treatment was promptly withdrawn after the onset of pulmonary symptoms, both patients died despite antibiotic and glucocorticoid pulse administration. As they also received chemotherapeutic agents known to cause interstitial pneumonitis, the exact role of interferon alfa cannot be determined, but there may be an increased risk of very severe pulmonary toxicity attributable to the drug combination.

- A 71-year-old woman developed reversible interstitial pneumonia after 5 months of peginterferon alfa and ribavirin treatment for chronic hepatitis C (58). Lung biopsy showed a granulomatous process.

In this patient there was convincing evidence of a causal relation, since her symptoms improved after withdrawal and recurred after readministration. The authors briefly reviewed the spectrum of pulmonary toxicity associated with interferon alfa from 32 other published cases in patients with chronic hepatitis C. Three cases involved peginterferon, but 10 patients had also received ribavirin. There was interstitial pneumonitis in 15 patients, sarcoid-like reactions in 10, bronchiolitis obliterans organizing pneumonia in three, asthma exacerbation in two, pleural effusion in one, and acute respiratory distress in one. All the patients survived after interferon alfa withdrawal, except one who was treated with peginterferon alfa and who died from an acute respiratory distress syndrome.

In a retrospective review of 70 patients with hepatitis C enrolled in four clinical trials, there were four cases of significant pulmonary toxicity (two of bronchiolitis obliterans and two of interstitial pneumonitis) (59). Three recovered completely, but one still required glucocorticoids for exertional dyspnea that persisted 17 months after interferon alfa withdrawal. The authors suggested that there was an increased risk with high-dose interferon, because three of these patients received high doses (5 MU/day) or pegylated interferon alfa. In contrast, they were unaware of any significant pulmonary toxicity in any of their approximately 500 patients with hepatitis C.

Interferon alfa can cause exacerbation of asthma (SED-14, 1248).

- Acute exacerbation of asthma has been reported in two men aged 27 and 57 years with a previous history of mild asthma (60). They developed progressive aggravation of asthma within 8–10 weeks of treatment for chronic hepatitis C, and finally required emergency treatment with systemic glucocorticoids and inhaled beta$_2$ adrenoceptor agonists. Severe asthma recurred 2–3 weeks after interferon alfa rechallenge in both patients.

Although these cases are anecdotal, they strongly suggest that interferon alfa should be regarded as a possible cause of asthma exacerbation in predisposed patients.

- Sustained and isolated dry cough has been attributed to interferon alfa in a 49-year-old woman (61). The symptoms disappeared after withdrawal, recurred on readministration, and again resolved after withdrawal. No other cause was found after thorough investigation.

Pleural effusion has been attributed to interferon alfa (62).

- A 54-year-old man received interferon alfa-2a, 9 MU/day, for chronic hepatitis C. He developed an asymptomatic right pleural effusion after 14 days. Although his serum titer of antinuclear antibodies was slightly increased, a more complete screening for autoimmune disease was negative. An infectious origin was also ruled out. The pleural effusion spontaneously disappeared after interferon alfa withdrawal and did not recur.

Although the mechanism of this adverse effect was purely speculative, it was suggested that interferon alfa might have induced a reaction similar to the immunopathological mechanism involved in serositis associated with systemic lupus erythematosus.

Pulmonary artery hypertension has been attributed to interferon alfa (63).

- A 23-year-old man taking hydroxycarbamide 1.5 g/day and interferon alfa-2b (less than 10 MU/day) for chronic lymphocytic leukemia had progressive dyspnea and a non-productive cough after about 5 months. The electrocardiogram showed axis deviation, incomplete right bundle-branch block, and right ventricular hypertrophy. The estimated pulmonary artery pressure by echocardiography was 80 mmHg and there were signs of right heart failure. Respiratory function tests showed a restrictive defect, and the chest X-ray showed pulmonary congestion without infiltrates. The patient's clinical status and respiratory function tests improved rapidly after withdrawal of interferon alfa while hydroxycarbamide was continued, and a mean pulmonary artery pressure of 34 mmHg was measured by right heart catheterization 1 month later. At 6 months, the systolic pulmonary artery pressure estimated by echocardiography had fallen to 35 mmHg and the electrocardiogram returned to normal after 1 year.

The authors mentioned that intravenous interferon alfa in sheep had caused an increase in pulmonary artery pressure.

Nervous system

Diffuse electroencephalographic slowing, which reversed after the completion of treatment, has previously been found in patients being treated for chronic hepatitis C (SEDA 24, 411). In a prospective study of the effect of age on the electroencephalogram in 98 patients, alterations became more marked as age increased (64).

Peripheral neuropathy
Dose-related distal paresthesia occurred in as many as 7% of patients (25), but new onset or worsening of neuropathy has rarely been attributed to interferon alfa. A sensorimotor polyneuropathy was the most frequent presentation (65). The symptoms usually developed after 2–28 weeks of treatment. Such reports were mostly described in patients who received high cumulative doses of interferon alfa, but induction or exacerbation of peripheral sensorimotor axonal neuropathy, particularly in patients with chronic hepatitis C and mixed cryoglobulinemia, was also observed after long-term or low-dose treatment (SED-13, 1092) (SEDA-20, 327) (SEDA-21, 370) (66). Nerve biopsy showed necrotizing vasculitis or axonal degeneration. Most patients stabilized or improved slowly over several months after interferon alfa withdrawal and/or treatment with corticosteroids. Several patients also required plasmapheresis or cyclophosphamide. Although several authors have suggested an autoimmune process, the underlying pathogenic mechanism is unclear.

Other rare forms of neuropathy (SED-13, 1092) include mononeuropathy multiplex (67), acute axonal polyneuropathy (68), anterior ischemic optic neuropathy (69), trigeminal sensory neuropathy (70), bilateral neuralgic amyotrophy (71), brachial plexopathy (72), and symptoms suggestive of leukoencephalopathy (73).

Interferon can cause multifocal leukoencephalopathy and sensorimotor polyneuropathy, in which case long-term interferon therapy should be reviewed.

- A 77-year-old man developed a progressive exercise-induced gait disturbance with right-sided predominance, non-systemic vertigo, muscle pain in both thighs after exercise, restless legs, stocking-type sensory disturbances bilaterally, and hypesthesia in the left arm (74). He had received intramuscular interferon alfa 2b 4.5–5 million units 3 times weekly for 18 years for hairy cell leukaemia. A brain MRI scan showed multiple non-enhancing T2-hyperintense lesions supratentorially, interpreted as vascular lesions. He was also taking pramipexole, levodopa, cabergoline, diltiazem, molsidomine, finasteride, and allopurinol. There was

bilateral hypacusis, bilateral dysmetria, exaggerated biceps tendon reflexes on the right side, reduced biceps reflexes on the left, discrete proximal weakness in the left arm, proximal weakness in both legs, reduced ankle reflexes on the right, extensor plantar reflexes bilaterally, left-sided hemihypesthesia, stocking-type sensory disturbances bilaterally, and a tendency to fall during Romberg's test and Unterberger's treadmill test. Serum IgG antibodies against *Borrelia burgdorferi* were positive, but the cerebrospinal fluid was negative.

Cranial nerve palsies
There have been reports of interferon alfa-induced cranial nerve palsies, including Bell's palsy.

Two patients, one of whom also received ribavirin, had facial nerve palsy after 5 and 8 months of interferon alfa therapy (75). The palsy resolved completely in one patient after withdrawal and the administration of prednisolone; however, in the other case, the palsy resolved without drug withdrawal, suggesting coincidence.

Two other cases of Bell's palsy, which reversed after interferon alfa withdrawal, have been reported (76). Although the delay in onset (7.5 and 8 weeks) suggested that interferon alfa might be the cause, one patient had no recurrence after rechallenge. A coincidental adverse event cannot therefore be completely ruled out.

Bell's palsy occurred in two patients aged 42 and 49 years with chronic hepatitis C (77). These patients, who also took ribavirin, developed Bell's palsy after 2-3 months of treatment. They recovered fully after withdrawal, but prednisone was also required in one. Whether interferon alfa played a direct role is questionable, because the first patient subsequently restarted the same treatment without recurrence, and the second patient had had an attack of *Herpes simplex* stomatitis shortly before the cranial nerve deficit occurred.

- Three men aged 43 to 55 years developed Bell's palsy after 4 days, 5 weeks, and 4 months of treatment with interferon alfa and ribavirin for chronic hepatitis C (78). The Bell's palsy spontaneously resolved within 6 weeks after withdrawal and none of the patients was rechallenged.

Although a coincidental adverse effect was possible, the authors suggested that immunotherapy may have produced a breakdown in peripheral tolerance to myelin sheath antigens.

Demyelination
Various forms of interferon alfa-induced neuropathy have been reported (SED-14, 1249), but chronic inflammatory demyelinating polyneuropathy has seldom been described (79,80).

- In two patients with chronic hepatitis C or malignant melanoma, paresthesia and tiredness occurred after 6 weeks and 9 months of treatment respectively. Despite withdrawal of interferon alfa, the neurological symptoms worsened initially and a diagnosis of chronic inflammatory demyelinating polyneuropathy was finally confirmed several weeks later. One patient improved after an extended course of plasma exchange and the other required immunoglobulins and prednisolone. Mild to moderate neurological abnormalities persisted at follow-up in both patients.

Multiple sclerosis has been attributed to interferon alfa (81).

- A 29-year-old woman received interferon alfa-2b (6–10 MU/day) for chronic myeloid leukemia, and about 3 years later developed headaches, back pain, progressive visual disturbance, and a sensory deficit in the legs. MRI scans of the brain and spinal cord suggested a first episode of multiple sclerosis. In addition, the myelin basic protein concentration was slightly raised, and perimetry showed bilateral optic neuritis. Most of her neurological symptoms, except central vision impairment, improved after interferon alfa withdrawal and treatment with high-dose methylprednisolone. As a major partial cytogenetic response had been obtained with interferon, she was given natural interferon alfa (3 MU/day) 2 months later. After 2 days of treatment, she complained of transient but severe pain in the back and the legs, and developed acute paraplegia and loss of micturition desire. Again, interferon was withdrawn and she was given high-dose methylprednisolone. There was no further neurological deterioration at follow-up.

This account is reminiscent of the various autoimmune diseases that can be unmasked or exacerbated by interferon alfa.

Features resembling multiple sclerosis have been reported in two women aged 29 and 62 years, treated with recombinant interferon alfa for chronic hepatitis C or chronic myelogenous leukemia (82). Both developed bilateral optic neuritis. Other symptoms included increased deep tendon reflexes and reduced sensation of vibration in one patient, and numbness of the legs and bowel and bladder dysfunction in the other. MRI findings and cerebrospinal fluid examination simulated multiple sclerosis. Similar manifestations were again noted in one patient after readministration of natural interferon alfa, and both partially recovered visual acuity after glucocorticoid treatment.

Other demyelinating events attributed to interferon alfa have included a report of Guillain–Barré syndrome 2 months after the end of a course of interferon alfa-2b and suramin sodium for metastatic pulmonary cancer in a 37-year-old woman (83). Although the concomitant use of suramin made the association of interferon alfa with the Guillain-Barré syndrome unclear, the authors suggested that the simultaneous occurrence of autoimmune liver disease and hematological changes in this patient argued for a causal role of interferon alfa, and they also suggested that this combination may have had a synergistic effect.

Movement disorders

Extrapyramidal effects have been occasionally reported as a manifestation of persistent neurotoxicity induced by interferon alfa. This issue has been addressed in a report of severe refractory akathisia (84).

- A 28-year-old man received interferon alfa (5 MU/day for 28 days) for chronic hepatitis B. At the end of treatment he developed a slight parkinsonian gait, and 8 days later had a fever with vomiting, insomnia, restlessness, and raised serum creatine kinase activity (4946 IU/l). He had severe akathisia with psychomotor excitement and parkinsonism. Despite treatment with clonazepam, thioridazine, propranolol, trihexyphenidyl, and bromocriptine, his condition progressively worsened. He was finally given intravenous levodopa for 8 days and recovered dramatically within the next few days.

This report, together with previous experimental data, suggests that levodopa might be useful in alleviating some of the manifestations of persistent interferon alfa-induced neurotoxicity.

In two other patients, akathisia occurred shortly after they had started to take interferon alfa; one improved after the frequency of injections was reduced (85). Unfortunately, this report did not provide sufficient convincing evidence for a causal relation; the development of akathisia may have been coincidental.

Chorea is a very rare manifestation of interferon alfa neurotoxicity (SEDA-20, 327).

- A 68-year-old woman developed progressive personality changes and 2 months later permanent choreic movements of the four limbs (86). She had taken interferon alfa-2b (3 MU/day) and hydroxyurea 50 mg/kg/day for chronic myeloid leukemia for 2 years and had no history of psychiatric disorders. Neuropsychological testing showed frontal subcortical dysfunction. There were no abnormalities in the Huntington disease gene. She progressively worsened over the next 6 months. The electroencephalogram was disorganized, with diffuse slow waves, and she was bedridden. Interferon alfa was withdrawn. The chorea ceased 1 month later and she completely recovered cognitive function. Electroencephalography was normal 6 and 12 months later.

The authors attributed these events to antidopaminergic effects of interferon alfa.

Severe and complex diffuse spinal myoclonus developed after 6 weeks of interferon alfa-2a (27 MU/week) in a 63-year-old man with metastatic renal cancer, but his symptoms were only partially improved after interferon withdrawal and anticonvulsant treatment (87). The exact role of interferon was therefore only speculative.

Involuntary facial movements have been more convincingly described (88).

- A 49-year-old woman progressively developed involuntary facial movements after taking interferon alfa-2a (6 MU/day) for 1.5 years for chronic myeloid leukemia. She had bilateral, involuntary, rhythmic, repetitive contractions of her frontalis, nasalis, orbicularis oris, and lower facial muscles. The movements worsened with anxiety and lack of sleep and during talking. Electromyography showed rhythmic synchronous muscle contractions compatible with myorhythmia.

Neurological examination was otherwise normal and no other causes were found. The facial movements completely resolved 1 month after interferon withdrawal, and recurred a few weeks after interferon rechallenge.

Interferon alfa-induced acute dystonia has been reported (89).

- A 24-year-old man with chronic hepatitis B who had tolerated interferon alfa for 6 months 9 years before, developed acute involuntary movements with fever and chills 2 hours after his first dose of peginterferon alfa. Electroencephalography and an MRI scan were normal. His symptoms resolved quickly after diazepam injection and he was given peginterferon again uneventfully.

In the context of an acute dystonia, the authors claimed that the development of tolerance to the second injection of interferon alfa further argued for a causal role of the drug.

Restless legs syndrome occurred 2 months after a 60-year-old man started to take interferon alfa and ribavirin for chronic hepatitis C (90). Evidence for a causal relation was obtained by complete resolution of the syndrome after withdrawal, careful elimination of other causes, and rapid recurrence of the symptoms on rechallenge with peginterferon alfa and ribavirin. Although both interferon alfa and ribavirin may have been the cause of the syndrome, the authors correctly argued that only interferon alfa is thought to affect dopaminergic neurotransmission, which is strongly suspected to be involved in the pathophysiology of restless legs syndrome.

Seizures

Interferon alfa can cause seizures in patients with no history of epilepsy, and one report of three patients suggested that seizures can occur at any time during treatment (91). Seizures were also reported in two of five children after intrathecal interferon alfa for subacute sclerosing panencephalitis, suggesting a much higher incidence in this condition (92). Although generalized tonic-clonic seizures have occasionally been described during trials of high doses of interferon alfa, they have also been reported after the use of intermediate or even low doses (93–95). There was a 1.3% incidence of generalized seizures in a retrospective study of 311 patients treated with low doses for chronic viral hepatitis (96). In another study, tonic-clonic seizures were identified in 4% of children treated for chronic hepatitis B (97). As seizures occurred only in children under 5 years of age with fever or potential perinatal nervous system injury, immaturity of the nervous system was suggested to be an additional factor for interferon alfa-induced neurotoxicity in children.

- In three patients, seizures occurred after a cumulative dose of 266–900 MU (98,99). Two were retreated with a lower dose and remained free of seizures, so that the strength of the causal relation was debatable (98). However, a dose-related effect was still possible.
- Another patient with a history of bipolar mood disorder experienced had his first four episodes of seizures

with a prolonged delirious state 1 week after withdrawal of interferon alfa (99).

Reversible photosensitive seizures have also been reported (100).

- A 62-year-old man without a personal or family history of epilepsy received interferon alfa (3 MU three times a week) for 2 years for multiple myeloma. He had frequent episodes of myoclonic jerks in the face, especially when the sun was shining while driving his car. He also had one generalized seizure. Electroencephalography showed a paroxysmal response to intermittent photic stimulation and magnetic resonance imaging was normal. The seizures disappeared and his electroencephalogram normalized after interferon alfa withdrawal.

In this case, the possible role of interferon alfa was suspected only late during treatment, indicating that patients should be regularly questioned about neurological symptoms, because more severe complications might have occurred.

In a prospective study of the effect of interferon alfa (56 MU/day for 4 weeks then 27 MU/week for 20 weeks) in 56 patients with chronic hepatitis C, there was diffuse electroencephalographic slowing at 2 and 4 weeks of treatment, suggesting mild encephalopathy (101). These changes completely reversed after withdrawal. However, the dosage used in this Japanese study was relatively high compared with the dosages currently used in Western countries. In addition, the clinical relevance of these electroencephalographic changes was not investigated.

Neuromuscular function

A number of reports have confirmed that interferon alfa can induce or unmask underlying silent myasthenia gravis (SED-13, 1097) (SEDA-20, 327) (SEDA-21, 370). The diagnostic criteria for myasthenia gravis were clearly fulfilled in these reports, and an autoimmune reaction was the most likely mechanism, as each patient had positive serum anti-acetylcholine receptor antibodies and required permanent anticholinesterase drugs long after interferon alfa withdrawal. Myasthenia gravis developed in two patients treated with interferon alfa-2b for chronic hepatitis C, one of whom also took ribavirin (102,103). Both had an increase in acetylcholine receptor antibody titers and required permanent pyridostigmine and immunosuppressant treatment. These findings suggest that interferon alfa does not cause myasthenia gravis but unmasks it.

Sensory systems

Eyes

Ophthalmic disorders can occur during interferon alfa treatment (104–110) and some of the literature has been reviewed (111). Retinopathy consisting of cotton-wool spots and/or superficial retinal hemorrhages has been reported with a variable incidence (18–86%), and the available data suggest that the increased incidence can be influenced by an initial high dose of interferon alfa. Whereas diabetes mellitus and systemic hypertension have been identified as possible susceptibility factors, the incidence of retinopathy was not significantly increased in 19 patients with chronic renal insufficiency, compared with 17 patients without chronic renal insufficiency (112). However, it was felt that renal insufficiency may be associated with the most severe cases, that is those requiring dosage reductions.

In one study, in which prospective ophthalmic examinations were made before and at regular 2-week intervals after the beginning and end of treatment, 28 of 81 patients who received a uniform total dose of natural interferon alfa (478 MU) for chronic hepatitis C developed the typical findings of interferon-induced asymptomatic retinopathy (cotton-wool spots and/or retinal hemorrhages) (110). In contrast, there were no lesions in the 25 patients with chronic hepatitis C who did not receive interferon alfa or in the 20 with diabetes mellitus and/or hypertension but without chronic hepatitis C. Most of the cases were observed within 4 months of treatment, and the lesions always abated after withdrawal or even despite continuation, suggesting that treatment can be continued unless patients develop symptoms. Indeed, most patients with retinopathy associated with interferon alfa remained asymptomatic.

Of 42 patients taking interferon alfa-2b and ribavirin for chronic hepatitis C and regularly followed for ocular complications, 27 developed a retinopathy with cotton-wool spots (n = 27), retinal hemorrhages (n = 6), subconjunctival hemorrhage (n = 2), and asymptomatic optic nerve edema (n = 1) (113). These lesions were usually transient, but some had recurrent episodes of retinopathy while continuing therapy. Although the retinopathy was considered benign in most cases, two patients complained of ocular symptoms consisting of blurring of vision in one and permanent peripheral monocular scotoma in the other. Three patients had to discontinue treatment because of more severe posterior segment complications. In contrast to several studies, the incidence of retinopathy was not dose related and hypertension was not identified as a significant risk factor.

- A 37-year-old man developed bilateral ischemic retinopathy and blurred vision and recovered promptly after peginterferon alfa-2b withdrawal (114).
- A 44-year-old woman developed visual difficulties in dim light, with abnormal electro-oculography; she slowly improved after peginterferon alfa withdrawal (115). She had previously tolerated standard interferon alfa for 6 months.
- A 46-year-old man developed visual complaints with predominant blurry vision and the classical cotton-wool spots of interferon-induced retinopathy after treatment with peginterferon alfa-2b; he recovered after 5 months despite continued treatment (116).

The last report also provided a complete detailed review of 27 reports of interferon alfa-related ocular abnormalities.

The pathogenesis of retinopathy associated with interferon alfa is unclear. In 45 patients with chronic hepatitis C (25 treated with interferon) there was an association

between retinal hemorrhages caused by interferon alfa (six patients) and a concomitant significant increase in plasma-activated complement (C5a), compared with baseline C5a serum concentrations (117). However, the signification and contribution of raised C5a concentrations in the pathogenesis of ocular complications needs to be clarified, although it has been suggested that retinal hemorrhage could be predicted on the basis of raised C5a concentrations (118).

Although most patients with interferon alfa retinopathy remain asymptomatic, ocular complications, such as reduced vision or complete visual loss due to occlusive vasculitis, central retinal artery occlusion, or anterior ischemic optic neuropathy, continue to be reported in a very few patients (SED-13, 1096) (108,111,119–123). However, subclinical but eventually long-lasting or even irreversible abnormally long visual evoked responses have been identified in 24% of patients (SEDA-22, 403). Regular ophthalmological monitoring to detect retinal changes, even though the patient is still asymptomatic, is therefore strongly recommended in patients receiving interferon alfa.

Owing to the possible risk of severe retinopathy, regular ophthalmic monitoring before treatment and at regular intervals during treatment is advocated by most authors. That interferon alfa-induced retinopathy is not always self-limiting and benign has been supported by the development of severe proliferative retinopathy in a 69-year-old woman with chronic myeloid leukemia who required extensive laser therapy and unilateral vitrectomy (124). Other severe posterior ocular complications recently described in patients with chronic hepatitis C treated with interferon alfa-2b have included:

- the development of an epiretinal membrane in a 47-year-old man (125)
- permanent visual loss attributed to combined retinal artery and central retinal vein obstruction in a 51-year-old woman (126);
- bilateral simultaneous anterior ischemic optic neuropathy in two patients (127).

Whether retinopathy is more frequent with peginterferon than with standard interferon is still unknown. In one prospective series of 23 patients co-infected with HIV and hepatitis C, who were treated with peginterferon alfa-2b and ribavirin, eight developed classical interferon retinopathy and two had impaired color vision, of whom one was sufficiently severely affected to require drug withdrawal (128). Although the incidence of retinopathy in this small study was within the expected range, the occurrence of two symptomatic cases is noteworthy.

Vogt–Koyanagi–Harada disease, an autoimmune syndrome with potentially severe ocular complications, has been reported (129).

- A 52-year-old woman developed visual blurring, eye pain, tinnitus, and alopecia after taking interferon alfa-2b 9 MU/week and ribavirin 1200 mg/day for 9 months. Further examination showed poliosis of the left eyelashes and reduced bilateral auditory acuity. Ophthalmic examination showed severe bilateral panuveitis with posterior synechiae, vitritis, and choroidal and scleral thickening. The antiviral treatment was withdrawn and she received local and systemic glucocorticoids.

The authors suggested that Vogt–Koyanagi–Harada disease, the pathophysiology of which involves a T cell-mediated autoimmune response to melanocytes, may have been exacerbated by interferon alfa in a genetically predisposed patient. Vogt–Koyanagi–Harada disease has also been observed after 4 months of interferon alfa-2b treatment in a 36-year-old man with chronic hepatitis C (130). Withdrawal and methylprednisolone therapy was followed by slow but complete recovery.

Other isolated reports of ophthalmic abnormalities refer to optic neuritis with blurring of vision, cortical blindness with fatal encephalopathy, mononeural abducent nerve paralysis, and complete but reversible bilateral oculomotor nerve paralysis (SED-13, 1096) (131–134).

- A 60-year-old smoker was treated with interferon alfa (100 MU/week for 2 months and 9 MU/week for 15 weeks) for cutaneous melanoma. Ocular examination was normal before treatment, but he developed acute loss of peripheral vision in his left eye after 23 weeks. Examination was consistent with anterior ischemic optic neuropathy, and there was optic disc edema, a pupillary defect, and circular visual field constriction in the left eye. There was renal artery constriction in both eyes. Despite treatment with aspirin, high-dose dexamethasone, heparin, and finally withdrawal of interferon alfa, loss of visual function progressed and affected both eyes. Ciclosporin was started, but he was considered to have irreversible loss of visual function.

This report shows that interferon alfa can be a potent precipitator of extremely severe ocular disorders and also argues for careful ocular surveillance in patients receiving adjuvant interferon alfa for high-risk resected melanoma.

Of 57 patients treated with interferon for renal cell carcinoma, two developed multiple retinal exudates associated with visual disturbance; both had taken vinblastine concurrently. The precise role of interferon in this reaction is unknown (135).

Ears

In 49 patients, there was reversible otological impairment with tinnitus, mild-to-moderate hearing loss or both in respectively 8, 16, and 20% of patients after interferon alfa or interferon beta administration (SEDA-19, 336). These disorders tended to occur more frequently in patients on high cumulative doses, but led to withdrawal of treatment in only two patients. Complete but reversible hearing loss, and acute unilateral vestibular dysfunction with spontaneous vertigo and nystagmus have each been reported in one patient receiving interferon alfa (SEDA-21, 372).

Of 27 patients treated with interferon alfa 30 MU/week for chronic hepatitis B and prospectively assessed for auditory function, hearing loss was found in nine after 7 days of treatment, with an increase in the degree of hearing loss until day 21 (136). However, auditory disability did not exceed 20 dB for any frequency despite continued treatment. Complete recovery was observed 1 month after withdrawal. More severe sudden hearing loss and tinnitus have been reported in six patients treated with pegylated interferon alfa and ribavirin (137). Symptoms occurred after 1 day or several weeks of treatment, suggesting that hearing loss can occur at any time. Although symptoms did not worsen on continued treatment in two patients, hearing loss persisted or improved minimally after withdrawal in four.

Sudden hearing loss has been reported (SEDA-21, 372) and in one case of promptly reversible hearing loss on interferon alfa withdrawal, the presence of anti-endothelial cell antibodies was suggested to have played a role in the development of autoimmune microvascular damage (138).

Sudden hearing loss and tinnitus occurred in six patients treated with peginterferon alfa (mostly alfa-2b); the onset of symptoms after the start of therapy was 1 day in one patient and 4–40 weeks in the other five (139). The authors estimated that about 1% of patients would experience this event. The ototoxicity was attributed to reversible biochemical and metabolic changes in the cochlea, but could also have been caused by an autoimmune mechanism with antiendothelial cell antibodies, which were found in nearly half of the patients. Autoimmune hearing loss occurs later in the course of treatment than biochemical and metabolic hearing loss, is mostly bilateral, and has a progressive course over the ensuing days and months. In contrast to direct ototoxicity caused by standard interferon, that caused by peginterferon does not recover fully after withdrawal. However, once symptoms appear they do not progress, and so the decision to continue or abandon treatment has to be individualized.

Taste

In a 40-year-old man taking interferon alfa for chronic hepatitis C ageusia and hyposmia resulted in loss of appetite and weight loss, but reversed on withdrawal (140).

Olfaction

Patients treated with interferon alfa sometimes complain of transient taste or smell alterations, but anosmia has been reported (SEDA-22, 403) (141).

- A 37-year-old man received interferon alfa for chronic hepatitis C. After 2 weeks he complained of smelling difficulties and subsequently developed complete anosmia. There were no other neurological symptoms and complete neurological examination was normal. Anosmia still persisted 13 months after drug withdrawal.
- In both patients, the persistence of anosmia late after interferon alfa was resumed indicates that a causal relation to treatment is purely speculative.

Psychological, psychiatric (*See also Interferons*)

Neuropsychiatric complications of interferon alfa were recognized in the early 1980s and represent one of the most disturbing adverse effects of interferon alfa (SED-13, 1091; SEDA-20, 327; SEDA-22, 400). Reviews have provided comprehensive analysis of the large amount of experimental and clinical data that have accumulated since 1979 (142,143).

DoTS classification (BMJ 2003; 327:1222–5)
Dose-relation: collateral effect
Time-course: intermediate or delayed
Susceptibility factors: pre-existing psychiatric disorders, organic brain injury, or addictive behavior

Presentation

Within a large spectrum of symptoms, complications are classified as acute, subacute, or chronic.

Acute neuropsychological disturbances are usually associated with the flu-like syndrome and include headache, fatigue, and weakness, drowsiness, somnolence, subtle impairment of memory or concentration, and lack of initiative (25). This pattern of cognitive impairment is similar to changes observed during influenza and has also been described in healthy patients who have received a single dose of interferon alfa (144). More severe acute manifestations (for example, marked somnolence or lethargy, frank encephalopathy with visual hallucinations, dementia or delirium, and sometimes coma) have been almost exclusively described in patients receiving more than 20–50 MU (25); vertigo, cramps, apraxia, tremor and dizziness were also reported.

The subacute or chronic neuropsychiatric effects of long-term therapy are usually non-specific, with cognitive impairment (for example visuospatial disorientation, attentional deficits, memory disturbances, slurred speech, difficulties in reading and writing), changes in emotion, mood, and behavior (for example psychomotor slowing, hypersomnia, loss of interest, affective disorders, irritability, agitation, delirium, paranoia, aggressiveness, and murderous impulses). Post-traumatic stress symptoms have also been reported (SEDA-22, 400). As a result, severe psychic distress can be observed during long-lasting treatment or in patients who are otherwise not severely affected (25,145). The most severe psychiatric complications of interferon alfa include rare cases of homicidal ideation, suicidal ideation, and attempted suicide (146).

The clinical features of mania have been described in four patients with malignant melanoma, with a detailed review of nine other published cases (147). Although seven suffered from depression during treatment, the onset of mania or hypomania was often associated with interferon alfa dosage fluctuation (withdrawal or dose reduction) or introduction of an antidepressant for interferon alfa-induced depression. In these patients, the risk of mood fluctuations persisted for several months after interferon alfa withdrawal, and low-dose gabapentin was considered useful in treating manic disorders and in preventing mood fluctuations. Interferon alfa was suggested as a possible cause of persistent manic-depressive illness for more than 4 years in a 40-year-old man (148). Although the manic episodes may have been coincidental,

the negative history and the age of onset are in keeping with a possible role of interferon alfa treatment.

The clinical features, management, and prognosis of psychiatric symptoms in patients with chronic hepatitis C have been reviewed using data from 943 patients treated with interferon alfa (85%) or interferon beta (15%) for 24 weeks (149). Interferon-induced psychiatric symptoms were identified in 40 patients (4.2%) of those referred for psychiatric examination. They were classified in three groups according to the clinical profile: 13 cases of generalized anxiety disorder (group A), 21 cases of mood disorders with depressive features (group B), and six cases of other psychiatric disorders, including psychotic disorders with delusions/hallucinations ($n = 4$), mood disorders with manic features ($n = 1$), and delirium ($n = 1$) (group C). The time to onset of the symptoms differed significantly between the three groups: 2 weeks in group A, 5 weeks in group B, and 11 weeks in group C. Women were more often affected than men. There was no difference in the incidence or nature of the disorder according to the type of interferon used. Whereas most patients who required psychotropic drugs were able to complete treatment, 10 had to discontinue interferon treatment because of severe psychiatric symptoms, 5 from group B and five from group C. Twelve patients still required psychiatric treatment for more than 6 months after interferon withdrawal. In addition, residual symptoms (anxiety, insomnia, and mild hypothymia) were still present at the end of the survey in seven patients. Delayed recovery was mostly observed in patients in group C and in patients treated with interferon beta. Although several patients with a previous history of psychiatric disorders are sometimes successfully treated with interferon alfa, severe decompensation with persistent psychosis should be regarded as a major possible complication (150).

The neuropsychological adverse effects of long-term treatment have been assessed in 14 patients with myeloproliferative disorders using a battery of psychometric and electroneurological tests before and after 3, 6, 9, and 12 months of treatment (median dose 25 MU/week) (151). In contrast to several previous studies, there was no significant impairment of neurological function, and attention and short-term memory improved during treatment. Despite the small number of patients, these results suggest that prolonged interferon alfa treatment did not cause severe cognitive dysfunction, at least in patients with cancer.

Diagnosis

Electroencephalographic (EEG) findings show reversible cerebral changes with slowing of dominant alpha wave activity, and occasional appearance of one and two activity in the frontal lobes, suggesting a direct effect on fronto-subcortical functions. Marked electroencephalographic abnormalities are sometimes observed in asymptomatic patients. The pattern of changes is identical whatever the dose, but the severity of symptoms is dose- and schedule-related. Most patients improve or recover after dosage reduction or withdrawal, and protracted toxicity, with impaired memory, deficits in motor coordination, persistent frontal lobe executive functions, Parkinson-like tremor, and mild dementia, have been occasionally reported (152).

In a study of 67 patients with chronic viral hepatitis, the self-administered Minnesota Multiphasic Personality Inventory (MMPI), which determines the patient's psychological profile, significantly correlated with the clinical evaluation and was a sensitive and reliable tool for identifying patients at risk of depressive symptoms before the start of interferon alfa therapy (153). It was also successfully used to monitor patients during treatment.

Frequency

The most typical psychiatric symptoms reported by patients taking interferon alfa are depressive symptoms, at rates of 10–40% in most studies (154–157). In four clinical studies in a total of 210 patients with chronic hepatitis C, the rate of major depressive disorders during interferon alfa treatment was 23–41% (158–161).

Suicidal ideation or suicidal attempts have been reported in 1.3–1.4% of patients during interferon alfa treatment for chronic viral hepatitis or even within the 6 months after withdrawal (162,163), but the excess risk related to interferon alfa is not known.

Time-course

Subacute or chronic neuropsychiatric manifestations are more typically identified after several weeks of treatment and are among the most frequent treatment-limiting adverse effects (145,156,164–166). The onset can be insidious in patients treated with low doses, or subacute in those who receive high doses. Most patients develop severe depressive symptoms within the first 3 months of treatment (158–161).

Although psychiatric manifestations usually appear during interferon alfa therapy, delayed reactions can occur.

- A 37-year-old man without a previous psychiatric history developed major depression with severe psychotic features within days after the discontinuation of a 1-year course of interferon alfa-2b (167).

In 10 patients with melanoma and no previous psychiatric disorders, depression scores measured on the Montgomery-Asberg Depression Rating Scale were significantly increased after 4 weeks of high-dose interferon alfa (168). Patients whose scores were higher before treatment developed the worst symptoms of depression during treatment. This positive correlation provides striking evidence that baseline and regular assessment of mood and cognitive functions are necessary to detect disorders as early as possible.

Mechanisms

Although very few studies have specifically investigated the role of the underlying disease, the findings of significant neuropsychiatric deterioration during interferon alfa treatment compared with placebo or no treatment in chronic hepatitis C, chronic myelogenous leukemia, or amyotrophic lateral disease strongly suggested a causal role of interferon alfa (169–171).

The mechanism by which the systemic administration of interferon alfa produces neurotoxicity is unclear, and might result from a complex of direct and indirect effects involving the brain vasculature, neuroendocrine system, neurotransmitters and the secondary cytokine cascade with cytokines which exert effects on the nervous system, for example interleukin-1, interleukin-2, or tumor necrosis factor alfa (172). Whether a clinical effect is directly mediated through the action of a given cytokine or results from a secondary pathway through the induction of other cytokines or second messengers is difficult to determine.

A study in 18 patients treated with interferon alfa for chronic hepatitis C has given insights into the possible pathophysiological mechanism of depression (173). Depression rating scales, plasma tryptophan concentrations, and serum kynurenine and serotonin concentrations were measured at baseline and after 2, 4, 16, and 24 weeks of treatment with interferon alfa 3–6 MU 3–6 times weekly. During treatment, tryptophan and serotonin concentrations fell significantly, while kynurenine concentrations rose significantly. Depression rating scales also rose from baseline after the first month of treatment, with continued increases thereafter. In addition, there was a relation between increased scores of depression and changes in serum kynurenine and serotonin concentrations. These changes suggested a predominant role of the serotonergic system in the pathophysiological mechanisms of interferon alfa-associated depression. Accordingly, 35 of 42 patients included in three open trials of antidepressant treatment responded to a selective serotonin reuptake inhibitor drug, such as citalopram or paroxetine, and were able to complete interferon treatment (160,174,175).

Susceptibility factors

Various possible susceptibility factors have been analysed in several studies (159–161,176). Sex, the dose or type of interferon alfa (natural or recombinant), a prior personal history of psychiatric disease, substance abuse, the extent of education, the duration and severity of the underlying chronic hepatitis, and scores of depression before interferon alfa treatment were not significantly different between patients with and without interferon alfa-induced depression. Advanced age was suggested to be a risk factor in only one study (161). Although a worsening of psychiatric symptoms was noted during treatment in 11 patients receiving psychiatric treatment before starting interferon, only one was unable to complete the expected 6-month course of interferon alfa and ribavirin therapy (159).

Of 91 patients treated with interferon alfa-2b and low-dose cytarabine for chronic myelogenous leukemia, 22 developed severe neuropsychiatric toxicity (177). Their symptoms consisted mostly of severe depression or psychotic behavior, which resolved on withdrawal in all patients. The time to toxicity ranged from as early as 2 weeks to as long as 184 weeks after the start of treatment. Five of six patients had recurrent or worse symptoms after re-administration of both drugs. Several baseline factors were analysed, but only a pretreatment history of neurological or psychiatric disorders was considered to be a reliable risk factor. Severe neuropsychiatric toxicity developed during treatment in 63% of patients with previous neuropsychiatric disorders compared with 10% in patients without. It is unlikely that the combination of interferon alfa-2b with low-dose cytarabine potentiated the neuropsychiatric adverse effects of interferon alfa in this study. Indeed, previous experience with this combination, but after exclusion of patients with a psychiatric history, was not associated with such a high incidence of neuropsychiatric toxicity or any significant difference in toxicity between interferon alfa alone and interferon alfa plus low-dose cytarabine.

Patients receiving high doses of interferon alfa or long-term treatment are more likely to develop pronounced symptoms (164). A previous history of psychiatric disorders, organic brain injury, or addictive behavior are among potential susceptibility factors, but worsening of an underlying psychiatric disease is not the rule, provided that strict psychiatric surveillance and continuation of psychotropic drugs are maintained (178). Other putative susceptibility factors include the intraventricular administration of interferon alfa, previous or concomitant cranial irradiation, asymptomatic brain metastases, and pre-existing intracerebral ateriosclerosis (SED-13, 1092) (156,179–183). Despite early findings, co-infection with HIV has not been confirmed to be a susceptibility factor (SEDA-20, 327).

The occurrence of psychiatric disorders has been prospectively investigated in 63 patients who received a 6-month course of interferon alfa (9 MU/week) for hepatitis C (184). All were assessed at baseline with the Structured Clinical Interview for DSM-III-R (SCID) and monitored monthly with the Hopkins Symptoms Checklist (SCL-90). Most had a history of alcohol or polysubstance dependence, and 12 had a lifetime diagnosis of major depression. There were no significant changes in the SCL-90 scores during the 6-month period of survey in the 49 patients who completed the study, even in those who had a lifetime history of major depression. At 6 months, there was probable minor depression in eight patients and major depression in one; none had attempted suicide.

In a prospective study, 50 patients with chronic hepatitis B or C who received 18–30 MU/week of natural or recombinant interferon alfa were followed for 12 months (185). The SCID before starting interferon alfa identified 16 patients with a current psychiatric diagnosis and eight with a previous psychiatric disorder; 26 patients free of any psychiatric history constituted the control group. Psychiatric manifestations during treatment occurred in 11 patients (five from the control group), major depression in five, depressive disorders in three, severe dysphoria in two, and generalized anxiety disorder in one. Most of them were successfully treated with psychological support and drug therapy. Overall, 20 patients interrupted interferon alfa (10 in each group), including three for psychiatric adverse effects, but patients with a pre-existing or recent psychiatric diagnosis were no more likely to withdraw from treatment than the controls.

Of 33 patients with chronic hepatitis C treated with interferon alfa, 9 MU/week for 3–12 months, prospectively evaluated using the Montgomery-Asberg Depression Rating Scale (MADRS) before and after 12 weeks of treatment, eight developed depressive symptoms, of whom four had major depression without a previous psychiatric history (186). All four recovered after treatment with antidepressants. This study confirmed that a high baseline MADRS is significantly associated with the occurrence of depressive symptoms.

These studies have confirmed that previous psychiatric disorders are not necessarily a contraindication to a potentially effective treatment. However, patients with depressive symptoms immediately before treatment are still regarded at risk of severe psychiatric deterioration with treatment (154).

Management
The management of the psychiatric complications of interferon alfa has not been carefully investigated, but multiple approaches are theoretically possible. Various pharmacological and non-pharmacological interventions have been discussed (187), and prompt intervention should be carefully considered in every patient who develops significant neuropsychiatric adverse effects while receiving interferon alfa. Depending on the clinical manifestations, proposed treatment options include antidepressants, psychostimulants, or antipsychotic drugs.

Based on a possible reduction in central dopaminergic activity mediated by the binding of interferon alfa to opioid receptors, naltrexone has been proposed as a means of improving cognitive dysfunction (164).

Selective serotonin re-uptake inhibitors have been advocated as the drugs of choice to allow completion of interferon alfa treatment (154), but that was based on very limited experience and the unproven assumption that SSRIs are safe in patients with underlying liver disease. The preliminary results of a double-blind, placebo-controlled study showed that 2 weeks of pretreatment with paroxetine significantly reduced the occurrence of major depression in 16 patients on high-dose interferon alfa for malignant melanoma (188). In a placebo-controlled trial, the preventive effects of paroxetine (mean maximal dose of 31 mg) were studied in 40 patients with high-risk malignant melanoma and interferon alfa-induced depression (189). Treatment started 2 weeks before adjuvant high-dose interferon alfa. Paroxetine significantly reduced the incidence of major depression (45% in the placebo group and 11% in the paroxetine group) and the rate of interferon alfa withdrawal (35 versus 5%). Although the number of patients was small and the duration of the survey short (12 weeks), this suggests that paroxetine effectively prevents the risk of depressive disorders in patients eligible for high-dose interferon alfa. However, these results are limited, because patients with melanoma who receive adjuvant high-dose interferon alfa are particularly likely to develop depression. The safety of prophylaxis with paroxetine also requires additional data, because three patients taking paroxetine developed retinal hemorrhages, including one with irreversible loss of vision.

- In contrast to this study, a 31-year-old woman with major depressive disorder, which responded to paroxetine and trazodone, had progressive recurrence of mood disorders after the introduction of interferon alfa for essential thrombocythemia (190).

This suggests that interferon alfa can also reverse the response to antidepressants.

Endocrine

- Autoimmune polyglandular syndrome with progressive thyroid autoimmunity, type 1 diabetes mellitus, amenorrhea, and adrenal insufficiency has been reported in a 51-year-old woman treated with interferon alfa for chronic hepatitis C (191). Pancreas and pituitary gland autoantibodies, which were undetectable before interferon alfa treatment, were present at the time of diagnosis. After withdrawal, she recovered normal thyroid function, but was still insulin dependent with amenorrhea and adrenal insufficiency.

Pituitary

Interferon alfa can stimulate the hypothalamic–pituitary–adrenal axis, with a marked increase in cortisol and adrenocorticotrophic hormone secretion after acute administration (SED-13, 1093). No further stimulation was observed after several weeks of treatment, pointing to possible down-regulation of the ACTH secretory system. As a result, long-term treatment with interferon alfa is not thought to influence pituitary hormones significantly, and the concentration of several hormones, for example calcitonin, LH, FSH, prolactin, growth hormone, ACTH, cortisol, testosterone, and estradiol, were not modified by prolonged interferon alfa treatment (192,193). No clinical endocrinopathies attributable to such disorders in the regulation of these hormones have yet been reported.

Although the rate of growth was significantly lower than predicted in 35% of children receiving long-term treatment for recurrent respiratory papillomatosis (194), only one case of growth retardation has been reported in other settings (SEDA-19, 335). A significant reduction in weight and nutritional status was observed during treatment with interferon alfa for 6 months for chronic viral hepatitis in children aged 4–16 years, but this was transient and not associated with growth impairment (195).

Hypopituitarism has rarely been reported in patients with chronic hepatitis C.

- A 44-year-old woman developed severe weakness after 2 months of interferon alfa 9 MU/week (196). Interferon was withdrawn and subsequent treatment with 4.5 MU/week produced similar symptoms within 3 months. Serum cortisol concentrations were dramatically reduced and plasma ACTH was undetectable. Other hormonal concentrations were within the reference ranges and adrenal cortex antibodies were negative. Biological anomalies normalized after interferon alfa withdrawal and 6-month substitutive hydrocortisone treatment.

- A 39-year-old man received interferon alfa-2b 9 MU/week and ribavirin 400 mg/day for 1 year, and amantadine 200 mg/day for 9 months (197). He had major weight gain (23 kg), reduced libido, and neuropsychiatric disturbances during treatment, and there was only partially improvement 1 year after the completion of treatment. There was testosterone, gonadotrophin, and growth hormone deficiency, but antipituitary antibodies were not detected. Although his symptoms markedly improved with recombinant human growth hormone, permanent pituitary impairment was suspected.

The authors suggested that in the second case pituitary dysfunction might have caused persistent symptoms after interferon alfa treatment. Although involvement of the immune system was suggested in both cases, autoantibodies were not detected.

Reversible hypopituitarism with antibodies to pituitary GH3 cells and exacerbation of Sheehan's syndrome have been reported (SED-13, 1093; SEDA-21, 371).

A syndrome resembling inappropriate antidiuretic hormone secretion has been described in a few patients receiving high-dose interferon alfa (SED-13, 1093) (198).

Thyroid

Since the original 1988 report of hypothyroidism in patients with breast cancer receiving leukocyte-derived interferon alfa (199), numerous investigators have provided clear clinical and biological data on thyroid disorders induced by different forms of interferon in patients with various diseases (26,200–203). Two of these reports also mentioned associated adverse effects that developed concomitantly, namely myelosuppression and severe proximal myopathy (Hoffmann's syndrome).

Presentation and outcomes

The spectrum of interferon alfa-induced thyroid disorders ranges from asymptomatic appearance or increase in antithyroid autoantibody titers to moderate or severe clinical features of hypothyroidism, hyperthyroidism, and acute biphasic thyroiditis. Antithyroid hormone antibodies have also been found in one patient, and this could have been the cause of erroneously raised thyroid hormone concentrations (204).

The clinical, biochemical, and thyroid imaging characteristics of thyrotoxicosis resulting from interferon alfa treatment have been retrospectively analysed from data on 10 of 321 patients with chronic hepatitis (75 with chronic hepatitis B and 246 with chronic hepatitis C) who developed biochemical thyrotoxicosis (205). Seven patients had symptomatic disorders, but none had ocular symptoms or a palpable goiter. Six had features of Graves' disease that required interferon alfa withdrawal in four and prolonged treatment with antithyroid drugs in all six. Three presented with transient thyrotoxicosis that subsequently progressed to hypothyroidism and required interferon withdrawal in one and thyroxine treatment in all three.

Although much work on thyroid autoimmunity associated with interferon alfa has accumulated, little is known about the very long-term outcome of this disorder. In 114 patients with chronic hepatitis C and no previous thyroid disease who were treated with interferon alfa-2a for 12 months, data on thyroid status were retrospectively obtained at the end of treatment, 6 months after withdrawal, and after a median of 6.2 years (206). Among 36 patients who had thyroid autoantibodies at the end of treatment, the authors identified three groups according to the long-term outcome: 16 had persistent thyroiditis, 10 had remitting/relapsing thyroiditis (that is antibodies became negative after 6 months of therapy and were again positive thereafter), and 10 had transient thyroiditis. Therefore, 72% of these patients had chronic thyroid autoimmunity at the end of follow-up and 12 developed subclinical hypothyroidism. In contrast, only one of 78 patients negative for thyroid autoantibodies developed thyroid autoantibodies. Although none of the patients had clinical thyroid dysfunction, this study suggests that long-term surveillance of thyroid disorders is useful in patients who have high autoantibody titers at the end of treatment with interferon alfa.

Although thyroid disorders in patients treated with interferon alfa generally follows a benign course after interferon alfa withdrawal or specific treatment, severe long-lasting ophthalmopathy resulting from Graves' disease has been described in a 49-year-old woman (207).

Time-course
Clinical symptoms usually occur after 2–6 months of treatment and occasionally after interferon alfa withdrawal.

- A middle-aged woman developed subacute thyroiditis by the sixth month of treatment with interferon alfa (208). She also had the classic symptoms of hyperthyroidism, although it is clear that these could easily have been mistaken for adverse effects of interferon alfa itself, for example weakness, weight loss, and palpitation.

After 6 months of treatment, 12% of patients with chronic hepatitis C had thyroid disorders, compared with 3% of patients with chronic hepatitis B. This study also suggested a possible relation between low free triiodothyronine serum concentrations before treatment and the subsequent occurrence of thyroid dysfunction. After a follow-up of 6 months after the end of interferon alfa treatment, 60% of affected patients with chronic hepatitis C still had persistent thyroid dysfunction; all had been positive for thyroid peroxidase antibodies before treatment. Long-term surveillance is therefore needed in these patients.

Frequency
In a prospective study, the overall incidence of biochemical thyroid disorders was 12% in 254 patients with chronic hepatitis C randomized to receive ribavirin plus high-dose interferon alfa (6 MU/day for 4 weeks then 9 MU/week for 22 weeks) or conventional treatment (9 MU/week for 26 weeks) (209). There was no difference in the incidence or the time to occurrence of thyroid disorders between the groups. Of the 30 affected patients,

11 (37%) had positive thyroid peroxidase autoantibodies (compared with 1% of patients without thyroid dysfunction), nine developed symptomatic thyroid dysfunction, and only three had to discontinue treatment. There was no correlation between the viral response and the occurrence of thyroid disorders, and only female sex and Asian origin were independent predictors of thyroid disorders.

Data on interferon alfa-associated thyroid disease have been comprehensively reviewed (210,211). There was a mean prevalence of 6% for incident thyroid dysfunction, and treatment for malignancies was associated with the highest prevalence (11%). Hypothyroidism occurs more often than hyperthyroidism, and spontaneous resolution is expected in almost 60% of patients with or without interferon alfa withdrawal. Finally, female sex and the presence of baseline thyroid autoimmunity were confirmed to be the most significant risk factors. The mechanisms of interferon alfa-induced thyroid dysfunction are not yet fully clarified. Although an autoimmune reaction or immune dysregulation are the most likely mechanisms, a direct inhibitory effect of interferon alfa on thyrocytes should be considered in patients without thyroid antibodies.

Data on the incidence of thyroid disorders in interferon alfa-treated patients vary, largely because the follow-up duration, the nature of the study (prospective or retrospective), biological monitoring, diagnostic criteria, and the underlying disease differ from study to study (200). The incidence of clinical or subclinical thyroid abnormalities is generally 5–12% in large prospective studies in patients with chronic hepatitis C treated for 6–12 months, but it reached 34% in one study (26,212). The incidence was far lower in patients with chronic hepatitis B, at 1–3%. A wider range in incidence was found in patients with cancer, with no clinical thyroid disorders in 54 patients treated during a mean of 16 months for hematological malignancies (213), whereas in many other studies there was a 10–45% incidence (200). Even more impressive was the escalating incidence of thyroid disorders in patients with cancer receiving both interferon alfa and interleukin-2 (qv).

Mechanisms
Possible mechanisms need to be clarified. Since thyroid autoantibodies are detected in most patients who develop thyroid disorders, the induction or exacerbation of pre-existing latent thyroid autoimmunity is the most attractive hypothesis. This is in accordance with the relatively frequent occurrence of other autoantibodies or clinical autoimmune disorders in patients who develop thyroid disorders (214). However, 20–30% of patients who develop thyroid diseases have no thyroid antibodies, and it is thus not yet proven that autoimmunity is the universal or primary mechanism. In fact, there were subtle and reversible defects in the intrathyroidal organification of iodine in 22% of antithyroid antibody-negative patients treated with interferon alfa (215). In addition, the acute systemic administration of interferon alfa in volunteers or chronic hepatitis patients reduces TSH concentrations (SED-13, 1093) (216), and in vitro studies have suggested

that interferon alfa directly inhibits thyrocyte function (SED-13, 1093) (217). Finally, the thyroid autoantibody pattern in patients who developed thyroid dysfunction during cytokine treatment was not different from that of patients without thyroid dysfunction, but differed significantly from that of patients suffering from various forms of spontaneous autoimmune thyroid disease (218).

Susceptibility factors

In addition to the underlying disease, there are many potential susceptibility factors (26,219). There is as yet no definitive evidence that age, sex, dose, and duration of treatment play an important role in the development of thyroid disorders. However, patients with previous thyroid abnormalities are predisposed to develop more severe thyroid disease (SEDA-20, 328). The incidence of thyroid disease was not different between natural and recombinant interferon alfa. Although this should be taken into account, a previous familial or personal history of thyroid disease was generally not considered a major risk factor. Finally, only pre-treatment positivity or the development of thyroid antibodies during treatment seem to be strongly associated with the occurrence of thyroid dysfunction.

In 175 patients with hepatitis B or C virus infections, women with chronic hepatitis C and patients with previously high titers of antithyroid autoantibodies were more likely to develop thyroid disorders (220).

The immunological predisposition to thyroid disorders has been studied in 17 of 439 Japanese patients who had symptomatic autoimmune thyroid disorders during interferon alfa treatment (221) There was a significantly higher incidence of the human leukocyte antigen (HLA)-A2 haplotype compared with the general Japanese population (88 versus 41%), suggesting that HLA-A2 is a possible additional risk factor for the development of interferon alfa-induced autoimmune thyroid disease.

Among other potential predisposing factors, treatment with iodine for 2 months in 21 patients with chronic hepatitis C receiving interferon alfa did not increase the likelihood of thyroid abnormalities compared with eight patients who received iodine alone, but abnormal thyroid tests were more frequent compared with 27 patients who received interferon alfa alone (222). This suggests that excess iodine had no synergistic effects on the occurrence of thyroid dysfunction induced by interferon alfa.

The occurrence of thyroid dysfunction in 72 patients treated with interferon alfa plus ribavirin (1.0–1.2 g/day) has been compared with that of 75 age- and sex-matched patients treated with interferon alfa alone for chronic hepatitis C (223). Of the former, 42 patients, and of the latter, 40 patients had received previous treatment with interferon alfa alone. There was no difference in the rate of thyroid autoimmunity (antithyroglobulin, antithyroid peroxidase, and thyroid-stimulating hormone receptor antibodies) between the two groups, but the patients who received interferon alfa plus ribavirin developed subclinical or overt hypothyroidism more often (15 versus 4%). Similarly, the incidence of hypothyroidism increased

to 19% in patients who underwent a second treatment with interferon alfa plus ribavirin compared with 4.8% after the first treatment with interferon alfa alone, while the incidence remained essentially the same in patients who had two consecutive treatments with interferon alfa alone (4.7 and 7.1% respectively). Furthermore, there was no higher incidence of thyroid autoimmunity or clinical disorders after a second course of interferon alfa whether alone or combined with ribavirin in patients who had no thyroid autoantibodies at the end of a first course of interferon alfa alone, suggesting that these patients are relatively protected against the development of thyroid autoimmunity.

Management

The management of clinical thyroid dysfunction depends on the expected benefit of interferon alfa. Assay of thyroid antibodies before treatment, and regular assessment of TSH concentrations in treated patients, even after interferon alfa withdrawal, are useful as a means of predicting and detecting the risk of thyroid disorders. Complete recovery of normal thyroid function is usually observed after thyroxine replacement but sometimes requires interferon alfa withdrawal. Sustained hypothyroidism requiring long-term substitution treatment has occasionally been observed (SED-13, 1092) (224), and is more likely in patients with initially severe hypothyroidism and raised thyroid antibody titers (225). By contrast, hyperthyroidism generally requires the prompt withdrawal of interferon alfa, and severe forms may require radical radioiodine therapy. Although not enough data are available on the long-term consequences of interferon alfa-induced thyroid dysfunction, the recurrence of thyroid abnormalities after the administration of pharmacological doses of iodine should be borne in mind (226).

Parathyroid

Exacerbation of secondary hyperparathyroidism occurred in a 20-year-old renal transplant patient who also developed psoriasis during interferon alfa treatment (227). Both disorders resolved after withdrawal.

Adrenal

Of 62 initially autoantibody-negative patients treated with interferon alfa for chronic hepatitis C for a mean of 8 months, three developed antibodies to 21b-hydroxylase, a sensitive assay of adrenocortical autoimmunity (228). However, there were no cases of Addison's disease or subclinical adrenal insufficiency. This study suggested that the adrenal cortex is another potential target organ of autoimmune effects of interferon alfa, along with thyroid and pancreatic islet cells.

Metabolism

Diabetes mellitus

The development or worsening of insulin-dependent diabetes mellitus is limited to isolated case reports in patients treated with interferon alfa or interferon alfa plus interleukin-2 (SEDA-20, 328) (SEDA-21, 371). In chronic hepatitis, diabetes mellitus was noted in only 10 of 11

241 treated patients (21). Although a relation between chronic hepatitis C and the occurrence of glucose metabolism disorders is possible (229), reports of diabetes mellitus in patients treated with interferon alfa were probably more than coincidental. Indeed, there have been reports of prompt amelioration or complete recovery after interferon alfa withdrawal (SED-13, 1092) (230–232) and of successive episodes of diabetes after each course of interferon alfa (SEDA-21, 371).

- In three middle-aged patients, diabetes was diagnosed after 3–7 months of treatment with interferon alfa-2b and ribavirin, and two presented with severe ketoacidosis (233,234). There was a family history of diabetes in one patient and two had high titers of glutamic acid decarboxylase antibodies before treatment. One patient never had diabetes-related serum autoantibodies before or after interferon alfa therapy. All three required permanent insulin treatment despite withdrawal of interferon alfa.
- Insulin-dependent diabetes mellitus has been reported after 2 weeks to 6 months of treatment with interferon alfa in four patients with chronic hepatitis C (235). All discontinued interferon alfa, and one woman who restarted treatment had a subsequent increase in insulin requirements.

A 75 g oral glucose tolerance test was performed before and after 3 months of interferon alfa treatment in 32 patients with chronic hepatitis C, of whom 15 also had an intravenous glucose test (236). Baseline evaluation showed that five patients had mild diabetes mellitus, three had impaired glucose tolerance, and 24 were normal. After 3 months of treatment, two patients with diabetes mellitus shifted to impaired glucose tolerance, and all patients with impaired glucose tolerance had normal glucose tolerance. Only three initially normal patients developed impaired glucose tolerance and none had newly diagnosed diabetes mellitus. From these results, and in contrast to previous reports (SED-14, 1250), it appears that interferon alfa did not have any adverse effects on insulin sensitivity and glucose tolerance after 3 months of treatment.

Interferon alfa may produce more severe changes than interferon beta (237).

- A 39-year-old man with diabetes, stabilized with insulin 22 U/day for 13 years, received interferon beta (6 MU/day) for chronic hepatitis C. His diabetes progressively worsened, necessitating insulin 50 U/day. After 4 weeks, interferon beta was replaced by interferon alfa (10 MU/day). Shortly afterwards he developed severe diabetic ketoacidosis and shock, which reversed after hemodynamic support and continuous hemodiafiltration.
- A 39-year-old man who had received peginterferon alfa-2b and ribavirin for 9 months developed diabetes mellitus 3 months after interferon withdrawal (238). He required oral hypoglycemic drugs and insulin, but fasting plasma glucose and glycosylated hemoglobin concentrations later normalized despite withdrawal of antidiabetic treatment.

Incidence

In patients with chronic hepatic B or C the respective prevalences of pancreatic autoantibodies increased from 2% and 3% at baseline to 5% and 7% after interferon (239). In all, 31 published cases of type 1 diabetes mellitus attributed to interferon alfa treatment were detailed, mostly in patients with hepatitis C. Irreversible diabetes required permanent insulin treatment in all but eight cases. At least one marker of pancreatic autoimmunity was positive in nine of 18 patients before treatment, and in 23 of 30 patients at the onset of diabetes. In accordance with these results and the likelihood of a genetic predisposition, the authors recommended screening for islet cell and glutamic acid decarboxylase autoantibodies before and during interferon alfa treatment. However, owing to the low number of reported cases and the paucity of studies that have examined the relation between pancreatic autoimmunity and the occurrence of diabetes, further research on the predictive potential of such a systematic investigation is warranted.

In a randomized trial in 74 patients with chronic hepatitis C treated with interferon alfa-2b and ribavirin, plus placebo or amantadine, two developed glutamic acid decarboxylase (GAD) autoantibodies, but none developed IA-2 or insulin autoantibodies (240). One had an increased titer of GAD autoantibodies during a first sequence of interferon alfa monotherapy, then a further rise during subsequent combination therapy, and finally developed diabetes mellitus after 5 months of treatment. The authors suggested that repetitive treatment with interferon alfa could increase the risk of type 1 diabetes in patients previously positive for islet antibodies.

Mechanisms

Autoimmunity was suggested as a likely mechanism, with HLA-DR4 haplotype and/or islet cell antibody (ICA) positivity at the time of diagnosis in several patients. Because the induction of ICA antibodies in patients treated with interferon alfa has never been otherwise demonstrated (26), the triggering, rather than the induction, of a latent autoimmune phenomenon in patients with a genetic susceptibility is probable (241).

More direct interference with glucose metabolism cannot be excluded. Interferon alfa can reduce the sensitivity of peripheral tissues or liver to insulin and accelerate the destruction of stimulated pancreatic beta-cells (242,243); this could be a possible mechanism in patients not exhibiting islet cell antibodies. This is also in keeping with rare instances of induction or exacerbation of type II non-insulin dependent diabetes mellitus (SEDA-19, 335).

Insulin antibodies were also found in six of 58 patients treated for chronic viral hepatitis (244) and that was associated with signs of insulin allergy in one patient (SEDA-19, 335).

Susceptibility factors

Patients with obesity and a family or previous history of glucose intolerance should be considered more predisposed to interferon alfa-induced diabetes, but the association is not consistently found (SEDA-20, 328).

Dyslipidemia

Dyslipidemia

Interferon alfa often affects lipid metabolism and produces a reversible reduction in cholesterol and, more consistently, increases in triglyceride concentrations (SEDA-20, 328; SEDA-21, 371). Meticulous blood lipid investigation showed a significant rise in serum triglyceride and lipoprotein(a) concentrations and reductions in total cholesterol, HDL cholesterol, LDL cholesterol, and apoprotein A1.

Marked hypertriglyceridemia (10–20 µg/ml), which abates when treatment is withdrawn, has sometimes been observed (SED-13, 1093) (245,246). Inhibitory effects of interferon alfa on lipoprotein lipase and triglyceride lipase or increased hepatic lipogenesis have been suggested (247,248). Diet and lipid-lowering drugs have been proposed as means of maintaining acceptable triglyceride concentrations during long-term interferon alfa therapy. Although the possibility of pancreatic or cardiovascular complications should be borne in mind, no secondary clinical consequences of interferon alfa-induced blood lipid disorders have been so far reported.

In a prospective study of lipid changes in 36 patients with chronic hepatitis C treated with interferon alfa for 6 months, the most prominent findings included increases in triglycerides, VLDL cholesterol, and apolipoprotein B, and falls in HLD cholesterol and apolipoprotein A1 (249). Three patients also developed chylomicronemia and two of those had severe hypertriglyceridemia. All three patients had triglycerides over 2 µg/ml before treatment, suggesting that patients with abnormal serum triglyceride concentrations at baseline are more likely to develop marked hypertriglyceridemia.

Severe hypertriglyceridemia (7.5–19 mmol/l; 653–1644 mg/dl) has been reported in three patients receiving adjuvant high-dose interferon alfa for malignant melanoma (250). The authors reviewed the available literature and proposed a detailed surveillance and management plan for this metabolic disorder in patients with melanoma treated with interferon.

Porphyria

Porphyria

- A severe acute flare of porphyria cutanea tarda has been reported in a 61-year-old man after 4 months of treatment with interferon alfa-2b plus ribavirin for chronic hepatitis C (251). No further relapse was observed after chloroquine treatment, despite continuation of the antiviral drugs.

This patient had previously had episodes of small blisters that spontaneously resolved, and hereditary porphyria cutanea tarda was demonstrated by chromatographic and mutation analysis.

Hematologic

Hematological toxicity due to interferon alfa commonly includes dose-related leukopenia, neutropenia, and thrombocytopenia, whereas anemia is rare and usually moderate (252).

Reductions in platelet count and leukocyte count were usually in the range of 30–50% compared with baseline, but severe and reversible thrombocytopenia ($<49 \times 10^9$/l) or neutropenia ($<0.9 \times 10^9$/l) were noted in 10 and 20% of patients (253). However, life-threatening neutropenia or thrombocytopenia were reported in only six of 11 241 patients with chronic viral hepatitis (21), and reports of Coombs'-negative hemolytic anemia or complete agranulocytosis are sparse (SED-13, 1094; SEDA-20, 329).

In a retrospective study of 158 patients with chronic viral hepatitis treated for 6–12 months, lymphoblastoid interferon alfa produced the largest fall in leukocyte and platelet counts (−38 and −32% versus baseline values), recombinant interferon alfa was associated with intermediate toxicity (−32 and −26%), and leukocyte interferon alfa produced the smallest reduction (−27 and −22%) (254). The lowest mean values were observed after an average of 4–5 months. However, the clinical relevance of these differences is probably minimal, because the overall reduction in leukocyte and platelet numbers was small, and no patients developed clinical symptoms of cytopenia.

- A 50-year-old woman with a liver transplant, who had taken interferon alfa-2b for post-transplant recurrence of hepatitis C infection, developed pancytopenia with a severe anemia (hemoglobin: 2.9 g/dl) after 5 months (255). The leukopenia and thrombocytopenia abated after withdrawal of interferon alfa, but the anemia persisted over the next 3 months. Pure red cell aplasia was confirmed on bone marrow biopsy and no other causes were found. She finally responded to a course of intravenous immunoglobulin.

Pernicious anemia has been associated with interferon alfa (256).

- During interferon alfa treatment for chronic hepatitis C, a 45-year-old man and a 58-year-old woman developed pernicious anemia, with anti-intrinsic factor antibodies and neurological symptoms. Interferon alfa was continued and both recovered with intramuscular cyanocobalamin.
- A 61-year-old man with macrocytosis received interferon alfa-2a 9 MU/week for chronic hepatitis C (257). His hemoglobin concentration progressively fell from 15.6 to 8.1 g/dl after 2 months of treatment and he had worsening of pre-existing paresthesia, tongue pain, ataxia, and mucosal pallor. There was hypergastrinemia, low serum vitamin B_{12} concentrations, and chronic atrophic gastritis. The anemia resolved after interferon alfa withdrawal and intramuscular administration of vitamin B_{12} and calcium folate. Anti-intrinsic factor antibodies were later detected.
- A 34-year-old woman developed aplastic anemia after 8 months of interferon alfa treatment for chronic myeloid leukemia (258). The pancytopenia was transfusion-dependent. There was no hemopoietic recovery 16 months after withdrawal of interferon alfa and no response to two courses of immunosuppressants. Only allogeneic bone marrow transplantation, performed because of recurrent chronic myeloid leukemia, finally restored normal hemopoiesis.

- Severe bone marrow hypoplasia occurred in a 48-year-old woman with multiple myeloma also taking thalidomide. She recovered after interferon alfa withdrawal and supportive treatment with G-CSF and erythropoietin (259).
- A 21-year-old woman developed severe bone marrow aplasia after about 3 years of interferon alfa treatment for chronic myelogenous leukemia (260). She recovered after interferon alfa withdrawal, G-CSF supportive treatment, and ciclosporin-based immunosuppressive treatment.

The authors of the last report identified only eight cases with complete recovery from the 18 previously published cases of bone marrow hypoplasia induced by interferon alfa in patients with chronic myelogenous leukemia. They suggested that interferon alfa-induced aplasia is immune mediated and speculated that immunosuppressive therapy may accelerate recovery.

Tolerance of peginterferon and ribavirin for recurrent hepatitis C after liver transplantation is poor. In a retrospective study of 31 patients, changes in the doses of interferon and peginterferon had to be made in 87% and of ribavirin in 81% of the patients (261). Hemopoietic growth factors were often used, for example G-CSF in 88% of patients treated with peginterferon and ribavirin.

Mechanisms

Interferon alfa has direct myelosuppressive effects and can also cause hematological disorders by immune blood cell destruction, as suggested by reports of immune-mediated thrombocytopenia, immune hemolytic anemia (262,263), or a positive direct Coombs' test, with or without hemolysis (264–266).

Anemia in the course of the combination treatment with interferon and ribavirin is mostly multifactorial, and is due to ribavirin-induced extravascular hemolysis and interferon-induced suppression of bone marrow. The response of erythropoietin is blunted compared with the severity of the anemia. It has been speculated that interferon might cause an inflammatory state severe enough to induce anemia similar that seen in chronic inflammatory disorders (267).

The kinetics of the hemotoxic effects of interferon alfa have been studied in 76 patients with chronic hepatitis C (268). There were significant falls in white blood cell count and platelet count within 12 hours after the first injection, and a second fall in platelet count after 2 weeks, but not further thereafter. This rapid time-course suggests that liver or spleen sequestration of blood cells, rather than direct bone marrow suppression or immune-mediated hematological toxicity, is the most likely explanation for this acute hemotoxic effect, which does not preclude continuation of treatment.

Susceptibility factors

Susceptibility factors for severe hematological toxicity include cirrhosis and hypersplenism.

Prior interferon alfa treatment lasting for more than 6 months and withdrawn for 2–3 months was also one of the most significant factors to explain a reduced yield of peripheral blood stem cells in 88 previously autografted patients with myeloma undergoing G-CSF stimulation for future autotransplantation (269). As suggested by this study, the myelosuppressive effects of interferon alfa may be prolonged to such an extent that a minimum delay of more than 2–3 months after interferon alfa withdrawal should be considered before harvesting bone marrow cells.

Pancytopenia or aplasia, sometimes fatal, have been reported only in patients who had received previous chemotherapy, as has severe and even fatal erythrocytosis in patients with hairy cell leukemia (SED-13, 1094) (270–274).

Anemia

Isolated anemia is not a common feature of the hemotoxic effects of interferon alfa, and pure red cell aplasia has been reported in two patients with chronic leukemia for several months (275,276). Both patients improved progressively after replacement of interferon alfa by hydroxyurea. However, one required erythrocyte transfusions for 14 months.

- Pernicious anemia with a low vitamin B_{12} concentration and positive intrinsic factor antibodies has been reported in a 54-year-old woman who was receiving interferon alfa as a maintenance treatment for relapsing chronic hepatitis C (277).

Rapid exacerbation (1–21 days) or delayed (3–38 months) de novo appearance of immune hemolytic anemia has been reported after initiation of interferon alfa treatment in nine patients with lymphoproliferative disorders (278). However, this rare event was identified in only 1% of 581 patients receiving interferon alfa alone or as part of a chemotherapeutic regimen for chronic myelogenous leukemia (279). A mechanism close to that observed with alpha-methyldopa has been thought to be involved (265). The direct antiglobulin test was positive in 32% of 28 chronic myeloid leukemia patients after a median of 1 year of treatment with interferon alfa (280).

Interferon alfa can also induce multiple antibody formation to transfused blood cell antigens, with subsequent massive hemolysis (281).

- A 33-year-old man developed a delayed hemolytic reaction 7 days after red cell transfusion (282). Additional investigations showed the presence of an anti-M antibody, the production of which was supposedly caused by chemoimmunotherapy (interferon alfa, interleukin-2, and 5-fluorouracil) which was begun 24 hours after transfusion.

Leukopenia

The dose of interferon alfa is usually halved when the neutrophil count falls below $0.75 \times 10^9/l$ or the drug is permanently withdrawn when it falls below $0.5 \times 10^9/l$. However, this issue has recently been challenged by a retrospective analysis of 11 patients with compensated cirrhosis, four of whom had severe neutropenia (that is 0.4 to $0.67 \times 10^9/l$) during the first 2 months of treatment

(283). They remained asymptomatic and the neutropenia spontaneously reversed despite continued treatment.

In one study in 119 patients treated with interferon alfa and ribavirin for chronic hepatitis C, in whom neutropenia was not considered as a cause for exclusion or dosage modification, the neutrophil count fell by an average of 34% (31–74%) (284). During the course of treatment, 32 patients had at least one neutrophil count below $1 \times 10^9/$l, 11 had a neutrophil count below $0.75 \times 10^9/$l, and 2 had a neutrophil count below $0.5 \times 10^9/$l; however, none of these patients required dosage modification because of neutropenia. None of the 22 patients who developed documented or suspected bacterial infections during or immediately after treatment withdrawal had concomitant neutropenia. The three black patients with constitutional neutropenia (pretreatment neutrophil counts below $1.5 \times 10^9/$l) had only minimal changes in their neutrophil counts during treatment and no infection, suggesting that these patients can be safely treated.

Thrombocytopenia

Although inhibition of stem-cell proliferation is the most likely mechanism of hematological toxicity, increased platelet hepatic uptake has been suggested to account for thrombocytopenia (285). Raised serum thrombopoietin concentrations were found in patients with interferon alfa-induced thrombocytopenia (286). However, there is evidence that serum thrombopoietin concentrations in patients who have had thrombocytopenia during interferon alfa treatment for chronic viral hepatitis C either do not increase (in patients with compensated cirrhosis) or increase only moderately and less than expected (in non-cirrhotic patients) (287). The authors proposed that interferon alfa impairs liver production of thrombopoietin, raising the possibility of testing thrombopoietin administration in patients with severe thrombocytopenia before or during treatment with interferon alfa (288).

- In a 45-year-old man treated with pegylated interferon alfa-2b for relapsing chronic hepatitis C, thrombocytopenia recovered over 2 months, despite initial treatment with glucocorticoids and immunoglobulin (289).

Acute thrombocytopenia has been reported with both standard and pegylated forms of interferon alfa in five patients with chronic hepatitis C or chronic myeloid leukemia (290,291,292,293). All resolved when interferon alfa was withdrawn and/or with intravenous immunoglobulin and glucocorticoid therapy.

- A 55-year-old woman with chronic myelogenous leukemia developed reversible autoimmune thrombocytopenia, which recurred after rechallenge with the same formulation of purified interferon alfa (294). Further administration with a different purified formulation was well tolerated, suggesting that subtle differences between two types of interferon explained the lack of cross-reactivity.

Interferon alfa-induced immune-mediated thrombocytopenia shares many features with idiopathic thrombocytopenic purpuras and may be therefore coincidental (SED-13, 1094) (SEDA-20, 328) (SEDA-21, 371), but recurrence of thrombocytopenia on interferon alfa readministration strongly supports a causal role of interferon alfa (295). Cross reaction with interferon beta was not found in an isolated report (SEDA-20, 329). Even though severe and even fatal worsening of idiopathic thrombocytopenic purpura has been observed after administration of interferon alfa (SED-13, 1094) (SEDA-20, 328), interferon alfa was not considered harmful in patients with chronic hepatitis C who were previously positive for platelet-associated immunoglobulin G (296).

Thrombotic thrombocytopenic purpura is a possible complication of interferon alfa in patients with chronic myelogenous leukemia, and can develop even after a successful prolonged (2–3 years) treatment (297). Complete recovery is expected after prompt medical management with plasma exchange and glucocorticoids.

Platelet aggregation

The effects of interferon alfa-2b on platelet aggregation have been studied in 29 patients with melanoma who received a low-dose regimen (9 MU/week in five patients) or a high-dose regimen (100 MU/m^2/week intravenously for 4 weeks, then 30 MU/m^2/week subcutaneously for 48 weeks in 24 patients) (298). Compared with pretreatment values and healthy controls, there was significant inhibition of platelet aggregation in the high-dose group, while the effects were minimal in the low-dose group. In the high-dose group, the inhibition was more prominent during the subcutaneous maintenance dose and was still detectable 8 weeks after interferon alfa withdrawal in 60% of the tested samples. An increased risk of bleeding should therefore be anticipated in patients who receive high-dose interferon alfa.

Clotting factors

Since 1994, there have been four case reports and one study of the development of inhibitors directed against factor VIII in patients receiving interferon alfa. However, all of the patients had additional risk factors for the development of factor VIII inhibitors—either a hematological malignancy or a history of hemophilia A. Thus, there has not been unequivocal evidence that interferon alfa alone is associated with the development of an autoantibodies to factor VIII.

- A 53-year-old black man with a 30-year history of asymptomatic hepatitis C infection developed cryoglobulinemic glomerulonephritis and was given peginterferon, ribavirin, and prednisone (299). After 6 months he noted scleral bleeding, easy bruising, and large ecchymoses after minor trauma. There was no history of a bleeding disorder and he had never had blood transfusions. Other medications included hydrochlorothiazide, labetalol, furosemide, amlodipine, and pantoprazole, and he had no known drug allergies. The hematocrit was 28%, platelet count $200 \times 10^9/$l, creatinine 186 µmol/l, partial thromboplastin time 74 seconds, and prothrombin time 12 seconds. Factor VIII activity was <1% and an inhibitor was detected in a titer of 24 Bethesda units. Peginterferon and ribavirin were withdrawn and he was given prednisone 1 mg/kg/day. After 3 weeks, there was no change in the inhibitor titer. Prednisone was gradually tapered and withdrawn. He was given rituximab 375 mg/m^2/week for 1 month, with little effect. One month later, he received another 4-week course of rituximab and the inhibitor titers normalized.

Exposure to interferon in the setting of cryoglobulinemia may be a risk factor for the development of autoantibodies to factor VIII. The use of interferon, and particularly peginterferon, which has a longer half-life, may be hazardous in patients with this disorder. Repeated examination of such patients for the presence of factor VIII autoantibodies seems warranted.

- Asymptomatic prolongation of the activated partial thromboplastin time associated with lupus anticoagulant and a reduction in the coagulation activity of factors IX, XI, and XII occurred after 10 weeks of interferon alfa-2b and ribavirin in a 60-year-old woman with chronic hepatitis C (300). There were no arguments in favor of an antiphospholipid syndrome, and all the abnormalities normalized after withdrawal.

Interferon alfa was suspected of having induced the development of anti-factor VIII autoantibodies in one patient with hemophilia who survived and one without hemophilia who subsequently died from acute hemorrhage (SED-13, 1094; (301,302).

- There was significant bleeding with hematomas in association with an inhibitor of factor VIII in a 58-year-old man who took interferon alfa for 1 year for chronic myelogenous leukemia (303). The factor VIII inhibitor, which was markedly raised, disappeared within 6 weeks of interferon alfa withdrawal and prednisone treatment.

By contrast, in a small uncontrolled study, there was no increase in antifactor VIII antibodies in patients with hemophilia A treated with interferon alfa (304).

Gastrointestinal

Mild and transient gastrointestinal disorders, namely nausea, vomiting, diarrhea or anorexia, were observed in 30–40% of patients, and their severity is typically dose-related (25). Dryness or inflammation of the oropharynx, and moderate stomatitis were sometimes noted, but severe painful oral ulcers recurring after interferon alfa re-administration have been reported (305). More severe forms of digestive disease have been described in isolated case histories with microscopic colitis and the occurrence or the exacerbation of ulcerative colitis (SED-13, 1094) (SEDA-21, 372).

Celiac disease

Celiac disease was observed after 2–3 months of interferon alfa treatment in two patients with chronic hepatitis C aged 34 and 38 years (306). Both had total villous atrophy on distal duodenal biopsy, were positive for antiendomysial antibodies, and responded to a gluten-free diet. Three other cases were reported after 1–5 months of treatment for chronic hepatitis C (46,307). The diagnosis was confirmed in all three patients, based on the presence of total villous atrophy on distal duodenal biopsy, positivity of antiendomysial antibodies, and recovery with a gluten-free diet. Pretreatment antiendomysial antibodies were positive in the two patients tested. As suggested in one patient, interferon alfa can be safely continued providing that a gluten-free diet is strictly respected.

In a retrospective study of 534 patients treated with various formulations of interferon alfa for chronic hepatitis C, 25 developed symptoms consistent with celiac disease and 13 required withdrawal or dosage reduction (308). A screen for celiac disease-specific anti-transglutaminase (anti-TG2) antibodies on serum samples obtained before treatment gave positive results in six of 25 patients with interferon alfa-induced symptoms of celiac disease compared with one of 170 patients also treated with interferon-alfa but free of symptoms and one of 225 controls.

Enteritis

Eosinophilic enteritis has been attributed to interferon alfa (309).

- A 23-year-old man with no previous history of digestive disorders took interferon alfa for chronic hepatitis C. After 2 weeks of treatment, he had severe abdominal pain and diarrhea. The absolute eosinophil count was 7.5×10^9/l, with 40% eosinophils on bone marrow aspiration and a markedly high IgE concentration. Radiological examination showed diffuse jejunal and ileal wall thickening and gross ascites with numerous eosinophils. Complete resolution was obtained after interferon alfa withdrawal and prednisolone treatment. There was no recurrence after prednisolone was withdrawn.
- A 23-year-old man with no previous history of digestive disorders took interferon alfa for chronic hepatitis C. After 2 weeks of treatment, he had severe abdominal pain and diarrhea. The absolute eosinophil count was 7.5×10^9/l, with 40% eosinophils on bone marrow aspiration and a markedly high IgE concentration. Radiological examination showed diffuse jejunal and ileal wall thickening and gross ascites with numerous eosinophils. Complete resolution was obtained after interferon alfa withdrawal and prednisolone treatment. There was no recurrence after prednisolone was withdrawn.

Colitis

Microscopic colitis and new or worsened ulcerative colitis have been attributed to interferon alfa (SED-13, 1094) (SEDA-21, 372). Ischemic colitis has been reported in two of 280 patients treated for chronic hepatitis C (310).

Liver

Asymptomatic and reversible rises in serum transaminases have been reported in 25–30% of patients receiving high-dose interferon alfa (25). Although direct hepatotoxicity has been suspected in isolated and unexplained cases of severe liver failure (SED-13, 1094), most data favored exacerbation of chronic viral hepatitis or latent autoimmuine hepatitis.

Exacerbation of chronic viral hepatitis
In the treatment of chronic hepatitis B, HBe seroconversion was sometimes preceded by transient and moderate worsening of serum transaminases, but severe exacerbation of chronic hepatitis B infection and fatal liver failure can occur. Such fatalities were reported in under 0.5% of patients with hepatitis B (311). Patients with active cirrhosis or a previous history of decompensated cirrhosis are particularly susceptible to these complications (312).

Acute exacerbation of hepatitis is an extremely rare complication of chronic hepatitis C treatment. An exaggerated immune response to hepatitis virus was supposedly the cause of acute icteric hepatitis in two patients (SEDA-21, 372) (313).

- A 43-year-old man had a moderate rise in hepatic transaminase activities after 4 weeks of interferon alfa treatment. His liver tests normalized after withdrawal, but the aspartate transaminase activity increased dramatically shortly after treatment was restarted. His condition rapidly deteriorated, with a diagnosis of hepatorenal failure, and he finally required liver transplantation. Histological examination of the liver showed advanced micronodular cirrhosis, a feature not found on pretreatment liver biopsy.

In another study, only four of 11 241 patients treated with interferon alfa died of fulminant liver failure (21).

Autoimmune hepatitis
More disturbing are reports of interferon alfa-induced acute exacerbation of latent autoimmune hepatitis (SED-13, 1094) (314-321). Further analysis showed that these patients were initially misdiagnosed as having hepatitis C, and autoimmune hepatitis reversible by glucocorticoid treatment was later proven to be the correct diagnosis. It was later found that latent chronic autoimmune hepatitis can be present in patients with unequivocal serological evidence of chronic hepatitis C (314,322). The co-existence of serological markers of autoimmune hepatitis and confirmed hepatitis C before treatment with interferon alfa in the same patient is very disturbing, because the distinction cannot readily be made on the basis of biological and histological data. As glucocorticoids can increase the extent of viremia, and since in addition interferon alfa can acutely exacerbate latent autoimmune liver disease, this has raised the question of how to deal with these patients. In those without serological markers of autoimmune liver disease, only close monitoring of liver function to detect any sudden increase in alanine transaminase activity is helpful, because the systematic detection of autoantibodies proves unable to predict the risk of overt autoimmune hepatitis (322). In those with both hepatitis C virus and autoantibodies, there is as yet no consensus on the therapeutic management. Glucocorticoids may actually increase the extent of viremia, while interferon alfa may exacerbate

autoimmune hepatitis. As a result, controversies have emerged, with several investigators advocating glucocorticoids as first line treatment and providing a safe option in patients with high antibody titers, whereas others have found interferon alfa to be more appropriate (SED-13, 1094) (SEDA-20, 329) (SEDA-21, 372).

De novo induction rather than exacerbation of autoimmune hepatitis is still possible, as indicated by anecdotal reports in patients with cancer or chronic viral hepatitis (SED-13, 1094) (SEDA-22, 402). Such a very rare event is in keeping with the usual absence of autoantibody specific for autoimmune liver disease after interferon alfa treatment.

Positive serological markers of autoimmune hepatitis before treatment in patients with concomitant chronic hepatitis C are sometimes associated with further exacerbation of an underlying autoimmune liver disease during interferon alfa treatment. Of three patients with raised antimitochondrial antibodies (over 1:160), only the two patients with M2 (with or without M4 or M8) subtypes had biochemical exacerbation of cholestasis and an unfavorable response to interferon alfa (320). Although very few patients were investigated, determination of antibodies against submitochondrial particles may help to identify patients who are likely to have no benefit and even exacerbation of liver disease with interferon alfa.

In 25 children with chronic hepatitis C, pretreatment positivity for liver/kidney microsomal type 1 (LKM-1) antibodies was associated with more frequent treatment-limiting increases in serum alanine transaminase activity (321). Withdrawal of interferon alfa-2b because of hypertransaminasemia was required in three of four LKM-1 positive children compared with two of the 21 LKM-1 negative children. Although none developed features of autoimmune hepatitis, careful surveillance of hepatic function is recommended in LKM-1-positive patients.

Other complications
Interferon alfa-associated macrovesicular steatosis has been reported (323).

- A 50-year-old woman without a history of liver disease, dyslipidemia, diabetes, obesity, or alcoholism started taking interferon alfa (7.5 MU/day) for chronic myelogenous leukemia together with allopurinol and hydroxyurea for 2 weeks. Her liver tests were normal before treatment but transaminase activities were greatly increased after 2 weeks. Serological tests for hepatitis B and C and HIV were negative, as was screening for serum antitissue antibodies. Liver biopsy showed severe macrovesicular steatosis (80% of hepatocytes) without steatohepatitis. Liver tests completely normalized on interferon alfa withdrawal. A few weeks after interferon alfa was restarted in a lower dose (3-5 MU/day), she again had a rise in liver enzymes, and a second liver biopsy showed unchanged findings. Liver tests remained stable despite treatment continuation.

The rapid growth of a pre-existing liver hemangioma was attributed to a 6-month course of interferon alfa-2b in a 16-year-old girl with chronic hepatitis C (324). However, liver enlargement was noted only 12 months after withdrawal of interferon. The authors also recognized that they were unable to determine which factors among puberty, hepatitis C infection, or interferon alfa treatment, could have influenced the rapid progression.

Other adverse liver effects reported with interferon alfa include primary biliary cirrhosis (SEDA-20, 329) and granulomatous hepatitis (SEDA-20, 329; SEDA-21, 372; 325).

Pancreas

Asymptomatic rises in pancreatic enzymes and reversible acute pancreatitis have been reported in isolated patients, with no mention of hypertriglyceridemia (SEDA-19, 336).

- A 54-year-old man developed abdominal pain from the beginning of interferon alfa treatment (326). Two weeks later his serum amylase and lipase peaked at about three times the upper limit of normal. Careful radiological investigations ruled out pancreatic calcification and biliary or pancreatic lithiasis and showed only pancreatic enlargement. Complete improvement occurred after treatment withdrawal. As in the very few previous cases, there was no hypertriglyceridemia in this patient.

A definite case of pancreatitis proven by positive rechallenge was also briefly cited in a review of drug-associated pancreatitis spontaneously reported to the Dutch adverse drug reactions system (327).

Two cases of interferon alfa-induced acute pancreatitis in patients with chronic hepatitis C were particularly convincing, because other causes were carefully ruled out and clinical symptoms or biological abnormalities recurred after rechallenge in both patients (328). Although one patient also took ribavirin, recurrence was observed after re-administration of interferon alfa alone. Lipid disorders were not found in these patients, confirming that interferon alfa-induced pancreatitis is not due to hypertriglyceridemia.

Among 1706 patients treated with interferon alfa and ribavirin for chronic hepatitis C, acute pancreatitis meeting the criteria of drug-induced pancreatitis was retrospectively found in seven after a median duration of 12 (4–21) weeks (329). All promptly recovered after withdrawal and none had recurrent pancreatic symptoms after a median follow-up of 18 months. Only one was rechallenged and did not develop evidence of recurrent pancreatitis, but re-treatment lasted only 1 month because of severe fatigue and diarrhea. Although hypertriglyceridemia is a well-known complication of interferon alfa treatment, only two patients had very mildly raised serum triglyceride concentrations. Because of the association of pancreatitis only with high-dose intravenous ribavirin and the previous experience of recurring pancreatitis after interferon alfa readministration, the authors strongly suspected that interferon alfa had been the cause.

- A 52-year-old man with no known risk factors rapidly developed acute pancreatitis after receiving a single injection of peginterferon alfa-2b and ribavirin for 7

days for chronic hepatitis C (330). He promptly improved after withdrawal of both interferon and ribavirin.

Urinary tract

Mild and usually asymptomatic proteinuria, leukocyturia, microscopic hematuria, or moderate increases in serum creatinine were observed in 15–25% of patients (25). There was moderate deterioration of glomerular and tubular renal function in most interferon alfa-treated patients assessed prospectively with a number of renal function markers (331).

In patients receiving high-dose interferon alfa, severe proteinuria and nephrotic syndrome have sometimes been noted (332).

- A 57-year-old woman developed severe nephrotic syndrome after 3 months of interferon alfa re-treatment, and renal biopsy showed minimal change nephrotic syndrome with T cell-predominant interstitial nephritis (333). Proteinuria persisted despite interferon alfa withdrawal and resolved only after glucocorticoid treatment.
- A 55-year-old woman was treated with interferon alfa and ribavirin for 1 year and developed asymptomatic nephrotic syndrome with focal segmental glomerulosclerosis on renal biopsy (334). Proteinuria slowly improved over the next 21 months.
- A 63-year-old man with Sézary syndrome developed renal thrombotic microangiopathy after 4 years of interferon alfa treatment (335).
- A 46-year-old woman developed acute irreversible renal insufficiency due to interstitial nephritis during treatment with peginterferon alfa-2a; she had had underlying hepatitis C-associated focal segmental glomerulosclerosis (336).
- A 54-year-old man developed acute tubular necrosis and worse IgA nephropathy on renal biopsy after treatment for 9 days with peginterferon alfa-2a and ribavirin for chronic hepatitis C (337).

Interferon alfa-induced acute renal insufficiency is rare and has mostly been reported in patients with underlying renal disease, in those receiving high dosages, or in those with varied malignancies (SED-13; 1095; SEDA-20, 329; SEDA-21, 372) (338,339). Very few cases have been described in patients with chronic hepatitis (SEDA-21, 372; SEDA-22, 402). It has also been reported after intravesical administration (SEDA-21, 372). When available, pathological findings have pointed variously to nephrotic syndrome with minimal-change nephropathy and acute tubulointerstitial nephritis, nephrotic syndrome with severe glomerular changes, membranoproliferative glomerulonephritis, extracapillary glomerulonephritis with crescents, focal segmental glomerulosclerosis, and acute tubular necrosis. Renal dysfunction usually resolves after withdrawal of interferon alfa, but irreversible alteration or incomplete resolution of renal function have occasionally been noted.

A review of 15 other available reports of renal insufficiency and proteinuria in patients with chronic myeloid leukemia or other malignancies confirmed that the histological spectrum of renal lesions associated with interferon alfa is varied, and includes membranous glomerulonephritis, minimal change glomerulonephritis, acute interstitial nephritis, hemolytic–uremic syndrome, and thrombotic microangiopathy. Renal complications were reversible in nine patients; three patients had persistent proteinuria, and four had persistent renal dysfunction, of whom three required chronic hemodialysis. Two-thirds of the patients developed renal complications within 1 month of treatment with interferon alfa, and one-third had received a relatively low dosage of interferon alfa (9–15 MU/week).

The mechanism of interferon alfa nephrotoxicity is probably multifactorial and may involve a direct nephrotoxic effect with a possible additive effect of concomitant NSAID therapy, a T cell-mediated immune effect, immune-complex renal disease, or an autoimmune etiology (SED-13, 1095) (340–345).

Acute renal insufficiency has also occurred as a consequence of interferon alfa-induced hemolytic–uremic syndrome, thrombotic thrombocytopenic purpura, or renal thrombotic microangiopathy (SED-13, 1095) (297,346–349). The diagnosis was made after 7 months to 10 years of treatment (median 50 months) with weekly doses of 15–70 MU. This rare but extremely severe complication has almost exclusively been reported in patients treated for chronic myelogenous leukemia, so that the respective roles of interferon alfa and the underlying hematological malignancy are undetermined. However, at least one case has been reported in a patient with hepatitis C (SEDA-22, 402). From a total of 15 cases, renal prognosis was poor, with early deaths in four patients, chronic hemodialysis in eight, and chronic renal insufficiency in three.

- A 29-year-old patient and a 57-year-old patient with chronic myeloid leukemia had received interferon alfa and hydroxycarbamide for 5 and 51 months (350). Thrombocytopenia and microangiopathic hemolytic anemia promptly resolved after interferon alfa withdrawal in both patients, whereas renal dysfunction progressed to dialysis-dependent renal insufficiency in one.

Of 11 other published cases of hemolytic–uremic syndrome associated with interferon alfa in patients with chronic myelogenous leukemia, five developed end-stage renal disease despite prednisolone and plasmapheresis in some.

- Focal segmental glomerulosclerosis has been attributed to interferon alfa in a 40-year-old man with chronic myelogenous leukemia and in a 56-year-old woman with chronic hepatitis C (351,352). The times to occurrence of the first symptoms were 1 and 2.5 months after starting interferon alfa. The interferon dosages were 9 MU/week and 21 MU/week. Although the first patient improved after interferon alfa withdrawal and prednisone treatment, the second patient progressed to end-stage renal disease and required long-term hemodialysis.

To determine the characteristics of thrombotic microangiopathy associated with interferon alfa, data from eight patients were carefully examined (353). All had chronic myeloid leukemia and had received high-dose interferon alfa (mean 39 MU/week) for a long time (mean 32 months) before diagnosis. Severe arterial hypertension was the most common sign before diagnosis. Five patients had distal ischemic lesions that required amputation in one, and all had typical lesions of renal thrombotic microangiopathy involving both glomeruli and small arterioles. After interferon alfa withdrawal, two recovered normal renal function, three had persistent renal insufficiency, one relapsed 17 months after treatment withdrawal, and two required chronic dialysis. One patient had already had reversible renal insufficiency during a previous course of interferon alfa. From a review of 21 other previously published similar cases, the authors confirmed that interferon alfa-induced thrombotic microangiopathy mostly occurred in patients with chronic myeloid leukemia, whereas only two cases were reported in patients with chronic hepatitis C and one in a patient with hairy cell leukemia. The delayed occurrence of renal toxicity was also suggested to be highly predictive of histological thrombotic microangiopathy.

Possible deleterious effects of interferon alfa on renal graft function are repeatedly reported (354).

- A 43-year-old man with stable renal graft function, taking ciclosporin, methylprednisolone, and azathioprine, developed chronic myelogenous leukemia and received interferon alfa (3 MU/day). Seven weeks later, he became tired and had increased proteinuria and a raised serum creatinine concentration (574 µmol/l). Interferon alfa was withdrawn and he received high-dose methylprednisolone for suspected acute graft rejection. This was unsuccessful and a first renal biopsy showed widespread interstitial edema that could not be correctly interpreted. Hemodialysis was restarted, and he finally developed a catheter infection and died from sepsis. Histology of the explanted renal graft showed severe, predominantly acute, vascular rejection.

The rapid occurrence of renal dysfunction after interferon alfa in this case suggested a causal relation.

Simultaneous thrombocytopenia and nephrotic syndrome has been reported (355).

- A 44-year-old man was given ribavirin and pegylated interferon alfa-2b for chronic hepatitis C. After 16 weeks he simultaneously developed severe thrombocytopenia and nephrotic syndrome due to focal segmental glomerulosclerosis. Both are rare adverse effects. Antiviral treatment was immediately interrupted. He received immunosuppressive therapy and promptly recovered from the thrombocytopenia and partially and slowly from the nephrotic syndrome.

Skin

A wide range of skin lesions has been reported, and most include skin dryness, rash, diffuse erythema, or urticaria, occurring in 5–12% of patients (25,356). However, severe dermatological complications are rare. In a prospective survey of 120 patients treated during 6–18 months for chronic viral hepatitis, only three developed lichen planus and one relapsing aphthous stomatitis (357).

The type and incidence of interferon alfa-induced cutaneous adverse effects has been systematically examined before and after every 3 months of treatment in 33 patients treated with interferon alfa-2a 9 MU/week for malignant melanoma (358). There were cutaneous adverse effects in 29, but they were usually mild and never required withdrawal. The most common symptoms were hair loss (n = 16), eczematous reactions at the injection site or at a distance (n = 13), hair discoloration (n = 6), xerostomia (n = 6), Raynaud's phenomenon (n = 4), and livedo reticularis (n = 4). Other effects included widespread and isolated pruritus, livedo reticularis, seborrheic dermatitis, recurrent herpes labialis, urticaria or angioedema, pytiriasis versicolor, aphthous ulcers, and vitiligo, but it was unclear whether they were fortuitous and how the causal relation with treatment was assessed in each case. In another briefly reported study in 52 patients treated with a combination of pegylated interferon alfa and ribavirin for chronic hepatitis C there were de novo skin lesions, mostly eczematous, in 12 patients (359). The lesions did not require withdrawal and regressed on completion of treatment.

Injection site reactions

Subcutaneous interferon alfa sometimes causes local erythema and skin induration, which can be prevented by regularly changing the site of injection. Isolated reports have described more severe local reactions, with inflammatory painful nodules, purpuric papules and vasculitis, local ulceration, and injection site necrosis (SED-13; 1095; SEDA-20, 330; SEDA-21, 372; SEDA-22, 402). Despite previous findings, even patients receiving low-dose interferon alfa can have severe injection site reactions. A localized intradermal bullous eruption, which recurred following each interferon alfa injection, was also reported (360).

Severe injection site reactions have been extensively detailed in six patients who had local cutaneous necrosis or indurated erythema after 1–10 months of treatment with low-dose interferon alfa (361). Four patients had concomitant risk factors known to reduce microcirculation, that is beta-blockers, dihydroergotamine, and cigarette smoking. The lesions healed after medical treatment in five patients, but one required surgical excision. The ulcers healed slowly and full recovery occurred only after a mean of 16 weeks after drug withdrawal. The lesions did not recur after interferon alfa re-administration at the other injection sites.

As with interferon alfa, pegylated interferon alfa has been associated with injection site skin necrosis (362). Severe local reactions after subcutaneous injections mostly consist of ulceration and skin necrosis, but a variety of reactions have been described. Prominent suppuration and granulomatous dermatitis at the injection sites of interferon alfa have been reported in two patients (363). Three patients who had severe rashes while receiving pegylated interferon alfa-2a or 2b had positive intracutaneous tests to both pegylated forms of interferon alfa but not to standard interferon alfa-2a or 2b (364). One of these patients subsequently tolerated standard interferon alfa. Cutaneous ulcers have also been reported in patients treated with peginterferon alfa-2b (365,366). In the latter

case, the lesions healed under local therapy and the same dose of interferon was maintained.

Lichen planus

The new occurrence or exacerbation of lichen planus is a well-known complication of interferon alfa, but this has been a source of a considerable debate (SED-13, 1095; SEDA-20, 330; SEDA-22, 402). Indeed, most patients have received interferon alfa for chronic hepatitis C, an underlying disease that is controversially thought to be associated with a spontaneously higher incidence of lichen planus (367). In addition, complete reversibility of previous lichen planus was sometimes observed in patients treated with interferon alfa (SEDA-20, 330). Whatever the truth of the matter, the recurrence of lesions after interferon alfa re-administration or reports of lichen planus in patients with cancer (368–370) argues strongly for a direct causal link with interferon alfa. Local treatment and PUVA were sometimes sufficient to alleviate symptoms, but withdrawal of interferon alfa was required in the most severe cases (SED-13, 1095).

Pemphigus and pemphigoid

Bullous pemphigus and pemphigoid with circulating pemphigus-like autoantibodies, pemphigus foliaceus with anti-intercellular IgG antibodies, and paraneoplastic pemphigus due to interferon alfa have rarely been reported (SED-13, 1095), as has extensive oral pemphigus (371).

- A 28-year-old woman developed oral ulcers after a 5-month course of interferon alfa-2a for chronic hepatitis C. She had multiple erosions on both lips, the tongue, the floor of the mouth, the soft palate, the pharyngeal walls, and the laryngeal mucosa, but there were no skin or genital lesions. Raised double-stranded DNA antibody titers were found. Histology showed pemphigus vulgaris, and complete resolution was obtained by withdrawal of interferon alfa and immunosuppressive and local treatment.

This case was thought to have been due to the immunomodulatory effects of interferon alfa.

Psoriasis

The first reports of psoriasis in cancer patients treated with high-dose interferon alfa were followed by a controversial debate (372,373). However, numerous cases have confirmed that interferon alfa can either induce typical psoriasis or worsen pre-existing psoriasis (SED-13, 1095) (374), an observation that is compatible with interferon alfa-induced imbalance toward an increased Th1 response. This was particularly exemplified by the reversibility of the lesions after withdrawal of treatment and the prompt recurrence of symptoms after interferon alfa re-administration. Exacerbation of psoriasis usually occurred within the first month, whereas a minimum of 2–3 months of treatment was required in patients without a past history of psoriasis (374). Psoriatic lesions at the sites of injection were suggested to be potential indicators for further generalization of psoriasis. In more severe cases, there was concomitant development of monoarticular or polyarticular joint symptoms (SED-13, 1095; SEDA-21, 372). Pustular psoriasis with balanitis and erosive monoarthritis, suggesting incomplete Reiter's syndrome, was also reported in one patient with HLA-B27 (375).

Vitiligo

Vitiligo has sometimes been reported in patients with malignant melanoma, for example in 17–25% of patients receiving interferon alfa alone (SEDA-21, 372). There have also been a few reports of vitiligo in patients with chronic hepatitis C (SEDA-20, 330), and one patient had both scleroderma and vitiligo (SEDA-19, 336).

Hyperpigmentation

Hyperpigmentation of the skin and tongue has been reported after several weeks of interferon alfa and ribavirin in two patients with chronic hepatitis C (376). The hyperpigmented regions disappeared slowly after withdrawal or diminished in intensity after dosage reduction of interferon alfa. Immunohistochemical analysis of a tongue biopsy in one patient showed melanin deposits, and the authors speculated that interferon alfa may have facilitated increased production of melatonin.

Other complications

A spectrum of cutaneous lesions has been described distant from sites of interferon alfa injection. The clinical and histological characteristics of inflammatory skin lesions that occurred away from injection sites have been investigated in 20 patients treated with interferon alfa 2a or 2b plus ribavirin for chronic hepatitis C (377). Cutaneous lesions developed between 2 weeks and 4 months and consisted of pruritic papular erythematous eruptions with occasional vesicles. These eczema-like skin lesions predominated on the distal limbs, the head, and the neck. Photosensitivity was also noted in four patients and mucous lesions in two. Skin biopsy mostly showed non-specific mononuclear infiltrates. The skin lesions were promptly reversible in 10 patients who required treatment withdrawal, while others improved after symptomatic treatment. Two of the three patients who again received the same or another type of interferon alfa had recurrence of their lesions. Skin tests performed in six patients were negative, including the two patients who relapsed after rechallenge with interferon alfa, and were therefore considered unhelpful.

Of 37 patients with chronic hepatitis C who received double or triple antiviral therapy, including peginterferon alfa-2b, nine developed skin reactions, sufficiently severe in five cases to require withdrawal of treatment (378). The skin lesions occurred after a mean of 2 months and consisted of diffuse, erythematous, eczematous, pruritic lesions (n = 2), erythematous papules and desquamative patches (n = 6), and diffuse exacerbation of pre-existing Darier's disease. Histological findings in five cases showed features resembling contact dermatitis. There was slow recovery of skin lesions in all but one case, and

all those who continued interferon required topical glucocorticoids for clinical improvement. Eight of these patients had injection site reactions to peginterferon, as did 14 patients with no cutaneous lesions. Apart from a possible synergistic action of amantadine, a pathogenic role of the pegylated formulation of interferon alfa was suggested.

- Radiation recall dermatitis developed in a 29-year-old woman after high-dose intravenous interferon alfa-2b was given 5 days after the completion of radiotherapy for malignant melanoma (379).

The authors suggested that interferon alfa can trigger an inflammatory reaction in patients whose inflammatory response threshold has been lowered by irradiation.

- A 46-year-old woman developed transient facial erythema with telangiectasia after each injection of interferon alfa, resolving within 1–2 days, and completely disappearing after definitive withdrawal of treatment 7 months later (380).
- Two patients treated with pegylated interferon alfa-2b and ribavirin developed cutaneous thrombotic microangiopathy (381).
- A 54-year-old woman had small bullous lesions mainly on the backs of her hands and feet after 6 months of treatment with pegylated interferon alfa-2b plus ribavirin for chronic hepatitis C. The lesions lasted 48 hours and healed rapidly after rupture of the bullae.
- A 62-year-old woman developed generalized pruritus and excoriated lesions after 5 months of treatment with pegylated interferon alfa-2b plus ribavirin. The lesions were maximal on the backs of her hands and feet after 8 months of treatment, but there were no bullae.

Skin biopsies in the last two patients showed microthrombi of the dermal capillaries and a necrotic epidermis. Immunofluorescence showed only fibrinogen. Some bullous lesions were still present 1 month after withdrawal in the first patient, while the second patient responded to local corticosteroids and a reduced dose of pegylated interferon alfa, but continued to have episodes of severe pruritus.

Other dermatological complications have in most instances been reported as single case histories, so that any causal relation with interferon alfa awaits confirmation. These reports have included worsening of lichen myxedematosus, injection site pyoderma gangrenosum, a polymorphous light eruption (SEDA-22, 372), and a fatal case of histiocytic cytophagic panniculitis (SEDA-20, 330).

- Meyerson's phenomenon, multiple focal and transient eczematous eruptions around melanocytic nevi, has been reported in a 24-year-old man when the dosage of interferon alfa for Behçet's disease was doubled (382).
- Bullous lesions with specific infiltrates of mycosis fungoides have been reported in a 67-year-old woman who took interferon alfa for 2 months for mycosis fungoides (383).

Although in the second case the syndrome could not be definitely attributed to interferon alfa, the authors noted that bullous mycosis fungoides is an extremely rare variant of this disease and withdrawal of interferon alfa led to healing of the blisters without further recurrence.

- A 50-year-old man developed severe, generalized, nummular eczema during treatment with interferon alfa-2b and ribavirin (384). His lesions persisted despite withdrawal of treatment and required maintenance therapy with glucocorticoids.
- A 41-year-old man treated with peginterferon alfa-2b and ribavirin developed generalized exfoliative dermatitis, with recurrence on peginterferon readministration (385).
- A 44-year-old man developed verrucous psoriasis with ichthyosis hystrix-like plaques after receiving peginterferon for 2 months, with rapid spontaneous healing of the lesions after withdrawal of treatment (386). He had a history of inverted psoriasis, but he had not suffered an exacerbation after previous treatment with standard interferon for 1 year.
- A 10-year-old girl developed vitiligo after having received interferon alfa-2a for 6 months and received topical glucocorticoids while interferon alfa was continued (387). She developed psoriatic lesions 4 months later of treatment and interferon was withdrawn after 1 year of treatment. The depigmented macules and the erythematous lesions continued to evolve and were still present 3 years later. The lesions were resistant to topical glucocorticoids during the next year.
- A 44-year-old man had an exacerbation of cryoglobulinemia-associated vasculitis and renal insufficiency 1 month after starting peginterferon alfa; he had tolerated three previous courses of standard interferon alfa (388).

Two of these patients developed cutaneous lesions while receiving peginterferon after uneventful previous treatment with standard interferon alfa. This again suggests that pegylated interferon may be associated with more frequent or more severe dermatological adverse effects.

- A 52-year-old woman developed dermatitis herpetiformis during treatment with interferon alfa-2b and ribavirin for chronic hepatitis C (389). Anti-endomysial and anti-gliadin IgA autoantibodies were retrospectively detected in serum samples collected before treatment. As she also had asymptomatic villous atrophy of the duodenum and recovered after a gluten-free diet and dapsone treatment, her skin lesions were attributed to latent gluten hypersensitivity, which could have been unmasked by interferon alfa.

Injection site reactions to subcutaneous interferon alfa can be severe.

- A 55-year-old diabetic man developed severe streptococcal cellulitis requiring surgical treatment, despite intensive antibiotic therapy, after receiving injections of peginterferon alfa-2b for 5 months (390).

Hair

Moderate and reversible alopecia secondary to telogen effluvium is common (7–30%), and sometimes recedes despite continued treatment (391). Alopecia areata has very occasionally been described (392).

- Injection-site alopecia has been reported in three patients, affecting the thighs in two patients and the abdomen in one (393). A reversible focal telogen effluvium secondary to high local concentrations of interferon alfa was the most likely cause, indicating that rotating injection sites are needed to prevent this adverse effect.
- Alopecia areata after 7 months of interferon alfa, slowly reversible on withdrawal, has also been reported in a 36-year-old woman (394).

In two patients with previously natural curly hairs, the combination of interferon alfa plus ribavirin was suggested to have triggered a rapid change in hair texture, with diffuse straightening hairs, eyelashes, and eyebrow hypertrichosis (395). In one patient, a causal role of treatment was supported by the spontaneous recovery of hair abnormalities after withdrawal and the recurrence of similar abnormalities on rechallenge.

Excessive growth of eyelashes and nail damage due *to Tinea unguium* have been occasionally reported in patients receiving interferon alfa (SED-13, 1095; SEDA-21, 372). Hypertrichosis of the eyelashes (trichomegaly) developed in two of 36 patients with chronic viral hepatitis who were examined for ocular complications during treatment with high-dose interferon alfa (18–30 MU/week) (112). These two patients had received the highest dose of interferon alfa.

Musculoskeletal

Arthralgia, myalgias, and muscle weakness are typically observed during the early influenza-like reaction (25). Direct muscle toxicity of interferon alfa can result in acute rhabdomyolysis, in some cases fatal after the dose of interferon alfa was increased (SEDA-19, 330; SEDA-21, 372).

- A 26-year-old man with a malignant melanoma had two episodes of acute severe rhabdomyolysis after each exposure to a chemotherapy regimen containing interferon alfa and dacarbazine (396). As a few cases of rhabdomyolysis have been previously reported after interferon alfa alone (SEDA-19, 336; SEDA-20, 330; SEDA-22, 403), interferon alfa was suggested as the most likely cause.
- Acute rhabdomyolysis occurred in a 34-year-old woman with a melanoma treated with interferon alfa 20 MU/m^2/day (397). There was no recurrence on retreatment with a lower dose (down to 6.6 MU/m^2/day), suggesting that this was a dose-related complication.
- A 34-year-old man with scleromyxedema had flu-like symptoms and muscle pain after the first injection of interferon alfa 6 MU (398). After three additional injections at 2-day intervals, his muscle symptoms worsened and were associated with mild quadriparesis, reduced deep tendon reflexes, dark urine, confusion, and agitation. Biological findings were consistent with acute rhabdomyolysis, and electromyography showed

rare denervation potentials. His symptoms resolved and the laboratory findings normalized within 15 days.

Delayed muscular adverse effects have been occasionally reported including the clinical exacerbation of a latent myopathy (SEDA-19, 336), delayed and severe myopathic changes (399), myositis, polymyositis, and a Lambert–Eaton-like syndrome (SEDA-19, 336; SEDA-20, 330; SEDA-22, 403).

- Three patients developed unilateral or bilateral avascular necrosis of the femoral head after 3–54 months of treatment with interferon alfa for chronic myelogenous leukemia (400). One required bilateral hip replacement and two significantly improved after interferon alfa withdrawal. One patient received further interferon alfa without exacerbation.

Although there were risk factors for avascular necrosis in two of the patients (a short course of methylprednisolone and moderate alcohol consumption), the authors did not consider them to be significant. They identified seven other reported cases of avascular necrosis in patients with chronic myeloid leukemia, including two patients who were taking interferon alfa at the time of the complication. One patient with pre-existing avascular necrosis had an acute exacerbation within 1 month of interferon alfa and required hip replacement. Although any causal relation with treatment remains purely speculative, the authors argued that the known antiangiogenic effects of interferon alfa could have predisposed patients to avascular necrosis.

Myopathy has been reported (401).

- A 33-year-old man without previous muscle disorders was treated for 24 weeks with a combination of peginterferon alfa-2b (1.5 micrograms/kg/week) and ribavirin (1 g/day) for acute hepatitis C. After 14 weeks he developed weakness and generalized myalgia, with an increase in creatine kinase activity to 904 UI/l and a small rise in CK-MB activity. There were no electromyographic signs of myositis and no inflammatory signs. The dose of peginterferon was halved for the next 10 weeks. The muscle symptoms were unchanged but the creatine kinase activity slowly fell. There was prompt resolution after the end of treatment.

This report suggests that muscle disorders associated with interferon alfa are likely to be collateral reactions.

There has been a further report of severe but reversible acute rhabdomyolysis, a very rare adverse effect of interferon alfa, in a 70-year-old man who had taken interferon alfa-2b 9MU/week alone for only 2 weeks for chronic hepatitis C (402).

Sexual function

Sexual complaints attributed to interferon alfa, namely decreased libido, impotence, or erectile failure, are usually concomitant to other neuropsychiatric symptoms, and cases of reversible impotence are anecdotal (403). The mechanisms accounting for these adverse effects are unclear, and changes in sex hormone concentrations have not been consistently reported. In one study in healthy women, interferon alfa produced falls in serum progesterone and estradiol concentrations (404), but neither impairment of libido nor impairment of fertility has apparently been reported in women. No evidence of gonadal toxicity or sexual dysfunction was found in 43 men with hairy cell leukemia who received interferon alfa for 2–12 months compared with 33 patients who received no systemic therapy (405). Finally, sexual complaints reported during interferon alfa treatment of chronic hepatitis C were presumably related to fatigue, anxiety, or psychological disorders rather than to endocrinological changes (SEDA-21, 373).

Immunologic

Hypersensitivity reactions

No IgE-mediated immediate-type allergic reactions to interferon alfa have ever been conclusively documented. A recurrent non-IgE-mediated anaphylactic reaction, possibly due to mast cell degranulation, has been described in a patient with mastocytosis (406).

- A 64-year-old man with no history of allergy had progressive fatigue, loss of appetite, and facial edema after 6 months of interferon alfa-2b treatment for chronic hepatitis C (407). Angioedema was diagnosed and it resolved after withdrawal of interferon alfa and a short course of prednisolone. Serum immunoglobulin E and plasma bradykinin concentrations were raised, but the C1 esterase inhibitor and serum complement concentrations were normal.
- A 47-year-old man, who had previously received a 2-month course of interferon alfacon-1 for chronic hepatitis C, started interferon alfa-2b 8 months later (408). He developed mild generalized pruritus the day after the second injection, and dyspnea with diffuse urticaria within a few hours after the third injection. Skin tests were not performed, and IgG but not IgE antibodies to interferon alfa were found.

These cases do not formally show a causal relation with interferon alfa, and at best they suggest that an IgE-mediated reaction is probably not the cause of hypersensitivity reactions to interferon alfa. In the first case, the mechanism may have been similar to that observed with angiotensin-converting enzyme inhibitors.

Cases of contact dermatitis have suggested that interferon alfa-2c can cause cell-mediated delayed hypersensitivity (409,410).

Interferon alfa antibodies

Both binding and neutralizing antibodies to interferon alfa can be detected in interferon alfa-treated patients, and the incidence or clinical significance of these antibodies is the subject of continuous controversy, which has been addressed in a number of studies or general reviews (SED-13, 1096; SEDA-20, 330; SEDA-21, 373) (200,411–413).

In various studies, the incidence of antibody formation has ranged from zero to more than 50% of patients. However, any comparison between studies is difficult, because the underlying treated disease, the type of interferon used, the route of administration, the dosage regimen, the schedule of treatment, and the method of assay have differed from one investigation to another. Antibodies to interferon alfa have been reported to be more frequent in patients receiving long-term, instead of short-term, treatment, in patients receiving subcutaneous rather than intravenous interferon alfa, and in patients receiving low rather than high doses (200). Complete disappearance of interferon alfa antibodies was usually observed after withdrawal. Interferon alfa formulations also differ in antigenicity. Using the same anti-interferon alfa antibody assay, a higher frequency of antibodies to recombinant interferon alfa-2a has repeatedly been reported compared with other recombinant or natural interferon alfa formulations in patients treated by the same route and with the same treatment schedule. In 296 patients with chronic hepatitis, binding and neutralizing antibodies were found in 45 and 20% respectively of patients receiving recombinant interferon alfa-2a compared with 15 and 6.9% of those receiving recombinant interferon alfa-2b, and 9.4 and 1.2% of those receiving interferon alfa-n1 (414,415). There were similar differences in the immunogenic potential of the two recombinant interferon alfa formulations in 159 patients with chronic myelogenous leukemia (416). Overall, the incidence of binding antibodies is in the range of 20–50% for recombinant interferon alfa-2a, 6–10% for recombinant interferon alfa-2b, and 1–6% for interferon alfa-n1 (412).

The clinical significance of binding antibodies appears to be limited to a possible change in interferon alfa pharmacokinetics. By contrast, neutralizing antibodies, which are usually detected within 2–4 months of treatment, can theoretically reduce the clinical response to interferon alfa and cause interpatient variability in response to treatment, but this is still debated (411–413). Although several studies have failed to detect any loss of therapeutic response, response failure or the reversal of an initial clinical response, simultaneously with (or soon after) the appearance of neutralizing antibodies, have been reported. In large-scale studies, the clinical response to recombinant interferon alfa was significantly less in patients with neutralizing antibodies, and it has therefore been suggested that the appearance of neutralizing antibodies provides the prime explanation for those instances in which there is a relapse or a breakthrough of the disease before the completion of treatment (417–423). In addition, in patients who cease to respond therapeutically after recombinant interferon alfa antibody formation, a change to natural interferon alfa has proved successful in restoring the response in some cases (423–426). This has led to the suggestion that the formation of neutralizing antibodies represents a specific immune response to the recombinant preparations, and that natural interferon alfa can overcome the neutralizing activity of antibodies to recombinant interferon alfa. On the other hand, neutralizing antibodies were not associated with immune complex-associated diseases or hypersensitivity reactions, and exerted no influence on interferon alfa-associated adverse effects. They can even be accompanied by improvement in the flu-like syndrome.

Autoantibodies

Collectively, several antibodies (mostly antinuclear, antithyroid, parietal cell, liver/kidney microsome, and smooth muscle antibodies, and rheumatoid factor) can be detected before interferon alfa treatment in about one-third of patients. Increased titers or the occurrence of various autoantibodies were observed in 4–30% of previously autoantibody-negative patients (26,200). These autoantibodies do not affect the response to interferon alfa treatment (427–429). Although it was initially felt that interferon alfa might facilitate the development of autoimmune disease in patients previously positive for non-organ-specific autoantibodies, the evidence is still limited and the clinical consequences of such autoantibodies are unclear. Except for thyroiditis, large studies in patients with chronic viral hepatitis C receiving interferon alfa did not show a significant increase in overt autoimmune diseases, despite the pre-existence or further positivity of several autoantibodies (SED-13, 1096; SEDA-20, 330; SEDA-21, 373; SEDA-22, 405). However, in 83 patients with chronic hepatitis C there were one or more pre-existing autoantibodies (mostly low-titer antinuclear antibodies) in 35 patients (group I), of whom seven had clinical evidence of immune-mediated disorders, whereas five of 48 patients without pre-existing autoantibodies (group II) had similar disorders (430). After 12–48 weeks of treatment with interferon alfa in 44 patients, there were new immune-mediated disorders in six of the 20 patients in group I (thyroid disorders in three, arthropathy in two, and psoriasis in one), but none in the 24 patients in group II. Patients who are positive for autoantibodies before interferon alfa may therefore be much more likely to develop autoimmune diseases, particularly thyroid disorders, during interferon alfa treatment.

As a result, there is no clear consensus about the management of patients previously positive for non-organ specific autoantibodies, but it is usually considered that low autoantibody titers or the absence of concomitant symptoms suggestive of autoimmune disease is not a contraindication to treatment.

Autoimmune disorders

The possibility of autoimmune disorders during interferon alfa treatment has been addressed by many authors. The spectrum of interferon alfa-induced immune diseases includes organ-specific and systemic autoimmune diseases, such as thyroiditis, diabetes, hematological disorders, systemic lupus erythematosus, rheumatoid arthritis, dermatological disease, and myasthenia gravis (200). Several have been discussed in appropriate sections elsewhere in this monograph. The exact role of interferon alfa is usually difficult to ascertain, because the underlying disease, that is chronic hepatitis C, can also be associated with immune-mediated disease.

Two studies have provided insights into the incidence and risk factors of the immune-mediated complications of interferon alfa in patients with chronic myeloid leukemia. In the first study, 13 of 46 patients had autoimmune manifestations consisting of a combination of autoimmune thyroiditis in four, a direct antiglobulin test without hemolysis in eight, cryoagglutinins in one, Raynaud's phenomenon in two, and chronic autoimmune hepatitis in one (431). Overall, six patients had clinically symptomatic manifestations after a median of 15 months of treatment. In the second study, there were autoimmune diseases in seven of 76 patients after a median of 19 months of treatment, including hypothyroidism in one, immune-mediated hemolysis in two, systemic lupus erythematosus in two, Raynaud's phenomenon in one, and mixed connective tissue disease in one (432). In both studies there was a strong association with female sex and it was confirmed that patients who developed clinical autoimmune complications had had relatively long exposures to interferon alfa.

The management of patients with chronic hepatitis C and associated features of autoimmune disease carries the risk of exacerbating the underlying disease. Different treatment strategies, including interferon alfa alone or combined with ribavirin or glucocorticoids, or no treatment, have been discussed (433).

Behçet's disease

Characteristic features of Behçet's disease or isolated positive skin tests were found in one study of patients with chronic myelogenous leukemia treated with interferon alfa (434). No other reports confirmed these findings.

Dermatomyositis

Occasional reports have suggested that interferon alfa can cause dermatomyositis (435).

- A 57-year-old woman received adjuvant high-dose interferon alfa 16 months after removal of a malignant melanoma. About 6 weeks later, she developed hand swelling, fatigue, myalgia, arthralgia, and weakness. Interferon alfa was withdrawn. She had multiple joint involvement, and radiological imaging showed bilateral interstitial pulmonary infiltrates. Anti-Jo antibodies were positive but other autoantibodies were negative. She also had violet eyelid discoloration with edema, tenderness in various muscle groups, and reduced strength in the shoulders. The muscle biopsy showed scattered necrotic fibers and basophilic regenerative fibers. She gradually recovered with methotrexate and corticosteroids, and the titer of anti-Jo antibodies fell dramatically.

Polyarteritis nodosa

Severe polyarteritis nodosa-like systemic vasculitis has been reported (SEDA-20, 330) and cutaneous polyarteritis nodosa has been attributed to interferon alfa (436).

- A 50-year-old woman was given interferon alfa for chronic hepatitis C and primary biliary cirrhosis, and within 2 months became febrile and developed a diffuse nodular erythematous rash. The skin biopsy showed typical features of necrotizing angiitis, and cutaneous periarteritis nodosa was diagnosed. Full recovery was obtained after interferon alfa withdrawal and prednisolone treatment.

According to the authors, it is not known whether this complication was directly due to interferon alfa, represented the triggering of latent periarteritis nodosa in a patient with primary biliary cirrhosis, or whether it was a coincidental adverse event.

Polymyositis

- Polymyositis has very rarely been associated with interferon alfa, but has been reported together with autoimmune thyroiditis in a 48-year-old woman after treatment for 5 months for malignant melanoma (437).

There have been two reports of polymyositis in association with interferon alfa treatment for hematological malignancies (164,438). In both cases, clinical and/or electrophysiological recovery occurred after drug withdrawal, spontaneously or after a short course of glucocorticoids.

Sarcoidosis

The early impression that interferon alfa, alone or in combination with ribavirin, could reactivate or cause new subcutaneous sarcoid nodules and pulmonary or generalized sarcoidosis, has been confirmed by several reports, with prompt recovery after interferon alfa withdrawal (SED-13, 1097; SEDA-20, 330; SEDA-22, 404). The incidence may have been underestimated; in one series, 3 patients out of 60 who received interferon alfa alone or combined with ribavirin developed pulmonary sarcoidosis (439). In a review of 27 cases, the time to onset was 15 days to 30 months, and there were dermatological signs in 50% (440). Five patients had also taken ribavirin, but an enhanced T cell immune reaction from the combination of interferon alfa plus ribavirin is speculative. However, the association of cutaneous or systemic sarcoidosis with interferon alfa, alone or in association with ribavirin, has been exemplified by various reports (441,442), including one patient whose sarcoidosis resolved with prednisone despite continued interferon alfa treatment (443).

Reports of cutaneous or systemic sarcoidosis in patients treated with interferon alfa alone or in association with ribavirin have been published (444,445,446). The course of the disease was usually benign and there was spontaneous recovery despite continuation of the antiviral treatment in one patient with cutaneous sarcoidosis. In another patient, asymptomatic liver and pulmonary sarcoidosis was diagnosed on a systematic liver biopsy 5 years after interferon alfa withdrawal, with diffuse sarcoid granulomas not found before treatment (447). These findings suggest that the incidence of interferon alfa-associated sarcoidosis may have been underestimated.

- A 60-year-old woman receiving interferon alfa developed cutaneous sarcoid foreign body granulomas at the sites of a previously childhood skin injury (448).

This suggests that interferon alfa may facilitate the development of cutaneous sarcoid granuloma from particulate foreign matter.

De novo sarcoidosis has been reported in six patients (449–453) and reactivation of pre-existing disease in one (454). One of these patients had chronic hepatitis B, suggesting that interferon alfa treatment rather than the underlying disease was the most probable triggering factor. Remission was observed in all patients after withdrawal, either spontaneously or after glucocorticoid treatment.

Sjögren's syndrome

One report of Sjögren's syndrome in a patient taking interferon alfa should be regarded with caution (SEDA-20, 330).

Systemic lupus erythematosus and rheumatoid arthritis

The possible role of interferon alfa in the development of rheumatoid arthritis or systemic lupus erythematosus has been described in isolated cases (455,456), and confirmed cases of systemic lupus erythematosus have very occasionally been reported (SED-13, 1096) (SEDA-20, 330). In most of these cases, the predominance of young patients and female sex, the presence of renal or skin involvement, the findings of positive antibodies to double-stranded DNA, and the rapid onset after the start of treatment, as well as persistence of symptoms after interferon alfa withdrawal, are more in keeping with unmasking by interferon alfa of idiopathic lupus rather than with a new drug-induced illness. The reactivation or appearance of inflammatory rheumatological disorders consistent with rheumatoid arthritis or lupus-like polyarthritis were also consistently reported (SEDA-20, 330) (457,458). In a review of 37 published cases of interferon alfa-induced arthritis, symmetrical polyarthritis was the most common feature, and antinuclear antibodies or rheumatoid factor were found in 72 and 34%, respectively (459). Although spontaneous improvement was sometimes observed after withdrawal of interferon alfa, more severe cases required anti-inflammatory, antimalarial, or immunosuppressive drugs. In five of eight patients rechallenged with interferon alfa, there was recurrence of arthritis.

There was an unexpectedly high incidence of rheumatoid and lupus-like symptoms (27 of 137 patients), namely arthralgia, arthritis, myalgia, and Raynaud's phenomenon, in patients with myeloproliferative disorders taking interferon alfa alone or combined with interferon gamma (457). However, only a minority of affected patients fulfilled the diagnostic criteria for systemic lupus erythematosus. By contrast, systemic autoimmune diseases appeared to be genuine but very rare complications of interferon alfa in chronic hepatitis C, with only one case of lupus-like syndrome and two cases of polyarthritis in a survey of 677 patients (23).

Other reports included seronegative polyarthritis, acute seronegative monoarthritis of the hip, and seropositive monoarthritis of the metatarsophalangeal of the right foot (SED-13, 1097).

Systemic lupus erythematosus has been reported in two patients given interferon alfa for chronic hepatitis C (455,460). However, it is not known whether this complication was coincidental or treatment-related.

Systemic sclerosis

Systemic sclerosis has been attributed to interferon alfa-2a (461).

- A 52-year-old woman received interferon alfa-2a for chronic myeloid leukemia and after about 2 years developed fever, dyspnea, and limb edema. The erythrocyte sedimentation rate was 50 mm/hour and pulmonary imaging showed pulmonary vascular congestion. She improved after interferon alfa withdrawal and administration of diuretics, but similar symptoms recurred 3 months later. She also had progressive thickening of the skin on the hands and wrists. There was diffuse parenchymal and interstitial fibrosis of the lungs, absence of peristalsis on esophagogastroduodenoscopy, renal impairment, and positive antisclero-70 antibodies. Capillaroscopy showed typical features of scleroderma. Based on these findings, a diagnosis of systemic sclerosis was suggested and she slowly improved over the next months with cyclophosphamide, prednisone, iloprost, and hydroxyurea.

This patient had the HLA-DR11 haplotype, which is associated with systemic sclerosis, and this suggests that interferon alfa may have triggered the autoimmune phenomenon.

- A 47-year-old woman without a family or personal history of autoimmune disease developed bilateral Raynaud's phenomenon after 6 months of treatment with interferon alfa-2b for chronic hepatitis C (462). Her symptoms worsened over the next 3 months and she had telangiectasiae, finger edema, diffuse pulmonary fibrosis, raised antinuclear autoantibodies, and a positive rheumatoid factor. There was rapid improvement after withdrawal of interferon and prednisone treatment, but she still had a limited form of systemic sclerosis with Raynaud's phenomenon, sclerodactyly, and calcinosis on follow-up after 6 years.

The favorable outcome after interferon alfa withdrawal and the absence of detectable hepatitis C-RNA or mixed cryoglobulinemia at the time of diagnosis suggested that interferon alfa was the most likely cause.

Vasculitis with cryoglobulinemia

Interferon alfa can sometimes aggravate hepatitis C-related cryoglobulinemia.

- A fatal exacerbation of hepatitis C-related cryoglobulinemia, preceded by rapid deterioration of neurological status, massive upper gastrointestinal bleeding, and diffuse hemorrhagic gastritis with vasculitic changes on gastroscopy, has been reported within the first 3 weeks of interferon alfa treatment in a 51-year-old woman (463).

However, cryoglobulin-associated vasculitis is a recognized manifestation of hepatitis C infection. Reports of vasculitis in interferon alfa-treated patients should therefore be interpreted with caution (SED-13, 1095) (464). In addition, several isolated reports suggested that exacerbation of cryoglobulinemia might also be the result of interferon alfa treatment (SEDA-19, 337).

Interferon alfa and transplantation

A possibly deleterious effect of pre- or post-transplant interferon alfa therapy in transplant recipients has been emphasized by several studies and case reports. These point to a higher incidence or greater severity of graft-versus-host reactions in patients who have received an allogeneic bone marrow transplant (SED-13, 1097) (465,466), significant deterioration of renal function or an increased risk of glucocorticoid-resistant rejection in renal transplant patients (SED-13, 1097) (467), and a possibly increased risk of acute or chronic rejection in liver transplant patients (468,469). These findings are still the subject of debate, as the available data consist mainly of retrospective or poorly controlled studies with a limited number of patients (470,471). In addition, several other investigators have failed to identify any deleterious influence of interferon alfa treatment in transplant recipients (SEDA-21, 373; SEDA-22, 403). This issue warrants further large scale, prospective, controlled studies.

In 23 liver transplant recipients treated with pegylated-interferon alfa-2b for 6 months, 35% had evidence of acute or chronic rejection in liver biopsies within 1–6 months after the start of interferon (472). Most of these patients had no history of previous rejection. Two patients had graft loss due to chronic rejection. Only a younger age at transplantation (44 versus 51 years) was a significant susceptibility factor. There was no relation between the time of rejection after transplantation, virological response, length of interferon exposure, history of rejection before interferon treatment, or pretreatment alanine transaminase activity. Persistent elevation of alanine transaminase activity can herald rejection. Unfortunately, rejection episodes can be unresponsive to withdrawal of interferon.

Immunosuppressive effects
Isolated reports of *Candida* esophagitis or *Pneumocystis jiroveci* (*Pneumocystis carinii*) infections in immunocompetent patients and the possible decrease in CD4+ T cells with or without opportunistic infections in several HIV-infected patients (SED-13, 1097) (473) suggest that unexpected immunosuppressive effects of interferon alfa can occur. An autoimmune destruction of CD4 cells in patients with a particular HLA haplotype has been proposed as a possible mechanism (474). One patient also had an acute and fatal acute precipitation of infection with *Entamoeba histolytica* (SEDA-22, 403). However, the available evidence is still very limited and no firm conclusion can be drawn on a possible association between interferon alfa treatment and a fall in CD4 cell count or an immunosuppressive effect.

Two patients aged 38 and 54 years with hemodialysis-dependent end-stage renal insufficiency developed severe bacterial infections, osteomyelitis, and prostatitis, within 3 months of interferon alfa-2b treatment for hepatitis C virus infection (475).

Possible exacerbation of latent parasitic infection by interferon alfa has been reported (SEDA-22, 403).

- Two patients receiving interferon alfa plus ribavirin for chronic hepatitis C developed symptomatic strongyloidiasis within 2–3 weeks of treatment (476).

Because both drugs have immunomodulatory effects, it was not determined which one was the more likely cause.

Graft-versus-host disease
There are still uncertainties about the possible relation between interferon alfa and an increased incidence or severity of acute graft-versus-host disease after bone marrow transplantation. Late-onset, severe, atypical chronic graft-versus-host disease has been attributed to interferon alfa (477).

- A 44-year-old woman received interferon alfa 6 MU/day for relapse of chronic myeloid leukemia 7 years after successful bone marrow transplantation. About 2 years later, interferon alfa was withdrawn because of diffuse erythematous skin lesions with discoid lupus erythematosus on skin biopsy and severe dysphagia with esophagitis and pseudomembranes at endoscopy. Fever, bilateral pulmonary infiltrates, and respiratory distress syndrome subsequently developed, and she required mechanical ventilation. An open lung biopsy showed features of chronic pulmonary graft-versus-host disease. All her symptoms completely resolved with ciclosporin and corticosteroids. An infectious cause was ruled out.

In this case, the clinical presentation was compatible with typical chronic graft-versus-host disease. Whether interferon alfa induced or aggravated chronic graft-versus-host disease in this patient was an open question.

Infection risk
Abscess formation has been attributed to interferon alfa in three patients with chronic hepatitis B or C (478). All three required surgical drainage. None had other underlying diseases, local ports of entry, or concomitant immunosuppressive treatment. Their neutrophil counts were normal. The infection recurred once on interferon alfa readministration in one patient, but all the patients later restarted their treatment uneventfully. Although the role of interferon alfa was debatable in these cases, the authors found three cases of abscesses among 68 patients treated with interferon alfa for chronic hepatitis, but no such cases among 132 chronic healthy carriers or 39 patients with chronic hepatitis B treated with lamivudine.

A higher risk of non-respiratory infections independent of neutropenia has been attributed to peginterferon alfa. In a retrospective study of patients treated with ribavirin plus peginterferon alfa-2a/2b (n = 152) or standard interferon alfa (n = 103), there were 26 infections in the first group and five in the second group (479). Although neutropenia (under 1.0×10^9/l) was significantly more frequent in the pegylated interferon group, this was associated only with respiratory infections (n = 8), without a statistical difference between the groups (five cases in the pegylated interferon group versus three in the standard interferon group). In contrast, no non-respiratory infection was observed in patients with neutropenia.

Death

There is a debate as to whether previous interferon alfa adversely affects the outcome of bone marrow transplantation in chronic myelogenous leukemia.

In a retrospective study of 153 patients who underwent bone-marrow transplantation for chronic myelogenous leukemia, pretransplant interferon alfa treatment for more than 12 months was associated with a significant increase in transplant-related mortality during the first 2 years when compared with patients who received pretransplant interferon alfa for less than 12 months (28 of 46 patients versus nine of 38) (480). This adverse outcome was also more frequent in patients who discontinued treatment less than 3 months before transplantation.

Of eight studies, five showed no harmful effect and three suggested an increased risk of post-transplant complications or mortality (SEDA-22, 403; SEDA-23, 395).

The outcome of bone marrow transplantation in 152 patients (86 on interferon alfa, 66 on chemotherapy) included in two consecutive randomized trials, has been analysed prospectively (481). Whereas the duration of interferon alfa treatment did not influence the outcome of transplantation, there was a significant reduction in survival: the 5-year survival was 45% in the 50 patients who were still receiving interferon alfa within 3 months before bone marrow transplantation and 71% in the 36 patients who were not. According to the authors, interferon alfa should not be prescribed in patients who are likely candidates for early bone marrow transplantation.

Long-Term Effects

Tumorigenicity

Compared with untreated patients and patients treated with busulfan or hydroxyurea, interferon alfa produced a significantly higher frequency of clonal aberrant cytogenetic abnormalities and chronic clonal evolution in patients with chronic myeloid leukemia (482). However, the possible role of interferon alfa in the secondary occurrence of hematological malignancies is purely speculative. Only isolated cases of myeloproliferative syndrome, leukemia, or lymphoma have been attributed to interferon alfa (SED-13, 1098; SEDA-20, 331; SEDA-21, 373). There was no increased incidence of second cancers in patients treated for hairy cell leukemia (SEDA-20, 331).

Second-Generation Effects

Teratogenicity

In experimental models there has been no evidence of mutagenic or teratogenic effects of interferon alfa, and placental transfer is unlikely or very low (483). Immediately after delivery, interferon alfa concentrations in the breast milk or in the sera of two newborns were very low compared with maternal serum concentrations (484). Uncomplicated and successful pregnancies have been detailed in several patients treated for hematological malignancies or chronic hepatitis C, with interferon alfa exposure during the first trimester or even the whole of pregnancy (SED-13, 1098; SEDA-20, 331) (485–488). In only three cases have premature delivery or moderate intrauterine growth retardation been observed, and any direct causal relation to interferon alfa treatment is doubtful. One report mentioned transient and moderate thrombocytopenia in a neonate born to a woman who had received interferon alfa throughout pregnancy (487).

Although no long-term follow-up is as yet available, clinical examination performed up to 2–3 years after delivery in at least seven babies has proved normal. Despite these reassuring data, the safety of interferon alfa during pregnancy still awaits further documentation, and it is advisable to delay therapy of non-life-threatening disease in pregnant women, especially during the first trimester.

Susceptibility Factors

Genetic

Genotype 4a hepatitis C virus is the most prevalent type in patients with Egyptian chronic active hepatitis, and confers poor sensitivity to standard interferon and interferon + ribavirin. Pegylated interferon alfa-2b (n = 30) and interferon alfa-2b (n = 31), both in combination with ribavirin 800–1200 mg/day, resulted in comparable safety and tolerability, but poor responses to antiviral therapy, which may be related to intrinsic resistance to the direct antiviral effect of interferon (489).

Age

There is little information on the use of interferon alfa in children with chronic hepatitis C. In a review of 19 studies published between 1990 and 2000, there were data on only 366 treated children (105 untreated) and they suggested a higher rate of sustained response than in adults (490). Besides flu-like symptoms, reversible weight loss, neutropenia, and alopecia were the most commonly reported adverse events, but adverse events were not systematically recorded in these studies.

Although the anti-angiogenic effects of interferon alfa have been successfully used to treat severe hemangiomas in infants, the possibility of spastic diplegia is a matter of concern (SEDA-22, 404). Spastic diplegia developed in five of 26 infants during treatment, with possibly significant functional sequelae (491), and in one of 53 infants treated for a median of 51 weeks (492). Persistent, severe, spastic diplegia occurred after 1 year of treatment in a 5-month-old boy (493). Because the immature central nervous system of infants may be more susceptible to interferon alfa toxicity, it has been stressed that this treatment should be reserved for infants with life-threatening hemangiomas (494). That interferon alfa can play a role in the occurrence of this acute neurological complication has been further substantiated by the finding of abnormally high interferon serum concentrations in 45% of neonates with spontaneous cerebral palsy compared with control children (495).

Renal disease

High daily dosage or serum concentrations of interferon alfa may be associated with more frequent and more severe adverse effects during the treatment of chronic hepatitis C in hemodialysis patients. In one study, three of 10 hemodialysis patients had severe neurological adverse effects (generalized seizures or posterior leukoencephalopathy) (496). In another study, three of six patients receiving daily injections had to discontinue treatment because of depression, loss of consciousness, and persistent high-grade fever, while no serious adverse effects were reported in three other patients who received interferon alfa three times a week (497). In both studies, there were significant changes in interferon alfa pharmacokinetics (higher C_{max}, AUC, and half-life) in hemodialysis patients compared with patients with normal renal function, consistent with altered clearance of interferon alfa.

Liver disease

Interferon alfa can cause an acute exacerbation of autoimmune hepatitis, but whether interferon can be used safely for patients with chronic hepatitis C who also have autoimmune hepatitis is still not established. Of 2342 patients treated with interferon alfa or beta for chronic hepatitis C, three developed jaundice and increased transaminases during or after interferon treatment, but HCV-RNA concentrations fell dramatically in all (498). Retrospective investigations suggested concomitant probable autoimmune hepatitis before treatment in these patients. Based on these findings, the authors suggested that interferon alfa should remain the frontline regimen in patients with autoimmune hepatitis-like manifestations if autoantibody titers are low, chronic hepatitis C is strongly suspected, and interferon alfa is expected to be effective.

The efficacy and safety of interferon alfa monotherapy in dialysis patients with chronic hepatitis C have been systematically reviewed (499). There were 14 published prospective studies (269 patients), but only two were controlled trials. Flu-like symptoms, neuropsychiatric disorders, and gastrointestinal abnormalities were the most frequent causes of withdrawal, and the overall rate of withdrawal because of adverse effects was 17%. Altered pharmacokinetics of interferon, older age, and frequent co-morbid conditions in hemodialysis patients may have accounted for poorer tolerance.

Liver transplantation

The safety of interferon alfa in liver transplant patients with recurrent hepatitis C is a matter of concern. In a retrospective study of 44 such patients treated with standard interferon alfa (n = 11) or pegylated interferon alfa (n = 33), with or without ribavirin, for a mean of 11 months, five developed biopsy-proven acute cellular rejection after a mean of 3.3 months (500). Four of 33 patients had received peginterferon alfa and one of 10 standard interferon alfa. Three patients responded to antirejection treatment, but two subsequently developed cirrhosis. The other two failed to respond to treatment: one underwent retransplantation while the other died from sepsis after graft loss. These five patients first received interferon alfa long after transplantation (mean of 42 months) and were correctly stabilized (only one had had a previous episode of rejection 5 years before). Of the remaining 39 patients, 13 had to discontinue interferon alfa prematurely because of another adverse effect (leukopenia in four, refractory anemia in three, thrombocytopenia in one, retinopathy in one, and depression in four). Overall, these data suggested poor tolerance of interferon in these patients.

There were three cases of late hepatic artery stenosis among 18 liver transplant patients treated with peginterferon alfa-2b and ribavirin for recurrent chronic hepatitis C (501). These events occurred after a mean treatment duration of 6 months and a median time of 28 months after transplantation. The authors noted that this rate was higher than expected from large studies in liver transplant patients not receiving interferon alfa and their own experience with non-pegylated interferon alfa. Thus, interferon may have a deleterious effect in transplant patients with recurrent hepatitis C. Any additional effect of pegylated interferon alfa needs to be confirmed.

Finally, one of 39 consecutive liver transplant patients treated with peginterferon alfa and ribavirin for recurrent hepatitis C had acute hepatic deterioration with massive ascites and renal insufficiency 1 month after the antiviral treatment had been started (502). The patient recovered after withdrawal and aggressive medical treatment.

HIV infection

The effectiveness and tolerability of interferon alfa in HIV-infected patients is probably lower than in HIV-negative patients. In an open randomized controlled trial in 180 HIV-infected patients with chronic hepatitis C, interferon alfa-2b 3 MU/day plus ribavirin was compared with interferon alfa-2b 3 MU 3 times weekly plus ribavirin; 42 patients had to discontinue interferon alfa because of an adverse effect related to treatment, with no significant between-group differences (503). Although seven adverse effects were serious, no patients died during treatment or developed opportunistic infections. There was no significantly adverse impact of interferon + ribavirin on markers of HIV disease or the effectiveness of antiretroviral treatment. The rate of sustained hepatitis C virus response was only 13%, compared with an expected 42–46% in non HIV-infected patients, with a non-significant higher response in the daily compared with the three-times weekly group.

Drug Administration

Drug formulations

The FDA has expanded the indications for a combination product to include patients with chronic hepatitis C who have not been treated with interferon alfa. This product, Rebetron Combination Therapy (Schering), contains recombinant interferon alfa-2b for injection (Intron A) plus ribavirin (Rebetol) in capsules, and was previously only approved for patients who had relapsed after treatment with interferon alone (504). Serious adverse effects, such as depression, suicidal ideation, and suicide, have occurred with this regimen and patients should be closely monitored.

The two currently available formulations of pegylated interferon alfa differ in their chemistry and pharmacokinetic and pharmacodynamic properties, but owing to a lack of direct comparisons, it is as yet unknown whether these differences are relevant in terms of efficacy or adverse effects (505).

Drug administration route

The use of intraspinal interferon alfa (1 MU three times a week for 4 weeks) in 22 patients with neoplastic meningitis was associated with frequent adverse effects that mostly manifested as chronic fatigue syndrome in 91% of patients (severe in 45%) and arachnoiditis in 73% (severe in 9%) (506). Interferon alfa-induced immune dysregulation in an immunologically predisposed patient was suggested to account for this complication.

Drug–Drug Interactions

Alcohol

Even moderate but continuing alcohol consumption needs to be taken seriously in patients receiving interferon alfa, and exacerbation of previous acute alcohol hepatitis has been reported in two patients with chronic hepatitis C, despite reduced alcohol consumption (507). Liver transaminases subsequently normalized after withdrawal of interferon alfa in both patients.

Angiotensin-converting enzyme inhibitors

An increased risk of severe and early but reversible neutropenia has been found in patients taking angiotensin-converting enzyme inhibitors (enalapril and captopril) with interferon alfa (508).

Carmustine

Owing to its antineoplastic properties, interferon alfa is sometimes used with cytostatic drugs. In 275 patients randomized to receive radiation and carmustine either alone or with interferon alfa for high-grade glioma, there was no significant improvement in the overall survival and time to disease progression in those given interferon alfa, but a higher incidence of adverse effects, namely fever, chills, myalgia, lethargy, headache, and seizures (509).

13-Cisretinoic acid

The combination of interferon alfa with 13-*cis*-retinoic acid may have potentiated the occurrence of fatal radiation pneumonitis (SEDA-21, 374).

Clozapine

Agranulocytosis was observed when interferon alfa was given to a patient taking long-term clozapine (SEDA-22, 404), but this is a known risk of the latter.

Coumarin anticoagulants

There have been two reports of increased prothrombin time in patients taking warfarin or acenocoumarol (SEDA-19, 337) (SEDA-22, 374).

Cyclophosphamide

Depending on the timing of exposure, interferon alfa may adversely affect the pharmacokinetic and hematological effects of cyclophosphamide. In 10 patients with multiple myeloma, interferon alfa given 2 hours before cyclophosphamide infusion significantly reduced cyclophosphamide clearance and produced less exposure to its metabolite 4-hydroxycyclophosphamide compared with interferon administration 24 hours after cyclophosphamide (510). This resulted in a significantly greater fall in white blood cell count in patients who received interferon alfa after cyclophosphamide.

Erythropoietin

Reduced efficacy of human erythropoietin, requiring increased erythropoietin dosages, has been clearly documented in several patients receiving interferon alfa (511–513), an effect that is probably mediated by interferon alfa-induced suppression of erythropoiesis.

5-Fluorouracil

One would expect drugs with myelosuppressive effects to exacerbate the hematological toxicity of interferon alfa. However, even though interferon alfa is increasingly used with other cytotoxic drugs, no specific and unexpected adverse effects have been reported, except for the combination of interferon alfa with 5-fluorouracil, which produced increased serum concentrations of fluorouracil and a significantly higher incidence of severe adverse effects, namely gastrointestinal and myelosuppressive adverse effects (514,515).

Melphalan

Interferon-induced fever has been thought to increase the cytotoxicity of melphalan (516).

Methadone

The safety of interferon alfa in opioid-dependent patients with chronic hepatitis C is of major concern. This issue has been explored in 50 stable patients taking methadone maintenance treatment for a median of 21 months (517). Compared with a prospectively matched group of 50 chronic controls with hepatitis C, who also received peginterferon alfa-2B and ribavirin for 24–48 weeks, more patients from the methadone group discontinued treatment during the first 8 weeks of the study (4% versus 22%). The discontinuation rates were similar in the two groups after 8 weeks of treatment. Poor compliance, patient request, or treatment failure were the major reasons for discontinuation, and only four patients in the methadone group withdrew because of an adverse effect versus two in the control group. There were no major psychiatric events in either group and no significant differences in the number of patients who received an antidepressant in the two arms. The sustained viral response at the end of treatment was similar in the methadone group compared with the control group (56% versus 42%), but only 25 patients taking methadone group versus 38 controls were able to complete the study. In addition, interferon alfa treatment did not appear to require changes in the daily dose of methadone.

Peginterferon alfa-2a treatment for 4 weeks produced only minimal changes in the pharmacokinetics of methadone maintenance treatment in 24 patients (518).

Oral hypoglycemic drugs

There has been one report of hypoglycemia in a patient treated with metformin and chlorpropamide (SEDA-21, 374).

Paroxetine

It has been suggested that concomitant treatment with paroxetine may be a susceptibility factor for retinal damage by interferon alfa (108).

Phenazone (antipyrine)

Interferon alfa inhibits several hepatic microsomal cytochrome P_{450} enzymes in vitro and in vivo (SED-13, 1099) (25,519). However, repeated injections of interferon alfa produced conflicting results, with no change in salivary phenazone clearance (SED-13, 1099) (520).

Ribavirin

The combination of interferon alfa with ribavirin is one of the most promising treatments for chronic hepatitis C. However, two patients developed rapid and particularly severe anemia within 4 and 6 weeks of combined treatment (521). One patient required erythrocyte transfusions, and both recovered after withdrawal. The combination of pure red cell aplasia due to interferon alfa and hemolytic anemia due to ribavirin was suggested to have accounted for this possible interaction.

There was an increased incidence of adverse skin effects, mostly eczema, malar erythema, and lichenoid eruptions, in 33 patients who received combination of interferon alfa with ribavirin compared with 35 patients treated with interferon alfa alone (522).

Thalidomide

In 13 patients with metastatic renal cell carcinoma, the combination of interferon alfa-2a (27 MU/week) and thalidomide produced severe neurological toxicity in four patients, an incidence that was considered to be far greater than would be expected with either drug alone (523).

Theophylline

Interferon alfa inhibits several hepatic microsomal cytochrome P_{450} enzymes in vitro and in vivo (25) and reduces theophylline clearance (SED-13, 1099) (524–526).

Zidovudine

Synergistic hemotoxicity has sometimes resulted from the combination of interferon alfa with zidovudine in AIDS-associated Kaposi's sarcoma, but this regimen is considered to be relatively safe (527,528).

Management of Adverse Drug Reactions

Antiviral pegylated interferon + ribavirin in patients with hepatitis C virus infection can result in thrombocytopenia, which has been successfully treated with rituximab (529).

- A 43-year-old Caucasian woman with hepatitis C infection (genotype 1b) received pegylated interferon + ribavirin for 11 months, during which time her platelet count remained above 150 x 10^9/l. After 1 year the platelet count fell to 31 x 10^9/l and she developed bruising and petechiae. Interferon + ribavirin was withdrawn without improvement in the platelet count. After 1 month, the platelet count fell to 13 x 10^9/l. She was given intravenous immunoglobulin, with transient improvement, but the platelet count fell again and she had bruising and petechiae. She was given intravenous anti-Rhesus D 50 mg/kg and four infusions of rituximab 375 mg/m^2/week. After the second dose her platelet count rose to 249 x 10^9/l and after the fourth dose it was 156 x 10^9/l. Pegylated interferon + ribavirin was restarted after the third infusion of rituximab.

References

1. Keeffe EB, Hollinger FBConsensus Interferon Study Group. Therapy of hepatitis C: consensus interferon trials. Hepatology 1997;26(3 Suppl 1):S101–7.
2. Aspinall RJ, Pockros PJ. The management of side-effects during therapy for hepatitis C. Aliment Pharmacol Ther 2004;20:917–29.
3. Davis GL. New schedules of interferon for chronic hepatitis C. J Hepatol 1999;31(Suppl 1):227–31.
4. Janssen HL, Gerken G, Carreno V, Marcellin P, Naoumov NV, Craxi A, Ring-Larsen H, Kitis G, van Hattum J, de Vries RA, Michielsen PP, ten Kate FJ, Hop WC, Heijtink RA, Honkoop P, Schalm SW. Interferon alfa for chronic hepatitis B infection: increased efficacy of prolonged treatment. The European Concerted Action on Viral Hepatitis (EUROHEP). Hepatology 1999;30(1):238–43.
5. Cotler SJ, Wartelle CF, Larson AM, Gretch DR, Jensen DM, Carithers RL Jr. Pretreatment symptoms and dosing regimen predict side-effects of interferon therapy for hepatitis C. J Viral Hepat 2000;7(3):211–7.
6. Kantarjian HM, O'Brien S, Smith TL, Rios MB, Cortes J, Beran M, Koller C, Giles FJ, Andreeff M, Kornblau S, Giralt S, Keating MJ, Talpaz M. Treatment of Philadelphia chromosome-positive early chronic phase chronic myelogenous leukemia with daily doses of interferon alpha and low-dose cytarabine. J Clin Oncol 1999;17(1):284–92.

7. Scott LJ, Perry CM. Interferon-alpha-2b plus ribavirin: a review of its use in the management of chronic hepatitis C. Drugs 2002;62(3):507–56.
8. San Miguel R, Guillen F, Cabases JM, Buti M. Meta-analysis: combination therapy with interferon-alpha 2a/2b and ribavirin for patients with chronic hepatitis C previously non-responsive to interferon. Aliment Pharmacol Ther 2002;16(9):1611–21.
9. Gaeta GB, Precone DF, Felaco FM, Bruno R, Spadaro A, Stornaiuolo G, Stanzione M, Ascione T, De Sena R, Campanone A, Filice G, Piccinino F. Premature discontinuation of interferon plus ribavirin for adverse effects: a multicentre survey in "real world" patients with chronic hepatitis C. Aliment Pharmacol Ther 2002;16(9):1633–9.
10. Barbaro G, Grisorio B, Fruttaldo L, Bacca D, Babudieri S, Torre D, Francavilla R, Rizzo G, Belloni G, Lucchini A, Annese M, Matarazzo F, Hazra C, Barbarini G. Good safety profile and efficacy of leucocyte interferon-alpha in combination with oral ribavirin in treatment-naive patients with chronic hepatitis C: a multicentre, randomised, controlled study. BioDrugs 2003;17:433–9.
11. Tripi S, Soresi M, Di Gaetano G, Carroccio A, Giannitrapani L, Vuturo O, Di Giovanni G, Montalto G. Leucocyte interferon-alpha for patients with chronic hepatitis C intolerant to other alpha-interferons. BioDrugs 2003;17:201–5.
12. Neri S, Pistone G, Saraceno B, Pennisi G, Luca S, Malaguarnera M. L-carnitine decreases severity and type of fatigue induced by interferon-alfa in the treatment of patients with hepatitis C. Neuropsychobiology 2003;47:94–7.
13. Bagheri H, Fouladi A, Barange K, Lapeyre-Mestre M, Payen JL, Montastruc JL, Vinel JP. Follow-up of adverse drug reactions from peginterferon alfa-2b-ribavirin therapy. Pharmacotherapy 2004;24:1546–53.
14. Ascione A, De Luca M, Di Costanzo GG, Picciotto FP, Lanza AG, Canestrini C, Morisco F, Tuccillo C, Caporaso N. Incidence of side effects during therapy with different types of alpha interferon: a randomised controlled trial comparing recombinant alpha 2b versus leukocyte interferon in the therapy of naive patients with chronic hepatitis C. Curr Pharm Des 2002;8(11):977–80.
15. Manns MP, McHutchison JG, Gordon SC, Rustgi VK, Shiffman M, Reindollar R, Goodman ZD, Koury K, Ling M, Albrecht JK. Peginterferon alfa-2b plus ribavirin compared with interferon alfa-2b plus ribavirin for initial treatment of chronic hepatitis C: a randomised trial. Lancet 2001;358(9286):958–65.
16. Heathcote EJ, Shiffman ML, Cooksley WG, Dusheiko GM, Lee SS, Balart L, Reindollar R, Reddy RK, Wright TL, Lin A, Hoffman J, De Pamphilis J. Peginterferon alfa-2a in patients with chronic hepatitis C and cirrhosis. N Engl J Med 2000;343(23):1673–80.
17. Zeuzem S, Feinman SV, Rasenack J, Heathcote EJ, Lai MY, Gane E, O'Grady J, Reichen J, Diago M, Lin A, Hoffman J, Brunda MJ. Peginterferon alfa-2a in patients with chronic hepatitis C. N Engl J Med 2000;343(23):1666–72.
18. Wiselka MJ, Nicholson KG, Kent J, Cookson JB, Tyrrell DA. Prophylactic intranasal alpha 2 interferon and viral exacerbations of chronic respiratory disease. Thorax 1991;46(10):706–11.
19. Saracco G, Rizzetto M. A practical guide to the use of interferons in the management of hepatitis virus infections. Drugs 1997;53(1):74–85.
20. Poynard T, Leroy V, Cohard M, Thevenot T, Mathurin P, Opolon P, Zarski JP. Meta-analysis of interferon randomized trials in the treatment of viral hepatitis C: effects of dose and duration. Hepatology 1996;24(4):778–89.
21. Fattovich G, Giustina G, Favarato S, Ruol A. A survey of adverse events in 11,241 patients with chronic viral hepatitis treated with alfa interferon. J Hepatol 1996;24(1):38–47.
22. De Sanctis GM, D'Errico DAF, Leonetti G, et al. Occurrence of major side effects in patients with chronic viral liver disease treated with interferons. Mediter J Infect Parasit Dis 1995;10:225–30.
23. Okanoue T, Sakamoto S, Itoh Y, Minami M, Yasui K, Sakamoto M, Nishioji K, Katagishi T, Nakagawa Y, Tada H, Sawa Y, Mizuno M, Kagawa K, Kashima K. Side effects of high-dose interferon therapy for chronic hepatitis C. J Hepatol 1996;25(3):283–91.
24. Iorio R, Pensati P, Botta S, Moschella S, Impagliazzo N, Vajro P, Vegnente A. Side effects of alpha-interferon therapy and impact on health-related quality of life in children with chronic viral hepatitis. Pediatr Infect Dis J 1997;16(10):984–90.
25. Vial T, Descotes J. Clinical toxicity of the interferons. Drug Saf 1994;10(2):115–50.
26. Vial T, Bailly F, Descotes J, Trepo C. Effets secondaires de l'interféron alpha. [Side effects of interferon-alpha.] Gastroenterol Clin Biol 1996;20(5):462–89.
27. Cacopardo B, Benanti F, Brancati G, Romano F, Nunnari A. Leucocyte interferon-alpha retreatment for chronic hepatitis C patients previously intolerant to other interferons. J Viral Hepat 1998;5(5):333–9.
28. Sonnenblick M, Rosin A. Cardiotoxicity of interferon. A review of 44 cases. Chest 1991;99(3):557–61.
29. Kruit WH, Punt KJ, Goey SH, de Mulder PH, van Hoogenhuyze DC, Henzen-Logmans SC, Stoter G. Cardiotoxicity as a dose-limiting factor in a schedule of high dose bolus therapy with interleukin-2 and alpha-interferon. An unexpectedly frequent complication. Cancer 1994;74(10):2850–6.
30. Angulo MP, Navajas A, Galdeano JM, Astigarraga I, Fernandez-Teijeiro A. Reversible cardiomyopathy secondary to alpha-interferon in an infant. Pediatr Cardiol 1999;20(4):293–4.
31. Kadayifci A, Aytemir K, Arslan M, Aksoyek S, Sivri B, Kabakci G. Interferon-alpha does not cause significant cardiac dysfunction in patients with chronic active hepatitis. Liver 1997;17(2):99–102.
32. Sartori M, Andorno S, La Terra G, Pozzoli G, Rudoni M, Sacchetti GM, Inglese E, Aglietta M. Assessment of interferon cardiotoxicity with quantitative radionuclide angiocardiography. Eur J Clin Invest 1995;25(1):68–70.
33. Kuwata A, Ohashi M, Sugiyama M, Ueda R, Dohi Y. A case of reversible dilated cardiomyopathy after alpha-interferon therapy in a patient with renal cell carcinoma. Am J Med Sci 2002;324(6):331–4.
34. Gupta SK, Glue P, Jacobs S, Belle D, Affrime M. Single-dose pharmacokinetics and tolerability of pegylated interferon-alfa2b in young and elderly healthy subjects. Br J Clin Pharmacol 2003;56:131–4.
35. Al-Zahrani H, Gupta V, Minden MD, Messner HA, Lipton JH. Vascular events associated with alfa interferon therapy. Leuk Lymphoma 2003;44:471–5.
36. Wisniewski B, Denis J, Fischer D, Labayle D. Péricardite induite par l'interféron alfa au cours d'une hépatite chronique C. Gastroentérol Clin Biol 2004;28:315–16.

37. Gressens B, Gohy P. Pericarditis due to interferon-alpha therapy during treatment for chronic hepatitis C. Acta Gastroenterol Belg 2004;67:301–2.

38. Colivicchi F, Magnanimi S, Sebastiani F, Silvestri R, Magnanimi R. Incidence of electrocardiographic abnormalities during treatment with human leukocyte interferon-alfa in patients with chronic hepatitis C but without pre-existing cardiovascular disease. Curr Ther Res Clin Exp 1998;59:692–6.

39. Parrens E, Chevalier JM, Rougier M, Douard H, Labbe L, Quiniou G, Broustet A, Broustet JP. Apparition d'un bloc auriculo-ventriculaire du troisième degré sous interféron alpha: à propos d'un cas. [Third degree atrio-ventricular block induced by interferon alpha. Report of a case.] Arch Mal Coeur Vaiss 1999;92(1):53–6.

40. Creutzig A, Caspary L, Freund M. The Raynaud phenomenon and interferon therapy. Ann Intern Med 1996;125(5):423.

41. Schapira D, Nahir AM, Hadad N. Interferon-induced Raynaud's syndrome. Semin Arthritis Rheum 2002;32(3):157–62.

42. Iorio R, Spagnuolo MI, Sepe A, Zoccali S, Alessio M, Vegnente A. Severe Raynaud's phenomenon with chronic hepatis C disease treated with interferon. Pediatr Infect Dis J 2003;22(2):195–7.

43. Cid MC, Hernandez-Rodriguez J, Robert J, del Rio A, Casademont J, Coll-Vinent B, Grau JM, Kleinman HK, Urbano-Marquez A, Cardellach F. Interferon-alpha may exacerbate cryoblobulinemia-related ischemic manifestations: an adverse effect potentially related to its anti-angiogenic activity. Arthritis Rheum 1999;42(5):1051–5.

44. Mirro J Jr, Kalwinsky D, Whisnant J, Weck P, Chesney C, Murphy S. Coagulopathy induced by continuous infusion of high doses of human lymphoblastoid interferon. Cancer Treat Rep 1985;69(3):315–7.

45. Durand JM, Quiles N, Kaplanski G, Soubeyrand J. Thrombosis and recombinant interferon-alpha. Am J Med 1993;95(1):115–6.

46. Becker JC, Winkler B, Klingert S, Brocker EB. Antiphospholipid syndrome associated with immunotherapy for patients with melanoma. Cancer 1994;73(6):1621–4.

47. Karim A, Ahmed S, Khan A, Steinberg H, Mattana J. Interstitial pneumonitis in a patient treated with alpha-interferon and ribavirin for hepatitis C infection. Am J Med Sci 2001;322(4):233–5.

48. Chin K, Tabata C, Sataka N, Nagai S, Moriyasu F, Kuno K. Pneumonitis associated with natural and recombinant interferon alfa therapy for chronic hepatitis C. Chest 1994;105(3):939–41.

49. Ishizaki T, Sasaki F, Ameshima S, Shiozaki K, Takahashi H, Abe Y, Ito S, Kuriyama M, Nakai T, Kitagawa M. Pneumonitis during interferon and/or herbal drug therapy in patients with chronic active hepatitis. Eur Respir J 1996;9(12):2691–6.

50. Ogata K, Koga T, Yagawa K. Interferon-related bronchiolitis obliterans organizing pneumonia. Chest 1994;106(2):612–3.

51. Kumar AS, Russo MW, Esposito S, Borczuk A, Jacobson I, Brown M, Brown RS. Severe pulmonary toxicity of interferon and ribavirin therapy in chronic hepatitis C. Am J Gastroenterol 2001;96(Suppl):127.

52. Patel M, Ezzat W, Pauw KL, Lowsky R. Bronchiolitis obliterans organizing pneumonia in a patient with chronic myelogenous leukemia developing after initiation of interferon and cytosine arabinoside. Eur J Haematol 2001;67(5–6):318–21.

53. Anderson P, Hoglund M, Rodjer S. Pulmonary side effects of interferon-alpha therapy in patients with hematological malignancies. Am J Hematol 2003;73(1):54–8.

54. Tokita H, Fukui H, Tanaka A, Kamitsukasa H, Yagura M, Harada H, Hebisawa A, Kurashima A, Okamoto H. Circulating KL-6 level at baseline is a predictive indicator for the occurrence of interstitial pneumonia during interferon treatment for chronic hepatitis C. Hepatol Res 2003;26:91–7.

55. Abi-Nassif S, Mark EJ, Fogel RB, Hallisey RK. Pegylated interferon and ribavirin-induced interstitial pneumonitis with ARDS. Chest 2003;124:406–10.

56. Kalambokis G, Stefanou D, Arkoumani E, Kitsanou M, Bourantas K, V Tsianos E. Fulminant bronchiolitis obliterans organizing pneumonia following 2 d of treatment with hydroxyurea, interferon-alpha and oral cytarabine ocfosfate for chronic myelogenous leukemia. Eur J Haematol 2004;73:67–70.

57. Yamamoto S, Tomita Y, Hoshida Y, Iizuka N, Marubashi S, Miyamoto A, Nagano H, Dono K, Umeshita K, Nakamori S, Sakon M, Aozasa K, Monden M. Interstitial pneumonia induced by combined intraarterial 5-fluorouracil and subcutaneous interferon-alpha therapy for advanced hepatocellular carcinoma. J Gastroenterol 2004;39:793–7.

58. Midturi J, Sierra-Hoffman M, Hurley D, Winn R, Beissner R, Carpenter J. Spectrum of pulmonary toxicity associated with the use of interferon therapy for hepatitis C: case report and review of the literature. Clin Infect Dis 2004;39:1724–9.

59. Kumar KS, Russo MW, Borczuk AC, Brown M, Esposito SP, Lobritto SJ, Jacobson IM, Brown RS Jr. Significant pulmonary toxicity associated with interferon and ribavirin therapy for hepatitis C. Am J Gastroenterol 2002;97(9):2432–40.

60. Bini EJ, Weinshel EH. Severe exacerbation of asthma: a new side effect of interferon-alpha in patients with asthma and chronic hepatitis C. Mayo Clin Proc 1999;74(4):367–70.

61. Pileire G, Leclerc P, Hermant P, Meeus E, Camus P. Toux chronique isolée pendant un traitement par interféron. [Isolated chronic cough during interferon therapy.] Presse Méd 1999;28(17):913.

62. Takeda A, Ikegame K, Kimura Y, Ogawa H, Kanazawa S, Nakamura H. Pleural effusion during interferon treatment for chronic hepatitis C. Hepatogastroenterology 2000;47(35):1431–5.

63. Fruehauf S, Steiger S, Topaly J, Ho AD. Pulmonary artery hypertension during interferon-alpha therapy for chronic myelogenous leukemia. Ann Hematol 2001;80(5):308–10.

64. Kamei S, Oga K, Matsuura M, Tanaka N, Kojima T, Arakawa Y, Matsukawa Y, Mizutani T, Sakai T, Ohkubo H, Matsumura H, Moriyama M, Hirayanagi K. Correlation between quantitative-EEG alterations and age in patients with interferon-alpha-treated hepatitis C. J Clin Neurophysiol 2005;22:49–52.

65. Boonyapisit K, Katirji B. Severe exacerbation of hepatitis C-associated vasculitic neuropathy following treatment with interferon alpha: a case report and literature review. Muscle Nerve 2002;25(6):909–13.

66. Tambini R, Quattrini A, Fracassetti O, Nemni R. Axonal neuropathy in a patient receiving interferon-alpha therapy for chronic hepatitis C. J Rheumatol 1997;24(8):1656–7.

67. Maeda M, Ohkoshi N, Hisahara S, Mizusawa H, Shoji S. [Mononeuropathy multiplex in a patient receiving interferon alpha therapy for chronic hepatitis C.]Rinsho Shinkeigaku 1995;35(9):1048–50.

68. Jaubert D, Hauteville D, Pelissier JF, Muzellec Y. Neuropathie periphérique au cours d'un traitement par interféron alpha. [Peripheral neuropathy during interferon alpha therapy.] Presse Méd 1991;20(5):221–2.

69. Purvin VA. Anterior ischemic optic neuropathy secondary to interferon alfa. Arch Ophthalmol 1995;113(8):1041–4.

70. Read SJ, Crawford DH, Pender MP. Trigeminal sensory neuropathy induced by interferon-alpha therapy. Aust NZ J Med 1995;25(1):54.

71. Bernsen PL, Wong Chung RE, Vingerhoets HM, Janssen JT. Bilateral neuralgic amyotrophy induced by interferon treatment. Arch Neurol 1988;45(4):449–51.

72. Loh FL, Herskovitz S, Berger AR, Swerdlow ML. Brachial plexopathy associated with interleukin-2 therapy. Neurology 1992;42(2):462–3.

73. Merimsky O, Reider I, Merimsky E, Chaitchik S. Interferon-related leukoencephalopathy in a patient with renal cell carcinoma. Tumori 1991;77(4):361–2.

74. Finsterer J, Sommer O, Stiskal M. Multifocal leukoencephalopathy and polyneuropathy after 18 years on interferon alpha. Leuk Lymphoma 2005;46(2):277–80.

75. Ogundipe O, Smith M. Bell's palsy during interferon therapy for chronic hepatitis C infection in patients with haemorrhagic disorders. Haemophilia 2000;6(2):110–2.

76. Hwang I, Calvit TB, Cash BD, Holtzmuller KC. Bell's palsy. A rare complication of interferon therapy for hepatitis C. Am J Gastroenterol 2002;87(Suppl):207–8.

77. Hwang I, Calvit TB, Cash BD, Holtzmuller KC. Bell's palsy: a rare complication of interferon therapy for hepatitis C. Dig Dis Sci 2004;49:619–20.

78. Hoare M, Woodall T, Alexander GJ. Bell's palsy associated with IFN-alpha and ribavirin therapy for hepatitis C virus infection. J Interferon Cytokine Re. 2005;25:174–6.

79. Anthoney DA, Bone I, Evans TR. Inflammatory demyelinating polyneuropathy: a complication of immunotherapy in malignant melanoma. Ann Oncol 2000;11(9):1197–200.

80. Meriggioli MN, Rowin J. Chronic inflammatory demyelinating polyneuropathy after treatment with interferon-alpha. Muscle Nerve 2000;23(3):433–5.

81. Kataoka I, Shinagawa K, Shiro Y, Okamoto S, Watanabe R, Mori T, Ito D, Harada M. Multiple sclerosis associated with interferon-alpha therapy for chronic myelogenous leukemia. Am J Hematol 2002;70(2):149–53.

82. Matsuo T, Takabatake R. Multiple sclerosis-like disease secondary to alfa interferon. Ocul Immunol Inflamm 2002;10:299–304.

83. Bachmann T, Koetter KP, Muhler J, Fuhrmeister U, Seidel G. Guillain-Barre syndrome after simultaneous therapy with suramin and interferon-alfa. Eur J Neurol 2003;10:599.

84. Sunami M, Nishikawa T, Yorogi A, Shimoda M. Intravenous administration of levodopa ameliorated a refractory akathisia case induced by interferon-alpha. Clin Neuropharmacol 2000;23(1):59–61.

85. Horikawa N, Yamazaki T, Sagawa M, Nagata T. A case of akathisia during interferon-alpha therapy for chronic hepatitis type C. Gen Hosp Psychiatry 1999;21(2):134–5.

86. Moulignier A, Allo S, Zittoun R, Gout O. Recombinant interferon-alpha-induced chorea and frontal subcortical dementia. Neurology 2002;58(2):328–30.

87. Benatru I, Thobois S, Andre-Obadia N, Gonnaud PM, Beaugendre Y, Berger C, Gonce M, Broussolle E. Atypical propriospinal myoclonus with possible relationship to alfa interferon therapy. Mov Disord 2003;18:1564–8.

88. Tan EK, Chan LL, Lo YL. "Myorhythmia" slow facial tremor from chronic interferon alfa-2a usage. Neurology 2003;61:1302–3.

89. Atasoy N, Ustundag Y, Konuk N, Atik L. Acute dystonia during pegylated interferon alpha therapy in a case with chronic hepatitis B infection. Clin Neuropharmacol 2004;27:105–7.

90. LaRochelle JS, Karp BI. Restless legs syndrome due to interferon-alpha. Mov Disord 2004;19:730–1.

91. Legroux-Crespel E, Lafaye S, Mahe E, Picard-Dahan C, Crickx B, Sassolas B, Descamps V. Crise épileptiques généralisées: un effet secondaire de l'interféron alfa: 3 cas. Ann Dermatol Venéréol 2003;130:202–4.

92. Caksen H, Odabas D, Anlar O, Atas B, Tuncer O. Onset of generalized seizures after intrathecal interferon therapy of SSPE. Pediatr Neurol 2003;29:78–9.

93. Hibi H, Itoh K, Kamiya T, Yamada Y, Shimoji T. [Grand mal like attack by interferon injection in case of renal cell carcinoma.]Hinyokika Kiyo 1991;37(1):69–72.

94. Janssen HL, Berk L, Vermeulen M, Schalm SW. Seizures associated with low-dose alpha-interferon. Lancet 1990;336(8730):1580.

95. Miller VS, Zwiener RJ, Fielman BA. Interferon-associated refractory status epilepticus. Pediatrics 1994;93(3):511–2.

96. Shakil AO, Di Bisceglie AM, Hoofnagle JH. Seizures during alpha interferon therapy. J Hepatol 1996;24(1):48–51.

97. Woynarowski M, Socha J. Seizures in children during interferon alpha therapy. J Hepatol 1997;26(4):956–7.

98. Ameen M, Russell-Jones R. Seizures associated with interferon-alpha treatment of cutaneous malignancies. Br J Dermatol 1999;141(2):386–7.

99. Seno H, Inagaki T, Itoga M, Miyaoka T, Ishino H. A case of seizures 1 week after the cessation of interferon-alpha therapy. Psychiatry Clin Neurosci 1999;53(3):417–20.

100. Brouwers PJ, Bosker RJ, Schaafsma MR, Wilts G, Neef C. Photosensitive seizures associated with interferon alfa-2a. Ann Pharmacother 1999;33(1):113–4.

101. Kamei S, Tanaka N, Mastuura M, Arakawa Y, Kojima T, Matsukawa Y, Takasu T, Moriyama M. Blinded, prospective, and serial evaluation by quantitative-EEG in interferon-alpha-treated hepatitis-C. Acta Neurol Scand 1999;100(1):25–33.

102. Borgia G, Reynaud L, Gentile I, Cerini R, Ciampi R, Dello Russo M, Piazza M. Myasthenia gravis during low-dose IFN-alpha therapy for chronic hepatitis C. J Interferon Cytokine Res 2001;21(7):469–70.

103. Weegink CJ, Chamuleau RA, Reesink HW, Molenaar DS. Development of myasthenia gravis during treatment of chronic hepatitis C with interferon-alpha and ribavirin. J Gastroenterol 2001;36(10):723–4.

104. Guyer DR, Tiedeman J, Yannuzzi LA, Slakter JS, Parke D, Kelley J, Tang RA, Marmor M, Abrams G, Miller JW, et al. Interferon-associated retinopathy. Arch Ophthalmol 1993;111(3):350–6.

105. Hayasaka S, Fujii M, Yamamoto Y, Noda S, Kurome H, Sasaki M. Retinopathy and subconjunctival haemorrhage in patients with chronic viral hepatitis receiving interferon alfa. Br J Ophthalmol 1995;79(2):150–2.

106. Kawano T, Shigehira M, Uto H, Nakama T, Kato J, Hayashi K, Maruyama T, Kuribayashi T, Chuman T, Futami T, Tsubouchi H. Retinal complications during interferon therapy for chronic hepatitis C. Am J Gastroenterol 1996;91(2):309–13.

107. Soushi S, Kobayashi F, Obazawa H, Kigasawa K, Shiraishi K, Itakura M, Matsuzaki S. [Evaluation of risk factors of interferon-associated retinopathy in patients with type C chronic active hepatitis.]Nippon Ganka Gakkai Zasshi 1996;100(1):69–76.

108. Hejny C, Sternberg P, Lawson DH, Greiner K, Aaberg TM Jr. Retinopathy associated with high-dose interferon alfa-2b therapy. Am J Ophthalmol 2001;131(6):782–7.

109. Jain K, Lam WC, Waheeb S, Thai Q, Heathcote J. Retinopathy in chronic hepatitis C patients during interferon treatment with ribavirin. Br J Ophthalmol 2001;85(10):1171–3.

110. Saito H, Ebinuma H, Nagata H, Inagaki Y, Saito Y, Wakabayashi K, Takagi T, Nakamura M, Katsura H, Oguchi Y, Ishii H. Interferon-associated retinopathy in a uniform regimen of natural interferon-alpha therapy for chronic hepatitis C. Liver 2001;21(3):192–7.

111. Hayasaka S, Nagaki Y, Matsumoto M, Sato S. Interferon associated retinopathy. Br J Ophthalmol 1998;82(3):323–5.

112. Kadayifcilar S, Boyacioglu S, Kart H, Gursoy M, Aydin P. Ocular complications with high-dose interferon alpha in chronic active hepatitis. Eye 1999;13(Pt 2):241–6.

113. Schulman JA, Liang C, Kooragayala LM, King J. Posterior segment complications in patients with hepatitis C treated with interferon and ribavirin. Ophthalmology 2003;110:437–42.

114. Sandner D, Pillunat LE. Reversible Perimakuläre Retinopathie bei Interferon-α-Langzeittherapie. Klin Monatsbl Augenheilkd 2004;221:63–6.

115. Crochet M, Ingster-Moati I, Even G, Dupuy P. Retinopathie à l'interféron-α associée à la ribavirine: atteinte de l'électro-oculogramme: à propos d'un cas. J Fr Ophtalmol 2004;27:257–62.

116. Willson RA. Visual side effects of pegylated interferon during therapy for chronic hepatitis C infection. J Clin Gastroenterol 2004;38:717–22.

117. Sugano S, Suzuki T, Watanabe M, Ohe K, Ishii K, Okajima T. Retinal complications and plasma C5a levels during interferon alpha therapy for chronic hepatitis C. Am J Gastroenterol 1998;93(12):2441–4.

118. Sugano S, Yanagimoto M, Suzuki T, Sato M, Onmura H, Aizawa H, Makino H. Retinal complications with elevated circulating plasma C5a associated with interferon-alpha therapy for chronic active hepatitis C. Am J Gastroenterol 1994;89(11):2054–6.

119. Ene L, Gehenot M, Horsmans Y, Detry-Morel M, Geubel AP. Transient blurred vision after interferon for chronic hepatitis C. Lancet 1994;344(8925):827–8.

120. Yamada H, Mizobuchi K, Isogai Y. Acute onset of ocular complications with interferon. Lancet 1994;343(8902):914.

121. Lohmann CP, Kroher G, Bogenrieder T, Spiegel D, Preuner J. Severe loss of vision during adjuvant interferon alfa-2b treatment for malignant melanoma. Lancet 1999;353(9161):1326.

122. Perlemuter G, Bodaghi B, Le Hoang P, Izem C, Buffet C, Wechsler B, Piette JC, Cacoub P. Visual loss during interferon-alpha therapy in hepatitis C virus infection. J Hepatol 2002;37(5):701–2.

123. Gupta R, Singh S, Tang R, Blackwell TA, Schiffman JS. Anterior ischemic optic neuropathy caused by interferon alpha therapy. Am J Med 2002;112(8):683–4.

124. Savant V, Gillow T. Interferon-associated retinopathy. Eye 2003;17:534–6.

125. Kargi SH, Oz O, Ustundag Y, Firat E. Epiretinal membrane development during interferon treatment. Can J Ophthalmol 2003;38:610–12.

126. Rubio JE, Charles S. Interferon-associated combined branch retinal artery and central retinal vein obstruction. Retina 2003;23:546–8.

127. Vardizer Y, Linhart Y, Loewenstein A, Garzozi H, Mazawi N, Kesler A. Interferon-alfa-associated bilateral simultaneous ischemic optic neuropathy. J Neuroophthalmol 2003;23:256–9.

128. Farel C, Suzman DL, McLaughlin M, Campbell C, Koratich C, Masur H, Metcalf JA, Robinson MR, Polis MA, Kottilil S. Serious ophthalmic pathology compromising vision in HCV/HIV co-infected patients treated with peginterferon alpha-2b and ribavirin. AIDS 2004;18:1805–9.

129. Sylvestre DL, Disston AR, Bui DP. Vogt-Koyanagi-Harada disease associated with interferon alfa-2b/ribavirin combination therapy. J Viral Hepat 2003;10:467–70.

130. Kasahara A, Hiraide A, Tomita N, Iwahashi H, Imagawa A, Ohguro N, Yamamoto S, Mita E, Hayashi N. Vogt-Koyanagi-Harada disease occurring during interferon alpha therapy for chronic hepatitis C. J Gastroenterol 2004;39:1106–9.

131. Manesis EK, Petrou C, Brouzas D, Hadziyannis S. Optic tract neuropathy complicating low-dose interferon treatment. J Hepatol 1994;21(3):474–7.

132. Merimsky O, Nisipeanu P, Loewenstein A, Reider-Groswasser I, Chaitchik S. Interferon-related cortical blindness. Cancer Chemother Pharmacol 1992;29(4):329–30.

133. Fukumoto Y, Shigemitsu T, Kajii N, Omura R, Harada T, Okita K. Abducent nerve paralysis during interferon alpha-2a therapy in a case of chronic active hepatitis C. Intern Med 1994;33(10):637–40.

134. Bauherz G, Soeur M, Lustman F. Oculomotor nerve paralysis induced by alpha II-interferon. Acta Neurol Belg 1990;90(2):111–4.

135. Fossa SD. Is interferon with or without vinblastine the "treatment of choice" in metastatic renal cell carcinoma? The Norwegian Radium Hospital's experience 1983–1986. Semin Surg Oncol 1988;4(3):178–83.

136. Gorur K, Kandemir O, Unal M, Ozcan C. The effect of recombinant interferon alfa treatment on hearing thresholds in patients with chronic viral hepatitis B. Auris Nasus Larynx 2003;30:41–4.

137. Formann E, Stauber R, Denk DM, Jessner W, Zollner G, Munda-Steindl P, Gangl A, Ferenci P. Sudden hearing loss in patients with chronic hepatitis C treated with pegylated interferon/ribavirin. Am J Gastroenterol 2004;99:873–7.

138. Cadoni G, Marinelli L, De Santis A, Romito A, Manna R, Ottaviani F. Sudden hearing loss in a patient hepatitis C virus (HCV) positive on therapy with alpha-interferon: a possible autoimmune-microvascular pathogenesis. J Laryngol Otol 1998;112(10):962–3.

139. Formann E, Stauber R, Denk DM, Jessner W, Zollner G, Munda-Steindl P, Gangl A, Ferenci P. Sudden hearing loss in patients with chronic hepatitis C treated with pegylated interferon/ribavirin. Am J Gastroenterol 2004;99(5):873–7.

140. Aleixandre SI, Castiella Eguzkira A, Fernandez Fernandez FJ. Reversible ageusia and hyposmia induced by interferon in a patient with chronic hepatitis C. Rev Esp Enferm Dig 2003;95:364–8.

141. Kraus I, Vitezic D. Anosmia induced with alpha interferon in a patient with chronic hepatitis C. Int J Clin Pharmacol Ther 2000;38(7):360–1.

142. Schaefer M, Engelbrecht MA, Gut O, Fiebich BL, Bauer J, Schmidt F, Grunze H, Lieb K. Interferon alpha (IFNalpha) and psychiatric syndromes: a review. Prog Neuropsychopharmacol Biol Psychiatry 2002;26(4):731–46.

143. Van Gool AR, Kruit WH, Engels FK, Stoter G, Bannink M, Eggermont AM. Neuropsychiatric side effects of interferon-alfa therapy. Pharm World Sci 2003;25(1):11–20.

144. Smith A, Tyrrell D, Coyle K, Higgins P. Effects of interferon alpha on performance in man: a preliminary report. Psychopharmacology (Berl) 1988;96(3):414–6.

145. Meyers CA, Valentine AD. Neurological and psychiatric adverse effects of immunological therapy. CNS Drugs 1995;3:56–68.

146. James CW, Savini CJ. Homicidal ideation secondary to interferon. Ann Pharmacother 2001;35(7–8):962–3.

147. Greenberg DB, Jonasch E, Gadd MA, Ryan BF, Everett JR, Sober AJ, Mihm MA, Tanabe KK, Ott M, Haluska FG. Adjuvant therapy of melanoma with interferon-alpha-2b is associated with mania and bipolar syndromes. Cancer 2000;89(2):356–62.

148. Monji A, Yoshida I, Tashiro K, Hayashi Y, Tashiro N. A case of persistent manic depressive illness induced by interferon-alfa in the treatment of chronic hepatitis C. Psychosomatics 1998;39(6):562–4.

149. Hosoda S, Takimura H, Shibayama M, Kanamura H, Ikeda K, Kumada H. Psychiatric symptoms related to interferon therapy for chronic hepatitis C: clinical features and prognosis. Psychiatry Clin Neurosci 2000;54(5):565–72.

150. Schafer M, Boetsch T, Laakmann G. Psychosis in a methadone-substituted patient during interferon-alpha treatment of hepatitis C. Addiction 2000;95(7):1101–4.

151. Mayr N, Zeitlhofer J, Deecke L, Fritz E, Ludwig H, Gisslinger H. Neurological function during long-term therapy with recombinant interferon alpha. J Neuropsychiatry Clin Neurosci 1999;11(3):343–8.

152. Rohatiner AZ, Prior PF, Burton AC, Smith AT, Balkwill FR, Lister TA. Central nervous system toxicity of interferon. Br J Cancer 1983;47(3):419–22.

153. Scalori A, Apale P, Panizzuti F, Mascoli N, Pioltelli P, Pozzi M, Redaelli A, Roffi L, Mancia G. Depression during interferon therapy for chronic viral hepatitis: early identification of patients at risk by means of a computerized test. Eur J Gastroenterol Hepatol 2000;12(5):505–9.

154. Dieperink E, Willenbring M, Ho SB. Neuropsychiatric symptoms associated with hepatitis C and interferon alpha: A review. Am J Psychiatry 2000;157(6):867–76.

155. Zdilar D, Franco-Bronson K, Buchler N, Locala JA, Younossi ZM. Hepatitis C, interferon alfa, and depression. Hepatology 2000;31(6):1207–11.

156. Renault PF, Hoofnagle JH, Park Y, Mullen KD, Peters M, Jones DB, Rustgi V, Jones EA. Psychiatric complications of long-term interferon alfa therapy. Arch Intern Med 1987;147(9):1577–80.

157. Prasad S, Waters B, Hill PB, et al. Psychiatric side effects of interferon alpha-2b in patients treated for hepatitis C. Clin Res 1992;40:840A.

158. Bonaccorso S, Marino V, Biondi M, Grimaldi F, Ippoliti F, Maes M. Depression induced by treatment with interferon-alpha in patients affected by hepatitis C virus. J Affect Disord 2002;72(3):237–41.

159. Dieperink E, Ho SB, Thuras P, Willenbring ML. A prospective study of neuropsychiatric symptoms associated with interferon-alpha-2b and ribavirin therapy for patients with chronic hepatitis C. Psychosomatics 2003;44(2):104–12.

160. Hauser P, Khosla J, Aurora H, Laurin J, Kling MA, Hill J, Gulati M, Thornton AJ, Schultz RL, Valentine AD, Meyers CA, Howell CD. A prospective study of the incidence and open-label treatment of interferon-induced major depressive disorder in patients with hepatitis C. Mol Psychiatry 2002;7(9):942–7.

161. Horikawa N, Yamazaki T, Izumi N, Uchihara M. Incidence and clinical course of major depression in patients with chronic hepatitis type C undergoing interferon-alpha therapy: a prospective study. Gen Hosp Psychiatry 2003;25(1):34–8.

162. Janssen HL, Brouwer JT, van der Mast RC, Schalm SW. Suicide associated with alfa-interferon therapy for chronic viral hepatitis. J Hepatol 1994;21(2):241–3.

163. Rifflet H, Vuillemin E, Oberti F, Duverger P, Laine P, Garre JB, Cales P. Pulsions suicidaires chez des malades atteints d'hépatite chronique C au cours ou au décours du traitement par l'interféron alpha. [Suicidal impulses in patients with chronic viral hepatitis C during or after therapy with interferon alpha.] Gastroenterol Clin Biol 1998;22(3):353–7.

164. Valentine AD, Meyers CA, Kling MA, Richelson E, Hauser P. Mood and cognitive side effects of interferon-alpha therapy. Semin Oncol 1998;25(1 Suppl 1):39–47.

165. Adams F, Quesada JR, Gutterman JU. Neuropsychiatric manifestations of human leukocyte interferon therapy in patients with cancer. JAMA 1984;252(7):938–41.

166. Bocci V. Central nervous system toxicity of interferons and other cytokines. J Biol Regul Homeost Agents 1988;2(3):107–18.

167. Prior TI, Chue PS. Psychotic depression occurring after stopping interferon-alpha. J Clin Psychopharmacol 1999;19(4):385–6.

168. Capuron L, Ravaud A. Prediction of the depressive effects of interferon alfa therapy by the patient's initial affective state. N Engl J Med 1999;340(17):1370.

169. McDonald EM, Mann AH, Thomas HC. Interferons as mediators of psychiatric morbidity. An investigation in a trial of recombinant alpha-interferon in hepatitis-B carriers. Lancet 1987;2(8569):1175–8.

170. Pavol MA, Meyers CA, Rexer JL, Valentine AD, Mattis PJ, Talpaz M. Pattern of neurobehavioral deficits associated with interferon alfa therapy for leukemia. Neurology 1995;45(5):947–50.

171. Poutiainen E, Hokkanen L, Niemi ML, Farkkila M. Reversible cognitive decline during high-dose alpha-interferon treatment. Pharmacol Biochem Behav 1994;47(4):901–5.

172. Licinio J, Kling MA, Hauser P. Cytokines and brain function: relevance to interferon-alpha-induced mood and cognitive changes. Semin Oncol 1998;25(1 Suppl 1):30–8.

173. Bonaccorso S, Marino V, Puzella A, Pasquini M, Biondi M, Artini M, Almerighi C, Verkerk R, Meltzer H, Maes M. Increased depressive ratings in patients with hepatitis C receiving interferon-alpha-based immunotherapy are related to interferon-alpha-induced changes in the serotonergic system. J Clin Psychopharmacol 2002;22(1):86–90.

174. Gleason OC, Yates WR, Isbell MD, Philipsen MA. An open-label trial of citalopram for major depression in patients with hepatitis C. J Clin Psychiatry 2002;63(3):194–8.

175. Kraus MR, Schafer A, Faller H, Csef H, Scheurlen M. Paroxetine for the treatment of interferon-alpha-induced depression in chronic hepatitis C. Aliment Pharmacol Ther 2002;16(6):1091–9.

176. Pariante CM, Landau S, Carpiniello BCagliari Group. Interferon alfa-induced adverse effects in patients with a psychiatric diagnosis. N Engl J Med 2002;347(2):148–9.

177. Hensley ML, Peterson B, Silver RT, Larson RA, Schiffer CA, Szatrowski TP. Risk factors for severe neuropsychiatric toxicity in patients receiving interferon alfa-

2b and low-dose cytarabine for chronic myelogenous leukemia: analysis of Cancer and Leukemia Group B 9013. J Clin Oncol 2000;18(6):1301–8.

178. Van Thiel DH, Friedlander L, Molloy PJ, Fagiuoli S, Kania RJ, Caraceni P. Interferon-alpha can be used successfully in patients with hepatitis C virus-positive chronic hepatitis who have a psychiatric illness. Eur J Gastroenterol Hepatol 1995;7(2):165–8.

179. Adams F, Fernandez F, Mavligit G. Interferon-induced organic mental disorders associated with unsuspected pre-existing neurologic abnormalities. J Neurooncol 1988;6(4):355–9.

180. Meyers CA, Obbens EA, Scheibel RS, Moser RP. Neurotoxicity of intraventricularly administered alpha-interferon for leptomeningeal disease. Cancer 1991;68(1):88–92.

181. Hagberg H, Blomkvist E, Ponten U, Persson L, Muhr C, Eriksson B, Oberg K, Olsson Y, Lilja A. Does alpha-interferon in conjunction with radiotherapy increase the risk of complications in the central nervous system? Ann Oncol 1990;1(6):449.

182. Laaksonen R, Niiranen A, Iivanainen M, Mattson K, Holsti L, Farkkila M, Cantell K. Dementia-like, largely reversible syndrome after cranial irradiation and prolonged interferon treatment. Ann Clin Res 1988;20(3):201–3.

183. Mitsuyama Y, Hashiguchi H, Murayama T, Koono M, Nishi S. An autopsied case of interferon encephalopathy. Jpn J Psychiatry Neurol 1992;46(3):741–8.

184. Mulder RT, Ang M, Chapman B, Ross A, Stevens IF, Edgar C. Interferon treatment is not associated with a worsening of psychiatric symptoms in patients with hepatitis C. J Gastroenterol Hepatol 2000;15(3):300–3.

185. Pariante CM, Orru MG, Baita A, Farci MG, Carpiniello B. Treatment with interferon-alpha in patients with chronic hepatitis and mood or anxiety disorders. Lancet 1999;354(9173):131–2.

186. Castera L, Zigante F, Bastie A, Buffet C, Dhumeaux D, Hardy P. Incidence of interferon alfa-induced depression in patients with chronic hepatitis C. Hepatology 2002;35(4):978–9.

187. Valentine AD. Managing the neuropsychiatric adverse effects of interferon treatment. BioDrugs 1999;11:229–37.

188. Miller A, Musselman D, Pena S, Su C, Pearce B, Nemeroff C. Pretreatment with the antidepressant paroxetine, prevents cytokine-induced depression during IFN-alpha therapy for malignant melanoma. Neuroimmunomodulation 1999;6:237.

189. Musselman DL, Lawson DH, Gumnick JF, Manatunga AK, Penna S, Goodkin RS, Greiner K, Nemeroff CB, Miller AH. Paroxetine for the prevention of depression induced by high-dose interferon alfa. N Engl J Med 2001;344(13):961–6.

190. McAllister-Williams RH, Young AH, Menkes DB. Antidepressant response reversed by interferon. Br J Psychiatry 2000;176:93.

191. Sasso FC, Carbonara O, Di Micco P, Coppola L, Torella R, Niglio A. A case of autoimmune polyglandular syndrome developed after interferon-alfa therapy. Br J Clin Pharmacol 2003;56:238–9.

192. Del Monte P, Bernasconi D, De Conca V, Randazzo M, Meozzi M, Badaracco B, Mesiti S, Marugo M. Endocrine evaluation in patients treated with interferon-alpha for chronic hepatitis C. Horm Res 1995;44(3):105–9.

193. Muller H, Hiemke C, Hammes E, Hess G. Sub-acute effects of interferon-alpha 2 on adrenocorticotrophic hormone, cortisol, growth hormone and prolactin in humans. Psychoneuroendocrinology 1992;17(5):459–65.

194. Crockett DM, McCabe BF, Lusk RP, Mixon JH. Side effects and toxicity of interferon in the treatment of recurrent respiratory papillomatosis. Ann Otol Rhinol Laryngol 1987;96(5):601–7.

195. Gottrand F, Michaud L, Guimber D, Ategbo S, Dubar G, Turck D, Farriaux JP. Influence of recombinant interferon alpha on nutritional status and growth pattern in children with chronic viral hepatitis. Eur J Pediatr 1996;155(12):1031–4.

196. Levy S, Abdelli N, Diebold MD, Gross A, Thiefin G. Insuffisance surrénale réversible au cours d'un traitement par interféron alfa d'une hépatite chronique C. Gastroentérol Clin Biol 2003;27:563–4.

197. Concha LB, Carlson HE, Heimann A, Lake-Bakaar GV, Paal AF. Interferon-induced hypopituitarism. Am J Med 2003;114:161–3.

198. Farkkila AM, Iivanainen MV, Farkkila MA. Disturbance of the water and electrolyte balance during high-dose interferon treatment. J Interferon Res 1990;10(2):221–7.

199. Fentiman IS, Balkwill FR, Thomas BS, Russell MJ, Todd I, Bottazzo GF. An autoimmune aetiology for hypothyroidism following interferon therapy for breast cancer. Eur J Cancer Clin Oncol 1988;24(8):1299–303.

200. Vial T, Descotes J. Immune-mediated side-effects of cytokines in humans. Toxicology 1995;105(1):31–57.

201. Fortis A, Christopoulos C, Chrysadakou E, Anevlavis E. De Quervain's thyroiditis associated with interferon-alpha-2b therapy for non-Hodgkin's lymphoma. Clin Drug Invest 1998;16:473–5.

202. Ghilardi G, Gonvers JJ, So A. Hypothyroid myopathy as a complication of interferon alpha therapy for chronic hepatitis C virus infection. Br J Rheumatol 1998;37(12):1349–51.

203. Schmitt K, Hompesch BC, Oeland K, von Staehr WG, Thurmann PA. Autoimmune thyroiditis and myelosuppression following treatment with interferon-alpha for hepatitis C. Int J Clin Pharmacol Ther 1999;37(4):165–7.

204. Papo T, Oksenhendler E, Izembart M, Leger A, Clauvel JP. Antithyroid hormone antibodies induced by interferon-alpha. J Clin Endocrinol Metab 1992;75(6):1484–6.

205. Wong V, Fu AX, George J, Cheung NW. Thyrotoxicosis induced by alpha-interferon therapy in chronic viral hepatitis. Clin Endocrinol (Oxf) 2002;56(6):793–8.

206. Carella C, Mazziotti G, Morisco F, Manganella G, Rotondi M, Tuccillo C, Sorvillo F, Caporaso N, Amato G. Long-term outcome of interferon-alpha-induced thyroid autoimmunity and prognostic influence of thyroid autoantibody pattern at the end of treatment. J Clin Endocrinol Metab 2001;86(5):1925–9.

207. Binaghi M, Levy C, Douvin C, Guittard M, Soubrane G, Coscas G. Ophtalmopathie de Basedow sévère liée a l'interféron alpha. [Severe thyroid ophthalmopathy related to interferon alpha therapy.] J Fr Ophtalmol 2002;25(4):412–5.

208. Sunbul M, Kahraman H, Eroglu C, Leblebicioglu H, Cinar T. Subacute thyroiditis in a patient with chronic hepatitis C during interferon treatment: a case report. Ondokuz Mayis Univ Tip Derg 1999;16:62–6.

209. Dalgard O, Bjoro K, Hellum K, Myrvang B, Bjoro T, Haug E, Bell H. Thyroid dysfunction during treatment of chronic hepatitis C with interferon alpha: no association with either interferon dosage or efficacy of therapy. J Intern Med 2002;251(5):400–6.

210. Monzani F, Caraccio N, Dardano A, Ferrannini E. Thyroid autoimmunity and dysfunction associated with type I interferon therapy. Clin Exp Med 2004;3:199–210.

211. Prummel MF, Laurberg P. Interferon-alfa and autoimmune thyroid disease. Thyroid 2003;13:547–51.

212. Preziati D, La Rosa L, Covini G, Marcelli R, Rescalli S, Persani L, Del Ninno E, Meroni PL, Colombo M, Beck-Peccoz P. Autoimmunity and thyroid function in patients with chronic active hepatitis treated with recombinant interferon alpha-2a. Eur J Endocrinol 1995;132(5):587–93.

213. Vallisa D, Cavanna L, Berte R, Merli F, Ghisoni F, Buscarini L. Autoimmune thyroid dysfunctions in hematologic malignancies treated with alpha-interferon. Acta Haematol 1995;93(1):31–5.

214. Marazuela M, Garcia-Buey L, Gonzalez-Fernandez B, Garcia-Monzon C, Arranz A, Borque MJ, Moreno-Otero R. Thyroid autoimmune disorders in patients with chronic hepatitis C before and during interferon-alpha therapy. Clin Endocrinol (Oxf) 1996;44(6):635–42.

215. Roti E, Minelli R, Giuberti T, Marchelli S, Schianchi C, Gardini E, Salvi M, Fiaccadori F, Ugolotti G, Neri TM, Braverman LE. Multiple changes in thyroid function in patients with chronic active HCV hepatitis treated with recombinant interferon-alpha. Am J Med 1996;101(5):482–7.

216. Barreca T, Picciotto A, Franceschini R, et al. Effects of acute administration of recombinant interferon alpha 2b on pituitary hormone secretion in patients with chronic active hepatitis. Curr Ther Res 1992;52:695–701.

217. Yamazaki K, Kanaji Y, Shizume K, Yamakawa Y, Demura H, Kanaji Y, Obara T, Sato K. Reversible inhibition by interferons alpha and beta of ^{125}I incorporation and thyroid hormone release by human thyroid follicles in vitro. J Clin Endocrinol Metab 1993;77(5):1439–41.

218. Schuppert F, Rambusch E, Kirchner H, Atzpodien J, Kohn LD, von zur Muhlen A. Patients treated with interferon-alpha, interferon-beta, and interleukin-2 have a different thyroid autoantibody pattern than patients suffering from endogenous autoimmune thyroid disease. Thyroid 1997;7(6):837–42.

219. Watanabe U, Hashimoto E, Hisamitsu T, Obata H, Hayashi N. The risk factor for development of thyroid disease during interferon-alpha therapy for chronic hepatitis C. Am J Gastroenterol 1994;89(3):399–403.

220. Fernandez-Soto L, Gonzalez A, Escobar-Jimenez F, Vazquez R, Ocete E, Olea N, Salmeron J. Increased risk of autoimmune thyroid disease in hepatitis C vs hepatitis B before, during, and after discontinuing interferon therapy. Arch Intern Med 1998;158(13):1445–8.

221. Kakizaki S, Takagi H, Murakami M, Takayama H, Mori M. HLA antigens in patients with interferon-alpha-induced autoimmune thyroid disorders in chronic hepatitis C. J Hepatol 1999;30(5):794–800.

222. Minelli R, Braverman LE, Valli MA, Schianchi C, Pedrazzoni M, Fiaccadori F, Salvi M, Magotti MG, Roti E. Recombinant interferon alpha (rIFN-alpha) does not potentiate the effect of iodine excess on the development of thyroid abnormalities in patients with HCV chronic active hepatitis. Clin Endocrinol (Oxf) 1999;50(1):95–100.

223. Carella C, Mazziotti G, Morisco F, Rotondi M, Cioffi M, Tuccillo C, Sorvillo F, Caporaso N, Amato G. The addition of ribavirin to interferon-alpha therapy in patients with hepatitis C virus-related chronic hepatitis does not modify the thyroid autoantibody pattern but increases the risk of developing hypothyroidism. Eur J Endocrinol 2002;146(6):743–9.

224. Marcellin P, Pouteau M, Renard P, Grynblat JM, Colas Linhart N, Bardet P, Bok B, Benhamou JP. Sustained hypothyroidism induced by recombinant alpha interferon in patients with chronic hepatitis C. Gut 1992;33(6):855–6.

225. Mekkakia-Benhabib C, Marcellin P, Colas-Linhart N, Castel-Nau C, Buyck D, Erlinger S, Bok B. Histoire naturelle des dysthyroïdies survenant sous interféron dans le traitement des hépatites chroniques C. [Natural history of dysthyroidism during interferon treatment of chronic hepatitis C.] Ann Endocrinol (Paris) 1996;57(5):419–27.

226. Minelli R, Braverman LE, Giuberti T, Schianchi C, Gardini E, Salvi M, Fiaccadori F, Ugolotti G, Roti E. Effects of excess iodine administration on thyroid function in euthyroid patients with a previous episode of thyroid dysfunction induced by interferon-alpha treatment. Clin Endocrinol (Oxf) 1997;47(3):357–61.

227. Calvino J, Romero R, Suarez-Penaranda JM, Arcocha V, Lens XM, Mardaras J, Novoa D, Sanchez-Guisande D. Secondary hyperparathyroidism exacerbation: a rare side-effect of interferon-alpha? Clin Nephrol 1999;51(4):248–51.

228. Wesche B, Jaeckel E, Trautwein C, Wedemeyer H, Falorni A, Frank H, von zur Muhlen A, Manns MP, Brabant G. Induction of autoantibodies to the adrenal cortex and pancreatic islet cells by interferon alpha therapy for chronic hepatitis C. Gut 2001;48(3):378–83.

229. Fraser GM, Harman I, Meller N, Niv Y, Porath A. Diabetes mellitus is associated with chronic hepatitis C but not chronic hepatitis B infection. Isr J Med Sci 1996;32(7):526–30.

230. Gori A, Caredda F, Franzetti F, Ridolfo A, Rusconi S, Moroni M. Reversible diabetes in patient with AIDS-related Kaposi's sarcoma treated with interferon alpha-2a. Lancet 1995;345(8962):1438–9.

231. Guerci AP, Guerci B, Levy-Marchal C, Ongagna J, Ziegler O, Candiloros H, Guerci O, Drouin P. Onset of insulin-dependent diabetes mellitus after interferon-alfa therapy for hairy cell leukaemia. Lancet 1994;343(8906):1167–8.

232. Mathieu E, Fain O, Sitbon M, Thomas M. Diabète autoimmun après traitement par interféron alpha. [Autoimmune diabetes after treatment with interferon-alpha.] Presse Méd 1995;24(4):238.

233. Eibl N, Gschwantler M, Ferenci P, Eibl MM, Weiss W, Schernthaner G. Development of insulin-dependent diabetes mellitus in a patient with chronic hepatitis C during therapy with interferon-alpha. Eur J Gastroenterol Hepatol 2001;13(3):295–8.

234. Recasens M, Aguilera E, Ampurdanes S, Sanchez Tapias JM, Simo O, Casamitjana R, Conget I. Abrupt onset of diabetes during interferon-alpha therapy in patients with chronic hepatitis C. Diabet Med 2001;18(9):764–7.

235. Mofredj A, Howaizi M, Grasset D, Licht H, Loison S, Devergie B, Demontis R, Cadranel JF. Diabetes mellitus during interferon therapy for chronic viral hepatitis. Dig Dis Sci 2002;47(7):1649–54.

236. Ito Y, Takeda N, Ishimori M, Akai A, Miura K, Yasuda K. Effects of long-term interferon-alpha treatment on glucose tolerance in patients with chronic hepatitis C. J Hepatol 1999;31(2):215–20.

237. Hayakawa M, Gando S, Morimoto Y, Kemmotsu O. Development of severe diabetic keto-acidosis with shock

after changing interferon-beta into interferon-alpha for chronic hepatitis C. Intensive Care Med 2000;26(7):1008.

238. Tasi TN, Hsieh CH. Development of reversible diabetes mellitus after cessation of interferon-alpha therapy for chronic hepatitis C infection. NZ Med J 2004;117:U1230.

239. Fabris P, Floreani A, Tositti G, Vergani D, De Lalla F, Betterle C. Type 1 diabetes mellitus in patients with chronic hepatitis C before and after interferon therapy. Aliment Pharmacol Ther 2003;18:549–58.

240. Schories M, Peters T, Rasenack J, Reincke M. Autoantikörper gegen Inselzellantigene und Diabetes mellitus Typ 1 unter Interferon alpha-Kombinationtherapie. Dtsch Med Wochenschr 2004;129:1120–4.

241. Fabris P, Betterle C, Greggio NA, Zanchetta R, Bosi E, Biasin MR, de Lalla F. Insulin-dependent diabetes mellitus during alpha-interferon therapy for chronic viral hepatitis. J Hepatol 1998;28(3):514–7.

242. Koivisto VA, Pelkonen R, Cantell K. Effect of interferon on glucose tolerance and insulin sensitivity. Diabetes 1989;38(5):641–7.

243. Imano E, Kanda T, Ishigami Y, Kubota M, Ikeda M, Matsuhisa M, Kawamori R, Yamasaki Y. Interferon induces insulin resistance in patients with chronic active hepatitis C. J Hepatol 1998;28(2):189–93.

244. di Cesare E, Previti M, Russo F, Brancatelli S, Ingemi MC, Scoglio R, Mazzu N, Cucinotta D, Raimondo G. Interferon-alpha therapy may induce insulin autoantibody development in patients with chronic viral hepatitis. Dig Dis Sci 1996;41(8):1672–7.

245. Elisaf M, Tsianos EV. Severe hypertriglyceridaemia in a non-diabetic patient after alpha-interferon. Eur J Gastroenterol Hepatol 1999;11(4):463.

246. Junghans V, Runger TM. Hypertriglyceridaemia following adjuvant interferon-alpha treatment in two patients with malignant melanoma. Br J Dermatol 1999;140(1):183–4.

247. Shinohara E, Yamashita S, Kihara S, Hirano K, Ishigami M, Arai T, Nozaki S, Kameda-Takemura K, Kawata S, Matsuzawa Y. Interferon alpha induces disorder of lipid metabolism by lowering postheparin lipases and cholesteryl ester transfer protein activities in patients with chronic hepatitis C. Hepatology 1997;25(6):1502–6.

248. Yamagishi S, Abe T, Sawada T. Human recombinant interferon alpha-2a (r IFN alpha-2a) therapy suppresses hepatic triglyceride lipase, leading to severe hypertriglyceridemia in a diabetic patient. Am J Gastroenterol 1994;89(12):2280.

249. Fernandez-Miranda C, Castellano G, Guijarro C, Fernandez I, Schoebel N, Larumbe S, Gomez-Izquierdo T, del Palacio A. Lipoprotein changes in patients with chronic hepatitis C treated with interferon-alpha. Am J Gastroenterol 1998;93(10):1901–4.

250. Wong SF, Jakowatz JG, Taheri R. Management of hypertriglyceridemia in patients receiving interferon for malignant melanoma. Ann Pharmacother 2004;38:1655–9.

251. Jessner W, Der-Petrossian M, Christiansen L, Maier H, Steindl-Munda P, Gangl A, Ferenci P. Porphyria cutanea tarda during interferon/ribavirin therapy for chronic hepatitis C. Hepatology 2002;36(5):1301–2.

252. Ernstoff MS, Kirkwood JM. Changes in the bone marrow of cancer patients treated with recombinant interferon alpha-2. Am J Med 1984;76(4):593–6.

253. Poynard T, Bedossa P, Chevallier M, Mathurin P, Lemonnier C, Trepo C, Couzigou P, Payen JL, Sajus M, Costa JMMulticenter Study Group. A comparison of three interferon alfa-2b regimens for the long-term treatment of chronic non-A, non-B hepatitis. N Engl J Med 1995;332(22):1457–62.

254. Toccaceli F, Rosati S, Scuderi M, Iacomi F, Picconi R, Laghi V. Leukocyte and platelet lowering by some interferon types during viral hepatitis treatment. Hepatogastroenterology 1998;45(23):1748–52.

255. Arcasoy MO, Rockey DC, Heneghan MA. Pure red cell aplasia following pegylated interferon alpha treatment. Am J Med. 2004;117:619–20.

256. Andres E, Loukili NH, Ben Abdelghani M, Noel E. Pernicious anemia associated with interferon-alpha therapy and chronic hepatitis C infection. J Clin Gastroenterol 2004;38:382.

257. Borgia G, Reynaud L, Gentile I, Borrelli F, Cerini R, Ciampi R, Piazza M. Pernicious anemia during IFN-alfa treatment for chronic hepatitis C. J Interferon Cytokine Res 2003;23:11–12.

258. Alabdulaaly A, Rifkind J, Solow H, Messner HA, Lipton JH. Rescue of interferon induced bone marrow aplasia in a patient with chronic myeloid leukemia by allogeneic bone marrow transplant. Leuk Lymphoma 2004;45:175–7.

259. Gomez-Rangel JD, Ruiz-Delgado GJ, Ruiz-Arguelles GJ. Pegylated-interferon induced severe bone marrow hypoplasia in a patient with multiple myeloma receiving thalidomide. Am J Hematol 2003;74:290–1.

260. Hishida A, Yamamoto K, Kato C, Yokozawa T, Emi N, Tanimoto M, Saito H. Recovery of normal hematopoiesis after severe bone marrow aplasia induced by interferon-alfa in a patient with chronic myelogenous leukemia. Int J Hematol 2003;77:55–9.

261. Ross AS, Bhan AK, Pascual M, Thiim M, Benedict CA, Chung RT. Pegylated interferon alpha-2b plus ribavirin in the treatment of post-liver transplant recurrent hepatitis C. Clin Transpl 2004;18(2):166–73.

262. de-la-Serna-Higuera C, Barcena-Marugan R, Sanz-de-Villalobos E. Hemolytic anemia secondary to alpha-interferon treatment in a patient with chronic C hepatitis. J Clin Gastroenterol 1999;28(4):358–9.

263. Landau A, Castera L, Buffet C, Tertian G, Tchernia G. Acute autoimmune hemolytic anemia during interferon-alpha therapy for chronic hepatitis C. Dig Dis Sci 1999;44(7):1366–7.

264. Akard LP, Hoffman R, Elias L, Saiers JH. Alpha-interferon and immune hemolytic anemia. Ann Intern Med 1986;105(2):306.

265. Barbolla L, Paniagua C, Outeirino J, Prieto E, Sanchez Fayos J. Haemolytic anaemia to the alpha-interferon treatment: a proposed mechanism. Vox Sang 1993;65(2):156–7.

266. Braathen LR, Stavem P. Autoimmune haemolytic anaemia associated with interferon alfa-2a in a patient with mycosis fungoides. BMJ 1989;298(6689):1713.

267. Trivedi HS, Trivedi M. Subnormal rise of erythropoietin in patients receiving interferon and ribavirin combination therapy for hepatitis C. J Clin Gastroenterol 2004;38(7):595–8.

268. Dormann H, Krebs S, Muth-Selbach U, Brune K, Schuppan D, Hahn EG, Schneider HT. Rapid onset of hematotoxic effects after interferon alpha in hepatitis C. J Hepatol 2000;32(6):1041–2.

269. Singhal S, Mehta J, Desikan K, Siegel D, Singh J, Munshi N, Spoon D, Anaissie E, Ayers D, Barlogie B. Collection of peripheral blood stem cells after a preceding autograft: unfavorable effect of prior interferon-alpha therapy. Bone Marrow Transplant 1999;24(1):13–7.

270. Harousseau JL, Milpied N, Bourhis JH, Guimbretiere L, Talmant P. Aplasie fatale après traitement par interféron alpha d'une leucémie myéloïde chronique après greffe de moelle osseuse allogénique. [Lethal aplasia after treatment

with alpha-interferon of recurrent chronic myeloid leukemia following allogeneic bone marrow graft.] Presse Méd 1988;17(2):80–1.

271. Hoffmann A, Kirn E, Krueger GR, Fischer R. Bone marrow hypoplasia and fibrosis following interferon treatment. In Vivo 1994;8(4):605–12.

272. Shepherd PC, Richards S, Allan NC. Severe cytopenias associated with the sequential use of busulphan and interferon-alpha in chronic myeloid leukaemia. Br J Haematol 1994;86(1):92–6.

273. Talpaz M, Kantarjian H, Kurzrock R, Gutterman JU. Bone marrow hypoplasia and aplasia complicating interferon therapy for chronic myelogenous leukemia. Cancer 1992;69(2):410–2.

274. Steis RG, VanderMolen LA, Lawrence J, Sing G, Ruscetti F, Smith JW 2nd, Urba WJ, Clark J, Longo DL. Erythrocytosis in hairy cell leukaemia following therapy with interferon alpha. Br J Haematol 1990;75(1):133–5.

275. Hirri HM, Green PJ. Pure red cell aplasia in a patient with chronic granulocytic leukaemia treated with interferon-alpha. Clin Lab Haematol 2000;22(1):53–4.

276. Tomita N, Motomura S, Ishigatsubo Y. Interferon-alpha-induced pure red cell aplasia following chronic myelogenous leukemia. Anticancer Drugs 2001;12(1):7–8.

277. Willson RA. Interferon alfa-induced pernicious anemia in chronic hepatitis C infection. J Clin Gastroenterol 2001;33(5):426–7.

278. Andriani A, Bibas M, Callea V, De Renzo A, Chiurazzi F, Marceno R, Musto P, Rotoli B. Autoimmune hemolytic anemia during alpha interferon treatment in nine patients with hematological diseases. Haematologica 1996;81(3):258–60.

279. Sacchi S, Kantarjian H, O'Brien S, Cohen PR, Pierce S, Talpaz M. Immune-mediated and unusual complications during interferon alfa therapy in chronic myelogenous leukemia. J Clin Oncol 1995;13(9):2401–7.

280. Steegmann JL, Pinilla I, Requena MJ, de la Camara R, Granados E, Fernandez Villalta MJ, Fernandez-Ranada JM. The direct antiglobulin test is frequently positive in chronic myeloid leukemia patients treated with interferon-alpha. Transfusion 1997;37(4):446–7.

281. McNair ANB, Jacyna MR, Thomas HC. Severe haemolytic transfusion reaction occurring during alpha-interferon therapy for chronic hepatitis. Eur J Gastroenterol Hepatol 1991;3:193–4.

282. Parry-Jones N, Gore ME, Taylor J, Treleaven JG. Delayed haemolytic transfusion reaction caused by anti-M antibody in a patient receiving interleukin-2 and interferon for metastatic renal cell cancer. Clin Lab Haematol 1999;21(6):407–8.

283. Renou C, Harafa A, Bouabdallah R, Demattei C, Cummins C, Rifflet H, Muller P, Ville E, Bertrand J, Benderitter T, Halfon P. Severe neutropenia and post-hepatitis C cirrhosis treatment: is interferon dose adaptation at once necessary? Am J Gastroenterol 2002;97(5):1260–3.

284. Soza A, Everhart JE, Ghany MG, Doo E, Heller T, Promrat K, Park Y, Liang TJ, Hoofnagle JH. Neutropenia during combination therapy of interferon alfa and ribavirin for chronic hepatitis C. Hepatology 2002;36(5):1273–9.

285. Sata M, Yano Y, Yoshiyama Y, Ide T, Kumashiro R, Suzuki H, Tanikawa K. Mechanisms of thrombocytopenia induced by interferon therapy for chronic hepatitis B. J Gastroenterol 1997;32(2):206–10.

286. Shiota G, Okubo M, Kawasaki H, Tahara T. Interferon increases serum thrombopoietin in patients with chronic hepatitis C. Br J Haematol 1997;97(2):340–2.

287. Peck-Radosavljevic M, Wichlas M, Pidlich J, Sims P, Meng G, Zacherl J, Garg S, Datz C, Gangl A, Ferenci P. Blunted thrombopoietin response to interferon alfa-induced thrombocytopenia during treatment for hepatitis C. Hepatology 1998;28(5):1424–9.

288. Martin TG, Shuman MA. Interferon-induced thrombocytopenia: is it time for thrombopoietin. Hepatology 1998;28(5):1430–2.

289. Sagir A, Wettstein M, Heintges T, Haussinger D. Autoimmune thrombocytopenia induced by PEG-IFN-alpha2b plus ribavirin in hepatitis C. Dig Dis Sci 2002;47(3):562–3.

290. Fujii H, Kitada T, Yamada T, Sakaguchi H, Seki S, Hino M. Life-threatening severe immune thrombocytopenia during alfa-interferon therapy for chronic hepatitis C. Hepatogastroenterology 2003;50:841–2.

291. Herishanu Y, Trestman S, Kirgner I, Rachmani R, Naparstek E. Autoimmune thrombocytopenia in chronic myeloid leukemia treated with interferon-alfa: differential diagnosis and possible pathogenesis. Leuk Lymphoma 2003;44:2103–8.

292. Medeiros BC, Seligman PA, Everson GT, Forman LM. Possible autoimmune thrombocytopenia associated with pegylated interferon-alfa2a plus ribavirin treatment for hepatitis C. J Clin Gastroenterol 2004;38:84–6.

293. Sevastianos VA, Deutsch M, Dourakis SP, Manesis EK. Pegylated interferon-2b-associated autoimmune thrombocytopenia in a patient with chronic hepatitis C. Am J Gastroenterol 2003;98:706–7.

294. Arimura K, Arima N, Ohtsubo H, Matsushita K, Kukita T, Ayukawa T, Kuroki T, Tei C. Severe autoimmune thrombocytopenic purpura during interferon-alpha therapy for chronic myelogenous leukemia. Acta Haematol 2004;112:217–8.

295. Zuffa E, Vianelli N, Martinelli G, Tazzari P, Cavo M, Tura S. Autoimmune mediated thrombocytopenia associated with the use of interferon-alpha in chronic myeloid leukemia. Haematologica 1996;81(6):533–5.

296. Taliani G, Duca F, Clementi C, De Bac C. Platelet-associated immunoglobulin G, thrombocytopenia and response to interferon treatment in chronic hepatitis C. J Hepatol 1996;25(6):999.

297. Rachmani R, Avigdor A, Youkla M, Raanani P, Zilber M, Ravid M, Ben-Bassat I. Thrombotic thrombocytopenic purpura complicating chronic myelogenous leukemia treated with interferon-alpha. A report of two successfully treated patients. Acta Haematol 1998;100(4):204–6.

298. Gutman H, Schachter J, Stopel E, Gutman R, Lahav J. Impaired platelet aggregation in melanoma patients treated with interferon-alpha-2b adjuvant therapy. Cancer 2002;94(3):780–5.

299. Herman C, Boggio L, Green D. Factor VIII inhibitor associated with peginterferon. Haemophilia 2005;11(4):408–10.

300. Carmona-Soria I, Jimenez-Saenz M, Gonzalez-Vilches J, Herrerias-Gutierrez JM. Development of lupic anticoagulant during combination therapy in a patient with chronic hepatitis C. J Hepatol 2001;34(6):965–7.

301. Castenskiold EC, Colvin BT, Kelsey SM. Acquired factor VIII inhibitor associated with chronic interferon-alpha therapy in a patient with haemophilia A. Br J Haematol 1994;87(2):434–6.

302. Stricker RB, Barlogie B, Kiprov DD. Acquired factor VIII inhibitor associated with chronic interferon-alpha therapy. J Rheumatol 1994;21(2):350–2.

303. English KE, Brien WF, Howson-Jan K, Kovacs MJ. Acquired factor VIII inhibitor in a patient with chronic myelogenous leukemia receiving interferon-alfa therapy. Ann Pharmacother 2000;34(6):737–9.

304. Mauser-Bunschoten EP, Damen M, Reesink HW, Roosendaal G, Chamuleau RA, van den Berg HM. Formation of antibodies to factor VIII in patients with hemophilia A who are treated with interferon for chronic hepatitis C. Ann Intern Med 1996;125(4):297–9.

305. Qaseem T, Jafri W, Abid S, Hamid S, Khan H. A case report of painful oral ulcerations associated with the use of alpha interferon in a patient with chronic hepatitis due to non-A non-B non-C virus. Mil Med 1993;158(2):126–7.

306. Bardella MT, Marino R, Meroni PL. Celiac disease during interferon treatment. Ann Intern Med 1999;131(2):157–8.

307. Cammarota G, Cuoco L, Cianci R, Pandolfi F, Gasbarrini G. Onset of coeliac disease during treatment with interferon for chronic hepatitis C. Lancet 2000;356(9240):1494–5.

308. Durante-Mangoni E, Iardino P, Resse M, Cesaro G, Sica A, Farzati B, Ruggiero G, Adinolfi LE. Silent celiac disease in chronic hepatitis C: impact of interferon treatment on the disease onset and clinical outcome. J Clin Gastroenterol 2004;38:901–5.

309. Kakumitsu S, Shijo H, Akiyoshi N, Seo M, Okada M. Eosinophilic enteritis observed during alpha-interferon therapy for chronic hepatitis C. J Gastroenterol 2000;35(7):548–51.

310. Tada H, Saitoh S, Nakagawa Y, Hirana H, Morimoto M, Shima T, Shimamoto K, Okanoue T, Kashima K. Ischemic colitis during interferon-alpha treatment for chronic active hepatitis C. J Gastroenterol 1996;31(4):582–4.

311. Janssen HL, Brouwer JT, Nevens F, Sanchez-Tapias JM, Craxi A, Hadziyannis S. Fatal hepatic decompensation associated with interferon alfa. European concerted action on viral hepatitis (Eurohep). BMJ 1993;306(6870):107–8.

312. Krogsgaard K, Marcellin P, Trepo C, Berthelot P, Sanchez-Tapias JM, Bassendine M, Tran A, Ouzan D, Ring-Larsen H, Lindberg J, Enriquez J, Benhamou JP, Bindslev N. Prednisolone withdrawal therapy enhances the effect of human lymphoblastoid interferon in chronic hepatitis B. INTERPRED Trial Group. J Hepatol 1996;25(6):803–13.

313. Lock G, Reng CM, Graeb C, Anthuber M, Wiedmann KH. Interferon-induced hepatic failure in a patient with hepatitis C. Am J Gastroenterol 1999;94(9):2570–1.

314. Farhat BA, Johnson PJ, Williams R. Hazards of interferon treatment in patients with autoimmune chronic active hepatitis. J Hepatol 1994;20(4):560–1.

315. Papo T, Marcellin P, Bernuau J, Durand F, Poynard T, Benhamou JP. Autoimmune chronic hepatitis exacerbated by alpha-interferon. Ann Intern Med 1992;116(1):51–3.

316. Payen JL, Rabbia I, Combis M, Voigt JJ, Vinel P, Pascal JP. Révélation d'une hépatite auto-immune par l'interféron. [Disclosure of autoimmune hepatitis by interferon.] Gastroenterol Clin Biol 1993;17(5):404–5.

317. Ruiz-Moreno M, Rua MJ, Carreno V, Quiroga JA, Manns M, Meyer zum Buschenfelde KH. Autoimmune chronic hepatitis type 2 manifested during interferon therapy in children. J Hepatol 1991;12(2):265–6.

318. Tran A, Beusnel C, Montoya ML, Lussiez V, Hebuterne X, Rampal P. Hépatite autoimmune de typ 1 révélée par un traitement par interféron. [Autoimmune hepatitis type 1 revealed during treatment with interferon.] Gastroenterol Clin Biol 1992;16(8–9):722–3.

319. Vento S, Di Perri G, Garofano T, Cosco L, Concia E, Ferraro T, Bassetti D. Hazards of interferon therapy for HBV-seronegative chronic hepatitis. Lancet 1989;2(8668):926.

320. Garrido Palma G, Sanchez Cuenca JM, Olaso V, Pina R, Urquijo JJ, Lopez Viedma B, Bustamante M, Berenguer M, Berenguer J. Response to treatment with interferon-alfa in patients with chronic hepatitis C and high titers of -M2, -M4 and -M8 antimitochondrial antibodies. Rev Esp Enferm Dig 1999;91(3):168–81.

321. Iorio R, Giannattasio A, Vespere G, Vegnente A. LKM1 antibody and interferon therapy in children with chronic hepatitis C. J Hepatol 2001;35(5):685–7.

322. Garcia-Buey L, Garcia-Monzon C, Rodriguez S, Borque MJ, Garcia-Sanchez A, Iglesias R, DeCastro M, Mateos FG, Vicario JL, Balas A, et al. Latent autoimmune hepatitis triggered during interferon therapy in patients with chronic hepatitis C. Gastroenterology 1995;108(6):1770–7.

323. Castera L, Kalinsky E, Bedossa P, Tertian G, Buffet C. Macrovesicular steatosis induced by interferon alfa therapy for chronic myelogenous leukaemia. Liver 1999;19(3):259–60.

324. Strzelczyk J, Bialkowska J, Loba J, Jablkowski M. Rapid growth of liver hemangioma following interferon treatment for hepatitis C in a young woman. Hepatogastroenterology 2004;51:1151–3.

325. Ryan BM, McDonald GS, Pilkington R, Kelleher D. The development of hepatic granulomas following interferon-alpha2b therapy for chronic hepatitis C infection. Eur J Gastroenterol Hepatol 1998;10(4):349–51.

326. Sevenet F, Sevenet C, Capron D, Descombes P. Pancréatite aiguë et interféron. [Acute pancreatitis and interferon.] Gastroenterol Clin Biol 1999;23(11):1256.

327. Eland IA, van Puijenbroek EP, Sturkenboom MJ, Wilson JH, Stricker BH. Drug associated acute pancreatitis: twenty-one years of spontaneous reporting in The Netherlands. Am J Gastroenterol 1999;94(9):2417–22.

328. Eland IA, Rasch MC, Sturkenboom MJ, Bekkering FC, Brouwer JT, Delwaide J, Belaiche J, Houbiers G, Stricker BH. Acute pancreatitis attributed to the use of interferon alfa-2b. Gastroenterology 2000;119(1):230–3.

329. Chaudhari S, Park J, Anand BS, Pimstone NR, Dieterich DT, Batash S, Bini EJ. Acute pancreatitis associated with interferon and ribavirin therapy in patients with chronic hepatitis C. Dig Dis Sci 2004;49:1000–6.

330. Cecchi E, Forte P, Cini E, Banchelli G, Ferlito C, Mugelli A. Pancreatitis induced by pegylated interferon alfa-2b in a patient affected by chronic hepatitis C. Emerg Med Australas 2004;16:473–5.

331. Kurschel E, Metz-Kurschel U, Niederle N, Aulbert E. Investigations on the subclinical and clinical nephrotoxicity of interferon alpha-2B in patients with myeloproliferative syndromes. Ren Fail 1991;13(2–3):87–93.

332. Selby P, Kohn J, Raymond J, Judson I, McElwain T. Nephrotic syndrome during treatment with interferon. BMJ (Clin Res Ed) 1985;290(6476):1180.

333. Nishimura S, Miura H, Yamada H, Shinoda T, Kitamura S, Miura Y. Acute onset of nephrotic syndrome during interferon-alpha retreatment for chronic active hepatitis C. J Gastroenterol 2002;37(10):854–8.

334. Willson RA. Nephrotoxicity of interferon alfa–ribavirin therapy for chronic hepatitis C. J Clin Gastroenterol 2002;35(1):89–92.

335. Politou M, Tsaftarides P, Vassiliades J, Siakantaris MP, Michail S, Nakopoulou L, Pangalis GA, Vaiopoulos G. Thrombotic microangiopathy in a patient with Sézary syndrome treated with interferon-alpha. Nephrol Dial Transplant 2004;19:733–5.

336. Fisher ME, Rossini M, Simmons E, Harris RC, Moeckel G, Zent R. A woman with chronic hepatitis C infection and nephrotic syndrome who developed multiple renal lesions after interferon alfa therapy. Am J Kidney Dis 2004;44:567–73.

337. Gordon A, Menahem S, Mitchell J, Jenkins P, Dowling J, Roberts SK. Combination pegylated interferon and ribavirin therapy precipitating acute renal failure and exacerbating IgA nephropathy. Nephrol Dial Transplant 2004;19:2155.

338. Nassar GM, Pedro P, Remmers RE, Mohanty LB, Smith W. Reversible renal failure in a patient with the hypereosinophilia syndrome during therapy with alpha interferon. Am J Kidney Dis 1998;31(1):121–6.

339. Dimitrov Y, Heibel F, Marcellin L, Chantrel F, Moulin B, Hannedouche T. Acute renal failure and nephrotic syndrome with alpha interferon therapy. Nephrol Dial Transplant 1997;12(1):200–3.

340. Averbuch SD, Austin HA 3rd, Sherwin SA, Antonovych T, Bunn PA Jr, Longo DL. Acute interstitial nephritis with the nephrotic syndrome following recombinant leukocyte a interferon therapy for mycosis fungoides. N Engl J Med 1984;310(1):32–5.

341. Kimmel PL, Abraham AA, Phillips TM. Membranoproliferative glomerulonephritis in a patient treated with interferon-alpha for human immunodeficiency virus infection. Am J Kidney Dis 1994;24(5):858–63.

342. Traynor A, Kuzel T, Samuelson E, Kanwar Y. Minimal-change glomerulopathy and glomerular visceral epithelial hyperplasia associated with alpha-interferon therapy for cutaneous T cell lymphoma. Nephron 1994;67(1):94–100.

343. Fahal IH, Murry N, Chu P, Bell GM. Acute renal failure during interferon treatment. BMJ 1993;306(6883):973.

344. Lederer E, Truong L. Unusual glomerular lesion in a patient receiving long-term interferon alpha. Am J Kidney Dis 1992;20(5):516–8.

345. Durand JM, Retornaz F, Cretel E, et al. Glomérulonéphrite extracapillaire au cours d'un traitement par interféron alpha. Rev Med Interne 1993;14:1138.

346. Jadoul M. Interferon-alpha-associated focal segmental glomerulosclerosis with massive proteinuria in patients with chronic myeloid leukemia following high dose chemotherapy. Cancer 1999;85(12):2669–70.

347. Honda K, Ando A, Endo M, Shimizu K, Higashihara M, Nitta K, Nihei H. Thrombotic microangiopathy associated with alpha-interferon therapy for chronic myelocytic leukemia. Am J Kidney Dis 1997;30(1):123–30.

348. Ravandi-Kashani F, Cortes J, Talpaz M, Kantarjian HM. Thrombotic microangiopathy associated with interferon therapy for patients with chronic myelogenous leukemia: coincidence or true side effect? Cancer 1999;85(12):2583–8.

349. Vacher-Coponat H, Opris A, Daniel L, Harle JR, Veit V, Olmer M. Thrombotic microangiopathy in a patient with chronic myelocytic leukaemia treated with alpha-interferon. Nephrol Dial Transplant 1999;14(10):2469–71.

350. Ohashi N, Yonemura K, Sugiura T, Isozaki T, Togawa A, Fujigaki Y, Yamamoto T, Hishida A. Withdrawal of interferon-alfa results in prompt resolution of thrombocytopenia and hemolysis but not renal failure in hemolytic uremic syndrome caused by interferon-alfa. Am J Kidney Dis 2003;41:E10.

351. Bremer CT, Lastrapes A, Alper AB, Mudad R. Interferon-alfa-induced focal segmental glomerulosclerosis in chronic myelogenous leukemia: a case report and review of the literature. Am J Clin Oncol 2003;26:262–4.

352. Saxena AK, Panhotra BR, Chopra R. FSGS and its progression to irreversible renal failure, despite cessation of interferon therapy for HCV infection: implications for renal transplantation. Dialysis Transplant 2003;32:619–23.

353. Zuber J, Martinez F, Droz D, Oksenhendler E, Legendre CGroupe D'Etude Des Nephrologues D'Ile-de-France (GENIF). Alpha-interferon-associated thrombotic microangiopathy: a clinicopathologic study of 8 patients and review of the literature. Medicine (Baltimore) 2002;81(4):321–31.

354. Bren A, Kandus A, Ferluga D. Rapidly progressive renal graft failure associated with interferon-alpha treatment in a patient with chronic myelogenous leukemia. Clin Nephrol 1998;50(4):266–7.

355. Alves Couto C, Costa Faria L, Dias Ribeiro D, de Paula Farah K, de Melo Couto OF, de Abreu Ferrari TC. Life-threatening thrombocytopenia and nephrotic syndrome due to focal segmental glomerulosclerosis associated with pegylated interferon alpha-2b and ribavirin treatment for hepatitis C. Liver Int 2006;26(10):1294–7.

356. Asnis LA, Gaspari AA. Cutaneous reactions to recombinant cytokine therapy. J Am Acad Dermatol 1995;33(3):393–410.

357. Dalekos GN, Hatzis J, Tsianos EV. Dermatologic disease during interferon-alpha therapy for chronic viral hepatitis. Ann Intern Med 1998;128(5):409–10.

358. Guillot B, Blazquez L, Bessis D, Dereure O, Guilhou JJ. A prospective study of cutaneous adverse events induced by low-dose alfa-interferon treatment for malignant melanoma. Dermatology 2004;208:49–54.

359. Kerl K, Negro F, Lübbe J. Cutaneous side-effects of treatment of chronic hepatitis C by interferon alfa and ribavirin. Br J Dermatol 2003;149:656.

360. Andry P, Weber-Buisset MJ, Fraitag S, Brechot C, De Prost Y. Toxidermie bulleuse à l'Introna. [Bullous drug eruption caused by Introna.] Ann Dermatol Venereol 1993;120(11):843–5.

361. Sparsa A, Loustaud-Ratti V, Liozon E, Denes E, Soria P, Bouyssou-Gauthier ML, Le Brun V, Boulinguez S, Bedane C, Scribbe-Outtas M, Outtas O, Labrousse F, Bonnetblanc JM, Bordessoule D, Vidal E. Réactions cutanées ou nécrose à l'interféron alpha: peut-on reprendre l'interféron? A propos de six cas. [Cutaneous reactions or necrosis from interferon alpha: can interferon be reintroduced after healing? Six case reports.] Rev Med Interne 2000;21(9):756–63.

362. Kurzen H, Petzoldt D, Hartschuh W, Jappe U. Cutaneous necrosis after subcutaneous injection of polyethylene-glycol-modified interferon alpha. Acta Dermatol Venereol 2002;82(4):310–2.

363. Sanders S, Busam K, Tahan SR, Johnson RA, Sachs D. Granulomatous and suppurative dermatitis at interferon alfa injection sites: report of 2 cases. J Am Acad Dermatol 2002;46(4):611–6.

364. Jessner W, Kinaciyan T, Formann E, Steindl-Munda P, Ferenci P. Severe skin reactions during therapy for chronic hepatitis C associated with delayed hypersensitivity to pegylated interferons. Hepatology 2002;36:361.

365. Heinzerling L, Dummer R, Wildberger H, Burg G. Cutaneous ulceration after injection of polyethylene-glycol-

modified interferon alpha associated with visual disturbances in a melanoma patient. Dermatology 2000;201(2):154–7.

366. Bessis D, Charron A, Rouzier-Panis R, Blatiere V, Guilhou JJ, Reynes J. Necrotizing cutaneous lesions complicating treatment with pegylated-interferon alfa in an HIV-infected patient. Eur J Dermatol 2002;12(1):99–102.

367. Cribier B, Garnier C, Laustriat D, Heid E. Lichen planus and hepatitis C virus infection: an epidemiologic study. J Am Acad Dermatol 1994;31(6):1070–2.

368. Dupin N, Chosidow O, Frances C, et al. Lichen planus after alpha-interferon therapy for chronic hepatitis C. Eur J Dermatol 1994;4:535–6.

369. Strumia R, Venturini D, Boccia S, Gamberini S, Gullini S. UVA and interferon-alfa therapy in a patient with lichen planus and chronic hepatitis C. Int J Dermatol 1993;32(5):386.

370. Aubin F, Bourezane Y, Blanc D, Voltz JM, Faivre B, Humbert PH. Severe lichen planus-like eruption induced by interferon-alpha therapy. Eur J Dermatol 1995;5:296–9.

371. Marinho RT, Johnson NW, Fatela NM, Serejo FS, Gloria H, Raimundo MO, Velosa JF, Ramalho FJ, Moura MC. Oropharyngeal pemphigus in a patient with chronic hepatitis C during interferon alpha-2a therapy. Eur J Gastroenterol Hepatol 2001;13(7):869–72.

372. Quesada JR, Gutterman JU. Psoriasis and alpha-interferon. Lancet 1986;1(8496):1466–8.

373. Harrison PV, Peat MJ. Effect of interferon on psoriasis. Lancet 1986;2(8504):457–8.

374. Nguyen C, Misery L, Tigaud JD, Petiot A, Fiere D, Faure M, Claudy A. Psoriasis induit par l'interféron alpha. A propos d'une observation. [Psoriasis induced by interferon-alpha. Apropos of a case.] Ann Med Interne (Paris) 1996;147(7):519–21.

375. Cleveland MG, Mallory SB. Incomplete Reiter's syndrome induced by systemic interferon alpha treatment. J Am Acad Dermatol 1993;29(5 Pt 1):788–9.

376. Willems M, Munte K, Vrolijk JM, Den Hollander JC, Bohm M, Kemmeren MH, De Man RA, Brouwer JT. Hyperpigmentation during interferon-alfa therapy for chronic hepatitis C virus infection. Br J Dermatol 2003;149:390–4.

377. Dereure O, Raison-Peyron N, Larrey D, Blanc F, Guilhou JJ. Diffuse inflammatory lesions in patients treated with interferon alfa and ribavirin for hepatitis C: a series of 20 patients. Br J Dermatol 2002;147(6):1142–6.

378. Cottoni F, Bolognini S, Deplano A, Garrucciu G, Manzoni NE, Careddu GF, Montesu MA, Tocco A, Lissia A, Solinas A. Skin reaction in antiviral therapy for chronic hepatitis C: a role for polyethylene glycol interferon? Acta Derm Venereol 2004;84:120–3.

379. Thomas R, Stea B. Radiation recall dermatitis from high-dose interferon alfa-2b. J Clin Oncol 2002;20(1):355–7.

380. Tursen U, Kaya TI, Ikizoglu G. Interferon-alpha 2b induced facial erythema in a woman with chronic hepatitis C infection. J Eur Acad Dermatol Venereol 2002;16(3):285–6.

381. Creput C, Auffret N, Samuel D, Jian R, Hill G, Nochy D. Cutaneous thrombotic microangiopathy during treatment with alpha-interferon for chronic hepatitis C. J Hepatol 2002;37(6):871–2.

382. Krischer J, Pechere M, Salomon D, Harms M, Chavaz P, Saurat JH. Interferon alfa-2b-induced Meyerson's nevi in a patient with dysplastic nevus syndrome. J Am Acad Dermatol 1999;40(1):105–6.

383. Pfohler C, Ugurel S, Seiter S, Wagner A, Tilgen W, Reinhold U. Interferon-alpha-associated development of

bullous lesions in mycosis fungoides. Dermatology 2000;200(1):51–3.

384. Moore MM, Elpern DJ, Carter DJ. Severe, generalized nummular eczema secondary to interferon alfa-2b plus ribavirin combination therapy in a patient with chronic hepatitis C virus infection. Arch Dermatol 2004;140:215–7.

385. Gallelli L, Ferraro M, Mauro GF, De Sarro G. Generalized exfoliative dermatitis induced by interferon alfa. Ann Pharmacother 2004;38:2173–4.

386. Scavo S, Gurrera A, Mazzaglia C, Magro G, Pulvirenti D, Gozzo E, Neri S. Verrucous psoriasis in a patient with chronic C hepatitis treated with interferon. Clin Drug Invest 2004;24:427–29.

387. Seckin D, Durusoy C, Sahin S. Concomitant vitiligo and psoriasis in a patient treated with interferon alfa-2a for chronic hepatitis B infection. Pediatr Dermatol 2004;21:577–9.

388. Batisse D, Karmochkine M, Jacquot C, Kazatchkine MD, Weiss L. Sustained exacerbation of cryoglobulinaemia-related vasculitis following treatment of hepatitis C with peginterferon alfa. Eur J Gastroenterol Hepatol 2004;16:701–3.

389. Borghi-Scoazec G, Merle P, Scoazec JY, Claudy A, Trepo C. Onset of dermatitis herpetiformis after treatment by interferon and ribavirin for chronic hepatitis C. J Hepatol 2004;40:871–2.

390. Lanternier F, Bon C, Memain N, Roulot D, Lortholary O. Severe streptococcal cellulitis after subcutaneous pegylated interferon alfa-2B in a diabetic patient. Gastroenterol Clin Biol 2004;28:1294–5.

391. Tosti A, Misciali C, Bardazzi F, Fanti PA, Varotti C. Telogen effluvium due to recombinant interferon alpha-2b. Dermatology 1992;184(2):124–5.

392. Agesta N, Zabala R, Diaz-Perez JL. Alopecia areata during interferon alpha-2b/ribavirin therapy. Dermatology 2002;205(3):300–1.

393. Lang AM, Norland AM, Schuneman RL, Tope WD. Localized interferon alfa-2b-induced alopecia. Arch Dermatol 1999;135(9):1126–8.

394. Kernland KH, Hunziker T. Alopecia areata induced by interferon alpha? Dermatology 1999;198(4):418–9.

395. Bessis D, Luong MS, Blanc P, Chapoutot C, Larrey D, Guilhou JJ, Guillot B. Straight hair associated with interferon-alfa plus ribavirin in hepatitis C infection. Br J Dermatol 2002;147(2):392–3.

396. Hauschild A, Moller M, Lischner S, Christophers E. Repeatable acute rhabdomyolysis with multiple organ dysfunction because of interferon alpha and dacarbazine treatment in metastatic melanoma. Br J Dermatol 2001;144(1):215–6.

397. van Londen GJ, Mascarenhas B, Kirkwood JM. Rhabdomyolysis, when observed with high-dose interferon-alfa (HDI) therapy, does not always exclude resumption of HDI. J Clin Oncol 2001;19(17):3794.

398. Ozdag F, Akar A, Eroglu E, Erbil H. Acute rhabdomyolysis during the treatment of scleromyxedema with interferon alfa. J Dermatolog Treat 2001;12(3):167–9.

399. Dippel E, Zouboulis CC, Tebbe B, Orfanos CE. Myopathic syndrome associated with long-term recombinant interferon alfa treatment in 4 patients with skin disorders. Arch Dermatol 1998;134(7):880–1.

400. Kozuch P, Talpaz M, Faderl S, O'Brien S, Freireich EJ, Kantarjian H. Avascular necrosis of the femoral head in chronic myeloid leukemia patients treated with interferon-alpha: a synergistic correlation? Cancer 2000;89(7):1482–9.

401. Golstein PE, Delforge ML, Deviere J, Marcellin P. Reversible myopathy during successful treatment with pegylated interferon and ribavirin for acute hepatitis C. J Viral Hepat 2004;11:183–6.

402. Gabrielli M, Santarelli L, Serricchio M, Leo D, Pola P, Gasbarrini A. Acute reversible rhabdomyolysis during interferon alfa 2B therapy for hepatitis C. Am J Gastroenterol 2003;98:940.

403. Alvarez JS, Sacristan JA, Alsar MJ. Interferon alpha-2a-induced impotence. Ann Pharmacother 1991;25:1397.

404. Kauppila A, Cantell K, Janne O, Kokko E, Vihko R. Serum sex steroid and peptide hormone concentrations, and endometrial estrogen and progestin receptor levels during administration of human leukocyte interferon. Int J Cancer 1982;29(3):291–4.

405. Schilsky RL, Davidson HS, Magid D, Daiter S, Golomb HM. Gonadal and sexual function in male patients with hairy cell leukemia: lack of adverse effects of recombinant alpha 2-interferon treatment. Cancer Treat Rep 1987;71(2):179–81.

406. Pardini S, Bosincu L, Bonfigli S, Dore F, Longinotti M. Anaphylactic-like syndrome in systemic mastocytosis treated with alpha-2-interferon. Acta Haematol 1991;85(4):220.

407. Ohmoto K, Yamamoto S. Angioedema after interferon therapy for chronic hepatitis C. Am J Gastroenterol 2001;96(4):1311–2.

408. Beckman DB, Mathisen TL, Harris KE, Boxer MB, Grammer LC. Hypersensitivity to IFN-alpha. Allergy 2001;56(8):806–7.

409. Detmar U, Agathos M, Nerl C. Allergy of delayed type to recombinant interferon alpha 2c. Contact Dermatitis 1989;20(2):149–50.

410. Pigatto PD, Bigardi A, Legori A, Altomare GF, Riboldi A. Allergic contact dermatitis from beta-interferon in eye-drops. Contact Dermatitis 1991;25(3):199–200.

411. Antonelli G. In vivo development of antibody to interferons: an update to 1996. J Interferon Cytokine Res 1997;17(Suppl 1):S39–46.

412. McKenna RM, Oberg KE. Antibodies to interferon-alpha in treated cancer patients: incidence and significance. J Interferon Cytokine Res 1997;17(3):141–3.

413. Bonino F, Baldi M, Negro F, Oliveri F, Colombatto P, Bellati G, Brunetto MR. Clinical relevance of anti-interferon antibodies in the serum of chronic hepatitis C patients treated with interferon-alpha. J Interferon Cytokine Res 1997;17(Suppl 1):S35–8.

414. Antonelli G, Currenti M, Turriziani O, Dianzani F. Neutralizing antibodies to interferon-alpha: relative frequency in patients treated with different interferon preparations. J Infect Dis 1991;163(4):882–5.

415. Antonelli G, Currenti M, Turriziani O, Riva E, Dianzani F. Relative frequency of nonneutralizing antibodies to interferon (IFN) in hepatitis patients treated with different IFN-alpha preparations. J Infect Dis 1992;165(3):593–4.

416. Von Wussow P, Hehlmann R, Hochhaus T, Jakschies D, Nolte KU, Prummer O, Ansari H, Hasford J, Heimpel H, Deicher H. Roferon (rIFN-alpha 2a) is more immunogenic than intron A (rIFN-alpha 2b) in patients with chronic myelogenous leukemia. J Interferon Res 1994;14(4):217–9.

417. Milella M, Antonelli G, Santantonio T, Currenti M, Monno L, Mariano N, Angarano G, Dianzani F, Pastore G. Neutralizing antibodies to recombinant alpha-interferon and response to therapy in chronic hepatitis C virus infection. Liver 1993;13(3):146–50.

418. Antonelli G, Giannelli G, Currenti M, Simeoni E, Del Vecchio S, Maggi F, Pistello M, Roffi L, Pastore G, Chemello L, Dianzani F. Antibodies to interferon (IFN) in hepatitis C patients relapsing while continuing recombinant IFN-alpha2 therapy. Clin Exp Immunol 1996;104(3):384–7.

419. Hanley JP, Jarvis LM, Simmonds P, Ludlam CA. Development of anti-interferon antibodies and breakthrough hepatitis during treatment for HCV infection in haemophiliacs. Br J Haematol 1996;94(3):551–6.

420. Roffi L, Mels GC, Antonelli G, Bellati G, Panizzuti F, Piperno A, Pozzi M, Ravizza D, Angeli G, Dianzani F, et al. Breakthrough during recombinant interferon alfa therapy in patients with chronic hepatitis C virus infection: prevalence, etiology, and management. Hepatology 1995;21(3):645–9.

421. Rajan GP, Seifert B, Prummer O, Joller-Jemelka HI, Burg G, Dummer R. Incidence and in-vivo relevance of anti-interferon antibodies during treatment of low-grade cutaneous T-cell lymphomas with interferon alpha-2a combined with acitretin or PUVA. Arch Dermatol Res 1996;288(9):543–8.

422. Tefferi A, Grendahl DC. Natural leukocyte interferon-alpha therapy in patients with chronic granulocytic leukemia who have antibody-mediated resistance to treatment with recombinant interferon-alpha. Am J Hematol 1996;52(3):231–3.

423. Russo D, Candoni A, Zuffa E, Minisini R, Silvestri F, Fanin R, Zaja F, Martinelli G, Tura S, Botta G, Baccarani M. Neutralizing anti-interferon-alpha antibodies and response to treatment in patients with Ph+ chronic myeloid leukaemia sequentially treated with recombinant (alpha 2a) and lymphoblastoid interferon-alpha. Br J Haematol 1996;94(2):300–5.

424. Milella M, Antonelli G, Santantonio T, Giannelli G, Currenti M, Monno L, Turriziani O, Pastore G, Dianzani F. Treatment with natural IFN of hepatitis C patients with or without antibodies to recombinant IFN. Hepatogastroenterology 1995;42(3):201–4.

425. Wussow PV, Jakschies D, Freund M, Hehlmann R, Brockhaus F, Hochkeppel H, Horisberger M, Deicher H. Treatment of anti-recombinant interferon-alpha 2 antibody positive CML patients with natural interferon-alpha. Br J Haematol 1991;78(2):210–6.

426. von Wussow P, Pralle H, Hochkeppel HK, Jakschies D, Sonnen S, Schmidt H, Muller-Rosenau D, Franke M, Haferlach T, Zwingers T, et al. Effective natural interferon-alpha therapy in recombinant interferon-alpha-resistant patients with hairy cell leukemia. Blood 1991;78(1):38–43.

427. Wada M, Kang KB, Kinugasa A, Shintani S, Sawada K, Nishigami T, Shimoyama T. Does the presence of serum autoantibodies influence the responsiveness to interferon-alpha 2a treatment in chronic hepatitis C? Intern Med 1997;36(4):248–54.

428. Cassani F, Cataleta M, Valentini P, Muratori P, Giostra F, Francesconi R, Muratori L, Lenzi M, Bianchi G, Zauli D, Bianchi FB. Serum autoantibodies in chronic hepatitis C: comparison with autoimmune hepatitis and impact on the disease profile. Hepatology 1997;26(3):561–6.

429. Noda K, Enomoto N, Arai K, Masuda E, Yamada Y, Suzuki K, Tanaka M, Yoshihara H. Induction of antinuclear antibody after interferon therapy in patients with type-C chronic hepatitis: its relation to the efficacy of therapy. Scand J Gastroenterol 1996;31(7):716–22.

430. Bell TM, Bansal AS, Shorthouse C, Sandford N, Powell EE. Low-titre auto-antibodies predict autoimmune disease during interferon-alpha treatment of chronic hepatitis C. J Gastroenterol Hepatol 1999;14(5):419–22.

431. Steegmann JL, Requena MJ, Martin-Regueira P, De La Camara R, Casado F, Salvanes FR, Fernandez Ranada JM. High incidence of autoimmune alterations in chronic myeloid leukemia patients treated with interferon-alpha. Am J Hematol 2003;72(3):170–6.

432. Tothova E, Kafkova A, Stecova N, Fricova M, Guman T, Svorcova E. Immune-mediated complications during interferon alpha therapy in chronic myelogenous leukemia. Neoplasma 2002;49(2):91–4.

433. Lunel F, Cacoub P. Treatment of autoimmune and extrahepatic manifestations of hepatitis C virus infection. J Hepatol 1999;31(Suppl 1):210–6.

434. Budak-Alpdogan T, Demircay Z, Alpdogan O, Direskeneli H, Ergun T, Bayik M, Akoglu T. Behçet's disease in patients with chronic myelogenous leukemia: possible role of interferon-alpha treatment in the occurrence of Behçet's symptoms. Ann Hematol 1997;74(1):45–8.

435. Dietrich LL, Bridges AJ, Albertini MR. Dermatomyositis after interferon alpha treatment. Med Oncol 2000;17(1):64–9.

436. Dohmen K, Miyamoto Y, Irie K, Takeshita T, Ishibashi H. Manifestation of cutaneous polyarteritis nodosa during interferon therapy for chronic hepatitis C associated with primary biliary cirrhosis. J Gastroenterol 2000;35(10):789–93.

437. Cirigliano G, Della Rossa A, Tavoni A, Viacava P, Bombardieri S. Polymyositis occurring during alpha-interferon treatment for malignant melanoma: a case report and review of the literature. Rheumatol Int 1999;19(1–2):65–7.

438. Hengstman GJ, Vogels OJ, ter Laak HJ, de Witte T, van Engelen BG. Myositis during long-term interferon-alpha treatment. Neurology 2000;54(11):2186.

439. Hoffmann RM, Jung MC, Motz R, Gossl C, Emslander HP, Zachoval R, Pape GR. Sarcoidosis associated with interferon-alpha therapy for chronic hepatitis C. J Hepatol 1998;28(6):1058–63.

440. Cogrel O, Doutre MS, Marliere V, Beylot-Barry M, Couzigou P, Beylot C. Cutaneous sarcoidosis during interferon alfa and ribavirin treatment of hepatitis C virus infection: two cases. Br J Dermatol 2002;146(2):320–4.

441. Savoye G, Goria O, Herve S, Riachi G, Noblesse I, Bastien L, Courville P, Lerebours E. Probable sarcoïdose cutanée après bi-thérapie associant ribavirine et interferon-alpha pour une hépatite chronique virale C. [Probable cutaneous sarcoidosis associated with combined ribavirin and interferon-alpha therapy for chronic hepatitis C.] Gastroenterol Clin Biol 2000;24(6–7):679.

442. Vander Els NJ, Gerdes H. Sarcoidosis and IFN-alpha treatment. Chest 2000;117(1):294.

443. Fiorani C, Sacchi S, Bonacorsi G, Cosenza M. Systemic sarcoidosis associated with interferon-alpha treatment for chronic myelogenous leukemia. Haematologica 2000;85(9):1006–7.

444. Marzouk K, Saleh S, Kannass M, Sharma OP. Interferon-induced granulomatous lung disease. Curr Opin Pulm Med 2004;10:435–40.

445. Rogers CJ, Romagosa R, Vincek V. Cutaneous sarcoidosis associated with pegylated interferon alfa and ribavirin therapy in a patient with chronic hepatitis C. J Am Acad Dermatol 2004;50:649–50.

446. Toulemonde A, Quereux G, Dreno B. Granulome sarcoïdosique sur tatouage induit par l'interféron alpha. Ann Dermatol Venereol 2004;131:49–51.

447. Tortorella C, Napoli N, Panella E, Antonaci A, Gentile A, Antonaci S. Asymptomatic systemic sarcoidosis arising 5 years after IFN-alpha treatment for chronic hepatitis C: a new challenge for clinicians. J Interferon Cytokine Res 2004;24:655–8.

448. Eberlein-Konig B, Hein R, Abeck D, Engst R, Ring J. Cutaneous sarcoid foreign body granulomas developing in sites of previous skin injury after systemic interferon-alpha treatment for chronic hepatitis C. Br J Dermatol 1999;140(2):370–2.

449. Gitlin N. Manifestation of sarcoidosis during interferon and ribavirin therapy for chronic hepatitis C: a report of two cases. Eur J Gastroenterol Hepatol 2002;14(8):883–5.

450. Husa P, Klusakova J, Jancikova J, Husova L, Horalek F. Sarcoidosis associated with interferon-alpha therapy for chronic hepatitis B. Eur J Intern Med 2002;13(2):129–31.

451. Nawras A, Alsolaiman MM, Mehboob S, Bartholomew C, Maliakkal B. Systemic sarcoidosis presenting as a granulomatous tattoo reaction secondary to interferon-alpha treatment for chronic hepatitis C and review of the literature. Dig Dis Sci 2002;47(7):1627–31.

452. Noguchi K, Enjoji M, Nakamuta M, Sugimoto R, Kotoh K, Nawata H. Various sarcoid lesions in a patient induced by interferon therapy for chronic hepatitis C. J Clin Gastroenterol 2002;35(3):282–4.

453. Tahan V, Ozseker F, Guneylioglu D, Baran A, Ozaras R, Mert A, Ucisik AC, Cagatay T, Yilmazbayhan D, Senturk H. Sarcoidosis after use of interferon for chronic hepatitis C: report of a case and review of the literature. Dig Dis Sci 2003;48(1):169–73.

454. Li SD, Yong S, Srinivas D, Van Thiel DH. Reactivation of sarcoidosis during interferon therapy. J Gastroenterol 2002;37(1):50–4.

455. Boonen A, Stockbrugger RW, van der Linden S. Pericarditis after therapy with interferon-alpha for chronic hepatitis C. Clin Rheumatol 1999;18(2):177–9.

456. Johnson DM, Hayat SQ, Burton GV. Rheumatoid arthritis complicating adjuvant interferon-alpha therapy for malignant melanoma. J Rheumatol 1999;26(4):1009–10.

457. Wandl UB, Nagel-Hiemke M, May D, Kreuzfelder E, Kloke O, Kranzhoff M, Seeber S, Niederle N. Lupus-like autoimmune disease induced by interferon therapy for myeloproliferative disorders. Clin Immunol Immunopathol 1992;65(1):70–4.

458. Conlon KC, Urba WJ, Smith JW 2nd, Steis RG, Longo DL, Clark JW. Exacerbation of symptoms of autoimmune disease in patients receiving alpha-interferon therapy. Cancer 1990;65(10):2237–42.

459. Nesher G, Ruchlemer R. Alpha-interferon-induced arthritis: clinical presentation treatment, and prevention. Semin Arthritis Rheum 1998;27(6):360–5.

460. Fukuyama S, Kajiwara E, Suzuki N, Miyazaki N, Sadoshima S, Onoyama K. Systemic lupus erythematosus after alpha-interferon therapy for chronic hepatitis C: a case report and review of the literature. Am J Gastroenterol 2000;95(1):310–2.

461. Beretta L, Caronni M, Vanoli M, Scorza R. Systemic sclerosis after interferon-alfa therapy for myeloproliferative disorders. Br J Dermatol 2002;147(2):385–6.

462. Solans R, Bosch JA, Esteban I, Vilardell M. Systemic sclerosis developing in association with the use of interferon alpha therapy for chronic viral hepatitis. Clin Exp Rheumatol 2004;22:625–8.

463. Friedman G, Mehta S, Sherker AH. Fatal exacerbation of hepatitis C-related cryoglobulinemia with interferon-alpha therapy. Dig Dis Sci 1999;44(7):1364–5.

464. Pateron D, Fain O, Sehonnou J, Trinchet JC, Beaugrand M. Severe necroziting vasculitis in a patient with hepatitis C virus infection treated by interferon. Clin Exp Rheumatol 1996;14(1):79–81.

465. Samson D, Volin L, Schanz U, Bosi A, Gahrtron G. Feasibility and toxicity of interferon maintenance therapy after allogeneic BMT for multiple myeloma: a pilot study of the EBMT. Bone Marrow Transplant 1996;17(5):759–762.

466. Morton AJ, Gooley T, Hansen JA, Appelbaum FR, Bruemmer B, Bjerke JW, Clift R, Martin PJ, Petersdorf EW, Sanders JE, Storb R, Sullivan KM, Woolfrey A, Anasetti C. Association between pretransplant interferon-alpha and outcome after unrelated donor marrow transplantation for chronic myelogenous leukemia in chronic phase. Blood 1998;92(2):394–401.

467. Rostaing L, Izopet J, Baron E, Duffaut M, Puel J, Durand D. Treatment of chronic hepatitis C with recombinant interferon alpha in kidney transplant recipients. Transplantation 1995;59(10):1426–31.

468. Dousset B, Conti F, Houssin D, Calmus Y. Acute vanishing bile duct syndrome after interferon therapy for recurrent HCV infection in liver-transplant recipients. N Engl J Med 1994;330(16):1160–1.

469. Féray C, Samuel D, Gigou M, Paradis V, David MF, Lemonnier C, Reynes M, Bismuth H. An open trial of interferon alfa recombinant for hepatitis C after liver transplantation: antiviral effects and risk of rejection. Hepatology 1995;22(4 Pt 1):1084–9.

470. Pohanka E, Kovarik J. Is treatment with interferon-alpha in renal transplant recipients still justified? Nephrol Dial Transplant 1996;11(6):1191–2.

471. Min AD, Bodenheimer HC Jr. Does interferon precipitate rejection of liver allografts? Hepatology 1995;22(4 Pt 1):1333–5.

472. Stravitz RT, Shiffman ML, Sanyal AJ, Luketic VA, Sterling RK, Heuman DM, Ashworth A, Mills AS, Contos M, Cotterell AH, Maluf D, Posner MP, Fisher RA. Effects of interferon treatment on liver histology and allograft rejection in patients with recurrent hepatitis C following liver transplantation. Liver Transpl 2004;10(7):850–8.

473. Soriano V, Bravo R, Samaniego JG, Gonzalez J, Odriozola PM, Arroyo E, Vicario JL, Castro A, Colmenero M, Carballo E, et alHIV-Hepatitis Spanish Study Group. CD4+ T-lymphocytopenia in HIV-infected patients receiving interferon therapy for chronic hepatitis C. AIDS 1994;8(11):1621–2.

474. Vento S, Di Perri G, Cruciani M, Garofano T, Concia E, Bassetti D. Rapid decline of CD4+ cells after IFN alpha treatment in HIV-1 infection. Lancet 1993;341(8850):958–9.

475. Marten D, Holtzmuller K, Julia F. Bacterial infections complicating hepatitis C infected hemodialysis dependent patients treated with interferon alfa. Am J Gastroenterol 2002;97(Suppl):163–4.

476. Parana R, Portugal M, Vitvitski L, Cotrim H, Lyra L, Trepo C. Severe strongyloidiasis during interferon plus ribavirin therapy for chronic HCV infection. Eur J Gastroenterol Hepatol 2000;12(2):245–6.

477. Serrano J, Prieto E, Mazarbeitia F, Roman A, Llamas P, Tomas JF. Atypical chronic graft-versus-host disease following interferon therapy for chronic myeloid leukaemia relapsing after allogeneic BMT. Bone Marrow Transplant 2001;27(1):85–7.

478. Gogos CA, Starakis JK, Bassaris HP, Skoutelis AT. Remote abscess formation during interferon-alfa therapy for viral hepatitis. Clin Microbiol Infect 2003;9:540–2.

479. Puoti M, Babudieri S, Rezza G, Viale P, Antonini MG, Maida I, Rossi S, Zanini B, Putzolu V, Fenu L, Baiguera C, Sassu S, Carosi G, Mura MS. Use of pegylated interferons is associated with an increased incidence of infections during combination treatment of chronic hepatitis C: a side effect of pegylation? Antivir Ther 2004;9:627–30.

480. Beelen DW, Elmaagacli AH, Schaefer UW. The adverse influence of pretransplant interferon-alpha (IFN-alpha) on transplant outcome after marrow transplantation for chronic phrase chronic myelogenous leukemia increases with the duration of IFN-alpha exposure. Blood 1999;93(5):1779–810.

481. Hehlmann R, Hochhaus A, Kolb HJ, Hasford J, Gratwohl A, Heimpel H, Siegert W, Finke J, Ehninger G, Holler E, Berger U, Pfirrmann M, Muth A, Zander A, Fauser AA, Heyll A, Nerl C, Hossfeld DK, Loffler H, Pralle H, Queisser W, Tobler A. Interferon-alpha before allogeneic bone marrow transplantation in chronic myelogenous leukemia does not affect outcome adversely, provided it is discontinued at least 90 days before the procedure. Blood 1999;94(11):3668–77.

482. Johansson B, Fioretos T, Billstrom R, Mitelman F. Aberrant cytogenetic evolution pattern of Philadelphia-positive chronic myeloid leukemia treated with interferon-alpha. Leukemia 1996;10(7):1134–8.

483. Waysbort A, Giroux M, Mansat V, Teixeira M, Dumas JC, Puel J. Experimental study of transplacental passage of alpha interferon by two assay techniques. Antimicrob Agents Chemother 1993;37(6):1232–7.

484. Haggstrom J, Adriansson M, Hybbinette T, Harnby E, Thorbert G. Two cases of CML treated with alpha-interferon during second and third trimester of pregnancy with analysis of the drug in the new-born immediately postpartum. Eur J Haematol 1996;57(1):101–2.

485. Delage R, Demers C, Cantin G, Roy J. Treatment of essential thrombocythemia during pregnancy with interferon-alpha. Obstet Gynecol 1996;87(5 Pt 2):814–7.

486. Hiratsuka M, Minakami H, Koshizuka S, Sato I. Administration of interferon-alpha during pregnancy: effects on fetus. J Perinat Med 2000;28(5):372–6.

487. Mubarak AA, Kakil IR, Awidi A, Al-Homsi U, Fawzi Z, Kelta M, Al-Hassan A. Normal outcome of pregnancy in chronic myeloid leukemia treated with interferon-alpha in 1st trimester: report of 3 cases and review of the literature. Am J Hematol 2002;69(2):115–8.

488. Trotter JF, Zygmunt AJ. Conception and pregnancy during interferon-alpha therapy for chronic hepatitis C. J Clin Gastroenterol 2001;32(1):76–8.

489. Derbala M, Amer A, Bener A, Lopez AC, Omar M, El Ghannam M. Pegylated interferon-alpha 2b-ribavirin combination in Egyptian patients with genotype 4 chronic hepatitis. J Viral Hepat 2005;12(4):380–5.

490. Jacobson KR, Murray K, Zellos A, Schwarz KB. An analysis of published trials of interferon monotherapy in children with chronic hepatitis C. J Pediatr Gastroenterol Nutr 2002;34(1):52–8.

491. Barlow CF, Priebe CJ, Mulliken JB, Barnes PD, Mac Donald D, Folkman J, Ezekowitz RA. Spastic diplegia as a complication of interferon Alfa-2a treatment of hemangiomas of infancy. J Pediatr 1998;132(3 Pt 1):527–30.

492. Dubois J, Hershon L, Carmant L, Belanger S, Leclerc JM, David M. Toxicity profile of interferon alfa-2b in children: a prospective evaluation. J Pediatr 1999;135(6):782–5.

493. Worle H, Maass E, Kohler B, Treuner J. Interferon alpha-2a therapy in haemangiomas of infancy: spastic diplegia as a severe complication. Eur J Pediatr 1999;158(4):344.

494. Enjolras O. Neurotoxicity of interferon alfa in children treated for hemangiomas. J Am Acad Dermatol 1998;39(6):1037–8.

495. Grether JK, Nelson KB, Dambrosia JM, Phillips TM. Interferons and cerebral palsy. J Pediatr 1999;134(3):324–32.

496. Rostaing L, Chatelut E, Payen JL, Izopet J, Thalamas C, Ton-That H, Pascal JP, Durand D, Canal P. Pharmacokinetics of alphaIFN-2b in chronic hepatitis C virus patients undergoing chronic hemodialysis or with normal renal function: clinical implications. J Am Soc Nephrol 1998;9(12):2344–8.

497. Uchihara M, Izumi N, Sakai Y, Yauchi T, Miyake S, Sakai T, Akiba T, Marumo F, Sato C. Interferon therapy for chronic hepatitis C in hemodialysis patients: increased serum levels of interferon. Nephron 1998;80(1):51–6.

498. Sezaki H, Arase Y, Tsubota A, Suzuki Y, Kobayashi M, Saitoh S, Suzuki F, Akuta N, Someya T, Ikeda K, Kumada H. Type C-chronic hepatitis patients who had autoimmune phenomenon and developed jaundice during interferon therapy. J Gastroenterol 2003;38:493–500.

499. Fabrizi F, Dulai G, Dixit V, Bunnapradist S, Martin P. Meta-analysis: interferon for the treatment of chronic hepatitis C in dialysis patients. Aliment Pharmacol Ther 2003;18:1071–81.

500. Saab S, Kalmaz D, Gajjar NA, Hiatt J, Durazo F, Han S, Farmer DG, Ghobrial RM, Yersiz H, Goldstein LI, Lassman CR, Busuttil RW. Outcomes of acute rejection after interferon therapy in liver transplant recipients. Liver Transpl 2004;10:859–67.

501. Biselli M, Lorenzini S, Gramenzi A, Andreone P, Bernardi M, Rossi C, Grazi GL. Hepatic artery stenosis in liver transplanted patients treated with pegylated interferon alpha-2b and ribavirin. Transplantation 2004;78:953.

502. Mukherjee S. Reversible decompensated liver disease as a possible complication of pegylated-interferon alfa 2b and ribavirin for recurrent hepatitis C. J Gastroenterol Hepatol 2004;19:723–4.

503. Sulkowski MS, Felizarta F, Smith C, Slim J, Berggren R, Goodman R, Ball L, Khalili M, Dieterich DT for the Hepatitis Resource Network Clinical Trials Group. Daily versus thrice-weekly interferon alfa-2b plus ribavirin for the treatment of chronic hepatitis C in HIV-infected persons: a multicenter randomized controlled trial. J Acquir Immune Defic Syndr 2004;35:464–72.

504. Anonymous. Interferon alfa-2b and ribavirin combination therapy—indications extended: previously untreated hepatitis C patients. WHO Newsletter 1999;1/2:9.

505. Foster GR. Pegylated interferons: chemical and clinical differences. Aliment Pharmacol Ther 2004;20:825–30.

506. Chamberlain MC. A phase II trial of intra-cerebrospinal fluid alpha interferon in the treatment of neoplastic meningitis. Cancer 2002;94(10):2675–80.

507. Zylberberg H, Fontaine H, Thepot V, Nalpas B, Brechot C, Pol S. Triggering of acute alcoholic hepatitis by alpha-interferon therapy. J Hepatol 1999;30(4):722–5.

508. Casato M, Pucillo LP, Leoni M, di Lullo L, Gabrielli A, Sansonno D, Dammacco F, Danieli G, Bonomo L. Granulocytopenia after combined therapy with interferon and angiotensin-converting enzyme inhibitors: evidence for a synergistic hematologic toxicity. Am J Med 1995;99(4):386–91.

509. Bauckner JC, Schomberg PJ, McGinnis WL, Cascino TL, Scheithauer BW, O'Fallon JR, Morton RF, Kuross SA, Mailliard JA, Hatfield AK, Cole JT, Steen PD, Bernath AM. A phase III study of radiation therapy plus carmustine with or without recombinant interferon-alpha in the treatment of patients with newly diagnosed high-grade glioma. Cancer 2001;92(2):420–33.

510. Hassan M, Nilsson C, Olsson H, Lundin J, Osterborg A. The influence of interferon-alpha on the pharmacokinetics of cyclophosphamide and its 4-hydroxy metabolite in patients with multiple myeloma. Eur J Haematol 1999;63(3):163–70.

511. Chan TM, Wu PC, Lau JY, Lok AS, Lai CL, Cheng IK. Interferon treatment for hepatitis C virus infection in patients on haemodialysis. Nephrol Dial Transplant 1997;12(7):1414–9.

512. Desai RG. Drug interaction between alpha interferon and erythropoietin. J Clin Oncol 1991;9(5):893.

513. Nordio M, Guarda L, Lorenzi S, Lombini C, Marchini P, Mirandoli F. Interaction between alpha-interferon and erythropoietin in antiviral and antineoplastic therapy in uraemic patients on haemodialysis. Nephrol Dial Transplant 1993;8(11):1308.

514. Czejka MJ, Schuller J, Jager W, Fogl U, Weiss C. Influence of different doses of interferon-alpha-2b on the blood plasma levels of 5-fluorouracil. Eur J Drug Metab Pharmacokinet 1993;18(3):247–50.

515. Greco FA, Figlin R, York M, Einhorn L, Schilsky R, Marshall EM, Buys SS, Froimtchuk MJ, Schuller J, Schuchter L, Buyse M, Ritter L, Man A, Yap AK. Phase III randomized study to compare interferon alfa-2a in combination with fluorouracil versus fluorouracil alone in patients with advanced colorectal cancer. J Clin Oncol 1996;14(10):2674–81.

516. Ehrsson H, Eksborg S, Wallin I, Osterborg A, Mellstedt H. Oral melphalan pharmacokinetics: influence of interferon-induced fever. Clin Pharmacol Ther 1990;47(1):86–90.

517. Mauss S, Berger F, Goelz J, Jacob B, Schmutz G. A prospective controlled study of interferon-based therapy of chronic hepatitis C in patients on methadone maintenance. Hepatology 2004;40:120–4.

518. Sulkowski M, Wright T, Rossi S, Arora S, Lamb M, Wang K, Gries JM, Yalamanchili S. Peginterferon alfa-2a does not alter the pharmacokinetics of methadone in patients with chronic hepatitis C undergoing methadone maintenance therapy. Clin Pharmacol Ther 2005;77:214–24.

519. Mannering GJ, Deloria LB. The pharmacology and toxicology of the interferons: an overview. Annu Rev Pharmacol Toxicol 1986;26:455–515.

520. Echizen H, Ohta Y, Shirataki H, Tsukamoto K, Umeda N, Oda T, Ishizaki T. Effects of subchronic treatment with natural human interferons on antipyrine clearance and liver function in patients with chronic hepatitis. J Clin Pharmacol 1990;30(6):562–7.

521. Tappero G, Ballare M, Farina M, Negro F. Severe anemia following combined alpha-interferon/ribavirin therapy of chronic hepatitis C. J Hepatol 1998;29(6):1033–4.

522. Sookoian S, Neglia V, Castano G, Frider B, Kien MC, Chohuela E. High prevalence of cutaneous reactions to interferon alfa plus ribavirin combination therapy in patients with chronic hepatitis C virus. Arch Dermatol 1999;135(8):1000–1.

523. Nathan PD, Gore ME, Eisen TG. Unexpected toxicity of combination thalidomide and interferon alpha-2a treatment in metastatic renal cell carcinoma. J Clin Oncol 2002;20(5):1429–30.

524. Israel BC, Blouin RA, McIntyre W, Shedlofsky SI. Effects of interferon-alpha monotherapy on hepatic drug metabolism in cancer patients. Br J Clin Pharmacol 1993;36(3):229–35.

525. Williams SJ, Baird-Lambert JA, Farrell GC. Inhibition of theophylline metabolism by interferon. Lancet 1987;2(8565):939–41.

526. Jonkman JH, Nicholson KG, Farrow PR, Eckert M, Grasmeijer G, Oosterhuis B, De Noord OE, Guentert TW. Effects of alpha-interferon on theophylline pharmacokinetics and metabolism. Br J Clin Pharmacol 1989;27(6):795–802.

527. Burger DM, Meenhorst PL, Koks CH, Beijnen JH. Drug interactions with zidovudine. AIDS 1993;7(4):445–60.

528. Krown SE, Gold JW, Niedzwiecki D, Bundow D, Flomenberg N, Gansbacher B, Brew BJ. Interferon-alpha with zidovudine: safety, tolerance, and clinical and virologic effects in patients with Kaposi sarcoma associated with the acquired immunodeficiency syndrome (AIDS). Ann Intern Med 1990;112(11):812–21.

529. Weitz IC. Treatment of immune thrombocytopenia associated with interferon therapy of hepatitis C with the anti-CD20 monoclonal antibody, rituximab. Am J Hematol 2005;78(2):138–41.

Interferon beta

See also Interferons

General Information

Interferon beta is used in the form of natural fibroblast or recombinant preparations (interferon beta-1a and interferon beta-1b) and exerts antiviral and antiproliferative properties similar to those of interferon alfa. Although its efficacy has been debated (1), interferon beta has been approved for the treatment of relapsing–remitting multiple sclerosis, and more recently for secondary progressive multiple sclerosis.

The general toxicity of interferon beta is very similar to that of interferon alfa (2), with no apparent differences between the two recombinant preparations with any route of injection (SEDA-20, 332) (3–6). In multiple sclerosis, fatigue and a transient flu-like syndrome responsive to paracetamol or the combination of paracetamol plus prednisone have been observed in about 60% of patients during the first weeks of treatment, and tachyphylaxis usually developed after several doses (7). Patients with chronic progressive disease are more likely to discontinue treatment because of adverse effects (8).

Clinically relevant adverse effects associated with interferon beta and their management have been lengthily reviewed (9). Interferon beta-1a and beta-1b, the two recombinant available forms of interferon beta, have not been directly compared. From the results of a randomized, crossover study in 12 healthy volunteers, a single injection of interferon beta-1a 6 MU (Rebif) was suggested to produce less frequent and less severe fever than interferon beta-1b 8 MU (Betaseron), but identical pharmacodynamic effects (10).

A flu-like illness is the most common adverse effect of interferon beta. In an open, randomized study of the effects of paracetamol 1 g or ibuprofen 400 mg before and 6 hours after interferon beta injection on interferon beta-induced flu-like symptoms in 104 patients, the two drugs were equally effective (11).

The therapeutic efficacy and adverse effects of subcutaneous interferon beta-1b in the management of relapsing–remitting and secondary progressive multiple sclerosis have been extensively reviewed (12). Interferon beta was considered to be a valuable first-line therapy in relapsing–remitting multiple sclerosis, and potentially useful in secondary progressive multiple sclerosis, although its effects on disease progression is uncertain.

Systematic reviews

In a meta-analysis of seven trials in 1215 patients with relapsing remitting multiple sclerosis, there was a modest effect of recombinant interferon beta on the rate of clinical exacerbation at 1 year of treatment, but no clear clinical benefit beyond 1 year (13). Compared with placebo, the most common adverse effects were flu-like symptoms (48% versus 28%), injection-site reactions (62% versus 14%), nausea and vomiting (32% versus 20%), hair loss (36% versus 2.5%), leukopenia (6% versus 0.6%), thrombocytopenia (3.5% versus 0.5%) and increased alanine transaminase activity (9% versus 3%).

Organs and Systems

Cardiovascular

Cardiovascular adverse effects of interferon beta include isolated reports of severe Raynaud's phenomenon (SEDA-22, 374) and acute myocarditis (SEDA-21, 374).

Fatal capillary leak syndrome has been reported (14).

- A 27-year-old woman had an 8-month history of relapsing–remitting neurological symptoms and a monoclonal gammopathy. She started to take interferon beta-1b for multiple sclerosis, but had marked somnolence 30 hours after a single injection. She rapidly became unresponsiveness, and hemodynamic tests showed low central venous and pulmonary capillary wedge pressures with generalized peripheral edema, ascites, and bilateral pleural effusions. She died within 80 hours after injection from multiple organ failure. At postmortem she was found to have C1 esterase inhibitor deficiency.

In the light of the possible effects of interferon beta on cytokine release and complement activation, a cytokine-mediated reaction was discussed as the cause of the capillary leak syndrome in this case.

Respiratory

Bronchiolitis obliterans with organizing pneumonia has been reported in a patient taking interferon beta (15).

- A 49-year-old man had a progressive unproductive cough and right hemithoracic pain after 3 months of interferon beta-1a 30 micrograms/week for multiple sclerosis. A CT scan showed a right basal pulmonary infiltrate and transbronchial biopsies showed features consistent with bronchiolitis obliterans with organizing pneumonia. The lesions resolved fully on interferon beta-1a withdrawal and prednisone treatment.

Nervous system

Although direct toxic effects of natural interferon beta on the nervous system have been regarded as a possible risk of intraventricular and/or intratumoral injection (16), interferon beta is considered to be markedly less neurotoxic than interferon alfa (17).

Although headache was not specifically identified as an adverse effect of interferon beta in pivotal trials, the frequency, duration, and intensity of headache increased during the first 6 months of treatment in 65 patients (18). There was a 35% probability of aggravated headaches in patients with pre-existing headaches.

The possible deleterious effects of interferon beta-1b on increased spasticity have been examined in 19 patients with primary progressive multiple sclerosis, 19 untreated matched patients, and 10 patients treated with interferon beta-1b for relapsing–remitting multiple sclerosis (19). Patients with primary progressive multiple sclerosis had frequent (68%) and clinically relevant increased spasticity (seven required oral baclofen), usually after about 2 months of treatment, whereas only two (11%) of the untreated patients and none of the patients with relapsing–remitting multiple sclerosis had similar disabling spasticity. Seven patients had to discontinue treatment after 6 months because of spasticity, and symptoms improved over several months after withdrawal. The authors suggested that this possible adverse effect should be taken into account in clinical trials because it could mask the positive clinical effects of interferon beta-1b.

Neurosarcoidosis has been reported in a patient with chronic hepatitis C who was treated with interferon beta.

- A 56-year-old woman developed numbness and difficulty in swallowing and in closing her left eye several weeks after starting interferon beta for chronic hepatitis C (20). She had facial paresthesia, a left facial nerve palsy, dysphagia, and signs of radiculopathy on the left side. Serum angiotensin-converting enzyme activity was raised. She had bilateral hilar lymphadenopathy without interstitial changes and increased radiogallium uptake in hilar lymph nodes and the parotid glands. Although the cerebrospinal fluid was normal, a diagnosis of neurosarcoidosis was considered, and she recovered completely after interferon beta withdrawal and glucocorticoid therapy.

Moderate exacerbation of multiple sclerosis sometimes occurs in the first 3 months of interferon beta treatment.

- A 21-year-old man had an acute and very severe clinical relapse, with multiple disseminated demyelinating lesions and axonal injury on MRI and cerebral biopsy, after the third injection of interferon beta-1a (21).

Whether this case was due to interferon beta or resulted from spontaneous exacerbation was open to question.

There is an association between multiple sclerosis and central nervous system tumors.

- A 19-year-old woman with multiple sclerosis had a concomitant right intraventricular tumor, consistent with meningioma on an MRI scan (22). Two years after the start of treatment with interferon beta, a brain MRI scan showed enlargement of the intraventricular mass and a relative increase in the number of white matter lesions without significant clinical deterioration. She underwent almost total resection of the mass. A papillary meningioma was confirmed histologically.

This association could have been coincidental. However, meningiomas have been previously reported in two patients with multiple sclerosis, with progression during treatment with interferon-beta (23). Based on immunohistochemistry, the meningioma was speculated to have resulted from enhanced platelet-derived growth factor receptors and/or down-regulated transforming growth factor receptors on the tumor itself.

Sensory systems

Retinal complications of interferon alfa in chronic viral hepatitis patients are well known, but few cases have been described with interferon beta.

- Bilateral retinopathy with similar features to those observed with interferon alfa has been reported in a 40-year-old woman treated with interferon beta-1b for multiple sclerosis (24).

In 49 patients, there was reversible otological impairment with tinnitus, mild-to-moderate hearing loss, or both in respectively 8, 16, and 20% of patients after administration of interferon alfa or interferon beta (SEDA-19, 336). These disorders tended to occur more frequently in patients on high cumulative doses, but led to withdrawal of treatment in only two patients.

Psychological, psychiatric (*See also Interferons*)

There have been reports of depression, suicidal ideation, and attempted suicide in patients receiving interferon beta (2,8,25). The lifetime risk of depression in patients with multiple sclerosis is high, and there has been a lively debate about whether interferon beta causes or exacerbates depression in such patients. Impressions of a possibly raised incidence of depression among patients treated with interferon beta for multiple sclerosis should be interpreted in the light of the spontaneous tendency to depressive disorders and suicidal ideation, which is encountered even in patients with untreated multiple sclerosis. Moreover, no raised incidence of these complications has been recorded in some studies (4,5). A critical review of the methodological limitations in studies that assessed mood disorders in patients on disease-modifying drugs for multiple sclerosis may help explain the widely divergent results from one study to another (26). Some results have argued against a specific role of interferon beta in the risk of depressive disorders.

Although isolated reports of psychotic delusional symptoms and depression continue to be published (27), recent controlled trials or longitudinal studies have not provided evidence of an increase in depression scores or in the rate of depression in patients treated with interferon beta (SEDA-27, 389). In a meta-analysis of seven trials in 1215 patients with relapsing remitting multiple sclerosis, the incidence of depression was 16% and did not differ between interferon-beta and controls, but the scales used to assess depression were specified in only three trials (83). Using a public reimbursement database for multiple sclerosis, the prevalence and incidence of depression and depression scores were not different in 163 patients treated with interferon beta or glatiramer, but the study was poorly controlled for potential biases (28). Overall, the current data suggest that interferon-beta is not substantially associated with depression.

A multicenter comparison of 44 and 22 micrograms of interferon beta-1a and placebo in 365 patients showed no significant differences in depression scores between the groups over a 3-year period of follow-up (29). In 106 patients with relapsing–remitting multiple sclerosis, depression status was evaluated before and after 12 months of interferon beta-1a treatment (30). According to the Beck Depression Inventory II scale, most of the patients had minimum (53%) or mild (32%) depression at baseline, and depression scores were not significantly increased after 1 year of treatment. There were no cases of suicidal ideation. In another study of 42 patients treated with interferon beta-1b, major depression at baseline was found in 21% of patients and was associated with a past history of psychiatric illness in most cases (31). Major depression was not considered as an exclusion criterion for interferon beta treatment when patients were on antidepressant therapy. There was a three-fold reduction in the prevalence of depression over the 1-year course of interferon treatment, suggesting a possible beneficial effect of treatment on mood. Finally, a single subcutaneous injection of interferon beta-1b did not alter cognitive performance and mood states in eight healthy volunteers (32).

The emotional state of 90 patients with relapsing–remitting multiple sclerosis has been carefully assessed with a battery of psychological tests at baseline and after 1 and 2 years of treatment with interferon beta-1b (33). In contrast to what was expected, and despite the lack of controls, there was significant improvement in emotional state, as shown by significant reductions in scores of anxiety and depression over time. In addition, there was no effect of low-dose oral glucocorticoids in a subgroup of 46 patients.

Depression has been quantified by telephone interview in 56 patients with relapsing multiple sclerosis 2 weeks before treatment, at the start of treatment, and after 8 weeks of treatment (34). Patients with a high depressive score 2 weeks before treatment significantly improved on starting treatment and returned to baseline within 8 weeks, whereas the depression score in non-depressed patients remained essentially unchanged. The investigators therefore suggested that patients' expectations had temporarily resulted in improvement of depression, and that increased depression during treatment is more likely to reflect pretreatment depression.

The clinical features, management, and prognosis of psychiatric symptoms in patients with chronic hepatitis C have been reviewed using data from 943 patients treated with interferon alfa (85%) or interferon beta (15%) for 24 weeks (35). Interferon-induced psychiatric symptoms were identified in 40 patients (4.2%) of those referred for psychiatric examination. They were classified in three groups according to the clinical profile: 13 cases of generalized anxiety disorder (group A), 21 cases of mood disorders with depressive features (group B), and six cases of other psychiatric disorders, including psychotic disorders with delusions/hallucinations ($n = 4$), mood disorders with manic features ($n = 1$), and delirium ($n = 1$) (group C). The time to onset of the symptoms differed significantly between the three groups: 2 weeks in group A, 5 weeks in group B, and 11 weeks in group C. Women were more often affected than men. There was no difference in the incidence or nature of the disorder according to the type of interferon used. Whereas most patients who required psychotropic drugs were able to complete treatment, 10 had to discontinue interferon treatment because of severe psychiatric symptoms, five from group B and five from group C. Twelve patients still required psychiatric treatment for more than 6 months after interferon withdrawal. In addition, residual symptoms (anxiety, insomnia, and mild hypothymia) were still present at the end of the survey in seven patients. Delayed recovery was mostly observed in patients in group C and in patients treated with interferon beta.

One debatable case of visual pseudo-hallucinations occurred only, but not reproducibly, within 30–60 minutes after interferon beta-1a injection in a 37-year-old woman with disseminated encephalomyelitis (36).

Endocrine

While no evidence of thyroid dysfunction or antithyroid antibodies was found in 20 patients receiving interferon beta during 24 weeks for hematological malignancies (37), antithyroid antibodies were detected in 29% of patients with multiple sclerosis after a prospective follow-up performed at 6, 12, and 18 months of treatment (38). Biological thyroid abnormalities without antithyroid antibodies have also been found (SEDA-20, 332). Overall, thyroid disorders with antithyroid antibodies were reported in only three patients on long-term interferon beta treatment for multiple sclerosis (38,39).

Thyroid disorders before and during the first 9 months of interferon beta-1b treatment have been systematically investigated in eight patients with relapsing–remitting multiple sclerosis (40). Before treatment, one patient had positive thyroperoxidase antibodies and one was taking thyroxine for multinodular goiter. After 3 months three other patients developed sustained positive titers of thyroperoxidase antibodies, of whom one developed hypothyroidism after 9 months. These results are in accordance with a previous similar study and isolated case reports (SEDA-21, 374) (SEDA-22, 405), and suggest that interferon beta, like interferon alfa, can cause thyroid autoimmunity.

As suggested in a more comprehensive long-term follow-up study, interferon beta-induced thyroid dysfunction is often transient or has limited clinical consequences (41). Of 31 patients with multiple sclerosis regularly assessed for 30–42 months for thyroid function, 13 developed thyroid disorders during treatment with interferon beta-1b. None withdrew because of thyroid disorders. Of the eight patients with no previous thyroid disorders, one had a persistent but isolated increase in antithyroglobulin titer, six developed transient signs of hypothyroidism or hyperthyroidism during the first year of therapy, and only one had overt hypothyroidism after 12 months of treatment and required thyroxine replacement. Of the five patients with baseline signs of Hashimoto's thyroiditis, one had a transiently positive antithyroglobulin titer, one developed transient hyperthyroidism, and the three patients who had previously had or who newly developed subclinical hypothyroidism remained stable throughout the study. Overall, thyroid disorders occurred only during the first 12 months of treatment and no additional cases were detected after the first year of therapy. In the authors' opinion, pre-existing or new thyroiditis is not a contraindication to continuing interferon beta-1b treatment. Two patients took thyroxine replacement and continued to receive interferon beta-1b (42).

Conflicting data have been reported on the association between interferon beta therapy in multiple sclerosis and thyroid disease. In 106 patients (76 women) with multiple sclerosis who received interferon beta-1a or beta-1b for up to 84 (median 42) months, there was baseline thyroid autoimmunity in 8.5% and hypothyroidism in 2.8% (43). Thyroid dysfunction (80% hypothyroidism, 92% subclinical, 56% transient) developed in 24% (68% with autoimmunity) and autoimmunity in 23% (46% with dysfunction), without a significant difference between the two cytokines; 68% of the cases of dysfunction occurred within the first year. Thyroid dysfunction was generally subclinical and was transient in over half of cases. Autoimmunity was the only predictive factor for the development of dysfunction (relative risk = 8.9), but sustained disease was also significantly associated with male sex.

Among 700 patients with multiple sclerosis treated with interferon beta-1a (n = 467) or beta-1b (n = 233), overt hyperthyroidism occurred in five patients treated with interferon-beta-1b, three of whom required withdrawal and long-term carbimazole, while there were two cases of hypothyroidism and one of goiter without thyroid dysfunction in patients treated with interferon beta-1a (44). Clinical abnormalities occurred after a mean of 14 months and there were thyroid antibodies in four of the eight patients. The frequency of clinical thyroid dysfunction was higher, but not statistically different for interferon-beta-1b compared with interferon beta-1a (2.15% versus 0.64%). A severe form of hypothyroidism with signs of Hashimoto's encephalopathy has also been attributed to interferon beta-1a treatment in a 54-year-old woman (45).

Metabolism

Severe hypertriglyceridemia, a well-known adverse effect of interferon beta, has been reported and fully investigated in a 39-year-old man receiving interferon beta for chronic hepatitis C (46).

Hematologic

Moderate reductions in white blood cell count (that is lymphopenia, leukopenia, and granulocytopenia) have been observed with recombinant interferon beta, and marked eosinophilia recorded in an atopic patient (SEDA-21, 374).

- A 42-year-old woman developed aplastic anemia after using interferon beta-1a for 1 year (47). There was hematological improvement after withdrawal and immunosuppressive therapy.

In an analysis of six controlled clinical trials of interferon beta-1a, there was grade 4 toxicity at similar rates in the placebo and treated patients (0.2% in each group) (48). Most of the abnormalities occurred within the first 6 months of treatment and resolved despite continued treatment. In the pooled safety database, which also included uncontrolled studies, only 12 of 3995 patients had to stop taking interferon beta-1a. The highest rate of treatment withdrawal was reported with the highest dose (0.8% in those given 44 micrograms 3 times weekly). Although the authors stated that symptomatic events were rare, uncommon infections were more frequent in the highest dose group compared with placebo (2.8% versus 1.6%), but unfortunately no in-depth analysis of these data was provided. Post-marketing surveillance data did not add significant additional information, but there was one convincing case of autoimmune hemolytic anemia, which resolved after interferon beta-1a withdrawal and recurred on readministration.

Acquired hemophilia A has been attributed to interferon beta-1a (49).

- A 41-year-old man with a history of autologous stem-cell transplantation received interferon beta-1a for multiple sclerosis. After 18 months of treatment, he developed persistent hematuria and 1 month later developed large ecchymoses, and bleeding from puncture sites. Prothrombin and activated partial thromboplastin times were prolonged. Factor VIII activity was only 2% and increased after administration of immunosuppressive drugs, gammaglobulins, and factor VIIa, but again fell to 4%. The patient died from severe hemorrhage.

This report of an acquired factor VIII inhibitor after administration of interferon beta is reminiscent of previous cases associated with interferon-alfa (SED-14, 1252; SEDA-25, 433).

Reversible autoimmune hemolytic anemia (SEDA-20, 332) has also been reported.

Liver

Increased transaminase activities are common in patients taking interferon-beta, and regular assessment of liver tests is usually recommended during the first 6-12 months. In a retrospective review of the charts of 844 patients with multiple sclerosis treated with one of the three commercially available forms of interferon-beta, there were de novo rises in alanine transaminase in 37% of patients (50). There were marked increases (5–20 times the upper limit of the reference range) in only 1.4% of the patients, and only two developed jaundice. There was a relation between increasing dose, the frequency of injections, and transaminase increases. Transaminase activity peaked within the first 6 months of treatment with subcutaneous administration, and at 6-12 months with intramuscular administration. The manufacturer of the subcutaneous form of interferon beta-1a (Rebif®) also specifically analysed the data from six controlled trials and concluded that raised alanine transaminase activity was mostly asymptomatic, dose-related, and resolved either spontaneously or after dosage adjustment (51). Only 0.4% of patients discontinued treatment because of liver damage. From post-marketing surveillance data, the estimated rate of serious symptomatic liver injury was 1/2300 patients, but it is not known how many cases were clearly related to interferon beta-1a. The authors also mentioned that the only reported case of fulminant hepatitis attributed to interferon beta-1a (SEDA-25, 436) was confounded by concomitant exposure to nefazodone, which is hepatotoxic.

The relevance of liver test abnormalities in patients treated with interferon beta has been discussed at length (52). Although rises in transaminase activities occur in about one-third of patients in clinical trials, they are mild and self-limiting in most cases (53). The clinical impact, if any, is therefore very small. However, an autoimmune mechanism may be involved in isolated cases, as suggested in a report of autoimmune hepatitis attributed to interferon beta-1a in a 52-year-old woman (54). Adverse hepatic effects of interferon beta were usually limited to a dose-dependent increase in transaminases, but transient autoimmune hepatitis has been described in one patient (55).

It has been suspected that interferon beta can accelerate hepatocellular carcinoma (56).

- A 62-year-old man with severe chronic hepatitis and positive serum anti-HCV, HBs, and HBc antibodies underwent unsuccessful treatment with interferon alfa for 3 months and then received interferon beta for 6 months with a partial response. During treatment, his alfa-fetoprotein (normal before treatment) progressively increased to seven-fold the upper limit of the reference range. There was also a slight increase in interleukin-6 serum concentration. A hepatocellular carcinoma was diagnosed 9 months later.

Liver carcinoma is an unexpected consequence of interferon beta in patients with chronic hepatitis C. However, the authors cited other published Japanese reports of hepatocellular carcinoma during or after interferon treatment. It is worth noting that interferon beta, but not interferon alfa, significantly increased serum interleukin-6 concentrations in patients with chronic hepatitis C (57), and that interleukin-6 has been suggested to promote the growth of hepatocellular carcinoma.

Fulminant liver failure has been attributed to interferon beta-1a (58), but it was subsequently confirmed that this case was confounded by concomitant exposure to nefazodone, which is hepatotoxic (59).

Urinary tract

Reversible hemolytic–uremic syndrome has been reported (SEDA-22, 405).

- Two women aged 24 and 44 years, whose primary diagnosis was relapsing multiple sclerosis, developed renal impairment and a thrombotic thrombocytopenic purpura-like syndrome within 2–4 weeks after starting interferon beta-1a (60). Thrombocytopenia and renal function normalized in the first patient, whereas the second patient had thrombotic angiopathy on renal biopsy and required dialysis while awaiting renal transplantation.

As interferon alfa has also been suggested to produce hemolytic–uremic syndrome with thrombotic thrombocytopenic purpura, it is tempting to speculate that either interferons or other cytokines may play a role in this syndrome.

Proteinuria, nephrotic syndrome, and various forms of renal lesions can be caused or exacerbated by interferon alfa (SED-14, 1252).

- Proteinuria with minimal–change nephrotic syndrome on renal biopsy has been attributed to interferon beta in a 64-year-old man with malignant melanoma (61). Although the proteinuria abated after withdrawal, the potential role of previous chemotherapy cannot be excluded.
- Nephrotic syndrome with segmental glomerulosclerosis has been reported in a 32-year-old woman with multiple sclerosis receiving interferon beta (62).
- A 39-year-old man given interferon beta-1a for relapsing-remitting multiple sclerosis developed nephrotic syndrome with minimal glomerular lesions on histological examination in (63).

The clinicopathological features of proteinuria have been investigated in 23 patients with chronic hepatitis C who had new or worsened proteinuria during interferon treatment (interferon alfa 6–10 MU/day in three patients and interferon beta 6 MU/day in 20 patients (64). Renal function and urinary findings were normal before treatment in 21 subjects. Proteinuria appeared after a mean of 12 (range 5–30) days after the start of treatment, and the mean value was 2.1 g/day. There was low selective proteinuria in 78% of the patients. Renal histopathology in 11 patients showed IgA glomerulonephritis in five, mesangial proliferative glomerulonephritis in four, membranoproliferative glomerulonephritis in one, and nephrosclerosis in one. There was only trace deposition of hepatitis C virus core antigen in three of nine patients, suggesting that hepatitis C was not the primary cause of these glomerulopathies.

Skin

Injection-site reactions are common after subcutaneous injection of interferon beta-1b, and are more frequent than with any other available interferons. In a multi-center placebo-controlled trial, 65% of patients receiving interferon beta-1b had reactions at the injection site compared with 6% in the placebo group (2). In contrast, only 5% of those who received interferon beta-1a had injection site reactions (3). The clinical features of injection site reactions to interferon beta-1b mostly consist of benign inflammatory reactions, but they can sometimes be more severe, with sclerotic dermal plaques, painful erythematous nodules, and deep cutaneous ulcers with skin necrosis (SEDA-21, 374) (65). Late severe reactions have included a case of squamous cell carcinoma at the injection site (SEDA-21, 375) and a case of panniculitis (66).

Since interferon beta-1b became available, 1443 instances of injection site reactions, 212 cases of injection site necrosis, and 10 cases of non-injection site necrosis were notified to the US Food and Drug Administration, and antibiotic therapy or surgery was required in 20–30% of patients (67). Severe necrotizing cutaneous lesions have also been attributed to subcutaneous interferon beta-1a (68).

In contrast to previous claims, even low-dose interferon beta-1b can produce severe local reactions and cutaneous necrosis, with no recurrence after interferon alfa injection and expected better tolerance to interferon beta-1a (69–72). The mechanisms of interferon beta-induced local skin reaction might involve a local vascular inflammatory process or platelet-dependent thrombosis, but positive intracutaneous tests to interferon beta have also been found (73).

There have been other isolated reports of skin lesions.

- Intravascular papillary endothelial hyperplasia with multiple lesions on both hands has been attributed to interferon beta-1b in a 50-year-old man with multiple sclerosis (74).
- Granulomatous dermatitis with disseminated pruritic papules and histological features resembling those of sarcoid granulomas has been described in a 57-year-old man who received interferon beta-1b (75). The first lesions were observed after 2 months of treatment, persisted for 2 years, and slowly improved after interferon beta withdrawal and treatment with hydroxychloroquine PUVA.
- Erythromelalgia has been attributed to interferon beta-1a in a 38-year-old woman (76). Complete recovery was obtained only after interferon withdrawal.
- In one patient, there was cutaneous mucinosis on skin biopsy, and skin lesions persisted for several months before healing spontaneously (77).

Other reports include exacerbation of quiescent psoriasis or induction of psoriatic lesions at the injection sites (SED-13, 1099; SEDA-20, 332), and the development or exacerbation of lichen planus (SEDA-21, 375).

Musculoskeletal

Rhabdomyolysis associated with interferon beta has been reported (78).

- A 39-year-old man developed acute generalized myalgia and weakness in all four limbs 3 months after starting interferon beta-1a (22 micrograms three times a week) for relapsing–remitting multiple sclerosis. Serum creatine kinase activity peaked at about 95 times the upper limit of the reference range. Infectious and metabolic causes were ruled out and there was no argument in favor of an underlying metabolic muscle disorder. He recovered fully after interferon beta withdrawal and supportive treatment.

Although rhabdomyolysis has not been previously attributed to interferon beta, this case is in keeping with those described with interferon alfa.

- A 39-year-old man developed monoarthritis in his right elbow after receiving interferon beta for 16 days for chronic hepatitis C (79).

Although the arthritis resolved after withdrawal, this report casts doubt on the causal relation, as there was no recurrence on re-administration.

Reproductive system

Mild to moderate menstrual disorders were twice as frequent in patients receiving interferon beta compared with placebo (17 versus 8%). Severe persistent vaginal bleeding has been reported in a 19-year-old woman (SEDA-21, 375).

Immunologic

Hypersensitivity reactions

Hypersensitivity reactions to interferon beta are rare, with only a few cases of immediate-type reactions (SEDA-20, 332) (80), and one case of urticaria associated with exacerbation of asthma (81).

- A 21-year-old woman had a severe anaphylactic-like reaction with laryngospasm and undetectable blood pressure within 10 minutes after interferon beta-1a injection (80). It is still uncertain that anaphylaxis was definitely attributable to interferon, because rechallenge and skin tests were not performed in this patient, who had tolerated the treatment for the 6 previous months.
- Urticaria developed after 9 months of treatment with interferon beta-1b in a 32-year-old woman with a previous history of penicillin allergy (81). She also had an exacerbation of asthma shortly after starting treatment. A positive intradermal test to interferon beta-1b, but not to interferon beta-1a or the diluents, suggested a specific IgE allergic reaction.

Another isolated case history has suggested that interferon beta-1b might have favored the development of a non-IgE-mediated anaphylactic reaction to previously well-tolerated injections of methylprednisolone (82).

- A 36-year-old woman with multiple sclerosis and positive serology for hepatitis C developed biopsy-proven non-specific lymphocytic cutaneous vasculitis, with renal involvement marked by proteinuria and hematuria, after 2 years of interferon beta-1a treatment (83). Immunological investigations were negative, except for a previously known positive antinuclear antibody, and serology did not indicate a recent infectious episode. There was complete recovery within 2 weeks after withdrawal.

Although the long delay in this case suggested a possible coincidental adverse event, this case is reminiscent of reports of vasculitis involving interferon alfa.

Autoimmune disorders
In contrast to interferon alfa, the autoimmune consequences of interferon beta treatment have been poorly evaluated. Interferon beta does not appear to be associated with the appearance or increased titres of several auto-antibodies, and no clinical features of autoimmune disease were observed in patients receiving a 6-month course of interferon beta-1a or interferon beta-1b (84,85).

Subcutaneous lupus erythematosus reversible on withdrawal of treatment (SEDA-22, 405) and myasthenia gravis (SEDA-21, 375) have been described, but each only in a single patient receiving interferon beta.

The possible involvement of interferon beta in the occurrence of sarcoidosis has been noted in two patients (86,87).

Interferon beta antibodies
In multiple sclerosis, neutralizing antibodies to recombinant interferon beta occurred in 12–38% of patients treated for 2–3 years (3–5,88). There were no adverse consequences in patients who developed antibodies to interferon beta-1a (4), but there was reduced therapeutic efficacy in terms of clinical relapse rate or magnetic resonance imaging in several patients who had neutralizing antibodies to interferon beta-1b (SEDA-20, 332) (5,89).

The systemic availability of interferon beta-1b, measured by a myxovirus protein A assay, was completely inhibited in patients with neutralizing antibodies (90). The presence of increased titers of serum-binding antibodies increased the likelihood of neutralizing antibodies. From an in vitro study of nine patients who developed neutralizing antibodies against the available formulations of recombinant interferon beta (three on interferon beta-1a and six on interferon beta-1b), it appears that these antibodies systematically cross-react in both binding and biological assays (91). Although the sample size was small and lacked clinical confirmation, these results suggest that clinical benefit will not be obtained by switching to an alternative formulation when the absence of response or relapse during treatment is due to neutralizing antibodies.

In an open study in 78 patients treated with various commercially available forms of interferon beta, the highest incidence of persistent neutralizing antibodies was in those who received subcutaneous interferon beta-1b (35%), with an intermediate rate in those who received subcutaneous interferon beta-1a (20%), and the lowest rate in those who received intramuscular interferon beta-1a (3%) (92). Patients with persistent neutralizing antibodies had a higher probability of worsening over the 3 years of the study and a shorter time to relapse, again suggesting that these antibodies have a negative influence on disease outcome. There were very similar findings in another study in 90 patients, which also suggested that the simultaneous presence of high titers of both binding and neutralizing antibodies correlated with higher degrees of disease activity and progression (93). Although both studies involved a limited number of patients, the minimal 3-year length of follow-up allowed accurate analysis of the clinical impact of antibodies.

An international panel specialized in the treatment of multiple sclerosis has summarized the question of whether neutralizing antibodies to interferon beta develop in patients with multiple sclerosis (94). They concluded that interferon beta can induce the development of antibodies that reduce or "neutralize" the biological activity of interferon beta. The extent of development of these antibodies differs amongst the available interferon beta products. Interferon beta-1b is associated with the highest incidence; of interferon beta-1a products, Avonex is associated with the lowest incidence. Neutralizing antibodies are clinically important, since the efficacy of interferon is reduced in patients with persistent and significant titers. Thus, optimal management of patients with multiple sclerosis should not only consider the relative efficacy and safety of interferon beta, but also their immunogenic potential. An international standardized assay for neutralizing antibodies is required and all patients with multiple sclerosis who receive interferon beta should be evaluated for their presence.

Interferon beta-1b induces neutralizing antibodies earlier and more often than interferon beta-1a, but whether the development of neutralizing antibodies is associated with reduced clinical efficacy in multiple sclerosis is a matter of continuing debate (95). In a large longitudinal study in which neutralizing antibodies were regularly assessed in 541 patients with relapsing remitting multiple sclerosis who were treated with various formulations of interferon beta, the presence of neutralizing antibodies was associated with a significantly higher rate of relapse and a shortened time to first relapse, but did not affect disease progression (96). No predictors of antibody formation were identified. However, these results were strongly disputed because of possible methodological flaws and the use of a non-standard assay (97). In another study the development of neutralizing antibodies was investigated in 718 patients with secondary progressive multiple sclerosis in a placebo-controlled trial, of whom 360 received interferon beta-1b (98). The incidence of neutralizing antibody, defined by at least two consecutive positive titers, was 28%. Again, there was no significant effect of the presence of neutralizing antibodies on the progression of disability. The impact of neutralizing antibodies on the relapse rate was inconsistent, but suggested an increased relapse rate during periods of high neutralizing antibody titers. The study also showed that 37% of patients with neutralizing antibodies durably reverted to antibody-negative status. Accordingly, most authors agreed that treatment decisions in patients with positive neutralizing antibodies should still be based on the clinical outcome.

The clinical significance of these antibodies is uncertain (99) and it has been proposed that decisions to discontinue treatment should be based on individual clinical responses and unequivocal demonstration of neutralizing antibodies with a reliable assay (88).

One study has shown that antibodies can spontaneously disappear in patients receiving long-term treatment. Of 24 of 51 patients who initially developed neutralizing antibodies, generally during the first year of treatment, only five still had antibodies after a mean treatment duration of 102 months (100). The mean time to antibody disappearance was 20 months.

A randomized study has been conducted in 161 patients to evaluate whether a monthly intravenous pulse of methylprednisolone reduces the frequency of neutralizing antibodies to interferon beta-1b (101). The patients who received both interferon beta and methylprednisolone had a 55% relative reduction in the development of neutralizing antibody and were significantly more likely to remain negative for neutralizing antibodies after 6 months of treatment compared with those who received interferon beta alone. The overall frequency of adverse effects and the number of withdrawals were similar between the groups, but headaches, fatigue, and myalgia were less frequent in the combination therapy group. A limitation of this study was that there was no difference in clinical outcome between the two groups.

Finally, possible differences in the immunogenic potential of recombinant and natural interferon beta preparations were found in a small study (102).

Second-Generation Effects

Teratogenicity

Data on outcomes in pregnancy after treatment with interferon beta-1a in patients with multiple sclerosis have been obtained from clinical trials (103). Of 29 pregnancies that occurred during or shortly after treatment withdrawal, 13 resulted in normal outcomes, two in premature births, one in fetal death, six in induced abortions, and seven in spontaneous abortions.

- A child whose mother had received interferon beta until 2.5 months before pregnancy had a right incomplete double renal pelvis and ureter (104). Although the authors discussed the possible role of interferon therapy, the timing of exposure was obviously not suggestive of a causal relation.

Although the data are still very limited, it is advisable to reassure exposed patients and to withdraw interferon beta up to the time of delivery.

Drug Administration

Drug dosage regimens

In 188 patients with relapsing–remitting multiple sclerosis assigned to receive interferon beta-1a 30 micrograms intramuscularly once a week or interferon beta-1b 44 micrograms subcutaneously three times a week, only injection site reactions and neutralizing antibodies to interferon beta were significantly more frequent in the interferon beta-1b group (105). These differences were probably related to the subcutaneous route of administration of interferon beta-1b. In contrast, there were significant differences in favor of interferon beta-1b for clinical outcomes after 2 years of treatment.

In a comparison of two regimens of interferon beta-1a (44 micrograms Rebif subcutaneously three times a week versus 30 micrograms Avonex intramuscularly once a week) in 677 patients with relapsing–remitting multiple sclerosis, Rebif was more effective on primary clinical outcomes (patients remaining relapse-free at 24 weeks), but produced significantly more frequent injection site reactions (88 versus 28%), asymptomatic and mild liver enzyme abnormalities (18 versus 9%), mild white cell abnormalities (11 versus 5%), and neutralizing antibodies to interferon beta (25 versus 2%) (106). However, the severity of adverse events and withdrawal due to an adverse event were similar in both groups.

Drug–Drug Interactions

Theophylline

Interferon beta reduced theophylline clearance by 29% in seven patients (SED-13, 1099).

Management of Adverse Drug Reactions

In a randomized double-blind trial in 84 patients, ibuprofen, paracetamol, and prednisone were equally effective in controlling the incidence and severity of the flu-like syndrome that occurs during the first month of intramuscular interferon beta-1a (107). However, patients who received ibuprofen had less severe symptoms on the day of interferon injection.

References

1. Anonymous. Euromedicines evaluation: the striptease begins. Lancet 1996;347(9000):483(see also Harvey P. Why interferon beta 1b was licensed is a mystery. BMJ 1996;313(7052):297–8 and Napier JC. Reputation of interferon beta-1b. Lancet 1996;347(9006):968).
2. The IFNB Multiple Sclerosis Study GroupThe University of British Columbia MS/MRI Analysis Group. Interferon beta-1b in the treatment of multiple sclerosis: final outcome of the randomized controlled trial. Neurology 1995;45(7):1277–85.
3. Jacobs LD, Cookfair DL, Rudick RA, Herndon RM, Richert JR, Salazar AM, Fischer JS, Goodkin DE,

Granger CV, Simon JH, Alam JJ, Bartoszak DM, Bourdette DN, Braiman J, Brownscheidle CM, Coats ME, Cohan SL, Dougherty DS, Kinkel RP, Mass MK, Munschauer FE 3rd, Priore RL, Pullicino PM, Scherokman BJ, Whitham RH, et al. Intramuscular interferon beta-1a for disease progression in relapsing multiple sclerosis. The Multiple Sclerosis Collaborative Research Group (MSCRG). Ann Neurol 1996;39(3):285–94.

4. Ebers GC, Hommes O, Hughes RAC, et alPRISMS (Prevention of Relapses and Disability by Interferon beta-1a Subcutaneously in Multiple Sclerosis) Study Group. Randomised double-blind placebo-controlled study of interferon beta-1a in relapsing/remitting multiple sclerosis. Lancet 1998;352(9139):1498–504.

5. European Study Group on Interferon Beta-1b in Secondary Progressive MS. Placebo-controlled multicentre randomised trial of interferon beta-1b in treatment of secondary progressive multiple sclerosis. Lancet 1998;352(9139):1491–7.

6. Weinstock-Guttman B, Rudick RA. Prescribing recommendations for interferon-beta in multiple sclerosis. CNS Drugs 1997;8:102–12.

7. Rio J, Nos C, Marzo ME, Tintore M, Montalban X. Low-dose steroids reduce flu-like symptoms at the initiation of IFNbeta-1b in relapsing-remitting MS. Neurology 1998;50(6):1910–2.

8. Neilley LK, Goodin DS, Goodkin DE, Hauser SL. Side effect profile of interferon beta-1b in MS: results of an open label trial. Neurology 1996;46(2):552–4.

9. Bayas A, Rieckmann P. Managing the adverse effects of interferon-beta therapy in multiple sclerosis. Drug Saf 2000;22(2):149–59.

10. Buraglio M, Trinchard-Lugan I, Munafo A, Macnamee M. Recombinant human interferon-beta-1a (Rebif) vs recombinant interferon-beta-1b (Betaseron) in healthy volunteers. A pharmacodynamic and tolerability study. Clin Drug Invest 1999;18:27–34.

11. Reess J, Haas J, Gabriel K, Fuhlrott A, Fiola M, Schicklmaier P. Both paracetamol and ibuprofen are equally effective in managing flu-like symptoms in relapsing-remitting multiple sclerosis patients during interferon beta-1a (AVONEX) therapy. Mult Scler 2002;8(1):15–8.

12. McCormack PL, Scott LJ. Interferon-beta-1b: a review of its use in relapsing-remitting and secondary progressive multiple sclerosis. CNS Drugs 2004;18:521–46.

13. Filippini G, Munari L, Incorvaia B, Ebers GC, Polman C, D'Amico R, Rice GP. Interferons in relapsing remitting multiple sclerosis: a systematic review. Lancet 2003;361:545–52.

14. Schmidt S, Hertfelder HJ, von Spiegel T, Hering R, Harzheim M, Lassmann H, Deckert-Schluter M, Schlegel U. Lethal capillary leak syndrome after a single administration of interferon beta-1b. Neurology 1999;53(1):220–2.

15. Ferriby D, Stojkovic T. Clinical picture: bronchiolitis obliterans with organising pneumonia during interferon beta-1a treatment. Lancet 2001;357(9258):751.

16. Matsumura S, Takamatsu H, Sato S, Ara S. [Central nervous system toxicity of local interferon-beta therapy. Report of three cases.]Neurol Med Chir (Tokyo) 1988;28(3):265–70.

17. Liberati AM, Biagini S, Perticoni G, Ricci S, D'Alessandro P, Senatore M, Cinieri S. Electrophysiological and neuropsychological functions in patients treated with interferon-beta. J Interferon Res 1990;10(6):613–9.

18. Pollmann W, Erasmus LP, Feneberg W, Then Bergh F, Straube A. Interferon beta but not glatiramer acetate therapy aggravates headaches in MS. Neurology 2002;59(4):636–9.

19. Bramanti P, Sessa E, Rifici C, D'Aleo G, Floridia D, Di Bella P, Lublin F. Enhanced spasticity in primary progressive MS patients treated with interferon beta-1b. Neurology 1998;51(6):1720–3.

20. Miwa H, Furuya T, Tanaka S, Mizuno Y. Neurosarcoidosis associated with interferon therapy. Eur Neurol 2001;45(4):288–9.

21. Von Raison F, Abboud H, Saint Val C, Brugieres P, Cesaro P. Acute demyelinating disease after interferon beta-1a treatment for multiple sclerosis. Neurology 2000;55(9):1416–7.

22. Drevelegas A, Xinou E, Karacostas D, Parissis D, Karkavelas G, Milonas I. Meningioma growth and interferon beta-1b treated multiple sclerosis: coincidence or relationship? Neuroradiology 2005;47(7):516–9.

23. Batay F, Al-Mefty O. Growth dynamics of meningiomas in patients with multiple sclerosis treated with interferon: report of two cases. Acta Neurochir (Wien) 2002;144(4):365–8.

24. Sommer S, Sablon JC, Zaoui M, Rozot P, Hosni A. Rétinopathie à l'interféron bêta au cours d'une sclérose en plaques. [Interferon beta-1b retinopathy during a treatment for multiple sclerosis.] J Fr Ophtalmol 2001;24(5):509–12.

25. Lublin FD, Whitaker JN, Eidelman BH, Miller AE, Arnason BG, Burks JS. Management of patients receiving interferon beta-1b for multiple sclerosis: report of a consensus conference. Neurology 1996;46(1):12–8.

26. Feinstein A. Multiple sclerosis, disease modifying treatments and depression: a critical methodological review. Mult Scler 2000;6(5):343–8.

27. Goeb JL, Cailleau A, Lainé P, Etcharry-Bouyx F, Maugin D, Duverger P, Gohier B, Rannou-Dubas K, Dubas F, Garre JB. Acute delirium, delusion, and depression during IFN-beta-1a therapy for multiple sclerosis: a case report. Clin Neuropharmacol 2003;26:5–7.

28. Patten SB, Fridhandler S, Beck CA, Metz LM. Depressive symptoms in a treated multiple sclerosis cohort. Mult Scler 2003;9:616–20.

29. Patten SB, Metz LMSPECTRIMS Study Group. Interferon beta1a and depression in secondary progressive MS: data from the SPECTRIMS Trial. Neurology 2002;59(5):744–6.

30. Zephir H, De Seze J, Stojkovic T, Delisse B, Ferriby D, Cabaret M, Vermersch P. Multiple sclerosis and depression: influence of interferon beta therapy. Mult Scler 2003;9(3):284–8.

31. Feinstein A, O'Connor P, Feinstein K. Multiple sclerosis, interferon beta-1b and depression. A prospective investigation. J Neurol 2002;249(7):815–20.

32. Exton MS, Baase J, Pithan V, Goebel MU, Limmroth V, Schedlowski M. Neuropsychological performance and mood states following acute interferon-beta-1b administration in healthy males. Neuropsychobiology 2002;45(4):199–204.

33. Borras C, Rio J, Porcel J, Barrios M, Tintore M, Montalban X. Emotional state of patients with relapsing-remitting MS treated with interferon beta-1b. Neurology 1999;52(8):1636–9.

34. Mohr DC, Likosky W, Dwyer P, Van Der Wende J, Boudewyn AC, Goodkin DE. Course of depression during

the initiation of interferon beta-1a treatment for multiple sclerosis. Arch Neurol 1999;56(10):1263–5.

35. Hosoda S, Takimura H, Shibayama M, Kanamura H, Ikeda K, Kumada H. Psychiatric symptoms related to interferon therapy for chronic hepatitis C: clinical features and prognosis. Psychiatry Clin Neurosci 2000;54(5):565–72.

36. Moor CC, Berwanger C, Welter FL. Visual pseudo-hallucinations in interferon-beta 1a therapy. Akt Neurol 2002;29:355–7.

37. Pagliacci MC, Pelicci G, Schippa M, Liberati AM, Nicoletti I. Does interferon-beta therapy induce thyroid autoimmune phenomena? Horm Metab Res 1991;23(4):196–7.

38. Martinelli V, Gironi M, Rodegher M, Martino G, Comi G. Occurrence of thyroid autoimmunity in relapsing remitting multiple sclerosis patients undergoing interferon-beta treatment. Ital J Neurol Sci 1998;19(2):65–7.

39. Schwid SR, Goodman AD, Mattson DH. Autoimmune hyperthyroidism in patients with multiple sclerosis treated with interferon beta-1b. Arch Neurol 1997;54(9):1169–90.

40. Rotondi M, Oliviero A, Profice P, Mone CM, Biondi B, Del Buono A, Mazziotti G, Sinisi AM, Bellastella A, Carella C. Occurrence of thyroid autoimmunity and dysfunction throughout a nine-month follow-up in patients undergoing interferon-beta therapy for multiple sclerosis. J Endocrinol Invest 1998;21(11):748–52.

41. Monzani F, Caraccio N, Casolaro A, Lombardo F, Moscato G, Murri L, Ferrannini E, Meucci G. Long-term interferon beta-1b therapy for MS: is routine thyroid assessment always useful? Neurology 2000;55(4):549–52.

42. McDonald ND, Pender MP. Autoimmune hypothyroidism associated with interferon beta-1b treatment in two patients with multiple sclerosis. Aust NZ J Med 2000;30(2):278–9.

43. Caraccio N, Dardano A, Manfredonia F, Manca L, Pasquali L, Iudice A, Murri L, Ferrannini E, Monzani F. Long term follow-up of 106 multiple sclerosis patients undergoing interferon-beta 1a or 1b therapy: predictive factors of thyroid disease development and duration. J Clin Endocrinol Metab 2005;90(7):4133–7.

44. Kreisler A, de Seze J, Stojkovic T, Delisse B, Combelles M, Verier A, Hautecoeur P, Vermersch P. Multiple sclerosis, interferon beta and clinical thyroid dysfunction. Acta Neurol Scand 2003;107:154–7.

45. Polman CH, Jansen PH, Jansen C, Uitdehaag BM. A rare, treatable cause of relapsing encephalopathy in an MS patient on interferon beta therapy. Neurology 2003;61:719.

46. Homma Y, Kawazoe K, Ito T, Ide H, Takahashi H, Ueno F, Matsuzaki S. Chronic hepatitis C beta-interferon-induced severe hypertriglyceridaemia with apolipoprotein E phenotype E3/2. Int J Clin Pract 2000;54(4):212–6.

47. Aslam AK, Singh T. Aplastic anemia associated with interferon beta-1a. Am J Ther 2002;9(6):522–3.

48. Rieckmann P, O'Connor P, Francis GS, Wetherill G, Alteri E. Haematological effects of interferon-beta-1a (Rebif) therapy in multiple sclerosis. Drug Saf 2004;27:745–56.

49. Kaloyannidis P, Sakellari I, Fassas A, Fragia T, Vakalopoulou S, Kartsios C, Garypidou B, Kimiskidis V, Anagnostopoulos A. Acquired hemophilia-A in a patient with multiple sclerosis treated with autologous hematopoietic stem cell transplantation and interferon beta-1a. Bone Marrow Transplant 2004;34:187–8.

50. Tremlett HL, Yoshida EM, Oger J. Liver injury associated with the beta-interferons for MS: a comparison between the three products. Neurology 2004;62:628–31.

51. Francis GS, Grumser Y, Alteri E, Micaleff A, O'Brien F, Alsop J, Stam Moraga M, Kaplowitz N. Hepatic reactions during treatment of multiple sclerosis with interferon-beta-1a: incidence and clinical significance. Drug Saf 2003;26:815–27.

52. Tremlett H, Oger J. Hepatic injury, liver monitoring and the beta-interferons for multiple sclerosis. J Neurol 2004;251:1297–303.

53. Tremlett HL, Oger J. Elevated aminotransferases during treatment with interferon-beta for multiple sclerosis: actions and outcomes. Mult Scler 2004;10:298–301.

54. Wallack EM, Callon R. Liver injury associated with the beta-interferons for MS. Neurology 2004;63:1142–3.

55. Durelli L, Bongioanni MR, Ferrero B, Oggero A, Marzano A, Rizzetto M. Interferon treatment for multiple sclerosis: autoimmune complications may be lethal. Neurology 1998;50(2):570–1.

56. Malaguarnera M, Restuccia S, Di Fazio I, Di Marco R, Pistone G, Trovato BA. Rapid evolution of chronic viral hepatitis into hepatocellular carcinoma after beta-interferon treatment. Panminerva Med 1999;41(1):59–61.

57. Furusyo N, Hayashi J, Ohmiya M, Sawayama Y, Kawakami Y, Ariyama I, Kinukawa N, Kashiwagi S. Differences between interferon-alpha and -beta treatment for patients with chronic hepatitis C virus infection. Dig Dis Sci 1999;44(3):608–17.

58. Yoshida EM, Rasmussen SL, Steinbrecher UP, Erb SR, Scudamore CH, Chung SW, Oger JJ, Hashimoto SA. Fulminant liver failure during interferon beta treatment of multiple sclerosis. Neurology 2001;56(10):1416.

59. Francis GS, Grumser Y, Alteri E, Micaleff A, O'Brien F, Alsop J, Stam Moraga M, Kaplowitz N. Hepatic reactions during treatment of multiple sclerosis with interferon-beta-1a: incidence and clinical significance. Drug Saf 2003;26(11):815–27.

60. Herrera WG, Balizet LB, Harberts SW, Brown ST. Occurrence of a TTP-like syndrome in two women receiving beta interferon therapy for relapsing multiple sclerosis. Neurology 1999;52:135.

61. Nakao K, Sugiyama H, Makino E, Matsuura H, Ohmoto A, Sugimoto T, Ichikawa H, Wada J, Yamasaki Y, Makino H. Minimal change nephrotic syndrome developing during postoperative interferon-beta therapy for malignant melanoma. Nephron 2002;90(4):498–500.

62. Gotsman I, Elhallel-Darnitski M, Friedlander Z, Haviv YS. Beta-interferon-induced nephrotic syndrome in a patient with multiple sclerosis. Clin Nephrol 2000;54(5):425–6.

63. Tola MR, Caniatti LM, Gragnaniello D, Russo M, Stabellini N, Granieri E. Recurrent nephrotic syndrome in patient with multiple sclerosis treated with interferon beta-1a. J Neurol 2003;250:768–9.

64. Ohta S, Yokoyama H, Wada T, Sakai N, Shimizu M, Kato T, Furuichi K, Segawa C, Hisada Y, Kobayashi K. Exacerbation of glomerulonephritis in subjects with chronic hepatitis C virus infection after interferon therapy. Am J Kidney Dis 1999;33(6):1040–8.

65. Elgart GW, Sheremata W, Ahn YS. Cutaneous reactions to recombinant human interferon beta-1b: the clinical and histologic spectrum. J Am Acad Dermatol 1997;37(4):553–8.

66. Heinzerling L, Dummer R, Burg G, Schmid-Grendelmeier P. Panniculitis after subcutaneous injection of interferon beta in a multiple sclerosis patient. Eur J Dermatol 2002;12(2):194–7.

67. Gaines AR, Varricchio F. Interferon beta-1b injection site reactions and necroses. Mult Scler 1998;4(2):70–3.

68. Radziwill AJ, Courvoisier S. Severe necrotising cutaneous lesions complicating treatment with interferon beta-1a. J Neurol Neurosurg Psychiatry 1999;67(1):115.

69. Berard F, Canillot S, Balme B, Perrot H. Nécrose cutanée locale après injections d'interféron béta. [Local cutaneous necrosis after injection of interferon beta.] Ann Dermatol Venereol 1995;122(3):105–7.

70. Sheremata WA, Taylor JR, Elgart GW. Severe necrotizing cutaneous lesions complicating treatment with interferon beta-1b. N Engl J Med 1995;332(23):1584.

71. Benincasa P, Bielory L. Necrotizing cutaneous lesions as a complication of subcutaneous interferon beta-1b. J Allergy Clin Immunol 1996;97:343.

72. Webster GF, Knobler RL, Lublin FD, Kramer EM, Hochman LR. Cutaneous ulcerations and pustular psoriasis flare caused by recombinant interferon beta injections in patients with multiple sclerosis. J Am Acad Dermatol 1996;34(2 Pt 2):365–7.

73. Feldmann R, Low-Weiser H, Duschet P, Gschnait F. Necrotizing cutaneous lesions caused by interferon beta injections in a patient with multiple sclerosis. Dermatology 1997;195(1):52–3.

74. Durieu C, Bayle-Lebey P, Gadroy A, Loche F, Bazex J. Hyperplasie endothéliale papillaire intravasculaire: multiples lésions apparues au cours d'un traitement par interféron béta. [Intravascular papillary endothelial hyperplasia: multiple lesions appearing in the course of treatment with interferon beta.] Ann Dermatol Venereol 2001;128(12):1336–8.

75. Mehta CL, Tyler RJ, Cripps DJ. Granulomatous dermatitis with focal sarcoidal features associated with recombinant interferon beta-1b injections. J Am Acad Dermatol 1998;39(6):1024–8.

76. Demirkaya S, Bulucu F, Odabasi Z, Vural O. An erythromelalgia case occurred during interferon beta treatment for multiple sclerosis. Eur J Neurol 2002;9(Suppl 2):220.

77. Benito-Leon J, Borbujo J, Cortes L. Cutaneous mucinoses complicating interferon beta-1b therapy. Eur Neurol 2002;47(2):123–4.

78. Lunemann JD, Schwarzenberger B, Kassim N, Zschenderlein R, Zipp F. Rhabdomyolysis during interferon-beta 1a treatment. J Neurol Neurosurg Psychiatry 2002;72(2):274.

79. Murata K, Shiraki K, Takase K, Nakano T, Tameda Y. Mono-arthritis following intensified interferon beta therapy for chronic hepatitis C. Hepatogastroenterology 2002;49(47):1418–9.

80. Corona T, Leon C, Ostrosky-Zeichner L. Severe anaphylaxis with recombinant interferon beta. Neurology 1999;52(2):425.

81. Brown DL, Login IS, Borish L, Powers PL. An urticarial IgE-mediated reaction to interferon beta-1b. Neurology 2001;56(10):1416–7.

82. Clear D. Anaphylactoid reaction to methyl prednisolone developing after starting treatment with interferon beta-1b. J Neurol Neurosurg Psychiatry 1999;66(5):690.

83. Debat Zoguereh D, Boucraut J, Beau-Salinas F, Bodiguel E, Lechapois D, Pomet E. Vascularite cutanée avec atteinte rénale compliquant un traitement par interféron-

84. beta 1a pour une sclérose en plaques. Rev Neurol 2004;160:1081–4.

84. Colosimo C, Pozzilli C, Frontoni M, Farina D, Koudriavtseva T, Gasperini C, Salvetti M, Valesini G. No increase of serum autoantibodies during therapy with recombinant human interferon-beta1a in relapsing-remitting multiple sclerosis. Acta Neurol Scand 1997;96(6):372–4.

85. Kivisakk P, Lundahl J, von Heigl Z, Fredrikson S. No evidence for increased frequency of autoantibodies during interferon-beta1b treatment of multiple sclerosis. Acta Neurol Scand 1998;97(5):320–3.

86. Abdi EA, Nguyen GK, Ludwig RN, Dickout WJ. Pulmonary sarcoidosis following interferon therapy for advanced renal cell carcinoma. Cancer 1987;59(5):896–900.

87. Bobbio-Pallavicini E, Valsecchi C, Tacconi F, Moroni M, Porta C. Sarcoidosis following beta-interferon therapy for multiple myeloma. Sarcoidosis 1995;12(2):140–2.

88. The IFNB Multiple Sclerosis Study Group and the University of British Columbia MS/MRI Analysis Group. Neutralizing antibodies during treatment of multiple sclerosis with interferon beta-1b: experience during the first three years. Neurology 1996;47(4):889–94.

89. Rudick RA, Simonian NA, Alam JA, Campion M, Scaramucci JO, Jones W, Coats ME, Goodkin DE, Weinstock-Guttman B, Herndon RM, Mass MK, Richert JR, Salazar AM, Munschauer FE 3rd, Cookfair DL, Simon JH, Jacobs LD. Incidence and significance of neutralizing antibodies to interferon beta-1a in multiple sclerosis. Multiple Sclerosis Collaborative Research Group (MSCRG). Neurology 1998;50(5):1266–72.

90. Deisenhammer F, Reindl M, Harvey J, Gasse T, Dilitz E, Berger T. Bioavailability of interferon beta 1b in MS patients with and without neutralizing antibodies. Neurology 1999;52(6):1239–43.

91. Khan OA, Dhib-Jalbut SS. Neutralizing antibodies to interferon beta-1a and interferon beta-1b in MS patients are cross-reactive. Neurology 1998;51(6):1698–702.

92. Malucchi S, Sala A, Gilli F, Bottero R, Di Sapio A, Capobianco M, Bertolotto A. Neutralizing antibodies reduce the efficacy of betaIFN during treatment of multiple sclerosis. Neurology 2004;62:2031–7.

93. Perini P, Calabrese M, Biasi. The clinical impact of interferon beta antibodies in relapsing-remitting MS. J Neurol 2004;251:305–9.

94. Hartung HP, Munschauer F, III, Schellekens H. Significance of neutralizing antibodies to interferon beta during treatment of multiple sclerosis: expert opinions based on the Proceedings of an International Consensus Conference. Eur J Neurol 2005;12(8):588–601.

95. Wolinsky JS, Toyka KV, Kappos L, Grossberg SE. Interferon-beta antibodies: implications for the treatment of MS. Lancet Neurol 2003;2:528.

96. Sorensen PS, Ross C, Clemmesen KM, Bendtzen K, Frederiksen JL, Jensen K, Kristensen O, Petersen T, Rasmussen S, Ravnborg M, Stenager E, Koch-Henriksen N. Clinical importance of neutralising antibodies against interferon beta in patients with relapsing-remitting multiple sclerosis. Lancet 2003;362:1184–91.

97. Giovannoni G, Deisenhammer F. Neutralising antibodies against interferon beta in multiple sclerosis. Lancet 2004;363:166–7.

98. Polman C, Kappos L, White R, Dahlke F, Beckmann K, Pozzilli C, Thompson A, Petkau J, Miller D. Neutralizing

antibodies during treatment of secondary progressive MS with interferon beta-1b. Neurology 2003;60:37–43.

99. Cross AH, Antel JP. Antibodies to beta-interferons in multiple sclerosis: can we neutralize the controversy? Neurology 1998;50(5):1206–8.

100. Rice GP, Paszner B, Oger J, Lesaux J, Paty D, Ebers G. The evolution of neutralizing antibodies in multiple sclerosis patients treated with interferon beta-1b. Neurology 1999;52(6):1277–9.

101. Pozzilli C, Antonini G, Bagnato F, Mainero C, Tomassini V, Onesti E, Fantozzi R, Galgani S, Pasqualetti P, Millefiorini E, Spadaro M, Dahlke F, Gasperini C. Monthly corticosteroids decrease neutralizing antibodies to IFNbeta1 b: a randomized trial in multiple sclerosis. J Neurol 2002;249(1):50–6.

102. Fierlbeck G, Schreiner T. Incidence and clinical significance of therapy-induced neutralizing antibodies against interferon-beta. J Interferon Res 1994;14(4):205–6.

103. Sanberg-Wollheim M. Outcome of pregnancy during treatment with interferon-beta-1a (Rebif) in patients with multiple sclerosis. Neurology 2002;58(Suppl):A445–6.

104. Watanabe M, Kohge N, Akagi S, Uchida Y, Sato S, Kinoshita Y. Congenital anomalies in a child born from a mother with interferon-treated chronic hepatitis B. Am J Gastroenterol 2001;96(5):1668–9.

105. Durelli L, Verdun E, Barbero P, Bergui M, Versino E, Ghezzi A, Montanari E, Zaffaroni MIndependent Comparison of Interferon (INCOMIN) Trial Study Group. Every-other-day interferon beta-1b versus once-weekly interferon beta-1a for multiple sclerosis: results of a 2-year prospective randomised multicentre study (INCOMIN). Lancet 2002;359(9316):1453–60.

106. Panitch H, Goodin DS, Francis G, Chang P, Coyle PK, O'Connor P, Monaghan E, Li D, Weinshenker BEVIDENCE Study Group. EVidence of Interferon Dose-response: European North American Compartative Efficacy; University of British Columbia MS/MRI Research Group. Randomized, comparative study of interferon beta-1a treatment regimens in MS: The EVIDENCE Trial. Neurology 2002;59(10):1496–506.

107. Rio J, Nos C, Bonaventura I, Arroyo R, Genis D, Sureda B, Ara JR, Brieva L, Martin J, Saiz A, Sanchez Lopez F, Prieto JM, Roquer J, Dorado JF, Montalban X. Corticosteroids, ibuprofen, and acetaminophen for IFNbeta-1a flu symptoms in MS: a randomized trial. Neurology 2004;63:525–8.

Interferon gamma

See also Interferons

General Information

Recombinant interferon gamma (interferon gamma-1b) is currently only approved as an adjunct to antibacterial therapy in chronic granulomatous disease (1), but its immunoregulatory potential has been investigated in other diseases. Clinical experience with interferon gamma is therefore limited and the most relevant information on long-term safety has been obtained from the ICGDSCG trial (2). In this study, adverse effects that were significantly more frequent with interferon gamma-

1b (1.5 micrograms/kg or 50 micrograms/m^2) than with placebo included mild fever and flu-like symptoms, headache, and moderate injection site reactions. There were no adverse consequences on growth and development in children followed for a mean of 2.5 years (3,4). Several other adverse effects have been reported to the manufacturers, but causal evaluation is lacking (1).

Interferon gamma has been investigated in 27 patients with systemic sclerosis randomized to receive interferon gamma for 12 months. Most of them complained of symptoms consistent with a flu-like syndrome, namely headache (85%), fever (81%), and arthralgia and myalgia (70%) (5). There were adverse events (one or more per patient) leading to treatment withdrawal in four cases, including arthralgia, cardiac pain, atrioventricular block, reversible loss of hearing, and impotence; however, a causal relation with interferon gamma was not documented.

Organs and Systems

Cardiovascular

Heart rate, ventricular or supraventricular extra beats, and asymptomatic cardiac events were not significantly different during treatment compared with baseline in 20 patients receiving interferon gamma (6). Interferon gamma rarely produced cardiovascular adverse effects. Hypotension, dysrhythmias, and possible coronary spasm were sometimes observed, mostly in patients receiving high doses or with previous cardiovascular disorders (SEDA-20, 333; SEDA-22, 405) (7,8).

Exacerbation of Raynaud's syndrome occurred in five of 20 patients with systemic sclerosis treated with interferon gamma (SEDA-20, 333).

Respiratory

Of 10 patients treated with interferon gamma-1b 200 micrograms three times a week for advanced idiopathic pulmonary fibrosis, four developed irreversible acute respiratory failure (9). All four patients had increasing dyspnea, fever, and rapidly progressive hypoxemia, and had new alveolar opacities on lung imaging. The symptoms occurred shortly after interferon gamma had been started in three patients, and after 35 injections in the fourth. Three patients died from refractory hypoxemia and the fourth underwent lung transplantation, but died a few weeks later. Pathological examination in two patients showed diffuse alveolar damage with pre-existing interstitial pneumonitis. Interferon gamma was suspected, as no other cause of abrupt pulmonary deterioration was found. Although the number of patients was small, the authors noted that before interferon beta pulmonary function tended to be worse in the four patients who developed acute respiratory failure than in the other six.

In a double-blind study in 330 patients with idiopathic pulmonary fibrosis randomized to receive interferon gamma-1b or placebo, there were no significant differences in progression-free survival or in conventional measures of lung function, gas exchange, or quality of life (10). As expected, patients who received interferon gamma-1b had more frequent constitutional symptoms, such as fever, rigors, and flu-like symptoms. They also had more frequent upper respiratory tract infections and non-fatal pneumonias.

Interferon gamma-associated lung toxicity has been further supported by a report of acute respiratory insufficiency after 4 months in a 68-year-old man with idiopathic pulmonary fibrosis (11).

Psychological, psychiatric (See also interfeous)

Neuropsychiatric disturbances have not been consistently found in patients receiving interferon gamma, despite electroencephalographic monitoring and psychometric tests (12). However, careful examination led to the impression that interferon gamma can cause neurophysiological changes similar to those of interferon alfa (13), and data from the manufacturers also point to rare cases of nervous system adverse effects in patients treated with high-dose interferon gamma (1).

Endocrine

Interferon gamma can increase serum cortisol concentrations (14).

Metabolism

Reversible dose-dependent hypertriglyceridemia has been attributed to interferon gamma (15).

Hyperglycemia, reversible on interferon gamma withdrawal and a short course of insulin, has been reported in one patient (SEDA-22, 406).

Hematologic

Interferon gamma was supposedly the cause of asymptomatic non-immune hemolytic anemia in one patient receiving both interleukin-2 and interferon gamma (16).

Only minimal effects of interferon gamma on white blood cell counts have been observed (17).

Auto-immune thrombocytopenia occurred in a patient receiving interferon gamma (SEDA-22, 406).

Gastrointestinal

A convincing case of severe aphthous stomatitis has been reported in a patient receiving interferon gamma (SEDA-20, 333).

Urinary tract

Dose-related asymptomatic proteinuria was sometimes observed, and severe proteinuria with nephrotic syndrome has been reported once after low-dose interferon gamma (SED-13, 1100). Acute renal insufficiency is extremely rare (SEDA-20, 333) (18).

Skin

Induction of psoriatic lesions at the injection site has been observed in 10 of 42 patients treated with interferon gamma for psoriatic arthritis, while the joint symptoms were improved (19).

Single or multiple lesions of erythema nodosum leprosum occurred in 60% of patients given intradermal interferon gamma for lepromatous leprosy, and severe systemic symptoms required thalidomide treatment in two patients (20).

Severe erythroderma was observed in five of 10 patients after interferon gamma was added to ciclosporin for autologous bone marrow transplantation (21).

Immunologic

An anaphylactoid reaction and severe bronchospasm have been reported once after the first injection of interferon gamma (12).

Although interferon gamma is mainly used for its immunoregulatory properties, the possibility of clinical immune adverse consequences has been addressed in a limited number of prospective studies. In two studies involving patients with chronic hepatitis B treated for 4–6 months, most developed a new autoantibody (22,23), but none developed clinical evidence of autoimmune disease. However, other reports suggested that interferon gamma can either improve or aggravate immune or inflammatory conditions. Although no change in antinuclear antibodies was reported in a trial of 54 patients with rheumatoid arthritis (24), increased or new antinuclear antibodies were observed in three of six patients with rheumatoid arthritis who received interferon gamma for 2–8 months, and two patients had clinical exacerbations of the disease (25). Isolated cases of systemic lupus erythematosus have been reported in patients receiving interferon gamma for rheumatoid arthritis (26,27). Rheumatoid or lupus-like symptoms associated with raised antinuclear antibodies titers were also noted in 17% of patients receiving interferon alfa and interferon gamma for myeloproliferative disorders, and in only 8.3% of patients treated with interferon alfa alone (28). Interferon gamma was also involved in the induction or reactivation of seronegative arthritis in patients with cutaneous psoriasis (29) and the unexpected exacerbation of multiple sclerosis in 39% of patients (30). Finally, neutralizing antibodies have exceptionally been found (SEDA-20, 333).

- A 10-year-old boy with chronic granulomatous disease developed systemic lupus erythematosus after treatment with interferon gamma 50 micrograms/m^2 three times weekly for 4 years (31). The boy was heterozygous for the allelic variants of FcγRIIa, which has been linked to an increased risk of autoimmune disorders in patients with chronic granulomatous disease. In addition, his mother had a disease similar to discoid lupus.

References

1. Todd PA, Goa KL. Interferon gamma-1b. A review of its pharmacology and therapeutic potential in chronic granulomatous disease. Drugs 1992;43(1):111–22.
2. The International Chronic Granulomatous Disease Cooperative Study Group. A controlled trial of interferon gamma to prevent infection in chronic granulomatous disease. N Engl J Med 1991;324(8):509–16.

3. Bemiller LS, Roberts DH, Starko KM, Curnutte JT. Safety and effectiveness of long-term interferon gamma therapy in patients with chronic granulomatous disease. Blood Cells Mol Dis 1995;21(3):239–47.

4. Weening RS, Leitz GJ, Seger RA. Recombinant human interferon-gamma in patients with chronic granulomatous disease—European follow up study. Eur J Pediatr 1995;154(4):295–8.

5. Grassegger A, Schuler G, Hessenberger G, Walder-Hantich B, Jabkowski J, MacHeiner W, Salmhofer W, Zahel B, Pinter G, Herold M, Klein G, Fritsch PO. Interferon-gamma in the treatment of systemic sclerosis: a randomized controlled multicentre trial. Br J Dermatol 1998;139(4):639–48.

6. Friess GG, Brown TD, Wrenn RC. Cardiovascular rhythm effects of gamma recombinant DNA interferon. Invest New Drugs 1989;7(2–3):275–80.

7. Sonnenblick M, Rosin A. Cardiotoxicity of interferon. A review of 44 cases. Chest 1991;99(3):557–61.

8. Yamamoto N, Nishigaki K, Ban Y, Kawada Y. Coronary vasospasm after interferon administration. Br J Urol 1998;81(6):916–7.

9. Honore I, Nunes H, Groussard O, Kambouchner M, Chambellan A, Aubier M, Valeyre D, Crestani B. Acute respiratory failure after interferon-gamma therapy of end-stage pulmonary fibrosis. Am J Respir Crit Care Med 2003;167(7):953–7.

10. Raghu G, Brown KK, Bradford WZ, Starko K, Noble PW, Schwartz DA, King TE Jr; Idiopathic Pulmonary Fibrosis Study Group. A placebo-controlled trial of interferon gamma-1b in patients with idiopathic pulmonary fibrosis. N Engl J Med 2004;350:125–33.

11. Carvalho CR, Kairalla RA, Schettino GP. Acute respiratory failure after interferon gamma therapy in IPF. Am J Respir Crit Care Med 2004;169:543–4.

12. Mattson K, Niiranen A, Pyrhonen S, Farkkila M, Cantell K. Recombinant interferon gamma treatment in non-small cell lung cancer. Antitumour effect and cardiotoxicity. Acta Oncol 1991;30(5):607–10.

13. Born J, Spath-Schwalbe E, Pietrowsky R, Porzsolt F, Fehm HL. Neurophysiological effects of recombinant interferon-gamma and -alpha in man. Clin Physiol Biochem 1989;7(3–4):119–27.

14. Krishnan R, Ellinwood EH Jr, Laszlo J, Hood L, Ritchie J. Effect of gamma interferon on the hypothalamic–pituitary–adrenal system. Biol Psychiatry 1987;22(9):1163–6.

15. Kurzrock R, Rohde MF, Quesada JR, Gianturco SH, Bradley WA, Sherwin SA, Gutterman JU. Recombinant gamma interferon induces hypertriglyceridemia and inhibits post-heparin lipase activity in cancer patients. J Exp Med 1986;164(4):1093–101.

16. Rabinowitz AP, Hu E, Watkins K, Mazumder A. Hemolytic anemia in a cancer patient treated with recombinant interferon-gamma. J Biol Response Mod 1990;9(2):256–9.

17. Aulitzky WE, Tilg H, Vogel W, Aulitzky W, Berger M, Gastl G, Herold M, Huber C. Acute hematologic effects of interferon alpha, interferon gamma, tumor necrosis factor alpha and interleukin 2. Ann Hematol 1991;62(1):25–31.

18. Ault BH, Stapleton FB, Gaber L, Martin A, Roy S 3rd, Murphy SB. Acute renal failure during therapy with recombinant human gamma interferon. N Engl J Med 1988;319(21):1397–400.

19. Fierlbeck G, Rassner G, Muller C. Psoriasis induced at the injection site of recombinant interferon gamma. Results of immunohistologic investigations. Arch Dermatol 1990;126(3):351–5.

20. Sampaio EP, Moreira AL, Sarno EN, Malta AM, Kaplan G. Prolonged treatment with recombinant interferon gamma induces erythema nodosum leprosum in lepromatous leprosy patients. J Exp Med 1992;175(6):1729–37.

21. Horn TD, Altomonte V, Vogelsang G, Kennedy MJ. Erythroderma after autologous bone marrow transplantation modified by administration of cyclosporine and interferon gamma for breast cancer. J Am Acad Dermatol 1996;34(3):413–7.

22. Weber P, Wiedmann KH, Klein R, Walter E, Blum HE, Berg PA. Induction of autoimmune phenomena in patients with chronic hepatitis B treated with gamma-interferon. J Hepatol 1994;20(3):321–8.

23. Kung AW, Jones BM, Lai CL. Effects of interferon-gamma therapy on thyroid function, T-lymphocyte subpopulations and induction of autoantibodies. J Clin Endocrinol Metab 1990;71(5):1230–4.

24. Cannon GW, Emkey RD, Denes A, Cohen SA, Saway PA, Wolfe F, Jaffer AM, Weaver AL, Manaster BJ, McCarthy KA. Prospective 5-year followup of recombinant interferon-gamma in rheumatoid arthritis. J Rheumatol 1993;20(11):1867–73.

25. Seitz M, Franke M, Kirchner H. Induction of antinuclear antibodies in patients with rheumatoid arthritis receiving treatment with human recombinant interferon gamma. Ann Rheum Dis 1988;47(8):642–4.

26. Graninger WB, Hassfeld W, Pesau BB, Machold KP, Zielinski CC, Smolen JS. Induction of systemic lupus erythematosus by interferon-gamma in a patient with rheumatoid arthritis. J Rheumatol 1991;18(10):1621–2.

27. Machold KP, Smolen JS. Interferon-gamma induced exacerbation of systemic lupus erythematosus. J Rheumatol 1990;17(6):831–2.

28. Wandl UB, Nagel-Hiemke M, May D, Kreuzfelder E, Kloke O, Kranzhoff M, Seeber S, Niederle N. Lupus-like autoimmune disease induced by interferon therapy for myeloproliferative disorders. Clin Immunol Immunopathol 1992;65(1):70–4.

29. O'Connell PG, Gerber LH, Digiovanna JJ, Peck GL. Arthritis in patients with psoriasis treated with gamma-interferon. J Rheumatol 1992;19(1):80–2.

30. Panitch HS, Hirsch RL, Schindler J, Johnson KP. Treatment of multiple sclerosis with gamma interferon: exacerbations associated with activation of the immune system. Neurology 1987;37(7):1097–102.

31. Badolato R, Notarangelo LD, Plebani A, Roos D. Development of systemic lupus erythematosus in a young child affected with chronic granulomatous disease following withdrawal of treatment with interferon-gamma. Rheumatology 2003;42:804–5.

Lamivudine

See also Nucleoside analogue reverse transcriptase inhibitors (NRTIs)

General Information

Lamivudine (3TC) is a nucleoside analogue reverse transcriptase inhibitor that has been widely used against HIV infection which also has antiviral effects against hepatitis B (1).

Observational studies

In a 24-week phase I/II study, 89 children aged 3 months to 17 years (median 7.3 years) were treated with lamivudine for 24 weeks in dosages of 1–20 mg/kg/day (2). Dosages over 20 mg/kg/day were not tested because of reported neutropenia in adults at this dosage (3). Lamivudine was generally well tolerated in these children. Ten children were withdrawn because of presumed adverse effects: three because of increased serum transaminase activities, three because of neutropenia, and two because of hyperactivity. One child became ataxic shortly after the start of therapy, and one developed pancreatitis during hospitalization for acute cryptosporidiosis. All of these events resolved with supportive care on withdrawal of lamivudine, except one case of hepatitis, which persisted up to 10 months after withdrawal. Treatment was discontinued temporarily in other patients because of pancreatitis ($n = 2$), rashes ($n = 2$), neutropenia ($n = 1$), anemia ($n = 1$), and increased serum transaminase activities ($n = 1$). On resolution of the presumed adverse reaction, the drug was reintroduced in all of these cases without further problems. There were no significant hematological or biochemical changes. Neither the incidence nor the severity of the observed adverse events was dose-related. The assignment of causality to lamivudine of most of the adverse events was complicated by intercurrent conditions and concomitant medications.

In a study of the efficacy of lamivudine (25, 100, or 300 mg/day for 12 weeks) in the treatment of chronic hepatitis B virus infections, lamivudine was similarly well tolerated in 32 patients (4). Only minor non-specific non-dose-related adverse reactions were observed. In addition, there were mild asymptomatic increases in serum activities of amylase, lipase, and creatine kinase, which in most cases resolved despite continued therapy.

Of 20 children and adolescents with chronic hepatitis B infection after failure of interferon given lamivudine 3 mg/kg (maximum 100 mg/day) for 1 year, 44% had sustained undetectable hepatitis B virus DNA and there were no adverse effects (5).

Lamivudine 100 mg/day was also used in a 6-year-old patient with nephritic syndrome associated with hepatitis B and resulted in complete resolution after 12 months without adverse effects (6).

Reactivation of hepatitis B virus infection in asymptomatic hepatitis B surface antigen carriers undergoing chemotherapy or immunosuppressive therapy was reduced by 4–7 times in patients who took lamivudine prophylaxis in a systematic review of nine prospective studies and one randomized-controlled trial (7). Adverse effects included rises in amylase, lipase and creatine kinase, nausea, headache, dizziness, and pancreatitis.

In 17 patients with severe acute or fulminant hepatitis B common adverse effects of lamivudine included headache, upper respiratory tract infection, nasopharyngitis, cough, pyrexia, dyspepsia, upper abdominal pain, fatigue, diarrhea, back pain, and myalgia, most of which were of mild-to-moderate intensity (8).

Comparative studies

The safety and efficacy of lamivudine (300–600 mg/day) in combination with zidovudine (600 mg/day) in the treatment of antiretroviral-naive and zidovudine-experienced HIV-infected persons has been compared with zidovudine monotherapy in two placebo-controlled studies of 129 and 223 patients (9,10). There were no significant differences in the incidence or severity of adverse effects between patients taking zidovudine alone or in combination with lamivudine. In both studies gastrointestinal symptoms, notably nausea, were the most commonly observed adverse reactions, occurring in 5–11% of zidovudine-experienced patients and 23–29% of antiretroviral drug-naive individuals. Although one antiretroviral drug-naive patient taking combined therapy had an asymptomatic rise in pancreatic amylase activity, acute pancreatitis was not observed in either study. Grade 1 peripheral neuropathy was reported in one zidovudine-experienced patient taking low-dosage lamivudine (150 mg bd) and zidovudine.

Long-term lamivudine 100 mg/day (n = 14) was as effective as lamivudine plus hepatitis B immune globulin (n = 15) in the prevention of recurrence of hepatitis B after liver transplantation (11). No major adverse effects were reported.

Oral entecavir 0.5 mg/day produced greater improvement in biochemical, histological, and virological outcomes than lamivudine 100 mg/day in HBeAg-positive and HBeAg-negative patients. Common adverse effects included headache, upper respiratory tract infections, nasopharyngitis, cough, pyrexia, dyspepsia, upper abdominal pain, fatigue, diarrhea, back pain, and myalgia; most were of mild-to-moderate intensity. More patients had rises in alanine transaminase activity and there were more withdrawals due to adverse effects in those who took lamivudine (12,13,14,15).

Of 136 patients given pegylated interferon alfa-2b monotherapy 49 (36%) had lost HBeAg at the end of follow-up at week 78, compared with 46 (35%) of 130 who were given interferon in combination with lamivudine (16). More of those in the combination group had cleared HBeAg at the end of treatment at week 52 (44% versus 29%) but they relapsed during follow-up. The patterns were similar when the response was assessed by suppression of serum hepatitis B virus (HBV) DNA or a change in alanine transaminase activity. Response rates (HBeAg loss) varied significantly by HBV genotype: genotype A 47%; genotype B 44%; genotype C 28%; genotype D 25%.

Lamivudine has been well tolerated in long term studies in hepatitis B positive patients; malaise and fatigue were the most common adverse effects (17).

It has been suggested that lamivudine be used prophylactically in only those patients with breast cancer who have active HBV viral replication (18).

Placebo-controlled studies

In 276 HBeAg-positive children the incidence of YMMD mutations (tyrosine, methionine, aspartate, aspartate) increased over time, resulting in lower response rates (19). Common adverse effects included infections of the ear, nose, and throat, headaches, cough, disturbances of temperature regulation, abdominal discomfort, nausea, and vomiting. During continued treatment with lamivudine for up to 3 years there were no new or unexpected adverse events or laboratory abnormalities.

Organs and Systems

Hematologic

A case report has provided strong evidence that lamivudine can on rare occasions cause severe anemia; the mechanism is unclear (20).

Gastrointestinal

In a 1-year trial of lamivudine in 10 children with vertically acquired chronic hepatitis B, lamivudine made serum hepatitis B viral DNA undetectable in all the patients within 24 weeks (21). Serum alanine transaminase activity returned to the reference range within 36 weeks. Although nausea and vomiting were reported in one child, it was not necessary to withhold treatment.

Liver

Lamivudine is used orally to treat chronic hepatitis B in adults and children (22). It increases the rate of loss of hepatitis B e antigen and seroconversion in compensated chronic carriers, with improvement of histology at a similar rate to interferon alfa. However, the tyrosine-methionine-aspartate-aspartate (YMDD) mutation prevents efficacy and can cause flares of hepatitis. The indications for treatment must therefore be established with care and only by those who have expert knowledge of the disease, the drug, and the YMDD mutation.

The reports of all suspected hepatic adverse drug reactions with a fatal outcome received by the Uppsala Monitoring Centre between 1968 and 2003 included 56 cases of lamivudine-associated hepatotoxicity (23). Lamivudine, as part of an antiretroviral regimen was reported to have caused liver failure in four HIV-infected patients (24). Lamivudine given alone in hepatitis B cirrhosis was reported to have caused hepatic decompensation, postulated to be due to an intense immune response, with associated vigorous cytotoxic T cell activity; however, the role of lamivudine in causing further decompensation of the liver when used in patients with already decompensated liver disease was impossible to assess (25). In some cases lamivudine was associated with transient leukopenia, which recovered on withdrawal.

Of 814 HBV eAg-positive patients two taking lamivudine monotherapy had hepatic decompensation after the end of treatment (26). One received a liver transplant and made a full recovery; the other died.

Nails

Paronychia has been reported in 12 HIV-infected patients who had taken lamivudine only for 3 months before the onset of symptoms (27). Microbiological investigations for fungi or bacteria were negative. All were treated with topical antiseptics; surgical procedures were performed in four. Five patients healed without recurrence, while the paronychia recurred in six. There was no mention of withdrawal of lamivudine. The causative role of lamivudine in the development of paronychia in these patients remains obscure.

Musculoskeletal

There were persistent rises in creatine kinase activity (478–1753 IU/l) with normal blood gases and lactate and no associated myalgia or other complaints in two of 20 HBe-Ag-positive children taking prolonged lamivudine therapy; the activities normalized on withdrawal of lamivudine (28).

Long-Term Effects

Drug tolerance

The incidence of viral resistance to lamivudine increases with the duration of therapy in patients with chronic hepatitis B (29). However, the effect of viral resistance on hepatic synthetic function has not been well defined. In 38 patients (26 with cirrhosis) in an open study there was an initial antiviral response in all patients (hepatitis B virus DNA became undetectable by a hybridization assay), and nine of 22 (41%) hepatitis B e antigen-positive patients underwent hepatitis B e antigen seroconversion. In 29 patients with undetectable serum hepatitis B viral DNA at the end of the study, the mean serum albumin concentration rose from 40 to 43 g/l, corresponding to a yearly increase of 1.85 g/l; this was largely attributable to an increase in the cirrhotic patients. Resistance to lamivudine developed in nine patients. Suppression of viral replication by lamivudine improves hepatic synthetic function in chronic hepatitis B patients, but emergence of drug resistance is associated with a rapid fall in serum albumin.

Drug resistance

Lamivudine is associated with a high rate of resistance (30). Baseline serum hepatitis B virus DNA > 106 copies/ml was a strong predictor for breakthrough disease because of drug-resistant mutations (31).

References

1. Jarvis B, Faulds D. Lamivudine. A review of its therapeutic potential in chronic hepatitis B. Drugs 1999;58(1):101–41.
2. Lewis LL, Venzon D, Church J, Farley M, Wheeler S, Keller A, Rubin M, Yuen G, Mueller B, Sloas M, Wood L, Balis F, Shearer GM, Brouwers P, Goldsmith J, Pizzo PA. Lamivudine in children with human immunodeficiency virus infection: a phase I/II study. The National Cancer Institute Pediatric Branch–Human Immunodeficiency Virus Working Group. J Infect Dis 1996;174(1):16–25.

3. McLean TW, Kurth S, Gee B. Pelvic osteomyelitis in a sickle-cell patient receiving deferoxamine. Am J Hematol 1996;53(4):284–5.

4. Dienstag JL, Perrillo RP, Schiff ER, Bartholomew M, Vicary C, Rubin M. A preliminary trial of lamivudine for chronic hepatitis B infection. N Engl J Med 1995;333(25):1657–61.

5. Hartman C, Berkowitz D, Shouval D, Eshach-Adiv O, Hino B, Rimon N, Satinger I, Kra-Oz T, Daudi N, Shamir R. Lamivudine treatment for chronic hepatitis B infection in children unresponsive to interferon. Pediatr Infect Dis J 2003;22:224–9.

6. Connor FL, Rosenberg AR, Kennedy SE, Bohane TD. HBV associated nephrotic syndrome: resolution with oral lamivudine. Arch Dis Child 2003;88:446–9.

7. Kohrt HE, Ouyang DL, Keeffe EB. Systematic review: lamivudine prophylaxis for chemotherapy-induced reactivation of chronic hepatitis B virus infection. Aliment Pharmacol Ther 2006;24(7):1003–16.

8. Tillmann HL, Hadem J, Leifeld L, Zachou K, Canbay A, Eisenbach C, Graziadei I, Encke J, Schmidt H, Vogel W, Schneider A, Spengler U, Gerken G, Dalekos GN, Wedemeyer H, Manns MP. Safety and efficacy of lamivudine in patients with severe acute or fulminant hepatitis B, a multicenter experience. J Viral Hepat 2006;13(4):256–63.

9. Katlama C, Ingrand D, Loveday C, Clumeck N, Mallolas J, Staszewski S, Johnson M, Hill AM, Pearce G, McDade H. Safety and efficacy of lamivudine–zidovudine combination therapy in antiretroviral-naive patients. A randomized controlled comparison with zidovudine monotherapy. Lamivudine European HIV Working Group. JAMA 1996;276(2):118–25.

10. Staszewski S, Loveday C, Picazo JJ, Dellarnonica P, Skinhoj P, Johnson MA, Danner SA, Harrigan PR, Hill AM, Verity L, McDade H. Safety and efficacy of lamivudine–zidovudine combination therapy in zidovudine-experienced patients. A randomized controlled comparison with zidovudine monotherapy. Lamivudine European HIV Working Group. JAMA 1996;276(2):111–7.

11. Buti M, Mas A, Prieto M, Casafont F, Gonzalez A, Miras M, Herrero JI, Jardi R, Cruz de Castro E, Garcia-Rey C. A randomized study comparing lamivudine monotherapy after a short course of hepatitis B immune globulin (HBIg) and lamivudine with long-term lamivudine plus HBIg in the prevention of hepatitis B virus recurrence after liver transplantation. J Hepatol 2003;38:811–17.

12. Chang TT, Gish RG, de Man R, Gadano A, Sollano J, Chao YC, Lok AS, Han KH, Goodman Z, Zhu J, Cross A, DeHertogh D, Wilber R, Colonno R, Apelian D; BEHoLD AI463022 Study Group. A comparison of entecavir and lamivudine for HBeAg-positive chronic hepatitis B. N Engl J Med 2006;354(10):1001–10.

13. Lai CL, Shouval D, Lok AS, Chang TT, Cheinquer H, Goodman Z, DeHertogh D, Wilber R, Zink RC, Cross A, Colonno R, Fernandes L; BEHoLD AI463027 Study Group. Entecavir versus lamivudine for patients with HBeAg-negative chronic hepatitis B. N Engl J Med 2006;354(10):1011–20.

14. Sherman M, Yurdaydin C, Sollano J, Silva M, Liaw YF, Cianciara J, Boron-Kaczmarska A, Martin P, Goodman Z, Colonno R, Cross A, Denisky G, Kreter B, Hindes R; AI463026 BEHoLD Study Group. Entecavir for treatment of lamivudine-refractory, HBeAg-positive chronic hepatitis B. Gastroenterology 2006;130(7):2039–49.

15. Han SH, Natural course, therapeutic options and economic evaluation of therapies for chronic hepatitis B. Drugs 2006;66(14):1831–51.

16. Janssen HL, van Zonneveld M, Senturk H, Zeuzem S, Akarca US, Cakaloglu Y, Simon C, So TM, Gerken G, de Man RA, Niesters HG, Zondervan P, Hansen B, Schalm SW; HBV 99-01 Study Group; Rotterdam Foundation for Liver Research. Pegylated interferon alfa-2b alone or in combination with lamivudine for HBeAg-positive chronic hepatitis B: a randomised trial. Lancet 2005;365(9454):123–9.

17. Buti M, Jardi R, Rodriguez-Frias F, Valdes A, Schaper M, Esteban R, Guardia J. Changes in different regions of hepatitis B virus gene in hepatitis B 'e' antigen-negative patients with chronic hepatitis B: the effect of long-term lamivudine therapy. Aliment Pharmacol Ther 2005;21(11):1349–56.

18. Dai MS, Chao TY. Lamivudine therapy in HBsAg-carrying breast cancer patients undergoing chemotherapy: prophylactic or preemptive? Breast Cancer Res Treat 2005;92(1):95–6.

19. Sokal EM, Kelly DA, Mizerski J, Badia IB, Areias JA, Schwarz KB, Vegnente A, Little NR, Gardener SD, Jonas MM. Long-term lamivudine therapy for children with HBeAg-positive chronic hepatitis B. Hepatology 2006;43(2):225–32.

20. Weitzel T, Plettenberg A, Albrecht D, Lorenzen T, Stoehr A. Severe anemia as a newly recognized side-effect caused by lamivudine. AIDS 1999;13(16):2309–11.

21. Zuccotti GV, Cucchi C, Gracchi V, D'Auria E, Riva E, Tagger A. A 1-year trial of lamivudine for chronic hepatitis B in children. J Int Med Res 2002;30(2):200–2.

22. Sokal E. Lamivudine for the treatment of chronic hepatitis B. Expert Opin Pharmacother 2002;3(3):329–39.

23. Bjornsson E, Olsson R. Suspected drug-induced liver fatalities reported to the WHO database. Dig Liver Dis 2006;38(1):33–8.

24. Wang SS, Chou NK, Chi NH, Hsu RB, Huang SC, Chen YS, Yu HY, Tsao CI, Ko WJ, Lai MY, Chu SH. Successful treatment of hepatitis B virus infection with Lamivudine after heart transplantation. Transplant Proc 2006;38(7):2138–40.

25. Murakami R, Amada N, Sato T, Orii T, Kikuchi H, Haga I, Ohashi Y, Okazaki H. Reactivation of hepatitis and lamivudine therapy in 11 HBsAg-positive renal allograft recipients: a single centre experience. Clin Transplant 2006;20(3):351–8.

26. Lau GK, Piratvisuth T, Luo KX, Marcellin P, Thongsawat S, Cooksley G, Gane E, Fried MW, Chow WC, Paik SW, Chang WY, Berg T, Flisiak R, McCloud P, Pluck N; Peginterferon Alfa-2a HBeAg-Positive Chronic Hepatitis B Study Group. Peginterferon alfa-2a, lamivudine, and the combination for HBeAg-positive chronic hepatitis B. N Engl J Med 2005;352(26):2682–95.

27. Zerboni R, Angius AG, Cusini M, Tarantini G, Carminati G. Lamivudine-induced paronychia. Lancet 1998;351(9111):1256.

28. Hartman C, Berkowitz D, Eshach-Adiv O, Hino B, Rimon N, Satinger I, Kra-Oz T, Shamir R. Long-term lamivudine therapy for chronic hepatitis B infection in children unresponsive to interferon. J Pediatr Gastroenterol Nutr 2006;43(4):494–8.

29. Hui JM, George J, Liddle C, Lin R, Samarasinghe D, Crewe E, Farrell GC. Changes in serum albumin during treatment of chronic hepatitis B with lamivudine: effects of response and emergence of drug resistance. Am J Gastroenterol 2002;97(4):1003–9.

30. Zoulim F, Poynard T, Degos F, Slama A, El Hasnaoui A, Blin P, Mercier F, Deny P, Landais P, Parvaz P, Trepo C; Lamivir Study Group. A prospective study of the evolution of lamivudine resistance mutations in patients with chronic hepatitis B treated with lamivudine. J Viral Hepat 2006;13(4):278–88.

31. Manolakopoulos S, Bethanis S, Elefsiniotis J, Karatapanis S, Triantos C, Sourvinos G, Touloumi G, Economou M, Vlachogiannakos J, Spandidos D, Avgerinos A, Tzourmakliotis D. Lamivudine monotherapy in HBeAg-negative chronic hepatitis B: prediction of response-breakthrough and long-term clinical outcome. Aliment Pharmacol Ther 2006;23(6):787–95.

Lopinavir and ritonavir

See also Protease inhibitors

General Information

Lopinavir and ritonavir are nucleoside analogue reverse transcriptase inhibitors that are used in combination in the treatment of AIDS.

The pharmacology, clinical pharmacology, uses, adverse effects, and interactions of lopinavir + ritonavir have been reviewed (1).

Observational studies

In studies of combining ritonavir with fluconazole (2) and ritonavir with mefloquine (3) there were no significant effects, and dosage adjustment is not warranted.

Placebo-controlled studies

In a randomized, double-blind study in 70 patients taking a regimen containing protease inhibitors, lopinavir + ritonavir 400/100 mg or 400/200 mg bd was substituted (4). On day 15 nevirapine 200 mg bd was added and NRTIs were changed to include at least one NRTI not previously taken. Despite a more than four-fold reduction in phenotypic susceptibility to the pre-entry protease inhibitor in 63% of the patients, mean plasma HIV-1 RNA concentrations fell by 1.14 log copies/ml after 2 weeks. At week 48, 86% had plasma HIV-1 RNA concentrations of under 400 copies/ml, and 76% under 50 HIV-1 RNA copies/ml. Mean CD4 cell counts increased by 125 cells/μl. The most common adverse events were diarrhea ($n = 16$) and weakness ($n = 4$). There were rises in gamma-glutamyl-transferase activity ($n = 18$), total cholesterol ($n = 17$), and triglycerides ($n = 17$), and transient rises in transaminase activities ($n = 11$). Three patients discontinued therapy because of drug-related adverse events.

General adverse effects

In early monotherapy studies, including 62 and 87 patients (5,6), in which the potent antiretroviral effect of ritonavir was first demonstrated, the most common adverse events were nausea, diarrhea, headache, circumoral paresthesia, and altered taste sensation. Nausea, vomiting, and diarrhea are common during the start of therapy and usually disappear over the first few weeks of treatment. These adverse effects can be markedly reduced by using a step-up approach, increasing to the full dose over six days. General weakness, circumoral paresthesia, and taste disturbance occur in 5–10% of patients and are seldom dose-limiting.

Ritonavir does not have a broad therapeutic margin, and patients with higher ritonavir concentrations are at a higher risk of neurological or gastrointestinal adverse effects. It is feasible to individualize the dosage regimen with the aid of plasma concentration measurement and close observation of adverse effects, and there is a close relation between the two; this may enable one to increase substantially the percentage of patients who tolerate ritonavir without risking underdosage (7).

An important consideration in the use of all HIV-1 protease inhibitors, but of ritonavir in particular, is their potential for drug interactions through their effects on cytochrome P450 isozymes. The various interactions of ritonavir with other anti-HIV drugs have recently been reviewed (8).

Organs and Systems

Cardiovascular

Two of 16 patients taking lopinavir + ritonavir developed so-called inflammatory edema, which resolved on withdrawal and recurred after rechallenge (9). In three of eight patients inflammatory edema occurred 1–4 weeks after they started to take regimens that contained lopinavir + ritonavir (10). The edema affected the feet, ankles, and calves and was associated in one case with fever and in another with a transient rash; in one case the left shoulder and groin were also affected. All three recovered completely within 1–4 weeks despite continued drug treatment, but 7 months later one had a relapse that required withdrawal of lopinavir + ritonavir.

From a case in which there was positive dechallenge and rechallenge it has been concluded that edema of the lower limbs can be an adverse effect of ritonavir in some HIV-positive patients (11). The authors suspected a relation to the drug's vasodilatory activity. However, it should also be borne in mind that ritonavir has caused reversible renal insufficiency, which should be looked for in any patient who develops edematous changes.

Nervous system

Myasthenia has been attributed to ritonavir (12).

- A 71-year-old man with an 8-year history of HIV infection developed slurred speech, difficulty in climbing stairs, bilateral ptosis, and lateral rectus weakness 3 weeks after having started to take ritonavir (1200 mg/day). All his signs worsened with prolonged testing, and edrophonium produced improvement in ptosis and speech. Specific electromyographic testing confirmed myasthenia gravis. Computed tomography of the chest was normal. After withdrawal of ritonavir the signs and symptoms partly resolved by 3 months.

A definite causal link with ritonavir could not be established, but the authors speculated that ritonavir may have unmasked myasthenia gravis in this patient.

Sensory systems

Ototoxicity has been attributed to lopinavir + ritonavir (13).

- A 46-year-old man took a regimen containing lopinavir + ritonavir, and 4 weeks later complained of reduced auditory acuity accompanied by lancinating pain. Audiology showed mild to moderate bilateral sensorineural hearing loss. On withdrawal and use of efavirenz his hearing recovered.

Based on the time-course and the effect of withdrawal, lopinavir/ritonavir appears to have been the causative agent in this case.

Endocrine

In five of 19 adolescents with HIV infection who took inhaled or intranasal fluticasone and low-dose ritonavir there was associated laboratory evidence of adrenal suppression (14). There were clinical features of Cushing syndrome, especially weight gain, in most of these patients, with laboratory evidence of adrenal suppression from exogenous corticosteroids. Both the laboratory abnormalities and the Cushingoid feature resolved when the fluticasone and/or ritonavir were withdrawn in four of the five cases.

Six patients with pre-existing HIV-lipodystrophy developed symptomatic Cushing's syndrome when treated with inhaled fluticasone at varying doses for asthma while concurrently taking low-dose ritonavir-boosted protease inhibitor antiretroviral regimens for HIV infection (15). Stimulation studies showed evidence of adrenal suppression in all patients. After withdrawal of inhaled fluticasone, four patients developed symptomatic hypoadrenalism, and three required oral glucocorticoid support for several months. Other complications included evidence of osteoporosis (n = 3), crush fractures (n = 1), and exacerbation of pre-existing type 2 diabetes mellitus (n = 1).

Metabolism

In 19 patients taking lopinavir + ropinavir, either 533/133 mg bd (in patients co-treated with NNRTIs) or 400/100 mg bd, there were increases in triglyceride and total cholesterol concentrations after 4 weeks; seven patients developed grade 2 or worse rises in lipids and three patients developing grade 3 hyperlipidemia. HDL cholesterol increased over the course of 48 weeks, but LDL did not change significantly (16). At baseline nine patients had lipodystrophy but that did not worsen over the next 48 weeks. However, CT-based standardized analysis

showed an increase in total abdominal fat (+14%) and a loss of limb fat (−8%) in the 16 patients who completed the study. There was a significant correlation between lopinavir trough concentrations and changes in limb fat, but no association between lopinavir concentration and changes in total abdominal fat, visceral fat, or subcutaneous abdominal fat. The authors cautioned against interpreting these data as showing a causative relation between lopinavir and lipodystrophy, as the study population was rather small and confounding elements could not be ruled out. However, they advised the use of plasma concentration measurement in order to avoid unnecessarily high lopinavir concentrations, which may be associated with lipodystrophy.

In contrast, in 55 patients taking lopinavir + ropinavir, even though there was a significant increase in triglyceride concentrations over 12 weeks, this did not correlate with higher plasma lopinavir concentrations (17). The authors cautioned against premature dosage adjustments pending larger studies to elucidate this phenomenon.

Gastrointestinal

The most common adverse effect associated with ritonavir boosted lopinavir is diarrhea (18).

In a prospective, randomized trial in 100 patients, the most common adverse effects of lopinavir + ritonavir included abnormal stools, diarrhea, and nausea (19).

Liver

The incidence of hepatotoxicity after treatment with lopinavir and the possible association with lopinavir plasma concentrations has been assessed in a retrospective analysis of 120 HIV-infected patients (52% with hepatitis C co-infection) (20). The cumulative incidence of severe liver toxicity (grade 3 or 4) was 4% over 12 months; 8% of patients with hepatitis C co-infection developed liver toxicity, suggesting a higher risk of liver-related complications of lopinavir in patients with pre-existing liver disease. Plasma concentrations of lopinavir and HbsAg positivity were not associated with hepatotoxicity. Compared with results with other protease inhibitors, these data suggest that lopinavir is relatively safe for the liver, even in the presence of hepatitis C co-infection. However, the overall frequency was low and a relation between lopinavir concentration and hepatotoxicity cannot be ruled out.

Urinary tract

In a retrospective analysis of 87 HIV-positive patients taking ritonavir in combination with two nucleoside analogues, serum creatinine increased in 12 cases by 66 (5–242)% from 66 (range 46–102) µmol/l, with a median glomerular filtration rate (GFR) of 116 ml/minute (60–202). Ten of the 12 patients had other risk factors for nephrotoxicity, such as dehydration or use of nephrotoxic drugs (21). It is prudent to monitor renal function in patients receiving ritonavir, particularly in the presence of other risk factors for renal dysfunction.

Three patients developed reversible renal insufficiency 10–12 days after starting treatment including ritonavir (22).

Reproductive system

Four women taking ritonavir developed hypermenorrhea, strongly suggesting that this can occur as a complication of ritonavir treatment (23).

Immunologic

Subcutaneous non-tuberculous granulomatous lesions developed in a 48-year-old HIV-positive man when he was given ritonavir (24).

Susceptibility Factors

Age

The adverse effects commonly observed with ritonavir in adults are reportedly similar in children (25). Of 51 children aged 6 months to 18 years who took escalating doses of a liquid formulation of ritonavir (from 250 up to 400 mg/m^2 every 12 hours), seven withdrew because of gastrointestinal toxicity and four because of grade three hepatic transaminase rises. Both serum triglyceride and cholesterol concentrations increased significantly from baseline within 12 weeks of treatment.

Drug–Drug Interactions

From 122 outpatients, 748 lopinavir and 748 ritonavir plasma concentrations were available for analysis (26). The interaction between the drugs was described by a time-independent inverse relation between exposure to ritonavir over a dosing interval and the apparent oral clearance (CL/F) of lopinavir. No patient characteristics, other than the use of NNRTIs had a significant effect on the pharmacokinetics of lopinavir combined with ritonavir.

From 186 patients in an Amsterdam clinic, 505 ritonavir plasma concentrations at a single time and 55 full pharmacokinetic profiles were available, resulting in a database of 1228 plasma ritonavir concentrations (27). The concomitant use of lopinavir resulted in a significant 2.7-fold increase in the clearance of ritonavir. No patient characteristics affected the pharmacokinetics of ritonavir.

In 45 HIV-positive patients taking lopinavir + ritonavir plus efavirenz and 24 patients taking lopinavir + ritonavir plus nucleoside/nucleotide reverse transcriptase inhibitors, lopinavir metabolism was induced by efavirenz by about 25% (28).

Alprazolam

The inhibitory effect of ritonavir (a viral protease inhibitor) on the metabolism of alprazolam, a CYP3A-mediated reaction, has been investigated in a double-blind study (29). Ten subjects took alprazolam 1.0 mg plus either low-dose ritonavir (four doses of 200 mg) or placebo. Ritonavir reduced alprazolam clearance by 60%, prolonged its half-life, and magnified its benzodiazepine agonist effects, such as sedation and impairment of performance.

Amprenavir

In attempts to lower the amprenavir capsule burden, low-dose ritonavir has been used as a pharmacokinetic booster. When ritonavir was added to amprenavir, the amprenavir AUC increased 3–4 times (30), which should allow the total daily capsule burden to be reduced. Adverse effects included diarrhea, nausea, paresthesia, rash, increased cholesterol, and increased triglycerides. The frequency of adverse events correlated with the dose of ritonavir.

Carbamazepine

Ritonavir can interact with carbamazepine (31).

- A 49-year-old woman with a long history of HIV infection developed worsening ataxia leading to two falls. Four days before admission she had had her antiretroviral drugs changed from zidovudine, lamivudine, and indinavir to ritonavir, saquinavir, and efavirenz. She was also taking carbamazepine to control generalized seizures resulting from a previous right thalamic infarction. The change in antiretroviral therapy resulted in an increase in serum carbamazepine concentration from 6.9 to 20 μg/ml. Serum concentrations in the usual target range were eventually achieved by a sixth of the dose of carbamazepine that had been required before starting ritonavir.

As saquinavir is only a mild inhibitor of CYP3A4 and efavirenz is an inducer, this effect was attributed to the potent inhibitor ritonavir.

Coumarin anticoagulants

Ritonavir can dangerously increase warfarin concentrations (32).

In contrast, ritonavir has been reported to reduce the effect of acenocoumarol (33).

- A 46-year-old man with prosthetic cardiac valves took acenocoumarol and later started to take stavudine, lamivudine, and ritonavir 600 mg bd. His INR fell markedly. Although the dose of acenocoumarol was progressively increased to three times the original dose, it was impossible to achieve the previous INR, and ritonavir was withdrawn.

Ecstasy

Ritonavir can interact with methylenedioxymetamfetamine (MDMA, ecstasy) (34).

- A 32-year old HIV-positive man who added ritonavir 600 mg bd to his existing antiretroviral regimen of zidovudine and lamivudine became unwell within hours after having ingested two and a half tablets of ecstasy, estimated to contain 180 mg of methylenedioxymetamfetamine. He was hypertonic, sweating profusely, tachypneic, tachycardic, and cyanosed. Shortly after, he had a tonic-clonic seizure and cardiorespiratory arrest. Attempts at resuscitation were unsuccessful. Blood concentrations obtained postmortem showed an concentration of 4.56 µg/ml, in the range of that reported in a patient with a life-threatening illness and symptoms similar to this patient after an overdose of 18 tablets of MDMA.

Ritonavir inhibits CYP2D6, which is the principal pathway by which MDMA is metabolized.

Efavirenz

Efavirenz reduced the trough concentration of lopinavir given with low-dose ritonavir by a median 39% and increased the inter-individual variability of lopinavir concentrations (35). The authors recommended that the dosage of lopinavir + ritonavir be increased to 533 mg + 133 mg bd if given with an NNRTI and that drug concentrations should be monitored.

Ergot alkaloids

Ergot alkaloids are among the many medications that are contraindicated in patients taking ritonavir. One patient developed severe ergotism after taking ritonavir and ergotamine for 13 days (36), and other cases of ischemia of the lower limbs or elsewhere in the periphery have been reported in patients taking both ritonavir and ergotamine (37–39).

Fentanyl

By inhibiting CYP3A4, ritonavir can significantly inhibit the metabolism of fentanyl, and considerable caution is needed (40).

Ritonavir is an inhibitor of HIV protease and a potent inhibitor of CYP3A4 and CYP2D6. The interaction between ritonavir and intravenous fentanyl has been investigated in 12 healthy volunteers in a double-blind, placebo-controlled, crossover study (41). The volunteers took ritonavir 600 mg on day 1, ritonavir 900 mg and intravenous fentanyl 5 µg/kg on day 2, and ritonavir 300 mg or placebo on day 3. Ritonavir reduced the clearance of fentanyl by 67% by inhibiting its metabolism. This could result in prolongation of fentanyl-induced respiratory depression in a patient with an already compromised cardiorespiratory system.

Indinavir

Using ritonavir as a booster may allow indinavir to be given twice daily and with food. In a cohort survey of 100 patients the combination of indinavir plus ritonavir (400 mg/400 mg or 800 mg/100 mg bd) was a safe and effective option to reduce the tablet burden and improve adherence (42).

Ketoconazole

In 12 patients taking ritonavir and saquinavir for HIV infection, ketoconazole significantly increased the AUC, the plasma concentration at 12 hours, and the half-life of ritonavir by 29, 62 and 31% respectively (43). Similar increases of 37, 94, and 38% were recorded for saquinavir. CSF concentrations of ritonavir were raised by 178% by ketoconazole, but there was no significant change in CSF concentrations of saquinavir.

Lamotrigine

The effect of lopinavir + ritonavir on the pharmacokinetics of lamotrigine and lamotrigine 2N-glucuronide and the effect of lamotrigine on the pharmacokinetics of lopinavir + ritonavir have been studied in 24 healthy subjects compared with historical controls (44). Lopinavir + ritonavir caused a mean 55% reduction in lamotrigine concentration on day 20, probably by induction of glucuronidation.

Methadone

Withdrawal symptoms, and the need to increase the dose of methadone, in a 51-year-old man previously stable on a maintenance dosage, have been attributed to ritonavir (45).

Methylenedioxymetamfetamine

An important consideration in the use of all HIV-1 protease inhibitors, but of ritonavir in particular, is their potential for drug interactions through their effects on cytochrome P450 isozymes. The various interactions of ritonavir with other antiretroviral drugs have been reviewed (46). Ritonavir, which inhibits CYP2D6, the principal pathway by which MDMA is metabolized, can also produce clinically relevant interactions with recreational drugs (47).

- A 32-year-old HIV-positive man, who added ritonavir 600 mg bd to his antiretroviral regimen of zidovudine and lamivudine, became unwell within hours after having taken two and a half tablets of ecstasy, estimated to contain 180 mg of MDMA. He was hypertonic, sweating profusely, tachypneic, tachycardic, and cyanosed. Shortly after he had a tonic-clonic seizure and a cardiorespiratory arrest. Attempts at resuscitation were unsuccessful. Blood concentrations obtained post-mortem showed an MDMA concentration of 4.56 mg/l, in the range of that reported in a patient with a life-threatening illness and symptoms similar to this patient after an overdose of 18 tablets of MDMA.

A patient infected with HIV-1 who was taking ritonavir and saquinavir had a prolonged effect from a small dose of MDMA and a near-fatal reaction from a small dose of gamma-hydroxybutyrate (48).

Nevirapine

Nevirapine reduced the trough concentration of lopinavir given with low-dose ritonavir by a median 39% and increased the inter-individual variability of lopinavir concentrations (35). The authors recommended that the dosage of lopinavir/ritonavir be increased to 533 mg + 133 mg bd if given with an NNRTI and that drug concentrations should be monitored.

Protease inhibitors

Ritonavir is mainly restricted to its use as a potent inhibitor of cytochrome P450 metabolism. When it is given in combination with the protease inhibitors amprenavir, atazanavir, indinavir, lopinavir, or saquinavir, ritonavir prolongs their half-lives and increases their trough concentrations and AUCs, allowing twice or even once-daily and administration. For this indication, ritonavir is given in very low doses (100 mg od or bd), which significantly reduces ritonavir-associated adverse effects (49).

Rifamycins

In a pilot, non-randomized study, HIV-infected patients with tuberculosis were treated with regimens containing rifampicin and ritonavir (50). Despite the effects on CYP3A4, there was no significant interaction; plasma ritonavir concentrations remained sufficiently high and rifampicin concentrations did not rise to the toxic range.

Statins

An interaction of ritonavir with simvastatin reportedly resulted in rhabdomyolysis (51).

- A 51-year-old woman taking zidovudine, lamivudine, indinavir, and simvastatin started to take ritonavir and after 1 week developed diffuse muscle weakness and body aches. Creatine kinase (total and the MB isozyme), lactate dehydrogenase, and transaminases were all raised. There were crystals and hemoglobin in the urine.

Presumably this interaction occurred through inhibition of CYP3A4 by ritonavir.

Rhabdomyolysis occurred in a 34-year-old man with AIDS and severe liver disease, because of a drug interaction involving clarithromycin 500 mg/day, atorvastatin 40 mg /day, and lopinavir/ritonavir (52).

Triazolam

The inhibitory effect of ritonavir on the biotransformation of triazolam and zolpidem has been investigated (53).

Short-term low-dose ritonavir produced a large and significant impairment of triazolam clearance and enhancement of its clinical effects. In contrast, ritonavir produced small and clinically unimportant reductions in zolpidem clearance. The findings are consistent with the complete dependence of triazolam clearance on CYP3A activity, compared with the partial dependence of zolpidem clearance on CYP3A.

Voriconazole

In a randomized double-blind study 20 healthy subjects, stratified according to CYP2C19 genotype, took oral ritonavir 300 mg bd or placebo for 2 days (54). A single dose of voriconazole 400 mg was given together with the first dose of ritonavir or placebo. Ritonavir significantly reduced voriconazole apparent oral clearance by about 43%. The authors concluded that co-administration of ritonavir, a potent inhibitor of CYP3A4, leads to increased exposure to voriconazole, which might increase the risk of adverse reactions.

Zolpidem

The inhibitory effect of ritonavir on the biotransformation of triazolam and zolpidem has been investigated (55). Short-term, low-dose ritonavir produced a large and significant impairment of triazolam clearance and enhancement of its clinical effects. In contrast, ritonavir produced small and clinically unimportant reductions in zolpidem clearance. The findings are consistent with the complete dependence of triazolam clearance on CYP3A activity, compared with the partial dependence of zolpidem clearance on CYP3A.

Monitoring Drug Therapy

In a study of 12-hour pharmacokinetic profiles of lopinavir in 100 patients, 12 models were used to predict AUC and Cmax; measurements at 2 and 6 hours after the dose gave acceptable estimated (56). The evening trough concentration gave the best estimate of the daily nadir.

References

1. Qazi NA, Morlese JF, Pozniak AL. Lopinavir/ritonavir (ABT-378/r). Expert Opin Pharmacother 2002;3(3):315–27.
2. Koks CH, Crommentuyn KM, Hoetelmans RM, Burger DM, Koopmans PP, Mathot RA, Mulder JW, Meenhorst PL, Beijnen JH. The effect of fluconazole on ritonavir and saquinavir pharmacokinetics in HIV-1-infected individuals. Br J Clin Pharmacol 2001;51(6):631–5.
3. Khaliq Y, Gallicano K, Tisdale C, Carignan G, Cooper C, McCarthy A. Pharmacokinetic interaction between mefloquine and ritonavir in healthy volunteers. Br J Clin Pharmacol 2001;51(6):591–600.
4. Benson CA, Deeks SG, Brun SC, Gulick RM, Eron JJ, Kessler HA, Murphy RL, Hicks C, King M, Wheeler D, Feinberg J, Stryker R, Sax PE, Riddler S, Thompson M, Real K, Hsu A, Kempf D, Japour AJ, Sun E. Safety and

antiviral activity at 48 weeks of lopinavir/ritonavir plus nevirapine and 2 nucleoside reverse-transcriptase inhibitors in human immunodeficiency virus type 1-infected protease inhibitor-experienced patients. J Infect Dis 2002;185(5):599–607.

5. Markowitz M, Saag M, Powderly WG, Hurley AM, Hsu A, Valdes JM, Henry D, Sattler F, La Marca A, Leonard JM, et al. A preliminary study of ritonavir, an inhibitor of HIV-1 protease, to treat HIV-1 infection. N Engl J Med 1995;333(23):1534–9.

6. Danner SA, Carr A, Leonard JM, Lehman LM, Gudiol F, Gonzales J, Raventos A, Rubio R, Bouza E, Pintado V, et alEuropean-Australian Collaborative Ritonavir Study Group. A short-term study of the safety, pharmacokinetics, and efficacy of ritonavir, an inhibitor of HIV-1 protease. N Engl J Med 1995;333(23):1528–33.

7. Gatti G, Di Biagio A, Casazza R, De Pascalis C, Bassetti M, Cruciani M, Vella S, Bassetti D. The relationship between ritonavir plasma levels and side-effects: implications for therapeutic drug monitoring. AIDS 1999;13(15):2083–9.

8. Hsu A, Granneman GR, Bertz RJ. Ritonavir. Clinical pharmacokinetics and interactions with other anti-HIV agents. Clin Pharmacokinet 1998;35(4):275–91.

9. Lascaux AS, Lesprit P, Bertocchi M, Levy Y. Inflammatory oedema of the legs: a new side-effect of lopinavir. AIDS 2001;15(6):819.

10. Eyer-Silva WA, Neves-Motta R, Pinto JF, Morais-De-Sa CA. Inflammatory oedema associated with lopinavir-including HAART regimens in advanced HIV-1 infection: report of 3 cases. AIDS 2002;16(4):673–4.

11. Dol L, Geffray L, el Khoury S, Cevallos R, Veyssier P. Oedèmes des membres inférieurs chez un patient seropositif pour le VIH: effet secondaire du ritonavir?. [Edema of the lower extremities in a HIV seropositive patient: secondary effect of ritonavir?.] Presse Méd 1999;28(2):75.

12. Saadat K, Kaminski HJ. Ritonavir-associated myasthenia gravis. Muscle Nerve 1998;21(5):680–1.

13. Williams B. Ototoxicity may be associated with protease inhibitor therapy. Clin Infect Dis 2001;33(12):2100–2.

14. Arrington-Sanders R, Hutton N, Siberry GK. Ritonavir–fluticasone interaction causing Cushing syndrome in HIV-infected children and adolescents. Pediatr Infect Dis J 2006;25(11):1044–8.

15. Samaras K, Pett S, Gowers A, McMurchie M, Cooper DA. Iatrogenic Cushing's syndrome with osteoporosis and secondary adrenal failure in human immunodeficiency virus-infected patients receiving inhaled corticosteroids and ritonavir-boosted protease inhibitors: six cases. J Clin Endocrinol Metab 2005;90(7):4394–8.

16. Gutierrez F, Padilla S, Masia M, Navarro A, Gallego J, Hernandez I, Ramos JM, Martin-Hidalgo A. Changes in body fat composition after 1 year of salvage therapy with lopinavir/ritonavir-containing regimens and its relationship with lopinavir plasma concentrations. Antivir Ther 2004;9(1):105–13.

17. Torti C, Quiros-Roldan E, Regazzi-Bonora M, De Luca A, Lo Caputo S, Di Giambenedetto S, Patroni A, Villani P, Micheli V, Carosi G; Resistance and Dosage Adapted Regimens Study Group of the MASTER Cohort. Lipid abnormalities in HIV-infected patients are not correlated with lopinavir plasma concentrations. J Acquir Immune Defic Syndr 2004;35(3):324–6.

18. Macassa E, Delaugerre C, Teglas JP, Jullien V, Tréluyer JM, Veber F, Rouzioux C, Blanche S. Change to a once-daily combination including boosted atazanavir in HIV-1-infected children. Pediatr Infect Dis J 2006;25(9):809–14.

19. Murphy RL, Brun S, Hicks C, Eron JJ, Gulick R, King M, White AC Jr, Benson C, Thompson M, Kessler HA, Hammer S, Bertz R, Hsu A, Japour A, Sun E. ABT-378/ritonavir plus stavudine and lamivudine for the treatment of antiretroviral-naive adults with HIV-1 infection: 48-week results. AIDS 2001;15(1):F1–9.

20. Gonzalez-Requena D, Nunez M, Jimenez-Nacher I, Gonzalez-Lahoz J, Soriano V. Short communication: liver toxicity of lopinavir-containing regimens in HIV-infected patients with or without hepatitis C coinfection. AIDS Res Hum Retroviruses 2004;20(7):698–700.

21. Bochet MV, Jacquiaud C, Valantin MA, Katlama C, Deray G. Renal insufficiency induced by ritonavir in HIV-infected patients. Am J Med 1998;105(5):457.

22. Duong M, Sgro C, Grappin M, Biron F, Boibieux A. Renal failure after treatment with ritonavir. Lancet 1996;348(9028):693.

23. Nielsen H. Hypermenorrhea associated with ritonavir. Lancet 1999;353(9155):811–2.

24. Kawsar M, El-Gadi S. Subcutaneous granulomatous lesions related to ritonavir therapy in a HIV infected patient. Int J STD AIDS 2002;13(4):273–4.

25. Mueller BU, Nelson RP Jr, Sleasman J, Zuckerman J, Heath-Chiozzi M, Steinberg SM, Balis FM, Brouwers P, Hsu A, Saulis R, Sei S, Wood LV, Zeichner S, Katz TT, Higham C, Aker D, Edgerly M, Jarosinski P, Serchuck L, Whitcup SM, Pizzuti D, Pizzo PA. A phase I/II study of the protease inhibitor ritonavir in children with human immunodeficiency virus infection. Pediatrics 1998;101(3 Pt 1):335–43.

26. Crommentuyn KM, Kappelhoff BS, Mulder JW, Mairuhu AT, van Gorp EC, Meenhorst PL, Huitema AD, Beijnen JH. Population pharmacokinetics of lopinavir in combination with ritonavir in HIV-1-infected patients. Br J Clin Pharmacol 2005;60(4):378–89.

27. Kappelhoff BS, Huitema AD, Crommentuyn KM, Mulder JW, Meenhorst PL, van Gorp EC, Mairuhu AT, Beijnen JH. Development and validation of a population pharmacokinetic model for ritonavir used as a booster or as an antiviral agent in HIV-1-infected patients. Br J Clin Pharmacol 2005;59(2):174–82.

28. Dailly E, Allavena C, Raffi F, Jolliet P. Pharmacokinetic evidence for the induction of lopinavir metabolism by efavirenz. Br J Clin Pharmacol 2005;60(1):32–4.

29. Greenblatt DJ, von Moltke LL, Harmatz JS, Durol AL, Daily JP, Graf JA, Mertzanis P, Hoffman JL, Shader RI. Alprazolam-ritonavir interaction: implications for product labeling. Clin Pharmacol Ther 2000;67(4):335–41.

30. Sadler BM, Gillotin C, Lou Y, Stein DS. Pharmacokinetic and pharmacodynamic study of the human immunodeficiency virus protease inhibitor amprenavir after multiple oral dosing. Antimicrob Agents Chemother 2001;45(1):30–37.

31. Burman W, Orr L. Carbamazepine toxicity after starting combination antiretroviral therapy including ritonavir and efavirenz. AIDS 2000;14(17):2793–4.

32. Newshan G, Tsang P. Ritonavir and warfarin interaction. AIDS 1999;13(13):1788–9.

33. Llibre JM, Romeu J, Lopez E, Sirera G. Severe interaction between ritonavir and acenocoumarol. Ann Pharmacother 2002;36(4):621–3.

34. Henry JA, Hill IR. Fatal interaction between ritonavir and MDMA. Lancet 1998;352(9142):1751–2.

35. Solas C, Poizot-Martin I, Drogoul MP, Ravaux I, Dhiver C, Lafeuillade A, Allegre T, Mokhtari M, Moreau J, Lepeu G, Petit N, Durand A, Lacarelle B. Therapeutic drug monitoring of lopinavir/ritonavir given alone or with a non-nucleoside reverse transcriptase inhibitor. Br J Clin Pharmacol 2004;57(4):436–40.

36. Vila A, Mykietiuk A, Bonvehi P, Temporiti E, Uruena A, Herrera F. Clinical ergotism induced by ritonavir. Scand J Infect Dis 2001;33(10):788–9.

37. Montero A, Giovannoni AG, Tvrde PL. Leg ischemia in a patient receiving ritonavir and ergotamine. Ann Intern Med 1999;130(4 Pt 1):329–30.

38. Liaudet L, Buclin T, Jaccard C, Eckert P. Drug points: severe ergotism associated with interaction between ritonavir and ergotamine. BMJ 1999;318(7186):771.

39. Rosenthal E, Sala F, Chichmanian RM, Batt M, Cassuto JP. Ergotism related to concurrent administration of ergotamine tartrate and indinavir. JAMA 1999;281(11):987.

40. Olkkola KT, Palkama VJ, Neuvonen PJ. Ritonavir's role in reducing fentanyl clearance and prolonging its half-life. Anesthesiology 1999;91(3):681–5.

41. Olkkola KT, Palkama VJ, Neuvonen PJ. Ritonavir's role in reducing fentanyl clearance and prolonging its half-life. Anesthesiology 1999;91(3):681–5.

42. Burger DM, Hugen PW, Aarnoutse RE, Dieleman JP, Prins JM, van der Poll T, ten Veen JH, Mulder JW, Meenhorst PL, Blok WL, van der Meer JT, Reiss P, Lange JM. A retrospective, cohort-based survey of patients using twice-daily indinavir + ritonavir combinations: pharmacokinetics, safety, and efficacy J Acquir Immune Defic Syndr 2001;26(3):218–24.

43. Khaliq Y, Gallicano K, Venance S, Kravcik S, Cameron DW. Effect of ketoconazole on ritonavir and saquinavir concentrations in plasma and cerebrospinal fluid from patients infected with human immunodeficiency virus. Clin Pharmacol Ther 2000;68(6):637–46.

44. van der Leur MR, Burger DM, la Porte CJ, Koopmans PP. A retrospective TDM database analysis of interpatient variability in the pharmacokinetics of lopinavir in HIV-infected adults. Ther Drug Monit 2006;28(5):650–3.

45. Geletko SM, Erickson AD. Decreased methadone effect after ritonavir initiation. Pharmacotherapy 2000;20(1):93–4.

46. Henry JA, Hill IR. Fatal interaction between ritonavir and MDMA. Lancet 1998;352(9142):1751–2.

47. Harrington RD, Woodward JA, Hooton TM, Horn JR. Life-threatening interactions between HIV-1 protease inhibitors and the illicit drugs MDMA and gamma-hydroxybutyrate. Arch Intern Med 1999;159(18):2221–4.

48. Lauerma H, Wuorela M, Halme M. Interaction of serotonin reuptake inhibitor and 3,4-methylenedioxymethamphetamine? Biol Psychiatry 1998;43(12):929.

49. Cooper CL, Van Heeswijk RP, Gallicano K, Cameron DW. A review of low-dose ritonavir in protease inhibitor combination therapy. Clin Infect Dis 2003;36:1585–92.

50. Moreno S, Podzamczer D, Blazquez R, Iribarren JA, Ferrer E, Reparaz J, Pena JM, Cabrero E, Usan L. Treatment of tuberculosis in HIV-infected patients: safety and antiretroviral efficacy of the concomitant use of ritonavir and rifampin. AIDS 2001;15(9):1185–7.

51. Cheng CH, Miller C, Lowe C, Pearson VE. Rhabdomyolysis due to probable interaction between simvastatin and ritonavir. Am J Health Syst Pharm 2002;59(8):728–30.

52. Mah Ming JB, Gill MJ. Drug-induced rhabdomyolysis after concomitant use of clarithromycin, atorvastatin, and lopinavir/ritonavir in a patient with HIV. AIDS Patient Care STDS 2003;17:207–10.

53. Greenblatt DJ, von Moltke LL, Harmatz JS, Durol AL, Daily JP, Graf JA, Mertzanis P, Hoffman JL, Shader RI. Differential impairment of triazolam and zolpidem clearance by ritonavir. J Acquir Immune Defic Syndr 2000;24(2):129–36.

54. Mikus G, Schöwel V, Drzewinska M, Rengelshausen J, Ding R, Riedel KD, Burhenne J, Weiss J, Thomsen T, Haefeli WE. Potent cytochrome P450 2C19 genotype-related interaction between voriconazole and the cytochrome P450 3A4 inhibitor ritonavir. Clin Pharmacol Ther 2006;80(2):126–35.

55. Greenblatt DJ, von Moltke LL, Harmatz JS, Durol AL, Daily JP, Graf JA, Mertzanis P, Hoffman JL, Shader RI. Differential impairment of triazolam and zolpidem clearance by ritonavir. J Acquir Immune Defic Syndr 2000;24(2):129–36.

56. Alexander CS, Montaner JS, Asselin JJ, Ting L, McNabb K, Harris M, Guillemi S, Harrigan PR. Simplification of therapeutic drug monitoring for twice-daily regimens of lopinavir/ritonavir for HIV infection. Ther Drug Monit 2004;26(5):516–23.

Nelfinavir

See also Protease inhibitors

General Information

Nelfinavir is a non-peptidic inhibitor of the HIV protease. Its most prominent adverse effect is diarrhea, which occurs in up to one-third of individuals. However, the diarrhea is usually mild and can be controlled, if necessary, by antidiarrheal agents (1). Other adverse effects, including rash, nausea, headache, and weakness, are reported in under 5%.

Observational studies

In a multicenter, open, uncontrolled trial of protease inhibitors in conjunction with NRTIs for at least 96 weeks in 32 children, the pharmacokinetics of nelfinavir showed large interindividual differences (2). In all, 17 children suffered adverse events, most of which were mild and occurred early in treatment. The rate of drug-related adverse effects was 0.16 per patient-year in those taking nelfinavir.

Organs and Systems

Nervous system

Peripheral neuropathy has been reported with nelfinavir (3).

- In a 40-year-old HIV-positive patient nelfinavir 750 mg tds was added to existing combination therapy with stavudine, lamivudine, and loviride. He had taken various drugs during the 4 years before, including zidovudine, zalcitabine, didanosine, stavudine, loviride, ritonavir, and saquinavir. Zalcitabine had been withdrawn 3 years before because of pain and paresthesia in both feet after 4 weeks. Thereafter, he had persisting dysesthesia in the feet. One week after having started nelfinavir, this extended in both legs to the knees. He also had sharp pains in the Achilles tendons and bones and burning in the legs, which made walking impossible. He reduced the dosage to 1500 mg/day and felt an improvement within a few days; nelfinavir was then withdrawn, and his symptoms abated.

Although circumoral paresthesia has been a common adverse effect of ritonavir, peripheral neuropathy has not been previously reported with HIV-1 protease inhibitors. There has been a report of painful neuropathy in two patients who took ritonavir and indinavir respectively (4).

Metabolism

The effect of replacing nelfinavir with atazanavir on lipid concentrations has been studied. Atazanavir in combination with stavudine and lamivudine was given to 139 and 144 patients previously treated with atazanavir 400 mg/day or 600 mg/day for 48 weeks who continued their regimen and in 63 patients who had taken nelfinavir 1250 mg bd for 48 weeks and were allowed to switch to atazanavir 400 mg/day in an open rollover/switch study (5). Those who took atazanavir during the entire study had no significant changes in lipid profiles compared to baseline. The patients who had previously taken nelfinavir had reductions in total cholesterol, fasting LDL, and triglyceride concentrations and an increase in HDL concentrations at weeks 12 and 24.

An insulin-modified frequent sampling intravenous glucose tolerance test was performed in HIV-infected children, of whom 33 were taking a protease inhibitor and 15 were not (6). The former were also taking ritonavir (n = 10), nelfinavir (n = 14), indinavir (n = 2), lopinavir + ritonavir (n = 5), ritonavir + nelfinavir (n = 1), and nelfinavir + saquinavir (n = 1). There were no differences between the two groups with respect to fasting serum insulin or C-peptide, homeostatic model assessment of insulin resistance, or a quantitative insulin sensitivity check index. In a multiple regression analysis, the insulin sensitivity index and disposition index of children taking a protease inhibitor were significantly lower than in children who were not. In those taking a protease inhibitor, insulin sensitivity correlated inversely with visceral adipose tissue area and visceral to subcutaneous adipose tissue ratio. There was mildly impaired glucose tolerance in four of 21 subjects taking a protease inhibitor. These results suggest that protease inhibitor therapy reduces insulin sensitivity in HIV-infected children but also that it impairs the beta-cell response to this reduction in insulin sensitivity and, in a subset of

children, leads to the development of impaired glucose tolerance.

In a cross-sectional analysis of existing databases, 17 children with HIV infection were identified as having taken protease inhibitors, either ritonavir 20–30 mg/kg/day (n = 9) or nelfinavir 60–90 mg/kg/day (n = 8) for an average of 711 days (7). They were matched with 112 apparently healthy children admitted for minor surgical procedures. Plasma concentrations of cholesterol, triglycerides, and insulin-like growth factor 1 (IGF-1) tended to be high in those who had taken a protease inhibitor. The plasma concentrations of omega-6 long-chain polyunsaturated fatty acids and in particular of the highly unsaturated 22:4 omega-6 and 22:5 omega-6, were significantly increased. Infected children also had increased delta-6 and delta-4 desaturase activities and decreased delta-5 desaturase activity. The authors concluded that these children have a metabolic syndrome associated with significant changes in plasma fatty acid composition, similar to that observed in insulin resistance.

Gastrointestinal

The main adverse event that has been consistently reported with nelfinavir is diarrhea and loose stools in 20–30% of patients. The frequency with which diarrhea is truly dose-limiting is not yet clear, but clinical experience suggests that in many cases treatment can be continued in conjunction with non-specific antidiarrheal medication (SEDA-22, 317) (8,9).

Stool abnormalities were the most common adverse effects in a 48-week, phase 2, open comparison of nelfinavir versus atazanavir + didanosine + stavudine; diarrhea was 2.5 times more frequent in those who took nelfinavir (61% versus 25%) (10).

Pancreas

Pancreatitis, if it occurs, can be serious; any doubt as to the causal link has been largely dispelled by a case in which there was positive rechallenge (11).

Urinary tract

Nelfinavir has been reported to cause urinary stones.

- A 37-year-old woman stopped taking indinavir and other antiretroviral drugs because of skin reactions and started to take nelfinavir and delavirdine (12). An unspecified time later she presented with right flank pain, due to nephrolithiasis with hydronephrosis. She was treated with lithotripsy, but 6 months later had another episode of flank pain with microscopic hematuria and crystalluria. There was a large urinary stone in the right renal pelvis and a smaller stone obstructing the ureter. The stone fragments contained nelfinavir 99% and indinavir 1%.

The authors pointed out that nelfinavir is normally only 1–2% excreted in the urine, and they postulated that the indinavir may have started a nidus on which the nelfinavir was subsequently deposited.

Skin

Three HIV-infected patients, who had all taken zidovudine, didanosine, and co-trimoxazole (for *Pneumocystis jiroveci* prophylaxis) for over 12 months, developed generalized urticaria 8–10 days after starting nelfinavir 750 mg tds. Withdrawal of nelfinavir and treatment with antihistamines resulted in the disappearance of the lesions within 4–6 days. One patient restarted nelfinavir 250 mg tds without consulting his physician and had similar symptoms within 5 days. Desensitization by using a dose escalation protocol was attempted in all three patients after withdrawal of nelfinavir. This was successful in two cases; the third patient prematurely discontinued desensitization on his own request. During desensitization two of the patients had mild and transient pruritus and rash, reinforcing the diagnosis of an allergic skin reaction (13).

Patients often develop rashes due to HIV infection or drug allergy. Sulfa desensitization is well documented, but desensitization to other medications is not as well studied. One patient taking nelfinavir developed a severe, pruritic rash; management included nelfinavir desensitization over 12 days (14).

Susceptibility Factors

Age

The pharmacokinetics of three doses of nelfinavir, 15, 30, and 45 mg/kg bd, have been studied in 22 neonates, who were also given stavudine and didanosine for 4 weeks after birth (15). Median values of C_{min} (µg/ml) at 1, 7, 14, and 28 days of age were respectively: 0.19, 1.21, 0.51, and 0.33 (15 mg/kg); 1.02, 3.18, 0.73, and 0.55 (30 mg/kg); and 0.67, 3.21, 0.70, and 0.73 (45 mg/kg). The median values of steady-state AUC (hours.µg/ml) at 14 and 28 days were respectively: 14 and 8.7 (15 mg/kg); 19 and 16 (30 mg/kg); and 23 and 19 (45 mg/kg). There were no serious adverse events. Thus, systemic exposure to nelfinavir fell after 7 days of age, perhaps because of hepatic enzyme maturation, autoinduction of nelfinavir metabolism, and/or changes in nelfinavir absorption. Because of the highly variable kinetics the authors suggested that plasma concentration monitoring was warranted to ensure adequate nelfinavir dosing in neonates.

In a multicenter trial, 128 children were randomly assigned to one of three regimens: zidovudine + lamivudine ($n = 36$), zidovudine + abacavir ($n = 45$), or lamivudine + abacavir ($n = 47$) (16). The children who were free of symptoms ($n = 55$) were also randomized to nelfinavir or placebo, while those with more advanced disease received open-label nelfinavir ($n = 73$). All 13 episodes of diarrhea occurred in those taking nelfinavir. Nelfinavir did not affect serum cholesterol or triglyceride concentrations and there were no cases of lipodystrophy. There were 24 serious adverse events; six were in the symptom-free children (all taking nelfinavir), but all were attributed to the NRTIs.

Drug–Drug Interactions

Cannabinoids

The effects of smoked marijuana (3.95% tetrahydrocannabinol; up to three cigarettes per day) and oral dronabinol (2.5 mg tds) on the pharmacokinetics of nelfinavir 750 mg tds ($n = 34$) have been evaluated in a randomised, placebo-controlled study in HIV-infected patients (17). At day 14, marijuana reduced the 8-hour AUC of nelfinavir by 10%, the C_{max} by 17%, and the C_{min} by 12%. However, none of these changes was significant. Dronabinol had no effects.

Methadone

Withdrawal symptoms in a 40-year-old man maintained on methadone, necessitating an increase in dosage, were attributed to nelfinavir (18).

Phenytoin

While one might expect nelfinavir to reduce blood concentrations of certain other drugs by inducing their metabolism, it has been suggested that it might actually increase circulating concentrations of simultaneously administered phenytoin (19). However, a Japanese case has suggested that phenytoin concentrations can indeed be reduced; when the two drugs were co-administered a patient previously stabilized on phenytoin had a generalized convulsion (20).

Statins

Since HMG CoA reductase inhibitors (statins) are partially or completely metabolized by CYP3A4, drug interactions with protease inhibitors can be expected. When nelfinavir was combined with atorvastatin and simvastatin, their AUCs increased by 74 and 505% respectively (21). As myopathy and rhabdomyolysis are concentration-related, simvastatin is contraindicated with any protease inhibitor, not just nelfinavir. Based on the data, the starting dose of atorvastatin should be reduced by about 50%.

In 14 healthy subjects who took pravastatin 40 mg/day and nelfinavir 1250 mg bd over 16 days, steady-state nelfinavir reduced exposure to pravastatin by 47%, suggesting that higher doses of pravastatin need to be prescribed to achieve maximal lipid lowering activity (22).

Tacrolimus

An interaction of nelfinavir with the macrolide immunosuppressant tacrolimus has been reported in a patient coinfected with HIV and hepatitis C virus who had undergone orthotopic liver transplantation (23). The dose of tacrolimus had to be reduced to a 70th of the normal dose to avoid adverse effects. Nelfinavir serum concentrations were not affected by tacrolimus. The authors suggested that this effect had resulted from inhibition of the metabolism of tacrolimus, because both compounds are substrates of CYP3A4.

Zidovudine

There was a clinically non-significant interaction between nelfinavir and zidovudine in a pharmacokinetic study in 46 patients (24).

References

1. Jarvis B, Faulds D. Nelfinavir. A review of its therapeutic efficacy in HIV infection. Drugs 1998;56(1):147–67.
2. van Rossum AM, Geelen SP, Hartwig NG, Wolfs TF, Weemaes CM, Scherpbier HJ, van Lochem EG, Hop WC, Schutten M, Osterhaus AD, Burger DM, de Groot R. Results of 2 years of treatment with protease-inhibitor-containing antiretroviral therapy in Dutch children infected with human immunodeficiency virus type 1. Clin Infect Dis 2002;34(7):1008–16.
3. Grunke M, Kraetsch HG, Low P, Rascu A, Kalden JR, Harrer T. Nelfinavir associated with peripheral neuropathy in an HIV-infected patient. Infection 1998;26(4):252.
4. Colebunders R, De Droogh E, Pelgrom Y, Depraetere K, De Jonghe P. Painful hyperaesthesia caused by protease inhibitors? Infection 1998;26(4):250–1.
5. Wood R, Phanuphak P, Cahn P, Pokrovskiy V, Rozenbaum W, Pantaleo G, Sension M, Murphy R, Mancini M, Kelleher T, Giordano M. Long-term efficacy and safety of atazanavir with stavudine and lamivudine in patients previously treated with nelfinavir or atazanavir. J Acquir Immune Defic Syndr 2004;36(2):684–92.
6. Bitnun A, Sochett E, Dick PT, To T, Jefferies C, Babyn P, Forbes J, Read S, King SM. Insulin sensitivity and beta-cell function in protease inhibitor-treated and -naive human immunodeficiency virus-infected children. J Clin Endocrinol Metab 2005;90(1):168–74.
7. Aldamiz-Echevarria L, Pocheville I, Sanjurjo P, Elorz J, Prieto JA, Rodriguez-Soriano J. Abnormalities in plasma fatty acid composition in human immunodeficiency virus-infected children treated with protease inhibitors. Acta Paediatr 2005;94(6):672–7.
8. Markowitz M, Conant M, Hurley A, Schluger R, Duran M, Peterkin J, Chapman S, Patick A, Hendricks A, Yuen GJ, Hoskins W, Clendeninn N, Ho DD. A preliminary evaluation of nelfinavir mesylate, an inhibitor of human immunodeficiency virus (HIV)-1 protease, to treat HIV infection. J Infect Dis 1998;177(6):1533–40.
9. Moyle GJ, Youle M, Higgs C, Monaghan J, Prince W, Chapman S, Clendeninn N, Nelson MR. Safety, pharmacokinetics, and antiretroviral activity of the potent, specific human immunodeficiency virus protease inhibitor nelfinavir: results of a phase I/II trial and extended follow-up in patients infected with human immunodeficiency virus. J Clin Pharmacol 1998;38(8):736–43.
10. Sanne I, Piliero P, Squires K, Thiry A, Schnittman S; AI424-007 Clinical Trial Group. Results of a phase 2 clinical trial at 48 weeks (AI424-007): a dose-ranging, safety, and efficacy comparative trial of atazanavir at three doses in combination with didanosine and stavudine in antiretroviral-naive subjects. J Acquir Immune Defic Syndr 2003;32:18–29.
11. Di Martino V, Ezenfis J, Benhamou Y, Bernard B, Opolon P, Bricaire F, Poynard T. Severe acute pancreatitis related to the use of nelfinavir in HIV infection: report of a case with positive rechallenge. AIDS 1999;13(11):1421–3.
12. Engeler DS, John H, Rentsch KM, Ruef C, Oertle D, Suter S. Nelfinavir urinary stones. J Urol 2002;167(3):1384–5.
13. Demoly P, Messaad D, Trylesinski A, Faucherre V, Fabre J, Reynes J, Delmas C, Dohin E, Godard P, Bousquet J. Nelfinavir-induced urticaria and successful desensitization. J Allergy Clin Immunol 1998;102(5):875–6.
14. Abraham PE, Sorensen SJ, Baker WH, Cushing HE. Nelfinavir desensitization. Ann Pharmacother 2001;35(5):553–6.
15. Rongkavilit C, van Heeswijk RP, Limpongsanurak S, Thaithumyanon P, Boonrod C, Hassink EA, Srigritsanapol A, Chuenyam T, Ubolyam S, Hoetelmans RM, Ruxrungtham K, Lange JM, Cooper DA, Phanuphak P. Dose-escalating study of the safety and pharmacokinetics of nelfinavir in HIV-exposed neonates. J Acquir Immune Defic Syndr 2002;29(5):455–63.
16. Paediatric European Network for Treatment of AIDS (PENTA). Comparison of dual nucleoside-analogue reverse-transcriptase inhibitor regimens with and without nelfinavir in children with HIV-1 who have not previously been treated: the PENTA 5 randomised trial. Lancet 2002;359(9308):733–40.
17. Damle BD, Mummaneni V, Kaul S, Knupp C. Lack of effect of simultaneously administered didanosine encapsulated enteric bead formulation (Videx EC) on oral absorption of indinavir, ketoconazole, or ciprofloxacin. Antimicrob Agents Chemother 2002;46(2):385–91.
18. McCance-Katz EF, Farber S, Selwyn PA, O'Connor A. Decrease in methodone levels with nelfinavir mesylate. Am J Psychiatry 2000;157(3):481.
19. Brooks J, Daily J, Schwamm L. Protease inhibitors and anticonvulsants. AIDS Clin Care 1997;9(11):87–90.
20. Honda M, Yasuoka A, Aoki M, Oka S. A generalized seizure following initiation of nelfinavir in a patient with human immunodeficiency virus type 1 infection, suspected due to interaction between nelfinavir and phenytoin. Intern Med 1999;38(3):302–3.
21. Hsyu PH, Schultz-Smith MD, Lillibridge JH, Lewis RH, Kerr BM. Pharmacokinetic interactions between nelfinavir and 3-hydroxy-3-methylglutaryl coenzyme A reductase inhibitors atorvastatin and simvastatin. Antimicrob Agents Chemother 2001;45(12):3445–50.
22. Aberg JA, Rosenkranz SL, Fichtenbaum CJ, Alston BL, Brobst SW, Segal Y, Gerber JG; ACTG A5108 team. Pharmacokinetic interaction between nelfinavir and pravastatin in HIV-seronegative volunteers: ACTG Study A5108. AIDS 2006;20(5):725–9.
23. Schvarcz R, Rudbeck G, Soderdahl G, Stahle L. Interaction between nelfinavir and tacrolimus after orthoptic liver transplantation in a patient coinfected with HIV and hepatitis C virus (HCV). Transplantation 2000;69(10):2194–5.
24. Panhard X, Goujard C, Legrand M, Taburet AM, Diquet B, Mentre F; COPHAR 1-ANRS study group. Population pharmacokinetic analysis for nelfinavir and its metabolite M8 in virologically controlled HIV-infected patients on HAART. Br J Clin Pharmacol 2005;60(4):390–403.

Neuraminidase inhibitors

General Information

The major drawbacks of older drugs for the treatment of influenza (amantadine and rimantadine) include lack of activity against influenza B, a considerable frequency of adverse effects, and probably a higher likelihood of resistance. With the recent development of inhibitors of the viral neuraminidase, a more specific class of drugs has been added to the antiviral armamentarium. After the description of the crystal structure of the viral neuraminidase, a surface glycoprotein, specific inhibitors were designed. Two derivatives of sialic acid, the natural substrate of the neuraminidase, are currently in use, oseltamivir (rINN) and zanamivir (rINN). These drugs were designed to bind strongly to a conserved region of the cleavage site of the virus, which resulted in a class of compounds with limited risk of viral resistance. In fact, resistance is rare during the use of these drugs (1).

Oseltamivir is administered as an oral capsule and zanamivir by inhalation. Inhaled zanamivir has fewer adverse effects than oral oseltamivir, but there are concerns about its use in young and elderly patients, who are generally unable to achieve the inspiratory flow rate needed to ensure adequate lung deposition (2). Oral oseltamivir is more convenient and easier to administer, but needs to be given with food to lower the frequency of gastrointestinal adverse effects.

Oseltamivir

Oseltamivir (GS4104) is an ester prodrug, whose active metabolite (GS4071) is a potent selective inhibitor of influenza virus neuraminidase, to which it is rapidly hydrolysed by hepatic carboxylesterases. The metabolite is then completely excreted by the kidneys by filtration and secretion. In animals oseltamivir had a wide safety margin, with no evidence of teratogenicity or adverse effects on fertility. Oseltamivir has also been studied in the treatment and prevention of influenza A and B, but published results are sparse. To date it has been well tolerated. Headache was the most frequent complaint and mild to moderate gastrointestinal adverse effects (nausea, diarrhea) have been described (3).

The safety and pharmacology of oseltamivir have been reviewed (4). Transient gastrointestinal disturbances are the major adverse effects and are reduced when the drug is taken with a light snack. In the clinical trials program, severe adverse events were reported at the same frequency as with placebo (1.3% with 75 mg bd, 0.7% with 150 mg bd, 1.2% with placebo).

Zanamivir

Zanamivir has poor oral systemic availability. Intranasal administration slightly increases its availability, but in mice strongly reduced viral replication (5). In vitro, zanamivir does not significantly inhibit human lysosomal neuraminidases, and so the potential for severe adverse effects is low (6).

Zanamivir has been evaluated in the prevention and treatment of influenza A and B, and was effective against both viruses. In placebo-controlled trials, the adverse events profile, including local nasal irritation, did not differ significantly between zanamivir and placebo (3,7–9). Phase II and III clinical studies have shown that zanamivir by inhalation, either orally or combined orally and intranasally, slightly shortened the duration and severity of influenza symptoms and in high-risk patients reduced the risk of complications (1,10). Typical doses are 10 mg by oral inhalation and 6.4 mg by intranasal inhalation. Dosing frequency is 2–6 times/day. Zanamivir is well tolerated. In placebo-controlled studies its toxicity profile was similar to placebo.

A review of the use of zanamivir in 6000 patients in clinical studies has confirmed its favorable safety profile (11). To date there is still an optimistic impression that the adverse reactions pattern is no different from that of placebo and that interactions are also unlikely, since it does not inhibit cytochrome P450 (12).

The safety and efficacy of zanamivir have been evaluated in hospitalized patients with serious influenza (13). Zanamivir + rimantadine was compared to rimantadine + placebo in a randomized blinded design in seven centers. The study was terminated prematurely after approval of zanamivir made enrolment untenable (41 patients, calculated sample size 100). There were no differences in the proportions of patients shedding virus by treatment day 3, or duration of hospitalization, or use of oxygen. More patients taking zanamivir had a slight cough on day 3.

Observational studies

In 730 residents of nursing homes who took oseltamivir prophylaxis for a median of 9 days (range 5–12), adverse effects were identified in 30 (4.1%), of whom 20 had one adverse effect and 10 had two (14). The most common adverse effects were diarrhea ($n = 12$), nausea or vomiting ($n = 7$), cough ($n = 5$), and confusion ($n = 4$).

In seven patients with influenza (three type A and four type B) after allografting, inhaled zanamivir (10 mg bd) was used until excretion of virus ceased (median duration 15, range 5–44, days); there was rapid resolution of the influenza and no adverse effects attributable to zanamivir (15).

Several large postmarketing studies have underscored the safety of oseltamivir. There were no increases in cardiovascular, neuropsychiatric, or respiratory complications (16), nor in skin reactions (17).

In a large Japanese prospective multicenter study during the influenza season of 2002–3 oseltamivir was given to 803 patients with influenza A and 684 patients with influenza B; amantadine was given to 676 patients with influenza A (18). In each group, the duration of fever (body temperature over 37.5oC) was significantly shorter in patients who were treated within 12 hours after the onset of symptoms than in those who were treated more than 12 hours after the onset. The type of influenza, the highest body temperature, and the time between the onset of symptoms and the start of treatment independently affected the duration of fever. Only minor adverse reactions were reported by 19 patients with influenza A given oseltamivir, eight patients given amantadine, and one patient with influenza B given oseltamivir.

In a model of cost effectiveness comparing annual influenza immunization against empirical amantadine and rapid testing followed by oseltamivir if the results are positive, antiviral therapy without immunization was associated with the lowest overall costs ($234 per person per year for amantadine, $237 for oseltamivir). The cost of annual immunization was $239 per person and was associated with 0.0409 quality-adjusted days saved, for a marginal cost-effectiveness ratio of $113 per quality-adjusted day gained or $41 000 per quality-adjusted life-year saved compared with antiviral therapy (19). Adverse effects were included in the model at estimated baseline probabilities from published work. The adverse effects of influenza vaccine included minor effects, such as local soreness at the injection site, estimated at a probability of 0.64, and rarely Guillain–Barré syndrome (10^{-6}). The authors estimated that adverse effects other than Guillain–Barré syndrome lasted 2 days. Minor adverse effects due to drugs were also included with a probability of 0.09 for amantadine and 0.1 for oseltamivir.

Placebo-controlled studies

In a large placebo-controlled efficacy study in adolescents, oseltamivir was well tolerated and mild gastrointestinal symptoms were the major adverse effects, in 14% (oseltamivir) versus 8% (placebo); vomiting was reported in 8% versus 3% (20)

Systematic reviews

Recent trials of the neuraminidase inhibitors for influenza have been reviewed (21). The percentage of patients with serious or minor adverse effects associated with the administration of neuraminidase inhibitors was as follows for zanamivir: serious or life-threatening reactions were allergic or allergic-like reactions, dysrhythmias, bronchospasm, dyspnea, facial edema, rash, seizure, syncope, and urticaria (<1.5%). Minor adverse effects included headache (2%), dizziness (2%), nausea (3%), diarrhea (adults, 3%; children, 2%), vomiting (adults, 1%; children, 2%), sinusitis (3%), bronchitis (2%), cough (2%), other nasal signs and symptoms (2%), and infections (ear, nose, and throat: adults, 2%; children, 5%).

For oseltamivir the serious or life-threatening events were aggravation of diabetes, dysrhythmias, confusion, hepatitis, pseudomembranous colitis, pyrexia, rash, seizures, swelling of the face or tongue, toxic epidermal necrolysis, and unstable angina (<1%) The minor effects were insomnia (adults, 1%), vertigo (1%), nausea (10%), and vomiting (9%). The adverse effects of oseltamivir prophylaxis were similar to those reported during treatment, but generally with lower incidences. More common with prophylactic use were headache (20%), fatigue (8%), cough (6%), and diarrhea (3%).

Organs and Systems

Respiratory

Zanamivir is administered by inhalation of a powder in a lactose vehicle. It can cause bronchospasm (22).

- A 63-year-old man with oxygen-dependent chronic obstructive pulmonary disease was given zanamivir because of an exacerbation attributed to influenza. Shortly after each inhalation of the drug he reported increasing respiratory difficulty and wheezing. After 3 days he developed respiratory distress, hypoxia, and wheezing, and was treated with bronchodilators, glucocorticoids, and antibiotics.

Patients taking zanamivir are advised to stop using it if bronchospasm occurs, and those with underlying lung disease are advised to have a fast-acting bronchodilator to hand. The bronchospasm can be reduced in frequency by using an inhaled β_2-adrenoceptor agonist before zanamivir. Proper drug deposition depends on flow rate and inhaler technique, so young and elderly patients may not be the best to treat in this way (2,23).

Psychiatric

Oseltamivir was associated with abnormal behavior in four children, including meaningless speech, disorientation, fearful responses, and running around the room; there were no neurological sequelae (24).

Gastrointestinal

Upper gastrointestinal effects (nausea or nausea with vomiting) have been reported more often in those taking oseltamivir in placebo-controlled studies (25,26). Despite this mild gastrointestinal intolerance, withdrawal rates have been low.

Nausea and vomiting have led to drug withdrawal in less than 1% of the subjects in clinical trials (27). While nausea was equally frequent with once-daily influenza prophylaxis (8%) and twice-daily treatment (7%), the incidence of vomiting increased from 4.5% during prophylaxis to 10% when the drug was taken twice a day (28). A similar pattern was found in the pediatric subgroup, with vomiting incidences of 10% and 20% respectively.

The oral capsule of oseltamivir can be used in any patient who can swallow. However, it must be given with food, to reduce the frequency of nausea and gastrointestinal discomfort. Of 695 children aged 1–12 years,

14.3% had vomiting compared with 8.5% of those who took placebo; otherwise oseltamivir was well tolerated (29).

In a meta-analysis of trials, there was a reduction in the duration of symptoms of about 1 day (30). Oseltamivir caused significantly more gastrointestinal symptoms (dyspepsia or nausea) than placebo; the effect increased with dose. Similarly, zanamivir caused significantly more gastrointestinal symptoms than placebo (OR = 2.6; 95% CI = 1.6, 4.2).

Susceptibility Factors

Age

Exposure (AUC) to both the parent drug and its active metabolite were more than 80% higher in five very elderly subjects compared with young volunteers, but oseltamivir was well tolerated by all subjects (31).

Renal disease

Gastrointestinal adverse effects and dizziness can occur with increasing doses of oseltamivir, particularly in patients with renal failure (32).

Liver disease

A single oral dose of oseltamivir 75 mg has been studied in 11 subjects with (all with cirrhosis, seven alcohol-induced) and paired controls. Mean BMI and estimated serum creatinine clearance were matched. In hepatic impairment the values of oseltamivir and oseltamivir carboxylate C_{max} were <6% and <19% lower and their AUCs 33% higher and <19% lower respectively. Thus, the metabolism of oseltamivir is not compromised in hepatic impairment and no dosage adjustment is required (33).

Drug–Drug Interactions

Amoxicillin

In healthy subjects amoxicillin had no effect on the pharmacokinetics of oral oseltamivir (34).

Cimetidine

In healthy subjects cimetidine had no effect on the pharmacokinetics of oral oseltamivir (34).

Probenecid

In healthy subjects probenecid completely blocked the renal secretion of the active metabolite of oseltamivir after oral administration, increasing its AUC 2.5 times (34). In vitro studies of the metabolite on the human renal organic anionic transporter I (hOAT1) were investigated in Chinese hamster ovary cells stably transfected with the transporter. The metabolite was a low-efficiency substrate for hOAT1 and a very weak inhibitor of hOAT1-mediated transport of para-aminohippuric acid. Probenecid inhibited the transport of the metabolite, para-aminohippuric acid, and amoxicillin via hOAT1.

References

1. Waghorn SL, Goa KL. Zanamivir. Drugs 1998;55(5):721–5.
2. McNicholl IR, McNicholl JJ. Neuraminidase inhibitors: zanamivir and oseltamivir. Ann Pharmacother 2001;35(1):57–70.
3. Calfee DP, Peng AW, Cass LM, Lobo M, Hayden FG. Safety and efficacy of intravenous zanamivir in preventing experimental human influenza A virus infection. Antimicrob Agents Chemother 1999;43(7):1616–20.
4. Dutkowski R, Thakrar B, Froehlich E, Suter P, Oo C, Ward P. Safety and pharmacology of oseltamivir in clinical use. Drug Saf 2003;26:787–801.
5. Ryan DM, Ticehurst J, Dempsey MH, Penn CR. Inhibition of influenza virus replication in mice by GG167 (4-guanidino-2,4-dideoxy-2,3-dehydro-N-acetylneuraminic acid) is consistent with extracellular activity of viral neuraminidase (sialidase). Antimicrob. Agents Chemother 1994;38(10):2270–5.
6. Woods JM, Bethell RC, Coates JA, Healy N, Hiscox SA, Pearson BA, Ryan DM, Ticehurst J, Tilling J, Walcott SM, et al. 4-Guanidino-2,4-dideoxy-2,3-dehydro-N-acetylneuraminic acid is a highly effective inhibitor both of the sialidase (neuraminidase) and of growth of a wide range of influenza A and B viruses in vitro. Antimicrob Agents Chemother 1993;37(7):1473–9.
7. Hayden FG, Osterhaus AD, Treanor JJ, Fleming DM, Aoki FY, Nicholson KG, Bohnen AM, Hirst HM, Keene O, Wightman KGG167 Influenza Study Group. Efficacy and safety of the neuraminidase inhibitor zanamivir in the treatment of influenzavirus infections. N Engl J Med 1997;337(13):874–80.
8. Monto AS, Robinson DP, Herlocher ML, Hinson JM Jr, Elliott MJ, Crisp A. Zanamivir in the prevention of influenza among healthy adults: a randomized controlled trial. JAMA 1999;282(1):31–5.
9. Monto AS, Fleming DM, Henry D, de Groot R, Makela M, Klein T, Elliott M, Keene ON, Man CY. Efficacy and safety of the neuraminidase inhibitor zanamivirin the treatment of influenza A and B virus infections. J Infect Dis 1999;180(2):254–61.
10. The MIST (Management of Influenza in the Southern Hemisphere Trialists) Study Group. Randomised trial of efficacy and safety of inhaled zanamivir in treatment of influenza A and B virus infections. Lancet 1998;352(9144):1877–81.
11. Freund B, Gravenstein S, Elliott M, Miller I. Zanamivir: a review of clinical safety. Drug Saf 1999;21(4):267–81.
12. Daniel MJ, Barnett JM, Pearson BA. The low potential for drug interactions with zanamivir. Clin Pharmacokinet 1999;36(Suppl 1):41–50.
13. Ison MG, Gnann JW Jr, Nagy-Agren S, Treannor J, Paya C, Steigbigel R, Elliott M, Weiss HL, Hayden FG; NIAID Collaborative Antiviral Study Group. Safety and efficacy of nebulized zanamivir in hospitalized patients with serious influenza. Antivir Ther 2003;8:183–90.
14. Bowles SK, Lee W, Simor AE, Vearncombe M, Loeb M, Tamblyn S, Fearon M, Li Y, McGeer A, Marguerite V, Baker D, Collins V, VanHorne E, Louie MA, Friedman P, Kam A, Goette M, Lam J, LaRue M, MacDonald M, McArthur MA, Mazzulli T, McDougall R, Satchell S,

Simpson M, Stanton S, Warshawsky B, Wier H. Oseltamivir Compassionate Use Program Group. Use of oseltamivir during influenza outbreaks in Ontario nursing homes, 1999–2000. J Am Geriatr Soc 2002;50(4):608–16.

15. Johny AA, Clark A, Price N, Carrington D, Oakhill A, Marks DI. The use of zanamivir to treat influenza A and B infection after allogeneic stem cell transplantation. Bone Marrow Transplant 2002;29(2):113–5.

16. Enger C, Nordstrom BL, Thakrar B, Sacks S, Rothman KJ. Health outcomes among patients receiving oseltamivir. Pharmacoepidemiol Drug Saf 2004;13(4):227–37.

17. Nordstrom BL, Oh K, Sacks ST, L'Italien GJ. Skin reactions in patients with influenza treated with oseltamivir: a retrospective cohort study. Antivir Ther 2004;9(2):187–95.

18. Kawai N, Ikematsu H, Iwaki N, Satoh I, Kawashima T, Maeda T, Miyachi K, Hirotsu N, Shigematsu T, Kashiwagi S. Factors influencing the effectiveness of oseltamivir and amantadine for the treatment of influenza: a multicenter study from Japan of the 2002–2003 influenza season. Clin Infect Dis 2005;40(9):1309–16.

19. Rothberg MB, Rose DN. Vaccination versus treatment of influenza in working adults: a cost-effectiveness analysis. Am J Med 2005;118(1):68–77.

20. Singh S, Barghoorn J, Bagdonas A, Adler J, Treanor J, Kinnersley N, Ward P. Clinical benefits with oseltamivir in treating influenza in adult populations: results of a pooled and subgroup analysis. Clin Drug Invest 2003;23:561–9.

21. Moscona A. Neuraminidase inhibitors for influenza. N Engl J Med 2005;353(13):1363–73.

22. Williamson JC, Pegram PS. Respiratory distress associated with zanamivir. N Engl J Med 2000;342(9):661–2.

23. Lalezari J, Campion K, Keene O, Silagy C. Zanamivir for the treatment of influenza A and B infection in high-risk patients: a pooled analysis of randomized controlled trials. Arch Intern Med 2001;161(2):212–7.

24. Okumura A, Kubota T, Kato T, Morishima T. Oseltamivir and delirious behavior in children with influenza. Pediatr Infect Dis J 2006;25(6):572.

25. Treanor JJ, Hayden FG, Vrooman PS, Barbarash R, Bettis R, Riff D, Singh S, Kinnersley N, Ward P, Mills RGUS Oral Neuraminidase Study Group. Efficacy and safety of the oral neuraminidase inhibitor oseltamivir in treating acute influenza: a randomized controlled trial. JAMA 2000;283(8):1016–24.

26. Nicholson KG, Aoki FY, Osterhaus AD, Trottier S, Carewicz O, Mercier CH, Rode A, Kinnersley N, Ward P. Efficacy and safety of oseltamivir in treatment of acute influenza: a randomised controlled trial. Neuraminidase Inhibitor Flu Treatment Investigator Group. Lancet 2000;355(9218):1845–50.

27. Roche Laboratories Inc. Tamiflu (oseltamivir phosphate) capsules [package insert]. Nutley, NJ: Roche Laboratories Inc, 2000.

28. Hayden FG, Belshe R, Villanueva C, Lanno R, Hughes C, Small I, Dutkowski R, Ward P, Carr J. Management of influenza in households: a prospective, randomized comparison of oseltamivir treatment with or without postexposure prophylaxis. J Infect Dis 2004;189(3):440–9.

29. Whitley RJ, Hayden FG, Reisinger KS, Young N, Dutkowski R, Ipe D, Mills RG, Ward P. Oral oseltamivir treatment of influenza in children. Pediatr Infect Dis J 2001;20(2):127–33.

30. Pitts SR. Evidence-based emergency medicine/systematic review abstract. Use of the neuraminidase inhibitor class of antiviral drugs for treatment of healthy adults with an acute influenza-like illness. Ann Emerg Med 2002;39(5):552–4.

31. Abe M, Smith J, Urae A, Barrett J, Kinoshita H, Rayner CR. Pharmacokinetics of oseltamivir in young and very elderly subjects. Ann Pharmacother 2006;40(10):1724–30.

32. Karie S, Launay-Vacher V, Janus N, Izzedine H, Deray G. Pharmacokinetics and dosage adjustment of oseltamivir and zanamivir in patients with renal failure. Nephrol Dial Transplant 2006;21(12):3606–8.

33. Snell P, Dave N, Wilson K, Rowell L, Weil A, Galitz L, Robson R. Lack of effect of moderate hepatic impairment on the pharmacokinetics of oral oseltamivir and its metabolite oseltamivir carboxylate. Br J Clin Pharmacol 2005;59(5):598–601.

34. Hill G, Cihlar T, Oo C, Ho ES, Prior K, Wiltshire H, Barrett J, Liu B, Ward P. The anti-influenza drug oseltamivir exhibits low potential to induce pharmacokinetic drug interactions via renal secretion-correlation of in vivo and in vitro studies. Drug Metab Dispos 2002;30(1):13–9.

Nevirapine

See also Non-nucleoside reverse transcriptase inhibitors (NNRTIs)

General Information

Nevirapine is a non-nucleoside reverse transcriptase inhibitor (1). Concerns about the adverse effects of nevirapine have delayed its implementation in preventing perinatal HIV. Decision analysis has been used to compare three strategies: a single dose of nevirapine, a short course of zidovudine, and no intervention (2). The authors concluded that nevirapine would prevent more deaths than either zidovudine and no intervention, as long as the rate of nevirapine toxicity did not respectively exceed nine and 42 times that observed in an earlier nevirapine clinical trial (HIVNET). Nevirapine would be economically preferable to zidovudine as long as the rate of toxicity did not exceed 22 times that observed in the clinical trial. They thought that implementation of nevirapine should not be delayed by concerns about its adverse effects.

Observational studies

In 77 HIV-positive subjects randomized to switch from protease inhibitors to nevirapine or efavirenz or to continue taking protease inhibitors, quality of life significantly improved among those who switched (3). Lipid profiles improved in those who took nevirapine but gamma-glutamyltransferase and alanine aminotransferase activities increased significantly and one patient interrupted treatment because of hepatotoxicity.

In 197 women exposed to nevirapine for more than 7 days, adverse effects occurred in 11 (5.6%), leading to drug withdrawal in seven (4). There was one case of Stevens–Johnson syndrome. There was no serious liver toxicity, except for one case of grade 4 cholestasis. Co-infection with hepatitis C virus was associated with toxicity.

In a retrospective analysis of 703 HIV-1-positive pregnant women treated with a nevirapine-containing regimen, severe adverse reactions—hepatotoxicity, rashes, and Stevens–Johnson syndrome—occurred in 6.5, 2.4, and 1.1% respectively (5).

In a Nigerian prospective, observational, cohort study, 50 antiretroviral drug-naive patients in stage 2 or stage 3 World Health Organization clinical classification were treated with generic brands of oral nevirapine (Nevimal, Cipla, Mumbai, India) 200 mg/day, lamivudine (Lamivir, Cipla) 150 mg bd, and stavudine (Stavir, Cipla) 40 mg bd (6). At week 48 the median CD4+ cell count increased by $186 \times 10^6/l$, the frequency of opportunistic infections fell by 82%, and the median body mass index increased by 4.8 kg/m^2. There were minor and transient adverse effects in 36%. The most comment adverse effect was a rash associated with nevirapine.

Organs and Systems

Psychological, psychiatric

Various psychiatric abnormalities (delirium, an affective state, and a psychosis) have been described in three patients who took nevirapine for 10–14 days (7):

- cognitive impairment, clouding of consciousness, and a paranoid episode in a 35-year-old man
- delusions of persecutory and depressive thoughts in a 42-year-old woman
- delusions of persecution and infestation and hallucinations in a 36-year-old woman

All three responded to withdrawal of nevirapine.

Metabolism

Lipodystrophy, well recognized with stavudine, has also been reported in nine of 56 patients taking combined HAART therapy including nevirapine, although there must be some doubt as to which drug or combination was responsible (8).

Hematologic

Two patients developed grade 4 thrombocytopenia (under $25 \times 10^9/l$) after 42 and 151 days of treatment with nevirapine (9). In the first patient, zidovudine and intravenous immunoglobulin were added to continued nevirapine, and the thrombocytopenia resolved by day 89. In the second, nevirapine was discontinued and alternative antiretroviral therapy was started, whereupon the platelet count returned to normal within 22 days.

Liver

The most frequent laboratory abnormality during nevirapine treatment is an increase in serum gamma-glutamyl transpeptidase activity, usually without changes in other measures of hepatic function. However, in a randomized, placebo-controlled comparison of zidovudine plus nevirapine, zidovudine plus didanosine, and the triple combination of zidovudine plus didanosine and nevirapine in 151 patients, there were abnormal liver function tests in increased frequency in the patients taking nevirapine (19% and 12% versus 6%) (10). Five of 98 patients taking nevirapine had to stop the drug permanently because of raised alanine aminotransferase activity, after which the laboratory abnormalities resolved completely.

Of 70 HIV-infected patients taking nevirapine 33 developed rises in transaminase activities (11). Higher nevirapine concentrations and hepatitis C virus infection were independent predictors of liver toxicity. In those with chronic hepatitis C, nevirapine concentrations over 6 µg/ml were associated with a 92% risk of liver toxicity. The authors concluded that monitoring nevirapine concentrations, especially in individuals with chronic hepatitis C, may be warranted.

In a prospective study of the incidence of severe hepatotoxicity in 312 patients taking nevirapine, hepatitis C and hepatitis B viruses were detected in 43% (12). There was severe hepatotoxicity in 16%, but only 32% of episodes were detected during the first 12 weeks of therapy. The risk was significantly greater among those with chronic viral hepatitis (69% of cases) and those taking concurrent protease inhibitors (82% of cases). However, 84% of patients with chronic hepatitis C or hepatitis B did not have severe hepatotoxicity.

Severe hepatitis has been attributed to nevirapine (13).

- A 36-year-old woman with HIV and hepatitis C infection developed fever, right upper quadrant pain, headache, confusion, and a generalized skin rash. Her antiretroviral treatment included lamivudine (150 mg bd), stavudine (40 mg bd), and nevirapine (200 mg bd). The dose of nevirapine had been doubled 2 weeks before. Her laboratory results were as follows: white cell count $4.1 \times 10^9/l$ with 18% eosinophils, aspartate aminotransferase 879 U/l, alanine aminotransferase 1424 U/l, lactate dehydrogenase 3268 U/l, activated prothrombin time 14 seconds, partial thromboplastin time 29 seconds. Abdominal ultrasound was normal. Her antiretroviral medications were stopped. She was given intravenous prednisolone (60 mg 6-hourly) and within 8 hours the signs of sepsis had abated and her liver function began to improve and resolved over the next few days. Several weeks later, she was rechallenged with lamivudine and zidovudine, without recurrence of liver toxicity or rash.

The authors argued that, although the syndrome could not be definitively attributed to nevirapine, several factors suggested that nevirapine was the most likely precipitant, including the fact that hepatic failure coincided with a systemic allergic response, which is commonly associated with nevirapine, especially after dose escalation. The rapid improvement with steroid therapy suggested that the hepatic failure was a manifestation of a systemic hypersensitivity reaction. However, since stavudine has also been associated with hepatitis, a causative role of stavudine cannot be excluded. The fact that lamivudine was given again without recurrence of hepatitis renders it an unlikely culprit.

In another case there was severe hepatic failure, which resolved on withdrawal and in which there was no other obvious cause than the use of nevirapine (14).

Significant hepatic deterioration, consistent with cholestasis, occurred 5 months into a HAART regimen including nevirapine (15).

- A 49-year-old man with severe factor VIII deficiency and stable chronic hepatitis C infection took stavudine, didanosine, and nevirapine. After 5 months of well-tolerated therapy, during which his viral load fell from 39 550 cpm to under 50 cpm, he developed anorexia, nausea, vomiting, fatigue, agitation, and biochemical evidence of deteriorating liver function. All drugs were withdrawn. There was no evidence of an infective cause and his hepatitis C status had not changed. Transvenous biopsy of the right hepatic lobe showed profound cholestasis and mild sinusoidal fibrosis, consistent with drug-induced cholestasis.

In four men aged 27, 41, 47, and 49 years, nevirapine was associated with a skin rash, malaise, and icteric hepatitis, 4–6 weeks after the start of therapy; resolution occurred after withdrawal of nevirapine (16).

Two health-care workers exposed to HIV developed severe hypersensitivity reactions to nevirapine; one required orthotopic liver transplantation to overcome the complications of acute hepatic failure and coma (17).

Hepatotoxicity was studied in 1731 nevirapine-treated patients enrolled in trials and compared with 1912 control patients (18). In this analysis, risk factors for asymptomatic rises in liver transaminases were increased baseline concentrations (over 2.5 times the upper limit of the reference ranges, RR = 4.3) and co-infection with hepatitis B (RR = 2.3) or hepatitis C (RR = 5.2). In men, but not in women, a CD4 cell-count above $400 \times 10^6/l$ was also marginally associated with an increased risk of asymptomatic rises in liver enzymes. The authors also summarized the results of large observational cohorts and found no significant differences in the rate of serious hepatic events among patients treated with nevirapine and other antiretroviral regimens.

The reports of all suspected hepatic adverse drug reactions with a fatal outcome received by the Uppsala Monitoring Centre between 1968 and 2003 included over 50 cases of nevirapine-associated hepatotoxicity (19).

In a multicenter, cross-sectional, observation study 1.1% of patients (7/613) developed hepatic toxicity that led to withdrawal of nevirapine (20). However, patients who had discontinued treatment within the first 2 years because of adverse events were excluded. The overall rate of hepatotoxicity may therefore have been underestimated.

In a retrospective study of 123 pregnant women who took nevirapine as part of combination antiretroviral therapy, eight developed significant hepatotoxicity, including two who died from fulminant hepatitis (21). Women who had more severe hepatotoxicity had higher pretreatment CD4 counts.

In another case, fulminant liver failure requiring liver transplantation was attributed to nevirapine (22).

There were grade 3 or 4 hepatobiliary laboratory adverse events in 30 of 220 patients (13.6%) taking once-daily nevirapine and 32 of 387 patients (8.3%) taking twice-daily nevirapine in a large randomized trial (23). This is the first report that hepatotoxicity may be more frequent with once-daily nevirapine. One patient taking twice-daily nevirapine died because of acute drug induced hepatitis with liver failure that started after 32 days of therapy.

In a non-randomized study of 256 patients taking nevirapine-containing regimens, the prevalence of liver enzyme activities above twice the upper limit of normal increased from 80 patients to 134 during a median follow-up of 18 months; in contrast, it fell in a concurrent cohort of 287 patients who took efavirenz (24).

Data from the FDA adverse events reporting system have shown that in 30 HIV uninfected persons who experienced EOCG grade 3 or 4 hepatotoxicity during nevirapine containing post-exposure prophylaxis, there was an accompanying fever in 11, eosinophilia in five, and a rash in seven (25). In 30 HIV uninfected patients with EOCG grade 3 or 4 hepatotoxicity, there was an accompanying fever in 11, eosinophilia in five, and a rash in seven. One had fulminant liver necrosis and needed liver transplantation.

In a comprehensive review of 17 randomized clinical trials about 10% of patients taking nevirapine develop transaminase activities five times above the upper limit of normal, which represents a risk ratio of 1.5 (26). In two-thirds of cases the rises were asymptomatic. However, the risk ratio for symptomatic hepatitis was 3.5 and the association of rash and hepatitis was found nearly exclusively in patients taking nevirapine (frequency 1.5%, RR = 11.2) and occurred during the first 60 days of therapy.

In other small studies there was no clear relation between nevirapine plasma concentrations and hepatotoxicity (27,28).

In a cohort analysis of 152 patients co-infected with HIV and hepatitis C nevirapine was associated with faster progression of fibrosis in liver biopsies (29).

Boehringer Ingelheim have conducted an analysis of *hepatotoxicity* in all of their past controlled and uncontrolled studies (30). This analysis resulted in warnings that female sex and a higher CD4 cell count at the start of therapy increases the risk of hepatotoxicity, particularly during the first 6 weeks of treatment. They recommended against starting nevirapine in women with CD4 cell counts over $250 \times 10^6/l$ or in men with CD4 cell counts over $400 \times 10^6/l$.

Skin

The principal, dose-limiting, adverse effect of nevirapine in adults is a rash, which occurs in 3–20% of patients, and which progresses to Stevens–Johnson syndrome in 0.5–1% of cases (31). Rash appears to be sex-specific; in 95 women and 263 men, women had a seven-fold higher risk of severe rash and were 3.5 times more likely to discontinue nevirapine than men (32).

- A pregnant woman had a febrile episode with a pruritic rash and eosinophilia at 32 weeks of gestation, 6 weeks after starting to take antiretroviral therapy including nevirapine (33). The symptoms resolved after nevirapine withdrawal and glucocorticoid treatment. Most other differential diagnoses were ruled out.

Of a retrospective case series of 74 children treated with nevirapine in the UK between 1997 and 1999, 20% developed a rash (34). Rash developed independent of drug dose after a median duration of treatment of 9 (1–44) days and lasted for a median of 10 (1–60) days. In 5%, the rash led to withdrawal of nevirapine. There were no cases of Stevens–Johnson syndrome.

Of 216 patients 18% developed a rash of some grade, leading to drug withdrawal in seven (35).

In contrast, none of 33 healthy subjects developed a rash when nevirapine was given in a pre-exposure prophylaxis trial once or twice a week or every other day for 12 weeks (36). Similarly, there were no adverse effects in three comparative trials of single or short administration of nevirapine for the prevention of mother-to-child transmission in 2080 subjects (37,38,39).

In a review of the medical records of HIV-positive patients who had taken nevirapine, delavirdine, or both, the frequency of skin reactions was determined, as were the consequences of rechallenge with the same or the alternative agent (40). The overall incidence of rash attributed to the use of one of the non-nucleoside reverse transcriptase inhibitors (NNRTIs) was 37%. While rash due to delavirdine was more common (8/20 versus 25/69), the rash due to nevirapine was more severe and necessitated more frequent hospitalization. Rash recurred in six of eight patients who were rechallenged with the same agent and in seven of 10 who were switched to the alternative agent. The conclusion was drawn that there is little value in attempting to re-treat patients who have had skin reactions to NNRTIs, except possibly those with limited treatment options.

In a multiple dose study, 21 children aged 3 months to 15 years were treated with nevirapine in dosages of 7.5–400 mg/m^2/day for up to 168 days (41). When it was intended to use dosages over 240 mg/m^2/day they were pretreated with a lower dosage (120 mg/m^2/day) for 28 days to reduce the risk of rash. However, a rash developed in one child after 2 weeks of treatment with nevirapine at a dosage of 240 mg/m^2/day, resolved on withdrawal, and recurred on rechallenge with a single dose of 120 mg/m^2.

An unusually high incidence of nevirapine-associated rash has been reported in Chinese HIV-infected patients (42). Of eight Chinese patients, five developed a rash within 4 weeks of treatment, resolving on withdrawal. Since the total number of patients in this report was small, it remains to be shown whether Chinese are indeed at increased risk of hypersensitivity reactions to nevirapine.

One patient out of 387 taking twice-daily nevirapine in the 2NN trial died after developing Stevens–Johnson syndrome 39 days after the start of therapy (23). Of twelve HIV uninfected people with severe skin reactions three had Stevens–Johnson syndrome; their symptoms started 7–12 days after the start of nevirapine therapy (43).

Prevention

While skin reactions to nevirapine cannot be prevented by the use of glucocorticoids (9,44), it is possible to induce tolerance to them, enabling treatment to continue (9). All the same, very severe skin reactions, with characteristics of both Stevens–Johnson syndrome and Lyell's syndrome, have occasionally been reported (45).

A 14-day lead-in period reduces the frequency of rash, and other effective interventions include antihistamines and a longer lead-in period (4 weeks) (46). In an attempt to reduce the rate of nevirapine-associated rash, 469 patients were randomly assigned to different schedules of induction therapy (47). Using a standard procedure, 19% developed a rash compared with 11, 8.6, and 7.7% in subjects assigned to a slowly escalating dose, concomitant administration of prednisone, or both. The rate of drug withdrawal was also reduced by a half using the new approaches.

Immunologic

The most important adverse effect of nevirapine is a potentially fatal systemic hypersensitivity reaction presenting with a rash and/or hepatitis. This reaction may be more frequent in HIV uninfected persons, as suggested by a case-series and review (25). The authors suggested that nevirapine should not be a component of prophylaxis after HIV exposure.

Hypersensitivity reactions occurred in 16.5% of 115 individuals taking antiretroviral regimens that included nevirapine, abacavir, lamivudine, and zidovudine, compared with 5.3% of 114 who were not taking nevirapine (43). It is not possible to differentiate hypersensitivity reactions to nevirapine from those to abacavir.

Second-Generation Effects

Pregnancy

A randomized trial of nevirapine versus nelfinavir combination antiretroviral therapy with a nucleoside analogue backbone with lamivudine and zidovudine in 17 pregnant women was stopped early because of greater than expected toxicity in the nevirapine arm that led to withdrawal in five of them (48). All these women started nevirapine with CD4 counts over $250 \times 10^6/l$. Adverse effects included one death due to fulminant hepatic failure and one case of Stevens–Johnson syndrome. According to the US Health Service Task Force, nevirapine should not be used in pregnant women with CD4 counts above $250 \times 10^6/l$; if it is used, liver toxicity should be monitored closely during the first 18 weeks (49).

Lactation

In 20 mother-infant pairs in a pharmacokinetic study in Botswana, maternal serum concentrations of nevirapine were high (median 9534 ng/ml) at a median of 4 hours after nevirapine ingestion, and the median breast-milk concentration of nevirapine was two-thirds of the serum concentration (50). The median infant serum nevirapine concentration was 971 ng/ml, which is at least 40 times the 50% inhibitory concentration and similar to peak concentrations after a single 2-mg/kg dose of nevirapine. One infant had a rash 3 days after birth (and had also received oral nevirapine at birth), and four of the infants had either severe or life-threatening neutropenia or anemia at some time during breast-feeding.

Susceptibility Factors

Liver disease

Viral hepatitis is a frequent co-morbidity in HIV infection. The effect of chronic hepatitis C on nevirapine drug concentrations has been examined in 70 patients taking a nevirapine-containing triple drug regimen (51). Nevirapine concentrations were similar in those with positive (n = 32) and negative (n = 38) titers of hepatitis C virus antibody.

Drug–Drug Interactions

Atazanavir

There was a significant increase in atazanavir clearance when either nevirapine was co administered with atazanavir + ritonavir compared with patients treated with atazanavir + ritonavir and nucleoside reverse transcriptase inhibitors (52).

Methadone

When methadone is used in HIV infection (for pain or treatment of opioid dependence) nevirapine given simultaneously can reduce methadone blood concentrations,

the effect being sufficient to cause methadone withdrawal symptoms (53,54).

Oral contraceptives

The mutual pharmacokinetic interaction of nevirapine with ethinylestradiol + norethindrone has been studied in 10 women (55). After a single dose of ethinylestradiol + norethindrone, they took oral nevirapine 200 mg/day (days 2–15), followed by 200 mg bd (days 16–29); on day 30 they took another dose of ethinylestradiol + norethindrone. Steady-state nevirapine reduced the AUC of ethinylestradiol by 29% and significantly reduced its mean residence time and half-life. The AUC of norethindrone was significantly reduced by 18%, but there was no change in C_{max}, mean residence time, or half-life. The kinetics of nevirapine were not affected by the oral contraceptive. The authors attributed this interaction to increased clearance of ethinylestradiol and concluded that oral contraceptives should not be the primary method of birth control in women of child-bearing potential who are taking nevirapine.

Protease inhibitors

Nevirapine induces CYP3A4 and therefore interacts with drugs such as protease inhibitors (56).

St John's wort

St John's wort reduced nevirapine concentrations, because of induction of CYP3A4 (57).

Warfarin

Nevirapine induces the metabolism of warfarin.

- A 38-year-old man with severe primary pulmonary hypertension and AIDS took warfarin 2.5 mg/day while he was taking zidovudine plus didanosine (58). His INR was 2.1–2.4. Later, because of virological failure, he was given a combination of stavudine, lamivudine, and nevirapine, and the INR promptly fell to 1.3. The dosage of warfarin was increased to 5 mg/day and anticoagulant activity was restored. All the antiretroviral drugs were then stopped because of an urticarial eruption, and a few days later stavudine, lamivudine, and saquinavir were given; warfarin 2.5 mg/day produced a stable INR.

References

1. De Wit S, Sternon J, Clumeck N. La nevirapine (viramune): un nouvel inhibiteur du VIH. [Nevirapine (Viramune): a new HIV inhibitor.] Rev Med Brux 1999;20(2):95–9.
2. Stringer JS, Sinkala M, Rouse DJ, Goldenberg RL, Vermund SH. Effect of nevirapine toxicity on choice of perinatal HIV prevention strategies. Am J Public Health 2002;92(3):365–6.

3. Negredo E, Cruz L, Paredes R, Ruiz L, Fumaz CR, Bonjoch A, Gel S, Tuldra A, Balague M, Johnston S, Arno A, Jou A, Tural C, Sirera G, Romeu J, Clotet B. Virological, immunological, and clinical impact of switching from protease inhibitors to nevirapine or to efavirenz in patients with human immunodeficiency virus infection and long-lasting viral suppression. Clin Infect Dis 2002;34(4):504–10.

4. João EC, Calvet GA, Menezes JA, D'Ippolito MM, Cruz ML, Salgado LA, Matos HJ. Nevirapine toxicity in a cohort of HIV-1-infected pregnant women. Am J Obstet Gynecol 2006;194(1):199–202.

5. Marazzi MC, Germano P, Liotta G, Guidotti G, Loureiro S, da Cruz Gomes A, Valls Blazquez MC, Narciso P, Perno CF, Mancinelli S, Palombi L. Safety of nevirapine-containing antiretroviral triple therapy regimens to prevent vertical transmission in an African cohort of HIV-1-infected pregnant women. HIV Med 2006;7(5):338–44.

6. Idigbe EO, Adewole TA, Eisen G, Kanki P, Odunukwe NN, Onwujekwe DI, Audu RA, Araoyinbo ID, Onyewuche JI, Salu OB, Adedoyin JA, Musa AZ. Management of HIV-1 infection with a combination of nevirapine, stavudine, and lamivudine: a preliminary report on the Nigerian antiretroviral program. J Acquir Immune Defic Syndr 2005;40(1):65–9.

7. Wise ME, Mistry K, Reid S. Drug points: Neuropsychiatric complications of nevirapine treatment. BMJ 2002;324(7342):879.

8. Aldeen T, Wells C, Hay P, Davidson F, Lau R. Lipodystrophy associated with nevirapine-containing anti-retroviral therapies. AIDS 1999;13(7):865–7.

9. Demoly P, Messaad D, Fabre J, Reynes J, Bousquet J. Nevirapine-induced cutaneous hypersensitivity reactions and successful tolerance induction. J Allergy Clin Immunol 1999;104(2 Pt 1):504–5.

10. Montaner JS, Reiss P, Cooper D, Vella S, Harris M, Conway B, Wainberg MA, Smith D, Robinson P, Hall D, Myers M, Lange JM. A randomized, double-blind trial comparing combinations of nevirapine, didanosine, and zidovudine for HIV-infected patients: the INCAS Trial. Italy, The Netherlands, Canada and Australia Study. JAMA 1998;279(12):930–7.

11. Gonzalez de Requena D, Nunez M, Jimenez-Nacher I, Soriano V. Liver toxicity caused by nevirapine. AIDS 2002;16(2):290–1.

12. Sulkowski MS, Thomas DL, Mehta SH, Chaisson RE, Moore RD. Hepatotoxicity associated with nevirapine or efavirenz-containing antiretroviral therapy: role of hepatitis C and B infections. Hepatology 2002;35(1):182–9.

13. Leitze Z, Nadeem A, Choudhary A, Saul Z, Roberts I, Manthous CA. Nevirapine-induced hepatitis treated with corticosteroids? AIDS 1998;12(9):1115–7.

14. Cattelan AM, Erne E, Salatino A, Trevenzoli M, Carretta G, Meneghetti F, Cadrobbi P. Severe hepatic failure related to nevirapine treatment. Clin Infect Dis 1999;29(2):455–6.

15. Clarke S, Harrington P, Condon C, Kelleher D, Smith OP, Mulcahy F. Late onset hepatitis and prolonged deterioration in hepatic function associated with nevirapine therapy. Int J STD AIDS 2000;11(5):336–7.

16. Prakash M, Poreddy V, Tiyyagura L, Bonacini M. Jaundice and hepatocellular damage associated with nevirapine therapy. Am J Gastroenterol 2001;96(5):1571–4.

17. Johnson S, Baraboutis JG, Sha BE, Proia LA, Kessler HA. Adverse effects associated with use of nevirapine in HIV postexposure prophylaxis for 2 health care workers. JAMA 2000;284(21):2722–3.

18. Stern JO, Robinson PA, Love J, Lanes S, Imperiale MS, Mayers DL. A comprehensive hepatic safety analysis of nevirapine in different populations of HIV infected patients. J Acquir Immune Defic Syndr 2003;34 Suppl 1:S21-33.

19. Bjornsson E, Olsson R. Suspected drug-induced liver fatalities reported to the WHO database. Dig Liver Dis 2006;38(1):33–8.

20. Bonjoch A, Paredes R, Domingo P, Cervantes M, Pedrol E, Ribera E, Force L, Llibre JM, Vilaró J, Dalmau D, Cucurull J, Mascaró J, Masabeu A, Pérez-Alvarez N, Puig J, Cinquegrana D, Clotet B. Long-term safety and efficacy of nevirapine-based approaches in HIV type 1-infected patients. AIDS Res Hum Retroviruses 2006;22(4):321–9.

21. Lyons F, Hopkins S, Kelleher B, McGeary A, Sheehan G, Geoghegan J, Bergin C, Mulcahy FM, McCormick PA. Maternal hepatotoxicity with nevirapine as part of combination antiretroviral therapy in pregnancy. HIV Med 2006;7(4):255–60.

22. Buyse S, Vibert E, Sebagh M, Antonini T, Ichai P, Castaing D, Samuel D, Duclos-Vallée JC. Liver transplantation for fulminant hepatitis related to nevirapine therapy. Liver Transpl 2006;12(12):1880–2.

23. van Leth F, Phanuphak P, Ruxrungtham K, Baraldi E, Miller S, Gazzard B, Cahn P, Lalloo UG, van der Westhuizen IP, Malan DR, Johnson MA, Santos BR, Mulcahy F, Wood R, Levi GC, Reboredo G, Squires K, Cassetti I, Petit D, Raffi F, Katlama C, Murphy RL, Horban A, Dam JP, Hassink E, van Leeuwen R, Robinson P, Wit FW, Lange JM; 2NN Study team. Comparison of first-line antiretroviral therapy with regimens including nevirapine, efavirenz, or both drugs, plus stavudine and lamivudine: a randomised open-label trial, the 2NN Study. Lancet 2004;363(9417):1253–63.

24. Manfredi R, Calza L, Chiodo F. Efavirenz versus nevirapine in current clinical practice: a prospective, open-label observational study. J Acquir Immune Defic Syndr 2004;35(5):492–502.

25. Patel SM, Johnson S, Belknap SM, Chan J, Sha BE, Bennett C. Serious adverse cutaneous and hepatic toxicities associated with nevirapine use by non-HIV-infected individuals. J Acquir Immune Defic Syndr 2004;35(2):120–5.

26. Dieterich DT, Robinson PA, Love J, Stern JO. Drug-induced liver injury associated with the use of nonnucleoside reverse-transcriptase inhibitors. Clin Infect Dis 2004;38 Suppl 2:S80-9.

27. Almond LM, Boffito M, Hoggard PG, Bonora S, Raiteri R, Reynolds HE, Garazzino S, Sinicco A, Khoo SH, Back DJ, Di Perri G. The relationship between nevirapine plasma concentrations and abnormal liver function tests. AIDS Res Hum Retroviruses 2004;20(7):716–22.

28. Dailly E, Billaud E, Reliquet V, Breurec S, Perre P, Leautez S, Jolliet P, Bourin M, Raffi F. No relationship between high nevirapine plasma concentration and hepatotoxicity in HIV-1-infected patients naive of antiretroviral treatment or switched from protease inhibitors. Eur J Clin Pharmacol 2004;60(5):343–8.

29. Macias J, Castellano V, Merchante N, Palacios RB, Mira JA, Saez C, Garcia-Garcia JA, Lozano F, Gomez-Mateos JM, Pineda JA. Effect of antiretroviral drugs on liver fibrosis in HIV-infected patients with chronic hepatitis C: harmful impact of nevirapine. AIDS 2004;18(5):767–74.

30. Leith J, Piliero P, Storfer S, Mayers D, Hinzmann R. Appropriate use of nevirapine for long-term therapy. J Infect Dis 2005;192(3):545-6; author reply 546.

31. Metry DW, Lahart CJ, Farmer KL, Hebert AA. Stevens–Johnson syndrome caused by the antiretroviral drug nevirapine. J Am Acad Dermatol 2001;44(Suppl 2):354–7.

32. Bersoff-Matcha SJ, Miller WC, Aberg JA, van Der Horst C, Hamrick HJ Jr, Powderly WG, Mundy LM. Sex differences in nevirapine rash. Clin Infect Dis 2001;32(1):124–9.

33. Knudtson E, Para M, Boswell H, Fan-Havard P. Drug rash with eosinophilia and systemic symptoms syndrome and renal toxicity with a nevirapine-containing regimen in a pregnant patient with human immunodeficiency virus. Obstet Gynecol 2003;101:1094–7.

34. Verweel G, Sharland M, Lyall H, Novelli V, Gibb DM, Dumont G, Ball C, Wilkins E, Walters S, Tudor-Williams G. Nevirapine use in HIV-1-infected children. AIDS 2003;17:1639–47.

35. De Maat MM, Ter Heine R, Mulder JW, Meenhorst PL, Mairuhu AT, Van Gorp EC, Huitema AD, Beijnen JH. Incidence and risk factors for nevirapine-associated rash. Eur J Clin Pharmacol 2003;59:457–62.

36. Jackson JB, Barnett S, Piwowar-Manning E, Apuzzo L, Raines C, Hendrix C, Hamzeh F, Gallant J. A phase I/II study of nevirapine for pre-exposure prophylaxis of HIV-1 transmission in uninfected subjects at high risk. AIDS 2003;17:547–53.

37. Moodley D, Moodley J, Coovadia H, Gray G, McIntyre J, Hofmyer J, Nikodem C, Hall D, Gigliotti M, Robinson P, Boshoff L, Sullivan JL; South African Intrapartum Nevirapine Trial (SAINT) Investigators. A multicenter randomized controlled trial of nevirapine versus a combination of zidovudine and lamivudine to reduce intrapartum and early postpartum mother-to-child transmission of human immunodeficiency virus type 1. J Infect Dis 2003;187:725–35.

38. Jackson JB, Musoke P, Fleming T, Guay LA, Bagenda D, Allen M, Nakabiito C, Sherman J, Bakaki P, Owor M, Ducar C, Deseyve M, Mwatha A, Emel L, Duefield C, Mirochnick M, Fowler MG, Mofenson L, Miotti P, Gigliotti M, Bray D, Mmiro F. Intrapartum and neonatal single-dose nevirapine compared with zidovudine for prevention of mother-to-child transmission of HIV-1 in Kampala, Uganda: 18-month follow-up of the HIVNET 012 randomised trial. Lancet 2003;362:859–68.

39. Taha TE, Kumwenda NI, Gibbons A, Broadhead RL, Fiscus S, Lema V, Liomba G, Nkhoma C, Miotti PG, Hoover DR. Short postexposure prophylaxis in newborn babies to reduce mother-to-child transmission of HIV-1: NVAZ randomised clinical trial. Lancet 2003;362:1171–7.

40. Gangar M, Arias G, O'Brien JG, Kemper CA. Frequency of cutaneous reactions on rechallenge with nevirapine and delavirdine. Ann Pharmacother 2000;34(7–8):839–42.

41. Luzuriaga K, Bryson Y, McSherry G, Robinson J, Stechenberg B, Scott G, Lamson M, Cort S, Sullivan JL. Pharmacokinetics, safety, and activity of nevirapine in human immunodeficiency virus type 1-infected children. J Infect Dis 1996;174(4):713–21.

42. Ho TT, Wong KH, Chan KC, Lee SS. High incidence of nevirapine-associated rash in HIV-infected Chinese. AIDS 1998;12(15):2082–3.

43. Blanckenberg DH, Wood R, Horban A, Beniowski M, Boron-Kaczmarska A, Trocha H, Halota W, Schmidt RE, Fatkenheuer G, Jessen H, Lange JM; CHARM Study Group. Evaluation of nevirapine and/or hydroxyurea with nucleoside reverse transcriptase inhibitors in treatment-naive HIV-1-infected subjects. AIDS 2004;18(4):631–40.

44. Rey D, Partisani M, Krantz V, Kempf G, Nicolle M, de Mautort E, Priester M, Bernard-Henry C, Lang JM.

Prednisolone does not prevent the occurrence of nevirapine-induced rashes. AIDS 1999;13(16):2307.

45. Wetterwald E, Le Cleach L, Michel C, David F, Revuz J. Nevirapine-induced overlap Stevens–Johnson syndrome/toxic epidermal necrolysis. Br J Dermatol 1999;140(5):980–2.

46. Barreiro P, Soriano V, Gonzalez-Lahoz J, Colebunders R, Schrooten W, Desmet P, De Roo A, Dreezen C. Prevention of nevirapine-associated rash. Lancet 2001;357(9253):392–3.

47. Barreiro P, Soriano V, Casas E, Estrada V, Tellez MJ, Hoetelmans R, de Requena DG, Jimenez-Nacher I, Gonzalez-Lahoz J. Prevention of nevirapine-associated exanthema using slow dose escalation and/or corticosteroids. AIDS 2000;14(14):2153–7.

48. Hitti J, Frenkel LM, Stek AM, Nachman SA, Baker D, Gonzalez-Garcia A, Provisor A, Thorpe EM, Paul ME, Foca M, Gandia J, Huang S, Wei LJ, Stevens LM, Watts DH, McNamara J; PACTG 1022 Study Team. Maternal toxicity with continuous nevirapine in pregnancy: results from PACTG 1022. J Acquir Immune Defic Syndr 2004;36(3):772–6.

49. Public Health Service Task Force. Recommendations for use of antiretroviral drugs in pregnant HIV-1-infected women for maternal health and interventions to reduce perinatal HIV-1 transmission in the United States. http://aidsinfo.nih.gov/ContentFiles/PerinatalGL.pdf.

50. Shapiro RL, Holland DT, Capparelli E, Lockman S, Thior I, Wester C, Stevens L, Peter T, Essex M, Connor JD, Mirochnick M. Antiretroviral concentrations in breast-feeding infants of women in Botswana receiving antiretroviral treatment. J Infect Dis 2005;192(5):720–7.

51. Nunez M, Gonzalez-Requena D, Gonzalez-Lahoz J, Soriano V. Short communication: interactions between nevirapine plasma levels, chronic hepatitis C, and the development of liver toxicity in HIV-infected patients. AIDS Res Hum Retroviruses 2003;19:187–8.

52. João EC, Calvet GA, Menezes JA, D'Ippolito MM, Cruz ML, Salgado LA, Matos HJ. Nevirapine toxicity in a cohort of HIV-1-infected pregnant women. Am J Obstet Gynecol 2006;194(1):199–202.

53. Heelon MW, Meade LB. Methadone withdrawal when starting an antiretroviral regimen including nevirapine. Pharmacotherapy 1999;19(4):471–2.

54. Clarke SM, Mulcahy FM, Tjia J, Reynolds HE, Gibbons SE, Barry MG, Back DJ. Pharmacokinetic interactions of nevirapine and methadone and guidelines for use of nevirapine to treat injection drug users. Clin Infect Dis 2001;33(9):1595–7.

55. Mildvan D, Yarrish R, Marshak A, Hutman HW, McDonough M, Lamson M, Robinson P. Pharmacokinetic interaction between nevirapine and ethinyl estradiol/norethindrone when administered concurrently to HIV-infected women. J Acquir Immune Defic Syndr 2002;29(5):471–7.

56. Back D, Gibbons S, Khoo S. Pharmacokinetic drug interactions with nevirapine. J Acquir Immune Defic Syndr 2003;34(Suppl 1):S8–S14.

57. de Maat MM, Hoetelmans RM, Math t RA, van Gorp EC, Meenhorst PL, Mulder JW, Beijnen JH. Drug interaction between St John's wort and nevirapine. AIDS 2001;15(3):420–1.

58. Dionisio D, Mininni S, Bartolozzi D, Esperti F, Vivarelli A, Leoncini F. Need for increased dose of warfarin in HIV patients taking nevirapine. AIDS 2001;15(2):277–8.

Non-nucleoside reverse transcriptase inhibitors (NNRTIs)

See also Delavirdine, Efavirenz, Nevirapine

General Information

The non-nucleoside analogue reverse transcriptase inhibitors (NNRTIs) include delavirdine, efavirenz, and nevirapine (all rINNs). The pharmaceutical chemistry and uses of the first- and second-generation NNRTIs have been reviewed (1).

In a non-randomized study of 694 patients taking two NRTIs, who were also given either nevirapine or efavirenz the reasons for discontinuation were as shown in Table 1 (2). In passing, it should be noted that although this paper contains details of adverse events, there is no way of knowing that from a perusal of the title or abstract. Thus, anyone searching online for data for a systematic review of the adverse effects of these drugs might miss this information. When adverse events data are included in accounts of trials, the fact should be mentioned in the title and/or abstract.

Observational studies

Combinations of NNRTIs

Efavirenz and nevirapine, have been compared in a large 48-week unblinded randomized trial that included once-daily and twice-daily arms for nevirapine (400 mg/day and 200 mg bd) (the 2NN trial) (3). This trial distinguished between adverse effects of the individual drug and adverse effects of the drug class. In patients taking double NNRTI treatment there was an excess of treatment interruptions due to adverse effects without evidence of better virological or immunological responses. Antiretroviral regimens that include two NNRTIs should therefore not be used.

Combinations of NRTIs and NNRTIs

The efficacy and safety of abacavir (NRTI) and efavirenz (NNRTI) plus background therapy have been retrospectively evaluated in 50 patients, who had previously been treated with HAART (4). There was some immunological benefit, albeit limited, in most of the patients. Adverse effects were not mentioned in detail, but the dropout rate during the first 4 weeks of treatment was high, owing to skin rashes and hypersensitivity reactions.

Of 80 injecting drug users who took two NRTIs + nevirapine (NNRTI), 20 discontinued treatment and 20 had treatment failures (5). Adverse events led to withdrawal of therapy in 10 patients: eight had a rash (in one case Stevens–Johnson syndrome), one had liver failure thought to be unrelated to the drugs; and one had "general intolerance." Two patients taking methadone had opioid withdrawal symptoms, presumably due to induction of methadone metabolism by nevirapine.

Combinations of NRTIs, NNRTIs, and protease inhibitors

In an open, 48-week, single-arm, multicenter phase II study in 99 patients abacavir 300 mg bd, amprenavir 1200 mg bd, and efavirenz 600 mg/day were associated with rashes in 50 patients, possibly because of abacavir hypersensitivity; 17 permanently discontinued one or more drugs as a result (6). Other adverse effects included nausea ($n = 41$), diarrhea ($n = 27$), sleep disorders ($n = 27$), dizziness ($n = 25$), fatigue ($n = 23$), hypertriglyceridemia ($n = 18$), neutropenia ($n = 8$), hyperamylasemia ($n = 4$), leukopenia ($n = 3$), hypercholesterolemia ($n = 2$), raised alkaline phosphatase ($n = 1$), and raised aspartate transaminase ($n = 1$).

In infants and children assigned to different combinations of one or two NRTIs plus an NNRTI and/or a protease inhibitor, the numbers of patients with moderate or severe adverse events were as shown in Table 2 (7). In cases of rash, the rash was worse in those who were taking nevirapine-containing regimens.

Organs and Systems

Liver

In 13 patients, studied retrospectively, who were given a combination of two NNRTIs, nevirapine + efavirenz, the most common adverse event was a change in liver function tests, to almost three times the upper end of the reference range in three cases; however, it was directly attributable to the antiretroviral drugs in only one case (8).

The incidence of NNRTI-induced hepatotoxicity in HIV-infected patients and the effect of co-infection with hepatitis B or hepatitis C virus have been studied in 272 patients (9). The NNRTIs used were delavirdine ($n = 40$), efavirenz ($n = 91$), and nevirapine ($n = 141$). There were mild to moderate rises in serum aspartate transaminase or alanine transaminase in 81 patients and large rises in three patients, one of whom was taking efavirenz and two nevirapine. Hepatitis B or hepatitis C virus infection did not significantly increase the risk of hepatotoxicity. The authors concluded that NNRTIs cause little hepatotoxicity, despite the presence of co-infection with hepatitis B or hepatitis C viruses.

Table 1 Numbers (%) of patients who discontinued therapy in a follow-up study of two NRTIs plus either nevirapine or efavirenz

Reason for withdrawal	Nevirapine (n = 460)	Efavirenz (n = 234)
Patient-related reasons*	47 (6.7)	16 (6.8)
Failure of therapy	31 (10.2)	5 (2.1)
Adverse events		
Nervous system	1 (0.2)	12 (5.1)
Metabolism (lipodystrophy)	3 (0.7)	0
Gastrointestinal	12 (2.6)	1 (0.4)
Immunologic	56 (12.2)	5 (2.1)
Laboratory adverse events		
Hematologic	3 (0.7)	1 (0.4)
Liver	9 (2.0)	1 (0.4)

*Poor adherence, patient's decision, or therapy simplification.

Table 2   Numbers (%) of infants and children with moderate or severe adverse events in a study of one or two NRTIs plus an NNRTI and/or a protease inhibitor

System affected by the adverse event	Regimen			
	Stavudine + nevirapine + ritonavir (n = 41)	Stavudine + nevirapine + nelfinavir (n = 44)	Stavudine + lamivudine + nevirapine + nelfinavir (n = 44)	Stavudine + lamivudine + nelfinavir (n = 52)
Respiratory	12 (29)	18 (41)	23 (52)	50 (96)
Hematologic (neutropenia)	17 (41)	9 (20)	14 (32)	23 (44)
Gastrointestinal				
Nausea/vomiting	29 (71)	32 (73)	18 (41)	15 (29)
Other	10 (24)	25 (57)	18 (41)	19 (37)
Liver	12 (29)	14 (32)	18 (41)	23 (44)
Skin	27 (66)	41 (93)*	32 (73)	17 (33)*
Body temperature (fever)	24 (59)	30 (68)*	20 (45)	10 (19)*

*Significantly different from the other groups.

References

1. Campiani G, Ramunno A, Maga G, Nacci V, Fattorusso C, Catalanotti B, Morelli E, Novellino E. Non-nucleoside HIV-1 reverse transcriptase (RT) inhibitors: past, present, and future perspectives. Curr Pharm Des 2002;8(8):615–57.

2. Cozzi-Lepri A, Phillips AN, d'Arminio Monforte A, Piersantelli N, Orani A, Petrosillo N, Leoncini F, Scerbo A, Tundo P, Abrescia N, Moroni M. Italian Cohort Naive Antiretrovirals (I.Co.N.A.) Study Group. Virologic and immunologic response to regimens containing nevirapine or efavirenz in combination with 2 nucleoside analogues in the Italian Cohort Naive Antiretrovirals (I.Co.N.A.) study J Infect Dis 2002;185(8):1062–9.

3. van Leth F, Phanuphak P, Ruxrungtham K, Baraldi E, Miller S, Gazzard B, Cahn P, Lalloo UG, van der Westhuizen IP, Malan DR, Johnson MA, Santos BR, Mulcahy F, Wood R, Levi GC, Reboredo G, Squires K, Cassetti I, Petit D, Raffi F, Katlama C, Murphy RL, Horban A, Dam JP, Hassink E, van Leeuwen R, Robinson P, Wit FW, Lange JM; 2NN Study team. Comparison of first-line antiretroviral therapy with regimens including nevirapine, efavirenz, or both drugs, plus stavudine and lamivudine: a randomised open-label trial, the 2NN Study. Lancet 2004;363(9417):1253-63.

4. Wasmuth JC, Herhaus C, Romer K, Salzberger B, Kaiser R, Schliefer K, Voigt E, Rockstroh JK. Efficacy and safety of abacavir plus efavirenz as a salvage regimen in HIV-infected individuals after 48 weeks. AIDS 2002;16(7):1077–8.

5. Zaccarelli M, Barracchini A, De Longis P, Perno CF, Soldani F, Liuzzi G, Serraino D, Ippolito G, Antinori A. Factors related to virologic failure among HIV-positive injecting drug users treated with combination antiretroviral therapy including two nucleoside reverse transcriptase inhibitors and nevirapine. AIDS Patient Care STDS 2002;16(2):67–73.

6. Falloon J, Ait-Khaled M, Thomas DA, Brosgart CL, Eron JJ Jr, Feinberg J, Flanigan TP, Hammer SM, Kraus PW, Murphy R, Torres R, Masur H, Manion DJ, Rogers M, Wolfram J, Amphlett GE, Rakik A, Tisdale M. CNA2007 Study Team. HIV-1 genotype and phenotype correlate with virological response to abacavir, amprenavir and efavirenz in treatment-experienced patients. AIDS 2002;16(3):387–96.

7. Krogstad P, Lee S, Johnson G, Stanley K, McNamara J, Moye J, Jackson JB, Aguayo R, Dieudonne A, Khoury M, Mendez H, Nachman S, Wiznia A, Ballow A, Aweeka F, Rosenblatt HM, Perdue L, Frasia A, Jeremy R, Anderson M, Japour A, Fields C, Farnsworth A, Lewis R, Schnittman S, Gigliotti M, Maldonaldo S, Lane B, Hernandez JE, et al. Pediatric AIDS Clinical Trials Group 377 Study Team. Nucleoside-analogue reverse-transcriptase inhibitors plus nevirapine, nelfinavir, or ritonavir for pre-treated children infected with human immunodeficiency virus type 1. Clin Infect Dis 2002;34(7):991–1001.

8. Olivieri J. Nevirapine + efavirenz based salvage therapy in heavily pretreated HIV infected patients Sex Transm Infect 2002;78(1):72–3.

9. Palmon R, Koo BC, Shoultz DA, Dieterich DT. Lack of hepatotoxicity associated with nonnucleoside reverse transcriptase inhibitors. J Acquir Immune Defic Syndr 2002;29(4):340–5.

Nucleoside analogue reverse transcriptase inhibitors (NRTIs)

See also Abacavir, Didanosine, Lamivudine, Tenofovir, Zalcitabine, Zidovudine

General Information

The nucleoside analogue reverse transcriptase inhibitors (NRTIs) include abacavir, didanosine, lamivudine, stavudine, tenofovir, zalcitabine, and zidovudine (all rINNs). The following abbreviations have been used and may still be encountered in published papers:

- DDI: didanosine (inosine analog)
- 3TC: lamivudine (cytidine analog)
- D4T: stavudine (thymidine analog)
- DDC: zalcitabine (cytidine analog)
- ZDV: zidovudine (thymidine analog)

The abbreviation AZT should not be used to designate zidovudine, since it can be, and has been, confused with azathioprine (1).

To be converted into an active compound, all the NRTIs have to be phosphorylated intracellularly to the triphosphate form via monophosphates and diphosphates. The triphosphates compete with cellular nucleotides and inhibit HIV reverse transcriptase by introducing a chain terminator into the growing complementary DNA strand during reverse transcription. However, this mechanism also occurs when human DNA is transcribed by the human DNA polymerase. In fact, the first NRTI in clinical use (zidovudine) was initially developed as an anticancer drug, targeting human DNA polymerase in cancer cells. However, all NRTIs have higher specificity for viral reverse transcriptase than for the human DNA polymerase.

Observational studies

Combinations of different NRTIs

In a 24-week open, single-arm trial, 108 antiretroviral therapy-naive, HIV-infected prisoners were given a combination tablet of lamivudine + zidovudine (150 mg/300 mg) and a tablet of abacavir 300 mg bd (2). The plasma HIV-1 RNA concentration remained at 400 copies/ml or less in 85% of the patients and at less than 50 copies/ml in 75%. Nausea was the most common adverse effect ($n = 40$). Four patients withdrew prematurely because of one or more of the following: abdominal discomfort and pain; abdominal distension; neutropenia; malaise or fatigue; nausea and vomiting. Two patients had a suspected hypersensitivity reaction to abacavir and were withdrawn.

A combination of nucleosides analogues, stavudine 2 mg/kg/day plus didanosine 180 mg/m^2/day for 48 weeks, has been assessed in 16 asymptomatic children, median age 6.5 years, with HIV infection (3). They had plasma HIV RNA concentrations below 50 000 copies/ml and CD4 counts above 15%. Plasma HIV RNA was reduced to 400 copies/ml or less in 43% and 44% of cases at 24 and 48 weeks respectively. There were minor adverse effects in two-thirds of the children; none led to drug withdrawal. There were no cases of lipoatrophy. There were no resistance mutations linked to didanosine (L74V or M184V) or stavudine (V75T) nor multinucleoside resistant genotypes (151 complex or 69 inserts). However, four children developed zidovudine-like mutations, T215Y and/or M41L.

In a retrospective observational cohort study of 730 patients who took abacavir + lamivudine + zidovudine, treatment was withdrawn in 104 patients because of severe adverse events; of these there were 30 hematological, 21 gastrointestinal, six hepatic, four due to malaise, three due to mitochondrial toxicity, one myopathy, one migraine, one fever, and one rash (4).

Combinations of NRTIs and NNRTIs

The efficacy and safety of abacavir (NRTI) and efavirenz (NNRTI) plus background therapy have been retrospectively evaluated in 50 patients, who had previously been treated with highly active antiretroviral therapy (HAART) (5). There was some immunological benefit, albeit limited, in most of the patients. Adverse effects were not mentioned in detail, but the dropout rate during the first 4 weeks of treatment was high, owing to skin rashes and hypersensitivity reactions.

Of 80 injecting drug users who took two NRTIs + nevirapine (NNRTI), 20 discontinued treatment and 20 had treatment failures (6). Adverse events led to withdrawal of therapy in 10 patients: eight had a rash (in one case Stevens–Johnson syndrome); one had liver failure thought to be unrelated to the drugs; and one had "general intolerance." Two patients taking methadone had opioid withdrawal symptoms, presumably due to induction of methadone metabolism by nevirapine.

Combinations of NRTIs and protease inhibitors

A variable-dose plasma concentration-controlled approach to combination antiretroviral therapy (zidovudine, lamivudine, and indinavir) has been compared with conventional fixed-dose therapy in 40 patients in a randomized, 52-week, open trial (7). Significantly more concentration-controlled recipients achieved the desired concentrations for all three drugs: there was a good response in 15 of 16 concentration-controlled recipients compared with nine of 17 conventional regimen recipients. However, there was no difference in the occurrence of drug-related clinical events or laboratory abnormalities between the two regimens. Three patients withdrew because of gastrointestinal adverse effects, one because of a peripheral neuropathy, and one with headache and anemia. There was nephrolithiasis in four cases.

The combination of indinavir + ritonavir 400/400 mg bd plus two NRTIs has been studied in 93 patients in an open, uncontrolled, multicenter trial (8). Raised triglycerides ($n = 78$) and cholesterol ($n = 63$) were the commonest adverse effects, followed by nausea ($n = 22$) and circumoral paresthesia ($n = 9$). Withdrawal was required in four cases of nausea, four of lipodystrophy, one of diarrhea, and one of osteonecrosis.

In a multicenter, open, uncontrolled trial of protease inhibitors in conjunction with NRTIs for at least 96 weeks in 32 children, the pharmacokinetics of indinavir and nelfinavir showed large interindividual differences (9). In all, 17 children suffered adverse events. The most common adverse events in those taking indinavir were diarrhea ($n = 6$), vomiting ($n = 6$), anorexia ($n = 5$), hematuria ($n = 5$), abdominal pain ($n = 4$), and headache ($n = 3$); most were mild and occurred early in treatment. The rates of drug-related adverse effects were 0.4 per patient-year in those taking indinavir and 0.16 per patient-year in those taking nelfinavir.

Combinations of NRTIs, NNRTIs, and protease inhibitors

In an open 48-week, single-arm, multicenter phase II study in 99 patients abacavir 300 mg bd, amprenavir 1200 mg bd, and efavirenz 600 mg/day were associated with rashes in 50 patients, possibly because of abacavir hypersensitivity; 17 permanently discontinued one or more drugs as a result (10). Other adverse effects included nausea ($n = 41$), diarrhea ($n = 27$), sleep disorders ($n = 27$), dizziness ($n = 25$), fatigue ($n = 23$), hypertriglyceridemia ($n = 18$), neutropenia ($n = 8$), hyperamylasemia ($n = 4$), leukopenia ($n = 3$), hypercholesterolemia ($n = 2$), raised alkaline phosphatase ($n = 1$), and raised aspartate transaminase ($n = 1$).

In infants and children assigned to different combinations of one or two NRTIs plus an NNRTI and/or a protease inhibitor, the numbers of patients with moderate or severe adverse events were as shown in Table 1 (11). In cases of rash, the rash was worse in those who were taking nevirapine-containing regimens.

General adverse effects

Almost all NRTIs cause non-specific gastrointestinal and general adverse effects (nausea, sleep disturbances, headache) that are difficult to attribute to a single agent and usually disappear after the first 2–4 weeks of therapy.

Mitochondrial toxicity associated with NRTIs

Initially, most of the adverse effects seen with zidovudine use (in particular hematological effects) were attributed to interference with cellular DNA replication. However, DNA replication also occurs in mitochondria. Mitochondrial DNA encodes some of the enzymes used for oxidative phosphorylation. Only recently has it been hypothesized that inhibition of this pathway could lead to mitochondrial toxicity and be responsible for most of the toxicity seen with NRTIs, including polyneuropathy, myopathy, cardiomyopathy, steatosis, lactic acidosis, exocrine pancreas failure, bone marrow failure, and proximal tubular dysfunction (12). These adverse effects are also a compilation of the clinical features seen in several genetic mitochondrial cytopathies.

In vitro, NRTIs reduce mitochondrial DNA, most likely by inhibiting mitochondrial DNA polymerase gamma. Heterogeneous toxicity profiles of different NRTIs in vivo may be related to variable tissue sensitivity, cell entry, and drug phosphorylation. Therefore, each NRTI has a specific adverse effect profile (Table 2 (12)), but any feature of mitochondrial toxicity must be considered with all NRTIs. A major problem of mitochondrial toxicity is its delayed onset, sometimes after several months of treatment.

Organs and Systems

Cardiovascular

Cardiomyopathy is a rare adverse effect that has been observed in patients treated with didanosine, zidovudine, and zalcitabine (13). In a retrospective, case-control study, cardiomyopathy was 8.4 (95% CI = 1.7, 42) times more likely to develop in children who had previously used zidovudine than in children who had never been exposed to it (14).

Nervous system

Peripheral neuropathy is often observed in patients treated with didanosine, stavudine, and zalcitabine (15,16).

Sensory systems

In a report of ototoxicity in a patient taking three antiretroviral drugs, the authors suggested that nucleoside analogues should be avoided in patients with known neurotoxicity.

- A 23-year-old woman who took stavudine, lamivudine, and nevirapine had nausea and diarrhea during the first few days and within 2 weeks developed sudden bilateral hearing loss, dizziness, and tinnitus (17). The hearing defect was sensorineural. CT and MRI scans were normal. Prednisolone was ineffective.

Table 1 Numbers (%) of infants and children with moderate or severe adverse events in a study of one or two NRTIs plus an NNRTI and/or a protease inhibitor

System affected by the adverse event	Regimen			
	Stavudine + nevirapine + ritonavir (n = 41)	Stavudine + nevirapine + nelfinavir (n = 44)	Stavudine + lamivudine + nevirapine + nelfinavir (n = 44)	Stavudine+lamivudine + nelfinavir (n = 52)
Respiratory	12 (29)	18 (41)	23 (52)	50 (96)
Hematologic (neutropenia)	17 (41)	9 (20)	14 (32)	23 (44)
Gastrointestinal				
Nausea/vomiting	29 (71)	32 (73)	18 (41)	15 (29)
Other	10 (24)	25 (57)	18 (41)	19 (37)
Liver	12 (29)	14 (32)	18 (41)	23 (44)
Skin	27 (66)	41 (93)*	32 (73)	17 (33)*
Body temperature (fever)	24 (59)	30 (68)*	20 (45)	10 (19)*

*Significantly different from the other groups

Table 2 Spectrum of adverse effects of nucleoside analogue reverse transcriptase inhibitors

Adverse effects	Didanosine	Lamivudine	Stavudine	Zalcitabine	Zidovudine
Cardiomyopathy	+	−	+	−	+
Neuropathy	++	−	++	++	−
Lactic acidosis	−	−	+	+	+
Bone marrow toxicity	+	−	+	−	++
Hepatic steatosis	−	±	+	+	+
Pancreatitis	−	±	++	+	−
Myopathy	−	−	−	−	++

++ most prominent adverse effect
+ observed adverse effect
± possible adverse effect
− adverse effect typically not observed

Metabolism

Lipodystrophy

The prevalence of lipodystrophy has been studied during follow-up for 30 months of previously untreated patients who had been randomized to receive different nucleoside analogue combinations for 6 months in the ALBI–ANRS 070 trial (18). After 30 months 37 of 120 patients who had used nucleoside analogues with or without other antiretroviral drugs had at least one morphological change, and 21 of those had isolated peripheral lipoatrophy; the corresponding values for the patients who used only nucleoside analogues during follow-up were 20 of 66 and 14 of 21 respectively. The factors associated with lipodystrophy were initial assignment to stavudine plus didanosine compared with zidovudine plus lamivudine (OR = 6.7), age below 10 years (OR = 3.6), and HIV RNA concentration at month 30 (OR = 0.4). There were no differences in cholesterol and glucose concentrations. Thus, exposure to stavudine and didanosine was associated with lipodystrophy (predominantly lipoatrophy).

In a well-controlled French comparison of changes in body composition, body fat distribution, and insulin secretion, patients taking stavudine ($n = 27$) or zidovudine ($n = 16$) were compared with controls ($n = 15$) (19). The zidovudine group and the control group had similar body composition and regional fat distribution. Stavudine was associated with a significantly lower percentage of body fat than zidovudine (13 versus 15%), a markedly lower ratio of subcutaneous to visceral fat, and a higher mean intake of fat and cholesterol. Triglyceride concentrations were significantly higher with stavudine than in the controls, but did not differ between stavudine and zidovudine or between zidovudine and controls. Free fatty acids tended to be higher with stavudine. Lipodystrophy was observed in 17 patients taking stavudine, and in three taking zidovudine after a median time of 14 months. The relative risk of fat wasting with stavudine compared with zidovudine was 1.95. Five of twelve patients had a major or mild improvement in their lipodystrophy after stavudine was withdrawn. The authors concluded that lipodystrophy may be related to long-term NRTI therapy, particularly if it includes stavudine.

In 39 HIV-1 infected children who started treatment with stavudine, lamivudine, and nevirapine, lipodystrophy occurred in 11 children after a median of 49 months (20).

Improvement of lipoatrophy can occur after switching from stavudine to abacavir. In the MITOX extension study in 85 patients, in which there was long-term follow up for 104 weeks after switching from stavudine or zidovudine to abacavir, limb fat continued to increase in the abacavir group (21). There was no improvement in visceral fat accumulation, buffalo hump, self-assessed lipodystrophy, or the lipodystrophy case definition score.

In the 48-week TARHEEL study (Trial to Assess the Regression of Hyperlactatemia and to Evaluate the Regression of Established Lipodystrophy), substitution of stavudine by abacavir or zidovudine in 118 patients who had developed lipoatrophy after taking a stavudine-based regimen for at least 6 months resulted in a median increase in arm fat of 35%, leg fat of 12%, and trunk fat of 12% compared with baseline (22).

Mechanism

Studies in vitro and in animals have suggested that depletion of mitochondrial DNA (mDNA) followed by mitochondrial damage represent an underlying mechanism that leads to both lipoatrophy and hyperlactatemia and have shown different severities of mitochondrial toxicity from drug to drug (zalcitabine Œ didanosine Œ stavudine > lamivudine > zidovudine > abacavir, tenofovir).

Clinical studies that have tested this hypothesis are not easy to interpret or to summarize, because of large variability in study designs, the patients studied (antiretroviral naïve up to heavily pre-treated patients), the treatments examined, different observation periods, different outcome measures (mDNA from different anatomical systems, lactate acidosis, measurement methods of lipoatrophy), and different combinations of measurements.

- A patient taking HAART including didanosine and zidovudine developed a syndrome that evolved over 4 months, starting with weight loss (25%), vomiting, polyuria, bone fracture due to osteoporosis, profound proximal weakness, and a stocking sensory neuropathy

(23). He had a Fanconi-type proximal renal tubular dysfunction, lactic acidosis, myopathy, and pancreatic dysfunction. All signs and symptoms improved markedly after withdrawal of the antiretroviral drugs.

This case demonstrates almost the full clinical spectrum of mitochondrial toxicity.

In a cross-sectional study in 45 patients who had been enrolled 4 years before in a randomized comparative trial of stavudine-based therapy versus zidovudine-based therapy, there was a significantly higher prevalence of lipoatrophy in the stavudine recipients (24). Analysing only those who continued to take randomly allocated NRTIs, mDNA in peripheral blood mononuclear cells fell after the start of treatment in both groups, resulting in significantly lower amounts in patients with lipoatrophy. Furthermore, the mDNA content of subcutaneous adipose tissue from the thigh, but not from the back, was significantly lower in patients taking stavudine.

In 60 patients taking a regimen containing stavudine + didanosine + nevirapine for 36 months, 10 developed lipoatrophy, mainly after 15–18 months of therapy. The authors attributed this to the mitochondrial toxicity of didanosine (25). There was one case of hyperlactatemia with peripheral neuropathy and diabetes mellitus, one case of cytolytic hepatitis, and one case of symptomatic pancreatitis.

In several studies mDNA in peripheral blood mononuclear cells and liver in HIV patients has been measured and treatment regimens containing stavudine, zalcitabine, and/or didanosine have been compared with regimens containing other NRTIs. There was a significant reduction in the number of mDNA copies from cryopreserved peripheral blood mononuclear cells over 48 weeks in patients who participated in the Delta trial and were treated with zidovudine + zalcitabine and zidovudine + didanosine, but not with zidovudine alone (26).

In 58 individuals with HIV infection, those taking stavudine-containing regimens and drug-naïve HIV-positive patients had significantly lower median amounts of mDNA than HIV-positive patients who were taking antiretroviral drugs other than stavudine, and this was associated with higher lactate concentrations (27).

The aim of another substudy from MITOX was to determine whether switching treatment to abacavir is associated with a significant increase in the mDNA content of peripheral blood mononuclear cells paralleling the consequent improvement in lipoatrophy (28). However, despite the improvement in peripheral lipoatrophy in patients whose treatment was switched to abacavir, there was no significant change in mDNA copy numbers in peripheral blood mononuclear cells over 24 weeks.

In summary, a reduced amount of mDNA seems to be associated with raised lactate, lipoatrophy, and medications such as stavudine, didanosine, and zalcitabine. Stavudine is associated with a higher risk of lipoatrophy, but the reduction in mDNA is not always associated with signs of lipoatrophy.

Given the central role of NRTIs in the development of lipoatrophy, switching to regimen containing other drugs

may lead to improvement, although reversibility seems only to be partial up to 2 years of follow up, anatomically restricted, and only detectable by techniques such as computed tomography or dual energy x-ray absorptiometry. However, there was neither an improvement in clinical parameters, such as waist–hip ratio, patient questionnaire, or lipoatrophy score, nor a correlation with increasing mDNA in peripheral blood mononuclear cells after switching treatment. This suggests either that other factors determine lipoatrophy or that there is persistent mitochondrial toxicity despite withdrawal of NRTIs.

Until now, no other options for treating lipoatrophy besides switching antiretroviral drug treatment has been found to be effective. Rosiglitazone, a thiazolidinedione hypoglycemic drug that promotes subcutaneous fat growth in 108 patients with type 2 diabetes did not improve lipoatrophy after 48 weeks in HIV-1 infected adults with lipoatrophy who were taking antiretroviral drugs, measured by changes in limb fat mass, body weight, body mass index, and waist and hip circumferences (29).

Depletion of mitochondrial DNA and morphological changes in adipocytes have been assessed in a small study of fat biopsies from HIV-negative patients (n = 6), HIV-positive but drug naïve patients (n = 11), and patients taking NRTIs (zidovudine (n = 9) or stavudine (n = 12)) (30). Drug-naïve HIV-infected patients had similar contents of mitochondrial DNA in adipocytes, while patients taking NRTIs had significantly reduced mean mitochondrial DNA content per cell. Compared with HIV-infected controls, mitochondrial DNA depletion was 45% or 87% in those taking zidovudine and stavudine respectively. These results support in vitro findings that stavudine causes more pronounced mitochondrial toxicity than zidovudine (31). .

Lactic acidosis

Lactic acidosis is a severe and potentially fatal form of mitochondrial toxicity. Metabolic stress or vitamin deficiencies (riboflavin, carnitine) might provoke it. There is suggestive evidence of clinical benefit with riboflavin therapy (32).

Antiretroviral nucleoside analogues have been associated with hepatic steatosis and lactic acidosis. These compounds require phosphorylation to active triphosphate derivatives by cellular phosphokinases. The triphosphate nucleotide inhibits the growing proviral DNA chain, but it also inhibits host DNA polymerases, and this can result in compensatory glycolysis and lactic acidosis. Abnormal mitochondrial oxidation of free fatty acids causes the accumulation of neutral fat in liver cells, and this manifests as hepatomegaly with macrovesicular steatosis. Hepatic steatosis and lactic acidosis have been reported previously with zidovudine, didanosine, zalcitabine, Combivir (zidovudine plus lamivudine), and lamivudine. Of 349 Australian patients studied for 18 months (516 patient-years) taking NRTIs only two had severe lactic acidosis (33).

Lactic acidosis is a consequence of long-term mitochondrial toxicity of NRTIs. In a systematic review of published cases of lactic acidosis, NRTI use and female sex were identified as significant risk factors (34). Among patients taking a triple drug regimen, all were taking stavudine as one of their NRTIs (52% in combination with didanosine). The most frequent clinical manifestations were gastrointestinal (nausea, vomiting, abdominal pain) in 50%, and 41% had dyspnea and tachypnea. The median lactate concentration in symptomatic patients was 11 mmol/l and liver enzymes were abnormal in 65%. Almost half of the patients died within a median period of 7 days.

Three cases of steatosis/lactic acidosis syndrome associated with stavudine plus lamivudine have been reported (35).

- A 37-year-old HIV-infected woman receiving stavudine, lamivudine, and indinavir developed epigastric pain, anorexia, and vomiting. She had lactic acidosis (serum lactate 4.9 mmol/l), raised liver enzymes, and an increased prothrombin time. She had hepatomegaly and tachypnea and required mechanical ventilation. Her progress was complicated by pancreatitis and acute respiratory distress syndrome. Antiviral medication was stopped and she was treated with co-enzyme Q, carnitine, and vitamin C. The serum lactic acid and transaminases returned to normal over 4 weeks and she was weaned off the ventilator after 4 months.
- A 40-year-old HIV-infected woman receiving stavudine, lamivudine, nelfinavir, and co-trimoxazole developed dyspnea, dysphagia, and vomiting with lactic acidosis (serum lactate 9.4 mmol/l) and hepatomegaly. Despite ventilation for respiratory failure she died after 5 days. Autopsy showed massive hepatomegaly with steatosis.
- A 36-year-old HIV-infected woman who had been receiving stavudine, saquinavir, ritonavir, and didanosine developed lactic acidosis (serum lactate 11.4 mmol/l) and hepatomegaly. She had acute pancreatitis and, despite ventilatory support for respiratory failure, died after 8 weeks.

There have been similar reports related to didanosine plus stavudine (36) and stavudine alone (37).

Routine monitoring of lactate is not recommended, since lactate concentrations do not correlate with symptoms and patients may have asymptomatic lactatemia. Furthermore, technical difficulties in blood collection and processing, and the lack of a standardized definition of lactic acidosis for patients taking NRTIs, prevent any routine monitoring recommendations in the absence of symptoms. In addition, it is possible that NRTIs also cause other forms of mitochondrial dysfunction.

In patients with stavudine-induced hyperlactatemia, withdrawal promptly resulted in a reduction in lactate concentrations. In 10 patients once hyperlactatemia resolved switching to either abacavir or zidovudine resulted in maintenance of normal lactate concentrations (38).

Hematologic

The NRTIs are toxic to hemopoietic progenitor cells in vitro (39) and can cause anemia (40,41), neutropenia (40), and thrombocytopenia (42,43).

Liver

In patients co-infected with HIV and hepatitis C, treatment with dideoxynucleoside analogues, such as didanosine and stavudine, was associated with hepatic steatosis (OR = 4.63; 95% CI = 1.55, 14) (44). The presence of steatosis was in turn associated with fibrosis (OR = 1.37; 95% CI = 1.03, 1.81).

Pancreas

Pancreatitis has been observed in patients treated with didanosine, lamivudine, stavudine, and zalcitabine (45), but its incidence is also increased in drug-naive patients with advanced HIV infection (46).

Immunologic

- A man developed several erythematous plaques on his face due to borderline tuberculoid leprosy with a reversal reaction (47). He had severe CD4 T cell lymphocytopenia due to HIV infection and had been given HAART. A fall in viral load and an increase in CD4 count preceded the development of the skin lesions, suggesting immune reconstitution as the underlying mechanism for the reversal reaction.

Paradoxical reactions are often observed in patients with pulmonary and extra-pulmonary tuberculosis being treated with HAART. Clinicians need to distinguish these from other adverse reactions related to drug therapy. Reversal reactions in leprosy are increasingly likely as more patients with HIV infection are treated with HAART in developing countries.

Second-Generation Effects

Pregnancy

Two HIV-1-positive women, both of whom had taken regimens containing stavudine and didanosine for at least 2 years, presented in the third trimester of pregnancy, one with acute lactic acidosis and one with acute pancreatitis and lactic acidosis (48). In the first case both mother and baby died. It is not known whether pregnancy is a risk factor for NRTI-induced lactic acidosis, perhaps in combination with riboflavin deficiency or a metabolic defect in the fetus, or whether NRTIs independently cause lactic acidosis through mitochondrial toxicity.

Drug–Drug Interactions

Methadone

In those who are opioid-dependent, methadone can facilitate adherence to HAART regimens. The pharmacokinetics of the tablet formulations of didanosine and stavudine have been studied in 17 individuals taking stable methadone therapy in comparison with 10 untreated controls (49). Methadone reduced the AUC_{0-6} by 63% for didanosine and by 25% for stavudine and the C_{max} by 66% and 44% respectively. These effects appeared to result primarily from reduced systemic availability. Trough concentrations of methadone were comparable to those seen in historical controls, suggesting that the nucleoside analogues did not affect methadone disposition. The authors concluded that larger doses of the tablet formulation (or another type of formulation) may be necessary to provide HAART in subjects taking methadone.

References

1. Ambrosini MT, Mandler HD, Wood CA. AZT: zidovudine or azathioprine? Lancet 1992;339(8798):935.
2. Kirkland LR, Fischl MA, Tashima KT, Paar D, Gensler T, Graham NM, Gao H, Rosenzweig JR, McClernon DR, Pittman G, Hessenthaler SM, Hernandez JE. NZTA4007 Study Team. Response to lamivudine–zidovudine plus abacavir twice daily in antiretroviral-naive, incarcerated patients with HIV infection taking directly observed treatment. Clin Infect Dis 2002;34(4):511–8.
3. de Mendoza C, Ramos JT, Ciria L, Fortuny C, Garcia FJ, de Jose MI, Asensi F, Soriano V. Efficacy and safety of stavudine plus didanosine in asymptomatic HIV-infected children with plasma HIV RNA below 50,000 copies per milliliter. HIV Clin Trials 2002;3(1):9–16.
4. Berenguer J, Pérez-Elías MJ, Bellón JM, Knobel H, Rivas-González P, Gatell JM, Miguélez M, Hernández-Quero J, Flores J, Soriano V, Santos I, Podzamczer D, Sala M, Camba M, Resino S; Spanish Abacavir, Lamivudine, and Zidovudine Cohort Study Group. Effectiveness and safety of abacavir, lamivudine, and zidovudine in antiretroviral therapy-naive HIV-infected patients: results from a large multicenter observational cohort. J Acquir Immune Defic Syndr 2006;41(2):154-9.
5. Wasmuth JC, Herhaus C, Romer K, Salzberger B, Kaiser R, Schliefer K, Voigt E, Rockstroh JK. Efficacy and safety of abacavir plus efavirenz as a salvage regimen in HIV-infected individuals after 48 weeks. AIDS 2002;16(7):1077–8.
6. Zaccarelli M, Barracchini A, De Longis P, Perno CF, Soldani F, Liuzzi G, Serraino D, Ippolito G, Antinori A. Factors related to virologic failure among HIV-positive injecting drug users treated with combination antiretroviral therapy including two nucleoside reverse transcriptase inhibitors and nevirapine. AIDS Patient Care STDS 2002;16(2):67–73.
7. Fletcher CV, Anderson PL, Kakuda TN, Schacker TW, Henry K, Gross CR, Brundage RC. Concentration-controlled compared with conventional antiretroviral therapy for HIV infection. AIDS 2002;16(4):551–60.
8. Lichterfeld M, Nischalke HD, Bergmann F, Wiesel W, Rieke A, Theisen A, Fatkenheuer G, Oette M, Carls H, Fenske S, Nadler M, Knechten H, Wasmuth JC, Rockstroh JK. Long-term efficacy and safety of ritonavir/indinavir at 400/400 mg twice a day in combination with two nucleoside reverse transcriptase inhibitors as first line antiretroviral therapy HIV Med 2002;3(1):37–43.
9. van Rossum AM, Geelen SP, Hartwig NG, Wolfs TF, Weemaes CM, Scherpbier HJ, van Lochem EG, Hop WC, Schutten M, Osterhaus AD, Burger DM, de Groot R. Results of 2 years of treatment with protease-inhibitor-containing antiretroviral therapy in Dutch children infected with human immunodeficiency virus type 1. Clin Infect Dis 2002;34(7):1008–16.
10. Falloon J, Ait-Khaled M, Thomas DA, Brosgart CL, Eron JJ Jr, Feinberg J, Flanigan TP, Hammer SM, Kraus PW, Murphy R, Torres R, Masur H, Manion DJ, Rogers M, Wolfram J, Amphlett GE, Rakik A, Tisdale M. CNA2007 Study Team. HIV-1 genotype and phenotype correlate with virological response to abacavir, amprenavir and efavirenz in treatment-experienced patients. AIDS 2002;16(3):387–96.
11. Krogstad P, Lee S, Johnson G, Stanley K, McNamara J, Moye J, Jackson JB, Aguayo R, Dieudonne A, Khoury M, Mendez H, Nachman S, Wiznia A, Ballow A, Aweeka F, Rosenblatt HM, Perdue L, Frasia A, Jeremy R, Anderson M, Japour A, Fields C, Farnsworth A, Lewis R, Schnittman S, Gigliotti M, Maldonaldo S, Lane B, Hernandez JE, et al. Pediatric AIDS Clinical Trials Group 377 Study Team. Nucleoside-analogue reverse-transcriptase inhibitors plus nevirapine, nelfinavir, or ritonavir for pre-treated children infected with human immunodeficiency virus type 1. Clin Infect Dis 2002;34(7):991–1001.
12. Brinkman K, ter Hofstede HJ, Burger DM, Smeitink JA, Koopmans PP. Adverse effects of reverse transcriptase inhibitors: mitochondrial toxicity as common pathway. AIDS 1998;12(14):1735–44.
13. Herskowitz A, Willoughby SB, Baughman KL, Schulman SP, Bartlett JD. Cardiomyopathy associated with antiretroviral therapy in patients with HIV infection: a report of six cases. Ann Intern Med 1992;116(4):311–3.
14. Domanski MJ, Sloas MM, Follmann DA, Scalise PP 3rd, Tucker EE, Egan D, Pizzo PA. Effect of zidovudine and didanosine treatment on heart function in children infected with human immunodeficiency virus. J Pediatr 1995;127(1):137–46.
15. Fichtenbaum CJ, Clifford DB, Powderly WG. Risk factors for dideoxynucleoside-induced toxic neuropathy in patients with the human immunodeficiency virus infection. J Acquir Immune Defic Syndr Hum Retrovirol 1995;10(2):169–74.
16. Simpson DM, Tagliati M. Nucleoside analog-associated peripheral neuropathy in human immunodeficiency virus infection. J Acquir Immune Defic Syndr Hum Retrovirol 1995;9(2):153–61.
17. Rey D, L'Heritier A, Lang JM. Severe ototoxicity in a health care worker who received postexposure prophylaxis with stavudine, lamivudine, and nevirapine after occupational exposure to HIV. Clin Infect Dis 2002;34(3):418–9.
18. Chene G, Angelini E, Cotte L, Lang JM, Morlat P, Rancinan C, May T, Journot V, Raffi F, Jarrousse B, Grappin M, Lepeu G, Molina JM. Role of long-term nucleoside-analogue therapy in lipodystrophy and metabolic disorders in human immunodeficiency virus-infected patients. Clin Infect Dis 2002;34(5):649–57.
19. Saint-Marc T, Partisani M, Poizot-Martin I, Bruno F, Rouviere O, Lang JM, Gastaut JA, Touraine JL. A syndrome of peripheral fat wasting (lipodystrophy) in patients receiving long-term nucleoside analogue therapy. AIDS 1999;13(13):1659–67.
20. Scherpbier HJ, Bekker V, van Leth F, Jurriaans S, Lange JM, Kuijpers TW. Long-term experience with combination antiretroviral therapy that contains nelfinavir for up to 7 years in a pediatric cohort. Pediatrics 2006;117(3):e528-36.
21. Martin A, Smith DE, Carr A, Ringland C, Amin J, Emery S, Hoy J, Workman C, Doong N, Freund J, Cooper DA; Mitochondrial Toxicity Study Group. Reversibility of lipoatrophy in HIV-infected patients 2 years after switching from a thymidine analogue to abacavir: the MITOX Extension Study. AIDS 2004;18(7):1029-36.
22. McComsey GA, Ward DJ, Hessenthaler SM, Sension MG, Shalit P, Lonergan JT, Fisher RL, Williams VC, Hernandez JE; Trial to Assess the Regression of Hyperlactatemia and

to Evaluate the Regression of Established Lipodystrophy in HIV-1-Positive Subjects (TARHEEL; ESS40010) Study Team. Improvement in lipoatrophy associated with highly active antiretroviral therapy in human immunodeficiency virus-infected patients switched from stavudine to abacavir or zidovudine: the results of the TARHEEL study. Clin Infect Dis 2004;38(2):263-70.

23. Miller RF, Shahmonesh M, Hanna MG, Unwin RJ, Schapira AH, Weller IV. Polyphenotypic expression of mitochondrial toxicity caused by nucleoside reverse transcriptase inhibitors. Antivir Ther 2003; 8: 253-7.

24. van der Valk M, Casula M, Weverlingz GJ, van Kuijk K, van Eck-Smit B, Hulsebosch HJ, Nieuwkerk P, van Eeden A, Brinkman K, Lange J, de Ronde A, Reiss P. Prevalence of lipoatrophy and mitochondrial DNA content of blood and subcutaneous fat in HIV-1-infected patients randomly allocated to zidovudine- or stavudine-based therapy. Antivir Ther 2004;9(3):385-93.

25. Reliquet V, Allavena C, François-Brunet C, Perré P, Bellein V, Garré M, May T, Souala F, Besnier JM, Raffi F. Long-term assessment of nevirapine-containing highly active antiretroviral therapy in antiretroviral-naive HIV-infected patients: 3-year follow-up of the VIRGO study. HIV Med 2006;7(7):431-6.

26. Reiss P, Casula M, de Ronde A, Weverling GJ, Goudsmit J, Lange JM. Greater and more rapid depletion of mitochondrial DNA in blood of patients treated with dual (zidovudine + didanosine or zidovudine + zalcitabine) vs. single (zidovudine) nucleoside reverse transcriptase inhibitors. HIV Med 2004;5(1):11-14.

27. de Mendoza C, de Ronde A, Smolders K, Blanco F, Garcia-Benayas T, de Baar M, Fernandez-Casas P, Gonzalez-Lahoz J, Soriano V. Changes in mitochondrial DNA copy number in blood cells from HIV-infected patients undergoing antiretroviral therapy. AIDS Res Hum Retroviruses 2004;20(3):271-3.

28. Hoy JF, Gahan ME, Carr A, Smith D, Lewin SR, Wesselingh S, Cooper DA. Changes in mitochondrial DNA in peripheral blood mononuclear cells from HIV-infected patients with lipoatrophy randomized to receive abacavir. J Infect Dis 2004;190(4):688-92.

29. Carr A, Workman C, Carey D, Rogers G, Martin A, Baker D, Wand H, Law M, Samaras K, Emery S, Cooper DA; Rosey investigators. No effect of rosiglitazone for treatment of HIV-1 lipoatrophy: randomised, double-blind, placebo-controlled trial. Lancet 2004;363(9407):429-38.

30. Nolan D, Hammond E, Martin A, Taylor L, Herrmann S, McKinnon E, Metcalf C, Latham B, Mallal S. Mitochondrial DNA depletion and morphologic changes in adipocytes associated with nucleoside reverse transcriptase inhibitor therapy. AIDS 2003; 17: 1329-38.

31. Birkus G, Hitchcock MJ, Cihlar T. Assessment of mitochondrial toxicity in human cells treated with tenofovir: comparison with other nucleoside reverse transcriptase inhibitors. Antimicrob Agents Chemother 2002; 46: 716-23.

32. Fouty B, Frerman F, Reves R. Riboflavin to treat nucleoside analog-induced lactic acidosis. Lancet 1998;352(9124):291-2.

33. John M, Moore CB, James IR, Nolan D, Upton RP, McKinnon EJ, Mallal SA. Chronic hyperlactatemia in HIV-infected patients taking antiretroviral therapy. AIDS 2001;15(6):717-23.

34. Arenas-Pinto A, Grant AD, Edwards S, Weller IV. Lactic acidosis in HIV infected patients: a systematic review of published cases. Sex Transm Infect 2003; 79: 340-3.

35. Johri S, Alkhuja S, Siviglia G, Soni A. Steatosis-lactic acidosis syndrome associated with stavudine and lamivudine therapy. AIDS 2000;14(9):1286-7.

36. Brivet FG, Nion I, Megarbane B, Slama A, Brivet M, Rustin P, Munnich A. Fatal lactic acidosis and liver steatosis associated with didanosine and stavudine treatment: a respiratory chain dysfunction? J Hepatol 2000;32(2):364-5.

37. Mokrzycki MH, Harris C, May H, Laut J, Palmisano J. Lactic acidosis associated with stavudine administration: a report of five cases. Clin Infect Dis 2000;30(1):198-200.

38. Lonergan JT, Barber RE, Mathews WC. Safety and efficacy of switching to alternative nucleoside analogues following symptomatic hyperlactatemia and lactic acidosis. AIDS 2003;17(17):2495-9.

39. Dornsife RE, Averett DR. In vitro potency of inhibition by antiviral drugs of hematopoietic progenitor colony formation correlates with exposure at hemotoxic levels in human immunodeficiency virus-positive humans. Antimicrob Agents Chemother 1996;40(2):514-9.

40. McLeod GX, Hammer SM. Zidovudine: five years later. Ann Intern Med 1992;117(6):487-501.

41. Blanche P, Silberman B, Barreto L, Gombert B, Sicard D. Reversible zidovudine-induced pure red cell aplasia. AIDS 1999;13(12):1586-7.

42. Yarchoan R, Perno CF, Thomas RV, Klecker RW, Allain JP, Wills RJ, McAtee N, Fischl MA, Dubinsky R, McNeely MC, et al. Phase I studies of 2',3'-dideoxycytidine in severe human immunodeficiency virus infection as a single agent and alternating with zidovudine (AZT). Lancet 1988;1(8577):76-81.

43. Dolin R, Lambert JS, Morse GD, Reichman RC, Plank CS, Reid J, Knupp C, McLaren C, Pettinelli C. 2',3'-Dideoxyinosine in patients with AIDS or AIDS-related complex. Rev Infect Dis 1990;12(Suppl 5):S540-9.

44. McGovern BH, Ditelberg JS, Taylor LE, Gandhi RT, Christopoulos KA, Chapman S, Schwartzapfel B, Rindler E, Fiorino AM, Zaman MT, Sax PE, Graeme-Cook F, Hibberd PL. Hepatic steatosis is associated with fibrosis, nucleoside analogue use, and hepatitis C virus genotype 3 infection in HIV-seropositive patients. Clin Infect Dis 2006;43(3):365-72.

45. Nguyen BY, Yarchoan R, Wyvill KM, Venzon DJ, Pluda JM, Mitsuya H, Broder S. Five-year follow-up of a phase I study of didanosine in patients with advanced human immunodeficiency virus infection. J Infect Dis 1995;171(5):1180-9.

46. Dassopoulos T, Ehrenpreis ED. Acute pancreatitis in human immunodeficiency virus-infected patients: a review. Am J Med 1999;107(1):78-84.

47. Lawn SD, Wood C, Lockwood DN. Borderline tuberculoid leprosy: an immune reconstitution phenomenon in a human immunodeficiency virus-infected person. Clin Infect Dis 2003;36(1):e5-6.

48. Sarner L, Fakoya A. Acute onset lactic acidosis and pancreatitis in the third trimester of pregnancy in HIV-1 positive women taking antiretroviral medication. Sex Transm Infect 2002;78(1):58-9.

49. Rainey PM, Friedland G, McCance-Katz EF, Andrews L, Mitchell SM, Charles C, Jatlow P. Interaction of methadone with didanosine and stavudine. J Acquir Immune Defic Syndr 2000;24(3):241-8.

Protease inhibitors

See also Amprenavir, Indinavir, Lopinavir and ritonavir, Nelfinavir, Saquinavir

General Information

All the HIV protease inhibitors have in common a specific effect against the aspartic HIV protease that cleaves viral proteins to yield structural proteins. Competitive inhibition of this process by the protease inhibitors results in the production of immature, non-infectious virus particles. These drugs are also characterized by their high specificity, being more than a thousand-fold more active against viral than human aspartic proteases.

Combination therapy including a protease inhibitor resulted in the first breakthrough of antiviral treatment in the mid-1990s, since when several protease inhibitors have been developed. However, results from in vitro and clinical studies clearly showed that these drugs share a cross-resistance pattern, probably due to secondary conformational changes of the protease outside the active binding site of the protease inhibitor. Nevertheless, some patients may benefit from a second protease inhibitor if therapy is promptly switched before multiple mutations have accumulated.

Drug interactions with protease inhibitors

Drug interactions with protease inhibitors, which are potent inhibitors of CYP3A4 and CYP2D6, have been comprehensively reviewed (1), and an updated summary of interactions is presented on the World Wide Web by the University of Liverpool (http://www.hiv-druginteractions.org); the drugs that should not be co-administered with protease inhibitors are summarized in Table 1.

Table 1　Drugs that should not be co-administered with protease inhibitors

Drug	APV	ATV	IDV	LPV	NFV	RTV	SQV
Alprazolam			x				
Amiodarone			x	x	x	x	x
Astemizole	x	x	x	x	x	x	x
Atazanavir			x				
Bepridil			x				
Cisapride	x	x	x	x		x	x
Clorazepate						x	
Diazepam	x					x	
Ergot derivatives	x	x	x	x	x	x	x
Estazolam						x	
Flecainide	x	x	x			x	x
Flurazepam						x	
Halofantrine	x		x		x	x	x
Indinavir		x					
Lovastatin	x	x	x	x	x	x	x
Lumefantrine	x	x	x	x		x	x
Midazolam	x	x	x	x	x	x	x
Pethidine						x	
Pimozide	x	x	x	x	x	x	x
Piroxicam						x	
Propafenone	x	x				x	
Quinidine		x				x	x
Rifabutin					x	x	x
Rifampicin	x	x	x	x			x
Simvastatin	x	x	x	x		x	x
St John's wort	x	x	x	x	x	x	x
Terfenadine	x	x	x	x	x	x	x
Triazolam	x	x	x	x	x	x	x
Vardenafil			x			x	
Vitamin E	x					x	
Voriconazole							
Zolpidem						x	

APV=amprenavir and fosamprenavir
ATV=atazanavir
IDV=indinavir
LPV=lopinavir
NFV=nelfinavir
RTV=ritonavir
SQV=saquinavir

Interactions between protease inhibitors

Evidence suggests that some protease inhibitors can inhibit the metabolism of others. For example, ritonavir inhibits CYP3A4 and CYP2D6 and has been used to enhance the actions of other protease inhibitors, by inhibiting their clearance (2).

In an open, randomized study of amprenavir combined with indinavir, nelfinavir, and saquinavir (3), saquinavir lowered the amprenavir AUC by 32%; amprenavir did not alter the pharmacokinetics of saquinavir. The amprenavir AUC increased by 35% when it was combined with indinavir, and indinavir concentrations also fell, suggesting that this protease inhibitor combination should be avoided. There was no significant interaction of amprenavir with nelfinavir.

In attempts to lower the amprenavir capsule burden, low-dose ritonavir has been used as a pharmacokinetic booster. When ritonavir was added to amprenavir, the amprenavir AUC increased 3–4 times (4), which should allow the total daily capsule burden to be reduced. Adverse effects included diarrhea, nausea, paresthesia, rash, increased cholesterol, and increased triglycerides. The frequency of adverse events correlated with the dose of ritonavir.

In nine patients who received amprenavir 750 mg bd for at least 3 weeks plus one of two doses of lopinavir + ritonavir, the trough concentration of amprenavir was 0.35–2.54 with lopinavir + ritonavir 400/100 mg bd ($n = 5$) and 1.92–2.83 with lopinavir + ritonavir 532/133 mg bd ($n = 4$); the corresponding trough concentrations of lopinavir were 0.35–2.54 and 4.74–6.71 (5).

A pharmacokinetic model has been developed to describe the interaction of amprenavir with ritonavir (6). A two-compartment linear model with first-order absorption fitted the amprenavir data best, while a one-compartment model best described the ritonavir data. Inhibition of the elimination of amprenavir by ritonavir was modelled with an E_{max} inhibition model. Simulation of drug regimens based on the model suggested that in patients who fail to respond to a traditional amprenavir regimen, amprenavir 600 mg plus ritonavir 100 mg bd would produce similar $C_{min}:IC_{50}$ ratios to amprenavir 1200 mg bd alone.

Using ritonavir as a booster allows indinavir to be given twice daily and with food. In a cohort survey of 100 patients the combination of indinavir plus ritonavir (400 mg/400 mg or 800 mg/100 mg bd) was a safe and effective option to reduce the tablet burden and improve adherence (7).

Observational studies

Combinations of NRTIs and protease inhibitors

A variable-dose plasma concentration-controlled approach to combination antiretroviral therapy (zidovudine, lamivudine, and indinavir) has been compared with conventional fixed-dose therapy in 40 patients in a randomized, 52-week, open trial (8). Significantly more concentration-controlled recipients achieved the desired concentrations for all three drugs: there was a good response in 15 of 16 concentration-controlled recipients compared with nine of 17 conventional regimen recipients. However, there was no difference in the occurrence of drug-related clinical events or laboratory abnormalities between the two regimens. Three patients withdrew because of gastrointestinal adverse effects, one because of a peripheral neuropathy, and one with headache and anemia. There was nephrolithiasis in four cases.

The combination of indinavir + ritonavir 400/400 mg bd plus two NRTIs has been studied in 93 patients in an open, uncontrolled, multicenter trial (9). Raised triglycerides ($n = 78$) and cholesterol ($n = 63$) were the commonest adverse effects, followed by nausea ($n = 22$) and circumoral paresthesia ($n = 9$). Withdrawal was required in four cases of nausea, four of lipodystrophy, one of diarrhea, and one of osteonecrosis.

Table 2 Numbers (%) of infants and children with moderate or severe adverse events in a study of one or two NRTIs plus an NNRTI and/or a protease inhibitor

System affected by the adverse event	Regimen			
	Stavudine + nevirapine + ritonavir (n = 41)	Stavudine + nevirapine + nelfinavir (n = 44)	Stavudine + lamivudine + nevirapine + nelfinavir (n = 44)	Stavudine + lamivudine + nelfinavir (n = 52)
Respiratory	12 (29)	18 (41)	23 (52)	50 (96)
Hematologic (neutropenia)	17 (41)	9 (20)	14 (32)	23 (44)
Gastrointestinal				
Nausea/vomiting	29 (71)	32 (73)	18 (41)	15 (29)
Other	10 (24)	25 (27)	18 (41)	19 (37)
Liver	12 (29)	14 (32)	18 (41)	23 (44)
Skin	27 (66)	41 (93)*	32 (73)	17 (33)*
Body temperature (fever)	24 (59)	30 (68)*	20 (45)	10 (19)*

*Significantly different from the other groups

Combinations of NRTIs, NNRTIs, and protease inhibitors

In an open, 48-week, single-arm, multicenter phase II study in 99 patients abacavir 300 mg bd, amprenavir 1200 mg bd, and efavirenz 600 mg/day were associated with rashes in 50 patients, possibly because of abacavir hypersensitivity; 17 permanently discontinued one or more drugs as a result (10). Other adverse effects included nausea ($n = 41$), diarrhea ($n = 27$), sleep disorders ($n = 27$), dizziness ($n = 25$), fatigue ($n = 23$), hypertriglyceridemia ($n = 18$), neutropenia ($n = 8$), hyperamylasemia ($n = 4$), leukopenia ($n = 3$), hypercholesterolemia ($n = 2$), raised alkaline phosphatase ($n = 1$), and raised aspartate transaminase ($n = 1$).

In infants and children assigned to different combinations of one or two NRTIs plus an NNRTI and/or a protease inhibitor, the numbers of patients with moderate or severe adverse events were as shown in Table 2 (11). In cases of rash, the rash was worse in those who were taking nevirapine-containing regimens.

Organs and Systems

Nervous system

In a randomized comparison of a protease inhibitor-containing regimen and an efavirenz-containing regimen, nervous system adverse effects were specifically sought (12). Patients were randomized to two NRTIs plus one or more protease inhibitors ($n = 49$) or two NRTIs plus efavirenz ($n = 51$). The patients who took the protease inhibitors reported the following at week 4: light-headedness (8%), dizziness (5%), difficulty in sleeping (4%), nervousness (4%), and headaches (3%). They reported the following at week 48: difficulty in sleeping (4%), nervousness (3%), headaches (3%), and light-headedness (2%). Three patients withdrew because of adverse events: diarrhea ($n = 2$) and nephrolithiasis ($n = 1$).

Sensory systems

- Lipemia retinalis and pancreatitis have been reported in a 39-year-old man with HIV infection associated with protease inhibitor therapy (13). He developed lipemia retinalis after switching to an antiretroviral regimen including ritonavir and saquinavir (together with zalcitabine and delavirdine). He had previously been taking zidovudine, lamivudine, and indinavir.

The ophthalmoscopic changes of lipemia retinalis include a milky-white discoloration of the retinal vessels, beginning at the periphery but progressing to involve the posterior pole as the triglyceride concentration rises. The fundus can appear salmon-colored, owing to the effect of triglycerides on the choroidal circulation. Experience from HIV-negative patients with hyperlipidemia has shown that plasma triglyceride concentrations must be at least 28 mmol/l (2500 mg/dl) for lipemia retinalis to occur (reference value below 1.52 mmol/l). This patient had a plasma triglyceride concentration of 53 mmol/l. On withdrawal of ritonavir and saquinavir the appearance of his retinal vessels returned to normal in parallel with a fall in his plasma triglycerides.

Endocrine

Two case reports have suggested that when a protease inhibitor is used with a glucocorticoid the tendency to adverse corticosteroid effects is potentiated (14,15). Two HIV-positive patients developed severely disfiguring skin striae within 3 months of starting indinavir therapy (16).

Metabolism

Metabolic changes that protease inhibitors can cause after prolonged therapy include raised serum lactate, hypogonadism, hypertension and accelerated cardiovascular disease, reduced bone density, and avascular necrosis of the hip. Two large prospective studies in 1207 patients (17) and 3191 patients (18) have clarified the spectrum and incidence of metabolic changes in HAART and have explored the relative importance of protease inhibitors. In addition, data on fat redistribution from a postmarketing review of HIV-infected individuals taking indinavir have been published (19).

Blood glucose concentration
Protease inhibitors can cause a rise in blood glucose concentration, although only a few cases have been reported. Patients with a family history of diabetes mellitus may be at a greater risk, and they demand especially close monitoring, for example with both baseline and quarterly glucose determinations, at least during the first 6–12 months of treatment (20,21).

Dyslipidemias and Lipodystrophy
While abnormal concentrations of circulating lipids are common in patients with HIV infection (usually hypercholesterolemia and moderate hypertriglyceridemia), there is no doubt that some members of this group of drugs can cause much more marked changes. The possible differences between the effects of various protease inhibitors on the lipid spectrum have been characterized in 93 HIV-infected adults taking ritonavir, indinavir, or nelfinavir, alone or in combination with saquinavir (22). There was a rise in plasma cholesterol concentration in all those who took a protease inhibitor, but it was more pronounced with ritonavir than with indinavir or nelfinavir. Plasma HDL cholesterol was unchanged. Ritonavir, but not indinavir or nelfinavir, was associated with a marked rise in plasma triglyceride concentrations. The combination of ritonavir or nelfinavir with saquinavir did not further alter plasma lipid concentrations. There was a 48% increase in plasma concentrations of lipoprotein(a) in those taking a protease inhibitor, with pretreatment values exceeding 200 µg/ml. There were similar changes in plasma lipid concentrations in six children taking ritonavir. The risk of pancreatitis and premature atherosclerosis as a consequence of such dyslipidemia remains to be established.

- A 35-year-old HIV-positive man developed a serum cholesterol concentration of 38 mmol/l and a fasting serum triglyceride concentration of 98 mmol/l after he started to take ritonavir, saquinavir, nevirapine, and didanosine (23). All other medications had been stable during this time; the condition resolved with antiretroviral drug withdrawal and lipid-lowering therapy. It was striking that the raised cholesterol and triglyceride concentrations did not recur when therapy was restarted in modified form with nelfinavir, saquinavir, nevirapine, and didanosine; the hyperlipidemia was therefore attributed to ritonavir.

In 19 consecutive HIV-positive men examined before and during treatment with a protease inhibitor (nelfinavir, ritonavir, or indinavir) and two nucleoside analogue reverse transcriptase inhibitors (NRTI), median treatment duration 22 (range 7–40) weeks, the predominant feature of dyslipidemia was an increase in triglyceride-containing lipoproteins (24). This observation is in accordance with the hypothesis of increased apoptosis of peripheral adipocytes, release of free fatty acids, and subsequent increased synthesis of VLDL cholesterol. The lipid profile, based on the ratio of total cholesterol to HDL cholesterol and the ratio HDL2 to HDL3, is significantly more atherogenic than normal.

Soon after the introduction of highly active antiretroviral combination treatments (HAART), lipodystrophy was associated with the use of protease inhibitors, and several reports have confirmed that a syndrome of peripheral lipodystrophy, central adiposity, breast hypertrophy in women, hyperlipidemia, and insulin resistance with hyperglycemia is an adverse event associated with the use of potent combination antiretroviral therapy, particularly including HIV-1 protease inhibitors (25–30).

Lipodystrophy is not limited to patients taking protease inhibitors (31,32). Nevertheless, protease inhibitors are strongly associated with metabolic alterations and with lipodystrophy, while NRTIs are associated with low-grade lactic acidosis and less markedly with lipodystrophy. Some reports have speculated a link between mitochondrial dysfunction and lipodystrophy. It is clear, however, that the syndrome is related to total duration of antiviral therapy and inversely related to viral load.

Presentation

Dyslipidemia is a common accompaniment of the lipodystrophy syndrome observed in HIV-infected patients. This syndrome presents as a combination of peripheral lipoatrophy and the metabolic syndrome (central adiposity, insulin resistance, and dyslipidemia). The term lipodystrophy syndrome was first used in two case reports to describe a clinical picture of subcutaneous fat wasting in the face and limbs of HIV infected patients treated with indinavir, reminiscent of the rare congenital lipodystrophy syndromes (33,34). In addition, benign symmetric lipomatoses on the trunk and neck were described. A systematic study of this syndrome in the Australian HIV cohort showed co-existence of peripheral lipoatrophy with abdominal visceral obesity, dyslipidemia, and insulin resistance in HIV-infected patients with or without treatment with protease inhibitors (35).

In a longitudinal study from an Aquitaine cohort of more than 1400 subjects, hypertriglyceridemia was significantly associated with age, low viral load, and protease inhibitors, but not NRTIs or NNRTIs. However, lipodystrophy also occurred in patients naive to protease inhibitors (36).

- "Buffalo neck" was described in a middle-aged man taking indinavir who developed a lipomatous formation in the retrocervical area; abdominal fat also increased in volume, while the subcutaneous fat on the lower limbs decreased (37).

Of 494 patients during median follow-up of 18 months, 17% developed lipodystrophy (38). Study limitations included the short time of follow-up and the lack of a standardized and accepted definition of lipodystrophy.

Peripheral lipodystrophy in patients is characterized by fat wasting of the face, limbs, buttocks, and upper trunk, while central adiposity may cause an increase in belly size and an increase in the dorsocervical fat pad, creating the appearance of a "buffalo hump" (39–41). The increase in belly size is often associated with symptoms of abdominal fullness, distension, and bloating. This is probably due to a change in body fat distribution, with selective accumulation of fat intra-abdominally (42).

In general, most of the morphological and metabolic changes appear to aggregate in the same population and resemble the so-called "metabolic syndrome", also called "syndrome X". The WHO describes this syndrome as an accumulation of three or more of the following clinical features:

- abdominal obesity;
- raised triglycerides;
- low HDL cholesterol;
- raised blood pressure;
- raised fasting glucose.

More specifically, the ATP III guidelines, from the third report of the National Cholesterol Education Program's expert Adult Treatment Panel on Detection, Evaluation, and Treatment of High Blood Cholesterol in Adults (43), define the metabolic syndrome as involving three or more of the following:

- central/abdominal obesity as measured by waist circumference (men: greater than 102 cm (40 inches); women: greater than 88 cm (35 inches));
- fasting triglycerides greater than or equal to 1.7 mmol/l (150 mg/dl);
- HDL cholesterol (men: less than 1 mmol/l (40 mg/dl); women: less than 1.3 mmol/l (50 mg/ dl));
- blood pressure greater than or equal to 130/85 mmHg;
- fasting glucose greater than or equal to 6.1 mmol/l (110 mg/dl).

In the HIV infected population, further evidence suggested that visceral fat accumulation, dyslipidemia, and insulin resistance are closely linked and associated with antiretroviral treatment, most pronounced with the use of protease inhibitors. In contrast, subcutaneous fat wasting is primarily determined by the choice of nucleoside reverse transcriptase inhibitor (NRTI). Switching studies have supported this notion, since substitution of stavudine has been associated with improvement in fat wasting, while switching a protease inhibitor had no beneficial effect in more than 30 clinical trials (44).

The effects of the protease inhibitor indinavir and the NNRTI efavirenz on lipid concentrations have been compared in a large comparative randomized study (45). Each of the two comparison drugs were used in one arm (with a zidovudine + lamivudine backbone) and the combination of the two drugs in a third arm. Zidovudine and lamivudine did not play a role in the lipid changes. However, both of the comparison drugs significantly increased cholesterol concentrations.

In a prospective, non-randomized analysis of 212 patients treated with a regimen containing a protease inhibitor, the overall incidences of hypertriglyceridemia and hypercholesterolemia at 12 months of treatment were 38% and 25% respectively (46). Increased concentrations of triglycerides and LDL cholesterol were more pronounced in patients taking ritonavir or lopinavir/ritonavir compared with other protease inhibitors.

In a small randomized, open, comparative study patients who switched to abacavir from either stavudine or stavudine plus a protease inhibitor or NNRTI, or a protease inhibitor + NNRTI had improved total and LDL cholesterol (47). Total arm and leg fat mass, measured by DEXA scan, rose significantly in those who switched from stavudine to abacavir, suggesting an important role of stavudine in the pathogenesis of lipodystrophy.

Support for the hypothesis of a casual relation between stavudine and lipodystrophy comes from another randomized study in which stavudine-containing HAART regimens were switched to a combination of zidovudine, lamivudine, and abacavir (48). Eight patients were randomized to continue stavudine and 14 patients switched to the triple combination. Imbalance in the treatment arms resulted from exclusion of patients who maintained treatment for a minimum of 6 months of follow up. Over 48 weeks after randomization, the average leg and arm fat mass fell in the continuation arm but increased in the switch arm. One patient in the switch arm, who had previously taken zidovudine and lamivudine, had a therapeutic failure.

In a retrospective analysis of 36 patients who switched from stavudine to tenofovir while HIV RNA was below 20 copies/ ml for more than 6 months, two switched because of peripheral neuropathy and 34 because of lipoatrophy; the median duration of observation was 36 weeks (49). There was a significant fall in cholesterol concentrations from 5.5 mmol/l to 5.0 mmol/l at week 4, and 4.7 mmol/l at week 36. There was also a non-significant trend toward a fall in triglyceride concentrations.

In patients who had taken their first antiretroviral regimen for more than 6 months, using either stavudine (n = 75) or zidovudine (n = 75) plus lamivudine and indinavir total fat was significantly lower in patients taking stavudine but the lean body mass was similar in the two groups (50). Fat redistribution was common: 20 patients

were classified as having lipoatrophy, 33 lipodystrophy, and 41 a mixed syndrome. However, there were no statistically significant differences between the two groups. Lack of physical activity was the only independent predictor of isolated or mixed lipoatrophy. Whether physical activity in fact improves lipodystrophy is not known.

Lipid abnormalities are a major adverse effect of HIV protease inhibitors (28). In a 48-week comparison with nelfinavir, atazanavir did not significantly increase total cholesterol, fasting LDL cholesterol or triglyceride concentrations (+6.8%, −7.1%, +1.5% respectively), while the respective concentrations rose by 28%, 31%, and 42% in those who took nelfinavir. The incidence of grade 1?4 lipodystrophy was infrequent in both groups, but this endpoint was poorly defined in this study.

The effect of the combination of lopinavir + ritonavir on the atherogenic lipid profile has been evaluated in 24 HIV infected patients (51). At baseline, there was an abnormally small LDL density. After 1 month lopinavir + ritonavir increased triglyceride and apolipoprotein CIII concentrations, and LDL size fell further.

Several pathophysiological mechanisms have been proposed, including adverse effects of protease inhibitors on hepatocyte and fat cell function (52), mitochondrial toxicity from nucleoside analogues (53), excess of reactive oxygen species (54), and cytokine-mediated events (55). In vitro data and studies in healthy volunteers suggest a role for protease inhibitors in insulin resistance (56) and dyslipidemia (57).

There was significant improvement in signs of mitochondrial toxicity in 49 patients who switched from stavudine to abacavir compared with 63 patients who continued to take stavudine in a non-randomized study for 12 months (58). Only patients who remained on their assigned treatment were included in the analysis. Lactate concentrations were assessed at baseline, week 24, and week 48, and electrical bioimpedance was performed in 22 cases and 12 controls at baseline and week 48. There were significant falls in serum lactate concentrations at weeks 24 and 48 in cases compared with controls. Patients who switched had a trend towards fat gain, while controls had significant reductions in total body fat and percentage of body fat.

Insulin resistance

The independent effect of protease inhibitors on insulin resistance has been investigated. A single dose of indinavir was sufficient to produce a significant reduction in insulin sensitivity, assessed by euglycemic hyperinsulinemic clamp testing, in both HIV-infected and HIV-negative individuals (59).

Diabetes mellitus

A retrospective analysis of the development of diabetes in 1011 patients has been summarized (60). All were non-diabetic when antiretroviral treatment was started. Over 10 months, diabetes was diagnosed in 16 patients (2.06 per 100 person-years). Older age (HR = 1.1; 1.06, 1.16) was associated with a higher risk. In multivariate analysis adjusted for age and sex, the onset of diabetes was not related to CD4 cell count, viral load, or type of antiviral therapy (with or without protease inhibitors). However, patients taking stavudine or indinavir were at significantly higher risks (stavudine HR = 16, 95% CI = 3, 84; indinavir HR = 4.0, 95% CI = 1.3, 13). The strong association of stavudine with diabetes is surprising and needs further confirmation.

In a cohort study (61) in 1785 women, 69 incident cases of diabetes mellitus were diagnosed, with an average incidence of 1.5 cases per 100 patient-years. In patients taking protease inhibitors, incidence rates were about twice as high (2.8 cases per 100 patient-years) as among users of NNRTIs or untreated patients (1.2%) and uninfected controls (1.4%). In a multivariate model use of protease inhibitors (HR = 2.9; 1.5, 5.6), age, and BMI were independent risk factors for diabetes.

Hematologic

There is still no clear explanation for a series of seven patients in whom treatment with various protease inhibitors was associated with early thrombosis (62). Hematological or vascular effects conducive to thrombosis are not recognized as a complication with this group of compounds; the only previous similar cases were associated with prolonged therapy.

In contrast, a detailed report on a large series of patients has provided impressive evidence that the use of protease inhibitors in HIV-infected patients with hereditary bleeding disorders can lead to an increased bleeding tendency (63). This effect, which is most likely to occur when ritonavir is used, is also unexplained.

Liver

In a retrospective study of patients taking one protease inhibitor ($n = 39$) or two ($n = 27$) who discontinued protease inhibitor therapy as a result of hepatotoxicity, the proportions of patients with raised alanine transaminase activity to at least five times the upper limit of the reference range (26 versus 19%) and hyperbilirubinemia (38 versus 30%) were similar (64). Rates of withdrawal due to hepatotoxicity were also similar.

Pancreas

- Lipemia retinalis and pancreatitis have been reported in a 39-year-old man with HIV infection associated with protease inhibitor therapy (13). His plasma triglyceride concentration was 53 mmol/l.

Urinary tract

In a randomized comparison of a protease inhibitor-containing regimen and an efavirenz-containing regimen in 100 patients, six patients taking protease inhibitors developed nephrolithiasis; one withdrew as a result (12).

Musculoskeletal

The results of a questionnaire survey of 878 people with HIV infection treated with antiretroviral drugs also suggested that other protease inhibitors can cause arthralgia;

indinavir and the combination of ritonavir + saquinavir were particularly implicated (65).

Immunologic

A man developed several erythematous plaques on his face due to borderline tuberculoid leprosy with a reversal reaction (66). He had severe CD4 T cell lymphocytopenia due to HIV infection and had been given highly active antiretroviral therapy (HAART). A fall in viral load and an increase in CD4 count preceded the development of the skin lesions, suggesting immune reconstitution as the underlying mechanism for the reversal reaction. Paradoxical reactions are often observed in patients with pulmonary and extra-pulmonary tuberculosis being treated with HAART. Clinicians need to distinguish these from other adverse reactions related to drug therapy. Reversal reactions in leprosy are increasingly likely as more patients with HIV infection are treated with HAART in developing countries.

Drug–Drug Interactions

See also General information and Table 1.

Ecstasy

A patient infected with HIV-1 who was taking ritonavir and saquinavir had a prolonged effect from a small dose of methylenedioxymetamfetamine (MDMA, ecstasy) and a near-fatal reaction from a small dose of gammahydroxybutyrate (67).

Gammahydroxybutyrate

A patient infected with HIV-1 who was taking ritonavir and saquinavir had a prolonged effect from a small dose of MDMA and a near-fatal reaction from a small dose of gammahydroxybutyrate (67).

Statins

There has been a study of the effects of ritonavir 400 mg bd plus saquinavir soft-gel capsules 400 mg bd on the pharmacokinetics of pravastatin, simvastatin, and atorvastatin (40 mg/day each), and of the effect of pravastatin on the pharmacokinetics of nelfinavir in a randomized, open study in 56 healthy HIV-negative adults (68). Ritonavir + saquinavir reduced the steady-state AUC of pravastatin, markedly increased the AUC of simvastatin, and slightly increased the AUC of total active atorvastatin. The AUCs of nelfinavir and its active M8 metabolite were not altered by pravastatin. The authors concluded that simvastatin (and by implication lovastatin, which in common with simvastatin inhibits CYP3A4) should be avoided and atorvastatin should be used with caution in people taking ritonavir and saquinavir, that dosage adjustment of pravastatin may be necessary with concomitant use of ritonavir + saquinavir, and that concomitant use of pravastatin with nelfinavir appears to be safe.

References

1. Malaty LI, Kuper JJ. Drug interactions of HIV protease inhibitors. Drug Saf 1999;20(2):147–69.

2. Rathbun RC, Rossi DR, Nazario M, Edouard B. Low-dose ritonavir for protease inhibitor pharmacokinetic enhancement. Ann Pharmacother 2002;36(4):702–6.

3. Sadler BM, Gillotin C, Lou Y, Eron JJ, Lang W, Haubrich R, Stein DS. Pharmacokinetic study of human immunodeficiency virus protease inhibitors used in combination with amprenavir. Antimicrob Agents Chemother 2001;45(12):3663–8.

4. Sadler BM, Gillotin C, Lou Y, Stein DS. Pharmacokinetic and pharmacodynamic study of the human immunodeficiency virus protease inhibitor amprenavir after multiple oral dosing. Antimicrob Agents Chemother 2001;45(1):30–7.

5. Khanlou H, Graham E, Brill M, Farthing C. Drug interaction between amprenavir and lopinavir/ritonavir in salvage therapy. AIDS 2002;16(5):797–8.

6. Sale M, Sadler BM, Stein DS. Pharmacokinetic modeling and simulations of interaction of amprenavir and ritonavir. Antimicrob Agents Chemother 2002;46(3):746–54.

7. Burger DM, Hugen PW, Aarnoutse RE, Dieleman JP, Prins JM, van der Poll T, ten Veen JH, Mulder JW, Meenhorst PL, Blok WL, van der Meer JT, Reiss P, Lange JM. A retrospective, cohort-based survey of patients using twice-daily indinavir + ritonavir combinations: pharmacokinetics, safety, and efficacy J Acquir Immune Defic Syndr 2001;26(3):218–24.

8. Fletcher CV, Anderson PL, Kakuda TN, Schacker TW, Henry K, Gross CR, Brundage RC. Concentration-controlled compared with conventional antiretroviral therapy for HIV infection. AIDS 2002;16(4):551–60.

9. Lichterfeld M, Nischalke HD, Bergmann F, Wiesel W, Rieke A, Theisen A, Fatkenheuer G, Oette M, Carls H, Fenske S, Nadler M, Knechten H, Wasmuth JC, Rockstroh JK. Long-term efficacy and safety of ritonavir/indinavir at 400/400 mg twice a day in combination with two nucleoside reverse transcriptase inhibitors as first line antiretroviral therapy HIV Med 2002;3(1):37–43.

10. Falloon J, Ait-Khaled M, Thomas DA, Brosgart CL, Eron JJ Jr, Feinberg J, Flanigan TP, Hammer SM, Kraus PW, Murphy R, Torres R, Masur H, Manion DJ, Rogers M, Wolfram J, Amphlett GE, Rakik A, Tisdale M. CNA2007 Study Team. HIV-1 genotype and phenotype correlate with virological response to abacavir, amprenavir and efavirenz in treatment-experienced patients. AIDS 2002;16(3):387–96.

11. Krogstad P, Lee S, Johnson G, Stanley K, McNamara J, Moye J, Jackson JB, Aguayo R, Dieudonne A, Khoury M, Mendez H, Nachman S, Wiznia A, Ballow A, Aweeka F, Rosenblatt HM, Perdue L, Frasia A, Jeremy R, Anderson M, Japour A, Fields C, Farnsworth A, Lewis R, Schnittman S, Gigliotti M, Maldonaldo S, Lane B, Hernandez JE, et al. Pediatric AIDS Clinical Trials Group 377 Study Team. Nucleoside-analogue reverse-transcriptase inhibitors plus nevirapine, nelfinavir, or ritonavir for pre-treated children infected with human immunodeficiency virus type 1. Clin Infect Dis 2002;34(7):991–1001.

12. Fumaz CR, Tuldra A, Ferrer MJ, Paredes R, Bonjoch A, Jou T, Negredo E, Romeu J, Sirera G, Tural C, Clotet B. Quality of life, emotional status, and adherence of HIV-1-infected patients treated with efavirenz versus protease inhibitor-containing regimens. J Acquir Immune Defic Syndr 2002;29(3):244–53.

13. Eng KT, Liu ES, Silverman MS, Berger AR. Lipemia retinalis in acquired immunodeficiency syndrome treated with protease inhibitors. Arch Ophthalmol 2000;118(3):425–6.

14. Chen F, Kearney T, Robinson S, Daley-Yates PT, Waldron S, Churchill DR. Cushing's syndrome and severe adrenal suppression in patients treated with ritonavir and inhaled nasal fluticasone. Sex Transm Infect 1999;75(4):274.

15. Hillebrand-Haverkort ME, Prummel MF, ten Veen JH. Ritonavir-induced Cushing's syndrome in a patient treated with nasal fluticasone. AIDS 1999;13(13):1803.

16. Darvay A, Acland K, Lynn W, Russell-Jones R. Striae formation in two HIV-positive persons receiving protease inhibitors. J Am Acad Dermatol 1999;41(3 Pt 1):467–9.

17. Bonfanti P, Valsecchi L, Parazzini F, Carradori S, Pusterla L, Fortuna P, Timillero L, Alessi F, Ghiselli G, Gabbuti A, Di Cintio E, Martinelli C, Faggion I, Landonio S, Quirino T. Incidence of adverse reactions in HIV patients treated with protease inhibitors: a cohort study. Coordinamento Italiano Studio Allergia e Infezione da HIV (CISAI) Group. J Acquir Immune Defic Syndr 2000;23(3):236–45.

18. Thiebaut R, Dabis F, Malvy D, Jacqmin-Gadda H, Mercie P, Valentin VD. Serum triglycerides, HIV infection, and highly active antiretroviral therapy, Aquitaine Cohort, France, 1996 to 1998. Groupe d'Epidemiologie Clinique du Sida en Aquitaine (GECSA). J Acquir Immune Defic Syndr 2000;23(3):261–5.

19. Benson JO, McGhee K, Coplan P, Grunfeld C, Robertson M, Brodovicz KG, Slater E. Fat redistribution in indinavir-treated patients with HIV infection: a review of postmarketing cases. J Acquir Immune Defic Syndr 2000;25(2):130–9.

20. Kaufman MB, Simionatto C. A review of protease inhibitor-induced hyperglycemia. Pharmacotherapy 1999;19(1):114–7.

21. Rodriguez-Rosado R, Soriano V, Blanco F, Dona C, Gonzalez-Lahoz J. Diabetes mellitus associated with protease inhibitor use. Eur J Clin Microbiol Infect Dis 1999;18(9):675–7.

22. Periard D, Telenti A, Sudre P, Cheseaux JJ, Halfon P, Reymond MJ, Marcovina SM, Glauser MP, Nicod P, Darioli R, Mooser V. Atherogenic dyslipidemia in HIV-infected individuals treated with protease inhibitors. The Swiss HIV Cohort Study. Circulation 1999;100(7):700–5.

23. Echevarria KL, Hardin TC, Smith JA. Hyperlipidemia associated with protease inhibitor therapy. Ann Pharmacother 1999;33(7–8):859–63.

24. Berthold HK, Parhofer KG, Ritter MM, Addo M, Wasmuth JC, Schliefer K, Spengler U, Rockstroh JK. Influence of protease inhibitor therapy on lipoprotein metabolism. J Intern Med 1999;246(6):567–75.

25. Roth VR, Kravcik S, Angel JB. Development of cervical fat pads following therapy with human immunodeficiency virus type 1 protease inhibitors. Clin Infect Dis 1998;27(1):65–7.

26. Striker R, Conlin D, Marx M, Wiviott L. Localized adipose tissue hypertrophy in patients receiving human immunodeficiency virus protease inhibitors. Clin Infect Dis 1998;27(1):218–20.

27. Viraben R, Aquilina C. Indinavir-associated lipodystrophy. AIDS 1998;12(6):F37–9.

28. Toma E, Therrien R. Gynecomastia during indinavir antiretroviral therapy in HIV infection. AIDS 1998;12(6):681–2.

29. Lui A, Karter D, Turett G. Another case of breast hypertrophy in a patient treated with indinavir. Clin Infect Dis 1998;26(6):1482.

30. Walli R, Herfort O, Michl GM, Demant T, Jager H, Dieterle C, Bogner JR, Landgraf R, Goebel FD. Treatment with protease inhibitors associated with periph-

eral insulin resistance and impaired oral glucose tolerance in HIV-1-infected patients. AIDS 1998;12(15):F167–73.

31. Carr A, Samaras K, Thorisdottir A, Kaufmann GR, Chisholm DJ, Cooper DA. Diagnosis, prediction, and natural course of HIV-1 protease-inhibitor-associated lipodystrophy, hyperlipidaemia, and diabetes mellitus: a cohort study. Lancet 1999;353(9170):2093–9.

32. Saint-Marc T, Touraine JL. Effects of metformin on insulin resistance and central adiposity in patients receiving effective protease inhibitor therapy. AIDS 1999;13(8):1000–2.

33. Massip P, Marchou B, Bonnet E, Cuzin L, Montastruc JL. Lipodystrophia with protease inhibitors in HIV patients. Therapie 1997;52:615.

34. Viraben R, Aquilina C. Indinavir-associated lipodystrophy. AIDS 1998;12:F37-9.

35. Carr A, Samaras K, Chisholm DJ, Cooper DA. Pathogenesis of HIV-1-protease inhibitor-associated peripheral lipodystrophy, hyperlipidaemia, and insulin resistance. Lancet 1998;351:1881-3.

36. Thiebaut R, Daucourt V, Malvy D. Lipodystrophy, glucose and lipid metabolism dysfunctions. Aquitaine Cohort. First International Workshop on Adverse Drug Reactions and Lipodystrophy in HIV, San Diego, 1999: Abstract 17.

37. Milpied-Homsi B, Krempf M, Gueglio B, Raffi F, Stalder JF. "Bosse de bison": un effet secondaire inattendu des traitements parinhibiteurs de protéases anti-VIH. ["Buffalo neck": an unintended secondary effect of treatment with anti-HIV protease inhibitors.] Ann Dermatol Venereol 1999;126(3):254–6.

38. Martinez E, Mocroft A, Garcia-Viejo MA, Perez-Cuevas JB, Blanco JL, Mallolas J, Bianchi L, Conget I, Blanch J, Phillips A, Gatell JM. Risk of lipodystrophy in HIV-1-infected patients treated with protease inhibitors: a prospective cohort study. Lancet 2001;357(9256):592–8.

39. Carr A, Cooper DA. Images in clinical medicine. Lipodystrophy associated with an HIV-protease inhibitor. N Engl J Med 1998;339(18):1296.

40. Carr A, Samaras K, Burton S, Law M, Freund J, Chisholm DJ, Cooper DA. A syndrome of peripheral lipodystrophy, hyperlipidaemia and insulin resistance in patients receiving HIV protease inhibitors. AIDS 1998;12(7):F51–8.

41. Lo JC, Mulligan K, Tai VW, Algren H, Schambelan M. "Buffalo hump" in men with HIV-1 infection. Lancet 1998;351(9106):867–70.

42. Miller KD, Jones E, Yanovski JA, Shankar R, Feuerstein I, Falloon J. Visceral abdominal-fat accumulation associated with use of indinavir. Lancet 1998;351(9106):871–5.

43. Third Report of the Expert Panel on Detection, Evaluation, and Treatment of High Blood Cholesterol in Adults (Adult Treatment Panel III) http://www.nhlbi.nih.gov/guidelines/cholesterol/(accessed 20 March 2005).

44. Nolan D. Metabolic complications associated with HIV protease inhibitor therapy. Drugs 2003;63:2555-74.

45. Tashima KT. Lipid changes in patients initiating efavirenz- and indinavir-based antiretroviral regimens. 2003;4:29-36.

46. Calza L, Manfredi R, Farneti B, Chiodo F. Incidence of hyperlipidaemia in a cohort of 212 HIV-infected patients receiving a protease inhibitor-based antiretroviral therapy. Int J Antimicrob Agents 2003;22:54-9.

47. Moyle GJ, Baldwin C, Langroudi B, Mandalia S, Gazzard BG. A 48-week, randomized, open-label comparison of three abacavir-based substitution approaches in the management of dyslipidemia and peripheral lipoatrophy. J Acquir Immune Defic Syndr 2003;33:22-8.

48. John M, McKinnon EJ, James IR, Nolan DA, Herrmann SE, Moore CB, White AJ, Mallal SA. Randomized, controlled, 48-week study of switching stavudine and/or protease inhibitors to combivir/abacavir to prevent or reverse lipoatrophy in HIV-infected patients. J Acquir Immune Defic Syndr 2003;33:29-33.

49. Lafeuillade A, Jolly P, Chadapaud S, Hittinger G, Lambry V, Philip G. Evolution of lipid abnormalities in patients switched from stavudine- to tenofovir-containing regimens. J Acquir Immune Defic Syndr 2003;33:544-6.

50. Domingo P, Sambeat MA, Perez A, Ordonez J, Rodriguez J, Vazquez G. Fat distribution and metabolic abnormalities in HIV-infected patients on first combination antiretroviral therapy including stavudine or zidovudine: role of physical activity as a protective factor. Antivir Ther 2003;8:223-31.

51. Badiou S, De Boever CM, Dupuy AM, Baillat V, Cristol JP, Reynes J. Small dense LDL and atherogenic lipid profile in HIV-positive adults: influence of lopinavir/ritonavir-containing regimen. AIDS 2003;17:772-4.

52. Carr A, Samaras K, Burton S, Law M, Freund J, Chisholm DJ, Cooper DA. A syndrome of peripheral lipodystrophy, hyperlipidaemia and insulin resistance in patients receiving HIV protease inhibitors. AIDS 1998;12:F51-8.

53. Brinkman K, Smeitink JA, Romijn JA, Reiss P. Mitochondrial toxicity induced by nucleoside-analogue reverse-transcriptase inhibitors is a key factor in the pathogenesis of antiretroviral-therapy-related lipodystrophy. Lancet 1999;354:1112-5.

54. Miserez AR, Muller PY, Barella L, Schwietert M, Erb P, Vernazza PL, Battegay M; Swiss HIV Cohort Study. A single-nucleotide polymorphism in the sterol-regulatory element-binding protein 1c gene is predictive of HIV-related hyperlipoproteinaemia. AIDS 2001;15:2045-9.

55. Moyle G. Mitochondrial toxicity hypothesis for lipoatrophy: a refutation. AIDS 2001;15:413-5.

56. Murata H, Hruz PW, Mueckler M. The mechanism of insulin resistance caused by HIV protease inhibitor therapy. J Biol Chem 2000;275:20251-4.

57. Liang JS, Distler O, Cooper DA, Jamil H, Deckelbaum RJ, Ginsberg HN, Sturley SL. HIV protease inhibitors protect apolipoprotein B from degradation by the proteasome: a potential mechanism for protease inhibitor-induced hyperlipidemia. Nature Med 2001;7:1327-31.

58. Garcia-Benayas T, Blanco F, de la Cruz JJ, Soriano V, Gonzalez-Lahoz J. Replacing stavudine by abacavir reduces lactate levels and may improve lipoatrophy. AIDS 2003;17:921-4.

59. Noor MA, Lo JC, Mulligan K, Schwarz JM, Halvorsen RA, Schambelan M, Grunfeld C. Metabolic effects of indinavir in healthy HIV-seronegative men. AIDS 2001;15:F11-18.

60. Brambilla AM, Novati R, Calori G, Mcneghini E, Vacchini D, Luzi L, Castagna A, Lazzarin A. Stavudine or indinavir-containing regimens are associated with an increased risk of diabetes mellitus in HIV-infected individuals. AIDS 2003;17:1993-5.

61. Justman JE, Benning L, Danoff A, Minkoff H, Levine A, Greenblatt RM, Weber K, Piessens E, Robison E, Anastos K. Protease inhibitor use and the incidence of diabetes mellitus in a large cohort of HIV-infected women. J Acquir Immune Defic Syndr 2003;32:298-302.

62. George SL, Swindells S, Knudson R, Stapleton JT. Unexplained thrombosis in HIV-infected patients receiving protease inhibitors: report of seven cases. Am J Med 1999;107(6):624–30.

63. Wilde JT, Lee C, Collins P, Giangrande PL, Winter M, Shiach CR. Increased bleeding associated with protease

inhibitor therapy in HIV-positive patients with bleeding disorders. Br J Haematol 1999;107(3):556–9.

64. Lawn SD, Wood C, Lockwood DN. Borderline tuberculoid leprosy: an immune reconstitution phenomenon in a human immunodeficiency virus-infected person. Clin Infect Dis 2003;36(1):e5–6.

65. Cooper CL, Parbhakar MA, Angel JB. Hepatotoxicity associated with antiretroviral therapy containing dual versus single protease inhibitors in individuals coinfected with hepatitis C virus and human immunodeficiency virus. Clin Infect Dis 2002;34(9):1259–63.

66. Florence E, Schrooten W, Verdonck K, Dreezen C, Colebunders R. Rheumatological complications associated with the use of indinavir and other protease inhibitors. Ann Rheum Dis 2002;61(1):82–4.

67. Harrington RD, Woodward JA, Hooton TM, Horn JR. Life-threatening interactions between HIV-1 protease inhibitors and the illicit drugs MDMA and gamma-hydroxybutyrate. Arch Intern Med 1999;159(18):2221–4.

68. Fichtenbaum CJ, Gerber JG, Rosenkranz SL, Segal Y, Aberg JA, Blaschke T, Alston B, Fang F, Kosel B, Aweeka F. NIAID AIDS Clinical Trials Group. Pharmacokinetic interactions between protease inhibitors and statins in HIV seronegative volunteers: ACTG Study A5047. AIDS 2002;16(4):569–77.

Ribavirin

General Information

The synthetic triazole nucleoside, ribavirin (1-beta-D-ribofuranosyl-1,2,4-triazole-3-carboxamide, tribavirin, virazole), has a broad spectrum of antiviral activity, including DNA as well as RNA viruses (1). Ribavirin closely resembles guanosine and is converted intracellularly to mono-, di-, and triphosphate derivatives, which inhibit the virally induced enzymes involved in viral nucleic acid synthesis by different mechanisms that are not fully understood (2). Of the DNA viruses, ribavirin is active against *Herpes simplex* virus and hepatitis B virus; among the RNA viruses, good activity has been observed against hepatitis C virus, orthomyxoviruses, paramyxoviruses, arenaviruses, and bunyaviruses. Although active against HIV in vitro and in vivo (3), ribavirin is not widely used in the treatment of HIV infection. So far, drug resistance has not been described.

Oral ribavirin has been successfully used in the treatment of Lassa fever (4), Crimean Congo hemorrhagic fever (5), and in combination therapy with interferon alfa for hepatitis C infection (6,7). Several publications have suggested enhanced efficacy of the combination of interferon alfa with ribavirin when compared with monotherapy with interferon alfa. There is also evidence that re-treatment with the combination may succeed in controlling or eliminating viremia when monotherapy has failed. Although the combination may lead to some increase in the adverse effects normally associated with interferon alfa (dyspnea, pharyngitis, pruritus, nausea, insomnia, and anorexia) (SEDA-23, 315), there is no doubt that oral ribavirin adds to the overall toxicity of the combination by causing hemolytic anemia, which is usually mild.

Ribavirin is well absorbed orally, but it can be given in aerosol form for the treatment of respiratory syncytial virus (RSV) infections in immunocompromised patients, and in those with cardiopulmonary abnormalities, or in infants receiving mechanical ventilation (8,9).

The adverse effects and other safety aspects of interferon and ribavirin in the treatment of hepatitis C infection have been reviewed (10).

Observational studies

Chronic hepatitis
The combination of interferon + ribavirin has the same adverse effects in patients who are co-infected with hepatitis C and HIV as in those who have hepatitis C only. However, in one series of 68 patients some unexpected adverse effects were recorded (11). One subject developed pancreatitis and four others developed asymptomatic hyperamylasemia, which disappeared after withdrawal. All of them were taking concomitant didanosine. Secondly, lactate concentrations increased slightly in two individuals, both of whom were taking stavudine. Significant weight loss (4.5 kg on average within 6 months) may be another adverse effect resulting from the interaction of ribavirin and HIV nucleoside analogues.

The mechanism of the beneficial effect of adding ribavirin to interferon is not fully understood. Ribavirin monotherapy is not effective in hepatitis C. However, adding ribavirin to interferon increases the number of patients with a virological response although it also increases the number of adverse events. Both the benefits and harms of adding ribavirin to interferon for patients with chronic hepatitis C should be considered before therapy is started.

Hepatitis C virus (HCV) RNA kinetics have been studied on day 1 in 15 patients (nine and six of genotypes 1 and non-1 respectively) and at weeks 1, 4, and 12 in 53 patients (19 and 34 of genotypes 1 and non-1 respectively) during treatment with ribavirin + pegylated interferon α-2a (12). Patients with a sustained virological response (SVR) had a significantly more pronounced mean $\log_{10}$ decline from baseline in HCV RNA amounts at weeks 1 and 4 compared with patients who failed to achieve a sustained response, whereas there was no difference after day 1. For patients with a 2 $\log_{10}$ reduction in HCV RNA amounts on day 7, the positive predictive value for a sustained virological response was 92%, whereas week 12 was the best time point for predicting a later non-response in patients who failed to achieve a 2 $\log_{10}$ fall. In patients with genotype non-1 and a 2 $\log_{10}$ fall in HCV RNA amounts the positive predictive value for a sustained virological response was 89% at week 1, and 79% at weeks 4 and 12. The corresponding negative predictive values for patients with genotype non-1 were 43, 40, and 100% respectively. Of the 60 patients, one withdrew from treatment after the second dose of pegylated interferon α-2a, four withdrew prematurely, one each at treatment weeks

8, 12, and 27 for unknown reasons, and one at week 16 for psychiatric reasons. One other patient withdrew at week 16 because of arthralgia. Dosage reduction was required in three patients because of thrombocytopenia or neutropenia, and in three others the dosage of ribavirin was reduced because of anemia.

Combination treatment of interferon-alfa + ribavirin for chronic hepatitis D does not induce virological responses at a sufficient rate, despite its partial effectiveness in improving biochemical responses, and is not superior to interferon-alfa monotherapy. Patients with chronic hepatitis D (n = 19) were treated with interferon-alfa2b (10 million U three times/week subcutaneously) and ribavirin (1000–1200 mg/day orally) for 24 months, with follow-up for at least 6 months (range 7–19) (13). All had compensated liver disease, raised transaminase activities, and hepatitis D virus RNA positivity at baseline. Genotypic analyses showed hepatitis D virus genotype I and hepatitis B virus genotype D. There were biochemical responses in eight patients (42%) at the end of treatment and in seven patients (37%) at the end of follow-up. Only eight patients at the end of treatment and four at the end of follow-up had sustained virological responses. There were flu-like symptoms, generally mild or moderate, in most of the patients. Two patients required a short-term dosage reduction from 10 to 5 MU because of leukopenia and thrombocytopenia and two patients had a drop in hemoglobin, which was managed with a reduction in the dosage of ribavirin.

Prevention of recurrence after liver transplantation
It has been postulated that there is a risk of increased severity of recurrent hepatitis C virus infection in living donor liver transplantation (LDLT) patients. Preventive therapy for this has been studied in 23 patients (14). All received interferon-alpha 2b and ribavirin 1 month after transplantation and for 12 months after the first negative HCV RNA test. They were then observed without therapy for 6 months (Group 1). Therapy was continued for at least 12 months when the HCV-RNA test remained positive (Group 2). They were removed from the protocol if they could not continue therapy for 12 months because of adverse effects or could not start therapy because of early death. Eight patients were removed from the protocol (three died and two could not start because of their poor general condition). Nine patients were assigned to Group 1 and the other six to Group 2. The sustained virological response ratio was 39% (9/23). There was a significant difference between the groups in the histological activity score 1 year after therapy. No details were given of the adverse effects in the eight patients who were withdrawn.

Fulminant hepatitis C
Pegylated interferon and ribavirin have been used successfully to treat fulminant hepatitis C infection; there were only mild self-limiting adverse effects, such as hemolytic anemia (15).

Comparative studies
Two, large, randomized, placebo-controlled comparisons of interferon alfa-2b alone with the combination of interferon alfa-2b plus ribavirin have been published. In the initial treatment of chronic hepatitis C, 912 patients were randomly assigned to receive standard-dose interferon alfa-2b alone or in combination with ribavirin (1000 or 1200 mg/day orally, depending on body weight) for 24 or 48 weeks (7). As expected, dosage reduction for anemia was necessary in 8% of patients taking the combination therapy and in none of those treated with interferon alone. Dyspnea, pharyngitis, pruritus, rash, nausea, insomnia, and anorexia were adverse effects that were reported more often during combination therapy with ribavirin (7). In patients whose chronic hepatitis had relapsed after therapy with interferon alfa-2b alone, 345 patients were randomized to receive standard-dose interferon alfa-2b alone or in combination with ribavirin (1000 or 1200 mg/day orally, depending on body weight) for 6 months (6). Dosage reduction for anemia was required in 12/173 patients assigned to combination therapy and in none assigned to interferon alone. As was the case in the initial therapy study, dyspnea, nausea, and rash were significantly more common in patients treated with the combination of interferon and ribavirin (6).

Ribavirin 15 mg/kg/day plus interferon alfa in 12 teenagers has been compared with interferon alone in 10 (16). There was no difference in dropout rate, but viral clearance was achieved in 50% of the patients who took the combination treatment versus 30% of those who took monotherapy. Adverse events were similar in the two groups. There was mild hemolytic anemia at the end of the first month in most of the children who took ribavirin, but four had moderate to severe hemolysis and two had to stop taking ribavirin. Severe hemolysis in a patient with thalassemia warranted withdrawal of ribavirin within 3 months.

Systematic reviews
In a meta-analysis of 72 trials with a total of 9991 patients, ribavirin plus interferon significantly reduced morbidity plus mortality (OR = 0.46; 95% CI = 0.22, 0.96) and significantly improved sustained viral clearance in treatment-naive patients (RR = 0.72; 95% CI = 0.68, 0.76), relapsers (RR = 0.63; 95% CI = 0.54, 0.73), and nonresponders (RR = 0.89; 95% CI = 0.84, 0.94) (17). This gave the following numbers needed to treat for beneficial effects (NNT$_B$): for reduction in mortality/morbidity 444 (302–5650); for clearing of HCV-RNA 4 (4–5); and for improving the histological response 8 (17–100).

This analysis also gave information about the increased toxicity of adding ribavirin, with the following numbers needed to harm (NNT$_H$): anemia 4 (4–5); leukopenia 4 (3–7); rash 11 (9–17); pruritus 13 (10–20); insomnia 14 (6–20); dosage reductions 14 (11–17); dermatitis 14 (8–50); dyspnea 17 (11–25); fatigue/weakness 17 (11–33); dry skin 20 (11–50); anorexia/nausea 20 (13–50); dyspepsia 20 (13–50); pharyngitis 20 (13–100); cough 20 (14–33); and stopping treatment 50 (33–100).

Treatment with peginterferon alfa-2b + ribavirin can achieve a complete clinical response in about 75% of patients with hepatitis C-related vasculitis. A complete clinical response correlates with the eradication of the virus and requires a shorter treatment period than that previously reported for interferon alfa plus ribavirin (14 months) (18). The short course was well tolerated, although one patient withdrew because of neutopenia (19).

Organs and Systems

Nervous system

Headache is a frequent adverse effect of combination therapy with ribavirin + interferon. However, headaches have not been reported in early controlled trials of ribavirin monotherapy for chronic hepatitis C, and the frequency of this adverse effect is comparable in patients treated with interferon and ribavirin and in interferon monotherapy. Of 452 patients treated with combination therapy for chronic hepatitis C, seven developed new severe migraine headaches and two had worsening of pre-existing migraine (20). The symptoms mostly started with a delay of several weeks to months. In seven patients, the migraine improved considerably or resolved when ribavirin was withheld or the dose was reduced. All of them had a recurrence when they were re-challenged with full-dose ribavirin. A causal link between ribavirin and migraine appears plausible, but has not been proven.

Progressive multifocal leukoencephalopathy in an HIV negative patient has been attributed to pegylated interferon-alfa 2a + ribavirin (21).

Hematologic

Ribavirin accumulates in erythrocytes, resulting in hemolysis by an unknown mechanism, perhaps related to oxidative damage to the erythrocyte membrane. Time-dependent and dose-dependent hemolytic anemia (eventually associated with hyperbilirubinemia and a high reticulocyte count) is the only major toxic effect associated with oral or intravenous ribavirin and is reversible on withdrawal. There was a fall in hemoglobin concentrations below 10.0 g/dl in 9% of patients with hepatitis C treated with ribavirin and interferon alfa (6,7).

In 140 patients with Nipah virus infection there was no difference in the incidence of adverse effects between those who elected to have ribavirin treatment and those who refused (22). Dosing was based on recommendations used to achieve the same approximate concentrations as those seen with 100–1200 mg/day in the treatment of hepatitis C. Anemia occurred in 37% of the ribavirin-treated patients and in the same number of controls, suggesting that ribavirin was equally well tolerated in the two groups.

In patients taking ribavirin plus interferon alfa-2b the average fall in hemoglobin is 2–3 g/dl. Of 57 patients taking ribavirin 800 mg/day 28 were randomized to a high dose of peginterferon alfa-2b once a week (3 micrograms/kg for 1 week, 1.5 micrograms/kg for 3 weeks, and 1.0 microgram/kg for 44 weeks) and 27 patients were randomized to receive a low dose (0.5 micrograms/kg) for 48 weeks; three patients required reduced doses of ribavirin because of anemia (23).

The anemia associated with peginterferon + ribavirin is thought to be a mixed form of ribavirin-induced hemolysis and interferon-induced myelosuppression. However in an 8- week study of 97 patients receiving peginterferon + ribavirin, while the mean hemoglobin fell significantly from 14.4 to 11.9 g/dl the serum erythropoietin responses were lower than seen in historical controls with iron deficiency (24). The mean dosage of ribavirin was reduced from 986 to 913 mg/day. Only 74% maintained their dosage of ribavirin.

Pure red cell aplasia can be associated with hepatitis A, B, or C infections and also with ribavirin and pegylated interferon alfa. Ribavirin-associated pure red cell aplasia is fully reversible after withdrawal (25).

- A 61-year-old man with chronic hepatitis C infection developed progressive anemia with a low reticulocyte count after 12 weeks of therapy despite reduction of the dose of ribavirin, but by 12 weeks after withdrawal of ribavirin the anemia had normalized.

Studies in rhesus monkeys have shown a dose-related reversible anemia, erythroid hypoplasia, and vacuolization of erythroid precursors (26).

In a randomized controlled trial of high-dose interferon alfa-2b plus oral ribavirin for 6 or 12 months in 50 patients with chronic hepatitis C, the sequential effects of treatment on hemoglobin, leukocytes, and platelets were recorded (27). There was a fall in hemoglobin, and the lowest concentrations were recorded after 6 months of treatment in both groups. All hematological measurements returned to normal after the end of treatment.

Detailed studies of the effects of ribavirin on erythrocyte ATP content and on the hexose monophosphate shunt have been conducted in vitro. ATP concentrations were significantly reduced and the hexose monophosphate shunt increased, suggesting erythrocyte susceptibility to oxidation. In vivo, ribavirin, alone or in combination with interferon, was associated with significant reductions in hemoglobin concentrations and a marked increase in absolute reticulocyte counts. Erythrocyte Na/K pump activity was significantly reduced, whereas K/Cl co-transport and its dithiothreitol-sensitive fraction and malondialdehyde and methemoglobin concentrations increased significantly. Ribavirin-treated patients showed an increase in aggregated band 3, which was associated with significantly increased binding of autologous antibodies and complement C3 fragments, suggesting erythrophagocytic removal by the reticuloendothelial system (28).

A low pretreatment platelet count, the dose of interferon alfa, and the haptoglobin phenotype are risk factors for ribavirin-induced anemia, and the fall in hemoglobin is independent of dose in the therapeutic range (29). In five patients with chronic hepatitis C on hemodialysis who received subcutaneous interferon alfa-2b and oral ribavirin for 40 weeks, the dose of ribavirin was titrated based on hemoglobin, with bone marrow support by erythropoietin (30). There was significant bone marrow toxicity in all five. A dose of 200 mg/day produced a steady-state AUC comparable to that obtained with 1000–1200 mg/day in historical controls with normal renal function. More severe anemia was possibly due to chronic renal insufficiency in addition to the prolonged effects of ribavirin.

Treatment of ribavirin-induced hemolytic anemia with recombinant human erythropoietin has been described in 13 patients (31). The hemoglobin concentration increased from a nadir of 10.2 g/dl to a median of 11.5 g/dl and ribavirin treatment did not have to be withdrawn.

Liver

As part of a multicenter, randomized, double-blind, placebo-controlled trial of ribavirin in 59 patients with hepatitis C virus infection, liver biopsies were studied for iron deposition (32). Increased total iron deposition, preferentially in hepatocytes, occurred during a 9-month course of ribavirin. The deposition had no apparent effect on the biochemical or histological response to ribavirin therapy.

Urinary tract

Ribavirin can cause hemoglobinuria, resulting in black urine (33).

Skin

Photosensitivity after administration of ribavirin has been described (34). A well-documented photoallergic reaction in a woman who was taking both ribavirin and interferon alfa provided evidence that ribavirin is a potential photosensitizer for UVB, a problem that may become increasingly relevant in patients with chronic hepatitis C taking combination therapy for 6–12 months with interferon alfa and ribavirin (34).

Pruritus, xerosis, and mild skin eruptions, such as eczema and lichen planus, are common (23%) during ribavirin plus interferon therapy (35). Control of these symptoms mostly requires sustained therapy with moderately potent to potent topical glucocorticoids, combined with baseline emollients throughout the combination treatment period. However, there are occasional reports of marked erythematous maculopapular eruptions starting 3–4 days after the start of combination therapy (36). Although this form of skin reaction (which is probably T cell mediated) is rare, it should be emphasized that it can occur early during treatment and can evolve into Stevens–Johnson syndrome.

Dermatitis occurred in 36 patients who were given ribavirin + pegylated interferon (37). Half of the patients had clinical symptoms within the first month of combination treatment, and the first signs typically appeared distant from the sites of peginterferon injection. All complained of generalized itch, and most had xerosis and erythemato-papulo-microvesicular lesions with a predilection for the extensor surfaces of the limbs and skin sites exposed to friction. Seven had skin biopsies with a superficial dermal perivascular inflammation with

spongiosis and parakeratosis; erythrocyte extravasation, sparse keratinocyte necrosis, and extension of the inflammation to the interface were variable; the last of these occurred in the clinically more severe cases. Two patients developed specific skin signs that differed from the eczema-like pattern described above. One patient with generalized eczematous skin changes eventually developed malar hypertrichosis lanuginosa and bullous skin lesions with milia on the backs of both hands, leading to a diagnosis of porphyria cutanea tarda; one patient developed a bullous eruption with histological features of acantholytic dermatitis with a non-specific immunohistological profile.

Five cases of Meyerson's syndrome (halo dermatitis), a benign eczematous rash around a pre-existing nevus, have been reported during treatment for hepatitis C (38,39). This syndrome has been reported with interferon-alfa-2b but not ribavirin in other conditions and resolved on withdrawal of therapy.

There is a well established association between hepatitis C virus infection and porphyria cutanea tarda. However it is thought that ribavirin increases the risk by increasing iron overload via hemolysis. Two cases of porphyria cutanea tarda have been reported after treatment with ribavirin and interferon (40).

There have been three reports of cutaneous sarcoidosis associated with pegylated interferon alfa plus ribavirin treatment (41).

Susceptibility Factors

Renal impairment

The clearance of ribavirin is impaired in patients with renal dysfunction, and it is not removed by hemodialysis. It is therefore not recommended for patients with a creatinine clearance under 50 ml/minute. However hepatitis C infection is associated with renal complications, such as membranoproliferative glomerulonephritis with or without cryoglobulinemia, membranous glomerulonephritis, and focal segmental glomerulosclerosis. Of seven patients treated with interferon six became HCV-RNA PCR negative, four maintained both virological and renal remission, and one maintained virological and partial renal remission (42). Ribavirin-induced anemia was managed in five patients with low-dose iron and erythropoietin. The authors concluded that ribavirin can be used, with reasonable safety, in HCV-related vasculitis and glomerulonephritis irrespective of renal function.

Transient acantholytic dermatosis (Grover's disease) was first described by Grover in 1970 as a pruritic, self-limiting, popular, or papulovesicular eruption, mainly distributed on the trunk of white middle-aged men. The histopathological hallmark is suprabasal acantholysis at different levels of the epidermis. Its origin is uncertain; most cases are related to sunlight, heat, or sweating. Grover's disease has been attributed to ribavirin (43).

- A 55-year-old man with chronic hepatitis C presented with a pruritic papular eruption on the trunk lasting 2 weeks. He had multiple, erythematous, excoriated papules on the neck, trunk, upper arms, and thighs. The lesions appeared 2 weeks after combination therapy with oral ribavirin and subcutaneous interferon alfa-2b. He had previously been treated with interferon alfa alone (in the same dosage). On withdrawal of ribavirin the lesions gradually faded, but they returned 1 week after reintroduction.

Second-Generation Effects

Teratogenicity

Ribavirin is teratogenic and embryotoxic in laboratory animals and should not be given to pregnant women. Concern has been expressed about the safety of people in the same room as patients being treated with ribavirin by aerosol, particularly women of child-bearing age. However, no ribavirin was detected in the urine, plasma, or erythrocytes of 19 nurses exposed to ribavirin administered via ventilator, oxygen tent, or oxygen hood over 3 days (44).

Drug–Drug Interactions

The intracellular triphosphorylation and pharmacokinetics of lamivudine (3TC), stavudine (d4T), and zidovudine (ZDV) have been assessed in 56 patients co-infected with human immunodeficiency virus and hepatitis C virus receiving peginterferon alfa-2a (40KD) 180 micrograms/week plus either placebo or ribavirin 800 mg/day; there was no difference (45).

Didanosine

Multisystem organ dysfunction and lactic acidemia occurred in two of 15 patients with HIV and hepatitis C infections who received interferon alfa, didanosine, and ribavirin (46). Co-administration of didanosine with ribavirin can lead to increased toxicity secondary to raised intracellular concentrations of phosphorylated didanosine (ddA-TP) (47,48). Thus, the evidence suggests that the combination of didanosine plus ribavirin increases the risk of lactic acidosis.

Interferon alfa-2b

In a randomized 48-week study, 107 patients co-infected with human immunodeficiency virus (HIV) and hepatitis C (HCV) were given interferon alfa-2b together with either a full course of ribavirin or placebo for 16 weeks, followed by ribavirin (49). More than 80% of the patients in both groups also took HAART and 25–28% took zidovudine. Significantly more patients in those who took the full 48-week course of ribavirin had to reduce the dose of ribavirin because of anemia (28% versus 11%). Of those who also took zidovudine, only those who took the full course of ribavirin had to reduce the dose of ribavirin for any reason (67% versus 24%) or for anemia (60% versus 16%). Zidovudine, but no other nucleoside analogue, was associated with a significantly lower hemoglobin concentration (10.1 versus 13.0 g/dl) and a significantly larger fall (−3.64 versus −2.08 g/dl) However, there was no pronounced association between leukopenia or neutropenia and zidovudine. The combination of zidovudine with interferons and ribavirin should be avoided if possible.

Warfarin

An interaction of warfarin with ribavirin has been reported (50).

- In a 61-year-old white man with chronic hepatitis C, who took interferon plus ribavirin, the dosage of warfarin had to be increased by about 40% (from 45 to 63 mg/week) in order to maintain the desired degree of anticoagulation. This effect was reproduced on rechallenge with ribavirin.

The mechanism of this supposed interaction is not known. For example, ribavirin is cleared by intracellular phosphorylation and its metabolites by the kidneys, warfarin by cytochrome P450 isozymes in the liver; warfarin is highly protein bound, ribavirin is not. However, an effect on warfarin absorption or its action on clotting factor synthesis is possible.

Management of Adverse Drug Reactions

The management with epoetin alfa and danazol of anemia during therapy with interferon and ribavirin has been reported (51).

- A 50-year-old African–American man with chronic hepatitis C was initially given subcutaneous interferon alfa-2b (3 mU three times/week) and oral ribavirin 1200 mg/day. The pre-treatment hemoglobin was 14.3 g/dl. There was a good therapeutic response, but the hemoglobin fell firstly to 11.2 g/dl and then to 9.4 g/dl by week 42. This prompted a reduction in the dosage of ribavirin to 800 mg/day, and the hemoglobin rose to 11.8 g/dl. The antiviral therapy was withdrawn at week 48 and reintroduced 3 months later for a relapse. He was given subcutaneous peginterferon alfa-2a (180 micrograms/week) and oral ribavirin 1200 mg/day plus subcutaneous epoetin alfa 40 00 U/week to prevent anemia and therefore the need to reduce the dose of ribavirin. Serum hemoglobin at the start of the second course of therapy was 14.7 g/dl and it remained stable throughout the first 12 weeks of therapy. However, at week 16, there was an abrupt fall in hemoglobin from 14.6 to 8.5 g/dl. The ribavirin was immediately withdrawn, the dosage of peginterferon was reduced, and the dosage of epoetin alfa was increased to 60 000 U/week. At week 18, the hemoglobin fell to 7.2 g/dl and the peginterferon was withdrawn. At week 20, the hemoglobin reached a nadir of 5.6 g/dl, requiring transfusion with 3 units of packed erythrocytes. The patient continued to require about 1 unit of blood every week despite continuing epoetin alfa, which was finally stopped at week 26. Erythropoietin antibodies became detectable by week 12 and peaked at week 24. Danazol 200 mg bd then 400 mg bd was started 8 weeks after the withdrawal of epoetin alfa. The hemoglobin then became stable at 9–10 g/dl for 24 weeks.

References

1. Stapleton T, et al. Studies with a Broad Spectrum Antiviral Agent. International Congress and Symposium SeriesLondon/New York: Royal Society of Medicine Services;. 1986.

2. Couch RB. Respiratory disease. In: Galasso GJ, Whitley RJ, Merigan TC, editors. Antiviral Agents and Viral Diseases of Man . New York: Raven Press, 1990:327.

3. Japour AJ, Lertora JJ, Meehan PM, Erice A, Connor JD, Griffith BP, Clax PA, Holden-Wiltse J, Hussey S, Walesky M, Cooney E, Pollard R, Timpone J, McLaren C, Johanneson N, Wood K, Booth D, Bassiakos Y, Crumpacker CS. A phase-I study of the safety, pharmacokinetics, and antiviral activity of combination didanosine and ribavirin in patients with HIV-1 disease. AIDS Clinical Trials Group 231 Protocol Team. J Acquir Immune Defic Syndr Hum Retrovirol 1996;13(3):235–46.

4. McCormick JB, King IJ, Webb PA, Scribner CL, Craven RB, Johnson KM, Elliott LH, Belmont-Williams R. Lassa fever. Effective therapy with ribavirin. N Engl J Med 1986;314(1):20–6.

5. Fisher-Hoch SP, Khan JA, Rehman S, Mirza S, Khurshid M, McCormick JB. Crimean Congo–haemorrhagic fever treated with oral ribavirin. Lancet 1995;346(8973):472–5.

6. Davis GL, Esteban-Mur R, Rustgi V, Hoefs J, Gordon SC, Trepo C, Shiffman ML, Zeuzem S, Craxi A, Ling MH, Albrecht J. Interferon alfa-2b alone or in combination with ribavirin for the treatment of relapse of chronic hepatitis C. International Hepatitis Interventional Therapy Group. N Engl J Med 1998;339(21):1493–9.

7. McHutchison JG, Gordon SC, Schiff ER, Shiffman ML, Lee WM, Rustgi VK, Goodman ZD, Ling MH, Cort S, Albrecht JK. Interferon alfa-2b alone or in combination with ribavirin as initial treatment for chronic hepatitis C. Hepatitis Interventional Therapy Group. N Engl J Med 1998;339(21):1485–92.

8. Hall CB, McBride JT, Walsh EE, Bell DM, Gala CL, Hildreth S, Ten Eyck LG, Hall WJ. Aerosolized ribavirin treatment of infants with respiratory syncytial viral infection. A randomized double-blind study. N Engl J Med 1983;308(24):1443–7.

9. Smith DW, Frankel LR, Mathers LH, Tang AT, Ariagno RL, Prober CG. A controlled trial of aerosolized ribavirin in infants receiving mechanical ventilation for severe respiratory syncytial virus infection. N Engl J Med 1991;325(1):24–9.

10. Chutaputti A. Adverse effects and other safety aspects of the hepatitis C antivirals. J Gastroenterol Hepatol 2000;15(Suppl):E156–63.

11. Perez-Olmeda M, Nunez M, Romero M, Gonzalez J, Castro A, Arribas JR, Pedreira J, Barreiro P, Garcia-Samaniego J, Martin-Carbonero L, Jimenez-Nacher I, Soriano V. Pegylated IFN-alpha2b plus ribavirin as therapy for chronic hepatitis C in HIV-infected patients. AIDS 2003;17:1023-8.

12. Carlsson T, Reichard O, Norkrans G, Blackberg J, Sangfelt P, Wallmark E, Weiland O. Hepatitis C virus RNA kinetics during the initial 12 weeks treatment with pegylated interferon-alpha 2a and ribavirin according to virological response. J Viral Hepat 2005;12(5):473-80.

13. Kaymakoglu S, Karaca C, Demir K, Poturoglu S, Danalioglu A, Badur S, Bozaci M, Besisik F, Cakaloglu Y, Okten A. Alpha interferon and ribavirin combination therapy of chronic hepatitis D. Antimicrob Agents Chemother 2005;49(3):1135-8.

14. Sugawara Y, Makuuchi M. Should living donor liver transplantation be offered to patients with hepatitis C virus cirrhosis? J Hepatol 2005;42(4):472-5.

15. Yu ML, Hou NJ, Dai CY, Chang WY, Chuang WL. Successful treatment of fulminant hepatitis C by therapy with alpha interferon and ribavirin. Antimicrob Agents Chemother 2005;49(9):3986-7.

16. Suoglu D OD, Elkabes B, Sokucu S, Saner G. Does interferon and ribavirin combination therapy increase the rate of treatment response in children with hepatitis C? J Pediatr Gastroenterol Nutr 2002;34(2):199–206.

17. Brok J, Gluud LL, Gluud C. Effects of adding ribavirin to interferon to treat chronic hepatitis C infection: a systematic review and meta-analysis of randomized trials. Arch Intern Med 2005;165(19):2206-12.

18. Cacoub P, Saadoun D, Sene D, Limal N, Piette JC. Treatment of hepatitis C virus-related systemic vasculitis. J Rheumatol 2005;32(11):2078-82.

19. Cacoub P, Saadoun D, Limal N, Sene D, Lidove O, Piette JC. PEGylated interferon alfa-2b and ribavirin treatment in patients with hepatitis C virus-related systemic vasculitis. Arthritis Rheum 2005;52(3):911-5.

20. Brau N, Bini EJ, Stancic S, Finch DA, Aytaman A. Severe migraine headaches are caused by ribavirin but not by interferon alpha-2B in combination therapy for chronic hepatitis C. J Hepatol 2003;38:871-2.

21. Lima MA, Auriel E, Wuthrich C, Borenstein NM, Koralnik IJ. Progressive multifocal leukoencephalopathy as a complication of hepatitis C virus treatment in an HIV-negative patient. Clin Infect Dis 2005;41(3):417-9.

22. Chong HT, Kamarulzaman A, Tan CT, Goh KJ, Thayaparan T, Kunjapan SR, Chew NK, Chua KB, Lam SK. Treatment of acute Nipah encephalitis with ribavirin. Ann Neurol 2001;49(6):810–3.

23. Buti M, Sanchez-Avila F, Lurie Y, Stalgis C, Valdes A, Martell M, Esteban R. Viral kinetics in genotype 1 chronic hepatitis C patients during therapy with 2 different doses of peginterferon alfa-2b plus ribavirin. Hepatology 2002;35(4):930–6.

24. Balan V, Schwartz D, Wu GY, Muir AJ, Ghalib R, Jackson J, Keeffe EB, Rossaro L, Burnett A, Goon BL, Bowers PJ, Leitz GJ; HCV Natural History Study Group. Erythropoietic response to anemia in chronic hepatitis C patients receiving combination pegylated interferon/ribavirin. Am J Gastroenterol 2005;100(2):299-307.

25. Tanaka N, Ishida F, Tanaka E. Ribavirin-induced pure red-cell aplasia during treatment of chronic hepatitis C. N Engl J Med 2004;350(12):1264-5.

26. Canonico PG, Kastello MD, Cosgriff TM, Donovan JC, Ross PE, Spears CT, Stephen EL. Hematological and bone marrow effects of ribavirin in rhesus monkeys. Toxicol Appl Pharmacol 1984;74(2):163-72.

27. Marco VD, Almasio P, Vaccaro A, Ferraro D, Parisi P, Cataldo MG, Di Stefano R, Craxi A. Combined treatment of relapse of chronic hepatitis C with high-dose alpha2b interferon plus ribavirin for 6 or 12 months. J Hepatol 2000;33(3):456–62.

28. De Franceschi L, Fattovich G, Turrini F, Ayi K, Brugnara C, Manzato F, Noventa F, Stanzial AM, Solero P, Corrocher R. Hemolytic anemia induced by ribavirin therapy in patients with chronic hepatitis C virus infection: role of membrane oxidative damage. Hepatology 2000;31(4):997–1004.

29. Van Vlierbergh H, Delanghe JR, De Vos M, Leroux-Roel G. BASL Steering Committee. Factors influencing ribavirin-induced hemolysis. J Hepatol 2001;34(6):911–6.

30. Tan AC, Brouwer JT, Glue P, van Leusen R, Kauffmann RH, Schalm SW, de Vries RA, Vroom B. Safety of interferon and ribavirin therapy in haemodialysis

patients with chronic hepatitis C: results of a pilot study. Nephrol Dial Transplant 2001;16(1):193-5.

31. Gergely AE, Lafarge P, Fouchard-Hubert I, Lunel-Fabiani F. Treatment of ribavirin/interferon-induced anemia with erythropoietin in patients with hepatitis C. Hepatology 2002;35(5):1281-2.

32. Fiel MI, Schiano TD, Guido M, Thung SN, Lindsay KL, Davis GL, Lewis JH, Seeff LB, Bodenheimer HC Jr. Increased hepatic iron deposition resulting from treatment of chronic hepatitis C with ribavirin. Am J Clin Pathol 2000;113(1):35-9.

33. Massoud OI, Yousef WI, Mullen KD. Hemoglobinuria with ribavirin treatment. J Clin Gastroenterol 2003;36:367-8.

34. Stryjek-Kaminska D, Ochsendorf F, Roder C, Wolter M, Zeuzem S. Photoallergic skin reaction to ribavirin. Am J Gastroenterol 1999;94(6):1686-8.

35. Kerl K, Negro F, Lubbe J. Cutaneous side-effects of treatment of chronic hepatitis C by interferon alfa and ribavirin. Br J Dermatol 2003;149:656.

36. Okai T, Shirasaki F, Sawabu N. Erythematous maculopapular eruption due to ribavirin administration in a patient with chronic hepatitis C. J Clin Gastroenterol 2003;36:283-4.

37. Lubbe J, Kerl K, Negro F, Saurat JH. Clinical and immunological features of hepatitis C treatment-associated dermatitis in 36 prospective cases. Br J Dermatol 2005;153(5):1088-90.

38. Girard C, Bessis D, Blatire V, Guilhou JJ, Guillot B. Meyerson's phenomenon induced by interferon-alfa plus ribavirin in hepatitis C infection. Br J Dermatol 2005;152(1):182-3.

39. Conde-Taboada A, de la Torre C, Feal C, Mayo E, Gonzalez-Sixto B, Cruces MJ. Meyerson's naevi induced by interferon alfa plus ribavirin combination therapy in hepatitis C infection. Br J Dermatol 2005;153(5):1070-2.

40. Thevenot T, Bachmeyer C, Hammi R, Dumouchel P, Ducamp-Posak I, Cadranel JF. Occurrence of porphyria cutanea tarda during peginterferon/ribavirin therapy for chronic viral hepatitis C. J Hepatol 2005;42(4):607-8.

41. Hurst EA, Mauro T. Sarcoidosis associated with pegylated interferon alfa and ribavirin treatment for chronic hepatitis C: a case report and review of the literature. Arch Dermatol 2005;141(7):865-8.

42. Bruchfeld A, Lindahl K, Stahle L, Soderberg M, Schvarcz R. Interferon and ribavirin treatment in patients with hepatitis C-associated renal disease and renal insufficiency. Nephrol Dial Transplant 2003;18:1573-80.

43. Antunes I, Azevedo F, Mesquita-Guimaraes J, Resende C, Fernandes N, MacEdo G. Grover's disease secondary to ribavirin. Br J Dermatol 2000;142(6):1257-8.

44. Rodriguez WJ, Bui RH, Connor JD, Kim HW, Brandt CD, Parrott RH, Burch B, Mace J. Environmental exposure of primary care personnel to ribavirin aerosol when supervising treatment of infants with respiratory syncytial virus infections. Antimicrob Agents Chemother 1987;31(7):1143-6.

45. Rodriguez-Torres M, Torriani FJ, Soriano V, Borucki MJ, Lissen E, Sulkowski M, Dieterich D, Wang K, Gries JM, Hoggard PG, Back D. Effect of ribavirin on intracellular and plasma pharmacokinetics of nucleoside reverse transcriptase inhibitors in patients with human immunodeficiency virus-hepatitis C virus coinfection: results of a randomized clinical study. Antimicrob Agents Chemother 2005;49(10):3997-4008.

46. Lafeuillade A, Hittinger G, Chadapaud S. Increased mitochondrial toxicity with ribavirin in HIV/HCV coinfection. Lancet 2001;357(9252):280-1.

47. Kakuda TN, Brinkman K. Mitochondrial toxic effects and ribavirin. Lancet 2001;357(9270):1802-3.

48. Salmon-Ceron D, Chauvelot-Moachon L, Abad S, Silbermann B, Sogni P. Mitochondrial toxic effects and ribavirin. Lancet 2001;357(9270):1803-4.

49. Brau N, Rodriguez-Torres M, Prokupek D, Bonacini M, Giffen CA, Smith JJ, Frost KR, Kostman JR. Treatment of chronic hepatitis C in HIV/HCV-coinfection with interferon alpha-2b+ full-course vs. 16-week delayed ribavirin. Hepatology 2004;39(4):989-98.

50. Schulman S. Inhibition of warfarin activity by ribavirin. Ann Pharmacother 2002;36(1):72-4.

51. Stravitz RT, Chung H, Sterling RK, Luketic VA, Sanyal AJ, Price AS, Purrington A, Shiffman ML. Antibody-mediated pure red cell aplasia due to epoetin alfa during antiviral therapy of chronic hepatitis C. Am J Gastroenterol 2005;100(6):1415-9.

Rimantadine

General Information

Rimantadine hydrochloride, an alpha-methyl derivative of amantadine (alpha-methyl-1-adamantane methylamine hydrochloride), is more active than amantadine against influenza A viruses in vitro and in laboratory animals. It is an alternative to amantadine for the prevention and treatment of influenza A virus infections in adults and for the prevention of influenza in children. Adverse effects have been considered to be less common with rimantadine (SEDA-8, 143), and it is generally tolerated better than amantadine, because it causes fewer nervous system adverse effects [1]. Unfortunately, rimantadine is more costly, which has led many institutions to develop influenza treatment guidelines. Both drugs work by blocking the M2 ion channel, which is needed to affect a pH change that helps to initiate viral uncoating.

In a systematic review of seven trials of amantadine versus placebo ($n = 1797$), three of rimantadine versus placebo ($n = 688$), and two of amantadine versus rimantadine ($n = 455$) in the prevention of influenza A illness, there was a relative odds reduction of illness of 64% with amantadine compared with placebo, a 75% reduction with rimantadine compared with placebo, and no significant difference between the two drugs [2]. There was a significantly higher risk of central nervous system adverse reactions and premature withdrawal with amantadine compared with placebo, but not with rimantadine compared with placebo. However, there was a significant increase in the risk of gastrointestinal adverse events with rimantadine compared with placebo (OR = 3.34, 95% CI = 1.17, 9.55).

Organs and Systems

Nervous system

The major advantage of rimantadine is a lower risk of central nervous system effects, such as light-headedness, difficulty in concentrating, nervousness, and insomnia, which can be a significant problem with amantadine, particularly in older patients. The English-language literature from 1966 to 1994 has been reviewed (3). In 598 elderly patients, nervous system adverse effects were more common with higher dosages (4.9% at 100 mg/day to 12.5% at 200 mg/day).

Gastrointestinal

The English-language literature from 1966 to 1994 has been reviewed (3). Gastrointestinal effects, including nausea, loss of appetite, diarrhea, and dry mouth, occurred in 8.4% of children under 10 years of age, in 3.1% of adults, and in 2.9% of elderly patients taking 100 mg/day and 17.0% taking 200 mg/day.

Long-Term Effects

Drug tolerance

Emergence of resistance of influenza A virus to rimantadine in vivo has been reported (4), but the incidence of resistance in field isolates appears to be extremely low (5).

References

1. Fleming DM. Managing influenza: amantadine, rimantadine and beyond. Int J Clin Pract 2001;55(3):189–95.
2. Marra F, Marra CA, Stiver HG. A case for rimantadine to be marketed in Canada for prophylaxis of influenza A virus infection. Can Respir J 2003;10(7):381–8.
3. Wintermeyer SM, Nahata MC. Rimantadine: a clinical perspective. Ann Pharmacother 1995;29(3):299–310.
4. Hayden FG, Belshe RB, Clover RD, Hay AJ, Oakes MG, Soo W. Emergence and apparent transmission of rimantadine-resistant influenza A virus in families. N Engl J Med 1989;321(25):1696–702.
5. Ziegler T, Hemphill ML, Ziegler ML, Perez-Oronoz G, Klimov AI, Hampson AW, Regnery HL, Cox NJ. Low incidence of rimantadine resistance in field isolates of influenza A viruses. J Infect Dis 1999;180(4):935–9.

Saquinavir

See also Protease inhibitors

General Information

Of all protease inhibitors, saquinavir is the most potent in vitro. However, owing to poor systemic availability (less than 4%), it is the least potent of all protease inhibitors in use, although a formulation with increased availability has been marketed. However, when saquinavir is given together with ritonavir, the strong inhibitory effect on CYP3A4 of the latter results in high plasma concentrations of saquinavir. This interaction has been exploited, with favorable results, both in first-line protease therapy and as salvage treatment in patients with virus resistant to a regimen containing a protease inhibitor. Using 400 mg/400 mg and 800 mg/200 mg, the saquinavir soft gelatin capsule AUC increased by 17–23 times compared with saquinavir alone (1). Saquinavir had no clinically significant effect on the pharmacokinetics of ritonavir. Ritonavir as a booster allows a much higher saquinavir concentration to be obtained, allowing twice-daily dosing and a reduced capsule burden.

General adverse effects

Hitherto, no particular or frequent adverse effects attributable to saquinavir have been reported from trials in which saquinavir was used at the licensed dosage of 600 mg tds (2). Diarrhea, usually of only moderate severity, occurring in 3–4% of patients, is the most common single adverse effect (SEDA-22, 310).

As with other HIV-1 protease inhibitors, saquinavir may be associated with drug interactions as a result of the effect of saquinavir on the hepatic cytochrome P450 oxidase system. Although compared with other HIV protease inhibitors, saquinavir has less of an inhibitory effect on cytochrome P450 isozymes; clinically relevant interactions can nevertheless occur. Drug interactions with saquinavir have been reviewed (3).

Organs and Systems

Psychological, psychiatric

Saquinavir can occasionally be associated with acute paranoid psychotic reactions (4).

- A 41-year-old woman took zidovudine and didanosine for HIV-1 infection after an acute seroconversion illness. Zidovudine and didanosine had to be withdrawn because of neutropenia and nausea, so she was given stavudine plus lamivudine without any adverse effects. Saquinavir (600 mg tds) was added 12 months later because of weight loss and a falling CD4 cell count. Within 24 hours of starting to take saquinavir she developed agitated depression with paranoid ideation. After drug withdrawal her mental health returned to normal over 5 days. Over the next 6 weeks stavudine and lamivudine were reintroduced without any adverse effects and continued for a further 11 months. Saquinavir was then reintroduced at the previous dosage, and within 2 days she again became extremely mentally agitated with paranoid ideation. Saquinavir was withdrawn and she recovered within 7 days. She was later given indinavir without problems. Her mother had a history of major depressive illness.

Endocrine

Gynecomastia has been reported in a series of men taking saquinavir (5). In these cases the association was clear (particularly since there was positive dechallenge), but this is a rare effect and has not previously been reported with either this or other protease inhibitors, although it has been associated with the nucleoside analogue reverse transcriptase inhibitor stavudine.

Metabolism

Triglycerides rose by 3.45 mmol/l (314 mg/dl) and total cholesterol by 1.1 mmol/l (43 mg/dl) in 11 patients taking amprenavir, saquinavir, and ritonavir (6).

Gastrointestinal

Diarrhea, usually of only moderate severity, which occurs in 3–4% of patients, is the most common single adverse effect of saquinavir (7–9).

All of 11 subjects in a randomized study had some degree of self-limiting diarrhea, which improved within 2 weeks of the start of triple therapy with protease inhibitors (6). Six had loose stools throughout the study, but the symptoms were controllable with loperamide or diphenoxylate + atropine.

Nausea and vomiting were the main adverse effects that led to drug withdrawal in 8.6% of patients taking saquinavir + ritonavir 1000 + 100mg bd. (10).

Liver

Of 28 patients who took rifampicin 600 mg/day together with ritonavir 100 mg bd and saquinavir 100 mg bd for 28 days, 11 had significant hepatocellular damage (11).

Urinary tract

A renal stone has been attributed to saquinavir (12).

• A 42-year-old HIV-positive man with a prior history of *Pneumocystis jiroveci* pneumonia who had been treated with zidovudine and dideoxycytidine started to take saquinavir 600 mg tds. His CD4 cell count rose from 28×10^3/l to 101×10^3/l and zidovudine and dideoxycytidine were replaced by stavudine and lamivudine, because of mild peripheral neuropathy. Saquinavir was continued unchanged. A few months later he developed left-sided loin pain and hematuria and a left renal calculus was seen on ultrasound. A month later the same signs and symptoms recurred and a few weeks later he passed a small black stone in the urine. Ultrasonic lithotripsy was performed, with a good result. Saquinavir was discontinued, after which he had no further renal problems.

This case suggests that saquinavir, like indinavir, may be associated with renal calculus formation in some individuals.

Skin

Adverse skin reactions to saquinavir are exceptional, but erythema multiforme has been reported (13).

• A 32-year-old HIV-positive man, who had been treated with didanosine and lamivudine, added saquinavir (600 mg tds) because of a rising plasma HIV-1 RNA viral load. Five days later he presented with a generalized maculopapular skin eruption, the lesions being centered on a bulla, and erosive lesions on the palate. Histological examination was compatible with erythema multiforme. Saquinavir was discontinued and all the mucocutaneous lesions healed within 15 days. Rechallenge was not attempted.

Long-Term Effects

Tumorigenicity

• Eruptive angiolipomata occurred in a 49-year-old woman after she had taken stavudine 30 mg bd, lamivudine 150 mg bd, and saquinavir 600 mg 8-hourly for 3 months (14). This has also been reported with other protease inhibitors (15,16) and the mechanism is not known. In one case lipomata regressed after the introduction of indinavir (17).

Drug Administration

Drug formulations

Saquinavir has a very low systemic availability (4%), and is used in combination with ritonavir. In an attempt to improve its systemic availability, saquinavir soft-gel capsules have been manufactured. However, when they were co-administered with ritonavir, the soft-gel and the hard-gel formulations resulted in similar drug exposures in healthy volunteers (18). Drug exposure was even slightly better for the hard-gel capsules, but diarrhea was far less frequent. Of 442 patients who used the soft gelatin capsule 1200 mg tds, for 48 weeks, 8% withdrew because of adverse events, which were not necessarily related to saquinavir. No new adverse effects or laboratory abnormalities emerged compared with those previously observed with the hard gelatin capsule. The most frequent adverse effects were gastrointestinal, diarrhea being the most common (19).

Drug–Drug Interactions

Amprenavir

In an open, randomized study of amprenavir combined with indinavir, nelfinavir, and saquinavir (20), saquinavir lowered the amprenavir AUC by 32%; amprenavir did not alter the pharmacokinetics of saquinavir.

Ciclosporin

In an HIV-1-positive kidney transplant recipient, saquinavir increased the trough concentration of ciclosporin three-fold, resulting in fatigue, headache, and gastrointestinal discomfort. Ciclosporin, like saquinavir, is metabolized by CYP3A. Saquinavir plasma concentrations were likewise increased by ciclosporin. All the symptoms disappeared after downward adjustment of the doses of both ciclosporin and saquinavir (21).

Cytotoxic drugs

In 37 patients with HIV-associated non-Hodgkin's lymphoma who were treated with a 96-hour continuous intravenous infusion of cyclophosphamide + doxorubicin + etoposide, severe (grade 3 or 4) mucositis occurred in eight of 12 patients who received concomitant saquinavir (600 mg tds) compared with three of 25 who did not receive saquinavir. Although the authors did not measure saquinavir plasma concentrations, they suggested that this finding may have been explained by inhibition of the metabolism of one or more of the cytotoxic drugs by saquinavir (22).

Erythromycin

In 11 healthy men, erythromycin 250 mg qds increased the AUC of saquinavir (given as a soft gel capsule 1200 mg tds) by 69% when both were given for 7 days (23).

Itraconazole

The HIV protease inhibitor saquinavir has limited and variable oral systemic availability and ritonavir, an inhibitor of CYP450 and P glycoprotein, is widely used to increase its systemic exposure. A small pilot study in three HIV-infected patients has suggested that oral itraconazole can have similar effects on the oral availability of saquinavir (24). Concomitant use of itraconazole 200 mg/day with a combination of saquinavir and two nucleoside reverse transcriptase inhibitors led to a 2.5- to 6.9-fold increase in the AUC of saquinavir, a 2.0- to 5.4-fold increase in peak plasma concentrations, and a 1.6- to 17-fold increase in trough plasma concentrations. The effect of itraconazole on saquinavir was comparable to that of ritonavir.

Ketoconazole

In 12 healthy men, ketoconazole 400 mg/day increased the AUC of saquinavir (given as a soft gel capsule 1200 mg tds) by 190% when both were given for 7 days (23).

Midazolam

Saquinavir substantially potentiates the effects of midazolam by raising its blood concentrations (25).

Saquinavir substantially potentiates the effects of midazolam by raising its blood concentrations, and this suggests that a parallel interaction could occur with other protease inhibitors (and no doubt certain other benzodiazepines) in similar combinations (25).

Protease inhibitors

In a randomized study, 11 patients, all of whom had had at least one failure on a regimen containing a protease inhibitor, were assigned to either amprenavir + ritonavir (600/100 mg bd) or saquinavir + ritonavir (1000/100 mg

bd) for 7–10 days (6). After pharmacokinetic evaluation of the initial regimen, a third protease inhibitor was prescribed (either saquinavir or amprenavir). After 14 days of triple therapy the effects of the third protease inhibitor on pharmacokinetics of the previously prescribed drugs were assessed. The addition of saquinavir did not change the concentration of amprenavir, but adding amprenavir to saquinavir/ritonavir caused a reduction in saquinavir/ritonavir exposure. This effect was overcome by increasing the dosages of saquinavir and ritonavir to 1400 mg bd and 200 mg bd respectively.

The combination of saquinavir plus ritonavir/lopinavir has been studied in 45 heavily pretreated patients with limited reverse transcriptase options and 32 mostly treatment-naïve patients with advanced HIV disease, who received saquinavir, ritonavir, and two or three different NRTIs (26). The pharmacokinetics of saquinavir were similar in the two groups, suggesting that lopinavir has little effect on saquinavir concentrations. However ritonavir concentrations were significantly lower with lopinavir/ritonavir, which is consistent with previous results. Importantly this did not affect the boosting potential of the drug.

Taken together these data suggest that saquinavir can be combined with lopinavir + ritonavir without dosage adjustments. However, plasma concentration monitoring should be encouraged because the pharmacokinetics of saquinavir are highly variable.

Rifamycins

In 14 healthy men, rifampicin 600 mg/day reduced the AUC of saquinavir (given as a soft gel capsule 1200 mg tds) by 46% when both were given for 14 days; this reduction can be counteracted by the addition of ritonavir (18).

References

1. Buss N, Snell P, Bock J, Hsu A, Jorga K. Saquinavir and ritonavir pharmacokinetics following combined ritonavir and saquinavir (soft gelatin capsules) administration. Br J Clin Pharmacol 2001;52(3):255–64.
2. Pollard RB. Use of proteinase inhibitors in clinical practice. Pharmacotherapy 1994;14(6 Pt 2):S21–9.
3. Vella S, Floridia M. Saquinavir. Clinical pharmacology and efficacy. Clin Pharmacokinet 1998;34(3):189–201.
4. Finlayson JA, Laing RB. Acute paranoid reaction to saquinavir. Am J Health Syst Pharm 1998;55(19):2016–7.
5. Donovan B, Bodsworth NJ, Mulhall BP, Allen D. Gynaecomastia associated with saquinavir therapy. Int J STD AIDS 1999;10(1):49–50.
6. Corbett AH, Eron JJ, Fiscus SA, Rezk NL, Kashuba AD. The pharmacokinetics, safety, and initial virologic response of a triple-protease inhibitor salvage regimen containing amprenavir, saquinavir, and ritonavir. J Acquir Immune Defic Syndr 2004;36(4):921-8.
7. Kitchen VS, Skinner C, Ariyoshi K, Lane EA, Duncan IB, Burckhardt J, Burger HU, Bragman K, Pinching AJ, Weber JN. Safety and activity of saquinavir in HIV infection. Lancet 1995;345(8955):952–5.
8. Collier AC, Coombs RW, Schoenfeld DA, Bassett RL, Timpone J, Baruch A, Jones M, Facey K, Whitacre C,

McAuliffe VJ, Friedman HM, Merigan TC, Reichman RC, Hooper C, Corey L. Treatment of human immunodeficiency virus infection with saquinavir, zidovudine, and zalcitabine. AIDS Clinical Trials Group. N Engl J Med 1996;334(16):1011–7.

9. Noble S, Faulds D. Saquinavir. A review of its pharmacology and clinical potential in the management of HIV infection. Drugs 1996;52(1):93–112.

10. Macassa E, Delaugerre C, Teglas JP, Jullien V, Tréluyer JM, Veber F, Rouzioux C, Blanche S. Change to a once-daily combination including boosted atazanavir in HIV-1-infected children. Pediatr Infect Dis J 2006;25(9):809-14.

11. Gray A, Abdool Karim SS, Gengiah TN. Ritonavir/saquinavir safety concerns curtail antiretroviral therapy options for tuberculosis-HIV-co-infected patients in resource-constrained settings. AIDS 2006;20(2):302-3.

12. Green ST, McKendrick MW, Schmid ML, Mohsen AH, Prakasam SF. Renal calculi developing de novo in a patient taking saquinavir. Int J STD AIDS 1998;9(9):555.

13. Garat H, el Sayed F, Obadia M, Bazex J. Erythème polymorphe au saquinavir. [Erythema multiforme caused by saquinavir.] Ann Dermatol Venereol 1998;125(1):42–3.

14. Dauden E, Alvarez S, Garcia-Diez A. Eruptive angiolipomas associated with antiretroviral therapy. AIDS 2002;16(5):805–6.

15. Dank JP, Colven R. Protease inhibitor-associated angiolipomatosis. J Am Acad Dermatol 2000;42(1 Pt 1):129–31.

16. Bornhovd E, Sakrauski AK, Bruhl H, Walli R, Plewig G, Rocken M. Multiple circumscribed subcutaneous lipomas associated with use of human immunodeficiency virus protease inhibitors? Br J Dermatol 2000;143(5):1113–4.

17. Bates D. Valacyclovir neurotoxicity: two case reports and a review of the literature. Can J Hosp Pharm 2002;55:123–7.

18. Kurowski M, Sternfeld T, Sawyer A, Hill A, Mocklinghoff C. Pharmacokinetic and tolerability profile of twice-daily saquinavir hard gelatin capsules and saquinavir soft gelatin capsules boosted with ritonavir in healthy volunteers. HIV Med 2003;4:94-100.

19. Gill MJ. Safety profile of soft gelatin formulation of saquinavir in combination with nucleosides in a broad patient population. NV15182 Study Team. AIDS 1998;12(11):1400–2.

20. Sadler BM, Gillotin C, Lou Y, Eron JJ, Lang W, Haubrich R, Stein DS. Pharmacokinetic study of human immunodeficiency virus protease inhibitors used in combination with amprenavir. Antimicrob Agents Chemother 2001;45(12):3663–8.

21. Brinkman K, Huysmans F, Burger DM. Pharmacokinetic interaction between saquinavir and cyclosporine. Ann Intern Med 1998;129(11):914–5.

22. Sparano JA, Wiernik PH, Hu X, Sarta C, Henry DH, Ratech H. Saquinavir enhances the mucosal toxicity of infusional cyclophosphamide, doxorubicin, and etoposide in patients with HIV-associated non-Hodgkin's lymphoma. Med Oncol 1998;15(1):50–7.

23. Grub S, Bryson H, Goggin T, Ludin E, Jorga K. The interaction of saquinavir (soft gelatin capsule) with ketoconazole, erythromycin and rifampicin: comparison of the effect in healthy volunteers and in HIV-infected patients. Eur J Clin Pharmacol 2001;57(2):115–21.

24. Koks CH, van Heeswijk RP, Veldkamp AI, Meenhorst PL, Mulder JW, van der Meer JT, Beijnen JH, Hoetelmans RM. Itraconazole as an alternative for ritonavir liquid formulation when combined with saquinavir. AIDS 2000;14(1):89–90.

25. Palkama VJ, Ahonen J, Neuvonen PJ, Olkkola KT. Effect of saquinavir on the pharmacokinetics and pharmacodynamics of oral and intravenous midazolam. Clin Pharmacol Ther 1999;66(1):33–9.

26. Stephan C, Hentig N, Kourbeti I, Dauer B, Mosch M, Lutz T, Klauke S, Harder S, Kurowski M, Staszewski S. Saquinavir drug exposure is not impaired by the boosted double protease inhibitor combination of lopinavir/ritonavir. AIDS 2004;18(3):503-8.

Stavudine

See also Nucleoside analogue reverse transcriptase inhibitors (NRTIs)

General Information

Stavudine is a nucleoside analogue reverse transcriptase inhibitor. Its most important adverse effects are peripheral neuropathy and increases in hepatic transaminases, both of which usually resolve on withdrawal.

Observational studies

In a subgroup analysis of 655 antiretroviral therapy-naive patients who took lamivudine, stavudine and nevirapine for 12 months, five had to stop taking stavudine because of neuropathy, two because of lipodystrophy, and two for unknown reasons (1).

Comparative studies

In a comparison of efavirenz, ritonavir-boosted amprenavir, or stavudine, with a backbone of abacavir and lamivudine in 98 patients, the following adverse events were reported in those who took stavudine: hypertriglyceridemia (7%), diarrhea (3%), nausea (6%), rashes (4%), peripheral neuropathy (6%, compared with 0–1% in the other arms), raised lactate (5% compared with 0% in the other arms) (2).

Organs and Systems

Nervous system

The principal toxic effect of stavudine is a peripheral neuropathy, with symptoms similar to the neuropathy associated with didanosine and zalcitabine (3,4). The development of this neuropathy depends on the duration of treatment, with an increasing risk after 12 weeks of treatment. A prior history of neuropathy increases the risk of stavudine-induced neuropathy. After withdrawal, the symptoms usually improve within 2 weeks, although they can persist for several months.

In a small prospective study serum lactate concentration discriminated between distal sensory polyneuropathy and stavudine-associated neuropathy (5). A lactate concentration above the upper end of the reference range

(2.2 mmol/l) had a sensitivity and a specificity of 90% for stavudine-associated neuropathy.

In a systematic review of NRTI-associated neuropathies current knowledge in the field has been summarized (6). At the usual dose of stavudine the 1-year rate of sensory neuropathy is 12–15% and synergistic neurotoxicity with didanosine is marked.

Metabolism

Lipodystrophy, a syndrome characterized by fat redistribution, hyperglycemia/insulin resistance, and dyslipidemia, can be associated with long-term HIV infection or with highly active antiretroviral therapy (HAART). In 1035 patients, those who took stavudine were 1.35 times more likely to report lipodystrophy (7). However, the study was retrospective, and other factors unrelated to specific drug therapy may have had a greater effect on the adjusted odds ratio.

Hematologic

A modest dose-related macrocytosis without an associated anemia can occur during treatment with stavudine (3,4).

Liver

Asymptomatic increases in hepatic transaminases, which are not clearly dose-related, can occur during treatment with stavudine, requiring dosage modification because of moderate or severe toxicity in about 10% of patients (3,4).

The reports of all suspected hepatic adverse drug reactions with a fatal outcome received by the Uppsala Monitoring Centre between 1968 and 2003 included 120 cases of stavudine-associated hepatotoxicity (8).

Urinary tract

Renal tubular dysfunction and hypophosphatemia occurred in a patient who was taking both stavudine and lamivudine (9).

Sexual function

Painful bilateral gynecomastia and hypersexuality, possibly related to the use of stavudine, have been reported (10).

- A 25-year-old HIV-infected man reported hot flushes and headaches during the first days of treatment with stavudine (40 mg bd), followed by swelling and tenderness under both nipples. He also reported increased libido, premature ejaculation, and persistent erections. He denied illicit drug use. He had bilateral gynecomastia. Luteinizing hormone and testosterone concentrations were within the reference ranges, excluding primary causes of gynecomastia. One month after withdrawal of stavudine, the swelling and tenderness had abated and his sexual symptoms had resolved. He was not rechallenged with the drug.

The authors argued that although idiopathic gynecomastia could not be ruled out, the temporal relation between withdrawal of the drug and improvement in his symptoms suggested a causative role of stavudine.

In a Swiss case-control study gynecomastia was associated with stavudine in 47 cases. Blood concentrations of luteinizing hormone were above the reference range in 40% of the cases and 29% had reduced testosterone concentrations (11).

Second-Generation Effects

Fetotoxicity

Concerns continue about the toxicity of stavudine on the fetus. Five of ten infants born to ten pregnant women who took the drug during pregnancy developed grade 3 neutropenia, one developed hypoglycemia, and three developed hyperkalemia (12).

Drug–Drug Interactions

Isoniazid

A distal sensory neuropathy is the commonest neurological complication in HIV-infected individuals, and has been documented in up to 30% of patients with AIDS. There is evidence from a retrospective case-note review in 30 individuals that co-administration of stavudine and isoniazid increases the incidence of distal sensory neuropathy. Of 22 patients taking stavudine in combination with other drugs, all took isoniazid for tuberculosis and 12 developed a distal sensory neuropathy, with a median time to onset of 5 months (13). Those taking stavudine alone had an incidence of 11%.

A five-fold increase in the risk of distal sensory neuropathy has been reported in patients taking stavudine plus isoniazid (55 versus 11%) compared with patients taking stavudine without isoniazid (13). In nine of 12 patients, the neuropathy resolved on changing antiretroviral drugs. Peripheral neuropathy is a distressing complication during treatment with stavudine and the risk is considerably increased with co-administration of isoniazid. This combination of drugs should be avoided, if possible, in patients with tuberculosis and AIDS.

References

1. Calmy A, Pinoges L, Szumilin E, Zachariah R, Ford N, Ferradini L. Generic fixed-dose combination antiretroviral treatment in resource-poor settings: multicentric observational cohort. AIDS 2006;20(8):1163-9.
2. Bartlett JA, Johnson J, Herrera G, Sosa N, Rodriguez A, Liao Q, Griffith S, Irlbeck D, Shaefer MS; Clinically Significant Long-Term Antiretroviral Sequential Sequencing Study (CLASS) Team. Long-term results of initial therapy with abacavir and lamivudine combined with efavirenz, amprenavir/ritonavir, or stavudine. J Acquir Immune Defic Syndr 2006;43(3):284-92.
3. Riddler SA, Anderson RE, Mellors JW. Antiretroviral activity of stavudine (2′,3′-didehydro-3′-deoxythymidine, D4T). Antiviral Res 1995;27(3):189–203.
4. Skowron G. Biologic effects and safety of stavudine: overview of phase I and II clinical trials. J Infect Dis 1995;171(Suppl 2):S113–7.

5. Brew BJ, Tisch S, Law M. Lactate concentrations distinguish between nucleoside neuropathy and HIV neuropathy. AIDS 2003;17:1094-6.

6. Cherry CL, McArthur JC, Hoy JF, Wesselingh SL. Nucleoside analogues and neuropathy in the era of HAART. J Clin Virol 2003;26:195-207.

7. Heath KV, Hogg RS, Chan KJ, Harris M, Montessori V, O'Shaughnessy MV, Montanera JS. Lipodystrophy-associated morphological, cholesterol and triglyceride abnormalities in a population-based HIV/AIDS treatment database. AIDS 2001;15(2):231-9.

8. Bjornsson E, Olsson R. Suspected drug-induced liver fatalities reported to the WHO database. Dig Liver Dis 2006;38(1):33-8.

9. Morris AA, Baudouin SV, Snow MH. Renal tubular acidosis and hypophosphataemia after treatment with nucleoside reverse transcriptase inhibitors. AIDS 2001;15(1):140-1.

10. Melbourne KM, Brown SL, Silverblatt FJ. Gynecomastia with stavudine treatment in an HIV-positive patient. Ann Pharmacother 1998;32(10):1108.

11. Strub C, Kaufmann GR, Flepp M, Egger M, Kahlert C, Cavassini M, Battegay M. Swiss HIV Cohort Study. Gynecomastia and potent antiretroviral therapy. AIDS 2004;18(9):1347-9.

12. Blanche S. Safety of stavudine during pregnancy. J Infect Dis 2005;191(9):1567-8; author reply 1568-9.

13. Breen RA, Lipman MC, Johnson MA. Increased incidence of peripheral neuropathy with co-administration of stavudine and isoniazid in HIV-infected individuals. AIDS 2000;14(5):615.

Tenofovir

See also Nucleoside analogue reverse transcriptase inhibitors (NRTIs)

General Information

Tenofovir is a nucleotide (nucleoside monophosphate) analogue reverse transcriptase inhibitor, given as a prodrug, tenofovir disoproxil fumarate. In contrast to the other members of this class, it only needs to be phosphorylated twice intracellularly before it is pharmacologically active (1). It has a long intracellular half-life, which permits once-daily dosing.

Metabolism

In vitro studies have shown that tenofovir has a favorable mitochondrial toxicity profile (26,2)). Fatal lactic acidosis was reported in a patient who had recently switched to tenofovir (3). However, this patient was also taking didanosine and stavudine, a combination that is typically associated with lactic acidosis. Given the excellent in vitro characteristics of tenofovir in all known mitochondrial test systems and the notorious association of didanosine/stavudine combination with lactic acidosis, a causal relation of lactic acidosis with tenofovir use seems highly unlikely.

In a multicenter, double-blind, randomized study tenofovir was compared with stavudine over 3 years, both being given in combination with lamivudine and efavirenz in 602 antiviral drug-naive patients (4). There were significantly more favorable lipid profiles with tenofovir and more patients required the addition of lipid-lowering agents with stavudine. The overall incidence of mitochondrial toxicity was significantly less among patients who took tenofovir and a higher incidence of lipodystrophy in those who took stavudine. Patients who took tenofovir had gained weight by 144 weeks, in contrast to the patients who took stavudine, who lost weight from weeks 24 to 144

Electrolyte balance

Hypokalemia in 40 patients taking tenofovir has been reviewed; susceptibility factors were treatment with ritonavir or didanosine, lower weight, and longer duration of tenofovir therapy (5).

Hematologic

Several reviews have focused on the fact that the combination of didanosine + tenofovir has been associated with a paradoxical depletion of CD4+ T cells in the face of complete viral suppression (6,7). It has been speculated that this is due to interference with the activation pathways of these drugs.

Pancreas

The combination of didanosine and tenofovir has been associated with unexpected reductions in CD4+ T cell counts and an increased risk of pancreatitis (8). A direct toxic effect of didanosine, particularly when given in combination with tenofovir, on the pancreas has been suggested as the cause for an increase in fasting blood glucose concentrations. In a retrospective comparison of patients taking didanosine + tenofovir and either didanosine or tenofovir in combination with other antiretroviral drugs, both hyperglycemia and diabetes mellitus were significantly more common in recipients of didanosine + tenofovir after 12 months compared with patients taking either tenofovir or didanosine alone; the effect was more pronounced in patients taking didanosine 400 mg/day. The recommended dose for didanosine when given in combination with tenofovir was reduced in 2003 to 250 mg/day for subjects weighing more than 60 kg and 200 mg/day in those weighing less than 60 kg.

Urinary tract

Tenofovir is extensively excreted by the kidneys by glomerular filtration and active tubular secretion. It is structurally closely related to adefovir and cidofovir, which are used in the treatment of hepatitis B and cytomegalovirus infection and are known to be involved in nephrotoxicity. The potential of tenofovir for nephrotoxicity has been reviewed (9). The similarity of tenofovir to other highly nephrotoxic drugs (cidofovir, adefovir) is a concern, but no conclusive study has yet defined its nephrotoxic potential.

Acute tubular necrosis occurred in 5 HIV-infected patients who were taking tenofovir and who were compared with 22 patients described in published reports (10). The mean serum creatinine concentration rose from 80 to 345 µmol/l, and fell to 106 µmol/l during recovery. The renal damage resolved in 22 of the 27 patients after withdrawal of tenofovir. The most common drugs given with tenofovir were ritonavir or lopinavir + ritonavir (n = 21), atazanavir (n = 5), and didanosine (n = 9). The authors suggested that multiple drug interactions with tenofovir can lead to an increased risk of acute tubular necrosis and that frequent monitoring of renal function is warranted in any patients who are taking these combinations.

Tenofovir has been associated with acute renal failure and Fanconi's syndrome (11,12). The risk seems to be increased in patients who are taking tenofovir concurrently with ritonavir, lopinavir + ritonavir, didanosine, or atazanavir.

In a cohort study the risk of nephrotoxicity related to tenofovir was around 0.4% and was associated with other susceptibility factors for renal dysfunction (13).

Of 160 patients who received tenofovir in the Hôpital Saint-Louis in Paris three developed a nephropathy (14). All three had a very long history of HIV infection and had taken several antiretroviral drugs before tenofovir. None had renal insufficiency nor was taking another nephrotoxic agent. All three had a CD4 count below 50 x 10^6/l and a slight initial rise in serum creatinine within a few weeks after tenofovir was started. After several months of uneventful treatment with tenofovir, two patients had a rapid decline in renal function, for no other discernible reason. There was hypokalemia with increased potassium excretion, metabolic acidosis, and hypophosphatemia. Kidney biopsies showed severe acute tubular necrosis associated with marked interstitial fibroedema. The proximal and distal tubules were equally affected by severe epithelial lesions, and there was profound karyomegaly and large vacuoles (indicative of hypokalemia) in the proximal tubular cells. Withdrawal of all antiviral drugs was associated with rapid recovery of renal function and normalization of electrolyte changes, but the serum creatinine remained raised in both cases, suggesting irreversible damage.

The third patient developed severe nephrogenic diabetes insipidus with proteinuria and normoglycemic glycosuria 6 months after starting to take tenofovir. All signs of tubular damage disappeared within 3 months after tenofovir was withdrawn. The authors presented four arguments for a causal relation between tubular damage and tenofovir:

1. the laboratory values normalized after withdrawal;
2. known nephrotoxicity from animal studies;
3. similar nephrotoxicity from other compounds of the same class (cidofovir, adefovir);
4. a similar previous case report.

Seven HIV-infected patients taking tenofovir, who had previously taken NRTIs, developed renal tubular damage (15). The first symptoms of renal dysfunction occurred 5–64 weeks after the start of treatment. Six patients had a low body weight (under 60 kg), five took low-dose ritonavir, and one took didanosine. Renal tubular dysfunction was accompanied by hypophosphatemia, normoglycemic glycosuria, proteinuria, and reduced creatinine clearance. A renal biopsy in one patient was consistent with tubulointerstitial damage. In three patients with in whom tenofovir plasma concentrations were measured there were high values 16 hours after drug administration. After withdrawal of tenofovir the abnormalities resolved in under 4 months.

Proximal tubulopathy is a rare adverse effect of tenofovir. Possible predisposing factors include dose, low body weight, and pre-existing latent tubular damage (due for example to mitochondrial defects caused by previous treatment with NRTIs). Regular monitoring of tubulopathy markers in these patients could lead to early detection of renal dysfunction.

Acute irreversible renal insufficiency has been reported with tenofovir (16).

- A 39-year-old HIV-positive white man, developed acute renal insufficiency after taking lamivudine, zidovudine, and nevirapine for 2 weeks. His serum creatinine was 237 µmol/l. Renal ultrasound was normal and renal histology showed acute tubular necrosis with vacuolation of the proximal tubular cells and no evidence of focal or global glomerulosclerosis. All drugs were withdrawn but he required permanent dialysis.

Skin

A lichenoid eruption has been attributed to tenofovir (17).

- A 60-year-old man with HIV and chronic hepatitis B taking lamivudine, nevirapine, and stavudine, developed lipoatrophy. The stavudine was withdrawn and tenofovir started. Soon afterwards he developed a pruritic erythematous lichenoid rash without a fever or other symptoms or abnormalities of blood tests. Skin biopsy showed a florid lichenoid reaction with eosinophils, consistent with a drug eruption. All HIV and other medications (at least treated for 2 years), were stopped and he was given topical glucocorticoids and oral antihistamines. The rash resolved and all medications except tenofovir were reintroduced without recurrence.

Musculoskeletal

In the study mentioned above (29) there were small and largely non-progressive reductions in bone mineral density, but the changes were significantly larger in the lumbar spine in those who took tenofovir.

Drug-Drug Interactions

Didanosine

If tenofovir is combined with didanosine, the standard dose of the latter needs to be reduced to 250 mg/day (34). The combination of didanosine with tenofovir requires dosage reduction of didanosine from 400 mg/

day to 250 mg/day because of increased didanosine plasma concentrations when tenofovir is co-administered. However, there have not previously been data to suggest that dosage reduction results in appropriate drug exposure in subjects weighing less than 60 kg.

- A patient weighing 48 kg with renal impairment (creatinine clearance 57 ml/minute) developed didanosine toxicity with lactic acidosis and liver failure 3 months after starting a regimen containing tenofovir and didanosine 200 mg/day [18].

In 21 patients weighing under 60 kg who took didanosine 250 mg/day plus tenofovir there was full viral suppression but the mean CD4 cell count fell to 120×10^6/l; no explanation for the reduction in CD4 count was found other than the combined treatment of tenofovir and didanosine [19].

Therefore, in patients with low body weight and/or renal impairment, this drug combination should be used only with extreme caution, even when low doses of didanosine are used, and close clinical and biochemical monitoring is required.

Inidinavir

Tenofovir reduces the C_{max} but not the AUC of indinavir [34].

Lamivudine

Tenofovir reduces the C_{max} but not the AUC of lamivudine [34].

Lopinavir

Tenofovir slightly reduces the AUC of lopinavir (by 15%) but no dosage adjustment is needed [34]. Tenofovir concentrations increased with concomitant use of lopinavir/ritonavir (AUC +30%) and indinavir (C_{max} +14%) but no dosage adjustments are needed [34].

In a population pharmacokinetic study of 192 HIV-infected patients a two compartment model fitted best [20]. This study also confirmed the increase in AUC over 24 hours for tenofovir induced by lopinavir + ritonavir and no observable effect of didanosine. Tenofovir plasma clearance was related to the body weight/serum creatinine ratio but not to serum creatinine or estimated GFR.

Probenecid

Probenecid inhibits the elimination of tenofovir and significantly increases drug concentrations [34].

Saquinavir

Tenofovir had no effect on the pharmacokinetics of saquinavir hard gel + ritonavir in 18 HIV-infected subjects [21].

Adverse effects have been reported as flatulence, raised transaminases, raised creatine kinase activity, and rarely a raised serum creatinine [22]. Tenofovir does not currently appear to be nephrotoxic.

References

1. Schooley RT, Ruane P, Myers RA, Beall G, Lampiris H, Berger D, Chen SS, Miller MD, Isaacson E, Cheng AK; Study 902 Team. Tenofovir DF in antiretroviral-experienced patients: results from a 48-week, randomized, double-blind study. AIDS 2002;16:1257-63.
2. Birkus G, Hitchcock MJ, Cihlar T. Assessment of mitochondrial toxicity in human cells treated with tenofovir: comparison with other nucleoside reverse transcriptase inhibitors. Antimicrob Agents Chemother 2002;46:716-23.
3. Rivas P, Polo J, De Gorgolas M, Fernandez-Guerrero ML. Fatal lactic acidosis associated with tenofovir. Br Med J 2003;327:711.
4. Gallant JE, Staszewski S, Pozniak AL, DeJesus E, Suleiman JM, Miller MD, Coakley DF, Lu B, Toole JJ, Cheng AK; 903 Study Group. Efficacy and safety of tenofovir DF vs stavudine in combination therapy in antiretroviral-naive patients: a 3-year randomized trial. JAMA 2004;292(2):191-201.
5. Cirino CM, Kan VL. Hypokalemia in HIV patients on tenofovir. AIDS 2006;20(12):1671-3.
6. Bongiovanni M, Tordato F. Tenofovir plus didanosine as NRTI backbone in HIV-infected subjects. Curr Med Chem 2006;13(23):2789-93.
7. Barreiro P, Soriano V. Suboptimal CD4 gains in HIV-infected patients receiving didanosine plus tenofovir. J Antimicrob Chemother 2006;57(5):806-9.
8. García-Benayas T, Rendón AL, Rodríguez-Nóvoa S, Barrios A, Maida I, Blanco F, Barreiro P, Rivas P, González-Lahoz J, Soriano V. Higher risk of hyperglycemia in HIV-infected patients treated with didanosine plus tenofovir. AIDS Res Hum Retroviruses 2006;22(4):333-7.
9. Grim SA, Romanelli F. Tenofovir disoproxil fumarate. Ann Pharmacother 2003;37:849-59.
10. Zimmermann AE, Pizzoferrato T, Bedford J, Morris A, Hoffman R, Braden G. Tenofovir-associated acute and chronic kidney disease: a case of multiple drug interactions. Clin Infect Dis 2006;42(2):283-90.
11. Mathew G. and S.J. Knaus, Acquired Fanconi's syndrome associated with tenofovir therapy. J Gen Intern Med 2006;21(11):C3-5.
12. Williams J, Chadwick DR. Tenofovir-induced renal tubular dysfunction presenting with hypocalcaemia. J Infect 2006;52(4):e107-8.
13. Moreno S, Domingo P, Palacios R, Santos J, Falcó V, Murillas J, Estrada V, Ena J, Alvarez ML; Recover Study Group. Renal safety of tenofovir disoproxil fumarate in HIV-1 treatment-experienced patients with adverse events related to prior NRTI use: data from a prospective, observational, multicenter study. J Acquir Immune Defic Syndr 2006;42(3):385-7.
14. Karras A, Lafaurie M, Furco A, Bourgarit A, Droz D, Sereni D, Legendre C, Martinez F, Molina JM. Tenofovir-related nephrotoxicity in human immunodeficiency virus-infected patients: three cases of renal failure, Fanconi syndrome, and nephrogenic diabetes insipidus. Clin Infect Dis 2003;36:1070-3.
15. Peyriere H, Reynes J, Rouanet I, Daniel N, de Boever CM, Mauboussin JM, Leray H, Moachon L, Vincent D, Salmon-Ceron D. Renal tubular dysfunction associated with tenofovir therapy: report of 7 cases. J Acquir Immune Defic Syndr 2004;35(3):269-73.
16. Krummel T, Parvez-Braun L, Frantzen L, Lalanne H, Marcellin L, Hannedouche T, Moulin B. Tenofovir-induced acute renal failure in an HIV patient with normal renal function. Nephrol Dial Transplant 2005;20(2):473-4.

17. Woolley IJ, Veitch AJ, Harangozo CS, Moyle M, Korman TM. Lichenoid drug eruption to tenofovir in an HIV/hepatitis B virus co-infected patient. AIDS 2004;18(13):1857-8.

18 Masia M, Gutierrez F, Padilla S, Ramos JM, Pascual J. Didanosine-associated toxicity: a predictable complication of therapy with tenofovir and didanosine? J Acquir Immune Defic Syndr 2004;35(4):427-8.

19. Negredo E, Molto J, Burger D, Viciana P, Ribera E, Paredes R, Juan M, Ruiz L, Puig J, Pruvost A, Grassi J, Masmitja E, Clotet B. Unexpected CD4 cell count decline in patients receiving didanosine and tenofovir-based regimens despite undetectable viral load. AIDS 2004;18(3):459-63.

20. Jullien V, Treluyer JM, Rey E, Jaffray P, Krivine A, Moachon L, Lillo-Le Louet A, Lescoat A, Dupin N, Salmon D, Pons G, Urien S. Population pharmacokinetics of tenofovir in human immunodeficiency virus-infected patients taking highly active antiretroviral therapy. Antimicrob Agents Chemother 2005;49(8):3361-6.

21. Boffito M, Pozniak A, Kearney BP, Higgs C, Mathias A, Zhong L, Shah J. of pharmacokinetic drug interaction between tenofovir disoproxil fumarate and nelfinavir mesylate. Antimicrob Agents Chemother 2005;49(10):4386-9.

22. Barditch-Crovo P, Deeks SG, Collier A, Safrin S, Coakley DF, Miller M, Kearney BP, Coleman RL, Lamy PD, Kahn JO, McGowan I, Lietman PS. Phase I/II trial of the pharmacokinetics, safety, and antiretroviral activity of tenofovir disoproxil fumarate in human immunodeficiency virus-infected adults. Antimicrob Agents Chemother 2001;45(10):2733-9.

Valaciclovir

General Information

Valaciclovir is the L-valyl ester of aciclovir. After oral administration, it is rapidly and extensively converted to aciclovir by first-pass metabolism, resulting in plasma aciclovir concentrations previously only attainable with intravenous administration. Like aciclovir, valaciclovir is generally well tolerated. Compared with oral aciclovir, the systemic availability of aciclovir from oral valaciclovir is markedly improved.

Valaciclovir is highly active against *Herpes simplex* and *Herpes zoster*. It is also effective in suppressing recurrent episodes of genital herpes (1). Prophylactic administration of high doses of valaciclovir to prevent CMV disease was effective in patients with AIDS and in liver transplant recipients (2,3).

Observational studies

In a double-blind comparison of two regimens of valaciclovir 500 mg bd for recurrent genital herpes, a 5-day course ($n = 398$) and a 3-day course ($n = 402$), there were no significant differences in therapeutic outcome or adverse events between the two regimens (4). The most common adverse events were headache (10%), nausea (4%), diarrhea (3%), and fatigue (1.5%).

Comparative studies

The effects of aciclovir and valaciclovir for anogenital herpes have been studied in HIV-infected individuals in two controlled trials (5). In the first study, 1062 patients with CD4+ counts over $100 \times 10^6/l$ received valaciclovir or aciclovir for 1 year and were assessed monthly. In the second study, 467 patients were treated episodically for at least 5 days with valaciclovir or aciclovir and were assessed daily. Valaciclovir was as effective as aciclovir for suppression and episodic treatment of herpesvirus infections. Hazard ratios for the time to recurrence with valaciclovir 500 mg bd and 1000 mg od compared with aciclovir were 0.73 (95% CI = 0.50, 1.06) and 1.31 (0.94, 1.82). Valaciclovir 1000 mg bd and aciclovir had similar effects on the duration of infective episodes (HR = 0.92; CI = 0.75, 1.14). The most common adverse events, which occurred at similar rates with all regimens, were diarrhea, headache, infections, rashes, nausea, rhinitis, pharyngitis, abdominal pain, fever, depression, and cough.

Placebo-controlled studies

In large, placebo-controlled comparisons of the efficacy of valaciclovir and aciclovir in treating or suppressing recurrent genital *Herpes simplex* infections in immunocompetent people, dosages up to 2 g/day were well tolerated, with safety profiles comparable to aciclovir (1,6). In a comparison of high-dose valaciclovir (8 g/day) with two doses of aciclovir (0.8 and 3.2 g/day) for prophylaxis of cytomegalovirus disease in patients with advanced human immunodeficiency virus infection, intention-to-treat analysis showed a trend toward earlier mortality in those who received valaciclovir. In those who actually received valaciclovir, survival was significantly shorter. In view of the unexplained trend toward earlier mortality, as well as higher frequencies of renal toxicity (see the section on Urinary tract) and premature treatment discontinuation, the authors concluded that the dose of valaciclovir was too high and that better tolerated doses, which maintain a protective effect on cytomegalovirus disease, need to be identified (2).

After oral administration valaciclovir, a prodrug of aciclovir, has 3–5 times higher systemic availability than aciclovir. In two placebo-controlled studies in the short-term treatment of herpes labialis, one-day treatment (two doses of 2 g) was compared with two-day treatment (two doses of 2 g on day 1 and two doses of 1 g on day 2) (7). Both regimens significantly reduced the duration of the episode and the time to healing. Despite the high doses of valaciclovir, adverse events (headache, nausea, diarrhea, dyspepsia) occurred at similar frequencies in the three groups.

In a randomized, placebo-controlled study of valaciclovir 500 mg bd for suppression of recurrent genital herpes in HIV-infected subjects (8) recurrence was significantly less common in patients who took valaciclovir (35% versus 74%) and the time to first herpes occurrence was significantly shorter in placebo recipients (median 59 days versus over 180 days). Adverse events were equally frequent in the two groups with daily rates of 2.2% (placebo) versus 2.0% (valaciclovir). The most common adverse events with valciclovir versus placebo were headache (13% versus 8%), diarrhea (12% versus 12%), upper respiratory tract infections (9% versus 7%), vomiting (3% versus 9%), fatigue (8% versus 5%), influenza (8% versus 3%), nasopharyngitis (8% versus 2%), nausea (8% versus 8%), and rash (8% versus 1%).

In a multicenter study there were similar rates of adverse effects in those who took placebo (n = 741) and those who took valaciclovir (n = 743) (9).

Organs and Systems

Nervous system

Valaciclovir is a prodrug of aciclovir and can therefore cause similar effects, as two cases of nervous system effects have demonstrated (10).

- A 65-year-old man was given valaciclovir 1 g bd for 36 hours and had reduced concentration and was incoherent. All investigations were normal or negative. He improved rapidly on withdrawal of valaciclovir.
- A 44-year-old man was given valaciclovir 1 g tds for 5 days and developed a fever, disorientation, confusion, ataxia, dysarthria, and photophobia. All investigations were normal or negative. He was given antimicrobial drugs, including aciclovir, but his symptoms did not improve until the aciclovir was withdrawn.

Elderly patients and people with chronic renal insufficiency are most susceptible to the neurotoxic effects of aciclovir: confusion, hallucinations, dizziness, irritability, ataxia, tremor, myoclonus, and seizures. The symptoms usually occur within 3 days of the start of therapy and resolve within 5 days after withdrawal. Plasma aciclovir concentrations do not correlate with symptoms. Lumbar puncture and CT scans of the head are essentially unremarkable. The most common electroencephalographic abnormality is diffuse generalized slowing of brain wave activity.

Psychological, psychiatric

At high doses (8 g/day) hallucinations and confusion were a significant concern (2,3), but similar symptoms have also occurred at lower doses and in patients with renal insufficiency.

- Ocular and auditory hallucinations have been reported in a 60-year-old female patient on CAPD (11).
- A 58-year-old man with chronic renal insufficiency, who was hemodialysed twice a week, was treated with valaciclovir (1 g tds) for Herpes zoster (12). Two days later he became disoriented, dizzy, dysarthric, and experienced hallucinations. The serum aciclovir concentration was 21 µg/ml. Treatment was discontinued and he was treated with hemodialysis for 6 hours, resulting in marked clinical improvement. The next day his symptoms of dysarthria recurred, but

immediately and completely resolved after a second hemodialysis.

Hematologic

In one study, high-dose valaciclovir was associated with an increased risk of a thrombotic microangiopathy-like syndrome, reported as thrombocytopenic purpura or hemolytic-uremic syndrome (2). This syndrome occurred in 14 of 523 patients who received valaciclovir and in only four of 704 patients who received aciclovir after a median of 54 (range 8–84) weeks of treatment. The precise relation to valaciclovir remains unclear, since eight of 14 patients who were treated with valaciclovir had stopped treatment at least 1 week before the onset of the syndrome. In addition, all patients with thrombotic microangiopathy-like syndromes had taken multiple concomitant medications, and most had other intercurrent illnesses, which could have explained the hematological and renal abnormalities. The authors concluded that additional data are required to understand the role of valaciclovir and other medications for thrombotic microangiopathy-like syndromes, which are recognized with increasing frequency in patients with advanced HIV disease.

Gastrointestinal

Nausea, vomiting, and abdominal pain were commonly reported in human volunteers, but only diarrhea was significantly associated with exposure (13).

Urinary tract

Aciclovir is excreted renally. High plasma concentrations of aciclovir can lead to precipitation in renal tubules, causing impaired renal function, which is generally reversible. Since oral valaciclovir can result in plasma aciclovir concentrations comparable to those attained with intravenous dosing, reversible impairment of renal function can also occur after prolonged use of high-dose valaciclovir. Indeed, in a study of high-dose valaciclovir for prevention of cytomegalovirus disease in HIV-infected people, there was an association between treatment with valaciclovir and moderate nephrotoxicity (serum creatinine more than 1.5 times the upper limit of normal; estimated creatinine clearance under 50 ml/minute) (2).

Herpes simplex mucositis occurs in more than two-thirds of patients receiving bone marrow transplants and can be prevented by aciclovir. In 60 patients who took valaciclovir 500 mg bd for prevention of Herpes simplex mucositis after bone marrow transplantation (14) treatment with valaciclovir was given from the start of transplant conditioning until resolution of neutropenia (over $1.0 \times 10^9/l$). The patients were compared with a historical control group of 60 patients who had taken aciclovir 600 mg 6-hourly. There were no serious adverse reactions, but 10% required dosage adjustment because of an increase in serum creatinine compared with 35% in the historical controls.

Second-Generation Effects

Fetotoxicity

In a phase I trial, valaciclovir administered in the third trimester of pregnancy was well tolerated (15).

Susceptibility Factors

Renal disease

Adverse effects of valaciclovir, the L-valyl ester of aciclovir, can be associated with increased drug concentrations when the dose is not adjusted for reduced renal function. For example, aseptic meningitis has been associated with valaciclovir in a patient with renal insufficiency (16).

• An 88-year-old man with renal insufficiency took valaciclovir 1000 mg tds. After the first dose, he became disoriented and incontinent. Valaciclovir was withdrawn, but the symptoms continued and progressed to drowsiness and nuchal rigidity. After an extensive work-up, aseptic meningitis was diagnosed.

Given the patient's age and renal dysfunction, it is likely that excessive valaciclovir accumulation was responsible for this presentation.

Drug–Drug Interactions

Cimetidine

In an open, single-dose study of the effects of probenecid and cimetidine on the pharmacokinetics of valaciclovir and its metabolite aciclovir in 12 healthy men, valaciclovir 1 g, valaciclovir plus probenecid 1 g, valaciclovir plus cimetidine 800 mg, and valaciclovir with a combination of probenecid and cimetidine were studied (17). At three subsequent administrations, drug regimens were alternated among groups so that each group received each regimen. Probenecid and cimetidine respectively increased the mean C_{max} of valaciclovir by 23 and 53% and its AUC by 22 and 73%. Probenecid and cimetidine also respectively increased the mean aciclovir C_{max} by 22 and 8% and its AUC by 48 and 27%. The combination had a greater effect than either drug alone. Neither cimetidine nor probenecid affected the absorption of valaciclovir.

Probenecid

In an open, single-dose study of the effects of probenecid and cimetidine on the pharmacokinetics of valaciclovir and its metabolite aciclovir in 12 healthy men, valaciclovir 1 g, valaciclovir plus probenecid 1 g, valaciclovir plus cimetidine 800 mg, and valaciclovir with a combination of probenecid and cimetidine were studied (17). At three subsequent administrations, drug regimens were alternated among groups so that each group received each regimen. Probenecid and cimetidine respectively increased the mean C_{max} of valaciclovir by 23 and 53% and its AUC by 22 and 73%. Probenecid and cimetidine also respectively increased the mean aciclovir C_{max} by 22 and 8% and its AUC by 48 and 27%. The combination had a greater effect than either drug alone. Neither cimetidine nor probenecid affected the absorption of valaciclovir.

References

1. Reitano M, Tyring S, Lang W, Thoming C, Worm AM, Borelli S, Chambers LO, Robinson JM, Corey LInternational Valaciclovir HSV Study Group. Valaciclovir for the suppression of recurrent genital *Herpes simplex* virus infection: a large-scale dose range-finding study. J Infect Dis 1998;178(3):603–10.

2. Feinberg JE, Hurwitz S, Cooper D, Sattler FR, MacGregor RR, Powderly W, Holland GN, Griffiths PD, Pollard RB, Youle M, Gill MJ, Holland FJ, Power ME, Owens S, Coakley D, Fry J, Jacobson MA. A randomized, double-blind trial of valaciclovir prophylaxis for cytomegalovirus disease in patients with advanced human immunodeficiency virus infection. AIDS Clinical Trials Group Protocol 204/Glaxo Wellcome 123-014 International CMV Prophylaxis Study Group. J Infect Dis 1998;177(1):48–56.

3. Lowance D, Neumayer HH, Legendre CM, Squifflet JP, Kovarik J, Brennan PJ, Norman D, Mendez R, Keating MR, Coggon GL, Crisp A, Lee ICInternational Valacyclovir Cytomegalovirus Prophylaxis Transplantation Study Group. Valacyclovir for the prevention of cytomegalovirus disease after renal transplantation. N Engl J Med 1999;340(19):1462–70.

4. Leone PA, Trottier S, Miller JM. Valacyclovir for episodic treatment of genital herpes: a shorter 3-day treatment course compared with 5-day treatment. Clin Infect Dis 2002;34(7):958–62.

5. Conant MA, Schacker TW, Murphy RL, Gold J, Crutchfield LT, Crooks RJ, Acebes LO, Aiuti F, Akil B, Anderson J, Melville RL, Ballesteros Martin J, Berry A, Weiner M, Black F, Anderson PL, Bockman W, Borelli S, Bradbeer CS, Braffman M, Brandon W, Clark R, Wisniewski T, Bruun JN, Burdge D, Caputo RM, Chateauvert M, LaLonde R, Chiodo F, et alInternational Valaciclovir HSV Study Group. Valaciclovir versus aciclovir for *Herpes simplex* virus infection in HIV-infected individuals: two randomized trials. Int J STD AIDS 2002;13(1):12–21.

6. Tyring SK, Douglas JM Jr, Corey L, Spruance SL, Esmann JThe Valaciclovir International Study Group. A randomized, placebo-controlled comparison of oral valacyclovir and acyclovir in immunocompetent patients with recurrent genital herpes infections. Arch Dermatol 1998;134(2):185–91.

7. Spruance SL, Jones TM, Blatter MM, Vargas-Cortes M, Barber J, Hill J, Goldstein D, Schultz M. High-dose, short-duration, early valacyclovir therapy for episodic treatment of cold sores: results of two randomized, placebo-controlled, multicenter studies. Antimicrob Agents Chemother 2003;47:1072-80.

8. DeJesus E, Wald A, Warren T, Schacker TW, Trottier S, Shahmanesh M, Hill JL, Brennan CA; Valacyclovir International HSV Study Group. Valacyclovir for the suppression of recurrent genital herpes in human immunodeficiency virus-infected subjects. J Infect Dis 2003;188:1009-16.

9. Corey L, Wald A, Patel R, Sacks SL, Tyring SK, Warren T, Douglas JM Jr, Paavonen J, Morrow RA, Beutner KR,

Stratchounsky LS, Mertz G, Keene ON, Watson HA, Tait D, Vargas-Cortes M; Valacyclovir HSV Transmission Study Group. Once-daily valacyclovir to reduce the risk of transmission of genital herpes. N Engl J Med 2004;350(1):11-20.

10. Tornero C, Sanchez P, Castejon P, Rull S. Regrassion de la lipomatosis multiple con la administration de indinavir. [Reversal of multiple lipomatosis after indinavir treatment.] Med Clin (Barc) 1999;113(7):278–9.

11. Izzedine H, Launay-Vacher V, Aymard G, Legrand M, Deray G. Pharmacokinetic of nevirapine in haemodialysis. Nephrol Dial Transplant 2001;16(1):192–3.

12. Linssen-Schuurmans CD, van Kan EJ, Feith GW, Uges DR. Neurotoxicity caused by valacyclovir in a patient on hemodialysis. Ther Drug Monit 1998;20(4):385–6.

13. Jacobson MA, Gallant J, Wang LH, Coakley D, Weller S, Gary D, Squires L, Smiley ML, Blum MR, Feinberg J. Phase I trial of valaciclovir, the L-valyl ester of acyclovir, in patients with advanced human immunodeficiency virus disease. Antimicrob Agents Chemother 1994;38(7):1534–40.

14. Eisen D, Essell J, Broun ER, Sigmund D, DeVoe M. Clinical utility of oral valacyclovir compared with oral acyclovir for the prevention of herpes simplex virus mucositis following autologous bone marrow transplantation or stem cell rescue therapy. Bone Marrow Transplant 2003;31:51-5.

15. Kimberlin DF, Weller S, Whitley RJ, Andrews WW, Hauth JC, Lakeman F, Miller G. Pharmacokinetics of oral valacyclovir and acyclovir in late pregnancy. Am J Obstet Gynecol 1998;179(4):846–51.

16. Fobelo MJ, Corzo Delgado JE, Romero Alonso A, Gomez-Bellver MJ. Aseptic meningitis related to valacyclovir. Ann Pharmacother 2001;35(1):128–9.

17. De Bony F, Tod M, Bidault R, On NT, Posner J, Rolan P. Multiple interactions of cimetidine and probenecid with valaciclovir and its metabolite acyclovir. Antimicrob Agents Chemother 2002;46(2):458–63.

Zalcitabine

See also Nucleoside analogue reverse transcriptase inhibitors (NRTIs)

General Information

Zalcitabine is a nucleoside analogue reverse transcriptase inhibitor. Because of the high incidence of nervous system adverse effects and the availability of less toxic alternatives, zalcitabine is no longer used.

Comparative studies

Several large-scale studies of the efficacy of combined antiretroviral treatment with zalcitabine and zidovudine in HIV-infected patients (compared with zidovudine monotherapy or a combination of zidovudine and didanosine) have not shown unexpected adverse effects (1–4). The most common adverse effects in patients taking zalcitabine were peripheral neuropathy and aphthous mouth ulcers.

Organs and Systems

Nervous system

A peripheral neuropathy is often observed in patients treated with didanosine, stavudine, and zalcitabine (5,6). It usually occurs after prolonged treatment (more than 4 months), most often with zalcitabine, and often requires drug withdrawal, but it is sometimes not fully reversible. Mitochondrial alterations have been demonstrated in Schwann cells of peripheral nerves and dorsal root ganglia in rabbits treated with zalcitabine (7).

Hematologic

Thrombocytopenia has been observed in patients using zalcitabine and didanosine (8,9).

References

1. Delta Coordinating Committee. Delta: a randomised double-blind controlled trial comparing combinations of zidovudine plus didanosine or zalcitabine with zidovudine alone in HIV-infected individuals. Lancet 1996;348(9023):283–91.

2. Hammer SM, Katzenstein DA, Hughes MD, Gundacker H, Schooley RT, Haubrich RH, Henry WK, Lederman MM, Phair JP, Niu M, Hirsch MS, Merigan TC. A trial comparing nucleoside monotherapy with combination therapy in HIV-infected adults with CD4 cell counts from 200 to 500 per cubic millimeter. AIDS Clinical Trials Group Study 175 Study Team. N Engl J Med 1996;335(15):1081–90.

3. Schooley RT, Ramirez-Ronda C, Lange JM, Cooper DA, Lavelle J, Lefkowitz L, Moore M, Larder BA, St Clair M, Mulder JW, McKinnis R, Pennington KN, Harrigan PR, Kinghorn I, Steel H, Rooney JF. Virologic and immunologic benefits of initial combination therapy with zidovudine and zalcitabine or didanosine compared with zidovudine monotherapy. Wellcome Resistance Study Collaborative Group. J Infect Dis 1996;173(6):1354–66.

4. Saravolatz LD, Winslow DL, Collins G, Hodges JS, Pettinelli C, Stein DS, Markowitz N, Reves R, Loveless MO, Crane L, Thompson M, Abrams D. Zidovudine alone or in combination with didanosine or zalcitabine in HIV-infected patients with the acquired immunodeficiency syndrome or fewer than 200 CD4 cells per cubic millimeter. Investigators for the Terry Beirn Community Programs for Clinical Research on AIDS. N Engl J Med 1996;335(15):1099–106.

5. Fichtenbaum CJ, Clifford DB, Powderly WG. Risk factors for dideoxynucleoside-induced toxic neuropathy in patients with the human immunodeficiency virus infection. J Acquir Immune Defic Syndr Hum Retrovirol 1995;10(2):169–74.

6. Simpson DM, Tagliati M. Nucleoside analogue-associated peripheral neuropathy in human immunodeficiency virus infection. J Acquir Immune Defic Syndr Hum Retrovirol 1995;9(2):153–61.

7. Feldman D, Anderson TD. Schwann cell mitochondrial alterations in peripheral nerves of rabbits treated with 2′,3′-dideoxycytidine. Acta Neuropathol (Berl) 1994;87(1):71–80.

8. Yarchoan R, Perno CF, Thomas RV, Klecker RW, Allain JP, Wills RJ, McAtee N, Fischl MA, Dubinsky R, McNeely MC, et al. Phase I studies of 2′,3′-dideoxycytidine in severe human immunodeficiency virus infection as a single agent and alternating with zidovudine (AZT). Lancet 1988;1(8577):76–81.

9. Dolin R, Lambert JS, Morse GD, Reichman RC, Plank CS, Reid J, Knupp C, McLaren C, Pettinelli C. 2′,3′-Dideoxyinosine in patients with AIDS or AIDS–related complex. Rev Infect Dis 1990;12(Suppl 5):S540–9.

Zidovudine

See also Nucleoside analogue reverse transcriptase inhibitors (NRTIs)

General Information

Zidovudine is a nucleoside analogue reverse transcriptase inhibitor. Its adverse effects include hematological complications, severe headache, insomnia, confusion, nausea, vomiting, abdominal discomfort, myalgia (myopathy), and nail pigmentation (1).

Observational studies

Oral zidovudine in a dosage of 200 mg every 4 hours for 42 days was used as prophylaxis in health-care workers after percutaneous exposure to blood or body fluids from HIV-infected patients. Adverse reactions occurred in 73%, the most frequent being nausea (47%), headache (35%), and fatigue (30%). Of selected hematological laboratory markers only platelet counts increased significantly over 4 weeks. Although adverse reactions were not very severe and none of the laboratory changes was considered clinically significant, treatment was poorly accepted and stopped prematurely by 30% (2). Current guidelines for postexposure prophylaxis recommend a much lower dosage (300 mg bd) with much better tolerance (3).

In a retrospective study of patients co-infected with HIV and hepatitis C who were given peginterferon + ribavirin and concurrent zidovudine, significant anemia, defined as a loss of >25% of hemoglobin concentration over the first 4 weeks of therapy, was associated with zidovudine (OR = 4.4; 95% CI = 2.3, 8.5) (4).

Comparative studies

The safety and efficacy of lamivudine (300–600 mg/day) in combination with zidovudine (600 mg/day) in the treatment of antiretroviral-naive and zidovudine-experienced HIV-infected persons has been compared with zidovudine monotherapy in two placebo-controlled studies of 129 and 223 patients (5,6). There were no significant differences in the incidence or severity of adverse effects between patients taking zidovudine alone or in combination with lamivudine. In both studies gastrointestinal symptoms, notably nausea, were the most commonly observed adverse reactions, occurring in 5–11% of zidovudine-experienced patients and 23–29% of antiretroviral drug-naive individuals. Although one antiretroviral drug-naive patient taking combined therapy had an asymptomatic rise in pancreatic amylase activity, acute pancreatitis was not observed in either study. Grade 1

peripheral neuropathy was reported in one zidovudine-experienced patient taking low-dosage lamivudine (150 mg bd) and zidovudine.

Several large-scale studies of the efficacy of combined antiretroviral treatment with zalcitabine and zidovudine in HIV-infected patients (compared with zidovudine monotherapy or a combination of zidovudine and didanosine) have not shown unexpected adverse effects (7–10). The most common adverse effects in patients taking zalcitabine were peripheral neuropathy and aphthous mouth ulcers.

Organs and Systems

Nervous system

Various nervous system adverse effects of zidovudine have been reported, which may or may not be directly related to the drug. These include seizures, confusion, and acute encephalopathy occurring after zidovudine dosage reduction (11).

Metabolism

Lipodystrophy is a common adverse effect of antiretroviral drugs, particularly the NRTIs and has been reported with zidovudine (12).

- A 42-year-old woman developed abdominal and dorsocervical fat enlargement after having taken zidovudine for over 10 years. Zidovudine was withdrawn and the lesions improved considerably over the next 26 months.

Hematologic

The main dose-limiting adverse reactions of zidovudine therapy in HIV-infected adults and children are hematological complications (13). When zidovudine was introduced it was given in about twice the dosage used today. Consequently, hematological adverse effects occur at a much lower frequency than previously reported (14,15).

Almost uniformly, zidovudine treatment results in a progressive increase in the erythrocyte mean cell volume, which cannot be prevented by supplementation with vitamin B_{12} and folinic acid (16). Zidovudine can cause anemia (17) and reversible pure red cell aplasia (18). While recombinant erythropoietin is useful in correcting zidovudine-induced anemia, some cases of anemia are associated with high serum erythropoietin concentrations and normocytic cells, indicating bone marrow unresponsiveness to erythropoietin (19). Measuring baseline serum erythropoietin concentrations may help to predict the response to this very costly hormone supplementation.

Zidovudine can cause neutropenia (17).

Thrombocytopenia has been observed in patients using zidovudine and didanosine (20,21).

In 740 adults who were given zidovudine + lamivudine + efavirenz, 28 interrupted treatment because of zidovudine-attributable adverse effects, including 25 with severe anemia, two with severe neutropenia, and one with non-obstructive cardiomyopathy (22).

The rate of anemia in infants born to mothers receiving zidovudine for prevention of mother-to-child transmission was slighter higher than in those born to mothers receiving other nucleoside analogues (23).

Hypoproliferative anemia is a typical, severe adverse effect of zidovudine; it usually resolves promptly after withdrawal of the drug. However, three patients developed zidovudine-induced anemia that could have been mistaken for hemolytic disease, especially in the first 2 weeks after withdrawal of the drug (24).

Skin

Fatal toxic epidermolysis has been attributed to zidovudine (25) as has cutaneous hypersensitivity (26).

Zidovudine can occasionally cause unusual pigmentation, probably depending on the individual's pigmentary pattern (27).

Nails

The presenting sign of neutropenia in a neonate treated with prophylactic zidovudine for reduction of perinatal transmission was, unusually, severe paronychia of the large toes as a result of *Candida albicans* and *Escherichia coli* infection (28). The paronychia resolved after treatment with oral fluconazole and topical antiseptics. Paronychia of the large toes has also been observed when filgrastim (recombinant granulocyte colony stimulating factor) was used alongside chemotherapy in poor-risk patients with myelodysplastic syndrome (29), and there is some evidence that paronychia (even in the absence of blood disorders) can occur as an independent adverse reaction to various antiretroviral drugs, including indinavir and lamivudine (30).

Closely similar is the strong evidence from case-control studies that ingrowing toenails are associated with the use of indinavir (31) and possibly other similar compounds, but not lamivudine.

Musculoskeletal

Myopathy has been well described with long-term zidovudine and is reversible after withdrawal (32). Phosphorus magnetic resonance spectroscopy has been used to study the changes in phosphorylated metabolites (ATP, phosphocreatine, and inorganic phosphate) during exercise in 19 healthy volunteers, 6 untreated HIV-positive individuals, and 9 zidovudine-treated patients with biopsy-proven myopathy (33). Zidovudine altered the normal muscle energy metabolism in the patients with myopathy, suggesting that it reduces maximal work output, and thus the maximal rate of mitochondrial ATP synthesis, in human muscle. So far, the syndrome has not been associated with any other NRTI (34) and it has been suggested that other factors might contribute to the development of zidovudine-associated myopathy (35). Mitochondrial abnormalities have also been observed in biopsies from untreated patients infected with HIV-1, suggesting that the virus itself can also cause myopathy.

To assess the contribution of zidovudine to the mitochondrial damage, the effects of zidovudine on non-infected co-cultures of spinal ganglia, spinal cord, and skeletal muscle in fetal rats have been studied (36). There were significant changes not only in the mitochondria but also in the nuclei of all cells tested. These changes depended less on the concentration of zidovudine than on the duration of exposure.

Multiorgan failure

A well-documented case has shown that zidovudine can cause type B lactic acidosis and acute respiratory and hepatic failure (37).

- A 34-year-old obese woman developed nausea, vomiting, and intermittent diarrhea. Her current medications included zidovudine. She had tachypnea and tender hepatomegaly, and a CT scan of the abdomen showed hepatomegaly with fatty infiltration. The serum bicarbonate concentration was low and the lactate concentration three times normal. The tachypnea and dyspnea worsened as the lactate concentration rapidly increased to 15 times normal, and she died in acute respiratory and hepatic failure with multiorgan dysfunction.

Second-Generation Effects

Pregnancy

Zidovudine during pregnancy and delivery, followed by treatment of the infant for 6 weeks, prevents maternofetal transmission of HIV, and is associated with minimal short-term toxicity to both mother and child and no increased incidence of neonatal structural abnormalities (SEDA-19, 279) (38).

Zidovudine is relatively well tolerated in pregnancy, with anemia, neutropenia, or thrombocytopenia occurring in 10% and abnormalities of serum electrolytes and liver function in 5% (38).

Teratogenicity

There has been a randomized cohort study in the USA of children from 122 pregnancies in which zidovudine was given and of children from 112 pregnancies in which only a placebo was used (39). The median age of the children at the last follow-up visit was 4.2 years. There were no significant differences between children exposed to zidovudine and those who received placebo in terms of sequential data on lymphocyte subsets, weight, height, head circumference, and cognitive/developmental function. There were no deaths or malignancies. Two children (both exposed to zidovudine) were still being followed for unexplained abnormal fundoscopy. One child exposed to zidovudine had a mild cardiomyopathy on echocardiography at the age of 48 months but was clinically asymptomatic.

Lactation

Since zidovudine seems to be relatively well tolerated both in pregnancy and in neonates, there is also much reason to consider its use during lactation in order to reduce vertical transmission of HIV. Indeed, many would regard it as highly preferable to abandoning breast-feeding by HIV-infected women. In a critical double-blind West African study the effects of a 6-month course of treatment in prenatal and lactating mothers were examined (40). Eligible participants were women aged 18 years or older who had confirmed HIV-1 infection, were 36–38 weeks pregnant, and gave written informed consent. Exclusion criteria were severe anemia, neutropenia, abnormal liver function, and sickle cell disease. They were randomly assigned to zidovudine ($n = 214$; 300 mg bd until labor, 600 mg at the start of labor, and 300 mg bd for 7 days) or to matching placebo ($n = 217$). The Kaplan-Meier probability of HIV infection in the infant at 6 months was 18% in the zidovudine group and 28% in the placebo group, a relative efficacy of 0.38. In current and follow-up observations over 6 months, no major adverse biological or clinical events were reported in excess among women or children in the zidovudine group. The authors concluded that a short course of oral zidovudine given during the peripartum period is well tolerated and provides significant reduction in early vertical transmission of HIV-1 infection despite breast-feeding.

A second related study showed similar results (41), and the two papers together provide impressive evidence that it is proper and defensible to use zidovudine in breast-feeding mothers.

Susceptibility Factors

Age

The only recognized toxic effect in infants is anemia within the first 6 weeks of life, which is not associated with premature delivery, duration of maternal treatment, degree of maternal immunosuppression, or maternal anemia. An 18-month follow-up of 342 children born to mothers who had taken zidovudine or placebo during pregnancy has recently been reported (42). There were no differences in growth parameters or immune function in uninfected children. In addition, no childhood neoplasias were reported in either group.

Drug–Drug Interactions

Atovaquone

Atovaquone can potentiate the activity of zidovudine by inhibiting its glucuronidation (43).

The metabolism of the antiviral nucleoside zidovudine to the inactive glucuronide form in vitro was inhibited by atovaquone (44,45).

Azithromycin

Zidovudine does not affect azithromycin concentrations and azithromycin does not affect zidovudine concentrations (46).

Clarithromycin

Clarithromycin has an unpredictable effect on the absorption of zidovudine; blood concentrations may rise or fall (47,48).

Cytotoxic drugs

In 13 HIV-infected patients with cancer, the mean pharmacokinetics of zidovudine (AUC, half-life, oral clearance, and oral apparent volume of distribution) were no different with or without chemotherapy (49). However, there was a 57% reduction in C_{max} and a 66% increase in t_{max} after chemotherapy. There were no differences in the urinary excretion of zidovudine or zidovudine glucuronide. The authors concluded that these minor changes did not warrant any change in the dosage of zidovudine during concurrent chemotherapy.

Fluconazole

Zidovudine glucuronidation in human hepatic microsomes in vitro was inhibited more by the combination of fluconazole with valproic acid than with other drugs, such as atovaquone and methadone (50).

Ganciclovir

Zidovudine and ganciclovir have overlapping toxicity profiles with respect to adverse hematological effects. Severe life-threatening hematological toxicity has been reported in 82% of patients treated with a combination of zidovudine and ganciclovir (51). The combination of ganciclovir with didanosine was much better tolerated (52).

Methadone

The metabolism of the antiviral nucleoside zidovudine to the inactive glucuronide form in vitro was inhibited by methadone (53). The concentration of methadone required for 50% inhibition was over 8 µg/ml, a supratherapeutic concentration, thus raising questions about the clinical significance of the effect. However, in eight recently detoxified heroin addicts, acute methadone treatment increased the AUC of oral zidovudine by 41% and of intravenous zidovudine by 19%, following the start of oral methadone (50 mg/day) (54). These effects resulted primarily from inhibition of zidovudine glucuronidation, but also from reduced renal clearance of zidovudine, and methadone concentrations remained in the target range throughout. It is recommended that increased toxicity surveillance, and possibly reduction in zidovudine dose, are indicated when the two drugs are co-administered.

Oxazepam

There was a striking incidence of headache in a small series of patients when zidovudine was given with oxazepam (55).

Paracetamol

Concomitant administration of paracetamol and zidovudine leads to inhibition of glucuronidation and to potentiation of the toxicity of each drug (56,57).

Paracetamol (acetaminophen)

Paracetamol (acetaminophen) increased the clearance (and possibly reduced the effects) of zidovudine (58,59).

Probenecid

In two healthy volunteers, co-administration of probenecid 500 mg every 6 hours altered the pharmacokinetics of a single oral dose of zidovudine 200 mg (60). There was an increase in the average AUC, with a corresponding reduction in oral clearance, attributed to an inhibitory effect of probenecid on the glucuronidation and renal excretion of zidovudine.

Eight subjects took zidovudine for 3 days with and without probenecid 500 mg every 8 hours for 3 days, and then additional quinine sulfate 260 mg every 8 hours (61). Probenecid increased the AUC of zidovudine by 80%. Quinine prevented the probenecid effect but had no effect on zidovudine kinetics when it was taken without probenecid by four other subjects. All of the effects were secondary to changes in zidovudine metabolism, since neither probenecid nor quinine changed the renal elimination of zidovudine.

Probenecid reduces the renal tubular secretion of zidovudine (62,63).

Rifamycins

Rifampicin, a well-known enzyme inducer, increased the metabolism of zidovudine, and the effect persisted for 2 weeks after rifampicin had been withdrawn (64).

Trimethoprim and co-trimoxazole

In one pharmacokinetic study in eight HIV-infected subjects, the renal clearance of zidovudine was significantly reduced by trimethoprim (65). The authors concluded that zidovudine dosages may need to be reduced if trimethoprim is given to patients with impairment of liver function or glucuronidation. Zidovudine, on the other hand, did not alter the pharmacokinetics of trimethoprim.

References

1. Neuzil KM. Pharmacologic therapy for human immunodeficiency virus infection: a review. Am J Med Sci 1994;307(5):368–73.
2. Forseter G, Joline C, Wormser GP. Tolerability, safety, and acceptability of zidovudine prophylaxis in health care workers. Arch Intern Med 1994;154(23):2745–9.
3. Gerberding JL. Prophylaxis for occupational exposure to HIV. Ann Intern Med 1996;125(6):497–501.
4. Alvarez D, Dieterich DT, Brau N, Moorehead L, Ball L, Sulkowski MS. Zidovudine use but not weight-based ribavirin dosing impacts anaemia during HCV treatment in HIV-infected persons. J Viral Hepat 2006;13(10):683-9.
5. Katlama C, Ingrand D, Loveday C, Clumeck N, Mallolas J, Staszewski S, Johnson M, Hill AM, Pearce G, McDade H. Safety and efficacy of lamivudine–zidovudine combination therapy in antiretroviral-naive patients. A randomized controlled comparison with zidovudine monotherapy. Lamivudine European HIV Working Group. JAMA 1996;276(2):118–25.
6. Staszewski S, Loveday C, Picazo JJ, Dellarnonica P, Skinhoj P, Johnson MA, Danner SA, Harrigan PR, Hill AM, Verity L, McDade H. Safety and efficacy of lamivudine–zidovudine combination therapy in zidovudine-experienced patients. A randomized controlled comparison with zidovudine monotherapy. Lamivudine European HIV Working Group. JAMA 1996;276(2):111–7.
7. Delta Coordinating Committee. Delta: a randomised double-blind controlled trial comparing combinations of zidovudine plus didanosine or zalcitabine with zidovudine alone in HIV-infected individuals. Lancet 1996;348(9023):283–91Erratum in 1996;348:834.
8. Hammer SM, Katzenstein DA, Hughes MD, Gundacker H, Schooley RT, Haubrich RH, Henry WK, Lederman MM, Phair JP, Niu M, Hirsch MS, Merigan TC. A trial comparing nucleoside monotherapy with combination therapy in HIV-infected adults with CD4 cell counts from 200 to 500 per cubic millimeter. AIDS Clinical Trials Group Study 175 Study Team. N Engl J Med 1996;335(15):1081–90.
9. Schooley RT, Ramirez-Ronda C, Lange JM, Cooper DA, Lavelle J, Lefkowitz L, Moore M, Larder BA, St Clair M, Mulder JW, McKinnis R, Pennington KN, Harrigan PR, Kinghorn I, Steel H, Rooney JF. Virologic and immunologic benefits of initial combination therapy with zidovudine and zalcitabine or didanosine compared with zidovudine monotherapy. Wellcome Resistance Study Collaborative Group. J Infect Dis 1996;173(6):1354–66.
10. Saravolatz LD, Winslow DL, Collins G, Hodges JS, Pettinelli C, Stein DS, Markowitz N, Reves R, Loveless MO, Crane L, Thompson M, Abrams D. Zidovudine alone or in combination with didanosine or zalcitabine in HIV-infected patients with the acquired immunodeficiency syndrome or fewer than 200 CD4 cells per cubic millimeter. Investigators for the Terry Beirn Community Programs for Clinical Research on AIDS. N Engl J Med 1996;335(15):1099–106.
11. Langtry HD, Campoli-Richards DM. Zidovudine. A review of its pharmacodynamic and pharmacokinetic properties, and therapeutic efficacy. Drugs 1989;37:408–50.
12. Garcia-Benayas T, Blanco F, Gomez-Viera JM, Barrios A, Soriano V, Gonzalez-Lahoz J. Lipodystrophy body-shape changes in a patient undergoing zidovudine monotherapy. AIDS 2002;16(7):1087–9.
13. Pizzo PA, Wilfert C. Antiretroviral therapy for infection due to human immunodeficiency virus in children. Clin Infect Dis 1994;19(1):177–96.
14. Fischl MA, Parker CB, Pettinelli C, Wulfsohn M, Hirsch MS, Collier AC, Antoniskis D, Ho M, Richman DD, Fuchs E, et al. A randomized controlled trial of a reduced daily dose of zidovudine in patients with the acquired immunodeficiency syndrome. The AIDS Clinical Trials Group. N Engl J Med 1990;323(15):1009–14.
15. Collier AC, Bozzette S, Coombs RW, Causey DM, Schoenfeld DA, Spector SA, Pettinelli CB, Davies G, Richman DD, Leedom JM, et al. A pilot study of low-dose zidovudine in human immunodeficiency virus infection. N Engl J Med 1990;323(15):1015–21.

16. Falguera M, Perez-Mur J, Puig T, Cao G. Study of the role of vitamin B12 and folinic acid supplementation in preventing hematologic toxicity of zidovudine. Eur J Haematol 1995;55(2):97–102.

17. McLeod GX, Hammer SM. Zidovudine: five years later. Ann Intern Med 1992;117(6):487–501.

18. Blanche P, Silberman B, Barreto L, Gombert B, Sicard D. Reversible zidovudine-induced pure red cell aplasia. AIDS 1999;13(12):1586–7.

19. Kuehl AK, Noormohamed SE. Recombinant erythropoietin for zidovudine-induced anemia in AIDS. Ann Pharmacother 1995;29(7–8):778–9.

20. Yarchoan R, Perno CF, Thomas RV, Klecker RW, Allain JP, Wills RJ, McAtee N, Fischl MA, Dubinsky R, McNeely MC, et al. Phase I studies of 2'3'-dideoxycytidine in severe human immunodeficiency virus infection as a single agent and alternating with zidovudine (AZT). Lancet 1988;1(8577):76–81.

21. Dolin R, Lambert JS, Morse GD, Reichman RC, Plank CS, Reid J, Knupp C, McLaren C, Pettinelli C. 2',3'-Dideoxyinosine in patients with AIDS or AIDS-related complex. Rev Infect Dis 1990;12(Suppl 5):S540–9.

22. Danel C, Moh R, Anzian A, Abo Y, Chenal H, Guehi C, Gabillard D, Sorho S, Rouet F, Eholié S, Anglaret X. Tolerance and acceptability of an efavirenz-based regimen in 740 adults (predominantly women) in West Africa. J Acquir Immune Defic Syndr 2006;42(1):29-35.

23. Young B, Weidle PJ, Baker RK, Armon C, Wood KC, Moorman AC, Holmberg SD; HIV Outpatient Study (HOPS) Investigators. Short-term safety and tolerability of didanosine combined with high- versus low-dose tenofovir disproxil fumarate in ambulatory HIV-1-infected persons. AIDS Patient Care STDS 2006;20(4):238-44.

24. Koduri PR, Parekh S. Zidovudine-related anemia with reticulocytosis. Ann Hematol 2003;82:184-5.

25. Murri R, Antinori A, Camilli G, Zannoni G, Patriarca G. Fatal toxic epidermolysis induced by zidovudine. Clin Infect Dis 1996;23(3):640–1.

26. Duque S, de la Puente J, Rodriguez F, Fernandez Pellon L, Maquiera E, Jerez J. Zidovudine-related erythroderma and successful desensitization: a case report. J Allergy Clin Immunol 1996;98(1):234–5.

27. Zazo-Hernanz V, Sanchez-Herreros C, Gonzalez-Beato-Merino MJ, Lazaro-Ochaita P. Zidovudine pigmentation. Med Oral 1999;4:441.

28. Russo F, Collantes C, Guerrero J. Severe paronychia due to zidovudine-induced neutropenia in a neonate. J Am Acad Dermatol 1999;40(2 Pt 2):322–4.

29. Kang-Birken SL, Prichard JG. Paronychia of the great toes associated with protease inhibitors. Am J Health Syst Pharm 1999;56(16):1674–5.

30. Tosti A, Piraccini BM, D'Antuono A, Marzaduri S, Bettoli V. Paronychia associated with antiretroviral therapy. Br J Dermatol 1999;140(6):1165–8.

31. Bourezane Y, Thalamy B, Viel JF, Bardonnet K, Drobacheff C, Gil H, Vuitton DA, Hoen B. Ingrown toenail and indinavir: case-control study demonstrates strong relationship. AIDS 1999;13(15):2181–2.

32. Peters BS, Winer J, Landon DN, Stotter A, Pinching AJ. Mitochondrial myopathy associated with chronic zidovudine therapy in AIDS. Q J Med 1993;86(1):5–15.

33. Sinnwell TM, Sivakumar K, Soueidan S, Jay C, Frank JA, McLaughlin AC, Dalakas MC. Metabolic abnormalities in skeletal muscle of patients receiving zidovudine therapy observed by 31P in vivo magnetic resonance spectroscopy. J Clin Invest 1995;96(1):126–31.

34. Pedrol E, Masanes F, Fernandez-Sola J, Cofan M, Casademont J, Grau JM, Urbano-Marquez A. Lack of muscle toxicity with didanosine (ddI). Clinical and experimental studies. J Neurol Sci 1996;138(1–2):42–8.

35. Benbrik E, Chariot P, Bonavaud S, Ammi-Said M, Frisdal E, Rey C, Gherardi R, Barlovatz-Meimon G. Cellular and mitochondrial toxicity of zidovudine (AZT), didanosine (ddI) and zalcitabine (ddC) on cultured human muscle cells. J Neurol Sci 1997;149(1):19–25.

36. Schroder JM, Kaldenbach T, Piroth W. Nuclear and mitochondrial changes of co-cultivated spinal cord, spinal ganglia and muscle fibers following treatment with various doses of zidovudine. Acta Neuropathol (Berl) 1996;92(2):138–49.

37. Acosta BS, Grimsley EW. Zidovudine-associated type B lactic acidosis and hepatic steatosis in an HIV-infected patient. South Med J 1999;92(4):421–3.

38. Connor EM, Sperling RS, Gelber R, Kiselev P, Scott G, O'Sullivan MJ, VanDyke R, Bey M, Shearer W, Jacobson RL, et al. Reduction of maternal-infant transmission of human immunodeficiency virus type 1 with zidovudine treatment. Pediatric AIDS Clinical Trials Group Protocol 076 Study Group. N Engl J Med 1994;331(18):1173–80.

39. Culnane M, Fowler M, Lee SS, McSherry G, Brady M, O'Donnell K, Mofenson L, Gortmaker SL, Shapiro DE, Scott G, Jimenez E, Moore EC, Diaz C, Flynn PM, Cunningham B, Oleske J. Lack of long-term effects of in utero exposure to zidovudine among uninfected children born to HIV-infected women. Pediatric AIDS Clinical Trials Group Protocol 219/076 Teams. JAMA 1999;281(2):151–7.

40. Dabis F, Msellati P, Meda N, Welffens-Ekra C, You B, Manigart O, Leroy V, Simonon A, Cartoux M, Combe P, Ouangre A, Ramon R, Ky-Zerbo O, Montcho C, Salamon R, Rouzioux C, Van de Perre P, Mandelbrot L. 6-month efficacy, tolerance, and acceptability of a short regimen of oral zidovudine to reduce vertical transmission of HIV in breastfed children in Cote d'Ivoire and Burkina Faso: a double-blind placebo-controlled multicentre trial. DITRAME Study Group. DIminution de la Transmission Mere-Enfant. Lancet 1999;353(9155):786–92.

41. Wiktor SZ, Ekpini E, Karon JM, Nkengasong J, Maurice C, Severin ST, Roels TH, Kouassi MK, Lackritz EM, Coulibaly IM, Greenberg AE. Short-course oral zidovudine for prevention of mother-to-child transmission of HIV-1 in Abidjan, Côte d'Ivoire: a randomised trial. Lancet 1999;353(9155):781–5.

42. Sperling RS, Shapiro DE, McSherry GD, Britto P, Cunningham BE, Culnane M, Coombs RW, Scott G, Van Dyke RB, Shearer WT, Jimenez E, Diaz C, Harrison DD, Delfraissy JF. Safety of the maternal-infant zidovudine regimen utilized in the Pediatric AIDS Clinical Trial Group 076 Study. AIDS 1998;12(14):1805–13.

43. Lee BL, Tauber MG, Sadler B, Goldstein D, Chambers HF. Atovaquone inhibits the glucuronidation and increases the plasma concentrations of zidovudine. Clin Pharmacol Ther 1996;59(1):14–21.

44. Trapnell CB, Klecker RW, Jamis-Dow C, Collins JM. Glucuronidation of 3'-azido-3'-deoxythymidine (zidovudine) by human liver microsomes: relevance to clinical pharmacokinetic interactions with atovaquone, fluconazole, methadone, and valproic acid. Antimicrob Agents Chemother 1998;42(7):1592–6.

45. Lee BL, Tauber MG, Sadler B, Goldstein D, Chambers HF. Atovaquone inhibits the glucuronidation and increases the plasma concentrations of zidovudine. Clin Pharmacol Ther 1996;59(1):14–21.

46. Chave JP, Munafo A, Chatton JY, Dayer P, Glauser MP, Biollaz J. Once-a-week azithromycin in AIDS patients: tolerability, kinetics, and effects on zidovudine disposition. Antimicrob Agents Chemother 1992;36(5):1013–8.

47. Gustavson LE, Chu SY, Mackenthun A, Gupta MS, Craft JC. Drug interaction between clarithromycin and oral zidovudine in HIV-1 infected patients. Clin Pharmacol Ther 1993;53:163.

48. Vance E, Watson-Bitar M, Gustavson L, Kazanjian P. Pharmacokinetics of clarithromycin and zidovudine in patients with AIDS. Antimicrob Agents Chemother 1995;39(6):1355–60.

49. Toffoli G, Errante D, Corona G, Vaccher E, Bertola A, Robieux I, Aita P, Sorio R, Tirelli U, Boiocchi M. Interactions of antineoplastic chemotherapy with zidovudine pharmacokinetics in patients with HIV-related neoplasms. Chemotherapy 1999;45(6):418–28.

50. Trapnell CB, Klecker RW, Jamis-Dow C, Collins JM. Glucuronidation of 3'-azido-3'-deoxythymidine (zidovudine) by human liver microsomes: relevance to clinical pharmacokinetic interactions with atovaquone, fluconazole, methadone, and valproic acid. Antimicrob Agents Chemother 1998;42(7):1592–6.

51. Hochster H, Dieterich D, Bozzette S, Reichman RC, Connor JD, Liebes L, Sonke RL, Spector SA, Valentine F, Pettinelli C, et al. Toxicity of combined ganciclovir and zidovudine for cytomegalovirus disease associated with AIDS. An AIDS Clinical Trials Group Study. Ann Intern Med 1990;113(2):111–7.

52. Jacobson MA, Owen W, Campbell J, Brosgart C, Abrams DI. Tolerability of combined ganciclovir and didanosine for the treatment of cytomegalovirus disease associated with AIDS. Clin Infect Dis 1993;16(Suppl 1):S69–73.

53. Trapnell CB, Klecker RW, Jamis-Dow C, Collins JM. Glucuronidation of 3'-azido-3'-deoxythymidine (zidovudine) by human liver microsomes: relevance to clinical pharmacokinetic interactions with atovaquone, fluconazole, methadone, and valproic acid. Antimicrob Agents Chemother 1998;42(7):1592–6.

54. McCance-Katz EF, Rainey PM, Jatlow P, Friedland G. Methadone effects on zidovudine disposition (AIDS Clinical Trials Group 262). J Acquir Immune Defic Syndr Hum Retrovirol 1998;18(5):435–43.

55. Mole L, Israelski D, Bubp J, O'Hanley P, Merigan T, Blaschke T. Pharmacokinetics of zidovudine alone and in combination with oxazepam in the HIV infected patient. J Acquir Immune Defic Syndr 1993;6(1):56–60.

56. Shriner K, Goetz MB. Severe hepatotoxicity in a patient receiving both acetaminophen and zidovudine. Am J Med 1992;93(1):94–6.

57. Ameer B. Acetaminophen hepatotoxicity augmented by zidovudine. Am J Med 1993;95(3):342.

58. Sattler FR, Ko R, Antoniskis D, Shields M, Cohen J, Nicoloff J, Leedom J, Koda R. Acetaminophen does not impair clearance of zidovudine. Ann Intern Med 1991;114(11):937–40.

59. Shriner K, Goetz MB. Severe hepatotoxicity in a patient receiving both acetaminophen and zidovudine. Am J Med 1992;93(1):94–6.

60. Hedaya MA, Elmquist WF, Sawchuk RJ. Probenecid inhibits the metabolic and renal clearances of zidovudine (AZT) in human volunteers. Pharm Res 1990;7(4):411–7.

61. Kornhauser DM, Petty BG, Hendrix CW, Woods AS, Nerhood LJ, Bartlett JG, Lietman PS. Probenecid and zidovudine metabolism. Lancet 1989;2(8661):473–5.

62. Hedaya MA, Elmquist WF, Sawchuk RJ. Probenecid inhibits the metabolic and renal clearances of zidovudine (AZT) in human volunteers. Pharm Res 1990;7(4):411–7.

63. Veal GJ, Back DJ. Metabolism of Zidovudine. Gen Pharmacol 1995;26(7):1469–75.

64. Gallicano KD, Sahai J, Shukla VK, Seguin I, Pakuts A, Kwok D, Foster BC, Cameron DW. Induction of zidovudine glucuronidation and amination pathways by rifampicin in HIV-infected patients. Br J Clin Pharmacol 1999;48(2):168–79.

65. Lee BL, Safrin S, Makrides V, Gambertoglio JG. Zidovudine, trimethoprim, and dapsone pharmacokinetic interactions in patients with human immunodeficiency virus infection. Antimicrob Agents Chemother 1996;40(5):1231–6.

ANTIFUNGAL DRUGS

Amphotericin

General Information

Having a broad-spectrum fungicidal activity, amphotericin remains the mainstay of treatment of most invasive fungal infections. Compared with conventional amphotericin B deoxycholate, other lipid formulations of amphotericin (amphotericin B colloidal dispersion, amphotericin B lipid complex, and liposomal amphotericin B) facilitate treatment in patients with suspected and proven invasive mycoses, who are intolerant of or refractory to conventional amphotericin.

Compared with conventional amphotericin B deoxycholate, lipid-based formulations (amphotericin B colloidal dispersion, amphotericin B lipid complex, and liposomal amphotericin B) are less nephrotoxic (1,2).

Mechanism of action

The principal mechanism of action of amphotericin is based on its binding to lipids of the cell membrane of target cells, particularly to ergosterol, the predominant lipid in fungal cells, and cholesterol, the predominant lipid of the vertebrate cell membrane. The principle of selectivity is based on a higher affinity of amphotericin to ergosterol than cholesterol, but peroxidation of the membrane appears to be of equal importance (3–5).

Different formulations of amphotericin

Because of nephrotoxicity from amphotericin, which is common when amphotericin is given as the deoxycholate, lipid formulations have been developed. The formulations that are currently available are:

- amphotericin B deoxycholate (DAMB);
- amphotericin B colloidal dispersion (ABCD);
- amphotericin B lipid complex (ABLC);
- liposomal amphotericin B (L-Amb, AmBisome).

Pharmacokinetics

The pharmacokinetics of amphotericin are highly variable and depend on the formulation used and the infusion rate (6,7). Amphotericin, when administered as the deoxycholate complex, is highly bound to lipoproteins, mainly LDL and VLDL, and to a lesser extent to HDL (8,9) as well as to cell membranes of circulating blood cells. Binding is so avid that after spiking human plasma, no unbound amphotericin is detectable (9). Concentrations in peritoneal, pleural, and synovial fluids are usually less than half of those in serum, while cerebrospinal fluid concentrations range from undetectable to some 4% of the serum concentration, but over 40% in neonates (10). For DAMB the half-life is 1–2 days. Concentrations in bile are detectable for up to 12 days and in urine for 27–35 days. Clearance is faster and the volume of distribution smaller in neonates and infants (11) There is marked tissue storage of amphotericin, again depending on the formulation and the rate of infusion. Liver, spleen,

kidneys, and lungs accumulate large amounts. Tissue storage plays a major role in the pharmacokinetics of amphotericin, which can be detected in tissues much more than a year after the completion of therapy (12–14). Up to 40% of amphotericin is ultimately excreted unchanged in urine. Elimination via the bile plays a lesser role, and metabolism appears to be unimportant. Elimination is so slow that dosages need not be altered in patients with renal insufficiency.

Lipid formulations of amphotericin have individually variable pharmacokinetics. The use of DAMB in 20% Intralipid results in marked changes, with lower antifungally active blood concentrations (15). Infusion of ABLC, ABCD, or L-Amb, AmBisome results in plasma amphotericin concentrations specific to the individual formulation. The half-life of amphotericin after lipid formulations is prolonged compared with the deoxycholate formulation (16): 4–10 days for ABCD (17) and about 5 days for ABLC (18). The importance of these differences is unknown, because they do not reflect the biologically active concentration of amphotericin, which also varies with formulation. On a weight for weight basis amphotericin in lipid formulations is less active than in the deoxycholate formulation, because of lower systemic availability (7). One factor that complicates the interpretation of blood concentrations is the sparse data discriminating between amphotericin bound to the original lipid formula and to plasma lipoproteins, and the minute amount of unbound amphotericin not detectable by available analytical methods (9).

The interaction of amphotericin with serum lipoproteins (8) suggests that manipulations of blood lipids and blood lipoproteins might affect the pharmacokinetics of amphotericin, and therefore also alter its activity, including toxic effects, as suggested in animal studies (19).

Observational studies

Amphotericin is highly effective in the treatment of visceral leishmaniasis (20). In a prospective study of 938 patients from Bihar, India, who received the drug in a dosage of 1 mg/kg/day infused over 2 hours for 20 days, serum creatinine values over 177 µmol/l were noted in 6.3%, and acute renal insufficiency developed in three patients. Two patients died, possibly related to amphotericin, one with renal insufficiency and one with hypokalemia and cardiac arrest. Infusion-related chills occurred in 92% and fever in 40% of patients. The parasitological cure rate (no relapse within 6 months) exceeded 99%.

Amphotericin deoxycholate (DAMB)

The adverse effects of amphotericin deoxycholate have been reviewed in a retrospective analysis of 102 adult patients (median age 61 years) with a variety of underlying conditions who were admitted to a small community hospital in Honolulu and who received the drug for treatment of presumed or proven fungal infections that were mostly due to *Candida* species (21). The average total dose of amphotericin deoxycholate was comparatively low at 162 (range 10–840) mg. The initial dose averaged 16 (range 1–50) mg and the total duration of therapy was

8.3 (range 1–46) days. Chills, fever, and/or nausea were noted in 25% of the patients. Hypokalemia (a serum potassium concentration below 3.5 mmol/l) occurred in 19%, and nephrotoxicity (defined as a serum creatinine concentration of at least 141 µmol/l (1.6 mg/dl) with an increase of at least 44 µmol/l (0.5 mg/dl) during amphotericin deoxycholate therapy) in 15% of the patients. Nephrotoxicity increased with increasing total dose of amphotericin, while infusion-associated toxicity decreased with advancing age. The overall response rate to therapy with amphotericin deoxycholate was 83%.

Amphotericin colloidal dispersion (ABCD)

The safety and efficacy of amphotericin colloidal dispersion have been evaluated in 148 immunocompromised patients with candidemia (22). ABCD was given intravenously in a median daily dose of 3.9 (range 0.1–9.1) mg/kg for a median of 12 (range 1–72) days. In the safety analysis ($n = 148$ patients), nephrotoxicity occurred in 16% of the patients, with either doubling of the baseline serum creatinine concentration or an increase of 88 µmol/l (1.0 mg/dl) or a 50% fall in calculated creatinine clearance. Severe adverse events were believed to be probably or possibly related to ABCD in 36 patients (24%), including chills and fever (9.5%), hypotension and abnormal kidney function (4%), tachycardia, asthma, hypotension (3%), and dyspnea (2%). ABCD was withdrawn in 12% because of toxicity. The overall response rate in 89 evaluable patients was 66% with candidemia alone and 14% with disseminated candidiasis.

The safety and efficacy of ABCD have been studied in 133 patients with invasive fungal infections and renal impairment due to either amphotericin deoxycholate or pre-existing renal disease (23). The mean daily dose of ABCD was 3.4 (range 0.1–5.5) mg/kg, and the mean duration of therapy was 21 (range 1–207) days. Although individual patients had increases in serum creatinine concentrations, ABCD did not have an adverse effect on renal function: the mean serum creatinine concentration tended to fall slightly with days on therapy, and increases were not dose-related. Six patients discontinued ABCD therapy because of nephrotoxicity. Infusion-related adverse events occurred at least once in 74 patients (56%); however, while 43% of patients had infusion-related toxic effects on day 1, only 18% reported these events by day 7. There were complete or partial responses in 50% of the intention-to-treat population and in 67% of the 58 evaluable patients.

The safety of ABCD has been reviewed using data from 572 immunocompromised patients refractory to or intolerant of standard therapies enrolled in five phase I/II clinical trials (24). The mean daily dose of ABCD was 3.85 (median 3.8, range 0.1–9.1) mg/kg and the mean duration of treatment was 25 (median 16, range 1–409) days. Overall, the principal adverse events associated with ABCD therapy were chills (52%), fever (39%), and hypotension (19%). These infusion-related reactions were dose-related and were the dose-limiting adverse events, defining the maximum tolerated dosage at 7.5 mg/kg. ABCD did not adversely affect renal function, as measured by overall changes in serum creatinine from baseline to the end of therapy, even in patients with pre-existing renal impairment; only 3.3% of patients discontinued therapy because of nephrotoxicity. Complete or partial responses to treatment were reported in 149 of 260 evaluable patients (57%).

The safety and efficacy of ABCD have been studied in 220 bone marrow transplant recipients enrolled in the same five phase I or phase II studies (25). The median dose in this population was 4 (range 0.4–8.0) mg/kg, and the median duration of treatment was 16 (range 1–409) days. Overall, 37 (19%) of the patients had nephrotoxicity, defined as a doubling of serum creatinine from baseline, an increase of 88 µmol/l from baseline, or at least a 50% fall in calculated creatinine clearance. There were no significant changes in hepatic transaminases, alkaline phosphatase, or total bilirubin. Fever and chills were reported by 12 and 11% of patients respectively. Other acute, severe, infusion-related adverse events were hypoxia (4.1%), hypertension (2.7%), and hypotension (2.7%).

Mucormycosis has an exceedingly high mortality rate in immunocompromised patients. In five phase I and phase II studies of ABCD, 21 patients were given ABCD (mean dose 4.8 mg/kg per infusion for a mean duration of 37 days) on the basis of pre-existing renal insufficiency, nephrotoxicity during amphotericin B therapy, or refractory infections (26). Of 20 evaluable patients, 12 responded to ABCD, and there was no renal or hepatic toxicity. However, a previous randomized, comparative trial showed an at least similar if not increased frequency and severity of infusion-related reactions compared with conventional amphotericin B (27).

Amphotericin B lipid complex (ABLC)

The safety and efficacy of ABLC have been evaluated in 556 cases of proven or presumptive invasive fungal infection treated in an open, single-patient, US emergency-use study of patients who were refractory to or intolerant of conventional antifungal therapy (28). The daily dosage was either 5 mg/kg (87%) or 3 mg/kg. The investigators had the option of reducing the daily dosage as clinically warranted. Treatment was for 7 days in 540 patients (97%). During the course of ABLC therapy, serum creatinine concentrations in all patients fell significantly from baseline. In 162 patients with serum creatinine concentrations of at least 221 µmol/l (2.5 mg/dl) at baseline, the mean serum creatinine concentration fell significantly from the first week to the sixth week. The serum creatinine concentration increased from baseline to the end of therapy in 132 patients (24%). Hypokalemia (serum potassium concentration of less than 3 mmol/l) developed in 4.6%, and hypomagnesemia (serum magnesium concentration of less than 0.75 mmol/l) in 18%. There was a rise in serum bilirubin in 142/284 patients (33%); the overall increase was from 79 to 112 µmol/l (4.66–6.59 mg/dl) at the end of therapy. The mean alkaline phosphatase activity rose from 273 to 320 IU/l. There was no significant change overall in alanine transaminase activity, but the activity increased by the end of treatment in 16% of patients with initially normal values. There were complete or partial responses to therapy with ABLC in 167 of 291 mycologically confirmed cases evaluable for therapeutic response (57%).

The safety and efficacy of ABLC 5 mg/kg/day in patients with neutropenia and intolerance or refractoriness to amphotericin deoxycholate have been reported in two smaller series of 25 treatment courses from the UK. In one, the mean serum creatinine at the start of therapy was 139 μmol/l and at the end of therapy 132 μmol/l; there were no infusion-related adverse events (29). There was an increase in alanine transaminase activity in 12 of the 22 analysed treatment courses. In the other, there was an increase in serum creatinine in 5 of 18 courses (28%), and hypokalemia (less than 2.5 mmol/l) in two courses (11%); premedication for infusion-associated reactions was required in three courses (17%) (30). There were modest increases in serum alanine transaminase activities in five patients (30%).

In contrast to these reports, there was a high prevalence of adverse events with ABLC in the treatment of suspected or documented invasive fungal infections in 19 Scandinavian patients with mostly hematological malignancies (31). The mean starting dose of ABLC was 4.1 mg/kg/day, given for a median of 3 (range 1–19) days. ABLC was withdrawn because of adverse events in 14/19 patients (74%). These included rising creatinine concentrations ($n = 12$), increased serum bilirubin ($n = 7$), erythema ($n = 6$), increased alanine transaminase ($n = 6$), fever and chills ($n = 5$), hypoxemia ($n = 3$), hemolysis ($n = 2$), and back pain and increased serum alkaline phosphatase activity ($n = 1$ each). In patients with renal adverse effects, there were significantly increased serum creatinine concentration (from 85 to 199 mmol/l) and increased bilirubin concentration (from 17 to 77 μmol/l) in seven patients. The authors stated that while all the patients were very ill at the time of the start of ABLC therapy, in all cases the adverse effects had a direct and obvious correlation with the administration of ABLC. However, the reason for this unusual high rate of adverse events remains unclear.

Safety data have been published in a retrospective analysis of 551 patients with invasive fungal infections intolerant of or refractory to conventional antifungal therapy, 73 of whom received ABLC initially at 3 mg/kg/day instead of 5 mg/kg/day, as recommended in the protocol (32). There were no notable differences in adverse events (increased serum creatinine, infusion-related chills) between the two groups. Serum creatinine values were improved or stable at the end of therapy in 78 and 70% of patients respectively.

Two smaller series have addressed the safety of ABLC in immunocompromised patients (33,34). Each included about 30 patients who were treated with median dosages of 4.8 and 5.0 mg/kg for a median of 8 and 14 days. In contrast to a previous retrospective analysis that showed a 74% withdrawal rate, mostly due to infusion-related reactions (31), ABLC was well tolerated, with withdrawal rates of 6 and 0% and an overall trend for stable or improved serum creatinine values at the end of therapy.

Similarly, 13 infusion-related reactions have been reported among 308 infusions in four of ten patients with hematological malignancies receiving ABLC 3 mg/kg/day (35). These reactions (fever, rigors, myalgias), occurred during the first infusions, were judged to be mild, and resolved during later infusions. ABLC was well tolerated by 30 persistently febrile neutropenic patients with hematological malignancies who received it in a low dosage of 1 mg/kg/day for a median of 7.5 (range 2–19) days (36). Seven patients (23%) had mild to moderate infusion-related reactions, and no patient had nephrotoxicity. In one patient, ABLC was discontinued owing to intolerable infusion-related fever and chills.

The safety and efficacy of low dose ABLC (1 mg/kg/day) for empirical treatment of fever and neutropenia have been studied in 69 episodes in 61 patients with hematological malignancies (37). The median duration of therapy was 8 (range 2–19) days and 13 patients had mild to moderate infusion-related adverse events. Creatinine concentrations remained stable in 42 cases, improved in 9, and deteriorated in 18. There were no other toxic effects. The response rate (resolution of fever during neutropenia and absence of invasive fungal infection) was 67%.

Fungal infection remains an important cause of morbidity and mortality in lung transplant patients. In a prospective non-comparative evaluation in lung or heart-lung transplant recipients, ventilated patients received undiluted aerosolized ABLC 100 mg, and extubated patients received 50 mg; in all, 381 treatments were given (98 in ventilated patients and 283 in extubated patients) (38). The treatment was administered by face mask jet nebulizer with compressed oxygen at a flow rate of 7–8 l/minute and inhaled over 15–30 minutes. Treatments were delivered once every day for four consecutive days, then once a week for 2 months. In all, 381 treatments were given to 51 patients, and ABLC was subjectively well tolerated in 98%. Pulmonary function worsened by 20% or more in under 5% of all treatments. There were no significant adverse events.

In a retrospective comparison of outcomes in liver transplant recipients with invasive aspergillosis who received amphotericin B lipid complex (ABLC) or conventional amphotericin B, the 60-day mortality rate was lower in the ABLC cohort: four of 12 patients versus 24 of 29 patients. Only one of four ABLC recipients with definite invasive aspergillosis died, compared with all 11 in the amphotericin B group. The 60-day survival probability curves was significantly lower in the amphotericin B group. ABLC therapy was the only independent mortality-protective variable (OR = 0.31; 95% CI = 0.07, 0.44) (39). In a prospective historical control study 131 patients with acute myclogenous leukemia or high-risk myelodysplastic syndrome undergoing induction chemotherapy were given ABLC 2.5 mg/kg intravenously 3 times weekly as antifungal prophylaxis and were compared with 70 who had previously received LAMB 3 mg/kg 3 times weekly (40). Grade 3 and 4 adverse events (hyperbilirubinemia, 3% versus 6%; infusion-related adverse events, 3% versus 7%) were statistically similar between the groups, as were rates of withdrawal because of adverse events (18% versus 15%) and survival rates (92% versus 86%).

The risk of hematological, renal, and hepatic toxicity associated with ABLC has been assessed in a multicenter, open, non-comparative study in 93 patients from 17 different hospitals who received ABLC because of proven or suspected systemic fungal infection or leishmaniasis (41). Optimum treatment with ABLC comprised a 2-hour infusion of 5 mg/kg/day for a minimum of 14 days; the mean dose was 235 (range 9–500) mg/day and the total cumulative dose was 2894 (range 30–10 200) mg. In the whole group, the mean serum creatinine concentration was similar before and after ABLC (1.00 versus 1.20 mg/dl). There were no significant changes in concentrations of hemoglobin, potassium, and bilirubin, or in hepatic transaminase activities. There was no significant correlation between the dose and the serum creatinine concentrations. There was no greater nephrotoxicity in the patients with previous renal insufficiency, or in those who had previously received amphotericin B. There were serious adverse events in five patients, but other alternative causes were found in three of them. There were fevers or chills in 23% of the patients during the infusion of ABLC, but only in one case did this necessitate withdrawal. The overall rate of withdrawals due to adverse events was 9.7%.

Liposomal amphotericin (LAmB)

The safety, tolerance, and pharmacokinetics of liposomal amphotericin have been evaluated in an open, sequential dose escalation, multiple-dose phase I/II study in 36 patients with neutropenia and persistent fever requiring empirical antifungal therapy (42). The patients received doses of 1, 2.5, 5.0, or 7.5 mg/kg/day of liposomal amphotericin for a mean of 9.2 days. Liposomal amphotericin was well tolerated: infusion-related adverse effects (fever, chills, rigor) occurred in 15 (5%) of all 331 infusions, and only two patients (5%) required premedication (dyspnea and generalized flushing; facial urticaria). Hypotension (one infusion) and hypertension (three infusions) were infrequent. One patient each had sharp flank pain and dyspnea during one infusion; these symptoms did not recur during subsequent infusions. Serum creatinine, potassium, and magnesium concentrations were not significantly changed from baseline, and there were no net increases in serum transaminases. There was, however, a significant increase in serum alkaline phosphatase activity and increase in bilirubin concentration in the overall population as well as in individual dosage groups. One patient who received concomitant L-asparaginase had increases in serum lipase and amylase activities in association with symptoms of pancreatitis while receiving liposomal amphotericin; however, as he continued to receive the drug, the serum lipase and amylase returned to baseline. Liposomal amphotericin had non-linear pharmacokinetics consistent with reticuloendothelial uptake and redistribution. There were no breakthrough fungal infections during therapy.

The efficacy of two dosages of liposomal amphotericin in the treatment of proven or probable invasive aspergillosis in neutropenic patients with cancer or those undergoing bone marrow transplantation has been studied in a prospective, randomized, open, multicenter trial in 120 patients randomized to receive either 1 mg/kg/day or 4 mg/kg/day of liposomal amphotericin; 87 patients were available for evaluation (43). There was at least one toxic event during treatment in 15 of 41 patients given 1 mg/kg/day and 25 of 46 given 4 mg/kg/day, but the numbers of events per patient were similar. These events included headache, nausea, diarrhea, rash, liver toxicity, myalgia, dyspnea, fever, chills, and back pain. Renal toxicity definitely related to liposomal amphotericin occurred in 1/41 patients treated with 1 mg/kg/day and 5/46 patients treated with 4 mg/kg/day. Only in one case was treatment permanently discontinued because of toxicity related to liposomal amphotericin (4 mg/kg/day). No patient died from liposomal amphotericin toxicity. Overall, liposomal amphotericin was effective in 50–60% of patients; however, the number of cases with proven invasive aspergillosis was too small to allow a meaningful comparison of the two dosages regarding efficacy in this life-threatening disease.

The safety and efficacy of liposomal amphotericin have been compared with that of ABLC in a retrospective analysis of 59 adult patients with hematological malignancies who received 68 courses of either liposomal amphotericin ($n = 32$) or ABLC ($n = 36$) for a variety of presumed or confirmed invasive fungal infections (44). The median daily dosages were 1.9 (range 0.7–4.0) mg/kg for liposomal amphotericin and 4.8 (range 1.9–5.8) mg/kg for ABLC. There was no statistically significant difference in the overall outcome; febrile reactions were significantly more common with ABLC (36 versus 6%), but there were no significant differences in the median creatinine concentrations at baseline and at the end of therapy or in the number of patients with urinary loss of potassium or magnesium.

In an open, sequential phase II clinical study of three different regimens of liposomal amphotericin for visceral leishmaniasis (2 mg/kg on days 1–6 and on day 10; 2 mg/kg on days 1–4 and on day 10; 2 mg/kg on days 1, 5, and 10) in Indian and Kenyan patients in three developing countries, there were few infusion-associated adverse effects (45). Of 32 Brazilian patients (15 of whom received 2 mg/kg on days 1–10 because of poor responses to the first regimen, 37% had a fever with one or more infusions, 9% had chills, and 6% had back pain; in addition, three patients had respiratory distress and/or cardiac dysrhythmias. There were different response rates to the three regimens in the different countries, leading to the recommendation of 2 mg/kg on days 1–4 and day 10 in India and Kenya, and 2 mg/kg on days 1–10 in Brazil.

In order to determine the maximum tolerated dosage of liposomal amphotericin B, a phase I/II study was conducted in 44 adult patients with proven ($n = 21$) or probable ($n = 23$) infections due to *Aspergillus* species and other filamentous fungi (46). The dosages were 7.5, 10, 12.5, and 15 mg/kg/day. The number of infusions was 1–83 with a median duration of 11 days. The maximum tolerated dosage was at least 15 mg/kg. Infusion-related reactions included fever in 8 and chills and rigors in 5 of 43 patients. Three patients developed a syndrome of substernal chest tightness, dyspnea, and flank pain, which was relieved by diphenhydramine. Serum creatinine increased two times above baseline in 32% of patients, but this increase was not dose-related. Hepatotoxicity developed in one patient. Altogether, the most common adverse events included fever (48%), increased creatinine concentration (46%), hypokalemia (39%), chills (32%), and abdominal pain (25%), with no obvious dose-dependency. Nine patients (20%) stopped taking the drug because of an adverse event. The reasons included raised serum creatinine, renal insufficiency, pancreatitis, hyperbilirubinemia, hypotension associated with the infusion, cardiorespiratory failure, multiorgan failure, and relapse of the primary malignancy. The last three events were attributed to the underlying disease process. Discontinuation was unrelated to dosage. Pharmacokinetic analysis showed dose-related non-linear kinetics at dosages of 7.5 mg/kg/day and over.

A dermatosis commonly known as post kala-azar dermal leishmaniasis can develop after treatment of human visceral leishmaniasis. In about 15% of cases the disfiguring lesions persist, sometimes for many years. The usefulness of LAMB 2.5 mg/kg/day for 20 days in the treatment of persistent post kala-azar dermal leishmaniasis has been evaluated in 12 Sudanese subjects, who were regularly screened for adverse effects, LAMB completely cleared the rash in 10 (83%) of the patients and caused no detectable adverse effects (47).

Liposomal amphotericin B (L-AmB) has been studied in an open study with a combination of fluconazole + itraconazole as prophylaxis in patients undergoing induction chemotherapy for acute myelogenous leukemia and myelodysplastic syndrome (48). Patients were randomized to receive either fluconazole 200 mg orally every 12 hours + itraconazole 200 mg orally every 12 hours (n = 67) or L-AmB 3 mg/kg intravenously 3 times a week (n = 72). Altogether, 47% of the patients completed antifungal prophylaxis without a change in therapy for proven or suspected fungal infection. Three patients in each arm developed a proven fungal infection. Because of persistent fever 23% of those treated with L-AmB and 24% of those treated with fluconazole + itraconazole were changed to alternative antifungal therapy. Increases in serum creatinine concentrations to over 177 µmol/l (2 mg/dl) (20% versus 6%) and increases in serum bilirubin concentrations to over 2 mg/dl (43% versus 22%) were more common with L-AmB. There were infusion-related reactions in five patients who received L-AmB. Responses to chemotherapy and induction mortality rates were similar in the two arms. Thus, while L-AmB and fluconazole + itraconazole appeared to have similar efficacy, L-AmB was associated with higher rates of increased serum bilirubin and creatinine concentrations.

In 164 HIV-negative children (median age 1.6 years; range 4 months to 14 years) with Mediterranean visceral leishmaniasis L-AmB 3 mg/kg given on days 1–5 and 10 was not associated with adverse events (49).

Aerosolized amphotericin

In a prospective, randomized, multicenter trial, inhalation of aerosolized amphotericin (10 mg bd) has been investigated as prophylaxis against invasive aspergillosis in 382 cancer patients with an anticipated duration of neutropenia of at least 10 days (50). While there was no difference in the incidence of invasive aspergillosis, infection-related mortality, and overall mortality, 31% of the patients discontinued amphotericin prophylaxis prematurely owing to adverse effects (55%; most commonly cough, bad taste, nausea), inability to cooperate further (30%), violation of the study protocol (11%), and non-adherence (4%).

Comparative studies

Comparisons of different formulations of amphotericin
Amphotericin deoxycholate in glucose versus amphotericin deoxycholate in Intralipid
The safety of two formulations of intravenous amphotericin deoxycholate has been investigated in a randomized, open comparison in neutropenic patients with refractory fever of unknown origin or pulmonary infiltrates (51). Amphotericin deoxycholate was given in a dose of 0.75 mg/kg/day either in 250 ml of a 5% glucose solution or mixed with 250 ml of a 20% lipid emulsion (Intralipid 20%) on eight consecutive days and then on alternate days as a 1–4 hour infusion. The mean number of days of treatment was 11.3 versus 9.9 days. There were no statistically significant differences between the two cohorts with respect to the incidence of infusion-related adverse events, such as fever and chills, renal impairment, or treatment failure. However, grade 3–4 acute dyspnea occurred slightly more often with the lipid emulsion formulation, and there were significantly more other severe respiratory events in patients receiving lipid emulsion, raising the possibility of a causal relation via fat overload or incompatibility between amphotericin deoxycholate and the lipid emulsion.

The efficacy and tolerability of amphotericin prepared in Intralipid 20% have been evaluated in 16 patients with HIV infection and esophageal candidiasis or cryptococcosis and compared with standard amphotericin in a matched group of 24 patients (52). While both formulations had apparently similar clinical and microbiological efficacy, fewer patients receiving the lipid emulsion formulation required premedication or symptomatic therapy for infusion-associated adverse events, and fewer patients were withdrawn because of adverse effects. Renal adverse effects (a rise in serum creatinine and/or electrolyte loss) were more common in patients who received the conventional formulation.

The efficacy and safety of amphotericin in Intralipid 20% or 5% glucose has been evaluated in a retrospective case analysis in 30 patients with AIDS and cryptococcal meningitis who received either formulation 1 mg/kg/day for 20 days with or without flucytosine (n = 20) or fluconazole (n = 4), followed by maintenance therapy with fluconazole 400 mg/day (53). Twenty patients received amphotericin deoxycholate in 500 ml 5% glucose over 5 hours, and 10 received amphotericin deoxycholate in

100 ml of 20% Intralipid given over 2 hours. Complete clinical resolution was obtained in 55 and 60% of the patients respectively. There were no differences regarding infusion-related adverse effects, nephrotoxicity, or anemia.

Amphotericin deoxycholate versus amphotericin B colloidal dispersion

Amphotericin colloidal dispersion has been compared with amphotericin deoxycholate in a prospective, randomized, double-blind study in the empirical treatment of fever and neutropenia in 213 patients (27). Patients were stratified by age and concomitant use of ciclosporin or tacrolimus and then randomized to receive ABCD (4 mg/kg/day) or amphotericin deoxycholate (0.8 mg/kg/day) for 14 days. Renal dysfunction was less likely to develop and occurred later with ABCD than with amphotericin deoxycholate. Likewise, the absolute and percentage fall in the serum potassium concentration from baseline to the end of therapy was greater with amphotericin deoxycholate than ABCD. However, probable or possible infusion-related hypoxia and chills were more common with ABCD than amphotericin deoxycholate. There was a therapeutic response in 50% of the patients who received ABCD and 43% of those who received amphotericin deoxycholate. Thus, ABCD was of comparable efficacy and less nephrotoxic than amphotericin deoxycholate, but infusion-related events were more common with ABCD.

Amphotericin deoxycholate versus liposomal amphotericin

Liposomal amphotericin 5 mg/kg/day and amphotericin deoxycholate 1 mg/kg/day have been compared in the treatment of proven or suspected invasive fungal infections in neutropenic patients in a randomized, multicenter study (54). Significantly more patients given amphotericin deoxycholate had a greater than 100% increase in baseline serum creatinine. Treatment was temporarily discontinued or the dosage reduced because of an increase in serum creatinine in 18/54 (33%) patients treated with amphotericin deoxycholate versus 2/51 (4%) treated with liposomal amphotericin. There was no statistically significant difference in the number of patients with infusion-related toxicity (fever/chills), hypokalemia, or increases in serum transaminases, alkaline phosphatase, or serum bilirubin. In 66 patients eligible for analysis of efficacy, there was a trend to an improved overall response rate and a significant difference in the rate of complete responses in favor of liposomal amphotericin; death rates were also lower in patients treated with liposomal amphotericin.

The results of a randomized, double-blind, multicenter comparison of liposomal amphotericin (3.0 mg/kg/day) with conventional amphotericin deoxycholate (0.6 mg/kg/day) for empirical antifungal therapy in patients with persistent fever and neutropenia have been reported (55). The mean duration of therapy was 10.8 days for liposomal amphotericin (343 patients) and 10.3 days for amphotericin deoxycholate (344 patients). While the composite rates of successful treatment were similar (50% for liposomal amphotericin and 49% for amphotericin deoxycholate), significantly fewer of the patients who received the liposomal preparation had infusion-related fever (17 versus 44%), chills or rigors (18 versus 54%), or other reactions, including hypotension, hypertension, and hypoxia. Nephrotoxicity (defined by a serum creatinine concentration twice the upper limit of normal) was significantly less frequent among patients treated with liposomal amphotericin (19%) than among those treated with conventional amphotericin deoxycholate (34%).

Aerosolized DAMB 50 mg and ABLC 25 mg have been compared in lung transplant recipients in a prospective, randomized, double-blind trial in 100 subjects (56). The study drug was withdrawn because of intolerance in six of 49 patients treated with DAMB and three of 51 treated with ABLC. Those who received DAMB were more likely to have had an adverse event (OR = 2.16, 95%CI = 1.10, 4.24).

Amphotericin lipid complex versus liposomal amphotericin

Liposomal amphotericin and ABLC have been compared in an open randomized study in 75 adults with leukemia and 82 episodes of suspected or documented mycosis (57). The median durations of treatment and dosages were 15 days at 4 mg/kg/day for liposomal amphotericin and 10 days at 3 mg/kg/day for ABLC. Acute but not dose-limiting infusion-related adverse events occurred in 36 versus 70%. Bilirubin increased to over 1.5 times baseline in 59 versus 38%. There was no difference in the effects of either agent on renal function and drug-related withdrawals. The overall response rate to therapy in documented fungal infections (29 and 30% respectively) was not different between the two drugs.

Amphotericin deoxycholate in glucose versus amphotericin in nutritional fat emulsion

The safety of DAMB prepared in nutritional fat emulsion (a non-approved mode of amphotericin administration) has been reviewed (SEDA-21, 282) (SEDA-22, 285). It is not clear whether it has a better therapeutic index than other formulations, and methods of preparing it have not been standardized. The adverse effects of amphotericin prepared in nutritional fat emulsion have been compared with those of amphotericin prepared in 5% dextrose in two studies. While one of the studies showed a significantly lower frequency of infusion-related reactions and hypokalemia in patients receiving the fat emulsion (58), there were no differences in safety and tolerance between the two formulations in the other study (59). The safety of amphotericin prepared in nutritional fat emulsions has been reviewed (SEDA-21, 282) (SEDA-22, 285). Because of stability concerns and lack of systematic safety data, this form of amphotericin cannot be recommended.

Comparisons of amphotericin with other antifungal drugs

Antifungal azoles

Fluconazole

There has been an open, randomized comparison of amphotericin deoxycholate 0.5 mg/kg/day intravenously versus fluconazole 400 mg/day orally for empirical antifungal therapy in neutropenic patients with cancer and fever refractory to broad-spectrum antibiotics (60). Patients with abnormal hepatic or renal function were excluded, as were those with proven or suspected invasive fungal infection. The mean duration of therapy was 8.3 days with amphotericin deoxycholate and 7.9 days with fluconazole. Altogether, 32/48 patients randomized to amphotericin deoxycholate and 19/52 randomized to fluconazole had adverse affects (67 versus 36%). Two patients developed immediate hypersensitivity reactions (flushing, hypotension, bronchospasm) to amphotericin deoxycholate and had to be withdrawn. Hypokalemia was noted in 25 patients (52%), and nephrotoxicity, defined as a rise in serum creatinine of 44 µmol/l (0.5 mg/dl) or more compared with the baseline value, in nine patients (19%). The corresponding frequencies with fluconazole were 23 and 6% respectively. Treatment success rates and mortality were similar (46 versus 56% and 33 versus 27% respectively).

Fluconazole and amphotericin as empirical antifungal drugs in febrile neutropenic patients have been investigated in a prospective, randomized, multicenter study in 317 patients randomized to either fluconazole (400 mg qds) or amphotericin deoxycholate (0.5 mg/kg qds) (61). Adverse events (fever, chills, renal insufficiency, electrolyte disturbances, and respiratory distress) occurred significantly more often in patients who were given amphotericin (128/151 patients, 81%) than in those given fluconazole (20/158 patients, 13%). Eleven patients treated with amphotericin, but only one treated with fluconazole, were withdrawn because of an adverse event. Overall mortality and mortality from fungal infections were similar in both groups. There was a satisfactory response in 68% of the patients treated with fluconazole and 67% of those treated with amphotericin. Thus, fluconazole may be a safe and effective alternative to amphotericin for empirical therapy of febrile neutropenic patients; however, since fluconazole is ineffective against opportunistic molds, the possibility of an invasive infection by a filamentous fungus should be excluded before starting empirical therapy. Similarly, patients who take azoles for prophylaxis are not candidates for empirical therapy with fluconazole.

Conventional amphotericin deoxycholate (0.2 mg/kg qds) and fluconazole (400 mg qds) have been compared in a prospective randomized study in 355 patients with allogeneic and autologous bone marrow transplantation (62). The drugs were given prophylactically from day 1 until engraftment. There was no difference in the occurrence of invasive fungal infections, but amphotericin was significantly more toxic than fluconazole, especially in related allogeneic transplantation, after which 19% of patients developed toxicity compared with none of those who received fluconazole.

Itraconazole

Amphotericin and itraconazole have been compared in a multicenter, open, randomized study in 277 adults with cancer and neutropenia (63). Itraconazole oral solution (100 mg bd, $n = 144$) was compared with a combination of amphotericin capsules and nystatin oral suspension ($n = 133$). Adverse events were reported in about 45% of patients in each group. The most frequent were vomiting (14 versus 12 patients), diarrhea (12 versus 9 patients), nausea (5 versus 12 patients), and rash (2 versus 13 patients). There were no differences in liver function test abnormalities. Treatment had to be withdrawn because of adverse events (including death) in 34 patients who took itraconazole and 33 of those who took amphotericin plus nystatin; there were 17 deaths in each group and death was recorded as adverse event in 13 and nine patients respectively.

Intravenous amphotericin deoxycholate (0.7–1.0 mg/kg) and itraconazole (400 mg intravenously for 2 days, 200 mg intravenously for up to 12 days, then 400 mg/day orally) have been compared in 384 granulocytopenic patients with persistent fever in a randomized, multicenter trial (64). The median duration of therapy was 8.5 days. The incidence of drug-related adverse events (54 versus 5%) and the rate of withdrawal due to toxicity (38 versus 19%) were significantly higher with amphotericin. The most frequent reasons for withdrawal in patients taking itraconazole were nausea and vomiting (5%), rash (3%), and abnormal liver function tests (3%). Significantly more of the patients who received amphotericin had nephrotoxicity (24 versus 5%); however, fewer had hyperbilirubinemia (5 versus 10%). There was no difference in gastrointestinal adverse events between the two groups.

Amphotericin in capsules 500 mg qds has been compared with itraconazole elixir 2.5 mg/kg bd for the prophylaxis of systemic and superficial fungal infections in a double-blind, randomized, placebo-controlled, multicenter trial for 1–59 days (65). While itraconazole significantly reduced the frequency of superficial fungal infections, it was not superior in reducing invasive fungal infections or in improving mortality. Adverse events were reported in 222 patients taking itraconazole (79%) and in 205 patients taking amphotericin (74%). The commonest adverse events were gastrointestinal, followed by rash and hypokalemia, with no differences between the two regimens. In both groups, 5% of the adverse events were considered to be definitely drug-related. Comparable numbers of patients in the two groups permanently stopped treatment because of adverse events (including death), 78 (28%) in the amphotericin group and 75 (27%) in the itraconazole group. Nausea (11 and 9%) and vomiting (7 and 8%) were the most frequently reported adverse events that led to withdrawal. Biochemical changes were comparable in the two groups.

Echinocandins

In a randomized, double-blind comparison of amphotericin (0.5 mg/kg intravenously) and caspofungin acetate (35, 50, or 70 mg) once daily for 7–14 days in 140 patients with oropharyngeal and/or esophageal candidiasis, 63% had esophageal involvement and 98% were infected with HIV (66). Response rates were 63% with amphotericin and 74–91% with caspofungin. More patients receiving amphotericin had drug-related adverse effects (fever, chills, nausea, vomiting) than those receiving any dose of caspofungin. Two patients who took caspofungin 35 mg and one who was given amphotericin withdrew because of adverse effects. Drug-related laboratory abnormalities were also more common in patients who received amphotericin. The most common drug-related laboratory abnormalities in patients who received caspofungin were raised alanine transaminase, aspartate transaminase, and alkaline phosphatase, which were typically less than five times the upper limit of normal and resolved despite continued treatment. None of the patients receiving caspofungin and nine of those who received amphotericin developed drug-related increases in serum creatinine concentrations. No patient withdrew because of drug-related laboratory adverse effects.

Amphotericin has been compared with caspofungin in a multicenter, double-blind, randomized trial in 128 adults with endoscopically documented symptomatic *Candida* esophagitis (67). There was endoscopically verified clinical success in 63% of patients given amphotericin deoxycholate 0.5 mg/kg/day and in 74 and 89% of the patients who received caspofungin 50 and 70 mg/day respectively. Therapy was withdrawn because of drug-related adverse events in 24% of the patients who were given amphotericin and in 4 and 7% of those who were given caspofungin 50 and 70 mg/day respectively. More patients who received amphotericin had drug-related fever, chills, or nausea than those who received caspofungin. More patients who received amphotericin (91%) than caspofungin (61 and 32%) developed drug-related laboratory abnormalities. There were drug-related increases in blood urea–nitrogen concentrations in 15% of the patients who received amphotericin but none of those who received caspofungin. Likewise, serum creatinine concentrations increased in 16 patients who received amphotericin but in only one who received caspofungin. In summary, caspofungin was as effective as amphotericin but better tolerated in the treatment of esophageal candidiasis.

In a double-blind, randomized trial, amphotericin deoxycholate was compared with caspofungin for the primary treatment of invasive candidiasis (68). Patients who had clinical evidence of infection and a positive culture for *Candida* species from blood or another site were enrolled. They were stratified according to the severity of disease, as indicated by the presence or absence of neutropenia and the Acute Physiology and Chronic Health Evaluation (APACHE II) score, and were randomly assigned to receive either amphotericin (0.6–0.7 mg/kg/day or 0.7–1.0 mg/kg/day for patients with neutropenia) or caspofungin (50 mg/day with a loading dose of 70 mg on day 1). Of the 239 patients enrolled, 224 were included in the modified intention-to-treat analysis. Baseline characteristics, including the percentage of patients with neutropenia and the mean APACHE II score, were similar in the two treatment groups. The efficacy of amphotericin was similar to that of caspofungin, with successful outcomes in 62% of the patients treated with amphotericin and in 73% of those treated with caspofungin. There were significantly more drug-related adverse events (fever, chills, and infusion-related events) associated with amphotericin. Amphotericin caused more nephrotoxicity, as defined by an increase in serum creatinine of at least twice the baseline value or an increase of at least 88 µmol/l) (8.4 versus 25%). Only 2.6% of those who were given caspofungin were withdrawn because of adverse events, compared with 23% of those who were given amphotericin. Thus, caspofungin was at least as effective as amphotericin for the treatment of mostly non-neutropenic patients with invasive candidiasis but significantly better tolerated.

Comparisons with antimonials

Amphotericin might be useful in the treatment of leishmaniasis, as suggested by a comparative study (amphotericin in 14 doses of 0.5 mg/kg infused in 5% glucose on alternate days) against sodium stibogluconate (20 mg/kg in two divided doses daily for 40 days). All 40 patients taking amphotericin were cured, whereas in the stibogluconate group 28 of the 40 showed an initial cure but only 25 a definite cure (69).

Placebo-controlled studies

In a small randomized, double-blind, placebo-controlled study, liposomal amphotericin (2 mg/kg three times weekly) was investigated as prophylaxis against fungal infections in 161 patients undergoing chemotherapy or bone marrow transplantation for hematological malignancies (70). There were no statistically significant differences between the two study arms in the incidences of the most frequently reported adverse events or in changes in renal and hepatic laboratory parameters. Despite a sizable rate of suspected or documented fungal infections in the placebo arm, prophylactic therapy with liposomal amphotericin did not lead to a significant reduction in fungal infections or the requirement for systemic antifungal therapy.

General adverse effects

Fever, rigors, nausea, vomiting, headaches, muscle pains, and joint pains are common. The incidence and severity of these reactions are highest with rapidly increasing blood concentrations, and are frequent during the start of therapy (71). Hypersensitivity reactions have been described in case reports. Reports of rashes have been rare. The UK Committee on Safety of Medicines received 20 reports of the occurrence of rash over a 17-year period (SEDA-16, 289). However, with increased use of lipid formulations this could change (72–74). Tumor-inducing effects have not been demonstrated in animals or humans

Organs and Systems

Cardiovascular

Electrolyte disturbances (hyperkalemia, hypomagnesemia, renal tubular acidosis) due to renal toxicity can be additional factors that precipitate cardiac reactions.

Effects on blood pressure
Changes in blood pressure (hypotension as well as hypertension) have been reported (75,76).

- A 67-year-old man with multiple intraperitoneal and urinary fungal pathogens and a history of well-controlled chronic hypertension developed severe hypertension associated with an infusion of ABLC (77). He received a 5 mg test dose, which was tolerated without incident. About 60 minutes into the infusion (5 mg/kg), his blood pressure rapidly increased to 262/110 mmHg from a baseline of 150/80 mmHg. His temperature increased to 39.8°C, and tachycardia developed (up to 121/minute). The infusion was stopped, and he was given morphine, propranolol, and paracetamol. His blood pressure returned to baseline over the next 2 hours. Rechallenge with ABLC on the next day resulted in an identical reaction despite premedication with pethidine, diphenhydramine, and morphine. ABLC was permanently withdrawn, and the infection was managed with high dosages of fluconazole.

The etiology of amphotericin-associated hypertension has not been elucidated, but it may be related to vasoconstriction. Of note, the traditional test dose appears not to identify individuals predisposed to hypertensive reactions; four of six cases of amphotericin-associated hypertension received test doses without incident.

Of eight cases of hypertension in patients receiving amphotericin; six occurred within 1 hour (78). All except one had received a non-lipid-containing formulation.

- A 19-year-old girl with acute lymphoblastic leukemia developed sustained severe arterial hypertension shortly after being given amphotericin and continuing for several hours after the infusion (79).

Cardiac dysrhythmias
In 6 of 90 children given intravenous amphotericin there was a significant fall in heart rate, and monitoring of heart rate was recommended in children with underlying heart disease (80). These immediate reactions follow intravenous administration and occur particularly with excessively rapid infusion of DAMB.

Ventricular dysrhythmias have been reported after rapid infusion of large doses of DAMB (81) in patients with hyperkalemia and renal insufficiency, but not in patients with normal serum creatinine and potassium concentrations, even if they have received the drug over a period of 1 hour. Slower infusion rates and infusion during hemodialysis have been advocated in patients with terminal kidney insufficiency, in order to avoid hyperkalemia.

Chest pain
Three cases of chest discomfort associated with infusion of L-AmB at a dosage of 3 mg/kg/hour for 1 hour have been reported (82).

- The first patient had chest tightness and difficulty in breathing and the second had dyspnea and acute hypoxia (PaO$_2$ 55 mmHg; 7.3 kPa), both within 10 minutes of the start of the infusion. The third complained of chest pain 5 minutes after the start of two infusions. In all cases the symptoms resolved on terminating therapy. Two patients were later rechallenged with slower infusions and tolerated the drug well.

A review of the literature showed that similar reactions had been reported anecdotally in several clinical trials of L-AmB, with all other formulations, and with liposomal daunorubicin and doxorubicin. While the pathophysiology of such reactions is yet unclear, the authors recommended infusing L-AmB over at least 2 hours with careful monitoring of adverse events.

Myocardial ischemia
Rare instances of cardiac arrest have been reported (83).

Cardiomyopathy
Reversible dilated cardiomyopathy secondary to DAMB has been reported.

- A 20-year-old man with fluconazole-refractory disseminated coccidioidomycosis without evidence of cardiac involvement developed dilated cardiomyopathy and clinical congestive heart failure after 2 months of therapy with amphotericin B (0.7 mg/kg/day of amphotericin B deoxycholate, switched after 1 month of treat ment, because of rising serum creatinine concentrations, to amphotericin B lipid complex 5mg/ kg/day (84). His echocardiographic abnormalities and heart failure resolved within 6 weeks, posaconazole having been substituted for amphoericin B after about 90 days. A 41-year-old woman with cryptococcal meningitis and no previous cardiac disease developed a fatal cardiac dysrhythmia, acute renal failure, and anemia after an acute overdose of amphotericin B deoxycholate (85). The intention had been to give liposomal amphotericin 5 mg/kg/day; however, amphotericin B deoxycholate 5 mg/kg was inadvertently given instead, the usual dose of the deoxycholate formulation being 0.5–0.8 mg/kg/day.
- A subacute cardiomyopathy occurred in a 37-year-old HIV-positive woman 48 hours after the start of therapy with liposomal amphotericin 200 mg/day for disseminated candidiasis (86). Amphotericin was replaced by caspofungin 50 mg/day; she was given dobutamine 10 micrograms/kg/minute and recovered in 9 days.

Vascular effects
Phlebitis occurs in over 5% of patients receiving amphotericin deoxycholate through peripheral veins, which limits the concentration advisable for this route of administration.

Raynaud's syndrome has been attributed to DAMB phenomenon (87).

Extravasation can cause severe local reactions, including tissue necrosis. Safe venous access, preferably via a central line is advisable. The recommendation to use sodium heparin or buffered dextrose is not supported by clinical data.

Respiratory

Inhaled DAMB, and to a lesser extent inhaled liposomal amphotericin, can provoke pulmonary reactions, including bronchospasm, cough, and dyspnea (88–91).

Original suggestions that DAMB in combination with granulocyte transfusions (92) result in pulmonary toxicity have subsequently not been confirmed in prospective observations (71,93). Pulmonary reactions, including dyspnea, hemoptysis, and new infiltrates, have also been suspected to be caused by the combined use of blood platelet transfusions and DAMB (SEDA-12, 227). It therefore appears advisable to space transfusions of blood products and amphotericin if possible (94).

Intravenous administration of DAMB has been associated with pulmonary reactions, including dyspnea, bronchospasm, fever, and chills; in contrast to rare reports of dyspnea after liposomal amphotericin (7), dyspnea was not associated with general toxic reactions. The possibility that liposome overload is the explanation of this reaction should be considered.

A life-threatening event has been reported after the use of ABLC in a patient previously treated with amphotericin deoxycholate (95).

- Tachycardia, tachypnea, dyspnea, and severe hypoxemia occurred 90 minutes after the start of the first dose of ABLC, with radiological evidence of bilateral interstitial infiltrates, and required transient mechanical ventilation. After the event, treatment was continued with amphotericin deoxycholate without undesirable effects.

The adverse effect of aerosolized amphotericin B lipid complex once daily for 4 days then once weekly for 13 weeks have been reported in 40 subjects undergoing allogeneic hemopoietic stem cell transplantation in an open non-comparative study (96). Cough, nausea, taste disturbance, or vomiting occurred after 2.2% of 458 total inhalations; 5.2% of inhalations were associated with a 20% or more fall in FEV_1 or forced vital capacity, but no patients required bronchodilators or withdrawal.

Fatal fat embolism has been attributed to ABLC (97).

- A 41-year-old Caucasian man with AIDS received amphotericin for cryptococcal meningitis but developed renal insufficiency and was switched to ABLC. After about 48 hours he developed a tachycardia, tachypnea, respiratory failure, a fall in hematocrit, thrombocytopenia, and altered mental status. Autopsy findings included fat emboli involving the heart, lungs, kidney, and brain.

This is the first report of fatal fat embolism in a patient receiving a lipid formulation of amphotericin B, although the causal relation was unresolved.

Nervous system

Headache is common during the immediate infusion reaction. Neuropathy, convulsions, tremor, and paresis have also been attributed to amphotericin. It is difficult to assess these reports, because in systemic fungal infections, with the possibility of central nervous system involvement, the symptoms may be due to the underlying disease.

Reversible parkinsonism has been attributed to ABLC (98).

- A 10-year-old bone marrow recipient was given ABLC 7 mg/kg/day for prolonged periods of time. Ablation therapy before transplantation included cytosine arabinoside, cyclophosphamide, and total body irradiation. He developed progressive parkinsonian features; an MRI scan showed non-specific frontal cortex white matter abnormalities, and brain MR spectroscopy was consistent with significant neuronal loss in the left insular cortex, left basal ganglia, and left frontal white matter. He was given co-careldopa (carbidopa + levodopa) and made a slow recovery within 4 months. A follow-up MRI scan again showed frontal white matter changes, but repeat MR spectroscopy showed marked improvement in the areas previously examined.

This case shows that, regardless of the formulation of amphotericin, severe neurological adverse effects can occur, in particular in patients who receive large dosages of amphotericin after cranial irradiation. A clinical syndrome of akinetic mutism, incontinence, and parkinsonism has been described in patients who received large doses of amphotericin deoxycholate in association with central nervous system irradiation or infection (76).

Electrolyte balance

Selective distal tubular epithelial toxicity by amphotericin can cause hypokalemia, and hypokalemia can cause further tubular damage. There is some evidence that hypokalemia due to amphotericin is mitigated by both spironolactone (99) and amiloride (100).

Infusion of amphotericin deoxycholate can cause hyperkalemia, in particular in the setting of renal insufficiency (76). The primary mechanism is not known.

- Fatal cardiopulmonary arrest occurred in a 4-year-old boy with acute leukemia and disseminated invasive candidiasis after the third infusion of ABLC 5 mg/kg/day, infused over 1 hour (101). During resuscitation he had a serum potassium concentration of 16 mmol/l; there was no evidence of hemolysis or rhabdomyolysis and serum creatinine and potassium concentrations had been within the reference ranges earlier in the day. Autopsy showed numerous fungal abscesses, including several in the myocardium.

Serum potassium concentrations were determined at the end of a 2-hour infusion of amphotericin deoxycholate (1 mg/kg/day) in a 2-year-old girl with systemic candidiasis receiving long-term hemodialysis for renal dysplasia (101). The potassium concentration was 6.7 mmol/l, despite dialysis against a 1.5 mmol/l potassium bath just before the infusion. The next dose was given during dialysis, and the serum potassium concentration was 2.6 mmol/l after the infusion.

When giving amphotericin to dialysed patients, it may be necessary to give it during dialysis in order to avoid hyperkalemia.

Hypokalemia severe enough to cause rhabdomyolysis has only occasionally been reported (102).

- A 10-year old boy receiving partial parenteral nutrition was given amphotericin B 1 mg/kg/day (formulation unspecified) for a catheter-related infection with Candida albicans. After 6 days he developed hypokalemia (serum potassium 2.2– 3.2 mmol/l) and was given daily potassium replacement. One week after a 2-week course of amphotericin B he developed fatigue, inability to walk, progressive weakness, and pain in the legs, particularly the calves. He was dehydrated and had abdominal distention and reduced bowel sounds. His deep tendon reflexes and muscle strength were markedly reduced, especially in the legs, and there was tenderness on palpation of the calves and thighs. His serum potassium concentration was 1.7 mmol/l, sodium 137 mmol/l, and magnesium 0.62 mmol/l; creatine kinase activity was 3937 U/l, lactate dehydrogenase 432 U/l, aspartate transaminase 105 U/l, and alanine transaminase 105 U/l; there was a metabolic alkalosis and myoglobinuria.

The authors discussed three mechanism whereby hypokalemia (below 2.0 mmol/l) can cause rhabdomyolysis: reduced blood flow to anerobic muscles; suppressed synthesis and storage of glycogen; and reduced transmembrane cation transport.

Mineral balance

Children with acute lymphoblastic leukemia are at risk of serious electrolyte abnormalities.

- A child with acute lymphoblastic leukemia and cerebral and paranasal sinus mould infections developed severe hyperphosphatemia (maximum 6.1 mmol/l) as a consequence of a large exogenous load of phosphorus from high-dose liposomal amphotericin B (25 mg/kg/day; approved dosage up to 5 mg/kg/day; maximum tolerated dosage administered in a dose-ranging study without dose-limiting adverse events 15 mg/kg/day) (103). Three days after withdrawal of liposomal amphotericin B, the serum phosphorus concentration had fallen to 2.6 mmol/l and was within the reference range 3 weeks later. There were no symptoms associated with the event.

It is unclear whether the phosphorus measured in the child's plasma was free phosphorus available for precipitation with calcium or stably bound to the phospholipid moiety of the infused liposomes. However, even considering the desperate clinical problem, the dosage of liposomal amphotericin B administered to this patient was difficult to justify.

Hematologic

A normochromic, normocytic, usually mild anemia develops regularly during therapy with DAMB. The erythropoietin response to anemia appears to be blunted during DAMB therapy, and survival of erythrocytes may be reduced by toxic effects of amphotericin on the cell membrane (104).

Frank hemolysis has also been reported in rare instances, including, rarely, immune-mediated hemolysis (105,106).

Leukopenia has been reported (SED-12, 673).

Thrombocytopenia has been reported in several instances and also occurs with lipid formulations (107,108). During amphotericin therapy, the response of platelet counts to thrombocyte transfusions was reduced, as was platelet survival (109,110).

Gastrointestinal

Anorexia, nausea, and vomiting are common effects of parenteral administration of amphotericin. Gastrointestinal complaints are markedly less common with liposomal amphotericin than with ABCD(7).

Liver

Cholestasis has been reported in infants treated with amphotericin for systemic Candida infections (SEDA-14, 230). Most of the above reports were incidental, and amphotericin cannot be regarded as a known cause of liver damage. This does not necessarily also apply to liposomal amphotericin and other lipid formulations. Therapy with L-Amb, AmBisome was associated with a rise in alkaline phosphatase in over a third of children treated with AmBisome (111) and with hepatic dysfunction in a little under 20% of adolescents and adults. In a small retrospective study, ABLC was withdrawn in 27% of patients because of rises in serum bilirubin and alkaline phosphatase, a finding confirmed in a larger prospective study. Cholestasis has also been observed with ABCD, in contrast to reports that L-AmB does not increase transaminases (7).

Acute hepatic damage has rarely been reported with DAMB. Asymptomatic increases in hepatic serum enzyme activities were seen in one case (112).

- A 26-year-old previously healthy man with life-threatening pulmonary blastomycosis developed increased hepatic transaminases to a maximum of ten times (aspartate transaminase) and 20 times (alanine transaminase), the upper limit of the reference ranges, 10 days after the addition of amphotericin (0.5 mg/kg) to his initial itraconazole therapy (200 mg bd) (113). The serum transaminase activities returned to normal within 4 days after withdrawal of amphotericin, and the blastomycosis was successfully treated with itraconazole alone. A liver biopsy showed mild focal fatty changes but no evidence of blastomycosis.

The authors speculated that amphotericin may have facilitated the uptake of itraconazole into mammalian cells by its membrane-damaging action, leading to increased interaction of itraconazole with CYP450 enzymes and hepatocellular damage.

In a retrospective matched case–control study, cases of hepatotoxicity among patients who underwent bone marrow transplantation were investigated using multivariable logistic regression modelling to evaluate the relation between hepatotoxicity and exposure to antifungal medications (114). The unadjusted incidence of hepatotoxicity was 1.50 for liposomal amphotericin B. In case–control analyses liposomal amphotericin B was associated with a substantial increase in the risk of hepatotoxicity in these patients (OR = 3.33; 95% CI = 1.61, 6.88); there was a smaller increase in risk for fluconazole (OR= 1.99; 95% CI= 1.21, 3.26). Patients had greater rises in serum transaminases associated with exposure to larger cumulative doses of liposomal amphotericin B. In the follow-up analysis of patients who developed hepatotoxicity and who continued to receive antifungal medication, one-third of those who received liposomal amphotericin B had marked increases in bilirubin concentrations, as opposed to 8% of patients treated with fluconazole.

Pancreas

In a retrospective analysis, 5 of 31 children with cancers, who had received liposomal amphotericin in dosages of 1–3 mg/kg/day, had an isolated transient rise in the serum lipase activity during or shortly after therapy with liposomal amphotericin (115). Three of these patients had signs of pancreatitis. While the exact pathogenesis is unclear, the authors proposed fat overload or toxic damage to the pancreas by the liposomes or amphotericin itself as potential mechanisms.

Urinary tract

Amphotericin can cause both glomerular and tubular damage. Lipid-based formulations (colloidal dispersion, lipid complex, and liposomal amphotericin) are less nephrotoxic than conventional amphotericin deoxycholate (116). However, several caveats have to be kept in mind in making such comparisons. For example, there are no defined equivalent doses for amphotericin deoxycholate and its lipid-based counterparts.

Presentation

The clinical and laboratory findings include reductions in glomerular filtration rate and renal plasma flow, proteinuria, cylindruria, and hematuria; the last three are frequent but usually discrete. The reductions in renal plasma flow and filtration fraction occur early. Changes in tubular function can cause increased excretion of uric acid (an effect that can be used to monitor the tubular damage). Excessive loss of potassium, magnesium, and bicarbonate result in hypokalemia, hypomagnesemia, and renal tubular acidosis (117,118). Severe hypokalemia is common,

and requires parenteral potassium replacement. Acidosis can also be severe, requiring bicarbonate. Hypomagnesemia, which may be symptomatic, can cause secondary hypocalcemia, resulting in tetany (119–121).

Amphotericin can cause an inability to form a concentrated urine (hyposthenuria) although this rarely becomes clinically important (76).

Nephrogenic diabetes insipidus, resistant to vasopressin, following damage to the distal renal tubule, may be more common than reported (122,123). Careful monitoring of electrolytes is therefore recommended in all instances of amphotericin therapy.

- A 43-year-old HIV-infected patient presented with nephrogenic diabetes insipidus associated with amphotericin deoxycholate therapy for ocular candidiasis; rechallenge was positive (124).

In a prospective observational study in 108 adults who received different formulations of amphotericin, nephrotoxicity was associated with accelerated mortality and, among those who survived, increased duration of hospital stay (125). There was at least one adverse event in 83 patients and 24% developed nephrotoxicity, defined as a 50% increase in the baseline serum creatinine and a peak of at least 178 mmol/l (2 mg/dl).

Mechanism

The exact mechanisms involved in amphotericin-induced uremia are not yet fully understood. Changes in tubular ion permeability have been demonstrated both in vitro and in vivo (14,119,126). The uremia can be caused by tubuloglomerular feedback, a mechanism whereby increased delivery and re-absorption of chloride ions in the distal tubule initiates a reduction in the glomerular filtration rate. Tubuloglomerular feedback is amplified by sodium deprivation and suppressed by sodium loading. Other possible mechanisms are renal arteriolar spasm, calcium deposition during periods of ischemia, and direct cellular toxicity (14). Yet other lines of research have looked at the roles of prostaglandins and TNFα, with evidence that indometacin may abate prostaglandin-mediated toxicity (127), an approach that is not practical in most patients requiring antifungal drugs.

- Renal damage due to amphotericin has also been reportedly caused by a tumor lysis-like syndrome in a 41-year-old woman with visceral leishmaniasis, with hyperkalemia, hyperphosphatemia, hyperuricemia, and acute renal insufficiency (128).

Differences among formulations

The nephrotoxicity of lipid formulations of amphotericin varies from formulation to formulation.

Amphotericin B deoxycholate

The epidemiology of the nephrotoxicity of conventional amphotericin B has been investigated in a retrospective study in 494 adult inpatients who received two or more doses (129). Nephrotoxicity was defined as a 50–100% increase in the baseline creatinine concentration. The median cumulative dose was 240 mg and most of the patients received it for empirical therapy. Overall, 139 patients (28%) had renal toxicity, including 58 (12%) with moderate to severe nephrotoxicity. For each 10 mg increase in the mean daily dose, the adjusted rate of renal toxicity increased by a factor of 1.13. Five risk factors were defined: a mean daily dose of 35 mg or more, male sex, weight 90 kg or more, chronic renal disease, and concurrent use of amikacin or ciclosporin. The incidence of moderate to severe nephrotoxicity was 4% in patients with none of these risk factors, 8% in those with one, 18% in those with two, and 29% in those with three or more. Nephrotoxicity rarely led to hemodialysis ($n = 3$). However, at the time of discharge or death, 70% of the patients with moderate to severe nephrotoxicity had a serum creatinine concentration that was at least 44 µmol/l above baseline. This study shows dose-dependency of nephrotoxicity related to DAMB, accentuated by other nephrotoxic drugs and patient risk factors. The authors suggested that in patients with more than two risk factors alternative antifungal drugs should be considered.

The nephrotoxicity of amphotericin deoxycholate has been investigated in a retrospective multicenter study in 239 immunosuppressed patients with suspected or proven aspergillosis for a median duration of 15 days (130). The serum creatinine concentration doubled in 53% of the patients and exceeded 221 µmol/l in 29%; 15% underwent dialysis, and 60% died. Multivariate Cox proportional hazards analysis showed that patients whose creatinine concentration exceeded 221 µmol/l, and patients with allogeneic and autologous bone marrow transplants were at greatest risk of requiring hemodialysis. The use of hemodialysis, the duration of amphotericin therapy, and the use of nephrotoxic agents were associated with a greater risk of death, whereas patients who underwent solid organ transplantation were at lowest risk. The findings of this study suggest that raised creatinine concentrations during therapy with amphotericin are associated with a substantial risk of hemodialysis and a higher mortality rate, but these risks vary in different patients.

Amphotericin-associated nephrotoxicity has been studied in a retrospective analysis of 69 recipients of blood stem-cell transplants with multiple myeloma who received at least two doses of amphotericin deoxycholate during 1992–95 (131). Nephrotoxicity occurred in 30 patients (43%) and developed rapidly. Patients who developed nephrotoxicity were similar to those who did not in many aspects associated with their treatment. However, baseline-estimated creatinine clearance, ciclosporin therapy, nephrotoxic drug therapy within 30 days of starting amphotericin, and the number of concomitant nephrotoxic drugs were significant predictors of amphotericin-associated nephrotoxicity. The authors concluded that recipients of bone marrow or peripheral blood stem-cell transplants who have multiple myeloma and are receiving ciclosporin or multiple nephrotoxic drugs at the start of amphotericin therapy should be considered at high risk of amphotericin-associated nephrotoxicity.

In a retrospective study, renal function was investigated in patients receiving ciclosporin alone or in combination with amphotericin (24-hour infusion) after allogeneic stem-cell transplantation (132). Of 84 patients, 22 were treated with amphotericin. There was a statistically significant reduction in renal function compared with the 62 patients who received ciclosporin alone. However, renal insufficiency in all patients remained in a clinically acceptable range and was reversible in patients who survived to 1 year after transplantation.

Continuous infusion of amphotericin has been assessed in an open study in six lung transplant recipients with invasive or semi-invasive bronchopulmonary azole-resistant candidal infections who were treated for 40 (17–73) days by 24-hour continuous infusions of amphotericin 1 mg/kg (133). They received at least 1000 ml/day of 0.9% saline intravenously. Apart from ciclosporin, five patients received aminoglycosides for at least 2 weeks, and four received ganciclovir. Calculated creatinine clearance fell from 57 (43–73) ml/minute to a nadir of 35 (28–39) and recovered to 52 (33–60) after the end of therapy. One patient needed temporary hemofiltration for 7 days. Besides three episodes of mild hypokalemia there were no adverse effects attributable to amphotericin. Asymptomatic colonization with *Candida* persisted for 10 months in one case, but the other five patients were cured.

Amphotericin B colloidal dispersion
During ABCD therapy 8.5% of patients developed nephrotoxicity compared with 21% in those given amphotericin deoxycholate (7).

In a randomized, double-blind, multicenter trial, ABCD (Amphotec; 6 mg/kg/day) was compared with liposomal amphotericin (1.0–1.5 mg/kg/day) for the first-line treatment of invasive aspergillosis in 174 patients (134). The median duration of therapy was 13 (1–357) days in those given ABCD, and 15 (1–87) days in those given liposomal amphotericin. For evaluable patients ($n = 103$) given ABCD or liposomal amphotericin, the respective rates of therapeutic response (52 versus 51%), mortality (36 versus 45%), and death due to fungal infection (32 versus 26%) were similar. Renal toxicity was significantly lower (25 versus 49%) and the median time to onset of nephrotoxicity longer (301 versus 22 days;) in the patients who received ABCD. Rates of drug-related toxicity in the patients who received ABCD and liposomal amphotericin were respectively 53 versus 30% (chills), 27 versus 16% (fever), 1 versus 4% (hypoxia), and 22 versus 24% (toxicity requiring study drug withdrawal). Based on the results of this trial, ABCD appears to have equivalent efficacy to liposomal amphotericin, and superior renal safety in the treatment of invasive aspergillosis. However, infusion-related chills and fever occurred more often with ABCD.

Amphotericin B lipid complex (ABLC)
A comparison of ABLC with amphotericin deoxycholate showed significant differences, with a doubling of baseline creatinine in 28% during ABLC compared with 47% during conventional therapy (7).

There has been a retrospective comparison of the renal effects of ABLC with amphotericin deoxycholate in the treatment of invasive candidiasis and cryptococcosis in dosages of 0.6–5 mg/kg/day; most patients received 5 mg/kg/day (135). Changes in serum creatinine were evaluated in three ways: doubling of the baseline value, an increase from below 132 µmol/l (1.5 mg/dl) at baseline to over 132 µmol/l, and an increase from below 132 µmol/l at baseline to at least 177 µmol/l (2.0 mg/dl). These endpoints were achieved significantly more often with amphotericin deoxycholate than with ABLC, and the time needed to reach each of the endpoints was significantly shorter with amphotericin deoxycholate. An increased serum creatinine concentration was reported as an adverse event more often in patients receiving amphotericin deoxycholate than in patients receiving ABLC (24 versus 43%).

In a randomized, controlled trial in 105 adults with hematological malignancies and fever of unknown origin after chemotherapy or autologous stem cell transplantation, amphotericin was used as empirical antifungal therapy (136). Patients were randomly allocated to receive ABLC 1 mg/kg/day or amphotericin deoxycholate 0.6 mg/kg/day. The incidence of renal toxicity, defined as a doubling of baseline serum creatinine or an increase to at least 132 µmol/l (1.5 mg/dl), was significantly lower with ABLC (8% versus 32%). The rates of infusion-related adverse events were similar (73% versus 77%). Two patients randomized to ABLC and 11 randomized to DAMB withdrew because of toxicity. The overall response rate was 72% for ABLC and 48% for DAMB; this difference was mainly due to the significantly higher renal toxicity with DAMB.

The renal effects of high-dosage/long-duration ABLC therapy (over 5 mg/kg/day for over 12 days; n = 309) have been compared with those of low-dosage/short-duration ABLC therapy (5 mg/kg/day or less for 12 days or less; n = 1417) (137). The median change in serum creatinine from baseline was 0.1 mg/dl in both groups.

To investigate the renal safety of ABLC, the records of 3514 ABLC-treated patients with fungal infections registered in the CLEAR database were reviewed (138). The median change in predicted creatinine clearance from baseline to the end of therapy was –3 (range –119 to 118) ml/minute; the serum creatinine concentration doubled in 13% of patients and new dialysis was needed for 3% of patients. Patients with underlying renal disease who had received prior antifungal therapy had a median creatinine clearance change of 0.5 (range –107 to 52) ml/minute. Despite an increased risk of renal impairment in recipients of allogeneic hemopoietic stem-cell transplants, only 17% had end-of-therapy doubling of serum creatinine concentrations, and the median change in creatinine clearance was –10 (range –107 to 108) ml/minute. In patients given ABLC concomitant treatment with potentially nephrotoxic agents and a baseline serum creatinine concentration of <176 µmol/l were predisposing factors for nephrotoxicity. These data support the notion that ABLC may be used safely to treat patients who are at increased risk of renal impairment.

The rates of ABLC-associated nephrotoxicity in various clinical settings at a university hospital have been estimated retrospectively and compared with previously reported rates of nephrotoxicity (139). Data from 33 adult patients (20 men, 13 women; mean age 49 years) with and without neutropenia receiving ABLC were collected, and the degree of nephrotoxicity was determined using two definitions: (1) doubling of baseline serum creatinine concentration using the peak value within the first 7 days and (2) end-of-therapy doubling of baseline serum concentration using the end-of-therapy value. Using the selected definitions of ABLC-associated nephrotoxicity, there were only two cases. This rate was significantly below the 42% rate reported in the only large published study (95% CI = 1.7, 19.6). The median change in serum creatinine concentration was 8.9 (–97 to 380) µmol/l. The concomitant use of nephrotoxic agents was not associated with significant changes in serum creatinine concentration. The authors concluded that ABLC infrequently causes clinically significant nephrotoxicity, and that earlier data derived from a single study in febrile patients with neutropenia should be interpreted cautiously.

Amphotericin deoxycholate

In an open randomized trial, an oral rehydration solution (3 liters; sodium 90 mmol/l, chloride 104 mmol/l, bicarbonate 22 mmol/l, and potassium 12 mmol/l, osmolarity 290 mosm/l) has been compared with intravenous saline (1 liter; sodium 153 mmol/l, chloride 153 mmol/l, osmolarity 306 mosm/l) in the prevention of nephrotoxicity from amphotericin B deoxycholate in 48 adults with mucosal leishmaniasis (140). There were no differences in serum creatinine, creatinine clearance, serum urea, or serum sodium during treatment. However, serum potassium concentrations were lower at cumulative doses of amphotericin of 9.6, 14.4, and 25.2 mg/kg with saline compared with oral rehydration. The authors concluded that oral rehydration was comparable to intravenous saline in preventing glomerular damage and was associated with less hypokalemia.

Hydration and sodium loading may reduce the glomerular nephrotoxicity that is associated with amphotericin deoxycholate (141). A standardized protocol of hydration and electrolyte supplementation was studied prospectively in patients with hematological malignancies receiving empirical treatment, in order to evaluate its effect on amphotericin deoxycholate -related renal toxicity (142). In all, 77 consecutive patients received amphotericin deoxycholate (1 mg/kg/day) in association with initial intravenous hydration of at least 1 l/m², containing at least 1 liter of 0.9% saline daily. Hydration was increased when serum creatinine concentrations increased by 20% from baseline. Electrolytes were replaced when indicated by serum concentration measurements. The median duration of therapy was 14 days. The mean intravenous hydration and the mean diuresis were respectively 1530 and 1970 ml/m²/day. Overall, 55 patients (71%) received a mean of 19 days of therapy without dose-limiting adverse events. Despite significant increases in mean serum creatinine concentrations and reductions in mean creatinine clearance, observed early in the whole population, in only six patients (7.8%) was therapy withdrawn because of renal insufficiency, which always recovered after withdrawal. In eight patients (10%) therapy was withdrawn because of infusion-related adverse effects. Seven patients died without evidence of amphotericin deoxycholate-associated toxicity. This series confirms that adequate hydration (about 1500 ml/m^2/day) and careful electrolyte supplementation are simple measures that contain nephrotoxicity and permit amphotericin deoxycholate therapy.

Liposomal amphotericin B

In a prospective double-blind study of more than 600 patients, 0.6 mg/kg of DAMB was compared with 3.0 mg/kg of L-Amb, a dose relation at the lower limit of equivalent doses determined in an animal model (143). At these dosages, in a large prospective double-blind study, there was a doubling of serum creatinine concentration in 19% of neutropenic patients receiving empirical therapy with L-AmB and 34% receiving conventional amphotericin (55).

Liposomal amphotericin has been given to an immunosuppressed renal transplant patient with cerebral aspergillosis for almost 10 months at a cumulative dose of 42 g with no apparent changes in the function of the renal allograft, as measured by serum creatinine, creatinine clearance, and potassium concentrations (144). Therapy was ultimately successful and was discontinued after surgical resection of a residual sclerotic lesion.

Liposomal amphotericin (3 mg/kg/day) has been compared with conventional amphotericin (0.7 mg/kg/day) for induction therapy of moderate to severe disseminated histoplasmosis in a randomized, double-blind, multicenter trial in 81 patients with AIDS (145). The duration of induction was 2 weeks, to be followed by 10 weeks of itraconazole in the case of a response. Clinical success was achieved in 14 of 22 patients treated with conventional amphotericin compared with 45 of 51 patients who received liposomal amphotericin (difference, 24%; 95% CI = 1%, 52%). Culture conversion rates were similar. Three patients treated with conventional amphotericin and one treated with liposomal amphotericin died during induction. Infusion-related adverse effects were more common with conventional amphotericin (63%) than with liposomal amphotericin (25%). Nephrotoxicity occurred in 37% of patients treated with conventional amphotericin and 9% of patients treated with liposomal amphotericin. The results of this study suggest that liposomal amphotericin is less toxic than conventional amphotericin and is associated with improved survival.

Amphotericin alters cell membrane permeability and tubular cell function, leading to various tubular transport defects. In two cases of nephrogenic diabetes insipidus there was a possible association with liposomal amphotericin.

- A highly febrile patient with cancer, granulocytopenia, and watery diarrhea receiving empirical amphotericin (1 mg/kg) developed complex metabolic disorders, including mild acute renal insufficiency, nephrogenic diabetes insipidus documented by a high serum antidiuretic hormone concentration (18 pg/ml), polyuria, a negative desmopressin test, and type I renal tubular acidosis (146). Amphotericin was withdrawn and replaced by liposomal amphotericin B (3 mg/kg), despite which the diabetes insipidus and distal tubulopathy persisted. Liposomal amphotericin was therefore withdrawn after 1 week and within 5 days the renal symptoms began to resolve.

- A 38-year-old man with acute myelogenous leukemia, who had received a matched unrelated donor allogeneic bone marrow transplant and had glucocorticoid-responsive graft-versus-host disease, developed a fungal pneumonia with *Torulopsis glabrata* and was given liposomal amphotericin (2.5 mg/kg/day; baseline serum creatinine 121–132 µmol/l) (147). The dose was increased to 7.5 mg/kg/day and subsequently to 10 mg/kg/day because of immunosuppression and poor response. He required mechanical ventilation for biopsy-proven bronchiolitis obliterans organizing pneumonia. He also developed diffuse alveolar hemorrhage and received intravenous desmopressin, with a reduction in bloody secretions. He then developed hypernatremia (serum sodium 155 mmol/l) and had an inappropriately increased urine output consistent with nephrogenic diabetes insipidus, associated with an increasing cumulative dose of liposomal amphotericin B, despite concurrent use of intravenous desmopressin. Aggressive water replacement was effective.

Although the second patient received several other nephrotoxic drugs, the Naranjo probability scale classified this as a possible adverse reaction because of the temporal sequence of nephrogenic diabetes insipidus after high-dose liposomal amphotericin B and previously reported cases associated with amphotericin B desoxycholate.

Liposomal amphotericin (n = 539) and caspofungin (n = 556) as empirical antifungal therapy have been compared in a randomized, double-blind, multinational trial (148). Fewer patients who received caspofungin had nephrotoxicity (2.6% versus 12%), an infusion-related events (35% versus 52%), or a drug-related adverse event, or withdrew because of drug-related adverse events.

Prevention

Avoidance of salt depletion and a salt load (500–1000 ml of 0.9% saline) reduce the renal toxicity of DAMB (149–155). Also the maintenance of adequate serum potassium concentrations by replacement therapy may be important and may contribute to "kidney sparing" (156). Other preventive measures, including dopamine infusion, are of no value (157).

It has been suggested that amphotericin-induced nephrotoxicity may be mitigated by increasing renal blood flow and glomerular filtration rate with low-dose dopamine (1–3 µg/kg/minute). The efficacy of low-dose dopamine in preventing nephrotoxicity associated with amphotericin deoxycholate has been evaluated in a prospective randomized study in 71 patients after antineoplastic chemotherapy for autologous bone marrow transplantation or acute leukemia (157). The patients were randomly assigned to receive low-dose dopamine by continuous infusion (3 μg/kg/minute) or no dopamine. Amphotericin deoxycholate 0.5 or 1.0 mg/kg/day was given for respectively 8 and 13 days on average. Nephrotoxicity, defined as a 1.5-fold or greater increase in baseline serum creatinine concentration, was slightly less common, but not significantly so, in those given dopamine (67 versus 80%). The grade of nephrotoxicity was the same. Ten patients developed grade IV nephrotoxicity and were withdrawn from the study. The authors concluded that dopamine offers little benefit in preventing amphotericin deoxycholate-associated nephrotoxicity.

In a randomized, controlled, single-center study, continuous infusion of amphotericin reduced nephrotoxicity and infusion-associated reactions compared with the standard infusion over 2–4 hours in patients with neutropenia, refractory fever, and suspected or proven invasive fungal infections (158). However, the concentration-dependent pharmacodynamics of antifungal polyenes raise concerns about the antifungal effectiveness of this mode of administration in particular, as its therapeutic efficacy has not been adequately studied in animals or in patients with documented infections.

Skin

An allergic skin reaction to amphotericin has been reported (159).

- A 3-year-old child had a severe allergic reaction during treatment with liposomal amphotericin for persistent neutropenic fever following unrelated allogeneic cord blood stem-cell transplantation. The patient developed an extensive maculopapular rash and severe itching that was unresponsive to antihistamine and glucocorticoid medication and resolved only after drug withdrawal. Continuation of therapy with conventional amphotericin for 20 days was well tolerated, suggesting that the lipid carrier was responsible for the adverse event.

Immunologic

A literature review found no support for the routine use of a test dose of amphotericin before the first therapeutic dose of amphotericin deoxycholate, as is still recommended by the manufacturers (160). The mechanism of common infusion-related adverse effects does not appear to be allergic in nature, and true allergic reactions are rare. Moreover, the absence of a reaction to a test dose does not necessarily indicate that patients will not have a severe infusion-related reaction later in the course of therapy, and the procedure of administering a test dose can lead to a detrimental delay in adequate antifungal therapy. The authors recommended starting therapy with amphotericin deoxycholate at the full therapeutic target dose, with careful bedside monitoring for infusion-related adverse events throughout therapy.

Anaphylaxis is rare with amphotericin (76). It is important to note that a patient may tolerate one formulation and respond with anaphylaxis to another.

- Anaphylaxis after ABCD occurred in a patient who had previously been treated with both amphotericin deoxycholate and ABLC without infusion-related

adverse effects (74). During the first infusion of ABCD he developed spontaneously reversible severe back pain and then swelling of his lips, respiratory distress, and left-sided hemiparesis, which resolved after 24 hours. An MRI scan suggested an ischemic event in the right putamen, lending support to the hypothesis that he had had an anaphylactic reaction to ABCD, hypoperfusion, and a subsequent stroke.

- In another patient, serious adverse events (fever, severe rigors, a fall in blood pressure, worsening mental status, increasing creatinine concentration, and leukocytosis) occurred after unrecognized substitution of one amphotericin formulation (ABLC) by another (ABCD) (161). After discovery of the switch, ABLC therapy was reinstituted and tolerated without incident.

These cases underscore the need to monitor patients closely when infusing the first dose of a different formulation of amphotericin.

The clinical characteristics and treatment of patients with a distinctive triad of acute infusion-related reactions to L-AmB have been analysed in patients who participated in trials of L-AmB (162). Acute infusion-related reactions occurred alone or in combination in one of three symptom complexes: (1) chest pain, dyspnea, and hypoxia; (2) severe abdominal, flank, or leg pain; and (3) flushing and urticaria. Of 84 patients in a single center, 29 had symptoms of acute infusion-related reactions during the first infusion of L-AmB. Most of the reactions (86%) occurred within the first 5 minutes of infusion. All patients had rapid resolution of symptoms after withdrawal of L-AmB infusion and the administration of intravenous diphenhydramine. The overwhelming majority of patients (93%) were rechallenged and tolerated the remainder of the infusion well; they also tolerated subsequent infusions when given diphenhydramine premedication. In a multicenter analysis there was a mean overall frequency of 20% (range 0–100%) of reactions in 64 centers. There was no effect of medication lot, type of infusion bag and tubing, or infusion rate. The authors concluded that since these reactions are distinctively different from those observed with conventional amphotericin B, the liposomal carrier is probably the key factor in the etiology of acute infusion-related reactions and that complement activation may play a pathogenic role.

Autacoids

The effects of amphotericin B on the release of histamine from human peripheral blood cells, mononuclear cells, and mast cells has been investigated in cell cultures in vitro. Cultured human mononuclear (THP-1) and mast (HMC-1) cells from five healthy volunteers were incubated with increasing concentrations of amphotericin B deoxycholate, diphenhydramine, amphotericin B deoxycholate plus diphenhydramine, and the calcium ionophore A23187 for up to 24 hours. Histamine concentrations and histamine N-methyltransferase activity were determined at various times. Cell viability was assessed by exclusion of erythrocin B. A23187 increased histamine concentrations from baseline in peripheral blood and HMC-1 cells. There was no change in histamine concentrations in response to amphotericin B deoxycholate. There was no change in histamine concentrations in THP-1 cells in response to any agent tested. Similarly, histamine N-methyltransferase activity in peripheral blood was not affected by amphotericin B deoxycholate. These results support the view that amphotericin B-induced infusion-related reaction is not a histamine-mediated event (163).

Second-Generation Effects

Pregnancy

Experience with amphotericin in pregnancy is limited. Amphotericin crosses the placenta and can increase creatinine concentrations in the neonate. High tissue concentrations of amphotericin persist weeks after treatment has been stopped (164).

Susceptibility Factors

Age

Neonates

The safety and efficacy of liposomal amphotericin in 40 preterm and 4 full-term neonates with invasive yeast infections have been studied retrospectively (165). The initial dosage was 1 mg/kg/day, and was increased stepwise by 1 mg/kg to a maximum of 5 mg/kg, depending on the clinical condition. There were no infusion-associated reactions. Blood pressure, hepatic, renal, and hematological indices were not altered. Hypokalemia was noted in 16 infants but was always transient and responsive to potassium supplementation. Treatment with liposomal amphotericin was successful in 72% of the children. However, 12 of the 40 preterm infants succumbed to the fungal infection; all had a birth weight of less than 1.5 kg.

Changes in serum creatinine and serum potassium have been measured in 21 neonates of very low birth weight who received amphotericin for presumed or documented yeast infections (166). The median dosage was 2.6 (range 1–5) mg/kg/day, and the median duration of therapy was 2 (11–79) days. Hypokalemia (below 3 mmol/l) was observed in 30% before treatment and in 15% during treatment. However, 21 days after the end of therapy, hypokalemia was not present in any patient. The maximum creatinine concentration fell from 121 (71–221) µmol/l to 68 (31–171) µmol/l during treatment and 46 (26–62) µmol/l at 21 days after the end of therapy. However, creatinine concentrations were available for only 10, 18, and 15 of the 21 patients respectively, and no information was provided on the number of patients who had an increase in serum creatinine during therapy. All patients responded to therapy with liposomal amphotericin, although the number of proven invasive fungal infections was small (7/21).

High-dose (5–7 mg/kg/day) liposomal amphotericin B has been evaluated prospectively in 41 episodes of systemic candidiasis in 37 neonates (median birth weight: 860 g, range 495–3785; median gestational age: 27 weeks, range 25–40; median age at onset of systemic candidiasis: 17 days, range 25–40). *Candida* species were isolated from the blood in all the patients and from urine (n = 6), skin abscesses (n = 5), and peritoneal fluid (n = 1). Candidiasis was due to *Candida parapsilosis* (n = 17), *Candida albicans* (n = 15), *Candida tropicalis* (n = 5), *Candida guilliermondii* (n = 2), *Candida glabrata* (n = 2), and an unidentified *Candida* (n = 1). The initial dosage was 1 mg/kg/day, and the final dosage was determined by the clinical course. High-dose liposomal amphotericin B was effective and safe in the treatment of neonatal candidiasis. Infusion-related adverse events were not recorded. Only one patient developed transient liver function disturbances that did not require drug withdrawal. None of the patients developed hypokalemia (potassium below 3 mmol/l). There were no changes in renal function during therapy (167). In contrast to the dosing scheme used in this study, treatment of invasive fungal infections in clinical practice should generally start at the target dosage, with careful observation during the first dose in order to provide maximum effective treatment immediately (1).

Infants

The safety and efficacy of ABLC in 11 neonates with systemic *Candida* infections have been reported (168). The infants were aged 3–14 (median 7) weeks and weighed 0.7–5 (median 1.4) kg. The median duration of ABLC treatment was 23 (range 4–41) days at an average dose of 4.9 (range 3.2–6.5) mg/kg/day. Nine of the eleven patients improved clinically, and eight of nine evaluable patients had a mycological cure. No infant discontinued treatment because of adverse drug reactions, and none had appreciable hepatic or hematological toxicity. Renal function improved or did not change in 8 of the 11. The median pretreatment serum creatinine concentration was 80 (range 35–522) µmol/l and the median creatinine concentration at the end of treatment was 44 (range 18–628) µmol/l.

Children

The disposition of liposomal amphotericin in children and adolescents is similar to that in adults. Liposomal amphotericin and conventional amphotericin have been compared in a randomized, multicenter study in 204 children with cancers, pyrexia, and neutropenia (169). There was a 2.6 times lower incidence of adverse effects with the liposomal formulation. There were severe adverse effects in 1% of those who received liposomal amphotericin and 12% of those who received the conventional formulation.

Data on the safety of amphotericin deoxycholate have been reported for 50 therapeutic courses in 44 children and adolescents with cancer and a median age of 6.8 years (range 9 months to 18 years) (59). Amphotericin deoxycholate was given in a dose of 1 mg/kg over 2 hours for a mean duration of 7.8 days. Most of the patients received the drug as empirical antifungal therapy in the setting of persistent fever and neutropenia. Nephrotoxicity, defined as a 100% increase in the serum creatinine from baseline, was observed in only one patient. Infusion-related reactions (fevers and/or rigors) occurred in 24% of treatment courses. Thus, amphotericin deoxycholate was relatively well tolerated in this population, although the mean duration of therapy was comparatively short.

The safety and efficacy of ABLC have been studied in an open, emergency-use, multicenter study in 111 treatment episodes in children with invasive mycoses refractory to or intolerant of conventional antifungal drugs (170). The mean daily dosage was 4.85 (range 1.1–9.5) mg/kg, and the mean duration of therapy was 33 (range 1–191) days. While the proportion of patients with deteriorating renal function was not mentioned, the mean serum creatinine concentration in the entire study population did not change significantly between baseline (109 µmol/l) and withdrawal of ABLC (117 µmol/l) over 6 weeks. Similarly, there were no significant differences between initial and end-of-therapy concentrations of serum potassium, magnesium, transaminases, alkaline phosphatase, and hemoglobin. However, there was a significant increase in mean total bilirubin (from 63 to 91 µmol/l) at the end of therapy. ABLC was withdrawn because of toxicity in 7 of the 111 children; adverse events leading to withdrawal included intolerable infusion-related reactions and allergic reactions. Among the 54 evaluable patients, a complete or partial therapeutic response was obtained in 38.

The effectiveness and tolerability of three antifungal formulations, amphotericin B deoxycholate (DAMB), liposomal amphotericin B (L-AmB), and amphotericin B colloidal dispersion (ABCD), in the treatment of neonatal *Candida* bloodstream infections have been investigated in a prospective study of all patients hospitalized in the neonatal intensive care unit from 1996 to 2000 with *Candida* bloodstream infections (171). Patients with a serum creatinine concentration under 106 µmol/l received DAMB (1 mg/kg), and those with a serum creatinine concentration of 1.2 mg/dl and over received L-AmB (5 mg/kg) or ABCD (3 mg/kg on day 1, followed by 5 mg/kg thereafter). Complete blood counts, and renal and hepatic function tests were obtained before, during, and after treatment; blood cultures were performed daily until three consecutive cultures were negative. If cultures were positive for more than 10 days with clinical signs of fungal infection and/or persistent thrombocytopenia, a second antifungal drug was added. Of 56 infants (four term and 52 preterm, including 36 extremely low birth weight infants) DAMB was the initial treatment in 34, L-AmB in six, and ABCD in 16. There were no differences in mortality between the three groups. Sterilization of the blood was achieved with amphotericin B in 68% of patients, L-AmB in 83%, and ABCD in 57%, when used as monotherapy; with the addition of a second antifungal agent, success rates were 100%, 83% and 93% respectively. There were no differences between the groups in the time to resolution of fungemia or in the total duration of therapy. No patients had immediate local or systemic adverse events and in none was renal function altered. Potassium supplementation during treatment was required by 16 infants (47%) in the DAMB

cohort and in none of the infants in the other groups. There were no differences in liver function tests, white blood cell counts, and platelet counts in the three groups.

The pharmacokinetics of ABLC have been investigated in 28 neonates (median weight 1.06, range 0.48–4.9 kg; median gestational age 27, range 24–41 weeks) with invasive candidiasis enrolled in a phase II multicenter trial (172). They received intravenous ABLC 2.5 mg/kg/day (n = 15) or 5 mg/kg/day (n = 13) over 1 or 2 hours for a median of 21 (range 4–47) days. Population-based pharmacokinetic modelling of concentration data showed that the disposition of ABLC in neonates was similar to that observed in other age groups: weight was the only factor that influenced clearance. Based on these results and documented safety and efficacy, a daily dosage of 2.5–5.0 mg/kg for treatment of invasive *Candida* infections in neonates was recommended.

ABLC has also been assessed in 548 children and adolescents 0–20 years of age who were enrolled into the CLEAR registry. All had a cancer or had received a bone marrow, cord blood, or solid organ transplant and were receiving amphotericin B lipid complex for documented or suspected fungal infections (173). Most were either intolerant of or refractory to conventional antifungal therapy, and almost one-half were neutropenic at the start of treatment. Of the 548 patients, 300 (55%) were transplant recipients and 393 (72%) had received one or more concomitant nephrotoxins. *Candida* and *Aspergillus* were the most commonly isolated species in patients with proven or probable infections. Response data were evaluable for 255 of the 285 patients with documented single or multiple pathogens. A complete response (cured) or partial response (improved) was achieved in 55% of patients, and an additional 17% of patients had a stable outcome. There was no significant difference between the rates of new hemodialysis versus baseline hemodialysis. There were rises in serum creatinine of over 1.5 times baseline and over 2.5 times baseline values in 25 and 8.8% of all patients respectively.

Elderly people

ABLC has been evaluated retrospectively using the CLEAR database in 572 elderly patients (over 65 years of age) and 2930 controls (65 years or under) (174). The patients were typically receiving ABLC for candidiasis, multiple fungal pathogen infections, and aspergillosis, or were being treated empirically. The median cumulative dose of ABLC in the two groups was similar. Despite higher median pretreatment serum creatinine concentrations among the elderly patients (150 µmol/l versus 123 µmol/l), both groups had only a median change from baseline of 9 µmol/l by the end of therapy.

Orthotopic liver transplantation

The prevalence of fungal infection after orthotopic liver transplantation is 5–42%. The most commonly isolated pathogens are *Candida* and *Aspergillus* species. High-risk liver transplant recipients are more susceptible to invasive fungal infections, with a prevalence of >over 40% and mortality rates of 78–100%; however, a strategy for fungal prophylaxis in this population has not been defined. Among 100 consecutive orthotopic liver transplantations followed for 28 months, 21 recipients (15 men, overall mean age of 49, range 23–65 years) were considered to be at high risk of fungal infections when they had at least one of the following criteria: acute liver failure, assisted ventilation for more than 7 days, re-transplantation, re-laparotomy, antibacterial therapy for more than 14 days, transfusion requirements of over 20 units of red blood cells, and/or biliary leakage (175). This group received LAMB (1 mg/kg/day for 7–10 days). The one-year survival in the high-risk group was 80%. The prevalence of invasive fungal infections was 9.5%. No *Candida* infection was observed. Two patients developed *Aspergillus* infection, in one case with a fatal outcome. Adverse events related to the drug were hypokalemia (n = 2), back pain (n = 3), and renal dysfunction (n = 2). None of these events required withdrawal of the prophylactic regimen.

Drug Administration

Drug formulations

The cellular toxicity of different amphotericin B deoxycholate formulations has been investigated in vitro (176). Human mononuclear THP-1 cells were exposed for 2 hours to the following deoxycholate formulations of amphotericin B in concentrations of 2.5 and 5 µg/ml: Apothecon, Pharmacia, Sigma, Gensia, Pharma Tek, and VHA. Toxicity was assessed by measuring interleukin (IL)-1 beta expression, amphotericin B content was measured by enzyme-linked immunosorbent assay (ELISA), and amphotericin A and B contents were assessed by spectrophotometry. Endotoxin contamination was evaluated in all reagents. Expression of IL-1 beta from Sigma, Pharmacia, and Pharma-Tek formulations was increased by about 250%, 50%, and 25% respectively compared with amphotericin A. The amphotericin B content of Sigma, Pharmacia, Pharma-Tek, and Gensia formulations, as measured by ELISA, was increased by about 450%, 200%, 200%, and 100% respectively compared with Apothecon. This variation could not be explained by differences in amphotericin A or B contents. As in previous clinical observations, the current in vitro evaluation showed significant differences among different formulations of amphotericin B deoxycholate. Probably other polyenes or pyrogenic toxins in differing amounts are present in these formulations and may explain the variability in toxicity.

The toxicity of amphotericin B deoxycholate has led to an increased preference for lipid formulations with more favorable safety profiles. However, many hospital formularies list both lipid and non-lipid formulations. A dispensing and administration error that caused amphotericin B deoxycholate to be given instead of liposomal amphotericin B resulted in death (177).

Drug dosage regimens

By lowering the infusion rate and using continuous infusion, fewer toxic reactions have been observed. In contrast, varying the infusion time of daily doses between 1 and 6 hours made no important change in regard to toxicity (178–183). Alternatively, smaller starting doses have been suggested to reduce toxicity, a strategy that is often not advisable in acute severe mycoses. The use of lipid formulations reduces the incidence and severity of general toxic reactions and of nephrotoxicity (27,55,58). It is currently unclear, however, whether lipid formulations of amphotericin have a broader therapeutic index than conventional DAMB. It has to be borne in mind that lipid formulations, while affecting the pharmacokinetics of amphotericin, have no targeting properties that discriminate between the cell membrane of the pathogen and that of host cells. It therefore appears possible that some, if not all, of their advantages are in the avoidance of high concentrations of reactive amphotericin, an effect that could also be obtained by continuous infusion of the conventional formula (71). Monitoring of blood concentrations of amphotericin B is not practical, and there are no validated recommendations for its use, neither for avoidance of toxicity nor for monitoring of efficacy.

Drug administration route

Intravenous

In a randomized, controlled study continuous infusion of amphotericin reduced nephrotoxicity and infusion-associated reactions compared with the standard infusion over 2–4 hours in neutropenic patients with refractory fever and suspected or proven invasive fungal infections (184). The same group of investigators has now evaluated dose-escalation using continuous administration of DAMB In 33 patients (31 of whom were neutropenic), who received an initial dosage of DAMB of 1 mg/kg/day that was gradually increased to 2.0 mg/kg/day, provided that renal function remained stable and the drug was tolerated (185). Dose escalation was possible without delay in 28 patients. The median duration of therapy was 16 days (range 7–72 days). Infusion-related reactions accompanied under 18% of DAMB infusions. Twenty-seven patients had a fall in creatinine clearance. There was a greater than two-fold reduction in creatinine clearance in five patients, but the reduction was dose-limiting in only one; dialysis was not required. The authors concluded that continuous infusion of DAMB up to 2.0 mg/kg/day seems not to cause additional impairment of vital organ functions and is well tolerated by most patients. However, the concentration-dependent pharmacodynamics of antifungal polyenes (SEDA-27, 276) raise concerns about the effectiveness of this mode of administration, as its therapeutic efficacy has not been studied adequately in animals or in patients with documented infections.

Amphotericin given intravenously does not lead to adequate CSF concentrations, and it can rarely be given intrathecally in cases of cerebral infection. Nevertheless, amphotericin is effective even in cryptococcal meningitis, in the absence of marked meningeal inflammation. Continuous infusion may be associated with less toxicity, but experience is limited.

Home-based infusion therapy

The types and frequencies of adverse events associated with community-based amphotericin B infusion therapy have been analysed in 105 patients who received amphotericin B from a home-care provider (186). A total of 113 courses of amphotericin B formulations were administered: liposomal amphotericin B, 41 courses (36%), amphotericin B deoxycholate, 31 courses (27%), amphotericin B lipid complex, 31 courses (27%), and amphotericin B colloidal dispersion, three courses (3%); an additional seven courses consisted of sequential therapy with two different formulations. Nephrotoxicity was associated with 46 (41%) courses, electrolyte abnormalities with 40 (35%) courses, venous access device complications with 12 (11%) courses, and infusion reactions with 13 (12%) courses. Nephrotoxicity occurred most often in those aged 60 years or older, solid organ transplant recipients, and those receiving concomitant ciclosporin. Only two (12%) of 17 courses in children under 13 years were associated with nephrotoxicity. Thirteen of all 113 courses resulted in patients requiring hospital admission owing to adverse events. Monitoring of electrolyte, serum creatinine, and blood urea nitrogen concentrations 2 or 3 times a week was adequate for identifying these events.

Intra-arterial

Mucormycosis is a highly lethal invasive mycotic infection that is characterized by angioinvasion, infarction, and tissue necrosis. A patient with soft tissue mucormycosis of the left thigh developed progressive disease, despite surgical debridement and appropriate systemic amphotericin therapy (187). In this difficult case, liposomal amphotericin B was infused directly into the left common iliac artery. The patient responded and was ultimately cured. Although intra-arterial infusion is far from being a standard approach in the treatment of mucormycosis, this and other case reports support the notion that intra-arterial infusion of liposomal amphotericin B can be used as adjunctive therapy in selected patients.

Intrathecal or intraventricular

Delirium (188) and parkinsonism (189) have been described after intrathecal or intraventricular therapy.

Drug–Drug Interactions

Amiloride

Amiloride is a therapeutic option in reducing potassium losses in patients receiving amphotericin. When it was given to 19 oncology patients with marked amphotericin-induced potassium depletion mean serum potassium concentrations increased in the 5 days before and after administration (from 3.4 to 3.9 mmol/l) (190). There was also a trend toward reduced potassium supplementation (48 versus 29 mmol/day). Adverse reactions were limited

to hyperkalemia in two patients who took amiloride 20 mg/day and a high potassium intake.

Aminoglycoside antibiotics

Amphotericin prolongs the half-life of aminoglycoside antibiotics (191).

Antifungal azoles

In evaluating possible antagonism between amphotericin and antifungal azoles, details of the experimental set-up are crucial. When filamentous fungi were exposed to subfungicidal concentrations of azoles, before exposure to an amphotericin + azole combination, antagonism could always be shown both in vitro and in vivo (192–194).

In vitro studies and experiments in animals have given conflicting results relating to potential antagonism between the effects of fluconazole and amphotericin on *Candida* species (193). However, large, randomized, double-blind comparisons of fluconazole with and without amphotericin for 5 days in non-neutropenic patients with candidemia showed no evidence of antagonism, but faster clearance of the organism from the blood and a trend toward an improved outcome in those who received the combination (195).

The effect of the combination of itraconazole with amphotericin on liver enzyme activities has been studied retrospectively in 20 patients with hematological malignancies or chronic lung disease complicated by fungal infection or colonization (196). They took itraconazole 200–600 mg/day for a median of 143 (range 44–455) days. Nine had no abnormal liver function tests, including periods of high concentrations of itraconazole (over 5000 ng/ml) and its active hydroxylated metabolite; only one had received concomitant amphotericin. All of the 11 patients with liver function abnormalities had received concomitant amphotericin. For each patient, liver function abnormalities were greatest during the time of concomitant therapy with both antifungal drugs. Although liver enzyme abnormalities are uncommon with amphotericin (197), and although this retrospective analysis was subject to several flaws and potential biases, it nevertheless suggests that hepatotoxicity should be carefully monitored if itraconazole and amphotericin are coadministered.

The combination of amphotericin with ketoconazole appears to lead to antagonism (192). A study of the effects of combinations of amphotericin with fluconazole, itraconazole, or ketoconazole against strains of *Aspergillus fumigatus* in vitro showed antagonistic effects in some strains, but different effects in other strains (198). In one group of mice infected with *Candida*, combinations of amphotericin with fluconazole were more effective than fluconazole alone; in another group the combination showed no interaction, but was not better than either drug given alone (199).

Although there are no clinical data, it can be expected that similar antagonism occurs between amphotericin and squalene oxidase inhibitors, which also eliminate the primary target ergosterol from the fungal cell membrane.

Antimony and antimonials

Amphotericin can worsen stibogluconate-induced cardiotoxicity; a gap of at least 10 days between sodium stibogluconate and amphotericin is recommended (200).

Ciclosporin

Because ciclosporin causes a reduction in renal function, there is increased nephrotoxicity if amphotericin is also given (132,191).

Cisplatin

Amphotericin-induced hypomagnesemia may be more profound in patients who develop a divalent cation-losing nephropathy associated with cisplatin (14).

Cytotoxic drugs

Amphotericin can enhance the risk of adverse effects from cytotoxic drugs (201).

Digitalis

Hypokalemia due to amphotericin can enhance the toxicity of digitalis (14,156).

Diuretics

The concurrent use of diuretics is associated with a higher risk of nephrotoxicity from amphotericin (202).

Echinocandins

There were no pharmacokinetic interactions of caspofungin with amphotericin deoxycholate in healthy volunteers (203).

Flucytosine

Amphotericin in combination with flucytosine results in an increased risk of hematological complications, because amphotericin often impairs renal function, causing retention of flucytosine (12). This combination also delays hematopoietic recovery after cytotoxic chemotherapy; it should not be used as empirical antifungal therapy in febrile patients with neutropenia (204).

Neuromuscular blocking drugs

Hypokalemia due to amphotericin can enhance the curariform effect of neuromuscular blocking agents (14,156,201).

Tacrolimus

Tacrolimus (FK 506), a macrolide immunosuppressant, has adverse effects similar to those of ciclosporin, including nephrotoxicity. Increased nephrotoxicity can be expected when it is given with amphotericin (205).

Monitoring Therapy

The results of the Collaborative Exchange of Antifungal Research (CLEAR), an industry-supported registry of patients receiving ABLC for invasive fungal infections, have been published. The CLEAR database provides data on the efficacy and renal safety of ABLC in 3514 patients at over 160 institutions in the USA and Canada after regulatory approval (206).

Within the CLEAR database, the efficacy and renal safety of ABLC were assessed in 398 patients with invasive aspergillosis (207). The most common underlying conditions were hemopoietic stem-cell transplantation (25%), hematological malignancies (25%), and solid-organ transplants (27%). The most common reason for administration of ABLC was lack of response to prior antifungal therapy. Overall, 65% of patients had a favorable clinical response: 44% were cured or improved and 21% were stabilized. Clinical responses were similar in patients who received ABLC as either first-line or second-line therapy. Changes in serum creatinine concentrations were not clinically significant in most patients; however, dialysis was initiated in seven, of whom six had had prior antifungal therapy or had pre-existing renal disease.

Similarly, in over 900 patients with invasive candidiasis, clinical responses (cured or improved) were similar in patients infected with invasive *Candida albicans* and non-albicans *Candida* species (63% and 62% respectively) (208). Compared with patients who received lower doses of ABLC, those who required higher doses of ABLC because of more severe infections did not develop significant renal impairment, as assessed by end-of-therapy changes in serum creatinine concentration from baseline (median 9, range −343 to 211 µmol/l), the incidence of serum creatinine doubling (16%), and the need for new dialysis (7%).

Intermediate-dose ABLC (3 mg/kg/day) as primary or salvage treatment of fungal infections has been assessed in 74 adults with hematological malignancies, of whom 45 received upfront therapy and 29 received salvage therapy for their infection (209). Of 71 evaluable patients 48 responded, with complete responses in 40 (56%) and partial responses in eight (11%), and 15 (21%) died as a consequence of the fungal infection. In 40 patients with neutropenia-associated infection, rapid neutropenic recovery (at less than 10 days from study entry) was essential for a response (90% versus 32%). Treatment was well tolerated; 15% of the infusions were followed by infusion-related adverse events; there was nephrotoxicity in 7% of patients and 11% of withdrawals were due to toxicity.

References

1. Groll AH, Glasmacher A, Just-Nuebling G, Maschmeyer G, Walsh TJ. Clinical pharmacology of antifungal compounds. Infect Dis Clin North Am 2003;17:159-91.
2. Boucher HW, Groll AH, Chiou CC, Walsh TJ. Newer systemic antifungal agents: pharmacokinetics, safety and efficacy. Drugs 2004;64(18):1997-2020.
3. Bolard J. How do the polyene macrolide antibiotics affect the cellular membrane properties? Biochim Biophys Acta 1986;864(3–4):257–304.

4. Vertut-Croquin A, Bolard J, Chabbert M, Gary-Bobo C. Differences in the interaction of the polyene antibiotic amphotericin B with cholesterol- or ergosterol-containing phospholipid vesicles. A circular dichroism and permeability study. Biochemistry 1983;22(12):2939–44.

5. Sokol-Anderson ML, Brajtburg J, Medoff G. Amphotericin B-induced oxidative damage and killing of *Candida albicans*. J Infect Dis 1986;154(1):76–83.

6. Edmonds LC, Davidson L, Bertino JS. Effect of variation in infusion time and macrophage blockade on organ uptake of amphotericin B-deoxycholate. J Antimicrob Chemother 1991;28(6):919–24.

7. Wong-Beringer A, Jacobs RA, Guglielmo BJ. Lipid formulations of amphotericin B: clinical efficacy and toxicities. Clin Infect Dis 1998;27(3):603–18.

8. Brajtburg J, Elberg S, Bolard J, Kobayashi GS, Levy RA, Ostlund RE Jr, Schlessinger D, Medoff G. Interaction of plasma proteins and lipoproteins with amphotericin B. J Infect Dis 1984;149(6):986–97.

9. Ridente Y, Aubard J, Bolard J. Absence in amphotericin B-spiked human plasma of the free monomeric drug, as detected by SERS. FEBS Lett 1999;446(2–3):283–6.

10. Baley JE, Meyers C, Kliegman RM, Jacobs MR, Blumer JL. Pharmacokinetics, outcome of treatment, and toxic effects of amphotericin B and 5-fluorocytosine in neonates. J Pediatr 1990;116(5):791–7.

11. Starke JR, Mason EO Jr, Kramer WG, Kaplan SL. Pharmacokinetics of amphotericin B in infants and children. J Infect Dis 1987;155(4):766–74.

12. Polak A. Pharmacokinetics of amphotericin B and flucytosine. Postgrad Med J 1979;55(647):667–70.

13. Atkinson AJ Jr, Bennett JE. Amphotericin B pharmacokinetics in humans. Antimicrob Agents Chemother 1978;13(2):271–6.

14. Lyman CA, Walsh TJ. Systemically administered antifungal agents. A review of their clinical pharmacology and therapeutic applications. Drugs 1992;44(1):9–35.

15. Chavanet PY, Garry I, Charlier N, Caillot D, Kisterman JP, D'Athis M, Portier H. Trial of glucose versus fat emulsion in preparation of amphotericin for use in HIV infected patients with candidiasis. BMJ 1992;305(6859):921–5.

16. Janknegt R, de Marie S, Bakker-Woudenberg IA, Crommelin DJ. Liposomal and lipid formulations of amphotericin B. Clinical pharmacokinetics. Clin Pharmacokinet 1992;23(4):279–91.

17. Sanders SW, Buchi KN, Goddard MS, Lang JK, Tolman KG. Single-dose pharmacokinetics and tolerance of a cholesteryl sulfate complex of amphotericin B administered to healthy volunteers. Antimicrob Agents Chemother 1991;35(6):1029–34.

18. Adedoyin A, Bernardo JF, Swenson CE, Bolsack LE, Horwith G, DeWit S, Kelly E, Klasterksy J, Sculier JP, DeValeriola D, Anaissie E, Lopez-Berestein G, Llanos-Cuentas A, Boyle A, Branch RA. Pharmacokinetic profile of ABELCET (amphotericin B lipid complex injection): combined experience from phase I and phase II studies. Antimicrob Agents Chemother 1997;41(10):2201–8.

19. Vita E, Schroeder DJ. Intralipid in prophylaxis of amphotericin B nephrotoxicity. Ann Pharmacother 1994;28(10):1182–3.

20. Thakur CP, Singh RK, Hassan SM, Kumar R, Narain S, Kumar A. Amphotericin B deoxycholate treatment of visceral leishmaniasis with newer modes of administration and precautions: a study of 938 cases. Trans R Soc Trop Med Hyg 1999;93(3):319–23.

21. Pathak A, Pien FD, Carvalho L. Amphotericin B use in a community hospital, with special emphasis on side effects. Clin Infect Dis 1998;26(2):334–8.

22. Noskin GA, Pietrelli L, Coffey G, Gurwith M, Liang LJ. Amphotericin B colloidal dispersion for treatment of candidemia in immunocompromised patients. Clin Infect Dis 1998;26(2):461–7.

23. Anaissie EJ, Mattiuzzi GN, Miller CB, Noskin GA, Gurwith MJ, Mamelok RD, Pietrelli LA. Treatment of invasive fungal infections in renally impaired patients with amphotericin B colloidal dispersion. Antimicrob Agents Chemother 1998;42(3):606–11.

24. Herbrecht R, Letscher V, Andres E, Cavalier A. Safety and efficacy of amphotericin B colloidal dispersion. An overview. Chemotherapy 1999;45(Suppl 1):67–76.

25. Noskin G, Pietrelli L, Gurwith M, Bowden R. Treatment of invasive fungal infections with amphotericin B colloidal dispersion in bone marrow transplant recipients. Bone Marrow Transplant 1999;23(7):697–703.

26. Herbrecht R, Letscher-Bru V, Bowden RA, Kusne S, Anaissie EJ, Graybill JR, Noskin GA, Oppenheim BA, Andrès E, Pietrelli LA. Treatment of 21 cases of invasive mucormycosis with amphotericin B colloidal dispersion. Eur J Clin Microbiol Infect Dis 2001;20(7):460–6.

27. White MH, Bowden RA, Sandler ES, Graham ML, Noskin GA, Wingard JR, Goldman M, van Burik JA, McCabe A, Lin JS, Gurwith M, Miller CB. Randomized, double-blind clinical trial of amphotericin B colloidal dispersion vs. amphotericin B in the empirical treatment of fever and neutropenia. Clin Infect Dis 1998;27(2):296–302.

28. Walsh TJ, Hiemenz JW, Seibel NL, Perfect JR, Horwith G, Lee L, Silber JL, DiNubile MJ, Reboli A, Bow E, Lister J, Anaissie EJ. Amphotericin B lipid complex for invasive fungal infections: analysis of safety efficacy in 556 cases. Clin Infect Dis 1998;26(6):1383–96.

29. Allsup D, Chu P. The use of amphotericin B lipid complex in 15 patients with presumed or proven fungal infection. Br J Haematol 1998;102(4):1109–10.

30. Myint H, Kyi AA, Winn RM. An open, non-comparative evaluation of the efficacy and safety of amphotericin B lipid complex as treatment of neutropenic patients with presumed or confirmed pulmonary fungal infections. J Antimicrob Chemother 1998;41(3):424–6.

31. Ringden O, Jonsson V, Hansen M, Tollemar J, Jacobsen N. Severe and common side-effects of amphotericin B lipid complex (Abelcet). Bone Marrow Transplant 1998;22(7):733–4.

32. Linden P, Lee L, Walsh TJ. Retrospective analysis of the dosage of amphotericin B lipid complex for the treatment of invasive fungal infections. Pharmacotherapy 1999;19(11):1261–8.

33. Singhal S, Hastings JG, Mutimer DJ. Safety of high-dose amphotericin B lipid complex. Bone Marrow Transplant 1999;24(1):116–7.

34. Cook G, Franklin IM. Adverse drug reactions associated with the administration of amphotericin B lipid complex (Abelcet). Bone Marrow Transplant 1999;23(12):1325–6.

35. Martino R, Subira M, Sureda A, Sierra J. Amphotericin B lipid complex at 3 mg/kg/day for treatment of invasive fungal infections in adults with haematological malignancies J Antimicrob Chemother 1999;44(4):569–72.

36. Martino R, Subira M, Domingo-Albos A, Sureda A, Brunet S, Sierra J. Low-dose amphotericin B lipid complex for the treatment of persistent fever of unknown origin in

patients with hematologic malignancies and prolonged neutropenia. Chemotherapy 1999;45(3):205–12.

37. Subira M, Martino R, Sureda A, Altes A, Briones J, Brunet S, Sierra J. Safety and efficacy of low-dose amphotericin B lipid complex for empirical antifungal therapy of neutropenic fever in patients with hematologic malignancies. Methods Find Exp Clin Pharmacol 2001;23(9):505–10.

38. Palmer SM, Drew RH, Whitehouse JD, Tapson VF, Davis RD, McConnell RR, Kanj SS, Perfect JR. Safety of aerosolized amphotericin B lipid complex in lung transplant recipients. Transplantation 2001;72(3):545–8.

39. Linden PK, Coley K, Fontes P, Fung JJ, Kusne S. Invasive aspergillosis in liver transplant recipients: outcome comparison of therapy with amphotericin B lipid complex and a historical cohort treated with conventional amphotericin B. Clin Infect Dis 2003;37:17-25.

40. Mattiuzzi GN, Kantarjian H, Faderl S, Lim J, Kontoyiannis D, Thomas D, Wierda W, Raad I, Garcia-Manero G, Zhou X, Ferrajoli A, Bekele N, Estey E. Amphotericin B lipid complex as prophylaxis of invasive fungal infections in patients with acute myelogenous leukemia and myelodysplastic syndrome undergoing induction chemotherapy. Cancer 2004;100:581-9.

41. Aguado JM, Lumbreras C, Gonzalez-Vidal D; Grupo de Farmacovigilancia de Abelcet. Assessment of nephrotoxicity in patients receiving amphotericin B lipid complex: a pharmacosurveillance study in Spain. Clin Microbiol Infect 2004;10:785-90.

42. Walsh TJ, Yeldandi V, McEvoy M, Gonzalez C, Chanock S, Freifeld A, Seibel NI, Whitcomb PO, Jarosinski P, Boswell G, Bekersky I, Alak A, Buell D, Barret J, Wilson W. Safety, tolerance, and pharmacokinetics of a small unilamellar liposomal formulation of amphotericin B (AmBisome) in neutropenic patients. Antimicrob Agents Chemother 1998;42(9):2391–8.

43. Ellis M, Spence D, de Pauw B, Meunier F, Marinus A, Collette L, Sylvester R, Meis J, Boogaerts M, Selleslag D, Kremery V, von Sinner W, MacDonald P, Doyen C, Vandercam B. An EORTC international multicenter randomized trial (EORTC number 19923) comparing two dosages of liposomal amphotericin B for treatment of invasive aspergillosis. Clin Infect Dis 1998;27(6):1406–12.

44. Clark AD, McKendrick S, Tansey PJ, Franklin IM, Chopra R. A comparative analysis of lipid-complexed and liposomal amphotericin B preparations in haematological oncology. Br J Haematol 1998;103(1):198–204.

45. Berman JD, Badaro R, Thakur CP, Wasunna KM, Behbehani K, Davidson R, Kuzoe F, Pang L, Weerasuriya K, Bryceson AD. Efficacy and safety of liposomal amphotericin B (AmBisome) for visceral leishmaniasis in endemic developing countries. Bull World Health Organ 1998;76(1):25–32.

46. Walsh TJ, Goodman JL, Pappas P, Bekersky I, Buell DN, Roden M, Barrett J, Anaissie EJ. Safety, tolerance, and pharmacokinetics of high-dose liposomal amphotericin B (AmBisome) in patients infected with *Aspergillus* species and other filamentous fungi: maximum tolerated dose study. Antimicrob Agents Chemother 2001;45(12):3487–96.

47. Musa AM, Khalil EA, Mahgoub FA, Hamad S, Elkadaru AM, El Hassan AM. Efficacy of liposomal amphotericin B (AmBisome) in the treatment of persistent post-kala-azar dermal leishmaniasis (PKDL). Ann Trop Med Parasitol 2005;99(6):563-9.

48. Mattiuzzi GN, Estey E, Raad I, Giles F, Cortes J, Shen Y, Kontoyiannis D, Koller C, Munsell M, Beran M, Kantarjian H. Liposomal amphotericin B versus the combination of fluconazole and itraconazole as prophylaxis for invasive fungal infections during induction chemotherapy for patients with acute myelogenous leukemia and myelodysplastic syndrome. Cancer 2003;97:450-6.

49. Cascio A, di Martino L, Occorsio P, Giacchino R, Catania S, Gigliotti AR, Aiassa C, Iaria C, Giordano S, Colomba C, Polara VF, Titone L, Gradoni L, Gramiccia M, Antinori S. A 6 day course of liposomal amphotericin B in the treatment of infantile visceral leishmaniasis: the Italian experience. J Antimicrob Chemother 2004;54:217-20.

50. Schwartz S, Behre G, Heinemann V, Wandt H, Schilling E, Arning M, Trittin A, Kern WV, Boenisch O, Bosse D, Lenz K, Ludwig WD, Hiddemann W, Siegert W, Beyer J. Aerosolized amphotericin B inhalations as prophylaxis of invasive *Aspergillus* infections during prolonged neutropenia: results of a prospective randomized multicenter trial. Blood 1999;93(11):3654–61.

51. Schoffski P, Freund M, Wunder R, Petersen D, Kohne CH, Hecker H, Schubert U, Ganser A. Safety and toxicity of amphotericin B in glucose 5% or Intralipid 20% in neutropenic patients with pneumonia or fever of unknown origin: randomised study. BMJ 1998;317(7155):379–84.

52. Manfredi R, Chiodo F. Case-control study of amphotericin B in a triglyceride fat emulsion versus conventional amphotericin B in patients with AIDS. Pharmacotherapy 1998;18(5):1087–92.

53. Torre D, Banfi G, Tambini R, Speranza F, Zeroli C, Martegani R, Airoldi M, Fiori G. A retrospective study on the efficacy and safety of amphotericin B in a lipid emulsion for the treatment of cryptococcal meningitis in AIDS patients. J Infect 1998;37(1):36–8.

54. Leenders AC, Daenen S, Jansen RL, Hop WC, Lowenberg B, Wijermans PW, Cornelissen J, Herbrecht R, van der Lelie H, Hoogsteden HC, Verbrugh HA, de Marie S. Liposomal amphotericin B compared with amphotericin B deoxycholate in the treatment of documented and suspected neutropenia-associated invasive fungal infections. Br J Haematol 1998;103(1):205–12.

55. Walsh TJ, Finberg RW, Arndt C, Hiemenz J, Schwartz C, Bodensteiner D, Pappas P, Seibel N, Greenberg RN, Dummer S, Schuster M, Holcenberg JSNational Institute of Allergy and Infectious Diseases Mycoses Study Group. Liposomal amphotericin B for empirical therapy in patients with persistent fever and neutropenia. N Engl J Med 1999;340(10):764–71.

56. Drew RH, Dodds Ashley E, Benjamin DK Jr, Duane Davis R, Palmer SM, Perfect JR. Comparative safety of amphotericin B lipid complex and amphotericin B deoxycholate as aerosolized antifungal prophylaxis in lung-transplant recipients. Transplantation 2004;77(2):232-7.

57. Fleming RV, Kantarjian HM, Husni R, Rolston K, Lim J, Raad I, Pierce S, Cortes J, Estey E. Comparison of amphotericin B lipid complex (ABLC) vs. ambisome in the treatment of suspected or documented fungal infections in patients with leukemia. Leuk Lymphoma 2001;40(5–6):511–20.

58. Nucci M, Loureiro M, Silveira F, Casali AR, Bouzas LF, Velasco E, Spector N, Pulcheri W. Comparison of the toxicity of amphotericin B in 5% dextrose with that of amphotericin B in fat emulsion in a randomized trial with cancer patients. Antimicrob Agents Chemother 1999;43(6):1445–8.

59. Nath CE, Shaw PJ, Gunning R, McLachlan AJ, Earl JW. Amphotericin B in children with malignant disease: a

comparison of the toxicities and pharmacokinetics of amphotericin B administered in dextrose versus lipid emulsion. Antimicrob Agents Chemother 1999;43(6):1417–23.

60. Malik IA, Moid I, Aziz Z, Khan S, Suleman M. A randomized comparison of fluconazole with amphotericin B as empiric anti-fungal agents in cancer patients with prolonged fever and neutropenia. Am J Med 1998;105(6):478–83.

61. Winston DJ, Hathorn JW, Schuster MG, Schiller GJ, Territo MC. A multicenter, randomized trial of fluconazole versus amphotericin B for empiric antifungal therapy of febrile neutropenic patients with cancer. Am J Med 2000;108(4):282–9.

62. Wolff SN, Fay J, Stevens D, Herzig RH, Pohlman B, Bolwell B, Lynch J, Ericson S, Freytes CO, LeMaistre F, Collins R, Pineiro L, Greer J, Stein R, Goodman SA, Dummer S. Fluconazole vs low-dose amphotericin B for the prevention of fungal infections in patients undergoing bone marrow transplantation: a study of the North American Marrow Transplant Group. Bone Marrow Transplant 2000;25(8):853–9.

63. Boogaerts M, Maertens J, van Hoof A, de Bock R, Fillet G, Peetermans M, Selleslag D, Vandercam B, Vandewoude K, Zachee P, De Beule K. Itraconazole versus amphotericin B plus nystatin in the prophylaxis of fungal infections in neutropenic cancer patients. J Antimicrob Chemother 2001;48(1):97–103.

64. Boogaerts M, Winston DJ, Bow EJ, Garber G, Reboli AC, Schwarer AP, Novitzky N, Boehme A, Chwetzoff E, De Beule KItraconazole Neutropenia Study Group. Intravenous and oral itraconazole versus intravenous amphotericin B deoxycholate as empirical antifungal therapy for persistent fever in neutropenic patients with cancer who are receiving broad-spectrum antibacterial therapy. A randomized, controlled trial. Ann Intern Med 2001;135(6):412–22.

65. Harousseau JL, Dekker AW, Stamatoullas-Bastard A, Fassas A, Linkesch W, Gouveia J, De Bock R, Rovira M, Seifert WF, Joosen H, Peeters M, De Beule K. Itraconazole oral solution for primary prophylaxis of fungal infections in patients with hematological malignancy and profound neutropenia: a randomized, double-blind, double-placebo, multicenter trial comparing itraconazole and amphotericin B. Antimicrob Agents Chemother 2000;44(7):1887–93.

66. Arathoon EG, Gotuzzo E, Noriega LM, Berman RS, DiNubile MJ, Sable CA. Randomized, double-blind, multicenter study of caspofungin versus amphotericin B for treatment of oropharyngeal and esophageal candidiases. Antimicrob Agents Chemother 2002;46(2):451–7.

67. Villanueva A, Arathoon EG, Gotuzzo E, Berman RS, DiNubile MJ, Sable CA. A randomized double-blind study of caspofungin versus amphotericin for the treatment of candidal esophagitis. Clin Infect Dis 2001;33(9):1529–35.

68. Mora-Duarte J, Betts R, Rotstein C, Colombo AL, Thompson-Moya L, Smietana J, Lupinacci R, Sable C, Kartsonis N, Perfect JCaspofungin Invasive Candidiasis Study Group. Comparison of caspofungin and amphotericin B for invasive candidiasis. N Engl J Med 2002;347(25):2020–9.

69. Mishra M, Biswas UK, Jha AM, Khan AB. Amphotericin versus sodium stibogluconate in first-line treatment of Indian kala-azar. Lancet 1994;344(8937):1599–600.

70. Kelsey SM, Goldman JM, McCann S, Newland AC, Scarffe JH, Oppenheim BA, Mufti GJ. Liposomal amphotericin (AmBisome) in the prophylaxis of fungal infections

71. Chabot GG, Pazdur R, Valeriote FA, Baker LH. Pharmacokinetics and toxicity of continuous infusion amphotericin B in cancer patients. J Pharm Sci 1989;78(4):307–10.

72. Ringden O, Andstrom E, Remberger M, Svahn BM, Tollemar J. Allergic reactions and other rare side-effects of liposomal amphotericin. Lancet 1994;344(8930):1156–7.

73. Laing RB, Milne LJ, Leen CL, Malcolm GP, Steers AJ. Anaphylactic reactions to liposomal amphotericin. Lancet 1994;344(8923):682.

74. Kauffman CA, Wiseman SW. Anaphylaxis upon switching lipid-containing amphotericin B formulations. Clin Infect Dis 1998;26(5):1237–8.

75. Le Y, Rana KZ, Dudley MN. Amphotericin B-associated hypertension. Ann Pharmacother 1996;30(7–8):765–7.

76. Groll AH, Piscitelli SC, Walsh TJ. Clinical pharmacology of systemic antifungal agents: a comprehensive review of agents in clinical use, current investigational compounds, and putative targets for antifungal drug development. Adv Pharmacol 1998;44:343–500.

77. Rowles DM, Fraser SL. Amphotericin B lipid complex (ABLC)-associated hypertension: case report and review. Clin Infect Dis 1999;29(6):1564–5.

78. Wiwanitkit V. Severe hypertension associated with the use of amphotericin B: an appraisal on the reported cases. J Hypertens 2006;24(7):1445.

79. Rodrigues CA, Yamamoto M, Aranters AM, Chauffaille ML, Colombo AL, Bordin JO. Amphotericin B-induced severe hypertension in a young patient: a case report and a review of the literature. Ren Fail 2006;28(2):185-7.

80. Levy M, Domaratzki J, Koren G. Amphotericin-induced heart-rate decrease in children. Clin Pediatr (Phila) 1995;34(7):358–64.

81. el-Dawlatly AA, Gomaa S, Takrouri MS, Seraj MA. Amphotericin B and cardiac toxicity—a case report. Middle East. J Anesthesiol 1999;15(1):107–12.

82. Johnson MD, Drew RH, Perfect JR. Chest discomfort associated with liposomal amphotericin B: report of three cases and review of the literature. Pharmacotherapy 1998;18(5):1053–61.

83. DeMonaco HJ, McGovern B. Transient asystole associated with amphotericin B infusion. Drug Intell Clin Pharm 1983;17(7–8):547–8.

84. Danaher PJ, Cao MK, Anstead GM, Dolan MJ, DeWitt CC. Reversible dilated cardiomyopathy related to amphotericin B therapy. J Antimicrob Chemother 2004;53:115-7.

85. Burke D, Lal R, Finkel KW, Samuels J, Foringer JR. Acute amphotericin B overdose. Ann Pharmacother 2006;40(12):2254-9.

86. Quer N, Soy D, Castro P, Nicolás JM. Miocardiopatía subaguda y anfotericina B liposomal. [Subacute cardiomyopathy and liposomal amphotericin B.] Farm Hosp 2006;30(4):260-1.

87. Zernikow B, Fleischhack G, Hasan C, Bode U. Cyanotic Raynaud's phenomenon with conventional but not with liposomal amphotericin B: three case reports. Mycoses 1997;40(9–10):359–61.

88. Griese M, Schams A, Lohmeier KP. Amphotericin B and pulmonary surfactant. Eur J Med Res 1998;3(8):383–6.

89. Gryn J, Goldberg J, Johnson E, Siegel J, Inzerillo J. The toxicity of daily inhaled amphotericin. B. Am J Clin Oncol 1993;16(1):43–6.

90. Erjavec Z, Woolthuis GM, de Vries-Hospers HG, Sluiter WJ, Daenen SM, de Pauw B, Halie MR.

in neutropenic patients: a randomised, double-blind, placebo-controlled study. Bone Marrow Transplant 1999;23(2):163–8.

Tolerance and efficacy of amphotericin B inhalations for prevention of invasive pulmonary aspergillosis in haematological patients. Eur J Clin Microbiol Infect Dis 1997;16(5):364–8.

91. Dubois J, Bartter T, Gryn J, Pratter MR. The physiologic effects of inhaled amphotericin. B. Chest 1995;108(3):750–3.

92. Wright DG, Robichaud KJ, Pizzo PA, Deisseroth AB. Lethal pulmonary reactions associated with the combined use of amphotericin B and leukocyte transfusions. N Engl J Med 1981;304(20):1185–9.

93. Dutcher JP, Kendall J, Norris D, Schiffer C, Aisner J, Wiernik PH. Granulocyte transfusion therapy and amphotericin B: adverse reactions? Am J Hematol 1989;31(2):102–8.

94. Hussein MA, Fletcher R, Long TJ, Zuccaro K, Bolwell BJ, Hoeltge A. Transfusing platelets 2 h after the completion of amphotericin-B decreases its detrimental effect on transfused platelet recovery and survival Transfus Med 1998;8(1):43–7.

95. Garnacho-Montero J, Ortiz-Leyba C, Garcia Garmendia JL, Jimenez Jimenez F. Life-threatening adverse event after amphotericin B lipid complex treatment in a patient treated previously with amphotericin B deoxycholate. Clin Infect Dis 1998;26(4):1016.

96. Alexander BD, Dodds Ashley ES, Addison RM, Alspaugh JA, Chao NJ, Perfect JR. Non-comparative evaluation of the safety of aerosolized amphotericin B lipid complex in patients undergoing allogeneic hematopoietic stem cell transplantation. Transpl Infect Dis 2006;8(1):13-20.

97. Tolentino LF, Tsai SF, Witt MD, French SW. Fatal fat embolism following amphotericin B lipid complex injection. Exp Mol Pathol 2004;77:246-8.

98. Manley TJ, Chusid MJ, Rand SD, Wells D, Margolis DA. Reversible parkinsonism in a child after bone marrow transplantation and lipid-based amphotericin B therapy. Pediatr Infect Dis J 1998;17(5):433–4.

99. Ural AU, Avcu F, Cetin T, Beyan C, Kaptan K, Nazaroglu NK, Yalcin A. Spironolactone: is it a novel drug for the prevention of amphotericin B-related hypokalemia in cancer patients? Eur J Clin Pharmacol 2002;57(11):771–3.

100. Bearden DT, Muncey LA. The effect of amiloride on amphotericin B-induced hypokalaemia. J Antimicrob Chemother 2001;48(1):109–11.

101. Barcia JP. Hyperkalemia associated with rapid infusion of conventional and lipid complex formulations of amphotericin. B. Pharmacotherapy 1998;18(4):874–6.

102. Lucas da Silva PS, Iglesias SB, Waisberg J. Hypokalemic rhabdomyolysis in a child due to amphotericin B therapy. Eur J Pediatr 2007;166(2):169-71.

103. Jain A, Butani L. Severe hyperphosphatemia resulting from high-dose liposomal amphotericin in a child with leukemia. J Pediatr Hematol Oncol 2003;25:324-6.

104. Blum SF, Shohet SB, Nathan DG, Gardner FH. The effect of amphotericin B on erythrocyte membrane cation permeability: its relation to in vivo erythrocyte survival. J Lab Clin Med 1969;73(6):980–7.

105. Salama A, Burger M, Mueller-Eckhardt C. Acute immune hemolysis induced by a degradation product of amphotericin B. Blut 1989;58(2):59–61.

106. Juliano RL, Grant CW, Barber KR, Kalp MA. Mechanism of the selective toxicity of amphotericin B incorporated into liposomes. Mol Pharmacol 1987;31(1):1–11.

107. Charak BS, Iyer RS, Rajoor BG, Saikia TK, Gopal R, Advani SH. Amphotericin B-related thrombocytopenia.

A report of two cases. J Assoc Physicians India 1990;38(3):235–6.

108. Chan CS, Tuazon CU, Lessin LS. Amphotericin-B-induced thrombocytopenia. Ann Intern Med 1982;96(3):332–3.

109. Bock M, Muggenthaler KH, Schmidt U, Heim MU. Influence of antibiotics on posttransfusion platelet increment. Transfusion 1996;36(11–12):952–4.

110. Kulpa J, Zaroulis CG, Good RA, Kutti J. Altered platelet function and circulation induced by amphotericin B in leukemic patients after platelet transfusion. Transfusion 1981;21(1):74–6.

111. Ringden O, Andstrom EE, Remberger M, Dahllof G, Svahn BM, Tollemar J. Prophylaxis and therapy using liposomal amphotericin B (AmBisome) for invasive fungal infections in children undergoing organ or allogeneic bone-marrow transplantation. Pediatr Transplant 1997;1(2):124–9.

112. Miller MA. Reversible hepatotoxicity related to amphotericin B. Can Med Assoc J 1984;131(10):1245–7.

113. Gill J, Sprenger HR, Ralph ED, Sharpe MD. Hepatotoxicity possibly caused by amphotericin B. Ann Pharmacother 1999;33(6):683–5.

114. Fischer MA, Winkelmayer WC, Rubin RH, Avorn J. The hepatotoxicity of antifungal medications in bone marrow transplant recipients. Clin Infect Dis 2005;41(3):301-7.

115. Stuecklin-Utsch A, Hasan C, Bode U, Fleischhack G. Pancreatic toxicity after liposomal amphotericin B. Mycoses 2002;45(5–6):170–3.

116. Groll AH, Gea-Banacloche JC, Glasmacher A, Just-Nuebling G, Maschmeyer G, Walsh TJ. Clinical pharmacology of antifungal compounds. Infect Dis Clin North Am 2003;17(1):159–91.

117. McCurdy DK. Distal tubule affected by amphotericin B. N Engl J Med 1969;280(4):220-1.

118. McCurdy DK, Frederic M, Elkinton JR. Renal tubular acidosis due to amphotericin B. N Engl J Med 1968;278(3):124–30.

119. Sabra R, Branch RA. Amphotericin B nephrotoxicity. Drug Saf 1990;5(2):94–108.

120. Barton CH, Pahl M, Vaziri ND, Cesario T. Renal magnesium wasting associated with amphotericin B therapy. Am J Med 1984;77(3):471–4.

121. Tsau YK, Tsai WY, Lu FL, Tsai WS, Chen CH. Symptomatic hypomagnesemia in children. Zhonghua Min Guo Xiao Er Ke Yi Xue Hui Za Zhi 1998;39(6):393–7.

122. Spath-Schwalbe E, Koschuth A, Dietzmann A, Schanz J, Possinger K. Successful use of liposomal amphotericin B in a case of amphotericin B-induced nephrogenic diabetes insipidus. Clin Infect Dis 1999;28(3):680–1.

123. Barbour GL, Straub KD, O'Neal BL, Leatherman JW. Vasopressin-resistant nephrogenic diabetes insipidus. A result of amphotericin B therapy. Arch Intern Med 1979;139(1):86–8.

124. Araujo JJ, Dominguez A, Bueno C, Rodriguez J, Rios MJ, Muniain MA, Perez R. Diabetes insipida nefrogenica secundaria a la administracion de anfotericina B y anfotericina B liposomal. [Nephrogenous diabetes insipidus secondary to the administration of amphotericin B and liposomal amphotericin B.] Enferm Infecc Microbiol Clin 1998;16(4):204–5.

125. Chen CY, Kumar RN, Feng YH, Ho CH, You JY, Liao CC, Tseng CH, Mavros P, Gerth WC, Chen YC. Treatment outcomes in patients receiving conventional amphotericin B therapy: a prospective multicentre study in Taiwan. J Antimicrob Chemother 2006;57(6):1181-8.

126. Burges JL, Birchall R. Nephrotoxicity of amphotericin B, with emphasis on changes in tubular function. Am J Med 1972;53(1):77–84.

127. Hohler T, Teuber G, Wanitschke R, Meyer zum Buschenfeld KH. Indomethacin treatment in amphotericin B induced nephrogenic diabetes insipidus. Clin Investig 1994;72(10):769–71.

128. Liberopoulos E, Alexandridis G, Elisaf M. A tumor lysis-like syndrome during therapy of visceral leishmaniasis. Ann Clin Lab Sci 2002;32(4):419–21.

129. Harbarth S, Pestotnik SL, Lloyd JF, Burke JP, Samore MH. The epidemiology of nephrotoxicity associated with conventional amphotericin B therapy. Am J Med 2001;111(7):528–34.

130. Wingard JR, Kubilis P, Lee L, Yee G, White M, Walshe L, Bowden R, Anaissie E, Hiemenz J, Lister J. Clinical significance of nephrotoxicity in patients treated with amphotericin B for suspected or proven aspergillosis. Clin Infect Dis 1999;29(6):1402–7.

131. Gubbins PO, Penzak SR, Polston S, McConnell SA, Anaissie E. Characterizing and predicting amphotericin B-associated nephrotoxicity in bone marrow or peripheral blood stem cell transplant recipients. Pharmacotherapy 2002;22(8):961–71.

132. Furrer K, Schaffner A, Vavricka SR, Halter J, Imhof A, Schanz U. Nephrotoxicity of cyclosporine A and amphotericin B-deoxycholate as continuous infusion in allogenic stem cell transplantation. Swiss Med Wkly 2002;132(23–24):316–20.

133. Speich R, Dutly A, Naef R, Russi EW, Weder W, Boehler A. Tolerability, safety and efficacy of conventional amphotericin B administered by 24-hour infusion to lung transplant recipients. Swiss Med Wkly 2002;132(31–32):455–8.

134. Bowden R, Chandrasekar P, White MH, Li X, Pietrelli L, Gurwith M, van Burik JA, Laverdiere M, Safrin S, Wingard JR. A double-blind, randomized, controlled trial of amphotericin B colloidal dispersion versus amphotericin B for treatment of invasive aspergillosis in immunocompromised patients. Clin Infect Dis 2002;35(4):359–66.

135. Luke RG, Boyle JA. Renal effects of amphotericin B lipid complex. Am J Kidney Dis 1998;31(5):780–5.

136. Subira M, Martino R, Gomez L, Marti JM, Estany C, Sierra J. Low-dose amphotericin B lipid complex vs. conventional amphotericin B for empirical antifungal therapy of neutropenic fever in patients with hematologic malignancies—a randomized, controlled trial. Eur J Haematol 2004;72:342–7.

137. Hooshmand-Rad R, Reed MD, Chu A, Gotz V, Morris JA, Weinberg J, Dominguez EA. Retrospective study of the renal effects of amphotericin B lipid complex when used at higher-than-recommended dosages and longer durations compared with lower dosages and shorter durations in patients with systemic fungal infections. Clin Ther 2004;26:1652-62.

138. Alexander BD, Wingard JR. Study of renal safety in amphotericin B lipid complex-treated patients. Clin Infect Dis 2005;40 Suppl 6:S414-21.

139. Slain D, Miller K, Khakoo R, Fisher M, Wierman T, Jozefczyk K. Infrequent occurrence of amphotericin B lipid complex-associated nephrotoxicity in various clinical settings at a university hospital: a retrospective study. Clin Ther 2002;24(10):1636–42.

140. Echevarria J, Seas C, Cruz M, Chávez E, Campos M, Cieza J, Gotuzzo E, Llanos A. Oral rehydration solution to prevent nephrotoxicity of amphotericin B. Am J Trop Med Hyg 2006;75(6):1108-12.

141. Groll AH, Glasmacher A, Just-Nuebling G, Maschmeyer G, Walsh TJ. Clinical pharmacology of antifungal compounds. Infect Dis Clin N Am 2003;17:159-91.

142. Girmenia C, Cimino G, Di Cristofano F, Micozzi A, Gentile G, Martino P. Effects of hydration with salt repletion on renal toxicity of conventional amphotericin B empirical therapy: a prospective study in patients with hematological malignances. Support Care Cancer 2005;13(12):987-92.

143. Pahls S, Schaffner A. Comparison of the activity of free and liposomal amphotericin B in vitro and in a model of systemic and localized murine candidiasis. J Infect Dis 1994;169(5):1057–61.

144. Carlini A, Angelini D, Burrows L, De Quirieo G, Antonelli A. Cerebral aspergillosis: long term efficacy and safety of liposomal amphotericin B in kidney transplant. Nephrol Dial Transplant 1998;13(10):2659–61.

145. Johnson PC, Wheat LJ, Cloud GA, Goldman M, Lancaster D, Bamberger DM, Powderly WG, Hafner R, Kauffman CA, Dismukes WEU.S. National Institute of Allergy and Infectious Diseases Mycoses Study Group. Safety and efficacy of liposomal amphotericin B compared with conventional amphotericin B for induction therapy of histoplasmosis in patients with AIDS. Ann Intern Med 2002;137(2):105–9.

146. Gerbaud E, Tamion F, Girault C, Clabault K, Lepretre S, Leroy J, Bonmarchand G. Persistent acute tubular toxicity after switch from conventional amphotericin B to liposomal amphotericin B (Ambisome). J Antimicrob Chemother 2003;51:473-5.

147. Canada TW, Weavind LM, Augustin KM. Possible liposomal amphotericin B-induced nephrogenic diabetes insipidus. Ann Pharmacother 2003;37:70-3.

148. Walsh TJ, Teppler H, Donowitz GR, Maertens JA, Baden LR, Dmoszynska A, Cornely OA, Bourque MR, Lupinacci RJ, Sable CA, dePauw BE. Caspofungin versus liposomal amphotericin B for empirical antifungal therapy in patients with persistent fever and neutropenia. N Engl J Med 2004;351:1391-402.

149. Branch RA. Prevention of amphotericin B-induced renal impairment. A review on the use of sodium supplementation. Arch Intern Med 1988;148(11):2389–94.

150. Gardner ML, Godley PJ, Wasan SM. Sodium loading treatment for amphotericin B-induced nephrotoxicity. Dicp 1990;24(10):940–6.

151. Llanos A, Cieza J, Bernardo J, Echevarria J, Biaggioni I, Sabra R, Branch RA. Effect of salt supplementation on amphotericin B nephrotoxicity. Kidney Int 1991;40(2):302–8.

152. Stein RS, Alexander JA. Sodium protects against nephrotoxicity in patients receiving amphotericin B. Am J Med Sci 1989;298(5):299–304.

153. Arning M, Scharf RE. Prevention of amphotericin-B-induced nephrotoxicity by loading with sodium chloride: a report of 1291 days of treatment with amphotericin B without renal failure. Klin Wochenschr 1989;67(20):1020–8.

154. Anderson CM. Sodium chloride treatment of amphotericin B nephrotoxicity. Standard of care? West J Med 1995;162(4):313–7.

155. Heidemann HT, Gerkens JF, Spickard WA, Jackson EK, Branch RA. Amphotericin B nephrotoxicity in humans decreased by salt repletion. Am J Med 1983;75(3):476–481.

156. Bernardo JF, Murakami S, Branch RA, Sabra R. Potassium depletion potentiates amphotericin-B-induced toxicity to renal tubules. Nephron 1995;70(2):235–41.

157. Camp MJ, Wingard JR, Gilmore CE, Lin LS, Dix SP, Davidson TG, Geller RB. Efficacy of low-dose dopamine in preventing amphotericin B nephrotoxicity in bone marrow transplant patients and leukemia patients. Antimicrob Agents Chemother 1998;42(12):3103–6.

158. Eriksson U, Seifert B, Schaffner A. Comparison of effects of amphotericin B deoxycholate infused over 4 or 24 hours: randomised controlled trial. Bmj 2001;322(7286):579–82.

159. Cesaro S, Calore E, Messina C, Zanesco L. Allergic reaction to the liposomal component of liposomal amphotericin B. Support Care Cancer 1999;7(4):284–6.

160. Griswold MW, Briceland LL, Stein DS. Is amphotericin B test dosing needed? Ann Pharmacother 1998;32(4):475–7.

161. Johnson JR, Kangas PJ, West M. Serious adverse event after unrecognized substitution of one amphotericin B lipid preparation for another. Clin Infect Dis 1998;27(5):1342–3.

162. Roden MM, Nelson LD, Knudsen TA, Jarosinski PF, Starling JM, Shiflett SE, Calis K, DeChristoforo R, Donowitz GR, Buell D, Walsh TJ. Triad of acute infusion-related reactions associated with liposomal amphotericin B: analysis of clinical and epidemiological characteristics. Clin Infect Dis 2003;36:1213–20.

163. Cleary JD, Schwartz M, Rogers PD, de Mestral J, Chapman SW. Effects of amphotericin B and caspofungin on histamine expression. Pharmacotherapy 2003;23:966–73.

164. Dean JL, Wolf JE, Ranzini AC, Laughlin MA. Use of amphotericin B during pregnancy: case report and review. Clin Infect Dis 1994;18(3):364–8.

165. Scarcella A, Pasquariello MB, Giugliano B, Vendemmia M, de Lucia A. Liposomal amphotericin B treatment for neonatal fungal infections. Pediatr Infect Dis J 1998;17(2):146–8.

166. Weitkamp JH, Poets CF, Sievers R, Musswessels E, Groneck P, Thomas P, Bartmann P. Candida infection in very low birth-weight infants: outcome and nephrotoxicity of treatment with liposomal amphotericin B (AmBisome). Infection 1998;26(1):11–5.

167. Juster-Reicher A, Flidel-Rimon O, Amitay M, Even-Tov S, Shinwell E, Leibovitz E. High-dose liposomal amphotericin B in the therapy of systemic candidiasis in neonates. Eur J Clin Microbiol Infect Dis 2003;22:603–7.

168. Adler-Shohet F, Waskin H, Lieberman JM. Amphotericin B lipid complex for neonatal invasive candidiasis. Arch Dis Child Fetal Neonatal Ed 2001;84(2):F131–3.

169. Prentice HG, Hann IM, Herbrecht R, Aoun M, Kvaloy S, Catovsky D, Pinkerton CR, Schey SA, Jacobs F, Oakhill A, Stevens RF, Darbyshire PJ, Gibson BE. A randomized comparison of liposomal versus conventional amphotericin B for the treatment of pyrexia of unknown origin in neutropenic patients. Br J Haematol 1997;98(3):711–8.

170. Walsh TJ, Seibel NL, Arndt C, Harris RE, Dinubile MJ, Reboli A, Hiemenz J, Chanock SJ. Amphotericin B lipid complex in pediatric patients with invasive fungal infections. Pediatr Infect Dis J 1999;18(8):702–8.

171. Linder N, Klinger G, Shalit I, Levy I, Ashkenazi S, Haski G, Levit O, Sirota L. Treatment of candidaemia in premature infants: comparison of three amphotericin B preparations. J Antimicrob Chemother 2003;52:663–7.

172. Wurthwein G, Groll AH, Hempel G, Adler-Shohet FC, Lieberman JM, Walsh TJ. Population pharmacokinetics of amphotericin B lipid complex in neonates. Antimicrob Agents Chemother 2005;49(12):5092-8.

173. Wiley JM, Seibel NL, Walsh TJ. Efficacy and safety of amphotericin B lipid complex in 548 children and adolescents with invasive fungal infections. Pediatr Infect Dis J 2005;24(2):167-74.

174. Hooshmand-Rad R, Chu A, Gotz V, Morris J, Batty S, Freifeld A. Use of amphotericin B lipid complex in elderly patients. J Infect 2005;50(4):277-87.

175. Castroagudin JF, Ponton C, Bustamante M, Otero E, Martinez J, Tome S, Conde R, Segade FR, Delgado M, Brage A, Galban C, Varo E. Prospective interventional study to evaluate the efficacy and safety of liposomal amphotericin B as prophylaxis of fungal infections in high-risk liver transplant recipients. Transplant Proc 2005;37(9):3965-7.

176. Cleary JD, Rogers PD, Chapman SW. Variability in polyene content and cellular toxicity among deoxycholate amphotericin B formulations. Pharmacotherapy 2003;23:572-8.

177. Mohr JF, Hall AC, Ericsson CD, Ostrosky-Zeichner L. Fatal amphotericin B overdose due to administration of nonlipid formulation instead of lipid formulation. Pharmacotherapy 2005;25(3):426-8.

178. Cruz JM, Peacock JE Jr, Loomer L, Holder LW, Evans GW, Powell BL, Lyerly ES, Capizzi RL. Rapid intravenous infusion of amphotericin B: a pilot study. Am J Med 1992;93(2):123–30.

179. Arning M, Dresen B, Aul C, Schneider W. Influence of infusion time on the acute toxicity of amphotericin B: results of a randomized double-blind study. Recent Results Cancer Res 1991;121:347–52.

180. Ellis ME, al-Hokail AA, Clink HM, Padmos MA, Ernst P, Spence DG, Tharpe WN, Hillier VF. Double-blind randomized study of the effect of infusion rates on toxicity of amphotericin B. Antimicrob Agents Chemother 1992;36(1):172–9.

181. Nicholl TA, Nimmo CR, Shepherd JD, Phillips P, Jewesson PJ. Amphotericin B infusion-related toxicity: comparison of two- and four-hour infusions. Ann Pharmacother 1995;29(11):1081–7.

182. Spitzer TR, Creger RJ, Fox RM, Lazarus HM. Rapid infusion amphotericin B: effective and well-tolerated therapy for neutropenic fever. Pharmatherapeutica 1989;5(5):305–11.

183. Gales MA, Gales BJ. Rapid infusion of amphotericin B in dextrose. Ann Pharmacother 1995;29(5):523–9.

184. Eriksson U, Seifert B, Schaffner A. Comparison of effects of amphotericin B deoxycholate infusied over 4 or 24 hours: randomised controlled trial. Br Med J 2001;322:579-82.

185. Imhof A, Walter RB, Schaffner A. Continuous infusion of escalated doses of amphotericin B deoxycholate: an open-label observational study. Clin Infect Dis 2003;36:943-51.

186. Malani PN, Depestel DD, Riddell J, Bickley S, Klein LR, Kauffman CA. Experience with community-based amphotericin B infusion therapy. Pharmacotherapy 2005;25(5):690-7.

187. Mansueto P, Rizzo M, Affronti M, Malta R, Carmina E, Mansueto S, Masellis M, Rini GB. Safe and successful endoarterial infusion of liposomal amphotericin B in treatment of mucormycosis. New Microbiol 2003;26:395-8.

188. Winn RE, Bower JH, Richards JF. Acute toxic delirium. Neurotoxicity of intrathecal administration of amphotericin B. Arch Intern Med 1979;139(6):706–7.

189. Fisher JF, Dewald J. Parkinsonism associated with intraventricular amphotericin B. J Antimicrob Chemother 1983;12(1):97–9.

190. Bearden DT, Muncey LA. The effect of amiloride on amphotericin B-induced hypokalaemia. J Antimicrob Chemother 2001;48(1):109–11.

191. Goren MP, Viar MJ, Shenep JL, Wright RK, Baker DK, Kalwinsky DK. Monitoring serum aminoglycoside concentrations in children with amphotericin B nephrotoxicity. Pediatr Infect Dis J 1988;7(10):698–703.

192. Schaffner A, Frick PG. The effect of ketoconazole on amphotericin B in a model of disseminated aspergillosis. J Infect Dis 1985;151(5):902–10.

193. Pahls S, Schaffner A. *Aspergillus fumigatus* pneumonia in neutropenic patients receiving fluconazole for infection due to *Candida* species: is amphotericin B combined with fluconazole the appropriate answer? Clin Infect Dis 1994;18(3):484–6.

194. Schaffner A, Bohler A. Amphotericin B refractory aspergillosis after itraconazole: evidence for significant antagonism. Mycoses 1993;36(11–12):421–4.

195. Rex JH, Pappas PG, Karchmer AW, Sobel J, Edwards JE, Hadley S, Brass C, Vazquez JA, Chapman SW, Horowitz HW, Zervos M, McKinsey D, Lee J, Babinchak T, Bradsher RW, Cleary JD, Cohen DM, Danziger L, Goldman M, Goodman J, Hilton E, Hyslop NE, Kett DH, Lutz J, Rubin RH, Scheld WM, Schuster M, Simmons B, Stein DK, Washburn RG, Mautner L, Chu TC, Panzer H, Rosenstein RB, Booth JNational Institute of Allergy and Infectious Diseases Mycoses Study Group. A randomized and blinded multicenter trial of high-dose fluconazole plus placebo versus fluconazole plus amphotericin B as therapy for candidemia and its consequences in nonneutropenic subjects. Clin Infect Dis 2003;36(10):1221–8.

196. Persat F, Schwartzbrod PE, Troncy J, Timour Q, Maul A, Piens MA, Picot S. Abnormalities in liver enzymes during simultaneous therapy with itraconazole and amphotericin B in leukaemic patients. J Antimicrob Chemother 2000;45(6):928–9.

197. Groll AH, Piscitelli SC, Walsh TJ. Clinical pharmacology of systemic antifungal agents: a comprehensive review of agents in clinical use, current investigational compounds, and putative targets for antifungal drug development. Adv Pharmacol 1998;44:343–500.

198. Maesaki S, Kohno S, Kaku M, Koga H, Hara K. Effects of antifungal agent combinations administered simultaneously and sequentially against *Aspergillus fumigatus*. Antimicrob Agents Chemother 1994;38(12):2843–5.

199. Sugar AM, Hitchcock CA, Troke PF, Picard M. Combination therapy of murine invasive candidiasis with fluconazole and amphotericin B. Antimicrob Agents Chemother 1995;39(3):598–601.

200. Thakur CP. Sodium antimony gluconate, amphotericin, and myocardial damage. Lancet 1998;351(9120):1928–9.

201. Bickers DR. Antifungal therapy: potential interactions with other classes of drugs. J Am Acad Dermatol 1994;31(3 Pt 2):S87–90.

202. Fisher MA, Talbot GH, Maislin G, McKeon BP, Tynan KP, Strom BL. Risk factors for amphotericin B-associated nephrotoxicity. Am J Med 1989;87(5):547–52.

203. Groll AH, Walsh TJ. Caspofungin: pharmacology, safety and therapeutic potential in superficial and invasive fungal infections. Expert Opin Investig Drugs 2001;10(8):1545–1558.

204. Hiddemann W, Essink ME, Fegeler W, Zuhlsdorf M, Sauerland C, Buchner T. Antifungal treatment by amphotericin B and 5-fluorocytosine delays the recovery of normal hematopoietic cells after intensive cytostatic therapy for acute myeloid leukemia. Cancer 1991;68(1):9–14.

205. Peters DH, Fitton A, Plosker GL, Faulds D. Tacrolimus. A review of its pharmacology, and therapeutic potential in hepatic and renal transplantation. Drugs 1993;46(4):746–94.

206. Pappas PG. Amphotericin B lipid complex in the treatment of invasive fungal infections: results of the Collaborative Exchange of Antifungal Research (CLEAR), an industry-supported patient registry. Clin Infect Dis 2005;40 Suppl 6:S379-83.

207. Chandrasekar PH, Ito JI. Amphotericin B lipid complex in the management of invasive aspergillosis in immunocompromised patients. Clin Infect Dis 2005;40 Suppl 6:S392-400.

208. Ito JI, Hooshmand-Rad R. Treatment of Candida infections with amphotericin B lipid complex. Clin Infect Dis 2005;40 Suppl 6:S384-91.

209. Martino R, Cortes M, Subira M, Parody R, Moreno E, Sierra J. Efficacy and toxicity of intermediate-dose amphotericin B lipid complex as a primary or salvage treatment of fungal infections in patients with hematological malignancies. Leuk Lymphoma 2005;46(10):1429-35.

Antifungal azoles

See also Antifungal azoles and other drugs for topical use, Genaconazole, Itraconazole, Ketoconazole, Miconazole, Posaconazole, Ravuconazole, Saperconazole, Voriconazole

General Information

The antifungal azoles are a class of synthetic compounds that have one or more azole rings and a more or less complex side chain attached to one of the nitrogen atoms. They are either imidazole or triazole derivatives. The imidazoles miconazole and ketoconazole were the first azoles developed for systemic treatment of human mycoses. However, severe adverse effects associated with the drug carrier (in the case of miconazole) and erratic absorption and significant interference with cytochrome P450 isozymes (in the case of ketoconazole) have limited their usefulness (1). However, the subsequently developed triazoles fluconazole, itraconazole, and voriconazole have become useful additions to the antifungal armamentarium. They have a wider spectrum of activity and greater target specificity and are generally well tolerated (1,2). Other azoles for topical use are reviewed in the monograph on Antifungal azoles and other drugs for topical use.

The azoles act by inhibiting the fungal enzyme lanosterol 14-α-demethylase, which is involved in the synthesis of ergosterol from lanosterol or 24-methylenedihydrolanosterol in the fungal cell membrane. The consequent inhibition of ergosterol synthesis originates from binding of the unsubstituted nitrogen (N-3 or N-4) of the imidazole or triazole moiety to the heme iron and from binding of their N-1 substituent to the apoprotein of a cytochrome P-450 (P-450(14)DM) of the endoplasmic reticulum (3). This inhibition interrupts the conversion of lanosterol to ergosterol, which alters cell membrane function.

Itraconazole has the highest affinity for the cytochrome and is about three and ten times more active in vitro than miconazole and fluconazole, respectively (4). They also inhibit the uptake of triglycerides and phospholipids through the cell membrane.

Drug–Drug Interactions

Mechanisms

Drug interactions with the antifungal azoles are common for several reasons:

- they are substrates of CYP3A4, but also interact with the heme moiety of CYP3A, resulting in non-competitive inhibition of oxidative metabolism of many CYP3A substrates; to a lesser extent they also inhibit other CYP450 isoforms;
- although fluconazole undergoes minimal CYP-mediated metabolism, it nevertheless inhibits CYP3A4 in vitro, albeit much more weakly than other azoles (5,6); however, fluconazole also inhibits several other CYP isoforms in vitro and interacts with enzymes involved in glucuronidation (7);
- interaction of antifungal azoles and other CYP3A substrates can also result from inhibition of P-glycoprotein-mediated efflux; P-glycoprotein is extensively co-localized and exhibits overlapping substrate specificity with CYP3A (7); in a cell line in which human P-glycoprotein was overexpressed, itraconazole and ketoconazole inhibited P-glycoprotein function, with 50% inhibitory concentrations of about 2 and 6 µmol/l respectively; however, fluconazole had no effect (8).
- the systemic availability of the antifungal azoles depends in part on an acidic gastric environment and the activity of intestinal CYP3A4 and P-glycoprotein.

For details of interactions with individual antifungal azoles, see individual monographs (fluconazole, itraconazole, ketoconazole, miconazole, and voriconazole).

Inhibition of metabolism by CYP3A4, and inhibition of transport by multidrug transporters Both were important in a boy with toxicity from a chemotherapeutic regimen containing drugs that are handled by these systems (9)

- A 14-year-old boy with Hodgkin's lymphoma was given vinblastine, doxorubicin, methotrexate, and prednisone chemotherapy and low-dose radiotherapy. When he was given itraconazole for a presumed fungal infection during an episode of neutropenia, unexpectedly severe bone marrow toxicity and neuropathy suggested toxicity from the chemotherapy due to enhancement by itraconazole. The itraconazole was withdrawn and the neutropenia and neuropathic pain improved.

The authors suggested that itraconazole had interfered with the metabolism of vinblastine, resulting in neurotoxicity, and with the metabolism of doxorubicin and methotrexate and the transport of doxorubicin, resulting in bone marrow suppression.

Posaconazole is an exception, since it is eliminated unchanged in the feces (10).

A novel mechanism whereby azoles may take part in drug interactions has been described (11). Drug metabolism is controlled by a class of orphan nuclear receptors that regulate the expression of genes such as CYP3A4 and MDR-1 (multi-drug resistance-1). Xenobiotic-mediated induction of CYP3A4 and MDR-1 gene transcription was inhibited by ketoconazole, which acted by inhibiting the activation of human pregnenolone X receptor and constitutive androstene receptor, which are involved in the regulation of CYP3A4 and MDR-1. The effect was specific to this group of nuclear receptors.

Frequency

To assess the frequency of potential drug interactions with azole derivatives and the consequences of interactions between fluconazole and other drugs in routine in-patient care, a retrospective cohort study of patients with systemic fungal infections treated with an oral or intravenous azole derivative was conducted in a tertiary-care hospital (12). Of the 4185 admissions in which azoles (fluconazole, itraconazole, or ketoconazole) were given, 2941 (70%) admissions involved potential drug interactions, and in 2716 (92%) there were potential interactions with fluconazole. The most frequent interactions that were potentially moderate or severe were co-administration of fluconazole with prednisone (25%), midazolam (18%), warfarin (15%), methylprednisolone (14%), ciclosporin (11%), and nifedipine (10%). Charts were reviewed for 199 admissions in which patients were exposed to potential fluconazole drug interactions. While four adverse events were attributed to fluconazole, none was thought to have been due to a drug-drug interaction, although in one instance fluconazole may have contributed. The authors concluded that although fluconazole drug interactions were very frequent they had few apparent clinical consequences.

Alfentanil

In a randomized crossover study in 12 healthy volunteers, oral voriconazole (400 mg twice on the first day and 200 mg twice on the second day) increased the AUC of intravenous alfentanil 20 micrograms/kg six-fold, reduced its mean plasma clearance by 85%, from 4.4 to 0.67 ml/minute/kg, and prolonged its half-life from 1.5 to 6.6 hours (13). Alfentanil caused nausea in five subjects and vomiting in two, all when they were taking voriconazole. The authors attributed this interaction to inhibition of CYP3A by voriconazole.

All-trans-retinoic acid

See Tretinoin.

Amphotericin

In evaluating possible antagonism between amphotericin and antifungal azoles, details of the experimental set-up are crucial. When filamentous fungi were exposed to subfungicidal concentrations of azoles, before exposure to an amphotericin + azole combination, antagonism could always be shown both in vitro and in vivo (14–16).

In vitro studies and experiments in animals have given conflicting results relating to potential antagonism between the effects of fluconazole and amphotericin on *Candida* species (15). However, large, randomized, double-blind comparisons of fluconazole with and without amphotericin for 5 days in non-neutropenic patients with candidemia showed no evidence of antagonism, but faster clearance of the organism from the blood and a trend toward an improved outcome in those who received the combination (17).

The combination of amphotericin with ketoconazole appears to lead to antagonism (14). A study of the effects of combinations of amphotericin with fluconazole, itraconazole, or ketoconazole against strains of *Aspergillus fumigatus* in vitro showed antagonistic effects in some strains, but different effects in other strains (18). In one group of mice infected with *Candida*, combinations of amphotericin with fluconazole were more effective than fluconazole alone; in another group the combination showed no interaction, but was not better than either drug given alone (19).

Although there are no clinical data, it can be expected that similar antagonism occurs between amphotericin and squalene oxidase inhibitors, which also eliminate the primary target ergosterol from the fungal cell membrane.

Anidulafungin

In a placebo-controlled study in 17 subjects anidulafungin (200 mg on day 1 then 100 mg/day on days 2–4) had no effect on the pharmacokinetics of voriconazole (400 mg every 12 hours on day 1 then 200 mg every 12 hours on days 2–4) (20). There were no dose-limiting or serious adverse events, and all adverse events were mild and consistent with the known safety profiles of the two drugs.

Antacids

The potential for a pH-dependent pharmacokinetic interaction between posaconazole 200 mg and the antacid Mylanta (co-magaldrox) 20 ml has been investigated under fasting and non-fasting conditions (21). in a randomized, four-period, crossover, single-dose study in 12 healthy men completed this. Food increased the relative systemic availability of posaconazole by 400%, but antacid co-administration had no statistically significant effect.

The effects of an antacid suspension (aluminium hydroxide 220 mg + magnesium hydroxide 120 mg in 240 ml) on the oral absorption of itraconazole 200 mg from capsules has been investigated in a randomized, open, two-period, crossover study in 12 healthy Thai men (22). The t_{max} of itraconazole was prolonged and its C_{max} and AUC were markedly reduced by the antacid, implying that the antacid markedly reduced the speed and extent of itraconazole absorption.

Antihistamines

The effects of co-administration of ketoconazole 400–450 mg/day on the pharmacokinetics of ebastine 20 mg/day and loratadine 10 mg/day and on the QT_c interval

have been evaluated in two placebo-controlled studies in healthy men (n = 55 and 62) (23). Neither ebastine nor loratadine alone altered the QT_c interval. Ketoconazole and placebo increased the mean QT_c by 6.96 ms in the ebastine study and by 7.52 ms in the loratadine study. Mean QT_c was statistically significantly increased during administration of both ebastine + ketoconazole administration (12.21 ms) and loratadine + ketoconazole (10.68 ms) but these changes were not statistically significantly different from the increases seen with placebo + ketoconazole (6.96 ms). Ketoconazole increased the mean AUC for ebastine 43-fold, and that of its metabolite carebastine 1.4-fold. It increased the mean AUC of loratadine 4.5-fold and that of its metabolite desloratadine 1.9-fold. No subjects withdrew because of electrocardiographic changes or drug-related adverse events. Thus, the larger effect of ketoconazole on the pharmacokinetics of ebastine was not accompanied by a correspondingly larger pharmacodynamic effect on cardiac repolarization.

Antiretroviral drugs

Indinavir is metabolized mainly by CYP3A4. There have been two randomized placebo-controlled studies in healthy men of the pharmacokinetic interactions, safety, and tolerance of voriconazole and indinavir (24). The first was an open parallel-group study of the effect of indinavir on the steady-state pharmacokinetics of voriconazole in 18 volunteers. The subjects took voriconazole 200 mg bd (days 1–7), then voriconazole 200 mg bd plus either indinavir 800 mg or placebo tds (days 8–17). The second was a double-blind, randomized, crossover study of the effect of voriconazole on the steady-state pharmacokinetics of indinavir in 14 volunteers, who took indinavir 800 mg tds + voriconazole 200 mg or placebo bd for two 7-day treatment periods separated by a washout period of at least 7 days. There was no important changes in the pharmacokinetics of either compound. Voriconazole co-administered with indinavir was well tolerated without serious adverse events.

However, voriconazole has reportedly interacted with other antiretroviral drugs.

- A 10-year-old girl (weight 21 kg: height 130 cm) with vertically acquired AIDS received antiretroviral combination therapy and died of liver failure after starting to take voriconazole (25). While taking amprenavir (22.5 mg/kg bd), didanosine (120 mg/m^2 bd), nevirapine (4 mg/kg bd), lopinavir (10 mg/kg bd), and ritonavir (2.5 mg/kg bd), she was given voriconazole 200 mg bd for refractory esophageal candidiasis. The next day her liver function tests rose slightly and rapidly deteriorated within 7 days, when voriconazole was withdrawn. Infectious causes were excluded. After 2 days the plasma concentrations of the antiretroviral drugs were increased (lopinavir, 10 μg/ml; nevirapine, 7.7 μg/ml; amprenavir, 10.9 μg/ml) compared with concentrations during the 6 months before admission (lopinavir, 3.9–6.0 μg/ml; nevirapine, 3.5–8.4 μg/ml; amprenavir, 3.5–7.7 μg/ml). There was no fever. She was alert and afebrile and neither had any neurological symptoms nor complained of pain. In the presence of progressive liver dysfunction, voriconazole and HAART were withdrawn. However, irreversible liver failure ensued, followed by hepatic coma. She dies 28 days after the start of voriconazole therapy. A postmortem was not performed.

The authors concluded that an interaction with HAART was the most likely explanation for the ultimately fatal liver failure.

Aripiprazole

Aripiprazole is mainly metabolized in vitro by CYP3A4 and CYP2D6. The effect of itraconazole 100 mg/day for 7 days on the pharmacokinetics of a single oral dose of aripiprazole 3 mg has been studied in 24 healthy adult men (26). Itraconazole increased the C_{max}, AUC, and terminal half-life of aripiprazole by 19%, 48%, and 19% respectively and of its main metabolite OPC-14857 by 19%, 39%, and 53%. Itraconazole reduced the oral clearance of aripiprazole in extensive metabolizers by 27%, with an even greater reduction (47%) in intermediate metabolizers. For C_{max}, there was no significant difference between extensive metabolizers and intermediate metabolizers, and the percent change by co-administration of itraconazole was less than 20% in both groups. For OPC-14857, the t_{max} in intermediate metabolizers was longer than that in extensive metabolizers, and the difference was amplified by itraconazole. The AUC was similarly affected by itraconazole in all genotypes. The urinary 6-beta-hydroxycortisol/cortisol concentration ratio was halved by itraconazole, consistent with inhibition of CYP3A4. However, the effect of CYP3A4 inhibition on the pharmacokinetics of aripiprazole was not thought to be clinically significant. On the other hand, there were definite differences in pharmacokinetics between CYP2D6 genotypes.

Atenolol

The effect of itraconazole 200 mg bd for 2 days on the pharmacokinetics of atenolol 50 mg has been investigated in 10 healthy volunteers in a randomized crossover study (27). Itraconazole increased the AUC of atenolol and the amount excreted in the urine by about 12%, suggesting a slight increase in systemic availability. However, it had no statistically significant effect on the pharmacodynamics of atenolol.

Benzodiazepines

Bromazepam has been reported to be metabolized by cytochrome P450, although the isozyme responsible has yet to be determined. The effects of itraconazole, an inhibitor of CYP3A4, on the pharmacokinetics and pharmacodynamics of bromazepam have been investigated in a double-blind, randomized, crossover study in eight healthy men who took itraconazole 200 mg/day for 6 days or placebo (28). On day 4 each subject took a single oral dose of bromazepam 3 mg and blood samples were taken for 70 hours. The time course of the pharmacodynamic effects of bromazepam on the central nervous system was assessed using a subjective rating of sedation, continuous number addition test, and electroencephalography up to 22 hours after bromazepam. Itraconazole caused no significant changes in the pharmacokinetics or pharmacodynamics of bromazepam, suggesting that CYP3A4 is not involved in the metabolism of bromazepam to a major extent and that bromazepam can be used in the usual doses in patients taking itraconazole.

Brotizolam

The effect of itraconazole on the single oral dose pharmacokinetics and pharmacodynamics of brotizolam has been investigated in a randomized, double-blind, crossover trial in 10 healthy men who had taken either itraconazole 200 mg/day or matched placebo for 4 days (29). Itraconazole significantly reduced the apparent oral clearance of brotizolam, increased its AUC, and prolonged its half-life. Itraconazole significantly increases plasma concentrations of brotizolam probably by inhibiting CYP3A4.

Carbamazepine

Fluconazole-induced carbamazepine toxicity has been reported (30).

- A 29-year-old woman taking carbamazepine 1600 mg/day, lamotrigine 400 mg/day, and barbexaclone 100 mg/day developed severe diplopia, oscillopsia, nausea, vomiting, and gait instability within several days after starting to take fluconazole 150 mg/day for tinea corporis. The carbamazepine concentration was 1.5 times above the usual target range. Within 24 hours after withdrawal of fluconazole, the neurological deficits had disappeared and the carbamazepine concentrations had returned to the target range.

This interaction was probably due to inhibition by fluconazole of CYP3A4 and/or CYP2C9, isoenzymes that are involved in the metabolism of carbamazepine.

Celiprolol

The effects of itraconazole on the pharmacokinetics of celiprolol has been investigated in a randomized crossover study in 12 healthy volunteers who took itraconazole 200 mg orally or placebo bd or grapefruit juice 200 ml tds for 2 days (31). On the morning of day 3, 1 hour after drug ingestion, each subject took celiprolol 100 mg with 200 ml of water (placebo and itraconazole phases) or grapefruit juice. During the itraconazole phase, the mean AUC from 0 to 33 hours of celiprolol was 80% greater than in the placebo phase. Cumulative urinary excretion of celiprolol was increased by itraconazole by 59%. Hemodynamic variables did not differ between the phases. Itraconazole almost doubles plasma celiprolol concentrations. This interaction probably results from increased availability of celiprolol, possibly as a result of inhibition of P glycoprotein in the intestine.

The effects of itraconazole 100 mg/day for 14 days on the pharmacokinetics of a single oral dose of quazepam and its two active metabolites have been studied in 10 healthy men in a double-blind, crossover, randomized, placebo-controlled study (32). Blood samplings and evaluation of psychomotor function by the Digit Symbol Substitution Test and Stanford Sleepiness Scale were conducted up to 240 hours after quazepam. Itraconazole did not change the kinetics of quazepam but significantly reduced the C_{max} and AUC of 2-oxoquazepam and N-desalkyl-2-oxoquazepam. Itraconazole did not affect psychomotor function.

Ciclosporin

The extent of the pharmacokinetic interaction between ciclosporin and itraconazole oral solution in eight renal transplant recipients and the effect on daily drug costs has been determined in a single-center, open, non-randomized study (33). After transplantation, renal transplant recipients received itraconazole solution 200 mg bd and ciclosporin to achieve target blood concentrations. At steady state blood samples were collected over 12 hours for pharmacokinetic evaluation of ciclosporin, itraconazole, and hydroxyitraconazole. Itraconazole was withdrawn after about 3 months. Ciclosporin doses were again titrated to achieve target blood concentrations and ciclosporin concentrations were once again determined at steady state. Mean peak and trough itraconazole concentrations were 1.64 and 1.23 µg/ml respectively. Mean peak and trough hydroxyitraconazole concentrations were 2.37 and 2.20 µg/ml respectively. Itraconazole caused a 48% reduction in the mean total daily dose of ciclosporin necessary to maintain target concentrations, 171 versus 329 mg). This reduction in ciclosporin dose resulted in a discounted itraconazole daily drug cost of about 30% while providing antifungal coverage with adequate itraconazole trough concentrations.

In a retrospective study of 102 children with steroid-dependent nephrotic syndrome, 78 received daily ketoconazole 50 mg dose and a reduced dose of ciclosporin and 24 received ciclosporin alone (34). The mean duration of treatment was 23 months. Co-administration of ketoconazole significantly reduced the mean doses of ciclosporin by 48%, with a net cost saving of 38%. It also resulted in significant improvement in the response to ciclosporin, increased success in the withdrawal of steroids, and a reduced frequency of renal impairment.

A 14-year-old girl with an allogeneic bone marrow transplant stopped taking voriconazole because of worsening liver function tests; the ciclosporin trough blood concentrations fell (35). This observation emphasizes the need for careful monitoring and dosage adjustments of ciclosporin in patients who take antifungal azoles.

The outcomes in renal transplant patients have been monitored using simultaneous ciclosporin C0 and C2 concentration measurements and in patients in whom only ciclosporin C2 concentrations were measured (36). The latter had higher ciclosporin C2 concentrations, AUCs, and drug doses during the immediate postsurgical period, and at 2 weeks and 4 and 6 months after transplantation. Six of the latter and none of the former had severe liver toxicity, characterized by jaundice and raised liver enzymes, with negative serological tests for CMV, HVC,

and HVB. There was a correlation between aspartate transaminase activity and ciclosporin C2 concentrations and both normalized at 15–55 days after ciclosporin dosage reduction. High ciclosporin C2 concentrations, which have been recommended when the drug is used alone in renal transplantation, cannot be used in patients taking ketoconazole, because C2 does not reflect drug exposure and high C2 concentrations can cause liver toxicity.

Cimetidine

The effect of itraconazole on the renal tubular secretion of cimetidine has been investigated in healthy volunteers who received intravenous cimetidine alone and after 3 days of oral itraconazole 400 mg/day (37). The cimetidine AUC increased by 25% after itraconazole. Glomerular filtration rate of cimetidine was unchanged, but secretory clearance was significantly reduced presumably due to inhibition of P glycoprotein.

Cyclophosphamide

Cyclophosphamide is a prodrug that is metabolized by CYP450 enzymes to cytotoxic alkylating species, and the extent of metabolism correlates with both efficacy and toxicity. In a randomized study of the safety and efficacy of itraconazole or fluconazole in preventing fungal infections in patients undergoing allogeneic stem cell transplantation, itraconazole (200 mg/day intravenously or 2.5 mg/kg orally tds) or fluconazole (400 mg/day intravenously or orally) were given with from start of conditioning therapy until at least 120 days after transplantation (38). After enrolment of the first 197 patients, a data and safety monitoring board reviewed the potentially drug-related adverse effects. Patients who had taken itraconazole had higher serum bilirubin and creatinine concentrations in the first 20 days after transplantation; the highest values were in patients who had taken itraconazole concurrently with cyclophosphamide conditioning. Analysis of cyclophosphamide metabolism in a subset of patients showed greater exposure to toxic metabolites (in particular 4-hydroxycyclophosphamide and 4-ketocyclophosphamide) among recipients of itraconazole compared with fluconazole. In contrast, those who took fluconazole had greater exposure to the unmetabolized drug. Adverse effects occurred preferentially in patients who had greater exposure to cyclophosphamide metabolites. These data suggest that azole antifungals, through differential inhibition of hepatic cytochrome P450 isozymes, affect cyclophosphamide metabolism and conditioning-related adverse effects after allogeneic stem cell transplantation.

In a randomized comparison of itraconazole (200 mg/day intravenously or 2.5 mg/kg tds orally) and fluconazole (400 mg/day intravenously or orally) in preventing fungal infections in patients undergoing allogeneic stem cell transplantation, those who received itraconazole developed higher serum bilirubin and creatinine concentrations in the first 20 days after transplantation, with the highest values in those who received concurrent cyclophosphamide (39). There was higher exposure to toxic metabolites (in particular 4-hydroxycyclophosphamide and 4-

ketocyclophosphamide) among recipients of itraconazole compared with fluconazole. In contrast, recipients of fluconazole had higher exposure to unmetabolized drug. Adverse effects occurred preferentially in those who had higher exposure to cyclophosphamide metabolites. These data suggest that azole antifungals, by differential inhibition of hepatic CYP isoenzymes, affect cyclophosphamide metabolism.

Agents that are frequently co-administered with cyclophosphamide in high-dose chemotherapy regimens were tested for inhibition of the activation of cyclophosphamide in human liver microsomes. The K_m and V_{max} of the conversion of cyclophosphamide to 4-hydroxycyclophosphamide were 93 μmol/l and 4.3 mg.nmol/hour respectively; itraconazole was inhibitory at an IC_{50} of 5 μmol/l, which is higher than the usual plasma itraconazole concentration and was thus considered of no clinical relevance (40).

Cytarabine

In vitro assays with itraconazole have shown that cytarabine is a substrate of CYP3A4 (41). Cytarabine and itraconazole inhibit CYP3A4. Cytarabine metabolism was significantly reduced when it was combined with itraconazole. Inhibition of cytarabine metabolism may have important clinical implications and warrants investigation in vivo.

Dapsone

The formation of dapsone hydroxylamine is thought to be the cause of the high rates of adverse reactions to dapsone in HIV-infected individuals. The effect of fluconazole on hydroxylamine formation in individuals with HIV infection has been investigated in 23 HIV-infected subjects (42). Fluconazole reduced the AUC, percent of dose excreted in urine in 24 hours, and formation clearance of the hydroxylamine by 49%, 53%, and 55% respectively. This inhibition of in vivo hydroxylamine formation was quantitatively consistent with that predicted from human liver microsomal experiments. Rifabutin had no effect on the plasma AUC of hydroxylamine or the percent excreted in the urine in 24 hours but increased formation clearance by 92%. Dapsone clearance was increased by rifabutin and rifabutin plus fluconazole (67% and 38% respectively) but was unaffected by fluconazole or clarithromycin. Hydroxylamine production was unaffected by clarithromycin. On the basis of these data, and assuming that exposure to dapsone hydroxylamine determines dapsone toxicity, we predict that co-administration of fluconazole should reduce the rate of adverse reactions to dapsone in people with HIV infection and that rifabutin and clarithromycin will have no effect. When dapsone is given in combination with rifabutin, dapsone dosage adjustment may be necessary.

Dexloxiglumide

Dexloxiglumide is a cholecystokinin CCK_1 receptor antagonist under investigation for functional gastrointestinal disorders; it is metabolized by CYP3A4 and CYP2C9.

The effect of steady-state ketoconazole on the pharmacokinetics of dexloxiglumide and its primary metabolite O-demethyldexloxiglumide has been studied in healthy subjects in a randomized, two-period, crossover study (43). Ketoconazole increased dexloxiglumide C_{max} by 32% without affecting the C_{max} of changed O-demethyldexloxiglumide and increased the AUC of dexloxiglumide and O-demethyldexloxiglumide by 36%. There were no changes in the half-lives of dexloxiglumide or O-demethyldexloxiglumide.

Digoxin

The effect of multiple-dose voriconazole on the steady-state pharmacokinetics of digoxin in healthy men has been studied in a double-blind, randomized, placebo-controlled study (44). All the subjects took oral digoxin for 22 days (0.5 mg bd on day 1, 0.25 mg bd on day 2 and 0.25 mg/day on days 3–22). On days 11–22 they were randomized to either voriconazole 200 mg bd or placebo. Voriconazole did not significantly alter the C_{max}, C_{min}, AUC, t_{max}, or clearance of digoxin at steady state. There were no significant differences in adverse events, all of which were classified as mild and transient.

Disopyramide

The effect of fluconazole on the heart, and the interaction of fluconazole with disopyramide has been investigated in chick White Leghorns embryos (45). The drugs were injected into the air sac of each fertilized egg: fluconazole 0.4, 0.8, and 1.2 mg/egg alone, disopyramide 0.3 mg/egg alone, or fluconazole 0.4 mg/egg + disopyramide 0.3 mg/egg. Fluconazole 0.4 mg/egg had no effect on heart rate, but heart rate fell significantly after 0.8 and 1.2 mg/egg. The heart rate also fell significantly after fluconazole 0.4 mg/egg + disopyramide 0.3 mg/egg and there was a cardiac dysrhythmia. These experiments suggest that concurrent administration of fluconazole and Class I antidysrhythmic drugs may increase the risk of cardiotoxicity by additive effects on QT prolongation.

Docetaxel

The effects of ketoconazole 200 mg/day for 3 days on the pharmacokinetics of docetaxel 100 mg/m^2, have been investigated in a randomized crossover study in seven patients with cancer (46). Ketoconazole co-administration resulted in a 49% reduction in the clearance of docetaxel. The docetaxel clearance ratio in the presence and absence of ketoconazole was weakly related to the AUC of ketoconazole. Inhibition of CYP3A4 by ketoconazole in vivo resulted in docetaxel clearance values that have previously been shown to be associated with a several-fold increase in the odds for febrile neutropenia at standard doses. Caution should be taken and substantial dosage reductions are required if docetaxel has to be administered together with potent inhibitors of CYP3A4.

Domperidone

Domperidone prolongs the QT interval and increases the risk of serious cardiac dysrhythmias. It is metabolized by

CYP3A4. In healthy volunteers ketoconazole increased domperidone C_{max} and AUC 3- to 10-fold (47). There was QT interval prolongation of about 10–20 msec when domperidone 10 mg qds was given with ketoconazole 200 mg bd but not with domperidone alone. This is a potentially dangerous combination.

Echinocandins

In rats, caspofungin did not alter the plasma pharmacokinetics of ketoconazole, a potent inhibitor of CYP3A4 (48). Co-administration of caspofungin 50 mg/day and itraconazole 200 mg/day to healthy subjects for 14 days did not alter the pharmacokinetics of either drug (49).

Etizolam

The effects of itraconazole 200 mg/day for 7 days on the single oral dose pharmacokinetics and pharmacodynamics of etizolam have been examined in 12 healthy men (50). Itraconazole significantly increased the total AUC and the half-life of etizolam, but had no effect on two pharmacodynamic measures, the Digit Symbol Substitution Test and Stanford Sleepiness Scale. The results suggested that itraconazole inhibits the metabolism of etizolam, providing evidence that CYP3A4 is at least partly involved.

Everolimus

The effect of ketoconazole 200 mg bd for 8 days on the pharmacokinetics of a single dose of everolimus 2 mg has been investigated in a two-period, single-sequence, crossover study in 12 healthy subjects (51). Ketoconazole increased the C_{max} of everolimus 3.9-fold and the AUC 15-fold and prolonged the half-life from 30 to 56 hours. Everolimus did not alter ketoconazole predose concentrations. Given the magnitude of this drug interaction, ketoconazole should be avoided if possible in patients taking everolimus.

Fentanyl

Unintentional overdose of fentanyl has been attributed to inhibition of its metabolism by ketoconazole (52).

- A 46-year-old man was given a Durogesic (fentanyl) transdermal patch 150 micrograms/hour for pain, plus morphine 10 mg/day, diclofenac 50 mg tds, paracetamol 1 g tds, oxazepam 15 mg bd, zolpidem 5 mg at night, nystatin, lidocaine oral spray, lactulose, and metoclopramide. Of these drugs, fentanyl, lidocaine, and paracetamol are partly metabolized by CYP3A4. He was then given fluconazole 50 mg/day and after 3 days died in his sleep. Forensic analysis of femoral blood showed a toxic concentration of fentanyl (0.017 µg/g), high concentrations of fluconazole (2.4 µg/g), lidocaine (1.6 µg/g), and metoclopramide (0.15 µg/g), and a therapeutic concentration of zolpidem (0.07 µg/g). No ethanol or drugs of abuse were identified. There were no pathological findings other than pulmonary congestion and brain edema.

The coroner's conclusion was that the cause of death was respiratory depression and circulatory failure due to fentanyl intoxication, since the concentration was within the range of other reported lethal intoxications, up to 48 ng/ml.

Fexofenadine

The effects of itraconazole on the pharmacokinetics and pharmacodynamics of a single oral dose of fexofenadine 180 mg have been investigated in relation to the multidrug resistance gene MDR1 in seven healthy subjects with the 2677GG/3435CC (G/C) haplotype and seven with the 2677TT/3435TT (T/T) haplotype (53). One hour before the dose of fexofenadine, either 200 mg itraconazole or placebo was given in a double-blind, randomized, crossover manner with a 2-week washout period. Histamine-induced wheal and flare reactions were measured to assess the effects on the antihistamine response. In the placebo phase there was no difference between the two MDR1 haplotypes in the pharmacokinetics of either fexofenadine or itraconazole. However, after itraconazole pretreatment the differences in fexofenadine pharmacokinetics became statistically significant; the mean fexofenadine AUC in the T/T group was significantly higher than that in the G/C group and the oral clearance in the T/T group was lower than in the G/C group. Itraconazole pretreatment caused more than a 3-fold increase in the peak concentration of fexofenadine and the AUC to 6 hours compared with placebo. This resulted in significantly greater suppression of the histamine-induced wheal and flare reactions in the itraconazole pretreatment phase compared with placebo. Thus, the effect of these MDR1 haplotypes on fexofenadine disposition is magnified in the presence of itraconazole. Itraconazole pretreatment significantly altered the disposition of fexofenadine and thus its peripheral antihistamine effects.

Flucytosine

Flucytosine has been successfully used in combination with ketoconazole, fluconazole, and itraconazole. Flucytosine and ketoconazole were synergistic in about 40% of yeast isolates resistant to flucytosine alone. The synergistic action of flucytosine with the triazoles against *Candida* species was seen both in vitro and in vivo (54–57).

Gefitinib

CYP3A4 is involved in the metabolism of gefitinib (Iressa, ZD1839). The in vitro metabolism of (14)-gefitinib 1–3 µmol/l has been investigated in human liver microsomes and a range of expressed human cytochrome P450 enzymes, with particular focus on the formation of O-desmethylgefitinib (M523595), the major metabolite in human plasma. Ketoconazole was used as a probe drug. While formation of M523595 was CYP2D6 mediated, the overall metabolism of gefitinib depended primarily on CYP3A4, and this was not obviously reduced in liver microsomes from CYP2D6 poor metabolizers (58).

When gefitinib 250 and 500 mg was administered in the presence of itraconazole, mean AUC increased significantly by 78% and 61% respectively (59). Although exposure to gefitinib is increased by co-administration with CYP3A4 inhibitors such as itraconazole, dosage reduction is not recommended due to the good tolerability profile of gefitinib.

Glucocorticoids

In allergic bronchopulmonary aspergillosis itraconazole and topical or systemic glucocorticoids are commonly co-administered. Itraconazole inhibits the metabolic clearance of glucocorticoids by inhibiting CYP3A4 and it also directly inhibits steroidogenesis, thereby causing serious adverse effects.

- A 4-year-old boy with cystic fibrosis developed Cushing's syndrome after taking itraconazole 100 mg bd and inhaled budenoside 200 micrograms bd for 2 weeks (60). Adrenal suppression was documented and persisted for 3 months after stopping this combined regimen.

This report is in line with a previous systematic assessment of the pituitary?adrenal axis in patients taking itraconazole and budenoside (61). In this study, an adrenocorticotrophic hormone (ACTH) test with tetracosactide 250 micrograms was performed in 25 patients with cystic fibrosis taking both itraconazole and budesonide, and in 12 patients taking itraconazole alone. ACTH tests performed as part of a pretransplantation program in another 30 patients with cystic fibrosis were used as controls. Of the 25 patients taking both itraconazole and budesonide, 11 had adrenal insufficiency. None of the patients taking itraconazole alone nor the control patients had an abnormal ACTH test. Furthermore, in a randomized, double-blind, crossover study in 10 healthy subjects (62), itraconazole increased the mean AUC of inhaled budesonide 4.2 (range 1.7–9.8) times and the C_{max} 1.6 times compared with placebo. The mean half-life of budesonide was prolonged from 1.6 to 6.2 hours by itraconazole. Suppression of cortisol production after inhalation of budesonide was significantly increased by itraconazole compared with placebo, with a 43% reduction in the plasma cortisol AUC from 0.5 to 10 hours and a 12% reduction in the cortisol concentration 23 hours after administration of budesonide. Thus, itraconazole markedly increases systemic exposure to inhaled budesonide. This interaction can result in enhanced systemic effects of budesonide, including Cushing syndrome.

- A 70-year-old white woman taking long-term high-dose inhaled budesonide for asthma developed Scedosporium apiospermum infection of the skin and subcutaneous tissues (63). She was given itraconazole for 2 months and developed Cushing's syndrome, probably due to a cytochrome P450-mediated interaction between itraconazole and budesonide. She also had secondary adrenal insufficiency requiring prolonged replacement with hydrocortisone.

The combination of itraconazole and inhaled glucocorticoids is increasingly being used to treat conditions such as allergic bronchopulmonary aspergillosis. Clinicians need to be aware of the potential for an interaction with such a combination.

Haloperidol

The combined effects of the CYP3A4 inhibitor itraconazole 200 mg/day for 10 days and the CYP2D6*10 genotype on the pharmacokinetics and pharmacodynamics of haloperidol 5 mg, a substrate of both CYP2D6 and CYP3A4, have been studied in 19 healthy subjects (nine CYP2D6*1/*1 and ten CYP2D6*10/*10) (64). Four subjects (one CYP2D6*1/*1 and three CYP2D6*10/*10) did not complete this randomized placebo-controlled crossover study because of adverse events. Itraconazole increased the mean AUC of haloperidol by 55%. The subjects with the CYP2D6*10/*10 genotype had 81% higher values of AUC than those with the CYP2D6*1/*1 genotype. In the presence of itraconazole, those with the CYP2D6*10/*10 genotype had a 3-fold higher AUC of haloperidol than placebo-treated subjects with the CYP2D6*1/*1 genotype. The CYP2D6*10/*10 genotype and itraconazole pretreatment reduced the oral clearance of haloperidol by 24% and 25% respectively and in combination by 58%. The Barnes Akathisia Rating Scale (BARS) in CYP2D6*10/*10 subjects during itraconazole treatment was significantly higher than in CYP2D6*1/*1 subjects during placebo. Thus, the moderate effect of the CYP2D6*10/*10 genotype on the pharmacokinetics and pharmacodynamics of haloperidol is augmented by the presence of itraconazole.

Histamine H_2 receptor antagonists

The effects of the histamine H_2 receptor antagonists cimetidine and ranitidine on the steady-state pharmacokinetics of voriconazole have been determined in an open, randomized, placebo-controlled, crossover study in 12 healthy men, who took oral voriconazole 200 mg + cimetidine 400 mg, voriconazole 200 mg + ranitidine 150 mg, and voriconazole 200 mg + placebo, all twice a day (65). Treatment periods were separated by at least 7 days. Co-administration of cimetidine increased the C_{max} and AUC of voriconazole by 18% (90% CI = 6, 32) and 23% (90% CI = 13, 33) respectively; ranitidine had no significant effect. Most of the adverse events were mild and transitory; two subjects withdrew because of adverse events (burning and pruritus of the scrotum during the placebo period and raised hepatic transaminases during the cimetidine period). Thus, co-administration of cimetidine or ranitidine does not affect the steady-state pharmacokinetics of voriconazole in an important manner.

HMG coenzyme-A reductase inhibitors

The antifungal azoles inhibit CYP isozymes and can therefore interact with some statins.

In a randomized, double-blind, crossover study in 12 healthy volunteers, fluconazole increased the plasma concentrations of fluvastatin and prolonged its elimination; the mechanism was probably inhibition of the CYP2C9-mediated metabolism of fluvastatin (66). Care should be taken if fluconazole or other potent inhibitors of CYP2C9 are given to patients using fluvastatin.

The effects of itraconazole, a potent inhibitor of CYP3A4, on the pharmacokinetics of atorvastatin, cerivastatin, and pravastatin have been evaluated in an open, randomized, crossover study in 18 healthy subjects who took single doses of atorvastatin 20 mg, cerivastatin 0.8 mg, or pravastatin 40 mg, with and without itraconazole 200 mg (67). Itraconazole markedly raised atorvastatin plasma concentrations (2.5-fold) and produced modest rises in the plasma concentrations of cerivastatin (1.3-fold) and pravastatin (1.5-fold). These results suggest that in patients taking itraconazole, cerivastatin or pravastatin may be preferable to atorvastatin.

Physicians should check for lipid-lowering drugs before treating elderly individuals with itraconazole (68). Susceptibility to this interaction varies from statin to statin, in that simvastatin is more affected than pravastatin (69). Concomitant use of simvastatin with itraconazole should be avoided, and the same holds true for atorvastatin (70). In another study, the blood concentration of fluvastatin was not significantly increased, whereas that of lovastatin was (71).

Ketoconazole can also cause rhabdomyolysis when taken with both lovastatin and simvastatin (72).

The effect of itraconazole on the pharmacokinetics of rosuvastatin has been studied in two double-blind, crossover, randomized, placebo-controlled studies in healthy men, who took itraconazole 200 mg/day for 5 days and on day 4 rosuvastatin 10 mg (n = 12) or 80 mg (n = 14) (73). After co-administration of itraconazole, the rosuvastatin AUC was increased by 28?39% and the C_{max} by 15?36%. These effects are unlikely to be of clinical relevance and support previous in vitro findings that CYP3A4 plays a minor role in the metabolism of rosuvastatin.

- An 83-year-old white man with a history of congestive heart failure and hyperlipidemia who was taking simvastatin 40 mg/day was given fluconazole. He developed severe muscle weakness and a markedly raised serum creatine kinase activity, which resolved after withdrawal of simvastatin and fluconazole (74).

Rhabdomyolysis in this case was probably caused by an interaction of simvastatin with fluconazole; alternative statins should be used if an antifungal triazoles is needed (SEDA-27, 281). Rhabdomyolysis has also been reported in a patient taking atorvastatin and fluconazole (75).

Ibuprofen

The effects of voriconazole and fluconazole on the pharmacokinetics of S(+)-ibuprofen and R(−)-ibuprofen have been studied in 12 healthy men, who took a single oral dose of racemic ibuprofen 400 mg in randomized order either alone or 1 hour after voriconazole or fluconazole 400 mg bd on day 1 and 200 mg bd on day 2 (76). Voriconazole increased the AUC of S-ibuprofen to 205% and the C_{max} to 122%; the half-life was prolonged from 2.4 to 3.2 hours. Fluconazole increased the AUC of S-ibuprofen to 183% and the C_{max} to 116%; the half-life was prolonged from 2.4 to 3.1 hours. These effects were attributed to inhibition of CYP2C9-mediated metabolism of S-ibuprofen. Voriconazole and fluconazole had minor effects on the pharmacokinetics of R-ibuprofen. The authors recommended that the dosage of ibuprofen should be reduced when it is co-administered with voriconazole or fluconazole, especially when the initial dose of ibuprofen is high.

Idarubicin

In vitro assays with itraconazole have shown that idarubicin is a substrate of CYP2D6 and CYP2C9 (54). Idarubicin inhibits CYP2D6, and itraconazole inhibits CYP3A4.

Imatinib

The effect of ketoconazole 400 mg on the pharmacokinetics of the tyrosine kinase inhibitor imatinib 200 mg has been investigated in a two-period, random, crossover study in 14 healthy subjects (13 men, 1 woman) (77). After ketoconazole co-administration, the mean imatinib C_{max} and AUC increased significantly by 26% and 40% respectively. There was a statistically significant reduction in the apparent clearance of imatinib, with a mean reduction of 29%. The mean C_{max} and AUC of the metabolite CGP74588 fell by 23% and 5% after ketoconazole. Co-administration of ketoconazole and imatinib caused a 40% increase in exposure to imatinib in healthy volunteers. Given its previously demonstrated safety profile, this increased exposure to imatinib is likely to be clinically significant only at high doses. This interaction should be considered when administering inhibitors of the CYP3A family in combination with imatinib.

A severe pustular eruption was associated with the concurrent use of voriconazole and imatinib in a patient with chronic myeloid leukemia (78). At the time of his skin eruption, the plasma concentrations of imatinib was raised. Imatinib is primarily metabolized by CYP3A4. Monitoring imatinib plasma concentrations may help in identifying patients at risk of severe toxicity.

Interleukin-6

Interleukin-6 can down-regulate the hepatic cytochrome P450 system and consequently alter drug disposition. The potential interaction of interleukin-6 (IL-6) with itraconazole has been studied using human hepatocytes in primary cultures from five adult men (mean age 42 years) who had not taken any medicines known to interact with CYP3A4 (79). The cultures were exposed to itraconazole 500 ng/ml, and the effects of cimetidine 120 µg/ml, human IL-6 50 ng/ml, or IL-6 plus IL-6 receptor antagonist were analysed for 2, 4, 8, and 12 hours. IL-6 did not inhibit hydroxyitraconazole formation.

Irinotecan

Ketoconazole inhibits the glucuronidation of the UGT2B7 substrates zidovudine and lorazepam, but its effect on UGT1A substrates is unclear. Co-administration of irinotecan and ketoconazole led to a significant increase in the formation of SN-38 (7-ethyl-10-hydroxy-camptothecin), a UGT1A substrate (26). The contribution of ketoconazole to SN-38 formation by inhibition of SN-38 glucuronidation has been studied in pooled human liver microsomes and cDNA-expressed UGT1A isoforms (1A1, 1A7, and 1A9). Indinavir, which inhibits UGT1A1, was used as a positive control (80). Ketoconazole competitively inhibited SN-38 glucuronidation. Among the UGT1A isoforms screened, ketoconazole showed the highest inhibitory effect on UGT1A1 and UGT1A9, with K_i values of 3.3 µmol/l for UGT1A1 and 32 µmol/l for UGT1A9. This may be the basis for increased exposure to SN-38 when ketoconazole is co-administered with irinotecan.

Lidocaine

Lidocaine is metabolized by CYP3A4 and CYP1A2 in vitro. However, their relative contributions to the elimination of lidocaine depend on the lidocaine concentration. The effect of itraconazole 200 mg/day on the pharmacokinetics of inhaled lidocaine 1.5 mg/kg has been investigated in 10 healthy volunteers using a randomized, two-phase, crossover design (81). The AUC of lidocaine and its major metabolite monoethylglycinexylidide were similar during both phases. There were no statistically significant differences in any of the pharmacokinetics of lidocaine: C_{max}, t_{max}, or half-lives of lidocaine or monoethylglycinexylidide. The clinical implication of this study is that no lidocaine dosage adjustments are necessary if it is used to prepare the airway before endoscopic procedures or intubation in patients using itraconazole or other inhibitors of CYP3A4.

Loperamide

Loperamide is metabolized by CYP2C8 and CYP3A4 and is a substrate of P glycoprotein. Itraconazole inhibits CYP3A4 and P-glycoprotein and gemfibrozil inhibits CYP2C8. In a randomized crossover study 12 healthy volunteers took itraconazole 100 mg bd, gemfibrozil 600 mg bd, both itraconazole and gemfibrozil, or placebo for 5 days (82). On day 3 they took a single dose of loperamide 4 mg. Itraconazole increased loperamide C_{max} 2.9-fold and the AUC 3.8-fold and prolonged the half-life from 12 to 19 hours. Gemfibrozil increased the C_{max} of loperamide 1.6-fold and its AUC 2.2-fold and prolonged its half-life to 17 hours. The combination of itraconazole and gemfibrozil increased the C_{max} of loperamide 4.2-fold and its AUC 13-fold and prolonged the half-life to 37 hours. The amount of loperamide excreted into urine within 48 hours was increased 1.4-fold, 3.0-fold, and 5.3-fold by gemfibrozil, itraconazole, and the combination respectively, and the plasma $AUC_{0\rightarrow72}$ ratio of N-desmethyl-loperamide to loperamide was reduced by 46, 65, and 88%. There were no significant differences in the Digit Symbol Substitution Test or subjective drowsiness between treatments.

Loratadine

The effects of ketoconazole 600 mg on the pharmacokinetics of loratadine in two oral formulations have been studied in 32 healthy volunteers in an open, randomized, two-period, crossover study (83). The speed and extent of absorption of loratadine was not affected by ketoconazole.

Loratadine and its metabolite desloratadine are metabolized not only by CYP3A4 but also by CYP2D6. Therefore, administration of loratadine with inhibitors of CYP3A4 does not cause such severe adverse effects as with terfenadine and astemizole. Nevertheless, severe hepatotoxicity after co-administration of desloratadine and fluconazole has been reported.

- A 38-year-old woman with cancer was given intravenous fluconazole 400 mg while taking desloratadine 10 mg/day, clemastine, allopurinol, ranitidine, lorazepam, levofloxacin, spironolactone, and filgrastim, and developed a sudden rise in hepatic transaminases (84). Her drugs were withdrawn and her hepatic transaminases normalized within 1 week. She had received both fluconazole and desloratadine on separate occasions and had tolerated both drugs well.

Lumiracoxib

In a two-way crossover study of the effect of fluconazole on the pharmacokinetics and action of lumiracoxib in 13 healthy subjects, fluconazole caused a small (18%) but not clinically relevant increase in lumiracoxib mean AUC and had no effect on lumiracoxib mean C_{max} (85). The fall in thromboxane B_2 from predose was not affected by fluconazole. As fluconazole is a potent inhibitor of CYP2C9, other CYP2C9 inhibitors are unlikely to affect the pharmacokinetics of lumiracoxib, making dosage adjustment unnecessary.

Macrolide antibiotics

The effects of multiple-dose erythromycin or azithromycin on the steady-state pharmacokinetics of voriconazole have been investigated in an open, randomized study in 30 healthy men aged 20–41 years, who took oral voriconazole 200 mg bd for 14 days plus erythromycin (1 g bd on days 8–14), or azithromycin (500 mg/day on days 12–14), or placebo (twice daily on days 8–14) (86). There were no significant interactions. The most common study drug-related adverse events were visual disturbances (17/30 patients), reported in all groups, and abdominal pain in the voriconazole + erythromycin group (5/10 patients).

Mefloquine

The effect of ketoconazole 400 mg/day for 10 days on the plasma concentrations of a single oral dose of mefloquine 500 mg has been studied in an open, randomized, two-phase, crossover study in eight healthy Thai men (87). Ketoconazole increased mefloquine AUC, half-life, and C_{max} by 79%, 39%, and 64% respectively. The AUC and

C_{max} of mefloquine's carboxylic acid metabolite were reduced by 28% and 31% respectively.

Meglitinides

The effects of fluconazole 200 mg/day for 4 days on the pharmacokinetics and pharmacodynamics of a single dose of nateglinide 30 mg have been investigated in a double-blind, randomized, crossover study in 10 healthy volunteers (88). Fluconazole increased the AUC of nateglinide by 48% (range 20?73%) and prolonged its half-life from 1.6 to 1.9 hours, but did not alter C_{max}. The C_{max} of the M7 metabolite of nateglinide was reduced by 34% by fluconazole and its half-life was prolonged from 2.2 to 3.5 hours. However, fluconazole did not alter the blood glucose responses to nateglinide.

Possible interactions of gemfibrozil, itraconazole, and their combination with repaglinide have been investigated in a randomized crossover study in 12 healthy volunteers (89). They took gemfibrozil 600 mg bd, itraconazole 100 mg bd (first dose 200 mg), both gemfibrozil and itraconazole, or placebo for 3 days and then took repaglinide 0.25 mg. Plasma drug and blood glucose concentrations were followed for 7 hours and serum insulin and C peptide concentrations for 3 hours. Gemfibrozil increased the AUC of repaglinide 8.1 (range 5.5?15) times and prolonged its half-life from 1.3 to 3.7 hours. Although itraconazole alone increased repaglinide AUC only 1.4 (1.1?1.9) times, the combination of gemfibrozil + itraconazole increased it 19 (13?25) times and prolonged the half-life of repaglinide to 6.1 hours. The plasma repaglinide concentration at 7 hours was increased 29 times by gemfibrozil and 70 times by the combination of gemfibrozil + itraconazole. Gemfibrozil alone and in combination with itraconazole considerably enhanced and prolonged the blood glucose-lowering effect of repaglinide. Concomitant use of gemfibrozil and repaglinide is therefore best avoided.

Methadone

The pharmacokinetic interaction of voriconazole with methadone 30–100 mg/day at steady state has been studied in 23 men (90). Voriconazole increased steady-state exposure to (R)-methadone, the pharmacologically active enantiomer: the mean $AUC_{0 \rightarrow 24}$ increased by 47% (90% CI = 38, 57%), and the mean C_{max} increased by 31% (90% CI = 22, 40%). The magnitude of increase in (S)-methadone exposure was even greater: the $AUC_{0 \rightarrow 24}$ increased by 103% (90% CI = 85, 124%) and the C_{max} by 65% (90% CI = 53, 79%). Methadone had no effect on the steady-state pharmacokinetics of voriconazole.

Moxifloxacin

The effect of itraconazole 200 mg/day on the pharmacokinetics of moxifloxacin has been studied in 12 healthy men (91). There was no effect in the systemic availability of moxifloxacin or concentrations of its sulfated metabolite, but there was a 30% reduction in the AUC of moxifloxacin glucuronide and an approximately 54% increase in renal excretion, which may have been due to changes in

phase 2 metabolism and/or transport mechanisms by itraconazole. Exposure (AUC) to itraconazole and its hydroxylated metabolite were not significantly altered by moxifloxacin.

Nevirapine

Adverse events that occurred after initiation of nevirapine-based antiretroviral therapy have been investigated in HIV-infected Thai patients who did not receive fluconazole (group A, n = 225) or who received fluconazole 400 mg/week (group B, n = 392) or 200 mg/day (group C, n = 69) in a retrospective 6-month cohort study (92). The incidences of hepatitis were 2/225 (0.9%), 4/392 (1.0%), and 0/69 respectively; there were no significant differences in the frequencies of raised transaminases across the groups. Fluconazole treatment did not predict hepatitis, raised transaminases, or skin rashes. At 6 months after initiating nevirapine, 77–84% of patients were still taking it.

Nitrofurantoin

Combined pulmonary and hepatic toxicity was reportedly precipitated by acute use of fluconazole concomitantly with chronic nitrofurantoin (93).

• A 73-year-old white man who had taken nitrofurantoin 50 mg/day for 5 years developed combined hepatic and pulmonary toxicity after taking fluconazole for onychomycosis. His hepatic enzymes rose to 5 times the upper limits of the reference range and he reported fatigue, dyspnea on exertion, pleuritic pain, burning tracheal pain, and a cough. Chest X-rays showed bilateral pulmonary disease consistent with nitrofurantoin toxicity. Both drugs were withdrawn and the hepatic and pulmonary toxicity resolved with inhaled glucocorticoids.

While either drug may have caused abnormal liver function tests, it is possible that pharmacokinetic changes induced by an interaction with fluconazole precipitated the nitrofurantoin-induced pulmonary toxicity by an unknown mechanism.

Omeprazole

Itraconazole oral solution has improved systemic availability and reduced pH dependency compared with the capsule formulation. The effects of pharmacologically induced gastric hypoacidity with omeprazole on the pharmacokinetics of the oral solution have been investigated in a randomized, open, prospective, crossover study in 15 healthy, non-pregnant adults, who took a single dose of itraconazole oral solution 400 mg on two occasions, at least 7 days apart, with omeprazole 40 mg nightly for 7 days before one of the doses of itraconazole (94). Omeprazole did not significantly affect the C_{max}, t_{max}, or $AUC_{0?8}$ of itraconazole or hydroxyitraconazole following the administration of the cyclodextrin solution of itraconazole. A more than 50% reduction in the $AUC_{0?24}$ of itraconazole has been observed when omeprazole was given concomitantly with the capsule formulation (95) Thus, when both drugs have to be given concomitantly, itraconazole has to be administered In from of the cyclodextrin solution.

Omeprazole is predominantly metabolized by CYP2C19 and CYP3A4. The effects of omeprazole on the steady-state pharmacokinetics of voriconazole have been investigated in an open, randomized, placebo-controlled, crossover study in 18 healthy men, who took oral voriconazole 400 mg bd on day 1 followed by 200 mg bd on days 2?9 and a single dose of 200 mg on day 10, with either omeprazole 40 mg/day or placebo for 10 days (96). Co-administration of omeprazole increased the mean C_{max} and AUC of voriconazole by 15% (90% CI = 5, 25) and 41% (90% CI = 29, 55) respectively, with no effect on t_{max}. One subject withdrew from the study during the voriconazole + omeprazole treatment period because of treatment-related abnormal liver function tests. All other treatment-related adverse events resolved without intervention. There were visual adverse events in 20 of 35 treatment episodes; the median times to onset for these events were 18 and 35 minutes, and the median durations were 28 and 15 minutes with and without omeprazole respectively. Omeprazole had no important effect on voriconazole exposure, suggesting that no voriconazole dosage adjustment is necessary for patients in whom omeprazole therapy is initiated.

Paclitaxel

In guinea-pigs ketoconazole reduced the cumulative biliary excretion of paclitaxel and its metabolites up to 6 hours by 62% (97).

Phenytoin

Phenytoin induces CYP3A4 activity and is a substrate and inducer of CYP2C9 and CYP2C19. There have been two placebo-controlled studies in healthy men of the pharmacokinetic interaction of voriconazole with phenytoin (98). The first was an open study of the effect of phenytoin 300 mg/day on the steady-state pharmacokinetics of voriconazole 200 mg bd and 400 mg bd. The second was a double-blind randomized study of the effects of voriconazole 400 mg bd on the steady-state pharmacokinetics of phenytoin 300 mg/day. Phenytoin reduced the mean steady-state C_{max} and AUC of voriconazole by about 50% and 70% respectively; increasing the dose of voriconazole from 200 mg to 400 mg bd compensated for this effect. Voriconazole 400 mg bd increased the mean steady-state C_{max} and AUC of phenytoin by about 70% and 80% respectively. Plasma phenytoin concentrations should therefore be monitored and the dose adjusted as appropriate when phenytoin is co-administered with voriconazole.

Quinolone antibiotics

Torsade de pointes has been associated with the use of fluconazole plus levofloxacin (99). While there have been reports that fluconazole and levofloxacin can cause QT interval prolongation when given alone, co-administration may further increase the risk.

Ranolazine

The interactions of ranolazine, a new antianginal compound, with inhibitors and substrates of the CYP3A isoenzyme family have been studied in an open study and in four double-blind, randomized, multiple-dose studies in healthy adults. Ketoconazole increased ranolazine plasma concentrations and reduced the CYP3A4-mediated metabolic transformation of ranolazine, confirming that CYP3A4 is the primary metabolic pathway for ranolazine (100).

Rifampicin

The effect of rifampicin on the pharmacokinetics of fluconazole and on clinical outcomes of fluconazole treatment in patients with AIDS-related cryptococcal meningitis have been studied in 40 Thai patients with AIDS and cryptococcal meningitis, of whom 20 had been taking oral rifampicin for at least 2 weeks to treat tuberculosis (101). Concomitant administration of rifampicin with fluconazole resulted in significant changes in the pharmacokinetics of fluconazole, including a 28% shorter half-life, a 22% reduction in AUC, a 17% reduction in C_{max}, and a 30% increase in clearance. Different fluconazole regimens did not affect the extent of change in half-life. Although serum concentrations of fluconazole during the time that patients took rifampicin + fluconazole 200 mg/day were generally lower than the minimum inhibitory concentration for Cryptococcus neoformans, there were no significant differences in clinical outcomes between the two groups. Co-administration of rifampicin with fluconazole caused significant changes in the pharmacokinetics of fluconazole, and long-term monitoring for recurrent cryptococcal meningitis is required to assess the clinical significance of this interaction.

Risperidone

The effects of itraconazole 200 mg/day for 1 week on the plasma concentrations of risperidone 2–8 mg/day and its active metabolite 9- hydroxyrisperidone have been investigated in 19 patients with schizophrenia in relation to CYP2D6 genotype (102). Dose-normalized plasma concentrations of risperidone and 9-hydroxyrisperidone were significantly increased by itraconazole and fell 1 week after withdrawal. However, the ratio of risperidone/9-hydroxyrisperidone, an index of CYP2D6 activity, was not altered. Itraconazole significantly increased the concentrations of risperidone by 69% and 75% in CYP2D6 extensive and poor metabolizers respectively; concentrations of risperidone plus 9-hydroxyrisperidone increased to a similar extent without a significant difference between CYP2D6 genotypes. There were no major pharmacodynamic effects. Thus, concentrations of both risperidone and 9-hydroxyrisperidone were significantly increased by the CYP3A inhibitor itraconazole, and this was independent of CYP2D6 activity, providing evidence that CYP3A is involved in the metabolism of risperidone and its metabolite.

Ritonavir

In a randomized, placebo-controlled crossover study in 20 healthy subjects, stratified according to CYP2C19 genotype, the apparent oral clearance of voriconazole after a single oral dose was 26% lower in CYP2C19*1/*2 individuals and 66% lower in CYP2C19 poor metabolizers (103). The addition of ritonavir reduced voriconazole apparent oral clearance (from 354 to 202 ml/minute); this occurred in all CYP2C19 genotypes (463 versus 305 ml/minute for CYP2C19*1/*1; 343 versus 190 ml/minute for CYP2C19*1/*2; 158 versus 22 ml/minute for CYP2C19*2/*2).

Saint John's wort

The short-term and long-term effects of Saint John's wort (300 mg of LI 160 tds) on the pharmacokinetics of a single oral dose of voriconazole 400 mg have been investigated in a controlled, open study in 16 healthy men stratified for CYP2C19 genotype (104). During the first 10 hours of the first day of administration of St John's wort, the AUC of voriconazole increased by 22% compared with control, but after 15 days the AUC was reduced by 59%, with a corresponding increase in oral voriconazole clearance. The baseline oral voriconazole clearance and the absolute increase in oral clearance were smaller in carriers of one or two deficient CYP2C19*2 alleles compared with wild-type individuals. Thus, co-administration of St John's wort leads to a short-term but clinically irrelevant increase followed by a prolonged extensive reduction in voriconazole exposure; CYP2C19 wild-type individuals may be at highest risk of potential voriconazole treatment failure.

Sildenafil

The effects of a single dose of sildenafil 3, 15, and 30 mg/kg and combined sildenafil + itraconazole 100 mg/kg on blood pressure, heart rate, and QT interval have been investigated in conscious beagle dogs (105). There were no changes in blood pressure. Sildenafil 15 and 30 mg/kg increased heart rate from 0.5 to 6 hours after the dose and shortened the QT interval; these effects were significantly enhanced by itraconazole. This was attributed to inhibition of CYP3A4. Caution should therefore be taken when sildenafil is co-administered with itraconazole.

Sirolimus

Sirolimus is metabolized by CYP3A4 and is a substrate of P glycoprotein, both of which are inhibited by drugs like voriconazole, itraconazole, and fluconazole (106).

- A 60-year-old patient taking sirolimus 2 mg/day after renal transplantation received fluconazole 100 mg/day, and after 22 days sirolimus trough plasma concentrations had increased four-fold without clinically overt adverse effects. The dosages of both drugs were reduced and tapered to achieve sirolimus trough concentrations within the recommended target range.

This case demonstrates that it is essential to monitor the blood sirolimus concentrations and to adjust the sirolimus doses before and after co-administration of fluconazole and other antifungal triazoles.

Concomitant treatment with voriconazole and sirolimus in two renal transplant recipients reduced sirolimus dosage requirements by 75–88% (107).

In a review of the medical records of all recipients of allogeneic hemopoietic stem cell, 11 patients had received voriconazole and sirolimus concomitantly for a median of 33 (range 3–100) days (108). In eight patients whose sirolimus dosage was initially reduced by 90%, trough sirolimus concentrations were similar to those obtained before the administration of voriconazole; there were no adverse effects attributable to either drug during co-administration, but there were serious adverse events in two patients in whom sirolimus dosages were not adjusted during voriconazole administration.

In a recipient of a stem cell transplant, a 20-year-old African–American man, co-administration of itraconazole 200 mg bd with sirolimus increased trough sirolimus concentrations to over 17.5 ng/ml (usual target range 5–15 ng/ml) (109). Sirolimus was withheld and the sirolimus trough concentration fell to 4.4 ng/ml.

An interaction between itraconazole and sirolimus has been reported in a primary renal allograft recipient (110).

Solifenacin

In an open crossover study in 17 healthy subjects aged 18–65 years, oral ketoconazole 200 mg/day prolonged the half-life of a single oral dose of solifenacin 10 mg from 49 to 78 hours and increased the C_{max} 1.43 times and the AUC about two-fold (111). Solifenacin is metabolized by CYP3A4, which is inhibited by ketoconazole.

Statins

See HMG coenzyme-A reductase inhibiters

Sulfamethoxazole

The formation of sulfamethoxazole hydroxylamine, in combination with long-term oxidative stress, is thought to be the cause of high rates of adverse drug reactions to sulfamethoxazole in subjects infected by HIV. The effect of fluconazole on sulfamethoxazole hydroxylamine formation in individuals with HIV-1 infection has been investigated in a two-part open drug interaction study in 21 subjects (112). Fluconazole reduced the AUC, the percent of the dose excreted in urine in 24 hours, and formation clearance of the hydroxylamine by 37%, 53%, and 61% respectively. Rifabutin increased the AUC, percent excreted, and formation clearance of the hydroxylamine by 55%, 45%, and 53% respectively. Fluconazole plus rifabutin reduced the AUC, percent excreted, and formation clearance of the hydroxylamine by 21%, 37%, and 46% respectively. Clarithromycin had no effect on hydroxylamine production. If exposure to sulfamethoxazole hydroxylamine predicts sulfamethoxazole toxicity, rifabutin will increase and clarithromycin plus fluconazole or rifabutin plus fluconazole should reduce the rates of adverse reactions to sulfamethoxazole in HIV-infected subjects.

Tacrolimus

Tacrolimus concentrations and dosage requirements have been compared before and during azole therapy (fluconazole or itraconazole) in 31 pediatric thoracic transplant patients (113). The dose of tacrolimus was empirically reduced by about one-third when azole therapy was begun. Mean tacrolimus dosage requirements fell by 68% within the first month of therapy (before azole therapy 0.27 mg/kg/day; 30 days after azole therapy 0.087 mg/kg/day). Despite mean reductions in tacrolimus dosage from baseline of 33%, 42%, and 55% on days 1, 2, and 4 of azole therapy respectively, there was still an unintended 38% increase in tacrolimus concentrations during the first month of azole therapy. There was no difference in tacrolimus dosage reduction between fluconazole and itraconazole. Azole antifungals markedly reduce tacrolimus requirements within the first few days of therapy. An initial reduction in tacrolimus dose by one-third may be insufficient, and a dosage reduction of at least 50% appears to be warranted. Once azole antifungal therapy is begun, frequent monitoring is required.

- A 40-year-old Asian woman received a cadaveric renal transplant for end-stage renal disease due to IgA nephropathy and was given tacrolimus, thymoglobulin, mycophenolate mofetil, and prednisone, along with diltiazem for hypertension (114). On postoperative day 5, donor bronchoalveolar lavage revealed active tuberculosis. She was given rifampicin 600 mg/day, and the dose of diltiazem was increased. Over the next 12 days, the dose of tacrolimus was increased to 32 mg/day to achieve a target trough concentration of 10–15 ng/ml. She then received a course of fluconazole 100 mg/day and clarithromycin 1000 mg/day. Despite this, there was no increase in tacrolimus concentrations. Rifampicin was withdrawn, after which therapeutic tacrolimus concentrations were finally reached with usual doses.

Rifampicin is a potent inducer of tacrolimus metabolism, sufficient to overcome the inhibitory effects of diltiazem, fluconazole, and clarithromycin.

The manufacturers of voriconazole recommend reducing the daily dosage of tacrolimus by one-third when it is co-administered with voriconazole.

- A 44-year-old liver transplant recipient taking a stable maintenance dosage of tacrolimus was given voriconazole for coccidioidomycosis. The dosage of tacrolimus was reduced by one-third, but over the next 10 days, the dosage was reduced to one-tenth of the starting dosage in order to maintain tacrolimus blood concentrations within the target range of 5–15 ng/ml.

This case emphasizes that blood tacrolimus concentrations should be carefully monitored when voriconazole is coadministered with tacrolimus or when voriconazole is discontinued in a patient receiving both drugs together (115).

In a randomized study in 70 live-donor kidney transplant recipients the addition of ketoconazole 100 mg/day to tacrolimus therapy reduced the effective dose of the latter by 54% and the cost by 53%; there was also significant improvement in graft function (116). No adverse effects of ketoconazole were noted.

The interaction of itraconazole with tacrolimus in lung transplant recipients and the efficacy of itraconazole prophylaxis has been analysed in 40 patients who took prophylactic itraconazole 200 mg bd for the first 6 months after transplantation (117). The mean dose of tacrolimus during itraconazole treatment was 3.26 mg/day compared with 5.74 mg/day (76% higher) after itraconazole was stopped. There were no differences in the rejection or fungal infection rates or in renal toxicity between the periods with and without itraconazole, although fewer positive fungal isolates were identified during itraconazole therapy.

Telithromycin

Itraconazole 200 mg/day increased the steady-state AUC of telithromycin 800 mg/day in a non-randomized, sequential, multiple-dose study in 34 healthy men (118).

The effect of ketoconazole 400 mg/day for 5 days on the pharmacokinetics and pharmacodynamics (effect on the QT_c interval) of telithromycin 800 mg/day have been investigated using clarithromycin as a comparator in 32 subjects aged 60 years or over with renal impairment (119). In those with creatinine clearances of 30–80 ml/minute ketoconazole increased telithromycin plasma concentrations to an extent similar to that for clarithromycin. There was no clinically significant prolongation of the QT_c interval.

Tretinoin (all-trans-retinoic acid, ATRA)

Tretinoin toxicity thought to be secondary to an interaction with fluconazole has been reported.

- A 4-year-old boy with acute promyelocytic leukemia underwent induction chemotherapy including all-trans-retinoic acid 45 mg/m^2/day divided into two doses. After an episode of fever and granulocytopenia on day 20 he was given fluconazole 100 mg/day for antifungal prophylaxis and 7 days later developed headache, vomiting, and papilledema. A CT scan of the brain was normal as was a lumbar puncture, except for a raised opening pressure of over 200 mm of fluid, and a diagnosis of pseudotumor cerebri was made. All-trans-retinoic acid was withdrawn and all his symptoms resolved within 24 hours. A few days later, he was rechallenged with all-trans-retinoic acid but only tolerated a dosage of 30% of the original dose until withdrawal of fluconazole, when he was able to tolerate the full target dosage of all-trans-retinoic acid of 45 mg/m^2/day.

All-trans-retinoic acid is hepatically metabolized by CYP2C8, CYP2C9, and CYP3A4. In this case, fluconazole, which inhibits CYP 2C9 and CYP 3A4, may increase exposure to all-trans-retinoic acid, resulting in cerebral adverse events (120).

Tretinoin rarely causes hypercalcemia, but another case has been reported and attributed to inhibition of CYP3A4-mediated metabolism of tretinoin by voriconazole (121).

Tricyclic antidepressants

Interactions between tricyclic antidepressants and fluconazole are rare; only five published reports can be found.

- A 44-year-old woman became progressively drowsy and unresponsive and then delirious (122). Her medications included metoprolol 50 mg bd, extended-release isosorbide mononitrate 60 mg/day, and amitriptyline 200 mg/day for fibromyalgia. Four days before admission, she was given fluconazole 100 mg/day for oral candidiasis. The combined serum concentration of amitriptyline + nortriptyline was 956 ng/ml (usual target range 150?250 ng/ml), and an electrocardiogram showed QT_c interval prolongation to 493 ms. A CT scan of the head was normal. Amitriptyline was withdrawn and her delirium resolved within 24 hours. Her serum amitriptyline concentration fell to 190 ng/ml, and her electrocardiogram became normal. She and her husband denied accidental or intentional overdose of amitriptyline.

Amitriptyline is oxidatively metabolized In the liver by CYP3A4, CYP2C9, CYP2C19, and CYP2D6; it is likely that fluconazole inhibited the demethylation of amitriptyline by CYP3A4 and CYP2C19, leading to central anticholinergic toxicity.

Vinca alkaloids

Concomitant use of itraconazole can cause unusually severe neurotoxicity of vincristine.

- An 8-year-old boy with acute lymphoblastic leukemia developed status epilepticus and inappropriate antidiuretic hormone secretion while taking both itraconazole and vincristine (123).
- A 2-year-old boy with acute lymphoblastic leukemia developed paraparesis associated with symmetrical bilateral demyelinating changes on an MRI scan of the brain after only three weekly doses of vincristine while taking itraconazole (123).
- A 3-year-old boy with acute lymphoblastic leukemia received induction chemotherapy (124). On day 14, itraconazole 5 mg/kg was begun and 10 days later he developed paralytic ileus, neurogenic bladder, mild left ptosis, and absence of deep reflexes, with severe paralysis of the legs and mild weakness of the arms. Itraconazole withdrawal was followed by rapid improvement to normality within 6 weeks.

The interaction between these two drugs is dose-related. The mechanisms of this interaction have not been formally elucidated, but probably include either competitive inhibition of the oxidative metabolism of vincristine, leading to increased systemic exposure, or alternatively inhibition of the transmembrane P glycoprotein efflux pump, leading to an increased intracellular concentration of vincristine. The concomitant use of itraconazole and all vinca alkaloids should be contraindicated (SEDA-26, 308).

Warfarin

The effect of voriconazole 300 mg bd on the pharmacodynamics of a single oral dose of warfarin 30 mg has been investigated in a double-blind, crossover, placebo-controlled study, in healthy men (125). Both the mean maximum change from baseline prothrombin time and the mean area under the effect curve for prothrombin time during co-administration with voriconazole (17 seconds and 3211 second.hours respectively) were statistically significantly greater than the mean values observed during the placebo period (8 seconds and 2282 second.hours). Prothrombin times were still prolonged by a mean value of 5.4 seconds 144 hours after warfarin dose following co-administration with voriconazole compared with a mean value of 0.6 seconds in the placebo treatment period. Coadministration of voriconazole potentiates warfarin-induced prothrombin time prolongation. Regular monitoring of the prothrombin time and appropriate adjustment of the dose of warfarin are recommended if these drugs are co-administered.

Food-Drug Interactions

Grapefruit juice

The effect of repeated ingestion of grapefruit juice 240 ml on the systemic availability of itraconazole and hydroxyitraconazole serum concentrations in subjects given hydroxypropyl-beta-cyclodextrin-itraconazole oral solution has been investigated in a randomized, two-period, crossover study in 20 healthy adults (10 men, 10 women) (126). There was no difference in itraconazole C_{max} or t_{max}. Co-administration of grapefruit juice reduced hydroxyitraconazole C_{max} by nearly 10%, but this difference was not statistically significant. There was a statistically significant increase in itraconazole AUC. The apparent oral clearance of itraconazole was significantly reduced. Grapefruit juice produced a significantly reduced mean hydroxyitraconazole:itraconazole AUC ratio; it also reduced the mean hydroxyitraconazole:itraconazole C_{max} ratio, but the difference was not statistically significant. Thus, repeated grapefruit juice consumption only moderately affects itraconazole systemic availability. Unlike previous findings with itraconazole capsules, changes in the disposition of itraconazole and hydroxyitraconazole after repeated grapefruit juice consumption are consistent with inhibition by grapefruit juice of intestinal cytochrome CYP3A4.

References

1. Groll AH, Piscitelli SC, Walsh TJ. Clinical pharmacology of systemic antifungal agents: a comprehensive review of agents in clinical use, current investigational compounds, and putative targets for antifungal drug development. Adv Pharmacol 1998;44:343–500.
2. Hoffman HL, Ernst EJ, Klepser ME. Novel triazole antifungal agents. Expert Opin Investig Drugs 2000;9(3):593–605.

3. Vanden Bossche H, Marichal P, Gorrens J, Coene MC, Willemsens G, Bellens D, Roels I, Moereels H, Janssen PA. Biochemical approaches to selective antifungal activity. Focus on azole antifungals. Mycoses 1989;32(Suppl 1):35–52.

4. Vanden Bossche H, Marichal P, Gorrens J, Coene MC. Biochemical basis for the activity and selectivity of oral antifungal drugs. Br J Clin Pract Suppl 1990;71:41–6.

5. Francis P, Walsh TJ. Evolving role of flucytosine in immunocompromised patients: new insights into safety, pharmacokinetics, and antifungal therapy. Clin Infect Dis 1992;15(6):1003–18.

6. Azon-Masoliver A, Vilaplana J. Fluconazole-induced toxic epidermal necrolysis in a patient with human immunodeficiency virus infection. Dermatology 1993;187(4):268–9.

7. Gubbins PO, McConnell SA, Penzak SR. Antifungal Agents. In: Piscitelli SC, Rodvold KA, editors. Drug Interactions in Infectious Diseases. Totowa, NJ: Humana Press Inc, 2001:185–217.

8. Wang EJ, Lew K, Casciano CN, Clement RP, Johnson WW. Interaction of common azole antifungals with P glycoprotein. Antimicrob Agents Chemother 2002;46(1):160–5.

9. Bashir H, Motl S, Metzger ML, Howard SC, Kaste S, Krasin MP, Hudson MM. Itraconazole-enhanced chemotherapy toxicity in a patient with Hodgkin lymphoma. J Pediatr Hematol Oncol 2006;28(1):33-5.

10. Krieter P, Flannery B, Musick T, Gohdes M, Martinho M, Courtney R. Disposition of posaconazole following single-dose oral administration in healthy subjects. Antimicrob Agents Chemother 2004;48:3543-51.

11. Huang H, Wang H, Sinz M, Zoeckler M, Staudinger J, Redinbo MR, Teotico DG, Locker J, Kalpana GV, Mani S. Inhibition of drug metabolism by blocking the activation of nuclear receptors by ketoconazole. Oncogene 2007;26(2):258-68.

12. Yu DT, Peterson JF, Seger DL, Gerth WC, Bates DW. Frequency of potential azole drug-drug interactions and consequences of potential fluconazole drug interactions. Pharmacoepidemiol Drug Saf 2005;14(11):755-67.

13. Saari TI, Laine K, Leino K, Valtonen M, Neuvonen PJ, Olkkola KT. Voriconazole, but not terbinafine, markedly reduces alfentanil clearance and prolongs its half-life. Clin Pharmacol Ther 2006;80(5):502-8.

14. Schaffner A, Frick PG. The effect of ketoconazole on amphotericin B in a model of disseminated aspergillosis. J Infect Dis 1985;151(5):902–10.

15. Pahls S, Schaffner A. *Aspergillus fumigatus* pneumonia in neutropenic patients receiving fluconazole for infection due to *Candida* species: is amphotericin B combined with fluconazole the appropriate answer? Clin Infect Dis 1994;18(3):484–6.

16. Schaffner A, Bohler A. Amphotericin B refractory aspergillosis after itraconazole: evidence for significant antagonism. Mycoses 1993;36(11–12):421–4.

17. Rex JH, Pappas PG, Karchmer AW, Sobel J, Edwards JE, Hadley S, Brass C, Vazquez JA, Chapman SW, Horowitz HW, Zervos M, McKinsey D, Lee J, Babinchak T, Bradsher RW, Cleary JD, Cohen DM, Danziger L, Goldman M, Goodman J, Hilton E, Hyslop NE, Kett DH, Lutz J, Rubin RH, Scheld WM, Schuster M, Simmons B, Stein DK, Washburn RG, Mautner L, Chu TC, Panzer H, Rosenstein RB, Booth JNational Institute of Allergy and Infectious Diseases Mycoses Study Group. A randomized and blinded multicenter trial of high-dose fluconazole plus placebo versus fluconazole plus amphotericin B as therapy for candidemia and its consequences in nonneutropenic subjects. Clin Infect Dis 2003;36(10):1221–8.

18. Maesaki S, Kohno S, Kaku M, Koga H, Hara K. Effects of antifungal agent combinations administered simultaneously and sequentially against *Aspergillus fumigatus*. Antimicrob Agents Chemother 1994;38(12):2843–5.

19. Sugar AM, Hitchcock CA, Troke PF, Picard M. Combination therapy of murine invasive candidiasis with fluconazole and amphotericin B. Antimicrob Agents Chemother 1995;39(3):598–601.

20. Dowell JA, Schranz J, Baruch A, Foster G. Safety and pharmacokinetics of coadministered voriconazole and anidulafungin. J Clin Pharmacol 2005;45(12):1373-82.

21. Courtney R, Radwanski E, Lim J, Laughlin M. Pharmacokinetics of posaconazole coadministered with antacid in fasting or nonfasting healthy men. Antimicrob Agents Chemother 2004;48:804-8.

22. Lohitnavy M, Lohitnavy O, Thangkeattiyanon O, Srichai W. Reduced oral itraconazole bioavailability by antacid suspension. J Clin Pharm Ther 2005;30(3):201-6.

23. Chaikin P, Gillen MS, Malik M, Pentikis H, Rhodes GR, Roberts DJ. Co-administration of ketoconazole with H1-antagonists ebastine and loratadine in healthy subjects: pharmacokinetic and pharmacodynamic effects. Br J Clin Pharmacol 2005;59(3):346-54.

24. Purkins L, Wood N, Kleinermans D, Love ER. No clinically significant pharmacokinetic interactions between voriconazole and indinavir in healthy volunteers. Br J Clin Pharmacol 2003;56 Suppl 1:62-8.

25. Scherpbier HJ, Hilhorst MI, Kuijpers TW. Liver failure in a child receiving highly active antiretroviral therapy and voriconazole. Clin Infect Dis 2003;37:828-30.

26. Kubo M, Koue T, Inaba A, Takeda H, Maune H, Fukuda T, Azuma J. Influence of itraconazole co-administration and CYP2D6 genotype on the pharmacokinetics of the new antipsychotic aripiprazole. Drug Metab Pharmacokinet 2005;20(1):55-64.

27. Lilja JJ, Backman JT, Neuvonen PJ. Effect of itraconazole on the pharmacokinetics of atenolol. Basic Clin Pharmacol Toxicol 2005;97(6):395-8.

28. Oda M, Kotegawa T, Tsutsumi K, Ohtani Y, Kuwatani K, Nakano S. The effect of itraconazole on the pharmacokinetics and pharmacodynamics of bromazepam in healthy volunteers. Eur J Clin Pharmacol 2003;59:615-9.

29. Osanai T, Ohkubo T, Yasui N, Kondo T, Kaneko S. Effect of itraconazole on the pharmacokinetics and pharmacodynamics of a single oral dose of brotizolam. Br J Clin Pharmacol 2004;58:476-81.

30. Ulivelli M, Rubegni P, Nuti D, Bartalini S, Giannini F, Rossi S. Clinical evidence of fluconazole-induced carbamazepine toxicity. J Neurol 2004;251:622-3.

31. Lilja JJ, Backman JT, Laitila J, Luurila H, Neuvonen PJ. Itraconazole increases but grapefruit juice greatly decreases plasma concentrations of celiprolol. Clin Pharmacol Ther 2003;73:192-8.

32. Kato K, Yasui-Furukori N, Fukasawa T, Aoshima T, Suzuki A, Kanno M, Otani K. Effects of itraconazole on the plasma kinetics of quazepam and its two active metabolites after a single oral dose of the drug. Ther Drug Monit 2003;25:473-7.

33. Florea NR, Capitano B, Nightingale CH, Hull D, Leitz GJ, Nicolau DP. Beneficial pharmacokinetic interaction between cyclosporine and itraconazole in renal transplant recipients. Transplant Proc 2003;35:2873-7.

34. El-Husseini A, El-Basuony F, Mahmoud I, Donia A, Sheashaa H, Sabry A, Hassan N, Sayed-Ahmad N, Sobh M. Impact of the cyclosporine-ketoconazole interaction in children with steroid-dependent idiopathic nephrotic syndrome. Eur J Clin Pharmacol 2006;62(1):3-8.

35. Groll AH, Kolve H, Ehlert K, Paulussen M, Vormoor J. Pharmacokinetic interaction between voriconazole and ciclosporin A following allogeneic bone marrow transplantation. J Antimicrob Chemother 2004;53:113-4.

36. Videla C, Vega J, Borja H. Hepatotoxicity associated with cyclosporine monitoring using C2 recommendations in adult renal recipients receiving ketoconazole. Transplant Proc 2005;37(3):1574-6.

37. Karyekar CS, Eddington ND, Briglia A, Gubbins PO, Dowling TC. Renal interaction between itraconazole and cimetidine. J Clin Pharmacol 2004;44:919-27.

38. Marr KA, Leisenring W, Crippa F, Slattery JT, Corey L, Boeckh M, McDonald GB. Cyclophosphamide metabolism is affected by azole antifungals. Blood 2004;103:1557-9.

39. Marr KA, Leisenring W, Crippa F, Slattery JT, Corey L, Boeckh M, McDonald GB. Cyclophosphamide metabolism is affected by azole antifungals. Blood 2004;103:1557-59.

40. de Jonge ME, Huitema AD, van Dam SM, Rodenhuis S, Beijnen JH. Effects of co-medicated drugs on cyclophosphamide bioactivation in human liver microsomes. Anticancer Drugs 2005;16(3):331-6.

41. Colburn DE, Giles FJ, Oladovich D, Smith JA. In vitro evaluation of cytochrome P450-mediated drug interactions between cytarabine, idarubicin, itraconazole and caspofungin. Hematology 2004;9:217-21.

42. Winter HR, Trapnell CB, Slattery JT, Jacobson M, Greenspan DL, Hooton TM, Unadkat JD. The effect of clarithromycin, fluconazole, and rifabutin on dapsone hydroxylamine formation in individuals with human immunodeficiency virus infection (AACTG 283). Clin Pharmacol Ther 2004;76:579-87.

43. Jakate AS, Roy P, Patel A, Abramowitz W, Persiani S, Wangsa J, Kapil R. Effect of azole antifungals ketoconazole and fluconazole on the pharmacokinetics of dexloxiglumide. Br J Clin Pharmacol 2005;60(5):498-507.

44. Purkins L, Wood N, Kleinermans D, Nichols D. Voriconazole does not affect the steady-state pharmacokinetics of digoxin. Br J Clin Pharmacol 2003;56 Suppl 1:45-50.

45. Yoshiyama Y, Kanke M. Toxic interactions between fluconazole and disopyramide in chick embryos. Biol Pharm Bull 2005;28(1):151-3.

46. Engels FK, Ten Tije AJ, Baker SD, Lee CK, Loos WJ, Vulto AG, Verweij J, Sparreboom A. Effect of cytochrome P450 3A4 inhibition on the pharmacokinetics of docetaxel. Clin Pharmacol Ther 2004;75:448-54.

47. Medicines Control Council. Interaction between ketoconazole and domperidone and the risk of QT prolongation-important safety information. S Afr Med J 2006;96(7):596.

48. Groll AH, Walsh TJ. Caspofungin: pharmacology, safety and therapeutic potential in superficial and invasive fungal infections. Expert Opin Investig Drugs 2001;10(8):1545-58.

49. Stone JA, McCrea J, Wickersham P, Holland S, Deutsch P, Bi S, Cicero T, Greenberg H, Waldman SA. Phase I study of caspofungin evaluating the potential for drug interactions with itraconazole, the effect of gender and the use of a loading dose. In: Abstracts of the 40th Interscience Conference on Antimicrobial Agents and Chemotherapy 854 2000:1545-58.

50. Araki K, Yasui-Furukori N, Fukasawa T, Aoshima T, Suzuki A, Inoue Y, Tateishi T, Otani K. Inhibition of the metabolism of etizolam by itraconazole in humans: evidence for the involvement of CYP3A4 in etizolam metabolism. Eur J Clin Pharmacol 2004;60:427-30.

51. Kovarik JM, Beyer D, Bizot MN, Jiang Q, Shenouda M, Schmouder RL. Blood concentrations of everolimus are markedly increased by ketoconazole. J Clin Pharmacol 2005;45(5):514-8.

52. Hallberg P, Mart n L, Wadelius M. Possible fluconazole-fentanyl interaction-a case report. Eur J Clin Pharmacol 2006;62(6):491-2.

53. Shon JH, Yoon YR, Hong WS, Nguyen PM, Lee SS, Choi YG, Cha IJ, Shin JG. Effect of itraconazole on the pharmacokinetics and pharmacodynamics of fexofenadine in relation to the MDR1 genetic polymorphism. Clin Pharmacol Ther 2005;78(2):191-201.

54. Viviani MA. Flucytosine—what is its future? J Antimicrob Chemother 1995;35(2):241–4.

55. Hospenthal DR, Bennett JE. Flucytosine monotherapy for cryptococcosis. Clin Infect Dis 1998;27(2):260–4.

56. Wise GJ, Kozinn PJ, Goldberg P. Flucytosine in the management of genitourinary candidiasis: 5 years of experience. J Urol 1980;124(1):70–2.

57. Francis P, Walsh TJ. Evolving role of flucytosine in immunocompromised patients: new insights into safety, pharmacokinetics, and antifungal therapy. Clin Infect Dis 1992;15(6):1003–18.

58. McKillop D, McCormick AD, Millar A, Miles GS, Phillips PJ, Hutchison M. Cytochrome P450-dependent metabolism of gefitinib. Xenobiotica 2005;35(1):39-50.

59. Swaisland HC, Ranson M, Smith RP, Leadbetter J, Laight A, McKillop D, Wild MJ. Pharmacokinetic drug interactions of gefitinib with rifampicin, itraconazole and metoprolol. Clin Pharmacokinet 2005;44(10):1067-81.

60. De Wachter E, Vanbesien J, De Schutter I, Malfroot A, De Schepper J. Rapidly developing Cushing syndrome in a 4-year-old patient during combined treatment with itraconazole and inhaled budesonide. Eur J Pediatr 2003;162:488-9.

61. Skov M, Main KM, Sillesen IB, Muller J, Koch C, Lanng S. Iatrogenic adrenal insufficiency as a side-effect of combined treatment of itraconazole and budesonide. Eur Respir J 2002;20:127-33.

62. Raaska K, Niemi M, Neuvonen M, Neuvonen PJ, Kivisto KT. Plasma concentrations of inhaled budesonide and its effects on plasma cortisol are increased by the cytochrome P4503A4 inhibitor itraconazole. Clin Pharmacol Ther 2002;72:362-9.

63. Bolland MJ, Bagg W, Thomas MG, Lucas JA, Ticehurst R, Black PN. Cushing's syndrome due to interaction between inhaled corticosteroids and itraconazole. Ann Pharmacother 2004;38:46-9.

64. Park JY, Shon JH, Kim KA, Jung HJ, Shim JC, Yoon YR, Cha IJ, Shin JG. Combined effects of itraconazole and CYP2D6*10 genetic polymorphism on the pharmacokinetics and pharmacodynamics of haloperidol in healthy subjects. J Clin Psychopharmacol 2006;26(2):135-42.

65. Purkins L, Wood N, Kleinermans D, Nichols D. Histamine H2-receptor antagonists have no clinically significant effect on the steady-state pharmacokinetics of voriconazole. Br J Clin Pharmacol 2003;56 Suppl 1:51-5.

66. Kantola T, Backman JT, Niemi M, Kivisto KT, Neuvonen PJ. Effect of fluconazole on plasma fluvastatin and pravastatin concentrations. Eur J Clin Pharmacol 2000;56(3):225–9.

67. Mazzu AL, Lasseter KC, Shamblen EC, Agarwal V, Lettieri J, Sundaresen P. Itraconazole alters the pharmacokinetics of atorvastatin to a greater extent than either cerivastatin or pravastatin. Clin Pharmacol Ther 2000;68(4):391–400.

68. Horn M. Coadministration of itraconazole with hypolipidemic agents may induce rhabdomyolysis in healthy individuals. Arch Dermatol 1996;132(10):1254.

69. Neuvonen PJ, Kantola T, Kivisto KT. Simvastatin but not pravastatin is very susceptible to interaction with the CYP3A4 inhibitor itraconazole. Clin Pharmacol Ther 1998;63(3):332–41.

70. Kantola T, Kivisto KT, Neuvonen PJ. Effect of itraconazole on the pharmacokinetics of atorvastatin. Clin Pharmacol Ther 1998;64(1):58–65.

71. Kivisto KT, Kantola T, Neuvonen PJ. Different effects of itraconazole on the pharmacokinetics of fluvastatin and lovastatin. Br J Clin Pharmacol 1998;46(1):49–53.

72. Gilad R, Lampl Y. Rhabdomyolysis induced by simvastatin and ketoconazole treatment. Clin Neuropharmacol 1999;22(5):295–7.

73. Cooper KJ, Martin PD, Dane AL, Warwick MJ, Schneck DW, Cantarini MV. Effect of itraconazole on the pharmacokinetics of rosuvastatin. Clin Pharmacol Ther 2003;73:322-9.

74. Shaukat A, Benekli M, Vladutiu GD, Slack JL, Wetzler M, Baer MR. Simvastatin-fluconazole causing rhabdomyolysis. Ann Pharmacother 2003;37:1032-5.

75. Kahri J, Valkonen M, Bcklund T, Vuoristo M, Kivist KT. Rhabdomyolysis in a patient receiving atorvastatin and fluconazole. Eur J Clin Pharmacol 2005;60(12):905-7.

76. Hynninen VV, Olkkola KT, Leino K, Lundgren S, Neuvonen PJ, Rane A, Valtonen M, Vyyrylinen H, Laine K. Effects of the antifungals voriconazole and fluconazole on the pharmacokinetics of S-(+)- and R-(-)-ibuprofen. Antimicrob Agents Chemother 2006;50(6):1967-72.

77. Dutreix C, Peng B, Mehring G, Hayes M, Capdeville R, Pokorny R, Seiberling M. Pharmacokinetic interaction between ketoconazole and imatinib mesylate (Glivec) in healthy subjects. Cancer Chemother Pharmacol 2004;54:290-4.

78. Gambillara E, Laffitte E, Widmer N, Decosterd LA, Duchosal MA, Kovacsovics T, Panizzon RG. Severe pustular eruption associated with imatinib and voriconazole in a patient with chronic myeloid leukemia. Dermatology 2005;211(4):363-5.

79. Gubbins PO, Melchert RB, McConnell SA, Franks AM, Penzak SR, Gurley BJ. Effect of interleukin 6 on the hepatic metabolism of itraconazole and its metabolite hydroxyitraconazole using primary human hepatocytes. Pharmacology 2003;67:195-201.

80. Yong WP, Ramirez J, Innocenti F, Ratain MJ. Effects of ketoconazole on glucuronidation by UDP-glucuronosyltransferase enzymes. Clin Cancer Res 2005;11(18):6699-704.

81. Isohanni MH, Neuvonen PJ, Olkkola KT. Effect of itraconazole on the pharmacokinetics of inhaled lidocaine. Basic Clin Pharmacol Toxicol 2004;95:120-3.

82. Niemi M, Tornio A, Pasanen MK, Fredrikson H, Neuvonen PJ, Backman JT. Itraconazole, gemfibrozil and their combination markedly raise the plasma concentrations of loperamide. Eur J Clin Pharmacol 2006;62(6):463-72.

83. Piñeyro-López A, Pineyro-Garza E, Torres-Alanís O, Reyes-Araiza R, Gómez Silva M, Wacksman N, Lujàn Rangel R, de Lago A, Trejo D, Gonzàlez-de la Parra M, Namur S. Bioavailability of two oral formulations of loratadine 20 mg with concomitant ketoconazole: an open-label, randomized, two-period crossover comparison in healthy Mexican adult volunteers. Clin Ther 2006;28(1):110-5.

84. Schottker B, Dosch A, Kraemer DM. Severe hepatotoxicity after application of desloratadine and fluconazole. Acta Haematol 2003;110:43-4.

85. Scott G, Yih L, Yeh CM, Milosavljev S, Laurent A, Rordorf C. Lumiracoxib: pharmacokinetic and pharmacodynamic profile when coadministered with fluconazole in healthy subjects. J Clin Pharmacol 2004;44:193-9.

86. Purkins L, Wood N, Ghahramani P, Kleinermans D, Layton G, Nichols D. No clinically significant effect of erythromycin or azithromycin on the pharmacokinetics of voriconazole in healthy male volunteers. Br J Clin Pharmacol 2003;56 Suppl 1:30-6.

87. Ridtitid W, Wongnawa M, Mahatthanatrakul W, Raungsri N, Sunbhanich M. Ketoconazole increases plasma concentrations of antimalarial mefloquine in healthy human volunteers. J Clin Pharm Ther 2005;30(3):285-90.

88. Niemi M, Neuvonen M, Juntti-Patinen L, Backman JT, Neuvonen PJ. Effect of fluconazole on the pharmacokinetics and pharmacodynamics of nateglinide. Clin Pharmacol Ther 2003;74:25-31.

89. Niemi M, Backman JT, Neuvonen M, Neuvonen PJ. Effects of gemfibrozil, itraconazole, and their combination on the pharmacokinetics and pharmacodynamics of repaglinide: potentially hazardous interaction between gemfibrozil and repaglinide. Diabetologia 2003;46:347-51.

90. Liu P, Foster G, Labadie R, Somoza E, Sharma A. Pharmacokinetic interaction between voriconazole and methadone at steady state in patients on methadone therapy. Antimicrob Agents Chemother 2007;51(1):110-8.

91. Stass H, Nagelschmitz J, Moeller JG, Delesen H. Pharmacokinetics of moxifloxacin are not influenced by a 7-day pretreatment with 200 mg oral itraconazole given once a day in healthy subjects. Int J Clin Pharmacol Ther 2004;42:23-9.

92. Manosuthi W, Chumpathat N, Chaovavanich A, Sungkanuparph S. Safety and tolerability of nevirapine-based antiretroviral therapy in HIV-infected patients receiving fluconazole for cryptococcal prophylaxis: a retrospective cohort study. BMC Infect Dis 2005;5:67.

93. Linnebur SA, Parnes BL. Pulmonary and hepatic toxicity due to nitrofurantoin and fluconazole treatment. Ann Pharmacother 2004;38:612-6.

94. Johnson MD, Hamilton CD, Drew RH, Sanders LL, Pennick GJ, Perfect JR. A randomized comparative study to determine the effect of omeprazole on the peak serum concentration of itraconazole oral solution. J Antimicrob Chemother 2003;51:453-7.

95. Jaruratanasirikul S, Sriwiriyajan S. Effect of omeprazole on the pharmacokinetics of itraconazole. Eur J Clin Pharmacol 1998;54:159-61.

96. Wood N, Tan K, Purkins L, Layton G, Hamlin J, Kleinermans D, Nichols D. Effect of omeprazole on the steady-state pharmacokinetics of voriconazole. Br J Clin Pharmacol 2003;56 Suppl 1:56-61.

97. Bun SS, Giacometti S, Fanciullino R, Ciccolini J, Bun H, Aubert C. Effect of several compounds on biliary excretion of paclitaxel and its metabolites in guinea-pigs. Anticancer Drugs 2005;16(6):675-82.

98. Purkins L, Wood N, Ghahramani P, Love ER, Eve MD, Fielding A. Coadministration of voriconazole and phenytoin: pharmacokinetic interaction, safety, and toleration. Br J Clin Pharmacol 2003;56 Suppl 1:37-44.

99. Gandhi PJ, Menezes PA, Vu HT, Rivera AL, Ramaswamy K. Fluconazole- and levofloxacin-induced torsades de pointes in an intensive care unit patient. Am J Health-Syst Pharm 2003;60:2479-83.

100. Jerling M, Huan BL, Leung K, Chu N, Abdallah H, Hussein Z. Studies to investigate the pharmacokinetic interactions between ranolazine and ketoconazole, diltiazem, or simvastatin during combined administration in healthy subjects. J Clin Pharmacol 2005;45(4):422-33.

101. Panomvana Na Ayudhya D, Thanompuangseree N, Tansuphaswadikul S. Effect of rifampicin on the pharmacokinetics of fluconazole in patients with AIDS. Clin Pharmacokinet 2004;43:725-32.

102. Jung SM, Kim KA, Cho HK, Jung IG, Park PW, Byun WT, Park JY. Cytochrome P450 3A inhibitor itraconazole affects plasma concentrations of risperidone and 9-hydroxyrisperidone in schizophrenic patients. Clin Pharmacol Ther 2005;78(5):520-8.

103. Mikus G, Schwel V, Drzewinska M, Rengelshausen J, Ding R, Riedel KD, Burhenne J, Weiss J, Thomsen T, Haefeli WE. Potent cytochrome P450 2C19 genotype-related interaction between voriconazole and the cytochrome P450 3A4 inhibitor ritonavir. Clin Pharmacol Ther 2006;80(2):126-35.

104. Rengelshausen J, Banfield M, Riedel KD, Burhenne J, Weiss J, Thomsen T, Walter-Sack I, Haefeli WE, Mikus G. Opposite effects of short-term and long-term St John's wort intake on voriconazole pharmacokinetics. Clin Pharmacol Ther 2005;78(1):25-33.

105. Kim EJ, Seo JW, Hwang JY, Han SS. Effects of combined treatment with sildenafil and itraconazole on the cardiovascular system in telemetered conscious dogs. Drug Chem Toxicol 2005;28(2):177-86.

106. Sadaba B, Campanero MA, Quetglas EG, Azanza JR. Clinical relevance of sirolimus drug interactions in transplant patients. Transplant Proc 2004;36:3226-8.

107. Mathis AS, Shah NK, Friedman GS. Combined use of sirolimus and voriconazole in renal transplantation: a report of two cases. Transplant Proc 2004;36:2708-9.

108. Marty FM, Lowry CM, Cutler CS, Campbell BJ, Fiumara K, Baden LR, Antin JH. Voriconazole and sirolimus coadministration after allogeneic hematopoietic stem cell transplantation. Biol Blood Marrow Transplant 2006;12(5):552-9.

109. Said A, Garnick JJ, Dieterle N, Peres E, Abidi MH, Ibrahim RB. Sirolimus-itraconazole interaction in a hematopoietic stem cell transplant recipient. Pharmacotherapy 2006;26(2):289-95.

110. Kuypers DR, Claes K, Evenepoel P, Maes B, Vandecasteele S, Vanrenterghem Y, Van Damme B, Desmet K. Drug interaction between itraconazole and sirolimus in a primary renal allograft recipient. Transplantation 2005;79(6):737.

111. Swart PJ, Krauwinkel WJ, Smulders RA, Smith NN. Pharmacokinetic effect of ketoconazole on solifenacin in healthy volunteers. Basic Clin Pharmacol Toxicol 2006;99(1):33-6.

112. Winter HR, Trapnell CB, Slattery JT, Jacobson M, Greenspan DL, Hooton TM, Unadkat JD. The effect of clarithromycin, fluconazole, and rifabutin on sulfamethoxazole hydroxylamine formation in individuals with human immunodeficiency virus infection (AACTG 283). Clin Pharmacol Ther 2004;76:313-22.

113. Mahnke CB, Sutton RM, Venkataramanan R, Michaels M, Kurland G, Boyle GJ, Law YM, Miller SA, Pigula FA, Gandhi S, Webber SA. Tacrolimus dosage requirements after initiation of azole antifungal therapy in pediatric thoracic organ transplantation. Pediatr Transplant 2003;7:474-8.

114. Bhaloo S, Prasad GV. Severe reduction in tacrolimus levels with rifampin despite multiple cytochrome P450 inhibitors: a case report. Transplant Proc 2003;35:2449-51.

115. Pai MP, Allen S. Voriconazole inhibition of tacrolimus metabolism. Clin Infect Dis 2003;36:1089-91.

116. El-Dahshan KF, Bakr MA, Donia AF, Badr A El-S, Sobh MA-K. Ketoconazole-tacrolimus coadministration in kidney transplant recipients: two-year results of a prospective randomized study. Am J Nephrol 2006;26(3):293-8.

117. Shitrit D, Ollech JE, Ollech A, Bakal I, Saute M, Sahar G, Kramer MR. Itraconazole prophylaxis in lung transplant recipients receiving tacrolimus (FK 506):efficacy and drug interaction. J Heart Lung Transplant 2005;24(12):2148-52.

118. Shi J, Montay G, Leroy B, Bhargava VO. Effects of itraconazole or grapefruit juice on the pharmacokinetics of telithromycin. Pharmacotherapy 2005;25(1):42-51.

119. Shi J, Chapel S, Montay G, Hardy P, Barrett JS, Sica D, Swan SK, Noveck R, Leroy B, Bhargava VO. Effect of ketoconazole on the pharmacokinetics and safety of telithromycin and clarithromycin in older subjects with renal impairment. Int J Clin Pharmacol Ther 2005;43(3):123-33.

120. Vanier KL, Mattiussi AJ, Johnston DL. Interaction of all-trans-retinoic acid with fluconazole in acute promyelocytic leukemia. J Pediatr Hematol Oncol 2003;25:403-4.

121. Bennett MT, Sirrs S, Yeung JK, Smith CA. Hypercalcemia due to all trans retinoic acid in the treatment of acute promyelocytic leukemia potentiated by voriconazole. Leuk Lymphoma 2005;46(12):1829-31.

122. Duggal HS. Delirium associated with amitriptyline/fluconazole drug. Gen Hosp Psychiatry 2003;25:297-8.

123. Ariffin H, Omar KZ, Ang EL, Shekhar K. Severe vincristine neurotoxicity with concomitant use of itraconazole. J Paediatr Child Health 2003;39:638-9.

124. Bermudez M, Fuster JL, Llinares E, Galera A, Gonzalez C. Itraconazole-related increased vincristine neurotoxicity: case report and review of literature. J Pediatr Hematol Oncol 2005;27(7):389-92.

125. Purkins L, Wood N, Kleinermans D, Nichols D. Voriconazole potentiates warfarin-induced prothrombin time prolongation. Br J Clin Pharmacol 2003;56 Suppl 1:24-9.

126. Gubbins PO, McConnell SA, Gurley BJ, Fincher TK, Franks AM, Williams DK, Penzak SR, Saccente M. Influence of grapefruit juice on the systemic availability of itraconazole oral solution in healthy adult volunteers. Pharmacotherapy 2004;24:460-7.

Antifungal azoles and other drugs for topical use

See also Antifungal azoles

General Information

All the topically used azoles can cause local irritation, burning, and, if used intravaginally, burning, swelling, and discomfort during micturition. There is cross-sensitivity between econazole, enilconazole, miconazole, and probably all other phenethylimidazoles.

Contact allergy to topical imidazoles is rare, considering how commonly they are used. The imidazole derivatives most often reported to be allergens are miconazole, econazole, tioconazole, and isoconazole. As far as cross-reactivity is concerned, in one review, there were statistically significant associations between miconazole, econazole, and isoconazole; between sulconazole, miconazole, and econazole; and between isoconazole and tioconazole (1).

Of 3049 outpatients who were patch-tested for contact dermatitis at the Department of Dermatology, Nippon Medical School Hospital from January 1984 to August 1994, 218 were patch-tested with topical antimycotic agents (2). There were 66 positive tests with imidazole derivatives, of whom 35 were allergic to the active ingredients: 16 were allergic to sulconazole, 11 to croconazole, 3 to tioconazole, 3 to miconazole, 1 to bifonazole, and 1 to clotrimazole. Exposure to croconazole occurred after a significantly shorter time with less drug than with sulconazole. Of the 35 patients who were allergic to an imidazole, 21 cross-reacted to other imidazoles.

Azoles

Bifonazole

Bifonazole has a broad spectrum of activity in vitro against dermatophytes, moulds, yeasts, dimorphic fungi, and some Gram-positive bacteria. It has been used in a strength of 1% in creams, gels, solutions, and powders, applied once a day to treat superficial fungal infections of the skin, such as dermatophytoses, cutaneous candidiasis, and pityriasis versicolor (3). In a multicenter, double-blind, randomized, parallel-group comparison with flutrimazole cream 1% in the treatment of dermatomycoses in 449 patients the overall incidence of adverse effects (mainly mild local effects such as irritation or a burning sensation) was 5% (4).

Clotrimazole

Clotrimazole was the first oral azole. While it was effective in deep mycoses, its limited absorption and induction of liver microsomal enzymes after a few days, leading to accelerated metabolism of the compound, as well as its toxicity, preclude its use for systemic therapy. Clotrimazole is therefore currently only used for topical therapy of mucocutaneous candidiasis.

Comparisons of fluconazole 200 mg/day with clotrimazole 10 mg 5 times/day in the prevention of thrush in patients with AIDS showed little difference in the occurrence of undesirable effects and abnormalities in laboratory measurements but less efficacy of clotrimazole (5,6).

Local problems can occur, including hypersensitivity reactions (1). In one case of contact allergy, patch-testing was positive with clotrimazole (5% in petroleum), itraconazole (1% in ether), and croconazole (1% in ether) (7). The authors reviewed the possible cross-reactions between the subgroups of imidazoles.

- A 71-year-old woman had a severe exacerbation of vulval dermatitis for which she had been using Canesten (clotrimazole) cream (8). There was a positive patch-test reaction with clotrimazole (1% in petrolatum) and patch tests with the other constituents of Canesten were negative.

Topical vaginal administration of even relatively high doses of clotrimazole did not result in systemic toxicity (9).

Croconazole

In one case of contact allergy, patch-testing was positive with clotrimazole (5% in petroleum), itraconazole (1% in ether), and croconazole (1% in ether) (7). The authors reviewed the possible cross-reactions between the subgroups of imidazoles.

Econazole

Econazole is used topically on the skin and also intravaginally, after which about 3–7% is absorbed. It can cause pruritus (10) and vaginal burning (11).

Enilconazole

Enilconazole is used in 10% solution/cream. Contact dermatitis has been reported (12,13).

Isoconazole

Isoconazole is mainly used for vaginal infections with *Candida albicans*. Contact dermatitis has been reported (14), including an unusual case with a papulo-pustular reaction (15).

Itraconazole

In one case of contact allergy, patch-testing was positive with clotrimazole (5% in petroleum), itraconazole (1% in ether), and croconazole (1% in ether) (7). The authors reviewed the possible cross-reactions between the subgroups of imidazoles.

Lanoconazole (latoconazole)

Used in a 1% cream, lanoconazole is effective against *Tinea*, and is more active than clotrimazole or bifonazole. Several cases of contact dermatitis have been reported (16–20).

Miconazole

See the monograph on Miconazole.

Nimorazole

Nimorazole is believed to be active against *Trichomonas vaginalis*. No specific adverse effects have been described after local use.

Ornidazole

Complaints of dizziness (21), mild gastrointestinal symptoms (22), and headache (23) during treatment with intravaginal ornidazole have been reported, since ornidazole is relatively well absorbed after rectal and vaginal administration (23).

Terconazole

Terconazole is prepared in creams and ovules for intra-vaginal use. Besides local irritation it causes systemic reactions. Headache was reported in over a quarter of patients. Other effects include hypotension, fever, and chills. Terconazole is absorbed to a greater extent than other topical azoles (SED-12, 684).

Tioconazole

Tioconazole is mainly used for vaginal or inguinal *Candida* infections. It has fewer local adverse effects than some of the older imidazoles. Local irritation, burning, rash, erythema, and pruritus have been reported. In a few women there was marked burning on micturition; these women all had signs of vaginal epithelial atrophy (SED-12, 684) (24).

Other topical antifungal drugs

5-Bromo-4-chlorosalicylamide (multifungin)

5-Bromo-4-chlorosalicylamide is one of a group of local antiseptics and fungistats that can cause photosensitization (25). There is cross-sensitization with bithionol, fenticlor, and tribromosalicylanide.

Buclosamide (N-butyl-4-chlorosalicylamide)

Photocontact dermatitis has been described with buclosamide (25). There is cross-reactivity with a number of other drugs, notably oral hypoglycemic drugs, diuretics, and sulfonamides. Because of these reactions, buclosamide is not recommended for topical use.

Captan (Orthocide-406)

Captan is one of the older fungicides. It is used for pityriasis versicolor and is included in some soaps and cosmetics to provide bactericidal and fungicidal effects. It is allergenic. It is carcinogenic in mice, and in several countries control agencies have taken steps to prohibit its use in cosmetics and non-drug products (SEDA-13, 236).

Ciclopirox

Ciclopirox, a substituted pyridone unrelated to the imidazoles, is effective against a wide variety of dermatophytes, yeasts, actinomycetes, molds, and other fungi. Ciclopirox olamine is generally well tolerated locally, and reactions occur in only 1–4% of cases (SEDA-12, 684).

Clodantoin

Contact dermatitis has been rarely reported with clodantoin.

Fluonilide (4-fluoro-3',5'-thiocarbanilide)

Contact dermatitis has been reported with fluonilide (26).

Gentian violet

Gentian violet (27) was at one time the treatment of choice for vaginal and oral candidiasis but is now obsolete. The main problem is staining and the messiness of the application, since the purple-colored fluid has to be brushed on to the skin.

Hachimycin (trichomycin)

Contact dermatitis has been reported with hachimycin.

KP-363

KP-363, a benzylamine derivative, is used in creams and solutions in concentrations of 0.1 and 0.6%. It is reported to cause less irritation than bifonazole and tolciclate (SEDA-14, 235).

Naftitine

Naftitine is one of a series of allylamine antifungal agents, derived from heterocyclic spironaphthalenes. It is usually sold in the form of a 1% cream for topical treatment of dermatomycoses, dermatophytes, and yeasts. It is claimed to be more effective than the imidazoles. Local irritation and a burning sensation, if they occur, are only mild (SED-12, 684) (28), (SEDA-16, 297).

Natamycin (pimaricin)

No cases of contact dermatitis were described in industrial workers in frequent contact with natamycin (29).

In the Hungarian Case–Control Surveillance of Congenital Abnormalities between 1980 and 1996, of 38 151 pregnant women who delivered infants without any defects (controls) and 22 843 who had fetuses or neonates with congenital abnormalities, 62 (0.27%) and 98 (0.26%) were treated with vaginal natamycin in the two groups respectively (crude OR = 1.1; 95% CI = 0.8, 1.5). There was thus no evidence of a teratogenic effect of natamycin.

Nifuratel

Contact dermatitis, with facial edema and a generalized erythema, has been described in the partner of a woman treated with nifuratel vaginal suppositories (SED-11, 578) (30).

Niphimycin

Niphimycin, an antimycotic antibiotic derived from *Actinomyces hygroscopicus*, is effective against both dermatomycosis and onychomycosis, with a 16–26% success rate in the latter. Tolerance is reportedly good, but there is a notable lack of recent data (SEDA-12, 236).

Nystatin

See the monograph on Nystatin.

Pecilocin (Variotin)

Skin irritation has been reported in 2–6.5% of patients treated with pecilocin. Contact dermatitis has been described in a few cases (31,32). Of 44 patients treated with pecilocin who were patch-tested with pecilocin, 7 were allergic to it; in three of them the skin disease had been caused or exacerbated by pecilocin (33).

Pyrrolnitrin (Miutrin, 3-chloro-4-(3-chloro-2-nitrophenyl) pyrrole)

Contact dermatitis with pyrrolnitrin and cross-reactivity with dinitrochlorobenzene has been reported in one case (34).

Salicylic acid 3% with benzoic acid 6% (Whitfield's ointment)

Whitfield's ointment, used for *Trichophyton rubrum*, has a keratolytic effect, and local irritation can occur (SED-12, 685) (35).

Sulbentine (dibenzthion)

Photoallergic contact dermatitis has been described with sulbentine, probably through a breakdown product, benzylisothiocyanate (36).

Tolciclate

Tolciclate, a thiocarbamate, is active against most common dermatophytes. Contact dermatitis has been reported (37).

Tolnaftate

Tolerance of tolnaftate is good. Local erythema has been described, as has allergic dermatitis (38).

References

1. Dooms-Goossens A, Matura M, Drieghe J, Degreef H. Contact allergy to imidazoles used as antimycotic agents. Contact Dermatitis 1995;33(2):73–7.
2. Yoneyama E. [Allergic contact dermatitis due to topical imidazole antimycotics. The sensitizing ability of active ingredients and cross-sensitivity.]Nippon Ika Daigaku Zasshi 1996;63(5):356–64.
3. Lackner TE, Clissold SP. Bifonazole. A review of its antimicrobial activity and therapeutic use in superficial mycoses. Drugs 1989;38(2):204–25.
4. Alomar A, Videla S, Delgadillo J, Gich I, Izquierdo I, Forn JCatalan Flutrimazole Study Group. Flutrimazole 1% dermal cream in the treatment of dermatomycoses: a multicentre, double-blind, randomized, comparative clinical trial with bifonazole 1% cream. Efficacy of flutrimazole 1% dermal cream in dermatomycoses. Dermatology 1995;190(4):295–300.
5. Powderly WG, Finkelstein D, Feinberg J, Frame P, He W, van der Horst C, Koletar SL, Eyster ME, Carey J, Waskin H, et al. A randomized trial comparing fluconazole with clotrimazole troches for the prevention of fungal infections in patients with advanced human immunodeficiency virus infection. NIAID AIDS Clinical Trials Group. N Engl J Med 1995;332(11):700–5.
6. Koletar SL, Russell JA, Fass RJ, Plouffe JF. Comparison of oral fluconazole and clotrimazole troches as treatment for oral candidiasis in patients infected with human immunodeficiency virus. Antimicrob Agents Chemother 1990;34(11):2267–8.
7. Erdmann S, Hertl M, Merk HF. Contact dermatitis from clotrimazole with positive patch-test reactions also to croconazole and itraconazole. Contact Dermatitis 1999;40(1):47–8.
8. Cooper SM, Shaw S. Contact allergy to clotrimazole: an unusual allergen. Contact Dermatitis 1999;41(3):168.
9. Wolfson N, Riley J, Samuels B, Singh JM. Clinical toxicology of clotrimazole when administered vaginally. Clin Toxicol 1981;18(1):41–5.
10. Grigoriu D, Grigoriu A. Double-blind comparison of the efficacy, toleration and safety of tioconazole base 1% and econazole nitrate 1% creams in the treatment of patients with fungal infections of the skin or erythrasma. Dermatologica 1983;166(Suppl 1):8–13.
11. Gouveia DC, Jones da Silva C. Oxiconazole in the treatment of vaginal candidiasis: single dose versus 3-day treatment with econazole. Pharmatherapeutica 1984;3(10):682–5.
12. Piebenga WP, van der Walle HB. Allergic contact dermatitis from 1-[2-(2,4-dichlorophenyl)-2-(2-propenyloxy) ethyl]-1H-imidazole in a water-based metalworking fluid. Contact Dermatitis 2003;48(5):285–6.
13. van Hecke E, de Vos L. Contact sensitivity to enilconazole. Contact Dermatitis 1983;9(2):144.
14. Frenzel UH, Gutekunst A. Contact dermatitis to isoconazole nitrate. Contact Dermatitis 1983;9(1):74.
15. Lazarov A, Ingber A. Pustular allergic contact dermatitis to isoconazole nitrate. Am J Contact Dermat 1997;8(4):229–30.
16. Soga F, Katoh N, Kishimoto S. Contact dermatitis due to lanoconazole, cetyl alcohol and diethyl sebacate in lanoconazole cream. Contact Dermatitis 2004;50(1):49–50.
17. Umebayashi Y, Ito S. Allergic contact dermatitis due to both lanoconazole and neticonazole ointments. Contact Dermatitis 2001;44(1):48–9.
18. Taniguchi S, Kono T. Allergic contact dermatitis due to lanoconazole with no cross-reactivity to other imidazoles. Dermatology 1998;196(3):366.
19. Tanaka N, Kawada A, Hiruma M, Tajima S, Ishibashi A. Contact dermatitis from lanoconazole. Contact Dermatitis 1996;35(4):256–7.
20. Nakano R, Miyoshi H, Kanzaki T. Allergic contact dermatitis from lanoconazole. Contact Dermatitis 1996;35(1):63.
21. Erkkola R, Jarvinen H. Single dose of ornidazole in the treatment of bacterial vaginosis. Ann Chir Gynaecol Suppl 1987;202:94–6.
22. Fugere P, Verschelden G, Caron M. Single oral dose of ornidazole in women with vaginal trichomoniasis. Obstet Gynecol 1983;62(4):502–5.
23. Andersson KE. Pharmacokinetics of nitroimidazoles. Spectrum of adverse reactions. Scand. J Infect Dis Suppl 1981;26:60–7.
24. Uyanwah PO. An open non-comparative evaluation of single-dose tioconazole (6%), vaginal ointment in vaginal candidosis. Curr Ther Res 1986;39:30.
25. Burry JN. Photoallergies to Fenticlor and Multifungin. Arch Dermatol 1967;95(3):287–91.
26. van Hecke E. Contact allergy to the topical antimycotic fluoro-4-dichloro-3'5'-thiocarbanilid. Dermatologica 1969;138(6):480–2.
27. Docampo R, Moreno SN. The metabolism and mode of action of gentian violet. Drug Metab Rev 1990;22(2–3):161–78.
28. Ganzinger U, Stutz A, Petranyi G, Stephen A. Allylamines: topical and oral treatment of dermatomycoses with a new class of antifungal agents. Acta Dermatol Venereol Suppl (Stockh) 1986;121:155–60.
29. Raab WP. Natamycin (Pimaricin). Its Properties and Possibilities in MedicineStuttgart: Georg Thieme Verlag;. 1972.
30. Bedello PG, Goitre M, Cane D, Fogliano MR. Contact dermatitis from nifuratel. Contact Dermatitis 1983;9(2):166.
31. Sundararajan V. Variotin sensitivity. Contact Dermatitis Newslett 1970;8:188.
32. Groen J, Bleumink E, Nater JP. Variotin sensitivity. Contact Dermatitis Newslett 1973;15:456.
33. Norgaard O. Pecilocinum–Allergie. [Pecilocin allergy.] Hautarzt 1977;28(1):35–6.

34. Meneghini CL, Angelini G. Contact dermatitis from pyrrolnitrin (an antimycotic agent). Contact Dermatitis 1975;1(5):288–92.
35. Odom R. A practical review of antifungals. Mod Med Can 1987;42:GP54.
36. Wurbach VG, Schubert H. Untersuchungen über die Afungin Allergie. [Studies on Afungin hypersensitivity.] Dermatol Monatsschr 1976;162(4):317–22.
37. Veraldi S, Schianchi-Veraldi R. Allergic contact dermatitis from tolciclate. Contact Dermatitis 1991;24(4):315.
38. Gellin GA, Maibach HI, Wachs GN. Contact allergy to tolnaftate. Arch Dermatol 1972;106(5):715–6.

Echinocandins

General Information

The echinocandins are a class of semisynthetic antifungal lipopeptides that are structurally characterized by a cyclic hexapeptide core linked to a variably configured lipid side chain. The echinocandins act by non-competitive inhibition of the synthesis of 1,3-α-D-glucan, a major polysaccharide component of the cell wall of many pathogenic fungi, which plays a key role in cell division and cell growth and is absent from mammalian cells. In concert with chitin, the rope-like glucan fibrils are important in maintaining the osmotic integrity of the fungal cell and play a key role in cell division and cell growth.

The currently available echinocandins are anidulafungin (rINN; Versicor Inc, Freemont, CA), caspofungin (rINN) (Merck & Co, Inc, Rahway, NJ), and micafungin (rINN; Fujisawa Inc, Deerfield, IL). They have relatively similar pharmacological properties. All three have potent and broad-spectrum antifungal activity against *Candida* species and *Aspergillus* species without cross-resistance to existing agents. They have prolonged post-antifungal effects and fungicidal activity against *Candida* and they cause severe damage to *Aspergillus* at the sites of hyphal growth. Their efficacy against these organisms in vivo has been demonstrated in animals (1,2). Their activity against other fungal pathogens in vitro is variable (3,4)

Pharmacokinetics

The echinocandins are currently available only for intravenous administration. They have dose-proportional plasma pharmacokinetics, with half-lives of 10–15 hours, which allows once-daily dosing. All echinocandins are highly protein-bound (over 95%) and distribute into all major tissues, including the brain; concentrations in non-inflammatory CSF are low. The echinocandins are eliminated by degradation and/or hepatic metabolism, and are slowly excreted as inactive metabolites in the urine and feces; only small fractions are excreted unchanged in the urine. As a class they are generally well tolerated and lack significant potential for drug interactions mediated by CYP450 isozymes (1–5). In vitro biotransformation studies of caspofungin have shown that it is not a substrate of P-glycoprotein and is a poor substrate and a weak inhibitor of cytochrome P450 enzymes (6). However, other in vitro studies have shown that it may inhibit CYP3A4 (7).

Observational studies

The clinical efficacy of anidulafungin, caspofungin, and micafungin against *Candida* species has been documented in phase II or phase III studies in immunocompromised patients with esophageal candidiasis. All achieved therapeutic efficacy at least comparable with standard agents. Phase III efficacy studies of caspofungin for esophageal candidiasis, invasive candidiasis, and empirical antifungal therapy in persistently febrile neutropenic patients have been completed. Caspofungin had no serious adverse effects and had therapeutic efficacy that was at least as good as standard agents (8–10). It is approved in the USA and the EU for second-line therapy of definite or probable invasive aspergillosis and for primary therapy in non-neutropenic patients with invasive *Candida* infections (3,6).

Anidulafungin

The safety and efficacy of intravenous anidulafungin 50, 75, or 100 mg/day has been investigated in 123 patients, 68 evaluable, with invasive candidiasis, including candidemia (11). A total of eligible patients were randomized to one of three regimens. Adverse events considered to be related to treatment were reported by under 5% of patients in each dosage group. The most common events were hypotension, vomiting, constipation, nausea, and pyrexia. Three serious adverse events were reported as either probably or possibly related to treatment (neutropenic fever, n = 1; seizures, n = 2).

Caspofungin

The use of caspofungin with other agents against certain types of invasive fungal infections is appealing, given the poor response rates to standard agents and its unique mechanism of action. Two retrospective analyses have explored the safety and potential benefits of a combination of caspofungin with liposomal amphotericin B (L-AmB).

In the first analysis the efficacy and safety of caspofungin in combination with L-AmB in 48 patients with hematological malignancies with definite or probable or possible invasive aspergillosis were evaluated (12). Caspofungin was given intravenously as a 70 mg loading dose on day 1, followed by a daily dose of 50 mg; L-AmB was started at an intravenous dose of 5 mg/day. The median duration of therapy with the combination was 20 (range 7–180) days. The combination of caspofungin and L-AmB was well tolerated: seven patients developed mild-to-moderate renal insufficiency that was attributed to the use of L-AmB and four required withdrawal. There was hypokalemia in three of the patients. One patient had a fever associated with caspofungin and another had hepatic dysfunction of multifactorial origin. In no patient was the combination withheld due to unanticipated adverse effects.

The second analysis included 30 patients with hematological malignancies and proven (n = 6), probable (n = 4), or possible (n = 20) invasive fungal lung infections refractory to L-AmB monotherapy 3–5 mg/kg/day (13). The dosage of caspofungin was 50 mg/day with a single loading dose of 70 mg on day 1. The median duration of combination therapy was 24 (range 3–74) days. In 18 patients there was a favorable antifungal response, defined as improvement in both clinical and radiographic signs of fungal pneumonia. There was mild to moderate nephrotoxicity in 15 patients, necessitating the substitution of liposomal amphotericin. There were mild rises in alkaline phosphatase *activity* in nine patients. Caspofungin was temporarily withheld from one patient who developed moderate but reversible biochemical hepatotoxicity.

These analyses suggest that caspofungin and L-AmB can be used safely together.

Caspofungin has been investigated in a non-comparative phase II clinical study in 90 patients with invasive aspergillosis (14). Two patients withdrew because of drug-related adverse events; 84 developed at least one clinical adverse event. However, only 11 had an adverse event that was possibly, probably, or definitely related to caspofungin. Of the 90 patients 53 developed at least one laboratory adverse event, but only 12 were considered to have had a drug-related adverse event. Drug-related nephrotoxicity (1.1%) and hepatotoxicity (under 5%) occurred infrequently.

The efficacy and safety of caspofungin as salvage therapy for invasive aspergillosis has been studied in patients enrolled in the Caspofungin Compassionate Use Study (15). There was a favorable response in 20/45 patients, including nine and 11 with complete and partial responses respectively. One serious drug-related adverse event was reported in a patient with acute biphenotypic leukemia, who had an *anaphylactic reaction* to caspofungin, characterized by stridor/dyspnea, facial swelling, and accentuation of a pre-existing skin rash about 10 minutes into the infusion; all the symptoms resolved within 15 minutes of withdrawal of caspofungin and the administration of diphenhydramine and hydrocortisone.

Caspofungin 50 mg/day for a median duration of 20 (range 8–64) days has been used as first-line therapy for proven or probable pulmonary fungal infection in 32 immunocompromised patients with hematological malignancies (median age 52 years) (16). The overall response rate was 56% (18/32), with 12/18 complete responses and 6/18 partial responses. Granulocyte recovery and status of disease (remission/onset versus refractory/relapsed) were significantly associated with a favorable outcome. There were no clinical adverse events and only grades I and II transient increases in serum alkaline phosphatase and/or transaminase activities in 4/32 patients.

Caspofungin has been studied in a multicenter, open, non-comparative phase II trial in 83 patients with definite or probable invasive aspergillosis refractory to or intolerant of standard therapies (17,18). Common underlying conditions included hematological malignancies (48%), allogeneic blood and marrow transplantation (25%), and

solid organ transplantation (11%). Caspofungin was administered in a dose of 70 mg on day 1, followed by 50 mg/day for a mean duration of 34 (range 1–162) days. There was a favorable response to caspofungin in 37 patients, including 32 of 64 with pulmonary aspergillosis and 3 of 13 with disseminated aspergillosis. Caspofungin was generally well tolerated. One serious adverse event was reported as possibly drug-related. Infusion-related reactions, nephrotoxicity, and hepatotoxicity were uncommon. Two patients discontinued caspofungin because of adverse effects.

The safety and tolerability of caspofungin have been studied in 623 patients, including 295 who received at least 50 mg/day for at least 1 week in clinical studies. In 263 patients who received caspofungin in randomized, double-blind, active-control trials, there were no serious clinical or laboratory drug-related adverse events; caspofungin was withdrawn in 2% of these patients because of drug-related adverse effects (19).

Micafungin

In an open, non-comparative study, neonates, children, and adults with new or refractory candidemia were given micafungin for up to 42 days (20). A total of 126 patients were evaluable and received at least five doses. There was a complete or partial response in 83% and serious adverse events related to micafungin were uncommon.

Comparative studies

Amphotericin

Caspofungin (Cancidas; 50 and 70 mg/day for 14 days; n = 74) has been compared with conventional amphotericin (0.5 mg/kg/day for 14 days; n − 54) in the treatment of esophageal candidiasis in a randomized, double-blind, multicenter trial in South America (21). Most of the patients (over 75%) were HIV-infected and about half of them had CD4+ lymphocyte counts of under 50×10^6/l. Caspofungin was well tolerated: eight patients in the amphotericin group and one patient in the combined caspofungin group developed a raised serum creatinine of over 176 μmol/l (2 mg/dl) during treatment. Of the patients who received caspofungin, 4.1% withdrew prematurely owing to drug-associated adverse effects, compared with 22% in the amphotericin arm. There was a clinical response (symptoms plus endoscopy) in 85% of the patients in the combined caspofungin group versus 67% in the amphotericin group.

Caspofungin has been compared with amphotericin in a multicenter, double-blind, randomized trial in 128 adults with endoscopically documented symptomatic *Candida* esophagitis (8). There was endoscopically verified clinical success in 74 and 89% of the patients who received caspofungin 50 and 70 mg/day respectively, and in 63% of patients given amphotericin deoxycholate 0.5 mg/kg/day. Therapy was withdrawn because of drug-related adverse events in 24% of the patients who were given amphotericin and in 4 and 7% of those who were given caspofungin 50 and 70 mg/day respectively. The most frequent adverse events with caspofungin were fever, phlebitis, headache, and rash. Fewer patients who received caspofungin had drug-related fever, chills, or nausea than those who received amphotericin. More patients who received amphotericin (91%) than caspofungin (61 and 32%) developed drug-related laboratory abnormalities, the most common in the caspofungin groups being hypoalbuminemia and increased serum activities of alkaline phosphatase and transaminases. There were drug-related increases in blood urea nitrogen concentrations in 15% of the patients who received amphotericin but in none of those who received caspofungin. Likewise, serum creatinine concentrations increased in 16 patients who received amphotericin but in only one who received caspofungin. In summary, caspofungin was as effective as amphotericin but better tolerated in the treatment of esophageal candidiasis.

In a double-blind, randomized trial, caspofungin was compared with amphotericin deoxycholate for the primary treatment of invasive candidiasis (10). Patients who had clinical evidence of infection and a positive culture for *Candida* species from blood or another site were enrolled. They were stratified according to the severity of disease, as indicated by the presence or absence of neutropenia and the Acute Physiology and Chronic Health Evaluation (APACHE II) score, and were randomly assigned to receive either caspofungin (50 mg/day with a loading dose of 70 mg on day 1) or amphotericin (0.6–0.7 mg/kg/day or 0.7–1.0 mg/kg/day for patients with neutropenia). Of the 239 patients enrolled, 224 were included in the modified intention-to-treat analysis. Baseline characteristics, including the percentage of patients with neutropenia and the mean APACHE II score, were similar in the two treatment groups. The efficacy of caspofungin was similar to that of amphotericin, with successful outcomes in 73% of the patients treated with caspofungin and in 62% of those treated with amphotericin. There were significantly fewer drug-related adverse events associated with caspofungin: fever, chills, and infusion-related events were less frequent with caspofungin. Caspofungin caused less nephrotoxicity (as defined by an increase in serum creatinine of at least twice the baseline value or an increase of at least 88.4 μmol/l: 8.4 versus 25%). Only 2.6% of those who were given caspofungin were withdrawn because of adverse events, compared with 23% of those who were given amphotericin. Thus, caspofungin was at least as effective as amphotericin for the treatment of mostly non-neutropenic patients with invasive candidiasis but significantly better tolerated.

The safety, tolerability, and efficacy of caspofungin in patients with oropharyngeal and/or esophageal candidiasis have been investigated in a phase II dose-ranging study (22). The patients were randomized, double-blind to either caspofungin acetate (35, 50, or 70 mg) or amphotericin (0.5 mg/kg intravenously) once daily for 7–14 days. Of 140 patients, 63% had esophageal involvement and 98% were infected with HIV. Response rates with caspofungin groups were 74–91%, and 63% with amphotericin. Fewer patients receiving any dose of caspofungin had drug-related adverse effects (fever, chills, nausea, vomiting). Two patients who took caspofungin 35 mg and one who was given amphotericin withdrew because of adverse effects. Drug-related laboratory abnormalities were also more common in

patients who received amphotericin. The most common drug-related laboratory abnormalities in patients who received caspofungin were raised alanine transaminase, aspartate transaminase, and alkaline phosphatase, which were typically less than five times the upper limit of normal and resolved despite continued treatment. None of the patients receiving caspofungin and nine of those who received amphotericin developed drug-related increases in serum creatinine concentrations. No patient withdrew because of drug-related laboratory adverse effects.

Caspofungin (n = 556) has been compared with liposomal amphotericin (n = 539) in a randomized, double-blind, multinational trial as empirical antifungal therapy (23). Patients were stratified according to risk and whether they had previously received antifungal prophylaxis. Premature withdrawal for any cause was less common with caspofungin than amphotericin (10 versus 15%). Fewer patients who received caspofungin sustained nephrotoxicity (2.6 versus 12%), an *infusion-related event* (35 versus 52%), or a systemic drug-related adverse event or discontinued therapy because of drug-related adverse events.

Antifungal azoles

Caspofungin and fluconazole have been compared in adults with *Candida* esophagitis in a double-blind, randomized trial (9). Eligible patients had symptoms compatible with esophagitis, endoscopic mucosal plaques, and microscopic *Candida*. They were randomized to receive caspofungin (50 mg) or fluconazole (200 mg) intravenously once a day for 7–21 days. Most of them (154/177) had HIV infection, with a median CD4 count of $30 \times 10^6/l$. Favorable response rates were achieved in 66 of the 81 patients in the caspofungin arm and in 80 of the 94 patients in the fluconazole arm; symptoms had resolved in over 50% of the patients in both groups by the fifth day of treatment. Drug-related adverse effects were reported in 41% of patients given caspofungin and 32% of those given fluconazole; the most common events in both groups were phlebitis, headache, fever, nausea, diarrhea, abdominal pain, and rashes. Drug-related laboratory abnormalities developed in 29% of patients given caspofungin and in 34% of those given fluconazole. The most frequent laboratory abnormalities included reduced white blood cell count, hemoglobin concentration, and serum albumin concentration, and increased alkaline phosphatase and transaminases. No patient given caspofungin developed a serious drug-related adverse effect; therapy was withdrawn in only one patient (who was receiving fluconazole), because of an unspecified adverse effect.

In a randomized, double-blind, double-dummy comparison of intravenous anidulafungin 50 mg/day and oral fluconazole 100 mg/day for 14–21 days in 601 patients with proven esophageal candidiasis (24). The safety profiles were similar. Treatment-related adverse events occurred in 9.3% and 12% of patients respectively. Laboratory parameters were similar between the treatments. Drug withdrawal was required in five patients (four versus one) because of adverse events.

The dose-response relation of micafungin (50, 100, or 150 mg) has been compared with that of standard fluconazole treatment for esophageal candidiasis in a randomized, double-blind study in 245 HIV-infected patients (25). The incidence of adverse events was 93% with micafungin and 89% with fluconazole. The most common treatment-related adverse events were fever, abdominal pain, nausea, diarrhea, leukopenia, injection-site inflammation, and headache. There were no clinically important changes in laboratory measures. There were changes in liver function tests in 13% of patients: micafungin 24/185 patients and fluconazole 7/60 patients. There were no apparent dose-dependent differences in safety and tolerance across the investigated dosages of micafungin, and no differences in safety and tolerance between micafungin and fluconazole.

In a randomized, double-blind, multi-institutional, comparative phase III trial in antifungal prophylaxis during neutropenia in patients undergoing hemopoietic stem cell transplantation (26) micafungin 1 mg/kg/day up to a maximum of 50 mg/day was compared with fluconazole 8 mg/kg/day up to a maximum of 400 mg/day in 882 adults and children (26). All the patients had at least one adverse event, but fewer micafungin-treated patients withdrew as a result (4.2% versus 7.2%). Adverse events that were considered to be drug-related occurred in 15% of micafungin-treated patients and in 17% of fluconazole-treated patients. There were no differences in liver function tests between patients who received micafungin + ciclosporin and those who received micafungin alone.

These two studies suggest a safety profile of micafungin that is at least similar to that of fluconazole.

In a large, randomized comparative study in 523 adults with esophageal candidiasis randomized to either intravenous micafungin 150 mg/day or fluconazole 200 mg/day for a median of 14 days, the incidences of drug-related adverse events were 28% for micafungin and 21% for fluconazole (57). Six patients taking micafungin and two taking fluconazole withdrew, most commonly because of rashes.

In a sequential dose escalation study, 74 adults with cancer undergoing bone marrow or peripheral blood stem cell transplantation were given fluconazole 400 mg/day and either isotonic saline (control, n = 12) or micafungin 12.5–200 mg/day (n = 62) for up to 4 weeks (27). The maximum tolerated dose of micafungin was not reached, based on the Southwest Oncology Group criteria for grade 3 toxicity; drug-related adverse events were rare. Common adverse events that were considered to be related to micafungin were headache (6.8%), arthralgia (6.8%), hypophosphatemia (4.1%), insomnia (4.1%), maculopapular rashes (4.1%), and other rashes (4.1%). There was no clinical or kinetic evidence of an interaction of micafungin with fluconazole.

Organs and Systems

Hematologic

A 68-year-old man developed reversible severe thrombocytopenia, possibly due to caspofungin, after being successfully treated for *Candida albicans* endocarditis (28).

Autacoids

The effects of caspofungin and amphotericin B on the release of histamine from human peripheral blood cells, mononuclear cells, and mast cells have been investigated in in vitro cell culture experiments (29). Cultured human mononuclear (THP-1) and mast (HMC-1) cells from five healthy volunteers were incubated with increasing concentrations of amphotericin B deoxycholate, diphenhydramine, amphotericin B deoxycholate plus diphenhydramine, caspofungin, caspofungin plus diphenhydramine, and the calcium ionophore A23187 for up to 24 hours. Histamine concentrations and histamine N-methyltransferase activity were determined at various times. Cell viability was assessed by exclusion of erythrocin B. A23187 increased histamine concentrations from baseline in peripheral blood and HMC-1 cells. There were no change in histamine concentrations in response to amphotericin B deoxycholate, whereas caspofungin caused a significant increase in histamine release in peripheral blood cells and HMC-1 cells. There were no changes in histamine concentrations in THP-1 cells in response to any agent. Similarly, histamine N-methyltransferase activity in peripheral blood was not affected by amphotericin B deoxycholate, but was significantly reduced by caspofungin. These results suggest that the amphotericin B-induced infusion-related reaction is not a histamine-mediated event. Conversely, caspofungin increased histamine concentrations in whole blood and HMC-1 cells and inhibited histamine N-methyltransferase activity. Thus, infusion-related reactions associated with caspofungin may be mediated by histamine release secondary to caspofungin.

Susceptibility Factors

Age

Children

In a retrospective analysis of the safety of off-label caspofungin in 25 immunocompromised children (median age 9.8 years; range 0.3–26) who had received at least one dose of caspofungin, 21 received It in combination with L-AmB, and four received caspofungin alone (30). The median duration of caspofungin therapy was 32 (range 1?116) days; patients weighing over 50 kg received 50?75 mg/day, and those weighing under 50 kg received 0.8?1.6 mg/kg/day. Three patients had at least one adverse event that was judged possibly related to caspofungin: hypokalemia (n = 3), raised serum bilirubin (n = 2), reduced hemoglobin (n = 1), and increased alanine transaminase (n = 1). No patient had an adverse event that was probably or definitely related to caspofungin, and there were no serious drug-related adverse events. Antifungal efficacy was not analysed.

In 39 children aged 2–11 years and adolescents aged 12–17 years with neutropenia, caspofungin, 1 mg/kg/day or 50 or 70 mg/m^2/day, was generally well tolerated (31). None developed a serious drug-related adverse event or were withdrawn because of toxicity.

In 13 infants in whom caspofungin was added to conventional antifungal drugs (amphotericin B and/or fluconazole or flucytosine) for refractory candidemia, 12 of whom were preterm, sterilization of blood cultures was achieved in 11 at a median time of 3 (range 1–21) days (32). Adverse events included thrombophlebitis (n = 1), hypokalemia (n = 2), and raised liver enzymes (n = 4). Three infants had a second episode of candidemia and seven died.

In a multicenter, phase I, open, sequential-group, dose-escalation study in 77 children with neutropenia micafungin was begun at a dosage of 0.5 mg/kg/day and increased to 1.0, 1.5, 2.0, 3.0, and 4.0 mg/kg/day (33). The most common adverse events were diarrhea (20%), epistaxis (18%), abdominal pain (17%), and headache (17%). Nine patients had adverse events that were considered by the investigator to be possibly related to micafungin; of these, the most common were diarrhea, vomiting, and headache, each of which occurred in two patients. There was an inverse relation between age and micafungin clearance: in those aged 2–8 years, clearance was about 1.35 times that of patients 9 years of age and older.

Liver disease and renal disease

The pharmacokinetics and plasma protein binding of micafungin have been studied in patients with moderate hepatic dysfunction (n = 8), patients with creatinine clearances below 30 ml/minute (n = 9), and matched controls (n = 8 and 9 respectively) (34). The AUC of intravenous micafungin was significantly lower in those with moderate hepatic dysfunction than in controls, but there was no difference in micafungin weight-adjusted clearance. The difference in AUC may have been due to differences in body weight. Renal dysfunction did not alter micafungin pharmacokinetics.

Drug–Drug Interactions

Micafungin had little or no effect on CYP3A4-related metabolism in human liver microsomes or MDR1 transport activity in MDR1-overexpressing LLC-GA5-COL150 cells (35). Thus, micafungin is unlikely to cause drug-drug interactions by inhibition of CYP3A4 or MDR1. In similar in vitro studies, micafungin neither inhibited nor stimulated the metabolic activities mediated by CYP1A2, CYP2D6, CYP2E1, CYP2C9, or CYP2C19. The IC_{50} of micafungin against CYP3A4-mediated nifedipine oxidation was comparable with that of voriconazole and fluconazole, compatible with the finding that micafungin is a mild inhibitor of CYP3A4 (36) (37).

Amphotericin

There were no pharmacokinetic interactions of caspofungin with amphotericin deoxycholate in healthy volunteers (6) and two retrospective analyses have suggested that caspofungin and liposomal amphotericin can be given safely in combination in the usual therapeutic dosages (38,39).

Antifungal azoles

In rats, caspofungin did not alter the plasma pharmacokinetics of ketoconazole, a potent inhibitor of CYP3A4 (6). Co-administration of caspofungin 50 mg/day and itraconazole 200 mg/day to healthy subjects for 14 days did not alter the pharmacokinetics of either drug (40).

Ciclosporin

Ciclosporin increased the AUC of caspofungin by about 35%, but caspofungin did not increase the plasma concentrations of ciclosporin. Because of transient rises in hepatic transaminases not exceeding 2–3 times the upper limit of the reference range in single-dose interaction studies, concomitant use of caspofungin with ciclosporin should be undertaken with care (6), although limited experience in patients studied by the manufacturers suggests that it may be safe.

The effect of anidulafungin on ciclosporin metabolism has been studied in vitro, in pooled human hepatic microsomal protein fractions, and in vivo, in a multiple-dose, open study in 12 healthy volunteers (41c). Anidulafungin 200 mg intravenously was followed by 100 mg/day intravenously on days 2–8. Ciclosporin 1.25 mg/kg bd was given orally on days 5–8. The in vitro addition of anidulafungin had no effect on ciclosporin metabolism by hepatic microsomes. In the clinical study, there were no dose-limiting or serious adverse events; there was a small increase in anidulafungin concentrations and drug exposure (22%) after 4 days of ciclosporin, but this was not considered to be clinically important.

In a retrospective chart review at four sites in 40 patients who had taken both drugs for 1–290 days there were rises in transaminases in 14 patients (42). Five had rises in aspartate transaminase activity at least possibly related to caspofungin + ciclosporin, but none was over 3.6 times the normal upper limit. No rises in alanine transaminase activity were related to caspofungin + ciclosporin. In two of four patients who discontinued therapy because of hepatotoxicity, there was a possible relationship to caspofungin + ciclosporin. No serious adverse events occurred because of caspofungin.

In a study of safety data in 14 patients with refractory invasive mycoses who took ciclosporin + caspofungin for 2–34 days before the latter was licensed there were no clinically significant rises in serum transaminases and no patient had concomitant therapy discontinued or interrupted because of a drug-related adverse event (43).

Of 31 patients given caspofungin after allogeneic stem cell transplantation as a second-line agent for treatment of invasive fungal infection (n = 15) or for fever of unknown origin (n = 16), 23 received caspofungin + ciclosporin without any major adverse effects; increases in aspartate transaminase activity were not clinically significant (44).

The effect of micafungin 100 mg/day on the pharmacokinetics of a single oral dose of ciclosporin 5 mg/kg have been studied in 27 subjects (45). Micafungin inhibited ciclosporin metabolism by only about 16%.

Cytarabine

A high throughput microtiter assay was used to determine whether cytarabine, idarubicin, itraconazole, or caspofungin were CYP450 isoenzyme substrates or inhibitors of CYP450 isoenzymes, and to determine potential CYP450 metabolism interactions between them (46). Idarubicin was a substrate for CYP2D6 and CYP2C9 and cytarabine was a substrate of CYP3A4. Idarubicin inhibited CYP2D6, and cytarabine, itraconazole, and caspofungin inhibited CYP3A4. Cytarabine metabolism was significantly reduced when it was combined with caspofungin or itraconazole. This in vitro inhibition of cytarabine metabolism may have important clinical implications that warrant investigation with in vivo pharmacokinetic studies.

Enzyme inducers

Regression analysis of pharmacokinetic data from patients has suggested that co-administration of caspofungin with inducers of drug metabolism and mixed inducer/ inhibitors, namely carbamazepine, dexamethasone, efavirenz, nelfinavir, nevirapine, phenytoin, and rifampicin, can cause clinically important reductions in caspofungin concentrations. However, no data are currently available from formal interaction studies, and it is not known which clearance mechanisms of caspofungin are inducible. The manufacturer currently recommends considering an increase in the daily dose of caspofungin to 70 mg in patients who are taking these drugs concurrently and who are not responding (6).

Indinavir

In rats, caspofungin did not alter the plasma pharmacokinetics of indinavir, a substrate and competitive inhibitor of CYP3A2 (6).

Nelfinavir

The potential for an interaction of caspofungin (50 mg/ day intravenously; n = 10) with nelfinavir (1,250 mg bd orally; n = 9) has been evaluated in two parallel-panel studies. In study A, healthy subjects received a 14-day course of caspofungin alone (47). Nelfinavir did not significantly alter the pharmacokinetics of caspofungin.

Rifampicin

The potential for an interaction of caspofungin (50 mg/ day intravenously; n = 10) with rifampicin (1250 mg bd orally; n = 10) has been evaluated (47). Rifampicin increased the initial AUC and trough concentration of caspofungin and reduced the steady-state trough concentration, consistent with initial transient inhibition of metabolism and net induction at steady state. Therefore, an increase in the dose of caspofungin to 70 mg/day should be considered when it is co-administered with rifampicin.

Caspofungin treatment failure has been reported in a patient with invasive candidiasis and concomitant rifampicin treatment (48). Rifampicin is a non-specific inducer/ activator of mixed-function oxygenases and may increase the metabolism of caspofungin, thereby leading to ineffective plasma concentrations.

Tacrolimus

While tacrolimus had no effect on the plasma pharmacokinetics of caspofungin, chronic caspofungin reduced the AUC of tacrolimus by about 20% (6).

Intravenous micafungin 100 mg had no effect on the pharmacokinetics of oral tacrolimus 5 mg in 26 healthy volunteers (49).

Voriconazole

Co-administration of anidulafungin and voriconazole has been investigated in a placebo-controlled study in 17 healthy subjects (24). Anidulafungin was administered intravenously (200 mg on day 1 then 100 mg/day on days 2–4) and voriconazole orally (400 mg every 12 hours on day 1 then 200 mg every 12 hours on days 2–4). There were no dose-limiting or serious adverse effects and all adverse events were mild and consistent with the known safety profiles of the two drugs. There was no pharmacokinetic interaction.

In a retrospective analysis the combination of caspofungin + voriconazole at currently approved dosages has been assessed (50). Compared with a historical control cohort of patients taking voriconazole alone there was no indication of increased toxicity of the combination with caspofungin.

Monitoring Therapy

The mean trough blood concentration of micafungin in patients with blood diseases in whom it was effective against pulmonary aspergillosis was 5.23 µg/ml in markedly improved cases, 4.08 µg/ml in improved cases, and 3.45 µg/ml in successfully prevented cases, but there was no significant difference among the three groups (51). The authors suggested a target trough blood concentration of 5 µg/ml or higher in treating aspergillosis in patients with blood diseases.

References

1. Kurtz MB, Douglas CM. Lipopeptide inhibitors of fungal glucan synthase. J Med Vet Mycol 1997;35(2):79–86.
2. Georgopapadakou NH. Update on antifungals targeted to the cell wall: focus on beta-1,3-glucan synthase inhibitors. Expert Opin Investig. Drugs 2001;10(2):269–80.
3. Groll AH, Glasmacher A, Just-Nuebling G, Maschmeyer G, Walsh TJ. Clinical pharmacology of antifungal compounds. Infect Dis Clin North Am 2003;17:159-91.
4. Groll AH, Walsh TJ. Antifungal chemotherapy: advances and perspectives. Swiss Med Wkly 2002;132(23–24):303–11.
5. Groll AH, Gea-Banacloche JC, Glasmacher A, Just-Nuebling G, Maschmeyer G, Walsh TJ. Clinical pharmacology of antifungal compounds. Infect Dis Clin North Am 2003;17(1):159–91.
6. Groll AH, Walsh TJ. Caspofungin: pharmacology, safety and therapeutic potential in superficial and invasive fungal infections. Expert Opin Investig Drugs 2001;10(8):1545–58.

7. Colburn DE, Giles FJ, Oladovich D, Smith JA. In vitro evaluation of cytochrome P450-mediated drug interactions between cytarabine, idarubicin, itraconazole and caspofungin. Hematology 2004;9(3):217–21.

8. Villanueva A, Arathoon EG, Gotuzzo E, Berman RS, DiNubile MJ, Sable CA. A randomized double-blind study of caspofungin versus amphotericin for the treatment of candidal esophagitis. Clin Infect Dis 2001;33(9):1529–35.

9. Villanueva A, Gotuzzo E, Arathoon EG, Noriega LM, Kartsonis NA, Lupinacci RJ, Smietana JM, DiNubile MJ, Sable CA. A randomized double-blind study of caspofungin versus fluconazole for the treatment of esophageal candidiasis. Am J Med 2002;113(4):294–9.

10. Mora-Duarte J, Betts R, Rotstein C, Colombo AL, Thompson-Moya L, Smietana J, Lupinacci R, Sable C, Kartsonis N, Perfect JCaspofungin Invasive Candidiasis Study Group. Comparison of caspofungin and amphotericin B for invasive candidiasis. N Engl J Med 2002;347(25):2020–9.

11. Krause DS, Reinhardt J, Vazquez JA, Reboli A, Goldstein BP, Wible M, Henkel T; Anidulafungin Invasive Candidiasis Study Group. Phase 2, randomized, dose-ranging study evaluating the safety and efficacy of anidulafungin in invasive candidiasis and candidemia. Antimicrob Agents Chemother 2004;48:2021-4.

12. Kontoyiannis DP, Hachem R, Lewis RE, Rivero GA, Torres HA, Thornby J, Champlin R, Kantarjian H, Bodey GP, Raad II. Efficacy and toxicity of caspofungin in combination with liposomal amphotericin B as primary or salvage treatment of invasive aspergillosis in patients with hematologic malignancies. Cancer 2003;98:292-9.

13. Aliff TB, Maslak PG, Jurcic JG, Heaney ML, Cathcart KN, Sepkowitz KA, Weiss MA. Refractory *Aspergillus pneumonia in patients with acute leukemia: successful therapy with combination caspofungin and liposomal amphotericin. Cancer 2003;97:1025-32.*

14. Maertens J, Raad I, Petrikkos G, Boogaerts M, Selleslag D, Petersen FB, Sable CA, Kartsonis NA, Ngai A, Taylor A, Patterson TF, Denning DW, Walsh TJ; Caspofungin Salvage Aspergillosis Study Group. Efficacy and safety of caspofungin for treatment of invasive aspergillosis in patients refractory to or intolerant of conventional antifungal therapy. Clin Infect Dis 2004;39:1563-71.

15. Kartsonis NA, Saah AJ, Joy Lipka C, Taylor AF, Sable CA. Salvage therapy with caspofungin for invasive aspergillosis: results from the Caspofungin Compassionate Use Study. J Infect 2005;50(3):196-205.

16. Candoni A, Mestroni R, Damiani D, Tiribelli M, Michelutti A, Silvestri F, Castelli M, Viale P, Fanin R. Caspofungin as first line therapy of pulmonary invasive fungal infections in 32 immunocompromised patients with hematologic malignancies. Eur J Haematol 2005;75(3):227-33.

17. Maertens J, Raad I, Sable CA, Ngui A, Berman R, Patterson TF, Denning D, Walsh TJ. Multicenter, noncomparative study to evaluate safety and efficacy of caspofungin in adults with invasive aspergillosis refractory or intolerant to amphotericin B, amphotericin B lipid formulations, or azoles. In: Abstracts of the 40th International Conference on Antimicrobial Agents and Chemotherapy 1103 2000:2020–9.

18. Maertens J, Raad I, Petrikkos G, Boogaerts M, Selleslag D, Petersen FB, Sable CA, Kartsonis NA, Ngai A, Taylor A, Patterson TF, Denning DW, Walsh TJCaspofungin Salvage Aspergillosis Study Group. Efficacy and safety of caspofungin for treatment of invasive aspergillosis in patients refractory to or intolerant of conventional antifungal therapy. Clin Infect Dis 2004;39(11):1563–71.

19. Sable CA, Nguyen BY, Chodakewitz JA, DiNubile MJ. Safety and tolerability of caspofungin acetate in the treatment of fungal infections. Transpl Infect Dis 2002;4(1):25–30.

20. Ostrosky-Zeichner L, Kontoyiannis D, Raffalli J, Mullane KM, Vazquez J, Anaissie EJ, Lipton J, Jacobs P, van Rensburg JH, Rex JH, Lau W, Facklam D, Buell DN. International, open-label, noncomparative, clinical trial of micafungin alone and in combination for treatment of newly diagnosed and refractory candidemia. Eur J Clin Microbiol Infect Dis 2005;24(10):654-61.

21. Sable CA, Villanueva A, Arathon E, Gotuzzo E, Turcato G, Uip D, Noriega L, Rivera C, Rojas E, Taylor V, Berman R, Calandra GB, Chodakewitz J. A randomized, double-blind, multicenter trial of MK-991 (L-743,872) vs. amphotericin B (AMB) in the treatment of *Candida* esophagitis in adults. In: Abstracts of the 37th Interscience Conference on Antimicrobial Agents and Chemotherapy LB-33 1997:25–30.

22. Arathoon EG, Gotuzzo E, Noriega LM, Berman RS, DiNubile MJ, Sable CA. Randomized, double-blind, multicenter study of caspofungin versus amphotericin B for treatment of oropharyngeal and esophageal candidiases. Antimicrob Agents Chemother 2002;46(2):451–7.

23. Walsh TJ, Teppler H, Donowitz GR, Maertens JA, Baden LR, Dmoszynska A, Cornely OA, Bourque MR, Lupinacci RJ, Sable CA, dePauw BE. Caspofungin versus liposomal amphotericin B for empirical antifungal therapy in patients with persistent fever and neutropenia. N Engl J Med 2004;351:1391-402.

24. Krause DS, Simjee AE, van Rensburg C, Viljoen J, Walsh TJ, Goldstein BP, Wible M, Henkel T. A randomized, double-blind trial of anidulafungin versus fluconazole for the treatment of esophageal candidiasis. Clin Infect Dis 2004;39:770-5.

25. de Wet N, Llanos-Cuentas A, Suleiman J, Baraldi E, Krantz EF, Della Negra M, Diekmann-Berndt H. A randomized, double-blind, parallel-group, dose-response study of micafungin compared with fluconazole for the treatment of esophageal candidiasis in HIV-positive patients. Clin Infect Dis 2004;39:842-9.

26. van Burik JA, Ratanatharathorn V, Stepan DE, Miller CB, Lipton JH, Vesole DH, Bunin N, Wall DA, Hiemenz JW, Satoi Y, Lee JM, Walsh TJ; National Institute of Allergy and Infectious Diseases Mycoses Study Group. Micafungin versus fluconazole for prophylaxis against invasive fungal infections during neutropenia in patients undergoing hematopoietic stem cell transplantation. Clin Infect Dis 2004;39:1407-16.

27. Hiemenz J, Cagnoni P, Simpson D, Devine S, Chao N, Keirns J, Lau W, Facklam D, Buell D. Pharmacokinetic and maximum tolerated dose study of micafungin in combination with fluconazole versus fluconazole alone for prophylaxis of fungal infections in adult patients undergoing a bone marrow or peripheral stem cell transplant. Antimicrob Agents Chemother 2005;49(4):1331-6.

28. Lynch J, Wong-Beringer A. Caspofungin: a potential cause of reversible severe thrombocytopenia. Pharmacotherapy 2004;24:1408-11.

29. Cleary JD, Schwartz M, Rogers PD, de Mestral J, Chapman SW. Effects of amphotericin B and caspofungin on histamine expression. Pharmacotherapy 2003;23:966-73.

30. Franklin JA, McCormick J, Flynn PM. Retrospective study of the safety of caspofungin in immunocompromised pediatric patients. Pediatr Infect Dis J 2003;22:747-9.

31. Walsh TJ, Adamson PC, Seibel NL, Flynn PM, Neely MN, Schwartz C, Shad A, Kaplan SL, Roden MM, Stone JA, Miller A, Bradshaw SK, Li SX, Sable CA, Kartsonis NA. Pharmacokinetics, safety, and tolerability of caspofungin in children and adolescents. Antimicrob Agents Chemother 2005;49(11):4536-45.

32. Natarajan G, Lulic-Botica M, Rongkavilit C, Pappas A, Bedard M. Experience with caspofungin in the treatment of persistent fungemia in neonates. J Perinatol 2005;25(12):770-7.

33. Seibel NL, Schwartz C, Arrieta A, Flynn P, Shad A, Albano E, Keirns J, Lau WM, Facklam DP, Buell DN, Walsh TJ. Safety, tolerability, and pharmacokinetics of Micafungin (FK463) in febrile neutropenic pediatric patients. Antimicrob Agents Chemother 2005;49(8):3317-24.

34. Hebert MF, Smith HE, Marbury TC, Swan SK, Smith WB, Townsend RW, Buell D, Keirns J, Bekersky I. Pharmacokinetics of micafungin in healthy volunteers, volunteers with moderate liver disease, and volunteers with renal dysfunction. J Clin Pharmacol 2005;45(10):1145-52.

35. Sakaeda T, Iwaki K, Kakumoto M, Nishikawa M, Niwa T, Jin JS, Nakamura T, Nishiguchi K, Okamura N, Okumura K. Effect of micafungin on cytochrome P450 3A4 and multidrug resistance protein 1 activities, and its comparison with azole antifungal drugs. J Pharm Pharmacol 2005;57(6):759-64.

36. Niwa T, Inoue-Yamamoto S, Shiraga T, Takagi A. Effect of antifungal drugs on cytochrome P450 (CYP) 1A2, CYP2D6, and CYP2E1 activities in human liver microsomes. Biol Pharm Bull 2005;28(9):1813-6.

37. Niwa T, Shiraga T, Takagi A. Effect of antifungal drugs on cytochrome P450 (CYP) 2C9, CYP2C19, and CYP3A4 activities in human liver microsomes. Biol Pharm Bull 2005;28(9):1805-8.

38. Kontoyiannis DP, Hachem R, Lewis RE, Rivero GA, Torres HA, Thornby J, Champlin R, Kantarjian H, Bodey GP, Raad II. Efficacy and toxicity of caspofungin in combination with liposomal amphotericin B as primary or salvage treatment of invasive aspergillosis in patients with hematologic malignancies. Cancer 2003;98:292-99.

39. Aliff TB, Maslak PG, Jurcic JG, Heaney ML, Cathcart KN, Sepkowitz KA, Weiss MA. Refractory Aspergillus pneumonia in patients with acute leukemia: successful therapy with combination caspofungin and liposomal amphotericin. Cancer 2003;97:1025-32.

40. Stone JA, McCrea J, Wickersham P, Holland S, Deutsch P, Bi S, Cicero T, Greenberg H, Waldman SA. Phase I study of caspofungin evaluating the potential for drug interactions with itraconazole, the effect of gender and the use of a loading dose. In: Abstracts of the 40th Interscience Conference on Antimicrobial Agents and Chemotherapy 854 2000:451-7.

41. Dowell JA, Stogniew M, Krause D, Henkel T, Weston IE. Assessment of the safety and pharmacokinetics of anidulafungin when administered with cyclosporine. J Clin Pharmacol 2005;45(2):227-33.

42. Marr KA, Hachem R, Papanicolaou G, Somani J, Arduino JM, Lipka CJ, Ngai AL, Kartsonis N, Chodakewitz J, Sable C. Retrospective study of the hepatic safety profile of patients concomitantly treated with caspofungin and cyclosporin A. Transpl Infect Dis 2004;6:110-6.

43. Sanz-Rodriguez C, Lopez-Duarte M, Jurado M, Lopez J, Arranz R, Cisneros JM, Martino ML, Garcia-Sanchez PJ, Morales P, Olive T, Rovira M, Solano C. Safety of the concomitant use of caspofungin and cyclosporin A in

44. Trenschel R, Ditschkowski M, Elmaagacli AH, Koldehoff M, Ottinger H, Steckel N, Hlinka M, Peceny R, Rath PM, Dermoumi H, Beelen DW. Caspofungin as second-line therapy for fever of unknown origin or invasive fungal infection following allogeneic stem cell transplantation. Bone Marrow Transplant 2005;35(6):583-6.

45. Hebert MF, Townsend RW, Austin S, Balan G, Blough DK, Buell D, Keirns J, Bekersky I. Concomitant cyclosporine and micafungin pharmacokinetics in healthy volunteers. J Clin Pharmacol 2005;45(8):954-60.

46. Colburn DE, Giles FJ, Oladovich D, Smith JA. In vitro evaluation of cytochrome P450-mediated drug interactions between cytarabine, idarubicin, itraconazole and caspofungin. Hematology 2004;9:217-21.

47. Stone JA, Migoya EM, Hickey L, Winchell GA, Deutsch PJ, Ghosh K, Freeman A, Bi S, Desai R, Dilzer SC, Lasseter KC, Kraft WK, Greenberg H, Waldman SA. Potential for interactions between caspofungin and nelfinavir or rifampin. Antimicrob Agents Chemother 2004;48:4306-14.

48. Belmares J, Colaizzi L, Parada JP, Johnson S. Caspofungin treatment failure in a patient with invasive candidiasis and concomitant rifampicin treatment. Int J Antimicrob Agents 2005;26(3):264-5.

49. Hebert MF, Blough DK, Townsend RW, Allison M, Buell D, Keirns J, Bekersky I. Concomitant tacrolimus and micafungin pharmacokinetics in healthy volunteers. J Clin Pharmacol 2005;45(9):1018-24.

50. Marr KA, Boeckh M, Carter RA, Kim HW, Corey L. Combination antifungal therapy for invasive aspergillosis. Clin Infect Dis 2004;39:797-802.

51. Shimoeda S, Ohta S, Kobayashi H, Yamato S, Sasaki M, Kawano K. Effective blood concentration of micafungin for pulmonary aspergillosis. Biol Pharm Bull 2006;29(9):1886-91.

Fluconazole

See also Antifungal azoles

General Information

Fluconazole is an antifungal triazole that was derived from the older imidazoles. It has a lower molecular weight and is soluble in water. It can be administered orally and parenterally.

Pharmacokinetics

After oral administration, its systemic availability is about 90%; maximum plasma concentrations are seen in 1–2 hours, its half-life is about 30 hours and steady state is reached within 5–7 days. Fluconazole has low protein binding (about 12%), is hardly metabolized, and about 80% of it is excreted in the urine. Hemodialysis reduces plasma concentrations by about half in 3 hours (1–3). Tissue and body fluid penetration is good (2,4–9). The CSF concentration ratio is 0.5–0.9, the resultant concentrations being adequate for the treatment of cryptococcal meningitis (10–12). Penetration into the eye is good (13), and fluconazole has been used successfully in fungal endophthalmitis (14–17). In a patient in whom the concentrations of fluconazole in bile were studied, the concentrations after the first dose were about the same as in serum, but 10–12 hours after the dose, bile concentrations were higher than serum concentrations (18). Sputum concentrations are similar to plasma concentrations. Concentrations in vaginal secretions are slightly lower than in plasma, but persist for longer (19,20).

Like ketoconazole, fluconazole is a potent inhibitor of cytochrome P450, but with much higher specificity for fungal enzymes compared with human enzymes (21,22). Clinical interaction studies and some in vitro studies have suggested that azole antifungal drugs inhibit P glycoprotein. In a cell line in which human P glycoprotein was overexpressed, itraconazole and ketoconazole inhibited P glycoprotein function, with 50% inhibitory concentrations of about 2 and 6 µmol/l respectively; however, fluconazole had no effect (23).

Observational studies

The prophylactic use of oral fluconazole to prevent invasive *Candida* infections in 260 critically ill surgical patients has been investigated in a prospective, randomized, placebo-controlled trial in a single-center, tertiary-care surgical intensive care unit (24). The patients were randomly assigned to receive either oral fluconazole 400 mg/day or placebo. The risk of presumed and proven *Candida* infections in the patients who received fluconazole was significantly less than the risk in those who received placebo. After adjusting for several potentially confounding effects, fluconazole reduced the risk of presumed and proven fungal infection by 55%. There was no difference in death rate between fluconazole and placebo. The authors concluded that enteral fluconazole safely reduced the incidence of fungal infections in this high-risk population.

Outcomes in 20 patients with solid tumors and candidemia treated with high-dose fluconazole (at least 600 mg/day) have been reported (25). There were no significant adverse effects.

Comparative studies

Amphotericin

Fluconazole and amphotericin as empirical antifungal drugs in febrile neutropenic patients have been investigated in a prospective, randomized, multicenter study in 317 patients randomized to either fluconazole (400 mg qds) or amphotericin deoxycholate (0.5 mg/kg qds) (26). Adverse events (fever, chills, renal insufficiency, electrolyte disturbances, and respiratory distress) occurred significantly more often in patients who were given amphotericin (128/151 patients, 81%) than in those given fluconazole (20/158 patients, 13%). Eleven patients treated with amphotericin, but only one treated with fluconazole, were withdrawn because of an adverse event. Overall mortality and mortality from fungal infections were similar in both groups. There was a satisfactory response in 68% of the patients treated with fluconazole

and 67% of those treated with amphotericin. Thus, fluconazole may be a safe and effective alternative to amphotericin for empirical therapy of febrile neutropenic patients; however, since fluconazole is ineffective against opportunistic molds, the possibility of an invasive infection by a filamentous fungus should be excluded before starting empirical therapy. Similarly, patients who take azoles for prophylaxis are not candidates for empirical therapy with fluconazole.

Conventional amphotericin deoxycholate (0.2 mg/kg qds) and fluconazole (400 mg qds) have been compared in a prospective, randomized study in 355 patients with allogeneic and autologous bone marrow transplantation (27). The drugs were given prophylactically from day –1 until engraftment. There was no difference in the occurrence of invasive fungal infections, but amphotericin was significantly more toxic than fluconazole, especially in related allogeneic transplantation, after which 19% of patients developed toxicity compared with none of those who received fluconazole.

Echinocandins

Caspofungin and fluconazole have been compared in adults with *Candida* esophagitis in a double-blind, randomized trial (28). Eligible patients had symptoms compatible with esophagitis, endoscopic mucosal plaques, and microscopic *Candida*. They were randomized to receive caspofungin (50 mg) or fluconazole (200 mg) intravenously once a day for 7–21 days. Most of them (154/177) had HIV infection, with a median CD4 count of 30×10^6/l. Favorable response rates were achieved in 66 of the 81 patients in the caspofungin arm and in 80 of the 94 patients in the fluconazole arm; symptoms had resolved in over 50% of the patients in both groups by the fifth day of treatment. Drug-related adverse effects were reported in 41% of patients given caspofungin and 32% of those given fluconazole; the most common events in both groups were phlebitis, headache, fever, nausea, diarrhea, abdominal pain, and rashes. Drug-related laboratory abnormalities developed in 29% of patients given caspofungin and in 34% of those given fluconazole. The most frequent laboratory abnormalities included reduced white blood cell count, hemoglobin concentration, and serum albumin concentration, and increased alkaline phosphatase and transaminases. No patient given caspofungin developed a serious drug-related adverse effect; therapy was withdrawn in only one patient (who was receiving fluconazole), because of an unspecified adverse effect.

Other antifungal azoles

In 304 patients randomized to fluconazole 400 mg/day or itraconazole either orally 2.5 mg/kg tds or intravenously 200 mg/day for 180 days after allogeneic stem cell transplantation, or until 4 weeks after discontinuation of graft-versus-host disease therapy, more patients stopped taking itraconazole because of adverse effects (36% versus 16%), in most cases because of gastrointestinal complaints (24% versus 4%) (29). More patients who took itraconazole had at least a tripling of the baseline total bilirubin concentration (95% versus 86%).

Fluconazole and itraconazole have been compared in 252 non-neutropenic cancer patients with oropharyngeal candidiasis (30). The safety and tolerance profiles of the two drugs were comparable, but the adverse effects were not listed in detail.

Placebo-controlled studies

In a randomized, double-blind, placebo-controlled study in Saudi Arabia of oral fluconazole (200 mg/day for 6 weeks) in the treatment of cutaneous leishmaniasis, 106 patients were assigned to fluconazole and 103 to placebo (31). Follow-up data were available for 80 and 65 patients respectively. At the 3-month follow-up, healing of lesions was complete in 63 of the 80 patients who took fluconazole and 22 of the 65 patients who took placebo (relative risk of complete healing, 2.33; 95% CI = 1.63, 3.33). Adverse effects were mild and similar in the two groups.

General adverse effects

Fluconazole is generally well tolerated. The most common adverse effects are nausea and vomiting. Abnormal liver function tests and slight increases in hepatic enzymes have been reported, and there have been anecdotal reports of hepatitis and hepatic failure. Early studies have shown no changes in testosterone concentrations or in the adrenal response to ACTH. Rashes and a few cases of exfoliative skin disorder have been reported (SED-12, 681) and have been seen more frequently in patients with AIDS (32). Alopecia has been reported in a few cases with the use of high doses for prolonged periods of time (SEDA-18, 281). Rare instances of anaphylactoid reactions have been reported (SED-12, 681) (33). Rare instances of hypersensitivity reactions have occurred in other individuals (SEDA-16, 293) (SEDA-17, 320). Tumor-inducing effects have not been reported.

In a study using the UK General Practice Research Database to determine rates of drug-induced, rare, serious adverse effects on the liver, kidneys, skin, or blood, occurring within 45 days of completing a prescription or refill in 54 803 users of either fluconazole or itraconazole, three had illnesses for which a fluconazole-induced cause could not be ruled out; one with thrombocytopenia, one with neutropenia, and one with an abnormal liver function test just after receiving fluconazole (34). The rates were 2.8/100 000 prescriptions (95% CI = 0.8, 10) for serious, adverse blood events and 1.4/100 000 prescriptions (95% CI = 0.25, 8.2) for serious, adverse liver events. These results suggest that fluconazole does not commonly have serious adverse effects on the liver, kidneys, skin, or blood.

Organs and Systems

Cardiovascular

Prolongation of the QT interval is a class effect of the antifungal azoles and has occasionally been reported with fluconazole, with a risk of torsade de pointes.

- A 68-year-old woman with *Candida glabrata* isolated from a presacral abscess developed torsade de pointes after 8 days treatment with oral fluconazole 150 mg/day (35). She had no other risk factors for torsade de pointes, including coronary artery disease, cardiomyopathy, congestive heart failure, or electrolyte abnormalities. The dysrhythmia resolved when fluconazole was withdrawn, but she continued to have ventricular extra beats and non-sustained ventricular tachycardia for 6 days.
- A 59-year-old woman with liver cirrhosis and *Candida* peritonitis developed long QT syndrome and torsade de pointes after intravenous therapy with 400–800 mg/day of fluconazole for 65 weeks, followed by intraperitoneal administration (150 mg/day) (36). One day after the second intraperitoneal administration, she developed palpitation, multifocal ventricular extra beats, and syncope. In contrast to a normal electrocardiogram on admission, electrocardiography showed polymorphic ventricular extra beats, T wave inversion, alternating T wave amplitude, and a prolonged QT_c interval of 606 ms. Torsade de pointes required cardiopulmonary resuscitation. The fluconazole plasma concentration was 216 µg/ml (usual target range at 400–800 mg/day: 18–28 µg/ml). Fluconazole was withdrawn and all conduction abnormalities reversed fully within 3 weeks.
- A 33-year-old woman with systemic lupus erythematosus was given intravenous fluconazole 200 mg/day for Candida albicans pneumonia (37). She developed prolongation of the QT_c interval and torsade de pointes. She had also been given domperidone, and both that and the fluconazole were withdrawn, the domperidone being suspected as the causative agent. However, torsade de pointes recurred several weeks later when she was given fluconazole again and resolved on withdrawal.
- A 55-year-old woman with acute myelomonocytic leukemia with bone marrow eosinophilia who was receiving consolidation chemotherapy had an episode of prolonged QT interval and torsade de pointes after being given fluconazole (38). Intravenous magnesium sulfate and multiple attempts at electrocardioversion led to recovery.
- A 25-year-old woman with worsening endocarditis had a prolonged QT interval at baseline and developed monomorphic ventricular dysrhythmias, which were managed successfully with pacing and antidysrhythmic therapy, including amiodarone (39). Several days later, she was given high-dose fluconazole (800 mg/day) for fungemia and after 3 days had episodes of torsade de pointes.

In this case torsade de pointes developed in the presence of known risk factors—hypokalemia, hypomagnesemia, female sex, baseline QT interval prolongation, and ventricular dysrhythmias.

Nervous system

Central nervous system abnormalities constitute the major dose-limiting adverse effects of fluconazole and are observed at dosages over 1200 mg/day (40).

Dizziness, headache, and seizures were seen in 2–5% of 232 patients with severe systemic fungal infections taking fluconazole (41). In the same group there were three cases each of delirium and dysesthesia (1.3%). A possible effect of the underlying illness has to be considered. In 14 patients treated with fluconazole for cryptococcal meningitis, dizziness was reported in 14% (SED-12, 681).

Two Japanese patients developed clonic convulsions while taking fluconazole 800 mg/day (42).

- A 66-year-old woman with complicated invasive *Candida tropicalis* infection but no renal impairment took fluconazole 800 mg/day. On the 21st day she developed clonic convulsions. The fluconazole trough concentration at the time of the event was 82 µg/ml.
- A 62-year-old man with deteriorating renal and hepatic function after coronary artery bypass surgery was given fluconazole 400 mg bd for a fungal sternal wound infection. On the 15th day he developed seizures. His trough plasma fluconazole concentration was 88 µg/ml. Nineteen days after dosage adjustment to 400 mg qds, he had another seizure. The trough fluconazole concentration was 103 µg/ml, probably because of deteriorating renal function.

In both cases, the seizures abated after dosage reduction. These case reports suggest an association between trough plasma fluconazole concentrations of 80 µg/ml and central nervous system toxicity; they re-emphasize the need for careful monitoring and dosage adjustment of fluconazole in patients with reduced renal function.

Endocrine

Preliminary studies concerning a possible effect on testosterone concentrations and the adrenal response to adrenocorticotropic hormone did not show any changes. However, determinations were performed after only 14 days of fluconazole administration. In fact, antifungal azoles inhibit the production of adrenal steroids and can cause acute adrenal insufficiency, as has been reported in a 38-year-old man who was treated with fluconazole and broad-spectrum antibiotics (43).

In a retrospective analysis of the effects of fluconazole 400 mg/day in 154 critically ill surgical patients the median plasma cortisol concentration was 158 µg/l in 79 patients randomized to fluconazole and 167 µg/l in 75 patients randomized to placebo (44). Patients randomized to fluconazole did not have significantly increased odds of adrenal dysfunction compared with patients randomized to placebo (OR = 0.98; 95%CI = 0.48, 2.01). Randomization to fluconazole was not associated with a significant difference in cortisol concentrations over time. Mortality was not different between patients with and without adrenal dysfunction, nor between patients with adrenal dysfunction who were randomized to fluconazole and those randomized to placebo.

- Two critically ill patients, a 77-year-old man with esophageal cancer and a 66-year-old woman with multiple organ failure, developed reversible adrenal insufficiency temporally related to the use of high-dose fluconazole (800 mg loading dose followed by 400 mg/day), as assessed by short stimulation tests with cosyntropin (ACTH) (45). Although anecdotal, these data suggest that the possibility that high-dose fluconazole can cause adrenal insufficiency in already compromised critically ill patients needs to be investigated further.
- A 63-year-old man received high-dose cyclophosphamide for peripheral blood stem-cell harvesting, having been taking fluconazole 200 mg/day (46). On day 3 he developed atrial fibrillation and his blood pressure fell to 78 mmHg. A rapid ACTH stimulation test showed a blunted adrenal response. He was suspected of having adrenal failure, and fluconazole was withdrawn. A rapid ACTH test was normal on day 14. To clarify the association between adrenal failure and fluconazole, he was rechallenged with fluconazole 400 mg/day from day 16 and a rapid ACTH test was performed on day 21; it showed a blunted adrenal response.

Mineral balance

Hypokalemia was observed in only a few patients taking fluconazole, which contrasts with experience with itraconazole (41,47). However, hyperkalemia was reported in one paper (SED-12, 681) (47).

Hematologic

Cytopenias occur but seem to be mild. Occasionally, more marked changes have been described, but these could have been connected with the underlying disease (SED-12, 681) (47,48). In a single placebo-controlled study of fluconazole prophylaxis using a relatively high dose of 400 mg/day, a post-hoc analysis suggested prolongation of granulocytopenia after intensive chemotherapy for hematological neoplasms in the fluconazole group. This may have been due to an interaction with the antineoplastic drug (49).

Leukopenia with eosinophilia has been attributed to fluconazole (50).

- A 75-year-old man with non-Hodgkin's lymphoma and cryptococcal meningoencephalitis developed neutropenia with eosinophilia associated with fluconazole. After 1 week of fluconazole 400 mg/day his total leukocyte count began to fall and his eosinophil count increased. Concurrent medications included levothyroxine, famotidine, and co-trimoxazole. The last two drugs were withdrawn and he was given G-CSF. However, his leukocyte count continued to fall and 4 days later reached a nadir of 700×10^6/l; the platelet count remained normal. The leukopenia and eosinophilia resolved promptly after withdrawal of fluconazole.

Since the leukopenia and eosinophilia did not resolve until fluconazole was withdrawn, an effect of the compound was plausible. This case and two other reported cases (51,52) emphasize the importance of recognizing fluconazole as a rare but potential cause of bone marrow suppression in patients in whom drug-induced agranulocytosis is suspected.

In one study in patients with AIDS taking prophylactic maintenance fluconazole for cryptococcal meningitis, there was a higher rate of hematological toxicity with fluconazole than with placebo, but this probably reflected the greater proportional and absolute amounts of zidovudine used in the fluconazole group; there was no serious hematotoxicity (SED-12, 682).

Mouth

An oral fixed drug eruption has been attributed to fluconazole (53).

- A 19-year-old boy developed redness and swelling over the hard palate 7–8 hours after a dose of fluconazole 150 mg. There was mild pain and hypersalivation and the lesion gradually progressed to superficial erosion. He had had a similar episode about 1 year before after taking fluconazole for tinea cruris; he had stopped taking fluconazole and the oral lesion had healed by itself. The new erosion healed in about 15 days after treatment with chlorhexidine mouth rinse 0.12% 10 ml bd, warm saline rinses, oral cetirizine 10 mg/day, and topical triamcinolone in Orabase. Four weeks after the oral lesion had healed, oral provocation with fluconazole 50 mg caused the lesion to reappear in about 3 hours.

Gastrointestinal

Nausea and vomiting are mentioned in most reports of patients taking fluconazole, with an incidence of 10–15% (SED-12, 682) (41,47,48,54,55). Anorexia, mild abdominal pain, and diarrhea have been reported, but none was severe.

Liver

Raised liver enzyme activities have been reported in most studies. In some articles this effect was described as transient, in others as disappearing after withdrawal. The incidence varies from a few percent of cases to 35–45%, but occasionally the effect has been recorded in all cases treated. A temporal relation between these liver function changes and fluconazole treatment has been shown in many cases. Severe liver toxicity has not been reported (SED-12, 681) (21). While asymptomatic rises in transaminases were noted in some children with neoplastic disease who were treated concomitantly with fluconazole in a small study (56), there were no significant changes in a larger study in cancer patients treated with placebo or fluconazole 400 mg (49).

- A 45-year-old woman with protracted cryptococcal meningoencephalitis developed fulminant hepatic failure secondary to high fluconazole serum concentrations, possibly precipitated by renal dysfunction

induced by concomitant amphotericin therapy or concomitant therapy with lisinopril, atenolol, or amlodipine (57). Four days after the withdrawal of fluconazole 400 mg/day the serum concentration of fluconazole was 40 µg/ml.

This case points to the potential risks of fluconazole therapy in the setting of renal insufficiency, in particular with higher dosages (400 mg/day and more).

The hepatotoxicity of antifungal medications in bone marrow transplant recipients has been analysed in a retrospective matched-control study (21). The unadjusted incidence of hepatotoxicity was 0.98 cases per 100 patient-days of exposure to fluconazole (OR = 1.99; 95% CI = 1.21, 3.26). In the follow-up analysis of patients who developed hepatotoxicity and who continued to take antifungal medications, 8% of those who took fluconazole developed marked increases in serum bilirubin concentration. Thus, fluconazole was associated with an increased risk of hepatotoxicity, independent of other treatments or patient characteristics. However, patients who develop hepatotoxicity appear to tolerate continued therapy with fluconazole.

Hepatotoxicity with azoles is not necessarily a class effect.c

- A 32-year-old Hispanic man with type I diabetes mellitus and coccidioidal meningitis developed increased hepatic transaminases when given fluconazole but subsequently tolerated voriconazole (58). Rechallenge with fluconazole again led to increased transaminases, which normalized when voriconazole was reinstated.

This report suggests that voriconazole may be cautiously substituted for fluconazole in patients with fluconazole-induced hepatatotoxicity who require azole therapy.

Biliary tract

The use of fluconazole prophylaxis for 6 weeks has been studied in a historical comparison in infants of extremely low birth weight (59). During prophylaxis, 60/140 infants developed conjugated hyperbilirubinemia compared with 12/137 who did not have prophylaxis. However, fluconazole prevented several cases of candidiasis (none versus nine in the two groups) and the benefit to harm balance is probably favorable.

Urinary tract

In a randomized, blind, multicenter comparison of fluconazole 800 mg/day plus placebo and fluconazole plus amphotericin B deoxycholate 0.7 mg/kg/day, with the placebo/amphotericin component given only for the first 5?6 days, as therapy for candidemia due to species other than *Candida krusei* in 219 adults without granulocytopenia, success rates on day 30 were 57% for fluconazole plus placebo and 69% for fluconazole plus amphotericin (60). Overall success rates were 56% (60 of 107 patients) and 69% (77 of 112 patients) respectively; the bloodstream infection failed to clear in 17% and 6% of subjects respectively. Renal dysfunction led to a reduction in drug dosage in 3% and 23% but was the primary cause of study failure in only 5% and 3% of subjects respectively.

There were no differences in mortality within 90 days of starting therapy. Thus, in non-neutropenic subjects, the combination of fluconazole plus amphotericin was not antagonistic compared with fluconazole alone, and the combination tended to produce improved success and more rapid clearance from the bloodstream. Nevertheless, the combination also was associated with a higher rate of nephrotoxicity.

Skin

Rashes of several types occur with fluconazole and are more frequent in immunocompromised patients.

The risk of serious skin disorders has been estimated in 61 858 users of oral antifungal drugs, aged 20–79 years, identified in the UK General Practice Research Database (61). They had received at least one prescription for oral fluconazole, griseofulvin, itraconazole, ketoconazole, or terbinafine. The background rate of serious cutaneous adverse reactions (corresponding to non-use of oral antifungal drugs) was 3.9 per 10 000 person-years (95% CI = 2.9, 5.2). Incidence rates for current use were 15 per 10 000 person-years (1.9, 56) for itraconazole, 11.1 (3.0, 29) for terbinafine, 10 (1.3, 38) for fluconazole, and 4.6 (0.1, 26) for griseofulvin. Cutaneous disorders associated with the use of oral antifungal drugs in this study were all mild.

Pruritus has also been reported (62). A few cases of Stevens–Johnson disease have been reported worldwide (SED-12, 682) (22,54,63), as well as a few instances of fixed-drug eruption (64). In cases of hypersensitivity, desensitization has reportedly been used with success (65).

Fixed drug eruption caused by systemic fluconazole has been reported (66).

- A 36-year-old woman with a history of atopy and recurrent *Candida* vaginitis developed a fixed drug eruption while taking fluconazole 150 mg/day. Local provocation with 10% fluconazole in petrolatum applied at the site of a previous site of fixed drug eruption reproduced the eruption clinically and histopathologically.

Erythema multiforme associated with fluconazole has again been reported (67).

Nails

A patient developed a longitudinal band of pigmentation in the diseased nail after fluconazole therapy for onychomycosis at a dosage of 150 mg once a week for 4 weeks (68).

Immunologic

Fluconazole hypersensitivity has been reported in a healthy man.

- A previously healthy 39-year-old man took two single doses of fluconazole 150 mg 7 days apart and 4 days later developed malaise, generalized weakness, low grade fever, jaundice, and a mildly pruritic, erythematous, generalized rash (69). He had a mild eosinophilia,

a marked increase in hepatic transaminases to over 2000 U/l, and a serum bilirubin of over 513 μmol/l. There was no evidence of viral or autoimmune hepatitis, and a liver biopsy showed portal and lobular inflammation with cholestasis and apoptosis; a skin biopsy showed numerous necrotic keratinocytes. He was treated with methylprednisolone and two doses of immunoglobulin and made a full recovery within 3 months.

Second-Generation Effects

Teratogenicity

The teratogenic activity of triazole and two triazole derivatives, flusilazole (an agricultural triazole monoderivative fungicide) and fluconazole, has been studied in vitro (70). Rat embryos 9.5 days old (1–3 somites) were exposed in vitro to triazole 500–5000 μmol/l, flusilazole 3.125–250 μmol/l, or fluconazole 62.5–500 μmol/l and examined after 48 hours in culture. There were similar teratogenic effects (abnormalities at the branchial apparatus level and cell death at the level of the branchial mesenchyme) at 6.25 μmol/l and higher for flusilazole and 125 μmol/l and higher for fluconazole. In contrast, there was little effect at the highest concentrations of triazole, suggesting no teratogenic activity. These investigations have confirmed the embryotoxic potential of antifungal triazole derivatives, specifically on the branchial apparatus.

The branchial apparatus is essential for the development of the facial skeleton. The branchial arch mesenchyme is formed by two different cellular populations: paraxial mesenchyme and ectomesenchyme, which originate from rhombencephalic neural crest cell migration. The possible pathogenic pathways involved in fluconazole-related branchial arch abnormalities have been investigated in rat embryos 9.5 days old exposed in vitro to fluconazole 0 or 500 μmol/l (71). After 24, 36, or 48 hours of culture, the embryos were examined for apoptosis and cell proliferation. Rhombencephalic neural crest cell migration and the extracellular matrix were analysed using immunostaining. The differentiating capability of the branchial mesenchymes was investigated using anti-endothelin and anti-endothelin receptor antibodies. During the whole culture period, there were no changes in physiological apoptosis, cell proliferation, or mesenchymal cell induction in fluconazole-exposed embryos, but in contrast there were major changes in neural crest cell migration pathways. These findings suggest that fluconazole produces teratogenic effects by interfering with the cellular and molecular mechanisms that control neural crest cell migration.

- A 9-month-old boy was born to a 30-year-old woman following a 37-week pregnancy, which was complicated by maternal HIV infection and multiple drug exposures, including fluconazole (400 mg/day) until the fifth month and then from 6 months to term, efavirenz, nevirapine, methadone, dapsone, pentamidine, and cotrimoxazole (72). At birth the infant had multiple congenital anomalies and at 9 months had craniosynostosis secondary to coronal and lambdoidal suture closures, a shallow orbital region, hypoplastic supraorbital ridges, hypertelorism, and mild ptosis. He had radioulnar synostosis and metacarpophalangeal–proximal interphalangeal symphalangism of D2-D5 bilaterally.

The potential of in utero exposure to fluconazole to initiate teratogenesis has been analysed in ICR (CD-1) mice (73). Developmental phase specificity was determined by treating mice with single oral doses of 700 mg/kg on gestational days 8, 9, 10, 11, or 12. Control animals received vehicle on days 8–12. Day 10 was identified as the phase of maximal sensitivity for induction of cleft palate, the predominant teratogenic effect induced by fluconazole, when 50% of exposed fetuses were affected. After treatments on days 8, 9, 11, or 12, cleft palate occurred with lower frequencies:12, 21, 29, and 2.7% respectively. There were anomalies of the middle ear apparatus in 15% of the fetuses that were exposed on day 8. A dysmorphic tympanic ring and absence of the incus were the more common ear anomalies recorded. Humeral length was reduced in 22% of fetuses that were exposed on day 10. The dose-response relation was investigated by treating animals with 0 (vehicle), 87.5, 175, or 350 mg/kg on day 10, coincident with the phase of peak teratogenic sensitivity. There was a clear dose-response relation and 175 mg/kg was the lowest dose at which cleft palate induction was observed, 7.6% of exposed fetuses being affected. The risk of malformations and other outcomes in children exposed to fluconazole in utero has been examined in 165 women who had taken fluconazole just before or during pregnancy, mostly in the form of a single dose of 150 mg to treat vaginal candidiasis (74). Birth outcomes (malformations, low birth weight, and preterm delivery) were compared with the outcomes among 13 327 women who did not receive any prescriptions during their pregnancies. The prevalence of malformation was 3.3% (four cases) among the 121 women who had used fluconazole in the first trimester, and 5.2% (697 cases) in offspring to controls (OR = 0.65; CI = 0.24, 1.77). The risks of preterm delivery (OR = 1.17; CI = 0.63, 2.17) and low birth weight (OR = 1.19; CI = 0.37, 3.79) were not significantly increased in association with fluconazole. Thus, the study showed no increased risk of congenital malformations, low birth weight, or preterm birth in offspring to women who had used single doses of fluconazole before conception or during pregnancy.

The potential ability of fluconazole to modulate phenytoin teratogenesis has been studied in Swiss mice (75). Pretreatment with a non-embryotoxic dosage of fluconazole (10 mg) potentiated phenytoin teratogenesis; combined treatment of fluconazole 50 mg with phenytoin resulted in a significant increase in embryo deaths. The mechanism of this teratological interaction remains to be established.

Lactation

Fluconazole is found in breast milk at concentrations comparable with those found in the blood after single or multiple doses (2); this may be of clinical relevance (76).

Susceptibility Factors

Age

High-risk very low birth weight infants

Daily versus twice-weekly fluconazole prophylaxis for up to 6 weeks have been investigated in a randomized, double-blind trial in 81 preterm infants who weighed under 1000 g at birth and with an endotracheal tube and/or central vascular catheter (77). There was *Candida* colonization in nine children and *Candida* sepsis in three, with no difference between treatments. All fungal isolates were sensitive to fluconazole and no adverse effects were documented.

Children

In a group of children with fever, neutropenia, and neoplastic disease, there was an increase in renal fluconazole clearance (52). In infants and children, the volume of distribution of fluconazole is significantly higher and falls with age. With the exception of infants, who have a slower clearance rate, children clear the compound more rapidly (78). However, a second larger study reported slower elimination in children under 1 year of age, requiring dosage adjustments (79). Low birth-weight neonates have a particularly low clearance rate, which increases within weeks (80).

The use of fluconazole in 726 children under 1 year of age, reported in 78 publications, has been reviewed (81). They received a wide range of dosages for up to 162 days. Fluconazole was well tolerated and efficacious in the therapy of systemic candidiasis and candidemia in children under 1 year of age, including neonates and very low birth-weight infants. The daily dosage recommended by the manufacturers is 6 mg/kg, to be reduced in patients with impaired renal function in accordance with the guidelines given for adults.

The efficacy and safety of fluconazole in neonates with *Candida* fungemia has been evaluated in a multicenter prospective study (82). Fluconazole was safe and effective even in complicated cases, including infants of very low birth weights. Two of 50 neonates developed raised liver enzymes during fluconazole therapy and two others had raised serum creatinine concentrations. In none of them did these abnormalities necessitate discontinuation of antifungal therapy.

The safety profile of fluconazole has been assessed in 562 children (aged 0–17 years; 323 boys and 239 girls), enrolled into 12 clinical studies of prophylactic or therapeutic fluconazole in predominantly immunocompromised patients (83). Most of the children received multiple doses of fluconazole 1–12 mg/kg, given as oral suspension or intravenously. Overall, 58 children reported 80 treatment-related adverse effects. The most common adverse effects were associated with the gastrointestinal tract (7.7%), the skin (1.2%), or the liver and biliary system (0.5% or three patients). Overall, 18 patients discontinued treatment owing to adverse effects, mainly gastrointestinal. Dosage and age did not affect the incidence and pattern of adverse effects. Treatment-related laboratory abnormalities included transiently raised alanine transaminase (4.9%), aspartate transaminase (2.7%), and alkaline phosphatase (2.3%). Although 99% of patients were taking concomitant drugs, there were no clinical or laboratory interactions. The safety profile of fluconazole was compared with those of other antifungal agents, mostly oral polyenes, by using a subset of data from five controlled studies. Adverse effects were reported by more patients treated with fluconazole (45 of 382; 12%) than by patients treated with comparator agents (25 of 381; 6.6%); vomiting and diarrhea were the most common events in both groups. The incidence and type of treatment-related laboratory abnormalities were similar in the two groups. Fluconazole was well tolerated, mirroring the favorable safety profile seen in adults.

In 34 otherwise healthy infants with oral candidiasis randomized to either nystatin oral suspension qds for 10 days or fluconazole suspension 3 mg/kg in a single daily dose for 7 days, 6 of 19 were cured by nystatin and all of 15 by fluconazole (84). Fluconazole was tolerated without apparent adverse events.

Renal disease

If the creatinine clearance is below 40 ml/minute fluconazole doses should be adjusted (SED-12, 682; 7).

Advanced HIV infection

In HIV-infected patients, fluconazole prophylaxis is associated with reductions in the rate of fungal infection. However, there are concerns about fluconazole prophylaxis and the risk of fluconazole-resistant infections. In a randomized, open comparison of oral fluconazole given continuously (200 mg 3 times weekly; the "continuous fluconazole arm"; n = 413) and fluconazole that was provided only for episodes of orophayngeal candidiasis or esophageal candidiasis (the "episodic fluconazole arm"; n = 416) in HIV-infected persons with CD4+ T cell counts of under $150 \times 10^6/l$ and a history of orophayngeal candidiasis (85). The primary end point was the time to development of fluconazole-resistant orophayngeal or esophageal candidiasis, which was defined as a lack of response to fluconazole 200 mg/day for 14 or 21 days respectively. After 42 months, 17 subjects in the continuous fluconazole arm (4.1%) developed fluconazole-resistant orophayngeal or esophageal candidiasis, compared with 18 (4.3%) in the episodic fluconazole arm. There was no difference between treatments with regard to the time to development of a fluconazole-resistant infection within 24 months or before the end of the study. Continuous fluconazole therapy was associated with fewer cases of orophayngeal or esophageal candidiasis (0.29 versus 1.08 episodes per patient-year) and fewer invasive fungal infections. This study shows that fluconazole is not associated with a significant risk of fluconazole-resistant orophayngeal or esophageal candidiasis, compared with episodic fluconazole therapy in HIV-infected patients with access to active antiretroviral therapy.

Immunocompromised patients with esophageal candidiasis

Micafungin 150 mg/day and fluconazole 200 mg/day have been compared in a large randomized study in 523 patients with documented esophageal candidiasis aged 16 years and over (86). The median duration of therapy was 14 days. For the primary end point of endoscopic cure, the treatment difference was negligible (88% each), and the overall therapeutic response rate was 87%. The incidences of drug-related adverse events were 28% for micafungin and 21% for fluconazole. Therapy was withdrawn in six patients taking micafungin and two taking fluconazole; rash was the most common event that led to withdrawal

Drug–Drug Interactions

See also Antifungal azoles

Alfentanil

In a randomized, double-blind, placebo-controlled, crossover study in nine subjects, fluconazole 400 mg reduced the clearance of alfentanil 20 micrograms/kg by 55% and increased alfentanil-induced subjective effects (87).

There has been a double-blind randomized control study of the effect of the antifungal drug fluconazole 400 mg on the pharmacokinetics and pharmacodynamics of intravenous alfentanil (137). Fluconazole given either orally or intravenously 1 hour before alfentanil 20 µg/kg intravenously caused a significant doubling of the half-life, by inhibition of CYP3A4, which metabolizes alfentanil. Intravenous and oral fluconazole both increased alfentanil-induced respiratory depression by reducing the respiratory rate by 10–15% compared with alfentanil alone. Alfentanil should therefore be given cautiously to patients taking fluconazole and the authors suggested that such patients require 60% less alfentanil for maintenance of analgesia, irrespective of the mode of administration of the antifungal drug.

Amitriptyline

An interaction of fluconazole with amitriptyline has been reported (88).

- A 12-year-old boy with prostatic rhabdomyosarcoma had episodes of syncope periodically over 7 months while taking fluconazole for chemotherapy-induced mucositis. He had taken fluconazole in the past without problems but had also taken a stable dose of amitriptyline for neuropathic pain. On withdrawal of amitriptyline he had no further episodes. The effect was confirmed by readministration.

Concurrent administration of fluconazole probably causes increased exposure to amitriptyline. Three reports of adults have shown increased amitriptyline plasma concentrations with concurrent administration of fluconazole; in one patient, a 57-year-old woman, the QT interval was prolonged and torsade de pointes occurred (89).

Amphotericin

In vitro studies and experiments in animals have given conflicting results relating to potential antagonism between the effects of fluconazole and amphotericin on *Candida* species (90). However, large, randomized, double-blind comparisons of fluconazole with and without amphotericin for 5 days in non-neutropenic patients with candidemia showed no evidence of antagonism, but faster clearance of the organism from the blood and a trend toward an improved outcome in those who received the combination (91).

Antacids

Fluconazole absorption after oral administration is not influenced by gastric pH; thus there is no effect of antacids such as co-magaldrox (SED-12, 682) (92).

Antihistamines

The concurrent use of terfenadine with fluconazole can lead to dangerously high terfenadine concentrations, with resulting cardiotoxicity (93). It is suspected that the same may happen with astemizole (94).

Benzodiazepines

The interaction of fluconazole with bromazepam has been studied in 12 healthy men in a randomized, double-blind, four-way, crossover study (95). They received single oral or rectal doses of bromazepam (3 mg) after 4-day pretreatment with oral fluconazole (100 mg/day) or placebo. Fluconazole caused no significant changes in the pharmacokinetics and pharmacodynamics of oral or rectal bromazepam.

Fluconazole increased blood concentrations of midazolam (96) and triazolam (97).

The effects of fluconazole (400 mg loading dose followed by 200 mg/day) on the kinetics of midazolam have been studied in 10 mechanically ventilated adults receiving a stable infusion of midazolam (98). Concentrations of midazolam were increased up to fourfold after the start of fluconazole therapy; these changes were most marked in patients with renal insufficiency. During the study, the ratio of α-hydroxymidazolam to midazolam progressively fell. The authors concluded that in ICU patients receiving fluconazole, reduction of the dose of midazolam should be considered if the degree of sedation is increasing.

In a study of the pharmacokinetics and pharmacodynamics of oral midazolam 7.5–15 mg, switching from inhibition of metabolism by itraconazole 200 mg/day to induction of metabolism by rifampicin 600 mg/day caused an up to 400-fold change in the AUC of oral midazolam (99).

Calcium channel blockers

A pharmacokinetic interaction between fluconazole and nifedipine has been reported (100).

- Fluconazole enhanced the blood pressure-lowering effects of nifedipine by increasing its plasma concentrations in a 16-year-old man with malignant pheochromocytoma taking chronic nifedipine for arterial hypertension who was given fluconazole for *Candida* septicemia (100).

After careful pharmacokinetic assessment of the effects of fluconazole withdrawal on 24-hour ambulatory blood pressure and nifedipine plasma concentration, the authors concluded that fluconazole can enhance the blood-pressure lowering effects of nifedipine by increasing its plasma concentrations, most likely by inhibiting CYP3A4.

Carbamazepime

Fluconazole can cause carbamazepine toxicity, presumably by inhibition of CYP3A4 (101).

- A 33-year-old man on stable therapy with carbamazepine (400 mg tds) for a seizure disorder became stuporose due to carbamazepine toxicity after taking fluconazole 150 mg/day for 3 days. Withdrawal of both drugs resulted in a fall in carbamazepine concentrations (maximum concentration 25 µg/ml) and return of the patient's baseline mental status. Carbamazepine was restarted and the patient had no further adverse events.
- Carbamazepine serum concentrations increased during concomitant fluconazole administration (400 mg/day) in a 38-year-old man (102).

Ciclosporin

Fluconazole can increase concentrations of ciclosporin by inhibiting CYP3A4. In some studies, minimal or no effects were recorded, but in others ciclosporin concentrations were increased by fluconazole. Differences in the dosage and duration of fluconazole treatment could have explained these discrepancies (SED-12, 682) (21,103–106). For example, there was no interaction at a fluconazole dosage of 100 mg/day, but high dosages of fluconazole (400 mg/day or more) increase blood ciclosporin and tacrolimus concentrations (107,108).

The interaction of ciclosporin with fluconazole has been retrospectively evaluated in 19 kidney and pancreas/kidney transplant recipients (109). Both intravenous and oral fluconazole altered the blood concentration of ciclosporin. Five subjects did not have a significant interaction and 15 did. No patient had nephrotoxicity or transplant rejection related to antifungal therapy.

The effects of higher dosages of fluconazole on ciclosporin immunosuppression have been investigated in six renal transplant patients in a prospective, unblinded, crossover study (110). Baseline renal function, ciclosporin AUC, C_{max}, C_{min}, t_{max}, and clearance were compared with those 2, 4, and 7 days after starting fluconazole orally in a dosage of 200 mg/day. From day 8 onwards the ciclosporin dose was reduced by 50% and the above parameters were repeated on day 14. The results are shown in Table 1. On repeated-measures ANOVA only the AUC and C_{max} on day 4 of fluconazole were significantly higher than on day 0. There were no significant changes in ciclosporin clearance and t_{max}. The authors concluded that changes in C_{min} may not be sensitive enough to detect the described interaction and suggested monitoring

Table 1 Changes in ciclosporin kinetics during co-administration of fluconazole

Parameter	Day 0	Day 4	Day 7	Day 14
AUC (hours.ng/ml)	2887	4750	4052	2330
C_{max} (ng/ml)	701	941	768	498
C_{min} (ng/ml)	207	274	293	174
Clearance (ml/minute/kg)	17	13	12	30
t_{max} (hours)	3.0	2.0	2.5	3.0

the AUC near day 4 of treatment to guide ciclosporin dosage adjustments in all patients taking concomitant fluconazole.

Cimetidine

Fluconazole absorption after oral administration is not influenced by gastric pH; thus there is no effect of cimetidine.

Clarithromycin

The effects of fluconazole and clarithromycin on the pharmacokinetics of rifabutin and 25-*O*-desacetylrifabutin have been studied in 10 HIV-infected patients who were given rifabutin 300 mg qds in addition to fluconazole 200 mg qds and clarithromycin 500 mg qds (111). There was a 76% increase in the plasma AUC of rifabutin when either fluconazole or clarithromycin was given alone and a 152% increase when both drugs were given together. The authors concluded that patients should be monitored for adverse effects of rifabutin when it is co-administered with fluconazole or clarithromycin.

Cyclophosphamide

Cyclophosphamide is a prodrug that is metabolized by CYP450 enzymes to produce alkylating species, which are cytotoxic, and the extent of cyclophosphamide metabolism correlates with both treatment efficacy and toxicity. In vitro studies in six human liver microsomes showed that the IC_{50} of fluconazole for reduction of 4-hydroxycyclophosphamide production was 9–80 µmol/l (112).

A retrospective study in 22 children with cancers addressed the potential interaction between fluconazole and cyclophosphamide. Children with an established profile of cyclophosphamide metabolism who were not receiving other drugs known to affect drug metabolism were selected; 9 were taking fluconazole and 13 were controls. The plasma clearance was significantly lower in patients taking concomitant fluconazole (2.4 versus 4.2 l/hour/m^2). It is unclear whether this interaction is associated with a reduction in the therapeutic efficacy of cyclophosphamide.

Doxorubicin

The effect of fluconazole on the plasma pharmacokinetics of doxorubicin has been investigated in a randomized, crossover study in non-human primates (113). Fluconazole (10 mg/kg/day) was given intravenously for

4 days before doxorubicin (2.0 mg/kg intravenously). Pretreatment with fluconazole had no effect on the pharmacokinetics of doxorubicin, and the incidence of severe neutropenia (absolute neutrophil count below 0.5×10^9/l) was higher with doxorubicin alone than with the combination of doxorubicin and fluconazole. Thus, fluconazole does not appear to contribute to the marrow-suppressive effects of doxorubicin.

Flucytosine

The concurrent use of fluconazole with flucytosine may have an additive effect (114). This combination could be useful in the treatment of cryptococcal meningitis (21).

HIV protease inhibitors

The pharmacokinetic interaction of fluconazole 400 mg od and indinavir 1000 mg tds has been evaluated in a placebo-controlled, crossover study for 8 days; there was no significant interaction (115).

The effect of fluconazole on the steady-state pharmacokinetics of ritonavir and saquinavir has been studied in patients infected with HIV-1 (116). They received the protease inhibitor (saquinavir 1200 mg tds, $n = 5$, or ritonavir 600 mg bd, $n = 3$) alone on day 1 and then with fluconazole 400 mg on day 2 and 200 mg on days 3–8. The median increase in saquinavir AUC was 50%, and the median increase in C_{max} was 56%. In contrast, fluconazole had no effect on the disposition of ritonavir.

Hormonal contraceptives—oral

Fluconazole did not significantly alter the pharmacokinetics of ethinylestradiol or norgestrel, and this finding was interpreted by the investigators as suggesting that treatment with fluconazole in a user of oral contraceptives would not increase the risk of pregnancy (121). However, the study was carried out using a 50 mg dose of fluconazole, and experience with higher dosages has shown different results (SED-12, 682).

In another study, fluconazole reduced the systemic availability of ethinylestradiol in an oral contraceptive, but there was no information about the doses used (94).

In an open, crossover study in 10 young healthy subjects, fluconazole 150 mg increased the serum concentrations of ethinylestradiol 30–35 µg/day (122).

The potential pharmacokinetic interaction between fluconazole 300 mg once weekly and an oral contraceptive containing ethinylestradiol and norethindrone (Ortho Novum 7/7/7; Ortho-McNeil Pharmaceutical Inc, Raritan, NJ) has been studied in a placebo-controlled, double-blind, randomized, two-way, crossover study in 26 healthy women aged 18–36 years (123). During the first cycle they took the oral contraceptive only. During the second cycle they were assigned randomly to oral contraceptive + fluconazole or oral contraceptive + placebo. In the third cycle they were given the other treatment. Fluconazole caused small but statistically significant increases in the AUC_{0-24} of both ethinylestradiol and norethindrone. There were no adverse events related to treatment in those given fluconazole.

It therefore appears that there is no threat of contraceptive failure because of concomitant fluconazole administration.

Losartan

The pharmacokinetic interaction of fluconazole 200 mg/day with losartan 100 mg/day has been investigated in 32 healthy subjects (138). Fluconazole significantly increased the steady-state concentration of losartan by 66% and inhibited the formation of the active metabolite EXP-3174 by 34%.

Another very similar study investigated interactions with itraconazole, an inhibitor of CYP3A4, and fluconazole, which is more specific for CYP2C9 (139). Fluconazole reduced the mean peak plasma concentrations of losartan and E3174 to 30% and 47% of their control concentrations respectively. The half-life of losartan was prolonged by 67%. Itraconazole had no significant effect. The possibility of a reduced effect of losartan when co-administered with fluconazole should be expected.

Methadone

An interaction between fluconazole and methadone (a substrate of CYP3A4, CYP2C9, and CYP2C19) has been reported.

- While taking a stable dose of methadone, a 60-year-old man with advanced cancer developed respiratory depression 2 days after receiving intravenous fluconazole for refractory oral candidiasis (117). Intravenous naloxone reversed the respiratory depression.

In a randomized, double-blind, placebo-controlled study in 25 patients, fluconazole 200 mg/day increased methadone concentrations, but patients treated with fluconazole did not have signs or symptoms of significant narcotic overdose (118).

In a randomized, double-blind, placebo-controlled trial, oral fluconazole increased the serum methadone AUC by 35% (140). Although renal clearance was not significantly affected, mean serum methadone peak and trough concentrations rose significantly, while renal clearance was not significantly altered.

Omeprazole

Fluconazole absorption after oral administration is not influenced by gastric pH; thus there is no effect of omeprazole (119). Omeprazole is extensively metabolized in the liver by 5-hydroxylation and sulfoxidation reactions, catalysed predominantly by CYP2C19 and CYP3A4 respectively. Fluconazole is a potent competitive inhibitor of CYP2C19 and a weak inhibitor of CYP3A4. The effect of fluconazole on the pharmacokinetics of a single oral dose of omeprazole 20 mg has been evaluated after a single oral dose of fluconazole 100 mg and after 4 days of oral administration of 100 mg/day in 18 healthy male volunteers (126). Fluconazole increased the C_{max} and the mean AUC of omeprazole and prolonged its half-life (2.59 versus 0.85 hours).

Phenytoin

Co-administration of fluconazole and phenytoin resulted in markedly higher phenytoin concentrations (SED-12, 682) (21,103,106,124).

Rifamycins

The combination of rifampicin with fluconazole has insignificant effects (SED-12, 682) (21,106,125).

Fluconazole increased the AUC of rifabutin by 76% (141).

Statins

The effects of fluconazole on plasma fluvastatin and pravastatin concentrations have been studied in two separate, randomized, double-blind, two-phase, crossover studies (126). Healthy volunteers were given oral fluconazole (400 mg on day 1 and 200 mg on days 2–4) or placebo. On day 4, they took a single oral dose of fluvastatin 40 mg or pravastatin 40 mg. Fluconazole increased the plasma AUC and the half-life of fluvastatin by 80% but had no significant effects on the pharmacokinetics of pravastatin. The mechanism of the prolonged elimination of fluvastatin was probably inhibition of CYP2C9. Pravastatin, in contrast, appears not to be susceptible to interactions with fluconazole and other CYP2C9 inhibitors.

The effect of fluconazole on the pharmacokinetics of rosuvastatin has been investigated in a randomized, double-blind, two-way, crossover, placebo-controlled study (127). Healthy male volunteers ($n = 14$) were given fluconazole 200 mg/day or matching placebo for 11 days; rosuvastatin 80 mg was co-administered on day 8. Plasma concentrations of rosuvastatin, N-desmethylrosuvastatin, and active and total HMG-CoA reductase inhibitors were measured up to 96 hours after the dose. Fluconazole increased the AUC and C_{max} of rosuvastatin by 14% and 9% respectively. Limited data available for the N-desmethylated metabolite showed that the C_{max} fell by about 25%. Fluconazole did not affect the proportion of circulating active or total HMG-CoA reductase inhibitors accounted for by circulating rosuvastatin. Thus, fluconazole produced only small changes in rosuvastatin kinetics, which were not considered to be of clinical relevance.

Sulfonylureas

The concurrent administration of fluconazole with tolbutamide resulted in increased tolbutamide concentrations (SED-12, 682) (90).

- A 56-year-old HIV-positive patient with diabetes mellitus taking gliclazide 160 mg/day developed severe hypoglycemia when treated with co-trimoxazole 480 mg/day and fluconazole 200 mg/day (128). The authors speculated that fluconazole might have inhibited gliclazide metabolism by inhibiting CYP2C9.

The effects of fluconazole and fluvoxamine on the pharmacokinetics and pharmacodynamics of glimepiride have been studied in a randomized, double-blind, crossover study in 12 healthy volunteers who took fluconazole 200 mg/day (400 mg on day 1), fluvoxamine 100 mg/day, or placebo once daily for 4 days (129). On day 4 they took a single oral dose of glimepiride 0.5 mg. Fluconazole increased the mean total AUC of glimepiride to 238% and the peak plasma concentration to 151% of control values, and the half-life of glimepiride was prolonged from 2.0 to 3.3 hours. This was probably due to inhibition of CYP2C9-mediated biotransformation of glimepiride by fluconazole. However, fluconazole did not cause statistically significant changes in the effects of glimepiride on blood glucose concentrations.

Tacrolimus

Since tacrolimus (FK506) is metabolized by intestinal and hepatic CYP3A4, drugs that inhibit CYP3A4 can reduce the metabolism of tacrolimus and increase tacrolimus blood concentrations (130). The effect of fluconazole on the blood concentrations of tacrolimus have been investigated in eight liver transplant patients in whom prophylactic fluconazole (200 mg/day) was withdrawn because of rises in hepatic transaminases ($n = 6$), renal dysfunction, or eosinophilia ($n = 1$ each) (131). Calculated tacrolimus concentrations fell by 13–81% (median 41%) between the fourth and ninth days after withdrawal of fluconazole. Tacrolimus blood concentrations should be carefully monitored and dosages increased as necessary after withdrawal of fluconazole.

The interaction of tacrolimus with fluconazole has been retrospectively evaluated in 19 kidney and pancreas/kidney transplant recipients (109). Both intravenous and oral fluconazole altered the blood concentration of tacrolimus. Five subjects did not have a significant interaction and 15 did. No patient had nephrotoxicity or transplant rejection related to antifungal therapy.

- A 17-year-old man with cystic fibrosis who took itraconazole after a lung-liver transplant had high trough concentrations of tacrolimus, despite the relatively low dosage (0.1–0.3 mg/kg/day) (132).
- A patient taking tacrolimus 0.085 mg/kg bd with itraconazole 200–400 mg/day developed ketoacidosis, neutropenia, and thrombocytopenia, requiring the withdrawal of both drugs (133).
- A 34-year-old renal transplant recipient taking a stable regimen of tacrolimus and methylprednisolone was given itraconazole 100 mg bd for a yeast infection of the urinary tract (134). Concomitant therapy with itraconazole led to a marked increase in tacrolimus trough concentrations on the second day of therapy (from 13 to 21 ng/ml) and an increase in serum creatinine

concentrations, necessitating dosage reduction of tacrolimus by 50%.

When itraconazole was withdrawn the effect of itraconazole on the kinetics of tacrolimus took 12 days to reverse.

The inhibitory effect of itraconazole occurred quickly, while the time of disappearance was much longer, which is important for clinical management. Thus, during co-administration of itraconazole with tacrolimus, close monitoring of tacrolimus blood concentrations and careful dosage adjustments are essential to avoid toxicity.

Tricyclic antidepressants

Fluconazole can increase plasma concentrations of amitriptyline, presumably by inhibiting cytochrome P450 isozymes (CYP3A4 and CYP2C19), preventing its demethylation. Two patients taking amitriptyline and fluconazole developed syncope and in one concomitant electrocardiographic monitoring showed a prolonged QT interval and torsade de pointes (142,143). In neither case were serum amitriptyline concentrations measured, but the symptoms were consistent with tricyclic antidepressant toxicity. These case reports suggest that this combination should be used with caution and probably with monitoring of amitriptyline concentrations.

Warfarin

Co-administration of fluconazole and warfarin has led in some cases to prolongation of the prothrombin time (21,106,125,135).

Zidovudine

Zidovudine glucuronidation in human hepatic microsomes in vitro was inhibited more by the combination of fluconazole with valproic acid than with other drugs, such as atovaquone and methadone (136).

References

1. Berl T, Wilner KD, Gardner M, Hansen RA, Farmer B, Baris BA, Henrich WL. Pharmacokinetics of fluconazole in renal failure. J Am Soc Nephrol 1995;6(2):242–7.
2. Debruyne D. Clinical pharmacokinetics of fluconazole in superficial and systemic mycoses. Clin Pharmacokinet 1997;33(1):52–77.
3. Pittrow L, Penk A. Dosage adjustment of fluconazole during continuous renal replacement therapy (CAVH, CVVH, CAVHD, CVVHD). Mycoses 1999;42(1–2):17–9.
4. Schafer-Korting M. Pharmacokinetic optimisation of oral antifungal therapy. Clin Pharmacokinet 1993;25(4):329–41.
5. Grant SM, Clissold SP. Fluconazole. A review of its pharmacodynamic and pharmacokinetic properties, and therapeutic potential in superficial and systemic mycoses. Drugs 1990;39(6):877–916Erratum in: Drugs 1990;40(6):862.
6. Lazar JD, Hilligoss DM. The clinical pharmacology of fluconazole. Semin Oncol 1990;17(3 Suppl 6):14–8.
7. Brammer KW, Farrow PR, Faulkner JK. Pharmacokinetics and tissue penetration of fluconazole in humans. Rev Infect Dis 1990;12(Suppl 3):S318–26.
8. Goa KL, Barradell LB. Fluconazole. An update of its pharmacodynamic and pharmacokinetic properties and

therapeutic use in major superficial and systemic mycoses in immunocompromised patients. Drugs 1995;50(4):658–90Erratum in: Drugs 1996;51(3):505.

9. Debruyne D, Ryckelynck JP. Clinical pharmacokinetics of fluconazole. Clin Pharmacokinet 1993;24(1):10–27.

10. Tucker RM, Williams PL, Arathoon EG, Levine BE, Hartstein AI, Hanson LH, Stevens DA. Pharmacokinetics of fluconazole in cerebrospinal fluid and serum in human coccidioidal meningitis. Antimicrob Agents Chemother 1988;32(3):369–73.

11. Arndt CA, Walsh TJ, McCully CL, Balis FM, Pizzo PA, Poplack DG. Fluconazole penetration into cerebrospinal fluid: implications for treating fungal infections of the central nervous system. J Infect Dis 1988;157(1):178–80.

12. Menichetti F, Fiorio M, Tosti A, Gatti G, Bruna Pasticci M, Miletich F, Marroni M, Bassetti D, Pauluzzi S. High-dose fluconazole therapy for cryptococcal meningitis in patients with AIDS. Clin Infect Dis 1996;22(5):838–40.

13. Mian UK, Mayers M, Garg Y, Liu QF, Newcomer G, Madu C, Liu W, Louie A, Miller MH. Comparison of fluconazole pharmacokinetics in serum, aqueous humor, vitreous humor, and cerebrospinal fluid following a single dose and at steady state. J Ocul Pharmacol Ther 1998;14(5):459–71.

14. Christmas NJ, Smiddy WE. Vitrectomy and systemic fluconazole for treatment of endogenous fungal endophthalmitis. Ophthalmic Surg Lasers 1996;27(12):1012–8.

15. Zarbin MA, Becker E, Witcher J, Yamani A, Irvine AR. Treatment of presumed fungal endophthalmitis with oral fluconazole. Ophthalmic Surg Lasers 1996;27(7):628–31.

16. Luttrull JK, Wan WL, Kubak BM, Smith MD, Oster HA. Treatment of ocular fungal infections with oral fluconazole. Am J Ophthalmol 1995;119(4):477–81.

17. Akler ME, Vellend H, McNeely DM, Walmsley SL, Gold WL. Use of fluconazole in the treatment of candidal endophthalmitis. Clin Infect Dis 1995;20(3):657–64.

18. Bozzette SA, Gordon RL, Yen A, Rinaldi M, Ito MK, Fierer J. Biliary concentrations of fluconazole in a patient with candidal cholecystitis: case report. Clin Infect Dis 1992;15(4):701–3.

19. Edelman DA, Grant S. One-day therapy for vaginal candidiasis. A review. J Reprod Med 1999;44(6):543–7.

20. Houang ET, Chappatte O, Byrne D, Macrae PV, Thorpe JE. Fluconazole levels in plasma and vaginal secretions of patients after a 150-milligram single oral dose and rate of eradication of infection in vaginal candidiasis. Antimicrob Agents Chemother 1990;34(5):909–10.

21. Francis P, Walsh TJ. Evolving role of flucytosine in immunocompromised patients: new insights into safety, pharmacokinetics, and antifungal therapy. Clin Infect Dis 1992;15(6):1003–18.

22. Azon-Masoliver A, Vilaplana J. Fluconazole-induced toxic epidermal necrolysis in a patient with human immunodeficiency virus infection. Dermatology 1993;187(4):268–9.

23. Wang EJ, Lew K, Casciano CN, Clement RP, Johnson WW. Interaction of common azole antifungals with P glycoprotein. Antimicrob Agents Chemother 2002;46(1):160–5.

24. Pelz RK, Hendrix CW, Swoboda SM, Diener-West M, Merz WG, Hammond J, Lipsett PA. Double-blind placebo-controlled trial of fluconazole to prevent candidal infections in critically ill surgical patients. Ann Surg 2001;233(4):542–8.

25. Torres HA, Kontoyiannis DP, Rolston KV. High-dose fluconazole therapy for cancer patients with solid tumors and candidemia: an observational, noncomparative retrospective study. Support Care Cancer 2004;12:511-6.

26. Winston DJ, Hathorn JW, Schuster MG, Schiller GJ, Territo MC. A multicenter, randomized trial of fluconazole versus amphotericin B for empiric antifungal therapy of febrile neutropenic patients with cancer. Am J Med 2000;108(4):282–9.

27. Wolff SN, Fay J, Stevens D, Herzig RH, Pohlman B, Bolwell B, Lynch J, Ericson S, Freytes CO, LeMaistre F, Collins R, Pineiro L, Greer J, Stein R, Goodman SA, Dummer S. Fluconazole vs low-dose amphotericin B for the prevention of fungal infections in patients undergoing bone marrow transplantation: a study of the North American Marrow Transplant Group. Bone Marrow Transplant 2000;25(8):853–9.

28. Villanueva A, Gotuzzo E, Arathoon EG, Noriega LM, Kartsonis NA, Lupinacci RJ, Smietana JM, DiNubile MJ, Sable CA. A randomized double-blind study of caspofungin versus fluconazole for the treatment of esophageal candidiasis. Am J Med 2002;113(4):294–9.

29. Marr KA, Crippa F, Leisenring W, Hoyle M, Boeckh M, Balajee SA, Nichols WG, Musher B, Corey L. Itraconazole versus fluconazole for prevention of fungal infections in patients receiving allogeneic stem cell transplants. Blood 2004;103:1527-33.

30. Oude Lashof AM, De Bock R, Herbrecht R, de Pauw BE, Kremery V, Aoun M, Akova M, Cohen J, Siffnerova H, Egyed M, Ellis M, Marinus A, Sylvester R, Kullberg BJ; EORTC Invasive Fungal Infections Group. An open multicentre comparative study of the efficacy, safety and tolerance of fluconazole and itraconazole in the treatment of cancer patients with oropharyngeal candidiasis. Eur J Cancer 2004;40:1314-9.

31. Alrajhi AA, Ibrahim EA, De Vol EB, Khairat M, Faris RM, Maguire JH. Fluconazole for the treatment of cutaneous leishmaniasis caused by Leishmania major. N Engl J Med 2002;346(12):891–5.

32. Gupta AK, Katz HI, Shear NH. Drug interactions with itraconazole, fluconazole, and terbinafine and their management. J Am Acad Dermatol 1999;41(2 Pt 1):237–49.

33. Neuhaus G, Pavic N, Pletscher M. Anaphylactic reaction after oral fluconazole. BMJ 1991;302(6788):1341.

34. Bradbury BD, Jick SS. Itraconazole and fluconazole and certain rare, serious adverse events. Pharmacotherapy 2002;22(6):697–700.

35. Tholakanahalli VN, Potti A, Hanley JF, Merliss AD. Fluconazole-induced torsade de pointes. Ann Pharmacother 2001;35(4):432–4.

36. Wassmann S, Nickenig G, Bohm M. Long QT syndrome and torsade de pointes in a patient receiving fluconazole. Ann Intern Med 1999;131(10):797.

37. Pham CP, de Feiter PW, van der Kuy PH, van Mook WN. Long QTc interval and torsade de pointes caused by fluconazole. Ann Pharmacother 2006;40(7-8):1456-61.

38. Tatetsu H, Asou N, Nakamura M, Hanaoka N, Matsuno F, Horikawa K, Mitsuya H. Torsades de pointes upon fluconazole administration in a patient with acute myeloblastic leukemia. Am J Hematol 2006;81(5):366-9.

39. Khazan M, Mathis AS. Probable case of torsades de pointes induced by fluconazole. Pharmacotherapy 2002;22(12):1632–7.

40. Anaissie EJ, Kontoyiannis DP, Huls C, Vartivarian SE, Karl C, Prince RA, Bosso J, Bodey GP. Safety, plasma concentrations, and efficacy of high-dose fluconazole in invasive mold infections. J Infect Dis 1995;172(2):599–602.

41. Robinson PA, Knirsch AK, Joseph JA. Fluconazole for life-threatening fungal infections in patients who cannot be treated with conventional antifungal agents. Rev Infect Dis 1990;12(Suppl 3):S349–63.

42. Matsumoto K, Ueno K, Yoshimura H, Morii M, Takada M, Sawai T, Mitsutake K, Shibakawa M. Fluconazole-induced convulsions at serum trough concentrations of approximately 80 microg/mL. Ther Drug Monit 2000;22(5):635–6.

43. Santhana Krishnan SG, Cobbs RK. Reversible acute adrenal insufficiency caused by fluconazole in a critically ill patient. Postgrad Med J 2006;82(971):e23.

44. Magill SS, Puthanakit T, Swoboda SM, Carson KA, Salvatori R, Lipsett PA, Hendrix CW. Impact of fluconazole prophylaxis on cortisol levels in critically ill surgical Patients. Antimicrob Agents Chemother 2004;48:2471-6.

45. Albert SG, DeLeon MJ, Silverberg AB. Possible association between high-dose fluconazole and adrenal insufficiency in critically ill patients. Crit Care Med 2001;29(3):668–70.

46. Shibata S, Kami M, Kanda Y, Machida U, Iwata H, Kishi Y, Takeshita A, Miyakoshi S, Ueyama J, Morinaga S, Mutou Y. Acute adrenal failure associated with fluconazole after administration of high-dose cyclophosphamide. Am J Hematol 2001;66(4):303–5.

47. Larsen RA, Leal MA, Chan LS. Fluconazole compared with amphotericin B plus flucytosine for cryptococcal meningitis in AIDS. A randomized trial. Ann Intern Med 1990;113(3):183–7.

48. Bozzette SA, Larsen RA, Chiu J, Leal MA, Jacobsen J, Rothman P, Robinson P, Gilbert G, McCutchan JA, Tilles J, et al. A placebo-controlled trial of maintenance therapy with fluconazole after treatment of cryptococcal meningitis in the acquired immunodeficiency syndrome. California Collaborative Treatment Group. N Engl J Med 1991;324(9):580–4.

49. Schaffner A, Schaffner M. Effect of prophylactic fluconazole on the frequency of fungal infections, amphotericin B use, and health care costs in patients undergoing intensive chemotherapy for hematologic neoplasias. J Infect Dis 1995;172(4):1035–41.

50. Wong-Beringer A, Shriner K. Fluconazole-induced agranulocytosis with eosinophilia. Pharmacotherapy 2000;20(4):484–6.

51. Chuncharunee S, Sathapatayavongs B, Singhasivanon P, Singhasivanon V. Fluconazole-induced agranulocytosis. Therapie 1994;49(6):517–8.

52. Murakami H, Katahira H, Matsushima T, Sakura T, Tamura J, Sawamura M, Tsuchiya J. Agranulocytosis during treatment with fluconazole. J Int Med Res 1992;20(6):492–4.

53. Mahendra A, Gupta S, Gupta S, Sood S, Kumar P. Oral fixed drug eruption due to fluconazole. Indian J Dermatol Venereol Leprol 2006;72(5):391.

54. Sugar AM, Stern JJ, Dupont B. Overview: treatment of cryptococcal meningitis. Rev Infect Dis 1990;12(Suppl 3):S338–48.

55. Chabot GG, Pazdur R, Valeriote FA, Baker LH. Pharmacokinetics and toxicity of continuous infusion amphotericin B in cancer patients. J Pharm Sci 1989;78(4):307–10.

56. Lee JW, Seibel NL, Amantea M, Whitcomb P, Pizzo PA, Walsh TJ. Safety and pharmacokinetics of fluconazole in children with neoplastic diseases. J Pediatr 1992;120(6):987–93.

57. Crerar-Gilbert A, Boots R, Fraenkel D, MacDonald GA. Survival following fulminant hepatic failure from fluconazole induced hepatitis. Anaesth Intensive Care 1999;27(6):650–2.

58. Spellberg B, Rieg G, Bayer A, Edwards JE Jr. Lack of cross-hepatotoxicity between fluconazole and voriconazole. Clin Infect Dis 2003;36:1091-3.

59. Aghai ZH, Mudduluru M, Nakhla TA, Amendolia B, Longo D, Kemble N, Kaki S, Sutsko R, Saslow JG, Stahl GE. Fluconazole prophylaxis in extremely low birth weight infants: association with cholestasis. J Perinatol 2006;26(9):550-5.

60. Rex JH, Pappas PG, Karchmer AW, Sobel J, Edwards JE, Hadley S, Brass C, Vazquez JA, Chapman SW, Horowitz HW, Zervos M, McKinsey D, Lee J, Babinchak T, Bradsher RW, Cleary JD, Cohen DM, Danziger L, Goldman M, Goodman J, Hilton E, Hyslop NE, Kett DH, Lutz J, Rubin RH, Scheld WM, Schuster M, Simmons B, Stein DK, Washburn RG, Mautner L, Chu TC, Panzer H, Rosenstein RB, Booth J; National Institute of Allergy and Infectious Diseases Mycoses Study Group. A randomized and blinded multicenter trial of high-dose fluconazole plus placebo versus fluconazole plus amphotericin B as therapy for candidemia and its consequences in nonneutropenic subjects. Clin Infect Dis 2003;36:1221-8.

61. Castellsague J, Garcia-Rodriguez LA, Duque A, Perez S. Risk of serious skin disorders among users of oral antifungals: a population-based study. BMC Dermatol 2002;2(1):14.

62. Haria M, Bryson HM, Goa KL. Itraconazole. A reappraisal of its pharmacological properties and therapeutic use in the management of superficial fungal infections. Drugs 1996;51(4):585–620Erratum in: Drugs. 1996;52(2):253.

63. Gussenhoven MJ, Haak A, Peereboom-Wynia JD, van 't Wout JW. Stevens–Johnson syndrome after fluconazole. Lancet 1991;338(8759):120.

64. Morgan JM, Carmichael AJ. Fixed drug eruption with fluconazole. BMJ 1994;308(6926):454.

65. Craig TJ, Peralta F, Boggavarapu J. Desensitization for fluconazole hypersensitivity. J Allergy Clin Immunol 1996;98(4):845–6.

66. Heikkila H, Timonen K, Stubb S. Fixed drug eruption due to fluconazole. J Am Acad Dermatol 2000;42(5 Pt 2):883–4.

67. Dalle S, Skowron F, Ronger-Savle S, Balme B, Thomas L. Erythema multiforme induced by fluconazole. Dermatology 2005;211(2):169.

68. Kar HK. Longitudinal melanonychia associated with fluconazole therapy. Int J Dermatol 1998;37(9):719–20.

69. Su FW, Perumalswami P, Grammer LC. Acute hepatitis and rash to fluconazole. Allergy 2003;58:1215-16.

70. Menegola E, Broccia ML, Di Renzo F, Giavini E. Antifungal triazoles induce malformations in vitro. Reprod Toxicol 2001;15(4):421–7.

71. Menegola E, Broccia ML, Di Renzo F, Giavini E. Pathogenic pathways in fluconazole-induced branchial arch malformations. Birth Defects Res Part A Clin Mol Teratol 2003;67:116-24.

72. Lopez-Rangel E, van Allen MI. Prenatal exposure to fluconazole: an identifiable dysmorphic phenotype. Birth Defects Res A Clin Mol Teratol 2005;73(11):919-23.

73. Tiboni GM, Giampietro F. Murine teratology of fluconazole: evaluation of developmental phase specificity and dose dependence. Pediatr Res 2005;58(1):94-9.

74. Sorensen HT, Nielsen GL, Olesen C, Larsen H, Steffensen FH, Schonheyder HC, Olsen J, Czeizel AE. Risk of malformations and other outcomes in children exposed to fluconazole in utero. Br J Clin Pharmacol 1999;48(2):234–8.

75. Tiboni GM, Iammarrone E, Giampietro F, Lamonaca D, Bellati U, Di Ilio C. Teratological interaction between the bis-triazole antifungal agent fluconazole and the

anticonvulsant drug phenytoin. Teratology 1999;59(2):81–7.

76. Force RW. Fluconazole concentrations in breast milk. Pediatr Infect Dis J 1995;14(3):235–6.

77. Kaufman D, Boyle R, Hazen KC, Patrie JT, Robinson M, Grossman LB. Twice weekly fluconazole prophylaxis for prevention of invasive Candida infection in high-risk infants of <1000 grams birth weight. J Pediatr 2005;147(2):172-9.

78. Brammer KW, Coates PE. Pharmacokinetics of fluconazole in pediatric patients. Eur J Clin Microbiol Infect Dis 1994;13(4):325–9.

79. Schwarze R, Penk A, Pittrow L. Administration of fluconazole in children below 1 year of age. Mycoses 1999;42(1–2):3–16.

80. Saxen H, Hoppu K, Pohjavuori M. Pharmacokinetics of fluconazole in very low birth weight infants during the first two weeks of life. Clin Pharmacol Ther 1993;54(3):269–77.

81. Schwarze R, Penk A, Pittrow L. Anwendung von Fluconazol bei Kinden <1 Jahr: Ubersicht. [Use of fluconazole in children less than 1 year old: review.] Mycoses 1998;41(Suppl 1):61–70.

82. Huttova M, Hartmanova I, Kralinsky K, Filka J, Uher J, Kurak J, Krizan S, Krcmery V Jr. Candida fungemia in neonates treated with fluconazole: report of forty cases, including eight with meningitis. Pediatr Infect Dis J 1998;17(11):1012–5.

83. Novelli V, Holzel H. Safety and tolerability of fluconazole in children. Antimicrob Agents Chemother 1999;43(8):1955–60.

84. Goins RA, Ascher D, Waecker N, Arnold J, Moorefield E. Comparison of fluconazole and nystatin oral suspensions for treatment of oral candidiasis in infants. Pediatr Infect Dis J 2002;21(12):1165–7.

85. Goldman M, Cloud GA, Wade KD, Reboli AC, Fichtenbaum CJ, Hafner R, Sobel JD, Powderly WG, Patterson TF, Wheat LJ, Stein DK, Dismukes WE, Filler SG; AIDS Clinical Trials Group Study Team 323; Mycoses Study Group Study Team 40. A randomized study of the use of fluconazole in continuous versus episodic therapy in patients with advanced HIV infection and a history of oropharyngeal candidiasis: AIDS Clinical Trials Group Study 323/Mycoses Study Group Study 40. Clin Infect Dis 2005;41(10):1473-80.

86. de Wet NT, Bester AJ, Viljoen JJ, Filho F, Suleiman JM, Ticona E, Llanos EA, Fisco C, Lau W, Buell D. A randomized, double blind, comparative trial of micafungin (FK463) vs. fluconazole for the treatment of oesophageal candidiasis. Aliment Pharmacol Ther 2005;21(7):899-907.

87. Palkama VJ, Isohanni MH, Neuvonen PJ, Olkkola KT. The effect of intravenous and oral fluconazole on the pharmacokinetics and pharmacodynamics of intravenous alfentanil. Anesth Analg 1998;87(1):190–4.

88. Robinson RF, Nahata MC, Olshefski RS. Syncope associated with concurrent amitriptyline and fluconazole therapy. Ann Pharmacother 2000;34(12):1406–9.

89. Dorsey ST, Biblo LA. Prolonged QT interval and torsades de pointes caused by the combination of fluconazole and amitriptyline. Am J Emerg Med 2000;18(2):227–9.

90. Pahls S, Schaffner A. Aspergillus fumigatus pneumonia in neutropenic patients receiving fluconazole for infection due to Candida species: is amphotericin B combined with fluconazole the appropriate answer? Clin Infect Dis 1994;18(3):484–6.

91. Rex JH, Pappas PG, Karchmer AW, Sobel J, Edwards JE, Hadley S, Brass C, Vazquez JA, Chapman SW, Horowitz HW, Zervos M, McKinsey D, Lee J, Babinchak T, Bradsher RW, Cleary JD, Cohen DM, Danziger L, Goldman M, Goodman J, Hilton E, Hyslop NE, Kett DH, Lutz J, Rubin RH, Scheld WM, Schuster M, Simmons B, Stein DK, Washburn RG, Mautner L, Chu TC, Panzer H, Rosenstein RB, Booth JNational Institute of Allergy and Infectious Diseases Mycoses Study Group. A randomized and blinded multicenter trial of high-dose fluconazole plus placebo versus fluconazole plus amphotericin B as therapy for candidemia and its consequences in nonneutropenic subjects. Clin Infect Dis 2003;36(10):1221–8.

92. Thorpe JE, Baker N, Bromet-Petit M. Effect of oral antacid administration on the pharmacokinetics of oral fluconazole. Antimicrob Agents Chemother 1990;34(10):2032–3.

93. Venkatakrishnan K, von Moltke LL, Greenblatt DJ. Effects of the antifungal agents on oxidative drug metabolism: clinical relevance. Clin Pharmacokinet 2000;38(2):111–80.

94. Bickers DR. Antifungal therapy: potential interactions with other classes of drugs. J Am Acad Dermatol 1994;31(3 Pt 2):S87–90.

95. Ohtani Y, Kotegawa T, Tsutsumi K, Morimoto T, Hirose Y, Nakano S. Effect of fluconazole on the pharmacokinetics and pharmacodynamics of oral and rectal bromazepam: an application of electroencephalography as the pharmacodynamic method. J Clin Pharmacol 2002;42(2):183–91.

96. Olkkola KT, Ahonen J, Neuvonen PJ. The effects of the systemic antimycotics, itraconazole and fluconazole, on the pharmacokinetics and pharmacodynamics of intravenous and oral midazolam. Anesth Analg 1996;82(3):511–6.

97. Varhe A, Olkkola KT, Neuvonen PJ. Effect of fluconazole dose on the extent of fluconazole–triazolam interaction. Br J Clin Pharmacol 1996;42(4):465 70.

98. Ahonen J, Olkkola KT, Takala A, Neuvonen PJ. Interaction between fluconazole and midazolam in intensive care patients. Acta Anaesthesiol Scand 1999;43(5):509–14.

99. Backman JT, Kivisto KT, Olkkola KT, Neuvonen PJ. The area under the plasma concentration-time curve for oral midazolam is 400-fold larger during treatment with itraconazole than with rifampicin. Eur J Clin Pharmacol 1998;54(1):53–8.

100. Kremens B, Brendel E, Bald M, Czyborra P, Michel MC. Loss of blood pressure control on withdrawal of fluconazole during nifedipine therapy. Br J Clin Pharmacol 1999;47(6):707–8.

101. Nair DR, Morris HH. Potential fluconazole-induced carbamazepine toxicity. Ann Pharmacother 1999;33(7–8):790–2.

102. Finch CK, Green CA, Self TH. Fluconazole–carbamazepine interaction. South Med J 2002;95(9):1099–100.

103. Lazar JD, Wilner KD. Drug interactions with fluconazole. Rev Infect Dis 1990;12(Suppl 3):S327–33.

104. Milliken ST, Powles RL. Antifungal prophylaxis in bone marrow transplantation. Rev Infect Dis 1990;12(Suppl 3):S374–9.

105. Sugar AM, Saunders C, Idelson BA, Bernard DB. Interaction of fluconazole and cyclosporine. Ann Intern Med 1989;110(10):844.

106. Lyman CA, Walsh TJ. Systemically administered antifungal agents. A review of their clinical pharmacology and therapeutic applications. Drugs 1992;44(1):9–35.

107. Osowski CL, Dix SP, Lin LS, Mullins RE, Geller RB, Wingard JR. Evaluation of the drug interaction between

intravenous high-dose fluconazole and cyclosporine or tacrolimus in bone marrow transplant patients. Transplantation 1996;61(8):1268–72.

108. Lopez-Gil JA. Fluconazole–cyclosporine interaction: a dose-dependent effect? Ann Pharmacother 1993;27(4):427–30.

109. Mathis AS, DiRenzo T, Friedman GS, Kaplan B, Adamson R. Sex and ethnicity may chiefly influence the interaction of fluconazole with calcineurin inhibitors. Transplantation 2001;71(8):1069–75.

110. Sud K, Singh B, Krishna VS, Thennarasu K, Kohli HS, Jha V, Gupta KL, Sakhuja V. Unpredictable cyclosporin–fluconazole interaction in renal transplant recipients. Nephrol Dial Transplant 1999;14(7):1698–703.

111. Jordan MK, Polis MA, Kelly G, Narang PK, Masur H, Piscitelli SC. Effects of fluconazole and clarithromycin on rifabutin and 25-O-desacetylrifabutin pharmacokinetics. Antimicrob Agents Chemother 2000;44(8):2170–2.

112. Yule SM, Walker D, Cole M, McSorley L, Cholerton S, Daly AK, Pearson AD, Boddy AV. The effect of fluconazole on cyclophosphamide metabolism in children. Drug Metab Dispos 1999;27(3):417–21.

113. Warren KE, McCully CM, Walsh TJ, Balis FM. Effect of fluconazole on the pharmacokinetics of doxorubicin in nonhuman primates. Antimicrob Agents Chemother 2000;44(4):1100–1.

114. Mikami Y, Scalarone GM, Kurita N, Yazawa K, Uno J, Miyaji M. Synergistic postantifungal effect of flucytosine and fluconazole on *Candida albicans*. J Med Vet Mycol 1992;30(3):197–206.

115. De Wit S, Debier M, De Smet M, McCrea J, Stone J, Carides A, Matthews C, Deutsch P, Clumeck N. Effect of fluconazole on indinavir pharmacokinetics in human immunodeficiency virus-infected patients. Antimicrob Agents Chemother 1998;42(2):223–7.

116. Koks CH, Crommentuyn KM, Hoetelmans RM, Burger DM, Koopmans PP, Mathot RA, Mulder JW, Meenhorst PL, Beijnen JH. The effect of fluconazole on ritonavir and saquinavir pharmacokinetics in HIV-1-infected individuals. Br J Clin Pharmacol 2001;51(6):631–5.

117. Tarumi Y, Pereira J, Watanabe S. Methadone and fluconazole: respiratory depression by drug interaction. J Pain Symptom Manage 2002;23(2):148–53.

118. Cobb MN, Desai J, Brown LS Jr, Zannikos PN, Rainey PM. The effect of fluconazole on the clinical pharmacokinetics of methadone. Clin Pharmacol Ther 1998;63(6):655–62.

119. Zimmermann T, Yeates RA, Riedel KD, Lach P, Laufen H. The influence of gastric pH on the pharmacokinetics of fluconazole: the effect of omeprazole. Int J Clin Pharmacol Ther 1994;32(9):491–6.

120. Kang BC, Yang CQ, Cho HK, Suh OK, Shin WG. Influence of fluconazole on the pharmacokinetics of omeprazole in healthy volunteers. Biopharm Drug Dispos 2002;23(2):77–81.

121. Devenport MH, Crook D, Wynn V, Lees LJ. Metabolic effects of low-dose fluconazole in healthy female users and non-users of oral contraceptives. Br J Clin Pharmacol 1989;27(6):851–9.

122. Sinofsky FE, Pasquale SA. The effect of fluconazole on circulating ethinyl estradiol levels in women taking oral contraceptives. Am J Obstet Gynecol 1998;178(2):300–4.

123. Hilbert J, Messig M, Kuye O, Friedman H. Evaluation of interaction between fluconazole and an oral contraceptive in healthy women. Obstet Gynecol 2001;98(2):218–23.

124. Mitchell AS, Holland JT. Fluconazole and phenytoin: a predictable interaction. BMJ 1989;298(6683):1315.

125. Gericke KR. Possible interaction between warfarin and fluconazole. Pharmacotherapy 1993;13(5):508–9.

126. Kantola T, Backman JT, Niemi M, Kivisto KT, Neuvonen PJ. Effect of fluconazole on plasma fluvastatin and pravastatin concentrations. Eur J Clin Pharmacol 2000;56(3):225–9.

127. Cooper KJ, Martin PD, Dane AL, Warwick MJ, Schneck DW, Cantarini MV. The effect of fluconazole on the pharmacokinetics of rosuvastatin. Eur J Clin Pharmacol 2002;58(8):527–31.

128. Abad S, Moachon L, Blanche P, Bavoux F, Sicard D, Salmon-Ceron D. Possible interaction between gliclazide, fluconazole and sulfamethoxazole resulting in severe hypoglycaemia. Br J Clin Pharmacol 2001;52(4):456–7.

129. Niemi M, Backman JT, Neuvonen M, Laitila J, Neuvonen PJ, Kivisto KT. Effects of fluconazole and fluvoxamine on the pharmacokinetics and pharmacodynamics of glimepiride. Clin Pharmacol Ther 2001;69(4):194–200.

130. Moreno M, Latorre A, Manzanares C, Morales E, Herrero JC, Dominguez-Gil B, Carreno A, Cubas A, Delgado M, Andres A, Morales JM. Clinical management of tacrolimus drug interactions in renal transplant patients. Transplant Proc 1999;31(6):2252–3.

131. Hairhara Y, Makuuchi M, Kawarasaki H, Takayama T, Kubota K, Ito M, Tanaka H, Yoshino H, Hirata M, Kita Y, Kusaka K, Sano K, Saiura A, Ijichi M, Matsukura A, Watanabe M, Hashizume K, Nakatsuka T. Effect of fluconazole on blood levels of tacrolimus. Transplant Proc 1999;31(7):2767.

132. Billaud EM, Guillemain R, Tacco F, Chevalier P. Evidence for a pharmacokinetic interaction between itraconazole and tacrolimus in organ transplant patients. Br J Clin Pharmacol 1998;46(3):271–2.

133. Furlan V, Parquin F, Penaud JF, Cerrina J, Ladurie FL, Dartevelle P, Taburet AM. Interaction between tacrolimus and itraconazole in a heart–lung transplant recipient. Transplant Proc 1998;30(1):187–8.

134. Capone D, Gentile A, Imperatore P, Palmiero G, Basile V. Effects of itraconazole on tacrolimus blood concentrations in a renal transplant recipient. Ann Pharmacother 1999;33(10):1124–5.

135. Black DJ, Kunze KL, Wienkers LC, Gidal BE, Seaton TL, McDonnell ND, Evans JS, Bauwens JE, Trager WF. Warfarin–fluconazole. II. A metabolically based drug interaction: in vivo studies. Drug Metab Dispos 1996;24(4):422–8.

136. Trapnell CB, Klecker RW, Jamis-Dow C, Collins JM. Glucuronidation of 3'-azido-3'-deoxythymidine (zidovudine) by human liver microsomes: relevance to clinical pharmacokinetic interactions with atovaquone, fluconazole, methadone, and valproic acid. Antimicrob Agents Chemother 1998;42(7):1592–6.

137. Palkama VJ, Isohanni MH, Neuvonen PJ, Olkkola KT. The effect of intravenous and oral fluconazole on the pharmacokinetics and pharmacodynamics of intravenous alfentanil. Anesth Analg 1998;87(1):190–4.

138. Kazierad DJ, Martin DE, Blum RA, Tenero DM, Ilson B, Boike SC, Etheredge R, Jorkasky DK. Effect of fluconazole on the pharmacokinetics of eprosartan and losartan in healthy male volunteers. Clin Pharmacol Ther 1997;62(4):417–25.

139. Kaukonen KM, Olkkola KT, Neuvonen PJ. Fluconazole but not itraconazole decreases the metabolism of losartan to E-3174. Eur J Clin Pharmacol 1998;53(6):445–9.

140. Cobb MN, Desai J, Brown LS Jr, Zannikos PN, Rainey PM. The effect of fluconazole on the clinical

pharmacokinetics of methadone. Clin Pharmacol Ther 1998;63(6):655–62.

141. Jordan MK, Polis MA, Kelly G, Narang PK, Masur H, Piscitelli SC. Effects of fluconazole and clarithromycin on rifabutin and 25-O-desacetylrifabutin pharmacokinetics. Antimicrob Agents Chemother 2000;44(8):2170–2.

142. Dorsey ST, Biblo LA. Prolonged QT interval and torsades de pointes caused by the combination of fluconazole and amitriptyline. Am J Emerg Med 2000;18(2):227–9.

143. Robinson RF, Nahata MC, Olshefski RS. Syncope associated with concurrent amitriptyline and fluconazole therapy. Ann Pharmacother 2000;34(12):1406–9.

Flucytosine

General Information

Flucytosine is an antimetabolite of the fluoropyrimidine type. The principle of selectivity of flucytosine for the fungal cell is dual: it depends on an enzyme in order to penetrate the cell (fungal cytosine permease) and a fungal enzyme that deaminates flucytosine to the active antimetabolite 5-fluorouracil, which is metabolized to 5-fluorouridine. Replacement of 5-fluorouracil in RNA results in the disruption of protein synthesis in the fungus. Flucytosine has selective activity against pathogenic yeasts such as Candida, but only moderate activity against Aspergillus and chromoblastomycosis. There is synergy between flucytosine and amphotericin, and the combination is effective against meningeal cryptococcosis and pheohyphomycosis of the central nervous system, specifically disease caused by Xylohypha bantiana (1–3). Flucytosine can be given orally as well as parenterally. For most fungal infections it should not be given as a single agent, because of the development of secondary drug resistance (4), with the possible exception of urinary tract infection (5), in which secondary resistance does however also occur.

The mechanism of toxicity of flucytosine is not fully understood. Conversion of flucytosine to certain metabolites, in particular 5-fluorouracil, in the liver or by the intestinal microflora after oral administration has been proposed. Toxicity may also occur through impurities in the raw material and the formation of fluorouracil from flucytosine after sterilization and storage.

Pharmacokinetics

Flucytosine can be given orally, and peak serum concentrations occur within 1–2 hours in patients with normal renal function. The absorption of flucytosine can be delayed by food or antacids. Flucytosine is minimally bound to plasma proteins. It penetrates into the CSF, vitreous humor, peritoneal fluid, inflamed joints, and other fluid compartments. There are several methods for determining serum concentrations, particularly the creatinine iminohydrolase assay, which makes use of the spurious creatinine increase in serum, as measured by the Kodak Ektachem analyser, an apparatus widely available and providing a low cost method compared with HPLC (6).

Flucytosine accumulates in patients with renal insufficiency, resulting in potentially toxic serum concentrations. About 90% of the dose is excreted unchanged in the urine. Dose adjustments in renal insufficiency can be made in proportion to the reduction in creatinine clearance (7). Metabolism occurs to a limited extent (8). There is evidence that some flucytosine (5-fluorocytosine) is converted into 5-fluorouracil by the intestinal flora (9,10); 5-fluorouracil may also be present as an impurity in the original formulation or after prolonged storage (11).

Increases in hepatic transaminases and depression of hemopoiesis are the principal adverse effects of flucytosine; conversion of flucytosine to fluorouracil is thought to be responsible in most cases (12). The conversion of flucytosine to fluorouracil by micro-organisms in the human intestinal microflora has been studied in vitro using viable and non-viable Escherichia coli at different concentrations of flucytosine (13). Flucytosine conversion was also studied in fecal specimens from three neutropenic patients at the start of antimicrobial/antifungal prophylaxis (C/A regimen) and 1 week later. Flucytosine concentrations fell by an average of 72, 71, and 72% after incubation for 48 hours with viable Escherichia coli organisms 10^{10}/ml in suspension in broth containing flucytosine 13, 130, and 1300 µg/ml respectively. There was a 44% reduction in flucytosine concentrations when non-viable Escherichia coli were used, showing that bacterial viability is not necessary for this conversion. When fecal specimens from two patients were investigated before the C/A regimen, there was significant flucytosine conversion, whereas there was no conversion in the corresponding fecal specimens after 1 week of the C/A regimen.

Observational studies

Experience with flucytosine monotherapy of cryptococcosis has been reviewed in 27 patients treated between 1968 and 1973 who were selected for this form of therapy on the basis of criteria associated with good prognosis (4). Flucytosine was given as primary therapy to 18 patients and as secondary therapy (following failure of amphotericin deoxycholate) to nine patients in dosages of 4–10 g/day in four divided doses for 8 weeks. Toxicity associated with flucytosine was uncommon and mild. Mild leukopenia (nadir 3–4 $\times$ 10^9/l) developed in three patients, and mild thrombocytopenia (101 $\times$ 10^9/l) and worsening anemia occurred in one patient each. Therapy was stopped early or changed in two patients. In the first, therapy was stopped after 31 days because of a white cell count of 4.1 $\times$ 10^9/l; despite the shortened course of therapy, the patient achieved a long-term cure, and the leukopenia was ultimately believed to be secondary to sarcoidosis. There was bone marrow suppression in the second patient shortly after the withdrawal of flucytosine (because of failure to respond); later resumption of flucytosine during amphotericin therapy for this critically ill patient was associated with severe bone marrow suppression and death.

Comparative studies

Combination therapy with fluconazole (200 mg/day for 2 months) and flucytosine (150 mg/kg/day for the first 2 weeks; $n = 30$) has been compared with fluconazole monotherapy (200 mg/day for 2 months; $n = 28$) in a randomized open trial in Ugandan patients with AIDS-associated cryptococcal meningitis (14). Patients in both groups who survived for 2 months received maintenance therapy with fluconazole (200 mg three times per week for 4 months). There were no serious adverse events in any of the patients. The combination therapy prevented death within 2 weeks and significantly increased the survival rate at 6 months (32 versus 12%). However, the rate of positive cryptococcal antigen titers remained high at 2 months after treatment in both groups.

General adverse effects

The toxicity and drug interactions of flucytosine have been reviewed (15). Nausea, vomiting, and diarrhea are common. Enterocolitis is infrequent. Hepatic dysfunction, hepatitis, and even hepatic necrosis and blood disorders, including fatal aplastic anemia, can occur (16). Severe reactions mainly occurred at a time when the importance of high serum concentrations (in excess of 100 µg/ml) of flucytosine in causing these adverse effects was not recognized. The importance of parenteral versus intravenous administration in the development of enterocolitis is unclear. Patients have been described who suffered adverse effects, such as hepatotoxicity or eosinophilia, that were idiosyncratic and not related to flucytosine concentrations. Hypersensitivity reactions can occur. Neither teratogenic nor tumor-inducing effects have been recorded.

Organs and Systems

Cardiovascular

Life-threatening fluorouracil-like cardiotoxicity has been attributed to flucytosine (17).

- A 34-year-old woman took flucytosine, 500 mg 12 times a day for 2 days, for vaginal candidiasis. After the last dose she complained of chest pain, which persisted for a week and was associated with ST segment elevation during exercise. Coronary angiography showed normal coronary arteries. One month later she was rechallenged with 500 mg 12 times a day for 2 days. The day after completion of this regimen, she developed severe chest pain. Electrocardiography showed widespread ST segment elevation and echocardiography showed apicolateral septal hypokinesia with a left ventricular ejection fraction of less than 15%. Her flucytosine plasma concentration 48 hours after the last dose was not high, but the fluorouracil concentration was similar to that found during a 5-day continuous infusion of 5-fluorouracil. Her lymphocytes showed no abnormalities of intracellular flucytosine clearance, and cytosine deaminase, the enzyme that converts flucytosine to fluorouracil, was not detectable.

Similar cardiotoxicity has been reported with 5-fluorouracil. The reported events were generally consistent with a drug- or metabolite-induced increase in coronary vasomotor tone and spasm, leading to myocardial ischemia. The

authors concluded that more attention should be given to the conversion of flucytosine to fluorouracil; however, it is not clear whether flucytosine should be contraindicated in patients with vasospastic or exertional angina.

Nervous system

Headaches, confusion, hallucinations, somnolence, and vertigo can occur. There have also been a few reports of peripheral neuropathy. However, it is difficult to establish the role of flucytosine in these adverse reactions (SED-12, 674). An acute cerebellopathy has been attributed to flucytosine (18).

Hematologic

Bone marrow suppression is a recognized toxic effect of flucytosine; anemia, leukopenia, and thrombocytopenia occur in about 5% of cases. The hematological effects are dose-related and occur after prolonged high blood concentrations of flucytosine (over 100 µg/ml). Hematotoxicity is seen more often in the presence of renal insufficiency and hence during the use of flucytosine in combination with amphotericin. If bone marrow reserve has already been depleted by underlying disease or by medications, the risk of hematotoxicity increases.

The relation between the toxicity and pharmacokinetics of flucytosine has been investigated in a retrospective study in 53 patients in an intensive care unit (19). Thrombocytopenia, as a marker of bone marrow depression, was associated with a reduced clearance of flucytosine; the lowest thrombocyte count was linearly related to the clearance of flucytosine. Patients with flucytosine concentrations over 100 µg/ml were at higher risk of thrombocytopenia and raised hepatic transaminases than those who did not exceed this threshold. In a second study, the authors corroborated their earlier findings and showed a significant relation between the lowest thrombocyte counts and thrombocyte counts predicted on the basis of the creatinine clearance in a new set of patients admitted to the intensive care unit (20).

In a pilot study in six patients receiving intravenous flucytosine, hematotoxicity was monitored by measuring platelet and leukocyte counts; flucytosine and 5-fluorouracil serum concentrations were measured using HPLC (21). The concentrations of 5-fluorouracil in the 34 available serum samples were below the limit of quantification (0.05 µg/ml), but flucytosine was detectable in all samples and the 5-fluorouracil metabolite, α-fluoro-β-alanine (FBAL), was detected at low concentrations in several samples. One patient developed thrombocytopenia (50×10^9/l) during therapy, and one developed leukopenia (2.6×10^9/l). The fact that 5-fluorouracil was not detected in the serum made it unlikely that the toxic effects of intravenous flucytosine resulted from exposure to 5-fluorouracil.

Patients with AIDS, treated for antifungal disease with flucytosine and amphotericin, are particularly susceptible to develop bone marrow depression (1,3). On the other hand, flucytosine can be safely administered to patients with cancer (6) or AIDS (14,22), if toxic concentrations are avoided and blood counts monitored.

Gastrointestinal

Nausea, vomiting, and diarrhea are the most common adverse effects of flucytosine.

Enterocolitis, usually sparing the rectosigmoid, has been described in some cases (23). This type of flucytosine toxicity may be associated with deamination of flucytosine to 5-fluorouracil in the gut (1). Potentially fatal ulcerative colitis has been suspected in a few patients; however, in most cases there were no diagnostic data to back up the diagnosis.

Liver

Abnormal liver function tests (above normal serum alkaline phosphatase and transaminases and, less often, raised serum bilirubin concentrations) have all been reported. Hepatic involvement is rarely serious (SED-12, 675; 24).

Liver cell necrosis, detected by means of liver biopsy, has been rarely described (SED-9, 477; 25).

The reason for antifungal treatment and the concurrent use of other medication may contribute to hepatotoxicity (SED-12, 675; 1,6).

Urinary tract

Crystalluria has been reported, with urinary gravel, consisting of a co-precipitate of flucytosine and uric acid (26).

Skin

Maculopapular and urticarial rashes severe enough to require drug withdrawal have been reported (SEDA-2, 242). The incidence is low.

Acquired photosensitivity has been described in two cases (27).

Immunologic

Anaphylaxis has been reported in a patient with AIDS (28).

Second-Generation Effects

Teratogenicity

Flucytosine is teratogenic in rats, and its close chemical relation to antimetabolites plus the fact that 5-fluorouracil is a metabolite make it inadvisable to administer flucytosine during pregnancy or to fertile women taking no contraceptive precautions.

Drug–Drug Interactions

Amphotericin

Amphotericin in combination with flucytosine results in an increased risk of hematological complications, because amphotericin often impairs renal function, causing retention of flucytosine (7). This combination also delays hematopoietic recovery after cytotoxic chemotherapy; it should not be used as empirical antifungal therapy in febrile patients with neutropenia (29).

Antifungal azoles

Flucytosine has been successfully used in combination with ketoconazole, fluconazole, and itraconazole. Flucytosine and ketoconazole were synergistic in about 40% of yeast isolates resistant to flucytosine alone. The synergistic action of flucytosine with the triazoles against *Candida* species was seen both in vitro and in vivo (3–6).

The concurrent use of fluconazole with flucytosine may have an additive effect (31). This combination could be useful in the treatment of cryptococcal meningitis (6).

The activity of itraconazole against black fungi can be augmented by combining it with flucytosine; the combination has prevented the development of flucytosine resistance (3).

Cytarabine

The suggested interaction of cytarabine with flucytosine, in which there is competitive inhibition of antifungal activity, has not been confirmed (30).

Monitoring Therapy

Monitoring blood concentrations of flucytosine is essential, so that the dosage can be altered according to changing renal function and to avoid toxicity. Peak plasma concentrations should be 40–60 µg/ml after an oral dose, and under 100 µg/ml after intravenous infusion. To avoid fungal resistance, trough concentrations below 20 µg/ml are recommended (7).

References

1. Lyman CA, Walsh TJ. Systemically administered antifungal agents. A review of their clinical pharmacology and therapeutic applications. Drugs 1992;44(1):9–35.
2. Vanden Bossche H, Dromer F, Improvisi I, Lozano-Chiu M, Rex JH, Sanglard D. Antifungal drug resistance in pathogenic fungi. Med Mycol 1998;36(Suppl 1):119–28.
3. Viviani MA. Flucytosine—what is its future? J Antimicrob Chemother 1995;35(2):241–4.
4. Hospenthal DR, Bennett JE. Flucytosine monotherapy for cryptococcosis. Clin Infect Dis 1998;27(2):260–4.
5. Wise GJ, Kozinn PJ, Goldberg P. Flucytosine in the management of genitourinary candidiasis: 5 years of experience. J Urol 1980;124(1):70–2.
6. Francis P, Walsh TJ. Evolving role of flucytosine in immunocompromised patients: new insights into safety, pharmacokinetics, and antifungal therapy. Clin Infect Dis 1992;15(6):1003–18.
7. Polak A. Pharmacokinetics of amphotericin B and flucytosine. Postgrad Med J 1979;55(647):667–70.
8. Chouini-Lalanne N, Malet-Martino MC, Gilard V, Ader JC, Martino R. Structural determination of a glucuronide conjugate of flucytosine in humans. Drug Metab Dispos 1995;23(8):813–7.
9. Malet-Martino MC, Martino R, de Forni M, Andremont A, Hartmann O, Armand JP. Flucytosine conversion to fluorouracil in humans: does a correlation with gut flora status exist? A report of two cases using fluorine-19 magnetic resonance spectroscopy. Infection 1991;19(3):178–80.

10. Harris BE, Manning BW, Federle TW, Diasio RB. Conversion of 5-fluorocytosine to 5-fluorouracil by human intestinal microflora. Antimicrob Agents Chemother 1986;29(1):44–8.

11. Vermes A, van der Sijs H, Guchelaar HJ. An accelerated stability study of 5-flucytosine in intravenous solution. Pharm World Sci 1999;21(1):35–9.

12. Groll AH, Piscitelli SC, Walsh TJ. Clinical pharmacology of systemic antifungal agents: a comprehensive review of agents in clinical use, current investigational compounds, and putative targets for antifungal drug development. Adv Pharmacol 1998;44:343-500.

13. Vermes A, Kuijper EJ, Guchelaar HJ, Dankert J. An in vitro study on the active conversion of flucytosine to fluorouracil by microorganisms in the human intestinal microflora. Chemotherapy 2003;49:17-23.

14. Mayanja-Kizza H, Oishi K, Mitarai S, Yamashita H, Nalongo K, Watanabe K, Izumi T, Ococi-Jungala K, Augustine K, Mugerwa R, Nagatake T, Matsumoto K. Combination therapy with fluconazole and flucytosine for cryptococcal meningitis in Ugandan patients with AIDS. Clin Infect Dis 1998;26(6):1362–6.

15. Vermes A, Guchelaar HJ, Dankert J. Flucytosine: a review of its pharmacology, clinical indications, pharmacokinetics, toxicity and drug interactions. J Antimicrob Chemother 2000;46(2):171–9.

16. Groll AH, Gea-Banacloche JC, Glasmacher A, Just-Nuebling G, Maschmeyer G, Walsh TJ. Clinical pharmacology of antifungal compounds. Infect Dis Clin North Am 2003;17(1):159–91.

17. Isetta C, Garaffo R, Bastian G, Jourdan J, Baudouy M, Milano G. Life-threatening 5-fluorouracil-like cardiac toxicity after treatment with 5-fluorocytosine. Clin Pharmacol Ther 2000;67(3):323–5.

18. Cubo Delgado E, Sanz Boza R, Garcia Urra D, Barquero Jimenez S, Vargas Castrillon E. Acute cerebellopathy as a probable toxic effect of flucytosine. Eur J Clin Pharmacol 1997;51(6):505–6.

19. Vermes A, van Der Sijs H, Guchelaar HJ. Flucytosine: correlation between toxicity and pharmacokinetic parameters. Chemotherapy 2000;46(2):86–94.

20. Vermes A, Guchelaar HJ, Dankert J. Prediction of flucytosine-induced thrombocytopenia using creatinine clearance. Chemotherapy 2000;46(5):335–41.

21. Vermes A, Guchelaar HJ, van Kuilenburg AB, Dankert J. 5-fluorocytosine-related bone-marrow depression and conversion to fluorouracil: a pilot study. Fundam Clin Pharmacol 2002;16(1):39–47.

22. van der Horst CM, Saag MS, Cloud GA, Hamill RJ, Graybill JR, Sobel JD, Johnson PC, Tuazon CU, Kerkering T, Moskovitz BL, Powderly WG, Dismukes WE. Treatment of cryptococcal meningitis associated with the acquired immunodeficiency syndrome. National Institute of Allergy and Infectious Diseases Mycoses Study Group and AIDS Clinical Trials Group. N Engl J Med 1997;337(1):15–21.

23. Cappell MS, Simon T. Colonic toxicity of administered medications and chemicals. Am J Gastroenterol 1993;88(10):1684–99.

24. Larsen RA, Leal MA, Chan LS. Fluconazole compared with amphotericin B plus flucytosine for cryptococcal meningitis in AIDS. A randomized trial. Ann Intern Med 1990;113(3):183–7.

25. Bennet JE. Flucytosine. Ann Intern Med 1977;86(3):319–21.

26. Williams KM, Chinwah PM, Cobcroft R. Crystalluria during flucytosine therapy. Med J Aust 1979;2(11):617.

27. Shelley WB, Sica PA Jr. Disseminate sporotrichosis of skin and bone cured with 5-fluorocytosine: Photosensitivity as a complication. J Am Acad Dermatol 1983;8(2):229–35.

28. Kotani S, Hirose S, Niiya K, Kubonishi I, Miyoshi I. Anaphylaxis to flucytosine in a patient with AIDS. JAMA 1988;260(22):3275–6.

29. Hiddemann W, Essink ME, Fegeler W, Zuhlsdorf M, Sauerland C, Buchner T. Antifungal treatment by amphotericin B and 5-fluorocytosine delays the recovery of normal hematopoietic cells after intensive cytostatic therapy for acute myeloid leukemia. Cancer 1991;68(1):9–14.

30. Wingfield HJ. Absence of fungistatic antagonism between flucytosine and cytarabine in vitro and in vivo. J Antimicrob Chemother 1987;20(4):523–7.

31. Mikami Y, Scalarone GM, Kurita N, Yazawa K, Uno J, Miyaji M. Synergistic postantifungal effect of flucytosine and fluconazole on *Candida albicans*. J Med Vet Mycol 1992;30(3):197–206.

Genaconazole

See also Antifungal azoles

General Information

Genaconazole is an *N*-substituted triazole with a wide antifungal spectrum. Its absorption is slow, with peak concentrations 2–4 hours after a single dose and mean peak concentrations markedly higher after 16 days of administration. The mean elimination half-life is about 90–100 hours (1). It has high tissue penetration, including the CNS, and good broad-spectrum antifungal activity. However, it causes hepatocellular carcinoma in animals and the manufacturers stopped developing it because of concerns about toxicity.

Reference

1. Lin C, Kim H, Radwanski E, Affrime M, Brannan M, Cayen MN. Pharmacokinetics and metabolism of genaconazole, a potent antifungal drug, in men. Antimicrob Agents Chemother 1996;40(1):92–6.

Griseofulvin

General Information

Griseofulvin was originally isolated in 1939 as a natural product of *Penicillium griseofulvum* (1). It interferes with fungal microtubule formation, disrupting the cell's mitotic spindle formation and arresting the metaphase of cell division. Griseofulvin is a fungistatic compound and is active against *Trichophyton*, *Microsporon*, and *Epidermophyton* species (2). With the development of newer safer azoles and the introduction of terbinafine, all of which are easy to administer, the indications for griseofulvin have dwindled (3).

Pharmacokinetics

Griseofulvin is commercially available for oral administration as griseofulvin microsize (4 μm particle size) and griseofulvin ultramicrosize (1 μm particle size). The oral availability of the micronized formulation is variable, 25–70%; ultramicronized griseofulvin, in contrast, is almost completely absorbed (4–6). Peak plasma concentrations occur about 4 hours after dosing. Griseofulvin distributes to keratin precursor cells and is concentrated in skin, hair, nails, liver, adipose tissue, and skeletal muscle. In skin, a concentration gradient is established over time, with the highest concentrations in the outermost stratum corneum (7,8). However, within 48–72 hours after withdrawal, plasma concentrations of griseofulvin are markedly reduced and it is no longer detectable in the stratum corneum (4–6). The half-life of griseofulvin is 9–21 hours (9). It is oxidatively demethylated and conjugated with glucuronic acid, primarily in the liver; its major metabolite, 6-desmethylgriseofulvin, is microbiologically inactive (10). Within 5 days, about one-third of a single dose of micronized griseofulvin is excreted in the feces, and 50% in the urine, predominantly as glucuronidated 6-desmethylgriseofulvin (10,11). The slow penetration rate of griseofulvin into tissues may explain difficulties and delays in eradication of infection in nails (SED-12, 676) (12).

General adverse effects

The more common adverse effects of griseofulvin include headache and a variety of gastrointestinal symptoms. Griseofulvin can cause photosensitivity and exacerbate lupus and porphyria. Cases of erythema multiforme-like reactions, toxic epidermal necrolysis, and a reaction resembling serum sickness have been reported. Proteinuria, nephrosis, hepatotoxicity, leukopenia, menstrual irregularities, estrogen-like effects, and reversible impairment of hearing have been reported rarely (4–6,13). Griseofulvin is teratogenic in animals and has mutagenic and carcinogenic potential, but the significance of these observations for humans is unclear (13).

Organs and Systems

Nervous system

Headache is the most common adverse effect of griseofulvin (14). It occurs in about 50% of patients and can be severe. Drowsiness, dizziness, fatigue, confusion, depression, irritability, and insomnia have also been observed (14). Impaired co-ordination and unsteadiness while walking have been reported in some cases when there was confusion (15). Peripheral neuritis has been attributed to griseofulvin, but with little proof of a causal relation (16).

Sensory systems

Eyes
Blurring of vision has been reported with griseofulvin (17).

Taste
Griseofulvin can cause dysgeusia; both taste and smell disturbances may occur more frequently than has been realized (SED-12, 676)(18).

Psychological, psychiatric

The psychiatric effects of griseofulvin can be very disturbing and are aggravated by alcohol (SED-12, 676) (14).

Endocrine

An estrogen-type effect has been reported in children, affecting the genitals and the breasts (SED-11, 567)(19).

Griseofulvin interferes with porphyrin metabolism. In man, transient increases in erythrocyte protoporphyrin concentrations have been demonstrated, and the production and excretion of porphyrins is increased. Acute intermittent porphyria is an absolute contraindication to griseofulvin. In patients with other forms of porphyria it should also be avoided, in view of the many alternatives (20–25).

Hematologic

There is no evidence that griseofulvin can cause serious blood disorders. However, leukopenia, neutropenia, and monocytosis have been reported (14,26).

Gastrointestinal

Anorexia, a feeling of bloating, mild nausea, and mild diarrhea are common (17). Vomiting, abdominal cramps, and more severe diarrhea are rare. Black, furry tongue, glossodynia, angular stomatitis, and taste disturbances have been described (18).

Liver

While there have been anecdotal reports of hepatitis (14) and cholestasis (27), a causal relation has never been shown.

Skin

Dermatological adverse effects are not uncommon with griseofulvin and are of considerable variety. The following have been described: urticaria (28,29), photosensitivity eruptions (30), erythema multiforme (31), morbilliform rashes (32), serum sickness-like reactions (33), fixed drug eruption (29,34,35), Stevens–Johnson syndrome (36), vasculitis (37), toxic epidermal necrolysis (38,39), and lupus erythematosus (40,41).

- A 40-year-old woman had a burning sensation and erythema of the lips, buccal mucosa, palate, and vulva, which recurred within 4 hours of oral rechallenge with griseofulvin 125 mg (42).

The risk of serious skin disorders has been estimated in 61 858 users, aged 20–79 years, of oral antifungal drugs identified in the UK General Practice Research Database (43). They had received at least one prescription for oral fluconazole, griseofulvin, itraconazole, ketoconazole, or terbinafine. The background rate of serious cutaneous adverse reactions (corresponding to non-use of oral antifungal drugs) was 3.9 per 10 000 person-years (95% CI = 2.9, 5.2). Incidence rates for current use were 15 per 10 000 person-years (1.9, 56) for itraconazole, 11.1 (3.0, 29) for terbinafine, 10 (1.3, 38) for fluconazole, and 4.6 (0.1, 26) for griseofulvin. Cutaneous disorders associated with the use of oral antifungal drugs in this study were all mild.

Immunologic

The triggering of a lupus-like syndrome by griseofulvin, by way of an allergic reaction, has been described, but is rare (44).

Second-Generation Effects

Pregnancy

Embryotoxicity and mutagenicity have been shown in animal experiments with high doses of griseofulvin. Although for this reason most handbooks and drug formularies warn against the use of griseofulvin during pregnancy, there are no reports of any adverse effects on the human fetus, despite the fact that griseofulvin has been on the market since the early 1960s and has been used extensively and, no doubt, during many pregnancies (SED-12, 676) (19,45–47).

Susceptibility Factors

Patients with porphyria are at risk; acute intermittent porphyria is an absolute contraindication to griseofulvin.

Exposure to intense natural or artificial sunlight should be avoided during treatment with griseofulvin.

Patients with pre-existing systemic lupus erythematosus may be more susceptible to the development of skin manifestations from griseofulvin (48).

Drug–Drug Interactions

Alcohol

The effects of alcohol are potentiated by griseofulvin, and the use of alcohol increases the risk and severity of psychiatric disturbances.

Ciclosporin

Ciclosporin blood concentrations halved in a patient who took griseofulvin 500 mg/day, despite an increase in ciclosporin dose by 70% (49). When griseofulvin was withdrawn, the ciclosporin concentration rose. This interaction is attributable to induction of cytochrome P450 by griseofulvin.

Coumarin anticoagulants

Griseofulvin is a potent inducer of cytochrome P450 and has a significant effect on P450 expression in hepatocytes (SEDA-12, 236). It therefore increases the rate of metabolism of coumarin anticoagulants (50). However, both increases and decreases in prothrombin time have been reported (SED-12, 676) (18).

Hormonal contraceptives—oral

Griseofulvin modifies hepatic enzyme activity in mice, and although there is no good evidence of a major enzyme-inducing effect in humans, several case reports of pregnancies in women taking both oral contraceptives and griseofulvin suggest an interaction.

The combination of griseofulvin with oral contraceptives can lead to oligomenorrhea, amenorrhea, and breakthrough bleeding; unintended pregnancies have been reported (51). In a case of oligomenorrhea after treatment with griseofulvin, the use of a higher estrogen oral contraceptive restored regularity of the menstrual cycle (SEDA-12, 237; 51). The fact that griseofulvin has an estrogen-like effect in children suggests that it may affect the rate of estrogen metabolism (SED-12, 676; 52). The authorities in several countries have warned that the contraceptive effect may be diminished (54).

Phenobarbital

The concomitant use of phenobarbital reduces griseofulvin concentrations, an effect that has been attributed to induction of liver enzymes by phenobarbital (53).

References

1. Oxford AE, Raistrick H, Simonart P. Studies in the biochemistry of microorganisms: Griseofulvin, C17H17O6Cl, a metabolic product of Penicillium griseofulvum. Biochem J 1939;33:240–8.
2. Gull K, Trinci AP. Griseofulvin inhibits fungal mitosis. Nature 1973;244(5414):292–4.
3. Graybill JR, Sharkey PK. Fungal infections and their management. Br J Clin Pract Suppl 1990;71:23–31.
4. Blumer JL. Pharmacologic basis for the treatment of tinea capitis. Pediatr Infect Dis J 1999;18(2):191–9.
5. Gupta AK, Sauder DN, Shear NH. Antifungal agents: an overview. Part I. J Am Acad Dermatol 1994;30(5 Pt 1):677–98.
6. Gupta AK, Sauder DN, Shear NH. Antifungal agents: an overview. Part II. J Am Acad Dermatol 1994;30(6):911–33.
7. Epstein WL, Shah VP, Riegelman S. Griseofulvin levels in stratum corneum. Study after oral administration in man. Arch Dermatol 1972;106(3):344–8.
8. Shah VP, Riegelman S, Epstein WL. Determination of griseofulvin in skin, plasma, and sweat. J Pharm Sci 1972;61(4):634–6.
9. Rowland M, Riegelman S, Epstein WL. Absorption kinetics of griseofulvin in man. J Pharm Sci 1968;57(6):984–9.
10. Lin CC, Magat J, Chang R, McGlotten J, Symchowicz S. Absorption, metabolism and excretion of 14C-griseofulvin in man. J Pharmacol Exp Ther 1973;187(2):415–22.
11. Lin C, Symchowicz S. Absorption, distribution, metabolism, and excretion of griseofulvin in man and animals. Drug Metab Rev 1975;4(1):75–95.

12. Shah VP, Riegelman S, Epstein WL. Griseofulvin absorption, metabolism and excretion. In: Robinson HM, editor. The Diagnosis and Treatment of Fungal Infections. Sprnigfield IL: Thomas, 1974:315.

13. Friedlander SF, Suarez S. Pediatric antifungal therapy. Dermatol Clin 1998;16(3):527–37.

14. Gotz H, Reichenberger M. Ergebnisse einer Fragebogenaktion bei 1670 Dermatologen der Bundesrepublik Deutschland uber Nebenwirkungen bei der Griseofulvin Therapie. [Results of questionnaires of 1670 dermatologists in West Germany concerning the side effects of griseofulvin therapy.] Hautarzt 1972;23(11):485–92.

15. Hillstrom L, Kjellin A. Centralnervosa symtom som biverkning vid myjosbehandling med griseofulvin. [Central nervous symptoms as side effects of mycosis treatment with griseofulvin.] Lakartidningen 1974;71(44):4310.

16. Livingood CS, Stewart RH, Webster SB. Cutaneous vasculitis caused by griseofulvin. Cutis (NY) 1970;6:1346.

17. Swartz JH. Infections caused by dermatophytes. N Engl J Med 1962;267:1359–61.

18. Cohen J. Antifungal chemotherapy. Lancet 1982;2(8297):532–7.

19. Walter AM, Heilmeyer L. Antibiotika Fibel. In: Otten H, Siegenthaler W, editors. Antimykotica. Stuttgart: Georg Thieme Verlag, 1975:676.

20. Ziprkowski L, Szeinberg A, Crispin M, Krakowski A, Zaidman J. The effect of griseofulvin in hereditary porphyria cutanea tarda. Investigation of porphyrins and blood lipids. Arch Dermatol 1966;93(1):21–7.

21. Bickers DR. Environmental and drug factors in hepatic porphyria. Acta Dermatol Venereol Suppl (Stockh) 1982;100:29–41.

22. Shimoyama T, Nonaka S. Biochemical studies on griseofulvin-induced protoporphyria. Ann NY Acad Sci 1987;514:160–9.

23. Knasmuller S, Parzefall W, Helma C, Kassie F, Ecker S, Schulte-Hermann R. Toxic effects of griseofulvin: disease models, mechanisms, and risk assessment. Crit Rev Toxicol 1997;27(5):495–537Erratum in: Crit Rev Toxicol. 1998;28(1):102.

24. Felsher BF, Redeker AG. Acute intermittent porphyria: effect of diet and griseofulvin. Medicine (Baltimore) 1967;46(2):217–23.

25. Smith AG, De Matteis F. Drugs and the hepatic porphyrias. Clin Haematol 1980;9(2):399–425.

26. Weinstein L. In: Goodman L, Gilman A, editors. The Pharmacological Basis of Therapeutics, Antifungal Agents. 4th ed.. London: The MacMmillan Press Ld, 1970:1299–305.

27. Chiprut RO, Viteri A, Jamroz C, Dyck WP. Intrahepatic cholestasis after griseofulvin administration. Gastroenterology 1976;70(6):1141–3.

28. Ahrens J, Graybill JR, Craven PC, Taylor RL. Treatment of experimental murine candidiasis with liposome-associated amphotericin B. Sabouraudia 1984;22(2):163–6.

29. Feinstein A, Sofer E, Trau H, Schewach-Millet M. Urticaria and fixed drug eruption in a patient treated with griseofulvin. J Am Acad Dermatol 1984;10(5 Pt 2):915–7.

30. Kojima T, Hasegawa T, Ishida H, Fujita M, Okamoto S. Griseofulvin-induced photodermatitis—report of six cases. J Dermatol 1988;15(1):76–82.

31. Rustin MH, Bunker CB, Dowd PM, Robinson TW. Erythema multiforme due to griseofulvin. Br J Dermatol 1989;120(3):455–8.

32. Gaudin JL, Bancel B, Vial T, Bel A. Hepatite aiguë cytolytique et eruption morbiliforme imputables a la prise de griseofulvine. [Acute cytolytic hepatitis and morbilliform eruption caused by ingestion of griseofulvin.] Gastroenterol Clin Biol 1993;17(2):145–6.

33. Colton RL, Amir J, Mimouni M, Zeharia A. Serum sickness-like reaction associated with griseofulvin. Ann Pharmacother 2004;38(4):609–11.

34. Savage J. Fixed drug eruption to griseofulvin. Br J Dermatol 1977;97(1):107–8.

35. Thyagarajan K, Kamalam A, Thambiah AS. Fixed drug eruption to griseofulvin. Mykosen 1981;24(8):482–4.

36. Walinga H, van Beugen L. Syndroom van Stevens–Johnson na gebruik van griseofulvine. [Stevens–Johnson syndrome following administration of griseofulvin.] Ned Tijdschr Geneeskd 1981;125(19):759–60.

37. Amita DB, Danon YL, Garty BZ. Kawasaki-like syndrome associated with griseofulvin treatment. Clin Exp Dermatol 1993;18(4):389.

38. Taylor B, Duffill M. Toxic epidermal necrolysis from griseofulvin. J Am Acad Dermatol 1988;19(3):565–7.

39. Mion G, Verdon R, Le Gulluche Y, Carsin H, Garcia A, Guilbaud J. Fatal toxic epidermal necrolysis after griseofulvin. Lancet 1989;2(8675):1331.

40. Miyagawa S, Okuchi T, Shiomi Y, Sakamoto K. Subacute cutaneous lupus erythematosus lesions precipitated by griseofulvin. J Am Acad Dermatol 1989;21(2 Pt 2):343–6Erratum in: J Am Acad Dermatol 1990;22(2 Pt 2):345.

41. Anderson WA, Torre D. Griseofulvin and lupus erythematosus. J Med Soc N J 1966;63(5):161–2.

42. Thami GP, Kaur S, Kanwar AJ. Erythema multiforme due to griseofulvin with positive re-exposure test. Dermatology 2001;203(1):84–5.

43. Castellsague J, Garcia-Rodriguez LA, Duque A, Perez S. Risk of serious skin disorders among users of oral antifungals: a population-based study. BMC Dermatol 2002;2(1):14.

44. Watsky MS, Lynfield YL. Lupus erythematosus exacerbated by griseofulvin. Cutis 1976;17(2):361–3.

45. Pohler H, Michalski H. Allergisches Exanthem nach Griseofulvin. [Allergic exanthema caused by griseofulvin.] Dermatol Monatsschr 1972;158(5):383–90.

46. Inoue H, Baba H, Awano K, Yoshikawa K. Genotoxic effect of griseofulvin in somatic cells of *Drosophila melanogaster*. Mutat Res 1995;343(4):229–34.

47. Marchetti F, Tiveron C, Bassani B, Pacchierotti F. Griseofulvin-induced aneuploidy and meiotic delay in female mouse germ cells. II. Cytogenetic analysis of one-cell zygotes. Mutat Res 1992;266(2):151–62.

48. Miyagawa S, Sakamoto K. Adverse reactions to griseofulvin in patients with circulating anti-SSA/Ro and SSB/La autoantibodies. Am J Med 1989;87(1):100–2.

49. Abu-Romeh SH, Rashed A. Cyclosporin A and griseofulvin: another drug interaction. Nephron 1991;58(2):237.

50. Okino K, Weibert RT. Warfarin–gnriseofulrin interaction. Drug Intell Clin Pharm 1986;20(4):291–3.

51. McDaniel PA, Caldroney RD. Griseofulvin-oral contraceptive. Drug Intell Clin Pharm 1986;20:291.

52. Bickers DR. Antifungal therapy: potential interactions with other classes of drugs. J Am Acad Dermatol 1994;31(3 Pt 2):S87–90.

53. Riegelman S, Rowland M, Epstein WL. Griseofulvin-phenobarbital interaction in man. JAMA 1970;213(3):426–31.

54. van Dijke CP, Weber JC. Interaction between oral contraceptives and griseofulvin. BMJ (Clin Res Ed) 1984;288(6424):1125–6.

Itraconazole

See also Antifungal azoles

General Information

Itraconazole is a triazole antifungal drug. It is used orally to treat oropharyngeal and vulvovaginal candidiasis, pityriasis versicolor, dermatophytoses unresponsive to topical treatment, and systemic infections, including aspergillosis, blastomycosis, chromoblastomycosis, coccidioidomycosis, cryptococcosis, histoplasmosis, paracoccidioidomycosis, and sporotrichosis. It is also used to prevent fungal infections in immunocompromised patients.

Pharmacokinetics

The systemic availability of itraconazole and the bioequivalence of single 200 mg doses of itraconazole solution and two capsule formulations have been evaluated in a crossover study in 30 male volunteers (1). Itraconazole and hydroxyitraconazole were 30–37% more available from the solution and were greater than from either capsule formulation. However, the values of C_{max}, t_{max}, and half-lives were comparable. There were no differences in safety and tolerance. The normal t_{max} of itraconazole is 1.5–4 hours and serum concentrations are dose-related. Steady-state concentrations are reached after about 10–14 days and are high in comparison with those attained after single doses. With single daily dose treatment the half-life is 20–30 hours. Itraconazole is highly protein bound. The tissue concentrations in lung, kidney, liver, bone, spleen, and muscle are 2–3 times higher than the corresponding serum concentrations. Concentrations in omentum and adipose tissue are particularly high, and higher concentrations are also found in various parts of the genital tract. Itraconazole is markedly keratinophilic; after withdrawal it will take 1–2 weeks before concentrations in the skin start to fall. Itraconazole concentrations in urine, saliva, eye fluids, and cerebrospinal fluid are low. Penetration of itraconazole into ocular tissues is low compared with those of ketoconazole and fluconazole (2,3). Itraconazole is degraded in the liver and excreted via the bile and to some extent in the urine. Its metabolism is not altered by renal dysfunction (4–6).

Observational studies

The pharmacokinetics, safety, and antifungal efficacy of intravenous itraconazole (400 mg for 2 days then 200 mg for 12 days), followed by 12 weeks of oral capsules (400 mg/day) have been investigated in 31 immunocompromised patients with invasive pulmonary aspergillosis (7). All received intravenous itraconazole and 26 then took oral itraconazole for a median of 79 days. Potentially therapeutic trough plasma itraconazole concentrations of 0.5 microgram/ml or more were achieved in 64% of the patients by day 2 and were generally maintained after switching to oral therapy. There was a complete or partial response in 15 patients. There were adverse events during intravenous therapy in 28 patients, and 13 had adverse events that were possibly related to the drug. The main events (at least 10% incidence) were fever, diarrhea, increased blood urea nitrogen, and nausea. Two of these 13 patients had intravenous therapy withdrawn. There were no consistent clinically relevant changes in laboratory parameters. During oral therapy, nine patients had similar adverse events that were possibly related to itraconazole. Treatment was withdrawn in seven patients because of adverse events during this phase. There were no deaths related to itraconazole.

The pharmacokinetics and safety of intravenous itraconazole for 7 days (200 mg bd for 2 days, then 200 mg od for 5 days), followed by itraconazole oral solution 200 mg od or bd for 14 days, have been assessed in 17 patients with hematological malignancies requiring antifungal prophylaxis (8). The mean trough plasma concentration at the end of the intravenous period was 0.54 microgram/ml. This concentration was not maintained during once-daily oral treatment but increased further during twice-daily treatment, with a trough itraconazole concentration of 1.12 micrograms/ml at the end of oral treatment. All patients had some adverse events, mainly gastrointestinal. The two patients who were withdrawn from the study during intravenous treatment both reported fever; one also had pneumonitis and died from pneumonia 2 weeks after withdrawal, but this was unrelated to the drug. Patients were withdrawn during oral treatment because of fever ($n = 3$), pneumonitis ($n = 2$), colitis ($n = 1$), and abdominal pain and diarrhea ($n = 1$). Biochemical and hematological abnormalities were frequent, but there were no consistent changes.

The efficacy and safety of intermittent itraconazole therapy have been investigated in 635 patients with onychomycosis (9). Intermittent itraconazole (400 mg/day for 1 week per month for 2 months) was effective and safe. Most adverse events were minor and occurred infrequently; there were no major changes in liver function tests.

Two dosages of itraconazole have been compared in the treatment of tinea corporis or tinea cruris in a multicenter, randomized, double-blind, parallel-group study, which showed that itraconazole 200 mg for 1 week (54 patients) is similarly effective, equally well tolerated, and at least as safe as the established regimen of itraconazole 100 mg for 2 weeks (60 patients) (10). In a similar study in tinea pedis or tinea manum, itraconazole 400 mg once a week (66 patients) and itraconazole 100 mg once every 4 weeks (69 patients) were both effective; the two schedules were equally well tolerated and safe (11).

In an open multicenter study in 156 Chinese patients who were given intravenous itraconazole for 2 weeks followed by 200 mg bd orally for 28 day, the most common adverse events were hypokalemia (14%), gastrointestinal disorders (13%), raised liver enzymes (11%), and raised bilirubin (8.3%) (12).

Comparative studies

Amphotericin

Data on the safety of itraconazole have been collected in an open, randomized, multicenter study in 277 adults with cancer and neutropenia (13). Itraconazole oral solution (100 mg bd, $n = 144$) was compared with a combination of amphotericin capsules and nystatin oral suspension ($n = 133$). Adverse events were reported in about 45% of patients in each group. The most frequent were vomiting (14 versus 12 patients), diarrhea (12 versus 9), nausea (5 versus 12), and rash (2 versus 13 patients). There were no differences in liver function test abnormalities. Treatment had to be withdrawn because of adverse events (including death) in 34 patients who took itraconazole and 33 of those who took amphotericin plus nystatin; there were 17 deaths in each group and death was recorded as adverse event in 13 and 9 patients, respectively.

Itraconazole (400 mg intravenously for 2 days, 200 mg intravenously for up to 12 days, then 400 mg/day orally) and intravenous amphotericin deoxycholate (0.7–1.0 mg/kg) have been compared in 384 granulocytopenic patients with persistent fever in a randomized, multicenter trial (14). The median duration of therapy was 8.5 days. The incidence of drug-related adverse events (5 versus 54%) and the rate of withdrawal due to toxicity (19 versus 38%) were significantly lower with itraconazole. The most frequent reasons for withdrawal in patients taking itraconazole were nausea and vomiting (5%), rash (3%), and abnormal liver function tests (3%). Significantly fewer of the patients who received itraconazole had nephrotoxicity (5 versus 24%); however, more had hyperbilirubinemia (10 versus 5%). There was no difference in gastrointestinal adverse events between the two groups.

Itraconazole elixir 2.5 mg/kg bd ($n = 281$) has been compared with amphotericin capsules 500 mg qds ($n = 276$) for the prophylaxis of systemic and superficial fungal infections in a double-blind, randomized, placebo-controlled, multicenter trial for 1–59 days (15). While itraconazole significantly reduced the frequency of superficial fungal infections, it was not superior in reducing invasive fungal infections or in improving mortality. Adverse events were reported in 222 patients taking itraconazole (79%) and in 205 patients taking amphotericin (74%). The commonest adverse events were gastrointestinal, followed by rash and hypokalemia, with no differences between the two regimens. In both groups, 5% of the adverse events were considered to be definitely drug-related. Comparable numbers of patients in the two groups permanently stopped treatment because of adverse events (including death), 75 (27%) in the itraconazole group and 78 (28%) in the amphotericin group. Nausea (9 and 11%) and vomiting (8 and 7%) were the most frequently reported adverse events that led to withdrawal. Biochemical changes were comparable in the two groups.

Other antifungal azoles

The safety of continuous itraconazole (50–200 mg/day for up to 3 months) in the treatment of onychomycosis and dermatomycosis has been reviewed, using published and unpublished data from clinical trials (16). The overall incidence of adverse events in patients who took continuous itraconazole (21%) differed little from that in patients who took either topical miconazole or oral placebo (18%). The most frequently reported adverse events were gastrointestinal disorders (6.7%), headache (4.2%), and skin disorders (2.7%). No data were given on the incidence of serious adverse events attributed to itraconazole. Among laboratory abnormalities, clinically significant rises in liver function tests occurred in 3.4% of 527 patients treated with itraconazole (2.6% in patients treated with 50–200 mg/day for dermatomycosis versus 6.6% in patients treated with 200 mg/day for 3 months for onychomycosis).

Oral fluconazole 400 mg qds and oral itraconazole 200 mg bd have been compared in a randomized, double blind, placebo-controlled trial in 198 patients with progressive non-meningeal coccidioidomycosis (17). Overall, 57% and 72% of patients responded to 12 months of therapy with fluconazole and itraconazole, respectively. Relapse rates after withdrawal did not differ significantly. Both drugs were well tolerated. Serious adverse events occurred in eight of 97 fluconazole-treated patients and six of 101 itraconazole-treated patients. They included raised liver enzymes, gastrointestinal disturbances, hypokalemia, and skin rash. Alopecia was reported in 15 of 97 patients taking fluconazole and in only four of 101 patients taking itraconazole. Similarly, dry lips were reported in 11 of 97 patients taking fluconazole and in none of 101 patients taking itraconazole. Both adverse events have previously been reported with fluconazole.

In a double-blind comparison in oropharyngeal candidiasis in 244 patients with AIDS, itraconazole oral solution and fluconazole capsules (each 100 mg/day for 14 days) were equally efficacious; there were no significant differences in adverse effects (18).

Itraconazole oral solution and fluconazole tablets have been compared in oropharyngeal candidiasis in HIV/AIDS patients in a prospective randomized, blind, multicenter trial (19). Both regimens of itraconazole oral solution (100 mg bd for 7 days or 100 mg od for 14 days) were equivalent to fluconazole (100 mg od for 14 days). Itraconazole oral solution was well tolerated.

Oral itraconazole solution has been compared with intravenous/oral fluconazole for the prevention of fungal infections in a randomized, controlled trial in adult liver transplant recipients, who were randomized to receive either oral itraconazole solution (200 mg bd) or intravenous/oral fluconazole (400 mg/day) (20). Prophylaxis was started immediately before transplant surgery and continued for 10 weeks after transplantation. Proven fungal infection developed in nine of 97 patients given itraconazole and in four of 91 patients given fluconazole. Mortality from fungal infection was very low and occurred in only one of the 188 patients. Except for more frequent gastrointestinal adverse effects (nausea, vomiting, diarrhea) with itraconazole, both drugs were well tolerated and neither was associated with hepatotoxicity. Mean trough plasma concentrations of itraconazole were over 250 ng/ml throughout the study and were not affected by H_2 histamine receptor antagonists or antacids.

In a single-center trial, 304 patients were randomized to receive fluconazole (400 mg/day) or itraconazole (orally 2.5 mg/kg tds or intravenously 200 mg/day) for 180 days after stem cell transplantation, or until 4 weeks after the end of therapy for graft-versus-host disease (21). Fluconazole was given for a median of 120 (range 1–183) days after transplantation, and itraconazole for a median of 89 (range 1–189) days after transplantation. More of those who were given itraconazole discontinued therapy because of adverse effects (36% versus 16%); most of the cases of itraconazole withdrawal were for gastrointestinal complaints (24% versus 4%). More of those who were given itraconazole had at least three times the baseline total bilirubin concentration (95% versus 86%). Intention-to-treat analysis showed no difference in the incidence of invasive fungal infections (fluconazole 16% versus itraconazole 13%); however, of those given itraconazole fewer developed invasive fungal infections (fluconazole 15% versus itraconazole 7%). Itraconazole provided better protection against invasive mould infections (fluconazole 12% versus itraconazole 5%), but similar protection against candidiasis (3% versus 2%). There was no difference in overall survival. Itraconazole appears to prevent invasive mould infections in patients who tolerate it; however, adverse effects and poor tolerability may limit its success as prophylactic therapy.

In 140 patients undergoing allogeneic hemopoietic stem cell transplantation at five selected transplantation centers in the USA who received itraconazole (200 mg intravenously every 12 hours for 2 days followed by 200 mg intravenously every 24 hours or a 200 mg oral solution every 12 hours) or fluconazole (400 mg intravenously or orally every 24 hours) from day 1 until day 100 after transplantation, proven invasive fungal infections occurred in six of 71 itraconazole recipients and in 17 of 67 fluconazole recipients during the first 180 days after transplantation (22). Prophylaxis with itraconazole was associated with fewer invasive infections caused by either yeasts or molds. Except for more frequent gastrointestinal adverse effects (nausea, vomiting, diarrhea, or abdominal pain) in patients given itraconazole (24% versus 9%), both itraconazole and fluconazole were well tolerated. The overall mortality rate was similar in the two groups (32 of 71 patients given itraconazole versus 28 of 67 patients given fluconazole).

Fluconazole and itraconazole have been compared in 252 non-neutropenic cancer patients with oropharyngeal candidiasis (23). The safety and tolerance profiles of the two drugs were comparable, but the adverse effects were not listed in detail.

Terbinafine

Itraconazole (28 patients) and terbinafine (27 patients) have been compared in a double-blind, randomized study in tinea capitis (24). The cure rates at week 12 were 86% and 78% respectively. Adverse events were mild and did not warrant discontinuation of therapy.

Placebo-controlled studies

In a double blind, randomized, placebo-controlled, multicenter trial in plantar or moccasin-type tinea pedis in 72 patients, itraconazole 200 mg bd was significantly more effective than placebo; its safety and tolerability were comparable with placebo (25).

In a double-blind, randomized trial in 71 adults undergoing orthotopic liver transplantation to investigate the role of itraconazole for prevention of invasive fungal infections, the patients were randomly assigned to either oral itraconazole (5.0 mg/kg preoperatively and 2.5 mg/kg bd postoperatively) or placebo (26). Therapy continued for a maximum of 56 days or until the patient was discharged from hospital or met a predefined end-point. Nine patients in the placebo group and one patient in the itraconazole group developed fungal end-points requiring therapy with amphotericin. Adverse events were reported by 97% and 100% of the patients given itraconazole and placebo respectively, and there were one and six deaths respectively.

The efficacy of itraconazole as prophylaxis against serious fungal infections has been investigated in a randomized, double-blind, placebo-controlled study in 39 patients (mean age 15 years) with chronic granulomatous disease, a rare disorder in which the phagocytes fail to produce hydrogen peroxide (27). After the initial treatment, each patient alternated between itraconazole and placebo annually. Patients aged 13 years or older and all patients weighing at least 50 kg took itraconazole 200 mg/day; those under 13 years or weighing less than 50 kg took 100 mg/day. One patient (who had not adhered properly to treatment) had a serious fungal infection while taking itraconazole, compared with seven who had a serious fungal infection while taking placebo. There were no serious adverse effects, although one patient had a *rash* and another had abnormal liver function tests, both of which resolved after withdrawal of itraconazole. Itraconazole prophylaxis appears to be effective and well-tolerated in chronic granulomatous disease, but monitoring for long-term adverse effects is warranted.

General adverse effects

Most of the reported adverse effects of itraconazole are transient. Gastrointestinal reactions, mild dyspepsia, pyrosis, nausea, vomiting, diarrhea, and epigastric pain are not uncommon. In many of the published reports mention is made of increases in serum liver enzyme activities and hypertriglyceridemia, and symptomatic liver toxicity has been reported. Itraconazole does not induce drug-metabolizing enzymes and is a weaker inhibitor of microsomal enzymes than ketoconazole (4,28). In rats given doses of up to 160 mg/kg, there was no induction or inhibition of the metabolism of xenobiotics (SEDA-16, 285). Hypokalemia has often been reported, without an explanation of the mechanism. The use of higher doses (400 or even 600 mg/day) causes an increased incidence of adverse effects; among those documented at these dosages are severe hypokalemia, reversible adrenal insufficiency, and (in one published case) arrhythmias, the latter being connected with an interaction with terfenadine (SEDA-16, 295) (SEDA-18, 198). Skin rashes and pruritus have been reported. Tumor-inducing effects have not been described.

In patients taking itraconazole capsules for prolonged periods the common adverse effects were nausea and vomiting (in under 10%), hypertriglyceridemia (9%), hypokalemia (6%), raised transaminases (5%), rashes and/or pruritus (2%), headache or dizziness (under 2%), and foot edema (1%) (29).

In a study using the UK General Practice Research Database to determine rates of rare, serious drug-induced, adverse effects on the liver, kidneys, skin, or blood, occurring within 45 days of completing a prescription or refill in 54 803 users of either fluconazole or itraconazole, one patient had an abnormal liver function test while taking itraconazole in whom a drug-induced etiology could not be ruled out, a rate of 3.2 per 100 000 prescriptions (95% CI = 0.6, 18) for serious adverse liver effects (30). Thus, itraconazole does not commonly have serious adverse effects on the liver, kidneys, skin, or blood.

Organs and Systems

Cardiovascular

Ventricular fibrillation has been attributed to itraconazole-induced hypokalemia (31).

- Pleural and subsequent pericardial effusion developed in a woman treated with itraconazole 200 mg bd for a localized pulmonary infection with *Aspergillus fumigatus* (SEDA-18, 282). After more than 9 weeks of treatment she developed a pericardial effusion, which necessitated drainage. Itraconazole was withdrawn. Six weeks later, and 2 weeks after the resumption of itraconazole, she developed signs of pulmonary edema and cardiac enlargement. These signs disappeared rapidly on discontinuation of itraconazole.

Studies in dogs and healthy human volunteers have suggested that itraconazole has a negative inotropic effect; the mechanism is unknown. A systematic analysis of data from the FDA's Adverse Event Reporting System (AERS) identified 58 cases suggestive of congestive heart failure in patients taking itraconazole (32). A simultaneous search did not identify any cases of congestive heart failure in patients taking fluconazole and ketoconazole, ruling out the possibility of a class effect. In consequence, the labeling of itraconazole has been revised. Itraconazole is now contraindicated for the treatment of onychomycosis in patients with evidence of ventricular dysfunction. For systemic fungal infections, the risks and benefits of itraconazole should be reassessed if signs or symptoms of congestive heart failure develop.

Nervous system

Headache due to itraconazole has been mentioned in some reports (33). Dizziness is an uncommon complaint, as are mood disturbances.

- A 74-year-old man without a previous psychiatric history developed delirium 1 day after starting to take itraconazole 200 mg bd for disseminated histoplasmosis (34). Extensive diagnostic work-up for potential causes was negative, and he continued to take itraconazole because of clinical improvement. Four days later, his delirium worsened and the dose of itraconazole was reduced to 200 mg/day. Despite the lower dosage, his delirium continued to worsen, and itraconazole was withdrawn, after which his mental status improved rapidly. However, the fungal infection recurred, and itraconazole 200 mg/day was restarted. One day later, the delirium recurred.

The mechanism of this adverse effect is unclear; hypothetically, release of prostaglandins and cytokines after successful therapy could have resulted in delirium, although it could simply have been due to the patient's inherent risk of delirium secondary to age, anemia, and disseminated histoplasmosis.

Painful neuropathy associated with itraconazole has been reported in a man with type 1 diabetes (35).

Psychological, psychiatric

Visual hallucinations with confusion have been reported in a 75-year-old woman, occurring on three separate occasions, each time about 2 hours after a 200 mg dose of itraconazole. Her symptoms abated spontaneously over about 8 hours (36).

Electrolyte balance

Hypokalemia, either in isolation or with hypertension, occurs in about 6% of patients taking long-term itraconazole (23,37); the exact mechanism is unknown.

- Severe hypokalemia followed by rhabdomyolysis occurred in a 19-year-old man with chronic granulomatous disease 20 days after he started to take oral itraconazole (200 mg bd) for a cutaneous abscess caused by *Aspergillus fumigatus*; hypokalemia may have been aggravated by concomitant treatment with intravenous amphotericin B during the first seven days of antifungal therapy (38).

Marked ankle edema with weight gain was seen in a patient taking itraconazole 400 mg/day, in whom there was no explanation other than the use of the drug; after withdrawal of the itraconazole the symptoms disappeared. Hypokalemia and edema have also been observed in a number of patients taking high-dose therapy (600 mg/day) (SED-12, 680; SEDA-16, 295; SEDA-17, 321; 4,31,39,40), associated with mildly depressed aldosterone concentrations (SED-12, 680; SEDA-16, 295).

Hematologic

Leukopenia has been attributed to itraconazole.

- A 14-year-old patient received itraconazole 1 mg/kg/day for dermatophytosis and had a fall in white blood cell count from 5 x 10^9/l at baseline (53% neutrophils) to 2.1 x 10^9/l (41% neutrophils) after 16 weeks of therapy with itraconazole. The white blood cell count returned to normal 9 weeks after drug withdrawal (41).

Hematologic adverse events have been reported with fluconazole and voriconazole, and are listed in the package inserts. However, the package insert for itraconazole is devoid of such a reference. The joint database of the German *Arzneimittelkomission der Deutschen Ärzteschaft* (AkDÄ) and the *Bundesinstitut für Arzneimittel und Medizinprodukte* (BfArM), based on voluntary reports, lists 683 reports of untoward drug effects of itraconazole, of which 4% are reduced erythrocyte counts, 3.2% reduced leukocyte counts, and 9.2 % reduced platelet counts. Although compared with fluconazole and voriconazole, hematological adverse events associated with itraconazole were less frequent in absolute and relative terms, monitoring of blood counts is recommended when itraconazole is prescribed for prolonged periods of time (42).

Gastrointestinal

Dyspepsia, pyrosis, nausea, vomiting, mild epigastric discomfort, and diarrhea can occur in patients taking itraconazole (43). These gastrointestinal complaints are generally mild, but they seem to be the most frequent adverse effects during treatment. The total incidence of adverse effects was 3–5% in patients treated for superficial mycosis and 8% in 99 patients treated for deep mycosis (SED-12, 680; 44). An incidence closer to 15% was reported in a multicenter trial (SEDA-17, 321).

In 50 women with acute vaginal candidiasis, adverse effects were reported in 17 (35%), nausea in seven, headache in six, dizziness in three, and bloating in three, while aspartate transaminase activity was raised in one (45).

Of 1108 patients with HIV treated for mucosal candidiasis, 239 reported gastrointestinal symptoms (46).

Pseudomembranous colitis has been reported in association with exposure to itraconazole (47).

- A 54-year-old man developed new abdominal pain and non-bloody diarrhea 1 month after exposure to a 7-day course of oral itraconazole 200 mg/day. He was taking stable chronic sertraline, valproic acid, and perphenazine, and had not taken antimicrobial drugs for 6 months. Flexible sigmoidoscopy after clinical progression showed pseudomembranes, and subsequent evaluation excluded other causes of diarrhea. Although *Clostridium difficile* culture and toxin assay were eventually negative, possibly because of delayed stool sampling, he responded to a 10-day course of anti-anaerobic drug therapy and was discharged with completely resolved symptoms.

The authors proposed that itraconazole had disrupted the resident fungal flora of the colon.

Liver

In most clinical reports, there were some cases of raised liver enzyme activities; the changes were transient or disappeared after withdrawal of itraconazole (48). More serious hepatotoxicity was not reported.

- Focal nodular hyperplasia of the liver has been reported in a 38-year-old woman who had taken itraconazole 200 mg/day for 4 months for a fungal infection of the fingernails (49). She had taken no other drugs in the year during which focal nodular hyperplasia developed.
- Of three patients, two women aged 62 and 57 and a man aged 75 years, who developed symptomatic hepatic injury 5–6 weeks after starting to take itraconazole, two had the biochemical pattern of cholestatic liver damage (50).

All itraconazole clinical trials sponsored by Janssen Research Foundation for the treatment of onychomycosis, in which there was an assessment of laboratory safety, have been analysed (51). There were no significant differences in the number of code 4 abnormalities (baseline value is in the reference range and at least two values, or the last testing in the observation period, exceed twice the upper limit of the reference range) in liver function tests (alanine transaminase, aspartate transaminase, alkaline phosphatase, and total bilirubin). The incidence of all the code 4 abnormalities was under 2%. Itraconazole pulse therapy for onychomycosis appears to be safe, especially from the perspective of potential liver damage. In the itraconazole package insert, liver function tests are recommended in patients receiving continuous itraconazole for over 1 month. There is no such monitoring requirement for the pulse regimen, unless the patient has a history of underlying hepatic disease, the liver function tests are abnormal at baseline, or signs or symptoms suggestive of liver dysfunction develop at any time.

In 49 patients with invasive fungal infections who were given intravenous itraconazole for 2–42 days, liver function tests were abnormal in 20 (52). Those with liver enzyme abnormalities before treatment were more likely to have liver damage during treatment. However, hepatic damage was associated with itraconazole in only two patients with mild liver function test abnormalities.

Liver failure requiring liver transplantation after itraconazole treatment for 3 weeks for toenail onychomycosis has been reported in a 25-year-old woman (53).

Skin

Different types of rash, including a case of acneiform rash, have been reported in patients taking itraconazole. In one case there were bloody bullae (SED-12, 680; 54–57).

- A 29-year-old man developed an infiltrative maculopapular eruption after 1 week of itraconazole 100 mg bd for tinea corporis (58). Itraconazole was withdrawn, and the lesions disappeared within 7 days. Scratch tests, patch tests, scratch-patch tests, and drug induced lymphocyte stimulation tests for itraconazole were negative; however, rechallenge with systemic itraconazole induced a maculopapular eruption on the face, hands, and the dorsa of the feet. Empty itraconazole capsules had no cutaneous effects, suggesting an allergic reaction to a metabolite of the compound.

Photosensitivity has been attributed to itraconazole (200 mg qds for 5 days), with reduced minimal erythema dose for both UVB (0.12 J/cm²) and UVA (20.1 J/cm²), negative photopatch testing, and a positive photochallenge (59). The authors proposed a photoallergic mechanism because earlier exposure to itraconazole had been uneventful. However, details about sun exposure during the first exposure and about the intensity of sun exposure during the oral photochallenge procedure were not given. The eruption responded to oral steroids, which is more typical of photoallergic than phototoxic reactions.

The risk of serious skin disorders has been estimated in 61 858 users of oral antifungal drugs, aged 20–79 years, identified in the UK General Practice Research Database (60). They had received at least one prescription for oral fluconazole, griseofulvin, itraconazole, ketoconazole, or terbinafine. The background rate of serious cutaneous adverse reactions (corresponding to non-use of oral antifungal drugs) was 3.9 per 10 000 person-years (95% CI = 2.9, 5.2). Incidence rates for current use were 15 per 10 000 person-years (1.9, 56) for itraconazole, 11.1 (3.0, 29) for terbinafine, 10 (1.3, 38) for fluconazole, and 4.6 (0.1, 26) for griseofulvin. Cutaneous disorders associated with the use of oral antifungal drugs in this study were all mild.

- A 45-year-old woman with autoimmune polyglandular syndrome type I taking long-term betamethasone developed fatal toxic epidermal necrolysis when she was given itraconazole (61). It was unclear, however, whether the polyglandular syndrome or any of the other drugs used in her care (canrenoate, co-amoxiclav, furosemide, levofloxacin, and pantoprazole) were causally involved.

Sexual function

There are inconsistent reports about the effects of itraconazole on sex steroids. Concentrations of testosterone, corticosterone, and progesterone were unchanged in rats and in six dogs in whom possible endocrine effects were studied (SED-12, 680). On the other hand, the administration of itraconazole to seven male volunteers for 2 weeks did not produce detectable changes in plasma testosterone or cortisol concentrations. There was a slightly reduced cortisol response to ACTH stimulation 2 weeks after the start of high-dose itraconazole therapy (600 mg/day) in one of eight patients with severe mycosis (39).

Erectile impotence, with normal steroid concentrations, has been reported, as has a reduction in libido (SEDA-17, 321; 62).

Immunologic

Itraconazole 200 mg bd for 2 weeks caused a serum sickness-like reaction in a 53-year-old woman with Ménière's disease (63).

Immediate hypersensitivity to itraconazole is extremely rare.

- Angioedema of the face and generalized urticaria occurred in a 65-year-old man on day 3 of oral treatment with itraconazole 100 mg bd for prophylaxis of candidiasis (64). The symptoms responded promptly to parenteral glucocorticoids. Oral rechallenge with itraconazole resulted in a maculopapular rash and angioedema within 2 hours after ingestion. The patient had no personal or family history of allergy.

Urticaria and angioedema attributed to itraconazole have been reported (65).

Second-Generation Effects

Teratogenicity

Since embryotoxicity and teratogenicity have been found in rats, albeit after the administration of high doses, itraconazole should be avoided during pregnancy (SED-12, 681) (66).

Susceptibility Factors

Age

The safety, tolerability, and pharmacokinetics of itraconazole and its active metabolite hydroxyitraconazole after administration of itraconazole solution in hydroxypropyl-β-cyclodextrin have been investigated in a multicenter study in 26 infants and children aged 6 months to 12 years with mucosal candidiasis or at risk of invasive fungal disease (67). There was a trend to lower minimum plasma concentrations in children aged 6 months to 2 years. The systemic absorption of the solubilizer hydroxypropyl-β-cyclodextrin was less than 1%. Given at 5 mg/kg/day, this formulation provided potentially therapeutic concentrations in plasma, somewhat lower than those attained in adults, and it was well tolerated and safe.

Itraconazole 100 mg/day has been studied in 24 children with *Trichophyton tonsurans* tinea capitis (68). Itraconazole was well tolerated, but 15 children required re-treatment due to persistent infection.

The safety, pharmacokinetics, and pharmacodynamics of an oral suspension of cyclodextrin itraconazole (2.5 mg /kg od or bd for 15 days) have been investigated in an open, sequential, dose-escalation study in 26 children and adolescents, 5–18 years old, infected with HIV (mean CD4 count 128 × 10⁶/l) with oropharyngeal candidiasis (69). Apart from mild to moderate gastrointestinal disturbances in three patients, cyclodextrin itraconazole was well tolerated. Two patients withdrew prematurely because of adverse events. The oropharyngeal candidiasis score fell significantly from a mean of 7.46 at baseline to 2.8 at the end of therapy, demonstrating antifungal efficacy in this setting. Based on these results, a dosage of 2.5 mg/kg bd was recommended for the treatment of oropharyngeal candidiasis in children aged 5 years and over.

The safety and efficacy of oral cyclodextrin itraconazole (5 mg/kg/day) as antifungal prophylaxis has been assessed in an open trial in 103 neutropenic children (median age 5 years; range 0–15 years) (70). Prophylaxis was started at least 7 days before the onset of neutropenia and continued until neutrophil recovery. Of the 103 patients, only 47 completed the course of prophylaxis; 27 withdrew because of poor compliance, 19 because of adverse events, and 10 for other reasons. Serious adverse events (other than death) occurred in 21 patients, including convulsions ($n = 7$), suspected drug interactions ($n = 6$), abdominal pain ($n = 4$), and constipation ($n = 4$). The most common adverse events considered definitely or possibly related to itraconazole were vomiting ($n = 12$), abnormal liver function ($n = 5$), and abdominal pain ($n = 3$). Tolerability of the study medication at end-point was rated as good (55%), moderate (11%), poor (17%), or unacceptable (17%). There were no unexpected problems of safety or tolerability.

Cystic fibrosis

The pharmacokinetics of itraconazole and hydroxyitraconazole have been assessed in 17 patients with cystic fibrosis (71). Steady-state concentrations were achieved after 8 days. On day 14 average C_{mat} was 404 ng/ml in those under 16 years of age (n = 5) and 779 ng/ml in those aged 16 years and over (n = 11), excluding one patient who was concurrently taking oral clarithromycin. All the younger patients and half of the older patients failed to achieve a plasma steady-state trough concentration of over 250 ng/ml. Adverse events were reported by nine of the subjects. Most were mild or moderate in intensity and were not considered related to treatment. One patient withdrew because of two severe adverse events (nausea and vomiting). There were ten significant laboratory abnormalities in seven of 16 patients with paired data. Six of these were clinically relevant (hyperuricemia in two and leukocytosis, anemia, hypokalemia, and a mild rise in alanine transaminase in one each).

Diabetes mellits

Adverse effects due to drug–drug interactions are not expected in diabetic patients using insulin and oral hypoglycemic drugs that are not metabolized by CYP3A4 (for example tolbutamide, gliclazide, glibenclamide, glipizide, and metformin). The pharmacokinetic and safety data from clinical trials and postmarketing surveillance have been reviewed to assess the safety of itraconazole in diabetic patients with onychomycosis or dermatomycosis (72). Postmarketing surveillance, including all adverse event reports in patients taking itraconazole concomitantly with insulin or an oral hypoglycemic drug, revealed 15 reports suggestive of hyperglycemia and nine reports suggestive of hypoglycemia. In most patients there was no change in antidiabetic effect. From clinical trials in 189 diabetic patients taking itraconazole for various infections, one itraconazole-related adverse event was recorded; this was a case of aggravated diabetes in a renal transplant patient who was also taking ciclosporin.

Drug Administration

Drug formulations

Itraconazole is poorly soluble in water and highly lipophilic. It is available as capsules, as oral and parenteral solutions that both contain hydroxypropyl-β-cyclodextrin as a solubilizer.

Absorption of itraconazole from capsules depends on a low gastric pH and is reduced by fasting and improved by the presence of food and acidic beverages; it is unpredictable in patients with hypochlorhydria. The t_{max} occurs at 1.5–2 hours. When polyethylene glycol is used as a solvent, the absorption is not as good. Inadequate plasma concentrations are often found in patients receiving cytotoxic drug therapy, which predisposes them to mucositis, poor food intake, and vomiting. The absorption of oral itraconazole seems to be reduced in patients with AIDS (2,3).

Systemic absorption of the cyclodextrin carrier after oral administration is negligible. After intravenous administration the cyclodextrin is not metabolized and is almost completely eliminated by glomerular filtration within 24 hours. Although the cyclodextrin enhances the systemic availability of itraconazole, it can have gastrointestinal adverse effects when used in escalating dosages exceeding 400 mg/day.

A dose-ranging study of itraconazole in antifungal prophylaxis in 123 neutropenic patients with hematological malignancies has been reported (73). The dosing regimens included itraconazole capsules 400, 600, or 800 mg/day and itraconazole cyclodextrin solution 400 and 800 mg/day (with an additional loading with 800 mg of the capsule formulation for 7 days). Ten of twenty-eight patients taking 800 mg/day as a solution withdrew after 1–6 days with severe nausea and vomiting in temporal relation to ingestion of the drug. All the patients who discontinued the solution continued medication with itraconazole capsules in the same dosage without gastrointestinal adverse effects.

Drug–Drug Interactions

See also Antifungal azoles

Amphotericin

The in vitro effects of the combination of amphotericin with itraconazole was tested using six strains of *A. fumigatus*, and an antagonistic effect was found after pretreatment for all strains in vitro and for one strain in a mouse model of aspergillosis (74).

The effect of the combination of itraconazole with amphotericin on liver enzyme activities has been studied retrospectively in 20 patients with hematological malignancies or chronic lung disease complicated by fungal infection or colonization (75). They took itraconazole 200–600 mg/day for a median of 143 (range 44–455) days. Nine had no abnormal liver function tests, including periods of high concentrations of itraconazole (over 5000 ng/ml) and its active hydroxylated metabolite; only one had received concomitant amphotericin. All of the 11 patients with liver function abnormalities had received concomitant amphotericin. For each patient, liver function abnormalities were greatest during the time of concomitant therapy with both antifungal drugs. Although liver enzyme abnormalities are uncommon with amphotericin (76), and although this retrospective analysis was subject to several flaws and potential biases, it nevertheless suggests that hepatotoxicity should be carefully monitored if itraconazole and amphotericin are co-administered.

Antihistamines

It seems likely that combining itraconazole with astemizole (77) and terfenadine (78,79) will lead to increased effects of these antihistamines (80).

Barbiturates

Barbiturates lower itraconazole concentrations (SED-12, 681) (4,28).

Benzodiazepines

The effect of itraconazole on the single oral dose pharmacokinetics and pharmacodynamics of estazolam has been studied in a double-blind, randomized, crossover study in 10 healthy male volunteers, who took oral itraconazole 100 mg/day or placebo for 7 days and on day 4 a single oral dose of estazolam 4 mg (81). Blood samplings and evaluation of psychomotor function by the Digit Symbol Substitution Test, Visual Analogue Scale, and Stanford Sleepiness Scale were conducted up to 72 hours after estazolam. There was no significant difference between the placebo and itraconazole phases in peak plasma concentration, clearance, and half-life. Similarly, psychomotor function was unaffected. These findings suggest that CYP3A4 is not involved to a major extent in the metabolism of estazolam.

In a study of the effects of itraconazole 200 mg/day and rifampicin 600 mg/day on the pharmacokinetics and pharmacodynamics of oral midazolam 7.5–15 mg during and 4 days after the end of the treatment, switching from inhibition to induction of metabolism caused an up to 400-fold change in the AUC of oral midazolam (82).

Bupivacaine

The interaction of itraconazole 200 mg orally od for 4 days with a single intravenous dose of racemic bupivacaine (0.3 mg /kg given over 60 minutes) has been examined in a placebo-controlled crossover study in 10 healthy volunteers (83). Itraconazole reduced the clearance of R-bupivacaine by 21% and that of S-bupivacaine by 25%, but had no other significant effects on the pharmacokinetics of the enantiomers. Reduction of bupivacaine clearance by itraconazole is likely to increase steady-state concentrations of bupivacaine enantiomers by 20–25%, and this should be taken into account in the concomitant use of itraconazole and bupivacaine.

Buspirone

The interaction of itraconazole with the active 1-(2-pyrimidinyl)-piperazine metabolite of buspirone has been studied after a single oral dose of buspirone 10 mg (84). Itraconazole reduced the mean AUC of the metabolite by 50% and the C_{max} by 57%, whereas the mean AUC and C_{max} of the parent drug were increased 14.5-fold and 10.5-fold respectively. Thus, itraconazole caused relatively minor changes in the plasma concentrations of the active piperazine metabolite of buspirone, although it had major effects on the concentrations of buspirone after a single oral dose.

Itraconazole, an inhibitor of CYP3A, can increase buspirone concentrations (SEDA-22, 39).

Busulfan

Reduced elimination and increased toxicity of busulfan co-administered with itraconazole has been postulated (85).

In 13 patients given bone marrow transplantation, the clearance of busulfan was reduced by an average of 20% in patients taking itraconazole compared with control patients and patients taking fluconazole (152).

Carbamazepine

Low and sometimes very low serum concentrations of itraconazole have been seen during concurrent therapy of itraconazole with carbamazepine (86).

Ciclosporin

The combination of itraconazole with ciclosporin leads to a marked increase in blood ciclosporin concentrations, and this can result in a rise in serum creatinine, clearly pointing to renal damage as a result of the high ciclosporin concentrations (SED-12, 681) (87,88–90). However, an interaction has not been demonstrated in all cases (92).

Two cases of rhabdomyolysis caused by itraconazole in heart transplant recipients taking long-term ciclosporin and simvastatin have been reported (92,93). To avoid severe myopathy, ciclosporin concentrations should be monitored frequently and statins should be withdrawn or the dosage should be reduced, as long as azoles need to be prescribed in transplant recipients. Patients need to be educated about signs and symptoms that require immediate physician intervention.

Citrate-phosphate buffer

The citrate-phosphate buffer used to facilitate the absorption of dideoxyinosine (didanosine), prescribed for the treatment of AIDS, may interfere with the absorption of itraconazole (94).

Clarithromycin

A report of three HIV-negative patients has suggested that concomitant therapy with itraconazole and clarithromycin can lead to increased clarithromycin exposure, with an increased metabolic ratio, possibly related to itraconazole's effect on CYP3A4 (95). Nevertheless, in none of the three reported individuals were there adverse effects from this presumed interaction.

Clozapine

Itraconazole 200 mg had no significant effect on serum concentrations of clozapine 200–550 mg/day or desmethylclozapine in 7 schizophrenic patients (97).

Digoxin

Itraconazole inhibits the elimination of digoxin, eventually leading to toxicity (4,97–100) (SEDA-18, 198).

Itraconazole increases the digoxin AUC_{0-72} by about 50%, and reduces its renal clearance by about 20% (101). Apart from inhibition of the renal secretion of digoxin, which is probably mediated by inhibition of P glycoprotein, a study in guinea pigs also showed significantly reduced biliary excretion of digoxin by itraconazole, suggesting that the interaction between itraconazole and digoxin may not only be due to a reduction in renal clearance, but also to a reduction in the metabolic clearance of digoxin by itraconazole (102).

The importance of this interaction has been emphasized by a report of two renal transplant patients who had digoxin toxicity when they took itraconazole concurrently (103).

Famotidine

Famotidine 40 mg/kg/day reduced the peak and trough concentrations of itraconazole 200 mg/kg/day by about 35% in 18 patients undergoing chemotherapy for hematological malignancies (104).

Fentanyl

Fentanyl is a substrate of CYP3A4, CYP2C9, and CYP2C19. However, in one study, the pharmacokinetics and pharmacodynamics of fentanyl 3 micrograms/kg were similar after itraconazole 200 mg and placebo in 10 healthy volunteers (105).

- An interaction of itraconazole with fentanyl has been reported in a 67-year-old man with cancer on a stable dose of transdermal fentanyl 50 micrograms/hour (106). He took itraconazole 200 mg bd for oropharyngeal candidiasis, and 24 hours later developed signs of opioid toxicity, which was reversed by withdrawal of fentanyl and replacement with short-acting opioids.

This may be an interaction to which only some individuals are susceptible.

Oral itraconazole 200 mg did not alter the pharmacokinetics of intravenous fentanyl 3 µg/kg, despite being a strong inhibitor of CYP3A4 in vitro (153). In vitro research suggests that itraconazole should inhibit the elimination of fentanyl, as it has been shown to do to alfentanil. This difference can be accounted for by the higher hepatic extraction ratio of fentanyl (0.8–1.0) compared with alfentanil (0.3–0.5), so that even large changes in the activity of enzymes that metabolize fentanyl significantly affect its pharmacokinetics.

Flucytosine

The activity of itraconazole against black fungi can be augmented by combining it with flucytosine; the combination has prevented the development of flucytosine resistance (107).

Glucocorticoids

Itraconazole can inhibit the metabolic clearance of glucocorticoids by interfering with CYP3A4 and can directly inhibit steroidogenesis, thereby causing serious adverse effects. The effects on different glucocorticoids differ.

Budesonide

Two patients with cystic fibrosis developed profound adrenal failure and impairment of inhaled steroid clearance, resulting in paradoxical Cushing's syndrome, after long-term treatment with itraconazole and inhaled budesonide (108,109). Pituitary–adrenal axis and gonadal function was then assessed in 37 patients treated with itraconazole with or without budesonide (110). An adrenocorticotropic hormone (ACTH) test (tetracosactide 250 micrograms) was performed in 25 patients with cystic fibrosis taking itraconazole and budesonide and in 12 patients taking itraconazole alone (6 with cystic fibrosis and 6 with chronic granulomatous disease). Mineralocorticoid and gonadal steroid function were evaluated by measurements of plasma renin activity and follicle-stimulating hormone, luteinizing hormone, progesterone, estradiol, testosterone, and serum inhibin A and B concentrations. ACTH tests performed as part of a pretransplantation program in a further 30 patients with cystic fibrosis were used as controls. Eleven of the twenty-five patients who took both itraconazole and budesonide had adrenal insufficiency. None of the patients taking itraconazole alone and none of the control patients with cystic fibrosis had an abnormal ACTH test. Mineralocorticoid or gonadal insufficiency was not observed in any patient. Only one patient with an initial pathological ACTH test subsequently normalized; the other 10 patients improved but had not achieved normal adrenal function 2–10 months after itraconazole had been withdrawn.

The effects of itraconazole on the pharmacokinetics and cortisol-suppressing activity of budesonide by inhalation were further investigated in a randomized, double-blind, two-phase, crossover study in 10 healthy subjects who took oral itraconazole 200 mg/day or placebo for 5 days (111). On day 5, 1 hour after the last dose of itraconazole or placebo, they took budesonide 1000 micrograms by inhalation. Plasma budesonide and cortisol concentrations were measured up to 23 hours. Itraconazole increased the mean total AUC of inhaled budesonide 4.2-fold (range 1.7–9.8) and the peak plasma concentration 1.6-fold compared with placebo. The mean half-life of budesonide was prolonged from 1.6 to 6.2 hours. The suppression of cortisol production after inhalation of budesonide was significantly increased by itraconazole compared with placebo, with a 43% reduction in the AUC of plasma cortisol from 0.5 to 10 hours and a

12% reduction in the cortisol concentration measured 23 hours after budesonide, at 8 a.m. Thus, itraconazole markedly increased the systemic exposure to inhaled budesonide. This interaction resulted in enhanced systemic effects of budesonide, as shown by suppression of cortisol production.

Dexamethasone
Itraconazole markedly increases both systemic exposure to dexamethasone and its effects. This interaction has been investigated in a randomized, double-blind, placebo-controlled, crossover study (112). Eight healthy volunteers took either oral itraconazole 200 mg od or placebo for 4 days. On day 4, each subject was given oral dexamethasone 4.5 mg or intravenous dexamethasone sodium phosphate 5.0 mg. Itraconazole reduced the systemic clearance of intravenous dexamethasone by 68%, and increased its AUC and prolonged its half-life more than three-fold; the AUC of oral dexamethasone was increased nearly four-fold and its half-life nearly three-fold. Morning plasma cortisol concentrations at 47 and 71 hours after dexamethasone were significantly lower after itraconazole than placebo.

Methylprednisolone
The interaction of itraconazole with oral methylprednisolone has been examined in a randomized, double-blind, crossover study in 10 healthy volunteers taking either oral itraconazole 200 mg/day or placebo for 4 days (113). On day 4 each subject took methylprednisolone 16 mg. Itraconazole increased the total AUC of methylprednisolone 3.9-fold compared with placebo, the peak plasma methylprednisolone concentration 1.9-fold, and the half-life 2.4-fold. This effect was probably through inhibition of CYP3A4.

Similar effects were found in a study of the effects of itraconazole on the kinetics of intravenous methylprednisolone in a double-blind, randomized, crossover study in nine healthy volunteers (114). Itraconazole (200 mg for 4 days) increased the AUC of methylprednisolone (16 mg on day 4) 2.6-fold, and the AUC_{12-24} 12-fold. The systemic clearance of methylprednisolone was reduced to 40% by itraconazole and the half-life was increased from 2.1 to 4.8 hours. The mean morning plasma cortisol concentration was only 9% of that during the placebo phase. Thus, concomitant itraconazole greatly increased exposure to methylprednisolone during the night-time and led to enhanced adrenal suppression.

The effects of oral itraconazole (400 mg on the first day, then 200 mg/day for 3 days) on the pharmacokinetics of a single oral dose of methylprednisolone 48 mg have been studied in 14 healthy men in a two-period, crossover study (115). Plasma cortisol concentrations were determined as a pharmacodynamic index. Itraconazole significantly increased the mean AUC of methylprednisolone from 2773 to 7011 hours.ng/ml and the half-life from 3.2 to 5.5 hours. Cortisol concentrations at 24 hours were significantly lower after the administration of methylprednisolone with itraconazole than after methylprednisolone alone (24 versus 109 ng/ml).

Prednisolone
The effects of oral itraconazole (400 mg on the first day then 200 mg/day for 3 days) on the pharmacokinetics of a single oral dose of prednisolone 60 mg or methylprednisolone 48 mg have been studied in 14 healthy men in a two-period, crossover study (117). Plasma cortisol concentrations were determined as a pharmacodynamic index. The disposition of prednisolone was unchanged.

The effects of itraconazole on the pharmacokinetics and pharmacodynamics of oral prednisolone have been investigated in a double-blind, randomized, crossover study (116). Ten healthy subjects took either oral itraconazole 200 mg od or placebo for 4 days. On day 4 they took oral prednisolone 20 mg. Itraconazole increased the plasma AUC of prednisolone by 24% and its half-life by 29% compared with placebo. The peak plasma concentration and time to the peak of prednisolone were not affected. Itraconazole reduced the mean morning plasma cortisol concentration, measured 23 hours after prednisolone, by 27%.

The minor interaction of itraconazole with oral prednisolone is probably of limited clinical significance. The susceptibility of prednisolone to interact with CYP3A4 inhibitors is considerably smaller than that of methylprednisolone, and itraconazole and probably also other inhibitors of CYP3A4 can be used concomitantly with prednisolone without marked interaction.

Grapefruit juice

The effect of grapefruit juice on the systemic availability of itraconazole capsules has been investigated in a randomized, two-way, crossover design in 11 healthy volunteers (117). Grapefruit juice reduced the mean itraconazole AUC_{0-48} by 43% and the mean hydroxyitraconazole AUC_{0-72} by 47%. Grapefruit juice also significantly delayed the mean itraconazole t_{max} from 4 to 5.5 hours. The mechanisms for the impaired absorption of itraconazole in the presence of grapefruit juice remain to be elucidated; however, grapefruit juice is acidic and an inhibitor of CYP3A4.

Unexpectedly, in another study, grapefruit juice did not affect the pharmacokinetics of itraconazole in 22 healthy men while orange juice reduced the C_{max}, t_{max}, and AUC (118).

Haloperidol

The effects of itraconazole 200 mg/day for 7 days on the steady-state plasma concentrations of haloperidol and its reduced metabolite have been investigated in schizophrenic patients receiving haloperidol 12 or 24 mg/day (119). Itraconazole significantly increased trough plasma concentrations of both haloperidol and reduced haloperidol (17 versus 13 ng/ml and 6.1 versus 4.9 ng/ml respectively). There was no change in clinical symptoms, but neurological adverse effects of haloperidol were significantly increased during itraconazole co-administration. Similar findings in a similar study were reported for bromperidol and its reduced metabolite, although there were no differences in clinical symptoms or neurological adverse effects during concomitant itraconazole therapy (120).

Adverse effects can result from increased plasma concentrations of haloperidol during itraconazole treatment. This has been observed in 13 schizophrenic patients treated with haloperidol 12 or 24 mg/day who took itraconazole 200 mg/day for 7 days (154). Plasma concentrations of haloperidol were significantly increased and neurological adverse effects were more common. Itraconazole is a potent inhibitor of CYP3A4.

Hormonal contraceptives-oral

Itraconazole can delay withdrawal bleeding in women using oral contraceptives (122). An analysis using data from a spontaneous adverse drug reaction reporting system in the Netherlands and logistic regression analysis showed an odds ratio of 85 (CI = 32, 230) for delayed withdrawal bleeding in women who used both drugs concomitantly compared with woman who used neither oral contraceptives nor itraconazole. Nine of the ten reports of delayed withdrawal bleeding concerned oral contraceptives containing desogestrel. In an open, crossover study in 10 young healthy subjects, fluconazole 150 mg increased the serum concentrations of ethinylestradiol at a close of 30–35 micrograms/day (123).

Lidocaine

The effects of itraconazole, an inhibitor of CYP3A4, on the pharmacokinetics of lidocaine have been studied in nine healthy volunteers. Steady-state oral itraconazole had no effect on the plasma concentration versus time curve of lidocaine after intravenous administration nor on the plasma concentrations of the major metabolite of lidocaine, MEGX (21).

Omeprazole

In 11 healthy volunteers, omeprazole 40 mg reduced the systemic availability of itraconazole 200 mg; these two drugs should therefore not be used together (121).

Phenytoin

Low and sometimes very low serum concentrations of itraconazole have been seen during concurrent therapy of itraconazole with phenytoin (87). At the same time, phenytoin concentrations may themselves be lowered when it is used with itraconazole (87).

Quinidine

In vitro studies have suggested that the oxidation of quinidine to 3-hydroxyquinidine is a specific marker reaction for CYP3A4 activity. In six healthy young men the pharmacokinetics of a single oral dose of quinidine 200 mg were studied before and during daily administration of itraconazole 100 mg (124). Itraconazole reduced quinidine total clearance, partial clearance by 3-hydroxylation, and partial clearance by N-oxidation by 61, 84, and 73% respectively.

Rifamycins

Rifampicin 600 mg/day for 14 days had a very strong inducing effect on the metabolism of a single dose of itraconazole 200 mg, indicating that these two drugs should not be used concomitantly (125).

Ropivacaine

The effects of clarithromycin (250 mg bd) and itraconazole (200 mg od) for 4 days on the pharmacokinetics of ropivacaine (given intravenously as a single dose of 0.6 mg/kg on day 4) have been studied in a double-blind, three-way, crossover study in eight healthy volunteers (126). There were no significant changes in the pharmacokinetics of ropivacaine after clarithromycin or itraconazole. Although clarithromycin and itraconazole inhibited formation of the (S)-2′,6′-pipelodoxylidide metabolite of ropivacaine, there were no significant changes in the pharmacokinetics of ropivacaine itself.

Saquinavir

The HIV protease inhibitor saquinavir has limited and variable oral systemic availability and ritonavir, an inhibitor of CYP450 and P glycoprotein, is widely used to increase its systemic exposure. A small pilot study in three HIV-infected patients has suggested that oral itraconazole can have similar effects on the oral availability of saquinavir (127). Concomitant use of itraconazole 200 mg/day with a combination of saquinavir and two nucleoside reverse transcriptase inhibitors led to a 2.5- to 6.9-fold increase in the AUC of saquinavir, a 2.0- to 5.4-fold increase in peak plasma concentrations, and a 1.6- to 17-fold increase in trough plasma concentrations. The effect of itraconazole on saquinavir was comparable to that of ritonavir.

Selegiline

The effects of itraconazole (200 mg/day for 4 days) on the pharmacokinetics of selegiline (10 mg on day 4) have been investigated in a randomized, placebo-controlled, crossover study (128). Itraconazole did not alter the pharmacokinetics of selegiline and its primary metabolites, desmethylselegiline and 1-metamfetamine. In human liver microsomes, itraconazole did not inhibit the formation of either metabolite. These findings suggest that selegiline is not susceptible to interactions with CYP3A4 inhibitors.

Simvastatin

Itraconazole probably inhibits simvastatin metabolism (SEDA 21, 459).

Statins

Itraconazole increases the risk of skeletal muscle toxicity of some statins by increasing their serum concentrations (155), but not all statins are equally affected. Concomitant use of atorvastatin, lovastatin, and simvastatin with itraconazole should be avoided or the doses should be reduced; fluvastatin and pravastatin have much less potential than other statins for clinically significant interactions with itraconazole and other CYP3A4 inhibitors; the effects of cerivastatin are intermediate.

In a randomized, open, three-way, crossover study, 18 healthy subjects took single doses of cerivastatin 0.8 mg, atorvastatin 20 mg, or pravastatin 40 mg without or with itraconazole 200 mg (129). Concomitant cerivastatin + itraconazole and pravastatin + itraconazole produced small increases in AUC, C_{max}, and half-life (up to 51%, 25%, and 23% respectively). However, itraconazole markedly increased atorvastatin AUC (150%), C_{max} (38%), and half-life (30%). Thus, itraconazole markedly increases systemic exposure to atorvastatin, but results in only modest increases in the plasma concentrations of cerivastatin and pravastatin.

Atorvastatin

Itraconazole increases serum concentrations of atorvastatin by inhibiting CYP3A4. In a randomized, double-blind, crossover study in 10 healthy volunteers, itraconazole 200 mg increased the AUC and half-life of atorvastatin 40 mg about three-fold, with a change in C_{max} (130). The AUC of atorvastatin lactone was increased about 4-fold, and the C_{max} and half-life were increased more than 2-fold. Itraconazole significantly reduced the C_{max} and AUC of 2-hydroxyatorvastatin acid and 2-hydroxyatorvastatin lactone and increased the half-life of 2-hydroxyatorvastatin lactone. The concomitant use of itraconazole and other potent inhibitors of CYP3A4 with atorvastatin should therefore be avoided, or the dose of atorvastatin should be reduced accordingly.

Cerivastatin

Cerivastatin uses a secondary CYP2C8-mediated metabolic pathway, which is unaffected by itraconazole (131). The effects of itraconazole on the pharmacokinetics of cerivastatin and its major metabolites have been investigated in a randomized, double-blind, crossover study (131). Inhibition of the CYP3A4-mediated M-1 pathway led to raised serum concentrations of cerivastatin, cerivastatin lactone, and metabolite M-23, resulting in increased concentrations of active HMG-CoA reductase inhibitors. However, the effect was modest.

Fluvastatin

The effects of itraconazole 100 mg on the pharmacokinetics of fluvastatin 40 mg have been studied in a randomized, placebo-controlled, crossover study in 10 healthy volunteers (131). Itraconazole had no significant effect on the C_{max} or total AUC of fluvastatin, but slightly prolonged its half-life.

Lovastatin

The effects of itraconazole 100 mg on the pharmacokinetics of lovastatin 40 mg have been studied in a randomized, placebo-controlled, crossover study in 10 healthy volunteers (135). Itraconazole, even in this low dosage, greatly increased plasma concentrations of lovastatin and its active metabolite, lovastatin acid, and increased the C_{max} of lovastatin about 15-fold and the total AUC by more than 15-fold; similarly, the C_{max} and total AUC of lovastatin acid were increased about 12-fold and 15-fold respectively.

Pravastatin

The effects of itraconazole 200 mg on the pharmacokinetics of pravastatin have been studied in a randomized, double-blind, crossover study in 10 healthy volunteers (138). Itraconazole slightly increased the AUC and C_{max} of pravastatin, but the changes were not statistically significant; the half-life was not altered.

Simvastatin

The effects of itraconazole 200 mg on the pharmacokinetics of simvastatin have been studied in a randomized, double-blind, crossover study in 10 healthy volunteers (136). Itraconazole increased the C_{max} and AUC of simvastatin and simvastatin acid at least 10-fold. The C_{max} and AUC of total simvastatin acid (naive simvastatin acid plus that derived by hydrolysis of the lactone) were increased 17-fold and 19-fold respectively, and the half-life was increased by 25%.

In two cases, rhabdomyolysis was caused by itraconazole in heart transplant recipients taking long-term ciclosporin and simvastatin (93,94). To avoid severe myopathy, ciclosporin concentrations should be monitored frequently and statins should be withdrawn or the dosage should be reduced, as long as azoles need to be prescribed in transplant recipients. Patients need to be educated about signs and symptoms that require immediate physician intervention.

Tacrolimus

Tacrolimus concentrations and toxicity are affected by itraconazole (137).

- In a 17-year-old man with cystic fibrosis who received a hepato-pulmonary transplant, there was an interaction of itraconazole 600 mg bd with tacrolimus (138). High trough concentrations of tacrolimus were noted, despite the relatively low dosage (0.1–0.3 mg/kg/day).
- Another patient experienced an interaction of tacrolimus 0.085 mg /kg bd with itraconazole 200–400 mg per day, with resulting ketoacidosis, neutropenia, and thrombocytopenia, requiring the withdrawal of both drugs (139).
- A 30-year-old man with a renal transplant had a more than two-fold increase in blood tacrolimus concentrations after starting to take itraconazole 200 mg/day, accompanied by a reduced glomerular filtration rate and biopsy-proven tacrolimus-associated tubulopathy (140).

Because of the narrow therapeutic index of tacrolimus, blood concentrations should be monitored particularly carefully when itraconazole is co-administered, and the dosage of tacrolimus may have to be altered (141).

The interaction of itraconazole (100 mg bd) with tacrolimus has been studied in 28 heart or lung transplant recipients (142). Tacrolimus blood concentrations were monitored on alternate days for up to 21 days after the start of itraconazole therapy ($n = 18$) or withdrawal

($n = 10$). The dose of tacrolimus was adjusted with the aim of keeping the 12-hour trough blood concentration at 7–12 micrograms/ml. The mean dose of tacrolimus during itraconazole therapy fell significantly from 8.4 to 2.9 mg/day. There was no significant change in serum creatinine or liver function tests. In patients in whom itraconazole was withdrawn, the mean dose of tacrolimus required increased significantly from 4.7 to 8.8 mg/day. Thus, substantial changes in the dose of tacrolimus were required both when itraconazole was begun and when it was withdrawn, and it was difficult to maintain tacrolimus blood concentrations within the target range during the first 2 weeks. However, major toxicity or rejection did not occur. Co-administration of itraconazole may reduce the cost of post-transplant immunosuppression. This interaction is probably due to inhibition of CYP3A4 by itraconazole.

Vinca alkaloids

Enhanced and potentially life-threatening neurotoxicity of vinca alkaloids through concomitant therapy with itraconazole has been the subject of several compelling reports (143–146). Enhancement of vincristine neurotoxicity results in polyneuropathy and paralytic ileus (99,147,148). The interaction is reversible, and readministration of vinca alkaloids may be safe after a prolonged washout (143). The mechanism has not been formally elucidated, but may be either competition for oxidative metabolism, leading to increased systemic exposure (149), or inhibition of the transmembrane P glycoprotein efflux pump (150), leading to increased intracellular concentrations of vinca alkaloids (150). The concomitant use of itraconazole and vinca alkaloids is therefore contraindicated.

Two adults with acute lymphoblastic leukemia developed unusually severe neurotoxicity caused by vincristine, which was probably the result of an interaction with itraconazole suspension (147).

The concomitant use of itraconazole with vincristine increased the incidence of neurotoxicity in children and adults with ALL (156,157).

- A 19-year-old woman developed severe abdominal pain and constipation 28 days after starting to take itraconazole as antifungal prophylaxis when receiving vincristine for ALL. She had hypertension, marked abdominal distension and tenderness, and absent bowel sounds. Withdrawal of itraconazole resulted in resolution of symptoms and vincristine was continued (157).
- A 72-year-old patient developed painful oral mucositis and constipation 3 days after being given vinorelbine and itraconazole (158). Further complications included neutropenia and hypoxia and he died.

Concomitant use of itraconazole 2.5 mg/kg/day and vincristine in five children receiving vincristine resulted in hypertension, paralytic ileus, and SIADH (132).

Warfarin

Itraconazole can alter warfarin concentrations.

- Following the addition of itraconazole to a treatment regimen comprising warfarin, ranitidine, and terfenadine, cardiac dysrhythmias developed in a 62-year-old man. The signs and symptoms included prolongation of the QT interval and ventricular fibrillation (SEDA-18, 198).

This particular regimen apparently resulted in a second interaction, since unexpectedly very high concentrations of terfenadine were found (SEDA-18, 283). The phenomenon has been described by others, with a marked rise in terfenadine serum concentrations, and increased toxicity of the drug during concurrent ingestion of itraconazole. The mechanism is not known, but it is likely to be related to inhibition of CYP3A4 (80).

Zolpidem

Zolpidem is mainly transformed by CYP3A4. However, itraconazole 200 mg did not alter the pharmacokinetics and pharmacodynamics of zolpidem 10 mg in 10 healthy volunteers (151). Therefore, zolpidem may be used in normal or nearly normal doses together with itraconazole.

References

1. Barone JA, Moskovitz BL, Guarnieri J, Hassell AE, Colaizzi JL, Bierman RH, Jessen L. Enhanced bioavailability of itraconazole in hydroxypropyl-beta-cyclodextrin solution versus capsules in healthy volunteers. Antimicrob Agents Chemother 1998;42(7):1862–5.
2. Lyman CA, Walsh TJ. Systemically administered antifungal agents. A review of their clinical pharmacology and therapeutic applications. Drugs 1992;44(1):9–35.
3. Schafer-Korting M. Pharmacokinetic optimisation of oral antifungal therapy. Clin Pharmacokinet 1993;25(4):329–41.
4. Francis P, Walsh TJ. Evolving role of flucytosine in immunocompromised patients: new insights into safety, pharmacokinetics, and antifungal therapy. Clin Infect Dis 1992;15(6):1003–18.
5. Haria M, Bryson HM, Goa KL. Itraconazole. A reappraisal of its pharmacological properties and therapeutic use in the management of superficial fungal infections. Drugs 1996;51(4):585–620Erratum in: Drugs 1996;52(2):253.
6. Dupont B, Drouhet E. Early experience with itraconazole in vitro and in patients: pharmacokinetic studies and clinical results. Rev Infect Dis 1987;9(Suppl 1):S71–6.
7. Caillot D, Bassaris H, McGeer A, Arthur C, Prentice HG, Seifert W, De Beule K. Intravenous itraconazole followed by oral itraconazole in the treatment of invasive pulmonary aspergillosis in patients with hematologic malignancies, chronic granulomatous disease, or AIDS. Clin Infect Dis 2001;33(8):e83–90.
8. Boogaerts MA, Maertens J, Van Der Geest R, Bosly A, Michaux JM, Van Hoof A, Cleeren M, Wostenborghs R, De Beule K. Pharmacokinetics and safety of a 7-day administration of intravenous itraconazole followed by a 14-day administration of itraconazole oral solution in patients with hematologic malignancy. Antimicrob Agents Chemother 2001;45(3):981–5.
9. Haneke E, Abeck D, Ring J. Safety and efficacy of intermittent therapy with itraconazole in finger- and toenail onychomycosis: a multicentre trial. Mycoses 1998;41(11–12):521–7.

10. Boonk W, de Geer D, de Kreek E, Remme J, van Huystee B. Itraconazole in the treatment of tinea corporis and tinea cruris: comparison of two treatment schedules. Mycoses 1998;41(11–12):509–14.

11. Schuller J, Remme JJ, Rampen FH, Van Neer FC. Itraconazole in the treatment of tinea pedis and tinea manuum: comparison of two treatment schedules. Mycoses 1998;41(11–12):515–20.

12. Wang A, Zhang Y, He L, Shen Z, Liao W, Han M, Li R, Liang D, Wu S, Hahn-Ast C, Glasmacher A, Liekwok WA. Clinical study on the efficacy and safety of intravenous itraconazole infusion for the treatment of invasive fungal infection in China. Jpn J Infect Dis 2006;59(6):370-6.

13. Boogaerts M, Maertens J, van Hoof A, de Bock R, Fillet G, Peetermans M, Selleslag D, Vandercam B, Vandewoude K, Zachee P, De Beule K. Itraconazole versus amphotericin B plus nystatin in the prophylaxis of fungal infections in neutropenic cancer patients. J Antimicrob Chemother 2001;48(1):97–103.

14. Boogaerts M, Winston DJ, Bow EJ, Garber G, Reboli AC, Schwarer AP, Novitzky N, Boehme A, Chwetzoff E, De Beule KItraconazole Neutropenia Study Group. Intravenous and oral itraconazole versus intravenous amphotericin B deoxycholate as empirical antifungal therapy for persistent fever in neutropenic patients with cancer who are receiving broad-spectrum antibacterial therapy. A randomized, controlled trial. Ann Intern Med 2001;135(6):412–22.

15. Harousseau JL, Dekker AW, Stamatoullas-Bastard A, Fassas A, Linkesch W, Gouveia J, De Bock R, Rovira M, Seifert WF, Joosen H, Peeters M, De Beule K. Itraconazole oral solution for primary prophylaxis of fungal infections in patients with hematological malignancy and profound neutropenia: a randomized, double-blind, double-placebo, multicenter trial comparing itraconazole and amphotericin B. Antimicrob Agents Chemother 2000;44(7):1887–93.

16. Nolting SK, Gupta A, Doncker PD, Jacko ML, Moskovitz BL. Continuous itraconazole treatment for onychomycosis and dermatomycosis: an overview of safety. Eur J Dermatol 1999;9(7):540–3.

17. Galgiani JN, Catanzaro A, Cloud GA, Johnson RH, Williams PL, Mirels LF, Nassar F, Lutz JE, Stevens DA, Sharkey PK, Singh VR, Larsen RA, Delgado KL, Flanigan C, Rinaldi MGMycoses Study Group. Comparison of oral fluconazole and itraconazole for progressive, nonmeningeal coccidioidomycosis. A randomized, double-blind trial. Ann Intern Med 2000;133(9):676–86.

18. Phillips P, De Beule K, Frechette G, Tchamouroff S, Vandercam B, Weitner L, Hoepelman A, Stingl G, Clotet B. A double-blind comparison of itraconazole oral solution and fluconazole capsules for the treatment of oropharyngeal candidiasis in patients with AIDS. Clin Infect Dis 1998;26(6):1368–73.

19. Graybill JR, Vazquez J, Darouiche RO, Morhart R, Greenspan D, Tuazon C, Wheat LJ, Carey J, Leviton I, Hewitt RG, MacGregor RR, Valenti W, Restrepo M, Moskovitz BL. Randomized trial of itraconazole oral solution for oropharyngeal candidiasis in HIV/AIDS patients. Am J Med 1998;104(1):33–9.

20. Winston DJ, Busuttil RW. Randomized controlled trial of oral itraconazole solution versus intravenous/oral fluconazole for prevention of fungal infections in liver transplant recipients. Transplantation 2002;74(5):688–95.

21. Marr KA, Crippa F, Leisenring W, Hoyle M, Boeckh M, Balajee SA, Nichols WG, Musher B, Corey L. Itraconazole versus fluconazole for prevention of fungal infections in patients receiving allogeneic stem cell transplants. Blood 2004;103:1527-33.

22. Winston DJ, Maziarz RT, Chandrasekar PH, Lazarus HM, Goldman M, Blumer JL, Leitz GJ, Territo MC. Intravenous and oral itraconazole versus intravenous and oral fluconazole for long-term antifungal prophylaxis in allogeneic hematopoietic stem-cell transplant recipients. A multicenter, randomized trial. Ann Intern Med 2003;138:705-13.

23. Oude Lashof AM, De Bock R, Herbrecht R, de Pauw BE, Krcmery V, Aoun M, Akova M, Cohen J, Siffnerova H, Egyed M, Ellis M, Marinus A, Sylvester R, Kullberg BJ; EORTC Invasive Fungal Infections Group. An open multicentre comparative study of the efficacy, safety and tolerance of fluconazole and itraconazole in the treatment of cancer patients with oropharyngeal candidiasis. Eur J Cancer 2004;40:1314-9.

24. Jahangir M, Hussain I, Ul Hasan M, Haroon TS. A double-blind, randomized, comparative trial of itraconazole versus terbinafine for 2 weeks in tinea capitis. Br J Dermatol 1998;139(4):672–4.

25. Svejgaard E, Avnstorp C, Wanscher B, Nilsson J, Heremans A. Efficacy and safety of short-term itraconazole in tinea pedis: a double-blind, randomized, placebo-controlled trial. Dermatology 1998;197(4):368–72.

26. Sharpe MD, Ghent C, Grant D, Horbay GL, McDougal J, David Colby W. Efficacy and safety of itraconazole prophylaxis for fungal infections after orthotopic liver transplantation: a prospective, randomized, double-blind study. Transplantation 2003;76:977-83.

27. Gallin JI, Alling DW, Malech HL, Wesley R, Koziol D, Marciano B, Eisenstein EM, Turner ML, DeCarlo ES, Starling JM, Holland SM. Itraconazole to prevent fungal infections in chronic granulomatous disease. New Engl J Med 2003;348:2416-22.

28. Heykants J, Van Peer A, Van de Velde V, Van Rooy P, Meuldermans W, Lavrijsen K, Woestenborghs R, Van Cutsem J, Cauwenbergh G. The clinical pharmacokinetics of itraconazole: an overview. Mycoses 1989;32(Suppl 1):67–87.

29. Tucker RM, Haq Y, Denning DW, Stevens DA. Adverse events associated with itraconazole in 189 patients on chronic therapy. J Antimicrob Chemother 1990;26(4):561–6.

30. Bradbury BD, Jick SS. Itraconazole and fluconazole and certain rare, serious adverse events. Pharmacotherapy 2002;22(6):697–700.

31. Nelson MR, Smith D, Erskine D, Gazzard BG. Ventricular fibrillation secondary to itraconazole induced hypokalaemia. J Infect 1993;26(3):348.

32. Ahmad SR, Singer SJ, Leissa BG. Congestive heart failure associated with itraconazole. Lancet 2001;357(9270):1766–7.

33. Odom RB, Aly R, Scher RK, Daniel CR 3rd, Elewski BE, Zaias N, DeVillez R, Jacko M, Oleka N, Moskovitz BL. A multicenter, placebo-controlled, double-blind study of intermittent therapy with itraconazole for the treatment of onychomycosis of the fingernail. J Am Acad Dermatol 1997;36(2 Pt 1):231–5.

34. Mittal D, Wikaitis J. Itraconazole-induced delirium. Psychosomatics 2003;44:260-1.

35. Singh R, Cundy T. Itraconazole-induced painful neuropathy in a man with type 1 diabetes. Diabetes Care 2005;28(1):225.

36. Cleveland KO, Campbell JW. Hallucinations associated with itraconazole therapy. Clin Infect Dis 1995;21(2):456.

37. Tucker RM, Haq Y, Denning DW, Stevens DA. Adverse events associated with itraconazole in 189 patients on chronic therapy. J Antimicrob Chemotherapy 1990;26:561-6.

38. Ruiz-Contreras J, Rodriguez R, Gomez de Quero P, Gonzalez Tome MI, Sanchez Diaz JI. Severe hypokalemia and rhabdomyolysis associated with itraconazole therapy. Pediatr Infect Dis J 2003;22:1024-5.

39. Sharkey PK, Rinaldi MG, Dunn JF, Hardin TC, Fetchick RJ, Graybill JR. High-dose itraconazole in the treatment of severe mycoses. Antimicrob Agents Chemother 1991;35(4):707-13.

40. Nelson MR, Smith D, Erskine D, Gazzara BG. Ventricular fibrillation secondary to itraconazole induced hypobalaemia. J Infect 1993;26(3):348.

41. Nagaoka Y, Okochi H, Tamaki K. Leukocytopenia after administration of itraconazole. Mycoses 2003;46:240-1.

42. Arzneimittelkommission der Deutschen Ärzteschaft. Blutbildstoerungen unter einer antimykotischen Therapie mit Itraconazol. Deutsches Ärzteblatt 2004;101:B2033-4.

43. Dismukes WE, Bradsher RW Jr, Cloud GC, Kauffman CA, Chapman SW, George RB, Stevens DA, Girard WM, Saag MS, Bowles-Patton CNIAID Mycoses Study Group. Itraconazole therapy for blastomycosis and histoplasmosis. Am J Med 1992;93(5):489-97.

44. Cauwenbergh G, De Doncker P, Stoops K, De Dier AM, Goyvaerts H, Schuermans V. Itraconazole in the treatment of human mycoses: review of three years of clinical experience. Rev Infect Dis 1987;9(Suppl 1):S146-52.

45. Gryn J, Goldberg J, Johnson E, Siegel J, Inzerillo J. The toxicity of daily inhaled amphotericin B. Am J Clin Oncol 1993;16(1):43-6.

46. Barbaro G, Barbarini G, Calderon W, Grisorio B, Alcini P, Di Lorenzo G. Fluconazole versus itraconazole for *Candida* esophagitis in acquired immunodeficiency syndrome. *Candida* esophagitis. Gastroenterology 1996;111(5):1169-77.

47. Nguyen AJ, Nelson DB, Thurn JR. Pseudomembranous colitis after itraconazole therapy. Am J Gastroenterol 1999;94(7):1971-3.

48. Lavrijsen AP, Balmus KJ, Nugteren-Huying WM, Roldaan AC, van't Wout JW, Stricker BH. Hepatic injury associated with itraconazole. Lancet 1992;340(8813):251-2.

49. Wolf R, Wolf D, Kuperman S. Focal nodular hyperplasia of the liver after intraconazole treatment. J Clin Gastroenterol 2001;33(5):418-20.

50. Lavrijsen AP, Balmus KJ, Nugteren-Huying WM, Roldaan AC, van 't Wout JW, Stricker BH. Lever be schadiging tijdens gebruik van itraconazo (Trisporal). [Liver damage during administration of itraconazole (Trisporal).] Ned Tijdschr Geneeskd 1993;137(1):38-41.

51. Gupta AK, Chwetzoff E, Del Rosso J, Baran R. Hepatic safety of itraconazole. J Cutan Med Surg 2002;6(3):210-3.

52. Zhu LP, Yang FF, Weng XH, Huang YX, Chen S, Shi GF, Lu Q, Zhang WH, Zhang YX. [Hepatic safety of itraconazole intravenous solution in treatment of invasive fungal infection.] Zhonghua Yi Xue Za Zhi 2006;86(29):2028-32.

53. Srebrnik A, Levtov S, Ben-Ami R, Brenner S. Liver failure and transplantation after itraconazole treatment for toenail onychomycosis. J Eur Acad Dermatol Venereol 2005;19(2):205-7.

54. Ganer A, Arathoon E, Stevens DA. Initial experience in therapy for progressive mycoses with itraconazole, the first clinically studied triazole. Rev Infect Dis 1987;9(Suppl 1):S77-86.

55. Marco F, Pfaller MA, Messer SA, Jones RN. Antifungal activity of a new triazole, voriconazole (UK-109,496), compared with three other antifungal agents tested against clinical isolates of filamentous fungi. Med Mycol 1998;36(6):433-6.

56. Kramer KE, Yaar M, Andersen W. Purpuric drug eruption secondary to itraconazole. J Am Acad Dermatol 1997;37(6):994-5.

57. Park YM, Kim JW, Kim CW. Acute generalized exanthematous pustulosis induced by itraconazole. J Am Acad Dermatol 1997;36(5 Pt 1):794-6.

58. Goto Y, Kono T, Teramae K, Ishii M. Itraconazole-induced drug eruption confirmed by challenge test. Acta Derm Venereol 2000;80(1):72.

59. Alvarez-Fernandez JG, Castano-Suarez E, Cornejo-Navarro P, de la Fuente EG, Ortiz de Frutos FJ, Iglesias-Diez L. Photosensitivity induced by oral itraconazole. J Eur Acad Dermatol Venereol 2000;14(6):501-3.

60. Castellsague J, Garcia-Rodriguez LA, Duque A, Perez S. Risk of serious skin disorders among users of oral antifungals: a population-based study. BMC Dermatol 2002;2(1):14.

61. Porzionato A, Zancaner S, Betterle C, Ferrara SD. Fatal toxic epidermal necrolysis in autoimmune polyglandular syndrome type I. J Endocrinol Invest 2004;27:475-9.

62. Denning DW, Van Wye JE, Lewiston NJ, Stevens DA. Adjunctive therapy of allergic bronchopulmonary aspergillosis with itraconazole. Chest 1991;100(3):813-9.

63. Park H, Knowles S, Shear NH. Serum sickness-like reaction to itraconazole. Ann Pharmacother 1998;32(11):1249.

64. Martinez-Alonso JC, Dominguez-Ortega FJ, Fuentes-Gonzalo MJ. Urticaria and angioedema due to itraconazole. Allergy 2003;58:1317-8.

65. Schmutz JL, Barbaud A, Trechot P. Urticaria and angiodema due to itraconazole. Ann Dermatol Venereol 2005;132(4):403.

66. Van Cauteren H, Heykants J, De Coster R, Cauwenbergh G. Itraconazole: pharmacologic studies in animals and humans. Rev Infect Dis 1987;9(Suppl 1):S43-6.

67. de Repentigny L, Ratelle J, Leclerc JM, Cornu G, Sokal EM, Jacqmin P, De Beule K. Repeated-dose pharmacokinetics of an oral solution of itraconazole in infants and children. Antimicrob Agents Chemother 1998;42(2):404-8.

68. Abdel-Rahman SM, Powell DA, Nahata MC. Efficacy of itraconazole in children with *Trichophyton tonsurans* tinea capitis. J Am Acad Dermatol 1998;38(3):443-6.

69. Groll AH, Wood L, Roden M, Mickiene D, Chiou CC, Townley E, Dad L, Piscitelli SC, Walsh TJ. Safety, pharmacokinetics, and pharmacodynamics of cyclodextrin itraconazole in pediatric patients with oropharyngeal candidiasis. Antimicrob Agents Chemother 2002;46(8):2554-63.

70. Foot AB, Veys PA, Gibson BE. Itraconazole oral solution as antifungal prophylaxis in children undergoing stem cell transplantation or intensive chemotherapy for haematological disorders. Bone Marrow Transplant 1999;24(10):1089-93.

71. Conway SP, Etherington C, Peckham DG, Brownlee KG, Whitehead A, Cunliffe H. Pharmacokinetics and safety of itraconazole in patients with cystic fibrosis. J Antimicrob Chemother 2004;53:841-7.

72. Verspeelt J, Marynissen G, Gupta AK, De Doncker P. Safety of itraconazole in diabetic patients. Dermatology 1999;198(4):382-4.

73. Glasmacher A, Hahn C, Molitor E, Marklein G, Sauerbruch T, Schmidt-Wolf IG. Itraconazole through concentrations in antifungal prophylaxis with six different dosing regimens using hydroxypropyl-beta-cyclodextrin oral solution or coated-pellet capsules. Mycoses 1999;42(11–12):591–600.

74. Schaffner A, Bohler A. Amphotericin B refractory aspergillosis after itraconazole: evidence for significant antagonism. Mycoses 1993;36(11–12):421–4.

75. Persat F, Schwartzbrod PE, Troncy J, Timour Q, Maul A, Piens MA, Picot S. Abnormalities in liver enzymes during simultaneous therapy with itraconazole and amphotericin B in leukaemic patients. J Antimicrob Chemother 2000;45(6):928–9.

76. Groll AH, Piscitelli SC, Walsh TJ. Clinical pharmacology of systemic antifungal agents: a comprehensive review of agents in clinical use, current investigational compounds, and putative targets for antifungal drug development. Adv Pharmacol 1998;44:343–500.

77. Lefebvre RA, Van Peer A, Woestenborghs R. Influence of itraconazole on the pharmacokinetics and electrocardiographic effects of astemizole. Br J Clin Pharmacol 1997;43(3):319–22.

78. Honig PK, Wortham DC, Hull R, Zamani K, Smith JE, Cantilena LR. Itraconazole affects single-dose terfenadine pharmacokinetics and cardiac repolarization pharmacodynamics. J Clin Pharmacol 1993;33(12):1201–6.

79. de Wildt SN, van den Anker JN. Wegrakingen tijdens simultaan gebruik van terfenadine en itraconazol. [Syncopes with simultaneous use of terfenadine and itraconazole.] Ned Tijdschr Geneeskd 1997;141(36):1752–3.

80. Bickers DR. Antifungal therapy: potential interactions with other classes of drugs. J Am Acad Dermatol 1994;31(3 Pt 2):S87–90.

81. Otsuji Y, Okuyama N, Aoshima T, Fukasawa T, Kato K, Gerstenberg G, Miura M, Ohkubo T, Sugawara K, Otani K. No effect of itraconazole on the single oral dose pharmacokinetics and pharmacodynamics of estazolam. Ther Drug Monit 2002;24(3):375–8.

82. Backman JT, Kivisto KT, Olkkola KT, Neuvonen PJ. The area under the plasma concentration-time curve for oral midazolam is 400-fold larger during treatment with itraconazole than with rifampicin. Eur J Clin Pharmacol 1998;54(1):53–8.

83. Palkama VJ, Neuvonen PJ, Olkkola KT. Effect of itraconazole on the pharmacokinetics of bupivacaine enantiomers in healthy volunteers. Br J Anaesth 1999;83(4):659–61.

84. Kivisto KT, Lamberg TS, Neuvonen PJ. Interactions of buspirone with itraconazole and rifampicin: effects on the pharmacokinetics of the active 1-(2-pyrimidinyl)-piperazine metabolite of buspirone. Pharmacol Toxicol 1999;84(2):94–7.

85. Buggia I, Zecca M, Alessandrino EP, Locatelli F, Rosti G, Bosi A, Pession A, Rotoli B, Majolino I, Dallorso A, Regazzi MB. Itraconazole can increase systemic exposure to busulfan in patients given bone marrow transplantation. GITMO (Gruppo Italiano Trapianto di Midollo Osseo). Anticancer Res 1996;16(4A):2083–8.

86. Buggia I, Zecca M, Alessandrino EP, Locatelli F, Rosti G, Bosi A, Pession A, Rotoli B, Majolino I, Dallorso A, Regazzi MB. Itraconazole can increase systemic exposure to busulfan in patients given bone marrow transplantation. GITMO (Gruppo Italiano Trapianto di Midollo Osseo). Anticancer Res 1996;16(4A):2083–8.

87. Tucker RM, Denning DW, Hanson LH, Rinaldi MG, Graybill JR, Sharkey PK, Pappagianis D, Stevens DA. Interaction of azoles with rifampin, phenytoin, and carbamazepine: in vitro and clinical observations. Clin Infect Dis 1992;14(1):165–74.

88. Kwan JT, Foxall PJ, Davidson DG, Bending MR, Eisinger AJ. Interaction of cyclosporin and itraconazole. Lancet 1987;2(8553):282.

89. Kramer MR, Marshall SE, Denning DW, Keogh AM, Tucker RM, Galgiani JN, Lewiston NJ, Stevens DA, Theodore J. Cyclosporine and itraconazole interaction in heart and lung transplant recipients. Ann Intern Med 1990;113(4):327–9.

90. Kwan JT, Foxall PJ, Davidson DG, Bending MR, Eisinger AJ. Interaction of cyclosporin and itraconazole. Lancet 1987;2(8553):282.

91. Trenk D, Brett W, Jahnchen E, Bixnbaum D. Time course of cyclosporin/itraconazole interaction. Lancet 1987;2(8571):1335–6.

92. Navakova I, Donnelly P, de Witte T, de Pauw B, Boezeman J, Veltman G. Itraconazole and cyclosporin nephrotoxicity. Lancet 1987;2(8564):920–1.

93. Vlahakos DV, Manginas A, Chilidou D, Zamanika C, Alivizatos PA. Itraconazole-induced rhabdomyolysis and acute renal failure in a heart transplant recipient treated with simvastatin and cyclosporine. Transplantation 2002;73(12):1962–4.

94. Maxa JL, Melton LB, Ogu CC, Sills MN, Limanni A. Rhabdomyolysis after concomitant use of cyclosporine, simvastatin, gemfibrozil, and itraconazole. Ann Pharmacother 2002;36(5):820–3.

95. May DR, Drew RH, Yedinak KC, Bartlett JA. Effect of simultaneous didanosine administration on itraconazole absorption in healthy volunteers. Pharmacotherapy 1994;14(5):509–13.

96. Auclair B, Berning SE, Huitt GA, Peloquin CA. Potential interaction between itraconazole and clarithromycin. Pharmacotherapy 1999;19(12):1439–44.

97. Raaska K, Neuvonen PJ. Serum concentrations of clozapine and N-desmethylclozapine are unaffected by the potent CYP3A4 inhibitor itraconazole. Eur J Clin Pharmacol 1998;54(2):167–70.

98. Sachs MK, Blanchard LM, Green PJ. Interaction of itraconazole and digoxin. Clin Infect Dis 1993;16(3):400–3.

99. Woodland C, Ito S, Koren G. A model for the prediction of digoxin–drug interactions at the renal tubular cell level. Ther Drug Monit 1998;20(2):134–8.

100. Koren G, Woodland C, Ito S. Toxic digoxin–drug interactions: the major role of renal P-glycoprotein. Vet Hum Toxicol 1998;40(1):45–6.

101. Meyboom RH, de Jonge K, Veentjer H, Dekens-Konter JA, de Koning GH. Potentiering van digoxine door itraconazol. [Potentiation of digoxin by itraconazole.] Ned Tijdschr Geneeskd 1994;138(47):2353–6.

102. Jalava KM, Partanen J, Neuvonen PJ. Itraconazole decreases renal clearance of digoxin. Ther Drug Monit 1997;19(6):609–13.

103. Nishihara K, Hibino J, Kotaki H, Sawada Y, Iga T. Effect of itraconazole on the pharmacokinetics of digoxin in guinea pigs. Biopharm Drug Dispos 1999;20(3):145–9.

104. Mathis AS, Friedman GS. Coadministration of digoxin with itraconazole in renal transplant recipients. Am J Kidney Dis 2001;37(2):E18.

105. Kanda Y, Kami M, Matsuyama T, Mitani K, Chiba S, Yazaki Y, Hirai H. Plasma concentration of itraconazole in patients receiving chemotherapy for hematological malignancies: the effect of famotidine on the absorption of itraconazole. Hematol Oncol 1998;16(1):33–7.

106. Palkama VJ, Neuvonen PJ, Olkkola KT. The CYP 3A4 inhibitor itraconazole has no effect on the pharmacokinetics of i.v. fentanyl Br J Anaesth 1998;81(4):598–600.

107. Mercadante S, Villari P, Ferrera P. Itraconazole–fentanyl interaction in a cancer patient. J Pain Symptom Manage 2002;24(3):284–6.

108. Palkama VJ, Neuvonen PJ, Olkkola KT. The CYP 3A4 inhibitor itraconazole has no effect on the pharmacokinetics of i.v. fentanyl Br J Anaesth 1998;81(4):598–600.

109. Viviani MA. Flucytosine—what is its future? J Antimicrob Chemother 1995;35(2):241–4.

110. Main KM, Skov M, Sillesen IB, Dige-Petersen H, Muller J, Koch C, Lanng S. Cushing's syndrome due to pharmacological interaction in a cystic fibrosis patient. Acta Paediatr 2002;91(9):1008–11.

111. Parmar JS, Howell T, Kelly J, Bilton D. Profound adrenal suppression secondary to treatment with low dose inhaled steroids and itraconazole in allergic bronchopulmonary aspergillosis in cystic fibrosis. Thorax 2002;57(8):749–50.

112. Skov M, Main KM, Sillesen IB, Muller J, Koch C, Lanng S. Iatrogenic adrenal insufficiency as a side-effect of combined treatment of itraconazole and budesonide. Eur Respir J 2002;20(1):127–33.

113. Raaska K, Niemi M, Neuvonen M, Neuvonen PJ, Kivisto KT. Plasma concentrations of inhaled budesonide and its effects on plasma cortisol are increased by the cytochrome P4503A4 inhibitor itraconazole. Clin Pharmacol Ther 2002;72(4):362–9.

114. Varis T, Kivisto KT, Backman JT, Neuvonen PJ. The cytochrome P450 3A4 inhibitor itraconazole markedly increases the plasma concentrations of dexamethasone and enhances its adrenal-suppressant effect. Clin Pharmacol Ther 2000;68(5):487–94.

115. Varis T, Kaukonen KM, Kivisto KT, Neuvonen PJ. Plasma concentrations and effects of oral methylprednisolone are considerably increased by itraconazole. Clin Pharmacol Ther 1998;64(4):363–8.

116. Varis T, Kivisto KT, Backman JT, Neuvonen PJ. Itraconazole decreases the clearance and enhances the effects of intravenously administered methylprednisolone in healthy volunteers. Pharmacol Toxicol 1999;85(1):29–32.

117. Lebrun-Vignes B, Archer VC, Diquet B, Levron JC, Chosidow O, Puech AJ, Warot D. Effect of itraconazole on the pharmacokinetics of prednisolone and methylprednisolone and cortisol secretion in healthy subjects. Br J Clin Pharmacol 2001;51(5):443–50.

118. Varis T, Kivisto KT, Neuvonen PJ. The effect of itraconazole on the pharmacokinetics and pharmacodynamics of oral prednisolone. Eur J Clin Pharmacol 2000;56(1):57–60.

119. Penzak SR, Gubbins PO, Gurley BJ, Wang PL, Saccente M. Grapefruit juice decreases the systemic availability of itraconazole capsules in healthy volunteers. Ther Drug Monit 1999;21(3):304–9.

120. Kawakami M, Suzuki K, Ishizuka T, Hidaka T, Matsuki Y, Nakamura H. Effect of grapefruit juice on pharmacokinetics of itraconazole in healthy subjects. Int J Clin Pharmacol Ther 1998;36(6):306–8.

121. Yasui N, Kondo T, Otani K, Furukori H, Mihara K, Suzuki A, Kaneko S, Inoue Y. Effects of itraconazole on the steady-state plasma concentrations of haloperidol and its reduced metabolite in schizophrenic patients: in vivo evidence of the involvement of CYP3A4 for haloperidol metabolism. J Clin Psychopharmacol 1999;19(2):149–54.

122. Furukori H, Kondo T, Yasui N, Otani K, Tokinaga N, Nagashima U, Kaneko S, Inoue Y. Effects of itraconazole on the steady-state plasma concentrations of bromperidol and reduced bromperidol in schizophrenic patients. Psychopharmacology (Berl) 1999;145(2):189–92.

123. Yasui N, Kondo T, Otani K, Furukori H, Mihara K, Suzuki A, Kaneko S, Inoue Y. Effects of itraconazole on the steady-state plasma concentrations of haloperidol and its reduced metabolite in schizophrenic patients: in vivo evidence of the involvement of CYP3A4 for haloperidol metabolism. J Clin Psychopharmacol 1999;19(2):149–54.

124. Van Puijenbroek EP, Egberts AC, Meyboom RH, Leufkens HG. Signalling possible drug–drug interactions in a spontaneous reporting system: delay of withdrawal bleeding during concomitant use of oral contraceptives and itraconazole. Br J Clin Pharmacol 1999;47(6):689–93.

125. Sinofsky FE, Pasquale SA. The effect of fluconazole on circulating ethinyl estradiol levels in women taking oral contraceptives. Am J Obstet Gynecol 1998;178(2):300–4.

126. Jaruratanasirikul S, Sriwiriyajan S. Effect of omeprazole on the pharmacokinetics of itraconazole. Eur J Clin Pharmacol 1998;54(2):159–61.

127. Damkier P, Hansen LL, Brosen K. Effect of diclofenac, disulfiram, itraconazole, grapefruit juice and erythromycin on the pharmacokinetics of quinidine. Br J Clin Pharmacol 1999;48(6):829–38.

128. Jaruratanasirikul S, Sriwiriyajan S. Effect of rifampicin on the pharmacokinetics of itraconazole in normal volunteers and AIDS patients. Eur J Clin Pharmacol 1998;54(2):155–8.

129. Jokinen MJ, Ahonen J, Neuvonen PJ, Olkkola KT. Effect of clarithromycin and itraconazole on the pharmacokinetics of ropivacaine. Pharmacol Toxicol 2001;88(4):187–91.

130. Koks CH, van Heeswijk RP, Veldkamp AI, Meenhorst PL, Mulder JW, van der Meer JT, Beijnen JH, Hoetelmans RM. Itraconazole as an alternative for ritonavir liquid formulation when combined with saquinavir. AIDS 2000;14(1):89–90.

131. Kivisto KT, Wang JS, Backman JT, Nyman L, Taavitsainen P, Anttila M, Neuvonen PJ. Selegiline pharmacokinetics are unaffected by the CYP3A4 inhibitor itraconazole. Eur J Clin Pharmacol 2001;57(1):37–42.

132. Horn M. Coadministration of itraconazole with hypolipidemic agents may induce rhabdomyolysis in healthy individuals. Arch Dermatol 1996;132(10):1254.

133. Mazzu AL, Lasseter KC, Shamblen EC, Agarwal V, Lettieri J, Sundaresen P. Itraconazole alters the pharmacokinetics of atorvastatin to a greater extent than either cerivastatin or pravastatin. Clin Pharmacol Ther 2000;68(4):391–400.

134. Kantola T, Kivisto KT, Neuvonen PJ. Effect of itraconazole on the pharmacokinetics of atorvastatin. Clin Pharmacol Ther 1998;64(1):58–65.

135. Gubbins PO, McConnell SA, Penzak SR. Antifungal Agents. In: Piscitelli SC, Rodvold KA, editors. Drug Interactions in Infectious Diseases. Totowa, NJ: Humana Press Inc, 2001:185–217.

136. Kantola T, Kivisto KT, Neuvonen PJ. Effect of itraconazole on cerivastatin pharmacokinetics. Eur J Clin Pharmacol 1999;54(11):851–5.

137. Kivisto KT, Kantola T, Neuvonen PJ. Different effects of itraconazole on the pharmacokinetics of fluvastatin and lovastatin. Br J Clin Pharmacol 1998;46(1):49–53.

138. Neuvonen PJ, Kantola T, Kivisto KT. Simvastatin but not pravastatin is very susceptible to interaction with the CYP3A4 inhibitor itraconazole. Clin Pharmacol Ther 1998;63(3):332–41.

139. Katari SR, Magnone M, Shapiro R, Jordan M, Scantlebury V, Vivas C, Gritsch A, McCauley J, Starzl T, Demetris AJ, Randhawa PS. Clinical features of acute

reversible tacrolimus (FK 506) nephrotoxicity in kidney transplant recipients. Clin Transplant 1997;11(3):237–42.

140. Billaud EM, Guillemain R, Tacco F, Chevalier P. Evidence for a pharmacokinetic interaction between itraconazole and tacrolimus in organ transplant patients. Br J Clin Pharmacol 1998;46(3):271–2.

141. Furlan V, Parquin F, Penaud JF, Cerrina J, Ladurie FL, Dartevelle P, Taburet AM. Interaction between tacrolimus and itraconazole in a heart-lung transplant recipient. Transplant Proc 1998;30(1):187–8.

142. Ideura T, Muramatsu T, Higuchi M, Tachibana N, Hora K, Kiyosawa K. Tacrolimus/itraconazole interactions: a case report of ABO-incompatible living-related renal transplantation. Nephrol Dial Transplant 2000;15(10):1721–3.

143. Outeda Macias M, Salvador P, Hurtado JL, Martin I. Tacrolimus–itraconazole interaction in a kidney transplant patient. Ann Pharmacother 2000;34(4):536.

144. Banerjee R, Leaver N, Lyster H, Banner NR. Coadministration of itraconazole and tacrolimus after thoracic organ transplantation. Transplant Proc 2001;33(1–2):1600–2.

145. Jeng MR, Feusner J. Itraconazole-enhanced vincristine neurotoxicity in a child with acute lymphoblastic leukemia. Pediatr Hematol Oncol 2001;18(2):137–42.

146. Bosque E. Possible drug interaction between itraconazole and vinorelbine tartrate leading to death after one dose of chemotherapy. Ann Intern Med 2001;134(5):427.

147. Kamaluddin M, McNally P, Breatnach F, O'Marcaigh A, Webb D, O'Dell E, Scanlon P, Butler K, O'Meara A. Potentiation of vincristine toxicity by itraconazole in children with lymphoid malignancies. Acta Paediatr 2001;90(10):1204–7.

148. Sathiapalan RK, El-Solh H. Enhanced vincristine neurotoxicity from drug interactions: case report and review of literature. Pediatr Hematol Oncol 2001;18(8):543–6.

149. Gillies J, Hung KA, Fitzsimons E, Soutar R. Severe vincristine toxicity in combination with itraconazole. Clin Lab Haematol 1998;20(2):123–4.

150. Bohme A, Ganser A, Hoelzer D. Aggravation of vincristine-induced neurotoxicity by itraconazole in the treatment of adult ALL. Ann Hematol 1995;71(6):311–2.

151. Zhou-Pan XR, Seree E, Zhou XJ, Placidi M, Maurel P, Barra Y, Rahmani R. Involvement of human liver cytochrome P450 3A in vinblastine metabolism: drug interactions. Cancer Res 1993;53(21):5121–6.

152. Gupta S, Kim J, Gollapudi S. Reversal of daunorubicin resistance in P388/ADR cells by itraconazole. J Clin Invest 1991;87(4):1467–9.

153. Bohme A, Ganser A, Hoelzer D. Aggravation of vincristine-induced neurotoxicity by itraconazole in the treatment of adult ALL. Ann Hematol 1995;71(6):311–2.

154. Gillies J, Hung KA, Fitzsimons E, Soutar R. Severe vincristine toxicity in combination with itraconazole. Clin Lab Haematol 1998;20(2):123–4.

155. Bosque E. Possible drug interaction between itraconazole and vinorelbine tartrate leading to death after one dose of chemotherapy. Ann Intern Med 2001;134(5):427.

156. Luurila H, Kivisto KT, Neuvonen PJ. Effect of itraconazole on the pharmacokinetics and pharmacodynamics of zolpidem. Eur J Clin Pharmacol 1998;54(2):163–6.

Ketoconazole

See also Antifungal azoles

General Information

Ketoconazole was the first azole available for oral administration. However, it has been supplanted by newer azoles with fewer adverse effects and more reliable absorption from the gastrointestinal tract. Even after years of use, new adverse effects (often related to high doses and/or newer indications) continue to be reported.

Ketoconazole is water-soluble at a pH of below 3. Its oral absorption is influenced by the acidity of the stomach contents, and the concomitant administration of histamine H_2 receptor antagonists, proton pump inhibitors, antacids, or food affects its absorption. A high carbohydrate meal ingested with ketoconazole reduces total drug absorption, while a high lipid meal increases it. Erratic absorption is particularly apparent in patients with AIDS. Peak serum concentrations are seen within 2–3 hours. The half-life is about 8 hours. CSF penetration is less than 10% (1). Ketoconazole is extensively metabolized in the liver and excreted in the bile in an inactive form; less than 1% of the active drug is excreted in the urine. Clearance is not significantly altered by renal dialysis (1).

In man, higher doses of ketoconazole affect cortisol/cortisone and androgen/testosterone substrates. This finding has led to the use of ketoconazole in Cushing's disease and prostate cancer, but the phenomenon is also responsible for some of the adverse effects, especially those associated with higher doses and prolonged use (SED-12, 677). The potency of ketoconazole in inhibiting P450 isozymes (such as CYP3A4) is a cause of interactions with several other drugs.

Use in non-infective conditions

The efficacy of ketoconazole 400 mg qds in the early treatment of acute lung injury and acute respiratory distress syndrome has been investigated in a randomized, double-blind, placebo-controlled trial in 234 patients (2). Ketoconazole was safe but had no effects on mortality, lung function, or the duration of mechanical ventilation.

Because of its potent inhibitory effects on adrenal steroidogenesis by interference with cytochrome CYP450, ketoconazole controls hypercortisolism when surgery is contraindicated or unsuccessful. The effects of oral ketoconazole 200–1200 mg qds for 65–83 months in three patients who had residual or recurrent Cushing's disease after surgical treatment have been reported (3). The dosage of ketoconazole was adjusted according to the clinical response and 24-hour urinary excretion of free cortisol. All three patients had good clinical and biochemical responses to therapy with ketoconazole and had no adverse effects.

General adverse effects

Gastrointestinal complaints, including anorexia, nausea, gastralgia, and constipation are the most frequent adverse effects of ketoconazole. Hepatotoxicity, varying in degree from mild disturbances of liver function tests to hepatitis and rare cases of fulminating hepatic necrosis, has been reported. Some cases have been reported in the first weeks of treatment, but duration of treatment is of importance and in prolonged courses of treatment, monitoring is advisable. With the use of high doses, especially for longer periods of time, the effects of interference with hormonal balance should be watched for. Adrenal insufficiency has been reported even with low-dose treatment. Pruritus and skin reactions have been reported, but do not in general cause major problems. Hypersensitivity reactions are rare. Tumor-inducing effects have not been reported.

Organs and Systems

Cardiovascular

Ketoconazole has not previously been thought to be pro-dysrhythmic without concomitant use of drugs that cause prolongation of the QT interval, but this has now been reported (4).

- A 63-year-old woman with coronary artery disease developed a markedly prolonged QT interval and torsade de pointes after taking ketoconazole for a fungal infection. Her QT interval returned to normal on withdrawal of ketoconazole. There were no mutations in her genes that encode cardiac IK_r channel proteins.

The authors concluded that because it blocks inward rectifier potassium channels (IK_r) channels, ketoconazole alone can prolong the QT interval and induce torsade de pointes. This calls for attention when ketoconazole is given to patients with risk factors for the long QT syndrome.

Respiratory

Patients undergoing esophagectomy are at increased risk of acute lung injury, perhaps related to increased concentrations of thromboxane in the postpulmonary circulation. In 38 consecutive patients undergoing esophagectomy, perioperative ketoconazole, which inhibits thromboxane synthesis, reduced the incidence of acute lung injury (5).

Nervous system

Headache, dizziness, nervousness, and somnolence have been reported (SED-11, 573). The incidences are low. Encephalopathy can occur as a result of severe liver damage.

Sensory systems

Eyes

Papilledema has been reported in one patient taking ketoconazole. The condition cleared on withdrawal of ketoconazole and recurred on resumption 2 months later (SEDA-18, 284).

Psychological, psychiatric

In one patient taking a high dosage for prostate cancer, weakness was associated with mental disturbances, notably confabulation and disorientation in time and space (SEDA-13, 233).

Endocrine

Gynecomastia was occasionally observed in men when ketoconazole first became available. Ketoconazole has a marked effect on steroid concentrations, including a change in the testosterone/estradiol ratio, and this is most likely to be the basis of the gynecomastia. A lowering of testosterone serum concentrations and a reduced response of testosterone concentrations to human gonadotropin have been shown (SED-11, 573) (6). Various studies have shown suppression of testosterone, androstenedione, and dehydroepiandrosterone, with reciprocal increases in gonadotrophins.

Reductions in serum and urinary cortisol concentrations have been reported in patients taking ketoconazole and signs of hypoadrenalism have been seen during high-dose treatment. However, it is not clear whether the asthenia syndrome described in the past (severe muscle weakness, most pronounced in the legs, fatigue, apathy, and anorexia) is related to hypoadrenalism. In some cases, hypoadrenalism has been described shortly after the start of low-dose treatment. Substitution therapy may be required, since simple withdrawal of ketoconazole may not redress hormonal balance quickly enough.

Various studies have shown that ketoconazole interferes with 17- and 20-hydroxylases and inhibits mitochondrial 11-α-hydroxylase and cytochrome P450-dependent steroid hydroxylase enzymes (SED-12, 677; 7 (SEDA-12, 228; (SEDA-14, 234; 6).

Because of its effects on the pituitary/adrenal system, ketoconazole has been used in the long-term control of hypercortisolism of either pituitary or adrenal origin (SED-12, 677). In seven patients with Cushing's disease and one with an adrenal adenoma, ketoconazole 600–800 mg/day for 3–13 months produced rapid persistent clinical improvement (8). Plasma dehydroepiandrosterone sulfate concentrations and urinary 17-ketosteroid and cortisol excretion fell soon after the start of treatment, and remained normal or nearly so throughout treatment. Urinary tetrahydro-11-deoxycortisol excretion rose significantly. Plasma cortisol concentrations fell. Plasma ACTH concentrations did not change and individual plasma ACTH and cortisol increments in response to CRH were comparable before and during treatment. The cortisol response to insulin-induced hypoglycemia improved in one patient and was restored to normal in another. The patients recovered normal adrenal suppressibility in response to a low dose of dexamethasone during ketoconazole treatment.

The effect of ketoconazole appears to be mediated by inhibition of adrenal 11-β-hydroxylase and 17,20-lyase, and in some unknown way it prevents the expected rise in ACTH secretion in patients with Cushing's disease (SEDA-12, 228; SEDA-17, 323). It may, however, cause such a rapid reduction in serum cortisol concentrations that a crisis is precipitated, and patients' adrenal function should be carefully monitored. While ketoconazole (400–800 mg/day) may be a good alternative to other adrenal steroid inhibitors, patients should be observed for signs of hepatotoxicity. Acute adrenal crisis occasionally occurs (9).

Because ketoconazole has antiandrogenic properties, it is particularly suitable for women, in whom it has few effects on menstruation and does not cause hirsutism. In men, however, long-term inhibition of androgen production can be disruptive, especially if it leads to gynecomastia and hypogonadism. Combination with aminoglutethimide and metyrapone has been advocated in order to avoid these effects (SEDA-17, 323).

Ketoconazole (400 mg/day) has been used for the treatment of hirsutism and acne in women, but adverse effects, such as headache, nausea, loss of scalp hair, hepatitis, and biochemical changes, were impressive (10–12).

Hematologic

Fatal aplastic anemia has been reported during ketoconazole therapy (13).

Gastrointestinal

Nausea, mild gastrointestinal symptoms, and vomiting can occur in patients taking ketoconazole; diarrhea has been reported, but the incidence is low (SED-12, 678). The incidence of gastrointestinal complaints is higher with the use of daily doses above 800 mg (14).

Liver

Mild and often transient rises in serum liver enzyme activities are not uncommon in patients taking ketoconazole; the incidence is reported to be about 10–15% (SEDA-12, 229). This figure is higher than originally thought (SED-11, 573; 15), the newer figures probably representing a greater awareness of the risk rather than a true increase; it is possible, however, that the use of higher doses plays a role (SED-12, 678).

The incidence of symptomatic hepatic injury associated with ketoconazole is estimated to be about one in 10 000 treated cases (SEDA-18, 284). Biochemically, the pattern was hepatocellular in 54%, cholestatic in 16%, and mixed cholestatic-hepatocellular in 30%. Histology (14 cases) showed a predominantly hepatocellular pattern in 57%, with extensive centrilobular necrosis and mild to moderate bridging (15,16). Lethal cases of toxic hepatitis and a case necessitating transplantation have been reported (15,17,18).

- A girl developed fatal liver failure while taking ketoconazole for Cushing's syndrome (19). The authors proposed using metyrapone when temporary control of hypercortisolism is required in childhood and adolescence.

A cohort study of the risk of acute liver injury among users of oral antifungal drugs has been performed in the general population of the General Practice Research Database in the UK (20). The cohort included 69 830 patients, free from liver and systemic diseases, who had received at least one prescription for oral griseofulvin, fluconazole, itraconazole, ketoconazole, or terbinafine between 1991 and 1996. Five cases of acute liver injury occurred during the use of oral antifungal drugs. Two of the patients were taking ketoconazole, another two itraconazole, and one terbinafine. The incidence rates of acute liver injury were 134 (CI = 37, 488) per 100 000 person-months for ketoconazole, 10 (CI = 2.9, 38) for itraconazole, and 2.5 (CI = 0.4, 14) for terbinafine. One case was associated with past use of fluconazole. Ketoconazole was the antifungal drug that was associated with the highest relative risk, 228 (CI = 34, 933) compared with the risk among non-users, followed by itraconazole and terbinafine, with relative risks of 18 (CI = 2.6, 73) and 4.2 (CI = 0.2, 25) respectively.

Skin

The risk of serious skin disorders has been estimated in 61 858 users, aged 20–79 years, of oral antifungal drugs identified in the UK General Practice Research Database (21). They had received at least one prescription for oral fluconazole, griseofulvin, itraconazole, ketoconazole, or terbinafine. The background rate of serious cutaneous adverse reactions (corresponding to non-use of oral antifungal drugs) was 3.9 per 10 000 person-years (95% CI = 2.9, 5.2). Incidence rates for current use were 15 per 10 000 person-years (1.9, 56) for itraconazole, 11.1 (3.0, 29) for terbinafine, 10 (1.3, 38) for fluconazole, and 4.6 (0.1, 26) for griseofulvin. Cutaneous disorders associated with the use of oral antifungal drugs in this study were all mild.

Pruritus and occasional rashes can occur in patients taking ketoconazole (22). Rare cases of a fixed drug eruption have been reported (SED-12, 678; SEDA-17, 323; 23).

Musculoskeletal

- Muscle weakness and diffuse myalgia were reported in a 17-year-old man with multiple endocrine neoplasia syndrome type 1 taking ketoconazole for oral candidiasis. The electromyogram showed a distinct myopathic pattern. Withdrawal of the ketoconazole was followed by rapid improvement (SEDA-17, 323).

In rats, high doses (80 mg/kg) caused syndactyly in one experiment (SED-11, 574). There are insufficient data to determine whether there might be a harmful effect in humans.

Second-Generation Effects

Lactation

Ketoconazole can be found in the milk of lactating dogs receiving ketoconazole; there are insufficient data to decide whether harm might ensue to the breast-fed child.

Susceptibility Factors

Age

Because ketoconazole interferes with steroid synthesis and vitamin D metabolism, ketoconazole should not be used in children. It is not approved for use in children and there is no pediatric dosage range based on pharmacokinetic information in this population.

Sex

Liver complications may be more common in elderly people, in women, or in subjects in whom liver function is already compromised for other reasons.

Androgen-independent prostate cancer

The combination of high-dose ketoconazole + hydrocortisone is effective in androgen-independent prostate cancer. The median duration of response tends to be brief but a significant minority of patients have extended responses. Well characterized information about response and survival, especially in patients who have more durable responses has not been previously reported. The medical records of 78 patients with androgen-independent prostate cancer treated with high-dose ketoconazole + hydrocortisone between March 1991 and February 1999 were retrospectively reviewed and baseline clinical and laboratory factors predictive of a prolonged response and survival were identified (24). The median baseline prostate specific antigen concentration before the start of therapy was 25 µg/l. The number of patients with none, 1–3, and more than three lesions on bone scan were 25, 35, and 18 respectively. The median and mean times to progression of prostate specific antigen were 6.7 and 15 months. Median and mean survival times were 38 and 42 months respectively. Response time and survival were highly correlated. A total of 34 (44%) men had a greater than 75% fall in prostate specific antigen. The median survival times in men with more versus less than a 75% reduction were 60 versus 24 months respectively. In a Cox proportional hazard regression, prolonged survival was predicted by percent fall in prostate specific antigen, extent of disease on bone scan, and baseline prostate specific antigen. Adverse effects were not reported.

Other features of the patient

Ketoconazole interferes with steroidogenesis, and those who are critically ill (25) or HIV-positive (26) are particularly susceptible to adrenal insufficiency from the use of ketoconazole.

Drug–Drug Interactions

Aciclovir

Ketoconazole seems to have a synergistic antiviral effect when it is taken with aciclovir (27).

Alcohol

A disulfiram-type reaction of ketoconazole with alcohol has been reported (SEDA-12, 231).

Amphotericin

Combination of ketoconazole with amphotericin reportedly leads to antagonism, particularly if ketoconazole therapy precedes amphotericin (28).

Antacids

Co-administration of antacids reduces the absorption of ketoconazole (14).

Antihistamines

Ketoconazole can increase the concentrations of astemizole and terfenadine by inhibition of CYP3A4. High concentrations of terfenadine can cause cardiac toxicity. Increased plasma concentrations of unmetabolized terfenadine prolong the QT interval and carry the risk of torsade de pointes and other fatal ventricular arrhythmias (14).

Ebastine (10 and 20 mg/day) had no clinically important effect on QT_c interval in adults, including elderly people, children, and patients with hepatic or renal impairment, and co-administration with ketoconazole or erythromycin did not lead to significant changes in the QT_c interval (29,30).

The pharmacokinetic and pharmacodynamic interactions of emedastine difumarate and ketoconazole have been investigated in 12 healthy volunteers (31). Emedastine difumarate 4 mg was given orally for 10 consecutive days, and on days 6–10 ketoconazole 200 mg bd was co-administered. Emedastine steady-state pharmacokinetics were slightly altered by ketoconazole: the AUC rose by about 33% and total clearance fell by about 30%, with no change in the half-life, suggesting that the volume of distribution also changed; this pattern could have arisen from protein binding displacement. However, there was no change in the QT_c interval after 5 days of co-treatment. The authors concluded that concomitant treatment with emedastine and ketoconazole in subjects with normal QT intervals could therefore be undertaken without special precautions.

When single doses of erythromycin (333 mg) or ketoconazole (200 mg) were given to healthy men who used levocabastine, two sprays per nostril (0.05 mg/spray) bd for 6 days, there were no changes in the pharmacokinetics of levocabastine or in the QT_c interval (32).

Ketoconazole affects plasma concentrations of loratadine, a non-sedating antihistamine, but appears to be devoid of any electrocardiographic effects (33). In a randomized, single-blind, multiple-dose, three-way, crossover study, concomitant administration of loratadine 10 mg qds and ketoconazole 200 mg bd resulted in significantly increased mean loratadine plasma concentration by 307% and desloratadine plasma concentrations by 73%; ketoconazole plasma concentrations were unaffected by loratadine. Despite increased concentrations of loratadine and its metabolite, there were no statistically significant differences in the electrocardiographic QT_c interval.

Benzodiazepines

In a double-blind, crossover kinetic and dynamic study of the interaction of ketoconazole with alprazolam and triazolam, two CYP3A4 substrate drugs with different kinetic profiles, impaired clearance by ketoconazole had more profound clinical consequences for triazolam than for alprazolam (34). By the same mechanism ketoconazole also inhibits the metabolism of midazolam (35).

Ketoconazole can increase the concentrations of triazolam through inhibition of CYP3A4 (66).

Calcium channel blockers

The effects of ketoconazole 200 mg on the pharmacokinetics of nisoldipine 5 mg have been investigated in a randomized, cross-over trial (36). Pretreatment with and concomitant administration of ketoconazole resulted in 24-fold and 11-fold increases in the AUC and C_{max} of nisoldipine, respectively. The ketoconazole-induced increase in plasma concentrations of the metabolite M9 was of similar magnitude. Thus, ketoconazole and other potent inhibitors of CYP3A should not be used concomitantly with nisoldipine.

In an intestinal perfusion study of the effect of ketoconazole 40 µg/ml on the jejunal permeability and first-pass metabolism of (R)- and (S)-verapamil 120 µg/ml in six healthy volunteers, ketoconazole did not alter the jejunal permeability of the isomers, suggesting that it had no effect on the P-glycoprotein mediated efflux. However, the rate of absorption increased, suggesting inhibition by ketoconazole of the gut wall metabolism of (R/S)-verapamil by CYP3A4 (37).

Carbamazepine

Serum concentrations of ketoconazole are reduced by concomitant use of drugs that induce hepatic microsomal enzymes, such as carbamazepine. There may at the same time be a change in serum carbamazepine concentration (14,27).

Ciclosporin

The fact that ketoconazole inhibits cytochrome P450 accounts for some of its interactions. Ketoconazole increases ciclosporin concentrations, enhancing the risk of renal impairment, as shown by a fall in creatinine clearance (SED-12, 678) (7,14,27,38).

The effect of ketoconazole in ciclosporin-treated kidney transplant recipients has been the subject of a prospective randomized study (39). In 51 ketoconazole-treated patients and 49 controls there was a similar frequency of acute rejection episodes. However, in the control group, rejection episodes were more recurrent, with a poorer response to treatment. Acute ciclosporin nephrotoxicity was more common in the ketoconazole group, but this was encountered more at induction and rapidly reversed on further reduction of the dose of ciclosporin. Chronic graft dysfunction was significantly less in the ketoconazole group during the first year, but by the end of the study the difference was not statistically significant. Hepatotoxicity was similar in the two groups. Serum concentrations of cholesterol, low-density lipoprotein, and triglycerides were lower in the ketoconazole group. The authors concluded that long-term low-dose ketoconazole in ciclosporin-treated kidney transplant recipients is safe and cost-saving.

Citalopram

The effect of ketoconazole on the pharmacokinetics of citalopram has been studied in a double-blind, three-way, crossover trial in 18 men and women (40). The subjects received three treatments with a 14-day washout period: a single dose of ketoconazole 200 mg plus placebo, a single dose of citalopram 40 mg plus placebo, and a single dose of ketoconazole 200 mg plus a single dose of citalopram 40 mg. There were no changes in the pharmacokinetics of citalopram after co-administration of ketoconazole, suggesting that ketoconazole and other CYP3A4 inhibitors can be safely co-administered with citalopram.

Clozapine

The interaction of ketoconazole (400 mg/day for 7 days) with clozapine has been evaluated in five patients with schizophrenia given a single dose of clozapine 50 mg at the end of ketoconazole therapy (41). Ketoconazole did not significantly change the disposition of clozapine or its metabolism to its principal metabolites, desmethylclozapine and clozapine-N-oxide.

Donepezil

Donepezil 5 mg produced no change in plasma concentrations of ketoconazole 200 mg (42).

Erythromycin

If erythromycin and ketoconazole, both CYP3A4 inhibitors, are taken in combination, there will be an even more dramatic effect on the metabolism of other drugs, such as terfenadine and astemizole, midazolam and triazolam, and ciclosporin.

Halofantrine

Halofantrine, a highly lipophilic antimalarial drug with poor and erratic absorption, is metabolized to its equipotent metabolite desbutylhalofantrine and this is inhibited by oral ketoconazole (43).

Histamine H$_2$ receptor antagonists

Co-administration of histamine H$_2$ receptor antagonists reduces the absorption of ketoconazole (14).

HIV protease inhibitors

The effects of co-administration of ketoconazole 400 mg and amprenavir 1200 mg, which is mostly metabolized by CYP3A4, have been studied in an open, randomized, balanced, single-dose, three-period, crossover study in 12 healthy men (44). Co-administration of the two drugs increased the amprenavir AUC by 31% and reduced its C_{max} by 16%. Amprenavir increased the AUC of ketoconazole by 44% and increased its half-life and C_{max} by 23% and 16% respectively. Thus, co-administration of amprenavir and ketoconazole result in statistically significant increases in the AUCs of both agents, but the clinical significance of these changes remains to be investigated.

The pharmacokinetic interaction of fluconazole 400 mg od and indinavir 1000 mg tds has been evaluated in a placebo-controlled, crossover study for 8 days; there was no significant interaction (45).

The effect of ketoconazole (200 and 400 mg qds) on plasma and cerebrospinal fluid concentrations of ritonavir (400 mg bd) and saquinavir (400 mg bd) have been investigated in a two-period, two-group, longitudinal pharmacokinetic study in 12 HIV-infected patients (46). Ketoconazole significantly increased the AUC and trough concentrations of ritonavir and prolonged its half-life (by 29%, 62%, and 31% respectively). It produced similar changes (37%, 94%, and 38% respectively) in the kinetics of saquinavir. Ketoconazole significantly increased the ritonavir CSF concentration at 4–5 hours after the dose by 178% (from 2.4 to 6.6 ng/ml), with no change in the paired unbound plasma concentration (26 ng/ml). The changes were not related to ketoconazole dose or plasma exposure. The corresponding changes in saquinavir CSF concentrations were not significant. The authors concluded that ketoconazole inhibited the systemic clearance of ritonavir and, because of the disproportionate increase in CSF concentrations compared with the increase in plasma concentrations, that there was greater inhibition of drug efflux from the CSF.

A study in seven HIV-infected men who took saquinavir 600 mg tds in addition to two other antiretroviral drugs and concomitant ketoconazole (200 mg qds for 7 days, followed by 400 mg qds for another 7 days) showed no significant differences in peak and trough concentrations of saquinavir after the addition of ketoconazole (47). There was substantial inter-subject variability in the study, and the authors concluded that saquinavir concentrations may be unpredictable in individual patients and that drug monitoring may be required for optimizing saquinavir treatment.

In 12 healthy men, ketoconazole 400 mg/day increased the AUC of saquinavir (given as a soft gel capsule 1200 mg tds) by 190% when both were given for 7 days (64).

HMG coenzyme-A reductase inhibitors

Ketoconazole can also cause rhabdomyolysis when taken with both lovastatin and simvastatin (55).

Hormonal contraceptives-oral

The reported reduction in the effect of oral contraceptives (SEDA-12, 231) concurs with the effect of ketoconazole on steroid metabolism; the effect seems to be mild and may be mainly of importance during the use of formulations with low estrogen content (SED-12, 678; (50).

Ifosfamide

The effect of ketoconazole on the CYP-mediated metabolism of ifosfamide to 4-hydroxyifosfamide and the ultimate cytotoxic ifosforamide mustard, and its deactivation to 2- and 3-dechloroethylifosfamide has been studied in a randomized, crossover study in 16 patients, who received intravenous ifosfamide 3 g/m^2/day, either alone or in combination with ketoconazole 200 mg bd 1 day before treatment and during 3 days of concomitant administration (48). Ketoconazole did not affect the fraction metabolized or exposure to the dechloroethylated metabolites and thus did not alter the pharmacokinetics of ifosfamide or its metabolism.

Isoniazid

Combined administration of ketoconazole with isoniazid can lead to increased concentrations of the latter, and there are possibly also alterations in ketoconazole concentrations (SED-12, 678) (27).

Lopinavir and ritonavir

In 12 patients taking ritonavir and saquinavir for HIV infection, ketoconazole significantly increased the AUC, the plasma concentration at 12 hours, and the half-life of ritonavir by 29, 62 and 31% respectively (48). Similar increases of 37, 94, and 38% were recorded for saquinavir. CSF concentrations of ritonavir were raised by 178% by ketoconazole, but there was no significant change in CSF concentrations of saquinavir.

Meglitinides

In healthy subjects, ketoconazole increased mean AUC of repaglinide by 15% and mean C_{max} by 7% (62).

Methylprednisolone

The effect of ketoconazole on steroid metabolism is reflected in other interactions. Ketoconazole increases the total amount of methylprednisolone in the body (SED-12, 678) (49).

Paclitaxel

In patients with ovarian cancer, ketoconazole, 100–1600 mg as a single oral dose 3 hours after paclitaxel 175 mg/m^2 as a 3-hour continuous intravenous infusion, did not alter plasma concentrations of paclitaxel or its principal metabolite, 6-alpha-hydroxypaclitaxel (63).

Phenazone

Ketoconazole reduces the clearance of phenazone (antipyrine) (SED-11, 574) (51).

Phenytoin

Serum concentrations of ketoconazole are reduced by concomitant use of drugs that induce hepatic microsomal enzymes, such as phenytoin. There may at the same time be a change in serum phenytoin concentration (SED-12, 678) (14,27,52).

Proton pump inhibitors

Co-administration of proton pump blockers reduces the absorption of ketoconazole (14).

Reboxetine

Reboxetine is metabolized by CYP3A4. In 11 healthy volunteers, ketoconazole increased the AUC of reboxetine by about 50% and increased its half-life (53). The adverse effects profile of reboxetine was not altered by ketoconazole, but the finding suggests that reboxetine should be used with caution in combination with drugs that inhibit CYP3A4, for example nefazodone and fluvoxamine.

Rifamycins

Serum concentrations of ketoconazole are reduced by concomitant use of drugs that induce hepatic microsomal enzymes, such as rifampicin. There may at the same time be a change in rifampicin serum concentration (SED-12, 678) (14,27,39).

Ketoconazole interacts with rifampicin, and the serum concentrations of both drugs are reduced (SEDA-9, 269).

Ropivacaine

Ketoconazole 400 mg caused a minor reduction in the clearance of ropivacaine, which is mostly metabolized by CYP1A2 (54).

Theophylline

The combination of ketoconazole with theophylline was reported to have reduced theophylline concentrations, suggesting increased metabolism of theophylline (SEDA-13, 234). This is surprising, since one would have expected ketoconazole to have inhibited the metabolism of theophylline, if at all. However, in 10 healthy, non-smoking men aged 18–40 years aminophylline 6 mg/kg intravenously before and after they had taken oral ketoconazole 200 mg/day for 7 days caused no change in the half-life or clearance of theophylline (56), so there is probably no important interaction between these two drugs.

Tolterodine

Tolterodine is eliminated by two different oxidative metabolic pathways: hydroxylation, catalysed by CYP2D6, and D-alkylation, catalysed by CYP3A. The pharmacokinetics and safety of tolterodine and its metabolites in the absence and presence of ketoconazole have been investigated in healthy volunteers with deficient CYP2D6 activity (poor metabolizers) (57). Clearance of tolterodine fell by 60% during co-administration of ketoconazole, resulting in a 2.1-fold increase in AUC. Thus, caution is needed when ketoconazole and other potent inhibitors of CYP3A are used concomitantly with tolterodine.

Tretinoin

A potential interaction of ketoconazole with tretinoin (all-trans retinoic acid), resulting in slowed metabolism, is probably not of importance (58).

Warfarin

Potentiation of the effects of warfarin has been reported in one case (SED-12, 678; 59), but absence of interference with anticoagulants was also claimed.

Zaleplon

Ketoconazole increases plasma concentrations of the Z drugs, zaleplon, zolpidem, and zopiclone, and increases their sedative effects (66); this occurs to a lesser extent than the similar effect on benzodiazepines that are exclusively metabolized by CYP3A4.

Ziprasidone

Ziprasidone is oxidatively metabolized by CYP3A4, but it does not inhibit CYP3A4 or other isoenzymes at clinically relevant concentrations. The effect of ketoconazole 400 mg qds for 6 days on the single-dose pharmacokinetics of ziprasidone 40 mg has been evaluated in an open, placebo-controlled, crossover study in healthy volunteers (60). Ketoconazole caused a modest increase in the mean AUC (33%) and the mean C_{max} (34%) of ziprasidone. This effect was not considered clinically relevant and suggests that other inhibitors of CYP3A4 are unlikely to affect the pharmacokinetics of ziprasidone significantly. Most of the reported adverse events were mild. The adverse events that were most commonly reported in subjects who took the drugs concomitantly were dizziness, weakness, and somnolence. There were no treatment-related laboratory abnormalities or abnormal vital signs during the study and at the 6-day follow-up evaluation.

Zolpidem

Zolpidem is metabolized by CYP3A. Ketoconazole 200 mg bd impaired the clearance of zolpidem 5 mg and enhanced its benzodiazepine-like pharmacodynamic effects. In contrast, itraconazole 100 mg bd and fluconazole 100 mg bd had small effects on zolpidem kinetics and dynamics (61).

References

1. Lyman CA, Walsh TJ. Systemically administered antifungal agents. A review of their clinical pharmacology and therapeutic applications. Drugs 1992;44(1):9–35.
2. The ARDS Network. Ketoconazole for early treatment of acute lung injury and acute respiratory distress syndrome: a randomized controlled trial. JAMA 2000;283(15):1995–2002.
3. Chou SC, Lin JD. Long-term effects of ketoconazole in the treatment of residual or recurrent Cushing's disease. Endocr J 2000;47(4):401–6.
4. Mok NS, Lo YK, Tsui PT, Lam CW. Ketoconazole induced torsades de pointes without concomitant use of QT interval-prolonging drug. J Cardiovasc Electrophysiol 2005;16(12):1375-7.
5. Schilling MK, Eichenberger M, Maurer CA, Sigurdsson G, Buchler MW. Ketoconazole and pulmonary failure after esophagectomy: a prospective clinical trial. Dis Esophagus 2001;14(1):37–40.
6. Krause W, Effendy I. Wie wirkt Ketoconazol auf den Testosteron-Stoffwechsel?. [How does ketoconazole affect testosterone metabolism?.] Z Hautkr 1985;60(14):1147–55.
7. Drouhet E, Dupont B. Evolution of antifungal agents: past, present, and future. Rev Infect Dis 1987;9(Suppl 1):S4–S14.
8. Loli P, Berselli ME, Tagliaferri M. Use of ketoconazole in the treatment of Cushing's syndrome. J Clin Endocrinol Metab 1986;63(6):1365–71.
9. Khosla S, Wolfson JS, Demerjian Z, Godine JE. Adrenal crisis in the setting of high-dose ketoconazole therapy. Arch Intern Med 1989;149(4):802–4.
10. Venturoli S, Fabbri R, Dal Prato L, Mantovani B, Capelli M, Magrini O, Flamigni C. Ketoconazole therapy for women with acne and/or hirsutism. J Clin Endocrinol Metab 1990;71(2):335–9.
11. De Pedrini P, Tommaselli A, Spano G, Montemurro G. Clinical and hormonal effects of ketoconazole on hirsutism in women. Int J Tissue React 1988;10(3):193–8.
12. Akalin S. Effects of ketoconazole in hirsute women. Acta Endocrinol (Copenh) 1991;124(1):19–22.
13. Duman D, Turhal NS, Duman DG. Fatal aplastic anemia during treatment with ketoconazole. Am J Med 2001;111(9):737.
14. Francis P, Walsh TJ. Evolving role of flucytosine in immunocompromised patients: new insights into safety, pharmacokinetics, and antifungal therapy. Clin Infect Dis 1992;15(6):1003–18.
15. Lewis JH, Zimmerman HJ, Benson GD, Ishak KG. Hepatic injury associated with ketoconazole therapy. Analysis of 33 cases. Gastroenterology 1984;86(3):503–13.
16. Stricker BH, Blok AP, Bronkhorst FB, Van Parys GE, Desmet VJ. Ketoconazole-associated hepatic injury. A clinicopathological study of 55 cases. J Hepatol 1986;3(3):399–406.
17. Knight TE, Shikuma CY, Knight J. Ketoconazole-induced fulminant hepatitis necessitating liver transplantation. J Am Acad Dermatol 1991;25(2 Pt 2):398–400.
18. Duarte PA, Chow CC, Simmons F, Ruskin J. Fatal hepatitis associated with ketoconazole therapy. Arch Intern Med 1984;144(5):1069–70.
19. Zollner E, Delport S, Bonnici F. Fatal liver failure due to ketoconazole treatment of a girl with Cushing's syndrome. J Pediatr Endocrinol Metab 2001;14(3):335–8.
20. Garcia Rodriguez LA, Duque A, Castellsague J, Perez-Gutthann S, Stricker BH. A cohort study on the risk of acute liver injury among users of ketoconazole and other antifungal drugs. Br J Clin Pharmacol 1999;48(6):847–52.
21. Castellsague J, Garcia-Rodriguez LA, Duque A, Perez S. Risk of serious skin disorders among users of oral antifungals: a population-based study. BMC Dermatol 2002;2(1):14.
22. Smith EB, Henry JC. Ketoconazole: an orally effective antifungal agent. Mechanism of action, pharmacology, clinical efficacy and adverse effects. Pharmacotherapy 1984;4(4):199–204.
23. Cox FW, Stiller RL, South DA, Stevens DA. Oral ketoconazole for dermatophyte infections. J Am Acad Dermatol 1982;6(4 Pt 1):455–62.
24. Scholz M, Jennrich R, Strum S, Brosman S, Johnson H, Lam R. Long-term outcome for men with androgen independent prostate cancer treated with ketoconazole and hydrocortisone. J Urol 2005;173(6):1947-52.
25. Albert SG, DeLeon MJ, Silverberg AB. Possible association between high-dose fluconazole and adrenal insufficiency in critically ill patients. Crit Care Med 2001;29(3):668–70.
26. Etzel JV, Brocavich JM, Torre M. Endocrine complications associated with human immunodeficiency virus infection. Clin Pharm 1992;11(8):705–13.
27. Bickers DR. Antifungal therapy: potential interactions with other classes of drugs. J Am Acad Dermatol 1994;31(3 Pt 2):S87–90.
28. Schaffner A, Frick PG. The effect of ketoconazole on amphotericin B in a model of disseminated aspergillosis. J Infect Dis 1985;151(5):902–10.
29. Moss AJ, Chaikin P, Garcia JD, Gillen M, Roberts DJ, Morganroth J. A review of the cardiac systemic side-effects of antihistamines: ebastine. Clin Exp Allergy 1999;29(Suppl 3):200–5.
30. Moss AJ, Morganroth J. Cardiac effects of ebastine and other antihistamines in humans. Drug Saf 1999;21(Suppl 1):69–80.
31. Herranz U, Rusca A, Assandri A. Emedastine–ketoconazole: pharmacokinetic and pharmacodynamic interactions in healthy volunteers. Int J Clin Pharmacol Ther 2001;39(3):102–9.
32. Pesco-Koplowitz L, Hassell A, Lee P, Zhou H, Hall N, Wiesinger B, Mechlinski W, Grover M, Hunt T, Smith R, Travers S. Lack of effect of erythromycin and ketoconazole on the pharmacokinetics and pharmacodynamics of steady-state intranasal levocabastine. J Clin Pharmacol 1999;39(1):76–85.
33. Kosoglou T, Salfi M, Lim JM, Batra VK, Cayen MN, Affrime MB. Evaluation of the pharmacokinetics and electrocardiographic pharmacodynamics of loratadine with concomitant administration of ketoconazole or cimetidine. Br J Clin Pharmacol 2000;50(6):581–9.
34. Greenblatt DJ, Wright CE, von Moltke LL, Harmatz JS, Ehrenberg BL, Harrel LM, Corbett K, Counihan M, Tobias S, Shader RI. Ketoconazole inhibition of triazolam

and alprazolam clearance: differential kinetic and dynamic consequences. Clin Pharmacol Ther 1998;64(3):237–47.

35. Lam YW, Alfaro CL, Ereshefsky L, Miller M. Pharmacokinetic and pharmacodynamic interactions of oral midazolam with ketoconazole, fluoxetine, fluvoxamine, and nefazodone. J Clin Pharmacol 2003;43(11):1274–82.

36. Heinig R, Adelmann HG, Ahr G. The effect of ketoconazole on the pharmacokinetics, pharmacodynamics and safety of nisoldipine. Eur J Clin Pharmacol 1999;55(1):57–60.

37. Sandstrom R, Knutson TW, Knutson L, Jansson B, Lennernas H. The effect of ketoconazole on the jejunal permeability and CYP3A metabolism of (R/S)-verapamil in humans. Br J Clin Pharmacol 1999;48(2):180–9.

38. Sugar AM, Stern JJ, Dupont B. Overview: treatment of cryptococcal meningitis. Rev Infect Dis 1990;12(Suppl 3):S338–48.

39. Sobh MA, Hamdy AF, El Agroudy AE, El Sayed K, El-Diasty T, Bakr MA, Ghoneim MA. Coadministration of ketoconazole and cyclosporine for kidney transplant recipients: long-term follow-up and study of metabolic consequences. Am J Kidney Dis 2001;37(3):510–7.

40. Gutierrez M, Abramowitz W. Lack of effect of a single dose of ketoconazole on the pharmacokinetics of citalopram. Pharmacotherapy 2001;21(2):163–8.

41. Lane HY, Chiu CC, Kazmi Y, Desai H, Lam YW, Jann MW, Chang WH. Lack of CYP3A4 inhibition by grapefruit juice and ketoconazole upon clozapine administration in vivo. Drug Metabol Drug Interact 2001;18(3–4):263–78.

42. Tiseo PJ, Perdomo CA, Friedhoff LT. Concurrent administration of donepezil HCl and ketoconazole: assessment of pharmacokinetic changes following single and multiple doses. Br J Clin Pharmacol 1998;46(Suppl 1):30–4.

43. Khoo SM, Porter JH, Edwards GA, Charman WN. Metabolism of halofantrine to its equipotent metabolite, desbutylhalofantrine, is decreased when orally administered with ketoconazole. J Pharm Sci 1998;87(12):1538–41.

44. Polk RE, Crouch MA, Israel DS, Pastor A, Sadler BM, Chittick GE, Symonds WT, Gouldin W, Lou Y. Pharmacokinetic interaction between ketoconazole and amprenavir after single doses in healthy men. Pharmacotherapy 1999;19(12):1378–84.

45. De Wit S, Debier M, De Smet M, McCrea J, Stone J, Carides A, Matthews C, Deutsch P, Clumeck N. Effect of fluconazole on indinavir pharmacokinetics in human immunodeficiency virus-infected patients. Antimicrob Agents Chemother 1998;42(2):223–7.

46. Khaliq Y, Gallicano K, Venance S, Kravcik S, Cameron DW. Effect of ketoconazole on ritonavir and saquinavir concentrations in plasma and cerebrospinal fluid from patients infected with human immunodeficiency virus. Clin Pharmacol Ther 2000;68(6):637–46.

47. Collazos J, Martinez E, Mayo J, Blanco MS. Effect of ketoconazole on plasma concentrations of saquinavir. J Antimicrob Chemother 2000;46(1):151–2.

48. Kerbusch T, Jansen RL, Mathot RA, Huitema AD, Jansen M, van Rijswijk RE, Beijnen JH. Modulation of the cytochrome P450-mediated metabolism of ifosfamide by ketoconazole and rifampin. Clin Pharmacol Ther 2001;70(2):132–41.

49. Glynn AM, Slaughter RL, Brass C, D'Ambrosio R, Jusko WJ. Effects of ketoconazole on methylprednisolone pharmacokinetics and cortisol secretion. Clin Pharmacol Ther 1986;39(6):654–9.

50. Kovacs L, Somos P, Hamori M. Examination of the potential interaction between ketoconazole (Nizoral) and oral contraceptives with special regard to products of low hormone content (Rigevidon, Anteovin). Ther Hung 1986;34:167.

51. D'Mello AP, D'Souza MJ, Bates TR. Pharmacokinetics of ketoconazole–antipyrine interaction. Lancet 1985;2(8448):209–10.

52. Food and Drug Administration. Ketoconazole labeling revised. FDA Drug Bull 1984;14(2):17–8.

53. Herman BD, Fleishaker JC, Brown MT. Ketoconazole inhibits the clearance of the enantiomers of the antidepressant reboxetine in humans. Clin Pharmacol Ther 1999;66(4):374–9.

54. Arlander E, Ekstrom G, Alm C, Carrillo JA, Bielenstein M, Bottiger Y, Bertilsson L, Gustafsson LL. Metabolism of ropivacaine in humans is mediated by CYP1A2 and to a minor extent by CYP3A4: an interaction study with fluvoxamine and ketoconazole as in vivo inhibitors. Clin Pharmacol Ther 1998;64(5):484–91.

55. Gilad R, Lampl Y. Rhabdomyolysis induced by simvastatin and ketoconazole treatment. Clin Neuropharmacol 1999;22(5):295–7.

56. Heusner JJ, Dukes GE, Rollins DE, Tolman KG, Galinsky RE. Effect of chronically administered ketoconazole on the elimination of theophylline in man. Drug Intell Clin Pharm 1987;21(6):514–7.

57. Brynne N, Forslund C, Hallen B, Gustafsson LL, Bertilsson L. Ketoconazole inhibits the metabolism of tolterodine in subjects with deficient CYP2D6 activity. Br J Clin Pharmacol 1999;48(4):564–72.

58. Lee JS, Newman RA, Lippman SM, Fossella FV, Calayag M, Raber MN, Krakoff IH, Hong WK. Phase I evaluation of all-trans retinoic acid with and without ketoconazole in adults with solid tumors. J Clin Oncol 1995;13(6):1501–8.

59. Smith AG. Potentiation of oral anticoagulants by ketoconazole. BMJ (Clin Res Ed) 1984;288(6412):188–9.

60. Miceli JJ, Smith M, Robarge L, Morse T, Laurent A. The effects of ketoconazole on ziprasidone pharmacokinetics—a placebo-controlled crossover study in healthy volunteers. Br J Clin Pharmacol 2000;49(Suppl 1):S71–6.

61. Greenblatt DJ, von Moltke LL, Harmatz JS, Mertzanis P, Graf JA, Durol AL, Counihan M, Roth-Schechter B, Shader RI. Kinetic and dynamic interaction study of zolpidem with ketoconazole, itraconazole, and fluconazole. Clin Pharmacol Ther 1998;64(6):661–71.

62. Hatorp V, Hansen KT, Thomsen MS. Influence of drugs interacting with CYP3A4 on the pharmacokinetics, pharmacodynamics, and safety of the prandial glucose regulator repaglinide. J Clin Pharmacol 2003;43(6):649–60.

63. Jamis-Dow CA, Pearl ML, Watkins PB, Blake DS, Klecker RW, Collins JM. Predicting drug interactions in vivo from experiments in vitro. Human studies with paclitaxel and ketoconazole. Am J Clin Oncol 1997;20(6):592–599.

64. Grub S, Bryson H, Goggin T, Ludin E, Jorga K. The interaction of saquinavir (soft gelatin capsule) with ketoconazole, erythromycin and rifampicin: comparison of the effect in healthy volunteers and in HIV-infected patients. Eur J Clin Pharmacol 2001;57(2):115–21.

65. Bickers DR. Antifungal therapy: potential interactions with other classes of drugs. J Am Acad Dermatol 1994;31(3 Pt 2):S87–90.

66. Hesse LM, von Moltke LL, Greenblatt DJ. Clinically important drug interactions with zopiclone, zolpidem and zaleplon. CNS Drugs 2003;17(7):513–32.

Miconazole

See also Antifungal azoles

General Information

Miconazole has been evaluated as a topical, oral, and intravenous agent. Its absorption is slightly better than that of clotrimazole, but is still insufficient for use as a systemic agent, particularly in view of its relatively short half-life of around 8–9 hours (1). Miconazole has also been given intravenously and intrathecally. It is highly protein-bound and does not diffuse well into the CSF; however, it does penetrate readily into synovial and vitreous fluids (2).

The indications for miconazole have diminished with the advent of newer antifungal azoles (3) and it has largely been superseded by other azoles; infection with *Pseudallescheria boydii* remains one of the few indications (4).

General adverse effects

Topical miconazole is well tolerated. Parenteral administration carries a higher frequency of adverse effects, some probably being caused by Cremophor (polyethoxylated castor oil, the carrier). Adverse effects include fever, chills, pruritus, rash, nausea, vomiting, diarrhea, hyponatremia, cardiac toxicity, phlebitis, hyperlipidemia, and central nervous system disturbances. Hypersensitivity reactions can occur. Tumor-inducing effects have not been reported.

Organs and Systems

Cardiovascular

Local phlebitis is not uncommon after intravenous miconazole (5); the type of intravenous solution used is of importance.

Collapse after rapid intravenous injection has been described, as have some cases of tachycardia, ventricular tachycardia, and even, in a few instances, cardiorespiratory arrest, attributable to the histamine-releasing properties of Cremophor (SED-12, 679; 2).

Nervous system

Arachnoiditis has been described after intrathecal injection (6).

Psychological, psychiatric

Acute toxic psychosis is a rare consequence of miconazole (SED-12, 679; 7).

Endocrine

Hyperlipidemia has been described in many patients given miconazole; this may be caused by the solvent (SED-12, 679; 2).

Skin

Pruritus and rashes have been reported with miconazole but do not seem to be frequent (5).

Drug–Drug Interactions

Anticoagulants

Miconazole can potentiate the effects of coumarin anticoagulants (SED-12, 679) (2). Potentiation of the anticoagulatory effects of acenocoumarol was noted after vaginal administration of miconazole capsules to two postmenopausal patients (8) and after oral administration of miconazole gel in three elderly patients with oral candidiasis (9). In both instances there were no major bleeding complications. However, it appears prudent to consider the use of non-azole topical antifungal agents in patients who are taking CYP3A4 metabolized drugs with a narrow therapeutic index.

Phenytoin

Miconazole can increase serum concentrations of phenytoin (SED-12, 679) (10).

Sulfonylureas

Enhancement of the effects of hypoglycemic sulfonamides has been reported (SED-12, 679; 2).

References

1. Mikamo H, Kawazoe K, Sato Y, Ito K, Tamaya T. Pharmacokinetics of miconazole in serum and exudate of pelvic retroperitoneal space after radical hysterectomy and pelvic lymphadenectomy. Int J Antimicrob Agents 1997;9(3):207–11.
2. Drouhet E, Dupont B. Evolution of antifungal agents: past, present, and future. Rev Infect Dis 1987;9(Suppl 1):S4–S14.
3. Walsh TJ, Pizzo A. Treatment of systemic fungal infections: recent progress and current problems. Eur J Clin Microbiol Infect Dis 1988;7(4):460–75.
4. Walsh TJ, Peter J, McGough DA, Fothergill AW, Rinaldi MG, Pizzo PA. Activities of amphotericin B and antifungal azoles alone and in combination against *Pseudallescheria boydii*. Antimicrob Agents Chemother 1995;39(6):1361–4.
5. Stevens DA. Miconazole in the treatment of coccidioidomycosis. Drugs 1983;26(4):347–54.
6. Sung JP, Campbell GD, Grendahl JG. Miconazole therapy for fungal meningitis. Arch Neurol 1978;35(7):443–7.
7. Cohen J. Antifungal chemotherapy. Lancet 1982;2(8297):532–7.
8. Lansdorp D, Bressers HP, Dekens-Konter JA, Meyboom RH. Potentiation of acenocoumarol during

vaginal administration of miconazole. Br J Clin Pharmacol 1999;47(2):225–6.

9. Ortin M, Olalla JI, Muruzabal MJ, Peralta FG, Gutierrez MA. Miconazole oral gel enhances acenocoumarol anticoagulant activity: a report of three cases. Ann Pharmacother 1999;33(2):175–7.
10. Rolan PE, Somogyi AA, Drew MJ, Cobain WG, South D, Bochner F. Phenytoin intoxication during treatment with parenteral miconazole. BMJ (Clin Res Ed) 1983;287(6407):1760.

Nystatin

General Information

Nystatin, the earliest antifungal drug, is used today primarily by the oral route to treat gastrointestinal mycosis and topically for mucocutaneous candidiasis (SED-11, 576). Absorption from the gut is insufficient to produce systemic activity.

An intravenous multilamellar liposomal formulation of nystatin has been developed. Its pharmacokinetics are unique among polyenes and are characterized by high peak plasma concentrations and a rapid clearance rate. In a phase I study it was tolerated well at doses up to 8 mg/kg. It was efficacious in candidemia, invasive aspergillosis, and as empirical antifungal therapy in patients with cancer and neutropenia (1). However, it was subsequently withdrawn by the manufacturers.

Observational studies

The activity and safety of multilamellar liposomal nystatin for invasive aspergillosis has been investigated in 26 patients refractory to or intolerant of amphotericin (2). There was at least one infusion-related adverse event in 22 patients. The most frequent adverse effects included chills (n = 21), shivering (n = 6), and fever (n = 7), leading to withdrawal in two patients. There was grade 1 impairment of renal function in 10 patients, and hypokalemia in 13. Liposomal nystatin can be effective in salvage therapy of invasive aspergillosis. However, infusion-related adverse events are frequent.

Organs and Systems

Gastrointestinal

Nausea, vomiting, and diarrhea have been reported after oral administration, possibly due to the drug but perhaps also ascribable to the underlying disease. These effects are more marked with doses over 5 MU a day (SED-11, 576) (3).

Immunologic

Allergic reactions to topical use are rare (SED-11, 576) (4).

Contact allergy due to topical nystatin has been reported in a woman using a combination of clobetasol and nystatin (5). The contact allergy was demonstrated with a positive patch test on day 4.

In another case, the patient developed a maculopapular rash over the trunk and limbs, associated with fever, arthralgia, malaise, and diarrhea; nystatin patch tests were positive on days 2 and 4 (6).

References

1. Offner F, Krcmery V, Boogaerts M, Doyen C, Engelhard D, Ribaud P, Cordonnier C, de Pauw B, Durrant S, Marie JP, Moreau P, Guiot H, Samonis G, Sylvester R, Herbrecht R. EORTC Invasive Fungal Infections Group. Liposomal nystatin in patients with invasive aspergillosis refractory to or intolerant of amphotericin B. Antimicrob Agents Chemother 2004;48(12):4808–12.
2. Offner F, Krcmery V, Boogaerts M, Doyen C, Engelhard D, Ribaud P, Cordonnier C, de Pauw B, Durrant S, Marie JP, Moreau P, Guiot H, Samonis G, Sylvester R, Herbrecht R; EORTC Invasive Fungal Infections Group. Liposomal nystatin in patients with invasive aspergillosis refractory to or intolerant of amphotericin B. Antimicrob Agents Chemother 2004;48:4808-12.
3. Cohen J. Antifungal chemotherapy. Lancet 1982;2(8297):532–7.
4. Pareek SS. Nystatin-induced fixed eruption. Br J Dermatol 1980;103(6):679–80.
5. Cooper SM, Shaw S. Contact allergy to nystatin: an unusual allergen. Contact Dermatitis 1999;41(2):120.
6. Cooper SM, Reed J, Shaw S. Systemic reaction to nystatin. Contact Dermatitis 1999;41(6):345–6.

Posaconazole

See also Antifungal azoles

General Information

Posaconazole is a lipophilic antifungal triazole. In vitro, it has broad-spectrum activity against opportunistic, endemic, and dermatophytic fungi, including organisms that are often resistant to existing agents, such as *C. glabrata*, *C. krusei*, *A. terreus*, *Fusarium* species, and the *Zygomycetes* (1,2). A large variety of animal models of invasive fungal infections have provided consistent evidence of efficacy against these organisms in vivo.

Posaconazole was well tolerated and as effective as fluconazole 100 mg in two large randomized comparisons in HIV-infected patients with oropharyngeal candidiasis. In a salvage study in patients with a variety of invasive fungal infections, there were response rates of 44–80% in patients with aspergillosis, fusariosis, cryptococcosis, candidiasis, and pheohyphomycoses after 4–8 weeks of therapy (3,4).

Posaconazole is available as an oral suspension and achieves optimal exposure when administered in two to four divided doses given with food or a nutritional supplement. Posaconazole has a large volume of distribution, in the order of 5 l/kg, and a half-life of about 20 hours. It is not metabolized by CYP450 but is primarily excreted unchanged in the feces. It inhibits CYP3A4, but has no effects on CYP1A2, CYP2C8, CYP2C9, CYP2D6, and CYP2E1, and a limited spectrum of drug–drug interactions can therefore be expected.

Posaconazole had strong antifungal efficacy in phase II and III clinical trials in immunocompromised patients with oropharyngeal and esophageal candidiasis. It also had promising efficacy as salvage therapy in a large phase II study. In the subset of 107 patients with invasive aspergillosis, 42% had a complete or partial response at the end of treatment compared with 26% of a contemporaneous control cohort. Posaconazole appears to be as well tolerated as fluconazole and it is currently under regulatory review in the USA; it has been approved in the European Union for salvage treatment of invasive fungal infections including invasive aspergillosis (3,4).

Posaconazole has been approved in the EU for treatment of aspergillosis, fusariosis, chromoblastomycosis, and coccidioidomycosis refractory to or in patients intolerant of standard therapies; in addition, it has been approved for prophylaxis in high-risk patients with acute myeloblastic leukemia/myelodysplastic syndrome and allogeneic hemopoietic stem cell transplantation and graft-versus-host disease in both the EU and the USA. The recommended daily dosage for salvage treatment is 400 mg bd given with food; for patients who cannot take solid food, a dosage of 200 mg qds is recommended, preferably with a nutritional supplement. The dosage for prophylaxis is 200 mg tds.

Observational studies

The pharmacokinetics, efficacy, and adverse effects of different dosing schedules of posaconazole oral suspension in patients with possible, probable, and proven refractory invasive fungal infections (n= 32) or febrile neutropenia (n = 66) have been evaluated in a multicenter, open, parallel-group study (5). Exposure in allogeneic bone marrow transplant recipients (n = 12) was 52% lower than in others. There were treatment-related adverse effects in 24% of patients, mostly gastrointestinal.

Susceptibility Factors

Renal disease

The pharmacokinetics and safety of a single oral dose of posaconazole 400 mg have been studied in healthy subjects and in those with mild, moderate, and severe chronic renal disease (n = 6 in each group) (6). In those on hemodialysis a dose was given on a non-hemodialysis day, and another 6 hours before hemodialysis. Mild to moderate renal disease had no effect on the pharmacokinetics of posaconazole. Mean oral clearances before and during hemodialysis were comparable. Furthermore, the difference in the pre-dialysis and post-dialysis posaconazole concentrations was only about 3%, suggesting that posaconazole is not removed by hemodialysis. Protein binding was similar in all groups (about 98%) and was unaffected by hemodialysis. Posaconazole was generally well

tolerated. These results show that dosage adjustments are not required in patients with disease.

Food-Drug Interactions

The effects of prandial status and the fat content of a meal on the systemic availability of posaconazole suspension and co-precipitate tablet formulations has been studied in a randomized, open-label, 4-way crossover, single-dose study in 20 healthy men (7). The posaconazole suspension had greater availability than the tablet and availability was markedly increased by a high-fat meal.

References

1. Groll AH, Glasmacher A, Just-Nuebling G, Maschmeyer G, Walsh TJ. Clinical pharmacology of antifungal compounds. Infect Dis Clin N Am 2003;17:159-91.
2. Groll AH, Walsh TJ. Posaconazole: clinical pharmacology and potential for management of fungal infections. Expert Rev Anti Infect Ther 2005;3:467-87.
3. Groll AH, Gea-Banacloche JC, Glasmacher A, Just-Nuebling G, Maschmeyer G, Walsh TJ. Clinical pharmacology of antifungal compounds. Infect Dis Clin North Am 2003;17(1):159–91.
4. Hoffman HL, Ernst EJ, Klepser ME. Novel triazole antifungal agents. Expert Opin Investig Drugs 2000;9(3):593–605.
5. Ullmann AJ, Cornely OA, Burchardt A, Hachem R, Kontoyiannis DP, Töpelt K, Courtney R, Wexler D, Krishna G, Martinho M, Corcoran G, Raad I. Pharmacokinetics, safety, and efficacy of posaconazole in patients with persistent febrile neutropenia or refractory invasive fungal infection. Antimicrob Agents Chemother 2006;50(2):658-66.
6. Courtney R, Sansone A, Smith W, Marbury T, Statkevich P, Martinho M, Laughlin M, Swan S. Posaconazole pharmacokinetics, safety, and tolerability in subjects with varying degrees of chronic renal disease. J Clin Pharmacol 2005;45(2):185-92.
7. Courtney R, Wexler D, Radwanski E, Lim J, Laughlin M. Effect of food on the relative bioavailability of two oral formulations of posaconazole in healthy adults. Br J Clin Pharmacol 2004;57:218-22.

Ravuconazole

See also Antifungal azoles

General Information

Ravuconazole is similar in structure to fluconazole. It has a long half-life suitable for once-daily dosing.

The safety and efficacy of ravuconazole have been studied in a randomized, double-blind comparison with fluconazole in 71 patients with esophageal candidiasis; ravuconazole was as effective as fluconazole and there were no apparent differences in adverse effects (1,2).

References

1. Groll AH, Gea-Banacloche JC, Glasmacher A, Just-Nuebling G, Maschmeyer G, Walsh TJ. Clinical pharmacology of antifungal compounds. Infect Dis Clin North Am 2003;17(1):159–91.
2. Hoffman HL, Ernst EJ, Klepser ME. Novel triazole antifungal agents. Expert Opin Investig Drugs 2000;9(3):593–605.

Saperconazole

See also Antifungal azoles

General Information

Saperconazole is an experimental, water-insoluble, lipophilic, fluorinated triazole. Its structure resembles that of itraconazole and it has a long half-life. It has a broad antifungal spectrum, including *Cryptococcus* Species and *Aspergillus* species. In early studies in cases of compassionate use, only a few adverse effects were described, including hepatotoxicity (1,2), and its adverse effects were expected to resemble those of itraconazole (3). However, the manufacturers stopped developing it because of concerns about toxicity.

References

1. Lin C, Kim H, Radwanski E, Affrime M, Brannan M, Cayen MN. Pharmacokinetics and metabolism of genaconazole, a potent antifungal drug, in men. Antimicrob Agents Chemother 1996;40(1):92–6.
2. Khoo SH, Denning DW. Cure of chronic invasive sinus aspergillosis with oral saperconazole. J Med Vet Mycol 1995;33(1):63–6.
3. Lyman CA, Walsh TJ. Systemically administered antifungal agents. A review of their clinical pharmacology and therapeutic applications. Drugs 1992;44(1):9–35.

Terbinafine

General Information

The allylamine derivative terbinafine can be used orally or topically. It is active against a broad range of fungi, including filamentous fungi and, to a lesser extent, yeast-like fungi. However, probably because of irreversible protein binding, its clinical usefulness is limited to the treatment of dermatophyte infections and perhaps lymphocutaneous sporotrichosis.

Terbinafine acts by inhibiting the synthesis of fungal ergosterol at the level of squalene oxidase, leading to depletion of ergosterol and accumulation of toxic squalenes in the fungal cell membrane.

Pharmacokinetics

Numerous reviews and studies of the pharmacokinetics of terbinafine have appeared (1–13). Independent of food, its oral availability is 70–80%. With a single dose of 250 mg, plasma concentrations reach around 0.97 µg/ml after two hours. Apparent steady-state plasma concentrations are reached after 10–14 days after only two-fold accumulation, although the half-life is up to 3 weeks and microbiologically active concentrations can be measured in plasma for weeks to months after the last dose, which is consistent with slow redistribution from the peripheral tissues and fat. Protein binding is over 99% and the apparent volume of distribution over 2000 liters, high concentrations reaching adipose tissues, the dermis, epidermis, and nails. Less than 0.2% reaches the breast milk. After several weeks, no further accumulation of the compound occurs (14,15). In renal or hepatic failure, elimination is slowed. Terbinafine undergoes extensive and complex hepatic biotransformation involving at least seven CYP450 enzymes (16); none of its metabolites is mycologically active. Urinary excretion accounts for more than 70% and fecal elimination for 10% of total excretion; the extent of enterohepatic recycling is unknown. Children have a shorter half-life, a lower mean AUC, and a higher volume of distribution, reflecting their higher proportion of lipophilic tissues (17,18).

Comparative studies

The safety and efficacy of terbinafine 250 mg/day and itraconazole 200 mg/day given for 12 weeks for toenail onychomycosis have been compared in a randomized, double-blind study in 372 patients (19). Adverse events were reported in 39% of the terbinafine-treated patients and in 35% of the itraconazole-treated patients. The mean values of biochemical parameters of liver and kidney function did not change significantly. Terbinafine produced higher rates of clinical cure (76 versus 58%) and mycological cure (73 versus 46%) than itraconazole.

In a randomized, double-blind comparison of terbinafine 250 mg/day (*n* = 146) with itraconazole 200 mg/day (*n* = 146), administered for 12 weeks for toenail onychomycosis, mycological cure rates at the 36-week follow-up end-point (67 versus 61%) and the proportion of patients with adverse effects (23 versus 22%) were similar in both study arms. However, more patients taking terbinafine stopped treatment permanently because of treatment-related adverse events (8 versus 1%) (20).

In a double-blind, randomized, multicenter comparison of terbinafine (250 mg/day for 12 or 16 weeks) or itraconazole capsules (200 mg bd for 1 week every 4 weeks for 12 or 16 weeks), 236 patients reported at least one adverse event. All were within the known safety profile of both drugs, and there were no significant differences among the four treatment regimens. Continuous terbinafine was significantly more effective than intermittent itraconazole (mycological cure rates at week 72: 76, 81, 38, and 49%; significant for all comparisons between terbinafine and itraconazole) (21,22).

General adverse effects

Terbinafine is usually well tolerated. Gastrointestinal complaints (dyspepsia, nausea, diarrhea) were the most common reasons for withdrawal. Abdominal pain and loss of taste were reported, as well as mild nervous system symptoms (headache and dizziness) (23).

Organs and Systems

Sensory systems

Terbinafine causes taste loss in 0.6–2.8% of patients. However, many so-called taste problems actually reflect olfactory problems, and the sole empirical study published on this topic, based on whole-mouth testing of a single subject, found no terbinafine-related deficit. Using well-validated taste and smell tests, chemosensory function in six patients complaining of taste disturbance after terbinafine were assessed and compared with six age-, race-, and sex-matched controls (24). Taste function for sweet-, sour-, and bitter-tasting stimuli was significantly reduced in both the anterior and posterior lingual regions. For sodium chloride, the decrements were confined to the posterior region. Olfactory function was normal. These findings support anecdotal case reports of taste loss from terbinafine, demonstrate that all four major taste qualities are affected, and suggest that olfactory dysfunction is not involved.

Taste disturbance due to terbinafine is usually reversible, with a median time to recovery of 42 days. However, a 46-year-old woman had complete loss of taste after taking oral terbinafine, with persistent taste disturbances for 3 years after stopping the drug (25).

Hematologic

The projected rate of all blood dyscrasias associated with terbinafine has been estimated to be 32 per million patient-years (26). Pancytopenia has been reported (27).

Leukocytes
Neutropenia has been reported in patients taking terbinafine (26–29).

- A 55-year-old woman who was taking terbinafine and paroxetine presented with fever, diarrhea, and vomiting. A bone marrow biopsy showed overall reduced cellularity, and the aspirate showed a profound shift toward the production of immature myeloid cells, consistent with maturation arrest. Treatment consisted of withdrawal of all outpatient medications, hydration, intravenous fluids, broad-spectrum antibiotics, and G-CSF 5 μg/kg for 5 days. Mature granulocytes appeared in the peripheral blood on the fifth day in hospital, and she was discharged on the seventh hospital day with an absolute neutrophil count of $6.2 \times 10^9/l$. Paroxetine was resumed weeks after discharge from hospital without hematological toxicity over 6 months.
- A 60-year-old man presented with fever, oral mucositis, pedal cellulitis, and bacteremia after a 6-week course of

terbinafine 250 mg. He was taking concurrent yohimbine for impotence. Bone marrow examination showed a hypocellular marrow with myeloid maturation arrest. Treatment consisted of withdrawal of outpatient medications, broad-spectrum antibiotics, hydration, and G-CSF, and was ultimately successful. Yohimbine was resumed later without any adverse effects.
- A 42-year-old man presented with fever and granulocytopenia (absolute neutrophil count: $340 \times 10^6/l$; temperature: 39.5°C) after a 30-day course of oral terbinafine 250 mg/day for presumed onychomycosis (28). The granulocyte count recovered promptly after withdrawal of the drug and administration of G-CSF for 2 days.
- Agranulocytosis occurred in a 15-year-old who took terbinafine 250 mg/day for toenail onychomycosis and tinea pedis (30). This effect was noted 4 weeks after starting terbinafine and resolved within 1 week after its withdrawal.

Platelets
Thrombocytopenia has been attributed to terbinafine (31), and the incidence has been estimated at 1 in 200 000 patients (26).

- A 25-year-old Yemeni woman with familial-ethnic leukopenia developed thrombocytopenia with epistaxis after taking terbinafine 250 mg for 4 weeks (31). The platelet count recovered from a nadir of $63 \times 10^9/l$ to $314 \times 10^9/l$ after drug withdrawal.
- A 53-year-old woman developed severe thrombocytopenia after a 6-week course of terbinafine (250 mg/day) for onychomycosis (32). A bone marrow aspirate showed a normocellular marrow. She received a platelet transfusion and recovered after a short course of prednisolone.

Salivary glands

- A 38-year-old man presented with acute right otitis media and unrelated painless bilateral enlargement of the parotid glands 15 days after taking oral terbinafine for tinea cruris (33). He stopped taking terbinafine and 12 days later the swelling had significantly abated and completely disappeared 4 weeks later.

Gastrointestinal

In a multicenter, randomized, double-blind, parallel-group study, 63 patients aged 14–85 years with *Sporothrix schenckii* infections were treated with terbinafine 500 mg/day (n = 28) or 1000 mg/day (n = 35) (34). There were no cases of relapse after 24 weeks of follow-up with 1000 mg/day, compared with six relapses with 500 mg/day. Terbinafine was well tolerated; the frequency of drug-related adverse effects was slightly higher with 1000 mg/day (17/35 vs. 10/28). Two patients taking 1000 mg/day withdrew because of an adverse effect (gastrointestinal hemorrhage; increasing abdominal pain).

Liver

Minor abnormalities in liver function tests have been reported in up to 4% of patients taking oral terbinafine (35), but have not generally been considered clinically important (SEDA-16, 297) (SEDA-18, 7) (SEDA-18, 25).

Terbinafine can cause hepatitis, with an estimated rate of about 1 in 50 000 (36). Idiosyncratic reactions can lead to liver cell necrosis as well as cholestasis. Prolonged cholestatic hepatitis and liver failure have been reported (37–41).

- In four patients with cholestatic hepatitis associated with terbinafine, all presented with jaundice and direct hyperbilirubinemia, various other clinical signs of hepatitis, and mild to moderate rises in alkaline phosphatase and hepatic transaminase activities (36,41). Biopsies in two patients showed cellular infiltrates in the portal tracts and hepatocellular and canalicular cholestasis ($n = 1$) and hepatocyte degeneration ($n = 1$). In the two cases with long-term follow-up, hepatitis was reversible after withdrawal of terbinafine and liver tests normalized within 6 months.
- A 41-year-old man developed severe hepatic dysfunction following a 3.5-week course of terbinafine (250 mg/day) (42). He had marked pruritus, jaundice, malaise, anorexia, and loin pain. His serum bilirubin rose to a peak of 718 µmol/l with alkaline phosphatase 569 U/l, alanine transaminase 90 U/l, aspartate transaminase 63 U/l, and a prolonged prothrombin time of 21 seconds, unresponsive to vitamin K. Liver biopsy showed canalicular cholestasis consistent with a drug reaction. His symptoms resolved 11 months after drug withdrawal, and his liver function tests normalized after 15 months.
- A previously healthy 46-year-old man developed acute fulminant hepatitis following treatment with rabeprazole, citalopram hydrobromide, terbinafine, and a multivitamin formulation (43). Liver biopsy showed submassive centrilobular necrosis and intrahepatic cholestasis with florid bile duct proliferation.

In the last case, because of the similarity of the clinical, laboratory, and histological effects of omeprazole and lansoprazole, as previously reported, the authors concluded that the reaction in the second patient might have been caused by the proton pump inhibitor rather than terbinafine.

Two cases of severe cholestatic hepatitis with prominent eosinophilia associated with terbinafine have been reported.

- In the first case, symptoms developed within 4 weeks of starting terbinafine; the patient was treated successfully with glucocorticoids after partial responses to ursodeoxycholic acid and colestyramine and made a full recovery within 6 weeks (44).
- In the second case, liver dysfunction developed after a 7-day course of terbinafine. Terbinafine was withdrawn and ursodeoxycholic acid and ademethionine were given. Liver tests normalized 6 months later (45).

The authors of the first report speculated that terbinafine hepatotoxicity could be more than just an idiosyncratic reaction and that 7,7-dimethylhept-2-ene-4-ynal (TBF-A), the allylic aldehyde metabolite of terbinafine, may play a role in its pathogenesis (81).

Histological changes resembling acute cellular rejection have been attributed to terbinafine in a liver transplant patient (46).

- A 51-year-old Hispanic man developed raised liver enzymes about 5 years after orthotopic liver transplantation. A biopsy sample was interpreted as acute cellular rejection, and he was treated with increased immunosuppression. He had started to take terbinafine about 4 weeks before for onychomycosis, and it was withdrawn. However, he developed progressive jaundice, malaise, and nausea, and a second liver biopsy sample showed marked centrilobular cholestasis and severe bile duct damage, consistent with terbinafine hepatotoxicity. He was managed with supportive therapy and discharged 1 month after admission, but continued to have slightly raised transaminases and moderately raised bilirubin concentrations. He died about 3 months after terbinafine was withdrawn after suddenly collapsing in his home. No autopsy was performed.

Fatal hepatic veno-occlusive disease associated with terbinafine was reported in another liver transplant recipient (47).

- A 35-year-old man with familial amyloid polyneuropathy and orthotopic liver transplantation took terbinafine 250 mg/day for onychomycosis in addition to a stable regimen of ciclosporin. Six weeks later, he developed jaundice, nausea, and vomiting without liver enlargement or ascites. He had a markedly raised serum bilirubin and moderately raised hepatic transaminases; viral causes were excluded. Terbinafine was withdrawn and a liver biopsy showed subacute venous effluent obstruction and occlusion of central veins with intraluminal deposition of fibrous tissue, consistent with veno-occlusive disease. He died 9 weeks later with progressive hepatic and secondary multiorgan failure. Repeat biopsies showed progressive central vein sclerosis.

The temporal relation between the start of terbinafine and the onset of symptoms suggested a drug-induced adverse effect. However, a drug interaction with ciclosporin or transplantation-induced veno-occlusive disease could not be entirely excluded.

- A 56-year-old woman developed chronic biliary ductopenia and portal fibrosis 2 years after a course of terbinafine (48). Terbinafine treatment at that time had resulted in jaundice and evidence of cholestasis. After withdrawal of terbinafine, she continued to have pruritus and persistently raised serum alkaline phosphatase activity. Investigations for various types of chronic liver disease were negative and so chronic bile duct loss and periportal fibrosis were attributed to terbinafine.

Because of the rare and unpredictable nature of hepato-biliary reactions to terbinafine, the mechanism of hepato-toxicity has been hypothesized to be either immunological or metabolically mediated. A potentially toxic reactive metabolite of terbinafine, 7,7-dimethylhept-2-ene-4-ynal (TBF-A), the *N*-dealkylation product of ter-binafine, has been identified in vitro (49). The authors speculated that this allylic aldehyde metabolite, formed by liver enzymes and conjugated with glutathione, would be transported across the canalicular membrane of hepa-tocytes and concentrated in the bile. The reactive mono-glutathione conjugate could bind to hepatobiliary proteins and cause direct toxicity. Alternatively, it could modify canalicular proteins and lead to an immune-mediated reaction, causing cholestatic dysfunction.

Skin

Cutaneous adverse effects reportedly occur in 1–3% of patients taking terbinafine. The overwhelming majority of these reactions consist of mild to moderate macular exanthemas.

More serious skin disorders, such as erythema multi-forme, toxic epidermal necrolysis, Stevens–Johnson syn-drome, toxic erythema, cutaneous lupus erythematosus, and generalized pustular eruptions are rare (SEDA-23, 299; SEDA-24, 314). Five patients developed severe rashes while taking oral terbinafine (50).

The risk of serious skin disorders has been estimated in 61 858 users of oral antifungal drugs, aged 20–79 years, identified in the UK General Practice Research Database (51). They had received at least one prescription for oral fluconazole, griseofulvin, itraconazole, ketoconazole, or terbinafine. The background rate of serious cutaneous adverse reactions (corresponding to non-use of oral anti-fungal drugs) was 3.9 per 10 000 person-years (95% CI = 2.9, 5.2). Incidence rates for current use were 15 per 10 000 person-years (1.9, 56) for itraconazole, 11.1 (3.0, 29) for terbinafine, 10 (1.3, 38) for fluconazole, and 4.6 (0.1, 26) for griseofulvin. Cutaneous disorders asso-ciated with the use of oral antifungal drugs in this study were all mild.

Generalized rashes, fixed drug eruptions, toxic epider-molysis, and erythema exudativum multiforme have all been reported in association with terbinafine (52,53).

Pustular eruptions can occur occasionally (52–56).

- Acute generalized exanthematous pustulosis associated with terbinafine has been described in two patients (54,56). Both presented within 7–10 days after starting to take terbinafine with generalized pustular dermato-sis and leukocytosis; fever was a presenting symptom in one patient. Treatment with systemic corticosteroids was successful in both cases.
- A 62-year-old diabetic man on stable oral medication with glibenclamide, metformin, Zestoretic (lisinopril + hydrochlorothiazide), gemfibrozil, and aspirin devel-oped febrile generalized pustular eruptions after 44 days of therapy with oral terbinafine 250 mg/day (57). Withdrawal of terbinafine and symptomatic treatment with hydrotherapy and topical and systemic steroids

resulted in complete resolution of fever and pustulosis within 4 days. The erythematous component responded more slowly, and mildly pruritic erythematous plaques persisted for more than 40 days.
- In another case of acute generalized exanthematous pustulosis attributed to terbinafine (250 mg/day), epi-cutaneous and intracutaneous skin tests were negative, and there was no evidence of viral infection (58).

Acute generalized exanthematous pustulosis (AGEP) is considered to be a clinical reaction pattern, 90% of cases of which are due to systemic drugs. It is a rare presentation of an adverse drug reaction most frequently triggered by antimi-crobial drugs, including terbinafine (59,60). Acute general-ized exanthematous pustulosis attributed to terbinafine has been described in two cases (61,62).

- A 62-year-old woman, who had been taking predniso-lone 10 mg/day for 18 months for bullous pemphigoid, was given terbinafine 250 mg/day for a toenail infec-tion (63). After 8 days she developed a symmetrical papular eruption that began on the face and neck. The papules rapidly coalesced and within 2 days the rash became generalized, with targetoid lesions on the arms and pustules on the back. There was never any mucosal involvement. She was febrile, with a leukocytosis, neu-trophilia, and a slightly raised creatinine of 162 μmol/l. She continued to take prednisolone 10 mg/day. A skin biopsy showed small subcorneal pustules, with neutro-phils, underlying spongiosis, and an upper dermal lym-phocytic infiltrate. Bacterial swabs and fungal scrapings from the pustules were negative. The eruption settled within 12 days of terbinafine withdrawal.

According to the scoring system devised by Sidoroff et al. (64) this was probably a case of acute generalized exanthematous pustulosis. This case suggests that steroids may not be effective in preventing the condition.

- Drug-induced bullous pemphigoid occurred in an otherwise healthy 78-year-old patient who took terbi-nafine 250 mg/day for 20 days for onychomycosis of the toenails. He had a widespread pruritic blistering erup-tion, more severe on the limbs. He had not taken terbinafine before and was not taking any other drugs. The diagnosis of bullous pemphigoid was con-firmed by histopathology. He was treated successfully with oral glucocorticoids.

Use of the Naranjo probability scale indicated a probable relation between bullous pemphigoid and terbinafine in this patient (65).

Splinter hemorrhages developed in fingernails that were affected by Trichophyton rubrum infection 2 months after the start of therapy with oral terbinafine 250 mg/day in a 45-year-old Chinese man; the fingernails that were not infected were not involved (66). Three of the five fingernails resolved 3 months after switching to itracona-zole.

Terbinafine can cause or exacerbate psoriatic lesions.

- Severe pustular psoriasis provoked de novo by oral terbinafine has been reported in a 65-year-old man 2 weeks after the start of therapy for onychomycosis (67). Treatment of psoriasis was complicated and ultimately required continuous systemic and topical antipsoriatic therapy.

- A 74-year-old woman developed inverse psoriasis after 14 days of therapy with terbinafine 250 mg/day for onychomycosis (68). The lesions resolved almost completely on withdrawal of terbinafine and topical therapy.

Probable psoriatic onychodystrophy, misdiagnosed as onychomycosis and treated with terbinafine, induced inverse psoriasis in the second case, underscoring the importance of mycological confirmation of onychomycosis before therapy.

Ten cases of severe skin reactions probably associated with terbinafine requiring drug withdrawal have been reported: erythema multiforme ($n = 5$), erythroderma ($n = 1$), severe urticaria ($n = 1$), pityriasis rosea ($n = 1$), and worsening of pre-existing psoriasis ($n = 2$) (69). All the patients made an uneventful recovery with appropriate therapy. The authors pointed out that patients should be counselled about discontinuing terbinafine at the onset of a skin eruption and about seeking medical advice about further management.

There have been further reports of lupus erythematosus-like eruptions (70) and erythema multiforme (71) associated with oral terbinafine.

- Cutaneous lupus erythematosus attributed to terbinafine has been reported in two previously healthy women (72,73). In the first patient, the lesions improved but did not resolve completely; in the second the symptoms resolved completely with appropriate therapy and the patient remained disease-free after withdrawal of all medication
- Another woman with a previous history suggestive of lupus erythematosus developed a widespread flare in her skin 1 week after starting oral terbinafine (74). The eruption ultimately responded to systemic treatment with corticosteroids.

Baboon syndrome has been attributed to terbinafine (75).

- A 26-year-old man developed a fixed drug eruption on his hands and inguinal and gluteal areas after oral treatment of onychomycosis with terbinafine 250 mg/day. The rash showed the characteristic distribution of the baboon syndrome. Although epicutaneous and intracutaneous tests were negative, the rash recurred 20 hours after oral rechallenge with terbinafine.

The underlying pathogenic mechanism for the baboon syndrome has been suggested to be a systemically induced allergic contact dermatitis.

Hair

Hair loss has been attributed to terbinafine (76).

- A 69-year-old woman took terbinafine 250 mg/day for 112 days for subungual hyperkeratosis and developed hair loss after 3 months. She was also taking hydrochlorothiazide, amiloride hydrochloride, and amlodipine besilate, all in the same dosage for more than 5 years. Clinical and laboratory investigations showed no other obvious causes, and hair loss completely reversed on withdrawal of terbinafine.

Musculoskeletal

Since its introduction, the Netherlands Pharmacovigilance Foundation Lareb has received eight reports of arthralgia in patients taking terbinafine (77). In four cases, skin reactions were also present and in two cases urticaria. Two patients who reported arthralgia also had a fever. Logistic regression modelling showed that both urticaria and arthralgia were statistically significantly associated with reports on terbinafine compared with all other reports in the database. These findings may point toward a clustering of these symptoms in patients using terbinafine, suggesting a shared immunological reaction.

Rhabdomyolysis has been attributed to terbinafine (78).

- A 24-year-old man took terbinafine 250 mg/day for sycosis barbae for 15 days and developed weakness, myalgia, and dark urine. His creatine kinase activity was 1120 U/l, and renal function was normal. He stopped taking terbinafine and the creatine kinase activity returned to normal.

Immunologic

Exacerbation of lupus erythematosus has been reported during terbinafine therapy (72–74,79). Of 21 consecutive patients with subacute cutaneous lupus erythematosus who attended an outpatient dermatology department in Germany during 1 year, 4 had terbinafine-associated disease (80). In addition to high titers of antinuclear antibodies with a homogeneous pattern, anti-Ro(SS-A) antibodies were present; in three of the four women, anti-La(SS-B) antibodies were also found. All the patients had antihistone antibodies, as in drug-induced lupus, and showed the characteristic genetic association with the HLA-B8,DR3 haplotype; moreover, in two cases HLA-DR2 was also present. After withdrawal of terbinafine, antinuclear antibody titers fell and antihistone antibodies became undetectable within 4.5 months in three patients.

- A 66-year-old man with giant cell arteritis and hypertension developed a hypersensitivity reaction 4.5 weeks after starting to take terbinafine, with a skin eruption, fever, lymphadenopathy, and hepatic dysfunction (81). Concomitant medications included prednisone, doxazosin, and aspirin. His symptoms and signs resolved within 6 weeks after withdrawal of terbinafine and continuation of all the other medications. The hypersensitivity syndrome reaction in this case was idiosyncratic, with no apparent predisposing factors.

Subacute lupus erythematosus-like eruptions due to terbinafine have been reported in three women with onychomycosis. Two patients had serological evidence of autoimmune disease predisposing to photosensitivity; the third had neutropenia. Histology in two cases was consistent with subacute lupus erythematosus. All three patients had strikingly similar eruptions that took several weeks to resolve with topical and systemic glucocorticoids (82).

- A 59-year-old man with a history of cutaneous lupus erythematosus, which had been in complete remission for 5 years, took oral terbinafine for onychomycosis and after 1 month developed cutaneous lupus erythematosus, with typical papulosquamous lesions and a raised titer of antinuclear antibodies (83). Terbinafine was withdrawn and systemic and topical steroids were given; the eruption resolved over several weeks.
- A 50-year-old woman had a severe cutaneous flare-up of pre-existing systemic lupus erythematosus 7 weeks after starting treatment with terbinafine (84).
- A 25-year-old woman with stable systemic lupus erythematosus, who was taking corticosteroids and chloroquine 200 mg/day, was given terbinafine for onychomycosis and 7days later developed cheilitis and bilateral conjunctivitis followed by epidermolysis involving 10% of the skin (85). There was massive hematuria and proteinuria, and a renal biopsy showed lupus glomerulonephritis. Antihistone antibodies were strongly positive. She was given systemic corticosteroids, chloroquine, and cyclophosphamide and recovered.
- A 39-year-old women with systemic lupus erythematosus took terbinafine 250 mg/day for onychomycosis and after 7 days developed a widespread severe erythematous eruption (86). Clinical, histological, and immunofluorescent investigations confirmed the diagnosis of co-existing subacute cutaneous and systemic lupus erythematosus. Terbinafine was withdrawn.
- Three patients with sicca syndrome, lung carcinoma, and Kikuchi disease were given terbinafine for suspected dermatophytic infections; each developed subacute cutaneous lupus erythematosus within 7 weeks (87).

A rash with eosinophilia and systemic symptoms induced by terbinafine was associated with severe sialadenitis and a complete sicca syndrome (88). The rash resolved slowly after drug withdrawal, but the sicca syndrome persisted with only very mild improvement at 6 months.

Susceptibility Factors

Age

The pharmacokinetics of terbinafine and five known metabolites have been investigated in 12 children, (mean age 8 years; range 5–11 years), who took terbinafine 125 mg/day for 6–8 weeks for tinea capitis (15). The metabolism of terbinafine was similar to that observed in adults, and there were comparable steady-state plasma concentrations after administration of the same oral dose. Steady state was reached by day 21 with no further accumulation up to day 56. Terbinafine was effective in all patients and safe and well tolerated over 56 days.

In a randomized, double-blind comparison of terbinafine ($n = 27$) with itraconazole ($n = 28$) for 2 weeks for tinea capitis in Pakistani children (mean age 8 years), fever, body ache, and vertigo were seen with terbinafine in one patient each, and urticaria with itraconazole in two patients (89). There were no significant changes in hematological and biochemical profiles.

In an open assessment of the efficacy, safety, and tolerability of oral terbinafine 125–250 mg/day for 1, 2, and 4 weeks for tinea capitis in 132 Brazilian children aged 1–14 years, adverse events were reported in 10 patients (90). The drug was prematurely withdrawn in one patient. In the post-treatment evaluation, two patients had abnormal bilirubin concentrations and eight patients had abnormal alkaline phosphatase activities; none was considered clinically relevant.

In an open, non-comparative study of the use of terbinafine for 14 days to treat tinea capitis in 50 children and adolescents (mean age 7.6 years; range 24 months to 18 years), the clinical and mycological cure rates were 86%. The drug appeared to be well tolerated. Two children had reversible neutropenia, thought to be due to a preceding viral illness; other adverse effects were not observed (18).

In 14 children, aged 1–15 years, with *Microsporum canis* tinea capitis given oral terbinafine for 4 weeks, the recommended daily dose produced no response by week 4; the dose was doubled (to 250 mg/day) for a further 4–8 weeks in six patients and continued at the original dose in six patients (91). Two patients withdrew. Four patients were cured after 8–12 weeks of treatment, and all had taken the doubled dose of terbinafine, except for one who had taken the usual adult dose of 250 mg/day from the start. Oral terbinafine was well tolerated by all but one patient, who had gastrointestinal disturbance and slightly raised transaminase activities during the first 4 weeks of treatment.

In an open, prospective, uncontrolled study in 81 immunocompetent young children (aged 2–13 years) with tinea capitis due to *Microsporum canis*, oral terbinafine was given in dosages based on weight (62.5 mg for those weighing 10–20 kg and 125 mg for 20–40 kg), and applied topically to affected areas (1% cream bd) (92). Treatment lasted for 4 weeks, followed by an 8-week observation (treatment-free) period. All the subjects were assessed for efficacy and tolerability at 12 weeks. At 12 weeks: 32 had completely recovered, with no evidence of relapse during the observation period, and 21 had mycological cure but residual signs of infection. The effective cure rate was 65%. Terbinafine was well tolerated by these children: 75 had no adverse effects; the other six had abdominal pain ($n = 2$), vomiting ($n = 1$), generalized itching ($n = 1$), local itching ($n = 1$), and localized erythema ($n = 1$). Hematological and biochemical parameters remained normal during the study.

In 100 children with tinea capitis caused by *Microsporum canis* treated with oral terbinafine 3.3–12.5 mg/kg/day for 8 weeks, five patients withdrew because of adverse events (93).

The safety of terbinafine in children has been reviewed, including 989 children reported in 20 studies. In all, 106 patients had adverse events (11%). Only eight patients (0.8%) discontinued terbinafine. Adverse events included the gastrointestinal system (2.8%), skin (1.2%), and nervous system (0.9%). There were hematological and hepatic enzyme abnormalities in 1.3 and 1.8% of children respectively. The adverse events in children were similar to those in adults, and most of the events were mild and transient (94).

Limited data suggest that the safety profile of terbinafine in children is not different from that observed in adults and that terbinafine is well tolerated in this population over short periods of time. Terbinafine has been approved for the treatment of tinea capitis in many countries worldwide, and provides good efficacy rates for *Trichophyton* tinea capitis using shorter regimens than griseofulvin.

The single dose and steady-state pharmacokinetics of terbinafine have been investigated in 22 otherwise healthy children aged 4–8 years with tinea capitis were comparable between children and adults for the administered dose; however, Children had significantly lower values C_{max} and $AUC_{0 \to 24}$ when dose was corrected for weight (95). Age accounted for about 50% of the variability in dose-normalized C_{max} and AUC. Adverse events consisted principally of headache (n = 3) and gastrointestinal complaints (altered eating habits, n = 3; loss of appetite, n = 3; stomach ache, n = 4; diarrhea, n – 2). There was a reduced neutrophil count in five children, thought to be related to terbinafine in two cases.

Drug Administration

Drug dosage regimens

Pulse-dose terbinafine has been compared with standard continuous-dose terbinafine for toenail onychomycosis in a double-blind, randomized, single-center, non-inferiority study in 306 volunteers randomized to terbinafine 250 mg/day for 3 months (continuous) or 500 mg/day for 1 week per month for 3 months (pulse) (96). Continuous-dose terbinafine was more effective. The two regimens were equally well tolerated.

Drug–Drug Interactions

In a study of drugs that are co-prescribed with CYP2D6 inhibitors (bupropion, fluoxetine, paroxetine, and terbinafine) lists of patients taking both inhibitors and substrates of CYP2D6 were drawn from the prescription databases of three Norwegian primary pharmacies (97). The highest frequencies of co-prescribed substrates were found for *paroxetine* (101 events per 267 patients, 38%), and fluoxetine (36 events per 110 patients, 33%). The drugs that were most often detected in combination with the inhibitors were *codeine* (116 events) and metoprolol (38 events). The frequency of co-prescribed substrates was comparatively low (3/96), codeine being co-prescribed with terbinafine in all three cases.

Alfentanil

In a randomized crossover study in 12 healthy volunteers terbinafine had no statistically significant effect on the pharmacokinetics of intravenous alfentanil 20 micrograms/kg (98).

Carbamazepine

An interaction of terbinafine with carbamazepine has been described (99).

- A 50-year-old man taking carbamazepine 900 mg/day, bupropion 300 mg/day, and quetiapine 300 mg/day for bipolar affective disorder, among other drugs, was given terbinafine 250 mg/day. Within 2–3 days he felt dizzy, but continued to take his medications. Within 2 weeks, his terbinafine tablets ran out and his dizziness abated as well. About 1 month later he started to take terbinafine again and within 3 days his dizziness recurred, causing him to fall on two occasions. He also had blurred vision and double vision. There was nystagmus and numbness of the left face. A CT scan of the head showed no abnormality. His serum carbamazepine concentration was 73 µmol/l (usual target range 17–42).

The authors pointed out that terbinafine and carbamazepine are both metabolized by CYP3A4, CYP2C9, CYP1A2, CYP2C19, and CYP2C8, and suggested that any of those isoenzymes could have been involved in this interaction.

Rifamycins

Rifampicin 600 mg/day reduced terbinafine concentrations by about 50% by enzyme induction (100).

Theophylline

Theophylline is largely metabolized by CYP1A2 and terbinafine increases its half-life (101). In a randomized, crossover study in 12 healthy volunteers, terbinafine increased theophylline exposure by 16%, with a 14% reduction in clearance and a 24% increase in half-life (101). These pharmacokinetic changes may predispose individuals to accumulation of theophylline and unwanted toxicity. Caution should be taken in prescribing terbinafine for patients taking long-term theophylline.

Tricyclic antidepressants

Amitriptyline

- A 37-year-old white woman with normal CYP2D6 metabolic capacity taking amitriptyline, valproate, and olanzapine was given terbinafine and shortly after developed extreme dryness of the mouth, nausea, and dizziness, accompanied by a large increase in the serum concentrations of amitriptyline and nortriptyline (102). Terbinafine was withdrawn and the dose of amitriptyline was reduced. Surprisingly, the serum concentrations of amitriptyline and nortriptyline did not return to baseline until about 6 months later. Terbinafine is a highly potent competitive inhibitor of CYP2D6, an important intermediate enzyme in the metabolism of amitriptyline to nortriptyline. Nortriptyline is further metabolized to 10-hydroxy metabolites, mainly by CYP2D6. It is therefore likely that the concomitant use of terbinafine was the major cause of the increased serum concentrations of amitriptyline and nortriptyline. Based on the data of this case report, there is a risk of clinically significant drug-drug interactions for at least 3 months after withdrawal of terbinafine.

Desipramine

Inhibition of CYP2D6 by terbinafine has been evaluated by assessing 48-hour concentration-time profiles of the tricyclic antidepressant desipramine in 12 healthy volunteers identified as extensive CYP2D6 metabolizers by genotyping and phenotyping (103). The pharmacokinetics were evaluated at baseline (50 mg oral desipramine given alone), steady state (after 250 mg oral terbinafine for 21 days), and 2 and 4 weeks after terbinafine withdrawal. The pharmacodynamics were evaluated before and 2 hours after each dose of desipramine, using Mini-Mental Status Examination and electroencephalography. Terbinafine inhibited CYP2D6 metabolism, as indicated by significant increases in desipramine C_{max} and AUC and reductions in the C_{max} and AUC of the CYP2D6-mediated metabolite, 2-hydroxydesipramine, both of which were still altered 4 weeks after terbinafine withdrawal. Caution should be exercised when co-prescribing terbinafine and drugs that are metabolized by CYP2D6, particularly those with a narrow therapeutic index.

Imipramine

- A 51-year-old patient developed imipramine toxicity and increased plasma concentrations associated with the introduction of terbinafine, possibly due to inhibition of CYP2D6 (104).

Nortriptyline

Metabolism by CYP2D6 is of major importance for the hydroxylation of nortriptyline, making it susceptible to competitive inhibition by terbinafine. Nortriptyline intoxication provoked by terbinafine has been reported (105).

- A 74-year-old man taking a stable dose of nortriptyline for depression developed signs of nortriptyline intoxication 14 days after he started to take terbinafine (105). Nortriptyline serum concentrations were several times higher than the usual target range and fell to baseline after withdrawal of terbinafine. Re-challenge led to the same clinical and laboratory findings.
- Nortriptyline intoxication secondary to terbinafine has been observed in a woman with a major depressive disorder (106). After rechallenge her serum nortriptyline concentration rose and the serum concentrations of its two hydroxylated metabolites fell. She had a normal genotype for CYP2D6, suggesting that this

interaction can occur even in people without reduced CYP2D6 activity.

Warfarin

While terbinafine had no effect on warfarin in healthy volunteers, it can prolong the prothrombin time in some individuals (107-110), prompting intensification of laboratory control during terbinafine therapy.

- A 71-year-old woman taking a stable dose of warfarin and cimetidine was treated with terbinafine, and 32 days later developed profuse intestinal bleeding associated with a prothrombin time of 120 seconds, suggestive of an interaction between warfarin and terbinafine, either directly or through the mediation of cimetidine (which can reduce terbinafine clearance by 33%) (107).

However, a contrasting case has also been reported.

- A 68-year-old woman taking warfarin, glibenclamide, metformin, furosemide, and spironolactone was given terbinafine 250 mg/day and 4 weeks later required progressive increases in the warfarin dosage to maintain a therapeutic INR; after withdrawal of terbinafine, her warfarin requirements returned to baseline over 4 weeks, supporting enzyme induction with gradual onset and offset (108).

Since a pharmacokinetic study of a single dose of warfarin in 26 healthy volunteers treated with terbinafine showed no significant interaction, and since a large postmarketing study of terbinafine did not find any cases of interaction of warfarin with terbinafine, the manufacturers (111) and others (110) have cautioned about any generalization regarding an interaction between terbinafine and warfarin.

References

1. Shear NH, Villars VV, Marsolais C. Terbinafine: an oral and topical antifungal agent. Clin Dermatol 1991;9(4):487–95.
2. Balfour JA, Faulds D. Terbinafine. A review of its pharmacodynamic and pharmacokinetic properties, and therapeutic potential in superficial mycoses. Drugs 1992;43(2):259–84Erratum in: Drugs 1992;43(5):699.
3. Finlay AY. Pharmacokinetics of terbinafine in the nail. Br J Dermatol 1992;126(Suppl 39):28–32.
4. Finlay AY. Global overview of Lamisil. Br J Dermatol 1994;130(Suppl 43):1–3.
5. Jones TC. Overview of the use of terbinafine (Lamisil) in children. Br J Dermatol 1995;132(5):683–9.
6. Gupta AK, Shear NH. Terbinafine: an update. J Am Acad Dermatol 1997;37(6):979–88.
7. Faergemann J, Zehender H, Jones T, Maibach I. Terbinafine levels in serum, stratum corneum, dermis–epidermis (without stratum corneum), hair, sebum and eccrine sweat. Acta Dermatol Venereol 1991;71(4):322–6.
8. Kovarik JM, Kirkesseli S, Humbert H, Grass P, Kutz K. Dose-proportional pharmacokinetics of terbinafine and its N-demethylated metabolite in healthy volunteers. Br J Dermatol 1992;126(Suppl 39):8–13.
9. Kovarik JM, Mueller EA, Zehender H, Denouel J, Caplain H, Millerioux L. Multiple-dose pharmacokinetics

and distribution in tissue of terbinafine and metabolites. Antimicrob Agents Chemother 1995;39(12):2738–41.

10. Nedelman JR, Gibiansky E, Robbins BA, Cramer JA, Riefler JF, Lin T, Meligeni JA. Pharmacokinetics and pharmacodynamics of multiple-dose terbinafine. J Clin Pharmacol 1996;36(5):452–61.

11. Wildfeuer A, Faergemann J, Laufen H, Pfaff G, Zimmermann T, Seidl HP, Lach P. Bioavailability of fluconazole in the skin after oral medication. Mycoses 1994;37(3–4):127–30.

12. Faergemann J. Pharmacokinetics of fluconazole in skin and nails. J Am Acad Dermatol 1999;40(6 Pt 2):S14–20.

13. Humbert H, Cabiac MD, Denouel J, Kirkesseli S. Pharmacokinetics of terbinafine and of its five main metabolites in plasma and urine, following a single oral dose in healthy subjects. Biopharm Drug Dispos 1995;16(8):685–94.

14. Matsumoto T, Tanuma H, Kaneko S, Takasu H, Nishiyama S. Clinical and pharmacokinetic investigations of oral terbinafine in patients with tinea unguium. Mycoses 1995;38(3–4):135–44.

15. Humbert H, Denouel J, Cabiac MD, Lakhdar H, Sioufi A. Pharmacokinetics of terbinafine and five known metabolites in children, after oral administration. Biopharm Drug Dispos 1998;19(7):417–23.

16. Vickers AE, Sinclair JR, Zollinger M, Heitz F, Glanzel U, Johanson L, Fischer V. Multiple cytochrome P-450s involved in the metabolism of terbinafine suggest a limited potential for drug–drug interactions. Drug Metab Dispos 1999;27(9):1029–38.

17. Nejjam F, Zagula M, Cabiac MD, Guessous N, Humbert H, Lakhdar H. Pilot study of terbinafine in children suffering from tinea capitis: evaluation of efficacy, safety and pharmacokinetics. Br J Dermatol 1995;132(1):98–105.

18. Krafchik B, Pelletier J. An open study of tinea capitis in 50 children treated with a 2-week course of oral terbinafine. J Am Acad Dermatol 1999;41(1):60–3.

19. De Backer M, De Vroey C, Lesaffre E, Scheys I, De Keyser P. Twelve weeks of continuous oral therapy for toenail onychomycosis caused by dermatophytes: a double-blind comparative trial of terbinafine 250 mg/day versus itraconazole 200 mg/day J Am Acad Dermatol 1998;38(5 Pt 3):S57–63.

20. Degreef H, del Palacio A, Mygind S, Ginter G, Pinto Soares A, Zuluaga de Cadena A. Randomized double-blind comparison of short-term itraconazole and terbinafine therapy for toenail onychomycosis. Acta Dermatol Venereol 1999;79(3):221–3.

21. Evans EG, Sigurgeirsson BThe LION Study Group. Double blind, randomised study of continuous terbinafine compared with intermittent itraconazole in treatment of toenail onychomycosis. BMJ 1999;318(7190):1031–5.

22. Sigurgeirsson B, Billstein S, Rantanen T, Ruzicka T, di Fonzo E, Vermeer BJ, Goodfield MJ, Evans EG. L.I.ON. Study: efficacy and tolerability of continuous terbinafine (Lamisil) compared to intermittent itraconazole in the treatment of toenail onychomycosis. Lamisil vs. Itraconazole in Onychomycosis Br J Dermatol 1999;141(Suppl 56):5–14.

23. O'Sullivan DP, Needham CA, Bangs A, Atkin K, Kendall FD. Postmarketing surveillance of oral terbinafine in the UK: report of a large cohort study. Br J Clin Pharmacol 1996;42(5):559–65.

24. Doty RL, Haxel BR. Objective assessment of terbinafine-induced taste loss. Laryngoscope 2005;115(11):2035–7.

25. Bong JL, Lucke TW, Evans CD. Persistent impairment of taste resulting from terbinafine. Br J Dermatol 1998;139(4):747–8.

26. Gupta AK, Soori GS, Del Rosso JQ, Bartos PB, Shear NH. Severe neutropenia associated with oral terbinafine therapy. J Am Acad Dermatol 1998;38(5 Pt 1):765–7.

27. Kovacs MJ, Alshammari S, Guenther L, Bourcier M. Neutropenia and pancytopenia associated with oral terbinafine. J Am Acad Dermatol 1994;31(5 Pt 1):806.

28. Shapiro M, Li LJ, Miller J. Terbinafine-induced neutropenia. Br J Dermatol 1999;140(6):1196–997.

29. Ornstein DL, Ely P. Reversible agranulocytosis associated with oral terbinafine for onychomycosis. J Am Acad Dermatol 1998;39(6):1023–4.

30. Aguilar C, Mueller KK. Reversible agranulocytosis associated with oral terbinafine in a pediatric patient. J Am Acad Dermatol 2001;45(4):632–4.

31. Grunwald MH. Thrombocytopenia associated with oral terbinafine. Int J Dermatol 1998;37(8):634.

32. Tsai HH, Lee WR, Hu CH. Isolated thrombocytopenia associated with oral terbinafine. Br J Dermatol 2002;147(3):627–8.

33. Torrens JK, McWhinney PH. Parotid swelling and terbinafine. BMJ 1998;316(7129):440–1.

34. Chapman SW, Pappas P, Kauffmann C, Smith EB, Dietze R, Tiraboschi-Foss N, Restrepo A, Bustamante AB, Opper C, Emady-Azar S, Bakshi R. Comparative evaluation of the efficacy and safety of two doses of terbinafine (500 and 1000 mg day-1) in the treatment of cutaneous or lymphocutaneous sporotrichosis. Mycoses 2004;47:62-8.

35. van der Schroeff JG, Cirkel PK, Crijns MB, Van Dijk TJ, Govaert FJ, Groeneweg DA, Tazelaar DJ, De Wit RF, Wuite J. A randomized treatment duration-finding study of terbinafine in onychomycosis. Br J Dermatol 1992;126(Suppl 39):36–9.

36. Gupta AK, del Rosso JQ, Lynde CW, Brown GH, Shear NH. Hepatitis associated with terbinafine therapy: three case reports and a review of the literature. Clin Exp Dermatol 1998;23(2):64–7.

37. van 't Wout JW, Herrmann WA, de Vries RA, Stricker BH. Terbinafine-associated hepatic injury. J Hepatol 1994;21(1):115–7.

38. Lazaros GA, Papatheodoridis GV, Delladetsima JK, Tassopoulos NC. Terbinafine-induced cholestatic liver disease. J Hepatol 1996;24(6):753–6.

39. Mallat A, Zafrani ES, Metreau JM, Dhumeaux D. Terbinafine-induced prolonged cholestasis with reduction of interlobular bile ducts. Dig Dis Sci 1997;42(7):1486–8.

40. Agarwal K, Manas DM, Hudson M. Terbinafine and fulminant hepatic failure. N Engl J Med 1999;340(16):1292–3.

41. Fernandes NF, Geller SA, Fong TL. Terbinafine hepatotoxicity: case report and review of the literature. Am J Gastroenterol 1998;93(3):459–60.

42. Chambers WM, Millar A, Jain S, Burroughs AK. Terbinafine-induced hepatic dysfunction. Eur J Gastroenterol Hepatol 2001;13(9):1115–8.

43. Johnstone D, Berger C, Fleckman P. Acute fulminant hepatitis after treatment with rabeprazole and terbinafine. Arch Intern Med 2001;161(13):1677–8.

44. Ajit C, Suvannasankha A, Zaeri N, Munoz SJ. Terbinafine-associated hepatotoxicity. Am J Med Sci 2003;325:292-5.

45. Zapata Garrido AJ, Romo AC, Padilla FB. Terbinafine hepatotoxicity. A case report and review of literature. Ann Hepatol 2003;2:47-51.

46. Lovell MO, Speeg KV, Havranek RD, Sharkey FE. Histologic changes resembling acute rejection in a liver transplant patient treated with terbinafine. Hum Pathol 2003;34:187-9.

47. Walter RB, Lukaschek J, Renner EL, Mullhaupt B, Bachli EB. Fatal hepatic veno-occlusive disease associated with terbinafine in a liver transplant recipient. J Hepatol 2003;38:373-4.

48. Anania FA, Rabin L. Terbinafine hepatotoxicity resulting in chronic biliary ductopenia and portal fibrosis. Am J Med 2002;112(9):741–2.

49. Iverson SL, Uetrecht JP. Identification of a reactive metabolite of terbinafine: insights into terbinafine-induced hepatotoxicity. Chem Res Toxicol 2001;14(2):175–81.

50. Danielsen AG, Thomsen JS, Svejgaard EL. Medikamentelt eksantem udlost af terbinafin. [Severe skin rash in patients treated with terbinafine.] Ugeskr Laeger. 2006;168(44):3825-6.

51. Castellsague J, Garcia-Rodriguez LA, Duque A, Perez S. Risk of serious skin disorders among users of oral antifungals: a population-based study. BMC Dermatol 2002;2(1):14.

52. Carstens J, Wendelboe P, Sogaard H, Thestrup-Pedersen K. Toxic epidermal necrolysis and erythema multiforme following therapy with terbinafine. Acta Dermatol Venereol 1994;74(5):391–2.

53. Munn SE, Russell Jones R. Terbinafine and fixed drug eruption. Br J Dermatol 1995;133(5):815–6.

54. Condon CA, Downs AM, Archer CB. Terbinafine-induced acute generalized exanthematous pustulosis. Br J Dermatol 1998;138(4):709–10.

55. Kempinaire A, De Raeve L, Merckx M, De Coninck A, Bauwens M, Roseeuw D. Terbinafine-induced acute generalized exanthematous pustulosis confirmed by a positive patch-test result. J Am Acad Dermatol 1997;37(4):653–5.

56. Papa CA, Miller OF. Pustular psoriasiform eruption with leukocytosis associated with terbinafine. J Am Acad Dermatol 1998;39(1):115–7.

57. Bennett ML, Jorizzo JL, White WL. Generalized pustular eruptions associated with oral terbinafine. Int J Dermatol 1999;38(8):596–600.

58. Rogalski C, Hurlimann A, Burg G, Wuthrich B, Kempf W. Arzneimittelreaktion auf Terbinafin unter dem bild einer akuten generalisierten exanthematischen pustulose (AGEP). [Drug reaction to terbinafine simulating acute generalized exanthematous pustulosis.] Hautarzt 2001;52(5):444–8.

59. Beltraminelli HS, Lerch M, Arnold A, Bircher AJ, Haeusermann P. Acute generalized exanthematous pustulosis induced by the antifungal terbinafine: case report and review of the literature. Br J Dermatol 2005;152(4):780-3.

60. Gréco M, Plantin P. Acute generalized exanthematous pustulosis (AGEP) induced by terbinafine with involuntary positive reintroduction. Eur J Dermatol 2005;15(2):116.

61. Lombardo M, Cerati M, Pazzaglia A. Acute generalized exanthematous pustulosis induced by terbinafine. J Am Acad Dermatol 2003;49:158-9.

62. Taberner R, Puig L, Gilaberte M, Alomar A. Acute generalized exanthematous pustulosis induced by terbinafine. Eur J Dermatol 2003;13:313-4.

63. Bajaj V, Simpson N. Oral corticosteroids did not prevent AGEP due to terbinafine. Acta Derm Venereol 2006;86(5):448-9.

64. Sidoroff A, Halevy S, Bavinck JNB, Vaillant L, Roujeau J-C. Acute generalized exanthematous pustulosis (AGEP)— a clinical reaction pattern. J Cutan Pathol 2001;28(3):113-9.

65. Aksakal BA, Ozsoy E, Arnavut O, Ali Gurer M. Oral terbinafine-induced bullous pemphigoid. Ann Pharmacother 2003;37:1625-7.

66. Tan C, Zhu WY. Splinter haemorrhages associated with oral terbinafine in a Chinese man. Clin Exp Dermatol 2006;31(1):153-4.

67. Wilson NJ, Evans S. Severe pustular psoriasis provoked by oral terbinafine. Br J Dermatol 1998;139(1):168.

68. Pauluzzi P, Boccucci N. Inverse psoriasis induced by terbinafine. Acta Derm Venereol 1999;79(5):389.

69. Gupta AK, Lynde CW, Lauzon GJ, Mehlmauer MA, Braddock SW, Miller CA, Del Rosso JQ, Shear NH. Cutaneous adverse effects associated with terbinafine therapy: 10 case reports and a review of the literature. Br J Dermatol 1998;138(3):529–32.

70. McKellar G, Porter D, Burden D Terbinafine as a cause of cutaneous lupus erythematosus. Rheumatology (Oxford) 2004;43:249.

71. Carducci M, Latini A, Acierno F, Amantea A, Capitanio B, Santucci B. Erythema multiforme during cytomegalovirus infection and oral therapy with terbinafine: a virus-drug interaction. J Eur Acad Dermatol Venereol 2004;18:201-3.

72. Brooke R, Coulson IH, al-Dawoud A. Terbinafine-induced subacute cutaneous lupus erythematosus. Br J Dermatol 1998;139(6):1132–3.

73. Murphy M, Barnes L. Terbinafine-induced lupus erythematosus. Br J Dermatol 1998;138(4):708–9.

74. Holmes S, Kemmett D. Exacerbation of systemic lupus erythematosus induced by terbinafine. Br J Dermatol 1998;139(6):1133.

75. Weiss JM, Mockenhaupt M, Schopf E, Simon JC. Fixes Arzneimittelexanthem auf Terbinafin mit charakteristischen Verteilungsmuster eines Baboon–Syndroms. [Reproducible drug exanthema to terbinafine with characteristic distribution of baboon syndrome.] Hautarzt 2001;52(12):1104–6.

76. Richert B, Uhoda I, De la Brassinne M. Hair loss after terbinafine treatment. Br J Dermatol 2001;145(5):842.

77. van Puijenbroek EP, Egberts AC, Meyboom RH, Leufkens HG. Association between terbinafine and arthralgia, fever and urticaria: symptoms or syndrome? Pharmacoepidemiol Drug Saf 2001;10(2):135–42.

78. Møller M, Bygum A. Terbinafinudlost subakut kutan lupus erythematosus. [Cutaneous lupus erythematosus induced by terbinafine.] Ugeskr Laeger 2006;168(50):4427-8.

79. Schilling MK, Eichenberger M, Maurer CA, Sigurdsson G, Buchler MW. Ketoconazole and pulmonary failure after esophagectomy: a prospective clinical trial. Dis Esophagus 2001;14(1):37–40.

80. Bonsmann G, Schiller M, Luger TA, Stander S. Terbinafine-induced subacute cutaneous lupus erythematosus. J Am Acad Dermatol 2001;44(6):925–31.

81. Gupta AK, Porges AJ. Hypersensitivity syndrome reaction to oral terbinafine. Australas J Dermatol 1998;39(3):171–2.

82. Hill VA, Chow J, Cowley N, Marsden RA. Subacute lupus erythematosus-like eruption due to terbinafine: report of three cases. Br J Dermatol 2003;148:1056.

83. Amitay-Layish I, Feuerman H, David M. [Subacute cutaneous lupus erythematosus induced by terbinafine.] Harefuah 2006;145(7):480-2, 552.

84. Gallego Peris A, Sanfélix Gimeno G, Palop Larrea V, Sanfélix Genovés J. Rabdomiolisis y terbinafina.

[Rhabdomyolisis and terbinafine.] Med Clin (Barc). 2006 Nov 25;127(20):799.

85. Terrab Z, El Ouazzani T, Zouhair K, El Kabli H, Lakhdar H. Syndrome de Stevens–Johnson et aggravation d'un lupus systémique induits par la terbinafine. [Terbinafine-induced Stevens–Johnson syndrome and aggravation of systemic lupus erythematosus.] Ann Dermatol Venereol. 2006 May;133(5 Pt 1):463-6.

86. Cetkovská P, Pizinger K. Coexisting subacute and systemic lupus erythematosus after terbinafine administration: successful treatment with mycophenolate mofetil. Int J Dermatol 2006;45(3):320-2.

87. Farhi D, Viguier M, Cosnes A, Reygagne P, Dubertret L, Revuz J, Roujeau JC, Bachelez H. Terbinafine-induced subacute cutaneous lupus erythematosus. Dermatology 2006;212(1):59-65.

88. Abecassis S, Roujeau JC, Bocquet H, Copie-Bergman C, Radier C, Revuz J, Cosnes A. Severe sialadenitis: a new complication of drug reaction with eosinophilia and systemic symptoms. J Am Acad Dermatol 2004;51:827-30.

89. Jahangir M, Hussain I, Ul Hasan M, Haroon TS. A double-blind, randomized, comparative trial of itraconazole versus terbinafine for 2 weeks in tinea capitis. Br J Dermatol 1998;139(4):672-4.

90. Filho ST, Cuce LC, Foss NT, Marques SA, Santamaria JR. Efficacy, safety and tolerability of terbinafine for tinea capitis in children: Brazilian multicentric study with daily oral tablets for 1, 2 and 4 weeks. J Eur Acad Dermatol Venereol 1998;11(2):141-6.

91. Koumantaki E, Kakourou T, Rallis E, Riga P, Georgalla S. Doubled dose of oral terbinafine is required for *Microsporum canis* tinea capitis. Pediatr Dermatol 2001;18(4):339-42.

92. Silm H, Karelson M. Terbinafine: efficacy and tolerability in young children with tinea capitis due to *Microsporum canis*. J Eur Acad Dermatol Venereol 2002;16(3):228-30.

93. Devliotou-Panagiotidou D, Koussidou-Eremondi TH. Efficacy and tolerability of 8 weeks' treatment with terbinafine in children with tinea capitis caused by Microsporum canis: a comparison of three doses. J Eur Acad Dermatol Venereol 2004;18:155-9.

94. Gupta AK, Adamiak A, Cooper EA. The efficacy and safety of terbinafine in children. J Eur Acad Dermatol Venereol 2003;17:627-40.

95. Abdel-Rahman SM, Herron J, Fallon-Friedlander S, Hauffe S, Horowitz A, Riviere GJ. Pharmacokinetics of terbinafine in young children treated for tinea capitis. Pediatr Infect Dis J 2005;24(10):886-91.

96. Warshaw EM, Fett DD, Bloomfield HE, Grill JP, Nelson DB, Quintero V, Carver SM, Zielke GR, Lederle FA. Pulse versus continuous terbinafine for onychomycosis: a randomized, double-blind, controlled trial. J Am Acad Dermatol 2005;53(4):578-84.

97. Molden E, Garcia BH, Braathen P, Eggen AE. Co-prescription of cytochrome P450 2D6/3A4 inhibitor-substrate pairs in clinical practice. A retrospective analysis of data from Norwegian primary pharmacies. Eur J Clin Pharmacol 2005;61(2):119-25.

98. Saari TI, Laine K, Leino K, Valtonen M, Neuvonen PJ, Olkkola KT. Voriconazole, but not terbinafine, markedly reduces alfentanil clearance and prolongs its half-life. Clin Pharmacol Ther 2006;80(5):502-8.

99. Baath NS, Hong J, Sattar SP. Possible carbamazepine toxicity with terbinafine. Can J Clin Pharmacol 2006;13(2):e228-31.

100. Jensen JC. Pharmacokinetics of Lamisil in humans. J Dermatol Treat 1990;1(Suppl 2):15.

101. Trepanier EF, Nafziger AN, Amsden GW. Effect of terbinafine on theophylline pharmacokinetics in healthy volunteers. Antimicrob Agents Chemother 1998;42(3):695-7.

102. Castberg I, Helle J, Aamo TO. Prolonged pharmacokinetic drug interaction between terbinafine and amitriptyline. Ther Drug Monit 2005;27(5):680-2.

103. Madani S, Barilla D, Cramer J, Wang Y, Paul C. Effect of terbinafine on the pharmacokinetics and pharmacodynamics of desipramine in healthy volunteers identified as cytochrome P450 2D6 (CYP2D6) extensive metabolizers. J Clin Pharmacol 2002;42(11):1211-8.

104. Teitelbaum ML, Pearson VE. Imipramine toxicity and terbinafine. Am J Psychiatry 2001;158(12):2086.

105. van der Kuy PH, Hooymans PM. Nortriptyline intoxication induced by terbinafine. BMJ 1998;316(7129):441.

106. Van Der Kuy PH, Van Den Heuvel HA, Kempen RW, Vanmolkot LM. Pharmacokinetic interaction between nortriptyline and terbinafine. Ann Pharmacother 2002;36(11):1712-4.

107. Gupta AK, Ross GS. Interaction between terbinafine and warfarin. Dermatology 1998;196(2):266-7.

108. Warwick JA, Corrall RJ. Serious interaction between warfarin and oral terbinafine. BMJ 1998;316(7129):440.

109. Guerret M, Francheteau P, Hubert M. Evaluation of effects of terbinafine on single oral dose pharmacokinetics and anticoagulant actions of warfarin in healthy volunteers. Pharmacotherapy 1997;17(4):767-73.

110. Clarke MF, Boardman HS. Interaction between warfarin and oral terbinafine. Systematic review of interaction profile of warfarin is needed. BMJ 1998;317(7152):205-6.

111. Gantmacher J, Mills-Bomford J, Williams T. Interaction between warfarin and oral terbinafine. Manufacturer does not agree that interaction was with terbinafine. BMJ 1998;317(7152):205.

Voriconazole

See also Antifungal azoles

General Information

Voriconazole is a triazole that is structurally related to fluconazole. It has two interesting properties, activity against *Aspergillus* species and sufficient solubility in water to be administered parenterally. It is active against a wide spectrum of clinically important fungi, including *Candida* species, *Trichosporon beigelii*, *Cryptococcus neoformans*, *Aspergillus* species, *Fusarium* species, and other hyaline, dematiaceous, and dimorphic molds (1,2). It has demonstrated efficacy in various animal models of invasive fungal infections (1,2,3,4). It undergoes complex hepatic metabolism and has the potential for drug-drug interactions mediated by CYP 3A4, CYP2C9, and CYP2C19.

Voriconazole causes more liver function test abnormalities than fluconazole, and photosensitization, hallucinations, and visual disturbances can occur (5).

Pharmacokinetics

Voriconazole has a half-life of about 6 hours and undergoes complex hepatic metabolism. There is wide intersubject variability in the disposition of voriconazole, which is at least partly related to a CYP2C19 polymorphism. Voriconazole has the potential for drug–drug interactions mediated by CYP3A4, CYP2C9, and CYP2C19.

The kinetics of voriconazole are non-linear, as shown in 42 healthy men (6). Two groups of subjects participated. Group 1 ($n = 28$) took part in two study periods, each consisting of 14 days separated by a minimum 7-day washout period. During one of the periods, 14 subjects received intravenous voriconazole 6 mg/kg bd on day 1 followed by 3 mg/kg bd on days 2–7 and were then switched to 200 mg bd orally on days 8–14. During the other period, they received 6 mg/kg bd intravenously on day 1 followed by 5 mg/kg bd on days 2–7 and were then switched to 400 mg orally bd on days 8–14. The other 14 subjects in group 1 received a matching placebo throughout the study. In group 2 ($n = 14$), 7 subjects received 6 mg/kg intravenously bd on day 1 followed by 4 mg/kg bd on days 2–7 and were then switched to 300 mg bd orally on days 8–14. The other seven received a matching placebo. Voriconazole had non-linear pharmacokinetics, attributed to saturable metabolism. For intravenous dosing, a 1.7-fold increase in dose resulted in 2.4- and 3.1-fold increases in C_{max} and AUC respectively; a two-fold increase in oral dosing resulted in 2.8- and 3.9-fold increases in C_{max} and AUC respectively. The mean C_{max} after oral dosing was 63–90% of the intravenous C_{max}. After the switch from intravenous to oral dosing, most subjects achieved steady state by day 4. Voriconazole was well tolerated; the most commonly reported adverse events were mild to moderate headache, rash, and abnormal vision. Visual function tests conducted on all subjects detected no further abnormalities during voriconazole treatment and one abnormality (abnormal color vision test) during placebo treatment. All visual disturbances were mild to moderate in intensity, and all resolved spontaneously within 2 days of onset. No subject in any treatment group had a serious adverse event.

Observational studies

The efficacy of voriconazole has been demonstrated in non-comparative phase 1/2 studies in patients with oropharyngeal and esophageal candidiasis and acute and chronic invasive aspergillosis (7). In phase III clinical trials, it was superior to conventional amphotericin for first-line therapy of invasive aspergillosis and yielded comparable success rates but less proven and probable breakthrough infections compared with liposomal amphotericin as empirical antifungal therapy in patients with persistent neutropenia (8).

Voriconazole had excellent clinical efficacy in a non-comparative phase 1/2 studies in patients with oropharyngeal and esophageal candidiasis and acute and chronic invasive aspergillosis (9).

In an open, non-comparative, multicenter study in immunocompromised patients with proven or probable invasive aspergillosis, 116 patients were treated with intravenous voriconazole 6 mg/kg bd twice and then 3 mg/kg bd for 6–27 days, followed by 200 mg bd orally for up to 24 weeks; voriconazole was given as primary therapy in 60 (10). There were good responses in 56; 16 had a complete response and 40 a partial response; there was a stable response in 24 patients. There were adverse events in 91% of the patients who received at least one dose of voriconazole, but only 15% were attributed to the drug. The most common adverse events attributed to voriconazole were skin rash (8.7%), reversible visual disturbances (11%), and raised liver function tests (15%). There was evidence of a concentration-dependent incidence of adverse events: six of seven patients with voriconazole plasma concentrations over 10 µg/ml developed adverse events requiring drug withdrawal.

The efficacy, tolerability, and safety of voriconazole have been studied in 301 patients (11). Intravenous voriconazole was given in a loading dose of 6 mg/kg bd for the first 24 hours, followed by 4 mg/kg bd for at least 3 days, after which patients could switch to oral voriconazole 200 mg bd. Oral voriconazole was given in a dose of 400 mg bd on the first day, followed by 200 mg bd. Voriconazole was given intravenously for a median of 18 (range 1–138) days and orally for a median of 69 (range, 1?326) days. The efficacy rates for voriconazole were 44% for aspergillosis, 58% for candidiasis, 39% for cryptococcosis, 46% for fusariosis, and 30% for scedosporiosis. The most common treatment-related adverse effects were visual abnormalities (26%), rash (7.5%), nausea (6.5%), vomiting (4.8%), and headache (4.6%). There were abnormal increases in liver function tests in up to 11% of patients with normal baseline values and in up to 29% of patients with abnormal baseline values. Treatment had to be withdrawn in 8.5% of patients.

In 52 patients with invasive candidiasis intolerant of other antifungal agents or with infection refractory to other antifungal agents the median duration of voriconazole therapy was 60 (range 1–314) days, and the median dosage was 400 mg/day (7.48 mg/kg/day). The overall rate of response was 56% (95% CI = 41?70; *Candida albicans* 44%; *Candida glabrata* 38%; *Candida krusei* 70%; *Candida tropicalis* 67%; and other *Candida* species 100%) (12). The response rate in patients who had failed previous azole therapy was 58%. Common adverse effects included nausea and vomiting (25%), abnormal liver enzymes (23%), visual disturbances (21%), rash (15%), dysrhythmias (13%), and abdominal pain (12%). There were serious adverse effects in four patients.

The antifungal efficacy and safety of voriconazole in patients with invasive fungal infections intolerant to or progressive despite standard therapy have been analysed in 45 patients, 38 of whom had invasive *Aspergillus*, three had *Fusarium*, and two had *Scedosporium*, affecting the lungs (n = 26), the central nervous system (n = 5), the sinuses (n = 3), or more than one body site (n = 9) (13). The most common underlying illnesses were solid organ transplantation (n = 13), bone marrow transplantation (n = 11), and hematological malignancies (n = 7). The median duration of voriconazole therapy was 79 days and nine patients took it for over 1 year. Four patients withdrew because of adverse effects: abnormal liver function tests (n = 4), rash (n = 1), and atrial fibrillation (n = 1). There were no significant visual abnormalities.

Comparative studies

Amphotericin

Voriconazole has been compared with liposomal amphotericin for empirical antifungal therapy in a randomized, international, multicenter trial in 837 patients (415 assigned to voriconazole, 4 mg/kg bd (6 mg/kg bd on day 1) and 422 to liposomal amphotericin, 3 mg/kg/day) (14). The overall success rates were similar (26% with voriconazole and 31% with liposomal amphotericin); however, there were significantly fewer documented breakthrough fungal infections in patients who received voriconazole (8 versus 21). Those who received voriconazole had fewer severe infusion-related reactions, less mild nephrotoxicity, as defined by increases in serum creatinine to over 1.5 times baseline, and less hypokalemia; however, there was no difference whatsoever in the proportion of patients with more profound renal compromise (increased serum creatinine to over 2.0 times baseline; 7.0 versus 7.6%). The incidence of hepatotoxicity, as measured by raised hepatic transaminases and alkaline phosphatase, was similar in the two groups; increased serum bilirubin to over 1.5 times baseline was more common in patients who took amphotericin. Patients who received voriconazole had more episodes of transient visual changes than those who received liposomal amphotericin (22 versus 1%) and more episodes of visual hallucinations (4.3 versus 0.5%). Parenteral voriconazole was changed to the oral formulation in 22%, with a reduction in the mean duration of hospitalization by 1 day in all patients but by 2 days in patients at high risk (with relapsed leukemia or after allogeneic bone marrow transplant). Toxicity or lack of efficacy caused 9.9% of those who received voriconazole and 6.6% of those who received amphotericin to withdraw.

In a comparative, randomized, unblinded trial for primary therapy of invasive aspergillosis, 144 patients received either intravenous voriconazole (6 mg/kg bd on day 1, then 4 mg/kg bd for at least 7 days) followed by 200 mg bd orally, and 133 received intravenous amphotericin deoxycholate (1–1.5 mg/kg/day); other licensed antifungal treatments were allowed if the initial therapy failed or if the patient had an intolerance to the first drug used (15). Most patients had allogeneic hemopoietic cell transplants, acute leukemia, or other hematological diseases. At week 12, 53% of the patients in the voriconazole group and 32% of those in the amphotericin group (difference = 21%; 95% CI = 10, 33%) had a successful outcome. The survival rate at 12 weeks was 71% in the voriconazole group and 58% in the amphotericin group (hazard ratio = 0.59; 95% CI = 0.40, 0.88). Transient visual disturbances were more common with voriconazole (45 versus 4.3%). The most frequent descriptions of such disturbances were blurred vision, altered visual perception, altered color perception, and photophobia. Of the patients who received voriconazole, 9% had hallucinations or confusion considered possibly related to the drug

compared with 3.7% of those who received amphotericin. Infusion-related reactions (fever, chills, or both) were more common in those who received amphotericin (3.1 versus 25%), as were severe, potentially related adverse events (13 versus 24%); the most frequent events were renal impairment (14%) in those who received amphotericin and liver function abnormalities (4.8%) in those who received voriconazole.

Other antifungal azoles

In a multicenter, randomized, double-blind, double-dummy, parallel-group, dose-escalation comparison with fluconazole, the safety, tolerability, and pharmacokinetics of oral voriconazole were investigated in 24 subjects at high risk of fungal infections (with hematological malignancies, solid tumors, or autologous bone marrow transplants) (16). The subjects were randomized to receive voriconazole 200 mg bd (n = 9), voriconazole 300 mg bd (n = 9), or fluconazole 400 mg od (n = 6) for 14 days. There was an approximate five-fold accumulation of voriconazole during the dosing period and evidence of non-linear pharmacokinetics. Voriconazole was generally safe and well tolerated. Mild, reversible visual disturbances were the most commonly reported adverse events but they were not associated with treatment withdrawal. No patient developed a breakthrough fungal infection.

The efficacy, safety, and tolerability of voriconazole and fluconazole have been compared in 391 immunocompromised adults with esophageal candidiasis in a randomized, double-blind, multicenter trial (17). Most of the patients (94%) had AIDS. Following randomization, they took either voriconazole (200 mg bd) or fluconazole (400 mg on day 1 followed by 200 mg od) for a median of 14 or 15 days respectively. Treatment was continued for 7 days after the resolution of all signs and symptoms but was not allowed to exceed 42 days. The two drugs achieved comparable success rates (98% voriconazole, 95% fluconazole), as assessed by esophagoscopy in the primary efficacy analysis in 256 patients. More patients discontinued voriconazole because of laboratory test abnormalities (3.5 versus 1%) or treatment-related adverse events (2.5 versus 0.5%). The most frequent adverse event with voriconazole was transient visual disturbances (23 versus 8%); other clinical adverse events were similar in frequency. Increases of over three times the upper limit of the reference range in aspartate transaminase (20 versus 8%), alanine transaminase (11 versus 7%), and alkaline phosphatase (10 versus 8%) were more frequent with voriconazole.

Amphotericin + fluconazole

In a multicenter, randomized, non-inferiority study, voriconazole (n = 283) was compared with a regimen of amphotericin followed by fluconazole (n = 139) for candidemia in non-neutropenic patients (18). Voriconazole was not inferior to amphotericin + fluconazole in the primary efficacy analysis, with successful outcomes in 41% of patients in both treatment groups. Withdrawals due to all-cause adverse events were more frequent with voriconazole, although most were due to non-drug-related events and there were significantly fewer serious adverse events and cases of renal toxicity than with amphotericin + fluconazole.

Placebo-controlled studies

The safety, tolerance, and pharmacokinetics of oral voriconazole after single and multiple dosing have been investigated in 64 healthy subjects (19). Groups of eight subjects each took voriconazole doses of 2 mg/kg bd, 4 mg/kg/day, 2 mg/kg tds, or 3 mg/kg bd, 11 took 1.5 mg/kg tds, and 21 took placebo. The pharmacokinetics of voriconazole were non-linear (dose- and time-related), and there was intersubject variability in C_{max} and AUC. Visual inspection of C_{min} values together with statistical analyses of C_{max} and AUC suggested that steady-state concentrations were achieved by 5?6 days of multiple dosing. There were treatment-related adverse events in 67% of those who took voriconazole compared with 81% of those who took placebo. Visual disorders, including abnormal vision, conjunctivitis, eye pain, lacrimation disturbance, and photophobia, were reported by 16 subjects taking active treatment, 14 of whom reported a change in brightness of vision. All visual adverse effects cleared without intervention or after withdrawal. Blood pressure, pulse rate, electrocardiograms, and Holter data did not show any effects of voriconazole. Five of those who took voriconazole withdrew from the study because of visual disturbances (n = 3), ventricular tachycardia (n = 1), or ST segment depression (n = 1); in the placebo group, two subjects withdrew. The oral dosage regimen selected for subsequent phase 2/3 clinical trials was 200 mg bd, equivalent to 3 mg/kg bd.

The pharmacokinetics and safety of intravenous voriconazole have been investigated in healthy men in two single-blind, placebo-controlled studies (20). In the first study 12 subjects were randomized to voriconazole (3 mg/kg) or placebo, administered once daily on days 1 and 12, and every 12 hours on days 3–11. In the second study, 18 subjects were randomized to voriconazole or placebo; voriconazole was given as a loading dose of 6 mg/kg twice on day 1, then 3 mg/kg bd on days 2–9, and 3 mg/kg once on day 10. The use of a loading dose in the second study resulted in a shorter time to steady-state C_{min} than in the first study. The final day pharmacokinetics in the two studies were similar. On multiple dosing, voriconazole accumulated to an extent that was not predictable from the single-dose data. Multiple doses of voriconazole were well tolerated and no subject withdrew in either study. There were seven cases of possibly drug-related visual disturbances in three subjects. There were no clinically significant electrocardiographic changes and no abnormalities on Holter recording in those who took voriconazole.

Organs and Systems

Cardiovascular

QT interval prolongation has been attributed to voriconazole.

- A 15-year-old girl with acute lymphoblastic leukemia, who was taking voriconazole for a *Fusarium* infection, developed asymptomatic bradycardia, QT interval prolongation, and non-sustained, polymorphic ventricular tachycardia, which recurred on rechallenge (21). Voriconazole concentrations and metabolism were within expected values.

This patient was also taking several potentially arrhythmogenic drugs and had hypokalemia during the first episode; however, during rechallenge the electrolytes were within normal values and the only co-administered drugs included vancomycin and amikacin. The authors concluded that this observation suggested that closer attention should be paid to cardiac rhythm monitoring during voriconazole treatment.

Nervous system

Painful peripheral neuropathy associated with voriconazole has been reported (22).

- A 43-year-old woman who had undergone liver transplantation received voriconazole for invasive deep sinus aspergillosis and developed intolerable pain in all limbs. Electromyography and nerve conduction studies suggested a demyelinating neuropathy. The symptoms and signs of neuropathy disappeared permanently soon after voriconazole withdrawal.

Sensory systems

Visual adverse events, in particular enhanced brightness of light and color, are common with voriconazole (17). These visual adverse effects are transient and reversible. So far, comprehensive ophthalmological investigations have not shown any morphological correlates or long-term visual sequelae in any patient.

In a multicenter study of a parenteral formulation of voriconazole in 39 immunocompromised children (aged 2–11 years) visual disturbances occurred in five and were the only drug-related adverse events that occurred more than once (23).

Psychiatric

Voriconazole can be associated with visual and other hallucinations.

- A 78-year-old man began to have auditory hallucinations, specifically of Christmas music, on the second day of voriconazole therapy (24). Psychiatric evaluation was otherwise unremarkable. After withdrawal of voriconazole the hallucinations reduced in intensity by 2 days and ceased altogether by the third day.

An extensive literature search, including Pfizer drug trial safety data, yielded no other reports of auditory hallucinations with voriconazole. Several other cases of musical hallucinations secondary to a variety of causes have been reported. They tend to occur secondary to temporal lobe insults and often have religious or patriotic themes.

Liver

Liver function test abnormalities in patients taking voriconazole are not unexpected for an azole and can be explained by its extensive hepatic metabolic clearance (25).

Concerns about voriconazole-related liver toxicity have led to suggestions for individual voriconazole dosage modification based on plasma concentrations; however, the manufacturers cited data from three large randomized trials of voriconazole (26). These data showed no differences in the frequency of liver function test abnormalities between voriconazole and amphotericin formulations in patients with invasive aspergillosis, fever, and granulocytopenia; only in patients with esophageal candidiasis was the frequency of abnormal liver function tests greater among patients taking voriconazole than among those taking fluconazole. According to the authors, these data suggest that among patients with multiple co-morbidities and multiple co-medications, the frequency of abnormal liver function tests may depend on factors other than the type of antifungal agent. Furthermore, they presented a summary of unpublished logistic regression analyses that they had performed to identify any possible relation between plasma concentrations and abnormal liver function tests (27). These analyses, performed on about 3000 samples from 1000 patients enrolled into 10 trials showed only a weak and inconsistent association of plasma concentrations of voriconazole with abnormal liver function tests.

Hepatotoxicity has been assessed in a retrospective observational study in 35 patients who took oral voriconazole compared with 11 patients who received intravenous voriconazole during the first week followed by oral treatment (28). The incidence of increased liver enzymes was comparable in the two groups. Voriconazole was withdrawn in two patients in the oral group and one patient in the intravenous group because of hepatotoxicity. Liver enzyme rises in all 46 patients were higher than previously reported in a comparable study population. However, clinically significant hepatotoxicity was infrequent (3/46, 6.5%).

Biliary tract

Cholelithiasis with cholecystitis has been attributed to voriconazole in three cases (29).

Skin

Seven cases of photosensitivity during treatment with voriconazole have been reported (30). All these patients had severe immunosuppression and were taking voriconazole for fungal infections. The photosensitivity reactions occurred within 5 weeks to 14 months after the start of treatment and in all cases followed exposure to the sun, occasionally at low levels. The lesions disappeared rapidly on withdrawal of voriconazole.

Toxic epidermal necrolysis has been attributed to voriconazole (31).

- A 39-year-old man with metastatic breast cancer, who was receiving cisplatin, epirubicin, and 5-fluorouracil, was given voriconazole for a fungal infection. After 5 days he developed a rash. All drugs were withdrawn. Within 3 days, he had a fever and diffuse erythema of the trunk, limbs, and face, with blistering skin lesions and a positive Nikolsky's sign on erythematous areas, characteristic of toxic epidermal necrolysis and involving 85% of the body surface. The conjunctivae were injected and there were multiple bullae and ulcers on the lips and oropharyngeal mucosa. Biopsy showed necrosis of cells from the basal layer and stratum spinosum, resulting in detachment of the epidermis from the dermis. He recovered after 2 weeks of intensive therapy.

Severe retinoid-like photosensitivity (erythema, desquamation, and ulceration of light-exposed skin) has been reported in two children with chronic granulomatous disease who took voriconazole 200 mg bd for chronic invasive aspergillosis (32). Histopathological examination in one of the patients showed superficial and deep perivascular dermatitis with epidermal necrosis, compatible with a photo-induced drug eruption. Although strict sun protection and sun avoidance led to resolution of the acute lesions while voriconazole was continued, dark pigmented lentigines developed over previously involved areas. In the second patient, the lesions completely resolved after withdrawal of the drug. There were similar phototoxic manifestations in two adults who took long-term voriconazole (33). In both patients, voriconazole was withdrawn and the lesions resolved within 2 weeks. While the exact mechanisms of voriconazole-associated phototoxicity are unknown, inhibition of retinoid metabolism or a direct phototoxic effect of voriconazole or one of its metabolites have been implicated.
While the exact mechanisms of this phototoxicity are unknown, inhibition of retinoid metabolism or a direct phototoxic effect of voriconazole or one of its metabolites have been implicated (SEDA-29, 287).

Photoageing caused by voriconazole has been reported (34).

- A 15-year-old girl developed cheilitis and erythema over the sun-exposed areas of her body after taking voriconazole for 5 weeks for a severe fungal infection. The lesions improved transiently before subsequent photodamage occurred to the backs of her forearms, the backs of her hands, and face. Voriconazole was withdrawn once the fungal infection had completely resolved and her blisters, erythema, and cheilitis resolved. However, she was left with solar elastosis, multiple lentigines, and ephelides on sun-exposed areas.

Pseudoporphyria is an uncommon blistering disorder. It has clinical and histological similarities to porphyria cutanea tarda but without changes in urine and serum porphyrin concentrations. Pseudoporphyria has many causes, including chronic renal insufficiency, ultraviolet radiation, and many medications. Pseudoporphyria has been attributed to voriconazole for the first time (35).

Susceptibility Factors

Age

Although the disposition of voriconazole in children aged 12 years and over is similar to that in adults, children aged 2–11 years need larger dosages to achieve the same exposure that was effective in interventional trials in adults (36).

The safety and efficacy of voriconazole have been studied in 58 children (aged 9 months to 15 years; median 7 years) who were treated in the manufacturer's compassionate release program (37). They received voriconazole for an invasive fungal infection if they were refractory to or intolerant of conventional antifungal drugs. Voriconazole was given intravenously as a loading dose of 6 mg/kg every 12 hours on day 1 followed by 4 mg/kg every 12 hours thereafter. When feasible, the route of administration was changed from intravenous to oral (100 or 200 mg bd for patients weighing under or over 40 kg, respectively). At the end of therapy (mean 93 days, range 1–800), 26 patients had a complete or partial response, 4 had a stable response, 25 failed therapy, and 4 were withdrawn because of intolerance of voriconazole. Two had treatment-related serious adverse events (ulcerated lips with rash and raised hepatic transaminases or bilirubin). A total of 23 patients had voriconazole-related adverse events, three of which required withdrawal of voriconazole. The most commonly reported adverse events included raised hepatic transaminases or bilirubin ($n = 8$), skin rashes ($n = 8$), abnormal vision ($n = 3$), and a photosensitivity reaction ($n = 3$). The authors concluded that these data support the use of voriconazole for invasive fungal infections in children who are intolerant of or refractory to conventional antifungal drugs.

In a multicenter study of a parenteral formulation of voriconazole in 39 immunocompromised children (aged 2–11 years) was reported body weight was more influential than age in accounting for the observed variability in voriconazole pharmacokinetics (27). Thus, children have a higher capacity for elimination of voriconazole per kilogram of body weight than healthy adults

In a retrospective case review of 21 children aged 5 to 16 years with cystic fibrosis and allergic bronchopulmonary aspergillosis, voriconazole, used as monotherapy or in combination with an immunomodulatory agent, resulted in significant improvement in pulmonary function and serology (38). There were adverse effects in seven children: photosensitivity reactions (n = 3), nausea (n = 2), a rise in hepatic enzymes (n = 1) and hair loss (n = 1).

Renal disease

The pharmacokinetics of a single oral dose of voriconazole 200 mg have been studied in five patients with end-stage renal disease undergoing peritoneal dialysis (39). The t_{max} occurred in plasma at 2.4 hours and in dialysate at 2.8 hours. The dialysate to plasma ratio was 0.66. Less than 1% of the dose was recovered in dialysate 24 hours after dosing. These results suggest that voriconazole penetrates peritoneal fluid well; there is minimal peritoneal clearance and therefore no dosage adjustment is needed for patients undergoing peritoneal dialysis.

Drug Administration

Drug formulations

The intravenous formulation of voriconazole includes the solvent sulfobutylether β-cyclodextrin sodium, whose clearance is reduced in patients undergoing dialysis. Concentrations of voriconazole and sulfobutylether β-cyclodextrin sodium have been measured in four patients with renal insufficiency undergoing intermittent dialysis (40). Voriconazole plasma concentrations were generally below 1.5 µg/ml and there was no evidence of accumulation, in contrast to sulfobutylether β-cyclodextrin sodium, which did accumulate.

- A 75-year-old woman initially had a maximal sulphobutylether β-cyclodextrin sodium plasma concentration of 145 µg/ml, but after a few days renal function recovered and the plasma concentration fell to under 20 µg/ml.
- Sulfobutylether β-cyclodextrin sodium accumulated in three patients with renal failure during intravenous administration of voriconazole; the maximum concentrations were 523 µg/ml in a 18-year-old man, 409 µg/ml in a 57-year-old man, and 581 µg/ml in a 47-year-old man.

No adverse effects were observed in these cases, although the degree of exposure was comparable to that used in toxicity studies in animals, and the clinical relevance is uncertain.

Monitoring Therapy

Individual voriconazole dosage modification based on plasma concentrations has been suggested because of concerns about voriconazole-related liver toxicity (41). In a reply, the manufacturers presented a summary of unpublished data of logistic regression analyses that they had performed in order to identify any possible relation between plasma concentrations and increases in liver function tests (42). These analyses, performed on about 3000 samples from 1000 patients enrolled in 10 clinical trials showed only a weak and inconsistent association between plasma concentrations of voriconazole and abnormal liver function tests. There was no association between plasma concentrations and antifungal efficacy in a subset of patients with defined fungal infections. The authors concluded that plasma concentration monitoring is unlikely to add any value over clinical judgment and diligent monitoring of liver function tests, as no definite guidance can be provided for the interpretation of plasma concentrations of voriconazole.

In another report, the issue of drug monitoring in patients taking voriconazole used the examples of three patients with diverse manifestations of toxicity that were possibly attributable to high voriconazole concentrations (43). Of 16 patients who were treated with voriconazole at the authors' center between 1993 and 2001 and in whom

© 2010 Elsevier B.V. All rights reserved.

voriconazole trough concentrations were measured, three had significant toxicity, all with high trough plasma concentrations (hypoglycemia, 9.7 µg/ml; altered mental status and hallucinations, 9.0 µg/ml; hyperkalemia, 18 µg/ml). The authors argued that a case could be made for plasma concentration monitoring, based on these results, polymorphism in the CYP2C19 isoenzyme, and the non-linear disposition of voriconazole. While this will remain controversial until more systematic post-marketing studies have been concluded, assessing trough concentrations is a reasonable approach, at least in patients with adverse events that might be due to voriconazole.

Of 26 patients six had neurological adverse events (visual and auditory hallucinations, encephalopathy, and visual disturbances) during treatment with voriconazole (44). There was a significant difference between serum voriconazole concentrations in those who did and did not have adverse effects (median 5.7 versus 1.4 µg/ml). The hazard ratio for adverse effects per 0.1 µg/ml increase in concentration was 2.27 (95% CI = 1.45, 3.56).

The relation between plasma voriconazole concentrations and the risk of visual adverse effects or abnormal liver function tests has been studied in 1053 patients (2925 plasma samples) (45). There was a relation between plasma voriconazole concentrations and the risk of visual adverse effects and a weaker, but still significant, association with the risk of abnormalities of bilirubin, alkaline phosphatase, or aspartate transaminase, but not alanine transaminase. However, receiver-operating characteristic curve analysis showed that individual plasma voriconazole concentrations cannot be used to predict abnormal liver function tests.

Steady-state trough plasma voriconazole concentrations were obtained in 25 recipients of allogeneic hemopoietic stem cell transplants, once (n = 13), twice (n = 10), or at least three times (n = 2), 5–18 (median 10) days after starting voriconazole or dosage modification (46). The 41 voriconazole concentrations were 0.2–6.8 µg/ml; six were below 0.5 (possibly below the in vitro MIC_{90} for *Aspergillus* spp.). Voriconazole concentrations correlated with aspartate transaminase and alkaline phosphatase activities, but not with creatinine, bilirubin, or alanine transaminase. Since liver dysfunction is common after HSCT, it was not possible to determine whether the increases in aspartate transaminase and alkaline phosphatase activities were due to higher voriconazole concentrations. The authors concluded that trough voriconazole concentrations vary considerably between patients and they suggested monitoring concentrations in patients taking voriconazole for confirmed fungal infections and in those with increased aspartate transaminase or alkaline phosphatase activities.

Drug–Drug Interactions

Ciclosporin

The interaction of voriconazole with ciclosporin has been investigated in a randomized, double-blind, placebo-controlled, crossover study in kidney transplant recipients with stable renal function (47). During the first study period (7.5 days), subjects taking ciclosporin 150 mg/day received either concomitant voriconazole (200 mg every 12 hours) or a matching placebo. After a washout period of at least 4 days, they were switched to the other treatment. In the seven subjects who completed both regimens, concomitant administration with voriconazole resulted in a 1.7-fold increase (90% CI = 1.47, 1.96) in mean ciclosporin AUC during a dosage interval. Ciclosporin C_{max} and t_{max} were not significantly affected, but C_{min} was increased by voriconazole by a mean of 2.48 (range 1.88–3.03) times. Seven subjects withdrew during voriconazole administration, six for reasons that were considered to be drug-related; most were attributable to increased ciclosporin concentrations. Although not serious, all causality-related adverse events were more frequent during voriconazole administration than during placebo administration. Thus, when voriconazole is initiated or withdrawn in patients who are already taking ciclosporin, blood ciclosporin concentrations should be carefully monitored and the dose of ciclosporin adjusted as necessary.

References

1. Hoffman HL, Ernst EJ, Klepser ME. Novel triazole antifungal agents. Expert Opin Investig Drugs 2000;9(3):593–605.
2. Chiou CC, Groll AH, Walsh TJ. New drugs and novel targets for treatment of invasive fungal infections in patients with cancer. Oncologist 2000;5(2):120–35.
3. Groll AH, Glasmacher A, Just-Nuebling G, Maschmeyer G, Walsh TJ. Clinical pharmacology of antifungal compounds. Infect Dis Clin North Am 2003;17.159-91.
4. Theuretzbacher U, Ihle F, Derendorf H. Pharmacokinetic/pharmacodynamic profile of voriconazole. Clin Pharmacokinet 2006;45(7):649-63.
5. Boucher HW, Groll AH, Chiou CC, Walsh TJ. Newer systemic antifungal agents: pharmacokinetics, safety and efficacy. Drugs 2004;64(18):1997-2020.
6. Purkins L, Wood N, Ghahramani P, Greenhalgh K, Allen MJ, Kleinermans D. Pharmacokinetics and safety of voriconazole following intravenous- to oral-dose escalation regimens. Antimicrob Agents Chemother 2002;46(8):2546–53.
7. Marco F, Pfaller MA, Messer SA, Jones RN. Antifungal activity of a new triazole, voriconazole (UK-109,496), compared with three other antifungal agents tested against clinical isolates of filamentous fungi. Med Mycol 1998;36(6):433–6.
8. Groll AH, Gea-Banacloche JC, Glasmacher A, Just-Nuebling G, Maschmeyer G, Walsh TJ. Clinical pharmacology of antifungal compounds. Infect Dis Clin North Am 2003;17(1):159–91.
9. Groll AH, Walsh TJ. Antifungal chemotherapy: advances and perspectives. Swiss Med Wkly 2002;132(23–24):303–11.
10. Denning DW, Ribaud P, Milpied N, Caillot D, Herbrecht R, Thiel E, Haas A, Ruhnke M, Lode H. Efficacy and safety of voriconazole in the treatment of acute invasive aspergillosis. Clin Infect Dis 2002;34(5):563–71.
11. Perfect JR, Marr KA, Walsh TJ, Greenberg RN, DuPont B, De la Torre-Cisneros J, Just-Nubling G, Schlamm HT, Lutsar I, Espinel-Ingroff A, Johnson E. Voriconazole treatment for less-common, emerging, or refractory fungal infections. Clin Infect Dis 2003;36:1122-31.
12. Ostrosky-Zeichner L, Oude Lashof AM, Kullberg BJ, Rex JH. Voriconazole salvage treatment of invasive candidiasis. Eur J Clin Microbiol Infect Dis 2003;22:651-5.

13. Baden LR, Katz JT, Fishman JA, Koziol C, DelVecchio A, Doran M, Rubin RH. Salvage therapy with voriconazole for invasive fungal infections in patients failing or intolerant to standard antifungal therapy. Transplantation 2003;76:1632-7.

14. Walsh TJ, Pappas P, Winston DJ, Lazarus HM, Petersen F, Raffalli J, Yanovich S, Stiff P, Greenberg R, Donowitz G, Schuster M, Reboli A, Wingard J, Arndt C, Reinhardt J, Hadley S, Finberg R, Laverdiere M, Perfect J, Garber G, Fioritoni G, Anaissie E, Lee JNational Institute of Allergy and Infectious Diseases Mycoses Study Group. Voriconazole compared with liposomal amphotericin B for empirical antifungal therapy in patients with neutropenia and persistent fever. N Engl J Med 2002;346(4):225–34.

15. Herbrecht R, Denning DW, Patterson TF, Bennett JE, Greene RE, Oestmann JW, Kern WV, Marr KA, Ribaud P, Lortholary O, Sylvester R, Rubin RH, Wingard JR, Stark P, Durand C, Caillot D, Thiel E, Chandrasekar PH, Hodges MR, Schlamm HT, Troke PF, de Pauw B. Invasive Fungal Infections Group of the European Organisation for Research and Treatment of Cancer and the Global Aspergillus Study Group. Voriconazole versus amphotericin B for primary therapy of invasive aspergillosis. N Engl J Med 2002;347(6):408–15.

16. Lazarus HM, Blumer JL, Yanovich S, Schlamm H, Romero A. Safety and pharmacokinetics of oral voriconazole in patients at risk of fungal infection: a dose escalation study. J Clin Pharmacol 2002;42(4):395–402.

17. Ally R, Schurmann D, Kreisel W, Carosi G, Aguirrebengoa K, Dupont B, Hodges M, Troke P, Romero AJEsophageal Candidiasis Study Group. A randomized, double-blind, double-dummy, multicenter trial of voriconazole and fluconazole in the treatment of esophageal candidiasis in immunocompromised patients. Clin Infect Dis 2001;33(9):1447–54.

18. Kullberg BJ, Sobel JD, Ruhnke M, Pappas PG, Viscoli C, Rex JH, Cleary JD, Rubinstein E, Church LW, Brown JM, Schlamm HT, Oborska IT, Hilton F, Hodges MR. Voriconazole versus a regimen of amphotericin B followed by fluconazole for candidaemia in non-neutropenic patients: a randomised non-inferiority trial. Lancet 2005;366(9495):1435-42.

19. Purkins L, Wood N, Greenhalgh K, Allen MJ, Oliver SD. Voriconazole, a novel wide-spectrum triazole: oral pharmacokinetics and safety. Br J Clin Pharmacol 2003;56 Suppl 1:10-16.

20. Purkins L, Wood N, Greenhalgh K, Eve MD, Oliver SD, Nichols D. The pharmacokinetics and safety of intravenous voriconazole—a novel wide-spectrum antifungal agent. Br J Clin Pharmacol 2003;56 Suppl 1:2-9.

21. Alkan Y, Haefeli WE, Burhenne J, Stein J, Yaniv I, Shalit I. Voriconazole-induced QT interval prolongation and ventricular tachycardia: a non-concentration-dependent adverse effect. Clin Infect Dis 2004;39:e49-52.

22. Tsiodras S, Zafiropoulou R, Kanta E, Demponeras C, Karandreas N, Manesis EK. Painful peripheral neuropathy associated with voriconazole use. Arch Neurol 2005;62(1):144-6.

23. Walsh TJ, Karlsson MO, Driscoll T, Arguedas AG, Adamson P, Saez-Llorens X, Vora AJ, Arrieta AC, Blumer J, Lutsar I, Milligan P, Wood N. Pharmacokinetics and safety of intravenous voriconazole in children after single- or multiple-dose administration. Antimicrob Agents Chemother 2004;48:2166-72.

24. Agrawal AK, Sherman LK. Voriconazole-induced musical hallucinations. Infection 2004;32:293-5.

25. Ullmann AJ. Review of the safety, tolerability, and drug interactions of the new antifungal agents caspofungin and voriconazole. Curr Med Res Opin 2003;19(4):263–71.

26. Potoski BA, Brown JR. The safety of voriconazole. Clin Infect Dis 2002;35:1273-5.

27. Lutsar I, Hodges MR, Tomaszewski K, Troke PF, Wood ND. Safety of voriconazole and dose individualization. Clin Infect Dis 2003;36:1087-8.

28. den Hollander JG, van Arkel C, Rijnders BJ, Lugtenburg PJ, de Marie S, Levin MD. Incidence of voriconazole hepatotoxicity during intravenous and oral treatment for invasive fungal infections. J Antimicrob Chemother 2006;57(6):1248-50.

29. Gérardin-Marais M, Allain-Veyrac G, Danner I, Jolliet P. Lithiase biliaire et cholecystite au voriconazole (VFEND):à propos de 3 cas. [Biliary lithiasis and cholecystitis with voriconazole: à propos 3 cases.] Thérapie 2006;61(4):367-9.

30. Auffret N, Janssen F, Chevalier P, Guillemain R, Amrein C, Le Beller C. Photosensibilisation au voriconazole: 7 cas. [Voriconazole photosensitivity: 7 cases.] Ann Dermatol Venereol 2006;133(4):330-2.

31. Curigliano G, Formica V, De Pas T, Spitaleri G, Pietri E, Fazio N, de Braud F, Goldhirsch A. Life-threatening toxic epidermal necrolysis during voriconazole therapy for invasive aspergillosis after chemotherapy. Ann Oncol 2006;17(7):1174-5.

32. Rubenstein M, Levy ML, Metry D. Voriconazole-induced retinoid-like photosensitivity in children. Pediatr Dermatol 2004;21:675-8.

33. Vandecasteele SJ, Van Wijngaerden E, Peetermans WE. Two cases of severe phototoxic reactions related to long-term outpatient treatment with voriconazole. Eur J Clin Microbiol Infect Dis 2004;23:656-7.

34. Racette AJ, Roenigk HH Jr, Hansen R, Mendelson D, Park A. Photoaging and phototoxicity from long-term voriconazole treatment in a 15-year-old girl. J Am Acad Dermatol 2005;52(5 Suppl 1):S81-5.

35. Sharp MT, Horn TD. Pseudoporphyria induced by voriconazole. J Am Acad Dermatol 2005;53(2):341-5.

36. Walsh TJ, Karlsson MO, Driscoll T, Arguedas AG, Adamson P, Saez-Llorens X, Vora AJ, Arrieta AC, Blumer J, Lutsar I, Milligan P, Wood N. Pharmacokinetics and safety of intravenous voriconazole in children after single- or multiple-dose administration. Antimicrob Agents Chemother 2004;48(6):2166–72.

37. Walsh TJ, Lutsar I, Driscoll T, Dupont B, Roden M, Ghahramani P, Hodges M, Groll AH, Perfect JR. Voriconazole in the treatment of aspergillosis, scedosporiosis and other invasive fungal infections in children. Pediatr Infect Dis J 2002;21(3):240–8.

38. Hilliard T, Edwards S, Buchdahl R, Francis J, Rosenthal M, Balfour-Lynn I, Bush A, Davies J. Voriconazole therapy in children with cystic fibrosis. J Cyst Fibros 2005;4(4):215-20.

39. Peng LW, Lien YH. Pharmacokinetics of single, oral-dose voriconazole in peritoneal dialysis patients. Am J Kidney Dis 2005;45(1):162-6.

40. von Mach MA, Burhenne J, Weilemann LS. Accumulation of the solvent vehicle sulphobutylether beta cyclodextrin sodium in critically ill patients treated with intravenous voriconazole under renal replacement therapy. BMC Clin Pharmacol 2006;6:6.

41. Potoski BA, Brown JR. The safety of voriconazole. Clin Infect Dis 2002;35:1273-75.

42. Lutsar I, Hodges MR, Tomaszewski K, Troke PF, Wood ND. Safety of voriconazole and dose individualization. Clin Infect Dis 2003;36:1087-88.

43. Boyd AE, Modi S, Howard SJ, Moore CB, Keevil BG, Denning DW. Adverse reactions to voriconazole. Clin Infect Dis 2004;39:1241-4.

44. Imhof A, Schaer DJ, Schanz U, Schwarz U. Neurological adverse events to voriconazole: evidence for therapeutic drug monitoring. Swiss Med Wkly 2006;136(45-46):739-42.

45. Tan K, Brayshaw N, Tomaszewski K, Troke P, Wood N. Investigation of the potential relationships between plasma voriconazole concentrations and visual adverse events or liver function test abnormalities. J Clin Pharmacol 2006;46(2):235-43.

46. Trifilio S, Ortiz R, Pennick G, Verma A, Pi J, Stosor V, Zembower T, Mehta J. Voriconazole therapeutic drug monitoring in allogeneic hematopoietic stem cell transplant recipients. Bone Marrow Transplant 2005;35(5):509-13.

47. Romero AJ, Pogamp PL, Nilsson LG, Wood N. Effect of voriconazole on the pharmacokinetics of cyclosporine in renal transplant patients. Clin Pharmacol Ther 2002;71(4):226–34.

ANTIPROTOZOAL AND ANTIHELMINTHIC DRUGS

Albendazole

See also Benzimidazoles

General Information

The benzimidazole antihelminthic drugs albendazole and mebendazole (qv) are commonly used to treat soil-transmitted helminth infections, such as gastrointestinal roundworms, hydatid disease, neurocysticercosis, larva migrans cutanea, and strongyloidiasis (1,2,3). They bind to nematode β tubulin and inhibit parasite microtubule polymerization, which causes death of adult worms after a few days. Although both albendazole and mebendazole are broad-spectrum antihelminthic drugs, their use differs substantially in clinical practice. Both are effective against ascariasis in a single dose. In hookworm infections, however, a single dose of mebendazole is associated with a low cure rate and albendazole is more effective. Conversely, a single dose of albendazole is not effective in many cases of trichuriasis. For both trichuriasis and hookworm infection, several doses of benzimidazoles are commonly needed. Provided that an adequate concentration is attained within the cyst, it is scolicidal. In high doses given for prolonged periods or cyclically, it is effective in echinococcosis, in which it is given in a dosage of 10 mg/kg/day for 4 weeks, repeated in six cycles with 2-week rest periods between each cycle, although even with this high dose only about one-third of patients enjoy a complete cure, some 70% having a partial response. Albendazole is also active against *Pneumocystis jiroveci*, and is effective in prophylaxis and treatment in immunosuppressed mice (4). In hydatid disease a combination of albendazole and praziquantel is effective when either agent has failed when used alone (SED-12, 707) (5).

Another important difference between albendazole and mebendazole is that mebendazole is poorly absorbed from the gastrointestinal tract. For this reason the therapeutic action is largely directed at intraluminal (adult) worms. Albendazole is better absorbed, especially when it is taken with a fatty meal. It is metabolized in the liver to a sulfoxide derivative, which is highly distributed to the tissues. It is therefore used to treat disorders caused by tissue-migrating larvae, such as visceral larva migrans caused by *Toxocara canis*. Systemic adverse effects, such as those on the liver and bone marrow, are usually rare with benzimidazoles in the doses used to treat soil-transmitted helminth infections. However, transient abdominal pain, diarrhea, nausea, dizziness, and headache can occur. Since deworming programmes in endemic regions clearly improve child health and education as well as reducing the burden of disease attributed to soil-transmitted helminths, the use of these drugs is not limited to treatment of symptomatic infections, but also for prevention by mass deworming programmes of morbidity in children living in endemic areas. Concerns about the sustainability of periodic deworming with benzimidazoles and the emergence of resistance have prompted efforts to develop and test new control tools (such as plant cysteine proteinases from various sources of latex-bearing plants and fruits).

Observational studies

Ankylostomiasis

Albendazole has been used in the treatment of human hookworm and trichuriasis. In a mass-treatment report from Western Australia 295 individuals in a remote rural area were treated with albendazole 400 mg/day for 5 days because of possible *Giardia lamblia* and hookworm infections (6). The 37% prevalence of *Giardia* fell to 12% between days 6 and 9, but rose again to 28% between days 18 and 30. The effect on hookworms (*Ankylostoma duodenale*) was more pronounced and more sustained with a reduction of the pretreatment prevalence of hookworm infections from 76% before treatment to 0% after 3–4 weeks. The tolerability of the drug was judged to be excellent by 89%, good by 1%, and moderately good by 1%, while 9% gave no response. Adverse effects were reported by five individuals and consisted of mild abdominal pain ($n = 2$), mild or moderate diarrhea ($n = 2$), moderate fever ($n = 1$), and weakness ($n = 1$).

Ascariasis

The efficacy of 2 years of mass chemotherapy against ascariasis has been evaluated in Iran (7). A single dose of albendazole 400 mg was given at 3-month intervals for 2 years to every person, except children under 2 years of age and pregnant women. After 2 years of treatment the prevalences, based on 2667 post-treatment samples, had fallen (Table 1). There were no adverse effects of mass treatment with albendazole.

Echinococcosis

Hydatid disease is a common zoonosis caused by the larval cysts of *Echinococcus granulosus*. Hydatid cysts most commonly form in the liver, but can occur in any organ. The management and operative complications in 70 patients with hydatid disease aged 10–78 years have been studied retrospectively to assess the impact of albendazole and praziquantel compared with surgery (8). In all, 39 patients received albendazole and praziquantel in combination and 19 received albendazole alone; none was treated with praziquantel alone. The combined use of albendazole and praziquantel preoperatively significantly reduced the number of cysts that contained viable

Table 1 Changes in prevalences of helminthic infections in patients treated with albendazole (5)

Helminth	Number (%) of positive tests before treatment ($n = 3098$)	Number (%) of positive tests after treatment ($n = 2667$)
Ascaris lumbricoides	1198 (39%)	196 (7.4%)
Trichuris trichiura	22 (0.7%)	5 (0.2%)
Hymenolepis nana	63 (2%)	49 (1.8%)

protoscolices. During the 12-year follow-up period an initial 3 months of drug treatment (albendazole throughout and praziquantel for 2 weeks), re-assessment, followed by either surgery or continuation with chemotherapy was found to be a rational treatment algorithm. In 11 patients albendazole, given for a median of 3 months at a dose of 400 mg bd, had adverse effects: five patients developed nausea and six had abnormal liver function tests. Therapy was withdrawn in two patients owing to altered liver function.

The efficacy of albendazole emulsion has been studied in 212 patients with hydatid disease of the liver, aged 4–82 years (9). Two regimens of albendazole were given for a variable period (3 months to more than 1 year); 67 adults received albendazole 10 mg/kg/day and 145 adults received 12.5 mg/kg/day. The overall cure rate was 75%. In the follow-up study the recurrence rate was 10%. The highest cure rate was observed in those who received albendazole 12.5 mg/kg/day for 9 months. At the start of therapy about 15% of the patients had mild pruritus, rash, and transient gastric pain, which resolved without specific therapy. Two patients had alopecia. There were frequent rises in serum transaminase activities in both groups but not to above 30–50 IU/l, except in six patients, who had values above 200 IU/l. In two patients albendazole was withdrawn because of vomiting. In one patient who took 12.5 mg/kg/day severe adverse effects, such as anorexia, jaundice, anemia, edema, and hypoproteinemia, developed, necessitating withdrawal. Reintroduction of albendazole 10 mg/kg/day was uneventful.

The use of albendazole and mebendazole in patients with hydatidosis has been evaluated in 448 patients with *E. granulosis* hydatid cysts who received continuous treatment with albendazole 10–12 mg/kg/day for 3–6 months daily orally in a total dose of (323 patients) twice or mebendazole 50 mg/kg/day (10). At the end of treatment, 82% of the cysts treated with albendazole and 56% of the cysts treated with mebendazole showed degenerative changes. During long-term follow-up 25% of these cysts showed relapse, which took place within 2 years in 78% of cases. Further treatment with albendazole induced degenerative changes in over 90% of the relapsed cysts, without induction of more frequent or more severe adverse effects, as observed during the first treatment period. Adverse effects during the first treatment period consisted of raised transaminases with albendazole (67 of 323 patients) and mebendazole (16 of 125 patients), and abdominal pain in 12 and 11% respectively. With both drugs, occasional patients experienced headache, abdominal distension, vertigo, urticaria, jaundice, thrombocytopenia, fever, or dyspepsia, but most of these are known manifestations of *Echinococcus* infection. Six of 323 patients taking albendazole withdrew because of adverse effects compared with eight of 125 patients taking mebendazole. It appears that albendazole is more effective than mebendazole in the treatment of hydatid cysts caused by *E. granulosis* and that both the intensity and frequency of the usually mild adverse effects are comparable.

Filariasis

Treatment of patients with high *Loa loa* microfilaraemia is sometimes complicated by an encephalopathy, suggested to be related to a rapid killing of large number of *L. loa* microfilariae. If the *L. loa* microfilarial count could be reduced more slowly, before ivermectin is distributed, ivermectin-related encephalopathy might be prevented. In 125 patients with *L. loa* microfilariasis the effect of albendazole (800 mg/day for three consecutive days) or multivitamin tablets on *L. loa* microfilarial load and the occurrence of encephalopathy were studied (11). *L. loa* microfilarial loads were followed for 9 months. There was no significant change in the overall microfilarial loads among those treated with albendazole, although the loads in patients with more than 8000 microfilariae/ml tended to fall more progressively during the first 3 months of follow-up. There were no cases of encephalopathy. The main adverse effects reported were itching (in eight patients taking albendazole and seven taking multivitamins), abdominal pain (two taking albendazole), and diarrhea (one taking albendazole, two taking multivitamins); overall analysis showed no significant difference in these events between the groups. Albendazole was associated with modest but significantly raised plasma transaminase activities.

Neurocysticercosis

In a report on the use of albendazole 15 mg/kg/day in two divided doses for 14 days in the treatment of persistent neurocysticercosis (12), adverse reactions were monitored in 43 patients with seizures and a solitary cysticercal cyst, who had not been treated before. In all patients CT scans confirmed the presence of a solitary cyst less than 2 cm in diameter. Antiepileptic treatment was continued. In seven patients dexamethasone 8 mg/day in four divided doses was given for the first 5–7 days after the start of treatment. Follow-up CT scans at 4–10 weeks after the start of treatment showed responses in 20 patients, with complete disappearance in seven patients and a reduction to 50% of the pretreatment size in the other 13. There were adverse effects in 15 patients, with a maximum on the fifth day after the start of treatment. Six patients had severe headaches, 11 had partial seizures, and 2 had epileptic seizures and severe postictal hemiparesis that persisted for a week or more. Because of these serious adverse effects treatment was discontinued in seven patients and dexamethasone was added in those patients who were not already taking it, although its use proved questionable. Adverse effects were seen in three of seven patients who took prophylactic steroid therapy and in 12 of 36 patients who did not.

Albendazole was effective in neurocysticercosis in an optimal dosage of 15 mg/kg/day divided in two doses every 12 hours for 8 days (13). It was generally well tolerated, although several patients had adverse reactions during the first few days after the start of treatment, consisting of headache, vomiting, and exacerbation of neurological symptoms caused by an inflammatory reaction to antigens from degenerating cysts, necessitating the concomitant use of glucocorticoids. In very large cysticerci, or cysticerci located in risky areas like the brainstem, these reactions may rarely be life-threatening.

Protozoal infections
The efficacy of albendazole 800 mg bd for 14 days for persistent diarrhea due to cryptosporidiosis ($n = 10$), isosporiasis ($n = 54$), or microsporidiosis ($n = 23$) has been studied in 153 HIV-positive patients (14). Albendazole reduced the burden of protozoal infection and promoted mucosal recovery in 87 patients who had a complete clinical response. Two patients reported nausea and vomiting. One patient developed leukopenia ($1.9 \times 10^9/l$) after treatment and four patients developed thrombocytopenia ($51-98 \times 10^9/l$).

Toxocariasis
The efficacy of albendazole plus prednisolone has been studied in five patients aged 11–72 years with ocular toxocariasis (15). All had uveitis and retinochoroidal granulomas. Their symptoms had persisted for a mean of 14 months (range 3 days to 24 months). The adults were treated with albendazole 800 mg bd for 2 weeks plus prednisolone starting at 1.5 mg/kg/day tapering over 3 months. The children were treated with 400 mg bd for 2 weeks plus prednisolone 1.0 mg/kg/day. All tolerated the therapy well without adverse effects. In particular, there were no significant hypersensitivity reactions to dying Toxocara larvae. The uveitis resolved in all cases and there were no relapses. After treatment, all the granulomas had disappeared, leaving heavily pigmented chorioretinal scars without loss of vision.

Comparative studies

Ascariasis
See below under Tricuriasis

Echinococcosis
Human hydatid disease is caused by the metacestode of Echinococcus granulosus. There are few data on the treatment of pulmonary hydatid disease in children. Mebendazole and albendazole have been evaluated in 82 children with a total of 102 pulmonary hydatid cysts (16). Mebendazole was given as 50 mg/kg/day in three divided doses and albendazole was given as 10 mg/kg/day in two divided doses continuously or in cycles consisting of 4 weeks of treatment alternating with 2-week drug-free intervals. The duration of treatment was 1–36 months. While taking benzimidazoles eight patients had raised liver enzymes, three had rash, and one had neutropenia; all were reversible on withdrawal.

Hydatidosis
The use of albendazole and mebendazole in patients with hydatidosis has been evaluated in 448 patients with E. granulosis hydatid cysts who received continuous treatment with albendazole 10–12 mg/kg/day for 3–6 months daily orally in a total dose of (323 patients) twice or mebendazole 50 mg/kg/day (10). At the end of treatment, 82% of the cysts treated with albendazole and 56% of the cysts treated with mebendazole showed degenerative changes. During long-term follow-up 25% of these cysts showed relapse, which took place within 2 years in 78% of cases. Further treatment with albendazole induced degenerative changes in over 90% of the relapsed cysts, without induction of more frequent or more severe adverse effects, as observed during the first treatment period. Adverse effects during the first treatment period consisted of raised transaminases with albendazole (67 of 323 patients) and mebendazole (16 of 125 patients), and abdominal pain in 12 and 11% respectively. Headache occurred in eight patients taking albendazole and three taking mebendazole, abdominal distension in seven and five patients, vertigo in five and one, urticaria in five and three, jaundice in one and one, thrombocytopenia in two and none, fever in three and none, dyspepsia in two and four, and tachycardia in two and none. Six of 323 patients taking albendazole withdrew because of adverse effects compared with eight of 125 patients taking mebendazole. It appears that albendazole is more effective than mebendazole in the treatment of hydatid cysts caused by E. granulosis and that both the intensity and frequency of the usually mild adverse effects are comparable.

Neurocysticercosis
The treatment of subarachnoid and intraventricular neurocysticercosis is controversial. In a randomized trial, 36 patients with subarachnoid and intraventricular cysticercosis were assigned to albendazole 15 mg/kg/day ($n = 16$) or 30 mg/kg/day ($n = 20$) plus dexamethasone for 8 days (17). The results were in favor of the higher dose, with larger cyst reduction on MRI scans at 90 and 180 days and higher albendazole sulfoxide concentrations in the plasma. A single dose was insufficient in intraventricular and giant cysts. Adverse effects were similar in the two groups. Three patients in each group had new headaches or an increase in headaches, one in each group had seizures, and one had focal paresthesia. Rash and hyperthermia occurred in two patients taking high-dose albendazole and each occurred in one patient taking low-dose albendazole. Six patients who took low-dose albendazole and nine who took the high dose developed a leukocytosis or increased alanine transaminase activities to over three times the upper limit of normal. These laboratory abnormalities had disappeared by 30 days. One patient developed glucocorticoid-related hyperglycemia.

Pediculosis capitis
Pediculosis capitis, caused by head lice, is one of the most common human ectoparasitic infections, infesting schoolchildren of all socioeconomic groups. In 150 children, the usefulness of albendazole was studied in the treatment of pediculosis capitis in combination with 1% permethrin or alone (18). Group 1 ($n = 30$) took a single dose of albendazole 400 mg; group 2 ($n = 30$) took albendazole 400 mg for 3 days; group 3 ($n = 30$) used permethrin 1%; group 4 ($n = 30$) used permethrin 1% and took a single dose of albendazole 400 mg; and group 5 ($n = 30$) used permethrin 1% and took albendazole 400 mg for 3 days. Those

who took albendazole also took a dose of albendazole 400 mg 1 week later. The success rates after 2 weeks were 62%, 67%, 80%, 85%, and 82% respectively. There were no significant differences among the five groups. The results suggested that albendazole is effective against pediculosis capitis and that there is no synergistic effect between albendazole and 1% permethrin. Adverse effects were not reported in this study.

Trichuriasis

In a randomized trial in Mexico (19) 622 children with *Trichuris* were randomized to either albendazole 400 mg/day for 3 days, one dose of albendazole 400 mg, or one dose of pyrantel 11 mg/kg. The aim was to study efficacy and the effects on growth. After three courses at 1 year the level of infection with *Trichuris* was reduced by 99% in the 3-day albendazole treatment group, by 87% in the single-dose albendazole treatment group, and by 67% in the pyrantel group. There were no significant differences in the increases in height, weight, or arm circumference, but contrary to expectations there was a lower increase in the thickness of the triceps skin fold in those given 3-day courses of albendazole. This was only found in the patients with lower pretreatment *Trichuris* stool egg counts. These findings suggest that although elimination of *Trichuris* may promote growth in children, albendazole in a dose of 1200 mg/kg every 4 months may have an independent negative effect on growth. In an accompanying commentary (20) it was concluded that the suggestion that relatively high doses of albendazole may affect growth deserves study, but that this possible effect must be weighed against the negative effect of prolonged helminthic infestation on children's health, growth, and cognitive function. However, it is unlikely that high-dose treatment will be standard in mass-treatment campaigns, and these results should not deter the use of single-dose albendazole in mass-treatment programs in high-risk populations.

In 110 children with ascariasis or trichuriasis the efficacy of a single dose of albendazole 400 mg has been compared with that of nitazoxanide 100 mg bd for 3 days in children aged 1–3 years and 200 mg bd for 3 days in children aged 4–11 years (21). Nitazoxanide cured 89 and 89% of the cases of ascariasis and trichuriasis respectively. Albendazole cured 91 and 58% of the cases of ascariasis and trichuriasis respectively. Abdominal pain ($n = 9$), nausea ($n = 1$), diarrhea ($n = 2$), and headache ($n = 1$) were reported as mild adverse effects in 105 patients who took nitazoxanide, and abdominal pain ($n = 1$), nausea ($n = 1$), and vomiting ($n = 1$) were reported as adverse effects in 54 patients who took albendazole. All the adverse events were mild and transient and drug withdrawal was not necessary.

Placebo-controlled studies

Albendazole has been used in the treatment and prophylaxis of microsporidiosis in patients with AIDS. In a small, double-blind, placebo-controlled trial from France (22) the efficacy and safety of treatment with albendazole was studied in four patients treated with albendazole 400 mg bd for 3 weeks and in four patients treated with placebo.

Microsporidia were cleared in all patients given albendazole but in none of those given placebo. Afterwards all eight patients were again randomized to receive either maintenance treatment with albendazole 400 mg bd or no treatment for the next 12 months; none of the three patients taking maintenance treatment had a recurrence, while three of the five who took no maintenance therapy developed a recurrence. During the double-blind part of the trial there were no serious adverse effects in the patients who took albendazole, although two complained of headache, one of abdominal pain, one had raised transaminase activities, and one had thrombocytopenia. However, half the patients were also taking anti-HIV triple therapy, which makes it difficult to assess these abnormalities. The authors concluded that the adverse effects were not serious and did not hinder maintenance therapy. The tentative conclusion derived from these findings is that albendazole may be useful in the treatment of microsporidiosis, which in patients with AIDS often leads to debilitating chronic diarrhea and is difficult to treat.

Use in non-infective conditions

The efficacy of albendazole has been evaluated in a few patients with either hepatocellular carcinoma ($n = 1$) or colorectal cancer and hepatic metastases refractory to other forms of treatment ($n = 8$) (23). Apart from hematological and biochemical indices, the tumor markers carcinoembryonic antigen (CEA) and alpha-fetoprotein (AFP) were measured to monitor treatment efficacy. One other patient with a neuroendocrine cancer and a mesothelioma was treated on a compassionate basis and only monitored for adverse effects. Albendazole was given orally in a dose of 10 mg/kg/day in two divided doses for 28 days. Albendazole reduced CEA in two patients and in the other five patients with measurable tumor markers, serum CEA or AFP was stabilized in three. In the seven patients who completed this pilot study, albendazole was well tolerated and there were no significant changes in any hematological, kidney, or liver function tests. However, three patients were withdrawn because of severe neutropenia, which resulted in the death of one. Neutropenia was more frequent than is usually experienced in the treatment of hydatid disease. The authors speculated that this may relate to reduced metabolism in patients with liver cancer or liver metastases, leading to the passage of unmetabolized drug into the circulation.

General adverse effects

As with other antihelminthic drugs, the general adverse effects of albendazole can reflect the destruction of the parasite rather than a direct action of the drug; pyrexia is likely to be seen, even in the absence of other problems. Albendazole was well tolerated in 30-day courses of 10–14 mg/kg/day separated by 2-week intervals.

Its adverse effects are similar to those of mebendazole and are possibly more common because of better and more reliable absorption.

The direct adverse effects of albendazole are few and usually minor, and consist of gastrointestinal upsets, dizziness, rash, and alopecia, which usually do not require drug withdrawal. Early pyrexia and neutropenia can also occur. Cyst rupture can also occur, as with mebendazole. About 15% of patients treated with albendazole at higher doses develop raised serum transaminases, necessitating careful monitoring and sometimes withdrawal of treatment after prolonged use. Careful monitoring of leukocyte and platelet counts is also indicated. The possibility of teratogenicity and embryotoxicity from animal studies suggests that the drug should be avoided in pregnancy.

Organs and Systems

Nervous system

Used in the treatment of neurocysticercosis, albendazole (like praziquantel) can cause a CSF syndrome characterized by fever, headaches, meningism, and exacerbation of some or many of the neurological signs of the disease; it is thought to be due to a local reaction to dying and dead larvae and can be attenuated by prednisone (SED-12, 707) (24,25).

Since neurocysticercosis is a neurological infection, it is not surprising that when treating it with any drug some of the neurological reactions to that drug (or to the death of the parasite) are particularly pronounced. For example, with a dose of 1.5 mg/kg continued for some time in cases of neurocysticercosis, a majority of patients initially develop intolerance in the form of headache, vomiting, fever, and occasionally diplopia and meningeal irritation (26). Even shorter and less intensive treatment has produced similar effects. However, all of these symptoms are probably due to the death of the parasite, and if therapy is continued they usually disappear within a few days. Nevertheless, they can be alarming and demand treatment. Data from large studies mention somnolence and even transient hemiparesis as incidental adverse effects.

- A 35-year-old Chinese woman with a 26-year history of persistent headache as the dominant symptom of neurocysticercosis, relieved only by diuretic treatment, took albendazole 800 mg/day for 3 weeks (27). After 2 days she reported increased headache plus vertigo, nausea, and vomiting. Analysis of the cerebrospinal fluid showed increased white cell pleocytosis. Her headache improved after the addition of a glucocorticoid and 3 months after withdrawal of albendazole her headache had not recurred.

Very rarely, in cases of neurocysticercosis, the reaction of the nervous system to the death of the parasite is extremely violent. In one case cerebral edema resulted in permanent neurological damage (28), while other patients have suffered hydrocephalus or acute intracranial hypertension requiring treatment, for example with glucocorticoids or mannitol (29).

Albendazole has sometimes aggravated extrapyramidal disorders or precipitated seizures in patients with prior epileptic symptoms. The risk of intracranial hypertension has led some to suggest that glucocorticoids should be given preventively when using albendazole in neurocysticercosis (30); however, dexamethasone can interact with

albendazole, increasing its plasma concentrations (31), and it is not clear whether this might produce new problems.

Encephalopathy is an adverse event related to the treatment of *L. loa* with diethylcarbamazine or ivermectin, and it has also been related to albendazole (13).

- A 55-year-old woman from Cameroon took oral albendazole 200 mg bd for a symptomatic *L. loa* infection with microfilaremia of 152 microfilariae/ml and a *Mansonella perstans* infection of 133 microfilariae/ml. Three days after the start of therapy she developed an encephalopathy. Albendazole was withdrawn and she recovered without any specific treatment within the next 16 hours. On day 4, the *L. loa* microfilarial count was 29 microfilariae/ml.

The clinical presentation, the interval after starting treatment, the evolution of the episode, and the results of cerebral spinal fluid analysis and electroencephalography in this case were similar to those seen in cases of encephalopathy following treatment of *L. loa* with ivermectin or diethylcarbamazine. However, pretreatment filaremia was relatively low and *L. loa* microfilariae were not detectable in the cerebral spinal fluid. Thus, pre-existing conditions might increase the susceptibility to encephalopathy.

Sensory systems

An allergic conjunctivitis was seen in cases of industrial occupational skin reactions to albendazole (32).

Eye pain has been attributed to albendazole in a patient with ocular cysticercosis (33).

- A 19-year-old Nepalese housewife with horizontal diplopia due to orbital cysticercosis was given albendazole 15 mg/kg/day for 8 days. After 3 days she developed nausea, vomiting, and distressing nocturnal left eye pain. She was reluctant to continue taking albendazole and her symptoms settled after a short course of oral analgesia. Later ocular examinations did not show any residual orbital cyst.

Hematologic

There have been various reports of bone marrow depression. In one study (34) two of 20 patients had a reversible drop in leukocyte count. Pancytopenia, reversible on withdrawal, has been documented in an elderly woman (35). Even with high doses neutropenia occurs in under 1% of cases. In the older literature an occasional hematological death was reported.

- A 68-year-old man with a large cystic lung mass due to echinococcosis was given albendazole and 2 weeks later developed septic shock with severe pancytopenia (36). He died after 10 days with no marrow recovery. Autopsy was consistent with albendazole-induced pancytopenia.

The clearance of albendazole sulfoxide is impaired in liver disease, which was relevant in this case. The authors suggested that frequent serial monitoring of blood counts is warranted in patients with liver disease.

A megakaryocytic thrombocytopenia attributed to albendazole has been reported (37).

- A 25-year-old woman who had been taking albendazole 13 mg/kg/day for 5 months for hepatic and pulmonary echinococcosis developed fatigue, bleeding gums, and prolonged menstrual bleeding. She had ecchymoses and petechiae over her legs, marked thrombocytopenia (10×10^9/l), a mild iron deficiency anemia, and a normal leukocyte count. There was no antiplatelet immunoglobulin. A bone marrow aspiration showed absent megakaryocytes with normal granulocytes and mild erythroid hyperplasia. A cytogenetic study of the bone marrow showed normal karyotype and immunophenotype. The albendazole was withdrawn and oral iron given. At follow-up 2 months later all laboratory abnormalities had resolved.

Gastrointestinal

With a single oral dose of albendazole 400 mg, there is usually little more in the way of adverse effects than mild gastrointestinal disturbances (notably epigastric pain or dry mouth), occurring only in about 6% of patients in some large series; a few patients have abdominal pain. With higher doses, irritation of the central nervous system can lead to nausea and vomiting.

Diarrhea occurs in a few patients taking albendazole and is usually mild. However, a typical case of pseudomembranous colitis has been documented, although the patient also had AIDS and intestinal microsporidiosis and had taken a number of other drugs; the complication responded to vancomycin (38).

Liver

Even in single low doses a transient increase in transaminase activities has been repeatedly reported, generally affecting up to 13–20% of patients taking albendazole (SEDA-18, 315; 39). At the higher doses some evidence of moderate hepatitis has been claimed to be present in almost all patients, but in one series with high doses of albendazole or mebendazole for echinococcosis only 17% had a (generally slight) increase in serum transaminases, and a fair number of these had pre-existent liver disorders (40). Like various other adverse effects, the increase in transaminases may be attributable to the breakdown of liver cysts; it is almost always reversible and is usually not a reason for withdrawal; it does not become more marked during long-term treatment. A very occasional individual develops jaundice (41) or some other manifestation of hepatitis (42).

- In a 17-year-old man severe jaundice, anemia, and edema occurred 20 days after starting albendazole emulsion 12.5 mg/kg/day (43). After recovery albendazole was re-instituted in a dose of 10 mg/kg/day for 6 months without problems.

Skin

A generalized rash has sometimes been seen in patients taking albendazole (SEDA-15, 334), and skin complications (including urticaria and contact dermatitis) are a potential problem in employees in the pharmaceutical industry if they undergo heavy exposure to the drug (32).

- A 38-year-old woman with cough, eosinophilia, and pulmonary infiltrates due to visceral larva migrans from *Toxocara canis* infection took albendazole 600 mg for 8 weeks and developed slight transient skin eruptions (44).

Stevens–Johnson syndrome was reported in a man who took albendazole 400 mg/day for toxocariasis (45).

Hair

There are various well-documented reports of reversible alopecia in patients taking albendazole (SEDA-17, 358; SEDA-22, 324), which in one study occurred in 2% of cases (SEDA-18, 315) and in another study in one case of 20 (SED-13, 913) (34).

- Severe alopecia has been described in an almost 3-year-old child who took albendazole 400 mg/d for 3 days; 2 months later alopecia developed and resolved within 1 month (46).
- When one woman took 400 mg bd for 10 months for hydatid disease, she lost much of her hair; no other likely cause could be identified, and her hair growth recovered when the drug was stopped (47).

Oddly, however, a fair proportion of patients when specifically questioned seem to remark that their hair growth has actually improved during treatment (SED-13, 913) (34).

Musculoskeletal

Myalgia and arthralgia can occur in patients taking albendazole (48). However, these symptoms are often features of the disease being treated.

Second-Generation Effects

Teratogenicity

It has been emphasized that albendazole is teratogenic in animals and should not be used in pregnancy (49).

Susceptibility Factors

Age

The pharmacokinetics of albendazole/albendazole sulfoxide and praziquantel have been investigated in 20 Thai school-age children with *Giardia intestinalis* infections (50). They were randomized to a single oral dose of albendazole (400 mg) with or without a single oral dose of praziquantel (20 mg/kg). There was no significant pharmacokinetic interaction and no adverse effects.

Drug Administration

Drug formulations

In a randomized crossover study (51) in 10 healthy volunteers (ages 18–41 years, weights 55–110 kg, body mass index 19–34 kg/m^2, four men) the systemic availability of the commercially available tablet of albendazole was compared with three newer formulations:

1. an oral suspension in arachis oil and the surfactant polysorbate 80 (aimed at increasing lipid solubility);
2. an oral solution incorporated into hydroxypropyl-β-cyclodextrin (aimed at increasing water solubility);
3. a macrogol suppository (aimed at avoiding intestinal degradation of albendazole).

Compared with the results with the tablet, the systemic availability was significantly increased by the oil-surfactant suspension (4.3-fold) and the cyclodextrin solution (9.7-fold). Rectal administration of albendazole-containing macrogol suppositories failed. Seven subjects had diarrhea 4–6 hours after the administration of the solution with hydroxypropyl-β-cyclodextrin.

Drug–Drug Interactions

Antiepileptic drugs

The pharmacological interactions of the antiepileptic drugs phenytoin, carbamazepine, and phenobarbital with albendazole have been studied in 32 adults with active intraparenchymatous neurocysticercosis (52):

- nine patients took phenytoin 3–4 mg/kg/day;
- nine patients took carbamazepine 10–20 mg/kg/day;
- five patients took phenobarbital 1.5–4.5 mg/kg/day;
- nine patients took no antiepileptic drugs.

All were treated with albendazole 7.5 mg/kg every 12 hours on 8 consecutive days. Phenytoin, carbamazepine, and phenobarbital all induced the oxidative metabolism of albendazole to a similar extent in a non-enantioselective manner. In consequence, there was a significant reduction in the plasma concentration of the active metabolite of albendazole, albendazole sulfoxide.

Cimetidine

The poor intestinal absorption of albendazole, which may be enhanced by a fatty meal, contributes to difficulties in predicting its therapeutic response in echinococcosis. The effect of cimetidine co-administration on the systemic availability of albendazole has been studied in six healthy men (53). After an overnight fast, a single oral dose of albendazole (10 mg/kg) was administered on an empty stomach with water, a fatty meal, grapefruit juice, or grapefruit juice plus cimetidine. The systemic availability of albendazole was reduced by cimetidine. There were no adverse events. These results are consistent with presystemic metabolism of albendazole by CYP3A4.

Protease inhibitors

- A 40-year-old man with HIV infection, taking HAART, developed alveolar echinococcosis (54). He was given albendazole 400 mg bd. Within 2 weeks he developed pancytopenia. Zidovudine, lamivudine, nelfinavir, and pyrimethamine/sulfamethoxazole were withdrawn. Full recovery of the bone marrow occurred within 10 weeks. HAART was changed to stavudine, abacavir, and lopinavir/ritonavir. Instead of albendazole, he was given mebendazole. Since a pharmacokinetic interaction between albendazole and the HIV protease inhibitors was suspected to have contributed to the development of pancytopenia, mebendazole plasma concentrations were carefully monitored. Therapeutic mebendazole plasma concentrations were reached after 6 weeks at a dose of 150 mg bd instead of the usual dosage of 500–1500 mg bd.

These data suggest that there may be a significant pharmacokinetic interaction between benzimidazoles and CYP3A4 inhibitors like protease inhibitors. This may result in significantly increased serum benzimidazole concentrations, when benzimidazole doses are not adjusted under concomitant treatment with HIV protease inhibitors.

References

1. Venkatesan P. Albendazole. J Antimicrob Chemother 1998;41(2):145–7.
2. Bethony J, Brooker S, Albonico M, Geiger SM, Loukas A, Diemert D, Hotez PJ. Soil-transmitted helminth infections: ascariasis, trichuriasis and hookworm. Lancet 2006;367:1521-32.
3. Stepek G, Buttle DJ, Duce IR, Behnke JM. Human gastrointestinal nematode infections: are new control methods required? Int J Exp Pathol 2006;87:325-41.
4. Bartlett MS, Edlind TD, Lee CH, Dean R, Queener SF, Shaw MM, Smith JW. Albendazole inhibits *Pneumocystis carinii* proliferation in inoculated immunosuppressed mice. Antimicrob Agents Chemother 1994;38(8):1834–7.
5. Cook GC. Tropical medicine. Postgrad Med J 1991;67(791):798–822.
6. Reynoldson JA, Behnke JM, Gracey M, Horton RJ, Spargo R, Hopkins RM, Constantine CC, Gilbert F, Stead C, Hobbs RP, Thompson RC. Efficacy of albendazole against *Giardia* and hookworm in a remote Aboriginal community in the north of Western Australia. Acta Trop 1998;71(1):27–44.
7. Fallah M, Mirarab A, Jamalian F, Ghaderi A. Evaluation of two years of mass chemotherapy against ascariasis in Hamadan, Islamic Republic of Iran. Bull World Health Organ 2002;80(5):399–402.
8. Ayles HM, Corbett EL, Taylor I, Cowie AG, Bligh J, Walmsley K, Bryceson AD. A combined medical and surgical approach to hydatid disease: 12 years' experience at the Hospital for Tropical Diseases, London. Ann R Coll Surg Engl 2002;84(2):100–5.
9. Chai J, Menghebat, Jiao W, Sun D, Liang B, Shi J, Fu C, Li X, Mao Y, Wang X, Dolikun, Guliber, Wang Y, Gao F, Xiao S. Clinical efficacy of albendazole emulsion in treatment of 212 cases of liver cystic hydatidosis. Chin Med J (Engl) 2002;115(12):1809–13.

10. Franchi C, Di Vico B, Teggi A. Long-term evaluation of patients with hydatidosis treated with benzimidazole carbamates. Clin Infect Dis 1999;29(2):304–9.

11. Tsague-Dongmo L, Kamgno J, Pion SD, Moyou-Somo R, Boussinesq M. Effects of a 3-day regimen of albendazole (800 mg daily) on *Loa loa* microfilaraemia Ann Trop Med Parasitol 2002;96(7):707–15.

12. Rajshekhar V. Incidence and significance of adverse effects of albendazole therapy in patients with a persistent solitary cysticercus granuloma. Acta Neurol Scand 1998;98(2):121–3.

13. Sotelo J, Jung H. Pharmacokinetic optimisation of the treatment of neurocysticercosis. Clin Pharmacokinet 1998;34(6):503–15.

14. Zulu I, Veitch A, Sianongo S, McPhail G, Feakins R, Farthing MJ, Kelly P. Albendazole chemotherapy for AIDS-related diarrhoea in Zambia—clinical, parasitological and mucosal responses. Aliment Pharmacol Ther 2002;16(3):595–601.

15. Barisani-Asenbauer T, Maca SM, Hauff W, Kaminski SL, Domanovits H, Theyer I, Auer H. Treatment of ocular toxocariasis with albendazole. J Ocul Pharmacol Ther 2001;17(3):287–94.

16. Dogru D, Kiper N, Ozcelik U, Yalcin E, Gocmen. Medical treatment of pulmonary hydatid disease: for which child? Parasitol Int 2005;54:135-8.

17. Gongora-Rivera F, Soto-Hernandez JL, Gonzalez Esquivel D, Cook HJ, Marquez-Caraveo C, Hernandez Davila R, Santos-Zambrano J. Albendazole trial at 15 or 30 mg/kg/day for subarachnoid and intraventricular cysticercosis. Neurology 2006;66:436-8.

18. Akisu C, Delibas SB, Aksoy U. Albendazole: single or combination therapy with premethrin against pediculosis capitis. Pediatr Dermatol 2006;23(2):179-82.

19. Forrester JE, Bailar JC 3rd, Esrey SA, Jose MV, Castillejos BT, Ocampo G. Randomised trial of albendazole and pyrantel in symptomless trichuriasis in children. Lancet 1998;352(9134):1103–8.

20. Winstanley P. Albendazole for mass treatment of asymptomatic trichuris infections. Lancet 1998;352(9134):1080–1.

21. Juan JO, Lopez Chegne N, Gargala G, Favennec L. Comparative clinical studies of nitazoxanide, albendazole and praziquantel in the treatment of ascariasis, trichuriasis and hymenolepiasis in children from Peru. Trans R Soc Trop Med Hyg 2002;96(2):193–6.

22. Molina JM, Chastang C, Goguel J, Michiels JF, Sarfati C, Desportes-Livage I, Horton J, Derouin F, Modai J. Albendazole for treatment and prophylaxis of microsporidiosis due to *Encephalitozoon intestinalis* in patients with AIDS: a randomized double-blind controlled trial. J Infect Dis 1998;177(5):1373–7.

23. Morris DL, Jourdan JL, Pourgholami MH. Pilot study of albendazole in patients with advanced malignancy. Effect on serum tumor markers/high incidence of neutropenia. Oncology 2001;61(1):42–6.

24. Teggi A, Lastilla MG, De Rosa F. Therapy of human hydatid disease with mebendazole and albendazole. Antimicrob Agents Chemother 1993;37(8):1679–84.

25. Desser KB, Baden M. Allergic reaction to pyrvinium pamoate. Am J Dis Child 1969;117(5):589.

26. Escobedo F, Penagos P, Rodriguez J, Sotelo J. Albendazole therapy for neurocysticercosis. Arch Intern Med 1987;147(4):738–41.

27. Finsterer J, Li M, Rasmkogeler K, Auer H. Chronic long-standing headache due to neurocysticercosis. Headache 2006;46:523-4.

28. Noboa C. Albendazole therapy for giant subarachnoid cysticerci. Arch Neurol 1993;50(4):347–8.

29. Garcia HH, Gilman RH, Horton J, Martinez M, Herrera G, Altamirano J, Cuba JM, Rios-Saavedra N, Verastegui M, Boero J, Gonzalez AE. Albendazole therapy for neurocysticercosis: a prospective double-blind trial comparing 7 versus 14 days of treatment. Cysticercosis Working Group in Peru. Neurology 1997;48(5):1421–7.

30. Del Brutto OH. Clues to prevent cerebrovascular hazards of cysticidal drug therapy. Stroke 1997;28(5):1088.

31. Takayanagui OM, Lanchote VL, Marques MP, Bonato PS. Therapy for neurocysticercosis: pharmacokinetic interaction of albendazole sulfoxide with dexamethasone. Ther Drug Monit 1997;19(1):51–5.

32. Macedo NA, Pineyro MI, Carmona C. Contact urticaria and contact dermatitis from albendazole. Contact Dermatitis 1991;25(1):73–5.

33 Wong YC, Goh KY, Choo CT, Seah LL, Rootman J. An unusual cause of acquired horizontal diplopia in a young adult. Br J Opthalmol 2005;89:390-1.

34. Steiger U, Cotting J, Reichen J. Albendazole treatment of echinococcosis in humans: effects on microsomal metabolism and drug tolerance. Clin Pharmacol Ther 1990;47(3):347–53.

35. Fernandez FJ, Rodriguez-Vidigal FF, Ledesma V, Cabanillas Y, Vagace JM. Aplastic anemia during treatment with albendazole. Am J Hematol 1996;53(1):53–4.

36 Opatrny L, Prichard R, Snell L, Maclean JD. Death related to albendazole-induced pancytopenia: case report and review. Am J Trop Med Hyg 2005;72(3):291-4.

37. Yildiz BO, Haznedaroglu IC, Coplu L. Albendazole-induced amegakaryocytic thrombocytopenic purpura. Ann Pharmacother 1998;32(7–8):842.

38. Shah V, Marino C, Altice FL. Albendazole-induced pseudomembranous colitis. Am J Gastroenterol 1996;91(7):1453–4.

39. Horton RJ. Albendazole in treatment of human cystic echinococcosis: 12 years of experience. Acta Trop 1997;64(1–2):79–93.

40. Teggi A, Lastilla MG, Grossi G, Franchi C, De Rosa F. Increase of serum glutamic-oxaloacetic and glutamic-pyruvic transaminases in patients with hydatid cysts treated with mebendazole and albendazole. Mediterr J Infect Parasit Dis 1995;10:85–90.

41. Choudhuri G, Prasad RN. Jaundice due to albendazole. Indian J Gastroenterol 1988;7(4):245–6.

42. Luchi S, Vincenti A, Messina F, Parenti M, Scasso A, Campatelli A. Albendazole treatment of human hydatid tissue. Scand J Infect Dis 1997;29(2):165–7.

43 Silva MA, Mirza DF, Bramhall SR, Mayer AD, McMaster P, Buckels JAC. Treatment of hydatid disease of the liver: evaluation of a UK experience. Dig Surg 2004;21:227-34.

44. Inoue K, Inoue Y, Arai T, Nawa Y, Kashiwa Y, Yamamoto S, Sakatani M. Chronic eosinophilic pneumonia due to visceral larva migrans. Intern Med 2002;41(6):478–82.

45. Dewerdt S, Machet L, Jan-Lamy V, Lorette G, Therizol-Ferly M, Vaillant L. Stevens–Johnson syndrome after albendazole. Acta Dermatol Venereol 1997;77(5):411.

46. Herdy R. Alopecia associated to albendazole: a case report. An Bras Dermatol 2000;75:715–9.

47. Al Karawi M, Kasawy MI, Mohamed AL. Hair loss as a complication of albendazole therapy. Saudi Med J 1988;9:530.

48. Supali T, Ismid IS, Ruckert P, Fischer P. Treatment of *Brugia timori* and *Wuchereria bancrofti* infections in Indonesia using DEC or a combination of DEC and

albendazole: adverse reactions and short-term effects on microfilariae. Trop Med Int Health 2002;7(10):894–901.

49. Bialek R, Knobloch J. Parasitare Infektionen in der Schwangerschaft und konnatale Parasitosen. II. Teil: Helmintheninfektionen. [Parasitic infections in pregnancy and congenital parasitoses. II. Helminth infections.] Z Geburtshilfe Neonatol 1999;203(3):128–33.
50. Pengsaa K, Na-Bangchang K, Limkittikul K, Kabkaew K, Lapphra K, Sirivichayakul, Wisetsing P, Pojjaroen-Anant C, Chanthavanich P, Subchareon A. Pharmacokinetic investigation of albendazole and praziquantel in Thai children infected with Giardia intestinalis. Ann Trop Med Parasitol 2004;98:349-57.
51. Rigter IM, Schipper HG, Koopmans RP, van Kan HJM, Frijlink HW, Kager PA, Guchelaar HJ. Relative bioavailability of three newly developed albendazole formulations: a randomized crossover study with healthy volunteers. Antimicrobial Agents Chemother 2004;48:1051-4.
52. Lanchote VL, Garcia FS, Dreossi SA, Takayanagui OM. Pharmacokinetic interaction between albendazole sulfoxide enantiomers and antiepileptic drugs in patients with neurocysticercosis. Ther Drug Monit 2002;24(3):338–45.
53. Nagy J, Schipper HG, Koopmans RP, Butter JJ, Van Boxtel CJ, Kager PA. Effect of grapefruit juice or cimetidine coadministration on albendazole bioavailability. Am J Trop Med Hyg 2002;66(3):260–3.
54. Zingg W, Renner-Schneiter EC, Pauli-Magnus C, Renner EL, van Overbeck J, Schlapfer E, Weber M, Weber R, Opravil M, Gottstein B, Speck RF, and the Swiss HIV Cohort Study. Alveolar echinococcosis of the liver in an adult with human immunodeficiency virus type-1 infection. Infection 2004;32:299-302.

Amocarzine

General Information

Amocarzine is an antifilarial antihelminthic drug, derived from amoscanate, that is active against the adult worms of *Onchocerca volvulus*.

Although ivermectin is very effective in the treatment of onchocerciasis, a fully effective and safe macrofilaricidal drug is still lacking. In animal studies of filarial infections, amocarzine showed promise as a macrofilaricidal drug and was afterward extensively tried in humans. Promising results were obtained in the early 1990s in Ecuador with amocarzine 3 mg/kg bd for 3 days and with acceptable adverse effects, including dizziness, itching, and rash. There were reversible neurological symptoms, such as impaired coordination and a positive Romberg's sign, in 4–12% of patients (1). Amocarzine was later tried in Africa in higher doses, with less good results. In a study from Ghana (2), the combination of ivermectin and amocarzine was not more effective than ivermectin alone. Adverse effects were more severe with amocarzine alone compared with ivermectin alone or with amocarzine preceded by ivermectin. Mazzotti-type reactions, such as itching, rash, peripheral sensory phenomena, and swellings, were all more severe or frequent after amocarzine than after ivermectin. Pretreatment with ivermectin markedly reduced these adverse reactions but did not affect other symptoms, such as dizziness and gaze-evoked nystagmus, suggesting that these adverse effects were probably directly drug-related and not a reaction to dying worms.

References

1. Guderian RH, Anselmi M, Proano R, Naranjo A, Poltera AA, Moran M, Lecaillon JB, Zak F, Cascante S. Onchocercacidal effect of three drug regimens of amocarzine in 148 patients of two races and both sexes from Esmeraldas, Ecuador. Trop Med Parasitol 1991;42(3):263–85.
2. Awadzi K, Opoku NO, Attah SK, Addy ET, Duke BO, Nyame PK, Kshirsagar NA. The safety and efficacy of amocarzine in African onchocerciasis and the influence of ivermectin on the clinical and parasitological response to treatment. Ann Trop Med Parasitol 1997;91(3):281–96.

Amodiaquine

General Information

Amodiaquine is a Mannich base derivative related to chloroquine. While it is generally considered equivalent to chloroquine, more recent studies have shown that amodiaquine is superior to chloroquine in tackling resistant strains of *Plasmodium falciparum*, although there may be cross-resistance to chloroquine (SEDA-20, 260).

Compared with chloroquine, amodiaquine was more effective and better tolerated in outpatients with uncomplicated malaria tropica in Kenya (SEDA-20, 260).

Because of its adverse effects, amodiaquine is no longer in use in the countries of the European Union or the USA, and was dropped from malaria control programs by the WHO in 1990 (SEDA-20, 260). However, it remains in use in other areas, including Africa and Oceania (SEDA-18, 287).

Observational studies

Amodiaquine has been suggested to be effective and safe in pregnancy, but its use has been limited because of previous reports of neutropenia and lymphopenia. Amodiaquine has been studied alone or in combination with sulfadoxine + pyrimethamine in 900 women with P. falciparum malaria in Ghana, who were randomly assigned to chloroquine (n = 225), sulfadoxine + pyrimethamine (n = 225), amodiaquine (n = 225), or amodiaquine plus sulfadoxine + pyrimethamine (n = 225) (1). The primary outcome measure was parasitological failure at day 28 of treatment. Parasitemia, hemoglobin concentration, white blood cell count, and liver function were assessed on days 3, 7, 14, and 28, and reports of adverse events were solicited at each visit. At day 28, parasitological failure was 14%, 11%, 3%, and 0% in the women assigned chloroquine, sulfadoxine + pyrimethamine, amodiaquine, and amodiaquine with sulfadoxine + pyrimethamine respectively. General weakness, vomiting, dizziness, and nausea were the most frequent adverse effects and were more common in those who took amodiaquine and amodiaquine with sulfadoxine + pyrimethamine than in those who took sulfadoxine. However, there were no major changes in the white blood cell, bilirubin, and liver enzyme profiles during or at the end of the study. Of the 900 women, 711 (79%), were followed up to delivery and for 6 weeks after. The proportion with peripheral parasitemia was significantly lower in those who took amodiaquine + sulfadoxine than in those who took chloroquine (2% versus 10%), but there was no significant difference in the occurrence of placental parasitemia across the four treatment groups. No maternal deaths were recorded and 7 babies had extra digits (one chloroquine, five amodiaquine, and one amodiaquine with sulfadoxine + pyrimethamine).

This suggests that amodiaquine, alone or in combination with sulfadoxine + pyrimethamine, is associated with minor adverse effects but is efficacious in malaria in pregnancy in places where falciparum malaria is still sensitive to these drugs, as is the case in much of West Africa.

Organs and Systems

Cardiovascular

Prolongation of the QT interval is a recognized effect of 4-aminoquinolines. In 20 adult Cameroonian patients with non-severe falciparum malaria treated with amodiaquine (total dose 30 mg/kg or 35 mg/kg over 3 days) there was asymptomatic sinus bradycardia ($n = 16$) and prolongation of the PQ, QRS, and QT intervals at the time of maximum cumulative concentration of drug (day 2 of treatment) [2].

Sensory systems

Corneal and conjunctival changes, which included intralysosomal membranous and amorphous inclusions in the epithelial cells, as well as abnormal retinal test responses, were reported in a man who took amodiaquine for 1 year. Follow-up over the years after withdrawal showed a reduction in the abnormalities. There are no data on possible retinal changes similar to those seen with chloroquine.

In 69 children given 35 mg/kg over 3 days (SEDA-16, 306), parasitemia was cleared in all but one case. Tolerance was good, except that there was a fairly high incidence of conjunctival hyperemia.

Hematologic

The principal reason against recommending amodiaquine for malaria prophylaxis is the reporting of agranulocytosis, occasionally associated with hepatitis (SEDA-12, 241). Since specific IgG antibodies, which lead to leukopenia, can be detected, all this suggests that the agranulocytosis is immune-mediated.

It takes substantial doses for a couple of weeks to cause the adverse hematological effects (SEDA-16, 692; SEDA-17, 327). If this is correct, amodiaquine could still be of use for short intensive courses of treatment in areas of chloroquine resistance.

The protective effect of intermittent amodiaquine during the high-risk seasons on anemia and malarial fever has been investigated in a double-blind, randomized, placebo-controlled study in 291 infants aged 12?16 weeks, of whom 146 received amodiaquine for 3 days (10, 10, and 5 mg/kg) every 2 months for 6 months, either alone or combined with an iron supplement [3]. The protective efficacy of intermittent amodiaquine for malarial fever and anemia compared with placebo was 65% (95% CI = 42, 77%) and 67% (34, 84) respectively. Except for one infant with a neutrophil count of $1.5 \times 10^9/l$, there were no adverse effects on neutrophils or the liver, and in particular no cases of agranulocytosis or liver failure. Furthermore, no patient died or stopped taking the drugs because of severe adverse effects. Thus, at least for this dosage regimen in infants, the incidence of severe hematological and hepatic adverse events is under 3%.

Gastrointestinal

Gastrointestinal complaints (nausea, vomiting, diarrhea, or constipation) are not uncommon with amodiaquine (SEDA-11, 587).

Skin

Abnormal pigmentation of the palate, nail beds, and the skin of the face and neck has been reported. The duration of such pigmentation after withdrawal is unknown. The effects on the nails resemble those seen with chloroquine (SEDA-12, 692; SEDA-12, 241).

Nails

Abnormal pigmentation of nail beds has been reported. The duration of such pigmentation after withdrawal is unknown. The effects on the nails resemble those seen with chloroquine (SEDA-12, 692; SEDA-12, 241).

References

1. Tagbor H, Bruce J, Browne E, Randal A, Greenwood B, Chandramohan D. Efficacy, safety, tolerability of amodiaquine plus sulphadoxine–pyrimethamine used alone or in combination for malaria treatment in pregnancy: a randomized trial. Lancet 2006;368:1349-56.
2. Ngouesse B, Basco LK, Ringwald P, Keundjian A, Blackett KN. Cardiac effects of amodiaquine and sulfadoxine-pyrimethamine in malaria-infected African patients. Am J Trop Med Hyg 2001;65(6):711–6.
3. Massaga JJ, Kitua AY, Lemnge MM, Akida JA, Malle LN, Ronn AM, Theander TG, Bygbjerg IC. Effect of intermittent treatment with amodiaquine on anaemia and malarial fevers in infants in Tanzania: a randomised placebo-controlled trial. Lancet 2003;361:1853-60.

Amopyroquine

General Information

Amopyroquine is a 4-aminoquinoline, structurally related to amodiaquine. It is not a new compound, but it is of renewed interest as a result of the extensive occurrence of resistance to chloroquine and the adverse effects of prophylactic amodiaquine. In a study in 152 patients with malaria, the efficacy of a 12 mg/kg, given as two intramuscular injections of 6 mg/kg 24 hours apart, was described as good (1). All the patients became apyrexial and there was clearance of parasites on day 7 in 143 cases; the nine who retained a low level of parasitemia were all children. In 50% of the cases, the parasite had been chloroquine-resistant. The drug was well tolerated, and there were no major adverse effects.

Reference

1. Gaudebout C, Pussard E, Clavier F, Gueret D, Le Bras J, Brandicourt O, Verdier F. Efficacy of intramuscular amopyroquin for treatment of Plasmodium falciparum malaria in the Gabon Republic. Antimicrob Agents Chemother 1993;37(5):970–4.

Antimony and antimonials

General Information

Antimony is a brittle, bluish-white, metallic element (symbol Sb; atomic no 51). The symbol Sb comes from the Latin word stibium.

Antimony is found in such minerals as dyscrasite, jamesonite, kermesite, pyrargyrite, stephanite, tetrahedrite, and zinkenite.

The Arabic word for antimony stibnite or antimony trisulfate was kohl, from which the word alcohol ultimately derives (1). Antimonious ores were sometimes confused with lead ores, and alquifou was the name of a Cornish lead ore that looked like antimony and was used by potters to give a green glaze to earthenware. The word that the Quechua Indians of Peru use for antimony is surúcht, which gives soroche, a synonym for mountain sickness, which antimony was thought to cause.

Antimony salts have in the past found many uses in medicine, and antimony compounds, especially pentavalent ones, are still used to treat Schistosoma japonicum infestation and leishmaniasis (2). Antimony is also used as an emetic. Attention is being paid to the anticancer potential of antimony compounds (3,4). As with many other metals, occupational and environmental exposure is possible and can act additively with medical exposure.

Common adverse effects of antimony treatment include anorexia, nausea, vomiting, muscle ache, headache, lethargy, and bone and joint pain.

Antimony has been suggested to be a causal factor in sudden infant death syndrome, since fungal transformation of fire retardants containing antimony in cot mattresses will lead to the formation of stibine (SbH_3). However, the involvement of stibine in cot death is most unlikely (5,6).

Salts of antimony

Meglumine antimoniate

Meglumine antimoniate is a pentavalent antimonial chemically similar to sodium stibogluconate and is considered to have similar efficacy and toxicity. Meglutamine antimoniate solution contains pentavalent antimony 8.5% and stibogluconate 10%.

Sodium stibogluconate

Sodium stibogluconate is a pentavalent antimonial that contains pentavalent antimony 10%.

Stibocaptate

Stibocaptate is a trivalent antimony compound, whose toxic effects, especially its acute adverse effects, are similar to those of the pentavalent compounds.

Uses

Antimony salts have in the past found many uses in medicine, and antimony compounds, especially pentavalent ones, are still used to treat Schistosoma japonicum infestation and leishmaniasis (2). The standard treatment of most South American cutaneous leishmaniasis is systemic, because of the propensity of the parasites to spread to mucous membranes. The drugs in common use remain parenteral, and are fairly toxic. Sodium stibogluconate, in a dose of 20 mg/kg for 30 days, remains the gold standard, with liposomal amphotericin a possible alternative. Both drugs are also the preferred choice for all strains of visceral leishmaniasis. Stibogluconate achieves a cure in certain cases (7,8) and is reasonably safe, although transient pancreatitis, musculoskeletal pains, and loss of appetite have been reported. Primary unresponsiveness (cure not obtained by the first course of treatment) is being increasingly reported (9) and is a particular problem in the Bihar region of North-East India, where a report documented primary unresponsiveness in 33% of cases (10). The new treatment options for visceral leishmaniasis have been reviewed (11).

Antimony is also used as an emetic. Attention is being paid to the anticancer potential of antimony compounds (3,4). As with many other metals, occupational and environmental exposure is possible and can act additively with medical exposure.

Pharmacokinetics

Antimony is excreted in the urine. Peak concentrations are seen at about 1–2 hours after an intramuscular injection of meglumine antimonate. Serum concentrations fall to about 10% of peak concentrations after about 8 hours. There is some accumulation of antimony during continued treatment. On a weight for weight basis children require a higher dose and tolerate antimony better. Toxicity is more likely in patients with impaired renal function, as would be expected for a drug that is mainly excreted in the urine.

Observational studies

The characteristics of 111 consecutive patients with visceral leishmaniasis in Sicily have been described (12). They were given intramuscular meglumine antimoniate (560 mg/m^2 of pentavalent antimony), generally for 21 days. There were adverse effects in 16 patients, including rash ($n = 3$) and dry cough ($n = 13$). All the adverse effects bar one (a severe urticarial rash) were transient and self-limiting and did not require drug withdrawal.

Comparative studies

In a prospective, open trial, conventional treatment with sodium stibogluconate ($n = 69$) was compared with meglumine antimoniate ($n = 58$) for cutaneous *Leishmania braziliensis* (13). The trial was too small and of too short a duration to compare efficacy reliably, but significantly fewer patients on meglumine antimoniate developed the myalgia/arthralgia, headache, and abdominal pain that are the most common adverse effects of the drug, and that tend to increase as treatment continues. About 30% developed significant adverse effects early in treatment and 70% late when stibogluconate was used, whereas 12% had early and 45% late adverse effects for antimoniate. Unfortunately, QT intervals were not monitored; a fatal dysrhythmia, usually preceded by increased QT dispersion, is the complication of sodium stibogluconate therapy most likely to lead to death.

There has been a further comparison of meglumine antimoniate ($n = 47$) and sodium stibogluconate ($n = 64$) (14). The trial was too small to examine the efficacy of the two drugs, but there were more adverse events with sodium stibogluconate, with a greater proportion with raised transaminase and amylase activities. There were no differences in electrocardiographic abnormalities between the two groups.

Allopurinol

Allopurinol and meglumine antimoniate (Glucantime™) have been evaluated in a randomized, controlled trial in 150 patients with cutaneous leishmaniasis (15). They received oral allopurinol (15 mg/kg/day) for 3 weeks, or intramuscular meglumine antimoniate (30 mg/kg/day, corresponding to 8 mg/kg/day of pentavalent antimony, for 2 weeks), or combined therapy. There were a few adverse effects in those who used allopurinol: nausea, heartburn ($n = 3$), and mild increases in transaminases ($n = 2$). These symptoms subsided on drug withdrawal.

In an open study, 72 patients each received meglumine antimoniate (60 mg/kg/day) or allopurinol (20 mg/kg/day) plus low-dose meglumine antimoniate (30 mg/kg/day) for 20 days, and each was followed for 30 days after the end of treatment (16). Only six patients in the combined treatment group complained of mild abdominal pain and nausea; however, one patient who received meglumine antimoniate developed a skin eruption.

Generalized muscle pain and weakness occurred in four patients.

Paromomycin

In patients with visceral leishmaniasis, paromomycin (12 or 18 mg/kg/d) plus a standard dose of sodium stibogluconate for 21 days was statistically more effective than sodium stibogluconate alone in producing a final cure (17). In an open, randomized comparison of sodium stibogluconate either alone ($n = 50$) or in combination with two regimens of paromomycin ($n = 52$ and $n = 48$), there was improved parasitological cure in both groups given combination therapy (18). There were no differences in adverse events or biochemical and hematological measurements between any of the treatment arms. There was one serious adverse event (myocarditis) in the sodium stibogluconate monotherapy group. It should be noted that there were insufficient auditory examinations performed to assess any ototoxic effects of paromomycin.

Placebo-controlled studies

A randomized, double-blind, placebo-controlled study of sodium stibogluconate for 10 and 20 days has been conducted in 38 US military personnel with cutaneous leishmaniasis; 19 received sodium stibogluconate for 10 days (and placebo for 10 days), and 19 received sodium stibogluconate for 20 days (19). Treatment withdrawal was necessary as a result of pancreatitis in seven patients (four in the 10-day treatment group and three in the 20-day group), and this occurred during the first 10 days of therapy in all seven patients. Myalgia occurred in 8 patients in the 10-day group and in 13 patients in the 20-day group. Patients in the 20-day group had myalgia on significantly more days than those in the 10-day group. Increases in amylase, lipase, and transaminases and falls in white blood cell count, hematocrit, and platelet count also differed significantly between the two groups.

General adverse effects

Common adverse effects of antimony treatment include anorexia, nausea, vomiting, muscle ache, headache, lethargy, and bone and joint pain.

Common adverse effects of meglumine antimonate are anorexia, nausea, vomiting, malaise, myalgia, headache, and lethargy. Muscle, bone, and joint pains have been described (SEDA-12, 710; SEDA-13, 838; 20). Cardiac toxicity and electrocardiographic changes are dose-related. The general condition of the patient with visceral leishmaniasis probably plays a crucial role in these and other adverse effects. Malnutrition is common, the immune status often severely impaired, and patients are susceptible to intercurrent infections (SEDA-12, 710).

In 96 patients with visceral, mucosal, or viscerotropic leishmaniasis, who were given sodium stibogluconate 20 mg/kg/day for 20–28 days, adverse effects were common and necessitated withdrawal of treatment in 28% of cases. They included arthralgias and myalgias (58%), pancreatitis (97%), increased transaminases (67%), headache (22%), bone marrow suppression (44%), and rash (9%) (8). Arthralgias are more likely to represent reactions to the tissue of the dead or dying parasite than true allergies to the drug.

In 53 patients with dermal leishmaniasis after kala-azar, who were given sodium stibogluconate 20 mg/kg/day intramuscularly, adverse effects were changes in electrocardiographic ST segments and T waves (7%), arthralgias (11%), allergic rashes (7%), swelling at the site of injection (5%), neuralgia (4%), and a metallic taste (6%) (21).

Organs and Systems

Cardiovascular

Electrocardiographic changes are common in patients taking antimony salts; in one group there was an incidence of 7%. The most common changes are ST segment changes, T wave inversion, and a prolonged QT interval. The role of conduction disturbances in cases of cardiac failure and sudden death is not known. Cases of sudden death have been seen early in treatment after a second injection (SEDA-13, 838; SEDA-16, 311).

Changes in the electrocardiogram depend on the cumulative dose of antimony, and sudden death can occur rarely (22).

- A 4-year-old boy with visceral leishmaniasis was given intravenous sodium stibogluconate 20 mg/kg/day (1200 mg/day) and oral allopurinol 16 mg/kg/day (100 mg tds). On day 3 he reported chest pain and a persistent cough. Electrocardiography was unremarkable. The drugs were withdrawn and 3 days later he developed a petechial rash on the legs. Sepsis and other causes of petechial rashes were ruled out. Three days after treatment was discontinued he developed ventricular fibrillation and died.

The authors suggested that patients taking antimony compounds should be observed cautiously for signs of cardiological and hematological changes.

Myocarditis with electrocardiographic changes has been well described, but the risk of dysrhythmias is usually small. There have been reports of severe cardiotoxicity, leading in some cases to death (23,24). This may largely be due to changes in physicochemical properties of the drug; one cluster of cases was associated with a high-osmolarity lot of sodium stibogluconate (23).

Because of concerns regarding the cardiac adverse effects of antimonials, it is good practice to admit patients for the duration of therapy whenever practicable. This may mean admitting otherwise fit young patients for several weeks for treatment of a non-healing ulcer. To address the safety of outpatient management, a recent small study of 13 marines in the UK showed that they could be safely managed as outpatients with daily stibogluconate injections, provided there was close monitoring of electrocardiograms and blood tests to provide early warning of bone marrow toxicity (25). Three patients developed minor electrocardiographic changes and one developed thrombocytopenia. All these adverse effects resolved when treatment was withdrawn. Patients with a predisposition to dysrhythmias (such as some with ischemic heart disease) are best treated with pentavalent antimonials as inpatients to identify and manage adverse effects early when resources allow.

The cardiac toxicity of antimony has been explored in cultured myocytes (26,27). Potassium antimony tartrate disrupted calcium handling, leading to a progressive increase in the resting or diastolic internal calcium concentration and eventual cessation of beating activity and cell death. An interaction with thiol homeostasis is also involved. Reduced cellular ATP concentrations paralleled toxicity but appeared to be secondary to other cellular changes initiated by exposure to antimony.

Even the normal dose of sodium stibogluconate can lead to both cardiotoxicity and hemotoxicity, because of its cumulative effects.

- Fatal accumulation of sodium stibogluconate occurred in a 4-year-old boy with visceral leishmaniasis treated with intravenous sodium stibogluconate 20 mg/kg (1200 mg/day) and oral allopurinol 16 mg/kg/day (100 mg tds) (22). On day 3 he reported chest pain and persistent cough, and the drugs were withdrawn. Three days later he developed a petechial rash on the legs and died with ventricular fibrillation.

Respiratory

Antimonate ore caused chronic bronchitis in 16 of 100 miners exposed (28). The chronic bronchitis was characterized by a mild slow course with ventilation disturbances. There were no cases of pneumoconiosis.

Nervous system

Headaches are common during treatment with antimony. Generalized neuralgia was reported in one study, with an incidence of 4% (SEDA-13, 838; SEDA-16, 311).

Peripheral sensory neuropathy has been described after the use of sodium stibogluconate for cutaneous leishmaniasis (SEDA-21, 300).

Sensory systems

A metallic taste due to antimonials is uncommon but probably under-reported (SEDA-16, 311).

Hematologic

Autoimmune hemolytic anemia has been described with meglumine antimoniate (29).

Thrombocytopenia has been reported in a patient with *Leishmania donovani* infection and AIDS after stibogluconate therapy for 7 days (SEDA-13, 838). There have been two further reports, one involving a patient with cutaneous leishmaniasis (occurring after 19 days of treatment), the second a man with kala-azar (who became thombocytopenic 11 days after starting therapy); in kala-azar a low platelet count is common and the count normally rises with treatment (SEDA-18, 294).

Gastrointestinal

Anorexia, nausea, and vomiting are common with antimonials (30).

Liver

Hepatotoxicity has been described, but the disease itself may play an overriding role. In 16 patients with mucosal leishmaniasis treated with meglumine antimoniate 20 mg/kg intravenously for 28 days, there were raised liver enzyme in conjunction with electrocardiographic abnormalities and/or musculoskeletal complaints in three subjects (SEDA-16, 311).

Pancreas

Pancreatitis has occasionally been reported with antimonials. In 1993, four cases were described in three reports; two of the patients were immunocompromised. One was asymptomatic, while the other three complained of abdominal pain. Rechallenge with half of the standard dose was carried out in one case and resulted in a renewed increase of serum amylase activity.

Acute pancreatitis developed during treatment with meglumine antimoniate for visceral leishmaniasis in a young boy (31).

- A 2-year-old boy with a history of intermittent high-grade fever, sweating, and abdominal distension developed visceral leishmaniasis. He was given meglumine antimoniate (Glucantime®, Rhone Poulenc, France) 5 mg/kg/day, and the dose was doubled every other day to reach 20 mg/kg/day. Two days after he had reached the full dose, his temperature returned to normal, his general condition improved, and his liver and spleen began to shrink. However, the serum amylase increased to 254 U/l. Because he was asymptomatic, treatment with meglumine antimoniate was continued. However, on day 10 he complained of vomiting and abdominal pain with rebound tenderness. Acute pancreatitis was confirmed by serum amylase and lipase values up to 1557 and 320 U/l respectively and by ultrasound findings of dilatation and edema of the pancreatic ducts. Meglumine antimoniate was withdrawn and the pancreatitis was managed conservatively. Two days later his fever increased and the spleen and liver began to enlarge. He was given allopurinol (20 mg/kg/day) and ketoconazole (5 mg/kg/day) and became afebrile; the spleen and liver began to shrink, his pancytopenia improved, and the albumin:globulin ratio and serum amylase and lipase activities returned to normal. The acute pancreatitis resolved uneventfully.

The mechanism and frequency of this adverse effect are unknown. It has been suggested that immunocompromised patients may be at a higher risk (SEDA-18, 294), but pancreatitis is in any case seen more often in patients with AIDS, irrespective of drug treatment.

Urinary tract

- Septic shock with oliguria developed soon after the first intramuscular administration of meglumine antimoniate 20 mg/kg (equivalent to 510 mg of antimony) to a patient with visceral leishmaniasis and normal renal function (32). Creatinine clearance fell to 23 ml/min.

Treatment was withdrawn, and antimony urinary excretion was measured. After the initial dose, 500 mg of antimony was recovered in the urine over 8 days (98% of the dose); 66% was eliminated within the first 48 hours. Nine days after the dose, meglumine antimoniate was reintroduced in a dosage of 11.7 mg/kg (equivalent to 300 mg of antimony) every 48 hours, with good tolerance. At that time creatinine clearance had returned to 88 ml/min. By day 14 of therapy the dosage interval was reduced to 24 hours and from day 17 to day 31 the dosage was increased to 16.6 mg/kg/day (equivalent to 425 mg of antimony). The patient eventually completely recovered, with normal renal function. Although there are no specific guidelines for dosage adjustment in renal insufficiency, monitoring antimony urinary excretion indicates that the kidneys are the almost exclusive route of elimination.

Musculoskeletal

Arthralgia is a common complaint during treatment with pentavalent antimonials and is usually dose-related (30). Muscle pain and bone pain have also been described.

- A palindromic arthropathy with effusion and pancreatitis occurred in association with stibogluconate treatment for kala-azar in a 30-year-old man on hemodialysis for chronic renal insufficiency (SEDA-16, 311).

Immunologic

As an industrial and environmental toxin, antimony trioxide can cause disturbances of immune homeostasis. Workers in antimony trioxide manufacture had reduced serum concentrations of cytokines (interleukin 2, gamma interferon) and immunoglobulin (IgG1, IgE; 33).

Death

Antimony has been suggested to be a causal factor in sudden infant death syndrome, since fungal transformation of fire retardants containing antimony in cot mattresses will lead to the formation of stibine (SbH_3). However, the involvement of stibine in cot death is most unlikely (5,6).

Sudden death has been reported during the use of stibocaptate (SEDA-11, 599) (34).

Long-Term Effects

Drug tolerance

The treatment of cutaneous leishmaniasis has been reviewed, including the use of pentavalent antimonials (35). Antimony-resistant strains continue to emerge (36), leading to the use of higher dosages of antimonials or combinations of antimonials with other compounds, such as paromomycin or gamma interferon (37).

Drug–Drug Interactions

Amiodarone

A pharmacodynamic interaction has been described between amiodarone and meglumine antimoniate, both of which prolong the QT interval; the interaction resulted in torsade de pointes (38).

- A 73-year-old man with visceral leishmaniasis was given meglumine antimoniate intramuscularly 75 mg/kg/day. At that time his QT_c interval was normal at 0.42 seconds. Three weeks later his QT_c interval was prolonged to 0.64 seconds and he was given metildigoxin 0.4 mg and amiodarone 450 mg intravenously over 8.25 hours; 12 hours later he had a cardiac arrest with torsade de pointes, which was cardioverted by two direct shocks of 300 J and lidocaine 100 mg in two bolus injections. Because he had frequent episodes of paroxysmal atrial fibrillation, he was given amiodarone 100 mg over the next 40 hours, and developed recurrent self-limiting episodes of torsade de pointes associated with QT_c interval prolongation, which responded to intravenous magnesium 1500 mg. After withdrawal of amiodarone there was no recurrence and a week later the QT_c interval was 0.48 seconds. The plasma potassium concentration was not abnormal in this case.

In view of this report it is probably wise to avoid co-administration of antimonials and amiodarone.

Amphotericin

Amphotericin can worsen stibogluconate-induced cardio-toxicity; a gap of at least 10 days between sodium stibogluconate and amphotericin is recommended (24).

References

1. Aronson JK. Here's mud in your eye. BMJ 1996;312:373.
2. Croft SL, Yardley V. Chemotherapy of leishmaniasis. Curr Pharm Des 2002;8(4):319–42.
3. Yi T, Pathak MK, Lindner DJ, Ketterer ME, Farver C, Borden EC. Anticancer activity of sodium stibogluconate in synergy with IFNs. J Immunol 2002;169(10):5978–85.
4. Tiekink ER. Antimony and bismuth compounds in oncology. Crit Rev Oncol Hematol 2002;42(3):217–24.
5. Gates PN, Pridham JB, Webber JA. Sudden infant death syndrome and volatile antimony compounds. Lancet 1995;345(8946):386–7.
6. De Wolff FA. Antimony and health. BMJ 1995;310(6989):1216–7.
7. Karki P, Koirala S, Parija SC, Hansdak SG, Das ML. A thirty day course of sodium stibogluconate for treatment of kala-azar in Nepal. Southeast Asian J Trop Med Public Health 1998;29(1):154–8.
8. Aronson NE, Wortmann GW, Johnson SC, Jackson JE, Gasser RA Jr, Magill AJ, Endy TP, Coyne PE, Grogl M, Benson PM, Beard JS, Tally JD, Gambel JM, Kreutzer RD, Oster CN. Safety and efficacy of intravenous sodium stibogluconate in the treatment of leishmaniasis: recent U.S. military experience Clin Infect Dis 1998;27(6):1457–64.
9. Khalil EA, el Hassan AM, Zijlstra EE, Hashim FA, Ibrahim ME, Ghalib HW, Ali MS. Treatment of visceral leishmaniasis with sodium stibogluconate in Sudan: management of those who do not respond. Ann Trop Med Parasitol 1998;92(2):151–8.
10. Thakur CP, Sinha GP, Pandey AK, Kumar N, Kumar P, Hassan SM, Narain S, Roy RK. Do the diminishing efficacy and increasing toxicity of sodium stibogluconate in the treatment of visceral leishmaniasis in Bihar, India, justify its continued use as a first-line drug? An observational study of 80 cases. Ann Trop Med Parasitol 1998;92(5):561–9.
11. Murray HW. Treatment of visceral leishmaniasis (kala-azar): a decade of progress and future approaches. Int J Infect Dis 2000;4(3):158–77.
12. Cascio A, Colomba C, Antinori S, Orobello M, Paterson D, Titone L. Pediatric visceral leishmaniasis in Western Sicily, Italy: a retrospective analysis of 111 cases. Eur J Clin Microbiol Infect Dis 2002;21(4):277–82.
13. Saldanha AC, Romero GA, Merchan-Hamann E, Magalhaes AV, Macedo V de O. Estudo comparativo entre estibogluconato de sodio BP 88R e· antimoniato de meglumina no tratamento da leishmaniose cutanea: I. Eficacia e seguranca. [A comparative study between sodium stibogluconate BP 88R and meglumine antimoniate in the treatment of cutaneous leishmaniasis. I. The efficacy and safety.] Rev Soc Bras Med Trop 1999;32(4):383–7.
14. Saldanha AC, Romero GA, Guerra C, Merchan-Hamann E, Macedo V de O. Estudo comparativo entre estibogluconato de sodio BP 88 e antimoniato de meglumina no tratamento da leishmaniose cutanea. II. Toxicidade bioquimica e cardiaca. [Comparative study between sodium stibogluconate BP 88 and meglumine antimoniate in cutaneous leishmaniasis treatment. II. Biochemical and cardiac toxicity.] Rev Soc Bras Med Trop 2000;33(4):383–8.
15. Esfandiarpour I, Alavi A. Evaluating the efficacy of allopurinol and meglumine antimoniate (Glucantime) in the treatment of cutaneous leishmaniasis. Int J Dermatol 2002;41(8):521–4.
16. Momeni AZ, Reiszadae MR, Aminjavaheri M. Treatment of cutaneous leishmaniasis with a combination of allopurinol and low-dose meglumine antimoniate. Int J Dermatol 2002;41(7):441–3.
17. Krause PJ, Lepore T, Sikand VK, Gadbaw J Jr, Burke G, Telford SR 3rd, Brassard P, Pearl D, Azlanzadeh J, Christianson D, McGrath D, Spielman A. Atovaquone and azithromycin for the treatment of babesiosis. N Engl J Med 2000;343(20):1454–8.
18. Thakur CP, Kanyok TP, Pandey AK, Sinha GP, Zaniewski AE, Houlihan HH, Olliaro P. A prospective randomized, comparative, open-label trial of the safety and efficacy of paromomycin (aminosidine) plus sodium stibogluconate versus sodium stibogluconate alone for the treatment of visceral leishmaniasis. Trans R Soc Trop Med Hyg 2000;94(4):429–31.
19. Wortmann G, Miller RS, Oster C, Jackson J, Aronson N. A randomized, double-blind study of the efficacy of a 10- or 20-day course of sodium stibogluconate for treatment of cutaneous leishmaniasis in United States military personnel. Clin Infect Dis 2002;35(3):261–7.
20. Castro C, Sampaio RN, Marsden PD. Severe arthralgia, not related to dose, associated with pentavalent antimonial therapy for mucosal leishmaniasis. Trans R Soc Trop Med Hyg 1990;84(3):362.

21. Thakur CP, Kumar K. Efficacy of prolonged therapy with stibogluconate in post kala-azar dermal leishmaniasis. Indian J Med Res 1990;91:144–8.

22. Cesur S, Bahar K, Erekul S. Death from cumulative sodium stibogluconate toxicity on kala-azar. Clin Microbiol Infect 2002;8(9):606.

23. Sundar S, Sinha PR, Agrawal NK, Srivastava R, Rainey PM, Berman JD, Murray HW, Singh VP. A cluster of cases of severe cardiotoxicity among kala-azar patients treated with a high-osmolarity lot of sodium antimony gluconate. Am J Trop Med Hyg 1998;59(1):139–43.

24. Thakur CP. Sodium antimony gluconate, amphotericin, and myocardial damage. Lancet 1998;351(9120):1928–9.

25. Seaton RA, Morrison J, Man I, Watson J, Nathwani D. Outpatient parenteral antimicrobial therapy—a viable option for the management of cutaneous leishmaniasis. QJM 1999;92(11):659–67.

26. Wey HE, Richards D, Tirmenstein MA, Mathias PI, Toraason M. The role of intracellular calcium in antimony-induced toxicity in cultured cardiac myocytes. Toxicol Appl Pharmacol 1997;145(1):202–10.

27. Tirmenstein MA, Mathias PI, Snawder JE, Wey HE, Toraason M. Antimony-induced alterations in thiol homeostasis and adenine nucleotide status in cultured cardiac myocytes. Toxicology 1997;119(3):203–11.

28. Lobanova EA, Ivanova LA, Pavlova TA, Prosina II. Kliniko-patogeneticheskie osobennosti pri vozdeistvii antimonitovykh rud na organizm rabotaiushchikh. [Clinical and pathogenetic features of exposure of workers to antimonate ore.] Med Tr Prom Ekol 1996;(4):12–5.

29. De Pablos Gallego JM, Cabrera Torres A, Almagro M, De Puerta S, Lopez Garrido P, Gomez Morales M, Esquivias JJ. Kala-azar y anemia hemolitica autoimmune; a propósito de un caso de evolución fatal por hepatotoxicidad del antimonato de N-metilglucamina. [Kala-azar and autoimmune hemolytic anemia. Apropos of a fatally developing case caused by hepatotoxicity from N-methylglucamine antimonate.] Rev Clin Esp 1982;164(6):417–20.

30. Arfaa F, Tohidi E, Ardelan A. Treatment of urinary bilharziasis in a small focus with sodium antimony dimercaptosuccinate (astiban). Am J Trop Med Hyg 1967;16(3):300–29.

31. Kuyucu N, Kara C, Bakirtac A, Tezic T. Successful treatment of visceral leishmaniasis with allopurinol plus ketoconazole in an infant who developed pancreatitis caused by meglumine antimoniate. Pediatr Infect Dis J 2001;20(4):455–7.

32. Hantson P, Luyasu S, Haufroid V, Lambert M. Antimony excretion in a patient with renal impairment during meglumine antimoniate therapy. Pharmacotherapy 2000;20(9):1141–3.

33. Kim HA, Heo Y, Oh SY, Lee KJ, Lawrence DA. Altered serum cytokine and immunoglobulin levels in the workers exposed to antimony. Hum Exp Toxicol 1999;18(10):607–13.

34. Rees PH, Kager PA, Ogada T, Eeftinck Schattenkerk JK. The treatment of kala-azar: a review with comments drawn from experience in Kenya. Trop Geogr Med 1985;37(1):37–46.

35. Moskowitz PF, Kurban AK. Treatment of cutaneous leishmaniasis: retrospectives and advances for the 21st century. Clin Dermatol 1999;17(3):305–15.

36. Lira R, Sundar S, Makharia A, Kenney R, Gam A, Saraiva E, Sacks D. Evidence that the high incidence of treatment failures in Indian kala-azar is due to the emergence of antimony-resistant strains of Leishmania donovani. J Infect Dis 1999;180(2):564–7.

37. Aggarwal P, Handa R, Singh S, Wali JP. Kala-azar—new developments in diagnosis and treatment. Indian J Pediatr 1999;66(1):63–71.

38. Segura I, Garcia-Bolao I. Meglumine antimoniate, amiodarone and torsades de pointes: a case report. Resuscitation 1999;42(1):65–8.

Arsenobenzol

General Information

Arsenobenzol has been used to treat syphilis (1,2) and as a topical antiseptic in combination with cortisone and silver colloid (3). Like other compounds with an arsenic base, it can cause gastrointestinal complaints, but also polyneuritis and encephalopathy. Adrenaline prevented the development of encephalopathy and was helpful in treating hemorrhagic encephalopathy (SEDA-11, 597).

References

1. Ebner H, Raab W. Nachbeobachtungen an arsenobenzol-schwermetall-behandelten faellen frischer lues. Hautarzt 1964;15:120–40.

2. Rossberg J. Katamnestische Untersuchungen an Arsenobenzol-Penizillin-Wismut behandelten Syphilitikern. [Catamnestic studies of syphilis patients treated with arseno-benzene-penicillin-bismuth.] Dermatol Wochenschr 1967;153(7):161–4.

3. Sacco S, Barlocco ME. Cortisone-arsenobenzolo-argento colloidale: la piu recente associazione antisettico-antiflogistica in parodontologia e stomatologia. [Cortisone-arsenobenzol-silver colloid: the newest antiseptic and anti-inflammatory combination in periodontology.] Riv Ital Stomatol 1970;25(12):1071–85.

Artemisinin derivatives

General Information

The herb Qinghaosu (*Artemisia annua*) has been known to Chinese medicine for centuries and was used in the treatment of fevers, in particular malaria fever; it is not clear why it did not become more widely used elsewhere. The plant can be grown in locations other than China, and field studies in propagating and growing the plant are being carried out in many parts of the world. In 1979 the Qinghaosu Antimalarial Coordinating Research Group reported their experience with four formulations of qinghaosu in both *Plasmodium vivax* and *Plasmodium falciparum* malaria (SEDA-13, 818; SEDA-17, 326; 1).

Artemisinin is an antimalarial constituent isolated from Qinghao. It is a sesquiterpene lactone with an endoperoxide bridge, structurally distinct from other classes of antimalarial agents. Several derivatives of the original compound have proved effective in the treatment of *Plasmodium falciparum* malaria and are currently available in a variety of formulations: artesunate (intravenous, rectal, oral), artelinate (oral), artemisinin (intravenous, rectal, oral), dihydroartemisinin (oral), artemether (intravenous, oral, rectal), and artemotil (intravenous). Artemisinic acid (qinghao acid), the precursor of artemisin, is present in the plant in a concentration up to 10 times that of artemisinin. Several semisynthetic derivatives have been developed from dihydroartemisinin (1).

Artemether

Artemether is a methyl ether derivative of dihydroartemisinin. It is dispensed in ampoules for intramuscular injection suspended in groundnut oil and in capsules for oral use. Like artesunate and artemisinin it has been used for both severe and uncomplicated malaria. Artesunate is probably faster-acting than the other two.

Artemotil (arteether)

Artemotil is the ethyl ether derivative of dihydroartemisinin. It was the choice of the WHO for development and was considered less toxic, because one would expect it to be metabolized to ethanol rather than methanol. It is also more lipophilic than artemether, a possible advantage for accumulation in brain tissues. The beta anomer was chosen since it is a crystalline solid and relatively easy to separate from the alpha anomer, which is liquid; it was necessary to choose a single anomer because of the more complex rules for the development of a drug with two anomers in the USA.

Artesunate

Artesunate is a water-soluble hemisuccinate derivative, available in parenteral and oral formulations. The parenteral drug is dispensed as powdered artesunic acid. Neutral aqueous solutions are unstable. Artesunate is effective by the intravenous, intramuscular, and oral routes in a dose of 10 mg/kg given for 5–7 days. The combination with mefloquine is very effective even against highly multiresistant strains of *Plasmodium falciparum*; the combination must be given for at least 3 days.

None of these medications has yet been registered for use in Europe or North America. In recent years there has been a substantial increase in our knowledge of their safety, efficacy, and pharmacokinetics. Higher cure rates are achieved when they are combined with longer-acting antimalarial drugs, such as mefloquine. After years of continued use, the sequential use of artesunate and mefloquine remains an effective treatment in areas of multidrug resistance in South-East Asia and provides an impetus for the evaluation of other artesunate-containing combination regimens for the treatment of uncomplicated malaria, such as artemether + benflumetol (2–6).

Mechanism of action

All three *Artemisia* derivatives are quickly hydrolysed to the active substance dihydroartemisinin. They produce a more rapid clinical and parasitological response than other antimalarial drugs. There are no reports of significant toxicity, and as late as 1994 there was no convincing evidence of specific resistance, but chloroquine-resistant *Plasmodium berghei* is resistant to artemisinin as well. The recrudescence rate is fairly high (1).

The mode of action is not fully known. There is strong evidence that the antimalarial activity of these drugs depends on the generation of free radical intermediates; free radical scavengers, such as ascorbic acid and vitamin E, therefore antagonize the antimalarial activity.

Drug activation by iron and heme may explain why endoperoxides are selectively toxic to malaria parasites. The malaria parasites live in a milieu of heme iron, which the parasite converts into insoluble hemozoin. Chloroquine, which binds heme, antagonizes the antimalarial activity of artemisinin.

Observational studies

Some impression of possible adverse effects in humans can be gained from a primarily pharmacokinetic study, in which artemotil solution in sesame oil was given intramuscularly. The half-life was 25–72 hours. Adverse effects in 23 subjects after the single dose included local pain in two, bitter taste and dryness of the mouth in one, and a mild but slightly itching papular rash that persisted for 14 days in another. There were no biochemical or electrocardiographic changes. Similar adverse effects were seen after 5 days of therapy in 14 of the 27 subjects; there was local pain in three, metallic taste in two, flu-like symptoms in three, and in two a maculopapular rash, which receded within 24 hours. One subject developed shivering, clammy hands and feet, dizziness, headache, nausea, and a metallic taste in the mouth, all lasting for about an hour, but the same reaction occurred after an injection of sesame oil only. Apart from some increase in eosinophil count in all groups, there were no significant hematological changes (7).

In 25 patients with acute uncomplicated *Plasmodium falciparum* malaria treated with artemotil (3.2 mg/kg followed by either 1.6 mg/kg or 0.8 mg/kg), the most frequent adverse events were headache, dizziness, nausea, vomiting, and abdominal pain; two patients complained of mild pain at the site of injection (8).

In a postmarketing surveillance study of artemotil in 300 patients, 294 (98%) were cured, five improved, and one did not show any change (9). The adverse effects were mild headache, nausea, vomiting, and giddiness.

Combination therapy

The artemisinin derivatives are limited by an unacceptable incidence of recrudescence with monotherapy, and they therefore need to be used in combination. A summary of prospective trials that looked specifically for adverse effects showed that artemisinins alone are very well tolerated (10). The same study showed no evidence of adverse interactions of artesunate with mefloquine, with an incidence of adverse effects similar to that expected from malaria and mefloquine (25 mg/kg) together. Reducing doses of mefloquine increases recrudescence rates to unacceptable levels (11). Combinations of artemisinins with quinine, co-trimoxazole, and doxycycline are well tolerated.

Artesunate + amodiaquine

Of 1017 patients aged 6 months to 10 years with uncomplicated malaria, 400 were randomized to chloroquine (25 mg/kg over 3 days) + a single dose of sulfadoxine–pyrimethamine (sulfadoxine 25 mg/kg, pyrimethamine 1.25 mg/kg); amodiaquine 25 mg/kg over 3 days + sulfadoxine–pyrimethamine; or amodiaquine + artesunate (4 mg/kg/day for 3 days); 396 were assessed for safety (12). There were no major adverse events.

Amodiaquine and artesunate have been compared alone and in combination for the treatment of uncomplicated malaria in 87 children aged 1?15 years, who were randomly assigned to amodiaquine 10 mg/kg (n = 27), artesunate 4 mg/kg (n = 27), or amodiaquine + artesunate (n = 33) (13). All the regimens were well tolerated and there were no major adverse events. Hematological profiles and serum transaminases were normal in all the groups. Only one child treated with artesunate developed a transient rise in alanine transaminase, which normalized after a few days.

Artemether + benflumetol

A large trial of the first fixed-dose combination of an artemisinin derivative likely to be licensed (artemether + benflumetol) had disappointing relapse rates compared with mefloquine monotherapy (6). In the 126 patients who took artemether + benflumetol there were no adverse effects attributed to drug treatment. However, less than 70% of patients were cured at 28 days. Benflumetol may be more effective at higher concentrations (14) but toxicity studies are lacking.

Artesunate + clindamycin

The combination of artesunate + clindamycin (2 mg/kg + 7 mg/kg 12 hourly for 3 days) has been compared with quinine + clindamycin (15 mg/kg + 7 mg/kg 12 hourly for 3 days) in 100 patients in a randomized comparison (15). Asexual parasite clearance time was faster with artesunate + clindamycin (29 versus 46 hours), and patients who took artesunate + clindamycin also experienced a shorter time to fever clearance (21 versus 30 hours). Both regimens were well tolerated and no severe adverse events were recorded. However, one patient who took artesunate + clindamycin had diarrhea and two who took quinine + clindamycin developed diarrhea and tinnitus.

Artesunate + lumefantrine

There has been a meta-analysis of 15 trials from Africa, Europe, and Asia of the use of varying doses of artesunate plus lumefantrine compared with several alternative antimalarial drugs in 1869 patients conducted by the manufacturers Novartis into its clinical safety and tolerability in the treatment of uncomplicated malaria (16). The most common adverse events were gastrointestinal—nausea (6.3%), abdominal pain (12%), vomiting (2.4%), anorexia (13%)—or central nervous—headache (21%) or dizziness (16%). There were 20 serious adverse events with artesunate plus lumefantrine, but only one (hemolytic anemia) was possibly due to artesunate plus lumefantrine. There was no QT prolongation associated with artesunate plus lumefantrine.

Artesunate + mefloquine

Artesunate has been combined with mefloquine in areas with a high prevalence of multiresistant *Plasmodium falciparum* in Thailand and the Thai-Burmese border. In a study reported in 1992 in 127 patients who were followed for 28 days, group A took artesunate 100 mg immediately and then 50 mg every 12 hours for 5 days (total 600 mg), group M took mefloquine 750 mg and another 500 mg 6 hours later, and group AM took first artesunate and then the two doses of mefloquine (17). Fever and parasite clearance time were significantly shorter in the two groups treated with artesunate. Table 1 gives the results and adverse effects. The combination of artesunate with mefloquine was more effective than either drug alone. However, the trial design was such that the patients who took both drugs were in fact treated twice, so the findings did not prove synergism between the two drugs.

In a second Thai study, 652 adults and children were treated with artesunate plus mefloquine (18). A single

Table 1 Adverse effects of artesunate with or without mefloquine in acute uncomplicated malaria tropica

	Mefloquine	Artesunate	Artesunate + Mefloquine
Total number of patients (M/F)	39/4	38/4	39/3
Number of patients with 18-day follow-up	37	40	39
Number (%) cured at 28 days	30 (81)	35 (88)	39 (100)
Fever clearance time (hours)	70	35	38
Parasite clearance time (hours)	64	36	38
Headache (%)	17 (39)	14 (33)	12 (28)
Dizziness (%)	8 (18)	6 (14)	5 (11)
Nausea (%)	9 (20)	6 (14)	9 (21)
Vomiting (%)	7 (16)	8 (19)	11 (26)
Abdominal pain (%)	1 (2)	2 (4)	1 (2)
Diarrhea (%)	1 (2)	3 (7)	1 (2)
Itching and rash (%)	0	3 (7)	1 (2)

dose of artesunate 4 mg/kg plus mefloquine 25 mg/kg gave a rapid response but did not improve cure rate. Artesunate given for 3 days in a total dose of 10 mg/kg plus mefloquine was 98% effective. The incidence of vomiting was significantly reduced by giving the mefloquine on day 2 of the treatment. There were no adverse effects attributed to artesunate.

Adults and children with uncomplicated malaria were randomized to either a combined formulation of artesunate + mefloquine (n = 251) or artesunate and mefloquine as separate tablets (n= 249) (19). The PCR cure rates after 63 days were 92% (95% CI = 88, 96) with the combined formulation and 89% (85, 93) with the separate tablets. Patients who took the separate tablets had more vomiting.

Artesunate+pyrimethamine+sulfadoxine

In Irian Jaya, a randomized, controlled trial ($n = 105$; 88 children) of oral artesunate (4 mg/kg od for 3 days) with pyrimethamine (1.25 mg/kg) + sulfadoxine or pyrimethamine + sulfadoxine alone (same dose) showed reduced gametocyte carriage and reduced treatment failure rates (RR = 0.3; 95% CI = 0.1, 1.3) in the combination group (20). Self-limiting adverse events of combination treatment were mild diarrhea (2.1%), rashes (4.3%), and itching (2.1%).

In a double-blind, randomized, placebo-controlled trial in Gambian children with uncomplicated falciparum malaria treated with pyrimethamine + sulfadoxine (25/500 mg; $n = 600$) or pyrimethamine + sulfadoxine combined with two regimens of oral artesunate (4 mg/kg, $n = 200$ or 4 mg/kg od for 3 days, $n = 200$), there were mild adverse events, such as headache, anorexia, nausea, vomiting, abdominal pain, and diarrhea, in a high proportion of children (56%) (21). Combination treatment with artesunate was associated with more rapid parasite clearance and less gametocytemia. Three-dose artesunate conferred no additional benefit over the one-dose regimen.

Artesunate + tetracycline

In a comparison between oral artesunate (700 mg over 5 days) plus tetracycline (250 mg at 6-hour intervals) and quinine (600 mg quinine sulfate at 8-hour intervals) for 7 days, artesunate was more effective and better tolerated in uncomplicated malaria (see Table 2) (22). Convulsions occurred in one case.

Table 2 Adverse effects in a comparison of artesunate plus tetracycline versus quinine

	Artesunate + tetracycline	Quinine
Number	31	33
Parasite clearance time (hours)	37 (24–52)	73 (26–135)
Cure rate on day 27 (%)	97	10
Nausea (%)	14 (45)	20 (60)
Dizziness (%)	16 (52)	16 (48)
Vomiting (%)	9 (26)	30 (91)
Tinnitus (%)	0	29 (88)
Convulsions (%)	1 (3)	0
Bradycardia (%)	7 (23)	Not done

Dihydroartemisinin + piperaquine

The novel combination (Artekin™) of dihydroartemisinin and piperaquine has been assessed in 106 patients (76 children and 30 adults) with uncomplicated *Plasmodium falciparum* malaria in Cambodia (23). The respective doses of dihydroartemisinin and piperaquine, which were given at 0, 8, 24, and 32 hours, were 9.1 mg/kg and 74 mg/kg in children and 6.6 and 53 mg/kg in adults. All the patients became aparasitemic within 72 hours. Excluding the results in one child who died on day 4, there was a 97% 28-day cure rate (99% in children and 92% in adults). Patients who had recrudescent infections used low doses of Artekin. Adverse effects, most commonly gastrointestinal complaints, were reported by 22 patients (21%) but did not necessitate premature withdrawal.

Observational studies

The efficacy and safety of a 3-day course of oral artesunate (4 mg/kg/day) for uncomplicated *Plasmodium falciparum* malaria have been evaluated in 50 Gabonese children (24). On day 14, cure rates were high (92%), but fell to 72% by day 28.

Optimal dosage regimens for several of the artemisinin derivatives still need to be established. If artemisinin derivatives are used in monotherapy, treatment for 5?7 days is probably acceptable. However, the combination of artemisinin derivatives with longer acting antimalarial drugs is generally recommended.

In 32 pregnant women (mean gestation 30 weeks) with uncomplicated falciparum malaria, who were given a standard dose of artesunate (two tablets of 100 mg each) with sulfadoxine + pyrimethamine (sulfadoxine 500 mg plus pyrimethamine 25 mg) all as one dose, the treatments were well tolerated, the parasitemia cleared, and the patients were symptom-free within 2 days (25). All delivered full-term live babies, although one baby died on the fourth day. None of the women died and there were no miscarriages, still births, or congenital anomalies in the neonates.

Artemether (480 mg in six injections intramuscularly) was given to 28 pregnant Sudanese women with falciparum malaria after treatment failure with chloroquine and quinine (26). One patient received artemether in the 10th week of gestation, 12 during the second trimester, and 15 during the third trimester. Artemether was well tolerated. One patient delivered at 32 weeks and the baby died 6 hours later. The other 27 delivered full-term live babies. None of the women died and there were no stillbirths or abortions.

Although artemisinin and its derivatives, such as artesunate, arteether, artemether, and dihydroartemisinin, were originally developed for the treatment of malaria, they have antischistosomal activity. In 87 Nigerian children, aged 5–18 years, with *Schistosoma haematobium* ova-positive urine samples (27) who took two oral doses of artesunate 6 mg/kg 2 weeks apart and provided four urine samples (two before and two after treatment). There were no serious adverse effects within 7 days of either dose. Six of the subjects complained of headache, three of abdominal pain, and four of fever.

Comparative studies

Three large clinical trials (in Kenya, the Gambia, and Vietnam) compared intramuscular artemether with intravenous quinine in severe malaria tropica (SEDA-20, 259). The treatments were similarly efficacious. There were no serious adverse effects.

Children with severe malaria were randomly assigned to either (a) artesunate suppositories (n = 41) 8–16 mg/kg at 0 and 12 hours and then daily or (b) intramuscular artemether (n = 38) 3.2 mg/kg followed by 1.6 mg/kg/day; one died with multiple complications within 2 hours of admission, but the other 78 recovered uneventfully (28). In a subset of 29 children, plasma concentrations of artemether, artesunate, and their common metabolite dihydroartemisinin were measured for the first 12 hours. Primary endpoints included time to 50% and 90% parasite clearance and time to per os status. In those who received suppositories the plasma concentrations of active metabolite were higher at 2 hours and there were significantly earlier parasite clearance times, mirroring the higher plasma concentrations, which may offer a considerable advantage in the treatment of children with severe malaria, particularly when parenteral therapy is not possible.

Artesunate has been compared with quinine for the treatment of complicated malaria in 80 children in India (29). Artesunate was better tolerated than quinine: no child who received artesunate had an adverse event; in contrast among those who received quinine nausea, vomiting, headache, tinnitus, circulatory failure, and blindness were reported.

Artemether + lumefantrine versus artemether + mefloquine

Artemether + lumefantrine (n = 53) and artemether + mefloquine (n = 55) have been compared for the treatment of uncomplicated malaria in Laos (30). There were no major adverse events in either group.

Artemether + lumefantrine versus artesunate + amodiaquine

Amodiaquine + artesunate (n = 153) has been compared with artemether + lumefantrine (co-lumefantrine) (n = 142) in 295 children with uncomplicated malaria (31). There were no major adverse effects in either group, although *vomiting* was more frequent with artesunate + lumefantrine on days 1 and 2.

Artesunate + azithromycin versus artesunate + quinine

Patients with acute, uncomplicated falciparum malaria were randomly assigned to one of 4 regimens for 3 days: Group 1 (n = 27) received azithromycin 750 mg bd + artesunate 100 mg bd; Group II (n = 27) azithromycin 1000 mg bd + artesunate 200 mg/day; Group III ((n = 16) azithromycin 750 mg bd + quinine 10 mg/kg bd; Group IV (n = 26) azithromycin 500 mg tds + quinine 10 mg/kg tds (32). The 28-day cure rate was similarly high in groups I, II, and IV). In Group III, there were

three treatment failures and recruitment was discontinued. Artesunate combinations had faster clinical and parasitological outcomes than quinine combinations. There were no deaths or drug-related adverse events.

Artesunate + mefloquine versus dihydroartemisinin + piperaquine

Patients with uncomplicated malaria were randomly assigned to receive either dihydroartemisinin + piperaquine (n = 327; 156 supervised treatment and 171 unsupervised), 6.3 + 50 mg/kg for 3 days or artesunate + mefloquine (n= 325; 162 and 163 respectively) 12 + 25 mg/kg (33). The primary endpoint was PCR-confirmed parasitological failure rate on day 42. One patient died 22 days after receiving dihydroartemisinin + piperaquine and 16 patients were lost to follow up. On day 42, the failure rate was 0.6% (95% CI = 0.2, 2.5) for dihydroartemisinin + piperaquine and 0 (0, 1.2) for artesunate + mefloquine. No major adverse events were reported, although dizziness was the most frequent complaint.

Placebo-controlled studies

In a double-blind, randomized study in Vietnam (*n* = 227), extending the duration of oral artemisinin monotherapy 500 mg/day from 5 to 7 days did not reduce recrudescence rates (total 23%) (34).

The antischistosomal properties of artemether have been investigated in a double-blind, placebo-controlled study in 322 schoolchildren randomized to receive either artemether 6 mg/kg once every 4 weeks to a total of 6 doses (n = 156) or placebo (n = 150) (35). Based on urinary egg counts 3 weeks after the last dose, artemether was efficacious in preventing *Schistosoma hematobium* infection. However, its protective efficacy against *Schistosoma hematobium* was considerably less than Its protective efficacy against *Schistosoma japonicum* and *Schistosoma mansoni* in previous studies. The adverse events reported within 72 hours of oral administration of artemether were headache, dizziness, abdominal pain, diarrhea, nausea, vomiting, fever, chills, cough, itching, and constipation. None of these symptoms was reported more often with artemether than placebo.

Systematic reviews

The efficacy of the addition of artesunate to standard regimens of *Plasmodium falciparum* malaria has been evaluated based on data from 5948 patients in 16 randomized trials (36). The addition of artesunate (4 mg/kg/day) resulted in lower therapy failure rates compared with standard regimens. The addition of artesunate resulted in significantly shorter parasite clearance times and significantly reduced gametocyte counts.

In a meta-analysis of 16 randomized controlled clinical trials (n = 5948) the effect of adding artesunate to standard treatment of *Plasmodium falciparum* malaria was analysed (37). The review compared treatment failures at day 14 and day 28 for standard drug alone against standard drug plus artesunate given over 3 days;

parasitological failure was lower with artesunate. Serious adverse events did not differ significantly between the groups.

General adverse effects

The safety of the peroxide antimalarial drugs has been reviewed (38). In animal studies, high doses of artemotil and artemether have been associated with hemopoietic, cardiac, and nervous system toxicity. Some subclinical neurotoxicity has been reported, with a discrete distribution in the brain stems of rats and dogs after multiple (high) doses (39). Dogs given high doses (15 mg/kg artemether for 28 days) had a progressive syndrome of clinical neurological defects with terminal cardiorespiratory collapse and death. There has been no evidence of neurotoxicity in man, but the human dosage of these ethers is of course lower. Reviews of clinical trials have reaffirmed the high tolerability of the artemisinins (11,40). Adverse effects have been chiefly limited to the GI tract. Most reported adverse events were described as mild and transient and none resulted in withdrawal of treatment.

Organs and Systems

Cardiovascular

Sinus bradycardia and a reversible prolongation of the QT interval have been reported (SEDA-21, 293).

A combination of artemether + lumefantrine (co-artemether, six doses over 3 days) followed by quinine (a 2-hour intravenous infusion of 10 mg/kg, not exceeding 600 mg in total, 2 hours after the last dose of co-artemether) was given to 42 healthy volunteers in a double-blind, parallel, three-group study (14 subjects per group) to examine the electrocardiographic effects of these drugs (41). Co-artemether had no effect on the QT_c interval. The infusion of quinine alone caused transient prolongation of the QT_c interval, and this effect was slightly but significantly greater when quinine was infused after co-artemether. Thus, the inherent risk of QT_c prolongation by intravenous quinine was enhanced by prior administration of co-artemether. Overlapping therapy with co-artemether and intravenous quinine in the treatment of patients with complicated or multidrug-resistant *Plasmodium falciparum* malaria may result in a modest increased risk of QT_c prolongation, but this is far outweighed by the potential therapeutic benefit.

The effects on the QT_c interval of single oral doses of halofantrine 500 mg and artemether 80 mg + lumefantrine 480 mg have been studied in 13 healthy men in a double-blind, randomized, crossover study (42). The length of the QT_c interval correlated positively with halofantrine exposure but was unchanged by co-artemether.

Nervous system

A study in mice suggested that intramuscular artemether is significantly more neurotoxic than intramuscular artesunate (43). The brains of 21 adults who died despite treatment with high doses of artemether (4 mg/kg

followed by 2 mg/kg every 8 hours, total dose 4?44 mg/kg; n = 6) or quinine (n = 15) for severe falciparum malaria have been examined (44). There was no histological evidence of neurotoxicity and in particular no evidence of either irreversible neuronal injury or selective distribution of neuropathological abnormalities confined to certain brain stem nuclei. The widespread neuronal stress responses and axonal injuries that were found were comparable in recipients of artemether and quinine. The authors therefore concluded that these injuries had resulted from the severe malaria itself and not from artemether. These data suggest that artemether does not acutely damage the nervous system in humans. However, the duration of artemether exposure was rather short (time to death only 8–331 hours) and so neurotoxic effects of artemether cannot be definitively ruled out.

In one case, an inadequately treated non-immune subject developed neurological symptoms after relapse and re-treatment; the symptoms were ascribed to artemether (45), although the well-recognized post-malaria neurological syndrome was much more likely, which the authors did not discuss (46).

One man developed ataxia and slurred speech after taking a 5-day course of oral artesunate (SEDA-21, 293).

A study of brain-stem auditory-evoked potentials showed no electrophysiological evidence of brain-stem damage in adults treated with artemisinin derivatives (47).

Brainstem encephalopathy has been attributed to artemisinin in a patient with cancer (48).

- A 42-year-old woman with early breast carcinoma developed diplopia, dysarthria, and an ataxic gait. Her medications included tamoxifen 20 mg/day, fluoxetine 10 mg/day, and 2 weeks of herbal therapy for breast cancer. The herbs consisted of artemisinin tablets 200 mg bd and a daily combination containing *Paeonia alba*, *Atractylodes alba*, *Momordica*, *Cudrania*, *Cochinchinensis*, *Sophora flavescens*, and *Dioscorea*. She had conjugate downward gaze, prominent vertical nystagmus, dysarthric speech, bilateral incoordination of both legs and arms, and an unsteady wide-based gait. Laboratory investigations were unremarkable, but brain MRI scanning showed symmetrical punctate foci. There was no evidence of stroke, demyelinating disease, or metastasis. After withdrawal of artemisinin, her neurological symptoms rapidly resolved. A repeat MRI scan on day 7 showed improvement. Tamoxifen and fluoxetine were restarted without recurrent symptoms.

Tamoxifen infrequently causes reversible neurotoxicity, but at much higher doses (Œ 160 mg/m²/day). There are no reports of similar adverse effects due to fluoxetine or the other herbal medications that the patient was taking. In animals, artemisinin derivatives cause degeneration and necrosis in the pons, medulla, and spinal cord, but not in cortical neurons or astrocytes. The MRI findings in this patient correlated closely with brainstem injury and mimicked that found in animal studies.

Neurotoxicity from artemether is related to drug accumulation due to slow and prolonged absorption from intramuscular injection sites. In mice, high doses of intramuscular artemether (50–100 mg/kg/day for 28 days) resulted in an unusual pattern of selective damage to certain brain-stem nuclei, especially those implicated in hearing and balance (49).

Metabolism

According to a randomized, open comparison in 113 adults with severe falciparum malaria in Thailand, hypoglycemia seems to occur less often in patients with malaria treated with artesunate (2.4 mg/kg intravenously followed by 1.2 mg/kg 12 hours later and then by 1.2 mg/kg/day intravenously or 12 mg/kg orally for 7 days) compared with those treated with quinine (20 mg/kg intravenously over 4 hours followed by 10 mg intravenously over 2 hours or orally tds for 7 days) (50). Artesunate and quinine had comparable efficacy, but hypoglycemia was only observed in 10% of the patients treated with artesunate whereas it occurred in 28% of those treated with quinine.

Fluid balance

Sodium artesunate inhibits sodium chloride transport in the thick ascending limb of the loop of Henle and therefore has a diuretic effect.

- In two men, aged 16 and 32 years, with falciparum malaria who were given four intravenous doses of sodium artesunate 60 mg, neither of whom had received diuretics or vasoactive drugs, there was a diuresis (6 l/day) accompanied by a natriuresis (51).

Hematologic

A transient dose-dependent reduction in reticulocyte count has been reported in healthy subjects (SEDA-21, 294).

Immunologic

A hypersensitivity reaction to artemether + lumefantrine has been reported (52).

- A 9-year-old boy with suspected malaria was given oral artemether + lumefantrine (three weight-adjusted doses of 80 + 480 mg over 24 hours). After the second dose, he developed severe coughing and vomiting and his eyelids became swollen. The coughing and vomiting worsened after the third dose and the swelling extended to his cheeks. An allergic reaction was suspected, artemether + lumefantrine was withdrawn, and the coughing and swelling resolved within 1–2 days. *Plasmodium falciparum* malaria was confirmed on the fourth day. However, parasitemia always remained low (160 asexual parasites/µl), and recovery was uneventful.

The coughing and swelling was thought to have resulted from a hypersensitivity reaction to artemether/lumefantrine. Since the boy had previously been exposed to

artesunate but not to lumefantrine, hypersensitivity to dihydroartemisinin, the active metabolite of both artesunate and artemether, was suspected.

Body temperature

Drug fever has been reported in healthy subjects taking artemether, artesunate, and artemisinin (SEDA-21, 294).

Because vomiting, prostration, and impaired consciousness often preclude oral administration and since parenteral therapy is generally not feasible in remote areas, rectal artesunate (10–15 mg/kg at 0 and 12 hours) was evaluated in 47 children in Papua New Guinea with uncomplicated *Plasmodium falciparum* (n = 42) or *Plasmodium vivax* (n = 5) malaria (53). The children were monitored during the first 24 hours and then chloroquine and sufadoxine + pyrimethamine were given and the children were discharged. Artesunate suppositories were well tolerated in all cases. After 24 hours, only one child had not defervesced and one other had persistent parasitemia. However, three children had a high body temperature, tachycardia, and vomiting 24 hours after having received their first dose of artesunate. None of those three children had a history of significant fever or chills and all had had rapid and sustained parasite clearance. The late fever therefore seemed unlikely to have resulted from active parasite replication or the parasiticidal effect of artesunate. The authors concluded that the late fever might have resulted from a mild intercurrent viral infection or a drug-induced or metabolic effect.

Second-Generation Effects

Pregnancy

Artemisinin derivatives (artesunate and artemether) for the treatment of multidrug-resistant *Plasmodium falciparum* malaria have been evaluated in 83 Karen pregnant women in Thailand; 55 women were treated for recrudescent infection after quinine or mefloquine, 12 for uncomplicated hyperparasitemic episodes, and 16 had not declared their pregnancy when treated (54). Artesunate and artemether were well tolerated and there was no drug-related adverse effect. Overall, 73 pregnancies resulted in live births, three in abortions and two in stillbirths; five women were lost to follow-up before delivery. There was no congenital abnormality in any of the neonates, and the 46 children followed for more than 1 year all developed normally.

Artemisinin derivatives have been studied in 461 pregnant women in a prospective cohort study in Thailand over 8 years (55). Oral artesunate monotherapy was associated with a treatment failure rate of 6.6%. Artesunate and artemether were well tolerated. The rates of abortions (including 44 first-trimester exposures), stillbirths, or congenital abnormalities were 4.8%, 1.58%, and 0.8% respectively, and were not significantly different from pregnant controls. These results are reassuring, but further information is needed before the safety of artesunate in pregnancy can be confirmed.

Drug Administration

Drug formulations

When the sesame oil vehicle in beta-artemotil was replaced by Cremophor (polyethoxylated castor oil) the total exposure in rats was 2.7-fold higher, owing to increased systemic availability (56). Anorexia and gastrointestinal toxicity from artemotil in sesame oil were significantly more severe than with artemotil in Cremophor. However, histological examination of the brain showed neurotoxic changes, which were worse with the castor oil formulation.

Drug administration route

A crossover pharmacokinetic study of single-dose rectal artesunate (10 or 20 mg/kg as a suppository) or intravenous artesunate (2.4 mg/kg) in moderate falciparum malaria in 34 Ghanaian children has been reported (57). The intravenous route gave much higher peak concentrations but more rapid elimination of artesunate and its active metabolite dihydroartemisinin than the rectal route. Rectal artesunate had higher systemic availability in the low-dose group than in the high-dose group (58% versus 23%). This is lower than published estimates of the systemic availability of oral artesunate (61–85%) (58). Parasite clearance kinetics were comparable in the two groups. There were no adverse events attributable to the drug. The results of Phase III and Phase IV studies of rectal artesunate as an alternative to parenteral antimalarial drugs in African children are awaited.

Drug–Drug Interactions

Artemisinin is metabolized in vitro by CYP2B6, CYP3A4, and CYP2A6. Since artemisinin induces CYP2C19, the question arises whether artemisinin also induces some of the cytochromes involved in its own metabolism and thus increases its own elimination (autoinduction) (59). During treatment with oral artemisinin for 10 days (250 mg/day for 9 days and 500 mg on the tenth day), artemisinin oral clearance increased 5.3 times in six poor CYP2C19 metabolizers and eight extensive metabolizers. The underlying mechanism was probably induction of CYP2B6. Induction of CYP2B6 by artemisinin could affect the metabolism of drugs given concomitantly and lead to suboptimal artemisinin concentrations towards the end of artemisinin treatment.

Chloroquine

Although an early report suggested antagonism between chloroquine and artemisinin (60), more recent evidence suggests the opposite, that is a synergistic effect between the two (61,62).

Deferoxamine

Co-administration of artemisinin derivatives, such as artesunate, with deferoxamine may be useful in patients with cerebral malaria, because of the combination of rapid parasite clearance by the former and central nervous system protection by the latter. Artesunate has been studied alone and in combination with deferoxamine in a single-blind comparison (63). Adverse effects were generally mild and there were no differences between the two regimens.

Omeprazole

Artemisinin induces its own elimination and that of omeprazole through an increase in CYP2C19 activity and that of another enzyme, as yet to be identified (64).

References

1. Ridley RG, Hudson AT. Chemotherapy of malaria. Curr Opin Infect Dis 1998;11:691–705.
2. von Seidlein L, Jaffar S, Pinder M, Haywood M, Snounou G, Gemperli B, Gathmann I, Royce C, Greenwood B. Treatment of African children with uncomplicated falciparum malaria with a new antimalarial drug, CGP 56697. J Infect Dis 1997;176(4):1113–6.
3. von Seidlein L, Bojang K, Jones P, Jaffar S, Pinder M, Obaro S, Doherty T, Haywood M, Snounou G, Gemperli B, Gathmann I, Royce C, McAdam K, Greenwood B. A randomized controlled trial of artemether/benflumetol, a new antimalarial and pyrimethamine/sulfadoxine in the treatment of uncomplicated falciparum malaria in African children. Am J Trop Med Hyg 1998;58(5):638–44.
4. Hatz C, Abdulla S, Mull R, Schellenberg D, Gathmann I, Kibatala P, Beck HP, Tanner M, Royce C. Efficacy and safety of CGP 56697 (artemether and benflumetol) compared with chloroquine to treat acute falciparum malaria in Tanzanian children aged 1–5 years. Trop Med Int Health 1998;3(6):498–504.
5. van Vugt M, Brockman A, Gemperli B, Luxemburger C, Gathmann I, Royce C, Slight T, Looareesuwan S, White NJ, Nosten F. Randomized comparison of artemether-benflumetol and artesunate-mefloquine in treatment of multidrug-resistant falciparum malaria. Antimicrob Agents Chemother 1998;42(1):135–9.
6. Looareesuwan S, Wilairatana P, Chokejindachai W, Chalermrut K, Wernsdorfer W, Gemperli B, Gathmann I, Royce C. A randomized, double-blind, comparative trial of a new oral combination of artemether and benflumetol (CGP 56697) with mefloquine in the treatment of acute Plasmodium falciparum malaria in Thailand. Am J Trop Med Hyg 1999;60(2):238–43.
7. Kager PA, Schultz MJ, Zijlstra EE, van den Berg B, van Boxtel CJ. Arteether administration in humans: preliminary studies of pharmacokinetics, safety and tolerance. Trans R Soc Trop Med Hyg 1994;88(Suppl 1):S53–4.
8. Looareesuwan S, Oosterhuis B, Schilizzi BM, Sollie FA, Wilairatana P, Krudsood S, Lugt ChB, Peeters PA, Peggins JO. Dose-finding and efficacy study for i.m. artemotil (beta-arteether) and comparison with i.m. artemether in acute uncomplicated P. falciparum malaria Br J Clin Pharmacol 2002;53(5):492–500.
9. Asthana OP, Srivastava JS, Das Gupta P. Post-marketing surveillance of arteether in malaria. J Assoc Physicians India 2002;50:539–45.
10. Price R, van Vugt M, Phaipun L, Luxemburger C, Simpson J, McGready R, ter Kuile F, Kham A, Chongsuphajaisiddhi T, White NJ, Nosten F. Adverse effects in patients with acute falciparum malaria treated

with artemisinin derivatives. Am J Trop Med Hyg 1999;60(4):547–55.

11. Na-Bangchang K, Tippanangkosol P, Ubalee R, Chaovanakawee S, Saenglertsilapachai S, Karbwang J. Comparative clinical trial of four regimens of dihydroartemisinin–mefloquine in multidrug-resistant falciparum malaria. Trop Med Int Health 1999;4(9):602–10.

12. Staedke S G, Mpimbaza A, Kamya MR, Nzarubara BK, Dorsey G, Rosenthal PJ. Combination treatments for uncomplicated falciparum malaria in Kampala, Uganda: randomised clinical trial. Lancet 2004;364(9449):1950-7.

13. Barennes H, Nagot N, Valea I, Koussoube- Balima T, Ouedrago A, Sanou T, Ye S. A randomized trial of amodiaquine and artesunate alone and in combination for the treatment of uncomplicated falciparum malaria in children from Burkina Faso. Tropical Medicine and International Health 2004;9:438-44.

14. Ezzet F, Mull R, Karbwang J. Population pharmacokinetics and therapeutic response of CGP 56697 (artemether + benflumetol) in malaria patients Br J Clin Pharmacol 1998;46(6):553–61.

15. Ramharter M, Oyakhirome S, Klouwenberg PK, Adégnika AA, Agnandji ST, Missinou MA, Matsiégui P-B, Mordmüller B, Borrmann S, Kun JF, Lell B, Krishna S, Graninger W, Issifou S, Kremsner P. Artesunate–clindamycin versus quinine–clindamycin in the treatment of Plasmodium falciparum malaria: a randomized controlled trial. Clin Infect Dis 2005;40:1777-84.

16. Bakshi R, Hermeling-Fritz I, Gathmann I, Alteri E. An integrated assessment of the clinical safety of artemether–lumefantrine: a new oral fixed-dose combination antimalarial drug. Trans R Soc Trop Med Hyg 2000;94(4):419–24.

17. Looareesuwan S, Viravan C, Vanijanonta S, Wilairatana P, Suntharasamai P, Charoenlarp P, Arnold K, Kyle D, Canfield C, Webster K. Randomised trial of artesunate and mefloquine alone and in sequence for acute uncomplicated falciparum malaria. Lancet 1992;339(8797):821–4.

18. Nosten F, Luxemburger C, ter Kuile FO, Woodrow C, Eh JP, Chongsuphajaisiddhi T, White NJ. Treatment of multidrug-resistant Plasmodium falciparum malaria with 3-day artesunate–mefloquine combination. J Infect Dis 1994;170(4):971–7.

19. Ashley EA, Lwin KM, McGready R, Simon WH, Phaiphun L, Proux S, Wangseang N, Taylor W, Stepniewska K, Nawamaneerat W, Thwai KL, Barends M, Leowattana W, Olliaro P, Singhasivanon P, White NJ, Nosten F. An open label randomized comparison of mefloquine–artesunate as separate tablets vs. a new co-formulated combination for the treatment of uncomplicated multi-drug resistant falciparum malaria in Thailand. Trop Med Int Health 2006;11:1653-60.

20. Tjitra E, Suprianto S, Currie BJ, Morris PS, Saunders JR, Anstey NM. Therapy of uncomplicated falciparum malaria: a randomized trial comparing artesunate plus sulfadoxine-pyrimethamine versus sulfadoxine-pyrimethamine alone in Irian Jaya, Indonesia. Am J Trop Med Hyg 2001;65(4):309–17.

21. von Seidlein L, Milligan P, Pinder M, Bojang K, Anyalebechi C, Gosling R, Coleman R, Ude JI, Sadiq A, Duraisingh M, Warhurst D, Alloueche A, Targett G, McAdam K, Greenwood B, Walraven G, Olliaro P, Doherty T. Efficacy of artesunate plus pyrimethamine–sulphadoxine for uncomplicated malaria in Gambian children: a double-blind, randomised, controlled trial. Lancet 2000;355(9201):352–7.

22. Karbwang J, Na-Bangchang K, Thanavibul A, Bunnag D, Chongsuphajaisiddhi T, Harinasuta T. Comparison of oral artesunate and quinine plus tetracycline in acute uncomplicated falciparum malaria. Bull World Health Organ 1994;72(2):233–8.

23. Denis MB, Davis TM, Hewitt S, Incardona S, Nimol K, Fandeur T, Poravuth Y, Lim C, Socheat D. Efficacy and safety of dihydroartemisinin-piperaquine (Artekin) in Cambodian children and adults with uncomplicated falciparum malaria. Clin Infect Dis 2002;35(12):1469–76.

24. Borrmann S, Adegnika AA, Missinou MA, Binder RK, Issifou S, Schindler A, Matsiegui PB, Kun JF, Krishna S, Lell B, Kremsner PG. Short-course artesunate treatment of uncomplicated Plasmodium falciparum malaria in Gabon. Antimicrob Agents Chemother 2003;47:901-4.

25. Adam I, Ali DM, Abdalla MA. Artesunate plus sulfadoxine–pyrimethamine in the treatment of uncomplicated Plasmodium falciparum malaria during pregnancy in eastern Sudan. Trans R Soc Trop Med Hyg 2006;100:632-5.

26. Adam I, Elwasila E, Mohamed Ali DA, Elansari E, Elbashir MI. Artemether in the treatment of falciparum malaria in pregnancy in Western Sudan. Trans R Soc Trop Med Hyg 2004;98(9):509-13.

27. Inyang-Etoh PC, Ejezie GC, Useh MF, Inyang-Etoh EC. Efficacy of artesunate in the treatment of urinary schistosomiasis, in an endemic community in Nigeria. Ann Trop Med Parasitol 2004;98:491-9.

28. Karunajeewa HA, Reeder J, Lorry K, Dabod E, Hamzah J, Page-Sharp M, Chiswell GM, Ilett KF, Davis TME. Artesunate suppositories versus intramuscular artemether for treatment of severe malaria in children in Papua New Guinea. Antimicrob Agents Chemother 2006;50(3):968-74.

29. Mohanty AK, Rath BK, Mohanty R, Samal AK, Mishra K. Randomised control trial of quinine and artesunate in complicated malaria. Indian J Paediatr 2004;71:291-5.

30. Stohrer JM, Dittrich S, Thongpaseuth V, Vanisaveth V, Phetsouvanh R, Phompida S, Monti F, Christophel EH, Lindegardh N, Annerberg A, Jelinex T. Therapeutic efficacy of artemether–lumefantrine and artesunate–mefloquine for treatment of uncomplicated malaria in Luang Namtha Province, Lao People's Democratic Republic. Trop Med Int Health 2004;9:1175-83.

31. Ndayiragije A, Niyungeko D, Karenzo J, Niyungeko E, Barutwanayo M, Ciza A, Bosman A, Moyou- Somo R, Nahimana A, Nyarushatsi JP, Barihuta T, Mizero L, Ndaruhutse J, Delacollette C, Ringwald P, Kamana J. Efficacite de combinaisons therapeutiques avec des derives de l'artemisinine dans le traitement de l'acess palustre non-complique au Burundi. Trop Med Int Health 2004;9(6):673-9.

32. Noedl H, Krudsood S, Chalermratana K, Silachamroon U, Leowattana W, Tangpukdee N, Looareesuwan S, Miller RS, Fukuda M, Jongsakul K, Sriwichai S, Rowan J, Bhattacharyya H, Ohrt C, Knirsch C. Azithromycin combination therapy with artesunate or quinine for the treatment of uncomplicated Plasmodium falciparum malaria in adults: a randomized, phase 2 clinical trial in Thailand. Clin Infect Dis. 2006;43(10):1264-71.

33. Smithuis F, Kyaw MK , Phe O, Aye KZ, Htet L, Barends M, Lindegardh N, Singtoroj T, Ashley E, Lwin S, Stepniewska K, White NJ. Efficacy and effectiveness of dihydroartemisinin–piperaquine versus artesunate–mefloquine in falciparum malaria: an open-label randomized comparison. Lancet 2006;367:2075-85.

34. Giao PT, Binh TQ, Kager PA, Long HP, Van Thang N, Van Nam N, de Vries PJ. Artemisinin for treatment of

uncomplicated falciparum malaria: is there a place for monotherapy? Am J Trop Med Hyg 2001;65(6):690–5.

35. N'Goran EK, Utzinger J, Gnaka HN, Yapi A, N'Guessan NA, Kigbafori SD, Lengeler C, Chollet J, Shuhua X, Tanner M. Randomized, double-blind, placebo-controlled trial of oral artemether for the prevention of patent Schistosoma haematobium infections. Am J Trop Med Hyg 2003;68:24-32.

36. Adjuik M, Babiker A, Garner P, Olliaro P, Taylor W, White N. Artesunate combinations for treatment of malaria: meta-analysis. Lancet 2004;363:9-17.

37. Adjuik M, Babiker A, Garner P, Olliaro P, Taylor W, White N; International Artemisinin Study Group. Artemisinin combination for treatment of malaria. Lancet 2004;363(9402):9-12.

38. Brewer TG, Peggins JO, Grate SJ, Petras JM, Levine BS, Weina PJ, Swearengen J, Heiffer MH, Schuster BG. Neurotoxicity in animals due to arteether and artemether. Trans R Soc Trop Med Hyg 1994;88(Suppl 1):S33–6.

39. Wesche DL, DeCoster MA, Tortella FC, Brewer TG. Neurotoxicity of artemisinin analogues in vitro. Antimicrob Agents Chemother 1994;38(8):1813–9.

40. Ribeiro IR, Olliaro P. Safety of artemisinin and its derivatives. A review of published and unpublished clinical trials. Med Trop (Mars) 1998;58(Suppl 3):50–3.

41. Lefevre G, Carpenter P, Souppart C, Schmidli H, Martin JM, Lane A, Ward C, Amakye D. Interaction trial between artemether–lumefantrine (Riamet) and quinine in healthy subjects. J Clin Pharmacol 2002;42(10):1147–58.

42. Bindschedler M, Lefevre G, Degen P, Sioufi A. Comparison of the cardiac effects of the antimalarials co-artemether and halofantrine in healthy participants. Am J Trop Med Hyg 2002;66(3):293–8.

43. Nontprasert A, Nosten-Bertrand M, Pukrittayakamee S, Vanijanonta S, Angus BJ, White NJ. Assessment of the neurotoxicity of parenteral artemisinin derivatives in mice. Am J Trop Med Hyg 1998;59(4):519–22.

44. Hien TT, Turner GD, Mai NT, Phu NH, Bethell D, Blakemore WF, Cavanagh JB, Dayan A, Medana I, Weller RO, Day NP, White NJ. Neuropathological assessment of artemether-treated severe malaria. Lancet 2003;362:295-6.

45. Elias Z, Bonnet E, Marchou B, Massip P. Neurotoxicity of artemisinin: possible counseling and treatment of side effects. Clin Infect Dis 1999;28(6):1330–1.

46. White NJ. Neurological dysfunction following malaria: disease- or drug-related? Clin Infect Dis 2000;30(5):836.

47. Van Vugt M, Angus BJ, Price RN, Mann C, Simpson JA, Poletto C, Htoo SE, Looareesuwan S, White NJ, Nosten F. A case-control auditory evaluation of patients treated with artemisinin derivatives for multidrug-resistant Plasmodium falciparum malaria. Am J Trop Med Hyg 2000;62(1):65–9.

48. Panossian LA, Garga NI, Pelletier D. Toxic brainstem encephalopathy after artemisinin treatment for breast cancer. Ann Neurol 2005;58 (5):812-3.

49. Nontprasert A, Pukrittayakamee S, Dondorp AM, Clemens R, Looareesuwan S, White NJ. Neuropathologic toxicity of artemisinin derivatives in a mouse model. Am J Trop Med Hyg 2002;67(4):423–9.

50. Newton PN, Angus BJ, Chierakul W, Dondorp A, Ruangveerayuth R, Silamut K, Teerapong P, Suputtamongkol Y, Looareesuwan S, White NJ. Randomized comparison of artesunate and quinine in the treatment of severe falciparum malaria. Clin Infect Dis 2003;37:7-16.

51. Seguro AC, Campos SB. Diuretic effect of sodium artesunate in patients with malaria. Am J Trop Med Hyg 2002;67(5):473–4.

52. Krippner R, Staples J. Suspected allergy to artemether–lumefantrine treatment of malaria. J Travel Med 2003;10:303-5.

53. Karunajeewa HA, Kemiki A, Alpers MP, Lorry K, Batty KT, Ilett KF, Davis TM. Safety and therapeutic efficacy of artesunate suppositories for treatment of malaria in children in Papua New Guinea. Pediatr Infect Dis J 2003;22:251-6.

54. McGready R, Cho T, Cho JJ, Simpson JA, Luxemburger C, Dubowitz L, Looareesuwan S, White NJ, Nosten F. Artemisinin derivatives in the treatment of falciparum malaria in pregnancy. Trans R Soc Trop Med Hyg 1998;92(4):430–3.

55. McGready R, Cho T, Keo NK, Thwai KL, Villegas L, Looareesuwan S, White NJ, Nosten F. Artemisinin antimalarials in pregnancy: a prospective treatment study of 539 episodes of multidrug-resistant Plasmodium falciparum. Clin Infect Dis 2001;33(12):2009–16.

56. Li QG, Mog SR, Si YZ, Kyle DE, Gettayacamin M, Milhous WK. Neurotoxicity and efficacy of arteether related to its exposure times and exposure levels in rodents. Am J Trop Med Hyg 2002;66(5):516–25.

57. Krishna S, Planche T, Agbenyega T, Woodrow C, Agranoff D, Bedu-Addo G, Owusu-Ofori AK, Appiah JA, Ramanathan S, Mansor SM, Navaratnam V. Bioavailability and preliminary clinical efficacy of intrarectal artesunate in Ghanaian children with moderate malaria. Antimicrob Agents Chemother 2001;45(2):509–16.

58. Newton P, Suputtamongkol Y, Teja-Isavadharm P, Pukrittayakamee S, Navaratnam V, Bates I, White N. Antimalarial bioavailability and disposition of artesunate in acute falciparum malaria. Antimicrob Agents Chemother 2000;44(4):972–7.

59. Simonsson US, Jansson B, Hai TN, Huong DX, Tybring G, Ashton M. Artemisinin autoinduction is caused by involvement of cytochrome P450 2B6 but not 2C9. Clin Pharmacol Ther 2003;74:32-43.

60. Stahel E, Druilhe P, Gentilini M. Antagonism of chloroquine with other antimalarials. Trans R Soc Trop Med Hyg 1988;82(2):221.

61. Hallett RL, Sutherland CJ, Alexander N, Ord R, Jawara M, Drakeley CJ, Pinder M, Walraven G, Targett GA, Alloueche A. Combination therapy counteracts the enhanced transmission of drug-resistant malaria parasites to mosquitoes. Antimicrob Agents Chemother 2004;48(10):3940–3.

62. Olliaro PL, Taylor WR. Developing artemisinin based drug combinations for the treatment of drug resistant falciparum malaria: A review. J Postgrad Med 2004;50(1):40–4.

63. Looareesuwan S, Wilairatana P, Vannaphan S, Gordeuk VR, Taylor TE, Meshnick SR, Brittenham GM. Co-administration of desferrioxamine B with artesunate in malaria: an assessment of safety and tolerance. Ann Trop Med Parasitol 1996;90(5):551–4.

64. Svensson US, Ashton M, Trinh NH, Bertilsson L, Dinh XH, Nguyen VH, Nguyen TN, Nguyen DS, Lykkesfeldt J, Le DC. Artemisinin induces omeprazole metabolism in human beings. Clin Pharmacol Ther 1998;64(2):160–7.

Atovaquone

General Information

Atovaquone is a hydroxynaphthaquinone that is effective in the prevention and treatment of murine *Pneumocystis jiroveci* pneumonitis. It also has effects against *Toxoplasma gondii* and *Plasmodium falciparum*. Food increases its absorption. The maximum serum concentration is dose-dependent, but absorption is reduced at doses above 750 mg. The maximum concentration occurs after 4–6 hours, with a second peak 24–96 hours later, suggesting enterohepatic cycling. The half-life is 77 hours.

Observational studies

In a 3-week study with test doses of 100–3000 mg/day, atovaquone was well tolerated. Three patients reported increased appetite; two of these had transient sinus arrhythmia. One of the 24 patients had a transient maculopapular rash that resolved without withdrawal. There were no abnormalities in hematological parameters or renal function. Two patients had slightly raised serum bilirubin concentrations and one each had raised transaminase activities. Two other patients had mildly increased transaminase activities, but both were known to have chronic hepatitis B (SEDA-13, 828).

Comparative studies

Atovaquone 250 mg tds has been compared with co-trimoxazole 320/1600 mg/day for 21 days in the treatment of *P. jiroveci* pneumonia in 408 patients. Therapeutic efficacy was similar, but atovaquone was much better tolerated, with a far lower incidence of rash, liver dysfunction, fever, nausea, and pruritus, and no neutropenia, chills, headache, renal impairment, or thrombocytopenia (1). However, pre-existing diarrhea was associated with an increased mortality in the atovaquone group.

Of 39 patients who had bone marrow transplants and who were randomized to receive either co-trimoxazole or atovaquone as prophylaxis in an open-label trial, eight taking co-trimoxazole withdrew because of presumed drug reactions, although in five of these the reported neutropenia and thrombocytopenia could have been a consequence of transplantation itself or of other drugs (2). None of 16 patients treated with atovaquone withdrew. This rate of reported adverse effects with co-trimoxazole is higher than usually reported in clinical practice with prophylactic dosages.

A study conducted by the AIDS Clinical Trials Group (ACTG) has shown that among patients who cannot tolerate treatment with co-trimoxazole, atovaquone and dapsone are similarly effective in preventing *P. jiroveci* pneumonia. Among patients who did not originally take dapsone, atovaquone was better tolerated and it might be the preferred choice for prophylaxis of *P. jiroveci* pneumonia in this setting (3). Inexplicably the rate of *P. jiroveci* pneumonia showed a greater fall in patients who discontinued the study drugs compared with those who continued to take them.

When atovaquone was compared with intravenous pentamidine in the treatment of mild and moderate *Pneumocystis jiroveci* pneumonia in an open trial, the success rates were similar. However, withdrawal of the original treatment was much more frequent with pentamidine (36%) than atovaquone (4%) (4). However, the authors' conclusion that the two approaches have a similar success rate has been challenged, and their series was small (5,6). Treatment-limited adverse effects occurred in only 7% of patients given atovaquone, compared with 41% given pentamidine. They included cases of rash and an increase in creatinine concentrations; atovaquone (unlike pentamidine) produced no vomiting, nausea, hypotension, leukopenia, acute renal insufficiency, or electrocardiographic abnormalities, but it did cause one case of dementia (4).

Combinations

Atovaquone + azithromycin

Human babesiosis has been traditionally treated with quinine plus clindamycin, a combination that has been compared with atovaquone plus azithromycin in a randomized, multicenter, unblinded study (7). The treatments were both completely effective. There were considerably fewer adverse events with azithromycin plus atovaquone than with quinine plus clindamycin.

Atovaquone + proguanil

Atovaquone acts synergistically with proguanil, and the combination of these two drugs (Malarone®) is highly efficacious in the treatment of uncomplicated malaria (8), including that against multidrug resistant forms, and in prophylaxis (9). However, at least one case of acquired *Plasmodium falciparum* resistance to atovaquone/proguanil has been reported in a non-immune female traveller to Kenya who had been treated for *Plasmodium falciparum* malaria (10).

An inpatient study of 79 patients given proguanil + atovaquone compared with 79 patients given mefloquine showed no malaria-independent adverse effects (11). Although there was a significant transient increase in liver enzymes, this was probably of limited clinical importance.

Prophylaxis with either one or two tablets containing atovaquone 250 mg plus proguanil hydrochloride 100 mg (one quarter or one half of the daily treatment dose), taken once-daily for 10 weeks, prevented *P. falciparum* malaria in 100% of semi-immune adults in a highly endemic area of Kenya (12). Children in Gabon taking daily Malarone at approximately one quarter of the treatment dose were similarly protected (13). Gastrointestinal adverse effects, including abdominal pain and vomiting, were relatively common in the initial parasite clearance phase of the pediatric study (when a full treatment course was given) and there was one case of repeated vomiting in the parasite clearance phase of the adult study. In both studies, the regimens were well tolerated in the prophylaxis phase (no difference from placebo). This efficacy and tolerability profile may be applicable to malaria prevention outside Africa. The use of the combination is predicted to reduce the development of resistance to each drug. Furthermore, atovaquone eliminates parasites during the hepatic phase of infection (causal prophylaxis), potentially removing the requirement to continue prophylaxis for several weeks after return from a malarious area, a period when compliance with current regimens is likely to be poor.

Proguanil plus atovaquone has been studied for chemoprophylaxis of malaria in African children (13) and in travelers (14) and is formulated as a combination of proguanil 100 mg plus atovaquone 250 mg for daily dosing. Atovaquone is a hydroxynaphthoquinone that inhibits the electron transport system (bc1 system) of parasites. Proguanil plus atovaquone is active against hepatic stages of *P. falciparum*, making it unnecessary to continue 4 weeks of prophylaxis after return from an endemic region. Current recommendations are that proguanil plus atovaquone should be continued for 1 week after returning from a malaria-endemic region.

In a comparison of proguanil plus atovaquone with proguanil (100 mg/day) plus chloroquine (155 mg base weekly) in travelers, proguanil plus atovaquone was 100% effective in the prevention of malaria (14). Those who took proguanil plus atovaquone ($n = 540$) had significantly fewer adverse events than those who took proguanil plus chloroquine ($n = 543$) (22% versus 28% respectively), particularly less diarrhea, abdominal pain, and vomiting. Only one person who took proguanil plus atovaquone had to discontinue prophylaxis owing to adverse events, as opposed to ten who had to discontinue proguanil plus chloroquine. There have been no other studies of similar size on the use of proguanil plus atovaquone. This combination is becoming established for the prophylaxis of malaria and the results of further phase IV studies are awaited.

The use of proguanil plus atovaquone has been reviewed (15). Neuropsychiatric adverse events were more frequent with mefloquine, whereas other adverse events occurred with similar frequencies as with chloroquine plus proguanil and mefloquine. Proguanil plus atovaquone is contraindicated in severe renal insufficiency. Co-administration of proguanil plus atovaquone with rifampicin is not recommended because of reductions in plasma atovaquone concentrations.

Observational studies

So far, eight randomized, open trials have shown that atovaquone + proguanil (1000 mg/day + 400 mg/day for 3 days) is effective in acute uncomplicated *Plasmodium falciparum* malaria (16). In these trials, atovaquone = proguanil produced higher or equal cure rates compared with previously approved antimalarial regimens.

In an open study in Gabon atovaquone + proguanil (20 + 8 mg/kg/day for 3 days) and amodiaquine (10 mg/kg/day for 3 days) were compared in 200 children weighing 5–11 kg (17). On day 27, atovaquone + proguanil produced a cure rate of 95%, whereas the cure rate was only 53% for amodiaquine. However, atovaquone +

proguanil has so far not been directly compared with regimens containing an artemisinin derivative.

The efficacy of atovaquone + proguanil in the prophylaxis of malaria has so far been investigated in six trials in semi-immune and immune populations (16). Success rates were 98?100%. Because atovaquone + proguanil is active against exoerythrocytic and erythrocytic forms of *Plasmodium* species, it provides causal and suppressive prophylaxis. Atovaquone + proguanil therefore needs to be taken for malaria prophylaxis in adults at a dosage of 250 + 100 mg mg/day starting 1–2 days before exposure until only 1 week after departure from the malarious area.

General adverse effects

Mild rashes are fairly common, and more serious rashes, like erythema multiforme, are rare. Gastrointestinal upsets, including abdominal pain, nausea, and diarrhea, are common. Mild nervous system disturbances have been mentioned (SEDA-18, 286; 18). Other adverse effects include fever.

The combination of atovaquone + proguanil is generally well tolerated. Adverse effects occur more often with the four-times higher doses required for malaria treatment than with the rather low doses used in prophylaxis. Common adverse effects include abdominal pain (17%), nausea (12%), vomiting (12%), headache (10%), diarrhea (8%), dizziness (5%), anorexia (5%), weakness (5%), pruritus (up to 6%), tinnitus (3–13%), dyspepsia, gastritis, insomnia, rash, urticaria, and rises in transaminases. However, most of these are relatively common in acute malaria and frequencies were similar in patients receiving placebo (16).

Two studies have confirmed that the above listed complaints are the most common adverse effects of atovaquone + proguanil. The safety of malaria prophylaxis with atovaquone + proguanil has been evaluated in an open study for 6 months in 300 Danish soldiers in Eritrea using a questionnaire (19). The most common complaints were diarrhea, abdominal pain, headache, cough, and loss of appetite. There were no serious adverse events and no cases of Plasmodium falciparum malaria. Furthermore, a post-marketing surveillance study recorded the following adverse event frequencies in 150 patients with mefloquine intolerance taking atovaquone + proguanil as malaria prophylaxis for 4.5?34 weeks: diarrhea (18%), abdominal pain (11%), headache (9%) dizziness (5%), and insomnia (6%) (20).

Organs and Systems

Psychological

The effects of primaquine and atovaquone + proguanil on psychomotor performance have been explored in a double-blind crossover study in 28 healthy volunteers (18–52 years old), who took atovaquone + proguanil 250 + 100 mg/day, or primaquine 30 mg/day, or placebo for 7 days separated by wash-out periods of 3 weeks (21). Neither primaquine nor atovaquone + proguanil caused any effects on psychomotor performance, mood, sleepiness, or fatigue.

Gastrointestinal

In a prospective efficacy trial of atovaquone suspension (750 mg od or 250 mg tds for 1 year) in *P. jiroveci* prophylaxis in 28 liver transplant recipients intolerant of co-trimoxazole, the adverse events reported included diarrhea ($n = 7$) and bloating or abdominal pain ($n = 3$) (22). No patient had developed *P. jiroveci* pneumonia by 37 months. This is a smaller dose than approved for *Pneumocystis* prophylaxis in HIV infection (1500 mg/day). Further studies in recipients of solid organ transplants are needed to confirm the efficacy of this prophylactic dose.

Atovaquone suspension (1500 mg orally bd) plus either pyrimethamine (75 mg/day after a 200 mg loading dose) or sulfadiazine (1500 mg qds), as treatment for acute *Toxoplasma* encephalitis (for 6 weeks) and as maintenance therapy (for 42 weeks), has been studied in a randomized phase II trial in HIV-positive patients (23). There were good responses in 21 of 28 patients who received pyrimethamine and nine of 11 who received sulfadiazine. Of 20 patients in the maintenance phase, only one relapsed. Of 40 eligible patients, 11 discontinued treatment as a result of adverse events, nine because of nausea and vomiting or intolerance of the taste of the atovaquone suspension.

Second-Generation Effects

Pregnancy

The pharmacokinetic properties of atovaquone + proguanil have been studied in 24 women with recrudescent multidrug resistant falciparum malaria during the second and third trimester of pregnancy (24). The clearance and volume of distribution were about twice those reported previously in non-pregnant women. Correspondingly, plasma concentrations in the pregnant women were only half of previously reported values, suggesting that pregnant women might need higher doses of atovaquone + proguanil. The pregnant women tolerated atovaquone + proguanil well and the 21 women with follow-up information gave birth to healthy infants.

Drug Administration

Drug formulations

Atovaquone suspension (750 mg bd; $n = 34$) or tablets (750 mg tds; $n = 20$) have been retrospectively compared in the treatment of *P. jiroveci* pneumonia in HIV-positive individuals (25). Efficacy was similar (74 and 70% successfully treated). Atovaquone suspension was associated with nausea in one patient and a rash in another.

Drug–Drug Interactions

Hormonal contraceptives

Proguanil is metabolized by CYP2C19 to its active metabolite cycloguanil, which interferes with folate synthesis in *Plasmodium* species. In 43 pregnant women, plasma proguanil and cycloguanil concentrations 6 hours after a single dose of proguanil (4 mg/kg) and urinary excretion were determined in the 3rd trimester and 2 months after delivery; the same was done in 40 women before and 3 weeks after starting an oral contraceptive, since estrogens inhibit CYP2C19 (26). Both in the third trimester and after administration of an oral contraceptive there was reduced formation of the active metabolite cycloguanil. The authors therefore recommended increasing proguanil doses by 50% in such patients.

Zidovudine

The metabolism of the antiviral nucleoside zidovudine to the inactive glucuronide form in vitro was inhibited by atovaquone (27,28).

Atovaquone can potentiate the activity of zidovudine by inhibiting its glucuronidation (28).

References

1. Hughes WT, LaFon SW, Scott JD, Masur H. Adverse events associated with trimethoprim–sulfamethoxazole and atovaquone during the treatment of AIDS-related *Pneumocystis carinii* pneumonia. J Infect Dis 1995;171(5):1295–301.
2. Colby C, McAfee S, Sackstein R, Finkelstein D, Fishman J, Spitzer T. A prospective randomized trial comparing the toxicity and safety of atovaquone with trimethoprim/sulfamethoxazole as *Pneumocystis carinii* pneumonia prophylaxis following autologous peripheral blood stem cell transplantation. Bone Marrow Transplant 1999;24(8):897–902.
3. El-Sadr WM, Murphy RL, Yurik TM, Luskin-Hawk R, Cheung TW, Balfour HH Jr, Eng R, Hooton TM, Kerkering TM, Schutz M, van der Horst C, Hafner R. Atovaquone compared with dapsone for the prevention of *Pneumocystis carinii* pneumonia in patients with HIV infection who cannot tolerate trimethoprim, sulfonamides, or both. Community Program for Clinical Research on AIDS and the AIDS Clinical Trials Group. N Engl J Med 1998;339(26):1889–95.
4. Dohn MN, Weinberg WG, Torres RA, Follansbee SE, Caldwell PT, Scott JD, Gathe JC Jr, Haghighat DP, Sampson JH, Spotkov J, Deresinski SC, Meyer RD, Lancaster DJAtovaquone Study Group. Oral atovaquone compared with intravenous pentamidine for *Pneumocystis carinii* pneumonia in patients with AIDS. Ann Intern Med 1994;121(3):174–80.
5. Lederman MM, van der Horst C. Atovaquone for *Pneumocystis carinii* pneumonia. Ann Intern Med 1995;122(4):314.
6. Stoeckle M, Tennenberg A. Atovaquone for *Pneumocystis carinii* pneumonia. Ann Intern Med 1995;122(4):314.
7. Krause PJ, Lepore T, Sikand VK, Gadbaw J Jr, Burke G, Telford SR 3rd, Brassard P, Pearl D, Azlanzadeh J, Christianson D, McGrath D, Spielman A. Atovaquone

and azithromycin for the treatment of babesiosis. N Engl J Med 2000;343(20):1454–8.

8. Farver DK, Lavin MN. Quinine-induced hepatotoxicity. Ann Pharmacother 1999;33(1):32–4.

9. Kedia RK, Wright AJ. Quinine-mediated disseminated intravascular coagulation. Postgrad Med J 1999;75(885):429–30.

10. Schwartz E, Bujanover S, Kain KC. Genetic confirmation of atovaquone-proguanil-resistant Plasmodium falciparum malaria acquired by a nonimmune traveler to East Africa. Clin Infect Dis 2003;37:450-1.

11. Newton P, Keeratithakul D, Teja-Isavadharm P, Pukrittayakamee S, Kyle D, White N. Pharmacokinetics of quinine and 3-hydroxyquinine in severe falciparum malaria with acute renal failure. Trans R Soc Trop Med Hyg 1999;93(1):69–72.

12. Shanks GD, Gordon DM, Klotz FW, Aleman GM, Oloo AJ, Sadie D, Scott TR. Efficacy and safety of atovaquone/proguanil as suppressive prophylaxis for Plasmodium falciparum malaria. Clin Infect Dis 1998;27(3):494–9.

13. Lell B, Luckner D, Ndjave M, Scott T, Kremsner PG. Randomised placebo-controlled study of atovaquone plus proguanil for malaria prophylaxis in children. Lancet 1998;351(9104):709–13.

14. Hogh B, Clarke PD, Camus D, Nothdurft HD, Overbosch D, Gunther M, Joubert I, Kain KC, Shaw D, Roskell NS, Chulay JD. Malarone International Study Team. Atovaquone–proguanil versus chloroquine–proguanil for malaria prophylaxis in non-immune travellers: a randomised, double-blind study. Malarone International Study Team. Lancet 2000;356(9245):1888–94.

15. Anonymous. Atovaquone + proguanil for malaria prophylaxis Drug Ther Bull 2001;39(10):73–5.

16. Marra F, Salzman JR, Ensom MH. Atovaquone–proguanil for prophylaxis and treatment of malaria. Ann Pharmacother 2003;37:1266-75.

17. Borrmann S, Faucher JF, Bagaphou T, Missinou MA, Binder RK, Pabisch S, Rezbach P, Matsiegui PB, Lell B, Miller G, Kremsner PG. Atovaquone and proguanil versus amodiaquine for the treatment of Plasmodium falciparum malaria in African infants and young children. Clin Infect Dis 2003;37:1441-7.

18. Masur H. Prevention and treatment of Pneumocystis pneumonia. N Engl J Med 1992;327(26):1853–60.

19. Petersen E. The safety of atovaquone/proguanil in long-term malaria prophylaxis of nonimmune adults. J Travel Med 2003;10 Suppl 1:S13-15; discussion S21.

20. Overbosch D. Post-marketing surveillance: adverse events during long-term use of atovaquone/proguanil for travelers to malaria-endemic countries. J Travel Med 2003;10 Suppl 1:S16-20; discussion S21-3.

21. Paul MA, McCarthy AE, Gibson N, Kenny G, Cook T, Gray G. The impact of Malarone and primaquine on psychomotor performance. Aviat Space Environ Med 2003;74:738-45.

22. Meyers B, Borrego F, Papanicolaou G. Pneumocystis carinii pneumonia prophylaxis with atovaquone in trimethoprim-sulfamethoxazole-intolerant orthotopic liver transplant patients: a preliminary study. Liver Transpl 2001;7(8):750–1.

23. Chirgwin K, Hafner R, Leport C, Remington J, Andersen J, Bosler EM, Roque C, Rajicic N, McAuliffe V, Morlat P, Jayaweera DT, Vilde JL, Luft BJ. Randomized phase II trial of atovaquone with pyrimethamine or sulfadiazine for treatment of toxoplasmic encephalitis in patients with acquired immunodeficiency syndrome: ACTG 237/ANRS 039 Study. AIDS Clinical Trials Group 237/Agence Nationale de Recherche sur le SIDA, Essai 039. Clin Infect Dis 2002;34(9):1243–50.

24. McGready R, Stepniewska K, Edstein MD, Cho T, Gilveray G, Looareesuwan S, White NJ, Nosten F. The pharmacokinetics of atovaquone and proguanil in pregnant women with acute falciparum malaria. Eur J Clin Pharmacol 2003;59:545-52.

25. Rosenberg DM, McCarthy W, Slavinsky J, Chan CK, Montaner J, Braun J, Dohn MN, Caldwell PT. Atovaquone suspension for treatment of Pneumocystis carinii pneumonia in HIV-infected patients. AIDS 2001;15(2):211–4.

26. McGready R, Stepniewska K, Seaton E, Cho T, Cho D, Ginsberg A, Edstein MD, Ashley E, Looareesuwan S, White NJ, Nosten F. Pregnancy and use of oral contraceptives reduces the biotransformation of proguanil to cycloguanil. Eur J Clin Pharmacol 2003;59:553-7.

27. Trapnell CB, Klecker RW, Jamis-Dow C, Collins JM. Glucuronidation of 3′-azido-3′-deoxythymidine (zidovudine) by human liver microsomes: relevance to clinical pharmacokinetic interactions with atovaquone, fluconazole, methadone, and valproic acid. Antimicrob Agents Chemother 1998;42(7):1592–6.

28. Lee BL, Tauber MG, Sadler B, Goldstein D, Chambers HF. Atovaquone inhibits the glucuronidation and increases the plasma concentrations of zidovudine. Clin Pharmacol Ther 1996;59(1):14–21.

Benzimidazoles

See also Albendazole, Carnidazole, Fenbendazole, Mebendazole and flubenazole, Metronidazole, Niridazole, Satranidazole, Secnidazole, Tiabendazole, Tinidazole, Triclabendazole

General Information

The benzimidazoles are a group of compounds that include albendazole, flubendazole, mebendazole, niridazole, triabendazole, and triclabendazole.

The newer broad-spectrum benzimidazoles have a wide spectrum of antihelminthic activity, killing larval and adult cestodes as well as intestinal nematodes, with generally low mammalian toxicity, apart from a potential for teratogenicity and embryotoxicity. The principal members of the group are mebendazole, its fluorine analogue flubendazole, and the better-absorbed albendazole. Mebendazole and albendazole are active orally in a single dose for a wide range of intestinal nematodes and are being used increasingly in the treatment of hydatid disease, in which experience is rapidly advancing. Flubendazole is very poorly absorbed and causes local tissue reactions at the site of injection when given parenterally.

None of the benzimidazoles is known to be safe in pregnancy and animal studies and their spectrum of toxicity suggest that they should be avoided. The absence of reports of harm in human pregnancy does not mean that no harm can occur.

The antihelminthic activity of the benzimidazoles is thought to result from selective blockade of glucose uptake by adult worms lodged in the intestine and their tissue-dwelling larvae, resulting in endogenous depletion of glycogen stores and reduced formation of adenosine triphosphate, which appears to be essential for parasite reproduction and survival. However, benzimidazole antihelminthics may also have antiparasitic activity by binding to free β-tubulin, thereby inhibiting the polymerization of tubulin and microtubule-dependent glucose uptake (1).

Uses

Ankylostomiasis

In 13 British soldiers with cutaneous larva migrans after a 2-week jungle training exercise in Belize the median incubation period was 10 (range 4–38) days; in 12 there were skin lesions on the calves or shins, and only two had foot or ankle lesions (2). Ten received oral thiabendazole, one oral mebendazole, one oral albendazole, and one topical thiabendazole. All those treated with oral thiabendazole complained of unpleasant reactions, predominantly nausea, vomiting, and dizziness. The one patient treated with topical thiabendazole returned with a new lesion 12 months later. He was then treated with systemic albendazole, with rapid resolution of symptoms. A 45-year-old woman developed larva migrans 20 days after lying on a beach in Singapore and was treated with thiabendazole 50 mg/kg in two doses for one day; she had no adverse effects (3).

The treatment of cutaneous larva migrans in 56 Italian patients aged 2–60 years has been retrospectively reviewed (4). All 13 patients treated with cryotherapy reported that it was painful, but none had recurrent disease or scarring. A further six patients were treated with oral thiabendazole 25–50 mg/kg/day for 2 days, and one had both thiabendazole and cryotherapy. In all cases there was regression of itching and skin lesions, but they had nausea, diarrhea, and dizziness while taking oral thiabendazole. No adverse effects were reported in 36 patients who were treated with albendazole 400 mg/day for 3 days (two were also treated with cryotherapy). Despite the low dose, larval migration was stopped in 1–2 days. Although a prompt and definitive cure was achieved in all 56 patients, albendazole was considered the treatment of choice given its minimal adverse effects. Until about 1980, surgery was the only treatment available for larval infections with *Echinococcus granulosus* or *Echinococcus multilocularis*. However, cysts are not always amenable to surgical removal, and operation is associated with the risk of rupture, leading to anaphylactic shock and re-infection; a proportion of cases are in any case not fit enough for surgery. The benzimidazoles have been used in varying high dosages over extended periods, initially to treat inoperable hydatid cysts and before surgery in attempts to sterilize cysts.

Echinococcosis

The epidemiology, clinical presentation, and treatment of alveolar echinococcosis of the liver have been described in French patients followed between 1972 and 1993 (5). From 1982 benzimidazoles were used. Of 117 patients, 72 took either albendazole or mebendazole for 4–134 months. The most common adverse effects were an increase in alanine transaminase activity to more than five times the top of the reference range (in six patients taking albendazole and in three taking mebendazole). Neutropenia (leukocyte count below 1.0×10^9/l) occurred in two patients taking albendazole. Alopecia occurred in four patients taking mebendazole. Minor adverse effects of albendazole included malaise, anorexia, and digestive intolerance in one patient each. In 13 patients treatment had to be withdrawn because of adverse effects ($n = 10$) or non-adherence to therapy ($n = 3$).

While mebendazole is used in continuous therapy of human alveolar echinococcosis, albendazole has been used in cyclic treatment. One treatment cycle consists of 28 days followed by a washout phase of 14 days without treatment, intended to reduce toxicity. Whether albendazole can also be used on a continuous basis has recently been studied in an open observational study in 35 patients with alveolar echinococcosis (in seven of 35 patients a curative operation was performed) (6). The outcome (lack of progression) was compared with the results obtained with continuous treatment with mebendazole or cyclic albendazole. Albendazole 10–15 mg/kg/day and mebendazole 40–50 mg/kg/day were equally effective. Seven patients were treated with continuous albendazole for an average of 28 (range 13–50) months. All patients taking continuous albendazole had stable or even regressive disease. The continuous dosing regimen was well tolerated without increased toxicity or higher rates of adverse reactions. Therefore, continuous dosing of albendazole is a promising alternative in cases of inoperable or progressive alveolar echinococcosis.

Prolonged cyclic albendazole treatment (for more than 9 years) was safe and effective in a patient with isolated cervical spine echinococcosis in whom surgery was performed without preoperative antihelminthic therapy because of a delay in diagnosis (7).

There have been several studies of the efficacy of albendazole in preventing recurrences of hydatid disease and cyst fluid spillage complications after surgery. In one Turkish study 22 of 36 patients with echinococcosis were treated with albendazole after surgical intervention (8). There was no significant benefit of perioperative albendazole over operation alone, although the recurrence rate of hepatic echinococcosis was lower than in historical controls. In contrast, in another study in 22 patients with hepatic echinococcosis there was a clear benefit of peri- and postoperative cyclic albendazole (12–15 mg/kg/day in four divided doses) (9). There were no cases of secondary hydatid disease or recurrence after a mean follow-up of 20 months. In two cases there were liver function abnormalities, which normalized after withdrawal.

Two regimens of albendazole emulsion were used in 264 patients with hepatic cystic echinococcosis (10). In 71 albendazole emulsion was given in a dose of 10 mg/kg/day by mouth for 6 months to over 1 year (group A). In 62 cases follow-up extended to 3–4 years after treatment. In 193 cases albendazole emulsion was given in a dose of 12.5 mg/kg/day for 3 months to over 1 year (group B).

The follow-up study in 139 cases extended for 2–4 years after treatment. In 38 there were mild, self-limiting reactions: mild pruritus (n = 20), rash (n = 14), transient liver pain (n = 9), gastric pain (n = 11), alopecia (n = 2), anorexia (n = 4), nausea (n = 3), vomiting (n = 3), and headache (n = 2). These adverse effects gradually resolved over 2 weeks without any treatment. In two patients with anorexia, nausea, and vomiting treatment was stopped. There were slight rises in serum transaminases in both groups. In group A there were increases in 43% (aspartate transaminase) and 49% (alanine transaminase) after 3 months. In group B increases occurred 2 weeks after starting treatment in 64% (aspartate transaminase) and 43% (alanine transaminase).

In a retrospective study of 30 patients with echinococcosis (11), four took albendazole. In two cases albendazole could not be tolerated because of gastrointestinal adverse effects in one case and abnormal liver function tests in the other. The liver function tests normalized after withdrawal of albendazole.

Fascioliasis

In 165 patients with fascioliasis (n = 35) or supposed fascioliasis (based on clinical and laboratory data but without detectable fasciola eggs in stools, n = 130), who were randomly allocated to oral triclabendazole 10 mg/kg for 1, 2, or 3 days, there were mild drug complications, such as nausea, vomiting, weakness, pruritus, epigastric pain, and liver enlargement in five, eight, and five patients who took one, two, and three doses respectively (12). In the triple dose group, there was an increase in one or both transaminases in seven patients, which normalized in four of five patients on day 60.

Neurocysticercosis

Cysticercosis is caused by the larval stage of the pork tapeworm *Tenia solium*. Neurocysticercosis is the most severe and common clinical manifestation in humans and probably the most frequent parasitic infection of the central nervous system. *T. solium* is endemic in Latin America, Asia, and sub-Saharan Africa. However, with the advent of computerized neuroradiology and improved serological tests, neurocysticercosis is increasingly being diagnosed throughout the world.

Controversies in the management of neurocysticercosis have been described (13–15). The management of the neurological complications of cysticercosis and in particular the role of antiparasitic drugs are issues of debate. It is commonly believed that the use of antiparasitic drugs and steroids should be individualized, based on the presence of active or inactive disease, the location of the cysts, and the presence or absence of complications such as hydrocephalus.

Some imidazoles have been used to treat parenchymal brain cysticerci. Initially, flubendazole (40 mg/kg for 10 days) was given to 13 patients with neurocysticercosis, with promising results. However, owing to its poor intestinal absorption, the use of flubendazole is limited. Albendazole is usually well absorbed and well tolerated, and albendazole serum concentrations are not significantly affected by glucocorticoids or anticonvulsants. Albendazole was given in daily doses of 15 mg/kg for 30 days. Further studies, however, showed that a treatment course could be shortened from 30 to 8 days without affecting efficacy. Direct comparative trials have shown that albendazole usually destroys 75–90% of parenchymal brain cysts, whereas praziquantel destroys 60–70%. The advantage of albendazole over praziquantel is limited not to its better efficacy, but also to better penetration of the subarachnoid space, allowing destruction of meningeal cysticerci. It also costs less than praziquantel.

Paragonimiasis

Five patients, aged 7–38 years, with *Paragonimus skrjabini* infections, were treated with oral triclabendazole (10 mg/kg bd for 3 consecutive days), an antihelminthic benzimidazole derivative used in the treatment of fascioliasis in sheep (16). One patient had cerebral involvement and received two courses. All five were cured. Blood eosinophilia completely disappeared. There were no adverse effects. Hepatic and renal function tests were unaffected. These data suggest that *Paragonimus skrjabini* infections can be safely treated with triclabendazole.

Trichuris trichiura

In a randomized trial in 168 patients the duration of albendazole therapy (400 mg/day) for 3, 5, or 7 days was studied in relation to its effectiveness in the treatment of *Trichuris trichiura* infection (17). Treatment with albendazole for 7 days resulted in a significantly higher cure rate, in particular in patients who had heavy infections (at least 1000 *Trichuris* eggs/g of feces). The authors therefore suggested that albendazole should be given for at least 3 days to those with light infections and for 5–7 days to patients with heavy infections. All reported adverse effects were mild. One patient (treated for 3 days) reported headache. Two patients (one treated for 3 days the other for 7 days) reported dizziness. Insomnia was reported in two patients treated for 7 days. Jaundice was not detected at any time.

To test the efficacy of albendazole against the for school-based deworming 150 children with whipworm (*Trichuris trichiura*) infections were randomized to albendazole 400, 800, or 1200 mg, each repeated four times, and 50 randomized to placebo; there were no adverse drug-related events (18).

Comparative studies

Echinococcosis

In a meta-analysis the clinical outcomes in 769 patients with hepatic cystic echinococcosis treated with percutaneous aspiration-injection-reaspiration (PAIR) plus albendazole or mebendazole (group 1) was compared with 952 era-matched historical control subjects undergoing surgical intervention (group 2) (19). The rates of clinical and parasitological cure were higher in patients receiving PAIR plus chemotherapy with albendazole or mebendazole. Disease recurrence, minor non-life-threatening complications, major complications such as anaphylaxis, biliary fistula, cyst infection, sepsis, liver/intra-abdominal abscess, and death occurred more often among surgical control subjects. Patients in this meta-analysis took antiparasitic drug therapy for 1 week before and 4 weeks after PAIR. Hepatic and hematological adverse effects were the most common, but detailed information was not given.

Giardiasis

In a comparison of mebendazole and secnidazole for giardiasis, 146 children aged 5–15 years were randomly assigned to mebendazole 200 mg tds for 3 days or secnidazole 30 mg/kg in a single dose (20). There was no difference in cure rates (78% versus 79%). Both treatment regimens were well tolerated. Transient abdominal discomfort was significantly more common with mebendazole than secnidazole (27% versus 8.2%). A bitter taste was reported in six patients who took secnidazole (8.2%) but not in patients who took mebendazole. The other reported adverse effects were nausea (in about 9.5% in each group) and vomiting (in 4–5% in each group).

In an open study, 57 patients were randomized to metronidazole 500 mg tds for 5 days (n = 29) or albendazole 400 mg/day for 5 days (n = 28) (21). Albendazole was better tolerated than metronidazole, especially with respect to anorexia (2 versus 18 patients) and metallic taste (0 versus 9 patients).

Neurocysticercosis

In an open, randomized, controlled trial, children with neurocysticercosis and seizures, aged 1–14 years, the efficacy of albendazole plus dexamethasone was studied (22). Of 123 children, 61 were given dexamethasone 0.15 mg/kg/day for 5 days plus albendazole 15 mg/kg/day for 28 days. The controls (n = 62) were given neither dexamethasone nor albendazole. Antiepileptic therapy was given to both groups. The cysticercal brain lesions resolved completely or partially in significantly more children in the treatment than the control group (79 versus 57%). The proportion of children who had seizures was significantly lower in the albendazole plus dexamethasone treatment group compared with the control group at 3 months (10% versus 32%) and at 6 months (13% versus 33%). In the 15 days follow-up after enrolment, there were no significant differences in the proportions of children with headache, vomiting, or visual problems. Thus, albendazole plus dexamethasone increased complete or partial resolution of cysticercal brain lesions and reduced the risk of subsequent recurrence of seizures among children with neurocysticercosis and seizures.

In another study, the appropriate duration of albendazole therapy in neurocysticercosis was established in a double-blind, randomized, placebo-controlled trial in 122 children with neurocysticercosis and seizures, who were randomized to albendazole (15 mg/kg/day) for 7 days followed by either albendazole (n = 60) or placebo (n = 62) for the following 21 days (23). It appeared that 1 week of therapy with albendazole was as effective as 4 weeks in children with neurocysticercosis. A minority

reported *nausea* or mild epigastric discomfort (seven children taking albenzazole, four taking placebo). Two developed headache not associated with raised intracranial pressure. One developed a transient rash. All adverse effects were mild and resolved spontaneously.

Pediculosis capitis

To test the potential effectiveness of thiabendazole in pediculosis capitis, girls aged 7?12 years took oral thiabendazole 20 mg/kg bd for 1 day, with repeat treatment after 10 days (24). Of 23 patients, 21 responded to treatment, 14 showing complete resolution of infestation. The only adverse reactions were nausea and mild *dizziness*, which occurred in four patients, three of whom took the drug on an empty stomach.

Placebo-controlled studies

Filariasis

The combination of albendazole + diethylcarbamazine has been compared with placebo + diethylcarbamazine in a double blind, randomized, parallel-group, field study in 1396 patients living in an area endemic for lymphatic filariasis in India (25). The combination of albendazole + diethylcarbamazine was as safe as diethylcarbamazine alone. There were 270 adverse events in 693 patients by the 5th day in the placebo arm compared with 238 adverse events in 703 patients with albendazole + Diethylcarbamazine. The most common reported adverse event on day 5 was fever, followed by myalgia. The other signs and symptoms that affected daily activity 5 days after administration were headache, nausea, and giddiness in the placebo arm, and abdominal pain, fatigue, and headache in the albendazole + diethylcarbamazine arm. There were no serious adverse events.

Loiasis

In a placebo-controlled, double-blind, crossover study in 99 subjects with *Loa loa* microfilaremia were given albendazole 400 mg/day or placebo for 3 days and were followed for 180 days, when the groups were crossed over and followed for a further 6 months (26). There were few adverse events, the most common being pruritus (30%, similar in the two groups), abdominal discomfort (12%), and urticaria (2%). One patient developed urticaria 14 days after placebo administration, and another developed urticaria and intense pruritus 1 month after albendazole administration. Both responded well to cetirizine.

Neurocysticercosis

In a double-blind, double-dummy, placebo-controlled trial 120 patients with neurocysticercosis were randomly assigned to either albendazole 800 mg/day plus dexamethasone 6 mg/day for 10 days (n = 60) or placebo (n = 60) (27). Significantly more of those who took albendazole had abdominal pain (8 versus 0).

General adverse effects

In a French pharmacovigilance study adverse drug reactions were reported in 31 patients who took albendazole, in 22 who took mebendazole and in 62 who took thiabendazole (28). In six patients who took albendazole the adverse events were classified as severe, leading to hospitalization: two cases of agranulocytosis, one case of hepatitis, one of retrobulbar neuritis, one of acute renal insufficiency, and one of rash. In two cases there were severe adverse events after mebendazole, leading to hospitalization (one with abdominal pain, one with bone marrow aplasia). In two patients who took thiabendazole there were severe bullous eruptions and bradycardia, which led to hospitalization. All the adverse effects due to albendazole or mebendazole had already been described in the literature, except for renal insufficiency, which occurred in three patients who took albendazole.

Organs and Systems

Skin

Many laborers take antihelminthic drugs, such as like mebendazole or metronidazole, to avoid a positive stool test that may exclude them from work overseas. In a case-control study 46 Filipino laborers with Stevens–Johnson syndrome or toxic epidermal necrolysis were matched with 92 controls according to age, sex, and month of arrival in Taiwan (29). The odds ratio for the skin rashes was 9.5 (95% CI = 3.9, 24) among workers who had used both metronidazole and mebendazole at some time in the preceding 6 weeks. There was an increasing risk with increasing level of exposure to metronidazole (in particular with doses over 2000 mg). There was a reverse dose-response relation between the risk of the rashes and the level of exposure to mebendazole (in particular with doses under 1000 mg). Combination therapy involving metronidazole and mebendazole should therefore be avoided because of the increased risk of developing Stevens?Johnson syndrome or toxic epidermal necrolysis.

Second-Generation Effects

Teratogenicity

The use of mebendazole in pregnancy gives reason for concern, because of the relative scarcity of data on its safety in pregnancy. The Israeli Teratogen Information Service followed 192 women exposed to mebendazole in pregnancy (30). Most of them were exposed to mebendazole during the first trimester (71.5%), 21.5% during the second trimester, and 7.0% during the third trimester. Similar proportions of women reported using mebendazole in a single dose of 100 mg (29%), a single dose of 100 mg repeated after an interval (36%) and 100 mg/day for 3 consecutive days (35%). There was no increase in the rate of major anomalies after exposure to mebendazole compared with controls. In addition, the incidence of major anomalies was not increased in the subgroup of patients who received mebendazole in the first trimester of pregnancy compared with controls. These data suggest that mebendazole does not represent a major teratogenic risk in humans when it is used in the doses commonly prescribed for pinworm infestation.

In another study inadvertent exposure of pregnant women to albendazole and ivermectin during a mass drug administration program for lymphatic filariasis was investigated (31). Of 2985 women of childbearing age who were interviewed, 343 were pregnant, of whom 293 were excluded from the programme. However, 50 pregnant women were inadvertently treated. Of the six children with some congenital malformations identified in these communities, one had been exposed to the drugs in utero. The relative risk for congenital malformations after exposure was 1.05. Two of nine women with spontaneous abortions had been exposed to the drugs (RR = 1.67). Thus, there seems to be no evidence of a higher risk of congenital malformations or abortions in pregnant women inadvertently exposed to albendazole and ivermectin.

Susceptibility Factors

Genetic

Although certain individuals have an increased susceptibility to adverse reactions to multiple pharmaceutical agents, familial or genetic predisposition has not been elucidated.

- A 4-year-old Mexican girl with ascariasis, who was treated with thiabendazole syrup 375 mg tds for 3 days, developed a severe skin rash compatible with Stevens–Johnson syndrome after 21 days (32). The pediatrician prescribed the same treatment for her four siblings (aged 2–7 years). Two of them developed a similar skin rash after 7 and 10 days.

These observations suggest that these severe skin rashes may be subject to genetic predisposition.

Age

Experience with albendazole and mebendazole in children under 24 months has been reviewed (33). In 17 studies, over 2189 children under 24 months received treatment with a benzimidazole derivative. In an epidemiological survey of 1209 courses of treatment no adverse effects were documented. In another 979 courses of treatment adverse effects were actively sought but were not found. There was only one episode of convulsion reported in a 7-week-old infant treated with mebendazole, but the symptoms were thought not to have been related to mebendazole.

In a double-blind, randomized, placebo-controlled trial in 212 children aged under 24 months there was no statistically significant difference in the incidence rate of adverse effects with mebendazole compared with placebo (34).

Thus, the evidence suggests that albendazole and mebendazole can be used to treat soil-transmitted helminthiasis in children aged 12 months and older, provided that the case for their use is established. Under 12 months of age, drug absorption may be increased, resulting in an increased risk of benzimidazole toxicity.

References

1. Georgiev VS. Necatoriasis: treatment and developmental therapeutics. Expert Opin Investig Drugs 2000;9(5):1065–78.

2. Green AD, Mason C, Spragg PM. Outbreak of cutaneous larva migrans among British military personnel in Belize. J Travel Med 2001;8(5):267–9.

3. Gourgiotou K, Nicolaidou E, Panagiotopoulos A, Hatziolou JE, Katsambast AD. Treatment of widespread cutaneous larva migrans with thiabendazole. J Eur Acad Dermatol Venereol 2001;15(6):578–80.

4. Albanese G, Venturi C, Galbiati G. Treatment of larva migrans cutanea (creeping eruption): a comparison between albendazole and traditional therapy. Int J Dermatol 2001;40(1):67–71.

5. Bresson-Hadni S, Vuitton DA, Bartholomot B, Heyd B, Godart D, Meyer JP, Hrusovsky S, Becker MC, Mantion G, Lenys D, Miguet JP. A twenty-year history of alveolar echinococcosis: analysis of a series of 117 patients from eastern France. Eur J Gastroenterol Hepatol 2000;12(3):327–36.

6. Reuter S, Jensen B, Buttenschoen K, Kratzer W, Kern P. Benzimidazoles in the treatment of alveolar echinococcosis: a comparative study and review of the literature. J Antimicrob Chemother 2000;46(3):451–6.

7. Garcia-Vicuna R, Carvajal I, Ortiz-Garcia A, Lopez-Robledillo JC, Laffon A, Sabando P. Primary solitary echinococcosis in cervical spine. Postsurgical successful outcome after long-term albendazole treatment. Spine 2000;25(4):520–3.

8. Mentes A, Yalaz S, Killi R, Altintas N. Radical treatment for hepatic echinococcosis. HPB 2000;2:49–54.

9. Erzurumlu K, Hokelek M, Gonlusen L, Tas K, Amanvermez R. The effect of albendazole on the prevention of secondary hydatidosis. Hepatogastroenterology 2000;47(31):247–50.

10. Chai J, Menghebat, Wei J, Deyu S, Bin L, Jincao S, Chen F, Xiong L, Yiding M, Xiuling W, Dolikun, Guliber, Yanchun W, Fanghua G, Shuhua X. Observations on clinical efficacy of albendazole emulsion in 264 cases of hepatic cystic echinococcosis. Parasitol Int 2004;53:3-10.

11. Silva MA, Mirza DF, Bramhall SR, Mayer AD, McMaster P, Buckels JAC. Treatment of hydatid disease of the liver: evaluation of a UK experience. Dig Surg 2004;21:227-34.

12. Talaie H, Emami H, Yadegarinia D, Nava-Ocampo AA, Massoud J, Azmoudeh M, Mas-Coma S. Randomized trial of singe, double and triple dose of 10 mg/kg of a human formulation of triclabendazole in patients with fascioliasis. Clin Exp Pharmacol Physiol 2004;31:777-82.

13. Di Pentima MC, White AC. Neurocysticercosis: controversies in management. Semin Pediatr Infect Dis 2000;11:261–8.

14. Del Brutto OH. Medical therapy for cysticercosis: indications, risks, and benefits. Rev Ecuat Neurol 2000;9:13–5.

15. Garg RK. Medical management of neurocysticercosis. Neurol India 2001;49(4):329–37.

16. Gao J, Liu Y, Wang X, Hu P. Triclabendazole in the treatment of *Paragonimiasis skrjabini*. *Chin Med J 2003;116:1683-6.*

17. Sirivichayakul C, Pojjaroen-Anant C, Wisetsing P, Praevanit R, Chanthavanich P, Limkittikul K. The effectiveness of 3, 5 or 7 days of albendazole for the treatment of *Trichuris trichiura* infection. Ann Trop Med Parasitol 2003;97:847-53.

18. Adams VJ, Lombard CJ, Dhansay MA, Markus MB, Fincham JE. Efficacy of albendazole against the whipworm Trichuris trichiura – a randomized, controlled trial. S Afr Med J 2004;94:972-6.

19. Smego RA, Bhatti S, Khaliq AA, Beg MA. Percutaneous aspiration–injection–reaspiration drainage plus albendazole or mebendazole for hepatic cystic echinococcosis: a meta-analysis. Clin Infect Dis 2003;37:1073-83.

20. Escobedo AA, Canete R, Gonzalez ME, Pareja A, Cimerman S, Almirall P. A randomized trial comparing mebendazole and secnidazole for the treatment of giardiasis. Ann Trop Med Parasitol 2003;97:499-504.

21. Karabay O, Tamer A, Gunduz H, Kayas D, Arinc H, Celebi H. Albendazole versus metronidazole treatment of adult giardiasis: an open randomized clinical study. World J Gastroenterol 2004;10:1215-7.

22. Kalra V, Dua T, Kumar V. Efficacy of albendazole and short-course dexamethasone treatment in children with 1 or 2 ring-enhancing lesions of neurocysticercosis: a randomized controlled trial. J Pediatr 2003;143:111-4.

23. Singhi P, Dayal D, Khanderwal N. One week versus four weeks of albendazole therapy for neurocysticercosis in children: a randomized, placebo-controlled double blind trial. Pediatr Infect Dis J 2003;22:268-72.

24. Namazi MR. Treatment of pediculosis capitis with thiabendazole: a pilot study. Int J Dermatol 2003;42:973-6.

25. Kshirsagar NA, Gogtay NJ, Garg BS, Deshmukh PR, Rajgor DD, Kadam VS, Kirodian BG, Ingole NS, Mehendale AM, Fleckenstein L, Karbwang J, Lazdins-Helds JK. Safety, tolerability, efficacy and plasma concentrations of diethylcarbamazine and albendazole co-administration in a field study in an area endemic for lymphatic filariasis in India. Trans R Soc Trop Med Hyg 2004;98:205-17.

26. Tabi TE, Befidi-Mengue R, Nutman TB, Horton J, Folefack A, Pensia E, Fualem R, Fogako J, Gwanmesia P, Quakyi I, Leke R. Human loiasis in a Cameroonian village: a double-blind, placebo-controlled, crossover clinical trial of a three-day albendazole regimen. Am J Trop Med Hyg 2004;71:211-5.

27. Garcia HH, Pretell EJ, Gilman RH, Martinez SM, Moulton LH, Del Brutto OH, Herrera G, Evans CA, Gonzalez AE, for the Cysticercosis Working Group in Peru. A trial of antiparasitic treatment to reduce the rate of seizures due to cerebral cysticercosis. N Engl J Med 2004;350:249-58.

28. Bagheri H, Simiand E, Montastruc JL, Magnaval JF. Adverse drug reactions to anthelmintics. Ann Pharmacother 2004;38:383-8.

29. Chen KT, Twu SJ, Chang HJ, Lin RS. Outbreak of Stevens–Johnson syndrome/toxic epidermal necrolysis associated with mebendazole and metronidazole use among Filipino laborers in Taiwan. Am J Public Health 2003;93:489-92.

30. Diav-Citrin O, Shechtman S, Arnon J, Lubart I, Ornoy A. Pregnancy outcome after gestational exposure to mebendazole: a prospective controlled cohort study. Am J Obstet Gynecol 2003;188:282-5.

31. Gyapong JO, Chinbuah MA, Gyapong M. Inadvertent exposure of pregnant women to ivermectin and albendazole during mass drug administration for lymphatic filariasis. Trop Med Int Health 2003;8:1093-101.

32. Johnson-Reagan L, Bahna SL. Severe drug rashes in three siblings simultaneously. Allergy 2003;58:445-7.
33. Montresor A, Awasthi S, Crompton DWT. Use of benzimidazoles in children younger than 24 months for the treatment of soil-transmitted helminthiasis. Acta Trop 2003;86:223-32.
34. Montresor A, Stolzfus RJ, Albonico M, Tielsch JM, Rice A, Chwaya HM, Savioli L. Is the exclusion of children under 24 months from anthelminthic treatment justifiable? Trans R Soc Trop Med Hyg 2002;96:197-9.

Benznidazole

General Information

Benznidazole, a nitroimidazole, is used in the treatment of *Trypanosoma cruzi* infections and Chagas' disease, and is recommended for use in urogenital trichomoniasis, all forms of amebiasis, giardiasis, and anaerobic infections. Its adverse effects are similar to those of metronidazole (SEDA-13, 833). Drowsiness, dizziness, headache, and ataxia occur occasionally. With prolonged and/or high doses, transient peripheral neuropathy and epileptiform seizures can be seen. Other frequently mentioned adverse effects are unpleasant taste, furred tongue, nausea, vomiting, and gastrointestinal disturbances. Skin rash and pruritus can occur. One case each of erythema multiforme and of toxic epidermolysis have been reported. Benznidazole causes a disulfiram-like effect with ethanol.

Placebo-controlled studies

In a double-blind, randomized, clinical trial, benznidazole 5 mg/kg/day for 60 days was compared with placebo in children in the indeterminate phase of infection by *Trypanosoma cruzi* (1). In general, treatment was well tolerated. The treated children had a significant reduction in mean titers of antibodies against *T. cruzi* measured by indirect hemagglutination, indirect immunofluorescence, and ELISA. At 4-year follow-up, 62% of the benznidazole-treated children and no placebo-treated child were seronegative for *T. cruzi*. Xenodiagnosis after 48 months was positive in 4.7% of the benznidazole-treated children and in 51% of the placebo-treated children.

General adverse effects

The adverse effects of benznidazole can be classified into three groups (2,3):

1. symptoms of hypersensitivity—dermatitis with skin eruptions (usually occurring at 7–10 days of treatment), generalized edema, fever, lymphadenopathy, and joint and muscle pains;
2. bone marrow suppression, thrombocytopenia, and agranulocytosis being the most severe manifestations;
3. peripheral polyneuropathy, paresthesia, and polyneuritis.

In a Cochrane systematic review the incidence of adverse effects was less than 20% (4). In one study, under 5% of participants complained of a variety of minor symptoms, but rash and pruritus were reported more commonly. In children the drug was well tolerated and there were no severe adverse effects. The only study in adults reported a non-quantified variety of mild adverse effects (skin reactions, peripheral neuropathy, digestive disturbances), but it was said that they were less intense than those seen with nifurtimox.

Long-Term Effects

Mutagenicity

Like metronidazole, benznidazole is mutagenic. In tests for chromosomal aberrations and induction of micronuclei in cultures of peripheral lymphocytes from children with Chagas' disease, there were increases in micronucleated interphase lymphocytes and of chromosomal aberrations after treatment with benznidazole (4).

Drug–Drug Interactions

Disulfiram

Like metronidazole, benznidazole has a disulfiram-like effect if alcohol is taken (5).

References

1. Sosa Estani S, Segura EL, Ruiz AM, Velazquez E, Porcel DM, Yampotis C. Efficacy of chemotherapy with benznidazole in children in the indeterminate phase of Chagas' disease. Am J Trop Med Hyg 1998;59(4):526–9.
2. Rodriques Coura J, de Castro SL. A critical review on Chagas disease chemotherapy. Mem Inst Oswaldo Cruz 2002;97(1):3–24.
3. Cancado JR. Long term evaluation of etiological treatment of Chagas' disease with benznidazole. Rev Inst Med Trop Sao Paulo 2002;44(1):29–37.
4. Villar JC, Marin-Neto JA, Ebrahim S, Yusuf S. Trypanocidal drugs for chronic asymptomatic *Trypanosoma cruzi* infection. Cochrane Database Syst Rev 2002;(1):CD003463.
5. Castro JA, Diaz de Toranzo EG. Toxic effects of nifurtimox and benznidazole, two drugs used against American trypanosomiasis (Chagas' disease). Biomed Environ Sci 1988;1(1):19–33.

Bephenium

General Information

Bephenium hydroxynaphthoate is an antihelminthic drug that has been used in the treatment of hookworm infections due to *Ancylostoma duodenale* in a single dose (1). It is well tolerated, and reactions are confined to mild gastrointestinal disturbances (unpleasant taste, nausea, abdominal pain, and sometimes also vomiting and diarrhea), headache, and dizziness. It is reputed to be safe in pregnancy but is better avoided in conditions in which purgation could be dangerous; these naturally include the last few months of pregnancy, because of the risk of miscarriage.

Reference

1. Botero D. Chemotherapy of human intestinal parasitic diseases. Annu Rev Pharmocol Toxicol 1978;18:1–15.

Bithionol

General Information

Bithionol is a chlorinated bisphenol with antihelminthic properties. It is active against most trematodes and has been recommended as preferable to praziquantel in fascioliasis and paragonimiasis, proving both effective and well tolerated. Bithionol also has antibacterial properties and was for this reason included for a time in cosmetic formulations. However, when given topically it caused photosensitivity reactions, and this form of application has been abandoned.

Organs and Systems

Gastrointestinal

An Egyptian report in 1991 on the use of bithionol in fascioliasis in doses of 30 mg/kg every other day for five doses noted that unpleasant symptoms were common, but some or most of them were certainly due to the underlying disease and actually abated with treatment (1). The only symptom the incidence of which actually increased as a result of treatment was diarrhea, which occurred in 12 of 14 users; one patient also developed urticaria with pruritus.

Liver

The effects of bithionol on the liver have been examined in an Egyptian study in 1990 (2). The pathology of human fascioliasis was studied before and after bithionol treatment using light and transmission electron microscopy. Fine needle biopsies were taken from five patients with established fascioliasis, before and after drug administration. By light microscopy the pathology of human fascioliasis was similar to that reported in experimental fascioliasis. The ultrastructural picture showed bile ductule hyperplasia, fibrosis of portal tracts, widening of the interhepatic spaces by many microvilli, and dilated Disse spaces with collagen fibers. Bile ductule hyperplasia may be the initial factor in fibrinogenesis, which subsequently enhances the development of microvilli on the surfaces of hepatocytes. Both light and electron microscopy showed regression of the picture of fascioliasis to normal after bithionol treatment, with no sign of adverse effects on the liver.

References

1. Bassiouny HK, Soliman NK, el-Daly SM, Badr NM. Human fascioliasis in Egypt: effect of infection and efficacy of bithionol treatment. J Trop Med Hyg 1991;94(5):333–7.
2. Abou Basha LM, Salem AI, Fadali GA. Human fascioliasis: ultrastructural study on the liver before and after bithionol treatment. J Egypt Soc Parasitol 1990;20(2):541–8.

Bitoscanate

General Information

Bitoscanate (phenylene 1,4-diisothiocyanate) enjoyed some clinical interest in the treatment of hookworm infection in around 1975 (1), but it proved highly dangerous, and deaths have occurred with as little as 300 mg. Most of the publicity that it has had in recent years relates to its listing as an "extremely hazardous substance," and the US Office of Homeland Security has actually regarded it as a possible tool for bioterrorism.

Reference

1. Samuel MR. Clinical experience with bitoscanate. Prog Drug Res 1975;19:96–107.

Carnidazole

General Information

Carnidazole is an imidazole derivative, less potent than most other similar compounds (1). In early studies, its adverse effects were mild; nausea and vomiting, abdominal discomfort, dry mouth, dizziness, headache, and tiredness were reported (SEDA-6, 264). Metronidazole-like adverse effects should be anticipated.

Reference

1. Jokipii L, Jokipii AM. Comparative evaluation of the 2-methyl-5-nitroimidazole compounds dimetridazole, metronidazole, secnidazole, ornidazole, tinidazole, carnidazole, and panidazole against *Bacteroides fragilis* and other bacteria of the *Bacteroids fragilis* group. Antimicrob Agents chemother 1985;28(4):561–4.

Chloroquine and hydroxychloroquine

General Information

Chloroquine is rapidly and almost completely absorbed from the intestinal tract, peak serum concentrations being reached in 1–6 hours (average 3 hours). It is extensively distributed and redistribution follows. It is slowly metabolized by side-chain de-ethylation. The half-life is 30–60 days. Elimination is mainly via the kidney. Malnutrition can slow down the rate of metabolism.

The use of chloroquine has diminished in recent years owing to widespread parasite drug resistance and it is now used mainly for prophylaxis in combination with proguanil. Hydroxychloroquine is being increasingly used to treat immune diseases, such as systemic lupus erythematosus, rheumatoid arthritis, and chronic graft-versus-host disease (GVHD). It commonly causes gastrointestinal disturbances and less commonly retinal toxicity, itching, intravascular hemolysis, rashes, and bone marrow suppression.

Comparative studies

Amodiaquine and chloroquine have been compared in an open, randomized trial in uncomplicated falciparum malaria in Nigerian children (1). The doses were amodiaquine (n = 104) 10 mg/kg/day for 3 days and chloroquine (n = 106) 10 mg/kg/day for 3 days. After 28 days, the cure rate was significantly higher with amodiaquine than chloroquine (95% versus 58%). The rates of adverse events, most commonly pruritus (10%) and gastrointestinal disturbances (3%), were similar in the two groups. Cross-resistance between the two aminoquinolines is common, and there are concerns regarding toxicity of amodiaquine with repeated use.

General adverse effects

There are relatively few adverse effects at the doses of chloroquine that are used for malaria prophylaxis and standard treatment doses. However, the use of higher doses than those recommended, for example because of problems with resistance, can cause problems. Infants are very easily overdosed (SEDA-16, 302). In the treatment of rheumatoid arthritis and lupus erythematosus, larger doses are used, often for long periods of time, and with this use the incidence of adverse effects is high. Neuromyopathy, neuritis, myopathy, and cardiac myopathy can cause serious problems. Retinopathy can lead to blindness. Chloroquine has a long half-life and accumulates in the tissues, including the brain. Concentrations in the brain can have a bearing on mental status and psychotic syndromes. Chloroquine interferes with the action of several enzymes, including alcohol dehydrogenase, and blocks the sulfhydryl–disulfide interchange reaction. Allergic reactions are generally limited to rashes and pruritus.

Organs and Systems

Cardiovascular

Electrocardiographic changes, comprising altered T waves and prolongation of the QT interval, are not uncommon during high-dose treatment with chloroquine. The clinical significance of this is uncertain. With chronic intoxication, a varying degree of atrioventricular block can be seen; first-degree right bundle branch block and total atrioventricular block have been described. Symptoms depend on the severity of the effects: syncope, Stokes–Adams attacks, and signs of cardiac failure can occur. Acute intoxication can cause cardiovascular collapse and/or respiratory failure. Cardiac complications can prove fatal in both chronic and acute intoxication.

Third-degree atrioventricular conduction defects have been reported in two patients with rheumatoid arthritis after prolonged administration of chloroquine (2,3).

Intravenous administration can result in dysrhythmias and cardiac arrest; the speed of administration is relevant, but also the concentration reached: deaths have been recorded with blood concentrations of 1 µg/ml; concentrations after a 300 mg dose are usually 50–100 µg/ml (SEDA-13, 803).

Long-term chloroquine can cause cardiac complications, such as conduction disorders and cardiomyopathy (restrictive or hypertrophic), by structural alteration of the interventricular septum (4). Thirteen cases of cardiac toxicity associated with long-term chloroquine and hydroxychloroquine have been reported in patients with systemic autoimmune diseases. The cumulative doses were 600–2281 g for chloroquine and 292–4380 g for hydroxychloroquine.

- A 64-year-old woman with systemic lupus erythematosus took chloroquine for 7 years (cumulative dose 1000 g). She developed syncope, and the electrocardiogram showed complete heart block; a permanent pacemaker was inserted. The next year she presented with biventricular cardiac failure, skin hyperpigmentation, proximal muscle weakness, and chloroquine retinopathy. Coronary angiography was normal. An echocardiogram showed a restrictive cardiomyopathy. A skeletal muscle biopsy was characteristic of chloroquine myopathy. Chloroquine was withdrawn and she improved rapidly with diuretic therapy.
- Chloroquine cardiomyopathy occurred during long-term (7 years) treatment for rheumatoid polyarthritis in a 42-year-old woman, who had an isolated acute severe conduction defect, confirmed by histological study with electron microscopy (5).
- A 50-year-old woman took chloroquine for 6 years for rheumatoid arthritis and developed a restrictive cardiomyopathy, which required heart transplantation (6).

Regular cardiac evaluation should be considered for those who have taken a cumulative chloroquine dose of 1000 g, particularly elderly patients.

More than one mechanism may underlie the cardiac adverse effects of chloroquine. Severe hypokalemia after a single large dose of chloroquine has been documented, and some studies show a correlation between plasma potassium concentrations and the severity of the cardiac effects (7).

Light and electron microscopic abnormalities were found on endomyocardial biopsy in two patients with cardiac failure. The first had taken hydroxychloroquine 200 mg/day for 10 years, then 400 mg/day for a further 6 years; the second had taken hydroxychloroquine 400 mg/day for 2 years (SEDA-13, 239). A similar case was reported after the use of 250 mg/day for 25 years (SEDA-18, 286).

Toxic cardiac effects following acute ingestion of large doses of hydroxychloroquine include QRS widening, QT interval prolongation, torsade de pointes, other ventricular dysrhythmias, hypokalemia, and hypotension. Torsade de pointes following chronic use of hydroxychloroquine has been reported (8).

- A 67-year-old woman with systemic lupus erythematosus and asthma who had taken prednisolone 15 mg/day, long-acting theophylline 200 mg/day, and hydroxychloroquine 200 mg/day for 1 year developed sudden loss of consciousness and generalized rigidity. Moments later she regained consciousness with no residual symptoms, but the episode recurred several times. She had a past history of cirrhosis and hepatitis B virus-related hepatoma with portal vein thrombosis, an old myocardial infarction, and a small ventricular septal defect. An electrocardiogram showed multiple ventricular extra beats and she had another syncopal attack, with documented torsade de pointes. After defibrillation and lidocaine 100 mg her cardiac rhythm reverted to normal with a prolonged QT interval (600 msec). Cardiac enzymes were not raised. Hydroxychloroquine was suspected as the cause of ventricular tachycardia and was withdrawn. She was given intravenous magnesium sulfate 1 g followed by isoprenaline 2 micrograms/minute. After 4 days the ventricular dysrhythmia subsided but the QT interval was still prolonged (500-530 msec). After 3 weeks the ventricular dysrhythmias abated.

It is not clear in this case whether congenital long QT interval could have contributed. Chronic use of hydroxychloroquine in rheumatic diseases should be weighed against the risk of potentially lethal cardiac dysrhythmias.

Respiratory

Respiratory collapse can occur with acute overdosage.

Acute pneumonitis probably due to chloroquine has been described (9).

- A 41-year-old man with chronic discoid lupus erythematosus was given chloroquine 150 mg bd for 10 days followed by 150 mg/day. After 2 weeks he developed fever, a diffuse papular rash, dyspnea, and sputum. A chest X-ray showed peripheral pulmonary infiltrates. He improved on withdrawal of chloroquine and treatment with cefpiramide and roxithromycin. No

organism was isolated. A subsequent oral challenge with chloroquine provoked a similar reaction.

Nervous system

The incidence of serious nervous system events among patients taking chloroquine for less than a year has been estimated as one in 13 600.

Chloroquine, especially in higher doses, can cause a marked neuromyopathy, characterized by slowly progressive weakness of insidious onset. In many cases this weakness first affects the proximal muscles of the legs. Reduction in nerve conduction time and electromyographic abnormalities typical of both neuropathic and myopathic changes can be found. Histologically there is a vacuolar myopathy. Neuromyopathy is a rare adverse effect and is usually limited to patients taking 250–750 mg/day for prolonged periods. The symptoms can be accompanied by other manifestations of chloroquine toxicity (SEDA-11, 583). An 80-year-old woman developed symptoms after taking chloroquine 300 mg/day for 6 months (10), once more demonstrating that a standard dosage can be too much for elderly people.

A spastic pyramidal tract syndrome of the legs has been reported. In young children the features of an extrapyramidal syndrome include abnormal eye movements, trismus, torticollis, and torsion dystonia.

Chloroquine can cause seizures in patients with epilepsy. The mechanism is uncertain, but it may include reductions in inhibitory neurotransmitters and pharmacokinetic interactions that alter anticonvulsant concentrations. Tonic–clonic convulsions were reported in four patients in whom chloroquine was part of a prophylactic regimen. Antiepileptic treatment was required to control the seizures. None had further seizures after withdrawal of the antimalarial drugs (11).

Chloroquine and desethylchloroquine concentrations have been studied in 109 Kenyan children during the first 24 hours of admission to hospital with cerebral malaria (12). Of the 109 children 100 had received chloroquine before admission. Blood chloroquine and desethylchloroquine concentrations were no higher in children who had seizures than in those who did not, suggesting that chloroquine does not play an important role in the development of seizures in malaria.

- A 59-year-old woman had a generalized convulsion 24 hours after returning from a trip to Vietnam (13). She had a history of partial complex seizures (controlled with carbamazepine) due to a previous ruptured cerebral aneurysm. For the preceding 3 weeks she had been taking chloroquine 100 mg/day and proguanil 200 mg/day. A blood film was negative for malaria. A CT scan of the brain showed changes compatible with the previous hemorrhage. She was successfully treated with clobazam (dose not stated) until withdrawal of chemoprophylaxis.

The interaction between chloroquine and carbamazepine was not examined. Chloroquine should not be given to adults with a history of epilepsy.

Neuromuscular function

Severe neuromyopathy has been reported in patients taking chloroquine (SEDA-21, 295).

Chloroquine-induced neuromyopathy is a complication of chloroquine treatment of autoimmune disorders or long-term use of chloroquine as a prophylactic antimalarial drug (14).

Sensory systems

Eyes

Chloroquine and its congeners can cause two typical effects in the eye, a keratopathy and a specific retinopathy. Both of these effects are associated with the administration of the drug over longer periods of time.

Keratopathy

Chloroquine-induced keratopathy is limited to the corneal epithelium, where high concentrations of the drug are readily demonstrable. Slit lamp examination shows a series of punctate opacities scattered diffusely over the cornea; these are sometimes seen as lines just below the center of the cornea, while thicker yellow lines may be seen in the stroma. The keratopathy is often asymptomatic, fewer than 50% of patients having complaints. The commonest symptoms are the appearance of halos around lights and photophobia. Keratopathy can appear after 1–2 months of treatment, but dosages of under 250 mg/day usually do not cause it. Dust exposure can lead to similar changes. The incidence of keratopathy is high, occurring in 30–70% of patients treated with higher dosages of chloroquine. The condition is usually reversible on withdrawal and does not seem to involve a threat to vision (SEDA-13, 805). There are differences in incidence between chloroquine and hydroxychloroquine. In a survey of 1500 patients, 95% of the patients taking chloroquine had corneal deposition of the drug, while less than 10% of patients taking hydroxychloroquine showed any corneal changes (SEDA-16, 303).

Retinopathy

The retinopathy encountered with the prolonged use of chloroquine or related drugs is a much more serious adverse effect and can lead to irreversible damage to the retina and loss of vision. However, it is not possible to predict in which patients and in what proportion of patients an early retinopathy will progress to blindness. The typical picture is that of the "bull's eye," an intact foveal area surrounded by a depigmented ring, the whole lesion being enclosed in a scattered hyperpigmented area. At this stage the retinal vessels are contracted, there are changes in the peripheral retinal pigment epithelium, and the optic disk is atrophic. In the early stages there are changes in the macular retinal pigment epithelium. However, the picture is not always clear, and peripheral retinal changes may appear as the first sign. Another sign may be unilateral paramacular retinal edema. The macular changes and the "bull's eye" are occasionally seen in patients who have never been treated with chloroquine or related drugs (SED-13, 805). Retinopathy can occur after

chloroquine antimalarial chemoprophylaxis for less than 10 years: the lowest reported total dose was 110 g (15). A case of hydroxychloroquine-induced retinopathy in a 45-year-old woman with systemic lupus erythematosus has illustrated that maculopathy can be associated with other 4-aminoquinolines (16).

The resulting functional defects are varied: difficulty in reading, scotomas, defective color vision, photophobia, light flashes, and a reduction in visual acuity. Symptoms do not parallel the retinal changes. By the time that visual acuity has become impaired, irreversible changes will have taken place.

Testing of visual acuity, central fields (with or without the use of red targets), contrast sensitivity, dark adaptation, and color vision provides no early indication of chloroquine retinopathy. Careful ophthalmoscopic examination of the macula can be a sensitive index when visual acuity remains intact. More sophisticated tests, such as the measurement of the critical flicker fusion frequency and the Amsler grid test (detection of small peripheral scotoma), can be useful. It is important to trace, if at all possible, the results of a pretreatment ophthalmological examination after dilatation of the pupils, thus reducing the possibility of confusing senile degenerative changes with chloroquine-induced abnormalities.

Despite the fact that the retinopathy has been known for many years, it is still not clear why certain patients develop these changes while others do not. There is a clear relation to daily dosage: the retinopathy is rarely seen with daily doses below 250 mg of chloroquine or 400 mg of hydroxychloroquine; the daily dose seems to be more important than the total dose. Nevertheless, cases of retinopathy have been described after the use of small doses for relatively short periods of time, while prolonged treatment and total doses of a kilogram or more have been used in many other patients without any evidence of macular changes. In the published cases there is usually no information about other treatments given previously or concomitantly. More cases are seen in older people. Patients with lupus erythematosus are more susceptible than patients with rheumatoid arthritis. The presence of nephropathy increases the likelihood of retinopathy, as does the concomitant use of probenecid. Exposure to sunlight may be of importance, since light amplifies the risk of retinopathy. The retinopathic changes are probably connected with the concentrating capacity of the melanin-containing epithelium. Chloroquine inhibits the incorporation of amino acids into the retinal pigment epithelium.

Little is yet known about the development of the retinopathy after withdrawal of treatment. Retinal changes in the early stages are probably reversible if the drug is withdrawn, and progression of a severe maculopathy to blindness seems to be less frequent than feared. In 1650 patients with 6/6 vision and relative scotomas there was no further decline in visual acuity after drug withdrawal, but 63% of patients who presented with absolute scotomas lost further vision over a median period of 6 years. This suggests that withdrawal of chloroquine at an early stage halts progression of the disease (SEDA-17, 327).

Three patients with chloroquine retinopathy have been studied with multifocal electroretinography (17). All three had been taking chloroquine for rheumatological diseases and all had electroretinographic changes that were more sensitive than full field electroretinography. It may be that multifocal electroretinography will be a useful technique in the assessment of suspected cases of subtle chloroquine retinopathy.

The need for routine ophthalmological testing of all patients who take chloroquine is under discussion, an obvious element being the cost/benefit ratio. The best current opinion seems to be that at doses not exceeding 6.5 mg/kg/day of hydroxychloroquine, given for not longer than 10 years and with periodic checking of renal and hepatic function, the likelihood of retinal damage is negligible and ophthalmological follow-up is not required (SEDA-16, 303) (SEDA-17, 327). However, patients taking chloroquine or higher doses of hydroxychloroquine should be checked.

Other adverse effects on the eyes

Rhegmatogenous retinal detachment and bitemporal hemianopsia have both been seen in association with chloroquine retinopathy. Bilateral edema of the optic nerve occurred in a woman who took chloroquine 200 mg/day for 2.5 months. Diplopia and impaired accommodation (characterized by difficulty in changing focus quickly from near to far vision and vice versa) also affect a minority of patients (SEDA-13, 806).

Ears

Ototoxicity has been mentioned occasionally over the years; tinnitus and deafness can occur in relation to high doses; symptoms described after injection of chloroquine phosphate include a case of cochlear vestibular dysfunction in a child (18). However, there is insufficient evidence to attribute ototoxicity to chloroquine in humans, except as a rare individualized phenomenon. In guinea pigs given chloroquine 25 mg/kg/day intraperitoneally, one of the first signs of intoxication was ototoxicity (SEDA-11, 586).

- Unilateral sensorineural hearing loss occurred in a 7-year-old girl with idiopathic pulmonary hemosiderosis after she had taken hydroxychloroquine 100 mg bd for 2 years (19).

Taste

Disturbances of taste and smell have been attributed to chloroquine (20).

Psychological, psychiatric

Many mental changes attributed to chloroquine have been described, notably agitation, aggressiveness, confusion, personality changes, psychotic symptoms, and depression. Acute mania has also been recognized (SEDA-18, 287). The mental changes can develop slowly and insidiously. Subtle symptoms, such as fluctuating impairment of thought, memory, and perception, can be early signs, but may also be the only signs. The symptoms may be connected with the long half-life of chloroquine and its accumulation, leading to high tissue concentrations (SEDA-11, 583). Chloroquine also inhibits glutamate dehydrogenase activity and can reduce concentrations of the inhibitory transmitter GABA.

In some cases with psychosis after the administration of recommended doses, symptoms developed after the patients had taken a total of 1.0–10.5 g of the drug, the time of onset of behavioral changes varying from 2 hours to 40 days. Most cases occurred during the first week and lasted from 2 days to 8 weeks (SEDA-11, 583).

Hallucinations have been reported after hydroxychloroquine treatment for erosive lichen planus (21).

- A 75-year-old woman was given hydroxychloroquine 400 mg/day for erosive lichen planus in conjunction with topical glucocorticoids and a short course of oral methylprednisolone 0.5 mg/kg/day. After 10 days she became disoriented in time and place, followed by feelings of depersonalization and kinesthetic hallucinations, preceded by nightmares. She stopped taking hydroxychloroquine 1 week later and the hallucinations progressively disappeared. She recovered her normal mental state within 1 month and had not relapsed 2 years later.

Transient global amnesia occurred in a healthy 62-year-old man, 3 hours after he took 300 mg chloroquine. Recovery was spontaneous after some hours (SEDA-16, 302).

In one center, toxic psychosis was reported in four children over a period of 18 months (SEDA-16, 302). The children presented with acute delirium, marked restlessness, outbursts of increased motor activity, mental inaccessibility, and insomnia. One child seemed to have visual hallucinations. In each case, chloroquine had been administered intramuscularly because of fever. The dosages were not recorded. The children returned to normal within 2 weeks.

Metabolism

Hypoglycemia was reported in a fatal chloroquine intoxication in a 32-year-old black Zambian male (SEDA-13, 240). Hypoglycemia has also been seen in patients, especially children, with cerebral malaria (SEDA-13, 240). Further studies have shown that the hypoglycemia in these African children was usually present before the antimalarial drugs had been started; in a study in Gambia hypoglycemia occurred after treatment with the drug had been started, although it was not necessarily connected with the treatment (SEDA-13, 240). Convulsions were more common in hypoglycemic children. This commonly unrecognized complication contributes to morbidity and mortality in cerebral *Plasmodium falciparum* malaria. Hypoglycemia is amenable to treatment with intravenous dextrose or glucose, which may help to prevent brain damage (SEDA-13, 804).

- A 16-year-old girl was treated empirically with chloroquine (total 450 mg of chloroquine base) for fever, had no malarial parasites in the peripheral blood smear, but had severe hypoglycemia of 1.5 mmol/l (27 mg/dl) (22).

This suggests that therapeutic doses of chloroquine can cause hypoglycemia even in the absence of malaria.

Although hydroxychloroquine has been used to treat porphyria cutanea tarda (23), there are reports that it can also worsen porphyria (24,25).

Electrolyte balance

Severe hypokalemia after a single large dose of chloroquine has been documented, and some studies show a correlation between plasma potassium concentrations and the severity of the cardiac effects. In a retrospective study of 191 consecutive patients who had taken an overdose of chloroquine (mean blood chloroquine concentration 20 μmol/l; usual target concentration up to 6 μmol/l), the mean plasma potassium concentration was 3.0 mmol/l (0.8) and was significantly lower in those who died than in those who survived (7). Plasma potassium varied directly with the systolic blood pressure and inversely with the QRS and QT intervals. Plasma potassium varied inversely with the blood chloroquine.

Hematologic

Chloroquine inhibits myelopoiesis in vitro at therapeutic concentrations and higher. In a special test procedure, a short-lasting anti-aggregating effect could be seen with chloroquine concentrations of 3.2–32 μg/ml (SEDA-16, 303). These effects have clinical consequences. Chloroquine and related aminoquinolines have reportedly caused blood dyscrasias at antimalarial doses. Leukopenia, agranulocytosis, and the occasional case of thrombocytopenia have been reported (SEDA-13, 804) (26). There is some evidence that myelosuppression is dose-dependent. This is in line with the hypothesis that 4-aminoquinoline therapy merely accentuates the cytopenia linked to other forms of bone marrow damage (SEDA-11, 584; SEDA-16, 302).

Some studies have pointed to inhibitory effects of chloroquine on platelet aggregability. In an investigation, this aspect of chloroquine was studied in vitro in a medium containing ADP, collagen, and ristocetin. There was a highly significant effect at chloroquine concentrations of 3.2–32 μg/ml. However, there were no significant differences in platelet responses to ADP or collagen 2 or 6 hours after adding chloroquine, compared with pre-drug values. The investigators believed that these data provided no cause for concern in using chloroquine for malaria prophylaxis in patients with impaired hemostasis (SEDA-16, 303).

Chloroquine can cause methemoglobinemia, especially in enzyme-deficient subjects. An exceptionally severe case of methemoglobinema has been reported.

- A 16-year-old girl treated empirically for fever with chloroquine (total 450 mg of chloroquine base)

developed cyanosis, jaundice, and altered consciousness (22). She had a moderate hemolytic anemia (hemoglobin 13.3 g/dl), severe methemoglobinema (70%), and hypoglycemia (1.5 mmol/l; 27 mg/dl). No malarial parasites were found, a Coomb's test was negative, and erythrocyte glucose-6-phosphate dehydrogenase (G6PD) activity was normal. NADPH methemoglobin reductase was not evaluated. Other causes, such as exposure to nitro-compounds, solvents, or drugs other than chloroquine were excluded. She was treated with methylthioninium (methylene blue) and the methemoglobinemia resolved over the next few days.

Mouth

Pigmentation of the palate can occur as a part of a more generalized pigmentation in patients taking chloroquine (27). It has been assiciated with retinopathy

Several patients seen with chloroquine retinopathy in Accra have been observed to present with depigmented patches in the skin of the face. This may be associated with a greyish pigmentation of the mucosa of the hard palate. Two such cases are reported here to illustrate the condition. Stomatitis with buccal ulceration has occasionally been mentioned (SEDA-11, 584).

Gastrointestinal

Gastrointestinal discomfort is not unusual in patients receiving chloroquine, and diarrhea can occur. Changes in intestinal motility may be to blame; intramuscular injection of chloroquine caused a shortened orotecal time in the five cases in which this was measured. Overdosage can cause vomiting.

Liver

Hepatotoxicity, which is uncommon with either chloroquine or proguanil, has been reported after the use of a fixed-dose combination of chloroquine and proguanil (28).

- A day before visiting the Indian subcontinent a 50-year-old French Caucasian woman began a course of a fixed-dose combination of proguanil (200 mg) + chloroquine (100 mg), one tablet daily for chemoprophylaxis. Four days later she developed vomiting, discolored stools, dark urine, and general fatigue. The chloroquine + proguanil was stopped immediately. A week later she developed severe nausea, headache, and conjunctival hemorrhages. There was no abdominal pain, fever, or rash. She had abnormal liver function tests, with aspartate transaminase activity of 335 U/l (reference range ≤35 U/l), alanine transaminase activity of 660 U/l (≤41 U/l), and alkaline phosphatase activity of 744 U/l (60–279 U/l). Total bilirubin was 616 (34–222) μmol/l and direct bilirubin was 393 (17–68) μmol/l. There were bile salts and increased bile pigments in the urine. An abdominal scan was unremarkable and serology for hepatitis A, B, and C was negative. Liver biopsy was not performed. She had no known susceptibility factors for liver disease. She had previously taken chloroquine + proguanil in 1998, 2002, and 2003 and recalled experiencing severe abdominal discomfort on the last occasion. Her symptoms improved after withdrawal of chloroquine + proguanil and her liver function tests gradually returned to normal.

The temporal relation between drug therapy and the appearance of symptoms suggested an adverse reaction to chloroquine + proguanil. The abnormal liver function tests, negative tests for hepatitis, the absence of fever, and the resolution of symptoms after withdrawal all suggested a causal relation. The exact mechanism is not known but it could have been an allergic reaction after sensitization from previous exposures.

Urinary tract

Chloroquine-induced kidney damage has occasionally been reported. A remarkably well-documented case report further elucidates this adverse drug reaction.

- A 46-year-old woman with impaired renal function (glomerular filtration rate of 23–26 ml/minute/$1.7m^2$, serum creatinine 186 µmol/l (2.1 mg/dl) took chloroquine 155 mg/day for Sjögren's syndrome (29). After 5 months, her serum creatinine had increased to 339 µmol/l (2.7 mg/dl) and after 11 months to 442 µmol/l (5 mg/dl) with a glomerular filtration rate of 8 ml/minute/$1.7m^2$. Light microscopy of a kidney biopsy showed vascular parenchymal atrophy with accumulation of colloid material in atrophic tubules. Electron microscopy showed osmiophilic lamellated bodies mainly in podocytes and to a lesser degree in glomerular and vascular endothelial and vascular smooth muscle cells. Fabry's disease was ruled out based on normal activity of alpha-galactosidase A. Because the histopathological findings resembled those seen in chloroquine-induced myopathy/cardiomyopathy, chloroquine was withdrawn and 9 months later, her glomerular filtration rate was 19 ml/minute/$1.7m^2$ and serum creatinine 221 µmol/l (2.5 mg/dl), close to baseline values.

Skin

The most common dermatological adverse event associated with chloroquine is skin discomfort (often called pruritus). It is much more common in people with darker skins and has been ascribed to chloroquine binding to increased melanin concentrations in the skin. In a pharmacokinetic study, the ratio of AUC_{0-48} for chloroquine and its major metabolite desethylchloroquine was significantly higher in the plasma and urine of 18 patients with chloroquine-induced pruritus than in that of 18 patients without (30). These results imply that differences in metabolism and higher chloroquine concentrations may be partly responsible for chloroquine-induced pruritus.

Pruritus begins about 10 hours after the start of treatment, with a maximum intensity at about 24 hours. These times correspond to maximum serum concentrations of chloroquine and its metabolites after oral ingestion. In many cases, the itch is confined to the palms of the hands and the soles of the feet. In a study in Nigeria, the incidence of pruritus was 60–75%; the itch was considered unbearable in 40%, and 30% refused further chloroquine (31). In a second study, there was an even higher incidence. In a study elsewhere, the incidence of pruritus was 27% (SEDA-16, 304). Not surprisingly, pruritus is a major cause of non-adherence to treatment, and it may contribute largely to the emergence and spread of resistant *P. falciparum* (SEDA-16, 304). Pruritus is more often seen in black-skinned than in white-skinned people in Africa, a difference that has been ascribed to the binding of chloroquine to melanin, and hence a racial predisposition. No such reports have come from America (SEDA-11, 584; SEDA-16, 303; SEDA-17, 327; SEDA-18, 288). Antihistamine treatment can have a preventive effect on pruritus. Other treatments that have been mentioned include prednisone and niacin, but the results were not impressive (32).

A few cases of psoriasis, or severe exacerbation of psoriasis shortly after the start of treatment, have been reported (SEDA-13, 804; SEDA-16, 304; SEDA-17, 327).

Photosensitivity and photo-allergic dermatitis have been seen, particularly during prolonged therapy with high doses.

Chloroquine can cause vitiligo (SEDA-17, 327).

- A 44-year-old Hispanic woman developed depigmented patches on her chest, shoulders, forearms, back, and shins 1 month after switching from hydroxychloroquine to chloroquine 500 mg/day for cutaneous discoid lupus erythematosus (33). Chloroquine was immediately withdrawn, and within 2 months spontaneous re-pigmentation occurred in most of the depigmented patches.

Chloroquine can cause Stevens–Johnson syndrome and toxic epidermal necrolysis (34), which can be fatal (35) .

- A 32-year old woman weighing 61kg developed painful skin blisters and erosions accompanied by fever and myalgia (36). The symptoms started on the third day after a course of oral chloroquine (Tablets Lariago, chloroquine phosphate 500 mg, Ipca Laboratories, Mumbai, India); two tablets initially, one tablet after 6 hours, and one tablet/day for 2 days. Her symptoms began with erythematous itchy papular eruptions on the trunk and then progressed to involve the face, limbs, and mucous membranes of the mouth. She had difficulty in swallowing because of painful erosions of the mouth and oropharynx. There was conjunctivitis but no visual impairment. Bullae continued to appear on the trunk and limbs and a purpuric rash on the trunk. She responded well to intravenous fluids, antibiotics to prevent secondary infection, and hydrocortisone.
- A 39-year-old woman with rheumatoid arthritis took hydroxychloroquine 200 mg bd for painful synovitis, in addition to meloxicam, co-dydramol, and Gaviscon. She inadvertently took twice the prescribed dose of hydroxychloroquine, but stopped it after 2 weeks because of nausea. The next day she developed a widespread blotchy erythema and 2 weeks later was admitted to hospital with clinical and histological toxic epidermal necrolysis and

deteriorated rapidly with multiorgan failure; she died 1 week later.

There have been only a few isolated reports of Stevens–Johnson syndrome associated with hydroxychloroquine. Recently, a clear temporal relation to the start of treatment with hydroxychloroquine has been documented in a patient with rheumatoid arthritis (37).

An increased frequency of skin reactions to hydroxychloroquine was noted in 11 patients (seven of whom had systemic lupus erythematosus, two discoid lupus, and two a lupus-like syndrome) when a coloring agent (sunshine yellow E110) was removed from the formulation; the authors were unable to explain this unexpected finding (38).

There have been four case reports of photosensitivity associated with hydroxychloroquine (39) which has an estimated incidence of about 10 per 1000 patient-years (40).

Pemphigus has been attributed to chloroquine and hydroxychloroquine (41).

- A 52-year-old woman abruptly developed generalized blisters 2 weeks after starting to take hydoxychloroquine for rheumatoid arthritis. The eruption consisted of pruritic bullae and erosions on the head, trunk, limbs, and oral mucosa. She also had scattered urticarial lesions. She had had a previous similar but mild reaction following chloroquine therapy, which had cleared when the drug was withdrawn. Biopsy confirmed the diagnosis of pemphigus vulgaris with suprabasal slitting.

Drug-induced pemphigus is mainly caused by drugs containing a thiol (-SH) group, such as penicillamine, captopril, piroxicam, and ampicillin, or drugs with an active amide group, such as penicillin. In contrast, chloroquine and hydroxychloroquine are 4-aminoquinolines.

A bullous skin eruption induced by radiotherapy after the use of chloroquine has been described (42).

- A 12-year-old girl with a diffuse pontine glioma was treated with radiotherapy and developed a high-grade fever with chills. Although the blood smear was negative for malaria she was treated empirically with chloroquine 500 mg, 250 mg 6 hours later, and 250 mg/day for 2 days. On day 3 she developed localized bullous eruptions over the site of irradiation, which peeled in 6-7 hours leaving a patch of fulminant moist desquamation. The skin surrounding the patch was severely erythematous. Radiotherapy was withheld and she was treated with topical amniotic membrane and gentian violet. One week later the fever had cleared and the desquamation had almost healed. Radiotherapy was restarted and completed uneventfully.

The combination of heat, chloroquine, and radiotherapy appears to have enhanced an aggressive skin reaction in this patient. Caution is recommended when using chloroquine in patients receiving radiotherapy.

Hydroxychloroquine causes skin reactions such as urticaria. There is some support for the contention that hydroxychloroquine causes skin reactions more often than chloroquine.

Nails

Chloroquine can turn the nail bed blue–brown and the nail itself can develop longitudinal stripes and show a blue–grey fluorescence (SEDA-11, 584).

Musculoskeletal

Chloroquine and hydroxychloroquine occasionally cause a myopathy associated with muscle weakness, reduced or absent tendon reflexes, and raised creatine kinase activity; it usually develops gradually after 5–7 months of treatment. The muscle biopsy findings have been described in detail: atrophic muscle fibers, muscle fiber necrosis with vacuolar degradation, vacuoles staining positive with acid phosphatase with a granular pattern, autophagic vacuoles, and cytosomes with electron-dense curvilinear profiles on electron microscopy (43).

Severe vacuolar myopathy has been reported (44).

- A 51-year-old man with a mantle cell carcinoma was initially treated with cyclophosphamide, doxorubicin, vincristine, and prednisolone and obtained remission for 2 years. His lymphoma recurred and he was given rituximab, with a poor response. He subsequently received a bone marrow graft from his son. Despite complete remission he developed graft-versus-host disease with scleroderma and fascial involvement. He was given various drugs for graft-versus-host disease, including mycophenolate mofetil, tacrolimus, prednisolone, 2'-deoxycoformycin, and hydroxychloroquine. His condition was moderately sensitive to prednisolone in doses of 60–120 mg/day. While taking prednisolone and tacrolimus he was given hydroxychloroquine 400 mg bd. He then developed progressive debilitating limb and respiratory muscle weakness. Glucocorticoids were suspected of causing this and were tapered, but without much effect, and he gradually became too weak to walk. The serum creatine kinase activity was normal and acetylcholine receptor antibodies were negative. Electromyography showed a severe, non-irritable myopathy and a sensory motor axonal polyneuropathy. Muscle biopsy showed a necrotizing, vacuolar myopathy, with many fibers containing autophagic and red-rimmed vacuoles, consistent with an amphiphilic drug-induced myopathy. Following withdrawal of hydroxychloroquine, his strength and function improved considerably. Later prednisolone and tacrolimus were reintroduced and he made a good recovery.

Immunologic

Chloroquine can cause cutaneous necrotizing vasculitis.

- A 49-year-old man developed severe cutaneous necrotizing vasculitis after 13 days of treatment with a combination of chloroquine 100 mg and proguanil 200 mg/day (45). He had taken no other drugs. The lesions consisted of diffuse painful purpuric and extensive necrotizing plaques on the upper and lower limbs, mainly on the hands and feet, maculopapular erythema on the trunk, and petechial maculae on the hard palate. His temperature was normal and he had no organomegaly. Laboratory findings were normal, except for a eosinophilia of $0.72 \times 10^9/l$ and a C-reactive protein of 133 mg/l. Chloroquine and proguanil were withdrawn and his condition normalized within 3 weeks. Skin tests 3 month later were positive for chloroquine, but negative for quinine and the biguanide derivatives proguanil and metformin.

- Allergic contact dermatitis, which progressed to generalized dermatitis and conjunctivitis, followed later by severe asthma, occurred in a 60-year-old worker in the pharmaceutical industry after exposure to hydroxychloroquine (46). Patch-testing showed delayed sensitivity to hydroxychloroquine. Equivalent tests in five healthy volunteers were negative. The patch test reactions were pustular, and a biopsy was interpreted as multiform contact dermatitis. Bronchial exposure to hydroxychloroquine dust produced delayed bronchial obstruction over the next 20 hours, progressing to fever and generalized erythema (hematogenous contact dermatitis).

Infection risk

An acute gluteal abscess after an injection of chloroquine has been reported (47).

- A 24 year-old woman who had been pregnant for 24 weeks developed a fever, sweating, headache, and pain in the region where she had received an injection of chloroquine 9 days before. She was febrile (38.9°C). Chest examination was normal and the spleen was not palpable. She had a right gluteal abscess, which was later drained under general anesthesia. She was given ampicillin/cloxacillin, 500 mg intravenously at first and then orally for 7 days. *Staphylococcus aureus* sensitive to ampicillin/cloxacillin was isolated. She made uneventful recovery and delivered a live 2.9 kg baby.

Long-Term Effects

Drug tolerance

Chloroquine-resistant falciparum malaria was first reported in 1960. As of 1996, chloroquine resistance became widespread throughout the world and in many areas there is multidrug resistance. Preventive administration of drugs such as chloroquine, primaquine, and pyrimethamine, as well as the use of various sulfonamide mixtures and combinations of sulfonamides with trimethoprim, has progressively lost its usefulness. Currently, hardly half a century after the therapeutic breakthroughs occurred, quinine is once more one of the most valuable drugs in the treatment of malaria and there is a desperate need for other effective drugs.

Alongside the well-known development of resistance by *P. falciparum* to chloroquine, the emergence of chloroquine-resistant *Plasmodium vivax* is now clear (SEDA-

13, 801). An increased frequency of cerebral malaria appears to coincide with the growing emergence of the chloroquine-resistant strains in Francophone Africa.

Drug resistance

Chloroquine was withdrawn in Malawi in 1993, but is once again an effective treatment for malaria in that country (48). In 210 children with uncomplicated malaria randomized to chloroquine or sulfadoxine + pyrimethamine for 28 days, only one of those who were given chloroquine group had treatment failure compared with 71/87 of those who were given sulfadoxine + pyrimethamine. Cumulative efficacy, defined using a proportional-hazards model, was 99% (95% CI = 93, 100) with chloroquine compared with 21% (95% CI = 13, 30) with sulfadoxine + pyrimethamine. There were no major adverse events related to the study drugs. This suggests that the advantage enjoyed by chloroquine-resistant malaria parasites was lost after the withdrawal of chloroquine. Drug pressure selectively favors the development of drug-resistant parasites. This promising result needs to be confirmed in larger studies and different locations.

A study in children has suggested that the antihistamine chlorphenamine might reverse chloroquine resistance (SEDA-24, 331). Furthermore, ketotifen, another antihistamine, reversed chloroquine resistance in *Plasmodium falciparum* in vitro. Chloroquine (10 mg chloroquine base/kg for 3 days) alone and in combination with ketotifen (0.25 mg/kg followed by 0.125 mg/kg tds for 4 days) have therefore been compared in an open, randomized study in 150 children aged 1–10 years (49). Although the fever lasted slightly shorter in patients who took the combination of chloroquine and ketotifen compared with those treated with chloroquine only, parasite clearance times and cure rates were comparable, suggesting no effect of ketotifen on chloroquine resistance.

Second-Generation Effects

Pregnancy

Chloroquine inactivates DNA, and crosses the placenta in animals. Caution has generally been advised with respect to the use of chloroquine and related compounds during pregnancy, but except for one (perhaps coincidental) case, there have been no reports of complications to mother or child from treatment with chloroquine during pregnancy (SEDA-14, 239; SEDA-17, 326).

An observational comparison in a rural Ghanaian hospital of 2083 pregnant women and 3084 historical controls showed no serious adverse events with chloroquine chemoprophylaxis (300 mg/week), but a high rate of pruritus (50). There was a decrease in anemia in pregnancy but no increase in perinatal mortality or birth weight in the chloroquine-treated mothers, although this was only in comparison with historical controls.

Teratogenicity

In 133 consecutive pregnancies in 90 women who took 200 mg hydroxychloroquine twice a day (n = 122) or once a day (n = 11) the same number of pregnancies resulted in live births as in 70 consecutive pregnancies in 53 women with similar disorders who did not take hydroxychloroquine (51). Pregnancy outcomes and the results of follow-up examination of the children were comparable.

Susceptibility Factors

Genetic factors

Mutations in the ABCR gene (a photoreceptor-specific ATP-binding cassette transporter gene) have been associated with Stargardt disease, which has some features similar to chloroquine-induced retinopathy. In a case-control study of eight cases of chloroquine-induced retinopathy, five of the eight cases had mis-sense mutations in the ABCR gene, two of which have been associated with Stargardt disease (52). It may be that polymorphisms in the ABCR gene predispose to chloroquine-induced retinopathy.

Age

Small children have usually been considered as being relatively more sensitive to the effects of overdosage, but it has been calculated that on a mg/kg body weight basis, adults are in fact equally sensitive. Young children seem to be truly more susceptible to gastric irritation. Patients with a history of mania or epilepsy should be careful in taking chloroquine (11). The hypoxemic effects of chloroquine, reflecting cardiac and respiratory toxicity, pose a particular problem in the newborn, in whom existing malarial infection may not become clinically manifest until some months after birth (SEDA-16, 302).

Compared with adults, mortality in children after acute chloroquine poisoning is extremely high. Although the clinical presentation is mostly similar to that in adults (apnea, seizures, cardiac dysrhythmias), a single 300 mg chloroquine tablet was enough to kill a 12-month-old female infant (SEDA-16, 302).

Other features of the patient

Skin reactions to hydroxychloroquine occur more often in patients with dermatomyositis than in patients with systemic lupus erythematosus, as has been shown in a retrospective, age-, sex-, and race-matched case-control study in 78 patients (53). Twelve of 39 patients with dermatomyositis developed a skin reaction to hydroxychloroquine, compared with only one of 39 patients with lupus erythematosus.

Drug Administration

Drug formulations

Chloroquine has a bitter taste, which can deter children from taking it, so a sweet effervescent formulation of chloroquine phosphate has been compared with chloroquine tablets in a pharmacodynamic study (54). However, sweet-tasting medications carry a risk of accidental overdose in children.

Drug administration route

If given intravenously, chloroquine should be diluted and infused slowly, since rapid injection causes toxic concentrations. Toxicity and even death have been reported after intramuscular administration of larger doses; this is probably connected with rapid absorption in such cases (SEDA-17, 327).

Drug overdose

Acute intoxication, either accidental or in attempted suicide, can cause headache, drowsiness, vision disturbance, vomiting and diarrhea, cardiovascular collapse, and respiratory failure. Deaths have been recorded at blood concentrations of 1 µg/ml (SEDA-11, 586; 55,56). Compared with adults, mortality in children after acute chloroquine poisoning is extremely high. Although the clinical presentation is mostly similar to that in adults (apnea, seizures, cardiac dysrhythmias), a single 300 mg chloroquine tablet was enough to kill a 12-month-old female infant (SEDA-16, 302).

Massive hydroxychloroquine overdose has been reported (57).

- A 17-year-old girl took 22 g of hydroxychloroquine in a suicide attempt. She developed hypotension, life-threatening ventricular dysrhythmias, and mild hypokalemia. She was managed with saline infusion and dopamine for hypotension, gastric lavage and activated charcoal for decontamination, and lidocaine, magnesium sulfate and defibrillation for pulseless ventricular tachycardia. Potassium replacement and bicarbonate replacement were performed. She survived without sequelae.

Quick treatment of hypotension, gastric decontamination, continuous cardiac monitoring, and the treatment of dysrhythmias are critical in the management of hydroxychloroquine and that of chloroquine intoxication.

Deaths from chloroquine overdose have been reported with doses as low as 2–3 g in adults, and the death rate is as high as 25%. The effects of chloroquine overdose include cardiac effects (such as dysrhythmias, reduced myocardial contractility, and hypotension) and central nervous system complications (such as confusion, coma, and seizures).

There have been three reports of chloroquine overdose, two from Oman (58) and one from the Netherlands (59). The two reports from Oman were similar to previously published reports of chloroquine overdose associated with cardiac dysfunction, confusion, and coma; both patients had standard treatment with activated charcoal, diazepam infusions, and positive inotropic drugs, and both survived. The single case report from the Netherlands gave pharmacokinetic measurements performed before, during, and after hemoperfusion. This showed that hemoperfusion extracted very little chloroquine and was unlikely to be of any use in chloroquine overdose, as would be expected from the high protein binding and large volume of distribution of chloroquine.

In Zimbabwe, 544 cases of poisoning by a single agent were identified in a retrospective hospital record review (60). Antimalarial drugs accounted for the largest proportion of admissions (53%), and chloroquine accounted for 96% of these (279 cases). The median length of hospital stay in those who took chloroquine was significantly shorter (1 versus 2 days) and more patients took chloroquine deliberately (80% versus 69%). The mortality rate from chloroquine poisoning was significantly higher than from poisoning with other drugs (5.7% versus 0.7%).

Overdose with hydroxychloroquine is far less common than with chloroquine. Three of eight patients died (61). Life-threatening symptoms, such as hypotension, conduction disturbances, and hypokalemia can occur within 30 minutes of ingestion and are similar to those seen in chloroquine overdose. The lethal plasma concentration of hydroxychloroquine is not well established. Therapeutic drug concentrations are usually less than 1 μmol/l. Serious toxicity has been reported at plasma concentrations of 2.1–29 μmol/l.

Management of hydroxychloroquine overdose is similar to that of chloroquine overdose, including the use of charcoal for drug adsorption, diazepam for seizures and sedation, early intubation and mechanical ventilation, and potassium replacement for severe hypokalemia.

Drug–Drug Interactions

Amlodipine

Syncope occurred in a hypertensive 48-year-old man who took oral chloroquine sulfate (total 600 mg base) while also taking amlodipine 5 mg/day (62). Chloroquine and amlodipine both cause vasodilatation, perhaps by release of nitric oxide, and the syncope in this case was probably due to a synergistic mechanism. Malaria itself can also provoke orthostatic reactions, which may be why syncope is not a reported adverse effect of chloroquine. However, in this patient malaria had been excluded.

A possible interaction of amlodipine with chloroquine has been reported (63).

- A 48-year-old hypertensive physician, who had optimal blood pressure control after taking oral amlodipine 5 mg/day for 3 months, developed a slight frontal headache and fever, thought that he had malaria, and took four tablets of chloroquine sulfate (total 600 mg base). Two hours later he became nauseated and dizzy and collapsed; his systolic blood pressure was 80 mmHg and his diastolic pressure was unrecordable, suggesting vasovagal syncope, which was corrected by dextrose–saline infusion.

There was no malaria parasitemia in this case, and hence the syncope may have resulted from the acute synergistic hypotensive, venodilator, and cardiac effects of chloroquine plus amlodipine, possibly acting via augmented nitric oxide production and calcium channel blockade. Since malaria fever is itself associated with orthostatic hypotension, this possible interaction may be unrecognized and unreported in these patients.

Antibiotics

Studies of chloroquine used in combination with antibiotics showed an antagonistic effect with penicillin but a synergistic effect with chlortetracycline. Urinary tests after single doses of ampicillin 1 g and chloroquine 1 g showed a significant reduction in the systemic availability of the ampicillin.

Artemisinin derivatives

Although an early report suggested antagonism between chloroquine and artemisinin (64), more recent evidence suggests the opposite, that is a synergistic effect between the two (65,66).

Chlorphenamine

Chlorphenamine enhances the efficacy of chloroquine in acute uncomplicated falciparum malaria, but the disposition of chloroquine in these circumstances is unpredictable. Chloroquine (25 mg/kg) was given orally over 3 days in combination with chlorphenamine to Nigerian children with parasitemia (67). The peak whole blood chloroquine concentration was increased and the time to peak concentration shortened. In small trials there seemed to be an increase in QT interval with this combination, but less than with halofantrine (68). However, in other studies, the addition of chlorphenamine to chloroquine did not amplify the cardiac effects of chloroquine (68).

Ciclosporin

Chloroquine can increase ciclosporin blood concentrations (69).

Cimetidine

Cimetidine enhanced the susceptibility of *P. falciparum* to chloroquine in vitro in 60% of isolates (70).

Digoxin

The pharmacokinetic interaction of quinidine with digoxin also occurs with quinine and hydroxychloroquine (71).

Fansidar (sulfadoxine + pyrimethamine)

The combined use of Fansidar (sulfadoxine + pyrimethamine) with chloroquine has been reported to result in more severe adverse reactions (72). However, an increased risk has not been reported in recent studies (73).

Halofantrine

There is an increased risk of dysrhythmias, including torsade de pointes, when halofantrine is combined with quinine/quinidine or chloroquine and any other drug that prolongs the QT interval (74).

Insulin

There may be an interaction of chloroquine with insulin. An oral glucose load given to healthy subjects and to patients with non-insulin-dependent diabetes mellitus, before and during a short course of chloroquine, showed a small but significant reduction in fasting blood glucose concentration in the control group and improvement in glucose tolerance in the patients (SEDA-12, 240). The response seems to reflect reduced degradation of insulin rather than increased pancreatic output.

Probenecid

Probenecid may increase the risk of chloroquine-induced retinal damage (75).

Pyrimethamine

The combined use of Fansidar (sulfadoxine + pyrimethamine) with chloroquine has been reported to result in more severe adverse reactions (76). However, an increased risk has not been reported in recent studies (77).

Quinine

Chloroquine antagonizes the action of quinine against *P. falciparum* in vivo (78). However, no such evidence of antagonism was found in a study in which Malawian children with cerebral malaria were treated with quinine. There was no difference in survival and rate of recovery in patients who had also been given chloroquine compared with those who had not (SEDA-13, 816).

Sulfonylureas

The addition of hydroxychloroquine to sulfonylureas has been investigated in a placebo-controlled study in 125 adipose patients whose diabetes was not well enough controlled with a sulfonylurea alone (79). During the first six months HbA_{1c} was significantly reduced by 1.02%. There were no significant differences in adverse effects, but those who took hydroxychloroquine had a greater incidence of minor corneal changes.

Thyroxine

A marked increase in serum TSH occurred in the same patient on two occasions after several weeks of antimalarial prophylaxis with chloroquine and proguanil, the likely mechanism being enzyme induction and increased thyroxine catabolism (SEDA-22, 469).

Vaccines

Chloroquine 300 mg/week adversely affected the antibody response to human diploid-cell rabies vaccine administered concurrently. The mean rabies-neutralizing antibody titer was significantly reduced on each day of testing (SEDA-13, 806) (10). In contrast, retrospective studies of the response to pneumococcal polysaccharide in patients with systemic lupus erythematosus taking chloroquine or hydroxychloroquine, and of the response to tetanus–measles–meningococcal vaccine in a region of Nigeria where malaria is endemic, did not show an effect on antibody production. However, it was pointed out that the altered immune status of patients with systemic lupus erythematosus makes it difficult to compare their response to that of young healthy adults receiving rabies vaccine. Illness and nutritional state could have influenced the findings in the Nigerian study (SEDA-13, 807) (80).

Verapamil

Verapamil completely reversed pre-existing in vitro resistance to chloroquine to below the cut-off point of 70 nmol/l (68).

Smoking

Antimalarial drugs (chloroquine, hydroxychloroquine, or quinacrine) were given to 36 patients with cutaneous lupus, of whom 17 were smokers and 19 non-smokers (10). The median number of cigarettes smoked was one pack/day, with a median duration of 12.5 years. There was a reduction in the efficacy of antimalarial therapy in the smokers. Patients with cutaneous lupus should therefore be encouraged to stop smoking and consideration may be given to increasing the doses of antimalarial drugs in smokers with refractory cutaneous lupus before starting a cytotoxic agent.

Monitoring Therapy

Susceptibility factors for the development of toxic retinopathy with hydroxychloroquine include high daily doses, long duration of treatment, concomitant liver or kidney disease, and age over 60 years. Constant monitoring for retinal toxicity is therefore vital for the prevention of this potentially irreversible severe adverse effect. However, most methods for monitoring retinal toxicity, such as Amsler grid testing, color vision testing, and static perimetry, are subjective. Multifocal electronic retinography is a more objective test and offers multidimensional visualization of the retina. In a longitudinal study of 12 patients who had multifocal electronic retinography at baseline and at 12–24 months, serial recordings of retinal amplitudes and peak latencies showed that patients taking hydroxychloroquine had reduced retinal function while those who stopped taking it had improved retinal function (79). Multifocal electronic retinography therefore offers the possibility of detecting early changes in retinal function in patients taking hydroxychloroquine.

References

1. Sowunmi A, Ayede AI, Falade AG, Ndikum VN, Sowunmi CO, Adedeji AA, Falade CO, Happi TC, Oduola AM. Randomized comparison of chloroquine and amodiaquine in the treatment of acute, uncomplicated, *Plasmodium falciparum* malaria in children. Ann Trop Med Parasitol 2001;95(6):549–58.
2. Veinot JP, Mai KT, Zarychanski R. Chloroquine related cardiac toxicity. J Rheumatol 1998;25(6):1221–5.

3. Guedira N, Hajjaj-Hassouni N, Srairi JE, el Hassani S, Fellat R, Benomar M. Third-degree atrioventricular block in a patient under chloroquine therapy. Rev Rhum Engl Ed 1998;65(1):58–62.

4. Cervera A, Espinosa G, Font J, Ingelmo M. Cardiac toxicity secondary to long term treatment with chloroquine. Ann Rheum Dis 2001;60(3):301.

5. Charlier P, Cochand-Priollet B, Polivka M, Goldgran-Toledano D, Leenhardt A. Cardiomyopathie a la chloroquine revelée par un bloc auriculo-ventriculaire complete. A propos d'une observation. [Chloroquine cardiomyopathy revealed by complete atrio-ventricular block. A case report.] Arch Mal Coeur Vaiss 2002;95(9):833–7.

6. Freihage JH, Patel NC, Jacobs WR, Picken M, Fresco R, Malinowska K, Pisani BA, Mendez JC, Lichtenberg RC, Foy BK, Bakhos M, Mullen GM. Heart transplantation in a patient with chloroquine-induced cardiomyopathy. J Heart Lung Transpl 2004;23(2):252-5.

7. Clemessy JL, Favier C, Borron SW, Hantson PE, Vicaut E, Baud FJ. Hypokalaemia related to acute chloroquine ingestion. Lancet 1995;346(8979):877–80.

8. Chen CY, Wang FL, Lin CC. Chronic hydroxychloroquine use associated with QT prolongation and refractory ventricular arrhythmia. Clin Toxicol 2006;44:173-5.

9. Mitja K, Izidor K, Music E. Chloroquin-induzierte arzneimitteltoxische Alveolitis. [Chloroquine-induced drug hypersensitivity alveolitis.] Pneumologie 2000;54(9):395–7.

10. Blaison G, Tranchant C, Mohr M, Roth T, Warter JM. Les complications neuromusculaires des traitements par la chloroquine. Sem Hop Paris 1990;66:2425–8.

11. Fish DR, Espir ML. Convulsions associated with prophylactic antimalarial drugs: implications for people with epilepsy. BMJ 1988;297(6647):526–7.

12. Crawley J, Kokwaro G, Ouma D, Watkins W, Marsh K. Chloroquine is not a risk factor for seizures in childhood cerebral malaria. Trop Med Int Health 2000;5(12):860–4.

13. Guilloton L, Burckard E, Fresse S, Drouet A, Felten D. Crise epileptique apres chimioprophylaxie antipalustre par chloroquine. [Epileptic crisis after antimalaria chemoprophylaxis with chloroquine.] Presse Méd 2001;30(35):1745.

14. Wasay M, Wolfe GI, Herrold JM, Burns DK, Barohn RJ. Chloroquine myopathy and neuropathy with elevated CSF protein. Neurology 1998;51(4):1226–7.

15. Bertagnolio S, Tacconelli E, Camilli G, Tumbarello M. Case report: retinopathy after malaria prophylaxis with chloroquine. Am J Trop Med Hyg 2001;65(5):637–8.

16. Warner AE. Early hydroxychloroquine macular toxicity. Arthritis Rheum 2001;44(8):1959–61.

17. Kellner U, Kraus H, Foerster MH. Multifocal ERG in chloroquine retinopathy: regional variance of retinal dysfunction. Graefes Arch Clin Exp Ophthalmol 2000;238(1):94–7.

18. Mukherjee DK. Chloroquine ototoxicity—a reversible phenomenon? J Laryngol Otol 1979;93(8):809–15.

19. Coutinho MB, Duarte I. Hydroxychloroquine ototoxicity in a child with idiopathic pulmonary haemosiderosis. Int J Pediatr Otorhinolaryngol 2002;62(1):53–7.

20. Weber JC, Alt M, Blaison G, Welsch M, Martin T, Pasquali JL. Modifications du gout et de l'odorat imputables a l'hydroxychloroquine. [Changes in taste and smell caused by hydroxychloroquine.] Presse Méd 1996;25(5):213.

21. Ferraro V, Mantoux F, Denis K, Lay-Macagno M-A, Ortonne J-P, Lacour J-P. Hallucinations au cours d'un traitement par hydroxychloroquine. Ann Dermatol Venereol 2004;131:471-3.

22. Sharma N, Varma S. Unusual life-threatening adverse drug effects with chloroquine in a young girl. J Postgrad Med 2003;49:187.

23. Petersen CS, Thomsen K. High-dose hydroxychloroquine treatment of porphyria cutanea tarda. J Am Acad Dermatol 1992;26(4):614–9.

24. Kutz DC, Bridges AJ. Bullous rash and brown urine in a systemic lupus erythematosus patient treated with hydroxychloroquine. Arthritis Rheum 1995;38(3):440–3.

25. Baler GR. Porphyria precipitated by hydroxychloroquine treatment of systemic lupus erythematosus. Cutis 1976;17(1):96–8.

26. Don PC, Kahn TA, Bickers DR. Chloroquine-induced neutropenia in a patient with dermatomyositis. J Am Acad Dermatol 1987;16(3 Pt 1):629–30.

27. Bentsi-Enchill KO. Pigmentary skin changes associated with ocular chloroquine toxicity in Ghana. Trop Geogr Med 1980;32(3):216–20.

28. Wielgo-Polanin R, Largace L, Gautron E, Diquet B, Lainé-Cessac P. Hepatotoxicity associated with the use of a fixed combination of chloroquine and proguanil. Int J Antimicrob Agents 2005;26 (2):176-8.

29. Muller Hocker J, Schmid H, Weiss M, Dendorfer U, Braun GS. Chloroquine-induced phospholipidosis of the kidney mimicking Fabry's disease: case report and review of the literature. Hum Pathol 2003;34:285-9.

30. Onyeji CO, Ogunbona FA. Pharmacokinetic aspects of chloroquine-induced pruritus: influence of dose and evidence for varied extent of metabolism of the drug. Eur J Pharm Sci 2001;13(2):195–201.

31. Osifo NG. Chloroquine-induced pruritus among patients with malaria. Arch Dermatol 1984;120(1):80–2.

32. Ajayi AA, Akinleye AO, Udoh SJ, Ajayi OO, Oyelese O, Ijaware CO. The effects of prednisolone and niacin on chloroquine-induced pruritus in malaria. Eur J Clin Pharmacol 1991;41(4):383–5.

33. Martin Garcia RF, del R Camacho N, Sanchez JL. Chloroquine-induced, vitiligo-like depigmentation. J Am Acad Dermatol 2003;48:981-3.

34. Boffa MJ, Chalmers RJ. Toxic epidermal necrolysis due to chloroquine phosphate. Br J Dermatol 1994;131(3):444–5.

35. Murphy M, Carmichael AJ. Fatal toxic epidermal necrolysis associated with hydroxychloroquine. Clin Exp Dermatol 2001;26(5):457–8.

36. Beedimani RS, Rambhimaiah S. Oral chloroquine induced Stevens–Johnson syndrome. Ind J Pharmacol 2004;36(2):101.

37. Leckie MJ, Rees RG. Stevens–Johnson syndrome in association with hydroxychloroquine treatment for rheumatoid arthritis. Rheumatology (Oxford) 2002;41(4):473–4.

38. Salido M, Joven B, D'Cruz DP, Khamashta MA, Hughes GR. Increased cutaneous reactions to hydroxychloroquine (Plaquenil) possibly associated with formulation change: comment on the letter by Alarcon. Arthritis Rheum 2002;46(12):3392–6.

39. Metayer I, Balguerie X, Courville P, Lauret P, Joly P. Toxidermies photo-induites par l'hydroxychloroquine: 4 cas. [Photodermatosis induced by hydroxychloroquine: 4 cases.] Ann Dermatol Venereol 2001;128(6–7):729–31.

40. Singh G, Fries JF, Williams CA, Zatarain E, Spitz P, Bloch DA. Toxicity profiles of disease modifying antirheumatic drugs in rheumatoid arthritis. J Rheumatol 1991;18(2):188–94.

41. Ghaffarpour G, Jalali MHA, Yaghmaii B, Mazloomi S, Soltani-Arabshahi R. Chloroquine/hydroxychloroquine-induced pemphigus. Int Soc Dermatol 2006;45:1261-3.

42. Rustogi A, Munshi A, Jalali R. Unexpected skin reaction induced by radiotherapy after chloroquine use. Lancet Oncol 2006;7:608-9.

43. Richter JG, Becker A, Ostendorf B, Specker C, Stoll G, Neuen-Jacob E, Schneider M. Differential diagnosis of high serum creatine kinase levels in systemic lupus erythematosus. Rheumatol Int 2003;23:319-23.

44. Bolaños-Meade J, Zhou L, Hoke A, Corse A, Vogelsang G, Wagner KR. Hydroxychloroquine causes severe vacuolar myopathy in a patient with chronic graft-versus-host disease. Am J Hematol 2005;78:306-9.

45. Luong MS, Bessis D, Raison Peyron N, Pinzani V, Guilhou JJ, Guillot B. Severe mucocutaneous necrotizing vasculitis associated with the combination of chloroquine and proguanil. Acta Dermatol Venereol 2003;83:141.

46. Meier H, Elsner P, Wuthrich B. Berufsbedingtes kontaktekzem und Asthma bronchiale bei ungewohnlicher allergischer Reaktion vona Spattyp auf Hydroxychloroquin. [Occupationally-induced contact dermatitis and bronchial asthma in a unusual delayed reaction to hydroxychloroquine.] Hautarzt 1999;50(9):665–9.

47. Adam I, Elbashir MI. Acute gluteal abscess due to chloroquine injection in Sudanese pregnant woman. Saudi Med J 2004;25(7):963-4.

48. Laufer MK, Thesing PC, Edington ND, Masonga R, Dzinjalamala FK, Takala SL, Taylor TE, Plowe CV. Return of chloroquine antimalarial efficacy in Malawi. N Engl J Med 2006;355:1960-6.

49. Sowunmi A. A randomized comparison of chloroquine and chloroquine plus ketotifen in the treatment of acute, uncomplicated, Plasmodium falciparum malaria in children. Ann Trop Med Parasitol 2003;97:103-17.

50. Geelhoed DW, Visser LE, Addae V, Asare K, Schagen van Leeuwen JH, van Roosmalen J. Malaria prophylaxis and the reduction of anemia at childbirth. Int J Gynaecol Obstet 2001;74(2):133–8.

51. Costedoat-Chalumeau N, Amoura Z, Duhaut P, Huong du LT, Sebbough D, Wechsler B, Vauthier D, Denjoy I, Lupoglazoff JM, Piette JC. Safety of hydroxychloroquine in pregnant patients with connective tissue diseases: a study of one hundred thirty-three cases compared with a control group. Arthritis Rheum 2003;48:3207-11.

52. Shroyer NF, Lewis RA, Lupski JR. Analysis of the ABCR (ABCA4) gene in 4-aminoquinoline retinopathy: is retinal toxicity by chloroquine and hydroxychloroquine related to Stargardt disease? Am J Ophthalmol 2001;131(6):761–6.

53. Pelle MT, Callen JP. Adverse cutaneous reactions to hydroxychloroquine are more common in patients with dermatomyositis than in patients with cutaneous lupus erythematosus. Arch Dermatol 2002;138(9):1231–3.

54. Yanze MF, Duru C, Jacob M, Bastide JM, Lankeuh M. Rapid therapeutic response onset of a new pharmaceutical form of chloroquine phosphate 300 mg: effervescent tablets Trop Med Int Health 2001;6(3):196–201.

55. Bochner F, Carruthers G, Kampmann J, Steiner J. Handbook of Clinical PharmacologyBoston, MA: Little Brown;. 1978.

56. Di Maio VJ, Henry LD. Chloroquine poisoning. South Med J 1974;67(9):1031–5.

57. Yanturali S, Aksay E, Demir OF, Atilla R. Massive hydroxychloroquine overdose. Acta Anaesthesiol Scand 2004;48:379-81.

58. Reddy VG, Sinna S. Chloroquine poisoning: report of two cases. Acta Anaesthesiol Scand 2000;44(8):1017–20.

59. Boereboom FT, Ververs FF, Meulenbelt J, van Dijk A. Hemoperfusion is ineffectual in severe chloroquine poisoning. Crit Care Med 2000;28(9):3346–50.

60. Ball DE, Tagwireyi D, Nhachi CF. Chloroquine poisoning in Zimbabwe: a toxicoepidemiological study. J Appl Toxicol 2002;22(5):311–5.

61. Marquardt K, Albertson TE. Treatment of hydroxychloroquine overdose. Am J Emerg Med 2001;19(5):420–4.

62. Ajayi AA, Adigun AQ. Syncope following oral chloroquine administration in a hypertensive patient controlled on amlodipine. Br J Clin Pharmacol 2002;53(4):404–5.

63. Ajayi AA, Adigun AQ. Syncope following oral chloroquine administration in a hypertensive patient controlled on amlodipine. Br J Clin Pharmacol 2002;53(4):404–5.

64. Stahel E, Druilhe P, Gentilini M. Antagonism of chloroquine with other antimalarials. Trans R Soc Trop Med Hyg 1988;82(2):221.

65. Hallett RL, Sutherland CJ, Alexander N, Ord R, Jawara M, Drakeley CJ, Pinder M, Walraven G, Targett GA, Alloueche A. Combination therapy counteracts the enhanced transmission of drug-resistant malaria parasites to mosquitoes. Antimicrob Agents Chemother 2004;48(10):3940–3.

66. Olliaro PL, Taylor WR. Developing artemisinin based drug combinations for the treatment of drug resistant falciparum malaria: A review. J Postgrad Med 2004;50(1):40–4.

67. Okonkwo CA, Coker HA, Agomo PU, Ogunbanwo JA, Mafe AG, Agomo CO, Afolabi BM. Effect of chlorpheniramine on the pharmacokinetics of and response to chloroquine of Nigerian children with falciparum malaria. Trans R Soc Trop Med Hyg 1999;93(3):306–11.

68. Sowunmi A, Fehintola FA, Ogundahunsi OA, Ofi AB, Happi TC, Oduola AM. Comparative cardiac effects of halofantrine and chloroquine plus chlorpheniramine in children with acute uncomplicated falciparum malaria. Trans R Soc Trop Med Hyg 1999;93(1):78–83.

69. Guiserix J, Aizel A. Interactions ciclosporine chloroquine. [Cyclosporine–chloroquine interactions.] Presse Méd 1996;25(26):1214.

70. Ndifor AM, Howells RE, Bray PG, Ngu JL, Ward SA. Enhancement of drug susceptibility in Plasmodium falciparum in vitro and Plasmodium berghei in vivo by mixed-function oxidase inhibitors. Antimicrob Agents Chemother 1993;37(6):1318–23.

71. Leden I. Digoxin–hydroxychloroquine interaction? Acta Med Scand 1982;211(5):411–2.

72. Rombo L, Stenbeck J, Lobel HO, Campbell CC, Papaioanou M, Miller KD. Does chloroquine contribute to the risk of serious adverse reactions to Fansidar? Lancet 1985;2(8467):1298–9.

73. Rahman M, Rahman R, Bangali M, Das S, Talukder MR, Ringwald P. Efficacy of combined chloroquine and sulfadoxine–pyrimethamine in uncomplicated Plasmodium falciparum malaria in Bangladesh. Trans R Soc Trop Med Hyg 2004;98(7):438–41.

74. Simooya OO, Sijumbil G, Lennard MS, Tucker GT. Halofantrine and chloroquine inhibit CYP2D6 activity in healthy Zambians. Br J Clin Pharmacol 1998;45(3):315–7.

75. Frankel EB. Visual defect from chloroquine phosphate. Arch Dermatol 1975;111(8):1069.

76. Rombo L, Stenbeck J, Lobel HO, Campbell CC, Papaioanou M, Miller KD. Does chloroquine contribute to the risk of serious adverse reactions to Fansidar? Lancet 1985;2(8467):1298–9.

77. Rahman M, Rahman R, Bangali M, Das S, Talukder MR, Ringwald P. Efficacy of combined chloroquine and

sulfadoxine–pyrimethamine in uncomplicated *Plasmodium falciparum* malaria in Bangladesh. Trans R Soc Trop Med Hyg 2004;98(7):438–41.

78. Ajana F, Fortier B, Martinot A, et al. Mefloquine prophylaxis and neurotoxicity. Report of a case. Sem Hop 1990;66:918.
79. Gerstein HC, Thorpe KE, Taylor DW, Haynes RB. The effectiveness of hydroxychloroquine in patients with type 2 diabetes mellitus who are refractory to sulfonylureas—a randomized trial. Diabetes Res Clin Pract 2002;55(3):209–19.
80. Van der Straeten C, Klippel JH. Antimalarials and pneumococcal immunization. N Engl J Med 1986;315(11):712–3.
81. Lai TYY, Chan W-M, Li H, Lai RYK, Lam DSC. Multifocal retinographic changes in patients receiving hydroxychloroquine therapy. Am J Ophthamol 2005;140(5):794-808.

Dabequine

General Information

Dabequine is a 4-aminoquinoline derivative, the adverse reaction pattern of which is unknown. The WHO Scientific Group's report concerning this drug dates from 1984 and there seems to be little or no later information on it in humans.

Dichlorophen

General Information

Dichlorophen is an antihelminthic drug that was used in the treatment of tapeworm infections but has been superseded by praziquantel and niclosamide. It also has antifungal and antibacterial activity and has been used topically in the treatment of fungal infections and as a germicide in soaps and cosmetics (1).

During the first few hours after taking a single oral dose one-third of patients have nausea, diarrhea, or abdominal pain, and some experience vomiting. Urticaria, contact allergic dermatitis, and photosensitivity can occur. In the past, with larger doses, jaundice and even hepatic necrosis have been described.

Reference

1. Yamarik TA. Safety assessment of dichlorophene and chlorophene. Int J Toxicol 2004;23(Suppl 1):1–27.

Diethylcarbamazine

General Information

Diethylcarbamazine is an antihelminthic drug used in the treatment of filarial infections, in particular for *Loa Loa* infections and lymphatic filariasis caused by *Wuchereria bancrofti* and *Brugia malayi*. With some infecting species it is effective in both the adult and microfilarial stages, whilst with others it is active only against the microfilarial stages and does not eradicate the infection. It can be associated with significant systemic adverse effects, which can compromise adherence to therapy. It is commonly believed that adverse reactions to diethylcarbamazine result from proinflammatory responses to antigens released from killed microfilariae rather than by direct drug or metabolite toxicity.

Pharmacokinetics

Diethylcarbamazine is extensively metabolized, the half-life being 6–12 hours; the remainder enters the urine within 48 hours. Over the initial period the dosage should be increased slowly to avoid or reduce allergic responses as a result of destruction of parasites and liberation of antigen, and then maintained at 3 mg/kg tds for 34 weeks. Not all of its adverse effects are necessarily due to destruction of the parasite; weakness, lethargy, anorexia, and nausea can be due to the drug itself.

The pharmacokinetics, safety, and tolerability of co-administered diethylcarbamazine and albendazole have been investigated in a double-blind, randomized, placebo-controlled trial in 42 subjects (aged 18–52 years, weighing 46–67 kg) living in a lymphatic filariasis endemic region but without detectable microfilariae (1). Three groups of 14 patients received diethylcarbamazine 6 mg/kg alone, albendazole 400 mg alone, or diethylcarbamazine 6 mg/kg plus albendazole 400 mg. Both diethylcarbamazine and albendazole were well tolerated alone and in combination. In contrast to a study in patients with lymphatic filariasis (2), there were no adverse events in amicrofilaremic individuals. In all three treatment groups the drugs were rapidly absorbed from the gastrointestinal tract, although there was marked interindividual variation. The pharmacokinetics of diethylcarbamazine, albendazole, and albendazole sulfoxide were similar.

Observational studies

Brugia malayi and Wuchereria bancrofti
When treating *Brugia malayi* and *Wuchereria bancrofti* infections, reactions include headache and fever, sometimes accompanied by malaise, nausea, and vomiting. Urticarial skin rashes can occur, and subsequently lymphangitis and lymphadenopathy often appear (SEDA-17,

357). Abscess formation can occur in association with adult worms. Major systemic complications of therapy include proteinuria and severe pruritus; proteinuria has even been seen in some patients using the drug topically. Circulating immune complexes have been found to be increased in many cases. Those with CIq binding greater than 3% are at significantly increased risk of developing visual field constriction and proteinuria.

In lymphatic filariasis diethylcarbamazine can cause systemic reactions, such as fever, arthralgia, headache, and malaise, and local reactions such as swollen and painful lymph nodes. All of these reactions are thought to be caused by an allergic reaction to antigens from the dying microfilaria and not to a toxic effect of the drug itself. This causal relation is further emphasized by a study from Indonesia, in which adverse reactions to treatment with diethylcarbamazine were studied in patients with *B. malayi* filariasis—26 microfilaria-positive patients (mean 235 mf/10 ml), 12 "endemic" controls (from the endemic area, but microfilaria counts negative), and 17 patients with elephantiasis, of whom three had with high microfilaria counts (3). Adverse effects, mainly fever, headache, and body aches, started 2–24 hours after the administration of diethylcarbamazine 6 mg/kg/day for 12 days. Of the patients with positive pretreatment microfilaria counts 15% had severe adverse reactions, 19% moderate reactions, and 65% mild reactions. There was a direct relation between the severity of the adverse effects and the height of pretreatment microfilaria counts, and the more severe adverse effects occurred in the patients with the highest pretreatment microfilaria counts. In the endemic controls there were no reactions or only mild ones. In the patients with elephantiasis there were no reactions or only mild ones in all but two patients, both of whom had moderate reactions and had high pretreatment microfilaria counts.

That diethylcarbamazine may still be the more effective drug in *B. malayi* infections, despite these adverse effects, is suggested in a study from India, in which the efficacy and safety of several single-dose drug combinations, including albendazole, diethylcarbamazine, and ivermectin were compared in 51 microfilaria-positive patients with *B. malayi* and in which diethylcarbamazine was more effective in reaching a sustained reduction in microfilaria counts after 1 year than ivermectin, although the study was small (4). Adverse reactions (fever, headache, myalgia, and chills) occurred in all of the 16 patients treated with a combination of ivermectin (200 micrograms/kg as a single dose) and diethylcarbamazine (6 mg/kg as a single dose), in 15 of the 16 patients treated with albendazole (400 mg as a single dose) and diethylcarbamazine (6 mg/kg as a single dose), and in 12 of the 16 patients treated with ivermectin and albendazole.

In order to assess the effects of re-treatment in *B. malayi* infections, 35 asymptomatic microfilaremic patients were re-treated at the end of the first year with an additional single dose of the combination they had previously received (5). Eleven patients received ivermectin 200 micrograms/kg plus diethylcarbamazine 6 mg/kg,

nine patients received ivermectin 200 micrograms/kg plus albendazole 400 mg, and 15 patients received diethylcarbamazine 6 mg/kg plus albendazole 400 mg. The best suppression of brugian microfilaremia 1 year after re-treatment was obtained with combinations that included diethylcarbamazine. Whatever the drug regimen, both the frequency and intensity of adverse reactions after re-treatment were less than after initial treatment. The greatest difference was in patients who received ivermectin plus diethylcarbamazine, who also had the lowest mean microfilarial counts immediately before re-treatment. None of the adverse reactions after re-treatment was severe. Most of them, including fever, headache, and myalgia, were easily controlled with paracetamol. Postural hypotension and the "string sign" (dilated painful and inflamed lymphatic channels) did not occur with re-treatment.

In another study the cost-effectiveness of a revised mass annual single-dose regimen of diethylcarbamazine 6 mg/kg was estimated for large-scale control of lymphatic filariasis in a pilot program launched in Tamil Nadu (6). This regimen gave good coverage (90% of the population studied) and high compliance of 82%. Adverse effects occurred in 22% of patients and most were non-specific (giddiness in 54%, vomiting in 11%, nausea in 1%, fever in 14%, and headache in 20%).

The frequency, severity, and costs of adverse reactions after mass treatment for lymphatic filariasis in 71 187 people in Leogane (Haiti) using diethylcarbamazine and albendazole have been reported in detail (7). They received diethylcarbamazine plus albendazole 400 mg (n = 38 655 men and boys and 16 482 women and girls) or diethylcarbamazine alone (n = 15 335 women and girls). Of those treated, 17 421 (24%) reported one or more adverse reactions. The reactions were considered minor in 15 916 (91%) and moderate in 1502 (9%). The most commonly reported minor adverse reactions were systemic: 9766 (61%) reported some combination of headache, fever, or body aches. Local problems attributed to worm death were reported in 2170 (14%). Moderate adverse reactions were reported primarily by adults. Men outnumbered women in reporting moderate adverse reactions. Among children under 15 years (n = 27 115), boys were less likely to report moderate adverse reactions than girls. Adverse reactions were systemic in 131. In 355 there were localized scrotal reactions only; 83 reported itching, 70 reported gastrointestinal problems, six reported dizziness, and 56 reported miscellaneous problems. A total of 801 persons reported multiple adverse reactions: these were combinations of systemic, localized scrotal, and other problems. Three patients were hospitalized with severe adverse reactions (0.02% of all reactions). All three reported multiple problems:

- an acute scrotal reaction, fever, headache, vomiting, and abdominal pain in a 25-year-old man;
- itching, body aches, abdominal pain, and heartburn in a 14-year-old girl;
- dizziness, body aches, and abdominal pain in an 18-year-old woman.

Of the patients who reported moderate adverse reactions, 1136 (76%) reported their problems during the 4 days of mass drug distribution. By the third day after the end of mass drug distribution, 98% of all moderate adverse reactions had been reported.

Loa loa

In loiasis both adult and microfilariae are susceptible to diethylcarbamazine. However, encephalitis is a major risk in patients with heavy infestation (SEDA-17, 356) (8), and ivermectin should be preferred. Severe allergic reactions can need treatment with antihistamines and glucocorticoids. The risk of encephalitis has led to the recommendation that prophylactic use of diethylcarbamazine against *Loa loa* should only be contemplated when the chance of infection is considerable.

Onchocerca volvulus

In onchocerciasis, severe reactions can occur in the initial stages of therapy, particularly since diethylcarbamazine only kills the microfilariae of *Onchocerca volvulus* (resulting in the release of toxins) and does not eradicate the infection. The Mazzotti reaction, a Herxheimer-like response, can be severe and even fatal; it comprises a pruritic papular dermatitis, urticaria, fever, malaise, and postural hypotension; asthma and respiratory distress can occur and the hypotension can be associated with irreversible collapse. There can be painful lymphadenopathy.

Ocular complications are of particular importance: iritis can be induced by dying microfilariae and may call for topical or systemic glucocorticoid treatment. Associated complications include chorioretinitis, anterior uveitis, and punctate keratitis. Changes can also occur in the posterior segment (9), with visual field defects; there can be transient retinal pigment epithelial lesions at the posterior pole, globular infiltrates at the limbus, and optic disk leakage (10) with visible pallor. Similar ocular lesions can develop after topical diethylcarbamazine (SEDA-7, 316). Visual field defects are not reversible and they limit the clinical value of diethylcarbamazine in this disease.

Mass treatment with diethylcarbamazine is a key measure for control of the transmission of bancroftian filariasis. However, severe adverse reactions can occur in patients with onchocerciasis treated with high doses of diethylcarbamazine, which may limit the prospects for the use of common salt medicated with diethylcarbamazine in many parts of Africa. However, the daily dose of diethylcarbamazine-medicated salt is considerably lower than that of conventional tablets (25–50 mg od for the first 1 or 2 days followed by 100 mg bd for 5–7 days).

The adverse effects of diethylcarbamazine-medicated salt in patients with *O. volvulus* has been assessed in a double-blind, placebo-controlled trial in four groups of ten men (11). Groups I and II had *O. volvulus* microfilariae only, group III had both *O. volvulus* and *W. bancrofti* microfilariae, and group IV had *W. bancrofti* microfilariae only. Groups I, III, and IV received diethylcarbamazine-medicated salt. Group II served as a control group and received cooking salt. The medicated salt (0.33% w/w) originated from a batch previously produced for control

trials. Each individual was given a total daily dose of 5.8 g of salt for 10 days, corresponding to the average daily salt intake for individuals aged over 15 years in the area. The salt supplement was spaced over three daily meals: 0.5 g, 2.5 g, and 2.8 g at breakfast, lunch, and dinner respectively. Hence, the daily dose of diethylcarbamazine was 19.1 mg. Diethylcarbamazine-medicated salt had no significant effect on *O. volvulus* microfilarial counts, but *W. bancrofti* microfilarial counts were significantly reduced in groups III and IV. The most pronounced adverse reactions occurred in groups I and III and were mild to moderate itching and rash. They were observed on days 3–4 and lasted for the remaining medication period, but did not interfere with normal daily activities. At day 30, all the reactions had abated. There were no severe adverse events, perhaps because of low pre-existing microfilarial counts and the short duration of therapy. There was no evidence that patients with *O. volvulus* and *W. bancrofti* double infection had a different adverse reaction pattern than individuals with *O. volvulus* infection only. Thus, diethylcarbamazine-medicated salt may be an important drug for the control of bancroftian filariasis in Africa. Salt with an even lower concentration of diethylcarbamazine may still have microfilaricidal properties in bancroftian filariasis without inducing microfilarial killing and adverse reactions in onchocerciasis, which may further improve treatment compliance and ease of use.

Wolbachia

In 15 Indonesian patients with *B. malayi* infection, the release of *Wolbachia* bacteria was studied in relation to adverse events after diethylcarbamazine treatment (6 mg/kg orally for 12 days) (12). Three patients had severe reactions and six patients had moderate reactions. In all samples from the three patients with severe reactions and in one of the six with moderate reactions, *Wolbachia* PCR products were detected from 4 hours after treatment, and persisted for 8–20 hours. These data suggest that release of *Wolbachia* bacteria into the blood may be associated with severe inflammatory reactions after diethylcarbamazine. Adverse reactions associated with increases in proinflammatory cytokines have also been reported in bancroftian filariasis and onchocerciasis, suggesting that similar events can also occur in these filarial infections.

Comparative studies

Filariasis

The efficacy of new treatment strategies for lymphatic filariasis using a single dose of diethylcarbamazine or a combination of diethylcarbamazine plus albendazole has been studied in 30 people (aged 11–52 years, weighing 25–63 kg) infected with *Brugia timori* and compared with the results of 27 people (aged 13–52 years, weighing 27–73 kg) infected with *W. bancrofti* (2). All were allocated at random to diethylcarbamazine (100 mg on day 1 and up to a total dose of 6 mg/kg on day 3) or diethylcarbamazine plus albendazole group (placebo on day 1 and diethylcarbamazine 6 mg/kg plus albendazole 400 mg on day 3). There was no difference in adverse reactions between diethylcarbamazine alone and diethylcarbamazine plus albendazole.

Headache ($n = 15$), myalgia ($n = 13$), itching ($n = 8$), and adenolymphangitis ($n = 8$) were the most common adverse effects; none were severe or life-threatening. The microfilaricidal effect of the drugs was achieved more rapidly for *B. timori*, which is associated with more adverse reactions than *W. bancrofti* filariasis. As previously shown, there was a strong correlation of microfilarial density with the frequency and severity of adverse reactions. The addition of albendazole resulted in no additional adverse reactions compared with diethylcarbamazine alone.

In 58 Egyptian adults with asymptomatic *Wuchereria bancrofti* microfilaremia single-dose combination therapy with diethylcarbamazine 6 mg/kg and albendazole 400 mg was compared with the same dose combination therapy for 7 consecutive days (13). Adverse events of mild-to-moderate severity were common after therapy in both groups, with no significant difference in frequencies between the two groups. These data suggest that the observed adverse events were more likely related to a Mazzotti-like reaction to microfilarial killing rather than direct drug or metabolite toxicity. The most frequent adverse events were fever (single dose versus multiple doses 29% versus 27%), headache (25% versus 33%), myalgia (25% versus 37%), and scrotal pain (11% versus 6%). These symptoms usually resolved within 2–3 days. Subjective fever, myalgia, and headache were more common in subjects with high microfilarial counts. Scrotal discomfort peaked at 1 week after treatment, but one man had mild persistent scrotal discomfort after 4 weeks. All adverse events were scored as grade 1 or 2 events. Adverse events in the week after re-treatment of 51 subjects were greatly reduced compared with those after the first round (fever 9%, headache 5%, myalgia 7%, scrotal pain 1 week after treatment 0%).

In 150 subjects in whom diethylcarbamazine was used to prevent recurrent attacks of acute lymphadenitis during bancroftian filariasis, divided into those with mild-to-moderate lymphedema (n = 100) and those with severe lymphedema (n = 50, . the following treatments were randomly allocated (daily treatment for 12 months):

- penicillin G potassium 800 000 U/day (n = 30);
- diethylcarbamazine 50 mg/day (n = 30);
- daily diethylcarbamazine + penicillin (n = 30);
- daily topical antibiotics (n = 30);
- daily placebo (n = 30).

There were 72 episodes of adverse effects: 25 with diethylcarbamazine, 17 with diethylcarbamazine + penicillin, 10 with local antibiotics, and nine with placebo. No subject withdrew because of adverse effects (14). Most subjects reported just one adverse effect each: burning epigastric pain (74%), fever (10%), or headache/drowsiness (10%). The adverse effects were usually mild (60 episodes) but some were moderate (seven episodes) or even severe (six episodes). Five of the six severe events (all epigastric pain) were reported by the subjects who received diethylcarbamazine. Based on these observations the authors recommended a combination of penicillin prophylaxis but not diethylcarbamazine to reduce the risk of acute lymphadenitis in bancroftian filariasis.

n 96 subjects with *Brugia timori* microfilariae, a single oral dose of diethylcarbamazine 6 mg/kg combined with albendazole 400 mg was not associated with adverse effects (15).

General adverse effects

Adverse reactions to treatment with diethylcarbamazine vary with the infecting filarial species and are most severe in onchocerciasis. Minor reactions include malaise, nausea, and headache, but diethylcarbamazine also depresses the central nervous system in some individuals, resulting in dizziness and somnolence; reversible coma has been reported in patients in poor physical condition. Nicotine-like properties can produce autonomic effects. A degree of eosinophilia during treatment is usual.

Although over 120 million individuals have lymphatic filariasis, it may be eradicable. Newer strategies for the elimination of lymphatic filariasis aim at transmission control through the use of annual doses of combinations of ivermectin, diethylcarbamazine, or albendazole, and disease control through individual patient management. Mass chemotherapy appears to be essential in the control of lymphatic filariasis. However, drug availability and the co-endemicity of onchocerciasis and loiasis play crucial roles. Although a single annual dose of diethylcarbamazine may be an effective approach toward long-term suppression of brugian and bancroftian microfilaremia, repeated multidrug chemotherapy is the preferred approach for control of lymphatic filariasis, as in other chronic infections, such as tuberculosis and leprosy. In addition, combining diethylcarbamazine with albendazole has the advantage of controlling intestinal parasites.

The Mazzotti reaction

Presentation

The Mazzotti reaction, first described in 1948 (16), was originally described after treatment of onchocerciasis with diethylcarbamazine. It is characterized by fever, urticaria, swollen tender lymph nodes, tachycardia, hypotension, arthralgia, edema, and abdominal pain within 7 days of treatment of microfilariasis. The Mazzotti reaction to diethylcarbamazine in onchocerciasis is so common that it is the basis of a skin patch test used to confirm the diagnosis (17,18). The drug patch is placed on the skin, and if the patient is infected with the microfilaria of *Onchocerca volvulus*, localized pruritus and urticaria occur at the site of application. Intense pruritus is most marked where microfilariae are concentrated. Hypotension, fever, adenitis, pruritus, and peripheral blood eosinopenia and neutrophilia all correlate with the intensity of the infection (19).

- A 13-year-old Liberian boy was given ivermectin, praziquantel, and albendazole for intestinal parasitosis and 6 days later began to have mid-epigastric pain, vomiting, and urticaria (20). He was treated with intravenous hydration and diphenhydramine and improved. however, on the next day his symptoms returned with the additional features of fever, generalized myalgia, swelling of his face, feet, and penis, and intense pruritus.

His temperature was 38.6° C, pulse 120/minute, respiratory rate 22/minute, and blood pressure 100/60 mmHg. There was a generalized urticarial was on the arms, legs, and trunk, bulbar and palpebral conjunctival injection, and angioedema of his face. He was given intravenous methylprednisolone (2 mg/kg bolus followed by 2 mg/kg/day) and diphenhydramine (1 mg/kg every 6 hours as needed for itching) and all his symptoms resolved completely in 12 hours.

The Mazzotti reaction in this case was ascribed to the presence of undiagnosed onchocerciasis or *Wuchereria bancrofti* infection.

Occurrence

The incidence of the Mazzotti reaction during treatment of onchocerciasis with ivermectin is low, at about 10% (21), and about 25% of patients have only fever or pruritus. Albendazole has not been reported to cause Mazzotti-like reactions. Praziquantel is associated with rare adverse effects in schistosomiasis, including fever, abdominal pain, and urticaria, as the following case shows, but it does not cause the typical Mazzotti reaction.

- An 11-year-old African refugee was given praziquantel for intestinal schistosomiasis and 2 hours after the first dose developed abdominal pain, high fever, and an intensely pruritic, urticarial rash on the arms and trunk. He looked unwell, was wheezy, but had no peripheral edema. He had tender hepatosplenomegaly, confirmed by ultrasonography. His stool was positive for *Schistosoma mansoni*. He had fluctuating high fevers and wheeze during the next three days but his symptoms improved with conservative management. His brother had a similar, although less severe, reaction to praziquantel.

Pathogenesis

The Mazzotti reaction is generally ascribed to an inflammatory response, associated with eosinophil migration to the skin and degranulation (22). It is precipitated by abrupt release of parasite-specific antigens during cell death. *Wolbachia* bacteria, free-living endosymbionts of *Onchocerca volvulus*, form degenerating parasites, which may contribute to the pathogenesis of severe Mazzotti reactions, as occur in patients with high microfilarial loads.

- A 29-year-old New Zealand woman travelling to Asia was exposed to Bancroftian filariasis and took ivermectin (200 micrograms/kg) (23). She had transient pruritus ani, especially at night, and took mebendazole for presumed *Enterobius vermicularis* infestation.

Management

In patients receiving diethylcarbamazine a low dose of dexamethasone (3 mg/day), begun after the start of the reaction, modified its progression without interfering with the microfilaricidal efficacy of diethylcarbamazine (24). Pretreatment with low-dose dexamethasone before diethylcarbamazine therapy prevented the reaction but also greatly reduced the microfilaricidal activity. Diphenhydramine, given after the start of the Mazzotti

reaction, had no effect on the course or intensity of the Mazzotti reaction nor on microfilaricidal activity.

Organs and Systems

Immunologic

Brugia malayi is more susceptible to diethylcarbamazine than *W. bancrofti*. A study of the former, undertaken to explain the very severe effects often associated with diethylcarbamazine treatment of lymphatic filariasis, provided evidence of the involvement of the cytokine interleukin-6 (IL-6), concentrations of which were raised during treatment (25).

The involvement of inflammatory mediators in the development of adverse events has recently been studied in 29 patients with *B. malayi* microfilaremia treated with diethylcarbamazine (26). Before and at serial time points after the start of treatment, plasma concentrations of the inflammatory mediators interleukin-6, interleukin-8, interleukin-10, tumor necrosis factor-alfa, and lipopolysaccharide-binding protein were measured in relation to diethylcarbamazine concentrations and adverse events. The adverse effects of diethylcarbamazine correlated well with pretreatment microfilariae counts, consistent with previous experience with diethylcarbamazine in lymphatic filariasis and onchocerciasis. Concurrent measurements of diethylcarbamazine concentrations failed to establish a clear relation between diethylcarbamazine concentrations and adverse events. Detailed kinetic studies showed the strongest association of the severity of symptoms with interleukin-6 and lipopolysaccharide-binding protein. Concentrations of interleukin-6 started to rise as early as 2–4 hours and reached a maximum after about 8 hours. Fever also occurred at 4–8 hours, consistent with the pyrogenic activity of interleukin-6. In addition, interleukin-6 plays a central role in the induction of the acute phase proteins involved in inflammatory reactions. Indeed, concentrations of the acute phase protein, lipopolysaccharide-binding protein, started to rise at 8 hours (that is after interleukin-6), and also peaked later than interleukin-6, at 24–48 hours after diethylcarbamazine. These observations suggest that the adverse effects of diethylcarbamazine result from an exaggerated host inflammatory response stimulated by a high load of antigen released from killed or degenerating microfilariae.

Susceptibility Factors

Renal disease

The clearance of diethylcarbamazine is reduced in renal insufficiency (27), and dosages should be reduced.

Drug Administration

Drug dosage regimens

The dose of diethylcarbamazine is adjusted according to age, and the following regimen is used in the treatment of carriers of *Wuchereria bancrofti* microfilaria: 50 mg (1–2 years), 100 mg (3–4 years), 150 mg (5–8 years), 200 mg (9–11 years), 250 mg (12–14 years), and 300 mg for over 14 years. In the hope of improving adherence to therapy a simpler schedule has been studied: 100 mg (2–4 years), 200 mg (5–14 years), and 300 mg for over 14 years (28). However, in asymptomatic carriers of microfilaria the incidence of adverse reactions in those aged 4–8 years was 50% with a dose of 150 mg and 67% with 200 mg. There were no life-threatening adverse reactions. Fever, headache, and myalgia, the most common adverse reactions, were mild and similar with both schedules.

References

1. Shenoy RK, Suma TK, John A, Arun SR, Kumaraswami V, Fleckenstein LL, Na-Bangchang K. The pharmacokinetics, safety and tolerability of the co-administration of diethylcarbamazine and albendazole. Ann Trop Med Parasitol 2002;96(6):603–14.

2. Supali T, Ismid IS, Ruckert P, Fischer P. Treatment of *Brugia timori* and *Wuchereria bancrofti* infections in Indonesia using DEC or a combination of DEC and albendazole: adverse reactions and short-term effects on microfilariae. Trop Med Int Health 2002;7(10):894–901.

3. Haarbrink M, Terhell AJ, Abadi GK, Mitsui Y, Yazdanbakhsh M. Adverse reactions following diethylcarbamazine (DEC) intake in "endemic normals", microfilaraemics and elephantiasis patients. Trans R Soc Trop Med Hyg 1999;93(1):91–6.

4. Shenoy RK, Dalia S, John A, Suma TK, Kumaraswami V. Treatment of the microfilaraemia of asymptomatic brugian filariasis with single doses of ivermectin, diethylcarbamazine or albendazole, in various combinations. Ann Trop Med Parasitol 1999;93(6):643–51.

5. Shenoy RK, John A, Babu BS, Suma TK, Kumaraswami V. Two-year follow-up of the microfilaraemia of asymptomatic brugian filariasis, after treatment with two, annual, single doses of ivermectin, diethylcarbamazine and albendazole, in various combinations. Ann Trop Med Parasitol 2000;94(6):607–14.

6. Krishnamoorthy K, Ramu K, Srividya A, Appavoo NC, Saxena NB, Lal S, Das PK. Cost of mass annual single dose diethylcarbamazine distribution for the large scale control of lymphatic filariasis. Indian J Med Res 2000;111:81–9.

7. McLaughlin SI, Radday J, Michel MC, Addiss DG, Beach MJ, Lammie PJ, Lammie L, Rheingans R, Lafontant J. Frequency, severity and costs of adverse reactions following mass treatment for lymphatic filariasis using diethylcarbamazine and albendazole in Leogane, Haiti, 2000. Am J Trop Med Hyg 2003;68:568–73.

8. Carme B, Boulesteix J, Boutes H, Puruehnce MF. Five cases of encephalitis during treatment of loiasis with diethylcarbamazine. Am J Trop Med Hyg 1991;44(6):684–90.

9. Bird AC, El-Sheikh H, Anderson J, et al. Changes in visual function and in the posterior segment of the eye during treatment of loiasis with diethylcarbamazine. Am J Trop Med Hyg 1991;44:684–90.

10. Bird AC, El-Sheikh H, Anderson J, Fuglsang H. Visual loss during oral diethylcarbamazine treatment for onchocerciasis. Lancet 1979;2(8132):46.

11. Meyrowitsch DW, Simonsen PE, Magnussen P. Tolerance to diethylcarbamazine-medicated salt in individuals infected with *Onchocerca volvulus*. Trans R Soc Trop Med Hyg 2000;94(4):444–8.

12. Cross HF, Haarbrink M, Egerton G, Yazdanbakhsh M, Taylor MJ. Severe reactions to filarial chemotherapy and release of *Wolbachia* endosymbionts into blood. Lancet 2001;358(9296):1873–5.

13. El Setouhy M, Ramzy RMR, Ahmed ES, Kandil AM, Hussain O, Farid HA, Helmy H, Weil GJ. A randomized clinical trial comparing single- and multi-dose combination therapy with diethylcarbazine and albendazole for treatment of bancroftian filariasis. Am J Trop Med Hyg 2004;70:191-6.

14. Joseph A, Mony P, Prasad M, John S, Srikanth, Mathai D. The efficacies of affected-limb care with penicillin, diethylcarbamazine, the combination of both drugs or antibiotic ointment, in the prevention of acute adenolymphangitis during bancroftian filariasis. Ann Trop Med Parasitol 2004;98:685-96.

15. Fischer P, Djuardi Y, Ismid IS, Ruckert P, Bradley M, Supali T. Long-lasting reduction of Brugia timori microfilariae following a single dose of diethylcarbamazine combined with albendazole. Trans R Soc Trop Med Hyg 2003;97:446-8.

16. Mazzotti L. Onchocerciasis in Mexico. In: Proceedings of the 4th International Congress of Tropical Medicine: Malarias (Session 1 of 6). Washington DC, 1948:948-56.

17. Stingl P, Ross M, Gibson DW, Ribas J, Connor DH. A diagnostic "patch test" for onchocerciasis using topical diethylcarbamazine. Trans R Soc Trop Med Hyg 1984;78(2):254-8.

18. Kilian HD. The use of a topical Mazzotti test in the diagnosis of onchocerciasis. Trop Med Parasitol 1988;39(3):235-8.

19. Francis H, Awadzi K, Ottesen EA. The Mazzotti reaction following treatment of onchocerciasis with diethylcarbamazine: clinical severity as a function of infection intensity. Am J Trop Med Hyg 1985;34(3):529-36.

20. Olson BG, Domachowske JB. Mazzotti reaction after presumptive treatment for schistosomiasis and strongyloidiasis in a Liberian refugee. Pediatr Infect Dis J 2006;25(5):466-8.

21. Chijioke CP, Okonkwo PO. Adverse events following mass ivermectin therapy for onchocerciasis. Trans R Soc Trop Med Hyg 1992;86(3):284-6.

22. Ackerman SJ, Kephart GM, Francis H, Awadzi K, Gleich GJ, Ottesen EA. Eosinophil degranulation. An immunologic determinant in the pathogenesis of the Mazzotti reaction in human onchocerciasis. J Immunol 1990;144(10):3961-9.

23. Shaw MTM, Leggat PA. A case of exposure to Bancroftian filariasis in a traveller to Thailand. Travel Med Infect Dis 2006;4:290-3.

24. Stingl P, Pierce PF, Connor DH, Gibson DW, Straessle T, Ross MA, Ribas JL. Does dexamethasone suppress the Mazzotti reaction in patients with onchocerciasis? Acta Trop 1988;45(1):77-85.

25. Yazdanbakhsh M, Duym L, Aarden L, Partono F. Serum interleukin-6 levels and adverse reactions to diethylcarbamazine in lymphatic filariasis. J Infect Dis 1992;166(2):453-4.

26. Reuben R, Rajendran R, Sunish IP, Mani TR, Tewari SC, Hiriyan J, Gajanana A. Annual single-dose diethylcarbamazine plus ivermectin for control of bancroftian filariasis: comparative efficacy with and without vector control. Ann Trop Med Parasitol 2001;95(4):361-78.

27. Adjepon-Yamoah KK, Edwards G, Breckenridge AM, Orme ML, Ward SA. The effect of renal disease on the pharmacokinetics of diethylcarbamazine in man. Br J Clin Pharmacol 1982;13(6):829–34.

28   Pani SP, Das LK, Vanamail P. Tolerability and efficacy of a three-age class dosage schedule of diethylcarbamazine citrate (DEC) in the treatment of microfilaria carriers of *Wuchereria bancrofti and its implications in mass drug administration (MDA) strategy for elimination of lymphatic filariasis (LF). J Commun Dis 2005;37(1):12-17.*

Difetarsone

General Information

Difetarsone is a pentavalent arsenical that often causes minor adverse effects, such as rashes, nausea, vomiting, and abdominal discomfort. Transient increases in transaminase activities can occur (SEDA-13, 834) (1).

Organs and Systems

Immunologic

Generalized angioedema occurred in a patient taking difetarsone 500 mg tds for *Entamoeba histolytica* infection (SEDA-11, 597) (2).

References

1. Committee on Antimicrobial Agents, Canadian Infectious Disease Society. Treatment of parasitic infections: Canadian versus US recommendations. Can Med Assoc J 1988;139:849.

2. McIntyre L, Krajden S, Keystone JS. Angioedema due to diphetarsone and a review of its toxicity. Trop Geogr Med 1983;35(1):49–51.

Diloxanide

General Information

Diloxanide often causes flatulence and, occasionally, nausea, vomiting, diarrhea, urticaria, and pruritus. It is an excellent luminal amebicide and is indicated after treatment with the 5-nitroimidazole compounds, which have relatively weak activity on the cyst stage. Experience over 14 years has been summarized by the Centers for Disease Control and Prevention (CDC, Atlanta), confirming the minimal toxicity of diloxanide. Fewer adverse effects were reported in patients aged 20 months to 10 years than in those aged over 10 years. There is no record of interactions between diloxanide and either metronidazole or tinidazole (SEDA-13, 830; SEDA-17, 333).

Eflornithine

General Information

Eflornithine is a specific irreversible inhibitor of ornithine decarboxylase, the enzyme involved in the first step of mammalian polyamine biosynthesis (1). Decarboxylation of ornithine is an obligatory and rate-limiting step in the biosynthesis of polyamines, such as putrescine, spermidine, and spermine. These low molecular weight polyamines play an essential role in the growth, differentiation, and replication of the cell by participating in nucleic acid and protein synthesis, and are needed in the process of decoding genetic messages. In vitro studies of different types of cell lines (including human malignant cells) exposed to eflornithine demonstrated inhibition of growth. Eflornithine added to human erythrocytes infected with *Plasmodium falciparum* reduced parasite growth and intracellular polyamine content. Polyamines play an important role in the cellular metabolism of trypanosomatids (SED-12, 708; 2,3). Eflornithine can arrest viral replication.

Eflornithine hydrochloride can be given intravenously and orally. Absorption after oral administration is adequate. After intravenous administration, 80% is excreted unchanged in the urine within 14 hours. It penetrates the nervous system, and cerebrospinal fluid concentrations are 10–45% of serum concentrations.

Eflornithine has been used in the chemotherapy and chemoprevention of some tumors, including glioblastoma and colorectal carcinoma in the presence of polyposis coli (4). It has been used in the treatment of malaria tropica, in AIDS, and in *Pneumocystis jiroveci* infections, with varied success. It has also been approved for use in *Trypanosoma gambiense* (SED-12, 708; 5,6).

The most important adverse effect of eflornithine is a natural consequence of its mode of action, myelosuppression, which is frequent and sometimes treatment-limiting. Gastrointestinal toxicity is also common, and is more marked with oral administration. Seizures, hearing loss, alterations in liver function tests, and rash have been described in the treatment of *P. jiroveci* infections in patients with AIDS (SED-13, 835). In 31 patients with AIDS and *P. jiroveci* pneumonia, intolerant of and/or unresponsive to co-trimoxazole or pentamidine, about 50% reacted favorably to eflornithine. The adverse effects were no different from those seen in patients without AIDS, but the frequency of adverse effects was higher. The most common effects in this group were myelosuppression, thrombocytopenia being the most serious, with hepatitis (3%) and hearing loss (9%) among the others (SEDA-17, 332).

Eflornithine in trypanosomiasis

Melarsoprol (an arsenical compound) is still the most effective compound against stage II (nervous system) disease in both East and West African trypanosomiasis. However, it is toxic and causes death in about 2–8% of subjects treated. Eflornithine is an alternative to melarsoprol for West African trypanosomiasis (both early and late). However, it is very expensive, and its usefulness in endemic areas may therefore be limited. The standard regimen is 100 mg/kg intravenously 6-hourly for 14 days. There have been some anecdotal reports that a shorter 7-day course may be equally effective, with obvious cost-saving advantages.

There has been a multicenter, randomized, open comparison of treatment with eflornithine for 7 or 14 days (*n* = 321) (7). The subjects were divided into new cases and relapses. The 14-day course of eflornithine was superior to the 7-day course for the new cases, but there was no difference in the relapsing cases. However, the numbers of patients who relapsed were small (*n* = 47) and this may not have allowed the detection of a small difference between the groups. The most common adverse events associated with eflornithine were convulsions, altered consciousness, diarrhea, vomiting, nausea, abdominal pain, and secondary infections. Diarrhea and secondary infection were more common in subjects who took the 14-day course.

In 42 patients with late-stage *Trypanosoma brucei gambiense* trypanosomiasis, who relapsed after initial treatment with melarsoprol, a sequential combination of intravenous eflornithine (100 mg/kg every 6 hours for 4 days) followed by three daily injections of melarsoprol (3.6 mg/kg, up to 180 mg) was used (8). They were followed for 24 months. In one case, the administration of eflornithine had to be interrupted for 48 hours because of convulsions, but treatment was then resumed without recurrence. Other adverse effects during treatment were abdominal pain or vomiting (*n* = 4 each), diarrhea (*n* = 1), and loss of hearing (*n* = 1). Two patients died during treatment:

- A 37-year-old man died of an acute cholera-like syndrome, with severe diarrhea, vomiting, and dehydration, after the last dose of eflornithine but before receiving his first dose of melarsoprol.
- A 34-year-old man died of an unknown cause after having received all 16 doses of eflornithine as well as the first injection of melarsoprol.

There has been a comparison of melarsoprol 2.2 mg/kg/day plus prednisolone 20 mg/day (n=708) and eflornithine (400 mg/kg given as a slow intravenous infusion in four 3-hour doses daily for 14 days; n=251) in patients with trypanosomiasis who presented with meningoencephalitis (9). Acute reactive encephalopathy occurred in 80 (11%) of those who took melarsoprol and in one (0.4%) who took eflornithine. Of those who took melarsoprol 25 (3.5%) died (23 because of acute reactive encephalopathy) while only two (0.8%) of those who took eflornithine died. Fever, hypertension, macular rash, severe headache, peripheral neuropathy, and tremors were more common in those who took melarsoprol, while diarrhea was more frequent in those who took eflornithine. After 12 months of follow-up there was no significant difference between the rates of relapse.

Organs and Systems

Nervous system

The frequency of seizures in patients with trypanosomiasis taking eflornithine is about 7% (SEDA-16, 316) (10).

Sensory systems

Hearing loss is the dose-limiting toxic effect for eflornithine in patients taking over 2 g/m^2/day (11) (SEDA-11, 394). In one study, it was found in 48% of patients. In some cases the hearing loss was characterized audiographically as bilateral, sensorineural, primarily high frequency, with a median loss of 25–30 dB. All patients recovered within 1–3 months of withdrawal of therapy. There was no clear association between the total dose of eflornithine and the degree of hearing loss (12). Tinnitus has been reported in some patients taking eflornithine (13).

Hearing loss due to eflornithine is usually reversible, but irreversible hearing loss has also been described.

- A patient in a Barrett's esophagus chemoprevention trial developed a hearing deficit of 15 dB at frequencies of 250, 2000, and 3000 Hz in the right ear and a deficit of 20 dB or more at 4000–6000 Hz in the left ear after taking eflornithine 0.5 g/m^2/day for about 13 weeks (cumulative dose 45 g/m^2); clinical hearing was not affected, but the threshold shifts persisted 7 months after eflornithine was withdrawn (14).

In a placebo-controlled study in 123 patients with colorectal polyps and normal hearing at frequencies of 250–2000 Hz, eflornithine 0.075–0.4 g/m^2/day for 12 months had no effect on auditory pure-tone thresholds or distortion product otoacoustic emission (15). There was no hearing loss, in contrast to studies with higher dosages.

In 58 patients with metastatic malignant melanoma, 179 sequential audiograms were obtained from patients treated with eflornithine 2–12 g/m^2/day alone (*n* = 16) or in combination with interferon alfa-2b (*n* = 42) for 2–50 weeks (16). Total doses of 60–1390 g/m^2 correlated with clinical effects and pure-tone audiometric changes at multiple frequencies (500, 1000, 2000, 4000, and 8000 Hz). Patients with normal baseline audiography had more hearing loss than those with abnormal baseline audiography at the higher frequencies. Of the patients with normal prestudy hearing thresholds 10% or less developed a demonstrable hearing deficit at cumulative doses below 150 g/m^2. Conversely, up to 75% of the patients who received more than 250 g/m^2 developed hearing loss. Other factors that adversely affected hearing included age, male sex, and the concomitant use of interferon alfa-2b.

Hematologic

Myelosuppression, as evidenced by anemia, leukopenia, and thrombocytopenia, is common in patients taking eflornithine. The manufacturers quote respective incidence rates of 55, 37, and 14%. However, other reports have suggested that thrombocytopenia is the most frequent problem; it is dose-dependent and occurred in 90% of patients with cancer on a dose of 6–8 g/m^2 with eflornithine concentrations over 400 mmol/l (SED-12, 709) (17).

- A 32-year-old Haitian man with AIDS developed complications of *Isospora belli* enteritis (18). He was given intravenous eflornithine and developed severe thrombocytopenia, nausea, and vomiting, which recurred on rechallenge with low-dose oral eflornithine.

There is an impression that patients with AIDS have a higher incidence of adverse effects. The myelosuppression is reversible on withdrawal (SED-12, 708; SEDA-16, 316).

Gastrointestinal

Nausea, vomiting, abdominal pain, and diarrhea, mild or severe, are seen with parenteral and oral administration, but more markedly after oral administration (SED-12, 708) (19).

Hair

Alopecia has been reported in 5–10% of patients with trypanosomiasis taking eflornithine (10).

Second-Generation Effects

Teratogenicity

Animal studies have shown arrest of the growth of malignant cells and decrease in tumor size, but as one might expect, there was also an arrest of embryonic growth if eflornithine was given in the first days of pregnancy (SED-12, 708). On theoretical grounds, one would expect an effect on the development of the fast-growing embryo; whether or not this could lead to defects and/or deformities is not known.

References

1. Pasic TR, Heisey D, Love RR. Alpha-difluoromethylornithine ototoxicity. Chemoprevention clinical trial results. Arch Otolaryngol Head Neck Surg 1997;123(12):1281–6.
2. Anonymous. d,l-alpha-Difluoromethyl ornithine. DFMO. Eflornithine. Drugs Future 1985;10:242.
3. Anonymous. DFMO. Ann Drug Data Rep 1982;71.
4. Courtney ED, Melville DM, Leicester RJ. Review article: chemoprevention of colorectal cancer. Aliment Pharmacol Ther 2004;19(1):1–24.
5. Notification New drug for trypanosomiasis. Lancet 1991;337(8732):42.
6. Anonymous. Sleeping sickness. Wake-up call. Economist 1990;110Dec 12.
7. Pepin J, Khonde N, Maiso F, Doua F, Jaffar S, Ngampo S, Mpia B, Mbulamberi D, Kuzoe F. Short-course eflornithine in Gambian trypanosomiasis: a multicentre randomized controlled trial. Bull World Health Organ 2000;78(11):1284–95.
8. Mpia B, Pepin J. Combination of eflornithine and melarsoprol for melarsoprol-resistant Gambian trypanosomiasis. Trop Med Int Health 2002;7(9):775–9.
9. Chappuis F, Udayraj N, Stietenroth K, Meussen A, Bovier PA. Eflornithine is safer than melarsoprol for the treatment of second stage *Trypanosoma brucei gambiense* human African trypanosomiasis. Clin Infect Dis 2005;1:748-51.
10. Burri C, Brun R. Eflornithine for the treatment of human African trypanosomiasis. Parasitol Res 2003;90(Suppl 1):S49–52.
11. van der Velden JW, van Putten WL, Guinee VF, Pfeiffer R, van Leeuwen FE, van der Linden EA, Vardomskaya I, Lane W, Durand M, Lagarde C, et al. Subsequent development of acute non-lymphocytic leukemia in patients treated for Hodgkin's disease. Int J Cancer 1988;42(2):252–5.
12. Meyskens FL, Kingsley EM, Glattke T, Loescher L, Booth A. A phase II study of alpha-difluoromethylornithine (DFMO) for the treatment of metastatic melanoma. Invest New Drugs 1986;4(3):257–62.
13. Levin VA, Chamberlain MC, Prados MD, Choucair AK, Berger MS, Silver P, Seager M, Gutin PH, Davis RL, Wilson CB. Phase I-II study of eflornithine and mitoguazone combined in the treatment of recurrent primary brain tumors. Cancer Treat Rep 1987;71(5):459–64.
14. Lao CD, Backoff P, Shotland LI, McCarty D, Eaton T, Ondrey FG, Viner JL, Spechler SJ, Hawk ET, Brenner DE. Irreversible ototoxicity associated with difluoromethylornithine. Cancer Epidemiol Biomarkers Prev 2004;13(7):1250–2.
15. Doyle KJ, McLaren CE, Shanks JE, Galus CM, Meyskens FL. Effects of difluoromethylornithine chemoprevention on audiometry thresholds and otoacoustic emissions. Arch Otolaryngol Head Neck Surg 2001;127(5):553–8.
16. Croghan MK, Aickin MG, Meyskens FL. Dose-related alpha-difluoromethylornithine ototoxicity. Am J Clin Oncol 1991;14(4):331–5.
17. Ajani JA, Ota DM, Grossie VB Jr, Levin B, Nishioka K. Alterations in polyamine metabolism during continuous intravenous infusion of alpha-difluoromethylornithine showing correlation of thrombocytopenia with alpha-difluoromethylornithine plasma levels. Cancer Res 1989;49(20):5761–5.
18. Tietze KJ, Gaska JA, Cosgrove EM. Thrombocytopenia and vomiting due to difluoromethylornithine. Drug Intell Clin Pharm 1987;21(7–8):627–30.
19. Anonymous. Eflornithine hydrochloride. d,l-alpha-Difluoromethyl ornithine. DFMO. Drugs Future 1986;11:220.

Fenbendazole

See also Benzimidazoles

General Information

The antiprotozoal activity of fenbendazole, a benzimidazole derivative, is similar to that of mebendazole; the target seems to be the microtubule protein beta-tubulin. Its adverse effects are comparable to those of mebendazole (1).

Reference

1. Katiyar SK, Gordon VR, McLaughlin GL, Edlind TD. Antiprotozoal activities of benzimidazoles and correlations with beta-tubulin sequence. Antimicrob Agents Chemother 1994;38(9):2086–90.

Furazolidone

General Information

Furazolidone is used in the treatment of giardiasis. Its adverse effects are usually mild and transient: abdominal discomfort, nausea, and vomiting. The urine may be dark-colored (SEDA-11, 597). Metabolites of furazolidone inhibit monoamine oxidase (1) and there is therefore the potential for interactions with foods containing tyramine (2) and with opioid analgesics; hyperpyrexia has been reported in rabbits that were given furazolidone and pethidine (3).

Organs and Systems

Skin

Non-pigmented fixed drug eruptions have previously been attributed to furazolidone; they usually affect the palms, trunk, inguinal folds, and buttocks. Another case has been reported (4).

- A 23-year-old man developed well circumscribed bright red macular patches on the palms and soles after taking furazolidone for an acute intestinal infection. There was no central blistering or telangiectasis. The lesions subsided within 7 days. On rechallenge with a quarter of the dose he developed acute edematous erythematous lesions at the same sites as before.

References

1. Timperio AM, Kuiper HA, Zolla L. Identification of a furazolidone metabolite responsible for the inhibition of amino oxidases. Xenobiotica 2003;33(2):153 67.
2. Pettinger WA, Oates JA. Supersensitivity to tyramine during monoamine oxidase inhibition in man. Mechanism at the level of the adrenergic neuron. Clin Pharmacol Ther 1968;9(3):341–4.
3. Eltayeb IB, Osman OH. Furazolidone–pethidine interaction in rabbits. Br J Pharmacol 1975;55(4):497–501.
4. Tan C, Zhu WY. Furazolidine induced nonpigmenting fixed drug eruptions affecting the palms and soles. Allergy 2005;60(7):972-3.

Halofantrine

General Information

Halofantrine is a phenanthrene-methanol derivative of an aminoalcohol, active against multidrug-resistant *Plasmodium falciparum* malaria. Halofantrine was known during World War II but was little used at that time. It is slowly and incompletely absorbed with peak concentrations 3.5–6 hours after dosing. Its absorption in its original formulation was unpredictable (SEDA-13, 820).

Dose-finding studies with halofantrine showed treatment failures with a single dose but cures with two doses, one of 1000 mg and one of 500 mg. A regimen of 500 mg at 6-hourly intervals was also effective. Three doses of 500 mg at 6-hourly intervals were also used in a French study and treatment resulted in cure in semi-immune subjects. The dosage was insufficient in non-immune Caucasian patients.

Halofantrine is not indicated for prophylactic use. However, of 480 army personnel who took two doses of 1500 mg each on the third and the tenth days after they had returned to a non-malarial area, only one had *P. falciparum* malaria during the next 5 months (SEDA-13, 820) (1). There were no adverse effects, except for one case of morbilliform rash possibly due to the halofantrine.

General adverse effects

Adverse effects with the dosages originally recommended have in general been mild, no more than nausea, diarrhea, headache, and pruritus (SEDA-13, 820) (2–4). Pruritus occurred markedly less often with halofantrine than with chloroquine (SEDA-16, 306). A comparison between high-dose chloroquine (35 mg/kg total in three daily doses) and halofantrine in the standard dose (total 25 mg/kg given at 6-hour intervals) in patients 4–14 years old showed a fairly similar frequency of adverse effects. Itching was a common adverse effect of chloroquine (4).

Organs and Systems

Cardiovascular

In 1993, the sudden cardiac death of a 37-year-old woman after her ninth dose of halofantrine was reported. A subsequent prospective study showed that halofantrine was associated with a dose-related lengthening of the QT interval by more than 25% (SEDA-17, 328). Mefloquine did not cause such changes, but the combination of mefloquine with halofantrine had a more pronounced effect on the electrocardiogram. However, in the region where this investigation was carried out (on the Thai–Burmese border area) thiamine deficiency is common, and patients in this area have longer baseline QT intervals than are usually reported (SEDA-17, 328). Two patients, mother and son, both with congenital prolongation of the QT interval, suffered sustained episodes of torsade de pointes after a total dose of 1000 mg of halofantrine, and there have been other reports of dysrhythmias, including death in a patient who took mefloquine and halofantrine (SEDA-20, 260) (SEDA-21, 295). Dysrhythmias due to halofantrine may respond to propranolol (5).

African children who received halofantrine (three doses of 8 mg/kg 6-hourly) for uncomplicated *P. falciparum* malaria had increases in both the PR interval and the QT_c interval; out of 42 children in the study, two children developed first-degree heart block and one child second-degree heart block; the QT_c interval either increased by more than 125% of baseline value or by more than 0.44 seconds (an effect that persisted for at least 48 hours) (6).

There have been recent reports in the French medical press of cases of significant QT$_c$ prolongation in children returning to France, and caution in its use has been urged (7). A small trial in non-immune adults in the Netherlands and France also showed increased QT$_c$ dispersion with halofantrine but not artemether + lumefantrine (8).

• Death due to a dysrhythmia was reported in a woman who had taken halofantrine for malaria (9). She had a normal electrocardiogram before treatment and no family history of heart disease.

Prolongation of the QT interval occurred in 10 of 25 children treated with halofantrine (24 mg/kg oral suspension in three divided doses) for acute falciparum malaria (10).

Electrocardiographic monitoring is recommended for children and adults taking halofantrine.

Gastrointestinal

Gastrointestinal adverse effects are more common with halofantrine than with other antimalarial drugs (11).

Immunologic

Anaphylactic shock has been attributed to halofantrine (12).

Long-Term Effects

Drug tolerance

Early laboratory studies suggested cross-resistance of halofantrine with mefloquine. In rats, parasites that are resistant to mefloquine, quinine, chloroquine, and amodiaquine are also markedly resistant to halofantrine (13).

Susceptibility Factors

Pre-existing cardiac conduction abnormalities, including those induced by other drugs, are a definite risk. Due consideration should be given not only to the half-life of such drugs, but also to the tissue concentrations and total clearances of the agents involved.

Drug Administration

Drug formulations

Halofantrine is available in a micronized form. Early information suggested that this improved absorption, but two later studies provided conflicting evidence: there was a wide range of serum concentrations of halofantrine and its main metabolite N-desbutylhalofantrine after the administration of half of the standard dose (2) or the usual dose of 500 mg given three times with 6-hour intervals (3). In general, concentrations were similar to those obtained with the older formulation. Taking the drug with food is thought to increase absorption.

Food–Drug Interactions

Grapefruit juice

Since halofantrine is metabolized to N-debutyl-halofantrine by CYP3A4, grapefruit juice increases its systemic availability (14). Twelve healthy men and women took halofantrine 500 mg with water, orange juice, or grapefruit juice (250 ml/day for 3 days and once 12 hours before halofantrine) in a crossover study. Compared with water, grapefruit juice significantly increased halofantrine AUC and C_{max} 2.8-fold and 3.2-fold respectively; there was no significant change in half-life. Maximum QT$_c$ interval prolongation increased significantly from 17 ms when halofantrine was taken with water to 31 ms when it was taken with grapefruit juice. Grapefruit juice should be avoided by patients taking halofantrine.

References

1. Baudon D, Bernard J, Moulia-Pelat JP, Martet G, Sarrouy J, Touze JE, Spiegel A, Lantrade P, Picq JJ. Halofantrine to prevent falciparum malaria on return from malarious areas. Lancet 1990;336(8711):377.

2. Fadat G, Louis FJ, Louis JP, Le Bras J. Efficacy of micronized halofantrine in semi-immune patients with acute uncomplicated falciparum malaria in Cameroon. Antimicrob Agents Chemother 1993;37(9):1955–7.

3. Bouchaud O, Basco LK, Gillotin C, Gimenez F, Ramiliarisoa O, Genissel B, Bouvet E, Farinotti R, Le Bras J, Coulaud JP. Clinical efficacy and pharmacokinetics of micronized halofantrine for the treatment of acute uncomplicated falciparum malaria in nonimmune patients. Am J Trop Med Hyg 1994;51(2):204–13.

4. Wildling E, Jenne L, Graninger W, Bienzle U, Kremsner PG. High dose chloroquine versus micronized halofantrine in chloroquine-resistant *Plasmodium falciparum* malaria. J Antimicrob Chemother 1994;33(4):871–5.

5. Toivonen L, Viitasalo M, Siikamaki H, Raatikka M, Pohjola-Sintonen S. Provocation of ventricular tachycardia by antimalarial drug halofantrine in congenital long QT syndrome. Clin Cardiol 1994;17(7):403–4.

6. Sowunmi A, Falade CO, Oduola AM, Ogundahunsi OA, Fehintola FA, Gbotosho GO, Larcier P, Salako LA. Cardiac effects of halofantrine in children suffering from acute uncomplicated falciparum malaria. Trans R Soc Trop Med Hyg 1998;92(4):446–8.

7. Olivier C, Rizk C, Zhang D, Jacqz-Aigrain E. Allongement de l'espace QT$_c$ compliquant la prescription d'halofantrine chez deux enfants presentant un accès palustre a *plasmodium falciparum*. [Long QT$_c$ interval complicating halofantrine therapy in 2 children with *Plasmodium falciparum* malaria.] Arch Pediatr 1999;6(9):966–70.

8. van Agtmael M, Bouchaud O, Malvy D, Delmont J, Danis M, Barette S, Gras C, Bernard J, Touze JE, Gathmann I, Mull R. The comparative efficacy and tolerability of CGP 56697 (artemether + lumefantrine) versus halofantrine in the treatment of uncomplicated falciparum malaria in travellers returning from the Tropics to The Netherlands and France Int J Antimicrob Agents 1999;12(2):159–69.

9. Malvy D, Receveur MC, Ozon P, Djossou F, Le Metayer P, Touze JE, Longy-Boursier M, Le Bras M. Fatal cardiac incident after use of halofantrine. J Travel Med 2000;7(4):215–6.

10. Herranz U, Rusca A, Assandri A. Emedastine–ketoconazole: pharmacokinetic and pharmacodynamic interactions in healthy volunteers. Int J Clin Pharmacol Ther 2001;39(3):102–9.
11. Anabwani G, Canfield CJ, Hutchinson DB. Combination atovaquone and proguanil hydrochloride vs. halofantrine for treatment of acute *Plasmodium falciparum* malaria in children. Pediatr Infect Dis J 1999;18(5):456–61.
12. Fourcade L, Gachot B, De Pina JJ, Heno P, Laurent G, Touze JE. Choc anaphylactique associé au traitement du paludisme par halofantrine. [Anaphylactic shock related to the treatment of malaria with halofantrine.] Presse Méd 1997;26(12):559.
13. Peters W, Robinson BL, Ellis DS. The chemotherapy of rodent malaria. XLII. Halofantrine and halofantrine resistance. Ann Trop Med Parasitol 1987;81(5):639–46.
14. Charbit B, Becquemont L, Lepere B, Peytavin G, Funck-Brentano C. Pharmacokinetic and pharmacodynamic interaction between grapefruit juice and halofantrine. Clin Pharmacol Ther 2002;72(5):514–23.

Hexachloroparaxylene

General Information

Hexachloroparaxylene has been used in China and Russia as an antihelminthic drug, principally to treat the liver fluke infections (clonorchiasis due to *Clonorchis sinensis*, schistosomiasis due to *Schistosoma japonicum*, and opisthorchiasis due to *Opisthorchiidae* (1,2). However, other treatments are preferred. It is also used very extensively in the veterinary field in Russia.

Hexachloroparaxylene causes gastrointestinal reactions, cardiac dysrhythmias (perhaps in over 50% of cases), and nephrotoxicity. Hemolysis can occur both early and late, and death can occur from the hemolytic–uremic syndrome (SEDA-4, 219). Late-onset hemolysis after treatment is associated with beta-thalassemia, while early-onset hemolysis is associated with hemoglobin H disease.

References

1. Wan ZR. [Preliminary clinical observations on the treatment of clonorchiasis sinensis with hexachloroparaxylene.] Zhonghua Nei Ke Za Zhi 1979;18(6):406–8.
2. Plotnikov NN, Karnaukov VK, Zal'nova NS, Alekseeva MI, Borisov IA, Stromskaia TF. Lechenie fastsioleza u cheloveka khloksilom (geksakhlorparaksilol). [Treatment of human fascioliasis with chloxyle (hexachloroparaxylene).] Med Parazitol (Mosk) 1965;34(6):725–9.

Hycanthone

General Information

Hycanthone is a derivative of lucanthone, but has less gastrointestinal and nervous system toxicity. It is effective against both *Schistosoma hematobium* and *Schistosoma mansoni* and is given as a single intramuscular injection in doses of 1.0–2.5 mg/kg (1,2). However, it has largely been superseded in the treatment of schistosomiasis by more recent, less toxic compounds. The most common adverse reactions to hycanthone, which occur in up to half of all patients treated with higher-dose regimens, is nausea and vomiting (3), often associated with abdominal colic and diarrhea. There can be muscle pain, and electrocardiographic changes can occur (4).

Organs and Systems

Liver

Hepatotoxicity occurs with hycanthone; serum transaminases are often raised and less commonly there is overt jaundice (5,6). Hepatitis was the major complication with hycanthone in the treatment of schistosomiasis, sometimes associated with pancreatitis (7,8). Hycanthone has also been evaluated as a potential antitumor agent, being used as a radiosensitizer; here too, hepatitis was a dose-limiting effect, occurring in some patients with doses of 100 mg/m2 per day (3). The lowest dosage that caused hepatitis was 70 mg/m2/day. In several cases the hepatotoxicity proved fatal (SED-11, 597) (7–9).

Long-Term Effects

Mutagenicity

Experimentally, hycanthone has been reported to be mutagenic in *Salmonella typhimurium* and *Escherichia coli*, but no human data are available (10,11).

Tumorigenicity

Experimentally, hycanthone has been reported to be carcinogenic in mice (12,13), but no human data are available.

Second-Generation Effects

Teratogenicity

Experimentally, hycanthone has been reported to be teratogenic (14,15), although no human data are available.

References

1. Dennis EW, Kobus W. A review of the clinical pharmacology of hycanthone. Egypt J Bilharz 1974;1(1):35–53.
2. Farah A, Berberian DA, Davison C, Dennis EW, Donikian MA, Drobeck HP, Ferrari RA, Yarinsky A. Hycanthone: a review of its experimental chemotherapy, pharmacology and toxicology. Egypt J Bilharz 1974;1(2):181–95.
3. Kovach JS, Moertel CG, Schutt AJ, Eagan RT. Phase I study of hycanthone. Cancer Treat Rep 1979;63(11–12):1965–9.
4. Takaoka L, Baldy JL, Passos JD, Soares EC, Zeitune JM, Siqueira JE. Alteracoes eletrocardiograficas em pacientes

com esquistossomose mansonica tratados com hicantone. [Electrocardiographic changes in patients with schistosomiasis mansoni treated with hycanthone.] Rev Inst Med Trop Sao Paulo 1976;18(5):378–86.

5. Oostburg BF. Clinical trial with hycanthone in schistosomiasis mansoni in Surinam. Trop Geogr Med 1972;24(2):148–51.
6. Farid Z, Smith JH, Bassily S, Sparks HA. Hepatotoxicity after treatment of schistosomiasis with hycanthone. BMJ 1972;2(805):88–9.
7. Buchanan N, Thatcher CJ, Cane RD, Bartolomeo B. Fatal hepatic necrosis in association with the use of hycanthone. A case report. S Afr Med J 1978;53(7):257–8.
8. Goncalves CS, Buaiz V, Zanandrea J, Zanotti WM, Boni ES, de Castro Filho AK, Pereira FE. Reacoes toxicas com o uso do hycanthone. [Toxic reactions with the use of hycanthone.] AMB Rev Assoc Med Bras 1977;23(9):305–8.
9. Mengistu M. Fatal liver toxicity due to hycanthone (Etrenol) in a patient with pre-existing liver disease: a case report. Ethiop Med J 1982;20(3):145–7.
10. Hartman PE. Early years of the *Salmonella* mutagen tester strains: lessons from hycanthone. Environ Mol Mutagen 1989;14(Suppl 16):39–45.
11. Cook TM, Goldman CK. Hycanthone and its congeners as bacterial mutagens. J Bacteriol 1975;122(2):549–56.
12. Botros SS. Effect of praziquantel versus hycanthone on deoxyribonucleic acid content of hepatocytes in murine schistosomiasis mansoni. Pharmacol Res 1990;22(2):219–29.
13. Bulay O, Urman H, Patil K, Clayson DB, Shubik P. Carcinogenic potential of hycanthone in mice and hamsters. Int J Cancer 1979;23(1):97–104.
14. Moore JA. Teratogenicity of hycanthone in mice. Nature 1972;239(5367):107–9.
15. Nishimura H, Tanimura T. Clinical Aspects of the Teratogenicity of Drugs. Amsterdam, Oxford: Excerpta Medica, 1976.

Ipecacuanha, emetine, and dehydroemetine

General Information

Ipecacuanha is an extract of the root of *Psychotria ipecacuanha*, also known as *Cephaelis ipecacuanha*, a member of the Rubiaceae. It contains the emetic alkaloids cephaeline and emetine. It has often been used as a home remedy for various purposes, and not only as an emetic. It is a traditional ingredient of some expectorants, since expectoration often accompanies vomiting. Misuse of ipecacuanha by patients with anorexia nervosa and bulimia has resulted in severe myopathy, lethargy, erythema, dysphagia, cardiotoxicity, and even death. Use in infancy generally seems safe.

Emetine, once the drug of choice for the treatment of amebiasis, despite marked cardiotoxicity, has largely been replaced by metronidazole and related compounds for this indication. Large doses of emetine can damage the heart, liver, kidneys, intestinal tract, and skeletal muscle. Allergic reactions and tumor-inducing effects have not been described. Dehydroemetine is a little less toxic but

also less effective than emetine; its adverse effects are similar (SED-11, 594).

Gastrointestinal decontamination in acute toxic ingestion has been reviewed (1).

- Although ipecac generally seems to have a good safety profile, it can be associated with protracted vomiting. Other reported adverse effects include drowsiness, agitation, abdominal cramps, diarrhea, aspiration pneumonia, cerebral hemorrhage, pneumoperitoneum, and pneumomediastinum. Its use is not currently recommended (2).

Gastric lavage can be useful in some patients who have taken life-threatening doses of highly toxic substances. However, toxin absorption can be enhanced by gastric lavage. Reported adverse effects mainly include laryngospasm, hypoxemia, aspiration pneumonia, bradycardia, electrocardiographic ST segment elevation, and rarely mechanical injury to the gastrointestinal tract.

- Activated charcoal has gained popularity as a first choice for gut decontamination, based on its efficacy and relative lack of adverse effects. Poor patient acceptance is a disadvantage. Frequent vomiting can rarely become a problem.
- Saline laxatives (magnesium citrate, magnesium sulfate, sodium sulfate, and disodium phosphate) or saccharide laxatives (sorbitol, mannitol, lactulose) are also used in poisoned patients. Common adverse effects are abdominal cramps, excessive diarrhea, and abdominal distension. Dehydration and electrolyte imbalance in children, and hypermagnesemia and magnesium toxicity (with magnesium-based cathartics) have also been reported.
- Whole bowel irrigation to wash the entire gastrointestinal tract rapidly and mechanically is similar to the methods used by gastroenterologists to prepare patients for colonic investigation or bowel surgery. It is safe, even in children, pregnant women, and patients with cardiac or respiratory failure. Polyethylene glycol isotonic electrolyte solution is commonly used. Complications are usually minor and include nausea, vomiting, abdominal distension and cramps, and anal irritation.

Organs and Systems

Cardiovascular

Cardiotoxicity is the most serious and dangerous adverse effect of emetine. The clinical signs are tachycardia, dysrhythmias, and hypotension. Deaths have been described. Electrocardiographic abnormalities occur in 60–70% of cases; increased T wave amplitude, prolongation of the PR interval, ST segment depression, and T wave changes are all common. It seems possible that emetine influences the cell permeability of sodium and calcium ions, and this could be the basis of its effect on cardiac automaticity and contractility and on the electrocardiogram (SED-11, 594). The symptoms of emetine toxicity suggest that an effect on intracellular magnesium concentrations could be another possible explanation, but there are no data to support this hypothesis (SED-11, 594).

Respiratory

Asthma can occasionally be induced by ipecacuanha; when the compound was more widely used in medicine this was a familiar problem for those compounding medicines (3).

Gastrointestinal

Nausea and vomiting are frequent, perhaps in as many as one-third of all cases. In about half the cases, diarrhea is induced or existing diarrhea aggravated. Melena can occur, but this seems unlikely to be drug-induced (SED-11, 594).

Skin

Dermatitis has been attributed to emetine (4), as has cellulitis at the site of injection (5).

Musculoskeletal

Complaints of weakness, tenderness, and stiffness of skeletal muscles, especially in the neck and shoulder, are common. Following emetine aversion therapy for alcohol abuse, muscle weakness and pathological changes in muscle biopsy specimens have been described (SED-11, 594).

A reversible myopathy secondary to abuse of ipecacuanha by individuals with eating disorders has been noted (SED-12, 945) (SEDA-17, 421); the active alkaloid may have been responsible.

Long-Term Effects

Drug abuse

Munchausen's syndrome by proxy involving syrup of ipecacuanha has been reported in an 18-month-old child who was brought by his mother with persistent vomiting for 4 weeks with generalized myopathy and pneumonia (6). Its over-the-counter availability, low cost, and effective emetic properties give this drug a high appeal for such abuse.

Drug Administration

Drug overdose

There is some suggestion of cardiac impairment with overdosage, presumably also myopathic. Forced emesis can lead to esophageal damage or even complete rupture; pneumomediastinum and pneumoperitoneum have therefore sometimes complicated induced emesis (SEDA-10, 326). A bizarre case of neonatal vomiting, irritability, and hypothermia was attributed to the mother having added ipecacuanha surreptitiously to her baby's feed; at any age, however, emetine in excess can be very irritant to the gastrointestinal tract, resulting, for example, in bloody diarrhea.

Drug–Drug Interactions

Activated charcoal

Activated charcoal, sometimes used to treat self-poisoning, binds the active ingredients of ipecacuanha and inactivates them, at least partly (7). The administration of ipecacuanha also delays the administration of charcoal (8). They should not be co-administered.

References

1. Lheureux P, Askenasi R, Paciorkowski F. Gastrointestinal decontamination in acute toxic ingestions. Acta Gastroenterol Belg 1998;61(4):461–7.
2. Anonymous. Position paper: Ipecac syrup. J Toxicol Clin Toxicol 2004;42(2):133–43.
3. Persson CG. Ipecacuanha asthma: more lessons. Thorax 1991;46(6):467–8.
4. Schwank R, Jirasek L. Kožni přecitlivelost na emetin. [Skin sensitivity to emetine.] Cesk Dermatol 1952;27(1–2):50–6.
5. Anonymous. Drugs for parasitic infections. Med Lett Drugs Ther 1986;28(706):17.
6. Cooper C, Kilham H, Ryan M. Ipecac—a substance of abuse. Med J Aust 1998;168(2):94–5.
7. Krenzelok EP, Freedman GE, Pasternak S. Preserving the emetic effect of syrup of ipecac with concurrent activated charcoal administration: a preliminary study. J Toxicol Clin Toxicol 1986;24(2):159–66.
8. Kornberg AE, Dolgin J. Pediatric ingestions: charcoal alone versus ipecac and charcoal. Ann Emerg Med 1991;20(6):648–51.

Ivermectin

General Information

Ivermectin, a dihydroavermectin B1, is an effective microfilaricide used in the treatment of strongyloides, scabies, and all types of filariasis except *Dipalonema (Mansonella) perstans* infections.

Over a number of years, ivermectin has shown excellent results in the treatment of onchocerciasis, both in controlled studies and in the field, including use in the WHO-sponsored program of treatment. This experience has provided a thorough picture of its adverse effects. The effective dosage is of the order of 50–200 micrograms/kg. After a single oral dose, skin microfilariae remain at low levels for up to 9 months.

As antihelminthic drugs go, ivermectin can be considered a reasonably safe drug, and it is generally better tolerated than diethylcarbamazine. Clinical experience has often shown relatively little toxicity, although mild adverse effects, presumably due to the killing of the microfilariae, involve at least one-third of patients; some work has suggested that neutrophil activation may play a role in the development of these reactions (1). It has also been well tolerated in combinations, for example when given with albendazole in order to kill adult worms (which cannot be achieved with ivermectin alone) or with diethylcarbamazine for bancroftian filariasis (SEDA-20, 281).

The principal reservation from the start was that ivermectin has a long half-life and that some late effects might occur in certain individuals. During the early phases it was recommended that in areas where the drug had been widely used the health workers involved should continue to observe patients for a time, in case problems did arise, but no late complications have in fact been documented.

Mode of action

The mode of action of ivermectin has been reviewed (2). It has tentatively been identified as agonism at GABA receptors, with inhibition of ion channels that control specific nerve cell connections. The functioning of chloride channels should thus be altered in most organisms, leading to paralysis and death of parasites. Several sites of action have been proposed:

- a postsynaptic agonist site either on the receptor or in its immediate neighborhood;
- a presynaptic site of activation of GABA release;
- potentiation of GABA binding to its receptor.

Another mechanism of action involves the binding of ivermectin to P glycoprotein.

Observational studies

Of 458 Brazilians with intestinal helminthiasis and parasitic skin diseases treated with ivermectin 200 micrograms/kg and a second dose after 10 days, 24 were also given albendazole and 15 mebendazole (3). There were adverse events in 9.4% of treatments (85 individuals). They were all mild to moderate in intensity and were transient. Abdominal pain, the most common adverse event (23 out of 458 first-dose treatments with ivermectin), was probably caused by worms dying and disintegrating, especially *Ascaris lumbricoides*. Nausea and loose stools were reported after 15 and 12 first-dose treatments respectively.

The incidence of serious adverse events after mass treatment with ivermectin in areas co-endemic for loiasis and onchocerciasis has been determined in a retrospective analysis (4). In addition, potential risk factors associated with these serious adverse events, including encephalopathy, were identified. In the period December 1998 to November 1999, a total of 784 653 people were given ivermectin in onchocerciasis mass treatment programmes in Cameroon; 47 serious adverse events were reported, resulting in an overall incidence of about six serious adverse events per 100 000 people treated. The five most frequent initial symptoms and signs were as follows:

- fatigue and/or weakness and/or difficulty or inability to stand (43%);
- confusion, obtundation, stupor, or unconsciousness (26%);
- nausea and/or vomiting and/or diarrhea and/or dehydration (17%);
- fever and/or chills (17%);
- dysarthria and/or aphasia (15%).

The median age of the cases was 35 (range 6–72) years. Male patients represented 75% of the caseload and 87% of the patients experienced ivermectin for the first time. Symptoms began within the first 24–48 hours of administration, but there was a delay of about 48–84 hours in seeking help after the onset of symptoms. There was a presumptive neurological diagnosis in 35 cases; 29 were considered to be so-called PLERM (Probable *Loa loa* Encephalopathy temporally Related to Mectizan treatment). Six of the 47 reported patients with a serious adverse event died; five fitted the diagnosis of PLERM. First-time exposure to ivermectin was primarily associated with PLERM.

Among the published studies some have specifically sought to define the pattern of adverse effects. In one such study (5), although a single dose of the drug was combined in some patients with diethylcarbamazine, the adverse effect pattern was similar to that when ivermectin was used alone (SEDA-18, 312). There now seem to be some circumstances in which a single low dose of ivermectin is sufficient to have a prolonged effect, for example in loaiasis a dose of 150 micrograms/kg resulted in a very much reduced level of microfilaria as much as a year later, and seemed to eliminate the infestation entirely in more than half the users (6,7). If further work confirms the validity of this approach, the adverse reactions problem may be lessened, since at these doses the few reactions experienced were limited to the skin and joints, although some calabar-like swellings were also noted.

Brugia malayi

In an open study from India 21 asymptomatic microfilaria carriers (with counts of 109–6934/ml of blood) were treated with a single oral dose of ivermectin 400 micrograms/kg and a single oral dose of diethylcarbamazine 6 mg/kg for infection with *Brugia malayi* (8). Twelve hours after treatment microfilaria counts fell by 96–100% in all patients and 12 patients had become afilaremic. All had an adverse reaction, lasting up to 48 hours after treatment: fever ($n = 20$), myalgia ($n = 19$), headache ($n = 17$), lethargy ($n = 15$), chills ($n = 13$), sweating ($n = 11$), anorexia ($n = 11$), sore throat and pharyngeal congestion ($n = 10$), arthralgia ($n = 6$), giddiness ($n = 4$), nausea and vomiting ($n = 3$), abdominal pain ($n = 2$), and cough ($n = 1$). Postural hypotension, lasting 1 day, was noted in two individuals. Transient dilated and painfully inflamed lymphatic channels, which stood out in cords, were seen in two individuals. Most adverse effects were mild and self-limiting.

Gnathostomiasis

In a prospective open study in 20 Thai patients with cutaneous gnathostomiasis oral ivermectin 50, 100, 150, or 200 micrograms/kg was associated with the following adverse events: malaise (n = 7), myalgia (6), drowsiness (6), pruritus (4), nausea/vomiting (4), dizziness (3), diarrhea (3), a feeling of shortness of breath (2), palpitation (2), constipation (1), anorexia (1), and headache (1) (9).

These adverse events were self-limiting and there were no serious adverse events. There were laboratory abnormalities in three patients. Transient microscopic hematuria, pyuria, and mildly raised liver enzymes were found in one patient each.

Loa loa

The use of ivermectin and its adverse effects in patients infected with *Loa loa* have been reviewed (10). It was concluded that ivermectin in a single dose of 150–300 micrograms/kg is effective in reducing microfilaria counts by over 90% with suppressed counts to 25% of pretreatment values after 1 year. An even more sustained effect can be reached by more frequent dosing. There is also some evidence that ivermectin in higher doses (400 micrograms/kg twice yearly) can affect the adult *Loa loa*. Tolerance to ivermectin is generally excellent, but serious adverse effects, especially encephalopathy, can occur, principally in more heavily infected individuals.

Onchocerciasis

The impact of 5 years of annual community treatment with ivermectin on the prevalence of onchocerciasis and onchocerciasis-associated morbidity in the village of Gami (Central African Republic) has been assessed (11). Pruritus, onchocercal nodules, and impaired vision were all significantly reduced by annual treatment with ivermectin.

In a study of the effect of ivermectin on adult *Onchocerca* worms (12) the following regimens were compared:

- 150 micrograms/kg yearly (reference group; $n = 166$)
- 400 micrograms/kg then 800 micrograms/kg yearly ($n = 172$)
- 150 micrograms/kg 3-monthly ($n = 161$)
- 400 micrograms/kg then 800 micrograms/kg 3-monthly ($n = 158$)

After 3 years of treatment more female worms had died in those who were treated every 3 months than in the reference group; female worms were also less fertile. There was no difference between the two groups of patients who were treated yearly. There were no serious adverse events, even at high doses. However, subjective complaints of visual disturbances, such as blurred vision, ocular pain, or dyschromatopsia, were more frequent in those who were given 800 micrograms/kg than in those who were given 150 micrograms/kg; the effects lasted less than 1 week. Detailed ocular examination showed no differences between patients from the reference group and the three other groups.

In another study, 890 subjects were interviewed to monitor adverse reactions after repeated ivermectin treatment of onchocerciasis in Nigeria (13). After the first treatment round with ivermectin, 202 subjects reported pains in joints, 108 reported fever, 30 reported headache, 18 reported itching, and four reported dizziness. There were no adverse reactions in 528 (59%). After the sixth treatment round, no reactions were reported in 756 subjects (85%). Pains in joints were reported by 76, itching by 26, fever by 24, and dizziness by eight subjects. The relatively mild adverse reactions observed during the first treatment round did not affect future participation in community treatment with ivermectin.

In a study in Uganda, 737 of 1246 patients (59%) with onchocerciasis developed adverse reactions after first treatment with a single oral dose of ivermectin 150 micrograms/kg (14). Pain, swelling, and cutaneous reactions were the three most dominant symptoms of adverse effects (in 57%, 50%, and 38% of the 737 symptomatic patients respectively). Ten patients had severe adverse reactions, including severe postural hypotension and high fever. In spite of the fact that many patients had adverse reactions to ivermectin, the drug was well accepted and appreciated by the population.

Severe adverse events (e.g. the Mazzotti reaction, see diethylcarbamazine) can occur after ivermectin treatment in patients with high *Loa loa* microfilarial densities. In the context of mass ivermectin distribution for onchocerciasis control in Africa, it is crucial to define precisely the geographical distribution of *Loa loa* in relation to that of *Onchocerca volvulus* and predict the prevalence of heavy infections. To that end, the distribution of *Loa loa* microfilarial densities were determined in 4183 individuals living in 36 villages in the Lekie Division in central Cameroon, to predict prevalences of *Loa loa*-related post-ivermectin adverse events (15). The value of k, an estimate of the degree of microfilarial overdispersion in *Loa loa*, varied around 0.3 independently of microfilarial intensity, host age, village, and endemicity. Based on these results a semi-empirical model was developed to predict the prevalence of heavy *Loa loa* microfilarial loads in a community, which could be useful in identifying areas and populations at risk of severe adverse events after ivermectin.

In a 3-year double-blind trial (16) in Cameroon the effect of ivermectin, given at 3-monthly intervals and/or in high doses (800 micrograms/kg) on adult *Onchocerca volvulus* was determined in 643 individuals and compared with the standard annual dose of 150 micrograms/kg. No patient developed a serious adverse reaction, none required hospitalization, and none withdrew because of an adverse reaction. Mild adverse reactions began soon after the start of treatment. Of 1129 consultations for adverse reactions, 367 took place in the a few hours after treatment (day 0); 449, 195, and 118 consultations took place on days 1, 2, and 3 respectively. The incidence of adverse reactions fell steadily over the course of the trial. After the first dose, 3-monthly treatment was associated with a reduced risk of reactions, especially edematous swelling, pruritus, and back pain. Edematous swelling and subjective ocular problems were associated with high doses of ivermectin.

Sarcoptes scabiei

The treatment of scabies has been reviewed (17,18,19). Topical treatments are poorly tolerated by some patients. An alternative approach is to use oral ivermectin, which has been used extensively for several parasitic infections, including onchocerciasis, lymphatic filariasis, and other nematode-related infestations. Ivermectin is thought to interrupt glutamate-induced and γ-aminobutyric acid-induced neurotransmission in parasites, leading to nematode paralysis and death. In humans ivermectin does not cross the intact blood–brain barrier. The efficacy and effectiveness of oral ivermectin is equivalent to, or better than, that of topical lindane, benzylbenzoate, and permethrin. Cure rates are 70–74% with a single dose of ivermectin and 95% when a second dose is taken 2 weeks later. The poorer efficacy of a single dose may reflect the fact that ivermectin is not ovicidal.

Two cases of scabies treated with oral ivermectin (200 µg/kg) have been reported (20).

- A 72-year-old man developed crusted scabies, having used an oral glucocorticoid, owing to a presumed misdiagnosis by an earlier physician. He was successfully treated with two oral doses of ivermectin 7 days apart with topical crotamiton 10% and a keratolytic ointment (5% salicylic acid in petrolatum). However, the nail scabies failed to respond. Live mites were detected from all his toenails 2 weeks after the second dose of ivermectin. Complete cure was achieved by occlusive dressings with lindane for 1 month. Follow-up observations did not reveal any adverse effects of ivermectin or signs of relapse.
- A 52-year-old woman, who had taken oral glucocorticoids for mesangial nephritis, developed common scabies, but a topical scabicide, crotamiton, was not effective, and 2 weeks after treatment with a single dose of oral ivermectin eggs were still detected from a burrow on her trunk. Her treatment was completed after a further two doses of oral ivermectin at 7 day intervals.

In both cases, oral ivermectin did not cause any clinical or laboratory adverse effects. Oral ivermectin is effective for crusted scabies, but not effective for nail scabies. A repeat treatment with ivermectin appears to result in the highest rate of cure.

In an uncontrolled open study 101 patients with scabies were treated with a single oral dose of ivermectin 200 micrograms/kg and then followed at 3 days and at 2 and 4 weeks (21). Two weeks after the start of treatment 89 patients were completely free of scabies, while another three had only mild lesions and pruritus with negative skin scrapings. The other nine patients had persistent pruritus and new lesions and were treated with a second dose, with a complete cure in all cases after 4 weeks. Twelve patients reported minor adverse effects, consisting of drowsiness ($n = 4$), arthralgia and bone aches ($n = 2$), dyspnea ($n = 3$), headache ($n = 1$), nausea ($n = 1$), and blurred vision ($n = 1$). The adverse effects were mostly reported at the first follow-up and were easily tolerated. Ivermectin appears to be a safe and effective treatment for scabies in a dose of 200 micrograms/kg, although a second dose is necessary for complete cure in a few patients.

- An 11-year-old girl developed severe crusted Norwegian scabies (22). Gamma-benzene hexachloride

lotion and topical keratolytics had no significant effect. She was given a single oral dose of ivermectin 6 mg/kg with dramatic effect. The pruritus subsided in 4 hours and the lesions started to clear 2 days later. A second dose of 6 mg was given after 3 weeks when no skin lesions were found anymore. The only adverse effect was some edema of the skin after the first dose, which did not occur after the second dose, suggesting that the reaction was more related to the intensity of the infection then to the effect of the drug itself.

Outbreaks of scabies in elderly people require special management for disease control. Owing to the frequent failure of repeated non-synchronized therapeutic efforts with conventional external antiscabie treatments, special eradication programs are required. The management of outbreaks of scabies with allethrin, permethrin, and ivermectin has been evaluated (23). Healthy infested people ($n = 240$) were treated once simultaneously with an external scabicide, such as allethrin or permethrin; this was effective in 99%. Those with crusted scabies ($n = 12$) were hospitalized and treated with systemic ivermectin or ivermectin plus permethrin; seven patients received ivermectin twice after an interval of 8 days and one received permethrin three times. Unfortunately, no details of adverse effects were given.

Strongyloidiasis

The efficacy and adverse effects of ivermectin 200 micrograms/kg, repeated 2 weeks later, have been studied in 50 patients with chronic strongyloidiasis, aged 30–79 years (24). The eradication rate was 96% at 2 weeks after the first dose and 98% after the second dose. There was no recurrence after follow-up of 4 months. One patient had nausea and vomiting 3 hours after the first dose and again after the second dose, but they were transient and required no therapy. In four patients there were mild laboratory abnormalities (slight increases in liver function tests in two, microscopic hematuria in one, and mild leukopenia and lymphocytosis in one). Of the 50 patients 12 were positive for human T lymphotropic virus type-I.

Wuchereria bancrofti

Early-stage elephantiasis caused by bancroftian filariasis in a 27-year-old traveller was treated with a single-dose oral combination of ivermectin 24 mg plus albendazole 400 mg, followed by albendazole 800 mg for 21 days (25). To avoid a severe Mazzotti-like reaction, he was given oral glucocorticoids and antihistamines for 3 days. He had a transient rash, pruritus, and mild hypotension on the days after the initial treatment, but otherwise remained well and the swelling subsided. Within 1 month he was free of symptoms. At the last follow-up examination, 3 years after treatment, there was no clinical or laboratory evidence of relapse. The authors thought that this type of treatment should be evaluated on a wider scale, given the minimal adverse events and apparent therapeutic efficacy.

The efficacy of annual mass chemotherapy with a combination of diethylcarbamazine and ivermectin on bancroftian filariasis in rural southern India has been studied, as has the supplementary role of controlling the vector mosquito *Culex quinquefasciatus* (26). Nine villages, topographically and ecologically similar but reasonably isolated from each other, were selected and split into three comparable groups of three villages each. Group A received chemotherapy with diethylcarbamazine at about 6 mg/kg and ivermectin at 400 micrograms/kg. Group B received chemotherapy and vector control. The most important vector-breeding sites were soakage pits, which were treated with expanded polystyrene beads. Minor vector-breeding sources, such as domestic or irrigation wells, were treated by adding larvae-eating Talapia fish or a commercial insecticide based on *Bacillus sphaericus*. Group C received no intervention. After the first round of treatment, combination chemotherapy alone caused a 60% drop in the annual filarial transmission potential, whereas the combined strategy reduced the transmission potential by 96%. After two rounds of treatment, the reduction in transmission potential was similar with the two strategies (about 91–96% reduction), whereas the prevalence of microfilaremia was reduced by 88–92%. Adverse events after combination therapy were reported in 20% of those who had taken diethylcarbamazine and ivermectin for the first time. The patients with adverse events had increased microfilarial counts. The most common adverse effects were headache (72% of adverse events), giddiness (67%), fever, and weakness. The incidence of adverse events among those taking combination therapy for a second time was relatively low (5.5%). The adverse events were also less severe in the second round than in the first. When antifilarial treatment was withdrawn in the third and final year of the study, transmission was resumed in the absence of vector control, whereas no infective female mosquitoes were detected in villages with vector control. Vector control, although obviously not cost-effective in the short term, could therefore play an important supplementary role in an integrated program, by preventing re-establishment of transmission after chemotherapy has been completed.

Comparative studies

Gnathostomiasis
In a randomized open study, a single dose of oral ivermectin 200 micrograms/kg (n = 17) was compared with oral albendazole 400 mg/day (n = 14) for 21 days in the treatment of cutaneous gnathostomiasis (27). There were no major adverse effects.

Sarcoptes scabiei
In a randomized trial a single oral dose of ivermectin (200 micrograms/kg) has been compared with 1% gamma-benzene hexachloride lotion for topical application overnight in 200 patients with scabies (28). The patients were assessed after 48 hours, 2 weeks, and 4 weeks. After 4 weeks, 83% showed marked improvement with ivermectin, compared with 44% of those treated with gamma-benzene hexachloride. There were no adverse events reported with gamma-benzene hexachloride. Headache was reported only once with ivermectin.

In 80 children aged 6 months to 14 years a single dose of ivermectin 200 micrograms/kg was compared with topical benzyl benzoate for the treatment of pediatric scabies in a randomized, controlled trial (29). Ivermectin cured 24 of 43 patients and topical benzyl benzoate cured 19 of 37 patients at 3 weeks after treatment. There were no serious adverse effects with either treatment, although benzyl benzoate was more likely to produce local skin reactions. These results are in line with those of another study, in which 18 children aged 14 months to 17 years with either scabies (*n* = 11) or cutaneous larva migrans (*n* = 7) were treated with a single dose of ivermectin 150–200 micrograms/kg (30). A single oral dose cured 15 patients, and three patients with crusted scabies required a second dose. None had significant adverse reactions.

No serious adverse effects were noted in a programme of mass treatment with ivermectin for children with scabies in the Solomon Islands (31). In one study, there was an excess risk of deaths among elderly patients who took ivermectin for scabies, but selection bias and confounding factors were possible explanations; this observation has not been confirmed in other studies, including studies of residents in nursing homes.

Wuchereria bancrofti
In a study in which doses of 200 or 400 micrograms were used, with or without diethylcarbamazine, for *Wuchereria bancrofti* infection, there was a higher than average incidence of reactions (and a higher incidence with ivermectin than with diethylcarbamazine), but this perhaps reflected an unusually high success rate or the severity of the original infection (32). For similar reasons, repeated courses of treatment tend to show a falling incidence of adverse effects. Normally, such general symptoms as fever, weakness, anorexia, malaise, and chills occur in a substantial minority of patients on a first course, while at least one-third have muscle and/or joint pains. Vertigo, dyspnea, diarrhea and abdominal disturbances affect a few patients. The severity of adverse effects is not related to serum concentrations of the drug (33), which again reflects the fact that they are largely a consequence of the parasitic breakdown, rather than toxic effects of ivermectin.

In a randomized, double-blind field trial in Tanzania, 1221 children with *Wuchereria bancrofti* lymphatic filariasis were given a single dose of ivermectin 150–200 micrograms/kg alone or in combination with albendazole 400 mg (34). Adverse reactions were few and mild in both groups, and mainly reported from pre-treatment microfilaria and children positive for circulating filarial antigen. The most common adverse events were headache (n = 25) and fever (n = 19). All adverse events abated within a few days.

Placebo-controlled studies

Emesis, ataxia, and mydriasis are cardinal signs of ivermectin toxicity. The safety, tolerability, and pharmacokinetics of escalating high-dose ivermectin have been studied in 68 healthy subjects in a randomized, double-blind, placebo-controlled study (35) in the following doses:

- 30 mg fasted ($n = 15$)
- 60 mg fasted ($n = 12$)
- 90 mg fasted ($n = 12$)
- 120 mg fasted ($n = 12$)
- 30 mg fed ($n = 11$)

Ivermectin was generally well tolerated. Quantitative pupillometry ruled out any mydriatic effect of ivermectin. There was no nervous system toxicity associated with oral ivermectin at any of the doses. There were no serious clinical or laboratory adverse events. Three of the fifty-one subjects who took ivermectin fasted reported minor adverse gastrointestinal events: fecal abnormality ($n = 1$), nausea ($n = 1$), and vomiting ($n = 1$); six reported minor neurological adverse events: headache ($n = 4$), anxiety ($n = 1$), and dizziness ($n = 1$). There were no adverse events in the subjects who took ivermectin 120 mg. The absorption of ivermectin was about 2.5 times higher when it was given after a high-fat meal.

Onchocerciasis

In a randomized, double-blind, placebo-controlled trial the efficacy and pharmacokinetic interaction of co-administration of ivermectin with albendazole in onchocerciasis was studied and compared with ivermectin alone in male patients, who were randomized to receive ivermectin 200 µg/kg ($n = 14$), albendazole 400 mg ($n = 14$), or both ($n = 14$) (36). The adverse effects were mild to moderate in intensity and there were no serious adverse effects. Comparison of the two ivermectin-treated groups showed no significant differences in the intensity or frequency of any type of adverse effect, or on the perceived need for additional medications, such as paracetamol. Although the combination, but not ivermectin alone, produced a significantly higher total Mazzotti reaction score than albendazole, the clinical significance of this difference was minor. Thus, the safety profile of ivermectin was not altered by co-administration of albendazole. In addition, albendazole did not alter systemic exposure to ivermectin or vice versa. However, the co-administration of ivermectin with albendazole offered no advantage over ivermectin alone in terms of efficacy against *Onchocerca volvulus*.

Sarcoptes scabiei

In a randomized, double-blind comparison of the efficacy of oral ivermectin and topical gamma-benzene hexachloride 53 patients were randomly allocated to either a single oral dose of ivermectin 150–200 micrograms/kg and a placebo topical solution, or a single dose of gamma-benzene hexachloride topical solution 1% and placebo tablets (37). Patients who did not fulfil the criteria for clinical cure within 15 days, defined as the absence of both pruritus and clinical lesions or a reduction in signs and symptoms to a mild degree, repeated the initial treatment. Of the 53 patients 43 completed the study (19 of those treated with ivermectin and 24 of those treated with gamma-benzene hexachloride). After 15 days 74% of the patients treated with ivermectin and 54% of the patients

treated with gamma-benzene hexachloride were considered to be cured. At 29 days both treatments were equally effective, with cure rates of 95% and 96% respectively. Adverse effects were mild and transient in both groups. One of the patients treated with ivermectin had hypotension, one had abdominal pain, one had vomiting, and one complained of headache. There were no abnormalities on routine laboratory testing.

Wuchereria bancrofti

In a double-blind, placebo-controlled study in Ghana single doses of ivermectin 150–200 micrograms/kg and albendazole 400 mg, either separately or in combination, were given to 1425 individuals for *Wuchereria bancrofti* infection (38). Of these, 340 were microfilariae-positive before treatment. Ivermectin and ivermectin plus albendazole both produced statistically significant reductions in mean microfilaria counts at follow-up; the effect of ivermectin was longer lasting. Albendazole produced a non-significant reduction. Adverse reactions were few and mostly mild, and there were no severe reactions.

General adverse effects

Acute symptoms, often flu-like or affecting the skin, are related almost entirely to the release of toxic products and allergens from the killed filariae, and can affect two-thirds of patients; in conditions in which this type of reaction does not occur one may suspect that the drug is ineffective. The mechanism of the effects also explains why they tend to occur early and sometimes briefly, that is immediately after the microfilariae die. For similar reasons, these effects are most severe in patients with a high microfilaria count (39).

Despite the sometimes transient and apparently tolerable nature of the skin effects, they can persist in patients requiring long-term treatment, for example for onchocerciasis, and under these conditions they are sufficient to impair compliance with treatment (SEDA-21, 317).

The effects of age, sex, dosing round, time of day, and distance from the nurse monitor on adverse event reporting during mass ivermectin administration in Achi in South-East Nigeria have been examined (40). There was a significant increase in adverse reporting with age, but not sex. Fewer adverse effects were reported after starting at night than after starting by day. There was no significant effect of distances up to 1 km on adverse events reporting. Both compliance and adverse reporting were less after the second dosing round than after the first. These variables should be included in the standardization of adverse events reporting.

Onchocerciasis

Although adverse reactions after ivermectin in onchocerciasis are usually less severe than after diethylcarbamazine, they still affect a significant number of patients with onchocerciasis after the first dose. With subsequent treatments, these reactions become less frequent and severe. The so-called Mazzotti reaction, which is often seen after treatment of *Onchocerca volvulus* with diethylcarbamazine or ivermectin, is characterized by fever, tachycardia, hypotension, adenitis, pruritus, arthralgia, a papular or urticarial rash, and lymphedema. It is ascribed to an inflammatory host response to microfilarial killing and tends to be more severe in those who have greater numbers of parasites. The roles of chemoattractants, such as eotaxin, RANTES, and MCP-3, in the recruitment of eosinophils to the site of parasite killing has been studied in 13 patients with onchocerciasis and two control subjects before and after ivermectin (41). There were adverse reactions in eight patients, but none were severe. The reactions were fever (54%), pruritus (62%), rash (46%), and lymphedema (46%). There was no significant postural hypotension. There was endothelial expression of both RANTES and eotaxin after ivermectin, suggesting that these chemoattractants have an important role in eosinophil recruitment into the skin during killing or degeneration of parasites after ivermectin.

A role for the release of *Wolbachia* bacterial endosymbionts has been suggested in the pathogenesis of the Mazzotti reaction (SEDA-26, 345). There was a good correlation between *Wolbachia* DNA, serum TNF-alfa, and the antibacterial peptides calprotectin and calgranulin after treatment with ivermectin or diethylcarbamazine, supporting a role for *Wolbachia* products in mediating these inflammatory responses (42).

There has been an epidemiological survey of the endemicity of human onchocerciasis and the effects of subsequent mass distribution of ivermectin in villages of the Nzerem-Ikpem community in Nigeria (43). Of 1126 people studied, 527 were positive for skin microfilariae, 329 had a leopard skin (characterized by focal skin depigmentation), 385 had nodules, and 167 had onchodermatitis. There were adverse effects in 362 patients (19%): pruritus in 13%, limb swelling in 8.5%, facial swelling in 2%, weakness in 4.8%, nausea and vomiting in 3.4%, headache in 5.8%, diarrhea in 3.4%, and rheumatism in 3.5%. There were no severe reactions.

Organs and Systems

Cardiovascular

Supine and postural tachycardia with postural hypotension can occur; in one large study, such effects were found in three of 40 patients (SEDA-14, 262) (SEDA-22, 327). In another there was hypotension in 13 of 69 cases (SEDA-20, 280), but in some series these effects have not been observed at all (SEDA-17, 356). A massive community study in Ghana noted hypotension in only 37 of nearly 15 000 patients treated (44). Transient electrocardiographic changes are sometimes seen.

Respiratory

In the treatment of *Wuchereria bancrofti* filariasis in 23 patients with single doses up to 200 micrograms/kg respiratory capacity was evaluated; there was a transient but significant fall in vital capacity some 24–30 hours after administration, apparently due to bronchodilatation (45). Frank dyspnea occurred in 2% of cases in the study cited above (5). In other studies, a few patients have developed a transient cough and in others pneumonitic patches have been seen in the chest X-ray (SEDA-18, 313).

Nervous system

Headache and vertigo are very common and even usual as part of the flu-like reaction to ivermectin.

A puzzling reaction was recorded in a small hospitalized Canadian population of elderly subjects treated for scabies with a single dose of ivermectin (150–200 micrograms/kg). Within 6 months, 15 of the 47 patients had died. All those who died had developed a sudden change in behavior, with lethargy, anorexia, and listlessness before death (46). The effect may have been an artefact with some extraneous cause, and it is notable that other groups using this treatment for scabies have not recorded similar reactions.

Loa loa encephalopathy

When treating *Loa loa* infections on a large scale with ivermectin, the encephalopathy that was a much-feared complication with diethylcarbamazine again seems to occur, especially with heavily infected or older individuals (47). For this reason, mass use of ivermectin in areas of endemic *Loa loa* infection is no longer recommended (48).

Ivermectin appears to promote the passage of *Loa loa* microfilaria into the cerebrospinal fluid, with a maximum after 3–5 days, followed by an intense allergic reaction to the dying microfilaria. The Mectizan Expert Committee defined a definite case of *Loa loa* encephalopathy related to ivermectin as having to satisfy two criteria:

1. encephalopathy in which there is microscopic evidence of vasculopathy in the brain associated with *Loa loa* microfilaria;
2. the onset of symptoms of disturbed nervous system function within 5 days after treatment with ivermectin, progressing to coma without remission.

A probable case of *Loa loa* encephalopathy was defined as having to satisfy four criteria:

1. coma in a previously healthy individual;
2. the onset of nervous system signs within 5 days of treatment with ivermectin progressing to coma;
3. an initial microfilaremia of over 10 000/ml, or 1000/ml in a blood sample taken within 2 months of treatment;
4. the presence of *Loa loa* microfilaria in the CSF.

Clinically common features of this condition are impaired consciousness appearing 3–4 days after treatment and lasting for 2–3 days.

There is no consensus on the proper management of ivermectin-associated *Loa loa* encephalopathy, and it is uncertain if co-administration of glucocorticoids is of any use. In several patients with more severe reactions, conjunctival hemorrhages were seen.

A systematic examination of the conjunctivae in 1682 patients complaining of any adverse reactions showed that these hemorrhages were closely correlated with the pretreatment microfilaria counts. This sign can be found 2 days after treatment and may thus single out patients susceptible to encephalopathy and needing closer follow-up. Although the incidence of such cases is very low (in the order of 1 in 10 000 treated patients), this serious adverse effect makes mass treatment of *Loa loa* infection problematic, and also mass treatment of onchocerciasis in areas in which *Loa loa* is endemic. To illustrate this point three probable cases of *Loa loa* encephalopathy after ivermectin treatment for onchocerciasis have been described (49). All three were young men treated with ivermectin 150 micrograms/kg in a mass-treatment campaign in onchocerciasis.

- A 26-year-old previously healthy man developed nervous system symptoms in the form of an inability to stand or eat and stiffness of the neck by the third day. On the fourth day he had difficulty swallowing and speaking. On the fifth day he could not speak and was incontinent of urine. He was given dexamethasone, diazepam, furosemide, and atropine. On the sixth day he became comatose. On the ninth day he developed a high fever, and was given penicillin and tube-feeding. His condition gradually worsened and he died on the 21st day. Serum microfilaria counts on day 13 after treatment were still high (3600/ml), and live *Loa loa* (10/ml) were found in the CSF.

- A 32-year-old man with alcoholism had a very high pretreatment serum microfilaria count (50 000/ml). After starting ivermectin he took to his bed and would not speak. On the third day he developed a fever, possibly attributed to malaria and treated with chloroquine. On the fourth day he was unable to stand, and alternately restless or somnolent; his CSF contained live *Loa loa* microfilaria. He became more incoherent and fidgety and had a marked grasp reflex. Later in the day he developed spastic hypertonia. On the fifth day he became incontinent and still would not speak. Over the following days he gradually improved and 4 months later had no neurological abnormalities, although his relatives found that his behavior had changed and that he was much calmer then in the past. An electroencephalogram on day 15 showed periodic diffuse discharges of large amplitude during hyperventilation and on day 146 an asymmetric tracing with focal activity in the right parieto-occipital area, which worsened during hyperventilation. On day 233 the electroencephalogram was normal.

- An 18-year-old previously healthy man was given ivermectin. On the second day he was unable to work and stayed at home. On the third day he was found unconscious in bed, incontinent of urine and feces. On the fourth day he did not move and had absent pain sensation. There was hypertonia in the arms with marked cogwheeling. On the fifth and sixth days there was a swinging horizontal movement of the eyeballs, but otherwise he appeared to improve. On the seventh day he could stay seated in bed with help and spoke several sentences. He could perform slow voluntary movements and his muscle strength and sensation returned to normal, although the cogwheel phenomenon still persisted. He gradually returned to normal over the following weeks. After 5 months the neurological examination was normal but he still complained of headaches and episodic amnesia. His pretreatment serum microfilaria counts were high (152 940/ml) and the CSF collected on the fourth day contained live *Loa loa* microfilaria. An electroencephalogram on the 19th day was slow with spontaneous, diffuse, paroxysmal, monomorphic theta activity, lasting 2–3 seconds. An electroencephalogram on the 105th day showed improvement, but focal abnormalities persisted in the left occipital region. On the 159th day all previously recorded abnormalities had disappeared.

Sensory systems

Careful ophthalmological examination shows a striking increase in the number of microfilariae in the anterior chamber of the eye in a significant minority of patients, and a new inflammatory infiltrate can appear during treatment in already damaged areas of the retina (SEDA-15, 334). However, no permanent ocular sequelae have been documented. Most of the other ophthalmic symptoms, including edema and local inflammation, are those of the primary infection.

Conjunctival hemorrhages have been recorded in patients living in areas in which loiasis is endemic. Although ivermectin is usually well tolerated, these patients had serious adverse reactions after taking ivermectin, including an encephalopathy similar to that seen after treatment with diethylcarbamazine. In retrospect, these cases all had high *Loa loa* microfilaremia and *Loa loa* microfilariae in the cerebrospinal fluid. The authors suggested that ivermectin might have provoked the passage of *Loa loa* microfilariae into the cerebrospinal fluid. In a subsequent study of 1682 patients with loiasis treated with ivermectin 150 micrograms/kg, conjunctival hemorrhages were found in 41, nine of whom had previously received a microfilaricidal drug (50). The initial mean *Loa loa* microfilaremia was 14 900 microfilariae/ml (range 0–182 400; median 37 500), compared with 14.5 microfilariae/ml (range 0–97 600; median 0) in those without conjunctival hemorrhage. In addition, male sex and *Dipalonema perstans* microfilaremia were associated with conjunctival hemorrhages. There was a close relation between conjunctival hemorrhages and retinal lesions. Based on observations in three patients who all developed coma after ivermectin the authors suggested that retinal lesions may reflect what occurs in the cerebral circulation in patients with high *Loa loa* microfilaremia and neurological problems after ivermectin.

Following the observation in a 3-year double-blind, randomized, controlled trial in Cameroon that transient visual problems can occur after treatment with ivermectin in onchocerciasis, ophthalmological examinations were carried out (51). The visual complaints were significantly more frequent in those who had received high doses of ivermectin (for example 800 versus 150 microgram/kg annually).

Hematologic

When 28 Sudanese patients were treated with a single dose of ivermectin for onchocerciasis they developed a prolonged prothrombin time, which continued to lengthen significantly during the next 4 weeks; there were no changes in other clotting parameters (52). After a month, two of them developed hematomas, which continued to enlarge for the next 3–4 days. Both of these patients had received ivermectin 150 micrograms/kg. One was given a transfusion; the swellings in both cases resolved within a week. For a time it was considered that the prothrombin changes observed in some such cases were a potential problem; more recent work suggests that the prolongation of prothrombin ratio is in fact hardly more than with placebo, and that in fact ivermectin merely has a mild effect on vitamin K metabolism and little effect on coagulation (53). However, lymphadenitis has been noted in a few patients. In one Guatemalan study of biannual treatment of the population to eradicate *Onchocerca volvulus* infection, upper limb edema was noted in nearly 20% of cases receiving the treatment for the first time (54).

Gastrointestinal

In the late stages of *Strongyloides* hyperinfection, ileus can develop and hamper the absorption of oral medication (55).

- A 39-year-old Afro-Caribbean man with stage IVB T cell lymphoma due to HTLV-1 infection had invasive Strongyloides hyperinfection that did not respond to oral ivermectin plus albendazole because of concurrent ileus. He was treated with two 6 mg doses of a veterinary formulation of ivermectin subcutaneously. There were no adverse effects, apart from pain at the injection site.

Urinary tract

Proteinuria is unusual but has been described; it was detected 14 days after a single dose and disappeared during follow-up (SEDA-18, 313).

Observations that proteinuria and hematuria may occur in patients with filariasis bancrofti and loiasis, which may exacerbate after treatment with diethylcarbamazine or ivermectin, led to the study of kidney function in patients with onchocerciasis before and after treatment with ivermectin (56). The occurrence of renal abnormalities was studied in a population-based study in a meso-endemic village (40% microfilaria carriers), in a group of patients with a generalized or hyper-reactive form of onchocerciasis, and in 46 patients treated with ivermectin in a single oral dose of 150 micrograms/kg. All individuals in all three study groups were examined clinically and had skin snips, serological testing for onchocerciasis, and nodulectomy (when relevant). Tests for malaria, schistosomiasis, intestinal nematodes, and hepatitis B, and serum glucose, creatinine, IgE, and electrophoresis were also performed. The urine was tested for erythrocytes, leukocytes, protein, nitrites, pH, glucose, ketone bodies, urobilinogen, and creatinine. All the patients underwent renal ultrasound examination. There was no difference in renal function and renal ultrasound between patients with and without onchocerciasis. A raised urinary protein concentration (over 70 mg/g of creatinine) was common and occurred in 47% of the patients with onchocerciasis and 63% of the patients without onchocerciasis. In the 46 patients treated for onchocerciasis with a single dose of 150 micrograms/kg there was a slight but statistically significant increase in total urine protein after 2 and 5 days, especially in 16 patients with high pretreatment skin microfilaria counts. The abnormalities were minor and insignificant. Neither onchocerciasis itself nor treatment with ivermectin was associated with abnormalities of renal function.

Skin

A degree of pruritus, soreness, or burning sensation is common with ivermectin, and rashes or skin edema can occur, while pre-existing conditions of this type can be aggravated (SEDA-17, 355; SEDA-22, 327). The skin over hematomas can be discolored. Swelling of the limbs and face, like the dermatological symptoms, is probably a reaction to breakdown products of the helminth (SEDA-20, 280).

Patients with severe skin involvement ("sowda") as a facet of their onchocerciasis can experience transient aggravation of the condition, but the course is favorable, and they are less likely to have the same problem if it is later necessary to repeat treatment (SEDA-20, 281).

Rashes and swelling of the lymph nodes seem to be more common in patients with AIDS who take ivermectin (57).

Musculoskeletal

Joint or bone pains are common but usually mild; in one study myalgia occurred in 33% of cases and arthralgia in 33% (5).

Reproductive system

Orchitis or epididymitis with scrotal tenderness occurs in a few patients as a manifestation of the acute reaction as the parasite succumbs (SEDA-17, 356; SEDA-22, 327).

Immunologic

Ivermectin eradicates the microfilariae of *Onchocerca volvulus*. The major drawback is that treatment is associated with adverse host inflammatory responses. The association of proinflammatory chemokines with the intensity of infection and clinical adverse reactions has been studied (58) by measurement of chemokine serum concentrations in patients with *Onchocerca volvulus* following a single dose of ivermectin or placebo (ivermectin 100 micrograms/kg, n = 13; ivermectin 150 micrograms/kg, n = 8; ivermectin 200 micrograms/kg, n = 24; placebo, n = 37). Adverse reactions scores in patients increased significantly on the third day after ivermectin treatment, and were unchanged in those who took placebo. The adverse reactions scores were significantly related to the microfilarial density but not the dose of ivermectin or serum concentrations of proinflammatory chemokines.

Long-Term Effects

Drug resistance

Resistance to oral ivermectin has been reported in two patients who had been treated over 50 times, which suggests that resistance can be induced by repetitive treatment (59).

In an open case-control study, an attempt was made to determine if persistence of *Onchocerca volvulus* microfilaridermia after multiple treatments with ivermectin was due to drug resistance in the parasite in 21 suboptimal responders matched by age, weight, number of treatments, locality, and skin microfilarial counts, with 7 amicrofilaridermic responders and 14 ivermectin-naïve subjects (60). The cases of persistent significant microfilaridermia despite multiple treatments were mainly attributable to non-response of the adult female worms and not to inadequate drug exposure or other factors. The possibility that some adult female worms might have developed resistance to ivermectin could not be excluded. Adverse effects of ivermectin were mild, and were most commonly body pain (n = 8), followed by headache (n = 7), arthralgia, pruritus, gland pain, or limb swelling (n = 6 each) and muscle aches (n = 5).

Second-Generation Effects

Teratogenicity

There is no evidence of second-generation injury from ivermectin, but it is prudent to avoid it in pregnant women.

Lactation

Only a small amount of ivermectin is excreted in the breast milk (61) and it has been suggested that it is unnecessary to exclude lactating mothers from mass chemotherapy with ivermectin.

Susceptibility Factors

HIV infection

The adverse effects of a single dose of ivermectin 150 micrograms/kg for onchocerciasis have been compared in 1256 Ugandan patients with and without infection with human immunodeficiency virus (HIV-1) (62). In those aged over 15 years, the frequency of adverse reactions was higher among those who were HIV-1 seropositive (53% versus 46%), but the difference was not statistically significant. However, the severity of the adverse reactions was significantly less in the HIV-1 positive patients.

HTLV co-infection

In Japan co-infection with human T lymphotropic virus type I (HTLV-1) occurs in about 38% of those who are infected with *Strongyloides stercoralis*, which may limit the efficacy of ivermectin. In a retrospective follow-up study 312 patients with strongyloidiasis were treated with a single dose of ivermectin 6 mg and a repeat dose of 6 mg 2 weeks after the first dose; 121 were HTLV-1 antibody-positive (63). Short-term antihelminthic treatment at 4 weeks was 97.9% efficacious in HTLV-1 antibody-negative patients (187 of 191 patients) and 90.1% in HTLV-1 antibody-positive patients (109 of 121 patients). Long-term antihelminthic drug efficacy fell dramatically in HTLV-1 antibody-positive patients (from 90.1% to 50%). Of 97 subsequent patients with strongyloidiasis who were treated with a higher dose of ivermectin (200 micrograms/kg on day 1 and a repeat dose after 2 weeks), 20 were HTLV-1 antibody-positive. Short-term antihelminthic efficacy was 100% in HTLV-1 antibody-positive and 97.1% in HTLV-1 antibody-negative patients. Long-term efficacy remained high (90%) in HTLV-1 antibody-positive patients. In those who were given 110 micrograms/kg dose there were subjective adverse effects in 27 patients. The most common adverse effects were dizziness (n = 7), diarrhea (n = 5), and nausea (n = 3). There were liver function test abnormalities in 19 patients. In one patient the second dose of ivermectin was withheld because of a large rise in aspartate transaminase activity. There were adverse effects in six of 93 patients who were given ivermectin 200 micrograms/kg. There were mild liver function test abnormalities in five patients. The frequency of adverse effects was not significantly different between the two treatment groups and neither was the frequency of liver function abnormalities.

References

1. Njoo FL, Hack CE, Oosting J, Stilma JS, Kijlstra A. Neutrophil activation in ivermectin-treated onchocerciasis patients. Clin Exp Immunol 1993;94(2):330–3.
2. Bounias M. Pragmatic efficacy against conceptual precaution in parasite control: the case of avermectins. J Environ Biol 2000;21:275–85.
3. Heukelbach J, Winter B, Wilcke T, Muehlen M, Albrecht S, Sales de Oliveira FA, Kerr-Pontes LRG, Liesenfeld O, Feldmeier H. Selective mass treatment with ivermectin to control intestinal helminthiasis and parasitic skin disease in a severely affected population. Bull WHO 2004;82:563-71.
4. Twum-Danso NA, Meredith SEO. Variation in incidence of serious adverse events after onchocerciasis treatment with ivermectin in areas of Cameroon co-endemic for loiasis. Trop Med Int Health 2003;8:820-31.
5. Moulia-Pelat JP, Nguyen LN, Glaziou P, Chanteau S, Gay VM, Martin PM, Cartel JL. Safety trial of single-dose treatments with a combination of ivermectin and diethylcarbamazine in bancroftian filariasis. Trop Med Parasitol 1993;44(2):79–82.
6. Gardon J, Kamgno J, Folefack G, Gardon-Wendel N, Bouchite B, Boussinesq M. Marked decrease in Loa loa

microfilaraemia six and twelve months after a single dose of ivermectin. Trans R Soc Trop Med Hyg 1997;91(5):593-4.

7. Duong TH, Kombila M, Ferrer A, Bureau P, Gaxotte P, Richard-Lenoble D. Reduced *Loa loa* microfilaria count ten to twelve months after a single dose of ivermectin. Trans R Soc Trop Med Hyg 1997;91(5):592-3.

8. Shenoy RK, George LM, John A, Suma TK, Kumaraswami V. Treatment of microfilaraemia in asymptomatic brugian filariasis: the efficacy and safety of the combination of single doses of ivermectin and diethylcarbamazine. Ann Trop Med Parasitol 1998;92(5):579-85.

9. Bussaratid V, Krudsood S, Silachamroon U, Looareesuwan S. Tolerability of ivermectin in gnathostomiasis. Southeast Asian J Trop Med Public Health 2005;36(3):644-9.

10. Boussinesq M, Gardon J. Challenges for the future: loiasis. Ann Trop Med Parasitol 1998;92(Suppl 1):S147-51.

11. Kennedy MH, Bertocchi I, Hopkins AD, Meredith SE. The effect of 5 years of annual treatment with ivermectin (Mectizan) on the prevalence and morbidity of onchocerciasis in the village of Gami in the Central African Republic. Ann Trop Med Parasitol 2002;96(3):297-307.

12. Gardon J, Boussinesq M, Kamgno J, Gardon-Wendel N, Demanga-Ngangue, Duke BO. Effects of standard and high doses of ivermectin on adult worms of *Onchocerca volvulus*: a randomised controlled trial. Lancet 2002;360(9328):203-10.

13. Oyibo WA, Fagbento-Beyioku AF. Adverse reactions following annual ivermectin treatment of onchocerciasis in Nigeria. Int J Infect Dis 2003;7:156-9.

14. Kipp W, Bamhuhiiga J, Rubaale T, Buttner DW. Adverse reactions to ivermectin treatment in *Simulium neavei*-transmitted onchocerciasis. Am J Trop Med Hyg 2003;69:621-3.

15. Olson BG, Domachowske JB. Mazzotti reaction after presumptive treatment for schistosomiasis and stronglyloidiasis in a Liberian refugee. Pediatric Infect Dis J 2006;25(5):466-8.

16. Kamgno J, Gardon J, Gardon-Wendel N, Demanga-Ngangue, Duke BOL, Boussinesq M. Adverse systemic reactions to treatment of onchocerciasis with ivermectin at normal and high doses given annually or three-monthly. Trans R Soc Trop Med Hyg 2004;98:496-504.

17. Guldbakke KK, Khachemoune A. Crusted scabies: a clinical review. J Drugs Dermatol. 2006;5(3):221-7.

18. Heukelbach J, Feldmeier H. Scabies. Lancet 2006;367:1767-74.

19. Chosidow O. Scabies. N Engl J Med 2006;354:1718-27.

20. Ohtaki N, Taniguchi H, Ohtomo H. Oral ivermectin treatment in two cases of scabies: effective in crusted scabies induced by corticosteroid but ineffective in nail scabies. J Dermatol 2003;30:411-6.

21. Molina JM, Chastang C, Goguel J, Michiels JF, Sarfati C, Desportes-Livage I, Horton J, Derouin F, Modai J. Albendazole for treatment and prophylaxis of microsporidiosis due to *Encephalitozoon intestinalis* in patients with AIDS: a randomized double-blind controlled trial. J Infect Dis 1998;177(5):1373-7.

22. Jaramillo-Ayerbe F, Berrio-Munoz J. Ivermectin for crusted Norwegian scabies induced by use of topical steroids. Arch Dermatol 1998;134(2):143-5.

23. Paasch U, Haustein UF. Management of endemic outbreaks of scabies with allethrin, permethrin, and ivermectin. Int J Dermatol 2000;39(6):463-70.

24. Zaha O, Hirata T, Kinjo F, Saito A, Fukuhara H. Efficacy of ivermectin for chronic strongyloidiasis: two single doses given 2 weeks apart. J Infect Chemother 2002;8(1):94-8.

25. Grobusch MP, Gobels K, Teichmann D, Bergmann F, Suttorp N. Early-stage elephantiasis in bancroftian filariasis. Eur J Clin Microbiol Infect Dis 2001;20(11):835-6.

26. Reuben R, Rajendran R, Sunish IP, Mani TR, Tewari SC, Hiriyan J, Gajanana A. Annual single-dose diethylcarbamazine plus ivermectin for control of bancroftian filariasis: comparative efficacy with and without vector control. Ann Trop Med Parasitol 2001;95(4):361-78.

27. Kraivichian K, Nuchprayoon S, Sitichalernchai P, Chaicumpa W, Yentakam S. Treatment of cutaneous gnathostomiasis with ivermectin. Am J Trop Med Hyg 2004;71:623-8.

28. Madan V, Jaskiran K, Gupta U, Gupta DK. Oral ivermectin in scabies patients: a comparison with 1% topical lindane lotion. J Dermatol 2001;28(9):481-4.

29. Brooks PA, Grace RF. Ivermectin is better than benzyl benzoate for childhood scabies in developing countries. J Paediatr Child Health 2002;38(4):401-4.

30. del Mar Saez-De-Ocariz M, McKinster CD, Orozco-Covarrubias L, Tamayo-Sanchez L, Ruiz-Maldonado R. Treatment of 18 children with scabies or cutaneous larva migrans using ivermectin. Clin Exp Dermatol 2002;27(4):264-7.

31. Lawrence G, Leafasia J, Sheridan J, Hills S, Wate J, Wate C, Montgomery J, Pandeya N, Purdie D. Control of scabies, skin sores and haematuria in children in the Solomon Islands: another role for ivermectin. Bull World Health Organ 2005;83(1):34-42.

32. Addiss DG, Eberhard ML, Lammie PJ, McNeeley MB, Lee SH, McNeeley DF, Spencer HC. Comparative efficacy of clearing-dose and single high-dose ivermectin and diethylcarbamazine against *Wuchereria bancrofti* microfilaremia. Am J Trop Med Hyg 1993;48(2):178-85.

33. Njoo FL, Beek WM, Keukens HJ, van Wilgenburg H, Oosting J, Stilma JS, Kijlstra A. Ivermectin detection in serum of onchocerciasis patients: relationship to adverse reactions. Am J Trop Med Hyg 1995;52(1):94-7.

34. Simonsen PE, Magesa SM, Dunyo SK, Malecela-Lazaro MN, Michael E. The effect of single dose ivermectin alone or in combination with albendazole on Wuchereria bancrofti infection in primary school children in Tanzania. Trans R Soc Trop Med Hyg 2004;98:462-72.

35. Guzzo CA, Furtek CI, Porras AG, Chen C, Tipping R, Clineschmidt CM, Sciberras DG, Hsieh JY, Lasseter KC. Safety, tolerability, and pharmacokinetics of escalating high doses of ivermectin in healthy adult subjects. J Clin Pharmacol 2002;42(10):1122-33.

36. Awadzi K, Edwards G, Duke BOL, Opoku NO, Attah SK, Addy ET, Ardrey AE, Quartey BT. The co-administration of ivermectin and albendazole—safety, pharmacokinetics and efficacy against *Onchocerca volvulus*. Ann Trop Med Parasitol 2003;97:165-78.

37. Winstanley P. Albendazole for mass treatment of asymptomatic trichuris infections. Lancet 1998;352(9134):1080-1.

38. Dunyo SK, Nkrumah FK, Simonsen PE. A randomized double-blind placebo-controlled field trial of ivermectin and albendazole alone and in combination for the treatment of lymphatic filariasis in Ghana. Trans R Soc Trop Med Hyg 2000;94(2):205-11.

39. Chippaux JP, Boussinesq M, Gardon J, Gardon-Wendel N, Ernould JC. Severe adverse reaction risks during mass treatment with ivermectin in loiasis-endemic areas. Parasitol Today 1996;12(11):448-50.

40. Chijioke CP. Factors affecting adverse event reporting during mass ivermectin treatment for onchocerciasis. Acta Trop 2000;76(2):169-73.

41. Cooper PJ, Beck LA, Espinel I, Deyampert NM, Hartnell A, Jose PJ, Paredes W, Guderian RH, Nutman TB. Eotaxin and RANTES expression by the dermal endothelium is associated with eosinophil infiltration after ivermectin treatment of onchocerciasis. Clin Immunol 2000;95(1 Pt 1):51–61.

42. Keiser PB, Reynolds SM, Awadzi K, Ottesen EA, Taylor MJ, Nutman TB. Bacterial endosymbionts of *Onchocerca volvulus* in the pathogenesis of posttreatment reactions. J Infect Dis 2002;185(6):805–11.

43. Abanobi OC, Anosike JC. Control of onchocerciasis in Nzerem-Ikpem, Nigeria: baseline prevalence and mass distribution of ivermectin. Public Health 2000;114(5):402–6.

44. De Sole G, Awadzi K, Remme J, Dadzie KY, Ba O, Giese J, Karam M, Keita FM, Opoku NO. A community trial of ivermectin in the onchocerciasis focus of Asubende, Ghana. II. Adverse reactions. Trop Med Parasitol 1989;40(3):375–82.

45. Kumaraswami V, Ottesen EA, Vijayasekaran V, Devi U, Swaminathan M, Aziz MA, Sarma GR, Prabhakar R, Tripathy SP. Ivermectin for the treatment of *Wuchereria bancrofti* filariasis. Efficacy and adverse reactions. JAMA 1988;259(21):3150–3.

46. Barkwell R, Shields S. Deaths associated with ivermectin treatment of scabies. Lancet 1997;349(9059):1144–5.

47. The Mectizan Expert Committee. Central nervous complications of loiasis and adverse CNS events following treatmentAtlanta: Mectizan Donation Program;. 1996.

48. Gardon J, Gardon-Wendel N, Demanga-Ngangue, Kamgno J, Chippaux JP, Boussinesq M. Serious reactions after mass treatment of onchocerciasis with ivermectin in an area endemic for Loa loa infection. Lancet 1997;350(9070):18–22.

49. Boussinesq M, Gardon J, Gardon-Wendel N, Kamgno J, Ngoumou P, Chippaux JP. Three probable cases of *Loa loa* encephalopathy following ivermectin treatment for onchocerciasis. Am J Trop Med Hyg 1998;58(4):461–9.

50. Fobi G, Gardon J, Santiago M, Ngangue D, Gardon-Wendel N, Boussinesq M. Ocular findings after ivermectin treatment of patients with high Loa loa microfilaremia. Ophthalmic Epidemiol 2000;7(1):27–39.

51. Fobi G, Gardon J, Kamgno J, Aimard-Favennec L, Lafleur C, Gardon-Wendel N, Duke BO, Boussinesq M. A randomized, double-blind, controlled trial of the effects of ivermectin at normal and high doses, given annually or three-monthly, against Onchocerca volvulus: ophthalmological results. Trans R Soc Trop Med Hyg 2005;99(4):279-89.

52. Homeida MM, Bagi IA, Ghalib HW, el Sheikh H, Ismail A, Yousif MA, Sulieman S, Ali HM, Bennett JL, Williams J. Prolongation of prothrombin time with ivermectin. Lancet 1988;1(8598):1346–7.

53. Whitworth JA, Hay CR, McNicholas AM, Morgan D, Maude GH, Taylor DW. Coagulation abnormalities and ivermectin. Ann Trop Med Parasitol 1992;86(3):301–5.

54. Collins RC, Gonzales-Peralta C, Castro J, Zea-Flores G, Cupp MS, Richards FO Jr, Cupp EW. Ivermectin: reduction in prevalence and infection intensity of *Onchocerca volvulus* following biannual treatments in five Guatemalan communities. Am J Trop Med Hyg 1992;47(2):156–69.

55. Chiodini PL, Reid AJ, Wiselka MJ, Firmin R, Foweraker J. Parenteral ivermectin in *Strongyloides* hyperinfection. Lancet 2000;355(9197):43–4.

56. Burchard GD, Kubica T, Tischendorf FW, Kruppa T, Brattig NW. Analysis of renal function in onchocerciasis patients before and after therapy. Am J Trop Med Hyg 1999;60(6):980–6.

57. Fischer P, Kipp W, Kabwa P, Buttner DW. Onchocerciasis and human immunodeficiency virus in western Uganda: prevalences and treatment with ivermectin. Am J Trop Med Hyg 1995;53(2):171–8.

58. Fendt J, Hamm DM, Banla M, Schulz-Key H, Wolf H, Helling-Giese G, Heuschkel C, Soboslay PT. Chemokines in onchocerciasis patients after a single dose of ivermectin. Clin Exp Immunol 2005;142:318-26.

59. Burkhart CG. Recent immunologic considerations regarding the itch and treatment of scabies. Dermatol Online J 2006;12(7):7.

60. Awadzi K, Boakye DA, Edwards G, Opoku NO, Attah SK, Osei-Atweneboana MY, Lazdins-Helds JK, Ardrey AE, Addy ET, Quartey BT, Ahmed K, Boatin BA, Soumbey-Alley EW. An investigation of persistent microfilaridermias despite multiple treatments with ivermectin, in two onchocerciasis-endemic foci in Ghana. Ann Trop Med Parasitol 2004;98:231-49.

61. Ogbuokiri JE, Ozumba BC, Okonkwo PO. Ivermectin levels in human breastmilk. Eur J Clin Pharmacol 1993;45(4):389–90.

62. Kipp W, Bamhuhiiga J, Rubaale T, Kabagambe G. Adverse reactions to the ivermectin treatment of onchocerciasis patients: does infection with the human immunodeficiency virus play a role? Ann Trop Med Parasitol 2005;99(4):395-402.

63. Zaha O, Hirata T, Uchima N, Kinjo F, Saito A. Comparison of anthelmintic effects of two doses of ivermectin on intestinal strongyloidiasis in patients negative or positive for anti-HTLV-1 antibody. J Infect Chemother 2004;10:348-51.

Levamisole

General Information

Levamisole is the levorotatory isomer of tetramisole. Originally used only as an antihelminthic drug, it acts by paralysing the musculature of susceptible nematodes so that they are expelled by peristalsis. It is rapidly metabolized and excreted, with a half-life of about 4 hours.

It is nowadays mainly used as an immunomodulating drug in adjuvant therapy for colon cancer, usually in combination with 5-fluorouracil. It is also used in other conditions, including nephrotic syndrome, and in some infections, such as pediculosis recurrent aphthous ulcerations.

Use in infective conditions

Treatment of ascariasis with a single oral dose of levamisole 2.5 mg/kg is effective, with evidence of toxicity in under 1% of patients.

Levamisole has been used experimentally in leprosy, particularly in combination with dapsone. This combination was used in a documented series of Indian patients, some currently lepromatous and others in the course of a leprosy reaction (1). When using doses sufficient to provide as good an effect as that obtained with clofazimine + dapsone in a comparison group, adverse effects were limited to gastrointestinal intolerance (which was usually mild), affecting only five of the 30 patients treated; an incidental case developed pyrexia.

Brucellosis

Adding levamisole to conventional antibiotic therapy may improve anergy against Brucella, bacteria that can survive in phagocytic cells. This hypothesis has been investigated in patients with chronic brucellosis in Turkey (2). A 6-week course of levamisole in addition to conventional antibiotic therapy in chronic brucellosis was not superior to conventional antibiotic treatment alone with respect to lymphocyte subgroup ratios and phagocytic function. Adverse effects were not reported.

Use in non-infective conditions

Levamisole has immunostimulatory activity by modulating the cell-mediated immune response and restoring T cell functions. It has therefore been used extensively and for extended periods of time in various rheumatic and other chronic diseases, in aphthous ulceration, nephrotic syndrome, warts, and malignancies, such as cancers of the head and neck and, in combination with 5-fluorouracil, colorectal cancer (SEDA-20, 348). Under these conditions, its adverse effects are more frequent and rather different because of the differing dosage scheme and presumably also the greater sensitivity of the individual, quite apart from the fact that it is often used in combination, for example with 5-fluorouracil; some 5% of patients fail to complete the course of treatment because of adverse effects. Most of the material in this record is necessarily derived from experience with long-term treatment; where possible a distinction will be drawn between adverse effects occurring under these conditions and those experienced during the acute treatment of tropical disorders.

Observational studies

Aphthous stomatitis

Levamisole has been used in the treatment of recurrent aphthous stomatitis (3) and its value reviewed (4). In four of seven placebo-controlled studies there was a reduction in the frequency and duration of aphthous ulcers during levamisole treatment. Efficacy did not differ whether levamisole was given routinely or started at the first sign of ulcers. In most patients levamisole was well tolerated. Of 128 patients who took levamisole, two withdrew as result of adverse effects (nausea and flu-like symptoms). The most frequent adverse effects were dysgeusia (21%) and nausea (16%). The other adverse effects occurred in less than 10% of the patients and included dysosmia, headaches, diarrhea, flu-like symptoms, and rash, but not all may have been due to levamisole. Levamisole rarely results in objective clinical improvement, and the associated adverse effects discourage its use.

Cancers

In 63 patients with stage III and stage IV squamous cell carcinomas of the oral cavity, oropharynx, hypopharynx, and larynx, with no distant metastases, randomized to either adjuvant oral chemotherapy with futraful, uracil, and levamisole ($n = 29$) or no treatment ($n = 34$), oral chemotherapy showed a trend of better control of distant tumor recurrence (5). However, there was no statistically significant improvement in overall long-term survival. Of the 29 patients who received adjuvant oral chemotherapy, 17 finished the 1-year course without withdrawing. In nine patients futraful, uracil, and levamisole was withdrawn because of local, regional, or distant metastases. Three patients withdrew after 4 months because of vomiting and mucositis. One developed a mild gastric upset and completed the course. There were no major hematological or nephrotoxic adverse effects.

Levamisole is used as an immunomodulating drug in colorectal cancer, usually in combination with 5-fluorouracil. The IGCS-COL multicenter randomized phase III study partly addressed the role of levamisole in the modulation of 5-fluorouracil as adjuvant systemic chemotherapy in patients with colorectal cancer (6). There was no evidence of improvement of disease-free survival or overall survival advantage by adding levamisole; nor did the addition of levamisole produce any statistically significant effect on the adverse effects profile of 5-fluorouracil.

In another study in 598 patients with stage III colon cancer the addition of levamisole to adjuvant fluorouracil significantly worsened the prognosis (7).

Candidiasis

In two patients with thymoma associated with myasthenia gravis, who both had recurrent oral candidiasis after thymectomy, radiotherapy, and chemotherapy levamisole was added as adjunctive therapy in combination with oral nystatin (8). Oral candidiasis responded favorably and substantial relief was obtained, with a concurrent increase in T cells and CD4/CD8 ratio, suggesting restoration of T cell immunity. Adverse effects were not mentioned.

Glomerulonephritis

In an extensive review of the treatment of minimal lesion glomerulonephritis the use of levamisole was briefly mentioned (9). The author concluded that levamisole has a beneficial effect in this disorder, although no new studies have appeared in recent years and well-controlled studies are scarce. Levamisole appears to be well tolerated in this condition. The adverse effects were neutropenia, rash, and raised liver transaminases.

Nephrotic syndrome

In 11 children with nephrotic syndrome, of whom five were glucocorticoid-sensitive, six glucocorticoid-resistant, and all resistant to other immunosuppressive drugs, levamisole 2.5 mg/kg was given every 48 hours for up to 18 months (10). Two patients were also given ciclosporin. All the patients in the steroid-sensitive group but none in the steroid-resistant group reacted favorably to levamisole, with disappearance of protein from the urine. There were serious adverse effects in two patients: one developed a transient leukopenia 2 months after the start of treatment and another developed a severe exacerbation of pre-existing psoriasis, although that may have been due to the withdrawal of cyclophosphamide.

Levamisole 2 mg/kg on alternate days was given to 25 glucocorticoid-dependent children with frequent relapses of idiopathic nephrotic syndrome (11). The steroid was tapered, and continued for 3–14 months. During treatment with levamisole the relapse frequency was reduced by 40%. Two patients developed mild transient leukopenia, which disappeared 2 weeks after withdrawal. One had a slight rash that disappeared while treatment was continued and one complained of epigastric pain, which led to drug withdrawal.

In a detailed review of the management of nephrotic syndrome in childhood, levamisole was advocated as a weak but effective glucocorticoid-sparing agent (12). In a prospective study in 20 children (aged 3–15 years; 16 boys, 4 girls) with steroid-dependent minimal-change nephrotic syndrome, there were no significant adverse effects during adjunctive therapy with levamisole, which led to successful withdrawal of glucocorticoids after 2 months in 11 children (13).

It has been stressed that levamisole is generally well tolerated by children with steroid-dependent nephrotic syndrome (14) (15). Adverse effects are uncommon but include neutropenia, vasculitis, liver toxicity, and convulsions; they are reversible after withdrawal of levamisole. In 40 children with idiopathic steroid-dependent minimal-change nephrotic syndrome levamisole 2.5 mg/kg on alternate days was compared with intravenous cyclophosphamide 500 mg/m2/month for 6 months (16). Prednisolone was gradually tapered. After withdrawal of treatment, five children in the levamisole and cyclophosphamide groups stayed in remission at 6 months, four versus two respectively at 1 year, three versus one at 2 years, and one in each group at 3 and 4 years of follow-up. Adverse effects were mild, and none of the patients withdrew because of adverse effects. In the 20 patients who took levamisole, infections occurred in 13 but only when glucocorticoids were used in conjunction. Nine patients had respiratory infections (acute bronchitis), two had scalp infections, and one had a urinary tract infection. One developed sialadenitis. One had personality changes characterized by aggression and nervousness after the start of glucocorticoid therapy. None developed neutropenia but there was a lower leukocyte count during levamisole treatment.

Pediculosis capitis

In 28 patients with pediculosis capitis (aged 7–12 years) levamisole was given in a dose of 3.5 mg/kg for 10 days; there were no adverse reactions (17).

Rheumatoid arthritis

The possible benefits of levamisole in rheumatoid arthritis are generally outweighed by its adverse effects (SEDA-11, 277; 18).

Comparative studies

Colorectal cancer

The efficacy of levamisole has been studied in several studies of patients with colorectal carcinoma (19–21). In a phase III trial 5-fluorouracil alone, 5-fluorouracil with levamisole, and 5-fluorouracil with hepatic irradiation have been compared in patients with residual, non-measurable, intra-abdominal metastases after resection of colorectal carcinoma (19). The adverse effects were as expected, and there were no differences between any of the treatments. The main adverse effects were hematological and gastrointestinal. However, analysis of life-threatening adverse effects showed some slight differences: there were fewer than expected in the 5-fluorouracil alone group, and more than expected in the 5-fluorouracil plus hepatic irradiation group. There was no treatment advantage for any of the combinations over 5-fluorouracil alone.

Levamisole combined with 5-fluorouracil in the adjuvant treatment of resected colon cancer has been studied in a prospective, randomized trial in which 891 patients were randomized to receive either intensive fluorouracil and leucovorin combined with levamisole, or a standard regimen of fluorouracil plus levamisole (22). The patients were then again randomized to receive either 6 or 12 months of treatment. Standard fluorouracil plus levamisole was not as effective as fluorouracil, levamisole, and leucovorin, and treatment for 12 months was not superior to treatment for 6 months. Unfortunately, there was no treatment arm with fluorouracil and leucovorin only, which is now widely considered to be the treatment of choice. Serious grade 3–4 adverse effects were more frequent in the three-drug treatment groups, and consisted of diarrhea (13 versus 3 patients in the 6-month groups, 17 versus 7 in the 12-month groups) and stomatitis (10 versus 3 in the 6-month groups, 11 versus 6 in the 12-month groups). Leukopenia occurred more frequently in the standard treatment groups (10 versus 18, one of whom died, in the 6-month groups, and 13 patients, one of whom died, versus 14, one of whom died). There were four treatment-associated deaths.

In another study combined intravenous and intraperitoneal fluorouracil plus leucovorin was compared with standard treatment with fluorouracil and levamisole in 241 patients with resected stage 3 or high-risk stage 2 colon cancers (23). In the combined treatment group there was an increased disease-free interval, an estimated 43% reduction in death rate, and a reduction in local tumor recurrence. Adverse effects were relatively uncommon and were generally judged to be mild to moderate; they were slightly more common in those treated with fluorouracil and levamisole, and consisted of nausea and vomiting (18 versus 14%), diarrhea (16 versus 10%), mucositis (17 versus 12%), granulocytopenia (29 versus 23%), and thrombocytopenia (5 versus 3%). Four cases of unspecified nervous system toxicity were noted in those given fluorouracil plus levamisole. There was abdominal pain during or shortly after intraperitoneal drug administration in 19% of patients. Overall 53% of the patients given fluorouracil plus levamisole and 56% of those given fluorouracil plus leucovorin had mild to moderate adverse effects. Severe reactions, requiring a 20% dosage reduction of fluorouracil, were more common in the fluorouracil plus levamisole arm (13 versus 3%). There were no deaths. Unfortunately, in this study no patients were treated with fluorouracil and leucovorin intravenously only.

It is likely that most of these reported adverse effects, although perhaps enhanced by levamisole, except for the nervous system toxicity noted in a few individuals, were caused by fluorouracil. This has been further emphasized by a dose-finding study to determine the maximum tolerated dose of levamisole in the treatment of colon cancer in 38 patients with advanced non-resectable colon cancer, treated with fluorouracil 450 mg/m^2 by rapid intravenous infusion for 5 days (24). Levamisole was given orally three times daily for 5 days every 5 weeks until disease progression. The main dose-limiting toxic effects were nausea and vomiting and an unpleasant metallic taste. The dose used was about five times the total amount of levamisole given in the standard fluorouracil plus levamisole regimen. Levamisole enhanced the gastrointestinal toxicity of fluorouracil, with anorexia, nausea, vomiting, and occasional diarrhea, but did not enhance the bone-marrow suppression associated with fluorouracil. Increasing the dose of levamisole to 150 mg/m^2 tds for 5 days resulted in significant nervous system toxicity, with confusion, vertigo, and severe vomiting. None of the patients treated with this dosage were able to complete the course.

Fluorouracil plus leucovorin has been compared with fluorouracil plus levamisole and combined fluorouracil plus leucovorin and levamisole in 2151 patients with Dukes B and C colon cancers (25). The regimens were as follows:

- fluorouracil plus leucovorin: six 8-week cycles of leucovorin 500 mg/m^2 as a 2-hour infusion repeated weekly for six doses and fluorouracil 500 mg/m^2, given as an intravenous bolus 1 hour after the start of the leucovorin infusion, also weekly for six doses; the cycle was repeated after a rest period of 2 weeks;
- fluorouracil plus levamisole: fluorouracil 350 mg/m^2 as an intravenous bolus daily for five consecutive days, then once weekly starting on day 29 and levamisole orally tds for 3 days and repeated every 14 days;
- fluorouracil plus leucovorin plus levamisole: the same fluorouracil plus leucovorin treatment as described above, with the addition of levamisole in the dose used in the fluorouracil plus levamisole group.

There was a small prolongation of the disease-free interval and overall survival in favor of fluorouracil plus leucovorin, although of borderline statistical significance. Information on toxicity was obtained in 98% of the patients. Eighteen died while on chemotherapy, four in the fluorouracil plus leucovorin group, three in the fluorouracil plus levamisole group, and 11 in the fluorouracil plus leucovorin and levamisole group. Grade 3–4 toxicity was reported equally in the three groups: fluorouracil plus leucovorin 35%, fluorouracil plus leucovorin and levamisole 36%, and fluorouracil plus levamisole 28%. They consisted mainly of adverse effects attributed to fluorouracil, such as diarrhea, vomiting, and stomatitis. Hematological toxicity was minimal (less than 2% in grades 3–4) and not significantly different across the groups. Neurotoxicity was rare. Ataxia was the most frequent neurological disorder, in 2% of the patients who received fluorouracil plus levamisole and in 1% of the patients in the other two groups combined.

QUASAR was a study of the effects of a higher dose of leucovorin or the addition of levamisole to 5-fluorouracil and leucovorin on survival in 4927 patients with colorectal cancer with no evidence of residual disease after resection (20). High-dose leucovorin was not associated with a survival or recurrence benefit compared with low-dose leucovorin. The addition of levamisole had no apparent survival benefit compared with placebo, with slightly more deaths in patients assigned to levamisole than placebo. Tumor recurrences were also higher in those who took levamisole. Dermatological adverse effects were significantly more frequent in those who took levamisole compared with placebo.

In 680 patients with curatively resected stage III colon cancer, adjuvant treatment with 5-fluorouracil plus leucovorin was significantly more effective than 5-fluorouracil plus levamisole in reducing tumor relapse and improving survival (21). There were fewer adverse effects in those given 5-fluorouracil plus levamisole compared with 5-fluorouracil plus leucovorin (820 versus 1190); the difference was mainly due to gastrointestinal toxicity. Only a few patients developed grade 3 or grade 4 adverse effects. There were no treatment-related deaths in either group.

The Gastrointestinal Intergroup has studied postoperative adjuvant chemotherapy and radiation therapy in 1659 patients with T3/4 and lymph-node-positive rectal cancer after potentially curative surgery to try to improve chemotherapy and to determine the risk of systemic and local failure (26). There was no advantage to regimens containing leucovorin or levamisole over bolus 5-fluorouracil alone in the adjuvant treatment of rectal cancer when combined with irradiation. Local and distant recurrence rates were still high, especially in T3 and T4 lymph node positive patients, even with full adjuvant chemoradiation therapy.

In a multicenter, phase 3, randomized comparison of fluorouracil + levamisole (n = 92) versus fluorouracil alone (n = 93) in 185 patients with stage III colon cancer the relative contribution of levamisole (50 mg tds for 3 consecutive days, repeated every 2 weeks for 1 year) was established (27). After a median follow-up time of 48 months, 80 patients had recurrent disease (40 in each arm) and there were no advantages in terms of disease-free survival and overall survival for fluorouracil + levamisole. However, leukopenia (18% versus 4.3%) and hepatic toxicity (16% versus 4.4%) were more frequent in patients receiving fluorouracil + levamisole compared with fluorouracil alone, whereas other adverse effects were equally distributed among both treatment arms. Some patients had neurological symptoms, consisting of mood-altering effects and disabling cerebellar ataxia, attributed to treatment with levamisole. They abated when therapy was withdrawn.

In a randomized trial in 218 patients with stage II–III resectable rectal cancer, adjuvant postoperative radiotherapy has been compared with sequential radiotherapy and chemotherapy with fluorouracil + levamisole (28).

Adherence to chemotherapy in patients undergoing sequential radiotherapy and chemotherapy was poor; 32% of the patients had to stop chemotherapy owing to severe toxicity, mostly gastrointestinal. The authors concluded that fluorouracil + levamisole is not effective in patients with resected rectal cancer.

Nephrotic syndrome

In a retrospective analysis in 51 children with glucocorticoid-dependent nephrotic syndrome, the ability of levamisole to reduce the relapse rate and to spare prednisone therapy was compared with that of cyclophosphamide (29). Apart from one patient who had a spontaneously resolving skin rash with levamisole and three patients who had transient neutropenia with cyclophosphamide, there were no other clinically significant adverse effects.

In a systematic review of randomized controlled trials levamisole significantly reduced the risk of relapse in nephrotic syndrome (30). Few adverse effects were reported in trials in children; however, important adverse effects include neutropenia, gastrointestinal effects, and rarely disseminated vasculitis.

Levamisole has been studied as an immunomodulator in the treatment of frequently relapsing steroid-dependent idiopathic nephrotic syndrome in a controlled study (31). Levamisole 2.5 mg/kg on alternate days for 1 year (n = 32) was compared with low-dose prednisolone <0.5 mg/kg on alternate days for 1 year (n = 24). The mean relapse rate was reduced more by levamisole as was the mean cumulative dose of steroids. Therapy failed in three levamisole-treated individuals compared with 12 controls. There were no major adverse effects of levamisole, and in particular significant leukopenia was not encountered. One patient taking levamisole reported a mild transient gastrointestinal upset.

Warts

In 44 patients with multiple recalcitrant warts randomized to either oral cimetidine 30 mg/kg/day in three doses for 12 weeks, or 30 mg/kg/day for 12 weeks plus levamisole 2.5 mg/kg for 2 days per week, cimetidine plus levamisole produced significant improvement (32). Adverse effects of levamisole were infrequent, except for a metallic taste in one patient and nausea in two. In one patient the nausea was severe enough to necessitate withdrawal. There were no significant changes in leukocyte count or differential counts.

Placebo-controlled studies

Colorectal cancer

Fluorouracil (370 mg/m^2) plus high-dose (175 mg) or low-dose (25 mg) folinic acid and either active or placebo levamisole has been evaluated in 4927 patients with colorectal cancer (33). Levamisole produced a significant excess of adverse dermatological events. Serious unexpected adverse events were rare. The authors concluded that the inclusion of levamisole in chemotherapy regimens for colorectal cancer does not delay recurrence or improve survival.

Onchocerciasis

In two randomized, double-blind, placebo-controlled trials, in which levamisole 2.5 mg/kg was given alone or with ivermectin 200 micrograms/kg or albendazole 400 mg in 42 healthy male volunteers and in 66 patients with onchocerciasis, the frequencies of adverse events were unrelated to the treatment regimen (34). The most common drug-related events were general weakness and headache (n = 9 each), followed by itching (n = 6), rash (n = 5), and joint pain (n = 5). The adverse events were considered mild. The most common drug-related events in the 66 patients with onchocerciasis were itching (n = 36) and rash (n = 31). Other adverse events were headache (n = 20), arthralgia (n = 11), and other body pains (n = 11). None of the adverse events was serious. They occurred more often in patients who took levamisole and ivermectin.

Systematic reviews

Melanoma

In a systematic review of the role of systemic adjuvant therapy in patients with high-risk resected primary melanoma the authors reported that based on four randomized controlled trials in which levamisole was studied (three placebo-controlled) there was no effect on survival when levamisole was used as an adjuvant (35). Although morbidity from levamisole is generally mild, it was severe enough for therapy to be withdrawn in 16–44% of the patients who took part in the various trials. Hematological abnormalities were rare and there were no treatment-related deaths.

Organs and Systems

Cardiovascular

Hypotension has been attributed to levamisole (36).

Nervous system

Several reports point to a condition that can best be described as multifocal inflammatory leukoencephalopathy, occurring in about four cases in every thousand, during treatment with levamisole either alone (37) or in combination with 5-fluorouracil (SEDA-21, 321; 38,39). It presents with confusion, ataxia, dysarthria, diplopia, focal neurological signs, and seizures, and has been extensively discussed and reviewed (40).

In a retrospective study in nine patients (three men, six women) aged 24–70 years with recurrent aphthous ulcers, the imaging and clinical findings of leukoencephalopathy induced by oral levamisole 700–2250 mg were described (41). Intervals from the initial dose to the onset of symptoms were 37–144 (mean 76) days. All patients recovered to their baseline neurological status within 12 months after withdrawal of levamisole. In murine models, levamisole is not directly toxic to myelin. However, when mice are infected with a demyelination-inducing virus, demyelination and inflammation are augmented and accelerated. Based on these observations, the authors speculated that levamisole may induce leukoencephalopathy in humans by stimulating a destructive immune response to a novel antigen that persists but does not cause symptoms, resulting in demyelination in susceptible individuals.

One large literature survey (42) concluded that 6% of patients taking long-term levamisole, mostly for malignancies, experienced "sensory stimulation," for example in the form of "hyperalert states" or insomnia, although depression is less commonly on record. Diplopia and tremor have been observed, whilst a number of children treated for juvenile rheumatoid arthritis or nephrotic syndrome have developed generalized convulsions and coma, with EEG abnormalities suggestive of encephalitis (SED-12, 771; 43); the condition recovers spontaneously, but for a time anticonvulsants may be needed.

- A 65-year-old man developed impaired cognition and a disturbed gait 6 months after the removal of a Dukes C colon cancer with adjuvant chemotherapy with fluorouracil and levamisole. Three months later he developed arthralgias in the hands, elbows, and knees, followed 1 month later by intermittent monocular diplopia, an ataxic gait, and deteriorating cognitive function. An MRI scan of the brain showed multiple small round and oval hyperintense lesions in the periventricular white matter, without surrounding edema or mass effect. Most of the lesions showed ring enhancement after intravenous gadolinium. The cerebrospinal fluid protein content was raised, but cytology was normal. A stereotactic biopsy of one of the lesions showed marked cellularity of the white matter, with mononuclear cells in both the parenchyma and perivascular areas, and severe demyelination of the white matter with relative preservation of the axons. Most of the parenchymal cells were macrophages, containing myelin debris. All the findings were consistent with a diagnosis of multifocal inflammatory leukoencephalopathy after treatment with levamisole. The adjuvant therapy was discontinued and he was treated with methylprednisolone 1 g/day intravenously for 3 days, followed by dexamethasone 4 mg qds orally. Within weeks he started to improve, with resolution of the ataxia and improved cognitive function. However, there was residual mild left hemiparesis and he remained moderately unsteady on his feet. The glucocorticoids were slowly tapered and an MRI scan several weeks later still showed multiple patchy areas of bright signal, corresponding to the previous demyelination, but without enhancement. A further MRI scan after 1 year showed complete resolution. His neurological condition and performance had further improved, but there was residual impairment of short-term memory.

The authors suggested that this syndrome may be more common than supposed, and that it is likely that levamisole is the main causal factor, since the same symptoms have been described after treatment with high-dose levamisole alone, but that the effect may be enhanced by the co-administration of fluorouracil, which potentiates the immunostimulatory action of levamisole. Although there is no conclusive evidence of the value of glucocorticoids

in this syndrome, an adequate course of glucocorticoids is advised, comparable to the treatment of multiple sclerosis.

- A 57-year-old man had a Dukes C colon cancer removed and was given adjuvant chemotherapy with fluorouracil and levamisole at 4-week intervals (44). Four months later he developed insomnia, diplopia, and a reduced level of consciousness. He was disoriented in time and place, but there were no focal neurological defects. The cerebrospinal fluid protein concentration was slightly raised. A provisional diagnosis of fluorouracil/levamisole-induced neural toxicity was made and he was treated with dexamethasone 4 mg qds. There was an improvement in orientation after 3 days of treatment, but he remained mentally dull. After 28 days of therapy the corticosteroids were tapered. An MRI scan showed multiple focal hyperintensities, with involvement of both deep and subcortical white matter. The lesions were not associated with a mass effect and were mildly enhanced by gadolinium. At 30 months follow-up the patient was well without evidence of relapse. An MRI scan at 24 months showed small, residual, hyperintense lesions, which were not further enhanced by gadolinium, consistent with gliotic scars.
- A 3-year-old girl who took levamisole 100 mg bd for 3 days for anorexia suddenly developed hyperkinesia, paresthesia, fidgetiness, an unstable gait, and frequent falling (45). An MRI scan of the brain did not show any demyelinating changes. She recovered fully within 48 hours without any specific intervention.

The authors suggested that levamisole overdose had contributed to the profound encephalitis-like effects.

Fatal viral encephalitis due to enterovirus type 71 has been reported during use of levamisole (46).

Psychological, psychiatric

Anxiety and depression have been associated with levamisole (47).

Psychosis has also been reported (48).

- A 28-year-old man, without a psychiatric history, developed a paranoid psychosis. He had been taking levamisole twice a week in an unspecified dose for 2 years for a stage 4 melanoma and metastatic lymph nodes in the axilla. Physical examination, a CT scan, an electroencephalogram, and standard laboratory tests were all normal. He was treated with perphenazine, with partial success, but after tapering of the dose his symptoms reappeared. It was thought likely that the psychosis had been caused by levamisole, which was discontinued. Three weeks later he had recovered completely. Levamisole was not reintroduced.

Hematologic

Agranulocytosis has often been reported during long-term treatment with levamisole. Most cases are reversible and transient, but deaths have occurred. In 3900 patients on whom data were available to the manufacturers, there were 88 cases of agranulocytosis and 43 of leukopenia; such dyscrasias occurred in 4% of patients with rheumatoid arthritis and 2% of oncological cases (42). In other published material the incidence has sometimes been higher; such differences do not appear to be related to the dose or duration of therapy. Agranulocytosis is more prevalent in rheumatoid patients with an HLA-B27 genotype (SEDA-7, 317). Children are also susceptible, and fatal outcomes because of hematological disorders have been described in cases of juvenile rheumatoid arthritis treated intermittently with levamisole. Other deaths have occurred in adults concurrently taking glucocorticoids for several years.

Agranulocytosis can be asymptomatic and, since it occurs unpredictably, regular monitoring of the leukocyte count is advisable, especially in patients concurrently receiving combination chemotherapy. Whilst the mechanism is not clear, granulocyte-agglutinating antibodies have been found, suggesting that levamisole acts as a hapten on the leukocyte membrane.

Thrombocytopenia has been reported in a woman with rheumatoid arthritis; it recurred after rechallenge (49).

The effects of daily levamisole 2–3 mg/kg have been assessed in 36 children with steroid-sensitive nephrotic syndrome with frequent relapses and/or steroid dependency (50). There was transient leukopenia in nine. White blood cell counts returned to normal 1–2 weeks after withdrawal of levamisole in seven cases. Two patients had leukopenia for more than 4 weeks. Leukopenia did not recur after re-institution of levamisole treatment once the white blood cell count had returned to normal.

Gastrointestinal

Nausea, vomiting, and diarrhea are very common in patients taking levamisole, and are sometimes accompanied by abdominal pain or constipation, although when such symptoms occur it is not always clear that they are due to the drug rather than to its interaction with the disease.

Exacerbation of peptic ulceration has been described, and mouth ulcers and abnormalities of taste sensation can be troublesome in patients taking long-term therapy.

Liver

Neither animal nor most human studies point to hepatotoxicity, but in a series of 11 patients with pyoderma treated with levamisole two had increases in aspartate transaminase activity. Liver toxicity has also been reported in a child with nephrosis (51).

Urinary tract

There has been one published case of uremia (52) and one of a reversible nephropathy in a patient with rheumatoid arthritis (53).

Skin

Type I allergic reactions have caused pruritic rashes and urticaria.

Ischemic necrosis of the skin, reversible on withdrawal, has been documented (54).

Single cases of erythema multiforme and erythema nodosum have been observed (SEDA-7, 317).

Two patients developed lichenoid skin eruptions, which subsided when the drug was stopped, although one of these was left with severe scarring, alopecia of the scalp, and widespread atrophic and hyperpigmented skin lesions.

A healed varicose ulcer has been observed to break down after treatment (55).

Fever and rash have been attributed to levamisole (56).

- A 33-year-old man with vitiligo was given oral betamethasone 5 mg/day and levamisole 150 mg/day on 2 consecutive days every week. After 3 months, topical fluocinolone acetonide cream (0.01%) was added. After 8 months, on one occasion, 12 hours after taking the oral drugs, he developed a fever (38.9°C) with chills and rigor, followed by itching and redness of the skin over the palms, soles, and both legs. The rash resolved in 8 days. He restarted betamethasone and levamisole after 1 month and developed similar symptoms within 4–5 hours. Rechallenge with oral levamisole 150 mg caused a fever (38.9°C) after 5 hours, followed by itching, redness and swelling of the lips, palms and soles.

Necrotizing vasculitis

Cutaneous necrotizing vasculitis with histological changes resembling a type III hypersensitivity reaction has been described in patients taking levamisole (57,58). In a well-documented case, a widespread vasculitic rash, chiefly affecting the limbs, appeared in a woman with rheumatoid arthritis treated for 2 months. In these cases, serum complement was normal and there were no circulating immune complexes, although a histamine skin wheal test produced a vasculitis at a clinically non-affected site. Both cases were reversible.

- Leukocytoclastic vasculitis has been attributed to levamisole in a 7-year-old boy with glucocorticoid-dependent nephrotic syndrome (59).

The authors estimated that about 0.5% of patients treated with levamisole develop cutaneous vasculitis with circulating autoantibodies.

Five of 160 children with nephrotic syndrome developed distinctive vascular purpura (60). They had taken levamisole for a mean of 24 months when they developed purpuric erythematous macules, which evolved to ecchymotic and necrotic purpura. The lesions were mostly on the external ear. Biopsies obtained from the ear lesions in four patients showed vasculopathic reaction patterns, ranging from leukocytoclastic and thrombotic vasculitis to vascular occlusive disease without true vasculitis. There were anticardiolipin, antinuclear, and/or antineutrophil cytoplasmic antibodies in four patients. The lesions resolved within 2–3 weeks after levamisole withdrawal, whereas anticardiolipin and antineutrophil cytoplasmic antibodies disappeared after 2–14 months only. A direct effect of levamisole on the endothelial cells or levamisole-induced or unmasked latent immunological abnormalities was suspected.

Musculoskeletal

Arthritis has occurred in patients with Crohn's disease or Behçet's disease treated with levamisole (61,62), although it is well known that this can occur with either disease irrespective of drug treatment.

Muscle pain can be severe when levamisole is given with 5-fluorouracil for colonic cancer (63), or there can be a painless rise in creatine kinase activity (64).

Immunologic

Allergic reactions often reflect the effects of the drug or of parasitic breakdown products. They include pruritic skin eruptions, arthritic pain and swelling, muscular pain and swelling, especially in patients already suffering from rheumatoid arthritis, Sjögren's syndrome (65), or psoriatic arthropathy. Skin reactions of various types can occur and type III reactions have been noted. Influenza-like symptoms might be an unusual form of type I allergy or a consequence of restoration of cellular immunity.

Disseminated autoimmune disease has been described during treatment with levamisole for nephrotic syndrome (66).

- An 8 year old boy had a 5-year history of steroid-dependent nephrotic syndrome. After half a year of glucocorticoid treatment he was given levamisole 2.5 mg/kg on alternate days for 1 year, with complete suppression of the proteinuria. The proteinuria reappeared after withdrawal of levamisole, and glucocorticoids and levamisole were reintroduced as before. Two years later, while still taking levamisole, he developed hepatosplenomegaly, a low-grade fever, and a Coombs' negative hemolytic anemia. The anticardiolipin IgM titer was high, p-ANCA antibodies were positive, C3 was moderately low and antinuclear anti-DNA antibodies were negative. Levamisole was withdrawn. Two weeks later the clinical parameters had normalized and 4 weeks later the liver enlargement had disappeared. Although p-ANCA antibodies persisted at 1 month the anticardiolipin IgM titer had returned to normal. After 6 months there still was splenomegaly but no other symptoms, and proteinuria was absent.

The same group of authors have also described a distinctive vasculitis with circulating antibodies in children with nephrotic syndrome associated with long-term levamisole, presenting with purpura of the ears (60).

- Four boys and one girl (mean age 10 years) had had nephrotic syndrome for 2–8 years and had taken levamisole orally in doses of 1.7–2.5 mg/kg/day for 16–44 months (mean 24 months). All had a sudden onset of rapidly enlarging purpuric and erythematous macules progressing to the formation of necrotic areas, purpuric plaques, and hemorrhagic bullae. The pinnae were involved in all five. In three there were also lesions on the cheek or lower limbs. One had fever and another complained of arthralgia. Routine laboratory tests were all normal. Antibodies to extractable nuclear antigen, cryoglobulin, rheumatoid factor, Coombs' test, circulating immune complexes, and complement and components were all negative or normal. However, antiphospholipid antibodies and/or anticardiolipin IgG and IgM were positive in three patients, p-ANCA was positive in three, and c-ANCA in one. Antinuclear antibodies were positive in two patients and anti-double stranded DNA antibodies were positive in one. Biopsies of the skin lesions in four patients showed vasculitis, ranging from a leukocytoclastic and thrombotic vasculitis to vascular occlusive disease without true vasculitis. Two patients had a hypersensitivity vasculitis in the superficial and deep dermis, with neutrophilic infiltration of the vessel walls and fibrinoid necrosis. Features of panniculitis with occlusion of deep and superficial blood vessels by fibrin-platelet thrombi were found in one patient. The lesions completely resolved in all patients within 2–3 weeks after the withdrawal of levamisole. The serum autoantibodies had disappeared in all cases after 2–14 months.

Although leucocytoclastic immune-complex vasculitis induced by levamisole is well known, this specific presentation with involvement of the ears has not been described before.

Second-Generation Effects

Teratogenicity

Animal studies do not point to a teratogenic effect, but there are insufficient human data to assess the safety of levamisole in pregnancy, and the WHO recommends delaying treatment until after pregnancy when possible.

In an analysis of a large population-based data set in the Hungarian Case-Control Surveillance of Congenital Abnormalities, 1980–96, there was no evidence of a higher rate of congenital abnormalities in children born to mothers who had taken oral levamisole during pregnancy (67).

Susceptibility Factors

Age

The adverse effects that occur after monotherapy with levamisole in children with nephrotic syndrome include taste disturbance (dysgeusia), arthralgia, myalgia, anxiety, sleep disturbances, depression, neutropenia, diarrhea, nausea, and vomiting.

Hepatic disease

Levamisole is mostly metabolized to *para*-hydroxylevamisole (68). It should not be used in severe hepatic disease, nor should it be combined with hepatotoxic antihelminthic drugs or other drugs presenting risks to the liver.

Other features of the patient

Both Sjögren's syndrome (65) and psoriatic arthropathy (69) are conditions in which levamisole is probably better avoided because of the risk of hypersensitivity reactions.

In the treatment of the hyperimmunoglobulin E recurrent infection syndrome (Job's syndrome), infectious complications are more serious when levamisole is given, even where there is normal chemotactic responsiveness (70).

The risk of severe adverse reactions is greater in lymphatic filariasis (71).

Drug Administration

Drug overdose

Acute overdose of levamisole has been described (72).

- A 43-year-old man treated himself with a levamisole enema of 10 g (33 times the therapeutic dose) for a gastrointestinal worm infestation. Soon after he developed malaise, tachycardia, nausea, vertigo, and profuse diarrhea. He lost consciousness and developed generalized seizures and a respiratory arrest. He was intubated, ventilated, and treated with clonazepam. His condition improved after 4 hours. He remained somnolent for 6 hours, and was nauseated and vomited for 24 hours. There was hypokalemia after the diarrhea, raised creatine kinase activity, and a leukocytosis. An electrocardiogram showed ST depression. By the fourth day all his symptoms had subsided.

The symptoms in this case were attributed to the cholinergic effect that levamisole has at this high dose.

Drug-Drug Interactions

Albendazole

In a pharmacokinetic study, the only clinically relevant drug-drug interactions were levamisole-attributable increases in the AUC and C_{max} of ivermectin, and reductions in the AUC and C_{max} of the active metabolite of albendazole, albendazole sulfoxide (73).

Salicylates

Levamisole increases serum concentrations of salicylates, but the effect does not appear to have any clinical consequences (74).

Warfarin

Levamisole and 5-fluorouracil interact with warfarin to produce bleeding (75). The mechanism of this effect is unclear, but the reports are sufficiently clear to point to the need for careful monitoring when this drug combination is used.

References

1. Sharma L, Thalliath GH, Girgia HS, Sen PC. A comparative evaluation of levamisole in leprosy. Indian J Lepr 1985;57(1):11–6.
2. Dizer U, Hayat L, Beker CM, Gorenek L, Özgüven V, Pahsa A. The effect of the doxycycline–rifampicin and levamisole combination on lymphocyte subgroups and functions of phagocytic cells in patients with chronic brucellosis. Chemotherapy 2005;51:27-31.
3. Porter SR, Hegarty A, Kaliakatsou F, Hodgson TA, Scully C. Recurrent aphthous stomatitis. Clin Dermatol 2000;18(5):569–78.
4. Barrons RW. Treatment strategies for recurrent oral aphthous ulcers. Am J Health Syst Pharm 2001;58(1):41–53.
5. Lam P, Yuen AP, Ho CM, Ho WK, Wei WI. Prospective randomized study of post-operative chemotherapy with levamisole and UFT for head and neck carcinoma. Eur J Surg Oncol 2001;27(8):750–3.
6. De Placido S, Lopez M, Carlomagno C, Paoletti G, Palazzo S, Manzione L, Iannace C, Ianniello GP, De Vita F, Ficorella C, Farris A, Pistillucci G, Gemini M, Cortesi E, Adamo V, Gebbia N, Palmeri S, Gallo C, Perrone F, Persico G, Bianco AR. Modulation of 5-fluorouracil as adjuvant systemic chemotherapy in colorectal cancer: the IGCS-COL multicentre, randomised, phase III study. Br J Cancer 2005;93:896-904.
7. Schippinger W, Jagoditsch M, Sorre C, Gnant M, Steger G, Hausmaninger H, Mlineritsch B, Schaberl-Moser R, Mischinger HJ, Hofbauer F, Holzberger P, Mittlbock M, Jakesz R; Austrian Breast and Colorectal Cancer Study Group. A prospective randomised trial to study the role of levamisole and interferon alfa in an adjuvant therapy with 5-FU for stage III colon cancer. Br J Cancer 2005;92(9):1655-62.
8. Lai WH, Lu SY, Eng HL. Levamisole aids in treatment of refractory oral candidiasis in two patients with thymoma associated with myasthenia gravis: report of two cases. Chang Gung Med J 2002;25(9):606–11.
9. Bargman JM. Management of minimal lesion glomerulonephritis: evidence-based recommendations. Kidney Int Suppl 1999;70:S3–S16.
10. Tenbrock K, Muller-Berghaus J, Fuchshuber A, Michalk D, Querfeld U. Levamisole treatment in steroid-sensitive and steroid-resistant nephrotic syndrome. Pediatr Nephrol 1998;12(6):459–62.
11. Kemper MJ, Amon O, Timmermann K, Altrogge H, Muller-Wiefel DE. Die Behandlung des häufig Rezidivierenden steroidsensiblen idiopathischen nephrotischen Syndroms im Kindersalter mit Levamisol. [The treatment with levamisole of frequently recurring steroid-sensitive idiopathic nephrotic syndrome in children.] Dtsch Med Wochenschr 1998;123(9):239–43.
12. Holt RCL, Webb NJA. Management of nephrotic syndrome in childhood. Curr Paediatr 2002;12:551–60.
13. Donia AF, Amer GM, Ahmed HA, Gazareen SH, Moustafa FE, Shoeib AA, Ismail AM, Khamis S, Sobh MA. Levamisole: adjunctive therapy in steroid dependent minimal change nephrotic children. Pediatr Nephrol 2002;17(5):355–8.
14. Davin JC, Merkus MP. Levamisole in steroid-sensitive nephrotic syndrome of childhood: the lost paradise? Pediatr Nephrol 2005;20:10-14.
15. Hodson EM, Craig JC, Willis NS. Evidence-based management of steroid-sensitive nephrotic syndrome. Pediatr Nephrol 2005;20:1523-30.

16. Donia AF, Ammar HM, El-Agroudy AE, Moustafa FE, Sobh MA. Long-term results of two unconventional agents in steroid-dependent nephrotic children. Pediatr Nephrol 2005;20:1420-5.

17. Namazi MR. Levamisole: a safe and economical weapon against pediculosis. Int J Dermatol 2001;40(4):292–4.

18. Pinals RS, Robertson F, Blechman WJ. A double-blind comparison of high and low doses of levamisole in rheumatoid arthritis. J Rheumatol 1981;8(6):949–51.

19. Witte RS, Cnaan A, Mansour EG, Barylak E, Harris JE, Schutt AJ. Comparison of 5-fluorouracil alone, 5-fluorouracil with levamisole, and 5-fluorouracil with hepatic irradiation in the treatment of patients with residual, nonmeasurable, intra-abdominal metastasis after undergoing resection for colorectal carcinoma. Cancer 2001;91(5):1020–8.

20. Kerr DJ. A United Kingdom Coordinating Committee on Cancer Research study of adjuvant chemotherapy for colorectal cancer: preliminary results. Semin Oncol 2001;28(1 Suppl 1):31–4.

21. Porschen R, Bermann A, Loffler T, Haack G, Rettig K, Anger Y, Strohmeyer GArbeitsgemeinschaft Gastrointestinale Onkologie. Fluorouracil plus leucovorin as effective adjuvant chemotherapy in curatively resected stage III colon cancer: results of the trial adjCCA-01. J Clin Oncol 2001;19(6):1787–94.

22. O'Connell MJ, Laurie JA, Kahn M, Fitzgibbons RJ Jr, Erlichman C, Shepherd L, Moertel CG, Kocha WI, Pazdur R, Wieand HS, Rubin J, Vukov AM, Donohue JH, Krook JE, Figueredo A. Prospectively randomized trial of postoperative adjuvant chemotherapy in patients with high-risk colon cancer. J Clin Oncol 1998;16(1):295–300.

23. Scheithauer W, Kornek GV, Marczell A, Karner J, Salem G, Greiner R, Burger D, Stoger F, Ritschel J, Kovats E, Vischer HM, Schneeweiss B, Depisch D. Combined intravenous and intraperitoneal chemotherapy with fluorouracil + leucovorin vs fluorouracil + levamisole for adjuvant therapy of resected colon carcinoma Br J Cancer 1998;77(8):1349–54.

24. Reid JM, Kovach JS, O'Connell MJ, Bagniewski PG, Moertel CG. Clinical and pharmacokinetic studies of high-dose levamisole in combination with 5-fluorouracil in patients with advanced cancer. Cancer Chemother Pharmacol 1998;41(6):477–84.

25. Wolmark N, Rockette H, Mamounas E, Jones J, Wieand S, Wickerham DL, Bear HD, Atkins JN, Dimitrov NV, Glass AG, Fisher ER, Fisher B. Clinical trial to assess the relative efficacy of fluorouracil and leucovorin, fluorouracil and levamisole, and fluorouracil, leucovorin, and levamisole in patients with Dukes' B and C carcinoma of the colon: results from National Surgical Adjuvant Breast and Bowel Project C-04. J Clin Oncol 1999;17(11):3553–9.

26. Tepper JE, O'Connell M, Niedzwiecki D, Hollis DR, Benson AB 3rd, Cummings B, Gunderson LL, Macdonald JS, Martenson JA, Mayer RJ. Adjuvant therapy in rectal cancer: analysis of stage, sex, and local control—final report of intergroup 0114. J Clin Oncol 2002;20(7):1744–50.

27. Cascinu S, Catalano V, Piga A, Mattioli R, Marcellini M, Pancotti A, Bascioni R, Torresi U, Silva RR, Pieroni V, Giorgi , Catalano G, Cellerino R. The role of levamisole in the adjuvant treatment of stage III colon cancer patients: a randomized trial of 5-fluorouracil and levamisole versus 5-fluorouracil alone. Cancer Invest 2003;21:701-7.

28. Cafiero F, Gipponi M, Lionetto R and the PAR Cooperative Study Group. Randomised clinical trial of adjuvant postoperative RT vs. sequential postoperative RT plus 5-FU and levamisole in patients with stage II?III resectable rectal cancer: a final report. J Surg Oncol 2003;83:140-6.

29. Alsaran K, Grisaru S, Stephens D, Arbus G. Levamisole vs. cyclophosphamide for frequently-relapsing steroid-dependent nephrotic syndrome. Clin Nephrol 2001;56(4):289–94.

30. Hodson EM. The management of idiopathic nephrotic syndrome in children. Pediatr Drugs 2003;5:335-49.

31. Al-Saran K, Mirza K, Al-Ghanam G, Abdelkarim M. Experience with levamisole in frequently relapsing, steroid-dependent nephrotic syndrome. Pediatr Nephrol 2006;21:201-5.

32. Parsad D, Pandhi R, Juneja A, Negi KS. Cimetidine and levamisole versus cimetidine alone for recalcitrant warts in children. Pediatr Dermatol 2001;18(4):349–52.

33. Pak CY. Correction of thiazide-induced hypomagnesemia by potassium-magnesium citrate from review of prior trials. Clin Nephrol 2000;54(4):271–5.

34. Awadzi K, Edwards G, Opoku NO, Ardrey AE, Favager S, Addy ET, Attah SK, Yamuah LK, Quartey BT. The safety, tolerability and pharmacokinetics of levamisole alone, levamisole plus ivermectin, and levamisole plus albendazole, and their efficacy against Onchocerca volvulus. Ann Trop Med Parasitol 2004;98:595-614.

35. Verma S, Quirt I, McCready D, Bak K, Charette M, Iscoe N, on behalf of the Melanoma Disease Site Group of Cancer Care Ontario's Program in Evidence-Based Care. Systematic review of systemic adjuvant therapy for patients at high risk for recurrent melanoma. Cancer 2006;106:1431-42.

36. Holcombe RF, Li A, Stewart RM. Levamisole and interleukin-2 for advanced malignancy. Biotherapy 1998;11(4):255–8.

37. Lucia P, Pocek M, Passacantando A, Sebastiani MI, De Martinis C. Multifocal leucoencephalopathy induced by levamisole. Lancet 1996;348(9039):1450.

38. Savarese DM, Gordon J, Smith TW, Litofsky NS, Licho R, Ragland R, Recht L. Cerebral demyelination syndrome in a patient treated with 5-fluorouracil and levamisole. The use of thallium SPECT imaging to assist in noninvasive diagnosis—a case report. Cancer 1996;77(2):387–94.

39. Vaughn DJ, Haller DG. The role of adjuvant chemotherapy in the treatment of colorectal cancer. Hematol Oncol Clin North Am 1997;11(4):699–719.

40. Recht LD, Primavera JM. Case records of the Massachusetts General Hospital. Weekly clinicopathological exercises. Case 24–1999. Neurologic disorder in a 65-year-old man after treatment of colon cancer. N Engl J Med 1999;341(7):512–9.

41. Liu HM, Hsieh WJ, Yang CC, Wu VC, Wu KD. Leukoencephalopathy induced by levamisole alone for the treatment of recurrent aphthous ulcers. Neurology 2006;67:1065-7.

42. Symoens J, Veys E, Mielants M, Pinals R. Adverse reactions to levamisole. Cancer Treat Rep 1978;62(11):1721–30.

43. Palcoux JB, Niaudet P, Goumy P. Side effects of levamisole in children with nephrosis. Pediatr Nephrol 1994;8(2):263–4.

44. Yeo W, Tong MM, Chan YL. Multifocal cerebral demyelination secondary to fluorouracil and levamisole therapy. J Clin Oncol 1999;17(1):431–3.

45. Dubey AK, Gupta RK, Sharma RK. Levamisole induced ataxia. Indian Pediatr 2001;38(4):417–9.

46. Mabin D, Castel Y, Le Fur JM, Alix D, Chastel C, Le Roy JP. Encéphalite aiguë virale mortelle au cours d'un traitement par le lévamisole. [Acute viral encephalitis with

fatal issue during treatment by levamisole.] Nouv Presse Med 1978;7(45):4143.

47. Hsu WH. Toxicity and drug interactions of levamisole. J Am Vet Med Assoc 1980;176(10 Spec No):1166–9.

48. Jeffries JJ, Cammisuli S. Psychosis secondary to long-term levamisole therapy. Ann Pharmacother 1998;32(1):134–5.

49. Parkinson DR, Cano PO, Jerry LM, Capek A, Shibata HR, Mansell PW, Lewis MG, Marquis G. Complications of cancer immunotherapy with levamisole. Lancet 1977;1(8022):1129–32.

50. Fu LS, Shien CY, Chi CS. Levamisole in steroid-sensitive nephrotic syndrome children with frequent relapses and/or steroid dependency: comparison of daily and every-other-day usage. Nephron Clin Pract 2004;97:c137-41.

51. Bulugahapitiya DT. Liver toxicity in a nephrotic patient treated with levamisole. Arch Dis Child 1997;76(3):289.

52. Lesquesne M, Floquet J. Les effets secondaires au cours des traitements prolongés par le lévamisole notamment dans le polyarthrites. [Side effects during prolonged treatment with levamisole, especially in polyarthritis.] Nouv Presse Med 1976;5(6):358–9.

53. Hansen TM, Petersen J, Halberg P, Permin H, Ullman S, Brun C, Larsen S. Levamisole-induced nephropathy. Lancet 1978;2(8092 Pt 1):737.

54. Menni S, Pistritto G, Gianotti R, Ghio L, Edefonti A. Ear lobe bilateral necrosis by levamisole-induced occlusive vasculitis in a pediatric patient. Pediatr Dermatol 1997;14(6):477–9.

55. El-Ghobarey AF, Mavrikakis M, Morgan I, Mathieu JP. Delayed healing of varicose ulcer with levamisole. BMJ 1977;1(6061):616.

56. Gupta R, Gupta S. Drug rash due to levamisole. Indian J Dermatol Venereol Leprol 2005;71(6):428-9.

57. Scheinberg MA, Bezerra JB, Almeida FA, Silveira LA. Cutaneous necrotising vasculitis induced by levamisole. BMJ 1978;1(6110):408.

58. Laux-End R, Inaebnit D, Gerber HA, Bianchetti MG. Vasculitis associated with levamisole and circulating auto-antibodies. Arch Dis Child 1996;75(4):355–6.

59. Bagga A, Hari P. Levamisole-induced vasculitis. Pediatr Nephrol 2000;14(10–11):1057–8.

60. Rongioletti F, Ghio L, Ginevri F, Bleidl D, Rinaldi S, Edefonti A, Gambini C, Rizzoni G, Rebora A. Purpura of the ears: a distinctive vasculopathy with circulating autoantibodies complicating long-term treatment with levamisole in children. Br J Dermatol 1999;140(5):948–51.

61. Segal AW, Pugh SF, Levi AJ, Loewi G. Levamisole-induced arthritis in Crohn's disease. BMJ 1977;2(6086):555.

62. Siklos P. Levamisole-induced arthritis. BMJ 1977;2(6089):773.

63. Buecher B, Blanc JF, Magnien F, Bechade D, Lapprand M, Oddes B. Des myalgies sévères: un effet indésirable inhabituel du lévamisole associé au 5-fluorouracile. [Severe myalgias: an unusual undesirable effect of levamisole combined with 5-fluorouracil.] Gastroenterol Clin Biol 1996;20(4):407–8.

64. Cersosimo RJ, Lee JM. Creatine kinase elevation associated with 5-fluorouracil and levamisole therapy for carcinoma of the colon. A case report. Cancer 1996;77(7):1250–3.

65. Balint G, el-Ghobary A, Capell H, Madkour M, Dick WC, Ferguson MM, Anwar-ul-haq M. Sjögren's syndrome: a contraindication to levamisole treatment? BMJ 1977;2(6099):1386–7.

66. Barbano G, Ginevri F, Ghiggeri GM, Gusmano R. Disseminated autoimmune disease during levamisole treatment of nephrotic syndrome. Pediatr Nephrol 1999;13(7):602–3.

67. Kazy Z, Pucho E, Czeizel E. Levamisol lehetseges terato-genitasanak vizsgalata terhessegben. Orv Hetil 2005;146(49):2499-500.

68. Kouassi E, Caille G, Lery L, Lariviere L, Vezina M. Novel assay and pharmacokinetics of levamisole and p-hydroxyle-vamisole in human plasma and urine. Biopharm Drug Dispos 1986;7(1):71–89.

69. Trabert U, Rosenthal M, Muller W. Therapie entzundlich-rheumatischer Krankheiten mit Levamisol, einer immunmo-dulierenden Substanz. [Therapy of inflammatory-rheumatic diseases with levamisol, an immunity modulating substance.] Schweiz Med Wochenschr 1976;106(39):1293–301.

70. Swim AT, Bradac C, Craddock PR. Levamisole in Job's syndrome. N Engl J Med 1982;307(24):1528–9.

71. Merlin M, Carme B, Kaeuffer H, Laigret J. Activité du lévamisole (Solaskil) dans la filariose lymphatique a Wuchereria bancrofti (varieté pacifica). [Activity of levami-sole (Solaskil) in lymphatic filariasis caused by *Wuchereria bancrofti* (variety *pacifica*).] Bull Soc Pathol Exot Filiales 1976;69(3):257–65.

72. Joly C, Palisse M, Ribbe D, De Calmes O, Genevey P. Intoxication aiguë au lévamisole. [Acute levamisole poisoning.] Presse Med 1998;27(15):717.

73. Awadzi K, Edwards G, Opoku NO, Ardrey AE, Favager S, Addy ET, Attah SK, Yamuah LK, Quartey BT. The safety, tolerability and pharmacokinetics of levamisole alone, levamisole plus ivermectin, and levamisole plus albendazole, and their efficacy against Onchocerca volvulus. Ann Trop Med Parasitol 2004;98(6):595-614.

74. Rumble RH, Brooks PM, Roberts MS. Interaction between levamisole and aspirin in man. Br J Clin Pharmacol 1979;7(6):631–3.

75. Wehbe TW, Warth JA. A case of bleeding requiring hospitalization that was likely caused by an interaction between warfarin and levamisole. Clin Pharmacol Ther 1996;59(3):360–2.

Lumefantrine

General Information

Lumefantrine is a synthetic aminoalcohol fluorene derivative, related to halofantrine and mefloquine (1). It was highly effective in uncomplicated chloroquine-resistant malaria tropica in an open, non-comparative trial in 102 patients in China when given in four oral doses over 48 hours (2). No significant adverse effects have been reported. It has also been marketed in a combination of artemether (20 mg) plus lumefantrine (120 mg).

In 60 children in the Gambia with uncomplicated malaria tropica, lumefantrine was safe and effective. No neurological, cardiac, or other adverse effects were reported (3).

Lumefantrine was inferior to artesunate + mefloquine in an open, randomized comparison in 617 patients in Thailand with uncomplicated multidrug-resistant malaria tropica, but produced two to four times fewer adverse effects, such as nausea, vomiting, dizziness, sleep disorders, or other neurological symptoms (4).

In another study in Thailand, lumefantrine was less efficacious than mefloquine in a double-blinded, randomized trial in 252 adults (5). The lower efficacy of lumefantrine in Thailand may be due to poorer absorption and more resistant parasites in these areas, and the manufacturers have recommended higher doses than were used in these studies (6).

In an open, randomized trial in 260 Tanzanian children, lumefantrine was superior to chloroquine and did not produce major adverse effects (7).

References

1. Ridley RG, Hudson AT. Chemotherapy of malaria. Curr Opin Infect Dis 1998;11:691–705.
2. Fadat G, Louis FJ, Louis JP, Le Bras J. Efficacy of micronized halofantrine in semi-immune patients with acute uncomplicated falciparum malaria in Cameroon. Antimicrob Agents Chemother 1993;37(9):1955–7.
3. Bouchaud O, Basco LK, Gillotin C, Gimenez F, Ramiliarisoa O, Genissel B, Bouvet E, Farinotti R, Le Bras J, Coulaud JP. Clinical efficacy and pharmacokinetics of micronized halofantrine for the treatment of acute uncomplicated falciparum malaria in nonimmune patients. Am J Trop Med Hyg 1994;51(2):204–13.
4. Wildling E, Jenne L, Graninger W, Bienzle U, Kremsner PG. High dose chloroquine versus micronized halofantrine in chloroquine-resistant *Plasmodium falciparum* malaria. J Antimicrob Chemother 1994;33(4):871–5.
5. Toivonen L, Viitasalo M, Siikamaki H, Raatikka M, Pohjola-Sintonen S. Provocation of ventricular tachycardia by antimalarial drug halofantrine in congenital long QT syndrome. Clin Cardiol 1994;17(7):403–4.
6. Fourcade L, Gachot B, De Pina JJ, Heno P, Laurent G, Touze JE. Choc anaphylactique associé au traitement du paludisme par halofantrine. [Anaphylactic shock related to the treatment of malaria with halofantrine.] Presse Méd 1997;26(12):559.
7. Di Perri G, Di Perri IG, Monteiro GB, Bonora S, Hennig C, Cassatella M, Micciolo R, Vento S, Dusi S, Bassetti D, et al. Pentoxifylline as a supportive agent in the treatment of cerebral malaria in children. J Infect Dis 1995;171(5):1317–22.

Mebendazole and flubendazole

See also Benzimidazoles

General Information

Mebendazole, a benzimidazole, is poorly absorbed from the gut, although it dependably enters cyst fluid; it is therefore most useful for treating intestinal infections and cyst-forming infestations. It is essentially an antihelminthic drug, being effective against hookworm, ascariasis, enterobiasis, and trichuriasis. Mebendazole is effective against enteric *Strongyloides* but since it is not absorbed it is ineffective against tissue forms (1). However, it is also effective against *Giardia lamblia*, while *Trichomonas vaginalis* is susceptible in vitro.

Mebendazole does not interfere with the normal intestinal flora. Mebendazole binds to nematode β tubulin and inhibits parasite microtubule polymerization, which causes death of adult worms after a few days.

Mebendazole has been assessed in a range of doses and durations of treatment. The most usual dose is 100 mg bd for 3 days; absorption is minimal, but there is considerable variation in plasma concentrations; the half-life is 2–9 hours. Much higher doses, up to 60 mg/kg/day, have been used in inoperable cases of cystic echinococcosis infestation, and then unwanted effects are more common.

Flubendazole is an analogue of mebendazole used in intestinal helminthiasis and hydatid disease. In trials of two-dose oral treatment for intestinal helminthiasis, reactions were mild and uncommon. They consisted of nausea, abdominal pain, dyspepsia, and sleepiness (SEDA-6, 281). Subsequent field experience has not suggested that flubendazole differs appreciably from other members of the class as regards adverse effects.

Comparative studies

Echinococcosis

The use of albendazole and mebendazole in patients with hydatidosis has been evaluated in 448 patients with *Echinococcus granulosis* hydatid cysts who received continuous treatment with albendazole 10–12 mg/kg/day for 3–6 months daily orally in a total dose of (323 patients) twice or mebendazole 50 mg/kg/day (2). At the end of treatment, 82% of the cysts treated with albendazole and 56% of the cysts treated with mebendazole showed degenerative changes. During long-term follow-up 25% of these cysts showed relapse, which took place within 2 years in 78% of cases. Further treatment with albendazole induced degenerative changes in over 90% of the relapsed cysts, without induction of more frequent or more severe adverse effects, as observed during the first treatment period. Adverse effects during the first treatment period consisted of raised transaminases with albendazole (67 of 323 patients) and mebendazole (16 of 125 patients), and abdominal pain in 12 and 11% respectively. With both drugs, occasional patients experienced headache, abdominal distension, vertigo, urticaria, jaundice, thrombocytopenia, fever, or dyspepsia, but most of these are known manifestations of echinococcus infection. Six of 323 patients taking albendazole withdrew because of adverse effects compared with eight of 125 patients taking mebendazole. It appears that albendazole is more effective than mebendazole in the treatment of hydatid cysts caused by *E. granulosis* and that both the intensity and the frequency of the usually mild adverse effects are comparable.

In 78 patients with hydatid disease there was a low recurrence rate of hydatid disease (below 3%) after a postoperative prophylactic course of mebendazole 20 mg/kg/day in three divided doses for 3 months (3). The only adverse effect of mebendazole was excessive loss of hair in two women. The unusual low recurrence rate of hydatid disease after treatment with mebendazole in this study was subsequently questioned and attributed to meticulously careful surgical procedures, with avoidance of spillage of hydatid fluid and complete removal of parasitic components (4).

There are few data on the treatment of pulmonary hydatid disease in children. Mebendazole and albendazole have been evaluated in 82 children with a total of 102 pulmonary hydatid cysts (5). Mebendazole was given as 50 mg/kg/day in three divided doses and albendazole was given as 10 mg/kg/day in two divided doses continuously or in cycles consisting of 4 weeks of treatment alternating with 2-week drug-free intervals. The duration of treatment was 1–36 months. While taking benzimidazoles eight patients had raised liver enzymes, three had rash, and one had neutropenia; all were reversible on withdrawal.

General adverse effects

With normal doses (100 mg bd for 3 days) very slight headache, dizziness, and nausea or diarrhea are common; in principle allergy can occur. Mild and reversible rises in transaminases can occur and need to be followed, but even in high doses withdrawal is justified in only a few patients (6). Neutropenia has been noted and can be severe and persistent. High doses (up to 50–60 mg/kg) can also cause alopecia and cough. Mebendazole has been associated with extra-intestinal migration of *Ascaris* in heavily infected patients. Apart from the poor therapeutic response obtained in some 25% of cases, drug toxicity (especially at high prolonged dosage) has led to withdrawal in a small proportion of patients. Adverse effects severe enough to lead to withdrawal have included worsening of pre-existing hyperlipidemia (type IV), progressive uremia, and a marked rise in liver enzymes (7). One individual developed a rash accompanied by a striking rise in serum transaminases, which recurred on subsequent re-exposure (8). Experience in the treatment of *Echinococcus multilocularis* infection is similar. One patient with fatal agranulocytosis also had severe, probably unrelated, liver disease (SEDA-8, 292). Some 3–4% develop fever, which can be persistent and accompanied by respiratory symptoms and eosinophilia. Other adverse effects include pain over the site of the cyst, allergic reactions, alopecia, glomerulonephritis, and rashes.

Various authors have reported spontaneous rupture of hydatid cysts with mebendazole, and this is probably more frequent than in untreated individuals. Pleural and peritoneal cysts are more likely to rupture.

Evidence of teratogenicity in rats has not been accompanied by reports that it causes harm in human pregnancy, but the WHO recommends avoidance during the first trimester. It is not known if mebendazole enters the breast milk; no adverse effects have been reported but the issue has not been specifically studied.

Organs and Systems

Liver

All benzimidazoles can cause mild and reversible rises in transaminases, but even in high doses withdrawal is justified in only a few patients (6).

Granulomatous hepatitis with eosinophilia has been attributed with mebendazole (9).

- A 52-year-old man with ascariasis took two 3-day cycles of mebendazole 100 mg bd with a 2-week interval. Within 48 hours of the second course he developed fever (39°C), diarrhea, anorexia, and prostration. Ten days later he had tender hepatomegaly. His liver function tests were abnormal (aspartate transaminase 466 IU/l, alanine transaminase 458 IU/l). The serum alkaline phosphatase and bilirubin were normal and the gamma-glutamyl transferase mildly raised. The white blood cell count was 12.7×10^9/l with 18% eosinophils. Coagulation was normal. Tests for Hepatitis A, B, and C, cytomegalovirus, and Epstein–Barr virus were all negative. Serum ACE was not raised. Antimitochondrial antibodies were negative but antinuclear antibodies (1:60) and antibodies against smooth muscle (1:160) were positive. Extensive tests to exclude other causes of granulomatous hepatitis were all negative. A liver biopsy showed multiple granulomata consisting of epithelioid cells, multinucleated giant cells, plasma cells, and lymphocytes. There was slight fibrosis around the granulomata. There was no evidence of cholestasis. No helminthic ova were found. Ziehl-Nielsen and periodic acid Schiff stains were both negative. After 2 days the fever had subsided without treatment and he felt better. The serum transaminases returned to normal over the next 10 weeks and the eosinophilia disappeared.

Liver damage has been described after treatment with most benzimidazoles, but it is usually cholestatic. The liver damage described in this case was granulomatous. Liver damage after mebendazole in the low dose used in this case is rare, probably because of its poor absorption. It is more frequent, although still rare, in the higher doses used in the treatment of human echinococcosis.

In 76 children *Trichinella britovi* infection was benign and milder than in adults who had consumed the same amount of infected meat (10). The children were treated with mebendazole 25 mg/kg divided into three doses for 14 days. Those with severe symptoms were also treated with oral prednisolone 20 mg/day for 7 days. No child reported adverse effects attributable to mebendazole. Liver enzymes rose in a 14-year-old child after 10 days of treatment with mebendazole and returned to normal 7 days after withdrawal.

Urinary tract

Glomerulonephritis has been observed in five patients from Kenya (11).

Skin

Two cases of exfoliative dermatitis occurred in a total of 131 patients treated (11). In one Indian case, a fixed drug eruption was attributed to the drug (12) and other forms of rash have been seen.

An outbreak of Stevens–Johnson syndrome has been reported in 52 Filipino overseas contract workers (aged 20–30 years, 50 women) working in China who used mebendazole for helminthic prophylaxis (13). All took mebendazole at least once after the appearance of rashes and fever. Three women eventually died, primarily due to septicemia.

Second-Generation Effects

Pregnancy

Women in hookworm-endemic areas may benefit from deworming during pregnancy by reducing hookworm-attributable anemia. Whether the use of albendazole and mebendazole is associated with adverse birth outcomes has been studied in small observational studies only. In Iquitos, Peru, a large double-blind, randomized, placebo-controlled trial was conducted to study the occurrence of adverse birth outcomes in 1042 pregnant women who were randomized to either a single 500 mg dose of mebendazole (n = 522) or placebo (n = 520) together with a 30-day supply of ferrous sulfate (60 mg elemental iron) (14). There were no statistically significant differences between the mebendazole and placebo group in numbers of miscarriages, malformations, stillbirths, early neonatal deaths, or premature deliveries. These data suggest that deworming with mebendazole in hookworm-endemic areas can be safely carried out during pregnancy.

Teratogenicity

The effects of mebendazole during pregnancy have been investigated in a case-control study in the mothers of babies born with congenital abnormalities and in matched control mothers of babies born without congenital abnormalities in the population-based data set of the Hungarian Case-Control Surveillance of Congenital Abnormalities between 1980 and 1996 (15). Of 38 151 women whose neonates had no defects, 14 had taken mebendazole during pregnancy; of 22 843 women whose neonates had congenital abnormalities, 14 had taken mebendazole for intestinal parasites during pregnancy (OR = 1.67; 95% CI = 0.7, 4.2). In six groups of different congenital abnormalities there was no higher prevalence of mebendazole use by the mothers. Mean gestational age was longer and mean birth weight higher in neonates born to mothers who had taken mebendazole. Thus, treatment with mebendazole during pregnancy was not significantly teratogenic or fetotoxic, although the numbers of treated cases and controls in this study were limited, which may have reduced the statistical power of this case-control study.

Susceptibility Factors

Age

Children aged under 2 years who are infected with helminths are currently excluded from treatment with mebendazole and other antihelminthic drugs on the basis of the manufacturer's instructions. In a double-blind, randomized trial in Tanzania 212 children aged under 2 years were given a total of 653 antihelminthic treatments (317 mebendazole 500 mg; 336 placebo) (16). There were no significant differences in adverse events in the two groups. In the light of the potential nutritional benefit achieved by regular deworming in this age group, the policy that excludes children aged under 2 years from treatment should probably be reconsidered.

Drug-Drug Interactions

Protease inhibitors

- A 40-year-old man with HIV infection, taking HAART, developed alveolar echinococcosis (17). He was given albendazole 400 mg bd. Within 2 weeks he developed pancytopenia. Zidovudine, lamivudine, nelfinavir, and pyrimethamine/sulfamethoxazole were withdrawn. Full recovery of the bone marrow occurred within 10 weeks. HAART was changed to stavudine, abacavir, and lopinavir/ritonavir. Instead of albendazole, he was given mebendazole. Since a pharmacokinetic interaction between albendazole and the HIV protease inhibitors was suspected to have contributed to the development of pancytopenia, mebendazole plasma concentrations were carefully monitored. Therapeutic mebendazole plasma concentrations were reached after 6 weeks at a dose of 150 mg bd instead of the usual dosage of 500–1500 mg bd.

These data suggest that there may be a significant pharmacokinetic interaction between benzimidazoles and CYP3A4 inhibitors like protease inhibitors. This may result in significantly increased serum benzimidazole concentrations, when benzimidazole doses are not adjusted under concomitant treatment with HIV protease inhibitors.

References

1. Boken DJ, Leoni PA, Preheim LC. Treatment of *Strongyloides stercoralis* hyperinfection syndrome with thiabendazole administered per rectum. Clin Infect Dis 1993;16(1):123–6.
2. Franchi C, Di Vico B, Teggi A. Long-term evaluation of patients with hydatidosis treated with benzimidazole carbamates. Clin Infect Dis 1999;29(2):304–9.
3. Ammari FF, Omari AK. Surgery and postoperative mebendazole in the treatment of hydatid disease. Saudi Med J 2002;23(5):568–71.
4. Meshikhes AW. Surgery and postoperative mebendazole in the treatment of hydatid disease. Saudi Med J 2002;23(11):1425.
5. Dogru D, Kiper N, Ozcelik U, Yalcin E, Gocmen. Medical treatment of pulmonary hydatid disease: for which child? Parasitol Int 2005;54:135-8.
6. Bartoloni C, Tricerri A, Guidi L, Gambassi G. The efficacy of chemotherapy with mebendazole in human cystic echinococcosis: long-term follow-up of 52 patients. Ann Trop Med Parasitol 1992;86(3):249–56.
7. Gil-Grande LA, Boixeda D, Garcia-Hoz F, Barcena R, Lledo A, Suarez E, Pascasio JM, Moreira V. Treatment of liver hydatid disease with mebendazole: a prospective study of thirteen cases. Am J Gastroenterol 1983;78(9):584–8.

8. Seitz R, Schwerk W, Arnold R. Hepatocelluläre Arzneimittelreaktion unter Mebendazoltherapie bei *Echinococcus cystis*. [Hepatocellular drug reaction caused by mebendazole therapy in cystic echinococcosis.] Z Gastroenterol 1983;21(7):324–9.

9. Colle I, Naegels S, Hoorens A, Hautekeete M. Granulomatous hepatitis due to mebendazole. J Clin Gastroenterol 1999;28(1):44–5.

10. Ozdemir D, Ozkan H, Akkoc N, Onen F, Gurler O, Sari I, Akar S, Birlik M, Kargi A, Ozer E, Pozio E. Acute trichinellosis in children compared with adults. Pediatr Infect Dis J 2005;24:897-900.

11. Kung'u A. Glomerulonephritis following chemotherapy of hydatid disease with mebendazole. East Afr Med J 1982;59(6):404–9.

12. Nair LV, Devi U. Mebendazole induced fixed drug eruption. Indian J Dermatol Venereol Leprol 1991;57:191.

13. Ajonuma LC, Chika LC. Outbreak of Stevens–Johnson syndrome among Filipino overseas contract workers using mebendazole for helminthiasis prophylaxis. Trop Doct 2000;30(1):57.

14. Gyorkos TW, Larocque R, Casapia M, Gotuzzo E. Lack of risk of adverse birth outcomes after deworming in pregnant women. Pediatr Infect Dis J 2006;25:791-4.

15. Acs N, Banhidy F, Puho E, Czeizel AE. Population-based case-control study of mebendazole in pregnant women for birth outcomes. Congenital Anomalies 2005;45:85-8.

16. Montresor A, Stoltzfus RJ, Albonico M, Tielsch JM, Rice AL, Chwaya HM, Savioli L. Is the exclusion of children under 24 months from anthelmintic treatment justifiable? Trans R Soc Trop Med Hyg 2002;96(2):197–9.

17. Zingg W, Renner-Schneiter EC, Pauli-Magnus C, Renner EL, van Overbeck J, Schlapfer E, Weber M, Weber R, Opravil M, Gottstein B, Speck RF, and the Swiss HIV Cohort Study. Alveolar echinococcosis of the liver in an adult with human immunodeficiency virus type-1 infection. Infection 2004;32:299-302.

Mefloquine

General Information

Mefloquine, a fluorinated derivative of 4-quinoline methanol, is a product of the US Army's antimalarial research program. It is active against chloroquine-resistant *Plasmodium falciparum*, and has an excellent schizonticidal effect in the blood in experimentally induced *Plasmodium vivax* infections in volunteers. It is not gametocidal. *P. vivax* infections can persist after successful treatment of the falciparum infection with other drugs; the fact that mefloquine is effective against both organisms is thus of practical importance (SEDA-13, 808).

Mefloquine is readily absorbed after oral administration; absorption is influenced by the formulation and is more rapid from an aqueous solution. Maximum serum concentrations occur after 1–4 hours. Absorption is reduced by diarrhea. The half-life varies considerably and has been variously reported to be 7–23 days, 15–30 days, and 8–18 days (SEDA-13, 808; SEDA-16, 308; 1, 2).

Plasma protein binding is high. Sick patients have a prolonged t_{max}.

Mefloquine has a high cure rate after a single dose of 750–1000 mg. The use of combinations of mefloquine with other antimalarial drugs has been advocated in order to reduce the development of resistance. Mefloquine is effective as a prophylactic. Using a weekly dosage schedule, a dose of 250 mg is appropriate in adults. Because early reports also suggested that the drug was without serious adverse effects, mefloquine became widely advocated by various advisory bodies for prophylactic use, starting a week before travel to an endemic area where chloroquine resistance is common, and continuing for 4 weeks after departure from the area (SEDA-16, 306). Regrettably, but as might have been expected, it has also been used for prophylaxis in areas where its use is unnecessary (SEDA-13, 808; 3–5).

Mefloquine remains useful in the treatment of uncomplicated malaria in areas of chloroquine resistance, but recommendations for mefloquine as prophylaxis in travellers are under constant review. In visitors to the Kruger National Park (South Africa), adverse effects were reported in 325 (25%) of 1300 subjects taking mefloquine; gastrointestinal and neuropsychiatric effects were dominant (6). Four subjects required hospital attention for particularly severe neuropsychiatric reactions and 53 changed from mefloquine prophylaxis because of adverse effects. However in the same study chloroquine + proguanil prophylaxis led to reported adverse effects in 720 (29%) of 2488 subjects: one had a convulsion and 69 altered their prophylaxis because of adverse events or the dosing schedule. In this population mefloquine was as well tolerated as chloroquine + proguanil prophylaxis in general terms. This is in contrast to previous studies, in which the use of mefloquine led to higher rates of intolerance and severe adverse effects.

Observational studies

Malaria prevention measures taken by 5626 returning North American and European travellers departing from Kenyan airports have been examined in a cross-sectional questionnaire study. Mefloquine (74%) and chloroquine and proguanil (15%) were the most common drugs used (7). There were adverse events in 20% of the travellers who took mefloquine and 16% of those who took chloroquine and proguanil. Neuropsychological adverse events were reported by 7.8% of those taking mefloquine and 1.9% of those taking chloroquine. Despite adverse events, adherence was better in the mefloquine group (95 versus 81%; OR = 0.25), which may have been due to a lower dosing frequency.

Comparative studies

Atovaquone + proguanil (250/100 mg/day; $n = 493$) and mefloquine (250 mg/week; $n = 483$) have been compared in non-immune subjects attending travel clinics in North America, Europe, and South Africa in a randomized, double-blind study (8). Adverse events were reported by an equivalent proportion of subjects who had taken either drug (71 versus 67%; difference 4.1%, 95% CI = 1.7, 9.9). Those who took atovaquone + proguanil had significantly fewer treatment-related neuropsychiatric adverse events (14 versus 29%), fewer adverse events of moderate or severe intensity (10 versus 19%), and fewer adverse events that caused prophylaxis to be withdrawn (1.2 versus 5%), compared with those who took mefloquine. Adherence was better in the atovaquone + proguanil group, which may have been due to the shorter duration of post-travel dosing (1 week versus 4 weeks for mefloquine). There were no confirmed cases of malaria.

Mefloquine (125 or 250 mg/week; n = 56) has been compared with proguanil (100 or 200 mg/day; n = 57) for short-term (6 months) malaria chemoprophylaxis in Nigerians with sickle cell anemia in a non-blinded, randomized study (9). Efficacy was similar (89% for mefloquine and 82% for proguanil). Adverse events were reported by 32% of those who took proguanil and 20% of those who took mefloquine. Surprisingly, only 3.6% of the mefloquine group reported neuropsychiatric adverse events.

Single-dose pyrimethamine + sulfadoxine (25 mg/kg; n = 54) has been compared with mefloquine (15 mg/kg; n = 48) in the treatment of uncomplicated *P. falciparum* malaria in an unblinded, randomized study in 102 Malawi children (10). Immediate vomiting was more common in those who took mefloquine (eight cases) than in those who took pyrimethamine + sulfadoxine (one case), with comparable parasite failure rates at 14 days (20 and 22% respectively).

Mefloquine alone has been compared with a combination of sulfalene + pyrimethamine (Metakelfin) plus quinine in 187 patients with uncomplicated malaria, randomized to either mefloquine 25 mg/kg (n=93) or sulphalene + pyrimethamine plus quinine (sulfalene 1.25 mg/kg + pyrimethamine 25 mg/kg once on the first day, quinine 30 mg/kg/day in three doses; n = 94) (11). There was no significant difference between the cure rates in the two groups during the early follow-up period and there were no cases of recrudescence in the 135 subjects who completed the extended follow-up. Similarly, there was no difference in the parasite clearance time between the two groups, but patients who were given mefloquine had a shorter mean fever resolution time (36 versus 44 hours) and a shorter mean hospital stay (3.9 versus 4.6 days). Overall, the proportions of reported adverse effects was the same in the two groups, but patients treated with mefloquine had more central nervous system effects (29 versus 9.6%), including sleep disturbances (27 versus 9.6%).

General adverse effects

Although mefloquine is generally well tolerated, particularly when used prophylactically, the list of adverse effects has grown with accumulated experience. With therapeutic doses (for example 750–1500 mg in adults; 20 mg/kg in children) adverse effects are usually mild, but with occasional severe neuropsychiatric derangement. The overall incidence of adverse effects is about the same as with chloroquine, about 40–50%. Events most commonly reported include nausea, diarrhea, abdominal pain, dizziness, strange dreams, and insomnia. Adverse effects are dose-related, with an increase in dizziness and gastrointestinal complaints and fatigue at higher doses. Extensive acute, subacute, and chronic studies of mefloquine in animals have shown that it is not phototoxic, like some of the quinolone-methanols studied, nor was it mutagenic, teratogenic, or carcinogenic in these studies (SEDA-13, 808).

The adverse effects of mefloquine have been extensively reviewed both for prophylaxis (when rare neuropsychiatric adverse effects make its use controversial) and in treatment doses, when it has been linked to an increased incidence of the postmalaria neurological syndrome. A retrospective review of 5120 Italian soldiers showed an overall chemoprophylaxis curtailment rate of less than 1%, which was not significantly different from the combination of chloroquine and proguanil (12). A semi-systematic review also suggested no significant difference in tolerability compared with other antimalarial drugs (13).

The frequency and spectrum of adverse events associated with mefloquine (750 and 500 mg 6 hours apart) has been assessed in 22 healthy volunteers who were monitored for 21 days after drug administration (14). More women than men reported severe adverse reactions. The most commonly reported adverse effects were vertigo (96%), nausea (82%), and headache (73%). The vertigo was severe (grade 3) in 73% and required bed rest and specific medication for 1–4 days. In most cases (17/22) the symptoms resolved within 3 weeks after drug administration. Biochemical and hematological measures stayed within the reference ranges, but there were significant rises in serum sodium, chloride, calcium, bilirubin, gammaglutamyl transpeptidase, and lactate dehydrogenase.

Organs and Systems

Cardiovascular

Sinus bradycardia was seen in 18% of patients taking mefloquine (SEDA-12, 693) (15), occurring some 4–7 days after administration; the bradycardia was asymptomatic and lasted about 3–4 days. Transient sinus arrhythmia was also reported, without a need for treatment (SEDA-12, 808). Asymptomatic dysrhythmias were also recorded in a dosage comparison trial (SEDA-16, 308).

Respiratory

Interstitial pneumonia following mefloquine prophylaxis has been described (16).

- A 60-year-old Caucasian woman took mefloquine prophylaxis 250 mg/week 3 weeks before a trip to Kenya. One day after the first dose, she developed a high fever with chills and was given empirical unspecified antibiotic therapy. Four days later her condition worsened and she developed a severe fever, progressive dyspnea, a productive cough, myalgia, and a headache. She had no history of pulmonary or allergic conditions and she had never taken malaria prophylaxis before. She was also taking low-dose aspirin, bisoprolol and ciprofibrate. Her temperature was 38.5°C and there were bilateral crackles in the lungs. She had a leukocytosis, a raised C- reactive protein concentration, and a raised lactate dehydrogenase activity. Blood and sputum cultures yielded no growth. Chest radiography showed bilateral interstitial infiltrates, consistent with diffuse interstitial pneumonia. Without additional medication she gradually improved and was discharged 20 days after admission. The trip to East Africa was cancelled. Four months later again she took mefloquine prophylaxis before travelling to Kenya and developed a similar illness, with high-grade fever and severe respiratory distress. The results of laboratory investigations and chest radiography were similar to those found in the first episode. A high-resolution CT scan confirmed the diagnosis of diffuse pulmonary infiltration. She was given glucocorticoids and made an uneventful recovery.

Mefloquine-induced respiratory pneumonia is rare.

Nervous system

Neurological and psychiatric reactions occur to such an extent, even during prophylactic use, that a general recommendation for the use of mefloquine as prophylaxis has been called into question (SEDA-20, 258), although disabling symptoms occur in under 1% of travellers (SEDA-21, 296). Headaches, dizziness, vertigo, and light-headedness are common (SEDA-12, 693; SEDA-16, 307; 15, 17), the incidence varying between 20 and 90%. Dizziness is to some extent dose-related (18,19). Tinnitus and vertigo are less frequent.

A so-called postmalaria neurological syndrome (convulsions, tremor, confusion) has been described in about 4–5% of patients treated for severe malaria tropica with mefloquine (SEDA-21, 296).

Two cases of polyneuropathy following mefloquine treatment have been documented (20).

- A 50-year-old man developed an intermittent fever of 3 weeks' duration associated with chills and rigors. A blood smear showed Plasmodium vivax malaria, for which he was treated with oral chloroquine for 4-5 days. He became apyrexial after 2 days, and 9 days later, while still fever-free, he was given mefloquine 1000 mg on day 1 followed by 500mg on the next day. Within 24 hours he developed tingling numbness and a burning sensation in both legs, followed by progressive weakness. Over the next 2 days these symptoms spread to both arms. He was unable to stand or perform fine motor activities, such as writing and eating. The next day an erythematous rash appeared on both legs, especially on the shins and the dorsa of the feet. There was no muscle atrophy, but power was reduced around the knees and ankles and all deep reflexes were absent. Conduction studies showed increased latency, diminished amplitudes, and decreased velocity, all suggestive of a sensory motor neuropathy in all four limbs. Within 3 weeks, he gradually improved and began walking; the

rash began to disappear and the numbness and tingling sensation abated. He recovered after 6 months.
- A 40-year-old housewife from Bihar State, India, with a fever and impaired consciousness, was given parenteral artemisinin for 3 days for Plasmodium vivax malaria. The fever subsided and her consciousness improved within 48 hours. A week later she was given mefloquine 1000 mg. Within 24 hours she developed a severe dermatitis and two large ulcers on the legs, one of which became infected. This was followed by paresthesia, progressive weakness, and inability to perform fine motor activities. Vital signs were normal and higher functions were intact. However, there was diminished tone and significantly reduced reflexes. Power was reduced and she could walk only with assistance. Nerve conduction studies confirmed a sensory motor polyneuropathy. She improved gradually with vitamins and a course of co-trimoxazole for the septic ulcer. She recovered within 3 months.

All reports have suggested that the neuropsychiatric reactions to mefloquine are transient. They may be precipitated by alcohol (21).

Penetration into the brain of the (+) enantiomer of mefloquine is much higher than that of the (–) enantiomer (22) whilst plasma concentrations are greater for the (–) enantiomer, potentially providing a way of minimizing neurological adverse effects, which are often significantly overstated (23), by using the less toxic enantiomer. However, chiral separation technology is not sufficiently well developed to be economically realistic for clinical practice in the foreseeable future.

Sensory systems

Three cases of previously healthy patients who developed persisting high-frequency sensorineural hearing loss and tinnitus whilst taking mefloquine have been reported (24).

Psychological, psychiatric

At first thought to occur only after therapeutic doses of mefloquine, it is now clear that neuropsychiatric reactions occur after prophylactic use as well. The incidence is estimated at about one in 13 000 with prophylactic use, but as high as one in 215 with therapeutic use (SEDA-17, 329). Combination with other drugs that affect the nervous system can result in unpredictable reactions. The symptoms vary in type and severity: non-cooperation, disorientation, mental confusion, hallucinations, agitation, and impaired consciousness. An acute psychiatric syndrome with attempted suicide was reported in one case. A single dose can be all that is needed to evoke a mental reaction. Convulsions have been reported, with or without psychiatric symptoms; it seems that mefloquine can aggravate and perhaps even provoke latent epilepsy (SEDA-13, 809; SEDA-16, 307; SEDA-17, 329; 25).

- A severe psychiatric and neurological syndrome, with agitation, progressive delirium, and generalized rigors, was seen in a 47-year-old man after he had taken mefloquine 1500 mg over 24 hours (26).

- A 7-year-old Indian boy was diagnosed as having "cerebral malaria" and received quinine followed by mefloquine (dose not given) (27). He developed hallucinations and removed his clothes and danced. His symptoms resolved within 24 hours of stopping mefloquine. This case highlights the fact that mefloquine should not be given after quinine in cases of severe malaria.
- A 42-year-old man with no previous psychiatric history suddenly developed visual symptoms after the third dose (total dose 750 mg) of prophylactic mefloquine (28). The symptoms consisted of an impression of focusing on two different planes and of perceiving his surroundings as very far from him. They were associated with slurred speech and altered comprehension. They occurred daily, lasting up to an hour, for 6 months. He had previously taken a course of mefloquine for 7 weeks without any adverse events.
- A 52-year-old woman with no psychiatric history developed anxiety, paranoia, visual hallucinations, confusion, and depressive symptoms after 3 doses of prophylactic mefloquine (250 mg/week) (29). She had previously taken mefloquine prophylaxis intermittently for 4 years with no adverse events.
- After 4 weeks of malaria prophylaxis with mefloquine 250 mg/week, a 25-year-old woman developed bizarre paranoid delusions with auditory and visual hallucinations (30). MRI and MRA scans of the brain were unremarkable, but electroencephalography showed diffuse cerebral dysfunction. A malaria smear was negative. Mefloquine was withdrawn and the psychotic symptoms resolved over 6 days with temporary risperidone treatment. The symptoms did not recur during a 2-year follow-up period.

These case reports illustrate important neuropsychiatric adverse effects of mefloquine in individuals who had previously taken mefloquine safely and had no psychiatric history.

A postal survey of 5446 returning Danish travellers examined the adverse effects of unstated doses of mefloquine, chloroquine, and chloroquine plus proguanil for malaria prophylaxis (31). There were 4158 responses (76%); 1223 travellers took chloroquine, 1827 took chloroquine plus proguanil, and 809 took mefloquine. Overall, although chloroquine and chloroquine plus proguanil were associated with a large number of mild (mainly gastrointestinal) adverse effects, 30–50% had diarrhea and about 20% had nausea or abdominal pain. There was a significantly larger number of reported "unacceptable symptoms" (not defined) with mefloquine: 2.7%, 1.0%, and 0.6% for mefloquine, chloroquine, and chloroquine plus proguanil respectively. Most of the more serious adverse events were in those who took mefloquine. Compared with chloroquine alone the relative risk (95% CI) of "depression," experiencing "strange thoughts," or having altered spatial perception were 5.1 (2.7, 9.5), 6.4 (2.5, 16.1), and 3.0 (1.4, 6.2) respectively. There was also a higher incidence of depression in women than in men. The relative risk of hospital admission or early termination of travel possibly related to prophylaxis was higher with mefloquine than with either chloroquine or chloroquine plus proguanil; the relative risks (95% CI) were 162 (69, 498), 612 (169, 5054), and 261 (127, 649) respectively.

A postal survey of the incidence of psychiatric disturbances in 2500 returning Israeli travellers (32) showed that travellers with this class of adverse effects were more likely to have taken mefloquine than other antimalarial drugs. Of 117 travellers with psychiatric adverse effects, 115 had taken mefloquine compared with 948/ 1340 for the entire cohort. This was a retrospective postal study with a response rate of 54% (1340 out of 2500), and of those who responded 71% had taken mefloquine, 5% had taken chloroquine, and 24% had taken no prophylaxis. In this study 11% (117) of the respondents reported psychiatric disturbances, mainly sleep disturbance, fatigue, vivid dreams, or "lack of mood." Only 16 of the respondents had symptoms lasting 2 months or more. Those who had had a psychiatric disturbance were also more likely to have been female and to have taken recreational drug use.

Although the above studies were limited by retrospective design, their results are in broad agreement with the results of other studies over the past few years that indicate that women have a higher incidence of psychiatric adverse effects from mefloquine than men (33–36).

In a prospective, double-blind, randomized, placebo-controlled study in 119 healthy volunteers (mean age 35 years), who took either atovaquone 250 mg/day + chloroguanide 100 mg/day or mefloquine 250 mg/week, depression, anger, and fatigue occurred during the use of mefloquine but not atovaquone + chloroguanide (37).

In a review of 10 trials (n = 2750 non-immune adult travellers) (38) the effects of mefloquine in adult travellers were compared with the effects of other regimens in relation to episodes of malaria, withdrawal from prophylaxis, and adverse effects. Five trials were field studies of male soldiers. One comparison of mefloquine with placebo showed that mefloquine was effective in an area of drug resistance (OR = 0.04; 95% CI = 0.02, 0.08) and withdrawals in the mefloquine group were consistently higher in four placebo-controlled trials (OR = 3.56; 95% CI = 1.67, 7.60).

In five comparisons of mefloquine with other chemoprophylaxis regimens, there was no difference in tolerability. The only consistent adverse effects consistently specific to mefloquine in the controlled trials were insomnia and fatigue, but there were also 516 reports of adverse effects of mefloquine, 63% of which were in tourists and travellers. Four deaths were attributed to mefloquine.

In another major review the risk of depression, psychosis, a panic attack, or a suicide attempt during current or previous use of mefloquine was compared with the risk during the use of proguanil and/or chloroquine or doxycycline (39). The study population (n = 35 370) was aged 17–79 years (45% men). There was no evidence that the risk of depression was increased during or after the use of mefloquine, but psychoses and panic attacks were more frequent in current users of mefloquine than in those using other antimalarial drugs.

Severe depression has been attributed to mefloquine (40).

- A 48-year-old woman developed anxiety, tremor, depression, dry mouth, nausea, and marked weight loss. Physical examination, electrocardiography, chest X-ray, CT scan, and laboratory investigations were unremarkable. The Hamilton D score was 44 for 17 items. She had taken mefloquine 250 mg/week for 8 weeks for malaria prophylaxis, and after 2 weeks had started to feel unwell, with dysphoria, depression, and weakness. She was given fluoxetine 20 mg/day and alprazolam 1.5 mg/day. Her condition continued to deteriorate. The dose of fluoxetine was increased to 40 mg/day and flunitrazepam was added. She was later instead given milnacipran, a serotonin and noradrenaline reuptake inhibitor. Five months after the first course of mefloquine she had recovered sufficiently to return to work. However, she relapsed and she was eventually stabilized on venlafaxine 75 mg/day.

Mefloquine has been associated with a number of neuropsychiatric adverse effects, which are often mild and of short duration.

Airline pilots should not take routine prophylaxis with mefloquine because of the small risk of neuropsychiatric reactions.

Endocrine

- A 30-year-old woman took mefloquine, 250 mg/week, and developed abdominal pain, palpitation, and tremor; thyroid function tests were abnormal; 1 month after withdrawal the tests had returned to normal (SEDA-18, 289).

Hematologic

Bleeding disorders following mefloquine are unusual (41).

- A 46-year-old nurse on a 2-week medical assignment in Sri Lanka was given mefloquine in a single dose weekly for malaria prophylaxis. She was not taking any other medication and had not used mefloquine before. Three days after the first dose of mefloquine she developed bruising over her trunk and waist and had occasional per rectal bleeding. The bruising extended from the trunk to the buttocks. A full blood count was normal, the platelet count was 524 x 10^9/l and the hemoglobin 12.4 g/dl. The coagulation profile and liver function tests were normal. Mefloquine was withdrawn and the bruising resolved. A repeat full blood count 6 months later was normal and the platelet count had normalized to 461 x 10^9/l.

Cutaneous reactions to mefloquine include pruritus and maculopapular rashes. The temporal relation between drug exposure and the bleeding episode in this case suggested a causal relation.

Severe thrombotic thrombocytopenic purpura has been attributed to mefloquine (42).

- A 56-year-old Caucasian on a trip to Thailand suddenly developed a fever, confusion, lethargy, and blurred vision 2 weeks after starting to take mefloquine prophylaxis 250 mg/week. He had leg petechiae and scleral jaundice. Chest radiography, electrocardiography, abdominal echocardiography, and a cerebral CT scan showed no abnormalities. Laboratory tests showed thrombocytopenia (platelets 26 000 x 10^9/l), a leukocytosis (18.5 x 10^9/l), anemia (hemoglobin 10.1 g/dl with a high reticulocyte count), and raised lactate dehydrogenase activity and bilirubin concentration (total 55 µmol/l, indirect 43 µmol/l). Coagulation profiles, creatinine, and urine examination were normal. Stool and blood cultures yielded no growth and serological tests were negative. After plasmapheresis the neurological symptoms improved and the hematological profile gradually returned to normal.

Thrombocytopenia after mefloquine has been described before, but the severe neurological symptoms accompanied by fever, thrombocytopenia, and a microangiopathic anemia were unusual.

- Agranulocytosis occurred in a 31-year-old man with *P. vivax* parasitemia given an initial dose of mefloquine followed by 500 mg 8 hours later (SEDA-16, 307).

Gastrointestinal

Gastrointestinal complaints, such as nausea, vomiting, abdominal discomfort, and (usually mild) diarrhea, have been mentioned in most reports, the incidence being 10–25% (SED-12, 693; SEDA-16, 307; 15, 17). The frequency increases with higher doses, for example 25 instead of 15 mg/kg (19).

Liver

Acute fatty liver has recently been reported after malaria prophylaxis with mefloquine (43).

- A 46-year-old woman took five, weekly, doses of mefloquine 250 mg before discontinuing treatment because of neuropsychiatric and gastrointestinal symptoms. Over the next month she had watery diarrhea, 11 kg weight loss, dependent edema, and abdominal fullness. On examination the liver was substantially enlarged; ultrasound imaging showed massive hepatomegaly with diffuse high-grade steatosis. Serological investigations for infective and autoimmune causes were negative. Her symptoms abated with fluid, electrolyte, and albumin replacement. A fine-needle liver biopsy showed features of diffuse macrovesicular hepatic steatosis. Clinical and radiological changes subsided without sequelae.

Acute fatty liver in this case may have been an idiosyncratic adverse effect of mefloquine.

Raised transaminases (up to 20 times normal activities) have been seen in a man taking mefloquine prophylaxis; they resolved after withdrawal of mefloquine (44).

Skin

Maculopapular rash, urticaria, and itching have been reported; itching may be more common with mefloquine than with chloroquine. However, there are isolated case reports concerning more serious skin conditions: exfoliative dermatitis (45); cutaneous vasculitis (SEDA-17, 3) and a bilateral facial rash, comprising raised red lesions and flat bullae (SEDA-17, 329).

Mefloquine has been associated with erythema multiforme and its variants Stevens–Johnson syndrome (SEDA-16, 307), and toxic epidermal necrolysis (46).

It has tentatively been suggested that mefloquine can exacerbate psoriasis (as can other antimalarial drugs, such as quinidine, chloroquine, and proguanil) (47).

Skin reactions to mefloquine have been reviewed, in relation to 74 case reports published between 1983 and 1997 (48). Pruritus and maculopapular rash were the most common skin reactions: in some studies, their approximate frequency was 4–10% for pruritus and up to 30% for non-specific maculopapular rashes. Adverse effects less commonly associated with mefloquine included urticaria, facial lesions, and cutaneous vasculitis. There was one case of Stevens–Johnson syndrome and one fatal case of toxic epidermal necrolysis.

Long-Term Effects

Drug tolerance

Instances of mefloquine resistance were reported in Tanzania in 1983, in Thailand in 1989, and in Africa (Malawi) in 1991. Resistance to combinations of mefloquine with sulfadoxine and pyrimethamine was reported in 1985 (SEDA-13, 808; 49–53). The possibility of cross-resistance between mefloquine and halofantrine was raised in 1990 (SEDA-13, 808; 54). Currently there are extensive areas, including Thailand, Cambodia, Laos, Papua New Guinea, and Myanmar, where *P. falciparum* is resistant to mefloquine (55). High-dose mefloquine has been tried in areas with mefloquine-resistant *P. falciparum*; and a dose of 25 mg/kg is effective, even in a multidrug-resistant area.

Early laboratory studies suggested cross-resistance between mefloquine and halofantrine, but in a later rodent model cross-resistance was not absolute (SEDA-12, 700).

Second-Generation Effects

Pregnancy

A review of the use of mefloquine in pregnancy (56) did not suggest that mefloquine has a worse effect in pregnancy than other antimalarial drugs, such as chloroquine and pyrimethamine + sulfadoxine.

A 7-day quinine regimen (10 mg/kg salt, 8-hourly for 7 days) has been compared with oral mefloquine 25 mg/kg plus artesunate 4 mg/kg/day for 3 days in 108 women on the Thai-Burmese border (57). The mefloquine plus artesunate regimen was significantly more effective than quinine (day 63, cure rate 98 versus 67%). There were more episodes of dizziness (RR = 1.93; 95% CI = 1.14, 3.25) and tinnitus (RR = 3.93; 95% CI = 1.98, 7.80) with

quinine, but no serious adverse events were attributable to either drug. There were also two mid-trimester abortions with mefloquine plus artesunate and none with quinine. There were no birth defects in either group. Although the numbers were very small, the authors concluded that despite a better parasitological cure rate, the increased risk of abortion associated with mefloquine in pregnancy precluded its routine use. Larger studies are needed to confirm this observation.

Teratogenicity

Mefloquine prophylaxis was studied in a group of 339 pregnant women on the Thai-Burmese border in a double-blind placebo-controlled trial. Infants in the mefloquine group had a lower mean birth weight; there was also a higher rate of stillbirths and congenital anomalies, though these differences were not statistically significant (58,59).

However, a further study of 208 pregnant women on the Thai-Burmese border showed a significantly increased incidence of still-births compared with 1565 women treated with other antimalarial drugs (60). Other adverse effects were no more common than with other antimalarial drugs. The study was performed during a period of emerging mefloquine-resistant malaria, and the findings may also reflect the effect of suboptimal malaria treatment.

In a postmarketing survey by Roche, spontaneous reports from 1267 women who had taken mefloquine during pregnancy showed that there was neither a specific pattern of malformations nor an overall increase in congenital malformations over the 4% prevalence observed among the general population (61). These data have been confirmed by a survey among 72 US Army soldiers who had taken mefloquine during pregnancy (62).

Susceptibility Factors

Sex

A study of the pharmacokinetics of oral mefloquine in 12 healthy adults (6 men, 6 women) over 10 weeks has given insights into sex differences in mefloquine pharmacokinetics (63). Five weekly doses of mefloquine 250 mg were given to healthy volunteers. After this, half the subjects took 5 weekly doses of mefloquine 125 mg and half continued to have 250 mg per week. By the second week, all the subjects had plasma mefloquine concentrations over 1.5 µmol/l (the effective prophylactic threshold), but it was only after the fourth dose that the trough concentrations reached this threshold. The women had significantly higher values of C_{oax} and $C_{min.ss}$ than the men. Although the dose of mefloquine was reduced to 125 mg, the plasma mefloquine concentration was maintained above 1.5 µmol/l in all subjects. In this small study, the most commonly reported adverse events were headache, insomnia, and vertigo, with most adverse events occurring between weeks 5 and 8, when plasma mefloquine concentrations were highest. Women had significantly more adverse events (number of days with adverse events/total number of days exposed 149/420) than men (43/420).

These results may explain why earlier studies in male military personnel failed to detect a higher proportion of neuropsychiatric problems with mefloquine compared with other prophylactic regimens.

Neuropsychiatric events have been extensively reviewed, but little is known about the sex-related incidence. Of 179 travellers (mean age 39 years) who took mefloquine for a 3-week prophylactic period before travelling, the women reported adverse events significantly more often than the men (64). There was an increase in fatigue exclusively in the women, especially in first-time users of mefloquine.

Drug Administration

Drug overdose

Mefloquine has a relatively wide therapeutic margin, but can cause predictable and sometimes long-lasting toxicity in overdose.

- Accidental ingestion of 5.25 g of mefloquine over 6 days by a 36-year-old woman caused vertigo, difficulty in visual accommodation, myalgia, hypotension, and tachycardia; most of the anomalies disappeared in 2 weeks (SEDA-16, 307).

Two case reports have described how the antifungal drug terbinafine (Lamasil) was confused with Lariam, leading to accidental mefloquine overdosage and neuropsychiatric adverse effects, including ataxia, high-frequency hearing loss, depression, and paresthesia (65).

Drug–Drug Interactions

Alcohol

One case history has suggested that the use of alcohol with mefloquine can precipitate a neuropsychiatric reaction (21).

Other antimalarial drugs

Mefloquine has been given in combination with other antimalarial drugs, with the aim of delaying drug resistance. Empirical combinations have been made with pyrimethamine and sulfa drugs. However, no drug has been proven to be synergistic with mefloquine, which, with its very long half-life, is difficult to match.

The chemotherapeutic response of *Plasmodium berghei* to various combinations of mefloquine with other drugs (sulfadoxine + pyrimethamine, primaquine, floxacrine) have shown that the desired effects are purely additive (SEDA-13, 809), so the adverse effects too are probably only those of the individual compounds. Adverse reactions occurred in 46% of 400 patients treated with Fanimef (mefloquine + pyrimethamine + sulfadoxine) (SEDA-12, 693). Of note were dizziness (29%), nausea (9.5%), vomiting (7.3%), weakness/lassitude (5.8%), abdominal discomfort or pain (5.5%), diarrhea (3.8%), pruritus (3.0%), insomnia (2.0%), and headache (2.0%).

The combination of mefloquine with artesunate improves tolerance to mefloquine and the therapeutic response is faster (SEDA-21, 296).

Rabies vaccine

Reports of a possible interaction between concurrent mefloquine administration and intradermal rabies immunization have not been substantiated. Four cases in which rabies vaccine and mefloquine were accidentally given concurrently all led to good antirabies antibody responses (66).

Valproate

Mefloquine has been reported to oppose the effect of valproate (67).

- A 20-year-old woman with bilateral myoclonus and generalized tonic–clonic seizures which had been controlled with valproic acid was given 2 prophylactic doses of mefloquine and developed generalized tonic–clonic seizures.

The authors suggested a causal relation to the administration of mefloquine.

References

1. White NJ. Clinical pharmacokinetics of antimalarial drugs. Clin Pharmacokinet 1985;10(3):187–215.
2. Boudreau EF, Fleckenstein L, Pang LW, Childs GE, Schroeder AC, Ratnaratorn B, Phintuyothin P. Mefloquine kinetics in cured and recrudescent patients with acute falciparum malaria and in healthy volunteers. Clin Pharmacol Ther 1990;48(4):399–409.
3. Lobel HO, Bernard KW, Williams SL, Hightower AW, Patchen LC, Campbell CC. Effectiveness and tolerance of long-term malaria prophylaxis with mefloquine. Need for a better dosing regimen. JAMA 1991;265(3):361–4.
4. Hoffman SL. Prevention of malaria. JAMA 1991;265(3):398–9.
5. UNDP/World Bank/WHO update. Development of mefloquine as an antimalarial drug. Bull World Health Organ 1983;61(2):169–78.
6. Durrheim DN, Gammon S, Waner S, Braack LE. Antimalarial prophylaxis—use and adverse events in visitors to the Kruger National Park. S Afr Med J 1999;89(2):170–5.
7. Lobel HO, Baker MA, Gras FA, Stennies GM, Meerburg P, Hiemstra E, Parise M, Odero M, Waiyaki P. Use of malaria prevention measures by North American and European travelers to East Africa. J Travel Med 2001;8(4):167–72.
8. Overbosch D, Schilthuis H, Bienzle U, Behrens RH, Kain KC, Clarke PD, Toovey S, Knobloch J, Nothdurft HD, Shaw D, Roskell NS, Chulay JDMalarone International Study Team. Atovaquone–proguanil versus mefloquine for malaria prophylaxis in nonimmune travelers: results from a randomized, double-blind study. Clin Infect Dis 2001;33(7):1015–21.
9. Nwokolo C, Wambebe C, Akinyanju O, Raji AA, Audu BS, Emodi IJ, Balogun MO, Chukwuani CM. Mefloquine

versus proguanil in short-term malaria chemoprophylaxis in sickle cell anaemia. Clin Drug Invest 2001;21:537–44.

10. MacArthur J, Stennies GM, Macheso A, Kolczak MS, Green MD, Ali D, Barat LM, Kazembe PN, Ruebush TK 2nd. Efficacy of mefloquine and sulfadoxine–pyrimethamine for the treatment of uncomplicated *Plasmodium falciparum* infection in Machinga District, Malawi, 1998. Am J Trop Med Hyg 2001;65(6):679–84.

11. Matteelli A, Saleri N, Bisoffi Z, Gregis G, Gaviera G, Visonà R, Tedoldi S, Scolari C, Marocco S, Gulletta M. Mefloquine versus quinine plus sulphalene–pyrimethamine (Metakelfin) for treatment of uncomplicated imported falciparum malaria acquired in Africa. Antimicrob Agents Chemother 2005;49(2):663-7.

12. Peragallo MS, Sabatinelli G, Sarnicola G. Compliance and tolerability of mefloquine and chloroquine plus proguanil for long-term malaria chemoprophylaxis in groups at particular risk (the military). Trans R Soc Trop Med Hyg 1999;93(1):73–7.

13. Schlagenhauf P. Mefloquine for malaria chemoprophylaxis 1992–1998: a review. J Travel Med 1999;6(2):122–33.

14. Rendi-Wagner P, Noedl H, Wernsdorfer WH, Wiedermann G, Mikolasek A, Kollaritsch H. Unexpected frequency, duration and spectrum of adverse events after therapeutic dose of mefloquine in healthy adults. Acta Trop 2002;81(2):167–73.

15. Kofi Ekue JM, Ulrich AM, Rwabwogo-Atenyi J, Sheth UK. A double-blind comparative clinical trial of mefloquine and chloroquine in symptomatic falciparum malaria. Bull World Health Organ 1983;61(4):713–8.

16. Soentjens P, Delanote M, Van Gompel A. Mefloquine induced pneumonitis. J Travel Med 2006;13(3):172-4.

17. Harinasuta T, Bunnag D, Lasserre R, Leimer R, Vinijanont S. Trials of mefloquine in vivax and of mefloquine plus "Fansidar" in falciparum malaria. Lancet 1985;1(8434):885–8.

18. Smithuis FM, van Woensel JB, Nordlander E, Vantha WS, ter Kuile FO. Comparison of two mefloquine regimens for treatment of *Plasmodium falciparum* malaria on the northeastern Thai–Cambodian border. Antimicrob Agents Chemother 1993;37(9):1977–81.

19. ter Kuile FO, Nosten F, Thieren M, Luxemburger C, Edstein MD, Chongsuphajaisiddhi T, Phaipun L, Webster HK, White NJ. High-dose mefloquine in the treatment of multidrug-resistant falciparum malaria. J Infect Dis 1992;166(6):1393–400.

20. Jha S, Kumar Rajesh, Kumar Raj. Mefloquine toxicity presenting with polyneuropathy—a report of two cases in India. Trans R Soc Trop Med Hyg 2006;100:594-6.

21. Wittes RC, Saginur R. Adverse reaction to mefloquine associated with ethanol ingestion. CMAJ 1995;152(4):515–7.

22. Pham YT, Nosten F, Farinotti R, White NJ, Gimenez F. Cerebral uptake of mefloquine enantiomers in fatal cerebral malaria. Int J Clin Pharmacol Ther 1999;37(1):58–61.

23. Reid AJ, Whitty CJ, Ayles HM, Jennings RM, Bovill BA, Felton JM, Behrens RH, Bryceson AD, Mabey DC. Malaria at Christmas: risks of prophylaxis versus risks of malaria. BMJ 1998;317(7171):1506–8.

24. Fusetti M, Eibenstein A, Corridore V, Hueck S, Chiti-Batelli S. [Mefloquine and ototoxicity: a report of 3 cases.]Clin Ter 1999;150(5):379–82.

25. Chamberland M, Duperval R, Marcoux JA, Dube P, Pigeon N. Severe falciparum malaria in nonendemic areas: an unrecognized medical emergency. CMAJ 1991;144(4):455–8.

26. Speich R, Haller A. Central anticholinergic syndrome with the antimalarial drug mefloquine. N Engl J Med 1994;331(1):57–8.

27. Havaldar PV, Mogale KD. Mefloquine-induced psychosis. Pediatr Infect Dis J 2000;19(2):166–7.

28. Borruat FX, Nater B, Robyn L, Genton B. Prolonged visual illusions induced by mefloquine (Lariam): a case report. J Travel Med 2001;8(3):148–9.

29. Javorsky DJ, Tremont G, Keitner GI, Parmentier AH. Cognitive and neuropsychiatric side effects of mefloquine. J Neuropsychiatry Clin Neurosci 2001;13(2):302.

30. Kukoyi O, Carney CP. Curses, madness, and mefloquine. Psychosomatics 2003;44:339-41.

31. Petersen E, Ronne T, Ronn A, Bygbjerg I, Larsen SO. Reported side effects to chloroquine, chloroquine plus proguanil, and mefloquine as chemoprophylaxis against malaria in Danish travelers. J Travel Med 2000;7(2):79–84.

32. Potasman I, Beny A, Seligmann H. Neuropsychiatric problems in 2,500 long-term young travelers to the tropics. J Travel Med 2000;7(1):5–9.

33. Schwartz E, Potasman I, Rotenberg M, Almog S, Sadetzki S. Serious adverse events of mefloquine in relation to blood level and gender. Am J Trop Med Hyg 2001;65(3):189–92.

34. Huzly D, Schonfeld C, Beuerle W, Bienzle U. Malaria chemoprophylaxis in German tourists: a prospective study on compliance and adverse reactions. J Travel Med 1996;3(3):148–55.

35. Phillips MA, Kass RB. User acceptability patterns for mefloquine and doxycycline malaria chemoprophylaxis. J Travel Med 1996;3(1):40–5.

36. Schlagenhauf P, Steffen R, Lobel H, Johnson R, Letz R, Tschopp A, Vranjes N, Bergqvist Y, Ericsson O, Hellgren U, Rombo L, Mannino S, Handschin J, Sturchler D. Mefloquine tolerability during chemoprophylaxis: focus on adverse event assessments, stereochemistry and compliance. Trop Med Int Health 1996;1(4):485–94.

37. van Riemsdijk MM, Sturkenboom MC, Ditters JM, Ligthelm RJ, Overbosch D, Stricker BH. Atovaquone plus chloroguanide versus mefloquine for malaria prophylaxis: a focus on neuropsychiatric adverse events. Clin Pharmacol Ther 2002;72(3):294–301.

38. Cayley WE Jr. Mefloquine for preventing malaria in non-immune adult travelers. Am Fam Phys 2004;69(3):521-2.

39. Meier CR, Wilcock K, Jick SS. The risk of severe depression, psychosis or panic attacks with prophylactic antimalarials. Drug Saf 2004;27(3):203-13.

40. Whitworth AB, Aichhorn W. First time diagnosis of depression. Induced by Mefloquine? J Clin Pyschopharmacol 2005;25(4):399-400.

41. Chin Chew H, Ponampalam R. An unusual cutaneous manifestation with mefloquine. Am J Emerg Med 2006;24:634-5.

42. Fiaccadori E, Maggiore U, Rotelli C, Giacosa R, Parenti E, Cabassi A, Ariya K, Wirote L. Thrombotic thrombocytopenic purpura following malaria prophylaxis with mefloquine. J Antimicrob Chemother 2006;57:160-1.

43. Miller KD, Jones E, Yanovski JA, Shankar R, Feuerstein I, Falloon J. Visceral abdominal-fat accumulation associated with use of indinavir. Lancet 1998;351(9106):871–5.

44. Gotsman I, Azaz-Livshits T, Fridlender Z, Muszkat M, Ben-Chetrit E. Mefloquine-induced acute hepatitis. Pharmacotherapy 2000;20(12):1517–9.

45. Martin GJ, Malone JL, Ross EV. Exfoliative dermatitis during malarial prophylaxis with mefloquine. Clin Infect Dis 1993;16(2):341–2.

46. McBride SR, Lawrence CM, Pape SA, Reid CA. Fatal toxic epidermal necrolysis associated with mefloquine antimalarial prophylaxis. Lancet 1997;349(9045):101.
47. Potasman I, Seligmann H. A unique case of mefloquine-induced psoriasis. J Travel Med 1998;5(3):156.
48. Antunes I, Azevedo F, Mesquita-Guimaraes J, Resende C, Fernandes N, MacEdo G. Grover's disease secondary to ribavirin. Br J Dermatol 2000;142(6):1257–8.
49. Van der Straeten C, Klippel JH. Antimalarials and pneumococcal immunization. N Engl J Med 1986;315(11):712–3.
50. Karwacki JJ, Webster HK, Limsomwong N, Shanks GD. Two cases of mefloquine resistant malaria in Thailand. Trans R Soc Trop Med Hyg 1989;83(2):152–3.
51. Nosten F, ter Kuile F, Chongsuphajaisiddhi T, Luxemburger C, Webster HK, Edstein M, Phaipun L, Thew KL, White NJ. Mefloquine-resistant falciparum malaria on the Thai–Burmese border. Lancet 1991;337(8750):1140–3.
52. Ooi WW. Failure of mefloquine prophylaxis in east Africa. N Engl J Med 1991;324(2):130.
53. Hoffman SL, Rustama D, Dimpudus AJ, Punjabi NH, Campbell JR, Oetomo HS, Marwoto HA, Harun S, Sukri N, Heizmann P, Laughlin LW. RII and RIII type resistance of *Plasmodium falciparum* to combination of mefloquine and sulfadoxine/pyrimethamine in Indonesia. Lancet 1985;2:1039–40.
54. Gay F, Bustos DG, Diquet B, Rojas Rivero L, Litaudon M, Pichet C, Danis M, Gentilini M. Cross-resistance between mefloquine and halofantrine. Lancet 1990;336(8725):1262.
55. Noticeboard Resistant malaria in Cambodia. Lancet 1992;339:735.
56. Phillips-Howard PA, Steffen R, Kerr L, Vanhauwere B, Schildknecht J, Fuchs E, Edwards R. Safety of mefloquine and other antimalarial agents in the first trimester of pregnancy. J Travel Med 1998;5(3):121–6.
57. McGready R, Brockman A, Cho T, Cho D, van Vugt M, Luxemburger C, Chongsuphajaisiddhi T, White NJ, Nosten F. Randomized comparison of mefloquine–artesunate versus quinine in the treatment of multidrug-resistant falciparum malaria in pregnancy. Trans R Soc Trop Med Hyg 2000;94(6):689–93.
58. Nosten F, ter Kuile F, Maelankiri L, Chongsuphajaisiddhi T, Nopdonrattakoon L, Tangkitchot S, Boudreau E, Bunnag D, White NJ. Mefloquine prophylaxis prevents malaria during pregnancy: a double-blind, placebo-controlled study. J Infect Dis 1994;169(3):595–603.
59. White AC Jr, Runnels JH. Mefloquine prophylaxis in pregnancy. J Infect Dis 1995;171(1):253.
60. Nosten F, Vincenti M, Simpson J, Yei P, Thwai KL, de Vries A, Chongsuphajaisiddhi T, White NJ. The effects of mefloquine treatment in pregnancy. Clin Infect Dis 1999;28(4):808–15.
61. Maguire RB, Stroncek DF, Campbell AC. Recurrent pancytopenia, coagulopathy, and renal failure associated with multiple quinine-dependent antibodies. Ann Intern Med 1993;119(3):215–7.
62. Metzger W, Mordmuller B, Graninger W, Bienzle U, Kremsner PG. High efficacy of short-term quinine-antibiotic combinations for treating adult malaria patients in an area in which malaria is hyperendemic. Antimicrob Agents Chemother 1995;39(1):245–6.
63. Kollaritsch H, Karbwang J, Wiedermann G, Mikolasek A, Na-Bangchang K, Wernsdorfer WH. Mefloquine concentration profiles during prophylactic dose regimens. Wien Klin Wochenschr 2000;112(10):441–7.
64. van Riemsdijk MM, Ditters JM, Sturkenboom MC, Tulen JH, Ligthelm RJ, Overbosch D, Stricker BH. Neuropsychiatric events during prophylactic use of mefloquine before travelling. Eur J Clin Pharmacol 2002;58(6):441–5.
65. Lobel HO, Coyne PE, Rosenthal PJ. Drug overdoses with antimalarial agents: prescribing and dispensing errors. JAMA 1998;280(17):1483.
66. Lau SC. Intradermal rabies vaccination and concurrent use of mefloquine. J Travel Med 1999;6(2):140–1.
67. Besser R, Kramer G. Verdacht auf anfallfordernde Wirkung von Mefloquin (Lariam). [Suspected convulsive side-effect of mefloquine (Lariam).] Nervenarzt 1991;62(12):760–1.

Melarsoprol

General Information

Melarsoprol is a trivalent arsenical with activity against East African and West African trypanosomiasis. It is the drug of choice in the case of *Trypanosoma rhodesiense* infection with nervous system involvement (stage II disease) and in stage I patients refractory or intolerant to suramin and pentamidine. Melarsoprol administered intravenously can cause a reactive encephalopathy, with a clinical picture consisting of high fever, headache, tremor, convulsions, and on occasion coma and death. The incidence of arsenic encephalopathy varies from 3 to 18% in various series (SEDA-12, 708; 1).

In a case-control study of physical growth, sexual maturity, and academic performance in 100 young subjects (aged 6–20 years) with and without a past history of sleeping sickness, melarsoprol-treated patients weighed less, were shorter, and had sexual maturity ratings significantly different from the corresponding controls (2).

The efficacy and safety of two regimens of melarsoprol have been compared in patients with nervous system involvement from *Trypanosoma brucei gambiense*: a conventional regimen lasting 26 days, starting at 1.2 mg/kg and rising to 3.6 mg/kg (n = 259), and a regimen of 2.2 mg/kg for 10 days (n = 250) (3). Parasitological cure 24 hours after treatment was 100% in both groups. Disappointingly, the rates of encephalopathy were no better (in fact marginally worse) with the new schedule, despite using 30% less drug overall, and there were six drug-related deaths in each arm. However, the new schedule is quicker and cheaper, and treatment deviations were significantly fewer in the new schedule arm. In a further detailed prospective study of the drug in eight patients with advanced leukaemia, three developed seizures attributable to the drug at doses broadly comparable to those used in trypanosomiasis.

In 42 patients with late-stage *Trypanosoma brucei gambiense* trypanosomiasis, who relapsed after initial treatment with melarsoprol, a sequential combination of intravenous eflornithine (100 mg/kg every 6 hours for 4 days) followed by three daily injections of melarsoprol (3.6 mg/kg, up to 180 mg) was used (4). They were followed for 24 months. In one case the administration of eflornithine had to be interrupted for 48 hours because of convulsions, but treatment was then resumed without recurrence. Other adverse effects during treatment were abdominal pain or vomiting ($n = 4$ each), diarrhea ($n = 1$), and loss of hearing ($n = 1$). Two patients died during treatment:

- A 37-year-old man died of an acute cholera-like syndrome, with severe diarrhea, vomiting, and dehydration, after the last dose of eflornithine but before receiving his first dose of melarsoprol;
- A 34-year-old man died of an unknown cause after having received all 16 doses of eflornithine as well as the first injection of melarsoprol.

In a randomized trial in 500 patients infected with *Trypanosoma brucei gambiense* treated with 10 daily consecutive doses of melarsoprol 2.2 mg/kg, the adverse effects were: encephalopathic syndrome 5.6%, death from encephalopathy 2.4%, polyneuropathy less than 1%, severe bullous dermatitis 1.2%, severe maculopapular rash 3.2%, severe pruritus 3.2% (5). Milder reactions were fever, headache, and diarrhea.

Organs and Systems

Nervous system

Melarsoprol given intravenously in patients with trypanosomiasis can cause a peripheral neuropathy within 2–5 weeks (SEDA-14, 243). It also causes a reactive arsenical encephalopathy in 3–5% of patients with trypanosomiasis (SEDA-13, 834) (6).

- Myalgia, distal paresthesia, and rapidly progressive weakness in all limbs developed in a young woman treated for 38 days with melarsoprol; there was massive distal Wallerian degeneration in the peripheral nerve, and abnormalities in the dorsal ganglia and spinal cord. Very high concentrations of arsenic were found in the spinal cord. All the findings were typical of toxic arsenic accumulation; in this case, renal dysfunction was probably at the root of the arsenic poisoning (SEDA-16, 316).

In patients with *Trypanosoma gambiense* sleeping sickness, the incidence of drug-induced encephalopathy was increased in patients with trypanosomes present in the nervous system, in patients with high CSF lymphocyte counts, and among those in whom no trypanosomes were found in either the blood or a lymph node aspirate. The authors of this report considered that aggressive therapeutic schemes may result in greater toxicity, especially in patients with impaired blood–brain barrier (SEDA-16, 316). Without data on dosage, renal function, and cerebral involvement before therapy it is impossible to assess this conclusion.

In a large-scale review it was estimated that encephalopathic syndromes occur in 5–10% of patients, and that 10–50% of these die as a result (7).

Encephalopathic syndromes complicating treatment of stage II human African trypanosomiasis with melarsoprol

in 588 patients have been reviewed (8). The overall rate of encephalopathy was 5.8% and presented in three ways: coma, convulsions, and psychotic reactions. The overall death rate was 38%. Comatose patients had a death rate of 52% and were commonly co-infected with malaria (14/16). Symptoms during treatment of fever (RR = 11.5), headache (RR = 2.5), bullous eruptions (RR = 4.5), and systolic hypotension (RR = 2.6) were associated with an increased risk of encephalopathic syndromes, especially coma.

In 56 patients with African trypanosomiasis, one treated with melarsoprol (total dose 26 mg/kg) developed a reactive arsenical encephalopathy (9).

Drug Administration

Drug dosage regimens

Melarsoprol dosage regimens vary; usually 3 or 4 series of 3 or 4 injections of increasing doses are given, with rest periods of 7–10 days. Shorter regimens are desirable. In 2020 patients with human African trypanosomiasis who were treated with the shortened melarsoprol treatment schedule (2.2 mg/kg/day for 10 days) the cure rate 24 hours after treatment was 94%; 2 years later it was 86% (10). However, 935 patients were lost to follow-up. The case fatality rate was 5.9%. Of the treated patients 8.7% had an encephalopathy that was fatal 46% of the time. The rates of severe bullous and maculopapular reactions were 0.8% and 6.8% respectively.

The 10-day melarsoprol regimen is effective and easier to administer than longer regimens. However, the high number of adverse events shows that the development of more effective and less toxic drugs remains a top priority.

References

1. Pepin J, Milord F. The treatment of human African trypanosomiasis. Adv Parasitol 1994;33:1–47.
2. Aroke AH, Asonganyi T, Mbonda E. Influence of a past history of Gambian sleeping sickness on physical growth, sexual maturity and academic performance of children in Fontem, Cameroon. Ann Trop Med Parasitol 1998;92(8):829–35.
3. Burri C, Nkunku S, Merolle A, Smith T, Blum J, Brun R. Efficacy of new, concise schedule for melarsoprol in treatment of sleeping sickness caused by *Trypanosoma brucei gambiense*: a randomised trial. Lancet 2000;355(9213):1419–25.
4. Mpia B, Pepin J. Combination of eflornithine and melarsoprol for melarsoprol-resistant Gambian trypanosomiasis. Trop Med Int Health 2002;7(9):775–9.
5. Blum J, Burri C. Treatment of late stage sleeping sickness caused by *T.b. gambiense*: a new approach to the use of an old drug Swiss Med Wkly 2002;132(5–6):51–6.
6. Nkanga NG, Mutombo L, Kazadi K, Kazyumba GL. Neuropathies arsénicales après traitement de la trypanosomiase humaine au mélarsoprol. Med Afr Noire 1988;35:73.
7. Anonymous. WHO Control and surveillance of African trypanosomiasis. WHO Tech Rep Ser 1998;881:1–114.
8. Blum J, Nkunku S, Burri C. Clinical description of encephalopathic syndromes and risk factors for their occurrence

and outcome during melarsoprol treatment of human African trypanosomiasis. Trop Med Int Health 2001;6(5):390–400.

9. Ruiz JA, Simarro PP, Josenando T. Control of human African trypanosomiasis in the Quicama focus, Angola. Bull World Health Organ 2002;80(9):738–45.

10. Schmid C, Richer M, Miaka Mia Bilenge C, Josenando T, Chappuis F, Manthelot CR, Nangouma A, Doua F, Asumu P, Simarro PP, Burri C. Effectiveness of a 10 day melarsoprol schedule for the treatment of late stage human African trypanosomiasis: confirmation from a multinational study (IMPANEL II). J Infect Dis 2005;195:1922-31.

Mepacrine

General Information

Following the administration of mepacrine quinccrine in normal doses, adverse effects are usually limited to gastrointestinal upsets, with nausea and sometimes vomiting. Yellow discoloration of the skin and conjunctivae is common, but is usually considered more of a cosmetic problem. The aplastic anemia seen with the use of mepacrine may be caused by a hypersensitivity reaction. The efficacy of mepacrine in the treatment of Creutzfeldt?Jakob disease has been investigated (1,2).

Organs and Systems

Cardiovascular

Mepacrine inhibits phospholipase A2 and subsequently leukotrienes, which are calcium ionophores. Mepacrine also has an inhibitory effect on phospholipase C and subsequent inositol phosphate formation, which mobilizes cytosolic calcium from intracellular stores. It has therefore been suggested that this might influence myocardial contractile function (SEDA-13, 241). However, mepacrine is not cardiotoxic, although problems could arise in the presence of a "sick" myocardium.

Nervous system

Headache and dizziness are not uncommon with ordinary doses of mepacrine and after transcervical administration for sterilization (3).

Signs of central nervous system excitation, including convulsions, have been seen with the use of large parenteral doses of mepacrine in the treatment of malignancies (SEDA-11, 592).

Sensory systems

Yellow discoloration of the conjunctivae and sclerae can occur in patients taking mepacrine. The corneal deposits and mild diffuse cornea edema, which can cause blurring of vision, are reversible.

There have been occasional reports of retinopathy resembling that seen with chloroquine (SEDA-11, 593;

4). However, as a rule mepacrine is non-toxic to the retina (SEDA-14, 241).

Psychological, psychiatric

Acute psychosis has been seen with the use of large parenteral doses of mepacrine in the treatment of malignancies (SEDA-11, 592).

- An acute transient psychosis (screaming, kicking, and hallucinations) was seen in an 11-year-old boy after 5 days' treatment with quinacrine 100 mg tds for *Giardia lamblia* infestation. Recovery occurred after withdrawal (SEDA-16, 309).

Hematologic

Mepacrine used in the prophylaxis of malaria can cause aplastic anemia. The incidence among soldiers during World War II was 2.84 per 100 000 compared with 0.66 per 100 000 in controls. A skin rash or lichenoid eruption often preceded the blood dyscrasia (SEDA-11, 593).

- Aplastic anemia has been reported in a 28-year-old woman given mepacrine for several months for the treatment of lupus erythematosus (SEDA-16, 309).

Liver

Rare cases of hepatitis and even hepatic necrosis have been reported in patients taking mepacrine (SEDA-11, 593).

Three patients with probable Creutzfeldt–Jakob disease were given quinacrine 1000 mg on the first day followed by 300 mg/day and liver enzymes were monitored (26). At 7–42 days after the start of therapy there were increases in liver enzymes: aspartate transaminase, alanine transaminase, gamma-glutamyltransferase, and alkaline phosphatase activities increased 3–30 times, 2–11 times, 2–4 times, and 2–8 times respectively. Since the neurological symptoms did not improve, quinacrine was withdrawn. In two patients, the liver enzymes normalized on days 21 and 30 after withdrawal of quinacrine. Although all three patients died shortly thereafter, no death was related to hepatic insufficiency. Autopsy confirmed Creutzfeldt–Jakob disease in all three, and histological examination of the liver showed cytolytic hepatitis and cholangitis. Since liver enzymes and liver histology were normal in five other patients who died from Creutzfeldt–Jakob disease, the hepatitis and cholangitis in the former three patients was probably unrelated to Creutzfeldt–Jakob disease, but caused by the experimental use of quinacrine.

Skin

Mepacrine causes a marked yellow discoloration of the skin and often also the conjunctivae. This is often combined with a blue–black discoloration of the palate and a curious discoloration of the nails, which can be brownish-black, yellowish-green, or sometimes white fluorescent in appearance. This phenomenon is related to the cumulative dose, though it is occasionally also seen with short-term use. The discoloration disappears after withdrawal. Mepacrine-induced discoloration shows up under Wood's lamp as a brilliant yellow–green fluorescence of the nails and palms of the hands and also of the urine.

Skin changes attributed to mepacrine include lichen planus nodules, wart-like growths, erosions, and ulcerations (5).

In allied soldiers during World War II, the most common adverse effect of mepacrine was a drug eruption, tropical lichenoid dermatitis, which sometimes led to permanent sequelae at an early stage (6). Late sequelae occurred 7–17 years after the war, and malignancies developed in two cases. Mepacrine can cause skin cancers as late as 34 years after administration.

A skin rash or lichenoid eruption often precedes mepacrine-induced aplastic anemia (7).

Sexual function

The local placing in the cornual portion of the fallopian tube of a 250 mg quinacrine pellet caused necrosis of the epithelial lining and an acute inflammatory reaction, with subsequent progressive fibrosis and occlusion of the lumen of the tube; in that particular study it was a desired effect (SEDA-16, 309).

Reproductive system

In three non-comparative phase I trials of transcervical administration of quinacrine pellets 250 mg in 21 women who were scheduled to undergo hysterectomy, five women reported pelvic/abdominal cramps (3).

Long-Term Effects

Tumorigenicity

The possibility of a relation of mepacrine to the incidence of squamous cell carcinoma was suspected in Australian ex-servicemen, but causality is not clear (SEDA-13, 818; SEDA-14, 241). Extensive sun exposure could have been a factor, and that would explain the difference in findings around the world.

Smoking

Antimalarial drugs (chloroquine, hydroxychloroquine, or quinacrine) were given to 36 patients with cutaneous lupus, of whom 17 were smokers and 19 non-smokers (8). The median number of cigarettes smoked was one pack/day, with a median duration of 12.5 years. There was a reduction in the efficacy of antimalarial therapy in the smokers. Patients with cutaneous lupus should therefore be encouraged to stop smoking and consideration may be given to increasing the doses of antimalarial drugs in smokers with refractory cutaneous lupus before starting a cytotoxic agent.

References

1. Scoazec JY, Krolak-Salmon P, Casez O, Besson G, Thobois S, Kopp N, Perret-Liaudet A, Streichenberger N. Quinacrine-induced cytolytic hepatitis in sporadic Creutzfeldt-Jakob disease. Ann Neurol 2003;53:546-7.
2. Kobayashi Y, Hirata K, Tanaka H, Yamada T. [Quinacrine administration to a patient with Creutzfeldt–Jakob disease who received a cadaveric dura mater graft—an EEG evaluation.] Rinsho Shinkeigaku 2003;43:403-8.
3. Laufe LE, Sokal DC, Cole LP, Shoupe D, Schenken RS. Phase I prehysterectomy studies of the transcervical administration of quinacrine pellets. Contraception 1996;54(3):181–6.
4. Zuehlke RL, et al. For lupus erythematosus quinacrine is less oculotoxic than chloroquine. Int J Dermat 1981;20:57.
5. Callaway JL. Late sequelae of quinacrine dermatitis, a new premalignant entity. J Am Acad Dermatol 1979;1(5):456.
6. Bauer F. Quinacrine hydrochloride drug eruption (tropical lichenoid dermatitis). Its early and late sequelae and its malignant potential: a review. J Am Acad Dermatol 1981;4(2):239–48.
7. Dollinger MR, Krakoff IH, Karnofsky DA. Quinacrine (Atabrine) in the treatment of neoplastic effusions. Ann Intern Med 1967;66(2):249–57.
8. Rahman P, Gladman DD, Urowitz MB. Smoking interferes with efficacy of antimalarial therapy in cutaneous lupus. J Rheumatol 1998;25(9):1716–9.

Metrifonate

General Information

Metrifonate, which is given orally, is effective in *Schistosoma hematobium* infections in three doses of 7.5–10 mg/kg 14 days apart. When metrifonate was used in daily doses, as in the treatment of *Onchocerca volvulus* infections, it produced muscarinic effects, and in one case there was proximal weakness due to a nicotinic effect. The combination of polyarthritis, fever, and a raised sedimentation rate was described in 11 of 34 patients treated. Metrifonate inhibits blood cholinesterase activity for up to 48 hours, and common reactions that probably result from this effect comprise nausea, vomiting, abdominal pain, diarrhea, dizziness, weakness, headache, and muscle cramps. Because of its prolonged inhibition of brain cholinesterase and increased steady-state concentrations of acetylcholine in the cortex and the hippocampus, it is now also increasingly used in the treatment of Alzheimer's disease.

Metrifonate has been used for the treatment of schistosomiasis for almost 40 years. Its identification as a cholinesterase inhibitor, together with recognition of the cholinergic deficit in Alzheimer's disease, has led to its use in Alzheimer's disease.

The pharmacology and pharmacokinetics of metrifonate and experience with its use in Alzheimer's disease have been reviewed (1).

Two reviews (2,3) of the use of metrifonate in Alzheimer's disease have been published. Both reported a positive effect of metrifonate, with generally mild and usually transient adverse effects, consisting of gastrointestinal symptoms (such as abdominal pain, diarrhea, flatulence, and nausea, probably reflecting cholinergic overactivation) and leg cramps, possibly caused by the overstimulation of nicotinic receptors at the neuromuscular junction. No laboratory abnormalities were reported.

In a randomized, double-blind, placebo-controlled trial in 408 patients with Alzheimer's disease, metrifonate (20 mg/kg/day for 2 weeks followed by 0.65 mg/kg/day for 24 weeks) significantly improved several mental performance scales (4). Of the 273 patients treated with metrifonate 12% discontinued treatment because of adverse effects, compared with 4% of the 134 patients treated with placebo. The adverse effects that led to withdrawal were mainly gastrointestinal in nature. Diarrhea occurred in 18% of the patients treated with metrifonate (leading to withdrawal in 3%) and in 8% of the patients treated with placebo; 2% of the patients treated with metrifonate discontinued treatment because of nausea and vomiting and 1% because of dyspepsia. Nausea occurred in 12% of the patients treated with metrifonate and in 10% of the patients treated with placebo. Vomiting occurred in 7% and 4% respectively.

In a second randomized, double-blind, placebo-controlled trial (5) 480 patients were randomized to receive placebo ($n = 120$), a low dose of metrifonate (0.5 mg/kg/day for 2 weeks followed by 0.2 mg/kg/day for 10 weeks) ($n = 121$), a moderate dose of metrifonate (0.9 mg/kg/day for 2 weeks followed by 0.3 mg/kg/day for 10 weeks) ($n = 121$), or a high dose (2.0 mg/kg/day for 2 weeks followed by 0.65 mg/kg/day for 10 weeks) ($n = 118$). These doses were selected to achieve steady-state erythrocyte acetylcholinesterase inhibition of 30%, 50%, and 70% respectively. There was a significant dose-related improvement in several mental performance scales with metrifonate. Most of the adverse events were mild and transient. Adverse events that occurred more often in the patients treated with metrifonate were abdominal pain (placebo 4%, low dose 3%, moderate dose 11%, high dose 12%), diarrhea (8%, 9%, 11%, 19%, respectively), flatulence (9%, 2%, 8%, 16%), and leg cramps (1%, 1%, 3%, 8%). Bradycardia, presumably related to the vagotonic effect of acetylcholinesterase inhibition, led to withdrawal of treatment in three patients with asymptomatic bradycardia. All three were in the loading-dose phase of the highest metrifonate dosage regimen.

In 16 patients with Alzheimer's disease, adverse effects were dose-related (6). In eight patients who took 2.5 mg/kg/day for 14 days, 4 mg/kg/day for 3 days, and then 2.0 mg/kg/day for 14 days (acetylcholinesterase inhibition 88–94%) treatment had to be withdrawn because of moderate to severe adverse effects in 6 patients on day 28 of the planned 31. In eight patients treated with 2.5 mg/kg/day for 14 days followed by 1.5 mg/kg/day for 35 days (acetylcholinesterase inhibition 89–91%), the frequency of more severe adverse effects was considerably lower despite similar acetylcholinesterase inhibition. The most frequent adverse events in the high-dose group were muscle cramps and abdominal discomfort during the loading-dose phase, followed by increasing gastrointestinal symptoms in the second loading phase, accompanied by headache and muscle aches. The adverse events profile initially improved during the maintenance phase. However, after 11 days of maintenance treatment, six patients complained of generalized moderate to severe muscle cramps, weakness, inability to resume daily activities, and difficulties with coordinations. The adverse events profile was much more favorable with the lower dose, with the same, but much less severe, range of adverse effects. Again, the most frequent adverse effects were gastrointestinal disturbances, muscle cramps, and light-headedness. One patient had increased sweating, dizziness, and palpitation on day 29 and another developed severe abdominal tenderness on day 31, which led to termination of treatment. There were no abnormalities in laboratory parameters. The authors proposed a maximum tolerated dose of 1.5 mg/kg/day of metrifonate for maintenance therapy in patients with Alzheimer's disease.

The Metrifonate Alzheimer's Trial (MALT) was designed to evaluate a wide range of symptoms and efficacy measures in four main clinical domains of Alzheimer's disease: cognition, psychiatric and behavioral features, activities of daily living, and global functioning (7). These are considered key targets for antidementia drugs. This prospective, multicenter, randomized, double-blind, parallel group study was conducted at 71 independent study centers, and 605 patients were randomized to placebo ($n = 208$), metrifonate 40 or 50 mg/day ($n = 200$), and metrifonate 60 or 80 mg/day ($n = 197$); within each treatment group the dose was determined according to body weight, and treatment lasted for 24 weeks. Metrifonate improved a wide range of symptoms across all four clinical domains of Alzheimer's disease in a dose-dependent manner, and was well tolerated at both doses. One patient withdrew from the study during the high-dose loading phase owing to cholinergic effects, resulting in muscle weakness. Metrifonate was also associated with a slight gradual lowering of hemoglobin, hematocrit, and erythrocyte count in both groups during the first 12 weeks; these were nonprogressive, stabilized over the 6-month treatment period, and hence were not considered to be clinically relevant. Metrifonate was associated with a higher incidence of bradycardia (under 50/minute); serious adverse events possibly related to bradycardia (hypotension, postural hypotension, syncope, dizziness, malaise, and accidental injury) occurred in three patients taking placebo, eight taking low-dose metrifonate (three of whom were withdrawn from the study), and two taking high-dose metrifonate.

In patients with mild to moderate Alzheimer's disease metrifonate significantly improved behavior as well as cognition, function in activities of daily living, and global functional status, as shown by a pooled analysis of four prospective, multicenter, randomized, double-blind, parallel-group, placebo-controlled trials, meeting FDA guidelines (8).

The safety and tolerability of once-daily oral metrifonate has been evaluated in patients with probable mild to moderate Alzheimer's disease in a randomized, double-blind, placebo-controlled, parallel-group study (9). Metrifonate was given to 29 patients as a loading dose (2.5 mg/kg) for 2 weeks, followed by maintenance dose (1 mg/kg) for 4 weeks; 10 patients received placebo. The proportion of patients who had at least one adverse event was comparable in the two groups: metrifonate 76%, placebo 80%. Selected adverse events, defined as those for which the incidence in the metrifonate and placebo group differed by at least 10%, were diarrhea, nausea, leg cramps, and accidental injury. The adverse events were predominantly mild and transient. Those who took metrifonate had a significantly lower heart rate. Metrifonate had no clinically important effect on laboratory tests, such as liver function tests, and did not affect exercise tolerance or pulmonary function.

Organs and Systems

Nervous system

Patients with Alzheimer's disease taking long-term treatment with metrifonate suffered seizures after abrupt withdrawal of antimuscarinic agents (10).

- A 58-year-old woman was given hyoscyamine for abdominal cramps by her local physician without the knowledge of her neurologists. She had a generalized seizure 36 hours after stopping the drug.
- A 66-year-old woman was treated for a skin allergy with doxepin cream in large doses. Withdrawal of this treatment led to two complex partial seizures.

The authors speculated that the antimuscarinic drugs impaired the cholinergic receptor down-regulation that would normally occur in the presence of the increased concentrations of acetylcholine caused by acetylcholinesterase inhibition. Withdrawal of the antagonist therefore abruptly exposed the receptors to high concentrations of the neurotransmitter, leading to seizures.

Second-Generation Effects

Teratogenicity

The birth of a hydrocephalic infant with a large meningomyelocele to a mother who had been treated with metrifonate for an infection with *Schistosoma hematobium* during the second month of pregnancy has been reported, but the report is not recent and the association was probably coincidental (SED-11, 598).

Drug Administration

Drug dosage regimens

The effects of a loading dose on the adverse events of metrifonate 50 mg/day have been studied in Alzheimer's disease. The regimen without a loading dose was better tolerated during the 4-week study period (11).

Drug–Drug Interactions

Organophosphorus insecticides

Theoretically, suxamethonium will be potentiated after administration of metrifonate and, in view of its ability to suppress enzyme activity, metrifonate should be used with great caution in areas where organophosphorus insecticides are used, since they have a similar effect.

References

1. Cummings JL, Ringman JM. Metrifonate (Trichlorfon): a review of the pharmacology, pharmacokinetics and clinical experience with a new acetylcholinesterase inhibitor for Alzheimer's disease. Expert Opin Investig Drugs 1999;8(4):463–71.
2. Cummings JL. Metrifonate: overview of safety and efficacy. Pharmacotherapy 1998;18(2 Part 2):43–6.
3. Mucke HAM. Metrifonate: treatment of Alzheimer's disease, acetylcholinesterase-inhibitor. Drugs Future 1998;23:491–7.
4. Morris JC, Cyrus PA, Orazem J, Mas J, Bieber F, Ruzicka BB, Gulanski B. Metrifonate benefits cognitive, behavioral, and global function in patients with Alzheimer's disease. Neurology 1998;50(5):1222–30.
5. Cummings JL, Cyrus PA, Bieber F, Mas J, Orazem J, Gulanski BMetrifonate Study Group. Metrifonate treatment of the cognitive deficits of Alzheimer's disease. Neurology 1998;50(5):1214–21.
6. Cutler NR, Jhee SS, Cyrus P, Bieber F, TanPiengco P, Sramek JJ, Gulanski B. Safety and tolerability of metrifonate in patients with Alzheimer's disease: results of a maximum tolerated dose study. Life Sci 1998;62(16):1433–41.
7. Dubois B, McKeith I, Orgogozo JM, Collins O, Meulien D. A multicentre, randomized, double-blind, placebo-controlled study to evaluate the efficacy, tolerability and safety of two doses of metrifonate in patients with mild-to-moderate Alzheimer's disease: the MALT study. Int J Geriatr Psychiatry 1999;14(11):973–82.
8. Farlow MR, Cyrus PA. Metrifonate therapy in Alzheimer's disease: a pooled analysis of four randomized, double-blind, placebo-controlled trials. Dement Geriatr Cogn Disord 2000;11(4):202–11.
9. Blass JP, Cyrus PA, Bieber F, Gulanski BThe Metrifonate Study Group. Randomized, double-blind, placebo-controlled, multicenter study to evaluate the safety and tolerability of metrifonate in patients with probable Alzheimer disease. Alzheimer Dis Assoc Disord 2000;14(1):39–45.
10. Piecoro LT, Wermeling DP, Schmitt FA, Ashford JW. Seizures in patients receiving concomitant antimuscarinics and acetylcholinesterase inhibitor. Pharmacotherapy 1998;18(5):1129–32.
11. Jann MW, Cyrus PA, Eisner LS, Margolin DI, Griffin T, Gulanski BMetrifonate Study Group. Efficacy and safety of a loading-dose regimen versus a no-loading-dose regimen of metrifonate in the symptomatic treatment of Alzheimer's disease: a randomized, double-masked, placebo-controlled trial. Clin Ther 1999;21(1):88–102.

Metronidazole

See also Benzimidazoles

General Information

Metronidazole is a benzimidazole derivative.

Uses

Metronidazole is effective in the treatment of many protozoal diseases, notably trichomoniasis, amebiasis, schistosomiasis, strongyloidiasis, and giardiasis, and has been in use for over 20 years.

For amebiasis there is still discussion about the use of a single high dose versus repeated lower doses, both as regards efficacy and adverse effects.

The use of metronidazole against infections with anerobic bacteria has increased over the years, and with this indication the use of metronidazole in combination with many other drugs used by patients with conditions likely to develop secondary anaerobic bacterial infections. With increased use there is also a widespread and increasing incidence of resistance of various strains of bacteria. The use of metronidazole as an added medication merely "to make assurance double sure" is to be discouraged. It is to be especially discouraged in immunocompromised patients, because of the risk of emergence of resistant bacterial strains. With increased use there is also an increased number of reports of some more unusual adverse effects. Overall, metronidazole can still be considered safe, if used in generally recommended doses.

Metronidazole has been formulated as a vaginal gel (0.75%) for the treatment of bacterial vaginosis. A single daily 5-day regimen has been approved by the FDA and has been shown to be as effective as oral metronidazole (1).

In the eradication of *Helicobacter pylori*, metronidazole plus bismuth is effective, but causes more adverse effects than omeprazole plus amoxicillin plus either clarithromycin or metronidazole.

Pharmacokinetics

Metronidazole has excellent systemic availability, and absorption after oral administration is not significantly affected by food. Peak serum concentrations occur about 1 hour after ingestion. Multiple doses every 6–8 hours result in some drug accumulation, the half-life averaging about 8 hours. There is a linear relation between dose and serum concentration. Rectal administration results in serum concentrations about half those seen after oral administration. Systemic absorption after local use in the vagina is slow, maximum serum concentrations being reached only after 8–24 hours; they are only about 20% of those attained after oral administration. Metronidazole is extensively metabolized, and only about 20% of the dose is excreted unchanged in the urine. Tissue concentrations are similar to serum concentrations, and metronidazole penetrates readily into the central nervous system. As with many metabolized drugs, the half-life is markedly prolonged in neonates; there is an inverse relation between gestational age and half-life. Prolonged elimination times are also seen in the presence of serious liver disease (SEDA-13, 831; SEDA-16, 310; SEDA-17, 332; SEDA-18, 294).

Preliminary case reports have suggested that azithromycin or metronidazole can improve ciclosporin-induced gingival hyperplasia (SEDA-19, 350).

General adverse effects

Metronidazole is generally well tolerated. With high doses and high-dose prolonged treatment, nausea, vomiting, and central nervous system symptoms ranging from headache and dizziness to neuritis can occur. The commonest reactions are nausea, a metallic taste in the mouth, furry tongue, and vulvovaginal irritation in patients who take metronidazole for bacterial vaginosis (SEDA-13, 831; SEDA-18, 295). Reports of pancreatitis, neuropathy, and optic neuritis call for caution. Hypersensitivity reactions are unusual, but rashes have been described.

Organs and Systems

Cardiovascular

Thrombophlebitis can occur after intravenous administration of metronidazole; an incidence of 6% has been cited (SEDA-7, 297) (2).

Respiratory

Pneumonitis has been attributed to metronidazole, with recurrence after rechallenge (SEDA-14, 242).

Nervous system

Central nervous system symptoms can occur with standard doses of metronidazole, but they are mainly seen with high doses and especially when such doses are given for a long time. Under the latter conditions, there was a 25% incidence of such symptoms as headaches, dizziness, tremor, ataxia, and confusion.

However, symptoms have also been reported in some cases after short-course metronidazole treatment.

- A middle-aged woman took metronidazole 400 mg tds for 9 days for suspected irritable bowel syndrome (3). On the ninth day, she developed profound weakness of the left arm and leg and an ascending sensory loss on the left side. Her reflexes were unaffected. The sensory hemisyndrome resolved after 6 days, but weakness of the left arm and lower leg, and sensory loss of the left distal leg partly persisted after 18 months. Except for metronidazole, no other cause for this sensorimotor neuropathy could be identified despite extensive further testing and examinations.
- A 74-year-old man took metronidazole 500 mg tds, levofloxacin, and amoxicillin for a purulent abdominal abscess (4). After a total dose of 75 g of metronidazole, he developed bilateral leg weakness, a mild peripheral neuropathy, dysarthria, dysmetria, ataxia, and mild left

gaze nystagmus. MRI scanning showed a symmetrical, non-enhancing increased signal intensity In the dentate nuclei on T2 weighted and FLAIR images, and subtle non-enhancing areas of increased T2 intensity in the subcortical and periventricular white matter. Metronidazole toxicity was suspected, and metronidazole was withdrawn. Thereafter, his condition improved dramatically, and an MRI scan 8 weeks later showed complete resolution.

- Similar findings of high signal intensity in the diffuse subcortical white matter, anterior commissure, splenium, midbrain, cerebellar white matter, and basal ganglia were seen in the T2 weighted and FLAIR MRI images of a 74-year-old woman who presented with dysarthria, dysphagia, and gait disturbances after taking metronidazole 1000 mg/day for 6 months (5).

A polyneuropathy is a well-recognized adverse effect of metronidazole (6). A 6% incidence of neuropathy has been quoted; polyneuropathy, mainly sensory in nature, has been recorded during the treatment of Crohn's disease, but again in connection with the prolonged use of high doses (SEDA-12, 705) (7,8). This complication is not restricted to patients with Crohn's disease; it is also seen when metronidazole is given for other purposes, such as radiosensitization. Electrophysiological studies have suggested distal sensory axonal degeneration, with loss of sensory nerve potential over the distal segment and normal motor nerve conduction. In some cases the severity of clinical and electrophysiological abnormalities is closely related to the total amount of metronidazole administered (SEDA-14, 242). Metronidazole is structurally similar to thiamine, and thiamine-synthesizing gut flora may synthesize a neurotoxic analogue of thiamine from ingested metronidazole. However, this hypothesis is weakened by the fact that nitroheterocyclic drugs other than metronidazole are also neurotoxic, despite a much weaker structural analogy to thiamine (SEDA-12, 706) (9).

Peripheral neuropathy was associated in one case with intermittent use of metronidazole (2 g/day for 5 days every other month) (10).

- A 65-year-old white woman with small intestine bacterial overgrowth developed persistent numbness and tingling of her upper and lower extremities. She had been taking alternating courses of tetracycline and metronidazole for 5 days every other month for about 1 year. Other medications included amitriptyline, lisinopril, digoxin, omeprazole, and tamoxifen. Serum vitamin B_{12} and folate concentrations were within the reference ranges. She had reduced sensation in a stocking-glove distribution, reduced sensation to touch and pin-prick, intact reflexes, and no weakness. Neuropathy was attributed to metronidazole, which was withdrawn. After 4 months she reported improvement. On follow-up at 5 months, there was no evidence of peripheral neuropathy.

Metronidazole-associated sensory neuropathy usually presents with pain and reduced thermal and pin-prick sensation, but normal strength, proprioception, and

tendon reflexes. Four patients with metronidazole-associated sensory symptoms had detailed electrodiagnostic studies and nerve or muscle biopsies (11). All had taken different doses of metronidazole and one (with a mitochondrial myopathy) had only used the drug topically. After withdrawal, sensory complaints persisted without progression in two patients and resolved in one. In the fourth, a lower dosage led to partial resolution. Nerve conduction studies were normal in all four cases, but quantitative sensory testing and quantitative sudomotor autonomic reflex testing showed abnormalities of small-fiber function. A sural nerve biopsy from one patient confirmed some loss of small myelinated axons. Metronidazole probably caused a small-fiber neuropathy in these four cases. The authors also reviewed nerve conduction studies in 30 reported cases of metronidazole-associated neuropathy; they were sometimes normal, suggesting that sensory symptoms related to metronidazole may be caused by a mixture of small-fiber and large-fiber sensory dysfunction.

Autonomic neuropathy has been attributed to metronidazole (12).

- A previously healthy 15-year-old girl developed burning feet after a short course of metronidazole for vaginitis. She could only obtain pain relief by submerging her feet in iced water. Her feet were cold and swollen and became erythematous and very warm when removed from the water. Temperature perception was reduced to the upper third of the shin bilaterally. Deep tendon reflexes and power were preserved. Nerve conduction studies showed a peripheral neuropathy with reduced sensory nerve and compound muscle action potentials. Reproducible sympathetic skin potential responses could not be obtained in the hands and feet, providing evidence of a concurrent autonomic neuropathy. No other possible cause of her condition was found and nerve conduction and sympathetic skin potentials gradually returned to normal over 6 months.

Aseptic meningitis is a rare, possibly allergic adverse effect of metronidazole; the one published case was well documented, with positive rechallenge (13).

There has been a report of visual loss and headache after metronidazole (14).

- A 68-year-old man with a tooth abscess had a tooth extraction and received amoxicillin. A few weeks later he developed toothache again and was given amoxicillin and metronidazole 400 mg tds; he took no other drugs. Six hours after the first dose he developed a headache. He continued with metronidazole for a total of three doses, and 6 hours after the last dose the headache resolved. Two days later he noticed flashing lights in both eyes. He then developed a central visual field defect and progressive visual loss. His blood pressure was 220/120 mmHg, but it settled spontaneously. Visual acuity was 6/12 in both eyes and fundoscopy showed marked disc swelling with hemorrhages, without other features of hypertensive retinopathy. Full blood count, plasma viscosity, routine biochemistry, vasculitis screen, anticardiolipin antibodies, angiotensin converting enzyme assay, chest X-ray, CT of the brain and orbits, MRI, and MRA were all normal. CSF examination showed an opening pressure of 24 cm of water and 13 white cells/μl. Over the next few months his visual symptoms slowly improved but he developed secondary optic atrophy.

In this case there was no other obvious cause for visual loss and it could have been caused by metronidazole. The exact mechanism was unclear, but it may have been related to raised intracranial pressure.

Convulsions occurred in an 87-year-old man who had taken metronidazole (15).

Encephalopathy associated with metronidazole has been reported in a patient with chronic renal insufficiency (16).

- A 58-year-old woman with end-stage renal insufficiency secondary to diabetic nephropathy developed abdominal wall cellulitis 4 days after insertion of a peritoneal dialysis catheter. She was given vancomycin, cefepime, and metronidazole in reduced doses (doses not stated) and 2 days later developed dysarthria, an intention tremor, dysmetria, and dysdiadochokinesia. "Routine" biochemical tests were unchanged and a CT scan of the brain was unremarkable, but an MRI scan showed cerebral and cerebellar atrophy with multifocal ischemic glial lesions. Metronidazole was withdrawn and 2 days later her symptoms and signs had completely resolved.

It is difficult to ascribe encephalopathy unequivocally to metronidazole in this case.

Nervous and sensory systems

- A 20-year-old man with ulcerative colitis who had taken metronidazole 1500 mg/day for 2 years developed reduced visual acuity (2/10 bilaterally) and major impairment of red color discrimination, painful distal paresthesia, and dysarthria and impaired coordination in both hands (17). Pattern visual evoked potentials were absent but low amplitude flash visual evoked potentials were elicited with markedly prolonged latencies. An MRI scan of the brain and optic nerves showed increased signal intensities in the splenium, truncus, and genu of the corpus callosum but normal optic nerves. Metronidazole was withdrawn and he gradually improved. The painful paresthesia resolved within 3 months. Repeat brain scans after 2 and 8 months showed moderate resolution of the increased signal in the corpus callosum. After 14 months all the other symptoms and signs had normalized and pattern evoked visual latencies had normalized.

Multiple sclerosis is the most common disease associated with corpus callosum hyperintensities. However, in this case the reversibility of the symptoms and signs, the reversible non-contrasting MRI images, and the absence of new neurological complications pointed to a causal effect of metronidazole.

Sensory systems

Eyes

Acute myopia occurred after 11 days of treatment with metronidazole for *Trichomonas* infection; it resolved within 4 days after withdrawal but recurred on rechallenge. The combined figures of two major American reporting systems for adverse reactions listed seven cases of retrobulbar or optic neuritis associated with oral metronidazole; two had a concurrent peripheral neuropathy. However, there is insufficient information to evaluate these reports (SEDA-17, 333).

Ears

Two reports of moderate to severe sensorineural deafness after metronidazole therapy, which resolved slowly after therapy ended suggest that this is an additional adverse effect (18). Both cases of deafness were preceded by tinnitus, which may be a warning to withdraw the drug.

Psychiatric

Metronidazole can cause delirium.

- A 75-year-old man took oral metronidazole 500 mg tds for *Clostridium difficile* colitis and 48 hours after the start of therapy he became withdrawn and less responsive; during the next 24 hours he developed hallucinations and confusion (19). Metronidazole was switched to oral vancomycin, and his symptoms resolved within 24 hours. One month later, he was re-challenged with metronidazole for recurrent *Clostridium difficile* diarrhea without knowledge of his prior adverse drug reaction. Soon after taking the first dose of metronidazole, he again developed hallucinations, which resolved after switching to vancomycin, confirming that metronidazole was the cause of his mental confusion.

Metabolism

Since anecdotal observations suggested a hypolipidemic effect of metronidazole, the effect of metronidazole 250 mg tds on serum lipids has been evaluated in 30 volunteers who twice took metronidazole for 14 days (20). On both occasions total serum cholesterol fell by 16% and LDL cholesterol by 21%.

Hematologic

Metronidazole can produce leukopenia and neutropenia, usually only associated with prolonged therapy and reversible on withdrawal (SEDA-13, 832). One case each of agranulocytosis and of aplastic anemia have been reported (SEDA-11, 595) and a single case of hemolytic–uremic syndrome in six children (SEDA-15, 298).

Gastrointestinal

Anorexia, nausea, vomiting, abdominal pain, and diarrhea have all been reported in patients taking metronidazole (21). The use of a large single dose most commonly leads to these complaints. A metallic taste also seems to be quite common (21), as is the occurrence of a black tongue (SEDA-13, 832) (22).

Although metronidazole is often used to treat pseudo-membranous colitis, it can also occasionally cause it (23,24).

Liver

Raised serum liver enzyme activities were reported in one case about 15 years ago (SEDA-11, 595) (25), but there has been no more recent confirmation.

Pancreas

Pancreatitis has been reported in various individual case histories (SEDA-15, 298), but there is reason to think that some of these cases at least were due to other factors, such as alcohol use (SEDA-13, 832).

- Pancreatitis has been described in a 63-year-old woman with Crohn's disease, coinciding with the administration of metronidazole and disappearing 1–2 days after withdrawal, but even here there was little support for a causal link (26–28).
- Pancreatitis was attributed to metronidazole in a 61-year-old woman given intravenous metronidazole 500 mg 6-hourly (29). The relation between pancreatitis and metronidazole in this case was less convincing than in previously reported cases, as there was no rechallenge.

This association has been investigated in a large population-based case control study (30). Computer-based prescription records of 3083 cases of acute pancreatitis were compared with the records of 30 083 matched controls. The odds ratios for acute pancreatitis in those who had redeemed a prescription for metronidazole were 3.0 (95% CI = 1.4, 4.6), 1.8 (1.2, 2.9), and 1.1 (0.6, 1.8) within 30, 31–180, and 181–365 days before hospitalization or the index date respectively. Among those with a concomitant prescription for proton pump inhibitors and/or amoxicillin, macrolides, or tetracyclines the respective adjusted odds ratios were 8.3 (2.6, 26), 2.7 (1.4, 5.5), and 1.7 (0.6, 4.8).

Thus, metronidazole increases the risk of acute pancreatitis, with an intermediate time-course, and the risk is even higher when it is used in combination with other drugs used for the treatment of *Helicobacter pylori*.

Urinary tract

Darkening of the urine can occur in patients taking metronidazole (31); this is a harmless discoloration, mostly seen during prolonged treatment.

Skin

Pruritus and rashes have been reported in patients taking metronidazole, including a fixed drug eruption (32) and a pityriasis rosea-like eruption. Urticaria after a single dose has been reported but could have been coincidental (SEDA-17, 333).

In a case of fixed drug eruption, a provocation test showed cross-reactivity with tinidazole but not with secnidazole (32).

Patients with Stevens?Johnson syndrome or toxic epidermal necrolysis (n = 46) and controls (n = 92) have been compared in a case-control study. There was an increased risk among individuals exposed to metronidazole + mebendazole during the previous 6 weeks (OR = 9.5; 95% CI = 3.9, 24) (33). However, exposure to metronidazole or mebendazole alone was not associated with an increased risk.

Stevens–Johnson syndrome after metronidazole has been reported (34).

- A 43-year-old woman developed bullous lesions and erosions of the lips, back, submammary and axilllary areas, and groins, with generalized erythema and edema of the distal extremities. She also complained of general malaise and chills. Before the development of the skin lesions she had taken amoxicillin and metronidazole for gingivitis. A day later she developed pruritus, and amoxicillin was withdrawn and replaced with cefuroxime. Histology of her skin showed a reactive epidermis with numerous apoptotic cells, vacuolar degeneration of the dermoepidermal junction, a moderate cellular infiltrate with marked eosinophilia, and pigment incontinence, consistent with Stevens–Johnson syndrome. Amoxicillin was withdrawn and she was given oral and topical steroids. The skin lesions regressed. Patch-testing with amoxicillin, cefuroxime, and metronidazole produced a vesicular reaction to metronidazole but no reaction to amoxicillin.

The positive patch test with metronidazole in this case suggests a causal relation.

Reproductive system

A genital mucosal erosion occurred in a 38-year-old woman who had taken metronidazole 400 mg tds for 10 days for bacterial vaginitis (SEDA-16, 310).

Long-Term Effects

Drug tolerance

Resistance of *H. pylori* to metronidazole was found in 30% of isolates in the Lebanon (35), 42% in Brazil (36), and 80–90% in Africa (37).

Mutagenicity

Mutagenicity of metronidazole has been demonstrated in some bacterial systems (SEDA-13, 832). Studies on breakages in single-stranded DNA in the lymphocytes of patients treated with metronidazole for *Trichomonas* vaginitis have suggested that such breakages were repaired after withdrawal. Another study reported chromosomal aberrations in the lymphocytes of ten volunteers taking metronidazole (SEDA-21, 301). A mutagenic effect would theoretically be possible in patients with a DNA repair defect (SEDA-16, 310).

There has been concern that metronidazole may be genotoxic, as there have been reports of mutagenicity in several bacterial species. The genotoxic effects of metronidazole (250 mg bd for 10 days) and nalidixic acid

(400 mg bd for 10 days) have been assessed in women with *Trichomonas vaginalis* infections (38). The genotoxic potential of these drugs was evaluated using a sister chromatid exchange test in peripheral blood lymphocytes. Metronidazole had no effect but nalidixic acid caused an increase in sister chromatid exchange frequency. This result confirms that there is little evidence of genotoxicity with metronidazole.

Tumorigenicity

Prolonged high-dose exposure of mice to metronidazole leads to an increased incidence of lung tumors, and in one study there was an increase in lymphoreticular neoplasia in female animals. These results, which caused much concern when first published, are probably non-specific and not relevant to humans; these and other neoplasms have also been induced in mice merely by varying the diet. Several long-term follow-up studies in man have failed to demonstrate an excess cancer risk (SEDA-13, 831). There has been a single report of cancers in three patients with Crohn's disease who had taken metronidazole for years (SEDA-11, 595) (39), but they had also taken sulfasalazine and glucocorticoids, and this cannot be regarded as constituting reasonable evidence of a causal link.

Second-Generation Effects

Teratogenicity

Tests for embryotoxicity and teratogenicity in different animal species have been negative, and there have been no reports of adverse effects on the fetus in pregnant women given metronidazole for trichomoniasis. Despite this, it is still wise to avoid metronidazole during the first trimester of pregnancy.

In a retrospective cohort study using the national birth registry in Denmark, comparing 124 women who took the drug with 13 327 who did not, there was no evidence of any increased risk to the unborn child (40).

In a prospective case-control study in Israel of 857 pregnant women seeking telephone advice regarding gestational exposure to prescribed drugs, 228 women who had taken metronidazole were compared with 629 controls exposed to non-teratogenic agents (41). The mean daily dose of metronidazole was 973 mg for a mean duration of 7.9 days; 90% had used the medication orally, 6% by suppository, and 4% intravenously. Most (86%) had been exposed to metronidazole in the first trimester of pregnancy. There was no difference in the rate of major congenital malformations between the groups (1.6 versus 1.4%), even after accounting for terminations due to prenatally diagnosed malformations. Neonatal birth weight was reduced in the metronidazole group (3.2 versus 3.3 kg) and this was not explained by an earlier gestational age at delivery or a higher prematurity rate but may have been due to the underlying conditions for which metronidazole was prescribed. These findings agree with previous meta-analyses showing that the use of metronidazole in pregnancy is not associated with an increased risk of fetal abnormality, despite in vitro evidence of mutagenesis and inconsistent animal evidence of fetal abnormalities caused by metronidazole.

However, another report has suggested a possible association of the use of vaginal metronidazole suppositories and hydrocephalus (42). The study was a population-based case-control comparison of 38 151 pregnant women who had newborn babies without congenital anomalies and 22 843 pregnant women who delivered babies or fetuses with congenital anomalies. Of the cases, 388 (1.7%) reported having used vaginal metronidazole compared with 570 (1.5%) in the control group (OR=1.1; 95% CI=1.0, 1.3). A further comparison of cases and controls showed an association between the use of vaginal metronidazole in the second and third months of gestation and hydrocephalus (OR=11; 95% CI=1.1, 105). However, this finding was based on only five cases and evaluation of medically recorded metronidazole treatment was non-confirmatory. This association has not previously been reported and requires further investigation.

Lactation

Metronidazole is excreted in the breast milk. There were no adverse effects in nursing infants (SEDA-13, 832), but one should still be cautious in using metronidazole in nursing mothers (SEDA-18, 295).

Susceptibility Factors

Age

In a randomized trial in 100 Iranian children, mebendazole (200 mg tds for 5 days, $n = 50$) was compared with metronidazole (5 mg/kg tds for 7 days, $n = 50$) in giardiasis (43). The two drugs were equally effective (over 85% cure rates). There were no adverse effects of mebendazole, whereas nausea, anorexia, and metallic taste were respectively observed in 4.9, 6, and 24% of those taking metronidazole.

Drug–Drug Interactions

The effect of 1 week of *Helicobacter pylori* eradication therapy with rabeprazole, clarithromycin, and metronidazole on CYP-dependent hepatic metabolism has been determined using the (^{13}) aminophenazone (aminopyrine) breath test (44). The test was performed before treatment, immediately after treatment, and 1 month after treatment. There was no change in hepatic metabolic function during and after treatment.

Alcohol

Metronidazole has a disulfiram-like effect in users of alcohol, sufficient to justify a warning (SEDA-13, 833).

An unusual Antabuse-type reaction reported on one occasion seems to have been due to an interaction of metronidazole with the alcohol present in X-Prep (SEDA-15, 398).

Antibiotics

In vitro, the combination of metronidazole with antibiotics has an additive effect against anaerobic bacteria (SEDA-13, 833).

Busulfan

In 24 patients with graft-versus-host disease, metronidazole significantly increased busulfan plasma concentrations from 452 to 807 ng/ml (45). The authors concluded that metronidazole should not be administered simultaneously with busulfan, because of the risk of severe toxicity and/or mortality.

The effect of metronidazole on busulfan pharmacokinetics was therefore studied in patients taking busulfan with and without metronidazole (46). Metronidazole coadministration significantly increased busulfan trough concentrations by about 80%. Patients who took metronidazole and busulfan concomitantly also had more frequent severe adverse effects. The authors concluded that patients taking high dose busulfan should use other drugs than metronidazole as prophylaxis for graft–versus–host disease.

Carbamazepine

Metronidazole can increase the toxicity of carbamazepine (47) by inhibiting its metabolism.

Cephalosporins

The use of high doses of metronidazole in combination with cefamandole and clindamycin has been associated with encephalopathy (SEDA-12, 705).

Ciclosporin

An interaction with metronidazole and ciclosporin, in which ciclosporin blood concentrations rise, has been suggested, though only in isolated case histories (SEDA-19, 351) (48).

It has been confirmed that metronidazole can produce a two-fold increase in blood concentrations of ciclosporin and tacrolimus, with a subsequent increase in serum creatinine in both cases (49).

The interaction of metronidazole with ciclosporin significantly increased blood ciclosporin and tacrolimus concentrations in two patients (48). Since both of these immunosuppressive drugs are toxic in overdosage, and since patients taking them are prone to infections, this is potentially a serious interaction.

Fabaceae

An unusual Antabuse-type reaction reported on one occasion seems to have been due to an interaction of metronidazole with the alcohol present in the high-dosage form X-Prep (SEDA-15, 398).

Fluorouracil

Pretreatment with metronidazole increased the toxicity of fluorouracil given by a daily bolus dose (50). The clinical significance of this is yet to be determined.

Gentamicin

In guinea pigs, metronidazole augmented gentamicin-induced ototoxicity, determined by the measurement of compound action potentials (51).

Lithium

Metronidazole can increase the toxicity of lithium (52).

Nephrotoxicity due to lithium was reportedly exacerbated by metronidazole (51).

Phenytoin

In a patient taking phenytoin, there was a disproportionate increase in the serum concentration of the hydroxylated metabolite of metronidazole, suggesting that phenytoin induces metronidazole-metabolizing enzymes (SEDA-12, 706) (53).

Quinidine

Serum quinidine concentrations rose during concomitant administration of metronidazole (54). The authors speculated that the mechanism was inhibition of cytochrome P_{450}.

Tacrolimus

A possible interaction with metronidazole recently reported is that it significantly increased blood ciclosporin and tacrolimus concentrations in two patients (48). Since both of these immunosuppressive drugs are toxic in overdosage, and since patients taking them are susceptible to infections, this is potentially a serious interaction.

- A 24-year-old man with a renal transplant, who had been stabilized on tacrolimus 4 mg bd (trough concentrations of 7–10 ng/ml) for 2 months and prednisolone 20 mg/day, developed severe diarrhea due to *Clostridium difficile* (55). He was given metronidazole 500 mg qds. Within 4–14 days his serum creatinine and tacrolimus trough concentrations rose to 292 µmol/l and 26.3 ng/ml. The dose of tacrolimus was reduced to 1 mg bd and after withdrawal of metronidazole the tacrolimus trough concentration fell to 9.4 ng/ml and the serum creatinine to 68 µmol/l.

Vecuronium bromide

Metronidazole can potentiate the effects of non-depolarizing muscle relaxants (56).

Serum concentrations of metronidazole rose during concomitant administration of ciprofloxacin and metronidazole (57). The authors speculated that the mechanism was inhibition of cytochrome P_{450} by ciprofloxacin.

Warfarin and other coumarins

The effect of warfarin is potentiated by metronidazole (58,59). The mechanism is stereoselective inhibition by metronidazole of the metabolism of *S*-warfarin, the more potent isomer (58). There is a similar interaction with acenocoumarol (60,61).

References

1. Wain AM. Metronidazole vaginal gel 0.75% (MetroGel-Vaginal): a brief review Infect Dis Obstet Gynecol 1998;6(1):3–7.

2. Stranz MH, Bradley WE. Metronidazole (Flagyl IV, Searle). Drug Intell Clin Pharm 1981;15(11):838–46.

3. Rustscheff S, Hulten S. An unexpected and severe neurological disorder with permanent disability acquired during short-course treatment with metronidazole. Scand J Infect Dis 2003;35:279-80.

4. Heaney CJ, Campeau NG, Lindell EP. MR imaging and diffusion-weighted imaging changes in metronidazole (Flagyl)-induced cerebellar toxicity. Am J Neuroradiol 2003;24:1615-7.

5. Seok JI, Yi H, Song YM, Lee WY. Metronidazole-induced encephalopathy and inferior olivary hypertrophy: lesion analysis with diffusion-weighted imaging and apparent diffusion coefficient maps. Arch Neurol 2003;60:1796-800.

6. Freedman B, Shah S, Lau A. Metronidazole-induced peripheral neuropathy. J Appl Ther Res 2000;3:49–54.

7. Urtasun RC, Rabin HR, Partington J. Human pharmacokinetics and toxicity of high-dose metronidazole administered orally and intravenously. Surgery 1983;93(1 Pt 2):145–8.

8. Roe FJ. Toxicologic evaluation of metronidazole with particular reference to carcinogenic, mutagenic, and teratogenic potential. Surgery 1983;93(1 Pt 2):158–64.

9. Alston TA. Neurotoxicity of metronidazole. Ann Intern Med 1985;103(1):161.

10. Dreger LM, Gleason PP, Chowdhry TK, Gazzuolo DJ. Intermittent-dose metronidazole-induced peripheral neuropathy. Ann Pharmacother 1998;32(2):267–8.

11. Zivkovic SA, Lacomis DL, Giuliani MJ. Sensory neuropathy associated with metronidazole: report of four cases and review of the literature. J Clin Neuromuscular Dis 2001;3:8–12.

12. Hobson-Webb LD, Roach ES, Donofrio PD. Metronidazole: newly recognized cause of autonomic neuropathy. J Child Neurol 2006;21(5):429-31.

13. Corson AP, Chretien JH. Metronidazole-associated aseptic meningitis. Clin Infect Dis 1994;19(5):974.

14. Allroggen H, Abbott RJ, Bibby K. Acute visual loss following administration of metronidazole: a case report. Neuro-Ophthalmology 2000;23:89–94.

15. Beloosesky Y, Grosman B, Marmelstein V, Grinblat J. Convulsions induced by metronidazole treatment for Clostridium difficile-associated disease in chronic renal failure. Am J Med Sci 2000;319(5):338–9.

16. Arik N, Cengiz N, Bilge A. Metronidazole-induced encephalopathy in a uremic patient: a case report. Nephron 2001;89(1):108–9.

17. De Bleecker JL, Leroy BP, Meire V. Reversible visual deficit and corpus callosum lesions due to metronidazole toxicity. Eur Neurol 2005;53(2):93-5.

18. Iqbal SM, Murthy JG, Banerjee PK, Vishwanathan KA. Metronidazole ototoxicity—report of two cases. J Laryngol Otol 1999;113(4):355–7.

19. Mahl TC, Ummadi S. Metronidazole and mental confusion. J Clin Gastroenterol 2003;36:373-4.

20. Shamkhani K, Azarpira M, Akbar MH. An open label crossover trial of effects of metronidazol on hyperlipidaemia. Int J Cardiol 2003;90:141-5.

21. Andersson KE. Pharmacokinetics of nitroimidazoles. Spectrum of adverse reactions. Scand J Infect Dis Suppl 1981;26:60–7.

22. Gugler R, Jensen JC, Schulte H, Vogel R. Verlauf des Morbus Crohn und Nebenwirkungs-profil unter Langzeittherapie mit Metronidazol. [The course of Crohn disease and side effect profile with long-term treatment using metronidazole.] Z Gastroenterol 1989;27(11):676–82.

23. Vivian AS, Stevenson JG. Metronidazole: cause as well as cure for colitis. Am J Hosp Pharm 1981;38(10):1442.

24. de Dios Garcia Diaz J, Moreno Sanchez D, Campos Cantero R, Medina Asensio J. Colitis seudomembranosa de curso prolongado tras uso de metronidazol intravaginal. [Long-standing pseudomembranous colitis ulcer after the use of intravaginal metronidazole.] Med Clin (Barc) 1987;88(16):652.

25. Appleby DH, Vogtland HD. Suspected metronidazole hepatotoxicity. Clin Pharm 1983;2(4):373–4.

26. Corey WA, Doebbeling BN, DeJong KJ, Britigan BE. Metronidazole-induced acute pancreatitis. Rev Infect Dis 1991;13(6):1213–5.

27. Britigan BE, Doebeling BN. Metronidazole and pancreatitis. Clin Infect Dis 1992;15:751.

28. Romero Y, Yebra M, Lacoma F, Manzano L. Metronidazole and pancreatitis. Clin Infect Dis 1992;15(4):750–1.

29. Sura ME, Heinrich KA, Suseno M. Metronidazole-associated pancreatitis. Ann Pharmacother 2000;34(10):1152–5.

30. Nørgaard M, Ratanajamit C, Jacobsen J, Skriver MV, Pedersen L, Sørensen HT. Metronidazole and risk of acute pancreatitis: a population based- case control study. Aliment Pharmacol Ther 2005;21:415-20.

31. Kapoor K, Chandra M, Nag D, Paliwal JK, Gupta RC, Saxena RC. Evaluation of metronidazole toxicity: a prospective study. Int J Clin Pharmacol Res 1999;19(3):83–8.

32. Thami GP, Kanwar AJ. Fixed drug eruption due to metronidazole and tinidazole without cross-sensitivity to secnidazole. Dermatology 1998;196(3):368.

33. Chen KT, Twu SJ, Chang HJ, Lin RS. Outbreak of Stevens–Johnson syndrome/toxic epidermal necrolysis associated with mebendazole and metronidazole use among Filipino laborers in Taiwan. Am J Public Health 2003;93:489-92.

34. Piskin G, Mekkes JR. Stevens–Johnson syndrome from metronidazole. Contact Dermatitis 2006;55:192-3.

35. Sharara AI, Chedid M, Araj GF, Barada KA, Mourad FH. Prevalence of *Helicobacter pylori* resistance to metronidazole, clarithromycin, amoxycillin and tetracycline in Lebanon. Int J Antimicrob Agents 2002;19(2):155–8.

36. Mendonca S, Ecclissato C, Sartori MS, Godoy AP, Guerzoni RA, Degger M, Pedrazzoli J Jr. Prevalence of *Helicobacter pylori* resistance to metronidazole, clarithromycin, amoxicillin, tetracycline, and furazolidone in Brazil. Helicobacter 2000;5(2):79–83.

37. Alarcon T, Domingo D, Lopez-Brea M. Antibiotic resistance problems with *Helicobacter pylori*. Int J Antimicrob Agents 1999;12(1):19–26.

38. Akyol D, Mungan T, Baltaci V. A comparative study of genotoxic effects in the treatment of *Trichomonas vaginalis* infection: metronidazole or nalidixic acid. Arch Gynecol Obstet 2000;264(1):20–3.

39. Krause JR, Ayuyang HQ, Ellis LD. Occurrence of three cases of carcinoma in individuals with Crohn's disease treated with metronidazole. Am J Gastroenterol 1985;80(12):978–82.

40. Sorensen HT, Larsen H, Jensen ES, Thulstrup AM, Schonheyder HC, Nielsen GL, Czeizel A. Safety of metronidazole during pregnancy: a cohort study of risk of congenital abnormalities, preterm delivery and low birth weight in 124 women. J Antimicrob Chemother 1999;44(6):854–6.

41. Diav-Citrin O, Shechtman S, Gotteiner T, Arnon J, Ornoy A. Pregnancy outcome after gestational exposure

to metronidazole: a prospective controlled cohort study. Teratology 2001;63(5):186–92.

42. Kazy J, Puhó E, Czeizel E. Teratogenic potential of metronidazole vaginal treatment during pregnancy. Eur J Obstet Gynaecol Reprod Biol 2005;123(2):174-8.

43. Sadjjadi SM, Alborzi AW, Mostovfi H. Comparative clinical trial of mebendazole and metronidazole in giardiasis of children. J Trop Pediatr 2001;47(3):176–8.

44. Giannini EG, Malfatti F, Botta F, Polegato S, Testa E, Fumagalli A, Mamone M, Savarino V, Testa R. Influence of 1-week *Helicobacter pylori* eradication therapy with rabeprazole, clarithromycin, and metronidazole on 13C-aminopyrine breath test. Dig Dis Sci 2005;50(7):1207-13.

45. Nilsson C, Aschan J, Hentschke P, Ringden O, Ljungman P, Hassan M. The effect of metronidazole on busulfan pharmacokinetics in patients undergoing hematopoietic stem cell transplantation. Bone Marrow Transplant 2003;31(6):429–35.

46. Nilsson C, Aschan J, Hentschke P, Ringden O, Ljungman P, Hassan M. The effect of metronidazole on busulfan pharmacokinetics in patients undergoing hematopoietic stem cell transplantation. Bone Marrow Transplant 2003;31:429-35.

47. Patterson BD. Possible interaction between metronidazole and carbamazepine. Ann Pharmacother 1994;28(11):1303–4.

48. Campana C, Regazzi MB, Buggia I, Molinaro M. Clinically significant drug interactions with cyclosporin. An update. Clin Pharmacokinet 1996;30(2):141–79.

49. Herzig K, Johnson DW. Marked elevation of blood cyclosporin and tacrolimus levels due to concurrent metronidazole therapy. Nephrol Dial Transplant 1999;14(2):521–3.

50. Bardakji Z, Jolivet J, Langelier Y, Besner JG, Ayoub J. 5-Fluorouracil–metronidazole combination therapy in metastatic colorectal cancer. Clinical, pharmacokinetic and in vitro cytotoxicity studies. Cancer Chemother Pharmacol 1986;18(2):140–4.

51. Riggs LC, Shofner WP, Shah AR, Young MR, Hain TC, Matz GJ. Ototoxicity resulting from combined administration of metronidazole and gentamicin. Am J Otol 1999;20(4):430–4.

52. Teicher MH, Altesman RI, Cole JO, Schatzberg AF. Possible nephrotoxic interaction of lithium and metronidazole. JAMA 1987;257(24):3365–6.

53. Ralph ED. Clinical pharmacokinetics of metronidazole. Clin Pharmacokinet 1983;8(1):43–62.

54. Cooke CE, Sklar GE, Nappi JM. Possible pharmacokinetic interaction with quinidine: ciprofloxacin or metronidazole? Ann-Pharmacother 1996;30(4):364–6.

55. Page II RL, Klem PM, Rogers C. Potential elevation of tacrolimus trough concentrations with concomitant metronidazole therapy. Ann Pharmacother 2005;39(6):1109-13.

56. McIndewar IC, Marshall RJ. Interactions between the neuromuscular blocking drug Org NC 45 and some anaesthetic, analgesic and antimicrobial agents. Br J Anaesth 1981;53(8):785–92.

57. Cooke CE, Sklar GE, Nappi JM. Possible pharmacokinetic interaction with quinidine: ciprofloxacin or metronidazole? Ann Pharmacother 1996;30(4):364–6.

58. O'Reilly RA. The stereoselective interaction of warfarin and metronidazole in man. N Engl J Med 1976;295(7):354–7.

59. Colquhoun MC, Daly M, Stewart P, Beeley L. Interaction between warfarin and miconazole oral gel. Lancet 1987;1(8534):695–6.

60. Marotel C, Cerisay D, Vasseur P, Rouvier B, Chabanne JP. Potentialisation des effets de l'acenocoumarol par le gel buccal de miconazole. [Potentiation of the effects of acenocoumarol by a buccal gel of miconazole.] Presse Méd 1986;15(33):1684–5.

61. Ortin M, Olalla JI, Muruzabal MJ, Peralta FG, Gutierrez MA. Miconazole oral gel enhances acenocoumarol anticoagulant activity: a report of three cases. Ann Pharmacother 1999;33(2):175–7.

Myrrh

General Information

Myrrh is an oleo gum resin obtained from the stem of *Commiphora molmol*, a tree that grows in north-east Africa and the Arabian Peninsula. In mice, myrrh showed no mutagenic effects and was a potent cytotoxic drug against solid tumor cells (1). The antitumor potential of *Commiphora molmol* was comparable with that of cyclophosphamide. Studies in hamsters suggested an antischistosomal activity of myrrh (2).

Observational studies

Facioliasis

The efficacy of myrrh has been studied in seven patients aged 10–41 years (five men, two women) with fascioliasis and 10 age- and sex-matched healthy volunteers (3). Myrrh was given orally in the morning on an empty stomach in a dosage of 12 mg/kg/day for 6 days. All the patients were passing *Fasciola* eggs in their stools (mean 36 eggs per gram of stool). The symptoms and signs of fascioliasis resolved during treatment with myrrh, and *Fasciola* eggs could not be demonstrated in the stools 3 weeks and 3 months after treatment. Antifasciola antibody titers became negative in six of the seven patients. There were no adverse effects.

Schistosomiasis

The efficacy and adverse effects of myrrh and the most effective dosage schedule have been studied in 204 patients (169 men and 35 women) with schistosomiasis aged 12–68 years and 20 healthy non-infected age- and sex-matched volunteers (2). The patients were divided into two groups: 86 patients with schistosomal colitis and 118 with hepatosplenic schistosomiasis, further divided into two subgroups—77 patients with compensated disease and 41 with decompensated disease. All but 12 had received one or more courses of praziquantel. The dosage of myrrh was 10 mg/kg/day for 3 days on an empty stomach 1 hour before breakfast. A second course of 10 mg/kg/day for 6 days was given to patients who still had living ova in rectal or colonic biopsy specimens. The response rate to a single course of myrrh was 92% in 187 patients. The cure rates were 91, 94, and 90% in patients with schistosomal colitis, compensated hepatosplenic schistosomiasis, and decompensated hepatosplenic schistosomiasis respectively. The cure rate was less in patients who had previously taken praziquantel and in patients with impaired liver function. *Schistosoma hematobium* infection was the most responsive (*n* = 4, cure rate 100%), followed by mixed infections (*n* = 29, cure rate 93%). Those infected with *Schistosoma mansoni* had the lowest cure rate (*n* = 171, cure rate 91%). There was no impairment of liver function after treatment with myrrh. In contrast, liver function tests significantly improved in patients with impaired liver function. There were no significant effects of myrrh on the electrocardiogram. Adverse effects of myrrh were reported in 24 of the 204 patients. Giddiness, somnolence, or mild fatigue were the most common (2.5%), and all other adverse effects were minor and less frequent. None of the healthy volunteers reported any adverse effects, nor were there any significant changes in liver or kidney function. A second course of myrrh resulted in a cure in 13 of the 17 patients who did not respond to a single course.

References

1. al-Harbi MM, Qureshi S, Ahmed MM, Rafatullah S, Shah AH. Effect of Commiphora molmol (oleo-gum-resin) on the cytological and biochemical changes induced by cyclophosphamide in mice. Am J Chin Med 1994;22(1):77–82.
2. Sheir Z, Nasr AA, Massoud A, Salama O, Badra GA, El-Shennawy H, Hassan N, Hammad SM. A safe, effective, herbal antischistosomal therapy derived from myrrh. Am J Trop Med Hyg 2001;65(6):700–4.
3. Massoud A, El Sisi S, Salama O, Massoud A. Preliminary study of therapeutic efficacy of a new fasciolicidal drug derived from Commiphora molmol (myrrh). Am J Trop Med Hyg 2001;65(2):96–9.

Niclosamide

General Information

Niclosamide, widely used in the treatment of tapeworm infestation, is well tolerated in doses of 2 g taken orally before breakfast and is the drug of choice for mass chemotherapy (1). It is has also been used as a molluscicide and in the control of *Schistosoma japonicum* (2).

Niclosamide is not absorbed and reactions consist of mild gastrointestinal disturbances. When treating *Tenia solium* infections an antiemetic is usually given before niclosamide, and patients are subsequently purged to reduce the theoretical risk of cysticercosis, because the dose of niclosamide active against *T. solium* does not destroy ova contained within the tapeworm. Alcohol is usually restricted during treatment, since niclosamide can interfere with its metabolism.

References

1. Sarti E, Rajshekhar V. Measures for the prevention and control of *Taenia solium* taeniosis and cysticercosis. Acta Trop 2003;87(1):137–43.
2. Lowe D, Xi J, Meng X, Wu Z, Qiu D, Spear R. Transport of *Schistosoma japonicum* cercariae and the feasibility of niclosamide for cercariae control. Parasitol Int 2005;54(1):83–9.

Nifurtimox

General Information

There are still few drugs available for the treatment of late-stage sleeping sickness due to *Trypanosoma gambiense*. Melarsoprol and eflornithine are equally efficacious, but both are toxic, and eflornithine is expensive. High-dose nifurtimox (30 mg/kg/day) has been used in Zaire for the treatment of arsenic-resistant trypanosomiasis; this dose was more effective and also more toxic. Of the 30 patients, one died after a period of confusion and coma and eight developed neurological effects, namely confusion, tremor, vertigo, and convulsions. One developed a rash and 12 had marked weight loss (SEDA-17, 333).

Acute polymyositis and toxic erythema purpura have been described in a young woman treated with nifurtimox for Chagas' disease; she died with respiratory and renal failure (SEDA-11, 598) (1).

Nifurtimox is the only drug approved for use in the USA (for Chagas' disease), but it is available exclusively on request from the Drug Service of the CDC. For treatment of the acute phase of Chagas' disease and in congenital cases the recommended dosage is 8–10 mg/kg/day, divided into two or three daily doses, for 30–60 consecutive days. In acute Chagas' disease, nifurtimox reduces both the duration of symptoms and parasitemia and mortality, but with limited efficacy in the eradication of parasites. The more frequent adverse effects are anorexia, loss of weight, psychic alterations, excitability or sleepiness, and digestive manifestations, such as nausea, vomiting, and occasionally intestinal colic and diarrhea (2). Other common adverse effects include neurological adverse effects (such as paresthesia, polyneuropathy, and convulsions), hematological adverse effects (such as thrombocytopenia and granulocytopenia), and allergic skin reactions (SEDA-13, 836) (3).

Long-Term Effects

Mutagenicity

Chromosomal aberrations were significantly increased in cultures of peripheral lymphocytes from a small group of children with Chagas' disease treated with nifurtimox. G-binding analysis of chromosomal aberration sites showed that treated patients presented coincidence in the chromosome regions affected (SEDA-15, 298).

References

1. Shaw M, Petroue J, Iglesias D, et al. Polimiositis aguda por nifurtimox. Arch Argent Dermatol 1982;32:191.
2. Rodriques Coura J, de Castro SL. A critical review on Chagas disease chemotherapy. Mem Inst Oswaldo Cruz 2002;97(1):3–24.
3. Pepin J, Milord F, Meurice F, Ethier L, Loko L, Mpia B. High-dose nifurtimox for arseno-resistant *Trypanosoma brucei gambiense* sleeping sickness: an open trial in central Zaire. Trans R Soc Trop Med Hyg 1992;86(3):254–6.

Niridazole

See also Benzimidazoles

General Information

Niridazole is used in the treatment of schistosomiasis and of guinea-worm (*Dracunculus medinensis*) infections; it is given in divided daily doses of 25 mg/kg for 5–10 days, depending on the infecting species. Niridazole is well tolerated in *Schistosoma hematobium* infections. It is metabolized in the liver, and its metabolites color the urine dark brown. In the treatment of schistosomiasis it has been largely superseded by newer drugs.

General adverse effects

In the treatment of schistosomiasis adverse effects were seen in 80% of inpatients and 33% of outpatients. Gastrointestinal effects were the most common, with vomiting in 50% of cases. Neurotoxic and psychiatric effects, convulsions, and cardiac effects are less common, except in patients with liver disease. Insomnia, somnolence, vertigo, nightmares, headache, weakness, jaundice, and muscle pains have all been reported (SEDA-12, 244). Toxic reactions are seen more often in *Schistosoma mansoni* infections, especially in patients with poor liver function or portocaval shunts, when neuropsychiatric complications can be expected. Some toxicity is directly related to parasite destruction and liberation of antigen rather than to the drug itself. Allergic reactions related to parasite destruction include urticaria, allergic conjunctivitis, and fever with peripheral eosinophilia. Niridazole is a potent carcinogen in mice, and tumorigenic in hamsters. It is not clear what the risks are in man, but it is of course usually given for only a brief period.

Organs and Systems

Cardiovascular

Minor electrocardiographic abnormalities, especially T wave changes, are common but are probably of no

functional significance of patients (1). Dysrhythmias with prolongation of the QT interval occur in a small minority of patients (2).

Respiratory

Cough, fever, and dyspnea with pulmonary infiltration have been reported in two cases (3).

Nervous system

Headache, drowsiness, and dizziness are common with niridazole (2,4,5). More severe neuropsychiatric symptoms are more frequent in patients with liver disease, especially those with portosystemic shunts, in whom the drug bypasses the liver (6). Symptoms in these cases include insomnia, anxiety, depression, confusion, hallucinations, and convulsions; the reactions may prove fatal. The electroencephalogram can show slowed alpha rhythms, beta waves, and theta waves, as well as sharp wave and spike forms with niridazole (7). A single case of acute cortical necrosis was recorded in the much older literature, but was probably coincidental (SED-8, 691) (8). Agitation can occur in patients with abnormal liver function.

Hematologic

A peripheral eosinophilia is usual (2). Hemolysis has been reported once in glucose-6-phosphate dehydrogenase (G6PD) deficiency (SED-8, 691) (9).

Gastrointestinal

Bad taste, anorexia, nausea, vomiting, diarrhea, and abdominal pain are common with niridazole (1,2,5). Gastrointestinal bleeding has been seen (10).

Liver

Prolongation of the prothrombin time can occur (11). Niridazole is contraindicated in liver disease.

Urinary tract

Metabolites of niridazole color the urine dark brown (12,13).

Musculoskeletal

Muscle pain, joint pain, and bone pain are commonly reported but may be related to parasite destruction rather than to direct toxicity (5).

Long-Term Effects

Mutagenicity

Niridazole should not be used in pregnancy; mutagenic effects have been seen in bacteria (14).

Second-Generation Effects

Fertility

Transient reduction in spermatogenesis, apparently reversible, can occur (15).

Susceptibility Factors

Genetic factors

Caution is required when using niridazole in individuals with G6PD deficiency (16).

Hepatic disease

Niridazole should not be used in the presence of hepatic disease (17).

References

1. Abdallah A, Saif M, Abdel-Meguid M, Badran A, Abdel-Fattah F, Aly IM. Treatment of urinary and intestinal bilharziasis with Ciba 32644-Ba (Ambilhar). A preliminary report. J Egypt Med Assoc 1966;49(2):145–63.
2. Katz N, et al. Clinical trials with CIBA 32644-Ba (Ambilhar) in schistosomiasis mansoni. Folha Med 1966;53(4):561.
3. Farid Z, Bassily S, Lehman JS Jr, Ayad N, Hassan A, Sparks HA. A comparative evaluation of the treatment of Schitosoma mansoni with niridazole and potassium antimony tartrate. Trans R Soc Trop Med Hyg 1972;66(1):119–24.
4. Abdallah A, Saif M, el-Mawla NG, Abdel-Fattah F, Aly IM. Spaced-dosage treatment of bilharziasis with niridazole. J Egypt Med Assoc 1968;51(9):823–30.
5. Ruas A, Almeido Franco LT. The effect of CIBA 32,644-Ba in the treatment of 1,059 cases of vesical and intestinal schistosomiasis. Ann Trop Med Parasitol 1966;60(3):288–92.
6. Basmy K, Shoeb SM, Mohran Y. The role of liver dysfunction in the occurrence of the neuropsychiatric side effects of Ambilhar. J Egypt Med Assoc 1969;52(2):196–204.
7. Davidson JC. Neuropsychiatric effects and E.E.G. changes in niridazole therapy Trans R Soc Trop Med Hyg 1969;63(5):579–81.
8. Emerit J, Saigot T, Escourolle R, Begue P. Un cas de nécrose massive de la substance blanche hémisphérique au décours d'un traitement par l'Ambilhar chez une cirrhotique. [A case of massive necrosis of the hemispheric white substance during treatment with Ambilhar in a cirrhotic female patient.] Ann Med Interne (Paris) 1974;125(1):65–9.
9. McCaffrey RP, Farid Z, Kent DC. Acute haemolysis with Ambilhar treatment in glucose-6-phosphate dehydrogenase deficiency. Trans R Soc Trop Med Hyg 1972;66(5):795–7.
10. de Almeida Junior N, Penha LA, Pereira N, de Oliveira CA, de Oliveira JV, Crosara A, Habib P, Jacob M, de Oliveira S. Hemorragia digestiva no decorrer do tratamento da esquistossomose mansonica com o niridazol (pesquisas realizadas em 28 pacientes no sentido de apurar o provavel mecanismo de sua producao). [Gastrointestinal hemorrhage during the management of schistosomiasis mansoni using niridazole (studies in 28 patients for the purpose of accelerating of proving its action mechanism).] Hospital (Rio J) 1968;74(5):1639–47.
11. Rodrigues LD, Vilela Mde P, Guimaraes RX, Jafferian PA, Miszputen SJ, Costa A. Estudo do comportamento do "tempo de protrombina" en pacientes esquisto-somoticos medicados com Ambilhar. [Study of the behavior of "prothrombin time" in schistosomatic patients using Ambilhar.] Hospital (Rio J) 1969;75(1):87–95.
12. Kelani YZ, Wilson P. Experiences with Ciba 32644 Ba (Ambilhar) in the treatment of schistosomiasis. J Kuwait Med Assoc 1969;2:151.
13. Mistry CJ, Mandanna KK. Ambilhar in amebic dysentery and hepatic amebiasis. Indian J Med Sci 1968;22(10):709–12.
14. McCalla DR, Voutsinos D, Olive PL. Mutagen screening with bacteria: niridazole and nitrofurans. Mutat Res 1975;31(1):31–7.
15. El-Beheiry AH, Kamel MN, Gad A. Niridazole and fertility in bilharzial men. Arch Androl 1982;8(4):297–300.
16. Gentillini M, Capron A, Imbert JC, Escande JP, Vernes A, Domart A. Essais thérapeutiques d'un dérivé du nitrothiazole (CIBA 32644) dans la bilharziose chronique. Etude clinique et sérologique portant sur 100 malades. [Therapeutic trials of a nitrothiazole derivative in chronic bilharziasis. Clinical and serological study of 100 patients.] Bull Mem Soc Med Hop Paris 1966;117(4):323–41.
17. Coutinho A, Barreto FT. Treatment of hepatosplenic schistosomiasis mansoni with niridazole: relationships among liver function, effective dose, and side effects. Ann NY Acad Sci 1969;160(2):612–28.

Ornidazole

See also Benzimidazoles

General Information

Used in single large doses for the treatment of *Trichomonas urogenitalis* or *Giardia* infections, ornidazole can cause gastrointestinal symptoms (SEDA-11, 597) (1).

Organs and Systems

Nervous system

- A meningeal syndrome with fever has been reported in a 65-year-old man on the fourth day of administration of ornidazole (500 mg tds) (2). Spontaneous recovery occurred within a few days and without sequelae.

Liver

Ornidazole can cause hepatotoxic damage resembling acute cholestatic hepatitis (3).

References

1. Chaisilwattana P, Bhiraleus P, Patanaparnich P, Bhadrakom C. Double blind comparative study of tinidazole and ornidazole as a single dose treatment of vaginal trichomoniasis. J Med Assoc Thai 1980;63(8):448–53.
2. Mondon M, Ollivier L, Daumont A. Méningite aseptique rapportée a l'ornidazole au cours d'une endocardite infectieuse. [Aseptic meningitis ornidazole-induced in the course

of infectious endocarditis.] Rev Med Interne 2002;23(9):784–7.

3. Tabak F, Ozaras R, Erzin Y, Celik AF, Ozbay G, Senturk H. Ornidazole-induced liver damage: report of three cases and review of the literature. Liver Int 2003;23(5):351–4.

Oxamniquine

General Information

Oxamniquine, in doses of 15 mg/kg or in a single oral dose of up to 15 mg/kg bd for 2 days, depending on the sensitivity of *Schistosoma mansoni* in the area concerned, is effective with minimal toxicity. It has no effect in *Schistosoma hematobium*. It is no longer given intramuscularly, because of severe pain at the injection site.

Reactions occur in up to one-third of patients and include dizziness, drowsiness, headache, amnesia, occasional behavioral disturbances (hallucinations, excitement), and even seizures (SEDA-14, 262); there is often some nausea, vomiting, and diarrhea (1). Allergic manifestations include fever and pruritic skin rashes. Minor abnormalities of liver function and creatine kinase, proteinuria, and hematuria have been reported.

Organs and Systems

Cardiovascular

Electrocardiographic and electroencephalographic changes have been reported as rare adverse effects of oxamniquine (SEDA-11, 598; 2).

Nervous system

Oxamniquine can cause neurological adverse effects (3). At high doses, somnolence was reported in 25% of cases. Somnolence is reportedly more common if the drug is given just before a meal. In a comparison of oxamniquine with praziquantel, there were adverse effects in 45% of patients treated with oxamniquine and 71% of patients given praziquantel; somnolence occurred less often (11% of cases) with praziquantel (SEDA-12, 709; 4).

Headache and dizziness have also been reported with oxamniquine (5).

Gastrointestinal

Gastrointestinal complaints are common with oxamniquine, and take the form of anorexia, abdominal discomfort, and diarrhea (4).

Liver

Hepatic enzyme changes, initially reported as rare with oxamniquine, may in fact be frequent if sought. In one study, there were changes in 79% of patients, with raised alkaline phosphatase in 36% and a raised eosinophil count in 52% (SEDA-13, 837; 6).

Immunologic

Fever occurs in about a quarter of patients taking oxamniquine (SEDA-11, 598) (7), and in some 15% of these, a Löffler-like syndrome with eosinophilia and pulmonary infiltrates was seen when it was specifically looked for.

References

1. de Carvalho SA, Shikanai-Yasuda MA, Amato Neto V, Shiroma M, Luccas FJ. Neurotoxicidade do oxamniquine no tratamento da infeccao humana pelo *Schistosoma mansoni*. [Neurotoxicity of oxamniquine in the treatment of human *Schistosoma mansoni* infection.] Rev Inst Med Trop Sao Paulo 1985;27(3):132–42.
2. Anonymous. Drugs for parasitic infections. Med Lett Drugs Ther 1986;28(706):9–16.
3. De Carvalhi SA, Shikanai-Yasuda MA, Amato Neto V, Shiroma M, Luccas FJ. Neurotoxicidade do oxamniquine no tratamento da infeccao humana pelo. Rev Inst Med Trop Sao Paula 1985;27:132–42.
4. Taddese K, Zein ZA. Comparison between the efficacy of oxamniquine and praziquantel in the treatment of *Schistosoma mansoni* infections on a sugar estate in Ethiopia. Ann Trop Med Parasitol 1988;82(2):175–80.
5. Walder Bezerra SA. Tratamento da esquitossomose de Mansno-Piraja da Silva pela oxamniquine oral. Ceara Medico 1980;2:31.
6. Kilpatrick ME, El Masry NA, Bassily S, Farid Z. Oxamniquine versus niridazole for treatment of uncomplicated *Schistosoma mansoni* infection. Am J Trop Med Hyg 1982;31(6):1164–7.
7. Higashi GI, Farid Z. Oxamniquine fever—drug-induced or immune-complex reaction? BMJ 1979;2(6194):830.

Pentamidine

General Information

Pentamidine, an aromatic diamine, has been known since the late 1930s as a treatment for trypanosomiasis and some forms of leishmaniasis. In recent times it has been extensively used in the treatment of *Pneumocystis jiroveci* pneumonia. Its mechanism of action is probably related to inhibition of dihydrofolate reductase and inhibition of oxidative phosphorylation and nucleic acid synthesis, as well as an effect on aerobic glycolysis.

Pharmacokinetics

The pharmacokinetics of pentamidine are incompletely known. It is not absorbed after oral administration and needs to be given parenterally or by aerosol. After intramuscular injection peak concentrations are seen after about 1 hour, and the serum concentration stays about the same for 24 hours. A study of multiple dosing over 2 weeks showed progressive accumulation in the plasma during that time. After multiple intravenous doses the half-life was 12.5 days. After a course of treatment decreasing amounts of pentamidine can be found in the urine for as long as 8 weeks. It seems that pentamidine is stored or bound in the tissues and excreted slowly in the urine and the amount excreted changes only marginally with repeated doses. The highest tissue concentrations have been found in the kidney, followed by the liver and then other tissues, but pentamidine was not found in the brain. Serum concentrations after aerosol therapy are markedly lower than after intravenous therapy. Uptake via the lungs is limited, which explains the lack of serious systemic toxicity seen with aerosol treatment, but also explains the reported occurrence of extrapulmonary *Pneumocystis* infections.

These pharmacokinetic data merit attention, because they may be helpful in preventing toxicity, which seems to be dictated by tissue accumulation rather than serum concentrations; major toxic reactions usually occur only after the first week of parenteral treatment. The pharmacokinetics may also explain the varied response to treatment with aerosolized pentamidine and the more satisfactory results of prophylaxis with the aerosol after initial parenteral treatment (SEDA-13, 824; SEDA-16, 312). A retrospective study showed efficacy of intravenous pentamidine in AIDS patients with *Pneumocystis jiroveci* pneumonia, the most frequent toxic effects being gastrointestinal, especially nausea (SEDA-21, 299).

Observational studies

In an uncontrolled study in French Guiana, intramuscular pentamidine isethionate (two 4 mg/kg injections 48 hours apart) in 198 patients with cutaneous leishmaniasis produced a cure rate of 87%; 80% of treatment failures responded to an identical second course (1). Compared with published studies, adverse events were relatively mild: pain on injection (54%), gastrointestinal effects (53%), and hypotension (8%). There were no dysrhythmias or glucose abnormalities. This may reflect the brief course of pentamidine used.

Pentamidine is the drug of choice for the treatment of cutaneous leishmaniasis in Surinam. Pentamidine mesylate in 235 patients and pentamidine isethionate in 80 patients have been compared in a retrospective study; the cure rate (healing without relapse) was nearly 90% in both groups (2). Relapses occurred in about 10% of patients in both groups. Minor adverse effects, such as pain at the injection site, bitter taste, and nausea, occurred with both drugs in about 65% of patients. Respiratory tract problems occurred in under 10% of patients who took pentamidine isethionate but were uncommon in those who took pentamidine mesylate.

General adverse effects

Pentamidine in therapeutic doses has a high rate of adverse effects (over 50%). Toxicity seems to occur more often in patients with AIDS. Hypotension subsequent to injection or infusion, hypoglycemia, and nephrotoxicity are the major adverse effects. Hepatotoxicity and neutropenia are not uncommon. Compared with the general population there is a high incidence of pancreatitis in patients with AIDS, particularly during aerosol treatment. Hypoglycemia has also been seen after aerosol treatment. Intermittent prophylactic aerosol use causes few adverse effects, apart from cough and bronchial irritation after inhalation. There is a higher incidence of spontaneous pneumothorax with aerosol prophylaxis. There is also a higher incidence of extrapulmonary infections with *P. jiroveci* in patients treated with aerosolized pentamidine. Another disturbing finding was the higher incidence of other opportunistic infections with pentamidine aerosol versus placebo, while pentamidine aerosol was associated with a markedly lower recurrence rate of *P. jiroveci* pneumonitis in a placebo-controlled pentamidine aerosol study of prophylaxis after initial treatment (SEDA-13, 824). Intramuscular administration of pentamidine brings its own adverse effects: localized pain, erythema, and sterile abscesses are frequent and troublesome (SEDA-13, 824) (3). The use of the Z-track technique for injection mitigates the local effects. Allergic reactions have been reported. There is no information about tumor-inducing effects.

Organs and Systems

Cardiovascular

Severe hypotension can occur after a single intramuscular injection of pentamidine or with rapid intravenous administration, but has been seen with slow infusion as well. Infusing the drug over 60 minutes or more may reduce this risk. Facial flushing, breathlessness, dizziness, and nausea and vomiting can occur at the same time.

Cardiac dysrhythmias, including ventricular tachycardia, have been reported during treatment (SEDA-13, 824; SEDA-16, 331) (4,5). Prolongation of the QT interval, which usually precedes the development of ventricular dysrhythmias with pentamidine, occurs in one-third of patients, usually within 2 weeks of starting therapy. Torsade de pointes has been described. Any dysrhythmia can recur many days after the pentamidine has been discontinued, which is not surprising, in view of the long half-life and tissue accumulation. Electrolyte abnormalities, including low serum magnesium concentrations, have been noticed at times of dysrhythmias (SEDA-16, 315; SEDA-17, 331) (4–6).

- Torsade de pointes has been reported in a 48-year-old HIV-positive woman treated with intravenous pentamidine (7).

Local thrombophlebitis can occur after injection of pentamidine, but problems are more often seen at the injection site after intramuscular injection.

Respiratory

Inhalation of pentamidine can cause intolerable coughing. Bronchospasm can occur, especially in cases of asthma; tolerance of inhaled pentamidine is increased in nearly all patients by pretreatment with inhaled beta$_2$-adrenoceptor agonists (SEDA-16, 313; SEDA-17, 330; SEDA-18, 290) (4).

In lung function tests, high-dose aerosolized pentamidine (600 mg/month) was associated with an increased pulmonary residual volume, reduced flow rates, and increased airway reactivity (SEDA-18, 291).

There is an increased incidence of spontaneous pneumothorax after the administration of pentamidine by aerosol, which may be connected with the effect on airway resistance. There was a particularly high frequency of spontaneous pneumothorax in people with hemophilia; the authors suggested that *P. jiroveci* infection and treatment resistance had played a role (SEDA-16, 313).

Acute eosinophilic pneumonia after one dose of inhaled pentamidine of 300 mg has been reported; the reaction subsided within 2 weeks but recurred on rechallenge (SEDA-18, 292).

Dissemination of lung infection is a potentially serious matter, especially when using pentamidine by the aerosol route. While high alveolar drug concentrations can be reached in the most accessible parts of the lung, systemic absorption is minimal and the organism can spread through the lung and beyond, despite containment of the initial pulmonary infection; in some cases there has been extensive spread of *P. jiroveci* into major organs and the bone marrow (SEDA-16, 313; SEDA-17, 332). Patients who have been treated with parenteral pentamidine are at a lower risk of disseminated *P. jiroveci* infection than those given aerosol prophylaxis only (SEDA-16, 313; SEDA-17, 322).

Nervous system

Mild dizziness can occur with pentamidine, but nervous system adverse effects are uncommon.

Psychological, psychiatric

Confusion and hallucinations have occasionally been reported with pentamidine. Magnesium deficiency may affect mental function; a flat affect, slow speech, and mental withdrawal are some of the typical effects (SEDA-13, 825) (6). The symptoms of hypomagnesemia can be ill defined; unexplained symptoms, despite improvement of the *P. jiroveci* infection, demand measurement of the serum magnesium concentration.

Metabolism

Hypoglycemia can be a serious and life-threatening effect of pentamidine and is seen in 10–30% of cases, mainly with parenteral use, although it can also occur with inhalation. In one 21-day study it was equally common with either form of therapy. Higher doses and longer durations of treatment increase the likelihood, as does prior treatment with pentamidine (SEDA-16, 314); uremia also increases the risk. In one study there was nephrotoxicity in all cases with hypoglycemia. The hypoglycemia is the result of a direct toxic effect on the pancreatic beta cells, resulting in insulin release and transient hypoglycemia, which is followed by beta cell destruction and insulin deficiency, which in turn can eventually lead to an irreversible state of diabetes mellitus (SEDA-13, 825; SEDA-16, 315; SEDA-17, 331) (4).

Mineral balance

Hypocalcemia has been attributed to pentamidine but not explained (8).

Hypomagnesemia (related to excess urinary excretion of magnesium due to pentamidine) and clinical signs of magnesium deficiency can have psychiatric consequences. Magnesium deficiency itself can affect mental function; a flat affect, slow speech, and mental withdrawal are some of the typical effects (SEDA-13, 825; 6). In one case (6) hypomagnesemia was still present 2 months after intravenous pentamidine treatment, although aerosolized pentamidine was being continued and could have caused hypomagnesemia. Some patients are particularly sensitive to this effect, and previous renal damage (for example by other drugs) can be a risk factor (SEDA-13, 825).

Hematologic

Anemia, leukopenia, or thrombocytopenia occur in less than 5% of cases (SEDA-13, 825). Thrombocytopenia is more likely to occur during prolonged therapy than initially.

Megaloblastic bone marrow changes can occur with prolonged therapy (SEDA-11, 598). Low blood cell counts have been reported even after aerosolized pentamidine.

- Intravenous pentamidine caused megaloblastic anemia in a 38-year-old woman with *Pneumocystis jiroveci* pneumonia (9).

In one patient with AIDS with severe but reversible thrombocytopenia after intravenous pentamidine, the serum during the acute phase contained antiplatelet antibodies that reacted with glycoprotein IIb/IIIa, similar to the reactions observed with quinine-induced thrombocytopenia (SEDA-18, 292). This suggests that even aerosol treatment or environmental exposure will need to be avoided in such patients.

Gastrointestinal

Gastrointestinal complaints due to pentamidine are usually minor. Nausea and vomiting can occur. Dysgeusia has been reported in a few cases with intravenous therapy, but one study of aerosol administration specifically noted its absence (SEDA-13, 825).

Liver

Occasionally, abnormal liver function tests have been seen in patients receiving pentamidine (10).

Pancreas

Episodes of acute pancreatitis and of hemorrhagic pancreatitis have been reported. This may or may not be combined with evidence of damage to pancreatic beta cells (SEDA-13, 825; SEDA-16, 315; SEDA-17, 331; 4). However, pancreatitis has also been seen in patients with AIDS who did not receive pentamidine. The risk of pancreatitis seems to be greater in children with CD4 counts under 100×10^6/l. In a case-control study 12 of 44 patients with AIDS and pancreatitis had used pentamidine (SEDA-20, 264).

Urinary tract

Nephrotoxicity is common with pentamidine. In one study, serum creatinine concentrations increased in nine of 10 patients with *P. jiroveci* infection during 2–3 weeks of treatment. The incidence is 20–35% (SEDA-13, 825), and in AIDS higher still, for example 50–65% (SEDA-17, 331; 4). Renal toxicity is more pronounced in patients with diarrhea and probably more severe with intramuscular use, perhaps because of dehydration, which is more readily corrected if the drug is given by infusion. In one study there was evidence of renal toxicity in all patients with pentamidine-induced hypoglycemia, in contrast to an incidence of 38% in the group who remained euglycemic. Because of the marked accumulation of pentamidine, it is conceivable that after an initial period of daily administration a modified treatment regimen, every other day or even twice a week, may prevent renal and pancreatic permanent damage (SEDA-13, 825).

Skin

Rashes can occur with pentamidine but are not common (11).

Local pain at the injection site, local infiltration, and sterile abscesses have been reported after intramuscular pentamidine (2,3).

There have been reports of a Herxheimer reaction and of skin lesions resembling toxic epidermal necrolysis, both in children (SEDA-13, 825).

Musculoskeletal

• Two patients (aged 31 and 38 years) with cutaneous leishmaniasis given intramuscular pentamidine 600 mg twice in 48 hours developed rhabdomyolysis (12). They recovered with fluid replacement and alkaline diuresis.

Drug Administration

Drug formulations

The administration of nebulized pentamidine has an environmental impact; handling the nebulizer, cleaning and preparing it for use, and assisting the patient exposes health-care workers to pentamidine (13). Adverse reactions, such as ocular and pulmonary irritation and irritation of exposed skin, have been reported in health care workers (14).

Drug–Drug Interactions

Amiloride

Pentamidine is structurally similar to amiloride and can cause severe hyperkalemia if co-prescribed with potassium-sparing diuretics (16). This is a particularly important interaction in patients with AIDS.

Potassium-sparing diuretics

Pentamidine is structurally similar to amiloride and can cause severe hyperkalemia if co-prescribed with potassium-sparing diuretic (15). This is a particularly important interaction in patients with AIDS.

References

1. Nacher M, Carme B, Sainte Marie D, Couppie P, Clyti E, Guibert P, Pradinaud R. Influence of clinical presentation on the efficacy of a short course of pentamidine in the treatment of cutaneous leishmaniasis in French Guiana. Ann Trop Med Parasitol 2001;95(4):331–6.
2. Lai A, Fat EJ, Vrede MA, Soetosenojo RM, Lai A Fat RF. Pentamidine, the drug of choice for the treatment of cutaneous leishmaniasis in Surinam. Int J Dermatol 2002;41(11):796–800.
3. Cheung TW, Matta R, Neibart E, Hammer G, Chusid E, Sacks HS, Szabo S, Rose D. Intramuscular pentamidine for the prevention of *Pneumocystis carinii* pneumonia in patients infected with human immunodeficiency virus. Clin Infect Dis 1993;16(1):22–5.
4. Masur H. Prevention and treatment of *Pneumocystis* pneumonia. N Engl J Med 1992;327(26):1853–60.
5. Ryan C, Madalon M, Wortham DW, Graziano FM. Sulfa hypersensitivity in patients with HIV infection: onset, treatment, critical review of the literature. WMJ 1998;97(5):23–7.
6. Gradon JD, Fricchione L, Sepkowitz D. Severe hypomagnesemia associated with pentamidine therapy. Rev Infect Dis 1991;13(3):511–2.
7. Kroll CR, Gettes LS. T wave alternans and torsades de Pointes after the use of intravenous pentamidine. J Cardiovasc Electrophysiol 2002;13(9):936–8.
8. Anonymous. Pentamine isethionate. Drugs Today 1985;21:315.
9. Au WY, Ma ES, Kwong YL. Intravenous pentamidine induced megaloblastic anaemia. Haematologica 2002;87(1):ECR06.
10. Bonacini M. Hepatobiliary complications in patients with human immunodeficiency virus infection. Am J Med 1992;92(4):404–11.

11. Leen CL, Mandal BK. Rash due to nebulised pentamidine. Lancet 1988;2(8622):1250–1.
12. Lieber-Mbomeyo A, Lipsker D, Milea M, Heid E. Rhabdomyolyse induite par l'isethionate de pentamidine (Pentacarinat) lors du traitement d'une leishmaniose cutanée. 2 cas. [Rhabdomyolysis induced by pentamidine (Pentacarinat) during treatment of cutaneous leishmaniasis: 2 cases.] Ann Dermatol Venereol 2002;129(1 Pt 1):50–2.
13. Beach JR, Campbell M, Andrews DJ. Exposure of health care workers to pentamidine isethionate. Occup Med (Lond) 1999;49(4):243–5.
14. McDiarmid MA, Fujikawa J, Schaefer J, Weinmann G, Chaisson RE, Hudson CA. Health effects and exposure assessment of aerosolized pentamidine handlers. Chest 1993;104(2):382–5.
15. Perazella MA. Drug-induced hyperkalemia: old culprits and new offenders. Am J Med 2000;109(4):307–14.

Pentaquine

General Information

Pentaquine is an 8-aminoquinoline that has been used to treat malaria (1,2) and trypanosomiasis (3,4). It has adverse effects very similar to those of primaquine.

References

1. Hall WH, Latts EM. Pentaquine and quinine in the treatment of Korean vivax malaria; a controlled study in 101 patients. J Lab Clin Med 1955;45(4):573–9.
2. Eldin GN, Morcos F. Pentaquine in the treatment of malaria. J Egypt Med Assoc 1952;35(5):330–4.
3. Rubio M, Pizzi T. Accion de la primaquina, pentaquina y pentaquina-quinina, sobre formas sanguaneas virulentas de *Trypanosoma cruzi*. [Effect of primaquine, pentaquine and pentaquine-quinine on virulent blood forms of Trypanosoma cruzi.] Bol Chil Parasitol 1954;9(3):75–9.
4. Neghme A, Agosin M, Christen R, Jarpa A, Atias AV. Ensayos de quimioterapia de la enfermedad de Chagas experimental. VIII. Accion de la cortisona sola y asociada al fosfato de pentaquina o al compuesto de sulfato de quinina-fosfato de pentaquina; estudio histopatologico. [Attempts at the chemotherapy of experimental Chagas' disease. VIII. Effect of cortisone alone and associated with pentaquine phosphate or with the combination of quinine sulfate and pentaquine phosphate; histopathological study.] Bol Inf Parasit Chil 1951;6(3):36.

Piperaquine

General Information

Piperaquine is a synthetic 4-aminoquinoline with high blood schizonticidal activity, similar to that of chloroquine (SEDA-13, 810). There is some evidence that piperaquine is active against chloroquine-resistant *Plasmodium falciparum*, but laboratory studies suggest a degree of cross-resistance. Piperaquine 600 mg/month was well tolerated. Its reported adverse effects included headache, dizziness, vomiting, and diarrhea.

Organs and Systems

Gastrointestinal

The novel combination (Artekin™) of dihydroartemisinin and piperaquine has been assessed in 106 patients (76 children and 30 adults) with uncomplicated *P. falciparum* malaria in Cambodia (1). The respective doses of dihydroartemisinin and piperaquine, which were given at 0, 8, 24, and 32 hours, were 9.1 and 74 mg/kg in children and 6.6 and 53 mg/kg in adults. All the patients became aparasitemic within 72 hours. Excluding the results in one child who died on day 4, there was a 97% 28-day cure rate (99% in children and 92% in adults). Patients who had recrudescent infections used low doses of Artekin. Adverse effects, most commonly gastrointestinal complaints, were reported by 22 patients (21%) but did not necessitate premature withdrawal.

Reference

1. Denis MB, Davis TM, Hewitt S, Incardona S, Nimol K, Fandeur T, Poravuth Y, Lim C, Socheat D. Efficacy and safety of dihydroartemisinin-piperaquine (Artekin) in Cambodian children and adults with uncomplicated falciparum malaria. Clin Infect Dis 2002;35(12):1469–76.

Piperazine

General Information

Piperazine is an antihelminthic drug that selectively blocks the neuromuscular cholinergic receptors of worms. It is readily absorbed, but has a highly variable half-life. The adult oral dose of 4 g of piperazine hydrate has been used extensively in the treatment of ascariasis. A very old drug, it is still considered sufficiently safe for use, although in most developed countries it has been abandoned, primarily because of concerns about possible carcinogenicity and electroencephalographic changes (SEDA-12, 267).

General adverse effects

In most patients, piperazine is free of adverse reactions. Mild gastrointestinal disturbances may occur; neurotoxicity is rare. Eczematous skin reactions, lacrimation, rhinorrhea, joint pains, productive cough, and bronchospasm can develop after sensitization, especially with occupational exposure. Urticaria has also been reported. When hypersensitivity reactions occur it should be withdrawn and not used again in the same patient. Mononitrosylation of piperazine can occur in the stomach, releasing the potential carcinogen N-mono-nitrosopiperazine, but there is no direct proof of risk in human subjects.

Organs and Systems

Cardiovascular

Cardiac conduction defects have been described in patients taking piperazine (1).

Respiratory

Allergic respiratory reactions can occur in patients taking piperazine, resulting in cough and bronchospasm (2).

Nervous system

Headache, dizziness, and somnolence occur in a small proportion of individuals who take piperazine (3). More serious neurological reactions occur rarely, but tend to be reported in young children, in people with neurological or renal disease, or after overdosage. Symptoms in such cases include ataxia, paresthesia, undue clumsiness, myoclonus, and nystagmus. Choreiform movements and an electroencephalogram with prominent slow waves have been reported as well as an exacerbation of petit mal (4) and absence seizures (5). In a child, horizontal nystagmus and hypotonia have been reported after a normal dose (6).

- A previously well 23-month-old girl was given piperazine 65 mg/kg/day for 7 days for a suspected worm infestation and developed cerebellar ataxia after 8 days (7). Over the next 48 hours her symptoms gradually settled. She made a complete recovery 5 days later.

Sensory systems

Piperazine can cause a range of visual effects, including dry eyes, bulging eyes, difficulty in focusing, and double vision (8).

Reports of cataract after piperazine have not been authenticated (9).

Hematologic

One suspected case of hemolysis after piperazine treatment was published but it appeared as long ago as 1971 and the patient had G6PD deficiency (10).

A case of temporary thrombocytopenia has been described (11), probably due to prior sensitization by ethylenediamide (a stabilizer in some creams), with which piperazine cross-reacts.

Gastrointestinal

Nausea, vomiting, abdominal pain, and diarrhea can occur occasionally with piperazine (12).

Liver

A single incident resembling viral hepatitis after piperazine and recurring after further dosage has been reported (13).

Skin

Erythema and rarely allergic reactions can occur in patients taking piperazine (12).

Second-Generation Effects

Teratogenicity

There is no evidence that piperazine has any second-generation effects; it has been used extensively in pregnancy without untoward incidents, but as a precaution WHO advises against administration in the first trimester.

Drug–Drug Interactions

Chlorpromazine

High doses of piperazine can enhance the adverse effects of chlorpromazine and other phenothiazines (14,15).

Pyrantel

Piperazine can antagonize the antihelminthic efficacy of pyrantel and vice versa (16).

References

1. Gouffault J, Van den Driessche J, Pony JC, Courgeon P, Thomas R. Les troubles de conduction induits par la pipérazine: étude clinique et expérimentale. [Conduction disorders induced by piperazine; clinical and experimental study.] Arch Mal Coeur Vaiss 1973;66(10):1289–95.
2. McCullagh SF. Allergenicity of piperazine: a study in environmental aetiology. Br J Ind Med 1968;25(4):319–25.
3. Onuaguluchi G, Mezue WC. Some effects of piperazine citrate on skeletal muscle and central nervous system. Arch Int Pharmacodyn Ther 1987;290(1):104–16.
4. Vallat JN, Vallat JM, Texier J, Leger J. Les signes neurologiques d'intoxication par la pipérazine (à propos de deux observations recentes). [Neurologic manifestations of piperazine poisoning (apropos of 2 cases).] Bord Med 1972;5(4):391–400.

5. Yohai D, Barnett SH. Absence and atonic seizures induced by piperazine. Pediatr Neurol 1989;5(6):393–4.

6. Bomb BS, Bedi HK. Neurotoxic side-effects of piperazine. Trans R Soc Trop Med Hyg 1976;70(4):358.

7. Shroff R, Houston B. Unusual cerebellar ataxia: "worm wobble" revisited. Arch Dis Child 2002;87(4):333–4.

8. In: Frauenfelder FT, editor. Drug-Induced Ocular Side Effects and Drug Interactions. 3rd ed.. Philadelphia: Lea & Febiger, 1989:494–580.

9. Radnot M, Varga M. Structure histologique de la cataracte causée par le pipérazine. [Histologic structure of cataracts caused by piperazine.] Ann Ocul (Paris) 1969;202(4):325–9.

10. Buchanan N, Cassel R, Jenkins T. G-6-PD deficieny and piperazine. BMJ 1971;2(753):110.

11. Cork MJ, Cooke NJ, Mellor E. Pruritus ani, piperazine, and thrombocytopenia. BMJ 1990;301(6765):1398.

12. Point G. Incidents neurologiques lors de l'utilisation de la pipérazine comme vermifuge. [Neurologic complications during the use of piperazine as a vermifuge.] Pediatrie 1965;20(5):600–4.

13. Hamlyn AN, Morris JS, Sarkany I, Sherlock S. Piperazine hepatitis. Gastroenterology 1976;70(6):1144–7.

14. Sturman G. Interaction between piperazine and chlorpromazine. Br J Pharmacol 1974;50(1):153–5.

15. Boulos BM, Davis LE. Hazard of simultaneous administration of phenothiazine and piperazine. N Engl J Med 1969;280(22):1245–6.

16. Aubry ML, Cowell P, Davey MJ, Shevde S. Aspects of the pharmacology of a new anthelmintic: pyrantel. Br J Pharmacol 1970;38(2):332–44.

Pneumocandins

General Information

The pneumocandins are a class of anti-*Pneumocystis* agents, thought to act by inhibiting the synthesis of beta-1,3-glucan, a component of the *Pneumocystis jiroveci* cyst wall. Chemical modification of the poorly water-soluble pneumocandins resulted in L-693,989, which can be given by aerosol. In rats, this compound prevents *P. jiroveci* pneumonia after daily or weekly administration, without evidence of toxicity. Since there is no counterpart for beta-1,3-glucan synthesis in humans, there are no obvious reasons to suspect mechanism-based toxicity with this class of compounds. More effective compounds are being studied (1).

Reference

1. Powles MA, McFadden DC, Liberator PA, Anderson JW, Vadas EB, Meisner D, Schmatz DM. Aerosolized L-693,989 for *Pneumocystis carinii* prophylaxis in rats. Antimicrob Agents Chemother 1994;38(6):1397–401.

Praziquantel

General Information

Praziquantel, initially introduced as a veterinary cesticidal drug, was found to have efficacy against all the human species of schistosomes (including cerebral forms), fasciolopsiasis, cysticercosis, paragonimiasis (lung fluke), *Clonorchis sinensis* (oriental liver fluke), and *Opisthorchis viverrini* infections. It is used in a single oral dose of 40 mg/kg, except in infestations with *Schistosoma japonicum*, *Clonorchis*, and *Paragonimus*, for which more prolonged administration is required, doses varying from 40 to 70 mg/kg/day depending on the infecting species. Evaluation of the result of praziquantel treatment has shown a lower laboratory rate of success than reported for clinical response (SEDA-16, 311).

Praziquantel is effective in human cysticercosis in doses of 10–100 mg/kg for 3–21 days (1). Initially, longer courses of praziquantel were advocated, but even shorter treatment regimens are equally effective: a complete course can be administered in a single day with comparable efficacy as conventional therapy of 15 days. Praziquantel was originally introduced as a racemic mixture; there is evidence that the levorotatory isomer is relatively more effective, but has the same incidence of adverse reactions (2).

Praziquantel is well absorbed and penetrates cyst walls, but it undergoes first-pass metabolism, especially when given together with glucocorticoids and anticonvulsants.

Observational studies

Echinococcosis

The management and operative complications in 70 patients with hydatid disease aged 10–78 years have been studied retrospectively to assess the impact of albendazole and praziquantel compared with surgery (3). In all, 39 patients received albendazole and praziquantel in combination and 19 received albendazole alone; none was treated with praziquantel alone. The combined use of albendazole and praziquantel preoperatively significantly reduced the number of cysts that contained viable protoscolices.

Neurocysticercosis

Praziquantel is generally used to treat neurocysticercosis in an oral dose of 50 mg/kg/day (divided into three doses) for 15 days. Sedation can be marked and driving should be avoided. Liver enzymes sometimes increase. Ocular cysticercosis should not be treated with praziquantel, because destruction of the parasite within the eye can cause irreparable lesions.

A one-day intensive course of praziquantel has been used experimentally, but results vary, and when there are multiple brain cysts present the outcome is poor (4). When used for this purpose in a dose of 25 mg/kg at 2-hour intervals, adverse effects included mild headache, dizziness, nausea, and vomiting. All the adverse effects remitted with analgesics or dexamethasone 0.2 mg/kg/day and continued for 2 days.

Paragonimiasis

Paragonimiasis is a food-borne parasitic disease common in Southeast Asia, especially in Japan, Korea, The Philippines, Taiwan, and parts of China. In Japan, paragonimiasis is caused by either *Paragonimus westermani* or *Paragonimus miyazakii*. Traditionally biothionol was used to treat paragonimiasis, a food-borne zoonosis that is endemic in limited areas of the world. However, owing to the need for long-term administration and moderate to severe adverse effects, such as nausea and diarrhea, biothionol has been replaced by praziquantel. At a dose of 75 mg/kg/day for only 2–3 days, praziquantel has the advantage of an easier dosing schedule in combination with excellent therapeutic efficacy. Adverse effects of praziquantel in paragonimiasis, if any, are mild and transient (5). In patients with pleural effusion, pleural fluid must be drained before starting chemotherapy; insufficient drainage often causes complications, such as chronic empyema or insufficient inflation of the lungs. Praziquantel is also effective for cutaneous, cerebral, or any form of extrapulmonary paragonimiasis.

The radiological features and treatment of paragonimiasis have been described in 13 patients (10 men, three women, aged 25–77 years) (6). All were treated with praziquantel 75 mg/kg/day for 2–3 days. One patient with empyema was also given bithionol. There was mild urticaria in two patients and no serious adverse effects.

Schistosomiasis

The current chemotherapeutic armory for schistosomiasis has been reviewed (7). Since the development of praziquantel in 1970, morbidity and mortality due to schistosomiasis has been significantly reduced. Nevertheless, praziquantel does not prevent reinfection, and repeated treatments are usually necessary in endemic areas. In addition, there are at least two threatening consequences of relying on single-drug therapy. One is the possibility of infection with a praziquantel-resistant strain and the other is a increase in the frequency of acute manifestations of the disease, for which praziquantel is not effective. Its most common adverse effects are usually mild and include headache, dizziness, nausea, diarrhea, abdominal pain, and anorexia.

The efficacy and adverse effects of two courses of praziquantel (40 mg/kg 4 weeks apart) for *Schistosoma haematobium* infection have been evaluated in 354 school children aged 5–15 years (8). The two doses of praziquantel were highly effective. Of the 354 children, 165 complained of one (33%), two (11%), or even three symptoms (2.5%). These adverse effects occurred within 1 hour of treatment, were mild, and gradually resolved without specific interventions; abdominal pain (18%), nausea (12%), headache (9.6%), and dizziness (9.6%) were the most common. None of the symptoms

appeared to be related to the intensity of the infection. Significantly more girls than boys complained of vomiting, dizziness, and abdominal pain and they were more frequent among older children. There were fewer adverse effects after the second dose. Only 43 (13%) of the children reported adverse effects after the second treatment dose, and headache was the most prevalent. The children who reported adverse effects after the first dose were no more likely to report adverse effects after the second dose than children who were asymptomatic after the first dose.

The effect of a single oral dose of praziquantel 40 mg/kg against *Schistosoma mansoni* has been studied in a rural community in Côte d'Ivoire (9). Among the 200 individuals treated, 25 reported one or more adverse effects within 24 hours after treatment. A significantly higher proportion of individuals with high infection intensities (>over 400 *Schistosoma mansoni* eggs/gram of feces) reported diarrhea and dizziness compared with those with light or moderate infections. None of the other adverse effects, such as itching (n = 5), vomiting (n = 2), headache (n = 2), nausea (n = 2), and urticaria (n = 1) was associated with the severity of the infection, nor with age. Abdominal pain was the only adverse effect associated with (female) sex.

Tapeworm infections

The efficacy and safety of praziquantel in *Diphyllobothrium nihonkaiense* infections has been studied in 14 Japanese men who took a single dose of praziquantel 5–10 mg/kg (10). All were cured and had not expelled proglottides after 1 year of follow-up. There were no adverse effects.

Tenia infections

In a report from India the efficacy and safety of treatment of niclosamide-resistant *Tenia saginata* infections with praziquantel 10 mg/kg orally has been confirmed in 185 consecutive patients (11). Follow-up stool examinations at 4 and 12 weeks showed a cure rate of 96%. Eleven patients were lost to follow-up, and eight still produced proglottides at the end of 12 weeks. None passed the worm in their stools, since praziquantel destructs the worm, after which the scolex and worm are digested. Thirty patients (16%) reported minimal adverse effects, such as nausea (n = 4), abdominal discomfort (n = 10), and giddiness (n = 16).

Comparative studies

Neurocysticercosis

Praziquantel 100 mg/kg in three divided doses has been compared with albendazole 15 mg/kg/day for 1 week in the treatment of neurocysticercosis in 20 patients (12). In the patients treated with albendazole the number of cysts fell from 64 to 7 and in the patients treated with praziquantel it fell from 59 to 24. The difference was not statistically significant. All the patients were concomitantly treated with high doses of glucocorticoids. Nine of the ten patients treated with praziquantel had seizures, headache, and dizziness, and two had hemiparesis before treatment. A few hours after the last dose of praziquantel six patients had adverse reactions, including headache and vomiting in five patients, seizures in one patient, and worsening of the pre-existing motor deficit in one. Analgesics and antiemetics improved the symptoms in four patients. In two patients treatment with mannitol was needed to relief symptoms of increased cranial pressure. The results suggested that single-day treatment with praziquantel of neurocysticercosis may be a useful option. The observed adverse effects were considered to be the result of the inflammatory reaction following the dying of worms and not a toxic effect of the drug itself.

Schistosomiasis

Oxamniquine and praziquantel have been compared in a triple-masked, randomized, controlled trial in 106 patients with *Schistosoma mansoni* infections (13). They were randomized to treatment with praziquantel 60 mg/kg/day on 3 consecutive days, oxamniquine 10 mg/kg twice on 1 day followed by placebo on days 2 and 3; starch for 3 consecutive days. When cure was evaluated by stool examination, oxamniquine and praziquantel had cure rates of 90% and 100% respectively. However, when the oogram was used as an indicator of sensitivity, the oxamniquine cure rate fell to 42% whereas the rate for praziquantel remained high, at 96%. The adverse effects of the two drugs were similar and mild; the most common adverse effects were headache, dizziness, drowsiness, and abdominal pain. Patients who took the placebo also had drowsiness and abdominal pain.

Placebo-controlled studies

In a double-blind, randomized, placebo controlled trial of praziquantel in 42 patients with clonorchiasis and an open study in 32 patients, the adverse effects of praziquantel were transient and included nausea and vomiting (15%), vertigo (12%), hepatomegaly (4.5%), headache (1.5%), rash (1.5%), and hypotension (1.5%) (14). Of 20 patients who received placebo, one developed a transient skin rash, fever, and chills. There were minor and transient, albeit statistically significant, changes in hemoglobin and serum concentrations of total protein, uric acid, cholesterol, and bilirubin.

General adverse effects

Although adverse effects are common they tend to be mild and it is rarely necessary to withdraw praziquantel for this reason; even when used in relatively high doses, as in schistosomiasis, praziquantel has, according to the *British National Formulary* (September 2005), "the most attractive combination of effectiveness, broad-spectrum activity, and low toxicity [of all antihelminthic drugs]." The most common problems relate not so much to direct adverse effects of the drug as to allergic and inflammatory reactions in the host to the presence of dying parasites. Fever, headache, meningism, and exacerbation of neurological symptoms have all been noted; these symptoms tend to be more severe when the pretreatment parasitic infection is widespread and intense (SEDA-20, 282; SEDA-21, 318). The main direct adverse effects of praziquantel are abdominal discomfort, diarrhea, nausea and vomiting, dizziness, and somnolence (SED-13, 837; SEDA-16, 310) (15). Among children somnolence was seen in 11%. Headaches, skin rashes, and fever are less common (SED-13, 837). Non-specific effects observed in a minority of patients taking praziquantel include generalized weakness, swelling of the legs, epigastric area, scrotum, or more generally, fatigue (SEDA-16, 354).

Organs and Systems

Respiratory

Acute respiratory failure with exudative polyserositis has been reported in a patient taking praziquantel (16).

Nervous system

An important aspect of drug treatment in neurocysticercosis is the simultaneous use of glucocorticoids with cysticidal drugs (17,18). This combination has been recommended to avoid the secondary effects of treatment due to destruction of parasites within the brain parenchyma. However, these reactions are usually mild and transient and may be ameliorated with analgesics or antiemetics, questioning the need for corticosteroids in every case. Glucocorticoids are currently indicated for patients who develop intracranial hypertension during treatment with cysticidal drugs. This can be anticipated in patients with multiple lesions. However, some forms of neurocysticercosis should not be treated with cysticidal drugs (17,18). Both albendazole and praziquantel can exacerbate the syndrome of intracranial hypertension observed in patients with cysticercotic encephalitis, and are contraindicated during the acute phase of the disease (1). In patients with mixed forms of neurocysticercosis, including hydrocephalus and parenchymal brain cysts, cysticidal drugs should only be used after prior ventricular shunt placement to avoid a further increase in intracranial pressure after treatment. When this precaution is not taken, a further increase in intracranial pressure precipitated by praziquantel can prove fatal (19).

- A 66-year-old man with neurocysticercosis treated with glucocorticoids and praziquantel developed headache and confusion. He did not have a ventricular shunt inserted. A contrast-enhanced CT scan showed multiple focal enhancing lesions with mild edema. An MRI scan of the head was reported as being most consistent with neurocysticercosis. He was given dexamethasone 2 mg bd and praziquantel 50 mg/kg/day. A few days later his headache worsened, with nausea and drowsiness. After 2 weeks he became stuporose and had to be ventilated. A CT scan showed multiple areas of deep subcortical focal edema near the areas of previously enhancing cysts, a striatocapsular stroke, and obstructive hydrocephalus. Two weeks after the last dose of praziquantel and despite a ventriculostomy tube he died.

The authors reported that deaths related to praziquantel in neurocysticercosis are rare. However, this case had characteristics suggestive of a high risk of post-treatment complications. Death was attributed to a sudden increase in intracranial pressure, with multiple foci of edema, meningeal inflammation, and stroke, and occurred despite the concomitant use of glucocorticoids.

In patients with neurocysticercosis treated with praziquantel in increasing doses of 10–50 mg/kg/day for the first week and maintenance therapy during the second week, 27 (60%) presented with adverse effects, three requiring interruption of therapy. Increased intracranial pressure occurred in two cases (one fatal). Exacerbation of CSF pleocytosis was recorded in 26 patients (57%) (SEDA-16, 311).

A delayed reaction, with central nervous system involvement, has been described in patients with cerebral cysticercosis; papilledema, hemorrhages, focal seizures, motor weakness (SEDA-13, 242), and in one case hemiplegia from a vasculitic infarct occurred (SEDA-14, 243). The possibility that this reaction is caused by a massive inflammatory response was discussed (SEDA-14, 243). If this hypothesis is correct, glucocorticoid treatment could be contemplated. A review has suggested that the use of glucocorticoids, although still controversial, is generally accepted as a means of alleviating inflammatory complications; however, simultaneous use of a glucocorticoid seems to reduce the plasma praziquantel concentration significantly. According to one group of investigators, no delayed reaction was seen in patients with cerebral schistosomiasis (SED-13, 837).

Dose-dependent dizziness is a recognized effect in some 14% of cases at higher doses of praziquantel (SEDA-12, 267). Patients predisposed to epilepsy may develop convulsions, probably as a reaction to the death of the parasite (20). Headache (40%) and drowsiness (25–40%) have been common in some studies and the manufacturers warn against driving or operating machinery while taking praziquantel.

Most patients treated for neurocysticercosis with praziquantel develop an early cerebrospinal fluid reaction; a similar late reaction, some 2 weeks after treatment has finished, has also been described (21). In both cases clinical signs and symptoms can include papilledema, headache, nausea, vomiting, neck stiffness, and even focal seizures. Glucocorticoids can usually prevent or relieve both the early and late reactions, but they can also reduce efficacy by lowering plasma concentrations of the drug by some 50% (22).

There is a risk that if patients treated with praziquantel for a disease other than neurocysticercosis do in fact also have neurocysticercosis, serious neurological reactions

(notably seizures) can occur as the parasite is killed and toxins are released (23).

Sensory systems

Praziquantel should not be used in ocular cysticercosis, because of the risk of inoperable lesions from destruction of the parasite within the eye (24).

Hematologic

Treatment of human schistosomiasis rapidly causes *eosinophilia* and this provides an opportunity to study the development of eosinophilia in relation to other immunological events induced by the release of worm antigens after treatment and the mechanisms that prevent systemic post-treatment hypersensitivity reactions, which might be expected to occur in the presence of high concentrations of circulating worm-specific IgE. This was studied in 69 fishermen (mean age 35, range 18–45 years), infected with *S. mansoni*, who had lived in Bugoigo, Uganda for at least 3 years (25). After praziquantel 40 mg/kg, blood eosinophil numbers fell within 24 hours, and significant eosinophilia developed after 3 weeks. Data based on measurements of cellular eosinophil cationic protein content, cellular eosinophil protein X, interleukins 5, 6, and 10, and tumor necrosis factor alpha and eotaxin concentrations before treatment and 24 hours and 3 weeks after treatment suggested that blood eosinophils are activated during *S. mansoni* infection and that treatment induces a burst of released antigens, causing increased production of IL-5, IL-6, IL-10, and eotaxin and a fall in TNF-alfa production, combined with transient sequestration of eosinophils. The data suggested that infection intensity-dependent concentrations of plasma IL-10, a major anti-inflammatory and regulatory cytokine that inhibits mast cell degranulation, may be involved in the prevention of treatment-induced anaphylactic reactions. It is certainly noteworthy that severe allergic reactions to praziquantel in schistosomiasis are not as frequent as they are after treatment of filariasis, in which such reactions are significantly associated with microfilarial density and fluctuations in circulating eosinophil numbers.

- A 46-year-old Korean woman developed left upper quadrant pain after taking praziquantel for suspected hepatic fascioliasis (26). Peripheral eosinophilia, hyperamylasemia, and hyperlipasemia were documented and ascribed to pancreatic fascioliasis. IgG antibodies to *Fasciola hepatica* were positive. Abdominal CT showed multiple hypodense foci, which had coalesced, forming irregular nodules in the left lobe of the liver, and similar lesions in the pancreas. Treatment with biothionol resulted in reversal of the eosinophilia and improvement in the patient's symptoms and laboratory findings. After 10 weeks imaging studies showed normal liver and pancreas.

Gastrointestinal

Praziquantel can cause dose-related nausea, vomiting, heartburn, and abdominal pain. For example, at doses of 30 mg/kg in schoolchildren there was stomachache in some 16% (SED-12, 775), while in a study with 40 mg/kg 35% of users had abdominal pain (27).

Bloody diarrhea occurs in some patients, but it can be difficult to distinguish this as an adverse effect from pretreatment symptoms; one Ethiopian study of the use of praziquantel in suspected schistosomiasis found that before treatment there was blood in the stool in 55% of cases, diarrhea in 61%, and abdominal discomfort in 80%, and the figures recorded the next day after treatment were not very different (28).

The efficacy and adverse effects of treatment with a single oral dose of praziquantel 40 mg/kg, in relation to egg counts and morbidity, have been studied in 611 primary schoolchildren infected with *Schistosoma mansoni* in Northeastern Ethiopia (29). Before treatment 40% of the patients had no symptoms and 30–40% complained of nausea, abdominal cramps, and/or bloody diarrhea. The symptoms before treatment were not related to nutritional status, intensity of *S. mansoni* egg excretion, or the presence of concomitant intestinal parasites. In the first 4–6 hours, 90 children (15%) developed severe gastrointestinal symptoms, with vomiting, abdominal cramps, and/or bloody diarrhea. They had higher mean pretreatment egg counts than the children who did not have these symptoms. The day after treatment 529 children (87%) were reviewed. Adverse effects were reported by 92% and consisted of abdominal cramps (87%), bloody diarrhea (50%), dizziness (31%), and vomiting (29%). Skin rashes and edema were observed in four individuals. The combination of abdominal cramps with vomiting, bloody diarrhea, and general weakness was significantly more common in the malnourished children and in the children with higher pretreatment egg counts. The overall cure rate after treatment with praziquantel was 83% after 5 weeks, but this rate fell with increasing pretreatment egg counts. These findings confirm that praziquantel is effective in the treatment of *S. mansoni* infections but that treatment may be associated with severe abdominal adverse effects, which may reduce drug compliance in population chemotherapy.

This point has been further evaluated in a larger double-blind, placebo-controlled study of the concurrent administration of albendazole and praziquantel in over 1500 children with high prevalences of schistosomiasis and other helminthic diseases in China and the Philippines, including two strains of *S. japonicum*, and two different areas of Kenya, one each with *S. mansoni* or *Schistosoma hematobium* (30). There was no difference in the rate of adverse effects after treatment with albendazole compared with placebo, but after treatment with praziquantel the children had significantly more nausea, abdominal pain, and headache. These adverse effects, although considered mild, were more common in children with schistosomiasis, which suggests a reaction to dying schistosomes rather than a toxic effect of the drug itself. There was a very high rate of complaints in Kenya, but both the history and the reactions after placebo suggested that many of the adverse effects reported after treatment reflected complaints before treatment.

Liver

It has been suggested that praziquantel can cause hepatomegaly and splenomegaly in children with schistosomiasis (30), but hepatomegaly and splenomegaly have also been described as complications of the disease in these young patients, and they regress with effective treatment (31).

Immunologic

Allergic reactions to praziquantel can be due to parasite death and include fever, urticaria, pruritic skin rashes, and eosinophilia. In one violent reaction there was marked eosinophilia, pleuritic chest pain, cardiac effusion, and ascites, pointing strongly to an exudative polyserositis (16).

Long-Term Effects

Drug resistance

Although praziquantel is currently the drug of choice for schistosomiasis, there is concern that schistosomes might become resistant.

- A previously healthy 26-year-old British man travelled to Kenya for 10 weeks and acquired *Schistosoma mansoni* (32). He received standard treatment with praziquantel 40 mg/kg, with no adverse effects. Evidence of active infection persisted after three courses of praziquantel in 4 years. Since oxamniquine as an alternative treatment agent was not available in the UK, he received a prolonged course of praziquantel 40 mg/kg/day for 3 consecutive days, with suggestion of a response to treatment. There was some response to treatment, suggesting tolerance to praziquantel or in part dose-dependent resistance.

Unfortunately, although this patient was followed for several years, the authors did not provide details about the follow-up nor did they mention polymerase chain reaction-based studies on this potential praziquantel-resistant *Schistosoma mansoni* strain. In general, resistance to praziquantel has occasionally been suggested but rarely or not at all observed in travellers. From studies in regions in which *Schitsosoma mansoni* is endemic, it usually appeared that although parasite resistance or tolerance to praziquantel could not be definitively excluded, poor cure rates after praziquantel were more likely to be due to high pre-treatment parasite burdens, high rates of re-infections, or schistosome immaturity.

Diminished susceptibility to praziquantel of *S. mansoni* was reported in an area in Northern Senegal (33).

Second-Generation Effects

Pregnancy

In a prospective study eastern Sudan in 25 pregnant women with *Schistosoma mansoni* infection were given a single oral dose of praziquantel at 40 mg/kg, six in the first trimester, 12 in the second trimester, and seven in the third trimester (34). There were no maternal deaths, stillbirths, or congenital abnormalities.

Teratogenicity

In a retrospective study, 88 women who took praziquantel during pregnancy were compared with a control group of 549 women who had not taken praziquantel (35). There were no significant differences between the groups in the rates of abortion or preterm deliveries. There were no congenital abnormalities in any of the babies born to either group. These data suggest that praziquantel during pregnancy is not associated with an increased risk of congenital abnormalities.

With the introduction of praziquantel in the early 1980s, a safe, single-dose antihelminthic drug became available for the treatment of schistosomiasis. In addition, several other parasitic diseases, such as clonorchiasis, paragonimiasis, and cysticercosis, could be cured. Unfortunately, praziquantel was not tested in pregnant and lactating women before marketing, despite its lack of known toxicity. It was therefore released as a pregnancy category B drug (presumed to be safe based on animal studies). In consequence, a large number of pregnant and lactating women living in endemic countries are currently not treated in targeted antihelminthic mass treatment programs or are treated after a significant delay. The available evidence on the toxicology of praziquantel, combined with over two decades of clinical experience has been reviewed with special emphasis on pregnancy and lactation (36). In contrast to most other antihelminthic drugs, praziquantel, given either acutely or chronically and in doses well above those routinely used in humans had no detrimental effect on experimental animals or in reproductive studies in rats, rabbits, and hamsters. Since its release, no cases have been reported that suggest adverse birth outcomes. In addition, several pregnant women have intentionally taken praziquantel for cysticercosis with no apparent adverse outcomes. In some mass treatment studies, several hundred pregnant women were inadvertently treated without adverse birth outcomes.

Lactation

Praziquantel is excreted in the breast milk, and mothers should not breastfeed for 72 hours after a dose (37).

Drug–Drug Interactions

Cimetidine

Praziquantel is well absorbed after oral administration and undergoes extensive first-pass metabolism, especially when it is given simultaneously with corticosteroids and anticonvulsants (1). Cimetidine significantly increases praziquantel serum concentrations by inhibiting its first-pass metabolism (38). The concurrent administration of cimetidine with praziquantel in neurocysticercosis allows simplification of the effective praziquantel regimen from 50 mg/kg/day for 2 weeks to a one-day regimen of three doses of 25 mg/kg at 2-hour intervals; this regimen increases the time over which the parasite is exposed to high drug concentrations and produces similar benefits, and also reduces the cost, length of treatment, and total dose used (39). It has also been proposed that simultaneous administration of praziquantel and cimetidine could improve the efficacy of single-day therapy with praziquantel for cysticercosis and other parasitic diseases, such as schistosomiasis (40).

Glucocorticoids

Serum concentrations of praziquantel fall when glucocorticoids are used simultaneously, which is usually the case in moderate to heavy infections; no mechanism has bee proposed for this effect (41).

Food–Drug Interactions

Serum praziquantel concentrations increase when a carbohydrate-rich diet is administered (1).

Grapefruit juice

A single dose of grapefruit juice 250 ml significantly increased the AUC and C_{max} of a single dose of praziquantel without changing the t_{max} or half-life, suggesting increased systemic availability (42).

References

1. Garg RK. Medical management of neurocysticercosis. Neurol India 2001;49(4):329–37.
2. Yue-Han L, Xiao-Gen W, Min-Xin Q, et al. A comparative trial of single dose treatment with praziquantel and levopraziquantel in human *Schistosomiasis japonica*. Jpn J Parasitol 1988;37:331.
3. Ayles HM, Corbett EL, Taylor I, Cowie AG, Bligh J, Walmsley K, Bryceson AD. A combined medical and surgical approach to hydatid disease: 12 years' experience at the Hospital for Tropical Diseases, London. Ann R Coll Surg Engl 2002;84(2):100–5.
4. Pretell EJ, Garcia HH, Gilman RH, Saavedra H, Martinez M. Cysticercosis Working Group in Peru. Failure of one-day praziquantel treatment in patients with multiple neurocysticercosis lesions. Clin Neurol Neurosurg 2001;103(3):175–7.
5. Nakamura-Uchiyama F, Mukae H, Nawa Y. Paragonimiasis: a Japanese perspective. Clin Chest Med 2002;23(2):409–20.
6. Mukae H, Taniguchi H, Matsumoto N, Iiboshi H, Ashitani J, Matsukura S, Nawa Y. Clinicoradiologic features of pleuropulmonary *Paragonimus westermani* on Kyusyu Island, Japan. Chest 2001;120(2):514–20.
7. Ribeiro dos Santos, Verjovski-Almeida S, Leite LCC. Schistosomiasis – a century searching for chemotherapeutic drugs. Parasitol Res 2006;99:505-21.

8. N'Goran EK, Gnaka HN, Tanner M, Utzinger J. Efficacy and side-effects of two praziquantel treatments against *Schistosoma haematobium infection, among schoolchildren from Côte d'Ivoire.* Ann Trop Med Parasitol 2003;97:37-51.

9. Raso G, N'Goran EK, Toty A, Luginbuhl A, Adjoua CA, Tian-Bi NT, Bogoch II, Vounatsou P, Tanner M, Utzinger J. Efficacy and side effects of praziquantel against Schistosoma mansoni in a community of western Côte d'Ivoire. Trans R Soc Trop Med Hyg 2004;98:18-27.

10. Ohnishi K, Kato Y. Single low-dose treatment with praziquantel for *Diphyllobothrium nihonkaiense infections.* Intern Med 2003;42:41-3.

11. Koul PA, Waheed A, Hayat M, Sofi BA. Praziquantel in niclosamide-resistant *Taenia saginata* infection. Scand J Infect Dis 1999;31(6):603-4.

12. Del Brutto OH, Campos X, Sanchez J, Mosquera A. Single-day praziquantel versus 1-week albendazole for neurocysticercosis. Neurology 1999;52(5):1079-81.

13. Ferrari MLA, Coelho PMZ, Antunes CMF, Tavares CAP, Da Cunha AS. Efficacy of oxamniquine and praziquantel in the treatment of *Schistosoma mansoni infection: a controlled trial.* Bull WHO 2003;81:190-6.

14. Yangco BG, De Lerma C, Lyman GH, Price DL. Clinical study evaluating efficacy of praziquantel in clonorchiasis. Antimicrob Agents Chemother 1987;31(2):135-8.

15. Pungpak S, Bunnag D, Harinasuta T. Studies on the chemotherapy of human opisthorchiasis: effective dose of praziquantel in heavy infection. Southeast Asian J Trop Med Public Health 1985;16(2):248-52.

16. Azher M, el-Kassimi FA, Wright SG, Mofti A. Exudative polyserositis and acute respiratory failure following praziquantel therapy. Chest 1990;98(1):241-3.

17. Di Pentima MC, White AC. Neurocysticercosis: controversies in management. Semin Pediatr Infect Dis 2000;11:261-8.

18. Del Brutto OH. Medical therapy for cysticercosis: indications, risks, and benefits. Rev Ecuat Neurol 2000;9:13-5.

19. Chang GY, Ko DY. Isolated Echinococcus granulosus hydatid cyst in the CNS with severe reaction to treatment. Neurology 2000;54(3):778-9.

20. Bada JL, Trevino B, Cabezos J. Convulsive seizures after treatment with praziquantel. BMJ (Clin Res Ed) 1988;296(6622):646.

21. Ciferri F. Delayed CSF reaction to praziquantel. Lancet 1988;1(8586):642-3.

22. Del Brutto OH. Delayed CSF reaction to praziquantel. Lancet 1988;2(8606):341.

23. Torres JR. Use of praziquantel in populations at risk of neurocysticercosis. Rev Inst Med Trop Sao Paulo 1989;31(4):290.

24. Auzemery A, Andriantsimahavandy A, Bernardin P, Queguiner P. La cysticercose intravitréenne. Evolution spontanée. A propos d'un cas. [Intravitreous cysticercosis. Spontaneous course. Apropos of a case.] J Fr Ophtalmol 1996;19(8-9):556-8.

25. Reimert CM, Fitzsimmons CM, Joseph S, Mwatha JK, Jones FM, Kimani G, Hoffmann KF, Booth M, Kabatereine NB, Dunne DW, Vennervald BJ. Eosinophil activity in schistosoma mansoni infections in vivo and in vitro in relation to plasma cytokine profile pre- and posttreatment with praziquantel. Clin Vacc Immunol 2006;13(5):584-93.

26. Lee OJ, Kim TH. Indirect evidence of ectopic pancreatic fascioliasis in a human. J Gastroenterol Hepatol 2006;21(10):1631-3.

27. Jaoko WG, Muchemi G, Oguya FO. Praziquantel side effects during treatment of *Schistosoma mansoni* infected pupils in Kibwezi, Kenya. East Afr Med J 1996;73(8):499-501.

28. Fletcher M, Teklehaimanot A. *Schistosoma mansoni* infection in a new settlement in Metekel district, north-western Ethiopia: morbidity and side effects of treatment with praziquantel in relation to intensity of infection. Trans R Soc Trop Med Hyg 1989;83(6):793-7.

29. Berhe N, Gundersen SG, Abebe F, Birrie H, Medhin G, Gemetchu T. Praziquantel side effects and efficacy related to *Schistosoma mansoni* egg loads and morbidity in primary school children in north-east Ethiopia. Acta Trop 1999;72(1):53-63.

30. Olds GR, King C, Hewlett J, Olveda R, Wu G, Ouma J, Peters P, McGarvey S, Odhiambo O, Koech D, Liu CY, Aligui G, Gachihi G, Kombe Y, Parraga I, Ramirez B, Whalen C, Horton RJ, Reeve P. Double-blind placebo-controlled study of concurrent administration of albendazole and praziquantel in schoolchildren with schistosomiasis and geohelminths. J Infect Dis 1999;179(4):996-1003.

31. Stephenson LS, Latham MC, Kinoti SN, Oduori ML. Regression of splenomegaly and hepatomegaly in children treated for *Schistosoma haematobium* infection. Am J Trop Med Hyg 1985;34(1):119-23.

32. Lawn SD, Lucas SB, Chiodini PL. Case report: *Schistosoma mansoni infection: failure of standard treatment with praziquantel in a returned traveller.* Trans R Soc Trop Med Hyg 2003;97:100-1.

33. Fallon PG, Sturrock RF, Niang AC, Doenhoff MJ. Short report: diminished susceptibility to praziquantel in a Senegal isolate of *Schistosoma mansoni.* Am J Trop Med Hyg 1995;53(1):61-2.

34. Adam I, Elwasila E, Homeida M. Praziquantel for the treatment of schistosomiasis mansoni during pregnancy. Ann Trop Med Parasitol 2005;99(1):37-40.

35. Adam I, Elwasila ET, Homeida M. Is praziquantel therapy safe during pregnancy? Trans R Soc Trop Med Hyg 2004;98:540-3.

36. Olds GR. Administration of praziquantel to pregnant and lactating women. Acta Trop 2003;86:185-95.

37. Putter J, Held F. Quantitative studies on the occurrence of praziquantel in milk and plasma of lactating women. Eur J Drug Metab Pharmacokinet 1979;4(4):193-8.

38. Castro N, Gonzalez-Esquivel D, Medina R, Sotelo J, Jung H. The influence of cimetidine on plasma levels of praziquantel after a single day therapeutic regimen. Proc West Pharmacol Soc 1997;40:33-4.

39. Sotelo J, Jung H. Pharmacokinetic optimisation of the treatment of neurocysticercosis. Clin Pharmacokinet 1998;34(6):503-15.

40. Jung H, Medina R, Castro N, Corona T, Sotelo J. Pharmacokinetic study of praziquantel administered alone and in combination with cimetidine in a single-day therapeutic regimen. Antimicrob Agents Chemother 1997;41(6):1256-9.

41. Garcia HH, Del Bruto, for the Cysticercosis Working Group in Peru. Neurocysticercosis: updated concepts about an old disease. Lancet Neurol 2005;4:653-61.

42. Castro N, Jung H, Medina R, Gonzalez-Esquivel D, Lopez M, Sotelo J. Interaction between grapefruit juice and praziquantel in humans. Antimicrob Agents Chemother 2002;46(5):1614-6.

Primaquine

General Information

The 8-aminoquinolines were the first synthetic antimalarial drugs to be introduced into medicine. Pamaquine (Plasmochin) was the first to be marketed in 1926, but primaquine proved to have the highest chemotherapeutic index of the many compounds tested.

Primaquine is rapidly absorbed, extensively distributed, and predominantly cleared by non-renal elimination. Its principal metabolite is carboxyprimaquine. While primaquine itself is rapidly eliminated from the plasma, the drug is effective when given once daily or even once weekly (SEDA-13, 810). The pharmacokinetics in children, pregnant women, and patients with renal or hepatic dysfunction are unknown.

Primaquine is mainly used to eradicate the exoerythrocytic stages of *Plasmodium vivax* and *Plasmodium ovale*, which if untreated cause late relapse (SEDA-18, 287). There is growing concern about primaquine resistance in *P. vivax* (SEDA-21, 296).

Observational studies

Primaquine base 0.5 mg/kg/day in the prophylaxis of *Plasmodium falciparum* and *P. vivax* malaria for a year did not cause noteworthy adverse effects. General complaints were less than in the placebo group but about the same as in those treated with chloroquine. None of the volunteers (smokers or non-smokers) had a methemoglobin concentration greater than 13% (1).

A randomized, placebo-controlled trial of supervised malaria prophylaxis with primaquine (30 mg/day for 20 weeks) in 97 non-immune adults with normal glucose-6-phosphate dehydrogenase (G6PD) concentrations in Papua New Guinea showed 93% protective efficacy of primaquine against malaria (95% CI = 71, 98%) (2). The most common adverse events were headache, abdominal pain, cough, and nausea, but these were not more frequent than with placebo. Transient rises in methemoglobin concentrations (mean 3.4% on the last day of prophylaxis, resolving by day 18) were asymptomatic.

Supervised treatment of *P. vivax* malaria with chloroquine (600 mg on day 1, 450 mg on days 2 and 3) and primaquine (15 mg/day for 14 days) has been studied in 50 patients in a non-endemic area of Brazil in a prospective open trial (3). G6PD status was not checked. The relapse-free cure rate at 6 months was 86%. There were no important adverse events. Risk factors for relapse included lower doses of primaquine. In patients over 60 kg in weight, the dose of primaquine can fall short of recommendations (0.25–0.3 mg/kg/day), and this can contribute to the risk of relapse.

Combination therapy

Primaquine + clindamycin

Primaquine on its own has some effect against *Pneumocystis jiroveci*, but in dosages of 0.25 or 0.5 mg/

kg, primaquine alone was ineffective in rats, confirming earlier clinical experience. Clindamycin alone was also ineffective. The combination was effective both in vitro and in animals for treatment and prophylaxis.

Intravenous clindamycin, 900 mg every 8 hours, with oral primaquine 26.3 mg, was effective in a number of patients with active disease. A maintenance oral dose of clindamycin 150 mg four times daily plus primaquine 26.3 mg/day was adequate. The drugs are thought to be synergistic. In a second study, clindamycin 600 mg qds was used intravenously, with a maintenance dose of 300–400 mg orally and primaquine base orally, 15 mg/day. Tolerance was reasonably good; the most frequent adverse effect, in half the patients, was a generalized maculopapular rash after 10–12 days. Rash was accompanied by fever in three cases. Other adverse effects were leukopenia ($n = 2$), nausea ($n = 2$), and diarrhea ($n = 1$) (SEDA-12, 703) (4). In a third study with higher doses of primaquine, methemoglobinemia was a major adverse effect. In a fourth study, clindamycin was given 900 mg tds with primaquine 30 mg/day, except in three of 28 episodes of *P. jiroveci* pneumonia treated in a total in 26 patients, who received primaquine 15 mg/day (two) or 30 mg on alternate days (one). In 11 episodes the patients had been intolerant of standard therapy, in 13 episodes conventional therapy had failed, and in four episodes there had been treatment failure and intolerance. Of the 28 episodes, 24 were successfully treated with clindamycin/primaquine, the most common adverse effect being rash (5). With the use of higher doses of clindamycin, gastrointestinal effects, especially diarrhea and colitis, can be expected. The primaquine component can cause hemolytic anemia and methemoglobinemia. Patients should be screened for G6PD deficiency.

Organs and Systems

Psychological, psychiatric

Severe mental depression and confusion was reported in one patient who had been treated with chloroquine beforehand; all the symptoms disappeared on withdrawal (SEDA-11, 588).

Hematologic

Mild anemia, methemoglobinemia, and leukocytosis have been mentioned occasionally, as well as a very occasional case of agranulocytosis, usually associated with overdosage (6).

Primaquine and its congeners can cause hemolytic anemia in people with G6PD deficiency. The effects are more pronounced in the B type (the Mediterranean type) than in the A type. In a 28-year-old Thai soldier, who developed a hemolytic anemia, G6PD deficiency was not mentioned (SEDA-16, 308), but four patients reported in Vanuatu in 1992 had G6PD deficiency; all developed acute intravascular hemolysis resulting in anemia, hemoglobinuria, and systemic illness after a single dose of the drug (SEDA-17, 328).

Gastrointestinal

The most common adverse effects of primaquine are gastrointestinal: mild to moderate abdominal cramps and occasional gastric distress. In those without G6PD deficiency, primaquine is well tolerated as a prophylactic at doses of 15 mg/day (7), with only 1/106 patients withdrawing because of gastrointestinal upset in one study.

Liver

Primaquine can cause dose-dependent hepatotoxicity, as illustrated by a case of acute liver failure (with spontaneous recovery) caused by accidental overdosage (1260 mg on the second day of treatment for *P. vivax*) (8).

Drug–Drug Interactions

Dapsone

The combination of primaquine with clindamycin is used as second choice in the treatment or prevention of *P. jiroveci* pneumonia. If the patient has been treated immediately beforehand with dapsone, methemoglobinemia can result, especially in patients infected with HIV (SEDA-21, 296).

References

1. Kremsner PG, Radloff P, Metzger W, Wildling E, Mordmuller B, Philipps J, Jenne L, Nkeyi M, Prada J, Bienzle U, et al. Quinine plus clindamycin improves chemotherapy of severe malaria in children. Antimicrob Agents Chemother 1995;39(7):1603–5.
2. Baird JK, Lacy MD, Basri H, Barcus MJ, Maguire JD, Bangs MJ, Gramzinski R, Sismadi P, Krisin, Ling J, Wiady I, Kusumaningsih M, Jones TR, Fryauff DJ, Hoffman SL. United States Naval Medical Research Unit 2 Clinical Trials Team. Randomized, parallel placebo-controlled trial of primaquine for malaria prophylaxis in Papua, Indonesia. Clin Infect Dis 2001;33(12):1990–7.
3. Duarte EC, Pang LW, Ribeiro LC, Fontes CJ. Association of subtherapeutic dosages of a standard drug regimen with failures in preventing relapses of vivax malaria. Am J Trop Med Hyg 2001;65(5):471–6.
4. Ruf B, Pohle HD. Clindamycin/primaquine for *Pneumocystis carinii* pneumonia. Lancet 1989;2(8663):626–7.
5. Noskin GA, Murphy RL, Black JR, Phair JP. Salvage therapy with clindamycin/primaquine for *Pneumocystis carinii* pneumonia. Clin Infect Dis 1992;14(1):183–8.
6. Jaremin B, Felczak-Korzybska I, Myjak P. Przypadek methemoglobinemii i agranulocytozy w przebiegu leczenia malarii (*Pl. ovale*) arechina I primachina. [A case of methemoglobinemia and agranulocytosis during the treatment of malaria (*Pl. ovale*) with arequine and primaquine.] Wiad Lek 1982;35(9):591–4.
7. Schwartz E, Regev-Yochay G. Primaquine as prophylaxis for malaria for nonimmune travelers: A comparison with mefloquine and doxycycline. Clin Infect Dis 1999;29(6):1502–6.
8. Lobel HO, Coyne PE, Rosenthal PJ. Drug overdoses with antimalarial agents: prescribing and dispensing errors. JAMA 1998;280(17):1483.

Proguanil and chlorproguanil

General Information

Proguanil is one of the antimalarial drugs most widely used for prophylactic purposes, usually in combination with chloroquine or atovaquone in malaria prophylaxis, and with atovaquone in malaria treatment (SEDA-21, 297). A biguanide, it is rapidly absorbed in standard doses and mainly excreted by the kidneys. Its antimalarial effect is due to its metabolite cycloguanil. However, its metabolism varies individually, and this is reflected in a variable degree of efficacy (SEDA-17, 328).

A derivative, chlorproguanil, is similarly effective in chemoprophylaxis of malaria tropica (SEDA-20, 260).

No serious adverse effects of proguanil have been reported in otherwise healthy patients (SEDA-13, 811; SEDA-17, 328). Skin rashes and hair loss can occur. Mouth ulcers have been mentioned, as have abdominal discomfort and vomiting. The incidence of mouth ulcers in a group of soldiers was 24% in those taking proguanil only and 37% in those taking proguanil 200 mg plus chloroquine either 300 or 150 mg weekly (SEDA-13, 811). With the use of large doses hematuria has been seen.

Organs and Systems

Hematologic

Since chlorproguanil + dapsone exerts lower resistance pressure on *Plasmodium falciparum* than does pyrimethamine + sulfadoxine, a randomized trial in outpatients with uncomplicated falciparum malaria was conducted in Africa in 910 children (1). Treatment failure was more common with pyrimethamine + sulfadoxine. Despite the rapid elimination of chlorproguanil + dapsone, children treated with this combination did not have more episodes of malaria than those who were treated with pyrimethamine + sulfadoxine. However, there was a higher incidence of anemia.

Liver

Hepatitis with mild jaundice has been attributed to proguanil (2).

Susceptibility Factors

Genetic factors

The major limitation of proguanil is that in most ethnic groups there are individuals who have limited ability to metabolize proguanil to the active metabolite cycloguanil. Consequently, poor proguanil metabolizers treated with combination drugs are effectively taking monotherapy. Poor metabolizers also have an increased incidence of adverse effects with proguanil, especially gastrointestinal effects (3).

Renal disease

The urinary excretion of proguanil may mean that caution is advisable when treating patients with renal disorders. Two patients with renal insufficiency became severely ill while taking standard doses of proguanil (SEDA-12, 694) (4). One developed anorexia, dizziness, vomiting, diarrhea, mouth ulcers, a low white cell count, and a low platelet count. The other developed extensive purpura, epistaxis, and vomiting. Bone marrow studies showed hypoplasia and gross megaloblastic changes. The relation of these findings to proguanil therapy is nevertheless questionable, and whether the drug was or was not causative is not known. While it is tempting to seek an explanation in the fact that these biguanides interfere with (plasmodial) folate synthesis, the serum folate and vitamin B_{12} concentrations were normal in both patients.

Drug–Drug Interactions

Cimetidine

Cimetidine co-administration increased the C_{max} of proguanil, and significantly reduced the C_{max} and AUC of cycloguanil, presumably by inhibiting the metabolism of proguanil; co-administration is probably inadvisable (5).

Coumarin

Proguanil potentiates the response to warfarin (6), perhaps by inhibiting CYP2C19 (7).

Levothyroxine

A rise in serum TSH has been described after antimalarial prophylaxis with chloroquine and proguanil in patients taking levothyroxine (SEDA-22, 469). In one case there was a marked increase in serum TSH in the same patient on two occasions after several weeks of antimalarial prophylaxis with chloroquine and proguanil, the likely mechanism being enzyme catabolism (8).

References

1. Sulo J, Chimpeni P, Hatcher J, Kublin JG, Plowe CV, Molyneux ME, Marsh K, Taylor TE, Watkins WM, Winstanley PA. Chlorproguanil–dapsone versus sulfadoxine–pyrimethamine for sequential episodes of uncomplicated falciparum malaria in Kenya and Malawi: a randomised clinical trial. Lancet 2002;360(9340):1136–43.
2. Oostweegel LM, Beijnen JH, Mulder JW. Hepatitis during chloroguanide prophylaxis. Ann Pharmacother 1998;32(10):1023–5.
3. Kaneko A, Bergqvist Y, Taleo G, Kobayakawa T, Ishizaki T, Bjorkman A. Proguanil disposition and toxicity in malaria patients from Vanuatu with high frequencies of CYP2C19 mutations. Pharmacogenetics 1999;9(3):317–26.
4. White NJ. Clinical pharmacokinetics of antimalarial drugs. Clin Pharmacokinet 1985;10(3):187–215.
5. Kolawole JA, Mustapha A, Abdul-Aguye I, Ochekpe N, Taylor RB. Effects of cimetidine on the pharmacokinetics of proguanil in healthy subjects and in peptic ulcer patients. J Pharm Biomed Anal 1999;20(5):737–43.
6. Jassal SV. Warfarin potentiated by proguanil. BMJ 1991;303(6805):789.
7. Goldstein JA. Clinical relevance of genetic polymorphisms in the human CYP2C subfamily. Br J Clin Pharmacol 2001;52(4):349–55.
8. Hassan Alin M, Ashton M, Kihamia CM, Mtey GJ, Bjorkman A. Multiple dose pharmacokinetics of oral artemisinin and comparison of its efficacy with that of oral artesunate in falciparum malaria patients. Trans R Soc Trop Med Hyg 1996;90(1):61–5.

Pyrantel

General Information

Pyrantel is an antihelminthic drug that is effective against intestinal nematodes, including roundworms (*Ascaris lumbricoides*), threadworms (*Enterobius vermicularis*), *Trichostrongylus* species, and the tissue nematode *Trichinella spiralis*. Although it is effective against hookworms, it is less effective against *Necator americanus* than against *Ancylostoma duodenale*.

Pyrantel is usually given as the embonate or pamoate in a single dose of 10 mg/kg, and adverse reactions rarely impede treatment. They include gastrointestinal disturbance (nausea, anorexia, abdominal pain, diarrhea), headache, and vomiting, which are usually mild but can occur in up to 20% of patients.

Organs and Systems

Nervous system

Myasthenia gravis has reportedly been aggravated by pyrantel pamoate (1).

Liver

In children aged 5–10 years pyrantel caused mild transient changes in liver function tests (2).

Drug–Drug Interactions

Piperazine

Piperazine can antagonize the antihelminthic efficacy of pyrantel and vice versa (4).

Theophylline

An interaction of pyrantel with theophylline has been reported (3).

- An 8-year-old boy with status asthmaticus was given intravenous aminophylline and then switched to modified-release oral theophylline on day 3, when his serum theophylline concentration was 15 µg/ml. On day 4 he was given a single dose of pyrantel 160 mg (for *A. lumbricoides* infection) at the same time as his second oral dose of theophylline. About 2.5 hours later his serum theophylline concentration was 24 µg/ml, and a further 1.5 hours later it had risen to 30 µg/ml. No further theophylline was given and no theophylline toxicity occurred.

References

1. Bescansa E, Nicolas M, Aguado C, Toledano M, Vinals M. Myasthenia gravis aggravated by pyrantel pamoate. J Neurol Neurosurg Psychiatry 1991;54(6):563.
2. Dotsenko VA, Ordyntseva AP, Makarova TA, Shirinian AA, Lysakova LA. [Experience with the use of nemocide (pyrantel pamoate) in nematodiases. Med Parazitol (Mosk) 1989;(5):36–9.
3. Hecht L, Mssurray WE. Theophylline–pyrantel pamoate interaction. DICP 1989;23(3):258.

Pyrimethamine

General Information

Pyrimethamine is the most active antimalarial of the 2–4-diaminopyrimidines, its effect being due to inhibition of the conversion of folic acid to its active form, folinic acid. It is also effective in toxoplasmosis. Its antiprotozoal and antimalarial activity is enhanced by the addition of sulfonamides.

Pyrimethamine is well absorbed in healthy subjects; the half-life is 80–95 hours (SEDA-13, 811). Absorption after intramuscular injection is slower; this may be of importance in patients with reduced muscle blood flow (SEDA-17, 328). Pyrimethamine penetrates the cerebrospinal fluid.

With the usual antimalarial prophylactic dosage of 25 mg/week, adverse reactions are generally slight or absent. With intensive treatment in high cumulative doses, as used in the treatment of toxoplasmosis, gastrointestinal intolerance, neurological symptoms, and depression of hemopoiesis can occur. Allergic reactions have not been reported. Tumor-inducing effects have also not been reported.

Pyrimethamine + azithromycin

Azithromycin is efficacious in animal models of toxoplasmic encephalitis. In a Phase I/II dose-escalation study of pyrimethamine (50 mg/day) plus azithromycin (900, 1200, or 1500 mg/day) for induction and maintenance treatment in 30 patients with AIDS and definite or suspected *Toxoplasma* encephalitis, the overall response rate was 67% after 6 weeks of induction therapy (1). However, maintenance therapy for 24 weeks with this combination was associated with a high relapse rate (47%); only six patients successfully completed induction and maintenance therapy. Adverse events were common (particularly in those taking azithromycin 1500 mg) and included hepatotoxicity, bone marrow suppression, ototoxicity, and gastrointestinal disturbances, which led 20%

of patients to withdraw. All adverse events resolved on withdrawal.

In a prospective, randomized, open, multicenter trial of pyrimethamine + azithromycin versus pyrimethamine + sulfadiazine for the treatment of ocular toxoplasmosis in 46 patients with sight-threatening ocular toxoplasmosis, the two regimens had similar efficacy; however, the adverse effects were significantly less common and severe with pyrimethamine + azithromycin (2).

Pyrimethamine + clarithromycin

Clarithromycin, a macrolide, and other macrolide and lincosamine antibiotics (azithromycin, clindamycin, spiramycin, and roxithromycin) have been used in combination with pyrimethamine in the treatment of *Toxoplasma gondii* infections, especially cases of *Toxoplasma* encephalitis.

Clarithromycin 2 g plus pyrimethamine 75 mg/day has been given for 6 weeks to a few AIDS patients with encephalitis (SEDA-13, 814). The adverse effects were many and severe: severe thrombocytopenia, anemia, neutropenia, liver toxicity of varying degree, nausea, vomiting, skin rashes, and hearing loss were found in two of three patients tested in a group of 13. The dose of clarithromycin in this study was the maximum dosage used in an earlier investigation of the treatment of mycobacterial infections in HIV-infected patients.

Pyrimethamine + clindamycin

Pyrimethamine 50 mg/day has been used in combination with clindamycin for the treatment of *Toxoplasma* encephalitis in AIDS. Adverse effects were common (rash, diarrhea, nausea), but the incidence of hematological reactions was lower than with the combination of sulfadiazine and pyrimethamine (SEDA-16, 309).

Pyrimethamine + dapsone

Pyrimethamine and dapsone are available in two fixed combinations:

1. 12.5 mg of pyrimethamine + 100 mg of dapsone = Maloprim.
2. 25 mg of pyrimethamine + 100 mg of dapsone = Deltaprim.

A "dapsone syndrome" was reported in a 30-year-old woman after 4 weeks of treatment with Maloprim and chloroquine base 300 mg/weekly. The symptoms comprised fever, joint and muscle pains, dry cough, and a diffuse red urticarial rash, followed by generalized lymphadenopathy, a painful exudative tonsillitis, and a prominent atypical lymphocytosis.

In Britain, the retrospective reported rate for serious reactions with Maloprim was one in 9100, the incidence of blood dyscrasias being one in 20 000. These figures are lower than those reported with Fansidar (pyrimethamine + sulfadoxine) (SEDA-16, 309).

Pyrimethamine + sulfadoxine

Pyrimethamine 25 mg plus sulfadoxine 500 mg is available in the combination formulation known as Fansidar. This combination, while effective in the prevention and treatment of *Plasmodium falciparum* malaria, carries a high frequency of adverse effects; hematological and serious skin reactions occur, but there are also reports of polyneuritis, vasculitis, and hepatotoxicity. Most of the severe skin reactions and the cases of vasculitis developed within under a month (SEDA-13, 812). The use of Fansidar for malaria prophylaxis has therefore been virtually abandoned (SEDA-21, 297). However, Fansidar is being increasingly used for the treatment of *P. falciparum* malaria in Africa. With the higher dosage used for that purpose, an increase in adverse effects, particularly hematological, can be expected. In Britain the retrospective reported rate for all serious reactions to Fansidar was one in 2100 prescriptions and for skin reactions one in 4900 prescriptions, the death rate being one in 11 100 (SEDA-16, 309).

Observational studies

In a single-arm, open, prospective study between 1990 and 1995 (before HAART) the prophylactic efficacy of Fansidar was evaluated in 95 HIV-infected patients with successfully treated *Pneumocystis jiroveci* pneumonia and no history of *Toxoplasma* encephalitis (3). Patients took Fansidar with folinic acid (15 mg) twice weekly and were followed for a median of 19 (range 1–72) months. Five patients had a *Pneumocystis* relapse, but three had not taken their therapy. Of the 69 patients positive for anti-*Toxoplasma* IgG antibodies, only one developed toxoplasma encephalitis after 50 months. A rash developed in 16 patients after a median of 3 weeks, and required withdrawal in six. Two developed Stevens–Johnson syndrome after three or four doses. There was no significantly increased risk of adverse reactions to Fansidar in patients with previous hypersensitivity reactions to co-trimoxazole. The results of this study are of particular relevance to areas in which HAART is unavailable and where the antimalarial activity of Fansidar may confer additional benefit.

Comparative studies

The efficacy of pyrimethamine + sulfadoxine seems to be unsatisfactory, at least in Laos, where 100 patients with uncomplicated falciparum malaria were randomized to either pyrimethamine + sulfadoxine (a single dose of pyrimethamine 1.25 mg/kg and sulfadoxine 25 mg/kg) or chloroquine (10 mg base/kg immediately, followed by 10 mg/kg 24 hours and 5 mg/kg 48 hours after the start of therapy) (4). There were treatment failures in 18% of those treated with pyrimethamine + sulfadoxine and in 36% of those treated with chloroquine. Thus, both regimens were considered inadequate.

Placebo-controlled studies

In a placebo-controlled study of chemoprophylaxis for malaria, 701 Tanzanian infants were assigned to intermittent pyrimethamine + sulfadoxine (under 5 kg, a quarter of a tablet; 5–10 kg, half a tablet; over 10 kg, one tablet; each tablet contained pyrimethamine 25 mg plus sulfadoxine 500 mg) alongside routine childhood immunizations and iron supplementation at 2, 3, and 9 months of age (5). The combination was well tolerated, with no reported adverse events. Episodes of clinical malaria fell by 59% (95% CI = 41, 72) and the incidence of severe anemia by 50% (95% CI = 8, 73%) in the first year of life. Contrary to previous studies involving continuous prophylaxis in infants, there was no increase in the frequency of rebound episodes of malaria up to 18 months of age, suggesting that the development of malaria-specific immunity was unimpaired. Responses to vaccines were unaffected.

Pyrimethamine + sulfadoxine + mefloquine

Pyrimethamine (25 mg), sulfadoxine (500 mg), and mefloquine (250 mg) are available in the combination formulation known as Fansimef. The adverse effects characteristic of all three components can be expected.

Pyrimethamine + trimethoprim

With the combination of pyrimethamine plus trimethoprim the risk of megaloblastic anemia is higher than with pyrimethamine alone, which on theoretical grounds might be expected. Concomitant administration of folic acid has been recommended (SEDA-11, 590), but the effect of folic acid on efficacy is not known.

Organs and Systems

Respiratory

Pyrimethamine

Non-cardiogenic pulmonary edema has been reported with pyrimethamine alone (SEDA-13, 811).

Pyrimethamine + sulfadoxine

Dyspnea and pleurisy have been described. Of 52 travellers with adverse reactions to Fansidar in Sweden, six had pulmonary infiltrates accompanied by fever (SEDA-13, 241). Such infiltrates have also been described in the past, and in one case a diagnosis of eosinophilic infiltration was made (SEDA-11, 590). A case of non-cardiogenic pulmonary edema was reported in 1989 (SEDA-13, 813).

Nervous system

High doses of pyrimethamine can cause rapid development of neurological symptoms such as ataxia, tremor, and convulsions, probably by a direct toxic effect (SEDA-11, 588).

An unusual case of extrapyramidal syndrome after sulfadoxine–pyrimethamine has been reported (6).

- A 39-year-old man with malaria was given three tablets of sulfadoxine–pyrimethamine. After 50 minutes, he developed extrapyramidal signs, with spasmodic torticollis, trismus, and akathisia. He was given intravenous diazepam 10 mg and made a quick recovery with no residual extrapyramidal signs.

As the extrapyramidal signs occurred 50 minutes after a single dose of sulfadoxine–pyrimethamine there is a strong possibility that it was causally related, although sulfadoxine–pyrimethamine has not previously been reported to cause extrapyramidal effects.

Hematologic

Pyrimethamine

Leukopenia, agranulocytosis, and thrombocytopenia have been reported and, as might be expected in view of folate antagonism, megaloblastic anemia. The latter is more common when high doses are used or when pyrimethamine is given in combination with a drug such as trimethoprim. The bone marrow depression can be reversed using folic acid (SEDA-18, 287).

Pancytopenia has been reported after the use of pyrimethamine alone, but is more often the result of its use in combination with dapsone or sulfonamides (SEDA-11, 589; SEDA-14, 241; SEDA-17, 328).

Although the hematological complications of pyrimethamine are generally a consequence of therapeutic use, long-term prophylactic treatment with pyrimethamine and dapsone in malaria could well involve an increased risk of megaloblastic anemia in patients whose nutritional state is not optimal (SEDA-13, 812).

Pyrimethamine + dapsone

The combination of pyrimethamine plus dapsone causes a higher incidence of blood dyscrasias than pyrimethamine alone; in particular, the occurrence of agranulocytosis increases and even deaths have been reported (SEDA-13, 812; SEDA-18, 287).

Dapsone (alone or in combination with pyrimethamine) can cause methemoglobinemia and hemolytic anemia. These complications tend to be dose-related and are more often encountered in G6PD-deficient subjects (SEDA-18, 287).

When the pyrimethamine + dapsone combination was used in the prophylaxis of *Pneumocystis jiroveci* pneumonia in 173 patients with AIDS, there was anemia in about 20, and in all 117 cases for which data were available, serum haptoglobin concentrations had fallen (SEDA-18, 287).

Pyrimethamine + sulfadoxine

The hematological adverse effects of the combination of pyrimethamine plus sulfadoxine are largely those known from pyrimethamine, that is leukopenia, agranulocytosis, thrombocytopenia, and pancytopenia, but the literature gives the impression that when using this combination these effects are more marked, with lower cell counts (7).

Intermittent iron supplementation with pyrimethamine + sulfadoxine has been investigated in 328 anemic but symptom-free Kenyan children, who were randomly given either iron (ferrous fumarate suspension 6.25 g/l twice a week) or placebo and pyrimethamine + sulfadoxine (25 mg and 1.25 mg/kg once every 4 weeks) or placebo (82 in each group) (8). After 12 weeks, those who took iron and pyrimethamine + sulfadoxine, iron alone, or pyrimethamine + sulfadoxine alone had higher hemoglobin concentrations than those who took the double placebo. No adverse effects were reported.

Severe megaloblastic anemia has also been described (9).

- Hemolytic–uremic syndrome was reported in a 24-year-old man with glucose-6-phosphate dehydrogenase (G6PD) deficiency (SEDA-18, 288).

Gastrointestinal

Gastric disturbances due to pyrimethamine are dose-related; the high doses used in the treatment of toxoplasmosis caused abdominal pains, vomiting, and dizziness (10). Gastrointestinal bleeding related to thrombocytopenia has also been reported (SEDA-11, 589).

Liver

Pyrimethamine + sulfadoxine

Liver function abnormalities with Fansidar (pyrimethamine + sulfadoxine) vary from raised serum transaminase activities to more marked disturbances, with jaundice and granulomatous hepatitis. An occasional case of fatal hepatic failure has been reported; this was the case in a young white American woman who had taken three doses of Fansidar with chloroquine (SEDA-12, 242). Hepatic symptoms may be part of a vasculitis syndrome or can be seen in association with skin reactions (SEDA-13, 241).

Skin

Pyrimethamine

Photosensitivity has been attributed to pyrimethamine (11).

Hyperpigmentation is a very rare adverse effect of pyrimethamine.

- A 29-year-old woman (HIV-negative) developed hyperpigmentation after taking pyrimethamine 25 mg bd for 40 days (12). The hyperpigmentation regressed 15 days after withdrawal of pyrimethamine.

Pyrimethamine + dapsone

Skin rashes are uncommon with the combination of pyrimethamine + dapsone. A lichen planus type of skin reaction has been described (13).

Pyrimethamine + sulfadoxine

Severe skin reactions have been reported with the combination of pyrimethamine + sulfadoxine (Fansidar) from various countries. These include erythema exudativum multiforme, Stevens–Johnson syndrome, toxic epidermal necrolysis, cutaneous vasculitis, lichen planus, a single case of ectodermosis pluriorificialis, and some cases of photosensitivity.

The incidence of skin reactions seems to vary regionally: there is a low incidence in Switzerland (14), but a high incidence has been reported by the US Centers for Disease Control. The latter reported skin reactions in 1:5000 to 1:8000 cases, with fatal reactions in 1:10 000 to 1:25 000 users (SEDA-12, 242). In Britain the retrospective reporting rate for all serious reactions was 1:2100 and for cutaneous reactions one in 4900 prescriptions, with a fatality rate of one in 11 100 (SEDA-16, 309).

In a spontaneous reporting system of individuals who had been exposed to Fansidar in 27 countries, an estimated 117 million users, there were 126 reports: 87 cases of erythema multiforme or Stevens–Johnson syndrome and 39 cases of toxic epidermal necrolysis; 86% of the cases were reported in Europe or North America (15). The fatality rates were 36% for toxic epidermal necrolysis (95% CI = 21, 53) and 9% for erythema multiforme/ Stevens–Johnson syndrome (4, 18). The overall SCAR risk was 1.1 (0.9, 1.3) per million. For developing countries with mainly single-dose use, the risk was estimated at 0.1 (0.0, 0.1) per million. For Europe and North America, with mainly prophylactic use, the risks were 10 (8, 12) and 36 (23, 48) per million respectively.

The prevalence of oral lichenoid reactions was 4.8% in 186 Malay army personnel who used Fansidar for 9 weeks, 0.5% in 186 army personnel who had stopped using Fansidar for 2 months, and 0% in 143 army personnel (control group) who had not used Fansidar for at least 4 months (16).

- Exudative erythema multiforme with chronic proliferation of the conjunctiva causing blindness was described in a 24-year-old man who had taken one tablet of Fansidar at 6-day intervals on three occasions (SEDA-16, 309).

Immunologic

Pyrimethamine + dapsone
Maloprim given for antimalarial prophylaxis was associated with immunosuppression: in military personnel in Singapore (17). The incidence of upper respiratory tract infections was 64% higher than in the non-treated group.

Three patients developed a hypersensitivity syndrome after taking pyrimethamine 12.5 mg + dapsone 100 mg weekly as malaria prophylaxis (18). The diagnosis was based on the presence of fever, lymphadenopathy, a maculopapular rash, and hepatitis. A mild Coombs'-positive hemolytic anemia was also observed in one of the patients. All the clinical, hematological, and biochemical abnormalities normalized within 3 months of tapering regimens of moderate-dose prednisolone.

Three cases of dapsone hypersensitivity syndrome have been reported in male military recruits taking Maloprim prophylaxis for malaria (19). They developed high-grade fever and rash after taking weekly doses of Maloprim for 6–8 weeks. The first patient had an erythematous, macular rash over the upper limbs and trunk, which looked like a viral exanthem, the second had a generalized maculopapular rash over the trunk, and the third had an erythematous rash on his trunk and limbs. All three responded well to topical and systemic glucocorticoids and oral antihistamines. Recovery was uneventful. Early and immediate institution of glucocorticoids and withdrawal of Maloprim is necessary for the treatment of this rare condition.

Pyrimethamine + sulfadoxine
Generalized vasculitis was reported in two cases in Sweden; one developed fever, jaundice, orchitis, and gastrointestinal bleeding after 2 weeks, the other cutaneous vasculitis and rapid progressive nephritis after 3 weeks (SEDA-13, 241).

An illness resembling Sézary syndrome was seen after combined treatment with chloroquine 150 mg every third day (total seven tablets) plus Fansidar one tablet weekly (total six tablets); the symptoms comprised fever, diarrhea, erythroderma, jaundice, lymphadenopathy, and hepatosplenomegaly (SEDA-12, 696).

The safety and efficacy of a fixed combination of pyrimethamine 25 mg + sulfadoxine 500 mg, supplemented with folinic acid 15 mg, both twice a week, as primary prophylaxis of *Pneumocystis* pneumonia and *Toxoplasma* encephalitis has been evaluated in 106 patients infected with HIV in a single-arm, open, prospective study (20). There were allergic reactions in 18 patients and permanent withdrawal was required in seven. One patient who took continued prophylaxis despite progressive hypersensitivity reactions developed a serious adverse reaction (Stevens–Johnson syndrome).

Long-Term Effects

Drug tolerance
Resistance to pyrimethamine has been reported from the Amazon region and Southeast Asia (SEDA-21, 297).

Second-Generation Effects

Pregnancy

Pyrimethamine + dapsone
A stillborn male child with a severe defect of the abdominal and thoracic wall and a missing left arm was delivered at 26 weeks to a woman who had used Maloprim on days 10, 20, and 30 after conception (SEDA-11, 589). The case was anecdotal and the relation to medication uncertain.

Teratogenicity

Pyrimethamine
Pyrimethamine is an inhibitor of dihydrofolate reductase and causes tetrahydrofolate deficiency. It is teratogenic in animals: in rats it produces limb defects, cleft palate, and brachygnathia, and in chick embryos micromelia. Fetal death has been seen in rats and hamsters. However, in toxoplasmosis, pyrimethamine, with or without a sulfonamide, has been given to pregnant women without evidence of subsequent abnormalities. Supplementation with folic acid has been advocated to prevent or reduce adverse effects (SEDA-13, 812), but it is not known if this could impair efficacy.

Pyrimethamine + sulfadoxine
Malaria during pregnancy is associated with an increased risk of severe anemia and babies of low birth weight. Effective intermittent therapy with

pyrimethamine + sulfadoxine reduces parasitemia and severe anemia and improves birth weight in areas in which *P. falciparum* is sensitive to this combination. In an open, prospective trial in 287 pregnant women in the Gambia who were exposed to a single dose of a combination of artesunate and pyrimethamine + sulfadoxine there was no evidence of a teratogenic or otherwise harmful effect (21).

Susceptibility Factors

Pyrimethamine

Pyrimethamine should not be given to patients with depleted folic acid reserves (22).

Pyrimethamine + sulfadoxine

Sulfonamide and/or pyrimethamine sensitivity, pregnancy, and G6PD deficiency are contraindications. Use in young infants is considered inadvisable; the history of an 8-month-old infant with *P. falciparum* malaria who developed high fever, tachycardia, hypotension, chills, jaundice, and splenomegaly 48 hours after a single parenteral dose of Fansidar (pyrimethamine + sulfadoxine) (SEDA-16, 309) seems to confirm the wisdom of this advice. It has been advocated that Fansidar should not be used prophylactically if exposure to malaria will last less then 3 weeks, in view of the incidence of severe skin reactions during the first month.

Drug–Drug Interactions

Chloroquine

The combined use of Fansidar (sulfadoxine + pyrimethamine) with chloroquine has been reported to result in more severe adverse reactions (23). However, an increased risk has not been reported in recent studies (24).

Liver metabolized drugs

Both pyrimethamine and sulfonamides are liver enzyme inhibitors, and can cause interactions with drugs that are normally metabolized in the liver.

References

1. Jacobson JM, Hafner R, Remington J, Farthing C, Holden-Wiltse J, Bosler EM, Harris C, Jayaweera DT, Roque C, Luft BJ. ACTG 156 Study Team. Dose-escalation, phase I/II study of azithromycin and pyrimethamine for the treatment of toxoplasmic encephalitis in AIDS. AIDS 2001;15(5):583–9.
2. Bosch-Driessen LH, Verbraak FD, Suttorp-Schulten MS, van Ruyven RL, Klok AM, Hoyng CB, Rothova A. A prospective, randomized trial of pyrimethamine and azithromycin vs pyrimethamine and sulfadiazine for the treatment of ocular toxoplasmosis. Am J Ophthalmol 2002;134(1):34–40.
3. Schurmann D, Bergmann F, Albrecht H, Padberg J, Grunewald T, Behnsch M, Grobusch M, Vallee M, Wunsche T, Ruf B, Suttorp N. Twice-weekly pyrimethamine–sulfadoxine effectively prevents *Pneumocystis carinii* pneumonia relapse and toxoplasmic encephalitis in patients with AIDS. J Infect 2001;42(1):8–15.
4. Mayxay M, Newton PN, Khanthavong M, Tiengkham P, Phetsouvanh R, Phompida S, Brockman A, White NJ. Chloroquine versus sulfadoxine-pyrimethamine for treatment of Plasmodium falciparum malaria in Savannakhet Province, Lao People's Democratic Republic: an assessment of national antimalarial drug recommendations. Clin Infect Dis 2003; 3: 1021–8.
5. Schellenberg D, Menendez C, Kahigwa E, Aponte J, Vidal J, Tanner M, Mshinda H, Alonso P. Intermittent treatment for malaria and anaemia control at time of routine vaccinations in Tanzanian infants: a randomised, placebo-controlled trial. Lancet 2001;357(9267):1471–7.
6. Adam I, Elbashir MI. Extrapyramidal syndrome after treatment with sulphadoxine–pyrimethamine. Saudi Med J 2004;25(9):1303–4.
7. Hellgren U, Rombo L, Berg B, Carlson J, Wiholm BE. Adverse reactions to sulphadoxine–pyrimethamine in Swedish travellers: implications for prophylaxis. BMJ (Clin Res Ed) 1987;295(6594):365–6.
8. Verhoef H, West CE, Nzyuko SM, de Vogel S, van der Valk R, Wanga MA, Kuijsten A, Veenemans J, Kok FJ. Intermittent administration of iron and sulfadoxine–pyrimethamine to control anaemia in Kenyan children: a randomised controlled trial. Lancet 2002;360(9337):908–14.
9. Chute JP, Decker CF, Cotelingam J. Severe megaloblastic anemia complicating pyrimethamine therapy. Ann Intern Med 1995;122(11):884–5.
10. Deron Z, Jablkowski M. Objawy uboczne w przebiegu leczenia toksoplazmozy. [Side effects of toxoplasmosis treatment.] Pol Tyg Lek 1980;35(23):857–9.
11. Craven SA. Letter: Photosensitivity to pyrimethamine? BMJ 1974;2(918):556.
12. Ozturk R, Engin A, Ozaras R, Mert A, Tabak F, Aktuglu Y. Hyperpigmentation due to pyrimethamine use. J Dermatol 2002;29(7):443–5.
13. Cutler TP. Lichen planus caused by pyrimethamine. Clin Exp Dermatol 1980;5(2):253–6.
14. Steffen R, Somaini B. Severe cutaneous adverse reactions to sulfadoxine–pyrimethamine in Switzerland. Lancet 1986;1(8481):610.
15. Sturchler D, Mittelholzer ML, Kerr L. How frequent are notified severe cutaneous adverse reactions to Fansidar? Drug Saf 1993;8(2):160–8.
16. Zain RB. Oral lichenoid reactions during antimalarial prophylaxis with sulphadoxine–pyrimethamine combination. Southeast Asian J Trop Med Public Health 1989;20(2):253–6.
17. Lee PS, Lau EY. Risk of acute non-specific upper respiratory tract infections in healthy men taking dapsone–pyrimethamine for prophylaxis against malaria. BMJ (Clin Res Ed) 1988;296(6626):893–5.
18. Thong BY, Leong KP, Chng HH. Hypersensitivity syndrome associated with dapsone/pyrimethamine (Maloprim) antimalaria chemoprophylaxis. Ann Allergy Asthma Immunol 2002;88(5):527–9.
19. Tee AKH, Oh HML, Wee IYJ, Khoo BP, Poh WT. Dapsone hypersensitivity syndrome masquerading as a viral exanthem: three cases and a mini- review. Ann Acad Med Singapore 2004;33(3):375–8.
20. Schurmann D, Bergmann F, Albrecht H, Padberg J, Wunsche T, Grunewald T, Schurmann M, Grobusch M, Vallee M, Ruf B, Suttorp N. Effectiveness of twice-weekly

pyrimethamine–sulfadoxine as primary prophylaxis of Pneumocystis carinii pneumonia and toxoplasmic encephalitis in patients with advanced HIV infection. Eur J Clin Microbiol Infect Dis 2002;21(5):353–61.

21. Fishman JA. Prevention of infection caused by Pneumocystis carinii in transplant recipients. Clin Infect Dis 2001;33(8):1397–405.
22. Akinyanju O, Goddell JC, Ahmed I. Pyrimethamine poisoning. BMJ 1973;4(5885):147–8.
23. Rombo L, Stenbeck J, Lobel HO, Campbell CC, Papaioanou M, Miller KD. Does chloroquine contribute to the risk of serious adverse reactions to Fansidar? Lancet 1985;2(8467):1298–9.
24. Rahman M, Rahman R, Bangali M, Das S, Talukder MR, Ringwald P. Efficacy of combined chloroquine and sulfadoxine–pyrimethamine in uncomplicated Plasmodium falciparum malaria in Bangladesh. Trans R Soc Trop Med Hyg 2004;98(7):438–41.

Pyrvinium embonate

General Information

Pyrvinium is a deep-red insoluble dye used as an antihelminthic drug. It is well tolerated in doses up to 5 mg/kg. Nausea, vomiting, diarrhea, and cramping abdominal pain are more frequent at higher doses. Feces and vomit are stained red. Isolated cases of severe allergy (1), transient photosensitivity, and Stevens–Johnson syndrome (2) have been reported.

References

1. Desser KB, Baden M. Allergic reaction to pyrvinium pamoate. Am J Dis Child 1969;117(5):589.
2. Coursin DB. Stevens–Johnson syndrome: nonspecific parasensitivity reaction? JAMA 1966;198(2):113–6.

Quinfamide

General Information

Quinfamide acts on the trophozoites of Entamoeba histolytica, making the trophozoite incapable of propagation. It is not active against amebic cysts. In doses of 100–1200 mg adverse effects have been frequent but mild, mainly comprising headaches and nausea (SEDA-12, 705) (1).

Reference

1. Robinson CP. Quinfamide. Drugs Today 1984;20:479.

Quinine

General Information

Quinine was originally extracted from the bark of the Cinchona tree (Peruvian bark or Jesuits' bark) and was used to treat ague, that is fever, usually due to malaria. It fell out of fashion with the advent of other antimalarial drugs, but has once again become the drug of first choice for malaria originating in areas with multiresistant Plasmodium falciparum. To be effective, quinine plasma concentrations greater than the minimal inhibitory concentration must be achieved and maintained.

Pharmacokinetics

Quinine given orally is well absorbed; the half-life is 11 hours. Clearance is predominantly by hepatic metabolism; urinary clearance accounts for only 20%. Information on pharmacokinetics in healthy volunteers can be misleading, since plasma quinine concentrations are higher in the presence of malaria infection than in healthy subjects given the same dose (SEDA-13, 814). The dosage regimen therefore needs to be adapted to the severity of the illness and amended as improvement occurs.

A population pharmacokinetic study of intramuscular quinine (loading dose 20 mg/kg salt diluted 1:1 in water) in 120 Ghanaian children with severe malaria showed predictable profiles, which were within the target range for quinine (15–20 µg/ml) and independent of clinical and laboratory variables (1). Adverse events included skin induration or abscesses at the injection site (12%), all of which resolved without surgical intervention, and hypoglycemia (10%), a special risk in children who were hypoglycemic at presentation.

Adverse effects of quinine are common at plasma concentrations over 10 µg/ml. The dose often recommended, 10 mg/kg intravenously over 10–20 minutes, may be too high in patients with cerebral malaria (SEDA-14, 240). In the USA, intravenous quinine has been discontinued in favor of quinidine (SEDA-17, 329). In some areas, a high rate of recrudescence is seen after short-term treatment with quinine. The addition of specific antibiotics may improve the cure rate.

Placebo-controlled studies

A double-blind, placebo-controlled trial of a 3-day combination regimen of quinine (8 mg/kg tds) and clindamycin (5 mg/kg tds) (n = 53) versus 7-day quinine (8 mg/kg tds intravenously for 3 days, then orally; n = 55) to treat uncomplicated imported falciparum malaria showed no significant differences in the parasite and fever clearance times or the 28-day cure rate (100 versus 96%) (2). The frequencies of mild adverse events (tinnitus and nausea) were similar in the two groups. There were two serious adverse events that necessitated treatment withdrawal: one patient taking quinine alone had a hemolytic episode and another a "severe toxic rash."

General adverse effects

Quinine is not pleasant to take, and adverse effects, including nausea, tinnitus, dizziness, and hypoglycemia, are well recognized and common compared with other antimalarial drugs (3) in the doses used to treat malaria, although not in the smaller doses used to prevent leg cramps, when classical allergic reactions can still occur (4). Most serious adverse effects are due to the prodysrhythmic properties of quinine, and its effects as a hypoglycemic agent, especially in pregnant women with severe malaria. The prolonged use of normal or low doses of quinine can lead to "cinchonism" in sensitive individuals; this in mild form consists of tinnitus, headache, nausea, and visual disturbances (SEDA-13, 814). Overdosage can cause marked gastrointestinal intolerance, central nervous system disturbances (especially vertigo), visual disorders (very occasionally involving sudden blindness), and cardiovascular problems related to impaired intracardiac conduction. A major risk with quinine is that of direct intravenous injection given too fast. Allergic reactions are not uncommon. They are usually limited to fever and rashes, but angioedema and asthma have been seen. Thrombocytopenic purpura and thrombocytopenia are, at least in some cases, caused by an allergic reaction, and the amounts of quinine present in some "tonics" are sufficient to trigger thrombocytopenia in such patients. Anaphylactic shock has been reported in rare cases (SEDA-13, 814).

Organs and Systems

Cardiovascular

Quinine can cause atrioventricular conduction disturbances. In sensitive patients, such changes can occur with normal dosages given over a prolonged period; however, in most cases cardiac effects are due to overdosage.

Electrocardiographic changes, such as prolongation of the QT interval, widening of the QRS complex, and T wave flattening, can be seen with plasma concentrations above 15 µg/ml (SEDA-11, 590; SEDA-14, 239).

Quinine, and more profoundly quinidine, its diastereomer, can cause ventricular tachycardia, torsade de pointes, and ventricular fibrillation by prolonging the QT interval (SEDA-20, 261).

- An 8-year-old child given an incorrect dose of quinine had ventricular tachycardia and status epilepticus after 48 hours; the plasma quinine concentration was 20 µg/ml (SEDA-18, 288), compared with the target range of 1.9–4.9 µg/ml.

Respiratory

Quinine can cause allergic asthmatic reactions.

- A 45-year-old woman with long-standing rheumatoid arthritis developed wheeze, severe anxiety, breathlessness, cough, orthopnea, mild fever, chills, and pleuritic chest discomfort after taking a single dose of quinine for nocturnal leg cramps (5). A chest X-ray showed diffuse, bilateral pulmonary infiltrates suggestive of pulmonary edema. No cause other than acute quinine ingestion was identified despite thorough cardiac and infectious disease evaluation.

Pulmonary edema is uncommon after treatment with quinine, but a case has been reported (6).

- A 57-year-old man with leg cramps developed transient acute pulmonary edema and hypotension 30–40 minutes after taking quinine sulfate 300 mg orally on two occasions. He was not taking any other drugs and there was no explanation for either event. Serial troponin T tests, an electrocardiogram, echocardiogram, and coronary angiogram were all normal.

Quinine poisoning can cause respiratory depression.

Nervous system

Tinnitus and vertigo are not uncommon with quinine, especially at higher dosages (7,8).

Headache and tinnitus occur in chronic toxicity (cinchonism) (9).

Acute intoxication can be followed by convulsions and coma (10).

- A general organic brain syndrome was observed in a 24-year-old woman with tropical malaria after the third day of treatment with quinine sulfate 500 mg tds; her symptoms were headache, blurred vision, vertigo, tinnitus, impaired hearing, increasing apathy, disorientation, speech changes, incoherent thinking, and disorientation with respect to time and place; recovery followed withdrawal (SEDA-16, 305).

A quinidine-induced myopathy has been reported (11).

Sensory systems

Eyes

Amaurosis connected with damage to the retina is most common with high plasma concentrations, and thus follows high dosages and especially overdosage. The outdated, but still practiced, use of quinine to induce abortion is probably the most common cause. Quinine initially affects the photoreceptor and ganglion cell layers; retinal vascular changes are secondary. The first sign may be widely dilated pupils that still respond to light; later, visual field contraction and loss of vision can occur. In milder cases, vision can return, but possibly with a residual disturbance of dark adaptation and/or restricted visual fields. However, loss of vision can be permanent. In cases of permanent damage, the classic late appearance of the fundus after quinine intoxication, with marked pallor and vascular narrowing, appears after some months (SEDA-13, 815). Loss of vision occurs mainly at serum concentrations over 10 µg/ml. Such concentrations are thought not to be toxic in patients with malaria, who have high circulating concentrations of alpha$_1$-acid glycoprotein, resulting in a lower fraction of unbound drug in the plasma (SEDA-17, 329).

Ocular toxicity with vasospasm has been described after poisoning with 4.5 g of quinine 24–36 hours before

(12). Therapy for vasospasm using nimodipine, hemodilution, and hypervolemia was instituted, with subsequent resolution of the symptoms.

Blindness has been attributed to quinine (13).

- A 43-year-old Ghanaian man with a mild attack of malaria due to *P. falciparum* received quinine 750 mg intravenously over 5 hours and another identical infusion 3.5 hours after the end of the first infusion. Half an hour after the start of the second infusion he wakened and reported complete blindness. The infusion was immediately stopped, and ocular examination showed fixed, dilated pupils, complete blindness, and normal fundi. A CT scan of the brain, electroencephalography, and retinal fluorescein angiography were all normal. Serum quinine concentrations continued to rise after the end of the infusion, but this effect was mitigated by an increase in the serum concentration of alpha$_1$-acid glycoprotein, to which quinine binds. Six hours after the end of the infusion of quinine his sight began to improve.

In a case of quinine poisoning, stellate ganglion block was performed immediately on the basis of the clinical history of visual disturbance without waiting for physical signs to develop. There was no residual field defect despite the presence of toxic concentrations of the drug. The authors suggested that stellate ganglion block may prevent development of visual field defects due to quinine toxicity (14). However, in other cases it was ineffective (15,16). The effectiveness of this treatment may be a function of the speed with which it is instituted.

Ears

Tinnitus is a fairly frequent complaint and is not only seen after quinine overdosage. Permanent impairment of hearing has long been thought to be a possible consequence of long-term use of quinine, but this belief has recently been challenged, and the original description of it is questionable (SEDA-13, 816); the complication is certainly rare (17).

Serial audiometry in 10 patients receiving quinine for acute falciparum malaria showed a reduction in high-tone auditory acuity in all patients, resulting in flattening of the audiogram. The onset of the effect was rapid and it resolved completely after the end of treatment; only seven of these patients reported tinnitus. Hearing impairment was investigated in six volunteers after single doses of quinine at 5, 10, and 15 mg/kg. A clear effect on hearing was found, but with high variability (SEDA-17, 306).

Psychiatric

Probable suicide after quinine treatment for chloroquine-resistant malaria has been reported (18).

- A 27-year-old man with falciparum malaria was given an infusion of quinine 600 mg in 5% dextrose tds until his vomiting stopped. Five hours later he was found dead by hanging using his turban. There was no adverse

social history and the patient was in general good health.

The cause of this man's suicide is not known and it was probably not related to quinine.

Metabolism

Clinical signs and symptoms of hypoglycemia are reported occasionally; most cases are subclinical, but severe cases have been described (SEDA-13, 815). A study of the effect of quinidine on glucose homeostasis in Thai patients with malaria showed a near doubling of plasma insulin concentrations and a corresponding fall in serum glucose concentrations. An additional factor may have been impaired nutritional status and the effects of parenteral quinine in severely ill patients not taking food (SEDA-13, 815; SEDA-14, 240; SEDA-18, 288).

Hematologic

Thrombocytopenia is often reported with quinine. It is probably due to hypersusceptibility rather than a toxic effect, since even the ingestion of minimal amounts of quinine, such as those present in commercial tonic waters, can cause it. A drug–antibody complex has been demonstrated (SEDA-11, 591). In some cases of quinine-induced thrombocytopenia, there was autoantibody-binding to glycoprotein Ib-IX, IIb, and IIIa complexes. In three published cases, quinine-dependent autoantibodies to glycoproteins Ib-IX and IIb/IIIa were associated with both thrombocytopenia and a hemolytic–uremic syndrome (SEDA-16, 305; SEDA-17, 329). Two other patients had recurrent febrile illnesses characterized by hypotension, pancytopenia, coagulopathy, and renal insufficiency, and both had high titers of quinine-dependent antibodies, which showed cross-sensitivity with quinidine. In one case there was a link with tonic water, while the other had taken quinine sulfate for leg cramps before each episode (19). A list of nine earlier published cases with antibody findings was added to these two case histories.

Isolated thrombocytopenia after the use of quinine for malaria or leg cramps has been described in isolated cases. The FDA's Center for Drug Evaluation and Research received 141 reports of isolated thrombocytopenia in association with quinine from 1974 to December 2000 (20). After elimination of cases that were confounded by acute or chronic disease or concomitant drug therapy, 64 reports of quinine-associated thrombocytopenia were analysed. Thrombocytopenia occurred soon after the start of therapy (median 7 days) and was often severe (hospitalization reported in 55 of the 64 cases).

Since 1972, the Australian Adverse Drug Reactions Advisory Committee has received 198 reports of thrombocytopenia associated with quinine, four of which had a fatal outcome (21). In 17 of the 20 reports received since the beginning of 2000, patients had platelet counts of $0–14 \times 10^9$/l; most of them required hospitalization and treatment with platelet transfusions, glucocorticoids, or immunoglobulin. In most cases the platelet count normalized within 1 week of quinine withdrawal. As quinine-induced thrombocytopenia has an immune-based mechanism, the Committee suggested that patients who develop this reaction should subsequently avoid all products that contain quinine, including drinks such as tonic water. They also reminded prescribers that quinine is no longer recommended for the treatment of nocturnal cramps; the FDA withdrew nocturnal cramps as an indication for all quinine products in 1995 because of lack of evidence of efficacy, and the Australian Medicines Handbook advises against its use for this indication.

Glycoprotein epitopes involved in quinine-induced thrombocytopenia have been characterized (22).

A well documented report has shown that re-exposure to a single dose of quinine for leg cramps about 40 years after a previous exposure can be sufficient to cause a severe episode of hemolytic?uremic syndrome + thrombotic thrombocytopenic purpura.

- A 67-year-old woman with hypertension, chronic kidney disease (serum creatinine 177 µmol/l (2 mg/dl) and a history of a short period of hemodialysis for acute renal insufficiency 10 years before developed nausea, vomiting, diffuse abdominal cramp, and blurred vision 1 hour after taking one tablet of quinine for leg cramps (23). She had a fever of 39.7°C, a raised blood pressure, and confusion, an anemia of 10.4 g/dl with schistocytes and burr cells, thrombocytopenia of 14 x 10⁹/l, acute renal insufficiency with a serum creatinine of 504 µmol/l (5.7 mg/dl), raised total and direct bilirubin, raised liver enzymes, and raised lactate dehydrogenase activity. She was given antibiotics for presumed sepsis, and her symptoms resolved within 24 hours. However, her urine output continued to fall and hemolytic?uremic syndrome + thrombotic thrombocytopenic purpura was diagnosed. After 7 sessions of plasmapheresis, the hematological parameters normalized. After 3 weeks of hemodialysis, kidney function returned to baseline. At a follow-up visit 9 months later, quinine-associated antiplatelet antibodies were detected. She then recalled that about 40 years before she had taken quinine for malaria prophylaxis.

Acute intravascular hemolysis with renal involvement and even renal insufficiency can occur with quinine and can follow relatively small doses. Quinine-induced hemolysis has probably played a role in the clinical syndrome of blackwater fever in the past.

- The combination of renal insufficiency with cortical necrosis, thrombocytopenia, intravascular coagulation, and deposition of fibrin was seen in a 63-year-old woman who had drunk tonic water. She had had two previous episodes of acute renal insufficiency also associated with quinine-containing drinks; this most certainly reflected a hypersensitivity reaction (SEDA-13, 815).
- A 65-year-old man, who had taken a single dose of quinine 300 mg for leg cramps, developed both acral necrosis and hemolytic–uremic syndrome, which resolved promptly after treatment with glucocorticoids (24).

Lupus anticoagulant has been reported with the use of quinine and quinidine, and an associated antiphospholipid syndrome has been described (25).

Disseminated intravascular coagulation has been attributed to quinine (26).

- A 79-year-old woman developed disseminated intravascular coagulation. She had taken a dose of quinine for leg cramps 3 months before and on the day before admission. Her only other medication was bendroflumethiazide. Investigations showed a platelet-associated immunoglobulin with positive immunofluorescence on exposure to quinine sulfate.
- Quinine-induced disseminated intravascular coagulation and hemolytic–uremic syndrome occurred in a 78-year-old woman who took quinine 150 mg for leg cramps; this is the first report of the two diseases occurring simultaneously after quinine (27).

There have been 16 cases of quinine-induced disseminated intravascular coagulation, most of them in women (28). The symptoms usually started within hours of taking quinine. Single small doses of quinine may be sufficient to provoke disseminated intravascular coagulation. Most (80%) of the cases presented with gastrointestinal complaints, such as abdominal pain, nausea, vomiting, diarrhea, and hematemesis and/or melena. Other common symptoms were petechial or ecchymotic rashes, back pain, myalgia, headache, fever, chills, and malaise. Typical laboratory results included thrombocytopenia, raised fibrin degradation products, raised D dimers, and coagulopathy. In addition, there can be uremia, raised plasma creatinine, lactate dehydrogenase, and bilirubin, lactic acidosis, reduced haptoglobin, and a urinary sediment. Quinine induces antiplatelet antibodies of the IgG or IgM classes and possibly also antibodies against erythrocytes, neutrophils, T lymphocytes, B lymphocytes, and endothelial cells. In some cases, antibodies were not detected during the initial days of the disease, but were detected after the patient had recovered. Patients with quinine-induced disseminated intravascular coagulation are treated with supportive care and plasmapheresis. Renal function recovered in most cases, but 31% developed chronic renal insufficiency and 19% required permanent hemodialysis. Patients with a history of quinine-induced disseminated intravascular coagulation must avoid quinine and all quinine-containing products.

In a retrospective survey of thrombotic thrombocytopenic purpura with hemolytic–uremic syndrome reported to the Oklahoma TTP/HUS Registry, 17 of 225 cases were associated with quinine (doses not stated) taken long-term for leg cramps (29). Patients typically presented with an acute onset of fever, chills, nausea, vomiting, diarrhea, and abdominal pain within hours of quinine ingestion. Laboratory findings included thrombocytopenia, microangiopathic hemolytic anemia, and liver and renal dysfunction. This is an immune-mediated reaction associated with quinine-dependent antiplatelet antibodies and a high mortality (three of the 17 patients). It can be triggered by small amounts of quinine, such as those present in tonic water.

Gastrointestinal

Nausea and abdominal pain can occur with quinine (30). With high doses, diarrhea has been seen (30).

Liver

Granulomatous hepatitis has been reported in a patient taking quinine sulfate for night cramps (31). Cholestatic jaundice was reported in another patient who had taken quinine for night cramps (SEDA-13, 815). Except for the occasional anecdotal case, there is no evidence that true hepatotoxicity occurs with quinine.

Urinary tract

Renal damage accompanies acute hemolysis due to quinine. Renal insufficiency in cases of quinine poisoning is probably due to circulatory collapse. Allergic reactions underlie at least some cases. The picture can be complex, with the renal insufficiency coming in association with cortical necrosis, thrombocytopenia, intravascular coagulation, and deposition of fibrin.

Skin

Rashes are common in allergic reactions. Photosensitivity induced by quinine has been reported; some of the cases occurred in elderly persons taking quinine for night cramps (32). Another form of hypersensitivity to quinine is cutaneous neutrophilic vasculitis, which is a form of photosensitivity (SEDA-16, 305). Local pigmentation has been described after intramuscular injection (SEDA-13, 815).

Immunologic

Quinine can cause a variety of immune-mediated syndromes, most commonly isolated thrombocytopenia, but rarely microangiopathic hemolytic anemia with thrombocytopenia and acute renal insufficiency (hemolytic–uremic syndrome). Two reports of immune-mediated syndromes following the use of quinine for leg cramps have helped to provide an immunopathological explanation for the diversity of such presentations (33,34).

- One patient presented with thrombocytopenic purpura, presumed to be idiopathic (which responded to glucocorticoids and intravenous immunoglobulin) and subsequently presented again with hemolytic–uremic syndrome, and required intensive renal replacement and immunosuppressive therapy. Analysis of serum samples from the isolated thrombocytopenic stage of the presentation showed the presence of quinine-dependent antibodies specific for platelet surface glycoprotein GPIb/IX. Quinine-dependent antibody targets widened to include glycoprotein IIb/IIIa during the hemolytic–uremic phase of the illness, with additional binding to neutrophils and lymphocytes.
- In another case of acute systemic allergy to quinine, which mimicked septic shock, with little hemolysis or renal involvement, the patient presented twice with a virtually identical clinical picture: sudden fever, rigors, and back pain, followed by hypotension, metabolic acidosis, granulocytopenia, and disseminated intravascular coagulation. On each occasion clinical and laboratory indices recovered spontaneously within 36 hours. A retrospective analysis of the patient's serum showed the presence of neutrophil-specific, quinine-dependent antibodies.

A quinine-induced lupus-like syndrome, including pericarditis and polyarthralgia, and positive antinuclear and anti-cardiolipin antibodies, and a polymyalgia rheumatica-like syndrome have been described with quinine or quinidine (SEDA-20, 261; SEDA-21, 298).

Second-Generation Effects

Teratogenicity

Quinine crosses the placenta and can be found in relatively high concentrations in cord blood; it is also excreted in the breast milk. Data about possible teratogenicity related to the therapeutic use of malaria are scanty (SEDA-14, 239). Hypoplasia of the optic nerve and deafness have been reliably described in children born after failure to induce abortion with the drug.

Susceptibility Factors

Age

Children under 2 years are more vulnerable to quinine toxicity as measured by prolongation of the QRS interval at 2–4 hours after intravenous quinine (SEDA-20, 261).

In 10 children with severe *Plasmodium falciparum* malaria aged 6 months to 11 years a loading dose of quinine dihydrochloride 20 mg/kg in 500 ml of 5% dextrose infused over 4 hours was followed by 10 mg/kg every 8 hours (35). Holter monitoring was performed for the first 24 hours. Mean heart rate was 100–156/minute and there were no major ventricular dysrhythmias. The high cardiac rate could have been due to pyrexia and anemia. There was no prolongation of the PQ or QT intervals and no other adverse effects were reported.

Renal disease

A pharmacokinetic study in adults with renal insufficiency showed that concentrations of the main metabolite, hydroxyquinine, rose to 45% of the concentrations of the parent compound, and may contribute up to 25% of the cardiac effects of quinine (36).

Other features of the patient

The use of quinine to treat cerebral malaria, and to a lesser extent severe malaria, has always been considered more risky than treatment of common cases of malaria. The changes in the pharmacokinetics of quinine caused by the malaria provide an explanation: the standard dose of 10 mg/kg is usually well tolerated in patients with uncomplicated malaria but causes markedly higher plasma serum concentrations in patients with cerebral malaria.

Total quinine clearance and total apparent volume of distribution are significantly lower in severe malaria, and after recovery, the pharmacokinetics return to normal. Probably the first loading dose should be as generally advised, but with a reduction in subsequent doses until the general condition has improved. Monitoring of plasma or red cell concentrations of quinine would of course be ideal, but this luxury is rarely available (SEDA-13, 816).

Drug Administration

Drug overdose

Early symptoms of intoxication with quinine are unremarkable, except for mild visual and auditory complaints (37,38). The principal sign is the sudden onset of bilateral pupil dilatation. Other symptoms are tinnitus, mild deafness, vertigo, mild drowsiness, vague disturbances of consciousness, headache, nausea, vomiting, and abdominal pain. The electrocardiogram may show prolongation of the QT interval and ST abnormalities. Deafness can occur within hours or days of the acute symptoms. Visual disturbances do not necessarily occur early on. Acute intoxication can be seen after ingestion of doses of 4–12 g, but a dose of 8 g can prove lethal. Death is commonly caused by respiratory arrest, but is sometimes due to renal insufficiency. In five children with quinine overdosage, there were serious neurological and cardiac effects, such as seizures, ventricular tachycardia, and cardiac arrest (SEDA-20, 261).

Repeated oral administration of activated charcoal reduces plasma quinine concentrations and mitigates toxicity (39).

Drug–Drug Interactions

Anticoagulants

Quinine can increase the action of anticoagulants, possibly by inhibition of prothrombin synthesis (40).

Atorvastatin

Quinine-induced acute renal failure has been described as a result of a combination of hemolytic–uremic syndrome and rhabdomyolysis with disseminated intravascular coagulation, thought to be secondary to an interaction with atorvastatin (41).

- A 54-year-old woman became acutely ill within 1 hour of taking quinine 300 mg. She had a history of hypertension, dyslipidemia, chronic back pain, and a previous cholecystectomy. She was currently taking atenolol 50 mg/day, aspirin 100 mg/day, glucosamine 750 mg bd, and atorvastatin 20 mg/day. Nausea, vomiting, diarrhea, and abdominal pains were followed by rigors, myalgia, dizziness, and peripheral paresthesia. She was febrile (38°C), with a tachycardia (101/minute), tachypnea (26/minute), and hypotension (100/60 mmHg). Electrocardiography showed widespread

T wave inversion. Hemoglobin was 13 g/dl, white cell count 2.6 x 10^9/l, and platelets 247 x 10^9/l. Urea, electrolytes, and liver function tests were normal and a chest X ray showed no active infection. She received intravenous fluids and ceftriaxone. Repeat biochemistry 18 hours later showed acute renal insufficiency (urea 13.5 mmol/l, creatinine 200 µmol/L), rhabdomyolysis (CK 9300 U/l). and acute hepatitis (bilirubin 40 mmol/l, alkaline phosphatase 70U/l, alanine transaminase 456 U/l, gamma glutamyl transferase 123 U/l). Urine microscopy was unhelpful, but urine dipstick showed protein 4+, blood 4+, and myoglobin. She later developed a petechial rash and anemia (hemoglobin 9.5 g/dl) and thrombocytopenia (61 x 10^9/l). There was a marked increase in lactate dehydrogenase, consistent with microangiopathic hemolysis. The international normalized ratio of 2.1, activated partial thromboplastin time of 42, decreased fibrinogen concentration of 1.3 g/L (normal 2.0-4.0) and increase D-dimers pointed to disseminated intravascular coagulation. Following aggressive hydration, urinary alkalinization, and diuresis, her hematology and biochemistry gradually returned to normal. However, flow cytometry showed the presence of antiplatelet antibodies, which returned to normal by 12 months. She had had a previous similar episode after taking a single quinine tablet along with atorvastatin.

This potentially fatal adverse event may have been triggered by quinine, a potent inhibitor of CYP3A4. Quinine may have increased the plasma concentrations of atorvastatin, by reducing its first-pass metabolism.

Carbamazepine

Quinine inhibits the metabolism of carbamazepine (42).

Chloroquine

Chloroquine antagonizes the action of quinine against *P. falciparum* in vivo (43). However, no such evidence of antagonism was found in a study in which Malawian children with cerebral malaria were treated with quinine. There was no difference in survival and rate of recovery in patients who had also been given chloroquine compared with those who had not (SEDA-13, 816).

Chloroquine and hydroxychloroquine

Chloroquine antagonizes the action of quinine against *P. falciparum* in vivo (43). However, no such evidence of antagonism was found in a study in which Malawian children with cerebral malaria were treated with quinine. There was no difference in survival and rate of recovery in patients who had also been given chloroquine compared with those who had not (SEDA-13, 816).

Clindamycin

The combination of quinine with clindamycin improved the cure rate after a short course of quinine (44). This combination enhances parasite clearance and recovery from fever and is of value in combating bacteremia (45).

Codeine

Quinidine inhibits the metabolism of codeine by CYP2D6 in extensive but not in poor metabolizers (46).

Digoxin

Quinine reduces the clearance of digoxin by inhibiting its biliary excretion (47). There may also be a small effect on renal digoxin clearance (48).

Enzyme inhibitors and inducers

Enzyme inhibitors, such as cimetidine and ketoconazole, increase the toxicity of quinine or quinidine (49), while enzyme inducers, such as rifampicin, reduce the toxic effects (50).

Erythromycin

Erythromycin is a competitive in-vitro inhibitor of quinine 3-hydroxylation and may therefore interact with quinine (51).

Flecainide

Flecainide is metabolized by CYP2D6, and is subject to polymorphic metabolism. In extensive metabolizers its clearance is reduced by quinine (52).

Halofantrine

Quinine should not be combined with halofantrine, since both drugs impair atrioventricular conduction and since quinine can enhance the cardiotoxic effects of halofantrine by inhibiting its metabolism (53).

Phenobarbital

Quinine inhibits the metabolism of phenobarbital (42).

Rifampicin

Since the use of antimalarial drugs in combination increases cure rates and can prevent drug resistance, and since rifampicin has antimalarial activity in experimental studies, the combination of quinine with rifampicin has been compared with standard quinine treatment in 59 adults with uncomplicated falciparum malaria (54). However, quinine + rifampicin failed, since its recrudescence rates were five times higher than with standard quinine. A pharmacokinetic evaluation showed that rifampicin co-therapy increases quinine clearance and thus lowers plasma concentrations of quinine and its partly active metabolite 3-hydroxyquinine. This effect is probably due to induction of intestinal and hepatic CYP3A4. Quinine and rifampicin should therefore not be combined in the treatment of malaria, and the dose of quinine should be increased in patients who are already taking rifampicin.

Tetracycline

The antimalarial effect of tetracycline is well known. The combination of quinine with doxycycline improved the cure rate (44,45). However, none of the patients taking quinine in combination with either clindamycin or tetracycline had high-grade resistant *P. falciparum* infections. The drawback of combining quinine with tetracycline is the longer time required for adequate treatment, which in most cases demands supervision. Furthermore, tetracycline should not be given to young children or pregnant women. The combination can be used in multiresistant and mefloquine-resistant *P. falciparum* malaria. When quinine is combined with doxycycline, the effects of the antibiotic itself may be enhanced.

Theophylline

Quinine raised serum theophylline concentrations in an elderly patient (55).

Troleandomycin

Troleandomycin is an in vitro competitive inhibitor of quinine 3-hydroxylation, which is mediated by CYP3A4, and may therefore interact when co-administered with quinine (56).

Food-Drug Interactions

Grapefruit juice

Grapefruit juice inhibits CYP3A4, which is involved in the metabolism of quinine. Thus, concomitant use of grapefruit juice with quinine might increase plasma quinine concentrations and increase the risk of adverse effects of quinine.

- A 31-year-old woman taking atenolol for asymptomatic long QT syndrome developed diabetes mellitus and took excessive amounts of tonic water containing quinine and grapefruit juice (57). Shortly after admission, she developed torsade de pointes with a QT_c interval of 0.58 seconds. Serum electrolytes, blood glucose, and thyroid function were normal. Two days later, after withdrawal of the drinks, her QT_c interval had shortened to 0.45 seconds and there were no dysrhythmias, even after programmed electrical stimulation.

The authors suggested that QT interval prolongation might have been caused by a pharmacokinetic interaction of quinine with grapefruit juice.

References

1. Krishna S, Nagaraja NV, Planche T, Agbenyega T, Bedo-Addo G, Ansong D, Owusu-Ofori A, Shroads AL, Henderson G, Hutson A, Derendorf H, Stacpoole PW. Population pharmacokinetics of intramuscular quinine in children with severe malaria. Antimicrob Agents Chemother 2001;45(6):1803–9.
2. Parola P, Ranque S, Badiaga S, Niang M, Blin O, Charbit JJ, Delmont J, Brouqui P. Controlled trial of 3-day quinine-clindamycin treatment versus 7-day quinine treatment for adult travelers with uncomplicated falciparum malaria imported from the tropics. Antimicrob Agents Chemother 2001;45(3):932–5.
3. Shapiro TA, Ranasinha CD, Kumar N, Barditch-Crovo P. Prophylactic activity of atovaquone against *Plasmodium*

falciparum in humans. Am J Trop Med Hyg 1999;60(5):831–6.

4. Bustos DG, Canfield CJ, Canete-Miguel E, Hutchinson DB. Atovaquone–proguanil compared with chloroquine and chloroquine–sulfadoxine–pyrimethamine for treatment of acute *Plasmodium falciparum* malaria in the Philippines. J Infect Dis 1999;179(6):1587–90.

5. Krantz MJ, Dart RC, Mehler PS. Transient pulmonary infiltrates possibly induced by quinine sulfate. Pharmacotherapy 2002;22(6):775–8.

6. Everts RJ, Hayhurst MD, Nona BP. Acute pulmonary edema caused by quinine. Pharmacotherapy 2004;24(9):1221–4.

7. Gopal KV, Gross GW. Unique responses of auditory cortex networks in vitro to low concentrations of quinine. Hear Res 2004;192(1–2):10–22.

8. Karbwang J, Sukontason K, Rimchala W, Namsiripongpun W, Tin T, Auprayoon P, Tumsupapong S, Bunnag D, Harinasuta T. Preliminary report: a comparative clinical trial of artemether and quinine in severe falciparum malaria. Southeast Asian J Trop Med Public Health 1992;23(4):768–72.

9. Anonymous. Quinine and cramp: uncertainty efficacy, major risks. Prescrire Int 2000;9(49):154–7.

10. Browne GF, Coppel DL. Management of quinine overdose. Hum Toxicol 1984;3(5):399–402.

11. Edwards DA, Johnson MC, O'Neill TJ. Quinine-induced myopathy. Lancet 1978;2:845.

12. Barrett NA, Solano T. Quinine ocular toxicity: treatment of blindness using therapy for vasospasm. Anaesth Intensive Care 2002;30(2):234–5.

13. Di Perri G, Allegranzi B, Bonora S. Quinine-induced blindness reversed by an increase in alpha1-acid glycoprotein level. Ann Intern Med 2002;136(4):339.

14. Boscoe MJ, Calver DM, Keyte C, Ayres JG. Quinine overdose. Prevention of visual damage by stellate ganglion block. Anaesthesia 1983;38(7):669–71.

15. Bacon P, Spalton DJ, Smith SE. Blindness from quinine toxicity. Br J Ophthalmol 1988;72(3):219–24.

16. Dyson EH, Proudfoot AT, Prescott LF, Heyworth R. Death and blindness due to overdose of quinine. BMJ (Clin Res Ed) 1985;291(6487):31–3.

17. Centers for Disease Control (CDC). Human rabies—Kenya. MMWR Morb Mortal Wkly Rep 1983;32(38):494–5.

18. Adam I, Elbashir MI. Suicide after treatment of chloroquine resistant falciparum malaria with quinine. Saudi Med J 2004;25(2):248–9.

19. Laughlin JC. Agricultural production of artemisinin—a review. Trans R Soc Trop Med Hyg 1994;88(Suppl 1):S21–2.

20. Brinker AD, Beitz J. Spontaneous reports of thrombocytopenia in association with quinine: clinical attributes and timing related to regulatory action. Am J Hematol 2002;70(4):313–7.

21. Anonymous. Quinine. Reports of thrombocytopenia. WHO Pharmaceuticals Newslett 2000;4:10.

22. Burgess JK, Lopez JA, Berndt MC, Dawes I, Chesterman CN, Chong BH. Quinine-dependent antibodies bind a restricted set of epitopes on the glycoprotein Ib-IX complex: characterization of the epitopes. Blood 1998;92(7):2366–73.

23. Baliga RS, Wingo CS. Quinine induced HUS-TTP: an unusual presentation. Am J Med Sci 2003;326:378–80.

24. Agarwal N, Cherascu B. Concomitant acral necrosis and haemolytic uraemic syndrome following ingestion of quinine. J Postgrad Med 2002;48(3):197–8.

25. Bird MR, O'Neill AI, Buchanan RR, Ibrahim KM, Des Parkin J. Lupus anticoagulant in the elderly may be associated with both quinine and quinidine usage. Pathology 1995;27(2):136–9.

26. Birku Y, Makonnen E, Bjorkman A. Comparison of rectal artemisinin with intravenous quinine in the treatment of severe malaria in Ethiopia. East Afr Med J 1999;76(3):154–9.

27. Morton AP. Quinine-induced disseminated intravascular coagulation and haemolytic–uraemic syndrome. Med J Aust 2002;176(7):351.

28. Knower MT, Bowton DL, Owen J, Dunagan DP. Quinine-induced disseminated intravascular coagulation: case report and review of the literature. Intensive Care Med 2003;29:1007–11.

29. Kojouri K, Vesely SK, George JN. Quinine-associated thrombotic thrombocytopenic purpura–hemolytic uremic syndrome: frequency, clinical features, and long-term outcomes. Ann Intern Med 2001;135(12):1047–51.

30. Bateman DN, Dyson EH. Quinine toxicity. Adverse Drug React Acute Poisoning Rev 1986;5(4):215–33.

31. Katz B, Weetch M, Chopra S. Quinine-induced granulomatous hepatitis. BMJ (Clin Res Ed) 1983;286(6361):264–5.

32. Meyrick Thomas RH, Munro DD. Lichen planus in a photosensitive distribution due to quinine. Clin Exp Dermatol 1986;11(1):97–101.

33. Glynne P, Salama A, Chaudhry A, Swirsky D, Lightstone L. Quinine-induced immune thrombocytopenic purpura followed by hemolytic uremic syndrome. Am J Kidney Dis 1999;33(1):133–7.

34. Schattner A. Quinine hypersensitivity simulating sepsis. Am J Med 1998;104(5):488–90.

35. Bregani ER, van Tien T, Cabibbe M, Figini G, Manetti F. Holter monitoring in children with severe Plasmodium falciparum malaria during i.v. quinine treatment. J Trop Pediatr 2004;50(1):61.

36. Beyens MN, Guy C, Ollagnier M. Effets indésirables de la quinine dans l'indication de crampes musculaires. [Adverse effects of quinine in the treatment of leg cramps.] Therapie 1999;54(1):59–62.

37. Townend BS, Sturm JW, Whyte S. Quinine associated blindness. Aust Fam Physician 2004;33(8):627–8.

38. Langford NJ, Good AM, Laing WJ, Bateman DN. Quinine intoxications reported to the Scottish Poisons Information Bureau 1997–2002: a continuing problem. Br J Clin Pharmacol 2003;56(5):576–8.

39. Guly U, Driscoll P. The management of quinine-induced blindness. Arch Emerg Med 1992;9(3):317–22.

40. Weser JK, Sellers E. Drug interactions with coumarin anticoagulants. N Engl J Med 1971;285(10):547–582.

41. Lim AKH, Ho L, Levidiotis V. Quinine-induced renal failure as a result of rhabdomyolysis, hemolytic uremic syndrome and disseminated intravascular coagulation. Intern Med J 2006;36:465–7.

42. Amabeoku GJ, Chikuni O, Akino C, Mutetwa S. Pharmacokinetic interaction of single doses of quinine and carbamazepine, phenobarbitone and phenytoin in healthy volunteers. East Afr Med J 1993;70(2):90–3.

43. Ajana F, Fortier B, Martinot A, et al. Mefloquine prophylaxis and neurotoxicity. Report of a case. Sem Hop 1990;66:918.

44. Shmuklarsky MJ, Klayman DL, Milhous WK, Kyle DE, Rossan RN, Ager AL Jr, Tang DB, Heiffer MH, Canfield CJ, Schuster BG. Comparison of beta-artemether and beta-arteether against malaria parasites in vitro and in vivo. Am J Trop Med Hyg 1993;48(3):377–84.

45. Meshnick SR. The mode of action of antimalarial endoperoxides. Trans R Soc Trop Med Hyg 1994;88(Suppl 1):S31–2.

46. Kirkwood LC, Nation RL, Somogyi AA. Characterization of the human cytochrome P450 enzymes involved in the metabolism of dihydrocodeine. Br J Clin Pharmacol 1997;44(6):549–55.

47. Hedman A, Angelin B, Arvidsson A, Dahlqvist R, Nilsson B. Interactions in the renal and biliary elimination of digoxin: stereoselective difference between quinine and quinidine. Clin Pharmacol Ther 1990;47(1):20–6.

48. Aronson JK, Carver JG. Interaction of digoxin with quinine. Lancet 1981;1(8235):1418.

49. Zhao XJ, Ishizaki T. A further interaction study of quinine with clinically important drugs by human liver microsomes: determinations of inhibition constant (Ki) and type of inhibition. Eur J Drug Metab Pharmacokinet 1999;24(3):272–8.

50. Pukrittayakamee S, Prakongpan S, Wanwimolruk S, Clemens R, Looareesuwan S, White NJ. Adverse effect of rifampin on quinine efficacy in uncomplicated falciparum malaria. Antimicrob Agents Chemother 2003;47(5):1509–13.

51. Gibson JR, Saunders NA, Burke B, Owen RJ. Novel method for rapid determination of clarithromycin sensitivity in *Helicobacter pylori*. J Clin Microbiol 1999;37(11):3746–8.

52. Munafo A, Reymond-Michel G, Biollaz J. Altered flecainide disposition in healthy volunteers taking quinine. Eur J Clin Pharmacol 1990;38(3):269–73.

53. Baune B, Furlan V, Taburet AM, Farinotti R. Effect of selected antimalarial drugs and inhibitors of cytochrome P-450 3A4 on halofantrine metabolism by human liver microsomes. Drug Metab Dispos 1999;27(5):565–8.

54. Pukrittayakamee S, Prakongpan S, Wanwimolruk S, Clemens R, Looareesuwan S, White NJ. Adverse effect of rifampin on quinine efficacy in uncomplicated falciparum malaria. Antimicrob Agents Chemother 2003;47:1509–13.

55. Shane R. Potential toxicity of theophylline in combination with Quinamm. Am J Hosp Pharm 1982;39(1):40.

56. Zhao XJ, Ishizaki T. A further interaction study of quinine with clinically important drugs by human liver microsomes: determinations of inhibition constant (Ki) and type of inhibition. Eur J Drug Metab Pharmacokinet 1999;24(3):272–8.

57. Hermans K, Stockman D, Van den Branden F. Grapefruit and tonic: a deadly combination in a patient with the long QT syndrome. Am J Med 2003;114:511–2.

Satranidazole

See also Benzimidazoles

General Information

Satranidazole is a highly active amebicidal agent with a slightly wider spectrum than metronidazole against micro-aerophilic and anaerobic bacteria; it is also effective against *Giardia* and *Trichomonas*. It is less toxic than metronidazole, nimorazole, secnidazole, and ornidazole. Gastrointestinal tolerability was good in early studies.

Therapeutic doses did not cause an adverse interaction with alcohol (SEDA-12, 707) (1,2).

References

1. Edlind TD, Hang TL, Chakraborty PR. Activity of the anthelmintic benzimidazoles against *Giardia lamblia* in vitro. J Infect Dis 1990;162(6):1408–11.

2. Arya VP. Satranidazole. Drugs Future 1983;8:797.

Secnidazole

See also Benzimidazoles

General Information

Secnidazole is more active and gives more prolonged blood concentrations than metronidazole; it is effective in hepatic amebiasis. Gastrointestinal adverse effects are less common (SEDA-11, 597) (1).

Reference

1. Andre LJ. Traitement de l'amibiase par le secnidazole. Ann Gastroenterol Hepatol 1979;15:221.

Suramin

General Information

Suramin is a trypanocide that has been used in the treatment of African trypanosomiasis and onchocerciasis, and has been studied in AIDS (SEDA-10, 277; SEDA-15, 335). However, the high doses that are required for a worthwhile effect are extraordinarily toxic. It is particularly likely to cause adverse effects in the malnourished and has therefore largely been abandoned for the treatment of trypanosomiasis and onchocerciasis, but it still seems uniquely capable of killing the adult onchocerciasis worm, although results are sometimes incomplete even in that respect (SEDA-20, 283).

Suramin is mainly used now in the treatment of hormone-refractory prostate cancer, in which it has shown some antitumor effect, although accompanied by extensive and sometimes severe adverse effects (1), and in the treatment of high-grade glioma. Since it suppresses adrenocortical function it is usually given in conjunction with glucocorticoids. The concomitant use of glucocorticoids and anti-androgen withdrawal usually makes it difficult to establish the true response to suramin.

Suramin inhibits the binding of a number of growth factors to their specific cell-surface receptors. Because angiogenesis is tightly regulated by many pro-angiogenic and anti-angiogenic growth factors, suramin can block transduction of extracellular signals in not only tumor cells but also endothelial cells. This may explain some of its antitumor effects, in particular in the treatment of urological malignancies. Its complex pharmacokinetics, narrow therapeutic margin, long half-life (55 days). and adverse effects (neuropathy, renal dysfunction, and myelosuppression), make it necessary to monitor blood suramin concentrations, aiming at concentrations of 200–250 μg/ml.

Experience with suramin in antiangiogenic therapy for urinary cancer has been reviewed (2). In advanced renal cell carcinoma, suramin yielded a response rate of only 0–4%. In 17 patients with hormone refractory prostate cancer, suramin produced a complete response in three and a partial response in three. In combination with glucocorticoids suramin produced a fall in prostate-specific antigen concentrations of at least 50% in 24–54% of eligible patients and yielded an objective response in 0–19% of patients with measurable disease. In patients with superficial bladder cancer, suramin can be administered intravesically which was tolerated well in phase I trials without serious toxicity. The paper did not contain further details of the adverse effects of suramin.

Observational studies

AIDS

For a time, unsuccessful attempts were made to treat AIDS with suramin (3), because of its proven ability to impair the in vitro infectivity (and inhibit the cytopathic effect) of human T cell lymphotropic virus type III (HTLV-III) or lymphadenopathy associated virus (LAV). However, it is no longer used for this indication. Its adverse effects were more severe and numerous than in its traditional field of use: two-thirds of patients had malaise, fever, and raised transaminases, and a quarter had adrenal insufficiency. Erythematous drug eruptions (particularly on sun-exposed surfaces) were common; some patients had a burning sensation of the skin, particularly on the limbs. The most common laboratory abnormalities were proteinuria, microscopic pyuria, trace hemoglobinuria, and occasional granular casts. There were rises in hepatic transaminase activities in some patients, usually during the second and third weeks, and others had eosinophilia (maximum 14%) during a drug eruption. There was a high incidence of keratopathy (SEDA-14, 263).

Bladder cancer

Intravesical suramin once weekly for 6 weeks has been studied in nine patients with histologically proven transitional cell carcinoma (4). The dose was slowly increased from 18 to 36 800 mg in 60 ml of fluid. Plasma suramin concentrations after treatment were 2–38 μg/ml and were not related to dose. Complications included self-limiting bladder irritation in four of 54 treatments, bladder spasm in four, and new or worsening vesicoureteral reflux in three. Another patient had bladder spasm, skin flushing, and fever. These symptoms resolved within 48 hours and did not recur after five subsequent treatments. An intravesical dose of 9180 mg/in 60 ml was defined as safe, with

acceptable plasma concentrations and minimal adverse effects.

In a phase I open, non-randomized dose-escalation study intravesical suramin 10, 50, 100, and 150 mg/ml was studied in 12 patients with recurrent transitional cell bladder carcinoma (5). Three patients had minor rises in fasting blood glucose and one had minor lymphopenia at the lowest dose. There were no serious adverse events and no patient complained of urinary symptoms related to treatment. There were no drug-related adverse events worse than grade 1. Systemic absorption of suramin was only found at the highest dose of 150 mg/ml.

Glioblastoma multiforme

In 55 patients with newly diagnosed glioblastoma multiforme (6) suramin was given in a conventional intermittent fixed-dosing regimen for 1 week before and during cranial radiotherapy (60 Gy in 30 fractions, weeks 2–7) aimed to maintain plasma concentrations in the 150–250 micrograms/ml range. Patients with stable or responsive disease at week 18 received an additional 4 weeks of twice-weekly intravenous infusions of suramin 275 mg/ m^2 (weeks 19–22). Two patients died of possibly related neurological events (stroke and raised intracranial pressure). Otherwise, adverse effects were generally transient and self-limiting. However, overall survival was not significantly improved when compared with other therapeutic regimens for glioblastoma.

Hormone-refractory prostate cancer

Much work relating to the use of suramin in prostatic cancer involves drug combinations, including aminoglutethimide, epirubicin, and hydrocortisone, with or without androgen deprivation, and in these studies it is hardly possible to determine which drug was responsible for a given adverse effect.

In 58 patients with hormone-refractory prostate carcinoma suramin was given in three 1-hour infusions each month (tapering the dose from 2400 mg/m^2 on day 1 to 1292 mg/m^2 on day 3) (7). Grade III fatigue (14%) was the predominant adverse effect. Suramin plasma concentrations were high even 3 months after withdrawal of therapy.

The adverse effects of suramin that were reported in a group of 69 patients with hormone-refractory prostate cancer were anorexia (19%), malaise and fatigue (40%), paresthesia (10%), weakness (9%), and skin rash (6%). In another group of patients there were higher frequencies of adverse effects: fatigue occurred in 70% and neuropathy in 16%. Hematological abnormalities occurred often, but were mostly mild and consisted of neutropenia (30%), anemia (74%), thrombocytopenia (26%), and coagulopathy (30%). Other common adverse effects have included uremia (21%), increased serum transaminase activities (19%), nausea and vomiting (30%), constipation (9%), edema (33%), dysrhythmias (7%), mild hyperglycemia (86%), and rash (60%) (SEDA-20, 283).

In 81 patients intravenous suramin (peak plasma concentration 300 μg/ml trough concentration 175 μg/ml) combined with aminoglutethimide 250 mg qds in patients with progressive androgen-refractory prostate cancer after antiandrogen treatment had been withdrawn, effectiveness was limited, whereas most adverse effects were attributed to suramin (8). There were 38 episodes of grade 3 and 4 toxic effects in 29 patients. Severe thrombocytopenia occurred in four patients. There were four episodes of atrial fibrillation. One patient developed uremia which required dialysis. One patient developed grade 3 neurosensory changes, but none had neuromotor changes. There was one episode of grade 4 rash, which was probably attributable to aminoglutethimide, consisting of diffuse erythematous exfoliating papules over the chest, back, arms, and face. All adverse effects were reversible.

The efficacy and adverse effects of treatment with suramin for hormone-refractory prostate cancer have been evaluated in 27 patients (9). The treatment regimen consisted of a loading phase, targeted to reach suramin serum concentrations of 180–250 µg/ml using a dose of 1.4 g/m^2 at 3-day intervals. Constant suramin concentrations were obtained with a dose of 0.5–1 g/m^2 every 7–10 days. Six patients did not complete the suramin loading phase because of adverse effects and were withdrawn. About one-third of the assessable patients had a more than 50% reduction in prostate specific antigen and/or serum alkaline phosphatase. Two of these also had a reduction in metastases on bone scan. Another 48% of the patients had unchanged prostate specific antigen or serum alkaline phosphatase during treatment with suramin, suggesting stable disease. The mean survival time was 41 weeks. Responders had a survival of 70 weeks compared with 12 weeks among non-responders. However, the adverse effects were substantial. The most common adverse effects were renal impairment (18 patients), 10 of whom had a mild increase in serum creatinine and seven had a moderate increase. One patient died of multiorgan failure, including renal shutdown. Suramin treatment was interrupted for 7–14 days when the creatinine clearance was under 40 ml/minute, which resulted in improvement in renal function. A sensimotor polyneuropathy occurred in 18 patients and typically presented as paresthesia involving the limbs, combined with reduced nerve conducting velocity. Mild and moderate sensorimotor polyneuropathy occurred in 14 patients. Severe polyneuropathy occurred in two cases and led to the withdrawal of suramin. These effects were only partially reversible. Eleven patients had no neurotoxic symptoms. Allergic rashes occurred in 30% of patients and consisted of a moderate, diffuse, morbilliform rash during the loading phase of suramin treatment, usually disappearing within a few days without further therapy. Two patients with more severe and prolonged rashes were treated with high-dose corticosteroids with a beneficial effect. Hematological toxicity consisted of anemia in 22% of the patients, who all needed blood transfusions, leukopenia in 15%, of whom one with a severe leukopenia was successfully treated by granulocyte-macrophage colony stimulating factor, and thrombocytopenia, which occurred in 15% of patients and led to spontaneous bleeding in two. The platelet count returned to normal in both cases after the withdrawal of suramin. Vortex keratopathy occurred in 15%

of the patients. Corneal changes were always minimal and vision was not affected. One patient had a retinal bleed, which led to withdrawal. Severe infections occurred in 26% of the patients, and also led to the withdrawal of suramin and required intravenous antibiotics. These results have further confirmed that suramin has a limited but statistically significant effect in the treatment of hormone-refractory prostate cancer. Although the severity of adverse effects were somewhat less than in previous studies they were still substantial. The toxicity of suramin seems to be closely related to the cumulative dose, peak concentration, and treatment regimen, and toxicity may be reduced by reducing daily doses and giving additional treatment-free intervals.

Suramin has been combined with epirubicin in 26 patients with hormone-refractory prostate cancer (10). No additional therapeutic effect was found compared with suramin or epirubicin alone. Suramin was given in an initial daily dose of 350 mg/m^2, with weekly infusions thereafter, targeted to maintain suramin plasma concentrations at 200–250 µg/ml for a maximum of 6 months. Cortisone acetate was added after 4 weeks in order to prevent adrenal insufficiency. Epirubicin was given as a weekly intravenous bolus from the start, also for a maximum of 6 months. The median duration of therapy with suramin and epirubicin was 9 (2–29) weeks. The toxic effects of this combined treatment included grade 1–2 nausea and vomiting (54%), fatigue (54%), anorexia (58%), stomatitis (52%), diarrhea (8%), mild rash (11%), neutropenia, usually mild (65%), low-grade fever (26%), mild increases in serum creatinine concentrations (27%), proteinuria (58%), and peripheral neurotoxicity (mild 23%, severe 4%); alopecia was caused by the epirubicin in 58%.

Combination chemotherapy with estramustine, docetaxel, and suramin has been studied in 42 patients with symptomatic progressive hormone-refractory prostate cancer (11). Estramustine was given at an oral dosage of 10 mg/kg/day on days 1–21 every 28 days, docetaxel 70 mg/m2 intravenously on day 2 every 28 days, and a total dose of 2150 mg of suramin was given in each cycle. Treatment was continued until disease progression or excessive toxicity. The median number of consecutive cycles was 8. The median time to progression was 57 weeks and median overall survival was 132 weeks. Most adverse events were moderate and were managed medically. Major adverse effects consisted of grade 3–4 anemia in 50%, leukopenia in 33%, and thrombocytopenia in 21%. The duration of leukopenia was generally less than 2 weeks and there were no deaths because of the sequelae of leukopenia. Severe neuropathy and gastrointestinal toxicity were uncommon, although there was moderate neuropathy in two patients. Six patients had a *rash*, a well-known adverse effect of suramin. Grade 3 or 4 non-hematological adverse effects included edema in three patients, malaise and/or fatigue in two, and dyspnea in three. Grade 1 or 2 adverse effects included hyperglycemia, malaise/fatigue, peripheral edema, nausea, and anorexia. Suramin dosage reduction was required in six patients because of adverse events (rash, weakness, and thrombocytopenia). There were no treatment-related deaths.

The use of suramin plus hydrocortisone and androgen deprivation and the use of multiple courses of suramin have been assessed in 59 patients with newly diagnosed metastatic prostate cancer (12). Suramin (doses aimed at plasma concentrations between a trough of 150 µg/ml and a peak of 250 µg/ml) was given in a 78-day fixed dosage schedule (one cycle) and suramin treatment cycles were repeated every 6 months to a total of four cycles. There was significant broad-spectrum toxicity throughout the study, leading to withdrawal of treatment in 33 patients. Cardiovascular events (dysrhythmias, hypotension, and congestive heart failure), neurotoxic effects, and respiratory effects were more frequent than expected. In consequence, repeated courses of suramin could be given in a minority of cases only. The authors felt that in the light of the relatively non-toxic palliation achieved with standard hormonal therapy, suramin in this dosage schedule has only limited use in patients with newly diagnosed metastatic prostate cancer.

The effects of fixed-dose suramin plus hydrocortisone have been studied in 50 patients with hormone-refractory prostate cancer (13). Suramin was initially given as a 30-minute test infusion of 200 mg. In the absence of allergic reactions, additional 24-hour intravenous infusions of 500 mg/m^2 were given daily for the next 5 days. Thereafter, 2-hour intravenous infusions (350 mg/m^2) were given weekly on an outpatient basis for 12 weeks or until disease progression. The median duration of response was 16 weeks and the median time to disease progression 13 weeks. Fatigue and lymphopenia were the most commonly reported adverse effects, in 27 patients (54%) and 39 patients (78%) respectively. Skin rash occurred in 12 patients (24%). Suramin was withdrawn in three patients because of acute renal insufficiency ($n = 2$) and Stevens–Johnson syndrome ($n = 1$).

In a randomized study in 390 patients suramin has been given in a fixed low dose (3.192 g/m^2), intermediate dose (5.320 g/m^2), or high dose (7.661 g/m^2) to determine whether its efficacy and toxicity in the treatment of patients with hormone-refractory prostate cancer is dose-dependent (14). There was no clear dose–response relation for survival or progression-free survival, but toxicity increased especially with the higher dose. There were neurological adverse effects in 40% of the patients and cardiac adverse effects in 15%. This raises questions about the usefulness of suramin, particularly in high doses, in advanced prostate cancer. However, in another phase I study of suramin with once- or twice-monthly dosing in patients with advanced cancer, suramin was relatively safely administered without using plasma concentrations to guide dosing (15). Dose-limiting toxic effects included fatigue, neuropathy, anorexia, and renal toxicity. Diffuse colitis, erythema multiforme, and hemolytic anemia were reported as unusual effects.

High-grade glioma

The efficacy, toxicity, and pharmacology of suramin have been studied in 12 patients with recurrent or progressive recurrent high-grade gliomas aged 26–67 years (16). Suramin was given in doses similar to those used in patients with hormone-refractory prostate cancer. Treatment-related adverse effects were usually mild and reversible. Three patients developed transient grade 3–4 toxicity (leukopenia, a rise in serum creatinine, and diarrhea). There was no coagulopathy or central nervous system bleeding. All patients reached target suramin concentrations. The pharmacology of suramin was not affected by anticonvulsant therapy. Median time to progression was 55 days (range 17–242) and median survival was 191 days (range 42–811). There were no partial or complete remissions at 12 weeks. However, the clinical outcome in three patients suggested that effect of suramin may be delayed. One patient who progressed after 12 weeks had a subsequent marked reduction in tumor size and maintained an excellent partial response for over 2 years without other therapy. The two others had disease stabilization and lived for 16 and 27 months respectively. Based on these observations, suramin and radiotherapy are now being used concurrently in patients with newly diagnosed glioblastoma multiforme to study survival as the primary outcome.

Human African trypanosomiasis

Human African trypanosomiasis is a fatal disease that has re-emerged in recent years; it is caused by *Trypanosoma brucei gambiense* or *Trypanosoma brucei rhodesiense*. However, little progress has been made in the development of new drugs; most of the drugs still in use were developed one or more decades ago and are generally toxic and of limited effectiveness. Suramin, a symmetrical polysulfonated naphthylamine polyanionic compound was introduced in the 1920s and to this day remains the drug of choice for the early phase of *Trypanosoma brucei rhodesiense* infections (17,18). The trypanocidal action of suramin is slow. It is given by slow intravenous injection. Subcutaneous or intramuscular injections are not recommended, because they cause local inflammation and necrosis. Because of its poor central nervous system penetration, suramin is not effective in late-stage trypanosomiasis, although it has been used in pre-treatment to reduce the toxicity of melarsoprol or to sterilize patients until they reach hospital, where melarsoprol will be given. Immediate life-threatening adverse effects include collapse, with nausea, vomiting and shock. Severe delayed reactions include *renal damage*, particularly in malnourished patients, exfoliative dermatitis, agranulocytosis, hemolytic anemia, jaundice, and severe diarrhea, all of which can be fatal. Polyneuropathy and stomatitis have also been described. A test dose of 200 mg is sometimes recommended to prevent idiosyncratic reactions. Trypanosomal resistance to suramin has not been a serious problem, even after 80 years of treating trypanosomiasis with this drug.

Lung cancer

The properties of low-dose suramin every 3 weeks as a chemosensitizer have been evaluated in a phase 1 study in 15 patients with advanced non-small cell lung cancer (19). The patients received 85 courses of suramin followed by paclitaxel (175–200 mg/m^2) and carboplatin (AUC of 6 minutes.mg/ml). The initial dose of suramin was 240 mg/ m^2, and the doses for subsequent cycles were calculated based on the 72-hour pretreatment plasma concentrations. The most common adverse effects were neutropenia (31 courses resulted in grade 3 neutropenia and 30 courses resulted in grade 4 neutropenia lasting for less than 5 days and never associated with neutropenic fever), nausea/vomiting grade 3 (after one course), malaise/fatigue grade 3 (after 13 courses), grade 3 peripheral neuropathy (after one course), grade 3 hypersensitivity/rash (after one course), and grade 3 diarrhea (after one course). There were no cases of adrenal dysfunction or episodes of sepsis. Dividing the suramin dose to be administered into two doses 24 hours apart yielded the target concentrations and avoided undesirable peak concentrations. There was discernible antitumor activity in seven of 10 patients with measurable disease, including two with prior chemotherapy. The median time to tumor progression was 8.5 months. The authors concluded that low-dose suramin does not increase the toxicity of the combination of paclitaxel + carboplatin.

Renal cell carcinoma

Suramin was not effective in one phase II study in advanced renal cell carcinoma, in which it was given in a fixed dose plus hydrocortisone to 22 patients (19 men, three women, aged 30–74 years) (20). Three patients had grade 4 toxicity (hypersensitivity, urethral obstruction, hypotension, and neutropenic sepsis). Eleven developed grade 3 toxicity, mainly abdominal pain, anemia, diarrhea, erythema, dyspnea, fatigue, and fever.

Placebo-controlled studies

The antitumor effect of suramin has been evaluated in a prospective randomized trial in 458 patients with hormone-refractory prostate cancer and significant opioid analgesic-dependent pain (21). Reduction of pain and opioid requirements served as surrogates for tumor responsiveness. The patients were given either suramin (aiming at sustained plasma concentrations of 100–300 µg/ml) plus hydrocortisone (40 mg/day) or placebo plus hydrocortisone. Patients treated with suramin plus hydrocortisone had greater reductions in combined pain and opioid analgesic intake. Suramin did not reduce the quality of life or performance status. However, overall survival was similar. Most of the adverse events were mild or moderate and were easily managed medically. Frequent adverse effects of suramin were rash, chills, fever, and taste disturbance. In contrast to the results of earlier studies, in which different suramin dosage regimens were used, neurological, renal, hepatic, and coagulation abnormalities were rare.

Organs and Systems

Nervous system

The neurological adverse effects of suramin have been reviewed in the context of a broad review of neuropathies associated with malignancy and chemotherapy (22).

Neurotoxicity is a dose-limiting adverse effect of suramin and there are two distinct types of neuropathy: a mild, length-dependent, axonal polyneuropathy, and a more serious subacute demyelinating, Guillain–Barré-like polyneuropathy. The reported incidence of neuropathy is 25–90% with a mean of about 50%. Various neuropathies were noted in four of 38 patients infused with suramin for various malignancies (23).

The milder axonal polyneuropathy is the most common neurological adverse effect of suramin and causes distal paresthesia, reduced pain and vibration sensation in the feet, weak toe extensors, and absent ankle jerks; this neuropathy is largely reversible. Milder neuropathies occurred in 50–70% of patients with plasma concentrations below 300 µg/ml, and severe motor neuropathy was rare in this category of patients.

The more severe demyelinating neuropathies appear to be dose-related and occur when peak suramin plasma concentrations are maintained above 350 µg/ml; the effective serum concentration in cancers is about 250 µg/ml (SEDA-20, 283) (24,25). A Guillain–Barré-like polyradiculoneuropathy occurs in 10–20% of patients after 1–5 months of treatment, with a maximum at 2–9 weeks after the start of treatment. The first symptoms are distal limb and or facial paresthesia, followed by diffuse, symmetrical, proximal weakness and areflexia. About 25% of these patients eventually require ventilation. The CSF protein content may be raised.

In 24 patients with hormone-refractory prostate cancer given suramin twice weekly intravenously targeted to reach plasma concentrations of 50–100, 101–150, 151–200, or 201–250 µg/ml plus doxorubicin, fatigue occurred in 18 and was dose-limiting in two (26). Eight developed neurological symptoms, of whom three, all receiving the highest dose, developed grade 3 toxicity. There were five cases of neuropathies. Two patients had evidence of a demyelinating neuropathy, one of whom developed a Guillain–Barré-like syndrome and inflammatory myopathy. A further patient had a mixed axonal and demyelinating peripheral neuropathy. Two patients developed a motor neuropathy that exacerbated pre-existing neurological defects. Other frequent adverse events were proteinuria, leukopenia, and alopecia. However, the respective roles of suramin and doxorubicin in causing these adverse effects were uncertain.

Evidence of an underlying distal axonal polyneuropathy can be detected by electromyelography (27). Electroencephalography shows slow motor conduction velocities and electromyography shows reduced recruitment in both proximal and distal muscles. In the more severe cases denervation emerges. Sural nerve biopsies have shown a reduced density of the large and small myelinated fibers, occasional axonal degeneration, and demyelination. Epineural and endoneural mononuclear inflammatory cell infiltrates are sometimes seen. After withdrawal of treatment symptoms may deteriorate further for several weeks with recovery, sometimes incomplete, after 1–2 months.

The precise mechanism of the neurotoxicity of suramin is unknown, although both inhibition of the effects of nerve growth factors and a possible immune-mediated effect, consistent with the many immunomodulating effects, have been suggested. In a recent experimental study in dorsal root ganglion cell cultures suramin disrupted the transport and metabolism of glycolipids, with accumulation of the GM1 ganglioside and ceramide, leading to cell death (28).

Guillain–Barré syndrome has been attributed to suramin.

- A 37-year-old woman with a poorly differentiated metastatic pulmonary adenocarcinoma developed Guillain–Barré syndrome after receiving suramin + interferon alfa (29). She received 2 cycles of intravenous immunoglobulin and her flaccid tetraparesis remitted over a few weeks.

Although treatment with suramin may be associated with a Guillain–Barré type of sensorimotor neuropathy, interferon was considered the most likely cause in this case, given the simultaneous existence of autoimmune liver disease, hematological changes, and autoantibodies.

Sensory systems

Photophobia, lacrimation, and palpebral edema are recognized late effects of suramin, and there is evidence of a late optic atrophy of suramin (30).

High-dose treatments for cancers seem to produce changes in most patients consistent with a vortex keratopathy (SEDA-21, 320), although it can be entirely symptomless. Other cases have a hyperoptic refractory shift leading to blurred vision. Of 114 patients, 19 developed ocular symptoms and signs while taking suramin sodium for metastatic cancer of the prostate (31). Of these, 13 developed bilateral corneal epithelial whorl-like deposits, in 10 cases associated with a foreign body sensation and lacrimation. Symptoms in all cases resolved with topical lubricants. Three patients developed asymptomatic corneal deposits. Seven had blurred vision and had a mean hyperopic shift in refractive error of 1.13 (range 0.75–2.00) diopters, which persisted throughout treatment. None of these patients had a reduction in best-corrected visual acuity. At the very least, ophthalmological surveillance is necessary when suramin is used.

Suramin has been used in the treatment of AIDS and adrenal carcinoma, in both of which there was a high incidence of keratopathy (SED-13, 836).

Endocrine

The adrenal glands are sensitive to the toxic effects of suramin; both glucocorticoid and mineralocorticoid functions can be impaired at doses normally used, necessitating replacement therapy (32).

Hematologic

Occasional cases of hemolytic anemia and agranulocytosis have been reported in patients taking suramin (SED-13, 836). Severe neutropenia has been attributed to suramin in six cases (33). Plasma concentrations of platelet-derived growth factor-AB (PDGF-AB) and fibroblast growth factor basic correlated with the time-course of the neutropenia, which was unpredictable and occurred both during and after withdrawal of suramin. Plasma concentrations of

suramin and G-CSF did not correlate with the neutropenia, but there was a rapid response to G-CSF.

Blood coagulation can be altered by suramin, which causes accumulation of glycosaminoglycans, which have heparin-like properties (34). In patients who had received suramin intravenously for 2 weeks there was inhibition of factors V, VIII, IX, X, XI, and XII, while thrombin, prothrombin, and factor VII were unaffected (35). The inhibition of factor V was virtually irreversible, although the effect of suramin on the other factors is readily reversed by dilution.

In one patient with cancer, immunological complement-mediated destruction of circulating platelets was induced by suramin, demanding urgent treatment (36).

Gastrointestinal

Suramin can cause vomiting (37).

Liver

In three men with severe chronic active hepatitis, suramin treatment prolonged the prothrombin time in all three, caused a rise in bilirubin in two, and may have led to hemorrhage from esophageal varices in one patient and to hepatic encephalopathy in another (38).

Urinary tract

Suramin can cause acute renal failure (39,40).

Skin

Skin effects were prominent in patients treated with the high doses needed in malignancies, including morbilliform rashes in 67%, UV recall in 35%, urticaria in 18%, and keratotic papules in 12% (41). Late reactions include various skin eruptions, hyperesthesia and paresthesia (particularly affecting the palms and soles of the feet); erythema multiforme as part of a generalized skin reaction has been well documented in some individual cases (42) and toxic epidermal necrolysis has been reported (43).

A single dose of intravenous hydrocortisone 200 mg was protective against skin complications in another small group of patients.

Second-Generation Effects

Teratogenicity

Suramin has generally been avoided in pregnancy because its safety is uncertain; it is teratogenic in experimental animals (44–46).

Drug–Drug Interactions

Furosemide

In 26 patients treated with suramin for hormone-refractory prostate cancer, furosemide reduced the total body clearance of suramin by 36% (47). In view of the increased risk of severe adverse effects after treatment with higher plasma concentrations of suramin, it would be prudent to alter dosage schemes in patients treated with both suramin and furosemide.

Warfarin

In 13 men with advanced hormone-refractory prostate cancer the interaction between suramin and warfarin was studied because of potential worries that suramin may affect blood coagulation (48). After initial stabilization to an International Normalized Ratio (INR) of about 2.0 suramin plus hydrocortisone was started, after which warfarin requirements fell by 0.50–0.78 mg/day. The difference did not reach statistical significance. There were no bleeding problems. These results suggest that suramin and warfarin can be safely co-administered, provided that coagulation status is monitored.

Monitoring Therapy

The pharmacokinetics of suramin have been studied in 62 patients with non-small-cell lung cancers with the aim of using it as a chemosensitizer for paclitaxel and carboplatin chemotherapy (49). For this purpose, plasma suramin concentrations should be kept within the effective range (10–50 µmol/l) duration the time that the chemotherapeutic agents are present at therapeutically significant concentrations (e.g. 48 hours for paclitaxel and carboplatin). A dosing nomogram was developed by using pharmacokinetic data that were derived from 15 patients who had participated in phase I trials and evaluated in 47 patients in phase II trials. The chemosensitizing dose of suramin had a terminal half-life of 202 hours and a total body clearance of 29 ml/hour/m^2. The dosing nomogram delivered the target concentration in more than 95% of treatments. There were no significant pharmacokinetic interactions between suramin, paclitaxel, and carboplatin, but the disposition of suramin was non-linear.

References

1. Knox JJ, Moore MJ. Treatment of hormone refractory prostate cancer. Semin Urol Oncol 2001;19(3):202–11.
2. Kanda S, Miyata Y, Kanetake H. Current status and perspective of antiangiogenic therapy for cancer: urinary cancer. Int J Clin Oncol 2006;11:90–107.
3. Broder S, Yarchoan R, Collins JM, Lane HC, Markham PD, Klecker RW, Redfield RR, Mitsuya H, Hoth DF, Gelmann E, et al. Effects of suramin on HTLV-III/LAV infection presenting as Kaposi's sarcoma or AIDS-related complex: clinical pharmacology and suppression of virus replication in vivo. Lancet 1985;2(8456):627–30.
4. Uchio EM, Linehan WM, Figg WD, Walther MM. A phase I study of intravesical suramin for the treatment of superficial transitional cell carcinoma of the bladder. J Urol 2003;169:357–60.
5. Ord JJ, Streeter E, Jones A, Le Monnier K, Cranston D, Crew J, Joel SP, Rogers MA, Banks RE, Roberts ISD, Harris AL. Phase I trial of intravesical suramin in recurrent superficial transitional cell bladder carcinoma. Br J Cancer 2005;92:2140–7.
6. Laterra JJ, Grossman SA, Carson KA, Lesser GJ, Hochberg FH, Gilbert MR. Suramin and radiotherapy in newly diagnosed glioblastoma: phase 2 NABTT CNS Consortium study. Neuro-Oncology 2004;6:15–20.
7. Vogelzang NJ, Karrison T, Stadler WM, Garcia J, Cohn H, Kugler J, Troeger T, Giannone L, Arrieta R, Ratain MJ, Vokes EE. A phase II trial of suramin monthly x 3 for hormone-refractory prostate carcinoma. Cancer 2004;100:65–71.
8. Dawson N, Figg WD, Brawley OW, Bergan R, Cooper MR, Senderowicz A, Headlee D, Steinberg SM, Sutherland M, Patronas N, Sausville E, Linehan WM, Reed E, Sartor O. Phase II study of suramin plus aminoglutethimide in two cohorts of patients with androgen-independent prostate cancer: simultaneous antiandrogen withdrawal and prior antiandrogen withdrawal. Clin Cancer Res 1998;4(1):37–44.
9. Garcia-Schurmann JM, Schulze H, Haupt G, Pastor J, Allolio B, Senge T. Suramin treatment in hormone- and chemotherapy-refractory prostate cancer. Urology 1999;53(3):535–41.
10. Falcone A, Antonuzzo A, Danesi R, Allegrini G, Monica L, Pfanner E, Masi G, Ricci S, Del Tacca M, Conte P. Suramin in combination with weekly epirubicin for patients with advanced hormone-refractory prostate carcinoma. Cancer 1999;86(3):470–6.
11. Safarinejad MR. Combination chemotherapy with docetaxel, estramustine and suramin for hormone refractory prostate cancer. Urol Oncol 2005;23:93–101.
12. Hussain M, Fisher EI, Petrylak DP, O'Connor J, Wood DP, Small EJ, Eisenberger MA, Crawford ED. Androgen deprivation and four courses of fixed-schedule suramin treatment in patients with newly diagnosed metastatic prostate cancer: a Southwest Oncology Group Study. J Clin Oncol 2000;18(5):1043–9.
13. Calvo E, Cortes J, Rodriguez J, Sureda M, Beltran C, Rebollo J, Martinez-Monge R, Berian JM, de Irala J, Brugarolas A. Fixed higher dose schedule of suramin plus hydrocortisone in patients with hormone refractory prostate carcinoma a multicenter Phase II study. Cancer 2001;92(9):2435–43.
14. Small EJ, Halabi S, Ratain MJ, Rosner G, Stadler W, Palchak D, Marshall E, Rago R, Hars V, Wilding G, Petrylak D, Vogelzang NJ. Randomized study of three different doses of suramin administered with a fixed dosing schedule in patients with advanced prostate cancer: results of intergroup 0159, cancer and leukemia group B 9480. J Clin Oncol 2002;20(16):3369–75.
15. Ryan CW, Vokes EE, Vogelzang NJ, Janisch L, Kobayashi K, Ratain MJ. A phase I study of suramin with once- or twice-monthly dosing in patients with advanced cancer. Cancer Chemother Pharmacol 2002;50(1):1–5.
16. Grossman SA, Phuphanich S, Lesser G, Rozental J, Grochow LB, Fisher J, Piantadosi S. New Approaches to Brain Tumor Therapy CNS Consortium. Toxicity, efficacy, and pharmacology of suramin in adults with recurrent high-grade gliomas. J Clin Oncol 2001;19(13):3260–6.
17. Fairlamb AH. Chemotherapy of human African trypanosomiasis: current and future prospects. Trends Parasitol 2003;19:488–94.
18. Docampo R, Moreno SNJ. Current chemotherapy of human African trypanosomiasis. Parasitol Res 2003;90:S10–13.
19. Villalona-Calero MA, Wientjes MG, Otterson GA, Kanter S, Young D, Murgo AJ, Fischer B, DeHoff C, Chen D. Yeh TK, Song SH, Grever M, Au JLS. Phase I study of low-dose suramin as a chemosensitizer in patients with advanced non-small cell lung cancer. Clin Cancer Res 2003;9:3303–11.

20. Schroder LE, Lew D, Flanigan RC, Eisenberger MA, Seay TE, Hammond N, Needles BM, Crawford ED. Phase II evaluation of suramin in advanced renal cell carcinoma. A Southwest Oncology Group study. Urol Oncol 2001;6(4):145–8.

21. Small EJ, Meyer M, Marshall ME, Reyno LM, Meyers FJ, Natale RB, Lenehan PF, Chen L, Slichenmyer WJ, Eisenberger M. Suramin therapy for patients with symptomatic hormone-refractory prostate cancer: results of a randomized phase III trial comparing suramin plus hydrocortisone to placebo plus hydrocortisone. J Clin Oncol 2000;18(7):1440–50.

22. Amato AA, Collins MP. Neuropathies associated with malignancy. Semin Neurol 1998;18(1):125–44.

23. La Rocca RV, Meer J, Gilliatt RW, Stein CA, Cassidy J, Myers CE, Dalakas MC. Suramin-induced polyneuropathy. Neurology 1990;40(6):954–60.

24. Chaudhry V, Eisenberger MA, Sinibaldi VJ, Sheikh K, Griffin JW, Cornblath DR. A prospective study of suramin-induced peripheral neuropathy. Brain 1996;119(Pt 6):2039–52.

25. Soliven B, Dhand UK, Kobayashi K, Arora R, Martin B, Petersen MV, Janisch L, Vogelzang NJ, Vokes EE, Ratain MJ. Evaluation of neuropathy in patients on suramin treatment. Muscle Nerve 1997;20(1):83–91.

26. Tu SM, Pagliaro LC, Banks ME, Amato RJ, Millikan RE, Bugazia NA, Madden T, Newman RA, Logothetis CJ. Phase I study of suramin combined with doxorubicin in the treatment of androgen-independent prostate cancer. Clin Cancer Res 1998;4(5):1193–201.

27. Rosen PJ, Mendoza EF, Landaw EM, Mondino B, Graves MC, McBride JH, Turcillo P, deKernion J, Belldegrun A. Suramin in hormone-refractory metastatic prostate cancer: a drug with limited efficacy. J Clin Oncol 1996;14(5):1626–36.

28. Gill JS, Windebank AJ. Suramin induced ceramide accumulation leads to apoptotic cell death in dorsal root ganglion neurons. Cell Death Differ 1998;5(10):876–83.

29. Bachmann T, Koetter KP, Muhler J, Fuhrmeister U, Seidel G. Guillain–Barré syndrome after simultaneous therapy with suramin and interferon-alpha. Eur J Neurol 2003;10:599.

30. Thylefors B, Rolland A. The risk of optic atrophy following suramin treatment of ocular onchocerciasis. Bull World Health Organ 1979;57(3):479–80.

31. Hemady RK, Sinibaldi VJ, Eisenberger MA. Ocular symptoms and signs associated with suramin sodium treatment for metastatic cancer of the prostate. Am J Ophthalmol 1996;121(3):291–6.

32. Kobayashi K, Weiss RE, Vogelzang NJ, Vokes EE, Janisch L, Ratain MJ. Mineralocorticoid insufficiency due to suramin therapy. Cancer 1996;78(11):2411–20.

33. Dawson NA, Lush RM, Steinberg SM, Tompkins AC, Headlee DJ, Figg WD. Suramin-induced neutropenia. Eur J Cancer 1996;32A(9):1534–9.

34. Horne MK 3rd, Stein CA, LaRocca RV, Myers CE. Circulating glycosaminoglycan anticoagulants associated with suramin treatment. Blood 1988;71(2):273–9.

35. Horne MK 3rd, Wilson OJ, Cooper M, Gralnick HR, Myers CE. The effect of suramin on laboratory tests of coagulation. Thromb Haemost 1992;67(4):434–9.

36. Seidman AD, Schwartz M, Reich L, Scher HI. Immune-mediated thrombocytopenia secondary to suramin. Cancer 1993;71(3):851–4.

37. Cheson BD, Levine AM, Mildvan D, Kaplan LD, Wolfe P, Rios A, Groopman JE, Gill P, Volberding PA, Poiesz BJ, et al.

38. Loke RH, Anderson MG, Coleman JC, Tsiquaye KN, Zuckerman AJ, Murray-Lyon IM. Suramin treatment for chronic active hepatitis B—toxic and ineffective. J Med Virol 1987;21(1):97–9.

39. Figg WD, Cooper MR, Thibault A, Headlee D, Humphrey J, Bergan RC, Reed E, Sartor O. Acute renal toxicity associated with suramin in the treatment of prostate cancer. Cancer 1994;74(5):1612–4.

40. Smith A, Harbour D, Liebmann J. Acute renal failure in a patient receiving treatment with suramin. Am J Clin Oncol 1997;20(4):433–4.

41. Lowitt MH, Eisenberger M, Sina B, Kao GF. Cutaneous eruptions from suramin. A clinical and histopathologic study of 60 patients. Arch Dermatol 1995;131(10):1147–53.

42. Katz SK, Medenica MM, Kobayashi K, Vogelzang NJ, Soltani K. Erythema multiforme induced by suramin. J Am Acad Dermatol 1995;32(2 Pt 1):292–3.

43. Falkson G, Rapoport BL. Lethal toxic epidermal necrolysis during suramin treatment. Eur J Cancer 1992;28A(6–7):1294.

44. Mercier-Parot L, Tuchmann-Duplessis H. Action abortive et tératogène d'un trypanocide, la suramine. [Abortifacient and teratogenic effect of suramin, a trypanocide.] C R Seances Soc Biol Fil 1973;167(11):1518–22.

45. Freeman SJ, Lloyd JB. Evidence that suramin and aurothiomalate are teratogenic in rat by disturbing yolk sac-mediated embryonic protein nutrition. Chem Biol Interact 1986;58(2):149–60.

46. Manner J, Seidl W, Heinicke F, Hesse H. Teratogenic effects of suramin on the chick embryo. Anat Embryol (Berl) 2003;206(3):229–37.

47. Piscitelli SC, Forrest A, Lush RM, Ryan N, Whitfield LR, Figg WD. Pharmacometric analysis of the effect of furosemide on suramin pharmacokinetics. Pharmacotherapy 1997;17(3):431–7.

48. Meyer M, Jeong E, Bolinger B, Chen L, Lenehan P, Slichenmyer W, Natale RB. Phase 1 drug interaction study of suramin and warfarin in patients with prostate cancer. Am J Clin Oncol 2001;24(2):167–71.

49. Chen D, Song SH, Wientjes MG, Yeh TK, Zhao L, Villalona-Calero M, Otterson GA, Jensen R, Grever M, Murgo AJ, Au JLS. Nontoxic suramin as a chemosensitizer in patients: dosing nomogram development. Pharm Res 2006;23(6):1265–74.

Suramin therapy in AIDS and related disorders. Report of the US Suramin Working Group. JAMA 1987;258(10):1347–51.

Tetrachloroethylene

General Information

Tetrachloroethylene is a chlorinated derivative of a simple hydrocarbon, $H_2C{=}CH_2$, in which each of the four hydrogen atoms is replaced by chlorine, $Cl_2C{=}CCl_2$. It is a heavy liquid that has been used to treat hookworm infection. It is given orally in a dose of 0.1 ml/kg up to a maximum of 5 ml as a single dose on an empty stomach. Usually formulated as capsules or emulsion, it is unstable, especially if exposed to light. Concurrent *Ascaris* infection should be treated first to avoid migration of worms and the risk of peritonitis.

General adverse effects

The adverse effects of tetrachloroethylene are similar to those of carbon tetrachloride (CCl_4), but less severe. Tetrachloroethylene is hepatotoxic and neurotoxic, but gastrointestinal disturbance is the only common adverse effect when it is used carefully. There is some risk of addiction to the inhaled vapor: inhalation can result in vascular reactions, loss of consciousness, pulmonary edema, and fatal hepatic and renal damage. Alcohol and fatty foods increase absorption and hepatic toxicity. Exposure to tetrachloroethylene has been known to lead to vinyl chloride disease. Allergic reactions have not been reported.

It is highly unlikely that single-dose use of tetrachloroethylene could have a tumorigenic effect; however, animal studies conducted in connection with exposure to tetrachloroethylene in the course of industrial use showed an increased occurrence of liver tumors.

Organs and Systems

Cardiovascular

In cases of poisoning with tetrachloroethylene, hypotension can occur; sympathomimetic drugs should not be used to treat it, since ventricular fibrillation can be precipitated.

Respiratory

Hypersensitivity pneumonitis has been attributed to occupational exposure to tetrachloroethylene in a 42-year-old dry cleaner; the diagnosis was confirmed by lung biopsy (1).

Nervous system

Headache and vertigo are common with tetrachloroethylene (2,3). Patients should remain at rest for 3 hours after administration and take only water.

Inhalation of tetrachloroethylene vapor can cause stupor (4).

Psychological, psychiatric

Reversible neuropsychiatric symptoms, readily resembling alcoholic intoxication, have occurred after a single dose of 5 ml. Sleep apnea causing neuropsychiatric abnormalities has been attributed to exposure to high concentrations of tetrachloroethylene and *N*-butanol vapors; however, the patient was obese and the association with the solvents was not clear (5).

Gastrointestinal

Nausea, vomiting, colicky abdominal pain, and diarrhea are common with tetrachloroethylene (2).

Liver

Hepatotoxicity from tetrachloroethylene is similar to that caused by carbon tetrachloride and can occur after oral treatment or even inhalation of the vapor (6).

Urinary tract

Tetrachloroethylene can cause renal damage, especially after inhalation (7).

Skin

If tetrachloroethylene comes into contact with the skin, either directly or after vomiting, burn-like reactions can occur (8).

Erythema multiforme has followed oral administration of tetrachloroethylene (9).

Second-Generation Effects

Pregnancy

It is unwise to use tetrachloroethylene in pregnancy in view of its potential for hepatotoxicity, but there is no specific information on the risks to the fetus as a result of therapeutic use; however, work relating to industrial use (exposure of women in the dry-cleaning industry) suggests the need for caution (10).

Susceptibility Factors

Age

It is dangerous to use tetrachloroethylene in young children (11).

References

1. Tanios MA, El Gamal H, Rosenberg BJ, Hassoun PM. Can we still miss tetrachloroethylene-induced lung disease? The emperor returns in new clothes. Respiration 2004;71(6):642–5.
2. Ng TP, Tsin TW, O'Kelly FJ. An outbreak of illness after occupational exposure to ozone and acid chlorides. Br J Ind Med 1985;42(10):686–90.
3. Sorensen S, Melgaard B. Treatment of hookworm anemia. Scand J Infect Dis 1971;3(1):65–9.
4. Lackore LK, Perkins HM. Accidental narcosis. Contamination of compressed air system. JAMA 1970;211(11):1846.
5. Muttray A, Randerath W, Ruhle KH, Gajsar H, Gerhardt P, Greulich W, Konietzko J. Obstruktives Schlafapnoesyndrom durch eine berufliche Losungsmittelexposition. [Obstructive sleep apnea syndrome caused by occupational exposure to solvents.] Dtsch Med Wochenschr 1999;124(10):279–81.
6. Brautbar N, Williams J 2nd. Industrial solvents and liver toxicity: risk assessment, risk factors and mechanisms. Int J Hyg Environ Health 2002;205(6):479–91.
7. Salahudeen AK. Perchloroethylene-induced nephrotoxicity in dry-cleaning workers: is there a role for free radicals? Nephrol Dial Transplant 1998;13(5):1122–4.

8. Hake CL, Stewart RD. Human exposure to tetrachloroethylene: inhalation and skin contact. Environ Health Perspect 1977;21:231–8.
9. Hisanaga N, Jonai H, Yu X, Ogawa Y, Mori I, Kamijima M, Ichihara G, Shibata E, Takeuchi Y. [Stevens–Johnson syndrome accompanied by acute hepatitis in workers exposed to trichloroethylene or tetrachloroethylene.]Sangyo Eiseigaku Zasshi 2002;44(2):33–49.
10. Ahlborg G Jr. Pregnancy outcome among women working in laundries and dry-cleaning shops using tetrachloroethylene. Am J Ind Med 1990;17(5):567–75.
11. Balmer S, Howells G, Wharton B. The effects of tetrachlorethylene in children with kwashiorkor and hookworm infestation. J Trop Pediatr 1970;16(1):20–3.

Tiabendazole

See also Benzimidazoles

General Information

Tiabendazole is a non-carbamate benzimidazole which inhibits cellular enzyme systems specific to some species of helminths. It is active against *Giardia lamblia*, but less effective than albendazole and mebendazole. It was at one time the drug of choice against *Strongyloides stercoralis*, both the enteric and tissue forms, but has been superseded by more modern drugs. It is also effective against *Enterobius*, hookworms, and *Trichuris trichiura*. Normally given orally, it is rapidly absorbed, with peak serum concentrations 1–2 hours after ingestion. It is rapidly metabolized and excreted; 40% of the drug and its metabolites are excreted during the first 4 hours and 80% during the first 24 hours. Its main metabolite, 5-hydroxythiabendazole, is inactive.

General adverse effects

Common toxic effects of tiabendazole include nausea, vomiting, and dizziness. Malaise and drowsiness are also common. Liver disorders can occur and are the most serious complications. Most systems can on occasion be affected. Allergic reactions are essentially due to parasite destruction rather than a direct effect of the drug itself. Chills, fever, lymphadenopathy, angioedema, and pruritic rashes all can occur; and treatment should in that case be stopped, since otherwise more serious reactions (for example Stevens–Johnson syndrome) can follow. Tumor-inducing effects have not been reported.

Organs and Systems

Cardiovascular

Bradycardia, hypotension, and syncope can occur with tiabendazole, even to the point of collapse (1).

Nervous system

Somnolence has been described with high doses of tiabendazole rectally or orally, and drowsiness, headache, malaise, and fatigue are common (1). Patients should be warned not to drive or carry out other potentially hazardous pursuits during treatment.

More severe symptoms of neurotoxicity are not unusual and include disorientation, confusional states, feelings of detachment, overt psychosis (2), and possibly epileptiform convulsions (3), although the latter have been observed only in a case of Down syndrome.

Sensory systems

Effects of tiabendazole on vision are not uncommon: abnormal sensations in the eyes, xanthopsia (a yellow tinge to objects), blurred vision, drying of mucous membranes, and sicca syndrome all have been described (4).

Allergic reactions to tiabendazole can cause keratoconjunctivitis sicca as part of Sjögren's syndrome (5).

Tinnitus has been attributed to tiabendazole (6).

Metabolism

Instances of both hypoglycemia and hyperglycemia have been recorded in patients taking tiabendazole.

Hematologic

Marked peripheral eosinophilia has also been described after rectal thiabendazole (7).

Gastrointestinal

Nausea, anorexia, vomiting, abdominal pain, and diarrhea are common and occur in a high proportion of patients taking tiabendazole (8).

Liver

Parenchymal liver damage can occur in patients taking tiabendazole and abnormal liver function tests have been documented (9). There have been well-studied cases of bile duct injury, which can lead to micronodular cirrhosis (10), and a case in which these various forms of liver disorder co-existed and liver transplantation proved necessary (11).

Persistent cholestasis can occur in patients taking tiabendazole (12).

- A 27-year-old patient from Surinam with beta-thalassemia took tiabendazole 1250 mg bd for 2 days for strongyloidiasis (13). One week later she became icteric, with raised total and conjugated bilirubin, alkaline phosphatase, gamma-glutamyltransferase, and transaminases. Tests for antinuclear antibodies, parietal cells antibodies, smooth muscle antibodies, mitochondrial antibodies, hepatitis A, B, C, cytomegalovirus, Epstein–Barr virus, mumps, and measles were negative. Ultrasonography showed normal intrahepatic and extrahepatic bile ducts. Liver biopsy showed intrahepatic cholestasis and a slightly increased infiltrate in the portal areas. One week later she developed a generalized urticarial rash. She had mildly abnormal liver tests for the next 7 years, at which time a liver biopsy showed a slight lymphatic infiltrate in the portal fields, without signs of cirrhosis, chronic hepatitis, or primary biliary cirrhosis.

- A 42-year-old woman, also from Surinam, with beta-thalassemia and non-insulin dependant diabetes mellitus took tiabendazole 1250 mg bd for 2 days for strongyloidiasis (13). Five weeks later she developed general malaise, anorexia, weight loss, icterus, and a tender liver. She had raised total and direct bilirubin, gamma-glutamyltransferase, and alkaline phosphatase, but only marginally raised transaminases. Tests for Hepatitis A, B, and C, cytomegalovirus, and schistosomiasis were negative. Tests for antinuclear antibodies and antibodies against liver cell membranes, smooth muscle, and mitochondria were negative, but there were parietal cell antibodies. Ultrasonography and ERCP showed normal intrahepatic and extrahepatic bile ducts. Liver biopsy showed severe centrally localized cholestasis. A year later all clinical and laboratory abnormalities had disappeared.

In view of these and previous cases of severe cholestasis after tiabendazole and the availability of less toxic equally effective drugs (albendazole or preferably ivermectin), tiabendazole must be considered obsolete in the treatment of strongyloidiasis.

Urinary tract

Crystalluria has been noted in patients taking tiabendazole, sometimes with hematuria (1).

Tiabendazole can give the urine an asparagus-like smell because of the presence of a mercaptan metabolite of tiabendazole (14).

Skin

Pruritus and skin rashes can occur (15). Much more rarely, toxic epidermal necrolysis has occurred after a total dose of 1800 mg, and Stevens–Johnson syndrome and erythema multiforme have also been reported (16).

Very occasionally a topical form of tiabendazole is used, for example to treat rosacea, and in one such case contact dermatitis aggravated by sunlight occurred as a complication, with positive tests for tiabendazole sensitivity (17).

Musculoskeletal

Tiabendazole can cause severe muscle pain on exercise (18).

Second-Generation Effects

Teratogenicity

There is no firm information on adverse reactions during pregnancy, but animal evidence points to teratogenicity and there have been official warnings against use by pregnant women (19).

Susceptibility Factors

Renal disease

Tiabendazole is metabolized almost completely by the liver, and its active metabolites are substantially excreted by the kidney; the risk of toxicity may therefore be greater in patients with impaired renal function (20).

Hepatic disease

Because tiabendazole is metabolized almost completely by the liver (20), it should be used with caution in hepatic impairment.

Drug–Drug Interactions

Theophylline

Tiabendazole can markedly increase serum concentrations of theophylline, with a prolonged half-life and a reduced clearance rate; concomitant administration of theophylline and thiabendazole resulted in severe nausea and vomiting (21).

References

1. Bagheri H, Simiand E, Montastruc JL, Magnaval JF. Adverse drug reactions to anthelmintics. Ann Pharmacother 2004;38(3):383–8.
2. Schantz PM, Van den Bossche H, Eckert J. Chemotherapy for larval echinococcosis in animals and humans: report of a workshop. Z Parasitenkd 1982;67(1):5–26.
3. Tchao P, Templeton T. Thiabendazole-associated grand mal seizures in a patient with Down syndrome. J Pediatr 1983;102(2):317–8.
4. Medwatch. Summary of safety-related drug labeling changes approved by FDA Center for Drug Evaluation and Research (CDER). Mintezole (Thiabendazole). 01/06/2001.
5. Fink AI, MacKay CJ, Cutler SS. Sicca complex and cholangiostatic jaundice in two members of a family probably caused by thiabendazole. Ophthalmology 1979;86(10):1892–6.
6. Council on Drugs. Evaluation of a broad-spectrum anthelmintic thiabendazole (Mintezol). JAMA 1968;205(3):172–3.
7. Boken DJ, Leoni PA, Preheim LC. Treatment of *Strongyloides stercoralis* hyperinfection syndrome with thiabendazole administered per rectum. Clin Infect Dis 1993;16(1):123–6.
8. Igual-Adell R, Oltra-Alcaraz C, Soler-Company E, Sanchez-Sanchez P, Matogo-Oyana J, Rodriguez-Calabuig D. Efficacy and safety of ivermectin and thiabendazole in the treatment of strongyloidiasis. Expert Opin Pharmacother 2004;5(12):2615–9.
9. Hennekeuser HH, Pabst K, Poeplau W, Gerok W. Thiabendazole for the treatment of trichinosis in humans. Tex Rep Biol Med 1969;27(Suppl 2):581.
10. Manivel JC, Bloomer JR, Snover DC. Progressive bile duct injury after thiabendazole administration. Gastroenterology 1987;93(2):245–9.
11. Skandrani K, Richardet JP, Duvoux C, Cherqui D, Zafrani ES, Dhumeaux D. Transplantation hépatique pour ductopénie sévère associée la prise de thiabendazole. [Hepatic transplantation for severe ductopenia related to

ingestion of thiabendazole.] Gastroenterol Clin Biol 1997;21(8–9):623–5.

12. Ishizaki T, Kamo E, Boehme K. Double-blind studies of tolerance to praziquantel in Japanese patients with *Schistosoma japonicum* infections. Bull World Health Organ 1979;57(5):787–91.
13. Eland IA, Kerkhof SC, Overbosch D, Wismans PJ, Stricker BH. Cholestatische hepatitis toegeschreven aan her gebruik van tiabendazole. [Cholestatic hepatitis ascribed to the use of thiabendazole.] Ned Tijdschr Geneeskd 1998;142(23):1331–4.
14. Morgan M. Practice management of helminth infections. Practitioner 2000;11(10):.
15. Sanchez del Rio J, Ramos Polo E, Nosti Martinez D, Rozado Fernandez S, Ribas Barcelo A. Exantema fijo generalizado por thiabendazol. [Fixed, generalized exanthema caused by thiabendazole.] Actas Dermosifiliogr 1982;73(3–4):125–8.
16. Humphreys F, Cox NH. Thiabendazole-induced erythema multiforme with lesions around melanocytic naevi. Br J Dermatol 1988;118(6):855–6.
17. Izu R, Aguirre A, Goicoechea A, Gardeazabal J, Diaz Perez JL. Photoaggravated allergic contact dermatitis due to topical thiabendazole. Contact Dermatitis 1993;28(4):243–4.
18. Parasitic infections. In: Merck Manual. 17th ed. Section 13, Chapter 161.
19. Anonymous. Communication from the Department of Health and Social Security. London, 22-02-1988.
20. Letter from the US Department of Health and Human Services to Merck Inc, 2000.
21. Schneider D, Gannon R, Sweeney K, Shore E. Theophylline and antiparasitic drug interactions. A case report and study of the influence of thiabendazole and mebendazole on theophylline pharmacokinetics in adults. Chest 1990;97(1):84–7.

Tinidazole

See also Benzimidazoles

General Information

The adverse effects of tinidazole resemble those of metronidazole. It can be given intravenously and is well tolerated, although thrombophlebitis has been reported. In one healthy volunteer, fainting with low blood pressure and nausea and tiredness for several hours was reported (SEDA-11, 597) (1).

Organs and Systems

Gastrointestinal

Drug-induced esophagitis is rare, accounting for about 1% of all cases of esophagitis. An incidence of 3.9 in 100 000 has been reported. After the first description, there have been more than 250 observations, with more than 50 different drugs. Among those, the principal antibiotics included tetracyclines (doxycycline, metacycline, minocycline, oxytetracycline, and tetracycline), penicillins (amoxicillin, cloxacillin, penicillin V, and pivmecillinam),

clindamycin, co-trimoxazole, erythromycin, lincomycin, spiramycin, and tinidazole. Doxycycline alone was involved in one-third of all cases. Risk factors included prolonged esophageal passage, due to motility disorders, stenosis, cardiomegaly, the formulation, supine position during drug ingestion, and failure to use liquid to wash down the tablet. Direct toxic effects of the drug (pH, accumulation in epithelial cells, non-uniform dispersion) also seem to contribute to the development of drug-induced esophagitis (2).

Skin

Skin reactions can occasionally occur in patients taking tinidazole.

- A fixed eruption with pruritus was observed in a 27-year-old man treated for *Entamoeba histolytica* with tinidazole 500 mg, four tablets in a single dose for 3 days.
- An erythematous patch with a sensation of burning appeared over the left buttock in a 32-year-old man while he was taking tinidazole for giardiasis; the allergic lesion was provoked by a challenge dose.

Both of these patients reacted with the same skin reaction on rechallenge with metronidazole (SEDA-16, 310).

In a patient with a fixed drug eruption to metronidazole, a provocation test showed cross-reactivity with tinidazole but not with secnidazole (3).

References

1. Aase S, Olsen AK, Roland M, Fagerhol MK, Liavag I, Bergan T, Leinebo O. Severe toxic reaction to tinidazole. Eur J Clin Pharmacol 1983;24(3):425–7.
2. Zerbib F. Les oesophagites médicamenteuses. Hepato-Gastro 1998;5:115–20.
3. Thami GP, Kanwar AJ. Fixed drug eruption due to metronidazole and tinidazole without cross-sensitivity to secnidazole. Dermatology 1998;196(3):368.

Triclabendazole

See also Benzimidazoles

General Information

Triclabendazole is a benzimidazole derivative primarily used in veterinary medicine but has also been used experimentally in man. Chills, fever, leukopenia, and upper abdominal colic have been described (SEDA-14, 263).

Several reports have suggested that triclabendazole may be of use in the treatment of *Fasciola hepatica* infection. In 20 patients with fascioliasis treated with two single doses of triclabendazole 10 mg/kg, the plasma concentrations of triclabendazole, its active metabolite triclabendazole-SO, and its sulfone metabolite were doubled by food (1). There were no serious adverse effects, except for somesided upper abdominal pain in several patients, which was relieved by oral spasmolytics. Triclabendazole should be administered with food.

Reference

1. Lecaillon JB, Godbillon J, Campestrini J, Naquira C, Miranda L, Pacheco R, Mull R, Poltera AA. Effect of food on the bioavailability of triclabendazole in patients with fascioliasis. Br J Clin Pharmacol 1998;45(6):601–4.

Trimetrexate

General Information

Trimetrexate is a lipid-soluble analogue of methotrexate that has been used in the management of *Pneumocystis jiroveci* in patients with AIDS when other therapy has proved ineffective. It has also been used as an antineoplastic drug in the management of various solid tumors. It is given with leucovorin (folinic acid) to minimize hematological toxicity. Trimetrexate can cause neutropenia and/or thrombocytopenia (SEDA-12, 704) (1,2). Fever and raised liver transaminases, while uncommon, have been noticed. The efficacy of trimetrexate is not as high as that of co-trimoxazole and the recurrence rate is markedly higher (3).

References

1. Hughes WT. *Pneumocystis carinii* pneumonitis. N Engl J Med 1987;317(16):1021–3.
2. Allegra CJ, Chabner BA, Tuazon CU, Ogata-Arakaki D, Baird B, Drake JC, Simmons JT, Lack EE, Shelhamer JH, Balis F, et al. Trimetrexate for the treatment of *Pneumocystis carinii* pneumonia in patients with the acquired immunodeficiency syndrome. N Engl J Med 1987;317(16):978–85.
3. Masur H. Prevention and treatment of *pneumocystis* pneumonia. N Engl J Med 1992;327(26):1853–60.

Triperaquine

General Information

Triperaquine is a 4-aminoquinoline derivative, the adverse reaction pattern of which is unknown. The WHO Scientific Group's report concerning this drug dates from 1984 and there seems to be little or no later information about it in humans.

WR-242511

General Information

WR-242511 is an 8-aminoquinoline that is being tested for activity against *Pneumocystis jiroveci* (1). In animals, it was more likely than primaquine to cause methemoglobinemia (SEDA-12, 703) (2).

References

1. Goheen MP, Bartlett MS, Shaw MM, Queener SF, Smith JW. Effects of 8-aminoquinolines on the ultrastructural morphology of *Pneumocystis carinii*. Int J Exp Pathol 1993;74(4):379–87.
2. Bartlett MS, Queener SF, Tidwell RR, Milhous WK, Berman JD, Ellis WY, Smith JW. 8-Aminoquinolines from Walter Reed Army Institute for Research for treatment and prophylaxis of *Pneumocystis* pneumonia in rat models. Antimicrob Agents Chemother 1991;35(2):277–82.

VACCINES

VACCINES

Editor's note: Abbreviations used in this section:aP: acellular pertussis

- BCG: Bacillus Calmette Guérin
- DTaP: Diphtheria + tetanus toxoids + acellular pertussis
- DTaP-Hib-IPV-HB: Diphtheria + tetanus toxoids + acellular pertussis + IPV + Hib + hepatitis B (hexavalent vaccine)
- DTwP: Diphtheria + tetanus toxoids + whole cell pertussis
- HA(V): Hepatitis A (virus)
- HbOC (also called PRP-CRM): conjugated Hib vaccine (Hib capsular antigen polyribosylphosphate covalently linked to the non-toxic diphtheria toxin variant CRM197)
- HB(V): Hepatitis B (virus)
- Hib: Hemophilus influenzae type b
- HPV: Human papilloma virus
- IPV: Inactivated polio vaccine
- JE vaccine: Japanese encephalitis vaccine
- MMR: measles + mumps + rubella
- MMRV: measles + mumps + rubella + varicella
- OPV: Oral polio vaccine
- PRP-D-Hib: conjugated Hib vaccine(Hib capsular antigen polyribosylphosphate covalently linked to a mutant polypeptide of diphtheria toxin)
- SV40: Simian virus 40
- Td: Diphtheria + tetanus toxoids (adult formulation)
- wP: whole cell pertussis
- YF vaccine: yellow fever vaccine

General Information

In the early days of vaccine development the majority of untoward effects after immunization were associated with faulty production, and the control of biological products as they exist today has been developed largely as a result of major accidents. For example, the Cutter incident in the USA in 1955, in which a batch of "inactivated" poliomyelitis vaccine containing live poliovirus was inadvertently released, had devastating consequences. The World Health Organization (WHO) subsequently took over the responsibility for international biological standardization. Currently, more than 50 WHO requirements for the manufacture and control of biological substances have been adopted and updated. As a result of the incorporation of WHO requirements, and their strict observance by manufacturers and control authorities, accidents due to faulty production of vaccines have become rare (SEDA-13, 271).

However, during the last two decades rapid increases in immunization coverage have been reported worldwide; globally, 5–10 billion injections are now given each year, mainly for treatment of illness but also for prophylactic purposes such as immunization. Unavoidably, the growth of WHO's Expanded Program on Immunization has sometimes resulted in extension of immunization work into areas of the developing world where logistic support and training programs have not been adequate. In such circumstances, avoidable faults have been made, involving variously improper or inadequate sterilization, incorrect doses and routes of vaccine administration, or substitution of drugs for diluents or vaccines (SEDA-15, 340). The use of unsterilized or improperly sterilized needles and syringes is particularly common in these regions and contributes largely to the spread of hepatitis B and C as well as to the spread of human immunodeficiency virus (HIV) and other blood-borne pathogens. The WHO recognizes this as a major public health problem and has initiated a program of activities to ensure safe injections (1).

In some highly developed countries that have had great success in reducing dangerous infectious diseases through immunization, there are controversial discussions regarding the benefits to harm balance of immunization. Because the incidence of vaccine-preventable diseases is reduced by increasing coverage with an efficacious vaccine, adverse events, both those caused by vaccines themselves (that is true adverse reactions) and those associated with immunization only by coincidence or as a result of faulty routines, become increasingly frequent. Not surprisingly, vaccine safety concerns have become increasingly prominent in successful immunization programs. Chronic illnesses recently claimed to be linked with immunizations include asthma, autism, diabetes, and multiple sclerosis. Given the current increasingly "anti-vaccine" environment that has developed in some countries, it is hard to imagine that the full potential of new vaccines will be harnessed. To avoid damage to current and future immunization programs considered as among the most successful and cost-effective public health interventions, we need to examine critically the factors that have influenced this change in public attitudes.

We have been relatively slow in appreciating the importance that the public now places on vaccine safety. In fact, much of our resource allocations still unfortunately reflect safety last rather than safety first. This reflects in part an unfortunate legacy, in which we have characterized this arena for years in narrow, negative terms of adverse events, instead of the more broad and positive terms of safety. Furthermore, it shows that we have not been as interested in preventing vaccine-induced illnesses as we are in vaccine-preventable diseases (2). Increasing interest in this problem can therefore be observed in many developed and developing countries, and international institutions and organizations, for example the World Health Organization (WHO), the United Nations Children's Fund (UNICEF), and vaccine manufacturers. The careful review of the current knowledge on adverse effects after immunization is a useful contribution to vaccine safety and helps to maintain confidence in immunization programs.

On the other hand, we must not forget that the ultimate goal of an immunization program is control of disease or even its regional elimination or worldwide eradication. A successful immunization program can lead to the eradication of disease and opens up the possibility of ultimately abandoning immunization (and with it the occurrence of adverse effects) completely. Smallpox eradication made it possible to stop smallpox immunization and poliomyelitis eradication, which is expected to be achieved within a few years, will mean the end of polio vaccine.

Surveillance of adverse events following immunization

Currently, in many highly developed countries in which the incidences of dangerous infectious diseases have been markedly reduced through immunization, there are controversial discussions about the balance of benefits and harms in immunization. Vaccine adverse events, both those caused by vaccines (that is true adverse reactions) and those associated with immunization only by coincidence, become more visible than the natural disease. Not surprisingly, vaccine safety concerns have become increasingly prominent in such successful immunization programs. Vaccines have been spuriously linked by various researchers to asthma, autism, diabetes, inflammatory bowel disease, multiple sclerosis, permanent brain damage, and sudden infant death syndrome (SIDS). Modern communication is providing even more penetrating ways of communicating messages on the subject through the World Wide Web. The result is a formidable challenge to immunization service providers. Neil Halsey, head of the Institute for Vaccine Safety at Johns Hopkins University, has summarized the features common to recent publications on vaccine adverse effects:

- a causal link is usually claimed with a disease or condition of unknown or unclear cause;
- the association is claimed by one investigator or a group of investigators;
- the association is not confirmed by peers or by subsequent research;
- the claims are made with no apparent concern for potential harm from public loss of confidence and refusal to immunize;
- findings of subsequent studies that fail to confirm the original claim never get the publicity given to the original finding, and so the public never gets a balanced view(2,3).

A critical examination (4) of a report (5) of several children whose chronic bowel and behavioral abnormalities were linked to measles, mumps, and rubella (MMR) immunization can be used as an example to underline Halsey's comments. Without effective and credible systems for the detection of vaccine-associated adverse events through pharmacovigilance, for distinguishing causal reactions from coincidental reactions by pharmacoepidemiological or other studies, and for risk communication, vaccine safety concerns may confuse the media and the public.

A concerted effort is needed to improve communications at all levels regarding the real risks associated with vaccines and immunization and helping to reassure the public of the overwhelming safety record of vaccines. The medical community is still preeminent as advice givers to the public on matters of immunization and should play the key role in improved communication (3,6).

The US Vaccine Adverse Events Reporting System (VAERS) has been described (SEDA-14, 919) and the pros and cons of the system have been discussed (7). About 1000 reports per month are submitted by manufacturers (39%), state health coordinators (34%), health care professionals (25%), and parents (2%). Manufacturers' reports are primarily based on information received from health care providers or the parents of vaccinees. As a passive reporting system, VAERS is subject to numerous well-known deficiencies, such as under-reporting, incomplete and missing data, and recall bias. Resources are not adequate to allow follow-up for complete or accurate data in all cases. However, VAERS is useful to detect early warning signals and to generate hypotheses. For example, the information from a parent that her daughter had lost her hair after the second and third doses of hepatitis B vaccine initiated a study that found 59 similar cases, including three cases of positive rechallenge. As a result of this study, alopecia was added to the list of adverse reactions in the hepatitis B vaccine package insert. Individual case reports can also trigger a complete review of the VAERS database. For example, three reports of idiopathic thrombocytopenic purpura after measles-containing vaccine were received in a short period of time. The study of idiopathic thrombocytopenic purpura reported in the database found 54 other reports during the period 1990–95, including one case of positive rechallenge, which increased the likelihood that the association was causal.

Reports on adverse events after the administration of the two hepatitis B vaccines licensed in the US (Engerix-B and Recombivax HB) have been compared in two different surveillance systems: VAERS and Vaccine Safety Datalink (VSD) (8). VSD is a computerized record linkage system designed to allow more rigorous evaluation of causality of adverse events after immunization. Since 1989, VSD has actively maintained medical files on over 500 000 children, aged from birth to 6 years enrolled at four West-coast health maintenance organizations. Immunization records, medical diagnoses, and diseases from clinic, hospitalization, and emergency room visits are coded. Whereas VAERS found that the reporting rate for events after brand 1 vaccine was at least three times higher than the reporting rate after brand 2 vaccine, VSD found no difference between rates of hospitalization or emergency room visits in recipients of the two brands. The authors concluded that the results of the VAERS database are subject to the inherent limitations of a passive surveillance system. This study underlined the importance of using other analytical studies, such as VSD, to evaluate preliminary results obtained by VAERS.

Adverse events reported to the VAERS during the period 1991–2001 have been summarized and evaluated (9). The main results were as follows:

- During 1991–2001, annual reports of deaths constituted 1.4–2.3% of all reports, and reports of life-threatening illness constituted 1.4–2.8% of all reports.
- A clinical research team followed up all deaths reported to VAERS; most of these deaths were ultimately classified as SIDS; analysis of the age

distribution and seasonality of infant deaths reported to VAERS showed that they matched the age distribution and seasonality of SIDS: both peaked at 2–4 months and during the winter; the reduction in the number of deaths reported to VAERS since 1992–93 parallels the overall reduction in the incidence of SIDS in the US population since the implementation of the "Back to sleep" campaign; the Food and Drug Administration (FDA) and the Institute of Medicine (IOM) reviewed 206 deaths reported to VAERS during 1990–91: only one death was believed to have resulted from a vaccine.

- Intussusception after rotavirus vaccine was described as an example to underline the value of passive surveillance systems as an indicator for carefully designed follow-up studies.
- An increase in the number of reports of Guillain–Barré syndrome after the receipt of influenza vaccine was noted in VAERS data by week 29 of the 1993–94 influenza season; the numbers of reports of Guillain–Barré syndrome increased from 23 during 1991-92 to 40 during 1992–93 and to 80 during 1993–94; a study of the VAERS signal showed that slightly more than one additional case of Guillain–Barré syndrome occurred per 1 million people immunized against influenza, a risk that is less than the risk from severe influenza, which can be prevented by the vaccine.
- A detailed review of VAERS reports received during the first 3 years after the licensure of the *Varicella* vaccine documented that the majority of reported adverse events were minor and serious adverse events were rare.

Terms and definitions used in surveillance programs of adverse events and adverse effects after immunization

Definitions are fundamental to the establishment of any functioning surveillance system. There must be different definitions for monitoring and evaluation. An international voluntary collaboration has been established to develop globally accepted standardized case definitions of adverse events following immunization (AEFI). The so-called Brighton Collaboration (initiated in 1999 during a vaccine meeting in Brighton, UK) took into consideration that there is a general lack of widely accepted and implemented case definitions, because only a limited set of case definitions elaborated by WHO has been available internationally. Working groups on fever, local reactions, intussusception, persistent crying, convulsion, and hypotonic-hyporesponsive episodes were already established. The website of the Brighton Collaboration contains information about the process and progress of the collaboration, the work statement for the working groups, a template format for draft definitions, and much more (10).

On behalf of the Brighton Collaboration, an electronic discussion on the wisdom of using the term "adverse event following immunization" (AEFI) was undertaken during the year 2001. Concern has sometimes been expressed that the word "following" could imply causality. At least 10 alternative terms have been proposed and discussed. Finally, the term "adverse event possibly related to immunization" (AEPRI) received the majority of votes. However, there were many participants in the discussion who believed that the term AEFI should be kept because it is used in many countries as well as in guidelines issued by the WHO and UNICEF and because a change will create new confusion. It therefore seems that the term AEFI will not be replaced (11).

The Brighton Collaboration took into consideration the fact that there is a general lack of widely accepted and implemented case definitions in order to allow comparability between studies. Based on a network of more than 300 participants from 34 countries and collaborating with many national and international organizations, institutions, and manufacturers (for example WHO, FDA, CDC, GlaxoSmithKline), case definitions and guidelines have been developed for intussusception, fever, prolonged crying, seizure, hypotonic-hyporesponsive episodes, and local reactions, such as nodules, swelling, cellulitis, and abscesses.

The case definitions of the Brighton Collaboration could be useful worldwide for clinical trials, epidemiological studies and post-marketing surveillance (12,13). The first six standardized case definitions were: fever (14), generalized convulsive seizure (15), hypotonic-hyporesponsive episodes (HHE) (16), intussusception (17), nodule at injection site (18), and persistent crying (19). New working groups have already been formed or will soon be formed on: allergic reactions, chronic fatigue syndrome, idiopathic thrombocytopenia, myalgia, paresthesia, rash, and smallpox vaccine-associated AEFI (12,13).

Subsequently, new working groups finalized and published the case definitions and guidelines for anaphylaxis (20), aseptic meningitis (21), encephalitis, myelitis, and acute disseminated encephalomyelitis (ADEM) (22), fatigue (23), local reactions (abscesses (24), cellulitis (25), induration (26), swelling (27)), rash (28), sudden infant death syndrome (SIDS) (29), idiopathic thrombocytopenia (30), and smallpox vaccine-associated adverse events following immunization (eczema (31), generalized vaccinia (32), vaccinia progressiva (33), inadvertent inoculation (34), and robust take (35)).

Monitoring

Data collection is the first step in surveillance. Terms and definitions developed for monitoring purposes should be useful both for practitioners and health workers. Monitoring definitions therefore have to cover a broad range. For monitoring purposes, the term "adverse event" has been introduced, defined as an untoward event temporally associated with immunization, which might or might not be caused by the vaccine or the immunization process. The term "adverse effect" does not seem useful for monitoring purposes, since it relates to an untoward effect after immunization that is linked to the vaccine (antigen or vaccine component) or to the immunization procedure. The term and its definition imply a distinct degree of causality. However, the term "suspected adverse effect" is acceptable.

Large linked databases

Combining administrative databases for pharmacovigilance purposes became possible in the 1980s with increased automation of pharmacy prescriptions and medical outcome records. These combined databases were referred to as "large linked databases" (LLDBs) because of their relatively large size (storing details about millions of patients) and the need for linkage of different data sets that were created separately from each other. Such databases became popular in vaccine safety surveillance. Their most obvious advantage is the ability to study rare events. The rarity of an event that can be studied depends on the size of population and the level of immunization coverage. Owing to the rarity of intussusception (25 per 100 000 infants per year) and the relatively low prevalence of rotavirus immunization (less than 20%), a cohort study of rotavirus vaccine and intussusception required the participation of 10 Health Maintenance Organizations, with a combined population of over 460 000 infants 1?11 months of age. The use of large linked databases is most advanced in the UK and the USA and the opportunities and hazards that they afford have been reviewed (36).

Evaluation

Statistical analysis and expert evaluation of the collected data should be carried out centrally. Expert evaluation of reports should be based on precise evaluation definitions, taking into account current scientific knowledge. One classification of events is that concerning causality, for example certain, probable, possible, unlikely, unclassified (when no attempt to classify has been made), and unclassifiable events (for example due to lack of data) (37). Alternatively one can classify reports into those for which there is no evidence of a causal relation, those for which evidence is insufficient to indicate a causal relation, those for which the evidence does not indicate a causal relation, those for which the evidence is consistent with a causal relation, and finally those for which the evidence indicates a causal relation (38,39). An evaluated adverse event classified as either "certain" or "probable" or "consistent with a causal relation" or "indicating a causal relation" fulfils the criteria for the recognition of an adverse effect after immunization.

Following WHO's proposal to distinguish four types of adverse events after immunization, the evaluation should furthermore try to distinguish:

- vaccine-induced adverse events (for example BCG lymphadenitis, BCG osteitis, vaccine-associated poliomyelitis, allergic reactions);
- vaccine-precipitated events (for example a simple febrile seizure following DPT immunization in a predisposed child);
- programmatic errors (for example an abscess due to improper sterilization);
- coincidental events (40).

Most commonly, passive surveillance systems are used for the surveillance of adverse events after immunization. Spontaneous reports from health care providers on temporally related adverse events suspected of being caused by the vaccine or the immunization procedure are collected and evaluated. However, the significant degree of under-reporting, together with the non-specific nature of most adverse events reports, highlights the limitations of passive surveillance systems of adverse events after immunization. A critical evaluation of the limitations of the US Vaccine Adverse Event Reporting System (VAERS) (SED-13, 919) and its predecessor, the Monitoring System for Adverse Events Following Immunization (MSAEFI) (SEDA-13, 273) (SEDA-16, 320) has been made by Rosenthal and Chen (41,42). The authors concluded that reports to VAERS and MSAEFI are essentially non-controlled clinical case reports or case series, useful for generating hypotheses but not for testing them. Other controlled studies are therefore often necessary to evaluate hypotheses raised by passive surveillance system reports: whether a given adverse event can be caused by a specific vaccine and if so how commonly it occurs. However, if reporting is reasonably consistent, it may be possible to detect changes in trends of known common adverse events. In addition, passive surveillance remains a potentially cost-effective way of monitoring rare events that cannot be detected in small prelicensing trials.

Cohort studies of rare adverse events would be useful, but are generally not feasible, particularly for economic reasons. One promising alternative approach is the use of large linked databases. The databases appropriate for such studies are derived from defined populations, such as members of health maintenance organizations, universal health care systems, and Medicaid programs. Information on both exposure (immunization records) and outcome (for example diagnoses recording potential adverse events) is usually computerized for members of such populations. These databases can be linked for epidemiological studies assessing potential associations between immunization and outcomes. Large linked databases have already been used by various groups to examine the association between DTP vaccine and SIDS, DTP vaccine and neurological events, and MMR vaccine and seizures. The Centers for Disease Control and Prevention (CDC) in Atlanta use large linked databases (now called Vaccine Safety Datalink) in addition to VAERS Vaccine Safety Datalink, covering a defined cohort of approximately 500 000 children up to the age of 5 years (42–44).

An epidemiological and statistical method based on linkage of routinely available computerized hospital admission records with vaccination records has been described by Farrington (45). This active surveillance method has been used to assess the attributable risk of convulsions after DTP and MMR immunization and to investigate the relation between MMR vaccine and idiopathic thrombocytopenic purpura in children under 2 years of age in five districts in England.

Causality assessment of adverse events following immunization (AEFI)

Adverse events following immunization can be causally related to the inherent properties of the vaccine, linked to errors in administration, quality, storage, and transport of the vaccine (programmatic errors), but also occurring coincidentally after immunization. It is therefore necessary to investigate and evaluate AEFI, particularly those that are serious or unknown.

The most reliable way of determining the causality of an AEFI is by randomized comparisons of events in an immunized group with events in a non-immunized group. However, such trials can never be large enough to detect very rare events; postmarket surveillance is required to identify rare events. A Global Advisory Committee on Vaccine Safety, constituted by the WHO in 1999, has developed criteria for AEFI causality assessment. Criteria to be considered are consistency (the findings should be replicable), strengths of association (in an epidemiological sense), specificity, temporal relation, and biological plausibility (according to what is known about the natural history and biology of the disease). Not all these criteria need to be present to establish causality (46).

In Canada, an advisory committee on causality assessment (ACCA) has been established to evaluate serious individual adverse event reports collected through active surveillance of pediatric hospitals or a passive voluntary reporting system. The ACCA is composed of specialists in epidemiology, immunology, infectious diseases, microbiology, neurology, pathology, and pediatrics. A causality assessment form has been developed (including criteria similar to the WHO criteria described above). For final classification of the evaluated adverse events the causality assessment criteria of the WHO (see Table 1 are used. The great majority of reports collected through the surveillance system describe minor or well-known reactions (over 95%); only the most serious and unusual adverse events after immunization requiring detailed review are submitted to ACCA. At each twice-yearly meeting of the ACCA 60–110 reports of severe and unusual reports are evaluated (47).

Surveillance systems in different countries and between-country collaboration

Some countries have long experience in monitoring adverse events after immunization, starting with smallpox vaccination.

Worldwide, under 10% of countries have implemented postlicensure surveillance systems. The withdrawal of rotavirus vaccine (Rotashield, Wyeth Laboratories) based on postmarketing surveillance data that showed an excess risk of intussusception after the use of the vaccine convincingly showed that such systems are essential components of vaccine program implementation. However, the event also showed the power of active surveillance systems. Active systems in Minnesota and the Northern California Kaiser Permanente Health Maintenance Organization have provided preliminary data that suggest an increased risk of intussusception after the administration of Rotashield. Further analysis of the data collected both from the passive US national AEFI reporting system (VAERS) and from various active postmarketing surveillance systems led to a decision to withdraw the vaccine from the market (48).

USA

Since March 21, 1988, health care providers and vaccine manufacturers have been required by law to report certain events following specific immunizations (SED-12, 792) to the US Department of Health and Human Services. Table 2 shows the updated table for reportable events after vaccination, which became effective on 1 February 2007, is shown in Table 2 (49). The VAERS accepts all reports of suspected adverse events after the administration of any vaccine, including but not limited to those certain events required by law to be reported.

A guide for evaluating vaccine safety concerns 'Do Vaccines Cause That?' has been published by Myers and Pineda (50). The book provides evidence-based information about vaccine safety questions, such as suspected links between various vaccines and asthma, autism, damage to the immune system, diabetes, Guillain–Barré syndrome, arthritis, multiple sclerosis, inflammatory bowel disease, febrile seizures, encephalopathy, sudden infant death, AIDS, birth defects, mad cow disease, and cancers.

On 19 April 2007 at a press conference two US representatives introduced a bill. the Vaccine Safety and Public Confidence Assurance Act, which would give responsibility for the nation's vaccine safety to an independent

Table 1 WHO causality assessment criteria for adverse reactions to vaccines

Probability	Criteria
Very likely/ certain	Clinical event with a plausible time relationship to vaccine administration, and which cannot be explained by concurrent disease or other drugs or chemicals
Probable	Clinical event with a reasonable time relationship to vaccine administration, and is unlikely to be attributed to concurrent disease or other drugs or chemicals
Possible	Clinical event with a reasonable time relationship to vaccine administration, but which could also be explained by concurrent disease or other drugs or chemicals
Unlikely	Clinical event whose time relationship to vaccine administration makes a causal connection improbable, but which could plausibly be explained by underlying disease or other drugs or chemicals
Unrelated	Clinical event with an incompatible time relationship to vaccine administration, and which could be explained by underlying disease or other drugs or chemicals
Unclassifiable	Clinical event with insufficient information to permit assessment and identification of the cause

Table 2   Vaccine injury table

Vaccine	Adverse event	Interval
I Tetanus toxoid-containing vaccines (for example DTaP, Tdap, DTP-Hib, DT, Td, TT)	A. Anaphylaxis or anaphylactic shock	0–4 hours
	B. Brachial neuritis	2–28 days
	C. Any acute complication or sequel (including death) of the above events	Not applicable
II Pertussis antigen-containing vaccines (for example DTaP, Tdap, DTP, P, DTP-Hib)	A. Anaphylaxis or anaphylactic shock	0–4 hours
	B. Encephalopathy or encephalitis	0–72 hours
	C. Any acute complication or sequel (including death) of the above events	Not applicable
III Measles, mumps, and rubella virus-containing vaccines in any combination (for example MMR, MR, M, R)	A. Anaphylaxis or anaphylactic shock	0–4 hours
	B. Encephalopathy or encephalitis	5–15 days
	C. Any acute complication or sequel (including death) of the above events	Not applicable
IV Rubella virus-containing vaccines (for example MMR, MR, R)	A. Chronic arthritis	7–42 days
	B. Any acute complication or sequel (including death) of the above event	Not applicable
V Measles virus-containing vaccines (for example MMR, MR, M)	A. Thrombocytopenic purpura	7–30 days
	B. Vaccine-Strain Measles Viral Infection in an immunodeficient recipient	0–6 months
	C. Any acute complication or sequel (including death) of the above events	Not applicable
VI Polio live virus-containing vaccines (OPV)	A. Paralytic polio	
	• in a non-immunodeficient recipient	0–30 days
	• in an immunodeficient recipient	0–6 months
	• in a vaccine associated community case	Not applicable
	B. Vaccine-strain polio viral infection	0–30 days
	• in a non-immunodeficient recipient	0–6 months
	• in an immunodeficient recipient	Not applicable
	• in a vaccine associated community case	Not applicable
	C. Any acute complication or sequel (including death) of the above events	
VIII nactivated polio virus-containing vaccines (for example IPV)	A. Anaphylaxis or anaphylactic shock	0–4 hours
	B. Any acute complication or sequel (including death) of the above event	Not applicable
VIII Hepatitis B antigen-containing vaccines	A. Anaphylaxis or anaphylactic shock	0–4 hours
	B. Any acute complication or sequel (including death) of the above event	Not applicable
IX *Hemophilus influenzae* (type b polysaccharide conjugate vaccines)	A. No condition specified for compensation	Not applicable
X Varicella vaccine	A. No condition specified for compensation	Not applicable
XI Rotavirus vaccine	A. No condition specified for compensation	Not applicable
XII Vaccines containing live, oral, rhesus-based rotavirus	A. Intussusception	0–30 days
	B. Any acute complication or sequel (including death) of the above event	Not applicable
XIII Pneumococcal conjugate vaccines	A. No condition specified for compensation	Not applicable
XIV Any new vaccine recommended by the Centers for Disease Control and Prevention for routine administration to children, after publication by the Secretary, HHS, of a notice of coverage	A. No condition specified for compensation	Not applicable

agency within the Department of Health and Human Services, removing most vaccine safety research from the Centers for Disease Control and Prevention (CDC). Currently, the CDC has responsibility for both vaccine safety and promotion, which is an inherent conflict of interest increasingly garnering public criticism. The initiative is joined by several groups advocating vaccine safety reform, including the National Autism Association. The proposed legislation came as the Senate considered legislation to reform the way that the Federal Government conducts drug safety at the FDA (51). The bill was updated in August 2009 (52).

Canada
The reporting of vaccine-associated adverse events temporally associated with immunization by health-care providers is voluntary in Canada, except in the province of Ontario, which has specific mandatory requirements.

However, there is no evidence of a higher reporting rate with the latter approach. This is partly explained by the fact that immunization in Ontario is usually provided by physicians, who have lower reporting rates than public-health nurses. In addition to its spontaneous voluntary reporting system, Canada also has an active surveillance system, IMPACT, for serious adverse events, vaccination failures, and selected infectious diseases. It involves a network of 11 pediatric centers across Canada. The results of the Canadian surveillance system have been reviewed (SEDA-16, 367) (SEDA-17, 363) (SEDA-21, 324) (SEDA-22, 333).

The Netherlands

A special committee of the Health Council of the Netherlands has been created (37), with the task of analysing, classifying, and interpreting the adverse effects which are reported to the National Institute of Health and Environmental Protection in Bilthoven (SEDA-12, 271) (SEDA-13, 278) (SEDA-14, 269) (SEDA-15, 341) (SEDA-16, 365) (SEDA-17, 363) (SEDA-18, 325).

New Zealand

Reported adverse events after immunization in New Zealand from 1990 to 1995 have been presented (53). Reactions at the injection site following adult tetanus + diphtheria vaccine were the most commonly reported (68 reports per 100 000 immunizations). The authors concluded that the picture confirmed the overall safety of vaccines and the value of an adverse events monitoring system.

Spain

In Spain, reports on adverse events after immunization are collected through the Spanish Pharmacovigilance System by means of a yellow card. In children, the most commonly involved pharmaceutical groups were antibiotics, respiratory drugs, and vaccines. A review of reports received over 10 years (1982–91) has been provided (54). In the framework of the Global Training Network, a WHO initiative to improve the quality of vaccines and their use, a model for a simple national system for dealing with vaccine safety and emergencies as they arise has been elaborated. The authors have described the model and have outlined a training program designed to help develop such a system (55).

Other countries

There are other national surveillance programs on adverse events after immunization, for example in Denmark, Germany, Sweden, the UK, Hungary; developing countries such as India (SEDA-17, 363) and Brazil (SEDA-16, 367) have also already started to collect data on severe adverse events.

International surveillance programs
The Program on International Drug Monitoring of the WHO has participation from over 50 member states. In all member countries, national Drug Reaction Monitoring Centers collect reports from health professionals and pass them on for entry into the international database housed at the WHO Collaborating Center for International Drug Monitoring in Uppsala, Sweden. Currently, there are some 2.5 million reports on file, and each year about 200 000 new reports are added. The database generates signals of potentially severe drug toxicity and provides confirmation of signals generated in specific countries. The Center also acts as the guardian of the standardized terminology of adverse drug reactions in computerized systems. Although vaccines are considered to be safe, the work of the Center underlines the importance of collecting information on the safety of vaccines to provide sound reference data in the event of a problem. Since vaccines are administered to healthy people, mostly children, the impact of a perceived problem can be enormous. Because of these considerations, many countries have already established specific programs for the monitoring of adverse events after immunization (56).

Clinical immunization safety assessment centers
The National Immunization Program of the Centers for Disease Control and Prevention (CDC), Atlanta, GA, USA, is trying to set up a network of Clinical Immunization Safety Assessment Centers (CISA). Based on standardized clinical evaluation protocols the centers will assist health care providers in evaluating patients who may have had an adverse reaction after immunization. Furthermore, the centers will evaluate newly hypothesized syndromes or events identified through the routine VAERS (57).

Guaranteeing vaccine quality
During the last 20–30 years, great progress has been achieved in both vaccine production technologies and testing technologies. In addition to sophisticated tests, vaccine regulation entails a number of procedures that ensure safety, including the characterization of starting material, cell banking, seed lot systems, principles of good manufacturing practices, independent release of vaccines on a lot-by-lot basis by national regulatory authorities, and enhanced premarketing and postmarketing surveillance for possible rare adverse events after immunization (58).

Perspectives on prelicensure trials
The withdrawal of tetravalent rhesus-based rotavirus vaccine from the market illustrates an important problem regarding prelicensure testing and its ability to identify

rare vaccine-related adverse events. A sample size of 10 000 volunteers may provide excellent estimates of reactogenicity (local and systemic reactions) and efficacy but be inadequate as a denominator for ruling out rare adverse events. Table 3 shows how trial size determines the ability to detect frequent and rare events. Plans for future large prelicensure trials, for example for newly developed rotavirus vaccines, therefore include 10 000–60 000 volunteers. These large trials can be seen as a bridge between prelicensure and improved postlicensure surveillance (59,60).

Improving monitoring of vaccine safety through postlicensure studies

Gaps in current vaccine safety monitoring methods have been analysed and the needs for improvements outlined (61,62). The well-known limitations associated with pre-licensure trials have led to the expectation that postli-censure studies would address safety issues better. To meet this expectation, steps that should be taken include bar-code labeling of vaccines to improve the accuracy and completeness of information about admi-nistered vaccines, establishing immunization registers to provide numerator data, developing generally agreed case definitions for adverse events following immuniza-tion (AEFI), to allow comparability between studies, to facilitate long-term follow-up, and to do linked database studies. Only by investing in vaccine safety infrastruc-ture will the high expectations of postlicensure studies be met.

Passive surveillance and reported fatalities

The fatalities reported to the federally administered (US) VAERS have been examined (63). A total of 1266 deaths were reported to VAERS between July 1990 and June 1997. Table 4 shows the numbers of deaths by age (total sample 1199 individuals) and Table 5 shows the causes of death (total sample 1244 individuals).

Nearly half of the deaths were attributed to SIDS. Since 1992/93, the trend of decreasing numbers of reported deaths follows that observed for SIDS overall in the US general population following implementation of the "Back to Sleep" program. Therefore, the VAERS data support the findings of controlled studies that show that

Table 3 Sample size necessary to have 80% power to detect an increase in the risk of a rare event

Background rate (1:1 randomization)	Detectable relative risk	Total study population required
1:100	5	1443
1:100	2	5916
1:1000	5	14 428
1:1000	2	59 160
1:10 000	5	144 280
1:10 000	2	591 600

Table 4 Deaths related in time to immunization reported to VAERS 1990–97, by age

Age (years)	Number (%)
Overall	1199 (100)
<1	808 (67)
1–4	117 (9.8)
5–9	18 (1.5)
10–17	17 (1.4)
18–45	43 (3.6)
46–64	55 (4.6)
≥65	141 (12)

Table 5 Causes of death related in time to immunization reported to VAERS 1990–97

Cause of death	Number (%)
Overall	1244 (100)
SIDS	592 (48)
Congenital	38 (3.1)
Infectious	164 (13)
Neoplastic	15 (1.2)
Other	261 (21)
Unknown	174 (14)

the association between infant immunization and SIDS is coincidental and not causal.

Adverse events after immunization—the need for improved research, surveillance, and communication

The success of vaccines is not only impressive and con-vincing but also makes immunization its own worst enemy. When a naturally occurring disease becomes less common, concerns about vaccine safety increase, particu-larly since no vaccine can be regarded as being completely safe, although some are very much safer than others. Progress in the development, manufacture, and control of modern vaccines has contributed to the safety of cur-rent vaccines. However, it is absolutely necessary to know what the true complications of immunization are, what the gaps in our current knowledge are, and when there is no evidence linking immunization and a suspected adverse effect.

Current knowledge about vaccine risks is incomplete, as was noted in extensive reviews in the early 1990s by the IOM in the USA (38,39). Two-thirds of the 76 vaccine-associated adverse events evaluated by the IOM had either no evidence or inadequate evidence to assess the causal role of the vaccine. Specifically, the reviews identi-fied the following limitations:

- inadequate understanding of the biological mechan-isms that underlie adverse events;
- insufficient or inconsistent information from case reports and case series;

- inadequate size or length of follow-up of many population-based epidemiological studies;
- limitations of existing surveillance systems to provide persuasive evidence of causation;
- few experimental studies published relative to the total number of epidemiological studies published.

A concerted effort is therefore needed to improve both research on and surveillance of the risks associated with immunization, as well as better communication aiming to reassure health-care providers, parents, and the public (including the media) about the steps taken to ensure vaccine safety and the overwhelming safety record of vaccines. The medical community still has preeminence as the purveyor of advice to the public on matters of immunization and should play a key role in improving communication. Nothing will strengthen our ability to communicate risk more than ensuring that there is adequate capacity and funding for vaccine safety infrastructure and research (2).

The response to a newly published adverse event associated with immunization must be rapid. First, if the reported association is correct, urgent re-evaluation of the immunization program is necessary. However, if the reported association is false, a credible countermessage must be sent, to minimize the negative impact on the immunization program (64).

The pros and cons of the (US) VAERS (SEDA-14, 919) have been critically reviewed (SEDA-23, 336) (65).

Among the many new developments in the communication of adverse events, the increasing role of the Internet should be mentioned. Many national bodies (for example the FDA, Rockville, Maryland, USA, www.fda.gov, or the CDC, Atlanta, Georgia, USA, www.vaers.org) and intergovernmental health authorities responsible for licensure of vaccines (for example the European Agency for the Evaluation of Medicinal Products (EMEA), www.eudra.org/emea.html), and/or vaccine safety (for example the World Health Organization, www.who.int/vaccines-diseases/safety/), as well as vaccine manufacturers (through postmarketing surveillance of their products, for example SmithKline Biologicals, www.worldwidevaccines.com), universities, and private organizations have also launched Websites providing information on vaccine safety and immunization risks.

National compensation programs for vaccine-related injuries

The need for some form of compensation when an individual is seriously injured by vaccination, particularly when the immunization has been compulsory or recommended by the health authority, has been accepted in many countries. Table 6 shows details of compensation programs in Canada (the province of Quebec), Denmark, France, Germany, Italy, Japan, New Zealand, Norway, Sweden, Switzerland, Taiwan, the UK, and the USA (66).

In the USA, the National Childhood Vaccine Injury Act of 1986 established the National Vaccine Injury Compensation Program as a federal no-fault compensation system for individuals who may have been injured by specific vaccines. This compensation program relies on a Vaccine Injury Table that lists the vaccines that are covered by the program, as well as injuries, disabilities, illnesses, and conditions (including death) for which compensation may be awarded. To better reflect current scientific knowledge about vaccine injuries, the Vaccine Injury Table was revised in 1995 and has been subsequently further modified. The latest modification, which became effective on December 1, 2004, is shown in Table 7. This revision took into account a review of the literature on specific adverse consequences of pertussis and rubella vaccines performed and published by the IOM (SED-12, 817) (SED-12, 825). In addition to the seven vaccines (diphtheria, pertussis, tetanus, measles, mumps, rubella, and poliomyelitis) included in the first Vaccine Injury Table, the 1997 revision includes hepatitis B, *Hemophilus influenzae* type b, and *Varicella* vaccines, as well as any future licensed vaccine recommended by the Advisory Committee on Immunization Practices (ACIP) for routine administration to children (67).

There are other compensation programs, for example in Japan, where the program covers damage caused by compulsory immunizations; provisions include medical allowance, care giver's allowance, disability pension, and a funeral grant; a national expert committee reviews applications.

Although not focusing primarily on vaccine-induced injury, attention should be paid to a contribution to the field of liability for drug-induced disease (68).

On December 1, 2004, the Secretary published a notice in the Federal Register announcing the addition of hepatitis A vaccines to the Vaccine Injury Table under Category XIV with an effective date of December 1, 2004.

The balance of benefits and harms in immunization
The risks of adverse effects after immunization against diphtheria, pertussis, tetanus, poliomyelitis, measles, and tuberculosis have been discussed in the framework of the WHO Expanded Program on Immunization and compared with the complication rate following natural disease (69). Table 8 presents a comparison of the estimated risks of adverse reactions after DTP immunization with the complication rates of natural whooping cough, while Table 9 shows a similar comparison for measles immunization and natural measles. The authors concluded that no vaccine is without adverse effects, but that the risks of serious complications from vaccines used in WHO's Expanded Program on Immunization are much lower than the risks from the natural disease.

Bioterrorism and prevention through immunization
Anthrax
Human anthrax is endemic in agricultural regions of the world where animal anthrax is common; sporadic cases occur in industrialized countries. Cutaneous, intestinal, and pulmonary anthrax can be distinguished clinically. Untreated cutaneous anthrax has a case fatality rate of 5–10%; death occurs only rarely in properly treated patients. Intestinal and pulmonary anthrax have much higher case fatality rates, even in properly treated patients. Transmission from person to person is very

Table 6 Vaccine injury compensation programs

	Denmark	France	Germany	Italy	Japan	New Zealand
Enacted	1972 (1978)	1964	1961	1992	1970 (1977)	1974
Administrative entity	National Social Security Office	Ministry of Solidarity, Health and Welfare	State (Länder) pension system (federal law guides outcome)	Ministry of Health	Ministry of Health and Welfare	Accident Rehabilitation and Compensation Insurance Corporation (semi-governmental)
Vaccines covered	Government-provided (free) (D, T, P, M, MR, polio, Hib, BCG)	Compulsory (D, T, polio, BCG, typhoid)	Recommended (most routine and specific use vaccines)	Compulsory (D, T, HBV, typhoid)	Recommended (polio, D, P, M, R, JE, and others for control of epidemics)	All that are administered
Filing deadlines	1 year after onset of symptoms	4 years after injury stabilization	None (states may have limitations)	Injury: 3 years; Death: no limit	No limit	1 year (exceptions if claim is incomplete)
Compensatable injuries	Injuries with reasonable probability caused by vaccine	All damage directly attributable to vaccination	Injuries exceeding normal extent of reactions—aggravation of pre-existing	Death or injury resulting in permanent physical or mental impairment	Disability or death resulting from vaccination	Vaccine-related injury that is rare and severe
Process and decision-making	Notice must be filed within 1 year of symptoms	Choice of ministerial commission or administrative tribunal	Claims evaluated by pension office; 1971 law gives more flexibility on causation	Medical Hospital Board for eligibility; compensation based on category level of injury (8 levels)	Committee on Public Health decides eligibility and damages	Patient files claim form, reviewed by ACC Medical Misadventure Advisory Committee
Proof needed	Reasonable probability	Clear and convincing evidence	Probable cause	Not specified	Not specified	Balance of probabilities
Elements of compensation	Medical costs, lost wages, death benefits, non-economic damages	Medical costs, disability pensions, death benefits, non-economic damages	Medical costs, disability pensions, lost wages, funeral costs, non-economic damages	Free medical care, medications, disability pension, death benefit	Medical costs, lost wages, disability pensions, death benefits	Medical costs, disability pensions, lost wages, death benefits
Funding source	National treasury	National treasury	General revenues of the state	National treasury	Treasury (50%), Municipal (25%), Prefecture (25%)	Employers, wage earners, auto licensing fees, government, investment income
Appeal rights	Yes	Yes	Yes	Yes	Yes	Yes
Litigation rights	Yes (with limits)	No	Yes (with limits)	Yes	Yes	No
Total number of claims filed to June 1999	55	51	4569	366	2982	211
Number compensated	5 (9%)	37 (73%)	1139 (25%)	260 (71%)	2720 (91%)	68 (32%)

Continued

Enacted	1995	1985	1978	1970	1988	1979	1988
Administrative entity	Ministry of Health and Social Affairs	Ministry of Health and Social Services	Pharmaceutical insurance (non-governmental)	State (Canton) (federal law guides outcome)	Department of Health	Department of Social Security	Department of Health and Human Services
Vaccines covered	Routine childhood (D, T, P, Hib, M, MR, polio, BCG)	Voluntary (none is compulsory) (D, T, P, MMR, BCG, HAV, HBV, typhoid)	All approved products marketed in Sweden	Recommended by cantons (D, T, P, HBV, Hib, M, MR, polio)	Compulsory (BCG, HBV, OPV, D, T, P, MMR, JE, D, T)	Routine childhood (D, T, P, M, MR, polio, BCG, Hib)	Routine childhood (D, T, P, MMR, polio, Hib, Varicella, HBV, rotavirus)
Filing deadlines	3 years after vaccination or death, or onset of chronic illness	3 years after being made first aware of injury	None (states may have limitations)	1 year after vaccination	Within 6 years of the later of either immunization date or age	Injury: 3 years; Death: 2 years (new vaccination or conditions—8 years)	
Compensatable injuries	Not specified	Any serious permanent damage, whether physical or mental, including death	Those noted as adverse effects in FASS or in medical literature	Injury has to exceed a normal post-vaccination reaction	Table of compensatable injuries	Severe disability to extent of 80% or more as a result of immunization	Injuries listed on the Vaccine Injury Table or by proving causation or significant aggravation
Process and decision-making	Minister of Health decides eligibility	Claim reviewed by three-member Medical Evaluation Committee; final decision by Minister of Health	Claims manager with Zurich insurance company makes decision with medical consultations as needed	Health dept. reviews claim and seeks supporting information	Vaccine Injury Comfort Fund Working Group, Department of Health	Claim evaluated by medical officer; makes recommendation to Secretary of State for Social Security	DHHS reviews claim and makes recommendation US Court of Federal Claims makes decision
Proof needed	Balance of probabilities	Balance of probabilities	Strong probability of cause and effect	Not specified	Balance of probabilities	Balance of probabilities	Balance of probabilities

Table 6 (Continued)

	Norway	Quebec	Sweden	Switzerland	Taiwan	UK	USA
Elements of compensation	No precedent (only award to date was a lump sum payment)	Unreimbursed medical costs, lost wages, rehabilitation, death benefits	Unreimbursed medical costs, lost wages, death benefits	Medical costs, lost wages, death benefits, non-economic damages	Medical costs, health care expenses, burial expenses	Lump sum of payment of statutory sum (40 000 pounds)	Unreimbursed past and future medical expenses, non-economic damage, attorneys' fees
Funding source	National treasury with contribution from manufacturer	National treasury	Manufacturers pay premiums into fund	General revenues of the states	Manufacturers and local society and community	Consolidated fund provided by Parliament	National treasury (pre-enactment); excise tax on covered immunization (postenactment)
Appeal rights	Yes	Yes	Yes	Yes	Yes	Yes	Yes
Litigation rights	Yes	Yes	No	Yes (with limits)	Yes	Yes (with limits)	Yes (with limits)
Total number of claims filed to June 1999	1	142	140	1	123	4012	5355
Number compensated	1 (100%)	17 (12%)	79 (56%)	1 (100%)	62 (50%)	890 (22%)	1390 (26%)

Table 7 National Childhood Vaccine Injury Act: Vaccine Injury Table

Vaccine	Illness, disability, injury, or condition covered	Time interval
I. Vaccines containing tetanus toxoid (for example DTaP, DTP-Hib, DT, Td, or TT)	A. Anaphylaxis	0–4 hours
	B. Brachial neuritis	2–28 days
	C. Any acute complication or sequel (including death) of the above events	Not applicable
II. Pertussis antigen-containing vaccines (for example DTaP, DTP, P, DTP-Hib)	A. Anaphylaxis or anaphylactic shock	0–4 hours
	B. Encephalopathy (or encephalitis)	0–72 hours
	C. Any acute complication or sequel (including death) of the above events	Not applicable
III. Vaccines containing measles, mumps, and rubella viruses in any combination (for example MMR, MR, M, R)	A. Anaphylaxis or anaphylactic shock	0–4 hours
	B. Encephalopathy (or encephalitis)	5–15 days
	C. Any acute complication or sequel (including death) of the above events	Not applicable
IV. Vaccines containing rubella virus (for example MMR, MR, R)	A. Chronic arthritis	7–42 days
	B. Any acute complication or sequel (including death) of the above event	Not applicable
V. Vaccines containing measles virus (for example MMR, MR, M)	A. Thrombocytopenic purpura	7–30 days
	B. Vaccine-strain measles viral infection in an immunodeficient recipient	0–6 months
	C. Any acute complication or sequel (including death) of the above events	Not applicable
VI. Vaccines containing live oral polio virus (OPV)	A. Paralytic polio	
	in a non-immunodeficient recipient	0–30 days
	in an immunodeficient recipient	0–6 months
	in a vaccine associated community case	Not applicable
	B. Vaccine-strain polio viral infection	
	in a non-immunodeficient recipient	0–30 days
	in an immunodeficient recipient	0–6 months
	in a vaccine associated community case	Not applicable
	C. Any acute complication or sequela (including death) of the above events	Not applicable
VII. Vaccines containing polio inactivated virus (for example IPV)	A. Anaphylaxis or anaphylactic shock	0–4 hours
	B. Any acute complication or sequela (including death) of the above event	Not applicable
VIII. Hepatitis B antigen-containing vaccines	A. Anaphylaxis or anaphylactic shock	0–4 hours
	B. Any acute complication or sequel (including death) of the above event	Not applicable
IX. Hemophilus influenzae type b polysaccharide conjugate vaccines	No condition specified for compensation	Not applicable
X. Varicella vaccine	No condition specified for compensation	Not applicable
XI. Rotavirus vaccine	No condition specified for compensation	Not applicable
XII. Vaccines containing live, oral, rhesus-based rotavirus	A. Intussusception	0–30 days
	B. Any acute complication or sequel (including death) of the above event	Not applicable
XIII. Pneumococcal conjugate vaccines	No condition specified for compensation	Not applicable
XIV. Any new vaccine recommended by the Centers for Disease Control and Prevention for routine administration to children, after publication of a notice of coverage by the Secretary of the HHS	No condition specified for compensation	Not applicable

Table 8 A comparison of the estimated risks of adverse reactions after DTP immunization with the complication rates of natural whooping cough

Adverse reaction	Whooping cough complication rates per 100 000 cases	DTP vaccine adverse reaction rates per 100 000 immunizations
Permanent brain damage	600–2000 (0.6–2.0%)	0.2–0.6
Death	100–400 (0.1–4.0%)	0.2
Encephalopathy/ encephalitis*	90–4000 (0.09–4.0%)	0.1–3.0
Convulsions	600–8000 (0.6–8.0%)	0.3–90
Shock		0.5–30

*Including seizures, focal neurological signs, coma, and Reye's syndrome

Table 9 A comparison of the estimated risks of adverse reactions after measles immunization with the complication rates of natural measles

Adverse reaction	Measles complication rates per 100 000 cases	Adverse reaction rates per 100 000 vaccines	Background illness rate per 100 000
Encephalopathy/ encephalitis	50–400 (0.05–0.4%)	0.1	0.1–0.3
Subacute sclerosing panencephalitis	500–2000 (0.5–2.0%)	0.05–0.1	—
Pneumonia	3800–7300 (3.8–7.3%)	—	—
Convulsions	500–1000 (0.5–1.0%)	0.02–190	30
Death	10–10 000 (0.01–10%)	0.02–0.3	—

rare. Recently, anthrax has been considered a leading potential agent in bioterrorism. As was first demonstrated in 1979 at Sverdlovsk, USSR (an environmental accident in a biological weapon manufacturing facility), inhalational anthrax accounts for most of the cases and for all deaths following the use of anthrax as an aerosolized biological weapon (70).

In 2001, the first bioterrorist anthrax attack occurred in the USA. From 3 October 2001 to 21 November 2001, the CDC had received reports of 23 human cases of anthrax, 18 confirmed and five suspected. There were five deaths from pulmonary anthrax (71).

Three human vaccines against anthrax are currently available commercially (produced in the UK, the USA, and Russia). Two are inactivated cell-free products, whereas the Russian product contains a live attenuated vaccine. Current vaccine supplies are limited; however, even if enough vaccine were available, mass immunization would not be recommended. Immunization could be considered for essential service personnel and, following a terroristic attack, would be combined with an antibiotic to protect against residual retained spores. In 1970, an anthrax vaccine, derived from a sterile filtrate of an avirulent non-encapsulated strain of *Bacillus anthracis* and adsorbed on to aluminium hydroxide, was licensed by the FDA to protect people who might be exposed to anthrax. The ACIP has recommended immunization with this vaccine for people working with *B. anthracis* in the laboratory,

those working with imported animal hides or furs in conditions that are inadequate to prevent exposure to anthrax spores, and military personnel deployed to areas with high risk for exposure. The vaccine was used to protect military personnel during the 1991 Gulf War. Although it has a safety record based on clinical trials carried out in the 1950s and 1960s and on the experience of over 30 years of use by thousands of military personnel, woolworkers, and veterinarians (72,73), the debate over its safety has come to the forefront in recent years, including the hypothesis of an alleged link between anthrax vaccine and disease reported by military personnel after the 1991 Gulf War. As a result of this concern, Congress ordered an in-depth investigation into the safety of the anthrax vaccine. The IOM was asked to initiate a comprehensive study and to report on the safety and efficacy of the vaccine in an effort to answer the questions raised by Congress, the Department of Defense, and the public. Because of immediate concerns over anthrax vaccine safety issues, the IOM provided on 30 March 2000 a letter report to the Department of Defense, summarizing the Institute's literature review on the safety of anthrax vaccine, prepared by its Committee on Health Effects Associated with Exposures during the Gulf War (74). The committee evaluated primary peer-reviewed literature and did not draw conclusions from secondary literature. It considered that there have been only a few published peer-reviewed studies of the safety of the anthrax vaccine in humans, and only one published series of studies discussed the long-term follow-up of individuals who received multiple vaccines including the anthrax vaccine. The committee concluded that (a) published studies have reported no significant adverse effects of the vaccine, but the literature is limited to a few short-term studies and (b) in peer-reviewed publications there is inadequate/insufficient evidence to determine whether there is or is not an association between anthrax immunization and long-term adverse health outcomes. The committee considered the findings and conclusions as an early step in the complex process of understanding the vaccine's safety, which began with its licensure in 1970 and includes the 1985 FDA advisory panel's finding that categorized the anthrax vaccine as safe and effective. The committee included in its evaluation the results of the VAERS, and considered the VAERS data as being useful as a sentinel for adverse events but of limited value for assessing the rate of causality of adverse events. The VAERS data are as follows:

(a) From 1 January 1990 to 31 August 2000, 1544 adverse events after anthrax immunization (nearly 2 million doses administered) were reported to VAERS.
(b) The most frequently reported adverse events were local reactions (864 reports).
(c) Other reports included headache (239 reports), arthralgia (232 reports), weakness (215 reports), pruritus (212 reports), and a few other systemic reactions.
(d) There were 76 (5%) serious adverse events (death or life-threatening disease, hospitalization, permanent disability); based on autopsy results, one of two reported deaths was due to aplastic anemia and the other was due to coronary arteritis (75).

- A 34-year-old man reported mild tenderness at the injection site after the first dose of anthrax vaccine; after the second dose he felt sweaty and weak and was pale; 20 hours after the third dose he had a life-threatening anaphylactic reaction (dyspnea, sweating, pallor, and urticarial wheals on the face, arms, and torso). After intensive care measures all his symptoms and signs resolved (76).

In 1997 manufacturers in South Korea started to develop a new anthrax vaccine. However, it will take a few years more to put the vaccine to practical use (71).

Botulism

Worldwide, sporadic cases and limited outbreaks of botulism can occur when food and food products are prepared or preserved by improper methods that do not destroy the spores of *Clostridium botulinum* and permit the formation of botulinum toxin. In industrially developed countries, the case fatality rate of food-borne botulism is 5–10%. Person-to-person transmission of botulism is not known. Botulinum toxin is the most poisonous substance known and poses a major bioweapon threat. In addition to the clinical forms of natural botulism (food-borne, wound, and intestinal), there is a fourth, man-made form of inhalational botulism that results from aerosolized botulinum toxin.

A pentavalent botulinum toxoid (botulinum toxin in different antigenic types) has been used for more than 30 years

in some countries to prevent the disease in laboratory workers and to protect troops against attack. Pre-exposure immunization for the general population is neither feasible nor desirable; the vaccine is ineffective for postexposure prophylaxis. Treatment of botulism consists of passive immunization and supportive care. Most licensed antitoxins contain antibodies against the most common toxin types A, B, and E. About 9% of recipients of equine antitoxin developed urticaria, serum sickness, or other hypersensitivity reactions. In 2% of recipients anaphylaxis occurred within 10 minutes of antitoxin administration. Before administering antitoxin the patient should be screened for hypersensitivity (77).

Plague

The large plague pandemics (the Justinian plague in the 6th century and pandemics in the 14th–15th centuries and 19th century) killed millions of people worldwide and were feared as the Black Death, a term that was actually not introduced until Victorian times. There is evidence that all three plagues were caused by the same variant of the plague bacterium, *Yersinia pestis*, var Orientalis (78). Progress in hygiene, public health, and antibiotic therapy make future pandemics improbable.

Wild rodent plague still exists in many parts of the world, including areas of Africa, Latin America, Asia, South-Eastern Europe, and the Western half of the USA. Domestic pets, for example house cats and dogs, can bring infected wild rodent fleas into homes, and the bite of infected fleas can result in human disease. Bubonic, pulmonary, and septicemic plague are life-threatening diseases with very high case fatality rates in untreated patients and in patients in whom treatment is delayed. Pneumonic plague is highly communicable, particularly in overcrowded facilities.

The potential use of *Y. pestis*, the causative agent of plague, as a biological weapon is of great concern. A killed whole bacillus plague vaccine (limited data on reactogenicity; see SED-14, 1086) is no longer available in the USA. This killed vaccine was efficacious in preventing or ameliorating bubonic plague but was ineffective against primary pneumonic plague. A live plague vaccine is manufactured in Russia, but no data on its efficacy and safety are available. Recommendations for antibiotic therapy of plague are available (79).

Smallpox

The last natural case of smallpox occurred in 1977 in Somalia. Because of the success of worldwide co-ordinated efforts, particularly the use of smallpox vaccine, natural smallpox has been eradicated. In 1989, the WHO declared the global eradication of this dangerous and often deadly disease (the case fatality rate in non-immunized patients was 20–40%). During natural smallpox outbreaks the secondary attack rate in non-immunized contacts was about 50%.

The smallpox virus belongs to the limited number of organisms that could be used as a biological weapon, causing one of the most serious diseases. There is no causal treatment of smallpox available. Immunization is the most effective measure for pre-exposure prevention and postexposure infection control. However, there were severe complications associated with the use of the old vaccines: encephalitis (mainly in primary vaccines; sometimes fatal or causing permanent neurological sequelae), progressive vaccinia (in both primary and secondary vaccinees), eczema vaccinatum (in vaccinees with either active or healed eczema), and generalized vaccinia resulting from viremia.

With the global eradication of smallpox, smallpox immunization was stopped in all countries of the world, and complications of smallpox vaccination became largely of historical interest. However, recently, the threat of bioterrorism has made it necessary again to consider immunization strategies and the potential hazards of immunization. The old vaccine used for prevention of smallpox contained vaccinia virus strains grown on scarified calves. Some countries and manufacturers have retained limited stocks of vaccine, and the WHO has 500 000 doses. However, there is no reserve large enough to meet more than very limited potential emergency needs (80). Strategies for smallpox immunization in emergencies have been elaborated, focusing on priority immunization of individuals at greatest risk and immunization to contain outbreaks (81). However, the old smallpox vaccine hardly meets modern safety requirements. Therefore, the development and licensure of a modern tissue cell culture vaccine and the establishment of new vaccine production facilities is necessary. Various manufacturers have initiated such developments. A British biotechnology company announced its intention to begin clinical trials on immunogenicity and reactogenicity of a newly developed tissue culture smallpox vaccine in early 2002 (82).

Tularemia

Tularemia is a zoonotic bacterial disease that occurs in North America, China, Japan, parts of Europe, and parts of the former USSR. Various wild animals are the reservoir of the causative agent, *Francisella tularensis*, which is transmitted through the bite of arthropods, including ticks and mosquitoes, by eating insufficiently cooked meat, by drinking contaminated water, or by inhalation of dust from a contaminated environment. Clinically, an ulceroglandular type can be distinguished from an oropharyngeal type; inhalation of infectious material can be followed by pneumonia or septicemia, with a 30–60% case fatality rate in untreated patients. Person-to-person transmission of tularemia is not known.

Francisella tularensis is considered to be a dangerous potential biological weapon. Live tularemia vaccines have been developed and used in the USSR (to protect millions of persons in endemic areas) and the USA (as an investigational vaccine to protect laboratory workers). After the administration of the US vaccine, symptoms of ulceroglandular disease were considered milder; however, the vaccine did not protect all recipients against aerosol challenge with *F. tularensis*. The live vaccine is currently under review by the US Food and Drug Administration (FDA), and its future availability is undetermined. Taking into account the short incubation period of

tularemia and incomplete protection through immunization with the current live vaccine, post-exposure immunization is not recommended. Treatment with antibiotics is recommended (83).

General adverse effects after immunization

The reports of the Institute of Medicine, National Academy of Sciences, Washington (on adverse events after pertussis and rubella immunization (SED-12, 817) (SED-12, 825) and on adverse events after immunization against tetanus, diphtheria, measles, mumps, poliomyelitis, *H. influenzae* type b, and hepatitis B (SEDA-18, 325) have provided useful reviews (38,39). The 1996 "Update on vaccine side effects, adverse reactions, contraindications, and precautions" elaborated and published by the US Advisory Committee on Immunization Practices includes summarized conclusions of evidence for possible associations between specific adverse effects and childhood vaccines (Table 10 and Table 11). The conclusions are based on the reports of the IOM mentioned above (84). The respective chapters of "Vaccines" (85) also provide useful information on adverse effects and immunization risks.

General risks of multiple immunizations

Some parents, particularly in industrialized countries, believe that infants get more vaccines than are good for them, and they fear that too many immunizations could overwhelm the infant's immune system. However, the actual number of antigens that children receive in the USA (the schedule includes 11 vaccines) or in some Western European countries (on average 9 vaccines) has declined when compared with immunization programs used during the 1960s or 1980s. Whereas smallpox vaccine contained about 200 proteins, the current routinely recommended vaccines for children in countries in which whole-cell pertussis vaccine was replaced by acellular pertussis vaccine (for example Austria, Germany, Japan, the Scandinavian countries, the USA) contain a total of 50–125 proteins and polysaccharides. The replacement of whole-cell pertussis vaccine by acellular pertussis vaccine reduced the content of immunogenic proteins and polysaccharides from about 3000 to a range of 2-5. (Table 12) (86).

If vaccines overwhelmed the immune system, smaller immune responses would be expected. However, the authors of many carefully designed studies have concluded that there is no evidence that adding vaccines to combination products increases the burden on the immune system. Young infants have a large capacity to respond to multiple vaccines, as well as to many other environmental challenges. Increased reactogenicity after the receipt of combination vaccines has not been a major issue. Combining antigens usually does not increase the risk of adverse effects, and can actually lead to an overall reduction in the numbers of adverse events (87) despite public concern (88).

The Safety Review Committee of the IOM has reviewed the evidence regarding multiple immunizations

Table 10 Evidence for possible associations between specific adverse effects and childhood vaccines

Evidence	DT/Td/tetanus toxoid	Measles vaccine	Mumps vaccine	OPV/IPV	Hepatitis B vaccine	Hemophilus influenzae type b (Hib) vaccine
None available to establish a causal relation	None	None	Neuropathy Residual seizure disorder	Transverse myelitis (IPV) Thrombocytopenia (IPV)	None	None
Inadequate to accept or reject a causal relation	Residual seizure disorder other than infantile spasms Demyelinating diseases of the central nervous system Mononeuropathy Arthritis Erythema multiforme	Encephalopathy Subacute sclerosing panencephalitis Residual seizure disorder Sensorineural deafness (MMR) Optic neuritis Transverse myelitis Guillain–Barré syndrome Thrombocytopenia Insulin-dependent diabetes mellitus	Encephalopathy Aseptic meningitis Sensorineural deafness (MMR) Insulin-dependent diabetes mellitus Sterility Thrombocytopenia Anaphylaxis	Transverse myelitis (OPV) Guillain–Barré syndrome (IPV) Death from SIDS	Guillain–Barré syndrome Demyelinating diseases of the central nervous system Arthritis Death from SIDS	Guillain–Barré syndrome Transverse myelitis Thrombocytopenia Anaphylaxis Death from SIDS
Favored rejection of a causal relation	Encephalopathy Infantile spasms (DT only) Death from SIDS (DT only)	None	None	None	None	Early-onset Hib disease (conjugate vaccines)
Favored acceptance of a causal relation	Guillain–Barré syndrome Brachial neuritis	Anaphylaxis	None	Guillain–Barré syndrome (OPV)	None	Early-onset Hib disease in children aged 18 months whose first Hib vaccination was with unconjugated PRP vaccine
Established a causal relation	Anaphylaxis	Thrombocytopenia (MMR) Anaphylaxis (MMR) Death from measles infection	None	Poliomyelitis in recipient or contact (OPV) Death from polio infection	Anaphylaxis	None

See the original paper for extensive notes to this table.
OPV—oral poliovirus vaccine
IPV—inactivated poliovirus vaccine
DT—diphtheria and tetanus toxoids for pediatric use
Td—diphtheria and tetanus toxoids for adult use
SIDS—sudden infant death syndrome

Table 11 Evidence for possible associations between specific adverse effects and childhood vaccines (DTP and MMR)

Evidence	Adverse effect	DTP vaccine–RA 27/3 MMR
Strong evidence against a causal relation	Autism	None
Inadequate to accept or reject a causal relation	Aseptic meningitis	Radiculoneuritis and other neuropathies
	Chronic neurological damage	Thrombocytopenic purpura
	Erythema multiforme or other rash	
	Guillain–Barré syndrome	
	Hemolytic anemia	
	Type 1 diabetes	
	Learning disabilities and attention-deficit disorder	
	Peripheral mononeuropathy	
	Thrombocytopenia	
Favored rejection of a causal relation	Infantile spasms	None
	Hypsarrhythmia	
	Reye's syndrome	
	Sudden infant death syndrome	
Favored acceptance of a causal relation	Acute encephalopathy	Chronic arthritis
	Shock and unusual shock-like state	
Causal relation established	Anaphylaxis	Acute arthritis
	Protracted inconsolable crying	

DTP—diphtheria toxoid, tetanus toxoid, and pertussis vaccine
MMR—measles, mumps, and rubella vaccine

Table 12 Number of immunogenic pathogens/polysaccharides in vaccines over the past 100 years

1900		1960s		1980s		2000	
Vaccine	Proteins	Vaccine	Proteins	Vaccine	Proteins	Vaccine	Proteins/polysaccharides
Smallpox	~200	Smallpox	~200	Diphtheria	1	Diphtheria	1
—	—	Diphtheria	1	Tetanus	1	Tetanus	1
—	—	Tetanus	1	WC-pertussis	~3000	AC-pertussis	2–5
—	—	WC-pertussis	~3000	Poliomyelitis	15	Poliomyelitis	15
—	—	Poliomyelitis	15	Measles	10	Measles	10
—	—	—	—	Mumps	9	Mumps	9
—	—	—	—	Rubella	5	Rubella	5
—	—	—	—	—	—	H. influenzae	2
—	—	—	—	—	—	Hepatitis B	1
—	—	—	—	—	—	Varicella*	69
—	—	—	—	—	—	Pneumococcus*	8
Total		200		3200		3040	46–49 (Germany) 123–126 (USA)

and immune dysfunction (89). The review did not support the hypothesis that an infant's immune system is inherently incapable of handling the number of antigens to which children in the USA are exposed during routine immunizations. The committee rejected a causal relation between multiple immunizations and increased risks of infections or type 1 diabetes mellitus. The evidence was inadequate to accept or reject a causal relation between multiple immunizations and allergic disease, particularly asthma.

Sudden infant deaths shortly after the administration of hexavalent vaccines

HEXAVAC® and Infanrix™hexa are combination vaccines that were authorized in the European Union on 23 October 2000. Both protect infants and children against diphtheria, tetanus, whooping cough (pertussis), hepatitis B virus, polio virus, and Hemophilus influenzae type b.

- A 3-month-old girl died suddenly after being given a hexavalent vaccine (90). Investigation of the brainstem on serial sections showed bilateral hypoplasia of the arcuate nucleus. The cardiac conduction system had persistent fetal dispersion and resorptive degeneration.

During its April 2003 meeting, the scientific Committee for Proprietary Medicinal Products (CPMP) of the European Agency for the Evaluation of Medicinal Products (EMEA) conducted a detailed review of five reports of unexplained deaths in children within 24 hours of immunization with a hexavalent vaccine (91). These reports were received as part of routine post-marketing safety monitoring over 2.5 years. During this time an estimated 8.7 million doses of the vaccines were used world wide, corresponding to the immunization of some 3 million children. Participants at the meetings included the pathologists who had conducted the autopsies, pediatricians with experience in vaccines and Sudden Infant Death Syndrome (SIDS), and epidemiologists. On the basis of the available information, the CPMP concluded that: the causes of the deaths remained unexplained, and on the basis of available data it is not possible to establish a cause-and-effect association with the hexavalent vaccines. In several cases SIDS, viral infection, metabolic disorders, allergic reactions, or airway obstruction were plausible, although none of these could be definitely proven to have caused the deaths. The Committee also considered possible risk factors, including a family history of epilepsy or convulsions at an early age, which was reported in three of the five cases. However, they concluded that the clinical description of these individual cases did not provide sufficient evidence to identify a family history of epilepsy as a possible risk factor. In summary, the Committee concluded that there was no change in the benefit/harm balance of these products and that the benefits of immunization far outweigh the possible risks of existing vaccines, including hexavalent vaccines. They recommended that immunization should be continued according to national immunization schedules without changes to the present conditions of use.

Since the EMEA Public Statement on 28 April 2003, EMEA's CPMP has continued to monitor all data submitted and at its November 2003 meeting reviewed the conclusions taken by a number of specialized expert groups and working parties with regard to HEXAVAC® and Infanrix™hexa. Since the products were authorized (in 2000) four sudden unexpected deaths in close temporal association with hexavalent immunization have been reported in children during the second year of life (three in Germany, one in Austria) over a period of 3 years,. A retrospective analysis based on an epidemiological approach to this problem was performed on data from Germany and showed that the observed number of cases of sudden unexpected death in the second year of life within 48 hours of HEXAVAC® administration exceeded the expected number of cases. This finding was based on three cases amongst more than 700 000 children who were given a booster in the period analysed (November 2000–June 2003).

The CPMP experts considered that this temporal relation raised a possible signal linking HEXAVAC®

immunization and sudden unexplained death, but they acknowledged some inevitable limitations of the data sources and methods used to calculate the expected numbers. In any case, a signal only raises a suspicion, and does not prove a cause-and-effect relation. The CPMP has agreed that a possible signal has been generated, but believes that this does not currently constitute a risk to public health. It is known that a very small number of children die in their early years from a number of causes, some of which are unknown. The fact that a child dies shortly after being immunized may lead to the reporting of this event to the responsible health authority as being temporally associated with immunization. Further studies are needed to establish whether or not there is a real risk. Being aware of this possibility, and taking into account the findings for Hexavac® as suggested by the statistical analysis, the following actions are being undertaken to find out whether the signal is real or not. Prospective and retrospective studies to investigate sudden unexpected deaths will be conducted by independent institutions and will be assessed by regulatory authorities. These active surveillance programs, which started in 2004, should facilitate the collection of further data and monitoring of vaccines by regulatory authorities and the marketing authorization holders. The results will be followed closely, so that timely regulatory action can be taken, if necessary.

Based on their review, the CPMP concluded that there was no change in the benefit/harm balance of these products and therefore recommended no changes to the present conditions of use (92).

In a letter to the editor, Zinka et al (93) described six cases of "sudden infant death" within 48 hours of administration of hexavalent DTaP–Hib–IPV–HBV combination vaccine. While they stated that the causality of immunization and death was uncertain, they aimed to inform physicians, pediatricians, and parents about possibly fatal complications after the use of hexavalent vaccines. In a reply, Schmitt et al (94) suggested that the data given in the letter fell short of providing any scientifically valid evidence. Basic information about the cases was missing; the descriptions of the cases lacked information on any special circumstances or potential risk factors, such as the sleeping position, the type of bed, bedding, or pillows, or the potential for foreign-body aspiration. There was no information about the autopsy protocol, about which tests and examinations had been performed on which of the cases, or whether histology was done on all organs. There was histological evidence of encephalitis, including necrosis, at least in some cases, but there was no information about microbiological tests for non-colonizing micro-organisms or tests done to look for intercurrent infections. Zinka et al. had stated that all the children had "extraordinary brain oedema", but it was not clear if this statement had been based on subjective personal impressions at autopsy, or the weight of the brains relative to a valid reference scale, or histological findings. They also pointed out that neither of the currently available hexavalent vaccines contains hepatitis B serum, but rather recombinant hepatitis B surface antigen protein.

Likewise, neither contains "influenza" antigen, but rather Hemophilus influenzae type b polysaccharide conjugated to tetanus toxoid. The recommended schedule for immunization of children in Germany, where the cases were observed, was not 2, 4, 6, and 12?14 months, as described in the letter.

In an other reply to the letter to the editor, von Kries (95) expressed doubt about the conclusions of Zinka et al and their recommendation to switch from a hexavalent combination vaccine to a pentavalent combination vaccine. The rationale was based on a statement in the discussion: 1/198 cases of sudden infant death shortly after immunization had been observed in 1994–2000 (when tetravalent or pentavalent vaccines were used) compared with six in 74 cases in 2001–2004 (since the launch of hexavalent vaccines). Five of these cases were in close temporal association ("shortly after vaccination") with the use of HEXAVAC® (four in the first year and one in the second year of life), one with the use of Infanrix™ hexa. Comparing the 1/198 cases and the 4/72 cases (excluding the case in the second year of life and the Infanrix™ hexa case) yields a two-sided Fischer's exact P value of 0.019. This does not, however, generate a signal for Hexavac® in the first year of life. Von Kries added that comparisons with historic control groups are always problematic, since changes in risk factors over time cannot be disentangled from the exposure of interest. Unless there are data to show that the proportion of cases of sudden infant death undergoing autopsy has been constant over time there is a possibility of selection bias. There had been 198 autopsies in cases of sudden infant death in 7 years (1994–2000) compared with only 74 in 4 years (2001–2004), which might be explained by the reduced incidence of sudden infant death in recent years. However, there is also a possibility that cases in temporal association with immunization were more likely to have an autopsy after 2001 than before.

Finally, in a systematic analysis of all cases of sudden infant death in temporal association with hexavalent vaccines before June 2003 there was no signal for HEXAVAC® or Infanrix™hexa in the first year of life. Given the limitations of the data, von Kries doubted that the observations in the Munich Institute for Legal Medicine provided support for an additional empirical signal for HEXAVAC® in the first year of life. He concluded that the observations reported by Zinka et al. justified neither an intimidating "information" policy for immunizing physicians and parents nor a recommendation to switch to pentavalent DTPaP–IPV–Hib vaccines.

In a press release (21 April, 2005) the European Agency for the Evaluation of Medicinal Products (EMEA) informed a meeting of the Committee for Medicinal Products for human use as follows: At the request of the German regulatory agency in February 2005 the committee again reviewed the safety of the hexavalent combination vaccine HEXAVAC® and verified its previous position (December 2003) that no regulatory action or changes to the product information are necessary (96).

During its twelfth meeting (June 9–10, 2005) the Global Advisory Committee on Vaccine Safety (GACVS), an expert clinical and scientific advisory body that reports to the WHO, reviewed the available data concerning a purported association between sudden unexplained death and hexavalent diphtheria, tetanus, acellular pertussis, Haemophilus influenzae type b, poliovirus, and hepatitis B (DTaP–Hib–IPV–HBV) combination vaccines. The Committee concluded that the evidence did not support a causal association, but recommended that additional studies be conducted to extend the observation period to the first 2 years of life. There have been no additional signals from Germany, where the first reports originated. The results from a population-based Italian study on sudden infant deaths and sudden unexplained deaths in five birth cohorts (between 1999 and 2003) have produced no evidence of increased risks in the 48 hours after any immunization in the first year of life. Nor has any increased risk been observed in relation to the number of doses of vaccine administered.

In Italy, between 1990 and 2001, infant mortality from any cause fell from 8.1 to 4.8 per 1000 births. Between 1991 and 2002, infant mortality from sudden infant deaths fell in Germany from 1.5 to 0.6 per 1000 births. This happened despite the introduction in both countries of several new vaccines in the infant schedule and increased immunization coverage in both countries. GACVS concluded that these data are inconsistent with any association between hexavalent vaccines and sudden infant death or sudden unexplained death (97).

A case-control study on SIDS (GeSID) was carried out in Germany between 1998 and 2001 in 18 forensic pathology institutes covering half of Germany (98). Infants who were immunized after the introduction of the hexavalent vaccines in October 2000 were analysed separately. There were 129 deaths from SIDS during the period October 2000 to October 2001. Of these, 22 (17%) had received a hexavalent vaccine compared with 100 of 378 controls (27%). If immunization increased the risk of SIDS one would expect a higher immunization rate among the SIDS cases than among the controls. In this study the opposite was the case. More controls were immunized and the control infants started their immunization schedule earlier. Even when the data were restricted to the 14 days before death/interview, there was no increased risk of SIDS in temporal relation to immunization. This study has provided further evidence that immunization is not a risk factor for SIDS.

Combination vaccines versus concomitant administration of single vaccines

The safety and reactogenicity of a booster dose of hexavalent DTaP–HBV–IPV/Hib vaccine (GSK Biologicals; n = 4725) has been compared with the separate administration of DTaP–IPV/Hib and HBV vaccines (GSK Biologicals; n = 4474) in two open, randomized, multicenter studies (99). In the first study (n = 1149), the incidences of symptoms were similar in the two groups; no serious adverse events were either reported within 4 days of immunization or considered to be causally related to immunization. In the second study (n = 8050), in which fever was the only solicited symptom, the rectal temperature was 39.5°C or over in 2.5% and 2.8% of the subjects respectively. Fever of 40.0°C or more was rare (0.6%), and only two cases of febrile convulsions were recorded during the 4 days after immunization, both in the control group. Extensive swelling (defined as local injection-site swelling with a diameter over 50 mm, noticeable diffuse injection-site swelling, or a noticeably increased circumference of the injected limb) was reported after 2.3% of the booster vaccine doses, regardless of the vaccine used. Extensive swelling involving an adjacent joint was reported in 0.1% of subjects.

The authors concluded that the hexavalent combination DTaP–HBV–IPV/Hib vaccine and the DTaP–IPV/Hib and HBV vaccines administered separately have similar good reactogenicity and safety profiles when given as booster doses in the second year of life.

Organs and Systems

Cardiovascular

Pentavalent and heptavalent vaccines and apnea or bradycardia

The incidence and clinical significance of apnea or bradycardia after immunization with pentavalent (DTaP–Hib–IPV) and heptavalent (DTaP–Hib–IPV–HB) combination vaccines have been evaluated in respiratory stable preterm infants (100). The medical records of 53 infants with a mean gestational age of 28 weeks, hospitalized in the neonatal intensive care unit of the University Children's Hospital in Basel from January 2000 to June 2003, were analysed. Clinical data were recorded for 72 hours before and after the first immunization. Of the 53 infants, seven had a transient recurrence or an increase in episodes of apnea or bradycardia after immunization. Five of these seven required interventions, ranging from tactile stimulation to bag-and-mask ventilation, but there were no serious consequences. The rate of fever over 38°C after immunization was higher in affected infants than in those without recurrence of or increase in apnea or bradycardia. The authors recommended monitoring of all preterm infants after immunization in neonatal intensive care units.

Respiratory

There is a hypothesis that the seeds of immune system problems, perhaps including asthma, may be sown in the early weeks of life and activated later by immunization. However, studies have given inconsistent results. A randomized controlled trial showed no effect of pertussis vaccine on the risk of asthma (101), which was suggested by a cohort study (102). A British cohort study also showed no association between immunization and wheeze, but there was a lower risk of eczema (103). Data from the Christchurch Health and Development Study showed a link between immunization and asthma

(104). However, the findings were based on only 23 children who had not received DTP or polio vaccine. There was no effect from measles vaccine. This is somewhat in contrast to data from Africa suggesting that measles may protect against asthma (SEDA-21, 323).

Nervous system

Guillain–Barré syndrome

The Global Advisory Committee on Vaccine Safety (GACVS) has discussed the possible relation between vaccines and Guillain–Barré syndrome (SEDA-21, 323), which is relatively rare (1–2 cases per 100 000 people annually) (105). Guillain–Barré syndrome has occasionally been observed in temporal association with immunization; this association has been considered as causal in cases after swine influenza vaccine (attributable risk 9.5 per million doses administered) as well as rabbit brain and other nervous-tissue derived rabies vaccines. With conflicting evidence, cases of Guillain–Barré syndrome have been reported in temporal association with other vaccines, including seasonal influenza, tetanus, meningococcal conjugates ,and diphtheria–tetanus–pertussis (DTP). The GACVS considers that investigation of a possible causal relation could best be achieved by large-scale studies of the incidence of Guillain–Barré syndrome before and after an immunization program. All incident cases would need to be carefully ascertained and documented to ensure as accurate a diagnosis as possible. Improved understanding of the pathogenesis of all forms of the disease would assist the investigation of possible associations between Guillain–Barré syndrome and immunization. In this context, the collection of serum samples from incident cases would contribute to the identification of the different forms of the disease and to understanding their possible relationship with vaccines.

Recent advances in understanding the mechanisms of some of the Guillain–Barré syndrome subtypes should be considered when evaluating the possible relation with vaccines. Guillain–Barré syndrome consists of at least four subtypes of acute peripheral neuropathy (106). The histological appearance of the acute inflammatory demyelinating polyradiculoneuropathy (AIDP) subtype resembles experimental autoimmune neuritis, which is predominantly caused by T cells directed against peptides from the myelin proteins P0, P2, and PMP22. The role of T-cell-mediated immunity in AIDP is unclear, and there is evidence for involvement of antibodies and complement. Strong evidence now exists that axonal subtypes of Guillain–Barré syndrome, acute motor axonal neuropathy (AMAN), and acute motor and sensory axonal neuropathy (AMSAN), are caused by antibodies to gangliosides on the axolemma that target macrophages to invade the axon at the node of Ranvier. About a quarter of patients with Guillain–Barré syndrome have had a recent *Campylobacter jejuni* infection, and axonal forms of the disease are especially common in these people. The Fisher's syndrome subtype is especially associated with antibodies to GQ1b, and similar cross-reactivity with ganglioside structures in the wall of *C. jejuni* has been discovered. Anti-GQ1b antibodies damage the motor nerve terminal in vitro by a complement-mediated mechanism.

In a prospective, multicenter study in 95 children with Guillain–Barré syndrome the frequency and causes of antecedent diseases were investigated (107). All infections and immunizations that occurred within 6 weeks before the onset of Guillain–Barré syndrome were documented. Preceding infections were reported in 82% of children; the most frequently involved agents were Coxsackie viruses (15%), *Chlamydia pneumoniae* (8%), cytomegalovirus (7%), and *Mycoplasma pneumoniae* (7%). There was serological evidence of a *Campylobacter jejuni* infection in six patients (7%). Eight children had been immunized during the 6 weeks preceding the onset of Guillain–Barré syndrome; in six of these children concomitant infectious diseases were reported, and in one child the time between immunization and Guillain–Barré syndrome was extremely short. The authors concluded that *Campylobacter* spp. do not seem to play a major role in childhood Guillain–Barré syndrome, in contrast to adults. Most of the children who had been immunized had also concomitant infectious diseases.

Haemophilus influenzae

Guillain–Barré syndrome has occasionally been reported after immunization with a conjugate vaccine against *H. influenzae* type b. In one case there were immunoglobulin antibodies against the *H. influenzae* type b polysaccharide (PRP) component of the vaccine, with a high titer of anti-PRP IgM antibody (108).

Hepatitis

Guillain–Barré syndrome has been reported after the administration of plasma-derived hepatitis B vaccines. In the 3 years between 1 June 1982 and 31 May 1985, an estimated 850 000 people received a new plasma-derived hepatitis B vaccine (Heptavax-B, Merck Sharp and Dohme) and there were 41 reports of the following neurological adverse events: Bell's palsy ($n = 10$), Guillain–Barré syndrome ($n = 9$), convulsions ($n = 5$), lumbar radiculopathy ($n = 5$), optic neuritis ($n = 5$), transverse myelitis ($n = 4$), and brachial plexus neuropathy ($n = 3$) (109). Half occurred after the first of three doses. There were no deaths.

Guillain–Barré syndrome has also occasionally been reported after the administration of recombinant hepatitis B vaccines (110,111).

Influenza

Following the swine-flu immunization campaign in 1976/77 in the USA there was a significant increase in the incidence of Guillain–Barré syndrome in immunized versus non-immunized people, from 2.6 per million to 13.3 per million (112). Peak time of onset was 2–3 weeks after receiving the vaccine, and cases among vaccinees were less likely to have a history of antecedent infection than were cases in unvaccinated persons. Since 1977 the risk of

influenza vaccine–induced syndrome appears to be the same as the risk in the non-immunized population.

However, whether influenza immunization (with the exception of the swine-flu immunization programme in 1976 in the USA) is associated with Guillain–Barré syndrome remains uncertain. (113) In two studies using population-based health-care data from the province of Ontario, Canada from 1 April 1992 to 31 March 2004, there were 1601 incident hospital admissions because of Guillain–Barré syndrome. In 269 patients, the syndrome was diagnosed within 43 weeks of immunization against influenza. The estimated relative incidence during the primary risk interval (weeks 2–7) compared with the control interval (weeks 20–43) was 1.45 (95% CI = 1.05, 1.99). This association persisted in several sensitivity analyses using risk and control intervals of different durations. However, a separate time-series analysis showed no evidence of seasonality and no statistically significant increase in hospital admissions because of Guillain–Barré syndrome after the introduction of the universal influenza immunization program. The authors concluded that influenza immunization is associated with a small but significantly increased risk of hospitalization because of Guillain–Barré syndrome.

The data from the Vaccine Adverse Event Reporting System included 54 reports of Guillain–Barré syndrome after immunization that occurred in the USA in 2004 (114). In 38 of the patients, Guillain–Barré syndrome occurred within 6 weeks, and the authors considered this suggestive of a causal association. The highest incidence within 6 weeks was observed in patients who received influenza vaccine ($n = 23$), followed by six cases after hepatitis vaccine (not distinguished into hepatitis A and hepatitis B). Two cases of Guillain–Barré syndrome were temporally related to Td vaccine (adult formulation). After other vaccines, including combination vaccines and concomitant administration of single vaccines, seven cases were reported. This study had many limitations, because it was based on data from a passive surveillance system and no attempt was made to determine the incidence of Guillain–Barré syndrome in vaccinees compared with the healthy population.

Measles–mumps–rubella

Guillain–Barré syndrome has been reported to be associated with measles, mumps, and rubella immunization (115).

Polio

Following a mass immunization campaign against poliomyelitis in Finland (1984/85), an analysis based on hospital records covering a population of 1.17 million and 6 years showed a significantly increased incidence of Guillain–Barré syndrome coinciding with the campaign (116).

In a retrospective analysis of the incidence of Guillain–Barré syndrome in Finland in 1981–86 by examination of medical records identified from the nationwide Hospital Discharge Register, based on a population of 5 million people, 247 patients fulfilled the accepted criteria of Guillain–Barré syndrome corresponding to a mean annual incidence of 0.82 per 100 000 (117). Monthly rates showed an increased incidence of Guillain–Barré syndrome in March 1985, a few weeks after the start of the nationwide oral poliovirus vaccine campaign and partly overlapping it. However, the increase had begun just before the onset of the campaign and may have been due to wild-type 3 poliovirus in the population immediately before the campaign and could have been due to that rather than the administration of the oral poliovirus vaccine.

Rabies

Guillain–Barré syndrome was a well-known complication in the era of brain tissue rabies vaccines, and there have been a few cases associated with the use of human diploid cell rabies vaccines (118,119).

Tetanus and diphtheria

Tetanus and diphtheria toxoid vaccines are reportedly associated with polyradiculitis, polyneuritis, and mononeuritis (120–122).

Immunization and Multiple sclerosis (encephalitis disseminata)

The possible relation between immunization and an attack of encephalitis disseminata has been investigated in an analysis of 16 confirmed cases and 24 suspected cases in connection with the distribution of approximately 100 million doses of vaccines manufactured by Behringwerke (1980–89) (123). The data did not support an increased risk of initial manifestation of encephalitis disseminata or of renewed attacks in patients with pre-existing disease.

Studies of the role of influenza or influenza immunization in multiple sclerosis have been reviewed by Jeffery (124). He considers that in patients with multiple sclerosis and advanced disability influenza can constitute a life-threatening illness. Even in patients who have minimal disability, severe influenza can be followed by secondary bacterial infection. The risk of relapse after influenza may be as high as 33%, whereas the risk of relapse after influenza immunization appears to be negligible. The only exception to the general rule that influenza immunization is beneficial for patients with multiple sclerosis is in patients with rapidly evolving neurological deficits due to active disease, in whom immunization should be withheld pending treatment with high-dose glucocorticoids to suppress inflammation or stabilization through immunomodulatory treatment. In the absence of active disease, Jeffery strongly recommends influenza immunization for patients with multiple sclerosis. Regarding hepatitis B immunization and multiple sclerosis, Jeffery recommends hepatitis B immunization only for patients with multiple sclerosis at high risk of hepatitis B infection. If a high-risk patient has active multiple sclerosis, immunization should be withheld until the disease activity has been adequately treated. With regard to other immunizations, Jeffery has made similar recommendations: immunization should be postponed if a patient with multiple sclerosis shows disease activity.

Psychiatric

Autism

Most studies, experts, and scientific committees have concluded that there is no proof of a causal tie between autism and thiomersal or the MMR vaccine (SED-15, 2207; SEDA-25, 387; SEDA-26, 350; SEDA-27, 338; SEDA-28, 363; SEDA-30, 375). Multiple independent lines of evidence all point in the same direction: vaccines in general, and thiomersal in particular, do not cause autism, which more probably has its roots in genetics. In the USA, 5 years after the removal of thiomersal from all childhood vaccines (except influenza vaccines), autism diagnosis rates have continued to increase.

In reality, the true reason for the huge increase in autism prevalence over the last 15–20 years is because of the broadening of the diagnostic criteria for a diagnosis of autism or autism spectrum disorder (ASD) beginning in the early 1990s, resulting in widespread.

However, the controversy, particularly in the USA, is not over. Some doctors and scientists, some groups representing families with autistic children, and many parents fervently believe that there is a connection. More than 4800 such families (Autism Omnibus) have petitioned the (US) Federal Vaccine Injury Compensation Program (VICP) for compensation, based on the claim that their children's autism was caused by vaccines. The VICP was adopted through legislation by the Congress in 1988.

The Poling case

The "Poling case" provided some support for the claiming families. Given the volume of more than 4800 cases, the VICP adjudicators called Special Masters (judges trained in vaccine problems) decided to hear a small number of "test cases". Why the case of Hannah Poling was settled separately is not known.

When Hannah Poling was 19 months old, she received five vaccines (diphtheria–tetanus–acellular pertussis, *Haemophilus influenzae* type b, measles–mumps–rubella, varicella, and inactivated polio). At that time, Hannah was interactive, playful, and communicative. Two days later, she was lethargic, irritable, and febrile. Ten days after immunization, she developed a rash consistent with vaccine-induced varicella. Months later, with delays in neurological and psychological development, encephalopathy caused by a mitochondrial enzyme deficit was diagnosed. Hannah's signs included problems with language, communication, and behaviour—all features of autism spectrum disorder. Although it is not unusual for children with mitochondrial enzyme deficiencies to develop neurological signs between their first and second years of life, Hannah's parents believed that vaccines had triggered her encephalopathy. They sued the Department of Health and Human Services (DHHS) for compensation under the Vaccine Injury Compensation Program and won (beginning of 2008). According to the leaked document posted online, the Special Masters, Division of Vaccine Injury Compensation, concluded that five shots that Hannah had received in July 2000, when she was 19 months old, "significantly aggravated an underlying mitochondrial disorder" and had resulted in a brain disorder "with features of autism spectrum disorder" (125).

The stakes for decision-makers are high, and not just for the 4800 families. If any of the petitioners win test cases based on hypothesis and despite the evidence, it will open the floodgates for the rest of the petitioners. This will probably bankrupt the Vaccine Injury Compensation Program and will also risk the US vaccine infrastructure. Pharmaceutical companies will be reluctant to subject themselves to the liability of selling vaccines if even the truth cannot protect them from lawsuits (126,127).

Statement of the Global Advisory Committee on Vaccine Safety (GAVCS): mitochondrial diseases and immunization

The Poling case offered advocates a new theory: that vaccines cause or contribute to an underlying mitochondrial disorder, which in turn causes autism. Although autism is common among children with mitochondrial disorders, several experts in the disorders dismissed the notion that vaccines may cause the disease, which is widely understood to have a genetic origin.

Mitochondrial diseases result from failures of the mitochondria, specialized compartments present in almost every type of cell in the body. Mitochondria are responsible for creating more than 90% of the energy needed by the body to sustain life and support growth. When they fail, less and less energy is generated within the cell. Cell injury and even cell death follow. If this process is repeated throughout the body, whole systems begin to fail, and the life of the person in whom this is happening is severely compromised. The disease primarily affects children, but adult onset is becoming more and more common. Diseases of the mitochondria appear to cause most damage to cells in the brain, heart, liver, skeletal muscle, kidneys, and the endocrine and respiratory systems. Depending on which cells are affected, symptoms may include loss of motor control, muscle weakness and pain, gastrointestinal disorders and swallowing difficulties, poor growth, cardiac disease, liver disease, diabetes, respiratory complications, seizures, visual/hearing problems, lactic acidosis, developmental delays, and susceptibility to infection.

At a meeting (18–19 June 2008) of the GAVCS, an expert clinical and scientific advisory body, established by the WHO to deal with vaccine safety problems of potential global importance independently from the WHO and with scientific rigor, discussed the hypothesis that immunization could affect patients with mitochondrial disease, leading to autism. The committee issued the following statement (128):

Mitochondrial diseases are inherited disorders of energy metabolism that tend to affect tissues with high energy requirements, such as the brain, heart, and liver. These diseases are associated with a variety of symptoms. They are often difficult to diagnose, and there is no effective treatment. Mitochondrial disorders can also lead to cognitive impairment and encephalopathy, resulting in blunted social interaction.

Physiological stress triggered by external factors (for example, fever, cold, heat, starvation, sleep deprivation) may result in a worsening of the metabolic situation which results in deterioration of affected organs. Additionally, inflammatory responses associated with most infectious diseases can precipitate a clinical deterioration in an underlying mitochondrial disease. While vaccines may cause fever, clinicians caring for children with mitochondrial disease recommend vaccinating their patients since the risk of developing an even more devastating clinical deterioration would be associated with natural infection. The GACVS concluded, on the basis of the limited data available from the United Kingdom and the United States, that there is no convincing evidence to support an association between vaccination and deterioration of mitochondrial disease. The topic will be reviewed further if new findings become available. GACVS supports the current practice standard: children with mitochondrial diseases should receive the immunizations recommended for healthy children.

Metabolism

Childhood immunization and diabetes mellitus

In a study of all children born in Denmark from 1 January 1990 to 31 December 2000, for whom detailed information on immunizations and type 1 diabetes was available, type 1 diabetes was diagnosed in 681 children during 4 720 517 person-years of follow-up (129). The rate ratio for type 1 diabetes among children who received at least one dose of vaccine, compared with unimmunized children, was 0.91 (95% CI = 0.74, 1.12) for *Hemophilus influenzae* type b vaccine; 1.02 (0.75, 1.37) for diphtheria, tetanus, and inactivated poliovirus vaccine; 0. 96 (0.71, 1.30) for diphtheria, tetanus, acellular pertussis, and inactivated poliovirus vaccine; 1.06 (0.80, 1.40) for whole-cell pertussis vaccine; 1.14 (0.90, 1.45) for measles, mumps, and rubella vaccine; and 1.08 (0.74, 1.57) for oral poliovirus vaccine. Thus, the development of type 1 diabetes in genetically predisposed children (defined as those who had siblings with type 1 diabetes) was not significantly associated with immunization. Furthermore, there was no evidence of any clustering of cases 2–4 years after immunization with any vaccine. The authors concluded that their results did not support a causal relation between childhood immunization and type 1 diabetes.

Liver

Hepatitis risk and suspension of marketing authorization for HEXAVAC®

At its meeting of 12?15 September 2005, the Committee for Medicinal Products for Human Use (CHMP) of the European Medicines Agency (EMEA) recommended as a precautionary measure the suspension of the marketing authorization for HEXAVAC® because of doubts about its ability to confer long-term protection against hepatitis B (130). The recommendation was made after identification of reduced immunogenicity of the hepatitis B component. This is supposed to be due to variability in the production process for the hepatitis B component. There is no immediate concern for children already immunized with HEXAVAC®. However, the Committee asked Sanofi Pasteur MSD, the marketing authorization holder, to design a specific surveillance program to investigate whether infants and children would need to be re-immunized at a later stage, for instance at adolescence, to ensure long-term protection against hepatitis B.

Musculoskeletal

A child with fibrodysplasia ossificans progressiva developed permanent heterotopic ossification at the injection site after intramuscular DTP immunization (131). The authors considered that intramuscular injections in children with fibrodysplasia ossificans progressiva were potentially risky, but that subcutaneous injections could be carried out.

Immunologic

The questions of whether early childhood immunization affects the development of atopy and whether it causes allergic reactions have been reviewed (132). The authors concluded that immunization programs do not explain the increasing incidence of allergic diseases, but that individual children may uncommonly develop an allergic reaction to a vaccine.

Cases reported to the post-marketing surveillance system of the Kitasato Institute have been examined and categorized into two groups: allergic reactions and severe systemic illnesses (133). Patients with anaphylaxis due to gelatin allergy after immunization with live measles, rubella, and mumps monovalent vaccines have been reported since 1993, but the number of reported cases with anaphylaxis dramatically fell after 1999, when gelatin was removed from all brands. The incidence of anaphylactic reactions was estimated to be 0.63 per million for Japanese encephalitis virus vaccine, 0.95 for DPT, and 0.68 for influenza vaccine, but the causative component has not yet been specified. Among 67.2 million immunizations, there have been six cases of encephalitis or encephalopathy, seven of acute disseminated encephalomyelitis (ADEM), 10 of Guillain–Barré syndrome, and 12 of idiopathic thrombocytopenic purpura (ITP). The wild-type measles virus genome was detected in a patient with encephalitis and in two of four bone marrow aspirates obtained from ITP after measles immunization. Enterovirus infection was identified in two patients after mumps immunization (one each with encephalitis and ADEM), one patient with encephalitis after immunization with JEV vaccine, and one with aseptic meningitis after immunization with influenza vaccine. The total estimated incidence of serious neurological illness after immunization was 0.1–0.2 per million immunizations. It is noteworthy that enterovirus infection or wild-type measles virus infection was coincidentally associated with immunization in several cases suspected of being vaccine adverse events.

There has been a systematic review of the risk of allergic disease after infant immunization (134). The authors searched MEDLINE from 1966 to March 2003 and bibliography lists from retrieved articles, and consulted experts in the field to identify all articles relating immunization (diphtheria, tetanus, and pertussis, measles, mumps, and rubella, and BCG vaccine) to allergy. The design and quality of the studies varied considerably. Many did not address possible confounders, such as the lifestyle factors, leaving them susceptible to bias. The studies that offered the strongest evidence, including the only randomized controlled trial published to date, suggested that infant immunization does not increase the risk of allergic disease. Furthermore, BCG does not seem to reduce the risk of allergies.

Non-sterile injection equipment can transmit HIV and other infectious agents, including hepatitis viruses. There is also a possibility of needle transmission from an HIV-infected person to a vaccinator. Data from the USA show that the risk of transmission of HIV through needle stick is very low, perhaps 20 times lower than in the case of hepatitis B, and in the order of one per 100 accidents. Furthermore, the types of injections given during immunization sessions do not as a rule cause bleeding. The risk of transmission is thus extremely low. No instances of immunization-related spread of HIV to other infants have been reported, and if proper sterilization of needles and syringes is performed and vaccines are administered correctly the risk of HIV transmission is zero (135). However, the use of unsterilized or improperly sterilized needles and syringes is common, particularly in many developing countries, and contributes largely to the spread of hepatitis B and C, as well to the spread of human immunodeficiency virus and other blood-borne pathogens. These risks are recognized by the WHO as major public health problems and led the Organization to initiate a broad program of activities to ensure safe injection techniques (1). The guidelines published by WHO and UNICEF in 1986/87 (136,137), which can be set out briefly as follows, are still valid. Essentially they state that:

(a) a single sterile needle and a single sterile syringe should be used with each injection;
(b) reusable needles and syringes are recommended for use in developing countries; they should be steam-sterilized between uses; boiling is an acceptable alternative procedure when steam sterilization is not available;
(c) disposable needles and syringes should only be used if an assurance can be obtained that they will be destroyed after a single use;
(d) disease transmission by use of jet injectors is theoretically possible and has been demonstrated in human beings in a single situation (SEDA-11, 296); until further studies clarify the risks of disease transmission with different types of injectors, their use should be restricted to special circumstances in which large numbers of persons need to be immunized within a short period of time.

Jet-gun-associated infections
The first outbreak of a disease in which a jet injector was implicated as the vehicle of transmission has been reported. Thirty-one attendees at a weight-reduction

clinic in Southern California experienced hepatitis B after daily parenteral injections of human chorionic gonadotrophin given by jet injectors; transmission appeared to have resulted from the multiple repeated jet injections (138). WHO and UNICEF have stated in their "Guidelines for selecting injection equipment for the Expanded Program on Immunization" that the use of jet injectors should be restricted to circumstances in which reusable or disposable equipment is not feasible because of the large number of persons to be immunized within a short period of time (136).

A new type of needleless jet-injector (Mini-Imojet) administers liquid vaccines from a single-use, prefilled cartridge ("imule"), thereby avoiding the risk of cross-contamination. Administration of various vaccines by jet-injector has been compared with standard syringe technique. All the jet-administered vaccines were of equivalent or superior immunogenicity. The most common reactions were mild (minor bleeding, superficial papules, erythema, induration). The technical and safety advantages of the Mini-Imojet reinforce the potential use of this technique for mass immunization (139).

Mixed bacterial vaccines

Starting in the 1920s, very many different mixed bacterial vaccine products (including inactivated bacteria such as *Staphylococcus aureus*, *Streptococcus* species, *Streptococcus pneumoniae*, *Moraxella catarrhalis*, *Klebsiella pneumoniae*, *H. influenzae*) were marketed worldwide. Currently, there are still several products available in European countries, and one product in the USA. Most vaccines have been used for treatment of recurrent and chronic infections of the respiratory tract. The efficacy of these products is doubtful. Delayed hypersensitivity to bacterial products is common. Delayed reactions, sometimes associated with vague malaise or myalgia, can occur after the administration of maintenance doses for months. If delayed skin reactions are accompanied by any systemic symptoms, administration of the mixed vaccine should be drastically reduced or stopped (140).

Immunization and autoimmune disease

Autoimmunity is characterized by the development of one or several immune responses, directed against antigenic components of the host. The detection of antibodies against host antigens (autoantibodies) or autoreactive T cells does not indicate current or future disease, and autoimmunity does not always result in autoimmune disease. An autoimmune disease results from autoimmunity when autoantibodies or autoreactive T cells reach the corresponding antigen in a target organ and become pathogenic, or when autoantibodies form pathogenic immune complexes with antigens released from host cells. An autoimmune disease can be clinically silent for months or years before the destruction of the tissues involved leads to clinical symptoms, for example the appearance of type 1 diabetes following the autoimmune destruction of pancreatic islets.

As many as 5% of individuals in Europe and North America have some form of autoimmune disease. There is clear evidence that genetic predisposition is necessary for the development of some autoimmune diseases. In genetically predisposed individuals an infection can induce or trigger an autoimmune disease. Of the many potential environmental factors, infections are the most likely cause. The causal linkage between some infections and autoimmune diseases has raised the question as to whether autoimmune diseases might also be triggered by vaccines (141).

Autoimmune diseases include autoimmune thrombocytopenia, Graves' disease, hemolytic anemia, Hashimoto's thyroiditis, insulin-dependent diabetes mellitus (diabetes type 1), multiple sclerosis, myasthenia gravis, rheumatoid arthritis, systemic lupus erythematosus, dermatomyositis, and Reiter's syndrome (141–143).

The number of publications that include claims and counterclaims about the risk of autoimmune diseases after immunization is increasing. However, only in rare instances has evidence established a causal relation or favored acceptance of a causal relation:

(a) Rabies vaccine produced in rabbit nervous tissue caused acute disseminated encephalomyelitis in some immunized individuals (144).
(b) During the US swine flu immunization campaign (1976–77) immunized individuals had a much higher risk of Guillain–Barré syndrome than non-immunized individuals (relative risk 7.60); currently, the risk of Guillain–Barré syndrome after immunization (one additional case per million people immunized) is estimated to be substantially lower than the risk of severe influenza and its complications (145).
(c) Idiopathic thrombocytopenia can occur after MMR immunization (about one case in 30 000 immunized children); however, the risks of thrombocytopenia after natural measles or rubella are respectively five and 10 times higher (141,146).

For other potential association between vaccines and autoimmune diseases, the evidence favors rejection of a causal relation:

(a) hepatitis B vaccine and multiple sclerosis (SEDA-24, 374) (SEDA-25, 386);
(b) vaccines and diabetes mellitus (147,148);
(c) Lyme disease vaccine and a treatment-resistant form of autoimmune arthritis; however, owing to public misperception and the promotion of false concerns about its safety, the demand for Lyme disease vaccine did not reach a sustainable level and the manufacturer withdrew it (SEDA-24, 366) (SEDA-26, 357).

How to assess a potential link between a vaccine and an autoimmune condition

Careful epidemiological studies should be the basis of conclusions about an association between a specific vaccine and a particular autoimmune disease. Wraith and colleagues took into account WHO recommendations for the assessment of adverse events after immunization and established the following four basic principles that apply to autoimmune diseases:

1. consistency of findings in various studies;
2. strength (in an epidemiological sense);
3. specificity;
4. temporal relation (148).

Death

Sudden infant death syndrome and multiple immunizations

Current recommendations call for infants and very young children to receive multiple doses of vaccines during their first year of life, and since SIDS is the most frequent cause of death in highly developed countries during that early period of life, it is important to take into account concerns that multiple immunization might play a role in SIDS. SIDS is the diagnosis most commonly used to explain unexpected sudden death of uncertain cause occurring mainly in infancy at 2–7 months of life, sometimes later. By definition, the cause or causes of SIDS are unknown, but risk factors (for example maternal characteristics, prenatal factors, and postnatal conditions, including the infant's sleeping position) have been identified. The results of the most comprehensive analysis "Vaccinations and sudden unexpected death in infancy" have been provided by the Immunization Safety Review Committee of the IOM (149). The committee reviewed an extensive collection of material, primarily from published peer-reviewed scientific and medical literature. The causality conclusions of the committee were as follows:

(a) The evidence favors rejection of a causal relation between exposure to multiple vaccines and SIDS.
(b) The evidence favors acceptance of a causal relation between diphtheria toxoid and whole cell pertussis vaccine and death due to anaphylaxis in infants; however, anaphylaxis of infants following vaccination is very rare, and a fatal outcome is extremely rare.
(c) The committee has not recommended a policy review of the recommended (US) childhood immunization schedule on the basis of concerns about SIDS.

The committee also noted that there are studies that show that vaccinated infants are at a reduced risk of SIDS (149,150).

Susceptibility Factors

It is well known that immune deficiency can play a distinct role in the development of adverse effects after immunization, particularly in connection with the administration of live virus vaccines. The first such experience was with smallpox vaccine; people with immune defects had a markedly increased risk of complications, for example generalized vaccinia. Disseminated BCG infection is usually associated with severe abnormalities of cellular immunity. The risk of vaccine-associated poliomyelitis is increased in immunodeficient children. There have been several reports of adverse events or death after measles immunization in immunodeficient children. Based on

these experiences, there is a general consensus that live vaccines should not be given to people with immune deficiency diseases or to those whose immune response is suppressed because of leukemia, lymphoma, generalized malignancy, or therapy with glucocorticoids, alkylating agents, antimetabolites, or radiation (SEDA-12, 268).

Various studies in both symptomatic and asymptomatic HIV-infected individuals have failed to show any special sensitivity to adverse effects after other immunizations, for example in children receiving live oral or inactivated polio vaccine, DPT or DT vaccine, or measles vaccine (151–154).

Recommendations by the US ACIP, WHO's Global Programme on Immunization, and the European Advisory Group on Expanded Program on Immunization have all been published in detail (SEDA-12, 270) (SEDA-13, 273). Based on the recommendations elaborated by WHO and ACIP, similar recommendations have been prepared in other countries. WHO recommends that non-immunized individuals with symptomatic HIV infection should not receive BCG, but should receive the other vaccines; whereas ACIP recommends that children with symptomatic HIV infection should not receive live virus or live bacterial vaccines. Taking into account the hazard of measles in children with AIDS and HIV-infection, ACIP additionally recommends measles vaccine for all (both symptomatic and asymptomatic) HIV-infected children.

Drug Administration

Drug additives

Aluminium

A vaccine adjuvant is defined as an agent that increases specific immune responses to an antigen (155). The only vaccine adjuvants currently licensed by the FDA are aluminum salts. All other adjuvants are considered experimental and must undergo special preclinical testing. Real and theoretical risks of vaccine adjuvants comprise various local acute or chronic inflammation with formation of abscesses and nodules; induction of hypersensitivity to the host's own tissues, producing autoimmune arthritis, amyloidosis, anterior uveitis; cross-reactions with human antigens, such as glomerular basement membranes or neurolemma, causing glomerulonephritis or meningoencephalomyelitis; sensitization to tuberculin or to other skin test antigens; carcinogenesis; pyogenesis; teratogenesis; abortion; and adverse pharmacological effects, such as hypoglycemia.

Macrophagic myofasciitis is an uncommon inflammatory disorder of muscle, believed to be due to persistence of vaccine-derived aluminium hydroxide at the site of injection. The condition is characterized by diffuse myalgia, arthralgia, and fatigue. In one case of histologically confirmed macrophagic myofasciitis left chest and upper limb pain developed more than 10 years after immunization (156). Treatment with steroids led to symptomatic improvement.

A meta-analysis has compared the reactogenicity of vaccines containing aluminium hydroxide versus vaccines without adjuvants in children aged up to 18 months, and vaccines containing different types of aluminium versus vaccines without adjuvants in children aged 10?16 years (157). In young children, vaccines containing adjuvants caused significantly more erythema and induration than plain vaccines (OR = 1.87; 95% CI = 1.57, 2.24) and significantly fewer reactions of all types (OR = 0.21; 0.15, 0.28). In older children, there was no association between exposure to aluminium-containing vaccines and the onset of local reactions or a raised temperature, but there was an association with local pain lasting up to 14 days (OR = 2.05; CI = 1.25, 3.38). The authors found no evidence that aluminium salts in vaccines cause any serious or long-lasting adverse effects.

During trials in Gothenburg, Sweden, of aluminium-adsorbed diphtheria–tetanus/acellular pertussis vaccines from a single producer, persistent itching nodules at the immunization site were observed in an unexpectedly high frequency: in 645 children out of about 76 000 immunized (0.8%) after both subcutaneous and intramuscular injection.. The itching was intense and long-lasting. After a median of 4 years 75% still had symptoms. There was contact hypersensitivity to aluminium in 77% of the children with itching nodules and in 8% of their symptomless siblings who had received the same vaccines (158). The authors suspected that the high incidence of itching nodules was related to the injection technique used. Post-marketing surveillance data from other regions in Sweden, Denmark, and Norway have suggested that the incidence of itching nodules is low after correct intramuscular administration of aluminium-adsorbed vaccines manufactured by Statens Seruminstitut in Copenhagen, Denmark (159).

The effect of reducing the aluminium content of a combined reduced-antigen-content Tdap vaccine on immunogenicity and safety has been evaluated in 647 healthy adolescents aged 10–18 years (160). Of those enrolled, 224 (35%) received a Tdap formulation with aluminium 0.5 mg, 209 (32%) a formulation with aluminium 0.3 mg, and 214 (33%) a formulation with aluminium 0.133 mg. One month after administration of the booster dose, all the subjects were seroprotected against diphtheria and tetanus toxoids. All were seropositive for anti-filamentous hemagglutinin and anti-pertactin antibodies, but 4% of those who were initially seronegative in both reduced aluminium groups did not seroconvert for anti-pertussis toxin. Booster responses did not differ significantly between the groups for any antibody, but geometric mean concentrations of anti-pertussis toxin after booster immunization differed significantly between groups and fell when vaccine aluminium content was reduced. There were no clear differences between the study groups in local or general adverse effects. The most frequently reported symptoms after immunization were injection site pain (90–91%), fatigue (42–47%) and headache (41–45%). This study showed that the aluminium content has a specific influence on the immunogenicity of this Tdap vaccine.

Thiomersal

Thiomersal (see the monograph on Mercury and mercurial salts) has been used as an additive to biologics and vaccinessince the 1930s because it is very effective in killing bacteria used in several vaccines and in preventing bacterial contamination, particularly in opened multidose containers. Some but not all vaccines contain thiomersal. Billions of children and adults have been immunized worldwide and there is no scientific data that thiomersal-containing products have caused problems of toxicity in humans. The main causes of concern with thiomersal are allergic reactions and the potential risk of neurotoxicity. The WHO has recommended a Permissible Total Weekly Intake (PTWI) of 200 µg for non-pregnant adults. Owing to the vulnerability of the developing brain, the PTWI for pregnant women and infants should be lower, but there is currently no international recommendation for maximum intake in infants. Considering the thiomersal content of some vaccines, intake of mercury in infants can exceed that which could be considered safe. The US Public Health Service, the American Academy of Pediatrics, and US vaccine manufacturers are in agreement that thiomersal should be eliminated from vaccine manufacture as soon as possible. Similar conclusions were reached in 1999 in a meeting attended by the European vaccine manufacturers, European regulatory agencies, and the FDA. Already, many new vaccines are thiomersal-free and the manufacturers are committed to the target of reducing as much as possible the thiomersal content in all vaccines. However, given that the known risks of not immunizing children far outweigh the unknown and much smaller risk of thiomersal, clinicians and parents are encouraged to continue immunizing all infants with the currently available vaccines despite their thiomersal content (161).

Questions and answers on thiomersal have been provided by the WHO, which has underlined the fact that the risk of adverse effects of thiomersal is theoretical, uncertain, and at most extremely small (162). However, the WHO aims to replace thiomersal with other preservatives in the long term. Combination vaccines can reduce the amount of mercury to an absolute minimum.

In 2001 the Immunization Safety Review Committee of the Institute of Medicine issued a report on thiomersal-containing vaccines and neurodevelopmental disorders (163). "The committee concluded that although the hypothesis that exposure to thiomersal-containing vaccines could be associated with neurodevelopmental disorders is not established and rests on indirect and incomplete information, primarily from analogies with methylmercury and levels of maximum mercury exposure from vaccines given in children, the hypothesis is biologically plausible. · · · [However,] the evidence is inadequate to accept or reject a causal relationship between thiomersal exposure from childhood vaccines and the neurodevelopmental disorders of autism, attention deficit/hyperactivity disorder, and speech or language delay." Hypersensitivity to thiomersal is a contraindication to immunization (164).

Gelatine

Since approval of live varicella vaccine (Oka strain) in 1986 in Japan, the effectiveness and safety of the vaccine has been investigated. From 1994, infants have been given acellular pertussis vaccine combined with diphtheria and tetanus toxoid (DTaP) until 12 months of age, before the administration of live vaccines, such as measles, rubella, mumps, and varicella vaccines. Increasing numbers of anaphylactic/allergic reactions to those vaccines have since been reported. Almost all of these subjects had previously been given three or four doses of DTaP containing gelatine. Gelatine-associated allergic reactions were also reported with varicella vaccine, and gelatine-free live varicella vaccine was introduced in 1999. Removal of gelatine from the live vaccine resulted in a dramatic reduction in anaphylactic/allergic reactions to this vaccine. The reported rates of anaphylactic/allergic reactions to gelatine-containing and gelatine-free varicella vaccines and titers of IgE antibodies to gelatine in those who developed anaphylactic/allergic reactions have been compared (165). After the use of gelatin-containing varicella vaccine (1994–9, 1 410 000 distributed doses), 28 serious anaphylactic reactions and 139 non-serious allergic reactions were reported; in contrast, there were no serious and only five non-serious reactions after the use of gelatin-free vaccine (1999–2000, 1 300 000 distributed doses). All nine sera available from children with serious reactions tested positive for gelatin-specific IgE, whereas 55 of the 70 available from those with non-serious reactions were positive, with one false positive. There was no correlation between gelatin-specific IgE antibody titers and the severity of allergic reaction. Anti-varicella antibody titers after immunization were comparable.

Squalene

The Global Advisory Committee on Vaccine Safety (GACVS) has discussed the safety of squalene (166). Squalene is commercially extracted from fish oil, in particular shark-liver oil, and is purified when used in pharmaceutical products and vaccines. Squalene alone is not an adjuvant, but emulsions of squalene with surfactants enhance the immune response when added to antigens. MF59, a proprietary adjuvant containing squalene, is included in some seasonal subunit influenza vaccines. The vaccine contains about 10 mg of squalene per dose. Since 1999, many million doses have been distributed since that time. This vaccine has been administered primarily to individuals aged 65 years and older, for whom the vaccine was licensed. Reported rates of adverse events and local reactogenicity are not in excess of those that would be expected with other inactivated seasonal flu vaccines, suggesting that squalene in this vaccine poses no significant risk. Several experimental vaccines, including some pandemic flu vaccines, malaria vaccines, and various viral and bacterial vaccines, are also being developed with squalene-containing adjuvants, with the intention of enhancing immunogenicity and thereby efficacy. Clinical studies of squalene-containing vaccines have been performed in infants and neonates without evidence of safety concerns. A link between the health problems of Gulf-War veterans and the possible presence of squalene in vaccines received by these soldiers has not been confirmed.

The Committee concurred that fears about squalene are unfounded. It did note, however, that the experience of squalene-containing vaccines has been primarily in older age groups and recommended that as squalene-containing vaccines are introduced in other age groups, careful post-marketing follow up to detect any vaccine-related adverse events needs to be performed.

Thiomersal

A WHO consultation on thiomersal (thimerosal) in vaccines from the regulatory perspective was held in 2002 (167). The main conclusions were as follows:

- recommendations for the removal of thiomersal developed by health authorities are mainly driven by public perceptions of risk and not by any scientific evidence of toxicity;
- limits for chronic exposure to methyl mercury derivatives from food should not be used to set limits for acute exposure to ethyl mercury derivatives (for example thiomersal) that can occur through immunization;
- making changes to the thiomersal content of vaccines already licensed to include thiomersal is a complex issue that requires careful consideration; any change in a formulation could have a serious impact on the quality, safety, and efficacy of vaccines and should be considered on a case-by-case basis; generally, products whose formulation changes are considered as new products and may require clinical trials.

In 2003, the Global Advisory Committee on Vaccine Safety (GACVS) noted that there is insufficient evidence to reach definite conclusions regarding the safety of thiomersal-containing vaccines in groups that may be at special risk, notably malnourished infants and preterm or low-birth-weight neonates. The GACVS reported to WHO that there is no scientific basis for changing current WHO recommendations for thiomersal-containing vaccines, including administration of a birth dose of hepatitis B vaccine and immunization of low-birth-weight infants when indicated (168).

In a review of the safety of thiomersal in vaccines Clements stated that generally thiomersal has been convincingly shown to be safe (169). However, the scientific evidence is not sufficiently strong to provide the same level of assurance for thiomersal-containing vaccines for use in pregnant women or premature or low-birth-weight infants. The fetal brain is more sensitive to mercury, whether ethyl mercury or methyl mercury, and it is at least possible that premature infants of very low birth weights may be at increased risk from thiomersal-containing vaccines. Until scientific evidence is available, thiomersal-free products of hepatitis B vaccine are to be preferred for the dose that is given at birth.

Yeast

The preparation of recombinant hepatitis B vaccines involves using cellular cultures of *Saccharomyces cerevisiae*, otherwise known as baker's yeast. Before vaccine licensure, clinical trials were performed to address whether residual yeast proteins in the vaccines could induce *anaphylaxis*, including testing for IgE anti-yeast antibody titers. There were anti-yeast IgE antibodies in 1–2% of subjects before immunization, but there was no significant rise in IgE after hepatitis B immunization. The authors of a study searched reports in the Vaccine Adverse Event Reporting System (VAERS) for those that mentioned a history of allergy to yeast and then reviewed the adverse events described in these reports for potential anaphylactic reactions (170). Probable anaphylaxis was defined as the presence of one or more skin symptoms and one or more respiratory, gastrointestinal, or cardiovascular symptoms with onset within 4 hours of hepatitis B immunization. Possible anaphylaxis was defined in one of two ways: (1) cases that described skin or respiratory symptoms (but not both) occurring within 4 hours of immunization; or (2) cases that described one or more skin and/or respiratory symptoms occurring 4–12 hours after immunization. Among the 107 reports of pre-existing "yeast allergies", 11 reports described probable or possible anaphylaxis after hepatitis B immunization. Four additional cases were described after other vaccines. Most of the vaccinees who met the case definitions and had a history of yeast allergies were female, aged 10–64, and symptom onset ranged from 15 minutes to 5 hours after immunization. No deaths were reported. The small number of reports to VAERS may be partly because health-care professionals observe current contraindications by not immunizing yeast-sensitive individuals. Nevertheless, yeast-associated anaphylaxis after hepatitis B immunization in sensitized patients appears to be rare.

Drug contamination

Since the problem of bovine spongiform encephalitis (BSE) emerged in Europe around 1987, the WHO and ministries of health have been alert to the danger that BSE-contaminated materials might be used in biological products, including vaccines. Guidelines were created that ensured that when bovine materials were needed in a biological product, the materials were obtained from countries that were free from BSE. Bovine products are used in the production of certain vaccines, in creating seed lots, or in amplifying the seed lot into working lots. The material is generally calf serum, which, even if it came from an infected animal, is extremely unlikely to transmit the prions that cause the disease. An exceptional situation arose on 20 October 2000, when the UK withdrew one manufacturer's OPV because of the possibility that the vaccine had been manufactured using serum from calves that were from a BSE-infected country. Even though the risk was extremely remote, the health authorities decided to have the vaccine withdrawn. This particular product was not the main OPV used in the UK. To clarify the situation in the UK and to reassure countries throughout the world that there were no implications for OPV used in the eradication of poliomyelitis, the WHO issued the following position statement (171): "On 20 October 2000, Evans/Medeva Oral Polio Vaccine

(OPV), which to the knowledge of the World Health Organization has only been used in the United Kingdom and Republic of Ireland, was recalled in the UK. This vaccine has never been used in the immunization campaigns that are ongoing as part of the Global Polio Eradication Initiative. The recall was prompted by evidence that the Evans/Medeva vaccine was manufactured in contravention of European Union guidelines, and was not based on any adverse events related to the vaccine. The specific concern is that fetal bovine calf serum (FCS) from the UK was used in the manufacture of the Evans/Medeva vaccine, at a time when there was a risk of bovine spongiform encephalopathy (BSE) in that country. WHO endorses the recall step as a precaution, even though FCS is removed in the manufacturing process of the vaccine and there is no evidence that FCS could transmit BSE. Due to this breach of EU guidelines and acting on a precautionary basis, the Health Departments in the UK have withdrawn remaining shelf stocks of the Evans/Medeva brand of polio vaccine, which had already ceased production. The World Health Organization imposes strict production and quality controls to ensure the safety and efficacy of OPV. WHO recommends that manufacturers use the safest source of materials from countries which have not reported indigenous BSE cases and have a compulsory BSE notification system, compulsory clinical and laboratory verification of suspected cases and a surveillance program. The Global Polio Eradication Initiative only distributes vaccine supplied by vaccine manufacturers who follow National Control Authorities' criteria provided by WHO."

Two advisory committees of the US FDA, the Transmissible Spongiform Encephalopathies Advisory Committee and the Vaccines and Related Biologicals Product Advisory Committee, said at a joint meeting on 3 August 2000 that vaccines made from bovine-derived materials from countries with a known or uncertain risk of BSE carry only an infinitesimal risk of new variant Creutzfeldt–Jakob disease, and that no change in US immunization practice is indicated (Evans G, personal communication, 3 August 2000).

References

1. Kane M. Unsafe injections. Bull World Health Organ 1998;76(1):99–100.
2. Chen R. Vaccine risks: real, perceived, and unknown. 4th European Conference on Vaccinology, March 17–19, 1999. Brighton, United Kingdom 1999;.
3. Clements CJ, Evans G, Dittman S, Reeler AV. Vaccine safety concerns everyone. Vaccine 1999;17(Suppl 3):S90–4.
4. Chen RT, DeStefano F. Vaccine adverse events: causal or coincidental? Lancet 1998;351(9103):611–2.
5. Wakefield AJ, Murch SH, Anthony A, Linnell J, Casson DM, Malik M, Berelowitz M, Dhillon AP, Thomson MA, Harvey P, Valentine A, Davies SE, Walker-Smith JA. Ileal-lymphoid-nodular hyperplasia, non-specific colitis, and pervasive developmental disorder in children. Lancet 1998;351(9103):637–41.
6. Jefferson T. Vaccination and its adverse effects: real or perceived. Society should think about means of linking

exposure to potential long term effect. BMJ 1998;317(7152):159–60.

7. Varricchio F. The vaccine adverse event reporting system. J Toxicol Clin Toxicol 1998;36(7):765–8.

8. Niu MT, Rhodes P, Salive M, Lively T, Davis DM, Black S, Shinefield H, Chen RT, Ellenberg SS, Braun M, Donlon J, Krueger C, Rastogi S, Varricchio F, Wise R, Haber P, Lloyd J, Terracciano G, Eltermann D, Gordon S, DeStefano F, Glasser J, Handler S, Kimsey D Jr, Swint E Jr, Fireman B Jr, Hiatt R Jr, Lewis N Jr, Lieu T Jrthe VAERS and VSD Working Group T Jr. Comparative safety of two recombinant hepatitis B vaccines in children: data from the Vaccine Adverse Event Reporting System (VAERS) and Vaccine Safety Datalink (VSD). J Clin Epidemiol 1998;51(6):503–10.

9. Surveillence for safety after immunization: Vaccine Adverse Event Reporting System (VAERS)-United States, 1991–2001. MMWR Morb Mortal Wkly Rep 2003;52(SS-1):1.

10. The Brighton Collaboration. http://brightoncollaboration.org/internet/en/index/html.

11. The term AEFI — a debate. VACSAF-L@LIST.NIH.GOV (accessed July 2001).

12. Bonhoeffer J, Heininger U. Standardized cases definitions of adverse events following immunization. Vaccine 2004;22:547–50.

13. Kohl KS, Bonhoeffer J, Chen R, Duclos P, Heijbel H, Heininger U, Loupi E. The Brighton Collaboration: enhancing comparability of vaccine safety data. Pharmacoepidemiol Drug Saf 2003;12:335–40.

14. Marcy SM, Kohl KS, Dagan R, Nalin D, Blum M, Jones MC, Hansen J, Labadie J, Lee L, Martin BL, O'Brien K, Rothstein E, Vermeer P, The Brighton Collaboration Fever Working Group. Fever as an adverse event following immunization: case definition and guidelines for data collection, analysis, and presentation. Vaccine 2004;22:551–6. Available online at www.sciencedirect.com.

15. Bonhoeffer J, Menkes J, Gold SM, de Souza-Brito G, Fisher MC, Halsey N, Vermeer P, The Brighton Collaboration Seizure Working Group. Generalized seizure as an adverse event following immunization: case definition and guidelines for data collection, analysis, and presentation. Vaccine 2004;22:557–62.

16. Bonhoeffer J, Gold SM, Heijbel H, Vermeer P, Blumberg D, Braun M, De Souza-Brito G, Davis RL, Halperin S, Heininger U, Khuri-Bulos N, Menkes J, Nokleby H, The Brighton Collaboration HHE Working Group. Hypotonic-hyporesponsive episode (HHE) as an adverse event following immunization: case definition and guidelines for data collection, analysis, and presentation. Vaccine 2004;22:563–568.

17. Bines JE, Kohl KS, Gorster J, Zanardi LR, Davis RL, Hansen J, Murphy TM, Music S, Niu M, Varricchio F, Vermeer P, Wong EJC, The Brighton Collaboration Intussusception Working Group. Acute intussusception in infants and children as an adverse event following immunization: case definition and guidelines for data collection, analysis, and presentation. Vaccine 2004;22:569–74.

18. Rothstein E, Kohl KS, Ball L, Halperin SA, Halsey N, Hammer SJ, Heath PT, Hennig R, Kleppinger C, Labadie J, Varricchio F, Vermeer P, Walop W, The Brighton Collaboration Local Reaction Working Group. Nodule at injection site as an adverse event following immunization: case definition and guidelines for data collection, analysis, and presentation. Vaccine 2004; 22:575–85.

19. Bonhöffer J, Vermeer P, Halperin S, Kempe A, Music S, Shindman J, Walop W, The Brighton Collaboration Persistent Crying Working Group. Persistent crying in infants and children as an adverse event following immunization: case definition and guidelines for data collection, analysis, and presentation. Vaccine 2004;22:586–91.

20. Rüggeberg JU, Gold MS, Bayas JM, Blum MD, Bonhoeffer J, Friedlander S, de Souza Brito G, Heininger U, Imoukhuede B, Khamesipour A, Erlewyn-Lajeunesse M, Martin S, Mäkelä M, Nell P, Pool V, Simpson N; The Brighton Collaboration Anaphylaxis Working Group. Anaphylaxis: case definition and guidelines for data collection, analysis, and presentation of immunization safety data. Vaccine 2007;25:5675–84.

21. Tapiainen T, Prevots R, Izurieta HS, Abramson J, Bilynsky R, Bonhoeffer J, Bonnet MC, Center K, Galama J, Gillard P, Griot M, Hartmann K, Heininger U, Hudson M, Koller A, Khetsuriani N, Khuri-Bulos N, Marcy SM, Matulionyte R, Schöndorf I, Sejvar J, Steele R; The Brighton Collaboration Aseptic Meningitis Working Group. Aseptic meningitis: case definition and guidelines for collection, analysis and presentation of immunization safety data. Vaccine 2007;25:5793–802.

22. Sejvar JJ, Kohl KS, Bilynsky R, Blumberg D, Cvetkovich T, Galama J, Gidudu J, Katikaneni L, Khuri-Bulos N, Oleske J, Tapiainen T, Wiznitzer M; The Brighton Collaboration Encephalitis Working Group. Encephalitis, myelitis, and acute disseminated encephalomyelitis (ADEM):case definitions and guidelines for collection, analysis, and presentation of immunization safety data. Vaccine 2007;25:5771–92.

23. Jones JF, Kohl KS, Ahmadipour N, Bleijenberg G, Buchwald D, Evengard B, Jason LA, Klimas NG, Lloyd A, McCleary K, Oleske JM, White PD; The Brighton Collaboration Fatigue Working Group. Fatigue: case definition and guidelines for collection, analysis, and presentation of immunization safety data. Vaccine 2007;25:5685–96.

24. Kohl KS, Ball L, Gidudu J, Hammer SJ, Halperin S, Heath P, Hennig R, Labadie J, Rothstein E, Schuind A, Varricchio F, Walop W; The Brighton Collaboration Local Reactions Working Group for Abscess at Injection Site. Abscess at injection site: case definition and guidelines for collection, analysis, and presentation of immunization safety data. Vaccine 2007;25:5821–38.

25. Halperin S, Kohl KS, Gidudu J, Ball L, Hammer SJ, Heath P, Hennig R, Labadie J, Rothstein E, Schuind A, Varricchio F, Walop W; The Brighton Collaboration Local Reaction Working Group for Cellulitis at Injection Site. Cellulitis at injection site: case definition and guidelines for collection, analysis, and presentation of immunization safety data. Vaccine 2007;25:5803–20.

26. Kohl KS, Walop W, Gidudu J, Ball L, Halperin S, Hammer SJ, Heath P, Hennig R, Rothstein E, Schuind A, Varricchio F; The Brighton Collaboration Local Reactions Working Group for Induration at or near Injection Site. Induration at or near injection site: case definition and guidelines for collection, analysis, and presentation of immunization safety data. Vaccine 2007;25:5839–57.

27. Kohl KS, Walop W, Gidudu J, Ball L, Halperin S, Hammer SJ, Heath P, Varricchio F, Rothstein E, Schuind A, Hennig R; The Brighton Collaboration Local Reaction Working Group for Swelling at or near Injection Site. Swelling at or near injection site: case definition and guidelines for collection, analysis and presentation of immunization safety data. Vaccine 2007;25:5858–74.

28. Beigel J, Kohl KS, Khuri-Bulos N, Bravo L, Nell P, Marcy SM, Warschaw K, Ong-Lim A, Poerschke G, Weston W, Lindstrom JA, Stoltman G, Maurer T; The Brighton Collaboration Rash Working Group. Rash including mucosal involvement: case definition and guidelines for collection, analysis, and presentation of immunization safety data. Vaccine 2007;25:5697–706.

29. Jorch G, Tapiainen T, Bonhoeffer J, Fischer TK, Heininger U, Hoet B, Kohl KS, Lewis EM, Meyer C, Nelson T, Sandbu S, Schlaud M, Schwartz A, Varricchio F, Wise RP; The Brighton Collaboration Unexplained Sudden Death Working Group. Unexplained sudden death, including sudden infant death syndrome (SIDS), in the first and second years of life: case definition and guidelines for collection, analysis, and presentation of immunization safety data. Vaccine 2007;25:5707–16.

30. Wise RP, Bonhoeffer J, Beeler J, Donato H, Downie P, Matthews D, Pool V, Riise-Bergsaker M, Tapiainen T, Varricchio F; The Brighton Collaboration Thrombocytopenia Working Group. Thrombocytopenia: case definition and guidelines for collection, analysis, and presentation of immunization safety data. Vaccine 2007;25:5717–24.

31. Nell P, Kohl KS, Graham PL, Larussa PS, Marcy SM, Fulginiti VA, Martin B, Trolin I, Norton SA, Neff JM; The Brighton Collaboration Vaccinia Virus Vaccine Adverse Event Working Group for Eczema Vaccinatum. Eczema vaccinatum as an adverse event following exposure to vaccinia virus: case definition & guidelines of data collection, analysis, and presentation of immunization safety data. Vaccine 2007;25:5725–34.

32. Beigel J, Kohl KS, Brinley F, Graham PL, Khuri-Bulos N, Larussa PS, Nell P, Norton S, Stoltman G, Tebaa A, Warschaw K; The Brighton Collaboration Vaccinia Virus Vaccine Adverse Event Working Group for Generalized Vaccinia. Generalized vaccinia as an adverse event following exposure to vaccinia virus: case definition and guidelines for data collection, analysis, and presentation of immunization safety data. Vaccine 2007;25:5745–53.

33. Nell P, Kohl KS, Graham PL, Larussa PS, Marcy SM, Fulginiti VA, Martin B, McMahon A, Norton SA, Trolin I; The Brighton Collaboration Vaccinia Virus Vaccine Adverse Event Working Group for Progressive Vaccinia. Progressive vaccinia as an adverse event following exposure to vaccinia virus: case definition and guidelines of data collection, analysis, and presentation of immunization safety data. Vaccine 2007;25:5735–44.

34. Wenger P, Oleske JM, Kohl KS, Fisher MC, Brien JH, Graham PL, Larussa PS, Lipton S, Tierney B; The Brighton Collaboration Vaccinia Virus Adverse Event Working Group for Inadvertent Inoculation. Inadvertent inoculation as an adverse event following exposure to vaccinia virus: case definition and guidelines for data collection, analysis, and presentation of immunization safety data. Vaccine 2007;25:5754–62.

35. Graham PL, Larussa PS, Kohl KS; The Brighton Collaboration Vaccinia Virus Adverse Event Working Group for Robust Take. Robust take following exposure to vaccinia virus: case definition and guidelines of data collection, analysis, and presentation of immunization safety data. Vaccine 2007;25:5763–70.

36. Verstraeten T, DeStefano F, Chen RT, Miller E. Vaccine safety surveillance using large linked databases: opportunities, hazards, and proposed guidelines. Expert Rev Vaccines 2003;2:21–9.

37. Health Council of the Netherlands. Adverse reactions to vaccines used in the national vaccination programme in 1989. 1999 Report, Gezondheidsraad, The Hague.

38. Adverse effects of pertussis and rubella vaccines. In: Howson CP, Howe CJ, Fineberg HV, editors. A report of the Committee to Review the Adverse Consequences of Pertussis and Rubella Vaccines. Washington, DC: National Academy Press, 1991:681–8.

39. In: Stratton KR, Howe CJ, Johnson Jr. RB Jr, editors. Adverse effects associated with childhood vaccines. Washington, DC: National Academy of Sciences, 1994:681–8.

40. Centers for Disease Control (CDC). Vaccine Adverse Event Reporting System—United States. MMWR Morb Mortal Wkly Rep 1990;39(41):730–3.

41. Rosenthal S, Chen R. The reporting sensitivities of two passive surveillance systems for vaccine adverse events. Am J Public Health 1995;85(12):1706–9.

42. Chen RT, Rastogi SC, Mullen JR, Hayes SW, Cochi SL, Donlon JA, Wassilak SG. The Vaccine Adverse Event Reporting System (VAERS). Vaccine 1994;12(6):542–50.

43. Wassilak SG, Glasser JW, Chen RT, Hadler SC. Utility of large-linked databases in vaccine safety, particularly in distinguishing independent and synergistic effects. The Vaccine Safety Datalink Investigators. Ann NY Acad Sci 1995;754:377–82.

44. Chen RT, Glasser JW, Rhodes PH, Davis RL, Barlow WE, Thompson RS, Mullooly JP, Black SB, Shinefield HR, Vadheim CM, Marcy SM, Ward JI, Wise RP, Wassilak SG, Hadler SC, Swint E, Hardy JR, Payne T, Benson P, Draket J, Drew L, Mendius B, Ray P, Lewis N, Fireman BH, Jing J, Wulfsohn M, Lugg MM, Osborne P, Rastogi S, Patriarca P, Caserta V. Vaccine Safety Datalink project: a new tool for improving vaccine safety monitoring in the United States. The Vaccine Safety Datalink Team. Pediatrics 1997;99(6):765–73.

45. Farrington P, Pugh S, Colville A, Flower A, Nash J, Morgan-Capner P, Rush M, Miller E. A new method for active surveillance of adverse events from diphtheria/tetanus/pertussis and measles/mumps/rubella vaccines. Lancet 1995;345(8949):567–9.

46. Anonymous. Causality assessment of adverse events following immunization. Wkly Epidemiol Rec 2001;76(12):85–9.

47. Collet JP, MacDonald N, Cashman N, Pless R, Halperin S, Landry M, Palkonyay L, Duclos P, Mootrey G, Ward B, LeSaux N, Caserta V. Monitoring signals for vaccine safety: the assessment of individual adverse event reports by an expert advisory committee. Advisory Committee on Causality Assessment. Bull World Health Organ 2000;78(2):178–85.

48. Delage G. Rotavirus vaccine withdrawal in the United States: the role of post-marketing surveillance. Can J Infect Dis 2000;11:10–2.

49. National childhood vaccine injury act. Vaccine injury table. <http:www.hrsa.gov/vaccinecompensation/table.htm> (last accessed 26 July 2007).

50. Myers MG, Pineda D. Do vaccines cause that? Galveston, Texas: Immunizations for Public Health, 2008.

51. Open Congress. Vaccine Safety and Public Confidence Assurance Act of 2007. http://www.opencongress.org/bill/110-h1973/show.

52. Govtrack.us. H.R. 2618: Vaccine Safety and Public Confidence Assurance Act of 2009. http://www.govtrack.us/congress/bill.xpd?bill=h111-2618.

53. Mansoor O, Pillans PI. Vaccine adverse events reported in New Zealand 1990–5. NZ Med J 1997;110(1048):270–2.

54. Morales-Olivas FJ, Martinez-Mir I, Ferrer JM, Rubio E, Palop V. Adverse drug reactions in children reported by means of the yellow card in Spain. J Clin Epidemiol 2000;53(10):1076–80.

55. Mehta U, Milstien JB, Duclos P, Folb PI. Developing a national system for dealing with adverse events following immunization. Bull World Health Organ 2000;78(2):170–7.

56. Adverse drug reaction monitoring: new issues. WHO Drug Inf 1997;11:1–4.

57. Announcing funding for clinical immunization safety assessment centers. www.cdc.gov/od/pgo/funding/01112.htm, 27/06/2001.

58. Dellepiane N, Griffiths E, Milstien JB. New challenges in assuring vaccine quality. Bull World Health Organ 2000;78(2):155–62.

59. Black S. Perspectives on the design and analysis of prelicensure trials: bridging the gap to postlicensure studies. Clin Infect Dis 2001;33(Suppl 4):S323–6.

60. Jacobson RM, Adegbenro A, Pankratz VS, Poland GA. Adverse events and vaccination—the lack of power and predictability of infrequent events in pre-licensure study. Vaccine 2001;19(17–19):2428–33.

61. Heijbel H, Jefferson T. Vaccine safety—improving monitoring. Vaccine 2001;19(17–19):2457–60.

62. Chen RT, Pool V, Takahashi H, Weniger BG, Patel B. Combination vaccines: postlicensure safety evaluation. Clin Infect Dis 2001;33(Suppl 4):S327–33.

63. Silvers LE, Ellenberg SS, Wise RP, Varricchio FE, Mootrey GT, Salive ME. The epidemiology of fatalities reported to the vaccine adverse event reporting system 1990–1997. Pharmacoepidemiol Drug Saf 2001;10(4): 279 85.

64. Duclos P, Ward BJ. Measles vaccines: a review of adverse events. Drug Saf 1998;19(6):435–54.

65. Singleton JA, Lloyd JC, Mootrey GT, Salive ME, Chen RT. An overview of the vaccine adverse event reporting system (VAERS) as a surveillance system. VAERS Working Group. Vaccine 1999;17(22):2908–17.

66. Evans G. Personal communication. 1999.

67. Red Book. Report of the Committee on Infectious Diseases. In: American Academy of Pediatrics. 24th ed. 1997:681–8.

68. Dukes MNG, Swartz B. Responsibility for Drug-induced InjuryAmsterdam: Elsevier;. 1988.

69. Galazka AM, Lauer BA, Henderson RH, Keja J. Indications and contraindications for vaccines used in the Expanded Programme on Immunization. Bull World Health Organ 1984;62(3):357–66.

70. Inglesby TV, Henderson DA, Bartlett JG, Ascher MS, Eitzen E, Friedlander AM, Hauer J, McDade J, Osterholm MT, O'Toole T, Parker G, Perl TM, Russell PK, Tonat K. Anthrax as a biological weapon: medical and public health management. Working Group on Civilian Biodefense. JAMA 1999;281(18):1735–45.

71. Anonymous. South Korea: development of anthrax vaccine close to completion. The Korea Herald 14/01/2000;.

72. Snyder JW. The anthrax vaccine: a question of safety. Clin Microbiol Newslett 2001;23:51–4.

73. Pittman PR, Gibbs PH, Cannon TL, Friedlander AM. Anthrax vaccine: short-term safety experience in humans. Vaccine 2001;20(5–6):972–8.

74. Committee on Health Effects Associated with Exposures during the Gulf War. An assessment of the safety of the anthrax vaccine. A letter report. Washington, DC: Institute of Medicine. http://www.nap.edu/html/anthrax_vaccine, 02/10/2027.

75. Advisory Committee on Immunization Practices. Use of anthrax vaccine in the United States. MMWR Recomm Rep 2000;49(RR-15):1–20.

76. Swanson-Biearman B, Krenzelok EP. Delayed life-threatening reaction to anthrax vaccine. J Toxicol Clin Toxicol 2001;39(1):81–4.

77. Arnon SS, Schechter R, Inglesby TV, Henderson DA, Bartlett JG, Ascher MS, Eitzen E, Fine AD, Hauer J, Layton M, Lillibridge S, Osterholm MT, O'Toole T, Parker G, Perl TM, Russell PK, Swerdlow DL, Tonat KWorking Group on Civilian Biodefense. Botulinum toxin as a biological weapon: medical and public health management. JAMA 2001;285(8):1059–70.

78. Drancourt M, Roux V, Dang LV, Tran-Hung L, Castex D, Chenal-Francisque V, Ogata H, Fournier PE, Crubezy E, Raoult D. Genotyping, Orientalis-like *Yersinia pestis*, and plague pandemics. Emerg Infect Dis 2004;10(9):1585–92.

79. Inglesby TV, Dennis DT, Henderson DA, Bartlett JG, Ascher MS, Eitzen E, Fine AD, Friedlander AM, Hauer J, Koerner JF, Layton M, McDade J, Osterholm MT, O'Toole T, Parker G, Perl TM, Russell PK, Schoch-Spana M, Tonat K. Plague as a biological weapon: medical and public health management. Working Group on Civilian Biodefense. JAMA 2000;283(17):2281–90.

80. Henderson DA, Inglesby TV, Bartlett JG, Ascher MS, Eitzen E, Jahrling PB, Hauer J, Layton M, McDade J, Osterholm MT, O'Toole T, Parker G, Perl T, Russell PK, Tonat K. Smallpox as a biological weapon: medical and public health management. Working Group on Civilian Biodefense. JAMA 1999;281(22):2127–37.

81. Centers for Disease Control and Prevention (CDC). Interim smallpox response plan and guidelines (draft 2.0—21 November 2001). http://www.bt.cdc.gov/DocumentsApp/Smallpox?RPG/plan.

82. Reuters Medical News. New smallpox vaccine nears clinical trial. http://primarycare.medscape.com/reuters/prof/2001/09/09.21/20010920drgd002.html.

83. Dennis DT, Inglesby TV, Henderson DA, Bartlett JG, Ascher MS, Eitzen E, Fine AD, Friedlander AM, Hauer J, Layton M, Lillibridge SR, McDade JE, Osterholm MT, O'Toole T, Parker G, Perl TM, Russell PK, Tonat KWorking Group on Civilian Biodefense. Tularemia as a biological weapon: medical and public health management. JAMA 2001;285(21):2763–73.

84. Advisory Committee on Immunization Practices (ACIP). Update: vaccine side effects, adverse reactions, contraindications, and precautions. MMWR Recomm Rep 1996;45(RR-12):1–35.

85. In: Plotkin SA, Orenstein WA, editors. Vaccines. 3rd ed.. Philadelphia: Saunders, 1999:681–8.

86. Offit PA, Quarles J, Gerber MA, Hackett CJ, Marcuse EK, Kollman TR, Gellin BG, Landry S. Addressing parents' concerns: do multiple vaccines overwhelm or weaken the infant's immune system? Pediatrics 2002;109(1):124–9.

87. Halsey NA. Safety of combination vaccines: perception versus reality. Pediatr Infect Dis J 2001;20(Suppl 11): S40–4.

88. Andreae MC, Freed GL, Katz SL. Safety concerns regarding combination vaccines: the experience in Japan. Vaccine 2004;22(29–30):3911–6.

89. Institute of Medicine (IOM) Report. Immunizations Safety Review. Multiple immunizations and immune dysfunction. http://www.cdc.gov/nip//vacsafe/concerns/gen/multiplevac_iom.htm, 28/10/2002.

90. Ottaviani G, Lavezzi AM, Matturri L. Sudden infant death syndrome (SIDS) shortly after hexavalent vaccination: another pathology in suspected SIDS? Virchows Archiv 2006;448:100–4.

91. EMEA Public statement. EMEA reviews hexavalent vaccines: Hexavac and Infanrix Hexa. London, April 28, 2003. htpp://www.emea.eu.int.

92. EMEA Public statement. EMEA update on hexavalent vaccines: Hexavac and Infanrix Hexa. London, December 1, 2003. htpp://www.emea.eu.int.

93. Zinka B, Rauch E, Buettner A, Ruëff F, Penning R. Unexplained cases of sudden infant death shortly after hexavalent vaccination. Vaccine 2006;24(31-32):5779–80.

94. Schmitt HJ, Siegrist CA, Salmaso S, Law B, Booy R. Comment on B. Zinka et al. Unexplained cases of sudden infant death shortly after hexavalent vaccination. Vaccine 2006;24(31-32):5781–2.

95. von Kries R. Comment on B. Zinka et al. Unexplained cases of sudden infant death shortly after hexavalent vaccination. Vaccine 2006;24(31-32):5783–4.

96. EMEA Public statement: Committee for Medicinal Products for Human Use: Extension of indications and recommendations. London, April 21, 2005. htpp://www.e-mea.eu.int.

97. Global Advisory Committee on Vaccine Safety. Safety of hexavalent vaccines. June 9–10, 2005. Weekly Epidemiol Rec 2005;80:245.

98. Vennemann MM, Butterfass-Bahloul T, Jorch G, Brinkmann B, Findeisen M, Sauerland C, Bajanowski T, Mitchell EA; the GeSID Group. Sudden infant death syndrome: no increased risk after immunisation. Vaccine 2007;25(2):336–40.

99. Saenger R, Maechler G, Potreck M, Zepp F, Knuf M, Habermehl P, Schuerman L. Booster vaccination with hexavalent DTPa–HBV–IPV/Hib vaccine in the second year of life is as safe as concomitant DTPa–IPV/Hib +HBV administered separately. Vaccine 2005;23:1135–43.

100. Schulzke S, Heininger U, Lücking-Famira M, Fahnenstich H. Apnea and bradycardia in preterm infants following immunization with penta- and heptavalent vaccines. Eur J Pediatr 2005;164:432–5.

101. Nilsson L, Kjellman NI, Storsaeter J, Gustafsson L, Olin P. Lack of association between pertussis vaccination and symptoms of asthma and allergy. JAMA 1996;275(10):760.

102. Odent MR, Culpin EE, Kimmel T. Pertussis vaccination and asthma: is there a link? JAMA 1994;272(8):592–3.

103. In: Butler NR, Golding J, editors. From birth to 5: a study of the health and behaviour of Britain's 5-year-olds. Oxford: Pergamon Press, 1986:681–8.

104. Kemp T, Pearce N, Fitzharris P, Crane J, Fergusson D, St George I, Wickens K, Beasley R. Is infant immunization a risk factor for childhood asthma or allergy? Epidemiology 1997;8(6):678–80.

105. Vaccines and Guillain–Barré syndrome. Wkly Epidemiol Rec 2008;83:37.

106. Hughes RAC, Cornblath DR. Guillain–Barré syndrome. Lancet 2005;366:1653–66.

107. Schessl J, Luther B, Kirschner J, Mauff G, Korinthenberg R. Infections and vaccinations preceding childhood Guillain–Barré syndrome: a prospective study. Eur J Pediatr 2006;165:605–12.

108. Gervaix A, Caflisch M, Suter S, Haenggeli CA. Guillain–Barré syndrome following immunisation with Haemophilus influenzae type b conjugate vaccine. Eur J Pediatr 1993;152(7):613–4.

109. Shaw FE Jr, Graham DJ, Guess HA, Milstien JB, Johnson JM, Schatz GC, Hadler SC, Kuritsky JN, Hiner EE, Bregman DJ, et al. Postmarketing surveillance for neurologic adverse events reported after hepatitis B vaccination. Experience of the first three years. Am J Epidemiol 1988;127(2):337–52.

110. Sinsawaiwong S, Thampanitchawong P. Guillain–Barré syndrome following recombinant hepatitis B vaccine and literature review. J Med Assoc Thai 2000;83(9):1124–6.

111. Khamaisi M, Shoenfeld Y, Orbach H. Guillain–Barré syndrome following hepatitis B vaccination. Clin Exp Rheumatol 2004;22(6):767–70.

112. Marks JS, Halpin TJ. Guillain–Barré syndrome in recipients of A/New Jersey influenza vaccine. JAMA 1980;243(24):2490–4.

113. Juurlin DN, Stukel TA,, Jeffrey Kwong J. Guillain–Barré syndrome after influenza vaccination in adults. A population-based study. Arch Intern Med 2006;166:2217–21.

114. Souayah N, Nasar A, Suri MF, Qureshi AI. Guillain–Barré syndrome following vaccination in the United States. Vaccine 2007;25:5253–5.

115. Morris K, Rylance G. Guillain–Barré syndrome after measles, mumps, and rubella vaccine. Lancet 1994;343(8888):60.

116. Kinnunen E, Farkkila M, Hovi T, Juntunen J, Weckstrom P. Incidence of Guillain–Barré syndrome during a nationwide oral poliovirus vaccine campaign. Neurology 1989;39(8):1034–6.

117. Kinnunen E, Junttila O, Haukka J, Hovi T. Nationwide oral poliovirus vaccination campaign and the incidence of Guillain–Barré syndrome. Am J Epidemiol 1998;147(1):69–73.

118. Knittel T, Ramadori G, Mayet WJ, Lohr H, Meyer zum Buschenfelde KH. Guillain–Barré syndrome and human diploid cell rabies vaccine. Lancet 1989; 1(8650):1334–5.

119. Courrier A, Stenbach G, Simonnet P, Rumilly P, Lopez D, Coquillat G, Scherer C, Chopin J. Peripheral neuropathy following fetal bovine cell rabies vaccine. Lancet 1986;1(8492):1273.

120. Bakshi R, Graves MC. Guillain–Barré syndrome after combined tetanus–diphtheria toxoid vaccination. J Neurol Sci 1997;147(2):201–2.

121. Dieckhofer K, Scholl R, Wolf R. Neurologische Storungen nach Tetanusschutzimpfung. Ein kasuistischer Beitrag. [Neurologic disorders following tetanus vaccination. A case report.] Med Welt 1978;29(44):1710–2.

122. Hamati-Haddad A, Fenichel GM. Brachial neuritis following routine childhood immunization for diphtheria, tetanus, and pertussis (DTP): report of two cases and review of the literature. Pediatrics 1997;99(4):602–3.

123. Quast U, Herder C, Zwisler O. Vaccination of patients with encephalomyelitis disseminata. Vaccine 1991;9(4):228–30.

124. Jeffery DR. The use of vaccinations in patients with multiple sclerosis. Infect Med 2002;19:73–9.

125. Chew K. The case of Hannah Poling. htpp://www.autism-vox.com/the-case-of-hannah-poling.

126. Novella S. The Antivaccine Movement. Skeptical Inquirer 2007;Nov-Dec:31.

127. Offit PA. Thimerosal and vaccines—a cautionary tale. N Engl J Med 2007;357:1278–9.

128. Mitochondrial disease and immunization. Wkly Epidemiol Rec 2008;83:291.

129. Hviid A, Stellfeld M, Wohlfahrt J, Melbye M. Childhood vaccination and type 1 diabetes. N Engl J Med 2004;350:1398–404.

130. EMEA Press release. EMEA recommends suspension of Hexavac. London, September 20, 2005. htpp://www.e-mea.eu.int.

131. Lanchoney TF, Cohen RB, Rocke DM, Zasloff MA, Kaplan FS. Permanent heterotopic ossification at the injection site after diphtheria–tetanus–pertussis immunizations in children who have fibrodysplasia ossificans progressiva. J Pediatr 1995;126(5 Pt 1):762–4.

132. Gruber C, Nilsson L, Bjorksten B. Do early childhood immunizations influence the development of atopy and do they cause allergic reactions? Pediatr Allergy Immunol 2001;12(6):296–311.

133. Nakayama T, Onoda K. Vaccine adverse events in post-marketing study of the Kitasato Institute from 1994–2004. Vaccine 2007;25:570–6.

134. Koppen S, de Groot R, Neijens HJ, Nagelkerke N, van Eden W, Rümke HC. No epidemiological evidence for infant vaccinations to cause allergic disease. Vaccine 2004;25-6:3375–85.

135. La Force FM. Immunization of children infected with human immunodeficiency virus. WHO/EPI/GEN/86.6 Rev 1, Geneva. 1986.

136. Expanded Programme on Immunization. Immunization policy. WHO/EPI/GEN/86.7 Rev 1, Geneva. 1986.

137. Expanded Programme on Immunization. Joint WHO/UNICEF statement on immunization and AIDS. Wkly Epidemiol Rec 1987;62(9):53.

138. Shah RH, Mackey K, Wallace H, Yawata K, Roberto R, Meissinger J, Ascher M, Hagens S, Chin JCenters for Disease Control (CDC). Hepatitis B associated with jet gun injection California. MMWR Morb Mortal Wkly Rep 1986;35(23):373–6.

139. Parent du Chatelet I, Lang J, Schlumberger M, Vidor E, Soula G, Genet A, Standaert SM, Saliou P, Gueye A, Julien H, Lafaix C, Lemardeley P, Monnereau A, Spiegel A, Soke M, Varichon JP. Clinical immunogenicity and tolerance studies of liquid vaccines delivered by jet-injector and a new single-use cartridge (Imule): comparison with standard syringe injection. Imule Investigators Group. Vaccine 1997;15(4):449–58.

140. Grabenstein JD. A miscellany of obscure vaccines: adenovirus, anthrax, mixed bacteria, and staphylococcus. Hosp Pharm 1993;28:259–66.

141. Wraith DC, Goldman M, Lambert PH. Vaccination and autoimmune disease: what is the evidence? Lancet 2003;362(9396):1659–66 http://image.thelancet.com/extras/02art9340web.pdf.

142. Offit PA, Hackett CJ. Addressing parents' concerns: do vaccines cause allergic or autoimmune diseases? Pediatrics 2003;111(3):653–9.

143. Shoenfeld Y, Aron-Maor A. Vaccination and autoimmunity-vaccinosis: a dangerous liaison? J Autoimmun 2000;14(1):1–10.

144. Stuart G, Krikorian KS. The neuroparalytic accidents of anti-rabies treatment. Ann Trop Med Parasitol 1928;22: 327–77.

145. Kilbourne ED, Arden NH. Inactivated influenza vaccines. In: Plotkin SA, Orenstein WA, editors. Vaccines. 3rd ed. Philadelphia: Saunders, 1999:542.

146. Miller E, Waight P, Farrington CP, Andrews N, Stowe J, Taylor B. Idiopathic thrombocytopenic purpura and MMR vaccine. Arch Dis Child 2001;84(3):227–9.

147. CDC. National Immunization program. Diabetes and vaccines. Questions and answers. http://www.cdc.gov/nip/vacsafe/concerns/diabetes/q&a.htm, 10/05/2003.

148. World Health Organization. Vaccines and Biologicals. Diabetes. http://www.who.int/vaccines-diseases/safety/infobank/diabetes.shtml, 29/11/2003.

149. Institute of Medicine (IOM) Report. Immunization Safety Review. Vaccinations and sudden unexpected death in infancy. The National Academies Press. http://www.nap.edu/nap-cgi, 20/06/2003.

150. Fleming PJ, Blair PS, Platt MW, Tripp J, Smith IJ, Golding J. The UK accelerated immunisation programme and sudden unexpected death in infancy: case-control study. BMJ 2001;322(7290):822.

151. Centers for Disease Control (CDC). Immunization of children infected with human T-lymphotropic virus type III/lymphadenopathy-associated virus. MMWR Morb Mortal Wkly Rep 1986;35(38):595–8603–6.

152. McLaughlin M, Thomas P, Onorato I, Rubinstein A, Oleske J, Nicholas S, Krasinski K, Guigli P, Orenstein W. Live virus vaccines in human immunodeficiency virus-infected children: a retrospective survey. Pediatrics 1988;82(2):229–33.

153. Gutfreund K, Cheatham-Speth D, Rossol S, Voth R, Clemens R, Hess G. The effect of vaccination against hepatitis B on the CD4 cell account in anti-HIV positive man. MontrealAbstracts, V International Conference on AIDS 1989;5:435.

154. Buchbinder SP, Hessol N, Lifson A, O'Malley P, Barnhart I, Rutherford G, Hadler S. The interaction of HIV and hepatitis B vaccination in a cohort of homosexual and bisexual men. MontrealAbstracts, V International Conference on AIDS 1989;5:259

155. Edelman R. An update on vaccine adjuvants in clinical trial. AIDS Res Hum Retroviruses 1992;8(8):1409–11.

156. Ryan AM, Bermingham N, Harrington HJ, Keohane C. Atypical presentation of macrophagic myofasciitis 10 years post vaccination. Neuromusc Disord 2006;16:867–9.

157. Jefferson T, Rudin M, Di Pietrantonj C. Adverse events after immunization with aluminium-containing DTP vaccines: systematic review of evidence. Lancet Infect Dis 2004;4:84.

158. Bergfors E, Trollfors B, Inerot A. Unexpectedly high incidence of persistent itching nodules and delayed hypersensitivity to aluminium in children after the use of adsorbed vaccines from a single manufacturer. Vaccine 2004;22:158.

159. Thierry-Carstensen B, Stellfeld M. Itching nodules and hypersensitivity to aluminium after the use of adsorbed vaccines from SSI. Vaccine 2004;22:1845.

160. Theeten H, Van Damme P, Hoppenbrouwers K, Vandermeulen C, Leback E, Sokal EM, Wolter J, Schuerman L. Effects of lowering the aluminium content of a DTPa vaccine on its immunogenicity and reactogenicity when given as a booster to adolescents. Vaccine 2005;23:1515–21.

161. Centers for Disease Control and Prevention (CDC). Thimerosal in vaccines: a joint statement of the American Academy of Pediatrics and the Public Health Service. MMWR Morb Mortal Wkly Rep 1999;48(26):563–5.

162. World Health Organization. Questions and answers on thiomersal. July 1999. http://www.who.int/vaccines-diseases/safety/hottop/thiomersal.htm, 15/11/2000.

163. Immunization Safety Review. Thimerosal-Containing Vaccines and Neurodevelopmental DisordersWashington, DC: National Academy Press;. 2001.

164. Advisory Committee on Immunization Practices (ACIP). Inactivated Japanese encephalitis virus vaccine. MMWR Recomm Rep 1993;42(RR-1):1–15.

165. Ozaki T, Nishimura N, Muto T, Sugata K, Kawabe S, Goto K, Koyama K, Fujita H, Takahashi Y, Akiyama M. Safety and immunogenicity of gelatine-free varicella vaccine in epidemiological and serological studies in Japan. Vaccine 2005, 23:1205–8.

166. Safety of squalene. Wkly Epidemiol Rec 2006;81:274–5.

167. Knezevic I, Griffiths E, Reigel F, Dobbelaer R. Thiomersal in vaccines: a regulatory perspective. Meeting report on a WHO consultation, Geneva, April 15-16, 2002. Vaccine 2004;22:1836–41.

168. Global Advisory Committee on Vaccine Safety, 11-12 June, 2003. Weekly Epidemiol Rec 2003;78:284.

169. Clements CJ. The evidence for the safety of thiomersal in newborns and infant vaccines. Vaccine 2004;22:1854–61.

170. DiMiceli L, Pool V, Kelso JM, Shadomy SV, Iskander J. Vaccination of yeast sensitive individuals: review of safety data in the US vaccine adverse event reporting system (VAERS). Vaccine 2006;24:703–7.

171. World Health Organization. Bovine spongiform encephalitis and oral polio vaccine. Position statement. October 2000. http://www.who.int/vaccines-diseases/safety/hottop/bse.htm, 11/10/2000.

Anthrax vaccine

General Information

Inactivated anthrax vaccine is mainly used for protection against occupational anthrax exposure. A complete vaccine series consists of three 0.5-ml subcutaneous doses at 2-week intervals, followed by three additional doses 6, 12, and 18 months after the first dose. Mild local reactions occur in 30% of vaccinees, including local erythema and tenderness, which occurs within 24 hours and begins to subside within 48 hours. The reactions tend to increase in severity by the fifth injection. Systemic reactions are rare and usually characterized by malaise and lassitude, chills, and fever (1).

The authors of an extremely controversial review of anthrax vaccine concluded that the vaccine has not been shown to be safe or effective and accused the US Department of Defense of having withheld reports on vaccine-related adverse events and criticized the Food and Drug Administration (FDA) for not properly performing many of its oversight duties (2).

Observational studies

Until recently, there has been little research into anthrax vaccines, other than that carried out for antibacteriological warfare purposes by the military. Currently, three human vaccines against *Bacillus anthracis* (produced in Russia, the UK, and the USA)

are commercially available. The results of two field trials of two vaccines produced in Russia and the USA have been analysed (3). The US killed vaccine was 93% effective in preventing cases of anthrax, and the Russian live attenuated vaccine afforded 75% protection when given by scarification and 84% when a jet-gun was used. The rates of local reactions (erythema, induration, and edema) and systemic reactions (fever, malaise, arthralgia, rash, headache) after the US vaccine were 5.75 and 0.4% respectively, compared with 0.54% local reactions and no systemic reactions after placebo. Adverse effects data on the Russian vaccine were not presented.

In a study by the Advisory Group of Medical Countermeasures of the UK Ministry of Defence only mild discomfort at the injection site was reported after the administration of a total of 55 000 doses of anthrax vaccine (4).

In 2000 the Institute of Medicine of the United States Academy of Sciences encouraged the evaluation of active long-term monitoring studies of large populations to further evaluate the relative safety of anthrax vaccine. The association of anthrax immunization with arthritic, immunological, and gastrointestinal adverse reactions has been evaluated, based on an analysis of the Vaccine Adverse Events Reporting System (VAERS) database (15 December 1997 to 12 April 2000) (5). Anthrax vaccine was one of the most reactogenic vaccines included in VAERS. The incidence of adverse reactions reported after anthrax vaccine was higher for every reaction analysed compared with the adult vaccine control groups. The authors concluded that the current anthrax vaccine may be acceptable in military populations in an impending threat of anthrax exposure. Civilian anthrax immunization will require a less reactogenic vaccine.

Organs and Systems

Sensory systems

Optic neuritis has been attributed to anthrax vaccine.

- Two patients, aged 23 and 39 years, developed acute optic neuritis 2 weeks and 1 month after anthrax booster immunizations. The first had excellent visual recovery, but the second required chronic immunosuppression to maintain his vision (6).

Second-Generation Effects

Pregnancy

A recent study was designed to determine whether military's women's pregnancy rates were affected by having been immunized with anthrax vaccine (7). The pregnancy rate ratio, birth odds ratio, and adverse birth outcomes ratio of 385 women was comparable with non-immunized women.

References

1. Centers for Disease Control and Prevention (CDC). Use of anthrax vaccine in response to terrorism: supplemental recommendations of the Advisory Committee on Immunization Practices. MMWR Morb Mortal Wkly Rep 2002;51(45):1024–6.
2. Nass M, Nicolson GL. The anthrax vaccine: historical review and current controversies. J Nutr Environ Med 2002;12:277–86.
3. Demicheli V, Rivetti D, Deeks JJ, Jefferson T, Pratt M. The effectiveness and safety of vaccines against human anthrax: a systematic review. Vaccine 1998;16(9–10):880–4.
4. Blain P, Lightfoot N, Bannister B. Practicalities of warfare required service personnel to be vaccinated against anthrax. BMJ 1998;317(7165):1077–8.
5. Geier MR, Geier DA. Gastrointestinal adverse reactions following anthrax vaccination. Hepato-Gastroenterology 2004;51:762–7.
6. Kerrison JB, Lounsbury D, Thirkill CE, Lane RG, Schatz MP, Engler RM. Optic neuritis after anthrax vaccination. Ophthalmology 2002;109:99–104.
7. Wiesen AR, Littell CT. Relationship between pre-pregnancy anthrax vaccination and pregnancy and birth outcomes among US army women. J Am Med Assoc 2002;287:1556–60.

Bacille Calmette–Guérin (BCG) vaccine

General Information

Bacille Calmette–Guérin (BCG) vaccine is a suspension of living tubercle bacilli of the Calmette–Guérin strain. It is used mainly prophylactically against tuberculosis, but also as a means of stimulating the immune response in malignant disease. There are variations in the characteristics of BCG vaccines, depending on the strain of BCG derived from the original BCG strain and employed for vaccine production. BCG is generally used intradermally, except for instillation in intravesical immunotherapy. The risk of adverse effects after BCG immunization is related to the BCG strain, the dose, the age of the vaccinee, the technique of immunization, and the skill of the vaccinator.

Therapeutic uses of BCG

In addition to its use in preventing tuberculosis, BCG has been used as an immunostimulant or immunomodulator. The degree of safety of this procedure differs with the technique and the purpose for which it is used. In most areas, the use of BCG to counter cancer has proved disappointing, although it is still used to some extent, generally as an adjunct to other forms of treatment (1–7). More encouraging is the use of intravesical instillation of BCG for recurrent superficial transitional cell carcinoma of the bladder, for which it now constitutes the treatment of choice.

BCG immunotherapy in bladder tumors

Intravesical instillation of BCG has been used to treat superficial bladder carcinoma and interstitial cystitis. Many reports have confirmed the efficacy of BCG in the treatment of transitional cell bladder cancers and have delineated its adverse effects (SEDA-12, 273) (SEDA-13, 278) (SEDA-15, 344) (SEDA-16, 375) (SEDA-17, 366) (SEDA-18, 328) (SEDA-20, 287) (SEDA-21, 328) (SEDA-22, 336). The exact mechanism of its antitumor activity is unknown, but live BCG provokes an inflammatory response that includes activation of macrophages, a delayed hypersensitivity reaction, and stimulation of T and B lymphocytes and natural killer cells.

In general, BCG immunotherapy of bladder cancer is considered to be relatively safe. However, it does have adverse effects, including fever, arthritis/arthralgia, bladder irritability, bladder contracture, cytopenias, cystitis, disseminated intravascular coagulation, respiratory failure, epididymitis, hepatitis, loss of bladder capacity, miliary tuberculosis, pneumonitis, polyarthritis, prostatitis, pyelonephritis, pseudotumoral granulomatous renal mass, rhabdomyolysis, renal granulomas, renal insufficiency, skin abscess, tuberculous aneurysm of the aorta and femoral artery, ureteral obstruction, vertebral osteomyelitis, and psoas abscess. A small number of reports of life-threatening adverse effects after BCG instillation have been published, including disseminated BCG infection (8–11), some fatal. These tragic cases illustrate many points of critical importance to all urologists using BCG. BCG should never be given at the same time as tumor resection or transurethral resection of the prostate. The dose of BCG given intravesically corresponds to a potentially lethal intravenous dose. Intravasation as a result of catheterization, tumor resection or biopsy, or cystitis has occurred in two-thirds of the reported cases of systemic BCG infection.

In a review of those observed among 195 patients with bladder cancer treated with various substrains of BCG, there were frequent but mild to moderate local adverse effects, with irritative cystitis leading to frequency and dysuria in 91% of patients and hematuria in 43% (12). Low-grade fever (24%), malaise (24%), and nausea (8%) also occurred. These symptoms usually occurred after two or three instillations and lasted for about 2 days. It has been stated that these frequent adverse effects did not seriously affect the quality of life of patients (13). Additional information was obtained from a multinational retrospective survey in order to cover the whole scope of severe and/or systemic complications associated with BCG immunotherapy, and to propose guidelines for management (14). Among 2602 patients in this survey, more than 95% had no serious adverse effects. Apart from fever higher than 39°C (2.9%) and major hematuria (1%) serious adverse effects comprised granulomatous prostatitis (0.9%), granulomatous pneumonitis and/or hepatitis (0.7%), arthritis and arthralgia (0.5%), epididymo-orchitis (0.4%), life-threatening BCG sepsis (0.4%), skin rashes (0.3%), ureteric obstruction (0.3%), bladder contracture (0.2%), renal abscesses (0.1%), and cytopenias (0.1%). There was no major difference in incidence among the different substrains used. Lowering the dose of BCG in an attempt to reduce the incidence of adverse effects produced somewhat contrasting results, with a reduced incidence of various adverse effects but no significant difference (or an apparently increased incidence) in the case of pollakiuria, hematuria, fever, and headache (SEDA-20, 287) (15).

Other rare complications have been seldom reported, namely cryoglobulinemia with evidence of disseminated BCG infection (16), ruptured mycotic aneurysm of the abdominal aorta (17), bladder wall calcification (18), rhabdomyolysis (19), iritis or conjunctivitis with arthritis or Reiter's syndrome (20,21), and severe acute renal insufficiency due to granulomatous interstitial nephritis, which can occur even in the absence of other systemic complications (22).

In 1990, the US Food and Drug Administration approved the marketing of BCG Live (intravesical) for use in the treatment of primary or relapsed carcinoma in situ of the urinary bladder, with or without associated papillary tumors. BCG is not recommended for treatment of papillary tumors that occur alone. The drug is marketed by Connaught Laboratories as TheraCys and by Organon as TiceBCG. The manufacturers recommend a 6-week induction course of weekly intravesical BCG, usually starting 1–2 weeks after biopsy or after transurethral resection of papillary tumors. Follow-up courses of treatment at 3, 6, 12, and 24 months or monthly for 6–12 months after initial treatment are recommended.

Influence of dosage

The degree of success of different doses of BCG vaccine (100–120 mg, 20–50 mg, or a much smaller dose of 1 mg) in preventing tumor relapse has been described in patients with superficial bladder cancers (23). Adverse reactions were dose-related. The authors considered that endovesical instillation of BCG vaccine 1 mg would be the optimal dosage for prevention of relapse.

In 108 patients with bladder cancer, tumor relapse was prevented by the use of BCG vaccine 1 mg (24). Inguinal lymphadenitis and dysuria have occurred (SEDA-20, 287).

Comparison of BCG strains

In a comparison of 56 patients who received BCG instillations using Berna strain BCG and 32 patients who received Pasteur strain BCG for treatment of superficial bladder cancer, the patients who received Pasteur strain BCG had the highest tumor-free rate but had significantly more toxicity (25). The answer to another difficult question connected with the use of intravesical BCG, that is whether the treatment increases the incidence of second primary malignancies, has been sought (26,27). It was suggested that BCG immunotherapy could accelerate the growth and cause metastatic spread of a growing second primary malignancy that had remained undetected at the start of BCG therapy, and that the time relation between the starting point of second primary tumor development and the starting point of BCG treatment might be crucial in determining whether BCG eradicates

the tumor or accelerates its growth. However, in 153 patients there was no evidence that intravesical BCG did increase the incidence of second primary malignancies (26). The matter is therefore still unresolved.

Comparison of different regimens for carcinoma in situ of the bladder

The efficacy and adverse effects of various alternative treatment regimens for carcinoma in situ of the bladder have been compared with those of instillation of BCG in 21 patients. All were treated initially with intravesicular instillations of Keyhole-Limpet Hemocyanin (first course: 20 mg weekly for six weeks; second course: 20 mg monthly for 1 year or bimonthly for 2 subsequent years). Patients who did not respond to two courses were treated with regular instillations of BCG Connaught strain 120 mg. Eleven patients were free from tumor tissue after the first or second course of Keyhole-Limpet Hemocyanin. Ten patients had to have a cystectomy because of persistence or progression of carcinoma after hemocyanin or hemocyanin with subsequent BCG. However, instillations of BCG caused severe dysuria in 60% and fever in 40% of patients, whereas hemocyanin treatment had only minor adverse effects (28). Combined therapy with mitomycin C and BCG was more effective in 28 patients with carcinoma in situ of the bladder than mitomycin alone (29). Compared with BCG monotherapy there were only a few adverse effects. The success rates were comparable.

Two different methods of treating superficial bladder cancer have also been compared in a randomized, multicenter trial, setting transurethral resection only against transurethral resection plus adjuvant mitomycin C and BCG instillation (30). The rate of progression was comparable in the two groups; at a medium follow-up of 20 months there was a reduction in recurrence rates with the combination therapy. Adverse effects occurred most often during or after BCG instillation. Other investigators have found the same degree of efficacy of the same regimen in reducing the incidence of recurrence of superficial urothelial cancer after transurethral bladder resection in 99 patients (31).

Treatment of adverse effects after intravesical instillation of BCG

The prompt recognition of risk factors for severe complications, namely traumatic catheterization or concurrent cystitis, that increase BCG absorption, and treatment of early adverse effects, is expected to reduce the incidence of severe adverse effects. Severe local and systemic adverse effects can be successfully treated with tuberculostatic drugs for up to 6 months (14).

Severe local and systemic adverse effects of BCG treatment can be treated successfully with tuberculostatic drugs, to most of which BCG is very susceptible, for up to 6 months (32). The effects of isoniazid on the incidence and severity of adverse effects of intravesical BCG therapy have been analysed in patients who received BCG with ($n = 289$) and without ($n = 190$) isoniazid (33). The authors concluded that prophylactic oral administration of isoniazid (300 mg/day with every BCG instillation) caused no reduction in any adverse effect of BCG. In contrast, transient liver function disturbances occurred slightly more often when isoniazid was used. The polymerase chain reaction has been used to monitor BCG in the blood after intravesical BCG instillation (22 patients) as well as after antituberculosis therapy (34). The early and fast diagnosis of BCG in the blood was considered to be potentially valuable in initiating specific early treatment of BCG complications.

General adverse effects of prophylactic BCG immunization

BCG immunization is generally well tolerated. Locally a small papule appears which scales and ultimately leaves a scar; however, abnormal reactions can occur. The most common adverse local reaction, suppurative lymphadenitis, has been reported in 0.1–10% of immunized children under 2 years of age. Faulty immunization technique is the most frequent cause of severe abnormal BCG primary reactions (35). The most serious generalized complications of BCG immunization involve disseminated infection with the BCG bacillus and BCG osteitis. Allergic reactions are unusual, but severe anaphylactic reactions can occur, especially when the product is used as an immunostimulant.

Tuberculin

Mammalian tuberculin purified protein derivative (tuberculin PPD) is the active principle of old tuberculin. A small test dose in a healthy individual, given intracutaneously, is likely to produce only a little local pain and pruritus. If tuberculous infection is present, the local reaction is more marked, with vesiculation, ulceration, and even granuloma annulare or necrosis.

If more than a minimal dose is used in cases of tuberculous infection, a severe and even fatal generalized anaphylactic reaction can develop within about 4 hours of the injection (36).

People who are engaged in manufacturing PPD can easily become sensitized to it, and severe allergic reactions can occur if they later inhale even small quantities (37).

Lymphangitis after tuberculin testing is rare (SED-11, 686).

Acute panuveitis has been reported (38). The episodes developed after each of two tests carried out at intervals of 8 years and responded well to glucocorticoid therapy.

Organs and Systems

Cardiovascular

A mycotic aneurysm, a rare complication of intravesical BCG therapy, has been reported (17).

- A 71-year-old man with bladder carcinoma in situ received six instillations of BCG at weekly intervals followed 3 months later by three booster instillations at weekly intervals. Four months later an inflammatory

aortic aneurysm, which had ruptured into a pseudoaneurysm, was diagnosed and excised. *Mycobacterium bovis* was found. After treatment with isoniazid and rifampicin he recovered. There was no sign of tumor in the bladder at cystocopy 8 months after the last BCG instillation.

Respiratory

There have been two reports of micronodular pulmonary infiltrates (BCG pneumonitis) associated with fever, chills, and night sweats following multiple instillations of intravesical BCG (39). Both patients were 71 years old. The reactions, including radiographic infiltrates, resolved spontaneously or after steroid therapy.

Nervous system

There have been two reports of tuberculous meningitis after BCG immunization in immunocompetent individuals, in two French children aged 4.5 and 5 years (40) and in a 22-year-old woman from Cambridge, UK (41).

Polyneuritis has been attributed to BCG immunization (42).

Sensory systems

Responding to the question of whether accidental inoculation of one drop of BCG vaccine into the eye of a health-care worker could be a risk, Pless (personal communication) has reported the case of a urologist who developed a corneal ulcer after a similar accident.

Endogenous endophthalmitis (SEDA-15, 344) and bilateral optic neuritis (43) have been reported after BCG immunization.

Hematologic

Clinical trials in BCG-vaccinated newborns in different countries have variously found a dose–effect relation for the risk of suppurative lymphadenitis:

- Croatia (SED-12, 799)
- French Guyana (SEDA-16, 373)
- Germany (SED-12, 798)
- Hong Kong (SED-12, 799)
- Hungary (SED-12, 798)
- India (SEDA-18, 328)

or a strain-dependent relation:

- Austria (SEDA-16, 374) (SEDA-17, 366)
- Germany (SED-12, 798)
- India (SEDA-18, 328)
- Saudi Arabia (SED-12, 799)
- Togo (SED-12, 799)
- Turkey (SED-12, 799)
- Zaire (SED-12, 799)

Since 1984, WHO's Expanded Programme on Immunization has received many reports from various countries of an increased incidence of suppurative lymphadenitis after BCG immunization. Careful investigations of risk factors have been carried out, particularly in Zimbabwe and Mozambique (1987, 1988). Those studies established a strong association between an increased risk of lymphadenitis within 6 months of immunization and both the use of the Pasteur BCG strain, and programmatic errors, such as poor injection technique, poor technique in reconstituting and mixing the freeze-dried vaccine with diluent, or an incorrectly administered dose of vaccine. The incidence of lymphadenitis was 9.9% with the suspect Pasteur strain versus 0% with two other strains used under the same conditions. Experience gained in other countries has suggested that when Pasteur strain BCG vaccine is administered properly, the rate of lymphadenitis in newborns should not exceed 1%. The increased occurrence of lymphadenitis may require both a reactogenic strain and poor technique (44). A rare case of BCG lymphangitis occurring 11 years and again 18 years after immunization has been reported (45).

Liver

Granulomatous neonatal hepatitis has been reported after BCG immunization (46).

Skin

Since the initial report of lupus vulgaris following BCG immunization in 1946, about 60 cases have been published, mostly following (multiple) revaccination. The risk of developing lupus vulgaris following primary immunization is extremely low.

- Lupus vulgaris occurred in a 7-year-old girl after only a single BCG immunization (47). She was treated with conventional antituberculosis therapy with an excellent response.
- Six months after BCG vaccination, an 18-year-old man developed lupus vulgaris on his right shoulder (48). He was successfully treated with rifampicin, isoniazid, and ethambutol. He had had lupus vulgaris after BCG vaccination on his left shoulder 8 years before.

Acute febrile neutrophilic dermatosis (49) and eczema vaccinatum (50) have been reported after BCG immunization.

Musculoskeletal

Arthritis and arthralgia are well-known adverse effects of intravesical BCG instillation as part of therapy of bladder cancer (SED-13, 925). The etiology and the different clinical pictures of BCG immunotherapy have been discussed (51). Considering that mycobacteria are potent stimulators of the immune system and especially of T cells, it is not surprising to observe T cell-mediated aseptic arthritis after BCG therapy. The authors suggested that the site of immune stimulation is critical, since intradermal injection produces a clinical presentation similar to reactive arthritis, and intravesical therapy causes a clinical picture identical to Reiter's syndrome.

In a large worldwide analysis of BCG adverse effects (1948–74) co-ordinated by the International Union Against Tuberculosis and Lung Disease (SED-12, 795) there were 272 cases of lesions of bones and joints, including synovial lesions. However, case reports of arthritis after BCG vaccination in healthy individuals are rare.

Polyarthritis has been reported in a 33-year-old healthy woman 3 weeks after BCG vaccination (52).

Osteitis

Osteitis occurred in <0.1–30/100 000 vaccinees and has been reported mainly among infants immunized with BCG in the neonatal period in the Scandinavian countries. A retrospective study showed that BCG osteitis was present in Sweden from 1949 onwards. The reported incidence was one per 40 000 in children born between 1960 and 1969 (53,54). In Sweden, the reported incidence of osteitis rose to one per 3000 and one per 4000 for children vaccinated in the neonatal period during 1972–75. Compulsory notification of BCG adverse effects to the Swedish Adverse Drug Reaction Committee was introduced at that time.

The incidence of BCG osteitis correlates closely with the BCG vaccine used. Between 1960 and 1970, when the vaccine based on the Gothenburg strain was prepared in Sweden, the incidence of BCG osteitis was 7.3 per 100 000 vaccinees. There was a significant increase to 37 per 100 000 in the early 1970s, when the vaccine was prepared in Copenhagen using the same Gothenburg strain. Because of the increased incidence, in 1978 the vaccine was replaced by a BCG vaccine made by Glaxo, UK. Since then, the incidence has been similar to that reported in the 1960s (6.4 per 100 000 vaccinees) (55).

Following a report of 10 cases of BCG osteitis in Finland (56), the medical records of 222 children with BCG osteitis registered from 1960 to 1988 in Finland were analysed (55). The most common sites of osteitis were the metaphyses of the long bones; the legs were affected more often (58%) than the arms (14%). There was also osteitis of the sternum (15%) and ribs (11%). With adequate treatment, the prognosis for children with BCG osteitis was good, but six children were left with sequelae, with abnormalities of the limbs in five cases and pronounced cheloid formation in the other.

In Czechoslovakia, BCG osteitis was not diagnosed before 1981. However, from March 1980 onwards another BCG vaccine (Moscow strain) was introduced (57), and after 1981 12 cases of BCG osteitis were diagnosed (58). Most of the cases developed between 7 and 24 months after immunization, but some occurred later. The risk of osteitis rose to 35 per million in the period 1982–85. There were 28 cases of BCG osteomyelitis during the period from 1980 to June 1985, when Russian BCG vaccine containing a higher amount of culturable particles was used, and only 11 cases during the period from July 1985–89, by which time the dose of vaccine had been halved (59). During the last 2 years of the second period there was only one case in each immunized birth cohort. Elsewhere, an infant developed BCG osteomyelitis of the upper spine, a very rare complication described only three times before (60).

Six cases of BCG osteitis in Switzerland occurred in 1980–85 (61). Reports from various countries, for example New Zealand (62) and India (63), have emphasized the increasing knowledge of BCG osteitis. BCG osteitis has never been reported in the UK with use of the Glaxo

Table 1 Rates of BCG osteitis in different countries

Country	Number of vaccinees (all age groups) (millions)	BCG osteitis Number	BCG osteitis Rate per million vaccinees
Finland	2.79	128	45.9
Sweden	3.42	121	35.4
Denmark	2.28	4	1.75
West Germany	9.03	14	1.55
Norway	2.32	2	0.86
East Germany	8.68	5	0.58
Switzerland	2.17	1	0.46
France	16.2	6	0.37
Austria	2.86	1	0.35
Yugoslavia	17.4	2	0.11
Europe (33 countries)	498	284	0.57
Israel	1.92	3	1.56
Algeria	8.05	2	0.25
Japan	166	2	0.01

vaccine. Table 1 shows rates of BCG osteitis in different countries (64).

Sexual function

- Following a 6-week course of intravesical BCG a 65-year-old man with carcinoma of the bladder developed a BCG-derived inflammatory infiltrate of the penis. The induration and lesions resolved after treatment with isoniazid and ethambutol (65).

Immunologic

The incidence of adverse effects after BCG immunization has been extensively investigated by the Committee on Prophylaxis of the International Union Against Tuberculosis and Lung Disease (IUATLD). Retrospective studies including 51 countries worldwide and collecting data from 1948–74, according to organ and system category, have been published (SED-12, 795) (64,66). The IUATLD carried out a second (prospective) 6-country study (1979–83) (67), using the classification system already used in the retrospective study. The mean risk of local complications and suppurative lymphadenitis was low: 0.387 per 1000 vaccinees or 0.093 per 1000 with positive bacteriological/histological findings, respectively. There were 21 cases of disseminated BCG infections and allergic manifestations recorded in four countries. The estimated risks of serious disseminated BCG infection were higher than calculated previously (except for bone and joint lesions), but very low when comparing the benefit and risks of BCG immunization, especially in infants (67).

Anaphylactic reactions

Allergy to BCG exceptionally occurs when intravesical BCG is used as an immunoenhancing agent. An anaphylactic reaction has been reported (68).

Three cases of anaphylactic reactions to BCG have been described in young children, one (in a 3-month-old girl) being fatal (69).

An acute shock-like syndrome developed 30 minutes after BCG immunization of a newborn girl.

Non-IgE-mediated anaphylactic (anaphylactoid) reactions suspected to be caused by dextran as used in BCG vaccines have been described (SEDA-16, 375).

Infection risk

Disseminated BCG infection

Dissemination of BCG infection occurs in under 0.1/100 000 vaccinees and is usually associated with severe abnormalities of cellular immunity. Data collected on cases immunized between 1948 and 1974 (from the retrospective IUATLD study) are shown in Table 2 (64,66).

Many authors have provided detailed reports of disseminated BCG infection after BCG immunization at birth in newborns with various underlying immunodeficiency syndromes (severe combined immunodeficiencies, cellular immunodeficiency syndromes, X-linked chronic granulomatous disease or autosomal recessive chronic granulomatous disease) (70–72) and include patients with AIDS. 108 cases of disseminated BCG infection reported worldwide since 1951, including 30 cases of disseminated infection during 1974 to 1994 in France, have been analysed (73,74). Four well-defined immunodeficiency conditions predispose to disseminated BCG infection: severe combined immunodeficiency, chronic granulomatous disease, Di George syndrome, and AIDS. About half the cases of disseminated BCG infection occur in the absence of any well-defined underlying immunological defect. However, there is little doubt that children with idiopathic BCG infection are immunodeficient, and there is good evidence that their immunodeficient status is inherited: four pairs of siblings and one pair of cousins were found among 60 children with idiopathic disseminated BCG infection; in addition, among the 50 single-case families, parental consanguinity was found in 7 of 24 families for whom information was available.

There have been reports of disseminated BCG infection in children with chronic granulomatous disease which can result in prolonged and relapsing local complications to BCG immunization (75–77). In a series of autopsies, 26 of 36 children who had been given BCG shortly after birth showed tuberculoid granulomas at various sites. None of the infants had histological evidence of immune deficiency (SEDA-8, 301) (78).

- Recovery from BCG sepsis has been documented in a 7-year-old girl with immunodeficiency (79).
- Two other infants with severe combined immunodeficiency who had developed BCG dissemination after

Table 2 Number of complications recorded until December 1977 in a retrospective study of adverse effects of BCG worldwide among cohorts vaccinated during 1948–74

Adverse effect	Patients in 187 countries (n = 1 470 208 160)		Patients in 51 countries in which some cases were recorded in any category (n = 1 053 402 835)		Cases proven bacteriologically and/ or histologically	
	Number	Rate	Number	Rate	Number	Rate
1 Abnormal BCG primary complex[a]	6602	4.49	6602	6.27	1100	1.04
2 and 3. Disseminated BCG infection: generalized and/or localized lesions (non-fatal and fatal)[b]	1072	0.73	1072	1.02	561	0.53
4. Syndromes or disease clinically associated with BCG immunization[c]	1838	1.25	1838	1.74	7	0.01
All categories	9512	6.47	9512	9.04	1668	1.58

[a]Such cases have often been proven bacteriologically and/or histologically.
[b]Until now, such cases have never been proven either bacteriologically or histologically.
[c]For these seven cases, TB lesions were seen at post-mortem examination, but BCG etiology was doubtful.

neonatal BCG immunization were treated successfully by bone marrow transplantation and tuberculostatic therapy (80).

However, recovery may not always be complete.

- One girl in her third year, who had been immunized against tuberculosis at birth, developed an abscess of the associated lymph nodes (which were extirpated) and some weeks later developed intestinal BCG dissemination, which appeared to be cured by tuberculostatic treatment. Despite this, at the age of 22 years she developed a left-sided hemiplegia due to aneurysms and thrombosis of cerebral arteries, and 4 years later an oculomotor nerve paralysis was diagnosed. She died at 26 from recurrent intestinal BCG dissemination, which developed at the end of a pregnancy (a healthy premature child was born).

The autopsy confirmed the diagnoses and showed acid-fast bacilli in the adventitia of the basilar artery; the paralysis of the oculomotor nerve was caused by the brain lesion. Defective function of macrophages was suggested as the possible cause of the underlying immunological abnormality (81).

- A 7-week-old infant died under circumstances reminiscent of SIDS; the histopathological examination revealed disseminated BCG infection, no abnormalities of the immune system were detected (82).
- X-ray and CT scan examinations suggested a teratoma in a 1-year-old girl who had received BCG vaccine at birth; however, histology and microbiology revealed the diagnosis of mediastinal BCGitis (83).

Histiocytosis

Cases of fatal histiocytosis have been reported in babies with immune defects immunized with BCG shortly after birth (84).

- One boy who had been immunized shortly after birth developed ipsilateral axillary lymphoma at the age of 6 months. Microscopically the picture was typical of BCG histiocytosis. The child survived. The underlying immunological disturbance was considered to be a temporary derangement of T lymphocyte function (SEDA-8, 300) (85).

Safe immunization with BCG: the WHO view

Taking into account the results of the investigations mentioned above, the WHO made the following recommendations to ensure safe and effective vaccination:

1. Pasteur BCG vaccine should be supplied only to countries using the product without problems. In no case should the vaccine be sent to a country which has been successfully using another product unless the country specifically requests this product.
2. UNICEF supplies of BCG vaccine should specify the doses to be given to infants (0.05 ml), and when possible 0.05 ml syringes should be supplied with the vaccine to reduce the likelihood of too large a dose being given.
3. WHO will supply to country programs a short protocol to assess the incidence of BCG-associated lymphadenitis using a standard case definition. Evidence of a rate greater than 1% would be the basis for

investigating the problem and for reviewing the vaccine strain being used.

Additionally it is recommended that training in the technique of BCG immunization should be re-emphasized (44).

Treatment

Erythromycin for a period of 2–4 weeks resolved troublesome post-BCG-lesions in six patients (86). The effect of erythromycin on the atypical mycobacteria was described as early as 1957 (87), but its use for treatment of BCG lesions has not previously been reported. Erythromycin has been used to treat cold abscesses (88).

Neonates with suppurative lymphadenitis have been treated with isoniazid (10 mg/kg/day for 3–9 months) (89). The treatment resulted in complete resolution of the adenitis. There was no significant difference between infants with ruptured nodes and those with intact nodes.

There was no statistical difference between patients who received different forms of treatment for suppurative lymphadenitis: 36 patients received erythromycin, 21 isoniazid, and 21 isoniazid + rifampicin (90). When lymphadenitis developed rapidly (within 2 months), the incidence of spontaneous drainage and suppuration was reported to be significantly higher than in patients with slowly developing processes. Total surgical excision is recommended in these rapidly evolving cases.

Long-Term Effects

Tumorigenicity

The effects of immunization with BCG have been studied extensively (91–94) (SEDA-7, 323) (95,96). On the whole the results are inconclusive, with little good evidence of either preventive or tumor-inducing effects.

Second-Generation Effects

Pregnancy

No harmful effects of BCG vaccine on the fetus have been seen. Nevertheless, it is prudent to avoid immunization of women during pregnancy, unless there is immediate excessive risk of unavoidable exposure to infective tuberculosis (97).

Susceptibility Factors

Disseminated BCG infection is usually associated with severe abnormalities of immunity. BCG for prevention of tuberculosis should therefore not be given to persons with impaired immune responses, such as occur in congenital immunodeficiency, leukemia, lymphoma, generalized malignancy, or AIDS, and when immunological responses have been suppressed by glucocorticoids, alkylating agents, antimetabolites, or radiation (97). The risks are greater in the debilitated or the very young.

Age

WHO has previously recommended that in countries with a high burden of tuberculosis, BCG vaccine should be given to all healthy infants as soon as possible after birth, unless the child presents with symptomatic HIV infection. However, recent evidence has shown that children who were HIV-infected when immunized with BCG at birth, and who later developed AIDS, were at increased risk of developing disseminated BCG disease. Further studies have contributed to updated WHO recommendations. In a prospective hospital-based surveillance study in the Western Cape Province, South Africa, the risk of disseminated BCG disease was found to be increased several hundred-fold in HIV-infected infants compared with the risk in HIV non-infected infants (98). In a retrospective study in the Hospital de Niños "R. Gutiérrez" in Buenos Aires, Argentina, in 310 infants perinatally infected with HIV and immunized with BCG after birth, 28/310 (9%) developed BCG disease, among them for cases of disseminated BCG (99).

HIV/AIDS

Reports on complications after BCG immunization in HIV-positive individuals and patients with AIDS have been published. Disseminated BCGitis has been described (100–103), for example involving the spleen and mediastinal and mesenteric lymph nodes in one case and the liver and the lung in another case (100) or leading to pneumonitis (104). In one case BCG lymphadenitis occurred 30 years after BCG immunization in a 36-year-old patient with AIDS (105). Surgical excision and biopsy revealed a puriform abscess. Pathological examination established the diagnosis of Kaposi's sarcoma; no granulomatous changes were found. BCG was grown. The authors believe that the BCG lymphadenitis was due to a late reactivation of the bacillus.

BCG reactions in HIV-positive and HIV-negative children have been compared in African children. The rates of local adenitis were equal (106,107). These results have been confirmed by a number of other studies, among them one in Haiti. The Haitian investigators found that the risk of complications after BCG vaccination in HIV-infected children is low and that the risk does not outweigh the benefits of BCG vaccination in populations at high risk of tuberculosis during infancy and childhood. Mild or moderate adverse effects occurred in 19 (9.6%) of 166 infants born to HIV-seronegative mothers as compared with 4 (31%) of 13 HIV-infected infants (108).

Organisms of the *Mycobacterium avium* complex (MAC) commonly cause disseminated bacterial infection among patients with AIDS. There is evidence that immunoprophylaxis against MAC infection may be possible. A heat-killed *Mycobacterium vaccae* vaccine was given in a three-dose schedule to 12 HIV-infected adults with CD4 cell counts below $300 \times 10^6/l$ (109). The vaccine was well tolerated and produced detectable immunological responses in 3 of 11 subjects who completed the trial.

The WHO has recommended that individuals with clinical (symptomatic) AIDS or other clinical manifestations of

HIV-infection should not receive BCG (110). The Global Advisory Committee on Vaccine Safety has noted that there has been repeated reference to local or disseminated BCG infection several years after BCG immunization in HIV-infected persons (111). However, the Committee has not recommended a change in immunization policy (BCG immunization recommended in asymptomatic HIV-infected persons; not recommended in symptomatic HIV-infected persons), but surveillance for BCG-immunized, HIV-infected persons should be continued for 5–7 years.

The updated WHO recommendations are as follows. Infants who are HIV-infected, with or without symptoms, and infants whose HIV infection status is unknown but who have signs suggestive of HIV infection and who are born to HIV-infected mothers should not be immunized (112).

Drug Administration

Drug administration route (see also Infection risk above)

The intravesical instillation of BCG is the treatment of choice for recurrent superficial transitional cell carcinoma of the bladder, and BCG instillation is said to be effective for carcinoma in situ. Overviews, case reports, and case series have confirmed the efficacy of BCG and have described the resulting adverse effects.

- In a 77-year-old patient BCG ileitis followed BCG instillation to treat recurrent bladder carcinoma (113). The abnormal part of the terminal ileum was resected, and 2 years after surgery the patient was well and with no evidence of neoplastic disease.

References

1. Crispen RG. BCG and cancer. Dev Biol Stand 1986;58(Pt A):371–7.
2. Schult C. Nebenwirkungen der BCG-Immunotherapie bei 511 Patienten mit malignem Melanom. [Side effects of BCG immune therapy in 511 patients with malignant melanoma.] Hautarzt 1984;35(2):78–83.
3. Grigorovich NA, Risina DIa, Nodel'son SE. [Treatment of the complications occurring in BCG vaccine immunotherapy of patients with malignant neoplasms.]Vopr Onkol 1984;30(7):102–6.
4. Hoover HC Jr, Surdyke MG, Dangel RB, Peters LC, Hanna MG Jr. Prospectively randomized trial of adjuvant active-specific immunotherapy for human colorectal cancer. Cancer 1985;55(6):1236–43.
5. Shea CR, Imber MJ, Cropley TG, Cosimi AB, Sober AJ. Granulomatous eruption after BCG vaccine immunotherapy for malignant melanoma. J Am Acad Dermatol 1989;21(5 Pt 2):1119–22.
6. Torisu M, Iwasaki K, Sakata M. Immunotherapy of cancer patients with BCG: summary of ten years experience in Japan. Dev Biol Stand 1986;58(Pt A):451–6.
7. The Ludwig Lung Cancer Study Group (LLCSG). Immunostimulation with intrapleural BCG as adjuvant therapy in resected non-small cell lung cancer. Cancer 1986;58(11):2411–6.
8. Steg A, Leleu C, Debre B, Boccon-Gibod L, Sicard D. Systemic bacillus Calmette–Guerin infection, "BCGitis", in patients treated by intravesical Bacillus Calmette–Guerin therapy for bladder cancer. Eur Urol 1989;16(3):161–4.
9. Deresiewicz RL, Stone RM, Aster JC. Fatal disseminated mycobacterial infection following intravesical bacillus Calmette–Guerin. J Urol 1990;144(6):1331–3.
10. Rawls WH, Lamm DL, Lowe BA, Crawford ED, Sarosdy MF, Montie JE, Grossman HB, Scardino PT. Fatal sepsis following intravesical Bacillus Calmette–Guerin administration for bladder cancer. J Urol 1990;144(6):1328–30.
11. Sakamoto GD, Burden J, Fisher D. Systemic Bacillus–Calmette Guerin infection after transurethral administration for superficial bladder carcinoma. J Urol 1989;142(4):1073–5.
12. Lamm DL, Stogdill VD, Stogdill BJ, Crispen RG. Complications of Bacillus Calmette–Guerin immunotherapy in 1,278 patients with bladder cancer. J Urol 1986;135(2):272–4.
13. Bohle A, Balck F, von Weitersheim J, Jocham D. The quality of life during intravesical Bacillus Calmette–Guerin therapy. J Urol 1996;155(4):1221–6.
14. Lamm DL, van der Meijden PM, Morales A, Brosman SA, Catalona WJ, Herr HW, Soloway MS, Steg A, Debruyne FM. Incidence and treatment of complications of Bacillus Calmette–Guerin intravesical therapy in superficial bladder cancer. J Urol 1992;147(3):596–600.
15. Galvan L, Ayani I, Arrizabalaga MJ, Rodriguez-Sasiain JM. Intravesical BCG therapy of superficial bladder cancer: study of adverse effects. J Clin Pharm Ther 1994;19(2):101–4.
16. Durand JM, Roubicek C, Retornaz F, Cretel E, Payan MJ, Bernard JP, Kaplanski G, Soubeyrand J. Cryoglobulinemia after intravesical administration of Bacille Calmette–Guerin. Clin Infect Dis 1998;26(2):497–8.
17. Damm O, Briheim G, Hagstrom T, Jonsson B, Skau T. Ruptured mycotic aneurysm of the abdominal aorta: a serious complication of intravesical instillation Bacillus Calmette–Guerin therapy. J Urol 1998;159(3):984.
18. Spirnak JP, Lubke WL, Thompson IM, Lopez M. Dystrophic bladder wall calcifications following intravesical BCG treatment for superficial transitional cell carcinoma of bladder. Urology 1993;42(1):89–92.
19. Armstrong RW. Complications after intravesical instillation of Bacillus Calmette–Guerin: rhabdomyolysis and metastatic infection. J Urol 1991;145(6):1264–6.
20. Nesher G. Syndrome de Reiter après BCG thérapie intravésicale. [Reiter syndrome after intravésical BCG therapy.] Rev Rhum Ed Fr 1993;60(12):941.
21. Price GE. Arthritis and iritis after BCG therapy for bladder cancer. J Rheumatol 1994;21(3):564–5.
22. Binaut R, Bridoux F, Provot F, Daniel N, Fleury D, Mougenot B, Vanhille P. Néphrite interstitielle granulomateuse avec insufifsance rénale aiguë, une complication potentielle de la BCG thérapie intravésicale. [Granulomatous interstitial nephritis with acute renal insufficiency, a potential complication of intravesicular bcg therapy.] Nephrologie 1997;18(5):187–91.
23. Corti Ortiz D, Rivera Garay P, Aviles Jasse J, Hidalgo Carmona F, MacMillan Soto G, Coz Canas LF, Vargas Delaunoy R, Susaeta Saenz de San Pedro R. Profilaxis del cancer vesical superficial con 1 mg de BCG endovesical: comparacion con otras dosis. [Prophylaxis of superficial bladder cancer with 1 mg of intravesical BCG: comparison with other doses.] Actas Urol Esp 1993;17(4):239–42.

24. Rivera P, Caffarena E, Cornejo H, Del Pino M, Foneron A, Haemmersli J, Sepulveda M, Ubilla A. Microdosis de vacuna BCG como profilaxis en cancer vesical etapa T1. [Microdoses of BCG vaccine for prophylaxis in bladder cancer stage T1.] Actas Urol Esp 1993;17(4):243–6.

25. Lo Cigno M, Emili E, Iraci F, Soli M, Bercovich E, Rusconi R. Confronto tra BCG Berna e Pasteur F nella profilassi delle recidive neoplastiche superficiali della vescica. Acta Urol Ital 1991;6(Suppl 1):145–8.

26. Guinan P, Brosman S, DeKernion J, Lamm D, Williams R, Richardson C, Reitsma D, Hanna M. Intravesical Bacillus Calmette–Guerin and second primary malignancies. Urology 1989;33(5):380–1.

27. Khanna OP. Intravesical BCG and second primary malignancies. Urology 1989;34(2):113.

28. Jurincic-Winkler C, Metz KA, Beuth J, Sippel J, Klippel KF. Effect of keyhole limpet hemocyanin (KLH) and Bacillus Calmette–Guerin (BCG) instillation on carcinoma in situ of the urinary bladder. Anticancer Res 1995;15(6B):2771–6.

29. Rintala E, Jauhiainen K, Rajala P, Ruutu M, Kaasinen E, Alfthan O, Hansson E, Juusela H, Kanerva K, Korhonen H, Nurmi M, Permi J, Petays P, Tainio H, Talja M, Tuhkanen K, Viitanen JThe Finnbladder Group. Alternating mitomycin C and Bacillus Calmette–Guerin instillation therapy for carcinoma in situ of the bladder. J Urol 1995;154(6):2050–3.

30. Krege S, Giani G, Meyer R, Otto T, Rubben H. A randomized multicenter trial of adjuvant therapy in superficial bladder cancer: transurethral resection only versus transurethral resection plus mitomycin C versus transurethral resection plus Bacillus Calmette–Guerin. Participating Clinics. J Urol 1996;156(3):962–6.

31. Nohales Taurines G, Cortadellas Angel R, Arango Toro O, Bielsa Gali O, Gelabert Mas A. Resultados de un estudio prospectivo de quimioprofilaxis con mitomycina–C y BGG alternades: respuesta completa, indice de recidives y de progresion. [Results of a prospective study of chemoprophylaxis with alternating mitomycin-C and BCG: complete response and recurrence and progression index.] Arch Esp Urol 1996;49(7):689–92.

32. van der Meijden AP. Practical approaches to the prevention and treatment of adverse reactions to BCG. Eur Urol 1995;27(Suppl 1):23–8.

33. Vegt PD, van der Meijden AP, Sylvester R, Brausi M, Holtl W, de Balincourt C, Andriole GL. Does isoniazid reduce side effects of intravesical Bacillus Calmette–Guerin therapy in superficial bladder cancer? Interim results of European Organization for Research and Treatment of Cancer Protocol 30911. J Urol 1997;157(4):1246–9.

34. Tuncer S, Tekin MI, Ozen H, Bilen C, Unal S, Remzi D, Lamm DL. Detection of Bacillus Calmette–Guerin in the blood by the polymerase chain reaction method of treated bladder cancer patients. J Urol 1997;158(6):2109–12.

35. Galazka AM, Lauer BA, Henderson RH, Keja J. Indications and contraindications for vaccines used in the Expanded Programme on Immunization. Bull World Health Organ 1984;62(3):357–66.

36. DiMaio VJ, Froeda RG. Allergic reactions to the tine test. JAMA 1975;233(7):769.

37. Radonic M. Systemic allergic reactions due to occupational inhalation of tuberculin aerosol. Ind Med Surg 1966;35(1):24–6.

38. Burgoyne CF, Verstraeten TC, Friberg TR. Tuberculin skin-test-induced uveitis in the absence of tuberculosis. Graefes Arch Clin Exp Ophthalmol 1991;229(3):232–6.

39. Namen AM, Grosvenor AR, Chin R, Daybell D, Adair N, Woodruff RD, Kavanagh PV, Haponik EF. Pulmonary infiltrates after intravesical Bacille Calmette–Guerin: two cases and review of the literature. Clin Pulm Med 2001;8:177–9.

40. Tardieu M, Truffot-Pernot C, Carriere JP, Dupic Y, Landrieu P. Tuberculous meningitis due to BCG in two previously healthy children. Lancet 1988;1(8583):440–1.

41. Morrison WL, Webb WJ, Aldred J, Rubenstein D. Meningitis after BCG vaccination. Lancet 1988;1(8586):654–5.

42. Katznelson D, Gross S, Sack J. Polyneuritis following BCG re-vaccination. Postgrad Med J 1982;58(682):496–7.

43. Yen MY, Liu JH. Bilateral optic neuritis following Bacille Calmette–Guerin (BCG) vaccination. J Clin Neuroophthalmol 1991;11(4):246–9.

44. Milstien JB, Gibson JJ. Quality control of BCG vaccine by WHO: a review of factors that may influence vaccine effectiveness and safety. Bull World Health Organ 1990;68(1):93–108.

45. Easton PA, Hershfield ES. Lymphadenitis as a late complication of BCG vaccination. Tubercle 1984;65(3):205–8.

46. Simma B, Dietze O, Vogel W, Ellemunter H, Guggenbichler JP. Bacille Calmette–Guerin-associated neonatal hepatitis. Eur J Pediatr 1991;150(6):423–4.

47. Kanwar AJ, Kaur S, Bansal R, Radotra BD, Sharma R. Lupus vulgaris following BCG vaccination. Int J Dermatol 1988;27(7):525–6.

48. Sasmaz R, Altinyazar HC, Tatlican S, Eskioglu F, Yurtsever P. Recurrent lupus vulgaris following repeated BCG (Bacillus Calmette Guerin) vaccination. J Dermatol 2001;28(12):762–4.

49. Radeff B, Harms M. Acute febrile neutrophilic dermatosis (Sweet's syndrome) following BCG vaccination. Acta Derm Venereol 1986;66(4):357–8.

50. Sadeghi E, Kumar PV. Eczema vaccinatum and postvaccinal BCG adenitis—case report. Tubercle 1990;71(2):145–6.

51. Buchs N, Chevrel G, Miossec P. Bacillus Calmette–Guerin induced aseptic arthritis: an experimental model of reactive arthritis. J Rheumatol 1998;25(9):1662–5.

52. Kodali VR, Clague RB. Arthritis after BCG vaccine in a healthy woman. J Intern Med 1998;244(2):183–4.

53. Bottiger M, Romanus V, de Verdier C, Boman G. Osteitis and other complications caused by generalized BCG-itis. Experiences in Sweden. Acta Paediatr Scand 1982;71(3):471–8.

54. Boman G, Sjogren I, Dahlstrom G. A follow-up study of BCG-induced osteo-articular lesions in children. Bull Int Union Tuberc 1984;59:198.

55. Kroger L, Korppi M, Brander E, Kroger H, Wasz-Hockert O, Backman A, Rapola J, Launiala K, Katila ML. Osteitis caused by Bacille Calmette–Guerin vaccination: a retrospective analysis of 222 cases. J Infect Dis 1995;172(2):574–6.

56. Peltola H, Salmi I, Vahvanen V, Ahlqvist J. BCG vaccination as a cause of osteomyelitis and subcutaneous abscess. Arch Dis Child 1984;59(2):157–61.

57. Krepela K, Galliova J, Sejdova E, Hajkova H, Maliniak J. Kostni komplikace po BCG vakcinaci. [Osseous complications after BCG vaccination.] Cesk Pediatr 1985;40(5):263–6.

58. Marik I, Kubat R, Slosarek M. BCG osteomyelitis et gonitis u batolete. [BCG osteomyelitis and gonitis in a small child.] Acta Chir Orthop Traumatol Cech 1984;51(6):495–503.

59. Krepela V, Galliova J, Kubec V, Marik J. Vliv snizene davky BCG vakciny na vyskyt kostnich komplikaci po kalmetizaci. [The effect of reduced doses of BCG vaccine on the occurrence of osseous complications after vaccination.] Cesk Pediatr 1992;47(3):134–6.

60. Geissler W, Pumberger W, Wurnig P, Stuhr O. BCG osteomyelitis as a rare cause of mediastinal tumor in a one-year-old child. Eur J Pediatr Surg 1992;2(2):118–21.

61. Hanimann B, Morger R, Baerlocher K, Brunner C, Giger T, Schopfer K. BCG Osteitis in der Schweiz. [BCG osteitis in Switzerland. A report of 6 cases.] Schweiz Med Wochenschr 1987;117(6):193–8.

62. Aftimos S, Nicol R. BCG osteitis: a case report. NZ Med J 1986;99(800):271–3.

63. Kolandaivelu G, Manohar K, Bose JC, Rajagopal P. Osteitis of humerus following BCG vaccination. J Indian Med Assoc 1986;84(6):184–5.

64. Lotte A, Wasz-Hockert O, Poisson N, Dumitrescu N, Verron M, Couvet E. BCG complications. Estimates of the risks among vaccinated subjects and statistical analysis of their main characteristics. Adv Tuberc Res 1984;21: 107–193.

65. Latini JM, Wang DS, Forgacs P, Bihrle W 3rd. Tuberculosis of the penis after intravesical Bacillus Calmette–Guerin treatment. J Urol 2000;163(6):1870.

66. Lotte A, Wasz-Hockert O, Poisson N, Dumitrescu N, Verron M, Couvet E. A bibliography of the complications of BCG vaccination. A comprehensive list of the world literature since the introduction of BCG up to July 1982, supplemented by over 100 personal communications. Adv Tuberc Res 1984;21:194–245.

67. Lotte A, Wasz-Hockert O, Poisson N, Engbaek H, Landmann H, Quast U, Andrasofszky B, Lugosi L, Vadasz I, Mihailescu P, et al. Second IUATLD study on complications induced by intradermal BCG-vaccination. Bull Int Union Tuberc Lung Dis 1988;63(2):47–59.

68. Proctor JW, Zidar B, Pomerantz M, Yamamura Y, Eng CP, Woodside D. Anaphylactic reaction to intralesional B.C.G Lancet 1978;2(8081):162.

69. Tshabalala RT. Anaphylactic reactions to BCG in Swaziland. Lancet 1983;1(8325):653.

70. Gonzalez B, Moreno S, Burdach R, Valenzuela MT, Henriquez A, Ramos MI, Sorensen RU. Clinical presentation of Bacillus Calmette–Guerin infections in patients with immunodeficiency syndromes. Pediatr Infect Dis J 1989;8(4):201–6.

71. Minegishi M, Tsuchiya S, Imaizumi M, Yamaguchi Y, Goto Y, Tamura M, Konno T, Tada K. Successful transplantation of soy bean agglutinin-fractionated, histoincompatible, maternal marrow in a patient with severe combined immunodeficiency and BCG infection. Eur J Pediatr 1985;143(4):291–4.

72. Lin CY, Hsu HC, Hsieh HC. Treatment of progressive Bacillus Calmette–Guerin infection in an immunodeficient infant with a specific bovine thymic extract (thymostimulin). Pediatr Infect Dis 1985;4(4):402–5.

73. Casanova JL, Jouanguy E, Lamhamedi S, Blanche S, Fischer A. Immunological conditions of children with BCG disseminated infection. Lancet 1995;346(8974):581.

74. Casanova JL, Blanche S, Emile JF, Jouanguy E, Lamhamedi S, Altare F, Stephan JL, Bernaudin F, Bordigoni P, Turck D, Lachaux A, Albertini M, Bourrillon A, Dommergues JP, Pocidalo MA, Le Deist F, Gaillard JL, Griscelli C, Fischer A. Idiopathic disseminated Bacillus Calmette–Guerin infection: a French national retrospective study. Pediatrics 1996;98(4 Pt 1):774–8.

75. Hodsagi M, Uhereczky G, Kiraly L, Pinter E. BCG dissemination in chronic granulomatous disease (CGD). Dev Biol Stand 1986;58(Pt A):339–46.

76. Kobayashi Y, Komazawa Y, Kobayashi M, Matsumoto T, Sakura N, Ishikawa K, Usui T. Presumed BCG infection in a boy with chronic granulomatous disease. A report of a case and a review of the literature. Clin Pediatr (Phila) 1984;23(10):586–9.

77. Smith PA, Wittenberg DF. Disseminated BCG infection in a child with chronic granulomatous disease. A case report. S Afr Med J 1984;65(20):821–2.

78. Trevenen CL, Pagtakhan RD. Disseminated tuberculoid lesions in infants following BCG vaccination. Can Med Assoc J 1982;127(6):502–4.

79. Erdos Z, Szabo I. Recovered case of BCG sepsis. Dev Biol Stand 1986;58:319.

80. Heyderman RS, Morgan G, Levinsky RJ, Strobel S. Successful bone marrow transplantation and treatment of BCG infection in two patients with severe combined immunodeficiency. Eur J Pediatr 1991;150(7):477–80.

81. Ehrengut W. BCG-itis während der Kindheit und in der Schwangerschaft. Zugleich ein Beitrag zu einer BCG-bedingten nekrotisierenden zerebralen Arteriitis. [BCG-induced inflammation during childhood and in pregnancy. Additionally a contribution to BCG-induced necrotising cerebral arteritis.] Klin Padiatr 1990; 202(5):303–7.

82. Molz G, Hartmann HP, Griesser HR. Generalisierte BCG-Infektion bei einem 7 Wochen alten, plötzlich gestorbenen Säugling. [Generalized BCG infection associated with the sudden death of a 7-week-old infant.] Pathologe 1986;7(4):216–21.

83. Wolff M, Dopfer R, Hassberg D, Niethammer D. BCGi tis als Ursache eines Mediastinaltumors. [BCGitis as a cause of mediastinal tumor.] Monatsschr Kinderheilkd 1993;141(5):409–11.

84. Baum WF, Wessel H, Exadaktylos P, et al. Die BCG-Histiocytose-eine Form der generalisierten BCG-Infektion. Dtsch Gesundheitswes 1983;37:1384.

85. Kunzel W, Frey G, Gunther J, et al. Geheilte BCG-Histiocytose bei isolierter temporärer Störung der T-Lymphocytenfunktion. Dtsch Gesundheitswes 1982;37:1384.

86. Power JT, Stewart IC, Ross JD. Erythromycin in the management of troublesome BCG lesions. Br J Dis Chest 1984;78(2):192–4.

87. Wolinsky E, Smith MM, Steenken W Jr. Drug susceptibilities of 20 atypical as compared with 19 selected strains of mycobacteria. Am Rev Tuberc 1957;76(3):497–502.

88. Singh G, Singh M. Erythromycin for BCG cold abscess. Lancet 1984;2(8409):979.

89. Akenzua GI, Sykes RM. Management of suppurative regional lymphadenitis complicating BCG vaccination in newborns. Niger J Pediatr 1986;13:65.

90. Caglayan S, Yegin O, Kayran K, Timocin N, Kasirga E, Gun M. Is medical therapy effective for regional lymphadenitis following BCG vaccination? Am J Dis Child 1987;141(11):1213–4.

91. Skegg DC. BCG vaccination and the incidence of lymphomas and leukaemia. Int J Cancer 1978;21(1):18–21.

92. Lilienfeld AM, Pedersen E, Dowd JE. In: Cancer Epidemiology: Methods of Study. Baltimore, MD: Johns Hopkins Press,, 1967:72.

93. Snider DE, Comstock GW, Martinez I, Caras GJ. Efficacy of BCG vaccination in prevention of cancer: an update. J Natl Cancer Inst 1978;60(4):785–8.

94. Kendrick MA, Comstock GW. BCG vaccination and the subsequent development of cancer in humans. J Natl Cancer Inst 1981;66(3):431–7.

95. Ambrosch F, Wiedermann G, Krepler P. Studies on the influence of BCG vaccination on infantile leukemia. Dev Biol Stand 1986;58(Pt A):419–24.

96. Haro AS. The effect of BCG-vaccination and tuberculosis on the risk of leukaemia. Dev Biol Stand 1986;58(Pt A):433–49.

97. Immunizations Practices Advisory Committee (ACIP). Recommendations on BCG vaccines. MMWR Morb Mortal Wkly Rep 1979;28(1):241.

98. Hesseling AC, Marais BJ, Gie RP, Schaaf HS, Fine PEM, Godfrey-Faussett P, Beyers N. The risk of disseminated Bacille Calmette-Guérin (BCG) disease in HIV-infected children. Vaccine 2007;25:14–18.

99. Fallo A, Torrado L, Sanchez A, Cerqueiro C, Shadgrosky L, Lopez EL. Delayed complications of BCG vaccination in HIV-infected children. Presented at the International AIDS Society meeting 2005 <http://www.who.int/vaccine_-safety/topics/bcg/immunocompromised/index.html.

100. Ninane J, Grymonprez A, Burtonby G, Francois A, Cornus G. Disseminated BCG in HIV infection. Arch Dis Child 1988;63:1268–9.

101. Clements CJ, von Reyn CF, Mann JM. HIV infection and routine childhood immunization: a review. Bull World Health Organ 1987;65(6):905–11.

102. Besnard M, Sauvion S, Offredo C, Gaudelus J, Gaillard JL, Veber F, Blanche S. Bacillus Calmette-Gucrin infection after vaccination of human immunodeficiency virus-infected children. Pediatr Infect Dis J 1993;12(12):993–7.

103. Borderon JC, Despert F, Le Touze A, Boscq M, Quentin R, Laugier J. Mesenteric adenitis due to BCG in an 8-year-old girl with AIDS vaccinated at the age of 1 month. Med Mal Infect 1996;26(Special Issue Jun):676–8.

104. von Reyn CF, Clements CJ, Mann JM. Human immuno-deficiency virus infection and routine childhood immunisation. Lancet 1987;2(8560):669–72.

105. Reynes J, Perez C, Lamaury I, Janbon F, Bertrand A. Bacille Calmette–Guerin adenitis 30 years after immuniza-tion in a patient with AIDS. J Infect Dis 1989;160(4):727.

106. Mvula M, Ryder R, Manzila T, et al. In: Response to childhood vaccination in African children with HIV infectionAbstracts, IV International Conference on AIDS, Stockholm, Sweden 12–16 June 1988:681–8.

107. Embree J, Datta P, Braddick M, et al. In: Vaccinations of infants of HIV-seropositive mothersAbstracts, IV International Conference on AIDS, Stockholm, Sweden 12–16 June 1988:681–8.

108. O'Brien KL, Ruff AJ, Louis MA, Desormeaux J, Joseph DJ, McBrien M, Coberly J, Boulos R, Halsey NA. Bacillus Calmette–Guerin complications in children born to HIV-1-infected women with a review of the literature. Pediatrics 1995;95(3):414–8.

109. Committee on Immunization. Guide for adult immunizationPhiladelphia: American College of Physicians;. 1985.

110. Global Advisory Group of the Expanded Programme on Immunization (EPI). Report on the meeting 13–17 October, 1986, New Delhi. Unedited document. WHO/EPI/Geneva/87/1, 1986.

111. Global Advisory Committee on Vaccine Safety. Wkly Epidemiol Rec 2003;78(32):282–411–12 June 2003.

112. Global Advisory Committee on Vaccine Safety (GACVS). Revised BCG vaccination guidelines for infants at risk for HIV infection. Wkly Epidemiol Rec 2007;82:193–6.

113. Satgé D, Pommepuy I, Goburdhun J, Flejou J-F. Ulcerative terminal ileitis after BCG therapy for bladder. Histopathology 2002;41:266–8.

Cholera vaccine

General Information

Cholera vaccines consist of live attenuated or heat-killed *Vibrio cholerae* organisms. They can be given orally or parenterally.

Following the administration of live oral cholera vac-cine (containing the attenuated strain *V. cholerae* CVD 103-HgR, prepared from *V. cholerae* 01 strain 569B), there were significant rises in serum antitoxin concen-trations and only few mild adverse effects (SED-13, 925). There was protective efficacy in 82–100% of healthy adult volunteers and no difference in adverse effects between 25 recipients of the vaccine and 26 controls (1).

A randomized, double-blind, placebo-controlled trial using a live oral cholera vaccine (strain CVD 103, derived from the *V. cholerae* 01 classical Inaba strain 569B by deletion of the genes encoding the A subunit of cholera toxin) was conducted in 50 healthy Swiss adults. There was a significant rise in serum antitoxin titers in 76% of the volunteers. Two vaccinees reported watery stools after immunization (2).

When an oral single-dose cholera vaccine against *V. cholerae* 0139 was given to ten volunteers there was 83% protective efficacy and the only adverse effect, in one volunteer, was mild diarrhea (3).

The results with an oral cholera vaccine consisting of the immunogenic but completely non-toxic B subunit of cholera toxin in combination with heat- and formalin-killed *V. cholerae* represent an improvement compared with previous parenteral vaccines (4). The frequency of adverse effects (pain at the injection site, nausea, diar-rhea) was low (5).

Severe complications connected with cholera (or com-bined) immunization are extremely rare and the causal relation is always doubtful. However, when they do occur they constitute a contraindication to further administration. There are occasional reports of neurological and psychiatric reactions (SED-8, 706) (SEDA-1, 246), Guillain–Barré syn-drome (SEDA-1, 246), myocarditis (6,7), myocardial infarc-tion (SEDA-3, 261), a syndrome similar to immune complex disease (8), acute renal insufficiency accompanied by hepa-titis (9), and pancreatitis (10).

Drug Administration

Drug formulations

A parenteral cholera vaccine consisting of a heat-killed, phenol-preserved, mixed suspension of the Inaba and Ogawa subtypes of *V. cholerae*, Serovar 01, has been used subcutaneously or intramuscularly, or for booster doses intradermally. About 1% of vaccinees develop mild local skin lesions comprising transitory soreness at the injection site within 5-7 days after injection, characterized by erythema, swelling, pain, and induration, and rarely resulting in ulceration. More general reactions are allergic: as a rule, these amount at most to slight pyrexia, headache, and malaise. However, considering the low protective efficacy and high reactogenicity of this vaccine, the World Health Organization and many national health authorities do not recommend it anymore and most manufacturers have stopped producing it.

References

1. Davis R, Spencer CM. Live oral cholera vaccine. A preliminary review of its pharmacology and clinical potential in providing protective immunity against cholera. Clin Immunother 1998;4:235–47.
2. Cryz SJ Jr, Levine MM, Kaper JB, Furer E, Althaus B. Randomized double-blind placebo controlled trial to evaluate the safety and immunogenicity of the live oral cholera vaccine strain CVD 103-HgR in Swiss adults. Vaccine 1990;8(6):577–80.
3. Coster TS, Killeen KP, Waldor MK, Beattie DT, Spriggs DR, Kenner JR, Trofa A, Sadoff JC, Mekalanos JJ, Taylor DN. Safety, immunogenicity, and efficacy of live attenuated *Vibrio cholerae* O139 vaccine prototype. Lancet 1995;345(8955):949–52.
4. Holmgren J, Svennerholm AM. New vaccines against bacterial enteric infections. Scand J Infect Dis Suppl 1990;70:149–56.
5. Markman B. Symptoms of reactogenicity in field trial of oral cholera vaccine. Lancet 1990;336(8710):320.
6. Gavrilesco S, Streian C, Constantinesco L. Tachycardie ventriculaire et fibrillation auriculaire associées après vaccination anticholériques. [Associated ventricular tachycardia and auricular fibrillation after anticholera vaccination.] Acta Cardiol 1973;28(1):89–94.
7. Driehorst J, Laubenthal F. Akute Myocarditis nach Choleraschutzimpfung. [Acute myocarditis after cholera vaccination.] Dtsch Med Wochenschr 1984;109(5):197–8.
8. Mall T, Gyr K. Episode resembling immune complex disease after cholera vaccination. Trans R Soc Trop Med Hyg 1984;78(1):106–7.
9. Eisinger AJ, Smith JG. Acute renal failure after TAB and cholera vaccination. BMJ 1979;1(6160):381–2.
10. Gatt DT. Pancreatitis following monovalent typhoid and cholera vaccinations. Br J Clin Pract 1986;40(7):300–1.

Diphtheria vaccine

General Information

Diphtheria vaccine contains diphtheria toxoid carried on aluminium hydroxide or calcium phosphate. Single antigen products are available only for cases in which combined antigens should not be used. The formulations used in most countries are a childhood formulation containing 25–30 Lf (flocculating units) of diphtheria toxoid (D) and

an adult formulation containing 2 Lf of diphtheria toxoid (d). The formulations of choice in routine immunization are DTP (diphtheria and tetanus toxoids combined with pertussis vaccine), DT (diphtheria and tetanus toxoids) for pediatric use, and Td (tetanus and diphtheria toxoids with a limited amount of diphtheria antigen) for use in older children and adults.

Combined diphtheria + tetanus immunization

In a prospective cohort study in Italy, 380 children aged 6 years were randomly assigned to receive either the DT or the Td vaccine as a booster dose in order to determine whether a booster dose of Td (diphtheria–tetanus vaccine with a reduced amount of diphtheria toxoid, adult formulation) would produce comparable diphtheria antibody titers but lower reactogenicity than DT (childhood formulation). The frequencies of symptoms within 3 days of vaccine administration were similar in the two groups, except for local redness and swelling, which were significantly more common in the children who received DT vaccine: redness 31% versus 16%, and swelling 36% versus 26%. The mean duration of local symptoms was 3.3 days in the diphtheria–tetanus group and 2.6 days in the Td group. After booster immunization, 97% of children in the DT group and 91% of those in the Td group had antibody concentrations of at least 1 IU/ml ("long-term" protection titer) (1).

There have been comparisons of the immunogenicity and reactogenicity of different diphtheria vaccines. They have involved single or combined administration of diphtheria and/or tetanus toxoids (SEDA-13, 279) (SEDA-15, 345), booster immunization using Td vaccines including either aluminium hydroxide or calcium phosphate as adjuvant (SEDA-20, 288), or either plain or adsorbed formulations (SEDA-21, 328).

Adverse events after diphtheria–tetanus vaccine in the USA in 1982–84 have been reviewed in detail (SEDA-13, 273). The usual types of local intolerance can be seen. For example, some 5% of schoolchildren develop redness and swelling, whilst some older children develop enlargement of the regional lymph nodes. Such reactions are much less common in young children, and much more common in children given combined vaccines.

General reactions (seen in some older children and adults) are usually limited to brief fever; sustained fever and other systemic reactions are uncommon unless the person has been hyperimmunized. After the administration of DT vaccine, local reactions, generally erythema and induration, with or without tenderness, can occur. In hyperimmunized cases, Arthus-type hypersensitivity reactions can occur. These characteristically severe local reactions generally start 2–8 hours after an injection. People who have such reactions usually have very high serum antitoxin concentrations and one should be careful not to administer a booster more than once every 10 years.

The occurrence of epidemic diphtheria in Eastern Europe led to the recommendation in the UK that those aged 15–18 years should receive a combined tetanus and low-dose diphtheria toxoid vaccine instead of a tetanus booster alone. In March 1995, 220 children aged 14–16 years were inadvertently given high-dose diphtheria and tetanus toxoid vaccine, and their parents were sent a questionnaire; 153 replied. A total of 141 (92%) of adolescents reported one or more reactions, most of which were classified as mild or moderate and lasted less than 1 week. However, 47 (31%) reported at least one severe local or systemic reaction (2).

In a study of adverse events after immunization in New Zealand in 1990–95 (3), reactions at the injection site after adult tetanus–diphtheria vaccine (68 reports per 100 000 immunizations) were reported five times more often than with tetanus vaccine.

Combined diphtheria + tetanus + pertussis immunization

An overview of clinical trials with a special diphtheria and tetanus toxoids and acellular pertussis (DTaP) vaccine has been published (4). The vaccine contains as pertussis components purified filamentous hemagglutinin, pertactin, and genetically engineered pertussis toxin. The vaccine induces high and long-lasting immunity and is at least as efficacious as most whole-cell pertussis vaccines and similar in efficacy to the most efficacious DTaP vaccines that contain three pertussis antigens. The vaccine is better tolerated than whole-cell vaccines and has a similar reactogenicity profile to other acellular vaccines.

A vaccine containing diphtheria and tetanus toxoids and DTaP with reduced antigen content for diphtheria and pertussis (TdaP) has been compared with a licensed reduced adult-type diphtheria–tetanus vaccine and with an experimental candidate monovalent DTaP vaccine with reduced antigen content (ap) (5). A total of 299 healthy adults (mean age 30 years) were randomized into three groups to receive one dose of the study vaccines. The antibody responses (antidiphtheria, antitetanus, antipertussis toxin, antipertactin, antifilamentous hemagglutinin) were similar in all groups. The most frequently reported local symptom was pain at the injection site (62–94%), but there were no reports of severe pain; redness and swelling with a diameter of 5 cm or more occurred in up to 13%. The incidence of local symptoms was similar after TdaP and Td immunization. The most frequently reported general symptoms were headache and fatigue (20–50%). The incidence of general symptoms was similar in the TdaP and Td groups. There were no reports of fever over 39°C. No serious adverse events were reported.

Data from the Third National Health and Nutrition Survey (1988–94) have been used to analyse the possible effects of DTP or tetanus immunization on allergies and allergy-related symptoms among 13 944 infants, children, and adolescents aged 2 months to 16 years in the USA (6). The authors concluded that DTP or tetanus immunization increases the risk of allergies and related respiratory symptoms in children and adolescents. However, the small number of non-immunized individuals and the study design limited their ability to make firm causal inferences about the true magnitude of effect.

Tetravalent, pentavalent, and hexavalent immunization

DTaP or DTwP vaccine can be combined with other antigens, such as *Haemophilus influenzae* type b (Hib), inactivated poliovaccine (IPV), and hepatitis B vaccine. In children DTaP or DTwP vaccines are the basis for such combinations, while in adults it is mostly Td or Tdap vaccine combined with inactivated poliovaccine (IPV). Current safety concerns regarding combination vaccines have been defined and reviewed (7). The author concluded that there is no evidence that adding vaccines to combination products increases the burden on the immune system, which can respond to many millions of antigens. Combining antigens usually does not increase adverse effects, but it can lead to an overall reduction in adverse events. Before licensure, combination vaccines undergo extensive testing to assure that the new products are safe and effective.

The frequency, severity, and types of adverse reactions after DTP-Hib immunization in very preterm babies have been studied (8). Adverse reactions were noted in 17 of 45 babies: nine had major events (apnea, bradycardia, or desaturation) and eight had minor reactions (increased oxygen requirements, temperature instability, poor handling, and feeding intolerance). Babies who had major adverse reactions were significantly younger at the time of immunization than the babies who did not have major reactions. Of 27 babies immunized at 70 days or less, nine developed major reactions compared with none of those who were immunized at over 70 days.

The Hexavalent Study Group has compared the immunogenicity and safety of a new liquid hexavalent vaccine against diphtheria, tetanus, pertussis, poliomyelitis, hepatitis B, and Hib (DTP + IPV + HB + Hib vaccine, manufactured by Aventis Pasteur MSD, Lyon, France) with two reference vaccines, the pentavalent DTP + IPV + Hib vaccine and the monovalent hepatitis B vaccine, administrated separately at the same visit (9). Infants were randomized to receive either the hexavalent vaccine (*n* = 423) or (administered at different local sites) the pentavalent and the HB vaccine (*n* = 425) at 2, 4, and 6 months of age. The hexavalent vaccine was well tolerated (for details, see the monograph on Pertussis vaccines). At least one local reaction was reported in 20% of injections with hexavalent vaccine compared with 16% after the receipt of pentavalent vaccine or 3.8% after the receipt of hepatitis B vaccine. These reactions were generally mild and transient. At least one systemic reaction was reported in 46% of injections with hexavalent vaccine, whereas the respective rate for the recipients of pentavalent and HB vaccine was 42%. No vaccine-related serious adverse event occurred during the study. The hexavalent vaccine provided immune responses adequate for protection against the six diseases.

Organs and Systems

Cardiovascular

Myopericarditis has been attributed to Td-IPV vaccine (10).

- A 31-year-old man developed arthralgia and chest pain 2 days after Td-IPV immunization and had an acute myopericarditis. He recovered within a few days with high-dose aspirin.

The authors discussed two possible causal mechanisms, natural infection or an immune complex-mediated mechanism. Infection was excluded by negative bacterial and viral serology and the favorable outcome within a few days without antimicrobial treatment.

Nervous system

Supposed neurological adverse effects of diphtheria immunization have been reported, but a causal connection was unclear (11). Of five cases of neurological complications after diphtheria or diphtheria-tetanus immunizations two were classified as vaccine-induced poliomyelitis; the other three could be traced back to a hyperergic reaction to diphtheria toxoids in the cerebral vessels (12).

Guillain–Barré syndrome (13) and polyradiculoneuritis (14,15) have rarely been reported. In a national surveillance study of Guillain–Barré syndrome in the USA, 31 of 998 cases developed the illness within 8 weeks after immunization. Of these 31 cases, 5 had been immunized with DT or DPT vaccine (16).

The bioelectric activity of the brain after DT immunization was studied in healthy children. Electroencephalography showed significant changes in 13 of 17 children, which resolved within 3 weeks (17). The question has arisen of whether this is of more than experimental significance: between 1980 and 1982 in the Campana region of Italy, several cases of encephalopathy in children who had been given DT immunization 1 week before were reported to the health authorities. However, summarizing the results of a case-control study, Greco pointed out that the statistical association that he found between the incidence of encephalopathy and DT administration did not imply a causal association (18).

Sensory systems

Optic neuritis has been attributed to Td-IPV vaccine (19).

- Ten days after receiving Td-IPV vaccine a 56-year-old woman developed acute unilateral optic neuritis. Complete remission occurred within 6 weeks of prednisolone treatment. No other causes were found.

Skin

Erythema multiforme developed 8 hours after diphtheria–tetanus immunization in a 9-month-old infant (20). There have also been reports of erythema multiforme after hepatitis B vaccine, MMR vaccine, and DPT vaccine.

Bullous pemphigoid has been attributed to DTP-IPV vaccine (21).

A previous healthy 3.5-month-old infant developed bullous pemphigoid 3 days after receiving a first dose of DTP-IPV vaccine. *Staphylococcus aureus* was isolated from purulent bullae. The lesions resolved rapidly after treatment with antibiotics and methylprednisolone.

The authors mentioned 12 other cases of bullous pemphigoid, reported during the last 5 years, that had possibly been triggered by vaccines (influenza, tetanus toxoid booster, and DTP-IPV vaccine).

Immunologic

- A six-year-old child had anaphylaxis 30 minutes after a fifth dose of DT vaccine (22). Skin tests, in vitro determination of specific IgE antibodies, and immunoblotting assays showed that the IgE response was directed against tetanus and diphtheria toxoids. Cross-reactivity between the two toxoids was not demonstrated, indicating the presence of co-existing but non-cross-reacting IgE and IgG antibodies.

Susceptibility Factors

The only contraindication to administering single diphtheria toxoid or combined diphtheria and tetanus toxoids is a history of a severe hypersensitivity or neurological reaction after a previous dose.

References

1. Ciofi degli Atti ML, Salmaso S, Cotter B, Gallo G, Alfarone G, Pinto A, Bella A, von Hunolstein C. Reactogenicity and immunogenicity of adult versus paediatric diphtheria and tetanus booster dose at 6 years of age. Vaccine 2001;20(1–2):74–9.
2. Sidebotham PD, Lenton SW. Incidence of adverse reactions after administration of high dose diphtheria with tetanus vaccine to school leavers: retrospective questionnaire study. BMJ 1996;313(7056):533–4.
3. Mansoor O, Pillans PI. Vaccine adverse events reported in New Zealand 1990–5. NZ Med J 1997;110(1048):270–2.
4. Matheson AJ, Goa KL. Diphtheria-tetanus-acellular pertussis vaccine adsorbed (Triacelluvax; DTaP3-CB): a review of its use in the prevention of *Bordetella pertussis* infection. Paediatr Drugs 2000;2(2):139–59.
5. Van der Wielen M, Van Damme P. Tetanus–diphtheria booster in non-responding tetanus–diphtheria vaccinees. Vaccine 2000;19(9–10):1005–6.
6. Hurwitz EL, Morgenstern H. Effects of diphtheria-tetanus-pertussis or tetanus vaccination on allergies and allergy-related respiratory symptoms among children and adolescents in the United States. J Manipulative Physiol Ther 2000;23(2):81–90.
7. Halsey NA. Combination vaccines: defining and addressing current safety concerns. Clin Infect Dis 2001;33(Suppl 4):S312–8.
8. Sen S, Cloete Y, Hassan K, Buss P. Adverse events following vaccination in premature infants. Acta Paediatr 2001;90(8):916–20.
9. Mallet E, Fabre P, Pines E, Salomon H, Staub T, Schodel F, Mendelman P, Hessel L, Chryssomalis G, Vidor E,

Hoffenbach A, Abeille A, Amar R, Arsene JP, Aurand JM, Azoulay L, Badescou E, Barrois S, Baudino N, Beal M, Beaude-Chervet V, Berlier P, Billard E, Billet L, Blanc B, Blanc JP, Bohu D, Bonardo C, Bossu CHexavalent Vaccine Trial Study Group. Immunogenicity and safety of a new liquid hexavalent combined vaccine compared with separate administration of reference licensed vaccines in infants. Pediatr Infect Dis J 2000;19(12):1119–27.

10. Boccara F, Benhaiem-Sigaux N, Cohen A. Acute myopericarditis after diphtheria, tetanus, and polio vaccination. Chest 2001;120(2):671–2.

11. Van Ramshorst JD, Ehrengut W. Die Diphtherieschutzimpfung. In: Herrlich A, editor. Handbuch der Schutzimpfungen. Berlin: Springer, 1965:394.

12. Ehrengut W. Komplikationen nach Diphtherieschutzimpfung und Impfungen mit Diphtherietoxoid-Mischimpfstoffen. [Neural complications after diphtheria vaccination and inoculations with diphtheria toxoid-mixed vaccines. Observations on their etiopathogenesis.] Dtsch Med Wochenschr 1986;111(24):939–42.

13. Onisawa S, Sekine I, Ichimura T, Homma N. Guillain–Barré syndrome secondary to immunization with diphtheria toxoid. Dokkyo J Med Sci 1985;12:227.

14. Holliday PL, Bauer RB. Polyradiculoneuritis secondary to immunization with tetanus and diphtheria toxoids. Arch Neurol 1983;40(1):56–7.

15. Immunization Practices Advisory committee (ACIP). Diphtheria, tetanus, and pertussis: recommendations for vaccine use and other preventive measures. MMWR Recomm Rep 1991;40(RR-10):1–28.

16. Hurwitz ES, Holman RC, Nelson DB, Schonberger LB. National surveillance for Guillain–Barré syndrome: January 1978 March 1979. Neurology 1983;33(2):150–7.

17. Wstepne D. Prophylactic vaccinations and seizure activity in EEG. Neurol Neurochir Pol 1981;5:553.

18. Greco D. Case-control study on encephalopathy associated with diphtheria–tetanus immunization in Campania, Italy. Bull World Health Organ 1985;63(5):919–25.

19. Burkhard C, Choi M, Wilhelm H. Optikusneuritis als Komplikaton einer Tetanus–Diphtherie–Poliomyelitis–Schutzimpfung: ein Fallbericht. [Optic neuritis as a complication in preventive tetanus–diphtheria–poliomyelitis vaccination: a case report.] Klin Monatsbl Augenheilkd 2001;218(1):51–4.

20. Griffith RD, Miller OF 3rd. Erythema multiforme following diphtheria and tetanus toxoid vaccination. J Am Acad Dermatol 1988;19(4):758–9.

21. Baykal C, Okan G, Sarica R. Childhood bullous pemphigoid developed after the first vaccination. J Am Acad Dermatol 2001;44(Suppl 2):348–50.

22. Martin-Munoz MF, Pereira MJ, Posadas S, Sanchez-Sabate E, Blanca M, Alvarez J. Anaphylactic reaction to diphtheria-tetanus vaccine in a child: specific IgE/IgG determinations and cross-reactivity studies. Vaccine 2002;20(27–28):3409–12.

Haemophilus influenzae type b (Hib) vaccine

General Information

Haemophilus influenzae causes several infectious diseases in man, the most serious being meningitis. Most cases of *Haemophilus influenzae* infection are due to type b of the organism (Hib). Two types of Hib vaccines have been developed: Hib capsular polysaccharide (PRP) vaccines are first-generation Hib vaccines, while Hib conjugate vaccines are second-generation vaccines. The capsular polysaccharide vaccines do not protect infants and children under 18 months, whereas Hib conjugate vaccines have greater immunogenicity and induce a high rate of protection in children under 18 months. The latter have therefore completely replaced the first-generation polysaccharide vaccines.

Comparisons of adverse reactions to different vaccines are difficult, because vaccines are virtually always administered together with other vaccines. However, the general experience is that adverse reactions to Hib vaccine are mild. Most reactions develop when the vaccine is given simultaneously with DTP vaccine. There have been no deaths or permanent sequelae attributable to Hib immunization.

Four different types of conjugated vaccines are commercially available:

1. PRP-D Hib vaccine (a mutant polypeptide of diphtheria toxin covalently linked to PRP), for example ProHIBit (produced by Connaught Laboratories, Philadelphia).
2. HbOC vaccine (Hib oligosaccharides linked to the non-toxic diphtheria toxin variant CRM197), for example HibTITER (produced by Lederle-Praxis Biologicals, Pearl River, NY).
3. PRP-OMC (PRP conjugated to outer membrane protein of Neisseria meningitidis group B), for example PedvaxHIB (produced by Merck Sharp & Dohme Research Laboratories).
4. PRP-T Hib vaccine (tetanus toxoid linked to PRP), for example ActHIB or OmniHIB (produced by Pasteur Merieux, Lyon, France) and PRP-T (produced by Pasteur Mérieux Connaught and Glaxo SmithKline).

PRP-D-Hib has been almost completely replaced by different conjugated Hib vaccines (1–3). The four types of conjugated Hib vaccines have been reviewed (SED-12, 803) (SEDA-16, 377) (SEDA-17, 367) (SEDA-18, 330). More than 100 million doses of conjugated vaccines have now been administered worldwide, and no reports of deaths, anaphylaxis, or residual neurological damage have been causally connected with them.

In a study of the tolerability and immunogenicity of Hib vaccines, 30 volunteers aged 69–84 years were immunized with either Pedvax-Hib (a conjugate of Hib polysaccharide and an outer membrane protein complex of *Neisseria meningitidis*-PRP-OMP) or Hib TITER (a conjugate of Hib oligosaccharide and a non-toxic mutant diphtheria toxin, CRM 197-HbOC) (4). The volunteers received a pediatric dose. Before immunization, 40% of the volunteers had serum anti-PRP antibody concentrations below 1.0 µg/ml. Four weeks after immunization, all the volunteers had concentrations over 1.0 µg/ml, which is generally considered to be protective. Adverse effects of immunization were mild, except in one volunteer given HbOC, who developed extensive erythema and swelling at the injection site.

Different vaccines and immunization strategies have been evaluated in Denmark, Finland, Iceland, Norway, and Sweden (5). Few places outside Scandinavia have collected data on Hib immunization programs for so long (more than a decade has elapsed since universal Hib immunization was initiated in Scandinavia) and with similar accuracy. Phase 3 studies with PRP-D-Hib vaccine were done in Finland in the late 1980s, and PRP-D-Hib vaccine has been the only vaccine used in Iceland. HbOC vaccine was first compared with PRP-D-Hib vaccine in Finland and then reintroduced to the primary health-care system as the only Hib vaccine used. Finally, PRP-T-Hib vaccine was first temporarily used in Finland, and then as almost the only vaccine in Denmark, Norway, and Sweden. Besides the different conjugate vaccines, the immunization programs have differed in other aspects, such as immunization schedule and administration of vaccines (separate versus simultaneous administration with other vaccines, such as DT, DTP, DTaP, IPV, or MMR).

Experience with PRP-D derives from Finland and Iceland. In Finland, 14.1 adverse reactions per 100 000 doses were reported in all, consisting respectively of 5.3, 6.3, 4.4, and 2.9 per 100 000 doses of local reactions, fever, rash, and irritability. These rates probably underestimate the true rate. In Iceland, adverse effects have not been monitored, but no serious events have been reported. For PRP-CRM, there were 17.8 per 100 000 doses in all, of which 7.7 per 100 000 doses were due to local reactions, 8.9 to fever, and 8.3 to rash. PRP-T currently enjoys the largest use in Scandinavia, being the routine choice in three countries. Of 115 reactions reported in Denmark, none was of serious concern; most were local reactions, fever, or rash. In Norway, the incidence of systemic reactions was 1 in 550 doses; fever and other symptoms and signs similar to those after DTP vaccination were the most common complaints. However, two findings were characteristic of Norway, where local reactions were reported with an overall frequency of 1/1500 doses:

- immediate mild allergic reactions were more common than if DTP or IPV was given alone; their incidence was 1 in 35 000 doses of PRP-T;
- large swellings at the vaccination site, with edema and sometimes bluish or other discoloration of the legs, were reported in 18 children, suggesting an incidence of about 1 in 30 000 doses; this reaction developed within hours after vaccination and subsided within 24 hours; all the vaccinees recovered spontaneously without treatment or sequelae.

About 1.5 million doses of PRP-T have been used in Sweden. Similar concerns about adverse reactions have been raised as in Norway, and further elucidation has recently begun. In Finland (where PRP-T was used in 1990–93), there were 15.9 reports per 100 000 doses; 6.2 per 100 000 doses were local reactions, 5.9 fever, 2.8 rash, and 2.4 irritability (5).

Notwithstanding the different approaches taken in the various Scandinavian countries, the results are similar: before vaccination against invasive Hib diseases, first introduced in Finland in 1986, the incidence of cases in the five Scandinavian countries was 49 per 100 000 per year in 0- to 4-year olds and 3.5 per 100 000 overall.

During the next decade, Hib conjugates given to young children had about 95% effectiveness, regardless of which conjugate was used, whether two or three primary doses were used, and at no matter what age in early infancy the first vaccination was given. Invasive diseases due to Hib have thus been nearly eliminated.

On the request of the French Drug Agency, the Regional Pharmacovigilance Center of Tours has analysed the adverse events that occurred in 1986–90 with the use of three different tetravalent DTwP-IPV vaccines produced by French vaccine manufacturers (6). The most frequent adverse events were local reactions (43% of the 631 events reported). Serious adverse events represented 25% of all reported events, including 23 reports of persistent crying, 12 febrile seizures, 14 apyretic seizures, and 3 reports of shock-like events.

Immunological interference, particularly to the Hib response, has been assessed in 135 infants at 2, 4, 6, and 18 months of age in studies of two different types of administration of DTaP and Hib vaccines: combined administration of the two vaccines mixed in the same syringe and simultaneous administration of separate injections at different sites (7). The vaccines were well tolerated and there were no differences in the rates of local and systemic reactions. Immune responses were also comparable between the two groups.

In 57 volunteers, 32 of whom had recently undergone splenectomy, who received Hib conjugate vaccine, antibodies to Hib were measured at 2, 6, 12, 24, and 36 months after immunization (8). All tolerated the vaccine well and reached protective antibody titers. The authors concluded that the vaccine is safe and protective in patients with thalassemia.

Combinations of vaccines

Since immunization against diphtheria, tetanus, and pertussis is recommended at the same age as immunization against *Haemophilus influenzae*, children must usually receive two intramuscular injections at separate sites during the same visit. Combined vaccines, for example DTP/Hib vaccines, requiring one injection, could be preferable and are commercially available. Results of clinical trials comparing the safety and efficacy of combined and simultaneously administrated vaccines (Hib, DTP, MMR, IPV) have been presented (SEDA-17, 369) (SEDA-18, 330). In general, the rates of local and systemic reactions and antibody responses did not differ significantly between the groups. The immunogenicity and safety of a diphtheria–tetanus–pertussis–Hib combination vaccine (tetanus-conjugated Hib vaccine) have been compared with those of the same combination obtained by the reconstitution of lyophilized Hib vaccine with liquid DTP vaccine in 262 healthy infants randomized to receive injections at 2, 4, and 6 months of age, a subgroup of 134 of whom received a booster dose at 12 months (9). Systemic and local reactions were generally mild and did not differ significantly between the two groups. With regard to Hib antibodies, the combination vaccine was at least as immunogenic as the lyophilized formulation.

Four-, five- and six-component combination vaccines based on DTaP or DTwP vaccine, and including other antigens such as hepatitis B or Hib or IPV, will play an important role in future worldwide immunization programs. The first hexavalent combination vaccines (DTaP-IPV-HB-Hib) were licensed in Germany; two hexavalent vaccines are available, manufactured by SmithKline Beecham and Aventis Pasteur.

The Hexavalent Study Group has compared the immunogenicity and safety of a new liquid hexavalent vaccine against diphtheria, tetanus, pertussis, poliomyelitis, hepatitis B, and Hib (DTP + IPV + HB + Hib vaccine, manufactured by Aventis Pasteur MSD, Lyon, France) with two reference vaccines, the pentavalent DTP + IPV + Hib vaccine and the monovalent hepatitis B vaccine, administrated separately at the same visit (10). Infants were randomized to receive either the hexavalent vaccine (n = 423) or (administered at different local sites) the pentavalent and the HB vaccine (n = 425) at 2, 4, and 6 months of age. The hexavalent vaccine was well tolerated (for details, see the monograph on Pertussis vaccines). At least one local reaction was reported in 20% of injections with hexavalent vaccine compared with 16% after the receipt of pentavalent vaccine or 3.8% after the receipt of hepatitis B vaccine. These reactions were generally mild and transient. At least one systemic reaction was reported in 46% of injections with hexavalent vaccine, whereas the respective rate for the recipients of pentavalent and HB vaccine was 42%. No vaccine-related serious adverse event occurred during the study. The hexavalent vaccine provided immune responses adequate for protection against the six diseases.

Regimens using different conjugated vaccines for primary series

Currently most advisory bodies for immunization practice recommend that the same Hib conjugate vaccine used to initiate a priming series should be continued for the entire series. However, interest in using different vaccines for sequential doses (mixed regimens) has been long-standing. Two- or three-dose mixed regimens of Hib conjugate vaccines have been compared in two randomized trials in 140 and 181 infants (11). In both trials, a group of infants received PRP-meningococcal protein conjugate vaccine (OMP, outer membrane protein) at 2 months of age, followed by either the same or another conjugate vaccine, oligosaccharide CRM197 Hib vaccine (HbOC, Hib oligosaccharides linked to the non-toxic diphtheria toxin variant CRM197) or PRP-tetanus toxoid Hib vaccine, either once (at 4 months of age) or twice (at 4 and 6 months of age). All the mixed regimens provided a mean anti-PRP antibody concentration that did not differ substantially compared with currently recommended regimens of proven or inferred efficacy. Recipients of a second dose of OMP vaccine had higher rates of swelling, erythema, and induration compared with the other recipients.

Organs and Systems

Nervous system

The neurological complications of conjugated Hib vaccines have been reviewed (12). A single convulsion occurred 12 hours after immunization in a 3-month-old child and a hyporesponsive episode occurred in another 3-month-old child.

Guillain–Barré syndrome has been reported after immunization with several different vaccines, including Hib conjugate vaccine (13,14). One patient had also received DTP and oral poliomyelitis vaccine.

Metabolism

The incidence of diabetes mellitus has been studied in Finnish children born between October 1983 and September 1985 compared with children born between October 1985 and 1987 (1). Of the children born in 1985–87, 50% received Hib vaccine (diphtheria conjugated PRP-D-Hib vaccine) at 3, 4, and 6 months as a primary course and a booster dose at 14–18 months; the other 50% received one dose of the same vaccine at the age of 24 months. Taking into account the documented increase in diabetes in Finnish children, the small difference between the incidence rate of diabetes in children born during the period 1985–87 and immunized at 18 months and the incidence rate of children born in 1983–85 and not immunized against Hib disease was expected, with a slightly higher non-significant incidence in immunized children. In children over 4 years of age primed during the first year of life the incidence rate of diabetes was also slightly but non-significantly higher than in children immunized at 18 months of age. The authors concluded that early PRP-D-Hib immunization does not increase the risk of diabetes during the first 10 years of life.

This study has been criticized on the grounds that the authors did not present data comparing the incidence rates of children born in 1985–87 and primed during the first year of life with children born in 1983–85 and not immunized (15). The critics presented significant differences in the incidence rate of diabetes between the two groups of children and concluded that immunization had increased the risk of diabetes. However, the relative risk of diabetes during the first 10 years of life in the immunized children born in 1985–87 was 1.19 compared with the non-immunized children born in 1983–85.

In another Finnish study, about 116 000 children born in Finland between 1 October 1985 and 31 August 1987 were randomized to receive four doses of Hib vaccine (PPR-D, Connaught) starting at 3 months of life or one dose starting after 24 months of life (16). A control cohort included all 128 500 children born in Finland in the 24 months before the study. The difference in cumulative incidence between those who received four doses of the vaccine and those who received none was 54 cases of diabetes per 100 000 at 7 years (relative risk = 1.26). Most of the extra cases of diabetes occurred in statistically significant clusters starting about 38 months after immunization and lasting about 6–8 months. The authors concluded that exposure to Hib immunization is associated with an increased risk of type 1 diabetes mellitus.

In 1997, US researchers suggested that immunization at 28 days after birth can cause type 1 diabetes mellitus in susceptible individuals. In May 1998, several institutions, including the National Institute of Allergy and Infectious Diseases, the Centers for Disease Control, the World Health Organization, and the UK's Department of Health, sponsored a workshop to assess the evidence of a possible link. Immunologists, diabetologists, epidemiologists, policymakers, and observers debated the available evidence and concluded that a causal link between immunization and type 1 diabetes is not supported. The results of a large, randomized, controlled trial of immunization against Hib carried out in Finland in 1985–87 (5) were also reanalysed and showed no association between the incidence of diabetes mellitus and the addition of another antigen to the schedule, irrespective of timing. Data reanalysis was made possible by prospective linking of individual information on exposure (in this case infant immunization or the administration of placebo) with the Finnish diabetes register (17).

Long-Term Effects

Tumorigenicity

The results of a large case-control study in the USA in 1999 raised the possibility that conjugate vaccine against *Hemophilus influenzae* type b was inversely associated with the risk of childhood leukemia (18). In 2000, reanalysis of data from an earlier trial in Finland suggested a non-significant protective effect of early versus late administration of a Hib conjugate vaccine (19). A further re-analysis of data from a nationwide immunization trial in Finland in the 1980s (20) was recently used to study the incidence of childhood leukemia in Hib-immunized children. In the 1980s, all 125 129 children born in Finland between 1 September 1987 and 31 August 1989 were enrolled, with a participation rate of 94%. All participating children received three doses of Hib vaccine at 4, 6, and 14-18 months of age, either PRP-D-Hib vaccine (children with an odd day of birth) or HbOC vaccine (children with an even day of birth). All cases of childhood leukemia diagnosed in Finland were taken from the Finnish Cancer Registry. A total of 80 cases of leukemia were diagnosed from birth to age 12 years among children born during the trial period. There were 35 cases among children born on an odd date and 45 among those born on an even date. This corresponds to a relative risk of 1.14 (95% CI = 0.63, 2.08) for subjects in the HbOC vaccine group. There were 69 cases of acute lymphoblastic leukemias, of which 30 were in the cohort of children born on an odd date, and 39 in the cohort born on an even date. There was no suggestion of different risks of childhood leukemia among Finnish children who received either PRP-D-Hib or HbOC vaccine. The authors concluded that the results of their study could be taken to suggest that the studies mentioned above did not show a causal relation, i.e. lack of protective effect of Hib immunization against childhood leukemia (21).

Susceptibility Factors

Of 23 patients with Hodgkin's disease and splenectomy immunized with Hib conjugate vaccine, most responded, although the antibody response was significantly lower than in healthy people (22). There were adverse effects in three of the vaccinees: in one case nausea, vertigo, and weakness occurred 2–4 days after administration of the vaccine, myalgias occurred in another case, and fever and myalgias (after primary immunization, with milder symptoms after the booster) in a third.

References

1. Karvonen M, Cepaitis Z, Tuomilehto J. Association between type 1 diabetes and *Haemophilus influenzae* type b vaccination: birth cohort study. BMJ 1999;318(7192):1169–72.
2. Evans G. Personal communication, 1999.
3. Dukes MNG, Swartz B. Responsibility for Drug-induced Injury. Amsterdam: Elsevier, 1988.
4. Kantor E, Luxenberg JS, Lucas AH, Granoff DM. Phase I study of the immunogenicity and safety of conjugated *Hemophilus influenzae* type b vaccines in the elderly. Vaccine 1997;15(2):129–32.
5. Peltola H, Aavitsland P, Hansen KG, Jonsdottir KE, Nokleby H, Romanus V. Perspective: a five-country analysis of the impact of four different *Haemophilus influenzae* type b conjugates and vaccination strategies in Scandinavia. J Infect Dis 1999;179(1):223–9.
6. Jonville-Bera AP, Autret-Leca E, Radal M. [Adverse effects of the vaccines Tetracoq, IPAD/DTCP and DTCP. A French study of regional drug monitoring centers.]Arch Pediatr 1999;6(5):510–5.
7. Lee CY, Thipphawong J, Huang LM, Lee PI, Chiu HH, Lin W, Debois H, Harrison D, Xie F, Barreto L. An evaluation of the safety and immunogenicity of a five-component acellular pertussis, diphtheria, and tetanus toxoid vaccine (DTaP) when combined with a *Haemophilus influenzae* type b–tetanus toxoid conjugate vaccine (PRP-T) in Taiwanese infants. Pediatrics 1999;103(1):25–30.
8. Cimaz R, Mensi C, D'Angelo E, Fantola E, Milone V, Biasio LR, Carnelli V, Zanetti AR. Safety and immunogenicity of a conjugate vaccine against *Haemophilus influenzae* type b in splenectomized and nonsplenectomized patients with Cooley anemia. J Infect Dis 2001;183(12):1819–21.
9. Amir J, Melamed R, Bader J, Ethevenaux C, Fritzell B, Cartier JR, Arminjon F, Dagan R. Immunogenicity and safety of a liquid combination of DTP-PRP-T vs lyophilized PRP-T reconstituted with DTP. Vaccine 1997;15(2):149–54.
10. Mallet E, Fabre P, Pines E, Salomon H, Staub T, Schodel F, Mendelman P, Hessel L, Chryssomalis G, Vidor E, Hoffenbach A, Abeille A, Amar R, Arsene JP, Aurand JM, Azoulay L, Badescou E, Barrois S, Baudino N, Beal M, Beaude-Chervet V, Berlier P, Billard E, Billet L, Blanc B, Blanc JP, Bohu D, Bonardo C, Bossu CHexavalent Vaccine Trial Study Group. Immunogenicity and safety of a new liquid hexavalent combined vaccine compared with separate

administration of reference licensed vaccines in infants. Pediatr Infect Dis J 2000;19(12):1119–27.

11. Bewley KM, Schwab JG, Ballanco GA, Daum RS. Interchangeability of *Haemophilus influenzae* type b vaccines in the primary series: evaluation of a two-dose mixed regimen. Pediatrics 1996;98(5):898–904.

12. Weinberg GA, Granoff DM. Polysaccharide–protein conjugate vaccines for the prevention of *Haemophilus influenzae* type b disease. J Pediatr 1988;113(4):621–31.

13. D'Cruz OF, Shapiro ED, Spiegelman KN, Leicher CR, Breningstall GN, Khatri BO, Dobyns WB. Acute inflammatory demyelinating polyradiculoneuropathy (Guillain–Barré syndrome) after immunization with *Haemophilus influenzae* type b conjugate vaccine. J Pediatr 1989;115(5 Pt 1):743–6.

14. Gervaix A, Caflisch M, Suter S, Haenggeli CA. Guillain–Barré syndrome following immunisation with *Haemophilus influenzae* type b conjugate vaccine. Eur J Pediatr 1993;152(7):613–4.

15. Classen JB, Classen DC. Association between type 1 diabetes and Hib vaccine. Causal relation is likely. BMJ 1999;319(7217):1133.

16. Classen JB, Classen DC. Clustering of cases of insulin dependent diabetes (IDDM) occurring three years after *Hemophilus influenza* B (HiB) immunization support causal relationship between immunization and IDDM. Autoimmunity 2002;35(4):247–53.

17. Jefferson T. Vaccination and its adverse effects: real or perceived. Society should think about means of linking exposure to potential long term effect. BMJ 1998;317(7152):159–60.

18. Groves FD, Gridley G, Wacholder S, Shu X-O, Robison LL, Linet MS. Infant vaccinations and risk of childhood acute lymphoblastic leukaemia in the United States. Br J Cancer 1999;81:175–8.

19. Auvinen A, Hakulinen T, Groves FD. *Haemophilus influenzae type b vaccination and risk of childhood leukaemia in a vaccine trial in Finland.* Br J Cancer 2000;83:956–8.

20. Peltola H, EskolaJ, Kayhty H, Takala AK, Makela PH. Clinical comparison of the *Haemophilus influenzae type b polysaccharide-diphtheria toxoid and the oligosaccharide-CRM197 protein vaccines in infancy.* Arch Pediatr Adolesc Med 1994;148:620–5.

21. Groves F, Sinha D, Auvinen A. *Haemophilus influenzae type b vaccine formulation and risk of childhood leukaemia.* Br J Cancer 2002;87:511–2.

22. Jakacki R, Luery N, McVerry P, Lange B. *Haemophilus influenzae* diphtheria protein conjugate immunization after therapy in splenectomized patients with Hodgkin disease. Ann Intern Med 1990;112(2):143–4.

Hepatitis vaccines

General Information

Hepatitis A

The successful propagation of hepatitis A virus in cell culture made the development of hepatitis A vaccines a realistic possibility. Various experimental hepatitis A vaccines have been tested in clinical trials. In December 1991, the first hepatitis A vaccine was licensed in Western European countries. Currently, two different hepatitis vaccines (prepared using different inactivation process) as well as vaccines for children (containing 720 enzyme-linked immunosorbent assay units per dose) and adults (containing 1440 units per dose) are commercially available and further products are under development. The results of safety and immunogenicity testing of hepatitis A vaccines developed by SmithKline-Beecham and by Merck Research Laboratories have been reviewed (SEDA-16, 384) (SEDA-17, 373) (SEDA-18, 333) (SEDA-21, 331). These vaccines were highly immunogenic. There were mild transient reactions at injection sites, but systemic reactions were minor and uncommon.

In a paper dealing mainly with indications for the use of hepatitis vaccine, the data on the hepatitis A vaccines most widely used, HAVRIX (manufactured by Glaxo SmithKline) and VAQTA (manufactured by Merck), have been summarized (1). The data are based on pre-licensure clinical trials and worldwide follow-up reports. No serious adverse effects have been attributed to hepatitis A vaccines. In children who received HAVRIX, soreness (15%) and induration (4%) at the injection site, feeding problems (8%), and headaches (4%) have been the most frequently observed adverse effects. In children who received VAQTA, the most common adverse effects were pain (19%), tenderness (17%), and warmth (9%) at the injection site. The reported frequencies were similar to the frequencies reported with hepatitis B vaccines.

Hepatitis B

Two types of hepatitis B vaccine are commercially available: plasma-derived hepatitis B vaccine and yeast recombinant hepatitis B virus vaccine. The two vaccines are equally immunogenic, protective, and safe. However, in most countries the recombinant vaccine is considered the vaccine of choice.

Plasma-derived hepatitis B virus vaccine

Plasma-derived hepatitis B virus vaccine is prepared from the plasma of chronic HBsAg carriers, and consists of purified, inactivated 20-nm HBsAg particles adsorbed on to an aluminium adjuvant. The use of a vaccine produced with plasma derived from infected individuals represented a major departure from conventional approaches, and safety testing has therefore been designed to cover all possibilities of risk and to ensure freedom from transmission of residual HBV and other blood-borne agents. Various clinical trials (SEDA-10, 289) (SEDA-11, 289) have confirmed the safety of plasma-derived hepatitis B virus vaccines produced by different manufacturers. Fears that plasma-derived vaccine may transmit AIDS can be considered unfounded.

Subsequent to the identification of HIV and the growing knowledge of its relatively easy inactivation with procedures such as heat or triple-step chemical inactivation, the complete safety and suitability of plasma-derived vaccines that meet WHO requirements has now been generally accepted and their safety demonstrated (SEDA-12, 279). This type of hepatitis B vaccine is generally well tolerated. The most common adverse effect has been local soreness at the injection site. Less common local reactions include erythema, swelling, and induration; all usually subside within 48 hours. Transient low-grade fever has occurred occasionally; but malaise, fatigue, vomiting, dizziness, myalgia, and arthralgia have been infrequent. Individual reports of more severe suspected adverse effects (erythema multiforme, hepatitis-like changes in liver function, hypersensitivity, lichen planus, menstrual abnormality, myasthenia gravis, neurological disorders including transverse myelitis and Guillain–Barré syndrome, reactive arthritis, Takayasu's arteritis, urticaria, uveitis) are very rare and are no more than could be expected by chance (SED-12, 804) (SEDA-16, 384) (SEDA-17, 373) (SEDA-18, 334). In one reported case with neurological symptoms, a preceding viral illness was the likely cause and in the patient with hepatitis-like symptoms mentioned above, the possibility of other causes was not excluded. In a case with urticaria, patch tests revealed a hypersensitivity to thiomersal, the preservative in the vaccine.

By 1988 it was possible to summarize the adverse effects reported after the distribution of over 1.8 million doses of plasma-derived hepatitis B vaccine (Table 1) (2). From 1982 onwards, the Centers for Disease Control, the Food and Drug Administration, and the manufacturers, Merck Sharp & Dohme, had supported a special surveillance system to monitor spontaneous reports of reactions to plasma-derived hepatitis vaccine. During the first 3 years, about 850 000 persons were immunized. In all, 41 reports were received for one of the following neurological adverse events: convulsion ($n = 5$), Bell's palsy ($n = 10$), Guillain–Barré syndrome ($n = 9$), lumbar radiculopathy ($n = 5$), brachial plexus neuropathy ($n = 3$), optic neuritis ($n = 5$), and transverse myelitis ($n = 4$). Half of these events occurred after the first vaccine dose. However, no conclusive causal association could be made between any neurological adverse event and the vaccine (3).

Yeast-derived recombinant hepatitis B vaccine
The yeast-derived recombinant vaccines first licensed in 1986 represented the first vaccine of any kind

manufactured by recombinant technology. The vaccines are prepared using antigen produced by recombinant technology in yeast (*Saccharomyces cerevisiae*). The recombinant vaccine produced by Merck, Sharp & Dohme was as immunogenic and protective against hepatitis B as plasma-derived vaccine (4). Clinical reactions in this series were mild and transient. About 17% of all recipients had pain, soreness, and tenderness at the injection site. A smaller proportion reported headache, weakness, nausea, or malaise. Between February 1984 and August 1986, 33 investigators in 19 countries carried out clinical trials with the yeast-derived recombinant hepatitis B vaccine produced by SmithKline Biologicals. Among other risk groups, neonates, patients with thalassemia and sickle cell anemia, and hemodialysis patients have been vaccinated. All the results point to the safety and acceptability of a yeast-derived vaccine. The incidence of reported reactions varied widely in different studies depending on the scrupulousness with which minor signs were reported. No serious, severe, or anaphylactic reactions occurred. The incidence of local and systemic reactions reported in each study tended to decrease after successive doses, suggesting that immunization did not induce hypersensitivity (SEDA-12, 280).

Reports of studies of the efficacy and safety of recombinant hepatitis vaccines have been published (SED-12, 805; SEDA-14, 282; SEDA-15, 351; SEDA-16, 385; SEDA-17, 373; SEDA-18, 334). Table 1 shows the adverse effects reported after the distribution of 205 000 doses of recombinant vaccine.

Third-generation hepatitis B vaccines
Between 5 and 15% of healthy immunocompetent individuals do not seroconvert after receipt of the currently licensed hepatitis B vaccines containing only the major surface protein HbsAg without pre-S epitopes. In a study of a hepatitis B vaccine containing pre-S1, pre-S2, and antigenic components of both viral subtypes adw and ayw, all three antigenic components were produced in a continuous mammalian cell line, after transfection of the cells with recombinant hepatitis B surface antigen DNA (5). The vaccine was manufactured as an aluminium

Table 1 Adverse effects after immunization with recombinant and plasma-derived hepatitis B vaccine

| Adverse effect | Recombinant vaccine[a] | | Plasma-derived vaccine[b] | |
	Number of adverse effects	Rate per dose	Number of adverse effects	Rate per dose
Pruritus	2	1:103 000	38	1:48 000
Urticaria	2	1:103 000	23	1:79 000
Exanthems	6	1:103 000	58	1:30 000
Angioedema	–	–	8	1:228 000
Facial edema	1	1:205 000	10	1:182 000
Eczema	–	–	2	1:912 000
Nodule formation	–	–	3	1:608 000
Erythema nodosum	1	1:205 000	1	1:1 823 000
Total	12	1:17 000	99	1:18 000

[a]Over 200 000 doses distributed (August 1986–April 1987).
[b]Over 1.8 million doses distributed (July 1982–December 1987).

hydroxide adjuvant formulation. The new vaccine (5, 10, 20, or 50 μg) was given to 68 individuals with HBs antibody titers below 10 IU/l. Seroconversion rates in the four groups were 60, 76, 64, and 80%. There were local or systemic reactions in 15%. No dose-related incidence was seen.

Surveillance of hepatitis B adverse events in the USA and Australia

Adverse events after hepatitis B vaccine reported between 1 January 1991 and 31 May 1995 to the US Vaccine Adverse Events Reporting System have been reviewed (6). The patients included 58 neonates and 192 infants who were immunized with hepatitis B vaccine alone and 1469 infants who had received hepatitis B vaccine in combination with DTP vaccine. The serious adverse events reported in neonates and infants included fever, agitation, and apnea. The events reported for infants who received hepatitis B vaccine and DTP vaccine simultaneously were compared with the events reported for infants who received either DTP vaccine or hepatitis B vaccine alone. The reports filed for the infants who received DTP vaccine alone or in combination with hepatitis B vaccine differed from the reports filed for infants who received hepatitis B vaccine alone, suggesting that these events may have been associated with the DTP vaccine. The reviewers concluded that no unexpected adverse events had occurred in neonates and infants

given hepatitis B vaccine, despite the administration of at least 12 million doses.

Between 1988 and 1996, the Australian Drug Evaluation Committee received some 600 reports of suspected adverse events after hepatitis B immunization. There were no serious events, and the overwhelming majority were well-known mild-to-moderate local or systemic reactions (7). Musculoskeletal symptoms, such as arthralgia, arthritis, and myalgia, were mentioned in 106 reports.

General adverse effects

The reports of major adverse reactions that have been published since the introduction of recombinant hepatitis B vaccine have been reviewed (8). In the clinical trials with hepatitis B vaccine, the most frequent adverse effects were soreness at the injection site, sometimes accompanied by erythema (3–29%), fatigue (15%), headache (9%), and a temperature increase higher than 37.7°C. The postmarketing surveillance literature (4.5 million doses) showed an overall rate of one adverse effect per 15 500 doses. Of these, local reactions were reported at a rate of 1 in 85 000 doses. Systemic reactions included nausea, rash, headache, fever, malaise, fatigue, flu-like symptoms, diarrhea, urticaria, paresthesia, and somnolence, all of which resolved, generally within 24–48 hours of vaccine administration. Reactions were less frequent with subsequent doses. Major adverse effects have

Table 2 Summary of important adverse effects after recombinant hepatitis B vaccination

Adverse effect	Sex	Age (years)	Number	Time after immunization	Duration of symptoms
Nervous system					
Cerebellar ataxia	F	25	2	10 days	4 months
Demyelination	F	26	3	6 weeks	3 weeks
Demyelination	F	28	2	6 weeks	3 months
Multiple sclerosis	F	43	1	7–10 weeks	4 weeks
Transverse myelitis	M	40	1	2 weeks	6 weeks
Hematologic					
Evans' syndrome	M	33	2	2 days	2 months
Thrombocytopenic purpura	F	15	3	4 weeks	4 months
Thrombocytopenic purpura	F	21	2	3 weeks	2 months+
Skin					
Erythema nodosum	F	43	1	4 days	Several weeks
Lichen planus	F	19	2	2 months	Not reported
Lichen planus	M	50	2	1 month	3 months
Acute posterior multifocal placoid pigment epitheliopathy	M	31	4	3 days	9 months
Urticaria	F	24	1	30 minutes	30 minutes
Musculoskeletal					
Polyarthritis + erythema nodosum	M	31	1	1 day	6 weeks
Polyarthritis	F	41	1	2 weeks	7 months
Reiter's syndrome	M	29	2	4 weeks	4 months
Rheumatoid arthritis	F	49	1	24 hours	Not reported
Immunologic					
Pulmonary and cutaneous vasculitis	F	45	1	2 days	1 week
Systemic lupus erythematosus	F	43	1	2 weeks	Not reported
Median		31		14 days	8 weeks
Average		32		19 days	11 weeks

been published as case reports: anaphylaxis; urticaria, erythema nodosum, lichen planus; arthritis, Reiter's syndrome; pulmonary and cutaneous vasculitis; systemic lupus erythematosus; glomerulonephritis; Evan's syndrome, thrombocytopenic purpura; acute posterior multifocal placoid pigment epitheliopathy; Guillain–Barré syndrome, transverse myelitis, multiple sclerosis, acute cerebellar ataxia; chronic fatigue syndrome. Table 2 summarizes the reports. Discussing the cause of the reported major adverse effects and a possible causal relation with vaccine administration, the authors considered that apart from anaphylaxis and urticaria, most of the reactions described were not allergic in nature and that the symptoms were those of immune-complex disease due to autoimmune mechanisms. Because of the extreme rarity of such serious adverse events, coincidence seems the simplest explanation, but an immune-complex mediated pathogenesis should not be excluded, given the close temporal relation between immunization and the onset of disease. Apart from immune-complex mechanisms, there may be reactions to other components of the vaccine, such as thiomersal, aluminium, or small quantities of yeast proteins.

Hexavalent immunization, including hepatitis B

The Hexavalent Study Group has compared the immunogenicity and safety of a new liquid hexavalent vaccine against diphtheria, tetanus, pertussis, poliomyelitis, hepatitis B, and *Haemophilus influenzae* type b (DTP + IPV + HB + Hib vaccine, manufactured by Aventis Pasteur MSD, Lyon, France) with two reference vaccines, the pentavalent DTP + IPV + Hib vaccine and the monovalent hepatitis B vaccine, administrated separately at the same visit (9). Infants were randomized to receive either the hexavalent vaccine ($n = 423$) or (administered at different local sites) the pentavalent and the HB vaccine ($n = 425$) at 2, 4, and 6 months of age. The hexavalent vaccine was well tolerated (for details, see the monograph Pertussis vaccines). At least one local reaction was reported in 20% of injections with hexavalent vaccine compared with 16% after the receipt of pentavalent vaccine or 3.8% after the receipt of hepatitis B vaccine. These reactions were generally mild and transient. At least one systemic reaction was reported in 46% of injections with hexavalent vaccine, whereas the respective rate for the recipients of pentavalent and HB vaccine was 42%. No vaccine-related serious adverse event occurred during the study. The hexavalent vaccine provided immune responses adequate for protection against the six diseases.

Hepatitis E

Hepatitis E is an important cause of morbidity and mortality in young adults in developing countries, and is particularly dangerous in pregnant women. Through recombinant technology a hepatitis E vaccine has been developed; in its first clinical trials the vaccine was found to be safe and immunogenic (10).

Organs and Systems

Nervous system

Hepatitis A

Probable posthepatitis A immunization encephalopathy has been reported (11).

During 1987–2001 there were 12 published cases of Guillain–Barré syndrome after hepatitis A immunization, mainly in adults, but also in a 3-year-old child.

- An 18-month-old child developed Guillain–Barré syndrome 10 days after a dose of hepatitis A vaccine (12). His neurological status improved slowly, and 4 months after admission the physical examination was normal.

Hepatitis B

Neuralgic amyotrophy (13), febrile convulsions (14), and Guillain–Barré syndrome (15) have been attributed to hepatitis B vaccine. Fatal inflammatory polyradiculoneuropathy has been reported in temporal relation to hepatitis B administration (16).

Demyelination

Two patients developed neurological symptoms and signs of central nervous system demyelination 6 weeks after administration of recombinant hepatitis B vaccine (17). One was a known case of pre-existent multiple sclerosis but the other had no history of neurological diseases. Both had HLA haplotype DR2 and B7, which are associated with multiple sclerosis. A causal link between immunization and demyelination cannot be established from these two case reports, but the time interval would fit a proposed immunological mechanism. In addition, the Centers for Disease Control (CDC) in Atlanta, Georgia, received reports on four cases of "chronic demyelinating disease." Several other reports describe related conditions or other forms of neurological disorder.

- Acute myelitis occurred in a 56-year-old man 3 weeks after hepatitis B immunization (18).
- Optic neuritis has been attributed to hepatitis B vaccine (19).
- Transverse myelitis developed 3 weeks after the first dose of hepatitis B vaccine in an 11-year-old girl (20).

Guillain–Barré syndrome occurred in a 7-year-old girl after the administration of recombinant hepatitis vaccine (21). The author noted that several other such incidents had been reported after the use of recombinant vaccines, two involving optic neuritis and one Guillain–Barré syndrome.

Multiple sclerosis

Eight cases of nervous system diseases suspected to be either recurrent disseminated encephalitis or multiple sclerosis after hepatitis B immunization have been reported (22). Symptoms started in four cases at 4–14 days after vaccination, and in the other four cases 42–70 days after vaccination. There was a family history of multiple sclerosis in two patients, one of whom had had symptoms compatible with optic neuritis before immunization. In a third patient, there was a history of an episode compatible with Lhermitte's sign. The authors concluded that the risk of demyelinating diseases is unknown. They recommended avoiding hepatitis B immunization in individuals with a personal or familial history of symptoms suggestive of an inflammatory or demyelinating disease.

- A 45-year-old woman with a history of epilepsy developed a lumbosacral acute demyelinating polyneuropathy 1 month after her second hepatitis B immunization (23). She had an uncommon syndrome that combined demyelinating and axonal features confined to the lumbosacral roots whose relation to Guillain–Barré syndrome was unclear. Viral or bacterial causes of the disease could not be found.

The authors concluded that a relation between immunization and disease could not be excluded.

- A 44-year-old female health worker developed bilateral optic neuritis 7 days after immunization against poliomyelitis (OPV) and hepatitis B (24). All hematological, biochemical, bacterial, and virological investigations were normal, and there was no evidence of demyelination on MRI scanning.

Although the current hepatitis B vaccine seems to be one of the safest ever produced, concerns are still sometimes expressed (SEDA-22, 346) (25), particularly since 1996 when a French neurologist announced that he had seen several cases of multiple sclerosis or demyelinating disease in women who had received hepatitis B vaccine. This report was taken up by the media and anti-immunization groups, resulting in further inquiry. In 1998 the French Health Authorities invited leading experts in the fields of immunization and possible adverse effects to meet in Paris and discuss scientific studies carried out in France on the possible relation between hepatitis B vaccine and multiple sclerosis. The conclusion of the meeting was that there was no evidence of a causal link between multiple sclerosis and hepatitis B vaccine. Nevertheless, the Minister of Health decided to stop hepatitis B immunization of schoolchildren aged 11–12 years, but to continue immunizing infants and adults at high risk. The rationale for this decision was the limited opportunity for the full discussion needed to attain informed consent during immunization sessions carried out in schools. It was recommended that immunization of school children and adolescents would in future be performed by general practitioners, who have a better opportunity to discuss the benefits and risks of immunization on an individual basis and to obtain informed consent.

The impact of this decision, or any other to curtail an immunization program, should not be underestimated. It raises concerns not only about the safety of the hepatitis B vaccine, but also about the safety of vaccines in general. In view of this, the WHO took strong action, and its Global Program on Immunization issued a statement in 1997 (after the French media had taken up the issue) that there was no evidence that hepatitis B vaccine causes multiple sclerosis (26). This conclusion was based on the following considerations:

- A comparison of the geographical incidence and prevalence of hepatitis B with that of multiple sclerosis, which shows large differences: Scandinavia and Northern Europe have the highest rates of multiple sclerosis and the lowest rates of hepatitis B infection, whereas in Africa and in Asia there are very low rates of multiple sclerosis and the highest rates of hepatitis B infection. If the virus does not cause multiple sclerosis, it is unlikely that the vaccine can do so.
- None of the postmarketing surveillance studies from the different vaccine manufacturers in North America showed any evidence of an increased risk of multiple sclerosis.
- Reanalysis by the WHO of the French data showed that the notification rate of demyelinating diseases following the administration of the hepatitis B vaccine was 0.6 cases/100 000 vaccinees, which is a lower rate than the expected incidence in the same population (estimated in France to be 1–3 cases/100 000 vaccinees).

The data and analyses were examined once more at a scientific conference organized by the Viral Hepatitis Prevention Board in collaboration with WHO, assisted by experts from many disciplines. Their final conclusion, published in a press release, was that the available scientific data did not demonstrate a causal association between hepatitis B vaccine and central nervous system diseases, including multiple sclerosis (27). It was pointed out moreover that since 1981 more than a billion doses of hepatitis B vaccine had been used worldwide, with an outstanding record of safety and efficacy. The WHO also warned of the consequences of stopping any immunization program on the basis of unfounded concerns, as had happened with pertussis vaccine in the UK, leading to major epidemics of natural pertussis in that country.

The National Multiple Sclerosis Society in the USA made a statement referring to anecdotal reports suggesting that immunization against hepatitis B may increase the risk of multiple sclerosis (28). They noted that:

- such reports had not been confirmed by any statistically significant scientific studies to date;
- because of the potential for public concern about this issue, further studies of the possibility of association of hepatitis B vaccine and demyelinating disease, including multiple sclerosis, were already under way in the USA and Europe;
- hepatitis B infection could result in serious, sometimes fatal, disease and immunization was effective in its prevention.

The official French position on hepatitis B immunization, expressed in January 1999, was that children should be vaccinated against hepatitis B, that adolescents should receive a dose of hepatitis B vaccine if they have not been vaccinated earlier, that adults at risk should be similarly vaccinated, and that the obligation is maintained for health professionals to be vaccinated against hepatitis B. The steps taken in 1998 had sought only to ensure that informed consent could be attained before children were vaccinated.

All available data that may throw light on the hypothesis that hepatitis B vaccine is causally linked to multiple sclerosis have been carefully reviewed (29). The authors concluded that the most plausible explanation for the observed temporal association between immunization and multiple sclerosis is coincidence.

A statement from the Viral Hepatitis Prevention Board (8 June 2000), working closely together with the WHO, is worth repeating: the data available to date do not show a causal association between hepatitis B immunization and central nervous system demyelinating disease, including multiple sclerosis. No evidence presented indicates a need to change public health policies with respect to hepatitis B immunization. Therefore, based on demonstrated important benefits (including prevention of liver cirrhosis and primary liver cancer) and a purely hypothetical risk, the Viral Hepatitis Prevention Board supports the recommendation of the WHO that all countries should have universal infant and/or adolescent immunization programs and should continue to immunize adults at increased risk of hepatitis B infection as appropriate (30).

However, the French Ministry of Health has already compensated 14 patients on the basis of a link between hepatitis B vaccine and neurological effects (multiple sclerosis, retrobulbar neuritis) or rheumatological symptoms. According to a press release from the French Ministry of Health, the decision was taken in the interest of the patients, and despite the fact that even the experts from the Agence française de sécurité sanitaire des produits, who were charged with regularly re-evaluating the safety profile of hepatitis B vaccine, have not yet been able to conclude whether there is an actual association between the vaccine and the development of multiple sclerosis or autoimmune disorders. It also stated that the decision in no way challenges the benefit:harm evaluation of hepatitis B vaccine and recommendations with respect to national immunization policies (31).

In the USA, following public discussions on the safety of hepatitis B vaccine, health officials, testifying in 1999 before the Government Reform Subcommittee on Criminal Justice, Drug Policy, and Human Resources, said that use of the vaccine has been monitored for 15 years, that it has not been proven to be the cause of deaths, and is only rarely linked to serious adverse effects (32).

Further studies have confirmed that there is no scientific evidence of a causal link between hepatitis B vaccine and multiple sclerosis (33–36). A European-wide study should be particularly mentioned. In 643 patients in various European countries during 1993–97, 15% of whom received various immunizations within 1 year before relapse, there was no increase in the risk in immunized patients compared with patients who did not receive any vaccine (37).

The results of a hospital-based case-control study in 121 patients with a first episode of central nervous demyelination occurring within 180 days after either hepatitis B vaccine or other vaccines have been reported (38). The results were compared with age- and sex-matched controls seen during the same period. No conclusion regarding a causal relation between hepatitis B vaccine and a first MS episode could be drawn, but the authors were not able to exclude such an association with certainty.

In summary the international consensus is that hepatitis B vaccine is still among the safest and most powerful vaccines in immunization programs and that it should be used worldwide.

Leukoencephalitis

Two episodes of leukoencephalitis occurred in a previously healthy patient after a second dose of HB vaccine and rechallenge with a third dose (39).

- A 39-year-old woman developed a complete right homonymous hemianopia and severe dyslexia 4 weeks after receiving a second dose of HB vaccine. Brain MRI showed a large lesion that occupied most of the left occipital lobe and extended into the splenium of the corpus callosum. Histological examination with immunoperoxidase staining, although not pathognomonic, was consistent with demyelinating disease. She underwent surgery and 1 week later there was marked improvement in her condition. Three months later (4.5 months after the second dose), she received a third dose of HB vaccine and 11 days later developed a left hemiparesis and acute progressive deterioration in vision. Brain MRI showed a new large lesion in the right parieto-occipital region, with the characteristics associated with the previous lesion. In comparison with previous findings, there was significant improvement in the left occipital lobe, in which there remained a proencephalic cyst. She was treated with dexamethasone and markedly improved. At 1 year and 2.5 years after the first episode she had residual dyslexia and a complete right homonymous hemianopia. An MRI scan showed almost complete resolution of the previous findings, with the exception of the proencephalic cyst.

The authors considered that acute leukoencephalitis is a rare but possible complication of HB vaccine.

Sensory systems

Eyes

Three days after recombinant hepatitis B booster immunization, a 31-year-old man developed acute posterior multifocal placoid pigment epitheliopathy (visual loss) and eosinophilia (40,41).

Papillitis (42) and retinal vein occlusion (43) have been attributed to hepatitis B vaccine.

Optic neuritis has been attributed to immunization with hepatitis A and B (44).

- A 21-year-old woman developed acute irreversible loss of vision to 0.05 and a nasal visual field defect in the left eye 2 weeks after immunization with hepatitis A and B and yellow fever vaccine. An MRI scan showed hyperintense thickening of the optic nerve, and a diagnosis of optic neuritis was made. Vision acuity did not recover but the scotoma disappeared within 6 weeks.

Ears

Two cases of hearing loss have been reported in patients given hepatitis B immunization (40). Acute tinnitus with permanent audiovestibular damage has been reported in temporal relation to hepatitis B administration (45).

Hematologic

There have been reports of individual cases of thrombocytopenia purpura (46) and pancytopenia (47) in temporal relation with hepatitis B vaccination.

- A healthy 7-year-old girl developed thrombocytopenic purpura after hepatitis B immunization, three doses of vaccine every month followed by a booster (48).
- A 45-day-old previously healthy girl developed thrombocytopenia purpura 40 days after receiving hepatitis B and BCG vaccines; she recovered completely within 2 weeks (49).

Pancytopenia has been reported in temporal relation to hepatitis B administration (50).

Mouth

Oral lichenoid lesions have been reported in temporal relation to hepatitis B administration (51).

Liver

Severe jaundice and raised serum liver enzyme activities have been reported after a dose of Twinrix® (combination HA/HB vaccine) (52).

- A 26-year-old man had a raised concentration of immunoglobulin G and an antinuclear antibody titer of 1:320. Liver biopsy showed marked bridging liver fibrosis and chronic inflammation, compatible with autoimmune hepatitis. Treatment led to complete normalization of liver function tests. The patient had never had jaundice or abnormal liver function tests.

The authors suggested that the vaccine had induced an acute exacerbation of an unrecognized autoimmune hepatitis.

Urinary tract

Acute glomerulonephritis (53) and minimal-change nephrotic syndrome (54) have been attributed to hepatitis B vaccine.

Skin

Erythema multiforme (55) and pityriasis rosea (56) have been reported after hepatitis B vaccination.

Alopecia (57), anetoderma (a disorder characterized by loss of dermal substance clinically and loss of elastic substance histologically) (58), lichen planus (59), urticaria (60), and white dot syndrome (61) have been attributed to hepatitis B vaccine.

Four cases of urticaria and one case of angioedema have been reported after hepatitis B immunization (62). The results emphasize that urticaria can be due to sensitization to the hepatitis Bs antigen itself, hyperimmunization, or a non-allergic reaction.

The first case of erythermalgia (paroxysmal attacks of bilateral pain in the extremities, with an acute increase in local heat in the affected parts, the production and aggravation of the distress by heat, relieved by cold) probably caused by hepatitis B vaccine has been reported (63).

- A 17-year-old boy developed the symptoms of erythermalgia two days after a third dose of hepatitis B vaccine. He was severely ill and recovered only 2 months after treatment with immunoglobulin and hypnotherapy. During hospitalization it was learned that he had had transient moderate pain of both feet for 3 nights after the second dose of vaccine.

The disease that followed the third dose of hepatitis B vaccine therefore met the criteria for positive rechallenge.

Musculoskeletal

In 31 cases of suspected chronic fatigue syndrome associated with hepatitis B immunization, no causal relation was determined (64).

Juvenile chronic polyarthropathy (65) and reactive arthritis (66) have been attributed to hepatitis B vaccine. Erythema nodosum and polyarthritis occurred the day after Engerix-B vaccine administration (67). The authors referred to reports of three other cases of polyarthritis in the literature.

The Global Advisory Committee on Vaccine Safety (GACVS) has considered the potential association of hepatitis B immunization and rheumatoid arthritis (68). The committee was presented with a preliminary analysis of the interaction between hepatitis B virus and HLA status in respect of the occurrence of rheumatoid arthritis. This is the relevant question if the adverse effect is limited to or occurs largely among a particular genetically determined subgroup. It has the advantage that analysis can be done using cases of rheumatoid arthritis alone. These are compared between the genetically determined groups, which will not differ in regard to the likelihood of hepatitis B virus exposure. There are various subtypes of HLA DRB1*04; nine were found in this study, although analyses were limited to two of the subtypes. Hepatitis B virus exposure in the 90, 180, and 365 days before the onset of rheumatoid arthritis symptoms was examined. There was no statistically significant evidence of an increased risk in the genetic subgroups examined, and point estimates were less than unity. However, an increased risk in a subgroup could not be excluded, because of the low power of the study, given the small numbers of immunized cases. This is an inevitable limitation, but it also makes it clear that hepatitis B virus makes at most very little impact on the incidence of rheumatoid arthritis. In addition, whether HLA DRB1*04 is the best genetic marker for increased risk of development of rheumatoid arthritis is still unknown. The GACVS concluded, based on a review of the limited data available, that there was no convincing evidence to support an association between HBV and rheumatoid arthritis.

Immunologic

Hepatitis A

Vasculitis suspected to be caused by the first dose of hepatitis A vaccine has been reported (SEDA-21, 331).

Hepatitis B

Polyarteritis nodosa (56), Sjögren's syndrome (69), lymphocytic vasculitis (70), vasculitis (71), and systemic lupus erythematosus (72) have been attributed to hepatitis B vaccine.

- After hepatitis B immunization, three doses of vaccine every month followed by a booster, a healthy 24-year-old woman lost weight and developed migratory arthralgia (48). Acute disseminated lupus erythematosus was diagnosed.

The authors discussed the possibility that immunization could have introduced an antigen that may have provoked an autoimmune reaction in genetically predisposed family members.

Churg–Strauss vasculitis (allergic angiitis and granulomatosis) has been attributed to hepatitis B immunization (73).

- A 20-year-old woman developed chronic rhinitis 1 month after the last dose of hepatitis B, followed about 1 year later by severe asthma, nasal polyposis, and petechial purpura in her fingernail beds and on her feet. A skin biopsy from the left leg showed infiltrates consistent with leukocytoclastic vasculitis.

The interval between immunization and the development of the vasculitis made it very difficult to establish a causal relation in this case.

Large artery vasculitis (two cases) (74) and polyarteritis nodosa (75) have been reported in temporal relation to hepatitis B administration.

A case of arthritis (76) and three cases of vasculitis (77) have been described 7–20 days after immunization with hepatitis B. A causal relation in such very rare cases is unclear, and probably they do not occur more than could be expected by chance.)

Second-Generation Effects

Teratogenicity

In the children of 10 women who had received hepatitis B vaccine during the first trimester of pregnancy there were no congenital abnormalities; at 2–12 months the infants were physically and developmentally normal for their ages (78).

Drug Administration

Drug additives

To increase immunogenicity, the hepatitis A vaccines commercially available are coupled to adjuvant aluminium phosphate or aluminium hydroxide. However, alum precipitates provoke inflammatory responses at the injection site. Immunostimulating reconstituted influenza virosomes have therefore been used as an alternative adjuvant. In 1994, a hepatitis A vaccine using the new adjuvant was licensed in Switzerland, and it was later approved for use in other countries: the vaccine was well tolerated and highly immunogenic (SEDA-20, 290) (SEDA-22, 344). Nine people with a history of ocular sensitivity were immunized with hepatitis B, without untoward reactions. However, this result in such a small series should not be overestimated (79). There have been reports of three cases of inflammatory nodular reactions after hepatitis B immunization; aluminium allergy was confirmed (80–82).

Four cases of reactions to thiomersal in hepatitis B vaccines (both plasma-derived and recombinant) have been reported from two centers (83,84). Although thiomersal is present in these vaccines at a concentration of only 1:20 000, it can cause severe cutaneous reactions of the delayed hypersensitivity type, and sometimes the reactions can be very long lasting.

To interpret strictly the package insert for hepatitis B vaccines would preclude its administration to persons with a history of ocular sensitivity to thiomersal.

Drug dosage regimens

Revaccination is sometimes necessary because only 50–70% of immunocompromised persons, especially dialysis patients, develop antibodies, and the anti-HBs titers in these cases are low. In revaccinated non-responders to primary hepatitis immunization using either 20 µg of plasma-derived vaccine or 10 µg of recombinant vaccine, depending on the vaccine used for previous doses, the revaccinations were well tolerated (85,86). Only 6.6% of the vaccinees reported slight irritation at the injection site, tenderness, minimal pain, or swelling lasting for a few hours up to 2 days.

Drug administration route

Aiming at cost reduction of hepatitis B immunization programs, the administration of low doses (2 micrograms) of vaccine given intradermally had by 1987 been evaluated in clinical trials in health-care workers (87) and children (88). The resulting seroconversion rates were 96 and over 90%, respectively. A minimum of local side effects occurred. In a comparison of antibody responses and adverse effects after intradermal or subcutaneous administration of 2 micrograms of a plasma-derived hepatitis B vaccine and intramuscular administration of 20 µg, the intradermal and intramuscular routes gave the highest seroconversion rates (100 and 96%, respectively) and the highest mean titers of anti-HBs (89). The aluminium adjuvant in the vaccine was assumed to cause a substantial number of local reactions (37% discoloration, 17% itching, and 13% nodule formation) after intradermal administration; other routes of administration showed adverse effects only rarely. Correct intradermal deposition of the vaccine is crucial.

Drug–Drug Interactions

Other vaccines

When recombinant hepatitis B vaccine was given together with DTP vaccine and oral poliomyelitis vaccine, there was no evidence of increased reactogenicity after simultaneous administration compared with hepatitis B vaccine alone (90).

In 40 children born to HBsAg-positive mothers who had received second and third doses of hepatitis B vaccine simultaneously with DTP vaccine and inactivated poliomyelitis vaccine, immunogenicity and reactogenicity were comparable with non-simultaneous administration of the different vaccines (91).

References

1. Rosenthal P. Hepatitis A vaccine: current indications. J Pediatr Gastroenterol Nutr 1998;27(1):111–3.
2. Quast U, Freiburg K. Zur Verträglichkeit gentechnisch hergestellter Impfstoffe. Die gelben Hefte. Immunbiol Inform 1988;28(1):41.
3. Shaw FE Jr, Graham DJ, Guess HA, Milstien JB, Johnson JM, Schatz GC, Hadler SC, Kuritsky JN, Hiner EE, Bregman DJ, et al. Postmarketing surveillance for neurologic adverse events reported after hepatitis B vaccination. Experience of the first three years. Am J Epidemiol 1988;127(2):337–52.
4. Hilleman MR. Yeast recombinant hepatitis B vaccine. Infection 1987;15(1):3–7.
5. Zuckerman JN. Hepatitis B third-generation vaccines: improved response and conventional vaccine non-response—third generation pre-S/S vaccines overcome non-response. J Viral Hepat 1998;5(Suppl 2):13–5.
6. Niu MT, Davis DM, Ellenberg S. Recombinant hepatitis B vaccination of neonates and infants: emerging safety data from the Vaccine Adverse Event Reporting System. Pediatr Infect Dis J 1996;15(9):771–6.
7. Anonymous. Adverse events following hepatitis B immunization. Aust Adv Drug React Bull 1996;15:6.
8. Grotto I, Mandel Y, Ephros M, Ashkenazi I, Shemer J. Major adverse reactions to yeast-derived hepatitis B vaccines—a review. Vaccine 1998;16(4):329–34.
9. Mallet E, Fabre P, Pines E, Salomon H, Staub T, Schodel F, Mendelman P, Hessel L, Chryssomalis G, Vidor E, Hoffenbach A, Abeille A, Amar R, Arsene JP, Aurand JM, Azoulay L, Badescou E, Barrois S, Baudino N, Beal M, Beaude-Chervet V, Berlier P, Billard E, Billet L, Blanc B, Blanc JP, Bohu D, Bonardo C, Bossu CHexavalent Vaccine Trial Study Group. Immunogenicity and safety of a new liquid hexavalent combined vaccine compared with separate administration of reference licensed vaccines in infants. Pediatr Infect Dis J 2000;19(12):1119–27.
10. Safary A. Perspectives of vaccination against hepatitis E. Intervirology 2001;44(2–3):162–6.
11. Hughes PJ, Saadeh IK, Cox JP, Illis LS. Probable post-hepatitis A vaccination encephalopathy. Lancet 1993;342(8866):302.
12. Blumenthal D, Prais D, Bron-Harlev E, Amir J. possible association of Guillain–Barré syndrome and hepatitis A vaccination. Ped Infect Dis J 2004;23:586–8.
13. Reutens DC, Dunne JW, Leather H. Neuralgic amyotrophy following recombinant DNA hepatitis B vaccination. Muscle Nerve 1990;13(5):461.
14. Hartman S. Convulsion associated with fever following hepatitis B vaccination. J Paediatr Child Health 1990;26(1):65.
15. Sinsawaiwong S, Thampanitchawong P. Guillain–Barré syndrome following recombinant hepatitis B vaccine and literature review. J Med Assoc Thai 2000;83(9):1124–6.
16. Sindern E, Schroder JM, Krismann M, Malin JP. Inflammatory polyradiculoneuropathy with spinal cord involvement and lethal outcome after hepatitis B vaccination. J Neurol Sci 2001;186(1–2):81–5.
17. Herroelen L, de Keyser J, Ebinger G. Central-nervous-system demyelination after immunisation with recombinant hepatitis B vaccine. Lancet 1991;338(8776):1174–5.
18. Mahassin F, Algayres JP, Valmary J, Bili H, Coutant G, Bequet D, Daly JP. Myelite aiguë après vaccination contre l'hépatite B. [Acute myelitis after vaccination against hepatitis B.] Presse Méd 1993;22(40):1997–8.
19. Albitar S, Bourgeon B, Genin R, Fen-Chong M, N'Guyen P, Serveaux MO, Atchia H, Schohn D. Bilateral retrobulbar optic neuritis with hepatitis B vaccination. Nephrol Dial Transplant 1997;12(10):2169–70.
20. Trevisani F, Gattinara GC, Caraceni P, Bernardi M, Albertoni F, D'Alessandro R, Elia L, Gasbarrini G. Transverse myelitis following hepatitis B vaccination. J Hepatol 1993;19(2):317–8.
21. Tuohy PG. Guillain-Barre syndrome following immunisation with synthetic hepatitis B vaccine. NZ Med J 1989;102(863).114–5.
22. Tourbah A, Gout O, Liblau R, Lyon-Caen O, Bougniot C, Iba-Zizen MT, Cabanis EA. Encephalitis after hepatitis B vaccination: recurrent disseminated encephalitis or MS? Neurology 1999;53(2):396–401.
23. Creange A, Temam G, Lefaucheur JP. Lumbosacral acute demyelinating polyneuropathy following hepatitis B vaccination. Autoimmunity 1999;30(3):143–6.
24. Stewart O, Chang B, Bradbury J. Simultaneous administration of hepatitis B and polio vaccines associated with bilateral optic neuritis. Br J Ophthalmol 1999;83(10):1200–1.
25. Marshall E. A shadow falls on hepatitis B vaccination effort. Science 1998;281(5377):630–1.
26. Expanded programme on immunization (EPI). Lack of evidence that hepatitis B vaccine causes multiple sclerosis. Wkly Epidemiol Rec 1997;72(21):149–52.
27. World Health Organization. Press Release. WHO/67, 2 October 1998.
28. Medical Advisory Board of the National Multiple Sclerosis Society: Statement from 14 August 1998. National Multiple Sclerosis Society: News Desk Research Bulletins, 3 September 1998.
29. Monteyne P, Andre FE. Is there a causal link between hepatitis B vaccination and multiple sclerosis? Vaccine 2000;18(19):1994–2001.
30. WHO. Press release. Hepatitis B and multiple sclerosis. Geneva, 8 June 2000. http://www.who.int/vaccines-diseases/safety/hottop/hepb.htm (accessed 20 August 2000).
31. French Ministry of Health. Press release, 23 May 2000. Compensation for hepatitis B vaccination. http:www.who.int/vaccines-diseases/safety/hottop/hepb.htm (accessed 2 August 2000).
32. Associated Press. Press release, 15 April 1999. Hepatitis vaccine safety questioned in the United States. http://

www.who.int/vaccines-diseases/safety/hottop/HBV_Press.html (accessed 2 August 2000).

33. Hostetler L. Vaccinations and multiple sclerosis. N Engl J Med 2001;344(23):1795.

34. Gellin BG, Schaffner W. The risk of vaccination—the importance of "negative" studies. N Engl J Med 2001;344(5):372–3.

35. Ascherio A, Zhang SM, Hernan MA, Olek MJ, Coplan PM, Brodovicz K, Walker AM. Hepatitis B vaccination and the risk of multiple sclerosis. N Engl J Med 2001;344(5):327–32.

36. Soubeyrand B, Boisnard F, Bruel M, Debois H, Delattre D, Gauthier A, Soum S, Thebault C. Pathologies démyélinisantes du système nerveux central rapportées après vaccination hépatite B par GenHevac B (1989–1998). [Central nervous system demyelinating disease following hepatitis B vaccination with GenHevac B. Review of ten years of spontaneous notifications (1989–1998).] Presse Méd 2000;29(14):775–80.

37. Confavreux C, Suissa S, Saddier P, Bourdes V, Vukusic SVaccines in Multiple Sclerosis Study Group. Vaccinations and the risk of relapse in multiple sclerosis. Vaccines in Multiple Sclerosis Study Group. N Engl J Med 2001;344(5):319–26.

38. Touze E, Gout O, Verdier-Taillefer MH, Lyon-Caen O, Alperovitch A. Premier épisode de démyelinisation du système nerveux central et vaccination contre l'hépatite B. [The first episode of central nervous system demyelinization and hepatitis B virus vaccination.] Rev Neurol (Paris) 2000;156(3):242–6.

39. Konstantinou D, Paschalis C, Maraziotis T, Dimopoulos P, Bassaris H, Skoutelis A. Two episodes of leukoencephalitis associated with recombinant hepatitis B vaccination in a single patient. Clin Infect Dis 2001;33(10):1772–3.

40. Brezin A, Lautier-Frau M, Hamedani M, Rogeaux O, Hoang PL. Visual loss and eosinophilia after recombinant hepatitis B vaccine. Lancet 1993;342(8870):563–4.

41. Biacabe B, Erminy M, Bonfils P. A case report of fluctuant sensorineural hearing loss after hepatitis B vaccination. Auris Nasus Larynx 1997;24(4):357–60.

42. Berkman N, Benzarti T, Dhaoui R, Mouly P. Neuro-papillite bilaterale au decours d'une vaccination contre l'hépatite B. [Bilateral neuro-papillitis after hepatitis B vaccination.] Presse Méd 1996;25(28):1301.

43. Devin F, Roques G, Disdier P, Rodor F, Weiller PJ. Occlusion of central retinal vein after hepatitis B vaccination. Lancet 1996;347(9015):1626.

44. Voigt U, Baum U, Behrendt W, Hegemann S, Terborg C, Strobel J. Optikusneuritis nach Impfung gegen Hepatitis A, B und Gelbfieber mit irreversiblem Visusverlust. [Neuritis of the optic nerve after vaccinations against hepatitis A, hepatitis B and yellow fever.] Klin Monatsbl Augenheilkd 2001;218(10):688–90.

45. DeJonckere PH, de Surgeres GG. Acute tinnitus and permanent audiovestibular damage after hepatitis B vaccination. Int Tinnitus J 2001;7(1):59–61.

46. Maezono R, Escobar AM. Purpura trombocitopenico apos vacina de hepatite B. [Thrombocytopenic purpura after hepatitis B vaccine.] J Pediatr (Rio J) 2000;76(5):395–8.

47. Viallard JF, Boiron JM, Parrens M, Moreau JF, Ranchin V, Reiffers J, Leng B, Pellegrin JL. Severe pancytopenia triggered by recombinant hepatitis B vaccine. Br J Haematol 2000;110(1):230–3.

48. Finielz P, Lam-Kam-Sang LF, Guiserix J. Systemic lupus erythematosus and thrombocytopenic purpura in two members of the same family following hepatitis B vaccine. Nephrol Dial Transplant 1998;13(9):2420–1.

49. Nascimento-Carvalho HNC, Athayde-Oliveira C, Lyra I, Moreira LM. Thrombocytopenia purpura after hepatitis B vaccine: case report and review of the literature. Ped Infect Dis J 2004;23:183–4.

50. Ashok Shenoy K, Prabha Adhikari MR, Chakrapani M, Shenoy D, Pillai A. Pancytopenia after recombinant hepatitis B vaccine—an Indian case report. Br J Haematol 2001;114(4):955.

51. Anonymous. Cutaneous drug reaction case reports: from the world literature. Am J Clin Dermatol 2001;2:49–56.

52. Csepregi A, Treiber G, Röcken C, Malfertheiner P. Acute exacerbation of autoimmune hepatitis induced by Twinrix. World J Gastroenterol 2005;11:4114–6.

53. Carmeli Y, Oren R. Hepatitis B vaccine side-effect. Lancet 1993;341(8839):250–1.

54. Islek I, Cengiz K, Cakir M, Kucukoduk S. Nephrotic syndrome following hepatitis B vaccination. Pediatr Nephrol 2000;14(1):89–90.

55. Loche F, Schwarze HP, Thedenat B, Carriere M, Bazex J. Erythema multiforme associated with hepatitis B immunization. Clin Exp Dermatol 2000;25(2):167–8.

56. De Keyser F, Naeyaert JM, Hindryckx P, Elewaut D, Verplancke P, Peene I, Praet M, Veys E. Immune-mediated pathology following hepatitis B vaccination. Two cases of polyarteritis nodosa and one case of pityriasis rosea-like drug eruption. Clin Exp Rheumatol 2000;18(1):81–5.

57. Wise RP, Kiminyo KP, Salive ME. Hair loss after routine immunizations. JAMA 1997;278(14):1176–8.

58. Daoud MS, Dicken CH. Anetoderma after hepatitis B immunization in two siblings. J Am Acad Dermatol 1997;36(5 Pt 1):779–80.

59. Trevisan G, Stinco G. HBV vaccination and lichen planus. G Ital Dermatol Venereol 1993;128:545–8.

60. Hudson TJ, Newkirk M, Gervais F, Shuster J. Adverse reaction to the recombinant hepatitis B vaccine. J Allergy Clin Immunol 1991;88(5):821–2.

61. Baglivo E, Safran AB, Borruat FX. Multiple evanescent white dot syndrome after hepatitis B vaccine. Am J Ophthalmol 1996;122(3):431–2.

62. Barbaud A, Trechot P, Reichert-Penetrat S, Weber M, Schmutz JL. Allergic mechanisms and urticaria/angioedema after hepatitis B immunization. Br J Dermatol 1998;139(5):925–6.

63. Rabaud C, Barbaud A, Trechot P. First case of erythermalgia related to hepatitis B vaccination. J Rheumatol 1999;26(1):233–4.

64. Anonymous. Alleged link between hepatitis B vaccine and chronic fatigue syndrome. CMAJ 1992;146(1):37–8.

65. Bracci M, Zoppini A. Polyarthritis associated with hepatitis B vaccination. Br J Rheumatol 1997;36(2):300–1.

66. Biasi D, De Sandre G, Bambara LM, Carletto A, Caramaschi P, Zanoni G, Tridente G. A new case of reactive arthritis after hepatitis B vaccination. Clin Exp Rheumatol 1993;11(2):215.

67. Rogerson SJ, Nye FJ. Hepatitis B vaccine associated with erythema nodosum and polyarthritis. BMJ 1990;301(6747):345.

68. Hepatitis B vaccination and rheumatoid arthritis. Wkly Epidemiol Rec 2008;83:39.

69. Toussirot E, Lohse A, Wendling D, Mougin C. Sjogren's syndrome occurring after hepatitis B vaccination. Arthritis Rheum 2000;43(9):2139–40.

70. Drucker Y, Prayson RA, Bagg A, Calabrese LH. Lymphocytic vasculitis presenting as diffuse subcutaneous edema after hepatitis B virus vaccine. J Clin Rheumatol 1997;3:158–61.

71. Mathieu E, Fain O, Krivitzky A. Cryoglobulinemia after hepatitis B vaccination. N Engl J Med 1996;335(5):355.

72. Tudela P, Marti S, Bonal J. Systemic lupus erythematosus and vaccination against hepatitis B. Nephron 1992;62(2):236.

73. Vanoli M, Gambini D, Scorza R. A case of Churg–Strauss vasculitis after hepatitis B vaccination. Ann Rheum Dis 1998;57(4):256–7.

74. Zaas A, Scheel P, Venbrux A, Hellmann DB. Large artery vasculitis following recombinant hepatitis B vaccination: 2 cases. J Rheumatol 2001;28(5):1116–20.

75. Saadoun D, Cacoub P, Mahoux D, Sbai A, Piette JC. Vascularites postvaccinales: à propos de trois observations. [Postvaccine vasculitis: a report of three cases.] Rev Med Interne 2001;22(2):172–6.

76. Casals JL, Vazquez MA. Artritis inducida por vacunacion antihepatitis B. [Arthritis induced by antihepatitis B vaccine.] An Med Interna 1999;16(11):601–2.

77. Le Hello C, Cohen P, Bousser MG, Letellier P, Guillevin L. Suspected hepatitis B vaccination related vasculitis. J Rheumatol 1999;26(1):191–4.

78. Levy M, Koren G. Hepatitis B vaccine in pregnancy: maternal and fetal safety. Am J Perinatol 1991;8(3):227–32.

79. Kirkland LR. Ocular sensitivity to thimerosal: a problem with hepatitis B vaccine? South Med J 1990;83(5):497–9.

80. Cosnes A, Flechet ML, Revuz J. Inflammatory nodular reactions after hepatitis B vaccination due to aluminium sensitization. Contact Dermatitis 1990;23(2):65–7.

81. Hutteroth TH, Quast U. Aluminiumhydroxid-Granulome nach Hepatitis-B-Impfung (Fragen aus der Praxis). [Aluminum hydroxide granuloma following hepatitis B vaccination.] Dtsch Med Wochenschr 1990;115(12):476.

82. Skowron F, Grezard P, Berard F, Balme B, Perrot H. Persistent nodules at sites of hepatitis B vaccination due to aluminium sensitization. Contact Dermatitis 1998;39(3):135–6.

83. Rietschel RL, Adams RM. Reactions to thimerosal in hepatitis B vaccines. Dermatol Clin 1990;8(1):161–4.

84. Jungkunz G, Kohler P, Holbach M, Schweisfurth H. Kasuistik: Zwei Fälle mit heftiger lokaler Reaktion nach aktiver Hepatitis-B-Impfung bei Sensibilisierung auf Thiomersal. Hyg Med 1990;15:418–20.

85. Jilg W, Schmidt M, Deinhardt F. Immune response to hepatitis B revaccination. J Med Virol 1988;24(4):377–84.

86. Jilg W, Schmidt M, Weinel B, Kuttler T, Brass H, Bommer J, Muller R, Schulte B, Schwarzbeck A, Deinhardt F. Immunogenicity of recombinant hepatitis B vaccine in dialysis patients. J Hepatol 1986;3(2):190–5.

87. Safary A, Andre F. Clinical development of a new recombinant DNA hepatitis B vaccine. Postgrad Med J 1987;63(Suppl 2):105–7.

88. Wiedermann G, Ambrosch F, Kremsner P, Kunz C, Hauser P, Simoen E, Andre F, Safary A. Reactogenicity and immunogenicity of different lots of a yeast-derived hepatitis B vaccine. Postgrad Med J 1987;63(Suppl 2):109–13.

89. Wahl M, Hermodsson S. Intradermal, subcutaneous or intramuscular administration of hepatitis B vaccine: side effects and antibody response. Scand J Infect Dis 1987;19(6):617–21.

90. Giammanco G, Li Volti S, Mauro L, Bilancia GG, Salemi I, Barone P, Musumeci S. Immune response to simultaneous administration of a recombinant DNA hepatitis B vaccine and multiple compulsory vaccines in infancy. Vaccine 1991;9(10):747–50.

91. Torres JM, Bruguera M, Vidal J, Artigas N. Immune response, efficacy and reactogenicity of hepatitis B vaccine administered simultaneously with DTP and poliomyelitis vaccine. Gastroenterol Hepatol 1993;16:470–3.

Herpes simplex virus vaccine

Genital herpes infection is caused by *Herpes simplex* virus type 2 (HSV-2). The infection can be asymptomatic or severe, with painful skin lesions and complications. There have been two double-blind, randomized trials in volunteers whose sexual partners had genital herpes (study 1: 847 volunteers seronegative for *Herpes simplex* virus type 1 and type 2; study 2: 1867 volunteers of any *Herpes simplex* virus status) [1]. The volunteers were randomized to receive a *Herpes simplex* virus type 2 glycoprotein-D subunit vaccine or placebo at 0, 1, and 6 months. There were follow-up visits, including serological analyses, over 19 months. The vaccine did not prevent infection with *Herpes simplex* virus type 2 in group 1, but was 74% effective in women who were seronegative for both *Herpes simplex* virus types 1 and 2. Adverse events were mainly limited to pain at the injection site. The use of condoms and antiviral agents to prevent *Herpes simplex* virus type 2 infection in sexual partners should still be emphasized.

Reference

1. Cole C. Vaccine prevents genital herpes in subgroup of women. J Fam Pract 2003;52:94–6.

Human immunodeficiency virus (HIV) vaccine

General Information

The difficult problems connected with clinical trials that have not been approved by independent authorities were highlighted in 1991 when Zagury published the first reports of immunization of humans using *Vaccinia* vaccine expressing HIV glycoprotein gp-160 (1,2). The first HIV vaccine approved for clinical trial status (1989) by the US Food and Drug Administration (FDA) was a recombinant gp-160 vaccine produced in a baculovirus-insect cell expression system by MicroGeneSys (3). Since then, various clinical trials using different HIV vaccines have been carried out. However, all HIV vaccines are still experimental. An overview of the current status of HIV vaccine development, with emphasis on efficacy and safety, has been provided by the AIDS Division of the National Institute of Allergy and Infectious Diseases (4).

References

1. Dorozynski A, Anderson A. Deaths in vaccine trials trigger French inquiry. Science 1991;252(5005):501–2.
2. Guillaume JC, Saiag P, Wechsler J, Lescs MC, Roujeau JC. Vaccinia from recombinant virus expressing HIV genes. Lancet 1991;337(8748):1034–5.

3. Midthun K, Garrison L, Gershman K. In: Cellular immunity in HIV-1 rgp 160 vaccinesAbstracts, V International Conference on AIDS, Montreal 1989:544.
4. National Institute of Allergy and Infectious Diseases (NIAID), National Institutes of Health (NIH). HIV vaccines. http://www.niaid.nih.gov/daids/vaccine/default.htm, 20/06/2005.

Human papilloma virus (HPV) vaccine

General Information

Cervical cancer is the second most common cause of cancer deaths in women worldwide and the number one cause in the developing world. It is almost invariably associated with HPV infection. HPV type 16 is found in about 50% of cervical cancers. About 70% of cervical cancers are associated with HPV types 16, 18, and 8 other HPV types. Types 18, 31, and 45 account for 25% of HVP-positive tumors. HPV types 6 and 11 can cause benign genital warts.

Other types of human papilloma virus, such as types 31, 33, and 45, are involved in the development of less common cancers (oropharyngeal, esophageal, penile, and anal cancers).

The development of a safe and effective HPV vaccine could prevent premalignant and malignant disease associated with HPV infection. Various HPV vaccines are under development or are undergoing clinical trials. Among others, GlaxoSmithKline (bivalent HPV types 16 and 18) and Merck (HPV types 6, 11, 16, 18) are both conducting expanded phase III trials and the researchers say the results so far have been promising. Merck plans to apply for approval to the US Food and Drug Administration in late 2005.

Clinical studies with a recombinant vaccine (using *Vaccinia* virus expressing HPV 16, 18, E6, and E7 proteins) in patients with preinvasive and invasive cancer have been reviewed (1).

Observational studies

There has been a trial of a papilloma (HPV-16) virus-like particle vaccine in 72 healthy volunteers, aged 18–27 years (2). The vaccine was well tolerated and highly immunogenic.

Placebo-controlled studies

In a double-blind study in 2392 young women (aged 16–23 years) randomly assigned to either three doses of placebo or HPV-16 vaccine at day 0, month 2, and month 6, the vaccine reduced the incidence of both HPV-16 infection (3.8 per 100 woman-years at risk in the placebo group versus 0 per 100 woman-years at risk in the vaccine group) and HPV-16-related cervical intraepithelial neoplasia (all nine cases of neoplasia occurred among the placebo recipients) (3).

Systematic reviews

The Global Advisory Committee on Vaccine Safety (GACVS) has reviewed the safety of human papilloma virus vaccines (both the tetravalent Gardasil® and the bivalent Cervarix®). Data from prelicensing randomized controlled trials and post-licensing surveillance reports from the two vaccine manufacturers and from the European Medicines Evaluation Agency (EMEA), the US Food and Drug Administration (FDA), and the US Centers for Disease Control and Prevention (CDC) were included (4). The current evidence on the safety of HPV vaccines is reassuring. The reviewed data covered short-term local and systemic events, and long-term events up to 6 years after immunization, including events in pregnancy. Injection-site reactions and muscle pain were common. During adolescent vaccine campaigns, some mass sociogenic illnesses, such as post-immunization dizziness and syncope, have been reported. These events have been prevented by observing adolescents for 15 minutes after immunization and encouraging good hydration. The Committee recommended good surveillance systems to identify possible rare adverse effects and specific adverse effects during pregnancy, as the target group includes women of reproductive age, although according to the Summary of Product Characteristics of the HPV vaccine Gardasil®, immunization of pregnant women is not recommended.

References

1. Adams M, Borysiewicz L, Fiander A, Man S, Jasani B, Navabi H, Lipetz C, Evans AS, Mason M. Clinical studies of human papilloma vaccines in pre-invasive and invasive cancer. Vaccine 2001;19(17–19):2549–56.
2. Harro CD, Pang YY, Roden RB, Hildesheim A, Wang Z, Reynolds MJ, Mast TC, Robinson R, Murphy BR, Karron RA, Dillner J, Schiller JT, Lowy DR. Safety and immunogenicity trial in adult volunteers of a human papillomavirus 16 L1 virus-like particle vaccine. J Natl Cancer Inst 2001;93(4):284–92.
3. Koutsky LA, Ault KA, Wheeler CM, Brown DR, Barr E, Alvarez FB, Chiacchierini LM, Jansen KU. Proof of Principle Study Investigators. A controlled trial of a human papillomavirus type 16 vaccine. N Engl J Med 2002;347(21):1645–51.
4. Global Advisory Committee on Vaccine Safety. Meeting on 12–13 June 2007. Safety of HPV vaccines. Wkly Epidemiol Rec 2007;82:255–6.

Influenza vaccine

General Information

Influenza vaccine viruses are propagated in embryonated chicken eggs. The virus-containing extra-embryonic fluid is harvested, purified, and inactivated with formalin. Inactivated flu vaccine is produced either as whole virus vaccine or ether-disrupted split or subunit preparations.

However, many other new or modified influenza vaccines are already available or are expected to appear in the near future, for example vaccines containing new adjuvants, live attenuated vaccines, and vaccines administered by alternative routes.

Comparisons of reactogenicity (and immunogenicity) of different vaccine types have been provided (SEDA-8, 299; SEDA-10, 290; SEDA-11, 290; SEDA-12, 282; SEDA-13, 286; SEDA-14, 284; SEDA-15, 351; SEDA-16, 386; SEDA-17, 375; SEDA-18, 335) (1). In most of these trials the investigators found little difference in reactogenicity between the various vaccines.

The safety and immunogenicity of an inactivated subunit influenza vaccine containing MF59 (oil-in-water emulsion, squalene, and Tween 80 and sorbitan trioleate as stabilizers) as an adjuvant has been described in two studies of 92 and 211 elderly persons (65 years of age and over) (2,3). Investigations were carried out during three consecutive influenza seasons. Compared with a commercial non-adjuvant subunit vaccine containing the same influenza strains recommended by the World Health Organization, geometric mean titers and seroconversion rates were higher after the use of the newly developed vaccine. The adjuvant vaccine caused more local reactions than the conventional vaccine. However, the reactions were mild and limited to the first 2–3 days after immunization. Systemic reactions were not significantly different, except for mild transient malaise. Considering the better immunogenicity of the adjuvant vaccine, the authors recommended it particularly for elderly people, who are at greatest risk of developing severe influenza disease. The vaccine manufactured by Chiron Behring has been already licensed for persons of 65 years of age and over in Italy and Germany.

The results of several clinical trials with a liposomal influenza vaccine have been reviewed (4). This trivalent liposomal influenza vaccine consists of purified influenza hemagglutinin inserted into a membrane of phosphatidyl-choline and phosphatidyl-ethanolamine; it contains 15 micrograms of hemagglutinin per viral strain per dose. The trials included two randomized studies (in 126 healthy nursing home residents aged 63–102 years and 72 elderly individuals aged 60–98 years) and four double-blind studies (in a total of 831 elderly persons aged 67–71 years and younger adults aged 28–38 years); further studies included 24 children and adults with cystic fibrosis and 49 children at high risk of influenza. In all of these studies, the liposomal vaccine was compared with commercially available whole and subunit influenza virus vaccines. In general, seroconversion rates were significantly higher with the liposomal vaccine than with the commercially available vaccines. Local adverse reactions, such as pain at the injection site (up to 62% of children and up to 11% of adults), local induration and swelling (33% of all vaccine recipients), or redness (5% of adults and 20% of children) were transient and usually mild. One child with cystic fibrosis and two elderly persons had severe pain. Between 68 and 100% of children with cystic fibrosis reported at least one systemic reaction

(fatigue, cough, coryza, headache) after the liposomal vaccine compared with 23–50% after the commercially available subunit vaccine. Fatigue (up to 19%), malaise (up to 14%), headache (up to 10%), and cough (up to 8%) were the most common systemic reactions reported by young adults and elderly people. There were single cases of severe fatigue, cough, and diarrhea in children with cystic fibrosis, young adults, and elderly people. Liposomal influenza vaccine did not induce a mean anti-phospholipid antibody response in the elderly volunteers.

In a randomized, double-blind study, trivalent, live, attenuated, cold-adapted intranasal influenza vaccine (FluMist) has been compared with intranasal placebo plus a trivalent injected inactivated influenza vaccine (5). The 200 patients were aged 65 years and over and had chronic cardiovascular or pulmonary conditions or diabetes mellitus. During the 7 days after immunization, sore throat was reported on at least one day by significantly more of the FluMist recipients (15 versus 2%). The increased frequency of sore throat may have been attributable to direct or indirect effects of vaccine virus replication. No other symptom was associated with FluMist. These findings were consistent with evaluations of other live, attenuated, cold-adapted influenza vaccine formulations in older adults. However, further studies of the safety of FluMist are warranted.

The immunogenicity and safety of inactivated intranasal influenza vaccine have been reviewed (6). The author concluded that the vaccine is highly immunogenic and well tolerated by most vaccinees, in terms of both local nasal symptoms and possible vaccine-mediated systemic symptoms. The symptoms were primarily mild, occasionally moderate, and in a few cases more severe; in most cases they lasted for only 1–2 days.

In 1997, an avian influenza A/Hongkong/97 (H5N1) virus emerged as a pandemic threat. A non-pathogenic variant influenza A/duck/Singapore/97 (H5N3) was identified as a leading vaccine candidate, but the non-adjuvanted antibody response was poor; however, the addition of the adjuvant MF59 (oil-in-water suspension) boosted the antibody responses to protective levels. In 65 volunteers who received either the non-adjuvanted or the adjuvanted vaccine, both vaccines were well tolerated and did not differ significantly. There was pain at the injection site of varying intensity in nine of the 32 volunteers who received the adjuvanted vaccine, and in none of the volunteers who received the non-adjuvanted vaccine (7).

General adverse effects

Local adverse reactions after flu immunization are few and infrequent. Slight to moderate tenderness, erythema, and induration at the injection site lasting 1–2 days occur in 15–30% of recipients. Fever, malaise, myalgia, and other symptoms of toxicity are rare (about 2%) and most often affect persons with no prior exposure to the flu antigens in the vaccine, for example young children. These reactions usually begin 6–12 hours after immunization and can last 1 or 2 days. They have been attributed to the vaccine, although the virus is inactivated. On the other hand, cases of respiratory diseases among vaccinees are coincidental. Although current flu vaccines are highly purified, they can cause hypersensitivity reactions such as hives or angioedema, perhaps due to residual egg protein (8,9). Notwithstanding the fact that the egg protein content is small, asthma or anaphylactic reactions with vascular purpura and encephalopathy can occur in those who are sensitive to the material (10,11).

Incidence

A nationwide surveillance system covering illness after flu immunization in the USA in 1976–77 among over 48 million persons immunized in 1976 with A/New Jersey/76 influenza vaccine (swine flu vaccine) resulted in a total of 4733 reports of illness, including reports of 223 deaths (12). Since most of the deaths occurred within 48 hours of immunization, the figures for deaths per 100 000 vaccinees (by diagnosis) were compared with the expected death rate (by the same diagnosis) per 100 000 population for a 2-day period. In general, the crude expected death rate was much higher than the death rate among vaccinees. Other than Guillain–Barré syndrome and rare cases of anaphylaxis, no serious illnesses seemed to be causally associated with flu immunization. However, widespread under-reporting of illness and death in the passive phase of this surveillance system impaired the validity of the study. Allergic skin reactions were reported at a rate of 0.3 per 100 000 vaccinees and severe anaphylaxis at a rate of 0.024 per 100 000. There was a cluster of four cases of encephalitis within 1 week of vaccine administration in one state. There were three deaths from cardiovascular disease in chronically ill persons over 70 years of age immunized in one clinic. It was not possible to establish a causative link between immunization and death. Persons immunized in the clinic died at rate of 5 per 100 000 per day, in contrast to the expected rate of 17 per 100 000 per day for people aged 65 years and older in the respective state.

Case reports of complications temporarily connected with the administration of flu vaccine have been published. They include reports of acute disseminated encephalitis (13), acute thrombocytopenic purpura (14), acute transverse myelitis (15), aseptic meningitis (16), bullous pemphigoid (SEDA-21, 334), encephalopathy (17), erythromelalgia (18), optic neuritis (19) with reversible blindness (SEDA-4, 226) (SEDA-21, 334), optic atrophy (SEDA-6, 287), pericarditis (SEDA-7, 324), polymyalgia rheumatica (20), microscopic polyangiitis involving the skin and joints (21), acute symmetrical polyarthropathy with orbital myositis and posterior scleritis (22), systemic vasculitis (23), a trigeminal neuralgia-like symptoms (24), and vascular purpura with histological features of cutaneous necrotizing vasculitis (25).

The current status of adjuvanted influenza vaccines has been reviewed (26). The authors concluded that the vaccine produces a higher titer of antibodies than non-adjuvanted or virosomal vaccines. Local reactions occur more often, but are mild and transient. The results of a trial with two doses of an intranasally administered inactivated virosome-formulated influenza vaccine containing *Escherichia coli* heat-labile toxin as a mucosal adjuvant in 106 volunteers aged 33–63 years have been reported (27). About 50% of vaccinees had local adverse reactions (44% after the first dose and 54% after the second dose) or systemic adverse reactions (48 and 46%) after administration of the vaccine. Rhinorrhea, sneezing, and headache were the most common reactions; they were mild and transient and resolved within 24–48 hours. No febrile reactions were associated with immunization. Between 77 and 92% of vaccinees developed protective hemagglutination inhibition antibody titers against the two influenzae A strains of the vaccine, whereas protective antibody titers against the B strain of the vaccine were achieved in only 49–58%.

Oculorespiratory syndrome

In 2000, oculorespiratory syndrome was identified as a new influenza vaccine-associated adverse effect in Canada. The case definition requires the presence of red eyes or respiratory symptoms or facial edema at 2–24 hours after immunization and lasting 48 hours (28,29). About 20% of vaccinees with oculorespiratory syndrome described the symptoms as mild and 42% described them as severe. The cause of oculorespiratory syndrome is debated, but the main hypothesis Is that It Is due to large viral aggregates in the vaccine. The reduction In the number of oculorespiratory syndrome reports in 2001–2 with the use of reformulated vaccines with lower aggregate content supports this hypothesis. The authors recommended that manufacturers should consider oculorespiratory syndrome in order to improve the acceptance of influenza vaccines through limitation of the aggregate and unsplit virion content.

Organs and Systems

Cardiovascular

Influenza infection has been a significant problem in cardiac transplant patients; immunization of such patients could therefore be beneficial. However, its use has been limited by concern that stimulation of the immune system might in principle cause an increased risk of cardiac rejection. In the renal transplant experience, influenza infection itself can trigger an immunological response to cause graft rejection, as well as predisposing to other infections. Another concern is whether an immunosuppressed cardiac transplant recipient could seroconvert sufficiently. In a case-control study in 18 cardiac transplant recipients and 18 control patients 6 months or more beyond transplant surgery, there were no differences in the incidence of cardiac rejection or immune responses (30).

There have been reports of pericarditis (31,32) in temporal relation to influenza vaccine.

Respiratory

Of 109 children with asthma aged 6 months to 18 years immunized with trivalent subvirion influenza vaccine, 59 vaccinees had no asthma symptoms on the day of immunization, but 50 had an exacerbation requiring prednisone (33). Antibody responses were not different in the two groups. Adverse effects, including local swelling at the injection site, fever, rash, and headache, were not different in the two groups.

Pulmonary edema is a rare complication of influenza immunization.

- Interstitial lung edema occurred 3 days after influenza immunization in a 67-year-old patient (34). An infectious cause was excluded and an allergic reaction was suspected. After antibiotic treatment and high-dosage glucocorticoids the patient

The safety of cold-adapted trivalent intranasal influenza virus vaccine (CAIV–FluMist) have been determined in 9689 children and adolescensts aged 1–17 years using vaccine or placebo (randomization 2:1) (35). Children under 9 years of age received a second dose of CAIV or placebo 28–42 days after the first dose. Of the four prespecified diagnostic categories (acute respiratory tract events, systemic bacterial infection, acute gastrointestinal tract events, and rare events potentially associated with wild-type influenza), none was associated with the vaccine. For reactive airway disease there was a significantly increased relative risk in children aged 18–35 months, with a relative risk of 4.06 (90% CI = 1.29, 17.86). The authors concluded that CAIV was generally safe in children and adolescents.

Nervous system

Adverse effects of flu immunization on the nervous system range from polyneuropathy to meningoencephalitis and Guillain–Barré syndrome (36).

Examination of new cases of multiple sclerosis among the 45 million swine flu vaccine recipients indicated no excess over the expected frequencies. Inactivated swine flu vaccine did not influence the onset or exacerbation of the disease (37).

The Vaccine Safety Committee of the Institute of Medicine (IOM), Academies of Science, reviewed the data on influenza vaccine and neurological conditions (38) and reached the following conclusions:

- Studies that examined the association between the swine influenza vaccine campaign in 1976 and Guillain–Barré syndrome, including analysis and reanalysis of nationwide data and state-based studies, consistently showed an increased risk of Guillain–Barré syndrome in the immunized population. The evidence therefore favors acceptance of a causal relation between the 1976 swine influenza vaccine and Guillain–Barré syndrome in adults.
- The Committee reviewed several population-based studies and a study of military personnel concerning the occurrence of Guillain–Barré syndrome after the use of influenza vaccines introduced after 1976. These studies differed in terms of their design, the case definitions for Guillain–Barré syndrome, the methods of ascertainment, the sizes of study populations, and the influenza

seasons covered. The findings were mixed. The Committee concluded that the evidence that influenza vaccines other than the 1976 swine flu vaccine caused Guillain–Barré syndrome is inadequate to accept or reject a causal relation.
- The Committee examined reports on epidemiological studies of relapses among patients with multiple sclerosis after influenza immunization; it separately examined a smaller set of reports concerning the risk of onset of multiple sclerosis. All the studies concerned influenza vaccines in various years, including the swine flu vaccines of 1976. The Committee concluded that the evidence favors rejection of a causal relation between influenza vaccines and relapse of Guillain–Barré syndrome in adults. Only one of the small set of studies on influenza immunization and the onset of multiple sclerosis provided a thorough description of the study methods and outcomes. This study found no increased risk for the onset of multiple sclerosis associated with influenza immunization, but in the absence of confirmation from other sources the Committee concluded that the evidence is inadequate to accept or reject a causal relation between influenza vaccines and multiple sclerosis in adults.
- However, the biological mechanisms involved in the onset of multiple sclerosis are presumed to be related to those involved in relapse. With the data favoring the rejection of a causal relation between influenza vaccines and relapse of multiple sclerosis, the committee saw no reason to suspect that there might be a causal relation between influenza vaccines and the onset of multiple sclerosis.
- With a single epidemiological study available (on optic neuritis) and several case reports mentioning the occurrence of other demyelinating neurological disorders (acute disseminated encephalomyelitis, transverse myelitis) after influenza immunization, the Committee concluded that the evidence is inadequate to accept or reject a causal relation between influenza vaccines and other demyelinating neurological disorders.
- Based on the lack of published evidence on influenza vaccines and demyelinating neurological disorders in children, especially those aged 6–23 months, the Committee concluded that there is no evidence to support a causal relation between influenza vaccines and demyelinating neurological disorders in children aged 6–23 months.

Guillain–Barré syndrome

Guillain–Barré syndrome was observed during the 1976/7 mass immunization campaign in the USA. The vaccine then used was A/New Jersey/76 (H1N1) flu vaccine (swine influenza). The overall incidence of cases of Guillain–Barré syndrome attributed to the use of vaccine at that time was 4.9–5.9 per million vaccinees (39). Various authors tried to settle the question of a cause and effect relation. Detailed reports have been published (SEDA-10, 289) (SEDA-11, 290) and the resulting litigation has been reviewed (40). In an analysis of computerized summaries of 1300 cases, immunized cases with extensive paresis or paralysis occurred in a characteristic epidemiological pattern, suggesting a causal relation between immunization and Guillain–Barré syndrome (39). Cases with limited motor involvement showed no such pattern. Unlike the 1976 swine flu vaccine, vaccines used subsequently have not been associated with an increased frequency of Guillain–Barré syndrome. It has been calculated that the risk of polyneuropathy following immunization is one in 200 000, compared with a population incidence of spontaneous Guillain–Barré syndrome of one in 1 000 000 (41).

The original Centers for Disease Control study of the relation between A/New Jersey/876 (swine flu) vaccine and Guillain–Barré syndrome showed a statistically significant association and suggested a causal association between the two events. In an evaluation of the medical records of all previously reported adult patients with Guillain–Barré syndrome in Michigan and Minnesota from 1 October 1976 to 31 January 1977, the relative risk during the 6 weeks after flu immunization in adults was 7.10 (excess cases attributed to the vaccine: 8.6 per million vaccinees in Michigan and 9.7 per million vaccinees in Minnesota), comparable to the relative risk of 7.60 found in the original study (42). There was no increase in the relative risk of Guillain–Barré syndrome beyond 6 weeks after immunization.

A retrospective study (1980–88) conducted to determine if the US Army's mass influenza immunization program was associated with an increased incidence of Guillain–Barré syndrome found no temporally related increase (43).

The number of reports of influenza vaccine-associated Guillain–Barré syndrome to the US Vaccine Adverse Event Reporting System increased from 37 in 1992–93 to 74 in 1993–94, raising concerns about a possible increase in vaccine-associated risk. Detailed data analyses showed that the relative risk of Guillain–Barré syndrome associated with influenza immunization, adjusted for age, sex, and vaccine season, was 2.0 for the 1992–93 season and 1.5 for the 1993–94 season. For the two seasons combined, the adjusted relative risk of 1.7 suggested that there was slightly more than one additional case of Guillain–Barré syndrome per million vaccinees. An accompanying editorial also referred to the occurrence of Guillain–Barré syndrome during the swine flu immunization campaign in 1976. The authors considered the results of this study as epidemiological evidence that immunization against strains of influenza other than swine flu may increase the risk of Guillain–Barré syndrome, albeit minimally (44).

Among 382 patients with Guillain–Barré syndrome after influenza vaccine to the Vaccine Adverse Events Reporting System (VAERS) database from 1991 through 1999 (45). there was a statistically significant increase in the incidence of Guillain–Barré syndrome after influenza immunization, compared with a Td vaccine control group. The overall mean incidence of Guillain–Barré syndrome was 0.95 per million

influenza immunizations compared with 0.22 per million Td immunizations. However, in the report of the Vaccine Safety Committee on neurological complications after influenza vaccine mentioned above (46) it was concluded that the information from VAERS added little to the Committee's ability to assess causality.

The relation of Guillain–Barré syndrome to pericarditis and nephrotic syndrome after influenza immunization has been discussed in the context of a case (47).

- A 68-year-old woman and a 72-year-old man developed distal weakness of the limbs and numbness within 2 weeks after influenza immunization. Guillain–Barré syndrome was diagnosed in both cases. The first patient also had pericarditis and the second had nephrotic syndrome.

Bell's palsy

After the introduction of an inactivated intranasal influenza vaccine, which was used only in Switzerland, 46 cases of Bell's palsy were reported. In a matched case-control study and a case-series analysis all primary-care physicians, ear, nose, and throat specialists, and neurologists in German-speaking regions of Switzerland were asked to identify cases of Bell's palsy in adults between 1 October 2000 and 30 April 2001 (48). Each physician was invited to select three control patients for each patient with Bell's palsy, matching by age, date of clinic visit, and physician. They identified 773 patients with Bell's palsy. Of the 412 who could be evaluated, 250 were enrolled and matched with 722 control patients; the other 162 patients had no controls. In the case-control study, 68 patients with Bell's palsy (27%) and eight controls (1.1%) had received the intranasal vaccine. In contrast to parenteral influenza vaccines, the intranasal vaccine significantly increased the risk of Bell's palsy (adjusted OR = 84; 95%CI = 20, 352). Even according to conservative assumptions, the relative risk of Bell's palsy was estimated to be 19 times the risk in the controls, corresponding to 13 excess cases per 10 000 vaccinees within 1–91 days after immunization. In the case-series analysis, the period of highest risk was 31–60 days after immunization. This vaccine is no longer used in Switzerland.

Sensory systems

Four patients with corneal transplants developed ocular manifestations (bilateral graft rejection in two cases, uveitis, and epithelial and stromal herpetic kerato-uveitis) at 3 days to 6 weeks after the receipt of inactivated influenza vaccine (49). Whereas case reports of ocular manifestations after influenza immunization are known, this is perhaps the first report of vaccine-related herpetic recurrence. The authors advised caution when influenza immunization is considered for patients who have had a corneal transplant.

Gastrointestinal

Pancolitis has been attributed to influenza immunization (50).

- Pancolitis occurred in a 70-year-old woman 2 hours after influenza immunization. She developed bloody diarrhea, abdominal cramps, and over the next several days arthralgia of the knees. The diagnosis was confirmed by abdominal CT scan; the patient refused endoscopy and recovered within 10 days.

The authors favored the hypothesis of vasculitis as a delayed hypersensitivity reaction after influenza immunization. The patient had already been given influenza vaccine six times before.

Urinary tract

There has been a report of minimal-change nephrotic syndrome (51) in temporal relation to influenza vaccine.

Skin

Influenza vaccine is one of at least 30 drugs believed to cause bullous pemphigoid or cicatricial pemphigoid (52).

- A 90-year-old woman developed a generalized bullous eruption resembling bullous pemphigoid 12 hours after influenza immunization (53).

Influenza vaccine can also cause pemphigus (54).

- Two women, one aged 28 the other 32 years, developed dusky red macules with pain and a burning sensation within 24 hours of receiving influenza vaccine (55). One had blisters within the lesion. The clinical feature and histopathology were consistent with the fixed drug eruptions. The lesions subsided within 2 weeks of topical glucocorticoid treatment. The diagnosis was confirmed by a topical provocation prick test with influenza vaccine.

Musculoskeletal

There has been a report of rhabdomyolysis in temporal relation to influenza vaccine (56).

Macrophagic myofasciitis has been observed in a patient who received annual injections of influenza vaccine for 4 years (57).

- A 59-year-old woman developed slowly progressive pain in the right thigh over 2 years. She had previously complained of diffuse myalgia and polyarthralgia for 2 years. She had been immunized with influenza vaccines annually after the diffuse myalgia had developed. There was severe focal muscle tenderness in the right thigh without muscle weakness or wasting. Clinical chemistry and serology were all normal. Open biopsy of the right vastus lateralis muscle and fascia showed characteristics of macrophagic myofasciitis, with conspicuous infiltration of macrophages, stained with PAS and CD68, but not desmin and smooth muscle actin or

CD1a and S100 protein. The macrophage infiltrates were multifocal, and lymphocytes were mostly CD8+ T cells. CD4+ T cells and CD45 cells were rare, and neither CD20+ B cells nor plasma cells were found. Prednisolone was given and the right thigh pain abated somewhat.

Immunologic

Changes in the lymphocyte population similar to those observed during virus infections occurred within the first 2 weeks after immunization (58). There were no reports describing more severe courses of infectious diseases during this period.

The question of whether egg allergy is a justified contraindication to influenza immunization has been studied in 80 individuals with egg allergy and 124 control subjects, who received influenza vaccine containing ovalbumin/ovomucoid 0.02, 0.1, or 1.2 µg/ml (59). The individuals with egg allergy received the vaccine in two doses 30 minutes apart; the first dose was one-tenth and the second dose nine-tenths of the recommended dose. The patients with egg allergy, even those with significant allergic reactions after egg ingestion, safely received influenza vaccine in this two-dose protocol with vaccine containing no more than 1.2 µg/ml of egg protein.

There have been reports of individual cases of giant cell arteritis (60) and polymyalgia rheumatica (61,62) in temporal relation to influenza vaccine.

- A 70-year-old man, previously healthy, developed giant cell arteritis 5 days after influenza immunization (63).

The authors mentioned another case reported in 1976.

Second-Generation Effects

Tumorigenicity

The cause of multiple myeloma is unknown. Exposure to chemicals, radiation, viruses, and other systemic factors are likely to interplay in the development of multiple myeloma in specific patients. In a case-control study there was no association between immunization and the development of light-chain multiple myeloma. Although there has been a report of multiple myeloma 3 months after influenza immunization (64), since multiple myeloma takes many years to develop from the first abnormal cell, this association can be discounted.

Susceptibility Factors

Age

Children

An update on Flumist®, a cold-adapted live attenuated influenza vaccine, was presented at the June 2007 meeting of the Global Advisory Committee on Vaccine Safety (GACVS) (65). Studies in young children showed that it is effective against circulating H1N1 and H3N2 strains, including H3N2 strains that are antigenically dissimilar to the strain that is included in the annual vaccine. Efficacy has also been demonstrated against circulating B strains. However, there was a significantly increased incidence of medically significant episodes of wheezing within 42 days of immunization among children aged 6–23 months. The manufacturer has applied to extend the indication to children under 5 years of age, and this is under review by the FDA.

Elderly people

Data on immunogenicity and safety from about 3500 individuals (healthy elderly and elderly with underlying chronic diseases) immunized in 13 clinical trials with either MF59-adjuvanted influenza vaccine (n = 1890) or non-adjuvanted influenza vaccine (n = 1374) were analysed (66). On day 28 the GMT ratios (adjuvanted versus non-adjuvanted vaccine) were as given in Table 2. Local and systemic reactions were more common in those who received the adjuvanted vaccine: pain at the injection site: 33% versus 13%, erythema 18% versus 13%, induration 15% versus 9%; malaise 6% versus 4%, myalgia 8% versus 3%, headache 6% versus 4% The reactions were mainly mild and transient, and there were no serious reaction.

Drug Administration

Drug dosage regimens

Current recommendations for the use of influenza vaccine in adults are based on a single injection. This may not be valid in case of a new pandemic caused by an antigenic shift of the influenza virus. Currently, the only group for whom a second dose is recommended comprises children who have never been immunized. However, when two-dose regimens in adults have been studied, the second dose of vaccine has not been associated with higher rates of reactions than the first. People who have a stronger local reaction after a first injection are more likely to have another such reaction after a second injection (33).

Drug administration route

A cold-adapted, live, attenuated, trivalent influenza vaccine to be administered by intranasal spray has been licensed by the Food and drug Administration (FDA) for children, adolescents, and adults aged 5–49 years. and its immunogenicity, efficacy, and safety have been

Table 2 GMT ratios healthy elderly vaccinees and in elderly vaccinees with chronic diseases

Antigen	Healthy elderly vaccinees	Elderly vaccinees with chronic diseases
A/H3N2 antigen	1.18	1.43
B antigen	1.17	1.37
A/H1N1 antigen	1.10	1.17

reviewed (67,68). Nasal congestion occurred in 7–11% of immunized children and fever in 4%. Vomiting, abdominal pain, and myalgia were rare. The reactions were usually self-limiting. In most published studies, there was no statistically significant difference between vaccine and placebo. The safety profile in adults is similar, except that in half of the studies a sore throat was reported in significantly more vaccine recipients (10–15%) than in controls.

Trivalent cold-adapted intranasal influenza vaccine was used to immunize 1602 healthy children aged 15–71 months in a randomized, double-blind, placebo-controlled trial (69). One year later 1358 were reimmunized. The vaccine provided efficacy of 92% during 2 years against virologically confirmed influenza. Transient, minor symptoms of respiratory illness (rhinorrhea, nasal congestion, low-grade fever) were reported more often in vaccinees than in controls; no significant differences were noted after dose 1 and dose 2.

Drug–Drug Interactions

Anticoagulants

In a case-control study 90 patients (88 taking warfarin and two taking acenocoumarol) were immunized with influenza vaccine; 45 non-immunized patients were used as controls (70). Influenza immunization caused significant prolongation of the prothrombin time within 7–10 days in 49 patients, and two had bleeding episodes. The authors concluded that the data supported the hypothesis of a potentially serious interaction between warfarin and influenza vaccine and recommended monitoring the International Normalized Ratio (INR) in patients taking anticoagulants during the immediate period after immunization.

In a prospective study of the effect of flu immunization on the prothrombin time in eight patients taking long-term anticoagulant treatment the prothrombin time was prolonged by 40%. In healthy subjects there was no significant effect on warfarin metabolism (SEDA-10, 289).

- An 81-year-old patient who had been well controlled by anticoagulants for 12 years had an episode of gastrointestinal bleeding associated with influenza immunization (71).

However, others did not confirm this effect (72), and the lack of a clinical interaction with warfarin has also been confirmed by a report of the US Immunization Practices Advisory Committee (9).

Anticonvulsants

In a study of serum concentrations of the anticonvulsants phenytoin, phenobarbital, and carbamazepine, before and after mentally retarded patients received flu vaccine, the authors concluded that serum concentrations of these drugs may increase as a result of flu immunization and that dosage adjustments may be necessary (73).

Benzodiazepines

The metabolism of lorazepam and chlordiazepoxide was not altered by flu immunization (74).

Pneumococcal vaccine

In a study of the interaction between 23-valent pneumococcal polysaccharide vaccine and influenza vaccine, 152 adults with chronic respiratory disease were randomized to receive both vaccines either simultaneously or at an interval of 1 month (77). There were no significant differences in serological responses between the groups. The incidence and severity of both local and systemic adverse effects were also similar: there were mild local reactions in 38 and 36% and systemic reactions in five and three of the vaccinees respectively.

Theophylline

After influenza immunization there was a reduction in blood theophylline concentrations in patients and healthy volunteers (75). The authors concluded that flu vaccine may influence the pharmacokinetics of several drugs, and a second group found that theophylline oxidation was significantly reduced at 1 day, but not at 7 days, after immunization (74). However, others did not confirm these effects (72,76). The lack of a clinical interaction with theophylline has also been confirmed by a report of the US Immunization Practices Advisory Committee (9).

References

1. Palache AM. Influenza vaccines. A reappraisal of their use. Drugs 1997;54(6):841–56.
2. Minutello M, Senatore F, Cecchinelli G, Bianchi M, Andreani T, Podda A, Crovari P. Safety and immunogenicity of an inactivated subunit influenza virus vaccine combined with MF59 adjuvant emulsion in elderly subjects, immunized for three consecutive influenza seasons. Vaccine 1999;17(2):99–104.
3. De Donato S, Granoff D, Minutello M, Lecchi G, Faccini M, Agnello M, Senatore F, Verweij P, Fritzell B, Podda A. Safety and immunogenicity of MF59-adjuvanted influenza vaccine in the elderly. Vaccine 1999;17(23–24):3094–101.
4. Holm KJ, Goa KL, Oxford JS, McElhaney JE. Liposomal influenza vaccine. Biodrugs 1999;11:137–46.
5. Jackson LA, Holmes SJ, Mendelman PM, Huggins L, Cho I, Rhorer J. Safety of a trivalent live attenuated intranasal influenza vaccine, FluMist, administered in addition to parenteral trivalent inactivated influenza vaccine to seniors with chronic medical conditions. Vaccine 1999;17(15–16):1905–9.
6. Glueck R. Review of intranasal influenza vaccine. Adv Drug Deliv Rev 2001;51(1–3):203–11.
7. Nicholson KG, Colegate AE, Podda A, Stephenson I, Wood J, Ypma E, Zambon MC. Safety and antigenicity of non-adjuvanted and MF59-adjuvanted influenza A/Duck/Singapore/97 (H5N3) vaccine: a randomised trial of two potential vaccines against H5N1 influenza. Lancet 2001;357(9272):1937–43.
8. Committee on Immunization. In: Guide for adult immunization. Philadelphia: American College of Physicians, 1985:58.
9. Centers for Disease Control (CDC). Prevention and control of influenza. MMWR Morb Mortal Wkly Rep 1987;36(24):373–80385–7.
10. Stefanini M, Piomelli S, Mele R, Ostroski JT, Colpoys WP. Acute vascular purpura following immunization with Asiatic-influenza vaccine. N Engl J Med 1958;259(1):9–12.
11. Yahr MD, Lobo-Antunes J. Relapsing encephalomyelitis following the use of influenza vaccine. Arch Neurol 1972;27(2):182–3.
12. Retailliau HF, Curtis AC, Storr G, Caesar G, Eddins DL, Hattwick MA. Illness after influenza vaccination reported through a nationwide surveillance system, 1976–1977. Am J Epidemiol 1980;111(3):270–8.
13. Nagano T, Mizuguchi M, Kurihara E, Mizuno Y, Tamagawa K, Komiya K. [A case of acute disseminated encephalomyelitis with convulsion, gait disturbance, facial palsy and multifocal CT lesions.]No To Hattatsu 1988;20(4):325–9.
14. Casoli P, Tumiati B. Porpora trombocitopenica idiopatica acuta dopo vaccinacione antinfluenzale. [Acute idiopathic thrombocytopenic purpura after anti-influenza vaccination.] Medicina (Firenze) 1989;9(4):417–8.
15. Bakshi R, Mazziotta JC. Acute transverse myelitis after influenza vaccination: magnetic resonance imaging findings. J Neuroimaging 1996;6(4):248–50.
16. Ichikawa N, Takase S, Kogure K. [Recurrent aseptic meningitis following influenza vaccination in a case of systemic lupus erythematosus.]Rinsho Shinkeigaku 1983;23(7):570–6.
17. Morimoto T, Oguni H, Awaya Y, Hayakawa T, Fukuyama Y. A case of a rapidly progressive central nervous system disorder manifesting as a pallidal posture and ocular motor apraxia. Brain Dev 1985;7(4):449–53.
18. Confino I, Passwell JH, Padeh S. Erythromelalgia following influenza vaccine in a child. Clin Exp Rheumatol 1997;15(1):111–3.
19. Hull TP, Bates JH. Optic neuritis after influenza vaccination. Am J Ophthalmol 1997;124(5):703–4.
20. Beijer WE, Sprenger MJ, Masurel N. Polymyalgia rheumatica und Grippa–Schutzimpfung. [Polymyalgia rheumatica and influenza vaccination.] Dtsch Med Wochenschr 1993;118(5):164–5.
21. Kelsall JT, Chalmers A, Sherlock CH, Tron VA, Kelsall AC. Microscopic polyangiitis after influenza vaccination. J Rheumatol 1997;24(6):1198–202.
22. Thurairajan G, Hope-Ross MW, Situnayake RD, Murray PI. Polyarthropathy, orbital myositis and posterior scleritis: an unusual adverse reaction to influenza vaccine. Br J Rheumatol 1997;36(1):120–3.
23. Mader R, Narendran A, Lewtas J, Bykerk V, Goodman RC, Dickson JR, Keystone EC. Systemic vasculitis following influenza vaccination—report of 3 cases and literature review. J Rheumatol 1993;20(8):1429–31.
24. Demmler M, Heidel G. Trigeminus-Affektion nach Influenza-Schutzimpfung. [Trigeminal involvement following preventive influenza vaccination.] Psychiatr Neurol Med Psychol (Leipz) 1985;37(7):428–33.
25. Vidal E, Gaches F, Berdah JF, Nadalon S, Lavignac C, Mitrea L, Loustaud-Ratti V, Liozon F. Vasculitis after influenza vaccination. Rev Med Intern 1993;14:1173.
26. Dooley M, Goa KL. Adjuvanted influenza vaccine. Biodrugs 2000;14:61–9.
27. Gluck R, Mischler R, Durrer P, Furer E, Lang AB, Herzog C, Cryz SJ Jr. Safety and immunogenicity of intranasally administered inactivated trivalent virosome-formulated influenza vaccine containing Escherichia coli heat-labile toxin as a mucosal adjuvant. J Infect Dis 2000;181(3):1129–32.

28. De Serres G, Grenier JL, Toth E, Menard S, Roussel R, Tremblay M, Douville Fradet M, Landry M, Robert Y, Skowronski DM. The clinical spectrum of the oculo-respiratory syndrome after influenza vaccination. Vaccine 2003;21:2354–61.

29. Skowronska DM, Strauss B, De Serres G, MacDonald D, Marion SA, Naus M, Patrick DM, Kendall P. Oculo-respiratory syndrome: a new influenza vaccine-associated adverse event? Clin Infect Dis 2003;36:705–13.

30. Kobashigawa JA, Warner-Stevenson L, Johnson BL, Moriguchi JD, Kawata N, Drinkwater DC, Laks H. Influenza vaccine does not cause rejection after cardiac transplantation. Transplant Proc 1993;25(4):2738–9.

31. Medearis DN Jr, Neill CA, Markowitz M. Influenza and cardiopulmonary disease. II. Med Concepts Cardiovasc Dis 1963;32:813–6.

32. Zanettini MT, Zanettini JO, Zanettini JP. Pericarditis. Series of 84 consecutive cases. Arquivos Brasileiros de Cardiologia 2004;82(4):360–9.

33. Park CL, Frank AL, Sullivan M, Jindal P, Baxter BD. Influenza vaccination of children during acute asthma exacerbation and concurrent prednisone therapy. Pediatrics 1996;98(2 Pt 1):196–200.

34. Franzen D, Schneider M, Karsten R, Von Eiff M. Interstitielles Lungenoedem nach Grippeschutzimpfung. Atemwegs Lungenkrankheiten 2003;29:86–8.

35. Bergen R, Black S, Shinefield H, Lewis E, Ray P, Hansen J, Walker R, Hessel C, Cordova J, Mendelman PM. Safety of cold-adapted live attenuated influenza vaccine in a large cohort of children and adolescents. Ped Infect Dis J 2004;23:138–44.

36. Hayase Y, Tobita K. Influenza virus and neurological diseases. Psychiatry Clin Neurosci 1997;51(4):181–4.

37. Kurland LT, Molgaard CA, Kurland EM, Wiederholt WC, Kirkpatrick JW. Swine flu vaccine and multiple sclerosis. JAMA 1984;251(20):2672–5.

38. Immunization Safety Review Committee, Institute of Medicine, National Academies. Immunization Safety Review. Influenza Vaccines and Neurological Complications. Washington, DC: The National Academies Press, 2004.

39. Langmuir AD, Bregman DJ, Kurland LT, Nathanson N, Victor M. An epidemiologic and clinical evaluation of Guillain–Barré syndrome reported in association with the administration of swine influenza vaccines. Am J Epidemiol 1984;119(6):841–79.

40. Dukes MNG, Swartz B. Responsibility for Drug-induced InjuryAmsterdam: Elsevier;. 1988.

41. Feschank R, Kunzel U, Quast U. Das Guillain–Barré syndrome-eine Impfkomplikation? Z Allg Med 1986;62:71.

42. Safranek TJ, Lawrence DN, Kurland LT, Culver DH, Wiederholt WC, Hayner NS, Osterholm MT, O'Brien P, Hughes JM. Reassessment of the association between Guillain–Barré syndrome and receipt of swine influenza vaccine in 1976–1977: results of a two-state study. Expert Neurology Group. Am J Epidemiol 1991;133(9):940–51.

43. Roscelli JD, Bass JW, Pang L. Guillain–Barré syndrome and influenza vaccination in the US Army, 1980–1988. Am J Epidemiol 1991;133(9):952–5.

44. Lasky T, Terracciano GJ, Magder L, Koski CL, Ballesteros M, Nash D, Clark S, Haber P, Stolley PD, Schonberger LB, Chen RT. The Guillain–Barré syndrome and the 1992–1993 and 1993–1994 influenza vaccines. N Engl J Med 1998;339(25):1797–802.

45. Geier MR, Geier DA, Zahalsky AC. Influenza vaccination and Guillain–Barré syndrome. Clin Immunol 2003;107:116–21.

46. Immunization Safety Review Committee, Institute of Medicine, National Academies. Immunization Safety Review. Influenza Vaccines and Neurological Complications. Washington, DC: The National Academies Press, 2004.

47. Kao CD, Chen JT, Lin KP, Shan DE, Wu ZA. Liao KK. Guillain–Barré syndrome coexisting with pericarditis or nephritic syndrome after influenza vaccination. Clin Neurol Neurosurg 2004;106:136–8.

48. Mutsch M, Zhou W, Rhodes P, Bopp M, Chen RT, Linder T, Spyr C, Steffen R. Use of the inactivated intranasal influenza vaccine and the risk of Bell's palsy in Switzerland. N Engl J Med 2004;350:896–903.

49. Solomon A, Siganos CS, Frucht-Pery J. Adverse ocular effects following influenza vaccination. Eye 1999;13(Pt 3a):381–2.

50. Luca L, Morisset M, Kanny G, Moneret-Vautrin DA. Pancolitis after influenza vaccination. Allergy (Copenhagen) 2004;59:367.

51. Kielstein JT, Termuhlen L, Sohn J, Kliem V. Minimal change nephrotic syndrome in a 65-year-old patient following influenza vaccination. Clin Nephrol 2000;54(3):246–8.

52. Vassileva S. Drug-induced pemphigoid: bullous and cicatricial. Clin Dermatol 1998;16(3):379–87.

53. Garcia-Doval I, Roson E, Feal C, De la Torre C, Rodriguez T, Cruces MJ. Generalized bullous fixed drug eruption after influenza vaccination, simulating bullous pemphigoid. Acta Dermatol Venereol 2001;81(6):450–1.

54. Mignogna MD, Lo Muzio L, Ruocco E. Pemphigus induction by influenza vaccination. Int J Dermatol 2000;39(10):800.

55. Al-Mutairi N, Al-Fouzan A, Nour-Eldin O. Fixed drug eruption due to influenza vaccine. J Cutan Med Surg 2004;8:16–18.

56. Plotkin E, Bernheim J, Ben-Chetrit S, Mor A, Korzets Z. Influenza vaccine—a possible trigger of rhabdomyolysis induced acute renal failure due to the combined use of cerivastatin and bezafibrate. Nephrol Dial Transplant 2000;15(5):740–1.

57. Park JH, Na KS, Park YW, Paik SS, Yoo DH. Macrophagic myofasciitis unrelated to vaccination. Scand J Rheumatol 2005;34:65–7.

58. Gerth HG. Grippeschutzimpfung. Dtsch Med Wochenschr 1989;114:180.

59. James JM, Zeiger RS, Lester MR, Fasano MB, Gern JE, Mansfield LE, Schwartz HJ, Sampson HA, Windom HH, Machtinger SB, Lensing S. Safe administration of influenza vaccine to patients with egg allergy. J Pediatr 1998;133(5):624–8.

60. Perez C, Loza E, Tinture T. Giant cell arteritis after influenza vaccination. Arch Intern Med 2000;160(17):2677.

61. Liozon E, Ittig R, Vogt N, Michel JP, Gold G. Polymyalgia rheumatica following influenza vaccination. J Am Geriatr Soc 2000;48(11):1533–4.

62. Perez C, Maravi E. Polymyalgia rheumatica following influenza vaccination. Muscle Nerve 2000;23(5):824–5.

63. Finsterer J, Artner C, Kladosek A, Kalchmayr R, Redtenbacher S. Cavernous sinus syndrome due to vaccination-induced giant cell arteritis. Arch Intern Med 2001;161(7):1008–9.

64. Schattner A, Berrebi A. Several possible causes for multiple myeloma including a vaccination in a single case study. Vaccine 2004;22:2509–10.

65. Global Advisory Committee on Vaccine Safety. Meeting on 12–13 June 2007. Influenza vaccines: update. Wkly Epidemiol Rec 2007;82:255–6.

66. Banzhoff A, Nacci P, Podda A. A new MF59-adjuvanted influenza vaccine enhances the immune response in the elderly with chronic diseases. Results from an immunogenicity meta-analysis. Gerontology 2003;49:177–84.

67. Eyles JE, Williamson ED, Alpar HO. Intranasal administration of influenza vaccines: current status. Biodrugs 2000;13:35–59.

68. Zangwill KM. Cold-adapted, live attenuated intranasal influenza virus vaccine. Pediatr Infect Dis 2003;22:273–4.

69. Belshe RB, Gruber WC. Prevention of otitis media in children with live attenuated influenza vaccine given intranasally. Pediatr Infect Dis J 2000;19(Suppl 5):S66–71.

70. Paliani U, Gresele P. Significant potentiation of anticoagulation by flu vaccine during the season 2001-2002. J Haematol 2003;88:539–40.

71. Kramer P, Tsuru M, Cook CE, McClain CJ, Holtzman JL. Effect of influenza vaccine on warfarin anticoagulation. Clin Pharmacol Ther 1984;35(3):416–8.

72. Gomolin IH, Chapron DJ, Luhan PA. Lack of effect of influenza vaccine on theophylline levels and warfarin anticoagulation in the elderly. J Am Geriatr Soc 1985;33(4):269–72.

73. Jann MW, Fidone GS. Effect of influenza vaccine on serum anticonvulsant concentrations. Clin Pharm 1986;5(10):817–20.

74. Meredith CG, Christian CD, Johnson RF, Troxell R, Davis GL, Schenker S. Effects of influenza virus vaccine on hepatic drug metabolism. Clin Pharmacol Ther 1985;37(4):396–401.

75. Kramer P, McClain CJ. Depression of aminopyrine metabolism by influenza vaccination. N Engl J Med 1981;305(21):1262–4.

76. Grabowski N, May JJ, Pratt DS, Richtsmeier WJ, Bertino JS Jr. The effect of split virus influenza vaccination on theophylline pharmacokinetics. Am Rev Respir Dis 1985;131(6):934–8.

77. Fletcher TJ, Tunnicliffe WS, Hammond K, Roberts K, Ayres JG. Simultaneous immunisation with influenza vaccine and pneumococcal polysaccharide vaccine in patients with chronic respiratory disease. BMJ 1997;314(7095):1663–5.

Japanese encephalitis vaccine

General Information

During the last half century, Japanese encephalitis has been recognized as an important arboviral disease in man in Japan, China, Korea, Thailand, India, Nepal, Sri Lanka, and Vietnam. In 1954, Japanese encephalitis vaccine of the mouse brain type for human use was licensed in Japan. However, there was strong criticism of mouse brain vaccine, which has continued for many years. Therefore, in 1965, the Nippon Institute of Biological Products and the Biken Foundation implemented more advanced purification procedures, such as alcohol precipitation and ultracentrifugation.

Three types of Japanese encephalitis vaccine are currently produced:

1. mouse brain-derived inactivated vaccine, commercially available;
2. cell culture-derived inactivated vaccine;
3. cell culture-derived attenuated vaccine, produced and used exclusively in China.

The Chinese cell culture-derived inactivated vaccine is produced in primary hamster kidney cells; a highly purified Vero cell culture-derived inactivated vaccine is under clinical development in France. A Chinese live-attenuated vaccine is also produced in primary hamster kidney cells. The efficacy of one dose is 80%, and the efficacy of two doses, given 1 year apart, is 97.5%; this vaccine was evaluated in a randomized trial in 26 239 children, half of whom received the vaccine and half served as controls, and its adverse effects have been reviewed (SEDA-22, 350).

In 1965, a special surveillance team was formed by the Japanese Ministry of Health and Welfare to investigate adverse events following the administration of Japanese encephalitis vaccine. No severe adverse event was reported among 21 396 vaccinees of whom 18 401 were adolescents and children under 18 years of age. Some mild reactions (fever, malaise, abdominal symptoms) were noted in 1.2% of the vaccinees. Using a countrywide hospital network, the surveillance team studied any severe neurological disease occurring within 1 month after receipt of the vaccine. During 1957–66, 26 cases (nine cases of meningitis, ten cases of convulsions, five cases of polyneuropathy, and two cases of demyelinization) were analysed. No evidence was provided showing a causative relation between these clinical syndromes and Japanese encephalitis immunization. The incidence of neurological disease was considered minor compared with the millions of doses distributed annually in Japan (1).

Since 1989, urticaria and/or angioedema of the extremities, face, and oropharynx, and respiratory distress have been reported from Europe, North America, and Australia as a new pattern of adverse effects. Collapse due to hypotension has required hospitalization in several cases, and erythema multiforme and erythema nodosum have also occurred; the reported rates of such adverse effects varied markedly in different countries (respective ranges 0.7–12 per 10 000 and 50–104 per 10 000). The vaccine constituents responsible for the adverse effects have not yet been identified.

Japanese encephalitis vaccine has been reviewed (2). A supplementary volume of the journal Vaccine has dealt with results presented at a WHO meeting held in Bangkok, Thailand, in 1998 (3–6). Comprehensive data were provided on the epidemiological and virological situation in southeast Asia and Australia, control measures, vaccine production capacities, and different vaccines against Japanese encephalitis. Adverse events after the use of inactivated mouse brain vaccine (the only vaccine that is currently licensed for international use) have been reviewed in detail (3).

Current and future Japanese encephalitis (JE) vaccines have been reviewed (7). With the exception of China, where a live attenuated vaccine is used, all countries practising immunization against Japanese encephalitis use formalin-inactivated mouse brain vaccines. The vaccine is generally well tolerated. Injection site reactions were reported in about 20%, and systemic adverse effects (fever, headache, malaise, rashes, chills, myalgia, gastrointestinal symptoms) in 5-10% of vaccinees. More serious adverse events fall into two categories: allergic and neurological. Hypersensitivity reactions (urticaria, angioedema, and bronchospasm) are reported at an incidence of 0.2-0.6 per 1000 vaccinees. Most patients respond well to anti-allergic treatment. Anaphylaxis can be life-threatening, and at least three deaths have been attributed to the vaccine.

The Global Advisory Committee on Vaccine Safety (GACVS) has considered a report from an Indian expert panel that assessed cases of serious adverse events after immunization campaigns (including about 9.3 million children aged 1–15 years) with the live attenuated SA-14-14-2 Japanese encephalitis vaccine (8). A total of 65 serious adverse events were reported, 22 of which were fatal. Most of the serious adverse events were considered to be unrelated to the vaccine. Two clusters of encephalitis-like syndromes were detected; the cases in one cluster probably represented cases of natural Japanese encephalitis, and the cases in the second cluster were classified as acute encephalopathy syndrome of unknown cause. A thorough investigation into possible alternative causes was not conducted. The committee concluded that the type of clustering of encephalopathy/encephalitis cases made it unlikely that they had been related to the vaccine. The committee recommended that future immunization campaigns should be accompanied by better adverse event monitoring and investigations.

Organs and Systems

Nervous system

Postmarketing surveillance data of adverse events after Japanese encephalitis immunization in Japan and the USA have been compared (9). The rates of total reported adverse events were 2.8 per 100 000 doses in Japan and 15.0 per 100 000 doses in the USA. In Japan, 17 neurological disorders were reported from April 1996 to October 1998 (0.2 per 100 000 doses), whereas in the USA there were no serious neurological adverse events temporally associated with Japanese encephalitis vaccine from January 1993 to June 1999. Rates for systemic hypersensitivity reactions were 0.8 and 6.3 per 100 000 doses in Japan and the USA respectively.

Acute disseminated encephalomyelitis after Japanese encephalitis immunization has been reported from Japan.

• A 6-year-old girl and a 5-year-old boy had drowsiness, paresthesia, and gait disturbance 14 and 17 days respectively after immunization with Japanese encephalitis vaccine. Treatment with prednisolone improved the clinical findings (10,11).

Another report included seven cases of acute disseminated encephalomyelitis after administration of Japanese encephalitis vaccine between 1968 and 1990 (12). Three other cases of neurological complications after the use of Japanese encephalitis vaccine have been reported in Denmark. In two cases the clinical picture was consistent with acute demyelinating encephalomyelitis (SEDA-21, 334).

The rate of neurological complications seen in Japan in 1965–73 was of the order of 1 per 2.3 million vaccinees, and in Denmark the rate was 3 per 175 000 immunized individuals (one of them with a predisposition to multiple sclerosis) (13).

Two deaths from acute anaphylaxis and four cases of acute encephalopathy or acute disseminated encephalomyelitis (two fatal), temporally related to Japanese encephalitis immunization, have been reported from the Republic of Korea (3). The live-attenuated vaccine developed and used in China had 98% efficacy in a two-dose schedule. Concern that a live vaccine derived from an encephalitogenic virus might lead to vaccine-associated encephalitis could not be addressed satisfactorily, even in a study of 26 000 children. During efficacy studies in different geographic areas of China and in different years, the observations were combinable: the average encephalitis risk after live-attenuated Japanese encephalitis vaccine was 1.59 per 100 000 vaccinees.

Immunologic

Hypersensitivity reactions (generalized urticaria or angioedema) after the use of Japanese encephalitis vaccine have been reported from some countries (see Table 1); the vaccine constituents responsible for these events have not been identified (14). There has been a detailed report of the adverse effects, mainly allergic

Table 1 Allergic reactions after Japanese encephalitis immunization

Country	Estimated number of vaccines	Number of reactions	Estimated rate per 100 000 vaccines
Denmark	42 000	21	50
Sweden	15 000	1	7
UK	1950	1	51
Australia			
Nationwide	3400	4	118
Fairfield Hospital	601	3	499
Canada			
University of Calgary	96	1	1042
USA			
Travellers	1328	2	151
Army	526	1	190
Army and dependents (Okinawa)	35 253	220	624
Total	100 154	254	254

mucocutaneous reactions, of Biken vaccine in Danish travellers and US Marine Corps personnel (SEDA-22, 351).

The Advisory Committee on Immunization Practices (ACIP) has recommended that vaccinees should be observed for 30 minutes after immunization and that medications to treat anaphylaxis should be available (14) [http://www.cdc.gov/mmwr/PDF/rr/rr4201.pdf]. A personal history of allergic disorders should be considered when weighing the risks and benefits of the vaccine for an individual. Japanese encephalitis vaccine should not be given to persons who had a previous adverse reaction after receiving Japanese encephalitis vaccine or a previous hypersensitivity reaction to other vaccines of neural origin.

In Japan, children who had immediate-type allergic reactions to Japanese encephalitis vaccine had antigelatin IgE in their sera. However, the immunological mechanism of non-immediate-type allergic reactions that consist of cutaneous signs developing several hours or more after Japanese encephalitis immunization is not yet clear. Serum samples taken from 28 children who had non-immediate-type allergic skin reactions have been compared with serum samples taken from 10 children who had immediate-type reactions (15). All the children who had had immediate-type reactions had antigelatin IgE and IgG. Of 28 children who had had non-immediate-type reactions, one had antigelatin IgE and nine had antigelatin IgG. These results suggest that some children who develop non-immediate-type allergic reactions have also been sensitized to gelatin.

Susceptibility Factors

Age

In children passive surveillance in Japan between 1965 and 1978 showed an incidence of neurological complications (encephalitis, encephalopathy, convulsions, peripheral neuropathy, transverse myelitis) of 1.0-2.3 per million vaccinees. Between 1992 and 1996, 16 cases of acute disseminated encephalomyelitis were recognized in Japan, South Korea, and Denmark. The incidence in Japan was estimated to be below 11 per million vaccinees, but much higher in Denmark at 1 per 50 000-75 000 vaccinees. Some researchers suspect that the illness originates from mouse brain tissue.

New inactivated JE vaccines manufactured in Vero cells are in advanced preclinical or early clinical development. A new live attenuated vaccine that uses a reliable flavivirus (yellow fever 17D) as a live vector is in early trials and a single dose appears to be immunogenic and well tolerated. Other approaches such as vaccinia and avipox vectored vaccines, recombinant subunit vaccines, have been taken into consideration (16).

Allergic disorders

An association between reactions to Japanese encephalitis vaccine and a history of urticaria or allergic rhinitis has been identified (17).

Drug Administration

Drug formulations

Mouse brain-derived Japanese encephalitis vaccine
On 30 May 2005 the Japanese Health, Labor, and Welfare Ministry called a halt to immunization with the mouse brain-derived Japanese encephalitis vaccine after a junior high school student developed acute disseminated encephalomyelitis following immunization (18). The ministry recognized that Japan now typically has less than 10 cases of Japanese encephalitis each year, and that the number of people who have adverse effects from immunization sometimes surpasses the number of people infected with the virus.

During its 12th meeting (9–10 June 2005) the Global Advisory Committee on Vaccine Safety (GACVS) considered the decision taken by the Japanese Government (19). It was advised that the Japanese national advisory committee on vaccine adverse events could not rule out a causal link between Japanese encephalitis immunization and the single case of acute disseminated encephalomyelitis, and that there is no definite evidence of an increased risk of acute disseminated encephalomyelitis. It concluded, on currently available information, that there is no good reason for the WHO and national immunization programs to change the current recommendations for Japanese encephalitis immunization for residents in and travellers to endemic regions. The Committee will review the problem if further information becomes available.

Live attenuated Japanese encephalitis vaccine
Researchers in China have attenuated the Japanese encephalitis virus in primary hamster kidney cells and have derived the strain SA 14-14-2, which has proved to be safe and immunogenic in animals and humans (20). The vaccine was licensed in China in 1988. The current production of live attenuated SA 14-14-2 vaccine exceeds 50 million doses annually, most of which are used in China. Neuroattenuation of the SA 14-14-2 strain is reported to be based on 57 nucleotide changes and 24 amino acid substitutions, suggesting that reversion to neurovirulence of the vaccine strain would be highly unlikely. Vaccine production is in accordance with WHO technical specifications, including detailed screening for adventitious viruses. Data reported from several studies have shown vaccine efficacy to be 80–99% following single-dose immunization and 98% or greater with two doses of the vaccine. Preliminary data on co-administration of the SA 14-14-2 vaccine with measles vaccine have been reassuring. SA 14-14-2 has been shown to be 99.3% effective in preventing encephalitis some weeks after immunization. A single dose of SA 14-14-2 was 98.5% effective 12–15 months after immunization. Although important questions about its protective efficacy up to 5 or more years after immunization still need to be answered, this study suggests that a single-dose strategy, which lowers the cost of immunization and increases the supply of the vaccine, should be evaluated.

More than 200 million doses of SA 14-14-2 vaccine have been given to Chinese children without reports of serious adverse events. In large safety studies in China, there were no cases of vaccine-related encephalitis, meningitis, or acute disseminated encephalomyelitis. In a post-licensure study of nearly 183 000 children aged 1–10 years, followed for at least 2 weeks after immunization, there were no serious adverse events.

In 26 000 children who were randomized to groups that immediately received SA 14-14-2 or had immunization delayed there were no cases of encephalitis, meningitis, or anaphylaxis during the 30 days after immunization (31). The rates of hospital admissions, seizures, and less serious adverse events (for example fever, rash) did not differ between the groups.

In a safety study in South Korea, where SA 14-14-2 vaccine was licensed in 2001, there were no serious adverse events in 522 children who were actively monitored for 4 weeks after immunization; about 10% developed a fever of 38°C or higher and under 1% had local injection site reactions (redness and swelling) (31).

During its 12th meeting the Global Advisory Committee on Vaccine Safety (GACVS) considered recent data on the safety profile of live attenuated SA 14-14-2 Japanese encephalitis vaccine (31). Data presented to GACVS, covering a 20-year period from 1979 to 1998, contained no reported cases of vaccine-associated Japanese encephalitis. GACVS acknowledged the excellent safety and efficacy profile of the SA 14-14-2 vaccine, but nevertheless recommended more detailed studies of the following:

- the safety profile in special risk groups, including immunocompromised people and pregnant women;
- whether viral shedding occurs in vaccinees and the potential implications of such shedding;
- further analysis of sequential or co-administration of Japanese encephalitis and measles vaccines;
- the interchangeability of inactivated and live Japanese encephalitis vaccines;
- the safety of vaccine administration to infants aged under 1 year;
- the implications for the efficacy and safety of the vaccine in infants with maternal antibodies against Japanese encephalitis virus.

References

1. Oya A. Japanese encephalitis vaccine. Acta Paediatr Jpn 1988;30(2):175–84.
2. Sabchareon A, Yoksan S. Japanese encephalitis. Ann Trop Paediatr 1998;18(Suppl):S67–71.
3. Tsai TF. New initiatives for the control of Japanese encephalitis by vaccination: minutes of a WHO/CVI meeting, Bangkok, Thailand, 13–15 October 1998. Vaccine 2000;18(Suppl 2):1–25.
4. Markoff L. Points to consider in the development of a surrogate for efficacy of novel Japanese encephalitis virus vaccines. Vaccine 2000;18(Suppl 2):26–32.
5. Kurane I, Takasaki T. Immunogenicity and protective efficacy of the current inactivated Japanese encephalitis

vaccine against different Japanese encephalitis virus strains. Vaccine 2000;18(Suppl 2):33–5.
6. Tsarev SA, Sanders ML, Vaughn DW, Innis BL. Phylogenetic analysis suggests only one serotype of Japanese encephalitis virus. Vaccine 2000;18(Suppl 2):36–43.
7. Monath TP. Japanese encephalitis vaccines: current vaccines and further prospects. Curr Top Microbiol Immunol 2002; 267: 105–38.
8. Global Advisory Committee on Vaccine Safety. Meeting on 29–30 November. Safety of Japanese encephalitis immunization in India. Wkly Epidemiol Rec 2007;82:23–4.
9. Takahashi H, Pool V, Tsai TF, Chen RT. Adverse events after Japanese encephalitis vaccination: review of post-marketing surveillance data from Japan and the United States. The VAERS Working Group. Vaccine 2000;18(26):2963–9.
10. Demmler M, Heidel G. Trigeminus-Affektion nach Influenza-Schutzimpfung. [Trigeminal involvement following preventive influenza vaccination.] Psychiatr Neurol Med Psychol (Leipz) 1985;37(7):428–33.
11. Ohtaki E, Murakami Y, Komori H, Yamashita Y, Matsuishi T. Acute disseminated encephalomyelitis after Japanese B encephalitis vaccination. Pediatr Neurol 1992;8(2):137–9.
12. Ohtaki E, Matsuishi T, Hirano Y, Maekawa K. Acute disseminated encephalomyelitis after treatment with Japanese B encephalitis vaccine (Nakayama–Yoken and Beijing strains). J Neurol Neurosurg Psychiatry 1995;59(3):316–7.
13. Anonymous. Vaccination against Japanese encephalitis for all travellers not currently recommended. Drugs Ther Perspect 1997;10:11–3.
14. Advisory Committee on Immunization Practices (ACIP). Inactivated Japanese encephalitis virus vaccine. MMWR Recomm Rep 1993;42(RR-1):1–15.
15. Sakaguchi M, Miyazawa H, Inouye S. Specific IgE and IgG to gelatin in children with systemic cutaneous reactions to Japanese encephalitis vaccines. Allergy 2001;56(6):536–9.
16. Dynport smallpox vaccine shows fewer side effects than Dryvax in phase I study. Clin Infect Dis 2003; 36: News.
17. Sakaguchi M, Yoshida M, Kuroda W, Harayama O, Matsunaga Y, Inouye S. Systemic immediate-type reactions to gelatin included in Japanese encephalitis vaccines. Vaccine 1997;15(2):121–2.
18. Xinhua News Agency (05/30/05).
19. Global Advisory Committee on Vaccine Safety. Mouse-brain-derived Japanese encephalitis vaccine. Weekly Epidemiol Rec 2005;80:242.
20. Global Advisory Committee on Vaccine Safety. Safety of SA-14-14-2 Japanese encephalitis vaccine. Weekly Epidemiol Rec 2005;80:242–3.

Lyme disease vaccine

General Information

Lyme disease is a tick-borne, spirochetal zoonosis, characterized by a distinctive skin lesion, systemic symptoms, and neurological, rheumatological, and cardiac involvement, occurring in varying combinations over a period of months to years. *Borrelia burgdorferi* is the causative agent in North America, whereas in Europe three genomic groups (named *Borrelia burgdorferi sensu stricto*, *Borrelia garinii*, and *Borrelia afzelii*) have been identified.

Endemic foci have been found in North America, Europe, the former USSR, Japan, and China. In many of these areas, Lyme disease is now the most common vector-borne disease. Because of this epidemiological problem and the severity of the disease, a vaccine was developed. High titers of antibody to outer surface protein A (OspA) of the spirochete prevented *B. burgdorferi* infection in mice and subsequently in immunized hamsters, dogs, and monkeys.

In December 1998, based on many prelicensure clinical trials, the first Lyme disease vaccine (LYMErix, manufactured by the then SmithKline Beecham) was licensed by the Food and Drug Administration (FDA) for individuals aged 15–70 years old (1) and subsequently became commercially available in the USA. In 1999, the Advisory Committee on Immunization Practices (ACIP) made recommendations for Lyme disease vaccine (2), including data on efficacy and safety (SEDA-23, 338). No convincing evidence was found that the vaccine caused serious problems, but discussion about its safety continued and the demand for the vaccine did not reach a sustainable level. Therefore, in February 2002 the vaccine was withdrawn by the manufacturers.

In two overviews the results of various clinical and efficacy trials were summarized (3,4) and the safety, immunogenicity, and efficacy of the vaccine were underlined. The authors considered that the intravector mode of action of the vaccine was unique and opened the door to a new method of preventing insect-borne illnesses in humans.

Dr Neal Halsey, head of the Institute for Vaccine Safety at John Hopkins University, explained that the poor sales had resulted from "public misperception and the promotion of false concerns" (5). Lyme disease researchers consider the Lyme vaccine story in the USA to have been a setback in Lyme disease prevention. Vaccines meeting the specific epidemiological situation in Europe are under development. Lyme disease prevention and prophylaxis has been reviewed (6).

Reports collected through the Vaccine Adverse Events Reporting System (VAERS) from December 1998 to October 2000 have been analysed to examine adverse reactions (arthritis, neuropathy, convulsions, thrombocytopenia, lymphadenopathy, flu-like syndrome, alopecia, gastrointestinal disease, and paralysis) after Lyme immunization in the adult population of the USA. Statistical methods were used to determine whether the increased incidence rates of serious adverse reactions achieved statistical significance over those reactions that were reported after Td and rubella immunizations in adults. Table 1 shows the increases in both acute and chronic adverse reactions after Lyme vaccine compared with Td vaccine (7). The authors also compared the incidence of arthritic reactions reported after Lyme immunization with those reported after adult rubella immunization, and found a statistical increase in the incidence of arthritic reactions after Lyme vaccine (RR = 3.4, attributable risk = 2.4, percent association = 77). There was also a statistical increase in the incidence of chronic arthritis after Lyme immunization compared with adult rubella immunization (RR = 4.8, attributable risk = 3.8, percent association = 83). Based on these results, the authors concluded that the withdrawal of Lyme vaccine from the market seems to have been justified.

Observational studies

LYMErix safety data reported to the Vaccine Adverse Event Reporting System (VAERS) from 21 December 1998 to 31 October 2000 mentioned reports of adverse events associated with Lyme vaccine in prelicensure trials, including injection site reactions, transient arthralgia and myalgia within 30 days of vaccination, fever, and a flu-like illness (8). Allergic reactions were reported to the VAERS and some could have plausibly been linked to the vaccine because of the short latency between vaccination and reaction onset. No clear patterns in age, sex, time to onset, or vaccine dose were identified, although the unexpected predominance of reports of arthrosis in men might warrant further consideration.

Placebo-controlled studies

The results of two efficacy and safety trials using Lyme disease vaccine with or without adjuvant have been reported. In a double-blind trial, 10 305 subjects at least 18 years old, recruited at 14 sites in areas of the USA where Lyme disease was endemic, were randomly assigned to receive either placebo (n = 5149) or OspA vaccine (n = 5156) (9). The first two injections were given 1 month apart and 7515 subjects also received a booster dose at 12 months. The efficacy of the vaccine was 68% in the first year of the study in the entire population and 92% in the second year among the 3745 subjects who received a third injection. The vaccine was well tolerated. There was a higher incidence of mild, self-limiting, local and systemic reactions in the vaccine group, but only during the 7 days after vaccination (Table 2). There was no significant increase in the frequency of arthritis or neurological events in vaccine recipients. The authors concluded that OspA vaccine was safe and effective in the prevention of Lyme disease.

In another randomized, double-blind trial in 10 936 subjects in areas of the USA in which Lyme disease is endemic, either recombinant *B. burgdorferi* OspA with adjuvant or placebo was given initially and at 1 and 12 months (10). After two injections, 22 subjects given vaccine and 43 given placebo contracted definite Lyme disease; vaccine efficacy was 49% (95% CI = 15, 69%). In the second year, after the third injection, 16 vaccine recipients and 66 placebo recipients contracted definite Lyme disease; vaccine efficacy was 76% (CI = 58, 86%). The efficacy of the vaccine in preventing asymptomatic infection was 83% in the first year and 100% in the second year. Injection of the vaccine was associated with mild to moderate local or systemic reactions lasting a median of 3 days (Table 3).

Table 1 Adverse reactions after Lyme vaccine compared with Td vaccine

Type of reaction	Incidence per million Lyme vaccine doses	Incidence per million Td doses	Relative risk	Attributable risk	Percent association	Statistical significance (P value)
Chronic arthritis	16	0.054	296	295	99	<0.0001
Arthritis	27	0.22	123	122	99	<0.0001
Gastrointestinal disease	3.6	0.039	92	91	99	<0.0001
Chronic gastrointestinal disease	2.1	0.023	91	90	99	<0.0001
Flu-like syndrome	64	1.0	64	63	98	<0.0001
Alopecia	1.4	0.39	36	35	97	NS
Thrombocytopenia	2.1	0.070	30	29	97	<0.0001
Chronic paralysis	1.4	0.054	30	29	97	NS
Paralysis	2.9	0.12	24	23	96	<0.0001
Chronic neuropathy	2.1	0.12	18	17	95	<0.0001
Neuropathy	5	0.36	14	13	93	<0.0001
Chronic lymphadenopathy	2.1	0.18	12	11	92	<0.0001
Lymphadenopathy	7.1	2.2	3.2	2.2	76	<0.00005
Convulsions	2.9	1.2	2.4	1.4	70	NS

Table 2 Percentage incidences of adverse effects within 7 days after injection

Adverse effect	Vaccine	Placebo
First injection		
Number of subjects	5156	5149
Any adverse effect	9.8	4.1
Musculoskeletal	6.4	1.3
Myalgia	5.5	0.6
General	1.8	0.9
Pain at injection site	0.3	0.04
Second injection		
Number of subjects	5050	5034
Any adverse effect	6.1	3.1
Musculoskeletal	3.3	1.1
Myalgia	2.5	0.4
General	1.7	0.8
Pain at injection site	0.8	0.1
Third injection		
Number of subjects	3745	3770
Any adverse effect	11.2	5.5
General	7.3	2.6
Tenderness	2.3	0.2
Pain at injection site	1.5	0.2
Unspecified pain	1.0	0.1
Reaction at injection site	0.8	0.2
Swelling	0.6	0.1
Pain in limb	0.5	0.02
Edema at injection site	0.5	0
Rigors	0.2	0
Skin or subcutaneous tissue	2.1	0.2
Erythematous rash	1.9	0.1

Table 3 Percentages of subjects with symptoms with an overall incidence of at least 1% that were classified as related or possibly related to immunization or unrelated to immunization

Symptom	Vaccine	Placebo	P-value
*Related or possibly related to immunization**			
Local at injection site			
Soreness	24.1	7.6	<0.001
Redness	1.8	0.5	<0.001
Swelling	0.9	0.2	<0.001
Systemic: early (<30 days)			
Total*	19.4	15.1	<0.001
Arthralgia	3.9	3.5	0.34
Headache	3.0	2.5	0.14
Myalgias	3.2	1.8	<0.001
Fatigue	2.3	2.0	0.37
Aching	2.0	1.4	0.01
Influenza-like illness	2.0	1.1	<0.001
Fever	2.0	0.8	<0.001
Chills	1.8	0.5	<0.001
Upper respiratory tract infection	1.0	1.1	0.69
Systemic: late (>30 days)			
Total*	4.1	3.4	0.06
Arthralgia	1.3	1.2	0.54
Unrelated to immunization			
Early (<30 days)	27.1	27.9	0.37
Late (>30 days)	53.3	52.6	0.48

* Totals include all early or late related or possibly related systemic events, not just those with a frequency of at least 1%.

Organs and Systems

Musculoskeletal

Suspicions were expressed in the Mealey Publication's Drug and Medical Device Report that the Lyme disease vaccine LYMErix could cause an incurable form of auto-immune arthritis. It was hypothesized that blood concentrations of OspA after three doses of vaccine place vaccinees classified by genetic type HLA-DR4+ at risk of developing treatment-resistant Lyme arthritis. The pre-market trials for the vaccine were assessed by an independent advisory committee, which found no link between Lyme disease immunization and autoimmune arthritis (11). However, the committee stressed the need for long-term surveillance and further studies in those over 70 years and in children, and the effect of the vaccine in patients with chronic arthritis; the possible development of autoimmunity deserves further study (12). After licensing of the vaccine, more than 1 million Americans received it and no unusual adverse effects were reported to the manufacturer (11).

However, subsequently the FDA received some reports of arthritis and Lyme disease after the use of the vaccine (13). Therefore, on 31 January 2001, the FDA's Vaccines and Related Biological Products Advisory Committee held a meeting to evaluate safety data of LYMErix. Among other information, the experts considered a manufacturer's briefing document as well as safety data reported to the VAERS. The Associated Press Report from 31 January 2001 on the FDA meeting on Lyme vaccine read (in part) as follows: "Vaccine Safety experts found no proof that the LYMErix vaccine is dangerous—the rare cases of arthritis and other symptoms could be coincidence. Many reports are of minor complaints or are not believed to have been caused by the vaccine, but the FDA is studying 133 reports of severe arthritis-like symptoms. That's because of a theory that the vaccine might set off an autoimmune reaction where the body attacks its own tissues, particularly in people who carry a certain gene called HLA-4." The panel said that ultimately "no convincing evidence exists that the vaccine causes serious problems", but the experts urged the FDA and the manufacturer to expedite new safety studies and demanded that the Government should act to ensure that patients are told about the possible risks before inoculation. They also urged that the CDC should aggressively distribute a patient-friendly safety fact sheet about LYMErix (14).

A postmarketing assessment cohort study using auto-mated record linkage was initiated at Harvard Pilgrim Health Care (HPHC) in order to address the theoretical concern that immunization with a vaccine containing OspA might cause an autoimmune arthritis and to evaluate whether exposure to the vaccine is a risk factor for Lyme disease, treatment-resistant Lyme disease, rheumatoid arthritis, certain neurological diseases, allergic events, hospitalization, and death (SEDA-24, 367) [http://www.fda.gov/ohrms/dockets/ac/01/slides/3680s2-05_platt.pdf]. The study involves 25 000 HPHC members who are expected to receive the vaccine and 75 000 non-immunized controls. For the most recent report, matched data were available for 2568 vaccinees and 7497 controls. At this stage, the available results do not suggest that the outcomes of interest were more frequent among vaccinees than among non-vaccinees. Having recognized that enrolment into the database is at a slower rate than anticipated, two additional Health Maintenance Organizations in countries where Lyme disease is endemic have been identified and will contribute.

On 30 November 2000, 1.4 million doses of vaccine had been distributed. From January 1999 to the time of the database query, there was no evidence that the incidence of these arthritic conditions was higher than reported in the general population or that they were associated with an autoimmune process (7).

To address further the question of a possible link between Lyme vaccine and arthritis, the FDA is conducting a telephone survey of individuals who have reported arthritic conditions to the VAERS after receiving the vaccine (7). This survey is a census of available and willing individuals who submitted reports that have been coded as arthritis, arthrosis, rheumatoid arthritis, joint disease, and arthralgia. The goals of the survey are to describe the characteristics of these adverse events, to identify concomitant factors that might influence the characteristics of these events, and to describe the relation of the events to immunization. After this survey phase is complete, the VAERS will identify cases of arthritis and will conduct a case-control study to examine the hypothesis that Lyme vaccine causes arthritis. It is planned to compare people who report arthritis after Lyme disease vaccine with two control groups: people who report arthritis to the VAERS after other vaccines and people who report adverse events to the VAERS other than arthritis after Lyme vaccine. All cases will be age-, sex-, and race-matched with controls. All three groups will be tested for DR HLA haplotypes at the allele level and for peripheral blood lymphocyte responses to OspA and leukocyte function-associated antigen-1 (LFA-1). This is an attempt to determine if people who report arthritis after Lyme vaccine have a higher prevalence of certain HLA alleles that are known to be associated with rheumatoid arthritis and have the same third common hypervariable region, while simultaneously having greater peripheral T cell reactivity to OspA and LFA-1. Given the relatively small number of arthritis cases reported after Lyme vaccine, probably only a very high risk will be detectable.

Four cases of arthritis associated with the administration of Lyme vaccine have been reported (15).

- 15 weeks after a third dose of Lyme vaccine a 9-year-old boy developed arthritis, including both knees, the right elbow, the left hip, the right ankle, and the right thumb; he was probably in an asymptomatic phase of natural Lyme infection.
- About 3 months after a third dose of Lyme vaccine a 16-year-old boy developed arthritis of both knees.
- 24 hours after a second dose of Lyme vaccine a 53-year-old man developed flu-like symptoms and arthralgia; he later developed swelling of the finger joints and toe joints.

- 24 hours after a second dose of Lyme vaccine a 43-year-old man developed multiple synovitis.

In all cases the disease was self-limiting and to the knowledge of the authors inconsequential in the long term. They considered that the findings supported postinfectious and mimicry models, by showing that OspA, an outer surface protein of the causative organism of Lyme disease, *B. burgdorferi*, could cause acute arthritis, the possibility that it was associated with a more protracted form of arthritis, and that it perhaps had a modulating effect in individuals with concurrent Lyme infection.

Susceptibility Factors

Genetic factors

From reports containing information on HLA types, clinical descriptions of adverse events in those given Lyme disease vaccine are similar in people with DR4 and non-DR4 HLA haplotypes and do not suggest more inflammatory arthritis in people of DR4 haplotype (16) [http://www.fda.gov/ohrms/dockets/ac/01/briefing/3680b2-06.pdf]. The characteristics of adverse events in people with a self-reported history of Lyme disease do not differ substantially from all adverse events after Lyme vaccine.

Drug Administration

Drug dosage regimens

In a randomized study in 956 volunteers aged 17–72 years a shortened immunization schedule of injections at 0, 1, and 2 months were compared with a schedule of injections at 0, 1, and 12 months (17). Adverse events were transient and mild to moderate. Soreness was the most frequently reported local symptom (82%), whereas fatigue (20–22%) was the most frequently reported general symptom. Two volunteers had more serious adverse events: severe chills and shaking in one and an episode of syncope (lasting a few minutes with complete recovery) on the day of the first dose in another. The authors concluded that doses at 0, 1, and 2 months would provide protection during a typical tick-transmission season.

References

1. Anonymous. Lyme disease vaccine. Med Lett Drugs Ther 1999;41(1049):29–30.
2. Advisory Committee on Immunization Practices (ACIP). Recommendations for the use of Lyme disease vaccine. MMWR Recomm Rep 1999;48(RR-7):1–1721–5.
3. Thanassi WT, Schoen RT. The Lyme disease vaccine: conception, development, and implementation. Ann Intern Med 2000;132(8):661–8.
4. Onrust SV, Goa KL. Adjuvanted Lyme disease vaccine: a review of its use in the management of Lyme disease. Drugs 2000;59(2):281–99.
5. Immunization News for April 26, 2002. http://www.immunizationinfo.org accessed.
6. Hayney MS, Grunske MM, Boh LE, Da Camada CC, Perreault MM. Lyme disease prevention and vaccine prophylaxis. Ann Pharmacother 1999;33(6):723–9.
7. Geier DA, Geier MR. Lyme vaccination safety. J Spirochetal Tick-borne Dis 2002;9:16–22.
8. FDAVaccines and Related Biological Products Advisory CommitteeLYMErix. Lyme Disease Vaccine Safety Update, 31 January 2001. http://www.fda.gov/ohrms/dockets/ac/01/briefing/368ob2.htm accessed.
9. Sigal LH, Zahradnik JM, Lavin P, Patella SJ, Bryant G, Haselby R, Hilton E, Kunkel M, Adler-Klein D, Doherty T, Evans J, Molloy PJ, Seidner AL, Sabetta JR, Simon HJ, Klempner MS, Mays J, Marks D, Malawista SE. A vaccine consisting of recombinant *Borrelia burgdorferi* outer-surface protein A to prevent Lyme disease. Recombinant Outer-Surface Protein A Lyme Disease Vaccine Study Consortium. N Engl J Med 1998;339(4):216–22.
10. Steere AC, Sikand VK, Meurice F, Parenti DL, Fikrig E, Schoen RT, Nowakowski J, Schmid CH, Laukamp S, Buscarino C, Krause DS, Cohen S, Boyer J, Hanrahan K, Dalgin P, Dalgin J, Garrett A, Petelaba M, Feder H, Good S, Green J, Miller K, Spiegel M, Daniel G, Jacob R, Maderazo E, Maiorano M, Seidner A, Bruno LLyme Disease Vaccine Study Group. Vaccination against Lyme disease with recombinant *Borrelia burgdorferi* outer-surface lipoprotein A with adjuvant. N Engl J Med 1998;339(4):209–15.
11. SmithKline to vigorously defend itself against class action. Mealey Publications Report. Press Release Newswire Association Baltimore, 23 December 1999.
12. Marwick C. Guarded endorsement for Lyme disease vaccine. JAMA 1998;279(24):1937–8.
13. Noble HB. Concerns over reactions to Lyme shots. NewYork Times, 21 November 2000.
14. Associated Press Report, 31 January 2001. Warnings urged for Lyme vaccine. http://www.fda.gov (accessed 1 February 2001).
15. Rose CD, Fawcett PT, Gibney KM. Arthritis following recombinant outer surface protein A vaccination for Lyme disease. J Rheumatol 2001;28(11):2555–7.
16. FDA. LYMErix Safety Data Reported to the Vaccine Adverse Event Reporting System; (VAERS) from December 21, 1998 through October 31, 2000. http://www.fda.gov/ohrms/dockets/ac/01/briefing/3680b2_06.pdf.
17. Schoen RT, Sikand VK, Caldwell MC, Van Hoecke C, Gillet M, Buscarino C, Parenti DL. Safety and immunogenicity profile of a recombinant outer-surface protein A Lyme disease vaccine: clinical trial of a 3-dose schedule at 0, 1, and 2 months. Clin Ther 2000;22(3):315–25.

Malaria vaccine

General Information

For centuries, attempts have been made to find a vaccine against malaria, especially malaria tropica (1). Most of these attempts have been disappointing (SEDA-13, 822) (SEDA-21, 298). However, the results of a number of field studies have been reported.

SPf66 vaccine is a synthetic polypeptide based on pre-erythrocyte and asexual blood stage proteins of *Plasmodium falciparum*. A trial in Tanzania, in an area in which malaria is not only endemic, but the parasite load heavy, showed an estimated efficacy of 30% (95% CI = 0, 52%). The vaccine was immunogenic. There was also a difference in mortality between the groups, with one death among the vaccinated children and five in the control group. Adverse effects were in general mild and similar to those in earlier work (2,3).

In Colombia, the vaccine brought about an immune response, and adverse effects were minimal. The first randomized, double-blind, controlled trial in Colombia, in which 1548 volunteers participated, showed an overall efficacy of 39%. In a small, selected group of volunteer soldiers there was a protective effect of 60–80%. In 96% of cases there were no adverse effects and the adverse effects in the remainder were mainly localized erythema, discomfort, and sometimes induration (4,5).

A study in Ecuador showed 67% mean prophylactic efficacy against *P. falciparum*, but the 95% CI were very wide (2.7, 89%). The vaccine was administered in three doses, at zero time and after 30 and 180 days. Adverse effects (local pain, erythema, and local induration) occurred after the first dose in 6.2%, after the second dose in 19%, and after the third dose in 14%. The placebo group had similar adverse effects, but in a lower proportion of subjects and with fewer cases of induration. There was no protection against *Plasmodium vivax* infections (6).

In a Venezuelan study (7), 1442 villagers were vaccinated and vaccination was completed with the three doses in 976 subjects, with a comparison group of 938 subjects from the same area. Antibodies were produced against the vaccine. The adverse effects were as in earlier work, but there was also some contralateral induration, mainly after the third dose of vaccine, which was given alternately into the left and right arms. Five women had generalized pruritus. Bronchospasm was observed in one case.

A new recombinant vaccine has shown promise in vitro and in rabbits. It combines segments of 21 different immunogenic peptides from *P. falciparum* into a single recombinant protein, eliciting a multilayered immune response, including B cell and T cell responses (1).

Organs and Systems

Immunologic

Immediate-type hypersensitivity reactions (acute, systemic urticaria after the third immunization) occurred in two of 39 volunteers immunized with a synthetic multi-antigen peptide vaccine (PfCS-MAP1NYU) against *P. falciparum* sporozoites, with detection of serum IgE MAP antibody (8). Immediate pain at the injection site was associated with the adjuvant QS-21, and delayed local inflammatory reactions were associated with high titers of circulating IgG anti-MAP antibody. Skin tests using intradermal injections of diluted MAP vaccine, to identify those who were sensitized to the vaccine, were negative in seven volunteers tested 27 days after the first vaccination, but six of these developed positive wheal and flare reactions when tested 14 or 83 days after the second vaccination; IgE MAP antibody was detected in only one of them. Skin tests may help in identifying individuals who have been sensitized to malaria peptides and who are at risk of developing systemic allergic reactions after revaccination.

On the other hand, delayed-type hypersensitivity testing has been used to test T cell functional activity in 27 volunteers immunized with a synthetic multi-antigen peptide vaccine (MAP) (PfCS-MAP1NYU) against *P. falciparum* sporozoites (9). Intradermal inoculations (0.02 ml) of several concentrations of the MAP vaccine and adjuvant control solutions were applied and induration measured 2 days after. Nine of 14 vaccinees with high serum titers of anti-MAP antibody developed positive skin tests (at least 5 mm induration), which first appeared by 29 days after immunization and persisted for at least 3–6 months after one or two more immunizations. In contrast, skin tests were negative in all of eight vaccinees with no or low antibody titers, and in all of five non-immunized volunteers. Biopsies of positive skin test sites were histologically compatible with a delayed hypersensitivity reaction. The authors concluded that the presence of T cell functional activity is reflected by a positive skin test response to the MAP antigen and may serve as another marker for vaccine immunogenicity.

References

1. Holder AA. Malaria vaccines. Proc Natl Acad Sci USA 1999;96(4):1167–9.
2. Alonso PL, Smith T, Schellenberg JR, Masanja H, Mwankusye S, Urassa H, Bastos de Azevedo I, Chongela J, Kobero S, Menendez C, et al. Randomised trial of efficacy of SPf66 vaccine against *Plasmodium falciparum* malaria in children in southern Tanzania. Lancet 1994;344(8931):1175–81.
3. Alonso PL, Tanner M, Smith T, Hayes RJ, Schellenberg JA, Lopez MC, Bastos de Azevedo I, Menendez C, Lyimo E, Weiss N, et al. A trial of the synthetic malaria vaccine SPf66 in Tanzania: rationale and design. Vaccine 1994;12(2):181–6.
4. Amador R, Moreno A, Murillo LA, Sierra O, Saavedra D, Rojas M, Mora AL, Rocha CL, Alvarado F, Falla JC, et al. Safety and immunogenicity of the synthetic malaria vaccine SPf66 in a large field trial. J Infect Dis 1992;166(1):139–44.
5. Valero MV, Amador LR, Galindo C, Figueroa J, Bello MS, Murillo LA, Mora AL, Patarroyo G, Rocha CL, Rojas M, et al. Vaccination with SPf66, a chemically synthesised vaccine, against Plasmodium falciparum malaria in Colombia. Lancet 1993;341(8847):705–10.
6. Sempertegui F, Estrella B, Moscoso J, Piedrahita L, Hernandez D, Gaybor J, Naranjo P, Mancero O, Arias S, Bernal R, et al. Safety, immunogenicity and protective effect of the SPf66 malaria synthetic vaccine against *Plasmodium falciparum* infection in a randomized double-blind placebo-controlled field trial in an endemic area of Ecuador. Vaccine 1994;12(4):337–42.
7. Noya O, Gabaldon Berti Y, Alarcon de Noya B, Borges R, Zerpa N, Urbaez JD, Madonna A, Garrido E, Jimenez MA, Borges RE, et al. A population-based clinical trial with the SPf66 synthetic Plasmodium falciparum malaria vaccine in Venezuela. J Infect Dis 1994;170(2):396–402.
8. Edelman R, Wasserman SS, Kublin JG, Bodison SA, Nardin EH, Oliveira GA, Ansari S, Diggs CL, Kashala OL, Schmeckpeper BJ, Hamilton RG. Immediate-type hypersensitivity and other clinical reactions in volunteers immunized with a synthetic multi-antigen peptide vaccine (PfCS-

MAP1NYU) against *Plasmodium falciparum* sporozoites. Vaccine 2002;21(3–4):269–80.

9. Kublin JG, Lowitt MH, Hamilton RG, Oliveira GA, Nardin EH, Nussenzweig RS, Schmeckpeper BJ, Diggs CL, Bodison SA, Edelman R. Delayed-type hypersensitivity in volunteers immunized with a synthetic multi-antigen peptide vaccine (PfCS-MAP1NYU) against *Plasmodium falciparum* sporozoites. Vaccine 2002;20(13–14):1853–61.

Measles, mumps, and rubella vaccines

General Information

Measles vaccine

Live measles vaccine

Live measles virus vaccine is available in monovalent (measles only) form and in combinations: measles–rubella (MR) and measles–mumps–rubella (MMR) vaccines. Measles vaccines based on further attenuated strains (beyond the level of the original strain, for example the Edmonston B strain or Schwarz strain) produces a mild or subclinical and non-communicable infection.

Inactivated measles vaccine

Inactivated measles vaccine was not distributed after 1967 (owing to lack of efficacy). Reports of the atypical measles syndrome that can occur after immunization with this vaccine type have been reviewed (SED-11, 679) (SEDA-8, 299) (SEDA-11, 291).

High-titer Edmonston–Zagreb measles vaccine

High-titer Edmonston–Zagreb measles vaccine is more immunogenic in young infants than measles vaccine based on other strains. It was hoped that the use of such vaccines could help to reduce the incidence of measles in infancy in developing countries. However, little was known about the long-term effects of high-titer measles vaccine given early in life. Subsequently, experience gained in Senegal, Guinea-Bissau, and other developing countries showed that child mortality after immunization was significantly higher in those who received high-titer Edmonston–Zagreb vaccine than in those given standard vaccine. The higher risk of death remained significant in multivariate analyses (1,2). There is a reason for concern about the long-term safety of high-titer measles vaccine. The World Health Organization has suspended the use of high-titer measles vaccine. Additionally, the mortality data in all West African trials have been analysed. All children were followed up to at least 3 years of age. Mortality in the high-titer group was significantly higher than in those given the standard vaccine. These results support the decision of the WHO that high-titer measles vaccines should no longer be used (3).

General adverse effects

About 5–15% of vaccinees develop a temperature of >39°C (beginning about the sixth day after immunization and lasting up to 5 days). Transient rashes have been reported in about 5% of vaccinees. Allergic reactions very rarely follow the immunization. Most of these reactions are considered minor and consist of wheal and flare or urticaria at the injection site. Very few cases of immediate allergic reactions in children who had histories of anaphylactoid reactions to egg ingestion have been reported, but these reactions could potentially have been life-threatening (4). This risk of egg allergy is excluded in modern vaccines, in which the virus is propagated in chicken or human fibroblast cell cultures.

In Japan, three further attenuated live measles vaccines are licensed for general use:

1. AIK-Cvaccine
2. Biken-CAM vaccine
3. Schwarz-FFB vaccine.No significant differences have been reported between the three vaccines regarding either immunogenicity or adverse reactions (5).

Data on adverse effects after measles immunization reported in the framework of the US Monitoring System on Adverse Events Following Immunization to the Centers for Disease Control have been published (SEDA-13, 274). Case reports on suspected encephalitis (including two cases with hearing loss) and convulsions have been reviewed (SEDA-8, 299) (SEDA-9, 284) (SEDA-11, 291) (SEDA-12, 283) (SEDA-16, 388); very rare individual case reports on cerebellar ataxia, diffuse retinopathy, optic neuritis, regional lymphadenitis, thrombocytopenic purpura (70% of cases of thrombocytopenic purpura occur following viral diseases; purpura has been reported also following the receipt of measles vaccine), paroxysmal cold hemoglobinuria, parkinsonism, pityriasis lichenoides et varioliformis acuta, nephrosis, and depression of the tuberculin skin test reaction were cited (SED-11, 679) (SED-12, 811) (SEDA-12, 283) (SEDA-16, 389) (SEDA-17, 376).

A review of the data generated in the last 4 years has amply described the continued efforts of the scientific community to monitor and understand true measles vaccine-associated adverse events (6). The rapidity and clarity of this same community's debunking of the spurious associations with Crohn's disease and autism suggests that those charged with vaccination programs have learned from past mistakes.

The immunogenicity and safety of two live attenuated measles vaccine strains (Schwarz strain and AIK-C strain) have been compared in 9-month-old Taiwanese infants in order to find a candidate for an effective measles immunization in the first year of life (7). Because of persisting maternal measles antibody, the commercially available measles vaccines do not produce a sufficient antibody response when given during the first year of life. However, early protection is particularly required in developing countries with high measles morbidity and mortality in infants. Attempts to find suitable candidates for early immunization, including the use of the high-titer Edmonston–Zagreb measles vaccine strain, failed (SED-14, 1078). The Japanese AIK-C measles vaccine strain (derived from the Edmonston strain) has been given to 67 infants and a measles vaccine based on the Schwarz strain was given to 68 infants. The AIK-C strain vaccine caused seroconversion in 65 infants (97%), whereas only 65 infants (70%) seroconverted with the Schwarz strain vaccine. Adverse effects of the two vaccines were comparable: fever 8.8% (Schwarz strain) versus 19% (AIK-C strain); skin rash 13 versus 13%; rhinorrhea 8.8 versus 12%; cough 7.4 versus 6%; diarrhea 2.9 versus 4.5%; poor appetite 2.9 versus 1.5%; irritability 2.9 versus 6%.

Mumps vaccine

Live mumps virus vaccine is available in monovalent (mumps only) form and in combination with measles (MM vaccine) and with measles and rubella (MMR vaccine). Vaccines based on the following mumps vaccine strains are in use:

- The Jeryl Lynn strain is used mainly in vaccines prepared in the USA. The virus was isolated from a female patient in 1963. Vaccines based on Jeryl Lynn strain are mostly used worldwide.
- The Urabe Am9 strain was attenuated in Japan and has been used for the preparation of MMR and other mumps-containing vaccines in Japan and Europe.
- The Rubini strain was isolated in Switzerland and was attenuated by passage in human diploid cells for use in vaccine production.
- The L-3 (Leningrad) strain was derived by combining five isolates of mumps virus from sick children. The attenuated strain is used for vaccine production, especially in Russia.
- The L-3 strain was further attenuated in Zagreb, Croatia, by adaptation and passage in SPF chick embryo fibroblast cell cultures (mumps vaccine strain L-Zagreb).The most commonly used vaccines have been based on the Urabe Am9 strain and the Jeryl Lynn strain. However, in September 1992 health authorities worldwide were informed by Smith Kline Beecham Biologicals that the company had decided to suspend the distribution of vaccines containing the Urabe Am9 strain, since alternative vaccines were available to maintain the immunization programs established in the various countries.

The Global Advisory Committee on Vaccine Safety (GACVS) has reviewed the data on the safety of mumps vaccine strains, particularly regarding the risk of vaccine-derived meningitis [8]. They noted that cases of aseptic meningitis and estimates of incidence rates have been reported after the use of Urabe, Leningrad–Zagreb, Hoshino, Torii, and Miyahara strains from various surveillance systems and epidemiological studies. The data up to now have shown low rates of aseptic meningitis and no cases of virologically proven meningitis after the use of the Jeryl–Lynn and RIT 4385 strains. Information about the Leningrad-3 strain is limited. No data were available to assess the safety of the S79 strain. There is still a lack of carefully designed studies to discriminate between potential strain variability and age-specific risk in different

populations. GACVS welcomed the establishment of a repository of mumps vaccine strains at the National Institute for Biological Standards and Control, Potters Bar, UK, and has urged the acceleration of work to gain insight into the biological determinants of risk from different strains.

General adverse effects

Various authors have investigated the immunogenicity and reactogenicity of different mumps vaccine strains. Among adverse reaction reports, episodes of parotitis and low-grade fever have been most prominent. Rash, pruritus, purpura, and other allergic reactions are uncommon, usually mild, and of brief duration (9). Data on adverse effects after immunization with mumps-containing vaccines collected in the framework of MSAEFI have been published (SEDA-13, 274). There have been reports of febrile convulsions, meningitis, orchitis, parotitis, swollen lymph nodes, and thrombocytopenia (SED-12, 813) (SEDA-16, 389) (SEDA-17, 377).

Rubella vaccine

Rubella vaccine is a live attenuated virus vaccine. Most of the vaccines currently manufactured outside Japan are produced in human diploid cells and are based either on the RA 27/3 strain (the most widely used) or the Cendehill strain. In Japan, five different vaccine strains (for example TO 336 and MEQ 11) are produced in two different non-human substrates. In China, another vaccine strain (BRD-2) has been developed and produced in human diploid cells. Its antigenicity and reactogenicity are comparable to those of the RA 27/3 strain.

General adverse effects

Subcutaneous injection is commonly followed by local soreness and induration. Children sometimes develop low grade fever, rash, and lymphadenopathy after immunization. Acute arthritis/arthralgia sometimes occurs.

Measles–mumps–rubella vaccine

In most industrialized countries, measles–mumps–rubella (MMR) vaccine has replaced the former use of single antigen vaccines against measles, mumps, and rubella in childhood immunization programs. Single antigen rubella vaccines are still used in postchildhood rubella prevention. Comparisons of the efficacy and safety of different MMR vaccines and monovalent versus bivalent or trivalent vaccines have been made in various clinical trials (SEDA-14, 285). Minor symptoms (fever, rash, malaise) occurred usually after 5–14 days and lasting for 2–3 days. Occasionally, febrile convulsions have been recorded within 3 weeks of immunization, and mild parotitis occurred rarely in the third week after immunization. On average, the seroconversion rates were between 95 and 99% for measles and rubella, and less for mumps.

In a crossover study among 581 twin pairs aged between 14 months and 6 years only 0.5–4% had adverse reactions to MMR vaccine. The difference between the reaction rate reported in the immunized members of the twin pairs and those reported in the placebo-injected twin sisters or brothers showed that most of the reactions were temporally and not causally related to immunization. Respiratory symptoms, nausea, and vomiting were more frequent in the placebo-injected group than in the MMR-immunized group (10).

General adverse effects

A Vaccine Safety Datalink project has been used to compare adverse events after MMR immunization either at 4–5 years or at 10–12 years (11). Information on events that are plausibly associated with MMR immunization (seizures, pyrexia, malaise/fatigue, musculoskeletal symptoms, rash, edema, induration, lymphadenopathy, thrombocytopenia, aseptic meningitis, joint pain) has been collected from 8514 children who received the vaccine at preschool age and from 18 036 schoolchildren. The results suggested that the risk of events is greater in those aged 10–12 years.

When the MMR immunization program was launched in Finland in 1982, a countrywide surveillance system, including all hospitals and health centers, was established to monitor serious adverse events after immunization. From 1982 to 1996 almost 3 million doses of MMR vaccine were distributed to 1.8 million individuals, mostly children. Most of the reported adverse events were minor or self-limiting events among 437 vaccinees, for example fever, rash, headache, fatigue, nausea, vomiting, transient arthralgia, and swelling of the parotid glands. In all, 173 potentially serious adverse events were evaluated in detail. The assessment of causality is shown in Table 1.

• One 13-month-old boy died 8 days after immunization. Autopsy showed that the cause of death was aspiration of vomit. The most commonly reported neurological adverse events were febrile seizures. Epilepsy was diagnosed in three children; symptoms manifested for the first time 1, 10, and 21 days after immunization. One child was later diagnosed as having severe Lennox–Gastaut syndrome; medical records subsequent to the acute phase were not available for the other two.In the four cases of encephalitis, a causal relation with MMR vaccine could not be excluded, because no other specific cause was detected. One child with acute lymphoblastic leukemia (diagnosed after immunization but with symptoms leading to the diagnosis already present at the time of immunization) developed, during immunosuppressive treatment, measles encephalopathy 54 days after immunization and interstitial pneumonia a few days later; 14 years later the leukemia had not relapsed, but she had developed severe epilepsy. The fourth child had *Herpes simplex* encephalitis, with a temporal association between MMR immunization and encephalitis (12). Idiopathic thrombocytopenic purpura was excluded from the analysis, because it has been analysed before (13).

Measles–mumps–rubella–varicella vaccine

A quadrivalent measles–mumps–rubella–varicella vaccine (ProQuad®, referred to as MMRV), which includes a

Table 1   Assessment of causality between MMR vaccination and 173 serious events

Event	Number of reports	Number not causally associated with MMR	Possibly causally associated with MMR		
			Number	%	Incidence per 100 000 doses
Respiratory					
Asthma	10	5	5	50	0.2
Pneumonia	12	7	5	42	0.2
Nervous system (n = 77)					
Febrile seizures	52	24	28	54	0.9
Epilepsy	3	2	1	33	0.03
Undefined seizure	4	2	2	50	0.07
Encephalitis	4	1	3	75	0.1
Meningitis	4	4	0	0	0
Guillain–Barré syndrome	2	0	2	100	0.07
Transient gait disturbance	5	0	5	100	0.2
Confusion during fever	3	1	2	67	0.07
Metabolism					
Diabetes	3	3	0	0	0
Reproductive function					
Orchitis	7	6	1	14	0.03
Immunologic (n = 74)					
Anaphylaxis	30	16	14	47	0.5
Henoch–Schönlein purpura	2	1	1	50	0.03
Urticaria	30	5	25	83	0.8
Stevens–Johnson syndrome	1	0	1	100	0.03
Death (n = 1)	1	1	0	0	0

varicella component of increased potency, has been evaluated in a blind multicenter study in 480 healthy children aged 12–23 months, who were randomized to either MMRV + placebo or MMR + monovalent varicella vaccine (14). Children who were randomized to MMRV + placebo received a second dose of MMRV 90 days later. Measles-like rash and fever during days 5–12 were more common after the first dose of combination MMRV (rash 5.9%; fever 28%) than after MMR + monovalent varicella vaccine (rash 1.9%; fever 19%). The incidences of other adverse events were similar between the groups. Response rates were over 90% to all vaccine components in both groups. Geometric mean titers to measles and mumps were significantly higher after one dose of MMRV than after MMR + varicella vaccine. The second dose of MMRV elicited slight to moderate increases in measles, mumps, and rubella antibody titers and a substantial increase in varicella antibody titer.

Organs and Systems

Respiratory

Measles
Measles giant-cell pneumonia after measles immunization has been described in a 21-year-old man with AIDS (15).

- A 21-year-old man developed AIDS followed by *Pneumocystis jiroveci* pneumonia. About a year after a booster immunization with MMR vaccine, he developed measles giant-cell pneumonia, confirmed by transbronchial and thoracoscopic lung biopsies. The entire genome of the isolated strain and that of the currently used vaccine strain Moraten were subsequently sequenced and were almost identical. Taking into consideration the long interval between measles immunization and pneumonia, the causal relation in this case was doubtful.

Nervous system

Measles
The long debate on the degree of risk to the nervous system presented by measles vaccine, especially encephalitis and encephalopathy, is not completely resolved (16). The problem has been analysed by the CDC, Atlanta, Georgia (4,17,18), the Ministry of Health and Welfare, Japan (5), the National Childhood Encephalopathy Study in the United Kingdom (SEDA-11, 285), and investigators in the Northwest Thames region of England (19). However, most investigators, having critically analysed the studies, have concluded that the incidence of suspected cases of encephalitis/encephalopathy after immunization is still much lower than that of natural infection, suggesting that some or most of the reported neurological disorders may be only temporally and not causally related to measles immunization.

Encephalopathy
Claims that cases of encephalopathy followed by permanent brain injury or death were due to measles immunization, submitted to the US National Vaccine Injury Compensation Program, have been reviewed (SED-13, 920) (20). A total of 403 claims of encephalopathy and/or seizure disorders after measles, measles–rubella, measles–mumps–rubella, mumps, or rubella immunization were identified during the period 1970–93. The medical records of these cases were reviewed by physicians in the compensation program to determine, if possible, the cause of injury and the classification of the findings. The inclusion criteria established by the compensation program were met by 48 claims by patients with acute encephalopathy of undetermined cause 2–15 days after immunization with attenuated measles virus. The clustering and peak onset of encephalopathy occurred in 17 patients on days 8 and 9, and the encephalopathy was followed by permanent brain impairment or death. The patients ranged in age from 10 months to 49 months, with a median age of 15 months. There were no cases of encephalopathy of undetermined cause within 15 days after the administration of mumps or rubella vaccine. Table 2 shows the clinical findings and sequelae among the 48 cases. The authors concluded that manifestations of acute encephalopathy among these 48 children were similar to the clinical features of acute encephalopathy described after natural measles. Vaccine-associated measles encephalopathy may be a rare complication of measles immunization. From 1970 to 1993 in the USA, about 75 million children received measles vaccine by age 4 years. The 48 cases of encephalopathy after measles immunization probably represented under-reporting to this passive compensation system. However, given the generous compensation offered in this program, it is reasonable to conclude that most serious cases temporally related to an immunization have been captured. The incidence of 48 cases of encephalopathy possibly caused by 75 million doses of vaccine can reasonably be described as low.

The clinical features of acute and chronic encephalopathy or death in these 48 patients were classified into three groups based on the initial findings of ataxia in 6, behavioral changes in 8, and seizures in 34. The onset of neurological findings varied in severity from ataxia or behavioral changes to prolonged seizures or coma. Fever preceded the onset of acute encephalopathy by several hours to several days in 43 of 48 children. There was a measles-like rash with a post-vaccination onset from day 6 to day 15 in 13 children.

The 1994 report of the Institute of Medicine concluded that the evidence was inadequate to accept or reject a causal relation between MMR and encephalopathy, and it is known that the incidence of encephalitis after measles immunization of healthy children tends to be lower than the observed incidence of encephalitis of unknown cause. Two large studies have been negative. In a study analogous to the British Childhood Encephalopathy Study there were no increased risks of either encephalopathy or neurological sequelae after measles immunization (21). A retrospective case-control study through the CDC Vaccine Safety Datalink assessing the risk for 300 000 doses of MMR found not a single case of encephalitis/encephalopathy within 30 days of the administration of MMR (22). In contrast, the review mentioned above (20) reported an association between measles vaccine and

Table 2 Clinical effects of acute encephalopathy in 48 patients 2–15 days after the first dose of measles, measles–rubella, or measles–mumps–rubella vaccine, and sequelae, 1970–93

Clinical onset	Number	Acute illness	Number	Neurological sequelae	Number
With initial ataxia (n = 6)					
Irritability	6	Ataxia	6	Ataxia (chronic)	4
Fever	5	Changed behavior	4	Mental retardation	3
Measles-like rash	3	Mental regression	3	Seizure disorder	1
		Hospitalization		Hearing loss	1
With initial behavioral changes (n = 8)					
Lethargy	3	Mental regression	8	Mental retardation	6
Irritability	2	Coma	5	Spastic paresis	5
Confusion	2	Hospitalization	6	Seizure disorder	1
Coma	1	Death	2	Choreoathetosis	1
Fever	6			Death (later)	1
Measles-like rash	1				
With initial seizures (n = 34)					
Fever	32	Hospitalization	33	Mental retardation	31
Status epilepticus	17	Mental regression	31	Seizure disorder	23
Generalized	14	Coma	29	Spastic paresis	10
Focal	3	Behavioral changes	5	Death (later)	3
Measles-like rash	9	Death	2		

encephalopathy. However, the conclusion of the report of the Institute of Medicine is still valid, namely that evidence is still inadequate to accept or reject a causal relation between measles vaccine and these diseases.

After the publication of the Institute of Medicine report, a literature search for adverse events after measles immunization, limited to publications published in 1994–98, unearthed a considerable amount of data that strengthened the rare association of measles-containing vaccines with postinfectious encephalomyelitis (6). The report has also been criticized as an attempt to establish an adverse event of a vaccine, without a specific laboratory finding, without a specific syndrome, without comparable data for non-immunized children, and by ignoring or minimizing years of descriptive epidemiology on reactions (23). In reply, the original authors reminded their critic that the US Congress had set up the Vaccine Injury Table to avoid disagreements over causation by making temporal association an important eligibility element for listed conditions (24). Although biologic plausibility is an important standard for assessing causation in medicine, it is less important in law. In the case of the National Vaccine Injury Compensation Program, the statute provided for a legal presumption of causation, if encephalopathy began 5–15 days after measles immunization.

Seizures

As with the administration of other agents that can produce fever, some children develop febrile seizures after receiving measles vaccine. Most of these convulsions are simple febrile seizures and do not in themselves increase the probability of subsequent epilepsy or other neurological disorders. There may be an increased risk of convulsions among children with a prior history of convulsions or those with a history of convulsions in siblings or parents. After analysing data on the increased risk of such

children, both the US Immunization Practices Advisory Committee (ACIP) and the Committee on Infectious Diseases of the American Academy of Pediatrics have recommended that children with a history of convulsions should be vaccinated because the benefits of immunization outweigh the risks. According to the ACIP, children at risk could receive antipyretics, starting before the expected onset of fever and continuing for 5–7 days. The Committee on Infectious Diseases was reluctant to recommend prevention with antipyretics, though after the onset of fever these are likely to be effective (18,25).

Epilepsy has been attributed to measles vaccine (26).

- A 23-month-old boy developed Lennox–Gastaut syndrome (an epileptic syndrome caused by an acquired organic brain insult) 14 days after measles immunization with Japanese FL-vaccine, based on AIK-C live attenuated measles vaccine strain. The child's prenatal and perinatal histories and his development until 23 months of life were uneventful. His monozygotic twin brother had no epileptic seizures, but it is not clear from the original paper whether or not the twin received the same vaccine.

Subacute sclerosing panencephalitis

Subacute sclerosing panencephalitis (SSPE) can occur after immunization with live attenuated measles vaccine, but the possible role of measles virus vaccine in the pathogenesis of SSPE has been neither proved nor disproved.

It is possible that cases reported without a history of natural measles may have resulted from a subclinical infection that occurred before administration of the vaccine. Even in cases in which the patient had not had measles in the past, wild-type measles sequences have been found in brain tissue of patients with subacute sclerosing panencephalitis (27,28,29). However, the frequency seems to be lower than after natural measles. A fall in the reported incidence of SSPE in the USA accompanied the reduction in the reported incidence of measles and following it by about 7 years (30). Similar experience has been reported from Israel (31), Japan (32,33), and the eastern part of Germany (34). There is no evidence of an increased risk of SSPE on revaccination. A different note is sounded from Romania, where epidemiological investigation of SSPE indicated a yearly incidence of 5–6 cases per million inhabitants without any change due to a measles immunization program implemented about 10 years earlier (35).

Using RNA-templated sequencing, vaccine-strain measles virus has been implicated as the cause of death in three immunocompromised children with inclusion body encephalitis (6). The authors referred to a case of measles vaccine virus-associated giant cell pneumonia in a patient with advanced HIV infection.

The WHO's Regional Office for Europe asked the Global Advisory Committee on Vaccine Safety (GACVS) to review the risk of measles vaccine strains causing subacute sclerosing panencephalitis. The US Institute of Medicine statements in its 1994 and 2001 reviews referred to absent or inadequate evidence either to reject or accept any causal relation between measles-containing vaccines and subacute sclerosing panencephalitis in immunocompetent individuals. It is uncertain whether there is enough evidence from viral RNA sequencing and classification to warrant any modification of this conclusion. However, GACVS noted that all reports published since the Institute of Medicine's review in 2001 containing information on measles virus classification in immunocompetent patients with subacute sclerosing panencephalitis showed the presence of wild (not vaccine) measles strains; and in countries in which measles has been controlled, subacute sclerosing panencephalitis has either declined substantially or no longer occurs. These findings do not suggest an association between measles vaccines and subacute sclerosing panencephalitis.

GACVS will commission a review of the epidemiology of subacute sclerosing panencephalitis in relation to measles vaccine, the results of which will be considered at its December 2005 meeting (36).

- A 21-month-old boy developed measles inclusion-body encephalitis 8.5 months after measles-mumps-rubella immunization (37). He had no history of immunodeficiency, measles exposure, or measles disease. He presented with status epilepticus, a fever of 39.5°C, but no signs of meningism. A cranial CT scan showed swelling of the left temporal lobe with narrowing of the ipsilateral ventricle. He had primary immunodeficiency, characterized by a profoundly depressed CD8 cell count and dysgammaglobulinemia. A brain biopsy showed histopathological changes consistent with measles inclusion-body encephalitis, and measles antigens were detected by immunohistochemical staining. The presence of measles virus in the brain tissue was confirmed by reverse transcription polymerase chain reaction. The nucleotide sequence in the nucleoprotein

and fusion gene regions was identical to that of the Moraten and Schwarz vaccine strains; the fusion gene differed from known genotype A wild-type measles viruses. Despite intensive medical interventions, neurological deterioration continued and he died.

This is an excellent example of the fourth ("fingerprint") type of definitive anecdotal adverse drug reaction (38)

Mumps
Meningitis
The risk of aseptic meningitis is increased after the administration of the Urabe mumps vaccine strain in mumps vaccine or MMR vaccine. Since 1992, most vaccine manufacturers have decided to suspend the distribution of vaccines containing the Urabe strain, provided that alternative vaccines were available to maintain the immunization programs established in the various countries. In 1997, a mass immunization campaign with an Urabe vaccine strain-containing MMR vaccine was carried out in the city of Salvador, in North-East Brazil (39). There was an increased risk of aseptic meningitis 3 weeks after mass immunization. The estimated risk of aseptic meningitis was 1 in 14 000 doses of MMR vaccine.

To assess the risk of mumps vaccine-associated meningitis, pediatricians were asked to report to the British Paediatric Surveillance Unit (BPSU) all confirmed and suspected cases during 1990–91. The risk based on confirmed cases was estimated to be 1 per 250 000 doses distributed (40). However, data from one district, based on two confirmed cases and one suspected case identified by the Nottingham Public Health Laboratory, suggested a much higher risk, about 1 in 4000 doses (41). To investigate whether the risk observed in Nottingham was atypical or indicative of substantial under-reporting elsewhere, additional studies were initiated. A laboratory study including four independent laboratories identified 13 cases of vaccine-associated meningitis after immunization with Urabe strain vaccines; in one-third of instances, mumps virus characterized as a vaccine-like virus was isolated. The risk estimate of 1 per 11 000 doses distributed was lower than in the Nottingham study but much higher than in the BPSU surveillance. Furthermore, children with a discharge diagnosis of viral meningitis were identified in the Oxford region and their immunization status ascertained. The estimated risk was 4.7 per 100 000 doses. Because there have been no cases of proven vaccine-associated mumps meningitis with isolation of the Jeryl Lynn vaccine virus in the UK, the decision was taken to change to Jeryl Lynn-containing vaccines (40).

Eight cases of mumps vaccine-associated meningitis have occurred in Canada. Mumps viruses have been isolated and characterized by nucleotide sequencing as Urabe Am9-like strains. Urabe mumps vaccine virus-containing vaccines are no longer licensed for sale in Canada (42).

Forsey and colleagues have examined over 80 mumps viruses from around the world; they included 20 isolates found in the cerebrospinal fluid of children with vaccine-associated meningitis in the UK and Ireland and isolates from parotitis and meningitis following mumps immunization from Australia, Belgium, Canada, France, Germany, and Japan (43). They were characterized by PCR and dideoxynucleotide sequencing for the differentiation of wild from attenuated mumps viruses. The isolates were characterized as Urabe Am9-like vaccine strains. The Jeryl Lynn mumps vaccine strain was found in one vaccinee from Germany, but no virus of this type was isolated from a patient with meningitis. Although one cannot conclude from this that Jeryl Lynn is free of such adverse effects, all the evidence points toward a considerable difference between the mumps vaccine strains in the degree of risk. Virological heterogeneity provides a clue for further investigation (44). Commercial mumps vaccine of the Jeryl Lynn strain contains at least two distinct mumps viruses; it has been suggested that this factor contributes significantly to the safety and efficacy of this vaccine (25).

The Global Advisory Committee on Vaccine Safety has considered a comprehensive review of the world literature on the safety of mumps immunization, with special attention to vaccine-associated meningitis. They found no cases of virologically proven aseptic meningitis after mumps immunization based on the Jeryl Lynn vaccine strain. If mumps vaccines based on vaccine strains such as Urabe Am9, Leningrad-3, or Leningrad–Zagreb are being used in mass immunization campaigns, the committee recommends that the potential for clustering of aseptic meningitis cases after the campaign should be taken into consideration. The available data are insufficient to distinguish between the safety profiles of the Urabe Am9, Leningrad-3, and Leningrad–Zagreb strains with regard to vaccine-associated meningitis (45).

Rubella

Adverse effects on the nervous system that have at least temporally been associated with rubella vaccination include myelitis, myeloradiculitis (SEDA-2, 268) (SEDA-20, 292), meningomyelitis (SEDA-10, 291), encephalitis (SEDA-5, 308), peripheral neuropathy (SEDA-12, 284), facial or peripheral paresthesia (SEDA-11, 295) (SEDA-12, 284), and carpal tunnel syndrome (SEDA-12, 284). In many of these cases the causal relation was doubtful. The authors of the report of the Institute of Medicine, National Academy of Sciences, Washington, DC (1991) entitled "Adverse Effects of Pertussis and Rubella Vaccines" (46) considered that there was insufficient evidence to indicate either the presence or absence of a causal relation between RA 27/3 rubella vaccine and radiculoneuritis and other neuropathies.

- Acute disseminated encephalomyelitis occurred in a 14-year-old boy 22 days after rubella immunization (47). The authors suggested that live rubella vaccine can occasionally trigger immunologically mediated demyelination within the nervous system.It has been suggested that there is insufficient evidence to indicate either the presence or absence of a causal relation between RA 27/3 rubella vaccine and radiculoneuritis and other neuropathies (46), although cases have been reported.
- A 23-year-old woman developed a mild distal demyelination neuropathy 4 weeks after rubella

immunization (48). Immunological studies showed the presence of antibodies to the rubella virus proteins and to the myelin basic protein.The authors suspected that a virus-induced immune response caused an auto-aggressive reaction responsible for demyelination.

Measles–mumps–rubella

A cranial nerve palsy has been attributed to MMR (49).

- A 13-month-old girl developed a recurrent sixth nerve palsy 1 week after MMR immunization. This resolved completely over 8 weeks and recurred 15 weeks after initial onset. Other causes for the sixth nerve palsy were excluded.

Meningitis

Vaccine-associated meningitis is a well-known adverse effect of mumps (or MMR) immunization. Until 1991 the reported incidence was low, but in that year Japanese researchers published the result of a nationwide survey, started in 1989, in which the incidence of vaccine-associated mumps meningitis varied from less than 1 case per 7000 to 1 per 405 from prefecture to prefecture. Meningitis was generally mild and there were no sequelae from the illness. The vaccine used contained the Urabe Am9 strain (50,51). In April 1993, the Ministry of Health and Welfare in Japan decided to interrupt the use of MMR vaccine (52).

A retrospective study was carried out to determine the incidence of vaccine-associated meningitis in Japan after the administration of different locally produced MMR vaccines (MHW MMR vaccine, Takeda MMR vaccine, Biken MMR vaccine, and Kitasato MMR vaccine) (53). Among the three MMR vaccines (Biken vaccine excepted) the incidence of meningitis was about 1 in 500–900 vaccinees. The criteria for inclusion of a case of meningitis were symptoms of meningitis and pleocytosis in the cerebrospinal fluid. In a complementary nationwide comparison of four MMR vaccines conducted in Japan by 1255 pediatricians the total number of vaccinees was 38 203 (54). All were arbitrarily given one of the MMR vaccines produced by three manufacturers (Biken, Kitasato, and Takeda) or the standard MMR vaccine made of designated strains (Biken's mumps Urabe Am9 strain, Kitasato's measles AIK-C strain, and Takeda's rubella To336 strain). The rates of virologically confirmed aseptic meningitis per 10 000 recipients were 16.6, 11.6, 3.2, and 0 for the standard MMR, Takeda MMR, Kitasato MMR, and Biken MMR vaccines respectively. The incidence of convulsions was the highest with the standard MMR vaccine, and the incidence of fever associated with vomiting at 15–35 days (symptoms relevant to aseptic meningitis) were also the highest with the standard MMR vaccine. The incidence of parotitis was the lowest with Takeda MMR vaccine.

The finding that the incidences of aseptic meningitis differed after administration of the standard MMR vaccine or Biken MMR vaccine has inevitably raised questions about the consistency of manufacture of the Urabe Am9 mumps virus vaccines. The National Institute of Health found that the biological characteristics of the

Urabe Am9 mumps virus contained in the standard MMR vaccine and in the Biken MMR vaccine were different. The Biken Company reported that the mumps vaccine in the standard MMR vaccine was a mixture of two Urabe Am9 mumps vaccine bulks, one identical to that contained in the Biken MMR vaccine and the other produced by a different manufacturing process.

In 1995, a somewhat different note was sounded from France, where all mumps vaccines produced and marketed at that time contained the Urabe vaccine strain. The incidence of vaccine-associated meningitis was estimated using two different data sources: the national network of hospital virology laboratories and the manufacturer's pharmacovigilance department (55). The risk of vaccine-associated meningitis was assessed at one case per 28 400 doses distributed when using the laboratory network data or one case per 13 000–67 200 doses distributed when using the pharmacovigilance data. The French vaccination committee recommended that the vaccine continue to be used whilst awaiting vaccine containing the Jeryl Lynn strain.

Cases of aseptic meningitis occurring 14–26 days after MMR immunization have been reviewed (56). All the patients had been immunized with the L-Zagreb mumps strain, Edmonston–Zagreb measles strain, and the RA 27/3 rubella strain. The incidence of vaccine-associated meningitis was 9 per 10 000 doses. This finding was similar to that reported for the same mumps vaccine strain by Cizman (57), 10 per 10 000 (SEDA-12, 815), but higher than that found by Kraigher (56).

Transmission of Urabe mumps vaccine strain between siblings has been reported (52). A girl developed parotitis after immunization with Urabe mumps strain; 19 days later her older sister developed mumps. The strain isolated showed molecular biological characteristics typical for the Urabe strain.

The reason for the high incidence of vaccine-associated mumps meningitis has been investigated by a comparison of the nucleotide sequence of the hemagglutinin-neuraminidase gene from vaccine virus with the genes of viruses isolated from patients with vaccine-associated meningitis (58). The analysis showed that the Urabe AM9 vaccine strain is a mixture of viruses that differ at nucleotide 1081: a wild type-like variant A and a variant G. Vaccinees who developed vaccine-associated meningitis or parotitis had predominantly variant A (98–100%), indicating strong selection of the wild type-like variant of the mumps virus.

The incidence of aseptic meningitis after immunization with MMR containing the Jeryl Lynn mumps vaccine strain has been assessed in a Vaccine Safety Datalink project (22). The overall rate of confirmed cases in the study population (children aged 1–2 years) was 172 cases per million children, very close to the background rate of 162 naturally occurring cases per million children aged 1–4 years in Olmsted County, MN. The authors concluded that there is no increased risk of aseptic meningitis after vaccination with MMR including the Jeryl Lynn mumps vaccine strain.

Gait disturbance

An analysis of 41 reports of gait disturbance in 15-month-old children in temporal relation with the first MMR immunization, collected in the framework of the Danish surveillance system for adverse events after immunization, has been reported (59). About 533 000 doses of MMR vaccine were administered to 15-month-old children in Denmark during the 10 years from 1987 to 1996. The number of reported cases of gait disturbance corresponded to a frequency of eight per 100 000 doses of MMR vaccine. The authors considered that the symptoms were characteristic of cerebellar ataxia. The high frequency and mainly mild course of gait disturbances might indicate a mumps-related reaction. The symptoms mainly occurred at 7–14 days after immunization, and the duration was on average 1–2 weeks (range 1 day to over 4 months). Most cases were mild and short-lasting and a longer duration of symptoms seems to be predictive of late sequelae. In the same period about the same number of doses were used for the second MMR immunization without similar reactions. Disturbance of gait has rarely been reported, and the authors listed in their paper two reports from Sweden and Germany in which symptoms were so mild that no invasive investigations were carried out.

Guillain–Barré syndrome

To test the hypothesis that MMR vaccine can cause Guillain–Barré syndrome, a retrospective study has been carried out in Finland, based on linkage of individual immunization records with nation-wide hospital discharge registers (60). MMR vaccine did not cause any increase in the incidence of Guillain–Barré syndrome over background and there was no clustering of cases at any time after the administration of the vaccine. The authors concluded that there is no causal association between MMR immunization and Guillain–Barré syndrome.

Sensory systems

Rubella

Bilateral optic neuritis (61) and retinal vasculitis (SEDA-20, 292) have been attributed to rubella immunization.

Psychological, psychiatric

Measles–mumps–rubella

The suggestion that measles/MMR immunization can cause autism has not been confirmed. Autism is characterized by absorption in self-centered subjective mental activities (such as day-dreams, fantasies, hallucinations), especially when accompanied by marked withdrawal from reality.

History

The scientific and public response to the 1998 publication of Wakefield and colleagues (62), and Wakefield's subsequent press conference, in which he suggested that immunization with MMR might be associated with Crohn's disease and autism was enormous and controversial. For example, Black and colleagues (63) stated that the publicity generated by this paper was out of proportion to the strength of the evidence it contained; Beale (65) suggested that the Lancet would bear a heavy responsibility for acting against the public health interest that the journal usually aims to promote; O'Brien and colleagues (65) considered that the substantial amount of evidence that contradicts the findings of Wakefield and colleagues did not achieve the same prominence in the popular press.

In replying to these letters, Wakefield (66) defended the clinician's duty to his patients and the researcher's obligation to test hypotheses. For his part, the editor of the Lancet pointed out that the paper had been presented with a commissioned commentary in the same issue; peer review had confirmed that the paper merited publication, with suitable revisions and editing, as an early report; finally, he considered that the press had presented the information in a balanced way (67). However, it was subsequently discovered that the work had been sponsored by a pharmaceutical company which had not been declared at the time (38). Furthermore, the children who were included in the study had been selected other than at random. Subsequently, 10 of Wakefield's colleagues withdrew the findings that they had initially reported and one other could not be traced; only Wakefield did not retract (69).

The Lancet then reviewed the developments in the discussion (70), with emphasis on evidence that contradicts the alleged association and new data presented by Wakefield. The last part of the editorial read as follows:

In a new twist, Wakefield's crusade fuelled further anxiety among parents when he and John O'Leary, director of pathology at Coombe Women's Hospital, Dublin, Ireland, presented unpublished data to the US Senate's congressional oversight committee in Washington on April 6 [2000]. The hearing was called by the chairman, Dan Burton, an Indiana Republican, whose grandson has autism and visited the Royal Free Hospital in November last year. At the hearing, six parents of children with autism gave moving testimonies of their children's illness. ··· Scientific evidence was presented by six chosen 'experts'. According to Wakefield's testimony, he has now studied more than 150 children with 'autistic enterocolitis'—an unproven association had become a disease—and a detailed analysis of the first 60 cases is to be published in the American Journal of Gastroenterology later this year.

Wakefield presented uninterpretable fragments of results only and concentrated on refuting studies that had contradicted his findings. His conclusions were surprisingly non-committal: 'the virological data indicate that this may be measles virus in some children'; he added that it would be imprudent to interpret the temporal relationship with MMR as a chance finding, in the absence of thorough investigation. O'Leary explained that gut-biopsy material from 24 of 25 children with autism was positive for measles virus compared with one of 15 controls, and that this material was presented to him by Wakefield using 'blinded protocols'. Since the controls are not further described and the details of these findings remain unpublished, this evidence raises far more questions than it answers.

"Autism is a poorly understood neurodevelopmental-disease spectrum with a heart-breaking personal story behind every case. But parents of such children have not been served well by these latest claims made well beyond the publically [sic] available evidence. A congressional hearing, like a press conference, is no place to make controversial scientific assessments. And if scientists question the safety of vaccines without making their evidence fully transparent, harm will be done to many more children than they purport to protect."

Wakefield and Montgomery subsequently raised doubts about the adequacy of the evidence that secured the license for MMR vaccine (71). Particularly in view of the immunosuppressive properties of the measles virus, they suggested that there is a potential for adverse interactions between the component live viruses. They therefore proposed that spaced monovalent measles, mumps, and rubella immunization should replace the use of the combined MMR vaccine. The continuing publications of Wakefield led to reduced MMR coverage in some parts of the UK and to well-publicized concerns about the potential for measles outbreaks among primary school entrants. In an editorial in the British Medical Journal, Elliman and Bedford replied to Wakefield's paper (72). They considered that the current concerns were idiosyncratic and presented reviews confirming the vaccine's safety. The Medicines Control Agency and the Department of Health in the UK rejected any suggestion by Wakefield and colleagues that combined MMR vaccines were licensed prematurely. A review of the licensing of MMR vaccines led to the assurance that the licensing procedure was normal and was based on robust studies (73). This position was shared by the Committee on Safety of Medicines and the Joint Committee on Vaccination and Immunisation. At the end of November 2001, Wakefield left his post at the Royal Free and University College Medical School in London. The college said: "Dr Wakefield's research was no longer in line with the Department of Medicine's research strategy and he left the university by mutual agreement" (74). The WHO strongly endorsed the use of MMR vaccine. The combination vaccine was recommended rather than monovalent presentations. There was no evidence to suggest impaired safety of MMR (75).

In April 2001, the Institute of Medicine's Immunization Safety Review Committee released its report "MMR vaccine and autism". Although scientists generally agreed that most cases of autism result from events that occur in the prenatal period or shortly after birth, there was concern because the symptoms of autism typically do not emerge until the child's second year, and this is the same time at which MMR vaccine is first administered in most developed countries. The committee took also into consideration the papers published by Wakefield and other groups and scientists suggesting evidence of a link between MMR vaccine and Crohn's disease and autism. Following review of the numerous research efforts on the MMR–autism hypothesis the committee concluded in its report "that the evidence favors rejection of a causal relationship at the population level between MMR vaccine and autistic spectrum disorders". Epidemiological evidence showed no association between MMR vaccine and autism, and the committee did not find a proven biological mechanism that would explain such a relation. Therefore, the committee did not recommend a policy review at this time of the licensure of MMR vaccine or of the current schedules and recommendations for MMR administration (76).

A conference of the American Academy of Pediatrics on "new challenges in childhood immunizations" was convened in Oak Brook, IL, on 12–13 June 2000 and reviewed data on what is known about the pathogenesis, epidemiology, and genetics of autism and the available data on hypothesized associations with Crohn's disease, measles, and MMR vaccine. The participants concluded that the available evidence did not support the hypothesis that MMR vaccine causes either Crohn's disease or autism or associated disorders. They recommended continued scientific efforts directed to the identification of the causes of autism (77).

Subsequently, it became known that Dr Wakefield was paid for a second study into whether children allegedly damaged by the MMR could sue. The MMR research was therefore compromised by a financial conflict of interests. The Editor of the Lancet concluded, "If we knew then what we know now, we certainly would not have published the part of the paper (78) that related to MMR" (79). In March 2004, 10 of the 12 authors of the article in question issued a "Retraction of an interpretation". The authors made it clear that in this paper no causal link was established between MMR vaccine and autism, as the data were insufficient to prove the hypothesis (80).

The response to a newly published adverse event due to immunization must be rapid. If the reported association is correct, urgent re-evaluation of the immunization program is necessary. Otherwise, if the reported association is false, a credible counter-message is necessary to minimize the negative impact on the immunization program (6). The rapidity of response to the 1998 publication of Wakefield and colleagues, including the convening of an independent review panel in the UK, was very useful.

Evidence

Two studies suggested a link between measles/MMR immunization and autism. Fudenberg reported that 15 of 40 patients with infantile autism developed symptoms within a week after MMR immunization (81). Wakefield and colleagues evaluated 12 children with chronic enterocolitis and regressive developmental disorders (62). The onset of behavioral symptoms was associated with MMR immunization in eight cases, as reported by the parents. Both reports were non-comparative and anecdotal. By chance alone some cases of autism will occur shortly after immunization, and most children in developed countries receive their first measles or MMR vaccination in the second year of life, when autism typically manifests. The imprecision of the interval between immunization and the onset of behavioral symptoms in the study by Wakefield

and colleagues made these data suspect, even before their retraction.

Inaccuracies in the study of Fudenberg, for example referring to hepatitis B vaccine as a live vaccine, cast some doubts on the carefulness of the entire report. Developmental delay is likely to be detected by a gradual awareness over a period of time, not on a particular day. Epidemiological studies in various countries (UK, Sweden, Finland), comparing the introduction and use of vaccines and the incidence of autism, have not supported a relation between measles/MMR vaccine and autism (82–85). Wing reviewed 16 studies in Europe, North America, and Japan and found no increase in autism with increasing use of measles or MMR vaccines (86). An analysis of two large European datasets produced similar results (87). In early 1998, experts in various medical disciplines reviewed the work of the Inflammatory Disease Study Group of the Royal Free Hospital in detail and concluded that there is no evidence for a link between measles/MMR vaccine and either Crohn's disease or autism (88).

Data from an earlier study have been reanalysed to test the hypothesis that MMR vaccine might cause autism but that the induction interval needs to be short (89). Evidence for an increased incidence was sought using the case-series method. The study used data on all MMR vaccines, including booster doses. The results of this study, combined with results obtained earlier by the same authors, provided powerful evidence against the hypothesis that MMR vaccine causes autism at any time after immunization.

In 2004, the Vaccine Safety Committee of the Institute of Medicine of the US National Academies of Sciences published the Immunization Safety Review on Vaccines and Autism. The Committee examined the hypothesis that vaccines, specifically the MMR vaccine and thiomersal-containing vaccines, are causally associated with autism. The Committee concluded that the body of epidemiological evidence favors rejection of a causal relation between the MMR vaccine and autism as well as a causal relation between thiomersal-containing vaccines and autism. The Committee further found that potential biological mechanisms for vaccine-induced autism that have been generated to date are theoretical only. In addition, the Committee recommended that available funding for autism research be channelled to the most promising areas (90).

The results of a retrospective study including 535 544 children aged 1-7 years who were immunized between November 1982 and June 1986 in Finland, based on linkage of individual MMR immunization data to a hospital discharge register, have been published (91). The numbers of cases of encephalitis and aseptic meningitis within a 3-month risk interval after immunization were compared with the estimated numbers of naturally occurring cases of encephalitis and aseptic meningitis during the subsequent 3-month interval. Changes in the overall number of hospitalizations because of autism and Crohn's disease were also checked throughout the study period. Of the 535 544 immunized children, 199 were hospitalized with encephalitis, 161 with aseptic meningitis, and 352 with autistic disorders. In nine children with encephalitis and 10 with meningitis, the disease developed within 3 months of immunization, revealing no increased occurrence within this designated risk period. Furthermore, in eight of nine cases of encephalitis, a very short interval of 2 days or an interval exceeding 1 month between immunization and hospitalization makes an association with immunization unlikely. As with encephalitis, an association between immunization and meningitis occurring on day 2 or over 1 month after immunization is unlikely. No clustering of hospitalization for autism has been detected after immunization; none of the autistic children went into hospital because of Crohn's disease.

A study on MMR immunization and autism has been carried out in Denmark in all 537 303 children born in Denmark from 1991 to 1998. Using national registry data on autistic disorders, the investigators found no association between MMR immunization and a subsequent diagnosis of autism (RR = 0.92; 95% CI = 0.68, 1.24) or a related disorder (RR = 0.83; 95% CI = 0.65, 1.07). This national cohort study obviated the problems of selection bias and misclassification bias (92).

In a case-control study of age at first MMR immunization in children with autism (n = 624) and children without (n = 1824), including children with regression in development, the overall distribution of ages at MMR immunization among children with autism was similar to that of matched control children (93). Most of the cases (71%) and control children (68%) were immunized at 12–17 months of age. Similar proportions of case and control children had been immunized before 18 or before 24 months of age. There were no significant associations for either of these age cutoffs in specific case subgroups, including those with evidence of developmental regression.

One of the major reasons that measles occurred among bone marrow transplant recipients during an outbreak of measles in Brazil was that they had not been immunized. In Brazil, a 2-year interval between bone marrow transplantation and measles immunization was recommended. Following a study in 79 patients the investigators concluded that MMR immunization between 12 and 24 months after bone marrow transplantation is safe and will therefore be recommended at 15 months after transplantation. However, because the seroconversion rate to measles is only 46% (rubella 91%), serology should be performed after immunization to confirm protection (94).

A survey commissioned by "Generation Rescue" (a parent-founded, parent-funded, and parent-led organization of more than 350 families) was published in June 2007. Data were gathered by Survey USA, a national market-research company, which carried out a telephone survey of the parents of more than 17 000 children, aged 4–17 years, in five counties in California (San Diego, Sonoma, Orange, Sacramento, and Marin) and four counties in Oregon (Multnomah, Marion, Jackson, and Lane) (95). They asked parents whether their child had been immunized, and whether that child had one or more of the following diagnoses: attention deficit disorder (ADD), attention deficit hyperactivity disorder (ADHD), Asperger's syndrome, pervasive development disorder not otherwise specified, or autism. Among more than 9000 boys aged 4–17 years, they found that immunized boys were 155% more likely to have neurological disorders compared with their non-immunized peers. Immunized boys were 224% more likely to have ADHD, and 61% more likely to have autism. Older immunized boys in the 11–17 age bracket were 158% more likely to have a neurological disorder, 317% more likely to have ADHD, and 112% more likely to have autism. It is open for discussion whether these results, obtained by this method, can really make a substantial contribution to a scientific question.

Metabolism

Measles–mumps–rubella

A total of 20 cases of type I diabetes mellitus suspected to be induced by MMR immunization have been reported to Behringwerke, Marburg, Germany, probably due to the mumps component (96). The earliest case occurred 3 days after receiving the vaccine and the latest 7 months after immunization. Twelve cases were diagnosed within 30 days of immunization. The investigators considered the cases of diabetes mellitus to have a temporal relation to the immunization. For every 5 million children immunized against mumps 50 spontaneous cases of diabetes mellitus are to be expected by random coincidence within a period of 30 days after immunization. In fact, only 12 cases were reported within 30 days after immunization. Mainly based on this analysis the Deutsche Vereinigung zur Bekämpfung der Viruskrankheiten (DVV) could not confirm the relation between mumps immunization and diabetes mellitus (97).

Hematologic

Measles

Anemia has been reported after measles immunization (98). The vaccine can cause a significant fall in hemoglobin that can persist for 14–30 days and can be difficult to distinguish from iron deficiency.

Rubella

Thrombocytopenic purpura has been attributed to rubella (99). However, the authors of the report of the Institute of Medicine, National Academy of Sciences, Washington, DC (1991) entitled "Adverse Effects of Pertussis and Rubella Vaccines" (46) considered that there was insufficient evidence to indicate either the presence or absence of a causal relation between RA 27/3 rubella vaccine and thrombocytopenic purpura.

Measles–mumps–rubella

There is a relation between MMR vaccination and thrombocytopenic purpura, but not with the measles component itself (20). Thrombocytopenic purpura after MMR has been reviewed, with discussion of pathogenesis and the vaccines and infections associated with this problem (100). Rubella vaccine is one of the most frequently reported causes of thrombocytopenia in Denmark (101). In France, a retrospective epidemiological survey (1984–92) showed that the rates of thrombocytopenic purpura per 100 000 vaccinees were 0.23 for measles vaccine, 0.17 for rubella vaccine, 0.87 for combination MR vaccine, and 0.95 for MMR vaccine (102). Thrombocytopenia was severe and always associated with purpura. Cases of recurrent thrombocytopenic purpura after repeated MMR immunization have been reported (103,104). The Advisory Committee on Immunization Practices (ACIP) of the Centers for Disease Control and Prevention (CDC) has recommended avoiding subsequent doses of MMR when a previous episode of thrombocytopenia occurred in close temporal proximity to the previous immunization, that is within 6 weeks (105,106).

The causal relation between MMR vaccine and idiopathic thrombocytopenic purpura has been confirmed using linkage of immunization records and hospital admission records; the absolute risk within 6 weeks of immunization was 1 per 22 300 doses (107).

Gastrointestinal

Measles–mumps–rubella
The suggestion that measles/MMR immunization can cause Crohn's disease has not been confirmed.

Inflammatory bowel disease (ulcerative colitis and Crohn's disease) is a general term for a group of chronic inflammatory disorders of unknown cause involving the gastrointestinal tract. Despite many attempts to confirm an infectious agent as the cause of disease, no bacterial, viral, or fungal agents have so far been isolated. There is strong evidence for a genetic predisposition.

In 1993 and 1994, researchers in the UK and Sweden suggested that Crohn's disease might be a late result of measles infection at a critical time during early childhood (108,109). In a study of the outcome of maternal measles infection in 25 Swedish babies, three of four children exposed to measles in utero subsequently developed Crohn's disease (110). Whereas wild measles virus was initially implicated, a controversial debate was initiated in 1995, when Thompson and colleagues suggested that attenuated measles vaccine virus might also cause inflammatory bowel disease, having found an increased risk of inflammatory bowel disease in about 3000 immunized individuals compared with about 11 000 non-immunized controls (111). Furthermore, the Inflammatory Bowel Disease Study Group of the Royal Free Hospital, London (Wakefield and colleagues), suggested that measles virus is present in the bowel of patients with Crohn's disease (with evidence from transmission electron microscopy, immunohistochemistry, in situ hybridization, and immunogold electron microscopy) (109,112,113).

Wakefield and colleagues then made two suggestions in a paper in the Lancet (62): that autism is linked to a form of inflammatory bowel disease and that this new syndrome is associated with MMR immunization. Their hypothesis was that MMR vaccine causes non-specific gut injury, allowing the absorption of normally non-permeable peptides, which in turn cause serious developmental disorders. The authors stated that they had not proved an association between measles, mumps, and rubella vaccines and either autism or inflammatory bowel disease. However, there were enough references in the text to lead the reader to the assumption that there is sufficient evidence provided by the study, and by other scientific publications, that there is a link. This paper of Wakefield and colleagues, publicized by a subsequent press conference, resulted in a heated debate and a huge number of letters to the editor of the Lancet, in turn severely criticizing both the article and its implications for immunization programs, blaming the editor for publishing the article, and defending the obligation of clinical researchers to publish provocative findings. Considerable evidence, mainly microbiological and epidemiological, has been collected by others to suggest that the association with MMR does not exist.

The alleged association of measles vaccination with Crohn's disease and autism has also been criticized as being based on poor science and as having been largely refuted by a large volume of stronger work, three types of evidence against the hypothesis (biological, microbiological, and epidemiological) being considered in detail (6).

Several groups have found no evidence of persistence of measles virus in the tissues of patients with Crohn's disease with a very sensitive test (polymerase chain reaction) (114–117). Furthermore, no viral genomic sequences of measles, mumps, and rubella viruses were found in intestinal specimens (117). The results of Iizuka and colleagues (115,118) were particularly interesting—they used the same monoclonal antibody in their immunohistochemical studies that Wakefield and colleagues had used. The antigen recognized could be a measles virus protein, but they considered it much more likely that the previous immunochemical observations were accounted for by antigen mimicry between measles virus and a host protein found in the intestinal tissue of patients with Crohn's disease. Serological studies have also shown lower measles complement fixation titers in patients with Crohn's disease than in controls, not supporting an association of measles vaccine with Crohn's disease (119,120).

The major weaknesses of the epidemiological studies reported by the Inflammatory Bowel Disease Study Group have been discussed in a number of letters to the editor (121,122) and editorials (122). On behalf of the World Health Organization's Expanded Program on Immunization, Lee and colleagues (82) questioned the conclusion that there was a temporal association between immunization and the onset of symptoms, because the study of Wakefield and colleagues had provided data on the interval between immunization and the onset of symptoms in only five of their 12 cases of so-called autism-bowel syndrome, and the age at which the vaccine was given was mentioned in only three. Furthermore, Lee and colleagues criticized the study on the grounds that no patient selection had been made by Wakefield and colleagues, other than the 12 patients referred to them; there were no controls and no blinding of the investigators (82). Payne and Mason (123) made the same comment, that the cases reported by Wakefield and colleagues were highly selected and that the underlying population was unclear.

Several epidemiological studies have failed to confirm an association between measles/measles vaccine and Crohn's disease. For example:

- *Copenhagen*: In a study of 25 mothers with measles during pregnancy there were no cases of Crohn's disease in their children (124).
- *Finland*: Peltola and colleagues (125) reported on over a decade's effort to detect all severe adverse events associated with MMR vaccine distributed in Finland. There was no evidence to support the hypothesis that the vaccine could cause pervasive developmental disorders or inflammatory bowel disease. Comparing the incidence of Crohn's disease and the Finnish data for measles and measles immunization, Pebody and colleagues (126) came to the same conclusion, namely that there is no association between Crohn's disease and measles vaccine.
- *Japan*: A nationwide survey of inflammatory bowel disease was carried out in 1979–93 in children under 16 years of age. From 1979 to 1992 the number of cases of inflammatory bowel disease was almost the same (ulcerative colitis rates 0.08–0.12 per 100 000, Crohn's disease 0.04–0.06 per 100 000). In 1992, the incidence of Crohn's disease rose to 0.10 and in 1993 to 0.12 per 100 000; the incidence rate of ulcerative colitis increased in 1993 to 0.18 per 100 000. Measles immunization (implemented in 1968) was 68% until 1993, and during the 1980s and early 1990s was relatively constant at about 70% (127).
- *UK*: In close to 7 million children who received MR vaccine in a catch-up campaign there was no increase in the number of new cases or exacerbation of existing cases of Crohn's disease (128).
- In an international case-control study of 499 patients with chronic inflammatory bowel disease and 998 control patients from nine countries, there was no difference in the risk of inflammatory bowel disease in association with either natural measles or measles immunization (129).

Pancreas

Mumps

Acute pancreatitis after mumps immunization is very rare and has been reported only sporadically in adolescents.

- A 13-month-old boy presented with an acute abdomen and surgery was performed for a suspected perforated appendicitis (130). The appendix was normal but the pancreas was enlarged, edematous, and covered with fibrin, with areas of superficial necrosis. The serum amylase activity was 528 IU (normal under 200 IU).Unfortunately the authors reported no data on the vaccine used or the interval between immunization and disease onset.

Measles–mumps–rubella

- A 19-year-old woman who received MMR vaccine developed pancreatitis 11 days later (131). Other causes of pancreatitis could not be implicated.

Skin

Measles

- An 11-month-old boy developed hypertrichosis at the injection site (2.5 × 3 cm) 1 month after measles immunization (132). The hypertrichosis and induration persisted during the follow-up period of 3 months.

Measles–mumps–rubella

Gianotti–Crosti syndrome has been reported in a child immunized with MMR (133).

- Three days after MMR immunization a 15-month-old boy developed a rash, initially on the arms but later involving the legs. Six weeks later he had an extensive, symmetrical, non-follicular, papular eruption on his face, arms, and legs, with striking sparing of his trunk. This was labeled Gianotti–Crosti syndrome, a self-healing non-recurrent erythematous or skin-colored papular eruption with symmetrical distribution on the face, buttocks, and extremities in children.A wide spectrum of infectious diseases has been associated with this syndrome, and preceding immunization against influenza, diphtheria, pertussis, and poliomyelitis has been reported.
- A 13-year-old girl developed toxic epidermal necrolysis 7 days after being immunized with live attenuated MMR vaccine (134).

Musculoskeletal

Rubella

Acute arthritis/arthralgia due to rubella vaccine occurs more often and tends to be more severe in susceptible women than in children, usually involving the small peripheral joints. Joint symptoms generally begin 3–25 days after immunization, persist for 1–11 days and rarely recur. Incidence rates are estimated to average 13–25% (135) among adult women after RA 27/3 immunization with much lower rates among children, adolescents, and adult men. The incidence of acute joint symptoms increases with increasing dose. There is also an analogous but chronic complication: in small studies of adult female vaccinees in Canada the incidence of persistent or recurrent arthritis was as high as 5–11%.

Radiological bone changes have also been reported after rubella immunization (135).

The committee responsible for the report of the Institute of Medicine, National Academy of Sciences, Washington, DC (1991) entitled "Adverse Effects of Pertussis and Rubella Vaccines" (46) found that the evidence indicates a causal relation between RA 27/3 rubella vaccine and acute arthritis and the available evidence was weaker but still consistent with a causal relation between RA 27/3 rubella vaccine and chronic arthritis. The section on rubella immunization and chronic arthritis included in the report has been reviewed critically (136). In the authors' opinion it is not justified to conclude on the basis of very limited experience that the evidence is consistent with a causal relation between the RA 27/3 rubella vaccine strain and chronic arthritis in adult women.

Table 3 Rubella immunization and arthritic reactions reported to the VAERS database

Arthritis reaction type	Number of reports	Women	Men	Mean onset (days)	Mean age (years)	Incidence per million immunizations
Arthralgia	191	170	17	11	40	78
Arthrosis	58	51	4	12	43	24
Arthritis	46	41	2	11	38	19
Joint disease	13	13	0	12	39	5

Other evidence comes from compensation claims. An analysis of 124 claims to the US National Vaccine Injury Compensation Program relating to chronic arthropathy showed that the onset of symptoms occurred at 1–6 weeks after rubella vaccination in 72 individuals (group 1) and at less than 1 week or more than 6 weeks after vaccination in 52 individuals (group 2). Various conditions developed in the two groups after immunization: unspecified arthritis (29 in group 1, one in group 2), specified arthritis (11 and 19), arthralgia (24 and 7), fibromyalgia (4 and 11), and multiple symptoms with minimal arthralgia or myalgia (4 and 14). The Program and the US Court of Federal Claims accepted that there is a causal relation between rubella vaccine and cases of chronic arthropathy that start at 1–6 weeks after vaccine administration (137).

There have also been reviews of this complication (138). Polyarthropathy associated with rubella vaccine predominantly affects the fingers and knees and usually resolves within 2–4 months. However, the duration of symptoms exceeded 18 months in about 5% of cases. Rubella virus RNA has been identified in the peripheral monocytes of such patients 8–10 months after vaccination. The association between adverse musculoskeletal and neurological events and rubella immunization has also been investigated prospectively in 546 healthy postpartum rubella-seronegative women, who received either monovalent rubella vaccine ($n = 270$) or saline placebo ($n = 276$). There was a significantly higher incidence of acute joint manifestations with rubella vaccine (30%) than with placebo (20%). The frequency of chronic or recurrent arthralgia or arthritis was only marginally significant (SEDA-22, 353).

Data from the VAERS on arthritic reactions (arthralgia, arthrosis, arthritis, and joint disease) after rubella immunization during 1991–98 have been analysed (Table 3) (139). Hepatitis A vaccine-associated arthritic reactions reported to VAERS during 1997–98 were used as controls. The analysis confirmed that rubella vaccine is associated with a large number of arthritic reactions. Among the female subjects given rubella vaccine, these reactions occurred primarily in the adult women. The incidence of arthritic reactions was 126 per million rubella immunizations, whereas the hepatitis A vaccine adult control group had an incidence rate of 3.2 per million immunizations.

Immunologic

Measles

It has been suggested that hydrolysed gelatin, rather than egg protein, is responsible for most episodes of anaphylaxis after measles immunization (6). Egg allergy should no longer be a contraindication to measles immunization. However, a previous anaphylactic reaction to measles or MMR vaccine remains a contraindication.

Toxic shock syndrome was reported to have developed within 3 hours of measles immunization in four children, three of whom died. It was initially seen in women who were using tampons in the presence of vaginal colonization and/or infection with toxin-producing strains of *Staphylococcus aureus*. The hallmarks of toxic shock syndrome are high fever, diarrhea, vomiting, tachycardia, hypertension, mucocutaneous ulceration, rash, conjunctival injection, red palms and soles, and a bleeding diathesis. There was some evidence that a used vial of measles vaccine had been kept in cold water and had thereby become contaminated (140). Similar events from various developing countries have been reported to the World Health Organization. Careful investigation pointed clearly to secondary bacterial contamination as the cause of disease and death (141).

Rubella

It has been suggested that RA 27/3 rubella strain could play a role in the etiology of chronic fatigue syndrome (142). Patients with chronic fatigue syndrome have raised IgG serum antibodies to multiple common viruses. Only IgG rubella antibodies correlate positively with the intensity of symptoms and reach titers that are significantly higher than in healthy controls.

Measles–mumps–rubella

There is a controversy about the safety of MMR vaccine in children with adverse reactions to egg. Investigating the vaccine produced by Merck Sharpe & Dohme (virus grown in chicken fibroblasts), some did not detect egg protein (143). However, using a competitive ELISA others detected 74 pg of ovalbumin per ml of the MMR vaccine produced by MSD (144). This extremely low amount of ovalbumin appears to be unlikely to provoke allergic reactions. The results of a study carried out in 15 children with egg-positive skin results confirmed the view that the MMR vaccine does not contain enough egg protein to cause reactions (145).

- Moraten vaccine, Berna (virus grown in human fibroblast cultures and therefore certainly ovalbumin-free) was given to a 2-year-old boy with atopic dermatitis (146). The skin tests were positive to cow's milk and egg. Within a few minutes after the receipt of the vaccine, the child developed severe dyspnea, rhinoconjunctivitis, and lip cyanosis; he was successfully treated for an acute anaphylactic reaction. The case seems to demonstrate that the very rare allergic reactions after the administration of measles and MMR vaccines could be due to causes other than egg protein.
- A 17-year-old woman who had an anaphylactic reaction to MMR vaccine was allergic to gelatin; after eating gelatin she developed ear and throat pruritus and tongue swelling (147). Prick skin tests were positive for MMR vaccine and gelatin. Immunoblotting confirmed the presence of IgE antibodies to multiple gelatin components from a variety of animal sources.MMR vaccine contains gelatin as a stabilizer. The authors concluded that the anaphylaxis was caused by the gelatin component of the vaccine.
- In 24 of 26 children aged 1–4 years who developed anaphylactic reactions within minutes after the administration of MMR vaccine there were IgE antibodies against gelatin (148).
- In 135 children with suspected or documented systemic allergy (atopic eczema, asthma, cow milk's allergy, severe systemic reactions after previous doses of different vaccines), who were prick-tested with undiluted MMR vaccine before immunization, 122 of 126 prick-test-negative children received the MMR vaccine. No untoward reactions developed, except for mild generalized urticaria or fever in two vaccinees. The author concluded that allergic diseases should not be considered as contraindications to MMR immunization (149).In a brief review of allergy and MMR vaccine the author concluded that anaphylactic reactions are very rare but potentially life-threatening (150). Immunization personnel must therefore be aware of this possibility and trained in its management. Most severe reactions occur within a few minutes after injection and it is extremely unlikely that a child who appears completely well after immunization will subsequently develop a severe reaction.

Infection risk

Mumps

Virus shedding after immunization with some live attenuated mumps vaccine strains has been well documented. However, it is generally accepted that horizontal transmission of vaccine strains is rare. Horizontal transmission of the Urabe AM9 vaccine strain from a symptomatic vaccinee to her younger sister was described in 1993 [151]. Horizontal transmission of the Leningrad-3 mumps vaccine strain from healthy vaccinees to six previously immunized contacts, resulting in symptomatic infection, has also been described [152]. In both reports,

parotitis developed as a unique consequence of horizontal transmission of mumps vaccine strains. All patients were of childhood age.

The first virologically confirmed cases of symptomatic mumps (causing parotitis as well as extrasalivary effects, including meningitis) has been described after horizontal transmission of the Leningrad–Zagreb mumps vaccine strain in adults [153]. The possible source of infection in the patients were their own children, who had been recently immunized with Leningrad–Zagreb mumps vaccine strain as a component of MMR vaccine. The first symptoms of disease in the parents appeared within 33–44 days after the immunization of their children.

Second-Generation Effects

Pregnancy

Measles

Live measles vaccine should not be given to pregnant women. This precaution is based on the purely theoretical risk of fetal infection (154).

Rubella

The risks of rubella infection to the unborn child are well known, and necessarily the question has arisen as to whether rubella vaccine is safer. Based on careful analyses by the CDC since 1979 and evaluation of reports from Germany and the UK showing that infants born to susceptible mothers did not develop the congenital rubella syndrome, the Immunization Practices Advisory Committee (155) has concluded that:

- pregnancy remains as a contraindication to rubella immunization because of the theoretical, albeit small, risk of congenital rubella syndrome
- reasonable precautions should be taken to preclude immunization of pregnant women
- rubella immunization of pregnant women should not ordinarily be a reason to consider interruption of pregnancy.Continued surveillance in the USA (data collected through the Vaccine in Pregnancy Registry) has not shown evidence that RA 27/3 rubella vaccine, administered during pregnancy, can cause the congenital rubella syndrome (156).

Susceptibility Factors

Measles

Persons who have experienced anaphylactic reactions to measles vaccine or neomycin (measles vaccine contains a small amount of neomycin) should not be immunized (157).

Replication of the measles vaccine virus may be potentiated in patients with immune deficiencies and by the suppressed immune responses that occur with leukemia, lymphoma or generalized malignancy, or during therapy with corticosteroids, alkylating drugs, antimetabolites or radiation; such patients should not be immunized (4).

Of 176 high-risk children with pre-existing damage to the central nervous system who were immunized to protect them against measles during an outbreak, only four developed adverse effects after immunization (convulsions, transient electroencephalographic changes) (158). Comparing the risks of immunization and of natural disease, the authors recommended measles immunization of neurologically high-risk children when an epidemic occurs.

Originally, the Committee on Infectious Diseases of the American Academy of Pediatrics and the Advisory Committee on Immunization Practices (ACIP) recommended that severely immunocompromised patients with HIV infection should not receive measles vaccine (159). However, in 1988, recognizing the severity of measles in immunodeficient individuals, it revised its measles immunization guidelines and recommended that children with asymptomatic HIV infection should be immunized and that immunization should be considered for symptomatic HIV-infected children (160). Health authorities in many countries have made similar recommendations. Since then, many HIV-infected children have been safely immunized, and the ACIP recommendation has been expanded to all age groups of HIV-infected persons.

Measles–mumps–rubella

Immunocompromise is a contraindication to the use of live virus vaccines, such as measles, mumps, rubella, and oral poliomyelitis vaccine. It is recommended that rubella or MMR vaccine should not be given to persons who are immunosuppressed because of AIDS or other clinical manifestations of HIV infection. The vaccine can be given to asymptomatic infected people (159).

Drug Administration

Drug additives

There has been a report of an allergic reaction after MMR vaccine (161). The author concluded that gelatin seems to be far more often the cause of allergic reactions to MMR vaccine than egg allergy. In vaccinees who receive DTaP vaccine (some brands contain gelatin) before MMR vaccine, the prior injection of gelatin-containing DTaP may be the cause of sensitization to gelatin and the subsequent reaction to other gelatin-containing vaccines.

References

1. Aaby P, Samb B, Simondon F, Whittle H, Seck AM, Knudsen K, Bennett J, Markowitz L, Rhodes P. Child mortality after high-titre measles vaccines in Senegal: the complete data set. Lancet 1991;338(8781):1518–9.
2. Garenne M, Leroy O, Beau JP, Sene I. Child mortality after high-titre measles vaccines: prospective study in Senegal. Lancet 1991;338(8772):903–7.
3. Knudsen KM, Aaby P, Whittle H, Rowe M, Samb B, Simondon F, Sterne J, Fine P. Child mortality following standard, medium or high titre measles immunization in West Africa. Int J Epidemiol 1996;25(3):665–73.
4. Centers for Disease Control (CDC). Recommendation of the Immunization Practices Advisory Committee (ACIP). Measles prevention. MMWR Morb Mortal Wkly Rep 1982;31(17):217–24229–31.
5. Isomura S. Measles and measles vaccine in Japan. Acta Paediatr Jpn 1988;30(2):154–62.
6. Duclos P, Ward BJ. Measles vaccines: a review of adverse events. Drug Saf 1998;19(6):435–54.
7. Tsai HY, Huang LM, Shih YT, Chen JM, Jiang TM, Tsai CH, Lee CY. Immunogenicity and safety of standard-titer AIK-C measles vaccine in nine-month-old infants. Viral Immunol 1999;12(4):343–8.
8. Global Advisory Committee on Vaccine Safety. Meeting on 29–30 November 2006. Safety of mumps vaccine strains. Wkly Epidemiol Rec 2007;82:20–2.
9. Centers for Disease Control (CDC). Mumps prevention. MMWR Morb Mortal Wkly Rep 1989;38(22):388–92397–400.
10. Peltola H, Heinonen OP. Frequency of true adverse reactions to measles–mumps–rubella vaccine. A double-blind placebo-controlled trial in twins. Lancet 1986;1(8487):939–42.
11. Davis RL, Marcuse E, Black S, Shinefield H, Givens B, Schwalbe J, Ray P, Thompson RS, Chen R, Glaser JW, Rhodes PH, Swint E, Jackson LA, Barlow WE, Immanuel VH, Benson PJ, Mullooly JP, Drew L, Mendius B, Lewis N, Fireman BH, Ward JI, Vadheim CM, Marcy SM, Jing J, Wulfson M, Lugg M, Osborne P, Wise RPThe Vaccine Safety Datalink Team. MMR2 immunization at 4 to 5 years and 10 to 12 years of age: a comparison of adverse clinical events after immunization in the Vaccine Safety Datalink project. Pediatrics 1997;100(5):767–71.
12. Patja A, Davidkin I, Kurki T, Kallio MJ, Valle M, Peltola H. Serious adverse events after measles–mumps–rubella vaccination during a fourteen-year prospective follow-up. Pediatr Infect Dis J 2000;19(12):1127–34.
13. Nieminen U, Peltola H, Syrjala MT, Makipernaa A, Kekomaki R. Acute thrombocytopenic purpura following measles, mumps and rubella vaccination. A report on 23 patients. Acta Paediatr 1993;82(3):267–70.
14. Shinefield H, Black S, Digilio L, Reisinger K, Blatter M, Gress JO, Hoffman Brown ML, Eves KA, Klopfer SO, Schadel F, Kuter BJ. Evaluation of a quadrivalent measles, mumps, rubella and varicella vaccine in healthy children. Pediatr Infect Dis J 2005;24:665–9.
15. Angel JB, Walpita P, Lerch RA, Sidhu MS, Masurekar M, DeLellis RA, Noble JT, Snydman DR, Udem SA. Vaccine-associated measles pneumonitis in an adult with AIDS. Ann Intern Med 1998;129(2):104–6.
16. Landrigan PJ, Witte JJ. Neurologic disorders following live measles-virus vaccination. JAMA 1973;223(13):1459–62.
17. Bloch AB, Orenstein WA, Stetler HC, Wassilak SG, Amler RW, Bart KJ, Kirby CD, Hinman AR. Health impact of measles vaccination in the United States. Pediatrics 1985;76(4):524–32.
18. Centers for Disease Control (CDC). Measles prevention. MMWR Morb Mortal Wkly Rep 1987;36(26):409–18423–5.
19. Pollock TM, Morris J. A 7-year survey of disorders attributed to vaccination in North West Thames region. Lancet 1983;1(8327):753–7.
20. Weibel RE, Caserta V, Benor DE, Evans G. Acute encephalopathy followed by permanent brain injury or death associated with further attenuated measles vaccines: a review of claims submitted to the National Vaccine Injury Compensation Program. Pediatrics 1998;101(3 Pt 1):383–7.

21. Miller D, Wadsworth J, Diamond J, Ross E. Measles vaccination and neurological events. Lancet 1997;349(9053):730–1.

22. Black S, Shinefield H, Ray P, Lewis E, Chen R, Glasser J, Hadler S, Hardy J, Rhodes P, Swint E, Davis R, Thompson R, Mullooly J, Marcy M, Vadheim C, Ward J, Rastogi S, Wise R. Risk of hospitalization because of aseptic meningitis after measles–mumps–rubella vaccination in one- to two-year-old children: an analysis of the Vaccine Safety Datalink (VSD) Project. Pediatr Infect Dis J 1997;16(5):500–3.

23. Sepkowitz S. Polio vaccine and polio. Pediatrics 1999;103(3):694–5.

24. Weibel RE, Caserta V, Benor DE, Evans G. Reply. Pediatrics 1999;103:695–6.

25. American Academy of Pediatrics Committee on Infectious Diseases. Personal and family history of seizures and measles immunization. Pediatrics 1987;80(5):741–2.

26. Ishikawa T, Ogino C, Chang S. Lennox–Gastaut syndrome after a further attenuated live measles vaccination. Brain Dev 1999;21(8):563–5.

27. Centers for Disease Control and Prevention. Epidemiology and Prevention of Vaccine-Preventable Diseases. 8th Edition. Chapter 10. Measles. Atlanta, GA: Centers for Disease Control and Prevention, 2004.

28. Barrero PR, Grippo J, Viegas M, Mistchenko AS. Wild-type measles virus in brain tissue of children with subacute sclerosing panencephalitis, Argentina. Emerg Infect Dis 2003;9:1333–6.

29. Jin L, Beard S, Brown DWG, Miller E. Characterization of measles virus strains causing SSPE: A study of 11 cases. J Neurovirol 2002;8:335–44.

30. Hinman AR, Orenstein WA, Bloch AB, Bart KJ, Eddins DL, Amler RW, Kirby CD. Impact of measles in the United States. Rev Infect Dis 1983;5(3):439–44.

31. Zilber N, Rannon L, Alter M, Kahana E. Measles, measles vaccination, and risk of subacute sclerosing panencephalitis (SSPE). Neurology 1983;33(12):1558–64.

32. Ueda S. SSPE-epidemiology in Japan and neurovirulence of SSPE virus. Brain Dev 1985;7:122.

33. Okuno Y, Nakao T, Ishida N, Konno T, Mizutani H, Fukuyama Y, Sato T, Isomura S, Ueda S, Kitamura I, et al. Incidence of subacute sclerosing panencephalitis following measles and measles vaccination in Japan. Int J Epidemiol 1989;18(3):684–9.

34. Gerike E, Dittmann S. Personal communication, 1990.

35. Cernescu C, Milea S. Epidemiology of subacute sclerosing panencephalitis (SSPE) in Romania between 1976–1982. Virologie 1983;34(4):239–50.

36. Global Advisory Committee on Vaccine Safety. Subacute sclerosing panencephalitis and measles vaccination. Weekly Epidemiol Rec 2005;80:241–9.

37. Bitnun A, Shannon P, Durward A, Rota PA, Bellini WJ, Graham C, Wang E, Ford-Jones EL, Cox P, Becker L, Fearon M, Petric M, Tellier R. Measles inclusion-body encephalitis caused by the vaccine strain of measles virus. Clin Infect Dis 1999;29:855–61.

38. Aronson JK, Hauben M. Anecdotes that provide definitive evidence. BMJ 2006;332:1267–9.

39. Dourado I, Cunha S, Teixeira MG, Farrington CP, Melo A, Lucena R, Barreto ML. Outbreak of aseptic meningitis associated with mass vaccination with a urabe-containing measles–mumps–rubella vaccine: implications for immunization programs. Am J Epidemiol 2000;151(5):524–30.

40. Miller E, Goldacre M, Pugh S, Colville A, Farrington P, Flower A, Nash J, MacFarlane L, Tettmar R. Risk of aseptic meningitis after measles, mumps, and rubella vaccine in UK children. Lancet 1993;341(8851):979–82.

41. Colville A, Pugh S. Mumps meningitis and measles, mumps, and rubella vaccine. Lancet 1992;340(8822):786.

42. Brown EG, Furesz J, Dimock K, Yarosh W, Contreras G. Nucleotide sequence analysis of Urabe mumps vaccine strain that caused meningitis in vaccine recipients. Vaccine 1991;9(11):840–2.

43. Forsey T, Bentley ML, Minor PD, Begg N. Mumps vaccines and meningitis. Lancet 1992;340(8825):980.

44. Peltola H. Mumps vaccination and meningitis. Lancet 1993;341(8851):994–5.

45. Global Advisory Committee on Vaccine Safety. Wkly Epidemiol Rec 2003;78(32):282–411–12 June 2003.

46. Adverse Effects of pertussis and rubella vaccines. In: Howson CP, Howe CJ, Fineberg HV, editors. A report of the Committee to Review the Adverse Consequences of Pertussis and Rubella Vaccines. Washington, DC: National Academy Press, 1991:681–8.

47. Tsuru T, Mizuguchi M, Ohkubo Y, Itonaga N, Momoi MY. Acute disseminated encephalomyelitis after live rubella vaccination. Brain Dev 2000;22(4):259–61.

48. Cusi MG, Bianchi S, Santini L, Donati D, Valassina M, Valensin PE, Cioe L, Mazzocchio R. Peripheral neuropathy associated with anti-myelin basic protein antibodies in a woman vaccinated with rubella virus vaccine. J Neurovirol 1999;5(2):209–14.

49. McCormick A, Dinakaran S, Bhola R, Rennie IG. Recurrent sixth nerve palsy following measles mumps rubella vaccination. Eye 2001;15(Pt 3):356–7.

50. Fujinaga T, Motegi Y, Tamura H, Kuroume T. A prefecture-wide survey of mumps meningitis associated with measles, mumps and rubella vaccine. Pediatr Infect Dis J 1991;10(3):204–9.

51. Sugiura A, Yamada A. Aseptic meningitis as a complication of mumps vaccination. Pediatr Infect Dis J 1991;10(3):209–13.

52. Sawada H, Yano S, Oka Y, Togashi T. Transmission of Urabe mumps vaccine between siblings. Lancet 1993;342(8867):371.

53. Ueda K, Miyazaki C, Hidaka Y, Okada K, Kusuhara K, Kadoya R. Aseptic meningitis caused by measles–mumps–rubella vaccine in Japan. Lancet 1995;346(8976):701–2.

54. Kimura M, Kuno-Sakai H, Yamazaki S, Yamada A, Hishiyama M, Kamiya H, Ueda K, Murase T, Hirayama M, Oya A, Nozaki S, Murata R. Adverse events associated with MMR vaccines in Japan. Acta Paediatr Jpn 1996;38(3):205–11.

55. Rebiere I, Galy-Eyraud C. Estimation of the risk of aseptic meningitis associated with mumps vaccination, France, 1991–1993. Int J Epidemiol 1995;24(6):1223–7.

56. Tesovic G, Begovac J, Bace A. Aseptic meningitis after measles, mumps, and rubella vaccine. Lancet 1993;341(8859):1541.

57. Cizman M, Mozetic M, Radescek-Rakar R, Pleterski-Rigler D, Susec-Michieli M. Aseptic meningitis after vaccination against measles and mumps. Pediatr Infect Dis J 1989;8(5):302–8.

58. Brown EG, Dimock K, Wright KE. The Urabe AM9 mumps vaccine is a mixture of viruses differing at amino acid 335 of the hemagglutinin–neuraminidase gene with one form associated with disease. J Infect Dis 1996;174(3):619–22.

59. Plesner AM, Hansen FJ, Taudorf K, Nielsen LH, Larsen CB, Pedersen E. Gait disturbance interpreted as

cerebellar ataxia after MMR vaccination at 15 months of age: a follow-up study. Acta Paediatr 2000;89(1):58–63.

60. Patja A, Paunio M, Kinnunen E, Junttila O, Hovi T, Peltola H. Risk of Guillain–Barré syndrome after measles–mumps–rubella vaccination. J Pediatr 2001;138(2):250–4.

61. Kazarian EL, Gager WE. Optic neuritis complicating measles, mumps, and rubella vaccination. Am J Ophthalmol 1978;86(4):544–7.

62. Wakefield AJ, Murch SH, Anthony A, Linnell J, Casson DM, Malik M, Berelowitz M, Dhillon AP, Thomson MA, Harvey P, Valentine A, Davies SE, Walker-Smith JA. Ileal-lymphoid-nodular hyperplasia, non-specific colitis, and pervasive developmental disorder in children. Lancet 1998;351(9103):637–41.

63. Black D, Prempeh H, Baxter T. Autism, inflammatory bowel disease, and MMR vaccine. Lancet 1998;351(9106):905–6.

64. Beale AJ. Autism, inflammatory bowel disease, and MMR vaccine. Lancet 1998;351(9106):906.

65. O'Brien SJ, Jones IG, Christie P. Autism, inflammatory bowel disease, and MMR vaccine. Lancet 1998;351(9106):906–7.

66. Author's reply Wakefield AJ. Lancet 1998;351(9106):908.

67. Editor's reply Horton R. Lancet 1998;351(9106):908.

68. Horton R. The lessons of MMR. Lancet 2004;363(9411):747–9.

69. Wakefield AJ, Harvey P, Linnell J. MMR—responding to retraction. Lancet 1998;351(9112):1356.

70. Anonymous. Measles, MMR, and autism: the confusion continues. Lancet 2000;355(9213):1379.

71. Wakefield AJ, Montgomery SM. Measles, mumps, rubella vaccine: through a glass, darkly. Adverse Drug React Toxicol Rev 2000;19(4):265–83.

72. Elliman D, Bedford H. MMR vaccine: the continuing saga. BMJ 2001;322(7280):183–4.

73. Letter from the Chief Medical Officer, the Chief Nursing Officer, and the Chief Pharmaceutical Officer, Department of Health, London. Current vaccine and immunisation issues. 1. MMR vaccine. http://www.doh.gov.uk/cmo/cmoh.htm.

74. BBC News. MMR research doctor resigns. http://news.bbc.co.uk/1/hi/health/1687967.stm.

75. WHO statement on MMR vaccine, 25 January 2001.

76. Immunization Safety Review Committee. Immunization safety review: measles–mumps–rubella vaccine and autism. In: Stratton K, Gable A, Shetty P, McCormick M, editors. Institute of MedicineNational Academy of Sciences, 2001:681–8.

77. Halsey NA, Hyman SLConference Writing Panel. Measles–mumps–rubella vaccine and autistic spectrum disorder: report from the New Challenges in Childhood Immunizations Conference convened in Oak Brook. Illinois, June 12–13, 2000Pediatrics 2001;107(5):E84.

78. Wakefield AJ, Murch SH, Anthony A, Linnell J, Casson DM, Malik M, Berelowitz M, Dhillon AP, Thomson MA, Harvey P, Valentine A, Davies SE, Walker-Smith JA. Ileal-lymphoid-nodular hyperplasia, non-specific colitis, and pervasive developmental disorder in children. Lancet 1998;351:637–41.

79. Horton R. The lessons of MMR. Lancet 2004;363:747–9.

80. Murch SH, Anthony A, Casson DH, Malik M, Berelowitz M, Dhillon AP, Thomson MA, Valentine A, Davies SE, Walker-Smith JA. Retraction of an interpretation. Lancet 2004;363:750.

81. Fudenberg HH. Dialysable lymphocyte extract (DLyE) in infantile onset autism: a pilot study. Biotherapy 1996;9(1–3):143–7.

82. Lee JW, Melgaard B, Clements CJ, Kane M, Mulholland EK, Olive JM. Autism, inflammatory bowel disease, and MMR vaccine. Lancet 1998;351(9106):905.

83. Immunizations Practices Advisory Committee (ACIP). Recommendations on BCG vaccines. MMWR Morb Mortal Wkly Rep 1979;28:241.

84. Global Advisory Group of the Expanded Programme on Immunization (EPI). Report on the meeting 13–17 October, 1986. New Delhi. Unedited document. WHO/87/1, 1986.

85. Corti Ortiz D, Rivera Garay P, Aviles Jasse J, Hidalgo Carmona F, MacMillan Soto G, Coz Canas LF, Vargas Delaunoy R, Susaeta Saenz de San Pedro R. Profilaxis del cancer vesical superficial con 1 mg de BCG endovesical: comparacion con otras dosis. [Prophylaxis of superficial bladder cancer with 1 mg of intravesical BCG: comparison with other doses.] Actas Urol Esp 1993;17(4):239–42.

86. Wing L. Autistic spectrum disorders. BMJ 1996;312(7027):327–8.

87. Fombonne E. Inflammatory bowel disease and autism. Lancet 1998;351(9107):955.

88. Department of Health, England and Wales. MMR vaccine is not linked to Crohn's disease or autism: conclusion of an expert scientific seminar. London: Press release 98/109, 24 March 1998.

89. Farrington CP, Miller E, Taylor B. MMR and autism: further evidence against a causal association. Vaccine 2001;19(27):3632–5.

90. Immunization Safety Review Committee. Institute of Medicine, National Academies of Sciences. Vaccines and autism. Washington, DC: National Academies Press, 2004. http://National-Academies.org (accessed 20 July, 2004).

91. Mäkelä A, Nuorti JP, Peltola H. Neurologic disorders after measles–mumps–rubella vaccination. Pediatrics 2002;110:957–63.

92. Madsen KM, Hviid A, Vestergaard M, Schendel D, Wohlfart J, Thorsen P, Olsen J, Melbye M. A population-based study of MMR and autism. New Engl J Med 2002;347:1477–82.

93. DeStefano F, Karapurkar Bhasin T, Thompson WW, Yeargin-Allsopp M, Boyle C. Age at first MMR vaccination in children with autism and school-matched control subjects: a population-based study in Metropolitan Atlanta. Pediatrics 2004;113:259–66.

94. Shaw PJ, Bleakley M Burges M. Safety of early immunization against measles/mumps/rubella after bone marrow transplantation. Blood 2002;99:3486.

95. The California–Oregon Unvaccinated Children Survey [www.GenerationRescue.org] (last accessed 26 July 2007).

96. Fescharek R, Quast U, Maass G, Merkle W, Schwarz S. Measles–mumps vaccination in the FRG: an empirical analysis after 14 years of use. II. Tolerability and analysis of spontaneously reported side effects. Vaccine 1990;8(5):446–56.

97. Deutsche Vereinigung zur Bekampfung der Viruskrankheiten (DVV). Mumpsschutzimpfung and Diabetes mellitus (Typ I). Bundesgesundhbl 1989;237:.

98. Olivares M, Walter T, Osorio M, Chadud P, Schlesinger L. Anemia of a mild viral infection: the measles vaccine as a model. Pediatrics 1989;84(5):851–5.

99. Tingle AJ, Chantler JK, Pot KH, Paty DW, Ford DK. Postpartum rubella immunization: association with development of prolonged arthritis, neurological sequelae, and chronic rubella viremia. J Infect Dis 1985;152(3):606–12.

100. Chang SK, Farrell DL, Dougan K, Kobayashi B. Acute idiopathic thrombocytopenic purpura following combined vaccination against measles, mumps, and rubella. J Am Board Fam Pract 1996;9(1):53–5.

101. Pedersen-Bjergaard U, Andersen M, Hansen PB. Thrombocytopenia induced by noncytotoxic drugs in Denmark 1968–91. J Intern Med 1996;239(6):509–15.

102. Autret E, Jonville-Bera AP, Galy-Eyraud C, Hessel L. Purpura thrombopenique après vaccination isolée on associée contre la rougeole, la rubeole et les oreillous. [Thrombocytopenic purpura after isolated or combined vaccination against measles, mumps and rubella.] Therapie 1996;51(6):677–80.

103. Drachman RA, Murphy S, Ettinger LJ. Exacerbation of chronic idiopathic thrombocytopenic purpura following measles–mumps–rubella immunization. Arch Pediatr Adolesc Med 1994;148(3):326–7.

104. Vlacha V, Forman EN, Miron D, Peter G. Recurrent thrombocytopenic purpura after repeated measles–mumps–rubella vaccination. Pediatrics 1996;97(5):738–9.

105. Rollan AR, Pool V, Chen R, Rhodes P. Indications for measles–mumps–rubella vaccination in a child with prior thrombocytopenia purpura. Pediatr Infect Dis J 1997;16(4):423–4.

106. Advisory Committee on Immunization Practices (ACIP). Update: vaccine side effects, adverse reactions, contraindications, and precautions. MMWR Recomm Rep 1996;45(RR-12):1–35.

107. Miller E, Waight P, Farrington CP, Andrews N, Stowe J, Taylor B. Idiopathic thrombocytopenic purpura and MMR vaccine. Arch Dis Child 2001;84(3):227–9.

108. Tardieu M, Truffot-Pernot C, Carriere JP, Dupic Y, Landrieu P. Tuberculous meningitis due to BCG in two previously healthy children. Lancet 1988;1(8583):440–1.

109. Wakefield AJ, Pittilo RM, Sim R, Cosby SL, Stephenson JR, Dhillon AP, Pounder RE. Evidence of persistent measles virus infection in Crohn's disease. J Med Virol 1993;39(4):345–53.

110. Ekbom A, Daszak P, Kraaz W, Wakefield AJ. Crohn's disease after in-utero measles virus exposure. Lancet 1996;348(9026):515–7.

111. Thompson NP, Montgomery SM, Pounder RE, Wakefield AJ. Is measles vaccination a risk factor for inflammatory bowel disease? Lancet 1995;345(8957):1071–4.

112. Radeff B, Harms M. Acute febrile neutrophilic dermatosis (Sweet's syndrome) following BCG vaccination. Acta Derm Venereol 1986;66(4):357–8.

113. Katznelson D, Gross S, Sack J. Polyneuritis following BCG re-vaccination. Postgrad Med J 1982;58(682):496–7.

114. Yen MY, Liu JH. Bilateral optic neuritis following Bacille Calmette–Guérin (BCG) vaccination. J Clin Neuroophthalmol 1991;11(4):246–9.

115. Simma B, Dietze O, Vogel W, Ellemunter H, Guggenbichler JP. Bacille Calmette–Guérin—associated neonatal hepatitis. Eur J Pediatr 1991;150(6):423–4.

116. Sadeghi E, Kumar PV. Eczema vaccinatum and postvaccinal BCG adenitis—case report. Tubercle 1990;71(2):145–8.

117. Haga Y, Funakoshi O, Kuroe K, Kanazawa K, Nakajima H, Saito H, Murata Y, Munakata A, Yoshida Y. Absence of measles viral genomic sequence in intestinal tissues from Crohn's disease by nested polymerase chain reaction. Gut 1996;38(2):211–5.

118. Iizuka M, Nakagomi O, Chiba M, Ueda S, Masamune O. Absence of measles virus in Crohn's disease. Lancet 1995;345(8943):199.

119. Fisher NC, Yee L, Nightingale P, McEwan R, Gibson JA. Measles virus serology in Crohn's disease. Gut 1997;41(1):66–9.

120. Power JT, Stewart IC, Ross JD. Erythromycin in the management of troublesome BCG lesions. Br J Dis Chest 1984;78(2):192–4.

121. Singh G, Singh M. Erythromycin for BCG cold abscess. Lancet 1984;2(8409):979.

122. Chen RT, DeStefano F. Vaccine adverse events: causal or coincidental? Lancet 1998;351(9103):611–2.

123. Payne C, Mason B. Autism, inflammatory bowel disease, and MMR vaccine. Lancet 1998;351(9106):907.

124. Nielsen LL, Nielsen NM, Melbye M, Sodermann M, Jacobsen M, Aaby P. Exposure to measles in utero and Crohn's disease: Danish register study. BMJ 1998;316(7126):196–7.

125. Peltola H, Patja A, Leinikki P, Valle M, Davidkin I, Paunio M. No evidence for measles, mumps, and rubella vaccine-associated inflammatory bowel disease or autism in a 14-year prospective study. Lancet 1998;351(9112):1327–8.

126. Pebody RG, Paunio M, Ruutu P. Measles, measles vaccination, and Crohn's disease. Crohn's disease has not increased in Finland. BMJ 1998;316(7146):1745–6.

127. Yamashiro Y, Walker-Smith JA, Shimizu T, Oguchi S, Ohtsuka Y. Measles vaccination and inflammatory bowel disease in Japanese children. J Pediatr Gastroenterol Nutr 1998;26(2):238.

128. Calman K. Measles, measles–mumps–rubella (MMR) vaccine, Crohn's disease and autism. Dear doctor letter from the Chief Medical Officer, March 1998, PL/CMO/98/2.

129. Gilat T, Hacohen D, Lilos P, Langman MJ. Childhood factors in ulcerative colitis and Crohn's disease. An international cooperative study. Scand J Gastroenterol 1987;22(8):1009–24.

130. Feldman G, Zer M. Infantile acute pancreatitis after mumps vaccination simulating an acute abdomen. Pediatr Surg Int 2000;16(7):488–9.

131. Adler JB, Mazzotta SA, Barkin JS. Pancreatitis caused by measles, mumps, and rubella vaccine. Pancreas 1991;6(4):489–90.

132. Ozkan H, Dundar NO, Ozkan S, Kumral A, Duman N, Gulcan H. Hypertrichosis following measles immunization. Pediatr Dermatol 2001;18(5):457–8.

133. Velangi SS, Tidman MJ. Gianotti–Crosti syndrome after measles, mumps and rubella vaccination. Br J Dermatol 1998;139(6):1122–3.

134. Dobrosavljevic D, Milinkovic MV, Nikolic MM. Toxic epidermal necrolysis following morbilli–parotitis–rubella vaccination. J Eur Acad Dermatol Venereol 1999;13(1):59–61.

135. Peters ME, Horowitz S. Bone changes after rubella vaccination. Am J Roentgenol 1984;143(1):27–8.

136. Bayer SR, Turksoy RN, Emmi AM, Reindollar RH. Rubella immunization. Fertil Steril 1992;57:229.

137. Weibel RE, Benor DE. Chronic arthropathy and musculoskeletal symptoms associated with rubella vaccines. A review of 124 claims submitted to the National Vaccine Injury Compensation Program. Arthritis Rheum 1996;39(9):1529–34.

138. Bannwarth B. Drug-induced rheumatic disorders. Rev Rhum Engl Ed 1996;63(10):639–47.

139. Geier DA, Geier MR. Rubella vaccine and arthritic adverse reactions: an analysis of the Vaccine Adverse Events Reporting System (VAERS) database from 1991 through 1998. Clin Exp Rheumatol 2001;19(6):724–6.

140. Phadke MA, Joshi BN, Warerkar UV, Diwan MP, Panse GA, Sokhey J, Bhate SM. Toxic shock syndrome: an unforeseen complication following measles vaccination. Indian Pediatr 1991;28(6):663–5.

141. Milstein J, Dittmann S. Personal communication, 1995.

142. Allen AD. Is RA27/3 rubella immunization a cause of chronic fatigue? Med Hypotheses 1988;27(3):217–20.

143. O'Brien TC, Maloney CJ, Tauraso NM. Quantitation of residual host protein in chicken embryo-derived vaccines by radial immunodiffusion. Appl Microbiol 1971;21(4):780–2.

144. Fasano MB, Wood RA, Cooke SK, Sampson HA. Egg hypersensitivity and adverse reactions to measles, mumps, and rubella vaccine. J Pediatr 1992;120(6): 878–81.

145. Greenberg MA, Birx DL. Safe administration of mumps–measles–rubella vaccine in egg-allergic children. J Pediatr 1988;113(3):504–6.

146. Giampietro PG, Bruno G, Grandolfo M, Businco L. Adverse reaction to measles immunization. Eur J Pediatr 1993;152(1):80.

147. Kelso JM, Jones RT, Yunginger JW. Anaphylaxis to measles, mumps, and rubella vaccine mediated by IgE to gelatin. J Allergy Clin Immunol 1993;91(4):867–72.

148. Sakaguchi M, Nakayama T, Inouye S. Food allergy to gelatin in children with systemic immediate-type reactions, including anaphylaxis, to vaccines. J Allergy Clin Immunol 1996;98(6 Pt 1):1058–61.

149. Juntunen-Backman K, Peltola H, Backman A, Salo OP. Safe immunization of allergic children against measles, mumps, and rubella. Am J Dis Child 1987;141(10):1103–5.

150. Lakshman R, Finn A. MMR vaccine and allergy. Arch Dis Child 2000;82(2):93–5.

151. Sawada H, Yano S, Oka Y and Togashi T. Transmission of Urabe mumps vaccine between siblings. Lancet 1993;342:371.

152. Atrashcuskaya AV, Neverov AA, Rubin S, Ignatyev GM. Horizontal transmission of the Leningrad-3 live attenuated mumps vaccine virus. Vaccine 2006;24:1530–6.

153. Tesović G, Poljak M, Lunar MM, Kocjan BJ, Seme K, Vukić BT, Sternak SL, Cajić V, Vince A. Horizontal transmission of the Leningrad–Zagreb mumps vaccine strain: a report of three cases. Vaccine 2008;22(16):1922–5.

154. Levine MM. Live-virus vaccines in pregnancy. Risks and recommendations. Lancet 1974;2(7871):34–8.

155. Centers for Disease Control (CDC). Rubella vaccination during pregnancy—United States, 1971–1986. MMWR Morb Mortal Wkly Rep 1987;36(28):457–61.

156. Centers for Disease Control (CDC). Rubella vaccination during pregnancy—United States, 1971–1988. MMWR Morb Mortal Wkly Rep 1989;38(17):289–93.

157. Kwittken PL, Rosen S, Sweinberg SK. MMR vaccine and neomycin allergy. Am J Dis Child 1993;147(2):128–9.

158. Hulsse C, Steffen W, Von Suchodeletz W. Masernausbruch in einem Heim für zerebral schwerstgeschädigte Patienten. Dtsch Gesundheitswes 1980;35:62.

159. Centers for Disease Control (CDC). Immunization of children infected with human T lymphotropic virus type III/ lymphadenopathy-associated virus. MMWR Morb Mortal Wkly Rep 1986;35(38):595–8603–6.

160. Watson JC, Hadler SC, Dykewicz CA, Reef S, Phillips L. Measles, mumps, and rubella—vaccine use and strategies for elimination of measles, rubella, and congenital rubella syndrome and control of mumps: recommendations of the Advisory Committee on Immunization Practices (ACIP). MMWR Recomm Rep 1998;47(RR-8):1–57.

161. Kelso JM. The gelatin story. J Allergy Clin Immunol 1999;103(2 Pt 1):200–2.

Meningococcal vaccine

General Information

Meningococcal vaccines have been comprehensively reviewed, distinguishing conventional polysaccharide vaccines, non-polysaccharide group B meningococcal vaccines, and conjugated meningococcal vaccines (1).

Reports of clinical trials in children, adults, and asplenic individuals have been provided (2) and surveyed (SEDA-11, 288; SEDA-16, 379; SEDA-17, 370; SED-12, 813).

The Global Advisory Committee on Vaccine Safety (GACVS) has been presented with data relating to the safety of outer membrane-vesicle-based meningococcal B vaccines, based on their use in Cuba, Norway, New Zealand, and parts of France (3). It has noted in particular the very carefully designed safety monitoring programme that was set up in New Zealand to ascertain possible serious adverse events following immunization of around 1 million people aged under 20 years from July 2004 onwards, including 200 000 vaccinees monitored through linkages to hospital admission records. The various surveillance methods used to detect potentially serious adverse events following immunization consistently found no evidence of such effects attributable to immunization.

Polysaccharide vaccines

Polysaccharide meningococcal vaccines in various combinations against meningococcal disease caused by meningococci of groups A, C, W135, and Y have been commercially available for many years. Studies mostly using a bivalent serogroup A + C vaccine carried out in about 15 countries, including some millions of people, have shown efficacy of 61–99%. Meningococcal group A vaccine is more immunogenic than group C vaccine in infants and small children. However, infants below 6 months of age produce a weak response, and meningococcal group C vaccine should not be used before the age of 2 years. Meningococcal group B polysaccharide vaccine is poorly immunogenic in humans and is therefore not available commercially.

Adverse reactions are infrequent and mild, consisting of local soreness or localized erythema at the injection site, and systemic reactions (transient fever, headache, fatigue), lasting 1–2 days (4,5). With the quadrivalent vaccine used in Canada, fever was reported in less than 1%, local reactions in 6.3%, and rash in 1.6% among those aged 11 years or older. Local reactions were also the most reported adverse effect in other reports.

Non-polysaccharide group B meningococcal vaccines

Considering the poor immunogenicity of polysaccharide group B vaccines, different vaccines have been developed—a Norwegian outer membrane complex group B vaccine and a Cuban vaccine in which the group C polysaccharide is added to a mixture of high molecular weight B outer membrane proteins and proteoliposomes. Both vaccines have been used in clinical trials, mainly in Latin America, but more conclusive studies with these two products are awaited. Adverse reactions with the Cuban vaccine have been studied. Among 16 700 vaccinees, mostly older than 4 years, local reactions were observed in 62%, and systemic reactions in 4.3%.

From studies in 370 infants and children and 171 adults aged 18–30 years, who received three doses of outer membrane protein meningococcal vaccine developed in either Cuba or Norway or a control vaccine (Hib vaccine), the vaccines are promising candidates for the control of epidemics caused by homologous epidemic strains (6). However, the vaccines would not confer protection during a heterologous epidemic. All vaccinees had more pain, induration, and erythema at the injection site than did Hib vaccine recipients; there were no serious adverse events, and 95% of individuals who reported symptoms said that the symptoms did not interfere with their normal activities.

A genetically engineered vaccine containing six meningococcal class I (PorA) outer membrane proteins, representing 80% of group B meningococcal strains prevalent in the UK, has been assessed in 103 infants who received the vaccine at ages 2, 3, and 4 months with routine infant immunization, and a fourth dose at 12–18 months (7). The vaccine was well tolerated, and after the fourth dose there were larger bactericidal responses to all six strains.

Conjugated meningococcal vaccines

Success with the protein-conjugate formulations of *Hemophilus influenzae* vaccines has facilitated research and development on conjugated meningococcal vaccines with preference for monovalent group C or bivalent group A/group C vaccines. Conjugated meningococcal vaccines of serogroup C made by various manufacturers were first licensed in 1999 in the UK and then (at the end of 2000) in many other countries, including some other member states of the European Union (Belgium, Germany, Greece, Luxembourg, Ireland, Portugal, Spain).

Field trials in the UK and the Gambia have shown immunogenicity not only in adults but also in toddlers and even infants. Except for local tenderness in 30–75% of vaccinees, no conjugate vaccine evaluated to date has been associated with significant adverse effects (1).

Vaccine containing oligosaccharides derived from group C meningococcal capsular polysaccharide coupled to CRM197, a non-toxic mutant diphtheria toxin (manufactured by Wyeth Lederle Vaccines and Pediatrics, Pearl River, NY) was given to 114 infants at 2, 3, and 4 months of age; 57 received a vaccine containing 2 µg of oligosaccharide and 5 µg of CRM197, and 57 received a vaccine containing 10 µg of oligosaccharide and 25 µg of CRM197 (8). The infants received DTP and Hib vaccine at the same time. In each group, 25 infants received a polysaccharide vaccine booster at median age of 83 weeks. Antibody concentrations required for protection are estimated to be 1–2 µg/ml. All the infants achieved a concentration of 2 µg/ml by age 4 months after two doses of conjugated meningococcal vaccine, and concentrations of bactericidal antibodies were much higher than after meningococcal polysaccharide vaccines. Antibody concentrations fell significantly by age 14 months. After boosting with polysaccharide vaccine all the children achieved antibody concentrations over 2 µg/ml. Local and systemic reactions were rare after meningococcal conjugate vaccines and were significantly less common at the site of DTP/Hib immunization for all three doses. The high-dose cohort had more systemic reactions than the low-dose cohort (Table 1). One infant required hospitalization for a viral illness soon after the second injection.

The results of a randomized, double-blind trial of safety, immunogenicity, and induction of immunological memory in 182 healthy infants has been published (9). The infants received either conjugated meningococcal vaccine (conjugated to CRM197, a non-toxic mutant of diphtheria toxin) of lot 1 (60 infants) or lot 2 (60 infants) or hepatitis B vaccine as a control vaccine. Diphtheria and tetanus toxoids and whole cell pertussis (DTP) vaccine reconstituted with Hib-tetanus conjugate was co-administered in the other leg. Polio vaccine was given orally. According to the UK immunization schedule, these vaccines were given at 2, 3, and 4 months of age. At 12 months the children received either meningococcal A and C polysaccharide vaccine or conjugated meningococcal serogroup C vaccine. The conjugated meningococcal vaccines were generally well tolerated and resulted in less tenderness and induration than the routine vaccines (DTP-Hib and hepatitis B) administered in the opposite leg. There was also no significant difference in systemic reactions between any of the vaccine groups. Parents of

Table 1 Adverse effects of a conjugated meningococcal group C vaccine

| Reaction | Conjugated meningococcal group C vaccine | | DTP/Hib vaccine |
	Low dose (2 µg/ml)	High dose (10 µg/ml)	
Erythema ≥2.5 cm	4/171 (2.3%)	5/173 (3.9%)	38/344 (12%)
Swelling ≥2.5 cm	2/171 (1.2%)	6/173 (3.5%)	64/344 (19%)
Fever ≥38°C for up to 3 days	7/171 (4.1%)	8/173 (4.6%)	No data
Systemic reactions in first 24 hours	26/171 (12%)	53/173 (31%)	No data
Crying for >1 hour in first 24 hours	3/156 (1.9%)	2/153 (1.3%)	No data

Table 2 Adverse effects of a purified group C meningococcal polysaccharide coupled to purified tetanus toxoid

Reaction	Number (n = 30)
Tenderness	15
Erythema	15
Pain	15
Induration	8
Fever ≥38°C	1
Headache	2
Myalgia	1
Arthralgia	0

the children who received meningococcal polysaccharide vaccine as a booster dose reported significantly more local tenderness, general irritability, and change in eating habits than those whose children received conjugate meningococcal vaccine. There was also an increased use of antipyretic drugs in children who received the polysaccharide vaccine. There were no differences in rash, sleepiness, unusual crying, vomiting, or diarrhea. The immunological results suggested that the conjugated vaccines were highly immunogenic and able to induce both a primary response in infants and immunological memory.

A somewhat different conjugated meningococcal vaccine (purified group C meningococcal polysaccharide coupled to purified tetanus toxoid, manufactured by North American Vaccine, Columbia, Maryland, USA) has been evaluated in 30 healthy adults (10). One dose containing 10 micrograms of polysaccharide and 15.5 micrograms of tetanus toxoid was given intramuscularly. All the volunteers developed a four-fold or greater increase in serum bactericidal antibody to group C meningococcus. Most of the local and systemic reactions were mild (Table 2). The only significant local erythema and swelling (over 3 cm) was related to an injection site hematoma, which resolved rapidly.

The results of a phase 1 study in 30 children aged 12–23 months, who received two doses of tetravalent A-C-W135-Y conjugated meningococcal vaccines (conjugated to diphtheria toxoid), have been reported (11). Three different doses were used: 1, 4, and 10 micrograms/ml polysaccharide of each serogroup. The 4 micrograms/ml dose appeared to be immunologically optimal. No children had a temperature of over 39.9°C. Local reactions were reported in 40–60% of children after each injection, with no apparent relation to dose. None of these local reactions was severe. The investigators considered the results as being sufficiently promising for further evaluation of this vaccine in a larger trial in toddlers and infants.

MCV-4 has been studied in three groups of 30 infants at 2, 4, and 6 months of age (12). The first group was given three doses of a quadrivalent polysaccharide meningococcal vaccine (group A, C, Y, W-135) conjugated to diphtheria toxoid (MCV-4) in a dosage of 1 microgram/ml of each serogroup polysaccharide. The second group was given MCV-4 in a dosage of 4 micrograms/ml, and the last group received 10 micrograms/ml. A subset of these children were immunized at 15–18 months with licensed meningococcal polysaccharide vaccine (A, C, Y, W-135). MCV-4 had a reactogenicity profile that was acceptable to parents and health-care providers. It was only modestly immunogenic in infants, but it primed the immune system of most of the infants who were given three doses in infancy. There was no statistically significant immunological advantage in increasing the dosage beyond 4 micrograms/ml, and local reactions were more frequent after 10 micrograms/ml.

Combined meningococcal and typhoid fever vaccine

A combined meningococcal and typhoid vaccine (Mérieux) was evaluated in 158 volunteers in a single-blind study (13). Comparing the effects with those in vaccinees receiving monocomponent vaccines or the combination vaccine, there was no significant difference in the reported frequency or duration of local and systemic reactions. However, vaccinees who received the monocomponent typhoid fever vaccine alone were less likely to complain of swelling or pain at the injection site.

Organs and Systems

Nervous system

During an immunization campaign with a polysaccharide vaccine in 130 000 children in Auckland, New Zealand, there were 92 reports of apparent peripheral nerve involvement, including 80 reports of unexplained weakness and 57 reports of paresthesia or dysesthesia (13).

Acute disseminated encephalomyelitis has been reported 4 weeks after immunization with a polysaccharide meningococcal A + C vaccine in a 23-year-old woman; she recovered with glucocorticoid treatment (15).

In Norway during the 1970s there was extensive use of the vaccine in trials among teenagers and young adults. In some age groups, up to 40% of the population had been immunized. Despite media reports of a possible increased risk of myalgic encephalomyelitis, also called chronic fatigue syndrome, the results from these trials provided no specific causes for concern with respect to serious adverse events following immunization.

During an efficacy trial using the Norwegian meningococcal group B vaccine carried out in Norwegian secondary schools from 1988 to 1991, seven of the reported events in vaccinees were classified as serious, among them one case was of acute transverse myelitis which was considered to be most probably a vaccine reaction. The remaining cases could have been triggered by the vaccine without being directly caused by the vaccine (16).

Data on adverse events reported to a passive provincial surveillance system have been evaluated after the mass immunization with a polysaccharide meningococcal vaccine of 1 198 751 people aged 6 months to 20 years in Quebec. A total of 118 reports of severe adverse events were selected. The most frequent were allergic reactions (9.2 per 100 000 doses), followed by few neurological reactions (0.5 per 100 000 doses) and very few anaphylactic reactions (0.1 per 100 000 doses). There were no reports of long-lasting sequelae or of encephalopathy, encephalitis, or meningitis (17).

As of 4 October 2005, the Vaccine Adverse Event Reporting System (VAERS) had received five reports of Guillain–Barré syndrome after the use of MCV4 vaccine (18). All occurred in people aged 17–18 years who were immunized with MCV4 during 10 June–25 July 2005 and had onset of symptoms 14–31 days after immunization. One patient reported another acute illness before the onset of the symptoms. The five patients received vaccine from four different lots. About 2.5 million doses of MCV4 had been distributed nationally since March 2005 (Sanofi–Pasteur, unpublished data, 2005). Data (1989–2001) from the Vaccine Safety Datalink in those aged 11–19 years show a background annual incidence of 1–2 cases per 100 000 person-years. This suggests that the rate of Guillain–Barré syndrome based on the number of cases reported within 6 weeks of administration of MCV4 is similar to what might have been expected to occur by chance alone. However, the timing of the onset of neurological symptoms (i.e. within 2–5 weeks of immunization) is of concern. Based on limited experience, it is not yet known whether these cases were caused by the vaccine or were coincidental.

Menactra®, a quadrivalent conjugated meningococcal vaccine (serogroups A,C,W$_{135}$,Y), is licensed in the USA and recommended for routine immunization at age 11–12 years (SEDA-29, 331). To date, more than 12 million doses have been delivered. As of 30 April 2007, a total of 19 cases of Guillain–Barré syndrome, occurring within 6 weeks of immunization, had been reported to the US Vaccine Adverse Event Reporting System (VAERS) (19). Analysis of the data could not exclude a slightly increased risk of Guillain–Barré syndrome after immunization, but this finding should be viewed with caution, given the limitations of the reporting system and the uncertainty of the background rate of Guillain–Barré syndrome and its potential for seasonal fluctuation. The manufacturer of the vaccine is planning further studies to evaluate the possible risk of Guillain–Barré syndrome after immunization with Menactra®.

Hematologic

The Norwegian-type vaccine had also been used in France in a three-dose schedule to immunize around 2700 children aged 12 months to 5 years in 1 administrative region (Department of Seine-Maritime), following an increased incidence of meningococcal disease in that department. Parents of vaccinees were sent a questionnaire to ascertain possible adverse events. A high proportion of these were returned, and nine serious adverse events were identified, eight of which were purpura (one idiopathic thrombocytopenic purpura, three Henoch–Schönlein purpura, four febrile purpura) and seven of which occurred after the second dose. All of these seven had received a third dose of vaccine with no reported ill effect. The significance of the cases of purpura following immunization was difficult to evaluate, because of the absence of data on background rates of purpura in the general population.

Urinary tract

The possible risk of relapse of nephrotic syndrome after administration of meningococcal serogroup C conjugate vaccine to children has been discussed (20). In 106 children with nephrotic syndrome, there were 63 relapses during the 12-month period before immunization and 96 relapses during the equivalent 12-month period after meningococcal immunization. The increase was statistically significant and the risk was especially higher in the first 6 months after immunization. Regarding the possible pathogenesis of the increased rate of relapses the authors discussed the hypothesis that disturbance of cytokines by the conjugated vaccine might mediate the onset of proteinuria. This is the first report of possible serious adverse effects of conjugated meningococcal immunization; further studies are needed before a real risk can be confirmed.

Musculoskeletal

A polyarthropathy has been reported after the administration of a conjugated meningococcal serogroup C vaccine (21).

- A 17-year-old boy developed a widespread rash on his back and the left side of his chest and painful swelling of his left elbow, right knee, and left ankle 4 days after receiving meningococcal serogroup C vaccine. The detection of meningococcal DNA in fluid from his knee made it probable that the disease was due to natural meningococcal infection, as the meningococcal vaccine does not contain nucleic acid. As the serogroup C vaccine protects only against group C infections, natural meningococcal infection of other serogroups, particularly the commonest serogroup B, can still occur.

Drug Administration

Drug formulations

Meningococcal diseases that occur worldwide are mainly caused by serogroups A, B, C, W$_{135}$, and Y. Bivalent and tetravalent polysaccharide vaccines against meningococcal diseases caused by *Neisseria meningitidis* serogroups A, C, W$_{135}$, and Y have been commercially available for several years, and have been considered to be safe and effective. However, the vaccines suffer from drawbacks common to polysaccharide vaccines. These drawbacks have been overcome by conjugation of the polysaccharide immunogen to a protein. In addition to conjugated Hib and pneumococcal vaccines, conjugated meningococcal vaccines of serogroup C have been developed, licensed, and very effectively introduced in some national immunization programs. Other conjugated meningococcal vaccines, such as bivalent A/C vaccines and tetravalent A/C/W$_{135}$/Y vaccines, are currently being developed or are undergoing clinical trials.

The development of a meningococcal serogroup B vaccine (either as polysaccharide or conjugated vaccine) is difficult and will take several more years.

The safety and immunogenicity of a tetravalent (A/C/W$_{135}$/Y) diphtheria toxoid conjugate meningococcal vaccine in 89 healthy adults, 18–55 years old have been evaluated (22). A single dose containing 1, 4, or 10 micrograms of each serogroup was acceptably tolerated and immunogenic. One volunteer reported a severe local and systemic reaction after receiving 4 micrograms. The proportion of vaccinees with local reactions (mainly swelling, erythema) and/or systemic reactions (mainly headache and malaise) increased with increasing dose. The symptoms were most often rated as mild and resolved by the third day after immunization.

Simultaneous administration of meningococcal C conjugate vaccine and other routine childhood vaccines

In a randomized, double-blind, controlled clinical trial 351 healthy 2-month-old infants were randomized to either meningococcal C conjugate vaccine or (as a control) hepatitis B vaccine (23). Simultaneously all infants received a pentavalent combination DTaP-Hib-IPV vaccine. DTaP-Hib-IPV vaccine was given at 2, 4, 6, and 15–18 months; meningitis C and hepatitis B vaccines were given at 2, 4, and 6 months or at 2, 4, and 15–18 months. The administration of meningitis C vaccine at the same time as the first four doses of DTaP-Hib-IPV vaccine did not adversely affect the antibody response to the antigens contained in either vaccine. All the vaccines were well tolerated. Local reactions at the injection site were significantly more commonly reported by recipients of meningitis C than recipients of hepatitis B. Injection site reactions in the limb injected with the pentavalent combination vaccine occurred at similar rates in those who were given meningitis C and hepatitis B, and generally at higher rates than in the limbs that had been injected with meningitis C and hepatitis B. Systemic reactions were also reported at similar rates; persistent crying was reported more often with the booster dose of meningitis C (in 3% of infants). Based on these results the Canadian National Advisory Committee on Immunization recommended that all Canadian infants be immunized simultaneously with meningitis C vaccine at 2, 4, and 6 months of age, at the same time that they receive DTaP-Hib-IPV.

Cardiorespiratory complications (apnea, bradycardia) after DTwP immunization of pre-term infants are well recognized. Similar events have been sought in 76 pre-term infants after simultaneous administration of conjugated meningitis C and DTaP-Hib vaccines (24,25). The infants (mean gestational age 27 weeks mean birth weight 990 g) were immunized at a median age of 65 days and monitored for cardiorespiratory events for 24 hours before and after immunization. There was no significant increase in the number of events and serious events were not seen.

References

1. Peltola H. Meningococcal vaccines. Current status and future possibilities. Drugs 1998;55(3):347–66.
2. Peltola H, Safary A, Kayhty H, Karanko V, Andre FE. Evaluation of two tetravalent (ACYW135) meningococcal vaccines in infants and small children: a clinical study comparing immunogenicity of O-acetyl-negative and O-acetyl-positive group C polysaccharides. Pediatrics 1985;76(1):91–6.
3. Safety of meningococcal B vaccines. Wkly Epidemiol Rec 2008;83:42.
4. Committee on Immunization. In: Guide for adult immunization. Philadelphia: American College of Physicians, 1985:63.
5. Roberts JS, Bryett KA. Incidence of reactions to meningococcal A&C vaccine among U.K. schoolchildren Public Health 1988;102(5):471–6.
6. Tappero JW, Lagos R, Ballesteros AM, Plikaytis B, Williams D, Dykes J, Gheesling LL, Carlone GM, Hoiby EA, Holst J, Nokleby H, Rosenqvist E, Sierra G, Campa C, Sotolongo F, Vega J, Garcia J, Herrera P, Poolman JT, Perkins BA. Immunogenicity of 2 serogroup B outer-membrane protein meningococcal vaccines: a randomized controlled trial in Chile. JAMA 1999;281(16):1520–7.
7. Cartwright K, Morris R, Rumke H, Fox A, Borrow R, Begg N, Richmond P, Poolman J. Immunogenicity and reactogenicity in UK infants of a novel meningococcal vesicle vaccine containing multiple class 1 (PorA) outer membrane proteins. Vaccine 1999;17(20–21):2612–9.
8. Richmond P, Borrow R, Miller E, Clark S, Sadler F, Fox A, Begg N, Morris R, Cartwright K. Meningococcal serogroup C conjugate vaccine is immunogenic in infancy and primes for memory. J Infect Dis 1999;179(6):1569–72.
9. MacLennan JM, Shackley F, Heath PT, Deeks JJ, Flamank C, Herbert M, Griffiths H, Hatzmann E, Goilav C, Moxon ER. Safety, immunogenicity, and induction of immunologic memory by a serogroup C meningococcal conjugate vaccine in infants: a randomized controlled trial. JAMA 2000;283(21):2795–801.
10. Richmond P, Goldblatt D, Fusco PC, Fusco JD, Heron I, Clark S, Borrow R, Michon F. Safety and immunogenicity of a new Neisseria meningitidis serogroup C-tetanus toxoid conjugate vaccine in healthy adults. Vaccine 1999;18(7–8):641–6.
11. Rennels M, King J Jr, Ryall R, Manoff S, Papa T, Weddle A, Froeschle J. Dose escalation, safety and immunogenicity study of a tetravalent meninogococcal polysaccharide diphtheria conjugate vaccine in toddlers. Pediatr Infect Dis J 2002;21(10):978–9.
12. Rennels M, King Jr. J, Ryall R, Papa T, Froeschle J. Dosage escalation, safety and immunogenicity study of four dosages of a tetravalent meninogococcal polysaccharide diphtheria toxoid conjugate vaccine in infants. Ped Infect Dis J 2004;23:429–35.
13. Khoo SH, St Clair Roberts J, Mandal BK. Safety and efficacy of combined meningococcal and typhoid vaccine. BMJ 1995;310(6984):908–9.
14. Hood DA, Edwards IR. Meningococcal vaccine—do some children experience side effects? NZ Med J 1989;102(862):65–7.
15. Py MO, Andre C. Encefalomielite disseminada aguda e vacinacao antimeningococica A e C. Relato de caso. [Acute disseminated encephalomyelitis and meningococcal A and C vaccine: case report.] Arq Neuropsiquiatr 1997;55(3B):632–5.
16. Halvorsen S. The meningococcal serogroup B vaccine protection trial in Norway 1988–1991: trial surveillance by an independent group. NIPH Ann 1991;14(2):135–7.

17. Yergeau A, Alain L, Pless R, Robert Y. Adverse events temporally associated with meningococcal vaccines. CMAJ 1996;154(4):503–7.

18. Guillain–Barré syndrome among recipients of Menactra® meningococcal conjugate vaccine—United States, June–July 2005. Morb Mortal Weekly Rep 2005;54:1–3.

19. Global Advisory Committee on Vaccine Safety. Meeting on 12–13 June 2007. Menactra and GBS. Wkly Epidemiol Rec 2007;82:256.

20. Abeyagunawardena AS, Goldblatt D, Andrews N, Trompeter RS. Risk of relapse after meningococcal C conjugate vaccine in nephrotic syndrome. Lancet 2003;362(9382):449–50.

21. Suresh E, Cox R, Morris I. A teenager with rash and joint swelling after meningococcal C conjugate vaccine. Lancet 2000;356(9240):1486.

22. Campbell JD, Edelman R, King Jr. JC, Papa T, Ryall R, Rennels MB. Safety, reactogenicity, and immunogenicity of a tetravalent meningococcal polysaccharide-diphtheria toxoid conjugate vaccine given to healthy adults. J Infect Dis 2002;186:1848–51.

23. Halperin SA, Mcdonald J, Samson L, Danzig L, Santos G, Izu A, Smith B, MacDonald N. Simultaneous administration of meningococcal C conjugate vaccine and diphtheria?-tetanus?acellular pertussis?inactivated poliovirus? *Haemophlus influenzae type b conjugate vaccine in children: a randomized double-blind study.* Clin Invest Med 2002;25:243–51.

24. Slack MH, Schapira C, Thwaites RJ, Andrews N, Schapira D. Acellular pertussis and meningococcal C vaccines: cardiorespiratory events in pre-term infants. Eur J Pediatr 2003;162:436–7.

25. Jackson LA, Carste BA, Malais D, Froeschle J. Retrospective population-based assessment of medically-attended injection site reactions, seizures, allergic responses and febrile episodes after acellular pertussis vaccine combined with diphtheria and tetanus toxoids. Pediatr Infect Dis J 2002;21:781–6.

Parvovirus vaccine

Parvovirus B19 usually causes a mild and self-limiting illness. However, it can cause a variety of more serious conditions in immune-deficient individuals and hydrops fetalis or abortion in pregnant women. A recombinant human parvovirus B19 vaccine has been developed and evaluated in a randomized, double-blind phase 1 trial In 24 parvovirus B19-seronegative adults, who received 2.5 or 25 micrograms at 0, 1, and 6 months. All developed neutralizing antibodies. Mild or moderate injection site pain was reported by 75–92%. Systemic reactions such as fever, headache, gastrointestinal symptoms, and fatigue occurred in 25–50% of volunteers. One volunteer reported severe fatigue for 6 days after the first immunization. The rate and intensity of adverse effects did not increase with vaccine dose or number of immunizations (1).

Reference

1. Ballou WR, Reed JL, Young NS, Koenig S. Safety and immunogenicity of a recombinant parvovirus B19 vaccine formulated with MF59. J Infect Dis 2003;187:675-8.

Pertussis vaccines

Note on abbreviations

DTP = Diphtheria + tetanus toxoids + pertussis
DTaP = Diphtheria + tetanus toxoids + acellular pertussis
DTwP = Diphtheria + tetanus toxoids + whole cell pertussis

General Information

Whole-cell and acellular pertussis vaccines have been reviewed, with emphasis on the protectivity of the various virulence factors and antigens (1). The authors summarized their review as follows: although *Bordetella pertussis* has at least five proteins required for virulence and an additional two "toxic" components, only serum neutralizing antibodies to pertussis toxin have been shown to confer immunity to pertussis.

Acellular pertussis vaccine

Quality control of acellular pertussis vaccines presents particular problems related to the various methods used for preparation of the active components, the different compositions of the final formulations, and different amounts of antigen. Researchers in the National Institute for Biological Standards and Control in the UK have presented a strategy capable of addressing the key problem areas likely to be encountered with all existing types of acellular pertussis vaccines and combinations (2). Their proposal could be considered as a starting point for improvement of quality control programs for these vaccines.

An overview of clinical trials with a special diphtheria and tetanus toxoids and acellular pertussis (DTaP) vaccine has been published (3). The vaccine contains as pertussis components purified filamentous hemagglutinin, pertactin, and genetically engineered pertussis toxin. The vaccine induces high and long-lasting immunity and is at least as efficacious as most whole-cell pertussis vaccines and similar in efficacy to the most efficacious acellular pertussis vaccines that contain three pertussis antigens. The vaccine is better tolerated than whole cell vaccines and has a similar reactogenicity profile to other acellular vaccines.

A vaccine containing diphtheria and tetanus toxoids and acellular pertussis with reduced antigen content for diphtheria and pertussis (TdaP) has been compared with a licensed reduced adult-type diphtheria–tetanus (Td) vaccine and with an experimental candidate monovalent acellular pertussis vaccine with reduced antigen content (ap) (4). A total of 299 healthy adults (mean age 30 years) were randomized into three groups to receive one dose of the study vaccines. The antibody responses (anti-diphtheria, anti-tetanus, anti-pertussis toxin, anti-pertactin, anti-filamentous hemagglutinin) were similar in all groups. The most frequently reported local symptom was pain at the injection site (62–94%), but there were no reports of severe pain; redness and swelling with a diameter of 5 cm or more occurred in up to 13%. The incidence of local symptoms was similar after TdaP and Td immunization. The most frequently reported general symptoms were headache and fatigue (20–50%). The incidence of general symptoms was similar in the TdaP and Td groups. There were no reports of fever over 39°C. No serious adverse events were reported.

A retrospective assessment among the population of the Group Health Cooperative from 1997 to 2000 confirmed the safety of DTaP vaccine (5). Administrative databases were used to identify medical visits linked with diagnostic codes indicative of injection site reactions, seizures, allergic responses, and febrile episodes after DTaP immunization. During the study 76 133 doses of DTaP were administered, mainly as the fourth or fifth immunization doses. There were 26 injection site reactions; four involved the entire upper arm but were self-limiting. Febrile seizures occurred in 1 per 19 496 immunizations. There was no evidence of allergic responses.

The immune system is partly immature at birth, and therefore immunization schedules for the majority of childhood vaccines including pertussis vaccine do not recommend immunization before 2 months of age. However, the pertussis case fatality rate is highest in infants below 6 months of age. For this reason 45 infants were immunized at birth and at 3, 5, and 11 months of age (group 1) and compared with 46 infants who were immunized at 3, 5, and 11 months of age (group 2) with a trivalent acellular pertussis vaccine (genetically detoxified pertussis toxin, filamentous hemagglutinin, pertactine) (6). There were no adverse effects in children in the two groups. After the second dose of vaccine (administered at 3 months) the antibody titers of infants in group 1 were already similar to the titers achieved after the second dose (administered at 5 months) in infants in group 2. The authors concluded that immunization at birth may lead to earlier prevention of pertussis in infants under 6 months of age.

Japanese studies of acellular vaccines
The first generation of the new acellular vaccines was developed in Japan in the late 1970s (Sato). Since late 1981, acellular vaccines have replaced the whole-cell pertussis vaccines for use in the Japanese immunization program. Two types of acellular pertussis vaccine have been produced by six Japanese manufacturers (7,8) (SEDA-12, 276; SEDA-13, 283; SEDA-14, 279). No doubt as a result of these measures the rate of reported serious reactions has decreased in Japan. During the period 1975–81, when whole-cell vaccines were given, the rate was 0.4 per million doses, compared with a rate of 0.25 per million doses during the period 1982–84 (9). In 1988, Kumura and Kuno-Sakei (10) summarized the experience gained in Japan since the introduction in 1981 of the new DTP vaccines containing acellular pertussis components. Acellular vaccines seem to be effective in Japan, since in parallel with their wider use the numbers of reported cases of pertussis and pertussis deaths have declined.

Reactogenicity (as expressed by fever or local reactions) was very low. Kimura and Kuno-Sakei cited reports of Quincke's edema-like swelling of the whole arm following the third injection (0.17% of vaccinees) and the booster injection (2.61%), respectively. Data on more severe adverse events have been collected from 1970 to 1986 in the framework of the National Adverse Reaction Compensation System (SED-12, 818). Kimura and Kuno-Sakei considered the vaccines to be safe and effective enough to eliminate pertussis in Japan in the future (10).

European studies of acellular vaccines
The epidemiological circumstances in Sweden and in parts of Germany and Italy, where the immunization programs against pertussis using whole-cell vaccine had been discontinued or reduced because of public concern about rare severe adverse events, offered good opportunities to assess the clinical efficacy and safety of acellular pertussis vaccines. In Germany, both controlled field trials and a household contact study were carried out. Different mono-component and multi-component acellular vaccines from European and US manufacturers were used in European trials. They comprised a mono-component vaccine composed only of detoxified pertussis toxin (toxoid); a two-component vaccine composed of pertussis toxin and filamentous hemagglutinin; a three-component vaccine composed of pertussis toxin, filamentous hemagglutinin, Pertactin; and another five-component vaccine composed of pertussis toxin, filamentous hemagglutinin, Pertactin, and fimbriae antigens 2 and 3. In some instances they were compared directly with whole-cell pertussis vaccines. In general, the acellular vaccines were found to be immunogenic, epidemiologically effective, and less reactogenic than the whole-cell pertussis-component vaccines as assessed in terms of fever, pain, fretfulness, and local reactions at the injection site. Based on the results, acellular pertussis vaccines have since been licensed in many countries of the world both for primary and booster immunization. Details regarding the results and conclusions of the large-scale field trials completed in 1994 and 1995 in Germany, Italy, and Sweden have been reviewed (SEDA-19, 298), as have reports on other clinical trials using acellular pertussis vaccines (SED-12, 818; SEDA-12, 277; SEDA-13, 283; SEDA-14, 279; SEDA-15, 350; SEDA-16, 382; SEDA-17, 370; SEDA-18, 332; SEDA-20, 289; SEDA-21, 230). The optimal composition of acellular vaccines has not yet been determined: the number of antigens still varies from one to five, and antigen amounts are also different. Post-licensing studies will therefore be of the utmost importance in studying the induction of herd immunity and possible rare events.

Local effects of acellular vaccines: effect of number of injections
Data on the use of a single DTaP vaccine for four-dose or five-dose series are limited, but the available data show a substantial increase in the frequency and magnitude of local reactions with successive doses. The accompanying tables show reactions after the fourth dose (Table 1) and

Table 1 Reactions after the fourth dose of DTaP given at 2 or 3 years of age

Local and systemic reactions	ACEL-IMUNE	Tripedia	Infanrix*	Certiva
Pain		19%	26%	19
Erythema	10% ≥2.4 cm	30% ≥2.54 cm	14% ≥2 cm	6% ≥3 cm
Swelling		29% ≥2.54 cm	11% ≥2 cm	5% ≥3 cm
Induration	9% ≥2.4 cm			
Tenderness				
Fever = 38°C	26%		26%	6.3%
Fever = 38.3°C		5.5%		

*To be compared with reactions after the first dose: pain 2%, erythema 0%, swelling 0%, fever ≥38°C 6.3%

Table 2 Reactions after the fifth dose of DTaP administered at entry to school

Local and systemic reactions	ACEL-IMUNE	Tripedia	Infanrix
Pain		2.1% severe pain	1.6% severe pain
Erythema	20% >2–2.4 cm	31% >5 cm	
Swelling		25% >5 cm	30% >5 cm
Induration	14%		21% >5 cm
Tenderness	38%		

fifth dose (Table 2). The original data and references are included in the supplementary recommendations of the Advisory Committee on Immunization Practices (ACIP) on the use of DTaP vaccines in a five-dose series (11). Reports from Alberta and British Columbia provinces, Canada, have suggested that the incidence rates of severe local adverse reactions may increase with each dose (third, fourth, fifth) in preschool children (12).

Limb swelling after booster doses of acellular vaccines
Swelling involving the entire thigh or upper arm has been reported after booster doses of different acellular pertussis vaccines, for example in a German study during April 1993 to November 1994 using a fourth dose of Infanrix. There was an increase in thigh circumference in 1.2–3.2% of children; swelling began within 48 hours of a booster dose and the mean duration was 3.9 (range 1–7) days; the mean increase in circumference was 2.2–5.0 cm; in a few children, the swelling interfered with walking.

Similar swelling and substantial local reactions have been observed in fourth-dose and fifth-dose follow-up studies from the Multicenter Acellular Pertussis Trial, which examined 12 different DTaP vaccines, and in recent studies of the fifth dose of Tripedia and Infanrix in Germany. The pathogenesis of both substantial local reactions and limb swelling is unknown. Associations with pertussis toxoid, diphtheria toxoid, or aluminium in the vaccine have been discussed. Because reports to date have suggested that the reactions are self-limited and resolve without sequelae, and in recognition of the benefits of a fifth dose of DTaP, the ACIP has recommended that a history of extensive swelling after the fourth dose should not be considered to be a contraindication to a fifth dose of DTaP. Parents or care-givers of children who

receive fourth and fifth doses of a DTaP series should be informed of the increases in reactogenicity that have been observed (11).

Whole-cell pertussis vaccine

Pertussis whole cell vaccine is an adsorbed suspension of inactivated pertussis bacteria. The vaccine is available in monovalent form or in combination with diphtheria and tetanus toxoids (DTP). DTP vaccine is the preparation of choice in routine immunization practice.

Local reactions are common after DTP immunization (40–70% of the vaccinees) but are usually self-limiting (12). A nodule may be palpable at the injection site of adsorbed products for several weeks. Abscess at the injection site has been reported (6–10 per million vaccinees). Mild to moderate fever (38.0–40.4°C) occurs frequently (about 50% of vaccinated infants), generally within several hours of administration, persisting for 1–2 days. Fever and other systemic symptoms are much less common following immunization with preparations not containing the pertussis component. Arthus-type hypersensitivity reactions occur, particularly after booster doses. Rarely, severe systemic reactions (urticaria, anaphylaxis) have been reported.

The non-neurological adverse effects of whole-cell immunization have been reviewed (SED-12, 815; SEDA-16, 379; SEDA-17, 370; SEDA-18, 330). The most frequent are mild fever, drowsiness, and reduction in appetite; a very small percentage of vaccinees experience more severe fever, redness, swelling, pain, "fussiness", or vomiting.

A follow-up study has been carried out in 105 children with collapse (a hypotonic-hyporesponsive episode or a shock-like syndrome) after their first immunization with

DTwP + IPV vaccine (13). Information about subsequent immunizations, health, and development in 101 of the children was supplied by child health-care units. The parents of one child refused further immunization, 16 children completed their schedule with the combination diphtheria + tetanus + poliomyelitis vaccine (DT-IPV), and the other 84 children received further pertussis vaccine (DTP-IPV), totalling 236 doses; 74 children received the complete series of three additional doses. None of the children had recurrent collapse, and other adverse events were only minor. About half were given paracetamol prophylactically for the first subsequent dose; most of them did not take it for further doses. The authors suggested that it is unnecessary to withhold further doses of pertussis vaccine in a child with collapse after a previous dose. It has been suggested that the threat of natural pertussis in non-immunized children should be taken much more into account than the fear of developing a collapse reaction (14). In another study (15) in the USA, one of the 14 children not completely immunized because of a hypotonic-hyporesponsive episode after a previous dose later developed natural pertussis, which lasted for 3 months and was transmitted to both her parents.

Acellular pertussis vaccine versus whole-cell pertussis vaccine

In a randomized, double-blind trial to determine the efficacy of vaccination against *Bordetella* infection, a multicomponent acellular pertussis has been compared with a whole-cell product and diphtheria + tetanus toxoids (DT) in 8532 infants aged 2–4 months, who received four doses of either DTwP or DTaP vaccine at 3, 4.5, 6, and 15–18 months of age, and 1739 controls, who received three doses of DT vaccine at 3, 4.5, 15–18 months of age (16). All the vaccines were generally well tolerated. However, adverse reactions were significantly less common after DTaP compared with DTwP vaccine. Persistent inconsolable crying was four times more common in DTwP recipients than in DTaP recipients. High fever, 40.5°C or over, was three times more common in DTwP vaccinees than in DTaP vaccinees. Only one DTaP recipient had a convulsion in temporal relation to immunization.

The 2000 Childhood Immunization Schedule, proposed by the Advisory Committee of Immunization Practices (ACIP), the American Academy of Pediatrics, and the American Academy of Family Physicians, recommended that acellular pertussis vaccines be exclusively chosen for routine use in the USA (17).

One evaluation of the results of two studies in a total of 182 children primed either with acellular or with whole-cell pertussis vaccines at 2, 4, and 6 months of age and boosted with an acellular vaccine has shown that booster doses of acellular vaccine are safe and immunogenic (18). Local adverse reactions after booster immunization with acellular vaccine were more common in children primed with acellular vaccine than in those primed with whole-cell pertussis vaccine (68 versus 33%). In another and similar study, children primed with acellular or whole

cell pertussis combined with DT vaccine have been boosted with a recombinant acellular pertussis vaccine combined with DT vaccine. The vaccine was highly immunogenic and safe (19).

From 1991 to 1993 about 27 million doses of DTP vaccine and 5 million doses of DTaP (that is acellular) vaccine were distributed in the USA. The results of a postmarketing comparison of the safety of acellular pertussis vaccines with whole-cell pertussis vaccines have been published. The rates of reported adverse events per 100 000 immunizations were significantly lower after the administration of DTaP vaccine than after DTP vaccine for the following outcomes: all reports (2.9 versus 9.8); fever (1.9 versus 7.5); seizures (0.5 versus 1.7); and hospitalizations (0.2 versus 0.9) (20).

The safety and immunogenicity of 12 acellular pertussis vaccines and one whole-cell pertussis vaccine given as a fourth dose have been compared in 1293 children aged 15–20 months. In general, DTaP vaccines were associated with fewer adverse events than a US-licensed DTwP vaccine (SEDA-22, 342).

A randomized, controlled comparison of two-component, three-component, and five-component acellular pertussis vaccines and a whole-cell pertussis vaccine has been carried out in 82 892 infants who were immunized either at age 3, 5, and 12 months, or at 2, 4, and 6 months. High fever and seizures occurred more often after whole-cell vaccine than after any of the acellular vaccines. Hypotonic-hyporesponsive episodes also occurred significantly more often in the whole-cell group and were more frequent in the acellular groups than previously reported (SEDA-22, 342).

An informal consultation of invited epidemiologists, infectious disease clinicians, immunologists, representatives of regulatory agencies for biological products, and other scientists and public health officials was held in Geneva on 18–19 May 1998 on the control of pertussis using whole-cell and acellular pertussis vaccines. The consultation was called jointly by the Children's Vaccine Initiative and the World Health Organization Global Program on Vaccines. The report of the meeting including the conclusions at full length has been summarized (SEDA-22, 341). The main conclusions read as follows. Whole-cell vaccines of documented quality have proved to be highly effective tools for preventing pertussis. Acellular pertussis vaccines are valuable alternatives to whole-cell vaccines for immunization in infancy. Because of their safety profiles, acellular vaccines may be preferred alternatives in industrialized countries, in which pertussis vaccination with whole-cell vaccines is not widely accepted. In each country, recommendations for the use of pertussis vaccines will be based on local risk–benefit and cost–benefit analyses. The use of acellular vaccines in many circumstances can be further considered for booster doses (fourth and fifth doses) in improving pertussis control after evaluation by local authorities of the epidemiological, cost, and programmatic issues. Additional data on the potential benefit of booster doses using acellular pertussis vaccines in adolescence or older ages are needed. The current control regulations and recommendations for pertussis vaccines (whole-cell and acellular) needs thorough review in the light of recently available scientific data, and appropriate revisions should be made (21).

Comparisons of DTP and DT

An older study in the USA compared the rates of both minor and more serious reactions in 15 752 children 0–6 years of age to DTP vaccine and in 784 children to DT vaccine. The frequencies of minor reactions associated with DTP and DT are shown in Table 3. The reactions after DT were not only less frequent, but also less severe. Convulsions and hypotonic-hyporesponsive episodes each occurred in one per 1750 immunizations (22).

Data from the Monitoring System for Adverse Events Following Immunization (MSAEFI) (SEDA-13, 274) show that the adverse effects rate for DTP vaccine was twice as high as that for DT vaccine. The most reported adverse effect was fever (59% of all reports) followed by local reactions (36%) (23). Similar results have been published by Dittmann (13.5 severe adverse effects per million vaccinees after DTP immunization and 4.8 reactions per million vaccinees after DT immunization) (24) and by Miller and colleagues (25).

Tetravalent, pentavalent, and hexavalent immunization

DTaP or DTwP vaccine can be combined with other antigens, such as *Haemophilus influenzae* type b (Hib), inactivated poliovirus (IPV), and hepatitis B vaccine. In children DTaP or DTwP vaccines form the basis for such combinations, while in adults it is mostly Td vaccine. Current safety concerns regarding combination vaccines have been defined and reviewed (26). The author concluded that there is no evidence that adding vaccines to combination products increases the burden on the immune system, which can respond to many millions of antigens. Combining antigens usually does not increase adverse effects, but it can lead to an overall reduction in adverse events. Before licensure, combination vaccines undergo extensive testing to assure that the new products are safe and effective.

Table 3 Frequencies of minor reactions associated with DTP and DT

Adverse effect	DTP (%)	DT (%)
Fitfulness	53	23
Pain	51	9.9
Fever	47	9.3
Local swelling	41	7.6
Local redness	37	7.6
Drowsiness	32	15
Anorexia	21	7.0
Vomiting	8.2	2.6
Persistent crying	3.1	0.7
High pitched unusual cry	0.1	0

The frequency, severity, and types of adverse reactions after DTP-Hib immunization in very pre-term babies have been studied (27). Adverse reactions were noted in 17 of 45 babies: nine had major events (apnea, bradycardia, or desaturation) and eight had minor reactions (increased oxygen requirements, temperature instability, poor handling, and feeding intolerance). Babies who had major adverse reactions were significantly younger at the time of immunization than the babies who did not have major reactions. Of 27 babies immunized at 70 days or less, nine developed major reactions compared with none of those who were immunized at over 70 days.

The Hexavalent Study Group has compared the immunogenicity and safety of a new liquid hexavalent vaccine against diphtheria, tetanus, pertussis, poliomyelitis, hepatitis B, and *Haemophilus influenzae* type b (DTP + IPV + HB + Hib vaccine, manufactured by Aventis Pasteur MSD, Lyon, France) with two reference vaccines, the pentavalent DTP + IPV + Hib vaccine and the monovalent hepatitis B vaccine, administrated separately at the same visit (28). Infants were randomized to receive either the hexavalent vaccine (*n* = 423) or (administered at different local sites) the pentavalent and the HB vaccine (*n* = 425) at 2, 4, and 6 months of age. The hexavalent vaccine was well tolerated (for details, see Table 4 and Table 5). At least one local reaction was reported in 20% of injections with hexavalent vaccine compared with 16% after the receipt of pentavalent vaccine or 3.8% after the receipt of hepatitis B vaccine. These reactions were generally mild and transient. At least one systemic reaction was reported in 46% of injections with hexavalent vaccine, whereas the respective rate for the recipients of pentavalent and HB vaccine was 42%. No vaccine-related serious adverse event occurred during the study. The hexavalent vaccine provided immune responses adequate for protection against the six diseases.

Organs and Systems

Nervous system

The question of nervous system complications and persistent brain damage after whole-cell pertussis vaccination has been argued for a long time. During the 1980s it was debated by physicians as well as by the general public in several countries. Reports, publications, and national evaluations of the matter led some countries to change their national pertussis immunization policy. In the autumn of 1981, Japan replaced the whole-cell vaccine with acellular pertussis vaccine developed by Sato in Japan. Sweden ended its pertussis (whole-cell vaccine) immunization program in 1979. As a result of the prominence accorded to the risks involved, the pertussis vaccine coverage rates in the UK fell sharply at the beginning of the 1980s, but were increasing again when some large pertussis epidemics occurred as a result of diminished population immunity. In the 1980s, in the erstwhile Federal Republic of Germany, pertussis immunization was only recommended for children at special risk but from 1991

Table 4   Percentage rates of local adverse events within 72 hours of immunization in infants given a hexavalent vaccine (Hexavac) or separate injections of the reference vaccines

Events	Hexavac				Pentavac				Hepatitis B Vax II			
	First dose	Second dose	Third dose	All[a]	First dose	Second dose	Third dose	All[a]	First dose	Second dose	Third dose	All[a]
Number of injections	423	420	418	1261	424	418	417	1259	424	418	417	1259
Any local reaction	23[b]	18	20	20	14[b]	16	18	16	3.3	2.6	5.5	3.8
Skin redness≥2 cm	10	12	14	12	3.5	7.9	9.1	6.8	0.7	1.0	2.9	1.5
Skin redness≤5 cm	2.8	1.2	1.0	1.6	0.5	0.7	1.0	0.7	0	0	0	0
Skin induration ≥2 cm	15	14	14	15	11	14	16	14	2.6	2.4	4.8	3.3
Skin induration ≥5 cm	2.6	1.5	0.7	1.6	0.3	1.0	0.7	0.6	0	0	0.3	0.1
Other local reactions[c]	2.8	0.5	1.2	1.5	0.9	0.2	0.7	0.6	0.5	0	0.2	0.2

[a]At least one reaction to any of the three primary doses.
[b]Statistically significant between the groups.
[c]Other local reactions: hematoma, injection site pain, local maculopapular rash, local heat.

Table 5   Percentage rates of systemic adverse events within 72 hours of immunization in infants given a hexavalent vaccine (Hexavac) or separate injections of the reference vaccines

Events	Hexavac				Pentavac and Hepatitis B Vax II			
	First dose	Second dose	Third dose	All[a]	First dose	Second dose	Third dose	All[a]
Any systemic events (%)	52[b]	47	38	46	45[b]	42	40	42
Fever >38°C	6.9	19	18	15	4.7	17	22	15
Fever 38.0–38.9°C	6.4	18	14	13	4.2	17	18	18
Fever 39.0–39.9°C	0.5	1.4	3.6	1.8	0.2	0.2	3.4	1.3
Fever >40°C	0	0	0	0	0.2	0	0.5	0.2
Drowsiness	17	9.5	7.7	11	14	9.3	6.0	9.9
Irritability/unusual crying	33	27	21	27	29	23	20	24
Inconsolable crying >3 hours	0.2	0.2	0.2	0.2	0	0	0	0
Vomiting/diarrhea	7.3	5.7	5.0	6.0	6.8	4.3	6.5	5.9
Insomnia	6.1	5.2	3.1	4.8	4.5	4.1	5.0	4.5
Loss of appetite	13	8.3	6.5	9.3	9.2	9.1	7.4	8.6
Other systemic events[c]	4.3	5.0	7.7	5.6	3.8	4.1	6.2	4.7

[a]At least one reaction to any of the three primary doses.
[b]Statistically significant between the groups.
[c]Minor childhood illnesses (for example respiratory or gastrointestinal disorders).

onwards the approach changed again and use of DTP was advised throughout what was now a united Germany. Severe neurological adverse effects of whole-cell pertussis immunization has been reviewed in detail, including reports on the legal cases brought in the English High Court pertussis vaccine trials (SED-12, 817).

It would now seem that the long debate has been resolved. The first step toward its resolution was taken in a review of all information available worldwide which was presented in a report produced in 1991 by the Institute of Medicine of the National Academy of Sciences, Washington SED-12, 817) (29). It was concluded that:

- there is evidence indicating a causal relation between DTP vaccine and anaphylaxis, febrile seizures and inconsolable crying;
- there is weaker evidence pointing to a relationship between DTP vaccine and acute encephalopathy and hypotonic-hyporesponsive episodes;
- there is no reason to believe in an association between vaccination and the occurrence of infantile spasms, afebrile seizures, hypsarrhythmia, Reye's syndrome, or sudden infant death syndrome;
- there is insufficient evidence to indicate either the presence or absence of a causal relation between DTP

vaccine and chronic neurological damage, epilepsy, aseptic meningitis, erythema multiforme or other rash, Guillain–Barré syndrome, hemolytic anemia, juvenile diabetes, learning disabilities and attention-deficit disorder, peripheral mononeuropathy, or thrombocytopenia.

In 1993, Miller and colleagues (30) and Madge and colleagues (31) presented the results of a 10-year follow-up study of the National Childhood Encephalopathy Study (NCES) carried out in 1976–79. The findings suggested a small excess risk of severe acute neurological events within 7 days of pertussis immunization, but the risk of permanent damage due to the vaccine, if any, was slight. Follow-up of cases and controls from this study for some years has shown that significantly more children with such illnesses die or suffer subsequent educational, behavioural, or neurological deficits than expected by comparison with controls, but the number of cases associated with pertussis vaccine was small and statistically vulnerable.

The committee responsible for the report of the Institute of Medicine concluded in 1994 that the recent findings from the NCES necessitated a review of the conclusion that the evidence is insufficient to indicate a causal relation between DTP and permanent neurological damage. After having reviewed the new NCES data, the committee declared that the balance of evidence was consistent with a causal relation between DTP vaccine and the forms of chronic nervous system dysfunction described in the NCES in children who have a serious acute neurological illness within 7 days after receiving DTP vaccine. This type of serious acute neurological response to DTP is rare. The estimated excess risk ranged from 0 to 10.5 per million immunizations. The evidence remains insufficient to indicate the presence or absence of a causal relation between DTP vaccine and chronic nervous system dysfunction under any other circumstances (32).

In a retrospective case-control study including four health maintenance organizations, records from 1 January 1981 to 31 December 1995 were examined to identify children aged 0–6 years old hospitalized with encephalopathy or related conditions (33). The cause of the encephalopathy was categorized as known, unknown or suspected but unconfirmed. Up to three controls were matched to each case. Conditional logistic regression was used to analyse the relative risk of encephalopathy after immunization with diphtheria–tetanus–pertussis (DTP) or measles–mumps–rubella (MMR) vaccines in the 90 days before the onset of the disease, as defined by chart review, compared with an equivalent period among controls indexed by matching on case onset date. In all 452 cases were identified. Cases were no more likely than controls to have received either vaccine during the 90 days before the onset of the disease. When encephalopathies of known cause were excluded, the odds ratio for case children having received DTP within 7 days before the onset of the disease was 1.22 (95% CI = 0.45, 3.31) compared with control children. For MMR in the 90 days before onset of encephalopathy, the odds ratio was 1.23

(95% CI = 0.51, 2.98). Thus, in this study of more than 2 million children, DTP and MMR vaccines were not associated with an increased risk of encephalopathy after immunization.

A 7-month-old boy developed acute transverse myelitis after diphtheria–tetanus–pertussis immunization with acellular pertussis vaccine (34). The diagnosis was based on the clinical course and MRI findings.

Finally, Gangarosa and colleagues have reviewed in detail the so-called pertussis vaccine encephalopathy. Seven epidemiological studies have been conducted; they show that neurological reactions are exceedingly rare with whole-cell pertussis vaccines (SEDA-21, 329).

Susceptibility factors

The (US) Immunization Practices Advisory Committee (ACIP) has recommended that a personal history of a prior convulsion should be evaluated before initiating or continuing immunization with vaccines containing a pertussis component. The presence of an evolving neurological disorder contraindicates pertussis immunization. Other contraindications to the receipt of pertussis vaccine are hypersensitivity to vaccine components or a history of a severe reaction following an earlier dose (35). Reviewing the data on the relation between a family history of convulsions and immunization with pertussis-containing vaccines, both the ACIP as well as the US Committee on Infectious Diseases considered that a family history of convulsions in parents and siblings was not a contraindication to pertussis immunization. The ACIP and other authors, believe that the use of an antipyretic (for example paracetamol given at a dose of 15 mg/kg at the time of DTP immunization and again 4 hours later) in conjunction with DTP vaccine may be reasonable in children with personal or family histories of convulsions since it will reduce the incidence of postimmunization fever (36,37).

Skin

Bullous pemphigoid has been attributed to DTP-IPV vaccine (38).

- A previous healthy 3.5-month-old infant developed bullous pemphigoid 3 days after receiving a first dose of DTP-IPV vaccine. *Staphylococcus aureus* was isolated from purulent bullae. The lesions resolved rapidly after treatment with antibiotics and methylprednisolone.

The authors mentioned 12 other cases of bullous pemphigoid, reported during the last 5 years, that had possibly been triggered by vaccines (influenza, tetanus toxoid booster, and DTP-IPV vaccine).

Immunologic

Data from the Third National Health and Nutrition Survey (1988–94) have been used to analyse the possible effects of DTP or tetanus immunization on allergies and allergy-related symptoms among 13 944 infants, children, and adolescents aged 2 months to 16 years in the USA

(39). The authors concluded that DTP or tetanus immunization increases the risk of allergies and related respiratory symptoms in children and adolescents. However, the small number of non-immunized individuals and the study design limited their ability to make firm causal inferences about the true magnitude of effect.

It has been suggested that the development of a sterile abscess represents an idiosyncratic reaction of some individuals, perhaps genetically determined, which causes a granulomatous response to antigens, irrespective of the location of the vaccine (40). Others maintain that it is caused by a contaminated needle track or to vaccine material coating the outside of the needle, resulting from the lack of a proper injection technique.

Susceptibility Factors

Age

Details regarding the age-related efficacy of the use of paracetamol and the inefficacy of prophylactic paracetamol given in a single dose have been published (SEDA-14, 277).

To evaluate the safety of DTaP–IPV–HIB immunization in premature infants, Pfister et al (41) In an observational study, 78 premature infants of very low birth weights (mean gestational age 28 weeks; mean birth weight 1045 were given DTaP–IPV–HIB vaccine before discharge. The vaccine elicited resurgent or increased cardiorespiratory events in 47% of cases (15% had apnea, 21% had bradycardia, 42% desaturated). Most of the vaccine-triggered events resolved spontaneously or after brief stimulation. The relative risk was 5–8 times higher in infants with a severe clinical course or persistence of cardiorespiratory symptoms at the time of immunization. Bag and mask respiratory support was given to five infants, and oxygen requirements increased transiently in four of 21 infants with chronic lung disease; none required re-ventilation. Reintroduction of oxygen, interruption of active oral feeding, or postponing of hospital discharge was not required. The authors concluded that cardiorespiratory events were often problematic after DTaP–IPV–HIB immunization, requiring monitoring and appropriate intervention. However, these episodes did not have a detrimental impact on the infants' clinical course. Timely immunization is warranted, even in the most vulnerable preterm infants.

Drug Administration

Drug formulations

In a study of local and systemic reactions following 9920 DTP immunizations there were significant differences relating to different manufacturers and different vaccine lots (42).

Drug additives

Eight children developed urticaria within 30 minutes after administration of a diphtheria + tetanus + acellular pertussis (DTaP) vaccine that contained gelatin as a stabilizer (43). None of the children had anti-gelatin IgE, and only two had detectable concentrations of anti-toxoid IgE to diphtheria and pertussis toxoids. No methods to measure anti-thiomersal and anti-alum IgE were available. The authors recommended the development of such methods, which could improve research into the causality of adverse effects of this sort.

Drug administration route

The "two-needle strategy," based on the hypothesis that changing the needle on the syringe after drawing up the DTP vaccine and before injecting reduces local reactions by eliminating deposition of aluminium adjuvant in the subcutaneous track of the needle, has been evaluated (44). When immunizing 223 children using this strategy and 200 others using the "one-needle strategy," there was no significant difference in the occurrence of local or systemic reaction.

References

1. Robbins JB, Schneerson R, Bryla DA, Trollfors B, Taranger J, Lagergard T. Immunity to pertussis. Not all virulence factors are protective antigens. Adv Exp Med Biol 1998;452:207–18.
2. Corbel MJ, Xing DK, Bolgiano B, Hockley DJ. Approaches to the control of acellular pertussis vaccines. Biologicals 1999;27(2):133–41.
3. Matheson AJ, Goa KL. Diphtheria–tetanus–acellular pertussis vaccine adsorbed (Triacelluvax; DTaP3-CB): a review of its use in the prevention of Bordetella pertussis infection. Paediatr Drugs 2000;2(2):139–59.
4. Van der Wielen M, Van Damme P. Tetanus–diphtheria booster in non-responding tetanus–diphtheria vaccinees. Vaccine 2000;19(9–10):1005–6.
5. Jackson LA, Carste BA, Malais D, Froeschle J. Retrospective population-based assessment of medically-attended injection site reactions, seizures, allergic responses and febrile episodes after acellular pertussis vaccine combined with diphtheria and tetanus toxoids. Pediatr Infect Dis J 2002;21:781–6.
6. Belloni C, De Silvestri A, Tinelli C, Avanzini MA, Marconi M, Strano F, Rondini G, Chirico G. Immunogenicity of a three-component acellular pertussis vaccine administered at birth. Pediatrics 2003;111:1042–5.
7. Galazka A. Update on acellular pertussis vaccine. WHO/EPI/GEN/88.4. Geneva: World Health Organization, 1988.
8. Aoyama T, Hagiwara S, Murase Y, Kato T, Iwata T. Adverse reactions and antibody responses to acellular pertussis vaccine. J Pediatr 1986;109(6):925–30.
9. Kimura M, Kuno-Sakai H. Pertussis vaccines in Japan. Acta Paediatr Jpn 1988;30(2):143–53.
10. Kimura M, Kuno-Sakai H. Reports on cases of neurological illnesses occurring after administration of acellular pertussis vaccines in Japan. Tokai J Exp Clin Med 1988;13(Suppl):165–70.
11. Advisory Committee on Immunization Practices (ACIP). Use of diphtheria toxoid–tetanus toxoid–acellular pertussis vaccine as a five-dose series. MMWR Recomm Rep 2000;49(RR-13):1–8.

12. Scheifele DW, Meekison W, Arcand T, Humphrey G. Local adverse reactions to DTP vaccine, adsorbed, in Surrey, BC. CMAJ 1989;141:312.

13. Vermeer-de Bondt PE, Labadie J, Rumke HC. Rate of recurrent collapse after vaccination with whole cell pertussis vaccine: follow up study. BMJ 1998;316(7135):902–3.

14. Miller E. Collapse reactions after whole cell pertussis vaccination. BMJ 1998;316(7135):876.

15. Baraff LJ, Shields WD, Beckwith L, Strome G, Marcy SM, Cherry JD, Manclark CR. Infants and children with convulsions and hypotonic-hyporesponsive episodes following diphtheria–tetanus–pertussis immunization: follow-up evaluation. Pediatrics 1988;81(6):789–94.

16. Stehr K, Cherry JD, Heininger U, Schmitt-Grohe S, uberall M, Laussucq S, Eckhardt T, Meyer M, Engelhardt R, Christenson P, Muller W, Neugebauer A, Sailer K, Keller H, Kircher U, Netzel B, Sachsenhauser-Kratzer H, Thelen M, Buck KE, Nath G, Clapier E, Gelius P, Graf zu Castell B, Hess H-J, Maas-Doyle E, Mayer HPR, Renner K, Seltsam I, Seuwen G. A comparative efficacy trial in Germany in infants who received either the Lederle/Takeda acellular pertussis component DTP (DTaP) vaccine, the Lederle whole-cell component DTP vaccine, or DT vaccine. Pediatrics 1998;101(1 Pt 1):1–11.

17. Recommendations of the Advisory Committee on Immunization Practices (ACIP). The 2000 Immunization Schedule. MMWR Morb Mortal Wkly Rep 2000;49:25.

18. Halperin SA, Mills E, Barreto L, Pim C, Eastwood BJ. Acellular pertussis vaccine as a booster dose for seventeen-to nineteen-month-old children immunized with either whole cell or acellular pertussis vaccine at two, four and six months of age. Pediatr Infect Dis J 1995;14(9):792–7.

19. Podda A, Bona G, Canciani G, Pistilli AM, Contu B, Furlan R, Meloni T, Stramare D, Titone L, Rappuoli R, Granoff DM, Bartalini M, Budroni M, De Luca EC, Cascio A, Cascio G, Cossu M, Orto PD, Di Leo G. Effect of priming with diphtheria and tetanus toxoids combined with whole-cell pertussis vaccine or with acellular pertussis vaccine on the safety and immunogenicity of a booster dose of an acellular pertussis vaccine containing a genetically inactivated pertussis toxin in fifteen- to twenty-one-month-old children. Italian Multicenter Group for the Study of Recombinant Acellular Pertussis Vaccine. J Pediatr 1995;127(2):238–43.

20. Rosenthal S, Chen R, Hadler S. The safety of acellular pertussis vaccine vs whole-cell pertussis vaccine. A postmarketing assessment. Arch Pediatr Adolesc Med 1996;150(5):457–60.

21. Children's Vaccine Initiative (CVI)Global Programme on Immunization of the World Health Organization (WHO). In: Informal Consultation on control of pertussis with whole cell and acellular vaccinesReport on a meeting May 18–19, 1998. Geneva: World Health Organization, 1999:681–8.

22. Cody CL, Baraff LJ, Cherry JD, Marcy SM, Manclark CR. Nature and rates of adverse reactions associated with DTP and DT immunizations in infants and children. Pediatrics 1981;68(5):650–60.

23. Centers for Disease Control. Adverse Events Following Immunization. Surveillance report No. 2, US Department of Health and Human Services, Public Health Service, CDC, Atlanta, GA, 1986.

24. Dittmann S. Atypische Impfverläufe nach SchutzimpfungenLeipzig: Barth;. 1981.

25. Miller DL, Ross EM, Alderslade R, Bellman MH, Rawson NSB. Pertussis immunisation and serious acute neurological illness in children. BMJ 1981;282:1595–9.

26. Halsey NA. Combination vaccines: defining and addressing current safety concerns. Clin Infect Dis 2001;33(Suppl 4):S312–8.

27. Sen S, Cloete Y, Hassan K, Buss P. Adverse events following vaccination in premature infants. Acta Paediatr 2001;90(8):916–20.

28. Mallet E, Fabre P, Pines E, Salomon H, Staub T, Schodel F, Mendelman P, Hessel L, Chryssomalis G, Vidor E, Hoffenbach A, Abeille A, Amar R, Arsene JP, Aurand JM, Azoulay L, Badescou E, Barrois S, Baudino N, Beal M, Beaude-Chervet V, Berlier P, Billard E, Billet L, Blanc B, Blanc JP, Bohu D, Bonardo C, Bossu CHexavalent Vaccine Trial Study Group. Immunogenicity and safety of a new liquid hexavalent combined vaccine compared with separate administration of reference licensed vaccines in infants. Pediatr Infect Dis J 2000;19(12):1119–27.

29. In: Howson CP, Howe CJ, Fineberg HV, editors. Adverse effects of pertussis and rubella vaccines. A report of the Committee to Review the Adverse Consequences of Pertussis and Rubella Vaccines. Washington, DC: National Academy Press, 1991:681–8.

30. Miller D, Madge N, Diamond J, Wadsworth J, Ross E. Pertussis immunisation and serious acute neurological illnesses in children. BMJ 1993;307(6913):1171–6.

31. Madge N, Diamond J, Miller D, Ross E, McManus C, Wadsworth J, Yule W, Frost B. The National Childhood Encephalopathy study: a 10-year follow-up. A report on the medical, social, behavioural and educational outcomes after serious, acute, neurological illness in early childhood. Dev Med Child Neurol Suppl 1993;68:1–118.

32. In: Stratton KR, Howe CJ, Johnston Jr. RB Jr, editors. DPT Vaccine and Chronic Nervous System Dysfunction: A New Analysis. Washington, DC: National Academy Press, 1994:681–8.

33. Ray P, Hayward J, Michelson D, Lewis E, Schwalbe J, Black S, Shinefield H, Marcy M, Huff K, Ward J, Mullooly J, Chen R, Davis R. Encephalopathy after whole-cell pertussis or measles vaccination: lack of evidence for a causal association in a retrospective case-control study. Pediatr Infect Dis J 2006;25:768–73.

34. Riel-Romero RMS. Acute transverse myelitis in a 7-month-old boy after diphtheria–tetanus–pertussis immunization. Spinal Cord 2006;44:688–91.

35. Centers for Disease Control (CDC). Diphtheria, tetanus, and pertussis: guidelines for vaccine prophylaxis and other preventive measures. Immunization Practices Advisory Committee. MMWR Morb Mortal Wkly Rep 1985;34(27):405–14419–26.

36. Lewis K, Cherry JD, Sachs MH, Woo DB, Hamilton RC, Tarle JM, Overturf GD. The effect of prophylactic acetaminophen administration on reactions to DTP vaccination. Am J Dis Child 1988;142(1):62–5.

37. Anonymous. Prophylactic paracetamol with childhood immunisation? Drug Ther Bull 1990;28(19):73–4.

38. Baykal C, Okan G, Sarica R. Childhood bullous pemphigoid developed after the first vaccination. J Am Acad Dermatol 2001;44(Suppl 2):348–50.

39. Hurwitz EL, Morgenstern H. Effects of diphtheria–tetanus–pertussis or tetanus vaccination on allergies and allergy-related respiratory symptoms among children and adolescents in the United States. J Manipulative Physiol Ther 2000;23(2):81–90.

40. Vulginity V. Sterile abscesses after diphtheria–tetanus toxoids–pertussis vaccination. Pediatr Infect Dis J 1987;6:497.

41. Pfister RE, Aeschbach V, Niksic-Stuber V, Martin BC, Siegrist C-A. Safety of DTaP-based combined

immunization in very-low-birth-weight premature infants: frequent but mostly benign cardiorespiratory events. J Ped 2004;145:58–66.

42. Baraff LJ, Manclark CR, Cherry JD, Christenson P, Marcy SM. Analyses of adverse reactions to diphtheria and tetanus toxoids and pertussis vaccine by vaccine lot, endotoxin content, pertussis vaccine potency and percentage of mouse weight gain. Pediatr Infect Dis J 1989;8(8):502–7.

43. Sakaguchi M, Nakayama T, Inouye S. Cases of systemic immediate-type urticaria associated with acellular diphtheria–tetanus–pertussis vaccination. Vaccine 1998;16(11–12):1138–40.

44. Salomon ME, Halperin R, Yee J. Evaluation of the two-needle strategy for reducing reactions to DPT vaccination. Am J Dis Child 1987;141(7):796–8.

Plague vaccine

General Information

Plague vaccine is a suspension of the formaldehyde-killed encapsulated form of *Yersinia pestis*. Primary immunization involves three doses given intramuscularly.

The Working Group on Civilian Biodefense has developed consensus-based recommendations for measures to be taken by medical and public health professionals following the use of plague as a biological weapon against a civilian population (1). They concluded that an aerosolized plague weapon could cause fever, cough, chest pain, and hemoptysis, with signs consistent with severe pneumonia 1–6 days after exposure. Rapid evolution of disease would occur in the 2–4 days after the onset of symptoms and would lead to septic shock with a high mortality if early treatment was not instituted. They advised early treatment and prophylaxis with streptomycin or gentamicin or one of the tetracycline or fluoroquinolone classes of antimicrobials.

General adverse effects

General malaise, headache, fever, mild lymphadenopathy, or erythema and induration at the injection site have been reported following the administration of plague vaccine (10% of vaccinees); these effects are more common with repeated injections. Sterile abscesses and hypersensitivity reactions (urticaria, asthma) occur rarely (2).

Organs and Systems

Immunologic

Known allergy to any of the plague vaccine constituents (beef protein, soya, casein, phenol) contraindicates immunization. Severe local or systemic reactions following previous doses contraindicate revaccination (2).

References

1. Inglesby TV, Dennis DT, Henderson DA, Bartlett JG, Ascher MS, Eitzen E, Fine AD, Friedlander AM, Hauer J, Koerner JF, Layton M, McDade J, Osterholm MT, O'Toole T, Parker G, Perl TM, Russell PK, Schoch-Spana M, Tonat K. Plague as a biological weapon: medical and public health management. Working Group on Civilian Biodefense. JAMA 2000;283(17):2281–90.

2. Committee on Immunization. In: Guide for Adult Immunization. Philadelphia: American College of Physicians, 1985:65.

Pneumococcal vaccine

General Information

Two types of pneumococcal vaccine are available, polysaccharide vaccine and conjugated polysaccharide vaccine.

Pneumococcal vaccine is composed of a saline solution containing the purified capsular polysaccharides of 23 types of *Streptococcus pneumoniae*. The improved 23-valent vaccine, which replaced a 14-valent vaccine at the beginning of the 1980s, contains antigens to pneumococcal types that are responsible for about 85% of bacteremic pneumococcal pneumonia.

Pneumococcal vaccines produced by different manufacturers are currently available, for example "Pneumovax 23" produced by Merck Sharp & Dohme and "Pnu-Imune 23" produced by Lederle Laboratories. Each vaccine dose (0.5 ml) contains 25 µg of each polysaccharide antigen. Immunization is recommended for people who are at increased risk of developing pneumococcal disease because of underlying chronic health conditions and for older people. About 50% of vaccinees develop mild adverse effects, such as erythema and pain at the injection site. Fever, myalgia, and severe local reactions have been reported in under 1% of vaccinees. Severe systemic reactions, such as anaphylaxis have been rarely reported.

Incidental case reports relate to small vessel vasculitis after combined pneumococcal-influenza immunization (1), severe febrile reactions with leukocytosis (2), Sweet's syndrome (3), thrombocytopenia (4,5), and keratoacanthoma at the injection site (6).

Patients with AIDS have an impaired antibody response to pneumococcal vaccine. Adverse effects in symptomatic and asymptomatic HIV-infected vaccinees were not different from those in HIV-negative persons (7).

Pneumococcal polysaccharide vaccine

The question of whether revaccination with 23-valent pneumococcal polysaccharide vaccine (PPV) at least 5 years after the first vaccination is associated with more frequent or more serious adverse events than those after the first vaccination has been studied in patients aged 50–74 years who had never been vaccinated with PPV (n = 901) or who had been vaccinated once at least 5 years before enrolment (n = 513) (8). After one dose of PPV, local injection site reactions and prevaccination concentrations of type-specific antibodies were measured. Those who were re-vaccinated were more likely than those who received their first vaccinations to report a local injection site reaction of at least 10.2 cm (4 in.) in diameter within 2 days of vaccination (55/513 versus 29/901, or 11 versus 3%). The reactions resolved by a median of 3 days after vaccination. The highest rate was among revaccinated patients who were immune competent and did not have chronic illnesses: 15% (33/228) compared with 3% (10/337) among comparable patients receiving their first vaccinations. The risk of these local reactions correlated significantly with prevaccination geometric mean antibody concentrations. The authors concluded that physicians and patients should be aware that self-limited local injection site reactions occur more often after revaccination compared with a first vaccination; however, this risk does not represent a contraindication to revaccination with PPV in recommended patients.

Pneumococcal conjugated polysaccharide vaccines

Pneumococcal polysaccharides are not immunogenic in infants, but improved immunogenicity of polysaccharide-protein conjugates has been demonstrated. One of the major problems in developing a successful vaccine against S. pneumoniae is the large number of different serotypes involved. More than 83 serotypes of the bacterium are known to cause disease, although about 10 of these account for up to 70% of disease in young children. The frequency of the serotypes can vary from year to year, from one age group to another, and on a geographical basis. Various conjugated vaccines that include different serotype variations are already licensed (the heptavalent conjugated vaccine Prevnar/Prevenar), under development or undergoing clinical trial (9- and 11-valent conjugated vaccines). If the vaccines prove to be successful, it is estimated that their use could reduce child deaths from pneumococcal pneumonia by up to 25%, saving over 250 000 lives a year worldwide. The development of conjugate pneumococcal vaccines has also been driven by the high incidence of inner ear infections and the severity of meningitis due to S. pneumoniae in industrialized countries. The results of some clinical trials with different vaccines of this type have been reviewed (SEDA-22, 344).

The immunogenicity of 7-valent pneumococcal-conjugate vaccine plus 23-valent pneumococcal vaccine in 11 children has been compared with the immunogenicity of 23-valent vaccine alone in 12 children up to 2 years of age with sickle cell disease (9). IgG pneumococcal antibody concentrations were higher with combined administration, with no increase in adverse effects after immunization with 23-valent vaccine.

In 2000, the first heptavalent conjugated pneumococcal vaccine, Prevnar, which contains polysaccharides of

pneumococcal serotypes 4, 6B, 9V, 14, 19F, and 23F, and oligosaccharide of serotype 18C, conjugated to the protein carrier CRM 197 (non-toxic variant of diphtheria toxin), was licensed in the USA (covering 90% of pneumococcal serotypes found in young children in the USA) and in all EU member states, as well as in selected other countries in 2001. In a randomized, double-blind study, 302 healthy infants in the Northern California Kaiser Permanente Health Plan received either the pneumococcal vaccine or meningococcal group C conjugate vaccine as a control at 2, 4, and 6 months of age and a booster at 12–15 months of age (10). The immunogenicity and safety of simultaneous administration of vaccines used in the routine immunization program of children (DTwP or DTaP, Hib and OPV or hepatitis B vaccine) were also evaluated. Local reactions after pneumococcal, DTwP-Hib, and DTaP plus Hib vaccine were statistically less severe at the pneumococcal vaccine site than at the DTwP-Hib sites (Table 1). There were 12 emergency room visits and 8 hospitalizations that occurred soon after immunization (otitis, febrile illness, urinary tract infection, burns), but none was thought to be related to the vaccine. One hypotensive-hyporesponsive episode 15 minutes after pneumococcal vaccine, DTwP-Hib/HB, and OPV, with complete recovery within 1 hour, was considered to have been vaccine-related. After the booster dose of pneumococcal vaccine, the geometric mean titers of all seven serotypes increased significantly (compared with the values after dose 3 and before dose 4) to antibodies considered as protective. When vaccine was administered at the same time as the booster dose of DTaP and Hib vaccines, there were lower antibody titers for some of the antigens than the antibody response when the

Table 1 Local reactions within 48 hours of injection in infants given conjugated pneumococcal, DTwP-Hib, and DtaP + Hib vaccines

		Infants		Toddlers
Reaction	Number of children	Pneumococcal vaccine (%)	DTwP-Hib (%)	DtaP + Hib (%)
Redness				
Dose 1	183	17	18	
Dose 2	159	18	29	
Dose 3	160	16	22	
Dose 4	110	9.1		6.1
Swelling				
Dose 1	183	10	19	
Dose 2	159	11	20	
Dose 3	160	9.0	21	
Dose 4	110	6.4		3.7
Tenderness				
Dose 1	183	24	27	
Dose 2	159	21	26	
Dose 3	160	23	28	
Dose 4	110	16		9.8

pneumococcal vaccine was given separately. Because the geometric mean titers of the booster responses were all generally high and all subjects achieved similar percentages above predefined antibody titers, these differences were considered to be probably not clinically significant. Summarizing their results the authors concluded that the pneumococcal vaccine was safe and immunogenic.

A comprehensive technical overview on the epidemiology and prevention of pneumococcal disease, including the use of polysaccharide and conjugate vaccines has been provided (11).

Further results of field trials on the efficacy, safety, and immunogenicity of heptavalent conjugated pneumococcal vaccine have appeared. Between October 1995 and August 1998, 37 868 infants were included in a double-blind trial (12). At 2, 4, 6, and 12–15 months of age they were randomly assigned to receive either the pneumococcal conjugate vaccine or meningococcal conjugate vaccine. More than 95% of pneumococcal vaccine recipients developed = 0.15 µg/ml antibodies against all serotypes included in the vaccine. As of April 1999, a vaccine efficacy of 97% (prevention of invasive pneumococcal disease caused by vaccine serotypes) was calculated; in addition, there was a significant impact on otitis media. Data on reactogenicity of the conjugate pneumococcal and meningococcal vaccines are provided in Table 2 and Table 3. Local reactions were analysed separately for children who had received DTaP and DTwP

vaccine simultaneously. Local and systemic reactions were generally relatively mild with either vaccine, and more severe local and systemic reactions were uncommon and self-limiting. There were significant differences in outpatient clinic visits for seizures (11 pneumococcal vaccine recipients versus 23 controls), but none of the subcategories of seizure (febrile seizures, epilepsy, afebrile seizures) was significantly different. There were four cases of sudden infant death syndrome (SIDS) (0.2/1000) in the pneumococcal vaccine group and eight in the controls (0.4/1000). This rate is similar to the rate of 0.5/1000 children observed in the general infant population of California.

Recent advances in conjugated pneumococcal vaccines selected from current literature have been reviewed, including studies with experimental tetravalent and pentavalent conjugated vaccines and vaccines conjugated to various proteins (13).

The first heptavalent conjugate pneumococcal vaccine, Prevnar (containing polysaccharides of pneumococcal serotypes 4, 6B, 9V, 14, 19F, and 23F, and oligosaccharide of serotype 18C, conjugated to the protein carrier CRM_{197}, a non-toxic variant of diphtheria toxin), was licensed in the USA in 2000, and in all EU member states as well as in selected other countries in 2001. The efficacy, immunogenicity, and safety of conjugated pneumococcal vaccine have been evaluated in 1756 low birth weight infants (under 2500 g) and 4340 preterm infants (before

Table 2 Local reactions comparing pneumococcal conjugate (PNCRM7) and meningococcal conjugate (MnCC) as well as each of these with DtaP

Reaction	PNCRM7 (%)	DTaP (%)	P value	MnCC (%)	DtaP (%)	P value	PNCRM7 versus MnCC (P value)
Redness							
Dose 1	10	6.7	<0.001	6.5	5.6	0.345	0.124
Dose 2	12	11	0.512	7.6	11	0.011	0.003
Dose 3	14	11	0.143	9.3	8.2	0.557	0.011
Dose 4	11	3	0.004	4.5	4.0	0.999	0.226
Redness >3 cm							
Dose 1	0.3	0.0	0.500	0.1	0.3		0.999
Dose 2	0.0	0.2	0.999	0.2	0.4	0.999	0.481
Dose 3	0.2	0.2		1.3	0.8	0.999	0.105
Dose 4	0.6	0.6	0.999	0.0	0.0	0.625	0.255
Swelling							
Dose 1	9.8	6.6	0.002	4.2	4.3	0.999	0.013
Dose 2	12	11	0.312	5.1	7.4	0.080	0.001
Dose 3	10	10	0.999	6.9	8.3	0.473	0.001
Dose 4	12	5.5	0.013	4.5	3.4	0.688	0.247
Swelling >3 cm							
Dose 1	0.1	0.1		0.0	0.0		0.999
Dose 2	0.4	0.6	0.999	0.2	0.0	0.999	0.999
Dose 3	0.5	1.0	0.500	0.3	0.5	0.999	0.999
Dose 4	0.6	0.6	0.999	0.0	0.0		0.224
Tenderness							
Dose 1	18	16	0.053	18	19	0.265	0.970
Dose 2	19	17	0.080	15	16	0.677	0.069
Dose 3	15	13	0.265	12	12	0.999	0.280
Dose 4	23	18	0.096	15	15	0.999	0.052

Table 3 Fever within 48 hours of vaccination among infants receiving PNCRM7 or MnCC vaccine*

Reaction	PNCRM7		MnCC		
	%	Number	%	Number	P value
Fever ≥ 38° C					
Dose 1	15	709	9.4	710	0.001
Dose 2	24	556	11	507	0.001
Dose 3	19	461	12	414	0.003
Dose 4	21	224	17	230	0.274
Fever > 39° C					
Dose 1	0.9	709	0.3	710	0.178
Dose 2	2.5	556	0.8	507	0.029
Dose 3	1.7	461	0.7	414	0.180
Dose 4	1.3	224	1.7	230	0.999

*Concomitantly with DTaP and other recommended vaccines; MnCC, meningococcal conjugate.

38 weeks) compared with infants of normal birth weight and infants born at full term (14). Vaccine efficacy was 100% in both groups. Fever and local events were similar when adjusted for clustering among multiple doses per child. After dose 3, there was more redness and swelling in infants of low birth weights and more swelling in preterm infants.

A hendecavalent pneumococcal conjugate vaccine (including serotypes 1, 3, 4, 5, 6B, 7F, 9V, 14, 18C, 19F, and 23F) was given to 117 Finnish and 135 Israeli infants at ages 2, 4, 6, and 12 months together with other vaccines used in the respective national immunization programs (15). The authors concluded that the vaccine is safe and immunogenic in infants. There were no severe adverse reactions. After each dose 30% of the vaccinees had local reactions, of which pain was the most common. Fever ≥38°C was reported in 33–53% and a high fever (≥40°C) was reported six times.

23-valent pneumococcal polysaccharide vaccine

The efficacy of polysaccharide vaccine in preventing invasive pneumococcal disease, pneumonia, and death has been assessed in a double-blind, randomized, placebo-controlled trial in 1392 HIV1-infected adults in Uganda (16). The vaccine was well tolerated. However, it was ineffective and is not recommended for use in HIV1-infected individuals. Reassessment of recommendations for polysaccharide vaccine immunization may be necessary in some countries. The authors suggested that the vaccine causes destruction of polysaccharide-responsive B cell clones.

In another clinical trial the immunogenicity and safety of polysaccharide vaccine has been assessed in 21 renal transplant recipients (17). Protective antibody titers were reached at 6 and 12 weeks after immunization in all recipients, bar one. No local or systemic adverse effects were observed.

Systematic reviews

The Global Advisory Committee on Vaccine Safety (GACVS) has reviewed safety data from 62 studies, including randomized controlled trials and post-marketing studies (18). Since the 7-valent pneumococcal conjugate vaccine was licensed in 2000, and following its widespread use in the USA and more recently in Canada and some European countries, major safety concern have not been identified. The evidence on the safety of the vaccine is reassuring. However, as with the introduction of any new vaccine, further careful surveillance for possible rare and unexpected events is recommended.

Organs and Systems

Immunologic

Arthus reactions and systemic reactions have commonly been reported after booster doses of polysaccharide vaccine and are thought to result from antigen–antibody reactions involving antibodies induced by the previous immunization (19). Data on revaccination of children are not yet sufficient to provide a basis for recommendation.

An allergic reaction has been described to 23-valent polysaccharide pneumococcal vaccine (20).

- A 2-year-old child developed bronchospasm and cutaneous and laryngeal edema immediately after the injection of a 23-valent polysaccharide pneumococcal vaccine. The symptoms resolved within 1 hour of treatment with antihistamines, glucocorticoids, and aerosols. Skin tests and specific IgE tests showed that the pneumococcal antigens were responsible for the anaphylaxis.

Susceptibility Factors

Age

In an evaluation of simultaneous immunization in 85 elderly subjects, pre-immunization pneumococcal antibodies were associated with more reactions, both local and systemic, after vaccination (21). A rise in temperature (9% of vaccinees), and pain at the injection site (5% of vaccinees) were significantly associated with raised pre-immunization pneumococcal polysaccharide antibody concentrations.

Other features of the patient

The immunogenicity and safety of pneumococcal polysaccharide vaccine have been studied in renal allograft recipients, dialysis patients (22), children and adolescents with sickle-cell anemia (SED-11, 682) (SEDA-12, 277), people with diabetes mellitus (SED-11, 682), and children with nephrotic syndrome (23). When comparing the results with healthy persons there were no significant differences.

The problems connected with pneumococcal disease and its prevention in HIV-infected individuals have been reviewed (24). Pneumococcal disease occurs significantly more often in HIV-infected individuals, with pneumococcal pneumonia rates 5.5–17.5 times greater than population-based estimates in the USA, and the increasing rate of penicillin-resistant strains of *S. pneumoniae* highlight the need for improved prevention strategies. Studies of pneumococcal disease in HIV infection have repeatedly shown that over 85% of the isolates from bacteremic patients, in both the USA and Africa, are of serotypes included in the 23-valent vaccine. However, the proportion of HIV-positive individuals who respond to 23-valent pneumococcal polysaccharide vaccine has been shown in some but not all studies to be slightly reduced compared with age-matched controls but comparable to other high-risk groups, such as elderly people, in whom clinical efficacy has been established. Some studies have suggested a trend toward a lower response rate as the CD4 cell count falls. The reason for concern about the safety of pneumococcal immunization in HIV-infected individuals is the reported association between immunization and increasing HIV virus load.

Much of this concern arises from extrapolation from published data on influenza immunization. However, there are some data on pneumococcal immunization alone. In 32 HIV-positive patients with a median CD4 cell count of 242×10^6/l, who received Pneumovax and tetanus toxoid there was no change in plasma HIV 1 RNA at 20–56 days after immunization (25). In contrast there were marked increases in plasma viral RNA (1.6–586 times) reported in 12 asymptomatic HIV-positive individuals (mean CD4 cell count 374×10^6/l). More recently, a study of patients with more advanced disease found that HIV-1 RNA and DNA were unaffected by either conjugate or polysaccharide pneumococcal vaccine up to 309 days after immunization (26). In summary, the authors of the review (24) recommended that HIV-infected individuals be immunized with pneumococcal vaccine and that immunization should be carried out as early as possible in the course of HIV infection.

Drug–Drug Interactions

Influenza vaccine

In a study of the interaction between 23-valent pneumococcal polysaccharide vaccine and influenza vaccine, 152 adults with chronic respiratory disease were randomized to receive both vaccines either simultaneously or at an interval of 1 month (27). There were no significant differences in serological responses between the groups. The incidence and severity of both local and systemic adverse effects were also similar: there were mild local reactions in 38 and 36% and systemic reactions in five and three of the vaccinees respectively.

References

1. Houston TP. Small-vessel vasculitis following simultaneous influenza and pneumococcal vaccination. NY State J Med 1983;83(11–12):1182–3.
2. Gabor EP, Seeman M. Acute febrile systemic reaction to polyvalent pneumococcal vaccine. JAMA 1979;242(20):2208–9.
3. Maddox PR, Motley RJ. Sweet's syndrome: a severe complication of pneumococcal vaccination following emergency splenectomy. Br J Surg 1990;77(7):809–10.
4. Citron ML, Moss BM. Pneumococcal-vaccine-induced thrombocytopenia. JAMA 1982;248(10):1178.
5. Kelton JG. Vaccination-Associated relapse of immune thrombocytopenia. JAMA 1981;245(4):369–70.
6. Bart RS, Lagin S. Keratoacanthoma following pneumococcal vaccination: a case report. J Dermatol Surg Oncol 1983;9(5):381–2.
7. Centers for Disease Control (CDC). Pneumococcal polysaccharide vaccine. MMWR Morb Mortal Wkly Rep 1989;38(5):64–873–6.
8. Jackson LA, Benson P, Sneller VP, Butler JC, Thompson RS, Chen RT, Lewis LS, Carlone G, DeStefano F, Holder P, Lezhava T, Williams WW. Safety of revaccination with pneumococcal polysaccharide vaccine. JAMA 1999;281(3):243–8.
9. Vernacchio L, Neufeld EJ, MacDonald K, Kurth S, Murakami S, Hohne C, King M, Molrine D. Combined schedule of 7-valent pneumococcal conjugate vaccine followed by 23-valent pneumococcal vaccine in children and young adults with sickle cell disease. J Pediatr 1998;133(2):275–8.
10. Shinefield HR, Black S, Ray P, Chang I, Lewis N, Fireman B, Hackell J, Paradiso PR, Siber G, Kohberger R, Madore DV, Malinoski FJ, Kimura A, Le C, Landaw I, Aguilar J, Hansen J. Safety and immunogenicity of heptavalent pneumococcal CRM197 conjugate vaccine in infants and toddlers. Pediatr Infect Dis J 1999;18(9):757–63.
11. Overturf GD, Peter G, Pickering LK, MacDonald NE, Chilton L, Jacobs RF, Delage G, Dowell SF, Orenstein WA, Patriarca PA, Myers MG, Ledbetter EO, Kim J. American Academy of Pediatrics. Committee on Infectious Diseases. Technical report: prevention of pneumococcal infections, including the use of pneumococcal conjugate and polysaccharide vaccines and antibiotic prophylaxis. Pediatrics 2000;106(2 Pt 1):367–76.
12. Black S, Shinefield H, Fireman B, Lewis E, Ray P, Hansen JR, Elvin L, Ensor KM, Hackell J, Siber G, Malinoski F, Madore D, Chang I, Kohberger R, Watson W, Austrian R, Edwards K, Aguilar J, Bartlett M, Bergeb R, Burman M, Dorfman S, Easter W, Finkel A, Froehlich H, Glauber J, Herz A, Honeychurch D, Kleinrock R. Efficacy, safety and immunogenicity of heptavalent pneumococcal conjugate vaccine in children. Northern California Kaiser Permanente Vaccine Study Center Group. Pediatr Infect Dis J 2000;19(3):187–95.
13. Dabelstein D, Cromer B. Selections from current literature. Recent advances in conjugated pneumococcal vaccination. Fam Pract 2000;17(5):435–41.
14. Shinefield H, Black S, Ray P, Fireman B, Schwalbe J, Lewis E. Efficacy, immunogenicity and safety of heptavalent pneumococcal conjugate vaccine in low birth weight and preterm infants. Pediatr Infect Dis J 2002;21:182–6.
15. Dagan R, Käyhty H, Wuorimaa T, Yaich M, Bailleux F, Zamir O, Eskola J. Tolerability and immunogenicity of an eleven-valent mixed carrier Streptococcus pneumoniae

capsular polysaccharide-diphtheria toxoid or tetanus protein conjugate vaccine in Finnish and Israeli infants. Ped Infect Dis J 2004;23:91–8.

16. French N, Nakiyingi J, Carpenter LM, Lugada E, Watera C, Moi K, Moore M, Antvelink D, Mulder D, Janoff EN, Whitworth J, Gilks CF. 23-valent pneumococcal polysaccharide vaccine in HIV-1-infected Ugandan adults: double-blind, randomised and placebo controlled trial. Lancet 2000;355(9221):2106–11.

17. Kazancioglu R, Sever MS, Yuksel-Onel D, Eraksoy H, Yildiz A, Celik AV, Kayacan SM, Badur S. Immunization of renal transplant recipients with pneumococcal polysaccharide vaccine. Clin Transplant 2000;14(1):61–5.

18. Global Advisory Committee on Vaccine Safety. Meeting on 29–30 November 2006. Safety of pneumococcal conjugate vaccines: update. Wkly Epidemiol Rec 2007;82:24.

19. Centers for Disease Control (CDC). Update: pneumococcal polysaccharide vaccine usage—United States. MMWR Morb Mortal Wkly Rep 1984;33(20):273–6281.

20. Ponvert C, Ardelean-Jaby D, Colin-Gorski AM, Soufflet B, Hamberger C, de Blic J, Scheinmann P. Anaphylaxis to the 23-valent pneumococcal vaccine in child: a case-control study based on immediate responses in skin tests and specific IgE determination. Vaccine 2001;19(32):4588–91.

21. Sankilampi U, Honkanen PO, Pyhala R, Leinonen M. Associations of prevaccination antibody levels with adverse reactions to pneumococcal and influenza vaccines administered simultaneously in the elderly. Vaccine 1997;15(10):1133–7.

22. Rytel MW, Dailey MP, Schiffman G, Hoffmann RG, Piering WF. Pneumococcal vaccine immunization of patients with renal impairment. Proc Soc Exp Biol Med 1986;182(4):468–73.

23. Halsey NA, Spika JS, Lum GM, Schiffman GS, Lauer BA. Adverse reactions to pneumococcal polysaccharide vaccine in children. Pediatr Infect Dis 1982;1(1):34–6.

24. Moore D, Nelson M, Henderson D. Pneumococcal vaccination and HIV infection. Int J STD AIDS 1998;9(1):1–7.

25. Katzenstein TL, Gerstoft J, Nielsen H. Assessments of plasma HIV RNA and CD4 cell counts after combined Pneumovax and tetanus toxoid vaccination: no detectable increase in HIV replication 6 weeks after immunization. Scand J Infect Dis 1996;28(3):239–41.

26. Kroon FP, Van Furth R, Bruisten SM. The effects of immunization in human immunodeficiency virus type 1 infection. N Engl J Med 1996;335(11):817–8.

27. Fletcher TJ, Tunnicliffe WS, Hammond K, Roberts K, Ayres JG. Simultaneous immunisation with influenza vaccine and pneumococcal polysaccharide vaccine in patients with chronic respiratory disease. BMJ 1997;314(7095):1663–5.

Poliomyelitis vaccine

General Information

There are two types of poliomyelitis vaccines available. One is prepared from polioviruses that as a rule have been inactivated by formaldehyde. Inactivated poliomyelitis vaccine (IPV) is given parenterally. The second group of polio vaccines comprises attenuated strains of live polioviruses (oral poliomyelitis vaccine, OPV), which are given orally; these live vaccines are the most widely used.

Inactivated poliomyelitis vaccine

Inactivated poliomyelitis vaccine produced by improvements in manufacturing technology (potency-enhanced IPV-eIPV was licensed in 1987) is used for routine immunization in an increasing number of countries (for example in Finland, France, Germany, Iceland, the Netherlands, Norway, Sweden, and certain provinces of Canada) and is recommended in other countries for certain specific purposes, for example for persons with underlying immunological disorders or non-immunized adults exposed to high risk. A few countries (for example Denmark, Hungary, Italy, Lithuania, and Israel) use a mixed schedule, starting primary immunization with IPV followed by OPV.

The US 2000 childhood immunization schedule, proposed by the Advisory Committee of Immunization Practices (ACIP), the American Academy of Pediatrics, and the American Academy of Family Physicians, recommended an all-IPV schedule for routine use in the USA, aimed at the elimination of the rare vaccine-associated paralytic poliomyelitis (1). Since 1 January 2000, all children have received four doses of IPV at ages 2 months, 4 months, 6–18 months, and 4–6 years.

Some other industrialized countries that use OPV in their routine immunization programs, and which had no wild poliomyelitis cases for many years but some vaccine-associated poliomyelitis, are reassessing their immunization strategy. They are considering new concepts of shifting from OPV to IPV or from OPV to mixed schedules. This change could help to prevent vaccine-associated poliomyelitis mostly occurring after the first dose of OPV immunization.

Serious adverse effects after IPV immunization have not been documented (2). Because IPV contains streptomycin and neomycin, there is a possibility of allergic reactions in those who are sensitive to these antibiotics. Although it has been postulated that IPV, like some viral infections, might trigger Guillain–Barré syndrome, there is inadequate evidence to accept or reject such an association (3); in this respect it may differ from the oral vaccine.

Oral polio vaccine

In contrast to IPV, OPV causes important problems, including vaccine-associated polio and prolonged polio virus excretion.

Combination vaccines

The safety, immunogenicity, and lot consistency of five-component pertussis combination vaccine (DtaP + IPV + PRP-T-Hib) in infants have been compared to those of a whole-cell pertussis combination vaccine (DTwP + IPV + PRP-T-Hib), as have separate and combined injections of DTP + IPV and Hib. The combination vaccine DtaP + IPV + Hib were comparable or superior regarding safety and immunogenicity to the combination vaccine containing the whole cell pertussis component. There was no interaction between acellular pertussis and PRP-T-Hib, a feature that distinguishes this combination vaccine from some others, which depress anti-PRP responses. The combination vaccine DtaP + IPV + Hib produced significantly lower rates of local and systemic reactions than did the combination vaccine containing the whole-cell pertussis component. Local reactions, such as redness, swelling, and tenderness occurred two to three times more often after combination vaccine containing whole cell pertussis than after combination vaccines with acellular pertussis components. Fever was three times more common after whole cell combination vaccine. Fever over 40°C was rare in all vaccinees, because of the use of paracetamol prophylaxis. Systemic reactions, such as fussiness, crying, reduced activity, and anorexia, were about twice as frequent with whole cell vaccine as with acellular pertussis vaccine. Both local and systemic reactions persisted longer after whole cell vaccine than after acellular pertussis vaccine. There were no significant differences between reaction rates among infants given DtaP + IPV vaccine combined with PRP-T-Hib vaccine in the same syringe compared with those given separate injections, except for local redness after the first dose (4).

Hexavalent immunization

The Hexavalent Study Group has compared the immunogenicity and safety of a new liquid hexavalent vaccine against diphtheria, tetanus, pertussis, poliomyelitis, hepatitis B and *Haemophilus influenzae* type b (DTP + IPV + HB + Hib vaccine, manufactured by Aventis Pasteur MSD, Lyon, France) with two reference vaccines, the pentavalent DTP + IPV + Hib vaccine and the monovalent hepatitis B vaccine, administered separately at the same visit (5). Infants were randomized to receive either the hexavalent vaccine (*n* = 423) or (administered at different local sites) the pentavalent and the HB vaccine (*n* = 425) at 2, 4, and 6 months of age. The hexavalent vaccine was well tolerated (for details see the monograph on Pertussis vaccines). At least one local reaction was reported in 20% of injections with hexavalent vaccine compared with 16% after the receipt of pentavalent vaccine or 3.8% after the receipt of hepatitis B vaccine. These reactions were generally mild and transient. At least one systemic reaction was reported in 46% of injections with hexavalent vaccine, whereas the respective rate for the recipients of pentavalent and HB vaccine was 42%. No vaccine-related serious adverse event occurred during the study. The hexavalent vaccine provided immune responses adequate for protection against the six diseases.

Polio vaccines as a possible cause of AIDS

There is a hypothesis that the HIV virus might have jumped the species barrier from monkey to people via a contaminated polio vaccine because the vaccine was manufactured in primary monkey kidney tissue known to be sometimes contaminated with monkey viruses. The existing evidence, including tests of poliovirus seed stocks, more than 20 vaccine lots, and serum samples from vaccine recipients makes this hypothesis highly improbable (6,7).

Contamination of polio vaccines with simian papovavirus 40 (SV40)

The problem of early polio vaccines produced in the 1950s and early 1960s, and, in some instances, contaminated with the monkey virus SV40 (simian virus 40) has been discussed (8). From 1954 to 1962, millions of people were immunized with polio vaccines, which during that period contained SV40 as an unrecognized contaminant. Some studies sought to investigate possible causation between the receipt of the vaccine and the development of tumors. The results were not convincing (SEDA-15, 355) (SED-12,821). At a workshop in 1997 some of the scientific questions concerned were again considered (9). The presentations by polio vaccine manufacturers and the UK National Institute for Biological Standards and Control provided convincing evidence that currently used polio vaccine was free of SV40.

Critical assessment of virological and epidemiological data suggests a probable causative role for SV40 in certain human cancers, but that additional studies are necessary to prove etiology (8). To help answer these issues, the World Health Organization has provided the following statement (10): "Investigations from several medical research institutions have detected the presence of simian virus 40 (SV40) genome in certain rare human tumors, notably mesotheliomas, osteosarcomas, and brain tumors. SV40, which is known to induce tumors in laboratory rodents, is a polyomavirus identified in 1960 as a contaminant of some batches of primary rhesus monkey kidney cells used to produce polio vaccine in the 1950s. Soon after its discovery, measures to exclude the virus from polio and other vaccines were rapidly introduced into WHO Requirements for the Manufacture and Quality Control of Polio Vaccines, and these have been rigorously applied by vaccine manufacturers. For over 30 years now, polio vaccines made in primary monkey kidney cells have been shown to be free of live SV40."

The use of new and highly sensitive polymerase chain reaction (PCR) techniques for the detection of the SV40 genome in batches of oral polio vaccines from several manufacturers has confirmed that the measures taken have been effective in excluding SV40 from vaccines. There is no doubt that the early batches of polio vaccine that were used between 1955 and 1963 contained SV40. Epidemiological studies between 1960 and 1974 failed to show an association between exposure to SV40-contaminated vaccines and human tumors. More recent epidemiological data from tumor registers, involving more than 60 million person years of observation, have likewise found no differences in tumor incidence that could be attributable to SV40-contaminated polio vaccines. The latest data published in the Journal of the National Cancer Institute, USA, suggest that SV40 may be present in humans more commonly than had previously been thought, and raises the possibility of transmission of the virus among humans. Whether the antibodies that have been detected are due to exposure to SV40 or to cross-reacting human polyomaviruses is not known. Neither is it certain that SV40 strains now present in the human population originated from the use of the early polio vaccines. Despite almost 40 years of observation, there is still no evidence that SV40 contamination of some early batches of polio vaccine has had any adverse effect on human health.

"The River" by Edward Hooper

The hypothesis that oral polio vaccine played a key role in the current AIDS epidemic was raised again by Edward Hooper, who has worked many years for the BBC and the UN in Africa and some years ago wrote a book called "Slim," in which he described the AIDS epidemic in East Africa. His book "The River. A journey to the source of HIV and AIDS," published in 1999 (11), raised great public attention and was discussed in BBC Press Releases and in the New York Times. Leading experts in virology and AIDS research published comments in Science and other scientific journals. "The River" is a thoroughly researched, well-written book and deserves to be taken seriously. Hooper carefully collected data and events describing the first phases of HIV infection and AIDS, as well as the early development and implementation of polio vaccines. His book reflects some hundreds of interviews, including the leading researchers in the related fields, and he has documented more than 4000 references.

The hypothesis is based on the following facts and assumptions. In 1957 and 1958, Koprowski, from the Wistar Institute in Philadelphia, was administering oral polio vaccine in Africa (pre-licensure field trials in Burundi, Rwanda, and the North-East Congo) near Stanleyville (now Kisangani) in Congo. Not far from the base, chimpanzees for use in medical research were housed in Camp Lindi and might have carried a primate immunodeficiency virus (PIV). Chimpanzee kidneys for hepatitis research were shipped from Camp Lindi to the Virological Department of the Children Hospital in Philadelphia in 1958 and 1959. Hooper suggests that "it could be that [kidneys from these chimpanzees] ended up at the Wistar", the laboratory in Philadelphia where polio vaccines were manufactured, where they contaminated vaccines with PIV. The polio vaccine that was supposed to be contaminated with PIV was then used in the Congo, transmitting the virus that evolved into HIV-1, the starting point of the worldwide HIV-1 epidemic. Over the next 20 years, infected humans progressed to AIDS, and the disease became visible in central Africa in the mid-to-late 1970s.

Does Hooper prove his hypothesis beyond a shadow of doubt? No, but he makes a powerful case for soberly and squarely addressing the issue.

There are also strong arguments against the hypothesis. Polio vaccines were first propagated in kidney cultures of rhesus and cynomolgus macaques, and later in African green monkeys. Plotkin and Koprowski categorically stated in a letter to the editor of the New York Times (7 December 1999) that no chimpanzee tissues were used in the Wistar Institute for polio vaccine production. They added that two independent analyses of the probable timing of the crossover of HIV from chimpanzees into humans give dates earlier than 1957–59, the years in which the Wistar polio vaccine was used in the Congo (12). It should also be mentioned that the vaccine manufactured in Wistar at that time was not only used in Africa but also in Sweden, Poland, and the US. One vial of Wistar's oral polio vaccine stored in Stockholm has already tested negative. Garrett and colleagues in England experimentally examined the survival of human and simian immunodeficiency viruses in oral polio vaccine formulations; no live retrovirus came through the procedure. Wistar declared that they would release lab specimens from a polio vaccine project carried out at the end of the 1950s in Africa for examination in two independent laboratories, in order to dispel claims that Wistar scientists inadvertently caused the AIDS epidemic.

It should be mentioned that the majority of scientists believe that the AIDS epidemic began after the simian immunodeficiency virus was transmitted from chimpanzees to humans during the slaughter of chimpanzees as early as the 1930s (13–16).

Finally, it is worth repeating the statement of the US Centers for Disease Control and Prevention, issued in 1992 but still valid (17): "The suggestion that HIV, the AIDS virus, originated as a result of inadvertent inoculation of an HIV-like virus present in monkey kidney cell cultures used to prepare polio vaccine is one of a number of unsubstantiated hypotheses. The weight of scientific evidence does not support this idea and there is no more reason to believe this hypothesis than many other which have been considered and rejected on scientific grounds."

Nevertheless, there are important lessons to be learned from Hooper's book. For many years, virologists and regulatory authorities have been worried that using permanent cell lines for vaccine virus propagation may somehow transfer cancer-causing properties and animal viruses. African green monkey kidneys are still used as the main cell substrate for oral polio vaccine. Millions of doses have been made from simian immunodeficiency virus (SIV)-positive monkeys before screening was introduced. Now there are well-tested non-oncogenic cell substrates, and it is time to reopen the debate on the use of primary cells versus cell lines for live attenuated virus vaccines. There is also a need to strengthen research on the sources of AIDS. However, the main focus of AIDS research should be prevention and treatment.

Global eradication of poliomyelitis and immunization

Using oral vaccine, the global campaign to eradicate polio achieved a more than 90% reduction in the number of polio cases worldwide in the 11 years since it was launched, and is on track to eradicate polio (18). It is to be expected that in a few years both poliomyelitis and polio vaccines will be confined to history books and science museums. Polio immunization will stop for good. When the final goal of global eradication is achieved, a decision on when and how to stop polio immunization will be necessary. Such a decision has not yet been prepared. Different scenarios are under discussion: continuation of universal immunization programs, sequential removal of one or two of the Sabin strains of OPV, change to an all-IPV program, and discontinuation of OPV immunization simultaneously worldwide or selectively country by country (19). The Technical Consultative Group for the World Health Organization on the Global Eradication of Poliomyelitis has the task of elaborating a proposal for final approval by the World Health Assembly (20).

Organs and Systems

Cardiovascular

Myopericarditis has been attributed to Td-IPV vaccine (21).

- A 31-year-old man developed arthralgia and chest pain 2 days after Td-IPV immunization and had an acute myopericarditis. He recovered within a few days with high-dose aspirin.

The authors discussed two possible causal mechanisms, natural infection or an immune complex-mediated mechanism. Infection was excluded by negative bacterial and viral serology and a favorable outcome resulted within a few days without antimicrobial drug treatment.

Nervous system

Vaccine-associated poliomyelitis

There is no doubt that OPV can cause poliomyelitis in a minority of recipients and their contacts. The risk is very low, but the seriousness of such a complication is evident. Various case reports and national surveys of vaccine-associated poliomyelitis, as well as studies quantifying this risk for OPV, have been published (SED-12, 820) (SEDA-16, 390) (SEDA-18, 336). The main conclusions from the field study that WHO conducted over a period of 15 years in 13 countries were as follows:

- The type 1 strain is almost never implicated in vaccine-associated poliomyelitis cases.
- The type 2 strain is an occasional cause of paralysis, commoner in contacts of the vaccine than in recipients.
- Most of the very small number of cases that do occur are due to type 3, both in vaccine recipients and in contacts.

In general, the risk of vaccine-associated poliomyelitis is less than one per million children immunized. The type 1 strain has been confirmed to be as safe and effective as any biological substance can be. The type 2 strain is safe

for recipients of the vaccine but on rare occasions can cause paralysis in contacts, so that any contact whose immunization status is doubtful should be immunized at the same time as the original vaccinee. The type 3 strain is much less stable genetically than the two strains and requires constant monitoring in the laboratory and in the field (22).

- A 22-year-old woman, who had never been immunized against poliomyelitis, was probably exposed during travel in Latin America through contact with an infant who had recently been immunized with OPV (23). She developed paralytic poliomyelitis. Stool specimens were positive for Sabin-strain poliovirus type 2 and 3. Sixty days after the onset of weakness, she had residual weakness in both legs.

This is the first known occurrence of imported vaccine-associated paralytic poliomyelitis in an unimmunized US adult who travelled abroad.

Frequency
The risk of vaccine-associated poliomyelitis has remained exceedingly low but stable since the mid-1960s. In all, 260 cases of vaccine-associated poliomyelitis were reported in the USA between 1961 and 1989. Cases of vaccine-associated poliomyelitis appeared to occur randomly in time and One potential cluster of vaccine-associated poliomyelitis consisting of six cases occurring over an 18-month period in Indiana was investigated during the period 1980–89. These cases were not shown to be epidemiologically related.

Using the total number of 80 vaccine-associated poliomyelitis cases in the USA from 1980 to 1989 as a numerator, the overall risk of vaccine-associated poliomyelitis was one case per 2.5 million doses of trivalent OPV distributed. The overall risk of vaccine-associated poliomyelitis was 9.7 times greater after the first dose of OPV than after all subsequent doses. The overall risk of recipient vaccine-associated poliomyelitis was one case per 6.8 million doses. The risk among OPV recipients was 29.0 times higher after the first dose than after subsequent doses. Comparing the first dose of OPV to subsequent doses, the lowest relative risk was found in immunologically abnormal persons. Whereas the average annual rate of vaccine-associated poliomyelitis for the period 1980–89 was 0.34 cases per million population, the annual incidence of vaccine-associated poliomyelitis among immunologically competent infants, who were used as the reference group, was 7.6 cases per 10 million population. Children under 1 year of age with a primary immunodeficiency were at highest risk of vaccine-associated poliomyelitis (annual rate of 16 216 cases per 10 million population, or 0.16%), which is more than 2000 times higher than the rate in the reference group. Among household contacts of children below 6 years of age, the annual rate of contact vaccine-associated poliomyelitis was 0.45 cases per 10 million population. For the remaining US population, the annual rate of vaccine-associated poliomyelitis was 0.14 cases per 10 million population. Summarizing their experience (24), the authors of the 1980–89 study distinguish three groups at risk of vaccine-associated poliomyelitis:

- recipients of OPV, especially infants receiving their first dose of OPV;
- persons in contact with OPV recipients, mostly unimmunized or inadequately immunized adults;
- immunologically abnormal individuals.

There has been a retrospective cohort study of cases of acute flaccid paralysis reported to the Ministry of Health in Brazil between 1989 and 1995 (25,26). For the first dose of OPV the estimated risk was one case of vaccine-associated paralytic poliomyelitis per 2.39 million doses; for total doses of OVP the risk was one case in 13.03 million doses. Most of the cases of vaccine-associated paralytic poliomyelitis were in children with a mean age of 1 year. Paralysis of the lower limbs caused by poliovirus type 2 was dominant.

Poliomyelitis caused by vaccine-derived polioviruses Between 12 July and 18 November 2000, a total of 19 people (aged between 9 months and 21 years) with acute flaccid paralysis were identified in the Dominican Republic; one case occurred in Haiti (August 2000) (27). The case in Haiti and three of the cases in the Dominican Republic were laboratory-confirmed with poliovirus type 1 isolates. All cases were either unimmunized or incompletely immunized. The outbreak was unusual, because the virus is derived from oral polio vaccine virus, with 97% genetic similarity to the parental OPV strain. Normally, vaccine-derived isolates are more than 99.5% similar to the parent strain. In contrast, wild polioviruses normally have less than 82% genetic similarity to OPV. The differences in nucleotide sequences suggest that the virus causing the outbreak has been circulating for about 2 years in the area in which immunization coverage is very low, and that the virus had accumulated genetic changes that restored the essential properties of wild poliovirus. A mass immunization with OPV brought the outbreak under control.

Causative strains
In an outbreak in Egypt during 1988–93, 32 cases of polio were associated with vaccine-derived poliovirus type 2 (28). Nucleotide sequence analysis performed during 1999 showed that all isolates were related (93–96% similarity) to the OPV 2 vaccine strain. The isolates were not related (less than 81% similarity) to the wild poliovirus type 2 that had been indigenous in Egypt. OPV was probably low in the affected communities.

Between 15 March and 26 July 2001, three cases of acute flaccid paralysis associated with vaccine-derived polioviruses type 1 were reported in the Philippines (29). There was a 3% genetic sequence difference between OPV type 1 virus and the vaccine-derived isolates.

Mechanism
The molecular biological findings connected with the occurrence of vaccine-associated poliomyelitis have been summarized (30). There is convincing evidence that the Sabin 2 and 3 viruses themselves can revert to a neuro-virulent phenotype on passage in man. The authors report that a point mutation in the 5′-non-coding region of the genome of the poliovirus type 3 vaccine consistently reverts to the wild type in viruses isolated from cases of vaccine-associated poliomyelitis.

Prolonged poliovirus excretion
Molecular studies of poliovirus isolates have suggested that viral replication of vaccine-related polioviruses may have persisted for as long as 7 years in a patient with vaccine-associated paralytic poliomyelitis, in whom common variable immunodeficiency syndrome had previously been diagnosed (SEDA-21, 336).

DTP injections
The risk of DTP injection in provoking paralytic poliomyelitis during a large poliomyelitis outbreak in Oman has been evaluated (31). Health center immunization records for 70 children aged 5–24 months with confirmed poliomyelitis and 692 control children were reviewed. A significantly higher proportion of case-patients received a DTP injection within 30 days before onset of paralysis compared with controls (43 versus 28%). All the patients for whom this information was available had paralysis of the injected limb. This study, which provided the first quantitative estimate in this respect, confirmed that injections are an important cause of provocative poliomyelitis and stressed the recommendation to avoid unnecessary injections during poliomyelitis outbreaks.

Multiple injections
Studies in 1994–95 in Romania produced evidence that multiple (unnecessary) injections may actually increase the risk of vaccine-associated poliomyelitis (SEDA-19, 302).

Between 1976 and 1985, seven cases of neurological disease were reported to have occurred in Germany among young children after simultaneous administration of oral poliovirus vaccine and diphtheria–tetanus toxoids or diphtheria–tetanus–pertussis vaccine (32). However, the virological data were incomplete; only one case was confirmed by the isolation of a vaccine-like polio virus, and in three cases the clinical symptoms did not correspond to poliomyelitis. The author concluded that in some cases the simultaneous administration of injectable vaccines cannot be excluded as a cause for paralysis.

Acute disseminated encephalomyelitis
Acute disseminated encephalomyelitis associated with polio vaccine has been reported (33).

- A 6-year-old girl developed acute disseminated encephalomyelitis, and polio vaccine virus type 2 was isolated from her cerebrospinal fluid and pharynx. The virus was sequenced throughout the 5′ non-coding region of the genome by polymerase chain reaction and was determined to have undergone various mutations at nucleotides 481, 500, 795, and 1195. The clinical signs of disease had completely disappeared 2 months later.

Guillain–Barré syndrome

Coincident with the mass nationwide campaign of OPV immunization that interrupted the transmission of wild poliovirus in Finland in 1985, with 4.5 million doses of OPV being administered, there was an unexpected rise in the occurrence of Guillain–Barré syndrome. In 10 cases of Guillain–Barré syndrome with onset of symptoms within 10 weeks after OPV immunization, no specific agents were found (34). The authors concluded that live attenuated polioviruses might trigger Guillain–Barré syndrome.

The authors of a report of the Institute of Medicine concluded that the evidence favored a causal relation between OPV and Guillain–Barré syndrome, but considered the evidence inadequate to accept or to reject a causal relation between OPV and transverse myelitis (SEDA-18, 325) (3).

The available reports on a possible association between polio vaccine and Guillain–Barré syndrome have been reviewed (35). The conclusion of a 1994 US Institute of Medicine committee that the evidence favored acceptance of a causal relation was mainly based on two reports from Finland (36,37). An earlier study of Guillain–Barré syndrome and oral polio vaccine has been extended to include hospital reports from the whole of Finland for 1981–6, during which time an oral polio vaccine campaign was carried out to control an outbreak of poliomyelitis (38). The rise in the numbers of cases of .Guillain–Barré syndrome started before the immunization campaign. Because there had also been an influenza epidemic during that time, the researchers acknowledged that the increase in the incidence of Guillain–Barré syndrome could also have been associated with influenza. Data from the Americas are also not supporting an association between oral polio immunization campaigns and Guillain–Barré syndrome (39).

Sensory systems

Optic neuritis has been attributed to Td-IPV vaccine (40).

- Ten days after receiving Td-IPV vaccine a 56-year-old woman developed acute unilateral optic neuritis. Complete remission occurred within 6 weeks of prednisolone treatment. No other causes were found.

Gastrointestinal

Intussusception probably causally related to rotavirus vaccine (SEDA-23, 354) prompted studies to answer the question of whether polio vaccine could also cause intussusception. A workshop held in Atlanta on 15–16 June 2000 brought together experts from various fields with the primary investigators of the studies. The participants concluded that the available evidence favored rejection of a causal relation between OPV and intussusception (41).

Skin

Bullous pemphigoid has been attributed to DTP-IPV vaccine (42).

- A previously healthy 3.5-month-old infant developed bullous pemphigoid 3 days after receiving a first dose of DTP-IPV vaccine. *Staphylococcus aureus* was isolated from purulent bullae. The lesions resolved rapidly after treatment with antibiotics and methylprednisolone.

The authors mentioned 12 other cases of bullous pemphigoid, reported during the last 5 years, that had possibly been triggered by vaccines (influenza, tetanus toxoid booster, and DTP-IPV vaccine).

Long-Term Effects

Tumorigenicity

Simian virus 40 (SV40) occurs naturally in some species of monkeys. Poliovaccine made in the 1950s could be contaminated, because rhesus monkey kidney cells were used in preparing the vaccine. After the discovery of SV40 in 1960 and the finding in 1961 that the virus can cause tumors in rodents, SV40 was completely removed from the seed strains of the vaccine virus in the early 1960s. However, possibly contaminated polio vaccines were used until at least 1963. Interest in SV40 has increased in recent years, because the virus was found in certain forms of cancer in humans, for instance mesotheliomas, brain and bone tumors, and more recently some types of non-Hodgkin's lymphoma. It has not been determined that SV40 causes these cancers (43).

In 2002, the Vaccine Safety Committee of the Institute of Medicine of the US Academies of Sciences, reviewed the epidemiological evidence of an association between exposure to polio vaccines containing SV40 and the subsequent development of cancer (44). The available studies on cancer incidence, cancer mortality, and cancers after prenatal exposure to SV40-containing vaccines were reviewed. The committee concluded that

- there is strong biological evidence that SV 40 is a transforming virus;
- there is moderate biological evidence that SV40 exposure could lead to cancer in humans under natural conditions;
- there is moderate biological evidence that SV40 exposure from the polio vaccine is related to SV40 in humans;
- the evidence is inadequate to accept or reject a causal relation between SV40-containing polio vaccines and cancer.

Second-Generation Effects

Pregnancy

Pregnancy is not a contraindication to the use of IPV. Although there is no convincing evidence documenting adverse effects of OPV on the developing fetus, and some evidence points to safety, it is widely considered prudent to avoid immunizing pregnant women, especially during the first 4 months of pregnancy. However, if immediate protection against poliomyelitis is needed, poliomyelitis immunization is recommended (45,46).

In order to interrupt poliovirus transmission during an outbreak of wild poliomyelitis in Israel in 1988, some 90% of the population, including pregnant women, were immunized with trivalent OPV. In a study of the abortions that occurred within 4 months of immunization compared with figures from a similar period in the previous year, the number of spontaneous abortions did not differ between women immunized during the first trimester of pregnancy and the controls (47). The authors concluded that OPV administered during early pregnancy has no adverse effect on the embryo or placenta that would cause increased fetal deaths or spontaneous abortions, nor does it seem to cause a higher rate of congenital anomalies.

An analysis of the OPV mass campaign in Finland suggested that OPV during early pregnancy had no harmful effects on fetal development (48). There were no significant deviations from the baseline prevalence for all malformations. However, when the vaccine is administered later in pregnancy the prospects may be different.

- Irreparable damage to the anterior horn cells of the cervical and thoracic cord occurred in a 20-week-old fetus whose mother was immune to poliomyelitis before conceiving but who was inadvertently given OPV at 18 weeks gestation (49).

Susceptibility Factors

There are no known contraindications to the use of IPV. OPV should not be given to persons who are immunocompromised due to immunodeficiency diseases, leukemia, lymphoma or generalized malignancy or who are immunosuppressed due to therapy with glucocorticoids, alkylating drugs, antimetabolites, or radiation. If poliomyelitis immunization is indicated in such persons, IPV should be used. OPV should also be avoided when immunizing household contacts of immunocompromised patients.

In the view of the WHO, live vaccines should in general not be given to immunocompromised individuals, but in developing countries, the risk of poliomyelitis in non-immunized infants is high and the risk from these vaccines, even in the presence of symptomatic HIV infection, appears to be lower (50).

The Immunization Advisory Committees of many developed countries recommend that OPV should not be given to children and young adults who are immunocompromised due to AIDS or other clinical manifestations of HIV infection. OPV can be given to asymptomatic infected persons. However, because family members may be immunocompromised due to AIDS or HIV infection, it seems prudent to use IPV (51).

References

1. Recommendations of the Advisory Committee (ACIP). The 2000 Immunization Schedule. MMWR Morb Mortal Wkly Rep 2000;49:25.
2. Centers for Disease Control. Adverse Events Following Immunization. Surveillance Report No. 2, US Department of Health and Human ServicesAtlanta, GA: Public Health Service, CDC;. 1986.
3. In: Stratton KR, Howe CJ, Johnson Jr. RB Jr, editors. Adverse Events Associated with Childhood Vaccines. Washington, DC: National Academy of Sciences, 1994:681–8.
4. Mills E, Gold R, Thipphawong J, Barreto L, Guasparini R, Meekison W, Cunning L, Russell M, Harrison D, Boyd M, Xie F. Safety and immunogenicity of a combined five-component pertussis-diphtheria-tetanus-inactivated poliomyelitis-Haemophilus B conjugate vaccine administered to infants at two, four and six months of age. Vaccine 1998;16(6):576–85.
5. Mallet E, Fabre P, Pines E, Salomon H, Staub T, Schodel F, Mendelman P, Hessel L, Chryssomalis G, Vidor E, Hoffenbach A, Abeille A, Amar R, Arsene JP, Aurand JM, Azoulay L, Badescou E, Barrois S, Baudino N, Beal M, Beaude-Chervet V, Berlier P, Billard E, Billet L, Blanc B, Blanc JP, Bohu D, Bonardo C, Bossu CHexavalent Vaccine Trial Study Group. Immunogenicity and safety of a new liquid hexavalent combined vaccine compared with separate administration of reference licensed vaccines in infants. Pediatr Infect Dis J 2000;19(12):1119–27.
6. Anonymous. Wkly Epidemiol Rec 1985;35:269.
7. Curtis T. Possible origins of AIDS. Science 1992;256(5061):1260–1.
8. Butel JS, Lednicky JA. Cell and molecular biology of simian virus 40: implications for human infections and disease. J Natl Cancer Inst 1999;91(2):119–34.
9. Griffiths E. Personal communication, 1997.
10. World Health Organization. Hot topics: statement on simian virus (SV40) and polio vaccine. http://www.who.int/vaccines-diseases/safety/hottop/sv40.htm, 14/09/2000.
11. Hooper E. The River: a journey to the source of HIV and AIDSBoston, New York, London: Little, Brown and Company;. 1999.
12. Plotkin SA, Koprowski H. New York Times 7 December 1999;.
13. Wakefield AJ, Pittilo RM, Sim R, Cosby SL, Stephenson JR, Dhillon AP, Pounder RE. Evidence of persistent measles virus infection in Crohn's disease. J Med Virol 1993;39(4):345–53.
14. van der Meijden AP. Practical approaches to the prevention and treatment of adverse reactions to BCG. Eur Urol 1995;27(Suppl 1):23–8.
15. Vegt PD, van der Meijden AP, Sylvester R, Brausi M, Holtl W, de Balincourt C, Andriole GL. Does isoniazid reduce side effects of intravesical bacillus Calmette–Guérin therapy in superficial bladder cancer? Interim results of European Organization for Research and Treatment of Cancer Protocol 30911. J Urol 1997;157(4):1246–9.

16. Tuncer S, Tekin MI, Ozen H, Bilen C, Unal S, Remzi D, Lamm DL. Detection of Bacillus Calmette–Guérin in the blood by the polymerase chain reaction method of treated bladder cancer patients. J Urol 1997;158(6):2109–12.

17. US Department of Health and Human Services, Public Health Service, Centers for Disease Control. Origin of HIV. Press release, 6 March 1992.

18. World Health Organization. Bovine spongiform encephalitis and oral polio vaccine. Position statement. October 2000. http://www.who.int/vaccines-diseases/safety/hottop/bse.htm, 11/10/2000.

19. Wood DJ, Sutter RW, Dowdle WR. Stopping poliovirus vaccination after eradication: issues and challenges. Bull World Health Organ 2000;78(3):347–57.

20. Technical Consultative Group to the World Health Organization on the Global Eradication of Poliomyelitis. "Endgame" issues for the global polio eradication initiative. Clin Infect Dis 2002;34(1):72–7.

21. Boccara F, Benhaiem-Sigaux N, Cohen A. Acute myopericarditis after diphtheria, tetanus, and polio vaccination. Chest 2001;120(2):671–2.

22. Cockburn WC. The work of the WHO Consultative Group on Poliomyelitis Vaccines. Bull World Health Organ 1988;66(2):143–54.

23. Imported vaccine-associated paralytic poliomyelitis (VAAP)—United States, 2005. MMWR Morb Mortal Wkly Rep 2006, 55:97–9.

24. Strebel PM, Sutter RW, Cochi SL, Biellik RJ, Brink EW, Kew OM, Pallansch MA, Orenstein WA, Hinman AR. Epidemiology of poliomyelitis in the United States one decade after the last reported case of indigenous wild virus-associated disease. Clin Infect Dis 1992;14(2):568–579.

25. de Oliveira LH, Struchiner CJ. Vaccine-associated paralytic poliomyelitis in Brazil, 1989–1995. Rev Panam Salud Publica 2000;7(4):219–24.

26. de Oliveira LH, Struchiner CJ. Vaccine-associated paralytic poliomyelitis: a retrospective cohort study of acute flaccid paralyses in Brazil. Int J Epidemiol 2000;29(4):757–63.

27. Anonymous. Poliomyelitis, Dominican Republic and Haiti. Wkly Epidemiol Rec 2000;75(49):397–9.

28. Anonymous. Acute flaccid paralysis associated with circulating vaccine-derived poliovirus, Philippines, 2001. Wkly Epidemiol Rec 2001;76(41):319–20.

29. Centers for Disease Control and Prevention (CDC). Circulation of a type 2 vaccine-derived poliovirus—Egypt, 1982–1993. MMWR Morb Mortal Wkly Rep 2001;50(3):41–251.

30. Evans DM, Dunn G, Minor PD, Schild GC, Cann AJ, Stanway G, Almond JW, Currey K, Maizel JV Jr. Increased neurovirulence associated with a single nucleotide change in a noncoding region of the Sabin type 3 poliovaccine genome. Nature 1985;314(6011):548–50.

31. Sutter RW, Patriarca PA, Suleiman AJ, Brogan S, Malankar PG, Cochi SL, Al-Ghassani AA, el-Bualy MS. Attributable risk of DTP (diphtheria and tetanus toxoids and pertussis vaccine) injection in provoking paralytic poliomyelitis during a large outbreak in Oman. J Infect Dis 1992;165(3):444–9.

32. Ehrengut W. Role of provocation poliomyelitis in vaccine-associated poliomyelitis. Acta Paediatr Jpn 1997;39(6):658–62.

33. Ozawa H, Noma S, Yoshida Y, Sekine H, Hashimoto T. Acute disseminated encephalomyelitis associated with poliomyelitis vaccine. Pediatr Neurol 2000;23(2):177–9.

34. Kinnunen E, Farkkila M, Hovi T, Juntunen J, Weckstrom P. Incidence of Guillain-Barré syndrome during a nationwide oral poliovirus vaccine campaign. Neurology 1989;39(8):1034–6.

35. Salisbury DM. Association between oral poliovaccine and Guillain-Barré syndrome? Lancet 1998;351(9096):79–80.

36. Uhari M, Rantala H, Niemela M. Cluster of childhood Guillain-Barré cases after an oral poliovaccine campaign. Lancet 1989;2(8660):440–1.

37. Uhari M, Rantala H, Niemela M. Cluster of childhood Guillain–Barré cases after an oral poliovaccine campaign. Lancet 1989;2(8660):440–1.

38. Kinnunen E, Junttila O, Haukka J, Hovi T. Nationwide oral poliovirus vaccination campaign and the incidence of Guillain–Barré syndrome. Am J Epidemiol 1998;147(1):69–73.

39. Olive JM, Castillo C, Castro RG, de Quadros CA. Epidemiologic study of Guillain–Barré syndrome in children <15 years of age in Latin America. J Infect Dis 1997;175(Suppl 1):S160–4.

40. Burkhard C, Choi M, Wilhelm H. Optikusneuritis als Komplikaton einer Tetanus–Diphtherie–Poliomyelitis–Schutzimpfung: ein Fallbericht. [Optic neuritis as a complication in preventive tetanus-diphtheria-poliomyelitis vaccination: a case report.] Klin Monatsbl Augenheilkd 2001;218(1):51–4.

41. Anonymous. Oral poliovirus vaccine (OPV) and intussusception. Wkly Epidemiol Rec 2000;75(43):345–7.

42. Baykal C, Okan G, Sarica R. Childhood bullous pemphigoid developed after the first vaccination. J Am Acad Dermatol 2001;44(Suppl 2):348–50.

43. Centers for Disease Control and Prevention (CDC), National Immunization Program. Simian virus 40 (SV40), poliovaccine, and cancer. Questions and Answers. htpp://www.cdc.gov/nip/vacsafe/concerns/cancer/default.htm (accessed July 24, 2004).

44. Immunization Safety Review Committee of the Institute of Medicine, National Academies. Immunization Safety Review: SV40 contamination of poliovaccine and cancer. Washington, DC: The National Academies Press, 2002.

45. Centers for Disease Control (CDC). Poliomyelitis prevention. MMWR Morb Mortal Wkly Rep 1982;31(3):22–631–4.

46. Department of Health and Social Security. Immunization against infectious diseaseLondon: Her Majesty's Stationery Office;. 1992.

47. Ornoy A, Arnon J, Feingold M, Ben Ishai P. Spontaneous abortions following oral poliovirus vaccination in first trimester. Lancet 1990;335(8692):800.

48. Harjulehto T, Aro T, Hovi T, Saxen L. Congenital malformations and oral poliovirus vaccination during pregnancy. Lancet 1989;1(8641):771–2.

49. Burton AE, Robinson ET, Harper WF, Bell EJ, Boyd JF. Fetal damage after accidental polio vaccination of an immune mother. J R Coll Gen Pract 1984;34(264):390–4.

50. Global Advisory Group of the Expanded Programme on Immunization (EPI). Report on the meeting 13–17 October, 1986, New Delhi. WHO/EPI/Geneva/87/1, 1986.

51. Centers for Disease Control (CDC). Immunization of children infected with human T-lymphotropic virus type III/lymphadenopathy-associated virus. MMWR Morb Mortal Wkly Rep 1986;35(38):595–8603–6.

Rabies vaccine

General Information

Rabies vaccine was for a long time prepared from infected brain tissue, and neurological complications were likely to occur in as many as one in 300 cases. A second-generation vaccine prepared from duck embryo tissue has been found to be better tolerated in this respect, and third-generation vaccines prepared in non-neural tissue cultures or in human diploid cell (HDC) vaccines with a progressive improvement in safety have now become the vaccines of choice where they are available.

Any rabies vaccine can apparently cause mild local discomfort and swelling, and either of the animal preparations can result in hypersensitivity reactions, for example pyrexia, serum sickness, or urticaria; sensitization can occur.

Third-generation rabies vaccines include a purified vero-cell rabies vaccine (PVRV) and HDC vaccine purified by zonal centrifugation. In clinical trials (SEDA-12, 284) (SEDA-15, 357) the improved vaccines were well tolerated. In 1988, Rabies Vaccine Adsorbed (RVA), a new cell culture-derived rabies vaccine (Kissling strain of rabies virus adapted to a diploid cell line of the fetal rhesus lung cells in medium-free of human albumin and adsorbed on aluminium phosphate) for human use was licensed in the USA. Reactions after primary immunization are similar to those observed with HDC. Systemic allergic reactions have been reported at a rate <1% (HDC 6%) (1).

Organs and Systems

Nervous system

Brain tissue rabies vaccine

There are many countries that still use first-generation brain tissue rabies vaccine because of limited financial resources and where neurological complications after immunization still occur. The neurological effects produced by this type of vaccine are very variable. Sometimes a peripheral neuropathy occurs, while other vaccinees develop encephalomyelitis, dorsolumbar myelitis, ascending myelitis, hemiplegia, or general subjective neurological symptoms, such as stiffness of the neck or physical weakness.

It has been argued that there may be a risk of transmitting scrapie (spongiform encephalopathy of sheep, similar to bovine spongiform encephalopathy of cattle) through blood donation from individuals who have been immunized against rabies with sheep brain rabies vaccine still used in developing countries (2).

Duck embryo rabies vaccine (DEV)

Nervous system complications after administration of DEV occurred, but in much lower frequency than after brain tissue rabies vaccines. Reports of transverse myelitis and other neurological complications have been published (SEDA-8, 300). There have been some efforts to develop a more purified version of DEV, but improvements in third-generation vaccines have made these the vaccines of choice when they can be afforded, much reducing the motivation to develop duck embryo vaccines further.

Human diploid cell (HDC) vaccines

The major advantage of third-generation rabies vaccines prepared in non-neural tissue cultures or in human diploid cells is the greatly reduced risk of neurological complications. Millions of doses of these vaccines have been administered worldwide since 1974, and only a few neurological complications have been reported. The overall risk of neurological complications is estimated to be less than 1 per 150 000 (SEDA-15, 356), and complications after the administration of third-generation rabies vaccines cultured in non-neural tissues have been very rarely reported. However, those that have been published are of various types.

Isolated case reports of Guillain–Barré syndrome after HDC vaccine have been published (SED-11, 684) (SEDA-15, 356) (3). Polyneuropathy and oculomotor nerve impairment have been reported after Russian cell-culture vaccine (SEDA-12, 284).

- Two weeks after the second injection of HDC vaccine a 45-year-old farmer developed meningoradiculitis. The symptoms regressed spontaneously (4).
- A 25-year-old veterinary practitioner developed an inflammatory demyelinating process affecting the central nervous system 8 days after the second HDC vaccine immunization (5).
- Acute disseminated encephalomyelitis developed in a 15-year-old boy after HDC vaccine immunization (SEDA-21, 337).
- A 14-year-old boy had a seizure after the administration of human diploid rabies cell vaccine simultaneously with rabies immune globulin. The symptoms developed within minutes after injection. After treatment and about 2 hours later, his mental status returned to normal (6).

Immunologic

In recipients given primary courses of third-generation rabies vaccines, only mild local and systemic reactions were reported by about 20%. Systemic allergic reactions have occurred in 11 per 10 000 vaccinees (7). However, 2–21 days after the administration of HDC vaccine, about 5% of patients receiving booster injections for pre-exposure prophylaxis and a few receiving postexposure primary immunization develop an immune complex (serum sickness-like) reaction, including urticaria, fever, malaise, arthralgias, arthritis, nausea, and vomiting. This syndrome may prove to be less common with RVA, but direct comparisons are lacking.

Anaphylaxis has been reported rarely after HDC vaccine prophylaxis (3).

References

1. Anonymous. Human rabies prophylaxis, 1987. Wkly Epidemiol Rec 1988;13:357.
2. Arya SC. Blood donated after vaccination with rabies vaccine derived from sheep brain cells might transmit CJD. BMJ 1996;13(7069):1405.
3. Anonymous. Rabies vaccine. Med Lett 1991;117–8.
4. Moulignier A, Richer A, Fritzell C, Foulon D, Khoubesserian P, de Recondo J. Méningo-radiculite secondaire à une vaccination antirabique. [Meningoradiculitis after injection of an antirabies vaccine. A vaccine from human diploid cell culture.] Presse Méd 1991;20(24):1121–3.
5. Tornatore CS, Richert JR. CNS demyelination associated with diploid cell rabies vaccine. Lancet 1990;335(8701):1346–7.
6. Mortiere MD, Falcone AL, Plotkin SA, Loupi E, Lang J. An acute neurologic syndrome temporally associated with post-exposure treatment of rabies. Pediatrics 1997;100(4):718–21.
7. Committee on Immunization. Guide for Adult Immunization. Philadelphia: American College of Physicians, 1985.

Rotavirus vaccine

General Information

Rotavirus is the most common cause of severe gastroenteritis in infants and young children aged under 5 years in the USA, resulting in approximately 500 000 physician visits, 50 000 hospitalizations, and 20 deaths each year. Worldwide, rotavirus is a major cause of childhood death, accounting for an estimated 600 000 deaths annually among children aged under 5 years. Rotavirus vaccines offer the opportunity to reduce substantially the occurrence of this disease.

Various oral rotavirus vaccines (live attenuated rotavirus vaccines, including rhesus rotavirus and bovine rotavirus strains; serotype 1 bovine-human rotavirus reassortant vaccine; and both rhesus rotavirus monovalent and tetravalent reassortant vaccines) have been evaluated in clinical trials (SEDA-21, 292).

On August 31, 1998, a tetravalent rhesus-based rotavirus vaccine (RRV-TV) (Rotashield) was licensed in the USA. The Advisory Committee on Immunization Practices (ACIP), the American Academy of Pediatrics, and the American Academy of Family Physicians all recommended routine use of the vaccine in healthy infants. Rotashield was the first rotavirus vaccine to be licensed.

Observational studies

The safety and immunogenicity of two human-bovine reassortant rotavirus candidate vaccines have been evaluated in infants, children, and adults (1). One candidate vaccine contained a single human rotavirus gene, while the other candidate vaccine contained two human rotavirus genes. The remaining genes for both vaccines were derived from bovine rotavirus strain UK. Each of these vaccines was well tolerated and immunogenic. This was also the case in the important group of infants under 6 months of age after a single dose of vaccine.

Placebo-controlled studies

The safety, immunogenicity, and efficacy of a live oral human rotavirus vaccine 89–12 has been assessed in 213 US infants in a randomized, placebo-controlled, double-blind, multicenter trial (2). The infants received two doses of vaccine or placebo and were followed up through one rotavirus season. There was an immune response to vaccine in 94.4% of vaccinees, and vaccine efficacy was 89%. Adverse reactions were mild. Low-grade fever after the first dose was the only adverse effect that was significantly more common with the vaccine than with placebo.

Kawasaki disease

The Global Advisory Committee on Vaccine Safety (GACVS) has been presented with the latest information from the USA and the European Union on the potential association between Kawasaki disease and the administration of rotavirus vaccines (3). The US Centers for Disease Control and Prevention have defined a case of Kawasaki disease as an illness in a patient with fever lasting 5 or more days associated with at least four of the following five clinical signs: rash, cervical lymphadenopathy (at least 1.5 cm in diameter), bilateral conjunctival injection, oral mucosal changes, and peripheral extremity changes.

The possibility of an association had previously been raised after the observation of a non-statistically significant excess of Kawasaki disease in pre-licensure trials of Rotateq™ vaccine (five cases in vaccinees versus one case in controls). In June 2007, the FDA amended the product information in the USA to note the occurrence of such cases, but stated that causality had not been established.

Since June 2007 and as of 14 October 2007, a total of 16 cases of confirmed cases of Kawasaki disease had been reported to the US Vaccine Adverse Events Reporting System (VAERS) in the context of stimulated reporting. The reported rate was 1.6 per 100 000 person-years compared with an estimated background rate of 17 per 100 000 person-years among children aged under 5 years.

Active surveillance during the first year of life after immunization with Rotateq™ was implemented in the Vaccine Safety Datalink. Preliminary results suggest that there has been no confirmed case of Kawasaki disease after the administration of over 125 000 doses. Further work is planned to extend the monitoring. In the European Union, where Rotarix™ vaccine is licensed and in use, there have been no spontaneous reports of Kawasaki disease, with about 12 million doses administered in 6 million individuals. In clinical trials there has been a slight excess of cases in the Rotarix™ groups, with a non-statistically significant difference between the Rotarix™ and the placebo arms. Delay between immunization and the onset of Kawasaki disease varied from 2 weeks to 19 months. The overall conclusion of the GACVS was that the evidence for a causal association was not strong and that there was no reason for concern.

The US Food and Drug Administration (FDA) has reported that five cases of Kawasaki disease (mucocutaneous lymph node syndrome of unknown etiology) have been identified in children under 1 year of age who received the Rotavirus (RotaTeq®) vaccine during clinical trials conducted before the vaccine was licensed (4). Three cases of Kawasaki disease were detected after the vaccine was approved in February 2006 through the Vaccine Adverse Event Reporting System (VAERS). After learning about these reports, CDC identified an additional unconfirmed case through its Vaccine Safety Datalink (VSD) Project. The number of reports of Kawasaki disease does not exceed the number of cases expected based on the usual occurrence of Kawasaki disease in children. There is not a known cause-and-effect relation between receiving RotaTeq® or any other vaccine and the occurrence of Kawasaki disease. The Global Advisory Committee on Vaccine Safety (GACVS) has recommended that current and future studies should incorporate surveillance for Kawasaki disease after immunization (5).

Organs and Systems

Gastrointestinal

Natural rotavirus itself is a causative agent for intussusception (6). In rotavirus disease in Japan, hyperplasia of the intestinal lymphoid tissue is a common pathological finding in necropsy cases, and is identical with that observed in patients who receive the tetravalent rhesus-based rotavirus vaccine.

During the period from 1 September 1998 to 7 July 1999, 15 cases of intussusception among infants who had received rotavirus vaccine were reported to the Vaccine Adverse Event Reporting System (VAERS). Of the 15 infants, 13 (87%) had developed intussusception after the first dose of the three-dose RRV-TV series, and 12 had developed symptoms within 1 week of receiving any dose of RRV-TV. Thirteen of the 15 had received other vaccines concurrently. Intussusception was confirmed radiographically in all 15 patients. Eight infants required surgical reduction, and one required resection of 18 cm of distal ileum and proximal colon. Histopathological examination of the distal ileum showed lymphoid hyperplasia and ischemic necrosis. All the infants recovered. The dates of onset ranged from 21 November 1998 to 24 June 1999. The median age of patients was 3 (range: 2–11) months. Ten were boys. Topographically the cases were scattered, being reported from seven states. Of the 15 cases reported to VAERS, 14 were spontaneous reports and one was identified through active postlicensing surveillance.

The rate of hospitalization for intussusception among infants aged less than 12 months during 1991–97 (before RRV-TV licensure) was 51 per 100 000 infant-years in New York. The manufacturer had distributed about 1.8 million doses of RRV-TV as of June 1 1999, and estimated that 1.5 million doses (83%) had been administered. Given this information, 14–16 intussusception

cases among infants would be expected by chance alone during the week after the administration of any dose of RRV-TV. Fourteen of the 15 case-patients were vaccinated before June 1 1999, and of those, 11 developed intussusception within 1 week of receiving RRV-TV.

In prelicensure studies of Rotashield, five cases of intussusception had occurred among 10 054 vaccine recipients and one of 4633 controls, a difference that was not statistically significant. Three of the five cases among vaccinated children occurred within 6–7 days of receiving rotavirus vaccine. On the basis of these data, intussusception was included in the package insert as a potential adverse reaction, and the ACIP recommended postlicensure surveillance for this adverse event following vaccination. In addition, because of concerns about the findings in prelicensing trials, VAERS data were analysed early in the postlicensure period (7).

Although the number of reported intussusception cases occurring within 1 week of receiving any dose of vaccine is in the expected range, it must be borne in mind that reporting of suspected adverse events to VAERS is far from complete, and the actual number of intussusception cases occurring among RRV-TV recipients may be substantially greater than that reported. The data available to date suggest but do not establish a causal association between receipt of rotavirus vaccine and intussusception. However, based on these results, use of the vaccine was suspended in July 1999, pending a review of the data by the Advisory Committee on Immunization Practice (ACIP) (8). In further studies more cases of intussusception (including two deaths) were associated with the administration of Rotashield. Based on the results of an expedited review of the scientific data (indicating a strong association between Rotashield and intussusception) presented to the ACIP, it was recommended in 22 October 1999 that Rotashield should no longer be used. The recommendation was made in cooperation with the FDA, NIH, and Public Health Service officials, along with the manufacturer, Wyeth-Lederle (9). In late 1999 the CDC recommended postponing administration of RRV-TV to children scheduled to receive the vaccine before November 1999, including those who had already have begun the RRV-TV series (7).

The epidemiology of hospitalizations and deaths associated with intussusception among US infants has been described (10). Such data could be useful for further clinical trials with newly developed rotavirus vaccines.

Further studies have confirmed that there is an increased risk of intussusception associated with the use of rhesus rotavirus tetravalent vaccine (RRV-TV). The association has been assessed in infants in 19 states of the USA (11). Each infant hospitalized with intussusception between 1 November 1998 and 30 June 1999 was matched according to age with four healthy controls who had been born at the same hospital. The authors estimated that one case of intussusception would occur for every 4670–9474 infants immunized.

In a retrospective cohort study in 10 managed care organizations there was an increased risk of intussusception in immunized children (12). The risk was greatest at 3–7 days after the first dose of RRV-TV.

In an editorial in the journal Emergency and Office Pediatrics the critical role of physicians in the early detection of adverse events was stressed. Despite the demands of busy clinical practice, all physicians are obliged to be alert to unusual circumstances and to seek answers when unexpected findings—especially related to new drugs, vaccines, or procedures—seem to be more than just coincidence. It is incumbent on them to report these findings and to learn if their experience can be confirmed by colleagues (13).

Current prelicensure trials of newly developed rotavirus vaccines are taking into consideration the experiences gained during the licensure process of Rotashield. Phase 3 clinical trials have included over 10 000 volunteers in order to exclude the potential risk of intussusception.

The two frontrunners in the race to replace RotaShield are GlaxoSmithKline and Merck & Co. Both are currently testing their oral vaccine candidates (GSK human strains, Merck a mixture of cow and human strains) for safety and efficacy. GSK is conducting clinical trials in 60 000 children in Latin America. Both companies say their vaccines do not appear to cause intussusception (14). Meanwhile, the US National Institutes of Health (NIH) has licensed the rights of RotaShield to Minneapolis-based BIOVIRx, which hopes to eliminate the risk of intussusception and re-introduce the vaccine into the US market (15).

The Global Advisory Committee on Vaccine Safety (GACVS) has reviewed data presented by the Centers for Disease Control and Prevention (CDC) and the FDA related to the risk of intussusception, which had been identified as associated with a previous rotavirus vaccine. With respect to Rotarix®, there was no evidence from any of the studies, which included over 30 000 vaccinees in trials and worldwide usage (about 5 million doses distributed), that there was an excess incidence of intussusception. The cases of intussusception that were reported did not show a pattern with regard to the time of onset after immunization consistent with a causal relation. The overall number of cases reported was much smaller than would have been expected based on applying rates for the normal incidence of intussusception to a population of the size immunized.

A study of RotaTeq® in over 30 000 individuals has been reported, and an observational cohort study of 44 000 immunized children is planned. Most of the data, especially from post-marketing studies and spontaneous reporting, relate to developed countries. There is no evidence that the rate of intussusception is raised above background, and it is certainly much less than the rate previously observed with the vaccine that was withdrawn after an association with intussusception was described. The VAERS spontaneous reporting data also showed a lower than expected rate of intussusception. GACVS concluded that the data, particularly those from developed countries, are reassuring (16). However, it was noted that the current data relate mainly to vaccines that are used in young children at the recommended age. It is important that intussusception should be monitored in developing countries as rotavirus vaccines are introduced, especially because infants are likely to present for their first dose of vaccine at slightly older ages on average than is the case in developed countries.

Body temperature

Administration of the first dose of rhesus rotavirus-based tetravalent vaccine in children aged 6 weeks and over is followed by a transient febrile reaction at 3–4 days in about one-third of vaccinees. It has been hypothesized that giving the first dose of the vaccine during the neonatal period might reduce its reactogenicity without compromising its use (17). In a double-blind placebo-controlled trial 90 infants received the vaccine at 0–4–6, 0–2–4, or 2–4–6 months of age. The authors concluded that infants who received the first dose of the vaccine during the neonatal period did not have febrile reactions. The immune response in a three-dose schedule initiated in the neonatal period is somewhat dampened but still acceptable. Neonatal immunization might also reduce the very small risk of intussusception, which has been associated with administration of rotavirus-based tetravalent vaccine to older infants.

References

1. Eichelberger MC, Sperber E, Wagner M, Hoshino Y, Dudas R, Hodgins V, Marron J, Nehring P, Casey R, Burns R, Karron R, Clements-Mann ML, Kapikian AZ. Clinical evaluation of a single oral dose of human-bovine (UK) reassortant rotavirus vaccines Wa × UK (P1A[8], G6) and Wa × (DS-1 × UK) (P1A[8], G2) J Med Virol 2002;66(3):407–16.

2. Bernstein DI, Sack DA, Rothstein E, Reisinger K, Smith VE, O'Sullivan D, Spriggs DR, Ward RL. Efficacy of live, attenuated, human rotavirus vaccine 89-12 in infants: a randomised placebo-controlled trial. Lancet 1999;354(9175):287–90.

3. Rotavirus vaccines and Kawasaki disease. Wkly Epidemiol Rec 2008;83:43.

4. US Food and Drug Administration. Information pertaining to labeling revision for Rotateq. June 15, 2007. <http://www.fda.gov/cber/label/rotateqLBinfo.htm> (accessed 16 August, 2007).

5. Kawasaki Disease and RotaTeq® Vaccine <http://www.cdc.gov/od/science/iso/concerns/kawasaki_disease_rotavirus.htm> (last accessed 26 July 2007).

6. Suzuki H, Katsushima N, Konno T. Rotavirus vaccine put on hold. Lancet 1999;354(9187):1390.

7. Centers for Disease Control and Prevention (CDC). Intussusception among recipients of rotavirus vaccine—United States, 1998–99. MMWR Morb Mortal Wkly Rep 1999;48(27):577–81.

8. Salisbury DM. Association between oral poliovaccine and Guillain-Barré syndrome? Lancet 1998;351(9096):79–80.

9. Recommendations of the Advisory Committee on Immunization Practices (ACIP). US Rotavirus Vaccine. CDC Media Relations 22 October 1999;.

10. Parashar UD, Holman RC, Cummings KC, Staggs NW, Curns AT, Zimmerman CM, Kaufman SF, Lewis JE, Vugia DJ, Powell KE, Glass RI. Trends in

intussusception-associated hospitalizations and deaths among US infants. Pediatrics 2000;106(6):1413–21.

11. Murphy TV, Gargiullo PM, Massoudi MS, Nelson DB, Jumaan AO, Okoro CA, Zanardi LR, Setia S, Fair E, LeBaron CW, Wharton M, Livengood JR. Rotavirus Intussusception Investigation Team. Intussusception among infants given an oral rotavirus vaccine. N Engl J Med 2001;344(8):564–72.

12. Kramarz P, France EK, Destefano F, Black SB, Shinefield H, Ward JI, Chang EJ, Chen RT, Shatin D, Hill J, Lieu T, Ogren JM. Population-based study of rotavirus vaccination and intussusception. Pediatr Infect Dis J 2001;20(4):410–6.

13. Liebert PS. Reporting vaccine complications. Emerg Off Pediatr 1999;12:125.

14. http://internationalrotavirus.com (accessed July 20, 2004).

15. Zimmerman R. Drug companies race to develop virus vaccine. Wall Street J 2004;June 30.

16. Global Advisory Committee on Vaccine Safety. Meeting on 12–13 June 2007. Safety of rotavirus vaccines. Wkly Epidemiol Rec 2007;82:256–7.

17. Vesikari T, Karvonen A, Forrest BD, Hoshino Y, Chanock RM, Kapikian AZ. Neonatal administration of rhesus rotavirus tetravalent vaccine. Pediatr Infect Dis J 2006;25:118–22.

Smallpox vaccine

General Information

The last case of smallpox occurred in 1977, and the eradication of smallpox was declared complete by the World Health Assembly in 1980. Since then, routine smallpox vaccination has ceased in all countries, because it is no longer required and because serious adverse reactions sometimes occur after both primary vaccination and revaccination (SED-8, 709; SED-11, 685; SEDA-1, 247; SEDA-3, 262; SEDA-4, 227; SEDA-6, 289; SEDA-13, 289; SEDA-15, 357) (1–5). However, the threat of bioterrorism has made it necessary to consider prevention and control strategies through vaccination and the potential hazards associated with the administration of smallpox vaccine.

Adverse reactions to smallpox vaccination vary from what may be called a "normal" reaction via anomalous reactions to real complications. These can be divided into two categories (6):

1. complications in which sequelae or a fatal outcome are rare (sensitivity, rashes, generalized vaccinia, and auto-inoculation vaccinia);
2. complications that may well be fatal or have permanent sequelae (postvaccinial encephalomyelitis, vaccinia necrosum, eczema vaccinatum).

Both categories of complications are much less common after revaccination than after primary vaccination. *Vaccinia* virus can also spread by contact to other subjects and cause adverse effects (7,8).

In an Australian survey of 5 000 000 vaccinations carried out between 1960 and 1976, the frequency of all complications was 188 per million and the death rate 1.5 per million. The ratio of reactions in women to reactions in men was 1.6, increasing with age. Paradoxically, of eight reports of cardiac complications seven concerned men (9).

The frequencies of some complications in 1968 in the USA per 1 000 000 smallpox vaccinations in primarily vaccinated and revaccinated subjects are shown in Table 1 (7).

In a phase 1 trial in 350 volunteers an investigational cell-cultured smallpox vaccine developed in a joint venture between Computer Science Corporation and Porton International has been evaluated. The vaccine induces fewer adverse effects than the historic Dryvax vaccine, and fever, lymphangitis, rash, fatigue, headache, and nausea were at least 8% less common (10).

Alternative smallpox vaccines have been developed because of concerns over biological weapons. Live, attenuated vaccinia virus vaccines derived from calf lymph were used to eradicate smallpox worldwide. However, well documented safety limitations prevent their widespread use in civilian in the absence of an outbreak. Of every million people immunized for smallpox, 14–52 had serious or life-threatening adverse reactions, and 1–2 per million primary vaccinees died. Furthermore, live attenuated vaccinia virus vaccine is contraindicated in up to 30% or more of the population, including infants, pregnant women, women who are breastfeeding, the immunocompromised, those with eczema or exfoliative skin disorders, people who live in the same house or are in intimate contact with people with the above conditions, and people with cardiovascular conditions.

A new smallpox vaccine derived from cell culture has been compared with a vaccine derived from calf lymph in 350 healthy adults (11). All but one participant developed pock lesions. Vaccine-associated adverse reactions were similar between the groups. The cell-cultured vaccine was as immunogenic and safe as the calf-lymph-derived vaccine.

Recombinant DNA technology using *Vaccinia* virus

Because of its great genetic potential, *Vaccinia* virus is an ideal medium for recombined genes originating from different organisms. In order to use *Vaccinia* virus, efforts have been made to attenuate it further, either by inactivating the genes responsible for virulence (SEDA-15,

Table 1 Frequencies of some complications per 1 000 000 smallpox vaccinations

Complication	Primarily vaccinated	Revaccinated
Vaccinia gangrenosa	0.9	0.7
Eczema vaccinatum	10.0	0.9
Generalized vaccinia	23.4	1.2
Accidental infections	25.3	0.8

357) or by introducing human lymphokine genes into its genome.

Vaccinia DNA can tolerate large insertions into non-essential regions of the genome, and this opens the door to the making of polyvalent live *Vaccinia* recombinations. A major obstacle to their use as vaccines is that severe complications can occur after vaccination, especially in immunodeficient individuals. On the other hand, there is evidence that recombinant *Vaccinia* viruses have reduced pathogenicity. A genetically engineered *Vaccinia* virus expressing murine interleukin-2 has been described (12), and it has been shown that athymic nude mice infected with the *Vaccinia* virus recover from the virus infection rapidly, whereas mice infected with a control virus develop progressive vaccinia.

One attempt to develop safe and efficacious live recombinant vaccines is the use of low neurovirulent strains of vaccinia virus: LC 16 m O (m O) or LC 16 m 8 (m 8). A recombinant *Vaccinia* virus vaccine (RVV) expressing hepatitis B surface antigen likely to form the basis of a safe live RV vaccine against hepatitis B has been constructed (13).

Little is known of the ways in which orthopoxviruses are maintained in nature (14). There is a possibility that strains of *Vaccinia* used as vaccines may become established in nature, as *Vaccinia* may have become established in Indian buffaloes, and/or undergo genetic hybridization with existing orthopoxviruses (14). Enthusiasm for these new prospects should not be allowed to compromise the requirement for obtaining additional scientific information essential to ensure safety, efficacy, and the exercise of all reasonable caution in mounting field investigations (15).

There have already been accusations that two patients with AIDS, treated with an experimental vaccine prepared using a *Vaccinia* virus that had apparently been inactivated and genetically engineered to express HIV proteins, may have died from vaccinia gangrenosa (16,17). On the other hand, there is evidence that recombinant *Vaccinia* viruses have reduced pathogenicity (SEDA-13, 289).

Protection of laboratory workers exposed to orthopoxviruses

In 1980, the US Public Health Service first recommended the use of *Vaccinia* (smallpox) vaccine to protect laboratory workers occupationally exposed to orthopoxviruses. In 1991, the Centers for Disease Control, Atlanta, Georgia, published recommendations on *Vaccinia* vaccine. From 1983 to 1991, 4649 doses of smallpox vaccine were administered, of which 57% were given in 1989–91. The proportion of primary vaccinations increased from 4% in 1983–88 to 14% in 1989–91. Of vaccinees 93% reported no signs or symptoms after vaccination. Reported adverse reactions were mild: lymphadenopathy, fever or chills, and tenderness at the site of vaccination. No severe adverse effects were reported. However, one vaccinee reported a spontaneous abortion 5 months after primary vaccination (18).

A somewhat different note has been sounded from the Committee on Occupational Medical Practice of the American College of Occupational and Environmental Medicine. The committee pointed out that risks to laboratory workers resulting from spontaneous infection are presumably similar to those involved in vaccination. The positive aspect of immunization is that it renders possible control over the time and initial site of entry of the virus. In the Committee's opinion, it is important for scientists and technicians to understand the US Public Health Service recommendations and to have the opportunity to receive *Vaccinia* immunization. They should also understand the possible drawbacks and have the opportunity to refuse vaccination (19). These different recommendations in the USA are also reflected in different national recommendations in other countries.

Accidental needle-stick inoculation of *Vaccinia* virus has been reported (20).

- A 26-year-old laboratory worker, who had been vaccinated against smallpox in childhood, developed a pustule and erythema on his left thumb 3 days after an accidental needle-stick while working with *Vaccinia* virus. Further pustules occurred on the fourth and fifth fingers of the same hand, accompanied by a large erythematous lesion on the left forearm, secondary bacterial infection, and axillary lymphadenopathy. After surgical excision of the necrotic tissue, he improved slowly and the lesions healed in about 3 weeks.

Smallpox and bioterrorism

The threat of bioterrorism has made it necessary to consider again prevention and control strategies through vaccination and the potential hazards associated with the administration of smallpox vaccine. Guidelines for prevention and control of smallpox have been elaborated in many countries. The guidelines distinguish between pre-event vaccination programs (worldwide no re-emergence of smallpox) and postevent vaccination programs (re-emergence of smallpox confirmed). Taking into account the risk of smallpox vaccination, most countries have elaborated plans for postevent vaccination programs. During the US pre-event vaccination program, in 1 per 20 000 members of the military who received primary smallpox vaccination, cardiac inflammation (myocarditis and/or pericarditis), including one death (myocardial infarction), has been reported. Compared with the rate reported in an unvaccinated military population during 1998–2000, the rate of myocarditis/pericarditis was substantially increased. However, there were no cases after revaccination. In 2003, among 38 257 civilian health-care and public health workers vaccinated against smallpox, 17 suspected and 5 probable cases of myocarditis/pericarditis were reported. The other adverse events include one case of suspected encephalitis, three of generalized *Vaccinia*, three of ocular *Vaccinia*, and 21 of inadvertent inoculation.

During the US pre-event smallpox vaccination program, there were no reports of eczema vaccinatum, progressive *Vaccinia*, or fetal *Vaccinia* (21,22).

During a meeting in June 2003, the Global Advisory Committee on Vaccine Safety considered two expert reports on the safety of smallpox vaccination in detail. They concluded that there is a real risk of serious adverse events, including safety issues that have not previously been recognized. Therefore, if the vaccine is being used in mass campaigns, it would be very important to include adverse events monitoring. The committee found that the available data were insufficient to define the risk after primary vaccination compared with the risk of revaccination after a long interval (23).

Organs and Systems

Cardiovascular

Acute myocarditis after vaccination against smallpox has been reported (24). Fatal myocarditis is rare, but electrocardiographic evidence of myocarditis has been found more frequently; this adverse effect is probably not always noticed (25–27). Pericarditis after smallpox vaccination has also been described (28).

All case reports of myocarditis/pericarditis after smallpox vaccination have been carefully evaluated. It was concluded that the data are consistent with a causal relation between myocarditis/pericarditis and smallpox vaccination; however, no causal association between ischemic cardiac events and smallpox has been identified (21,22).

- A 57-year-old woman with a history of hypertension, a transient ischemic attack, and carotid endarterectomy died 22 days after smallpox vaccination. Histopathological evaluation showed no evidence of cardiac inflammation.

Pericarditis after smallpox vaccination has been described (28).

The Advisory Committee on Immunization Practices (ACIP) has recommended that people who have underlying heart disease, with or without symptoms, or who have three or more known major cardiac risk factors (that is hypertension, diabetes, hypercholesterolemia, heart disease at age 50 years in a first-degree relative, and smoking) should be excluded from the pre-event smallpox vaccination program (29).

During the period of routine smallpox vaccination, only rare reports of cardiac inflammation (pericarditis, fatal myocarditis, and electrocardiographic evidence of myocarditis) were published in the world literature (SED-8, 709). To determine the risk of cardiac death after smallpox vaccination, death certificates were analysed from a period in 1947 when 6 million New York City residents were vaccinated after a smallpox outbreak; the incidence of cardiac deaths did not increase after the vaccination campaign (30).

Respiratory

Pneumonia has been observed after smallpox vaccination (2).

Nervous system

Neurological complications of smallpox vaccination can cause paralysis; of 26 patients with such symptoms, most were children under 2 years (31).

The most dreaded complication of smallpox vaccination is postvaccinial encephalitis or encephalomyelitis, which is said to occur even without a cutaneous vaccination reaction (32), although this occurs rarely, if at all (2,33). It is mainly a complication of primary vaccination. There is increased morbidity with increasing age, especially around puberty. It is rare after revaccination.

Two conditions must be distinguished: a histopathological picture characterized by diffuse perivenous focal encephalitis and a condition based on a disturbance of the blood–brain barrier, that is an encephalopathy that characteristically occurs in infants. The incubation period of postvaccinial encephalomyelitis is 9–13 days. Its onset is mostly sudden. The clinical picture varies. Mortality is high (30–50%). Recovery may be complete, but there are often neurological sequelae, such as paresis and extrapyramidal disturbances.

The frequency of postvaccinial encephalomyelitis in different countries has varied considerably. Per million primary vaccinations 68 cases were recorded in Bavaria, 48 in the Netherlands, 15 in the UK, and 1.8 in the USA. The frequency increases with age beyond the first year (5). In 53 034 primarily vaccinated army recruits in the Netherlands there were 11 cases (1:5000) (34). In 1968 the frequency in the USA was 2.8 per million primarily vaccinated. There were no cases among 8.5 million revaccinations (8). In very young infants sudden death can occur (35), but information about frequency is difficult to assess, because of the occurrence of sudden unexplained deaths ("cot" deaths).

The risk of postvaccinial encephalomyelitis is reduced by simultaneous administration of hyperimmune *Vaccinia* gamma globulin (34) or by pre-immunization with an antigen from formol-inactivated *Vaccinia* virus (36,37). However, cases of postvaccinial encephalomyelitis have been described even after these procedures (34,38).

- A 13-year-old girl died of neuromyelitis optica. By inoculation of rabbits it was shown that the brain tissue contained *Vaccinia* antigen. Moreover, there were high titres against *Vaccinia* virus in both sera and spinal fluid. In infancy this patient had been vaccinated against smallpox (39).
- Accidental administration of 10 doses of smallpox vaccine to an already vaccinated girl resulted in a clinical picture characterized by neuraxitis with general tonic-clonic convulsions. Complete cure was attained after treatment with hyperimmune antivaccinia globulins and methisazone (40).

Polyneuropathy (41) and bacterial meningitis (42) have been described after smallpox vaccination.

Sensory systems

Ears

Disturbances of hearing and balance are described after smallpox vaccination (43).

Eyes

Vaccinial lesions on the eyelids and the conjunctivae are seen after secondary infection with *Vaccinia* virus by scratching (44,45). From these lesions a keratitis can develop, which sometimes extends to deeper layers of the cornea, with concomitant iridocyclitis. Papillitis with myelitis has been described after revaccination (46).

Metabolism

Diabetes mellitus occurred in a 1-year-old child about 4 weeks after smallpox vaccination (47).

Hematologic

Thrombocytopenic purpura is a rare complication of smallpox vaccination (48). Postvaccinial lymphadenitis can occur (49).

Mouth and teeth

Of 2568 people who were given oral vaccination, five bit open the vaccine-containing capsule or tablet, with release of the virus, and developed vaccinia lesions on the tongue, gums, or nares (50).

Urinary tract

Nephritis is a very rare complication of smallpox vaccination (51).

Skin

A 20-year-old man had active eczematous lesions on both wrists. After vaccination widespread *Vaccinia* developed in non-eczematous skin around the eyes and mouth, while the eczematous regions were unaffected. The possible reason was the use of a glucocorticoid ointment (52).

Skin reactions after smallpox vaccination can cause the following complications (2,3):

- Ectopic pocks near the vaccination site (vaccinia serpiginosa).
- Secondary vaccinia, which is caused by spreading of Vaccinia virus from the vaccination site to other body sites by scratching.
- Generalized vaccinia (vaccinia generalisata). Here the *Vaccinia* virus is spread by the blood stream and vaccinial lesions can develop all over the body some 9–10 days after vaccination.
- Eczema vaccinatum. This serious complication occurs most often in infants with active infantile eczema and occasionally in adults who have eczema or some other skin disease. It is seen after vaccination or contact with vaccinated persons with vaccinial lesions. The first symptoms are observed at the eczematous spots 3–4 days after vaccination. In many cases generalization of the vaccinial lesions follows. The estimated mortality is 25–30%.
- Progressive vaccinia (vaccinia gangrenosa). About 7 days after a normal vaccinial reaction the lesion extends, penetrates the underlying tissue, and becomes necrotic. Ulcers are formed, and the process spreads to adjacent areas of the skin. Secondary bacterial

infection is common. This rare complication may be due to some abnormality of serum proteins, interfering with immunity, such as agammaglobulinemia or hypogammaglobulinemia. Mortality is high (53–55).

- Keloid from vaccinial scars.

Abnormal hair growth (56), herpes virus infections (57), and malignant changes in smallpox vaccination scars have also been observed (58,59). Allergic skin reactions sometimes occur, for example urticaria and purpura (60) and possibly photosensitivity reactions.

Musculoskeletal

Osteomyelitis and periosteitis are exceedingly rare sequels to smallpox vaccination (61,62).

Acute arthritis has been described after smallpox vaccination (63).

Death

In France 4 113 109 primary vaccinations during 1968–77 have been surveyed. There were 30 deaths that could have been associated with the vaccine, 23 in the first year of life (64). The relation was considered certain in one case, probable in 5, possible in 6, and doubtful in 18. In the certain case there may have been immunodeficiency. In the second group there were three cases of acute encephalitis. Among the doubtful cases were 12 patients who had neurological symptoms.

Second-Generation Effects

Pregnancy

The incidence of abortions is said to be higher in women who have recently received smallpox vaccination; in some cases the aborted fetus showed evidence of *Vaccinia* infection (65).

Fetotoxicity

Although routine smallpox vaccination of infants was discontinued in the UK in 1971, some 20–28 cases of complications of vaccination continue to be reported yearly to the Committee on Safety of Medicines (66), including both cross-infection and fetal infection (66).

Susceptibility Factors

HIV infection is a risk factor for vaccinia generalisata, making it important to have *Vaccinia* immune globulin available for outbreak control (67).

- Vaccinia generalisata developed after smallpox vaccination (co-administrated with other vaccines) of an army recruit with asymptomatic HIV infection (68). After 2–3 weeks he developed cryptococcal meningitis, and a diagnosis of AIDS was made. While being treated for the meningitis he developed generalized vaccinia. He was treated with *Vaccinia* immune globulin and recovered from his vaccinia generalisata.

References

1. Moss B, Fuerst TR, Flexner C, Hugin A. Roles of vaccinia virus in the development of new vaccines. Vaccine 1988;6(2):161–3.

2. Herrlich A, Ehrengut W, Schleussing H. Die Pockenschutzimpfung. Der Impfschaden. In: Herrlich A, editor. Handbuch der Schutzimpfungen. 1st ed.. Berlin: Springer-Verlag, 1965:60.

3. Treatment and Nursing. Sequelae Complications. In: Dixon CW, editor. Smallpox. 1st ed.. London: J and A Churchill Ltd, 1962:143.

4. Copeman PW, Banatvala JE. The skin and vaccination against smallpox. Br J Dermatol 1971;84(2):169–73.

5. Dick G. Routine smallpox vaccination. BMJ 1971;3(767):163–6.

6. WHO Scientific Group. Human viral and rickettsial vaccines. World Health Organ Tech Rep Ser 1966;325:1–79.

7. Lane JM, Ruben FL, Neff JM, Millar JD. Complications of smallpox vaccination, 1968: results of ten statewide surveys. J Infect Dis 1970;122(4):303–9.

8. Lane JM, Ruben FL, Neff JM, Millar JD. Complications of smallpox vaccination, 1968. N Engl J Med 1969;281(22):1201–8.

9. Feery BJ. Adverse reactions after smallpox vaccination. Med J Aust 1977;2(6):180–3.

10. Dynport smallpox vaccine shows fewer side effects than Dryvax in phase I study. Clin Infect Dis 2003;36:News.

11. Greenberg RN, Kennedy JS, Clanton DJ, Plummer EA, Hague L, Cruz J, Ennis FA, Blackwelder WC, Hopkins RJ. Safety and immunogenicity of new cell-cultured smallpox vaccine compared with calf-lymph derived vaccine: a blind, single-centre, randomised controlled trial. Lancet 2005;365:398–409.

12. Ramshaw IA, Andrew ME, Phillips SM, Boyle DB, Coupar BE. Recovery of immunodeficient mice from a vaccinia virus/IL-2 recombinant infection. Nature 1987;329(6139):545–6.

13. Watanabe K, Kobayashi H, Kajiyama K, Morita M, Yasuda A, Gotoh H, Saeki S, Sugimoto M, Saito H, Kojima A. Improved recombinant LC16m0 or LC16m8 vaccinia virus successfully expressing hepatitis B surface antigen. Vaccine 1989;7(1):53–9.

14. Baxby D, Gaskell RM, Gaskell CJ, Bennett M. Ecology of orthopoxviruses and use of recombinant vaccinia vaccines. Lancet 1986;2(8511):850–1.

15. Brown F, Schild GC, Ada GL. Recombinant vaccinia viruses as vaccines. Nature 1986;319(6054):549–50.

16. Dorozynski A, Anderson A. Deaths in vaccine trials trigger French inquiry. Science 1991;252(5005):501–2.

17. Guillaume JC, Saiag P, Wechsler J, Lescs MC, Roujeau JC. Vaccinia from recombinant virus expressing HIV genes. Lancet 1991;337(8748):1034–5.

18. Stokes SL, Atkinson WL, Becher JA, Williams WW. Vaccination against orthopoxvirus infection and adverse events among laboratory personnel, United States, 1983–1991. Personal communication, 1992.

19. Perry GF. Occupational Medicine Forum: Pro and cons of vaccinia immunization. J Occup Med 1992;34:757.

20. Moussatche N, Tuyama M, Kato SE, Castro AP, Njaine B, Peralta RH, Peralta JM, Damaso CR, Barroso PF. Accidental infection of laboratory worker with vaccinia virus. Emerg Infect Dis 2003;9(6):724–6.

21. Centers for Disease Control and Prevention (CDC). Update: adverse events following civilian smallpox vaccination—United States, 2003. MMWR Morb Mortal Wkly Rep 2003;52(34):819–20.

22. Centers for Disease Control and Prevention (CDC). Update: adverse events following smallpox vaccination—United States, 2003. MMWR Morb Mortal Wkly Rep 2003;52(13):278–82.

23. Global Advisory Committee on Vaccine Safety. Wkly Epidemiol Rec 2003;78(32):282–411–12 June 2003.

24. Baldini G, Bani E. Sulle complicanze cardiache in corso di vaccinazione jenneriana. (Contributo clinico ed ec-grafie.). [Cardiac complications in Jennerian vaccination (Clinical and electrocardiographic studies).] Minerva Pediatr 1979;31(1):35–9.

25. Finlay-Jones LR. Fatal myocarditis after vaccination against smallpox. Report of a case. N Engl J Med 1964;270:41–2.

26. Mead J. Serum transaminase and electrocardiographic findings after smallpox vaccination: case report. J Am Geriatr Soc 1966;14(7):754–6.

27. Bessard G, Marchal A, Avezou F, Pont J, Rambaud P. Un nouveau cas de myocardite après vaccination anti-variolique. [A new case of myocarditis following smallpox vaccination.] Pediatrie 1974;29(2):179–84.

28. Price MA, Alpers JH. Acute pericarditis following smallpox vaccination. Papua N Guinea Med J 1968;11:30.

29. Centers for Disease Control and Prevention (CDC). Supplemental recommendations on adverse events following smallpox vaccine in the pre-event vaccination program: recommendations of the Advisory Committee on Immunization Practices. MMWR Morb Mortal Wkly Rep 2003;52(13):282–4.

30. Centers for Disease Control and Prevention (CDC). Cardiac deaths after a mass smallpox vaccination campaign—New York City, 1947. MMWR Morb Mortal Wkly Rep 2003;52(39):933–6.

31. Koen M. Paralyses following smallpox vaccination and revaccination. Probl Infect Parasit Dis 1978;6:64.

32. Rockoff A, Spigland I, Lorenstein B, Rose AL. Postvaccinal encephalomyelitis without cutaneous vaccination reaction. Ann Neurol 1979;5(1):99–101.

33. In: De Vries E, editor. Postvaccinal Perivenous Encephailitis. Amsterdam: Elsevier, 1960:60.

34. Nanning W. Prophylactic effect of antivaccinia gamma-globulin against post-vaccinal encephalitis. Bull World Health Organ 1962;27:317–24.

35. De Vries E. Plotselinge dood na pokkenvaccinatie bij zeer jonge kinderen. [Sudden death following smallpox vaccination in very young children.] Ned Tijdschr Geneeskd 1964;108:2061–3.

36. Herrlich A. Welchen Nutzen hat die Prophylaxe der postvakzinalen Enzephalitis?. [What is the advantage of a prevention of postvaccinal encephalitis? A comparative evaluation of present methods.] Dtsch Med Wochenschr 1964;89:968–74.

37. Dietzsch HJ. Kasuistik: Enzephalitis nach Pocken Erstimpfung trotz Vorbehandlung mit Vakzineantigen. [Encephalitis following 1st smallpox vaccination despite pretreatment with vaccine antigens.] Kinderarztl Prax 1966;34(9):425–8.

38. Eggers C. Die postvakzinale Polyneuritis als Komplikation nach Pockenschützimpfung. [The post-vaccinal polyneuritis as complication following smallpox-vaccination.] Monatsschr Kinderheilkd 1974;122(4):169–71.

39. Adams JM, Brown WJ, Eberle ED, Vorlty A. Neuromyelitis optica: severe demyelination occurring years after primary smallpox vaccination. Rev Roum Neurol 1973;10(3):227–31.

40. Bertaggia A. Nevrassite da inoculazione accidentale in dose massiva di vaccino antivaioloso per via sottocutanea in soggetto vaccinato e rivaccinate contro il vaiolo. Trattamento e

guarigione con immunoglobuline iperimmuni antivacciniche e methisazone. [Neuraxitis due to accidental inoculation of a massive subcutaneous dose of smallpox vaccine in a subject vaccinated and revaccinated against smallpox. Treatment and recovery with antivaccinal hyperimmune immunoglobulins and methisazone.] Minerva Pediatr 1975;27(29):1586–91.

41. Herrlich A. Über Vakzineantigen: Versuch einer Prophylaxe neuraler Impfschäden. [Vaccine antigen; an experiment in prevention of neural vaccinal complications.] Munch Med Wochenschr 1959;101(1):12–4.

42. Stickl H, Helming M. Eitrige Meningitiden nach der Pockenschutzimpfung. [Purulent meningitides following smallpox vaccination. On the problem of post-vaccinal decrease of resistance.] Dtsch Med Wochenschr 1966;91(29):1307–10.

43. Wirth G. Schädigung des Hör- und Gleichgewichtsorganes nach Wiederimpfung gegen Pocken ohne Impfenzephalitis. [Labyrinthine lesion following smallpox revaccination without vaccinia encephalitis.] Z Laryngol Rhinol Otol 1973;52(7):526–32.

44. Paufique L, Durand L, Magnard G, Dorne PA. Complications oculopalpébrales de la vaccination antivariolique: vaccine palpébrale. [Oculo-palpebral complications of smallpox vaccination: palpebral vaccinia.] Bull Soc Ophtalmol Fr 1968;68(7):673–7.

45. Ross J, Gorin M. Vaccinia infection of the eyelids: two case reports. Eye Ear Nose Throat Mon 1969;48(6):363–5.

46. Mathur SP, Makhija JM, Mehta MC. Papillitis with myelitis after revaccination. Indian J Med Sci 1967;21(7):469–71.

47. Schneider H. Diabetesmanifestation nach Pockenimpfung. [Manifestation of diabetes after smallpox vaccination.] Kinderarztl Prax 1975;43(3):101–7.

48. Burke PJ, Shah NR. Thrombocytopenic purpura after smallpox vaccine. Pa Med 1981;84(9):49–50.

49. Hartsock RJ. Postvaccinial lymphadenitis. Hyperplasia of lymphoid tissue that simulates malignant lymphomas. Cancer 1968;21(4):632–49.

50. Stickl H, Jung EG. Störungen des Impfverlaufes bei der Oral Impfung gegen Pocken. [Abnormal postvaccination course after oral smallpox immunisation.] Dtsch Med Wochenschr 1977;102(31):1118–9.

51. von Vacano D. Akute diffuse Glomerulonephritis nach Pockenschutzimpfung. [Acute diffuse glomerulonephritis following smallpox vaccination.] Monatsschr Kinderheilkd 1968;116(11):596–8.

52. Gundersen SG, Bjorvatn B. Vaccinia and topical steroids: a case report. Acta Derm Venereol 1980;60(5):445–7.

53. Stoop JW. Een patiëntje met vaccinia progressive et generalisata. [A patient with progressive and generalized vaccinia.] Ned Tijdschr Geneeskd 1972;116(44):1981–4.

54. Ziegler HK, Schock V. [Vaccinia generalisata progressiva in a case of alymphocytosis.] Z Kinderheilkd 1969;106(3):206–212.

55. Chandra RK, Kaveramma B, Soothill JF. Generalised non-progressive vaccinia associated with IgM deficiency. Lancet 1969;1(7597):687–9.

56. Kumar LR, Goyal BG. Pigmented hairy scar following smallpox vaccination. Indian J Pediatr 1968;35(245):283–4.

57. Warren WS, Salvatore MA. Herpesvirus hominis infection at a smallpox vaccination site. JAMA 1968;205(13):931–3.

58. Gordon HH. Complications of smallpox vaccination: Basal cell carcinoma, keloids, acute bullons reaction. Cutis 1974;13:444.

59. Haider S. Keratoacanthoma in a smallpox vaccination site. Br J Dermatol 1974;90(6):689–90.

60. Coskey RJ, Bryan HG. Photosensitivity secondary to smallpox vaccination. Cutis (NY) 1970;6:761.

61. Bennett NM, Yung AP, Lehmann NI. Periostitis following smallpox vaccination. Med J Aust 1968;1(24):1052–3.

62. Singhal RK. Osteo-articular complications of smallpox vaccination. J Indian Med Assoc 1970;55(1):20–2.

63. Silby HM, Farber R, O'Connell CJ, Ascher J, Marine EJ. Acute monarticular arthritis after vaccination. Report of a case with isolation of vaccinia virus from synovial fluid. Ann Intern Med 1965;62:347–50.

64. Martin-Bouyer C, Foulon G, de Solan M, Torgal J, N'Guyen K, Martin-Bouyer G. Etude des décès imputés à la vaccination antivariolique en France, de 1968 à 1977. [Deaths due to smallpox vaccination in France 1968–1977.] Arch Fr Pediatr 1980;37(3):199–206.

65. Tondury G, Kistler G. Die Gefährdung des ungeborenen bei Pockenschutzimpfung in graviditate. Prav-Med 1973;18:45.

66. Du Mont GC, Beach RC. Continuing mortality and morbidity from smallpox vaccination. BMJ 1979;1(6175):1398–9.

67. Heymann DL. Smallpox containment updated: considerations for the 21st century. Int J Infect Dis 2004;8(Suppl 2):S15–20.

68. La Force FM. Immunization of children infected with human immunodeficiency virus. 1986 WHO/EPI/GEN/86.6 Rev 1, Geneva.

Snakebite antivenom

General Information

Snakebite is an important medical emergency in some parts of the rural tropics. In most tropical countries it is an occupational disease of farmers, plantation workers, herders, and hunters. Every year thousands of people die in Africa, Central America, South Asia, and South-East Asia because of envenomation after snakebite. In India alone, an estimated 15–20 000 people die each year due to snakebite. In Sri Lanka, the overall incidence of snakebite exceeds 400 per 100 000 population per year, one of the highest in the world.

Antivenom (also known as antivenin, antivenene, and antisnakebite serum) is the concentrated enzyme-refined immunoglobulin of animals, usually horses or sheep, that have been exposed to venom. It is the only specific treatment currently available for the management of snakebite envenoming and has proved effective against many of the lethal and damaging effects of venoms. The most widely used antivenoms are F(ab')2-equine polyspecific antivenoms, raised against the venoms of many poisonous snakes. In the management of snakebite, the most important clinical decision is whether or not to give antivenom, because only some snake-bitten patients need it, it can produce severe reactions, and it is expensive and often in short supply.

Antivenom is most effective by intravenous injection. The range of venoms neutralized by an antivenom is usually stated in the package insert. If the biting species of snake is known, the appropriate monospecific antivenom should be used. In countries where several species produce similar signs, snakebite victims are treated with polyspecific antivenom, which contains a lower concentration of specific antibody to each species than the monospecific antivenom.

Organs and Systems

Immunologic

Antivenom treatment can be complicated by early reactions (anaphylaxis), pyrogenic reactions, or late reactions (serum sickness-type). The incidence and severity of early reactions is proportional to the dose of antivenom and the speed with which it enters the blood stream (1,2). These reactions usually develop within 10–180 minutes of starting antivenom therapy. The reported incidence of early reactions after intravenous antivenom in snakebite patients, which ranges from 43% (3) to 81% (4), appears to increase with the dose and decrease when refined antivenom is used and administration is by intramuscular rather than intravenous injection. Unless patients are watched carefully for 3 hours after treatment, mild reactions can be missed and deaths misattributed to the envenoming itself. In most cases symptoms are mild: urticaria, nausea, vomiting, diarrhea, headache, and fever; however, in up to 40% of cases severe systemic anaphylaxis develops, with bronchospasm, hypotension, or angioedema. However, deaths are rare (5).

Early reactions respond well to adrenaline given by intramuscular injection of 0.5–1 ml of a 0.1% solution (1:1000, 1 mg/ml) in adults (children 0.01 ml/kg) at the first sign of trouble. Antihistamines also should be given by intravenous injection to counteract the effects of histamine released during the reaction.

Pyrogenic reactions result from contamination of antivenom by endotoxin-like compounds. High fever develops 1–2 hours after treatment and is associated with rigors, followed by vasodilatation and a fall in blood pressure. Febrile convulsions can occur in children. Patients should be cooled and given antipyretic drugs by mouth, powdered and washed down a nasogastric tube, or by suppository.

Late (serum sickness-type) reactions develop 5–24 days after treatment. Symptoms include fever, itching, urticaria, arthralgia (which can involve the temporomandibular joint), lymphadenopathy, periarticular swellings, mononeuritis multiplex, albuminuria, and rarely encephalopathy. This is an immune complex disease which responds to antihistamines or, in more severe cases, to glucocorticoids.

Early antivenom reactions are not usually Type I IgE-mediated reactions to equine serum proteins and are not predicted by hypersensitivity tests. Several methods have been used to reduce acute adverse reactions to antivenom. A small test dose of antivenom to detect patients who may develop acute adverse reactions to the antivenom has no predictive value, can itself cause anaphylaxis,

and is no longer recommended (5). Prophylactic use of hydrocortisone and antihistamines before infusion with antivenom is also practiced widely, although the theoretical basis for their use is unclear. Antihistamines counter only the effects of histamine after its release and do not prevent further release; one small randomized controlled trial showed no benefit from the routine use of antihistamines (6). Hydrocortisone takes time to act and may be ineffective as a prophylactic against acute adverse reactions that can develop almost immediately after antivenom treatment, which is very often administered urgently to snakebite victims. One study has suggested that intravenous hydrocortisone is ineffective in preventing acute adverse reactions to antivenom, but if given together with intravenous chlorphenamine it can reduce these reactions (7). However, this trial recruited only 52 patients and was not designed to study the efficacy of chlorphenamine alone, making it difficult to give a clear interpretation of the results and recommendations on pretreatment with glucocorticoids and antihistamines to prevent acute reactions to antivenom. In one study of 105 patients, low-dose adrenaline given subcutaneously immediately before administration of antivenom to snakebite victims significantly reduced the incidence of acute adverse reactions to the serum (3). However, this trial did not enroll sufficient participants to establish safety adequately, a major concern regarding the use of adrenaline in a prophylactic role, particularly the risk of intra-cerebral haemorrhage. Therefore, further studies on the safety of this treatment are required before it can be recommended routinely. For the present, the only available alternative to prevention is the early detection of adverse reactions to antivenom and the ready availability of drugs such as adrenaline for their prompt treatment.

References

1. Anonymous. Antivenom therapy and reactions. Lancet 1980;1(8176):1009–10.
2. Reid HA. Antivenom reactions and efficacy. Lancet 1980;1(8176):1024–5.
3. Premawardhena AP, de Silva CE, Fonseka MM, Gunatilake SB, de Silva HJ. Low dose subcutaneous adrenaline to prevent acute adverse reactions to antivenom serum in people bitten by snakes: randomised, placebo controlled trial. BMJ 1999;318(7190):1041–3.
4. Ariaratnam CA, Sjostrom L, Raziek Z, Kularatne SA, Arachchi RW, Sheriff MH, Theakston RD, Warrell DA. An open, randomized comparative trial of two antivenoms for the treatment of envenoming by Sri Lankan Russell's viper *(Daboia russelii russelii)*. Trans R Soc Trop Med Hyg 2001;95(1):74–80.
5. Malasit P, Warrell DA, Chanthavanich P, Viravan C, Mongkolsapaya J, Singhthong B, Supich C. Prediction, prevention, and mechanism of early (anaphylactic) antivenom reactions in victims of snake bites. BMJ (Clin Res Ed) 1986;292(6512):17–20.
6. Fan HW, Marcopito LF, Cardoso JL, Franca FO, Malaque CM, Ferrari RA, Theakston RD, Warrell DA. Sequential randomised and double blind trial of promethazine prophylaxis against early anaphylactic reactions to antivenom for bothrops snake bites. BMJ 1999;318(7196):1451–2.

7. Gawarammana IB, Kularatne SA, Dissanayake WP, Kumarasiri RP, Senanayake N, Ariyasena H. Parallel infusion of hydrocortisone +/− chlorpheniramine bolus injection to prevent acute adverse reactions to antivenom for snakebites. Med J Aust 2004;180(1):20–3.

Tetanus toxoid

General Information

Tetanus toxoid is prepared from *Clostridium tetani* and can be given either in a fluid form (plain) or an adsorbed form. The slight local reactions that tend to occur (induration, erythema, tenderness) are more common with the adsorbed type. Intramuscular injection of tetanus toxoid is generally the route of choice, and the vaccine is best injected into a large muscle.

General adverse effects

Data on adverse events after tetanus immunization have been collected by Behringwerke, Germany (SED-12, 826) and in the framework of the US Monitoring System for Adverse Events Following Immunization (MSAEFI) (SEDA-13, 274). Some figures for particular types of complication are cited below.

Isolated reports of adverse effects have included: abscess formation (SEDA-11, 288); dermatomyositis, neuralgic amyotrophy, polyradiculoneuritis with paresis of the urinary bladder and bowel (SEDA-9, 283); asymmetrical polyneuropathy, demyelinating polyneuropathy, and Guillain–Barré syndrome (SEDA-14, 281); and subcutaneous nodules. Polyvinylpyrrolidone thesaurismosis, revealed by inflammatory manifestations after tetanus booster injection, has also been reported (SEDA-10, 288).

Combined diphtheria + tetanus + pertussis immunization

An overview of clinical trials with a special diphtheria and tetanus toxoids and acellular pertussis (DTaP) vaccine has been published (1). The vaccine contains as pertussis components purified filamentous hemagglutinin, pertactin, and genetically engineered pertussis toxin. The vaccine induces high and long-lasting immunity and is at least as efficacious as most whole-cell pertussis vaccines and similar in efficacy to the most efficacious acellular pertussis vaccines that contain three pertussis antigens. The vaccine is better tolerated than whole cell vaccines and has a similar reactogenicity profile to other acellular vaccines.

A vaccine containing diphtheria and tetanus toxoids and acellular pertussis with reduced antigen content for diphtheria and pertussis (TdaP) has been compared with a licensed reduced adult-type diphtheria–tetanus (Td) vaccine and with an experimental candidate monovalent acellular pertussis vaccine with reduced antigen content (ap) (2). A total of 299 healthy adults (mean age 30 years) were randomized into three groups to receive one dose of the study vaccines. The antibody responses (anti-diphtheria, anti-tetanus, anti-pertussis toxin, anti-pertactin, anti-filamentous hemagglutinin) were similar in all groups. The most frequently reported local symptom was pain at the injection site (62–94%), but there were no reports of severe pain; redness and swelling with a diameter of 5 cm or more occurred in up to 13%. The incidence of local symptoms was similar after TdaP and Td immunization. The most frequently reported general symptoms were headache and fatigue (20–50%). The incidence of general symptoms was similar in the TdaP and Td groups. There were no reports of fever over 39°C. No serious adverse events were reported.

Data from the Third National Health and Nutrition Survey (1988–94) have been used to analyse the possible effects of DTP or tetanus immunization on allergies and allergy-related symptoms among 13 944 infants, children, and adolescents aged 2 months to 16 years in the USA (3). The authors concluded that DTP or tetanus immunization increases the risk of allergies and related respiratory symptoms in children and adolescents. However, the small number of non-immunized individuals and the study design limited their ability to make firm causal inferences about the true magnitude of effect.

Tetravalent, pentavalent, and hexavalent immunization

DTaP or DTwP vaccine can be combined with other antigens, such as *Haemophilus influenzae* type b (Hib), inactivated poliovirus (IPV), and hepatitis B vaccine. In children DTaP or DTwP vaccines are the basis for such combinations, while in adults it is mostly Td vaccine. Current safety concerns regarding combination vaccines have been defined and reviewed (4). The author concluded that there is no evidence that adding vaccines to combination products increases the burden on the immune system, which can respond to many millions of antigens. Combining antigens usually does not increase adverse effects, but it can lead to an overall reduction in adverse events. Before licensure, combination vaccines undergo extensive testing to assure that the new products are safe and effective.

The frequency, severity, and types of adverse reactions after DTP-Hib immunization in very preterm babies have been studied (5). Adverse reactions were noted in 17 of 45 babies: nine had major events (apnea, bradycardia, or desaturation) and eight had minor reactions (increased oxygen requirements, temperature instability, poor handling, and feeding intolerance). Babies who had major adverse reactions were significantly younger at the time of immunization than the babies who did not have major reactions. Of 27 babies immunized at 70 days or less, 9 developed major reactions compared with none of those who were immunized at over 70 days.

The Hexavalent Study Group has compared the immunogenicity and safety of a new liquid hexavalent vaccine against diphtheria, tetanus, pertussis, poliomyelitis, hepatitis B and *H. influenzae* type b (DTP + IPV + HB + Hib vaccine, manufactured by Aventis Pasteur MSD, Lyon, France) with two reference vaccines, the pentavalent

DTP + IPV + Hib vaccine and the monovalent hepatitis B vaccine, administrated separately at the same visit (6). Infants were randomized to receive either the hexavalent vaccine ($n = 423$) or (administered at different local sites) the pentavalent and the HB vaccine ($n = 425$) at 2, 4, and 6 months of age. The hexavalent vaccine was well tolerated (for details see the monograph on Pertussis vaccines). At least one local reaction was reported in 20% of injections with hexavalent vaccine compared with 16% after the receipt of pentavalent vaccine or 3.8% after the receipt of hepatitis B vaccine. These reactions were generally mild and transient. At least one systemic reaction was reported in 46% of injections with hexavalent vaccine, whereas the respective rate for the recipients of pentavalent and HB vaccine was 42%. No vaccine-related serious adverse event occurred during the study. The hexavalent vaccine provided immune responses adequate for protection against the six diseases.

Organs and Systems

Cardiovascular

Myopericarditis has been attributed to Td-IPV vaccine (7).

- A 31-year-old man developed arthralgia and chest pain 2 days after Td-IPV immunization and had an acute myopericarditis. He recovered within a few days with high-dose aspirin.

The authors discussed two possible causal mechanisms, natural infection or an immune complex-mediated mechanism. Infection was excluded by negative bacterial and viral serology and a favorable outcome occurred within a few days without antimicrobial treatment.

Nervous system

Reports of neurological adverse effects after tetanus immunization have appeared (8). The most common reported complication is a polyneuropathy. In the majority of cases the onset occurred within 14 days of the last injection, and ranged in severity from a single nerve palsy to profound sensorimotor involvement of the nervous system, including cord and cortex. Recovery was usually complete (eight of 10 patients with onset at less than 14 days after injection) but three patients with onset at more than 14 days from injection had only partial recovery.

Mononeuritis and polyneuritis after tetanus immunization during 1970 and 1977 have been reviewed (9). The frequency of this adverse effect was 0.4 per one million distributed vaccine doses.

Sensory systems

Optic neuritis has been attributed to Td-IPV vaccine (10).

- Ten days after receiving Td-IPV vaccine a 56-year-old woman developed acute unilateral optic neuritis. Complete remission occurred within 6 weeks of prednisolone treatment. No other causes were found.

Skin

Bullous pemphigoid has been attributed to DTP-IPV vaccine (11).

- A previous healthy 3.5-month-old infant developed bullous pemphigoid 3 days after receiving a first dose of DTP-IPV vaccine. *Staphylococcus aureus* was isolated from purulent bullae. The lesions resolved rapidly after treatment with antibiotics and methylprednisolone.

The authors mentioned 12 other cases of bullous pemphigoid, reported during the last 5 years, that had possibly been triggered by vaccines (influenza, tetanus toxoid booster, and DTP-IPV vaccine).

- About 2 weeks after receiving a second dose of adsorbed tetanus toxoid a 50-year-old woman developed generalized morphea, a rare condition, in which multiple patches of skin sclerosis occur over much larger areas than in the localized variant. The patient denied taking any drugs. After prednisone therapy, a month later the lesions had dramatically improved (12).

The cause of this condition is unknown, but an autoimmune mechanism triggered by endogenous and exogenous factors has been suggested.

Immunologic

Allergic reactions to reinforcing doses of tetanus toxoid have been described by different investigators (SEDA-8, 300) (SEDA-11, 288). The association between high titers of antitoxin produced by active immunization and reactions is well established (13,14). Booster doses of tetanus toxoid are being given with unnecessary and indeed excessive frequency. Continuing to do this will produce a more highly toxoid-sensitive population without adding significantly to the already high protection that this immunized population has against tetanus. It is therefore recommended that routine boosters in individuals known to have had primary immunization including a reinforcing dose be given only at 10-year intervals, and that emergency boosters be given no less than 1 year apart (15). Allergic reactions may be due to an allergy to the toxoid or the proteins of *C. tetani* that co-purify with toxoid during the precipitation process used in its conventional preparation (16).

Susceptibility Factors

Age

In children aged 15–16 years receiving routine reinforcement tetanus immunization, adsorbed vaccine caused more intense and more frequent local reactions than did plain tetanus toxoid, and a higher incidence of pyrexia. The incidence of swelling and erythema at the inoculation site increased with serum antitoxin titre at the time of administration, whereas pain and tenderness were related to the presence of the aluminium hydroxide adjuvant (17). Based on similar experiences it has been widely recommended that plain and not adsorbed tetanus toxoid should be used when reinforcement of immunity to tetanus alone is desired.

Other features of the patient

The only contraindication to tetanus toxoid is a history of a severe allergic or neurological reaction after a previous dose. AIDS and HIV infections are not contraindications. Data on the immune response to tetanus in children with AIDS showed defective responses (18).

References

1. Matheson AJ, Goa KL. Diphtheria–tetanus–acellular pertussis vaccine adsorbed (Triacelluvax; DTaP3-CB): a review of its use in the prevention of Bordetella pertussis infection. Paediatr Drugs 2000;2(2):139–59.
2. Van der Wielen M, Van Damme P. Tetanus–diphtheria booster in non-responding tetanus–diphtheria vaccines. Vaccine 2000;19(9–10):1005–6.
3. Hurwitz EL, Morgenstern H. Effects of diphtheria–tetanus–pertussis or tetanus vaccination on allergies and allergy-related respiratory symptoms among children and adolescents in the United States. J Manipulative Physiol Ther 2000;23(2):81–90.
4. Halsey NA. Combination vaccines: defining and addressing current safety concerns. Clin Infect Dis 2001;33(Suppl 4):S312–8.
5. Sen S, Cloete Y, Hassan K, Buss P. Adverse events following vaccination in premature infants. Acta Paediatr 2001;90(8):916–20.
6. Mallet E, Fabre P, Pines E, Salomon H, Staub T, Schodel F, Mendelman P, Hessel L, Chryssomalis G, Vidor E, Hoffenbach A, Abeille A, Amar R, Arsene JP, Aurand JM, Azoulay L, Badescou F, Barrois S, Baudino N, Beal M, Beaude-Chervet V, Berlier P, Billard E, Billet L, Blanc B, Blanc JP, Bohu D, Bonardo C, Bossu CHexavalent Vaccine Trial Study Group. Immunogenicity and safety of a new liquid hexavalent combined vaccine compared with separate administration of reference licensed vaccines in infants. Pediatr Infect Dis J 2000;19(12):1119–27.
7. Boccara F, Benhaiem-Sigaux N, Cohen A. Acute myopericarditis after diphtheria, tetanus, and polio vaccination. Chest 2001;120(2):671–2.
8. Rutledge SL, Snead OC 3rd. Neurologic complications of immunizations. J Pediatr 1986;109(6):917–24.
9. Quast U, Hennessen W, Widmark RM. Mono- and polyneuritis after tetanus vaccination (1970–1977). Dev Biol Stand 1979;43:25–32.
10. Burkhard C, Choi M, Wilhelm H. Optikusneuritis als Komplikaton einer Tetanus–Diphtherie–Poliomyelitis–Schutzimpfung: ein Fallbericht. [Optic neuritis as a complication in preventive tetanus–diphtheria–poliomyelitis vaccination: a case report.] Klin Monatsbl Augenheilkd 2001;218(1):51–4.
11. Baykal C, Okan G, Sarica R. Childhood bullous pemphigoid developed after the first vaccination. J Am Acad Dermatol 2001;44(Suppl 2):348–50.
12. Drago F, Rampini P, Lugani C, Rebora A. Generalized morphoea after antitetanus vaccination. Clin Exp Dermatol 1998;23(3):142.
13. Levine L, Ipsen J Jr, McComb JA. Adult immunization. Preparation and evaluation of combined fluid tetanus and diphtheria toxoids for adult use. Am J Hyg 1961;73:20–35.
14. Relihan M. Reactions to tetanus toxoid. J Ir Med Assoc 1969;62(390):430–4.
15. Edsall G, Elliott MW, Peebles TC, Eldred MC. Excessive use of tetanus toxoid boosters. JAMA 1967;202(1):111–3.
16. Leen CL, Barclay GR, McClelland DB, Shepherd WM, Langford DT. Double-blind comparative trial of standard (commercial) and antibody-affinity-purified tetanus toxoid vaccines. J Infect 1987;14(2):119–24.
17. Collier LH, Polakoff S, Mortimer J. Reactions and antibody responses to reinforcing doses of adsorbed and plain tetanus vaccines. Lancet 1979;1(8131):1364–8.
18. Chen RT, Spira TJ. Tetanus prophylaxis in AIDS patients. JAMA 1986;255:1063.

Tick-borne meningoencephalitis vaccine

General Information

Tick-borne meningoencephalitis vaccine is an adsorbed formalin-inactivated virus vaccine prepared on chicken embryo tissue. Mild local reactions (soreness, redness, swelling) have been reported in 10–25% of vaccinees; systemic reactions occur at similar rates. Fever over 39°C is rare.

Positive merthiolate tests were found in eight of 30 patients with suspected adverse reactions to tetanus or tick-borne encephalitis vaccine (local inflammatory reactions at the injection site, fever, lymphadenopathy, urticarial or lichenoid exanthemas) (1).

An improved tick-borne encephalitis vaccine for children, produced by Baxter, has been licensed and is being used without safety concerns in many countries.

Organs and Systems

Nervous system

In 1989, 172 reports on adverse effects after tick-borne meningoencephalitis immunization were collected in the Federal Republic of Germany, among them 72 reports of suspect neurological complications. Among the 72 reports, there were only three cases of peripheral neuropathy that suggested a causal link with immunization (2). Three cases of mild meningitis, encephalitis, and convulsions were, because of incomplete diagnosis, difficult to evaluate.

The database of the Swiss Drug Monitoring Center included 20 spontaneous reports (1987 to July 1992) of adverse events after the administration of tick-borne meningoencephalitis vaccine, among them 11 reports of cases with neurological symptoms (for example meningism, polyradiculitis, ataxia, vestibulopathy, facialis paresis). Most recovered completely within a few days. The incompleteness of the data did not allow conclusions regarding the causality of immunization (3).

Individual reports of neuropathy (SEDA-13, 290), multifocal cerebral vasculitis and infarction (4), cervical myelitis (5), myelopolyradiculitis (5), and suspected encephalitis (SEDA-17, 363) have appeared.

During the period 1987–2000, the Swiss Center for Adverse Events Drug Monitoring (Schweizerische Arzneimittel-Nebenwirkungszentrale) received 33 reports, including 39 neurological adverse events, following the receipt of tick-borne encephalitis vaccine: headache (36% of reports), neuropathy (18%), and meningeal signs (13%); 12 of 33 patients were hospitalized and all recovered (6).

Immunologic

A case of facial edema and pain and swelling of the left knee following the receipt of tick-borne meningoencephalitis vaccine was suspected to be caused by thiomersal allergy (7).

Body temperature

Following the use of tick-borne encephalitis vaccine ("Ticovac," produced by Baxter) in German children, there was a sudden increase in the number of cases of high fever, and 20–30% of vaccinees, primarily those 3 years of age and younger, reacted with a high fever (39°C or more) after the first dose (8). The increased number of cases of high fever was observed after changes in the production of tick-borne encephalitis vaccine were made: the vaccine no longer contains thiomersal and human albumin. Currently, according to the German licensing authority, the Paul Ehrlich institute, the vaccine is no longer recommended for children under 3 years; children aged 3–15 years should only receive half a dose of Ticovac, and the vaccine should only be used when strongly indicated. The reason for the changed reactogenicity is still unclear.

References

1. Lindemayr H, Drobil M, Ebner H. Impfreaktionen nach Tetanus- und Frühsommer-meningoenzephalitisschutzimpfung durch Merthiolat (Thiomersal). [Reactions to vaccinations against tetanus and tick-borne encephalitis caused by merthiolate (thiomersal).] Hautarzt 1984;35(4):192–6.
2. Kappos L. Mogliche neurologische Nebenwirkungen nach FSME-ImpfungFachpressegedprach, Frankfurter Press Club;. 20 February 1990.
3. Bohus M, Glocker FX, Jost S, Deuschl G, Lucking CH. Myelitis after immunisation against tick-borne encephalitis. Lancet 1993;342(8865):239–40.
4. Schabet M, Wiethoelter H, Grodd W, Vallbracht A, Dichgans J, Becker W, Berg PA. Neurological complications after simultaneous immunisation against tick-borne encephalitis and tetanus. Lancet 1989;1(8644):959–60.
5. Goerre S, Kesselring J, Hartmann K, Kuhn M, Reinhart WH. Neurologische Nebenwirkungen nach Impfung gegen die Fruhsommer–Mening–Enzephalitis. Fahlbericht und Erfahrungen der Schweizerischen Arzneimittel–Nebenwirkungs–Zentrale (SANZ). [Neurological side effects following vaccination of early-summer meningoencephalitis. Case report and experiences of the Swiss Center for Adverse Drug Effects.] Schweiz Med Wochenschr 1993;123(14):654–7.
6. Doser AK, Hartmann K, Fleisch F, Kuhn M. Vermutete neurologische Nebenwirkungen der FSME–Impfung: Erfahrung Schweizerischen Arzneimittel–Nebenwirkungs–Zentrale (SANZ). [Suspected neurological side-effects of tick-borne meningoencephalitis vaccination: experiences of the Swiss Adverse Drug Reaction Reporting Center.] Schweiz Rundsch Med Prax 2002;91(5):159–62.
7. Ackermann R. Allergische Reaktion nach FSME-Auffrischimpfung (Anfrage und Antwort). Dtsch Med Wochenschr 1990;115:1213.
8. Statement of Paul Ehrlich Institute. Use of Ticovac TBE vaccineLangen: Germany;. 14/11/2000 http://www.pei.de/ticova_info.htm.

Typhoid vaccine (including typhoid-paratyphoid vaccine)

General Information

The different typhoid vaccines commercially available for civilian use are:

- an oral live attenuated vaccine based on the Ty21a strain of *Salmonella typhi*;
- a parenteral heat-phenol-inactivated vaccine, containing either killed *Salmonella typhi* or killed *Salmonella typhi* and *Salmonella paratyphi* A, B organisms; this preparation is no longer recommended by the WHO, although it is still produced in some countries;
- Vi capsular polysaccharide vaccine and vaccines based on auxotrophic mutants of Vi-positive and Vi-negative *Salmonella typhi* (SED-11, 687) (SEDA-12, 278) (SEDA-13, 284) (SEDA-14, 281).

Oral live vaccine

During volunteer studies and field trials using Ty21a vaccine, adverse reactions were rare and consisted of abdominal discomfort, nausea, vomiting, and rash or urticaria. Adverse reactions occurred with equal frequencies among groups receiving vaccine or placebo. There has been a report of a typhoid-fever-like syndrome after the use of Ty21a live oral vaccine (SEDA-15, 350).

Inactivated vaccine

Vaccines administered parenterally are not well tolerated. Severe local pain and/or swelling occurred in 6–40% of vaccinees; systemic reactions have been reported in 9–30% (headache), and in 14–29% (fever); 13–24% of vaccinees missed work or school due to adverse effects. More severe reactions (hypotension, shock) have been reported sporadically (1).

General adverse effects

There have been individual reports of fatal angioimmunoblastic lymphadenopathy, hemolytic uremic syndrome (after typhoid/paratyphoid/diphtheria vaccination), fatal hyperpyrexia (SED-11, 687), transverse myelitis (SEDA-10, 288), erythema nodosum (SEDA-11, 289) (SEDA-14, 281), and Reiter's syndrome (SEDA-15, 350).

The efficacy and safety of typhoid fever vaccine have been estimated in a meta-analysis of about 1.8 million vaccinees in efficacy trials and about 11 000 vaccinees in safety studies (2). The 3-year cumulative efficacy was 73% (65–80%) for two doses of whole cell vaccines, 51% (35–63%) for three doses of Ty21 live attenuated vaccine, and 55% (30–71%) for one dose of Vi vaccine.

After immunization, fever occurred in 16% (12–21%) of whole cell vaccine recipients, 2% (0.7–5.3%) of Ty21a vaccine recipients, and 1.1% (0.1–12.3%) of Vi vaccine recipients (Table 1 (3).

A meta-analysis of studies of the efficacy and toxicity of typhoid fever vaccines (SEDA-23, 343) has been criticized as an interesting mathematical exercise but of little practical relevance (4). The meta-analysis had lumped together different parenteral whole-cell vaccines (alcohol-inactivated, formalin-inactivated, acetone-inactivated, and dried whole-cell vaccines), but the only parenteral killed whole-cell vaccine that is available is the heat-inactivated phenol-preserved vaccine. The analysis should therefore have been limited to heat-phenolized vaccine. It was also mentioned that the quoted data on the occurrence of adverse reactions failed to include the largest clinical trial of Ty21a vaccine. Last but not least, the authors of the meta-analysis failed to relate the severity of reactions connected with the use of parenteral whole-cell vaccine. These adverse reactions, including high fever, malaise, and nausea, make the killed whole-cell vaccine among the most reactogenic vaccine ever

licensed. In contrast, oral Ty21a and parenteral Vi vaccines are well tolerated.

Combined meningococcal and typhoid fever vaccine

A combined meningococcal and typhoid vaccine (Mérieux) was evaluated in 158 volunteers in a single-blind study (5). Comparing the effects with those in vaccinees receiving monocomponent vaccines or the combination vaccine, there was no significant difference in the reported frequency or duration of local and systemic reactions. However, vaccinees who received the monocomponent typhoid fever vaccine alone were less likely to complain of swelling or pain at the injection site.

Organs and Systems

Musculoskeletal

Two cases of arthritis temporally connected with the administration of oral typhoid vaccine have been reported (6).

Table 1 Local and systemic reactions after the administration of typhoid fever vaccines

Study	Age (years)	Type of study	Number immunized	Fever (%)	Swelling (%)	Vomiting (%)	Diarrhea (%)	Missed school or work (%)
Ty21a vaccine								
Gilman et al.	Adults	Clinic	155	1	NA	3	10	ND
Murphy et al.	0.5–2	Clinic	18	11	NA	17	11	ND
Rahman et al.	3–70	Clinic	157	2	NA	0	1	ND
Cryz et al.	2–6	Clinic	317	<1	NA	1	<1	ND
Cryz et al.	16–56	Clinic	30	2	NA	0	20	ND
Pooled estimate (95% CI)				2.0 (0.7–5.3)		2.1 (0.6–7.8)	5.1 (1.7–15)	
Vi vaccine								
Levin et al.	ND	Clinic	21	24	ND	NA	NA	ND
Tacket et al.	20–24	Clinic	19	0	ND	NA	NA	0
Klugman et al.	5–15	Field	253	<1	4	NA	NA	ND
Cumberland et al.	18–22	Clinic	388	<1	1	NA	NA	ND
Mirza et al.	5–15	Field	435	0	8	NA	NA	ND
Pooled estimate (95% CI)				1.1 (0.1–1.2)	3.7 (1.3–9.6)			
Whole-cell vaccine (heat inactivated)								
YTC	5-50	Field	214	9	5	NA	NA	11
Ashcroft et al.	5–15	Field	193	13	61	NA	NA	14
YTC	NA	Field	66	29	ND	NA	NA	17
Hejfez et al.	7–18	Field	2621	30	19	NA	NA	ND
Hejfez et al.	ND	Field	3463	26	21	NA	NA	ND
Hejfez et al.	7–20	Field	2157	13	13	NA	NA	ND
Dimache et al.	16–18	Field	94	27	ND	NA	NA	ND
Dimache et al.	21	Clinic	113	1	ND	NA	NA	ND
Dimache	20	Field	100	34	ND	NA	NA	2
Cumberland et al.	18–22	Clinic	390	2	20	NA	NA	ND
Pooled estimate (95% CI)					16 (21–21)	20 (13–30)		10 (6–16)

For references, see 3
YTC=Yugoslav Typhoid Commission
NA=Not applicable
ND=Not described in study

- Arthritis of the knees, ankles, and hands occurred 8 weeks after immunization in a 27-year-old woman.
- Bilateral sacroiliitis was observed in a 66-year-old woman 1 day after she had completed the four-capsule series.

A rheumatologist evaluated both patients and felt that the diagnosis was reactive arthritis, in his opinion most probably vaccine-related. The two cases are the first reports of reactive arthritis following oral typhoid immunization. However, the time-courses (1 day, 8 weeks) make a causal relation doubtful.

References

1. Immunization Practices Advisory Committee (ACIP). Typhoid immunization. MMWR Recomm Rep 1990;39(RR-10):1–5.
2. Hodsagi M, Uhereczky G, Kiraly L, Pinter E. BCG dissemination in chronic granulomatous disease (CGD). Dev Biol Stand 1986;58(Pt A):339–46.
3. Engels EA, Falagas ME, Lau J, Bennish ML. Typhoid fever vaccines: a meta-analysis of studies on efficacy and toxicity. BMJ 1998;316(7125):110–6.
4. Clemens J, Hoffman S, Ivanoff B, Klugman K, Levine MM, Neira M, Pang T. Typhoid fever vaccines. Vaccine 1999;17(20–21):2476–8.
5. Khoo SH, St Clair Roberts J, Mandal BK. Safety and efficacy of combined meningococcal and typhoid vaccine. BMJ 1995;310(6984):908–9.
6. Adachi JA, D'Alessio FR, Ericsson CD. Reactive arthritis associated with typhoid vaccination in travelers: report of two cases with negative HLA-B27. J Travel Med 2000;7(1):35–6.

Varicella vaccine

General Information

A live attenuated *Varicella* vaccine was developed in 1973 by Takahashi using the Oka strain, which was isolated from a boy with chickenpox and named after the boy. Several producers use this live vaccine strain, for example Biken Institute, Merck, Sharp & Dohme, and SmithKline Beecham. Whereas the first Oka strain vaccines needed to be stored at $-20\,°C$, subsequent reformulation of the vaccine provided a shelf-life of up to 2 years at $+2\,°C$, with immunogenicity and safety comparable to other Oka strain *Varicella* vaccines (1). The safety and efficacy of *Varicella* vaccine have been critically reviewed (2).

Measles–mumps–rubella–varicella vaccine

A quadrivalent measles–mumps–rubella–varicella vaccine (ProQuad®, referred to as MMRV), which includes a varicella component of increased potency, has been evaluated in a blind multicenter study in 480 healthy children aged 12–23 months, who were randomized to either MMRV + placebo or MMR + monovalent varicella vaccine (3). Children who were randomized to MMRV + placebo received a second dose of MMRV 90 days later. Measles-like rash and fever during days 5–12 were more common after the first dose of combination MMRV (rash 5.9%; fever 28%) than after MMR + monovalent varicella vaccine (rash 1.9%; fever 19%). The incidences of other adverse events were similar between the groups. Response rates were over 90% to all vaccine components in both groups. Geometric mean titers to measles and mumps were significantly higher after one dose of MMRV than after MMR + varicella vaccine. The second dose of MMRV elicited slight to moderate increases in measles, mumps, and rubella antibody titers and a substantial increase in varicella antibody titer.

Non-infective uses

In an experimental model of multiple sclerosis, repeated high doses of antigen (myelin basic protein) deleted both the clinical and pathological manifestations of the disease. The effects of *Varicella* vaccine on 50 patients with chronic progressive multiple sclerosis have therefore been studied (4). The patients were immunized with *Varicella* vaccine and followed for 1 year. All were seropositive for *Varicella* before immunization and all had rises in *Varicella* antibodies after being given the vaccine. There was improvement in 14 patients, 4 became worse, and 29 were unchanged. Four patients developed mild chickenpox after immunization. No other untoward adverse effects occurred.

Observational studies

A subcutaneously administered, live, high-titer (18 700 to 60 000 plaque-forming units per dose) varicella zoster virus vaccine (zoster vaccine) of the Oka/Merck strain has been evaluated for the prevention of herpes zoster and the reduction of zoster-associated pain in adults aged 60 years and over (5). Zoster vaccine reduced the burden of herpes zoster illness by 61% compared with placebo, the incidence of herpes zoster by 51%, and the incidence of postherpetic neuralgia by 67% during more than 3 years of surveillance. The most frequently reported adverse reactions after immunization were injection-site reactions; the only systemic adverse event with zoster vaccine that differed significantly in incidence from that with placebo was headache.

Systematic reviews

The Global Advisory Committee on Vaccine Safety (GACVS) has discussed the safety of varicella vaccines (6). Some potential safety problems are common to other live attenuated viral vaccines, including the effect of immunization on pregnant women inadvertently immunized and the risk of secondary transmission of the attenuated virus vaccine strain from vaccinees to their contacts. To date, the data provide no cause for concern. An important safety question related to varicella vaccine is the impact on the epidemiology of herpes zoster, both in vaccinees and in individuals previously infected with wild-type varicella. Mathematical models have predicted that the limited circulation of wild virus as a result of universal varicella immunization programmes could lead to an increased incidence of herpes zoster for many years before an eventual decline over decades. The US data do not show signs of such an increase after 11 years of observation. However, given the natural history of reactivation, it is likely that decades of observation will be required before any conclusions can be drawn regarding the long-term impact on herpes zoster epidemiology. In view of the potential impact of these long-term effects, it is recommended that any countries planning to introduce varicella vaccine programmes should collect baseline data on the age-specific incidence of herpes zoster. These observations should be shared and continuously reviewed.

General adverse effects

There have been many clinical trials to assess the immunogenicity and safety of *Varicella* vaccine, in both healthy and immunocompromised individuals (SEDA-12, 285; SEDA-13, 290; SEDA-14, 285; SEDA-15, 289; SEDA-17, 379). The vaccine was safe and immunogenic. The reports of different investigators were similar. Mild local reactions at the injection site were the most commonly reported adverse effects, occurring in about 10% after the first and second dose. Vaccine-associated rashes about 1 month after the first dose were reported in about 6% of vaccinees, sore throat in 8%, and fever over 37.8°C in 2%. Rash and fever after the second dose of vaccine were reported by less than 1% of vaccinees. Maculopapular or papulovesicular rashes occurred much more often in vaccinees with suspended chemotherapy (42%) (7,8).

Oka strain *Varicella* vaccine (Merck) has been evaluated in immunocompromised children with leukemia in remission (9,10). Most children had chemotherapy stopped 1 week before and 1 week after immunization; glucocorticoids were also stopped for 3 weeks (from 1 week before to 2 weeks after immunization). *Varicella* vaccine was safe, immunogenic, and effective in leukemic children at risk of serious disease or death from chickenpox. The major adverse effect was a mild rash in 50% of the children within 1 month of immunization, about 40% of whom were treated with aciclovir. A mild form of *Varicella* developed in 14% of immunized children exposed to *Varicella* (household contacts). The vaccine protected completely against severe *Varicella*. Leukemic vaccinees were less likely to develop zoster than were comparable children with leukemia who had wild type *Varicella*.

Between 1987 and 1993, *Varicella* vaccine developed by the Biken Institute was given to 1.39 million individuals in Japan and 1.93 million individuals in South Korea. Adverse effects occurred in 6.9% of vaccinees. Despite more than 100 well-documented contacts with *Varicella* patients, only 2% of those immunized developed breakthrough *Varicella*, with very mild clinical features (11).

In 1995, live virus *Varicella* vaccine (Oka strain) was licensed in the USA. Of the 6574 reported adverse events 4% (67.5 per 100 000 doses) after immunization were serious, while the majority were minor reactions, such as rash, redness, or injection site pain (12). About 10% of immunized children developed mild chickenpox. A total 193

vaccinees reported neurological symptoms, including Bell's palsy, convulsions, and demyelinating syndromes; febrile seizures accounted for half of the cases of convulsions. There were 14 deaths, but a role for the vaccine was not proven.

Organs and Systems

Hematologic

Acute thrombocytopenic purpura developed 3 weeks after *Varicella* immunization (13).

Skin

Extensive *Varicella* vaccine-associated rashes occurred in four leukemic children; there was a relation between rash and glucocorticoid treatment (14).

Immunologic

In one case a hypersensitivity vasculitis developed 2 weeks after *Varicella* immunization (15).

Infection risk

Spread of infection to the siblings of immunized leukemic children has been observed. Between 18 and 36 days after the receipt of *Varicella* vaccine 2–10% of exposed children developed mild *Varicella* and/or seroconversion to *Varicella* virus.

Transmission of the *Varicella* vaccine virus from a toddler (with a history of allergic diathesis) to his pregnant mother has been reported and discussed (SEDA-22, 354).

Complicated wild *Varicella* in immunized individuals has been reported (16); among the two cases the first reported case of *Varicella* meningitis occurring in a child with documented immunization and seroconversion (17).

Cases of zoster both in healthy and in immunosuppressed persons have been reported (18–20).

A 19-month-old girl was immunized against *Varicella* at 15 months of age and later developed zoster infection (21). Viral cultures from various lesions isolated *Varicella zoster* virus. The Oka vaccine strain was revealed by polymerase chain reaction.

There have been three case reports of suspected reactivation of *Varicella zoster* through hepatitis A vaccine, influenza vaccine, and simultaneous administration of rabies and Japanese encephalitis vaccine (22).

- A 53-year-old woman without signs of immunodeficiency developed zoster in the left T10 dermatome 14 days after influenza immunization. The rash resolved without sequelae. Five months later she received another injection of influenza vaccine and 12 days later developed zoster in the left T1 dermatome. Recovery was prolonged.
- An 80-year-old woman with carcinoma developed long-lasting left thoracic zoster 6 days after influenza immunization. Influenza vaccination before and after the event had no adverse effect.

- A 27-year-old man developed zoster in the second and third branches of the trigeminal nerve 1 day after immunization against rabies and Japanese encephalitis.

Zoster in childhood is unusual and probably even less common after *Varicella* immunization.

- A 6-year-old boy developed a zoster infection (with a vesicular rash in a left second thoracic dermatome pattern on his back extending to the back of his left arm) 15 days after the receipt of *Varicella* vaccine (Oka strain) (23). Molecular biological analysis of the virus isolated from the vesicles showed a pattern consistent with wild-type *Varicella zoster* virus.

The authors felt that this case mandated a careful review of all cases of zoster after *Varicella* immunization. Zoster induced by *Varicella* immunization could have implications for the use of immunization to prevent zoster in the elderly, a population with almost uniform *Varicella zoster* latent virus and at higher risk of zoster.

Susceptibility Factors

Age

In 29 children aged 1–12 years with chronic liver disease who received one dose of *Varicella* vaccine, seroconversion rates at 8 weeks after immunization were 100% (24). The geometric mean titers tended to relate to the severity of liver disease. Local and systemic reactions did not differ from reactions reported in healthy children.

- A 5-year-old boy developed zoster-like vesicular lesions 4 years after *Varicella* immunization. Virological examination showed *Herpes simplex* virus type 1, and so the vesicular lesions could not be attributed to the *Varicella zoster* virus vaccine strain, demonstrating the difficulty in confirming causality between time-related events (25).

Other features of the patient

To determine the safety and immunogenicity of *Varicella* immunization in HIV-infected children, 41 children (aged 1–8 years) who were asymptomatic or mildly affected (according to CDC stages) received two doses of live attenuated *Varicella* vaccine 12 weeks apart (26). Two months after the second dose, 60% of the recipients had anti-*Varicella* antibody in their serum. A minority of recipients developed mild local or systemic reactions. The immunization had no effect on the clinical stage of HIV infection or the HIV RNA plasma load.

Live virus vaccines, such as *Varicella*, are contraindicated in pregnancy.

Drug–Drug Interactions

Other vaccines

Reaction rates were not increased if the administration of *Varicella* vaccine was combined with other vaccines, for example MMR vaccine (SEDA-15, 358).

References

1. Tan AY, Connett CJ, Connett GJ, Quek SC, Yap HK, Meurice F, Lee BW. Use of a reformulated Oka strain varicella vaccine (SmithKline Beecham Biologicals/Oka) in healthy children. Eur J Pediatr 1996;155(8):706–11.
2. Skull SA, Wang EE. Varicella vaccination—a critical review of the evidence. Arch Dis Child 2001;85(2):83–90.
3. Shinefield H, Black S, Digilio L, Reisinger K, Blatter M, Gress JO, Hoffman Brown ML, Eves KA, Klopfer SO, Schadel F, Kuter BJ. Evaluation of a quadrivalent measles, mumps, rubella and varicella vaccine in healthy children. Pediatr Infect Dis J 2005;24:665–9.
4. Ross RT, Nicolle LE, Cheang M. The *Varicella zoster* virus: a pilot trial of a potential therapeutic agent in multiple sclerosis. J Clin Epidemiol 1997;50(1):63–8.
5. Robinson DM, Perry CM. Zoster vaccine live (Oka/Merck). Drugs Aging 2006;23:525–31.
6. Varicella vaccine. Wkly Epidemiol Rec 2006;81:277.
7. Gershon AA, Steinberg SP, LaRussa P, Ferrara A, Hammerschlag M, Gelb L. Immunization of healthy adults with live attenuated *Varicella* vaccine. J Infect Dis 1988;158(1):132–7.
8. Gershon AA, Steinberg SP, Gelb L. Live attenuated *Varicella* vaccine use in immunocompromised children and adults. Pediatrics 1986;78(4 Pt 2):757–62.
9. Gershon AA, LaRussa P, Steinberg S. The *Varicella* vaccine. Clinical trials in immunocompromised individuals. Infect Dis Clin North Am 1996;10(3):583–94.
10. LaRussa P, Steinberg S, Gershon AA. *Varicella* vaccine for immunocompromised children: results of collaborative studies in the United States and Canada. J Infect Dis 1996;174(Suppl 3):S320–3.
11. Asano Y. *Varicella* vaccine: the Japanese experience. J Infect Dis 1996;174(Suppl 3):S310–3.
12. Wise RP, Salive ME, Braun MM, Mootrey GT, Seward JF, Rider LG, Krause PR. Postlicensure safety surveillance for *Varicella* vaccine. JAMA 2000;284(10):1271–9.
13. Lee SY, Komp DM, Andiman W. Thrombocytopenic purpura following *Varicella-zoster* vaccination. Am J Pediatr Hematol Oncol 1986;8(1):78–80.
14. Lydick E, Kuter BJ, Zajac BA, Guess HANIAID Varicella Vaccine Collaborative Study Group. Association of steroid therapy with vaccine-associated rashes in children with acute lymphocytic leukaemia who received Oka/Merck *Varicella* vaccine. Vaccine 1989;7(6):549–53.
15. Fraunfelder FW, Rosenbaum JT. Drug-induced uveitis. Incidence, prevention and treatment. Drug Saf 1997;17(3):197–207.
16. Pillai JJ, Gaughan WJ, Watson B, Sivalingam JJ, Murphey SA. Renal involvement in association with post-vaccination *Varicella*. Clin Infect Dis 1993;17(6):1079–80.
17. Naruse H, Miwata H, Ozaki T, Asano Y, Namazue J, Yamanishi K. *Varicella* infection complicated with meningitis after immunization. Acta Paediatr Jpn 1993;35(4):345–7.
18. Plotkin SA, Starr SE, Connor K, Morton D. Zoster in normal children after *Varicella* vaccine. J Infect Dis 1989;159(5):1000–1.
19. Hammerschlag MR, Gershon AA, Steinberg SP, Clarke L, Gelb LD. Herpes zoster in an adult recipient of live attenuated *Varicella* vaccine. J Infect Dis 1989;160(3):535–7.
20. Magrath DI. Prospective vaccines for national immunization programmes. In: Proceedings, 3rd Meeting of National Programme Managers on Expanded Programme on ImmunizationSt Vincent, Italy 22–25 May 1990:60 ICP/EPI 023/31, 1990.

21. Liang MG, Heidelberg KA, Jacobson RM, McEvoy MT. Herpes zoster after *Varicella* immunization. J Am Acad Dermatol 1998;38(5 Pt 1):761–3.
22. Walter R, Hartmann K, Fleisch F, Reinhart WH, Kuhn M. Reactivation of herpesvirus infections after vaccinations? Lancet 1999;353(9155):810.
23. Kohl S, Rapp J, La Russa P, Gershon AA, Steinberg SP. Natural *Varicella-zoster* virus reactivation shortly after varicella immunization in a child. Pediatr Infect Dis J 1999;18(12):1112–3.
24. Nithichaiyo C, Chongsrisawat V, Hutagalung Y, Bock HL, Poovorawa Y. Immunogenicity and adverse effects of live attenuated *Varicella* vaccine (Oka-strain) in children with chronic liver disease. Asian Pac J Allergy Immunol 2001;19(2):101–5.
25. Takayama N, Takayama M, Takita J. *Herpes simplex* mimicking *Herpes zoster* in a child immunized with *Varicella* vaccine. Pediatr Infect Dis J 2001;20(2):226–8.
26. Levin MJ, Gershon AA, Weinberg A, Blanchard S, Nowak B, Palumbo P, Chan CY. AIDS Clinical Trials Group 265 Team. Immunization of HIV-infected children with *Varicella* vaccine. J Pediatr 2001;139(2):305–10.

Yellow fever vaccine

General Information

Yellow fever vaccine contains the 17D virus strain grown in chick embryo tissue. The older (Dakar) yellow fever vaccine was prepared from more virulent material and often caused encephalitis, the risk in children being particularly high (SED-8, 712).

General adverse effects

The adverse effects of yellow fever vaccine have been documented by an expert group of the WHO (1).

Apart from minor postimmunization reactions the most common adverse effects are on the nervous system and allergic reactions.

On about the sixth day after immunization, under 5% of vaccinees develop fever, headache, and backache, lasting for 1–2 days.

Organs and Systems

Nervous system

About 20 cases of encephalitis have been recorded over a period of 40 years (SEDA-14, 1097). They all occurred in children: 12 in infants under 4 months old, two at 4 months, one at 6 months, one at 7 months, and one at 3 years of age. The last mentioned case was the only fatal one (17D virus was isolated from the brain), all the others recovered fully.

Encephalitis occurred after yellow fever immunization in a child older than 3 years of age and one over 9 months; a 13-year-old boy developed the disease 1 week after receipt of vaccine (2). The patient recovered after 1 month. There have been reports of encephalitis in a 29-year-old man and meningoencephalitis in two adults, suspected to be caused by the 17D yellow fever vaccine (3,4).

A case report of provocation of multiple sclerosis was published in 1967 (5), but this has not been confirmed.

Immunologic

Rash, erythema multiforme, urticaria, angioedema, and asthma occur infrequently, predominantly in people with a history of allergy, especially to eggs (1).

Severe immediate hypersensitivity reactions (type 1), sometimes accompanied by anaphylactic shock and circulatory collapse, have been described very rarely (1). Allergic reactions of the Arthus phenomenon type, characterized by local swelling and necrosis following less than 24 hours after immunization, have occurred in rare instances. Some of these cases have been fatal.

Two episodes have been reported from the Ivory Coast (1974) and Ghana (1982) (1). In the Ivory Coast, there were 39 cases of severe reactions with eight deaths following a mass campaign, in which 730 000 persons were immunized. The clinical features were uniform: a few hours after immunization the vaccinees developed signs of local inflammation. In severe cases, edema and inflammation were followed by cardiovascular collapse. Bacterial contamination could have been the cause; during the campaign, five-dose vaccine ampoules were pooled to prepare 50 and 100 doses for use in jet injectors. In Ghana, six vaccinees developed fulminant reactions 2–6 hours after immunization, including two deaths. The clinical features resembled those in the Ivory Coast episode. In 2001, during a mass vaccination campaign against yellow fever in Abidjan, the Ivory Coast, more than 2.6 million doses were administered and 87 adverse events were notified, of which 41 were considered to be vaccine-related. There was one case of anaphylaxis and 26 cases of urticaria, five of which were generalized (6).

People who are known to be suffering from allergy must be tested intradermally before immunization.

Multiorgan failure

There have been reports of seven cases of serious adverse events (including six deaths) after yellow fever immunization (7–11). The cases occurred from 1996 to 2001 in Australia (n = 1), Brazil (n = 2), and the USA (n = 4). The two people in Brazil were immunized with vaccine containing the live attenuated 17DD yellow fever strain and the others received vaccine containing the live attenuated 17D-204 strain; both strains are derived from the original 17D vaccine strain. All seven became ill within 2–5 days after immunization and required intensive care. Illness was characterized by fever, lymphocytopenia, thrombocytopenia, raised hepatocellular enzymes, hypotension, and respiratory failure. Most also had headache, vomiting, myalgia, hyperbilirubinemia, and renal insufficiency requiring hemodialysis. In some aspects the disease was similar to natural yellow fever. The causal association between multiorgan failure and the receipt of yellow fever vaccine is supported in most cases by isolation of the vaccine virus and histopathological changes; in cases with lack of specimens the temporal association and the similarity of the clinical presentations makes a causal association likely.

This has become known as viscerotropic disease, and further suspected cases have been reported to the Vaccine Adverse Event Reporting System (VAERS) between 20 June 2001 and 31 August 2002. Since 1996, 12 cases have been reported worldwide, among them five occurred in people under 50 years of age. During the same period four new cases of suspected yellow fever vaccine-associated neurotropic disease (previously called post-vaccinal encephalitis) have been reported to the VAERS. All four patients recovered. Encephalitis after yellow fever immunization has long been recognized as a vaccine-associated adverse event, but the incidence fell substantially with implementation of the seed-lot standardization process in 1945. Since then, 27 cases of yellow fever vaccine-associated neurotropic disease have been reported worldwide (12).

A yellow fever immunization campaign was conducted in the Ica Region of Peru following the earthquake in September 2007. Four cases of vaccine-associated viscerotropic disease, all fatal, were reported among about 40 000 individuals who received one particular lot of yellow fever vaccine. As a result, the national authorities suspended the campaign. The rate of vaccine-associated viscerotropic disease was about 10 per 100 000 doses administered for this vaccine lot, compared with an expected rate of approximately 0.3 per 100 000 based on previous experience with yellow fever vaccines (13).

Two fatal cases of yellow fever vaccine-associated viscerotropic disease were temporally associated with yellow fever immunization during a mass immunization program to control an outbreak of sylvatic yellow fever in Minas Gerais (2001), in the Southeast region of Brazil (14). Virus recovered from blood and post-mortem samples in both cases was identified as yellow fever virus. Partial nucleotide sequence of parts of prM/E and the non-structural 5 genes and 3' non-coding region (3' NCR) was used to characterize the origin of yellow fever virus involved in both cases. Wild-type yellow fever virus was identified as the causative agent. This report underlines the need for careful molecular biological analysis of isolated viruses in case of suspected complication after yellow fever immunization.

Kitchener (15) carried out an investigation of vaccine surveillance reports from Europe after the distribution of more than 3 million doses of ARILVAX. During 1991–2003 he found four cases each of yellow fever vaccine-associated viscerotropic disease (one death) and yellow fever vaccine-associated neurotropic disease.

Susceptibility Factors

Age

Infants under 9 months old are not generally immunized, except if they live in rural areas with a history of yellow fever epidemics (immunization at 6 months) or in an active epidemic focus (immunization at 4 months) (1).

References

1. World Heath Organization. Prevention and Control of Yellow Fever in AfricaGeneva: WHO;. 1986.
2. Schoub BD, Dommann CJ, Johnson S, Downie C, Patel PL. Encephalitis in a 13-year-old boy following 17D yellow fever vaccine. J Infect 1990;21(1):105–6.
3. Merlo C, Steffen R, Landis T, Tsai T, Karabatsos N. Possible association of encephalitis and 17D yellow fever vaccination in a 29-year-old traveller. Vaccine 1993;11(6):691.
4. Drouet A, Chagnon A, Valance J, Carli P, Muzellec Y, Paris JF. Meningo-encephalite après vaccination anti-amanile par la souche 17 D: deux observations. [Meningoencephalitis after vaccination against yellow fever with the 17D strain: 2 cases.] Rev Med Interne 1993;14(4):257–9.
5. Miller H, Cendrowski W, Shapira K. Multiple sclerosis and vaccination. BMJ 1967;2(546):210–3.
6. Fitzner J, Coulibaly D, Kouadio DE, Yavo JC, Loukou YG, Koudou PO, Coulombier D. Safety of the yellow fever vaccine during the September 2001 mass vaccination campaign in Abidjan, Ivory Coast. Vaccine 2004;23(2):156–62.
7. Centers for Disease Control and Prevention (CDC). Fever, jaundice, and multiple organ system failure associated with 17D-derived yellow fever vaccination, 1996–2001. MMWR Morb Mortal Wkly Rep 2001;50(30):643–5.
8. Anonymous. Adverse events following yellow fever vaccination. Wkly Epidemiol Rec 2001;76(29):217–8.
9. Vasconcelos PF, Luna EJ, Galler R, Silva LJ, Coimbra TL, Barros VL, Monath TP, Rodigues SG, Laval C, Costa ZG, Vilela MF, Santos CL, Papaiordanou PM, Alves VA, Andrade LD, Sato HK, Rosa ES, Froguas GB, Lacava E, Almeida LM, Cruz AC, Rocco IM, Santos RT, Oliva OF. Brazilian Yellow Fever Vaccine Evaluation Group. Serious adverse events associated with yellow fever 17DD vaccine in Brazil: a report of two cases. Lancet 2001;358(9276):91–7.
10. Chan RC, Penney DJ, Little D, Carter IW, Roberts JA, Rawlinson WD. Hepatitis and death following vaccination with 17D-204 yellow fever vaccine. Lancet 2001;358(9276):121–2.
11. Martin M, Tsai TF, Cropp B, Chang GJ, Holmes DA, Tseng J, Shieh W, Zaki SR, Al-Sanouri I, Cutrona AF, Ray G, Weld LH, Cetron MS. Fever and multisystem organ failure associated with 17D-204 yellow fever vaccination: a report of four cases. Lancet 2001;358(9276):98–104.
12. Adverse events associated with 17D-derived yellow fever vaccination—United States, 2001-2002. Morbid Mortal Weekly Rep 2002;51:989–93.
13. Safety of yellow fever vaccines. Wkly Epidemiol Rec 2008;83:39.
14. De Filippis AMB, Nogueira RMR, Jabor AV, Schatzmayr HG, Oliveira JC, Dinis SCM, Galler R. Isolation and characterization of wild type yellow fever virus in cases temporally associated with 17DD vaccination during an outbreak of yellow fever in Brazil. Vaccine 2003;22:1073–8.
15. Kitchener S. Viscerotropic and neurotropic disease following vaccination with the 17D yellow fever vaccine, ARILVAX®. Vaccine 2004;22:2103–5.

DISINFECTANTS AND ANTISEPTICS

Disinfectants and antiseptics

General Information

Antimicrobial drugs are widely used in topical medicaments, cosmetics, household products, and industrial biocides. Depending on their concentrations, they can function as disinfectants, antiseptics, or preservatives. The prevalence and rank order of sensitization to antimicrobial allergens in Europe have been reviewed (1,2). The most frequent antimicrobial allergens in 8521 patients who were patch-tested between 1985 and 1997 in Belgium are given in Table 1 (2).

In the multicenter study of the Information Network of Departments of Dermatology, sensitization rates of preservatives in the standard series were all over 1% in the test population of 11 485 patients. Thiomersal was rating highest (5.3%), chloromethyl-isothiazolinone/methyisothiazolinone, formaldehyde, and methyl-dibromo-glutaronitrile/phenoxyethanol were next at about 2%, and parabens rating lowest at 1.6%. Glutaral, a biocide mainly used as a disinfectant, showed a remarkable increase in sensitization from less than 1% in 1990 up to more than 4% at the end of 1994. Health personnel and cleaning personnel were often affected and showed a sensitization rate of 10% (1,3).

References

1. Schnuch A, Geier J, Uter W, Frosch PJ. Patch testing with preservatives, antimicrobials and industrial biocides. Results from a multicentre study. Br J Dermatol 1998;138(3):467–76.

Table 1 Most frequent antimicrobial allergens in Belgium out of 8521 patients in 1985–97

Rank	Allergen	Number
1	Methyl(chloro)isothiazolinone	143*
2	Thiomersal	136
3	Merbromine	94
4	Iodine	89
5	Cetrimide	88
6	Formaldehyde	80
7	Parabens	71
8	Chloramine	43
9	Quaternium-15	32
10	Nitrofurazone	29
11	Quinoline mix	28
12	Benzyl alcohol	25
	Benzoic acid	25
	Thiocyanomethylbenzothiazole	25
	Chlorhexidine	25
13	Glutaral (glutaraldehyde)	22
	Methyldibromoglutaronitrile + phenoxyethanol	22
14	Chloroacetamide	20
	Diazolidinyl urea	20

*Methyl (chloro) isothiazolinone was not tested until 1987

2. Goossens A, Claes L, Drieghe J, Put E. Antimicrobials: preservatives, antiseptics and disinfectants. Contact Dermatitis 1998;39(3):133–4.
3. Schnuch A, Uter W, Geier J, Frosch PJ, Rustemeyer T. Contact allergies in healthcare workers. Results from the IVDK. Acta Dermatol Venereol 1998;78(5):358–63.

Acrisorcin

General Information

Acrisorcin (aminoacridine 4-hexylresorcinolate) has been used for induction of abortion in mid-trimester pregnancies. Abortion was produced when a 0.1% solution of acrisorcin was introduced into the extra-amniotic space in 23 women. All patients aborted after a mean induction-delivery interval of 59 hours (SEDA-11, 474) (1).

Reference

1. Lewis BV, Pybus A, Stilwell JH. The oxytocic effect of acridine dyes and their use in terminating mid-trimester pregnancies. J Obstet Gynaecol Br Commonw 1971;78(9):838–42.

Aliphatic alcohols

General Information

The lower aliphatic alcohols (ethanol, isopropanol, 2-propanol, N-propanol, and 1-propanol) are widely used for skin antisepsis. In appropriate concentrations, these alcohols are bactericidal to most of the common pathogenic bacteria, but some rare species survive and can grow, especially since these alcohols are inactive against dried spores.

Seven cases of gaseous edema were observed in the former German Democratic Republic (GDR) after intramuscular injections following rubbing of the skin with ethanol (SEDA-11, 474); this led to a national recommendation that alcohols should not be used to cleanse the skin before intramuscular injection, injection of vasoconstrictors, injections in patients with disturbed peripheral circulation, or before lumbar, paraneural, or intra-articular injection and puncture (SEDA-11, 474) (1).

Organs and Systems

Cardiovascular

Skin disinfection before insertion of peripheral infusion catheters is standard practice. Ethanol 70% has been compared with 2% iodine dissolved in 70% ethanol in a prospective, randomized trial in 109 patients who were given infusions of prednisone and theophylline (2). Phlebitis occurred six times in the ethanol group and 12 times in the iodine group. The relative risk reduction of 53% failed to reach significance, but the power of the study was only 0.55, so there was a 45% chance of missing a true difference. As vast numbers of catheters are inserted each year, a small difference in phlebitis rate could save many patients discomfort.

Skin

Skin reactions to ethanol are extremely rare, although allergic contact dermatitis due to lower aliphatic alcohols has been described (SEDA-11, 474) (3). However, in premature infants of very low birth weight, second-degree and third-degree chemical skin burns were reported after the use of isopropanol, either for conduction in electro-cardiography, for the preparation of the umbilical stump for arterial catheterization, or for cleansing before her-niotomy (SEDA-11, 474) (4,5). Possible causes are hypo-perfusion of the skin by local pressure and general hypoperfusion derived from hypoxia, hypothermia, and acidosis. Immediate drying of the skin of small premature infants after antisepsis with alcoholic formulations and carefully avoiding that the alcohol is absorbed by the diapers is recommended.

Drug Administration

Drug administration route

When it is not possible to perform surgical treatment for omphalocele, application of ethanol is probably the safest method. However, dosage and frequency of application should be as limited, since absorption and intoxication can occur (SEDA-11, 474) (6).

References

1. Spengler W. Stellungnahme der Hauptabteilung Hygiene und Staatlichen Hygieninspektion im Ministerium für Gesundheitswesen zur Diskussion über die Hautdesinfektion vor Injektionen und Punktionen. Med Aktuell 1977;3:51.
2. de Vries JH, van Dorp WT, van Barneveld PW. A rando-mized trial of alcohol 70% versus alcoholic iodine 2% in skin disinfection before insertion of peripheral infusion catheters. J Hosp Infect 1997;36(4):317–20.
3. Ludwig E, Hausen BM. Sensitivity to isopropyl alcohol. Contact Dermatitis 1977;3(5):240–4.
4. Schick JB, Milstein JM. Burn hazard of isopropyl alcohol in the neonate. Pediatrics 1981;68(4):587–8.
5. Klein BR, Leape LL. Skin burn from Freon preparation. Surgery 1976;79(1):122.
6. Schroder CH, Severijnen RS, Monnens LA. Vergiftiging door desinfectans bij conservatieve behandeling van twee patiënten met omfalokele. [Poisoning by disinfectants in the conservative treatment of 2 patients with omphalocele.] Tijdschr Kindergeneeskd 1985;53(2):76–9.

Benzalkonium chloride

General Information

Quaternary ammonium compounds are surface-active agents. Some of them precipitate or denature proteins and destroy microorganisms. The most important disin-fectants in this group are cationic surface-active agents, such as benzalkonium chloride, benzethonium chloride and methylbenzethonium chloride, and cetylpyridinium chloride; the problems that they cause are similar.

Benzalkonium chloride is composed of a mixture of alkyldimethylbenzylammonium chlorides. The hydropho-bic alkyl residues are paraffinic chains with 8–18 carbon atoms. Benzalkonium chloride is used as a preservative in suspensions and solutions for nasal sprays and in eye-drops. Depending on the concentration of the solution, local irritant effects can occur. In nasal sprays it can exacerbate rhinitis (1) and in eye-drops it can cause irrita-tion or keratitis (2).

A total of 125 ophthalmologists in private practice located throughout France examined 919 glaucomatous patients treated with eye-drops which either did or did not contain a preservative; the proportion of patients who experienced discomfort or pain during instillation was 58% for eye-drops containing a preservative and 30% for eye-drops with no preservative (2). Moreover, the proportion of patients presenting at least one symptom of eye irritation (sensation of itching or burning, sensation of a foreign body in the eye, and flow of tears) was greater with preservative-containing eye-drops (53 versus 34%). The experience of discomfort during instillation was more often associated with problems later on. The patient's complaints were correlated with objective signs of con-junctival damage (conjunctival redness, conjunctival folli-cles), or corneal damage (superficial punctate keratitis). A higher proportion of patients treated with eye-drops containing a preservative had at least one conjunctival sign (52 versus 35%) or superficial punctate keratitis (12 versus 4%). In 164 patients whose treatment was changed from eye-drops containing a preservative to eye-drops with no preservative and who were examined a second time (mean interval between visits 3.3 months) the fre-quency of all symptoms and objective signs fell by a factor of 3 to 4.

Organs and Systems

Respiratory

An asthmatic patient whose salbutamol formulation was replaced by another containing benzalkonium chloride as an excipient developed bronchospasm as a result (3).

Benzalkonium chloride is used as a preservative in nebulizer solutions and can cause secondary paradoxical bronchoconstriction in patients with bronchial asthma. Although nebulizers containing beta-adrenoceptor ago-nists may also contain benzalkonium chloride, reports are more common with anticholinergic drugs. This is probably because of the more rapid onset and larger effect of sympathomimetic-induced bronchodilatation compared with anticholinergic drugs (4).

Bronchiolitis obliterans organizing pneumonia has been reported (5).

- Bronchiolitis obliterans organizing pneumonia (BOOP) was diagnosed in a 46-year-old cleaning lady who had severe dyspnea, cough, and fever 2 weeks after spilling a large amount of cleaning agent and inhaling its vapor. The components of the cleaning agent were benzalkonium compounds.

BOOP is an inflammatory lung disease that simultaneously involves the terminal bronchioles and alveoli. It is regarded as idiopathic in most cases, but it can be secondary to drugs, infections, organ transplantation, radiotherapy, and rarely occupational exposure to hazardous agents.

Ear, nose, throat

Benzalkonium chloride accentuated the severity of rhinitis medicamentosa and increased histamine sensitivity in a 30-day study with oxymetazoline nasal spray in healthy volunteers (6,7).

Sensory systems

Benzalkonium chloride is used in eye-drops in concentrations of 0.033 or 0.025%. At a dilution of 1:1000 (0.1%), a drop applied to the human cornea causes mild discomfort that persists for 2 or 3 hours. Slit-lamp examination within 90 seconds shows fine grey clots (epithelial keratitis) in the corneal epithelium. Within 10 minutes, a grey haze can be seen on the corneal surface; superficial desquamation of the conjunctival epithelium can follow. The superficial irritation and disturbances disappear in a day or less.

Patients with glaucoma, dry eyes, infections, or iritis, sometimes use solutions containing benzalkonium chloride often enough and for long enough to cause damage. In these patients, there is a higher incidence of endothelial damage, epithelial edema, and bullous keratopathy, and because of the severity of the disease, additional damage from the medication can be overlooked. This is especially true in patients with defective epithelium or corneal ulcers, in whom the medication can penetrate well and who can be most vulnerable. There have been analogous results in investigations of benzethonium chloride, cetrimonium bromide, cetylpyridinium chloride, decyldodecylbromide, hexadecyl, and tetradecyltrimethylammonium bromide (SEDA-11, 490).

Immunologic

Life-threatening anaphylactic reactions that rarely occur during general anesthesia are mostly due to neuromuscular blockers. They may be due to cross-allergy mediated by drug-specific IgE antibodies to the quaternary ammonium moiety of the neuromuscular blocker molecule, perhaps with a contribution from IgE-independent mechanisms. Quaternary ammonium compounds, such as benzalkonium, in cosmetics and toiletries may play a role in sensitization (8).

Allergic reactions can occur after topical use, but are fairly rare. Allergic contact dermatitis has been reported in some cases. Allergic rhinitis on contact has also been reported.

In a study of the efficacy and acceptability of benzalkonium chloride-containing contraceptives (vaginal sponges, pessaries, and creams) in 56 women, one developed an allergic reaction with edema of the vulva (4). Non-allergic local irritation, itching, and a burning sensation were reported in nine women and nine husbands.

Infection risk

The bactericidal activity of benzalkonium chloride is limited to the Gram-positive and some of the Gram-negative bacteria, but *Pseudomonas* species are especially resistant and can cause severe infection. Too often it is not realized that the disinfectant can be contaminated with active multiplying resistant organisms.

Pseudomonas bacteremia has been attributed to the use of material in open-heart surgery that was stored in accidentally contaminated benzalkonium solutions, and after cardiac catheterization caused by inadequate disinfection of the catheters with benzalkonium solutions. In 1961, about 15 patients were reported with *Pseudomonas* infections caused by cotton pledgets kept in a contaminated aqueous solution used for skin antisepsis before intravenous and intramuscular injection (9). In 1976 there were outbreaks of *Pseudomonas cepacia* infections in two American general hospitals (10) and pseudobacteremia (*Pseudomonas cepacia* or *Enterobacter*) caused by contamination of blood cultures in 79 patients in whom contaminated aqueous benzalkonium solutions were used for skin and antisepsis before venepuncture and due to contamination of the samples (SEDA-11, 490) (11).

References

1. Hillerdal G. Adverse reaction to locally applied preservatives in nose drops. ORL J Otorhinolaryngol Relat Spec 1985;47(5):278–9.
2. Levrat F, Pisella PJ, Baudouin C. Tolérance clinique des collyres antiglaucomateux conservés et non conservés. Résultats d'une enquête inédite en Europe. [Clinical tolerance of antiglaucoma eyedrops with and without a preservative. Results of an unpublished survey in Europe.] J Fr Ophtalmol 1999;22(2):186–91.
3. Ontario Medical Association's Committee on Drugs and Pharmacotherapy. Preservatives: bronchospasm. The Drug Report 1987;24:.
4. Meyer U, Gerhard I, Runnebaum B. Benzalkonium-chlorid zur vaginaten Kontrazeption—der Scheidenschwamm. [Benzalkonium chloride for vaginal contraception—the vaginal sponge.] Geburtshilfe Frauenheilkd 1990;50(7):542–7.
5. DiStefano F, Verna N, DiGiampaolo L, Boscolo P, DiGioacchino M. Cavitating BOOP associated with myeloperoxidase deficiency in a floor cleaner with incidental heavy exposure to benzalkonium compounds. J Occup Health 2003;45:182–4.

6. Graf P, Hallen H, Juto JE. Benzalkonium chloride in a decongestant nasal spray aggravates rhinitis medicamentosa in healthy volunteers. Clin Exp Allergy 1995;25(5):395–400.

7. Hallen H, Graf P. Benzalkonium chloride in nasal decongestive sprays has a long-lasting adverse effect on the nasal mucosa of healthy volunteers. Clin Exp Allergy 1995;25(5):401–5.

8. Weston A, Assem ES. Possible link between anaphylactoid reactions to anaesthetics and chemicals in cosmetics and biocides. Agents Actions 1994;41(Spec No):C138–9.

9. Lee JC, Fialkow PJ. Benzalkonium chloride-source of hospital infection with Gram-negative bacteria. JAMA 1961;177:708–10.

10. Dixon RE, Kaslow RA, Mackel DC, Fulkerson CC, Mallison GF. Aqueous quaternary ammonium antiseptics and disinfectants. Use and misuse. JAMA 1976;236(21):2415–7.

11. Kaslow RA, Mackel DC, Mallison GF. Nosocomial pseudobacteremia. Positive blood cultures due to contaminated benzalkonium antiseptic. JAMA 1976;236(21):2407–9.

Benzethonium chloride and methylbenzethonium chloride

General Information

Quaternary ammonium compounds are surface-active agents. Some of them precipitate or denature proteins and destroy microorganisms. The most important disinfectants in this group are cationic surface-active agents, such as benzalkonium chloride, benzethonium chloride and methylbenzethonium chloride, and cetylpyridinium chloride; the problems that they cause are similar.

In an extensive report, the Expert Panel of the American College of Toxicology (1) has concluded that both benzethonium chloride and methylbenzethonium chloride can be regarded as safe when applied to the skin at a concentration of 0.5% or when used around the eye in cosmetics at a maximum concentration of 0.02%. In clinical studies, benzethonium chloride produced mild skin irritation at 5%, but not at lower concentrations. Neither ingredient is considered to be a sensitizer.

Drug Administration

Drug overdose

The Paris Poison Center has received reports on 45 cases of acute accidental poisoning, with 18 deaths (2). All the victims were mentally disturbed patients who had ingested Airsane HP 800, a water-soluble powder packed in a sachet; it contains a mixture of quaternary ammonium compounds and was left in the patients' rooms by hospital workers. Symptoms were corrosive burns of the mouth, pharynx, esophagus, and sometimes of the respiratory tract.

References

1. Anonymous. Final report on the safety assessment of benzethonium chloride and methylbenzethonium chloride. J Am Coll Toxicol 1985;4:65.

2. Chataigner D, Garnier R, Sans S, Efthymiou ML. Intoxication aiguë accidentelle par un désinfectant hospitalier. 45 cas dont 13 d'évolution mortelle. [Acute accidental poisoning with hospital disinfectant. 45 cases of which 13 with fatal outcome.] Presse Méd 1991;20(16):741–3.

Benzoxonium chloride

General Information

Benzoxonium is a quaternary ammonium compound with antibacterial, antiviral, and antimycotic activity. It can be used in topical disinfection, disinfection of surgical instruments, inhibition of plaque formation, and in veterinary products.

Organs and Systems

Immunologic

Contact allergic reactions have been rarely reported, with potential cross-reactivity with benzalkonium chloride and domiphen bromide (1,2).

- A 37-year-old woman developed intense burning and pruritic eczema where she had applied a cream containing benzoxonium for seborrheic dermatitis for 5 months (3). The reaction disappeared on withdrawal of the cream. Patch tests were positive to benzoxonium chloride 0.1% aqueous on days 2 and 4. Patch tests with benzalkonium chloride and benzoxonium chloride in 20 controls were negative.

References

1. de Groot AC, Conemans J, Liem DH. Contact allergy to benzoxonium chloride (Bradophen). Contact Dermatitis 1984;11(5):324–5.

2. Bruynzeel DP, de Groot AC, Weyland JW. Contact dermatitis to lauryl pyridinium chloride and benzoxonium chloride. Contact Dermatitis 1987;17(1):41–2.

3. Diaz-Ramon L, Aguirre A, Raton-Nieto JA, de Miguel M. Contact dermatitis from benzoxonium chloride. Contact Dermatitis 1999;41(1):53–4.

Benzyl alcohol

General Information

Benzyl alcohol is commonly used as a preservative in multidose injectable pharmaceutical formulations. For this purpose, concentrations in the range of 0.5–2.0% are used and the whole amount of benzyl alcohol injected is generally very well tolerated. Concentrations of 0.9% are used in Bacteriostatic Sodium Chlorine (USP), which is often used in the management of critically ill patients to flush intravascular catheters after the addition of medications or the withdrawal of blood, and in Sterile Bacteriostatic Water for injection (USP), used to dilute or reconstitute medications for intravenous use. The content of benzyl alcohol in a lot of injectable pharmaceutical formulations needs to be considered carefully. The view still taken in many countries that the additives and excipients in medicines are trade secrets must be deplored. The duty to declare them is only realized in some countries.

The toxic effects of benzyl alcohol include respiratory vasodilatation, hypertension, convulsions, and paralysis.

Organs and Systems

Nervous system

The data on reported cases of neurological disorders after intrathecal chemotherapy with methotrexate or cytosine arabinoside that could be attributed to benzyl alcohol or to other preservatives have been reviewed in the context of a case of flaccid paraplegia after intrathecal administration of cytosine arabinoside diluted in bacteriostatic water containing 1.5% benzyl alcohol (1). Most commonly, flaccid paraparesis, with absent reflexes, developed rapidly, often with pain and anesthesia. Very often there was full recovery. The prognosis depended mainly on the concentration of the preservative and on the time of exposure. In some cases, the paralysis ascended to cause respiratory distress, cardiac arrest, and death. Only preservative-free sterile CSF substitute or saline, or preferably the patient's own CSF, should be used to dilute chemotherapeutic agents (SEDA-11, 475).

Skin

Allergic contact dermatitis, characterized by erythema, palpable edema, and raised borders, was attributed to benzyl alcohol (2). In this case, the benzyl alcohol was present as a preservative in an injectable solution of sodium tetradecyl sulfate, a sclerosing agent used for the treatment of varicose veins. The author provided a list of 151 injectable formulations (48 for subcutaneous administration) that contained benzyl alcohol as a preservative in the range 0.5–2.0%. The list included hormones and steroids, antihypertensive drugs (reserpine), vitamin formulations (vitamins B_{12} and B_6), ammonium sulfate, antihistamines, antibiotics, heparin (17 brands), tranquillizers, and sclerosing agents (sodium morrhuate and sodium tetradecyl sulfate).

- A patient with a contact allergic reaction to a topical antimycotic drug formulation that contained benzoyl alcohol had positive patch tests on day 4 and a positive repeated open application test to benzoyl alcohol 5% in petroleum jelly (3).

The authors noted that although contact allergic reactions to benzoyl alcohol are rarely reported, they can be responsible for contact allergy to topical glucocorticoid formulations.

Immunologic

Various allergic reactions have been attributed to benzyl alcohol.

- A 55-year-old man developed fatigue, nausea, and diffuse angioedema shortly after an intramuscular injection of vitamin B12 containing benzyl alcohol (4).
- In another male patient, fever developed, and a maculopapular rash occurred on his chest and arms after an injection of cytarabine, vincristine, and heparin in a dilution solution containing benzyl alcohol (5).

Susceptibility Factors

Age

A gasping syndrome in small premature infants who had been exposed to intravenous formulations containing benzyl alcohol 0.9% as a preservative has been described (SEDA-10, 421) (SEDA-11, 475) (6–8). The affected infants presented with a metabolic acidosis, seizures, neurological deterioration, hepatic and renal dysfunction, and cardiovascular collapse. Death was reported in 16 children who received a minimum of 99 mg/kg/day of benzyl alcohol. This metabolic acidosis is caused by accumulation of the metabolite benzoic acid and is mainly related to an excessive body burden relative to body weight, so that the load of the metabolite may exceed the capacity of the immature liver and kidney for detoxification. The FDA has recommended that neither intramuscular flushing solutions containing benzyl alcohol nor dilutions with this preservative should be used in newborn infants.

In a review of the hospital and autopsy records of infants admitted to a nursery during the previous 18 months, 218 patients had been given fluids containing benzyl alcohol as flush solutions and they were compared with 218 neonates admitted during the following 18 months (9). Withdrawal of benzyl alcohol as a preservative had no demonstrable effect on mortality, but the development of kernicterus was significantly associated with benzyl alcohol in 15 of 49 exposed patients, and no cases occurred after withdrawal of the preservative. However, this apparent association was not confirmed in a 5-year study of the use of benzyl alcohol as a preservative in intravenous medications in a neonatal intensive care unit (10). In 129 neonates who died between the ages of 2 and 28 days, there was no difference in the rate of kernicterus and the exposure to benzyl alcohol between neonates who developed kernicterus and the control group of unaffected infants who were born during the same period and who were of the same birth weight and gestation age. In this study, only estimates of the extent of exposure to benzyl alcohol were given, rather than exact doses and serum concentrations.

References

1. Hahn AF, Feasby TE, Gilbert JJ. Paraparesis following intrathecal chemotherapy. Neurology 1983;33(8):1032–8.
2. Shmunes E. Allergic dermatitis to benzyl alcohol in an injectable solution. Arch Dermatol 1984;120(9):1200–1.
3. Podda M, Zollner T, Grundmann-Kollmann M, Kaufmann R, Boehncke WH. Allergic contact dermatitis from benzyl alcohol during topical antimycotic treatment. Contact Dermatitis 1999;41(5):302–3.
4. Grant JA, Bilodeau PA, Guernsey BG, Gardner FH. Unsuspected benzyl alcohol hypersensitivity. N Engl J Med 1982;306(2):108.
5. Wilson JP, Solimando DA Jr, Edwards MS. Parenteral benzyl alcohol-induced hypersensitivity reaction. Drug Intell Clin Pharm 1986;20(9):689–91.
6. Gershanik J, Boecler B, George W, et al. Gasping syndrome: benzyl alcohol poisoning. Clin Res 1981;29:895a.
7. Gershanik J, Boecler B, Ensley H, McCloskey S, George W. The gasping syndrome and benzyl alcohol poisoning. N Engl J Med 1982;307(22):1384–8.
8. Gershanik J, Boecler B, George W, et al. Neonatal deaths associated with use of benzyl alcohol—United States. Munch Med Wochenschr 1982;31:290.
9. Jardine DS, Rogers K. Relationship of benzyl alcohol to kernicterus, intraventricular hemorrhage, and mortality in preterm infants. Pediatrics 1989;83(2):153–60.
10. Cronin CM, Brown DR, Ahdab-Barmada M. Risk factors associated with kernicterus in the newborn infant: importance of benzyl alcohol exposure. Am J Perinatol 1991;8(2):80–5.

Boric acid

General Information

In the past, boric acid was falsely considered to be relatively non-toxic, and had an unwarranted reputation as a germicide. However, it is only bacteriostatic, even in a saturated aqueous solution, and can cause adverse reactions. Boric acid has often proved poisonous, either by ingestion or after local use. Cases from the world literature have been reviewed (1,2). In 172 cases of boric acid intoxication, including 83 deaths, 37 deaths occurred after external use, including 23 children with nappy rash. From 1974 to 1984, the Poison Centre in Paris recorded 134 cases of intoxication with boric acid or borates, 88 of which were accidental and 31 iatrogenic.

Boric acid penetrates even intact skin, but it is readily absorbed through inflamed or otherwise damaged skin and through mucous membranes. After the application of wet compresses of boric acid to intact and eczematous skin in 21 patients over several days, blood concentrations of boric acid were generally not raised (3). One patient, however, did have a significant rise in blood boric acid concentration, which the authors ascribed to pre-existing kidney insufficiency.

Prolonged absorption of boric acid causes anorexia, weight loss, vomiting, mild diarrhea, skin rash, diffuse alopecia, convulsions, and anemia.

On account of the high toxicity and limited therapeutic value of boric acid, the borates have been abandoned as obsolete in many countries.

Organs and Systems

Skin

The skin, which is unlikely to react unfavorably to topical boric acid, can react strongly if it is absorbed systemically; resultant redness can mimic scarlet fever, and psoriasiform lesions, bullae, and alopecia can occur.

Drug Administration

Drug overdose

In acute intoxication, symptoms develop progressively, typically beginning with persistent vomiting and diarrhea (mucous and blood, with a bluish-green color). Shortly after the onset of the gastrointestinal symptoms, in some cases even earlier, a rash appears (with macules and/or papules), beginning on the abdomen, genitalia, and head, and rapidly spreading, followed by excoriation and intensive desquamation after 1–2 days. Mucous membranes are often involved, especially in young infants, in whom the mouth, pharynx, and conjunctivae are often inflamed. Later, there are central nervous system symptoms, with headache and mental confusion followed by convulsions. In children there is meningism and twitching of the facial muscles and limbs followed by convulsions. Acute tubular necrosis can occur, with oliguria or anuria, hypernatremia, hyperchloremia and hyperkalemia, proteinuria, erythrocyturia, and cylindruria. Finally, hyperthermia, a fall in blood pressure, tachycardia, and shock occur.

Fourteen cases of acute boric acid ingestion were reported in New York City over 30 months. In these patients excretion of urinary riboflavin (vitamin B_2) was determined, and in about two-thirds it was significantly increased. This is not surprising, since riboflavin and boric acid are known to form a water-soluble complex. The range of lethal doses is 1–3 g for babies, 5 g for infants, and 15–20 g for adults.

- Severe acute boric acid poisoning occurred in a 3-year-old boy after the application of a boric acid-containing talcum powder (4).
- A similar case occurred in a 27-day-old baby girl (5).
- Generalized erythema and skin desquamation, fever convulsions, and diarrhea followed repeated boric acid application to a denuded umbilical stump (6).
- Fatal intoxication has followed prostatectomy and subsequent bladder irrigation with 3% boric acid (7).

Of 172 published cases of boric acid intoxication, 83 proved fatal; the series included 37 deaths after external use of boric acid, 23 of these being children who had developed diaper rashes. The actual number of cases, mostly unpublished, must have been much greater: from 1974 to March 1984, 134 cases of intoxication by boric acid or borates were recorded in France alone, 88 being accidental and 31 associated with medicinal use.

References

1. Goldbloom RB, Goldbloom A. Boric acid poisoning; report of four cases and a review of 109 cases from the world literature. J Pediatr 1953;43(6):631–43.
2. Valdes-Dapena MA, Arey JB. Boric acid poisoning: three fatal cases with pancreatic inclusions and a review of the literature. J Pediatr 1962;61:531.
3. Schuppli R, Seiler H, Schneeberger R, Niggli H, Hoffmann K. Über die Toxizität der Borsäure. [Toxicity of boric acid.] Dermatologica 1971;143(4):227–34.
4. Skipworth GB, Goldstein N, McBride WP. Boric acid intoxication from "medicated talcum powder". Arch Dermatol 1967;95(1):83–6.
5. Baliah T, MacLeish H, Drummond KN. Acute boric acid poisoning: report of an infant successfully treated by peritoneal dialysis. Can Med Assoc J 1969;101(3):166–8.
6. Hillman DA, Hillman ES, Hall J. Boric acid poisoning. East Afr Med J 1970;47(11):572–5.
7. Schmid F, Zbinden J, Schlatter C. Zwei Fälle von letaler Borsäurevergiftung nach Blasenspülung. [Two cases of fatal boric acid poisoning after bladder irrigation.] Schweiz Med Wochenschr 1972;102(3):83–8.

Cetrimonium bromide and cetrimide

General Information

Cetrimonium bromide and cetrimide are quaternary ammonium antiseptics, trimethylammonium derivatives, with similar structures. Cetrimonium bromide is hexadecyltrimethylammonium. Cetrimide is a mixture of tetradecyltrimethylammonium (mostly), dodecyltrimethylammonium, and hexadecyltrimethylammonium. They dissociate in aqueous solution, forming a relatively large and complex cation, which is responsible for their surface activity, and a smaller inactive anion. They are emulsifiers and detergents and have bactericidal activity against Gram-positive and, at higher concentrations, some Gram-negative bacteria.

In three hospitals, 378 patients with hydatid cysts were treated surgically, including irrigation with cetrimide solutions in concentrations between 0.05 and 1% (1). No adverse effects were observed.

Organs and Systems

Sensory systems

Accidental use of cetrimide solution 0.1% as an irrigation solution during cataract surgery in two cases resulted in immediate corneal edema, which in turn resulted in a severe bullous keratopathy (2).

Acid–base balance

Severe metabolic acidosis without signs of peritonitis occurred after more than 1 liter of 1% cetrimide was instilled into hydatid cysts (SEDA-11, 490) (3).

Hematologic

Severe methemoglobinemia has been reported after excision of hydatid cysts and liberal irrigation of the cysts in the liver with a 0.1% solution of cetrimide (4).

Skin

Eighteen cases of an ichthyosiform contact dermatitis caused by antiseptic solutions containing 3% cetrimide and 0.3% chlorhexidine have been reported (5,6). Biopsies showed hyperkeratosis with striking vesiculation of lamellar bodies in the granular cells and upper spinous cells, premature secretion of lamellar bodies, and abundant remnants of lamellar bodies and retention of desmosomes between corneocytes. Cetrimide, and not chlorhexidine, was said to be the cause of the dermatitis.

References

1. Frayha GJ, Bikhazi KJ, Kachachi TA. Treatment of hydatid cysts (*Echinococcus granulosus*) by Cetrimide®. Trans R Soc Trop Med Hyg 1981;75(3):447–50.
2. van Rij G, Beekhuis WH, Eggink CA, Geerards AJ, Remeijer L, Pels EL. Toxic keratopathy due to the accidental use of chlorhexidine, cetrimide and cialit. Doc Ophthalmol 1995;90(1):7–14.
3. Momblano P, Pradere B, Jarrige N, Concina D, Bloom E. Metabolic acidosis induced by cetrimonium bromide. Lancet 1984;2(8410):1045.
4. Baraka A, Yamut F, Wakid N. Cetrimide-induced methaemoglobinaemia after surgical excision of hydatid cyst. Lancet 1980;2(8185):88–9.
5. Lee JY, Wang BJ. Contact dermatitis caused by cetrimide in antiseptics. Contact Dermatitis 1995;33(3):168–71.
6. Lee JY. Pathogenesis of abnormal keratinization in ichthyosiform cetrimide dermatitis: an ultrastructural study. Am J Dermatopathol 1997;19(2):162–7.

Chlorhexidine

General Information

Chlorhexidine (1,1'-hexamethylene-*bis*[5-(*p*-chlorophenyl)biguanide]) is a widely used antibacterial agent with activity against Gram-positive bacteria, Gram-negative bacteria (less against *Pseudomonas* species), and yeasts. It was introduced as an antiseptic in the early 1950s. It has been primarily used for topical antisepsis, for example in preoperative skin disinfection, and for disinfection of materials, mainly in combination with cetrimide. Long-term experience has shown a low incidence of sensitization and a low irritant potential (SEDA-11, 480).

Unwanted effects have resulted from undue reliance on the disinfecting properties of chlorhexidine. Hospital-acquired infections have been caused by infected chlorhexidine used for bladder irrigation and for storage or disinfection of catheter spigots and needles (SEDA-11, 480) (1,2). A microbiological analysis of chlorhexidine-cream tubes, repeatedly used by patients with indwelling urethral catheters, showed high contamination with potential pathogens in 32% of cream samples and in 35% of swabs taken from the outside of the tubes beneath the screw cap (SEDA-11, 480) (3).

Use of chlorhexidine in dentistry

Chlorhexidine has been used as an adjuvant for plaque control and in the treatment of gingival inflammation. It is generally considered to be effective in the control of plaque and can be helpful in the treatment of gingivitis. It can be applied in the form of a solution, used as a mouth rinse or with a toothbrush, in dentifrice or as a gel. The concentrations used are 0.05–2%.

It is very difficult to summarize the effect of chlorhexidine on oral hygiene, since studies differ markedly as regards the population studied, the occurrence of gingival lesions, the use of other oral hygiene regimens, previous scaling, and polishing of the teeth. The most frequently reported adverse effects of oral use are discoloration of the teeth, tongue, and buccal mucosa, taste disorders, and desquamation of the oral mucosa. A mild increase in gingival bleeding was reported after the use of chlorhexidine mouthwash compared with mechanical cleaning methods.

Use of chlorhexidine in neonatal skin care

The withdrawal of hexachlorophene-containing products for routine neonatal skin care stimulated investigations into the possible use of chlorhexidine in this field. Its activity range includes effectiveness of high dilutions against Gram-positive and Gram-negative bacteria, yeasts, and molds. In studies of nursery populations, chlorhexidine appears to be as effective as hexachlorophene in preventing staphylococcal colonization and infection. However, there is no evidence that chlorhexidine promotes Gram-negative colonization in neonates bathed in water-containing chlorhexidine (SEDA-11, 482).

Data presented in three studies have provided substantial evidence that there is very low percutaneous absorption in full-term infants and also in excessively exposed newborn rhesus monkeys. However, traces of chlorhexidine were found in adipose tissue (two of five monkeys), kidneys (five of five), and liver (one of five), suggesting some absorption percutaneously or by oral ingestion, following the rigorous bathing procedure in the above study. The grooming habits of the monkeys could have played a role (SEDA-12, 578) (4).

Use of chlorhexidine in routine neonatal cord care

About 0.2 ml of undiluted 4% chlorhexidine solution with a detergent was included in the daily routine of rubbing the dry cord stump and the surrounding skin, which were then rinsed and dried. No chlorhexidine was detectable in the blood samples of the neonates, taken on the fifth day (5).

Use of chlorhexidine in spermicides

The use of chlorhexidine in spermicides has been promoted as a strategy for protecting against sexually transmitted diseases, including HIV infection. However, both the claim of protection and the cytotoxicity of chlorhexidine, with a risk of damage to the epithelia of the vagina, cervix, and glans penis due to chronic exposure, have to be further validated (6).

Use of chlorhexidine in venesection

To minimize the bacterial contamination rate in blood collected from donors a study was designed to evaluate the suitability of a single-use chlorhexidine-alcohol antiseptic for donor arm preparation at all blood collection venues in Australia (7). The tolerability of an antiseptic for blood donor disinfection is important to minimize any factor that might cause

ors to stay away from future donation. A prospective study of bacterial load on the skin was performed in 616 blood donor arms before and after disinfection. Disinfection was achieved with a swab containing 1% chlorhexidine gluconate with 75% alcohol. Feedback from blood donors and staff was obtained using questionnaires. After disinfection, there was a marked reduction in skin bacterial counts, well under the target of the Australian Red Cross Blood Services. Sixteen donors reported skin irritation at the site of application; most of the reactions were self-limiting itchiness, with or without erythema. The majority of donors either preferred or did not object to the use of the chlorhexidine antiseptic.

Use of chlorhexidine to prevent HIV transmission during labor

An estimated 600 000 children world-wide were infected with HIV type 1 in 2001. Most of these infections occurred through mother-to-child transmission of HIV during pregnancy, around the time of labor and delivery. In developing countries, where the resources to prevent and manage these infections are limited, peripartum cleansing with chlorhexidine is a potentially simple and low-cost strategy for the prevention of mother-to-child transmission. However, low concentrations of chlorhexidine have not been proved to be effective in reducing mother-to-child transmission. Before assessing the effectiveness of higher chlorhexidine concentrations on mother-to-child transmission the highest tolerated concentration needed to be established. Three concentrations of chlorhexidine (0.25%, 1%, and 2%) have been evaluated as perinatal maternal and infant washes, to identify the maximum tolerated concentration for this intervention (8). Women were enrolled during their third trimester at a maternity unit in Soweto, South Africa. Subjective maternal symptoms and infant examinations were used to assess tolerability of the washes. The 0.25% concentration of chlorhexidine was well tolerated by the mothers. Ten of 79 complained of mild vaginal burning or itching from the 1% chlorhexidine washes, and washes were stopped in five. Of the 75 women in the 2% chlorhexidine group, 23 had subjective complaints and the washes were stopped in 12. There were no indications of toxicity from any of the chlorhexidine washes in the infants. The authors concluded that a 1% solution of chlorhexidine is safe and well tolerated and could be considered for a trial in the prevention of mother-to-child transmission.

Five studies of the prophylactic intravaginal use of chlorhexidine vaginal suppositories before delivery and in obstetrics have been reviewed (9). No severe adverse reactions were reported.

Accidental ingestion

A newborn had cyanotic spells associated with sinus bradycardia but not with apnea on the third day of life(10). The mother had used a chlorhexidine spray on her breasts from the third feed onwards. Bradycardia became less frequent and less severe from day 4 after the spray was stopped, and had abated completely by day 6.

Five healthy newborn breast-fed babies were accidentally fed a dilute antiseptic solution containing chlorhexidine 0.05% with cetrimide 1% instead of sterile water (11). They developed caustic burns of the lips, mouth, and tongue within minutes. One baby became severely ill due to acute pulmonary edema, but all survived without sequelae.

Systematic reviews

A review of the literature on chlorhexidine interventions (vaginal, newborn skin, and umbilical cord cleansing) focused on neonatal outcomes and safety. In summary, tens of thousands of neonates have received these chlorhexidine-based cleansing interventions without reported adverse effects. However, the data on safety are incomplete. Although the chlorhexidine concentrations used and reported thus far appear to be safe, the upper level of chlorhexidine that can be considered safe is not known and further research is required to inform public health safety in developing countries (12).

Placebo-controlled studies

In a multicenter, double-blind, randomized, placebo-controlled study, two decontamination regimens were assessed in the prevention of acquired infections in 515 high-risk intubated patients in intensive care (13):

- topical polymyxin + tobramycin (n = 130);
- nasal mupirocin ointment with chlorhexidine body washing (n = 130);
- the two regimens combined (n = 129);
- matching placebo (topical placebo and/or nasal placebo + liquid soap; n = 126).

There were fewer acquired infections with the combined regimens than with either regimen alone or placebo. There were no differences between either regimen alone and placebo. There were *allergic reactions* in six patients who received chlorhexidine and six patients who received the liquid soap. There was no intolerance to the nasal ointment. The polymyxin + tobramycin regimen was withdrawn in 37 patients, mostly because the serum tobramycin concentration exceeded 2 mg/l. Body washing was discontinued in nine patients receiving chlorhexidine and in eight patients receiving the liquid soap.

Organs and Systems

Respiratory

Chlorhexidine gluconate when ingested usually causes relatively mild symptoms, with poor gastrointestinal absorption, and is considered relatively safe. However, a rare fatality, with acute respiratory distress syndrome, has been reported (14).

- An 80-year-old woman with dementia accidentally took about 200 ml of chlorhexidine gluconate 5%. She aspirated her gastric contents and despite intensive treatment died of acute respiratory distress syndrome 12 hours later. The serum concentration of chlorhexidine gluconate was markedly high (25 µg/ml).

It is possible that in this case, although chlorhexidine was poorly absorbed from the gastrointestinal tract, absorption occurred through the pulmonary alveoli.

Occupational asthma caused by chlorhexidine alcohol disinfectant spray was well documented in three nurses (15).

Ear, nose, throat

Reversible hyposmia occurred after Hardy's operation for pituitary adenomas following preoperative disinfection of the nasal cavity with chlorhexidine (16). The olfactory disturbance improved after 3–7 weeks. Degeneration of the olfactory epithelium was also seen in guinea pigs when the nasal cavities were irrigated 3 times with 5 ml 0.5% chlorhexidine solution; regeneration of the epithelium started after 14 days and the surfaces appeared normal after 1 month.

Nervous system

The middle ear should be carefully protected against chlorhexidine solutions in preoperative skin disinfection in otolaryngology. Severe sensorineural deafness occurred in 14 of 97 patients who underwent myringoplasty (17). The only common factor in all these patients was the preoperative skin disinfection of the ear with 0.5% chlorhexidine in a 70% alcoholic solution.

In extensive investigations of the ototoxicity caused by chlorhexidine after introducing it into the middle ear of guinea pigs, the extent of severe vestibular and cochlear damage was related to the concentration of chlorhexidine, to the duration of exposure, and to the time-lapse after exposure (18–20).

Sensory systems

Eyes
Accidental contamination of the eye with chlorhexidine can cause adverse effects. Care should be taken during preoperative skin preparation to keep chlorhexidine out of the eye and to flush copiously with sterile saline solution or sterile water if contact accidentally occurs.

- Accidental use of chlorhexidine solutions (1:666 and 1:1000) as irrigation during cataract surgery resulted in immediate corneal edema, which resulted in a bullous keratopathy (21).

- Chlorhexidine disinfectant was accidentally used to irrigate the eyes of four patients; despite immediate treatment, corneal burns occurred in all (22).
- In four patients, accidental corneal exposure to Hibiclens (4% chlorhexidine formulated with a detergent) resulted in keratitis, with severe and permanent corneal opacification (23).
- Epithelial and stromal edema of the cornea and a diffuse bullous keratopathy developed in a 39-year-old woman 2 weeks after a preoperative disinfection of the face with an alcoholic chlorhexidine solution. This led to penetrating keratoplasty 10 months later (24).

The histopathological findings in the cornea have been described in such cases; they include epithelial edema with bullous changes, marked loss of keratocytes, thickening of Descemet's membrane, and an attenuated disrupted cell layer (25).

Chlorhexidine has also been accidentally irrigated into the anterior chamber of the eye, instead of balanced salt solution, during cataract surgery (26). Later in the operation, a decrease in corneal clarity was noted and an epithelial abrasion had to be performed. The inadvertent use of chlorhexidine in this patient resulted in reduced endothelial function and loss of corneal clarity.

Progressive ulcerative keratitis related to the use of chlorhexidine gluconate 0.02% eye-drops has been reported (27).

- A 45-year-old woman was treated for presumed *Acanthameba* keratitis with chlorhexidine gluconate 0.02% and propamidine 0.1% eye-drops. After using the eye-drops for 8 weeks she developed a near total loss of the corneal epithelium and progressive ulcerative keratitis, which eventually required penetrating keratoplasty. Histopathological examination of the corneal button showed ulceration and loss of Bowman's membrane, massive loss of keratocytes with apparent apoptosis, and loss of endothelial cells, with inflammatory cells adherent to the remaining cells. These findings were similar to those seen in chlorhexidine 4% keratopathy. No organisms were seen in stained sections and immune histochemistry showed no significant findings.

Two almost identical cases of cataract and iris atrophy have been reported as complications of the use of chlorhexidine for acanthameba keratitis (28). Both cases involved prolonged treatment with chlorhexidine 0.02% and propamidine isethionate 0.1%, but progressed to deep marginal ulceration, necessitating a preventive graft. In both cases perioperative findings showed a totally dense membrane cataract, which had developed over a few months, together with an atrophic iris and a maximally dilated pupil. A complicating cataract in uveitis usually takes more than a few months to reach maturity. The intraocular inflammation in these two cases was not as intense as might have been expected if the amoebae had spread across the aqueous and entered the iris and the lens through an intact capsule. The iris atrophy

was extreme, and the surface was greyish with little spots of clustered pigment. This appearance is almost never encountered in uveitis lasting only for a few months and this led the authors to suggest that a toxic mechanism was involved.

Taste

Taste impairment and/or disturbance is especially seen with chlorhexidine mouthwashes (SEDA-11, 481) (SEDA-12, 577) (29). It occurs in about one-third of subjects and is perhaps the main complaint. Symptoms include a burning sensation, a feeling of soreness, a dry mouth, and a bitter aftertaste lasting from a few minutes up to several hours. However, these types of symptoms also occur, albeit to a lesser extent, in placebo or control groups.

Mouth and teeth

Chlorhexidine is the most commonly used mouthwash for chemical plaque control. Its use is becoming widespread as an adjuvant treatment of mechanical control, particularly in individuals with compromised oral hygiene. However, several adverse effects that limit patients' acceptance of mouth rinsing have been identified—brown staining of the teeth, an unpleasant taste, and rarely painful desquamation of the oral mucosa. Alternative formulations of chlorhexidine and other chemical plaque controls have therefore been developed to minimize adverse effects. Desquamation and ulceration of the oral mucosa after the use of chlorhexidine mouthwashes. must be very infrequence. Histological and histochemical examination of mucosal biopsies taken after 18 months of daily exposure to chlorhexidine did not show any adverse effect on the oral mucosa. There was increased keratinization of human gingival cells in vitro in cell cultures if the chlorhexidine concentration exceeded 25 µg/ml, and the same acceleration of keratinization of human gingival cells in gingival swabs occurred after rinsing with 0.025, 0.05, and 0.1% chlorhexidine solutions (30). In one case there was excessive impairment of wound healing after daily rinses with a 0.1% chlorhexidine solution after oral surgery.

Staining of the teeth was the first and principal adverse effect observed with the dental use of chlorhexidine. In a 4-month study of soldiers who used chlorhexidine mouthwashes in concentrations of 0.1 and 0.2%, 15% of the interproximal surface and 62% of the fillings, especially the old and porous ones, were discoloured (31,32). The stain intensity seems to be directly correlated to the concentration of the chlorhexidine and to the frequency and duration of use. The type of administration (0.2% mouthwash, 0.2% spray, or 1% gel) does not seem to influence the amount of tooth-staining. The initial discoloration is yellow-brown, but prolonged use and stronger concentrations result in a dark-brown color. Extensive investigations have been performed to evaluate factors that influence tooth-staining and the possibility of avoiding it. The cause of extrinsic tooth discoloration is not fully understood. However, the available evidence suggests that browning and formation of pigmented metal sulfides are the most likely causes, while dietary factors (such as beverages or red wine) and smoking may play an aggravating role only (SEDA-10, 427).

A discoloration of the dorsum of the tongue occurs in up to one-third of subjects using chlorhexidine mouth rinses. It does not occur during the use of chlorhexidine-containing dentifrices or gels (SEDA-11, 481) (32,33).

The effects of three oral sprays containing chlorhexidine, benzydamine hydrochloride, and chlorhexidine plus benzydamine hydrochloride on plaque and gingivitis have been compared (34). Chlorhexidine and chlorhexidine + benzydamine hydrochloride sprays were equally effective and the benzydamine hydrochloride spray alone was less effective. The chlorhexidine/benzydamine hydrochloride spray caused a burning sensation.

Commercial formulations of chlorhexidine mouthwash include other active ingredients, in an attempt to improve their effectiveness and reduce adverse effects. Using a double-blind crossover design, three different commercial formulations of 0.12% chlorhexidine digluconate mouthwash were compared for their effects on dental plaque, supragingival calculus, and dental staining. Changes in gingival and dental staining indices were not affected differently by the three products, but tongue staining was more frequent with the product that contained cetylpyridinium chloride (35).

Dental flossing is effective in reducing interproximal gingivitis and preventing caries on proximal tooth surfaces. However, representative surveys have shown that dental flossing is used by only a small part of the population. The results of a trial designed to assess if chlorhexidine mouth rinse was a suitable alternative to flossing suggested that daily use of the tested mouth rinses may result in greater interproximal plaque reduction than daily flossing. However adverse effects were reported with chlorhexidine mouthwash, mainly staining of the teeth and tongue. Although from a medical viewpoint these were not considered severe, they may be esthetically unacceptable and can lead to reduced compliance (36).

The use of a chlorhexidine oral spray has been proposed as an alternative to reduce the adverse effects associated with chlorhexidine mouthwashes. The plaque inhibitory effects of chlorhexidine, cetypyridinium chloride, and triclosan delivered by sprays and mouth rinses have been investigated in 15 volunteers (37). Although the effects on plaque regrowth observed with chlorhexidine rinses was superior to that of chlorhexidine sprays, the latter did not cause adverse effects.

Reversible swelling of the parotid glands has been reported occasionally after the use of chlorhexidine mouthwashes; this is probably due to mechanical obstruction of the parotid duct by over-rigorous rinsing (SEDA-11, 481) (38).

Gastrointestinal

- Atrophic gastritis was reported in a 72-year-old man with Parkinson's disease who had used a 4% chlorhexidine solution as a daily mouthwash but had also swallowed it. Gastroscopy showed multiple erosions in the lower part of his stomach and the first part of the duodenum (39).

Urinary tract

In a prospective randomized single-blind study of the effect of chlorhexidine in urethral local anesthetic gel, 141 patients undergoing flexile cystoscopy were randomized to either 2% lidocaine gel with 0.05% chlorhexidine gluconate or 2% lidocaine aqueous gel (40). Pain scores were recorded on a numerical visual analogue scale form 0 to 10. The pain scores were not significantly different between the groups at the time of insertion of the scope or immediately after cystoscopy; however, there were significant differences in the mean pain scores between groups during the first void (1.8 versus 1.0) and after the first void (2.4 versus 1.2). There was also a significant increase in the levels of urgency after cystoscopy with chlorhexidine. There was no difference in the level of culture-proven symptomatic infection. Thus, chlorhexidine appears to contribute to significant pain and urgency after out-patient flexible cystoscopy and its presence in urethral local anesthetic preparations should be questioned.

Skin

Of 551 patients with venous or traumatic ulcers of the leg who were patch-tested with chlorhexidine, 10 developed severe dermatitis and 4 developed skin infections on the face and/or scalp (41).

Serosae

Sclerosing peritoneal disease occurred in peritoneal dialysis patients in whom the tubing connection had been disinfected with chlorhexidine (42–44); 214 cases were reported from 112 centers in European countries up to 1984 (SEDA-15, 250).

Musculoskeletal

Chlorhexidine is not normally used in arthroscopy, but it is a common irrigating fluid for surgical wounds. In three of five patients with pain, swelling, crepitus, and loss of range of movement following arthroscopy of the knee, there had been accidental irrigation with 1% aqueous chlorhexidine. Histological examination showed partial necrosis of the cartilage, with slight non-specific inflammation and fibrosis of synovial specimens (45). This shows that particular care is needed in checking irrigation fluids.

Even very dilute solutions of chlorhexidine can cause marked chondrolysis of articular cartilage, leading to severe permanent damage of the knee (46).

- A 20-year-old woman, a 30-year-old man, and a 62-year-old woman all had arthroscopic reconstruction of the anterior cruciate ligament performed by the same surgeon. The only difference in technique in these three cases, compared with the rest of his cases, was the use of chlorhexidine 0.02% for irrigation throughout the procedure. All had good immediate postoperative recovery with no sign of infection, and none had preceding rheumatoid or other inflammatory joint disease, systemic disease, or chronic use of drugs. However, they all developed pain, swelling, stiffness, and loud crepitus at 2–4 months after the procedure and had radiological evidence of loss of joint space, especially in the medial compartment. Arthroscopy in all three cases showed a large amount of loose chondral material, a "snowstorm" appearance, which could be washed out. Severe erosion of the articular cartilage and a mild synovitis were also demonstrated. The ligaments were all intact and culture of the synovial fluid was sterile. Histopathology of the fragments of cartilage showed non-viable chondrocytes, with an absence of acute inflammatory cells and very few chronic inflammatory cells. The synovial biopsies showed evidence of fibrosis. All three patients had severely damaged knees which required total knee replacement.

These cases of chondrolysis did not result from accidental use of relatively concentrated solutions of chlorhexidine, but from the use of a very low concentration, 0.02%, which is widely used as an irrigation solution during surgical procedures. Chlorhexidine has a damaging effect on the articular cartilage of the knee, and should not be used, even in low concentrations, to irrigate exposed articular cartilage.

Immunologic

Allergic reactions, including anaphylaxis, from chlorhexidine are reported with all types of use and are well documented. However, chlorhexidine may still not be suspected as a possible cause of anaphylaxis when several agents are used in the anesthetized surgical patient, and hypersensitivity to chlorhexidine may not be tested for (47). If a reaction occurs during anesthesia, there is often doubt about the exact agent responsible; patch-testing will help if there is doubt about causality.

Anaphylaxis occurred in a patient known to be allergic to chlorhexidine (48).

- A 79-year-old man, with a history of acute anaphylaxis to chlorhexidine given in a urethral gel 1 year before, underwent laparotomy for suspected small bowel obstruction. Before general anesthesia, an arterial catheter, a peripheral venous catheter, and an epidural catheter were inserted using povidone iodine as an antiseptic. A four-lumen central line was also inserted using local anesthetic while the patient was awake. Immediately on insertion the patient stated, "I'm not feeling well". Within 30 seconds his systolic blood pressure fell to about 50 mmHg and his heart rate rose to over 120/minute. There were no electrocardiographic ST segment changes. Anaphylaxis was suspected and the skin antiseptic and catheters were reviewed. On closer inspection of the central line packaging it became clear that it was impregnated with chlorhexidine. The catheter was immediately withdrawn, adrenaline, glucocorticoids, and fluids were administered, and he was successfully resuscitated. After a period of stabilization, surgery proceeded and he made an uneventful recovery.

The presence of chlorhexidine in the central venous catheter was mentioned in small writing on the outside of the package, but this was given much less emphasis than the latex-free statement in larger font on the main label.

Four other cases of allergy to chlorhexidine in urethral gel have been reviewed with reference to previous reports (49).

Mechanism
The molecular basis of the recognition of chlorhexidine in a sensitive patient has been examined (50).

- A 75-year-old man had three anaphylactic events after the use of chlorhexidine. The first occurred in September 1995 during general anesthesia for coronary artery bypass grafts. Ten minutes after induction he developed a marked fall in blood pressure, bronchospasm, tachycardia, and increased pulmonary artery pressure. In July 1996, a transurethral resection of the prostate was performed under spinal anesthetic. At cystoscopy he developed a headache, a rash, and bronchospasm, which settled after treatment. He had a further cystoscopy in February 1998, during which he became flushed, wheezy, and hypotensive, and had a cardiac arrest. He was successfully resuscitated. He had raised serum tryptase activities (60.4 and 26.6 µg/l at 3.5 and 9.5 hours after the event), suggesting a true anaphylactic reaction. Since the only pharmacological agent common to all three procedures was urethral jelly containing lidocaine 2% and chlorhexidine 0.05%, he subsequently had skin prick tests, intradermal tests, and sequential subcutaneous challenges to lidocaine without any positive or adverse effects. Because he had developed profound anaphylaxis with cardiac arrest after the topical administration of chlorhexidine, skin tests were deemed unethical, and an in vitro method for detecting sensitivity to chlorhexidine was pursued. Detailed quantitative hapten inhibition studies were carried out with chlorhexidine-reactive IgE antibodies identified in the serum of the patient.

The authors concluded that unlike most drug allergic determinants the whole chlorhexidine molecule is complementary to the IgE antibody combining sites and that the 4-chlorophenol, biguanide, and hexamethylene structures together comprise the allergenic component.

Frequency and susceptibility factors
In a collaborative Danish study, 2061 patients were patch-tested with chlorhexidine gluconate 1% in water. There was a positive reaction in 2.3% of the patients. This was more common in patients with leg eczema (6.8%) or leg ulcer (10.6%) than in those with eczema of the hands (1.9%) or at other sites (1.6%). Of the 14 patients who were retested with chlorhexidine, only one was positive to the 1% solution and none to a solution of 0.01%. This apparent loss of sensitivity may be due to irritable skin at the initial testing, the so-called excited skin syndrome. This study suggests that the sensitizing potential of

chlorhexidine is very low, but that it should be used with caution in dressings for leg ulcers.

In 1063 consecutive eczema patients tested with the ICDRG standard series, supplemented with chlorhexidine gluconate 1% in water and 1% in petrolatum, the frequency of positive reactions was similar to that in the collaborative study (51).

Application to the skin

In a report of generalized urticaria after skin cleansing with and urethral instillation of chlorhexidine-containing products, the authors suggest that there is under-reporting of such reactions and that alternative antiseptics should be considered in urological and gynecological procedures (52).

- Life-threatening anaphylactic reactions with generalized urticaria, dyspnea, and shock occurred in six patients after the use of 0.05, 0.5, or 1% chlorhexidine (53), and in one patient after topical use of an alcoholic 0.5% chlorhexidine digluconate solution (54). A severe systemic allergic response was observed in a patient in whom the large donor area of skin graft was dressed with Bactigras, which contains 0.5% chlorhexidine acetate (55).
- An acute anaphylactic reaction occurred in a 19-year-old man after cleaning of burns on the left arm (56). A scratch test with chlorhexidine acetate 0.05% was positive.
- Anaphylaxis occurred in four surgical patients after the use of chlorhexidine as a skin disinfectant. All four had a history of minor symptoms, such as rashes or faints, in connection with previous surgery or invasive procedures (57).
- Following disinfection of a drain insertion site with chlorhexidine digluconate 2% solution, a 43-year-old man had severe anaphylaxis, manifest as dyspnea, shock, and ST segment elevation (58). In the past he had had two episodes of contact dermatitis with chlorhexidine antiseptics.
- A 14-year-old girl had combined delayed and immediate types of allergy, with urticarial rash and syncope, after long-term use of an antiacne formulation containing chlorhexidine in an unknown concentration.
- When chlorhexidine was used as a skin disinfectant in a 53-year-old man undergoing lung resection for adenocarcinoma, anaphylaxis was complicated by coronary artery spasm (59). He had two anaphylactic reactions accompanied by severe myocardial ischemia. Immunological testing indicated chlorhexidine as the causative substance.

In the last two cases, epicutaneous tests with 1% chlorhexidine gluconate and acetate, and prick tests with 0.05 and 0.01% of the acetate solution were positive (60).

Local hypersensitivity reactions to chlorhexidine-impregnated patches occur in neonates (61). In a randomized comparison of povidone-iodine and a chlorhexidine gluconate impregnated dressing for the prevention of central venous catheter infections in neonates, the risk of local contact dermatitis limited the use of the chlorhexidine dressing (62).

Application to mucous membranes

Anaphylaxis has been reported after nasal application of chlorhexidine.

- Anaphylactic circulatory arrest occurred in a 53-year-old man with acromegaly when his nasal mucosa was cleaned with an aqueous solution of chlorhexidine gluconate 0.05% (63).

A report to the Japanese Ministry of Welfare about reactions observed between 1967 and 1984 included 22 cases with a fall in blood pressure, 13 with dyspnea, 9 with anaphylactic shock, 4 with cyanosis, 19 with erythema, 11 with urticaria, 9 with pruritus, and 7 with facial wheals; following this report, in 1984, the Japanese Ministry of Welfare recommended that the use of chlorhexidine on mucous membranes be prohibited because of the evidence of the risk of anaphylactic shock (53).

Application to wounds

Application of chlorhexidine to wounds is generally safe, but it should be used at the lowest bactericidal concentration of 0.05% (53).

Application to the urethra

Anaphylaxis after cystoscopy or urinary catheterization has been reported repeatedly (SEDA-18, 255) (SEDA-19, 235) (SEDA-20, 225) (64).

- A 64-year-old man had severe anaphylaxis induced by intraurethral chlorhexidine gel 0.05% (65). Previous hypersensitivity reactions during urethral dilatation had been thought to be due to lidocaine. Chlorhexidine-specific IgE antibodies were demonstrated, and the chlorhexidine gel had been used in all the preceding urological procedures the patient had undergone.

Severe allergic reactions occurred in five patients after the insertion of an intra-urethral lidocaine jelly containing chlorhexidine gluconate and instrumentation of the urethra (66). All had a positive response to chlorhexidine gluconate (0.0005%) and a negative response to lidocaine (0.2%).

Impregnated intravenous catheters

Chlorhexidine-coated catheters have been developed in the hope of reducing the incidence of central venous line sepsis. Package inserts warn that these should not be used in individuals who are thought to be sensitive to chlorhexidine.

It is possible that the potential benefit of reducing the incidence of central venous sepsis by using chlorhexidine-coated catheters is outweighed by the risk of sudden and profound anaphylaxis. Certainly a high degree of suspicion of chlorhexidine allergy should be exercised and skin tests performed.

In a randomized clinical study of the efficacy of catheters impregnated with antiseptics for the prevention of central venous catheter-related infections in intensive care units in 204 patients with 235 central venous catheters between November 1998 and June 1999, a standard triple-lumen polyurethane catheter and a catheter impregnated with chlorhexidine and silver sulfadiazine were indistinguishable from each other (67). Compared with standard polyurethane catheters, antiseptic catheters were less likely to be colonized by micro-organisms when they were cultured at removal (8 versus 20 colonized catheters per 100 catheters; relative risk 0.34 (95% CI = 0.15, 0.74). There were no significant differences between the groups in catheter-related infections (0.9 versus 4.9 infections per 100 catheters; relative risk 0.17 (95% CI = 0.03, 1.15)). Gram-positive cocci and fungi were more likely to colonize the standard polyurethane catheters than antiseptic catheters. Two of the cases in the control group died because of catheter-related candidemia. There were no adverse reactions such as hypersensitivity or leukopenia with the antiseptic catheters. The authors concluded that central venous catheters with antiseptic coating are safe and carry less risk of colonization of bacteria and fungi than standard catheters in critically ill patients.

However, although randomized studies have failed to show an association between hypersensitivity reactions and chlorhexidine-impregnated central venous catheters, there have been reports of anaphylaxis after insertion of these catheters (SEDA-22, 262) (68,69), and two life-threatening episodes of anaphylaxis in the same patient were attributed to a central venous catheter that had been impregnated with chlorhexidine and sulfadiazine (70).

- In a 51-year-old man, two episodes of pronounced, refractory cardiovascular collapse accompanied the insertion of a chlorhexidine-coated central venous catheter (71). Sensitivity to chlorhexidine was not at first suspected, but 5 months later, a skin prick test with chlorhexidine resulted in a characteristic sustained wheal and flare response, strongly suggesting IgE-mediated sensitivity. The patient subsequently underwent uneventful surgery following strict avoidance of chlorhexidine exposure.

Bronchospasm was not a feature in any of these cases.

The FDA has issued a public health notice to inform health-care professionals about the potential for serious hypersensitivity reactions to medical devices impregnated with chlorhexidine. The Agency is also seeking information and reports to better evaluate the potential health hazard these products might pose, and to decide on what action, if any, should be taken. Devices that incorporate chlorhexidine that the FDA has cleared for marketing include intravenous catheters, topical antimicrobial skin dressings, and implanted antimicrobial surgical mesh. The notice describes non-US reports of systemic reactions to chlorhexidine-impregnated gels or lubricants used during urological procedures and similarly impregnated central venous catheters. It also describes other types of reactions that have been reported in the USA, including localized reactions to impregnated patches in neonates and occupational asthma in nurses exposed to chlorhexidine and alcohol aerosols (61).

A meta-analysis of the clinical and economic effects of chlorhexidine and silver sulfadiazine antiseptic-impregnated catheters has been undertaken (72). The costs of hypersensitivity reactions were considered as part of the analysis, and the use of catheters impregnated with antiseptics resulted in reduced costs. The analysis used the higher estimated incidence of hypersensitivity reactions occurring in Japan, where the use of chlorhexidine-impregnated catheters is still banned (73).

Polyhexanide
Caution also seems to be warranted concerning polyhexanide, a chlorhexidine polymer, used in disinfectants for a relatively short time.

- Polyhexanide caused severe anaphylaxis in two young patients (74). Both were exposed to the disinfectant Lavasept®, containing polyhexanide, on surgical wounds during orthopedic interventions. They had never been exposed to polyhexanide before, but had been exposed to chlorhexidine. However, skin prick tests were positive for polyhexanide in both cases, while chlorhexidine was negative. Negative skin prick tests to polyhexanide were obtained from controls.

References

1. Mitchell RG, Hayward AC. Postoperative urinary-tract infections caused by contaminated irrigating fluid. Lancet 1966;1(7441):793–5.
2. Speller DC, Stephens ME, Viant AC. Hospital infection by *Pseudomonas cepacia*. Lancet 1971;1(7703):798–9.
3. Salveson A, Bergan T. Contamination of chlorhexidine cream used to prevent ascending urinary tract infections. J Hyg (Lond) 1981;86(3):295–301.
4. Gongwer LE, Hubben K, Lenkiewicz RS, Hart ER, Cockrell BY. The effects of daily bathing of neonatal rhesus monkeys with an antimicrobial skin cleanser containing chlorhexidine gluconate. Toxicol Appl Pharmacol 1980;52(2):255–61.
5. Johnsson J, Seeberg S, Kjellmer I. Blood concentrations of chlorhexidine in neonates undergoing routine cord care with 4% chlorhexidine gluconate solution. Acta Paediatr Scand 1987;76(4):675–6.
6. Salole EG, Shepherd AJ. Spermicides: anti-HIV activity and cytotoxicity in vitro. AIDS 1993;7(2):293–5.
7. Wong PY, Colville VL, White V, Walker HM, Morris RA. Validation and assessment of a blood-donor arm disinfectant containing chlorhexidine and alcohol. Transfusion 2004;44:1238–42.
8. Wilson CM, Gray G, Read JS, Mwatha A, Lala S, Johnson S, Violari A, Sibiya PM, Fleming TR, Koonce A, Vermund SH, McIntyre J. Tolerance and safety of different concentrations of chlorhexidine for peripartum vaginal and infant washes. Acquir Immune Defic Syndr 2004;35:138–43.
9. Weidinger H, Passloer HJ, Kovacs L, Berle B. Nutzen der prophylaktischen Vaginalantiseptik mit Hexetidin in Geburtshilfe und Gynäkologie. [The advantage of preventive vaginal antisepsis with hexetidine in obstetrics and gynecology.] Geburtshilfe Frauenheilkd 1991;51(11):929–35.
10. Quinn MW, Bini RM. Bradycardia associated with chlorhexidine spray. Arch Dis Child 1989;64(6):892–3.

11. Mucklow ES. Accidental feeding of a dilute antiseptic solution (chlorhexidine 0.05% with cetrimide 1%) to five babies Hum Toxicol 1988;7(6):567–9.
12. Mullany LC, Darmstadt GL, Tielsch JM. Safety and impact of chlorhexidine antisepsis interventions for improving neonatal health in developing countries. Ped Infect Dis J 2006;8:665–79.
13. Camus C, Bellissant E, Sebille V, Perrotin D, Garo B, Legras A, Renault A, Le Carre P. Prevention of acquired infections in intubated patients with the combination of two decontamination regimens. Crit Care Med 2005;33:307–14.
14. Hirata K, Kurokawa A. Chlorhexidine gluconate ingestion resulting in fatal respiratory distress syndrome. Vet Hum Toxicol 2002;44(2):89–91.
15. Waclawski ER, McAlpine LG, Thomson NC. Occupational asthma in nurses caused by chlorhexidine and alcohol aerosols. BMJ 1989;298(6678):929–30.
16. Yamagishi M, Kawana M, Hasegawa S, et al. Impairment of olfactory epithelium treated with chlorhexidine digluconate (Hibitane): postoperative olfactory disturbances after Hardy's operation and results of experimental study. Pract Otol 1985;78:399.
17. Bicknell PG. Sensorineural deafness following myringoplasty operations. J Laryngol Otol 1971;85(9):957–61.
18. Parker FL, James GW. The effect of various topical antibiotic and antibacterial agents on the middle and inner ear of the guinea-pig. J Pharm Pharmacol 1978;30(4):236–9.
19. Aursnes J. Vestibular damage from chlorhexidine in guinea pigs. Acta Otolaryngol 1981;92(1–2):89–100.
20. Aursnes J. Cochlear damage from chlorhexidine in guinea pigs. Acta Otolaryngol 1981;92(3–4):259–71.
21. van Rij G, Beekhuis WH, Eggink CA, Geerards AJ, Remeijer L, Pels EL. Toxic keratopathy due to the accidental use of chlorhexidine, cetrimide and cialit. Doc Ophthalmol 1995;90(1):7–14.
22. Nakamura Y, Inatomi T, Nishida K, Sotozono C, Kinoshita S. Four cases of chemical corneal burns by misuse of disinfectant. Jpn J Clin Ophthalmol 1998;52:786–8.
23. Tabor E, Bostwick DC, Evans CC. Corneal damage due to eye contact with chlorhexidine gluconate. JAMA 1989;261(4):557–8.
24. Phinney RB, Mondino BJ, Hofbauer JD, Meisler DM, Langston RH, Forstot SL, Benes SC. Corneal edema related to accidental Hibiclens exposure. Am J Ophthalmol 1988;106(2):210–5.
25. Varley GA, Meisler DM, Benes SC, McMahon JT, Zakov ZN, Fryczkowski A. Hibiclens keratopathy. A clinicopathologic case report. Cornea 1990;9(4):341–6.
26. Klebe S, Anders N, Wollensak J. Inadvertent use of chlorhexidine as intraocular irrigation solution. J Cataract Refract Surg 1998;24(6):729–30.
27. Murthy S, Hawksworth NR, Cree I. Progressive ulcerative keratitis related to the use of topical chlorhexidine gluconate (0.02%) Cornea 2002;21(2):237–9.
28. Ehlers N, Hjortdal J. Are cataract and iris atrophy toxic complications of medical treatment of acanthamoeba keratitis? Acta Ohthalmologica Scand 2004;82:228–31.
29. Bain MJ. Chlorhexidine in dentistry—a review. NZ Dent J 1980;76(344):49–54.
30. Heidemann D. Wundheiliungsstörungen nach Chlorhexidin-Anwedung: ein Fallbericht. [Disturbances in the wound healing process after chlorhexidine use—a case report.] ZWR 1981;90(9):68–70.
31. Flotra L, Gjermo P, Rolla G, Waerhaug J. Side effects of chlorhexidine mouth washes. Scand J Dent Res 1971;79(2):119–25.
32. Flotra L. Different modes of chlorhexidine application and related local side effects. J Periodontal Res Suppl 1973;12:41–4.
33. Prayitno S, Taylor L, Cadogan S, Addy M. An in vivo study of dietary factors in the aetiology of tooth staining associated with the use of chlorhexidine. J Periodontal Res 1979;14(5):411–7.
34. Bozkurt FY, Ozturk M, Tetkin Z. The effects of three oral sprays on plaque and gingival inflammation. J Periodontol 2005;76:1654–60.
35. Bascones A, Morante S, Mateos L, Mata M, Poblet J. Influence of additional ingredients on the effectiveness of non-alcoholic chlorhexidine mouthwashes: a randomized controlled trial. J Periodontol 2005;76:1469–75.
36. Zimmer S, Kolbe C, Kaiser G, Krage T, Ommerborn M, Barthel C. Clinical efficacy of flossing versus use of antimicrobial rinses. J Periodontol 2006;77:1380–5.
37. Pizzo G, Guiglia R, Imburgia M, Pizzo I, D'Angelo M, Giuliana G. The effects of antimicrobial sprays and mouthrinses on supragingival plaque regrowth: a comparative study. J Periodontol 2006;77:248–56.
38. Rushton A. Safety of Hibitane. II. Human experience. J Clin Periodontol 1977;4(5):73–9.
39. Roche S, Chinn R, Webb S. Chlorhexidine-induced gastritis. Postgrad Med J 1991;67(784):210–1.
40. Jayathillake A, Mason DFC, Broome K, Tan G. Chlorhexidine in urethral gel: does it cause pain at flexible cystoscopy? Urology 2006;67:670–3.
41. Osmundsen PE. Contact dermatitis to chlorhexidine. Contact Dermatitis 1982;8(2):81–3.
42. Junor BJ, Briggs JD, Forwell MA, et al. Sclerosing peritonitis—the contribution of chlorhexidine in alcohol. Periton Dial Bull 1985;5:101.
43. Oules R, Challah S, Brunner FP. Case-control study to determine the cause of sclerosing peritoneal disease. Nephrol Dial Transplant 1988;3(1):66–9.
44. Lo WK, Chan KT, Leung AC, Pang SW, Tse CY. Sclerosing peritonitis complicating prolonged use of chlorhexidine in alcohol in the connection procedure for continuous ambulatory peritoneal dialysis. Perit Dial Int 1991;11(2):166–72.
45. Douw CM, Bulstra SK, Vandenbroucke J, Geesink RG, Vermeulen A. Clinical and pathological changes in the knee after accidental chlorhexidine irrigation during arthroscopy. Case reports and review of the literature. J Bone Joint Surg Br 1998;80(3):437–40.
46 Schroder CH, Severijnen RS, Monnens LA. Vergiftiging door desinfectans bij conservatieve behandeling van twee patiënten met omfalokele. [Poisoning by disinfectants in the conservative treatment of 2 patients with omphalocele.] Tijdschr Kindergeneeskd 1985;53(2):76–9.
47. Evans P, Foxell RM. Chlorhexidine as a cause of anaphylaxis. Int J Obstet Anaesth 2002;11:145–6.
48. Kluger M. Anaphylaxis to chlorhexidine-impregnated central venous catheter. Anaesth Intens Care 2003;31:697–8.
49. Jayathallake A, Mason DFC, Broome K. Allergy to chlorhexidine gluconate in urethral gel. Report of four cases and review of the literature. Urology 2003;61:837.
50. Pham NH, Weiner JM, Reisner GS, Baldo BA. Anaphylaxis to chlorhexidine. Case report. Implication of immunoglobulin E antibodies and identification of an allergenic determinant. Clin Exp Allergy 2000;30(7):1001–7.
51. Lasthein Andersen B, Brandrup F. Contact dermatitis from chlorhexidine. Contact Dermatitis 1985;13(5):307–9.
52. Stables GI, Turner WH, Prescott S, Wilkinson SM. Generalized urticaria after skin cleansing and urethral

instillation with chlorhexidine-containing products. Br J Urol 1998;82(5):756–7.

53. Okano M, Nomura M, Hata S, Okada N, Sato K, Kitano Y, Tashiro M, Yoshimoto Y, Hama R, Aoki T. Anaphylactic symptoms due to chlorhexidine gluconate. Arch Dermatol 1989;125(1):50–2.

54. Ohtoshi T, Yamauchi N, Tadokoro K, Miyachi S, Suzuki S, Miyamoto T, Muranaka M. IgE antibody-mediated shock reaction caused by topical application of chlorhexidine. Clin Allergy 1986;16(2):155–61.

55. Cheung J, O'Leary JJ. Allergic reaction to chlorhexidine in an anaesthetised patient. Anaesth Intensive Care 1985;13(4):429–30.

56. Evans RJ. Acute anaphylaxis due to topical chlorhexidine acetate. BMJ 1992;304(6828):686.

57. Garvey LH, Roed-Petersen J, Husum B. Anaphylactic reactions in anaesthetised patients — four cases of chlorhexidine allergy. Acta Anaesthesiol Scand 2001;45(10):1290–4.

58. Ebo DG, Stevens WJ, Bridts CH, Matthieu L. Contact allergic dermatitis and life-threatening anaphylaxis to chlorhexidine. J Allergy Clin Immunol 1998;101(1 Pt 1):128–9.

59. Conraads VM, Jorens PG, Ebo DG, Claeys MJ, Bosmans JM, Vrints CJ. Coronary artery spasm complicating anaphylaxis secondary to skin disinfectant. Chest 1998;113(5):1417–9.

60. Thune P. To pasienter med klorheksidinallergi–anafylaktiske reaksjoner og eksem. [Two patients with chlorhexidine allergy—anaphylactic reactions and eczema.] Tidsskr Nor Laegeforen 1998;118(21):3295–6.

61. Nightingale SL. Hypersensitivity to chlorhexidine-impregnated medical devices. JAMA 1998;279:1684.

62. Garland JS, Alex CP, Mueller CD, Otten D, Shivpuri C, Harris MC, Naples M, Pellegrini J, Buck RK, McAuliffe TL, Goldmann DA, Maki DG. A randomized trial comparing povidone-iodine to a chlorhexidine gluconate-impregnated dressing for prevention of central venous catheter infections in neonates. Pediatrics 2001;107(6):1431–6.

63. Chisholm DG, Calder I, Peterson D, Powell M, Moult P. Intranasal chlorhexidine resulting in anaphylactic circulatory arrest. BMJ 1997;315(7111):785.

64. Leuer J, Mayser P, Schill WB. Anaphylaktischer schock durch intraoperative Anwendung von Chlorhexidin. HGZ Hautkr 2001;76:160–3.

65. Wicki J, Deluze C, Cirafici L, Desmeules J. Anaphylactic shock induced by intraurethral use of chlorhexidine. Allergy 1999;54(7):768–9.

66. Yong D, Parker FC, Foran SM. Severe allergic reactions and intra-urethral chlorhexidine gluconate. Med J Aust 1995;162(5):257–8.

67. Sheng WH, Ko WJ, Wang JT, Chang SC, Hsueh PR, Luh KT. Evaluation of antiseptic-impregnated central venous catheters for prevention of catheter-related infection in intensive care unit patients. Diagn Microbiol Infect Dis 2000;38(1):1–5.

68. Nikaido S, Tanaka M, Yamoto M, Minami T, Akatsuka M, Mori H. [Anaphylactoid shock caused by chlorhexidine gluconate.]Masui 1998;47(3):330–4.

69. Terazawa E, Shimonaka H, Nagase K, Masue T, Dohi S. Severe anaphylactic reaction due to a chlorhexidine-impregnated central venous catheter. Anesthesiology 1998;89(5):1296–8.

70. Stephens R, Mythen M, Kallis P, Davies DW, Egner W, Rickards A. Two episodes of life-threatening anaphylaxis in the same patient to a chlorhexidine-sulphadiazine-coated central venous catheter. Br J Anaesth 2001;87(2):306–8.

71. Pittaway A, Ford S. Allergy to chlorhexidine-coated central venous catheters revisited. Br J Anaesth 2002;88(2):304–5.

72. Veenstra DL, Saint S, Sullivan SD. Cost-effectiveness of antiseptic-impregnated central venous catheters for the prevention of catheter-related bloodstream infection. JAMA 1999;282(6):554–60.

73. Raad I, Hanna H. Intravascular catheters impregnated with antimicrobial agents: a milestone in the prevention of bloodstream infections. Support Care Cancer 1999;7(6):386–90.

74. Olivieri J, Eigenmann PA, Hauser C. Severe anaphylaxis to a new disinfectant: polyhexanide, a chlorhexidine polymer. Schweiz Med Wochenschr 1998;128(40):1508–11.

Chloroxylenol

General Information

Since the withdrawal of hexachlorophene from the non-prescription drug market in the USA, the antiseptic chloroxylenol, which is less antigenic, has been a substitute in a large number of products.

Organs and Systems

Skin

Topical exposure can cause rashes.

- A 10-day-old baby who was given a bath in a 25% solution of Dettol containing 1.2% chloroxylenol became dehydrated and developed diffuse erythema and vesicles, and afterwards exfoliative dermatitis (1). There was a good response to therapy with systemic glucocorticoids and supportive measures.

Immunologic

Allergic contact dermatitis can occur after sensitization to chloroxylenol, for example in medicated Vaseline or in electrocardiographic paste (2).

In a retrospective analysis of patch tests in 951 patients 1.8% had positive reactions to chloroxylenol (3). Most of the patients had been sensitized by popular proprietary formulations containing chloroxylenol (SEDA-11, 221).

Drug Administration

Drug overdose

There have been many cases of intoxication with oral Dettol liquid, a widespread household disinfectant that contains chloroxylenol 4.8%, pine oil, and isopropyl alcohol (4–8). Dettol was involved in 10% of hospital admissions related to self-poisoning in Hong Kong. In a retrospective study of 67 cases, serious complications were relatively common (8%) and these included aspiration of Dettol with gastric contents, resulting in pneumonia, cardiopulmonary arrest, bronchospasm, adult respiratory distress syndrome, and severe laryngeal edema with upper airway obstruction. Of 89 patients, five developed minor hematemesis, in the form of coffee-colored or blood-stained vomitus (6). One patient had a gastroscopy performed on the day after admission, which showed signs of chemical burns in the esophagus and stomach. Gastroscopy in another patient on day 11, done to rule out an esophageal stricture, showed no abnormality. All patients with hematemesis recovered completely. The authors suggest that upper gastrointestinal hemorrhage after Dettol ingestion tends to be mild and self-limiting. Gastroscopy, which may increase the risk of aspiration in patients with impaired consciousness, is not required unless other causes of gastrointestinal bleeding are suspected. Furthermore, Dettol poisoning can be associated with an increased risk of aspiration, possibly caused by the use of gastrointestinal lavage in 88% of the patients and vomiting in 62% (6,7).

Of 121 patients who ingested Dettol 200–500 ml, three developed renal impairment, as evidenced by raised plasma urea and creatinine (7). Two of these patients also had serious complications, including aspiration leading to pneumonia and adult respiratory distress syndrome; one died. Renal impairment only appears to be observed when relatively large amounts of Dettol are ingested (7).

References

1. Kumar B, Singh G, Roy SN. Dettol induced irritant dermatitis. Indian J Dermatol Venereol Leprol 1981;47:128.
2. Storrs FJ. Para-chloro-meta-xylenol allergic contact dermatitis in seven individuals. Contact Dermatitis 1975;1(4):211–3.
3. Myatt AE, Beck MH. Contact sensitivity to parachlorometaxylenol (PCMX). Clin Exp Dermatol 1985;10(5):491–4.
4. Chan TY, Lau MS, Critchley JA. Serious complications associated with Dettol poisoning. Q J Med 1993;86(11):735–8.
5. Chan TY, Critchley JA. Is chloroxylenol nephrotoxic like phenol? A study of patients with DETTOL poisoning. Vet Hum Toxicol 1994;36(3):250–1.
6. Chan TY, Critchley JA, Lau JT. The risk of aspiration in Dettol poisoning: a retrospective cohort study. Hum Exp Toxicol 1995;14(2):190–1.
7. Chan TY, Sung JJ, Critchley JA. Chemical gastro-oesophagitis, upper gastrointestinal haemorrhage and gastroscopic findings following Dettol poisoning. Hum Exp Toxicol 1995;14(1):18–9.
8. Chan TY, Critchley JA. Pulmonary aspiration following Dettol poisoning: the scope for prevention. Hum Exp Toxicol 1996;15(10):843–6.

Chlorphenesin

Chlorphenesin (3-(4-chlorophenoxy)-1,2-propane-diol) is an antimicrobial agent with antifungal, antibacterial, and anticandidal activity. There have been two cases of contact allergy attributed to chlorphenesin because of positive skin tests (1). The authors estimated the incidence of such reactions, based on their overall experience, at 0.02% per year.

Reference

1. Brown VL, Orton DI. Two cases of facial dermatitis due to chlorphenesin in cosmetics. Contact Dermatitis 2005;52(1):48–9.

Ethacridine

General Information

Ethacridine (6,9-diamino-2-ethoxyacridine) is widely used in the local treatment of inflammatory or ulcerative conditions of the skin, particularly crural eczema due to venous stasis. Many publications point to the high frequency of exacerbation involving local allergic reactions and also generalized eczematous reactions (SEDA-11, 474).

Ethacridine has also been given by slow extra-amniotic instillation of 150 ml of a 0.1% solution to induce abortion or delivery in patients with missed abortion or fetal death, and during the second and third trimester without or in combination with drip infusion of oxytocin or dinoprost (prostaglandin $F_{2\alpha}$) (SEDA-11, 474).

In 56 women (18–20 weeks gestation), treated at the Marie Stopes Clinic in Jodhpur, India, who underwent termination of pregnancy with 0.1% ethacridine lactate 150 ml injected into the intrauterine extra-amniotic space and in whom intravenous oxytocin was used to expedite the delivery of the abortus, ethacridine lactate induced successful abortion in 52 cases (1). Abortion failure occurred in the other four cases because of transverse lie of fetus ($n = 2$), cervical dystocia ($n = 1$), and uterine inertia ($n = 1$). In 41 women the abortion occurred at 12–24 hours after induction (mean 20 hours) which was shorter than that of previous reports (29.5–38 hours). There were complications in six cases: three women had cervical tears and three had incomplete expulsion. There was one case each of severe bleeding and vaginal laceration. There were no cases of sepsis. The authors concluded that ethacridine lactate performed better than other instillation abortion methods.

In a prospective, placebo-controlled study, tannin albuminate 500 mg plus ethacridine lactate 50 mg was given to 30 patients with Crohn's disease and chronic diarrhea for 5 days, followed by a 5-day break, and then a further 5 days treatment (2). Stool frequency was significantly reduced, albeit by only a small amount, with an improvement in the consistency of the stools.

References

1. Gupta S, Sachdeva L, Gupta R. Ethacridine lactate—a safe and effective drug for termination of pregnancy. Indian J Matern Child Health 1993;4(2):59–61.
2. Plein K, Burkard G, Hotz J. Behandlung der chronischen Diarrhoe beim Morbus Crohn. Eine Pilotstudie zur klinischen Wirkung von Tanninalbuminat und

Ethacridinlactat. [Treatment of chronic diarrhea in Crohn disease. A pilot study of the clinical effect of tannin albuminate and ethacridine lactate.] Fortschr Med 1993;111(7):114–8.

Ethylene oxide

General Information

Ethylene oxide is a gas that is used in the sterilization of equipment too large for other techniques, and for sterilizing rubber, plastic goods, and other materials that are damaged by heat and not adequately disinfected by other cold methods.

Ethylene oxide is highly toxic. If it is not eliminated after sterilization it can produce severe irritation and burns. In addition, it forms ethylene glycol with moisture and ethylene chlorhydrine with free chlorine atoms. Both products, also irritants, are absorbed by the sterilized object, from which they elute very slowly. The rate of elimination of ethylene oxide and its irritant reaction products depends on a variety of factors, such as the nature and thickness of the material, the duration and temperature of aeration, and the material used for wrapping the sterilized item. Ethylene oxide is also an alkylating agent, a directly acting mutagen and carcinogen.

Exposure

Exposure to ethylene oxide has been reported predominantly in workers in sterilization units, and should be kept as low as feasible. Health personnel working in close proximity to ethylene oxide should be given information about its dangers, and should be informed of the known and uncertain risks of exposure. Sterilizing equipment should be regularly checked, proper ventilation established, and alarm systems installed. These procedures must guarantee that the content of ethylene oxide is lower than 1 ppm, and that the content of halogenated ethylene hydrines is lower than 150 ppm on materials that have been sterilized by ethylene oxide at the time of use. The manufacturer of each device or instrument that should be sterilized by ethylene oxide must declare the conditions necessary for sterilization and decontamination (SEDA-11, 479).

Occupational exposure of sterilizing staff to ethylene oxide in hospitals, tissue banks, and research facilities can result during any of the following operations and conditions (1):

- changing pressurized ethylene oxide gas cylinders
- leaking valves, firings, and piping
- leaking sterilizer-door gaskets
- opening sterilizer doors at the end of a cycle
- improper ventilation at the sterilizer door
- improperly ventilated or unventilated air gap between the discharge line and the sewer drain
- removal of items from sterilizers and transfer of sterilized loads to aerators

- improper ventilation of aerators and aeration areas
- incomplete aeration items
- inadequate general room ventilation
- passing near sterilizers and aerators during operation.

In most studies, exposure appears to result mostly from peak emissions during such operations as opening the door of a sterilizer and unloading and transferring sterilized material. Although much smaller amounts are used in sterilizing medicinal instruments and supplies in hospitals and industrially, it is during these uses that the highest occupational exposures have been measured. On the other hand, proper engineering controls and work practices are reported to result in full-shift exposure of less than 0.1 ppm ($0.18 \, \text{mg/m}^3$) and short-term exposure of less than 2 ppm ($3.6 \, \text{mg/m}^3$) (1). Regular medical follow-up is advisable for sterilizing staff.

Monitoring exposure

Measurement of the concentration of ethylene oxide in workplace air is commonly used for exposure control. A standard of 1 ppm for workplace air is currently accepted as a threshold limit (SEDA-21, 254).

In 12 workers who were occupationally exposed to ethylene oxide during the sterilization of medical equipment, concentrations of 0.2–8.5 ppm were detected (2). This study also confirmed the relation between the ethylene oxide concentration in ambient air and the amount of N-2-hydroxyethylvaline in human globin, which has been used as a biological marker of carcinogenicity.

Disposition

Ethylene oxide is readily taken up by the lung. At steady state, 20–25% of inhaled ethylene oxide reaching the alveolar space is exhaled as unchanged compound and 75–80% is taken up by the body and metabolized. Aqueous ethylene oxide solutions can penetrate human skin.

Ethylene oxide is rapidly and uniformly distributed throughout the body. It is eliminated metabolically by hydrolysis and by conjugation with glutathione and glycol. The half-life has been estimated at 14 minutes to 3.3 hours (3–6). It is excreted mainly in the urine as thioethers; at higher doses, the proportion of thioethers is reduced, while the proportion of ethylene glycol increases.

Ethylene oxide alkylates nucleophilic groups in biological macromolecules. Hemoglobin adducts have been used to monitor tissue doses of ethylene oxide (7).

Local irritant effects

Some communications report irritating effects caused by liberation of residue of ethylene oxide and its reaction products in industrial materials, for example (SEDA-11, 479):

- tracheal stenosis by tracheotomy cannulae
- severe bronchospasm and asthmatic attacks after endotracheal anesthesia by endotracheal tubes
- anaphylaxis in a hemodialysis patient from plastic and rubber connecting tubes in an arteriovenous shunt

- postoperative inflammatory reactions to intraocular lenses
- allergic contact dermatitis
- hemolysis after exposure to plastic tubing sterilized with ethylene oxide.

Organs and Systems

Nervous system

In 12 men who were exposed to ethylene oxide after a sterilizer developed a leak, although the concentrations of ethylene oxide were not monitored, all four operators intermittently smelled the ethylene oxide gas, roughly indicating a concentration of more than 700 ppm (8). All four operators developed neurological disorders. One operator who had been working for only 3 weeks developed headache, nausea, vomiting, and lethargy, followed by major motor seizures. The other three had all been working for more than 2 years and had headache, limb numbness and weakness, increased fatigue, trouble with memory, and slurred speech. Three of them developed cataracts and one required bilateral cataract extractions. Four men, two of whom had not worked directly with the leaking sterilizer, had increased central corneal thickness with normal endothelial cell counts (SEDA-11, 479).

Other reports of neurological dysfunction related to acute and chronic exposure to ethylene oxide have appeared (9–15).

Sensory systems

Of 16 hospital sterilization operators who had been exposed to ethylene oxide and underwent medical and ocular examinations, 14 had lens opacities and 12 had abnormal contrast vision (an abnormality that is non-specific to cataracts but which often supports the diagnosis) (16).

Hematologic

Epidemiological studies have associated ethylene oxide with hematological diseases (mainly anemia, leukopenia, and leukemia) (SEDA-24, 271). To determine whether occupational exposure to low concentrations of ethylene oxide can cause hematological abnormalities and whether blood monitoring could be used as health surveillance, a cross-sectional study was undertaken (17). Blood samples were collected from 47 hospital workers who were exposed to ethylene oxide during a mean period of 6.6 years. Ethylene oxide concentrations were in the range <0.01–0.06 ppm. The control group, individually matched by age, sex, and smoking habits, consisted of 88 workers from the administrative sector who had never been occupationally exposed to ethylene oxide. There were significant differences between the exposed and the control groups in the frequency of workers with low leukocyte counts. There was no significant difference in the absolute mean number of total leukocytes, but there was an increase in the mean number of monocytes and eosinophils and a reduction in the absolute mean number of lymphocytes in the exposed group compared with the

controls. There was an increase in the percentage hematocrit and the mean absolute number of erythrocytes and a fall in the mean absolute number of platelets in the exposed group compared with the controls. The mean absolute numbers of eosinophils and erythrocytes were significantly higher, as was the hematocrit, and the mean absolute numbers of lymphocytes and platelets were significantly lower in the subgroups with a higher cumulative dose of exposure. There was a dose relation between cumulative exposure and the absolute mean number of eosinophils. The results of this study suggest that the total leukocyte count and the eosinophil count could be used to monitor for early detection of health problems in ethylene oxide workers.

Immunologic

Dialyser hypersensitivity syndrome (SEDA-11, 219) (SEDA-11, 479) presents as an acute anaphylactic reaction, the symptoms of which range from mild to life-threatening. The cause of the syndrome is unknown, but affected patients appear to have a high incidence of positive radioabsorbent tests to a conjugate of human serum albumin and ethylene oxide used to sterilize artificial kidneys. This conjugate may be the allergen responsible.

Immediate hypersensitivity reactions occurred in six of 600 donors who underwent automated platelet pheresis; skin-prick testing in four of them (but in none of 40 controls) was positive when an ethylene oxide human serum albumin reagent was used (18). Radioallergosorbent testing showed that serum from four of the six donors, but only one of 145 controls, contained IgE antibodies to ethylene oxide-albumin.

Long-Term Effects

Mutagenicity

There is overwhelming evidence that ethylene oxide produces genetic damage in a wide range of organisms and cells, including somatic cells of exposed humans (7,19). Ethylene oxide is a germ-cell mutagen in rodents. In male mice it induces chromosome breakage, leading to dominant lethal mutations and heritable translocations. Sensitive stages appear to be restricted to late spermatocytes and early spermatozoa.

In females, ethylene oxide also induces presumed, dominant, lethal mutations. When females are exposed shortly after mating or during the early pronuclear stage of the zygote, high frequencies of fetal anomalies are induced.

Ethylene oxide was ineffective in inducing morphological-specific locus mutations in spermatogonial stem cells; however, it produced dominant visible and electrophoresis-specific locus mutants in male mice, assumed to be derived from poststem cells.

The effectiveness of ethylene oxide in inducing chromosome breakage in germ cells of male mice is strongly influenced by varying degrees or rates of exposure. Since there was a dose-rate effect for ethylene oxide-induced, dominant, lethal mutations at high concentrations and over long exposure periods, considering the short burst exposure and low TWA exposure in humans, the question of whether significant dose-rate effects also exist at low exposures is unanswered.

Cytogenetic studies have been performed in the peripheral lymphocytes of humans exposed to ethylene oxide (SEDA-12, 574). Several studies in workers in hospital ethylene oxide sterilization units or plants have suggested that these workers have increased frequencies of chromosomal aberrations, including micronuclei and sister chromatid exchanges in peripheral lymphocytes.

The International Agency for Research on Cancer (IARC) Working Group reviewed the published studies of workers exposed to ethylene oxide in hospital and factory sterilization units and in ethylene manufacturing and processing plants (9). The studies consistently showed chromosomal damage in peripheral blood lymphocytes, including chromosomal aberrations in 11 of 14 studies, sister chromatid exchange in 20 of 23 studies, micronuclei in three of eight studies, and gene mutation in one study. In general, the degree of damage is correlated with the degree and duration of exposure. The induction of sister chromatid exchange appears to be more sensitive to exposure to ethylene oxide than is induction of either chromosomal aberrations or micronuclei. In one study, chromosomal aberrations were observed in the peripheral lymphocytes of workers 2 years after cessation of exposure to ethylene oxide, and sister chromatid exchanges 6 months after cessation of exposure. However, in one study, incidental exposure to high concentrations of ethylene oxide did not cause any measurable permanent mutational/cytogenetic damage in lymphocytes of exposed persons (20).

The effects of glutathione-S-transferase T1 and M1 genotypes on hemoglobin adducts in erythrocytes and sister chromatid exchange in lymphocytes have been examined in 58 hospital operators of sterilizers that used ethylene oxide and non-exposed workers (21). The results suggested that the glutathione-S-transferase T1 null genotype was associated with increased formation of ethylene oxide-hemoglobin adducts in relation to occupational exposure. This suggests that individuals with the glutathione-S-transferase T1 null genotype may be more susceptible to the genotoxic effects of ethylene oxide.

Tumorigenicity

Ethylene oxide is carcinogenic, assigned to Group 1 of the IARC (7). There is limited evidence of carcinogenicity in humans, but much evidence in experimental animals, and the IARC has classified ethylene oxide in category 1 ("carcinogenic in humans"), based primarily on evidence in animals and genotoxic considerations. The overall evaluation of the Working group of the IARC, updated in 1995, is based on the following supporting evidence.

Ethylene oxide is a directly acting alkylating agent that:

- induces a sensitive, persistent, dose-related increase in the frequency of chromosomal aberrations and sister chromatid exchange in peripheral lymphocytes and micronuclei in bone marrow cells of exposed workers;
- has been associated with malignancies of the lymphatic and hemopoietic systems in both humans and experimental animals;
- induces a dose-related increase in the frequency of hemoglobin adducts in exposed humans and a dose-related increase in the numbers of adducts in both DNA and hemoglobin in exposed rodents;
- induces gene mutations and heritable translocations in germ cells of exposed rodents;
- is a powerful mutagen and clastogen at all phylogenetic levels.

In 1979, three cases of hemopoietic cancer that had occurred between 1972 and 1977 were reported in workers at a Swedish factory where 50% ethylene oxide and 50% methyl formate had been used since 1968 to sterilize hospital equipment (22).

In epidemiological studies of exposure to ethylene oxide, the most frequently reported association has been with lymphatic and hemopoietic cancer. Two populations were studied: people using ethylene oxide as a sterilizing agent and chemical workers manufacturing or using ethylene oxide. Of studies of sterilization personnel, the largest and most informative is that conducted in the USA (23,24). Overall, mortality from lymphatic and hemopoietic cancers was only marginally raised, but there was a significant trend, especially for lymphatic leukemia and non-Hodgkin's lymphoma, in relation to estimated cumulative exposure. For exposure to 1 ppm (1.8 mg/m^3) over a working lifetime of 45 years, a ratio of 1.2 was estimated for lymphatic and hemopoietic cancers. The other studies of workers involved in sterilization in Sweden (25–27) and in the UK (28) each showed non-significant excesses of lymphatic and hemopoietic cancers. An assessment based on epidemiological data showed no increase in leukemia in those who had been exposed to ethylene oxide (29).

Because of the possibility of confounding occupational exposure in the studies of chemical workers exposed to ethylene oxide (20,25,26,28,30–38), less weight can be given to the positive findings. Nevertheless, they are compatible with a small but consistent excess of lymphatic and hemopoietic cancers found in studies of sterilization personnel. Some of the epidemiological studies have shown an additional risk of cancer of the stomach, which was significant only in one study from Sweden (25,26,30).

The incidence of breast cancer has been estimated in a cohort study in 7576 women employed for at least 1 year in commercial sterilization facilities and exposed to ethylene oxide for an average of 10.7 years (39). Compared with the non-exposed population, those in the upper quintile of cumulative exposure had a 27% increase in breast cancer, with a 15-year lag. The authors concluded that that ethylene oxide may be associated with breast cancer, but a causal interpretation was weakened owing to inconsistencies in exposure-reported trends and possible biases from non-responses and incomplete cancer ascertainment.

In 1994 the International Agency for Research on Cancer determined that the gas ethylene oxide was a definite (group 1) human carcinogen, based on limited evidence from epidemiological studies showing increased hemopoietic cancers and supported by positive human cytogenetic evidence.

In the mid-1980s the National Institute for Occupational Safety and Health assembled a cohort of 18 235 men and women workers who had been exposed to ethylene oxide. The results of this study were reviewed in 1987 and showed no overall excess of hemopoietic cancers, but did show a significant excess amongst men, and particularly non-Hodgkin's lymphoma.

Mortality reporting in this cohort has been extended from 1987 to 1998 (40). There were 2852 deaths, compared with 1177 in the earlier 1987 follow-up. There was no overall excess of hemopoietic cancers combined or of non-Hodgkin's lymphoma. However, internal exposure response analysis showed positive trends for hemopoietic cancers, limited to men. The reasons for this sex specificity are not known.

Second-Generation Effects

Teratogenicity

In animals, ethylene oxide has toxic effects on reproduction and is teratogenic, but the relevance of animal and epidemiological studies to occupational exposure of humans in the environment of sterilization units is unclear (SEDA-12, 576) (41).

References

1. Mortimer VD Jr, Kercher SL. Control Technology for Ethylene Oxide Sterilization in hospitalsCincinnati, OH: National Institute for Occupational Safety and Health;. 1989.
2. Leonardos G, Kendak D, Barnard N. Odor threshold determinations of 53 odorant chemicals. J Air Pollut Control Assoc 1969;19:51.
3. Brugnone F, Perbellini L, Faccini G, Pasini F. Concentration of ethylene oxide in the alveolar air of occupationally exposed workers. Am J Ind Med 1985;8(1):67–72.
4. Osterman-Golkar S, Bergmark E. Occupational exposure to ethylene oxide. Relation between in vivo dose and exposure dose. Scand J Work Environ Health 1988;14(6):372–7.
5. Osterman-Golkar S. Dosimetry of ethylene oxide. In: Bartsch H, Hemminki K, O'Neill IK, editors. Methods for detecting DNA Damaging Agents in Humans: Applications in Cancer Epidemiology and Prevention 89. Lyon: IARC, 1988:249–57.
6. Filser JG, Denk B, Tornqvist M, Kessler W, Ehrenberg L. Pharmacokinetics of ethylene in man; body burden with ethylene oxide and hydroxyethylation of hemoglobin due to endogenous and environmental ethylene. Arch Toxicol 1992;66(3):157–63.
7. WHO International Agency for Research of Cancer. Ethylene oxide. IARC Monogr Eval Carcinog Risks Hum 1994;60:73–159.
8. Jay WM, Swift TR, Hull DS. Possible relationship of ethylene oxide exposure to cataract formation. Am J Ophthalmol 1982;93(6):727–32.
9. Schroder JM, Hoheneck M, Weis J, Deist H. Ethylene oxide polyneuropathy: clinical follow-up study with morphometric and electron microscopic findings in a sural nerve biopsy. J Neurol 1985;232(2):83–90.
10. Fukushima T, Abe K, Nakagawa A, Osaki Y, Yoshida N, Yamane Y. Chronic ethylene oxide poisoning in a factory manufacturing medical appliances. J Soc Occup Med 1986;36(4):118–23.
11. Estrin WJ, Cavalieri SA, Wald P, Becker CE, Jones JR, Cone JE. Evidence of neurologic dysfunction related to long-term ethylene oxide exposure. Arch Neurol 1987;44(12):1283–6.
12. Estrin WJ, Bowler RM, Lash A, Becker CE. Neurotoxicological evaluation of hospital sterilizer workers exposed to ethylene oxide. J Toxicol Clin Toxicol 1990;28(1):1–20.
13. Crystal HA, Schaumburg HH, Grober E, Fuld PA, Lipton RB. Cognitive impairment and sensory loss associated with chronic low-level ethylene oxide exposure. Neurology 1988;38(4):567–9.
14. Klees JE, Lash A, Bowler RM, Shore M, Becker CE. Neuropsychologic "impairment" in a cohort of hospital workers chronically exposed to ethylene oxide. J Toxicol Clin Toxicol 1990;28(1):21–8.
15. Grober E, Crystal H, Lipton RB, Schaumburg H. Authors' commentary. EtO is associated with cognitive dysfunction. J Occup Med 1992;34(11):1114–6.
16. Sobaszek A, Hache JC, Frimat P, Akakpo V, Victoire G, Furon D. Working conditions and health effects of ethylene oxide exposure at hospital sterilization sites. J Occup Environ Med 1999;41(6):492–9.
17. Shaham J, Levi Z, Gurvich R, Shain R, Ribak J. Hematological changes in hospital workers due to chronic exposure to low levels of ethylene oxide. J Occup Environ Med 2000;42(8):843–50.
18. Leitman SF, Boltansky H, Alter HJ, Pearson FC, Kaliner MA. Allergic reactions in healthy plateletpheresis donors caused by sensitization to ethylene oxide gas. N Engl J Med 1986;315(19):1192–6.
19. Dellarco VL, Generoso WM, Sega GA, Fowle JR 3rd, Jacobson-Kram D. Review of the mutagenicity of ethylene oxide. Environ Mol Mutagen 1990;16(2):85–103.
20. Tates AD, Boogaard PJ, Darroudi F, Natarajan AT, Caubo ME, van Sittert NJ. Biological effect monitoring in industrial workers following incidental exposure to high concentrations of ethylene oxide. Mutat Res 1995;329(1):63–77.
21. Yong LC, Schulte PA, Wiencke JK, Boeniger MF, Connally LB, Walker JT, Whelan EA, Ward EM. Hemoglobin adducts and sister chromatid exchanges in hospital workers exposed to ethylene oxide: effects of glutathione S-transferase T1 and M1 genotypes. Cancer Epidemiol Biomarkers Prev 2001;10(5):539–50.
22. Hogstedt C, Malmqvist N, Wadman B. Leukemia in workers exposed to ethylene oxide. JAMA 1979;241(11):1132–3.
23. Stayner L, Steenland K, Greife A, Hornung R, Hayes RB, Nowlin S, Morawetz J, Ringenburg V, Elliot L, Halperin W. Exposure-response analysis of cancer mortality in a cohort of workers exposed to ethylene oxide. Am J Epidemiol 1993;138(10):787–98.
24. Steenland K, Stayner L, Greife A, Halperin W, Hayes R, Hornung R, Nowlin S. Mortality among workers exposed to ethylene oxide. N Engl J Med 1991;324(20):1402–7.
25. Hogstedt C, Aringer L, Gustavsson A. Epidemiologic support for ethylene oxide as a cancer-causing agent. JAMA 1986;255(12):1575–8.
26. Hogstedt C. Epidemiological studies on ethylene oxide and cancer: an updating. In: Bartsch H, Hemminki K, O'Neill, editors. Methods for Detecting DNA Damaging Agents in

Humans: Applications in Cancer Epidemiology and Prevention 89. Lyon: IARC, 1988:265–70.

27. Hagmar L, Welinder H, Linden K, Attewell R, Osterman-Golkar S, Tornqvist M. An epidemiological study of cancer risk among workers exposed to ethylene oxide using hemoglobin adducts to validate environmental exposure assessments. Int Arch Occup Environ Health 1991;63(4):271–7.

28. Gardner MJ, Coggon D, Pannett B, Harris EC. Workers exposed to ethylene oxide: a follow up study. Br J Ind Med 1989;46(12):860–5.

29. Baca D, Drexler C, Cullen E. Obstructive laryngotracheitis secondary to gentian violet exposure. Clin Pediatr (Phila) 2001;40(4):233–5.

30. Hogstedt C, Rohlen O, Berndtsson BS, Axelson O, Ehrenberg L. A cohort study of mortality and cancer incidence in ethylene oxide production workers. Br J Ind Med 1979;36(4):276–80.

31. Morgan RW, Claxton KW, Divine BJ, Kaplan SD, Harris VB. Mortality among ethylene oxide workers. J Occup Med 1981;23(11):767–70.

32. Shore RE, Gardner MJ, Pannett B. Ethylene oxide: an assessment of the epidemiological evidence on carcinogenicity. Br J Ind Med 1993;50(11):971–97.

33. Kiesselbach N, Ulm K, Lange HJ, Korallus U. A multicentre mortality study of workers exposed to ethylene oxide. Br J Ind Med 1990;47(3):182–8.

34. Benson LO, Teta MJ. Mortality due to pancreatic and lymphopoietic cancers in chlorohydrin production workers. Br J Ind Med 1993;50(8):710–6.

35. Teta MJ, Benson LO, Vitale JN. Mortality study of ethylene oxide workers in chemical manufacturing: a 10 year update. Br J Ind Med 1993;50(8):704–9.

36. Bisanti L, Maggini M, Raschetti R, Alegiani SS, Ippolito FM, Caffari B, Segnan N, Ponti A. Cancer mortality in ethylene oxide workers. Br J Ind Med 1993;50(4):317–24.

37. Thiess AM, Schwegler H, Fleig I, Stocker WG. Mutagenicity study of workers exposed to alkylene oxides (ethylene oxide/propylene oxide) and derivatives. J Occup Med 1981;23(5):343–7.

38. Greenberg HL, Ott MG, Shore RE. Men assigned to ethylene oxide production or other ethylene oxide related chemical manufacturing: a mortality study. Br J Ind Med 1990;47(4):221–30.

39. Steenland K, Whelan E, Deddens J, Stayner L, Ward E. Ethylene oxide and breast cancer incidence in a cohort study of 7576 women. Cancer Causes Control 2003;14:531–9.

40. Steenland K, Stayner L, Deddens J. Mortality analyses in a cohort of 18235 ethylene oxide exposed workers: follow up extended from 1987 to 1998. Occup Environ Med 2004;61:2–7.

41. Florack EI, Zielhuis GA. Occupational ethylene oxide exposure and reproduction. Int Arch Occup Environ Health 1990;62(4):273–7.

Formaldehyde

General Information

Aldehydes such as formaldehyde, glyoxal, and glutaral (glutaraldehyde) are used as solutions and vapours for disinfection and sterilization. They are irritating and sensitizing and cause contact dermatitis in health-care workers (SEDA-21, 254).

Formulations

Formaldehyde is released by numerous agents, such as paraformaldehyde, dichlorophene, Dowicil 75 and Dowicil 200 (cis-1-(3-chloroalkyl)-3,5,7-triaza-l-azonia-adamantane chloride), bronopol, Biocide DS 52–49 (1,2-benzoisothiazoline-3-one plus a formaldehyde releaser), and Bakzid (cyclic aminoacetal). Formalin is an alternative name for an aqueous solution of formaldehyde, but the latter name is preferred, since formalin is also used as a brand name in some countries.

Free formaldehyde is used in cosmetics, especially in hair shampoos, and in many disinfectants and antiseptics. The solid paraformaldehyde is used as a source of formaldehyde vapor for the disinfection of rooms. Noxythiolin, polynoxylin, hexamidine, and taurolidine act by slow release of formaldehyde. Formaldehyde solution contains 34–38% of formaldehyde methanol as a stabilizing agent to delay polymerization of the formaldehyde. Formaldehyde gel contains 0.75% of formaldehyde and is used to treat warts.

Formaldehyde cannot be applied safely to the skin or the mucous membranes in the concentration necessary to rapidly kill microbes, and formaldehyde solutions have to be diluted before use to a 2–8% solution to disinfect inanimate objects and to a 1–2% solution for disinfection by scrubbing. For fumigation of air a concentration of 1–2% is used.

General adverse effects

Discussion about the toxicity, mutagenicity, and potential carcinogenicity of formaldehyde relates more to occupational and environmental exposure, caused by its release from urea formaldehyde resins used for wood products and from foams for cavity-wall insulation, than its use in disinfection and sterilization (SEDA-11, 476) (SEDA-12, 569). However, the concentrations of formaldehyde that are found in the air after scrubbing with formaldehyde-containing disinfectants can be several hundred percentage higher than the maximum safe workplace concentration, even when scrubbing is carried out properly (1). Very high concentrations have also been found in pathologists' workrooms. In the pathology departments of two Italian hospitals the highest values of 2.6 and 6.0 ppm were measured in the dissection laboratories; in the histology and cytology laboratories, concentrations were less than 1 ppm, except when technicians handled formaldehyde solutions (2).

Primary irritant effects of formaldehyde

The minimum amount of formaldehyde that can be detected by odor varies considerably between individuals and ranges from 0.1 to 1.0 ppm (0.12–1.2 mg/m^3), close to the concentration at which minimal irritant effects are felt in the eyes and in the pulmonary airways (3). Thus, the fundamental toxicity of formaldehyde lies in primary irritation to the eyes, nose, and throat when the subject is exposed to concentrations in the range of 1–5 ppm. Concentrations above 2–5 ppm cause irritation of the pharynx, lungs, and eyes, and some erythema of vaporized areas of the skin, such as the face and neck. Acute exposure to concentrations of formaldehyde of the order of three times the maximum threshold of detection of the odor will most likely produce severe acute pulmonary edema after only a few minutes.

Acute toxicity after local administration of formaldehyde-containing solutions

Dilute solutions of 1–10% formalin have been instilled into the bladder to treat inoperable profusely bleeding tumors or intractable hemorrhagic cystitis. Anuria was a severe complication. This was due either to edematous obstruction of the ureter or to tubular or papillary necrosis, probably caused by systemic absorption. Bladder perforation with intraperitoneal spillage, peritonitis, and finally death was described in an elderly patient with a carcinoma of the uterine cervix (SEDA-11, 476).

In 1983, Godec and Gleich reviewed all published results of treatment of intractable hematuria with formalin. Dilutions of 1–10% formalin (containing 0.37–3.7% formaldehyde) were used; the most commonly used concentration of formalin was 10%. The authors concluded that formalin was probably the most effective tool for controlling massive hematuria, but also probably the most dangerous. The review covered 23 articles and 118 patients; in 104 cases, treatment was successful. However, in only 10 reports had the treatment been used without serious adverse effects; the other 13 articles listed four deaths and many serious local and systemic complications. The complication rate increased when the formalin concentration was higher, but the contact time and the volume instilled did not influence the occurrence of adverse effects. The most frequent local complications were reflux and hydronephrosis. Fibrosis of the bladder with reduced capacity was the usual clinical outcome. A systemic effect was tubular necrosis with anuria, with two deaths. Another complication was ureteric obstruction, which was not related to ureteric fibrosis or bladder wall fibrosis obstructing the intramural ureter; in two cases the obstruction appeared to be due to retroperitoneal fibrosis (SEDA-11, 476) (4).

In 1989, Donahue and Frank (SEDA-12, 569) (SEDA-14, 205) (5) published a systematic review of 235 cases of intravesical hemorrhagic cystitis treated with intravesical instillations of diluted formalin in concentrations of 1, 5, or 10%. Complete response rates were 71, 78, and 83% respectively. The average duration of a complete response was 3–4 months; the recurrence rate fell gradually with the use of higher concentrations. Complications were divided into two groups: "minor complications" included all mild or transient problems not requiring surgical intervention (fever, tachycardia, transient or minor rises in blood urea nitrogen or creatinine, mild hydronephrosis, grades I and II uricoure-thral reflux, increased urinary frequency, urgency, incontinence, suprapubic pain, or a reduction in bladder capacity not requiring urinary diversion); "major complications" were those that required surgical intervention, resulting in loss of renal function or causing damage to the supravesical urinary tracts (stricture formation), including anuria, acute tubular necrosis, papillary necrosis, ureteric or retroperitoneal fibrosis, uterovesical or uteropelvic junction obstruction, severe hydronephrosis, grades III or IV vesicoureteric

reflux, any vesical fistula, or a reduction in bladder capacity requiring urinary diversion. Major complications occurred in all treatment groups, including those treated with 1% formalin. The higher rate observed with 10% formalin was not significantly different from the rates associated with the use of 1 or 5% formalin. The mortality rates were 2.2% in all the formalin groups, but the rates were not significantly different. Formalin 10% resulted in a higher and favorable response rate, a lower recurrence rate, equal numbers of major complications and mortality rate, and a three-fold higher rate of minor complications than 5% formalin in patients with hemorrhagic cystitis due to radiotherapy for bladder tumors. In contrast, formalin 5% was more effective than formalin 10% in treating patients with intractable hematuria due to unresectable bladder tumors or cyclophosphamide-induced cystitis.

The use of 2 or 4% formaldehyde as a scolecidal agent for injection into hydatid cysts and for peritoneal lavage was followed by shock in seven cases and resulted in death in three. All three patients who died had undergone a peritoneal lavage with 2–8 liters of 2 or 4% formaldehyde (SEDA-11, 476).

Immunogenicity

Formaldehyde is one of the most frequent contact allergens. If the responses in some individuals to acute exposure to formaldehyde are immunogenically mediated, they are of serious clinical significance. These individuals should not be exposed to any concentration of the agent. However, the intensity and nature of the immunogenic response can so resemble a primary irritant effect that a different diagnosis is not possible.

Mutagenicity and carcinogenicity

Formaldehyde is mutagenic in vitro and although there is conflicting evidence about its carcinogenicity, this should be taken seriously.

Organs and Systems

Respiratory

Formaldehyde-induced bronchial asthma occurred in a laboratory technician and in a neurologist preparing brain specimens for a student demonstration (6).

In a renal dialysis unit, five of 28 members of the staff had respiratory symptoms associated with formaldehyde, and in two cases, attacks of wheezing could be provoked by exposure to formaldehyde. One nurse was particularly affected, the asthmatic wheeze persisting for 8 days after exposure (7).

Of 230 people who had been exposed to formaldehyde and had asthma-like respiratory symptoms with a bronchial provocation test, 12 were considered to be caused by specific sensitization to formaldehyde, 11 were triggered by a concentration of 2.5 mg/m^3, and one by 1.2 mg/m^3 (SEDA-11, 477) (8). The authors concluded that formaldehyde-induced asthma, although apparently rare, is under-reported.

When four aldehydes were entered into an Asthma Hazard Assessment Program using a quantitative structure-activity relationship model, all four had a common structural basis for increasing the risk of asthma (9).

Biliary tract

Three cases of secondary sclerosing cholangitis developed during the early postoperative phase of surgical treatment of hydatid liver cysts in which formaldehyde was used; one patient died within 3 months and the remaining two underwent liver transplantation following biliary sclerosis. Another group also reported three cases of sclerosing cholangitis. Both groups of authors concluded that because of the risk of complication and the unproven efficacy of intracystic injection of a scolecidal solution in preventing dissemination of the parasite, this technique should be abandoned in the surgical treatment of hydatid disease of the liver. However, in 560 patients, there were no cases of sclerosing cholangitis when the concentration was not more than 2% and the solution was injected only into cysts that contained clear fluid content (SEDA-11, 476) (10).

Skin

Formaldehyde-releasing preservatives, such as quaternium-15, diazolidinyl urea, and imidazolidinyl urea, are widely used in cosmetics and topical medications and are well-known contact sensitizers. In spite of positive patch test reactions to these preservatives in a number of patients, only some of these patients will react when they use the corresponding commercial formulations. This is because the concentrations of preservatives in the commercial products are often below the threshold necessary to produce a clinical reaction. This finding confirms the importance of using commercial formulations of topical agents in estimating the clinical relevance of patch test results (11).

The formaldehyde concentration in patch-testing, which has been lowered to 1% in the standard series in recent years, has been studied in a comparison of concentrations of 1% and 2% in 3734 consecutively patch-tested patients (12). Since there was no significant difference between 1 and 2% formaldehyde with respect to the frequency of positive patch test reactions, while there were more irritant reactions with 2%, a 1% patch test concentration can still be recommended.

Allergic dermatitis has been demonstrated from direct skin contact and from exposure to gaseous formaldehyde in the air. Various forms of reaction occur, from simple erythema to maculopapular lesions, hyperesthesia, and angioedema. Five patients developed an allergic contact dermatitis to plaster casts, caused by free formaldehyde released by a melamine–formaldehyde resin incorporated in the plaster.

Urticaria with acute Quincke's edema was reported following the use of a formaldehyde-containing dental paste (SEDA-11, 477) (13).

The first published case of photosensitivity to formaldehyde was reported in 1982 in a man who developed pruritus, burning, and redness of the skin within minutes of exposure to sunlight; photopatch tests showed specific photosensitivity to formaldehyde (14).

Immunologic

Aldehydes are irritating and sensitizing and cause contact dermatitis in health-care workers (SEDA-21, 254). The incidence of allergy to aldehydes has been examined in 280 health-care workers with skin lesions (15). Allergy was diagnosed in 64 (23%). Most (86%) were sensitive to only one aldehyde. Formaldehyde caused allergy slightly more often (14%) than glutaraldehyde (12%). Only five (1.9%) were sensitive to glyoxal. This hierarchy of sensitivity was also confirmed in animal testing.

Immediate-type allergy to formaldehyde mediated by ice occurred during the use of a formaldehyde reconditioned dialyser in a 20-year-old woman without a personal or familiar history of atopy.

A specific cold agglutinin cross-reacting with anti-N was detected in the sera of 68 (21%) of 325 hemodialysis patients; each had used a dialyser that had been sterilized with formaldehyde. The results of transfusion experiments suggested in vivo hemolytic activity of this antibody. The authors postulated that such in vivo exposure to formaldehyde might make the MN-receptor on erythrocytes immunogenic, thus inducing the formation of the anti-N-like antibody.

The commonly used Clinitest reaction for residual formaldehyde in reused dialysers fails to detect concentrations below 50 ppm (16). The use of Schiff reagent in ratios of 1:1 to 3:1, which can detect formaldehyde at concentrations of 3.6–5.0 ppm, has therefore been recommended (SEDA-12, 571) (17). It can also be used in combination with a glucose-containing dialysate.

Long-Term Effects

Mutagenicity

Formaldehyde is mutagenic in many laboratory test systems, for example fruit flies (Drosophila), grasshoppers, flowering plants, fungi, bacteria, and cultured human bronchial fibroblasts.

Formaldehyde may be genotoxic by a dual mechanism: direct damage to DNA and inhibition of repair of mutagenic and carcinogenic DNA lesions by other chemical and physical carcinogens.

Marked chromosomal abnormalities and chromosomal breaks were found in metaphases in direct bone marrow preparations from 40 patients undergoing maintenance hemodialysis (SEDA-11, 477) (18). During the period of these cytogenetic studies, the dialysers were reused after sterilization with formaldehyde, and each patient may have received residual amounts of as much as 127 (sd 51) mg of formaldehyde during each dialysis.

Tumorigenicity

There is evidence of possible carcinogenicity of formaldehyde from two inhalation studies on rats and mice (SEDA-10, 423; SEDA-11, 477; SEDA-12, 571) (13).

In man, a number of epidemiological studies using different designs have been conducted (18) on the health risks of non-medical exposure to formaldehyde and also in health-care professionals (19–24), with contradictory results. Cancers in excess in more than one study were: Hodgkin's disease (25,26), leukemia (19,20,23,24,27), cancers of the buccal cavity and pharynx (particular the nasopharynx) (19,20,26,28,29), lung (19,25,28,30–32), nose (33–37), prostate (20,25,27), bladder (20,24,27), brain (21), colon (19–21,26,28), skin (19,26), and kidney (28).

There was no association between formaldehyde exposure and lung cancer in a case-referral study among Danish physicians working in departments of pathology, forensic medicine, and anatomy (22).

Mortality from prostatic cancer was increased among embalmers (20) and industrial workers (25,27), but the excess was statistically significant only among embalmers (20). A slight excess of mortality from bladder cancer (20,23,27), a significant excess of colon cancers (19,20,28), and a significant excess mortality from skin cancer (19,26) were noted among British pathologists (23), embalmers (19,20), and industrial workers (26–28).

Excess mortality from leukemia and cancer of the brain was generally not seen among industrial workers, which suggests that the increased rates of these cancers among professionals (anatomists (21), pathologists (23), embalmers (19,20), and undertakers (24)) is due to factors other than just formaldehyde.

It is of course possible that the studies that have provided positive evidence of a link between formaldehyde and cancer related to more intensive exposure; for example, reports on the risk associated with chronic exposure to low concentrations of formaldehyde suggest that formaldehyde cannot be a potent carcinogen; if it were, the high degree of environmental exposure would result in much clearer evidence of risk. However, any compound that produces cancer in experimental animals or mutagenicity in several test systems should be considered as a potential cancer risk to human subjects, even though humans and animals may differ in their susceptibility to formaldehyde. The contradictory evidence from human studies should therefore be taken seriously and efforts should be made to reduce exposure.

Tumorigenicity

In 2004 the International Agency for Research on Cancer reclassified formaldehyde as a proven human carcinogen based on "sufficient evidence" of nasopharyngeal cancer. It also concluded that there was "strong but not sufficient evidence for a causal association between leukemia and occupational exposure to formaldehyde". This was primarily based on the observation that a mechanism for leukemia induction had not been identified. In a subsequent review of the data to assess the biological plausibility of formaldehyde as a potential human carcinogen there was inadequate evidence to conclude that there is a causal relation between formaldehyde and the risk of leukemia (38).

The need to re-evaluate the rationale underlying the use of formaldehyde, formocresol, and paraformaldehyde in dentistry has been stressed, since the clinical use and delivery of these products are considered to be arbitrary and unscientific (39).

References

1. Ponsold B, Schulze B, Kirsch H. Hygienische Probleme bei des Anwendung formaldehydheltiger Desinfektionsmitted. [Hygienic problems in the use of formaldehyde containing disinfectants.] Z Gesamte Hyg 1977;23(6):408–11.
2. De Zotti R, Petronio L, Negro C, Gabelli A. Inquinamento da formaldeide in anatomia patologica: esperienze in due ospedali regionali. [Formaldehyde pollution in pathologic anatomy: experiences at 2 regional hospitals.] Med Lav 1986;77(5):523–8.
3. Leonardos G, Kendak D, Barnard N. Odor threshold déterminations of 53 odorant chemicals. J Air Pollut Control Assoc 1969;19:51.
4. Ferrie BG, Smith PJ, Kirk D. Retroperitoneal fibrosis complicating intravesical formalin therapy. J R Soc Med 1983;76(10):831–2.
5. Donahue LA, Frank IN. Intravesical formalin for hemorrhagic cystitis: analysis of therapy. J Urol 1989;141(4):809–12.
6. Sakula A. Formalin asthma in hospital laboratory staff. Lancet 1975;2(7939):816.
7. Hendrick DJ, Lane DJ. Formalin asthma in hospital staff. BMJ 1975;1(5958):607–8.
8. Nordman H, Keskinen H, Tuppurainen M. Formaldehyde asthma—rare or overlooked? J Allergy Clin Immunol 1985;75(1 Pt 1):91–9.
9. Seed MJ, Hussey LJ, Lines SK, Turner S, Agius RM. Prediction of asthma hazard of glutaraldehyde substitutes. Occup Med (Lond) 2006;56:284–5.
10. Bourgeon R. Cholangite sclérosante due au formol dans le traitement du kyste hydatique du foie. [Sclerosing cholangitis caused by formol in the treatment of hydatid cyst of the liver.] Gastroenterol Clin Biol 1985;9(8–9):644–5.
11. Skinner SL, Marks JG. Allergic contact dermatitis to preservatives in topical medicaments. Am J Contact Dermat 1998;9(4):199–201.
12. Trattner A, Johansen JD, Menne T. Formaldehyde concentration in diagnostic patch testing: comparison of 1% with 2%. Contact Dermatitis 1998;38(1):9–13.
13. Burri C, Wüthrich B. Quincke-Ödem mit Urtikaria nach Zahnwurzelbehandlung mit einem Paraformaldehyd-haltigen Dentalantiseptikum bei Spättyp-Sensibilisierung auf Paraformaldehyd. Allergologie 1985;8:264.
14. Shelley WB. Immediate sunburn-like reaction in a patient with formaldehyde photosensitivity. Arch Dermatol 1982;118(2):117–8.
15. Kiec-Swierczynska M, Krecisz B, Krysiak B, Kuchowicz E, Rydzynski K. Occupational allergy to aldehydes in health care workers. Clinical observations. Experiments. Int J Occup Med Environ Health 1998;11(4):349–58.
16. Friedman EA, Lundin AP 3rd. Environmental and iatrogenic obstacles to long life on hemodialysis. N Engl J Med 1982;306(3):167–9.
17. Zasuwa G, Levin NW. Problems in hemodialysis. N Engl J Med 1982;306(25):1550.
18. IARC Monographs updating of Vol. 1 to 42, Supplement, 1987:211–16.
19. Walrath J, Fraumeni JF Jr. Mortality patterns among embalmers. Int J Cancer 1983;31(4):407–11.

20. Walrath J, Fraumeni JF Jr. Cancer and other causes of death among embalmers. Cancer Res 1984;44(10):4638–41.

21. Stroup NE, Blair A, Erikson GE. Brain cancer and other causes of death in anatomists. J Natl Cancer Inst 1986;77(6):1217–24.

22. Jensen OM, Andersen SK. Lung cancer risk from formaldehyde. Lancet 1982;1(8277):913.

23. Harrington JM, Oakes D. Mortality study of British pathologists 1974–80. Br J Ind Med 1984;41(2):188–91.

24. Levine RJ, Andjelkovich DA, Shaw LK. The mortality of Ontario undertakers and a review of formaldehyde-related mortality studies. J Occup Med 1984;26(10):740–6.

25. Blair A, Stewart P, O'Berg M, Gaffey W, Walrath J, Ward J, Bales R, Kaplan S, Cubit D. Mortality among industrial workers exposed to formaldehyde. J Natl Cancer Inst 1986;76(6):1071–84.

26. Stayner L, Smith AB, Reeve G, Blade L, Elliott L, Keenlyside R, Halperin W. Proportionate mortality study of workers in the garment industry exposed to formaldehyde. Am J Ind Med 1985;7(3):229–40.

27. Fayerweather WE, Pell S, Bender JB. Case-control study of cancer deaths in DuPont workers with potential exposure to formaldehyde. In: Clary JJ, Gibson JE, Waritz RS, editors. Formaldehyde: Toxicology, Epidemiology, Mechanisms. New York: Marcel Dekker, 1983:47–125.

28. Liebling T, Rosenman KD, Pastides H, Griffith RG, Lemeshow S. Cancer mortality among workers exposed to formaldehyde. Am J Ind Med 1984;5(6):423–8.

29. Vaughan TL, Strader C, Davis S, Daling JR. Formaldehyde and cancers of the pharynx, sinus and nasal cavity: I. Occupational exposures. Int J Cancer 1986;38(5):677–83.

30. Partanen T, Kauppinen T, Nurminen M, Nickels J, Hernberg S, Hakulinen T, Pukkala E, Savonen E. Formaldehyde exposure and respiratory and related cancers. A case-referent study among Finnish woodworkers. Scand J Work Environ Health 1985;11(6):409–15.

31. Coggon D, Pannett B, Acheson ED. Use of job-exposure matrix in an occupational analysis of lung and bladder cancers on the basis of death certificates. J Natl Cancer Inst 1984;72(1):61–5.

32. Bertazzi PA, Pesatori AC, Radice L, Zocchetti C, Vai T. Exposure to formaldehyde and cancer mortality in a cohort of workers producing resins. Scand J Work Environ Health 1986;12(5):461–8.

33. Hayes RB, Raatgever JW, de Bruyn A, Gerin M. Cancer of the nasal cavity and paranasal sinuses, and formaldehyde exposure. Int J Cancer 1986;37(4):487–92.

34. Olsen JH, Jensen SP, Hink M, Faurbo K, Breum NO, Jensen OM. Occupational formaldehyde exposure and increased nasal cancer risk in man. Int J Cancer 1984;34(5):639–44.

35. Olsen JH, Asnaes S. Formaldehyde and the risk of squamous cell carcinoma of the sinonasal cavities. Br J Ind Med 1986;43(11):769–74.

36. Acheson ED, Barnes HR, Gardner MJ, Osmond C, Pannett B, Taylor CP. Formaldehyde in the British chemical industry. An occupational cohort study. Lancet 1984;1(8377):611–6.

37. Vaughan TL, Strader C, Davis S, Daling JR. Formaldehyde and cancers of the pharynx, sinus and nasal cavity: II. Residential exposures. Int J Cancer 1986;38(5):685–8.

38. Golden R, Pyatt D, Shields PG. Formaldehyde as a potential human leukemogen: an assessment of biological plausibility. Clin Rev Toxicol 2006;36:135–53.

39. Lewis BB, Chestner SB. Formaldehyde in dentistry: a review of mutagenic and carcinogenic potential. J Am Dent Assoc 1981;103(3):429–34.

Glutaral

General Information

Aldehydes such as glutaral, formaldehyde, and glyoxal are used as solutions and vapors for disinfection and sterilization.

The safety and biocidal efficacy of glutaral has led to its endorsement by the CDC and WHO as a substitute for formaldehyde in high-level disinfection and cold sterilization (1,2). Glutaral is used in a 2% aqueous solution buffered to a pH of about 8 for sterilization of endoscopic and dental equipment and for other equipment that cannot be sterilized by heat.

Occupational safety considerations for glutaral largely relate to its volatility, and stringent precautions in handling are especially needed in tropical climates (3). The lack of precautionary details for glutaral fumes, especially in warm climates is inadequate. Manufacturers should provide details of possible adverse effects that could arise from glutaral vapor, and of the precautions that should be taken to keep the air concentration below the recommended limit. The odor threshold for glutaral vapor is about 0.04 ppm, the irritation threshold about 0.3 ppm, and the recommended Ceiling Threshold Limit Value 0.2 ppm. It should not be exceeded at any time during the working day (4–6). It is likely that glutaral vapor will be smelt before reaching overexposure concentration, but there is an urgent need to develop affordable and effective methods of containing glutaral fumes.

Of 169 nurses working in 17 hospitals, especially in endoscope units, 68% had symptoms, 38% two or more. The major complaints were eye irritation in 49%, skin discoloration or irritation in 41%, and cough or shortness of breath in 34% (7). Complaints were not related to habits, atopic status, or duration of exposure. In two hospitals, the time-weighted average concentrations were estimated. The 10-minute time-weighted average was below the UK occupational exposure standard of 0.2 ppm. In a similar survey of 150 staff in two Middlesex hospitals who were exposed to glutaral, the rate of complaints was in the same range (8).

Organs and Systems

Cardiovascular

Nine of 184 implantations of glutaral-preserved mitral valves became incompetent (9). This incidence of dehiscence substantially exceeded that previously noted with synthetic valves, and the authors suggested that incomplete removal of glutaral from the prosthesis might have contributed to failure of healing. Collagen ultrastructure investigations of glutaral-treated porcine aortic valve tissue showed that the long-term mechanical durability of treated aortic valves can be substantially increased if careful consideration is given to the pressure at which initial fixation of glutaral is carried out.

Respiratory

Nurses working in endoscopy units complained of respiratory symptoms on exposure to glutaral (10). The provocation test was positive in two of these patients.

Sensory systems

The eye is particularly sensitive to glutaral. The threshold for conjunctival irritation is 0.2–0.5% and for corneal injury 0.5–1.0%. At 2%, eye injury is moderate, whereas at 5% and above it is severe (6,11). In one case, acute eye injury was caused by a leakage of retained glutaral in an anesthesia mask (SEDA-13, 648).

Gastrointestinal

Residues of glutaral in endoscopes can cause significant mucosal injury resulting in acute colitis within 12–48 hours. Symptoms include hematochezia, fever, and tenesmus. Histological findings in biopsy specimens are similar to those of ischemic colitis, with crypt dilatation followed by epithelial cell dropout and subsequent acute inflammation. It cannot be diagnosed by histological analysis alone (12,13). Distinct from the type of colitis that is caused by hydrogen peroxide, the detrimental effects of glutaral are not immediately recognizable during the endoscopic examination. It does not cause raised white mucosal plaques ("pseudolipomatosis") (13).

Colitis was observed during endoscopy of the lower gastrointestinal tract in 21 patients (14). Some patients developed rectal bleeding, tenesmus, and increased frequency of stools, lasting 12 days.

The histological findings in 12 patients with suspected glutaral-induced hemorrhagic colitis were similar to those in rats after colonic instillation of the sterilizing solution (15).

- During uncomplicated right laparoscopic orchidopexy in a 3-year-old boy a few millilitres of 2% glutaral was deposited inadvertently into the peritoneal cavity from insufflation tubing that had been sterilized with glutaral. This resulted in multiple small bowel fistulae, poor bowel healing, fibrosis, and impaired gastrointestinal motility caused by segmental necrosis of the muscularis propria (16).

Skin

Solutions of glutaral can cause skin irritation. With sustained contact under occlusion or on sensitive skin, the threshold for producing irritation is 0.2–0.5% (6). However, concentrations up to 5% may not be irritant if applied only briefly to non-occluded skin. The site of skin contact, the length of contact time, and whether the area is occluded are factors that determine the likelihood and severity of irritation.

A few people develop allergic contact dermatitis to glutaral (SEDA-11, 478; SEDA-12, 572) (6,17).

Immunologic

Aldehydes are irritating and sensitizing and cause contact dermatitis in health-care workers (SEDA-21, 254). The incidence of allergy to aldehydes has been examined in 280 health-care workers with skin lesions (18). Allergy was diagnosed in 64 (23%). Most (86%) were sensitive to only one aldehyde. Formaldehyde caused allergy slightly more often (14%) than glutaral (12%). Only 5 (1.9%) were sensitive to glyoxal. This hierarchy of sensitivity was also confirmed in animal testing.

Anaphylaxis occurred in a woman after the fourth intramuscular injection of a glutaral-containing pollen formulation (SEDA-11, 478) (19).

Long-Term Effects

Tumorigenicity

Although glutaral represents a substantial proportion of the human exposure to aldehydes in medicine and industry, the epidemiology of cancer risk in exposed workers has not been extensively studied. Data on 188 male workers from a glutaral production unit have been evaluated to assess the risk of cancer. The rates of respiratory tract cancer or leukemia related to glutaraldehyde exposure were not increased, although the population was small and there was a low prevalence of smoking (20).

Drug Administration

Drug overdose

The symptoms of overexposure to glutaral vapor include watery and/or burning eyes, nose and throat irritation, and some respiratory discomfort. Glutaral also causes asthma-like symptoms in a few hypersensitive individuals. However, in none of these cases was there evidence that an immune-mediated process was present; these cases probably represent bronchial hyper-reactivity rather than respiratory sensitization (6).

References

1. Guidelines on sterilization and desinfection effective against human immunodeficiency virus (HIV). WHO AIDS series Geneva: WHO. 1989.
2. Centers for Disease Control (CDC). Recommendations for preventing transmission of infection with human T-lymphotropic virus type III/lymphadenopathy-associated virus in the workplace. MMWR Morb Mortal Wkly Rep 1985;34(45):681–6691–5.
3. Mwaniki DL, Guthua SW. Occupational exposure to glutaraldehyde in tropical climates. Lancet 1992;340(8833):1476–7.
4. American Conference of Governmental Industrial Hygienists (1991), 1991.
5. Ballantyne B. Toxicology of glutaraldehyde. Review of studies and human health effectsDanbury, CT: Union Carbide Corporation;. 1995.
6. Jordan SL. The correct use of glutaraldehyde in the health-care environment. Gastroenterol Nurs 1995;18(4):143–5.
7. Calder IM, Wright LP, Grimstone D. Glutaraldehyde allergy in endoscopy units. Lancet 1992;339(8790):433.
8. Waldron HA. Glutaraldehyde allergy in hospital workers. Lancet 1992;339(8797):880.

9. Wright JS, Newman DC. Complications with glutaraldehyde-preserved bioprostheses. Med J Aust 1980;1(11):542–3.

10. Corrado OJ, Osman J, Davies RJ. Asthma and rhinitis after exposure to glutaraldehyde in endoscopy units. Hum Toxicol 1986;5(5):325–8.

11. Ballantyne B, Berman B. Dermal sensitization potential of glutaraldehyde: a review and recent observations. J Toxicol Cut Ocul Toxicol 1984;3:251–62.

12. West AB, Kuan SF, Bennick M, Lagarde S. Glutaraldehyde colitis following endoscopy: clinical and pathological features and investigation of an outbreak. Gastroenterology 1995;108(4):1250–5.

13. Ryan CK, Potter GD. Disinfectant colitis. Rinse as well as you wash. J Clin Gastroenterol 1995;21(1):6–9.

14. Jonas G, Mahoney A, Murray J, Gertler S. Chemical colitis due to endoscope cleaning solutions: a mimic of pseudomembranous colitis. Gastroenterology 1988;95(5):1403–8.

15. Durante L, Zulty JC, Israel E, Powers PJ, Russell RG, Qizilbash AH, Morris JG Jr. Investigation of an outbreak of bloody diarrhea: association with endoscopic cleaning solution and demonstration of lesions in an animal model. Am J Med 1992;92(5):476–80.

16. Karpelowsky JJ, Maske CP, Sinclair-Smith C, Rode H. Glutaraldehyde-induced bowel injury after laparoscopy. J Pediatr Surg 2006;41:e24-5.

17. Rahi AH, Hungerford JL, Ahmed AI. Ocular toxicity of desferrioxamine: light microscopic histochemical and ultrastructural findings. Br J Ophthalmol 1986;70(5):373–81.

18. Kiec-Swierczynska M, Krecisz B, Krysiak B, Kuchowicz E, Rydzynski K. Occupational allergy to aldehydes in health care workers. Clinical observations. Experiments. Int J Occup Med Environ Health 1998;11(4):349–58.

19. Small P. Modified ragweed extract. J Allergy Clin Immunol 1982;69(6):547.

20. Collins JJ, Burns C, Spencer P, Bodnar CM, Calhoun T. Respiratory cancer risks amongst workers with glutaraldehyde exposure. J Occup Environ Med 2006:48:199–203.

Hexachlorophene

General Information

Hexachlorophene has been extensively used as an ingredient of innumerable kinds of consumer goods and medical formulations. Since 1961, when it was reported that daily bathing of newborn infants with a 3% hexachlorophene suspension prevented colonization of the skin by coagulase-positive staphylococci, hexachlorophene has been widely used in hospital nurseries.

However, hexachlorophene readily penetrates excoriated or otherwise damaged skin and absorption through intact skin has also been described. The most dramatic complication reported was due to accidental use in talcum powder in neonates, with neurological and other features and many deaths (1). Since then there has been reticence to use hexachlorophene in young infants at all, and certainly the customary 3% emulsion is too strong (2).

As a result of investigations of the toxicity of hexachlorophene in animals and reports of accidental intoxication in France, the FDA in 1972 banned all non-prescription uses of this drug, restricting hexachlorophene to prescription use only, as a surgical scrub and hand-wash product for health-care personnel. Hexachlorophene was excluded from cosmetics, except as a preservative in concentrations not exceeding 0.1%. Other countries followed suit. An extensive critical review of hexachlorophene is given by Delcour-Firquet (3).

Organs and Systems

Respiratory

- Occupational asthma occurred in a pediatric nurse who had worked with hexachlorophene for 15 years (4). The initial symptom was rhinitis but at the time of diagnosis she was also suffering from attacks of asthma.

Nervous system

Neurotoxicity has been observed after dermal application of hexachlorophene to large areas of burned or otherwise excoriated skin, after accidental application of extremely high concentrations on intact skin, and after ingestion. If hexachlorophene is applied in high concentrations or at frequent intervals to the intact skin, excoriation will result, increasing the risk of systemic effects.

In animals, there is a clear relation between dosage, blood concentrations, duration of treatment, and morphological and functional disturbances of the nervous system. In the lower dose range, there is unequivocal histological evidence of neurotoxicity, but without symptoms of neurotoxicity (SEDA-9, 397) (SEDA-11, 486).

Five Chinese patients aged 14–39 years were treated for *Conorchiasis sinensis* infection with oral hexachlorophene 20 mg/kg for 5–6 days (5). They developed nausea, vomiting, diarrhea, abdominal pain, general muscle weakness, and soreness in the eyeballs and legs. Of eight children treated similarly, one became comatose on the fourth day, with temporary loss of light reflex, alternating dilatation and contraction of the pupils, and positive cerebrospinal tract signs (6). The fundus showed papillary edema. Recovery followed symptomatic treatment.

In experimental animals, hexachlorophene can cause cerebral edema (7), and occasional cases have also been reported in humans (8). It affects exclusively the white matter of the brain and spinal cord and produces a spongiform encephalopathy, transforming the white matter into an extensive network of cystic spaces lined by fragments of myelin. Electron microscopy shows intramyelinic edema, with splitting and separation of the myelin lamellae. Nerve damage due to hexachlorophene appears to be reversible, although it takes many weeks for all the holes in the white matter to disappear. However, extensive edema occurring within a rigid structure such as the spinal canal can result in infarction of nervous tissue. In one study of all premature infants who weighed under 1400 g at birth and who survived at least 4 days, there was a significant association between repeated whole-body bathing in 3% hexachlorophene soap (undiluted pHisoHex) and a vacuolar encephalopathy of the brainstem reticular formation (9).

Sensory systems

Optic atrophy has been described after oral or topical use of hexachlorophene (10).

Second-Generation Effects

Teratogenicity

Teratogenicity has been reported (11), but was not confirmed in an expert review (12). In a 1977 study of Swedish medical personnel it was suggested that repeated hand-washing with hexachlorophene-containing detergents during the first trimester of pregnancy could be associated with a greatly increased incidence of both major and minor birth defects in the offspring (11). This publication was the subject of extensive discussion, but it had several serious methodological deficiencies, and the hypothesis that hexachlorophene was teratogenic has never been further examined or confirmed. One reason why it was initially taken seriously was the fact that hexachlorophene crosses the placenta and accumulates in fetal neural tissue in mice and rats; the administration of toxic doses in these species is associated with birth defects, including cleft palate, hydrocephalus, anophthalmia, and microphthalmia in rats.

Topical exposure of male neonatal rats to a commercial hexachlorophene formulation produced significantly reduced fertility, resulting in inability to ejaculate. It was possibly caused by dioxin, an "androgenic" contaminant of hexachlorophene, responsible for permanent disruption of the integrated ejaculatory reflex. However, in subneurotoxic concentrations, hexachlorophene did not seem to cause significant impairment of spermatogenesis in rats or dogs.

Lactation

Hexachlorophene is secreted into breast milk (13).

Susceptibility Factors

Age

The blood concentrations of hexachlorophene, determined during newborn skin care with hexachlorophene bathing, cover a range that varies between studies (14–17) but are similar to the concentrations associated with neurotoxicity in monkeys (18), rats (19), and mice (SEDA-9, 397). In interpreting hexachlorophene plasma concentrations, it must be remembered that most of the hexachlorophene is probably distributed very rapidly into lipophilic tissues and that low plasma concentrations may correspond to high tissue concentrations.

A critical review of the available data on the risk of hexachlorophene bathing or other kinds of hexachlorophene use in neonatal skin care has suggested that there is ample evidence of the toxic potential of this disinfectant. It is absolutely contraindicated in infants with a low birth weight (under 2000 g), excoriated areas of the skin, and raised serum bilirubin. The use of a dusting powder with a maximal concentration of 0.3–0.5% seems to be connected with a lower risk of toxicity, but there was marked absorption of hexachlorophene after the application of 0.33% hexachlorophene dusting powder (20) and further information is needed about the pharmacokinetics of hexachlorophene in neonates.

The view that a small degree of edematous change in the central nervous system caused by hexachlorophene use in neonates should be reversible and is very probably without influence on the further development of the child cannot be accepted, because these changes are certainly the first signs of central nervous system toxicity.

- In a neonate treated with a 3% hexachlorophene lotion which had not been rinsed off, the skin became excoriated 4 days later and there was muscular twitching, which progressed to convulsions (21). Four days after withdrawal of hexachlorophene, the convulsions disappeared; recovery was complete.

In assessing the use of hexachlorophene in the nursery, one also has to consider the fact that infections with highly infective strains of staphylococci, producing serious life-threatening diseases, do not appear to be prevented or aborted by hexachlorophene bathing, and the problem that reduced staphylococcal colonization of infants by hexachlorophene may lead to an increased number of infections with Gram-negative bacteria.

Other features of the patient

Owing to the high absorption rate of hexachlorophene through damaged skin and the risk of fatal intoxication, the use of hexachlorophene in the treatment of burns or on otherwise excoriated skin is strongly contraindicated.

- A 10-year-old boy who had sustained 25% first-degree and second-degree burns and was treated with frequent daily applications of a 3% hexachlorophene emulsion (diluted and undiluted) developed a fatal encephalopathy (7).

In six of eight burned children, including four with a history of convulsive seizures after hexachlorophene use, serum concentrations of hexachlorophene were 4–74 ng/ml (approximately equivalent to blood concentrations of 2–37 ng/ml) (22).

In 1972, six deaths related to the use of hexachlorophene in patients with burns were reported to the FDA (23).

In four children, two of whom were being treated with 3% hexachlorophene baths for burns and two for severe congenital ichthyosis, the interval between exposure and symptoms ranged from 6 hours to 10 days (24). All showed severe vacuolation of the white matter in different areas of the cerebrum and cerebellum.

Drug Administration

Drug administration route

Hexachlorophene in vaginal lubricants is variably absorbed from the vaginal mucosa, and hexachlorophene can be identified in maternal and cord serum in an appreciable number of women in whom vaginal examinations during labor were carried out with a hexachlorophene-containing antiseptic lubricant (25). Because of the potential for neonatal hexachlorophene toxicity, the use of alternative lubricants for pelvic examinations is recommended.

Drug overdose

In France, in 1972, 6% hexachlorophene was accidentally included in certain batches of a baby talcum powder. French law prohibited publication of the detailed report on the consequences of this contamination, until litigation was concluded. In 1977, a detailed report of 18 children involved in this accidental poisoning appeared (26) and in 1982 a report described 204 children (SEDA-10, 433) (1). Follow-up investigations of 14 surviving children have also appeared (26). There were no obvious cerebral sequelae, but follow-up was not long enough to exclude the possibility of more subtle damage, which may be manifest by learning difficulties or behavior disorders.

In four cases of accidental ingestion of hexachlorophene in human subjects, anorexia, nausea, vomiting, abdominal cramps, and diarrhea occurred (27). No neurological symptoms were reported.

- Bilateral atrophy of the optic nerve occurred in a 31-year-old woman who had ingested 10–15 ml of a 3% hexachlorophene emulsion orally each day for 10–11 months. She had also applied large amounts of hexachlorophene solution to her face every day as self-treatment for pimples. She was depressed during this period, noted no headache, diplopia, or dizziness, but may have had intermittent numbness of the left foot.
- Acute bilateral optic nerve necrosis occurred in a 7-year-old boy who accidentally received 3% hexachlorophene emulsion (about 12 g) over 52 hours. After the last dose he complained of intermittent blindness. Peritoneal dialysis was ineffective, and he died 98 hours after the last dose.

References

1. Martin-Bouyer G, Lebreton R, Toga M, Stolley PD, Lockhart J. Outbreak of accidental hexachlorophene poisoning in France. Lancet 1982;1(8263):91–5.
2. Garcia-Bunuel L. Toxicity of hexachlorophene. Lancet 1982;1(8282):1190.
3. Delcour-Firquet MP. La toxicité de l'hexachlorophène. [Toxicity of hexachlorophene.] Arch Belg Med Soc 1980;38(1):1–43.
4. Nagy L, Orosz M. Occupational asthma due to hexachlorophene. Thorax 1984;39(8):630–1.
5. Chung HL, Tsao WC, Hsue HC, Kuo CH, Ko HY, Mo PS, Chang HY, Chuo HT, Chou WH. Hexachlorophene (G-11) as a new specific drug against clonorchiasis sinensis; its efficacy and toxicity in experimental and human infection. Chin Med J (Engl) 1963;82:691–701.
6. John L, Wang CN, Yue JH, Wang MN, Chang CF, Chens S. Hexachlorophene in the treatment of clonorchiasis sinensis. Chin Med J (Engl) 1963;82:702–11.
7. Kinoshita Y, Matsumura H, Igisu H, Yokota A. Hexachlorophene-induced brain edema in rat observed by proton magnetic resonance. Brain Res 2000;873(1):127–30.
8. Chilcote R, Curley A, Loughlin HH, Jupin JA. Hexachlorophene storage in a burn patient associated with encephalopathy. Pediatrics 1977;59(3):457–9.
9. Shuman RM, Leech RW, Alvord EC Jr. Neurotoxicity of hexachlorophene in humans. II. A clinicopathological study of 46 premature infants. Arch Neurol 1975;32(5):320–5.
10. Slamovits TL, Burde RM, Klingele TG. Bilateral optic atrophy caused by chronic oral ingestion and topical application of hexachlorophene. Am J Ophthalmol 1980;89(5):676–9.
11. Halling H. Suspected link between exposure to hexachlorophene and malformed infants. Ann NY Acad Sci 1979;320:426–35.
12. Baltzar B, Ericson A, Kallen B. Delivery outcome in women employed in medical occupations in Sweden. J Occup Med 1979;21(8):543–8.
13. West RW, Wilson DJ, Schaffner W. Hexachlorophene concentrations in human milk. Bull Environ Contam Toxicol 1975;13(2):167–9.
14. Curley A, Kimbrough RD, Hawk RE, Nathenson G, Finberg L. Dermal absorption of hexochlorophane in infants. Lancet 1971;2(7719):296–7.
15. Alder VG, Burman D, Corner BD, Gillespie WA. Absorption of hexachlorophane from infants' skin. Lancet 1972;2(7773):384–5.
16. Plueckhahn VD. Hexachlorophane and skin care of newborn infants. Drugs 1973;5(2):97–107.
17. Tyrala EE, Hillman LS, Hillman RE, Dodson WE. Clinical pharmacology of hexachlorophene in newborn infants. J Pediatr 1977;91(3):481–6.
18. Bressler R, Walson PD, Fulginitti VA. Hexachlorophene in the newborn nursery. A risk-benefit analysis and review. Clin Pediatr (Phila) 1977;16(4):342–51.
19. Kennedy GL Jr, Dressler IA, Richter WR, Keplinger ML, Calandra JC. Effects of hexachlorophene in the rat and their reversibility. Toxicol Appl Pharmacol 1976;35(1):137–45.
20. Alder VG, Burman D, Simpson RA, Fysh J, Gillespie WA. Comparison of hexachlorophane and chlorhexidine powders in prevention of neonatal infection. Arch Dis Child 1980;55(4):277–80.
21. Herter WB. Hexachlorophene poisoning. Kaiser Found Med Bull 1959;7:228.
22. Larson DL. Studies show hexachlorophene causes burn syndrome. Hospitals 1968;42(24):63–4.
23. Lockart JD. How toxic is hexachlorophene? Pediatrics 1972;50:229.
24. Mullick FG. Hexachlorophene toxicity. Human experience at the Armed Forces Institute of Pathology. Pediatrics 1973;51(2):395–9.
25. Strickland DM, Leonard RG, Stavchansky S, Benoit T, Wilson RT. Vaginal absorption of hexachlorophene during labor. Am J Obstet Gynecol 1983;147(7):769–72.
26. Goutieres F, Aicardi J. Accidental percutaneous hexachlorophane intoxication in children. BMJ 1977;2(6088):663–5.
27. Wear JB Jr, Shanahan R, Ratliff RK. Toxicity of ingested hexachlorophene. JAMA 1962;181:587–9.

Hexetidine

General Information

Hexetidine has been used as an oral cavity antiseptic. At the concentration normally used (1 mg/ml), it is effective in vitro against Gram-positive bacteria and *Candida albicans*, but insufficiently active against most Gram-negative bacteria. However, the duration of the reduction in germ count is not longer than 1 hour. When applied to mucosae it also appears to have a mild local anesthetic property. Very little critical work on hexetidine has been published (SEDA-11, 482).

Organs and Systems

Sensory systems

Of 61 patients in whom hexetidine was used as a buccal and pharyngeal antiseptic, a minority, generally those who had just undergone tonsillectomy, complained of a burning sensation and a salty taste (1).

Skin

Allergic contact dermatitis has been reported (SEDA-11, 482) (2).

Drug Administration

Drug administration route

It is not clear whether the use of hexetidine as a spray, of which some will certainly be inhaled, affects its clinical tolerance. Adverse effects were not recorded in 81 patients treated in this way, but the report was not sufficiently detailed for a full assessment (3).

References

1. Platt P, Otten E. Untersuchungen über die Wirksamkeit von Hexidine bei akuten Erkrankungen des Rachens und der Mundhöhle sowie nach Tonsillektomie. Therapiewoche 1969;19:1565.
2. Merk H, Ebert L, Goerz G. Allergic contact dermatitis due to the fungicide hexetidine. Contact Dermatitis 1982;8(3):216.
3. Mann HJ, Wagner B. Klinische Erfahrugen bei der Behandlung der Tonsilitis acuta und der Pharyngitis acuta mit Hexoral Spray. Therapiewoche 1972;22:4316.

Parabens

General Information

Parabens (methyl, ethyl, propyl, and butyl esters of para-hydroxybenzoic acid) are used as preservatives in concentrations of 0.1–0.3% in pharmaceutical formulations and in concentrations of 0.01–0.1% in cosmetics and foods. In such concentrations they are devoid of systemic toxic effects, but allergic reactions have been reported.

Organs and Systems

Immunologic

Parabens can cause allergic contact dermatitis that can run an insidious course, especially when the parabens are in glucocorticoid ointments. In such cases, treatment leads to a protracted dermatitis without acute exacerbation, so that neither the patient nor the physician suspects parabens as a possible cause. A sensitization index of 0.8% was found in 273 patients with chronic dermatitis (1).

In 1973, in a multicenter study of 1200 individuals carried out by the North American Contact Dermatitis Group, there was a 3% incidence of delayed hypersensitivity reactions to parabens. General allergic reactions have also been reported after the injection of parabens-containing formulations of lidocaine and hydrocortisone and after oral use of barium sulfate contrast suspension, haloperidol syrup, and an antitussive syrup, all of which contained parabens (SEDA-11, 484).

Reference

1. Schorr WF. Paraben allergy. A cause of intractable dermatitis. JAMA 1968;204(10):859–62.

Peroxides

General Information

Solutions of hydrogen peroxide containing 5–7% H_2O_2 are used for cleaning wounds and ulcers. The disinfectant and deodorant actions of hydrogen peroxide occur by oxidation of cell materials during the rapid release of oxygen while hydrogen peroxide is in contact with the tissues. The solution does not penetrate well, but the effervescence provides a mechanical means for detaching necrotic tissue from inaccessible parts of wounds. It should not be used in closed body cavities. The germicidal action of hydrogen peroxide is relatively weak and of short duration.

Residues on insufficiently rinsed equipment disinfected by hydrogen peroxide can provoke local irritation, burns, and general reactions. Non-specific inflammation has been reported, with instantaneous blanching and effervescence on the surfaces of the intestinal mucosa during endoscopy (1).

Organs and Systems

Respiratory

Pneumomediastinum caused by subcutaneous emphysema has been reported in a 30-year-old man after the application of hydrogen peroxide solution to a root canal. The patient had acute pulpitis of the lower right third molar, treated by extirpation followed by irrigation with

3% hydrogen peroxide solution. Soon after irrigation, subcutaneous emphysema developed (2).

Hematologic

Hemolysis was reported after hemodialysis therapy with a dialysis fluid inadvertently contaminated with hydrogen peroxide (3).

References

1. Bilotta JJ, Waye JD. Hydrogen peroxide enteritis: the "snow white" sign. Gastrointest Endosc 1989;35(5):428–30.
2. Nahlieli O, Neder A. Iatrogenic pneumomediastinum after endodontic therapy. Oral Surg Oral Med Oral Pathol 1991;71(5):618–9.
3. Gordon SM, Bland LA, Alexander SR, Newman HF, Arduino MJ, Jarvis WR. Hemolysis associated with hydrogen peroxide at a pediatric dialysis center. Am J Nephrol 1990;10(2):123–7.

Phenols

General Information

Phenol is a benzyl alcohol and a major oxidized metabolite of benzene that was introduced into medicine as an antiseptic (1). Although it can be prepared in an aqueous solution or in glycerine, it appears to be more effective when mixed in aqueous compounds. At a concentration of 0.2% it is bacteriostatic and at over 1% bactericidal (2). In addition to its uses as an antiseptic and disinfectant, phenol is also used as a sclerosant, as a local anesthetic on the skin, and as an analgesic, by injection into nerves or spinally, but its use was limited by severe adverse effects. Current medical uses include cosmetic face peeling, nerve injections, and topical anesthesia. It is also an ingredient of various topical formulations, and is used as an environmental disinfectant.

Systemic adverse effects can occur through absorption from intact skin or wounds, by ingestion, or by absorption of vapor through the skin or via the lungs. They include central nervous stimulation followed by depression, seizures, coma, tachycardia, hypotension, dysrhythmias, pulmonary edema, metabolic acidosis, and hepatic and renal injury. Serious adverse reactions due to percutaneous absorption can occur and death has been described several times. The signs and symptoms of phenol toxicity have been reviewed (3–7) and are listed in Table 1.

Nonoxynols are ethoxylated alkyl phenols that are synthesized from alkylbenzene nonoxynol by reacting it with ethylene oxide to produce ethylene oxide polymers of various lengths. Each nonoxynol is followed by a number that indicates the approximate number of ethylene oxide groups it contains. In cosmetic products, nonoxynols are used as emulsifying, wetting, foaming, and solubilizing agents. They are used in hair and skin products, and in bath, shaving, and fragrance formulations (8). The

Table 1 The signs and symptoms of phenol toxicity

System	Symptoms
Cardiovascular	Cyanosis, cardiac dysrhythmias, electrocardiographic abnormalities, circulatory failure, collapse
Respiratory	Respiratory failure
Nervous system	Dizziness, coma
Sensory systems	Darkening of the cornea
Hematologic	Methemoglobinemia
Gastrointestinal	Abdominal pain
Urinary tract	Hemoglobinuria
Skin	Darkening of the face and hands

non-ionic surfactant properties of nonoxynols allow them to be used in a wide variety of industrial, household, agricultural, and pharmaceutical products. Nonoxynol-9 and nonoxynol-10 are surface-active agents used in antiseptic formulations, such as Hibitane solution, Betadine solution, Hexomedine transcutanée, and Hexomedine ointment. They are the most commonly used spermicidal contraceptives and have been recommended in the prevention of sexually transmitted diseases and in human immunodeficiency virus prophylaxis (9).

Pentachlorphenol is a disinfectant that is used in commercial laundries. In 1967 a hospital laundry accidentally used a product containing sodium pentachlorphenolate for a final rinse in the laundering of diapers and infants' bed linen. Twenty newborn infants developed sweating, tachycardia, tachypnea, hepatomegaly, and metabolic acidosis. Six children with severe reactions were subjected to exchange blood transfusion and in each instance there was a dramatic improvement. Two babies died before exchange transfusion could be carried out; postmortem examination showed fatty change in the liver and hydropic and fatty degeneration in the renal tubules and myocardium. There were toxic concentrations of pentachlorphenol in the serum of one patient and autopsy tissue of another.

The phenol derivatives paratertiary butylphenol and amylphenol are used in proprietary germicides.

The permeability of the human epidermis to many phenolic compounds correlates with their lipophilic pattern. However, phenolic compounds appear to produce denaturation in the skin, and an additional increase in permeability is attributed to the resulting damage to the epidermis. Complications of topical phenol, notably cardiac dysrhythmias, including death, can be caused by phenol face peels (10).

Phenol is so rapidly absorbed through the skin that severe systemic effects and even death can result within minutes to hours.

- A 90% solution of phenol spilled over the left sole and shoe of a 47-year-old tanker driver (11). He neither removed his soaked shoe nor attempted to decontaminate himself. He continued driving for 4.5 hours, after which he had vertigo and faintness. Fire fighters removed him from the vehicle, took off his shoes and clothes, and thoroughly washed his leg with copious amount of water. On admission to hospital he was alert but confused, his heart rate was 146/minute, his blood pressure 160/100 mmHg, and there was tense swelling and blue-black discoloration of his left foot, ankle, and distal part of the leg (3% total body surface area), with hypalgesia and hypesthesia over the affected area. His leg was irrigated with large amounts of water, longitudinal incisions of his left foot were performed, and he was transferred to the intensive care unit. Shortly afterwards, he developed rapid atrial fibrillation, ventricular extra beats, reduced blood pressure (90/60 mmHg), and fever of 38.3°C. He was treated with intravenous crystalloids, verapamil, dopamine, and phenylephrine. All the systemic symptoms resolved in 24 hours. Blood and urine cultures were negative. Over the next 3 weeks he was treated with 0.25% troclosene dressings and 1% micronized silver sulfadiazine, until the swelling resolved and the wound had healed. On discharge from hospital and at a 4-month follow up, only blue-black discoloration was noted.

Phenol can be applied percutaneously by the use of a monopolar needle electrode or by open injection when the nerve is exposed surgically. In addition, main nerve trunks or motor branches can be injected, depending on the clinical indication.

Organs and Systems

Cardiovascular

Phenol is cardiotoxic, and various cardiac dysrhythmias have been noted after application to the skin, or less commonly when it has been used for neurolysis. Ventricular extra beats occurred during topical application of phenol and croton oil in hexachlorophene soap and water for chemical peeling of a giant hairy nevus (12). Three of sixteen children treated with motor point blocks for cerebral palsy with a phenolic solution under halothane anesthesia developed cardiac dysrhythmias (13). Severe cardiac dysrhythmias followed by circulatory arrest occurred in an elderly patient with pancreatic cancer, injected with a phenolic solution to produce splanchnic neurolysis (14). The authors recommended that ethanol should replace phenol for this purpose.

In New Zealand, a patient died with brain damage after a cardiac arrest after being exposed to a chemical face peeling solution containing 64% phenol, Exoderm (15). The New Zealand Ministry of Health issued a public statement concerning the safety of phenol solutions.

Respiratory

Acute life-threatening epiglottitis developed in one patient after the use of a throat spray containing the equivalent of 1.4% phenol. The reaction may have been anaphylactic or a direct toxic effect (16).

A flu-like illness has been attributed to phenol.

- A 44-year-old man with hemorrhoids underwent submucosal injection sclerotherapy with 5% phenol in almond oil on two occasions 3 weeks apart (17). After the second injection he developed a flu-like illness with cough and dyspnea. He had no history of significance and was not taking any medications. He had a core temperature of 37.2°C, with a pulse of 115/minute, sinus rhythm, an initial blood pressure of 160/100 mmHg, a respiratory alkalosis, and mild hypoxia. His erythrocyte sedimentation rate was raised (38 mm/hour). An electrocardiogram was normal. A chest X-ray showed diffuse bilateral parenchymal changes with mottled shadowing. A high-resolution CT scan of the lungs showed peripheral interstitial involvement with limited subpleural parenchymal consolidation. He was given oxygen, intravenous fluids, and antibiotics (cefuroxime and levofloxacin) and recovered within 48 hours.

Nervous system

Early reports suggested that phenol caused selective damage only to small sensory nerve fibers (18). However, later studies showed that in concentrations of 1–7% it caused indiscriminate damage to efferent and afferent nerve fibers (19). At concentrations under 1%, phenol appears to have a local anesthetic affect, which is fully reversible (20). However, at higher concentrations it causes both Wallerian degeneration and axonal demyelination, leading to muscle denervation. After injection of 2% aqueous phenol there can be damage to the microcirculation around nerves, leading to occlusion of small blood vessels and fibrosis in the injected area (21).

The most common adverse effect of phenol injection is pain during injection, often described as a burning or stinging sensation. It can be associated with edema several hours after injection. The application of ice and the use of non-steroidal anti-inflammatory drugs often help to minimize the discomfort. In some studies, lidocaine mixed with phenol has been used to help diminish the local pain response associated with injection (22).

The most worrying adverse effect related to phenol injection is dysesthesia, caused by involvement of sensory nerve axons when perineural injection is attempted. Dysesthesia has been reported from a few days to about 2 weeks after injection; most patients describe a neuropathic component, including a burning pain with light tactile stimulation and involvement of only a small portion of the sensory distribution of the nerve that was blocked (23).

Sensory loss or loss of voluntary motor strength are not uncommon in the first few days after chemical neurolysis with phenol (24), but permanent loss, other than motor changes associated with the reduction in hypertonia, is less frequent (25,26). A muscle that is already weak is more susceptible to further weakening with chemical neurolysis (27).

- Loss of all sensation and strength after chemical neurolysis occurred in the distribution of the posterior tibial nerve after mixed sensorimotor block of this nerve, in this case after five blocks of the nerve or its motor branches over several years. Some strength and partial sensation returned after surgical lysis of excessive fibrous tissue at the site where the injection had been performed (28).

Chemoneurectomy with aqueous phenol injection in 116 selected patients with spastic cerebral palsy, in whom 246 peripheral nerves were blocked, caused complications in 11 patients (29). Five patients, in whom the posterior tibial nerve was blocked, developed paresthesia: one had complete loss of sensation, which recovered spontaneously after a couple of days; and three had pain at the site of injection or in the distribution of the injected nerve, lasting for a few days to a month. In another study there was a 3% complication rate in 98 blocks (30), while adverse effects occurred in nine of 150 blocks, with muscle weakness in eight cases and painful paresthesia in one (31).

Liver

Neonatal jaundice has been associated with the use of phenolic disinfectants in nurseries, not only when used in excessive concentrations, but also when applied in the recommended dilution (SEDA-5, 258) (SEDA-8, 246) (SEDA-11, 484).

Chronic liver disease has been attributed to long-term household poisoning with pentachlorphenol (32).

Skin

Phenol-related contact pemphigus has been described (23).

- A 32-year-old otherwise healthy woman developed oral erosive lesions of 6 weeks duration and a bullous eruption on her legs, back, and chest of 2 weeks duration. She had bullae and erosions on the left thigh and bullae on the legs, back, abdomen, and chest. There was no family history of pemphigus. Pemphigus vulgaris was diagnosed, and she was given prednisone 100 mg/day. She went home for weekends during hospitalization, and returned to hospital each time with worse lesions. She reported working with a cleaning agent that she had begun using before the onset of the disease. The cleaning agent contained nonyl phenol, and she challenged herself at her next visit home by putting the same agent on her hands for a full day. As a result, new bullae appeared in the mouth. She was advised to stop using the product and was discharged a few days later taking prednisone 80 mg/day, which was tapered as the disease subsided. No new lesions appeared after 1 year of follow-up.

Depigmentation has been attributed to germicides (O-Syl® and Ves-Phene®), containing a mixture of the phenol derivatives paratertiary butylphenol and amylphenol, in five hospital workers and seven other patients who had used them as household disinfectants. Patch tests with the phenolic components of the disinfectants on the patients and controls showed that virtually any moderately irritating phenolic compound can depigment the skin, but of those tested, paratertiary butylphenol and amylphenol most often depigmented the skin without producing toxic inflammation. Initial signs of depigmentation appeared 6 months after using the phenolic mixtures. Within 1 year, two of five patients noted a spontaneous return of pigment; another continued to use the product and there was no evidence of repigmentation (SEDA-11, 485) (33).

Musculoskeletal

Intramuscular injection of phenol can cause pain and swelling in the muscle (34,35). Sometimes, a firm nodular swelling develops in the calf 1–3 weeks after intramuscular neurolysis (36), particularly when larger quantities of phenol are injected into the intramuscular branches of the tibial nerve. This can usually be avoided by limiting the quantity of phenol injected to the minimum necessary and by applying cold packs to the injected area after the procedure.

Muscle necrosis and round cell infiltrates have been seen in histological studies of animal muscle recently injected with phenol, but not after a few months have passed (37).

Immunologic

Contact dermatitis in patients exposed to nonoxynols was initially considered to result from irritation, but allergic reactions have also been reported (38). Nonoxynol contact allergy has been described in two patients who developed contact photosensitivity to nonoxynol-l0 in the antiseptic product Hexomedine transcutanée (39). Among 32 control subjects, 13 had positive photopatch tests to Hexomedine transcutanée and four had positive photopatch tests to nonoxynol-10. Surprisingly, the authors observed that only undiluted nonoxynol was phototoxic. In another study, nonoxynol-9 was found to be rarely sensitizing and compatible with latex and silicone lubricants used in condoms (40).

There is evidence of immunosuppressive effects due to interference by dibenzo-p-dioxin and/or dibenzofuran with the chemical properties of pentachlorphenol (SEDA-11, 485) (41).

Death

Phenol can be fatal in the newborn.

- A 1-day-old child died 11 hours after 2% phenol had been applied to the umbilicus. The postmortem blood concentration of phenol was 125 µg/ml (SEDA-11, 485)(7).
- A 6-day-old child developed cerebral symptoms, circulatory failure, and methemoglobinemia after application of a phenol camphor solution (30% phenol, 60% camphor) to a skin ulcer. The child recovered after exchange transfusion (SEDA-11, 485)(7).

Drug Administration

Drug administration route

A phenol solution of 89% was mistakenly sprayed into the nostrils of a 79-year-old man (42). Immediately, blanching and local erythema developed. It was expected that the patient would develop a significant local burn and possibly systemic toxicity, but neither developed. This may have been due to the amount sprayed or the relatively small body surface area covered by the phenol.

References

1. Glenn MB, Elovic E. Chemical denervation for the treatment of hypertonia and related motor disordrers: phenol and botulinum toxin. J Head Trauma Rehabil 1997;12:40–62.
2. Felsenthal G. Pharmacology of phenol in peripheral nerve blocks: a review. Arch Phys Med Rehabil 1974;55(1):13–6.
3. Truppman ES, Ellenby JD. Major electrocardiographic changes during chemical face peeling. Plast Reconstr Surg 1979;63(1):44–8.
4. Del Pizzo A, Tanski A. Chemical face peeling—malignant therapy for benign disease? Plast Reconstr Surg 1980;66(1):121–3.
5. Ruedemann R, Deichmann WB. Blood phenol level after topical application of phenol-containing preparations. JAMA 1953;152(6):506–9.
6. Deichmann WB. Local and systemic effects following skin contact with phenol—a review of the literature. J Ind 1949;31:146.
7. Hinkel GK, Kintzel HW. Phenolvergiftungen bei Neugeborenen durch kutane Resorption. [Phenol poisoning of a newborn through skin resorption.] Dtsch Gesundheitsw 1968;23(51):2420–2.
8. Christian MS. Cosmetic ingredient review: final report on the safety assessment of nonoxynols 2, 4, 8, 9, 10, 12, 14, 15, 30, 40 and 50. J Am Coll Toxicol 1983;2:35–60.
9. Bird KD. The use of spermicide containing nonoxynol-9 in the prevention of HIV infection. AIDS 1991;5(7):791–6.
10. Botta SA, Straith RE, Goodwin HH. Cardiac arrhythmias in phenol face peeling: a suggested protocol for prevention. Aesthetic Plast Surg 1988;12(2):115–7.
11. Bentur Y, Shoshani O, Tabak A, Bin-Nun A, Ramon Y, Ulman Y, Berger Y, Nachlieli T, Peled YJ. Prolonged elimination half-life of phenol after dermal exposure. J Toxicol Clin Toxicol 1998;36(7):707–11.
12. Warner MA, Harper JV. Cardiac dysrhythmias associated with chemical peeling with phenol. Anesthesiology 1985;62(3):366–7.
13. Morrison JE Jr, Matthews D, Washington R, Fennessey PV, Harrison LM. Phenol motor point blocks in children: plasma concentrations and cardiac dysrhythmias. Anesthesiology 1991;75(2):359–62.
14. Gaudy JH, Tricot C, Sezeur A. Troubles du rythme cardiaque graves après phénolisation splanchnique peropératoire. [Serious heart rate disorders following perioperative splanchnic nerve phenol nerve block.] Can J Anaesth 1993;40(4):357–9.
15. Anonymous. Failure to provide the necessaries of life. NZ Med J 2003;116(1168):1.
16. Ho SL, Hollinrake K. Acute epiglottitis and Chloraseptic. BMJ 1989;298(6687):1584.
17. Lattuneddu A, Farneti F, Lucci E, Colinelli C. A pulmonary allergic reaction after injection sclerotherapy for hemorrhoids. Int J Colorectal Dis 2003;18:459–460.
18. Moller JE, Helweg-Larsen J, Jacobsen E. Histopathological lesions in the sciatic nerve of the rat following perineural application of phenol and alcohol solutions. Dan Med Bull 1969;16(4):116–9.
19. Bodine-Fowler SC, Allsing S, Botte MJ. Time course of muscle atrophy and recovery following a phenol-induced nerve block. Muscle Nerve 1996;19(4):497–504.
20. Burkel WE, McPhee M. Effect of phenol injection into peripheral nerve of rat: electron microscope studies. Arch Phys Med Rehabil 1970;51(7):391–7.
21. Glenn MB. Nerve blocks. In: Glenn M, Whyte J, editors. The Practical Management of Spasticity in Children and Adults. Philadelphia: Lea & Febiger, 1990:227–58.
22. Petrillo CR, Knoploch S. Phenol block of the tibial nerve for spasticity: a long-term follow-up study. Int Disabil Stud 1988;10(3):97–100.
23. Goldberg I, Sasson O, Brenner S. A case of phenol-related contact pemphigus. Dermatology 2001;203(4):355–6.
24. Khalili AA, Betts HB. Peripheral nerve block with phenol in the management of spasticity. Indications and complications. JAMA 1967;200(13):1155–7.
25. Tardieu G, Tardieu C, Hariga J, Gagnard L. Treatment of spasticity in injection of dilute alcohol at the motor point or by epidural route. Clinical extension of an experiment on the decerebrate cat. Dev Med Child Neurol 1968;10(5):555–68.
26. Copp EP, Harris R, Keenan J. Peripheral nerve block and motor point block with phenol in the management of spasticity. Proc R Soc Med 1970;63(9):937–8.
27. Khalili AA, Benton JG. A physiologic approach to the evaluation and the management of spasticity with procaine and phenol nerve block: including a review of the physiology of the stretch reflex. Clin Orthop Relat Res 1966;47:97–104.
28. Glenn MB. Nerve blocks for the treatment of spasticity. Phys Med Rehabil State of the Art Rev 1994;3:481–505.
29. Yadav SL, Singh U, Dureja GP, Singh KK, Chaturvedi S. Phenol block in the management of spastic cerebral palsy. Indian J Pediatr 1994;61(3):249–55.
30. Copp EP, Keenan J. Phenol nerve and motor point block in spasticity. Rheumatol Phys Med 1972;11(6):287–92.
31. Helweg-Larsen J, Jacobsen E. Treatment of spasticity in cerebral palsy by means of phenol nerve block of peripheral nerves. Dan Med Bull 1969;16(1):20–5.
32. Brandt M, Schmidt E, Schmidt FW. Chronische Lebererkrankung durch laugjahrige Intoxikation in Haushalt mit Pentachlorphenol. [Chronic liver disease caused by long term household poisoning with pentachlorophenol.] Verh Dtsch Ges Inn Med 1977;83:1609–11.
33. Kahn G. Depigmentation caused by phenolic detergent germicides. Arch Dermatol 1970;102(2):177–87.
34. Halpern D, Meelhuysen FE. Phenol motor point block in the management of muscular hypertonia. Arch Phys Med Rehabil 1966;47(10):659–64.
35. Garland DE, Lilling M, Keenan MA. Percutaneous phenol blocks to motor points of spastic forearm muscles in head-injured adults. Arch Phys Med Rehabil 1984;65(5):243–5.
36. Mullins RJ, Richards C, Walker T. Allergic reactions to oral, surgical and topical bovine collagen. Anaphylactic risk for surgeons. Aust NZ J Ophthalmol 1996;24(3):257–60.
37. Halpern D, Meelhuysen FE. Duration of relaxation after intramuscular neurolysis with phenol. JAMA 1967;200(13):1152–4.
38. Dooms-Goossens A, Deveylder H, de Alam AG, Lachapelle JM, Tennstedt D, Degreef H. Contact sensitivity

to nonoxynols as a cause of intolerance to antiseptic pre-parations. J Am Acad Dermatol 1989;21(4 Pt 1):723–7.

39. Michel M, Dompmartin A, Moreau A, Leroy D. Contact photosensitivity to nonoxynol used in antiseptic prepara-tions. Photodermatol Photoimmunol Photomed 1994;10(5):198–201.

40. Fisher AA. Allergic contact dermatitis to nonoxynol-9 in a condom. Cutis 1994;53(3):110–1.

41. Dickson D. PCP dioxins found to pose health risks. Nature 1980;283(5746):418.

42. Durback-Morris LF, Scharman EJ. Accidental intranasal administration of phenol. Vet Hum Toxicol 1999;41(3):157.

Polyhexanide

General Information

Polyhexanide is a polymerized form of chlorhexidine, used as a disinfectant (1).

Organs and Systems

Immunologic

Severe anaphylaxis occurred in an 18-year-old woman and a 15-year-old man when polyhexanide was used to clean surgical wounds (2). Immediate-type hypersensitiv-ity to polyhexanide was suggested by positive skin prick tests. Both patients had previously been exposed to chlor-hexidine, but skin tests with chlorhexidine were negative.

References

1. Kramer A, Behrens-Baumann W. Prophylactic use of topical anti-infectives in ophthalmology. Ophthalmologica 1997;211(Suppl 1):68–76.

2. Olivieri J, Eigenmann PA, Hauser C. Severe anaphylaxis to a new disinfectant: polyhexanide, a chlorhexidine polymer. Schweiz Med Wochenschr 1998;128(40):1508–11.

Polyvidone

General Information

Polyvidone (polyvinylpyrrolidone, povidone) is a vari-able-weight polymer of the monomer N-vinylpyrrolidi-none. When it enters the body, it causes histologically characteristic reactions in tissues with which it comes into contact (1,2).

Polyvidone co-polymers are used in cosmetics as anti-microbials, antistatics, binding compounds, stabilizers of emulsions, and film-forming, viscosity-controlling, hair-fixing, skin-conditioning, and skin-protective agents. It is used as a component of hair sprays and as a retardant for subcutaneous injections. It was formerly used as a plasma expander (3) and has been inappropriately used for intra-venous injection as a "blood tonic," especially in Asian societies. Some products intended for parenteral admin-istration contain polyvidone as an excipient. Polyvidone is widely used as a suspending and coating agent in tablets, for its film-forming properties in eye drops, and as a carrier molecule for iodine in disinfectants. About 20% of all tablets on the market contain polyvidone. It is also used in the cosmetics industry as a dispersing agent and as a lubricant in ointments.

Povidone-iodine

Povidone-iodine is a macromolecular complex (poly-I(I-vinyl-2-pyrrolidinone)) that is used as an iodophor. It is formulated as a 10% applicator solution, a 2% cleansing solution, and in many topical formulations, for example aerosol sprays, aerosol foams, vaginal gels, ointments, and mouthwashes. Because it contains very little free iodine (less than 1 ppm in a 10% solution), its antibacterial effectiveness is only moderate compared with that of a pure solution of iodine.

Systemic absorption

The extent of systemic absorption of povidone-iodine depends on the localization and the conditions of its use (area, skin surface, mucous membranes, wounds, body cavities).

Healthy skin

Repeated surgical skin antisepsis and hand washing did not increase serum iodine concentrations, but produced a small increase in iodine content in the 24-hour urine (4).

Burns

The use of povidone-iodine for the treatment of burns, for peritoneal lavage in the treatment of purulent peritonitis, or as a rinsing solution for body cavities can increase serum iodine concentrations associated with increased urinary excretion of iodine. In people with burns, the extent of iodine absorption depends on the extent of the burned body surface. It is not uncommon for serum iodine concentrations to rise to more than 1000 µg/ml. If renal function is intact, iodine elimination in the urine can be adequate. The serum iodine concentration returns to nor-mal about 1 week after the last application.

The penetration of povidone-iodine has been studied in vivo in rabbits (5). The penetration from third-degree burns on the back was measured autoradiographically in tissues, blood, urine, and bandages. The results showed that about 20% of iodine is absorbed through fresh necro-sis, whereas only 5% is absorbed through a clean wound or 24-hour-old necrosis. The passage through burn necro-sis was faster than through vital tissue.

In repeated topical use on burns, the extent of absorp-tion seems to decrease with the treatment time.

Wounds

Povidone-iodine inhibits leukocyte migration and fibro-blast aggregation in wounds. The effect on the wound healing process has been studied in 294 children undergoing surgery, 283 of whom had undergone appendectomy (6). In a first series using 5% povidone-iodine aerosol for preoperative disinfection, the postoperative wound infection rate was 19% in the test group and only 8% in the controls. When a 1% povidone-iodine solution was used, only 2.6% of the patients were infected (control group 8.5%). Using a drain with a cellulose viscose sponge, 5% povidone-iodine by aerosol inhibited leukocyte migration, but no cell aggregates or fibroblasts were detected. A 5% solution allowed better cellular movement and attachment to the framework, polymorphonuclear leukocytes predominating. The excipients in the aerosol formula must be more toxic to the cell than those in the solution. If a 1% povidone-iodine solution was absorbed by the sponge, the aggregation phenomenon was only slightly averted and cell morphology was similar to that of the saline control.

Povidone-iodine reduced the number of wound infections only in patients with appendicitis in whom neither peritonitis nor a periappendicular abscess had yet developed (SEDA-11, 489).

Mucous membranes

The effect of a povidone-iodine mouthwash on thyroid function has been studied in 16 medically healthy volunteers. After they had used the mouthwash four times daily for a period of 14 days, all thyroid tests were significantly changed, but there was no suppression of thyroid function. However, this was not to be expected, considering the short test period.

Body cavities

When povidone-iodine is used as a rinsing solution in body cavities, absorption of the whole macromolecular complex is possible. The complex has a molecular weight of about 60 000 and cannot be eliminated by the kidneys or metabolically. It is filtered by the reticuloendothelial system (4,7,8).

Although povidone-iodine is no longer used in dialysates, a povidone-iodine-containing cap is used to seal the Tenckhoff catheter during the day. Iodine-induced hypothyroidism occurred in a 3-year-old boy and an 18-month-old girl, in both cases due to the sealing cap (9). The povidone-iodine inside the cap diffused into the catheter and flushed into the peritoneal cavity at the next dialysis session.

Intravaginal administration

Systemic iodine absorption can occur after intravaginal administration of povidone-iodine (10). There were increases in serum iodine, protein-bound iodine, and inorganic iodine, but not serum thyroxine, after a 2-minute vaginal administration of povidone-iodine in non-pregnant women (11).

Guidelines for the safe use of povidoneiodine complexes

In 1985, a working group of the Federal German Medical Association issued a number of recommendations for the safe use of povidone-iodine complexes (12). They remain valid and can be summarized as follows:

1. The application of povidone-iodine formulations cannot be recommended for surgical hand disinfection, since active iodine-free formulations are available.
2. The activity of povidone-iodine in preoperative skin disinfection in adults is well proven.
3. Povidone-iodine is appropriate for skin disinfection before an incision, a puncture, with use of intravenous or arterial catheters, and for the prophylaxis of iatrogenic *Clostridia* infections.
4. In the case of superficial wounds, povidone-iodine can be applied occasionally or repeatedly in spite of increased iodine absorption through the broken skin surfaces.
5. Lavage of wound and body cavities with povidone-iodine or its instillation is not indicated because of increased iodine absorption.
6. Routine body washing of patients in intensive care units is not cost-beneficial.
7. Vaginal administration of povidone-iodine is not recommended.
8. Povidone-iodine is contraindicated in premature babies and neonates; this also applies to prophylactic disinfection of the umbilical stump.
9. The clinical usefulness of povidone-iodine in the treatment of burns is well proven.
10. Local mouth antiseptics serve no therapeutic purpose; this is also true for povidone-iodine.

Observational studies

In a meta-analysis of observational studies of iodopovidone for chemical pleurodesis, the only significant complication was *chest pain* of varying degree (13). Systemic hypotension was reported in three patients in only one study. There were no deaths related to chemical pleurodesis. Overall, the review supported the safety and efficacy of iodopovidone as an agent for chemical pleurodesis.

Comparative studies

In a randomized trial of povidone-iodine compared with standard care to reduce visual impairment from corneal ulcers in rural Nepal, although additional benefit could not be demonstrated, a 2.5% solution of povidone-iodine was well tolerated and produced no more adverse effects than the standard therapy in this population (topical chloramphenicol, gentamicin, or tetracycline) (14).

Organs and Systems

Nervous system

- A 62-year-old man, treated with continuous mediastinal irrigation with a 1:10 solution of povidone-iodine, developed seizures on the fifth day of drainage (15). After the seizure, his serum iodine concentration was raised (120 µg/ml). Renal insufficiency developed at the same time. The electroencephalogram showed no evidence of epileptic activity or other abnormalities. The povidone-iodine irrigation was replaced by continuous irrigation with a solution of neomycin and polymyxin B. Renal function improved and the creatinine concentration returned to normal 3 days after the seizure.

Seizures, which can occur with iodinated contrast media, have only rarely been associated with the use of povidone iodine. Partial complex seizures with secondary generalization occurred after intrapleural povidone iodine irrigation (16).

- A 67-year-old healthy white man developed a parapneumonic effusion, which was treated with chest drainage and systemic antibiotics. One week later 100 ml of povidone iodine 10%, equivalent to iodine 1%, was instilled and drained after 10 minutes via the chest tube. Ten minutes after drainage, he reported headache, especially on the right side, vertigo, deviation of the head and eyes toward the right side, and uncontrolled rapid movements of the right hand. This episode lasted several seconds. Soon after, he became unconscious for a few minutes; his blood pressure was 120/80 mm Hg, heart rate 88/minute, and temperature 37°C. On the same night he had a similar seizure and recovered spontaneously after 10 minutes, with complaints of fatigue, headache, somnolence, amnesia, nausea, and vomiting. Thereafter he had no further seizures or neurological disturbances. He was not rechallenged with povidone iodine.

Endocrine

Hypothyroidism
Extensive iodine absorption can cause transient hypothyroidism, or, in patients with latent hypothyroidism, the risk of destabilization and thyrotoxic crisis (SEDA-20, 226) (SEDA-22, 263). Especially at risk are patients with an autonomous adenoma, localized diffuse autonomy of the thyroid gland, nodular goiter, latent hyperthyroidism of autoimmune origin, or endemic iodine deficiency (4).

Altered metabolism of thyroid hormones occurs in children undergoing open heart surgery. The changes typically last from the onset of cardiopulmonary bypass until 5?7 days after surgery. Studies of the effects of perioperative disinfection of the skin with povidone iodine have suggested that cardiopulmonary bypass has a more profound effect on the hypothalamic–pituitary axis than povidone iodine. In a prospective study of the effect of perioperative topical povidone iodine on thyroid hormone status in 20 infants with delayed sternal closure, there was transient hypothyroidism in four of the infants (17). The authors concluded that although changes in thyroid hormone metabolism in critically ill infants are difficult to interpret, hypothyroidism in the late postoperative period can be caused by exposure to iodine from povidone iodine.

Severe hypothyroidism requiring levothyroxine replacement therapy has been reported in a neonate after prolonged use of an iodinated skin disinfectant (18).

- In a boy born at 25 weeks gestation, neonatal screening for hypothyroidism on day 4 after delivery was normal (thyroid-stimulating hormone, TSH, <20 mU/l). On day 10, septic thrombophlebitis of the scalp with abscess formation, probably secondary to an intravenous catheter, was treated with intravenous antibiotics and topical povidone iodine, polysporin, and saline compresses. The povidone iodine was discontinued after 20 days and thyroid function tests showed a marked increase in TSH (455 mU/l) and a reduction in free thyroxine concentration (0.51 ng/dl). Transient hypothyroidism secondary to iodine overload was diagnosed. There were no clinical signs of hypothyroidism. Ultrasonography of the thyroid gland showed enlargement of the two lobes and the isthmus, with increased vascularity. Levothyroxine 25 micrograms/day was started on day 39. Thyroid-stimulating hormone and free thyroxine concentrations returned to the reference range after 10 days of levothyroxine therapy and withdrawal of povidone iodine. The levothyroxine was withdrawn on day 65.

Thyroid hormones play a major role in postnatal brain development and linear growth. This case of severe transient hypothyroidism after the use of povidone iodine skin disinfectant suggests that neonatal iodine exposure should be minimized whenever possible and that TSH should be routinely measured soon after exposure to iodinated skin disinfectants. This is especially important in preterm neonates, in whom skin permeability is high, the thyroid gland is particularly sensitive to the antithyroid effects of iodine excess, and the clinical signs of hypothyroidism are difficult to recognize.

A report of hypothyroidism in a neonate in whom povidone iodine was used as an operative and postoperative skin disinfectant has drawn attention to the risk of transient hypothyroidism in neonates through the Wolff–Chaikoff effect, the prevention of iodine organification by iodine-containing drugs or solutions. It has been suggested that in three neonates with end-stage renal insufficiency hypothyroidism developed as a consequence of the Wolff–Chaikoff effect (19). Iodine is normally cleared from the circulation by the kidneys. Neonates or infants with renal insufficiency are therefore less likely to clear any iodine to which they are exposed and are potentially at increased risk of hypothyroidism. In all three cases the infants received renal replacement therapy using peritoneal dialysis. Iodine concentrations were higher in the peritoneal dialysis fluid than in the plasma, and it was suggested that the source of iodine had been the povidone iodine in the cap used at the end of the peritoneal dialysis catheter, which helps reduce the incidence of local infection and peritonitis (20).

Hyperthyroidism
Povidone-iodine-induced hyperthyroidism is rarer than hypothyroidism (SEDA-20, 226), but a history of long-term use of iodine-containing medications should be considered when investigating the cause of hyperthyroidism (21).

- A 48-year-old woman developed palpitation and insomnia. The clinical history, physical examination, and laboratory tests supported hyperthyroidism. Since July 1994, she had been combating constipation by improper use of an iodine-containing antiseptic cream for external use only. She had inserted povidone-iodine into her rectum by means of a cannula. The iodine-containing cream was withdrawn and she was given a beta-blocker. The palpitation resolved within 2 weeks and her plasma thyroid hormone concentrations normalized within 1 month.

Hyperthyroidism in this patient was probably triggered by improper long-term use of an over-the-counter iodine-containing cream.

Thyrotoxicosis related to iodine toxicity in a child with burns occurred after alternate-day povidone-iodine washes (22).

- A 22-month-old boy was admitted to a pediatric intensive care unit after partial and full thickness burns over 80% of his body surface area. After debridement he was given alternate-day povidone-iodine (Betadine) washes. He became increasingly tachycardic, hypertensive, and hyperpyrexic, with sweating, agitation, and diarrhea and developed neutropenia. Daily sepsis screens were negative. Thyroid function tests showed evidence of iodine-induced thyrotoxicosis. He was given propranolol and carbimazole, and chlorhexidine was substituted for povidone-iodine. His tachycardia, hypertension, and diarrhea slowly improved and his neutropenia resolved. By day 42 his free thyroxine concentration was normal and he was given thyroxine. On day 63 the carbimazole, propranolol, and thyroxine were withheld. However thyroid function tests 1 week later showed hypothyroidism, and thyroxine was restarted. Repeat plasma and urine iodine concentrations on day 57 (a month after withdrawal of povidone-iodine) continued to show marked urinary excretion of iodine (urine concentration 65 μmol/l, reference range 0.39–1.97) with high but falling plasma iodine concentrations (5.9 μmol/l, reference range 0.32–0.63).

Hematologic

Severe neutropenia occurred in a patient in whom deep, second-degree burns, involving about 50% of the body surface, were being treated with Betadine Helafoam twice a day (23).

Liver

There have been reports of liver damage from polyvidone (24).

Urinary tract

Povidone-iodine sclerosis has been suggested to be safe and effective in treating lymphoceles after renal transplantation, with only minor complications of the procedure, such as pericatheter cutaneous infections. However, a case of acute renal tubular necrosis has been reported (25).

- In a 23-year-old woman, a kidney allograft recipient with recurrent lymphoceles treated with povidone-iodine irrigations (50 ml of a 1% solution bd for 6 days), a metabolic acidosis occurred and renal function

deteriorated. After a few days, despite suspension of irrigation, the patient developed oliguria, and dialysis was needed. A renal biopsy showed acute tubular necrosis.

Iodine-induced renal insufficiency has also been reported after the use of topical povidone-iodine on the skin and after intracavity irrigation.

- A 65-year-old man with second- and third-degree burns covering 26% of his body was given intravenous lactated Ringer solution and topical silver sulfadiazine in addition to debridement and skin grafting (26). However, he developed a wound infection with *Pseudomonas aeruginosa*, which was treated successfully with topical povidone-iodine gel. Persistent nodal bradycardia with hypotension, metabolic acidosis, and renal insufficiency occurred 16 days later. Iodine toxicosis was suspected and the serum iodine concentration was 206 μg/ml (reference range 20–90 μg/ml). The povidone-iodine gel was therefore withdrawn immediately. His family refused hemodialysis and he died 44 days after admission.
- A 57-year-old man developed renal insufficiency after triple coronary bypass grafting, 7 days after povidone-iodine mediastinal irrigation and required 3 days of renal replacement therapy (27). Complete resolution occurred within 8 days and followed a short non-oliguric phase (4 days).

Although other common causes of acute renal insufficiency were present in the second case, the only significant change in management at the time of onset of renal insufficiency was the use of povidone-iodine.

In a patient who developed hepatic and renal dysfunction, with toxic iodine concentrations during continuous mediastinal irrigation with povidone iodine, hemodialysis and hemofiltration were used for iodine clearance (28). Hepatic function and renal function improved with falling plasma iodine concentrations.

Skin

Povidone-iodine causes concentration-dependent damage to cells and clusters. The effect is most pronounced for isolated cells, but it is also detectable in more complex tissues. Clinical experience with burn victims cannot rule out the possibility that the healing process may be slightly retarded. However, this deficiency may be balanced by an appropriate microbicidal effect on the healing edge (29).

Contact dermatitis from povidone iodine is not rare, and the responsible antigens are iodine (30) and polyoxyethylene nonylphenyl ether (31). However, immediate-type allergy is uncommon.

- A 59-year-old woman who had had several episodes of contact urticaria after hair treatment developed anaphylaxis after vaginal application of povidone iodine solution for disinfection (32). Prick tests showed wheal-and-flare responses to both povidone iodine (0.1% aqueous) and polyvinylpyrrolidone (0.001% aqueous) but not to iodine or polyoxyethylenenonylphenyl ether, both of which are also contained in povidone iodine solution. Basophils from her peripheral blood released considerable amounts of histamine on stimulation by polyvinylpyrrolidone. She was recommended to avoid

Both the shampoo and the permanent wave solution contained polyvinylpyrrolidone styrene-copolymer emulsion and polyvinylpyrrolidone N,N-dimethylaminoethylmethacrylic acid copolymer diethyl sulfate solution. Both of these agents provoked immediate skin responses on prick testing.

Chemical burns are a rare but potentially serious complication of povidone iodine used as a topical antiseptic (33).

- A 38-year-old woman underwent laparoscopic right ovarian cystoscopy and endometrial ablation. The antiseptic skin preparation used was 1% povidone-iodine solution. She had no allergies and was not taking any drugs. The morning after the procedure she developed burning, itching, pain, and marked redness on her back and buttocks. On the next day blisters developed. There was superficial skin loss on both sides of her back, some medial tenderness, but no evidence of cellulitis. The partial thickness chemical burns were treated with polymyxin B sulfate + bacitracin zinc ointment, and at 1 week the affected area was healed, with no permanent sequelae. There was some superficial skin loss, but no loss of pigmentation.

Polyvidone storage disease

Polyvidone molecules that weigh less than 20 kD can be excreted by a normally functioning kidney, whereas larger polymers are phagocytosed and permanently stored in the mononuclear phagocytic system, causing so-called polyvidone storage disease. Polyvidone storage disease occurs in patients who have received polyvidone for prolonged periods of time. The large polymers deposit in the histiocytes and cause them to proliferate and infiltrate histiocytes in the reticuloendothelial system, including osteocytes. There is generally no significant damage to these organs, except that prolonged administration can cause bone destruction, skin lesions, arthritis, and polyneuropathy.

The first cutaneous case of polyvidone storage disease, reported in 1964 (34), was caused by local injection of polyvidone-containing posterior pituitary extracts for the treatment of diabetes insipidus. Similar cases, including those following local injection of porcine polyvidone to treat neuralgia, were documented, mostly in European reports (35,36). Localized cutaneous polyvidone storage disease was then known as Dupont-Lachapelle disease (37).

Five cases of polyvidone storage disease with cutaneous involvement have been documented (38). Two patients presented with skin eruptions mimicking collagen vascular disease and chronic pigmented purpuric dermatosis. In one, polyvidone was found in a metastatic tumor and in the other in a pemphigus lesion. The fifth case was seen in a blind skin biopsy specimen taken to exclude Niemann-Pick disease after

examination of a bone marrow smear. The latter patient and the patient with a collagen vascular-like disease also had severe anemia and serious orthopedic and neurological complications due to massive infiltration of polyvidone-containing cells in the bone marrow, with destruction of the bone.

Polyvidone storage disease can easily be diagnosed by its histopathological features. Skin biopsy specimens show a variable number of characteristic blue–gray vacuolated cells around blood vessels and adnexal structures, and stain positively with mucicarmine, colloidal iron, and alkaline Congo red and negatively with periodic acid Schiff and Alcian blue.

Musculoskeletal

Pathological fractures of several bones and destructive lesions seen radiologically in other bones have been reported in patients who had received repeated intravenous injections of polyvidone for many years (39,40). Biopsies of the fracture sites showed both intracellular deposits of polyvidone and mucoid changes in the affected cells. If of sufficient severity, this may cause a virtual "melt down" of osseous tissue.

Immunologic

Polyvidone has been reported to cause anaphylaxis (41).

- A 32-year-old man took paracetamol (in Doregrippin) for flu-like symptoms and about 10 minutes later developed generalized urticaria, angioedema, hypotonia, and tachycardia, and became semiconscious. His symptoms were rapidly relieved by intravenous antihistamines and steroids. This was the first time he had taken Doregrippin, but he had previously taken paracetamol-containing formulations, which had been well tolerated. He was not taking any regular medications. Subsequent testing of the various constituents of the analgesic tablets identified polyvidone as the cause of the anaphylactic reaction.

This report demonstrates a rare case of a type I allergic reaction toward a commonly used ingredient of tablets and widely used disinfectants.

In principle, all forms of the well-known iodine-induced allergic reactions, such as iododerma tuberosum, dermatitis, petechiae, and sialadenitis are possible with povidone-iodine, but the incidence seems to be very low (SEDA-11, 489) (SEDA-12, 586) (4,42,43).

- A severe anaphylactoid reaction occurred immediately after the instillation of a 10% solution of povidone-iodine into a hydatid cyst cavity during surgery. Severe bronchospasm developed immediately and was followed by a coagulopathy and subsequent liver and renal insufficiency (44).

There have been only a few reports of contact allergy to povidone-iodine, despite its widespread use. In two cases there were positive patch test reactions on days 2, 3, and 7 to povidone-iodine (5% aqueous) and iodine (0.5% in petrolatum), but negative reactions to povidone itself (45).

Second-Generation Effects

Pregnancy and fetotoxicity

Routine vaginal douching with povidone-iodine during pregnancy causes maternal iodine overload and markedly increases the iodine content in amniotic fluid and of the fetal thyroid, as soon as the trapping mechanism of iodine by the thyroid has started to develop. Vaginal use of povidone-iodine is therefore not recommended during pregnancy (46,47) and labor (48).

The fetal thyroid starts to store iodine between the 10th and 13th weeks of gestation, and to secrete thyroid hormone between the 18th and 24th weeks. Especially after intravaginal administration during pregnancy, povidone-iodine can cause congenital goiter and hypothyroidism in newborn infants. However, hyperthyroidism can also occur.

In premature twins who had markedly enhanced thyroid stimulating hormone concentrations the mother had used povidone iodine pessaries for 7 weeks during pregnancy to prevent vaginal infection (49).

In 99 of 9320 newborns, TSH concentrations were above the reference range (20 mU/ml) on the fifth day of life, but between the 10th and 21st day, all these infants had normal TSH concentrations and normal thyroid function (50). In 76 of the newborns with hyperthyrotropinemia, urinary iodine excretion was significantly raised (above 16 µg/ml). Most of them were born in obstetric departments where iodophores were routinely used for disinfection during labor.

In 66 mothers and their infants, povidone-iodine was given during labor and delivery as a 1 or 2% solution pumped intravaginally through a plastic catheter until delivery (for 5–30 hours), urinary iodide concentrations on the first and the fifth day, and serum iodine concentrations at birth were significantly raised in the mothers as well as in the neonates (51). At birth, the TSH concentrations in the mothers and infants were no different from those in the controls, but on the third and fifth days they were significantly higher. Thyroxine concentrations were significantly lower in the exposed mothers and infants (at birth and on the third and the fifth days). One-fifth of the infants had high TSH concentrations (above 20 µU/ml) and low thyroxine values (below 7 µg/ml), which is suggestive of hypothyroidism. However, none of the infants developed clinical symptoms, and on the 14th day the values were normal again. In the iodine-exposed mothers and infants, tri-iodothyronine (T3) concentrations were significantly reduced at birth, but not thereafter. The concentrations of reverse T3 did not differ from the controls at birth, but were significantly lower on succeeding days. This reduction in reverse T3 in the iodine-exposed infants was probably due to reduced thyroxine concentrations, causing a lack of substrates for monodeiodination to reverse T3.

Iodine concentrations in breast milk and in random urine in neonates and the serum concentrations of neonatal TSH and free thyroxine on day 5 after delivery were measured after the use of povidone-iodine for disinfection after delivery (48). Iodine concentrations in the breast milk and neonatal TSH were significantly raised.

Perinatal iodine exposure causes transient hypothyroidism in a significant number of neonates, in whom careful monitoring and follow-up of thyroid gland function are needed. It is better to avoid the use of iodine-containing antiseptics in pregnancy and neonates, especially if follow-up cannot be guaranteed.

Susceptibility Factors

Age

Neonates

Hypothyroidism in neonates has been related to the use of small doses of iodine as an antiseptic. The high vulnerability of the neonatal thyroid is a reason for avoiding povidone-iodine for care of the umbilical stump or omphaloceles (SEDA-11, 488) (SEDA-12, 585) (52).

Serum TSH and thyroxine concentrations have been measured 57 days after birth in 365 healthy newborns whose umbilical stump had been treated with 10% povidone-iodine (53). The prevalence of high TSH concentrations was significantly higher in this group than in the general population (3.1 versus 0.4%), as was the rate of transient hypothyroidism (2.7 versus 0.25%). All the children were normal when retested 1 week later.

Transient hypothyroidism due to skin contamination with povidone-iodine occurred in a neonate with an omphalocele (54).

The postnatal iodine overload, measured as urinary iodine concentration, has been studied in ill neonates after the cutaneous application of povidone-iodine (0.96% I_2; Betadine) (SEDA-11, 488) (52). The mean iodine overload was 1297 µg/day in one povidone-iodine group and 1253 µg/day in a second group; in the control group, 64% of the newborns had iodinuria of less than 100 µg/day, and of the 10 others (mean ioduria 1212 has been studied in ill neonates) three were born by cesarean section: in these cases the mothers received an iodine-containing curariform agent. There were 12 cases of hypothyroidism among the neonates exposed to iodine-containing antiseptics, but none in the control group.

Very low birth weight infants admitted to a neonatal intensive care unit who had been given chlorhexidine-containing antiseptics ($n = 29$) were compared with infants in a comparable unit who had been given iodinated antiseptic agents ($n = 54$) (55). The latter had an up to 50-fold higher increase in urinary iodine excretion than the controls. The median serum TSH concentration was significantly higher in the iodine-exposed infants (4.6 mU/ml) than in the control infants (2.4 mU/ml). On day 14, TSH concentrations in nine of the 36 iodine-exposed infants were above 20 mU/ml, their mean thyroxine concentration was significantly lower (44 nmol/1) than the mean thyroxine concentrations (83 nmol/1) in both the exposed infants with normal TSH concentrations and the controls.

Renal disease

Since iodine is eliminated by the kidneys, renal insufficiency increases the risk of toxicity, and the risk may be further increased by metabolic acidosis (56,57).

Drug Administration

Drug contamination

Bacterial contamination of povidone-iodine formulations has been reported. *Pseudomonas cepacia* was discovered in the blood cultures of 52 patients in four hospitals in New York over 7 months, and of 16 patients in a Boston hospital over a 10-week period in 1980 (58). A contaminated povidone-iodine solution produced by one manufacturer was implicated as the source of the bacteria. It is not clear why this solution was contaminated, whereas other marketed povidone-iodine solutions containing equivalent amounts of available or free iodine remained sterile.

Drug administration route

Vaginal use of povidone-iodine is not recommended in pregnancy (SEDA-25, 277). However, it continues to be used prophylactically for abdominal hysterectomy, although there is a lack of research in this area. In a randomized trial in 150 women to assess whether infectious morbidity after total abdominal hysterectomy is reduced by the addition of povidone-iodine gel at the vaginal apex after the usual vaginal preparation, there was a reduced risk of pelvic abscess and no severe allergic reactions (59).

Silver nitrate is commonly used for renal pelvic installation sclerotherapy for chyluria. However, it can cause severe adverse effects. In one study silver nitrate and povidone-iodine had equivalent efficacy and povidone-iodine was well tolerated; the most significant adverse effect was severe flank pain, with an incidence of 11% in the silver nitrate group and 2% in the povidone-iodine group (60).

Interference with Diagnostic Tests

Povidone-iodine gives a positive reaction with an ortho-toluidine reagent used to detect blood in the urine, for example Hematest reagent tablets or dipsticks (SEDA-11, 488) (61).

Povidone-iodine used for skin disinfection before skin puncture blood was taken altered serum concentrations of potassium, phosphate, and uric acid.

References

1. Bergman M, Flance IJ, Blumenthal HT. Thesaurosis following inhalation of hair spray; a clinical and experimental study. N Engl J Med 1958;258(10):471–6.
2. Bergman M, Flance IJ, Cruz PT, Klam N, Aronson PR, Joshi RA, Blumenthal HT. Thesaurosis due to inhalation of hair spray. Report of twelve new cases, including three autopsies. Nord Hyg Tidskr 1962;266:750–5.
3. Weese HG, Periston H. Ein never Blutluessigkeitsersatz. Münch Med Wochenschr 1943;90:11–5.
4. Gortz G, Haring R. Wirkung und Nebenwirkung von Polyvinylpyrrolidon-Jod (PVP-Jod). Therapiewoche 1981;31:4364.
5. Colcleuth RG. Distribution protein binding of betadine ointment in burn wounds. In: Altemeier WA, editor. II World Congress/Antisepsis Proceedings. New York: HP Publishing Co, 1980:122–3.
6. Viljanto J. Disinfection of surgical wounds without inhibition of normal wound healing. Arch Surg 1980;115(3):253–6.
7. Glick PL, Guglielmo BJ, Tranbaugh RF, Turley K. Iodine toxicity in a patient treated by continuous povidone-iodine mediastinal irrigation. Ann Thorac Surg 1985;39(5):478–80.
8. Campistol JM, Abad C, Nogue S, Bertran A. Acute renal failure in a patient treated by continuous povidone-iodine mediastinal irrigation. J Cardiovasc Surg (Torino) 1988;29(4):410–2.
9. Vulsma T, Menzel D, Abbad FC, Gons MH, de Vijlder JJ. Iodine-induced hypothyroidism in infants treated with continuous cyclic peritoneal dialysis. Lancet 1990;336(8718):812.
10. Jacobson JM, Hankins GV, Murray JM, Young RL. Self-limited hyperthyroidism following intravaginal iodine administration. Am J Obstet Gynecol 1981;140(4):472–3.
11. Vorherr H, Vorherr UF, Mehta P, Ulrich JA, Messer RH. Vaginal absorption of povidone-iodine. JAMA 1980;244(23):2628–9.
12. Wissenschaftlicher Berat der Bundesärztäkammer. Für Anwendung von Polyvinylpyrolidon-Jod Komplexen. Dtsch Ärztebl 1985;82:1434.
13. Agarwal R, Aggarwal AN, Gupta D, Jindal SK. Efficacy and safety of iodopovidone in chemical pleurodesis: a meta-analysis of observational studies. Respir Med 2006;100:2043–7.
14. Katz J, Khatry SK, Thapa MD, Schein OD, Kimbroug L, Pradham K, le Clerq SC, West KP. A randomised trial of povidone-iodine to reduce visual impairment from corneal ulcers in rural Nepal. Br J Ophthalmol 2004;88:1487–92.
15. Zec N, Donovan JW, Aufiero TX, Kincaid RL, Demers LM. Seizures in a patient treated with continuous povidone-iodine mediastinal irrigation. N Engl J Med 1992;326(26):1784.
16. Azzam ZS, Farhat D, Braun E, Krivoy N. Seizures: an unusual complication of intrapleural povidone-iodine irrigation. J Pharm Technol 2003;19:94–6.
17. Kovacikova L, Kunovsky P, Lakomy M, Skrak P, Miskova Z, Siman J, Kostalova L, Tomeckova E. Thyroid hormone status after cardiac surgery in infants with delayed sternal closure and continued use of cutaneous povidone-iodine. Endocrine Regul 2003;37:3–9.
18. Khashu M, Chesex P, Chanoine J-P. Iodine overload and severe hypothyroidism in a premature neonate. J Pediatr Surg 2005;40:E1-E4.
19. Aliefendioğlu D, Sanli C, Cakmak M, Ağar A, Albayrak M, Evliyaoğlu O. Wolff–Chaikoff effect in a newborn: is it an overlooked problem? J Pediatr Surg 2006;41:e1-3.
20. Brough R, Jones C. Iatrogenic iodine as a cause of hypothyroidism in infants with end stage renal failure. Pediatr Nephrol 2006;21:400–2.
21. Grant JA, Bilodeau PA, Guernsey BG, Gardner FH. Unsuspected benzyl alcohol hypersensitivity. N Engl J Med 1982;306(2):108.

22. Robertson P, Fraser J, Sheild J, Weir P. Thyrotoxicosis related to iodine toxicity in a paediatric burn patient. Intensive Care Med 2002;28(9):1369.

23. Alvarez E. Neutropenia in a burned patient being treated topically with povidone-iodine foam. Plast Reconstr Surg 1979;63(6):839–40.

24. Golightly LK, Smolinske SS, Bennett ML, Sutherland EW 3rd, Rumack BH. Pharmaceutical excipients. Adverse effects associated with 'inactive' ingredients in drug products (Part II). Med Toxicol Adverse Drug Exp 1988;3(3):209–40.

25. Manfro RC, Comerlato L, Berdichevski RH, Ribeiro AR, Denicol NT, Berger M, Saitovitch D, Koff WJ, Goncalves LF. Nephrotoxic acute renal failure in a renal transplant patient with recurrent lymphocele treated with povidone-iodine irrigation. Am J Kidney Dis 2002;40(3):655–7.

26. Aiba M, Ninomiya J, Furuya K, Arai H, Ishikawa H, Asaumi S, Takagi A, Ohwada S, Morishita Y. Induction of a critical elevation of povidone-iodine absorption in the treatment of a burn patient: report of a case. Surg Today 1999;29(2):157–9.

27. Ryan M, Al-Sammak Z, Phelan D. Povidone-iodine mediastinal irrigation: a cause of acute renal failure. J Cardiothorac Vasc Anesth 1999;13(6):729–31.

28. Kanakiriya S, DeChazal I, Nath KA, Haugen EN, Albright RC, Juncos LA. Iodine toxicity treated with haemodialysis and continuous venovenous hemodiafiltration. Am J Kidney Dis 2003;41:702–8.

29. Kobayashi H. Review of the use of povidone-iodine (PVP-I) in the treatment of burns. Postgrad Med J 1993;69:584–92.

30. Erdmann S, Hertl M, Merk HF. Allergic contact dermatitis from povidone-iodine. Contact Dermatitis 1999;40:331–2.

31. Nishioka K, Seguchi T, Yasuno H, Yamamoto T, Tominaga K. The results of ingredients patch testing in contact dermatitis elicited by povidone-iodine preparations. Contact Dermatitis 2000;42:90–4.

32. Adachi A, Fukunaga A, Hayashi K, Kunisada M, Horikawa T, Anaphylaxis to polyvinylpyrrolidone after vaginal application of povidone-iodine. Contact Dermatitis 2003;48:133–6.

33. Lowe DO, Knowles SR, Weber EA, Railton CJ, Shear NH. Povidone-iodine-induced burn: case report and review of the literature. Pharmacotherapy 2006;26:1641–5.

34. Dupont A, Lachapelle JM. Dermite due à un depot medicamenteux au cours du traitement d'un diabète insipide. [Dermatitis due to a medicamentous deposit during the treatment of diabetes insipidus.] Bull Soc Fr Dermatol Syphiligr 1964;71:508–9.

35. Lachapelle JM. Thesaurismose cutanée par polyvinylpyrrolidone. [Cutaneous thesaurismosis due to polyvinylpyrrolidone.] Dermatologica 1966;132(6):476–89.

36. Mensing H, Koster W, Schaeg G, Nasemann T. Zur klinischen Varianz der Polyvinylpyrrolidon-Dermatose. [Clinical variability of polyvinylpyrrolidone dermatosis.] Z Hautkr 1984;59(15):1027–37.

37. Bazex A, Geraud J, Guilhem A, Dupre A, Rascol A, Cantala P. Maladie de Dupont et Lachapelle (thesaurismose cutanée par polyvinylpyrrolidone). [Dupont-Lachapelle disease (cutaneous thesaurismosis due to polyvinylpyrrolidone).] Arch Belg Dermatol Syphiligr 1966;22(4):227–33.

38. Kuo TT, Hu S, Huang CL, Chan HL, Chang MJ, Dunn P, Chen YJ. Cutaneous involvement in polyvinylpyrrolidone storage disease: a clinicopathologic study of five patients, including two patients with severe anemia. Am J Surg Pathol 1997;21(11):1361–7.

39. Kepes JJ, Chen WY, Jim YF. 'Mucoid dissolution' of bones and multiple pathologic fractures in a patient with past history of intravenous administration of polyvinylpyrrolidone (PVP). A case report. Bone Miner 1993;22(1):33–41.

40. Dunn P, Kuo T, Shih LY, Wang PN, Sun CF, Chang MJ. Bone marrow failure and myelofibrosis in a case of PVP storage disease. Am J Hematol 1998;57(1):68–71.

41. Ronnau AC, Wulferink M, Gleichmann E, Unver E, Ruzicka T, Krutmann J, Grewe M. Anaphylaxis to polyvinylpyrrolidone in an analgesic preparation. Br J Dermatol 2000;143(5):1055–8.

42. Zamora JL. Chemical and microbiologic characteristics and toxicity of povidone-iodine solutions. Am J Surg 1986;151(3):400–6.

43. Ancona A, Suarez de la Torre R, Macotela E. Allergic contact dermatitis from povidone-iodine. Contact Dermatitis 1985;13(2):66–8.

44. Okten F, Oral M, Canakici N, et al. An anaphylactoid induced with polyvinylpyrrolidone iodine. A case report. Turk Anesteziyol Reanim 1993;21:118–22.

45. Erdmann S, Hertl M, Merk HF. Allergic contact dermatitis from povidone-iodine. Contact Dermatitis 1999;40(6):331–2.

46. Melvin GR, Aceto T Jr, Barlow J, Munson D, Wierda D. Iatrogenic congenital goiter and hypothyroidism with respiratory distress in a newborn. S D J Med 1978;31(10):15–9.

47. Mahillon I, Peers W, Bourdoux P, Ermans AM, Delange F. Effect of vaginal douching with povidone-iodine during early pregnancy on the iodine supply to mother and fetus. Biol Neonate 1989;56(4):210–7.

48. Koga Y, Sano H, Kikukawa Y, Ishigouoka T, Kawamura M. Effect on neonatal thyroid function of povidone-iodine used on mothers during perinatal period. J Obstet Gynaecol 1995;21(6):581–5.

49. Muthers S, Krude H, Jager R, Rhode W, Graters A, Rossi R. Hypothyroidism in dizygotic premature twins due to excessive prepartal vaginal iodine application. Zentralbl Gynakol 2003;125:226–8.

50. Gruters A, l'Allemand D, Heidemann PH, Schurnbrand P. Incidence of iodine contamination in neonatal transient hyperthyrotropinemia. Eur J Pediatr 1983;140(4):299–300.

51. l'Allemand D, Gruters A, Heidemann P, Schurnbrand P. Iodine-induced alterations of thyroid function in newborn infants after prenatal and perinatal exposure to povidone-iodine. J Pediatr 1983;102(6):935–8.

52. Castaing H, Fournet JP, Leger FA, Kiesgen F, Piette C, Dupard MC, Savoie JC. Thyroide du nouveau-né et surcharge en iode après la naissance. [The thyroid gland of the newborn infant and postnatal iodine overload.] Arch Fr Pediatr 1979;36(4):356–68.

53. Arena J, Eguileor I, Emparanza J. Repercusion sobre la funcion tiroidea del RN a termino de la aplicacion de povidona iodada en el munon umbilical. [Repercussion of the application of povidone-iodine to the umbilical stump on thyroid function of the neonate at term.] An Esp Pediatr 1985;23(8):562–8.

54. Tummers RF, Krul EJ, Bakker HD. Passagere hypothereoidie ten gevolge van huidinfectie met jodium bij een pasgeborene met een omfalokele. [Transient hypothyroidism due to skin contamination with iodine in a newborn infant with an omphalocele.] Ned Tijdschr Geneeskd 1985;129(20):958–9.

55. Smerdely P, Lim A, Boyages SC, Waite K, Wu D, Roberts V, Leslie G, Arnold J, John E, Eastman CJ. Topical iodine-containing antiseptics and neonatal

hypothyroidism in very-low-birthweight infants. Lancet 1989;2(8664):661–4.

56. Wilson JP, Solimando DA Jr, Edwards MS. Parenteral benzyl alcohol-induced hypersensitivity reaction. Drug Intell Clin Pharm 1986;20(9):689–91.
57. Shmunes E. Allergic dermatitis to benzyl alcohol in an injectable solution. Arch Dermatol 1984;120(9):1200–1.
58. Craven DE, Moody B, Connolly MG, Kollisch NR, Stottmeier KD, McCabe WR. Pseudobacteremia caused by povidone-iodine solution contaminated with *Pseudomonas cepacia*. N Engl J Med 1981;305(11):621–3.
59. Eason E, Wells G, Garber G, Hemmings R, Luskey G, Gillett P, Martin M. Antisepsis for abdominal hysterectomy: a randomised controlled trial of povidone-iodine gel. BJOG 2004;111:695–9.
60. Goel S, Mandhani A, Srivastava A, Kapoor R, Gogoi S, Kumar A, Bhandari M. Is povidone-iodine an alternative to silver nitrate in renal pelvic installation in chyluria? BJU Int 2004;94:1082–5.
61. Van Steirteghem AC, Young DS. Povidone-iodine ("Betadine") disinfectant as a source of error. Clin Chem 1977;23(8):1512.

Salicylanilides

General Information

The halogenated salicylanilides, which include closantel, niclosamide, oxyclozanide, rafoxanide, and resorantel, have antiparasitic activity in animals (1). Closantel and rafoxanide are widely used for the control of infestation with *Hemonchus* species and *Fasciola* species in sheep and cattle and *Estrus ovis* in sheep. Niclosamide is used as an anticestode in a wide range of animals. Other parasites that are susceptible include hematophagous helminths and external parasites such as ticks and mites. Many halogenated salicylanilides, including dibromosalicylanilide, tribromosalicylanilide, and tetrachlorosalicylanilide, have been used in disinfectants.

Organs and Systems

Skin

Photocontact dermatitis can occur after the use of halogenated salicylanilides, such as tribromosalicylanilide in soaps (2) and bithionol in first-aid creams (3). Tetrachlorosalicylanilide caused many cases in the 1960s and was withdrawn, after which there was a striking reduction in the numbers of patients with positive photopatch tests (4). The authors concluded that these results were most likely to have been due to removal from the market of the more potent photosensitizing chemicals and increased familiarity of physicians with this effect. Photosensitivity has also been reported to other analogues in widespread use, such as the dibromo and tribromo derivatives. The photoallergy is localized, but transient generalized reactions can occur.

References

1. Swan GE. The pharmacology of halogenated salicylanilides and their anthelmintic use in animals. J S Afr Vet Assoc 1999;70(2):61–70.
2. Osmundsen PE. Fotokontaktdermatitis forarsaget af tribromsalicylanilid i toiletsaebe. [Photocontact dermatitis caused by tribromosalicylanilide in toilet soap.] Ugeskr Laeger 1967;129(48):1607–10.
3. O'Quinn SE, Kennedy CB, Isbell KH. Contact photodermatitis due to bithionol and related compounds. JAMA 1967;199(2):89–92.
4. Smith SZ, Epstein JH. Photocontact dermatitis to halogenated salicylanilides and related compounds. Our experience between 1967 and 1975. Arch Dermatol 1977;113(10):1372–4.

Sodium hypochlorite

General Information

Sodium hypochlorite is widely used as a cleaning agent and to deal with blood spillages. Chlorine gas is released during the use of hypochlorite. Alkaline hypochlorite solutions with 0.25% "available chlorine" have been used to clean and disinfect wounds.

In the USA 2.95 billion pounds of disinfectants were used in 1999. Most consisted of chlorine/hypochlorites used in disinfecting portable waste and in recreational water. Unlike work regulations regarding agriculture, no specific US laws regulate the exposure of young people to pesticides or disinfectants in non-agricultural industries. An epidemiological assessment of acute occupational disinfectant-related illnesses amongst young people in 1993?8 showed that hypochlorites were responsible for 45% of the illnesses (1). Most of the illnesses were mild (78%); there were no deaths.

Accidental injection

Local injection

A patient developed ulceration and paresthesia of the lip with facial swelling after accidental injection of Milton solution (sodium hypochlorite 1%, sodium chloride 16.5%) into the upper lip when a sharp needle was used for irrigation (2). The ulceration healed within 6 weeks, the paresthesia within 3 months. In similar cases hypochlorite solution (25 mg NaClO) was accidentally injected into periapical tissue (3), and into the mandibular branch of the facial nerve (90 mg NaClO) (4).

Intraperitoneal infusion

Accidental intraperitoneal infusion of hypochlorite solution on two occasions (the first time about 10 ml of a 5% solution the second time more diluted) has been reported (5). After the infusion the patient felt severe abdominal pain accompanied by nausea but not by vomiting. Raised peritoneal solute transport rates and reduced ultrafiltration gradually subsided, but they did not return to pre-infusion values.

Intravenous infusion

Accidental intravenous infusion has been reported (6–8).

- A 68-year-old man received an accidental infusion of a 1% sodium hypochlorite solution. After 1 hour and an infusion volume of 150 ml the infusion was stopped. By this time he had a slow heart rate, mild hypotension, and an increased respiratory rate. The blood cell count, hemoglobin, serum electrolyte, creatinine, and urea concentrations, and transaminase activities were normal and hemolysis did not occur. Treatment with NaCl 0.9% and dextrose at an infusion rate of 300 ml/hour was started immediately with the goal of maintaining an adequate diuresis. Furosemide and dopamine hydrochloride were also administered. His blood pressure and respiratory rate promptly returned to normal, but the bradycardia persisted for 3 days despite atropine sulfate (6).
- During dialysis a patient received an accidental infusion of 30 ml of a 5.25% solution of hypochlorite. This led to cardiorespiratory arrest and massive hemolysis with hyperkalemia, although the patient eventually recovered (8).
- In a third case, intravenous infusion of 60 ml of an oral and 0.3 ml of a parenteral 5.25% solution in a suicide attempt did not result in any serious effects (7).

Organs and Systems

Respiratory

Chlorine gas, released during the use of hypochlorite, can cause mucous membrane irritation, bronchospasm, pneumonia, and pulmonary edema. It is believed that when hypochlorite is used as a cleaning substance in low concentrations it does not cause respiratory damage, but in a comparison of pulmonary function tests in 23 cleaning workers and 14 technical personnel, as a control group, even low concentrations of hypochlorite affected pulmonary function, causing irritation in the airways (9).

References

1. Brevard TA, Calvert GM, Blondell JM, Mehler LN. Acute occupational disinfectant related illness among youth 1993?1998. Environ Health Perspect 2003;111:1654–9.
2. Linn JL, Messer HH. Hypochlorite injury to the lip following injection via a labial perforation. Case report. Aust Dent J 1993;38(4):280–2.
3. Becker GL, Cohen S, Borer R. The sequelae of accidentally injecting sodium hypochlorite beyond the root apex. Report of a case. Oral Surg Oral Med Oral Pathol 1974;38(4):633–8.
4. Herrmann JW, Heicht RC. Complications in therapeutic use of sodium hypochlorite. J Endod 1979;5(5):160.
5. Dedhia NM, Schmidt LM, Twardowski ZJ, Khanna R, Nolph KD. Long-term increase in peritoneal membrane transport rates following incidental intraperitoneal sodium hypochlorite infusion. Int J Artif Organs 1989;12(11):711–4.
6. Marroni M, Menichetti F. Accidental intravenous infusion of sodium hypochlorite. DICP 1991;25(9):1008–9.
7. Froner GA, Rutherford GW, Rokeach M. Injection of sodium hypochlorite by intravenous drug users. JAMA 1987;258(3):325.
8. Hoy RH. Accidental systemic exposure to sodium hypochlorite (Chlorox) during hemodialysis. Am J Hosp Pharm 1981;38(10):1512–4.
9. Demiralay R. Effects of the use of hypochlorite as a cleaning substance on pulmonary functions. Turk J Med Sci 2001;31:51–7.

Sulfites and bisulfites

General Information

Sulfites and bisulfites have been used extensively as preservatives in the food industry and also in drugs and bronchodilator inhalant solutions (1,2) as preservatives. Sodium metabisulfite is used commonly as an antioxidant in foods and drugs. As an additive in various pharmaceutical products, metabisulfite can cause unpleasant adverse reactions.

Organs and Systems

Respiratory

Status asthmaticus and acute bronchospasm have been linked to the use of metabisulfites (3).

Hematologic

Preservatives in subconjunctival gentamicin were identified as the cause of conjunctival chemosis and capillary closure (4). Patients undergoing cataract surgery were divided into three groups: one was given a subconjunctival injection of a preservative-free solution of gentamicin at the end of the cataract procedure, another received a subconjunctival injection of gentamicin containing sodium metabisulfite and disodium edetate as preservatives, and the third was the control group not given a subconjunctival injection. There was a significant difference in the severity of conjunctival chemosis between patients who received gentamicin with and without preservatives (4).

Immunologic

The major symptoms of an adverse reaction to a sulfite are flushing, acute bronchospasm, and hypotension (SED-11, 492) (SEDA-10, 232) (SEDA-11, 221) (5). The incidence of sulfite sensitivity in an asthmatic population is estimated at about 10%. Sulfites have therefore been withdrawn from the composition of several medicines intended for asthmatic patients.

Metabisulfite-induced anaphylaxis through an IgE-mediated mechanism has been described in a patient who developed urticaria, angioedema, and nasal congestion following provocative challenge with sodium metabisulfite (6).

The presence of sodium metabisulfite as an antioxidant in commercial lidocaine with adrenaline significantly increased discomfort during injection (7).

Anaphylactic shock occurring during epidural anesthesia for cesarean section has been attributed to sodium metabisulfite (8).

Reports of contact allergy to topical medicaments containing sodium metabisulfite are rare (9). In two cases, a topical corticosteroid formulation that contained sodium metabisulfite (Trimovate cream) caused contact allergy; patch tests were positive with both sodium metabisulfite and Trimovate cream (10).

References

1. Gunnison AF, Jacobsen DW. Sulfite hypersensitivity. A critical review. CRC Crit Rev Toxicol 1987;17(3):185–214.
2. Monafo WW, West MA. Current treatment recommendations for topical burn therapy. Drugs 1990;40(3):364–73.
3. Maria Y, Vaillant P, Delorme N, Moneret-Vautrin DA. Les accidents graves liés aux metabisulfites. [Severe complications related to metabisulfites.] Rev Med Interne 1989;10(1):36–40.
4. Pande M, Ghanchi F. The role of preservatives in the conjunctival toxicity of subconjunctival gentamicin injection. Br J Ophthalmol 1992;76(4):235–7.
5. Chan TY, Critchley JA. Is chloroxylenol nephrotoxic like phenol? A study of patients with DETTOL poisoning. Vet Hum Toxicol 1994;36(3):250–1.
6. Sokol WN, Hydick IB. Nasal congestion, urticaria, and angioedema caused by an IgE-mediated reaction to sodium metabisulfite. Ann Allergy 1990;65(3):233–8.
7. Long CC, Motley RJ, Holt PJ. Taking the "sting" out of local anaesthetics. Br J Dermatol 1991;125(5):452–5.
8. Soulat JM, Bouju P, Oxeda C, Amiot JF. Choc anaphylactoide aux metabisulfites au cours d'une césarienne sous anesthésie péridurale. [Anaphylactoid shock due to metabisulfites during cesarean section under peridural anesthesia.] Cah Anesthesiol 1991;39(4):257–9.
9. Heshmati S, Maibach HI. Active sensitization to sodium metabisulfite in hydrocortisone cream. Contact Dermatitis 1999;41(3):166–7.
10. Tucker SC, Yell JA, Beck MH. Allergic contact dermatitis from sodium metabisulfite in Trimovate cream. Contact Dermatitis 1999;40(3):164.

Tosylchloramide sodium

General Information

Tosylchloramide has been used as a wound disinfectant, a general surgical antiseptic, and for the disinfection of water (1).

Organs and Systems

Respiratory

Asthma has been attributed to tosylchloramide (2).

Immunologic

Contact sensitization to tosylchloramide has been reported (3).

- Urticaria, rhinitis, dyspnea, and edema of the face were reported in a female nurse after contact (SEDA-11, 492) (4). Specific IgE antibodies to tosylchloramide were demonstrated.

References

1. Reybrouck G. The bactericidal activity of aqueous disinfectants applied on living tissues. Pharm Weekbl Sci 1985;7(3):100–3.
2. Romeo L, Gobbi M, Pezzini A, Caruso B, Costa G. Asma da tosilcloramide sodica: descrizione di un caso. [Tosylchloramide-induced asthma: description of a case.] Med Lav 1988;79(3):237–40.
3. Metzner HH. Kontaktsensibilisierungen durch Tosylchloramidnatrium (Chloramin) und Hydroxychinolin (Sulfachin). [Contact sensitization caused by tosylchloramide sodium (chloramine) and hydroxyquinoline (Sulfachin).] Dermatol Monatsschr 1987;173(11):674–7.
4. Dooms-Goossens A, Gevers D, Mertens A, Vanderheyden D. Allergic contact urticaria due to chloramine. Contact Dermatitis 1983;9(4):319–20.

Triclocarban

General Information

Triclocarban is mainly used as an antibacterial agent in soaps and antiperspirants (1).

Organs and Systems

Hematologic

Methemoglobinemia was attributed to triclocarban in seven neonates (2). There was no history of familial methemoglobinemia, but in one case the mother had been taking phenacetin before labor and this was most probably the cause. In five of the six other cases the methemoglobinemia was thought to be due to the use of disinfecting solutions or ointments containing triclocarban that had been used as a vaginal disinfectant in four cases and as a powder for the umbilicus in one case. The authors advised against the use of triclocarban in maternity units and by those dealing with neonates.

Skin

Several patients developed relatively minor dermatological problems after treatment with triclocarban (3). Three developed very extensive erosive cutaneomucosal lesions after using triclocarban for periods ranging from 20 days to 3 months.

Immunologic

There have been several cases of allergic contact dermatitis after the use of antiperspirants containing triclocarban and propylene glycol (4).

References

1. Bodey GP, Arnett J, De Salva S. Comparative trial of bacteriostatic soap preparations: Hexachlorophene versus triclosan and triclocarbon. Curr Ther Res 1978;24:542.
2. Ponte G, Richard J, Bonte C, et al. Méthémoglobinèmes chez le nouveau-né: discussion du rôle étiologique du trichlorcarbanilide. Ann Pédiatr 1974;21:359.
3. Barriere H. La dermite cutanéomuqueuse caustique du trichloro carbanilide. [Caustic cutaneo-mucosal dermatitis caused by trichlorcarbanilide.] Therapeutique 1973;49(10):685–7.
4. Osmundsen PE. Concomitant contact allergy to propantheline bromide and TCC. Contact Dermatitis 1975;1(4):251–2.

Triphenylmethane dyes

Triphenylmethane is a hydrocarbon, $(C_6H_5)_3CH$, from which synthetic dyestuffs are derived, including bromocresol green, malachite green, brilliant green, crioglaucine, and crystal (gentian) violet. They are intensely coloured and poorly resistance to light and chemical bleaches. They are mainly used in copying papers and printing inks, and in textile applications for which light-fastness is not important.

Gentian violet is a purple dye, so called because its color resembles that of the flower; it has nothing to do with *Gentiana* species. It was used in the past by local application to treat oral and vaginal candidiasis and to prepare the vagina for gynecological operations and as an antihelminthic agent by oral administration. However, its use is now restricted in many countries, owing to concern over its mutagenic and carcinogenic effects.

Oral administration can cause gastrointestinal irritation and intravenous injection can cause a reduced white blood cell count [1]. Gentian violet is demethylated in the liver and is reduced to leucogentian violet by intestinal microflora. Complete demethylation produces leucopararosaniline, which is carcinogenic in rats. A free-radical derivative is also formed in the liver, but its toxicity is not clear. N-demethylation by peroxidases and cyclo-oxygenase are other routes of metabolism.

Respiratory

Mucosal ulceration and airways obstruction can occur with application of gentian violet, and occlusive laryngotracheitis requiring orotracheal intubation has been reported [2].

- Mucosal lesions consistent with oral candidiasis developed in a previously healthy, full-term, exclusively breast-fed, 2-week-old girl. She was treated with oral nystatin, resulting in an initial reduction in the severity of the lesions. After a few days, the thrush became more prominent. At 4 weeks of age, 1% aqueous gentian violet was prescribed and the day after she developed a cough and difficulty in feeding. There was no nasal congestion, fever, or rhinorrhea. Over the next 7 days her cough and feeding difficulties became progressively worse, and she developed a hoarse cry and stridor. Nasal washings for respiratory syncytial virus were negative. Intravenous fluconazole and ceftriaxone were given for presumed sepsis and fungal tracheitis. Lateral neck radiographs showed an absence of air in the cervical trachea. he was intubated with a 3.5 mm oral endotracheal tube for airway management, and was then taken for direct laryngoscopy under general anesthesia. The supraglottic, glottic, and subglottic structures were very edematous, but the vocal cords were mobile. Blood cultures on the day of admission failed to grow bacteria or fungi. There were no fungi in the supraglottic exudate. Nasopharyngeal samples obtained for viral culture were negative.

The obstructive chemical laryngotracheitis was thought to be secondary to gentian violet.

Urinary tract

Chemical cystitis due to intravesical installation of gentian violet is rare. Cases have occurred in adult women when an undiluted solution was used. Cystitis has been reported in a child after bladder instillation of diluted gentian violet [3].

- A 16 month-old boy developed painful gross hematuria after a herniorrhaphy. During the operation, gentian violet solution diluted to 0.1% had been instilled into the bladder to rule out bladder injury, and hematuria developed several hours later. There was no pyuria. Ultrasonography showed multiseptate structures resulting from edema and hematoma in the bladder and bilateral hydronephrosis. The hematuria responded to intravenous hydration, and follow-up ultrasonography showed bladder wall thickening with resolution of the strictures and less hydronephrosis.

Skin

Contact sensitization to gentian violet, brilliant green, and malachite green has been described [4].

Skin necrosis has been described in a child after the application of 2% gentian violet to the gluteal fold [5]. The authors recommended using concentrations below 1% for the treatment of intertrigo.

Immunologic

An allergic reaction, with urticaria, edema of the eyelids and lips, and hypotension, occurred aftert the use of Patent Blue Violet dye for lymphangiography [6].

References

1. Docampo R, Moreno SN. The metabolism and mode of action of gentian violet. Drug Metab Rev 1990;22(2-3):161–78.
2. Baca DJ, Drexler C, Cullen E. Obstructive laryngotracheitis secondary to gentian violet exposure. Clin Pediatr 2001;40:233–5.
3. Kim SJ, Koh H, Park JS, Ahn HS, Choi JB, Kim YS. Hemorrhagic cystitis due to intravesical instillation of gentian violet completely recovered with conservative management. Yonsei Med J 2003;44:163–5.
4. Bielicky T, Novák M. Contact-group sensitization to triphenylmethane dyes. Gentian violet, brilliant green, and malachite green. Arch Dermatol 1969;100(5):540–3.
5. Meurer M, Konz B. Hautnekrosen nach Anwendung 2%iger Pyoktaninlosung. [Skin necrosis following the use of a 2-per-cent Pyoctanin solution.] Hautarzt 1977;28(2):94–5.
6. Hietala SO, Hirsch JI, Faunce HF. Allergic reaction to Patent Blue Violet during lymphography. Lymphology 1977;10(3):158–60.

Index of drug names

Note: The letter '*t*' with the locater refers to tables.

Printed and bound by CPI Group (UK) Ltd, Croydon, CR0 4YY

03/10/2024

01040333-0019